INDUSTRIAL HYGIENE AND TOXICOLOGY

Second Revised Edition   *In Three Volumes*

VOLUME I:
General Principles

VOLUME II:
Toxicology

VOLUME III:
Industrial Environmental Analysis

INDUSTRIAL HYGIENE
AND
TOXICOLOGY

Second Revised Edition

FRANK A. PATTY, *Editor*

VOLUME II

TOXICOLOGY

David W. Fassett and Don D. Irish, Editors

Index by Mrs. Kathleen Kumler

Authors

W. B. Deichmann	F. W. Heyroth	V. K. Rowe
D. W. Fassett	C. H. Hine	H. E. Stokinger
H. W. Gerarde	D. D. Irish	W. L. Sutton
C. L. Hake	R. A. Kehoe	J. F. Treon
D. O. Hamblin	M. L. Keplinger	R. J. Weir
L. W. Hazleton	F. A. Patty	M. A. Wolf

INTERSCIENCE PUBLISHERS

a division of John Wiley & Sons, Inc., New York · London · Sydney

ISBN 0 470 67188 2

PRINTED IN THE UNITED STATES OF AMERICA

PREFACE

to the Second Edition, Volume II

Prefaces to the previous edition, possibly of historical interest to students of industrial health, are to be found in Volume I and are omitted from Volume II in the interest of avoiding unnecessary verbosity.

This book was planned as a ready, practical reference for persons interested in or responsible for safeguarding the health of others working with the chemical elements and compounds used in industry today. Although guide lines for selecting those chemical compounds of sufficient industrial importance for inclusion are not clearly drawn, those chemicals found in carload price lists seem to warrant first consideration.

Where available information is bountiful, an attempt has been made to limit the material presented to that of a practical nature, useful in recognizing, evaluating, and controlling possibly harmful exposures. Where the information is scanty, every fragment of significance, whether negative or positive, is offered the reader. The manufacturing chemist, who assumes responsibility for the safe use of his product in industry and who employs a competent staff to this end, as well as the large industry having competent industrial hygiene and medical staffs, are in strategic positions to recognize early and possibly harmful exposures in time to avoid any harmful effects by appropriate and timely action. Plant studies of individuals and their exposures regardless of whether or not the conditions caused recognized ill effects offer valuable experience. Information gleaned in this manner, though it may be fragmentary, is highly important when interpreted in terms of the practical health problem.

The very rapid expansion of the number of useful chemical products and the rapidly mounting stockpiles of toxicological data (frequently from studies of animals) have made the present volume timely but most difficult to compile and edit.

The last chapter is devoted to the recognition and control of exposures to be looked for in specific industries and occupations.

The efforts of the contributors, and their patience through seemingly unending delays, are recognized and appreciated. This volume would not have been possible without the competent and time-consuming work of the associate editors. No reference book is better than its index and this one will not be found wanting in that respect.

For the benefit of the reader who does not have Volume I immediately available we have listed on page xiii the contents of that volume, which deals with the broad aspects of industrial hygiene.

FRANK A. PATTY

Marine City, Michigan
September, 1962

AUTHORS of Volume II

William B. Deichmann, Ph.D., *Professor and Chairman, Department of Pharmacology, School of Medicine, University of Miami, Coral Gables, Florida.*

David W. Fassett, M.D., *Director, Laboratory of Industrial Medicine, Eastman Kodak Company, Rochester, New York.*

Horace W. Gerarde, M.D., Ph.D., *Senior Research Associate, Medical Research Division, Esso Research and Engineering Company, Linden, New Jersey.*

C. L. Hake, Ph.D., *Head, Chemistry Research Department, Pitman-Moore Company, Division of The Dow Chemical Company, Indianapolis, Indiana.*

Donald O. Hamblin, M.D., *Formerly Medical Director, American Cyanamid Company, New York, New York.*

Lloyd W. Hazleton, Ph.D., *President, Hazleton Laboratories, Falls Church, Virginia.*

Francis F. Heyroth, M.D., *(Deceased), Assistant Director, Kettering Laboratory of Applied Physiology, College of Medicine, University of Cincinnati, Cincinnati, Ohio.*

Charles H. Hine, M.D., *Associate Clinical Professor of Occupational Medicine and Toxicology, School of Medicine, University of California, San Francisco, California.*

Don D. Irish, Ph.D., *Biochemical Research Laboratory, The Dow Chemical Company, Midland, Michigan.*

Robert A. Kehoe, M.D., *Director, Department of Preventive Medicine and Industrial Health, College of Medicine, University of Cincinnati, Cincinnati, Ohio.*

M. L. Keplinger, Ph.D., *Toxicologist, Medical Department, Hercules Powder Company, Wilmington, Delaware.*

Frank A. Patty, *Formerly Director, Industrial Hygiene Department, Personnel Staff, General Motors Corporation, Detroit, Michigan.*

V. K. Rowe, M.S., *Biochemical Research Laboratory, The Dow Chemical Company, Midland, Michigan.*

Herbert E. Stokinger, Ph.D., *Chief, Toxicology Section, Division of Occupational Health, United States Public Health Service, Cincinnati, Ohio.*

William L. Sutton, M.D., *Laboratory of Industrial Medicine, Eastman Kodak Company, Rochester, New York.*

Joseph F. Treon, Ph.D., *Manager, Biological Sciences Section, Chemical Research Department, Atlas Chemical Industries, Wilmington, Delaware.*

Robert J. Weir, Ph.D., *Toxicologist, Hazleton Laboratories, Falls Church, Virginia.*

M. A. Wolf, M.S., *Biochemical Research Laboratory, The Dow Chemical Company, Midland, Michigan.*

USEFUL EQUIVALENTS AND CONVERSION FACTORS

1 kilometer = 0.6214 mile

1 meter = 3.281 feet

1 centimeter = 0.3937 inch

1 micron = 1/25,400 inch = 40 microinches = 10,000 Angstrom units

1 foot = 30.48 centimeters

1 inch = 25.40 millimeters

1 square kilometer = 0.3861 square mile (U.S.)

1 square foot = 0.0929 square meter

1 square inch = 6.452 square centimeters

1 square mile (U.S.) = 2,589,998 square meters = 640 acres

1 acre = 43,560 square feet = 4,047 square meters

1 cubic meter = 35.315 cubic feet

1 cubic centimeter = 0.0610 cubic inch

1 cubic foot = 28.32 liters = 0.0283 cubic meter = 7.481 gallons (U.S.)

1 cubic inch = 16.39 cubic centimeters

1 U. S. gallon = 3.7853 liters = 231 cubic inches = 0.13368 cubic foot

1 liter = 0.9081 quart (dry), 1.057 quarts (U.S., liquid)

1 cubic foot of water = 62.43 pounds (4°C.)

1 U.S. gallon of water = 8.345 pounds (4°C.)

1 kilogram = 2.205 pounds

1 gram = 15.43 grains

1 pound = 453.59 grams

1 ounce (avoir.) = 28.35 grams

1 gram mole of a perfect gas ≈ 24.45 liters (at 25°C. and 760 mm. Hg barometric pressure)

1 atmosphere = 14.7 pounds per square inch

1 foot of water pressure = 0.4335 pound per square inch

1 inch of mercury pressure = 0.4912 pound per square inch

1 dyne per square centimeter = 0.0021 pound per square foot

1 gram calorie = 0.00397 B.t.u.

1 B.t.u. = 778 foot-pounds

1 B.t.u. per minute = 12.96 foot-pounds per second

1 hp. = 0.707 B.t.u. per second = 550 foot-pounds per second

1 centimeter per second = 1.97 feet per minute = 0.0224 mile per hour

1 foot candle = 1 lumen incident per square foot = 10.764 lumens incident per square meter

1 grain per cubic foot = 2.29 grams per cubic meter

1 milligram per cubic meter = 0.000437 grains per cubic foot

To convert degrees centigrade to degrees Fahrenheit: °C. (9/5) + 32 = °F.

To convert degrees Fahrenheit to degrees centigrade: (5/9) (°F. − 32) = °C.

For solutes in water: 1 mg./liter ≈ 1 p.p.m. (by weight)

Atmospheric contamination: 1 mg./liter ≈ 1 oz./1000 cu. ft. (approx.)

For gases or vapors in air at 25°C. and 760 mm. Hg. pressure:

    To convert mg./liter to p.p.m. (by volume): mg.liter (24,450/mol. wt.) = p.p.m.

    To convert p.p.m. to mg./liter: p.p.m. (mol. wt./24,450) = mg./liter

CONVERSION TABLE FOR GASES AND VAPORS[a]

(Milligrams per Liter to Parts per Million and Vice Versa;
25°C. and 760 mm. Mercury Barometric Pressure)

Molecular Weight	1 mg./l. p.p.m.	1 p.p.m. mg./l.	Molecular p.p.m.	1 mg./l. p.p.m.	1 p.p.m. mg./l.	Molecular Weight	1 mg./l. p.p.m.	1 p.p.m. mg./.
1	24,450	0.0000409	51	479	0.002086	101	242.1	0.00413
2	12,230	.0000818	52	470	.002127	102	239.7	.00417
3	8,150	.0001227	53	461	.002168	103	237.4	.00421
4	6,113	.0001636	54	453	.002209	104	235.1	.00425
5	4,890	.0002045	55	445	.002250	105	232.9	.00429
6	4,075	.0002454	56	437	.002290	106	230.7	.00434
7	3,493	.0002863	57	429	.002331	107	228.5	.00438
8	3,056	.000327	58	422	.002372	108	226.4	.00442
9	2,717	.000368	59	414	.002413	109	224.3	.00446
10	2,445	.000409	60	408	.002454	110	222.3	.00450
11	2,223	.000450	61	401	.002495	111	220.3	.00454
12	2,038	.000491	62	394	.00254	112	218.3	.00458
13	1,881	.000532	63	388	.00258	113	216.4	.00462
14	1,746	.000573	64	382	.00262	114	214.5	.00466
15	1,630	.000614	65	376	.00266	115	212.6	.00470
16	1,528	.000654	66	370	.00270	116	210.8	.00474
17	1,438	.000695	67	365	.00274	117	209.0	.00479
18	1,358	.000736	68	360	.00278	118	207.2	.00483
19	1,287	.000777	69	354	.00282	119	205.5	.00487
20	1,223	.000818	70	349	.00286	120	203.8	.00491
21	1,164	.000859	71	344	.00290	121	202.1	.00495
22	1,111	.000900	72	340	.00294	122	200.4	.00499
23	1,063	.000941	73	335	.00299	123	198.8	.00503
24	1,019	.000982	74	330	.00303	124	197.2	.00507
25	978	.001022	75	326	.00307	125	195.6	.00511
26	940	.001063	76	322	.00311	126	194.0	.00515
27	906	.001104	77	318	.00315	127	192.5	.00519
28	873	.001145	78	313	.00319	128	191.0	.00524
29	843	.001186	79	309	.00323	129	189.5	.00528
30	815	.001227	80	306	.00327	130	188.1	.00532
31	789	.001268	81	302	.00331	131	186.6	.00536
32	764	.001309	82	298	.00335	132	185.2	.00540
33	741	.001350	83	295	.00339	133	183.8	.00544
34	719	.001391	84	291	.00344	134	182.5	.00548
35	699	.001432	85	288	.00348	135	181.1	.00552
36	679	.001472	86	284	.00352	136	179.8	.00556
37	661	.001513	87	281	.00356	137	178.5	.00560
38	643	.001554	88	278	.00360	138	177.2	.00564
39	627	.001595	89	275	.00364	139	175.9	.00569
40	611	.001636	90	272	.00368	140	174.6	.00573
41	596	.001677	91	269	.00372	141	173.4	.00577
42	582	.001718	92	266	.00376	142	172.2	.00581
43	569	.001759	93	263	.00380	143	171.0	.00585
44	556	.001800	94	260	.00384	144	169.8	.00589
45	543	.001840	95	257	.00389	145	168.6	.00593
46	532	.001881	96	255	.00393	146	167.5	.00597
47	520	.001922	97	252	.00397	147	166.3	.00601
48	509	.001963	98	249.5	.00401	148	165.2	.00605
49	499	.002004	99	247.0	.00405	149	164.1	.00609
50	489	.002045	100	244.5	.00409	150	163.0	.00613

Molecular Weight	1 mg./l. p.p.m.	1 p.p.m. mg./l.	Molecular Weight	1 mg./l. p.p.m.	1 p.p.m. mg./l.	Molecular Weight	1 mg./l. p.p.m.	1 p.p.m. mg./l.
151	161.9	0.00618	201	121.6	0.00822	251	97.4	0.01027
152	160.9	.00622	202	121.0	.00826	252	97.0	.01031
153	159.8	.00626	203	120.4	.00830	253	96.6	.01035
154	158.8	.00630	204	119.9	.00834	254	96.3	.01039
155	157.7	.00634	205	119.3	.00838	255	95.9	.01043
156	156.7	.00638	206	118.7	.00843	256	95.5	.01047
157	155.7	.00642	207	118.1	.00847	257	95.1	.01051
158	154.7	.00646	208	117.5	.00851	258	94.8	.01055
159	153.7	.00650	209	117.0	.00855	259	94.4	.01059
160	152.8	.00654	210	116.4	.00859	260	94.0	.01063
161	151.9	.00658	211	115.9	.00863	261	93.7	.01067
162	150.9	.00663	212	115.3	.00867	262	93.3	.01072
163	150.0	.00667	213	114.8	.00871	263	93.0	.01076
164	149.1	.00671	214	114.3	.00875	264	92.6	.01080
165	148.2	.00675	215	113.7	.00879	265	92.3	.01084
166	147.3	.00679	216	113.2	.00883	266	91.9	.01088
167	146.4	.00683	217	112.7	.00888	267	91.6	.01092
168	145.5	.00687	218	112.1	.00892	268	91.2	.01096
169	144.7	.00691	219	111.6	.00896	269	90.9	.01100
170	143.8	.00695	220	111.1	.00900	270	90.6	.01104
171	143.0	.00699	221	110.6	.00904	271	90.2	.01108
172	142.2	.00703	222	110.1	.00908	272	89.9	.01112
173	141.3	.00708	223	109.6	.00912	273	89.6	.01117
174	140.5	.00712	224	109.2	.00916	274	89.2	.01121
175	139.7	.00716	225	108.7	.00920	275	88.9	.01125
176	138.9	.00720	226	108.2	.00924	276	88.6	.01129
177	138.1	.00724	227	107.7	.00928	277	88.3	.01133
178	137.4	.00728	228	107.2	.00933	278	87.9	.01137
179	136.6	.00732	229	106.8	.00937	279	87.6	.01141
180	135.8	.00736	230	106.3	.00941	280	87.3	.01145
181	135.1	.00740	231	105.8	.00945	281	87.0	.01149
182	134.3	.00744	232	105.4	.00949	282	86.7	.01153
183	133.6	.00748	233	104.9	.00953	283	86.4	.01157
184	132.9	.00753	234	104.5	.00957	284	86.1	.01162
185	132.2	.00757	235	104.0	.00961	285	85.8	.01166
186	131.5	.00761	236	103.6	.00965	286	85.5	.01170
187	130.7	.00765	237	103.2	.00969	287	85.2	.01174
188	130.1	.00769	238	102.7	.00973	288	84.9	.01178
189	129.4	.00773	239	102.3	.00978	289	84.6	.01182
190	128.7	.00777	240	101.9	.00982	290	84.3	.01186
191	128.0	.00781	241	101.5	.00986	291	84.0	.01190
192	127.3	.00785	242	101.0	.00990	292	83.7	.01194
193	126.7	.00789	243	100.6	.00994	293	83.4	.01198
194	126.0	.00793	244	100.2	.00998	294	83.2	.01202
195	125.4	.00798	245	99.8	.01002	295	82.9	.01207
196	124.7	.00802	246	99.4	.01006	296	82.6	.01211
197	124.1	.00806	247	99.0	.01010	297	82.3	.01215
198	123.5	.00810	248	98.6	.01014	298	82.0	.01219
199	122.9	.00814	249	98.2	.01018	299	81.8	.01223
200	122.3	.00818	250	97.8	.01022	300	81.5	.01227

[a] A. C. Fieldner, S. H. Katz, and S. P. Kinney, "Gas Masks for Gases Met in Fighting Fires," U.S. Bur. Mines, Tech. Paper No. 248 (1921).

CONTENTS

Volume I

Volume II

CHAPTER XXII

Halogens

FRANCIS F. HEYROTH,* M.D.

FLUORINE, F_2

1. Source, Uses, Properties, and Industrial Exposures

Fluorine is a yellow gas that does not occur free in nature because of its great reactivity. Before World War II it was generated only in gram quantities, but the need for stable fluorocarbons in the atomic energy program stimulated the chemical industry to devise electrolyte cells for the preparation of elemental fluorine. The present availability of the compressed gas in nickel or steel cylinders has led to its use for the preparation of a host of new organic fluorine compounds, many of them of commercial value.[1]

The gas can be liquefied at low temperatures. Its odor, which differs from that of chlorine can be compared to that of relatively concentrated ozone.

It can be piped through standard steel pipe or copper tubing equipped with Monel metal or nickel valves with Teflon packing. With metals, fluorine reacts only slowly, except at sufficiently elevated temperatures. Under suitable conditions, it reacts spontaneously with most materials at room temperature except the inert gases, metal fluorides in their highest valance states, and carbon tetrachloride. Under some conditions, fluorine at atmospheric pressure can burn even steel equipment. For reasons unknown, fluorine does not always react with water, but at times the reaction may occur explosively. With the moisture of the air, it forms hydrogen fluoride and, possibly, oxygen fluoride, OF_2.

The industrial handling of fluorine has been described thoroughly by Landau and Rosen,[2] and the special precautions described in their article, including those for the disposal of unused gas, should be scrupulously enforced wherever fluorine is handled.

* This chapter was revised by Dr. Heyroth before his death. The very few additions pertinent to bring this up to date are as indicated. There has been very little new information on the halogens published in the last few years. A brief bibliography of recent literature has been added. EDITORS

[1] Symposium on Fluorine Chemistry, Ind. Eng. Chem., 39, 236 (1947).

[2] R. Landau and R. Rosen, Ind. Eng. Chem., 39, 281 (1947).

2. *Physiological Response*

Inhalation. The only studies of the effects of exposure of animals to metered dilutions of fluorine in nitrogen were those made at the University of Rochester by a team led by Stokinger.[3]

The gas was uniformly fatal to rabbits, guinea pigs, rats, and mice in exposures ranging from 5 minutes at 10,000 p.p.m. to 3 hours at 200 p.p.m. Guinea pigs survived an exposure of 7 hours to 100 p.p.m., but the over-all mortality among the various species was 60 per cent. Respiratory damage, with pulmonary edema, was the cause of death. More prolonged exposure, up to 35 days, was made to lesser concentrations. Irritation of the eyes and nasal and buccal mucosa was noted at concentrations of 5 to 10 p.p.m., and dogs exhibited irrational seizures, many of which were fatal. Moderate to severe pulmonary irritation occured at all levels down to 3 mg./cu. meter, and rats showed also a high degree of testicular degeneration at 25-mg. level. The tolerated exposure was taken as 1 p.p.m. (1.7 mg./cu. meter), and hydrogen fluoride and fluorine were regarded as independently toxic.

Cutaneous Burns. The Rochester group also exposed the skin of the back of anesthetized rabbits for periods of 0.2 to 0.6 sec. at a distance of 1 in. to fluorine under 40 lb. pressure. The briefest exposure led to the appearance of a small ischemic area $1/4$ in. in diameter, surrounded by an erythematous area. This became a superficial eschar that sloughed off by the fourth day, disclosing normal epidermis. The longer exposures were accompanied by a flash of flame, burning the hair and causing coagulation necrosis of the burned area and charring of the epidermis. The thermal flash burns resembled those induced by an oxyacetylene flame.

3. *Threshold Limit*

The American Conference of Governmental Industrial Hygienists (1961) has recommended 0.1 p.p.m. as the threshold limit for fluorine.

FLUORIDES OF THE ALKALIES AND ALKALINE EARTHS

1. *Source, Uses, and Industrial Exposures*

The chief source of fluorine compounds is fluorspar (calcium fluoride), with cryolite (sodium aluminum fluoride, Na_3AlF_6) imported from Greenland being of lesser importance.

The largest use of calcium fluoride is in steelmaking, where it first was used at Massillon, Ohio, in 1927; 20 years later it was in use in fifty steel plants producing 7 million tons of steel per year.[4]

[3] H. E. Stokinger, N. Eriksen, B. Andur, A. L. Shannon, N. Glover, M. Schlamowitz, P. Hoch, N. Murphy, C. W. LaBelle, J. A. Orcutt, N. Smith, E. W. Same, and I. Slotnik, *Natl. Nuclear Energy Ser., Div. VI,* **1,** Pt. II, 1021 (1949).

[4] K. E. Markuson, *Ind. Med.,* **16,** 434 (1947).

Added calcium or sodium fluorides prolong the fluidity of the ingot, thus facilitating the release of gaseous products. Fluorides are also used as a flux in the smelting of nickel, copper, gold, and silver. Other applications are in the opacifying of glass and enamels and in coatings for welding rods. Considerable quantities of calcium fluoride are treated with sulfuric acid for the preparation of hydrogen fluoride. Sodium fluoride is used as a rodenticide; antimony fluoride serves as a catalyst for certain organic reactions; and barium fluoride is used in baths for the electrolytic isolation of beryllium. Lesser uses of fluorides are found in the bleaching of cane for seats, in the disinfection of hides and skins, in the preservation of timbers, in the coagulation of latex, in cleaning graphite, metals, windows, and glassware, and variously in the optical, brewing, and dyeing industries.[5]

Cryolite is employed as an insecticide and, in molten form, as an electrolyte in the production of aluminum from alumina. In Danish plants for the grinding and packing of cryolite, the presence of large quantities of dust caused the fluoride content of the air to be about 35 mg./cu. meter.[6] Exposure to cryolite dusts also occurs in the manufacture of beryllium.

The use of fluoride-coated welding rods offers a definite hazard.[7] In magnesium founding, fluorides inhibit oxidation when sprayed upon cores or mixed with the core sand to the extent of 4 to 10 per cent. They also act as fluxes when added to the melting pots.[8] Williams[9] found the following quantities, expressed as milligrams of fluoride per cubic meter of air, near various operations: core spraying, 0.7; melting, 1.26; molding, 1.88; shakeout, 8.77. Analyses by the Illinois Department of Labor in other foundries showed the presence of fluorine equivalent to 2 p.p.m. of hydrogen fluoride at the molding operation, less than 3 p.p.m. in the shakeout areas, but as much as 7.2 p.p.m. near the pouring operation.[10] Largent and Ferneau,[11] who used the urinary excretion of fluorine by workers as a measure of the severity of exposure, found the greatest exposure in the core-spraying and pouring areas. Very similar results have been obtained in British magnesium foundries.[12]

During sulfuric acid treatment in the preparation of fertilizer from fluorine-bearing phosphate rock, it has been estimated that well over 70,000 tons of fluoride per year, largely as silicon tetrafluoride, escape into the atmosphere where it

[5] D. A. Greenwood, *Physiol. Revs.*, **20**, 582 (1940); S. J. Davenport and G. G. Morges, *U. S. Bur. Mines Inform. Circ.* No. **7687** (1954).

[6] K. Roholm, *Arch. Gewerbepathol. Gewerbehyg.*, **7**, 255 (1936).

[7] P. Drinker and K. W. Nelson, *Ind. Med.*, **13**, 673 (1944).

[8] M. E. Brooks and A. W. Winston, *Trans. Am. Foundrymen's Assoc.*, **49**, 165 (1941).

[9] C. R. Williams, *J. Ind. Hyg. Toxicol.*, **24**, 277 (1942).

[10] Anon., *Illinois Ind. Hyg. Div. Lab. Bull.*, **4**, No. 7, 4, 12 (1944); *J. Ind. Hyg. Toxicol.*, **26**, 103A (1944).

[11] E. J. Largent and I. Ferneau, *J. Ind. Hyg. Toxicol.*, **26**, 113 (1944).

[12] R. G. Bowler, M. Buckell, J. Garrad, A. B. Hill, D. Hunter, K. M. A. Perry, and R. S. F. Schilling, *Brit. J. Ind. Med.*, **4**, 216, 231 (1947)

reacts with water to form hydrogen fluoride and hydrofluosilicic acid.[13] Large quantities of fluorides are also emitted in the thermal processing of rock phosphate. Much effort has been expended in attempts to control the fluoride emission within the working areas of the plants or to design methods for recovering fluorides from the stack gases.[14] Atmospheric pollution with fluoride has caused actual economic loss from damage to vegetation, to cattle and other animals foraging in the vicinity of fertilizer, steelmaking, and aluminum-producing plants.

Gaseous fluorine-containing pollutants have caused relatively little damage to farm crops, but considerable damage has resulted where commercial florists have attempted to grow in these areas certain plants, notably gladioli, which are highly susceptible to low concentrations of hydrogen fluoride.[15]

The presence on forage of fluoride-bearing dusts derived from the soil of contaminated areas may be harmful to cattle and other animals. The effect observed on men in a British aluminum factory and on animals in its vicinity, at Invernessshire, have been described in a survey conducted by the Fluorosis Committee of the Medical Research Council.[16] Concentrations of fluorine within the furnace room of the plant ranged from 0.14 to 3.43 mg./cu. meter. Outside the plant the concentrations had decreased at 200 yards to 15 per cent of those found in the furnace room and at 1 mile to 3 per cent of these values. In the immediate vicinity of the factory, samples of dried soil had the very high fluorine content of 1010 p.p.m., and, in the direction of prevailing winds, the values were still 66 p.p.m. at the distance of 7 miles. The results of analyses of grass samples declined in a parallel fashion. Chemical analysis of the teeth and bones of animals grazing on such herbage showed excessive values, and some damage to the animals in this area was encountered.

2. Determination in the Atmosphere

The sample of air may be collected in an evacuated bulb and shaken with a 1 per cent solution of sodium hydroxide, or it may be passed through an impinger. If hydrogen fluoride is the only fluorine compound present, the remaining free alkali in the absorbing solution may be titrated with acid. If other volatile compounds or fluorine-containing dusts are present, the liquid is concentrated to a small volume, after the addition of calcium or magnesium oxide to prevent loss of fluorine. It is then distilled with perchloric acid and the fluorine determined

[13] W. H. MacIntyre and co-workers, *Ind. Eng. Chem.*, **41**, 2466 (1949).

[14] T. P. Hignett and M. R. Siegel, *Ind. Eng. Chem.*, **41**, 2493 (1949); A. B. Pettit, *Air Pollution and Smoke Prevention Assoc. Am., Proc.*, **44**, 98 (1951).

[15] F. Johnson, D. F. Allmendinger, V. L. Miller, and C. J. Gould, *Phytopathology*, **40**, 239 (1950); V. L. Miller, D. F. Allmendinger, F. Johnson, and D. Polley, *J. Agr. Food Chem.*, **1**, 526 (1953).

[16] G. H. Agate, D. Hunter, K. M. A. Perry, J. Gareod, and E. A. Cheeseman, *Med. Research Council Brit., Mem.* No. **22** (1948).

in the distillate by the back-titration method of Dahle, Bonnar, and Wichmann.[17] In efforts to devise more sensitive, or more easily performed methods, applicable as well to fluoride determinations in biological materials, a number of proposals have been made to use spectrophotometric,[18] fluorometric,[19] polarographic,[20] and even enzymic[21] methods. The determination of the small amounts present in blood has been described by Smith and co-workers[22] and by Epars.[23]

3. Physiological Response

Acute effects. The lethal dose of sodium fluoride when administered orally to rabbits is 200 mg./kg. of body weight[24] and, when given intraperitoneally, is 250 mg/kg., according to Handler.[25] Because of its lesser solubility, cryolite is much less toxic; rats are not killed by the largest dose that can be administered orally.[26]

The symptoms that follow the ingestion of soluble fluorides by man are listed in the order of diminishing frequency of occurrence: vomiting, abdominal pain, diarrhea, convulsions, generalized and muscular weakness, collapse, dyspnea, paresis, difficulty in articulation, thirst, weakness of the pulse, disturbed color vision, loss of consciousness, and motor unrest.[27] Albuminuria is frequently present. Acute toxic nephritis, hemorrhagic gastroenteritis, and more or less definite pathologic damage to other organs are found on examination. The calcium content of the blood is reduced following the ingestion of large amounts of fluorides[28] Fluoride acts as an inhibitor of certain intracellular enzymes concerned in the anaerobic glucolysis of many types of cells, plant as well as mammalian.[29] It interferes with enzymes concerned with the conversion of phosphoglyceric to phosphopyruvic

[17] D. Dahle, R. U. Bonnar, and H. J. Wichmann, *J. Assoc. Offic. Agr. Chemists,* **21,** 459 (1938).

[18] H. E. Bumsted and J. C. Wells, *Anal. Chem.,* **24,** 1595 (1952); A. D. Horton, P. F. Thomason, and F. J. Miller, *Anal. Chem.,* **24,** 548 (1952); D. Revinson and J. H. Harley, *Anal. Chem.,* **25,** 794 (1953).

[19] W. A. Powell and J. H. Saylor, *Anal. Chem.,* **25,** 960 (1953).

[20] B. J. MacNulty, G. F. Reynolds, and E. A. Terry, *Nature,* **169,** 888 (1952).

[21] H. Stetter, *Chem. Ber.,* **81,** 532 (1948).

[22] F. A. Smith, D. E. Gardner, M. J. Voss, D. Wing, P. Dunn, J. Figueres, and J. W. Keating, *J. Dental Research,* **30,** 182 (1951).

[23] L. Epars, *Bull. schweiz. Akad. med. Wiss.,* **8,** 360 (1952); *Chem. Abstr.,* **47,** 3385 (1953).

[24] C. W. Muehlberger, *J. Pharmacol. Exptl. Therap.,* **39,** 346 (1930).

[25] P. Handler, *J. Biol. Chem.,* **161,** 55 (1945).

[26] E. J. Largent, *J. Ind. Hyg. Toxicol.,* **30,** 92 (1948).

[27] K. Roholm, in Heffter-Huebner, *Handbuch exptl. Pharmakol. Erg.,* **7,** 1 (1938); *Z. ges. gerichtl. Med.,* **27,** 174 (1936); F. McClure, *Physiol. Revs.,* **13,** 277 (1933); D. A. Greenwood, *Physiol. Revs.,* **20,** 582 (1940).

[28] H. Wieland and G. Kurtzahn, *Arch. exptl. Pathol. u. Pharmakol.,* **97,** 488 (1923); A. Jodlbauer, *Arch. exptl. Pathol. u. Pharmakol.,* **164,** 464 (1931).

[29] F. Lipmann, *Biochem. J.,* **206,** 171 (1929); K. Lohmann, *Biochem. J.,* **222,** 324 (1930); F. Lipmann and K. Lohmann, *Biochem. J.,* **222,** 389 (1930); F. Dickens and F. Simer, *Biochem. J.,* **23,** 936 (1929).

acids, an essential link in the chain of reactions.[30] A number of other enzymes, particularly those concerned with processes of phosphorylation, are also affected by the fluoride ion.

The therapy of acute fluoride poisoning has been reviewed by Peters.[31]

Immediate intoxication by the ingestion of fluorides is rare in industry, but various degrees of respiratory irritation may result from the inhalation of fluorides in the form of dusts. Some magnesium founders complain of a severe biting sensation in the nose when the concentration of fluorides in the air exceeds 10 mg./cu meter. This is accompanied after a few minutes by a discharge from the nose or by nosebleed. No such effects are noted when the concentration does not exceed 2.5 mg./cu. meter.[32]

Fluorosis or Chronic Fluorine Intoxication. Chronic fluorosis was first described as occurring among animals. *Darmous,* an abnormality of the teeth and bones, occurs among sheep and cattle pastured in North Africa on land where the vegetation is contaminated by fluoride-bearing dusts derived from phosphate deposits.[33] In Iceland, sheep grazing on pastures contaminated with fluoride-containing volcanic ash suffer from *Gaddur,* a condition in which the teeth are spotted, the bones are thickened, and the animal becomes lame and weak. Severe cases among sheep, goats, and cattle have occurred both in Europe and in the United States in the vicinity of factories that emit fluorine-containing compounds into the atmosphere. Such cattle have a stiff-legged gait, swollen hock joints, and palpable lumps on their bones, as well as irregularly worn teeth.[34]

Many experiments have been performed in the effort to reproduce the condition by incorporating fluorides in the food of rats, dogs, sheep, swine, cattle, and chickens. They have been reviewed by Schmidt and Rand.[35] Estimates by various investigators of the maximum daily intake that may be tolerated by animals of various species, as recorded by Schmidt and Rand, fall within the following ranges (mg./kg. of body weight): dairy cattle, 1 to 3; swine, 5 to 12; rats, 10 to 20; guinea pigs, 12 to 20; chickens, 35 to 70. Analysis of the urinary fluoride of cattle grazing in contaminated areas has been suggested as a means of estimating the severity of the intoxication. Contamination of pasture plants to the extent of 25 to 50 p.p.m. (dry weight basis) may result in fluorosis in grazing animals.[36]

The continued daily ingestion by rats of diets containing from 7 to 12 p.p.m. by weight induces signs of dental fluorosis, detectable only by the aid of a lens as

[30] C. W. Bishop and E. Roberts, *Natl. Nuclear Energy Ser., Div. VI,* **1,** Book 4, 1867 (1953).

[31] J. H. Peters, *Am. J. Med. Sci.,* **216,** 278 (1948).

[32] C. R. Williams, *J. Ind. Hyg. Toxicol.,* **24,** 277 (1942).

[33] H. Velu, *Arch. inst. Pasteur Algérie,* **10,** 41 (1932).

[34] E. J. Largent, *Proc. Natl. Air Pollution Symposium, 1st Symposium, Pasadena, Calif.,* 1949.

[35] H. J. Schmidt and W. E. Rand, *Am. J. Vet. Research,* **13,** 38 (1952).

[36] H. H. Mitchell and M. Edman, *Nutrition Abstr. & Revs.,* **21,** 787 (1952).

fine lines of impaired calcification of the teeth. When the diet contains some-what greater amounts, the incisors become chalky, pitted, and corroded, and changes in the bony system may result.[37] Dietary levels of from 100 to 230 p.p.m. by weight inhibit growth. Rats that consume diets with 900 or more p.p.m. by weight die after a few weeks. According to Roholm,[38] who expressed the intake in relation to body weight, incipient dental fluorosis may be included by a daily intake of 1 mg./kg. Five times that intake gave rise to incipient changes in the bones and kidneys, and an intake of from 10 to 15 mg./kg. brought about recognizable signs of ill-health. Severe degenerative changes in the organs appeared when the daily intake was from 20 to 25 mg./kg., and death occurred within a few days or weeks following a period of inanition, lethargy, and weakness when it was 50 to 100 mg./kg. Such impairment of fertility as has been found in experiments on animals[39,40] may have resulted from the general state of inanition.[41] In pigs and dogs, nephritis has been induced in feeding experiments, but liver damage is exceptional. Anemia has been observed in some of these experiments but is not a uniform feature of chronic fluoride intoxication.[42]

Lesions of the skeletal system are of significance in industrial toxicology. In the attempt to elicit them, Largent[43] gave 65 mg. of fluoride daily to each of 2 dogs which were 11-month-old littermates. The third littermate served as control. The dose was given to one dog as sodium fluoride and to the other as cryolite. During the second month, the dog given sodium fluoride lost appetite and vomited occasionally, but this did not persist. The administration of fluorides was continued for 5 years and 5 months when it was stopped because of the death of the dog given none. During life, no bone changes could be detected roentgenographically. The ash of the bones of the dog given sodium fluoride contained 10 times the amount of fluorine found in that of the dog used as control. Among the soft tissues, increased amounts were found only in the lungs and kidneys which had 1.77 and 3.01 mg/kg., respectively. No noteworthy histopathological changes were found in the organs; and the bones, although chalky in appearance, exhibited no very striking abnormality, even though the daily intake had been of the order of 3 to 5 mg. of fluoride per kilogram, and the total amount of sodium fluoride consumed by one of the dogs has been about 250 g.

In 1892, however, Brandl and Tappeiner[44] had described crystalline deposits in the marrow spaces and Haversian canals in the bones of a dog that had been fed

[37] M. C. Smith and R. M. Leverton, *Ind. Eng. Chem.*, **26**, 791 (1934).

[38] K. Roholm, *Fluorine Intoxication*. London, 1937.

[39] A. R. Lamb, P. H. Phillips, E. B. Hart, and G. Bohstedt, *Am. J. Physiol.*, **106**, 350 (1933).

[40] H. M. Hauck, H. Steenbock. and H. T. Parsons, *Am. J. Physiol.*, **103**, 480 (1939).

[41] J. A. Schulz and A. R. Lamb, *Science*, **61**, 93 (1925).

[42] J. T. Ginn and J. T. Volker, *Proc. Soc. Exptl. Biol. Med.*, **57**, 189 (1944).

[43] E. J. Largent, *Proc. Natl. Air Pollution Symposium, 1st Symposium, Pasadena, Calif., 1949*.

[44] J. Brandl and H. Tappeiner, *Z. Biol.*, **28**, 518 (1892).

over 400 g. of sodium fluoride over a period of 21 months. During the last 5 months of this period the animal, in running or walking, held its spine, especially the sacral portion, in an unusually stiff position. On sacrificing the dog, the investigators noted that the tensile strength of the bones, which had an unusually white appearance was apparently increased.

Similar changes were noted by Bardelli and Menzani[45] in the bones of cattle poisoned by fluorides; and Hauck, Steenbock, and Parsons[46] found that the bones of rats similarly poisoned were 18 per cent heavier than normal and had a thickened cortex that in some instances encroached upon the medullary cavity. The skeletons of rabbits that were given 763 p.p.m. (by weight) fluorine in the diet as sodium fluoride for 92 days, or 2.980 p.p.m. for 47 days, were chalky, and small areas of porous bone, varying in diameter from 1 mm. to 1 cm., were scattered throughout the vertebra, cranium, zygoma, and maxillae, with larger plaques on a markedly bulked mandible.[47]

Whereas the bony alterations in man are of an osteosclerotic type, in the earlier descriptions those of animals which had grazed on contaminated foilage tended to be of an osteoporotic nature and the condition was sometimes confused with osteomalacia. The histological changes in the bones of experimental animals have been described by Sutro,[48] de Senarclens,[49] Pachaly,[50] and others,[51] but much remains to be learned regarding the factors that cause either absorption or deposition of bone to predominate.

Much attention has been devoted to the possible role of calcium in relation to the development of bony abnormalities. In chronic poisoning of animals, no significant diminution in the calcium content of the blood has been found.[52] Fluorides must be fed at a level that produces general intoxication in order to cause a diminution in the retention of calcium,[53,54] or to alter the calcium content or composition of the ash in the bones. There is little evidence that the phosphate content of the bone is affected.[55] The fluoride ion appears to be deposited in place of the hydroxyl ion in hydroxyapatite in the bone.[56] Evidence has been presented that two types of stores of fluoride may exist in the bones of rats. One is readily

[45] P. Bardelli and C. Menzani, *Atti ist. veneto sci.*, Pt. II, **97**, 623 (1937–38).
[46] H. M. Hauck, H. Steenbock, and H. T. Parsons, *Am. J. Physiol.*, **103**, 489 (1939).
[47] E. J. Largent, W. Machle, and I. F. Ferneau, *J. Ind. Hyg. Toxicol.*, **25**, 3936 (1943).
[48] C. J. Sutro, *A.M.A. Arch. Pathol.*, **19**, 159 (1935).
[49] F. de Senarclens, *Helv. Med. Acta*, **8**, 379 (1941).
[50] W. I. Pachaly, *Arch. exptl. Pathol. u. Pharmakol.*, **166**, 1 (1932).
[51] W. Dittrich, *Arch. exptl. Pathol. u. Pharmakol.*, **168**, 319 (1932); E. Rost. *Arch. Gewerbepathol. Gewerbehyg.*, **8**, 256 (1937).
[52] H. M. Hauck, H. Steenbock, and H. T. Parsons, *Am. J. Physiol.*, **103**, 489 (1939).
[53] F. J. McClure and H. H. Mitchell, *J. Biol. Chem.*, **90**, 297 (1931).
[54] E. M. Lantz and M. C. Smith, *Am. J. Physiol.*, **109**, 645 (1934).
[55] P. H. Phillips, *Science*, **76**, 239 (1932); M. C. Smith and E. M. Lantz, *J. Biol. Chem.*, **112**, 303 (1935).
[56] W. F. Neuman, M. W. Neuman, E. R. Main, J. O'Leary, and F. A. Smith, *J. Biol. Chem.*, **187**, 655 (1950); H. G. McCann, *J. Biol. Chem.*, **201**, 247 (1953).

eliminated following withdrawal of the dietary supplement while the other is more firmly held.[57]

Absorption and Storage of Fluoride in Man. Although the soft tissues contain fluorine to the extent of 20 to 60 μg. per 100 g.,[58] ingested fluorine is stored chiefly in the bones and teeth. Roholm,[38] who found from 0.048 to 0.21 per cent of fluoride in the ash of human ribs, estimated that the total amount in the bones of a man is of the order of 1.5 to 6 g. Machle and Scott[59] found 0.026 per cent in the ash of human bones.

In the absence of industrial exposure the chief source of fluorine is in drinking water, the fluorine content of which differs greatly in various localities. The daily intake by one resident of a low-fluoride area remained remarkably constant over a period of 16 weeks of unrestricted diet, the mean value being 0.457 mg. of fluorine. Of this 0.115 mg. was derived from the food and 0.299 mg. from the fluid constitutes of the diet. The mean daily fecal output was 0.039 mg, while the mean

TABLE 1

Community	Fluoride in water, p.p.m.	Urinary mean fluoride, p.p.m.	Mean daily intake, mg.	Mean daily output, mg.
Galesburg, Ill.	2	2.09	3.54	3.27
Ennis, Texas	5.5	5.46	7.30	7.02
Lake Preston, S. D.	6	7.80	8.30	8.48
Bartlett, Texas	8	8.71	15.30	13.17

daily urinary output was 0.377 mg. Over the entire period, no significant storage occurred.[60] In 30 Danish hospital patients, the daily urinary fluoride excretion varied from 0.18 to 1.85 mg., and the mean urinary concentration was 0.92 mg./liter.[61] The content of fluorine of many foods has been tabulated by McClure.[62] Largent and Heyroth[63] have also made metabolic observations on the fluoride balance of 8 persons long resident in four United States communities where the drinking water contains known concentrations of fluoride. The results, presented in Table 1, indicate that when the concentration in the water does not exceed 6 p.p.m. by weight, retention is slight although the daily intake increases in proportion to concentration in the water. Definite radiological signs of skeletal

[57] W. B. Savchuck and W. D. Armstrong, *J. Biol. Chem.*, **193**, 575 (1951); R. F. Miller and P. H. Phillips, *J. Nutrition,* **51**, 273 (1953).

[58] A. O. Gettler and L. Ellerbrook, *Am. J. Med. Sci.*, **197**, 625 (1939).

[59] W. Machle and E. W. Scott, *J. Ind. Hyg. Toxicol.*, **17**, 230 (1935).

[60] W. Machle and E. J. Largent, *J. Ind. Hyg. Toxicol.*, **25**, 112 (1943).

[61] G. C. Brun, H. Buchwald, and K. Roholm, *Acta Med. Scand.*, **106**, 261 (1941).

[62] F. J. McClure, *Public Health Repts. U. S.*, **64**, 1061 (1949).

[63] E. J. Largent and F. F. Heyroth, unpublished work.

retention occurred in only 11 to 12 per cent of the population at Barlett, Texas, where the mean daily intake approximated 15 mg.[64]

In metabolic studies, Machle and Largent[60] found that when the daily intake of fluoride was 6.47 mg., of which 6.28 mg. was absorbed, 63 per cent of the amount absorbed was retained in the body. McClure and associates[65] from brief experiments in which account was taken of the amounts excreted in the perspiration, found retention does not begin until the daily intake is greater than 4 mg.

There are some indications that the accumulation of fluorine in the bones may exert a retarding influence upon the rate at which it is stored in prolonged experiments.[66] In general, however, the urinary fluoride excretion may be used as a rough measure of the rate at which fluoride is being stored. Once a man has ceased the prolonged intake of fluoride at excessive rates, the stored fluoride is slowly eliminated from his tissues, the excess of his daily output over his intake decreasing exponentially.[67]

A remarkable parallelism between the fluoride content of the drinking water and concentration of fluoride in the urine of a large number of recruits has been demonstrated by McClure,[68] but no relation to their susceptibility to fractures of the bones could be found.

Collings[69] et al. have demonstrated an unexpectedly prompt increase of the urinary fluoride following brief periods of inhalations of rock phosphate dust.

Chronic Industrial Fluorosis. In the course of radiographic examinations of workers who had been exposed for several years to dust in amounts sufficient to supply from 0.2 to 0.35 mg. of fluorine per kilogram of body weight daily, Møller and Gudjonsson[70] discovered a generalized osteosclerosis in many. Abnormal density and indistinct structure of the bones were most striking in the spinal column pelvis, and ribs. Over half of those who had been exposed for about 10 years presented definite evidence of changes in the bone structure, varying from a fleecy thickening of the lamina or increased opacity of the bones to the appearance of exostoses and calcification of the ligamentous attachments. In some workers, the abnormalities were so pronounced as to cause stiffness of the spinal column and thorax. In a few cases, ankylosis of the joints of the vertebral column caused serious disability.

In the United States, a worker in the phosphate fertilizer industry[71,72] was

[64] Testimony of Dr. Francis A. Arnold, Jr., Hearings before the House Select Committee to Investigate the Use of Chemicals in Foods and Cosmetics. H. Rep. 82nd Congress, 2nd Session. Pursuant to H. Res. 74 and H. Res. 447, Part 3, 1655.

[65] F. J. McClure, H. H. Mitchell, T. S. Hamilton, and C. A. Kinser, J. Ind. Hyg. Toxicol., 27, 159 (1945).

[66] M. Lawrenz, H. H. Mitchell, and W. A. Ruth, J. Nutrition, 19, 531 (1948); F. F. Heyroth, Am. J. Public Health, 42, 1571 (1952).

[67] E. J. Largent, Arch. Ind. Hyg. Occupational Med., 6, 37 (1952).

[68] F. J. McClure, Public Health Repts. (U. S.), 59, 1543, 1575 (1944).

[69] G. H. Collings, Jr., R. B. L. Fleming, R. Mav, and W. O. Bianconi, Arch. Ind. Hyg. Occupational Med., 6, 368 (1951); G. H. Collings, Jr. R. B. L. Fleming, and R. May, Arch. Ind. Hyg. Occupational Med., 4, 585 (1951).

[70] F. Møller and S. V. Gudjonsson, Acta Radiol., 13, 269 (1932).

found to have bony changes that might have been related to the greatly increased fluoride concentration in his bones.

About half of the Danish cryolite workers complained of lack of appetite, nausea, and shortness of breath; and a smaller proportion mentioned constipation, localized pain in the region of the liver, and other symptoms.[38] Some degree of anemia was found in half of the workers with fluorosis. The mean concentration of fluorine in the urine of these workers was 16.05 mg./liter, the range being 2.41 to 43.41 mg./liter. In those with less severe exposure, the mean urinary concentration was 4.81 mg./ liter, with a range of 1.78 to 11.67 mg./liter.[61] A slight degree of fluorosis was found in workers exposed for 2 to $2^{1}/_{2}$ years, while more definite signs were found in those exposed nearly 5 years, and signs of moderate fluorosis appeared in those with more than 11 years of exposure. The most severe cases were those of men who had 21 years of exposure. However, not all workers developed fluorosis, no abnormalities being detected in one man after 24 years of work. From analysis of the bones of 2 workers, Roholm estimated that their skeletal systems contained 50 and 90 g. of fluorine, respectively. The latter amount had been deposited during 7500 working days, corresponding to an average deposition of 12 mg./day.[38]

A striking feature of the examination of men employed in an aluminum factory was the absence of disabling symptoms, although a few of the men employed in the furnace room exhibited radiological appearances consistent with those of skeletal fluorosis. The mean daily urinary output of fluorides in the most heavily exposed group was 9.03 mg.[73] In two factories in the United States, increases in the radiographic density of the bones have appeared in men whose urine was known to have contained 10 or more mg./liter.[74] More recent data offer some indications that such changes may occur to a slight degree in a few men with urinary concentrations of but 6 mg./liter.[75] To provide some factor of safety, some industrial physicians regard the maximum permissible exposure as one that will not cause the mean daily urinary output to exceed 4 mg.

4. Threshold Limit

The threshold limit for total fluoride published in 1959 by the American Conference of Governmental Industrial Hygienists is 2.5 mg./cu. meter of air.

HYDROGEN FLUORIDE, HF

1. Uses, Properties, and Industrial Exposures

Hydrogen fluoride is a colorless liquid, with molecular weight 20.01. It boils at 19.4°C. Measurements of the density of its vapor show that it may be asso-

[71] P. A. Bishop, Am. J. Roentgenol. Radium Therapy, **35**, 577 (1936).

[72] W. A. Wolff and E. G. Kerr, Am. J. Med. Sci., **195**, 493 (1938).

[73] G. H. Agate, D. Hunter, K. M. A. Perry, J. Gareod, and E. A. Cheeseman, Med. Research Council Brit., Mem. No. **22**, 1948.

[74] E. J. Largent, P. G. Bovard, and F. F. Heyroth, Am. J. Roentgenol. Radium Therapy, **65**, 42 (1951).

[75] F. F. Heyroth, Am. J. Public Health, **42**, 1568 (1952).

ciated to predominantly linear polymers.[76] Its great solubility in water causes it
to fume strongly in moist air. Partial-pressure measurements on the system
hydrogen-fluoride-water have been recorded.[77] Its aqueous solution dissolves
glass, reacting with the silica to form gaseous silicon tetrafluoride. Because of its
solvent properties, it is used in various concentrations for frosting, etching, and
polishing glass, and for removing sand from metal castings. Recently, large quantities have been used in refineries as a catalyst for the production of certain hydrocarbons for high-octane gasoline. Only occasionally does an operator of this
process exhibit an abnormally increased urinary fluoride concentration.[78] Because
of uncertainty in the degree of association of hydrogen fluoride vapor, and therefore, of its molecular weight, atmospheric concentrations may be more precisely
stated in terms of milligrams per cubic meter, rather than as parts per million. In
practice, this is frequently disregarded.

2. Physiological Response

Toxicity When Inhaled by Animals. When inhaled by rabbits and guinea pigs
in a concentration of not more than 50 mg./cu meter of air,[79] hydrogen fluoride
induced signs of mild irritation, such as coughing and sneezing, which appeared to
lessen after 5 to 15 minutes. Inhaled in greater concentrations, it acted as a severe irritant: the eyes were kept closed, paroxysms of coughing and sneezing
were frequent, the respirations were slowed, and there was a copious discharge
from the nose and eyes.

Animals died within 5 minutes when they inhaled air containing 1500 mg./cu.
meter.[79] The inhalation of air containing 1000 mg./cu. meter for 30 minutes killed
no animals but did cause damage to the tissues. All animals exposed to the concentration of 500 mg./cu. meter for 15 minutes or more showed signs of weakness
and ill-health; concentrations below 100 mg./cu. meter could be tolerated for 5
hours without causing death; and a concentration of 24 mg./cu. meter was tolerated for a total of 41 hours without fatality, although the animals subsequently
lost in weight. The concentration of 15 mg./cu. meter was found tolerable.

Necropsies on animals that survived after repeated exposure to hydrogen fluoride gave evidence of damage to the lung, liver, and kidney, of a nature that suggested that some process was involved beyond that usually induced by irritant
gases. Machle's experiments demonstrated that storage of fluoride in the bones
occurs as a result of repeated exposures to concentrations that are but slightly, if
at all, irritant.

Injury from Industrial Use. The highest concentration of hydrogen fluoride
that can be tolerated by man for 1 minute is 100 mg/cu. meter of air. This causes

[76] H. A. Berni and C. P. Smyth, *J. Chem. Phys.*, **15**, 337 (1947).

[77] P. A. Munter, O. T. Alphi, and R. A. Kossatz, *Ind. Eng. Chem.*, **41**, 1504 (1949).

[78] E. J. Largent, *J. Ind. Hyg. Toxicol.*. **29**, 53 (1947).

[79] W. F. Machle, F. Thamann, K. Kitzmiller and J. Cholak, *J. Ind. Hyg. Toxicol.*. **16**,
129 (1934); W. Machle and K. Kitzmiller, *J. Ind. Hyg. Toxicol.*. **17**, 223 (1935).

a definite smarting of the skin, a definite sour taste, and some degree of conjunctival and respiratory irritation.[79] In air containing 50 mg./cu meter, the taste is apparent and there is irritation of the eyes and nose, but no smarting of the skin. The concentration of 26 mg./cu. meter can be tolerated for several minutes, but the taste becomes evident after a short time, and there is mild smarting of the nose and eyes. No apparent habituation results from repeated brief exposures.

Contact of the skin with the anhydrous liquid produces severe burns that are felt immediately. Concentrated aqueous solutions also cause an early sensation of pain, but more dilute solutions may give no warning of injury. If the solution is not promptly removed, the skin may be penetrated by the fluoride ion, leading to the later development of painful ulcers, which heal slowly. A 0.03 per cent solution of sodium fluoride will destroy epithelium, according to Stanton and Kahn.[80] This observation suggests the necessity for preventing the diffusion of the fluoride ion in the treatment of all burns from hydrogen fluoride. After any contact with a solution of the acid, even though there be no immediate pain, the area should be flushed with copious amounts of water, after which it may be swabbed with cotton moistened with 10 per cent solution of 28 per cent aqueous ammonia and than allowed to remain immersed in a bath of water for a prolonged period. Prolonged washing, when carried out without any delay following contact, may succeed in restoring a pink color to any initially blanched area, thereby saving the tissue. Finally an ointment containing 20 per cent magnesium oxide in glycerin should be applied.[81] In the case of more serious burns, other procedures have been suggested,[82] such as setting up a barrier to the spread of the fluoride ion by injecting a 10 per cent solution of calcium gluconate subcutaneously around and underneath the affected area.

Precautions necessary for the safe handling of liquid hydrogen fluoride have been described by Flowers[83] and the Manufacturing Chemists' Association.[84]

3. Threshold Limit

The American Conference of Governmental Industrial Hygienists has adopted 3 p.p.m. or 2 mg./cu. meter as the threshold limit for hydrogen fluoride.

AMMONIUM BIFLUORIDE, $(NH_4)HF_2$, AND SODIUM BIFLUORIDE, $NaHF_2$

Sodium bifluoride is used in the refining of beryllium and, sometimes as welding flux or as a coating on welding rods. In order to reduce the porosity of steel,

---

[80] D. N. Stanton and M. Kahn, *J. Am. Med. Assoc.*, **64**, 1985 (1915).

[81] Booklet issued by the Universal Oil Products Company.

[82] H. Haar, *Zentr. Chir.*, **74**, 467 (1949); H. Bade, *Angew. Chem.*, **64**, 166 (1952); F. Flury, *J. Ind. Hyg. Toxicol.*, **24**, 92A (1942); A. Paley and J. Seifter, *Proc. Soc. Exptl. Biol. Med.*, **46**, 190 (1941).

[83] F. E. Flowers, *Proc. Am. Petro. Inst., Sect. I*, **28**, 95 (1948).

[84] Manufacturing Chemists' Assoc., *Chem. Safety Data Sheet, SD-25*, 1948; *Manual Sheet H-10*, 1948.

$NaHF_2$ may be added to the mold during pouring. Ammonium acid fluoride is also used as a flux in magnesium foundries. The hazards arising from the use of these salts are those of fluorides in general (see Fluorine and Fluorides of the Alkalies and Alkaline Earths sections p. 831).

BORON TRIFLUORIDE, BF_3, AND AMMONIUM BOROFLUORIDE, NH_4BF_4

Boron fluoride is a colorless gas at ordinary temperatures. It boils at $-101°C$. It has been used as a polymerization catalyst for certain organic reactions.[85] Ammonium borofluoride (fluoborate) is a solid which has been used as flux in magnesium founding. The fluoborates have found some use in the electroplating industry. When ingested, they do not appear to give rise to the retention of fluoride in the skeleton.

EDITOR'S ADDITION

Laboratory toxicological studies on boron trifluoride have been carried out at the Biochemical Research Laboratory of The Dow Chemical Co. This information should appear in publication soon.

Rats, guinea pigs, and rabbits were exposed 7 hours/day, 5 days/week, to an atmosphere to which BF_3 had been added.

At a concentration of 12 p.p.m. (based on analysis for fluoride) marked lung irritation was noted in animals sacrificed after 2 months of repeated exposures. Lung irritation decreased with concentration. At 1 to 2 p.p.m. for 6 months, some lung irritation was still indicated by a greater-than-normal incidence of bronchiolitis. Otherwise the animals were in good condition. A definite increase in fluoride content of the teeth was observed though the teeth appeared sound.

It should be recognized that the nature of the material breathed is uncertain due to the reaction of BF_3 with moisture. Concentrations were determined as fluoride and expressed as BF_3 equivalent.

SILICON TETRAFLUORIDE, SiF_4

Silicon tetrafluoride is a heavy gas with molecular weight 104.06 and a density of 3.6 referred to air $= 1$. It is not employed in industry, but it may be discharged into the air in smelting operations as the result of the interaction of calcium fluoride with sand or with the silica present in ores. Its toxicity as such has not been investigated, but products formed from it may contribute to the injury of plants and animals in the areas surrounding industrial operations.

HYDROFLUOSILICIC ACID, H_2SiF_6, AND ITS SALTS

Hydrofluosilicic acid, molecular weight 144.08, has been employed in dilute solution as a disinfectant in breweries. The barium salt has been used as an insec-

[85] *Oil Gas J.*, **43**, 81 (1944).

ticide. Sodium and other silicofluorides (fluosilicates) are employed as fluxes in magnesium founding,[86] in the extraction of beryllium, in disinfecting hides and skins, in hardening cement, in coagulating latex, and in cleaning windows.[87] In fluoridating communal water supplies for preventing dental caries, sodium silicofluoride is becoming the choice as the source of the fluoride ion, at least in the larger installations. The charging of the hoppers is a dusty operation, and while so engaged the operators should wear dust respirators approved by the Bureau of Mines. Dust collectors consisting of bag filters operating under positive air pressure vented to the outside have been recommended. Safe limits for the contamination of the air in the neighborhood should not be exceeded.[88] The observations recorded in this paper were made in a water works in which sodium fluoride, rather than sodium fluosilicate, was employed; direct observations and exposures to the latter do not appear to have been recorded.

Weber and Engelhardt[89] exposed guinea pigs to air bearing sodium silicofluoride as a dust in concentrations ranging from 13 to 55 mg./cu. meter and found the dust capable of causing pulmonary irritation. They concluded that the least concentration that caused death when inhaled for a period of 6 hours was 33 mg./cu. meter.

The sodium salt is highly toxic when ingested; numerous deaths have been recorded. The signs of poisoning resemble those seen in intoxication by fluorides.

When in contact with the skin, the acid and its salts cause redness and a burning sensation, sometimes followed by the formation of ulcers. A pustular rash has been observed among men who worked with the sodium salt.[90]

CHLORINE, Cl_2

1. *Source, Uses, and Industrial Exposures*

Chlorine is made at present by the electrolysis of chlorides, either fused or in aqueous solution.

Large quantities of chlorine are used for the sterilization of water supplies and swimming pools. Chlorine, as such, or in the form of "chloride of lime," is used as a bleaching agent in laundries. It is employed in the manufacture of chlorates, perchlorates, and numerous other inorganic compounds, and has a variety of uses in the organic chemical industry.

Chlorine is stored and shipped either in steel cylinders or in tank cars. Steel cylinders should be capable of resisting a pressure of 22 atm. and should not contain more than 1 kg. of liquid per liter of capacity. Precautions in the handling

[86] C. R. Williams, *J. Ind. Hyg. Toxicol.*, **24**, 277 (1942).

[87] D. A. Greenwood, *Physiol. Revs.*, **20**, 582 (1940).

[88] J. C. Zufelt and L. A. Smith, *J. Am. Water Works Assoc.*, **42**, 839 (1950).

[89] H. H. Weber and W. E. Engelhardt, *Zentr. Gewerbehyg. Unfallverhüt.*, **10**, 41 (1933).

[90] R. P. White, *The Dermatergoses, or Occupational Affections of the Skin.* 4th ed.. Lewis, London, 1934.

of chlorine have been described by Eichenhofer.[91] The development of a $1/8$-in. hole in the bottom of a 100 lb. tank while being transported along the streets of Brooklyn occurred near a subway ventilator grating. As a result, 418 persons had to be examined and 208 were admitted into eight hospitals where some had to remain for one or two weeks.[92] No deaths occurred. An explosion of a tank car containing 15 tons of chlorine caused injury to 85 persons, 3 of whom died.[93]

2. Physical and Chemical Properties

Chlorine is a greenish-yellow gas with an irritating odor. Its molecular weight is 70.91. It is 2.49 times as heavy as air and can be condensed to a liquid, which boils at $-33.6°C$. The latter has a high coefficient of expansion, its volume increasing by 21.9 per cent when it is heated from $-35°C$. to 60°C. One liter of liquid, when vaporized, forms 463.8 liters of gas at 0°C. and 760 mm. pressure. The vapor pressure of liquid chlorine is 3.66 atm. at 0°C., 6.62 atm at 20°C., 8.75 atm. at 30°C., and 41.7 atm. at 100°C. A liter of water dissolves 2.260 liters of chlorine at 20°C. The gas is readily absorbed upon charcoal and is soluble in carbon tetrachloride, other halogenated hydrocarbons, and sulfuryl chloride. Chlorine reacts readily with metals and, by substitution or addition, with a wide variety of organic compounds.

1 mg./liter $\approx$ 344 p.p.m. and 1 p.p.m. $\approx$ 2.9 mg./cu. meter at 25°C., 760 mm. Hg.

3. Detection and Determination in the Atmosphere

When present in the air to the extent of 5 p.p.m., chlorine has but a slight odor. In one-tenth of that concentration, it is odorless, but can be detected by its ability to give a blue color to paper impregnated with starch and potassium iodide.[94]

Improvements in the sampling and determination of chlorine in air, using either the potassium iodide or the o-toluidine method, have been described by Wallach and McQuarry.[95] In the latter method, a measured volume of air is passed through a solution of o-toluidine in hydrochloric acid, and the resultant yellow color is compared with that of standard color solutions of potassium dichromate.[96] An instrument has been devised for automatically giving a warning signal when-

[91] H. J. Eichenhofer, *Power*, **97**, No. 6, 112, 113 (1953).

[92] H. Chasis, J. A. Zapp, J. H. Bannon, J. S. Whittenberger, J. Helm, J. J. Doheny, and C. M. MacLeod, *Occupational Med.*, **4**, 152 (1947).

[93] A. Römcke and O. K. Evensen, *Nord. Med.* **7**, 1224 (1940); *Chem Abstr.*, **35**, 1541 (1941).

[94] E. Smolczyk and H. Cobler, *Gasmaske*, **2**, 27 (1930), cited in F. Flury and F. Zernik, *Schädliche Gase*. Springer, Berlin, 1931, p. 121.

[95] A. Wallach and W. A. McQuary, *Am. Ind. Hyg. Assoc. Quart.*, **9**, 63 (1948).

[96] L. E. Porter, *Ind. Eng. Chem.*, **18**, 730 (1926). Am. Public Health Assoc., *Standard Methods of Water Analysis*, 10th ed., 1955.

ever the chlorine content of the air exceeds 0.005 vol.% (50 p.p.m.).[97] It operates by measuring the conductivity of water through which the air is passed.

4. Physiological Response

Acute Effects. Chlorine acts as a sterilizing and deodorizing agent by virtue of its ability to oxidize organic matter, and to this destructive action is probably due its irritant effect.

Brief exposure to air containing 1000 p.p.m. (3000 mg./cu. meter) kills even large animals.[98] The mechanism that induces almost immediate death is unknown, but if the animals survive for a longer period, death results from pulmonary edema. The exposure of cats to a concentration of 900 mg./cu. meter (300 p.p.m.) for 1 hour may cause death after a period during which the conjunctiva is inflamed and there is coughing and dyspnea. On the other hand, it has been asserted that dogs rarely die following a 30-minute exposure to a concentration not greater than 1900 mg./cu. meter (650 p.p.m.), and never after a corresponding period of exposure to a concentration less than 800 mg./cu. meter (280 p.p.m.).

The respiratory rate of animals is increasing during exposure to air containing from 200 to 1000 p.p.m.,[99] but when the concentration is 10,000 or more p.p.m., the inspirations occur more slowly and deeply and are finally arrested. Artificial respiration was ineffective in the case of animals exposed to concentrations greater than 50,000 p.p.m.[100] The pulse rate of dogs is retarded during exposure to concentrations of 180 to 200 p.p.m. or more.[101]

The sensitivity of men to chlorine varies greatly. Men may work without interruption in air containing 3 to 6 mg. of chlorine/cu. meter (1 or 2 p.p.m), according to Matt.[102] Exposures to low concentrations, 10 to 20, mg./cu. meter or 3 to 6 p.p.m.,[98] cause a stinging or burning sensation in the eyes, nose and throat, and sometimes headache due to irritation of the accessory nasal sinuses. There may be redness and watering of the eyes, sneezing, coughing, and huskiness or loss of the voice. Bleeding of the nose may occur and sputum from the pharynx and trachea may be blood-tinged. There is little or no chest pain other than the muscular soreness associated with excessive coughing. Exposure for $^1/_2$ to 1 hour to a concentration of 40 to 60 mg./cu. meter (14 to 21 p.p.m.) is dangerous, and the concentration of 290 mg./cu. meter (100 p.p.m.) cannot be borne for longer than 1 minute.[103]

[97] T. A. H. Jeffrey, *Power*, **87**, 514 (1943); *Chem. Abstracts*, **38**, 1812 (1943).

[98] F. Flury and F. Zernik, *Schädliche Gase*. Springer, Berlin, 1931, p. 118.

[99] J. A. Gunn, *Quart. J. Med.*, **13**, 121 (1920); *Chem. Abstr.*, **14**, 1148 (1920).

[100] E. Schäfer, *Brit. Med. J.*, **1915**, II, 245.

[101] H. G. Barbour, *J. Pharmacol. Exptl. Therap.*, **14**, 47, 65 (1919).

[102] L. Matt, Dissertation, Würzburg, 1881, cited in F. Flury and F. Zernik, *Schädliche Gase*. Springer, Berlin, 1931, p. 120.

[103] E. B. Vedder, *The Medical Aspects of Chemical Warfare*. Williams & Wilkins, Baltimore, 1925, p. 70.

Damage from exposures to low concentrations is limited to the nose and throat because of the solubility of chlorine in water. Greater concentrations damage the lungs and give rise to pulmonary congestion and edema, the seriousness of which is increased by bronchocongestion.[104] Many factors, which have been discussed in detail by Kehoe and Kitzmiller,[105] contribute to the development of a state of anoxia. The initial response to anoxia is an increase in the cardiac rate and output and in arterial and venous pressure. The victim becomes prostrated, and the skin may be warm and purplish in color. Damage to the intrapulmonary vascular beds, and the decreased volume and increased viscosity of the blood resulting from the loss of water to the lung, put a severe strain upon the heart. If failure of the heart is imminent, the pulse becomes rapid and thready and the skin, especially that of the extremities, becomes cold and clammy and assumes a grayish pallor.[105,106]

The significance of the absorption of chlorine, and of hydrogen chloride formed by the reaction of chlorine with water, is difficult to appraise. In dogs that inhaled air containing 800 p.p.m. chlorine for 2 to 7 hours, a rapidly increasing acidosis occurred.[107] Among the occasional sequellae found in men gassed by chlorine in warfare are chronic bronchitis and emphysema.[108]

Effects of Repeated Inhalation of Low Concentrations. Lehmann,[109] believed that repeated exposure of animals to chlorine may lead to the development of some tolerance. Men rapidly lose their ability to detect the odor of chlorine in small concentrations.[110] Although Ronzani[121] found no damage in animals repeatedly exposed to the concentration of 2 mg./cu. meter, Skljanskaya and Rapport[111] found more recently that the repeated exposure of rabbits to concentrations ranging from 2 to 5 mg./cu. meter (0.7 to 1.7 p.p.m.) over periods up to 9 months caused a loss of weight and an increased incidence of respiratory disease. According to Ronzani,[121] men exposed in bleaching rooms to concentrations of the order of 15 mg./cu. meter (5 p.p.m.) age prematurely, suffer from disease of the bronchi, and become predisposed to tuberculosis. The teeth were corroded by hydrochloric acid formed in the mouth, and inflammation or ulceration of the mucous membrane

[104] H. G. Barbour, *J. Pharmacol. Exptl. Therap.*, **14**, 47, 65 (1919).

[105] R. A. Kehoe and K. V. Kitzmiller, *Cincinnati J. Med.*, **23**, 423 (1942).

[106] O. Klotz, *J. Lab. Clin. Med.*, **2**, 889 (1917); W. H. Schulz and H. R. Hunt, *J. Pharmacol. Exptl. Therap.*, **11**, 179 (1918).

[107] A. M. Hjort and F. A. Taylor, *J. Pharmacol. Exptl. Therap.*, **13**, 407 (1919).

[108] H. S. Gilchrist and P. B. Matz, *Med. Bull. Veterans Admin.*, **9**, 229 (1933); J. C. Meakins and J. G. Priestley, *Can. Med. Assoc. J.*, **9**, 968 (1919); R. G. Pearce, *J. Lab. Clin. Med.*, **5**, 441 (1920).

[109] K. S. Lehmann, *Arch. Hyg.*, **34**, 272 (1899), cited in F. Flury and F. Zernik, *Schädliche Gase*. Springer, Berlin, 1931, p. 119.

[110] G. Lutz, *Zentr. Gewerbehyg. Unfallverhüt.*, **14**, 175 (1927), cited in F. Flury and F. Zernick, Schädliche Gase, Springer, Berlin, 1931, p. 119.

[111] R. M. Skljanskaja and J. L. Rappoport, *Arch. exptl. Pathol. u. Pharmakol.*, **117**, 276 (1935).

of the nose occurred. In guinea pigs the inhalation of small quantities of chlorine accelerates the course of experimental tuberculosis, according to Arloing, Berthet, and Vallier.[112] These experimental clinical observations tend to controvert the belief, frequently expressed several years ago, that the presence of chlorine in small quantities in the air of workrooms lessens the incidence of respiratory diseases among workmen presumably by sterilizing the air.[113]

Effects on the Skin. In high concentrations chlorine irritates the skin, causing sensations of burning or pricking, inflammation, or even blister formation. Workers in electrolytic chlorine plants may develop a form of acne, particularly evident around the ears and on the face. It resembles acne due to contact with chlorinated naphthalenes and has been attributed to the action of some compound formed by the reaction of chlorine with agents used in the composition of the electrodes.[114]

5. Threshold Limit

The American Conference of Governmental Industrial Hygienists has adopted as the threshold concentration of chlorine 1 p.p.m. or 3 mg./cu. meter. Detailed safety measures for the handling of liquid chlorine for water chlorination have been described by Hedgepeth and Riggs.[115]

HYDROGEN CHLORIDE, HCl

1. Properties, Source, Uses, and Industrial Exposures

Hydrogen chloride is a colorless gas with a molecular weight of 36.47 and a density of 1.6397 g./liter[116] at 0°C. and 760 mm. Hg (about 1.3 times as heavy as air). Because of its great solubility in water, HCl fumes in moist air. It may be prepared from sodium chloride by the action of sulfuric acid or sodium bisulfate. Aqueous solutions containing about 35 per cent hydrogen chloride have a wide variety of uses. The impure commercial product is called muriatic acid. When employed in industries in which there is little familiarity with the handling of chemicals, the use of hydrochloric acid gives rise to numerous minor accidents. Serious intoxication from inhalation is rare because its irritant nature is such as to prevent uninterrupted work in air containing the gas in dangerous concentrations.

[112] Arloing, Berthet, and Vallier, *Presse méd.,* **48,** 361 (1940; *Chem. Abstr.,* **34,** 5540 (1940).

[113] C. Baskerville, *Science,* **50,** 50 (1919); *J. Ind. Eng. Chem.,* **12,** 293 (1920), **13,** 568 (1921), **15,** 746 (1923).

[114] K. Herxheimer, *Münch. med. Wochschr.,* **46,** 278 (1899); G. Thibierge and P. Pagniez, *Ann. dermatol. syphilig.,* 4th Ser., **1,** 815 (1900); J. Nicholas and M. Pillon, *Bull. soc. franç. dermatol. syphilig.,* **32,** 33 (1925), cited in R. P. White, *The Dermatergoses, or Occupational Affections of the Skin.* 4th ed., Lewis, London, 1934, pp. 222, 223.

[115] L. L. Hedgepeth and W. S. Riggs, *J. Am. Water Works Assoc.,* **30,** 1671 (1938).

[116] R. W. Gray, *Proc. Chem. Soc.,* **23,** 119 (1907).

1 mg./liter $\backsim$ 670 p.p.m. and 1 p.p.m. $\backsim$ 1.47 mg./cu. meter at 25°C., 760 mm. Hg.

2. Determination in the Atmosphere

Samples of air may be absorbed in a known quantity of an alkaline solution and determined by titration of the remaining alkali or by titration of the chloride ion with silver nitrate. Apparatus for the continuous determination of hydrogen chloride in air has been devised.[117] In the absence of interfering acids, hydrogen chloride can be determined in the field by the method described under Sulfuric Acid.

3. Physiological Response

Effects on Animals. When inhaled in sufficiently high concentration, hydrogen chloride acts as an irritant to the respiratory tract. The inhalation of air containing 6400 mg./cu. meter for 30 minutes by rabbits and guinea pigs resulted in death,[118] in many instances from laryngeal spasm, laryngeal edema, or rapidly developing pulmonary edema. Lehmann[119] found that exposure of cats, rabbits, and guinea pigs for $1^1/_2$ hours to the concentration of 5000 mg./cu. meter (34000 p.p.m.) caused death after 2 to 6 days, while a slightly shorter exposure to the concentration of 2000 mg./cu. meter (1350 p.p.m.) caused severe irritation, dyspnea, and clouding of the cornea. When the duration of exposure was 2 to 6 hours, the concentration of 1000 mg./cu. meter (675 p.p.m.) caused some fatalities.[118] A 6 hour exposure to the concentration of 450 mg./cu meter. (300 p.p.m.) caused slight corrosion of the cornea and upper respiratory irritation. This response was but slight after a similar exposure to the concentration of 150 to 210 mg./cu. meter (100 to 140 p.p.m.).[119] The increase in respiratory rate associated with elevation of environmental temperature increases absorption and thus adds to the danger of exposures to low concentrations.[120]

An exposure of 6 hours daily to the concentration of 150 mg./cu. meter (100 p.p.m.), repeated on 50 days, caused only slight unrest and irritation of the eyes and nose of rabbits, guinea pigs, and pigeons.[121] The hemoglobin content of the blood was only slightly diminished. Twenty exposures, each of 6 hours, to the concentration of 50 mg./cu. meter (33 p.p.m.) caused no harm to a monkey or to smaller animals.[118] Repeated exposure to higher concentrations resulted in a loss of weight that paralleled the severity of the exposure.

[117] E. C. White, *J. Am. Chem. Soc.*, **50**, 2148 (1928).

[118] W. Machle, K. V. Kitzmiller, E. W. Scott, and J. F. Treon, *J. Ind. Hyg. Toxicol.*, **24**, 222 (1942).

[119] K. B. Lehman, *Arch. Hyg.*, **5**, 16 (1886), **67**, 57 (1908); K. B. Lehmann and A. Burch, *Arch. Hyg.*, **72**, 343 (1910), cited in F. Flury and F. Zernik, *Schädliche Gase*. Springer, Berlin, 1931, p. 126.

[120] Leitz, *Arch. Hyg.*, **102**, 91 (1929), cited in F. Flury and F. Zernik, *Schädliche Gase*. Springer, Berlin, 1931, p. 126.

[121] E. Ronzani, *Arch. Hyg.*, **70**, 217 (1909), cited in F. Flury and F. Zernik, *Schädliche Gase*. Springer, Berlin, 1931, p. 127.

When inhaled in high concentrations, the gas causes necrosis of the tracheal and bronchial epithelium as well as pulmonary edema, atelectasis, and emphysema, and damage to the pulmonary blood vessels. Detailed descriptions of the lesions in the liver and other organs have been given by Machle and co-workers.[118] The insufflation of weak aqueous solutions of hydrochloric acid into the bronchi of rabbits sets up inflammatory processes resembling those seen in influenza or after exposure to certain chemical warfare agents.[122]

The repeated oral administration of dilute hydrochloric acid to dogs induces acute and chronic gastritis and duodenitis, and leads to the appearance of ulcers of the pylorus,[123] while toxic doses lower the alkaline reserve of the blood.[124]

Effects on Man. According to Matt,[125] work is impossible when one inhales air containing hydrogen chloride in concentrations of 75 to 150 mg./cu. meter (50 to 100 p.p.m.); work is difficult but possible when the air contains concentrations of 15 to 75 mg./cu. meter (10 to 50 p.p.m.); and work is undisturbed at the concentration of 15 mg./cu. meter (10 p.p.m.). Prolonged exposure to low concentrations causes erosion of the teeth. Mists of heated metal-pickling solutions may cause bleeding of the nose and gums, ulceration of the nasal and oral mucosa, and render the skin of the face so tender that shaving becomes painful.[126] Exposure of the skin to gaseous hydrogen chloride, escaping from leaks in apparatus or piping, has caused severe burns. Contact with concentrated solutions of hydrogen chloride (muriatic acid) in cleaning metal gives rise to small burns and ulcerations of the hands.[127]

4. Threshold Limit

The American Conference of Governmental Industrial Hygienists has adopted 5 p.p.m. or 7 mg. per cubic meter as the threshold concentration. According to Hirt,[128] a concentration seven times as great as this cannot be detected by taste or odor. (Editor's Note: Most people can detect 1–5 p.p.m.; 5–10 p.p.m. is disagreeable.)

5. Odor and Warning Properties

Sharp odor and an acid taste.

[122] M. C. Winternitz, G. H. Smith, and F. P. McNamara, *J. Exptl. Med.*, **32**, 199, 205 (1920).

[123] W. J. Gallagher, *A.M.A. Arch. Surg.*, **17**, 613 (1928).

[124] A. Loewy and F. Munzer, *Biochem. Z.*, **134**, 437 (1923).

[125] L. Matt, Dissertation, Würzburg, 1889, cited in F. Flury and F. Zernik, *Schädliche Gase.* Springer, Berlin, 1931, p. 128.

[126] W. Ludewig, *Arch. Gewerbepathol. Gewerbehyg.*, **11**, 296 (1942); L. Carozzi, *Occupation and Health.*

[127] R. P. White, *The Dermatergoses, or Occupational Affections of the Skin.* 4th ed., Lewis, London, 1934.

[128] Hirt, cited in F. Flury and F. Zernik, *Schädliche Gase.* Springer, Berlin, 1931, p. 127.

BLEACHING POWDER (Chloride of Lime) AND HYPOCHLORITES

Bleaching powder is made by passing chlorine over finely divided slaked lime in closed chambers. It is a hygroscopic white powder with an odor of chlorine. Its exact chemical nature is controversial, some chemists regarding it as a double calcium salt of hydrochloric and hypochlorous acids, $CaCl(ClO)$, and others as a mixture of calcium chloride, $CaCl_2$, and calcium hypochlorite, $Ca(ClO)_2$. It evolves chlorine, especially in moist air. The evolution of oxygen under the catalytic influence of impurities such as iron and manganese has led to the building up of dangerous pressures in closed containers, especially when heated.[129] This was more common when chlorine was made by the Weldon process. Bleaching powder is explosive when heated suddenly above 100°C. and deflagration occurs when it is mixed with combustible substances.[130]

Bleaching powder is used as a disinfecting, oxidizing, and chlorinating agent. As a bleaching agent, it is employed in the making of paper and textiles. The inhalation of dusts may damage the teeth, conjunctiva, and respiratory tract. It has been much employed by dyers to remove stains from the hands. If used as a paste with excessive rubbing for more than 3 to 4 minutes, it sometimes causes a moderate to severe degree of damage to the skin.[131]

Javel water, a solution of sodium hypochlorite, used for the removal of stains from textiles, has been reported as a cause of dermatitis among laundresses.[132] Hypochlorites have sensitizing properties.[133]

BROMINE, Br_2

1. Source, Properties, and Uses

Although formerly prepared by the action of chlorine upon bromides present in the mother liquors in salt manufacture, bromine is now obtained in large quantities from sea water.[134] It is a deep reddish brown liquid, with molecular weight 159.83 and a density of 3.12 at 15°C. It solidifies at −7.30°C. and boils at 58.8°C. Its vapor is about 5.5 times as heavy as air. The vapor pressure is equivalent to 77.3 mm. Hg at 4°C., 172 mm. at 20.6°C., and 378 mm. at 30.6°C. Bromine is soluble in water to the extent of 3.21 parts per 100 parts at 20°C. It is very soluble

[129] A. H. Gill, *Ind. Eng. Chem.*, **16**, 577 (1924).

[130] J. Weichherz, *Chemiker Ztg.*, **52**, 729 (1928).

[131] J. Lebduvska, J. Pidra, and F. Pokorny, *Arch. exptl. Pathol. u. Pharmakol.*, **193**, 629 (1939).

[132] H. Rabeau and Mlle Ukrainczyk, *Ann. dermatol. syphilig.*, **10**, 656 (1939); *Chem. Abstr.*, **33**, 8773 (1939).

[133] P. Ravaut and Koang, *Bull. soc. franç. dermatol. syphilig.*, **37**, 655 (1930), cited in R. P. White, *The Dermatergoses or Occupational Affections of the Skin.* 4th ed., Lewis, London, 1934, p. 243.

[134] C. M. A. Stine, *Ind. Eng. Chem.*, **21**, 434 (1929); L. C. Stewart, *Ind. Eng. Chem.*, **26**, 361 (1934).

in alcohol, ether, chloroform, and carbon disulfide. Liquid bromine, or its aqueous solution, is widely used as an oxidizing agent in chemical laboratories. It causes burns when spilled upon the skin. Large quantities are used in brominating hydrocarbons, and in the manufacture of fumigants, dyestuffs, drugs, and war gases. Bromides are widely used in photography.

1 mg./liter $\backsim$ 152 p.p.m. and 1 p.p.m. $\backsim$ 6.53 mg./cu. meter at 25°C., 760 mm. Hg.

2. Determination in the Atmosphere

The bromine content of air may be determined by passing a measured volume through a solution of potassium iodide, and titrating the liberated iodine with a standard solution of sodium thiosulfate.[135] It also may be absorbed in alkali, oxidized to bromate by sodium hypochlorite, and estimated iodometrically.[136] After absorption in alkali, bromine may also be liberated by chlorine water and determined colorimetrically, either in aqueous solution or after extraction with carbon tetrachloride.[137]

3. Physiological Response

Toxicity for Animals. Like chlorine, bromine exerts a strong irritant action upon the respiratory tract, the concentration of 3500 mg./cu. meter of air (about 550 p.p.m.) being immediately fatal to animals.[138] An exposure of 7 hours to air containing 150 mg./cu. meter (23 p.p.m.) provoked only irritation of the respiratory tract and slight dyspnea in cats, rabbits, and guinea pigs, while a similar exposure to air containing 1200 mg. of bromine/cu. meter (180 p.p.m.) caused clouding of the cornea (rabbits and guinea pigs), severe irritation of the respiratory tract, dyspnea, reduction of the respiratory rate, and some fatalities (rabbits). An exposure of 3 hours to air containing 2000 mg. of bromine/cu. meter (300 p.p.m.) caused, in addition, definite disturbances of function of the central nervous system. Guinea pigs died during the exposure and rabbits died a few hours later. The hair of rabbits became yellow and brittle. Necropsy revealed the presence of edema of the lungs, a pseudomembranous deposit on the trachea and bronchi, and hemorrhages of the gastric mucosa. Foci of bronchopneumonia were found in animals that died several days after exposure.

The effect of repeated exposure to air containing bromine in low concentrations has not been adequately investigated. Henderson and Haggard[139] concluded that the depressant action of the bromine absorbed and converted to the bromide

---

[135] F. H. Goldman and J. M. DallaValle, *Public Health Repts. U. S.,* **54,** 1728 (1930).

[136] I. M. Kolthoff and H. Yutzy, *Ind. Eng. Chem. Anal. Ed.,* **9,** 75 (1937).

[137] M. Lane, *Ind. Eng. Chem. Anal. Ed.,* **14,** 149 (1942).

[138] K. B. Lehmann and R. Hess, *Arch. Hyg.,* **7,** 335 (1887), cited in F. Flury and F. Zernik, *Schädliche Gase.* Springer, Berlin, 1931, p. 122.

[139] Y. Henderson and H. Haggard, *Noxious Gases.* 2nd ed., Reinhold, New York, 1944, p. 133.

ion was insignificant, even under the conditions of prolonged exposure to air containing bromine vapor. Flury[140] believes that bromine can be absorbed through the skin during exposure to the vapor.

Effects on Man. The symptoms arising in man following the inhalation of bromine in small amounts include: coughing, nosebleed, a feeling of oppression, dizziness, and headache, followed after some hours by abdominal pain and diarrhea, and sometimes by a measleslike eruption on the trunk and extremities. Oppenheim[141] mentioned the frequency with which discharging pustules and furuncles appear in exposed areas of the skin of men who handle bromine. Brief contact of the liquid with the skin leads to the formation of vesicles and pustules. If not removed at once, it induces deep, painful ulcers.

4. Safe Concentrations of Bromine Vapor in Air

Undisturbed work is possible when the respired air contains 1 to 2 mg. bromine/cu. meter (0.15 to 0.3 p.p.m.); work becomes difficult in the presence of 2 to 3 mg./cu. meter (0.3 to 0.45 p.p.m.) and impossible at 4 mg./cu. meter (0.6 p.p.m.), according to Matt.[142] Zederbaum[143] found no serious effects resulted from an exposure to 26 mg/cu. meter (about 4 p.p.m.); Henderson and Haggard[139] considered this the maximum concentration allowable for an exposure of less than 1 hour, and believed that even brief exposures to air containing 40 to 60 p.p.m. would be dangerous. The threshold concentration listed in 1961 by the American Conference of Governmental Industrial Hygienists is 0.1 p.p.m. or 0.7 mg./cu. meter. The least concentration that gives a detectable odor is of the order of 3.5 p.p.m., according to Fieldner, Katz, and Kinney.[144]

IODINE, I_2, AND IODIDES

1. Properties, Source, and Uses

Iodine, a crystalline, blackish violet element with a metallic luster and a characteristic odor, has a molecular weight of 253.84. Iodine melts at 114°C. and boils at 184°C. It sublimes readily, giving off a violet vapor that colors the skin. Its vapor pressure is equivalent to 0.03 mm. Hg at 0°C., 0.131 mm. at 15°C., 0.309 mm. at 25°C., 0.467 mm. at 30°C., 0.699 mm. at 35°C., 1.027 mm. at 40°C., and 2.144 mm. at 50°C.[145] It is slightly soluble in water, easily soluble in chloroform,

[140] F. Flury, *Festsch, Zangger*, **2**, 836 (1935); *Chem. Abstr.*, **31**, 6711 (1937).
[141] M. Oppenheim, *Wien. klin. Wochschr.*, **28**, 1273, cited in R. P. White, *The Dermatergoses, or Occupational Affections of the Skin.* 4th ed., Lewis, London, 1934, p. 178.
[142] L. Matt, Dissertation, Würzburg, 1889, cited in F. Flury and F. Zernik, *Schädliche Gase.* Springer, Berlin, 1931, p. 123.
[143] D. Zederbaum, *Gigiena Truda i Tekh Bezopasnosti*, 1927, p. 68, cited in F. Flury and F. Zernik, *Schädliche Gase.* Springer, Berlin, 1931, p. 123.
[144] A. C. Fieldner, S. H. Katz, and S. P. Kinney, *U. S. Bur. Mines Tech. Paper,* No. **248** (1921).
[145] L. J. Gillespie and L. H. D. Fraser, *J. Am. Chem. Soc.*, **58**, 2260 (1936).

carbon disulfide, or alcohol. Formerly obtained from the ash of seaweed, it is now obtained chiefly from the mother liquors remaining in the preparation of Chile saltpeter or from natural-occurring brines in the United States. It is liberated from its salts by the action of oxidizing agents and purified by sublimation. Iodine and its compounds are used in analytical chemistry, medicine, photography, and in the making of dyestuffs and numerous organic compounds.

1 mg./liter $\approx$ 96.5 p.p.m. and 1 p.p.m. $\approx$ 10.38 mg./meter at 25°C., 760 mm. Hg.

2. *Determination in the Atmosphere*

Fritted-glass scrubbers have been recommended for the absorption of iodine from air samples.[146] Standard volumetric procedures are available for its determination by titration with sodium thiosulfate, using starch-iodide indicator.

3. *Physiological Response*

Iodine is an essential element in nutrition, being required by the thyroid for the elaboration of its hormone, thyroxin. Because of its use in medicine in the treatment of hypothyroidism and other diseases, and because of the addition of iodides to salt for the prevention of endemic goiter, the pharmacology and metabolism of iodine have been the subjects of many investigations, which are described in texts on pharmacology and in reviews.[147] Industrial poisoning by iodine is uncommon.

Iodine vapor acts as an irritant when inhaled and is capable of causing a rapidly developing pulmonary edema.[148] Systemic poisoning may result from its absorption through various portions of the respiratory tract.

Lachrymation and a burning sensation in the eyes, blepharitis, rhinitis, catarrhal stomatitis, and chronic pharyngitis have been noted following industrial exposure.

In an experiment upon himself, Matt[149] found that work is undisturbed when the concentration of iodine vapor in the air is 0.1 p.p.m., difficult when it is 0.15 to 0.2 p.p.m., and not possible when it is 0.3 p.p.m. In one recent case of fatal poisoning from the ingestion of iodine,[150] nitrogen retention and anuria occurred, but no significant renal lesions were found at necropsy.

Experiments have not been performed to determine the effects upon animals of the repeated inhalation of iodine.

Chronic intoxication resulting from the ingestion of excessive amounts of

[146] *Am. Public Health Assoc. Year Book*, 92 (1939–40).

[147] R. E. Remington, *J. Chem. Educ.*, **7**, 2590 (1930) ; T. von Fellenberg, *Ergeb. Physiol.*, **25**, 176 (1926).

[148] A. B. Luckhardt, F. C. Koch, W. F. Schroeder, and A. H. Weiland, *J. Pharmacol., Exptl. Therap.* **15**, 1 (1920).

[149] L. Matt, Dissertation, Würzburg, 1889, cited in F. Flury and F. Zernik, *Schädliche Gase.* Springer, Berlin, 1931, p. 124.

[150] R. Finkelstein and M. Jacobi, *Ann. Internal Med.*, **10**, 1283 (1937).

iodides is characterized by signs of irritation and nervousness. Determination of the basal metabolic rate of workers exposed to iodine does not appear to have been made. Von Fellenberg[151] found that the average daily urinary excretion of iodine by 5 normal persons was 173 μg. Administered iodine is rapidly excreted in the urine and in smaller quantities in saliva, milk, sweat, bile, and other secretions. The storage of iodine in the thyroid depends upon the functional state of the gland.

Although iodine is an irritant to the skin, the repeated application of its aqueous or alcoholic solution upon the skin of white mice during a period of 18 months did not induce hyperkeratotic changes.[152] A vesicular or impetiginous eruption of the face, with acne and folliculitis, occurring among photographers, physicians, and nurses, has been ascribed to iodine or iodides.

4. Threshold Limit

The American Conference of Governmental Industrial Hygienists in 1961 adopted 0.1 p.p.m. or 1 mg./cu. meter as the threshold concentration for iodine.

HYDROGEN IODIDE, HI (Hydriodic Acid)

Hydrogen iodide is a colorless gas with molecular weight of 127.93. It is 4.4 times as heavy as air. Hydrogen iodide melts at $-50.8°$C. and boils at $-35.5°$C. It is soluble in alcohol and very soluble in water. It has a limited use in industry and may cause injury by virtue of its acid nature (see Hydrogen Chloride). Quantitative data are not available.

References to Recent Literature

E. J. King, M. Yoganathan, and G. Nagelschmidt, Tissue reaction produced by calcium fluoride in the lungs of rats, *Brit. J. Ind. Med.*, **15**, 168, 1958; abstracted in *A.M.A. Arch. Ind. Health*, **20**, 335, 1959.

Fluoride control at Geneva Works, *Intermountain Indust. and Mining Rev.*, 22, 1957.

E. A. Pfitzer, P. P. Yevich, E. A. Green, and K. H. Jacobson, Acute toxicity of red fuming nitric acid–hydrofluoric acid mixture, *A.M.A. Arch. Ind. Health*, **18**, 218, 1958.

T. Dalhamn, Chlorine dioxide, *A.M.A. Arch. Ind. Health*, **15**, 101, 1957.

H. J. Horn and R. J. Weir, Inhalation toxicology of chlorine trifluoride, *A.M.A. Arch. Ind. Health*, **12**, 515, 1955.

E. J. Largent and K. W. Largent, Hygienic aspects of fluorine and its compounds, *Am. J. Public Health*, **45**, 197, 1955.

Methods for the detection of toxic gases in industry. Chlorine, Dept. of Sci. and Ind. Res., Leaflet 10, H. M. Stationary Off. London, 1955, *A.M.A. Arch. Ind. Health*, **13**, 516, 1956.

J. V. Klauder, L. Schelansky, and K. Gabriel, Industrial uses of compounds of fluorine and oxalic acid. Cutaneous reaction and calcium therapy, *A.M.A. Arch. Ind. Health*, **12**, 412, 1955.

F. A. Smith, D. E. Gardner, and H. C. Hodge, Investigations on the metabolism of fluoride, *A.M.A. Arch. Ind. Health*, **11**, 2, 1955.

[151] T. von Fellenberg, *Biochem. Z.*, **184**, 85 (1927).

[152] J. Rosenstirn, *J. Cancer Research*, **10**, 61 (1926); K. Fritzler, *Arch. exptl. Pathol. u. Pharmakol.*, **114**, 6 (1926).

J. P. Nielson and A. D. Dangerfield, Use of ion exchange resins for determination of atmospheric fluorides, *A.M.A. Arch. Ind. Health,* **11,** 61, 1955.

J. J. Davenport and G. G. Morgis, Review of literature on health hazards of fluorine and its compounds in the mining and allied industries, *Inform. Circ. 7687, U. S. Bur. Mines,* June, 1954; abstracted in *A.M.A. Arch. Ind. Health,* **11,** 442, 1955.

K. T. Semrau, Emission of fluorides from industrial processes. A review, *J. Air Pollution Control Assoc.,* **7,** 92, 1957; abstracted in *A.M.A. Arch. Ind. Health,* **17,** 308, 1958.

J. Gloemme and K. D. Lungren, Health hazards from chlorine dioxide, *A.M.A. Arch. Ind. Health,* **16,** 169, 1957.

CHAPTER XXIII

Alkaline Materials

FRANK A. PATTY

The caustic alkalies, in solid form or concentrated liquid solution, have a more corrosive local action on the tissues than do most acids. The free caustic dusts, mists, and sprays may cause irritation of the eyes and respiratory tract and erosion of the nasal septum. Strong alkalies combine with tissue to form albuminates, and with natural fats to form soaps. They gelatinize tissue to form soluble compounds and by doing so may produce deep and painful destruction of tissue. Potassium and sodium hydroxides and oxides are the most active, while less active members of the group are calcium oxide and hydroxide, ammonia gas and ammonium hydroxide, sodium and potassium carbonates, trisodium phosphate, and sodium metasilicate. Even dilute solutions of the stronger alkalies tend to soften the epidermis and emulsify or dissolve the skin fats. Soaps readily dissociate and, when unbuffered, may act as free alkalies to irritate the skin.

When first encountered, atmospheres slightly contaminated with alkalies may be quite irritant, but the effect soon becomes less noticeable. Workmen frequently are found working unconcernedly in an atmosphere that causes coughing and painful throat and nasal irritation in a person unaccustomed to the exposure.

By far the greatest hazard of working with alkaline materials is from the splash or splatter of particles or solutions of the stronger alkalies entering the eyes of workmen. This can be prevented by the use of eye protection that is effective at all angles. Proper provisions should also be available for immediate and prolonged washing with water should such eye contamination occur.

AMMONIA, NH_3, AND AMMONIUM HYDROXIDE, NH_4OH

1. *Source, Uses, and Industrial Exposures*

Ammonia is produced as a by-product in the distillation of coal, by the action of steam on cyanamide, and by the catalytic combination of nitrogen and hydrogen gases at high temperature and pressure. Ammonia is used extensively in refrigeration, in petroleum refining, and in the manufacture of fertilizers, nitric acid, explosives, dyes, plastics, and other chemicals. It is also encountered in the silvering of mirrors, in gluemaking, in tannery work, and around nitriding furnaces. Ammonia also results from the decomposition of nitrogenous materials.

2. Physical and Chemical Properties

Physical state: colorless gas
Molecular weight: 17.03
Melting point: $-77.7°C.$
Boiling point: $-33.35°C.$
Vapor density: 0.59 (air = 1) at 25°C., 760 mm. Hg
Solubility: 90 g. in 100 ml. water at 0°C., 13.2 g. in 100 ml. alcohol at 20°C.
Alkalinity: the pH of a 1 per cent solution in water is about 11.7

The gas is compressed to a liquid and stored or transported in steel cylinders or tank cars. A 28 per cent solution of ammonia in water (sp.g. 0.90 at 25/25°C.)—called ammonium hydroxide or stronger ammonia water—is the common form in which ammonia is supplied and used. It is also available in 10 per cent solution (household ammonia). Ammonium carbonate and ammonium carbamate are crystalline solids containing 34 and 44 per cent ammonia, respectively. They decompose in the air to release ammonia. Compressed ammonia gas is obviously the greatest inhalation hazard because high atmospheric concentrations can be developed more rapidly by compressed gases, if released, than by vaporization from an aqueous solution or by gradual decomposition of the salts exposed to the air.

1 mg./liter $\approx$ 1438 p.p.m. and 1 p.p.m. $\approx$ 0.7 mg./cu. meter at 25°C., 760 mm. Hg.

3. Determination in the Atmosphere

In the method of determination most applicable to field use in low concentrations of ammonia and most other alkalies, the atmosphere is aspirated at a measured rate through a scrubber containing 0.01N sulfuric acid and methyl red or other suitable indicator (see Vol. I, page 183). The amount of standard acid that is to be used will depend upon the concentration of ammonia anticipated, 1 or 2 ml. made up to 10 ml. with water being sufficient for concentrations on the order of 100 p.p.m. or less, if the rate of sampling is on the order of 2 liters per minute. Since 1 ml. 0.01N sulfuric acid is equivalent to 0.17 mg. or 0.2445 ml. of ammonia gas at 25°C. and 760 mm. Hg pressure, results may be computed as follows:

$$C = (\text{ml. } 0.01N \text{ } H_2SO_4 \times 0.2445 \times 1000)/(\text{rate of sampling} \times \text{time})$$

where C = concentration of ammonia in the atmosphere expressed in parts per million, rate of sampling is expressed in liters per minute, time is the minutes of sampling required to change the color of the indicator in the scrubber, and 1000 is a factor that converts the milliliters of ammonia per liter of air (parts per thousand) to parts per million. When 1 ml. of 0.01N sulfuric acid is used and the sampling rate is 2.83 liters per minute (as with the use of a midget impinger) the equation becomes:

$$C = 86/\text{sampling time}$$

With reasonable care this method yields results that are within 5 per cent of the amount of ammonia present. If fritted scrubbers are used, an antifoaming agent may be required.

If preferred, the sample may be collected by scrubbing through an excess of standard sulfuric acid or 4 per cent boric acid solution and titrated,[1] or be collected in 0.1N sulfuric acid and nesslerized.

4. Physiological Response

Response in Animals. Boyd, MacLachlan, and Perry[2] exposed rabbits and cats to static atmospheres with initial concentrations on the order of 5000 to 10,000 p.p.m. ammonia, for 1 hour, and found this exposure to be approximately the LC_{50}. They found that the noses, mouths, and throats of the animals were severely affected, but that because the ammonia was absorbed upon these mucous surfaces the tracheae and bronchi were partially protected. Weatherby[3] found little evidence of chronic effects from prolonged daily exposure of animals to concentrations below those causing acute effects.

Pathology in Animals. In acute exposures the epithelial lining of the less resistant bronchioles was damaged; congestion, edema, atelectasis, hemorrhage, and emphysema were found in the alveoli; and there was an increase in respiratory-tract fluid.[2]

Absorption, Excretion, and Metabolism. Although the alkaline properties of ammonia might be expected to upset the normal pH of the human system after prolonged exposure to low concentrations, no data have yet been presented to show that such is the case. On the contrary, according to Sollmann,[4] ammonia in the body is rapidly converted to urea and ceases to act as ammonia. Concentrations below the amount that causes irritation are not known to have any adverse effect regardless of the length of exposure.

Part of any ammonia reaching the alveoli is neutralized by the carbon dioxide normally present, and part may be absorbed unchanged into the circulation. After exposure, traces of ammonia have been found in the sweat, urine, and exhaled air. Since ammonia is a normal constituent of blood and urine, analysis of these fluids does not offer satisfactory tests to indicate the degree of exposure.

Mode of Action. The irritation to mucous membranes becomes noticeable at about 100 p.p.m. Concentrations above 400 p.p.m. may destroy mucous surfaces upon prolonged contact by dissolving or emulsifying keratin, fat, and cholesterol. Ammonia may cause sensitization (see Vol. I, page 449).

Cause of Death. The most frequent cause of death in man from exposure to ammonia is pulmonary edema.[5]

[1] M. B. Jacobs, *The Analytical Chemistry of Industrial Poisons, Hazards, and Solvents.* 2nd ed., Interscience Publishers, New York, 1949.

[2] E. M. Boyd, M. L. MacLachlan, and W. F. Perry, *J. Ind. Hyg. Toxicol.*, **26**, 29 (1944).

[3] J. H. Weatherby, *Proc. Soc. Exptl. Biol. Med.*, **81**, 300 (1952).

[4] T. Sollmann, *A Manual of Pharmacology.* 6th ed., Saunders, Philadelphia, 1944.

[5] M. Coplin, *Lancet*, **241**, 95 (1941).

Effects Observed in Man. The effects of various concentrations of ammonia are given in Table 1.[6]

TABLE 1

Physiological Response to Ammonia[6]

Response	Concentration, p.p.m.
Maximum concentration for prolonged exposure	100
Maximum amount for 1 hr.	300–500
Least amount causing immediate irritation of eyes, nose, and throat	400–700
Dangerous for as little as ¹/₂ hr.	2,500–6,500
Rapidly fatal for short exposures	5,000–10,000

Silverman, Whittenberger, and Muller[7] found no significant changes in nitrogen metabolism due to exposure to 500 p.p.m. but reported such concentration to be physiologically undesirable.

High concentrations of ammonia, in addition to their corrosive action on mucous surfaces, which can cause permanent injury to the cornea, extensive damage to the throat and the upper respiratory tract, chronic bronchial catarrh, and edema, may affect heart action or cause cessation of respiration by reflex action.

During the approval testing of respiratory protective devices, the author has observed that atmospheres of 1 per cent ammonia are mildly irritant to the moist skin, those of 2 per cent have a more pronounced action, and concentrations of 3 per cent or greater cause a stinging sensation and may produce chemical burns with blistering after a few minutes exposure.

5. Hygienic Standard of Permissible Exposure

The maximum permissible concentration is accepted to be 100 p.p.m. (70 mg./cu. meter). A standard based upon comfort would be somewhat less than 100 p.p.m.

6. Flammability

The flammable range of ammonia is 15.5 to 26.6 per cent by volume in air (see Vol. I, Chapter XVI).

Odor and Warning Properties

Ammonia is detectable by odor in concentrations of less than 5 p.p.m. A concentration of 20 p.p.m. is easily noticeable and 100 p.p.m. has a moderately strong odor and is moderately irritant to the nose.

[6] Y. Henderson and H. W. Haggard, *Noxious Gases,* Reinhold, New York, 1943.

[7] L. Silverman, J. L. Whittenberger, and J. Muller, *J. Ind. Hyg. Toxicol.,* **31,** 74 (1949).

CALCIUM HYDROXIDE, Ca(OH)$_2$ (Slacked Lime, Hydrated Lime)

1. Source, Uses, and Industrial Exposures

The source and uses of calcium hydroxide are similar to those of calcium oxide, from which it is formed by the addition of water.

2. Physical and Chemical Properties

Physical state: colorless to white crystalline solid that changes to carbonate (CaCO$_3$) upon exposure to the air

Molecular weight: 74.10

Specific gravity: 2.34

Solubility: 1 g. in 590 ml. of water at 25°C. and in 1300 ml. at 100°C.

Alkalinity: the pH of a saturated solution in water at 25°C. is about 12.8

3. Physiological Response

Calcium hydroxide is a moderately caustic irritant.

4. Hygienic Standard of Permissible Exposure

No permissible level of air contamination has been published.

CALCIUM OXIDE, CaO (Lime, Burnt Lime, Quicklime)

1. Source, Uses, and Industrial Exposures

Lime is produced by burning limestone in kilns. It combines with water, with evolution of considerable heat, to form calcium hydroxide. Calcium oxide is used in making mortar, plaster, and chlorinated lime, in dehairing hides, in deodorizing vegetable oils, as a dehydrating agent, and in many other ways. Upon exposure to air it absorbs water and carbon dioxide to form calcium carbonate, becoming air-slaked and losing most of its causticity.

2. Physical and Chemical Properties

Physical state: white lumps

Molecular weight: 56.08

Specific gravity: 3.35 (20°C.)

Melting point: 2580°C.

Boiling point: 2850°C.

Solubility: 1 g. in 835 ml. of water at 25°C. and in 1670 ml. at 100°C.

Alkalinity: the pH of a saturated solution in water is about 12.5

3. Physiological Response

Lime, a moderately caustic irritant, is reported to have caused chemical pneumonia as a result of dust inhalation, but the severe irritation of the upper respiratory tract ordinarily causes persons to avoid serious inhalation exposure.

4. *Hygienic Standard of Permissible Exposure*

No permissible dustiness has been proposed.

NaK ("Nack")

NaK, an alloy of sodium and potassium in any proportions, has been used as a heat-transfer fluid. See Sodium; Potassium.

POTASSIUM, K

1. *Source, Uses, and Industrial Exposures*

Potassium salts are obtained from waste liquor from molasses treating and other molasses demineralizing processes, from feldspars, from salt deposits and naturally occurring brines. Metallic potassium is made by the thermal reduction of potassium chloride with sodium. One of the few commercial uses of potassium metal is in regenerative oxygen-breathing apparatus in the form of the superoxide (KO_2) which is produced from the metal by oxidation. Metallic potassium is similar to, but more reactive than, sodium. It must be used, handled, and stored with great care.

2. *Physical and Chemical Properties*

Physical state: soft, silvery, ductile metal
Molecular weight: 39.10
Specific gravity: 0.86 (20°C.)
Melting point: 62.3°C.
Boiling point: 760°C.
Vapor pressure: 8 mm. Hg, 432°C.
Potassium reacts with most gases, liquids, and solids. It is inert in argon, helium, and nitrogen.

POTASSIUM HYDROXIDE, KOH (Caustic Potash, Potassium Hydrate)

1. *Source, Uses, and Industrial Exposures*

Potassium hydroxide is manufactured by the electrolysis of potassium chloride solution. It is used in the manufacture of some other potassium compounds, including the carbonate, and in soaps. The carbonate is used extensively in glass manufacture.

2. *Physical and Chemical Properties*

Physical state: white deliquescent pellets, sticks, or cake
Molecular weight: 56.10
Specific gravity: 2.044 (20°C.)
Melting point: 360°C.

Boiling point: 1320°C.

Vapor pressure: 1 mm. Hg, 719°C.

Solubility: 100 g. in 90 ml. of water at 25°C.; 100 g. in 375 ml. of alcohol at 25°C.; 100 g. in 200 ml. of glycerine at 25°C.

Alkalinity: the pH of a 1 per cent solution in water is about 13

3. Determination in the Atmosphere

Similar to sodium hydroxide.

4. Physiological Response

Similar to that of sodium hydroxide.

5. Hygienic Standard of Permissible Exposure

Because of the similarity to sodium hydroxide, 2 mg./cu. meter is suggested.

6. Flammability

Potassium hydroxide is not flammable.

SODIUM, Na

1. Source, Uses, and Industrial Exposures

Sodium is manufactured in the United States by the electrolysis (Downs cell) of a mixture of molten sodium chloride and calcium chloride. It is available in bricks weighing 1 to 24 lb. as well as in drums and tank cars. The principal use of sodium is as an intermediate in the manufacture of tetraethyl-lead. It is used as a heat-transfer medium in nuclear power plants, in many chemical reactions, and in sodium–sodium hydride metal descaling baths. The handling and use of sodium require close observance of rules for safe handling, scrap disposal, and fire fighting. These are available from many sources, one of which is the A.E.C. Liquid Metals Handbook.[8]

2. Physical and Chemical Properties

Physical state: light ductile silvery metal

Molecular weight: 23.00

Specific gravity: 0.9684 (20°C.)

Melting point: 97.8°C.

Boiling point: 886°C.

Vapor pressure: 1 mm. Hg, 432°C.

Solubility: reacts violently with water and most common solvents

Fire hazard: the metal or its vapor may ignite spontaneously in air at

[8] C. B. Jackson, *Liquid Metals Handbook, Sodium-NaK Supplement*. Atomic Energy Commission, July, 1955.

temperatures above 115°C.; the resulting smoke, chiefly sodium oxide, is a serious respiratory hazard.

3. Physiological Response

Sodium fume or smoke resulting from the escape of hot sodium vapor to the atmosphere with or without combustion may contain micron and even submicron particles of sodium, sodium oxide, and sodium hydroxide, all of which may penetrate the respiratory tract. The result is a corrosive action and extreme irritation largely in the nose, throat, and upper passages. The irritant action serves as an effective warning.

4. Hygienic Standard of Permissible Exposure

Permissible limits of exposure have not been determined.

SODIUM CARBONATE, Na_2CO_3 (Soda Ash)

Crystallized with 10 moles of water ($Na_2CO_3 \cdot 10H_2O$), it is called decahydrate, sal soda, and washing soda.

1. Source, Uses, and Industrial Exposures

Sodium carbonate is manufactured by several processes, chief of which is the Solvay process in which carbon dioxide is passed through a strongly ammoniacal solution of common salt thereby precipitating sodium bicarbonate, which is then converted to the carbonate by calcining. It is used in the manufacture of sodium compounds including glass and soap, in washing and cleaning, in bleaching cotton and linen fabrics, in washing wool, in softening water, in dyeing and dye manufacture, in photography, and as a reagent chemical.

2. Physical and Chemical Properties

Physical state: white hygroscopic powder
Molecular weight: 106.0
Specific gravity: 2.53 (20°C.)
Melting point: 851°C.
Boiling point: decomposes
Solubility: 7.1 g. in 100 ml. of water at 0°C. and 45.5 g. in 100 ml. at 100°C.
Alkalinity: the pH of a 1 per cent solution in water is about 11.5

3. Determination in the Atmosphere

Similar to method for sodium hydroxide.

4. Physiological Response

Sodium carbonate is a primary skin irritant (Vol. I, page 449), and dusts or mists are moderately irritating to the nasal membranes.

5. Hygienic Standard of Permissible Exposure

None has been proposed, but it should be higher than that permissible for sodium hydroxide.

6. Flammability

Sodium carbonate is not flammable.

SODIUM HYDROXIDE, NaOH (Caustic Soda, Caustic Flake, Lye, Liquid Caustic)

1. Source, Uses, and Industrial Exposures

Sodium hydroxide is made chiefly from the electrolysis of brine, though some is manufactured in conjunction with the ammonia–soda process for the manufacture of sodium carbonate by the causticization of sodium carbonate with lime, after which it is evaporated to solid caustic, 98 per cent sodium hydroxide, or liquid caustic, 45 to 75 per cent sodium hydroxide.

Caustic soda is used in the manufacture of rayon, mercerized cotton, soap, paper, explosives, and dyestuffs. It is also used in the chemical industries, in metal cleaning, electrolytic extraction of zinc, tin plating, oxide coating, laundering, bleaching, and dishwashing.

2. Physical and Chemical Properties

Physical state: white deliquescent flakes, pellets, sticks, or cake. Liquid caustic is a solution of 45 to 75 per cent sodium hydroxide in water.

Molecular weight: 40.01

Specific gravity: 2.130 (20°C.)

Melting point: 318.4°C.

Boiling point: 1390°C.

Vapor pressure: 1 mm. Hg, 739°C.

Solubility: 42 g. in 100 ml. of water at 0°C. and 347 g. at 100°C.; freely soluble in glycerine and alcohol

Refractive index: 1.3576

Alkalinity: the pH of a 1 per cent solution is about 13

3. Determination in the Atmosphere

Caustics and other alkaline materials in the atmosphere, whether gases, mists, or dusts, may be determined by scrubbing a measured volume of air through a measured amount of standard sulfuric acid and titrating the excess acid with standard alkali. The analysis may be completed in the field by using a suitable quantity of standard sulfuric acid scrubbing agent containing methyl red or other appropriate indicator, and noting the time required for scrubbing a measured rate of air until the indicator changes color.

4. Physiological Response

Characteristic irritation of nasal tissue frequently causes sneezing. Tissue response to strong alkalies is described at the beginning of this chapter. The greatest hazard is that of rapid destruction of any tissue upon contact with the solid or concentrated solutions. The inhalation of dust or mist is of secondary industrial importance. Dermatitis resulting from contact with dilute solutions is a common problem (Vol. I, page 450).

5. Hygienic Standard of Permissible Exposure

The hygienic standard of permissible exposure has been established at 2 mg./cu. meter.

6. Flammability

Sodium hydroxide is not flammable.

SODIUM PEROXIDE, Na_2O_2 (Sodium Dioxide, Sodium Superoxide)

1. Source, Uses, and Industrial Exposures

Sodium peroxide is made by oxidizing sodium with dry air, first to the monoxide, then to the dioxide. It is used as an oxidant and bleaching agent, carbon dioxide absorbent, and in the laboratory as a fusion reagent.

2. Physical and Chemical Properties

Physical state: white powder, turning yellow
Molecular weight: 77.99
Specific gravity: 2.805 (20°C.)
Melting point: 460°C. (decomposes)
Reacts violently with water, forming sodium hydroxide and hydrogen peroxide, which in turn generates heat and liberates oxygen

3. Determination in the Atmosphere

(See under Sodium Hydroxide)

4. Physiological Response

Similar to that of sodium hydroxide except for a greater hazard of air-borne dust.

5. Hygienic Standard of Permissible Exposure

None has been proposed. Should not be greater than 2 mg./cu. meter.

TRISODIUM PHOSPHATE, $Na_3PO_4 \cdot 12 H_2O$ (TSP, Sodium Orthophosphate)

1. Source, Uses, and Industrial Exposures

Trisodium phosphate is used in detergents, in clarifying sugar, in photographic developers, in removing boiler scale, in softening water, in manufacturing paper, and in tanning leather.

2. Physical and Chemical Properties

Physical state: colorless crystals
Molecular weight: 380.21
Specific gravity: 1.62 (20°C.)
Melting point: 75°C.
Solubility: 25.8 g. in 100 g. of water at 20°C. and 157 g. in 100 g. at 70°C.¯
Alkalinity: the pH of a 1 per cent solution in water is about 11.6

3. Physiological Response

Physiological response is that of a moderately strong alkali.

SODIUM SILICATES

There are several sodium silicates with different formulas and various amounts of water of crystallization. They are variously soluble in water to form alkaline solutions ranging in pH between 11.0 and 12.0. They are used for preserving eggs, in cements, in water softeners, as detergents, in fireproofing fabrics, in waterproofing, and as adhesives.

CHAPTER XXIV

Arsenic, Phosphorus, Selenium, Sulfur, and Tellurium

FRANK A. PATTY

ARSENIC, As$_4$

1. Source, Uses, and Industrial Exposures

Elemental arsenic occurs to a limited extent in nature as a steel-gray brittle metal. It can be obtained by reduction of arsenic trioxide with carbon and by sublimation from metal arsenopyrites. The metal has limited industrial use as an alloying agent to harden lead for shot making and in lead-base bearing materials. It also is alloyed with copper to enhance its toughness and corrosion resistance. It has not been recognized as a noteworthy industrial hazard.

2. Physical and Chemical Properties

Physical state: gray, lustrous, crystalline mass, also black amorphous powder (density 3.7) and yellow crystals (density 2.0)

Molecular weight: 299.64

Specific gravity: 5.73

Melting point: sublimes without melting at 610°C.

Vapor pressure: 1 mm. Hg (372°C.)

ARSENIC COMPOUNDS (Except Arsine)

1. Source, Uses, and Industrial Exposures

The major source of arsenic compounds is the by-product arsenic trioxide recovered from the roasting of arsenic-containing ores.

The compounds of arsenic are used in medicine, in glass manufacture, in pigment production, in rodent poisons, in insecticides and fungicides, in weed killers, in textile printing, in tanning and taxidermy preservatives, in antifouling paints, and to control sludge formation in lubricating oils. The most serious exposures to fumes and dusts occur in connection with the smelting of copper, lead, zinc, iron, and other ores, and in the manufacture and use of insecticides. Non-industrial absorption of arsenic from eating sea food, sprayed fruits and vegetables, and from the use of medicinal arsenicals should always be considered when tracing exposures. In insecticide manufacture exposures to various arsenic compounds, such as calcium arsenate, arsenic trioxide, sodium arsenite, Paris green, Scheele's green, and others, occur. The greatest exposures in this industry are usually in mixing, screening, drying, bagging, and drum-filling operations, concentrations ranging from 0.5 to 45 mg. of arsenic per cubic meter of air being not uncommon. Respirators are frequently, but not universally, worn in such an environment.

In the smelting of arsenical ores many opportunities for exposures within, and even above, this range exist. Among the highest exposures are those encountered in the cleaning of dust collectors and flues, and in loading and transporting the flue dust. The dust is very fine and disperses readily wherever it is agitated, as in grinding, screening, shoveling, sweeping, transferring from or to wheelbarrows, cars, hoppers, bins, settling rooms, and collectors. The repairing and cleaning of furnaces and other equipment at intervals give rise to high exposures. The effluent from smelter stacks when not properly cleaned poses an atmospheric pollution problem, and vegetation for a considerable distance may have a high arsenic content, up to 350 times normal. Samples of soil in the province of Quebec, Canada, 2 miles distant from smelter stacks have been found to contain as much as 0.06 per cent arsenic. Respirators are sometimes used in the industry, but contamination frequently involves areas where no protection is provided. The experience under these conditions, however, has not been as adverse as might be expected. Many orchardists exposed for 2 or 3 months to concentrations of spray mists up to 4.8 mg. of arsenic per 10 cu. meters of air have not evidenced significant intoxication.

In occupations where exposures to arsenic compounds exist, workmen prefer respirators in which no rubber touches the face, because of the serious arsenic irritation resulting on the moist surface under and alongside the area covered by rubber. Cotton batting $1/2$ to 1 in. thick is popular, and somewhat effective, as a filter when properly adjusted to the face.

Men exposed to arsenic trioxide dust should wear protective clothing over the entire body and take a shower at the end of the day's work.[1]

Organic Arsenic Compounds. There are a number of trivalent and pentavalent arsenic compounds produced for medicinal use in the control of protozoan parasites (trypanosomes, amebae, and plasmodia, as well as spirochetes). In the manufacture of these compounds there exist exposures not only to arsenical raw materials and by-products, but also to the organic, finished products.

These organic arsenicals are nonionized and do not immediately produce arsenic effects when inhaled or ingested. They are only slowly broken down in the body and the ingested portion may be largely excreted unchanged. Inhaled dusts

[1] *Chemical Data Sheet, SD-60, Arsenic Trioxide.* Manufacturing Chemists' Assoc., Washington, D. C., 1956.

may be more or less readily absorbed in relation to their solubilities, and only slowly produce the toxic effects of inorganic arsenic. The threshold of harm from such exposures is even more obscure than that from inorganic arsenic compounds, but permissible concentrations probably are greater for the organic compounds. It is possible that the amount of arsenic in the urine could be directly compared with that resulting from exposures to other arsenic compounds, as a measure of excessive exposure.

Many organic arsenicals of varied, toxic properties have been investigated for their possible use in chemical warfare; but since they are not of industrial importance except in their manufacture, need not be discussed here.

o-ARSENIC ACID, $H_3AsO_4 \cdot {}^1/_2H_2O$

2. Physical and Chemical Properties

Physical state: white translucent hygroscopic crystals
Molecular weight: 150.94
Specific gravity: 2.0 to 2.5
Melting point: 35.5°C.
Boiling point: 160°C.
Solubility: 16.7 g. in 100 ml. of water at 20°C.
Per cent arsenic: 50

ARSENIC TRICHLORIDE, $AsCl_3$

2. Physical and Chemical Properties

Physical state: yellowish oily liquid (needle-shaped crystals)
Molecular weight: 181.28
Specific gravity: 2.163 (14°C.)
Melting point: −18°C.
Boiling point: 130.2°C.
Vapor density: 6.25 (air = 1)
Vapor pressure: 10 mm. Hg, 23.5°C.
Per cent in "saturated" air: 1.3 (23.5°C.)
Density of "saturated" air: 1.07 (air = 1)
Dissolves in water and decomposes to arsenic trioxide and hydrochloric acid
Per cent arsenic: 76
1 mg./liter ≎ 135 p.p.m., 1 p.p.m. ≎ 7.4 mg./cu. meter at 25°C., 760 mm. Hg

ARSENIC TRIOXIDE, As_2O_3 (White Arsenic)

2. Physical and Chemical Properties

Physical state: transparent crystals or amorphous white powder
Molecular weight: 197.82

Specific gravity: 3.74 to 4.15
Sublimes without melting at 193°C.
Solubility: 2 g. in 100 ml. of water at 25°C. and 11.5 g. in 100 ml. at 100°C.
Not flammable
Per cent arsenic: 76

ARSENIC PENTOXIDE, As_2O_5 (Anhydride of Arsenic Acid)

2. Physical and Chemical Properties

Physical state: white amorphous powder
Molecular weight: 229.82
Specific gravity: 4.086
Melting point: 315°C. (decomposes)
Solubility: 150 g. in 100 ml. of water at 16°C.; increasingly soluble with
temperature rise
Not flammable
Per cent arsenic: 65

CALCIUM ARSENATE, $Ca_3(AsO_4)_2$

2. Physical and Chemical Properties

Physical state: white amorphous powder
Molecular weight: 398.06
Solubility: slightly soluble in water, soluble in dilute acids
Per cent arsenic: 38; also occurs with 3 moles of water, in which case the
molecular weight is 452.11 (per cent arsenic: 33)

COPPER ACETOARSENITE, $3Cu(AsO_2)_2 \cdot Cu(COOCH_3)_2$ (approx.) (Copper Acetate Metarsenate, Imperial, Schweinfurth, Vienna, Parrot or Paris Green)

2. Physical and Chemical Properties

Physical state: emerald green powder
Molecular weight: 1013.7
Solubility: insoluble in water, soluble in dilute acids
Per cent arsenic: 44.3

CUPRIC ARSENITE, $CuHAsO_3$ (approx.) (Scheele's Green, Swedish Green)

2. Physical and Chemical Properties

Physical state: yellowish green powder
Molecular weight: 187.5

Solubility: insoluble in water and alcohol, soluble in dilute acids and in ammonia

Per cent arsenic: 40

LEAD ARSENATE, PbHAsO₄ (approx.)

2. Physical and Chemical Properties

Physical state: white monoclinic crystals

Molecular weight: 347.14

Specific gravity: 5.79

Solubility: insoluble in water, soluble in dilute nitric acid and in caustic alkalies

Per cent arsenic: 21.6

Per cent lead: 60 (see Chapter XXVI, Industrial Lead Poisoning)

LEAD ARSENITE, Pb(AsO₂)₂ (approx.)

2. Physical and Chemical Properties

Physical state: white powder

Molecular weight: 421.03

Specific gravity: 5.85

Solubility: insoluble in water, soluble in dilute nitric acid

Per cent arsenic: 35

Per cent lead: 49 (see Chapter XXVI, Industrial Lead Poisoning)

SODIUM ARSENITE, NaAsO₂ (Sodium Metarsenite)

2. Physical and Chemical Properties

Physical state: hygroscopic gray powder

Molecular weight: 129.9

Solubility: very soluble in hot or cold water, slightly soluble in alcohol

Per cent arsenic: 57.6

3. Determination in the Atmosphere

Dusts and mists containing arsenical compounds may be collected by impingment into dilute alkali or other suitable media, by filter paper, or by the electrostatic precipitator. Fumes are perhaps more reliably collected by the electrostatic precipitator. Estimation can then be made by the Gutzeit,[2] the silver diethyldithiocarbamate,[3] or the molybdenum blue[4] methods.

[2] M. B. Jacobs, *Analytical Chemistry of Industrial Poisons, Hazards, and Solvents.* 2nd ed., Interscience Publishers, New York, 1949.

[3] *Manual of Analytical Methods,* American Conference of Governmental Industrial Hygienists, 1958.

[4] E. B. Sandell, *Colormetric Determination of Traces of Metals.* Interscience Publishers, New York, 1950.

4. Physiological Response

Absorption, Excretion, and Metabolism. Compounds of arsenic may be absorbed industrially by inhalation, ingestion, and through the skin. Arsenic is largely eliminated in the urine, some in the feces, hair, epithelium, nails, and possibly small amounts in the exhaled breath. Excretion is slow, requiring up to 10 days after an acute absorption; sometimes more than a year after prolonged absorption. After subcutaneous injection of potassium arsenite, human beings and all animals except rats showed very small amounts of arsenic in the blood[5] and no evidence of accumulation in rapidly growing tissues. Arsenic is retained and stored in all the tissues, the bones, and especially the hair. It is said that the arsenic content of long hair, in relation to that of close-cropped hair, can be used to give an indication of the time and extent of absorption. Arsenic content of the hair several years after death has been presented as evidence of nonindustrial arsenic poisoning.

The arsenic content of the urine of a group of exposed persons is significant in evaluating their exposure. The normal excretion of arsenic has been found to be about 0.015 mg. of As_4/liter[6] with occasional values up to 0.06 mg./liter. The average urinary arsenic excretion among orchardists during periods of relatively high exposures averaged 0.22 to 0.24 mg./liter, with many values above 0.5 and one value as high as 2.0 mg., although no signs of adverse effects from the arsenic were exhibited. Values as high as 3 mg. of arsenic/liter of urine have been reported elsewhere without detectable signs of intoxication. Just where on this scale a mark can be chosen to indicate excessive exposure will require further study. Kunkele[7] considers amounts above 0.1 mg. of arsenic/liter of urine, and 3 μg. arsenic/g. of head hair, to indicate poisonings, but reports results as high as 1.5 mg./liter of urine and 120 μg./g. of hair. Young and Rice[8] were unable to distinguish between internally and externally deposited arsenic in the hair and question the value of tests for arsenic in the hair.

In isolated instances, men exposed to up to 0.6 mg./cu. meter of arsenicals (computed as arsenic trioxide, 76 per cent As_4) in the atmosphere, during the manufacture of arsphenamine and related compounds over an extended period, excreted from 1 to 5 mg. of arsenic (computed as the trioxide) per liter of urine, and exhibited minor symptoms of arsenic poisoning.[9]

Schrenk and Schreibis[10] found that the urinary arsenic level in the majority

[5] F. T. Hunter, A. F. Kip, and J. W. Irvine, Jr., *J. Pharmacol., Exptl. Therap.,* **76,** 207 (1942).

[6] P. A. Neal, W. C. Dreessen, T. I. Edwards, W. H. Reinhart, S. H. Webster, H. T. Castberg, and L. T. Fairhall, *U. S. Public Health Service Publ. No.* **267** (1941).

[7] E. F. Kunkele, *Chem. Ztg.,* **64,** 29 (1940).

[8] E. G. Young and F. A. H. Rice, *J. Lab. Clin. Med.,* **29,** 439 (1944).

[9] R. M. Watrous and M. B. McCaughey, *Ind. Med.,* **14,** 639 (1945).

[10] H. H. Schrenk and L. Schreibis, *Am. Ind. Hyg. Assoc. J.,* **19,** 225 (1958).

of unexposed persons is less than 0.1 mg./liter with a few values in excess of 0.2 mg./liter unless there is some specific arsenic intake, as for instance, sea food in the diet. Results on a single individual are of questionable value in assessing exposure, probably of less value as a diagnostic aid, and in no case are they indicative of industrial arsenic poisoning. Urinary arsenic levels in a group of exposed persons are a useful index of exposure and absorption.

Effects Observed in Man. Acute or subacute industrial poisoning from ingestion, skin contact, or inhalation of arsenical dusts is not common. Symptoms include gastrointestinal disturbances, irritation of the nose and conjunctiva, skin eruptions, and inflammation. Laryngitis, bronchitis, and huskiness of the voice can result from relatively brief, massive doses of arsenicals by inhalation. Systemically, arsenic relaxes the capillaries[11] and increases their permeability, simulating inflammation, and this dilation introduces changes in the circulation that cause disturbances in organic function.

After chronic exposure to arsenic compounds perforation of the nasal septum is a common occurrence, as well as irritation, and occasionally ulceration, of the skin. Pigmentation also may be produced. Peripheral neuritis[12] is said to occur in less than 5 per cent of the cases, tremors in 10 per cent, gastric symptoms in one third, and rashes in the majority of cases of chronic poisoning. Loss of nails and hair may result. An abnormal incidence of carcinogenic effects involving both skin and respiratory tract has been reported[13] among English factory employees exposed to arsenic compounds, but this has not been confirmed by experience[14] in the United States.

Small amounts of arsenic are thought to have a beneficial effect in treating leukemias, anemias, and some skin diseases. Arsenic in organic forms is used in the treatment of syphilis: for example, Mapharsen or the arsphenamines. Arsenic was formerly used extensively, even indiscriminately,[11] as an alterative and tonic in all kinds of blood disturbances. Small amounts of arsenic have been found to have a beneficial effect upon animals, especially animals on a diet containing selenium.[15]

5. Hygienic Standard of Permissible Exposure

The ACGIH Threshold Limit Values and the AIHA Hygienic Guide each place the maximum at 0.5 mg. of arsenic per cubic meter of air. Each compound must be considered in relation to its arsenic content. This standard if applied to arsenic trichloride would be equivalent to 0.067 p.p.m. Because of the irritant and

[11] T. Sollmann, *A Manual of Pharmacology.* 6th ed., Saunders, Philadelphia, 1944.

[12] T. Legge, *Industrial Maladies.* Oxford Univ. Press, London, 1934.

[13] A. B. Hill and E. L. Faning, *Brit. J. Ind. Med.,* **5,** 1 (1948).

[14] L. S. Snegireff and O. M. Lombard, *Arch. Ind. Hyg. Occupational Med.,* **4,** 199 (1951).

[15] A. L. Moxon, C. R. Paynter, and A. W. Halverson, *J. Pharmacol. Exptl. Therap.,* **84,** 115 (1945).

vesicant properties of arsenic trichloride, further information is necessary to establish its permissible amount, which probably should be lower than that.

6. Odor and Warning Properties

With the exceptions of arsine (which see) and arsenic trichloride the compounds of arsenic do not have odor or warning properties. Arsenic trichloride is strongly irritant to the nasal passages and respiratory tract. It has a vesicant action upon contact with the skin. One exposure of 20 minutes in 0.2 mg./liter of air was fatal to cats in 4 days.[16]

ARSINE, AsH_3

1. Source, Uses, and Industrial Exposures

Arsine results whenever nascent hydrogen comes in contact with a solution containing an inorganic arsenic compound. Exposure to arsine gas may result from the action of acids on metals containing arsenic, from the use of impure sulfuric acid made from pyrites containing arsenic, or from the use of hydrochloric acid made from impure sulfuric acid that contained arsenic. Arsine poisoning has resulted from slushing out steel tanks that had previously contained a commercial grade of sulfuric acid, the diluted acid acting upon the metal tank to generate hydrogen, which combined with arsenic impurities in the acid. Arsine may arise from the pickling of any metal containing arsenic; it has been formed from the action of water on metallic arsenides or hot dross containing arsenic and aluminum. Arsine may occur as an impurity in acetylene and present an exposure either in its manufacture or use. It may occur in soldering, etching, lead plating, electrolysis of arsenious solutions, by the action of moisture on ferrosilicon, or from the use of impure or inhibited acids[17] for scale removal. It has not been a hazard associated with the manufacture, maintenance, or use of lead storage batteries in the United States, where arsenic-free lead and sulfuric acid are used.

In England,[18] 13 of 15 cases of toxic jaundice in industry reported during 1949 to 1956 were attributed to arsine poisoning, 6 among smelters, 3 among persons entering and cleaning sulfuric acid tanks without respiratory protection, and 4 from miscellaneous occupations. Four of the 13 died from the exposure.

2. Physical and Chemical Properties

Physical state: colorless gas
Molecular weight: 77.9
Melting point: −116.3°C.
Boiling point: −55°C.

[16] F. Flury, Z. ges. exptl. Med., **13**, 523 (1921).
[17] G. F. Hawlick and E. B. Ley, Occupational Med., **1**, 388 (1946).
[18] A. T. Doig, Lancet, No. 7036, **88** (July 12, 1958).

Vapor density: 2.7 (air = 1)

Solubility: 20 ml. in 100 ml. of water at 20°C.

1 mg./cu. meter ⊃⊂ 0.313 p.p.m. and 1 p.p.m. ⊃⊂ 3.2 mg./cu. meter at 25°C., 760 mm. Hg

3. Determination in the Atmosphere

Ampoules for the rapid estimation of arsine in the field are now available but their dependability has not been appraised in the American literature. Modifications of methods discussed under arsenic are applicable to the determination of arsine.

4. Physiological Response

Acute Effects in Man. It is thought that arsine first combines with the hemoglobin in the blood corpuscles[19] in some manner, and soon thereafter hemolysis of the cells occurs, resulting in the destruction of the cells and solution of the hemoglobin in the serum. More arsenic has been demonstrated in the corpuscles than in the serum. During the ensuing anemia the blood count falls rapidly and hemoglobin is excreted in the urine. Within a few days destruction of the red cells may progress to the point where death occurs from anoxemia, and may be considered a form of chemical asphyxia. It is said that death frequently occurs from pulmonary edema[20] resulting either from primary irritation or secondary to failure of the circulation. Symptoms are similar to those observed in other forms of anoxemia: headache, weakness, vertigo, and nausea. Pain in the kidneys and liver develops,[11] and the kidneys may be blocked. Jaundice, both from disintegration of red blood cells and from disorder of the liver, appears and blends its yellow with the cyanotic pallor to create a peculiar bronze tint of the skin. The symptoms of arsenic poisoning (as opposed to those of arsine) may be present in addition to the ones described. Death may occur 2 to 9 days following exposure or, if due to nephritis from arsenic, may be delayed. Recovery, even from severe poisoning, is possible in the majority of instances where the cause is recognized early[21] and prompt measures taken. Table 1 lists responses to various concentrations of arsine.

Chronic Effects in Man. In prolonged exposure to low concentrations of arsine the symptoms bear more relation to effects produced by dusts and fumes of arsenic compounds, and albumin may appear along with hemoglobin in the urine. Prolonged exposure of animals to low concentrations of arsine[22] produced a compensated destruction of red blood cells, which gradually deteriorated to a stationary level of anemia. There was little injury to other organs. Chronic arsine

[19] F. Flury and F. Zernik, *Schädliche Gase.* Springer, Berlin, 1931.

[20] Y. Henderson and H. W. Haggard, *Noxious Gases.* 2nd ed., Reinhold, New York, 1943.

[21] C. A. Nau, W. Anderson, and R. E. Cone, *Ind. Med.,* **13,** 308 (1944).

[22] M. Kiese, *Arch. exptl. Pathol. u. Pharmakol.,* **186,** 337 (1937).

TABLE 1

Physiological Response to Various Concentrations of Arsine[20]

Response	Concentration, p.p.m.
Maximum concentration allowable for prolonged exposure	1
Slight symptoms after exposure of several hours	3–10
Maximum concentration that can be inhaled 1 hr. without serious consequences	6–30
Dangerous after exposure of 30 to 60 min.	16–60
Fatal after exposure of 30 min.	250

poisoning[23] occurred in a group of 9 men exposed several months to undetermined amounts of arsine in the cyanide extraction of gold. These men complained of headache, weakness, nausea, and vomiting, the attacks increasing in frequency and severity. Puffiness of face and eyelids, tingling sensation in the toes, lumbar and epigastric pains, garliclike odor of the breath, and change of complexion were also common symptoms. The red blood cells were markedly reduced, 8 of 9 cases having lows of 0.4 to 2.4 million per cubic millimeter. The arsenic content of the urine ranged from 0.3 to 3.3 mg./liter (0.37 to 4.3 mg. of As_2O_3). The maximum arsenic excretions in the urine of 7 of the 9 men ranged from 1.0 to 3.3 mg. of As/liter, with an average of 2 mg./liter. Three months later one man was excreting slightly above 1 mg./liter, one 0.4 mg./liter, and the others were approaching normal levels. All of the men recovered.

5. Hygenic Standards and Warning Properties

The AIHA Hygienic Standard (8-hour) and the 1961 ACGIH Threshold are each 0.05 p.p.m. The faint garliclike odor of arsine in concentrations below 1 p.p.m. cannot be considered a suitable warming property.

PHOSPHORUS, P₄

1. Source, Uses, and Industrial Exposures

Yellow phosphorus is one of the elements early to be recognized as a cause of occupational disease. Its use in the match industry in the United States was eliminated in 1912 by a prohibitive tax on matches made from yellow phosphorus; and for a decade it was more or less a laboratory curiosity. It was revived in fireworks manufacture, where it caused injuries and several deaths. Consequently, an agreement was reached in 1926 to discontinue the use of yellow phosphorus in fireworks. Other forms of phosphorus, such as the sesquisulfide, have been substituted safely in match manufacture.

Phosphorus is a by-product and intermediate in the smelting of phosphate rock for the production of phosphate fertilizer (see Chapter XLIX, Fertilizer

[23] F. M. R. Bulmer, H. E. Rothwell, S. S. Pollack, and D. W. Stewart, J. Ind. Hyg. Toxicol., 22, 111 (1940).

Manufacture). In this process phosphate rock, sand, and coke are heated in an electric furnace to produce elemental yellow phosphorous or phosphorus pentoxide, depending upon which product is in demand. Phosphorus is used in the manufacture of phosphor-bronze, the manufacture of tracer bullets, incendiaries, and smokes, and as an ingredient of rodent poison. Yellow phosphorus is stored and handled under water and wherever significant quantities of its vapor escape into the air well-engineered ventilation is required to prevent inhalation exposures. Pre-employment and periodic dental x-rays are vital factors in the control program.[24] Experience has been sufficient to indicate that present control practices are successful in combating the rather terrifying phosphorus necrosis of the jaw.

2. Physical and Chemical Properties

Physical state: yellow (white) waxlike solid
Molecular weight: 123.92
Specific gravity: 1.82 (20°C.)
Melting point: 44.1°C.
Boiling point: 280°C.
Solubility: 0.0003 g. in 100 ml. of water; very soluble in carbon disulfide, ether, chloroform, and benzene

3. Determination in the Atmosphere

Phosphorus and phosphine in the air may be determined by aspirating the atmosphere through a series of three scrubbers, each containing 10 ml. of $0.01N$ potassium permanganate solution plus 1 ml. of 5 per cent sulfuric acid solution.[25] The solutions are combined, and decolorized by heating with 10 ml. of $0.01N$ oxalic acid solution. Phosphoric acid in the resultant solution is determined colorimetrically by the Bell-Doisy-Briggs method. Another method utilizes bromine water as the scrubbing medium, and the excess bromine is expelled by boiling before the phosphoric acid is determined colorimetrically, as before.

4. Physiological Response

Absorption and Excretion. Phosphorus can be absorbed through the skin, by ingestion, and through the respiratory tract, but the latter is the chief industrial mode. Phosphorus burns on the skin may be deep and painful and unless the resultant phosphoric acid is removed, or neutralized, the corrosive action continues. A 2 to 5 per cent copper sulfate solution has been used to wash the skin where particles of phosphorus may be in contact with it; the theory is that any elemental phosphorus present would be coated with metallic copper and absorption or reaction prevented. Phosphorus is excreted chiefly in the urine in the combined forms, such as phosphates. The normal daily excretion is 1.5 to 1.75 g. of

[24] H. Heimann, *J. Ind. Hyg. Toxicol.*, **28**, 142 (1946).
[25] W. Muller, *Arch. Hyg. u. Bakteriol.*, **129**, 286 (1943).

phosphorus. In phosphorus poisoning, insignificant amounts of phosphorus may be excreted in the exhaled breath as well as in the sweat. Elemental phosphorus may be found in the breath, blood, and feces, but not in the urine. The odor and phosphorescent property of phosphorus offer two excellent means of identifying its presence in the air or biological materials. An increase in nitrogen excretion and the amount of amino acids[26] in the urine are aids in diagnosing phosphorus absorption.

Acute Effects in Man. The acute effects of phosphorus absorption have not been demonstrated industrially. This is probably due to the relatively low concentrations ordinarily encountered, the slow absorption, and delayed effects.

Chronic Effects in Man. Gastrointestinal upsets, jaundice, and sometimes a phosphorus odor of the breath are said to be early signs of phosphorus poisoning. There is loss of appetite, and natural metabolism is slowed. The formation of glycogen is inhibited[27] and the normal enzymic liver function is thought to be paralyzed, while autolytic processes continue and lead to toxic decomposition products. Cachexia results and jaundice is common and intense. Prolonged intake of phosphorus causes densification and changes of bone but does not cause calcification of cartilage. The bones become fragile and their resistance to infection is diminished. Necrosis of the jaw, which is thought to be fostered by defective teeth, has occurred in a small percentage of the persons exposed for a prolonged period; therefore, persons with defective teeth should not be subjected to exposure. The first symptom of necrosis may be a toothache occurring after prolonged exposure or even after termination of exposure. Suppurative ulceration of the gums around carious teeth, or abscesses that fail to heal after the extraction of teeth, may develop and progress into suppurating fistulas and necrosis of the jaw bone.

5. Hygienic Standard of Permissible Exposure

The threshold limit (ACGIH, 1961) is 0.1 mg./cu. meter.

6. Flammability

Phosphorus ignites spontaneously in air at 34°C., and must be kept under water.

Yellow phosphorus and its vapor in contact with air glow in the dark (phosphoresce).

Red phosphorus, a reddish brown powder, is more stable and less toxic. It does not ignite in air at temperatures below 240°C. Two other forms of phosphorus, violet and black, are not of industrial importance.

7. Odor and Warning Properties

Yellow phosphorus and its vapor have a characteristic, distinct odor and in the presence of air they glow in the dark.

[26] T. Sollmann, *A Manual of Pharmacology.* 6th ed., Saunders, Philadelphia, 1944.
[27] F. Fischler, *Münch. med. Wochschr.,* **621** (1941).

PHOSPHINE, PH_3 (Phosphoreted Hydrogen, Hydrogen Phosphide)

1. *Source and Industrial Exposures*

Phosphine is formed whenever phosphorus is dissolved in hot alkalies and when calcium phosphide comes in contact with water, as in the generation of acetylene from calcium carbide containing calcium phosphide as an impurity. It may be formed during the quenching of alloys or sludges containing phosphides and from the action of water on ferrosilicon containing phosphorus.

2. *Physical and Chemical Properties*

Physical state: colorless gas
Molecular weight: 34.04
Specific gravity: 1.17 (air = 1)
Melting point: —133.5°C.
Boiling point: —87.4°C.
Solubility: 26 ml. of gas dissolves in 100 ml. of water at 17°C.
1 p.p.m. ≎ 1.39 mg./cu. meter and 1 mg./cu. meter ≎ 0.72 p.p.m.

3. *Determination in the Atmosphere*

Phosphine can be determined in the same manner as arsine, and the two gases occurring together may be absorbed by scrubbing through bromine water, and the combined arsenic and phosphorus determined in an aliquot after expelling the bromine. The Deniges[28] (molybdenum blue) method is satisfactory. Phosphine also can be estimated by exposing mercuric chloride test paper to a measured volume of phosphine—air mixture and comparing the resultant color with similarly prepared standards. Also see method given under determination of phosphorus.

4. *Physiological Response*

Acute Effects in Man. Phosphine differs in its action from that of arsine in that it does not hemolyze the red blood cells. Early symptoms of poisoning are a feeling of coldness and a pain in the region of the diaphragm. Dyspnea, weakness, vertigo, bronchitis, edema, convulsions, and death may result from exposure to phosphine. Table 2 lists responses of humans to various concentrations of inhaled phosphine.

Chronic Effects in Man. The chronic effects produced by phosphine are essentially the same as those produced by phosphorus.

5. *Hygienic Standard of Permissible Exposure*

The threshold limit (ACGIH, 1961) is 0.05 p.p.m.

6. *Flammability*

Phosphine ignites spontaneously in air.

[28] M. B. Jacobs, *Analytical Chemistry of Industrial Poisons, Hazards, and Solvents.* 2nd ed., Interscience Publishers, New York, 1949.

TABLE 2

Response of Men to Inhalation of Phosphine[20]

Response	Concentration, p.p.m.
Rapidly fatal	2000
Death occurs following $1/2$ to 1 hr. exposure	400–600
Dangerous to life after 1 hr.	290–430
Maximum amount for $1/2$ to 1 hr. without serious effects	100–200
Serious effects after several hours	7
Limit of perceptibility	1.5–3

7. Odor and Warning Properties

Phosphine has a foul odor resembling that of decayed fish, but this gives no warning of its threshold concentration.

PHOSPHORUS SESQUISULFIDE, P_4S_3 (Tetraphosphorus Trisulfide)

1. Uses and Industrial Exposures

Tetraphosphorus trisulfide is used in making matches and friction strips for safety-match boxes.

2. Physical and Chemical Properties

Physical state: yellow crystals
Molecular weight: 220.12
Specific gravity: 2.03
Melting point: 172.5°C.
Boiling point: 407.5°C.
Solubility: insoluble in cold water, decomposes in hot water

3. Physiological Response

Phosphorus sesquisulfide is not sufficiently volatile to present a vapor hazard at the temperatures of occupied spaces. The dust or fume is irritant to the eyes, the respiratory tract, and the skin. Its toxicity, however, is minor when compared to that of yellow phosphorus and serious ill effects other than eczema have not been reported.

4. Hygienic Standard of Permissible Exposure

No hygienic standard has been accepted.

5. Flammability

The ignition temperature is about 100°C.

PHOSPHORUS PENTACHLORIDE, PCl$_5$

1. Source, Uses and Industrial Exposures

Phosphorus pentachloride is made by reacting yellow phosphorus with chlorine. It is used in chemical manufacturing. It produces phosphorus trichloride and chlorine when heated and phosphorus oxychloride, phosphoric acid, and hydrochloric acid when decomposed in water.

2. Physical and Chemical Properties

Physical state: pale yellow, fuming solid
Molecular weight: 208.31
Melting point: 148°C. under pressure
Boiling point: 160°C. under pressure
Sublimes at 100°C.

3. Physiological Response

Phosphorus pentachloride has a pungent, unpleasant odor and its vapor or fume is very irritant to all mucous surfaces including the lungs.

4. Hygienic Standard of Permissible Exposure

The threshold of permissible exposure has been set by the ACGIH at 1 mg./cu. meter.

PHOSPHORUS TRICHLORIDE, PCl$_3$

1. Source, Uses and Industrial Exposures

Phosphorus trichloride is made by reacting yellow phosphorus with chlorine. It is used in chemical manufacturing. It hydrolyzes to phosphorus acid and hydrochloric acid.

2. Physical and Chemical Properties

Physical state: colorless liquid
Molecular weight: 137.39
Specific gravity: 1.574(21°C.)
Melting point: —91°C.
Boiling point: 75.5°C.
1 p.p.m. $\backsimeq$ 5.6 mg./cu. meter and 1 mg./cu. meter $\backsimeq$ 0.18 p.p.m. at 25°C., 760 mm. Hg

3. Physiological Response

The vapors of phosphorus trichloride are very irritant to the respiratory tract.

4. Hygienic Standard of Permissible Exposure

The threshold limit is 0.5 p.p.m. (ACGIH 1961).

PHOSPHORIC ANHYDRIDE, P_2O_5 (PHOSPHORUS PENTOXIDE)

1. Source, Uses, and Industrial Exposures

Phosphorus pentoxide is formed by burning yellow phosphorus in dry air or oxygen. It has a great affinity for water and is used as a dessicating agent.

2. Physical and Chemical Properties

Physical state: white fluffy powder
Molecular weight: 142
Specific gravity: 2.38
Melting point: 347°C.
Solubility: very soluble in water; reacts violently with evolution of heat to form phosphoric acid, H_3PO_4, which in its most concentrated commercially available form contains 85 per cent H_3PO_4

3. Physiological Response

Corrosive irritant to mucous surfaces, eyes, and skin. The resulting phosphoric acid is less harmful than sulfuric acid.

4. Hygienic Standard of Permissible Exposure

The hygienic standard recommended in the AIHA Guide is 1 mg. of P_2O_5/cu. meter of air.

SELENIUM, Se_8, AND SELENIUM COMPOUNDS

1. Source, Uses, and Industrial Exposures

Selenium, a nonmetallic element of the sulfur group, although relatively scarce in quantity, is widely distributed in nature. It is recovered from flue dust collected during the burning of pyrites for the manufacture of sulfuric acid. Selenium, together with tellurium, occurs as an impurity in most sulfide ores of copper, gold, nickel, and silver, and during the course of refining these ores the selenium and tellurium must be removed. As a result these by-products, although relatively rare materials, are sufficiently moderately priced to be used commercially to a considerable extent. The annual production of selenium is about 1,000,000 pounds.

Selenium is used in the manufacture of pigments, in insecticides, in rubber compounding, in the manufacture of rectifiers and photoelectric cells, to remove the green (iron) tint of glass, to produce pink, ruby, and black glass glaze, to improve the machinability of copper alloys and stainless steel, to improve the grain, structure, and ductility of cast steel, to increase the depth of chill in cast iron, as a flameproofing agent for textiles and wire-cable coverings, and in chemical manufacture. Exposures to selenium may result during the smelting and refining of ores containing selenium, in the refining of copper, silver, and gold to remove the selenium, or from the use of selenium compounds.

SELENIUM

2. *Physical and Chemical Properties*

Physical state: red amorphous powder turning black upon standing and vitreous upon heating
Atomic weight: 78.96
Specific gravity: 4.3 to 4.8 (20°C.)
Melting point: 217°C.
Boiling point: 688°C.

SELENIUM DIOXIDE, SeO_2

2. *Physical and Chemical Properties*

Physical state: white crystalline powder
Molecular weight: 110.96
Specific gravity: 3.95 (15°/15°C.)
Melting point: 340°C. under pressure, sublimes at 315°C.
Readily soluble in hot or cold water to form selenious acid, H_2SeO_3

SELENIUM TRIOXIDE, SeO_3

2. *Physical and Chemical Properties*

Physical state: yellowish white hygroscopic powder
Molecular weight: 126.96
Specific gravity: 3.6
Solubility: dissolves readily in water to form selenic acid, H_2SeO_4, a hygroscopic corrosive acid similar in some respects to sulfuric acid

SODIUM SELENITE, $Na_2SeO_3 \cdot 5H_2O$

2. *Physical and Chemical Properties*

Physical state: white powder
Molecular weight: 263.04
Solubility: dissolves freely in water to form a slightly alkaline solution

SODIUM SELENATE, Na_2SeO_4

2. *Physical and Chemical Properties*

Physical state: colorless crystals with or without 10 moles of water
Molecular weight: 188.95
Solubility: very soluble in water

3. Determination in the Atmosphere

Dusts and fumes containing selenium or its compounds can be sampled with the electrostatic precipitator, and gases or vapors may be scrubbed through 40 to 48 per cent hydrobromic acid containing 5 to 10 per cent free bromine. The effluent side of such a scrubber should be provided with a soda-lime tube to absorb the corrosive vapors of hydrobromic acid and bromine. Soda lime has also been used to collect hydrogen selenide, and it is probable that silica gel could be used for any selenium compound in the gaseous or vapor state. After the sample has been collected the selenium compounds may be separated by distillation[29-31] with an excess of hydrobromic acid and bromine, and the selenium content estimated by gravimetric, volumetric, colorimetric, or spectrophotometric methods. Selenium dioxide can be collected in 10 ml. of water with the midget impinger and determined by a modification of Chernyi's method,[32] 5 ml. of c.p. hydrochloric acid being added and the sample being made up to 20 ml. Take 5 ml. in a Nessler tube, add 1 ml. of 5 per cent gum arabic solution and 1 ml. of a 10 per cent stannous chloride solution. Make up to 20 ml. with a mixture of 1 part hydrochloric acid and 3 parts water, shake, and compare the resultant color with similarly prepared standards that have stood a corresponding length of time. Amounts in the range of 0.01 to 0.1 mg. are satisfactory for comparison and it is possible to compare amounts up to 0.5 mg. The remainder of the sample can be used if desired to repeat the test with a greater or less amount as indicated by the intensity of the color. For sampling concentrations below 3 mg. per 10 cu. meters it is necessary to sample more than ten minutes and to use the entire sample. Dilution may be omitted and a smaller volume used for comparison as long as the same ratio of hydrochloric acid is maintained. The standard impinger is more satisfactory than the midget for concentrations below 1 mg. per 10 cu. meters.

4. Physiological Response

Acute Effects. Selenium somewhat resembles arsenic and tellurium in its physiological action. The relative toxicities of compounds of selenium, tellurium, and arsenic have been determined[33] in animals by intraperitoneal injection and the minimum fatal doses in mg./kg. (75 per cent of animals dying within 48 hours) found to be as follows: Na_2SeO_3, 3.25–3.5; Na_2SeO_4, 5.35–5.75; Na_2TeO_3, 2.25–2.5; Na_2TeO_4, 20.0–30.0; Na_2HAsO_3, 4.25–4.75; Na_2HAsO_4, 14.0–18.0. Selenium can be acquired from inhalation of the dust, vapors, gases, and fumes of the metal or its compounds, by ingestion, and to some extent by absorption through the skin.

[29] W. O. Robinson, H. C. Dudley, K. T. Williams, and H. G. Byers, *Ind. Eng. Chem. Anal. Ed.,* **6,** 274 (1934).

[30] H. C. Dudley, *Am. J. Hyg.,* **24,** 227 (1936).

[31] G. Wernimont and F. J. Hopkinson, *Ind. Eng. Chem. Anal. Ed.,* **12,** 308 (1940).

[32] M. E. Chernyi, *Chem. Abstr.,* **37,** 5925 (1943).

[33] K. W. Franke and A. L. Moxon, *J. Pharmacol. Exptl. Therap.,* **58,** 454 (1936); **61,** 89 (1937).

The first signs of acute selenium poisoning[34] are nervousness and fear followed by vomiting, then quietness and somnolence. Respiration becomes difficult, dyspnea develops, followed by opisthotonos, tetanic spasm, clonic spasm, falling blood pressure, and respiratory failure resulting from action on the central nervous system. Acute effects other than garlicky breath have not been reported from industrial exposure. The odor of the breath may be due to tellurium accompanying the selenium.

Chronic Effects. According to Dudley,[35] prolonged exposure to the inhalation of selenium compounds during extraction, purification, and processing of selenium-bearing ores has given rise to marked pallor, coated tongues, gastrointestinal disorders, nervousness, and a pronounced garlickly odor of the breath. Moxon and Rhian[34] report a very odoriferous selenium breath resulting from chronic selenium poisoning and enumerate such effects as nervous disorders, small local hemorrhages, severe ascites, liver and splenic damage, emaciation, apathy, and progressive anemia. Hamilton[36] also mentions the garlic odor as well as the irritation of nose, throat, and bronchi, pain in the lumbar region, nasal inflammation resembling that accompanying a cold, and night sweats. Waitkins, Bearse, and Shutt[37] state that the garlickly breath, which is known to result from absorption of tellurium and has been observed after exposure to selenium, is believed not to result from absorption of selenium but to be due to tellurium impurities present in the selenium. Commercial selenium now, however, is being refined in the United States to a tellurium content of less than 0.1 per cent. This amount is said not to have been known to cause garlicky breath, and selenium workers, not also exposed to tellurium, henceforth would not be expected to develop this symptom.

Naturally occurring compounds in the soil of sections of the northwestern plain states have produced grains and vegetation of sufficient selenium content to poison livestock. Foodstuffs such as cereals, vegetables, eggs, meats, and milk in such areas contain from 0.16 mg. to 18.0 mg. of selenium/kg. of foodstuff. Members of the rural population in these areas excrete from 0.10 mg. to 2.00 mg. of selenium/liter of urine,[38] a condition indicating absorption of selenium, but they do not evidence definite symptoms of selenium poisoning. High protein diets are protective against selenium poisoning, and sodium arsenite or arsenate prevents[39] normally toxic amounts of selenium from causing hepatic injury and destruction of hemoglobin. This action is independent of the route of administration of the arsenite[40] and therefore is not due to an inhibition of absorption of the selenium.

Absorption and Excretion. Industrially, absorption results essentially from

[34] A. L. Moxon and M. Rhian, *Physiol. Revs.,* **23,** 305 (1943).

[35] H. C. Dudley, *Am. J. Hyg.,* **23,** 181 (1936).

[36] A. Hamilton, *Industrial Toxicology.* Harper, New York, 1934.

[37] G. R. Waitkins, A. E. Bearse, and R. Shutt, *Ind. Eng. Chem.,* **34,** 899 (1942).

[38] M. I. Smith, *J. Am. Med. Assoc.,* **116,** 562 (1941).

[39] M. Rhian and A. L. Moxon, *J. Pharmacol. Exptl. Therap.,* **78,** 249 (1943).

[40] A. L. Moxon, C. R. Paynter, and A. W. Halverson, *J. Pharmacol. Exptl. Therap.,* **84,** 115 (1945).

inhalation, although ingestion occurs to some extent, and skin absorption may occur to a lesser degree. Fifty to eighty per cent of ingested[38] selenium is excreted through the kidneys. Three to ten per cent of subcutaneously injected[41] selenium is exhaled within 24 hours (half in the first 3 hours) in the form of an unidentified volatile compound of selenium. Farm animals with "alkali disease," as selenium poisoning of animals is known in seleniferous areas, excrete 0.6 mg. to 3.0 mg. of selenium/liter of urine. Selenium is stored in the liver, kidneys, spleen, pancreas, muscle, heart, lungs, and other tissues. As in arsenic poisoning, the amount stored in the hair may serve as a criterion of length of exposure and degree of storage. Urinary excretion closely parallels intake. The normal excretion[42] of persons outside seleniferous areas and not exposed to selenium, except through the normal ingestion with cereals, has been found to be from 0.01 mg. to 0.15 mg./liter of urine. Dudley[35] determined the selenium content of the urine of a number of men engaged in extracting and processing large quantities of selenium. These men were exposed to undetermined amounts of selenium dioxide, selenium dust, and presumably some hydrogen selenide. He reported a maximum finding of 0.069 mg./liter of urine with only minor ill effects, none of which could be definitely ascribed to selenium.

Tests Indicating Exposure. The most conclusive test of exposure is that of selenium in the urine;[31, 35, 43] when amounts in excess of 0.1 mg./liter are found, it may be accepted as evidence of unusual absorption. Amounts greater than 0.2 mg./liter indicate potentially harmful exposures, and amounts of 0.5 mg. or more per liter of urine warrant immediate consideration of corrective or control measures.

5. *Hygienic Standard of Permissible Exposure*

The threshold limit (ACGIH, 1961, and the AIHA Hygienic Guide) are both 0.1 mg. of selenium/cu. meter of air. This appears to have a liberal safety factor.

HYDROGEN SELENIDE, H_2Se

1. *Source and Industrial Exposures*

Hydrogen selenide may be formed by the reaction of acids or water with metal selenides or wherever nascent hydrogen is in contact with soluble selenium compounds. It has no commercial use.

2. *Physical and Chemical Properties*

Physical state: colorless gas
Molecular weight: 80.98
Specific gravity: 2.79 (air = 1)

[41] K. P. McConnell, *J. Biol. Chem.,* **145**, 55 (1942).
[42] J. H. Sterner and V. Lidfeldt, *J. Pharmacol. Exptl. Therap.,* **73**, 205 (1941).
[43] M. I. Smith and R. D. Lillie, *Natl. Insts. Health Bull.,* No. **174** (1940).

Melting point: —64°C.
Boiling point: —42°C.
Solubility: 270 ml. in 100 ml. of water at 22.5°C.; more soluble in alkaline water
1 mg./cu. meter ⌇ 0.3 p.p.m. and 1 p.p.m. ⌇ 3.3 mg./cu. meter at 25°C., 760 mm Hg

3. *Physiological Response*

Response in Animals. Guinea pigs showed no response during an 8-hour exposure to 0.3 to 2.1 p.p.m.[44] Fifty per cent of the animals died within one month following exposure. A high percentage of fatalities resulted from exposure of 2 hours to 10 p.p.m. Eye and nasal irritation was immediately evident. Deaths appeared to be due primarily to irritation of the pulmonary tissue resulting in pneumonitis.[44]

Effects Observed in Man. Garlic odor of breath, nausea, metallic taste, dizziness, and extreme lassitude. Less than 1-month exposure to less than 0.2 p.p.m. is reported to have caused symptoms.[45]

4. *Hygienic Standard of Permissible Exposure*

The permissible exposure for 8 hours (ACGIH, 1961 and AIHA Hygienic Standard) is 0.05 p.p.m.

5. *Odor and Warning Properties*

Hydrogen selenide has an offensive odor somewhat resembling that of decayed horseradish In the lower toxic range this odor is not a dependable warning. Eye and nasal irritation is moderate.

SELENIUM OXYCHLORIDE, $SeOCl_2$

1. *Uses and Industrial Exposures*

Selenium oxychloride is a powerful solvent, chlorinating agent, and resin plasticizer used in the chemical industry.

2. *Physical and Chemical Properties*

Physical state: clear pale yellow liquid
Molecular weight: 165.87
Specific gravity: 2.42 (22°C.)
Melting point: 8.5°C.
Boiling point: 176.4°C.
Vapor pressure: about 0.05 mm. Hg, 20°C.
Solubility: hydrolyzes in water to hydrochloric and selenious acids

[44] H. C. Dudley and J. W. Miller, *J. Ind. Hyg. Toxical.*, **23**, 470 (1941).
[45] R. F. Buchan, *Occupational Med.* **3**, 439 (1947).

3. Physiological Response

Selenium oxychloride is strongly vesicant[46] and will rapidly destroy the skin upon contact unless immediately removed by washing. Less than 0.01 ml. on the skin of rabbits has proved fatal within 24 hours, and selenium could be demonstrated in the blood and liver. Dudley[46] found the minimum lethal dose, when applied to the skin of a rabbit, to be 7 mg./kg. This would be equivalent to approximately 0.2 ml. applied to a man of average size. The application of less than 0.005 ml. to the arm of a man caused a painful burn with swelling, and its healing required a month. The vapors of selenium oxychloride are toxic, but their irritant and corrosive action on the respiratory tract is not as great as might be supposed, because the vapor readily decomposes in air and also its low vapor pressure limits the concentration possible in air.

4. Hygienic Standard of Permissible Exposure

No hygienic standard has been published.

SULFUR, S (Brimstone)

1. Source, Uses, and Industrial Exposures

Elemental sulfur is mined in great quantities in Texas and Louisiana. By-product sulfur is produced from hydrogen sulfide and sulfur dioxide. Sulfur is widely used and there are numerous operations, such as shoveling, grinding, screening, and bagging, where sulfur dust in considerable amount is found in the atmosphere.

2. Physiological Response

Sulfur dust is a very mild irritant and may irritate sensitive skins. No injurious effects have been ascribed to the inhalation of the dust in the United States,[47] but cases of "thiopneumoconiosis"[48] and bronchitis with emphysema[49] have been reported in literature from overseas.

3. Flammability

Sulfur and its dust suspension are flammable; the lower flammable limit for sulfur dust in air is 35 mg./liter and the ignition temperature is 190°C.

SULFUR DIOXIDE, SO₂

1. Source and Occurrence

Sulfur dioxide is formed whenever sulfur is burned in the air, and its odor is frequently described as that of burning sulfur. It is perhaps the most widely en-

[46] H. C. Dudley, *Public Health Repts., U. S.,* **53,** 94 (1938).
[47] S. S. Pinto, R. A. Brown, and B. H. Carlton, *J. Ind. Hyg. Toxicol.,* **25,** 149 (1943).
[48] G. P. Schiavina, *Rass. med. ind.,* **12,** 173, 244 (1941).
[49] G. Frada, G. Mentesana, and V. Azzaro, *Folia Med., Naples,* **40,** 525 (1957).

countered and best known irritant gas, not only because of its wide usage but also because of its frequent occurrence as an undesired by-product in the smelting of sulfide ores, in paper manufacture, in the combustion of sulfur-bearing coals and petroleum fuels, and in the action of sulfuric acid on reducing agents. Sulfur dioxide is one of the most prominent gases contributing to atmospheric pollution in large cities and in areas surrounding smelters. Sweetening plants for petroleum products sometimes dispose of sulfide gases by burning them to sulfur dioxide and discharging it from high stacks. The terrain, height of stack, rate of gas discharge, and atmospheric conditions present variable factors that have made the success of this dilution method unpredictable and frequently disappointing. The gas rises vertically for some distance above the stack, then spreads out laterally. The important factors in its dispersion are fog, wind direction, velocity, inversion, and turbulence. Sulfur dioxide in moist air or fogs combines with the water to form sulfurous acid, and is slowly oxidized to sulfuric acid.[50]

2. Uses and Industrial Exposures

Sulfur dioxide is an intermediate in the manufacture of sulfuric acid. It is also used in the manufacture of sodium sulfite and in other chemical processes. Large quantities are used in refrigeration, in bleaching, fumigating, and preserving. It is used as an antioxidant in melting and pouring magnesium where its is applied as the gas or generated by adding powdered sulfur to the surface of the molten metal in the ladle and on the surface of the poured casting. Sulfur dioxide up to 0.5 per cent is also used for the prevention of oxidation in controlled-atmosphere heat-treat ovens for magnesium. Breathing zone concentrations of sulfur dioxide in some magnesium foundries are highly variable and range from fractions of a part per million to over 10 p.p.m. as an average concentration with occasional peak concentrations of short duration in excess of 50 p.p.m.

3. Physical and Chemical Properties

Physical state: colorless irritant gas having a characteristic odor and taste
Molecular weight: 64.07
Specific gravity: 1.434 (liquid) at 0°C.
Melting point: —72.7°C.
Boiling point: —10°C.
Vapor density: 2.3 (air = 1)
Solubility: 0.6 g. in 100 g. of water at 90°C., 8.5 g. in 100 g. at 25°C., and 22.8 g. in 100 g. at 0°C.; more soluble in methyl and ethyl alcohol; soluble in acetic and sulfuric acids, chloroform, and ethyl ether
1 p.p.m. $\backsim$ 2.62 mg./cu. meter and 1 mg./cu. meter $\backsim$ 0.38 p.p.m. at 25°C., 760 mm. Hg

[50] H. F. Johnstone and D. R. Coughanowr, *Ind. Eng. Chem.*, **50**, 1169 (1958).

4. Determination in the Atmosphere

Aspirate the sample through two scrubbers, the first of which contains a measured quantity of 0.01N standard iodine in potassium iodide solution, and the second about one half as much 0.01N sodium thiosulfate solution. Wash the contents of the two scrubbers into a beaker and titrate with 0.01N sodium thiosulfate, using starch indicator. One milliliter of 0.01N iodine solution $\backsim$ 0.1223 ml. of sulfur dioxide gas at 25°C., 760 mm. Hg pressure. Determination may also be made by aspiration of the atmosphere at known rate through standard 0.002N iodine-starch solution until the color is just discharged, or better, to a very faint blue. The amount of iodine required to produce a faint blue in the scrubber must be considered. Hydrogen sulfide, hydrogen cyanide, and other reducing agents give the same reaction. A blank should be run parallel on air free from reducing gases, or air filtered through silica gel or activated carbon.

The atmosphere also may be scrubbed through an oxidizing solution such as 5 per cent potassium chlorate[51] and evaluated by acidifying with hydrochloric acid, adding barium chloride solution, and comparing with standards of potassium sulfate treated in a similar manner. Automatic recorders[52-54] are available and are widely used.

5. Physiological Response

Acute Effects. Sulfur dioxide is an irritant gas: 6 to 12 p.p.m. causes immediate irritation to nose and throat. Three tenths to 1 p.p.m. can be detected by the average individual, probably by taste rather than by odor, and 3 p.p.m. has an easily noticeable odor. About 20 p.p.m. is the least amount irritating to the eyes. One per cent is irritant to moist areas of the skin within a few minutes. Although sulfur dioxide dissolves readily and its inhalation affects chiefly the upper respiratory tract and bronchi, it may cause edema of the lungs or glottis and can produce respiratory paralysis.[55]

Chronic Effects. A period of human exposure of over 2 years to variable concentrations on the order of 30 p.p.m. with occasional peaks of up to 100 p.p.m. was found to have produced a significantly higher than normal incidence of nasopharyngitis,[56] an alteration of the senses of smell and taste, high urinary acidity, and increased fatigue. There was also an extension of the duration of colds, but not a significant change in their incidence.

Exposure of mice and guinea pigs to concentrations of 10, 25, 33, 65, 100, 150,

[51] S. Plisetskaya, *Lab. Prakt. USSR*, No. **12**, 25 (1939).

[52] M. D. Thomas, O. J. Ivie, J. N. Abersold, and R. H. Hendricks, *Ind. Eng. Chem. Anal. Ed.*, **15**, 287 (1943).

[53] H. L. Helwig and C. L. Gordon, *Anal. Chem.*, **30**, 1810 (1958).

[54] W. G. Cummings and M. W. Redfearn, *J. Inst. Fuel*, **30**, 628 (1957).

[55] R. T. Johnstone, *Occupational Diseases*. Saunders, Philadelphia, 1942.

[56] R. A. Kehoe, W. F. Machle, K. Kitzmiller, and T. J. LeBlanc, *J. Ind. Hyg.*, **14**, 159 (1932).

300, and 1000 p.p.m.[57] of sulfur dioxide revealed no significant effects in concentrations of 33 p.p.m. or less. In 65 p.p.m. one third of the animals evidenced acute distention of the stomach on the ninth day, and in 100 p.p.m. on the fourth day, at which time perforations of the stomach began to appear. The median lethal concentration for mice was 130 p.p.m. for 24 hours, 340 for 6 hours, 610 for 1 hour, and 1350 p.p.m. for 10 minutes. Less than $1/2$ p.p.m. is believed to be injurious to plant foliage.

6. Maximum Permissible Concentration and Warning Properties

The accepted permissible limit for prolonged exposure is 5 p.p.m. The irritating effects of this concentration are not sufficient to provide ample warning. Fifty to one hundred p.p.m. is considered the maximum permissible amount for 30 to 60 minutes' exposure, while 400 to 500 p.p.m. is immediately dangerous to life. Men are not likely to voluntarily enter concentrations high enough to be immediately harmful. The gas is not flammable.

SULFUR TRIOXIDE, SO₃ (Sulfuric Anhydride)

1. Physical and Chemical Properties

Physical state: colorless liquid or crystals
Molecular weight: 80.06
Melting point: 16.83°C.
Boiling point: 44.8°C.
Density: 2.8 (air = 1)
Solubility: up to 100 per cent sulfuric acid in water

SULFURIC ACID, H₂SO₄ (Oil of Vitriol)

1. Source, Uses, and Industrial Exposures

Sulfuric acid is made by oxidation of sulfur dioxide in the lead chamber and the contact processes. The acid is used in great quantities in fertilizer manufacture, chemical manufacture, petroleum refining, the production of rayon and film, iron and steel, explosives, textiles, and in pickling and anodizing. The fume of SO_3, or mist of H_2SO_4, may result also wherever sulfuric acid is heated in the open air or gas bubbles are released from a liquid surface containing sulfuric acid.

2. Physical and Chemical Properties

Physical state: colorless liquid (oil of vitriol)
Molecular weight: 98.06
Specific gravity: 1.834 (20°C.)
Melting point: 10.5°C.

[57] F. R. Weedon, A. Hartzell, and C. Setterstrom, contribs. *Boyce Thompson Inst.*, **10**, 281 (1939).

Boiling point: 330°C. (98.3 per cent)
Vapor pressure: 1 mm. Hg (146°C.)
Miscible with water
1 mg. SO_3/cu. meter $\backsimeq$ 0.3 p.p.m. and 1 p.p.m. $\backsimeq$ 3.2 mg./cu. meter at 25°C., 760 mm. Hg (1.2 mg. of H_2SO_4 contains 1 mg. of SO_3)

3. Determination in the Atmosphere

Sulfuric acid mists may be collected either in scrubbers or impingers and titrated with $0.01N$ alkali with a suitable indicator such as methyl red. Standard alkali may also be used with methyl red in a scrubber and air scrubbed at a measured rate through it until a color change of the indicator shows that neutralization has occurred, after which the atmospheric concentration is computed from the sampling rate and time. Sampling of the minute particles in sulfuric acid "fumes" or fog requires either very efficient scrubbing, electrostatic collectors employing acid-resistant linings or glass tubes, or collection on a suitable filter. The SO_4 may also be precipitated by barium chloride.

4. Physiological Response

Amdur[58] exposed guinea pigs to concentrations of 2 to 200 mg. of sulfuric acid mist/cu. meter of atmosphere for 1-hour periods. Using particle sizes of 0.8, 2.5, and 7 μ, Amdur found the smallest size was the most effective at the lowest level (2 mg./cu. meter). The 2.5 μ droplets caused the greatest response in the higher levels of concentration and at 200 mg./cu. meter caused death within one hour to all 4 animals exposed to this concentration. The 7 μ particles in concentrations up to 30 mg./cu. meter produced only slight response because they did not penetrate beyond the upper respiratory tract.

Sulfur trioxide and sulfuric acid mist are strongly irritant and the inhalation of concentrations of around 3 mg./cu. meter causes a choking sensation in the uninitiated. Persons accustomed to the exposure are unable to notice concentrations of this order of magnitude. Sulfur trioxide is irritant and corrosive to all mucous surfaces, causing inflammation of the upper respiratory tract, and possible lung injury. Sulfuric acid also attacks the enamel of the teeth.

5. Hygienic Standard of Permissible Exposure

The threshold limit of ACGIH and the AIHA Hygienic Standard for 8-hour exposure are both 1 mg./cu. meter.

HYDROGEN SULFIDE, H_2S

1. Source, Uses, and Industrial Exposures

Exposure to hydrogen sulfide results occasionally from its use as an industrial chemical, and frequently, from its occurrence as a by-product in industrial or

[58] M. O. Amdur, *A.M.A. Arch. Ind. Health,* **18,** 407 (1958).

natural processes wherever proteins decompose. It is encountered in mining, especially where sulfide ores are found; in excavating in swampy or filled ground, and hence sometimes in wells, caissons, and tunnels; in natural gas; in the production and refining of petroleum; in the waters of certain natural springs; in volcanic gases; in the low temperature carbonization of coal; in the manufacture of chemicals, dyes, and pigments; in the rayon industry; in the rubber industry; in tanneries; in the manufacture of glue; in the washings from sugar beets; and in sewer gases. Since hydrogen sulfide is soluble in water and oil, it may flow for a considerable distance from its place of origin to escape at unexpected areas. Many budding chemists have developed a casual disregard for the toxicity of hydrogen sulfide because of its general, and sometimes careless, use in the teaching of qualitative and quantitative analysis, and it is with great surprise that they later learn that the gas they used so consistently has a toxicity comparable to that of hydrogen cyanide. It is detectable by odor at about $1/_{400}$ of the lowest amount that can cause injurious effects.

2. Physical and Chemical Properties

Physical state: colorless gas with rotten egg odor
Molecular weight: 34.08
Specific gravity: 1.19 (air = 1)
Melting point: —82.9°C.
Boiling point: —61.8°C.
Solubility: 437 ml. in 100 ml. of water at 0°C. and 186 ml. in 100 ml. at 40°C.; also soluble in alcohol, petroleum solvents, and crude petroleum

3. Determination in the Atmosphere

Hydrogen sulfide in the atmosphere is readily detected by its blackening of moistened lead acetate paper. It may be determined quantitatively by scrubbing the air through a cadmium chloride solution and estimating iodometrically[59] or colorimetrically.[60] In the absence of interfering oxidizing or reducing gases, hydrogen sulfide may be determined, in the field, by scrubbing through water containing a measured amount of 0.01N standard iodine solution[61] and starch indicator. The midget impinger pump with a calibrated, fritted glass, or other suitable scrubber, is satisfactory for either the cadmium chloride or the iodine method. In the case of iodine a 0.01N solution may be carried into the field in a glass-stoppered bottle, measured into the scrubber with a pipet, and the starch added. Air can be drawn through the scrubber at a measured rate to a predeter-

[59] M. B. Jacobs, *Analytical Chemistry of Industrial Poisons, Hazards, and Solvents.* 2nd ed., Interscience Publishers, New York, 1949.

[60] F. H. Goldman, A. A. Coleman, H. B. Elkins, and H. H. Schrenk, *Am. J. Public Health,* **33,** 862 (1943).

[61] A. C. Fieldner, G. G. Oberfell, M. C. Teague, and J. N. Lawrence, *Ind. Eng. Chem.,* **11,** 523 (1919).

mined fading or almost complete loss of color and, since 1 ml. of $0.01N$ solution $\backsim 0.1223$ ml. of hydrogen sulfide at 25°C. and 760 mm. Hg pressure, the concentration of hydrogen sulfide in the air may be estimated by the following equation:

[ml. $0.01N$ iodine $\times$ 0.1223 $\times$ 1000]/
$$[\text{sampling time (min.)} \times \text{sampling rate (liters/min.)}] = \text{p.p.m.}$$

or where the rate is 1 liter per minute:

[ml. $0.01N$ iodine $\times$ 122.3]/[sampling time (min.)] = p.p.m. hydrogen sulfide

When a control test is made in hydrogen sulfide–free air and allowance made for the amount of iodine necessary to produce a detectable color in the scrubbing liquid, this method can give results of at least 95 per cent accuracy. The M.S.A. hydrogen sulfide detector[62] offers another rapid, convenient, and fairly accurate method. An automatic detection and control system has been described by Clough.[63]

4. Physiological Response

Acute Effects. By far the greatest danger from the inhalation of hydrogen sulfide is from its acute effects. Whether the effects are to be acute, or subacute and chronic, depends upon the concentration of the gas in the atmosphere. Concentrations of 700 p.p.m. (0.07 per cent by volume) and above may result in acute poisoning and, although the gas is an irritant, the systemic effects from absorption of hydrogen sulfide into the blood stream[64] overshadow the irritant effects. These acute systemic effects result from the action of free hydrogen sulfide in the blood stream and occur whenever the gas is absorbed faster than it can be oxidized to pharmacologically inert compounds, such as thiosulfate or sulfate. Such oxidation occurs rapidly in man or animals and, even following inhalation exposure to concentrations up to 700 p.p.m. hydrogen sulfide in the atmosphere, hydrogen sulfide does not appear in the exhaled breath. Relatively massive doses are required to overcome this protective activity of the body. Sodium sulfhydrate, NaHS, solution injected intravenously into dogs rapidly disappears from the circulating blood[65] when a rate equivalent to 0.1 to 0.2 mg. of hydrogen sulfide per kilogram of body weight per minute is not exceeded.

When the amount absorbed into the blood stream exceeds that which is readily oxidized, systemic poisoning results, with a general action on the nervous system, hyperpnea occurs shortly, and respiratory paralysis may follow immediately. This condition may be reached almost without warning as the originally detected odor of hydrogen sulfide may have disappeared, as a result of olfactory

[62] J. B. Littlefield, W. P. Yant, and L. B. Berger, *U. S. Bur. Mines Rept. Invest.*, No. **3276** (1935).

[63] J. Clough, *J. Soc. Chem. Ind. London*, **63**, 210 (1944).

[64] Y. Henderson and H. W. Haggard, *Noxious Gases*. 2nd ed., Reinhold, New York, 1943.

[65] V. A. Tichonravov, *Farmakol. i Toksikol.*, **6**, 36 (1943).

fatigue. Unless the victim is removed to fresh air within a very few minutes, and breathing stimulated or induced by artificial respiration, death occurs. Unconsciousness and collapse occur within seconds in high concentrations; for that reason many persons have lost their lives attempting to save a victim who has collapsed from exposure. In such a case, holding the breath will permit a brief stay in the atmosphere, while to inhale it would cause almost immediate collapse.

As an example of the rapidity of these effects and recovery may be cited the experimental results observed with a dog exposed to a concentration on the order of 1000 p.p.m. When first placed in the atmosphere, the animal frisked playfully for a short time, stopped and stood still momentarily, breathing laboriously, fell on his side, gasped once or twice, then remained motionless with legs extended. At the end of 1 minute the dog was removed from the chamber and given artificial respiration; within 1 or 2 minutes he resumed his frisking as though nothing had occurred. Had breathing not been induced by artificial respiration, the heart would have stopped within a very few minutes: death would have resulted from asphyxiation. The mechanics of this respiratory paralysis formerly were thought to involve a chemical reaction with the respiratory enzymes or with the hemoglobin or both; but, in high concentrations of the gas, are now believed to be due to

TABLE 3

Physiological Response to Various Concentrations of Hydrogen Sulfide[64]

Response	Concentration, p.p.m.
Maximum allowable concentration for prolonged exposure	20
Slight symptoms after several hours	70–150
Maximum concentration for 1 hr. without serious consequences	170–300
Dangerous after exposure of $1/_2$ to 1 hr.	400–700

reflexes resulting from irritation of the carotid sinus.[66] Moderately high concentrations cause apne vera after overstimulation of the respiratory center.[64] Regardless of the exact mechanism of the cessation of respiration, if the victim is removed to uncontaminated air and respiration set in motion by any means before heart action has ceased, rapid recovery may be expected.

Subacute Effects. Hydrogen sulfide is an irritant gas and exposure to concentrations between 70 and 700 p.p.m. may irritate the mucous membranes of the eyes and of the respiratory tract. Pulmonary edema[66] or bronchial pneumonia is likely to follow prolonged exposure to concentrations on the order of 250 to 600 p.p.m. These levels of exposure may cause such symptoms as headache, dizziness, excitement, nausea or gastrointestinal disturbances, dryness and sensation of pain in the nose and throat and chest, and coughing. Table 3 indicates responses to various concentrations of hydrogen sulfide in the atmosphere.

[66] T. Sollmann, *A Manual of Pharmacology.* 6th ed., Saunders, Philadelphia, 1944.

Among the subacute and chronic effects of exposure to hydrogen sulfide, eye irritation resulting in conjunctivitis or "gas eyes"[67] is the most common and, ranging from mild to severe with extent and intensity of exposure, may include itching and smarting, a feeling of sand in the eyes, marked inflammation and swelling, cloudy cornea, destruction of the epithelial layer with scaling resulting in blurring of vision. Exposure to light may increase the painful effect. Atmospheric concentrations above 50 p.p.m. and up to 300 p.p.m. are conducive to this condition.

Absorption and Excretion. The absorption of hydrogen sulfide is almost exclusively through the respiratory tract. Absorption through the skin has been demonstrated and discoloration of the skin reported, but that this is not a significant source of systemic poisoning is indicated by the fact that the routine for gas mask approval testing by the United States Bureau of Mines has included the wearing of gas masks for 30-minute periods in atmospheres containing 2 per cent (20,000 p.p.m.) hydrogen sulfide. During these tests, which include strenuous exercise, the subjects have noted slight skin irritation but no systemic effects indicative of hydrogen sulfide absorption and no discoloration of the skin. When free sulfide exists in the circulating blood a certain amount of hydrogen sulfide is excreted in the exhaled breath. This is sufficient to be detected by odor. The greater portion, however, is excreted in the urine, chiefly as sulfate, but some also as sulfide.

There are no known reliable tests, applicable to the exposed individual, that are indicative of the degree of exposure.

5. Maximum Permissible Concentration

The accepted maximum permissible concentration for prolonged exposure is 20 p.p.m.

6. Flammability

Hydrogen sulfide is flammable within the range of 4.30 to 45.5 per cent by volume. The ignition temperature is 558°F.

7. Odor and Warning Properties

Although the characteristic odor of the gas is detectable in concentrations as low as 0.025 p.p.m., is distinct at 0.3 p.p.m., is offensive and moderately intense at 3 to 5 p.p.m., is strong and marked but not intolerable at 20 to 30 p.p.m., the odor of higher concentrations does not become more intense, and above about 200 p.p.m. the disagreeable odor appears less intense. These perceptions are based upon initial inhalations, and with continuous inhalation the olfactory sense fatigues rapidly.

[67] W. P. Yant, *Am. J. Public Health*, **22**, 598 (1930).

ALKALINE SULFIDES

The sulfides of potassium, sodium, calcium, and barium are solids that, in the presence of water or acids, liberate hydrogen sulfide into the air. They are caustic because of formation of free alkali by hydrolysis.

CARBON DISULFIDE, CS_2 (Carbon Bisulfide)

1. Uses and Industrial Exposures

Carbon disulfide is used in the xanthation of cellulose in the preparation of viscose, and exposures exist not only in the xanthating process but also during spinning and washing of the viscose. In the rubber industry carbon disulfide has been used as a solvent for sulfur or as a diluent for sulfur chloride in vulcanizing and as a solvent for rubber cement. It has been used as an insecticide and in the chemical industry as a solvent for phosphorus, fats, oils, resins, and waxes. It is used in the manufacture of optical glass, to fill glass prisms. Carbon disulfide is also encountered in the destructive distillation of coal.

2. Physical and Chemical Properties

Physical state: colorless liquid
Molecular weight: 76.13
Specific gravity: 1.2626 (20°C.)
Melting point: —108.6°C.
Boiling point: 46.3°C.
Vapor density: 2.63 (air = 1)
Vapor pressure: 360 mm. Hg (25°C.)
Refractive index: 1.6232 (25°C.)
Per cent in "saturated" air: 47.4 (25°C.)
Density of "saturated" air: 1.74 (air = 1)
Solubility: 0.22 g. in 100 ml. of water at 22°C.; miscible with alcohol, ether and benzene
Flash point: —22°F. (closed cup)
1 mg./cu. meter $\approx$ 0.32 p.p.m. and 1 p.p.m. $\approx$ 3.12 mg./cu. meter at 25°C., 760 mm. Hg

3. Determination in the Atmosphere

Scrub the atmosphere through two fritted glass scrubbers containing a solution of 0.5 per cent each diethylamine and triethanolamine and 0.001 per cent cupric acetate in 95 per cent ethyl alcohol.[68] The sampling rate should be 0.5 to 1 liter per minute. One to fifty micrograms of carbon disulfide develops sufficient color and comparison can be made with prepared standards. The determination is more satisfactorily made by means of a photometer. The vapor of carbon disulfide

[68] R. W. McKee, J. Ind. Hyg. Toxicol., 23, 151 (1941).

may also be determined directly in the air by means of an ultraviolet photometer. Satisfactory results can be obtained by the methods of Viles[69] or Matuzak.[70] The interferometer may also be used successfully (see Vol. I, page 179).

4. Physiological Response

Acute Effects. Table 4 lists six representative levels of effect upon man, with corresponding ranges of concentration of inhaled carbon disulfide.[71] The predominant effect of high concentrations of carbon disulfide is narcosis, and death may result from respiratory failure. Less severe exposures may result in headache, giddiness, respiratory disturbances, precordial distress, and gastrointestinal disturbances.[72] The possibility of injury to the central nervous system from a single severe acute exposure is reported by Lewy.[73]

TABLE 4

Effects of Various Concentrations of Carbon Disulfide on Man[71]

Effects	Concentration	
	mg./liter	p.p.m.
Slight or no effect	0.5–0.7	160–230
Slight symptoms after several hours	1.0–1.2	320–390
Symptoms after $1/2$ hr.	1.5–1.6	420–510
Serious symptoms after $1/2$ hr.	3.6	1150
Dangerous to life after $1/2$ hr.	10.0–12.0	3210–3850
Fatal in $1/2$ hr.	15.0	4815

Chronic Effects. Repeated brief exposures to high concentrations or prolonged exposures to low concentrations are of much greater industrial importance than are the single acute exposures. Among the subjective complaints that characterize chronic carbon disulfide poisoning are: fatigue, loss of memory, insomnia, listlessness, headache, excessive irritability, melancholia, vertigo, weakness, loss of appetite, gastrointestinal disturbances, and impairment of sexual functions. Visual disturbances, loss of reflexes, hallucinations, mania, or chronic dementia may occur. Lung irritation has been reported. Degenerative changes in the blood and blood-forming organs are reported to occur sometimes after poisoning has progressed. Dermatitis, and even blistering, may result from contact of vapor or liquid with the skin or mucous surfaces. Rubin and Arieff[74] studied 100 workers exposed for 4 years to average concentrations of 1.0 to 5.5 p.p.m. hydrogen sulfide

[69] F. J. Viles, *J. Ind. Hyg. Toxicol.*, **22**, 188 (1940).
[70] M. P. Matuzak, *Ind. Eng. Chem. Anal. Ed.*, **4**, 98 (1932).
[71] F. Flury and F. Zernik, *Schädliche Gase.* Springer, Berlin, 1931.
[72] *Public Health Repts.*, *U. S.*, **56**, 574 (1941).
[73] F. H. Lewy, *Penna. Dept. Labor Industry Bull.*, No. **46** (1938).
[74] H. H. Rubin and A. J. Arieff, *J. Ind. Hyg. Toxicol.*, **27**, 123 (1945)

and 1.9 to 26.4 p.p.m. carbon disulfide, or a combined sulfide gas and vapor concentration of 2.9 to 31.9 p.p.m., and found no indication of intoxication. Improvement or complete recovery is to be expected if the exposure is discontinued before severe damage results. Barthelemy,[75] reporting on 10 years of experience in the manufacture of viscose rayon, cites 3 cases of poisoning due to excessive exposure to carbon disulfide: 1 of mental derangement, and 2 with impaired motor nerves adversely affecting the leg muscles. All 3 workmen recovered completely within a few months after termination of exposure. He further states that when the carbon disulfide in the air was kept below 30 p.p.m. and the hydrogen sulfide below 20 p.p.m. no trouble whatsoever was experienced.

Mice and rats exposed 8 hours per day for 20 weeks to an average concentration of 37 p.p.m. (0.114 mg./liter) carbon disulfide showed evidence of toxic effects.[76]

Absorption and Excretion. The absorption of carbon disulfide is mainly through the lungs, where it enters the blood stream and is distributed throughout the body. Limited absorption can occur through the skin, and if swallowed, it is absorbed from the gastrointestinal tract. Saturation of the body occurs rapidly (see Vol. I, page 157). Eighty-five to ninety per cent of the carbon disulfide is metabolized[77] and eliminated in the urine as inorganic sulfates and other sulfur compounds; the balance is eliminated unchanged: 8 to 13 per cent in the exhaled breath, $1/2$ per cent in the urine, and none in the feces. It is probable that a trace is also eliminated in the sweat, but this has not been demonstrated.

Tests Indicating Exposure. Although the total sulfur in the urine of exposed persons is considerably elevated, there are many other factors that cause that same result. Carbon disulfide in the blood or urine, however, is indicative of exposure, and its concentration can give some indication of the severity of the exposure.

5. Maximum Permissible Concentration

The present accepted maximum permissible concentration for 8-hour exposure is 20 p.p.m. (0.0624 mg./liter).

6. Flammability

Carbon disulfide has a range of flammability of 1.25 to 50.0 per cent by volume in air. The ignition temperature is 248°F., a temperature commonly encountered in steam pipes, electric light bulbs, and elsewhere. The flash point by the closed cup method is −22°F.

[75] H. L. Barthelemy, *J. Ind. Hyg. Toxicol.*, **21**, 141 (1939).

[76] F. H. Wiley, W. C. Hueper, and W. F. von Oettingen, *J. Ind. Hyg. Toxicol.*, **18**, 733 (1936).

[77] R. W. McKee, C. Kipper, J. H. Fountain, A. M. Riskin, and P. Drinker, *J. Am. Med. Assoc.*, **122**, 217 (1943).

7. Odor and Warning Properties

Carbon disulfide has a foul, slightly ethereal odor that, however, does not offer adequate warning in the lower harmful concentrations.

CARBONYL SULFIDE, COS (Carbon Oxysulfide)

1. Source and Industrial Exposures

Carbonyl sulfide is encountered in the destructive distillation of coal and in the purification of petroleum and it frequently is found with hydrogen sulfide and carbon disulfide.

2. Physical and Chemical Properties

Physical state: colorless gas
Molecular weight: 60.07
Melting point: $-138.2°C.$
Boiling point: $-50.2°C.$
Solubility: in water, 0.54 ml./ml. at 20°C.; in alcohol, 8 ml./ml. at 22°C.; and in toluene, 15 ml./ml. at 22°C.; it decomposes in moist air to carbon dioxide and hydrogen sulfide
    1 mg./cu. meter ≎ 0.41 p.p.m. and 1 p.p.m. ≎ 2.5 mg./cu. meter at 25°C., 760 mm. Hg

3. Determination in the Atmosphere

Carbonyl sulfide in the air can be determined by absorption in alcoholic potassium hydroxide solution, and precipitation of sulfide with cadmium chloride or other suitable means; or it may be oxidized with hydrogen peroxide or bromine and the sulfate determined by precipitation with barium chloride. The gas can also be determined readily with the gas interferometer.

4. Physiological Response

The gas carbonyl sulfide is only slightly irritant to the lungs. It acts principally upon the central nervous system with death resulting mainly from respiratory paralysis. Rabbits[71] showed some ill effects after an exposure of $1/2$ hour to 1300 p.p.m., convulsions and death following an exposure of 1 hour to 3200 p.p.m. With mice[78] death occurred in $3/4$ minute when they were exposed to 8900 p.p.m., $1^1/_2$ minutes to 2900 p.p.m., and 35 minutes to 1200 p.p.m. Sixteen minutes' exposure to 900 p.p.m. caused no perceptible effects.

Experience with exposure of human beings has not been recorded. It is probable that the effects can be assigned to the action of the hydrogen sulfide resulting from partial decomposition in the lungs and after absorption into the blood

[78] A. Klemenc, Ber. deut. chem. Ges., 76, 299 (1943).

stream. It is also probable that the permissible limit should be somewhat more liberal than for hydrogen sulfide.

5. Flammability

Carbonyl sulfide is flammable between the range of 11.90 to 28.50 per cent by volume in air.

6. Odor and Warning Properties

Pure carbonyl sulfide has inadequate warning properties.

SULFUR MONOCHLORIDE, S_2Cl_2

1. Uses and Industrial Exposures

Sulfur monochloride is used in vulcanizing and in curing rubber. In the manufacture of rubber-coated fabrics, sulfur chloride has been used in oven "curing" atmospheres, and in some such operations has been poured into open containers and placed on steam coils on the floor of the curing oven, having little or no ventilation. The leakage into the room in such an instance causes pronounced irritation to the eyes and nose of anyone working in the room. The distribution and collection of open containers involve brief exposures to relatively high concentrations. Men who do this sometimes inadvisedly rely upon holding their breath during the period of exposure to high vapor concentration, rather than wearing a gas mask. Sulfur monochloride is also used in the manufacture of organic chemicals, printer's inks, varnishes, and cements; in hardening soft woods; and as an agricultural insecticide.

2. Physical and Chemical Properties

Physical state: yellowish red viscous liquid
Molecular weight: 135.05
Specific gravity: 1.678 (20°C.)
Melting point: —80°C.
Boiling point: 135.6°C.
Solubility: soluble in carbon disulfide, benzene, and ether; the liquid and its vapor decompose in water (or humid air) to form sulfur, hydrochloric acid, and sulfur dioxide
1 mg./cu. meter $\cong$ 0.2 p.p.m. and 1 p.p.m. $\cong$ 5.5 mg./cu. meter at 25°C., 760 mm. Hg

3. Determination in the Atmosphere

Sulfur monochloride vapor in the air may be determined by scrubbing through two scrubbers containing a measured quantity of $0.1N$ silver nitrate acidified with nitric acid. When sampling has been completed, add $0.1N$ sodium chloride solution equivalent to the silver nitrate used, and titrate the excess chloride with additional 0.1N silver nitrate. The reaction is as follows:

$$2\,S_2Cl_2 + 2\,H_2O + 4\,AgNO_3 \rightarrow 4\,AgCl + 3\,S + SO_2 + 4\,HNO_3$$

4. Physiological Response

Sulfur chloride has a suffocating odor and it is strongly irritant to the eyes, nose, and throat. A concentration of 150 p.p.m. has been stated[71] to be fatal to mice after an exposure of 1 minute, but the degree of toxicity has not yet been well established. The irritant effects are due to the sulfur dioxide and hydrochloric acid liberated by hydrolysis. Since this occurs rather readily, most of the irritant action likely is expended upon the upper respiratory tract. However, if the hydrolysis should not be completed in the upper respiratory tract, injury to the bronchioles and alveoli would result.

5. Hygienic Standard of Permissible Exposure

ACGIH threshold for 8 hours is 1 p.p.m.

SULFURYL CHLORIDE, SO_2Cl_2

1. Uses and Industrial Exposures

Sulfuryl chloride is used as a chlorinating and sulfonating agent in chemical manufacturing and for treating wool to prevent subsequent shrinking.

2. Physical and Chemical Properties

Physical state: colorless liquid
Molecular weight: 134.97
Specific gravity: 1.667 (20°C.)
Melting point: −54.1°C.
Boiling point: 69.1°C.
Vapor density: 4.7 (air = 1)
Refractive index: 1.444 (20°C.)
Solubility: decomposes in water and moist air, soluble in benzene and in acetic acid
1 mg./cu. meter ≎ 0.2 p.p.m. and 1 p.p.m. ≎ 5.5 mg./cu. meter at 25°C., 760 mm. Hg

3. Determination in the Atmosphere

Collect in alkaline medium and determine as chloride or sulfate.

4. Physiological Response

Sulfuryl chloride is a respiratory irritant. No level of permissible exposure has been published. See sulfur monochloride and thionyl chloride.

THIONYL CHLORIDE, SOCl₂

1. *Uses and Industrial Exposures*

Thionyl chloride is used as a chlorinating agent in chemical manufacturing.

2. *Physical and Chemical Properties*

Physical state: colorless liquid
Molecular weight: 118.97
Specific gravity: 1.655 (10°C.)
Melting point: −105°C.
Boiling point: 75.5°C.
Vapor density: 4.1 (air = 1)
Vapor pressure: 110 mm. Hg, 26°C.
Per cent in "saturated" air: 14.5 (26°C.)
Density of "saturated" air: 1.5 (air = 1) at 26°C.

Decomposes in water or moist atmosphere, yielding sulfur dioxide, sulfur, chlorine, sulfur monochloride, and hydrochloric acid

1 mg./cu. meter ≎ 0.2 p.p.m. and 1 p.p.m. ≎ 4.87 mg./cu. meter at 25°C., 760 mm. Hg

3. *Determination in the Atmosphere*

Thionyl chloride may be determined by the same method as sulfur monochloride.

4. *Physiological Response*

Thionyl chloride is a respiratory irritant. A 20-minute exposure to 17.5 p.p.m. is said to have proved fatal to cats.[71] No hygienic standard of permissible exposure has been published. It probably should be of the same order of magnitude as that for sulfur monochloride.

TELLURIUM, Te₂

1. *Uses and Industrial Exposures*

The world's production of tellurium since 1940 has been about 200,000 lb. annually, or one fifth that of selenium, along with which it is found in sulfide ores. The principal source of tellurium (and selenium) is the anode mud resulting during the electrolytic refining of copper. Tellurium is used as a rubber improver; in tellurium vapor "daylight" lamps; in cast iron, where minute amounts stabilize the iron carbide and appreciably increase the depth of the chill. The gray iron industry uses hundreds of tons annually, a considerable amount being for hardening the surface of car wheels. It is also used in malleable iron to improve ductility and in stainless steel for machinability. A fraction of 1 per cent alloyed with lead improves the corrosion resistance, strength, and hardening properties of the lead.

Tellurium is used to increase the machinability of copper and bronze, and to improve other metals and alloys. It is also used in several chemical processes, including use as a catalyst, and it is a semiconductor.

2. Physical and Chemical Properties

Physical state: hexagonal, brittle metallic, silvery semiconductor or an amorphous brownish black powder

Atomic weight: 127.61

Specific gravity: metallic, 6.24 at 20°C. and amorphous powder, 6.00 at 20°C.

Melting point: 450°C.

Boiling point: approx. 990°C.

Solubility: insoluble in water and carbon disulfide; soluble in oxidizing acids and alkalies

3. Determination in the Atmosphere

Dusts and fumes are satisfactorily collected by the electrostatic precipitator, while hydrogen telluride or methyl telluride can be absorbed by scrubbing through 48 per cent hydrobromic acid containing 5 to 10 per cent bromine, as in the collection of hydrogen selenide. The amount of sample needed will depend upon the collection apparatus available and the concentration in the air. Determination may be made by the method of Steinberg and associates.[79] Best results are obtained when the aliquot used for comparison with standards contains between 5 and 50 μg. of tellurium.

4. Physiological Response

Acute Effects. The actions of tellurium and selenium are similar to those of inorganic arsenic, especially the injurious effect on the capillaries.[80] A garlic odor is imparted to the breath by brief and minor exposures to tellurium compounds. This may result from a single short inhalation exposure, or from skin absorption from handling tellurium compounds. The garlic odor may persist for months if the amount of tellurium absorbed was significant. Suppression of the sweat, nausea, metallic taste, and somnolence also may result from significant acute inhalation exposures. More serious effects have not been reported from acute exposures.

Chronic Effects. The toxic effects of tellurium on animals are similar to those of arsenic. Steinberg and his associates[79] report that workmen, exposed 2 years to amounts ranging from 0.1 mg. to 7.4 mg. of tellurium and tellurium oxide per 10 cu. meters of air, exhibited in decreasing frequency: garlic odor of the breath, dryness of the mouth, metallic taste, somnolence, garlic odor of the sweat, loss of appetite, nausea; but there was no suppression of sweat, nor evidence of intoxication. Approximately 90 per cent of the air samples ranged between 0.1 mg.

---

[79] H. H. Steinberg, S. C. Massari, A. C. Miner, and R. Rink, *J. Ind. Hyg. Toxicol.,* **24,** 183 (1942).

[80] T. Sollmann, *A Manual of Pharmacology.* 6th ed., Saunders, Philadelphia, 1944.

and 1.0 mg. of tellurium per 10 cu. meters of air. However, since the symptoms recorded over the 2-year period could also occur from brief exposure to higher concentrations, it is not possible to say that concentrations below 1 mg. per 10 cu. meters would produce all of these symptoms. It is reasonable to think that where concentrations do not rise above 1 mg. per 10 cu. meters, serious illness is not to be expected.

Absorption and Excretion. Tellurium compounds can be absorbed through the skin, by ingestion, and by inhalation. They are excreted in the exhaled breath, sweat, urine, and feces.

Tests Indicating Exposure. Exposure is indicated by a garlic odor of the breath, or by the presence of amounts of 0.01 mg. or more of tellurium/liter of urine. Concentrations up to 0.06 mg./liter have been found in the absence of any evidence of intoxication from tellurium.

5. *Hygienic Standard of Permissible Exposure*

The maximum permissible concentration in the atmosphere depends upon the criterion. If a foul breath is to be avoided, the maximum permissible concentration is on the order of 0.1 mg. or 0.2 mg. of tellurium per 10 cu. meters of air. If the manifestation of toxic effects is to be the criterion, 1 mg. per 10 cu. meters appears quite safe, and indications are that the toxic level is probably in excess of 8 mg. per 10 cu. meters. It seems somewhat logical that for practical purposes the presence or absence of the characteristic tellurium breath can be relied upon to indicate exposure, and if this is heeded as an indication of the necessity for control measures, injurious exposures are not likely to occur. The ACGIH threshold for 8-hour exposure is 0.1 mg./cu. meter.

6. *Flammability*

Tellurium burns slowly in air to form tellurium dioxide.

HYDROGEN TELLURIDE, H_2Te

1. *Uses and Industrial Exposures*

Hydrogen telluride has no industrial uses.

2. *Physical and Chemical Properties*

Physical state: colorless gas
Molecular weight: 129.63
Melting point: —51°C.
Boiling point: —4°C.
Vapor density: 4.5 (air = 1)
Solubility: dissolves and decomposes in water, precipitating elemental tellurium

1 mg./cu. meter $\backsim$ 0.2 p.p.m. and 1 p.p.m. $\backsim$ 5.3 mg./cu. meter at 25°C., 760 mm. Hg

3. *Physiological Response*

Hydrogen telluride has an odor somewhat resembling hydrogen sulfide, and is irritant in relatively low concentrations. Its actions on the human system are believed to be similar to those of other tellurium compounds. Its degree of toxicity has not been established but, owing to its ready decomposition, it is believed to be less toxic than arsine or hydrogen selenide.

CHAPTER XXV

Inorganic Compounds of Oxygen, Nitrogen, and Carbon

FRANK A. PATTY

OXYGEN, O_2

1. *Occurrence and Uses*

Oxygen (20.95 per cent by volume) is a normal constituent of the air we breathe. Men and animals are dependent upon the presence of oxygen and can live only a few minutes in its absence. Oxygen is required by the body for combustion in the tissues in amounts proportional to energy expenditures. Henderson and Haggard[1] give the approximate energy expenditures, oxygen consumption, and volume of breathing of an average (154 lb.) man (Table 1).

TABLE 1

Energy Expenditure, Oxygen Consumption, and Volume of Breathing of Man

Activity	Energy used, cal./min.	O_2 consumption, liter/min. (0°C., 760 mm.)	Vol. of air breathed, liter/min. (20°C.)
Rest in bed, fasting	1.15	0.240	6
Sitting	1.44	0.300	7
Standing	1.72	0.360	8
Walking, 2 m.p.h.	3.12	0.650	14
Walking, 4 m.p.h.	5.76	1.200	26
Slow run	9.60	2.000	43
Maximum exertion	14–20	3.000–4.000	65–100

The capacity to supply oxygen to the tissues declines with increasing age as indicated in Table 2.

Oxygen inhalation during decompression has been used to prevent compressed-air illness.[3]

[1] Y. Henderson and H. W. Haggard, *Noxious Gases*. Reinhold, New York, 1943.

[2] S. Robinson, *Arbeitsphysiologie*, **10**, 251 (1939).

[3] R. R. Jones, J. W. Crosson, F. E. Griffith, R. R. Sayers, H. H. Schrenk, and E. Levy, *J. Ind. Hyg. Toxicol.*, **22**, 427 (1940).

TABLE 2

Age and Maximal Oxygen Intake[2]

Mean age, years	Number of subjects	Mean maximal O_2 intake, liter/min.	O_2, cc./kg. of body wt./min.
17.4	11	3.61	52.8
24.5	11	3.53	48.7
35.1	10	3.42	43.1
44.3	9	2.92	39.5
51.0	7	2.63	38.4
63.1	8	2.35	34.5
75.0	3	1.71	25.5

2. Physical and Chemical Properties

Physical state: Colorless, odorless gas
Molecular weight: 32
Melting point: $-218.4°C$.
Boiling point: $-183°C$.
Solubility in water: 4.89 cc. per 100 ml. at $0°C$.

3. Physiological Response to Increased Concentrations

Effects on Man. The inhalation of 100 per cent oxygen for periods up to 16 hours per day for many days at atmospheric pressure has caused no observed injury to man. It is believed to have no serious adverse effect for a continuous exposure of 24 to 48 hours.[4] Pure oxygen for 24 hours at atmospheric pressure causes some substernal distress, but none at $1/2$ atm. (equivalent of 18,000 ft. altitude).[5] Mixtures of up to 65 per cent oxygen in air may be inhaled for extended periods with no known ill effects. The inhalation of pure oxygen at 3 atm. pressure (30 lb. gage) is safe for man for a period of 30 minutes.[6] Longer periods or higher pressures may produce oxygen poisoning. Convulsions have occurred in man after oxygen has been breathed for 45 minutes at 4 atm. pressure,[7] while, after 1 to 3 hours at 1 atm. pressure, neuromuscular coordination and the power of attention were adversely affected. In most subjects they were impaired or at least increased effort was required to maintain them (see also Vol. I, page 599).

Oxygen[8] at 3 atm. pressure can be breathed by young, healthy men for 3 hours without distressing symptoms. During the fourth hour, a progressive contraction of

[4] T. Sollman, *A Manual of Pharmacology*. 6th ed., Saunders, Philadelphia, 1944.

[5] J. H. Comroe, Jr., R. D. Dripps, P. R. Dumke, and M. Deming, *J. Am. Med. Assoc.*, **128**, 710 (1945).

[6] A. R. Behnke, L. A. Shaw, C. W. Shilling, R. M. Thomson, and A. C. Messer, *Am. J. Physiol.*, **107**, 13 (1934).

[7] A. R. Behnke, F. S. Johnson, J. R. Poppen, and E. P. Motley, *Am. J. Physiol.*, **110**, 565 (1935).

[8] A. R. Behnke, H. S. Forbes, and E. P. Motley, *Am. J. Physiol.*, **114**, 436 (1936).

the visual field with dilatation of the pupils and some impairment in central vision is the most constant criterion of oxygen toxicity. Circulatory changes indicative of peripheral vascular constriction are associated with the visual impairment and culminate during the fourth hour in an abrupt rise in systolic and diastolic blood pressure, increase in pulse rate, and extreme pallor. At this stage the subjects experience dizziness and a feeling of impending collapse, with partial stupefaction. Rapid and complete recovery, attended by a feeling of alertness and stimulation, results within an hour after air is substituted for oxygen.

Effects on Dogs. Among dogs[9] inhaling pure oxygen at atmospheric pressure, oxygen poisoning begins to develop after 36 hours, causes distress within 48 hours, and death in 60 hours. Ninety per cent oxygen in air requires double the exposure period for similar results; in 80 per cent oxygen in air the animals did not die but were ill at the end of a continuous exposure of 1 week. A decline in oxygen saturation of the blood, rise in hemoglobin, lung congestion and edema, right heart failure, and liver congestion were frequent findings in oxygen poisoning.

4. *Fire Hazards of Increased Concentrations*

In the use of air containing oxygen in concentrations above 21 per cent, or air at increased pressures, all materials are more readily oxidized than in the normal atmosphere and therefore will ignite more easily and burn more rapidly, thus presenting a greater fire hazard. Great care must be exercised to assure that cylinders, gages, valves, or lines to be used with compressed oxygen do not become contaminated with traces of oil or other readily oxidizable material. If such contamination exists, explosive oxidation may occur when contact is made with the compressed oxygen.

Oxygen Deficiency

1. *Occurrence*

Oxygen deficiency is of much more concern in industry than are high concentrations or high pressures of oxygen. It may be encountered in numerous situations such as in tanks, vats, holds of ships, silos, mines, or in any poorly ventilated area where the air may be diluted or displaced by gases or vapors of volatile materials, or where the oxygen may be consumed by chemical or biological reaction processes.

2. *Determination in the Atmosphere*

The analysis of the atmosphere for its oxygen content may be done with the Haldane or Orsat[10] gas-analysis apparatus. Safety lamps[11] are used for detecting

[9] J. R. Paine, A. Keys, and D. Lynn, *Am. J. Physiol.,* **133,** 406 (1941).

[10] G. A. Burrell and F. M. Seibert, revised by G. W. Jones, *U. S. Bur. Mines Bull.* No. **197** (1926).

[11] A. B. Hooker, E. J. Coggeshall, and G. W. Jones, *U. S. Bur. Mines Rept. Invest.* No. **3327** (1937).

oxygen deficiency, and where there was no danger of encountering flammable gases a lighted candle has been used to indicate unsafe atmospheres (less than 16.5 per cent oxygen). If the candle burns, there is believed to be sufficient oxygen to support life. This, however, has been demonstrated to be unreliable (see Vol. I, page 591). Oxygen-recording devices covering the range of 0 to 15 per cent oxygen are available.

3. *Physiological Response*

When the concentration of oxygen in the atmosphere falls below 16 per cent, symptoms of anoxia begin to appear. These may be classified by stages[1] as given in Table 3.

TABLE 3

Response of Man to the Inhalation of Atmospheres Deficient in Oxygen

Stage	Oxygen vol. %	Symptoms or phenomena
1	12–16	Breathing and pulse rate increased, muscular coordination slightly disturbed
2	10–14	Consciousness continues, emotional upsets, abnormal fatigue upon exertion, disturbed respiration
3	6–10	Nausea and vomiting, inability to move freely, loss of consciousness may occur; may collapse and although aware of circumstances be unable to move or cry out
4	Below 6	Convulsive movements, gasping respiration; respiration stops and a few minutes later heart action ceases

Mixtures of 2 per cent oxygen with nitrogen have been administered for 3 or 4 minutes in the treatment of certain forms of insanity[12] with only an occasional respiratory failure. There is almost immediate loss of consciousness, progressive stimulation of respiration, tachycardia and irregularity of heart action, muscular twitching, and opisthotonos. One or two inflations with oxygen are said to produce complete recovery. The suddenness with which oxygen-deficient atmospheres can cause unconsciousness and death may be explained as follows:

During the inhalation of normal air the arterial blood leaves the lungs about 95 per cent saturated with oxygen, and, with a subject standing at rest, the venous blood returns to the lungs about 60 to 70 per cent saturated. During 1 minute approximately 360 cc. of oxygen are used up. After a forced deep inspiration the normal lung volume is about 5 to 5.5 liters, 1 liter of which is oxygen, or nearly a 3-minute oxygen supply should the breath be held. However, when a subject inhales normally, the lung volume is only about 2.5 to 3 liters and if an oxygen-depleted atmosphere is inhaled at the rate of about 8 liters per minute, the 2.5 to 3 liters of atmosphere in the lungs is depleted of its oxygen by ventilation much more

[12] F. A. D. Alexander and H. E. Himwick, *Am. J. Physiol.,* **126,** 418 (1939).

rapidly than by absorption. Since the initial effects of anoxia are increased rate of breathing and circulation, these processes are speeded up, the oxygen percentage in the lung atmosphere falls below 10 per cent, and the arterial blood supply to the brain very quickly (seconds, rather than minutes) becomes insufficiently oxygenated to maintain consciousness. Respiratory failure and cessation of heart action soon follow unless the subject is returned to respirable air.

There is considerable personal variation in susceptibility to anoxia. Because of impaired compensation, persons with cardiac and pulmonary deficiencies are more susceptible. Hyperthyroid persons normally consume more oxygen and are therefore more susceptible, while the reverse is true of hypothyroid persons. Whenever anoxia is prolonged, recovery is slow and there are apt to be sequelae such as hallucinations, excitement, headache, nausea, and apathy extending over several hours; these are thought to result from pressure of cerebral edema. When the anoxia is severe and prolonged, with unconsciousness, irreversible degenerative changes[13] in the nervous system, especially in the cerebral cortex and basal ganglia, may occur. These result in paralyses, amnesia, and other manifestations of permanent injury. The condition is perhaps more often seen following prolonged unconsciousness due to carbon monoxide poisoning. There is a marked difference in survival time[14] of different nerve tissues when cut off from their blood supply, the cerebrum and cerebellum having the shortest, 8 and 13 minutes, and the myenteric plexus longest, 180 minutes (see Table 10). Likewise, the cortical center of the brain can revive after an interruption of blood supply not to exceed 5 minutes, but the cardioregulatory, vasomotor, and respiratory centers may survive up to 30 minutes' interruption. It is evident, then, why severe anoxia, whether from the inhalation of an oxygen-deficient atmosphere or from any other cause, may result in cerebral damage even though respiration may not have stopped.

4. *Warning Properties*

The warning properties of an atmosphere deficient in oxygen are completely inadequate and, although a trained observer may, when alert, recognize the increase in the pulse and rate of breathing in time to return to good air, the average individual fails to recognize the danger until he is too weak to save himself, especially where the return to good air involves climbing stairs or a ladder. The fire hazard of oxygen-deficient atmospheres is below normal and where the oxygen content of the air is below 16 per cent many common materials will not burn.

OZONE, O_3

1. *Occurrence and Uses*

Ozone in very small amounts is a frequent and variable constituent of the atmosphere we breathe. During and following electrical storms it may reach

[13] A. T. Steegmann, *A.M.A. Arch. Neurol. Psychiat.*, **41**, 955 (1939).
[14] C. K. Drinker, *Carbon Monoxide Asphyxia.* Oxford Univ. Press, New York, 1938.

sufficient concentrations to be readily recognizable by odor—on the order of 0.01 to 0.05 p.p.m. by volume. Reports of concentrations up to 100 times that amount in the outdoor air must be regarded with some skepticism. Ozone can be generated by a high-tension nonsparking discharge in air or oxygen. It is used for the sterilization of water; for bleaching oils, paper, and flour; for aging liquor; and in combating odors, in lieu of proper ventilation. Claims that ozone will detoxify exhaust gases, freshen or purify the air, kill bacteria in the atmosphere of inhabited spaces, and other equally fantastic feats have not been substantiated.

2. Physical and Chemical Properties

Physical state: colorless gas
Molecular weight: 48
Melting point: −192.1°C.
Boiling point: −112°C.
Vapor density: 1:65 (air = 1)
Solubility in water: 49 cc. per 100 ml. at 25°C.
1 mg./cu meter ⊃⊂ 0.51 p.p.m. and 1 p.p.m. ⊃⊂ 2 mg./cu. meter at 25°C., 760 mm. Hg

3. Determination in the Atmosphere

In the absence of other oxidizing gases such as nitrogen oxides and hydrogen peroxide, the atmosphere to be sampled may be scrubbed through an acidified solution of potassium iodide and titrated with $0.1N$ sodium thiosulfate.[15] In the presence of interfering gases, the sample should first be scrubbed through chromic acid and potassium permanganate. A similar method[16,17] utilizes alkaline potassium iodide as an absorbent. This solution is then neutralized and the intensity of the color measured. A method for total oxidants was proposed by Haagen-Smit.[18]

4. Physiological Response

Ozone is very irritant to all mucous membranes, and significant exposures may cause pulmonary edema. Its prolonged inhalation in concentrations above 0.05 p.p.m. is inadvisable because of the danger of pulmonary irritation. The effects of various concentrations as tabulated by Witheridge and Yaglou[19] are given in Table 4. The 1961 A.C.G.I.H. threshold is 0.1 p.p.m.

These investigations confirmed the findings of others, that ozone does not destroy odors, but masks the odor and appears to "freshen" the air by lowering

[15] M. B. Jacobs, The Analytical Chemistry of Industrial Poisons, Hazards, and Solvents. Interscience Publishers, New York–London, 1949.
[16] R. G. Smith and P. Diamond, Am. Ind. Hyg. Assoc. Quart., 13, 252 (1952).
[17] D. H. Byers, B. E. Saltzman and F. L. Hyslop, Rept. Environmental Research Laboratory. Dept. Public Health, Univ. of Washington, May 10, 1955.
[18] A. J. Haagen-Smit and M. F. Brunelle, Intern. J. Air Pollution, 1, 51 (1958).
[19] W. N. Witheridge and C. P. Yaglou, Trans. ASHVE, 45, 509 (1939).

TABLE 4

Effects of Ozone in Various Concentrations

Observed effect	Concentration, p.p.m.
Threshold of odor, normal person	0.01–0.015
Maximum allowable concentration	0.04
Objectionable to all normal persons, irritates the nose and throat of most persons	0.10
Disorders breathing, reduces oxygen consumption, and shortens lives of guinea pigs	0.5–1.0
Inhibits fungus and mold growth in cold storage rooms	0.3–1.5
Headache, respiratory irritation, and possible coma	1–10
Lethal to small animals within 2 hrs.	15–20
Lethal in a few minutes	>1700
Germicidal for air-borne organisms	6500

the perceptibility of odors. Odors of high intensity cannot be successfully masked by concentrations of ozone regarded as safe for prolonged inhalation. Flury and Zernik[20] state: (1) that men, when exposed to ozone, suffer eye, nasal, and throat irritation, cramps in the chest, frontal headache and vertigo, increasing fatigue, and lowering of blood pressure, owing to the centrally conditioned dilatation of the peripheral blood vessels; (2) that 0.5 p.p.m. causes distinct irritation; (3) that 1 p.p.m. for an hour causes cough and serious fatigue; and (4) that a brief period of inhalation of 5 to 10 p.p.m. accelerates the pulse and causes stupefaction and continuous body pain.

In a more recent study[21] in which animals were exposed repeatedly to the inhalation of 1 p.p.m. ozone over a prolonged period, the terminal airways of the lungs were thickened, the air passages narrowed, with fibrotic tissue formation, resulting in diminished capacity to move air through the lungs.

NITROGEN, N_2

Nitrogen is a colorless, odorless, physiologically inert gas that normally constitutes about 78 per cent of the atmosphere by volume. Its molecular weight is 28.02. It melts at −209.86 and boils at −195.8°C. Its solubility in water is 2.33 cc. at 0°C. and 1.02 cc. at 45°C. per 100 ml. water.

The only physiological effects due to inhalation of nitrogen result from oxygen dilution, and nitrogen must be present in sufficient amount to reduce the partial pressure of oxygen below about 100 mm. Hg equivalent before serious effects of anoxia result (see Oxygen Deficiency, page 914). Saturation of the body with nitrogen occurs rapidly (Vol. I, pages 157, 588). During rapid lowering of the pressure of the environmental atmosphere, nitrogen may separate as bubbles in

[20] F. Flury and F. Zernik, *Schädliche Gase.* Springer, Berlin, 1931.

[21] H. E. Stokinger, W. D. Wagner, and O. J. Dobrogorski, *A.M.A. Arch. Ind. Health,* **16,** 514 (1957).

the tissue, capillaries, and veins, causing "bends" (see Chapter XVII). Helium, because of its higher rate of diffusion, has been found useful[22,23] as an oxygen diluent in the prevention of "bends." Undiluted oxygen has likewise been used for this purpose.[24]

NITROUS OXIDE, N₂O

Nitrous oxide, molecular weight 44.02, is a colorless gas having a sweetish taste. Its melting point is —102.4°, and its boiling point is —89.5°C. Its solubility in water is 60 cc. per 100 ml. of water at 25°C. It is weakly narcotic and is used as an anesthetic: when it is mixed with air, it acts chiefly as an asphyxiant by lowering the oxygen percentage; when mixed with oxygen, the percentage of nitrous oxide must be more than 80 in order to produce deep anesthesia. Nitrous oxide is of little interest to industrial hygienists.

NITRIC OXIDE, NO

Nitric oxide, molecular weight 30.01, is a colorless gas slightly heavier than air. Its boiling point is —151.8° and its melting point is —163.6°C. Its solubility in water is 7.34 cc. in 100 ml. water at 0°C.

1 mg./cu. meter $\backsimeq$ 0.815 p.p.m. and 1 p.p.m. $\backsimeq$ 1.23 mg./cu. meter at 25°C., 760 mm. Hg.

Nitric oxide is oxidized in the air to nitrogen dioxide and so, for practical purposes, its toxicity need not be given marked attention, because the resulting nitrogen dioxide is much more insidious. Nitric oxide is not an irritant, but in animals it has been found to act upon the central nervous system to produce paralysis phenomena and convulsions.[20] It combines with hemoglobin and this is oxidized by oxygen in the blood to methemoglobin, with resultant anoxia.

Mice exposed to 2500 p.p.m. for 6 or 7 minutes were narcotized and death occurred within 12 minutes, but if the narcotized animals were returned to fresh air within 4 to 6 minutes, recovery was rapid.[20] Poisoning of man has not been reported.

The significant gases contributing to the toxicity of "nitrous fumes" are NO and NO₂. NO₂ is about 4 or 5 times as toxic as NO.[25]

The combustion products of internal combustion engines contain significant quantities of each of the above gases. See also under Carbon Monoxide, Industrial Exposures, page 926.

[22] R. R. Sayers, W. P. Yant, and C. Hildebrand, U. S. Bur. Mines Rept. Invest. No. 2670 (1925).

[23] R. R. Sayers and W. P. Yant, Anesthesia & Analgesia, 5, 127 (1926).

[24] R. R. Jones, J. W. Crossen, F. E. Griffith, R. R. Sayers, H. H. Schrenk, and E. Levy. J. Ind. Hyg. Toxicol., 22, 427 (1940).

[25] E. LeB. Gray, A.M.A. Arch. Ind. Health, 19, 479 (1959).

NITROGEN DIOXIDE, NO₂ AND NITROGEN TETROXIDE, N₂O₄

1. Uses and Industrial Exposures

Nitrogen dioxide and its polymer, nitrogen tetroxide, are always found together at normal environmental temperatures. Nitrogen dioxide can be used to nitrate benzene, anthracene, and naphthalene at 20 to 60°C. but has not been extensively applied. It is a by-product of many operations and results whenever nitric acid acts upon metals, as in bright dipping, pickling, and etching, or upon organic material, as in the nitration of cotton or other cellulose. It is a by-product of the manufacture of many chemicals including explosives, dyes, lacquers, and celluloid. It also results, in significant amounts, from the slow burning of explosives or the detonation of explosives having a high oxygen balance; from electric arcs or electric- and gas-welding or gas-shrinking operations in confined and unventilated areas; from the burning of nitrocellulose; from the accidental spillage of nitric acid; during operations incidental to the manufacture or recovery of nitric acid; from the reduction of nitrates as, for instance, in the accidental pollution of a molten nitrate salt, heat-treat bath with some readily oxidizable matter. Nitrogen dioxide has been used as a component of fuel for jet propulsion.

The brown mixture arising from the bright dipping of copper or brass or from nitration reactions is essentially a pure mixture of NO_2 and N_2O_4. So far as industrial exposures are concerned, it matters little whether the nitrogen oxides enter the air as NO, NO_2, or N_2O_4, since the NO gradually changes to NO_2 and the NO_2–N_2O_4 balance mentioned in the next paragraph then comes into play. This mixture has frequently been erroneously referred to as nitrous fumes, but it is not nitrous oxide, N_2O, and it is a gaseous mixture not a fume. It may, for convenience, more properly be called nitrogen dioxide. Fatal concentrations have been reported in farm silos shortly after filling.[26]

2. Physical and Chemical Properties

Nitrogen dioxide is a dark chocolate-brown gas, with molecular weight 46.01. Its polymer, nitrogen tetroxide, N_2O_4, with molecular weight 92.02, is colorless. At −9.3°C. and below, the oxides are a colorless solid composed completely of nitrogen tetroxide. At temperatures of 135°C. and above, the gas is a very dark chocolate-brown, composed essentially of NO_2. At temperatures within this range, the gases are always present as a mixture of the two. At 37.5°C., the temperature of the body and therefore the temperature at which the gases react upon the human lung, the ratio of NO_2 to N_2O_4 is 30:70.[27] Both gases are frequently referred to as the dioxide, and computations such as the following conversion factors are almost always made on the same basis as though the mixture were all NO_2. The gas reacts with water to form a mixture of nitrous acid, HNO_2, and nitric acid, HNO_3. The

[26] T. Lowry and L. M. Schuman, J. Am. Med. Assoc., 162, 153 (Sept. 15, 1956).
[27] Y. Henderson and H. W. Haggard, Noxious Gases. 2nd ed., Reinhold, New York, 1943.

relative proportions are frequently assumed to be equal, but may vary with circumstances.

1 mg./cu. meter $\backsimeq$ 0.53 p.p.m. and 1 p.p.m. $\backsimeq$ 1.88 mg./cu. meter at 25°C., 760 mm. Hg.

3. Determination in the Atmosphere

Perhaps the most satisfactory laboratory method of determining the concentration of nitrogen dioxide in the air is the phenol–disulfonic acid method described by Beatty, Berger, and Schrenk.[28] The method, however, is exacting and time-consuming, and unless meticulous care is employed it frequently gives low results. It does not differentiate between the dusts of nitrates and nitrites, or between the brown oxide gases and nitric acid, and its result is in terms of the total as nitrates. A satisfactory field method[29] for the determination of nitrogen dioxide has been developed from the α-naphthylamine–nitrite reaction. A sampling kit, with permanent dyed cellophane standards, Luer syringes, and reagent bottles, is available commercially.[30] The total time required for sampling and determination in the field is less than 10 minutes. Results obtained under controlled conditions in the laboratory, as well as in nitration plants, acid manufacture, bright dipping, welding, and mining operations, indicate that the method is reliable and convenient. By substituting α-naphthylamine hydrochloride for α-naphthylamine, using a 2-liter sample bottle and a spectrophotometer for comparison of standards with samples, the method is satisfactory for fractions of a part per million.

4. Physiological Response

The odor of nitrogen dioxide is characteristic and distinct in concentrations as low as 5 p.p.m. In concentrations of 10 to 20 p.p.m. the gas is mildly irritant to the eyes, nose, and upper respiratory mucosa. There is very little difference in intensity of odor and irritation, however, between concentrations of 20 and 100 p.p.m. In well-lighted areas nitrogen dioxide–air mixtures of 100 p.p.m. or more nitrogen dioxide have a visible, reddish brown tint. These properties of the gas can in no way be considered adequate warning. *It should be pointed out with emphasis that the ordinary type A (acid gas) or type AB (acid gas and organic vapor) canister gas masks, with soda lime or soda lime-activated carbon fills, do not offer satisfactory protection against nitrogen dioxide gas.*

Acute Effects. Nitrogen dioxide, in concentrations from 100 to 1000 p.p.m. or more, caused death in five species of animals—cats, guinea pigs, mice, rats, and rabbits—by asphyxia resulting from pulmonary edema induced by irritation of the lung tissue.[31] As related to the concentrations of gas, the average durations of

---

[28] R. L. Beatty, L. B. Berger, and H. H. Schrenk, *U. S. Bur. Mines Rept. Invest.* No. **3687** (1943).

[29] F. A. Patty and G. M. Petty, *J. Ind. Hyg. Toxicol.*, **25**, 361 (1943).

[30] Mine Safety Appliances Co., Pittsburgh, Pa.

[31] L. W. Latowsky, E. L. MacQuiddy, and J. P. Tollman, *J. Ind. Hyg. Toxicol.*, **23**, 129 (1941).

exposure causing death were found to be as tabulated in Table 5. This indicates a higher order of toxicity, but is not in serious disagreement with findings reported by Flury and Zernik[20] for mice, rabbits, and cats as tabulated in Table 6.

With man, concentrations considered dangerous for short exposures, above 50 p.p.m., are moderately irritating to the eyes and nasal passages. Higher concentrations, up to 150 p.p.m., cause an acid taste but are not painfully irritant. There have been many deaths resulting from acute exposure to nitrogen dioxide and, although there is little information on attendant atmospheric concentrations, there is reason to believe that the results of exposure of man are similar to those of animals, where death has been found to be due to asphyxia resulting from a pulmonary edema and not due to the effects of nitrite.

TABLE 5

Concentrations of Nitrogen Dioxide and Average Time to Produce Death in Animals

Concentration, p.p.m.	Time, min.
30	No deaths
100	318
150	90
400	58
600	32
800	29
1000	19

TABLE 6

Concentrations of Nitrogen Dioxide and Time of Exposure Causing Death of Animals within 24 Hours

Concentration, p.p.m.	Time, min.
110–125	360–420 (no effects)
225–230	315–420
340–410	60–105
1000	30–50
3350–7500	8–10

Most water-soluble, irritant gases exert their strongest effects at the earliest point of contact with moist mucous surfaces, but not so nitrogen dioxide. This difference has been accounted for by the fact that nitrogen dioxide hydrolyzes slowly in water or humid air to form nitrous and nitric acids. The theory is that during inhalation the relatively dry gas–air mixture reacts little with the slightly moist surfaces of the respiratory passages, whereas after reaching the alveoli the humid air, moist surfaces, and extended time promote almost complete hydrolysis in intimate contact with the alveolar tissue. According to Sollmann,[32] this always

[32] T. Sollmann, A Manual of Pharmacology. 6th ed., Saunders, Philadelphia, 1944.

results in edema, and a person who has been exposed to lethal concentrations of nitrogen dioxide may feel no discomfort for several hours after the end of exposure, but as long as 8 hours later may become distressed by the accumulation of fluid in his lungs. Whenever fluid collects in the lungs, it interferes with oxygen exchange and asphyxia may result. Symptoms may include weakness, a cold feeling, nausea, abdominal pain, coughing with a foamy yellow or brown expectorate, accelerated heart action, severe cyanosis with convulsions. Sollmann[32] points out that the slightest exertion under such circumstances may produce dyspnea, cyanosis, cardiac dilatation, and collapse; and, when this situation terminates fatally, death occurs in most cases within 8 to 48 hours following exposure. Where the pulmonary edema is survived, infectious pneumonia is a probable sequela that may cause death some weeks later. It has been pointed out by von Oettingen[33] that acute nitrogen dioxide poisoning may not always follow the usual pattern, but may cause a reversible type, characterized by dyspnea, cyanosis, vomiting, vertigo, somnolence, loss of consciousness, and methemoglobinemia, without pulmonary edema, from which the victim may recover completely if removed from exposure early. Another type, termed "shock type," is described in which a person exposed to a sudden, high concentration of nitrogen dioxide (and possibly nitric oxide) suffers asphyxiation, convulsions, and respiratory arrest. The similarity between nitrogen dioxide poisoning and phosgene poisoning is noteworthy.

Chronic Effects. The nitrite effect, resulting from absorption of nitrous acid hydrolyzed from nitrogen dioxide, is a factor to be considered along with pulmonary irritation in prolonged exposure to concentrations between 25 and 100 p.p.m. In concentrations above 100 p.p.m. it is probable that the effects of irritation outweigh any others. Adverse effects of exposures well below 25 p.p.m. have been reported by the Institute of Hygiene of Labor and Industrial Diseases, Leningrad,[34] but our experience in the United States does not lend support to these findings.

Men observed by the author working 6 to 8 hours daily in nitric acid recovery and fortification plants, where exposures ranged from 5 to 30 p.p.m. and averaged 10 to 20 p.p.m., for periods up to 18 months, evidenced no significant ill health nor were any characteristic adverse effects detected by periodic medical examinations.

From experimental exposures of guinea pigs and rats to filtered carbon-arc fumes, Tollman, MacQuiddy, and Schonberger[35] concluded that nitrogen dioxide inhaled in concentrations in excess of 100 p.p.m., 4 hours per day, will lead, in time, to fatal results.

McCord, Harrold, and Meek,[36] studying the effects of welding fumes on rabbits and rats, found that exposure to fumes containing up to 24 p.p.m. nitrogen

[33] W. F. von Oettingen, *U. S. Public Health Bull.* No. **272** (1941).

[34] N. A. Vigdortschik, E. C. Andreeva, I. Z. Matussevitsch, M. M. Nikulina, L. M. Frumina, and V. A. Striter, *J. Ind. Hyg. Toxicol.*, **19**, 469 (1937).

[35] J. P. Tollman, E. L. MacQuiddy, and S. Schonberger, *J. Ind. Hyg. Toxicol.*, **23**, 269 (1941).

[36] C. P. McCord, G. C. Harrold, and S. F. Meek, *J. Ind. Hyg. Toxicol.*, **23**, 200 (1941)

dioxide for 6 hours per day, 5 days a week, to a total of 45 days, produced an average of 2.9 per cent methemoglobin in rabbits and 15 per cent in rats; but they concluded that no permanent, harmful effects were demonstrated by prolonged exposure of rabbits and rats to atmospheres containing up to 70 p.p.m. nitrogen dioxide. They also reported 2.3 to 2.6 per cent methemoglobin in the blood of welders exposed to from 3.9 to 5.4 p.p.m. nitrogen dioxide, and suggest that methemoglobin might be useful as a measure of degree of exposure to welding fumes.

Other harmful effects, which are more or less common to all irritant acid gases, have been described by von Oettingen.[33]

Gray[25] has pointed out the disparity of reports dealing with the toxicity of the oxides of nitrogen. The relative amounts of nitric oxide and dioxide and the admixture of ozone are the most significant items in estimating the hazards of an inhalation exposure to these contaminants.

It has been stated that proof is lacking that nitrogen oxides, as such, are irritant. In support of the general opinion that they are irritant, it may be of interest to mention a personal experience concerning the action of the gas on normal dry skin. During the breaking of many glass ampoules of pure nitrate-free $NO_2-N_2O_4$ (purity attested by analysis at two different laboratories), whenever the liquid or the concentrated gas came in contact with the dry skin corrosion resulted. The corroded area had the same appearance that results from contact with nitric acid or its concentrated vapors except that the action was not as intense.

5. Hygienic Standard of Permissible Exposure

The AIHA Hygienic Standard for maximum atmospheric concentration for 8 hours and the ACGIH threshold for 8 hours are 5 p.p.m.

NITROGEN CHLORIDE, NCl_3

1. Uses and Industrial Exposures

Nitrogen chloride is used in the bleaching of flour and in the fumigation of citrus fruit.

2. Physical and Chemical Properties

Physical state: yellowish, oily liquid
Molecular weight: 120.38
Melting point: $<-40°C.$
Boiling point: $<70°C.$
Vapor pressure: 150 mm. Hg at 20°C.
Not soluble in water; dissolves readily in carbon tetrachloride, chloroform, benzene, and carbon disulfide
Flammability: subject to explosive decomposition from temperatures above 60°C.. from impact, or from supersonic sound waves

1 mg./cu. meter ≎ 0.2 p.p.m. and 1 p.p.m. ≎ 4.9 mg./cu. meter at 25°C., 760 mm. Hg

3. Physiological Response

The vapor–air mixtures of nitrogen chloride have a characteristic odor and are irritating to the mucous membranes, but less irritating than chlorine.[20] No permissible concentration has been published.

NITROSYL CHLORIDE, NOCl

Nitrosyl chloride may be encountered wherever aqua regia is made or used.

1. Physical and Chemical Properties

Physical state: yellowish gas
Molecular weight: 65.47
Melting point: −64.5°C.
Boiling point: −5.5°C.
Solubility in water: decomposes
1 mg./cu. meter ≎ 0.37 p.p.m. and 1 p.p.m. ≎ 2.7 mg./cu. meter at 25°C., 760 mm. Hg

2. Physiological Response

Nitrosyl chloride is a dangerous respiratory irritant. One exposure of 20 minutes to the inhalation of 100 p.p.m. proved fatal to cats, apparently from lung hemorrhage.[20] No permissible concentration has been published.

CARBON MONOXIDE, CO

Of all the gases that have poisonous effects upon man and animals, carbon monoxide is the most widely encountered. It exerts its effects by combining with the hemoglobin of the blood and interrupting the normal oxygen supply to the body tissues. Although this resultant oxygen deficiency is a reversible chemical asphyxia, nevertheless damage done by severe asphyxia from any cause may not be reversible.

1. Industrial Exposures

Exposure to carbon monoxide in industry, or in private life, may occur whenever carbonaceous matter, such as coal, wood, paper, oil, gas, gasoline, or any other organic material, is burned. Carbon monoxide is a product of incomplete combustion, and is not likely to result where a flame burns in an abundant air supply without contacting any surface. Whenever a flame touches a surface that is cooler than the ignition temperature of the gaseous part of the flame, carbon monoxide may result. Notorious in this respect are water heaters: the temperature of the water-filled coils cannot rise appreciably above the boiling point of water at the pressure involved; if a flame is allowed to play on the coils, a substantial amount of carbon monoxide is produced; and where the heater is not effectively vented to the exterior, contamination of the room atmosphere results.

Gas or coal heaters in the home and gas space heaters in industry have been frequent sources of carbon monoxide when not provided with effective vents. Gas heaters, although they may be properly adjusted when installed, may become hazardous sources of carbon monoxide if not correctly maintained. Automobile exhaust gas in garages, especially small private garages, is perhaps the most familiar source of carbon monoxide exposures.

Additional potential exposures occur in: the manufacture and use of illuminating gas, or "manufactured gas"; the manufacture of synthetic methanol or other organics from carbon monoxide; carbide manufacture; the distillation of coal or wood; operations near furnaces, ovens, stoves, forges, and kilns, which are especially likely to produce excessive carbon monoxide during the period in which they are being brought to normal operating temperature after a period of idleness; controlled atmosphere heat-treatment of metals; fire fighting; mines, following fires or the use of explosives; testing internal combustion engines; and many other exposure sources. Salamander stoves are dangerous sources of carbon monoxide in poorly ventilated areas. Among the situations less commonly warranting concern are: faulty exhaust discharge equipment on automobiles, buses, airplanes, and cabin cruisers; improperly located air inlets for automobile interiors; and compressed air for respiratory protective devices when supplied by reciprocating compressors (rarely, carbon monoxide may occur because of excessive overheating of the compressor).

The extent of room atmospheric pollution by carbon monoxide from any source, as for instance an automobile, can be computed from the rate of production of carbon monoxide, where constant, and the amount of general ventilation by use of the formula,[37]

$$C = 100K \, (1 - e^{-Rt})/RV$$

where $C =$ per cent carbon monoxide in a room after a given time, t; $R =$ air changes per hour; $t =$ time in hours; $V =$ volume of room in cubic feet; $K =$ cubic feet of carbon monoxide liberated per hour; and $e =$ the base of the natural system of logarithms.

It is obvious from inspection of the formula that as R or t increases, the factor $(1 - e^{-Rt})$ approaches 1 and equilibrium is reached—more quickly for larger values of R. Where V is relatively small the concentration builds up rapidly. $100K/RV =$ the equilibrium concentration. This formula can be used in computing general ventilation rates, the concentration of any gas or vapor in the air, or the rate of admitting any gas or vapor to the room, providing all the other factors are known.

[37] G. W. Jones, L. B. Berger, and W. F. Holbrook, *U. S. Bur. Mines Tech. Paper* No. **337** (1923).

The quantity of carbon monoxide produced by an automobile gasoline engine varies with the air/fuel ratio; speed; temperature of the combustion chamber, cylinder, and cylinder walls; compression ratio; spark advance; piston displacement; and other factors. An idling engine with a rich fuel/air ratio may produce an exhaust with upward of 7 per cent carbon monoxide, while the same engine with carburetor adjusted for efficient operation and at a speed of 35 or more miles per hour, may produce less than 0.5 per cent carbon monoxide. In general, combustion products containing a high percentage of carbon monoxide will have a low percentage of nitrogen oxides, and vice versa.

Engineering improvements in automobile engines have lowered the carbon monoxide content of vehicle combustion products appreciably over the years. Recent studies[38,39] have indicated that carbon monoxide in city streets is not a serious problem even with today's congested motor traffic.

The relative proportions of carbon monoxide and nitrogen oxides and their relative toxicities become of prime importance in the control of vehicle combustion products in tunnels and mines or in fact wherever quantities of ventilating air are regulated according to the concentration of carbon monoxide. Where the traffic is moving at less than 25 miles per hour the ratio of carbon monoxide to nitrogen oxides emitted by automobiles may be as much as 1000:1, but, where the traffic moves at not less than 35 miles per hour, the ratio may approach 1:1.

2. Physical and Chemical Properties

Physical state: colorless, odorless gas
Molecular weight: 28
Specific gravity: essentially the same as that of air
Melting point: —207°C.
Boiling point: —190°C.
Solubility: 3.5 cc. per 100 ml. water at 0°C. and 1.5 cc. at 60°C.; 20 cc. in 100 ml. alcohol at 20°C.
Flammability: flammable range 12.50 to 74.20 per cent; ignition temperature, 1130°F.

3. Determination in the Atmosphere

As with so many other compounds, the choice of a method of determining carbon monoxide will depend upon many factors, and no one method is satisfactory for all circumstances. Perhaps the most widely used is the catalytic oxidation indicator, which may be of the recorder type[40] or one of the portable types described

[38] G. D. Clayton, W. A. Cook, and W. G. Fredrick, *Am. Ind. Hyg. Assoc. J.*, **21**, 46 (1960).
[39] V. J. Castrop, J. F. Stephens, and F. A. Patty, *Am. Ind. Hyg. Assoc. Quart.*, **16**, 225 (1955).
[40] S. H. Katz, D. A. Reynolds, H. W. Frevert, and J. J. Bloomfield, *U. S. Bur. Mines Tech. Paper* No. **355** (1926).

in Volume I (page 181). The pyrotannic acid method[41] is fairly reliable, accurate within less than 0.03 per cent when good standards are used. Standards supplied with a commercial device utilizing the pyrotannic acid method have been found to vary as much as 25 per cent. The iodine pentoxide method is slow and requires meticulous attention to details and possible interfering gases. It is reliable and has been widely used as a laboratory method.

The Haldane or Orsat gas absorption devices may be used in certain instances, and when only small samples are available. The carbon monoxide can be burned and the resultant carbon dioxide absorbed; or the carbon monoxide can be absorbed as such in a mixture of cuprous sulfate, β-naphthol, and sulfuric acid. The limit of accuracy of the Orsat is 0.2 to 0.4 per cent, and the Haldane 0.02 to 0.04 per cent.

The use of sealed palladium chloride ampules, containing palladium chloride in a water–acetone mixture, is a very convenient method of roughly estimating the carbon monoxide content of the air. When the ampule is crushed, the contents moisten its cotton covering and, when suspended in the air for 10 minutes, it blackens in relation to the concentration of carbon monoxide. The method is not satisfactory in temperatures approaching $0°F$.

The National Bureau of Standards has developed a simple and practical indicator,[42-44] silica gel treated with sulfuric acid, ammonium molybdate, and palladium sulfate; these silica gel granules, contained in tubes, change color characteristically when air containing even less than 10 p.p.m. carbon monoxide is passed through the tube. The equipment is commercially available, and if checked and approved in conformity with the National Bureau of Standards, is reliable. Satisfactory infrared analyzers[38] are available commercially.

4. Determination in the Blood

The pyrotannic acid method[41,45] discussed under Determination in the Atmosphere is equally applicable to the determination of carbon monoxide in the blood. The Van Slyke[46] gasometric method, employing a manometric gas-analysis apparatus, requires a 2-ml. sample and is accurate to about ±0.025 volume per cent carbon monoxide. The Scholander[47] gasometric method is convenient and

[41] R. R. Sayers, W. P. Yant, and G. W. Jones, *Public Health Repts. U. S.*, **38**, 2311 (1923) (Reprint No. **872**).

[42] M. Shepard, "A Preliminary Report on the NBS Colorimetric Indicating Gel for the Rapid Determination of Small Amounts of Carbon Monoxide (mimeographed publication)," *Natl. Bur. Standards* (June 29, 1946).

[43] M. Shepard, *Anal. Chem.*, **19**, 77 (1947).

[44] R. L. Beatty, Methods for detecting and determining carbon monoxide, *U. S. Bur. Mines Bull.* No. **557**, 1955.

[45] R. R. Sayers and W. P. Yant, *U. S. Bur. Mines Tech. Paper* No. **373** (1927).

[46] P. B. Hawk and O. Bergeim, *Practical Physiological Chemistry*. 11th ed., Blakiston, Philadelphia, 1937.

[47] P. F. Scholander and F. J. Roughton, *J. Ind. Hyg. Toxicol.*, **24**, 218 (1942).

sufficiently accurate.[38] Spectrophotometric methods[48-50] are rapid and of various degrees of accuracy.

Blood for CO determinations must be kept from contact with the air or the carbon monoxide will soon be displaced. Blood out of contact with air will retain its carbon monoxide for indefinite periods and is best preserved by adding 0.3 per cent sodium fluoride as anticoagulant. Carbon monoxide may be demonstrated in exhumed bodies, even months after death but, since the plasma and cells may not be in their original proportions, all determinations of CO in blood taken from bodies after death should be referred to the same sample after complete saturation with CO.

Any sample of blood to be transported or stored before the determination of CO is accomplished should be kept in well-filled and sealed containers. If blood is obtained from a puncture wound, it must be collected promptly into a measuring pipet or other container to exclude air or else part of the carbon monoxide will be displaced by oxygen.

5. Physiological Response

Acute Effects. The symptoms caused by various percentages of CO hemoglobin in the blood, as tabulated by Sayers and Yant,[51] are given in Table 7.

TABLE 7

Symptoms Caused by Various Amounts of Carbon Monoxide Hemoglobin in the Blood

Blood saturation, % CO hemoglobin	Symptoms
0–10	No symptoms
10–20	Tightness across forehead; possibly slight headache, dilation of cutaneous blood vessels
20–30	Headache and throbbing in temples
30–40	Severe headache, weakness, dizziness, dimness of vision, nausea, vomiting, and collapse
40–50	Same as previous item with more possibility of collapse and syncope, and increased respiration and pulse
50–60	Syncope, increased respiration and pulse, coma with intermittent convulsions, and Chenye-Stokes respiration
60–70	Coma with intermittent convulsions, depressed heart action and respiration, and possibly death
70–80	Weak pulse and slow respiration, respiratory failure, and death

In carbon monoxide poisoning, as in any other form of asphyxia, there are many factors that may cause a greater susceptibility than the average. Notable

[48] B. L. Horecker and F. S. Brackett, *J. Biol. Chem.*, **152**, 669 (1944).

[49] H. Hartman, *Ergeb. Physiol. biol Chem. u. exptl. Pharmakol.*, **39**, 413 (1937).

[50] R. R. Sayers, F. V. Meriwether, and W. P. Yant, *Public Health Repts. U. S.*, **37**, 1127 (1922) (Reprint No. 748).

[51] R. R. Sayers and W. P. Yant, *U. S. Bur. Mines Rept. Invest.* No. **2476** (1923).

among these factors as pointed out by Drinker[52] are: any impairment in circulation, heart disease in any form, anemia, asthma, lung impairment, any condition that speeds metabolism, any increase in activity, high temperature and high humidity, and low barometric pressure (high altitude).

TABLE 8

Time for Various Concentrations of Carbon Monoxide to Produce 80 Per Cent Equilibrium Value of Blood Saturation[51]

CO in air (inclusive) vol. %	Blood satn., % (80% of approx. equil. values)	Time, hr.
0.02–0.03	23–30	5–6
0.04–0.06	36–44	4–5
0.07–0.10	47–53	3–4
0.11–0.15	55–60	$1^{1}/_{2}$–3
0.16–0.20	61–64	1–$1^{1}/_{2}$
0.21–0.30	64–68	$^{1}/_{2}$–$^{3}/_{4}$
0.31–0.50	68–73	20–30 (min.)
0.51–1.00	73–76	2–15 (min.)

Carbon monoxide combines with hemoglobin and reaches a state of equilibrium more slowly in low concentrations as shown in Table 8. The rate of combining is more rapid at first and slows as equilibrium is approached as indicated graphically in Figure 1. Both the rate of combining and the symptoms of poisoning are increased by exercise.

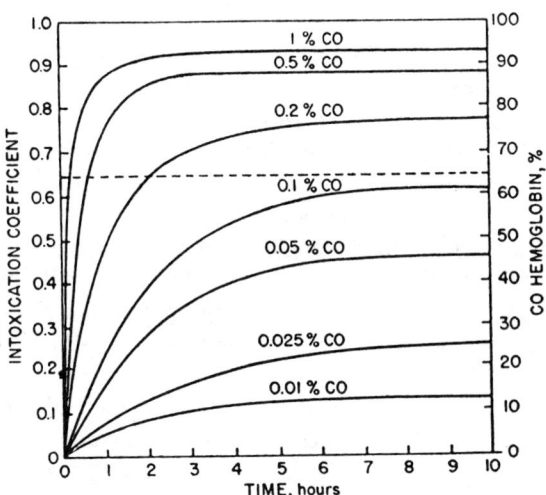

Fig. 1. Speed of saturation of hemoglobin with different concentrations of carbon monoxide until equilibrium between the concentration of carbon monoxide in air and blood is produced.[53]

[52] C. K. Drinker, *Carbon Monoxide Asphyxia*. Oxford Univ. Press, New York, 1938.
[53] W. F. von Oettingen, *U. S. Public Health Bull.* No. **290** (1944).

As with many other gases, the degree of harm from carbon monoxide is a product of concentration times the length of exposure. Henderson and Haggard[54] have proposed the following equations as a rough guide in estimating probable effects—obviously it does not apply to exposures longer than a few hours:

$$\text{Hours} \times \text{p.p.m.} = 300 \quad \text{(no perceptible effect)}$$
$$\text{Hours} \times \text{p.p.m.} = 600 \quad \text{(just perceptible effect)}$$
$$\text{Hours} \times \text{p.p.m.} = 900 \quad \text{(headache and nausea)}$$
$$\text{Hours} \times \text{p.p.m.} = 1500 \quad \text{(dangerous to life)}$$

The more precise figures (Table 9) of Henderson, Haggard, Teague, Prince, and Wünderlich[55] are also of interest.

TABLE 9

Physiological Response to Various Concentrations of Carbon Monoxide

Response	CO in air	
	p.p.m. by vol.	Vol. %
Concentration allowable for an exposure of several hours	100	0.01
Concentration inhaled for 1 hr. without appreciable effect	400–500	0.04–0.05
Concentration causing just appreciable effects after 1 hr. of exposure	600–700	0.06–0.07
Concentration causing unpleasant, but not dangerous, symtoms after 1 hr. of exposure	1000–1200	0.1–0.12
Dangerous concentration for exposure of 1 hr.	1500–2000	0.15–0.2
Concentrations fatal in exposures of less than 1 hr.	4000 and above	0.4 and above

Persons suffering prolonged unconsciousness from exposure to carbon monoxide may have permanent ill effects. As described by Drinker,[52] these include, rarely, damage to the heart, blood vessels, and various visceral organs, but more frequently to the brain and the nervous system. The most common neurological sequela of carbon monoxide poisoning is the "basal ganglia syndrome" as a result of injury to the brain tissue from anoxemia.

The dissociation curve of oxyhemoglobin in the presence of various quantities of CO hemoglobin (Fig. 2) illustrates why a normal person with 50 per cent saturation with carbon monoxide is unable to undergo physical exertion comparable to that of an anemic person with only 50 per cent of the normal amount of hemoglobin.

If the oxygen saturation of the available hemoglobin of the venous blood of each person is reduced to, say, 75 per cent at rest, the anemic person has a remaining tension of oxygen of 44 mm., while the person with 50 per cent carbon monoxide

[54] Y. Henderson and H. W. Haggard, *Noxious Gases*. 2nd ed., Reinhold, New York, 1943.

[55] Y. Henderson, H. W. Haggard, M. C. Teague, A. L. Prince, and R. M. Wünderlich, *J. Ind. Hyg.*, **3,** 79, 137 (1921)

saturation will have only 28 mm. of oxygen tension. Then if each person exercises, the venous blood becomes less saturated with oxygen—say, 60 per cent—and the anemic person retains a partial pressure of around 36 mm. of oxygen, which is still ample to oxidize tissue but the person with 50 per cent CO hemoglobin has suffered a drop in oxygen tension to less than 20 mm., which is probably insufficient to prevent fainting.[56] Other factors influence the dissociation of both oxyhemoglobin and CO hemoglobin: carbon dioxide, for instance, enhances the dissociation of each.

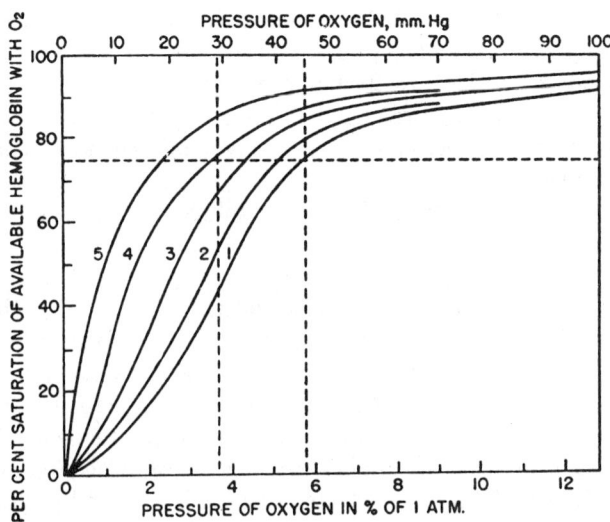

Fig. 2. Dissociation of oxyhemoglobin in the presence of various quantities of carbon monoxide hemoglobin: 1 = 0 per cent, 2 = 10 per cent, 3 = 25 per cent, 4 = 50 per cent, 5 = 75 per cent.[53]

Chronic Effects. The well-known effects of prolonged exposure to carbon monoxide are no different from the acute effects: headache, nausea, impaired senses, general debility, weakness, vertigo, and ataxia. Increase in hemoglobin and red cells as well as many more obscure effects have been attributed to chronic poisoning, some of them being reputed sequelae of acute poisoning as well. In evaluating such statements as far as acute poisoning is concerned, proof of prolonged unconsciousness should be an important factor; and, in the case of chronic poisoning, the certainty and degree of exposure and of carbon monoxide absorption are of vital concern.

Claims have been made of permanent, harmful effects from prolonged exposure to low concentrations of carbon monoxide, but they are not readily substantiated. Such possibilities have been discussed by Grut[57] and von Oettingen[53]

[56] For a more complete discussion of the physiology of this situation, consult C. K. Drinker's *Carbon Monoxide Asphyxia*. Oxford Univ. Press, New York, 1938, p. 18.

[57] A. Grut, *Chronic CO Poisoning*. Ejner Munksgard, Copenhagen, 1949.

and have been demonstrated[58] in dogs. Every smoker, as well as everyone present in a room filled with tobacco smoke, is exposed to small amounts of carbon monoxide. Cigarette smoke, as inhaled, contains 200 to 800 p.p.m. carbon monoxide, while cigar or pipe tobacco smoke has considerably more, and smokers' blood may become saturated with carbon monoxide to the extent of 5 per cent or greater within a period of 2 hours.[59]

That repeated exposures of persons to low but significant amounts of carbon monoxide do not ordinarily cause permanent ill effects has been demonstrated many times by carefully controlled experiments. Sayers, Meriwether, and Yant[50] exposed men to 200, 300, and 400 p.p.m. for periods of 6 hours, several times, during which their blood reached 15 to 28 per cent saturation with carbon monoxide. There were no indications of prolonged ill effects at the time, and any one who knows the subjects would feel certain they have suffered no cerebral damage or other permanent ill effects.

Again, Sayers, Yant, Levy, and Fulton[60] subjected 6 men, 4 to 7 hours daily for 68 days, to exhaust gas containing concentrations ranging from 200 to 400 p.p.m. carbon monoxide. The blood of these men acquired 20 to 40 per cent CO hemoglobin upon each of the 68 days, and the men suffered the usual acute symptoms. Although these men were subjected to exhaustive examinations and tests, no symptoms of a permanent or semipermanent nature were found during or following the exposures, other than a significant increase in hemoglobin and red cells, a few instances of urinary sugar, and a slight tendency toward poorer performance on a prolonged steadiness test. There were no apparent signs that the exposures produced deleterious effects upon the health and physical well-being of the subjects at the time or in the years following these experiments.

Considerably more convincing evidence of the absence of any signs of chronic carbon monoxide poisoning, especially where exposures are too low to cause acute symptoms, is the report by Sievers, Edwards, Murray, and Schrenk[61] on the results of clinical and laboratory examinations of 156 traffic officers stationed in the Holland Tunnel in New York. These men had been on duty 13 years in an exposure which averaged 65 to 85 p.p.m. carbon monoxide from exhaust gas, and the CO hemoglobin in their blood ranged from 0.5 to 13.1 per cent. They were found to be in exceptionally good physical condition.

Pathology. Carbon monoxide poisoning has been variously reported to have caused a vast assortment of ailments involving injury to practically all visceral organs.[53] No very satisfactory evidence has ever been presented, however, to

[58] F. H. Lewey and D. L. Drabkin, *Am. J. Med. Sci.*, **208**, 502 (1944).

[59] G. W. Jones, W. P. Yant, and L. B. Berger, *U. S. Bur. Mines Rept. Invest.* No. **2539** (1923)

[60] R. R. Sayers, W. P. Yant, E. Levy, and W. B. Fulton, *U. S. Public Health Bull.* No. **186** (1929).

[61] R. F. Sievers, T. I. Edwards, A. L. Murray, and H. H. Schrenk, *J. Am. Med. Assoc.*, **118**, 585 (1942).

indicate that permanent ill effects in men or animals are to be expected from a single acute exposure to carbon monoxide where the exposed person or animal remains conscious throughout.

Where poisoning is severe enough to cause unconsciousness, however, some damage to the brain, central nervous system, and circulation may occur, related in degree to the length and severity of the asphyxia. Table 10[52] shows the survival time of different nerve tissues when completely deprived of blood, and readily explains the often observed loss of cerebration in individuals severely poisoned by carbon monoxide.

TABLE 10

Survival Time of Nerve Tissues When Deprived of Blood

Tissue	Survival time, min.
Cerebrum, small pyramidal cells	8
Cerebellum, Purkinje's cells	13
Medullary centers	20–30
Spinal cord	45–60
Sympathetic ganglia	60
Myenteric plexus	180

Single exposures of dogs to concentrations of carbon monoxide causing unconsciousness and death within 30 minutes produced circulatory changes characterized by dilatation, stasis, perivascular hemorrhage, and edema. There were diffuse degenerative changes throughout the entire brain.[62] The neuropathology was quite similar to that resulting from asphyxiation by nitrogen in a similar time. Dogs dying in 11 to 15 minutes showed considerably less damage; while dogs kept in a state of unconsciousness and near death for several hours, by exposure to carbon monoxide, evidenced much more extensive damage; and dogs surviving this exposure for periods up to 165 days supplied evidence that much of the damage to the brain was of a permanent nature.

6. Absorption and Elimination

Absorption. Carbon monoxide is absorbed only through the lungs, where it enters the blood stream in the same manner as does oxygen. Carbon monoxide exerts its acute harmful effects by displacing oxygen in the blood. It has a greater affinity for hemoglobin than has oxygen and therefore forms a more stable compound. Douglas and Haldane[63] found that hemoglobin in equilibrium with atmospheres containing carbon monoxide and not less than 14 per cent oxygen was all

[62] W. P. Yant, J. Chorynak, H. H. Schrenk, F. A. Patty, and R. R. Sayers, *U. S. Public Health Bull. No.* **211** (1934).

[63] C. G. Douglas, J. S. Haldane, and J. B. S. Haldane, *J. Physiol,* **44**, 275 (1912).

in combination with one or the other of these gases and that the affinity of carbon monoxide for hemoglobin was approximately 300 times as great as that of oxygen. The relation[54] between the partial pressures of oxygen and carbon monoxide in the lungs and their combinations with hemoglobin can be expressed by the equation:

$$PCO \times 300/PO_2 = COHb/O_2Hb$$

and

$$\text{Per cent COHb} = PCO \times 300 \times 100/[PO_2 + (PCO \times 300)]$$

where Hb = hemoglobin; PCO and PO_2 = the partial pressures of carbon monoxide and oxygen, respectively. .

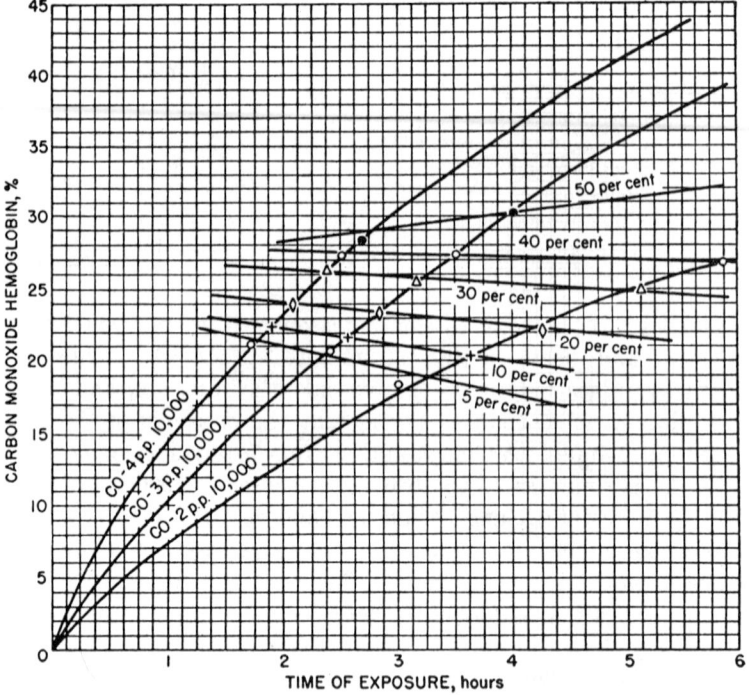

Fig. 3. Rate of saturation of blood with carbon monoxide, subjects exercising mildly during exposures to 200, 300, and 400 p.p.m. carbon monoxide. The per cent lines refer to the frequency of frontal headaches.[60]

Prince[64] pointed out that this relation made it easy to compute the concentration of carbon monoxide in an atmosphere that is in equilibrium with blood, where the partial pressure of oxygen and the amount of hemoglobin combined with either

[64] A. L. Prince, *Report of New York State Bridge and Tunnel Commission*, Sec. 9, Appendix 4, p. 188 (1921).

oxygen or carbon monoxide is known. Sayers, Yant, and Jones[65] developed a convenient method for the determination of CO hemoglobin and applied it to the determination of carbon monoxide in the air. Henderson and Haggard[54] have pointed out that in using this formula to compute concentrations of CO hemoglobin in the body, or concentrations of carbon monoxide inhaled, the partial pressure of the gases in the lungs must be used, and this is given as 15 per cent for oxygen. At equilibrium the partial pressure of carbon monoxide upon inhalation would be reduced somewhat, but not to such an extent as that of oxygen, the reduction being primarily due to increase in the partial pressure of water vapor. More recent information[66,67] indicates that the ratio of CO hemoglobin to oxyhemoglobin lies somewhere between 210 and 300 and is influenced by such factors as pH of the blood and partial pressure of carbon dioxide.

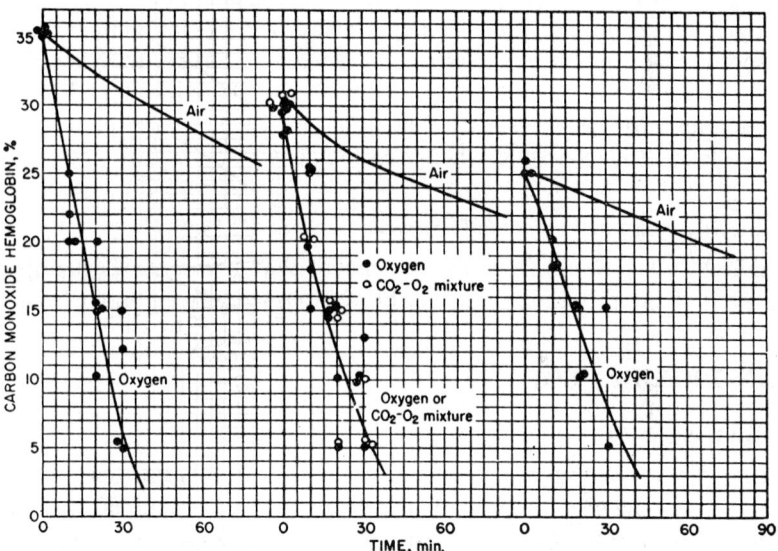

Fig. 4. Elimination of carbon monoxide from the blood as effected by breathing oxygen or a mixture of 5 per cent carbon dioxide in oxygen.[60]

Henderson and Haggard[54] determined that an average adult man at rest inhaling carbon monoxide absorbs enough in one hour to reach a maximum of 50 per cent of the equilibrium figure. Experiments conducted at the United States Bureau of Mines[60,68] with carbon monoxide–air mixtures support this finding,

[65] R. R. Sayers, W. P. Yant, and G. W. Jones, *Public Health Repts. U. S.,* **38**, 2311 (1923) (Reprint No. **872**).

[66] J. L. Lilienthal, Jr., R. L. Riley, D. D. Proemmeland, and R. E. Franke, *Am. J. Physiol.,* **145**, 351 (1946).

[67] N. Joels and L. G. C. E. Pugh, *J. Physiol, London,* **148**, 63 (1958).

[68] R. R. Sayers, F. V. Meriwether, and W. P. Yant, *Public Health Repts. U. S.,* **37**, 1127 (1922) (Reprint No. **748**).

in that men exposed to concentrations of 200 to 400 p.p.m. acquired in 1 hour a maximum of 45 per cent of the expected equilibrium value at rest, only slightly over 50 per cent of the equilibrium value at strenuous exercise, and somewhat less than 50 per cent with mild exercise. This last is indicated by the absorption curves shown in Figure 3.

In the development of safety standards the figure of 50 per cent of equilibrium as being the maximum attainment in 1 hour for adults at rest or doing light work may be utilized in connection with atmospheric pollution with CO below 0.1 per cent.

Elimination. The elimination of carbon monoxide is solely through the lungs and similar in many ways to the absorption. Although elimination is rapid at first, the last traces are eliminated very slowly. Exercise speeds elimination but is inadvisable because it may cause collapse. Increasing the partial pressure of oxygen by inhaling pure oxygen, or oxygen with 5 to 7 per cent carbon dioxide added, speeds the elimination. Curves presented by Sayers and associates[60] as portraying the elimination of carbon monoxide are reproduced in Figure 4. For each percentage saturation with CO, the upper curve shows the elimination when breathing air, and the lower one, when breathing oxygen.

7. *Hygienic Standard of Permissible Exposure*

AIHA, ACGIH, and ASA are in agreement that 100 p.p.m. is the upper limit of exposure for 8 hours. Other values for brief exposure may be taken from Table 9.

CARBON DIOXIDE, CO_2

1. *Uses and Industrial Exposures*

Carbon dioxide is a normal constituent of the atmosphere, about 0.03 per cent by volume above the ocean up to 0.06 per cent in urban areas; the exhaled breath contains up to 5.6 per cent; the gas is also widely encountered in industry in harmless concentrations. Carbon dioxide may be recovered from lime or cement kilns, from flue gases, from fermentation processes, and from some natural gas wells. It is first purified, then dehydrated, and compressed. If solid carbon dioxide, Dry Ice, is desired, it is then manufactured from the compressed liquid. When the pressure upon this compressed, liquefied CO_2 is suddenly released, a portion of it solidifies to "snow" as the balance expands to a gas again and is drawn off and recompressed. The snow is then pressed into 220-lb. blocks of solid Dry Ice. The compressed and bottled gas is used for carbonating beverages and the Dry Ice is used for preserving foods, especially during transportation. Other incidental uses are for chilling aluminum rivets and shrinking cylinder liners or bearing inserts. Industrial exposures may occur in mines, caves, tunnels, wells, the holds of ships, as well as tanks, vats, or any place where fermentation processes may have depleted the oxygen with formation of carbon dioxide. The manufacture, storage, and use of

Dry Ice also offer exposures, as do carbon dioxide fire extinguishers when operated in confined areas.

2. Physical and Chemical Properties

Physical state: colorless gas

Molecular weight: 44.01

Melting point: —56.6°C. at 5.2 atm. pressure

Boiling point: sublimes at —78.5°C. at atmospheric pressure

Vapor density: 1.5 (air = 1)

Solubility in water: 100 ml. water at 0°C. dissolves 180 cc. carbon dioxide, about $1/2$ that quantity at 20°C.; the solution is weakly acid

1 mg/liter ⪴ 556 p.p.m. and 1 p.p.m. ⪴ 1.8 mg./cu. meter at 25°C., 760 mm. Hg

3. Determination in the Atmosphere

From the viewpoint of industrial hygiene there is seldom occasion for making an analysis of the air for carbon dioxide. The Haldane or Orsat gas absorption devices (see Vol. I, page 180) are convenient means of estimation. In the absence of interfering alkaline or acid materials, air to be analyzed may be aspirated through a scrubber containing standard bicarbonate solution and phenolsulfon-phthalein indicator, and the air concentration of carbon dioxide determined by comparison with prepared color standards.[69] The color developed is not influenced by volume or rate of sampling but depends entirely upon the percentage of carbon dioxide in the air. Higher concentrations may be analyzed and recorded by utilizing an infrared analyzer. Circumstances requiring the analysis will dictate the choice of method, from among these or other methods, of which there are many.

4. Physiological Response

Acute Effects. Except as a contributor to oxygen deficiency, a very real danger, carbon dioxide does not offer serious industrial exposures. The initial effect of inhalation of excessive carbon dioxide is noticed in concentrations of about 2 per cent, 20,000 p.p.m., when the breathing becomes deeper and the tidal volume is increased.[70] The depth of respiration is markedly increased at 4 per cent; at 4.5 to 5 per cent breathing becomes labored, and distressing to some individuals. Concentrations of 8 to 10 per cent have been inhaled by men for periods up to 1 hour with no evident, harmful effects. The role of carbon dioxide in oxygen deficiency need not be elaborated upon, as it acts similarly to any other diluent gas. It is worth noting that in many instances the carbon dioxide may have been formed by processes, such as combustion or fermentation, that were at the same time depleting the oxygen supply in the air. Up to 10 per cent of carbon dioxide

[69] H. L. Higgins and W. M. Marriott, *J. Am. Chem. Soc.*, **39**, 68 (1917).

[70] T. Sollman, *A. Manual of Pharmacology*. 6th ed., Saunders, Philadelphia, 1944.

mixed with oxygen is given therapeutically[71] to improve respiration and lung ventilation and hasten the elimination of anesthetic gases or carbon monoxide.

Chronic Effects. Although repeated, daily 1-hour exposures to 8 per cent carbon dioxide increased the hemoglobin and red cells, and improved the gaseous exchange in the blood, no marked, deleterious effects were observed.[72] Cohn, Tannenbaum, Thalimer, and Hastings,[73] however, warn against such a procedure in the presence of pneumonia or other pulmonary or cardiovascular diseases. In such diseases excessively rapid or deep respirations are an added burden to an already overburdened respiratory and cardiovascular system.

5. *Hygienic Standard of Permissible Exposure*

The ACGIH threshold for 8-hours inhalation exposure is $1/2$ per cent.

PHOSGENE, $COCl_2$

1. *Uses and Industrial Exposures*

Phosgene may be encountered in certain chemical manufacturing operations, where it is used in many organic syntheses. It is also used in metallurgy to separate ores by chlorination of the oxides and volatilization. It is produced commercially by the catalytic chlorination of carbon monoxide and supplied in liquid form in steel cylinders. It was an effective combat gas used in chemical warfare from 1915 to 1918. Its chief importance in industrial hygiene, however, lies in its occurrence as one of the products of combustion whenever a volatile chlorine compound, such as a chlorinated solvent or its vapor, comes in contact with a flame or very hot metal. This ordinarily does not produce a serious threat to health except where ventilation is not satisfactory, the area is confined, or considerable quantities of chlorinated vapors are involved. It may be encountered in the use of carbon tetrachloride[74] for extinguishing fires in confined spaces.

2. *Physical and Chemical Properties*

Physical state: colorless gas
Molecular weight: 98.92
Melting point: —104°C.
Boiling point: 8.3°C.
Vapor density: 3.4 (air = 1)
Solubility: decomposes in water to carbon dioxide and hydrogen chloride; soluble in many solvents

[71] Y. Henderson and H. W. Haggard, *Noxious Gases.* 2nd ed., Reinhold, New York, 1943.

[72] W. Tomaszewski, J. Oszacki, and E. Dumoulin, *J. Am. Med. Assoc.,* **108,** 1016 (1937).

[73] D. J. Cohn, A. Tannenbaum, W. Thalimer, and A. B. Hastings, *J. Biol. Chem.,* **128,** 109 (1939).

[74] W. P. Yant, J. C. Olsen, H. H. Storch, J. B. Littlefield, and L. Scheflan, *Ind. Eng. Chem.,* *Anal. Ed.,* **8,** 20 (1936).

1 mg./liter $\backsim$ 247 p.p.m. and 1 p.p.m. $\backsim$ 4.05 mg./cu. meter at 25°C., 760 mm. Hg

3. Determination in the Atmosphere

The best method for the determination of phosgene in concentrations of 2 p.p.m. or greater is the diphenylurea[74] method, but for concentrations below this amount the method requires more than ordinarily careful technique and is not reliable. These low concentrations can be determined more satisfactorily, perhaps, by the use of phosgene test paper.[75,76]

4. Physiological Response

Phosgene is mildly irritant to mucous surfaces in concentrations below 10 p.p.m., and very irritant to the entire respiratory tract in considerably higher concentrations. A single, shallow respiration of a moderately high concentration causes a rasping, burning sensation in the nose, pharynx, and larynx that is not readily forgotten. One-half part per million by volume of phosgene can be recognized in air, through the sense of smell, by normal persons acquainted with its odor, and 1 p.p.m. is easily noticeable. At 2 p.p.m., the odor is moderately strong and the irritant action on eyes, nose, and throat is barely detectable.

The principal action of phosgene is that of a lung irritant. Only a relatively small portion of the inhaled gas hydrolyzes in the respiratory passages, but in the moist atmosphere of the terminal spaces of the lungs complete hydrolysis occurs with irritant effects upon the alveolar walls and blood capillaries.[77] The result of this action is a gradually increasing edema, until as much as 30 to 50 per cent of the total blood plasma has accumulated in the lungs, causing "dry land drowning." The air spaces grow less and less; the blood is thickened by loss of plasma, which results in slowed circulation; oxygen exchange is slowed; and the overworked heart, with insufficient oxygen, weakens. The end result may be either asphyxiation or heart failure, and this may be delayed. High concentrations of phosgene are immediately corrosive to lung tissue and result in sudden death by suffocation.

The normal responses to slight gassing are, besides a dryness or burning sensation in the throat, numbness, vomiting, pain in the chest, bronchitis, and possibly dyspnea. There is sometimes a latent effect: the period between inhaling low concentrations of the gas and the appearance of dyspnea may be several hours, almost free of symptoms. The action of phosgene and its sequelae resemble in some respects those of nitrogen dioxide (see Nitrogen Dioxide, page 922). The responses to various concentrations of phosgene in air as given by Flury and Zernik are shown in Table 11.

[75] A. C. Fieldner, G. G. Oberfeld, M. C. Teague, and J. N. Lawrence, *Ind. Eng. Chem.*, **11**, 519 (1919).

[76] F. A. Patty, *Am. J. Public Health*, **30**, 1191 (1940).

[77] F. Flury and F. Zernik, *Schädliche Gase*. Springer, Berlin, 1931.

TABLE 11

Physiological Response to Phosgene Gas[77]

Response	Concentration, p.p.m.
Maximum amount for prolonged exposure	1
Dangerous to life, for prolonged exposure	1.25–2.5
Cough or other subjective symptoms within 1 min.	5
Irritation of eyes and respiratory tract in less than 1 min.	10
Dangerous to life in 30 to 60 min.	12.5
Severe lung injury within 1 to 2 min.	20
Dangerous to life for as little as 30 min.	25
Rapidly fatal (30 min. or less)	90

5. *Hygienic Standard of Permissible Exposure*

The threshold limit of ACGIH is 1 p.p.m. for 8 hours.

CHAPTER XXVI

Industrial Lead Poisoning

ROBERT A. KEHOE, M.D.

I. The Occurrence of Lead Poisoning in Industry

There are no reliable statistics on the frequency of occurrence of occupational lead poisoning in the United States of America as a whole, by localities, or by occupations. This situation arises out of a number of conditions, among which must be mentioned the frequency of errors of diagnosis, the lack of adequate medical supervision of many industrial plants in which there is hazardous occupational exposure to lead, and the failure on the part of certain of the states to set up mechanisms for reporting accurately cases of industrial lead poisoning. As occupational medicine achieves a more complete coverage of industrial operations in the lead-using industries, and the application of state compensation laws becomes more effective and more precise in relation to this occupational disease, the statistical data will approach adequacy, but for the present the industrial hygienist will be well advised to rely only upon those recorded in areas in which the mechanisms for collecting and recording the facts are known by him to be reliable. On this account any statements concerning the incidence of industrial lead poisoning must of necessity be based in large part upon the observations and experience of an individual writer, without benefit of the collective records of his professional colleagues.

Such evidence as is recorded would seem to demonstrate that fatal lead poisoning has decreased in frequency more or less regularly and progressively over the period of the past three or four decades. Just how much of this decrease has resulted from changes in medical viewpoints, with reference to the *primary* cause of death of persons with an apparently significant history of occupational exposure to lead, cannot be determined. It is quite certain, however, that the more severe forms of lead poisoning are seen relatively infrequently now, as compared with earlier periods in the United States, and that primary, fatal, industrial lead poisoning, is infrequent in its occurrence. These facts argue that the more hazardous types of occupational exposure to lead have disappeared, or have been brought under some degree of control, and the experience of industrial physicians and hygienists validates this argument.

It is by no means so certain that there has been any very significant decrease in the incidence of the milder forms of industrial lead poisoning. The experience of the writer leads him to believe that occupational lead exposure has been brought under some measure of control in the large proportion of American industries within recent years, and that an entirely satisfactory quality of control has been achieved in many of these. On the other hand, the opportunities for hazardous exposure have not been controlled to the degree that is required to eliminate lead poisoning as the cause of partial and temporary disability among a large number of workmen in a wide variety of industrial occupations. Moreover, in some industries some men continue uninterruptedly at work after they have developed mild symptoms and signs of lead intoxication which are not disabling and which, if noticed, are taken to be the deviations from well-being that beset most people from time to time under a variety of circumstances. Regardless of statistics or the lack of them, it is apparent to any informed observer that industrial lead poisoning occurs with a frequency that bears no reasonable relationships to the accuracy of present criteria for differentiating safe from dangerous occupational exposure to lead, or to the adequacy of presently available medical and engineering means for controlling such exposure within entirely safe limits.

The causes for the present unnecessarily high incidence of industrial lead poisoning are numerous, but certain outstanding factors should be understood by those who would attempt to eliminate this disease from an industry or from the industrial community. First in importance among these causes is the prevailing lack of understanding on the part of industrial management, and many of its medical and technical advisers, of the primary factors that determine the frequency and severity of lead poisoning in a given occupation, and of the means now available for recognizing and appraising these factors. Certain confused and confusing ideas have persisted for a long time in connection with the lead trades. Allegedly wide variations in the "susceptibility" of individuals have given rise to the erroneous belief that there are no dependable means by which wholly safe occupational conditions of exposure to lead can be differentiated from those that are dangerous. Conversely, the comparative mildness of lead intoxication as it is seen commonly in American industry, and the frequency and promptness with which the more troublesome symptoms tend to subside after the termination of the occupational exposure to lead, contribute to the persistence of the irresponsible or illiterate view that there is no need to achieve complete control of the hazard. This view, it seems, is especially prevalent in some of the long-established lead trades in which the conditions are "much better than they used to be" while failing to meet adequate hygienic standards.

A second factor that exerts an important influence upon the incidence of intoxication is the continually changing pattern of modern industrial technology whereby new processes and products create new and sometimes unrecognized opportunities for occupational exposure to lead. By this means lead hazards are introduced into industrial plants in which there is no understanding of such hazards and no experience in coping with them. In other plants, new lead hazards may be added to recognized pre-existing hazards that were well under control, thereby, at times, increasing the exposure to lead above threshold levels and giving rise to cases of saturnism.

A third factor, and one that came into prominence during World War II and may operate during any period of rapid industrial expansion, is to be found in a large and abrupt increase in the number and volume of lead-bearing materials and commodities. Under such circumstances production schedules are likely to outrun precautionary measures, and the exposure of many workmen to lead may well increase by an increment that is disproportionate to the increase in production. Such increased exposure comes about through actual increase in the variety and volume of materials handled, through increased frequency of failures of plant equipment under the stress of production schedules, through the mistakes and faults of inadequately trained personnel, through inadequacies in supervision, through lengthened hours of work (overtime), and through deficiencies in plant housekeeping and in facilities and training in relation to personal hygiene.

Whatever may be the cause or combination of causes that gives rise to lead poisoning in a specific plant or industry, the means of prevention can be found in the careful study of the origin, nature, and magnitude of the exposure to lead by methods now readily available, and in the application of certain procedures that have demonstrated their capacity to reduce the exposure to a level compatible with the safety of plant personnel. This is not to say that there are no difficult problems to be met in certain industries. It is merely to say that the specifications for the maintenance of safety can be prescribed with adequate accuracy, and that the technical skill required to meet such specifications need not be of a higher order than that so generously expended upon problems of production. There are few types of occupational exposure to lead that offer insurmountable or even serious obstacles to modern engineering methods of attack, and there are none that can elude properly applied procedures for determining the extent of the hazard associated with them. What is required is a sound understanding of the problem that is to be solved, and a genuine determination to solve it, not by prophylactic medical therapy, but by the application of orthodox engineering principles and equipment coupled with satisfactory medical supervision. It is the purpose of the following paragraphs to outline the practical means by which the hazards of industrial exposure to lead may be understood, measured, and eliminated.

II. Types of Exposure to Lead in Industry

Generally speaking, there are two means for the entrance of inorganic lead compounds (cf. later discussion of organic compounds) into the human body under industrial conditions: (a) by way of the respiratory tract, through inhalation of vapor, fume, dust, or mist, and (b) by way of the gastroenteric tract, through swallowing lead compounds trapped in the upper respiratory tract, or introduced into the mouth on food, tobacco, tools, fingers, or other objects. The inorganic compounds of lead, in solid form or in solutions or dispersions in various

solvents or dispersion media, do not penetrate the normal or abraded human skin
to any practically important extent. Certain of these compounds may injure or
irritate the skin under suitable conditions, but even under these conditions their
percutaneous absorption is of only theoretical, rather than of practical, significance. These portals of entry are mentioned for the purpose of calling attention to
the forms in which inorganic lead compounds must exist if they are to gain entrance into the body. The presence of the vapors, fume, or fine dust of inorganic
lead compounds in the air breathed by workmen is the most generally important
factor in occupational exposure to lead. However, lead compounds that contaminate the hands, food, tobacco or other objects taken into the mouth may not be
ignored as sources of exposure. In this connection the not uncommon practice
among workmen in certain shops and factories of keeping food or bottles of milk
in convenient places in dusty workrooms or in locker rooms inadequately protected
against factory dust is a source of significant hazard. Likewise, the rooms in which
workmen eat lunches or "snacks," or make coffee or tea, are contaminated all too
frequently with dust from plant operations and materials.

The quantities of lead that are given off as vapor from pots containing molten
lead at temperatures under 1000°C. are insufficient in themselves to create an important lead hazard, but alloys of high lead content, prepared and handled at
higher temperatures—often near and sometimes above the boiling point of lead
(1525°C., 2777°F., see Table 1)—give rise to dangerous concentrations of vaporized lead in the air. Even at lower temperatures, however, lesser contributions
made by vaporized lead to the total exposure of workmen to lead may have
sufficient importance to warrant their elimination. Molten lead is easily oxidized
at its surface, and when it is skimmed, stirred, poured, or otherwise agitated in the
presence of air, variable quantities of finely divided lead oxide may be thrown into
the air. For these reasons the handling of molten lead or molten alloys of high lead
content, in all plant processes that are not fully enclosed or adequately ventilated,
is always associated with opportunities for some exposure to lead on the part of
workmen. The hazard of such exposure is especially prominent in certain foundry
operations in which castings of lead-containing alloys are made. The latter include
a wide variety of bearings for motor vehicles, ships, and railroad equipment. The
alloys are heated in furnaces to appropriate temperatures, which vary with their
composition and use, and in a common type of foundry are conveyed in the molten
state in pots on overhead rails to the desired site and poured into molds. In transit,
vaporized metal and fume escape into the atmosphere over and around the molds,
thereby increasing the respiratory exposure of molders and numerous other employees, whose absorption of lead, in connection with their work, otherwise might
well be slight or perhaps even negligible. A wide variety of industrial procedures
create exposure to lead in this form, including, as a few of the many examples:
soldering in the manufacture of automobile radiators, tin cans, and other receptacles; the production and reclamation of Babbitt metal; the manufacture of lead
shot and bullets; the preparation of granular metallic surfaces by means of a finely

TABLE 1

Relevant Physical Properties of Metallic Lead and Certain of Its Alloys and Compounds[a]

Material	Melting point, °C.	Pouring tempera-ture, °C.	Boiling point, °C.	Solubility, in 100 ml.			
				Cold water	Hot water	Acid	Alkali
Metallic lead (common)	327.4	382	1525	Insol.	Insol.	HNO3 Acetic	
"Babbitt metal" Pb-Sn-Sb		329–438					
Electrotype {94-3-3 metal \Pb-Sn-Sb		327					
Stereotype {80.5-6-13.5 metal \Pb-Sn-Sb		327					
Soft solder {38-62 \Pb-Sn	183						
Bearing {10-75-12-3 alloy \Pb-Sn-Sb-Cu	306	377					
Bearing {80-5-15 alloy \Pb-Sn-Sb	260						
Lead acetate (anhydrous)	280			44.3	221.20		
Lead arsenate (insecticide) mono, di, meta, and ortho	Decomp.			Insol.	Sl. sol.	Sol.	Sol.
Lead azide	Explodes at 350			0.023	0.09	Acetic	
Lead bromide	373		916	0.455	4.71	Sol.	Sol.
Lead carbonate (white lead)	Decomp.			Insol.	Insol.	Sol.	Sol.
Lead chloride	501		954	0.673	3.34	Sol.	Sol.
Lead chromate (chrome yellow)	844		Decomp.	0.00005		Sol. (Insol. acetic)	Sol.
Lead fluoride	855		1290	0.064		Sol. HNO3-H2SO4	Insol.
Lead nitrate	Decomp. at 470		Decomp.	38.8	138.8	Sol.	Sol.
Lead sesquioxide	Decomp. at 360		Decomp.	Insol.	Decomp.	Sol.	Sol.
Lead oxide (red lead)	Decomp. at 500		Decomp.	Insol.	Insol.	Sol.	Insol.
Lead monoxide	888			0.0017			Sol.
Lead suboxide	Decomp.			Insol.		Sol.	Sol.
Lead dioxide (peroxide)	Decomp. at 290			Insol.	Insol.	Sol.	Sol.
Lead silicate (glazes)	766			Insol.		Decomp.	Sol.
Lead sulfate	1170			0.0028	0.0056	Sol. conc., insol. acetic	
Lead sulfate (basic)	977			0.0044		Sl. sol. H2SO4	
Lead sulfide	1114			Essentially insol.		Sol.	Insol.
Lead tetraethyl	−130		±200	Insol.	Insol.	Insol.	Insol.
Lead tetramethyl	−27.5		±110	Insol.	Insol.	Insol.	Insol.

[a] Taken in part from *Lead in Modern Industries*. Lead Industries Assoc., New York, 1952.

divided spray of molten alloy; and the casting of type in the various processes such as linotyping and electrotyping. One common procedure in many types of chemical plants—that of lining tanks and reaction vessels with metallic lead and repairing these linings from time to time—is a source of significant exposure to lead fume, especially when the work has to be done on the inside of extensively corroded receptacles. Indeed, lead burning in general, which includes the operations just referred to and many others, may be a highly hazardous occupation under unfavorable conditions involving prolonged or regularly repeated exposure to vapor and fume. Somewhat similar techniques, from the aspect of respiratory exposure to the vapor or fume of lead compounds, are those involving the cutting, welding, or reshaping of metal surfaces heavily coated with lead-bearing paints or with other corrosion-resisting layers of high lead content such as that of terneplate. The application of a cutting flame to lead-containing alloys or to leaded steel likewise results in the release of lead fume, which under suitable conditions may give rise to hazardous exposure to lead.

Among the most frequently occurring types of industrial exposure to lead are those that arise from handling or processing lead compounds in such a way as to introduce respirable particles of such compounds into the surrounding air. Metallic lead and all of its compounds, when present in finely divided form in the atmosphere breathed by workmen, must be regarded as dangerous, unless the quantities present in the air remain within limits known to be safe. The industrial processes responsible for the dispersion of lead-containing dust into the atmosphere are too numerous and too variable from one time to another to lend themselves to classification with respect to hazard. The latter may be very great or negligible in a named occupation or industry, dependent upon its methods of operation or upon the quality of the hygienic control achieved in the operations. The most dangerous of operations, in terms of its potentialities, may be entirely safe, while the most innocent, disregarded or unrecognized, may be highly hazardous.

Lead compounds are also thrown into the atmosphere within droplets of mist in a number of operations in which paints, enamels, and glazes are applied as a spray. The inhalation of such mists constitutes a potential hazard that is not different in principle from that involved in the case of fume and dust. It is necessary here, however, to differentiate the effects induced by respiratory exposure to the solvent or dispersion medium from those due to the absorption of the lead compound, and while this, as a rule, is not very difficult, the failure to do so is not infrequent.

Certain organic lead compounds, of which only tetraethyl lead is manufactured and used in any considerable quantity at present, are liquids, which are sufficiently volatile at ordinary temperatures to give rise to dangerous concentrations of vapor. (Petroleum technologists have been interested for many years and have developed a greatly heightened interest in recent years in the use in motor gasoline of a more volatile compound of lead than tetraethyl lead, and it may be that tetramethyl lead or some other suitable lead alkyl will be substituted for tetraethyl lead in whole or in part.) Liquid lead compounds when handled in undiluted form or in concentrated solutions, as in their manufacture and transportation, and in plants in which they are mixed with gasoline, also give rise to exposure to lead by contact with the skin, through which they are absorbed. Any open receptacle that contains these liquids in high concentration, and any container, article of clothing, floor, or other surface that has not been decontaminated thoroughly after contact with them, may give rise to serious exposure to lead on the part of persons who are nearby or whose skin may come in contact with them. The acute and frequently fatal character of lead intoxication from absorption of organic lead compounds of this type justifies special precautions for the avoidance of exposure to them.

Many problems of industrial hygiene in connection with the manufacture and handling of tetraethyl lead (and other lead alkyls), differ so strikingly, both qualitatively and quantitatively, from those involved in the production and handling of metallic lead and its compounds, as to require separate treatment. Brief references to certain aspects of these problems will appear here and there in the text that follows, so as not to neglect altogether this important subject. The techniques of industrial hygiene in this industry cannot be covered satisfactorily, however, within the scope of this chapter, and for the most part, hereafter, unless tetraethyl lead (or other lead alkyl) is mentioned specifically, it does not enter into the discussion.

Differentiation should be made between the occupations involving opportunities for exposure to tetraethyl lead in concentrated form and those concerned only with the handling of gasoline containing tetraethyl lead in the concentrations in which it is employed in commercial motor fuels. Dilution of tetraethyl lead with gasoline to such an extent that there are one thousand parts or more of gasoline by volume to one part of tetraethyl lead, effectively prevents the absorption of appreciable quantities of tetraethyl lead through human skin. At the same time the vaporization of tetraethyl lead out of such dilute solutions in gasoline is so slight as to be practically insignificant under the normally prevailing conditions of handling and use that are required to avoid the risks of fire and explosion. For these reasons, the normal commercial distribution and use of leaded gasoline as a motor fuel involves no hazardous exposure to lead. (Attention is called to the fact that under the unique conditions that exist within tanks in which leaded gasoline has been stored, there may be highly dangerous exposure to tetraethyl lead and its decomposition products, as well as to obviously toxic gasoline vapors, on the part of men who enter the tanks to clean or repair them. Moreover, under any set of conditions in which considerable quantities of leaded gasoline are spilled or sprayed or otherwise vaporized into an enclosed, unventilated, or inadequately ventilated space, the concentration of lead in the air may exceed safe levels. Leaded gasoline is intended for use as a motor fuel only. It should not be used for other purposes, excepting under conditions that will certainly prevent hazardous exposure to lead.)

Certain other compounds of lead, such as the salts of complex organic acids, and especially fatty acids, have been regarded with distrust and apprehension from the aspect of their absorption through human skin. In particular, the use of lead soaps as lubricants and as technological adjuncts to lubricants, under conditions that bring these into frequent or regular contact with the skin of workmen in machining operations, has been under suspicion in this regard. It cannot be said that all of such compounds of lead have been investigated as to their absorbability through the skin, whether in their concentrated commercial form or in the industrial formulations in which they are used, often in high concentrations. Certain representatives of this group of compounds have been found to be essentially unabsorbable, however, through the skin of animals, and others have been found to be unabsorbed under the conditions of their use in industry. While definitive information is incomplete, the preponderance of the evidence, in our opinion, justifies the belief that these compounds are absorbed into the body in potentially dangerous quantities only through inhalation or ingestion.

Certain relevant physical properties of a number of products and compounds of lead are listed in Table 1. The melting points, pouring temperatures, and boiling points of metallic lead and a few of the more commonly used alloys will be of some use in pointing out conditions under which the ambient atmosphere may be expected to be contaminated with vaporized lead. Of some importance, but of considerably less direct significance, are the data with respect to the solubilities of various compounds of lead. While the facts in this regard are of great technological importance, their relation to toxicology and to the occupational hazard of exposure to lead should not be taken too seriously. For example, the solubility of a specific lead compound in water often has little direct bearing upon its absorbability from either the pulmonary or the intestinal surface. Moreover, in the case of a finely divided compound that is inhaled and retained to some extent within the pulmonary alveoli, relative solubility is of less significance than size of particle, since, if the particles are soluble at all in or on the pulmonary membrane (this is not necessarily a matter of solubility in water), they will be absorbed sooner or later, depending upon the surface relationships. The solubility of lead compounds has some significance in relation to their absorption in the alimentary tract. Even here, however, the highly soluble compounds are poorly absorbed, and the only compounds that are not absorbed at all are those that are incapable of conversion to soluble compounds in the complex medium of the alimentary contents within the time required to traverse the tract. Lead sulfide, for example, appears to be absorbed but little, if at all, from the lung, but some of it may be converted in the stomach to the slightly soluble chloride and so absorbed, to some extent, before reaching the colon, in which it may be converted again to the sulfide.

III. The Detection of Exposure to Lead in Industry

The *existence* of occupational exposure to lead can usually be recognized by careful observance of the activities of an industrial plant. Some experience is required before one can attempt to estimate the *severity* of the exposure, and even then it is best to reserve judgment until the results of more precise methods of estimation are available. Nevertheless, systematic examination of processes and equipment, while following the flow of materials through manufacturing operations, will give valuable information concerning conditions that require further study; therefore, frequent trips of inspection should be made throughout a potentially hazardous plant as part of a general program of hygienic supervision. Attention should be paid to the cleanliness and orderliness of the workrooms and to the condition of sanitary and clothes-changing facilities, as indices of the quality of the care being given to matters of health and safety.

In lieu of precise measurements of the extent of the potential exposure to lead in an industrial establishment, or of the actual absorption of lead by the workmen, both of which will be referred to extensively later, it may sometimes be desirable to arrive at a prompt conclusion concerning the severity of the exposure to which a group of men is being subjected. Here, resort may be had to an immediate examination into the hygienic status of the workmen. This may be done in several ways, depending upon the information and facilities that are available. An experienced physician may find it possible to establish the facts by examining a limited number of carefully selected and representative workmen. Examination of employment records, if such are available, will reveal the extent of the labor turnover, and if the latter is excessive, the records may suggest the reasons, especially if there are wide differences in this respect among the different occupations. Records of absenteeism and illness may also contribute highly significant information. The recorded incidence of lead poisoning among the workmen carries its own significance, but an unduly high rate of illness of any type, regardless of its classification, calls for interpretation. A survey conducted along these lines can be made quickly, and will usually provide a fairly definitive answer as to the satisfactory or unsatisfactory quality of industrial hygiene in the operations.

IV. The Measurement of Industrial Exposure to Lead

A. DETERMINATION OF LEAD IN AIR AS A MEANS OF ESTIMATING THE RESPIRATORY EXPOSURE OF WORKMEN

Poisoning solely or mainly from the ingestion of lead compounds is of relatively infrequent occurrence in American industry, partly because of the usually good quality of washroom and lunchroom facilities in industrial plants and the resultant promotion of cleanly habits among the workmen, but chiefly because of the virtual inevitability of the contamination of the atmosphere of workrooms in the lead-using industries with lead compounds, and the difficulties that attend the avoidance of some degree of respiratory exposure to, and absorption of, these compounds. In other words, ingestion of lead compounds in the course of the day's work can and should always be prevented through the exercise of regular, attentive care in matters of personal cleanliness and good housekeeping, while prevention of the inhalation of lead is much more difficult. Then, too, almost half of a somewhat soluble, particulate, *inhaled* lead compound (35 per cent or more of particles of 0.75 μ or less in diameter) is retained in the lungs and absorbed promptly or slowly while most of even the soluble lead that is *swallowed* (85 to 92 per cent) passes through the alimentary tract unabsorbed. As might be expected, therefore, much the largest proportion of the lead absorbed by the usual victim of industrial lead poisoning is derived from the respired air. For these reasons determinations of the respirable lead in the atmosphere of workrooms may yield fairly adequate information as to the magnitude of the occupational lead hazard in such rooms. Satisfactory methods for sampling and estimating the lead content of the atmosphere have been devised and employed extensively in a variety of industries, not only in hygienic surveys but also in the routine periodic investigation of occupational conditions. The student of this subject should consult standard texts and official reports, some of which are listed at the end of this chapter, for detailed information. The development of these methods, their practical application to the study of industrial processes, and the correlation of data obtained by their use with the results of clinical and physiological observations on workmen have provided a practical means for the appraisal of industrial lead hazards. As experience has been gained it has become increasingly apparent that it is feasible to express the general magnitude of the lead hazards of a plant and of its individual operations in terms of the lead content of sufficient numbers of properly selected air samples, and to anticipate with reasonable accuracy the results of exposure thereto.

The satisfactory application of the methods of sampling the air and of analyzing such samples requires a sound understanding of the principles involved in their use, as well as experience in handling the equipment and in selecting the sites for sampling. As to the actual analysis, a number of methods are available. (Cf. references at end of chapter.) The choice of method must be based upon the quantities of lead that are to be dealt with, and upon the presence or absence of other interfering substances in the atmosphere of the plant. Highly specific and sensitive methods must be used if the quantities of lead are small and if other metals are present in significant quantities. These matters should be considered carefully, and the choice and application of methods should be made by competent and experienced chemists.

In the case of a single analytical survey of a plant, a sufficient number of samples of air must be taken to give a comprehensive picture of all parts of the plant. Moreover, samples must be collected at appropriate intervals of time and with sufficient frequency in the various locations to yield representative results. If analyses of the air are employed as part of a regular program for controlling the exposure to lead, they should be applied either continuously at selected representative sites or at such intervals as will reveal the effects of any changes in environmental conditions, whether due to changing weather or to operating conditions within the plant. (It should be recognized that a sample collected over a period of hours may not reveal the existence of episodes of intense exposure of short duration, thereby failing to yield a true picture of the existing hazard.) Especially should the effects of all alterations in plant processes be checked by air analyses.

B. ANALYSIS OF THE BLOOD, URINE, AND FECES OF EXPOSED WORKMEN AS A MEANS OF MEASURING THE LEAD HAZARDS OF THEIR OCCUPATIONS

A second general method that is available for measuring the lead hazards of an individual, an occupational group, or an industrial population is based upon the facts pertaining to the physiological behavior of lead in the human body. Thus, the periodic determination of the lead content of the blood or urine of exposed workmen will demonstrate the general magnitude of their current absorption of lead in the complete absence of any or all other signs of lead absorption, and will disclose the severity of their hazard or the relative margin of their freedom from hazard. Under certain circumstances, but usually not as a regular procedure, the determination of the total lead content of samples of feces, so obtained as to comprise the total alimentary evacuation of each of a representative group of workmen after one or more typical workdays, will give an approximate but very useful measurement of the amount of lead swallowed from the upper respiratory tract, as well as that taken with food, beverages, and other objects introduced into the mouth, on the day or two immediately preceding the date of the collection of the samples. Such data are especially useful in relation to air-borne dust, in the appraisal of the use and adequacy of respirators (which may be required at times, in emergencies or under unusual circumstances, although dependence upon them is not the usual or recommended procedure), or in assessing the faithfulness and care of employees in their avoidance of the ingestion of lead in connection with their work.

The general physiological background on which the interpretation of the results of analyses of the lead in the blood and excreta must be based will be presented later in this chapter. Methods for collecting and analyzing samples are referred to in the bibliography at the end of this chapter. These matters play an all-important part in the medical supervision of workmen in certain lead-using industries, and, therefore, they require careful and expert consideration. They are referred to here only in relation to the measurement of exposure of lead, and the differentiation of safe from dangerous exposure.

V. Safe Occupational Exposure to Lead (Permissible Limits)

Inasmuch as lead is a normal constituent of human tissues, and the regular absorption of lead is a universal human experience, it is evident that there must be some line of demarcation between harmless and dangerous conditions and degrees of exposure to, and absorption of, lead. It is not surprising, therefore, that careful and prolonged medical supervision of workmen in a variety of potentially hazardous occupations should have demonstrated the existence of such a line or, more properly, such a zone, many years ago. The first expression of this differentiation was formulated in 1912 through the work of Legge and Duckering in British industry. More recent investigations have served to provide a sound physiological background for this concept, and at the same time, have yielded well-defined criteria for safety in the lead-using industries. In this connection, both clinical and physiological evidence refute the suspicion, often felt and sometimes expressed, that some slight exposure to, and absorption of, lead, above that which is "normal," should result in some slight or eventual harm, perhaps too insidious to be recognized as such. Experience has shown that when occupational exposure to lead, as measured technically or physiologically, is insufficient to cause occasional signs or episodes of toxemia resembling those characteristic of plumbism, no evidence is found of vague general disorders or obscure types of impairment of good health that differ in frequency or degree from those encountered in any comparable group of persons whose work environment is free of specific stress or hazard. On the basis of measurements of environmental exposure to lead, or of those concerned with the absorption of lead by workmen carried out in parallel with clinical observations that involve interpretations of human deviations from health and well-being, certain specifications have been advanced from time to time as hygienic standards in industrial practice. Of those worthy of consideration and application at present, there are two general types and three specific examples, which, by reason of their sound theoretical and practical backgrounds, are recommended.

A. LEAD IN THE ATMOSPHERE OF THE WORK SITE

Expressed in terms of air analyses, the upper limit of safety for industrial exposure to the more readily soluble, particulate, inorganic compounds of lead is represented in American practice to be the concentration of 0.2 mg. of lead (Pb) per cubic meter of air. This standard, put in plain and simple language, specifies that, when the air of workrooms contains regularly not more than 0.2 mg. of lead (in inorganic form) per cubic meter of air, as measured by prescribed methods, cases of disabling lead intoxication will not occur among the men who work for many years on the usual schedule in such workrooms, and cases of questionable or mild intoxication will occur only rarely, if at all. In practice, the attempt is made to maintain the lead content of the air within such limits as will yield an average of not more than 0.2 mg. of lead per cubic meter throughout the working day, while preventing the occurrence of materially higher concentrations (0.5 mg. or more per cubic meter). Evidence of the validity of this standard has been provided elsewhere and need not be enlarged upon here. It should be said, however, that its adequacy depends upon the limitation of the occupational exposure to lead dispersed in the air. It assumes, for its applicability, that the unnecessary and more easily avoided ingestion of lead, because of faulty personal habits and unclean surroundings in connection with the day's work, is kept at a negligible level.

The upper limit of safety in the range of the concentration of certain highly insoluble compounds of lead appears to be appreciably higher than that indicated above. Since it is difficult, on both practical and theoretical grounds, to give quantitative expression to differences of this type at this time, no attempt is made here in that direction.

B. LEAD IN THE BLOOD AND URINE OF WORKMEN

1. In Blood

The extreme upper limit of safety for the concentration of lead in the blood of men under the conditions associated with the absorption of inorganic compounds of lead in industry (or elsewhere), is just short of 80 μg. per 100 g. of whole blood. Since the analytical error is rarely less than ± 10 per cent in this general range (when dealing with samples of blood of 10 to 15 ml., containing somewhat more or less than 1 μg. of lead), one can be sure that the concentration of lead in the blood, at or near this threshold value, is slightly under, rather than somewhat over, this concentration only by analyzing a series of samples of blood by a method of high precision. The threshold value is actually so sharp that the utmost in precision is necessary to define it in the individual person, and in practice it is necessary for the analyst either to employ the procedure of analyzing multiple samples, or to adopt a threshold value sufficiently low to make allowance for the known error associated with his specific procedure and facilities. In any case, whether the analytical result is interpreted in its application by a physician, or by someone other than a physician, who accepts thereby a considerable share of the responsibility for the safety of workmen, the deviation of the analytical result must be defined precisely by the analyst, since otherwise a result near the threshold lacks its true significance.

With the foregoing factors in mind, this standard for industrial safety can best be stated (to be modified in the specific situation, as the precision of the analytical method requires or permits) as follows: When the concentration of lead in the whole blood of an individual workman is 70 μg. or less per 100 g., the individual, while he may have little margin of safety for further or more severe occupational exposure to lead, is in no danger from his exposure to, or absorption of, lead, as of the time of sampling.

In an occupational group any individual whose lead level in the blood is 80 μg. or more, while not necessarily ill in any way, is in danger of developing an episode of lead intoxication without warning, and should not incur further occupational exposure to lead until the concentration of lead in his blood has diminished significantly. Individuals in such a group, whose lead levels are below 70 μg. per 100 g. of whole blood, have margins of safety that are the greater the more nearly they approach the mean normal level of approximately 30 μg. per 100 g. The occupational hazard of a group of men whose work is subject to little variability in nature or duration is defined in fully predictable terms by the mean level of the lead concentration reached and maintained in their blood after a period of months (up to 8 or 9) during which they have engaged regularly in the occupation. Unless there is a significant change in the severity of their exposure, there will be no further significant change in the concentration of lead in their blood. Hence, a significant change in the latter is indicative of a change in the former. Such changes parallel each other, being comparable in direction, in rate, and in extent.

This standard, following the development and increasing use of precise analytical procedures, and with the test of time and experience, has become much the best and most reliable single criterion for the clean-cut differentiation of safe from dangerous types and conditions of industrial exposure to the inorganic compounds of lead. On the one hand, if the sampling of the personnel is adequate, the criterion can be counted on to ensure that there will be no cases of lead poisoning under conditions which it defines as safe, and, on the other, it can be trusted to point out the conditions that will be productive of lead poisoning, and the precise personnel among whom such cases will occur. In Table 2 characteristic analytical results obtained from a group of persons devoid of occupational exposure to lead are

TABLE 2

Frequencies of Lead Concentrations in Blood of Workmen
Correlated with Severity of Exposure to Lead

(Comparison of various occupational groups from various industries,[a] according to the frequencies of occurrence of various levels of the concentration, and according to the mean concentration, of lead in their blood, in association with varying degrees of severity of exposure, below and above the threshold of safety, as judged by the nonoccurrence or occurrence of lead poisoning within the individual groups)

Lead in blood, mg. per 100 g.	Severity of exposure to lead							
	Normal	Demonstrable		Slight to severe	Moderate to severe	Near threshold		Very severe, very dangerous
		Barely	Readily			Safe	Unsafe	
Total examined	86	36	19	272	59	62	54	75
0.00–0.019								
0.02–0.039	85	11	6	85	8	4	1	
0.04–0.059	1	24	8	122	35	30	18	9
0.06–0.079		1	5	50	10	22	21	12
0.08–0.099				7	5	5	6	17
0.10–0.119				6	1	1	6	14
0.12–0.139				2			1	5
0.14–0.159								4
0.16–0.179								1
0.18–0.199								1
0.20–0.219								2
0.22 and over							1	10
Mean	0.027[b]	0 042[b]	0.051[b]	0.050	0.057	0.060	0.075	0.130[b]
P.E.	0.003	0.001	±0.002	±0.001	0.002	±0.001	±0.004	±0.007
S.D.	0.004	0.008	±0.013	±0.019	0.022	±0.016	±0.041	±0.091

[a] The samples of blood repeated in this table were obtained at the same time from the same groups arranged in the same order as the samples of urine in Table 3.

[b] Calculated from a different arrangement of the frequencies.

compared with those obtained from certain groups of persons from a variety of industries whose occupations have subjected them to different degrees of severity of exposure to inorganic lead compounds. The gradation from safe to dangerous types of exposure is illustrated. It is important to note that certain individuals, within groups of men whose exposure to lead was predominantly nonhazardous, had absorbed dangerous quantities of lead. These individuals would require special investigation to determine wherein their exposure had deviated from the others in the group, unless the special hazard of their specific jobs should be readily apparent. In the instances illustrated here, the latter was true, for there were certain special hazards that had not been brought under complete control at the time of making the observations. It is important to realize, and to have the means of demonstrating that, in a generally safe series of industrial operations, there may be specific opportunities for excessive exposure to lead. The failure to recognize this simple fact in many a lead-using industry is the primary reason for the frequent belief and statement that an "unusually susceptible" workman contracted lead poisoning under conditions that were "well controlled" and safe for other workmen.

This standard cannot be employed successfully in any situation in which a significant exposure to tetraethyl lead occurs. The absorption of tetraethyl lead does not produce an increase in the lead content of the blood that bears any apparent or proportional relationship to the quantity of lead absorbed. Analysis of the blood, under these circumstances, for the purposes of either diagnosis or industrial hygiene, will yield a grossly misleading result.

2. In Urine

The upper limit of safety for the concentration of lead in the urine cannot be expressed in the simple and invariable terms that are applicable to the blood, for the reason that, under the same conditions in the same individual, the variability of the concentration of lead in the urine is many times greater than that in the blood. This greater variability is illustrated in Table 3, as compared with Table 2. The "spot" samples of urine tabulated for each group in Table 3 were obtained at the time the samples of blood (Table 2) were taken and so are generally comparable in relation to the physiological state of each individual in each group. This is not strictly true, however, since there is an appreciable delay in the occurrence of a change, as well as a leveling out of the magnitude of the change, in the concentration of lead in the blood in response to exposure to lead under ordinary occupational conditions. Not so in the case of the urine, which undergoes relatively prompt change in response to such exposure. In addition, however, to this response, there are other factors that influence the rate of the urinary excretion of lead, such as the amount of water available for excretion and the amount of physical exertion put forth by the individual during the excretory period. A seasonal factor, which does not lend itself to precise interpretation, is also discernible. These are not

TABLE 3

Frequencies of Lead Concentrations in Urine of Workmen
Correlated with Severity of Exposure to Lead

(Comparison of occupational groups from various industries,[a] according to the frequencies of occurrence of various levels of the concentration, and according to the mean concentration, of lead in their urine, in association with varying degrees of severity of exposure, below and above the threshold of safety, as judged by the nonoccurrence or occurrence of lead poisoning within the individual groups)

Lead in urine, mg. per liter	Severity of exposure to lead							
	Normal	Demonstrable		Slight to severe	Moder- ate to severe	Near threshold		Very severe, very danger- ous
		Barely	Readily			Safe	Unsafe	
Total examined	86	206	19	273	55	63	57	79
0.00–0.039	66	41	4	35	1		3	2
0.04–0.079	18	156	13	98	14	11	7	2
0.08–0.119	2	9	2	56	24	22	10	13
0.12–0.159				35	11	16	14	14
0.16–0.199				27	2	7	11	9
0.20–0.239				8	1	2	3	9
0.24–0.279				7	1	2	4	6
0.28–0.319				4		3	3	5
0.32 and over				3	1		2	19
Mean	0.033[b]	0.051[b]	0.067[b]	0.102[b]	0.108[b]	0.130[b]	0.152[b]	0.238
P.E.	±0.001	±0.007	±0.006	±0.003	±0.005	±0.005	±0.002	±0.011
S.D.	±0.015	±0.014	±0.037	±0.069	±0.057	±0.059	±0.024	±0.151

[a] The samples of urine represented in this table were obtained at the same time from the same groups arranged in the same order, as the samples of blood in Table 2.

[b] Calculated from a different arrangement of the frequencies.

necessarily minor variations, and the only way by which one can determine the current general rate of urinary excretion of lead of an individual or group, in its relation to the current rate of lead absorption, is that of obtaining an appropriate representation of the many variables, in a sample of urine of large volume (from an individual) or in a series of samples of small volume (from an individual or a representative group) in sufficient numbers for statistical treatment. The attempt has been made to reduce the variability of individual samples of urine to satisfactory limits for practical purposes by relating the lead content of a sample of urine to the time over which it was excreted, rather than to its volume, on the hypothesis that the output of lead from the body within a unit of time is essentially independent of the throughput of water. Other attempts have involved corrections of the concentration of lead in the urine in terms of the concentration of creatinine therein, or in terms of a standard "normal" specific gravity, on the theory that the true rate of excretion of lead in the urine may be gauged by the rate of excretion of a relatively constant metabolite or of the total solids. All of these procedures fail in practice at the very points at which they are most needed (when unusually low or high concentrations of lead are encountered in the urine and require interpretation), and they fail always, in the physiological sense, because they are based upon false premises. Instead of resorting to rules of thumb which reduce the physiological variability by purely manipulative devices, it is necessary to attack the problem directly by the methods indicated above. This is especially true in dealing with a diagnostic problem in an individual, in which a "short cut" usually provides misinformation and leads often to needless controversy and fruitless litigation. For the purpose of appraising the safety or hazard associated with industrial exposure to lead, the problem is soluble, however, in that one makes use of the results obtained by the periodic sampling of *groups* of employees whose exposure to lead can be visualized in relation to their participation in specific industrial operations. The information obtained in this manner under a wide variety of conditions in a large number of industries has provided a wealth of data for the differentiation of safe from dangerous exposure. The differentiation is made on a statistical basis, in terms of the mean concentration of lead (and the standard deviation) found in samples of urine collected in a carefully selected and uniform manner, at or within a designated time. The upper limit of safety for the concentration of lead in the urine varies significantly with the type of sample obtained for analysis. Thus composite samples of large volume (2 to 3 liters) collected from workmen after the hours of work or over the weekend are lower in their lead concentration than are individual samples of small volume collected (with corresponding care for the avoidance of contamination) during the workday or workshift. There is also a slight but statistically significant difference between the levels of lead concentration found in the urine on the last day of the work week and at the end of the long weekend.

The differences which arise out of variations in the volume of samples, in the time of sampling, and in the manner of sampling, with specific reference to the means of reducing or avoiding contamination of the samples, when added to the variations in the precision of the methods employed by various analysts, and when augmented further by the lack of uniformity in the interpretation (statistical or otherwise) of the results obtained, account, in part, for the variety of the standards that are advocated by observers in a number of American industries in which the results of analyses of the urine of workmen are being employed for purposes of hygienic control. A larger factor in this lack of uniformity is contributed, however, by the variation in the points of view of physicians or others responsible for industrial hygiene, as to the degree of freedom from hazard—that is, the margin of safety—that should be incorporated in a standard of this type. There are valid differences in professional opinion in such matters, and when these differences cannot be resolved on physiological or clinical grounds, they can sometimes be brought to a point of understanding or of practical compromise. The margin of safety that can be built into the specification, at a reasonable cost for compliance therewith, may well be a matter of policy, to be decided on economic or other practical grounds, so long as the interpretation of the nature and characteristics of lead intoxication as a medical entity is not obscured thereby. The standards promulgated in these paragraphs in physiological terms provide a margin of safety sufficient to include the variations in the susceptibility of workmen, as they are selected for their work in ordinary industrial medical practice, and so to prevent even the earliest and mildest signs of lead intoxication that are now recognizable.

In view of the fact that samples of urine of small volume, comprising a single voiding, are obtained for analysis more frequently than are other types of samples, the standard set herein is based on the analytical findings in such samples.

With the foregoing considerations in mind, the upper limit of safety, in terms of the mean concentration of lead in the urine of a group of workmen whose occupational exposure to lead is fairly uniform, is 0.13 mg. per liter for samples that exceed 0.18 mg. per liter only infrequently, and exceed 0.24 mg. per liter only rarely. This standard can be applied to an individual, or to a small number of individuals, only after the collection of multiple samples from the individual or individuals in sufficient number to provide a statistically stable group of observations. (In the individual case, in industries in which only inorganic compounds of lead are handled, it is both simpler and more dependable to rely upon the results of analyses of the blood, either exclusively or in combination with analyses of the urine.)

As indicated previously, the concentration of lead in the urine provides the only analytical criterion for the estimation of the significance of exposure to, and absorption of, tetraethyl lead, and the standard expressed above has demonstrated its efficacy in defining the upper limit of safety in the industries concerned with such exposure. It is equally applicable, however, to occupational conditions associated with exposure to the inorganic compounds of lead, where it provides a somewhat, but not excessively, larger margin of safety, as a rule, because of the lesser rates of both the absorption of and the excretion of inorganic, as compared to organic, compounds of lead.

3. Interpretation of Analytical Findings in Blood and Urine

Some discussion, along with comments on certain of the data in Table 3, may be useful in providing the most satisfactory interpretation of the analytical information from certain of the industrial groups. Some comparisons of these data with those of Table 2 may also be made with profit. It will be noted that there are a few excessively high levels of urinary lead concentration (i.e., in excess of 0.24 mg. per liter) in each group to the right of the third column of numbers in Table 3, despite the fact that the mean concentration of lead in the urine reaches the threshold level (as here defined) only at the sixth column. Reference to Table 2 will show that some individuals in each of the groups represented in the fourth, fifth, and sixth columns had excessive concentrations (0.08 mg. or more per 100 g.) of lead in their blood. The data in Table 3 do not *prove* that there was any real danger in the occupational conditions represented in the fourth, fifth, and sixth columns, although individual results in excess of 0.24 mg. per liter justify some concern on that score. Values in excess of 0.08 mg. per 100 g. of whole blood of themselves, however, denote the actual occurrence of dangerous absorption of lead. Thus it is clear that, while most of the individuals in the groups represented in the fourth, fifth, and sixth columns of both Table 2 and Table 3 had been working under safe conditions, a few in each group had not. Such findings are in full accordance with the observed facts concerning the several industries and occupations represented in Tables 2 and 3. While the observations recorded in the fourth column were being made the industry concerned was undergoing changes in their operations, which resulted a little later in the elimination of hazardous conditions. After these changes had been accomplished, the *mean* levels of lead concentration in the urine and blood of the workmen diminished but little, but the excessively high values disappeared completely from the blood (and thereafter none of the men were transferred to other operations), and those in the urine diminished nearly to the vanishing point.

Somewhat similar situations accounted for the excessively high results recorded in the fifth and sixth columns in Tables 2 and 3. There is nothing unusual in such a pattern of results. Only rarely are the composite operations of a lead-using industry so completely controlled and so nearly uniform in type that all of them are equally free of hazard; and only rarely can complete freedom from hazardous exposure be maintained indefinitely in the operations of these industries without frequent precise monitoring of both the operations and personnel. There were no cases of lead poisoning within any of these groups (i.e., up to the seventh column in Tables 2 and 3) during the period of several years of our surveillance, presumably for the reason that any man who gave evidence of excessive absorption, as defined herein, was transferred promptly to work involving great reduction in his exposure to lead or a termination of his exposure for a considerable period of time. This presumption, borne out amply by other experience, is supported by the positive clinical evidence associated with the findings tabulated in the seventh and eighth columns of Tables 2 and 3. In the plant represented by the data in the seventh column there were infrequent, irregularly spaced cases of mild lead intoxication. The findings recorded in the eighth column of both tables, on the other hand, are illustrative of highly dangerous conditions. Only a few of the persons in this industry (office staff) worked under even approximately safe conditions. Lead poisoning was frequent in its occurrence and all persons whose employment had persisted for as long as three weeks were found to have absorbed potentially dangerous quantities of lead. Nevertheless (and this is a fact that has contributed to the complacency of some employers, and has frustrated some industrial physicians who otherwise might have had more interest in good industrial hygiene), there were persons in this group who had worked steadily for several years without illness, and who, at the time of this survey, were wholly free of any sign of lead intoxication.

In the practice of industrial hygiene in the industries concerned with the inorganic compounds of lead, in which, despite generally effectual methods of control, there are opportunities for hazardous exposure, it is advantageous to obtain analytical information periodically and in parallel on both the blood and the urine of representative employees (and of all employees occasionally). The results of analyses of the blood reveal the range and the general level of the *total quantity of lead absorbed* by workmen individually and in occupational groups, in consequence of their more or less prolonged occupational exposure. The range of concentration, and the mean concentration, of lead in the urine of the group, in comparison with prior results on the same or on similar occupational groups portray the relative magnitude of the *current occupational exposure*. Significant variations in the mean concentration of lead in the urine of the group, from time to time, will serve, better and more quickly than any other means, to demonstrate improvement or deterioration in the efficacy of measures of hygienic control. (As indicated previously, the concentration of lead in the blood of an individual indicates clearly his status with reference to safety or danger, and provides a basis for action, if need be, on his individual behalf.)

C. CLINICAL CRITERIA FOR THE DIFFERENTIATION OF SAFE FROM HAZARDOUS ABSORPTION OF LEAD

Brief mention may be made here of the determination of basophilic granulation of the erythrocytes in the blood of workmen, and of the analysis of their urine for porphyrins, as medical techniques for the estimation of the extent of the hazard of occupational exposure to lead. The first of these procedures is an unsatisfactory and outmoded clinical method for this purpose. Since the phenomenon of "stippling" of the erythrocytes is not a specific sign of the absorption of lead, and since it does not vary quantitatively in the individual with either the rate of absorption or the quantity of lead absorbed, it is but a poor and makeshift substitute for the specific procedures recommended above. It serves no useful purpose in the absence of actual danger, and when danger exists it is not adequate to cope with it. It cannot be recommended as a means of differentiating, in any other than a very crude manner, harmless from hazardous absorption of lead, and no semiquantitative expression of the results of the application of this method or its modification can be recommended as a hygienic standard in industry. The characteristics of the phenomenon, and the extent of its usefulness for diagnostic purposes will be dealt with later in this text.

The significance of porphyrinuria, in relation to the diagnosis of lead intoxication, is considerable. Suffice it to say, in the present relationship, however, that abnormal quantities of porphyrins, chiefly coproporphyrin III, appear in the urine as an expression of a derangement of the synthesis of hemoglobin in a number of diseases, including lead poisoning, of which, at present, it seems to be among the first, if not the first of all, evidences of intoxication. Since the goal of industrial hygiene in the lead-using industries is the *prevention* of lead intoxication, reliance upon this incipient sign of intoxication, as a means of recognizing and terminating hazardous exposure to lead prior to the appearance of more serious manifestations, while advantageous to that extent, would seem to be considerably less than satisfactory from the medical viewpoint, since there are methods by which even the slightest manifestations of lead intoxication can be anticipated and avoided. There may be unusual situations or emergencies in which its use can be justified, as, for example, during brief periods when facilities for the performance of the more precise and specific analytical procedures are not available. In general, however, the use of this technical procedure lies in diagnostic, rather than in preventive, medicine, and it will not be dealt with further in this chapter, although references to it are listed at the end. It should be pointed out that this hematological derangement is not observed in poisoning with tetraethyl lead, in which the central nervous system is the primary, and usually the sole, point of attack.

VI. The Control of Industrial Exposure to Lead

A. ENGINEERING METHODS OF CONTROL

The elimination of hazardous exposure to lead from the many industrial operations that involve the use of lead compounds is primarily a technical problem. Its satisfactory solution in any instance depends upon adequate knowledge of how the specific industrial product can be made satisfactorily, economically, and safely. The choice of the materials (with respect to lead content or physical state, i.e., whether in the form of solid, powder, or paste), the processes, and the equipment will determine the general character of the plant that is to be built or adapted for the purpose. The potential health hazards should influence all of these choices, and in addition will introduce their own requirements into the project. For these reasons a new plant should be designed and built, or an old one should be remodeled, only after taking into account all the factors that have to do with the safety and health of the employees.

Certain general principles of procedure have been found to be worthy of attention in the construction and arrangement of industrial plants, and in the remodeling of old ones.

1. Hazardous types of exposure should not be distributed throughout a plant, if it is possible to localize them in specific areas where they can be subjected to concentrated procedures of control. The application of this principle has a bearing upon the movement of materials through a plant, the arrangement of equipment within it, and the manner of employing ventilating equipment (whether local or general, or both). These, along with other considerations, determine the layout of the plant, its structural characteristics and, thereby, to a considerable extent, its cost.

2. The equipment should be arranged with reference to the hazards and the means of their control, as well as to convenience and speed in production. For example, ventilation can rarely be introduced with completely satisfactory results as an afterthought. There must be space for fans and ducts and for the heating or cooling equipment that may be required to temper the air.

3. The matter of ventilation in general is dealt with in Volume I, and therefore, little need be said here. Three points, however, merit specific mention in connection with particulate lead compounds: (1) the necessity for the use of carefully designed local exhaust systems to remove vapor, fume, or dust at their points of origin; (2) the need to exercise great care to prevent the reintroduction of exhausted, contaminated, air into workrooms; and (3) the desirability of collecting exhausted dust and fume so as not to contribute to the pollution of the atmosphere in the neighborhood. The importance of these points will be appreciated only by those who have had broad experience, or by those who, lacking experience, have undertaken to solve these practical problems by methods of trial and error. It is of considerable importance also to so design the ventilation system and the ventilated areas as to ensure satisfactory results in independence of weather conditions and the personal predilections of workmen.

4. The construction and arrangement of the plant must be such as to promote cleanliness and good housekeeping. Floors, window ledges, walls, and working space in general must be designed with specific reference to the materials to be used and the methods of housekeeping that will be employed. For example, floors that are to be hosed down must have unobstructed surfaces and adequate drainage, and must not be slippery when wet. Similar purposes may be served by metal grills overlaying a secondary floor equipped with adequate drains. Areas and equipment that are to be cleaned by vacuum must be smooth, regular, and free of all unnecessary obstructions and sharp angles.

5. Facilities for medical services must be designed on the basis of the medical program that is contemplated, and not on what remains of space and location after other requirements have been satisfied.

6. Toilets, washrooms, locker rooms, lunchrooms, and sanitary equipment in general, including a satisfactory supply of drinking water, should be so planned as to encourage their use. Convenience, comfort, and adequacy in relation to the number of employees are first considerations. Almost without exception in lead-using industries, certain groups of workmen must change clothing completely when going to work, and change and bathe at the end of the day's work. Therefore, there must be double lockers for the two sets of clothing, and good bathing facilities. Workmen must not eat in certain workrooms; consequently, a clean and comfortable lunchroom must be provided, in which hot or cold beverages and perhaps hot food should be available. Good washroom facilities are required to enable the men to cleanse themselves properly before entering the lunchroom. All of these rooms and spaces must be located or otherwise designed so as to prevent their contamination with dust-laden or fume-laden air, and to facilitate frequent and thorough cleaning.

The problem of coping with exposure to lead in a plant already built and in operation resolves itself into two distinct parts. First, it is necessary to determine the origin and the extent of the exposure to lead associated with the various occupations and operations of the plant. Second, come the choice and application of the methods for control. The simplest means by which the first step can be achieved has been mentioned previously, but certain practical points are worthy of some elaboration. Adequate systematic sampling of the air of workrooms while work is in progress, and the use of accurate methods for the analysis of such samples, will serve to portray the general distribution and the order of severity of the lead hazards. In a plant in which the quality of sanitary facilities and of the general hygienic performance is good, the results of air analysis will demonstrate the hazard or the lack of it with considerable accuracy and will indicate clearly what is to be done and where to begin. It may happen, however, that hygienic conditions in general are unsatisfactory, that the instruction and supervision of the workmen in precautionary measures are defective, and that as a consequence the men are inadequately informed and indifferent in matters of safety. Under these conditions the extent of their occupational absorption of lead can be expected to be considerably greater than that which would naturally follow from the magnitude of the contamination of the air. In addition there is an educational problem to be met, both among the workmen and the management, as a first step toward the accomplishment of the task that lies ahead. Both must be brought to recognize the actuality of the hazards, while at the same time they are informed of their respective responsibilities in a comprehensive plan for the control of such hazards.

The methods for controlling the exposure to lead will depend upon the circumstances and the nature of the plant processes. In a generally well-designed and well-managed plant the problem is straightforward and comparatively simple. Mechanical measures of control can be applied where and as they are needed, the results being checked by air-sampling and analysis until the desired conditions have been achieved. Under the unsatisfactory conditions referred to above, however, and even under the best of conditions, considerable time may be required to obtain and install the necessary equipment. It may be highly important, therefore, to do all that can be done to reduce existing lead hazards, pending their elimination by more adequate or permanent means. In such instances the use of respirators is fully justified as a temporary expedient for the control of otherwise unavoidable respiratory exposure to lead. Moreover, every effort should be made to prevent the wholly unnecessary ingestion of lead that may result from inadequacies in the instruction of personnel and in plant practices with respect to personal cleanliness and the handling of food, beverages, and tobacco during working hours. Improvement of sanitary facilities, instruction of workmen through talks, formal regulations, and placards, as well as provision for adequate hygienic supervision in the performance of the day's work, may be highly necessary and helpful in alleviating the situation.

It has been implied above that respirators should be regarded as temporary means for preventing the inhalation of lead compounds. This implication was intentional. Occasional situations arise that necessitate the use of respiratory equipment of some type, and certain intermittent operations of brief duration may best be dealt with in this manner; but in the main, exposure to lead should be prevented by other means, not only for the sake of the comfort and efficiency of the workmen but also on account of the continual care and supervision that are required to see that this type of equipment is used properly and effectively. As to the selection of respirators or masks in relation to specific lead hazards, as well as the necessities for proper use and maintenance, the reader is referred to Volume I, Chapter XI. It should be emphasized, however, that no aspect of this matter should be left to the workmen. The care of such equipment, including inspection, cleansing, maintenance, and replacement, as well as all decisions as to when and how it is to be used, must be set up under the control of trained and responsible personnel.

In a potentially hazardous lead-using industry, all of the operations of production, repair and replacement of equipment, the investigative work on processes, the laboratory control of materials and products, and even general housekeeping, involve opportunities for exposure to lead, and, therefore, come within the province of industrial hygiene. What is often forgotten in dealing with pounds and tons of material is that one is concerned, in matters of human health, with mere milligrams of lead compounds. The technical background and experience required to grasp and retain this quantitative concept, as an effective principle of daily work, are not common among production personnel; therefore, the industrial hygienist must supply and implement this viewpoint. The continuous search for the small departures from satisfactory practice, for the minor sources of exposure to lead that otherwise pass unnoticed is his responsibility. In setting these upon a firm and understandable quantitative basis, in terms of accepted standards of air contamination, he eliminates mystery and conjecture and thereby may gain the effective assistance of technical and production personnel. A single demonstration of the efficacy of a procedure recommended for the elimination of air-borne contamination, by means of "before and after" analytical results, will have a more salutary effect than volumes of verbal propaganda.

The orderly and effective application of safe methods for the performance of many items of plant procedure, especially those involving repair and maintenance of equipment, often calls for written instructions and regulations. The preparation of such instructions requires a sound understanding of operating processes, as well as of hygienic and physiological principles. Here, and at many other points of contact, there must be effective collaboration between the industrial hygiene engineer, the physician, and production personnel. It is highly important, therefore, that the organization of the activities of a plant, in relation to industrial health in general, should be such as to facilitate teamwork of this type.

B. MEDICAL METHODS OF CONTROL

As has been indicated previously, the control of occupational exposure to lead through the maintenance of a safe environment is essentially a mechanical problem, which therefore lies within the province of the engineer. If the maintenance of a safe level of atmospheric contamination with lead were accepted as a necessary condition of plant hygiene in the lead trades generally, there would seem to be no medical problem. The situation is not quite so simple, however. There is a great deal of misunderstanding on this score on the part of industrial management, engineers, and even physicians. Actually safe conditions with respect to occupational exposure to lead are not generally achieved in industry, and there is an unfortunate belief that it is the function of the industrial physician, by some special clinical skill or legerdemain, to convert hazardous work into safe, or at least to avert the more serious consequences of hazardous exposure to lead. Let us clear the air by admitting that this cannot be done except to a limited degree, and only then through the use of highly specialized physiological tools. The physician may find it possible to recognize the early clinical manifestations of saturnism and thereby prevent the development of the more serious and disabling forms of lead poisoning among industrial workers. Under actually hazardous conditions of exposure to lead, however, he will not always succeed in this attempt, as experience has amply demonstrated. He may also alleviate suffering and shorten the period of disability through proper care and treatment of poisoned workmen, but by no stretch of the imagination can this be regarded as preventive medicine. Indeed to the extent that such medical practice tends to obscure the hazardous character of an occupation, it is an obstacle rather than an aid to the achievement of industrial health. There are, however, certain functions in the control of industrial lead hazards that can best be served by the industrial physician. There is great need at present for a greater understanding and a wider application of these medical functions.

The first and foremost function of the physician in the lead trades is to recognize and record the earliest manifestations of lead intoxication. The proper performance of this function serves a threefold purpose. It enables him to recognize and document the existence of hazardous exposure, thereby demonstrating that measures being taken for its control are inadequate. The recognized occurrence of lead poisoning, when related to the work location and the duration of the exposure of affected individuals, often serves to reveal the source and nature of the lead hazard, thereby pointing the way for its elimination. The existence of plumbism in one or more members of an occupational group will demonstrate the need for prompt protection of the others in the group, whether by special means or by temporary discontinuance of the work until the hazard has been eliminated.

It should be apparent to all students of the subject that criteria for the safety of any type of occupational exposure to lead are based, ultimately, upon the sensitivity of clinical methods for the recognition of the signs of lead intoxication. Such present criteria are wholly pragmatic and practical, since they have been derived from the correlation of analytical data and clinical observations. If new and better clinical methods based upon more complete understanding of the toxicology of lead compounds should be developed, these criteria may well be modified in the future as they have been in the past. The importance of continued clinical investigation in the lead industries, therefore, is quite obvious.

There is also, of course, the matter of nonoccupational illnesses among industrial workmen. The differentiation of these, especially the chronic diseases of obscure origin, from saturnism is a problem the complexity of which varies considerably with conditions that have little relation to clinical diagnosis as such. The inevitable tendencies of some workmen, members of their families, their legal advisors, and their sympathetic and sometimes ill-informed family physicians to attribute any illness to the effects of lead absorption are heightened tremendously by unsatisfactory labor-management relations, by stringent general, local, and individual economic conditions, and by local professional friction. Nevertheless the clinical problem itself is sometimes difficult to solve, requiring a high degree of technical expertness and general professional judgment, as well as complete freedom from prejudice. It is not enough for the physician to determine to his own satisfaction that an illness is or is not due to lead absorption. Proper professional standards as well as medicolegal considerations demand that he put forth every effort consonant with current medical knowledge, not only to convict or exclude lead as an etiological or contributory factor in an illness, but also to demonstrate the actual nature of the disease in question.

A second function of the physician in the lead trades lies in the utilization of physiological methods for the measurement of lead exposure, and the interpretation of the data obtained for this purpose under his supervision, when such information is required. Reference has been made previously to criteria for the differentiation of safe from dangerous lead exposure in terms of the concentration of lead in the blood or urine of exposed persons. The analysis of these materials calls for the services of a competent analytical chemist. The collection of suitable samples of such biological materials, on the other hand, should be done by or under the guidance of a physician whose familiarity with the workmen and their working conditions is complete, and whose understanding of the physiological and technical background of the sampling procedures is comprehensive. There appears to be considerable uncertainty as to the proper role of analytical data of this type in the control of industrial lead exposure. The specific meaning of various types of analytical data is dealt with elsewhere in this chapter, but questions concerning the desirability and the need for such data in the given instance can be answered briefly and simply. In general, when the exposure to lead in an occupation, a plant, or an industry is of such type and character as to be substantially uniform and readily controlled by mechanical means, so that essentially safe occupational conditions are maintained regularly, there may be no real need for systematic determination of the lead content of the blood or urine of workmen. In comparison with the sampling and analysis of the air of workrooms, the sampling and analysis of the blood and urine of workmen are difficult, time-consuming, and expensive, while, under the conditions indicated, the former procedures may well be entirely adequate to maintain safe working conditions. In such circumstances the analysis of biological materials is required only occasionally, if at all, in connection with medical or medicolegal differential diagnoses.

On the other hand, when the exposure to lead in an industry is highly variable for technical or other reasons, or when the control of such exposure is difficult to achieve and maintain, precise analytical data, especially through analyses of suitable samples of the blood or urine of the exposed personnel, become a necessity. Only by such means can one protect those who are actually involved in hazardous exposure. The medical problem in such a situation is to determine how long an individual or a group may safely continue to work. When workmen reach a level of absorption at which it is no longer safe for them to accept the ordinary or recurring opportunities for exposure associated with their day's work, they must be removed and transferred to work involving less exposure or perhaps none at all. The quantity of lead that is retained in the tissues of workmen under a given set of occupational conditions is determined largely by the factor of time, that is, by the duration of continuous employment in the specific occupation. The influence of the time factor in relation to a known intensity of exposure to lead can be estimated roughly, but in a number of industries and occupations it can be ascertained with adequate assurance only by the analyses of blood or urine or both blood and urine. For this reason in the medical supervision of occupations in which the exposure to lead is uncertain or near the threshold of danger, the use of such methods is generally advantageous and often indispensable.

A third function of the industrial physician is the maintenance of adequate records of preplacement and periodic examinations, and the use of such records to determine the hygienic status of individuals and occupational groups of plant personnel. The necessity and propriety of comprehensive examinations of this type in hazardous occupations is beyond argument and requires no discussion here. It is not so commonly recognized that the information that can be gained thereby in matters of general industrial health or in relation to specific hazards is of such importance as to justify considerable expenditure of time and effort in the accumulation of complete and precise records and in periodic systematic study of the data contained therein. Obviously, much of the information in such records must be held in proper professional confidence, but the essential data related to the main problems of industrial hygiene rarely have individual or personal implications, and can be employed to demonstrate the facts to the management.

1. Specific Medical Procedures

The preceding paragraphs are intended merely to outline the general professional methods of the physician in the lead industries. Certain specific procedures and practices merit detailed discussion, however, for the reason that their significance is not always fully understood nor are they always applied in a satisfactory manner. These are concerned primarily with the detection of dangerous levels of lead absorption prior to the development of toxic manifestations. The evidences of lead absorption, as distinguished from those of intoxication, consist of: (a) certain blood changes of which the most important are apparently the result of some degree of stimulation of the blood-forming tissues with the release of increased numbers of juvenile forms of erythrocytes (reticulocytes and erythrocytes with basophilic granulation); (b) the development of punctate deposits of lead sulfide in the gum tissues and in the mucosa of the colon and anus; and (c) an elevation of the lead content of the blood and urine (or tissues) above the normal level. The first two of these cannot be regarded as specific evidence of lead absorption, since many stimuli affect the hematopoietic tissues, and since any metal that will form a black sulfide may cause a blue line in the gum tissues or a bluish black deposit in the anal mucosa. The third, on the contrary, is entirely specific and reliable in indicating the extent of the absorption of lead.

(a) *The Significance of "Stippling" of the Erythrocytes.* The limits of normal numerical variation of the several formed elements of the blood that are specifically involved in lead absorption, have been observed and reported upon by numerous clinicians and investigators. Discussion herein will be limited to punctate basophilia or "stippling" of the erythrocytes. (References are given at the end of this chapter to articles on these and other blood elements and on methods.) The latter occur normally in human blood in small numbers. The numbers of stippled erythrocytes in the blood show little or no change with small increments of increased lead intake, but at higher levels of exposure and absorption their number in the circulating blood shows a somewhat irregular tendency toward an increase. Considerable variation in the degree of individual response is observed, and fairly wide variations occur from day to day, but with increasingly severe conditions of exposure to lead there is a trend toward increasing numbers of stippled erythrocytes in the blood of exposed workmen, if comparisons are based on groups rather than on individuals. Unfortunately the trend in this direction is neither striking nor dependable, and a sizable proportion of the results of any series of such observations is always within the range of normal values.

The above facts stated briefly and somewhat too simply show clearly why it is impossible to rely upon this and related hematological tests for estimating the extent of the absorption of lead by individuals and occupational groups with sufficient promptness and exactitude for practical purposes. And yet some modification of this procedure is the common, and in many instances the only, method employed for this purpose. The time has passed long since when industrial physicians may justify reliance upon these procedures as more than clinical adjuncts to other more specific and sensitive methods.

Certain additional facts demonstrate the value of blood examinations of this type in relation to clinical diagnosis, when properly employed and interpreted. Under conditions of hazardous exposure to lead, an abrupt and *progressive* increase in stippled erythrocytes above the previous normal level of the individual is strongly suggestive of an increased rate of absorption of lead, and should be so regarded unless there is adequate evidence to the contrary. Such an increase cannot be taken as evidence of the existence or imminence of lead intoxication, for large variations occur without symptoms or signs of illness, but it should be considered a harbinger of danger, somewhat in proportion to its magnitude and speed of development. On the other hand, the absence of stippled erythrocytes in the blood in a suspected or alleged case of lead poisoning, during a period of active intoxication, constitutes very nearly conclusive evidence that the toxemia is due to some factor other than lead. The only exceptions, in our experience, are found in cases of acute overwhelming intoxication from a single or very brief massive exposure, and in cases of poisoning by tetraethyl lead. The absorption of the latter compound has never been found to result in hematological abnormalities.

(b) *The Significance of the "Lead Line."* A lead line may appear at the margin of a tooth (or teeth) in gums that are somewhat inflamed by bacterial invasion—that is, when hydrogen sulfide is produced locally—whenever the lead content of the involved tissues is sufficiently elevated. We have observed the appearance of a punctate blue line in the tissue of an infected pocket overlying a single tooth in a person whose blood contained lead only to the extent of 0.05 mg. per 100 g. Easily identified lead lines usually signify somewhat greater absorption of lead than that indicated in this instance. Obvious deposits of lead sulfide are seen in the gums of persons with no demonstrable symptoms or signs of intoxication, but the intensification and extension of a lead line may be accepted as an indication of progressive absorption of lead and thereby as a warning of possible danger. The lead line must be differentiated from similar deposits resulting from the precipitation of other metallic sulfides, notably of bismuth; and it must not be confused with the normal pigment of Negroes and correspondingly dark-skinned peoples. The former differentiation is sometimes provided by the medical history of the person in question, but may require analysis of the blood or urine or both; the latter can usually be made by careful study of the locus and appearance of the pigment. The natural pigment is rarely found in the gum tissue on the lingual side, while despite the scant attention given the fact by clinicians in general, a favored site for the first appearance of a lead line is in the extreme lingual edge of the gum opposite the bicuspids and lower molars. The bluish purple line of gingival congestion may be taken for a lead line by the inexperienced physician, especially if examination is not made after expression of the blood by means of a transparent applicator (such as a glass slide). More commonly, the stained or discolored surface of a tooth just visible beneath a thin layer of gum tissue is called a lead line. The differentiation is not always easy, and resort must sometimes be had to the use of a hand lens, or even to biopsy followed by microchemical or, better, spectrographic analysis.

(c) *The Significance of the Results of Analyses of the Blood, the Urine, and (Post-Mortem) Various Tissues for Their Lead Content.* Information concerning the lead content of the blood and urine in relation to standards of industrial safety has been presented in some detail previously. In achieving an understanding of the physiological background on which such standards are based, the normal pattern of the metabolism of lead in the human organism and the quantitative changes in this pattern in response to abnormal quantities and types of lead absorption should be summarized briefly. The student of this subject as it relates to industrial and public health should seek more extensive information in the literature, some of which is indicated in the references at the end of this chapter.

VII. The Metabolism of Lead in Man

A. NORMAL LEAD METABOLISM

The general character of the metabolism of lead in the body of the average normal, healthy, North American adult may be described with considerable quantitative accuracy. He ingests with his food and beverages quantities of lead varying from 0.05 to somewhat more than 2.00 mg. (rarely in excess of 4 mg.) per day, the mean daily quantity over a period of months being approximately 0.30 mg., varying somewhat from individual to individual in accordance with the gross quantity and the variety of food consumed. He inhales air containing a small quantity, which rarely, if ever, exceeds 0.10 mg. per 24 hours, of which only a portion—for example, 35 to 50 per cent of particles 0.75 μ or less in diameter—is retained in the lungs and, if not utterly insoluble, is absorbed, rapidly or slowly, as the case may be. Fifty to sixty-five per cent of particles of such small size are exhaled. A large proportion of appreciably larger particles is trapped in the upper respiratory tract and subsequently swallowed, while still larger particles, 10 μ or more in diameter, fail to remain air-borne for long and thus occur infrequently in the general ambient atmosphere. (The quantity of lead absorbed from the "normal" atmosphere is so small as to be barely apparent in prolonged experiments in which the total output of lead in the feces and urine is balanced against the total intake of lead by ingestion. In Cincinnati it appears to be of the order of ±0.02 mg. per day.) Only a small proportion (8 to 10 per cent or less) of normally ingested lead is actually absorbed into the body. Most of it traverses the alimentary tract unabsorbed and appears in the feces. The portion absorbed is distributed, mainly into the skeleton, but also into other tissues, including the liver, from which a small quantity is secreted back into the alimentary tract with the bile. The effect of the generally poor alimentary absorption of lead, together with the biliary secretion (and perhaps other lead-containing secretions into the alimentary tract), is to make the daily fecal lead output almost equivalent to the total intake by ingestion. The small quantity that escapes elimination in the feces finds its way into the tissues (including the skeleton), the blood, the body fluids, and secretions. The concentration of lead in the blood at any specific time depends upon two main factors: the rate of absorption from the intestine and lung, and the magnitude of the countercurrent of lead entering the circulation from the tissues of the body collectively. Thus the whole blood of normal healthy North American adults usually contains from 0.01 to 0.05 mg. per 100 g., the mean concentration being approximately 0.03 mg. per 100 g. (see Table 2). Concentrations as high as 0.06 mg. per 100 g. are found occasionally in a large series of analyses, probably as an expression of a chance contamination or an unusually large analytical error. Only traces of the lead in the blood are found in the plasma, 95 per cent or more being found in the erythrocytes. Lead is found normally in physiologically representative volumes of urine (large samples of 2 to 3 liters) ranging from 0.01 to 0.08 mg. per liter. The mean concentration in the urine approximates 0.03 mg. per liter (sweat contains lead in concentrations comparable to those in the urine). The lead in the tissues of the average normal man is distributed in accordance with a definite pattern. This pattern is displayed partially in Table 4, but the other tissues and organs participate in it. The major portion is found in the skeleton, the long bones, such as the femur, containing higher concentrations than the flat. (This relationship is reversed during and for some time after a period of abnormal absorption (Table 4).) Measurable and fairly uniform concentrations are found in the other tissues. The average gross quantity of lead present in the body of the normal human adult in the United States of America is not less than 100 mg. nor much more than three times this quantity, with due regard to variations in the body weight, and to the quantity and variety of food and beverages consumed. Both direct and indirect types of evidence indicate that there is little or no progressive accumulation of lead in the human body under the average conditions of lead intake and absorption described above.

TABLE 4

The Levels of Concentration of Lead in Certain Human Tissues and Organs
under Varying Circumstances: Random and Inadequate Sampling
(Lead in milligrams per 100 g. of wet—fresh or fixed—tissue)

Tissue	Persons with no occupational exposure			Persons deceased (irrelevant cause) in course of prolonged occupational exposure			Fatal lead poisoning (children)		
	Mini-mum	Maxi-mum	Mean	Mini-mum	Maxi-mum	Aver-age	Mini-mum	Maxi-mum	Aver-age
Brain	0.01	0.09	0.02	Few results		0.24	0.24	0.58	0.40
Bone (long)	0.67	3.59	1.88				Few results		10.07
Bone (flat)	0.21	1.11	0.47	Few results		4.53	10.65	26.80	17.00
Kidney	0.02	0.16	0.027	0.07	0.32	0.16	0.22	2.59	1.71
Liver	0.04	0.28	0.13	0.60	1.75	0.92	2.96	4.40	3.71

Note: From the levels of the concentration of lead in these and other tissues, including striated muscle because of its gross weight, the lead content of the entire body can be calculated with fair accuracy. The total lead content of the body of the average adult in each of the three categories above, assuming an average weight of 75 kg., may be estimated at approximately 200, 460, and 1800 mg., respectively.

B. ABNORMAL LEAD METABOLISM

When there is an increase in the exposure of an individual to more than the usual quantities of lead in the "normal" environment, as defined in the foregoing paragraphs, whether by reason of an increase in the lead content of his food and drink, or of the air he breathes, the evidence of such an increase may be found promptly, although such increase may be of small magnitude, as well as regularly or irregularly intermittent in its occurrence, and of brief or prolonged duration. The means of finding the evidence will differ somewhat according to the nature of the increase. Since the primary concern here is that of occupational exposure to lead, attention should be focused mainly on the types and characteristics of such exposure. However, it must not be forgotten entirely that opportunities for abnormal exposure to lead other than those that occur in a specific industry present themselves to some industrial employees, since some of them have more than one job, and since, occasionally, the environment of the home may provide a significant exposure to lead, because of a faulty supply of water, or for some other strictly local reason. Moreover, persons have been known to take various lead compounds by mouth in order to induce or to prolong an episode of lead poisoning. (The writer has had but three such cases in the past twenty years, but such cases may be expected to occur from time to time, as an expression of the quest for compensation, or in pursuit of other economic or personal ends.) When abnormal quantities of lead are taken by mouth on a single occasion, there is a corresponding increase in the lead content of the feces for one to three days thereafter (until the alimentary tract has been completely evacuated). If the ingestion of an abnormal quantity of lead is repeated one or more times, at intervals, the fecal output of lead will undergo corresponding increases at approximately corresponding intervals, and if the ingestion is continuous (daily), the level of the lead output of the feces will be elevated correspondingly and regularly. The lead content of the feces of workmen in lead-using industries will also be increased when they work regularly (with one or two days of freedom at the weekend) without respiratory protection in an atmosphere containing lead-bearing dust of which some of the particles exceed 1 μ in diameter. The increase under these circumstances will vary with the severity of the atmospheric contamination with lead-containing particles of relatively large size (and mass). If the atmospheric contamination is limited to minute lead-bearing particles (vapor, fume, or extremely fine dust) under 1 μ in diameter, there will be little or no demonstrable increase in the lead content of the feces (only to the extent that there is an increase in the true alimentary excretion of lead).

Under any set of conditions through which greater than "normal" (usual) quantities of somewhat absorbable lead compounds (it is impossible at present to be certain that there is any particulate compound of lead that is completely unabsorbed for practical purposes by the pulmonary or alimentary membranes) are introduced into the body by the respiratory or alimentary routes, or by both routes as is usually the case in industry, there is a prompt increase in the rate of absorption of lead, as shown first by a proportional elevation of the concentration and output of lead in the urine. This is followed a little later (if the increase in absorption is sufficient) by a lesser increment of increase in the concentration of lead in the blood. There is also an increase in the distribution of lead into the tissues of the body, and for a time during the period of the elevated intake and absorption of lead there is a proportional increase in the retention and accumulation of lead in the body. If the increase in the intake and absorption of lead is maintained continuously (24 hours/day, 7 days/week) for months or years, there is a corresponding and steady increase (proportional to the daily dose) in the concentration of lead in the urine and blood and in the body as a whole. However, if the elevated absorption is intermittent, according to the usual industrial schedule of 40 hours, more or less, of exposure per week, the levels of lead concentration in the urine and blood reach a plateau after some months (the actual time required varies with the severity of the exposure), after which they remain essentially constant so long as the severity and duration of the exposure remain constant. After this plateau has been reached, and so long as it is maintained, there is little progressive increase in the accumulation of lead in the tissues. It cannot be said that there is none, for the reason that industrial exposure to lead is not only intermittent by reason of the hours and days of work, but it also varies somewhat in magnitude from day to day and from time to time. As it increases and decreases, the metabolic equilibrium is disturbed and the lead content of the body increases and decreases accordingly, with some delay, especially on the decreasing side of the cycle (a longer time is required for the elimination than for the absorption and retention of lead). The evidence indicates that there is also a shift of the lead within the skeleton from its more active (in terms of metabolism) to its less active parts during a period of elevated absorption, and this feature of the skeletal metabolism tends to result in a slow accumulation of lead within this great mass of tissue (roughly one-seventh of the body weight in the adult) during such a period. Such factors as these introduce serious difficulties into observations designed to determine the precise relationships between the intake and output of lead by man over prolonged periods of time, even under precise experimental conditions in the laboratory. Such experiments have shown, however, that during any period within which the concentrations of lead in the blood and urine remain essentially constant (i.e., subject only to ordinary and inevitable variations above and below a mean or medium level), progressive accumulation of lead within the body is slight to the point of insignificance.

After the discontinuance of occupational exposure to lead, the lead retained in the body during the period of increased absorption maintains the concentration of lead in the tissues, the blood, and the urine above normal levels for a period of time that depends upon both the quantity of lead retained and the time over which the retention was occurring. Thus the concentration of lead in the blood and urine decreases progressively, reaching the normal range in a period that varies from a few weeks to considerably more than one year. The alimentary tract, on the contrary, to the extent that particulate lead compounds from the nasopharynx were being swallowed, is emptied of its abnormal lead content to a very large extent within a few days after the termination of the occupational exposure. From then on the alimentary lead output is so largely composed of lead ingested with food and beverages as to be indistinguishable therefrom, except under carefully controlled conditions in which the total lead content of the ingested food and beverages is ascertained by the analysis of duplicate samples. The lead content of the tissues of the body diminishes in accordance with the elevated rate of the excretion of lead, there being some lag, but by no means immobility in this regard, on the part of the skeleton. Available analytical data indicate that approximately normal concentrations of lead in the tissues are commonly reached within 12 to 18 months after the termination of a prolonged period of occupational exposure, the skeleton being the last to return to the normal range.

VIII. Recommended Procedures of Preventive Medicine

The importance of the facts outlined in the foregoing paragraphs cannot be overemphasized in their relation to the problem of occupational exposure to lead. They demonstrate clearly that lead occurs in the human body without prejudice to normal health, and that a toxic effect on the part of this element is a matter of concentration. Combined with other experimental and clinical evidence, they also indicate that the absorption and excretion of quantities of lead considerably in excess of the usual and normal levels, are wholly compatible with normal and healthy human life and activity, and that lead intoxication is not to be anticipated if certain limits of exposure and absorption are not exceeded. They also point the way to the proper medical handling of workmen who are found to have absorbed quantities of lead that are potentially dangerous, as judged by the criteria recommended herein. The removal of such a man from the occupation in which he is engaged, prior to his development of any symptom or sign of intoxication, becomes an ideal procedure of preventive medicine, in that an occupational illness can certainly be avoided and the man can be returned to the same work, after he has excreted the potentially dangerous lead from his tissues, without prejudice to his health. If, in the meantime, while he works at a nonhazardous job, the hazardous conditions that were responsible for his excessive absorption of lead shall have been eliminated or brought under control, the entire problem will have been solved. On the same physiological basis, the rotation of men into and out of occupations that are unusually resistant to hygienic control becomes a rational procedure of preventive medicine. There are problems here with respect to the handling of personnel, admittedly, but the validity of such procedures in the strict terms of personal health, when properly implemented as medical and personnel policies, removes these problems from the realm of makeshift and compromise.

Considering all of the facts in relation to the problems of sampling and analysis and the interpretation of analytical data, there are several types and combinations of practical preventive procedures that can be employed in the medical supervision of workmen in hazardous occupations in lead-using industries. Certain of these that have proved their efficacy over a period of many years may be worthy of presentation. Any one of them may be the procedure of choice in a given situation.

As a means of demonstrating the severity of the exposure to lead associated with each of a series of different occupations in a plant, periodic analyses have been carried out after collecting samples of urine of large volume (2 liters or more) from each of a group of workmen chosen to represent the several operations and operating areas. These samples are obtained by the workmen during time spent at their homes, after having bathed and changed clothes at the plant. Verbal and written instructions are provided for their guidance so as to cultivate their informed cooperation and to avoid errors of sampling, so far as possible. Each series of analytical data is subjected to careful study and to statistical analysis. Since the occupational groups are kept unchanged for as long as possible and are modified in composition and augmented or diminished in number only after careful consideration, so as to maintain their representative character, the status of the hazard of the daily work, from the aspect of the absorption of lead associated with specific occupations, is made evident on each occasion. The results will reveal the status of individual workmen from time to time, but a comparison of the successive series of results is especially valuable in demonstrating the satisfactory or unsatisfactory quality of the hygienic control in all parts of the plant. The interval between successive series of analysis can be varied in accordance with the extent of the hazard associated with the operations, after this has been established. Inasmuch as this regimen fails to provide observations on all of the workmen who are subjected to exposure to lead, it is usually advisable to supplement it, periodically, by a complete analytical survey of all such workmen. This may be done in the manner described above, but analysis of the blood (except in connection with exposure to tetraethyl lead) is the method of choice in determining the status of individual workmen.

The foregoing regimen, while being highly useful and satisfactory, has met with difficulties in some instances, and has then been discarded in favor of the use of samples of urine of small volume (spot samples) collected, with sufficient frequency to make up for the greater variability of their lead content, from a similarly representative group of workmen. The results of these analyses are assembled periodically in relation to the occupations represented, and are subjected to statistical analysis, so as to reveal the hygienic status of the occupations. In addition, everyone engaged in operations that give rise to exposure to lead is sampled at least twice per year in the same manner. Any unusual result on the high side is checked by the analysis of such further samples of urine as may be required to determine the status of the individual concerned. Here again, the sampling and analysis of the blood usually prove, with the exception indicated above, to be the method of choice in investigating the absorption of lead by individuals, at infrequent intervals.

Another regimen of sampling and analysis—one that is employed frequently and advantageously—is that of obtaining samples of blood and "spot" samples of urine from all of the individuals who are subjected to occupational exposure to lead, on a schedule that is adjusted to the severity of the exposure to lead associated with the various operations and operating areas. Some of the operations and the personnel concerned with them may require little attention, while others may call for frequent analytical surveillance. The advantages of this regimen have been pointed out previously. An alternative that retains most of these advantages is that of limiting the analyses to the blood alone, and in some situations this limitation is fully justified. It is sometimes necessary, and it is usually wise, to keep the cost of all such services at the lowest level that is compatible with satisfactory performance. Here considerations of economy, tinctured with medical judgment, will dictate the choice of procedures that will best meet the requirements of the given situation.

The basis for the interpretation of the results obtained by the analysis of the blood and urine has been presented previously in some detail. It has been found to be useful, however, to set down certain practical applications of such interpretations as guides for uniform medical and administrative action. The presentation of these in the form of a directive might convey the impression that undue importance is attributed to the analytical regimen and its findings, and that the medical examination of workmen and the exercise of sound clinical judgment in their behalf are to some extent disregarded. Nothing could be further removed from the substance and intent of these statements concerning procedure. Rather, they stem from the basic consideration that the workmen in potentially hazardous industries should be kept under a medical surveillance that is competent and conscientious; that thorough physical examinations, supplemented by appropriate procedures of the clinical laboratory, are employed capably and with sufficient frequency to provide satisfactory information on the general health of the workmen, as well as on specific environmental hazards. Nothing said here or elsewhere in this chapter should be thought of as condoning the practice, often seen some years ago in the lead-using industries, of limiting the medical supervision of workmen to the examination of blood films for "stippled" erythrocytes by a technician. That the presently available analytical techniques are infinitely superior, in their specificity and sensitivity, to those based on hematological findings, cannot justify their substitution for medical knowledge, judgment, and skill. Despite the fact that analyses of the blood and urine for lead call for greater precision and skill, on the part of a good chemist, than are required in most of the procedures of the medical laboratory, they are clinical procedures and they should be regarded as a part of the armamentarium of the physician in industrial hygiene. In this context the following criteria for the interpretation and application of analytical results as a part of the medical handling of personnel are advanced.

A. CRITERIA FOR THE INTERPRETATION OF THE RESULTS OF ANALYSES OF BLOOD OR URINE, IN THE RECONSIDERATION OF THE CLINICAL DIAGNOSIS ARRIVED AT IN INDEPENDENCE OF SUCH RESULTS

1. A cause-and-effect or a contributory relationship between the absorption of lead and any type of illness presented by a workman whose current exposure to lead continued uninterruptedly up to the onset of his illness and in direct temporal relationship to the collection of the samples for analysis, and who has not

been subjected meanwhile to therapy of a type that would disturb the lead metabolism of the body, may be excluded, when (a) the levels of concentration of lead in duplicate samples of blood and a "spot" sample of urine are well within safe limits—not more than 0.06 mg. per 100 g. of blood and under 0.13 mg./liter of urine; or when (b) the levels of concentration of lead in repeated "spot" samples of urine are within safe limits as defined above; or when (c) the concentration of lead in a sample of urine of large volume is not in excess of 0.10 mg. per liter.

2. A cause-and-effect relationship between the absorption of lead and the existence of nonspecific or obscure symptoms or signs of illness cannot certainly be excluded and so may be assumed (not proved) to exist, when (a) other causes of illness have been excluded beyond reasonable doubt, and the levels of concentration of lead in duplicate samples of blood and a "spot" sample of urine, uninfluenced by eliminative therapy, are indicative of potentially dangerous absorption of lead—more than 0.08 mg. per 100 g. of whole blood, and more than 0.18 mg./liter of urine; or when (b) in the same clinical situation, the results of the analysis of multiple "spot" samples of urine (or of a sample of urine of large volume, collected with appropriate care to avoid contamination) are in agreement in demonstrating the occurrence of potentially dangerous absorption of lead—more than 0.18 mg./liter.

3. A cause-and-effect relationship between the absorption of lead and the existence of strongly suggestive or characteristic symptoms and signs of lead intoxication may be assumed to exist, when (a) the levels of concentration of lead in the blood and urine are within the dangerous range, as previously defined; or when (b) it is known that the levels of concentration were well above safe limits during a period of exposure to lead that has now been discontinued for some weeks.

B. CRITERIA FOR THE REMOVAL OF WORKMEN FROM OCCUPATIONS IN WHICH THEY MAY INCUR SIGNIFICANT EXPOSURE TO LEAD, DESPITE THEIR FAILURE TO DISPLAY ANY SYMPTOMS OR SIGNS SUGGESTIVE OR CHARACTERISTIC OF LEAD INTOXICATION, OR ANY OTHER OBSCURE OR ILL-DEFINED SYMPTOMS OR SIGNS OF ILLNESS

1. The exposure of a workman to lead should be terminated abruptly (a) when the concentration of lead in his blood, as demonstrated by multiple analyses, is found, more than fleetingly, to be at or in excess of 0.08 mg. per 100 g., or when (b) the level of concentration of lead in multiple "spot" samples of urine, or in a properly collected sample of urine of large volume, is found to be in excess of 0.18 mg./liter; or when (c) the concentration of lead in the blood is found to be increasing and approaching the threshold of danger, and the level of lead concentration in the urine remains constantly in the lower third of the normal range and thus gives presumptive evidence of impairment of renal function with respect to the excretion of lead.

C. CRITERIA FOR THE RESTORATION OF WORKMEN TO OCCUPATIONS FROM WHICH THEY WERE REMOVED BECAUSE OF POTENTIALLY DANGEROUS ABSORPTION OF LEAD

1. Workmen who are free of illness may be returned to their former occupations involving exposure to lead, with no more than the medical reservations of judgment that apply to the original selection of workmen for employment in such occupations, when (a) the concentration of lead in their blood has diminished to a point at which it is consistently well within the safe range, i.e., less than 0.06 mg. per 100 g.; or when (b) the general level of their urinary excretion of lead, as judged by the analysis of multiple samples of small or large volume, is well within the safe range, that is, less than 0.13 mg. per liter; or when (c) following the correction of the occupational conditions (whether personal or environmental) that gave rise to their excessive absorption of lead, the levels of lead concentration in the blood or urine (or both preferably) have diminished and are within the range indicated above.

2. The severity and the duration of the exposure that was responsible for the excessive absorption of lead are important factors in the consideration of the wisdom of returning such a workman to his former occupation, since his probable body burden of lead, in relation to the severity of the exposure to which he will be subjected on his return, will be a significant factor in the length of time he will be able to continue in his occupation without a further return to a dangerous level of lead absorption. His return to a safe work environment involves little or no risk, but if the environment is dangerous or only dubiously safe, the risk will have been multiplied by his previous experience. In any case, the duration of the period of his freedom from exposure should be extended in proportion to the duration of the period of exposure that resulted in his absorption of excessive quantities of lead. The elimination of lead accumulated in the body, during a period of exposure, requires at least twice as long as the period over which the accumulation occurred.

IX. The Problem of Industrial Lead Poisoning in Relation to Industrial Hygiene

The diagnosis and treatment of lead poisoning are in the realm of clinical medicine, and can scarcely be dealt with adequately in a discussion devoted primarily to the control of occupational exposure to lead and the prevention of occupational illness. Accordingly, attention is called to adequate sources of information in the references at the end of this chapter, and only a few points that impinge upon industrial hygiene will be touched upon.

A. PROPHYLAXIS

When incipient or well-defined lead intoxication has appeared in an individual case, certain steps should be taken at once (see criteria for disposition of men in relation to analytical results in preceding section). The first requirement is the elimination of further exposure to lead on the part of the affected person. The means by which this is done will depend upon the circumstances. If the exposure to lead has resulted from accidental and unforeseen causes, and if the resultant illness is slight and fleeting, it may be sufficient to see that the opportunities for subsequent exposure are reduced or dispensed with, after which the workman may resume his work. If the cause was carelessness, ineptitude, or inattention to instructions on the part of the workman, it may be necessary to transfer him indefinitely or permanently to other less exacting work.

When cases of lead intoxication occur within a group of employees, the other employees are likely to be concerned, and the physician will be concerned for them, and will consider what can be done to reduce their present danger. Dietary supplements for the workmen, such as acid beverages, milk, and vitamin preparations are not to be recommended as prophylactic measures. If they are employed as means of improving the quality of the food and correcting dietary deficiencies that often exist, they are not open to criticism on this score. Even then, however, it seems probable that advice and instruction in matters of diet, such as may be given by the industrial physician to individuals and groups of workmen, may better serve the purpose. A still more effective method for meeting nutritional needs is that of providing a satisfactory restaurant in which a good midday or mid-shift meal is served.

The regular administration of cathartic medication is to be condemned unreservedly, and this practice, often resorted to by the workmen themselves, should be combated. Constipation is a frequent human ailment, and for the workman in the lead-using trades it is especially disadvantageous, since it delays the evacuation of the alimentary tract and thereby promotes the absorption of lead. To the degree that it is caused by the absorption of lead, it is a sign of lead intoxication, and should be dealt with by the control of the exposure to lead, not by cathartics; but to the degree that it is due to other causes, such as carelessness and faulty diet, it should be dealt with by the recognition and elimination of these causes. The physician in industry has an important advisory and educational function to perform in this and many other hygienic matters.

The use of specific chelating agents, prophylactically, as a means of hastening the excretion of lead by men who are working in potentially dangerous occupations, regardless of the specificity or harmlessness of such agents of this type as may be made available in the future (no such specific or harmless agent is available as yet) is unworthy of favorable consideration by the members of a humane and socially sensitive profession. A physician should make clear to an employer the necessity of providing a safe and healthful work place for his employees, rather than give him support and comfort in his complacency and his preoccupation with other concerns. A physician can properly accept some responsibility for unsafe conditions and for employing means of mitigating their effects for a time only when every effort is being made to eliminate hazards as promptly as possible after they have been disclosed.

There are, however, times of stress in many industries, when the hazards of the day's work tend to be augmented or multiplied, and this situation arises from time to time in those concerned with the production or use of lead compounds. The physician should be especially vigilant in periods of accelerated industrial production when hours of work per day and per week tend to increase. At such times, as mentioned previously, it may be advantageous or even necessary to resort to the rotation of men and jobs so as to reduce the time spent by individuals in the more hazardous occupations. No specific rule can be given by which to adapt such rotation to the requirements of the situation, but any regularly occurring interruption of the exposure to lead will result in some reduction in absorption. The desired end can be accomplished satisfactorily under the guidance of an analytical surveillance of the men involved.

B. THE DISPOSITION OF WORKMEN AFTER RECOVERY

For reasons indicated previously, the therapy of lead poisoning is not dealt with herein. However, the recovery of a workman from an episode of lead poisoning creates a new problem of industrial hygiene, in that the question of re-employment then arises. Should the workman return to his former job, and if not, what work should he undertake? In general, it may be said that if he has regained good health and if the conditions that were responsible for his illness have been identified and corrected, he may return to his previous work. Otherwise, provision should be made for his employment in a position that will not subject him to potentially hazardous exposure to lead. This will not be difficult to do in an industry that is engaged in varied types of production, and it is not likely to be an insoluble problem in a well-managed lead-using industry. It should not be undertaken lightly, however, for if the outcome is to be successful, there must be fairly precise information on the status of the workman, with reference to his body burden, and adequate information as to the intensity of the exposure to lead, if any, to which he is to be subjected. It will not suffice to put such a man to work in an occupation or area that is considered, without evidence, to be essentially or largely free of exposure to lead, nor will it be satisfactory merely to provide a job, unsuited as it may be to the workman and his capabilities. All too frequently, the odd job or the "outside work" is potentially hazardous in terms of exposure to lead, often because it did not seem important enough or continuous enough to justify careful appraisal of its hazard. It is also worth pondering that a job that requires little effort or skill, or one manufactured for the occasion, may well be a means of converting a once skillful and energetic workman into a frustrated loafer or perhaps into a chronic invalid. Satisfactory work in favorable environment, on the other hand, often provides one of the necessary stimuli for prompt and complete rehabilitation.

General References

Methods for Sampling and Analyzing Air and Biological Materials for Lead

Air Hygiene Foundation of America, Inc., Determination of lead in the air, *Preventive Eng. Ser. Bull.* No. **2**, Part 6 (1938).

Amdur, M. O., and L. Silverman, Direct field determination of lead in air, *A.M.A. Arch. Ind. Hyg. Occupational Med.*, **10**, 152 (1954).

Bambach, K., and R. E. Burkey, Microdetermination of lead by dithizone with an ˙mproved lead-bismuth separation, *Ind. Eng. Chem. Anal. Ed.*, **14**, 904 (1942).

Bloomfield, J. J., and J. M. Dallavalle, The determination and control of industrial du.st, *U. S. Public Health Bull.* No. **217** (1935).

Cadle, R. D., P. L. Magill. A. A. Nichol, H. C. Ehrmantraut, and G. W. Newell, Sampling procedures, in P. L. Magill, F. R. Holden, C. Ackley, eds., *Air Pollution Handbook*. McGraw-Hill, New York, 1956, p. 10.

Chambers, L. A., M. J. Foter, and J. Cholak, A comparison of particulate loadings in the atmospheres of certain American cities, *Proc. Natl. Air Pollution Symposium, 3rd Symposium, Pasadena, Calif.*, **1955**, p. 24.

Cholak, J., Analytical methods, in P. L. Magill, F. R. Holden, C. Ackley, eds., *Air Pollution Handbook*. McGraw Hill, New York, 1956, p. 11.

Cholak, J., and K. Bambach, Determination of lead in biological material: a polarographic method, *Ind. Eng. Chem. Anal. Ed.*, **13**, 583 (1941).

Cholak, J., D. M. Hubbard, and R. E. Burkey, Microdetermination of lead in biological material with dithizone extraction at high pH, *Anal. Chem.*, **20**, 671 (1948).

Cholak, J., L. J. Schafer, and R. F. Hoffer, Results of a five-year investigation of air pollution in Cincinnati, *A.M.A. Arch. Ind. Hyg. Occupational Med..* **6**, 314 (1952).

Cholak, J., and R. V. Story, Spectrographic analysis of biological material. Lead, tin, aluminum, copper, and silver, *Ind. Eng. Chem. Anal. Ed.*, **10**, 619 (1938).

Clifford, P. A., and H. J. Wichmann, Dithizone methods for the determination of lead, *J. Assoc. Official Agr. Chemists,* **19**, 130 (1936).

Dixon, B. E., and P. Metson, A field method for determination of lead fume, *Analyst,* **84**, 46 (1959).

Drinker, P., and T. Hatch, *Industrial Dust. Hygienic Significance, Measurement and Control.* 2d ed., McGraw-Hill, New York, 1954.

Quino, E. A., Field method for the determination of inorganic lead fumes in air, *Am. Ind. Hyg. Assoc. J.*, **20**, 134 (1959).

Report of Subcommittee on Determination of Lead in Air, Occupational Health Section of the American Public Health Association, *Methods for Determining Lead in Air and in Biological Materials*. New York, 1955.

Wilkins, E. S., Jr., C. E. Willoughby, E. O. Kraemer, and F. L. Smith, 2nd, Determination of minute amounts of lead in biological materials: a titrimetric-extraction method, *Ind. Eng. Chem. Anal. Ed.*, **7**, 33 (1935).

Willoughby, C. E., E. S. Wilkins, Jr., and E. O. Kraemer, Determination of lead: removal of bismuth interference in the dithizone method, *Ind. Eng. Chem. Anal. Ed.*, **7**, 285 (1935).

Nonspecific Examination of the Blood and Urine in Relation to Lead Absorption

Microscopic Examination

Jones, R. R., Estimation of basophilic cells (reticulocytes) in blood by examination of ordinary blood film, *Public Health Repts. U. S.*, **48**, 1011 (1933).

Key, J. A., Lead studies. IV. Blood changes in lead poisoning in rabbits; stippled cells, *Am. J. Physiol.*, **70**, 86 (1924).

Kitzmiller, K. V., The organ systems in relation to occupational medicine. The hemopoietic system, *Proc. 8th Ann. Meeting, Am. Acad. Occupational Med.*, Cincinnati, Ohio, 1956, p. 20.

McCord, C. P., F. R. Holden, and J. Johnston, The basophilic aggregation test for lead poisoning and lead absorption, ten years after its first use, *Ind. Med.*, **4**, 180 (1935).

Sanders, L. W., Measurement of industrial lead exposure by determination of stippling of the erythrocytes, *J. Ind. Hyg. Toxicol.*, **25**, 38 (1943).

Biochemical Examinations

Chisolm, J. J., Jr., and H. E. Harrison, Quantitative urinary coproporphyrin excretion and its relation to edathamil calcium disodium administration in children with acute lead intoxication. *J. Clin. Invest.*, **35**, 1131 (1956).

Parkinson, E. S., and J. Cholak, Problems in the analysis of urinary coproporphyrin III, *Am. Ind. Hyg. Assoc. Quart.*, **13**, 158 (1952).

Pinto, S. S., C. Einert, E. J. Roberts, G. S. Winn, and K. W. Nelson, Coproporphyrinuria; study of its usefulness in evaluating lead exposure, *A.M.A. Arch. Ind. Hyg. Occupational Med.*, **6**, 496 (1952).

Wyllie, J., Urinary porphyrins in lead absorption, *A.M.A. Arch. Ind. Health*, **12**, 396 (1955).

Lead in Human Tissues and Excreta

Bagchi, K. N., H. D. Ganguly, and J. N. Sirdar, Lead in human tissues, *Indian J. Med. Research*, **26**, 935 (1939).

Brudevold, F., and L. T. Steadman, The distribution of lead in human enamel, *J. Dental Research*, **35**, 430 (1956).

Butt, E. M., and D. G. Simonsen, Mercury and lead storage in human tissues with special reference to thrombocytopenic purpura, *Am. J. Clin. Pathol.*, **20**, 716 (1950).

Gant, V. A., Lead poisoning. The "normal" occurrence of lead in tissues of the human body. The "normal" concentration of lead in body fluids, *Ind. Med.*, **7**, 616, 621 (1938).

Griffith, G. C., E. M. Butt, and J. Walker, The inorganic element content of certain human tissues, *Ann. Internal Med.*, **41**, 501 (1954).

Harris, W. J., J. A. Beauchemin, H. M. Hershenson, S. H. Roberts, and G. Matsuyama, Metal ions in the central nervous system, *J. Neuropathol. Exptl. Neurol.*, **13**, 427 (1954).

Henderson, D. A., and J. A. Inglis, The lead content of bone in chronic Bright's disease, *Australasian Ann. Med.*, **6**, 145 (1957).

Kehoe, R. A., J. Cholak, D. M. Hubbard, K. Bambach, and R. R. McNary, Experimental studies on lead absorption and excretion and their relation to the diagnosis and treatment of lead poisoning, *J. Ind. Hyg. Toxicol.*, **25**, 71 (1943).

Kehoe, R. A., J. Cholak, D. M. Hubbard, K. Bambach, R. R. McNary, and R. V. Story, Experimental studies on the ingestion of lead compounds, *J. Ind. Hyg. Toxicol.*, **22**, 381 (1940).

Kehoe, R. A., J. Cholak, and R. V. Story, A spectrochemical study of the normal ranges of concentration of certain trace metals in biological materials, *J. Nutrition*, **19**, 579 (1940).

Kehoe, R. A., J. Cholak, and R. V. Story, Editorial review: manganese, lead, tin, aluminum, copper and silver in normal biological material, *J. Nutrition*, **20**, 85 (1940).

Kehoe, R. A., F. Thamann, and J. Cholak, On the normal absorption and excretion of lead. I. Lead absorption and excretion in primitive life. II. Lead absorption and excretion in modern American life. IV. Lead absorption and excretion in infants and children, *J. Ind. Hyg.*, **15**, 257, 273, 301 (1933).

Koch, H. J., Jr., E. R. Smith, and J. McNeely, Analysis of trace elements in human tissues. II. The lymphomatous diseases, *Cancer*, **10**, 151 (1957).

Koch, H. J., Jr., E. R. Smith, N. F. Shimp, and J. Connor, Analysis of trace elements in human tissues. I. Normal tissues, *Cancer*, **9**, 499 (1956).

Morris, H. P., Age and the lead content of certain human bones: a compilation and statistical analysis of recently published data, *J. Ind. Hyg. Toxicol.*, **22**, 100 (1940).

Tompsett, S. L., The distribution of lead in human bones, *Biochem. J.*, **30**, 345 (1936).

Tompsett, S. L., and A. B. Anderson, The lead content of human tissues and excreta, *Biochem. J.*, **29**, 1851 (1935).

Vouk, V. B., K. Voloder, O. A. Weber, and L. Purec, Normal values of lead concentration in human blood, *Arhiv hig. rada*, **6**, 277 (1955).

Webster, S. H., The lead and arsenic content of urines from 46 persons with no known exposure to lead or arsenic, *Public Health Repts. U. S.*, **56**, 1953 (1941).

Willoughby, C. E., and E. S. Wilkins, Jr., The lead content of human blood, *J. Biol. Chem.*, **124**, 639 (1938).

Lead in Food and Beverages

Bibliography on heavy metals in food and biological material (from the beginning of the year 1921 to date). II. Lead, *Analyst*, **57**, 775 (1932).

Kehoe, R. A., J. Cholak, D. M. Hubbard, K. Bambach, R. R. McNary, and R. V. Story, Experimental studies on the ingestion of lead compounds, *J. Ind. Hyg. Toxicol.*, **22**, 381 (1940).

Kehoe, R. A., J. Cholak, and R. V. Story, Editorial review: manganese, lead, tin, aluminum, copper and silver in normal biological material, *J. Nutrition*, **20**, 85 (1940).

Kehoe, R. A., F. Thamann, and J. Cholak, On the normal absorption and excretion of lead. III. The sources of normal lead absorption, *J. Ind. Hyg.*, **15**, 290 (1933).

Larkin, D., M. Page, J. C. Bartlet, and R. A. Chapman, The lead, zinc, and copper content of foods, *Food Research*, **19**, 211 (1954).

Lockwood, H. C., Determination of lead in foodstuffs, *Analyst*, **79**, 143 (1954).

Monier-Williams, G. W., Reports on Public Health and Medical Subjects, No. **88**, H. M. Stationery Office, London, 1938.

Neal, P. A., W. C. Dreessen, T. I. Edwards, W. H. Reinhart, S. H. Webster, H. T. Castberg, and L. T. Fairhall, A study of the effect of lead arsenate exposure on orchardists and consumers of sprayed fruit, *U. S. Public Health Bull. No.* **267** (1941).

Observations on Industrial Exposure to Lead

Baumann, W. H., W. Smith, and D. J. Lauer, Environmental health studies associated with lead-bearing steel production, *Am. Ind. Hyg. Assoc. Quart.*, **18**, 29 (1957).

Bloomfield, J. J., V. M. Trasko, R. R. Sayers, R. T. Page, and M. F. Peyton, A preliminary survey of the industrial hygiene problem in the United States, *U. S. Public Health Bull.* No. **258** (1940).

Cholak, J., and K. Bambach, Measurement of industrial lead exposure by analyses of blood and excreta of workmen, *J. Ind. Hyg. Toxicol.*, **25**, 47 (1943).

Dreessen, W. C., T. I. Edwards, W. H. Reinhart, R. T. Page, S. H. Webster, D. W. Armstrong, and R. R. Sayers, The control of the lead hazard in the storage battery industry, *U. S. Public Health Bull.* No. **262** (1941).

Giel, C. P., M. Kleinfeld, and J. Messite, Lead toxicity in a storage-battery plant, *A.M.A. Arch. Ind. Health*, **13**, 321 (1956).

Goldblatt, M. W., Research in industrial health in the chemical industry, *Brit. J. Ind. Med.*, **12**, 1 (1955).

Goss, A. E., and A. M. Ross, Jr., Effective control of lead dust in the manufacture of vinyl plastics, *Am. Ind. Hyg. Assoc. Quart.*, **14**, 41 (1953).

Halperin, H. J., and G. S. Reichenbach, Jr., Engineering and medical control of lead exposure in patent annealing and galvanizing of wire, *Am. Ind. Hyg. Assoc. Quart.,* **18**, 55 (1957).

Holtaway, J. S., Evaluation and control of lead exposures in powder metallurgy operations, *Am. Ind. Hyg. Assoc. J.,* **19**, 481 (1958).

Kehoe, R. A., F. Thamann, and J. Cholak, Lead absorption and excretion in certain lead trades, *J. Ind. Hyg.,* **15**, 306 (1933).

Lauer, D. J., Clinical lead intoxication from brass-foundry operations, *A.M.A. Arch. Ind. Health,* **11**, 107 (1955).

Maloof, C. C., H. Bavley, and G. W. Boylen, The engineering and medical control of a lead hazard. A plant study, *Am. Ind. Hyg. Assoc. Quart.,* **15**, 64 (1954).

Manufacturing Chemists' Assoc., Properties and essential information for safe handling and use of lead oxides, *Chemical Safety Data Sheet* SD-64, 1956.

Meyers, G. B., Lead absorption experiences in the manufacture of electric storage batteries, *Ind. Med. and Surg.,* **25**, 4 (1956).

Moskowitz, S., B. Feiner, W. J. Burke, and A. E. Perina, Lead exposures from oxy-acetylene flame cleaning and cutting of lead-painted steel, *N. Y. State Dept. Labor Monthly Rev., Div. Ind. Hyg.,* **33**, 29 (1954).

Pagnotto, L. D., and H. Bavley, The control of a lead hazard in the silversmithing industry, *Am. Ind. Hyg. Assoc. J.,* **19**, 73 (1958).

Russell, A. E., R. R. Jones, J. J. Bloomfield, R. H. Britten, and L. R. Thompson, Lead poisoning in a storage battery plant, *U. S. Public Health Bull. No.* **205** (1933).

Shiels, D. O., Industrial lead poisoning in relation to climate, *Australasian Ann. Med.,* **4**, 178 (1955).

Storlazzi, E. D., Hygiene of welding in U. S. naval shipyards, *A.M.A. Arch. Ind. Health,* **19**, 307 (1959).

Lead Poisoning—Occurrence, Diagnosis, and Therapy

Chisolm, J. J., Jr., and H. E. Harrison, The treatment of acute lead encephalopathy in children, *Pediatrics,* **19**, 1 (1957).

Johnstone, R. T., Common errors in the diagnosis of plumbism, *Ind. Med. and Surg.,* **28**, 126 (1959).

Johnstone, R. T., A re-examination of the picture of plumbism, *Ind. Med. and Surg.,* **26**, 323 (1957).

Kehoe, R. A., A critical appraisal of current practices in the clinical diagnosis of lead intoxication, *Ind. Med. and Surg.,* **20**, 253 (1951).

Kehoe, R. A., Misuse of edathamil calcium-disodium for prophylaxis of lead poisoning. Report to the Council on Industrial Health and Council on Pharmacy and Chemistry, *J. Am. Med. Assoc.,* **157**, 341 (1955).

Kehoe, R. A., Lead poisoning, in R. L. Cecil and R. F. Loeb, eds., *A Textbook of Medicine.* 10th ed., Saunders, Philadelphia, 1959, p. 498.

Miller, L. H., EDTA therapy in persons with excessive lead absorption from industrial exposure, *Ind. Med. and Surg.,* **28**, 144 (1959).

Report of Committee on Lead Poisoning, Industrial Hygiene Section, American Public Health Association, *Occupational Lead Exposure and Lead Poisoning.* New York, 1943.

Smith, H. D., Lead poisoning in children and its therapy with EDTA, *Ind. Med. and Surg.,* **28**, 148 (1959).

Tetraethyl Lead—Certain Aspects of Toxicology, Occupational and General Exposure, and Characteristics of Poisoning

Advisory Committee on Tetraethyl Lead to Surgeon General of Public Health Service (H. J. Magnuson, M. D., Chairman), Public health aspects of increasing tetraethyl lead content in motor fuel, *U. S. Public Health Serv. Publ. No.* **712** (1959).

Barry, P. S. I., A recent case of mild tetra-ethyl lead intoxication. *Trans. Assoc. Ind. Med. Officers,* **7,** 71 (1957).

Boyd, P. R., G. Walker, and I. N. Henderson, The treatment of tetraethyl lead poisoning, *Lancet,* **1,** 181 (1957).

Kehoe, R. A., Tetra-ethyl lead poisoning: clinical analysis of a series of nonfatal cases, *J. Am. Med. Assoc.,* **85,** 108 (1925).

Kehoe, R. A., and F. Thamann, The behavior of lead in the animal organism. II. Tetraethyl lead, *Am. J. Hyg.,* **13,** 478 (1931).

Kehoe, R. A., F. Thamann, and J. Cholak, An appraisal of the lead hazards associated with the distribution and use of gasoline containing tetraethyl lead. Part I, *J. Ind. Hyg.,* **16,** 100 (1934).

Kehoe, R. A., F. Thamann, and J. Cholak, An appraisal of the lead hazards associated with the distribution and use of gasoline containing tetraethyl lead. II. The occupational lead exposure of filling station attendants and garage mechanics, *J. Ind. Hyg. Toxicol.,* **18,** 42 (1936).

Kitzmiller, K. V., J. Cholak, and R. A. Kehoe, Treatment of organic lead (tetraethyl) intoxication with edathamil calcium-disodium, *A.M.A. Arch. Ind. Hyg. Occupational Med.,* **10,** 312 (1954).

Machle, W. F., Tetra-ethyl lead intoxication and poisoning by related compounds of lead, *J. Am. Med. Assoc.,* **105,** 578 (1935).

Permissible Limits of Lead in the Atmosphere

Hygienic guide series: Lead and its inorganic compounds, *Am. Ind. Hyg. Assoc. J.,* **19,** 154 (1958).

Legge, T. M., and K. W. Goadby, *Lead Poisoning and Lead Absorption.* Longmans, Green, New York, 1912, p. 207.

Schrenk, H. H., Hygienic lead standards, *Ind. Med. and Surg.,* **28,** 106 (1959).

Threshold limit values for 1959, American Conference of Governmental Industrial Hygienists, *A.M.A. Arch. Ind. Health,* **20,** 266 (1959).

CHAPTER XXVII

The Metals (Excluding Lead)

HERBERT E. STOKINGER

ALUMINUM, Al

1. *Source and Production*

Primary Al production in the United States in 1957 was almost 1.65 million tons, a decrease of 2 per cent from the preceding year. Because of technological advances in smelting, the industry began to advertise pig of 99.5 per cent minimal purity instead of 99 per cent minimal average purity. For differing reasons most plants operated under capacity; plant expansion, however, continued so that primary Al production capacity in the United States would be about 2.4 million short tons by 1959. In Canada a second firm began producing primary Al; in French Cameroon the first primary ingot was produced in 1957.

Domestic recovery of secondary Al was 360,000 short tons in 1957; some of the recovery was from Cu-, Zn-, and Mg-base alloys.

Al is produced by the electrolysis of bauxite ($Al_2O_3 \cdot 2H_2O$) in a bath of molten cryolite ($3NaF \cdot AlF_3$). In addition to these sources Al exists in minerals such as spinel ($MgO \cdot Al_2O_3$), feldspars such as orthoclase ($K_2O \cdot Al_2O_3 \cdot 6SiO_2$), and micas such as muscovite ($K_2O \cdot 3Al_2O_3 \cdot 6SiO_2 \cdot 2H_2O$).

2. *Uses and Industrial Exposures*

The major use of Al is structural in the building industry; it is also used in consumer durables, and in containers and packaging. The automotive industry is an important and growing consumer, as is the canning industry. It is anticipated that large quantities will be used in highway programs in such applications as signs, fencing, lighting, and signal supports. Far smaller quantities of Al as powder, flake, and paste are used in the paint industry. Natural Al minerals (bentonite and clays) are used in water and sugar purification and in the brewing and paper industries; aluminum compounds [$Na_3AlO_3, NaAl(SO_4)_2$] are used in the food-processing and baking industries. Various forms of Al_2O_3 are used widely as abrasives and refractories and as catalysts and catalyst carrier; activated alumina as adsorbents; hydrated aluminas as pigments and fillers.

Hazardous exposures are confined to the production of Al and the making of Al abrasives; dusts and fumes in the reduction plants contain alumina (Al_2O_3), cryolite, NaOH, SiO_2, and fluorides of Na, Al, and Ca. The reason for this unusual limitation of hazard becomes apparent in Section 5, Physiological Responses.

3. *Physical and Chemical Properties*

The physical and chemical properties of Al and some of its compounds are listed in Table 1.

TABLE 1

Physical and Chemical Properties of Al and Some of Its Compounds

	Chemical symbol	At. or mol. wt.	Sp. gr.	M.p., °C.	B.p., °C.	Solubility
Aluminum	Al	26.98	2.6989[a] 2.703[b]	659.7	2057	Insol. hot and cold H_2O Sol. alkalies, HCl, H_2SO_4, conc. HNO_3, acetic acid
Alumina	Al_2O_3	101.94	3.5	2050	2250	0.98 mg./liter(20°) Insol. hot H_2O V. sl. sol. acids and alkalies
Aluminum chloride	$AlCl_3$	133.34	2.44(25°) Liq. 1.31(200°)	190[c]	182.7[d] Subl. 177.8	69.9 mg./liter(15°) Sol. with viol. V. sol. abs. alc. Sol. CCl_4 Sl. sol. chloroform

[a] 99.99+ per cent pure.

[b] 99.5 per cent pure, aluminum conductor grade.

[c] 2.5 atm.

[d] 752 mm. Hg.

Commercially pure Al may contain up to 1 per cent Fe, Si, and Cu. Other elements may be added as alloying components. Al strongly resists the action of air by forming a resistant oxide film that checks further oxidation and minimizes absorption of gases. Al electrical conductivity is 60 per cent that of Cu per area of cross section. Al is strongly electropositive so that it corrodes rapidly in contact with many other metals. The oxide occurs naturally as ruby, sapphire, corundum, and emery.

4. *Analytical Determination*

Following conventional methods of collecting dusts and fumes, Al is best determined spectrographically[1] in the amounts collected from air; the various colorimetric methods suitable for amounts in excess of 1 mg. are subject to numerous interfering elements at smaller amounts. Intermediate quantities in excess of 1 mg. may be determined chemically by precipitation as potassium cryolite

[1] I. R. Campbell, J. S. Cass, J. Cholak, and R. A. Kehoe, *A.M.A. Arch. Ind. Health.* **15**, 359 (1957).

followed by acidimetric titration.[2] The aluminum industry has standardized on photographic spectroscopy;[3] instrumentation is so elaborate as to be justified only in large-scale operations. The spectrographic method is also the one of choice for determining Al in biological material; concentrations as low as 0.05 p.p.m. may be determined reliably. Aluminon has been used to stain tissue sections containing hydrated Al_2O_3.[4]

5. Physiological Responses

An exhaustive and critical review (90 pages) by Kehoe and associates of the environmental aspects of the effects of Al on the health of man appeared in 1957; it contains 503 literature references.[1]

Acute Toxicity. Insoluble forms of Al appear to be without demonstrable toxicity by any route provided their administration is not associated with substances that are tissue irritants. Colloidal Al_2O_3 has been repeatedly administered intravenously to rabbits without inducing intoxication. When inhaled, powdered Al produces no injurious response. A mixture of insoluble Al phosphates administered to rabbits by stomach tube to the limit of feasibility resulted in no observable effects. More soluble forms of Al, however, have measurable toxicity. Single, subcutaneously injected doses of 100 mg. Al/kg. as $Al_2(SO_4)_3$ are acutely toxic and even fatal to rabbits; the acute intraperitoneal toxicity is somewhat greater. Solutions of Al salts more concentrated than 1 per cent precipitate in the blood stream, causing emboli. The lethal oral doses of the acetate and chloride of Al for experimental animals appear to lie between 5 and 15 g./kg. for the acetate and between 1 and 3 g./kg. for the chloride. For anhydrous sodium alum the lethal oral dose for the rabbit is about 2 g./kg. The interpretation placed on the contrasting toxicities of the soluble and insoluble forms of Al is that the anionic, acidic component of the soluble Al salts is responsible in some undetermined degree for the observed toxicity.[1]

Chronic Toxicity. No instance of this form of injury from Al or its salts has been demonstrated for either animal or man, again provided the exposure was not in association with other tissue irritants (cf. Shaver's disease page 990). Countless studies have been made in man and in animals in which an Al compound was added to the diet; all showed no measurable effect on health.[1] Clinically, no deleterious effects were seen in patients that had received 16 g./day $Al(OH)_3$ for 6 months or longer. Of equal significance is the lack of reports of any injury resulting from repeated inhalation of various forms of Al for the prevention of human silicosis.

Metabolism. The physiological basis for the innocuousness of Al per se would appear to reside, in part at least, in its outstanding lack of significant absorption from either the alimentary or the respiratory tracts.[1] Numerous studies

[2] H. L. Walts and D. W. Utley, *Anal. Chem.,* **25,** 864 (1953).

[3] J. R. Churchhill, *Iron Age,* **168,** 97 (1951).

[4] D. A. Irwin, *A.M.A. Arch. Ind. Health,* **12,** 218 (1955).

have revealed clinical and experimental evidence of (a) no hygienically significant retention of Al in the body tissues despite four generations of Al feeding in rats, (b) no significant increases in Al content of blood or urine following repeated Al administration. Consistent with the negative urinary findings have been repeated demonstrations in careful balance experiments accounting for all of the administered Al in the fecal excretion. The metabolic means by which measurable amounts of Al reach the tissues have not been explained.

Mode of Action. Because a substance may be systemically injurious without being absorbed, by such means as interferences with food assimilation, many studies on the mode of action of Al compounds have been made.[1] When amounts of Al considerably larger (2 to 3 times) more than those normally taken in the diet (10 to 100 mg./day) were administered, 10 to 25 per cent of the urinary P was deflected to the feces, but the total P excreted was unchanged. Amounts of Al compounds required to decrease P absorption from the alimentary tract are roughly 5 to 10 times greater than those that occur in the daily diet. Organic P may combine with Al in the digestive tract. Excessive levels (1400 p.p.m.) of Al in the diet lower the P content of blood and bone and lead to severe "rickets" in chicks, but similar, large quantities of Al in the human diet show no evidence of interfering with the assimilation of protein, carbohydrate, or fat. Numerous in vitro studies of the effect of Al compounds on digestive enzymes have shown pepsin to be inactivated, stomach acidity decreased without appreciable disturbance of acid base balance, gastric secretion decreased, and mucin secretion increased. The activity of pancreatic juice was unchanged by $Al(OH)_3$.

Industrial Experience. A few new hazards associated with the preparation and uses of Al and its compounds have been reported since the first edition in 1949. The decade has given opportunity for a better evaluation of the use of Al in the prevention of silicosis, and Shaver's disease has become a recognized entity.

Pulmonary changes in Al workers. A German report has appeared[5] of two types of pulmonary change seen in workers exposed to Al dust in mines, foundries, and factories. In more than 1000 given x-ray study, 3.5 per cent of those exposed to bauxite in mines showed pulmonary changes, and 4.9 per cent of the factory workers exposed to Al_2O_3 and cryolite dust in foundries. The tendency of the affection to be confined to the middle and lower lobes, the strandlike densities that were frequently unilateral, and its nonprogressive nature, would seem to differentiate these conditions from Shaver's disease.

Shaver's disease. This often fatal and rapidly progressive condition of interstitial fibrosis of the lung of the nonnodular type was first described in the United States in workers in the Al abrasives industry by Shaver and Riddell in 1947.[6] Goralewski[7] and others had previously called attention to the "aluminum lung"

[5] F. Geher, *Forschr. Gebiete Röntgenstrahlen,* **82,** 598 (1955).

[6] C. G. Shaver and A. R. Riddell, *J. Ind. Hyg. Toxicol.,* **29,** 145 (1947).

[7] G. Goralewski, *Arch. Gewerbepathol. Gewerbehyg.,* **9,** 676 (1939), **10,** 384 (1940), **11,** 116 (1941).

which Shaver's disease resembles, particularly with regard to spontaneous and frequently bilateral pneumothorax, but from which it differs somewhat in the type of exposure. The disease described by Shaver had a 10 per cent incidence of "well-established" radiographic changes and 3 per cent of "doubtful" changes among a worker population of 344 exposed for 6 to 8 years. The exact nature of the etiological agent is not known; exposure is to fumes of Al_2O_3, SiO_2, Fe, and other substances derived from a mixture fused at 2000°C., consisting of bauxite, Fe, coke, and silica. At least three other reports of lung affection like Shaver's disease have since appeared.[8-10] Because the great mass of evidence demonstrates the comparative harmlessness of Al for man, the role of Al in the etiology of Shaver's disease is certainly unclear at this time. See also page 400, Volume I.

Al powder treatment of silicosis. In 1939, Denny, Robson, and Irwin[11] showed that the addition of 1 per cent powdered Al to the dust in mines tends to protect miners from silicosis. The protective action was believed to result from a layer of crystalline α-monohydrate of Al adsorbed on the silica particles, which reduced SiO_2 solubility and reduced the capacity of SiO_2 to form fibrotic lesions. In the two decades that followed it was repeatedly demonstrated experimentally in vitro[12] and in vivo[13,14] that colloidal Al hydroxide or powdered Al hydrate acts to (a) prevent development of the fibrogenic action of quartz (b) cause resolution and healing of immature silicotic tissue reactions and prevention of enlargement of the fibrous nodules, (c) stimulate the tuberculous process, and (d) neutralize the histotoxic action of subcutaneously implanted quartz.[15] Moreover, reports giving an impression of clinical improvement in silicotic patients treated with Al dust appeared from time to time, without, however, clear, objective evidence that Al retarded the further development of silicosis or caused regression of the established disease. A carefully planned and controlled 4-year investigation was made by Kennedy[16] upon the recommendation of the Medical Research Council of Great Britain and endorsed by the Al Subcommittee of the Industrial Pulmonary Disease Committee. It was the conclusion of this investigator, on the basis of a study of 120 selected patients, that pure metallic Al dust provided no change in the radiological picture nor any objective evidence of improvement in functional capacity of the patients. Whatever symptomatic improvement occurred was mainly psychological. There was no evidence that Al dust favored the development of pulmonary tuberculosis in the group studied.

[8] J. Hagen, *Z. ges. inn. Med. u. ihre Grenzebiete*, **5**, 31 (1950).

[9] H. Gartner, *Arch. Ind. Hyg. Occupational Med.*, **6**, 339 (1952).

[10] J. R. Ruttner, P. Bovet, and H. Aufdermaur, *Deut. med. Wochnschr.*, **77**, 1413 (1952).

[11] J. J. W. Denny, W. D. Robson, and D. A. Irwin, *Can. Med. Assoc. J.*, **40**, 213 (1939).

[12] A. L. Godbert and R. K. Halpin, *Ministry Fuel Power Brit. Safety in Mines Research Estab., Research Rept.* No. **116** (1955).

[13] M. Dworski, *A.M.A. Arch. Ind. Health*, **12**, 229 (1955).

[14] E. J. King, C. V. Harrison, G. P. Mohanty and G. Nagelschmidt, *J. Pathol. Bacteriol.*, **69**, 81 (1955).

[15] G. W. H. Schepers, *9th Conf. McIntyre Research Foundation*, Jan., 1958.

[16] M. C. S. Kennedy, *Britt. J. Ind. Med.*, **13**, 85 (1956).

A case was reported of pulmonary fibrosis, in a worker making fine Al powder, that resembled Shaver's disease in all its characteristics except spontaneous pneumothorax.[17] The Al dust was mixed with 0.5 per cent by weight of stearine, which forms Al stearate. If other cases validate the etiology of this disease it would appear on present evidence that for Al to produce pulmonary disease in man it must be in some other form than metallic Al. It would appear, however, that Al metal powder can cause pulmonary fibrosis in certain animals (rats) but not others (rabbits), according to King and co-workers.[18]

Dermal effects of Al. Al filings rubbed into the skin of man or animals and Al splinters embedded into the skin do not induce a hypersensitive state.[19] Repeated contact of the skin with certain salts of Al results in irritation due to acid liberated by hydrolysis. A congestive, anesthetic condition of the fingers (acroanesthesia) occurs among wet bobbin winders in cotton mills from long contact of the skin with alum.[20]

Toxicity and hazards of alkylaluminum compounds.[21] These organometallic substances are being increasingly used as catalysts for "low-pressure polyethylene." They are extremely hazardous because of their reactivity and must be protected from air, moisture, and compounds containing active H by blanketing with an inert gas during use. Exothermic addition reactions present a fire and personnel-burn hazard. Fume toxicity exposures to diisobutyl aluminum chloride under conditions simulating a spill of 1 gal. per 2000 cu. ft. killed 2 of 6 rats after 1 hour's exposure. Death occurred from pulmonary hemorrhage. The fume toxicity of the triisobutyl derivative was somewhat less. Tests showed that the skin is easily burned by the undiluted compounds, even if protected by cloth or polyethylene sheeting, but 20 per cent dilutions were without reaction. The fumes are white, possess a characteristic musty odor, and are detectable at low concentrations.

6. Hygienic Standards of Exposure

No threshold limit for Al has been set by any official agency in this country because of the lack of hygienic significance associated with Al exposures. A level of 15 mg./cu. meter has been recommended by Elkins;[22] this represents an upper limit of dustiness set for all innocuous dusts.

[17] J. Mitchell, *Brit. J. Ind. Med.,* **16**, 123 (1959).

[18] F. M. Engelbrecht, P. D. Byers, B. D. Stacy, C. V. Harrison and E. J. King, *J. Pathol. Bacteriol.,* **77**, 407 (1959).

[19] E. Sedlacek, *Arch. Gewerbepathol. Gewerbehyg.,* **10**, 445 (1941); F. Marquardt, *Med. Welt.,* **1939**, 1317, *J. Ind. Hyg. Toxicol.,* **24**, 158A (1942).

[20] R. P. White, *The Dermatogoses, or Occupational Diseases of the Skin.* 4th ed., Lewis, London, 1934.

[21] J. E. Knap, R. E. Leech, A. J. Reid, and W. S. Tamplin, *Ind. Eng. Chem.,* **49**, 874 (1957).

[22] H. B. Elkins, *Chemistry of Industrial Toxicology.* 2nd ed., Wiley, New York, 1959.

7. Flammable Limits

The ease of ignition of powdered Al has caused many fires and explosions. When the mean particle diameter is 0.14 μ, the lower limiting explosive concentration is 40 to 50 mg./liter of air.[23] For details regarding ignition temperatures, minimum explosive concentrations, minimum energy required for ignition, and maximum permissible oxygen content to prevent ignition of atomized or stamped aluminum dust see Volume I, Chapter XVI, Section Two. Large fires must be isolated and allowed to burn out, but small ones should be controlled by sand, talc, or sodium chloride. Anhydrous $AlCl_3$ presents a fire hazard when exposed to moisture in metal containers from the H_2 liberated by HCl from hydrolysis of $AlCl_3$. The fire hazard from alkyl Al derivatives has been mentioned above.

ANTIMONY, Sb

1. Source and Production

Sources of free-world supply of primary Sb are chiefly from Mexico, Bolivia, and the Union of South Africa. Total world production of Sb decreased from a high in 1956 of 53,000 short tons to 44,000 in 1958, largely due to declines in industrial consumption; of this amount primary United States production (mine and smelter) was 9300 tons, secondary, about 20,000 tons. Primary Sb is recovered as impure cathode metal from complex Ag–Cu–Pb ore (Idaho); smelter recoveries were 47 per cent oxide, 43 per cent metal, 9 per cent primary residue, and 1 per cent sulfide. All secondary Sb was recovered from Pb and Sn alloys.

Although more than one hundred minerals contain Sb, commercial ores are limited to a dozen, chiefly metallic Sb, Sb oxides and sulfides, and complex Cu-, Pb-, Ag-, and HgSb sulfides, the most important of which is stibnite (Sb_2S_3).

2. Uses and Industrial Exposures

Use of Sb is about equally divided between metal and nonmetal products. Chief among the metal uses are alloys of Pb, bearing metals and bearings, battery parts, type metal, sheet and pipe, cable covering, castings, tubes and foil, solder, and ammunition. Among the more prominent users of Sb in nonmetal products are flameproof textiles, paints and lacquers, rubber compounds, ceramic enamels, glass and pottery, abrasives, phosphors, and matches. Recently intermetallic compounds of Sb such as indium antimonide (InSb) were produced for the electronics industry.

Hazardous exposures are still being reported[24] in the mining of Sb in some parts of the world (Morocco) and exposures to dusts and fumes of Sb and its oxides and sulfides are experienced by those who crush the ores, tend or clean the extraction chambers, or collect the oxide dusts from the roasting chambers in the

[23] R. B. Mason and C. S. Taylor, *Ind. Eng. Chem.*, **29**, 626 (1937), **32**, 67 (1940).

[24] J. Rodier and G. Souchere, *Bull. inst. hyg. Maroc.*, **15**, 93 (1955).

United States.[25] Foundry workers, linotypers, monotypers, and stereotypers may
be exposed to fumes containing Sb; rubber compounders may suffer exposure to
Sb compounds. Glassworkers using Sb in powdered form represent an exposed
group.[26] Machining, welding, or soldering such compounds as InSb could present
an exposure hazard, although this has not been evaluated as yet.

Stibine, the volatile hydride of Sb, presents a serious exposure risk in the
processing of metals containing Sb with reducing acids, or in overcharging of
storage batteries.[27] Possibly because SbH_3 is less readily formed than AsH_3 and is
less stable, or because exposures are unrecognized, reports of serious injury from
SbH_3 are uncommon. Minor uses of the metal and its compounds in the chemical
industry,[28] in metal polishing and decoration, and in the pharmaceutical industry
have not given rise to serious hygienic problems.

TABLE 2

Physical and Chemical Properties of Sb and Some of Its Compounds

	Chemical symbol	At. or mol. wt.	Sp. gr.	M.p., °C.	B.p., °C.	Solubility
Antimony	Sb	121.76	6.68(25°)	630.5	1440	Sol. hot conc. H_2SO_4, aq. reg.
Stibine	SbH_3	124.78	4.344(15°) (air = 1) Liq. 2.26(−25°) 530 g./liter	−88	−17	20 cm.³ per 100 g. H_2O 4 cm.³ per 100 g. $H_2O(100°)$
Antimony trioxide	Sb_2O_3	291.52	5.2	656	1550 (subl.)	Sl sol. hot H_2O Sol. HCl, KOH
Antimony pentoxide	Sb_2O_5	323.52	3.78	(−0)ᵃ 380	(−20)ᵃ 930	Insol. hot or cold H_2O Sol. HCl, KOH
Antimony trisulfide	Sb_2S_3	339.72	4.64	550		Dec. hot H_2O Sol. alkalies
Antimony pentasulfide	Sb_3S_5	403.85	4.12	Dec.		Insol. hot H_2O Sol. alkalies

ᵃ O = oxygen

3. Physical and Chemical Properties

The physical and chemical properties of Sb and some of its compounds are
listed in Table 2.

Sb alloys readily with many metals, Pb in particular. Sb is only slightly oxidized in air but when melted forms a volatile oxide. It may be oxidized by steam
or strongly oxidizing salts. Sb is converted to an oxide by nitric acid. Cl_2 and Br_2

[25] L. E. Renes, *Arch. Ind. Hyg. Occupational Med.*, **7**, 99 (1953).
[26] R. Gallina, and R. Luvoni, *Rass. med. ind.*, **27**, 28 (1958).
[27] H. F. Haring, and K. G. Compton, *Trans. Electrochem. Soc.*, **68**, 283 (1935).
[28] F. M. R. Bulmer and J. H. Johnston, *J. Ind. Hyg. Toxicol.*, **30**, 26 (1948).

attack heated Sb. Sb forms compounds with As, Te, Se, P and certain other elements, but not with Si, B, and C. Pb, As, and S are the chief impurities in Sb.

4. *Analytical Determination*

Following conventional collecting methods for dusts and fumes, Sb may be determined colorimetrically, using Rhodamine B[29] or 9-methyl-2,3,7-trihydroxy-fluor-6-one,[30] or spectrographically.[31] A potassium iodoantimonite complex has been used[32] to determine Sb in biological material. Maren's method[33] using Rhodamine B has been employed to determine Sb in animal tissues[34] but was not considered entirely satisfactory for Sb in urine.[35] Low concentrations of SbH_3 can be estimated by the degree of darkening of $AgNO_3$ paper;[30] greater accuracy is claimed if SbH_3 is first absorbed in $HgCl_2$ solution followed by colorimetric determination with Rhodamine B.

5. *Physiological Responses*

Fairhall and Hyslop have exhaustively reviewed (41 pages) the physiological action of Sb and its compounds in animals and man to 1947 and have summarized information on all other aspects of Sb of industrial hygiene significance;[36] the review contains 175 references.

Animals. The intraperitoneal LD_{50} of Sb and 5 of its industrially important compounds for rats ranged from 11 mg./kg. for potassium antimonyl tartrate to 4 g./kg. for Sb_2O_5; elemental Sb was more toxic (100 mg./kg.) than the inorganic salts tested; the dusts passed a 325-mesh screen and were injected in corn oil. It was Fairhall's conclusion[36] that the relative order of intraperitoneal toxicity, beginning with the most toxic, was: Sb, Sb_2S_3, Sb_2S_5, Sb_2O_3, and Sb_2O_5. The dust of the free metal was more toxic on inhalation than any of the oxides or sulfides. Generally trivalent Sb compounds were more toxic than the corresponding pentavalent forms, which are less irritating. Orally, Sb oxides in small doses accelerated the growth of rats; larger doses (200 mg.) retarded growth slightly when given daily. Cats lost weight when given daily 450 mg. of either oxide and showed diminished numbers of leucocytes but a relative increase in certain of the white cells. Rabbits fed daily up to 150 mg./kg. of Sb_2O_3 for 4 weeks showed no pathological changes. Chronic daily inhalation of Sb_2O_3 by guinea pigs, 2 to 3 hours daily to a mean dust level of 45.4 mg./cu. meter, produced extensive pneu-

[29] W. G. Fredrick, *Ind. Eng. Chem. Anal. Ed.*, **13**, 922 (1941).

[30] S. H. Webster, and L. T. Fairhall, *J. Ind. Hyg. Toxicol.*, **27**, 183 (1945); P. Wenger, R. Duckert, and C. P. Blancpain, *Mikrochim. Acta*, **3**, 13 (1938).

[31] J. Cholak and D. M. Hubbard, *J. Ind. Hyg. Toxicol.*, **28**, 121 (1946).

[32] E. W. McChesney, *Ind. Eng. Chem. Anal. Ed.*, **18**, 146 (1946).

[33] T. H. Maren, *Anal. Chem.*, **19**, 487 (1947).

[34] M. L. Westrick, *Proc. Soc. Exptl. Biol. Med.*, **82**, 56 (1953).

[35] H. Brieger, C. W. Semisch, J. Stasney, and D. A. Piatnek, *Ind. Med. Surg.*, **23**, 521 (1954).

[36] L. T. Fairhall and F. Hyslop, *Public Health Repts. Suppl.* No. **195** (1947).

monitis in all animals and fatty degeneration of liver in most. Electrocardiograms were normal and kidneys showed no changes. Hypertrophy of splenic follicles occurred along with a decreased polymorphonuclear leucocyte, eosinophile, and white-blood-cell count. The effects required exposures of 33 to 609 hours.

Rats, rabbits, and dogs exposed 7 hours daily by inhalation to the same preparation of Sb_2S_3 as were workmen in whom cardiac lesions had been demonstrated, in 6 weeks developed definite and consistent functional disorders of the heart and parenchymatous degeneration of the myocardium at an air concentration of 3.07 to 5.6 mg./cu. meter.[35] All animals survived exposure and had normal gain in weight. Blood sugar and blood urea nitrogen remained within normal range; liver function tests were unsuggestive of disorder and the numbers of formed elements of the blood were normal. Slight to moderate myocardial damage with a tendency to progression was seen; T-waves were especially altered in the rabbits, but not in the dogs.

Stibine. Production of SbH_3 occurs whenever an acid reacts with a metal containing Sb as an impurity, or in the case of highly electropositive metals such as Al, with water.[37] SbH_3 has been generated in overcharged storage batteries or whenever nascent H comes in contact with metallic Sb or a soluble Sb compound. Thus the chief sources of exposure are, (a) metallurgy, (b) welding or cutting with blow torches, (c) soldering, (d) filling of hydrogen balloons, (e) etching of Zn, (f) chemical laboratories.

The physiological action of this gas greatly resembles that of arsine; death occurs rapidly at 10,000 p.p.m., in a few hours at 100 p.p.m.[36] Stibine attacks the blood and central nervous system. In acute poisoning, the signs in man are headache, nausea, weakness, slow breathing, and weak and irregular pulse. Like arsine, SbH_3 is hemolytic and produces, in addition, characteristic "spine cells" in animals. One of the early signs of overexposure to SbH_3 in man may be hemoglobinuria.[37] No clear-cut case of fatal poisoning in man has been reported because of the usual contamination with arsine, hydrogen sulfide, phosphine, or lead. Chronic SbH_3 poisoning in man appears not to have been reported.

Metabolism. Most of the information on the metabolism of Sb concerns organic derivatives (antimonials) and may thus be of doubtful significance for inorganic derivatives of industrial interest. Antimonials showed very little tendency to store Sb in the tissues, but rather it is distributed nonspecifically in all, with the liver an exception. Thyroid and parathyroid contained the next largest amounts. There appears to be no evidence that Sb in tissues is more than transitory.[36]

Tissue distribution of Sb in rats fed 2 per cent Sb_2O_3 in casein diet for 49 days showed mean amounts of Sb from 6.7 to 88 μg./g. of tissue of 7 analyzed. The largest amounts were in thyroid and adrenal; spleen, liver, lung, heart, and kidney, in that order, had about 0.1 the concentration of Sb, as did the thyroid. Thyroxin administration increased slightly the Sb in all tissues.[34]

[37] C. U. Dernehl, F. M. Stead, C. A. Nau, *Ind. Med.*, **13**, 361 (1943).

Urinary excretion of Sb from inhaled Sb_2S_3 by dogs increased from 10 to 150-fold normal values during a 6-week exposure to about 5.5 mg. of Sb_2S_3/cu. meter.[35]

Mode of Action. Older studies in this area showing increased glutathione and nonprotein nitrogen content of the blood, increased epinephrine content of adrenals, and increased number of red cells in animals in which organic Sb derivatives were used[36] have possible pertinence here. Similarly, more recent studies by Barron and Kalintsky[38] with fuadin and a catechol derivative of Sb, showing 50 per cent inhibition of succinoxidase at $2 \times 10^{-3}M$ Sb, the effect on pyruvate brain oxidase of pigeons,[39] and the studies of Chen *et al.*[40] indicate the interference of Sb with cellular metabolism by combination with sulfhydryl groups in respiratory enzymes. The in vivo work of Westrich[34] with 2 per cent Sb_2O_3 in the diet of animals, however, showed little effect on O_2 uptake or body weight of rats. When thyroxin was given with Sb_2O_3, marked loss in body-weight gain occurred with increase in O_2 uptake; the relative distribution of Sb in the tissues was increased slightly over that of Sb alone.

2,3-Dimercaptopropanol (BAL) has been demonstrated experimentally in rabbits[41] and in pigeon-brain tissue slices[42] to detoxify Sb compounds (not SbH_3). BAL significantly increased urinary excretion of Sb, which was considered to account in part for BAL's protective action.

Industrial Experience. Four studies of adverse effects in workers from exposure to various Sb compounds have appeared since the first edition. Renes[43] has reported 69 cases of Sb poisoning among smelter workers in a 5-month period. Although As and NaOH were present in the atmosphere, Sb was shown by analysis to be the predominant contaminant and was felt by the author to account for the observed illness, on the basis that it resembled the signs and symptoms of Sb poisoning, whereas many of the characteristic signs of As poisoning were lacking. The mean Sb levels ranged from 4.69 to 11.81 mg./cu. meter, whereas the As levels were between 0.39 and 1.1 mg./cu. meter. Among the 13 pathological conditions diagnosed, dermatitis and rhinitis were most frequent (20 per cent), inflammation of the upper and lower respiratory tract including pneumonitis was next in order, with a few cases (3 to 4 per cent) of gastritis, conjunctivitis, and septal perforations. Penicillin provided relief and healing followed removal from exposure.

Brieger *et al.*[35] have reported on cases of cardiac injury and deaths in an abrasives industry in which Sb_2S_3 was used. Of 125 workers employed for 8 months

[38] E. S. G. Barron and G. Kalintsky, *Biochem. J.,* **41,** 346 (1947).
[39] R. H. S. Thompson and V. P. Whittaker, *Biochem. J.,* **41,** 346 (1947).
[40] G. Chen, E. M. K. Geiling, and R. M. MacHatton, *J. Infectious Diseases,* **76,** 152 (1936).
[41] H. Eagle, F. G. Germuth, H. J. Magnuson, R. Fleischman, J. C. Grossberg, and C. E. Tucker, *J. Pharmacol.,* **89,** 196 (1947).
[42] R. H. S. Thompson and V. P. Whittaker, *Biochem. J.,* **41,** 346 (1947).
[43] L. E. Renes, *Arch. Ind. Hyg.,* **7,** 99 (1953).

to 2 years, 6 died suddenly, in addition to 2 others who died of chronic heart disease; 37 of 75 examined showed EKG changes, mostly of the T-waves, 14 a blood pressure over 150/90, and 7 of 111 had ulcers. Despite an air-conditioned room, mean Sb values ranged mostly from 3 to 5.5 mg./cu. meter; urinary Sb values 0.8 to 9.6 mg./liter. When use of Sb_2S_3 was abandoned, no further instances of cardiovascular disease or deaths occurred. More information on the cardiac effects of Sb compounds seems desirable.

Two other reports of occupational exposures have appeared, one among the Sb mines in Morocco[24] in which chronic poisoning only was detected. Symptoms consisted of headache, sleeplessness, vertigo, loss of appetite, and muscular pains; the blood-cell picture was altered. All responses were mild. Sb was found in the urine and hair, but rarely in the blood. The ore was chiefly the sulfide. The other report[26] of Sb poisoning, which occurred in a Milan glass factory, concerned exposure to the pentasulfide. Symptoms consisted of nausea, vomiting, and diarrhea, a bitter taste in the mouth, and the characteristic shift in the leucocyte count. All patients showed Sb in the urine in amounts of 3 to 21 μg. per cent.

6. Hygienic Standards of Exposure

The threshold limit of exposure to Sb and its compounds (except stibine) is 0.5 mg. of Sb per cubic meter of air, according to the Threshold Limits Committee of the ACGIH. From the results of studies of workers and animals exposed to Sb_2S_3 Brieger considers the limit to be satisfactory, at least for this Sb compound.[35]

The threshold limit for stibine (SbH_3) is 0.1 p.p.m. or approximately 0.5 mg. of stibine per cubic meter of air. On the basis of a study of stibine in telephone offices at less than 0.1 p.p.m. in which no effects occurred, Haring and Compton[27] concluded that stibine below 1 p.p.m. was of no hygienic significance.

7. Flammable Limits

For details regarding antimony dust see Vol. I, Chapter XVI, Section Two.

8. Odor and Warning Properties

Stibine has an extremely unpleasant odor that can serve as warning; whether the threshold of odor detection is sufficiently low to prevent injury is uncertain.[27]

BARIUM, Ba

1. Source and Production

Despite a declining domestic production of 1.3 million short tons of barite ($BaSO_4$) less in 1957 than in 1956, imports established a new high by more than 240,000 tons over the preceding year. World production continues to increase. Several new production facilities began operation. A barite deposit near Austin, Nevada, discovered in 1956 was reported to contain 5 million tons 98 per cent barite. Arkansas is the leading producing state, Missouri second, Georgia third, and Nevada fourth. Lithopone (ZnS and BaSO$_4$) and Ba chemical output in 1957 was below 1956. Ba metal was produced in 1957 in small quantities by two companies.

Barite is produced from high-grade ore (75 to 98 per cent), often in association with granite and shale, crushed, beneficiated by froth flotation or by jigging, and dried. A new process makes 99 per cent pure Ba(OH)$_2$ directly from barite ore.

2. *Uses and Industrial Exposures*

About 90 per cent crude barite is currently used in manufacture of ground barite, the remainder for lithopone and Ba chemicals. Oil- and gas-well drilling, the largest use of ground barite, accounts for 95 per cent of the total output. The glass, paint, and rubber industries consume crushed and ground barite. The amount of barite used for lithopone is decreasing owing to competition from TiO$_2$ as a paint pigment. A barium–titanate ceramic has been patented, as well as a process for barium–titanate crystals. A rubber–barite mixture for road construction, vehicle undercoats, and roofing paints has been patented, using particles 90 per cent less than 10μ diameter. Ba chemicals of domestic industrial importance are the carbonate, chloride, hydroxide, sulfate, and oxide; sales of the oxide are increasing. Ba metal is used as a "getter" to remove gases from vacuum tubes.

TABLE 3

Physical and Chemical Properties of Ba and Some of Its Compounds

	Chemical symbol	At or mol. wt.	Sp. gr.	M.p., °C.	B p., °C.	Solubility
Barium	Ba	137.36	3.5(20°)	850	1140	Dec. ev. H$_2$ Sol. alc. Insol. bz.
Barium sulfate	BaSO$_4$	233.43	4.5(15°)	1580	Tr.[a] 1149– monocl.	2.46 mg./liter(25°) 4.13 mg./liter(100°) Sl. sol. HCl, H$_2$SO$_4$
Barium hydroxide	Ba(OH)$_2$	171.38	4.495			16.7 g./liter (0°) 1.014 kg./liter(100°)
Barium carbonate	BaCO$_3$	197.37	4.43	1740[b]	Dec.	20 mg./liter(20°) 60 mg./liter(100°) Sol. acid, NH$_4$Cl Insol. alc.

[a] Tr. = transition.
[b] 90 atm.

3. *Physical and Chemical Properties*

The physical and chemical properties of Ba and some of its compounds are listed in Table 3.

Ba, a silver-white metal obtained from the electrolysis of the molten chloride, is one of the less expensive metals that have the distinctive properties of absorbing gases. Ba is highly electronegative, liberating H_2 actively from cold water. $BaSO_4$ is the source of most Ba compounds. When $BaSO_4$ is reduced to the sulfide and treated with $ZnSO_4$ an insoluble mixture of $BaSO_4$ and ZnS, lithopone, is formed; for use as a paint pigment it is given a heat treatment.

4. Analytical Determination

Collection of samples from the air may be made by conventional methods. Determination of Ba by chemical methods (the chromate and the Yagoda method,[44] using mixed crystals of $BaSO_4$–$KMnO_4$) is rather unsatisfactory for microquantities because of either solubility relationships or significant interferences by metals (Ca and Pb) commonly present in industrial atmospheres. Accordingly, the spectrographic method of Grabowski and Unice[45] is recommended both for samples of air and of biological origin. It is applicable to any sample soluble in HCl following appropriate fusion or acid digestion.

5. Physiological Responses

Most of the information on the toxicity of Ba derives from physiological and pharmacological studies on animals made in the early 1900's. Unlike other elements in its group, Ca and Sr, Ba is highly toxic. $BaCO_3$ has found use as a rat poison, and the sulfide as a depilatory. The fatal dose of $BaCl_2$ for man is reported[46] to be about 0.8 to 0.9g. (0.55 to 0.6g. of Ba); that of the sulfide is considerably greater.

Fasekas et al.[47] have reported that a subcutaneous injection of $BaCl_2$ at 5 mg./kg. caused acute toxicity and death of rabbits after 2 to 2.5 hours. Chronic poisoning was achieved by the repeated injection of 10, 5, and 2 mg./kg.; the rabbits in this series were killed at 98 to 193 days. Effects on the central nervous system were described. Bronchogenic carcinoma (squamous-cell type) developed in rats intratracheally injected with radioactive particles of $BaSO_4$ (S^{35}).[48]

The symptoms of Ba poisoning are excessive salivation, vomiting, colic, diarrhea, convulsive tremors, slow, hard pulse, and elevated blood pressure. Hemorrhages may occur in the stomach, intestines, and kidneys. Muscular paralysis may follow. According to the dose and solubility of the Ba salt, death may occur in a few hours or a few days.

Metabolism. No study appears to have been made of the absorption of natural Ba, but a study of the metabolism of Ba^{140} in rats showed 24-hr. urinary and fecal excretions to be 7 and 20 per cent, respectively; Ba was irreversibly deposited in the skeleton in trace amounts.[49]

[44] H. Yagoda, J. Ind. Hyg. Toxicol., **26**, 224 (1944).

[45] R. J. Gabrowski and R. C. Unice, Anal. Chem., **30**, 1374 (1958).

[46] T. A. Sollmann, Manual of Pharmacology. 7th ed., Saunders, Philadelphia, 1953.

[47] I. G. Fasekas, R. Felkai, and B. Melegh, Arch. pathol. Anat. u. Physiol. Virchow's, **324**, 110 (1953).

[48] H. Comber and J. A. Watson, A.M.A. Arch. Ind. Health, **17**, 230 (1958).

[49] G. C. H. Bauser, A. Carlsson, and B. Lindquist, Biochem. J., **63**, 535 (1956)

The determination of Ba in human bone by radioactivation analysis showed no accumulation throughout a lifetime (7 p.p.m., ashed tissue) nor any accumulation in soft tissue as determined on 37 cadavars (11 to 24 p.p.m., ashed tissue). The pigmented parts of the eye, iris and choroid, contained more (25 p.p.m., wet tissue) than other parts of the eye.[50] Thus Ba differs strikingly from a heavier element in the group, Ra, which is a pronounced bone seeker.

Mode of Action. Ba stimulates all muscle, striped, unstriped, and cardiac,. irrespective of innervation. Ba is mutually antagonistic to all muscular depressants (Ca). Ba produces strong vasoconstriction by direct stimulation of arterial muscle, violent stimulation of smooth muscle (peristalsis), and stimulation followed by paralysis of the central nervous system. Hematopoiesis is reported stimulated in rabbits; toxic doses are hemolytic.[51] The intimate mechanisms whereby these varied effects are produced are not understood.

6. Industrial Experience

Although Ba poisoning has not been of much significance in industry, the potential from the more soluble forms exists. Pneumoconiosis has been reported among workers exposed to BaO[52] and a case of baritosis occurred in a worker exposed to finely divided $BaSO_4$.[53] Bronchial irritation was reported in a factory where bomb casings were heated in $BaCO_3$.[54] Recent reports in the European literature continue to add instances of pneumoconiosis production in workers inhaling barium-containing dusts (baryta and talc mixtures)[55] and lithopone.[56] The pneumoconioses were identified by x-ray examination, which revealed very fine nodulation on a reticulated background. Clinically there were no symptoms of emphysema or bronchitis, and lung-function tests indicated no pulmonary impairment, although there were complaints of dyspnea on exertion. Aspirated $BaSO_4$, however, has been reported to result in granuloma of the lung[57] and other sites in man.[58] BaO dust is considered a potential agent of dermal and nasal irritation.[59]

7. Hygienic Standards of Exposure

The threshold limit of 0.5 mg. Ba/cu. meter of air was set by American Conference of Governmental Industrial Hygienists on the advice of Edwin C. Hyatt,

[50] E. M. Snowden and S. R. Stitch, *Biochem. J.*, **67**, 104 (1957), **70**, 712 (1958); E. M. Snowden and A. Pirie, *Biochem. J.*, **70**, 716 (1958).

[51] G. G. Zolessi and G. Stazzi, *Med. lavoro*, **30**, 44 (1939).

[52] F. Spedini and P. L. Valdini, *Radiol. Med.*, **26**, 1 (1939).

[53] E. P. Pendergrass and R. R. Greening, *Arch. Ind. Hyg. Occupational Med.*, **7**, 43 (1953).

[54] Anon., *J. Am. Med. Assoc.*, **117**, 122 (1941).

[55] B. Gombos, *Pracovní lékařství*, **9**, 399 (1957).

[56] E. Wende, *Arch. Gewerbepathol. Gewerbehyg.*, **15**, 171 (1956).

[57] F. Fite, *A.M.A. Arch. Pathol.*, **59**, 673 (1955).

[58] S. Kay, *A.M.A. Arch. Pathol.*, **57**, 279 (1954), **59**, 388 (1955).

[59] H. B. Elkins, *Chemistry of Industrial Toxicology*. 1st ed., Wiley, New York, 1950.

who had employed this limit without mishap for a few years in the control of exposures to $Ba(NO_3)_2$ at Los Alamos. It is not known what degree of safety this limit incorporates for dust exposures to more insoluble Ba compounds.

8. Flammability

The free metal presents a fire and explosion hazard if exposed to moist air owing to liberation of H_2.

BERYLLIUM, Be[60]

1. Source and Production

Be is estimated to be equal to B and Co in abundance (0.001 per cent in the earth's crust). Beryl ($3BeO \cdot Al_2O_3 \cdot 6SiO_2$) is at present the chief source of Be, although there are about thirty recognized minerals containing Be, chief of which are chrysoberyl ($BeAl_2O_4$) and phenacite (Be_2SiO_4). Precious forms of beryl, emerald and aquamarine, approximate the pure form composition, of 14 per cent BeO. A form of phenacite found in certain regions of Utah is being exploited as a new local source of Be ore. Its distribution in minute particles relates to its solubility in weak acids.

Principal beryl-producing countries of the free world are Brazil, India, South Rhodesia, Argentina, and the United States. The Black Hills of South Dakota and Colorado, and now Utah, produce most of the Be ore in the United States; New Mexico and New Hampshire supply some. World production of beryl was 7000 short tons in 1958, declining from 12,000 tons in the 2 previous years. Domestic production has fluctuated around 500 tons since 1950; but domestic consumption reached an all-time high in 1958 of 6000 tons. Hand-gathering of the ore is still practiced for lack of mechanical methods. Interest in domestic production is increasing owing to the discovery of widespread deposits of beryl throughout many of the western states. Domestic demand for Be is increasing; production costs alone hinder radio development of the vast deposits.

The fluoride and the sulfate processes are used to convert beryl to BeO. The fluoride process involves sintering briquettes of a berylsodium ferric fluoride mixture, leaching the sintered briquettes with water, precipitating $Be(OH)_2$ with NaOH, filtering, and igniting the precipitate to form BeO. Some small operations in the western United States presently make Be concentrates and $Be(OH)_2$. The sulfate process involves melting beryl at 1625°C., quenching the melt in cold water to obtain a glass, leaching with strong H_2SO_4, extracting $BeSO_4$ and $Al_2(SO_4)_3$ with water, and precipitation of the latter with $(NH_4)_2SO_4$. The $BeSO_4$ is precipitated as the hydroxide and ignited to form BeO. Be metal is obtained by reducing BeF_2 with a stoichiometric deficiency of Mg. Excess BeF_2 acts as a flux permitting reaction at lower temperatures. At 1300°C. Be floats on the molten slag, from

[60] The author wishes to acknowledge with appreciation the help of Dr. Harriet Hardy in reviewing the section on beryllium.

which it is poured; the excess BeF_2 is recycled. $(NH_4)_2BeF_4$ is an intermediate in the preparation of BeF_2 from NH_4F and impure BeO.

2. Uses and Industrial Exposures

A large part of the Be produced is used as a hardening agent in alloys, mostly Cu. Be–Cu alloys are used in parts subject to abnormal wear, such as bushings, or extreme vibration or shock-loading, such as current-carrying springs, electric contacts and switches, and radio and radar devices. Be–Cu is also used in non-sparking tools. Be added to Zn with other metals improves creep and corrosion resistance and tensile strength. Be–Ni has high tensile strength, greater hardness than Be–Cu alloy, and age-hardening characteristics; for example, Be–Ni is used for diamond drill-bit matrix, watch-balance wheels, and certain airplane parts. Be is also used in a number of other alloys with Mg, Fe, Al, Ag, Au, and Pt. The light weight and high-temperature and corrosion resistance of Be make it attractive as a structural material for space vehicles, for which expanded use is probable. Wider use for this purpose hinges on development of ductile Be; brittleness of Be metal at present limits its use.

Be, however, has attractive nuclear properties that are finding increasing use in the atomic energy industry, as a moderator to retard neutrons in certain types of reactors because of its low neutron absorption properties; and its high-scattering cross section also increases its usefulness in atomic reactors. Specific uses of Be being given engineering attention are as a cladding material for U, as a combined moderator and fuel element in conjunction with fissionable material as a UBe_{13} alloy, and as a compact fuel element in reactors to propel rockets. A report is available on sources, extraction, and properties of Be.[61]

Much technological research effort is being given to developing better methods of extracting and purifying Be, of flotation and recovery, and production of high-purity Be. Much effort is also being given such metallurgic operations as purification, electron-beam melting, casting, forging, joining, sheetrolling, surface effects, and alloy development.

Chief hazards from exposure to Be and its compounds come wholly from its milling and use; past mining and handling of beryl appear not to be associated with a Be hazard. The now general and widespread recognition of the serious health hazards in Be plant operations and in Be metal handling, as well as the community air-pollution problem, have resulted in the redesign of plant equipment that virtually has eliminated injury to health as previously experienced. The insidious nature of the physiological action of Be even from minute exposures, however, requires constant vigilance to minimize exposure, particularly to Be metal fume and dust, certain forms of BeO, $Be(OH)_2$, BeF_2 fume, and to mists and dusts of soluble Be compounds such as $BeSO_4$. A comprehensive discussion of health protection in Be facilities, by Breslin and Harris, has been published,[62] and an

[61] W. Hodge, U. S. Govt. Research Repts., P. B. **121648** (Aug. 15, 1958).

[62] A. J. Breslin and W. B. Harris, U. S. Atomic Energy Comm., N. Y. Operations Office, Rept. HASL-36 (May, 1958).

evaluation of the hazards and their control in Be machining operations has been made by Mitchell and Hyatt.[63]

3. *Physical and Chemical Properties*

The physical and chemical properties of Be and some of its compounds are listed in Table 4.

TABLE 4

Physical and Chemical Properties of Be and Some of Its Compounds

	Chemical symbol	At. or mol. wt.	Sp. gr.	M.p., °C.	B.p., °C.	Solubility
Beryllium	Be	9.013	1.85	1278 ±	2970[a]	Insol. cold H_2O Sl. sol. dec. hot H_2O Sol. dil. acids and alkalies Insol. Hg
Beryllium oxide	BeO	25.01	3.01	2530 ± 30	ca. 3900	0.2 mg./liter(30°) Sol. conc. H_2SO_4 Fus. KOH
Beryllium hydroxide	$BeO \cdot xH_2O$			Dec.		Sl. sol. cold H_2O Dec. hot H_2O Sol. acids, alc., ether
Beryllium fluoride	BeF_2	47.01	1.986(25°)	800		Inf. sol. cold and hot H_2O Sol. alc., H_2SO_4
Beryllium sulfate	$BeSO_4$	105.08	2.443	Dec. 550– 600		425 g./liter(25°) 1 kg./liter(100°) Sl. sol. conc. H_2SO_4

[a] 760 mm. Hg.

Be metal has many unique properties; it is the only stable, lightweight metal with a high melting point; it has an especially high strength-to-weight ratio; and its alloying property confers to metals special properties of resistance to corrosion, vibration, and shock. The extreme hardness attributable to Be is considered to reside in a thin film of BeO on the surface. Be, as ordinarily prepared, is brittle, presumably because of impurities at grain boundaries. Ductility is sufficient at 1000°C. to permit Be to be swaged. Its useful nuclear properties are referred to in the section on Uses (page 1003). Be yields about 30 neutrons per million alpha particles.

Be reacts with other elements to a significant degree only at elevated temperatures; at 700°C. oxidation is noticeable; at 1000°C., rapid. Finely divided Be burns in air. Similar effects occur with N and C. Be is divalent in all its important compounds; despite this, Be resembles trivalent Al so closely as to render its

[63] R. N. Mitchell and E. C. Hyatt, *Am. Ind. Hyg. Assoc. Quart.*, **18**, 207 (1957).

separation difficult. Solutions of the water-soluble salts of Be are strongly acidic and dissolve large amounts of Be hydroxide. They are hydrolyzed to a great extent, however. The reason for this behavior is unknown.

4. Analytical Determination

Useful methods for the analysis of Be must be capable of detecting with accuracy 0.1 μg. of Be or less. Three general procedures are used for traces of Be: colorimetric, fluorimetric, and spectrographic. Colorimetric methods providing a sensitivity of from 0.05 to 0.2 μg. of Be/ml. are numerous; none of the dyes is specific for Be.[64] The fluorimetric methods possess a sensitivity in the range of 0.001 to 0.025 mg./liter for pure solutions, which is rarely attained in practice.[65] The method using morin appears to have the highest sensitivity. The spectrographic method is the most satisfactory for determining traces of Be because of the lack of interferences, in biological specimens, that are commonly a problem in the other methods. As little as 0.0005 μg. of Be can be readily detected. Fe must be removed if present in large amounts because it enhances the intensity of the Be line of 2348 A.[66] More recently, a rapid and sensitive spectrographic method of monitoring Be aerosols has been proposed by Fitzgerald.[67]

5. Physiological Response

Toxicity for Animals. Soluble forms of Be are highly toxic by all routes of administration. The LD_{50} of $BeSO_4 \cdot H_2O$ when administered intravenously to rats is 0.36 mg. of Be/kg.[68] Histological changes consisted of mid-zonal focal necrosis of the liver, epithelial necroses of the distal third of the proximal convoluted tubules of the kidney, and some degenerative changes in the hemopoietic system. Of considerable interest, but as yet unobserved in man, is the development of osteosarcoma after many months in the rabbit, first observed by Gardner[69] and later reproduced by others, by injecting ZnBe silicate.

Oral administration of soluble Be compounds, $BeSO_4$ and Be oxyfluoride, induced hemorrhagic necroses of the gastric mucosa, but the metal is poorly ab-

[64] M. W. Cucci, W. F. Neuman, and B. J. Mulryan, *Anal. Chem.*, **21**, 1358 (1949); E. B. Sandell, *Colorimetric Determination of Traces of Metals*. N. Y. Interscience Publishers, New York–London, 1950; A. L. Underwood and W. F. Neuman, *Anal. Chem.*, **21**, 1348 (1949); W. N. Aldridge and H. F. Liddell, *Analyst*, **73**, 607 (1948); F. A. Vinci, *Anal. Chem.*, **25**, 1580 (1953); J. H. Wood, *Mikrochim. Acta*, 11 (1955).

[65] T. Y. Toribara and R. E. Sherman, *Anal. Chem.*, **25**, 1594 (1953); F. W. Klemperer and A. P. Martin, *Anal. Chem.*, **22**, 828 (1950); G. Welford and J. Harley, *Am. Ind. Hyg. Assoc. Quart.*, **25**, 232 (1952); H. A. Laitinen and P. Kivalo, *Anal. Chem.*, **24**, 1467 (1952); J. M. Riley, *U. S. Bur. Mines Rept. Invest.* No. **5282**, 1956.

[66] J. Cholak and D. M. Hubbard, *Anal. Chem.*, **20**, 73, 970 (1948), *Am. Ind. Hyg. Assoc. Quart.*, **13**, 125 (1952).

[67] J. J. Fitzgerald, *A.M.A. Arch. Ind. Health*, **15**, 68 (1957).

[68] J. K. Scott, *U. S. Atomic Energy Comm. Rept.* MDDC-1237 (1947), *Arch. Pathol.*, **45**, 354 (1948).

[69] L. U. Gardner, *J. Ind. Hyg. Toxicol.*, **29**, 71A (1947).

sorbed from the gastrointestinal tract. Be metal or BeO can be incorporated to the extent of 5 per cent in the diet without affecting growth over long periods, but 5 per cent $BeCO_3$ in the diet completely inhibited growth. $BeSO_4$ interfered with growth at a dietary level of 1.4 per cent and Be oxyfluoride was toxic at 0.1 per cent. Repeated administration of Be compounds in the diet leads to a rachitic condition in rats.

By inhalation, BeF_2 was acutely more toxic than either the sulfate or the oxide;[70] 10 mg./cu. meter of the fluoride was lethal to several species in 15 days, whereas the same concentration of the sulfate was lethal to only 1 species (rat) in 95 days, and the oxide at this concentration caused no deaths to any species in 40 days. Exposure to 1 mg./cu. meter of BeF_2 (190 μg. of Be/cu. meter) for 100 days produced lung damage in animals comparable in extent to that produced by 10 mg./cu. meter of $BeSO_4 \cdot 6H_2O$ (420 μg. of Be/cu. meter). Characteristic pulmonary changes occurred, however, from subchronic exposures of $BeSO_4$ as low as 40 μg. of Be/cu. meter.

Acute pneumonitis closely resembling that in industrial Be workers was observed in animals exposed to certain grades of BeO dust; fluorescent grade BeO, characterized by fine particle size of large surface area, resulted in lung injury, whereas BeO of larger size particles and low surface area failed to produce lung injury.[71]

The toxicity of inhaled $BeSO_4$ mist was shown to be potentiated in animals by alternating the daily inhalations with low levels (8 mg./cu. meter) of HF gas.[72]

Anemia, of the macrocytic type, was shown to develop in animals exposed to Be compounds;[73] recovery occurred spontaneously after 3.5 to 4 months in dogs despite the presence of measurable amounts of Be in the tissues (bone and liver), indicating a secondary response to the Be. Be was shown by radiotracer technique to slow the rate of synthesis of both hemin and globin in the erythrocyte.[74]

Be granulomata have been reported by Davies and Harding[75] to form in the lungs of rats from intratracheal injection of amorphous BeO. The formation of the lesion, which was a similar though limited type of reaction to that of chronic granulomatosis in man, was aided by skim milk or by the presence of MnO_2. Granulomata have since been produced with Be compounds, in the rat, by others.

Experimental Be granulomata of the skin have been produced in pigs by Dutra[76] following implantation of a Be phosphor. In conformity with the findings of Hall et al.,[71] the more extensive and protracted reaction occurred with the Be dust that had been calcined least.

[70] H. E. Stokinger et al., Arch. Ind. Hyg. Occupational Med., **8**, 493 (1953).

[71] R. H. Hall et al., Arch. Ind. Hyg. Occupational Med., **2**, 25 (1950).

[72] H. E. Stokinger et al., Arch. Ind. Hyg. Occupational Med., **1**, 398 (1950).

[73] H. E. Stokinger, C. A. Stroud, and R. E. Root, J. Lab. Clin. Med., **38**, 173 (1951).

[74] H. E. Stokinger, K. I. Altman, and K. Salomon, Biochem. et Biophys, Acta, **12**, 439 (1953).

[75] T. A. L. Davies and H. E. Harding, Brit. J. Ind. Med., **7**, 70 (1950).

[76] F. R. Dutra, Arch. Ind. Hyg. Occupational Med., **3**, 81 (1951).

Prolonged inhalation of $BeSO_4$ at low levels (a few μg. of Be/cu. meter) by rats was reported to result in pulmonary tumors that metastasized to other organs and were transplantable.[77]

Metabolism. Like U, Be shows a sharp difference in distribution pattern between soluble and insoluble forms. The lungs of dogs exposed for 100 days to 40 μg. of Be/cu. meter as $BeSO_4 \cdot 6H_2O$ contained 200 times that in kidneys, 20 times that in the bone, and all other tissues except the pulmonary lymph nodes contained small fractions of that in the lung.[72] The distribution pattern of the water-soluble $BeF_2 \cdot 2H_2O$ was similar; dogs exposed for 87 days to about 100 μg. of Be/cu. meter showed approximately 40 times as much Be in the lungs as in the kidneys, 4 times as much as in the bone (femoral epiphyses).[70] In comparison, dogs exposed to the insoluble fluorescent-grade BeO at high levels (25 mg. of Be/cu. meter) for 17 days showed 27,500 times as much Be in the lungs as in the kidney, 2700 times that in bone. Again, pulmonary lymph node was the only site with Be content higher than that of lung; all other tissues contained but a minute fraction as much Be as that in lung. Lung Be content was of the order of several hundred μg. of Be/g. of wet tissue.[71] The important finding revealed by these distribution studies is that but fractions of a p.p.m. Be occur in tissues of severely poisoned animals.

The absorption of ingested Be compounds is of a low order: related only very roughly to the water solubility, absorption from the intestinal tract ranges from a few hundredths to a few thousandths per cent, depending on dose primarily and solubility secondarily.

The distribution pattern of intravenously injected Be as determined by tracer amounts of Be^7 differed markedly, depending on whether carrier was present;[78] differences from the inhalation route were also apparent.

It is important to note, in applying the experimental findings of tissue distribution of Be to cases of chronic Be poisoning in man, that Be may or may not be present in the lung or other body site, depending on the length of time since exposure. Moreover, Be may be found in tissues without Be disease being present.

Mechanism. A number of studies have been made to elucidate the mode of action of Be in acute poisoning in the hope that the results might be applicable in part to the more important chronic disease. Early studies made in vitro indicated that Be exerts a strong inhibitory action on Mg-activated alkaline phosphatase.[79] A Ca-activated enzyme, adenosine triphosphatase, was also shown by DuBois et al.[79] to be inhibited by Be. DuBois and co-workers[80] have extended their ex-

[77] G. W. Schepers, T. M. Durkan, A. B. Delahant, and F. T. Creedon, *A.M.A. Arch. Ind. Health,* **15,** 32 (1957).

[78] H. E. Stokinger, *Nucleonics,* **11,** 24 (1953).

[79] K. P. DuBois, K. W. Cochran, and M. Mazur, *Science,* **110,** 420 (1949); R. S. Grier, M. B. Hood, and M. B. Hoagland, *J. Biol. Chem.,* **180,** 289 (1949); F. W. Klemperer, J. M. Miller, and C. J. Hill, *J. Biol. Chem.,* **180,** 281 (1949); W. N. Aldridge, *Nature,* **165,** 772 (1950).

[80] K. W. Cochran, M. M. Zerwic, and K. P. DuBois, *J. Pharmacol. Exptl. Therap.,* **102,** 165 (1951).

amination of phosphatases and showed that Be in extremely low concentrations (1.8×10^{-6} — $1.2 \times 10^{-3}M$) produced 50 per cent inhibition of the phosphatase of serum and adrenal cortex of the rat; this result, moreover, was duplicated in the intact animal. Similarly adenosine triphosphatase was inhibited in vivo as was phosphoglucomutase of liver and of muscle. The extent and duration of the inhibition paralleled the dose, thus suggesting that these inhibitory actions of Be are associated with the acute toxic action of Be.

In addition to these enzymes, Vorwald and Reeves[81] have shown that hyaluronidase is inhibited by Be, but that succinic dehydrogenase is activated. The latter effect is believed related to the finding of increased activity of this enzyme in experimental tumors. Carbonic anhydrase, cytochrome oxidase, arginase, carboxylase, and uricase were reported to be unaffected by Be.

In addition to the rather general effect on phosphatases in different body sites, Spiegl et al.[82] showed Be to decrease the ratio of phospholipid to free cholesterol of the red blood cells in acutely poisoned dogs, and to increase markedly the ratio of uric acid to creatinine in the urine. The altered ratios paralleled closely the course of Be exposure.

Experimental Therapeutic Measures. A variety of chemical therapeutic agents of the metal-chelating type have been tested in animals and men against the acute and chronic forms of Be poisoning. Schubert and collaborators have found aurintricarboxylic acid (ATA) to increase the survival of mice and rats given acute lethal doses of Be provided ATA is given within a few hours following the Be dose.[83] The means by which ATA exerts its action against Be toxicity is through lake formation within Be. The salicylate portion of ATA was likewise found protective against acute Be poisoning. Subsequent work by others has found the toxicity of ATA to be a seriously limiting factor in its use. Ethylenediaminetetraacetic acid (EDTA) salts were shown to increase renal excretion of Be in 2 patients with Be disease, but clinical improvement was not demonstrated.[84] EDTA could, however, be used as an aid in diagnosis in early cases of Be poisoning.

Industrial and Clinical Experience. The various aspects of Be disease and its control are summarized in a report (172 pages) of a conference held in Boston in September, 1958, under the direction of Dr. Harriet L. Hardy.[81] Since 1952, 606 cases of acute and chronic Be poisoning in the United States have been collected

[81] Beryllium Disease and Its Control, Conf. Mass. Inst. Technol., Sept., 1959. *A.M.A. Arch. Ind. Health*, **19**, 91 (1958).

[82] C. J. Spiegl, L. LaFrance, and B. J. Ashworth, *Arch. Ind. Hyg. Occupational Med.*, **7**, 319 (1953).

[83] M. R. White, A. J. Finkel, and J. Schubert, *J. Pharmacol. Exp. Therap.*, **102**, 88 (1951); J. Schubert, M. R. White, and A. Lindenbaum, *J. Biol. Chem.*, **196**, 273, 279 (1952); A. J. Finkel and M. R. White, *Proc. Soc. Exptl. Biol. Med.*, **79**, 672 (1952).

[84] R. Cash, R. I. Shapiro, S. H. Levy, and S. M. Hopkins, *New Engl. J. Med.*, **260**, 683 (1959).

in a Case Registry. The report gives industrial sources of the disease as ore extraction, processing of Be and BeO, fluorescent powder production, and neon tubes, and nonoccupational. "Neighborhood" cases are still occurring in Pennsylvania.[85] Arrangement of the data in the Case Registry shows a marked reduction in Be cases after 1949, when controls were introduced. The attack rate was between 5.8 and 7.5 per cent; the case fatality between 5.8 and 23 per cent for the 13- to 20-year period studied. Over 60 per cent of the cases had chronic Be disease. The delay between exposure and symptoms of chronic disease ranged from less than 1 month to 20 years. The clinical impression that the chronic form is less disabling following a long induction or latent period is substantiated by the reduced mortality rate after a 5-year latent period. There appears to be no relation of age to onset of Be disease, but the morbidity rate for women appears higher than that for men from the data at hand (1958). Lack of knowledge of total exposed populations is a serious handicap to evaluation, however.

Symptoms of the chronic disease in decreasing frequency are: dyspnea, weight loss, cough, fatigue, chest pain, anorexia, and weakness; signs referable to the chest lead all others, but cyanosis and clubbing are prominent; complications in a significant percentage of cases (10 to 17 per cent) are: cardiac failure, pneumothorax, and renal stones. The stones, when tested, contain Be. Noted with increasing frequency, but still of low incidence, is renal injury related to one pole of the kidney. Be produces hypercalsuria without recorded hypercalcemia in a number of cases when looked for, a possible aid in differential diagnosis between sarcoidosis in which both hypercalcemia and hypercalsuria occur.[86] Differentiation of the two diseases has in the past rested on epidemiological data, spectrographic demonstration of Be, skin tests, and clinical and roentgenographic changes. Sarcoidosis reveals x-ray evidence of hilar adenopathy without pulmonary densities and cystic changes in the bones of hands and feet not characteristic of Be disease.

Respiratory pathophysiology of chronic Be disease is characterized by low arterial O_2 tension with resulting incomplete saturation of arterial blood at rest. The alveolar–arterial O_2 pressure difference is markedly elevated at rest and during exercise in room air and low O_2-breathing, which indicates an alveolar-capillary block as the basic physiopathology of Be disease.

Histologically, Be disease is not confined to the lungs, but may occur in many tissues, including the lymph nodes, liver, spleen, kidney, myocardium, skin, skeletal muscle, and pleura. The chronic reaction includes not only granulomas but a variable infiltration of lymphocytes and plasma cells that may be responsible for many of the symptoms.

Present treatment of Be disease consists of administration of the corticosteroids, cortisone or prednisone, for both initial and maintenance therapy. In general, initial dosage of cortisone is 100 to 300 mg.; the maintenance dose is 12.5 to 100

[85] J. Lieben, Third Air Pollution Seminar, New Orleans, March, 1960.
[86] H. L. Hardy, personal communication.

mg. daily, with wide variation, depending both on individual case and physician's judgment. One fifth this dosage is used for prednisone. Despite the risks of serious side effects (obesity, diabetes, gastric ulcer, capillary fragility, osteoporosis) and the realization that true therapy is not achieved, steroid administration is used because it alleviates the disability and prolongs the life of the patient.

Various problems associated with the control of operations in Be-processing plants, in Be uses, in machining operations, as well as the control of neighborhood air contamination near Be plants, are discussed in the report of the Be conference.[81]

6. Hygienic Standards of Permissible Exposures

The American Conference of Governmental Industrial Hygienists adopted a threshold limit of 2 μg. of Be/cu. meter on the basis of its apparently satisfactory prevention of disease in plants that had adopted this limit. It is not yet known with certainty, however, whether this limit will ensure complete freedom from disease in all instances, or whether some higher level such as 5 μg. of Be/cu. meter might be equally suitable; no plant data exist to show an unequivocal relation between dosage and response, despite air-monitoring programs for more than a decade. Moreover, suggested hypersusceptibility[81,85-87] to Be, which may depend on lack of knowledge of the prevalence of Be disease, makes extremely difficult the setting of any hygienic standard for Be. Recognition of this fact has resulted in the use of supplementary air standards for community air in the vicinity of Be plants of 0.01 μg. of Be/cu. meter averaged over 1 month. In addition a limit of 25 μg. of Be/cu. meter has been used in plants; this limit refers to the concentration, obtained from a single sample, that should not be exceeded for any period however short.

Among the practicing industrial hygienists for the industry there is increasing feeling that the present limit is unnecessarily low, that the community level of 0.01 μg. of Be/cu. meter is unrealistic, and that a single level of 5 μg. of Be/cu. meter should serve for control of all types of exposures. Against these suggestions are the facts that reliable exposure data for 15 years or so of the latent period of the disease have not yet been accumulated, that the difference in engineering requirements for relaxation of the present limit to 2.5 times its value constitutes unimportant savings in comparison with the possible risk, and finally, that an 8-hour daily threshold limit for adult industrial workers in comparatively good health should not be directly carried over to community air. Some consideration, however, should be given to revising the community air limit to 10- or 15-fold its present value.

[87] J. H. Sterner and M. Eisenbud, Arch. Ind. Hyg. Occupational Med., 4, 123 (1951).

CADMIUM, Cd

1. Source and Production

No ore of commercial importance is mined solely for Cd production; the most common mineral, greenockite (CdS, 77.8 per cent Cd), occurs as a coating on the ZnS ore sphalerite. Hence Cd is derived chiefly from this or other sulfide ores. The United States is by far the world's leading producer of Cd; other sources (Southwest Africa, Mexico, Canada, and Australia) are also important Cd (and Zn) producers.

All domestic supply of primary Cd is recovered from flue dusts of Zn-blende roasting and sintering furnaces and from Cu- and Pb-smelting plants, from Zn dust in early stages of distillation, and from electrolytic purification of Zn as a Cd precipitate.

World production of Cd amounted to about 10,000 tons in 1958 of which the United States produced about half, 11 per cent being obtained from foreign flue dust. About 5000 tons went for Cd metal; most of the remainder for alloys and Cd compounds.

2. Uses and Industrial Exposures

The largest use (about 50 per cent) of Cd (also oxide, hydrate, and chloride) is for electroplating; the next largest use is for Cd-bearing alloys. Cd-bearing alloys are of two types: Cd–Ni, 98.65 per cent: 1.35 per cent, and Cd–Ag–Cu with Ag ranging from 0.2 to 2.25 per cent, Cu 0.25 to 2 per cent, and the balance Cd. Other alloys of Cd are with Au that is used in jewelry. Solders, the next largest use, employ Cd with varying amounts of Cu, Pb, Sn, Zn, and Ag, of which the last is most widely used. Another large and growing use is in Cd–Ni batteries, which are used in civil and military aircraft, guided missiles, and refrigerated railway cars. Electroplated Cd items are used in automobile engine parts, aircraft parts, radio and television parts, and as nuts and bolts. Cd metal and $Cd(NO_3)_2$ are used in reactors to control the rate of nuclear fission and as reactor poisons. CdS and CdSeS are used as pigments. $CdBr_2$ and CdI_2 are used in photography. Diethyl Cd is used in a new process for making tetraethyl Pb.

Industrial exposures to Cd may occur from fumes emitted in smelting of impure Zn and in the distillation of Cd-sponge. Molten Zn that is the feed for fractionating Cd and Zn also contains Pb. Most common cases of accidental industrial exposure to Cd occur in remelting scrap not suspected to contain Cd. Fatal cases of Cd poisoning have occurred among workers who flanged Cd-plated steel pipe with the aid of a blow torch.[88] Welding Cd-plated stock and heating Cd-plated rivets also have led to poisoning.

[88] L. W. Spolyar, J. T. Keppler, and H. G. Porter, *J. Ind. Hyg. Toxicol.*, **26**, 232 (1944).

TABLE 5

Physical and Chemical Properties of Cd and Some of Its Compounds

	Chemical symbol	At. or mol. wt.	Sp. gr.	M.p., °C.	B.p. °C.	Solubility
Cadmium	Cd	112.41	8.642	320.9	767 ± 2	Insol. cold or hot H_2O Sol. a. NH_4NO_3 Sol. hot H_2SO_4
Cadmium oxide	CdO	128.41	6.95	>1426	Dec. 900–1000	Insol. cold or hot H_2O Sol. a. NH_4 salts Insol. alkalies
Cadmium chloride	$CdCl_2$	183.32	4.047(25°)	568	960	1.4 kg./liter(20°) 1.5 kg./liter(100°) 15.2 g./liter alc. Insol. acetone and ether
Cadmium nitrate	$Cd(NO_3)_2$	236.43		350		1.09 kg./liter(0°) 3.26 kg./liter(60°) V. sol. acid Sol ether and acetone
Cadmium sulfide	CdS	144.48	4.82	1750[a]	Subl. in N_2	1.3 mg./liter(18°) Colloidal in hot H_2O V. sl. sol. NH_4OH Sol. acid

[a] 100 atm.

3. Physical and Chemical Properties

The physical and chemical properties of Cd and some of its compounds are listed in Table 5.

Cd is a soft, ductile, silver-white, electropositive metal. Spontaneous annealing and recrystallization of chill-cast Cd, as well as recrystallization of the cold-worked Cd, occurs at room temperature. The electrode potential of Cd is 0.40 volt; Zn, 0.76 volt; and Ni, 0.250 volt.

The most usual valence of Cd is +2, the most common compound, CdS. A few compounds of Cd exist in which the valence is +1. Cd ions are precipitated from solution by OH ions; the precipitates do not dissolve in excess OH. Cd ions form insoluble white compounds, usually hydrated, with carbonates, phosphates, arsenates, oxalates, and ferrocyanides; ferricyanides give yellow to orange precipitates. All are soluble in NH_4OH with the formation of complex cations of Cd and NH_3.

4. Analytical Determination

A satisfactory colorimetric method for the determination of small amounts of Cd in biological specimens employs dithizone. As modified for biological specimens by Saltzman,[89] controlled amounts of CN permit accurate measurement of Cd in the presence of from 1000 to 10,000 times the amounts of interfering substances; conditions for improved color stability are given, as well as a procedure to

[89] B. E. Saltzman, *Anal. Chem.*, **25**, 493 (1953).

separate interfering Tl. For air samples or relatively uncontaminated samples, a polarographic method may be used;[90] concentrations of 1 μg. of Cd/ml. may be determined with an accuracy of 5 per cent. A spectrographic method for quantities of from 0.4 to 200 μg. of Cd employs the Cd line of 3261 A., and as internal standard, the Mo line of 3209 A.[91]

5. Physiological Response

Acute Toxicity. $CdCl_2$ is a powerful emetic; the minimal effective dose for cats is 4 mg./kg.[92] As little as 14.5 mg. of Cd taken orally by man has caused nausea and vomiting, but as much as 326 mg. was not fatal. Thirteen to fifteen p.p.m. Cd in popsicles has sickened children, as has 67 p.p.m. in punch, and 530 p.p.m. in gelatin.[93] The lethal oral dose for rabbits has been given as 150 to 300 mg./kg. The ingestion of Cd salts induces salivation, choking attacks, persistent vomiting, abdominal pains, tenesmus and diarrhea, vertigo, and loss of consciousness. Catarrhal and ulcerative gastroenteritis, congestion, pulmonary infarcts, and subdural hemorrhages are found at necropsy.

When injected subcutaneously, Cd salts induce inflammation and coagulation necrosis at the site of injection. The irritant action is caused by Cd rather than the acid radical, because cats undergo fatal pulmonary irritation following inhalation of CdO fume.[94] LC_{50} values for inhaled fumes from a Cd arc expressed as CT, product of minutes of exposure and concentration in mg./cu. meter are: rats, 500; mice, <700; rabbits, 2500; guinea pigs, 3500; dogs, 4000; monkeys, 15,000.[95] The estimated CT value for 2 fatal human exposures was between 2500 and 2900 mg. min./cu. meter. The corresponding CT value for dogs exposed to $CdCl_2$ aerosol was 9600;[96] death in all cases resulted from pulmonary injury.

Chronic Toxicity. When incorporated in the diet of cats, the largest dose of Cd tolerated without production of vomiting was 20 p.p.m.[92] Cd prevented the growth of rats and induced fatalities at a level of 250 p.p.m. in the diet.[97] Dietary levels as low as 15 p.p.m. Cd were toxic to rats in a series of studies at 45, 75, and 135 p.p.m. Cd; anemia was a consistent finding at the lowest level; at 135 Cd early deaths occurred.[98] The toxicity increased on low protein diets.

[90] F. L. Feicht, H. H. Schrenk, and G. E. Brown, *U. S. Bur. Mines Rept. Invest.* No. 3639, 1942; N. Ya. Khlopin, *Analyst,* **73,** 120 (1948) (abstr.); H. H. Bauer and P. J. Elving, *Anal. Chem.,* **30,** 334 (1958).

[91] J. Cholak and D. Hubbard, *Ind. Eng. Chem. Anal. Ed.,* **16,** 333 (1944).

[92] C. L. Alsberg and E. W. Schwartze, *J. Pharmacol.,* **13,** Proc. 504 (1919).

[93] L. V. Garrity, *J. Am. Water Works Assoc.,* **40,** 1194 (1948); Ohio River Valley Water Sanitary Comm., Subcomm. Toxicol., Metal-Finishing Industries Action Comm. Rept. No. **3,** 1950.

[94] L. Prodan, *J. Ind. Hyg.,* **14,** 132, 151 (1932).

[95] H. M. Barrett, D. A. Irwin, and E. Semmons, *J. Ind. Hyg. Toxicol.,* **29,** 279 (1947); H. M. Barrett and B. Y. Card, *J. Ind. Hyg. Toxicol.,* **29,** 286 (1947).

[96] H. E. Harrison, H. Bunting, N. K. Ordway, and W. S. Albrink, *J. Ind. Hyg. Toxicol.,* **29,** 302 (1947).

[97] A. D. Johns, C. O. Finks, and C. L. Alsberg, *J. Pharmacol.,* **21,** 59 (1923).

[98] O. G. Fitzhugh and F. H. Meiller, *J. Pharmacol. Exptl. Therap.,* **72,** 15 (1941).

Five groups of rats given water containing from 0.1 to 10 p.p.m. Cd for 1 year showed no toxic effects, nor histological changes in blood or any other tissue.[98] The Cd content of the liver and kidney, however, increased at *all* levels fed in proportion to the Cd intake, indicating the possibility of eventual toxicity.

Repeated subcutaneous administration of Cd, as sulfate, to rabbits resulted in pronounced damage to the renal tubules, no changes in glomeruli, but proteinuria after 5 or 6 weeks on a dose of 0.65 mg./kg.;[99] anemia, liver cirrhosis, and splenic hyperplasia also resulted.

Prolonged, 1 year, inhalation of CdO and CdS, separately, by dogs at levels of 3 to 7 mg./cu. meter daily resulted in no damage, although blood Cd concentration averaged 0.22 mg. per cent and urine values were 0.36 mg./liter.[100] Most of the Cd was retained in the lungs, considerable amounts in the kidney and liver, with lesser amounts in bone and teeth. The lack of effect was attributed to the low rate of solubility and can only be explained in the face of appreciable blood Cd levels as due to tightly complexed and thus inactive Cd.

Metabolism. The distribution of Cd in the rat administered $CdCl_2$ in the water for 1 year, at six levels from 0.1 to 50 p.p.m. Cd, showed less than 1 μg./g. in the bone at the highest level, but levels of 50 μg./g. or greater in the kidney and 40 μg./g. in the liver.[101] Retention of Cd was estimated to be about 0.17 per cent by the kidneys at all levels from 2.5 to 50 p.p.m. Cd; somewhat greater percentage retention (0.3) was found for the 0.1 p.p.m. Cd level. The percentage retention of Cd by the liver increased with the dose being of the order of 0.4 per cent at 50 p.p.m., about half this value at 0.1 p.p.m. level. CdO fed to pigs for 3 days at 68 and 91 mg./lb. of feed resulted in a similar distribution but in far lower concentrations than is found in "unexposed" human organs[102] in some countries.

A number of studies of metabolism of Cd have been made using Cd^{115} with carrier in rats, orally, intravenously, and subcutaneously,[103,104] and in rabbits after subcutaneous injection.[105] Following ingestion in rats, 88 per cent of the Cd^{115} was fecally excreted; about 1 per cent was in the kidney and the liver each. Little or no Cd^{115} was found in the bone or in the soft tissues. Following injection, the largest fraction of the dose, both relatively and absolutely, went to the liver and the level was maintained for 5 weeks; similar distribution was noted for the kidney. The lung showed measurable amounts of Cd^{115} but these decreased with time. About 20 per cent of the intravenous dose in rats was excreted via the feces

[99] T. Dalhamn and L. Friberg, *Acta Pathol. Microbiol. Scand.*, **40**, 475 (1957); L. Friberg, *Acta Pharmacol. Toxicol.*, **11**, 168 (1955).

[100] F. Princi, *Arch. Ind. Hyg. Occupational Med.*, **1**, 651 (1950); A. M. Potts *et al.*, *Arch. Ind. Hyg. Occupational Med.*, **2**, 175 (1950).

[101] L. E. Decker, R. U. Byerrum *et al.*, *A.M.A. Arch. Ind. Health*, **18**, 228 (1958).

[102] R. B. Forney, C. A. Bunde, and G. R. Burch, *Proc. Soc. Exptl. Biol. Med.*, **90**, 13 (1955).

[103] C. F. Decker, R. U. Byerrum, and C. A. Hoppert, *Arch. Biochem. Biophys.*, **66**, 140 (1957).

[104] L. Friberg and E. Oderblad, *Acta Pathol. Microbiol. Scand.*, **41**, 96 (1957).

[105] L. A. Carlson and L. Friberg, *Scand. J. Clin. Lab. Invest.*, **9**, 1 (1957).

in 72 hours and little or no Cd^{115} was found in the urine; the distribution in the liver of Cd^{115} as the sulfate was uniform,[104] whereas in the kidney the cortex showed high accumulation and the medulla content was low. The pancreas accumulated large amounts in uniform distribution. The spleen contained small amounts of Cd^{115} concentrated at the capsular or subcapsular regions. Distribution of radioactive Cd^{115} in the blood of rabbits showed more than 90 per cent in this tissue to be in the red cells and it quantitatively migrated with the hemoglobin.[105] In 4 individuals who had presented symptoms of Cd poisoning, Cd followed the distribution seen in animals[106] and agreed with previous findings in German and British workers; highest Cd content, 145 µg./g. of wet tissue, was in the kidney with high content also in the liver, pancreas, and thyroid. The retention of Cd was remarkable because exposure to Cd had ceased 3 to 9 years before organ analysis. Cd was below the detectable limit (10 µg./specimen) in unexposed individuals (Sweden). "Unexposed" British subjects showed 12 to 18 µg. of Cd/g. of wet tissue;[107] 1 exposed individual showed amounts of Cd very similar to those found in Sweden. Kidney content of 8 unexposed Americans (7 to 56 years of age) averaged 24 µg. of Cd/g. of wet tissue.[108]

A unique protein, called metallothionein, has been isolated from equine kidney and contains 4.8 per cent Cd, 1.1 per cent Zn, and 7.1 per cent S (molecular weight, 10,200).[109] Whether metallothionein is a "natural" protein, or one formed by combination with exogenously derived metal, is unclear.

Mechanism. A study of the change in motor chronaxie of men (Russians) exposed to CdO fume pointed to the direct action of CdO on the cortical cells of the motor analyzer; defense inhibition occurred in cases with functional changes in the cerebral cortex.[110] Inactivation by Cd of certain, but not all, enzymes bearing —SH groups has been demonstrated in vitro;[111] and uncoupling of oxidative phosphorylation by extremely low concentrations of Cd^{2+} ($5 \times 10^{-6}M$) in connection with the oxidation of succinate and of citrate in mitochondria has been shown;[112] EDTA and certain divalent ions (Mn, Co, Ni) reverse the Cd effect. Thus it would appear from these isolated pieces of information that part of the toxicity of Cd may be related to its combination with SH-containing enzyme systems, be they part of the nervous system or other tissues.

Industrial Experience. A number of cases of serious consequence resulting from the inhalation of Cd fumes arising after heating metal containing Cd were reported in 1938[113] and after. Early symptoms of acute intoxication include:

[106] L. Friberg, *A.M.A. Arch. Ind. Health,* **16,** 27 (1957).

[107] J. C. Smith, J. E. Kench and J. P. Smith, *Brit J. Ind. Med.,* **14,** 246 (1957).

[108] H. J. Koch, E. R. Smith, N. F. Shimp, and J. Connor, *Cancer,* **9,** 499 (1956).

[109] M. Margoshes and B. L. Vallee, *J. Am. Chem. Soc.,* **79,** 4813 (1957); J. H. R. Kägi, Federation Meeting, Chicago, 1960.

[110] R. S. Vorob'eva, *Abstr. Ind. Hyg. Digest,* **21,** 12 (1957).

[111] F. P. Simon, A. M. Potts, and R. W. Gerarde, *Arch. Biochem.,* **12,** 283 (1947).

[112] E. E. Jacobs, *et al., J. Biol. Chem.,* **223,** 157 (1956).

[113] F. M. R. Bulmer, N. E. Rothwell, and E. R. Frankish, *Can. Public Health J.,* **29,** (January 1938).

dryness of the throat, coughing, feeling of constriction in the chest, shivering, headache, and less commonly nausea and vomiting. Later a pneumonitis develops with excruciating pain in the chest, severe dyspnea, and prostration. In nonfatal cases, intensity of the respiratory distress may increase for a few days, followed by recovery in 8 to 14 days. Chronic Cd poisoning is characterized by emphysema and renal injury[107,114] in which the urine contains a protein of molecular weight of 20,000 to 30,000. The protein does not respond to routine tests of urinary proteins. The two conditions may exist together or separately. Urinary Cd excretion may exceed 30 μg./day in cases of chronic poisoning. Respiratory function tests revealed significant changes only in those exposed to Cd for more than 10 years.[115] A follow-up study 5 years later of chronically exposed Cd workers showed greater deterioration in performance by respiratory function tests among exposed groups as well as deterioration in those with emphysema despite no further Cd exposure in the majority.[116] Eighteen of 24 new cases showed proteinuria only. An unusual case of fatal Cd poisoning is reported following explosion of Cd propionate while drying.[117]

Treatment. Indications to date from experiences with BAL and EDTA in the treatment of acute and chronic Cd poisoning in man and animals are that EDTA may be of some benefit in acute poisoning but may actually worsen the effect in chronic poisoning.[118] Administration of Fe experimentally to Cd-poisoned rabbits is reported to prevent the anemia to some extent.[119]

6. Hygienic Standards of Permissible Exposure

A threshold limit of 0.1 mg. of Cd/cu. meter has been set by the American Conference of Governmental Industrial Hygienists to provide a safe daily repeated exposure to Cd dust and fume. There are no reports to indicate whether this limit actually ensures freedom from response. Compared to the proposed drinking water standard of 0.01 p.p.m. Cd, which is based on a 10-fold reduction of the lowest level (0.1 p.p.m.) producing Cd accumulation in the kidney, the air limit of 0.1 mg./cu. meter might not provide the desired factor of safety. More evidence is needed on this point.

[114] J. A. Bonnell, *Brit. J. Ind. Med.,* **12,** 181 (1955); L. Friberg, *A.M.A. Arch. Ind. Health,* **16,** 30 (1957).

[115] R. S. J. Buxton, *Brit. J. Ind. Med.,* **13,** 36 (1956); G. Kazantis, *Brit. J. Ind. Med.,* **13,** 30 (1956).

[116] J. A. Bonnell, G. Kazantis, and E. King, *Brit. J. Ind. Med.,* **16,** 135 (1959).

[117] C. H. Manley and R. A. Dailey, *Analyst,* **82,** 287 (1957).

[118] L. Friberg, *A.M.A. Arch. Ind. Health,* **13,** 18 (1956); C. P. Odescaichi, and U. Scudier, *Med. lavoro,* **47,** 103 (1956); J. Dalhamn and L. Friberg, *Acta Pharmacol. Toxicol.,* **11,** 168 (1955); L. H. Cotter, *J. Am. Med. Assoc.,* **166,** 735 (1958).

[119] L. Friberg, *Acta. Pharmacol. Toxicol.,* **11,** 168 (1955).

CHROMIUM, Cr

1. Source, Production, and Uses

The American chromium industry mainly uses chromite ore ($FeO \cdot Cr_2O_3$) for the production of Cr alloys, refractories, and chemicals. Chromite ore varies widely in composition, being basically chromium oxide in association with oxides of Mg and Al, in addition to Fe. More than 50 per cent is used for alloys, one-third for refractories, and the remainder for chemicals. Production is chiefly from California, Montana, and Utah; some ore is imported from the Transvaal.

Chromite ore is mined from both open-cut and underground operations. For ores requiring concentration, gravity methods using crushers, jigs, classifiers, and wet shaking tables are used. Some flotation and magnetic concentration methods are employed.

For the production of Cr-ferro alloys, as Cr additives to steel, Cr ore is reduced in an electric furnace yielding low-C ferrochromium as the chief product. This low-C alloy is usually made from chrome silicide (ferrochrome-silicon) and chromite; high-C ferrochromium and Cr silicide are made directly from the ore. Exothermic ferrochromium contains in addition, $NaNO_3$. Some chromite is added directly to the steel charge. Cr metal is made by aluminothermic and electrolytic methods. Cr refractories are bricks with a wide range of Cr content mixed with ground materials, cements, for lining furnaces. Cr chemicals (chromates) are made by roasting finely ground chromite ore with soda ash and lime at 900 to 1000°C. followed by water leaching, acidification to obtain pure Na_2CrO_4 or $Na_2Cr_2O_7$, from which all other Cr compounds are derived.

Improvements in technology have resulted in production of ultrapure Cr (10 p.p.m. impurities) from an iodide process; electroplating of Cr directly on Al is used by the chemical, pharmaceutical, food, electronic, aviation, and astronautic industries.

Chromates and dichromates have many applications in lithography, textile printing, tanning, dyeing, photography, and in the manufacture of dyes, pigments, wallpaper, electric cells, explosives, matches, and rubber goods. Zinc chromate is widely used in metal primers. Chromous salts have fewer uses.

2. Industrial Exposures

The chief exposure to hazardous Cr substances in American industry is believed to be to an acid-soluble, water-insoluble chromate–chromite mixture produced in the preparation of chromate.[120] Before recognition and correction of this source of worker exposure, a 29-fold increase in the incidence of bronchogenic carcinoma occurred in workers in the United States chromate industry over that among workers in other chemical operations. Perforation of the nasal septa from CrO_3 also occurs. Other sources of exposure are to fume and dust of chromite ore

[120] W. M. Gafafer et al., Health of workers in the chromate industry, U. S. Public Health Serv. Publ. No. **192**, 1953.

and chrome alloys about electric furnaces and in ore-crushing operations. Chromous salts provide little industrial hazard.

In the cement industry Cr-containing cements have led to dermatitis,[121] as have chromate and chromic acid solutions and mists, such as in Mg foundries where the castings are treated with chromic acid solutions for weather resistance. Contact dermatitis has been reported from handling timber in which Cr salts were impregnated.[122] Electrolytic plating baths may carry a mist of Cr into the air; of 233 Cr platers examined by Schwartz and Seike[123] in 1930, 42.6 per cent had dermatitis, ulceration, or scars; in 52 per cent of these, the nasal membrane was damaged.

3. Physical and Chemical Properties

The physical and chemical properties of Cr and some of its compounds are listed in Table 6.

TABLE 6

Physical and Chemical Properties of Cr and Some of Its Compounds

	Chemical symbol	At. or mol. wt.	Sp. gr.	M.p., °C.	B.p., °C.	Solubility
Chromium	Cr	52.01	7.20(28°)	1890	2480	Insol. hot or cold H_2O Sol. HCl, dil. H_2SO_4 Insol. HNO_3
Chromic oxide (chromium sesquioxide)	Cr_2O_3	152.02	5.21	1990		Insol. hot or cold H_2O Insol. acids, alc., alkalies
Chromium trioxide	CrO_3	100.01	2.70	196	Dec.	1.6 kg./liter(15°) 2.067 kg./liter(100°) Sol. ether, alc., H_2SO_4
Chromous chloride	$CrCl_2$	122.92	2.75	824		V. sol. cold or hot H_2O Sl. sol. alc. Insol. ether
Sodium chromate	Na_2CrO_4	162	2.71			873 g./liter(30°) Sol. methyl alc. Sl. sol. alc.

Cr exhibits a valence of $+2$, $+3$, and $+6$ in its compounds. Cr is a blue-white hard metal that is not oxidized in moist air and even when heated oxidizes to but a slight extent. In an atmosphere of CO_2 it oxidizes to Cr_2O_3; in HCl, to $CrCl_2$. Cr combines directly with N, C, Si, and B. A passive form of the metal is conferred by oxidizing acids, attributable to a film of Cr_2O_3.

[121] C. R. Denton, R. G. Keenan, and D. J. Birmingham, J. Invest. Dermatol., **23,** 189 (1954); E. N. Walsh, J. Am. Med. Assoc., **153,** 1305 (1953); L. E. Gaul, Ann. Allergy, **11,** 758 (1953).

[122] P. Behrbohm, Berufsdermatosen, **5,** 271 (1957).

[123] L. Schwartz and F. Seike, Zentr. Gewerbehyg. Unfallverhüt., **17,** 232 (1930).

Bivalent Cr compounds are basic; trivalent, amphoteric; and hexavalent, acidic. The chromate ion in acidic solution is a powerful oxidizing agent. Cr^{2+} compounds resemble closely Fe^{++} compounds; Cr^{3+} compounds, those of Al^{3+}. Cr forms a series of isopoly acids and salts (K_2CrO_4, $K_2Cr_2O_7$, $K_2Cr_3O_{10}$, $K_2Cr_4O_{13}$); a group of chrome alums (e.g., $NH_4Cr(SO_4)_2 \cdot 12H_2O$), and complex ammines (e.g. $[Cr(NH_3)_6]Cl_3 \cdot H_2O$), in which Cr is $+3$.

4. Analytical Determination

Samples of air containing Cr compounds may be collected in an impinger using a NNaOH solution. Upon addition of KI and acidification, the liberated I_2 is titrated with standardized $Na_2S_2O_3$ for a determination of Cr^{6+}. Microgram quantities may be determined colorimetrically by diphenylcarbazide following permanganate oxidation. The procedure is applicable to samples of air, water, and urine, with good recovery and a sensitivity of 0.03 μg. of Cr in 25 ml.[124] Tissues have been analyzed for water- and acid-soluble Cr by a modified carbazide test[125] and for acid insoluble Cr (chromite) by the method of Cahnmann.[126] Procedures exist for the separation and analysis of soluble Cr^{6+} and soluble total Cr in Portland cement, and for the determination of total Cr.[127]

5. Physiological Response

Cr^{3+} salts, including Cr_2O_3, are ascribed a low order of toxicity and have caused no significant industrial illness according to Akatsuka and Fairhall,[128] who also reviewed the industrial toxicology literature of Cr to 1934. On the other hand, chromates and dichromates act as protein precipitants and are irritant in action. Administered subcutaneously to rabbits and guinea pigs, chromates damage the kidney, with the production of albumin and casts.[129] Fatal nephritis has occurred in man as a result of cauterizing a wound with chromic acid.[130] Chromate ion is reported to be absorbed through the skin of rabbits in sufficient amounts to be systemically toxic.[131] Inhalation of chromic acid dust by rabbits resulted in pulmonary hyperemia and inflammation, and guinea pigs exposed to the mist of Cr-plating baths for from 0.5 to 3 hours daily over a 45-day period developed lesions of the mucosa and submucosa of the respiratory tract as well as changes in the spleen and kidney.[132] Exposure of mice and rats, however, to a "mixed" chromate dust at 1 to 3 mg. of CrO_3/cu. meter for 4 hours daily throughout the major

[124] B. E. Saltzman, Anal. Chem., 34, 1016 (1952).
[125] A. M. Baetjer, C. Damron, and V. Budacz, A.M.A. Arch. Ind. Health, 20, 136 (1959).
[126] H. J. Cahnmann and R. Bisen, Anal. Chem., 24, 1341 (1952).
[127] R. G. Keenan and V. B. Perone, Am. Ind. Hyg. Assoc. Quart., 18, 231 (1957).
[128] K. Akatsuka and L. T. Fairhall, J. Ind. Hyg., 16, 1 (1934).
[129] W. Ophuls, Proc. Soc. Exptl. Biol. Med., 9, 11, 13 (1911); W. C. Hunter and J. M. Roberts, Am. J. Pathol., 8, 665 (1932).
[130] R. H. Major, Bull. Johns Hopkins Hosp., 33, 56 (1922).
[131] L. Lewin, Chem. Ztg., 31, 1076 (1907).
[132] S. Gallero, Folia Med. Naples, 24, 1256 (1938), J. Ind. Hyg. Toxicol., 21, 98A (1939).

part of their lifetime resulted in no bronchogenic cancers, nor did intratracheal or intravenous injection of basic potassium zinc chromate or $BaCrO_4$ cause carcinoma of the lung.[133] On the other hand, $CaCrO_4$, and sintered CrO_3 produced cancers, mainly sarcomas, when introduced in pellet form in the thigh muscle and plural cavity of rats,[134] but irradiated Cr^{51} gave no gross evidence of injury as a metal implant in rats for 18 months.[135] Orally, neither hexavalent nor trivalent Cr was toxic to rats when administered in the drinking water for 1 year at levels from 0.45 to 25 p.p.m. Cr.[136]

Metabolism. Na_2CrO_4 or K_2CrO_4 injected intratracheally into guinea pigs is removed rapidly (a few days) from lungs except for a small amount still present at 140 days.[125] In 10 minutes, 20 per cent was in the blood and only 5 per cent in the liver, kidneys, and spleen together. Considerable amounts of Cr were attached to the erythrocytes. Water-soluble $CrCl_3$ remains fixed for long periods at the injection site. Water-soluble Cr forms retained by the tissues are saline-insoluble. Lung slices bound Cr^{3+} at a greater rate than Cr^{6+}. Little Cr was found in bones of either animals or men exposed to Cr salts and no relation was found between the amounts of the soluble or insoluble forms of Cr in cancerous lung tissue. The lung has no specific affinity for Cr^{6+} or Cr^{3+}.[137]

Distribution of Cr in rats fed Cr^{6+} in drinking water at various low levels (0.45 to 11 p.p.m.) for 1 year showed highest amounts in spleen, then bone, kidney, and liver. The p.p.m. Cr in the drinking water resulted in closely similar values in the spleen at 1 year. At 25 p.p.m. Cr in the water, Cr^{3+} retained was about $1/5$ to $1/10$ that of Cr^{6+}, in the rat.[136] Intravenous Cr^{3+} behaves like a small-sized colloid going to the reticuloendothelial system (bone marrow), according to Kraintz, *et al.*[138]

Mean urinary Cr value of white United States chromate workers was 43 μg./liter, for colored workers 71 μg./liter; white workers from this group had a lung cancer rate 14 times the expected number, colored 80 times the expected number.[120] Mean value for blood Cr of white males was 4 μg. per cent, for colored males, 6 μg. per cent; 12 per cent of the samples showed values of 10 μg. per cent or greater.

Mechanism. Both Cr^{3+} and Cr^{6+} have been shown[138] to (*a*) denature proteins (albumins) albeit at high concentrations (2 per cent), and at low pH values

[133] A. M. Baetjer, J. F. Lowney, H. Steffee, and V. Budacz, *A.M.A. Arch. Ind. Health,* **20,** 124 (1959).

[134] W. C. Hueper and W. W. Payne, *J. Am. Ind. Hyg. Assoc.,* **20,** 274 (1959); W. W. Payne, *Arch. Envir. Health,* **1,** 20 (1960).

[135] W. G. Myers, *Am. J. Roentgenol. Radium Therapy Nuclear Med.,* **81,** 99 (1959).

[136] R. D. MacKenzie, R. U. Byerrum, G. F. Decker, C. A. Hoppert, and R. F. Laugham, *A.M.A. Arch. Ind. Health,* **18,** 232 (1958).

[137] A. M. Baetjer, C. M. Damron, J. H. Clark, and V. Budacz, *A.M.A. Arch. Ind. Health,* **12,** 258 (1955).

[138] J. H. Clark, *A.M.A. Arch. Ind. Health,* **20,** 117 (1959); L. Kraintz and R. V. Talmadge, *Proc. Soc. Exptl. Biol. Med.,* **81,** 490 (1952).

(<5.4), and (b) to precipitate nucleic acids.[137] Cr catalyzes phosphoglucomutase reaction,[139] the succinate cytochrome C reductase system,[140] and influences favorably plant growth and many other biological reactions.[141] Recently Cr^{3+} has been identified as the glucose tolerance factor (GTF)[142] (GTF deficiency is characterized by delayed removal of glucose from the blood). This finding appears to indicate that Cr^{3+} is an essential bioelement for mammals.

Industrial Experience. The existence of a serious lung-cancer hazard to workers producing chromates from chromite ore and possibly during the handling of certain Cr pigments ($ZnCrO_4$, $BaCrO_4$, and $PbCrO_4$) has been realized for many years, first in Germany,[143] and later in the United States by Machle and Gregorios[143] and from an industry-wide survey by the Division of Occupational Health, USPHS.[120] At the time (1951), investigation of chromate workers in England revealed but 1 case of lung cancer in 724 employed in three plants. Within the following 6 years, however, 14 new cases of carcinoma of the lung were found by Bidstrup and Case.[143] These new cases resulted in a death rate from lung cancer of about 4 times normal, compared with a rate 29 times normal in the United States.[120] However, "normal" mortality from all causes in Britain was about twice that in the United States, and the normal mortality from lung cancer in the United States was less than half the rate in Britain. The lung cancer deaths in the chromate workers of the United States and Britain were 21 and 15 per cent, respectively, of the total deaths in these groups, excluding those expected from nonindustrial causes.

Although the precise form of Cr responsible for the production of lung cancer among chromate workers is not known with absolute certainty, experimental production of sarcomas by Hueper and Payne[134] relates the tumorigenesis to the degree of solubility of the compounds: sintered CrO_3, (Cr_5O_{12}) with highest water solubility gave the largest percentage of tumors, whereas $BaCrO_4$ with lowest solubility resulted in none under the same conditions.

Chrome ulcer is another characteristic injury resulting from exposure to chromates, which consists of a penetrating ulcer of hands and forearms. Painful only if on the knuckles, they heal in a few weeks with permanent scarring if further exposure is prevented early.[144] Over 50 per cent of a group of chromate workers had evidence of either active or healed chrome ulcers.[120] Only 2 per cent of the group had a contact dermatitis, however. Characteristic lesions also occur

[139] L. H. Strickland, *Biochem. J.,* **44,** 190 (1949).

[140] B. L. Horecker, E. Stotz, and T. R. Hogness, *J. Biol. Chem.,* **128,** 251 (1939).

[141] A. M. Baetjer, in M. J. Udy, ed., *Chromium,* Vol. 1. Reinhold, New York, 1956, p. 80.

[142] K. Schwartz and W. Mertz, *Arch. Biochem. Biophys.,* **85,** 292 (1959).

[143] E. Pfeil, *Deut. med. Wochschr.,* **61,** 1197 (1935); W. Alwens and W. Jonas, *Acta Univ. Intern. contra Cancrum,* **3,** 103 (1938); W. Machle and F. Gregorios, Public Health Repts. U. S., **63,** 1114 (1948); T. F. Mancuso and W. C. Hueper, *Ind. Med. and Surg.,* **20,** 358 (1951); P. L. Bidstrup and R. A. M. Case, *Brit. J. Ind.,* **13,** 260 (1956).

[144] C. P. McCord, H. G. Higginbotham, and J. C. McGuire, *J. Am. Med. Assoc.,* **94,** 1043 (1930).

in the nose, and may lead to atrophy of the mucous membranes, and ulceration or perforation of the nasal septum; detailed descriptions are given by Schwartz.[145] These lesions were found in over 50 per cent of chromate workers.[120] Small, multiple ulcers occur in the soft palate, posterior surface of the tongue, and floor of the mouth. Generalized dermatitis is common, may occur among lithographers. Dermatitis of the hands may occur among cement workers,[121] and on the wrists from wrist watches and bracelets.

A syndrome "chromium enteropathy" has been named for a general and enteral injury as a chronic condition from exposure to $CaCr_2O_7$.[146]

Control and Therapy. Absolute control of chromic acid mist exposure is reported by Hama, Fredrick *et al.*[147] by the use of a surface-active fluorocarbon. EDTA combined in an ointment has been reported[148] to be effective in healing Cr ulcers. The control of hazards in chrome plating and anodizing has been discussed by Riley and Goldman.[149]

6. *Hygienic Standards of Permissible Exposure*

A threshold limit of 0.1 mg. of CrO_3/cu. meter has been set for chromic acid mist and chromates by the American Conference of Governmental Industrial Hygienists. No reports to date indicate the suitability of this limit; the extensive study by the USPHS[120] of the chromate industry, in which a high incidence of lung cancer was shown, frequently recorded values for Cr in the workroom air in excess of the recommended limit. The absence of reported cases of lung cancer from chromate plants rebuilt with recognition of the Cr air standards augurs well for their suitability. A drinking water standard of 0.05 p.p.m. Cr^{6+} has been set by the USPHS.

COBALT, Co

1. *Source and Production*

Co is a relatively rare element, composing but 0.001 per cent of the earth's crust, as compared with Ni at 0.02 per cent. Important minerals are the arsenides, sulfides, and oxidized forms. The principal arsenides of Co are smaltite ($CoAs_2$), safflorite ($CoAs_2$), skuterudite ($CoAs_3$), and cobaltite (CoAsS). The principal sulfide minerals are carrolite ($CuCo_2S_4$) and linnaeite (Co_3S_4). The principal oxide minerals are asbolite (an impure mixture of Mn and other oxides), neterogenite (a hydrated oxide usually containing Cu and occasionally Ni and Fe),

[145] L. Schwartz, *U. S. Public Health Bull.* No. **249**, 44 (1939).

[146] H. Buess, *Helv. Med. Acta,* **17**, 104 (1950).

[147] G. Hama, W. Fredrick, D. Millage, and H. Brown, *Am. Ind. Hyg. Assoc. Quart.,* **15**, 211 (1954).

[148] C. C. Maloof, *A.M.A. Arch. Ind. Health,* **11**, 123 (1955).

[149] E. C. Riley and F. H. Goldman, *Public Health Repts. U. S.,* **52**, 172 (1937).

sphaerocobaltite ($CoCO_3$), and erythrite ($3CoO, As_2O_5 \cdot 8H_2O$). A number of other less known materials of Co exist but in insufficiently quantity to be mined.

The Belgian Congo is the largest producer of Co ore and metal, the United States is second. World production in 1958 totaled 14,600 short tons of which the United States produced more than 2,000; Canada, Morocco, and Rhodesia each produced somewhat less. World output of Co is progressing steadily as new deposits are brought into production. The United States consumes (domestic production and import) about 8000 tons, more than half of the world production.

Practically all Co produced is a by- or coproduct of other metals, chiefly Cu; accordingly description of mining process is omitted. A brief description of the Co mining of the leading United States producer may be found in Huttl.[150]

The metallurgy of Co is highly diversified, perhaps more than that of any other metal, because of the variety of ores that have to be treated. A new process[151] for recovery of Co by the largest United States producer entails the following principal steps: wet chemical autoxidation at high temperature using air under high pressure, acid leaching under pressure, filtration of the tailings, purification of the solution, neutralization with NH_3, H_2 reduction of the ammoniacal solution, electric furnacing for S removal, and granulation of the metal. An electrolytic unit producing high-purity Co may replace the H_2 reduction step. The U. S. Bureau of Mines has reported the experimental production of Co metal of 99.99 per cent purity.[152]

2. Uses and Industrial Exposures

Permanent-magnet alloys and high-temperature, high-strength alloys comprise the bulk uses (4.5 million pounds) of metallic Co. Other metal uses are for cemented tungsten carbides, high-speed steels, and alloyed hard-facing rods. Salts and driers for lacquers, varnishes, paints, inks, pigments, enamels, glazes, electroplating, and feed consume more than 1 million pounds. Nonmetallic uses, exclusive of salts and driers, consume somewhat less than 1 million pounds for pigments, ground-coat frits, etc. Co oxides offer good prospects as catalysts in afterburners for motor exhaust gases.

Exposures sustained in the milling of Co by the leading United States producer have changed owing to the development of a new wet-chemical, high-temperature, high-pressure oxidation process. As far as it has been studied, chief exposures are to Co fume and dust from powder falls in the electric furnace, and fume from melting and pouring of Co metal prior to pelleting.

In the production of cemented tungsten carbides (carballoy) exposures are to dust and fume of Co, in combination with dusts of WC, TiC, and TaC. Weighed

[150] J. B. Huttl, *Eng. Mining J.*, **151**, 91 (1950).

[151] J. S. Mitchell, *Mining Eng.*, **8**, 1093 (1956).

[152] K. K. Kershner, F. W. Hoertel, and J. C. Stahl, *U. S. Bur. Mines Rept. Invest.* No. **5175** (1956).

charges of Co metal powder, tungsten metal powder, and lampblack, together with small additions of Ta and Ti, are ground in ball mills. The charging and emptying of the containers present dust exposures. After pressing, the material is put through a presintering process, following which it is cut and ground. This also presents a dust exposure. The material is given a final sintering and the tips are brazed into holders (drills, lathe tools, saw teeth, etc.). Some fume may be produced at these operations. The tools are given a final (wet) grinding.

3. *Physical and Chemical Properties*

The physical and chemical properties of Co and some of its compounds are listed in Table 7.

TABLE 7

Physical and Chemical Properties of Co and Some of Its Compound

	Chemical symbol	At. or mol. wt.	Sp. gr.	M.p., °C.	B.p., °C.	Solubility
Cobalt	Co	58.94	8.9	1495	2900	Insol. hot or cold H_2O Sol. acid
Cobaltous oxide	CoO	74.94	5.7–6.7	1800(dec.)		Insol. hot or cold H_2O, NH_4OH, alc. Sol. acid
Cobaltic oxide	Co_2O_3	165.88	5.18	895(dec.)		Insol. hot or cold H_2O, alc. Sol. acid
Cobalto-cobaltic oxide	Co_3O_4	240.82	6.07			Insol. hot or cold H_2O, HCl, HNO_3, aq. reg. Sol. H_2SO_4
Cobaltous sulfide	CoS	91.01	5.45	>1116		3.8 mg./liter(18°) Sol. acid, alc.
Cobaltous chloride	$CoCl_2 \cdot 6H_2O$	237.95	1.924	86 $(-6H_2O)$110		767 g./liter(0°) 499 g./liter(20°) 1.90 kg./liter(100°) V. sol. alc. 2.9 g./liter ether Sol. acetone

Co is a hard magnetic metal, resembling Ni in appearance but with a pinkish tinge. The metal crystallizes in two allotropic forms: alpha, a close-packed hexagonal, and beta, a face-centered cubic. The magnetic permeability of Co averages less than two thirds that of Fe, but when alloyed with Fe and Ni, exceptional magnetic properties have been developed. Co is a relatively unreactive metal; it does not oxidize in dry or moist air at ordinary temperatures, and at red heat oxidation is superficial. Co reacts with most acids but becomes passive in concentrated nitric acid. Co is not attacked by alkalies, either in solution or fused, but combines with the halogens when heated. When reduced from the oxide in a fine powder form, Co is pyrophoric. At red heat, the metal decomposes steam. In NH_3, a nitride forms, which decomposes at higher temperatures. Under pressure at 150°C. Co forms characteristic orange crystals of the tetra carbonyl $[Co(CO)_4]_2$ (see section on Metal Carbonyls, page 1108).

Co forms three types of complexes: ammines $[Co(NH_3)_6]Cl_3$ $[Co(NH_3)_6-Cl]Cl_2$, complex nitrites $K_3[Co(NO_2)_6]$, and complex cyanides, $K_4[Co(CN)_6]$, $K_3[Co(CN)_6]$. Solutions of the Co ammines show none of the reactions of Co. Co also chelates with certain organic molecules such as ethylenediamine that possess O_2-carrying properties.

CoO varies in color from olive green to red, depending on particle size; usually however, it is dark gray. It is the principal constituent of the "gray" cobalt oxide of commerce; the other chief compound is Co_3O_4.

Co_2O_3 forms when Co compounds are heated at low temperatures in excess air; higher temperatures convert Co_2O_3 to Co_3O_4. Oxidation of cobaltous (Co^{++}) salts in acid or alkaline solutions gives hydrated cobaltic oxide. This oxide is amphoteric and forms complexes with metal oxides, for example, $ZnO \cdot Co_2O_3$. The "black" cobaltic oxide of commerce consists chiefly of this oxide together with a small amount of Co_3O_4.

Co_3O_4 is the stable oxide of Co. It is reducible, however, to Co by C, CO, and H_2.

The chlorides of Co and their hydrates comprise a large group of Co compounds because of their capacity to complex with NH_3 to form complex ammines. $CoCl_2$ hydrolyzes in aqueous solution to the extent of 0.11 per cent at $0.062M$, 0.17 per cent at $0.031M$. Co^{++} salts are pink or red; the corresponding Co^{+++} salts are commonly so unstable as not to exist under normal conditions. CoF_3, however, is used as a catalyst in the production of fluocarbons and in the cracking of gasoline.

4. Analytical Determination

A method for the determination of atmospheric samples of Co, particularly those from cemented tungsten carbide sources, has been developed by Keenan and Flick.[153] A certain oxidative procedure is incorporated to convert the metallic carbides to sulfates prior to complexing Co with nitroso-R salt. For samples containing 100 μg. of Co or greater, an accuracy of at least 95 per cent is reported. For removal of Cu and Ni interferences a preliminary extraction with 1-nitroso-2-naphthol is recommended by Saltzman and Keenan in a compilation of methods of biochemical analysis of Co.[154] Because colorimetric methods do not have the requisite sensitivity for quantitative determination of Co in normal biological

---

[153] R. G. Keenan and B. M. Flick, *Anal. Chem.*, **20**, 1238 (1948).

[154] B. E. Saltzman and R. G. Keenan, in D. Glick, ed., *Methods of Biochemical Analysis*, Vol. V. Interscience Publishers, New York–London, 1957, p. 181; R. G. Keenan and J. F. Kopp, *Anal. Chem.*, **28**, 185 (1956).

specimens, two spectrochemical procedures are recommended. Preliminary concentration is required for normal, unexposed tissues as well as a prior extraction with 1-nitroso-2-naphthol, to rid of interferences. By these procedures, recoveries of 90 to 98 per cent for a working range of 0.006 to 0.09 μg. of Co are reported.[154]

5. Physiological Response

Toxicity in Animals. The approximate acute lethal doses, LD_{100} of Co and certain of its compounds by several routes of administration and for several laboratory animal species are given in Table 8. Co metal powders, particle size not

TABLE 8

Approximate Acute Lethal Doses of Cobalt and Its Compounds for Laboratory Animals

Substance	Approx. LD_{100} mg. Co/kg.	Route	Species
Co[a]	1,500	Oral	Rat[a]
Co	<700	Oral	Rabbit[c]
Co	250–500	I.P.	Rat[d]
Co	100–200	I.P.	Rat[b]
Co	100	I.V.	Rabbit[e]
CoO	<70	Oral	Dog[f]
Co2O3	5000	I.P.	Rat[b]
CoSO4	3.4[g]	I.V.	Dog[f]
CoCl2·6H2O	7.5	I.V.	Dog[h]
CoCl2·6H2O	6.3?	I.V.	Rabbit[e]
CoCO3	1,000	I.P.	Rat[b]
Co lactate	10	I.P.	Rat[b]
Co albuminate	9	Subcut.	Mouse[i]

[a] An unestimated but probably large fraction of the Co metal and its oxides was unabsorbed.
[b] W. Fredrick and W. Bradley, 7th Ann. Meeting Am. Ind. Hyg. Assoc., Chicago, 1946.
[c] M. Simesin, *Arch. intern. pharmacodynamie,* **62,** 347 (1939).
[d] H. E. Harding, *Brit. J. Ind. Med.,* **7,** 76 (1950).
[e] P. Mascherpa, *Boll. soc. ita*c. *biol. sper.,* **15,** 226 (1940).
[f] W. B. Woodman and C. M. Tidy, Forensic Medicine and Toxicology. 1877, pp. 171, 214.
[g] Dog weight assumed to be 10 kg.
[h] F. Caujolle, *J. Pharmacol.,* **29,** 410 (1939).
[i] P. Mascherpa and R. Perito, *Arch. intern. pharmacodynamie,* **40,** 471 (1931).

defined, show an approximate LD_{100} parenterally in the rat and rabbit of from 100 to 500 mg. of Co/kg. of body weight. Orally, Co metal powder appears to be about 2 or 3 times less toxic for these species. CoO orally in the dog is relatively toxic acutely, but Co_2O_3 intraperitoneally in the rat is relatively nontoxic. The soluble Co salts, however, are highly toxic by vein in the dog and rabbit; the water-insoluble carbonate is only mildly toxic. Organic salts and protein complexes appear to have the same order of toxicity as the soluble inorganic salts independent of route of administration.

Co and its salts would appear to have a cumulative toxic action under conditions in which elimination cannot keep pace with absorption. Schepers[155] found repetition of a 5-mg. dose of Co metal (particle size not specified), intratracheally administered, to be lethal to rats, whereas a single 5-mg. dose was not. Similarly, Fredrick and Bradley[156] found young rats unable to survive repeated 30-mg. doses of Co metal in the diet longer than 1 month, whereas 1500 mg. of Co in a single dose was the LD_{100}. Stanley et al.[157] found doses of 9 mg. of Co as $CoCl_2 \cdot 6H_2O$ to be lethal subcutaneously to rats in 4 repeated daily doses. Apparently, dietary components can greatly modify the toxicity of ingested Co, as Underhill et al.[158] reported weanling rats died on daily doses of 1 and 0.5 mg. of Co for 3.5 months when fed a milk diet. Rats, however, were reported to tolerate a daily dose of 1 mg. of Co in their drinking water for 14 weeks when fed the usual laboratory food.

On the other hand, there appears to be experimental evidence that a tolerance may be developed to Co when initial doses are sufficiently low to be well tolerated; following daily subcutaneous injections of 10 mg. of $CoCl_2 \cdot 6H_2O$ (2.5 mg. Co) during a 13-day period, a relatively huge dose of 1 gram $CoCl_2 \cdot 6H_2O$ was required to be lethal.[159] Schepers[155] mentions a tolerance to Co metal in the lung of animals surviving milligram doses. Moreover, certain anions such as citrate may affect Co toxicity; rabbits survived injections of 100 mg. of Co citrate (20 per cent Co).[160]

A polycythemic level of soluble Co compounds for rats is considered to be 40 mg./kg. orally, 2.5 mg./kg. by injection.[161]

The toxicity of an As-containing Co ore (smaltite) has been reported[162] to cause a slowly developing corneal reaction, but cobaltite caused only a slight reaction. Co_2O_3 has a low order of toxicity for the guinea pig lung.[163]

Chronic inhalation by animals of a Co-metal blend as used in industry containing 46 per cent WC, 28 per cent TiC, 8 per cent TaC, 6 per cent Co, and 2.5 per cent SiO_2 at a level of 20 mg. of Co/cu. meter resulted, after 3 years, in focal fibrotic lesions, hyperplasia of the bronchial epithelium, and developing granulomas in areas of dust deposition, which appeared to simulate those reported in industrial workers.[164] Daily inhalation of Co metal fume, which was approximately equal parts of Co, CoO, and Co_3O_4, at 1 mg. of Co/cu. meter for 2 years failed to elicit these pulmonary reactions.

[155] G. W. H., Schepers, A.M.A. Arch. Ind. Health, 12, 127 (1955).

[156] W. Fredrick and W. Bradley, 7th Ann. Meeting Am. Ind. Hyg. Assoc., Chicago, 1946.

[157] A. J. Stanley, H. C. Hopps, and A. M. Shideler, Proc. Soc. Exptl. Biol. Med., 66, 19 (1947).

[158] F. A. Underhill, J. M. Orten, and R. C. Lewis, J. Biol. Chem., 91, 13 (1931).

[159] C. Seghini, Clin. med. ital., 71, 355 (1940).

[160] J. M. LeGoff, Compt. rend. soc. biol., 96, 455 (1927).

[161] E. J. Underwood and C. A. Elvehjem, J. Biol. Chem., 124, 419 (1938).

[162] A. Policard and J. Rollet, Compt. rend. soc. biol., 132, 192 (1939).

[163] G. W. H. Schepers, A.M.A. Arch. Ind. Health, 12, 124 (1955).

[164] H. E. Stokinger, W. D. Wagner, D. A. Fraser, and O. J. Dobrogorski, unpublished results.

For the effects of inhaling Co hydrocarbonyl, see Metal Carbonyls, page 1108.

Much study has been given the effects of dietary administration of Co compounds because Co has been found essential in nutrition of foraging animals, and because of its capacity to induce polycythemia in man and animals. The use of Co as an erythrogenic agent in man is extremely limited, as few anemias respond to Co therapy, and its purpose for combating anoxia is rather remote. The great amount of work on the erythropoietic action of Co, however, has led to the discovery of a cobalt-activated erythropoietic factor, erythropoietin,[165] The relation of Co deficiency to a nutritional disorder in ruminants is given in a review of the pertinent literature.[166]

Signs of acute poisoning in animals fed Co salts consist of diarrhea, loss of appetite, paralysis of the hind legs, and lowering of body temperature prior to death.[167] With high doses anuria occurred; with smaller doses, albuminuria. One of the immediate signs is cutaneous vasodilatation, especially of the nose and ear, within 3 minutes after administration and persisting for about 1 hour. The blood pressure may fall. Microscopically, all organs are congested, with small focal hemorrhages on the serosal surfaces and large hemorrhages in the liver and adrenals; bones show hyperplastic marrow. Lungs show alveolar thickening; the kidneys, tubular degeneration. Fibers of the myocardium are pale and shrunken and the pancreas shows degenerative changes. The detailed microscopic changes in acute Co poisoning, including the results of Perl's reaction in critical organs, have been summarized by Valerio.[168] Perl's stain indicated that the Co^{++} is the prevailing ion form in the tissues. Schepers[169] found that Co mixed with WC at high doses produces a transient inflammatory reaction in the guinea pig lung with residual papillary hypertrophy of the bronchial mucosa. The acute inflammatory reaction is thus replaced by a focal pneumonitis and residual hyperplasia and metaplasia of the bronchial epithelium. Co metal instilled in the lungs of guinea pigs, in large doses, produced obliterative bronchiolitis.[155] Intravenously injected $CoCl_2$ in rabbits produces severe and selective damage to the alpha cells of the islets of Langerhans.[170]

Two reports, one claiming production of an adenocarcinoma in the lung and a spindle-cell sarcoma in the bone of rabbits,[171] the other a rabdomyofibrosarcoma and other sarcomas in the rat,[172] seem to find no parallelism as yet in man; high neoplasm rates have not been reported from Co-mining areas; Canada (Cobalt City), Belgian Congo (Katanga), Norway (Skuterud), France (Allemont), or Czechoslovakia (Dobscheina).[155]

[165] E. Goldwasser, L. O. Jacobson, W. Fried, and L. Plzak, *Science,* **125,** 1085 (1958).

[166] K. C. Beeson, U. S. Dept. Agr., Agr. Infor. Bull. No. **7,** 1950.

[167] J. M. LeGoff, *Compt. rend. soc. biol.,* **101,** 797 (1929).

[168] V. Valerio, *Folia Med.* Naples, **32,** 104 (1949).

[169] G. W. H. Schepers, *A.M.A. Arch. Ind. Health,* **12,** 140 (1955).

[170] M. G. Goldner, B. W. Volk, and S. Lazarus, *Metabolism Clin. and Exptl.,* **1,** 544 (1952).

[171] H. R. Schinz, *Schweiz. med. Wochschr.,* **39,** 1070 (1942).

[172] J. C. Heath, *Brit. J. Cancer,* **10,** 668 (1956).

Metabolism. Co is an essential trace element for man and animals; it is an important constituent of vitamin B_{12} and certain enzymes, and is associated with the production of erythropoietin, and red cell stimulating factor. Common plants such as lettuce, beets, cabbage, spinach, and sweet potatoes act as sources of dietary Co containing from a few hundredths p.p.m. (sweet potatoes) to 0.7 p.p.m. (spinach) on a moisture-free basis.[166] Co content of plants varies somewhat with the region in which they are grown. Co content of fresh tissues of normal, unexposed dogs, rabbits, and rats ranged from a few $m\mu g$. (bone) to a few tenths μg. (thyroid and adrenals) when determined by a highly sensitive spectrochemical procedure developed by Keenan and Kopp.[154] The pancreas, kidney, and lung showed intermediate amounts. The spleens of rabbits and rats contained far higher amounts than did this organ in the dogs, although the order of tissue content of Co showed little consistency from species to species. Forbes *et al.*[173] reported the Co values for 10 tissues of 1 human cadaver to range from 0.01 p.p.m. Co for fat, nerve, muscle and gastrointestinal tract to 0.06 p.p.m. for liver; skin, skeleton, and heart were next highest; endocrine organs were not analyzed.

The degree of gastrointestinal absorption of Co and its salts depends on the dose; very small doses of the order of a few $\mu g./kg.$ are absorbed almost completely; larger doses are less well absorbed. Copp and Greenberg,[174] administering 10 $\mu g.$ of Co^{60} to rats, found more than 30 per cent excreted in the urine when given orally, if injected, more than 90 per cent. Although their urinary values appear somewhat high, their tissue distribution data are in conformity with others; the glandular organs, particularly the pancreas, accumulated largest amounts, as did the liver, spleen, and kidneys. Of the injected Co approximately equal quantities appeared in the bile and feces. Sheline and Chaïkoff,[175] using intravenous radiocobalt in dogs, found 0.1 to 0.3 per cent in the pancreatic juice in 2 or 3 days, whereas 5 per cent was found in the bile at the same time. The previously cited, later work of Goldner *et al.*[170] would indicate that possibly the Co in the pancreas is bound to cellular components with lesser amounts in the pancreatic juice. The metabolic fate of 13 mg. of Co, as $CoCl_2$, intravenously injected in man, determined spectrographically, resulted in a 10-fold increase in urinary output but only a 17-fold increase in fecal excretion during the first week after injection.[176] Slightly less than 3 mg. of the 13 mg. of Co injected was recovered in the excreta during one week, indicating a rather slow elimination of injected Co. The normal weekly urinary output amounted to 0.21 mg. of Co; fecal output, 1.04 mg.; normal urinary excretion of Co thus representing 17 per cent of the total output. Co injected in rabbits as the citrate is more rapidly eliminated; 50 and 65 per cent urinary elimination of the total dose of 10 and 13 mg., respectively, was eliminated in 24 hours.[160] The distribution of Co in rabbits given Co metal orally and

[173] R. M. Forbes, A. R. Cooper, and H. H. Mitchell, *J. Biol. Chem.,* **209,** 857 (1954).
[174] D. H. Copp and D. M. Greenberg, *Proc. Natl. Acad. Sci. U. S.,* **27,** 153 (1941).
[175] G. E. Sheline and I. L. Chaikoff, *Am. J. Physiol.,* **145,** 285 (1946).
[176] N. L. Kent and R. L. McCance, *Biochem. J.,* **35,** 877 (1941).

$[CoCO_3(NH_3)_4]Cl_2$ subcutaneously is reported by Simesin,[177] and the fate of $CoCl_2$ and cobaltic proteinate in guinea pigs by Untersteiner.[178] The retention by rats of low (0.003 and 0.08 p.p.m.) dietary levels of Co with and without Fe and Cu is reported by Houk et al.,[179] Comar et al.[180] have shown, in agreement with others, that Co is excreted mainly in the urine and that a smaller fraction may be recovered from the bile after intravenous injection; when given orally, a large fraction appears in the feces.

Mechanism. Present evidence would appear to indicate that (a) Co can exert a variety of physiological activities, (b) these activities are manifested at various levels of tissue Co concentration. Aside from the indirect action of Co as part of vitamin B_{12} one of whose actions is believed to be the regulation of tissue sulfhydryl concentration,[181] the intimate mechanism of the other activities of Co is as yet undisclosed. The over-all effect of many of these activities of Co is to stimulate rather than depress physiological action. The earliest sign of increased intake of Co in animals is the production of increased amounts of serum alpha globulins.[182] Because neuraminic acid is principally associated with the alpha globulins of the serum, their increased production is a direct result of Co action. This action of Co is not uniform in all animals, indicating that a sensitization mechanism is involved, and the action tends to regress to normal despite continued Co intake. At slightly higher levels of Co intake, 1 to 5 mg. of Co/kg. as soluble salt by mouth, polycythemia develops[183] by a mechanism believed to be a direct stimulating action on red bone marrow and possibly extramedullary hemopoietic tissue in other organs. The polycythemia is regularly accompanied by hemoglobinogenesis[158] provided Fe and Cu are present in adequate amounts. It now appears probable that the Co-stimulated erythropoietic factor[165,184] is part of the mechanism by which increased red blood cells are formed following the administration of Co. At higher doses Co specifically destroys the alpha cells in the pancreatic islets of the rabbit[170] and is accompanied by temporary elevation of blood-sugar levels.

At the enzyme level, certain Co ammine complexes are reported[185] to activate the serum inhibitor of hyaluronidase, and Co^{++} accelerates the hydrolytic rate of certain enzymes for peptide derivatives.[186]

[177] M. Simesin, *Arch. intern. pharmacodynamie,* **62,** 347 (1939).

[178] L. Untersteiner, *Arch. intern. pharmacodynamie,* **41,** 411 (1931).

[179] A. E. Houk, A. W. Thomas, and H. C. Sherman, *J. Nutrition,* **31,** 609 (1946).

[180] C. L. Comar and G. K. Davis, *J. Biol. Chem.,* **170,** 379 (1947); C. L. Comar, G. K. Davis, and R. F. Taylor, *Arch. Biochem.,* **9,** 149 (1946); C. L. Comar, G. K. Davis, and R. F. Taylor, *J. Nutrition,* **32,** 61 (1946).

[181] D. K. Kasbekar, W. V. Lavate, D. V. Rege, and A. Sreenivasan, *Biochem. J.,* **72,** 374 (1959).

[182] H. E. Stokinger and W. D. Wagner, *A.M.A. Arch. Ind. Health,* **17,** 273 (1958).

[183] G. Brewer, *Am. J. Physiol.,* **128,** 345 (1940).

[184] T. E. Brown and H. A. Meineke, *Proc. Soc. Exptl. Biol. Med.,* **99,** 435 (1958).

[185] M. B. Mathews et al., *Arch. Biochem. Biophys.,* **35,** 93 (1952).

[186] K. R. Rao et al., *J. Biol. Chem.,* **198,** 507 (1952).

Co is reported to act synergistically with antibiotics,[187] both in vitro and in vivo (mice). If antibiotics also synergize the action of Co, it could possibly explain the enhanced sensitivity to Co seen in some exposed individuals.

Industrial Experience. From present published and unpublished reports, it appears that Co is capable of giving rise to at least three clinical manifestations of industrial Co poisoning. A dermatitis of the allergic sensitivity type, has been described by Schwartz et al.[188] Distribution of the eruption was most marked at points of friction and seemed to be related to the abrasive nature of the dust (Co-cemented tungsten carbide (WC)). The same type of allergic dermatitis was described by Schwartz to occur among Co-alloy workers and also among Finnish pottery workers handling Co clay.[188] Fairhall has described a "carboloy itch" of rather low prevalence among workers in the cemented WC industry.

Associated with the Co-cemented WC industry both in the United States and in Europe is a pneumoconiosis. Miller et al.[189] have reported 3 cases of "peculiar pulmonary reactions with hyperglobulinemia" among workers of the United States tungsten-carbide tool industry, who became asymptomatic upon removal from the dust exposure. Lundgren et al.[189] have reported a fatal case, which they describe on x-ray and pathological evidence as a chronic interstitial pneumonitis with pulmonary insufficiency, in a worker exposed to Co-cemented WC and TiC, and have referred to other European cases in this industry based on x-ray findings.

An allergic basis has been considered for the respiratory difficulties found in 25 per cent of workers in a Co ore-reduction plant using a new ore-treatment process (see Uses and Industrial Exposures, page 1023). In a recent plant survey by the USPHS[189] an asthmalike respiratory disease was seen in addition to other respiratory symptoms. Exposures were not only to Co, however, but to varying amounts of Ni, As, NH_3, SO_2, and sulfuric acid mist. Neither the physical examination data, pulmonary function measurements, chest roentgenograms, nor clinical laboratory results correlated well with symptoms described. Accordingly, no specific chemical substance(s) could be incriminated.

It would appear in view of the unusual and presently unexplained manifestations of Co that consideration should be given to the possibility of synergistic action of Co with certain antibiotics now in widespread use.[187]

Clinical Experience. A relatively large literature exists on the hemopoietic effects of Co and its use in the therapy of certain of the more unusual anemias, particularly certain of the anemias of pregnancy and infancy in which response to

[187] R. Pratt, J. Dufrenoy, and L. A. Strait, *J. Bacteriol.*, **55**, 75 (1948).

[188] L. Schwartz, S. M. Peck, K. E. Blair, and K. E. Markuson, *J. Allergy*, **16**, 51 (1945); L. Schwartz, L. Tulipan, and D. J. Birmingham, *Occupational Diseases of the Skin.* 3rd ed., Lea & Febiger, Philadelphia, 1957.

[189] C. W. Miller, M. W. Davis, A. Goldman, and J. P. Wyatt, *Arch. Ind. Hyg. Occupational Med.*, **8**, 453 (1953); K. D. Lundgren and H. Ohman, *Arch. pathol. Anat. u. Physiol. Virchou's* **325**, 259 (1954); K. D. Lundgren and A. Swensson, *Acta Med. Scand.*, **145**, 20 (1953).

Fe is poor or ineffective.[190] On the other hand, Co had a goitrogenic effect in a 17-month-old girl,[191] and produced pronounced activation of acne (folliculitis) in patients taking vitamin–Fe–mineral supplements containing Co;[192] the dermal manifestations disappeared when Co was discontinued. Rostenberg and Perkins[193] cite evidence for believing Co may be as important a sensitizer as is Ni; when workers' skin is irritated by use of solvents or by skin cleansers, the development of Co sensitization is favored. It is currently believed that the eczematous sensitization by Co results from a combination of Co with some protein constituent, which leaves the site of its formation by way of the lymphatics. Both Ni and Co are cross-reactive and it is not presently possible to purify either Co or Ni sufficiently to eliminate one from the other. Accordingly it is not presently possible to tell whether this is a true cross-sensitization or whether the two metals cross-react because of contained traces of each other. The more probable view, however, is that because of similar chemical reactivities, the Co–Ni cross-sensitization represents a true cross-reaction.

Co has long been known to enhance the antibiotic activity of penicillin[187] both in vitro and in vivo. The action occurs at microgram levels of inorganic Co. The effect appears to be highly specific as salts of Ni, Mn, Pt, Ir, Fe, Zn, Sr, Cd, Li, Cu, Ag, Au, and Bi had no effect and Co alone exerted no protective action. More recently the Co complex of hydrocortisone was more effective on a dosage basis as an antiinflammatory agent in rats than was hydrocortisone alone.[194]

Therapy. The depressing effect on blood pressure, observed as $Co(NO_3)_2$ is given by vein to mice, can be diminished to a slight extent by BAL, according to Dalhamn.[195]

6. Hygienic Standards of Permissible Exposure

No threshold limit for Co has been recommended by an official agency in the United States. Elkins [196] has suggested a limit of 0.5 mg. of Co/cu. meter. This value appears to have considerable evidence to substantiate its choice both from animal and human experience; animals exposed to Co metal fume at 2 mg. of Co/cu. meter daily for 3.5 months revealed no pulmonary or other type lesions specifically referable to Co, whereas longer exposure to Co-metal blend at 20 mg. of Co/cu. meter resulted in pulmonary lesions ascribable to Co. On the other hand, Elkins quotes a case of pulmonary involvement resulting from exposure to 2 mg. of

[190] R. G. Holly, *J. Am. Med. Assoc.,* **158,** 1349 (1955); *Obst. Gynecol. Survey,* **5,** 562 (1955); B. L. Coles, and U. James, *Arch. Disease Childhood,* **29,** 85 (1954); ibid, *Lancet,* **75,** 79 (1955).

[191] J. S. Robey, P. M. Veazey, and J. D. Crawford, *New Engl. J. Med.,* **255,** 955 (1956).

[192] C. M. Sidell, J. G. Erickson, and J. E. McCleary, *Calif. Med.,* **88,** 20 (1958).

[193] A. Rostenberg and A. J. Perkins, *J. Allergy,* **22,** 466 (1951).

[194] J. W. Fisher, *Federation Proc.,* **19,** 158 (1960).

[195] T. Dalhamn, *Acta Pharmacol. Toxicol.,* **55,** 75 (1953).

[196] H. B. Elkins, *Chemistry of Industrial Toxicology,* Wiley, New York, 1959.

Co/cu. meter. In view of the suspected allergic sensitization response associated with exposure to Co, its compounds, and related industrial products, a threshold limit of 0.1 mg. of Co/cu. meter would appear to offer a larger factor of safety. From the data available at this time, no assurance can be given that 0.1 mg. of Co/cu. meter will provide freedom from unfavorable response to Co and its compounds for all persons.

7. Flammability

Only certain forms of Co metal are pyrophoric. The form prepared by reducing the oxides in H_2 is pyrophoric; when cobalt oxide is reduced with NH_3 so that it contains 14 to 16 per cent O_2, it glows when exposed to air. Pyrophoric Co is a black powder that burns brilliantly when in contact with O_2 or air.

COPPER, Cu

1. Source and Production

Cu occurs free in nature as native copper, as well as in several oxide ores, carbonate, and sulfide ores the last predominating. The processes employed for its extraction vary with the nature of the ore. In general they involve the operations of crushing, concentration by flotation, roasting at 600 to 800°C. to remove part of the sulfur, and smelting at 1100 to 1600°C. for the production of matte. The latter may contain 30 to 35 per cent copper as sulfide, together with iron sulfide. Treatment of molten matte with air in a converter removes the sulfur, and the addition of a siliceous flux permits the iron to be removed as an oxide slag. The product is brittle blister copper, which contains some sulfur, iron, and at times precious metals. It may be further refined with charcoal or coke in a reverberatory furnace, or cast into anodes and subjected to electrolytic refining, using as a bath an acid solution of copper sulfate. In the latter process, precious metals are recovered from the slime that collects at the bottom of the tanks.

2. Uses and Industrial Exposures

Because of its outstanding qualities as a conductor of electricity, Cu is widely used in the manufacture of such equipment. Marketed as castings, sheets, rods, tubing, and wire, it is used in chemical apparatus and equipment, cooking utensils, roofing, coinage, and for the formation of many Cu compounds. Cu forms many important alloys, which modify its properties for special purposes. *Be–Cu alloys* (1 to 2 per cent Be) find use in parts where toughness and resistance to vibration are needed. *Brass* containing 18 to 40 per cent Zn melts below Cu. *Bronze*, 3 to 8 per cent Sn, 11+ per cent Zn, and some Pb, is used for works of art because of its color, fusibility, and "patina." *Gun metal* contains 10 per cent, and *Bell metal* 20 to 24 per cent Sn. *German silver* contains 19 to 44 per cent Zn, 6 to 22 per cent Ni, and shows none of the copper hue. *Aluminum-bronze,* containing 5 to 10 per cent

Al, resembles gold in color. *Silicon-bronze* contains not more than 5 per cent Si and Cu_2Si, is less conductive than Cu, but more tenacious. *Phosphor-bronze* contains 9 per cent Sn with 0.5 to 1 per cent P and is used for machine parts. *Manganese-bronze* contains 30 per cent Mn. In many of these alloys the metals are partly in the form of chemical compounds such as Cu_2Zn_3 and Cu_3Sn. Cu is used in anti-fouling paints, in insecticides and fungicides, as an algicide, and as a rot-proofing agent (Cu naphthenate, Cu dimethyl gloxime) for fabrics.

Mining and extraction processes are attended by the hazard of silicosis and exposure to dusts and fumes, such as those of Sb, As, Bi, Pb, Se, Te, Zn, Hg, Ag, and Au. SO_2 is encountered in charging the roasting furnaces and acid fumes are present near the electrolytic cells. Welders may be exposed to Cu fumes, and producers of Cu-containing paint and economic poisons may be exposed to toxic dusts.

3. Physical and Chemical Properties

Cu has three distinctive properties that largely determine its use: toughness, high electric conductivity (second to Ag), and its toxicity for lower forms of life. The presence of a few tenths per cent impurities, especially arsenic, greatly decreases electric conductivity, but alloying with 1 to 2 per cent Be reduces conductivity only a small proportion. Cu forms two series of salts, Cu^+ and Cu^{++}; both valence types form complex ions that are stable.

TABLE 9

Physical and Chemical Properties of Cu and Some of Its Compounds

	Chemical symbol	At. or mol. wt.	Sp. gr.	M.p., °C.	B.p., °C.	Solubility
Copper	Cu	63.57	8.92	1083	2336	Insol. hot or cold H_2O Sol. HNO_3, hot H_2SO_4 V. sl. sol. HCl, NH_4OH
Copper sulfate	$CuSO_4 \cdot 5H_2O$	249.69	2.284	$(-4H_2O)110$	$(-5H_2O)150$	31.6 g./100 g.(0°) 203.3 g./100 g.(100°) 15.6 g./100 g. methyl alc.(18°)
Copper oxide	CuO	79.54	6.4	1026(dec.)		Insol. hot or cold H_2O Sol. a. NH_4Cl, KCN

4. Analytical Determination

Following collection of air samples by conventional methods, Cu may be determined chemically by a dithizone procedure.[197] This procedure is particularly applicable when Cu is in the presence of Pb, Zn, or other dithizone-complexing metals. Another chemical method both sensitive and highly specific that has

[197] G. H. Bendix and D. Grabenstetter, *Ind. Eng. Chem. Anal. Ed.*, **15**, 649 (1943).

proved satisfactory for air containing Cu fumes and dust employs Neo-cuproine (2,9-dimethyl-1,10-phenanthroline).[198] The method is also applicable to amounts of Cu found in biological samples following suitable preparation of the specimen. The spectrographic method is particularly suited to the determination of Cu;[199] indium is recommended as the ideal internal standard.

5. Physiological Responses

Cu is widely distributed and has diverse functions in plants and animals. Despite a concentration in certain plants and animals severalfold greater than that of other trace metals, minute amounts in water are toxic for algae, bacteria, and other unicellular forms. Cu is an essential trace metal for animals and man because it is required for the formation of erythrocytes and hemoglobin,[200-202] as well as oxidative enzymes such as catalase, peroxidase, cytochrome oxidase, and many others.[203]

Symptoms of Poisoning. Inhalation of dusts, fumes, and mists of Cu salts results in congestion of the nasal mucous membranes, ulceration with perforation of the nasal septum on occasion, and sometimes pharyngeal congestion. Cu metal fumes or salts leave a sweetish, metallic taste, cause salivation, nausea, vomiting, gastric pain, hemorrhagic gastritis, diarrhea, cramps in the calves, and, terminally, muscular rigor and prostration. In chronic exposures the liver, kidneys, and spleen may be injured and anemia may develop, although chronic poisoning like that from lead is unknown. Contact of the skin with Cu salts may result in an itching eczema of papulovesicular nature, which may be due to sensitization; contact of the eye will result in conjunctivitis, edema of the eyelids, and ulceration and turbidity of the cornea.

Acute Effects. Cu is a gastrointestinal-tract irritant, irritating the nerve endings in the stomach and initiating the vomiting reflex in the higher animals. The emetic dose for man is 500 mg. of the sulfate in a single administration. The sulfate in 50 mg./kg. dose was toxic for rabbits, a species that cannot vomit; 159 mg./kg. of the carbonate was lethal; 420 mg./kg. of the carbonate was lethal for goats.[204] Cu salts are highly toxic by vein; 25 mg. of $CuCl_2$ injected subcutaneously in a 1 per cent solution in rabbits caused local necrosis and loss in weight;[205] 4 to 5 mg./kg. is the lethal intravenous dose of the sulfate for rabbits and 2 mg./kg. is lethal for guinea pigs.[206]

[198] A. R. Gahler, *Anal. Chem.,* **26**, 577 (1954).

[199] G. E. Heggen and L. W. Strock, *Anal. Chem.,* **25**, 859 (1953).

[200] E. B. Hart, H. Steenbock, C. A. Elvehjem, and J. Waddell, *J. Biol. Chem.,* **65**, 67, (1925).

[201] C. A. Elvehjem, *Physiol. Revs.,* **15**, 471 (1935).

[202] G. E. Cartwright, *Blood,* **2**, 111 (1947).

[203] C. R. Dawson and M. F. Mallette, *Advances in Protein Chem.,* **2**, 179 (1945).

[204] R. Attia, Egypt Ministry Agr., Tech. Sci. Serv. Bull. No. **105** (1931); *Chem. Abstr.,* **26**, 243 (1932).

[205] H. Eggers, *Beitr. Klin. Tuberk.,* **47**, 373, (1921); *Chem. Abstr.,* **16**, 2359 (1922).

[206] Spector, ed., *Handbook of Toxicology,* Vol. I. Nat'l Res. Council, Apr. 1955.

Chronic Effects. The sulfate fed ad libitum in the diet of rats at a level of 500 p.p.m. caused retarded growth; 4000 p.p.m. caused starvation and death.[207] Copper poisoning in animals leads to injury of the liver, kidneys, and spleen;[208] access of sheep to salt licks containing 5 to 9 per cent $CuSO_4$ caused a sudden onset of anorexia, hemolytic anemia, icterus, and hemoglobinuria followed by death in a day or two; 40 to 49 g. or less was the estimated dose over a 25- to 86-day period. At necropsy the liver, kidneys, and spleen showed severe degenerative changes. The livers of 3 sheep contained 0.28, 0.53, and 1.47 mg./g. of Cu in comparison with normal Cu content of from 0.005 to 0.042 mg./g. Pulmonary edema is reported to result in dogs following short exposures to Cu acetate dusts.[209] Dusts of Cu stearate cause an acute reaction but the deposits are subsequently absorbed and result in no changes of chronic nature.[210]

Metabolism. The absorption of Cu from the gastrointestinal tract, as with all essential elements, is limited. The average total daily intake of Cu in the adult is 2 to 2.5 mg. of which 0.25 mg. (range, 0.08–0.7 mg.) is excreted in the urine. Increased Cu intake does not appreciably affect urinary excretion. Fecal excretion of Cu is 10 to 20 times that of urine. Cu retention is about 0.8 mg. from a daily intake of 2.5 mg.

Kehoe, Cholak, and Story,[211] after reviewing the work of others on the distribution of copper in the tissues of men and animals, reported the content of copper in various human organs, in mg. per 100 g. of fresh tissue, as follows: liver, 0.710; kidney, 0.166; heart, 0.190; spleen, 0.085; lung, 0.110; muscle, 0.125; stomach, 0.107; intestines, 0.110; rib, 0.37 to 0.47; and long bone, 1.19. The content in the liver is increased (2.40 mg. per 100 g.) in infancy. It is moderately increased in hemochromatosis and Wilson's disease, while in Laennec's cirrhosis as much as 27.4 mg. of copper per 100 g. of liver has been found. Relatively large quantities (0.22 to 0.68 mg. per 100 g.) are found in the brain. Blood contains 0.114 mg. per 100 g., with slightly more in the cells than in the plasma.

Cu has been shown to exhibit a number of interesting antagonisms: to have a beneficial effect in combating Mo toxicity in cattle,[212] to provide better growth of chicks and poults fed a diet containing 300 p.p.m. Mo,[213] and to reverse the toxicity of Zn in rats.[214]

Divalent Cu has been shown to bind histamine to serum albumin.[215] For some time metal fume fever has been placed on an allergic basis; in view of the physiological role of histamine in allergy, it is intriguing to postulate that the Cu-binding

[207] R. Boyden, V. R. Potter, and C. A. Elvehjem, *J. Nutrition,* **15,** 397 (1938).
[208] J. B. Boughton and W. T. Hardy, *Texas Agr. Expt. Sta. Bull.* No. **499** (1934).
[209] I. Brodskii, *Arch. Gewerbepathol. Gewerbehyg.,* **5,** 91 (1933).
[210] P. T. Knies, *J. Lab. Clin. Med.,* **25,** 726 (1940).
[211] R. A. Kehoe, J. Cholak, and R. V. Storey, *J. Nutrition,* **19,** 582 (1940), **20,** 85 (1940).
[212] R. W. Muir, *Vet. J.,* **97,** 387 (1941).
[213] F. H. Kratzer, *Proc. Soc. Exptl. Biol. Med.,* **80,** 483 (1952).
[214] S. E. Smith and E. J. Larson, *J. Biol. Chem.,* **163,** 29 (1946).
[215] A. C. Andrews and T. D. Lyons, *Science,* **126,** 561 (1957).

of histamine to a serum protein provides a means whereby a local action (fume inhalation) may become systemic (fume fever).

Industrial Experience. Few instances of illness from exposure to copper and its compounds have been reported. Schiötz[216] has described 7 cases of copper fever in a paint factory where red CuO was being pulverized. Men exposed to dusts of CuAc have complained of sneezing, coughing, digestive disorders, and fever.[209] Welders exposed to Cu fume gave similar complaints.[217] Metal fume fever was reported among workers in a Norwegian Cu plant.[218] Of 20 cases, 17 worked with Cu; 3 worked with Zn or brass. Typical fume fever was seen in 13; 7 had mild or indistinct symptoms. In 8 cases, serum Cu exceeded 160 µg. per 100 cc.; average serum Cu was found to be 126 ± 11 µg. per 100 cc. Normal serum Cu values were found among the 3 Zn workers who had fume fever. An instance in which Cu metal in a finely divided state caused nasal ulceration and bleeding is reported from Germany.[219]

6. *Hygienic Standard of Permissible Exposure*

No threshold limit for exposure to Cu fume or any Cu compound has been set by an official agency in this country. A limit of 0.1 mg./cu. meter for Cu fume has been tentatively suggested[217] on the basis that repeated exposures to fume levels up to 0.4 mg./cu. meter caused no complaints, and brief exposures to concentrations of 1 to 3 mg. of Cu/cu. meter resulted in no other discomfort than a sweet taste. The Public Health Service Drinking Water Standard for Cu is 1 mg./liter. Considering the toxicity of Cu by inhalation to be 10-fold that by mouth (see Section 5, Physiological Response) it can be calculated that 0.1 mg./cu. meter provides at least a 5-fold factor of safety before discomforting effects of Cu become manifest.

GALLIUM, Ga

1. *Source and Production*

Although Ga was discovered by a French chemist in 1875 from spectroscopic examination of sphalerite (ZnS ore), interest in Ga in the United States did not begin until 1915 when it was accidentally discovered as Ga–In beads in Pb slabs that had been exposed to sunlight. Because Ga exists as no concentrated ore body but rather associated in small amounts in zinc blende (ZnS ore), bauxite (0.002 to 0.008 per cent Ga), some Sn ores (0.01 to 0.05 per cent Ga), and coals from North England, commercial production was slow in developing. It was not until 1943 that the commercial production of Ga (and In) began, using the residues of leached Zn concentrates. In another process, Ga derived from the Bayer Al process liquors.

[216] E. H. Schiötz, *Proc. Intern. Congr. Ind. Med. 9th Congr., London,* **1948,** p. 798 (1949).
[217] N. E. Whitman, Bethlehem Steel Co., personal communication, 1957.
[218] J. R. Woldgren and O. Gorbatow, *Nord. Med.,* **41,** 764 (1949).
[219] R. Luchsinger, *Z. Unfalluerhüt. med. Berufskrankh.,* **44,** 274 (1951).

In England processes have been developed for extracting Ga from flue dust. It is estimated that 1000 tons of Ga per year are lost to the air in the British Isles from the burning of Ga-rich British coals. A new process of recovering Ga from ores (such as Al) treats subdivided ore with a current of HCl or HBr in substantially anhydrous conditions at 700 to 950°C.

2. Uses and Industrial Exposures

No large-scale use of Ga has yet been developed. Ga has been used in high-temperature thermometers (500 to 1200°C.), in plating optical mirrors, in replacement of Hg-quartz lamps in atomic and astrophysical research, in semiconducting materials as antimonides and arsenides, as a substitute for Hg in rectifiers, as a liquid metal, and as a sealant for glass joints and valves in vacuum equipment. Although Ga alloys with most metals to give improved structural characteristics, in general such improvement can be obtained more economically with other metals. Recently some proposed uses of GaP are: solar-cell power plants for space stations, and small rugged electronic parts for missiles, satellites, and future space probes.

Exposures are limited at present to small operations and few personnel. Hazards would appear to be related to substances associated with the production of Ga and to constituents with which Ga is combined, such as As, P, and Te. Welding with these Ga compounds would appear to present risks, but as yet no determination of the hazard has been made.

3. Physical and Chemical Properties

The physical and chemical properties of Ga and some of its compounds are listed in Table 10.

TABLE 10

Physical and Chemical Properties of Ga and Some of Its Compounds

	Chemical symbol	At. or mol. wt.	Sp. gr.	M.p., °C.	B.p., °C.	Solubility	Vapor pressure, mm. Hg
Gallium	Ga	69.72	Solid, 5.904 (29.6°) Liq., 6.095 (29.8°)	29.78	1983	Insol. hot or cold H_2O Sol. alkalies	0.004 (1000°)
Gallium oxide	Ga_2O_3	187.44	6.48	Tr. 600		Insol. hot or cold H_2O Sl. sol. hot acids and alkalies	
Gallium monotelluride	GaTe	197.33	5.44	824			

Ga normally forms trivalent salts, but under reducing conditions exhibits a valence of 1+ or 2+. Ga has a normal electrode potential of —0.56, thus occupying a position between Zn and In. Ga resembles Al with respect to apparent resistance to oxidation; a globule will retain its metallic luster at red heat, apparently due to a thin, impervious film of oxide, which protects the metal against further oxidation. Ga is relatively resistant to alkalies and acids but when treated with aqua regia it forms the chloride slowly. A sesquioxide, Ga_2O_3, exists in three forms, γ, β, and a hydrate; a suboxide, Ga_2O, is also known. Ga is amphoteric, forming soluble metal gallates[220] in which Ga acts as a nonmetal, and soluble Ga salts, as a metal. Recent industrial interest in Ga compounds has centered on GaAs, GaP, GaSb, and GaTe (sphalerite-type compounds); preparation and properties of single crystals have been described by Wolff et al.[221] There is some evidence of commercial research interest in alkyl and aryl Ga compounds such as trimethyl- and triphenyl Ga, and dimethyl Ga hydride.

4. Analytical Determination

Separation of Ga from normally associated elements is an essential preliminary step in analysis. The only method tested critically for quantitative determination of Ga in tissues is that of Dudley,[222] which depends upon the fluorescence of 8-quinolinate. The method was found satisfactory by Brucer,[223] who recommended 10-minute mechanical shaking of the $CHCl_3$ extractions. A promising method that bears further study is that of Onishi,[224] which utilizes preliminary concentration by ether extraction in $6M$ HCl followed by estimation with Rhodamine B; sensitivity to 0.01 μg. of Ga is claimed for qualitative detection. The spectrochemical properties of Ga are discussed by Ahrens.[225] Other proposed methods are either of limited application or sensitivity.

5. Physiological Response

A comprehensive study of the physiological and pharmacological actions of Ga, which included all necessary work for the proper understanding of its possible therapeutic applications as radioisotope Ga^{72}, was reported by Brucer in 1953[223] and should be consulted for details. Ga citrate was the compound used because of its stability upon injection; Ga lactate had been previously used by Dudley for a similar reason in a number of toxicological studies performed by him and his associates. The usual inorganic salts proved unsatisfactory for toxicological work because of hydrolysis to the hydroxide, which in a colloidal form is immediately lethal by vein.[226]

[220] Not to be confused with the same term that refers to the salts of the organic compound, gallic acid, 3,4,5-trihydroxybenzoic acid.

[221] G. Wolff, P. H. Keck, and J. D. Broder, Phys. Revs., 94, 755 (1954).

[222] H. C. Dudley, J. Pharm. Exptl. Therap., 95, 482 (1949).

[223] M. Brucer, Radiology, 61, 534 (1953).

[224] H. Onishi, Anal. Chem., 27, 832 (1955).

[225] L. H. Ahrens, Spectrochemical Analysis. Addison-Wesley, Cambridge, Mass., 1954.

[226] H. C. Dudley and M. D. Levine, J. Pharm. Exptl. Therap., 95, 487 (1949).

Toxicity in Animals. Feeding of $GaCl_3$ at dietary levels of 10, 100, 500, and 1000 p.p.m. as Ga for 26 weeks, however, resulted in no discernible toxic response in rats; no Ga could be found in the soft tissues; traces in bone indicating negligible absorption. Although rats exposed to nebulized solutions of $GaCl_3$ (25 to 125 mg./cu. meter) for from 0.5 to 4 hours experienced pulmonary consolidation bilaterally, it was felt to be no different than the response to inhaled NaCl acidified to an equivalent degree; no Ga passed from the lung. The LD_{50} of *Ga lactate* for rats intravenously was 47 mg. of Ga/kg.; subcutaneously, 121 mg. of Ga/kg.; for rabbits intravenously, 43 mg. of Ga/kg.; subcutaneously, 97 mg. of Ga/kg., indicating a moderate to high toxicity of Ga lactate by injection for rodents.

Ga *lactate* was considerably more toxic for larger than for smaller animal species on a mg. of Ga/kg. of body weight basis;[227] doses greater than 20 mg. of Ga/kg. were uniformly lethal to dogs and goats. Subcutaneous edema commonly occurred at the injection site. Ga *citrate*, also, was more toxic for the larger species. LD_{50} (10-day) in mg. of Ga/kg. were: mouse, 600; rat, 220 to 240; rabbit, 45; dogs and goats, 10 to 15. Brucer[223] found the LD_{50} (10-day) of single doses of *Ga citrate* by vein in dogs to be 18.2 mg. of Ga/kg.; for rats, however, the LD_{50} by vein was greater than 220 mg. of Ga/kg. A summative toxicity was noted in dogs injected with repeated small doses, (2.5 to 5 mg. of Ga/kg.); frequently dogs died from a total dose less than that from a single intravenous dose; Dudley had noted a similar summative effect in rabbits at a somewhat lower level (1 mg. of Ga/kg.) as well as in dogs.

Lethal or near lethal doses in dogs provoked vomiting soon after injection, diarrhea, bloody stools, food refusal, and weight loss. Urine samples contained many erythrocytes, granular and hyaline casts, and 1 to 4+ albumin, but no sugar. Hemoglobin values were reduced as much as 40 per cent in some dogs, unchanged in others, generally paralleling the degree of debility. Renal damage resembled that from Hg poisoning microscopically. Blood urea nitrogen values in dogs that died were between 190 and 300 mg. per cent. Lymph nodes swollen to 3 times normal showed marked nuclear fragmentation and necrosis of the lymphoid elements, with polymorphonuclear infiltration so dense as to suggest a purulent reaction (this histologic picture has no counterpart in human pathology from Ga). Aplastic changes occurred in the bone marrow. Dudley *et al.*[226] reported also photophobia and blindness in rats; other toxic signs resembled those given by Brucer. Both concluded that renal injury was the cause of death after 4 days but that other unknown changes were the cause of earlier deaths from Ga.

A Ga alloy consisting of 26.5 per cent Ga, 43.5 per cent Cu, and 26.5 per cent Sn, subcutaneously implanted in the abdominal wall of guinea pigs, produced caseation necrosis.[228]

[227] H. C. Dudley, K. E. Henry, and B. F. Lindsley, *J. Pharm. Exptl. Therap.*, **98**, 409, (1950).

[228] J. L. Hartley and N. O. Harris, *Air Univ. School of Aviation Med.*, Rept. No. **58-148** Randolph AFB, Nov., 1958.

Toxic Manifestations in Man. Clinical trials of Ga[72] mixed with stable Ga have been made in man; because of low specific activity, Ga[72] was given with amounts of stable Ga believed to be somewhat toxic. Ga[72] given in doses of 10 to 100 microcuries intravenously were usually accompanied by drowsiness, and later doses produced anorexia, nausea, vomiting. A frequent symptom was a peculiar metallic taste during treatment. Mild degrees of folliculitis or maculopapular rash developed, usually after the fifth or sixth injection; generalized exfoliative dermatitis occurred in some. Itching was severe and edema of the skin and subcutaneous tissue occurred in some. All dermatological manifestations cleared within a few days after the last of the Ga injections. Hematological changes, consisting of relative lymphopenia followed by decrease in total leucocytes, were frequently so severe as to require discontinuance of treatment. Later, platelets decreased in number. Anemia frequently followed early, progressive drop in erythrocytes. Recovery from bone-marrow depression usually occurred, but not in all cases. It is believed that (a) although radiation was the chief cause of bone-marrow depression, stable Ga toxicity was contributory, (b) the skin rash is due to stable Ga, and (c) the gastrointestinal symptoms are due to both forms.

Metabolism. Distribution and excretion patterns have been determined in animals[223,229] and man[223] both for stable Ga and for the radioisotopic Ga[72]. Preliminary data are available for Ga[67].[223] Stable Ga as the *lactate* injected intravenously in rabbits is transported wholly by the plasma, whence it is cleared in 1 day to be fixed in appreciable quantities in the liver and kidneys for more than 1 month. Ga goes to the bone in less than 4 hours, where it is retained for longer than 3 months. Excretion of Ga is by the urine; more than 85 per cent of a near toxic dose is thus eliminated. In the rabbit 100 mg. of Ga/kg. is retained more than 10 days. By subcutaneous injection only low blood levels are attained. Subcutaneous injection of Ga[72] *citrate*[230] is cleared more rapidly from the blood than the lactate and is found chiefly in the bone and kidney, other soft tissues receiving negligible amounts; rat, rabbit, and dog followed the same distribution pattern. While 30 to 48 per cent is absorbed by the skeleton, 30 to 55 per cent is being excreted in the urine; 1–2 per cent of the dose goes to the gastrointestinal tract. A similar distribution-excretion pattern for Ga[72]-labeled $GaCl_3$ was found by Brucer[223] in the rat; a greater amount of data on tissue deposition sites and time intervals are reported. The total 96-hour fecal excretion was 7.8 per cent for the *chloride*, 3.3 per cent for the *citrate*. The amount of Ga deposited in bone from Ga citrate is independent of route of administration (subcutaneous or intravenous) but is dependent on dose between 1 and 15 mg. of Ga/kg.[231] Ga content of bone decreased little from 1 day to 6 months. No significant change in urinary excretion of Ga was occasioned by giving NH_4Cl or BAL.

[229] H. C. Dudley, G. E. Maddox, and H. C. LaRue, *J. Pharm. Exptl. Therap.*, **96**, 135 (1949).

[230] H. C. Dudley, J. I. Munn, and K. E. Henry, *J. Pharm. Exptl. Therap.*, **98**, 105 (1950).

[231] H. C. Dudley and H. H. Marrer, *J. Pharm. Exptl. Therap.*, **106**, 129 (1952).

Ga alizarinate, a particulate lake or chelate, gave a distribution unlike that of the chloride:[223] about 50 per cent was retained by the liver, 20 per cent by bone; excretion was about equal between urine and feces, comprising at 10 days about 50 per cent of the dose.

Mechanism of Action. Study of the intimate mechanism by which Ga exerts its injurious action has not been made, but studies by Dudley *et al.*[227] of changes in constituents of blood and urine in rabbits given Ga citrate indicated renal poisoning, which ultimately resulted in an uncompensated acidosis. Nonprotein nitrogen, urea nitrogen, and blood sugar increased, accompanied by albuminuria and glycosuria, reduction of CO_2-combining power of the plasma, decrease in CO_2 content, and a decrease in pH of the whole blood; serum sodium and potassium decreased without increase in urinary sodium.

6. Hygienic Standards of Permissible Exposure

No limit has been set by any official agency for Ga or any of its compounds.

GERMANIUM, Ge

1. Source and Production

Ge is not found in the free state, but always in combination with other elements as a mineral, such as in the principal minerals, argyrodite (Ag_8GeS_6, 6 to 7 per cent Ge) and germanite ($7CuS \cdot FeS \cdot GeS_2$, 8.7 per cent Ge). Ge is most commonly distributed within the structure of other minerals, most abundantly in the sphalerite (ZnS) in quantities of less than 1 per cent. Some Ag and Sn ores contain Ge as do some coals; a germaniferous lignite in the United States has been reported. Oak and beech humus in one locality in Germany contains 70 p.p.m. Ge. Ge is taken up by cereals, especially oats, from Ge-bearing soils. The principal domestic source of Ge is from the residues of cadmium derived from Zn ores.

In the domestic recovery of Ge from ZnS concentrate, S is first removed by roasting which converts the Ge, Cd, and Zn to oxides. These are sintered with C and NaCl producing volatile chlorides of Ge, Cd, and Pb, which are collected in an electrostatic precipitator. Following removal of most of the Cd and Pb, the Ge is further purified by distillation as $GeCl_4$. For production of high purity GeO_2, crude $GeCl_4$ is multiple-distilled in the presence of excess Cl_2 and HCl. The pure $GeCl_4$ is hydrolyzed in cold water, rinsed free of chloride, and dried as GeO_2. Reduction to the metal is done by heating in an electric furnace with H_2 at 650 to 675°C. until H_2O ceases to form. Following complete reduction, the temperature is raised to 1000°C. to melt the Ge metal powder. When extreme metal purity is desired, multiple-zone melting and fractional crystallization are used, by which the impurities concentrate at one end of the bar.

2. Uses and Industrial Exposures

Most of the uses for which Ge is noted arise from its property as a semiconductor: in transistors, diodes, and rectifiers, replacing and exceeding the functions of vacuum tubes. More than 45 million Ge diodes, transistors, and rectifiers were produced in the United States in 1958. Their development in World War II was responsible to a large extent for the "revolution in electronics." Diodes are used in various types of electronic circuits, including radio, television, telegraphy, and telephony, multiposition switching, and in voltage-multiplier circuits. Small-sized diodes are made by the millions; large sizes for power generation are also produced. Transistors similarly find use in radio and television, as amplifiers and oscillators. Because Ge is only one of several substances with semiconductor properties, other substances such as Si are competing with Ge as transistor diode material. Various combinations, As and Sb alloyed with Al, Ga, and In also are being developed for these purposes. Lenses of Ge have been made for industrial infrared work. $MgGeO_3$ is used as a phosphor. A glass, in which Ge replaces Si, possesses a high refractive index and is used in wide-angle camera lenses and in microscope objectives. Ge is used as a catalyst, in the hydrogenation of coal, and particularly as an extremely low-temperature catalyst.

Industrial exposures are to the dusts and fumes of the metal and its oxide during the separation and purification from the ore concentrate. Cd, Pb, and Zn are concomitant exposures during this process. In the electronics industry, exposures are confined to metal fume from welding or multiple-zone melting.

3. Physical and Chemical Properties

The physical and chemical properties of Ge and some of its compounds are listed in Table 11.

TABLE 11

Physical and Chemical Properties of Ge and Some of Its Compounds

	Chemical symbol	At. or mol. wt.	Sp. gr.	M.p., °C.	B.p., °C.	Solubility
Germanium	Ge	72.60	5.35(20/20°)	958.5	(2700)	Insol. hot or cold H_2O, alkalies Sol. hot H_2SO_4, aq. reg.
Germanium dioxide	GeO₂	104.6	4.703(18°)	1115		4.47 g./liter(20°) 10 g./liter(100°) Sol. acids, alkalies
Germanium tetrachloride	GeCl₄	214.4	1.879(20°)	−49.5	83.1	Dec. hot or cold H_2O V. sol. dil. HCl Sol. alc., ether
Germane	GeH₄	76.63	3.43 g./liter (−142°) Liq., 1.532	−165	−90	Insol. hot or cold H_2O Sol. NaClO Sl. sol. hot HCl

Ge metal has the unusual property of being highly transparent to infrared light. This, coupled with a high index of refraction, is used to advantage in making unique optical elements. Crystalline Ge is a true semiconductor; traces of impurities of 1 part in 10 million such as Sb, Sn, and P will lower its electric resistivity from 50 ohm-cm. to 0.1 to 10 ohm-cm. Ge forms alloys with many metals, producing solid solutions or intermetallic compounds, which expand on solidification. Ge is not attacked by O_2 of the air or by HF (aq.) or HCl (aq.), but reacts slowly with concentrated nitric and sulfuric acids. Alkaline H_2O_2 or aqueous NaClO rapidly oxidize Ge. Ge forms two series of salts, divalent and tetravalent, which resemble stannous and stannic compounds in many ways. The tetravalent is the more common, stable form. Ge does not form salts with the common inorganic oxy acids. In its tetravalent compounds Ge is in coordinate valence with neighboring atoms. Ge forms a series of hydrides, germanes, by the reaction of HCl (aq.) and Mg_2Ge. They possess pungent odors, are combustible, and often explosive. Water reacts with NaGe to give $(GeH)_x$, a dark brown powder that explodes when dried. A series of halogenated germanes are known, which are either colorless gases or liquids. Although Ge does not form a carbide, it forms strong bonds with C atoms to form a variety of organo compounds such as tetraalkyl- and tetraarylgermanes. They are volatile, stable, colorless substances, insoluble in H_2O, but soluble in organic solvents.

Germanium Hydrides (Germanes). Ge forms a series of hydrides, which correspond chemically to the methane series of hydrocarbons and to silanes (silicon series of hydrides). The germanes, prepared by the action of a reducing acid on a Ge alloy such as Mg_2Ge, have properties given in Table 12.

TABLE 12

Properties of Germanium Hydrides

Ge hydride	Synonym	Chemical symbol	At. or mol. wt.	M.p., °C.	B.p., °C.
Germane	Germanomethane Germanium tetrahydride	GeH_4	76.63	−165	− 90
Digermane	Germanoethane	Ge_2H_6	151.25	−109	29
Trigermane	Germanopropane	Ge_3H_8	225.86	−105.6	110.5

The germanes are insoluble in hot and cold water but soluble in sodium hypochlorite (aq.) and slightly soluble in hot HCl (aq.).

No recent studies of the toxicity of the germanes appear to have reached the open literature, but early, limited toxicity tests (30 and 60 minutes) of GeH_4 indicated 610 mg./cu. meter for 30 minutes was ultimately lethal to a mouse, whereas 480 mg./cu. meter for 60 minutes was lethal to a mouse but not a guinea pig, and 310 mg./cu. meter for 60 minutes produced only dyspnea in a rabbit, not death.[232] Hemoglobinuria was noted in the animals, thus resembling AsH_3 in this respect. Thus, on an increasing toxicity scale would appear to be GeH_4, AsH_3, and SnH_4,

[232] F. Paneth. and G. Joachimoglu, *Ber. deut. chem. Ges.,* **57**, 1925 (1924).

but such comparisons on limited data are very uncertain. The higher hydrides of Ge are noted[233] to be less toxic than GeH_4 but no supporting evidence is given.

4. Analytical Determination

A molybdenum-blue method has been used by Rosenfeld[234] for the determination of microgram amounts of Ge in biological samples. Although the procedure does not appear to have been reported in detail, a critical examination of the optimal conditions for analysis of microgram quantities with an error of 5 per cent is given by Shaw and Corwin.[235] A spectrographic procedure, adapted for the analysis of Ge in Si alloys, with a standard deviation of ± 8 per cent has been reported.[236]

5. Physiological Response

A brief, but critical, review of various controversial aspects of the toxicology of Ge, that were current up to 1953, is included in an extensive report of acute and chronic toxicity of Ge in rats by Rosenfeld and Wallace.[237]

Acute Toxicity in Animals. The maximal tolerated single intraperitoneal dose of neutralized (pH 7.3) GeO_2[238] approximated 600 mg./kg. for the rat; the corresponding minimal lethal and "absolute" lethal doses were 700 and 1200 mg./kg., respectively. The estimated intraperitoneal LD_{50} was 750 mg./kg. Similarly administered Ge-mannitol and Ge-glycerin were tolerated at 140 mg./kg. Repeated intraperitoneal weekly doses for 8 weeks of GeO_2, Ge-mannitol, and Ge-glycerin at 100 mg. of compound/kg. were without manifest effect on rat growth, blood picture, or on critical organs. Particulate GeO_2 repeatedly injected into the peritoneal cavity of rats in 200 mg./kg. doses failed to produce fibrotic nodules or necrosis. Similarly, single subcutaneous injections of neutralized GeO_2 at doses of 150 to 600 mg./kg. produced no evidence of local tissue damage or systemic toxicity; nor did repeated weekly doses of 50 mg./kg. in young rats.

Most of the deaths from GeO_2 were rapid, occurring within 12 hours; all occurred within 24 hours. Gross tissue changes consisted of edematous, hemorrhagic lungs, multiple petechiae in walls of small intestine, and peritoneal effusion. The peritoneal fluid was voluminous, and rich in protein. Death was usually preceded by hypothermic shock (rectal temperatures sometimes below 80°F.), and almost complete respiratory depression. No effects on the nervous system or on the blood were found. Loss of retroperitoneal fat was shown due to alkalinity of Na_2GeO_3 and not to Ge per se.

[233] H. C. Dudley, *Arch. Ind. Hyg. Occupational Med.*, **8**, 528 (1953).

[234] G. Rosenfeld, *Arch. Biochem. Biophys.*, **48**, 84 (1954).

[235] E. R. Shaw and J. F. Corwin, *Anal. Chem.*, **30**, 134 (1958).

[236] M. C. Gardels and H. H. Whitaker, *Anal. Chem.*, **30**, 1496 (1958).

[237] G. Rosenfeld and E. J. Wallace, *Arch. Ind. Hyg. Occupational Med.*, **8**, 466 (1953).

[238] When GeO_2 dissolves in water, it forms germanic acid, H_2GeO_3,pH3.7; a "neutralized solution of GeO_2,pH7.3," is a mixture of H_2GeO_3 and Na_2GeO_3; Na_2GeO_3 has a pH value of 11.7.[235]

Chronic Toxicity in Animals. Neutralized GeO_2 incorporated in the diets of rats for 14 weeks stimulated growth at 10 p.p.m. to the same extent that 1000 p.p.m. depressed growth, but the latter level caused a 50 per cent mortality; 100 p.p.m. was without effect. GeO_2 at 100 p.p.m. in the drinking water resulted in a similar 50 per cent mortality in 4 weeks, indicating a high toxicity of soluble Ge salt by mouth. No hematological or gross pathological changes were found to account for the mortality. Thus, Ge compounds are more toxic by mouth than parenterally, a rather unusual phenomenon for non-volatile substances. The lack of gross tissue change in animals dying from oral intake of Ge would indicate interference with normal utilization of food in some manner, and to which some rats can subsequently adapt.

An acquired tolerance occurred from repeated sublethal doses of neutralized GeO_2, Ge-mannitol, or Ge-glycerin that protected all rats from an otherwise LD_{75}; the tolerance lasted 6 weeks. Muller[239] had reported a similar tolerance previously.

Topically, neutralized GeO_2 in 0.67 per cent solution produced no irritation or other cutaneous systemic effects following twice daily applications for 2 weeks on the shaved skin of the backs of rats.

Rosenfeld and Wallace[237] could not confirm the hemapoietic effect of Ge nor an increased oxidative metabolism (fat).

Metabolism. Inorganic Ge as neutralized GeO_2 is readily absorbed following small (a few milligrams) oral, subcutaneous, intramuscular, or intraperitoneal doses in the rat.[234] Blood levels attained somewhat less than 10 μg./g. within a few hours after administration; following a single small dose, Ge leaves the blood stream within a few hours. It may be true, as claimed, that Ge is transported via the blood unbound to protein, but the evidence is not convincing (Ge unbound to protein precipitated by 10 per cent trichloroacetic acid). Ge is widely distributed throughout the body and is not selectively retained in any tissue; after 1 week following a moderate intraperitoneal dose (40 mg. of Na_2GeO_3/kg.) all Ge that had gone to the tissues had disappeared. Ge was not found as a microconstituent in the 15 tissues examined of untreated rats, thus confirming its disappearance after administration, despite its presence in cereals and grains (the analytic method used had a sensitivity of 0.5 p.p.m.). Ge is excreted by both the kidneys and the gastrointestinal tract, but the kidney is the chief excretory organ (80 vs. 13 per cent).

Studies of excretion of elemental Ge^{71} and $Ge^{71}O_2$ in the rabbit and dog showed elimination chiefly via the urine[240] as was found in the rat.

Essentially the same type of metabolic activity occurred with neutron-activated Ge^{71} and $Ge^{71}O_2$ dusts by inhalation.[233] Rats exposed to (unmeasured) atmospheres of Ge^{71} for 1 hour at respirable particle sizes (1.7 and 0.45 μ mean diameter resp.) showed moderately fast elimination of the Ge^{71} from the lungs; 52 per cent disappeared in 1 day, 82 per cent in 7 days, the longest period of ob-

[239] J. H. Muller, *J. Pharm. Exptl. Therap.*, **42**, 277 (1931).
[240] H. C. Dudley and E. J. Wallace, *Arch. Ind. Hyg. Occupational Med.*, **6**, 263 (1952).

servation. The kidneys and liver contained only tracer amounts at 7 days. $Ge^{71}O_2$ was cleared somewhat more rapidly from the lungs, kidneys, and liver (79 per cent in 1 day), and only tracer amounts remained in these organs at 4 and 7 days.

Mechanism of Action. Only very general impressions of the manner in which Ge acts in the animal body exist; no studies using modern biochemical techniques have been made. The main action of Ge on which there is no dispute is its profound disturbance in water balance at high exposure levels leading to dehydration, hemoconcentration, fall in blood pressure, and hypothermia.[237] The means by which tolerance to Ge is acquired has not been studied, nor has the cause of the high toxicity of neutralized GeO_2 by mouth.

It is important to note that with the exception of Hueper[241] and Bailey *et al.*[242] no description of the microscopic changes in tissues has been made, although the Ge studies[233,234,237,240] would have led to a greater understanding of its mechanism of action, particularly as to its unexpectedly high toxicity by mouth. Bailey, *et al.*, in recording lesions from Ge, described histological changes from GeO_2 in rabbits subcutaneously injected with doses 1 to 5 per cent of that maximally tolerated. Hueper's histological descriptions concerned guinea pigs that died shortly after a single intraperitoneal injection (325 to 600 mg.), the rabbits dying from acute or subacute subcutaneous injections (repeated, 8 to 12 mg. of GeO_2/kg.). Bailey *et al.* found effects on the heart muscle with edematous or atrophied cells; similar edematous changes in the parenchymal cells of the liver and kidneys; the spleen, hyperplastic; bone marrow, fatty. Hueper found similar changes in these organs but noted no striking changes in the digestive tract with the exception of deposits of a greenish brown pigment. Hueper called attention to the marked cathartic action of toxic amounts of GeO_2 resulting in pronounced disturbance of water metabolism.

6. Hygienic Standards of Permissible Exposure

Despite considerable toxicological investigation, no air standards for Ge or any of its compounds have been set by an official agency. Information has been limited to experimental study of the element and GeO_2 chiefly; no worker exposure experience has been reported.

7. Flammability

Ge hydrides (germanes) are thermally unstable, much more so than the corresponding silanes (SiH_x). $(GeH)_x$, a dark brown powder from the reaction of sodium–germanium alloy (NaGe), decomposes explosively if dried in air. GeH_4 does not react with O_2 below 230°C., nor explode below 330°C.; digermane (Ge_2H_6) is more explosive.

[241] W. C. Hueper, *Occupational Med.*, **4**, 209 (1947).
[242] G. H. Bailey, P. B. Davidson, and C. H. Bunting, *J. Am. Med. Assoc.*, **84**, 1722 (1925).

INDIUM, In

1. *Source and Production*

In is widely distributed, but in minute quantities (generally 0.1 per cent and lower), in the lithosphere; it is also found in the hydrosphere. Mineral sources are most commonly dark sphalerite (ZnS), marmatite, and christophite (FeS:ZnS). In also occurs in small quantities in Sn ores, siderite, and in Mn and W ores. Ga is often associated with In in Zn and Sn ores. Many sulfide ores of Cu, Fe, Pb, Co, and Bi contain small quantities of In. Flue dusts of Zn smelters constitute the largest commercial source of In, in some cases amounting to more than 1 per cent In. Other commercial sources are derived from plant residues and dross from the refining of Zn, Pb, and Cd.

Current annual production is something over 10 tons and could be increased to 100 tons if all residues and dusts now discarded were processed for In. In is extracted metallurgically by several methods, depending upon the source and nature of the associated metals. One of the In recovery processes[243] starts with a lead-smelter charge such as concentrates, ores, dusts, and residues bearing In (a few thousandths per cent), and a small amount of Zn mixed with crude Pb bullion, which is likewise very low in In, but contains about 0.25 per cent Sn. The Sn oxidizes the In, which becomes concentrated with the Sn in the dross and removed; it is melted with $PbCl_2$, which converts In to $InCl_3$, in which form it is recovered as a slag. $ZnCl_2$ is then added to provide better solvent action for $InCl_3$, to reduce the melting point of the mix (300°C.), and to minimize volatilization of the In. The crude mix is wet-ground with H_2SO_4 and HCl, filtered to remove $PbSO_4$ and other insoluble sulfates and chlorides, yielding a filtrate that contains about 25 g./liter of In. The solution is purified by adding metallic In, which displaces Pb, Sn, and other metals. Purification of In is done electrolytically on Zn rods. The metallic-sponge In is washed, briquetted, melted under paraffin, and cast into bars of 99.99 per cent pure In.

2. *Uses and Industrial Exposures*

One of the principal uses of In is in surface protection of metals, particularly for prevention of corrosion of Cd- and Pb-types of bearing alloys, which also utilize the property of surface "wettability" of In. In is used in many alloys (dental and bearings) because of its capacity to increase hardness. It may be added to Be and Cu alloys to increase hardness. In increases the resistance to alkaline attack of solders (e.g., 25 per cent In, 37.5 per cent Pb, and 37.5 per cent Sn). In is used in making glass-to-glass or glass-to-metal seals; added to graphite as lubricant; used in motion picture screens, cathode oscillographs, and mirrors, because of property of equal reflectance of all spectral wavelengths for color. In oxides and sulfides are used in the electric industry because of fluorescence and

---

[243] T. R. Wright, *Metal. Research Min. Met.*, **26**, 559 (1945).

photoconductivity properties. In_2O_3 (sesquioxide) is used with sulfur to make yellow-colored glass. An increasing use of In is in transistors and infrared detectors; here In is combined with some metalloids such as Sb, As, Ge. The metals are zone-refined so that the intermetallic compounds are essentially 100 per cent pure (99.999999+ per cent).

Industrial exposures to In and its compounds, with the exception of those associated with its extraction and purification, are concerned chiefly with its use in plating and in making transistors, transistor materials (metal compounds with In), glasses, and other minor uses involving relatively small quantities. The hazards from the cyanide-plating baths are not well defined.[244] The hazard from fume in welding and soldering transistor parts containing In and associated metals has not been studied, nor that connected with glassmaking. The metallurgic extraction of In from its various sources has associated hazards of Pb, Sn, and Zn.

3. *Physical and Chemical Properties*

The Indium Corp. of America has compiled an exhaustive, annotated, but unindexed, bibliography of the metal dealing with all aspects of its technology.

The physical and chemical properties of In and some of its compounds are listed in Table 13.

TABLE 13

Physical and Chemical Properties of In and Some of Its Compounds

	Chemical symbol	At. or mol. wt.	Sp. gr.	M.p., °C.	B.p., °C.	Solubility
Indium	In	114.82	7.30	156.4	2000 ± 10	Insol. hot or cold H_2O Sol. acids V. sl. sol. NaOH
Indium sesquioxide	In_2O_3	277.52	7.179	Dec. 850	Volatile	Insol. cold H_2O Amor. s. a. Cr. i. a.
Indium trichloride	$InCl_3$	221.13	4(3.46)	586 Subl. <400	Volatile 600	V. sol. cold H_2O Sl. sol. alc., ether
Indium sulfate	$In_2(SO_4)_3 \cdot 9H_2O$	679.86			Dec. 250	V. sol. cold H_2O

In resembles Sn. It is extremely soft and malleable with a Brinell hardness of less than 1. In the electromotive series it appears between Fe and Sn, and does not decompose water at boiling temperature. It is stable in air but when heated it burns with a non-luminous blue-red flame yielding In_2O_3. The surface of In remains bright up to its melting point; above this, it forms a film of oxide. Indium

---

[244] C. P. McCord, S. F. Meek, G. C. Harrold, and C. E. Heussner, *J. Ind. Hyg. Toxicol.*, **24**, 243 (1942).

forms three series of compounds, mono-, di-, and trivalent, of which the last is the most common. The trichloride is deliquescent; the sulfate is hygroscopic. Indium compounds of greatest present industrial interest are the intermetallic compounds, the antimonides, arsenides, and germanides.

4. *Analytical Determination*

In has been determined in biological sources by direct analysis in the medium quartz spectrograph without prior chemical separation,[245] following wet- or dry-ashing at 450°C. Th is used as the internal standard, RbCl the spectroscopic buffer. When the spectral line 3079 A is used, a sensitivity on the electrode of 0.05 mg. of In with the standard error of an analysis of ±10 per cent is reported. Harrold and co-workers[246] describe a procedure for ashing, isolation, and determination of In in excreta and food applicable to the 1- to 25-μg. range, based on colorimetry of a quinalizarin lake. No discussion is given to the possible interfering action of other ions.

5. *Physiological Response*

Parenterally administered, In is one of the most toxic of elements; doses of fractions of a mg./kg. of certain In salts intravenously given to rodents are lethal. This was found by McCord *et al.*[244] in 1942 and confirmed later by Downs *et al.*[245] McCord *et al.* provide a brief review of the literature on In toxicity to 1942 in addition to supplying toxicity data on $InCl_3$ and $In_2(SO)_3$ in rabbits, rats, guinea pigs, and on skin tests in man. These latter produced no evidence of skin irritation. $InCl_3$ administered parenterally to rats, rabbits, and dogs as a partial citrate complex had an acute lethal dose range of from 0.33 to 3.6 mg. of In/kg.[245] Parenterally given In_2O_3 was far less toxic: single intraperitoneal doses of 955 mg./kg. or greater were required to be uniformly fatal to rats 9 days after injection, but doses of 546 mg./kg. were fatal to an occasional animal. In_2O_3 intravenously given to rabbits was moderately toxic, 100 mg./kg. being slowly fatal (1 to 3.5 months); dosages of 35 to 68 mg./kg. resulted in delayed severe reactions beginning at 3 weeks. On the other hand, multiple intravenous injections cumulative to a total as high as 424 mg./kg. resulted in no deaths of rabbits in 35 days or 209 mg./kg. at 103 days, the time of sacrifice. In_2O_3 when ingested by rats was practically non-toxic; 8 per cent In_2O_3 incorporated in the diet for 3 months had no discernible effect on growth, mortality, or tissue morphology; by contrast, $InCl_3$ caused a slight growth depression at 2.4 per cent and a marked depression at 4 per cent in the diet over a 3-month period.

Inhalation of the insoluble In_2O_3 dust (particle diameter 0.5 μ MMD)[247] 4

[245] W. L. Downs, J. K. Scott, L. T. Steadman, and E. A. Maynard, *Univ. Rochester Atomic Energy Rept.* **UR-558**, Nov. 1959.

[246] G. C. Harrold, S. F. Meek, N. Whitman, and C. P. McCord, *J. Ind. Hyg. Toxicol.,* **25,** 233 (1943).

[247] MMD = Mass median diameter.

hours daily for 3 months by rats resulted in an atypical inflammatory reaction, characterized by a paucity of cellular exudate, that persisted for 12 weeks beyond the termination of exposure (the longest period of observation). The lesion was further characterized by the absence of any change in severity and of any healing (fibrosis).[248]

In-treated Ag disks implanted subcutaneously, intramuscularly, and intraperitoneally induced only foreign-body reactions up to 156 days.[246]

A study by Harrold et al.[246] of possible In absorption in foods stored in roughly simulated In-plated ware indicated significant amounts of In (fractions of mg./ml.) were extracted by acidic foods, representing most of the plated In.

Gross signs of In poisoning on which there is agreement[244,245] are reduced food and water consumption, with accompanying weight loss, pulmonary edema, necrotizing pneumonia, blood damage, and degenerative changes in liver and kidneys. McCord et al.[244] described in addition, hind-leg paralysis in rabbits from $InCl_3$, degenerative changes in the genitourinary tract, localized convulsive motions, nosebleed, and increased reflexes. Histologically, most critical organs showed some degree of damage including the brain and heart, adrenals, and spleen. Downs et al.[245] noted a small and transient decrease in hemoglobin and hematocrit, and an elevated neutrophil and decreased lymphocyte count that persisted (rats). Renal histology, 24 hours postinjection, showed necrotic changes in the convoluted tubules surrounding the glomeruli but restricted to the proximal part of the proximal tubule; 1 week later tubular regeneration had occurred with typical and "atypical" epithelium.

Metabolism. Marked differences in distribution and excretion of In occur, depending on route of administration.[245] Orally, in rats, In_2O_3 is absorbed to the extent of a few p.p.m. in the critical tissues (lung, liver, kidneys, spleen, and bone); only from 0.2 to 0.4 per cent is absorbed from the gastrointestinal tract of dogs. On the other hand, repeated intravenous injections of In_2O_3 in rabbits resulted in the deposition of from several tens to several thousands p.p.m. in the tissues in conformity with the toxicity. Also, in agreement with the histological findings, were distributions in the tissues of In^{114m}; the liver content of In in rats given intraperitoneal doses ($InCl_3$) was more than twice that from intravenous doses; in contrast, In contents of kidney and pelts were half those by intravenous injection. Fecal excretion was about the same regardless of the route, but urinary excretion was 4-fold greater intravenously than intraperitoneally. Subcutaneous injection in rats of In citrate complex labeled with In^{114m} showed rapid (2 days) absorption of most (85 to 90 per cent) of the dose.[249] These findings are consistent with the high toxicity (2 mg./kg.) of $InCl_3$ and $In_2(SO_4)_3$ found by McCord et al.[244] upon subcutaneous injection. The metabolism of In^{114m} in the rat by intratracheal instillation was the same whether administered in the form of the hydroxide or the

    [248] L. J. Leach, J. K. Scott, L. T. Steadman, E. A. Maynard, and R. D. Armstrong, Univ. Rochester Atomic Energy Project, unpublished results, Feb., 1960.
    [249] G. A. Smith and J. K. Scott, *Univ. Rochester A.E. Rept.* **UR-507**, Oct., 1957.

citrate.[250] Tissue distribution was generally uniform: skin, muscle, and bone contained a large part of the dose, whereas the tracheobronchial lymph nodes, spleen, adrenals, kidneys, and liver contain higher concentrations of In. About 60 per cent of the dose had left the lung in about 2 weeks; thereafter the elimination rate slowed so that at 9 weeks only 82 per cent had left the lung. At this time, fecal excretion accounted for 50 per cent of the dose; the urine, 8 per cent.

No industrial experience has been reported of exposures to In or its compounds and no air standard has been set by any official agency.

IRON, Fe

1. Source and Production

About 80 per cent of past iron-ore production in the United States has come from the Lake Superior region; Alabama, New York, Utah, and Texas contributed the remainder. There is growing dependence on foreign ores and beneficiated domestic ore. Of foreign sources, Canada and Venezuela are the largest suppliers. About 110 million long tons of iron ore was mined in the United States in 1958, down about 50 million tons from 1957. The oxides, magnetite (Fe_3O_4), hematite (Fe_2O_3), and limonite ($2Fe_2O_3 \cdot 3H_2O$), containing 32 to 57 per cent Fe comprise the important United States ores. Other Fe-bearing minerals occur as carbonates, silicates, and sulfides.

Mining of Fe ore is mainly a problem of handling and transportation. Open-pit operations supply a large part of the iron ore. Underground mining, however, is practiced in year-round operations, whereas open-pit work is seasonal. Beneficiation of Fe ore varies from simple screening operations and washing to complex grinding, roasting, magnetic separation, or flotation. The metallurgy of Fe consists essentially of the passage of the ore, coke, and limestone through a blast furnace. As the mixture passes downward, C combines with O_2 of the ore, freeing Fe, which is tapped as a liquid from the furnace.

Present indications are that by 1975 about 37 per cent of domestic ore requirements will be met by foreign imports; Canada is expected to supply about 20 per cent, the remainder coming chiefly from Venezuela and Liberia.

Advances in technology consisted of new drilling methods, drillhole patterns, and explosive charges, which resulted in one of the largest blasts on record (1.1 million tons taconite with 99 per cent fragmentation). Research in Fe ore beneficiation resulted in pilot-plant tests of a high-tension separator. Study is being given self-fluxing sintering operations, pelletizing processes, and exploitation of titaniferous Fe ores. Commercial production of sponge Fe was begun in Mexico in 1958.

[250] G. A. Smith, R. G. Thomas, B. Black, and J. K. Scott, *Univ. Rochester Atomic Energy Rept.*, **UR-500**, July, 1957.

2. Industrial Exposures

Mining and handling of Fe ores provide exposures to dusts of SiO_2 and Fe oxides. CO is a hazard in the operation of blast furnaces for the production of pig iron. The use of fluorspar (CaF_2) in steelmaking gives rise to gases containing SiF_4 and other F-containing substances, and the manufacture of alloy steels introduces hazards attendant on the use of metals such as Cr, Mn, Ni, V, W, Mo, and Cu. "Pickling" of Fe containing As and P liberates arsine (AsH_3) and phosphine (PH_3). Certain grades of ferrosilicon used in steelmaking decompose with explosive violence on contact with moist air evolving various toxic gases, such as acetylene, H_2S, SiH_4, AsH_3, and PH_3. Fatal intoxications have occurred from such accidents during transportation, particularly at sea.[251] (See also Industrial Experience, page 1056.)

TABLE 14

Physical and Chemical Properties of Fe and Some of Its Compounds

	Chemical symbol	At. or mol. wt.	Sp. gr.	M.p., °C.	B.p., °C.	Solubility
Iron	Fe	55.85	7.86	1535	3000	Insol. hot or cold H_2O, alkalies, alc., ether Sol. acids
Ferric oxide	Fe_2O_3	159.70	5.24	1565		Insol. hot or cold H_2O Sol. HCl
Ferrous carbonate	$FeCO_3$	115.86	3.8	Dec.		67 mg./liter(25°) Sol. CO_2 soln.
Iron di-sulfide (pyrite)	FeS_2	119.98	5.0	1171	Dec.	4.9 mg./liter Sol. HNO_3 Insol. dilute acids
Ferrous sulfate	$FeSO_4 \cdot 7H_2O$	278.03	1.898	64 ($-6H_2O$)100 ($-7H_2O$)300		156.5 g./liter 486 g./liter(50°) Insol. alc.
Ferric sulfate	$Fe_2(SO_4)_3 \cdot$ $9H_2O$	562.04	2.1	Dec.		4.4 kg./liter Dec. hot H_2O Sol. abs. alc.
Ferric chloride	$FeCl_3 \cdot 6H_2O$	270.32		37	280–285	191 g./liter(20°) Inf. sol. hot H_2O Sol. alc., ether

3. Physical and Chemical Properties

The physical properties of Fe (see Table 14) are profoundly affected by impurities and by changes in temperature and treatment. Fe is superior to all other elements in magnetic properties. Fe, in almost pure state, loses its magnetism when removed from an electric field; when Fe contains small amounts of C, Co, or Ni

[251] C. E. Pellew, *J. Soc. Chem. Ind.*, **33**, 779 (1914); H. Hognested, *Med. Rev.*, **48**, 409 (1931); A. Jerwell and M. Haaland, *Med. Rev.*, **47**, 145 (1930).

the retention of magnetism is increased. When heated to 770°C., Fe loses its magnetism; on cooling, it regains this property. Fe undergoes a variety of structural changes (transformations) on heating that form the basis of the heat treatment of ferrous metals.

The principal compounds of Fe are ferrous (Fe^{2+}) and ferric (Fe^{3+}). In general, ferrous and ferric forms are mutually interconvertible. The oxidation potential against the normal hydrogen electrode for the ferrous form is —.43v., for the ferric form +.77v. Ferrous compounds are more stable than ferric when ionized, less stable when covalent. A large proportion of Fe salts are water-soluble; exceptions are carbonates, oxides, hydroxides, phosphates, sulfides, and ferrous fluoride. Fe of both valences tends to form complexes in which the commonest coordination number is 6. Fe has a strong tendency to combine with O, as in the form of OH groups, with resultant stable compounds, especially as chelates. Fe compounds exhibit marked catalytic activity in the promotion of oxidations, both of chemical and biologic importance. Fe forms several carbonyls; their properties and uses are discussed on pages 1106–1108.

4. Analytical Determination

α,α-Dipyridyl or 1,10-phenanthroline are the reagents of choice for the colorimetric determination of small amounts of Fe.[252] Keenan and Minderman[253] have determined the conditions necessary for accurate analysis of welding-fume samples containing high concentration of Fe.

Separation of small amounts of Fe is often desirable; this is best done by means of cupferron (ammonium salt of nitrosophenyl hydroxylamine).[252] Spectrographic determination of Fe is not particularly well suited for quantitative estimation of Fe in biological specimens.

5. Physiological Response

Orally, Fe salts of both valence forms are not strikingly toxic; on the other hand, when introduced directly into the blood stream Fe salts are highly and instantaneously toxic, particularly ferric salts. The acute lethal intravenous dose of $FeCl_3 \cdot 6H_2O$ for the rabbit is reported[254] to be 7.2 mg./kg. By comparison, the lethal dose of $FeSO_4 \cdot 7H_2O$ in this same species was 99 mg./kg.; both salts have about the same molecular weight. The corresponding lethal dose for ferrous lactate was 287 mg./kg. Subcutaneous lethal doses for these ferrous salts in the rabbit ranged from 189 to 2778 mg./kg. Oral lethal doses of these ferrous compounds for the rat ranged from 984 to 2778 mg./kg.

[252] E. B. Sandell, *Colorimetric Determination of Traces of Metals.* Interscience Publishers, New York–London, 1959, pp. 221, 550.

[253] R. G. Keenan and B. L. Minderman, *J. Ind. Hyg. Toxicol.,* **28,** 32 (1946).

[254] W. Starkenstein, in A. Heffter and W. Heubner, *Handbuch der experimentellen Pharmakologie,* Vol. 3, Sect. 2. Springer, Berlin, 1930, p. 1278.

Tne bulk of the toxicological information on Fe compounds is related to long-term exposures, and has been determined for the most part in man (see section on Industrial Experience, page 1056). The radiographic stippling seen in the lungs of rouge polishers, however, has been reproduced by Harding in rats[255] following intratracheal injection of rouge. Highly significant was the finding by Gardner and McCrum[256] showing that even severe pulmonary irritation from welding fumes (and gases) does not favor infection with tubercle bacilli nor occasion progressive tuberculosis. The daily administration of from 0.2 to 0.8 g. of ferrous chloride to dogs produced no noteworthy physiological changes.[257] For the toxicity of iron carbonyls see page 1108.

Metabolism. An excellent review of the absorption and metabolism of Fe has been made by Gubler.[258] The amount of Fe used daily by an adult for hemoglobin synthesis is 26 or 27 mg. Owing to extensive Fe reuse, this is far in excess of the daily dietary Fe requirement, which is about 1.2 mg. of Fe/day or the amount of Fe excretion, barring blood loss. Allowing a factor of 10 for low absorption of food Fe, 10 mg./day is considered a normal daily allowance for a man; 12 mg. for a woman. Ferrous Fe is generally absorbed from the gastrointestinal tract more readily than ferric Fe, presumably because of greater solubility of ferrous compounds. The amount of Fe absorbed is inversely proportional to the intake. A number of factors influence absorption; among them are acidity of the gastric juice, composition of diet (high phosphate and phytic acid reduce absorption), vitamin B6, Ca, Cu, and others. Regulation of intestinal Fe absorption is not completely understood but it appears to depend on body stores and requirements. An intestinal "acceptor" is postulated, because a single, large dose of Fe can block absorption for several days. The protein apoferritin is believed to be the Fe acceptor and is synthesized as need occurs. The steps in Fe absorption and storage are pictured by Granick[259] in the following way. Fe passes as ferrous Fe into the intestinal mucosal cell. In the cell Fe^{++} is converted to Fe^{+++} in ferritin. No Fe absorption as ferritin occurs until cell is physiologically "depleted." Fe, however, is withdrawn from ferritin as Fe^{++}, as need arises. Fe released directly into the blood stream is quickly oxidized by dissolved O_2 to Fe^{+++} whence it complexes with the specific Fe-transport B_1-globulin (transferrin). Fe in the plasma in excess of the binding capacity of this B_1-globulin is loosely combined in a nonspecific way with plasma proteins. It is this fraction that causes toxic reactions, and is rapidly removed from the plasma. Mean, normal, human plasma Fe levels are 129 μg. per 100 ml. for men, 110 μg. for women. This means that the B_1-globulin is normally about $1/3$ saturated with Fe on the basis of its constituting 3 per cent of the serum proteins, its

[255] H. E. Harding, *Brit. J. Ind. Med.*, **2**, 32 (1946).

[256] L. V. Gardner and D. S. McCrum, *J. Ind. Hyg. Toxicol.*, **24**, 173 (1942).

[257] R. Sanders, *Arch. exptl. Pathol. Pharmakol.*, **151**, 1 (1930); F. Hendych and K. Klinaesch, *Arch. exptl. Pathol. Pharmakol.*, **178**, 178 (1935).

[258] C. J. Gubler, *Science*, **123**, 87 (1956).

[259] S. Granick, *Ann. N. Y. Acad. Sci.*, **48**, 657 (1957).

molecular weight of 90,000, and its 2 atoms/molecule Fe-binding capacity. It is a very important fraction, however, although it comprises only 0.1 per cent of total circulating Fe and serves as a sensitive index of Fe metabolism. Possibly more use should be made of plasma-bound Fe as an index of exposure in workers exposed to excessive amounts of Fe dust (see Industrial Experience, below).

In addition to the above metabolic aspects, Fe functions in the transport of O_2 in hemoglobin in the blood, in myoglobin in the muscle, where it delivers its O_2 to the cytochrome system of the cells (a 4-Fe-porphyrin-containing complex of cytochrome oxidase, and the cytochromes a, b, c). Catalase and peroxidase are two other Fe-porphyrin enzymes present in nearly all tissues for the decomposition of peroxide oxygen.

Storage of Fe is divided into four main compartments. The normal human body contains 4.5 g. of Fe; of this, hemoglobin, which is almost entirely in the blood, comprises 72.9 per cent of total Fe; myoglobin, 3.3 per cent; parenchymal Fe (oxidative enzymes) 0.2 per cent; and storage Fe (ferritin, hemosiderin, and unaccounted Fe) 23.5 per cent. Most of the storage Fe is found in the liver, bone marrow, and spleen.

Industrial Experience. The hygienic significance of mottling of the lungs, siderosis, or as preferred by Sander[260] "iron pigmentation," is now considered that of a benign pneumoconiosis (see Vol. I, page 375) because it does not lead to fibrous proliferation, is of low order of severity, and usually requires 6 to 10 years of exposure before diagnosable roentgenographic changes occur.[261] The condition commonly occurs in electric-arc welders after years of exposure, but may occur in silver polishers or rouge users, according to McLaughlin et al.[262] Buckell et al.[263] x-rayed the lungs of 171 iron turners, found reticulation from Fe oxide present in 15 men. In 5 cases, workers had been at the trade for 20 years. The lung changes were moderate, symptoms were few, and only 1 man complained of shortness of breath, although 6 noted a tendency to cough. Healing of tuberculous lesions despite continuous exposure to Fe-oxide fumes, first noted by Sander,[260] has been repeatedly confirmed in man[264] and animals.[257] Physical examinations and tests of work capacity of welders with Fe pigmentation show that it causes little or no disability.[264] Gardner[265] regarded Fe oxide as a retardant of the development of conglomerate silicotic fibrosis. (See, however, discussion of lung fibrosis of steel

[260] O. A. Sander, *Am. J. Roentgenol.*, **58**, 277 (1947).

[261] O. A. Sander, *J. Ind. Hyg. Toxicol.*, **26**, 79 (1944); J. A. Groh, *Ind. Med.*, **13**, 598 (1944); W. Dreesen, H. P. Brinton, R. G. Keenan, et al., *U. S. Public Health Bull.* No. **298** (1947); E. P. Prendergrass and S. S. Leopold, *J. Am. Med. Assoc.*, **127**, 701 (1945).

[262] A. I. G. McLaughlin, J. L. A. Grout, H. J. Barrie and H. E. Harding, *Lancet*, **1**, 337 (1945).

[263] M. Buckell, J. Garrad, et al., *Brit. J. Ind. Med.*, **3**, 78 (1946).

[264] R. Fawcitt, *Brit. J. Radiol.*, **16**, 323 (1943); W. E. Fleischer, K. W. Nelson and P. Drinker, *J. Ind. Hyg. Toxicol.*, **27**, 94A (1945); K. Humperdinck, *Deut. med. Wochschr.*, **68**, 16 (1942).

[265] L. U. Gardner, *J. Ind. Hyg. Toxicol.*, **26**, 48A (1944).

foundrymen and hematite miners below.) Chemical data on the Fe content of the lungs of workers in the dusty trades are given by Gerstel.[266]

There is accumulating evidence, however, that Fe oxides may not act to prevent silicosis and tuberculosis in all types of exposure. Iron and steel foundrymen (England) subjected to high-temperature Fe-oxide fume along with silica were reported by McLaughlin and Harding[267] to develop "siderosis, silicosis, and mixed dust fibrosis." Moreover, a high incidence of bronchiogenic carcinoma was noted among the foundrymen. It would seem, in the face of apparent experimental evidence to the contrary,[268] that freshly formed, high-temperature-produced Fe oxides may act much the same way that Al_2O_3 and SiO_2 fumes act to produce Shaver's disease, a fibrotic condition of the lung.

Hematite pneumoconiosis in Cumberland Fe-ore miners has been reopened with a report by Faulds.[269] It is a progressive, massive fibrosis, appearing as a modified form of infective pneumoconiosis; it was considered to be tuberculosis when first described, but may not be so invariably. However, tuberculosis is the terminal event in most cases, probably because of a relative increase in the degree of silica exposure. An increased incidence of lung tumors in the Fe-ore miners was noted; of 238 necropsies, there were 24 cases of carcinoma, compared with less than $1/3$ this number in nonminers.

A definite correlation between serum Fe levels and radiographic findings in Fe workers in Bavaria has been reported;[270] serum Fe levels are considered as a reflection of partial release of Fe in the lungs. Serum Fe levels of 47 pneumoconiotic workers averaged 160 μg. per cent compared with a normal of 127 μg. per cent in healthy nonexposed workers. The increase was proportional to the degree of exposure to the ore dust and the lung changes.

Phosphine liberation was reported[271] in the ambient air about the machining of spheroidal graphite Fe; 6 p.p.m. PH_3 was measured near the tool point; at 20 cm., 0.8 p.p.m. No coolant was used in machining.

EDTA salts have been reported[272] to increase serum Fe levels immediately following injection into normal healthy individuals. On the same day about 12 times more Fe appeared in the urine than was normally present. After repeated administration of EDTA salts, Fe excretion decreased, indicating exhaustion of available Fe stores.

6. Hygienic Standards of Permissible Exposures

The American Conference of Governmental Industrial Hygienists has set a threshold limit of 15 mg./cu. meter for Fe-oxide fume. Weber,[273] however, is of the

[266] G. Gerstel, Arch. Gewerbepathol. Gewerbehyg., 10, 616 (1941).

[267] A. I. G. McLaughlin and H. E. Harding, A.M.A. Arch. Ind. Health, 14, 350 (1956).

[268] P. Gross, M. L. Westrick and J. M. McNerney, Diseases of Chest, 37, 35 (1960).

[269] J. S. Faulds, J. Clin. Pathol., 10, 187 (1957).

[270] H. J. Diesfeld, Arch. Gewerbepathol. Gewerbehyg., 15, 611 (1957).

[271] J. R. Bowker, Trans. Assoc. Ind. Med. Officers, 8, 50 (1958).

[272] J. Teisinger and V. Fiserova-Bergerova, Casopis lékáru českych, 96, 1605 (1957), (seen in abstract only).

[273] H. Weber, Am. Ind. Hyg. Assoc. Quart., 16, 38 (1955).

opinion that iron pigmentation occurs at exposure levels above 10 mg./cu. meter, and has used a limit of 5 mg./cu. meter in the control of fume exposures. This is in line with the earlier suggestion of Enzer and Sander[274] to reduce fume exposure below the point at which Fe accumulates in the lungs. Drinker, Warren, and Page[275] had previously suggested 10 mg./cu. meter as a limit, but later revised their opinion and suggested a limit of 30 mg./cu. meter on the basis that the lower limit is difficult to maintain in certain operations.

THE LANTHANONS (Rare Earth Elements)

The lanthanons comprise a group of fifteen elements of atomic numbers from 57 through 71 in which yttrium (At. No. 39) is usually included (see Table 15).

TABLE 15

The Lanthanons

Cerium subgroup, "light" lanthanons	Yttrium subgroup, "heavy" lanthanons	
Lanthanum, La	Samarium, Sm	Holmium, Ho
Cerium, Ce	Europium, Eu	Erbium, Er
Praseodymium, Pr	Gadolinium, Gd	Thulium, Tm
Neodymium, Nd	Terbium, Tb	Ytterbium, Yb
Prometheum (Illinium), Pm	Dysprosium, Dy	Lutetium, Lu
	(Yttrium), Y	

Because of their similar chemical and toxicological properties and because little or nothing is known of the toxicity or industrial hygiene significance of several of the members, they may best be considered as a group at the present time.

1. Source and Preparation

The lanthanons occur in monazite sand, which is a phosphate mineral of thorium and the lanthanons. After the lanthanons are separated from the thoria, they are recovered by precipitation to rid them of phosphate liberated during the monazite treatment. Recovery of lanthanons may be obtained as oxalates or as lanthanon sulfate double salts. Oxalate precipitation is complete, but double-sulfate precipitate leaves some of the Ce earths and most of the Y earths in the liquor. If hydrated oxides are desired the precipitates are boiled with NaOH which yields a granular hydroxide. Drying the hydroxides gives hydrated oxides. If oxides are desired, the oxalates or hydroxides are calcined. Separation of Ce, one of the lanthanons of greatest industrial use, depends on oxidation of Ce to the tetra-

[274] N. Enzer and O. A. Sander, *J. Ind. Hyg. Toxicol.*, **26**, 79 (1949).
[275] P. Drinker, H. Warren and R. Page., *J. Ind. Hyg.*, **17**, 133 (1935); P. Drinker, *U.S. Labor, Standards Bur.*, Spec. Bull. No. **5** (1941).

valent state, which has solubility properties that differ from most all of the lanthanons. Various methods are available for the isolation of Ce.[276]

The "heavy" metal lanthanons, Sm, 62, through Lutetium, 71, are separated by ion-exchange methods, the "light" elements La, 57, through Pm, 61, are separated by conventional crystallization procedures.

2. *Uses and Industrial Exposures*

It is only within the last few years that sufficient amounts of reasonably priced pure lanthanons have been available for industrial use. As a result, only a few of the heavy lanthanons are being used commercially, although others are being explored for industrial use. Among the lighter lanthanons, Ce, of course, has been used for many years in the gas-mantle industry; it has found use in lighter flints as the pyrophoric ferroalloy, as a component in tracer bullets and other pyrotechniques, and as a getter in thermionic tubes. Ce is also used alloyed with Al and Mg to improve mechanical properties of moving parts; in the ceramic field; the oxide is used as an abrasive superior to the best rouge; La is used similarly to some extent.

One of the uses for which the heavy lanthanons are being tried depends upon their neutron-absorbing capacity, such as for control rods in atomic reactors (lanthanons 62, 63, 64, and 66 have particularly high cross-section for neutron absorption). As control rod materials, they may be used in three ways: (*a*) as cermet (a dispersion of metal in a ceramic) control rods made by dispersing a lanthanon oxide in a metal matrix, in amounts as much as 40 per cent Gd_2O_3, in stainless steel, and Ti, (*b*) as metallic control rods by alloying, and (*c*) as solutions of lanthanon sulfates and nitrates in a reactor core. Dy has interest as a burnable "poison" in atomic fuel elements. Metallurgy is another area of use in spacecraft parts. Because of their capacity to withstand stresses and high temperatures, their use in ceramics and refractories is being explored. Their use as selective dehydration and dehydrogenation catalysts is being studied. The electronics industry is interested in the particular property of microwave transparency of certain lanthanon oxide–iron oxide combinations, as phosphor activators and other applications.

3. *Physical and Chemical Properties*

The physical and chemical properties of the lanthanon metals and their oxides are listed in Table 16.

The lanthanon metals have a silver-grey luster but tarnish quickly in air. The Ce-group metals are soft, but hardness increases with atomic number. Oxide-free metals are malleable, but oxide inclusions reduce this property. The metals are good heat conductors and moderate electric conductors. The metals react with hot and cold water and with dilute acids, but not readily with concentrated H_2SO_4.

[276] R. E. Kirk and D. F. Othmer eds., *Encyclopedia of Chemical Technologies*, Vol. 3. Interscience Publishers, New York, 1949, p. 640.

TABLE 16

Physical and Chemical Properties of Lanthanons and Their Oxides

Element	At. no.	At. wt.	Density	M.p., °C.	Oxidation states	Formula product of ignition in air	Formula wt.	Normal color	Approx. density	M.p., °C.
Lanthanons						*Oxides*				
Yttrium	39	88.92	4.472	1552	III	Y_2O_3	225.8	White	4.8–5.0	2410
Lanthanum	57	138.92	6.162	920	III	La_2O_3	325.8	White	6.51	2315
Cerium	58	140.13	6.768	804	III, IV	CeO_2	172.1	Yellow-white	7.3	2600
Praseodymium	59	140.92	6.769	935	III (IV)	Pr_6O_{11}	1021.5	Black		
Neodymium	60	144.27	7.007	1024	III	Nd_2O_3	336.5	Light blue	7.24	
Promethium	61	145.0	No stable isotopes							
Samarium	62	150.35	7.54	1052	(II), III	Sm_2O_3	348.7	Yellowish	7.43	
Europium	63	152.0	5.166	<900	(II), III	Eu_2O_3	352.0	White	7.42	
Gadolinium	64	157.26	7.868	1350	III	Gd_2O_3	362.5	White	7.4	
Terbium	65	158.93	8.253	1360	III (IV)	Tb_4O_7	747.7	Dark brown		
Dysprosium	66	162.51	8.556	>1400	III	Dy_2O_3	373.0	Off-white		
Holmium	67	164.94	8.799	>1400	III	Ho_2O_3	377.9	Yellow-green	7.8	
Erbium	68	167.27	9.058	1500–1550	III	Er_2O_3	382.5	Light yellow	8.6	
Thulium	69	168.94	9.318	1550–1600	III	Tm_2O_3	385.9	Rose-red	8.6	
Ytterbium	70	173.04	6.959	824	(II), III	Yb_2O_3	394.1	White (greenish)	9.2	
Lutetium	71	174.99	9.849	1650–1750	III	Lu_2O_3	398.0	White		

The metals are active reducing agents, ignite in air, are pyrophoric on filing (mixed metals), combine with halogens above 200°C., and with N_2 above 1000°C. Sulfides, carbides, silicides, and phosphides are formed by direct union on heating. The metals absorb H_2 forming interstitial hydrides of the approximate composition $RH_{2.8}$. Lanthanons alloy with almost all metals. Although all elements of the group are available in pure form, 99.99 per cent, Ce, La, Gd, Nd, Er, and Y are at present most used industrially.

The lanthanon salts precipitate readily at physiologic pH range because their isoelectric point is below pH7. La and other rare earths have been shown to form insoluble complexes with nucleic acids.[277] The stabilities of lanthanon complexes with EDTA have been determined at 20°C. and at an ionic strength of $\mu = 0.1$.[277]

4. Analytical Determination

The arc spectrographic method may be used to estimate the lanthanons provided a diffraction grating with adequate dispersion is used to separate the complex spectra of lanthanons.[278] More recently, flame photometry has been proposed as a more suitable method[279] because it is claimed to offer greater precision, speed, and reproducibility than the spectrographic method; use of an organic medium, hexone, for solution of the 2-thenoyltrifluoroacetone complex of the lanthanons increases the emissibility 20- to 100-fold. Claims are made for the quantitative estimation of a single element in complex mixtures of the lanthanons. Various chemical and physical methods of analysis of the lanthanons as well as separation techniques have been described in a recent report.[280]

5. Physiological Response

Much of the recent research investigations on the biological and medical aspects of the lanthanons through 1955 is recorded in the proceedings of a conference sponsored by the Medical Division of the Oak Ridge Institute of Nuclear Studies.[280] On available evidence, which is incomplete, the acute toxicity of the lanthanons, including Y, is low orally; high, parenterally. No sound judgment can be made regarding the toxicity by inhalation. Moreover, the parenteral toxicity varies greatly according to the associated nonmetallic component. An early report[281] on the acute parenteral LD_{50} in mice of organic salts of Ce, La, Nd, Pr, and Sm showed a relatively low toxicity: acetates, 10 to 13 g./kg. of body weight; propionates, 3.5 to 4.5 g./kg.; lactates, 5 to 10 g./kg.; maleates, 0.7 to 1 g./kg.;

[277] D. Laszlo, D. M. Ekstein, R. Lewin and K. G. Stern, J. Natl. Cancer Inst., 13, 559 (1952); E. J. Wheelwright, F. H. Spedding and G. Schwarzenbach, J. Am. Chem. Soc., 75, 4196 (1953).

[278] V. A. Fassel, J. Opt. Soc. Am., 39, 187 (1949).

[279] Chem. Eng. News, 41 (1959).

[280] Rare Earths in Biochemical and Medical Research, ORINS-12. Tech. Inform. Serv. Ext., Oak Ridge, Tenn., 1955.

[281] E. Vincke, Z. ges. exptl. Med., 113, 536 (1944).

the nicotinate salts of La, Pr, and Nd were said to be nontoxic at the limit of solubility (0.8 g./kg.). Later comparative acute intraperitoneal toxicity tests in mice and guinea pigs by Graca[282] of citrates and chlorides of Ce, La, Nd, and Pr showed the citrate complex to be more toxic by a large factor than the corresponding chloride; in mice, LD_{50} values for the citrates of the four lanthanons ranged from 78 to 147 mg./kg., whereas values for the chlorides were from 348 to 372 mg. of the salt/kg. of body weight; guinea pigs showed about a 2-fold difference in toxicity between the two types of salts, and were more susceptible to the lanthanons than were the mice. The intraperitoneal LD_{50} of $LaCl_3$ for rats, as determined by Kyker and Cress[283] was about half that found some years previously by Cochran, DuBois, et al.;[284] acute toxicity values obtained of Ce, Gd, Ho, Lu, Nd, Pr, and Y were all felt by these workers to be lower than previously found. As pointed out by Kyker and also by Graca, however, precise acute toxicity determinations of the lanthanons are difficult because of their protein-precipitating capacity and the unusually great influence of the nonmetallic components; toxicity values could be greatly modified, therefore, according to concentration and rate of the injected dose; animal strain differences may be another factor. Hart et al.[285] found that when La and Y were administered to man and animals, the uptake of La and Y by the tissues varied according to the compound administered, that is, whether ionized or as an EDTA complex.

The acute toxic response is similar for all lanthanons provided the nonmetallic moiety is the same. Marked differences in the acute response occur, however, between the citrate complex and the chlorides of the lanthanons.[281] The former is characterized by dyspnea and pulmonary edema; the latter, by liver edema and portal congestion, pleural effusion, and pulmonary hyperemia. Intraperitoneally injected chlorides show progressive local inflammatory changes ending in peritonitis with serous or hemorrhagic ascites. These findings have been confirmed in animals in a histopathologic study using Ce, La, Nd, Pr, Y, and a commercial lanthanon mixture, but were not seen in patients treated at lower levels.[286]

Rats administered either $Y(NO_3)_3$ alone or a lanthanon mixture as the nitrates and subsequently given whole-body irradiation showed mortalities that were interpreted as a result of synergistic interaction.[287] Wound healing was slowed in the presence of the lanthanons.

[282] J. G. Graca, E. L. Garst and W. E. Lowry, A.M.A. Arch. Ind. Health, 15, 9 (1957).

[283] G. C. Kyker and E. A. Cress, A.M.A. Arch. Ind. Health, 16, 475, (1957).

[284] K. W. Cochran, J. Doull, M. Mazur and K. P. DuBois, Arch. Ind. Hyg. Occupational Med., 1, 637 (1950).

[285] H. E. Hart, J. Greenburg, R. Lewin, H. Spencer, K. G. Stern and D. Laszlo, J. Lab. Clin. Med., 46, 182 (1955).

[286] C. H. Steffee, A.M.A. Arch. Ind. Health, 20, 414 (1959).

[287] Q. L. Hartwig, T. P. Leffingwell and G. S. Melville, A.M.A. Arch. Ind. Health, 18, 505 (1958); G. S. Melville, T. P. Leffingwell and Q. L. Hartwig, A.M.A. Arch. Ind. Health, 20, 15 (1959).

Although all of the physiological studies made thus far on the lanthanons have dealt with acute toxicity, one chronic study of the effects of repeated intraperitoneal injections of YCl_3 (20 mg./liter) has been made by MacDonald *et al.*[288] The fact that the rats tolerated without manifest systemic toxic effects a dose of 60 mg./kg. would indicate a low order of chronic toxicity for YCl_3 at least every 2 days for 5 months.

Y and six lanthanons, when applied to the rabbits' cornea denuded of epithelium, resulted in permanent opacification.[280] The intact epithelium, however, provided a complete barrier and protection for the cornea against these substances. Under industrial conditions in which fragments might penetrate the corneal epithelium severe damage could result if immediate removal of the lanthanons were not done. Table 17 shows the injuriousness of the lanthanons for corneal stroma in relation to other metals of interest.

TABLE 17

Comparison of Injuriousness of Certain Lanthanons and Other Metals for Corneal Stroma at Neutrality[280]

Metal	Opacification	Affinity for Corneal Stroma (Ca = 1)
Yttrium	4+[a]	62
Lanthanum	4+[a]	42
Cerium	4+[a]	
Praseodymium	4+[a]	55
Neodymium	4+[a]	78
Samarium	4+[a]	88
Gadolinium	4+[a]	70
Scandium	4+[a]	
Beryllium	4+[a]	4.3
Magnesium	0[b]	0.8
Calcium	0[b]	1.0
Strontium	0[b]	1.0
Barium	0[b]	0.5

[a] Permanent opacity.
[b] No permanent injury.

Preliminary study of EDTA treatment of corneas injured with Gd gave disappointing results, but more work is desirable.

Metabolism. The distribution of all fifteen lanthanons[280,289] as their radioisotopes in a dozen tissues and body sites of rats for periods of from 1 to 256 days

[288] W. S. MacDonald, R. E. Nusbaum, G. V. Alexander, F. Erzmirlian, P. Spain, and D. E. Rounds, *J. Biol. Chem.*, **195**, 837 (1952).

[289] P. W. Durbin, M. H. Williams, M. Gee, R. N. Newman, and J. G. Hamilton, *U. S. Atomic Energy Comm. Rept.* **UCRL-3066**, July, 1955.

by intramuscular injection showed: (a) deposition in the skeleton to increase progressively with atomic number from 20 per cent of the injected dose for La[140] to 70 per cent for Lu[177], (b) liver deposition to decrease correspondingly, and (c) urinary excretion to increase gradually but less regularly from 5 per cent for La[140] to 17 per cent for Lu.[177] Ce was marked by lowest per cent absorbed dose retained in skeleton, highest in liver, and highest fecal excretion of the fifteen isotopes. It is important to note that care must be taken in relating deposition data obtained on carrier-free radioisotopes, in which trivial amounts of substances are used, to amounts of industrial and toxicological interest. Y was classed as a bone-seeker on the basis of Y[91] studies[290] but was definitely found to have no selective affinity for bone when 60 mg./kg. was repeatedly administered for long periods.[288]

Distribution in mice of stable and radioisotopic La[140] as $LaCl_3$ (0.06 mg.) intravenously administered, was highest (24 per cent of dose) in liver and spleen (13 per cent) and low (2.8 per cent) in the skeleton; uptake by the tissues from intraperitoneally or subcutaneously injected $LaCl_3$ was very small, but when given intraperitoneally as the unionized versenate salt, better and more uniform distribution resulted.[291] Later distribution studies of V and La by these workers, in man and animals, showed that excretion and tissue distribution can be greatly modified by combination with complexing agents. Although Dudley[292] found EDTA-complexed Y[90] to go preferentially to bone, later work by Foreman and Finnegan[293] showed no such effect for Y[91] or indeed for Ce[144], Pm[147], Tb[160], and Tm[170] when CaEDTA was given *subsequent* to intravenous injections of isotopes.

Mode of Action. The complexing capacity of the lanthanons, particularly with proteins, is unquestionably the basis for their characteristic distribution and excretion pattern, varied toxicity by different routes and according to compound type, lack of mobility from an injection site, and their effect on the denuded cornea previously noted. This property is, in turn, a consequence of their high ionic charge, tri- or tetravalence. La, for example, forms insoluble complexes with nucleic acids at physiological pH values,[294] of which use has been made to analyze and purify nucleic acids and to electron-stain chromatin fibrils. Similarly, all salts of La, Pr, Nd, and Sm precipitate fibrinogen at concentrations of 0.01 to 1 per cent; higher salt concentrations partially resolve the precipitates and the chlorides of

[290] J. G. Hamilton, *Radiology,* **49,** 325 (1947) ; J. Schubert, M. P. Finkel, M. R. White, and G. M. Hirsch, *J. Biol. Chem.,* **182,** 635 (1950).

[291] D. Laszlo, M. Ekstein, R. Lewin, and K. G. Stern, *J. Natl. Cancer Inst.,* **13,** 559 (1952) ; H. E. Hart, J. Greenberg, R. Lewin, H. Spencer, K. G. Stern, and D. Laszlo, *J. Lab. Clin. Med.,* **46,** 182 (1955).

[292] H. C. Dudley, *J. Lab. Clin. Med.,* **45,** 792 (1955).

[293] H. Foreman and C. Finnegan, *J. Biol. Chem.,* **226,** 745 (1957).

[294] T. Cassperson, E. Hammarsten, and H. Hammarsten, *Trans. Faraday Soc.,* **31,** 367 (1935) ; J. N. Davidson and C. Waymouth, *Biochem. J.,* **38,** 39 (1944) ; E. Chargaff, E. Vischer, R. Doninger, C. Green, and F. Misani, *J. Biol. Chem.,* **177,** 405 (1949) ; F. Calvert, B. M. Siegel, and K. G. Stern, *Nature,* **162,** 305 (1948).

Pr, Nd, and Sm completely resolve them. Human serum proteins are incompletely precipitated at 0.25 to 0.7 per cent lanthanon salt, dissolve at concentrations greater than 1.5 per cent.[281] Bamann,[295] however, believes that the well-studied anticoagulant action of the lanthanons on blood is mediated through their capacity to act as phosphate acceptors and indirectly reduce prothrombin content of blood. The reasons for these actions are open to other interpretations.

In studying the means by which the lanthanons injure denuded cornea, Grant and Kern[280] found that the amount of lanthanon bound to the corneal stroma appeared to be governed by the amount of anion supplied by the tissue; methylation of the carboxyl groups of the tissue reduced the bonding of La (and Ca) to a very small fraction of that in normal corneas. Similarly, an increase in lanthanon-binding occurred when the basicity of the corneal components was reduced by treatment with formaldehyde.

Intravenous injection of the "light" lanthanons (Ce) produces fatty infiltration in the livers of rats,[296] characterized by an increase in neutral fat esters; total cholesterol and phospholipid of liver were unchanged. Ce was found associated almost wholly with the acid-soluble fraction of the liver and not with the lipid. The action of Ce was sex based, because it appeared most pronounced in the female, and testosterone-treated and ovariectomized females showed marked reduction in the response. Neither choline nor methionine exerted a protective effect against Ce.

At the enzyme level, Cochran, DuBois, et al.[284] found La to stimulate the activity of the succinic dehydrogenase system, as did Al, a trivalent element; but both La and Y inhibited adenosine triphosphatase activity at $10^{-3}M$. Horecker et al.[297] had previously shown that, like polyvalent Al and Cr, certain lanthanons had a promoting effect on the succinic dehydrogenase-cytochrome oxidase system.

Industrial Experience. Because of the recentness of availability in quantity and purity of the lanthanons, their handling and use have been limited almost entirely to amounts necessary for research and development of four or five of the group, including Y. Exposures are accordingly confined to relatively small numbers of persons and on an intermittent basis. On this basis, however, no indications of injury have been forthcoming when air concentrations were controlled within the threshold limit tentatively set for Y in 1956.

Medical Experience. Considerable study has been given in the past to the anticoagulant action of the lanthanons on blood for the prevention of thrombosis. More recently, the lanthanons have been explored, as their stable or radioisotopic forms, either to remove other deposited metals or to localize metals in order to achieve therapeutic effects. As a result, much useful clinical information has been obtained.

[295] E. Bamann, *Klin. Wochschr.,* **32,** 588 (1954).
[296] F. Snyder, E. A. Cress, and G. C. Kyker, *J. Lipid Research,* **1,** 125 (1959).
[297] B. L. Horecker, E. Stotz, and T. R. Hogness, *J. Biol. Chem.,* **128,** 251 (1939).

The anticoagulant effect of various lanthanons given intravenously has been demonstrated in the lower animals.[298] Lanthanons in doses of from 40 to 100 mg. had been injected into human subjects as therapeutic agents in tuberculosis.[299] Nd salts injected intravenously in total dosages of 250 to 500 mg. has been reported to prolong for 6 hours the coagulation time of the blood of man;[300] no ill effects were noted but oral or subcutaneous administration was without anticoagulant effect. Beaser et al.,[301] after having confirmed in rabbits the innocuousness of small anticoagulant doses of Nd (10 mg./kg.), critically evaluated the anticoagulant effects of Nd, La, and Ce in 18 patients with single and repeated daily doses of from 3 to 12.5 mg./kg. All the lanthanons tested increased the clotting time of blood to the point of incoagulability, which persisted in diminishing degree for 8 hours. The toxic side reactions of chills, fever, headache, muscle pains, abdominal cramps, hemoglobinemia, and hemoglobinuria were such as to exclude the use of lanthanons as anticoagulant agents. Determination of blood levels of Nd showed that it was still present in considerable amounts during decline of the anticoagulant effect. Nd was not found in the urine of the treated patients despite repeated analysis; available evidence suggested removal by the reticuloendothelial system of the liver and spleen. Local thrombophlebitis was observed in man, as reported in dogs, after injection of the lanthanon chlorides. Hemoglobinemia was unexpected; it had not been previously seen in animals and does not occur when the lanthanons are added to blood in vitro. These findings provide guides in the medical control of exposures to lanthanons. Blood-serum hemoglobin values would be the most sensitive indicator, values above 9 per cent being indicative of overexposure. Appearance of hematuria and altered prothrombin time would be less sensitive indicators. Continued or repeated headache, malaise, chills and fever, and nausea would be indicative symptoms.

In the field of radiation therapy the radioisotopes of the lanthanons are destined to play a large part because they comprise a large percentage (about $1/3$) of the total radioisotopes available with a practical half-life. Their relative immobility upon injection provides a degree of control that is desirable for localization in tumors, or they can be incorporated in ceramic beads and implanted, or included in a thread and sewn in the desired location.

6. Hygienic Standards of Exposure

A tentative threshold limit of exposure has been set for Y of 5 mg./cu. meter of air; no standards have been set for the lanthanons. Because no inhalation

[298] A. Heffter and W. Heubner, Handbuch der experimentallen Pharmakologie, Vol. 3, Sect. 3. Springer, Berlin, 1930; G. Guidi, Arch. interh. pharmacodynamic, 37, 305 (1930); E. Vincke and H. A. Oelkers, Arch. exptl. Pathol. Pharmakol., 187, 594 (1937).

[299] H. Grenet and H. Drouin, Gaz. hôp., 93, 789 (1920); M. Esnault and M. Brou, Bull. mém. soc. méd. hôp. Paris, 44, 606 (1920).

[300] H. Dyckerhoff and N. Goosens, Z. ges. exptl. Med., 106, 181 (1939).

[301] S. B. Beaser, A. Segel, and L. Vandam, J. Clin. Invest., 21, 447 (1942).

toxicity data were available, the Y standard was a calculated air concentration based on intravenous toxicity data. Limited experience for a few years in the use of this value has indicated apparent suitability.

7. Flammable Limits

Ce flashes at 150 to 180°C. and mixed lanthanons are pyrophoric on filing.

LITHIUM, Li

1. Source and Production

Li is widely distributed throughout the world in a variety of minerals, the chief five of which are listed in Table 18 with their lithia (Li_2O) content.

TABLE 18

Lithia Content of Some Minerals

Mineral	Chemical composition, approx.	Range of lithia content, %
Pentalite	$LiAlSi_4O_{10}$	2–4
Lepidolite	$K_2Li_3Al_4Si_7O_{21}$	3–4
Spodumene	$LiAlSi_2O_6$	4–7
Amblygonite	$LiAlFPO_4$	8–9
Dilithium sodium phosphate	Li_2NaPO_4	19–21

Li also occurs in the hydrosphere in low concentrations (11 p.p.m. in sea water) and in certain mineral waters. Some foods contain Li, the daily intake of which has been calculated to be 2 mg.

The world Li production increased from 4,000 short tons in 1925 to 122,000 tons in 1958, the number of producing countries from five to twelve, and United States production to around 50,000 tons (estimated).

Some Li mineral is used directly (ceramics). The recovery of Li, usually as the carbonate, varies with the mineral source, the phosphate ores being decomposed with mineral acids, whereas the silicate ores are usually calcined with NH_4Cl–$CaCO_3$ and subsequently leached. A recent process for the recovery of Li_2CO_3 from spodumene involves sulfating, leaching the Li_2SO_4, and treating with NH_4F to precipitate LiF. LiF is heated with $(NH_4)_2SO_4$, the resulting Li_2SO_4 is mixed with water and treated with CO_2 and NH_3, forming Li_2CO_3.

2. Uses and Industrial Exposures

Li minerals and compounds find uses in a wide variety of industries including metallurgy, ceramics, air-conditioning, chemical and pharmaceutical, and lubricating (grease). Major consumers are the grease and ceramics industries; Li stearate or other fatty-acid forms find use as all- or multipurpose greases, which retain their lubricating properties throughout extremes of temperature and, in addition, have good water-resistant qualities. A survey estimated that by 1965 about 42 per cent of the domestic grease used in the automotive industry will be Li-based, amounting to about 250 million lb./year. Certain natural Li minerals are used directly in ceramics. Li and its compounds are used in glasses, glazes, and enamels when high gloss, and superior scratch- and chemical resistance are required. Li is also added to porcelain enamels for Al and steel coatings to increase the fluidity of the coating and thus reduce the temperature of firing. LiCl and LiBr are used in welding, brazing, and in air-conditioning equipment; LiOH is used in alkaline storage batteries to increase output and cell life. Li metal and hydrides of Li, both simple (LiH) and complex (LiAlH₄), are finding increasing use in atomic

TABLE 19

Physical and Chemical Properties of Li and Some of Its Compounds

	Chemical symbol	At. or mol. wt.	Sp. gr.	M.p., °C.	B.p., °C.	Solubility
Lithium	Li	6.94	0.534	186	1336 ± 5	Dec. to LiOH + H₂ Dec. alc. Sol. acids
Lithium hydride	LiH	7.95	0.82	680		Dec. to LiOH + H₂
Lithium hydroxide hydrate	LiOH·H₂O	41.96				223 g./liter(10°) 268 g./liter(80°) Sl. sol. alc.
Lithium chloride	LiCl	42.40	2.068(25°)	613	1353	454 g./liter(20°) 1275 g./liter(100°) Sol. alc., ether, acetone
Lithium carbonate	Li₂CO₃	73.89	2.111(17.5°)	618	Dec.	13.3 g./liter(20°) 7.2 g./liter(100°) Sol. acids Insol. acetone, NH₃, alc.
Lithium stearate	LiC₁₈H₃₅O₂	290.40			220.5	0.1 g./liter(18°) Sl. sol. alc., ether, acetone

energy and in the chemical industries: Li metal as a scavenger in metal manufacture, and in organic synthesis; LiH for Li⁶ isotope extraction process; and LiAlH₄ as a selective, versatile, reducing agent in organic synthesis.

Industrial exposures are confined to extraction of Li from its ores, preparation of various Li compounds, welding, brazing, enameling, and more recently to Li hydrides. The newer extraction processes, however, use aqueous systems that reduce exposures materially in comparison with the calcining processes of the past. Li fumes comprise the potential exposures in welding and brazing, particularly from accidents or leaks in the use of the Li hydrides.

The physical and chemical properties of Li and some of its compounds are listed in Table 19.

3. Physical and Chemical Properties

Li is the lightest metal, with two stable isotopes, Li^6 and Li^7, and three radio-isotopes Li^5, Li^8, and Li^9 with short half-lives, 10^{-21}, 0.83, and 0.17 sec., respectively. Although Li belongs to the alkali metals, in the following respects it resembles the alkaline earths: (a) the relative insolubility of the carbonate, fluoride, phosphate, and hydroxide in water, (b) the greater solubility of the bicarbonate than the normal carbonate, and (c) the marked tendency of the chloride to form hydrates. LiCl can be extracted from other alkali-metal chlorides with amyl alcohol. Li colors the flame red, with a dominant line 6707 A, which is used for its photometric determination.

4. Analytical Determination

Li may be determined in tissues and plasma in a direct-reading flame photometer following ashing and solution according to a method by Hald.[302] Unknown solutions are compared with standard solutions of LiCl containing appropriate amounts of NaCl and KCl. A spectrographic method has been applied to human bone by Alexander et al.;[303] however, the lower limit of detection of Li in bone was 11 p.p.m. A concentration method to separate Li from the bulk of bone ash prior to spectrographic analysis is needed.

5. Physiological Response

An extensive review (58 pages, 342 references) of the biology and pharmacology of Li was made by Schou in 1957.[304] Li can be relatively toxic to man and animals, a fact often unappreciated, because its toxicity, dependent upon accumulation, is determined not only by the amount of Li given but on the amount of Na intake; the lower the Na intake, the more toxic is Li. Accordingly, statements on toxicity of Li are uninterpretable unless accompanied by statements on the Na intake. Sufficient work on the toxicity of Li has been done on varying known levels of Na intake to show that the primary toxic action of Li is on kidney function.[305] Rats given small (0.5 meq./kg./day) or moderate (3 meq./kg./day) doses of Li with a high Na intake suffer no accumulation of Li after the first day. If higher Li doses are given, or if Na intake is lowered, Li excretion lags behind intake, and Li accumulation occurs. The rise of serum Li is moderate at first but later renal function collapses, there is a transition from polyuria to oliguria and development of azotemia. The toxicity is no longer reversible and the animal dies even if admin-

[302] V. D. Davenport, Am. J. Physiol., 163, 633 (1950); P. M. Hald, J. Biol. Chem., 167, 499 (1947).
[303] G. V. Alexander, J. Biochem., 192, 489 (1951).
[304] M. Schou, Pharmacol. Revs., 9, 17, 1957.
[305] M. Schou, Acta Pharmacol, Toxicol., 15, 70, 85 (1958).

istration of Li is stopped. High Na intake prevents toxic injury. The interrelated actions of Li and Na are presently explained as a competition between Li and Na for reabsorption, in the renal tubules, into the blood stream.[305] A dose of LiCl of 60 mg. (1.4 mM)/kg./day administered orally or subcutaneously to cats and dogs will eventually be fatal from gastroenteritis. Cats appear to be more susceptible, and dogs less susceptible, than rats to gastrointestinal irritation produced by LiCl; a severe diarrhea was produced in rats that received 1.1 mM/kg./day for 8 days;[302] when given 6 mM of LiCl/kg. intraperitoneally, cerebral excitability was increased for 24 hours as measured by electroshock seizure threshold, caused in part by loss of plasma Na and low concentrations of Li in the brain.

Rats given chronic administration of 0.6 mM of LiCl/kg./day showed loss of appetite in the first and fifth and sixth weeks, lower seizure thresholds, and reduced plasma Na levels, but no indication of tissue accumulation of Li.[302] Lethal plasma Li levels for the rat are just in excess of 8 mM/liter; plasma levels are about 3 mM/liter at the start of behavioral changes, but there are no early irreversible changes if Li is withdrawn.[306] Prolonged ingestion by rats of LiCl at 20 mM/liter in the drinking water was just subtoxic, and apart from slight, transitory initial disturbances, caused no effects on health or behavior in a 2-year period.[306]

Lithium Hydride. The toxicity of LiH by inhalation for animals is marked by irritancy and corrosiveness.[307] All exposures of mice, rats, rabbits, and guinea pigs limited to either single, 4-hour periods or small multiples thereof, and air concentrations of from 5 to 55 mg. of LiH/cu. meter were intensely irritating, caused coughing and sneezing, inflammation of the conjunctivas, partial sloughing of mucosal epithelium of trachea, and some pulmonary emphysema, believed to be secondary to dyspnea and coughing. Corrosiveness of LiH was manifested by lesions of the forepaws and nose amounting to erosion of the septum in some instances. These actions are attributed to the alkalinity of the hydrolysis product (LiOH) and are believed to be characteristic of Li+. No mortality occurred as a result of any level of exposure. No chronic responses were noted up to 5 months, the length of the postexposure observation period. A level of 25 μg. of LiH/cu. meter was suggested as being free of toxicity.

Signs and Symptoms of Toxicity. The pharmacology of Li has been studied extensively in man as well as in animals because of its long history of medical use (120 years), first in gout (because of the solubility of the Li salt of uric acid), later as a dietary salt substitute in prescribed conditions of low-Na intake, and recently in psychiatry, particularly manic states, because of its sedative-like action. The clinical picture of Li poisoning is complex, involving many organs and systems (see Metabolism), and is thus often vague and ill-defined; but usually prominent are symptoms referable to the gastrointestinal tract, the central nervous system, but

[306] E. M. Trautner, R. R. Pennycink, R. J. H. Morris, S. Gershon, and K. H. Shankly, *Australian J. Exptl. Biol. Med. Sci.,* **36** (1958).

[307] C. J. Spiegl, J. K. Scott, H. Steinhardt, L. J. Leach, and H. C. Hodge, *A.M.A. Arch. Ind. Health,* **14,** 468 (1956).

particularly the kidneys. It has not been possible to establish the cause of death with certainty either in animal studies or in fatal human cases, but the kidney is considered the critical target organ.[305,308] Signs associated with Li poisoning are anorexia, extreme weight loss, general weakness and fatigue, dehydration, thirst, and dryness of the mouth. Animals show in addition reduced body temperature. Among the most common toxic signs referable to the gastrointestinal tract are: salivation, nausea, vomiting, diarrhea, frequently with loose, watery, or bloody stools. Signs referable to the nervous system are: fine tremor of hands, sometimes lips and jaws from moderate doses; in more severe poisoning muscular weakness, ataxia, positive Romberg sign, giddiness, tinnitis, drowsiness, slurred speech, and blurred vision; in more advanced stages, hyperactive deep reflexes, muscular hyperirritability, muscular fasciculations, particularly of the face, nystagmus, lethargy, and stupor; in most severe cases, coma and epileptic seizures. Electroencephalographic changes have been noted before clinical signs of toxicity and have persisted for about 1 week after ending Li treatment. Signs referable to the kidneys are first polyuria; later, elevation of nonprotein nitrogen; and oliguria in the terminal stages of poisoning. Histological changes occur in the epithelium of the proximal convoluted tubules of the kidney.[305,308]

Metabolism. Ionic Li is readily absorbed from the gastrointestinal tract, from subcutaneous, intramuscular, and intraperitoneal depots, but not through the skin (human). Absorption from the lung does not appear to have been studied, but there is no reason to suspect Li retention by this organ. Li ion is distributed widely and rather uniformly throughout the aqueous phase of the body, thus differing from Na and K distribution. Li does not appear to be transported bound to plasma protein.[308] Distribution studies in rats and dogs by intraperitoneal injection show that Li rapidly goes to liver, kidneys, and skin, more slowly to muscle and bone, and very slowly to brain (as do Na and K).

Li is usually eliminated mostly in the urine except when losses occur through vomiting and diarrhea; some is lost in the sweat. Reduced elimination of Li occurs on low salt intake when accumulation may reach toxic levels. The rate of Li elimination, therefore, depends on the Na intake; with normal salt intake, a large part of the administered dose is eliminated in the first few days, but small amounts of Li continue to be eliminated after 1 to 2 weeks. Psychotic patients are claimed to retain unusual amounts of Li during a manic phase; reduction in manic symptoms was accompanied by a profuse release of Li.[309] Li elimination is not modified by K administration.

It is important to note that although Li is distributed generally in the body following dosage, no Li was detected in bones of individuals from 5 months to 75 years of age by a spectrographic method[303] that was sensitive to 11 p.p.m. Li in bone.

[308] J. L. Radomski, H. N. Fuyat, A. A. Nelson, and P. K. Smith, *J. Pharmacol.,* **100,** 429 (1950).

[309] C. H. Noack and E. M. Trautner, *Med. J. Australia,* **38,** 219 (1951); **42,** 280 (1955).

Clinical Experience. Since Cade[310] accidentally observed a sedative-like action of Li in guinea pigs and subsequently found manic patients improved by its use, Li treatment of mania has been tried in a number of mental hospitals in Australia, France, and Denmark, mainly with good results.[304] Li salts, as carbonate or citrate, are given in tablet form in dosages usually of 40 to 60 meq./day, corresponding to 1.5 to 2.3g. of Li_2CO_3/day, although somewhat higher doses have been used. Atropine prevents the gastrointestinal irritation produced by Li. Frequently the dosage is reduced after 1 or 2 weeks, and a maintenance dose of 20 to 30 mM of Li (98 to 1.1g. of Li_2CO_3/day), suffices to keep the patient in a normal state. The counteracting effect of Li ion is rather specific as it differs from treatment with barbiturates and tranquilizers in that the patients do not act drowsy or drugged but are quiet and cooperative. Caution in the use of Li has been voiced, however, by Schou[311] because of the possibility of renal injury.

Treatment. Treatment of Li-poisoning in cases of psychomotor overactivity aims at promotion of cellular K uptake and release of Li for renal excretion.[312] It consists of: (*a*) cessation of intake, (*b*) promotion of diuresis, (*c*) promotion of cellular K uptake, (*d*) provision of calories, and (*e*) correction of acidosis. For promotion of K uptake, an oral dose of 1 g. of KCl daily is used; for correction of acidosis, usually glucose administration is sufficient, but in severe cases, infusion of $NaHCO_3$ solution may be necessary. The application of these treatment principles has produced a satisfactory reversal of the Li toxicity without any known residual or irreversible changes.

Industrial Experience. Concern for hazardous exposures to Li and its compounds has been confined to the hydrides. Spiegl *et al.*[307] have called attention to potential irritation from LiH, based on a report by Banus.[313] The principal physical property of Li hydrides important to their safe handling is the large amount of H_2 contained in a relatively small volume. The alloy hydrides because of their high relative density are largest reservoirs of H_2; the saline hydrides, both primary and complex, generally have much lower H_2 content per unit volume, but their reaction with H_2O makes them potentially larger sources of H_2. In the handling of Li hydrides, H_2O in any form must be eliminated. Fire hazards from H_2 liberation are increased with $LiAlH_4$, which is used in flammable solvents such as diethyl ether and tetrahydrofuran. Alloy hydrides are used as fine powders. Chief hazards in handling are dust production, static electricity, and fire. A fire once started cannot be extinguished by ordinary methods; smothering by dolomite powder is recommended. Detailed discussions of the process points at which particular care must be employed are given by Gaylord.[314] Experience has shown, however, that

[310] J. F. L. Cade, *Med. J. Australia*, **36**, 349 (1949).

[311] M. Schou, *Psychopharmocologia*, **1**, 65 (1959).

[312] D. A. Coats, E. M. Trautner, and S. Gershon, *Australasian Ann. Med.*, **6**, 11 (1957).

[313] M. D. Banus, *Chem. Eng. News*, **32**, 2424 (1954).

[314] N. Gaylord, *Reduction with Complex Metal Hydrides*. Interscience Publishers, New York–London, 1956.

Li hydrides can be handled safely if adequate precautions and careful engineering and process development are employed.

6. Hygienic Standards of Permissible Exposure

The only Li compound, based on apparent need, for which a limiting concentration in air has been set is LiH. The threshold limit of 25 μg. of LiH/cu. meter was adopted by the American Conference of Governmental Industrial Hygienists on the basis of the results of short-term exposures of animals by Spiegl *et al.*[307] The limit represents a level $1/_{200}$th that of the lowest level tested that produced irritant and corrosive manifestations. Because of the lack of development of any chronic changes or any morphological alterations other than those ascribable to acute irritation, the threshold limit is based on an acute response.

7. Flammability

LiH is pyrophoric and should be maintained and handled out of contact with air and moisture. LiAlH$_4$ has the additional fire hazard in that it is used in the flammable solvents, diethyl ether and tetrahydrofuran.

MAGNESIUM, Mg

1. Source and Production

Of the sixty or so Mg-bearing minerals, only magnesite (MgCO$_3$), brucite (MgO·H$_2$O), dolomite (MgCO$_3$·CaCO$_3$) a sea water (0.13 per cent Mg), sea-water bitterns, and well brines account for all of the Mg production. Minor quantities of olivine (MgFe)$_2$SiO$_4$, and serpentine (3MgO·2SiO$_2$·2H$_2$O) are used. So widespread and available are Mg sources, it has been estimated that if 100 million tons were taken from the sea annually for 1 million years, the Mg concentration in the sea would drop only 0.12 per cent. Both domestic and world production of Mg metal continued to fall in 1958, owing to accumulated stocks beyond annual average consumption. Domestic consumption, however, was greater than production although still lower than any year since 1951. The industry is divided into production of Mg metal and Mg compounds. Production of Mg metal in the U. S. is made by two reduction processes: electrolytic and silicothermic. In the electrolytic process, sea water is mixed with a slurry of Ca(OH)$_2$, precipitating Mg(OH)$_2$, which, on settling, separates out about 98 per cent of the sea water. The Mg(OH)$_2$ slurry is converted to MgCl$_2$, evaporated, and dehydrated to MgCl$_2$·1/2H$_2$O. The electrolytic cell further dehydrates and decomposes the chloride to Mg metal, Cl$_2$, and HCl. The silicothermic process starts with raw dolomite, briquetted (calcined) with 75 per cent grade ferrosilicon in a ratio 4:1. The briquettes are heated and evacuated, producing Mg vapor and dicalcium silicate. The Mg vapor, condensed to Mg crystals, may be melted and refined and cast into ingots. Many Mg-base alloys are produced, chief of which are those with Mn, Al, and Zn.

Caustic calcined magnesia (MgO) is produced from $MgCO_3$ (magnesite) by heating it below 1560°C. to retain 2 to 10 per cent CO_2; refractory magnesia is made by calcining at temperatures between 1560 and 1760°C. The products from calcining below 1700°C. are: (a) brick grade (83 to 90 per cent MgO), (b) maintenance grade (60 to 82 per cent MgO). Periclase is a product from calcining a purer grade of magnesite, or sea-water MgO above 1700°C. Iron oxide or silica may be added before heating to produce varying types of refractory magnesia. For the production of MgO from dolomite and sea water, dolomite is crushed, washed, and beneficiated, recrushed, screened, and calcined at 1100 to 1200°C., driving off CO_2 and producing dolomitic lime (MgO·CaO). This is reacted with sea water to yield $Mg(OH)_2$, filtered, and heated in rotary kilns to 1850°C. producing approximately 97 per cent MgO. MgO in turn may be combined with chromite (FeO·Cr$_2$O$_3$) or SiO_2 to produce refractories.

2. Uses and Industrial Exposures

Mg metal may be used directly for structural parts in airplanes, autos, boats, portable tools, handling equipment, machinery, etc. Mg is also used for cathodic protection of iron and steel; for alloys with a variety of metals, more recently, Th for jet engines and missiles, and Be for atomic reactors; and as a reducing agent in the production of metals (e.g., Ti). Research and development have produced new techniques to reduce impurities to less than 0.2 per cent, new alloys for withstanding vibrational stresses, new protective coatings of P and Ni, and new armor plate from Mg–Li alloys. Uses of Mg compounds and ores as refractories, except specified magnesias, $Mg(OH)_2$, and magnesium trisilicate, continued to expand for all-basic open hearth and all-basic roof construction. Caustic calcined magnesias were used for: oxychloride and oxysulfate cement, 50 per cent; insulation, 10 per cent; rubber, 4 per cent; Rayon, 2 per cent; fertilizer, uranium and chemical processing, and the paper industry consumed the bulk of the remainder. These same industries and the pharmaceutical utilized technical grade magnesias.

Industrial exposures within the Mg industry itself provide few hazards. In casting Mg, fluoride fluxes and certain inhibitors of oxidation are employed;[315] during the melting of the metal in steel crucibles, fluxes such as NH_4HF_2, NH_4BF_4, NH_4SiF_5 are sprinkled on the metal and stirred in when metal is molten. The charge is then heated to about 900°C., removed from the furnace, the dross skimmed, and a mixture of S and $Na_2B_4O_7$ added, producing a brief but violent evolution of gas and fumes. After the metal has been cooled to 700 or 800°C., more inhibitor is added, and the metal is poured. Inhibitor is also added to the metal in the pouring basins and risers, so that a blanket of SO_2 may prevent the Mg from burning.[316] The molding sand is also treated with an inhibitor and the cores are sprayed with F compounds, and sometimes with S dust. In some foundries the castings are heat-treated in an atmosphere of SO_2.

[315] M. E. Brooks and A. W. Winston, *Trans. Am. Foundrymen's Assoc.*, **49**, 165 (1941).
[316] A. B. Guise, C. V. Mars, and K. S. Wilson, *Light Metal Age*, **1**, 10, 20 (1943).

The hazards arising from the use of F in Mg-founding are discussed under Fluorine and Fluorides, page 833. The hazards from the use of S dust in coremaking are satisfactorily controlled by the precautions required to keep SiO_2 dust within safe limits.[317] Because SO_2 concentrations can reach undesirable levels in the shake-out area, at the core knockout, and near the pouring operation, ventilation control is required.

Concentrations of MgO sufficiently great to cause metal-fume fever have not been encountered in the air of the melting room, which has been reported to contain between 0.08 and 0.7 mg./cu. meter.[318] Fine particles of Mg dispersed in the air during trimming, filing, or buffing of castings may cause irritation to the mucous membranes.[319] The treatment of finished castings with chromic acid to increase weather resistance introduces hazards to the nasal mucosa and skin, discussed under Chromium, page 1017.

3. Physical and Chemical Properties

The physical and chemical properties of Mg and some of its compounds are listed in Table 20.

TABLE 20

Physical and Chemical Properties of Mg and Some of Its Compounds

	Chemical symbol	At. or mol. wt.	Sp. gr.	M.p., °C.	B.p., °C.	Solubility
Magnesium	Mg	24.32	1.74(5°)	651	1107	Insol. cold H_2O, alkalies Sl. sol. d. hot H_2O Sol. min. acids exc. CrO_3, conc. HF, NH_4 salts
Magnesium oxide (periclase)	MgO	40.32	3.58	2800		6.2 mg./liter(20°) 86 mg./liter(30°) Sol. acids, NH_4 salts Insol. alc.
Magnesium hydroxide (brucite)	$Mg(OH)_2$	58.34	2.38	$(-H_2O)350$		9 mg./liter(18°) 40 mg./liter(100°) Sol. acids, NH_4 salts
Magnesium chloride	$MgCl_2$	95.23	2.316	708	1412	542.5 g./liter(20°) 727 g./liter(100°) 500 g./liter alc.
Magnesium carbonate (magnesite)	$MgCO_3$	84.33	3.037	Dec. 350	$(-CO_2)900$	106 mg./liter(20°) Sol. acids, aq. CO_2 Insol. acetone, NH_3
Magnesium silicate	$MgSiO_3$	100.38	3.28	Dec. 1557		

[317] Illinois Labor Bull., **3**, 8, (1942).
[318] C. R. Williams, J. Ind. Hyg. Toxicol., **24**, 277 (1942).
[319] H. H. Gay, Iron Age, **150**, 60 (1942).

Mg, a silvery white metal, is the lightest structural, metallic-base material. Commercial Mg is about 99.9 per cent pure; chief contaminants are Al, Cu, Fe, Mn, Ni, and Si. Mg of high purity is obtained by distillation of impure metal *in vacuo*. Be in trace amounts (0.0002 per cent) will provide a protective film on Mg metal. Mg metal powder will ignite readily in air, and may explode. Mg alloys with Al, Ce, Mn, and Zr to produce some of the highest strength-to-weight ratios of any structural metallic material. Mg alloys have excellent machinability and adaptability to most processes of fabrication and assembly. Mg alloys have good thermal and electric conductivity, good stability to atmospheric exposure, and resistance to alkalies, certain acids, and many organic chemicals.

Mg will react with water but the reaction is self-limiting because of formation of a film of insoluble hydroxide. With an electro-negative potential of -1.63v. it can displace Zn from solution as well as Fe^{3+}, H^+, and Hg^{2+}. Mg reactions proceed at significant rate only in relatively concentrated acid solutions. Mg combines directly with free halogens if O_2 and N_2 are excluded. Other reactions of importance occur at elevated temperatures: with S, P, FeS, and $TiCl_4$ if O_2 and N_2 are not present. The reaction with FeS probably accounts for the production of tough ductile cast iron from brittle cast iron by addition of small amounts of Mg. The reaction with $TiCl_4$ is the basis of the Kroll process for the preparation of Ti. Mg forms an important group of organometallic derivatives soluble in organic solvents, which are the Grignard reagents of organic synthesis.

4. Analytical Determination

Few suitable methods exist for the accurate determination of Mg in quantities of industrial hygiene interest. Sandell[320] in reviewing colorimetric methods to 1959 finds the Titan-yellow method used by Williams[318] to have the disadvantages of colloidal lake formation, instability, and other difficulties. Eriochrome Black T, which avoids these difficulties, is highly sensitive to Ca, a common contaminant of Mg. If, however, 8-hydroquinolinol complex of Mg is extracted with an organic solvent, as little as 10 μg. of Mg may be determined. A number of industrial hygiene laboratories have found the spectrographic method suitable for the determination of microgram quantities of Mg; for larger amounts a gravimetric procedure may be used. A multichannel flame spectrometer has been used for the determination of serum Mg levels by Wacker and Vallee.[321]

5. Physiological Response

A comprehensive review (14 pages, 232 references) of the metabolism of Mg has been made by Wacker and Vallee.[322] Several reviews have appeared on the syndromes of Mg depletion and retention.[323]

[320] E. B. Sandell, *Colorimetric Determination of Traces of Metals.* Interscience Publishers, New York–London, 1959, p. 587 et seq.

[321] W. E. C. Wacker and B. L. Vallee, *New Engl. J. Med.*, **257**, 1254 (1957).

[322] W. E. C. Wacker and B. L. Vallee, *New Engl. J. Med.*, **259**, 431, 475 (1958).

[323] *Nutrition Revs.*, **17**, 112 (1959), **18**, 72, 101 (1960).

When administered intravenously, Mg is toxic, producing local and general anesthesia and narcosis,[324] which can be counteracted by intravenous Ca; muscular paralysis is also produced. There is a fall in blood pressure, and at 27 to 44 meq./liter, heart beat is arrested. At serum levels of 5 to 10 meq./liter in dogs, cardiac conduction is modified as shown by electrocardiographic changes.[325] The lethal dose for dogs is from 230 to 280 mg./kg. orally. $MgSO_4$ acts as a purgative by virtue of its poor absorption from the gut and consequent osmotic withdrawal of water from the gut wall. Fatal intoxication, however, can occur under circumstances that increase absorption of Mg, or when very high dietary levels (15,000 to 25,000 p.p.m.) are fed.[326]

Gardner and Delahant[327] observed no damage to the lungs of cats and guinea pigs following inhalation of finely divided Mg metal. It appears from their work and the similar responses of animals and man to Mg that the possibility is very remote that sufficient Mg dust would be inhaled during manufacturing processes to sustain injury. The effects of inhaling MgO for protracted periods do not appear to have been determined, however.

Metabolism. The normal adult body contains about 21 g. of Mg, 11 g. of which resides in the skeleton, 9.5 g. in the cells, and 0.5 g. in extracellular water, corresponding to about 2 meq./liter of plasma.[322] Next to K, Mg is the most abundant intracellular cation comprising 6 to 20 meq./liter compared to about 120 meq./liter for K. At ordinary levels of intake and absorption, renal conservation of Mg is high. As a consequence, plasma Mg levels usually bear little resemblance to intracellular levels.

Gastrointestinal absorption of Mg has been studied at physiologic doses (3 to 10 meq.) with labeled Mg^{28} in man;[328] at these doses, less than 10 per cent appeared in the urine in 72 hours, whereas 59 to 88 per cent was measured in the feces in 120 hours. A more detailed study of oral Mg absorption in rabbits, from about microcurie amounts of tracer Mg^{28} contained in 5 meq. of Mg, showed similar results; 10 to 12.5 per cent of the Mg was renally excreted in 48 hours when fed in highly diluted aqueous solution to fasted rabbits.[329] No absorption appeared to occur from the large intestine. Larger doses (160 meq.) resulted in

[324] J. Auer and S. J. Meltzer, *J. Exptl. Med.,* **23,** 643 (1916); C. J. Peth and S. J. Meltzer, *J. Am. Med. Assoc.,* **67,** 1131 (1916); J. T. Gwathney, *J. Am. Med. Assoc.,* **85,** 1482 (1925).

[325] S. A. Mathews and W. S. Austin, *Am. J. Physiol.,* **79,** 708 (1927); S. J. Meltzer and D. R. Lucas, *J. Exptl. Med.,* **9,** 298 (1907); G. Crisler, *Am. J. Physiol.,* **86,** 552 (1928); D. R. Joseph and S. J. Meltzer, *J. Pharmacol.,* **1,** 1 (1909); J. R. Miller and T. R. Van Dellen, *J. Lab. Clin. Med.,* **23,** 914 (1938).

[326] V. G. Heller and C. H. Larkwood, *Science,* **71,** 233 (1930); C. W. Barlow and M. S. Biskind, *Am. J. Physiol.,* **86,** 594 (1928).

[327] L. U. Gardner and A. B. Delahant, *Am. J. Public Health,* **33,** 153 (1943).

[328] J. K. Aikawa, E. I. Rhoades, and G. S. Gordon, *Proc. Soc. Exptl. Biol. Med.,* **98,** 29 (1958).

[329] J. K. Aikawa, *Proc. Soc. Exptl. Biol. Med.,* **100,** 293 (1959).

urinary recoveries of 43 per cent of the Mg in 24 hours in man;[330] similar results have been reported for normal dogs, rabbits, and rats.[330,331]

Mechanism. Mg is important in the neuromuscular conduction of skeletal and cardiac muscle, activates numerous enzymes such as the phosphatases, and catalyzes reactions involved with adenosine triphosphate (ATP). ATP is required in most of the important enzyme functions of the body, such as muscle contraction; carbohydrate utilization; protein, fat, nucleic acid, and coenzyme synthesis; methyl-group transfer; acetate, formate, and sulfate activation; and oxidative phosphorylation. In addition, Mg serves as a cofactor in carboxylase, along with thiamine pyrophosphate, for decarboxylation reactions. Certain peptidases require Mg. In heart muscle, there is considerable evidence that Mg is essential for the integrity of the cell mitochondria; a dietary deficiency of Mg leads to the uncoupling of oxidative phosphorylation,[332] which interferes with normal carbohydrate metabolism. The mitochondrial structures are thought to be responsible for the maintenance of normal gradients of Na and K in and out of the cell, and for active transport of ions.[333] This latter finding may explain the experimental observation that K accentuates the neuromuscular manifestations of Mg deficiency. The thyroxin interference with oxidative phosphorylation in mitochondria can be overcome by Mg.[334]

Industrial Experience. Industrial poisoning by Mg has not been reported. Examination of 95 men exposed to MgO dust revealed only slight irritation of the eyes and nose, although the Mg level in the serum of 60 per cent of those examined was above the normal upper limit of 3.5 mg. per cent.[319,335] Experimental metal-fume fever in man has however, been produced by exposure to excessive concentration of fresh MgO fume.[336] Mg slivers beneath the skin have been shown experimentally to produce gaseous blebs with delayed healing;[337] such injuries have not been a serious problem industrially. Dermatitis from Mg compounds is rare.

Clinical Experience. A syndrome of Mg deficiency in man seems fairly well-established, although convincing evidence for many of its complex aspects has yet to be demonstrated.[323] The syndrome is virtually identical to that seen in animals,[383] is a "true Mg tetany" characterized by "semicoma, severe neuromuscular hyperirritability, including Chvostek's sign and carpopedal spasm, athetoid movements, marked susceptibility to auditory, mechanical, and visual stimuli, a de-

[330] A. D. Hirschfelder, *J. Biol. Chem.*, **104**, 647 (1934).

[331] J. C. Forbes and F. P. Pitts, *J. Am. Pharm. Assoc.*, **24**, 450 (1935).

[332] J. J. Vitale, M. Nakamura, and D. M. Hegsted, *J. Biol. Chem.*, **228**, 573 (1957)

[333] W. Bartley and R. E. Davies, *Biochem. J.*, **57**, 37 (1954).

[334] J. J. Vitale, D. M. Hegsted, M. Nakamura, and P. Connors, *J. Biol. Chem.*, **226**, 597 (1957).

[335] A. Pleschtizer, *Arch. Gewerbepathol. Gewerbehyg.*, **7**, 8 (1936).

[336] P. Drinker, R. M. Thomson, and J. L. Finn, *J. Ind. Hyg.*, **9**, 187 (1927); K. R. Drinker and P. Drinker, *J. Ind. Hyg.*, **10**, 56 (1928).

[337] C. P. McCord, J. J. Prendergrast, S. F. Meek, and G. C. Harrold, *Ind. Med.*, **11**, 71 (1942); R. Z. Schultz and C. W. Walter, *J. Ind. Hyg. Toxicol.*, **24**, 142 (1942).

creased Mg and a normal serum Ca concentration." According to Vallee and co-workers,[338] the syndrome can only be distinguished from hypocalcemic tetany by determining serum Mg and Ca levels. Help in identifying clinical Mg deficiency is now offered by Wacker and Vallee's multichannel flame spectrometer,[321] which provides relative ease and increased accuracy in Mg analysis, and by Smith and Hammarsten,[339] who have utilized erythrocyte Mg as indicator of deficiency, recognizing the poor correlation of serum Mg levels and the deficiency state.

The syndrome is presently recognized in chronic alcoholism, delerium tremens, malnutrition, and in postoperative patients, but conceivably could occur in industry as a result of exposure to some agent such as carbon disulfide that mobilizes metals from tissue sites, particularly nerves. Moreover, K, Ca, and glucose administration aggravates Mg deficiency in ways that are not well understood; similarly the relationships with vitamin D, phosphate, and parathyroid activity are far from clear. On the other hand, man does not readily develop Mg deficiency because of relatively large stores of Mg in bone.

6. Hygienic Standards of Permissible Exposure

A threshold limit of 15 mg./cu. meter of air for MgO fume has been adopted by the American Conference of Governmental Industrial Hygienists. From evidence quoted herein[336] the limit seems appropriate. No limits have been set for Mg metal or other Mg compounds for lack of ostensible need.

7. Flammability and Explosibility

Mg metal in the powder or dust form is an extreme fire and explosion hazard. The fire hazard is high, particularly where ignition from outside sources is possible; this applies also to Mg turnings and chips. The ignition temperature of a dust cloud of pure atomized Mg is 600°C., milled Mg is 540°C. and stamped Mg, 520°C. (see pages 560, 562, 565, 567, 570 and 575 in Vol. I for other details regarding flammability of magnesium dust). Molten Mg must be protected by an atmosphere of some noncombustible gas such as SO_2 to prevent ignition. Molten Mg reacts violently with many hot metal oxides such as Fe and Ti oxides.

MANGANESE, Mn

1. Source and Production

Mn occurs in a great variety of minerals widely scattered over the earth, but the most important commercial source is the oxide, pyrolusite (MnO_2), from which the United States derives over 90 per cent of its supplies. Other Mn minerals of importance are manganite (MnOOH), hausmannite (Mn_3O_4), rhodochrosite ($MnCO_3$), and psilomelane ($4MnO_2$) (Mn,BaK)O·nH_2O. USSR, French Morocco, Cuba, Union of South Africa, Gold Coast, and India are the chief producers of Mn

---

[338] D. D. Ulmer, W. E. C. Wacker, and B. L. Vallee, *J. Clin. Invest.*, **38**, 1049 (1959).
[339] W. O. Smith and J. F. Hammarsten, *Am. J. Med. Sci.*, **237**, 413 (1959).

ore outside the United States. Domestic supplies of oxides are supplemented with $MnCO_3$ from Montana. Lately (1958) Nevada and Minnesota exceeded all other states in producing Mn and manganiferous ore (448,000 tons as against 884,000 tons from ten other states). By comparison 12,460,000 tons of Mn ore was produced by fifty other countries throughout the world. Manganiferous ore from Minnesota and Michigan contains but 5 to 10 per cent Mn; Mn ores from other states contain 10 to 35 per cent Mn. Pyrolusite and manganite contain 60 to 63 per cent Mn.

Mn mining methods range from primitive to most modern. Open-pit or shallow underground workings are the rule; an exception, mines at Butte, Montana, where $MnCO_3$, in quartz and granite is mined at depth. The large deposits contributing to world supply require only simple concentrating processes, such as washing to remove clay, and primary crushing. Low-grade ores are beneficiated by sulfuric acid, SO_2 leach, or by NO_2, nitric acid leach. Smelting of Mn ore is similar to that of Fe ore because of their basic similarity. The principal metallurgic form of Mn is ferromanganese, which contains 80 per cent or more of Mn, the remainder mainly Fe. Spiegeleisen is a high-carbon steel pig iron containing up to 20 per cent Mn and is similarly made in a blast furnace. Silicomanganese, containing 65 to 70 per cent Mn and 15 to 20 per cent Si, is made in an electric furnace. There are fume losses in the ferromanganese operation but they are recoverable. To minimize Mn dust losses hard, coarse ores are mixed with fine ores. The amount of P in ferromanganese is governed almost wholly by the content of the charge and does not usually exceed 0.35 per cent. Bases, consisting of CaO, MgO, and BaO, are added to the ferromanganese charge to combine with SiO_2 and liberate MnO_2 for reduction. There are similar (5 to 10 per cent) fume and dust losses in the electric furnace. High-purity Mn is made either electrolytically or by electric-furnace reduction of Mn ore with Si or Al. The ore may be first given a preliminary roast to reduce the higher oxides to MnO, which is then leached with H_2SO_4 to form $MnSO_4$. The solution is neutralized with NH_3, thickened, and the residue washed free of soluble salts. As, Cu, Co, Pb, Ni, and Mo are precipitated as the sulfides with H_2S. The Mn electrolized from this solution is 99.9+ per cent pure, S and SO_4 being the chief impurities; the remaining impurities are chiefly those metals mentioned above, not completely precipitated as the sulfide.

2. Uses and Industrial Exposures

Mn has three principal uses: (a) as a reagent in steelmaking (to reduce O_2 and S) and as an ingredient in special alloy steels, (b) in the manufacture of dry-cell batteries as MnO_2, and (c) in the chemical industry, as an oxidizing agent, for the production of $KMnO_4$ and other Mn chemicals. Manganin, an alloy of 13 per cent Mn, 83 per cent Cu, and 4 per cent Ni, is used in electric resistance coils. Mn also is one of the components of Mn-bronze and certain alloys with desirable magnetic properties. In addition to its use as MnO_2 in dry cells for depolarization it is employed in glassmaking and in ceramics. As a chemical, several salts are used as driers for linseed oil; the manganates and permanganates are oxidizing agents used for disinfection, bleaching, and as laboratory reagents. Manganous acetate is used in dyeing, tanning of leather, in fertilizers, and as a chemical catalyst. Manganese sulfate is used as a trace element in poultry and animal feeds.

Industrial exposures to Mn dusts occur in the mining, transporting, crushing, and sieving of the ore. Mn fumes and dusts occur near the reduction furnaces.[340] Most of the many reported cases of exposure have been acquired during these operations where Mn concentrations in the air have been recorded as high as 40 to 173 mg./cu. meter. Exposures to smaller amounts of associated metals such as As, Pb, Co, and Ni also present themselves. Relatively few cases of intoxication have occurred from the use of Mn in the metal trades, which use 90 to 95 per cent of the total metal output. The use of Mn in coated welding rods is a source of exposure; air concentrations have been found to contain from 2 to 23 mg. of Mn/cu. meter.[341]

3. Physical and Chemical Properties

The physical and chemical properties of Mn and some of its compounds are listed in Table 21.

TABLE 21

Physical and Chemical Properties of Mn and Some of Its Compounds

	Chemical symbol	At. or mol. wt.	Sp. gr.	M.p., °C.	B.p., °C.	Solubility
Manganese	Mn	54.94	7.2	1260	1900	Dec. hot or cold H_2O Sol. dil. acid
Manganese dioxide	MnO_2	86.93	5.026	(−0)535		Insol. hot or cold H_2O, HNO_3, acetone Sol. HCl
Manganous carbonate	$MnCO_3$	114.94	3.125	Dec.		65 mg./liter(25°) Sol. dil. acid Insol. NH_3, alc.
Manganous chloride	$MnCl_2$	125.84	2.997(25°)	650	1190	622 g./liter(10°) 1.238 kg./liter (100°) Sol. alc. Insol. ether, NH_3
Manganous acetate	$Mn(C_2H_3O_2)_2 \cdot 4H_2O$	245.08	1.589			Sol. cold H_2O, alc.
Potassium permanganate	$KMnO_4$	158.03	2.703	Dec. <240		28.3 g./liter(0°) 250 g./liter(65°) Dec. alc. Sol. H_2SO_4 V. sol. methyl alc., acetone

[340] R. H. Flinn, P. A. Neal, and W. B. Fulton, *J. Ind. Hyg. Toxicol.*, **23**, 374 (1941); R. F. Gayle, *J. Am. Med. Assoc.*, **85**, 2008 (1925); R. H. Flinn, P. A. Neal, et al., U. S. Public Health Serv. Bull. No. **247** (1940).

[341] C. P. McCord, G. C. Harrold, and S. F. Meek, *J. Ind. Hyg. Toxicol.*, **23**, 204 (1941).

Mn is a grey-white metal resembling Fe, but it is harder and more brittle, and is especially noted for alloying with metals and imparting hardness.

Mn exists in seven oxidation states, as shown in Table 22 along with typical compounds.

TABLE 22

Oxidation States of Mn and Some of Its Compounds

Compound	Oxidation state	Corresponding oxide	Typical compound
Univalent	Mn^+		$Na_3(Mn(CN)_4)$
Manganous	Mn^{2+}	MnO(basic)	$MnSO_4$
Manganic, manganates III	Mn^{3+}	Mn_2O_3(basic)	$MnCl_3$ or $K_3Mn(CN)_6$
Manganates IV	Mn^{4+}	MnO_2(amphoteric)	$Mn(SO_4)_2$ or $CaMnO_3$
Manganates V	Mn^{5+}		K_3MnO_4
Manganates VI	Mn^{6+}	(acidic)	K_2MnO_4
Permanganates VII	Mn^{7+}	Mn_2O_7(acidic)	$KMnO_4$

Mn is a highly reactive metal. The bivalent form is the most stable, the ionized form being more stable than ferrous Fe. Bivalent Mn can be oxidized by strong oxidizers; $Mn(OH)_2$ is readily oxidized to Mn^{3+}. There is mutual interaction of Mn^{2+} and $KMnO_4$ with the formation of MnO_2. Mn^{2+} is pink and its common salts are water-soluble; exceptions are carbonate, phosphate, and sulfide.

4. Analytical Determination

Richards' modification of the Willard-Greathouse periodate method[342] is the most suitable method for the determination of less than milligram amounts of Mn in specimens encountered in industrial hygiene practice. Richards examined the method critically and established optimal conditions for precision and accuracy. A spectrochemical method also may be used and is particularly applicable for determination of Mn in biological materials.[343]

5. Physiological Response

Three reviews of Mn poisoning and its various biological manifestations have been published. von Oettingen[344] reviewed the literature on its distribution, pharmacology, and health hazards to 1935; Fairhall and Neal[345] summarized information on industrial Mn poisoning with a review of pertinent toxicological and

---

[342] M. B. Richards, Analyst, 55, 554 (1930); F. H. Goldman and M. B. Jacobs, Chemical Methods in Industrial Hygne. Interscience Publishers, New York 1953; Manual of Analytical Methods. Comm. on Recent Analytical Methods, Am. Conf. Govtl. Ind. Hygienists.

[343] R. A. Kehoe, J. Cholak, and R. V. Storey, J. Nutrition, 19, 581, (1940).

[344] W. F. von Oettingen, Physiol. Revs., 15, 175 (1935).

[345] L. T. Fairhall and P. A. Neal, Natl. Insts. Health Bull. No. 182, (1943).

analytical information to 1943; and Cotzias[346] has critically presented Mn in health and disease in the light of biochemical concepts of 1958.

Animal Toxicity. Information on experimental Mn poisoning from the older literature indicated the toxicity to vary according to the salt administered; Mn^{2+} was shown to be 2.5 to 3 times as toxic as Mn^{3+} on the basis that doses causing death when Mn^{2+} was given were only slightly toxic when Mn^{3+} was substituted. This older finding has possible interesting significance for the present view[347] that Mn^{3+} is the truly biologically active form in mammals. Subcutaneous doses of 50 mg./kg. as $MnCl_2$ were fatal to mice, guinea pigs, and rabbits; whereas 18 mg./kg. of the same preparation was lethal to rabbits by the intravenous route; 56 mg./kg. was required for dogs.[348] The nature of the associated anion is reported to affect the toxicity of Mn^{2+};[349] manganous citrate injected subcutaneously is more rapidly fatal than $MnCl_2$. Particle size is reportedly implicated in the toxicity of Mn salts; particles above a certain size result in toxicity and death, not seen on administration of multilethal doses of smaller sized particles.[350]

MnO_2 and $MnCl_2$ separately injected intratracheally into rats, in an effort to simulate manganese pneumonitis seen in man, produced characteristic histological changes in the lungs.[351] Within minutes epithelial cells of the bronchi discharged mucus, the epithelium became ragged, and cells became loosened from the basal membrane. An intense mononuclear-cell infiltration of the alveolar walls and cells occurred; large hydropic cells appeared. Late and inconstant changes were granulomatous reaction and giant-cell formation. The injection of $MnCl_2$ caused intense congestion and pulmonary edema, which was often fatal. Thus, pneumonitis as seen in man was felt to have been reproduced in animals.

Similar pulmonary effects were reported by Russian workers[352] from intratracheally injected suspensions of MnO, MnO_2, Mn_2O_3, and Mn_3O_4 of particle sizes less than 3 μ in young rats. Important additional findings were: (a) the higher oxides were more toxic, and (b) greater toxicity was exhibited by freshly prepared oxides than by those stored 6 and 12 months.

Many attempts[353] to induce characteristic Mn brain damage by feeding Mn compounds to animals have been only partially successful, but have shown what von Oettingen[344] concluded long ago, that Mn given in inorganic form by mouth is slowly and incompletely absorbed in the blood stream.

[346] G. C. Cotzias, *Physiol. Revs.*, **38**, 503 (1958).

[347] D. C. Borg and G. G. Cotzias, *Federation Proc. Abstr.*, **19**, 248 (1960).

[348] F. Cervinka, *Compt. rend. soc. biol.*, **102**, 262 (1929).

[349] H. Handovsky, H. Schulz, and M. Staemmler, *Arch. exptl. Pathol. Pharmakol.*, **110**, 265 (1925).

[350] R. Nissen, *Klin. Wochschr.*, **1**, 1986 (1922).

[351] T. A. L. Davies and H. E. Harding, *Brit. J. Ind. Med.*, **6**, 82 (1949).

[352] E. N. Levina and E. G. Robachevsky, *Gigiena i Sanit.*, **1**, 25 (1955); in abstract, *Ind. Hyg. Digest*, **19**, No. 948 (1955).

[353] L. van Bogaert and M. J. Dallemagne, *Monatsschr. Psychiat. Neurol.*, **111**, 60 (1945); J. T. Skinner, *J. Nutrition*, **5**, 451 (1932); W. D. Gallup, L. E. Walter, and D. E. McOsker, *Proc. Oklahoma Acad. Sci.*, **32**, 71 (1951); B. M. Richards, *Biochem. J.*, **24**, 1572 (1930).

Many studies in animals have also been made to elucidate the effect of Mn on the formed elements of the blood[354] as reported by Flinn et al.[340] and Kesic and Hausler.[355] $MnSO_4$ injections stimulate hematopoiesis in animals; and permanganate, hemolysis. Polycythemia has also been reported. High Mn feeding (in lambs), however, caused reduction in hemoglobin, indicating that massive amounts of Mn interfere with dietary Fe absorption. There is some evidence of dietary interplay of Mn and Cu (Gubler et al.[354]), and the level of dietary Mn and low blood Mg in cows (Fain et al.[354]).

Chronic Animal Toxicity. Repeated oral administration of Mn to animals as a metal or its salts for prolonged periods gave no evidence of injury in moderate doses;[356] Mn stimulated growth when present in the diet up to 100 p.p.m.[357] but proved deleterious at 600 p.p.m.[358] It is certain, however, that the Mn levels only have significance relative to the composition of the diets; the relative amounts of other mineral and amino-acid constituents exert considerable effect on the biological activity of Mn.

Repeated intraperitoneal injections of $MnCl_2$ into monkeys for 18 months resulted in characteristic lesions of the basal nucleus; the animals exhibited choreiform movements; later rigidity of the muscles, disturbances in motility, tremors, and contractions of the hands.[359] Histological examination revealed marked changes in the corpus striatium, in the globus pallidus of the lenticular nucleus, degenerative changes in the caudate and striate nuclei, and in the large ganglion cells of the cortex and basal ganglia, in addition to acute hepatitis and liver necrosis. Similar changes have been observed in the brains of dogs following repeated subcutaneous injections and in rabbits after repeated, large oral doses.[344]

Inhalation exposures of rabbits to MnO_2 dust 4 hours daily for from 3 to 6 months at levels of 10 to 20 mg./cu. meter resulted in decreased hemoglobin and erythrocytes in the blood;[357] Mn pneumonitis did not occur, but fibrotic changes in the lung resembling those in silicosis were observed.

Metabolism. Mn salts are but slowly and poorly absorbed from the gastrointestinal tract.[344] This has been shown by both natural and radioisotopic Mn^{54};[360] 1 mg. of $MnCl_2$ containing 0.1 microcurie of Mn^{54}/mg. Mn was excreted

[354] L. Paterni, *Folia Med. Naples,* **37,** 994 (1954); R. H. Hartman, G. Matrone, and G. H. Wise, *J. Nutrition,* **57,** 429 (1955); C. J. D. Gubler, S. Taylor, G. E. Eichwald, G. E. Cartwright, and M. M. Wintrobe, *Proc. Soc. Exptl. Biol. Med.,* **86,** 223 (1954); P. Fain, J. Dennis, and F. G. Harbough, *Am. J. Vet. Research,* **13,** 348 (1952).

[355] B. Kesic and V. Hausler, *Arch. Ind. Hyg. Occupational Med.,* **10,** 336 (1954).

[356] W. F. von Oettingen and T. Sollman, *J. Ind. Hyg.,* **9,** 48 (1927); F. Hendrych and K. Klimesch, *Arch. exptl. Pathol. Pharmakol.,* **178,** 178 (1935).

[357] O. Ehrismann, *Z. Hyg. Infectionskrankh.,* **117,** 662 (1935); R. E. Carratala and C. L. Carboneschi, *Chem. Abstr.,* **31,** 8697 (1937).

[358] V. E. Nelson, J. M. Evvard, and W. E. Sewel, *Proc. Iowa Acad. Sci.,* **36,** 267 (1929), *Chem. Abstr.,* **25,** 1565 (1931).

[359] H. Mella, *Arch. Neurol. Psychiat.,* **11,** 405 (1924).

[360] D. M. Greenburg and W. W. Campbell, *Proc. Natl. Acad. Sci.,* **26,** 448 (1940); D. M. Greenberg, H. D. Copp, and E. M. Cuthbertson, *J. Biol. Chem.,* **147,** 749 (1943).

to the extent of 97.2 per cent in the feces at the end of 75 hours when given orally to a rat; following intraperitoneal injection, 90 per cent was similarly excreted. Only 28 per cent of the oral dose was distributed to the tissues; a detailed list of tissue Mn levels is given. The liver is a preferential site of accumulation. Even less Mn is absorbed when insoluble powdered ores are fed.[361] Moreover, as pointed out by Cotzias,[346] the rate of absorption from large amounts of inorganic Mn salts finds no parallel in the trace amounts found in food in which Mn probably exists chelated with organic constituents; the properties of the individual complex determine the rate of transport across the cell wall. Both distribution and excretion are markedly affected by chelating agents. Moderate levels (10 to 100 μg. of Mn) are absorbed to about 3 to 4 per cent of the oral dose, the Mn appearing rapidly in the bile, which probably constitutes the main excretory route; little or no Mn appears in the urine.[360] Bertinchamps and Cotzias[362] have reported electrophoretic evidence for three different kinds of Mn compounds in the bile; part of these fractions are reabsorbed.

In man, on a daily average intake of 3 to 9 mg. of Mn, the blood contains from 12 to 15 μg.% Mn, the urine somewhat less than 10 μg./liter, most of the intake appearing in the feces. Small quantities are to be found in most tissues, the bone, liver, and lymph nodes containing highest concentrations.

Mn is an essential element for normal metabolism. Rabbits require about 300 μg. of Mn in organic form daily for satisfactory growth; rats, 0.2 mg./day. Human requirements are only rough estimates but Kent and McCance[363] estimated between 3 and 9 mg. of Mn/day depending on diet selection to be the daily intake; Kehoe, Cholak, and Storey suggest 4 mg./day as an average value.[364]

Mechanism. Chronic Mn poisoning and Mn deficiency show a high degree of specificity, which in turn should be explicable in terms of biologic mechanisms of Mn action. Although there is an enormous literature on the biochemical action of Mn, its almost wholly in vitro character makes it impossible at present to relate the findings to meaningful interpretations of action in the intact animal. For example, Mn is essential in the mammal for the activity of the enzyme arginase, for the production of urea, but the fact that birds have a high requirement for Mn, produce no urea, and possess no arginase, inclines one to the view that the more important functions of Mn have yet to be found. Attempts in this direction made by Maynard and Cotzias[365] showed that Mn^{56} disappeared rapidly from the blood of man and rat to accumulate preferentially in organs rich in mitochondria—the liver, pancreas, kidneys, and brain, in which the rate of Mn turnover was high. Moreover, newly deposited radio Mn was easily removed

[361] C. K. Reimann and A. G. Minot, *J. Biol. Chem.*, **42**, 133 (1920).

[362] A. Bertinchamps and G. C. Cotzias, *Federation Proc.*, **17**, 428 (1958).

[363] N. L. Kent and R. A. McCance, *Biochem. J.*, **35**, 877 (1941).

[364] R. A. Kehoe, J. Cholak, and R. V. Storey, *J. Nutrition*, **19**, 579 (1940).

[365] L. S. Maynard and G. C. Cotzias, *J. Biol. Chem.*, **214**, 489 (1955); G. C. Cotzias and J. J. Greenough, *Physiol. Revs.*, **38**, 503 (1958).

from mitochondria, whereas older, stable, Mn was not, indicating different types of combinations of Mn with mitochondria. From this and other work on the specificity of displacement of Mn, Cotzias[346] has concluded that there is a highly specific step in the pathway of Mn through the body that is intracellularly located, probably providing chelated Mn^{3+}. Adrenal glucocorticoids, acting as metal ligands, have been found by Cotzias[366] to accelerate the turnover of Mn^{54}.

Mn deficiency states studied in animals and characterized by retardation of growth, bone abnormalities, dysfunction of reproduction, and central nervous system pathology, have been associated with lowered liver arginase activity, which could be raised by administering Mn. Lowered bone alkaline phosphatase has also been shown associated with Mn deficiency in rabbits.[367] The metabolism of choline appears intimately linked with that of Mn; both decrease liver fat in rats and in birds; Mn supplements require high intake of choline to prevent perosis (bone deformities and slipped tendons) in birds.[368] There is an interrelation between the storage of Mn in the liver and thiamine in the diet, and vice versa.[369] Cu assimilation and storage are affected by the level of dietary Mn (Gubler et al.);[354] high Mn intake changed the normal distribution of Cu among various organs but not total body Cu content. Curran[370] has reported that Mn increases cholesterol synthesis in rats. Mn has been suggested as being a cofactor in oxidative phosphorylation;[371] to be involved in human lupus erythematosus disseminatus in which Mn inhibits the formation of L.E. cells;[372] as having a favorable role in carcinogenesis by inhibition of certain types of tumor cell growth;[373] and possibly being a key factor in the development of Parkinsonism,[346] an extrapyramidal disease following childhood jaundice or following poisoning with CO or CN. These and a number of other proposed biochemical functions of Mn, such as involvement with antibody formation,[374] and chelation with the B vitamin inositol,[375] indicate varied and important roles for Mn whose interrelationships and significance await further clarification.

Industrial Experience. Manganese pneumonitis. "Manganese pneumonia" had been recognized before 1937 by Baader,[376] who collected evidence from various parts of the world of proneness to pulmonary disease, pleuritis, bronchopneumonia, and severe and often fatal pneumonia. This condition had been

[366] R. R. Hughes and G. C. Cotzias, *Federation Proc. Abstr.,* **19,** 249 (1960).

[367] S. E. Smith, M. Medlecott, and G. H. Ellis, *Arch. Biochem.,* **4,** 281 (1944).

[368] T. H. Jukes and A. D. Welch, *J. Biol. Chem.,* **146,** 19 (1942).

[369] R. M. Hill and D. E. Holtkamp, *J. Nutrition,* **53,** 73 (1954).

[370] G. L. Curran, *J. Biol. Chem.,* **170,** 765 (1954), *Proc. Soc. Exptl. Biol. Med.,* **88,** 101 (1955).

[371] O. Lindberg and L. Ernster, *Nature,* **173,** 1038 (1954).

[372] P. Comens, *Am. J. Med.,* **20,** 944 (1956).

[373] J. Balo and I. Banga, *Acta Unio Intern. contra Cancrum,* **13,** 463 (1957).

[374] L. E. Walbum and S. Schmidt, *Z. Immunitätsforsch.,* **13,** 32 (1925).

[375] A. C. Wiese, B. C. Johnson, C. A. Elvehjem, and E. B. Hart, *Science,* **88,** 383 (1938).

[376] E. W. Baader, *Aerztl. Schverst. Zty.,* **43,** 75 (1937).

recognized by others to occur among workers in German pyrolusite mines and mills. In addition to the pneumonia, fibrotic lesions formed about deposits rich in Mn. Stevedores regularly employed in handling Mn ores developed pneumonia from which 31 per cent died,[345] and Elstad reported pneumonia in Norwegian workers following introduction of an electric smelting furnace for Mn ore. Rodier[377] has noticed pneumonias from drilling and blasting in underground Moroccan mines and is inclined to ascribe them to associated factors with Mn the aggravating cause; Penalver[378] in Cuba has not reported Mn pneumonitis. The reason for this difference is not clear.

Chronic manganese poisoning. Chronic effects of Mn, recognized for more than 100 years, continue to be the subject of reports from several parts of the world. The diagnostic characteristics of manganism have been detailed by Flinn et al.,[340] Rodier et al.,[377] and Penalver.[378]

The onset is insidious, with apathy, anorexia, and asthenia. The Mn psychosis that follows has certain definitive features: unaccountable laughter, euphoria, impulsiveness, and insomnia, followed by overpowering somnolence. Headache is often present; recurring leg cramps; and sexual excitement, followed by impotence. Following or concomitantly with these manifestations are speech disturbances with slow and difficult articulation, incoherence, even complete muteness. Mask-like facies sets in, general clumsiness of movement, noticeable in altered gait and balance, which may develop into severe propulsive and retropulsive movements, and ultimate development of "hen's gait." Micrographia is a consistent finding. As the poisoning progresses, rigidity is marked, frequent falls occur, and tremor sets in, which becomes exaggerated by stresses such as fatigue or emotion. Absolute detachment, broken by sporadic and spasmodic laughter, ensues and, as in extra-pyramidal affections, salivation and excessive sweating. Despite the severe incapacitations imposed by the disease, the patient survives, although permanently disabled unless treated; chronic Mn poisoning is not a fatal disease.

Onset of the chronic disease occurs between 6 months and 2 years according to Rodier, although earlier cases have been reported from Chile (49 days); average exposure period has been estimated to be 178 days, but some cases have appeared after 16 years.[379] Individual susceptibility is marked; a high percentage of Spaniards, according to Penalver, work for years as if immune. Factors possibly influencing sensitivity are alcoholism, chronic infections, such as syphilis, malaria, tuberculosis, and avitaminosis, and liver dysfunction. The difference in time of onset and severity of the chronic disease seen throughout the world, although recognizably influenced by all these factors, still might be most importantly in-

[377] J. Rodier, *Brit. J. Ind. Med.,* **12,** 21 (1955); J. Rodier, R. Mallet, and L. Rodi, *Arch. maladies profess. med. travail et sécurité sociale,* **15,** 210 (1954); J. Boyer and J. Rodier, *Rev. neurol.,* **90,** 13 (1954).

[378] R. Penalver, *Ind. Med. and Surg.,* **24,** 1, (1955), *A.M.A. Arch. Ind. Health,* **16,** 64 (1957).

[379] P. Schuler, et al., *Ind. Med. and Surg.,* **26,** 167 (1957).

fluenced by nature of the Mn ore; Rodier has stated that braunite, a mixture of Mn_2O_3 and $MnSiO_3$, is much more injurious than pyrolusite (MnO_2).

No indisputable test of Mn poisoning has been developed, but Rodier[377] finds diminished excretion of 17-ketosteroids in 81 per cent of patients, a relative increase in lymphocytes, and a decrease in the number of polymorphonuclear cells in 52 per cent, and an increased basal metabolic rate. Increased urinary Mn is evidence of exposure, but the correlation with symptoms is not clearly established. Kesic and Hansler[355] found slightly higher (4.5×10^6 vs. 4.337×10^6) erythrocyte counts and decreased monocyte values (640 vs. 780) among Jugoslavian miners during the first phase of the disease, that later returned to normal; in only a few cases (4 of 60) was there a decreased leucocyte count. Fairhall, in his review,[345] mentions early rises in red blood cell count as the first symptom of the disease. At the present time, blood Mn levels are not used; Flinn et al.[340] found no Mn in the blood of the specimens they examined with the methods of the time (volumetric bismuthate); Penalver[378] states, without presenting evidence, that blood Mn levels are too variable to be of value. Possibly, application of newer analytical methods would prove of value.

The characteristic pathological lesion in man is destruction of the ganglion cells of the basal ganglia, followed by scarring and shrinking. Perivascular degenerative areas in the striatum and pallidum occur and to lesser extent in the frontal and parietal cortex. Others[380] believe Mn effects are not confined to the corpus striatum but involve the vegetative and vasomotor centers of the midbrain and also the medulla and cord. Unfortunately, owing to the customs in the countries in which manganism is currently most prevalent, autopsies cannot be performed and thus no recent histological examinations have been reported.

The prognosis of manganism depends on the duration of the disease; many of the symptoms will regress or disappear if the worker is removed from exposure shortly after the appearance of symptoms, although there may be some residual disturbances in speech and gait. Well-established manganism, however, is a crippling disease with permanent disability particularly in respect to the use of the legs. Even some of these patients may improve slowly with occasional remissions of tremor, weakness, and muscular cramps. Seriously poisoned individuals have been considered lifelong cripples. Recently, however, Penalver[381] has reported marked improvement in a chronic case of Mn poisoning on prolonged treatment with EDTA salt plus other drugs offering supportive treatment. Prior work with EDTA in animals[376,382] given single or repeated doses of Mn salts had shown EDTA effective in removing recent Mn from body tissues, possibly preventing the disease from advancing, but not in curing the nervous effects of Mn. Reports of more human cases treated with EDTA would be welcome. In a comparative test of three EDTA-type compounds against acute experimental Mn poisoning in

---

[380] H. Voss, Arch. Gewerbepathol. Gewerbehyg., **9**, 464 (1939).

[381] R. Penalver, A.M.A. Arch. Ind. Health, **16**, 64 (1957).

[382] M. F. Kosai and A. J. Boyle, Ind. Med. and Surg., **25**, 1 (1956).

rats, diethylene triaminepentaacetic acid was found to be somewhat more effective in preventing Mn intoxication than EDTA.[383]

A case of chronic manganese poisoning, believed to be the first reported from the German fertilizer industry, occurred in an attendant of the ore-roasting ovens preparatory to making $MnSO_4$ as a trace addition to the fertilizer.[384] Blood analyses revealed 2.3 and 2.6 μg.% Mn, extremely low values even for unexposed individuals in other countries, although the author claims 0.3 μg.% is normal for his area.

A Czech report[385] revealed that 5 of 54 employees engaged in the manufacture of permanganate for more than 10 years showed toxic damage to the central nervous system characteristic of Mn poisoning. Blood Mn values of exposed workers were 45 μg.% as against controls with an average of 11 μg.%. The atmosphere contained up to 17 mg./cu. meter of Mn. No respiratory symptoms were seen.

6. Hygienic Standards of Permissible Exposure

The threshold limit of 6 mg. of Mn/cu. meter of air has been adopted by the American Conference of Governmental Industrial Hygienists following publication of this value by the American Standards Association.[386] Despite the many years that have elapsed since its promulgation, no rigorous test of its validity has been made. From such information as exists, 6 mg./cu. meter is probably not sufficiently low to prevent chronic poisoning in all persons for an indefinitely long exposure. Surveys conducted by Flinn et al.[340] revealed that 1 person exhibited symptoms of Mn poisoning at 5 mg./cu. meter, 2 of 11 others were affected when exposed to between 30 and 59 mg./cu. meter, and 8 of 10 exposed to more than 9 mg./cu. meter. More recently, a Chilean report[379] showed chronic poisoning occurred in miners when but one third of the air concentration values were above the threshold limit. It would appear, therefore, that 6 mg. of Mn/cu. meter provides little or no factor of safety and indeed should be regarded as a maximal upper limit.

7. Explosivity

Mn dust clouds have a minimal ignition temperature of 450°C. The minimal explosive concentration is 125 oz. per 1000 cu. ft. producing a maximal explosion pressure of 50 p.s.i.; the limiting O_2 percentage preventing ignition of the dust cloud is 15 (see also Vol. I, Chapter XVI, Section Two).

[383] J. F. Fried, A. Lindenbaum, and J. Schubert, Proc. Soc. Exptl. Biol. Med., **100**, 570 (1959).

[384] D. Schurmann, Zentr. Arbeitmed. Arbeitsschutz, **6**, 106 (1956).

[385] V. Pekarek, E. Ponca, and Z. Jizera, Precovní lékařstuí, **9**, 104 (1957).

[386] Am. Standards Assoc., Pamphlet No. **237 B**, July 16, 1942.

MERCURY, Hg[387]

1. *Source and Production*

Hg ore is found in rocks of all classes. The common host rocks are limestone; calcareous shales; sandstone; serpentine ($3MgO \cdot 2SiO_2 \cdot 2H_2O$) chert; andesite (soda-lime feldspar) basalt; and rhyolite (alkaline feldspar and quartz). Hg is recovered almost entirely from cinnabar (α-HgS) 86.2 per cent Hg, although elemental Hg occurs in some ores. Other less important sources are livingstonite ($HgS \cdot 2Sb_2S_3$), metacinnabarite (βHgS), and about 25 other Hg-containing minerals. The dominant world producers of Hg are Spain and Italy; Mexico, Chile, Peru, Yugoslavia, the Philippines, and the United States are also significant producers. World output of Hg in 1958 rose for the tenth consecutive year, totaling 248,000 flasks; domestic production rose 10 per cent above 1957, totaling 38,000 flasks; California was the leading producer with 59 per cent, Nevada second, with 19 per cent, and Alaska, third.

In the United States, where Hg deposits are small and irregular, both surface and underground mining methods are used, but the latter preponderates. The lowest mine depth is about 2500 feet. Hg-mining practice differs from other mining in that very little ore is blocked out ahead of stoping; to offset this, more working faces are required and thus more Hg surface is exposed.

Hg is recovered from ores by heating in furnaces and retorts. Preliminary treatment involves sorting and screening and in a few instances milling. Retorts are used for high-grade ore where the deposit is small, or they are used in conjunction with furnaces to treat the soot. In the United States, ores are treated chiefly in rotary-kiln or multiple-hearth furnaces. According to Schuette,[388] furnaces for treating Hg ores are of two kinds, direct-fired furnaces and retorts with indirect firing. Unless HgS is heated above 300°C. little Hg is liberated; this is the limiting temperature below which the ore can be dried without danger of loss of Hg or poisoning of the workers. The firing of HgS is done at about red heat (700°C.). In direct-fired furnaces O_2 combines with the S to yield SO_2 and the Hg is set free as a vapor, which is condensed. In retorts, where there is no O_2, lime is added to combine with the S and thus to liberate Hg. Because of the small scale of the deposits almost every mine has its own treatment facilities. The metal thus obtained is generally pure enough for most commercial purposes. Schuette[389] claims that with ample condensing system, stack losses should not exceed 2 or 3 per cent of the metal in the furnace charge, and that a well-designed plant should recover 95 per cent of the input. Recoveries at efficient plants meet these conditions, but recoveries as low as 60 per cent of input have occurred in some locations.

---

[387] The author wishes to acknowledge with appreciation the help of Dr. Bertram Dinman and Dr. Leonard Goldwater in reviewing the section on mercury.

[388] C. N. Schuette, *U. S. Bur. Mines Bull.* No. **335** (1931).

[389] C. N. Schuette, *State of Ore. Bull.* No. **4** (1938).

2. Uses and Industrial Exposures

The largest present use of Hg is in electric apparatus (9335 flasks of 76-lb.); the two next largest uses, for industrial control instruments and agricultural and industrial poisons, insecticides, fungicides, and bactericides (6000 flasks each). Electrolytic cells for the preparation of Cl_2 and NaOH take nearly 5000 flasks annually, and pharmaceutical and dental preparations combined, a similar amount. Significant uses of Hg are in antifouling paint, as catalysts, for amalgamation, and general laboratory use.

Users of mercury-filled instruments, Hg-vapor lamps, thermometers, barometers, and gages are not always aware of the health hazard that may arise from Hg spilled on floors, tables, and sinks. A stream of air passed at 1 liter/min. over a 10-sq. cm. surface of Hg at 20°C. becomes about 15 per cent saturated, acquiring about 3 mg. of Hg vapor/cu. meter of air.[390] When Hg is spilled on floors and dirty tables evaporation is facilitated by the dust and grease, which maintain minute globules and provide large air-exposed surface areas. Grease, however, tends to retard Hg evaporation. In an investigation of 61 laboratories in the United States, the air in 10 contained from 0.02 to 0.07 mg. of Hg/cu. meter; in 28, less than 0.004 mg./cu. meter,[391] and in most of the others the concentration was from 0.04 to 0.4 mg./cu. meter (McCarroll[391]). The use of hot Hg in induction furnaces has caused acute intoxication; a concentration of 7000 mg./cu. meter was recorded in the surrounding air in one instance.[392] The use of Hg as a heat-transfer liquid in boilers for power generation is a potential source of danger. Large quantities of Hg used in metallurgy, for the extraction of Au and Ag, and the use of Hg amalgams, offer sources of exposure. In one instance, the air of a room in which Cu amalgam was heated contained 8.5 mg. of Hg/cu. meter.[393] Dental amalgams may be responsible for a significant part of the body burden of Hg in individuals who have had no industrial or therapeutic absorption.[394]

The danger of poisoning in mining and extraction of Hg has long been recognized, and studies continue to show excessive air concentrations of Hg in mines and mills. Similarly, in various countries reports continue to appear of excessive exposures in thermometer factories and in large chemical plants.

3. Physical and Chemical Properties

The physical and chemical properties of Hg and some of its compounds are listed in Table 23.

[390] A. C. Giese, Science, 91, 476 (1940).

[391] M. Shepherd, S. Schuhmann, R. H. Flinn, J. W. Hough, and P. A. Neal, J. Research Natl. Bur. Standards, 26, 357 (1941); C. F. McCarroll, U. S. Bur. Mines Rept. Invest. No. 3475 (1939); J. Gillis, J. Ind. Hyg. Toxicol., 26, 137A (1946); L. E. Renes and H. E. Seifert, Ind. Hyg. Quart., 7, 21 (1946).

[392] L. Jordon and W. P. Barrows, Ind. Eng. Chem., 16, 898 (1924).

[393] P. A. Neal and A. S. Gray, et al., J. Ind. Hyg. Toxicol., 22, 144A (1940).

[394] L. J. Goldwater, School Public Health, Columbia Univ., personal communication, 1959.

TABLE 23

Physical and Chemical Properties of Hg and Some of Its Compounds

	Chemical symbol	At. or mol. wt.	Sp. gr.	M.p., °C.	B.p., °C.	Solubility
Mercury	Hg	200.61	13.546	−38.87	356.58	Insol. hot or cold H_2O, dil. HCl, HBr, HI Sol. HNO_3
Mercuric oxide (montroydite)	HgO	216.61	11.14	Dec. 500		52 mg./liter(25°) 395 mg./liter(100°) Sol. acids Insol. alc., ether, acetone, alkalies, NH_3
Mercuric sulfide (cinnabar)	HgS	232.68	8.10	Subl. 583.5		10 μg./liter(18°) Sol. Na_2S, aq. reg. Insol. HNO_3, alc.
Mercuric chloride	$HgCl_2$	271.52	5.44(25°)	276	302	36 g./liter(0°) 69 g./liter(20°) 613 g./liter(100°) 330 g./liter alc. (25°) 250 g./liter ether Sol. acetic acid, pyridine
Mercurous sulfate	Hg_2SO_4	497.29	7.56	Dec.	Dec.	600 mg./liter(25°) 900 mg./liter(100°) Sol. H_2SO_4, HNO_3
Mercuric acetate	$Hg(C_2H_3O_2)_2$	318.7	3.27	Dec.		250 g./liter(10°) 1 kg./liter(100°) Sol. alc., acetic acid
Mercuric fulminate	$Hg(NCO)_2$	284.65	4.42	Exp!		Sl. sol. cold H_2O Sol. hot H_2O, alc., NH_4OH
Dimethyl mercury	$Hg(CH_3)_2$	230.68	3.069		96	Sol. alc., ether
Ethyl mercuric chloride	C_2H_5HgCl	265.13	3.482	193		Insol. cold H_2O V. sol. hot alc. Sl. sol. ether
Phenyl mercuric acetate	$C_6H_5HgO_2C_2H_3$	336.75		149		Sl. sol. hot or cold H_2O Sol. glacial acetic acid, benzene, alc.

Hg is the only metal that is liquid throughout usual temperature ranges. When solidified, Hg contracts and is highly ductile. Frozen Hg is used in the investment-casting industry because it has a low volume change on melting and has the property of self-welding. The vapor pressure of Hg is somewhat irregular as are some of its other properties; for approximate industrial calculations the following formula will serve from 0 to 150°C.:

$$\log P = -3212.5/T + 8.025$$

where P is pressure in mm. and T is absolute temperature. Hg has a relatively high surface tension, 480.3 dynes/sq. cm. vs. 75.6 for H_2O, and thus exhibits a reversed meniscus in a glass tube.

Chemically pure Hg is stable at ordinary temperatures, not reacting with air, O_2, CO_2, N_2O, or NH_3. In moist air Hg may oxidize slowly, forming a film of mercurous oxide (Hg_2O). In general, however, the film that forms is that of the oxides of traces of contained impurities. When subjected to prolonged heating, red HgO forms, which decomposes to Hg if heated above 500°C. Mercury combines with S and the halogens readily, but is relatively inert toward mineral acids except nitric acid. When Hg is in excess, or if unheated, the mercurous salts form, whereas the mercuric salts form with excess acid or with heat. Mercuric Hg forms basic salts of the type $Hg(OH)Cl$, $HgNH_2Cl$, $Hg=N—HgNO_2$, $Hg_2NI\cdot H_2O$. These salts are either infusible or insoluble and are formed when a base (HOH,NH_3,- NH_4OH) acts on the corresponding Hg salt. A solution of potassium mercuriiodide in KOH is Nessler's reagent, which becomes yellow upon contact with traces of NH_3 making it a valuable reagent for detecting traces of NH_3. A large series of simple organic Hg derivatives are known both of the alkyl and aryl series of the formula R_2Hg, as well as salts of the mono alkyl and aryl Hg ($RHgX$), and a vast number of more complex organic derivatives that have found use in medicine.

4. Analytical Determination

When Hg is present in the air wholly as a vapor, an Hg-vapor detector may be used as a preliminary survey instrument. The detector utilizes the ultraviolet resonance absorption line of 2537A., where Hg absorbs most strongly providing sensitivities of the order of a few tenths part per billion; interfering substances such as ozone, nitrobenzene, or nickel carbonyl absorb respectively from 800 to 500 times less strongly;[395] most other common industrial vapors absorb far less strongly, for example, sulfur dioxide, 33,000, and nitrogen dioxide, 80,000 times less. Various simple methods are available for detecting leaks of Hg vapor; for example, gold chloride on silica gel bleaches or turns gray when the air concentration of Hg exceeds 1 μg./liter.[396] Similarly, paper coated with selenium sulfide blackens in Hg vapor because of the formation of HgS.[397]

[395] D. J. Troy, *Anal. Chem.*, **27**, 1217 (1955).
[396] K. Grosskopf, *J. Ind. Hyg. Toxicol.*, **20**, 21A (1938).
[397] L. R. Biggs, *J. Ind. Hyg. Toxicol.*, **20**, 161 (1938).

Hg-containing dusts in the air may be trapped in an impinger, and dissolved in aqua regia, acidified $KMnO_4$, or KI_3 solution. When considerable dust is associated with Hg vapor, it is recommended that this sampling procedure be used in conjunction with a colorimetric determination, as the total Hg exposure may not be detected by the vapor detector alone. The colorimetric procedure of choice is the dithizone method recommended by the Committee on Analytical Methods of ACGIH.[398] Procedures are given also for its application to biological specimens. A rapid, sensitive method for the determination of Hg in urine is given by Miller and Swanberg;[399] using an acid peroxide digestion, as little as 1 μg. of Hg per 100 ml. of urine may be determined. A spectrographic method suitable for the quantitative determination of microgram quantities of Hg in types of specimens of industrial hygiene interest has not been reported.

5. Physiological Response

Hg and its salts have long been recognized to be general cellular poisons and effective protein precipitants because of their release of Hg^{2+} ion.

Acute Toxicity. The lethal oral dose of $HgCl_2$ for dogs is 10 to 15 mg./kg.;[400] intravenously 4 to 5 mg./kg.;[401] .for cats, the lethal dose is somewhat larger.[402] The oral dose of calomel (HgCl) producing systemic effects in the dog is 210 mg./kg.[403] Poisoning can follow inunction of Hg by absorption through the skin.[404] Certain Hg salts that do not ionize to the extent of $HgCl_2$, such as $Hg(CN)_2$, HgI_2, and mercuric salicylate, are less irritating to the tissues and less toxic than $HgCl_2$. In general, organic mercurials by vein are less irritant and less toxic than the inorganic salts.

Acute Hg intoxication following oral intake of ionizable salts of Hg is characterized by pharyngitis, dysphagia, abdominal pain, nausea and vomiting, bloody diarrhea, and shock. Later, swelling of the salivary glands, stomatitis, loosening of the teeth, nephritis, anuria, and hepatitis occur. Severe damage has been produced by Ashe *et al.*[405] in the kidneys, liver, brain, heart, lungs, and colon of rabbits exposed for a single 4-hour period to Hg vapor at an average concentration of 28.8 mg./cu. meter; mild damage to most of these organs occurred from a single hour of exposure. Acute intoxication from inhaling Hg vapor in high concentration used to be common among those who extracted Hg from its ores;

[398] *Manual of Analytical Methods.* Secy. Am. Conf. Govtl. Ind. Hygienists.

[399] V. L. Miller and F. Swanberg, *Anal. Chem.,* **29,** 391 (1957).

[400] R. E. Carratala and C. Guerra, *Rev. assoc. méd. arg.,* **49,** 926 (1936).

[401] W. D. Sansum, *J. Am. Med. Assoc.,* **70,** 824 (1918).

[402] W. Modell and S. Krop, *Proc. Soc. Exptl. Biol. Med.,* **55,** 80 (1944).

[403] G. B. Valeri, *Arch. intern. pharmacodynamie,* **19,** 315 (1910).

[404] J. F. Schamberg, G. W. Raiziss, and G. L. Garvan, *J. Am. Med. Assoc.,* **70,** 142 (1918); C. Stein, *Deut. med. Wochschr.,* **34,** 2126 (1908).

[405] W. F. Ashe, E. J. Largent, F. R. Dutra, D. M. Hubbard, and M. Blackstone, *Arch. Ind. Hyg. Occupational Med.,* **7,** 19 (1953).

now it is relatively infrequent, although cases still occur from time to time.[406] The condition is characterized by a metallic taste, nausea, abdominal pain, vomiting, diarrhea, headache, and sometimes albuminuria. After a few days, the salivary glands swell, stomatitis and gingivitis develop, and a dark line of HgS forms on the inflamed gums. The teeth may loosen, and ulcers may form on the lips and cheeks. In milder cases, recovery occurs within 10 to 14 days, but in others poisoning of the chronic type may ensue, accompanied by muscular tremors and psychic disturbances. Some of the acute cases have resulted from exposure concentrations of from 1.2 to 8.5 mg. of Hg/cu. meter.

Chronic Toxicity. The chronic form of mercurialism in man has been so long observed and well documented that it has tended to discourage the study of long-term effects in animals. At least two important contributions have appeared on this aspect, however, since the first publication of this volume.

Animals. Fitzhugh et al.[407] in studying the comparative effects of mercuric acetate, Hg (Ac)$_2$, and phenylmercury acetate in the diet of male and female rats for as long as 2 years, found that the latter compound was considerably more toxic than (HgAc)$_2$. Phenylmercury acetate produced renal lesions in the female at a dietary level of 0.5 p.p.m. Hg, whereas from 10 to 20 times as much Hg was required to produce similar effects with Hg(Ac)$_2$. Slight renal injury was produced even at a level of 0.1 p.p.m. Hg from phenylmercury acetate. Increased retention of Hg in the kidney was related in a general way to increased injury in this organ, although it should be noted that there was no greater retention of Hg in the female kidney, which showed greater injury, than in the male. Apparently, it is greater absorption and not renal retention of Hg that determines toxicity in this instance (see also section on Mechanism, page 1099).

Ashe et al.,[405] in studying the responses of animals to Hg vapor in repeated daily exposures for as long as 83 weeks and at levels that included from 6 to 0.1 mg. of Hg/cu. meter, found severe damage to kidney, heart, lung, and brain of rabbits after 6 weeks at 6 mg. of Hg/cu. meter, but no microscopic indication of tissue damage or of altered kidney function in dogs after 83 weeks' exposure at 0.1 mg. of Hg/cu. meter. Although the intermediate level, 0.86 mg./cu. meter, produced significant amounts of brain and kidney injury at 6 weeks this disappeared on cessation of exposure. Ashe et al. point out the greater susceptibility of renal tissue of their animals compared to that of man and thus the demonstration in the case of Hg-vapor exposure that the results in animals cannot be applied quantitatively to man. It is unfortunate that in this and other animal studies, the cat, which mimics most closely man's response to metals affecting the nervous system, was not studied.

Man. The insidious chronic form of mercurialism continues to occur from both inorganic and organic Hg compounds. It may appear after a few weeks of

[406] B. A. Warren, *U. S. Veterans, Bur. Med. Bull.* No. **6,** 39 (1930).

[407] O. G. Fitzhugh, A. A. Nelson, E. P. Lang, and F. M. Kunze, *Arch. Ind. Hyg. Occupational Med.*, **2,** 433 (1950).

exposure, or it may be delayed for much longer periods. Psychic and emotional disturbances are characteristic; the victim becomes excitable and irascible, especially when criticized. He loses the ability to concentrate mentally and becomes fearful, indecisive, or depressed, and may complain of headache, fatigue, weakness, loss of memory, and either drowsiness or insomnia. Objectively, he exhibits a fine tremor, and is unsteady in attempts to perform fine motions. The tremor may affect the hands, head, lips, tongue, or jaw. His writing is affected, with letters omitted, or even becomes illegible. Other neurological disturbances include paresthesias, affections of taste or smell, neuralgia, and dermographism. Other signs of systemic disease occur with less regularity. Signs of renal disease, however, are common; chronic nasal catarrh and epistaxis are not unusual. Salivation, gingivitis, and digestive disturbances are common. Stomatitis is sometimes severe. Ocular lesions occur, such as amblyopia, scotomas, and, particularly from organic mercurials, narrowing of the visual fields. In general, symptoms from organic Hg exposure are confined more specifically to the nervous system. Most patients show slow recovery when removed from exposure.

Metabolism. Estimates in the 1930's of the average daily amount of Hg entering the body from food vary from 5 μg., according to Stock in Germany,[408] to 20 μg. in the United States, according to Gibbs et al.[409] Little consideration seems to have been given to dental amalgams which, according to Goldwater,[394] are major contributors to Hg body-burden, at least in individuals in the United States who have had dental amalgam fillings. Other external sources, such as food, beverages, and packaging, contribute unknown but small amounts, a few μg./day at most. Blood Hg levels of 7.5 to 9 μg.% are not uncommon among otherwise nonexposed New Yorkers; urine levels may average from 6 to 16 μg./liter; significant elevations of urinary Hg values follow new fillings of dental amalgams. According to Stock[408] and Borinski,[410] normal individuals (Europeans) excrete 0.5 μg. of Hg in the urine, and 10 μg. in the feces daily. Organs of unexposed persons may contain up to 0.05 μg./g. of nerve and heart (Forbes et al.)[410] although the kidney may contain as much as 50 times this amount. Striated muscle contains on the order of 0.02 μg./gram.

Before proper inferences concerning man may be drawn from data on Hg metabolism in animals, it should be noted at the outset that not only do animal species vary considerably from one to another in respect to their absorption of inhaled Hg and excretion in the urine, but they vary widely in these respects from man. Moreover, the data on animals have been obtained largely from much shorter exposures, often before effects of renal injury from Hg on excretion make themselves felt. The study of Ashe et al.,[405] which probably provides the best data to date on Hg excretion, showed that the U/A ratio of excreted Hg in the urine to

[408] A. Stock, *Arch. Gewerbepathol. Gewerbehyg.*, **7**, 388 (1936).

[409] O. S. Gibbs, H. Pond, and G. A. Hausmann, *J. Pharmacol. Exptl. Therap.*, **72**, 16 (1941).

[410] P. Borinski, *Klin. Wochschr.*, **10**, 149 (1931); R. M. Forbes, A. R. Cooper, and H. H. Mitchell, *J. Biol. Chem.*, **209**, 857 (1954).

the inhaled Hg vapor concentration at the threshold limit (0.1 mg./cu. meter) was 0.14 for the rabbit, 0.4 for the dog after the animals had attained a steady state (several weeks). The U/A ratio for man, exposed for months and years to combined Hg vapor and Hg dusts at the same Hg level, was 7, a striking difference, indicating a far greater capacity either for pulmonary deposition, absorption, or urinary excretion of Hg in man than animals. As the Hg exposure concentration increased the ratio approached 3;[411] smaller decreases were noted in animals on increased exposure (See Section on organic Hg for U/A ratio, page 1102). It is the general conclusion of Goldwater,[394] Moskowitz,[411] and Grandjean[412] that there may be no correlation between urinary Hg excretion and clinical evidence of Hg poisoning, chiefly because, as indicated above, prolonged exposure may lead to diminished Hg excretion in the urine because of renal injury and possibly other factors. These conclusions do not deny the possibility of urinary Hg values serving as useful guides in early periods of exposure. Storlazzi and Elkins[413] believe that the U/A ratio affords a measure of the extent to which retention of Hg occurs. This is true as long as the type of Hg exposure is identical in the comparable case. In the past,[414] variations in the type of Hg exposure (vapor or dust) and the variations in the resulting pulmonary deposition, blood absorption, tissue retention, and renal excretion have given rise to apparent and disturbing inequalities in the U/A ratio, among worker groups that have been studied, that have not permitted the assignment of a urinary threshold limit for Hg, as for Pb. The early suggestion of Koelsch of a urinary threshold limit of 0.1 mg. of Hg/liter has not been borne out in this country from studies of Neal, Goldwater, Moskowitz, or Grandjean; clinical evidence of poisoning has been seen in persons with both lower and considerably higher urinary levels than 0.1 mg./liter when exposed to inorganic Hg. Friberg[415] finds no definite case of poisoning among Scandinavian workers exposed for long periods to inorganic Hg below 150 µg./liter of urine, and feels that 200 to 300 µg. of Hg/liter of urine can be tolerated for 10 years without manifest poisoning. On the other hand, Lundgren and Swensson,[416] in a study of 8 cases of organic (alkyl) Hg poisoning, report that symptoms may appear when the urine contains 50 to 100 µg. of Hg/liter, and that urinary Hg is a more sensitive test of exposure than clinical symptoms. Obviously there are not only wide differences in resultant urinary Hg levels among workers exposed to various inorganic Hg compounds, but exposure to alkyl Hg compounds presents a still further difference in this relation. Hg and Hg vapor may be found in the urine and feces for months after exposure has ceased, the average urinary concentration decreasing logarithmically

---

[411] S. Moskowitz, *N. Y. State Dept. Labor Monthly Rev.*, **29**, No. 5, 17 (1950).

[412] von H. Turrian, E. Grandjean, and V. Turrian, *Schweiz. Med. Wochschr.*, **86**, 1091 (1956).

[413] E. D. Storlazzi and H. B. Elkins, *J. Ind. Hyg. Toxicol.*, **23**, 459 (1941).

[414] F. F. Heyroth in F. A. Patty, ed., *Industrial Hygiene and Toxicology*, Vol. II. 1st ed., Interscience Publishers, New York, 1949, p. 720.

[415] L. Friberg, *Nord. Hyg. Tidskr.*, **1951**, 9 (abstract).

[416] K. D. Lundgren and A. Swensson, *J. Ind. Hyg. Toxicol.*, **31**, 190 (1949).

with time.[393] Urine samples collected from 23 workers during the fourth, ninth, eighteenth, and thirty-first week following cessation of exposure to high concentrations of Hg vapor, contained the following respective average concentrations, 0.54, 0.32, 0.17, and 0.07 mg./liter, but the range of variation was wide. On the other hand, excretion of Hg from alkyl Hg compounds is relatively rapid, being complete in from 1 to 3 months.[416]

More recent experimental studies in animals (rats and dogs) using $Hg(NO_3)_2$, phenyl mercuric acetate, and methyl mercuric hydroxide tagged with Hg^{203} by Swensson et al.[417] showed considerable differences in metabolic activity after intravenous administration. Both organic mercurials became bound to erythrocytes, but Hg from $Hg(NO_3)_2$ was bound to plasma proteins. Initially high blood levels of Hg fell rapidly in 5 to 10 minutes for all three compounds, and Hg appeared immediately in the urine. The methyl Hg compound was the slowest to be excreted. Deposition was greatest in the kidneys except for CH_3HgOH, which was more uniformly distributed in the tissues. Deposition of Hg in the central nervous system was small for all three compounds and showed no systematic differences. When $HgCl_2$, $Hg(NO_3)_2$, phenyl mercuric acetate, CH_3HgOH, and cyanomethyl mercuriguanidine were similarly labeled and fed in the drinking water to rats for periods up to 3 weeks at a level of 2 p.p.m. Hg, higher blood levels were attained from both types of the organic mercurials, and were bound to the erythrocytes; similarly higher values of Hg were found in the brain, liver, and kidney from the organic compounds by a factor of severalfold. A constancy was believed found for the ratio of the concentrations in the blood and brain, indicating a distribution equilibrium for each compound. Friberg,[418] in a comparative study of the metabolism of $HgCl_2$ and methyl mercury dicyandiamide $[CH_3HgNHC-(=NH)NHCN]$, found nearly 100-fold greater amounts Hg from the organic compound in the blood and 10-fold greater amounts in the cerebrum, cerebellum, brain stem, and spleen than were found from $HgCl_2$, upon repeated subcutaneous injections. Hg from $HgCl_2$, however, was twice as concentrated in the kidney. Strikingly, Hg bound in the brain from $HgCl_2$ could not be exchanged by subsequent Hg administration as it could from other organs.

Body clearance of inorganic Hg from $Hg(NO_3)_2$ by rats given small, essentially nontoxic intramuscular and intravenous injections (0.2 mg. of Hg/kg.), both singly and repeatedly, occurred in three phases, according to Rothstein and Hayes.[419] In the first, rapid phase, 35 per cent of the dose was cleared in a few days; in the second phase, 50 per cent of the dose cleared had a half-time of 30 days, and in the third phase, 15 per cent required 100 days for clearance. Tissue

[417] A. Swensson, K. D. Lundgren, and O. Lindstrom, A.M.A. Arch. Ind. Health, 20, 432, (1959).

[418] L. Friberg, A.M.A. Arch. Ind. Health, 20, 42 (1959); Acta Pharmacol. Toxicol., 12, 411 (1956).

[419] A. Rothstein and A. Hayes, Univ. Rochester Atomic Energy Comm. Rept. UR-556, 1959.

distribution was about the same regardless of dose except for the kidney, which accumulated less at smaller doses and with lessened urinary excretion. No special localization of Hg^{203} from the inorganic nitrate salt was seen in the brain.

A comparative study of the localization of Hg from $HgCl_2$ and phenyl mercury acetate following subcutaneous administration in the rabbit showed Hg to accumulate in the collecting tubules, the distal portions of the proximal convoluted tubules, and the wide parts of Henle's loop. No mercury was demonstrated in the glomeruli, and no differences due to compound were seen.[420]

Skin Absorption. Hg compounds, particularly the unionized or slightly ionized types, and metallic Hg itself may be absorbed through the skin in small amounts. The presence of Hg in the urine has been demonstrated following application of Hg antiseptics.[413] One gram of 10 per cent ammoniated Hg ointment ($HgNH_2Cl$) applied daily for 1 month to man caused a total increase of 500 μg. of Hg in the excreta for that time.[409] Erythema did not increase absorption.

It becomes clear from the foregoing that the metabolism and physiological behavior of Hg vapor, inorganic Hg salts, organic Hg salts, and true organic Hg compounds differ considerably depending on their chemical configuration and ionizing potential. It is not possible to generalize on the absorption, distribution, retention, and excretion of Hg from these varied structures.

Direct experimental evidence[421] now confirms this view at least in respect to mercurial diuretics. The action, and therefore the metabolism, of organomercurials are affected by pH; these Hg compounds are more unstable in acid medium, the affinity of Hg for —SH group is influenced by an acid medium. Thus, the one fact alone, that Hg is excreted complexed with cysteine,[422] the degree of which is in turn determined by pH, would seem to provide explanation for the first time for the varied excretion patterns of Hg not only among the various compounds, but among individuals.

Mechanism. There are at least three types of action of Hg for which it would be desirable to know the mechanism: the renal injury, nerve or brain injury, and diuretic action. Inhibition of succinic dehydrogenase by Hg has been histochemically localized in the distal segment of the rat kidney tubule.[423] Although formerly thought to demonstrate the site of action of mercurial diuretics, it has now been shown that certain Hg compounds can inhibit the dehydrogenase, but are nondiuretic.[424] Accordingly, this SH enzyme inhibition is now believed to be related to toxic injury of the kidney. This is also the conclusion of Bickers *et al.*[425] after having found an indiscriminate inhibition by a mercurial diuretic on 5 renal tubu-

[420] A. Bergstrand, L. Friberg, and E. Odeblad, *A.M.A. Arch. Ind. Health,* **17,** 253 (1958).

[421] C. T. Ray, *A.M.A. Arch. Internal Med.,* **102,** 1016 (1958).

[422] G. H. Mudge and I. M. Wiener, *Ann. N. Y. Acad. Sci.,* **71,** 344 (1958).

[423] K. K. Mustakallio and A. Tellka, *Science,* **118,** 320 (1953).

[424] R. H. Kessler, R. Lozano, and R. F. Pitts, *J. Pharmacol. Exptl. Therap.,* **121,** 432 (1957).

[425] J. N. Bickers, E. H. Bresler, and R. Weinberger, *J. Pharmacol. Exptl. Therap.,* **128,** 283 (1960).

lar enzyme systems, which was correlated in each instance with mercurial nephropathy. The 5 enzymes inhibited were succinic dehydrogenase, diphospho- and triphosphopyridine nucleotide diaphorase, glucose-6-phosphatase, and β-glycerylphosphatase. Morphological changes occurred earlier or at lower doses (3 to 4 mg. of Hg/kg.) than enzyme inhibition, and diuresis occurred at lower levels than were required for either. The precise means by which mercurials inhibit tubular resorption of Na is unknown, but there is some indirect evidence that extrarenal action of Hg may play a part.[421] In the kidney, not all of the Hg exerts diuretic action.[425] This has led to the unsupported view that trace amounts of Hg are locally liberated.

Organic mercurials have been shown by Cook and associates[426] to inhibit other enzyme systems, catalase, cytochrome oxidase, but the significance of these inhibitions for Hg action is unclear.

Industrial Experience. Reports of present-day industrial exposures concern for the most part vapors of Hg, either as elemental Hg or as alkyl Hg compounds; exposures to dusts of certain new organic structures occur, in which Hg is bonded to nitrogen as in methyl mercuridicyandiamide.[420] A relatively new type of exposure to Hg vapor in the frozen-Hg investment-casting process has been pointed out by Kramer and Goldwater,[427] with a description of the process and the requirements for health protection. Goldwater, Kleinfeld, and Berger[428] report a case of poisoning from Hg vapor in a university laboratory assistant. An outbreak of severe Hg poisoning among 22 of 36 employees who worked with an Hg–Cu–amalgam tamping compound was reported from Ohio by Benning;[429] a 3-year follow-up revealed improvement in the mildly affected cases, and reduction of symptoms in those more severely affected. An investigation by Dalhamn[430] of Hg levels in offices of dentists and assistants in Norway showed values to be below the recommended limit of 0.1 mg./cu. meter except in a few special instances. Application of antifouling paint, which requires preliminary heating and spraying, presents a danger of poisoning from its content of Hg.[431]

The irritant nature of $HgCl_2$, $Hg(NO_3)_2$, and HgI_2 to the skin among handlers of these substances is well known.

Signs of Poisoning. Despite the long clinical history of Hg poisoning there is still little emphasis or agreement on early signs of poisoning. Tara et al.[432] consider alveolar destruction of the maxillae as an early sign of mercurialism, as

[426] M. R. Sohler, M. A. Seibert, C. W. Kreke, and E. S. Cook, *J. Biol. Chem.,* **198,** 281 (1952); M. A. Seibert, C. W. Kreke, and E. S. Cook, *Science,* **112,** 649 (1950).

[427] I. R. Kramer and L. J. Goldwater, *A.M.A. Arch. Ind. Health,* **13,** 29 (1956).

[428] L. J. Goldwater, M. Kleinfeld, and A. R. Berger, *N. Y. State Dept. Labor Monthly Rev.,* **35,** No. 4, 13 (1956).

[429] D. Benning, *Ind. Med. and Surg.,* **27,** 354 (1958).

[430] T. Dalhamn, *Arch. Ind. Hyg. Occupational Med.,* **7,** 358 (1953) (abstract).

[431] L. J. Goldwater and C. P. Jeffers, *J. Ind. Hyg. Toxicol.,* **24,** 21 (1942).

[432] S. Tara, Y. Deplace, and A. Cavigneaux, *Arch. maladies process. méd. travail et sécurité sociale,* **13,** 478 (1952).

detected by x-ray, although it is recognized alveolysis is not unique to mercurialism. It is detectable, however, when there are no other signs except slight muscular tremors, loss of appetite, nausea, and diarrhea. Friberg[415] considers tremors the best early sign. Psychic and emotional disturbances are later signs, with an incidence of under 20 per cent; renal involvement is found in about one third of the cases; a somewhat higher incidence of cardiovascular disturbances, a tendency to be underweight, and ocular effects, consisting of restriction of the visual fields, also occur. Hunter and Lister[433] describe "mercurialentis" as a colored reflex from the anterior capsule of the lens, believed to result from deposition of Hg in a person with protracted exposures. Severe disturbances, such as mania, paralytic dementia, paralysis, and peripheral neuropathy, common in the past, are now rarely seen because of improved occupational procedures. A tendency toward lowered red-cell count, increased cell size and hemoglobin values, and absolute monocytosis, reported by Shoib et al.,[434] has also been noted by Grandjean.[412]

Organic Mercury Exposures. Organic mercury compounds, particularly alkyl derivatives, such as dimethyl Hg, and ethyl Hg halides and phosphate, are highly toxic, resulting in ataxia, dysarthria, constricted visual fields, and altered plantar reflexes. There is an intense and widespread degeneration of certain sensory nerve paths, the peripheral nerves and posterior spinal roots being affected first, the spinal cord later. There is also degeneration of certain neurons of the middle lobe of the cerebellum. The alkyl Hg halides are also irritating to the skin and may give rise to severe dermatitis.[435]

Dimethyl Hg is extremely toxic, producing a condition that combines the characteristics just described with those of acute mercurialism.[436] Small amounts cause weakness that lasts for weeks or months. Two chemists who participated in the early development of this substance were fatally poisoned, and 2 stenographers working in the vicinity of 20,000 lbs. of stored diethyl Hg were fatally poisoned.

Hg fulminate can give rise to dermatitis, with erythema, intense itching, edema, papules, pustules, and deep ulcers, "powder holes," especially on the tops of the fingers.[437] Fatigue, irritation of the conjunctiva and respiratory tract, and headache occur, along with allergic manifestations in some persons. Sensitized persons may show a fall in blood pressure, leucopenia, albuminuria, and edema of the face. Preventive measures have been described.[438]

The use of phenyl mercury oleate as a fungicide and wood preservative presents a dermatological hazard.[439]

[433] D. Hunter and A. Lister, *Brit. J. Ophthalmol.,* **37,** 234 (1953).
[434] M. O. Shoib, L. J. Goldwater, and M. Sass, *Am. Ind. Hyg. Assoc. Quart.,* **10,** 29 (1949).
[435] F. J. Vintinner, *J. Ind. Hyg. Toxicol.,* **22,** 297 (1940).
[436] W. H. Hill, *Can. Public Health J.,* **34,** 158 (1943).
[437] L. Naro, *Acta Med. Scand. Suppl.,* **120,** 95 (1941); Swanston, *Proc. Roy. Soc. Med.,* **36,** 633 (1943); A. Jordi, *Schweiz. med. Wochschr.,* **77,** 621 (1947).
[438] *Natl. Safety News,* **44,** 38 (1941); *J. Ind. Hyg. Toxicol.,* **24,** 93A (1942).
[439] C. P. McCord, S. F. Meek, and P. A. Neal, *J. Ind. Hyg. Toxicol.,* **23,** 466 (1941).

Lundgren and Swensson[416] classified their cases of organic poisoning derived from seed-treating and wood-preserving operations into (a) local action on skin and mucous membranes, (b) acute, systemic poisoning, and (c) chronic poisoning from protracted, low-grade exposures. Early dermal reactions developing in a few days to several weeks after exposure are warmth, redness, swelling, and burning, followed often by blister formation, which on breaking presents a sodden, grayish appearance. Irritating dryness of the mouth and blisters of mouth and throat may occur. These reactions usually disappear rapidly upon termination of exposure. In addition to symptoms above described for alkyl mercurial derivatives, Lundgren and Swensson quote German reports of respiratory- and alimentary-tract involvement, myalgia, and a transient albuminuria. In cases of severe poisoning, the physical defects and the mental deterioration may remain. Hg excretion in the urine of their 8 cases was so variable as not to permit of an estimate of an allowable urinary Hg limit, although 1 poisoning case showed only 50 to 100 μg. of Hg/liter. Unfortunately, their excretion curves are determined on relatively few points. The Hg exposure levels are not given nor are the structures of the Hg compounds in all cases stated. Moreover, in view of the ease with which losses occur during analysis for organic Hg, the values for urinary Hg would have been more assuring had the authors[416] given the details of their method.

Ahlmark,[440] in reporting on cases of exposure to methyl Hg compounds, concludes that the threshold limit for organic Hg should be less than one tenth the present limit used for inorganic Hg vapor, and that urinary Hg levels should be kept below 10 to 15 μg./liter. A careful study for $5^{1}/_{2}$ years of 20 workers exposed to four different organic Hg compounds, ethyl Hg chloride or phosphate, and ethyl or phenyl Hg acetate, by Dinman et al.[441] failed to reveal any consistent symptoms of Hg poisoning at Hg concentrations in the air between 0.01 and 0.1 mg. of Hg/cu. meter. On occasion, Hg levels considerably exceeded this range for brief periods. Workers studied were exposed 3 or more consecutive weeks or exhibited urinary levels greater than 100 μg./liter. Particular attention was given to accurate sampling and analysis for organic mercury. As characteristically found in other studies, the urinary excretion of Hg was individualized and irregular; if any rough U/A ratio could be derived from their combined data it approximated 1, a value far below that (3 to 7) derived by others on workers exposed to Hg vapor, but closer to that obtained by Ashe[405] for dogs similarly exposed. The only statistically significant correlation found was that between cumulative exposure time and urinary excretion, a considerable period (2 months) after exposure. Dinman et al.[441] feel that air concentrations of Hg express but inadequately the total exposure and that other portals of entry contribute to the exposure. The conclusion was reached that the present level set for inorganic Hg vapor of 0.1 mg./cu. meter offers sufficient protection from repeated daily exposures to ethyl Hg acetate, chloride, phosphate, and phenyl Hg acetate; and that the conclusion of the Scan-

[440] A. Ahlmark, Brit. J. Ind. Med., 5, 177 (1948).
[441] B. D. Dinman, E. E. Evans, and A. L. Linch, A.M.A. Arch. Ind. Health, 18, 248 (1958).

dinavians was erroneously based on recommending 0.01 mg. of Hg/cu. meter as a limiting exposure for these compounds; errors were considered to stem from analytical losses of volatile Hg, which would yield apparent low values for Hg for both air and urine samples. Losses of up to 100 per cent commonly occur in all but the most scrupulously designed analytical methods.[442]

Therapy and Protection. BAL has been found therapeutically effective in cases of acute and chronic mercurialism in which it has been tried. Longscope[443] found BAL to be effective in acute poisoning by $HgCl_2$. To be effective in these cases BAL must be given promptly. Adults were given 300 mg. of BAL on admission, followed by 150 mg. within a few hours. Subsequent doses were given up to 3 or 4 days if the seriousness of the poisoning warranted it. No symptoms of overdosage were seen. Gastric lavage of 5 to 10 per cent Na formaldehyde sulfoxylate solution was given directly on admission. BAL was also effective in increasing urinary Hg excretion with good clinical response in chronic poisoning cases, in a single instance in a worker poisoned by exposure to Hg fulminate[444] and in a case of combined Pb- and Hg-vapor poisoning.[445] In none of the 3 reports was $CaNa_2EDTA$ found effective. Woodcock[446] more recently has reported $CaNa_2$-EDTA effective in treating a case of chronic Hg-vapor poisoning, however. Earlier treatment with BAL did not increase urinary Hg excretion nor improve the patient's condition. The mechanism of the effect of BAL on Hg excretion in the urine, in acute poisoning in dogs with an organic mercurial diuretic (mercuric cysteine), was found to be pH-dependent,[447] increasing in alkaline urine, remaining unchanged in acid urine. Urinary pH had no effect, however, on removal of Hg from renal parenchyma. With other organic mercurials BAL either had no effect on excretion or the rate of urinary Hg excretion was independent of urinary pH.

Acetyl *dl*-penicillamine was found, by Elkins,[448] not to be particularly effective in treating chronic Hg poisoning in man; although the chelating agent increased urinary Hg (and Pb) excretion, there were no signs of clinical improvement.

Respirators offering protection against Hg vapor are described by Vouk.[449]

A striking demonstration of the capacity of the body to counteract the toxic effects of inorganic Hg in rats has been made by Surtshin *et al.*[450] Rats on high

[442] D. E. Davis and K. Linke, *Proc. Australian Pulp & Paper Ind. Tech. Assoc.*, **8**, 237 (1954).

[443] W. T. Longscope, *Bull. Ayer Clin. Lab. Penn. Hosp.*, **4**, 61 (1952).

[444] A. Hadengue, *et al.*, *Arch. maladies profess. méd. travail et sécurité sociale*, **18**, 561 (1957).

[445] R. F. Bell, J. C. Gilliland, and W. S. Dunn, *A.M.A. Arch. Ind. Health*, **11**, 231 (1955).

[446] S. M. Woodcock, *Brit. J. Ind. Med.*, **15**, 207 (1958).

[447] I. M. Weiner, K. Garlid, D. Sapir, and G. H. Mudge, *J. Pharmacol. Exptl. Therap.*, **127**, 325 (1959).

[448] H. B. Elkins, *Ann. Meeting Am. Ind. Hyg. Assoc.*, Rochester, N. Y., April, 1960.

[449] V. B. Vouk, Z. Topolnik, and M. Fugas, *Brit. J. Ind. Med.*, **10**, 69 (1953).

[450] A. Surtshin, et al., *Am. J. Physiol.*, **190**, 271, 278 (1957); K. Yagi and H. L. White, *Am. J. Physiol.*, **194**, 547 (1958); A. Surtshin and K. Yagi, *Am. J. Physiol.*, **192**, 405 (1958); A. Surtshin, M. Audia, and H. L. White, *Am. J. Physiol.*, **195**, 150 (1958).

sucrose diets fortified with vitamins can tolerate otherwise acute lethal doses of HgCl₂. The results are believed explained by the development of adaptive (inducible) enzymes containing —SH groups in amounts sufficient to bind added amounts of Hg. Higher amounts of Hg were associated with a soluble albuminlike fraction. Serum albumin injected into rats reduced renal toxicity. Moreover, renal pickup of Hg from HgCl₂ was reduced following sucrose diets, indicating a true increase in renal tolerance for Hg.

6. Hygienic Standards of Permissible Exposures

The limit of 0.1 mg. of Hg/cu. meter of air adopted by the American Conference of Governmental Industrial Hygienists would appear, from the ample evidence quoted in the foregoing sections, to provide freedom from effects of exposure to metallic Hg vapor with a relatively small margin of safety, and from exposure to inorganic Hg salts with possibly a somewhat larger factor of safety. Recently Dinman et al.[441] have presented evidence from an evaluation of 20 men exposed for 6 years to air concentrations of organic Hg exceeding 0.01 but below 0.1 mg./cu. meter, that the present limit of 0.1 mg. of Hg/cu. meter is also satisfactory for exposure to certain organic Hg compounds. The organic compounds concerned were ethyl Hg-chloride and phosphate adsorbed on clay, and the acetate salts of ethyl and phenyl Hg used in a solvent. Detailed data are lacking to determine whether the present limit for Hg vapor may also apply to such organic compounds of Hg as dimethyl and diethyl Hg. Accordingly, the recommended limit of 0.01 mg. of Hg/cu. meter would appear to provide a 10-fold factor of safety from certain organic compounds commonly used at present industrially and occupationally.

METAL CARBONYLS, $Me_x(CO)_y$

These substances represent a large group of compounds of the general formula shown above in which Me designates a metal and x and y are whole numbers. They differ from all other metallic compounds.

1. Preparation and Source

The metal carbonyls, with the exception of Cr, Re, Os, and Ir, are prepared by direct combination of metal, generally in finely divided form, with CO. This is the basis of the Mond process used since 1890 in industry; $Ni(CO)_4$ is produced in tonnage quantities by this process yearly. If a metal forms several carbonyls (see Physical Properties, page 1105), that one with the highest CO content is formed directly; the others form from this by subsequent reactions. Metal hydrocarbonyls may be prepared by acidification of a suitable salt of an organic base of the metal carbonyl. Thus, cobalt hydrocarbonyl forms by adding sulfuric acid to pyridinium cobalt carbonyl. A Grignard method is used to prepare carbonyls of Cr, Mo, and W; metal carbonyls of the Pt and Fe groups and Re form from the sulfide or halides. Nitrosyl carbonyls result from the action of NO on carbonyls of selected metals (Fe, Co).

Formation of metal carbonyls from methods other than industrial synthesis are the following: (a) in the steel industry in the Bessemer converter; amounts of carbonyl surviving, however, are small; (b) inadvertent introduction of CO onto Fe or other metal catalyst beds; (c) storage of CO in metal cylinders may result in concentrations of $Fe(CO)_5$ as high as 2000 p.p.m.; (d) slowly flowing water gas in iron pipe may result in significant amount of $Fe(CO)_5$; and (e) carbonyl probably exists in small quantities in the Fischer-Tropsch process for the liquefaction of coal.

2. Uses and Industrial Exposures

Metal carbonyls are used for the preparation of metals of high purity, and more recently as catalysts for organic reactions. The Mond process itself is a means of isolating Ni from its ore contaminants. $HCo(CO)_4$ is the catalyst for the OXO process that converts olefins to oxidized products. $Fe(CO)_5$ is used as a gasoline additive (0.2 per cent) in Europe similar to tetraalkyl Pb in the United States. The electronics industry utilizes pure metal powders from the carbonyls for radiofrequency transformers. In the research stage is the use of metal carbonyls for metal-filming. Extremely hazardous exposures exist about the Mond process because of the insidious nature of the exposure to carbonyls; in addition to the carbonyl itself exposure to free CO also occurs.

3. Physical and Chemical Properties

The properties and reactions of metal carbonyls (see Table 24) are presented in a number of reviews.[451-453] A bibliography to 1949 is also available.[454] The metal carbonyls are generally colorless or colored crystalline substances, except $Fe(CO)_5$, $Ni(CO)_4$, $Rh(CO)_5$, and $Os(CO)_5$, which are liquids. The volatility, conductivity, solubility, and other properties are consistent with their nonpolar structure. They are very reactive to oxygen and some are spontaneously flammable. They are thermally unstable. The chemical properties vary widely with the carbonyl, but certain generalities may be mentioned. The halides of carbonyl are known $(CO)_5X_2[X = halogen]$ formed directly from the halogen and, for example, $Fe(CO)_5$. The hydrocarbonyls, $HCo(CO)_4$, for example, and the nitrosyl derivatives, $Fe(CO)_2(NO)_2$, $Co(CO)_3NO$, have been mentioned, as well as a large variety of metal carbonyl complexes with ammonia or organic bases. The volatile metal carbonyls have a characteristic odor which for $Ni(CO)_4$ is detectable between 1 and 3 p.p.m. Properties of the more important carbonyls are given below.

[451] A. A. Blanchard, *Chem. Revs.*, **21**, 3 (1937).
[452] W. E. Trout, Jr., *J. Chem. Educ.*, **14**, 453, 575 (1937); **15**, 77, 113 (1938).
[453] J. W. Smith, *Sci. Progr.*, **35**, 283 (1947).
[454] F. E. Croxton, *U. S. Atomic Energy Comm. Rept.*, **AECU-171** (K-365 Pt. II), Mar., 1949.

TABLE 24

Physical and Chemical Properties of Volatile Metal Carbonyls

Metal carbonyl	Vapor pressure, mm. Hg	Sp. gr., 25°C.	M.p., °C.	B.p., °C.	Synthesis temp., °C.	Initial dec. temp., °C.	Appearance
$Ni(CO)_4$	261(15°)	1.31	−25	43.2	30–50 (1 atm.)	50	Colorless liq.
$H.Co(CO)_4$	High		−26.2		−20	−26	Gas
$[Co(CO)_4]_2$	0.7(15°)	1.73	50		150(40 atm.)	50	Orange-red crystals
$[Co(CO)_4]_3$	Very low						Solid
$Co(CO)_3NO$	91(20°)	147	−15	78.6	−10	55	Cherry red liquid
$Fe(CO)_5$	35(25°)	1.45	−21	104.6	173(200 atm.)	150	Colorless liq.
$H_2Fe(CO)_4$	11(−10°)		−70		−20	−10	Gas
$Fe_3(CO)_9$		$2.085^{18°}$				80	Yellow-orange crystals
$Fe(CO)_2(NO)_2$	21(20°)	1.56	18.4	110	−10	50	Dark-red crystals
$Cr(CO)_6$	1(48°)		Sinters 90°	Subl. room temp.	0–4	130	Colorless crystals
$Mo(CO)_6$	2.3(55°)	1.96	Sinters 120°	Subl. 30–40°	0–4	150	Colorless crystals
$W(CO)_6$	1.2(67°)		Sinters 125°	Subl. 50°	0–4	100	Colorless crystals
$Ru(CO)_6$	Very volatile		−22		300(400 atm.)	−15	Colorless liq.

Nickel Carbonyl. Ni(CO)$_4$ is a colorless liquid insoluble in water, unreactive with aqueous acids and alkalies, but soluble in organic liquids. Air mixtures may explode at 20°C. and at partial pressures of 15 mm. Hg, O$_2$ rapidly decomposes the vapor to an amorphous hydrous nickel carbonate oxide, H$_2$SO$_4$ liberates CO and nickelous salts; CS$_2$ yields NiS and C. Thermodynamic data reviewed by Spice *et al.*[455] show that unless the pressure of CO approximates 1 atmosphere Ni(CO)$_4$ is almost completely dissociated; at a partial pressure of 1000 p.p.m. CO at 25°C. the equilibrium concentration of Ni(CO)$_4$ is 0.02 p.p.m. Decomposition is not instantaneous, however, because Kincaid, Strong, and Sunderman,[456] working with air mixtures of Ni(CO)$_4$ between 2 and 350 p.p.m., found about 5 per cent dissociated in 50 seconds, and 30 per cent dissociated in a chamber with 1 air change every 4 minutes.

Cobalt Hydrocarbonyl. HCo(CO)$_4$, containing a readily dissociable H, behaves like a strong acid, titratable with alkali, forming salts. Although it decomposes rapidly in air, pyridine solutions out of contact with air may be maintained at refrigerator temperatures for months, according to Palmes *et al.*[457] They also reported 20 per cent decomposition after 2.5 minutes of a 1.5 per cent carbonyl-N$_2$ mixture; in air complete decomposition in less than 2 minutes; in an animal exposure chamber with 1 air change per minute, 90 per cent decomposition.

Iron Pentacarbonyl. Fe(CO)$_5$, like Ni(CO)$_4$, is insoluble in water and unreactive in dilute acids. It may ignite spontaneously in air. Concentrated reducing acids yield ferrous salts, as do gaseous halogens; CCl$_4$ yields COCl$_2$, CO, and FeCl$_3$. Fe(CO)$_5$ is a strong reducing agent reducing ketones to alcohols, benzil to benzoin, and nitrobenzene to aniline (see also Vol. I, Chapter XVI, Section Two).

4. Analytical Procedure

The only justifiable method for the analysis of air-borne metal carbonyls is an instantaneous procedure because of the serious health hazard. A recording air analyzer for Ni(CO)$_4$, but which is applicable to all metallorganic vapors, has been described;[458] concentrations in the range of 0.05 to 4 p.p.m. may be recorded. Kincaid, Strong, and Sunderman[459] have described another type of recording air analyzer with a sensitivity of a few tenths p.p.m., as well as detailed procedure for chemical sampling and analyses of air for Ni(CO)$_4$ with a sensitivity of 2 p.p.b., but the sampling requires 50 hours at this level. By a combination of collection of the sample in an evacuated bottle containing halogen, and the Billion-Aire

[455] J. E. Spice, L. A. K. Stavely, and G. A. Harrow, *J. Chem. Soc.*, **1955**, 100.

[456] J. F. Kincaid, J. S. Strong, and F. W. Sunderman, *Arch. Ind. Hyg. Occupational Med.*, **8**, 48 (1953).

[457] E. D. Palmes, N. Nelson, S. Laskin, and M. Kuschner, *Am. Ind. Hyg. Assoc. J.*, **20**, 453 (1959).

[458] J. E. McCarley, R. S. Saltzman, and R. H. Osborn, *Anal. Chem.*, **28**, 880 (1956).

[459] J. F. Kincaid, J. S. Strong, and F. W. Sunderman, *Arch. Ind. Hyg. Occupational Med.*, **8**, 48 (1953).

Analyzer, it should be possible to measure metal carbonyls in the p.p.m. to p.p.b. range.

5. Physiological Response

With present toxicological information limited to 4 metal carbonyls, health hazards resulting from exposure to them range from extremely serious for $Ni(CO)_4$, moderately serious for $HCo(CO)_4$, to mild for $Co_2(CO)_8$ and $Fe(CO)_5$. The basis for this wide range of toxic behavior appears to reside in part on the toxic character of the metal component and in part on the volatility and stability; the more toxic the metal and the more volatile and stable the carbonyl, the more hazardous. Although the metal carbonyls exhibit widely different degrees of toxicity their toxicological behavior is sufficiently similar to allow general comparisons.

Acute Toxicity. According to recent studies,[457,459] the acute toxicity of $Ni(CO)_4$ for laboratory animals is about twice that of $HCo(CO)_4$ as far as could be determined with the limited stability of the latter; the 30-minute LC_{50} of $Ni(CO)_4$ for the rat is given in Table 25 along with corresponding data for other metal carbonyls.

TABLE 25

Metal Carbonyls by Inhalation—30-Minute LC_{50} Values for Rats

Metal carbonyl	30-min. LC_{50}		Toxicity relative to $Ni(CO_4)$
	Mg. carbonyl/ cu. meter	Mg. metal/ cu. meter	
$Ni(CO)_4$	240	85	1[a]
$HCo(CO)_4$	560	165	0.52[a]
$Fe(CO)_5$	910	260	0.33[b]
	2190	625 (mice)	0.14[b]
$Co_2(CO)_8$	1400[c]	480[c]	$\ll 0.17$[d]

[a] E. D. Palmes, N. Nelson, S. Laskin, and M. Kuschner, *Am. Ind. Hyg. Assoc. J.*, **20**, 453 (1953).

[b] F. W. Sunderman, B. West, and J. F. Kincaid, *A.M.A. Arch. Ind. Health*, **19**, 11 (1959).

[c] A concentration that for 60 minutes resulted in no toxic signs, not the LC_{50}.

[d] H. M. Armit, *J. Hyg.*, **9**, 249 (1909).

It is seen from Table 25 that the 30-minute LC_{50} for rodents of 3 other metal carbonyls is lower than that of $Ni(CO)_4$; moreover, the hazard from certain carbonyls [$Co_2(CO)_8$, $Fe(CO)_5$] is considerably less than $Ni(CO)_4$ because of their lower volatility.

Both the cutaneous and the subcutaneous toxicities of $Co_2(CO)_8$ appear to be low.[460] A 5 per cent solution of this carbonyl in ethanol injected subcutaneously in

[460] J. F. Kincaid, J. S. Strong, and F. W. Sunderman, *Arch. Ind. Hyg. Occupational Med.*, **10**, 210 (1954).

rats in doses up to 0.4 mg./rat (130 mg./kg.) showed no effects other than ulceration at the site of injection. Repeated daily application of a 7 per cent solution of $Co_2(CO)_8$ in methyl ethyl ketone for 18 applications produced only eschar formation, no evidence of systemic effect.

Appreciable absorption through the skin has been claimed for liquid $Ni(CO)_4$.

All reports indicate a similarity of acute symptoms and lesions resulting from exposure to the metal carbonyls. Immediately after exposure giddiness and headache occur, accompanied at times with dyspnea and vomiting. Exposure to fresh air brings relief of symptoms. From 12 to 36 hours later dyspnea returns, cyanosis and leucocytosis appear, and the temperature begins to rise. Cough with more or less bloodstained sputum occurs on the second day or later. The pulse rate increases, but not in proportion to the respiratory rate. Delirium and other signs of disturbance of the central nervous system appear. Death occurs in fatal cases between 4 and 11 days.

Pathological findings consist of hepatization of the lung resembling changes produced by phosgene. Alveoli are filled with fibrin with very few cells. Changes in capillaries and arterioles explain the presence of multiple petechiae in lungs and degenerative changes in brain medulla and upper spinal cord. In man and animals $Ni(CO)_4$ has been considered to produce acute chemical pneumonitis.[457,461] The work of Barnes and Denz[462] confirms and clarifies the pathological changes seen in man.

Chronic Effects. Cancer of the lung and nose has long been considered the long-term effect in Ni refinery (Mond Process) workers. Sunderman *et al.*,[463] in an attempt to validate these effects in animals, found that chronic exposure of rats to $Ni(CO)_4$ resulted in extensive pulmonary changes including a remarkable degree of squamous metaplasia of the bronchial epithelium. Although the incidence and degree of squamous metaplasia was insufficient to incriminate $Ni(CO)_4$ as a carcinogenic agent in rats, the evidence was considered suspicious. In addition to these effects the long-term (1 year) symptoms were impaired health, reduced growth, and higher mortality rates than among unexposed rats. Exposure concentrations of the various groups were between 30 and 60 mg. of $Ni(CO)_4$/cu. meter. Ni content of the liver between 6 months and 1 year was only slightly greater than that of the controls.

Metabolism. Information in this area has thus far been confined to (*a*) determining blood levels (absorption) of the metal in man following exposure to $Ni(CO)_4$,[461] (*b*) limited determinations of metal in tissues of man and animal,[457,461] and (*c*) urinary and fecal excretion of the metal during and following exposure.[457,464] Thus, interest in carbonyl metabolism has been directed chiefly

[461] F. W. Sunderman and J. F. Kincaid, *J. Am. Med. Assoc.*, **155,** 889 (1954).

[462] J. M. Barnes and F. A. Denz, *Brit. J. Ind. Med.*, **8,** 117 (1951).

[463] F. W. Sunderman, J. F. Kincaid, A. J. Donnelly, and B. West, *A.M.A. Arch. Ind. Health*, **16,** 480 (1957).

[464] R. E. Tedeschi and F. W. Sunderman, *A.M.A. Arch. Ind. Health*, **16,** 486 (1957).

to determining the usefulness of these measures as aids in detecting the degree of the exposure. Nickel balance studies in dogs, in which Ni content of food as well as that from Ni(CO)$_4$ was measured, indicated within the limits of the study that 90 per cent of the ingested Ni is excreted via the gastrointestinal tract, 10 per cent via the urinary tract. Exposure to Ni(CO)$_4$ markedly increased urinary Ni excretion, not fecal. The balance study indicated all inhaled Ni was excreted in 6 days and it was concluded that no significant retention of inhaled Ni occurs; no analyses of tissues (bone) were made, however.

Absorption of the metal carbonyls from the lung appears to be rapid; urinary concentration of Co from inhaled HCo(CO)$_4$ in animals is reduced to 50 per cent the day following exposure;[457] in man Ni concentration in urine had fallen to almost normal values in 3 days following a mild exposure.[461] Both blood and urine metal values increase following exposure to carbonyls; values in blood of exposed individuals vary widely from the normal mean of 11 μg. per 100 ml., ranging from 16 to 225 μg. per 100 ml. with a mean of 63 μg. (13 cases). Normal urine values average 1.1 μg. per 100 ml. Analysis of organs from fatal exposures to Ni(CO)$_4$ showed no Ni in the liver, however; small amounts in the lung (0.4 to 22 μg. per 100 g. of wet tissue); and variable amounts in the kidney. Thus Sunderman and Kincaid[461] have emphasized the practical usefulness of measuring urinary Ni of workers exposed to nickel carbonyl.

Mode of Action. Various conjectures have been proposed from time to time as to how Ni(CO)$_4$ exerts its toxic action, but none have been given experimental study. For example, Flury and Zernik[465] are of the opinion that the action is that of a catalytic poison influencing particularly the central nervous system and the metabolic processes, and also suggest that Ni separates as a colloid in the lung and other tissues. This suggestion finds no support from the work of Sunderman and Kincaid,[461] who reported no Ni in the livers of exposed animals, an organ that acts as a colloid trap. Studies are needed in this area.

Industrial Experience. The Mond Process workers constitute the only group on which reports of serious injury have appeared. One hundred and sixty cases of acute chemical pneumonitis that occurred in the United States have come to the attention of Dr. Sunderman.[466] Other reports of similar acute effects have appeared.[467] Acute poisoning is of two types, initial and delayed.[461] Exposures may be fatal. In exposures with survival, any pulmonary fibrosis that occurs eventually resolves and the disease is without sequelae.[467] A review of 354 cases of poisoning by the Mond Nickel Co. showed in order of frequency: 56 per cent, frontal headache; 42 per cent, giddiness; 33 per cent, tightness of chest; 28 per cent, nausea; 22 per cent, weakness of limbs; 16 per cent, perspiring; 15 per cent, cough; 14 per cent, vomiting; 9 per cent, cold and clammy skin; and 9 per cent, shortness of

[465] F. Flury and F. Zernik, *Schädliche Gases.* Springer, Berlin, 1931.

[466] F. W. Sunderman, personal communication, 1959.

[467] J. L. Carmichael, *Arch. Ind. Hyg. Occupational Med.,* **8,** 143 (1953); W. W. Brandes, *J. Am. Med. Assoc.,* **102,** 1204 (1934).

breath. Free, excess CO may be a common complicating agent in industrial $Ni(CO)_4$ exposures.

Cancer of the lung and nose has occurred with high incidence (5 times normal) among Mond Process workers after long exposures.[468] From 1923 to 1948, 49 cases of cancer of the nose with 46 fatalities, and 82 cases of cancer of the lung with 72 fatalities, occurred among Ni workers in England. The average number of work years before cancer of the nose developed was 23, and 25 for the lung. Doll and Morgan,[469] on the basis of statistical-epidemiological studies, see no clear evidence that supports the view that $Ni(CO)_4$ is the etiological agent; and the latter is inclined to incriminate As as the causal agent, a view not accepted by informed observers in this country. Sunderman's studies indicate cancer of the lung in rats may follow a single heavy exposure to $Ni(CO)_4$ 2 years previously.[466]

Treatment of Industrial Poisoning. A new therapeutic agent for the treatment of persons exposed to $Ni(CO)_4$, Na diethyldithiocarbamate trihydrate (Dithiocarb) has been reported by Sunderman and Sunderman[470] following lack of success with Na_2EDTA or partial success with BAL. Use of Dithiocarb orally in 11 cases of accidental $Ni(CO)_4$ exposures relieved the symptoms of poisoning, delayed reactions were minimal, and convalescence uneventful. Although mobilization, translocation, and excretion may have been factors in the drug's efficacy, other beneficial actions appear possible; 16 days were required to eliminate the inhaled Ni, a period longer than normal. A plan of drug administration and case management is proposed.[470]

6. Hygienic Standards of Permissible Exposure

Kincaid, Sunderman, et al.[471] have suggested an air concentration of $Ni(CO)_4$ vapor of 0.04 p.p.m. by volume as the maximal limit for avoidance of acute effects in man. No limits for avoidance of acute effects have been suggested for other metal carbonyls but approximate limits could be derived from comparison of the relative acute LC_{50} values given in Table 25. Because of the possible relation of cancer to $Ni(CO)_4$, an attention-calling threshold limit of 0.001 p.p.m. $Ni(CO)_4$ vapor has been set by the American Conference of Governmental Industrial Hygienists for repeated 8-hour daily exposures.

Suggested limits for control of $Ni(CO)_4$ exposure hazards, both for in-plant and community air are proposed in Table 26, based on the recommendations and experiences of Kincaid, Sunderman, et al.[471]

[468] G. P. Barnett, *Annual Report, Chief Inspector of Factories for 1948.* H.M.S.O., London, 1949; W. C. Hueper, *Occupational Tumors and Allied Diseases.* Thomas, Springfield, Ill., 1942; A. C. Loken, *Tidsskr. Norske Laegeforen.* **7**, 376 (1950).

[469] R. Doll, *Brit. J. Ind. Med.,* **15**, 217 (1958); J. G. Morgan, *Brit. J. Ind. Med.,* **15**, 224 (1958).

[470] F. W. Sunderman and F. W. Sunderman, Jr., *Am. J. Med. Sci.,* **236**, 26 (1958).

[471] J. F. Kincaid, E. L. Stanley, C. H. Beckworth, and F. W. Sunderman, *Am. J. Clin. Pathol.,* **26**, 107 (1956).

TABLE 26

Suggested limits for Control of Ni(CO)$_4$ Exposure Hazards[a]

| | In-plant | | |
	Single air sample	Daily average concentration	Out-plant[b]
Target value	0.04[c]	0.001	0.0003
Discontinue operation (require respiration protection)	0.2–2.0	0.001–0.005	
Shutdown operation	2.0	0.005	0.001

[a] p.p.m. Ni(CO)$_4$ in air.

[b] Concentration averaged over 1 month.

[c] J. F. Kincaid, E. L. Stanley, C. H. Beckworth, and F. W. Sunderman, *Am. J. Clin. Pathol.*, 26, 107 (1956).

7. Flammability

Nickel carbonyl is flammable and burns with a yellow flame. It may decompose violently when heated at 60°C. in the presence of air or oxygen. Ten parts per million in the atmosphere is sufficient to impart luminosity to alcohol or carbon monoxide flames: this may be used as a semiquantitative test. Fe(CO)$_5$ has an ignition temperature of 320°C.; the minimal explosive concentration is 105 oz./cu. ft.; 10 per cent oxygen is the limiting concentration to prevent ignition[472] (see also Vol. I, Chapter XVI, Section Two).

8. Odor and Warning Properties

Although nickel carbonyl has a characteristic odor in low concentrations, this is considered to give inadequate warning. The minimal level of Ni(CO)$_4$ detectable by odor is reported[471] to be 1 to 3 p.p.m. under the most favorable circumstances; it is not sufficiently low to provide warning for prevention of either acute or possible chronic effect. On the other hand, the odor of Co$_2$(CO)$_8$ is so characteristic and unpleasant as to be felt and so provide warning against toxic exposures.[460]

MOLYBDENUM, Mo

1. Source and Production

United States production of Mo, by far the largest share of world production, has been rising steadily since 1953 to meet emergency needs in missiles and aircraft; world production in 1957 was 38,100 tons, United States 30,320 tons.

Molybdenite (MoS$_2$) is the only important mineral source at present, but small quantities of powellite (CaMoO$_4$) are mined from time to time; in the past, deposits of wulfenite (PbMoO$_4$) were worked. Molybdenite is commonly associated with Cu ores, thus production of Mo is regulated to a considerable degree by demand for Cu.

[472] Martis, *Mechanical Engineering Handbook*, 6th ed., 1958, p. 41.

Molybdenite concentrates are produced by flotation resulting in a rough concentrate of 10 or 12 to 1; it is reground and refloated yielding a final concentrate of 190:1. When Mo is a by-product of Cu mining, a concentrate of Cu and Mo is first produced and the two ores are later separated by differential flotation. Government specifications of molybdenite concentrates require: 80 per cent MoS_2 (min.), 1 per cent Cu (max.), 0.3 per cent Pb (max.), and 0.2 per cent (max.) As, P, and Sn combined.

Molybdenite concentrates are roasted to produce technical-grade oxide, considerable amounts of which are used directly in steel; the rest is converted to other Mo products. MoO_3 of higher purity is made by sublimation of the technical grade oxide or from $(NH_4)_2MoO_4$. Ferromolybdenum is made from the oxide by ignition with Al, iron ore, ferrosilicon, lime, and fluorspar.

Mo metal powder is made from MoO_3 by reduction; research on preparation of ductile Mo and its alloys by bomb reduction of Mo oxides is continuing (U. S. Bureau of Mines).

Mo cannot be fusion-welded or resistance-welded without becoming embrittled. Shielded W-arc welding can be done successfully if an adequate gas blanket is formed on all sides of the weld. Spot and seam-welding can be done if first etched.

Important by-products of Mo are W and Rh, the latter from the associated Cu ore.

2. Uses and Industrial Exposures

The largest use of Mo is in steel, alloys, and castings. Considerable and increasing amounts are used as wire, rod, and sheet, particularly Mo shapes in manufacturing parts for aircraft and missiles. Small but relatively important amounts of Mo metal are used in electronic parts, induction heating elements, as electrodes for glass melting, and in metal-spray applications for steel and other metals; Mo film provides excellent bonding for subsequent sprayed metal deposits when sprayed on as a thin layer on ferritic steel. Use of Mo compounds as a lubricant (MoS_2), catalysts (Mo, MoO_3 MoS_2), and colors (Mo inorganic and organic complexes), is small but growing; MoS_2 use as a lubricant rose 12 per cent in 1958 over the preceding year, but the quantity used in friction materials, brake linings, and rubber products remained about stationary. Unlike graphite, which it resembles, MoS_2 does not require water film for lubrication and thus can be used in vacuum. As catalysts, Mo compounds are used in hydrogenation cracking, alkylation, and reforming of petroleum fractions, in the Fischer-Tropsch synthesis, and in various oxidation-reduction reactions, and organic cracking reactions, such as cracking of acetone to ketone (250 million lb./yr.). Because traces of S poison other catalysts, but not Mo, this property extends the use of Mo particularly in reactions involving petroleum and natural gas. An exhaustive bibliography on the use of Mo catalysts has been prepared.[473] Mo finds use in colors as (a) molybdate

[473] S. H. Killeffer and A. Linz, *Molybdenum Compounds.* Interscience Publishers, New York–London, 1952.

pigments, (b) Mo colored complexes with animal fibers, (c) insoluble dye complexes with phosphomolybdic acid, and (d) mordants. In ceramics, molybdates permit application of vitreous enamels to steel, by virtue of their property of lowering the surface tension of silicate melts, increasing their fluidity, improving welting power, and increasing adherence to metal. Lead molybdate is used in labeling glass containers. About 1 ton Mo was used as a trace element in fertilizers in 1958, 3 tons each in making electric contacts and in Mo-bearing Ti alloys. Relatively small but wide use is made of a large variety of Mo compounds in various chemical uses. $Mo(OH)_3$ is used in electroplating to give black (moly-black) protective coatings; trivalent Mo compounds may be used to tan skins.

Industrial exposures related to the production and fabrication of Mo products are to dusts and fume of Mo, its oxides, and sulfides chiefly from electric furnace or other high-temperature treatment. In its applications, MoS_2 as a lubricant may be applied to metal surfaces at 700°F.; spraying of Mo may provide a hazard; and loss of Mo catalysts to the air adds to the metal burden of contaminated atmospheres. The sublimation characteristics of MoO_3 (above 800°C.) present a fume hazard.

3. Physical and Chemical Properties

The physical and chemical properties of Mo and some of its compounds are listed in Table 27.

The mechanical properties of Mo are dependent on its history of processing; as produced by powder-metallurgy methods or by arc melting, Mo is brittle, but

TABLE 27

Physical and Chemical Properties of Mo and Some of Its Compounds

	Chemical symbol	At. or mol. wt.	Sp. gr.	M.p., °C.	B.p., °C.	Solubility
Molybdenum	Mo	95.95	10.2	2620 ± 10	4800	Insol. hot or cold H_2O Sol. hot. conc. HNO_3, H_2SO_4 Sl. sol. HCl Insol. HF, dil. H_2SO_4
Molybdic oxide	MoO_3	143.95	4.50(19.5°)	795	Subl.	1.066 g./liter(18°) 20.55 g./liter(70°) Sol. acids, alkalies, NH_4OH
Molybdenum disulfide	MoS_2	160.08	4.8(14°)	1185	Dec. in air	Insol. hot or cold H_2O Sol. hot H_2SO_4, aq. reg., HNO_3 Insol. dil. acids
Ammonium molybdate	$(NH_4)_2MoO_4$	196.03	2.27	Dec.		400 g./liter(20°) Dec. hot H_2O Sol. acids, alkalies

can be made ductile by heating at 1000 to 1300°C. Mo can maintain its hardness at reasonably high levels to extremely high temperatures and is superior to best "super-alloys" of Ni for heat-resisting applications. Mo and W form a series of solid solutions with melting points that are higher than that of Mo. Mo also alloys with many lighter alloying substances, Co, Fe, Al, Cr, and Si, which effectively increase its strength in lesser amounts than W. Mo is completely miscible with Nb and Ta and with the high-temperature form of Ti.

Mo is a typical transition element exhibiting a variety of valence forms from 0 to 6, of which the last is the most stable, but several industrially important compounds exist, for example, MoS_2 and MoO, with intermediate valences. Mo metal is stable to O_2 at room temperatures but above 500°C. reacts rapidly to form MoO_3, which is volatile. Lower oxides MoO and Mo_2O_3, which form MoO_2 by partial reduction of MoO_3, are basic anhydrides; MoO_3 is an acid anhydride. Water vapor also oxidizes Mo forming a series of oxides and H_2. Protection against this form of corrosion is made either by coating with molybdenum disilicide by vapor-phase deposition, or by spraying with Si powder, or by Iconel cladding. Mo is inert in H_2, is not embrittled even by moist H_2. Mo reacts at elevated temperatures with S, C, N, and the halogens except I_2. Mo is resistant to acid attack of HF and HCl.

Mo forms an extremely complex series of compounds; with the exception of the halides, sulfides, and oxides, few simple salts of Mo are known. There are several reasons for the complexity. Mo compounds readily disproportionate to yield mixtures in which Mo occurs in various valence states. There is a strong tendency to form complex compounds. Shifts between different coordination numbers (4,6,8) result from relatively minor differences in conditions. Mo^{6+} has the striking tendency to form isopoly- and heteropoly acids and salts (e.g., $(NH_4)_6H_8$-Mo_7O_{28}, $12MoO_3 \cdot H_3PO_4 xH_2O$). Mo also forms a series of oxy acids and salts (e.g., $MoOCl_3$, MoO_2Cl_2, $MoOF_4$).

4. *Analytical Determination*

The methods that have been developed for the quantitative determination of Mo have been for its estimation in metallurgic samples. Fairhall *et al.* have, however, adapted a thiocyanate method to the analysis of Mo in soft and hard tissues, urine, and feces of experimental animals exposed to a variety of Mo compounds.[474] For determination of Mo in air (in exposure chambers) Fairhall *et al.* converted the sampled Mo to molybdic acid, dissolved it in H_2SO_4, reduced it in a Jones reductor, and titrated it with standard potassium permanganate.

Mo can be determined in the p.p.m. range by the emission spectrograph.[475] With enrichment and buffer techniques and use of In as internal standard a

[474] L. I. Fairhall, R. C. Dunn, N. E. Sharpless, and E. A. Pritchard, U. S. Public Health Bull. No. **293**, (1945); A. T. Dick and J. B. Bingley, *Australian J. Exptl. Biol. Med. Sci.*, **29**, 459 (1951).
[475] G. E. Heggen and L. W. Strock, *Anal. Chem.*, **25**, 859 (1953).

highly acceptable sensitivity can be obtained. The CN band head interferes with one of the most sensitive lines, 3798, but Mo 3170 is relatively sensitive and generally free from interference.[476] Simple techniques for the suppression of the CN bands, however, permit the use of the Mo 3798 line with a sensitivity of 0.25 μg. on the electrode.[477]

5. *Physiological Response*

In addition to its industrial hygiene significance, Mo is of considerable biological importance because it is necessary for the fixation of nitrogen in the soil by bacteria; it can poison cattle and sheep feeding on herbage that has taken up Mo in sufficient quantities. A review (84 pages, 392 references) of Mo in soil, plants, and animal nutrition to 1956 has been made.[478] Mo is an essential trace element as it functions as a Mo-flavoprotein in electron transport in the body.[479] The only definitive study of Mo toxicity of industrial hygiene interest is that of Fairhall et al.,[474] who reviewed the physiological responses of Mo, to 1945, and determined its toxicity for laboratory animals by various routes.

Orally, MoS_2 is well tolerated by rats in daily repeated doses of from 10 to 500 mg./rat/day. On the other hand, all hexavalent compounds tested, MoO_3, $(NH_4)_2MoO_4$, and $CaMoO_4$, were increasingly fatal over the same dose range; all daily doses in excess of 100 mg./rat/day were uniformly fatal. The approximate LD_{50} of daily repeated doses in mg. of Mo/kg./day for $CaMoO_4$ was 100; for MoO_3, 125; and for $(NH_4)_2MoO_4$, 333.

By inhalation, MoS_2 was similarly noninjurious to guinea pigs exposed 1 hour daily to 287 mg. of Mo/cu. meter; MoO_3 was highly irritating and lethal (about 70 per cent mortality in 30 days) under similar conditions; $CaMoO_4$ at 160 mg. of Mo/cu. meter produced no obvious effects. These exposure levels are exceedingly high. Inhalation of Mo metal (oxide) fume was reported to have no effect in guinea pigs at particle size of 1.6 μ and concentrations up to 190 mg. of Mo/cu. meter. Sodium molybdate was shown to be less toxic than either the corresponding salt of Cr or of W upon intraperitoneal injection in mice.

Sodium molybdate ($Na_2MoO_4 \cdot 2H_2O$) was uniformly fatal to the rabbit at all dietary levels of 0.1 per cent and higher within a few weeks.[480] Addition of Cu to the diet prevented the development of toxicity, and was therapeutically effective in combating Mo poisoning once established.

Signs of Mo poisoning are loss of appetite, listlessness, diarrhea, and reduced growth rate. Death from injected doses occurs in guinea pigs in 2 hours to 4 days depending on the dose. Exposure to MoO_3 is irritating to eyes and mucous membranes of nose and throat.[474] Anemia is characteristic of Mo toxicity, with low

[476] L. H. Ahrens, *Spectrochemical Analysis*. Addison-Wesley, Cambridge, Mass., 1950.
[477] R. G. Keenan and C. E. White, *Anal. Chem.*, **25**, 887 (1953).
[478] F. F. Bear, et al., *Soil Science*, **81**, 159 (1956).
[479] H. R. Mahler and D. E. Green, *Science*, **120**, 7 (1954).
[480] L. R. Arrington and G. K. Davis, *J. Nutrition*, **51**, 295 (1953).

hemoglobin concentration and reduced red-cell counts.[481] Cattle, rabbits, and chicks on high dietary levels of Mo show deformities of joints of the extremities.[482] Histopathologically, livers and kidneys of severely poisoned animals showed fatty degeneration. Bronchial and alveolar exudate, in moderate amounts, was present in animals exposed by inhalation.

Metabolism. Mo as MoO_3 is rapidly absorbed from the gastrointestinal tract (guinea pig) and deposited rather uniformly in the critical organs within 4 hours; the blood and bile contained relatively high levels of Mo.[474] Rabbits showed a similar rapid absorption of Mo following ingestion, with rapidly rising blood levels. Mo is rapidly eliminated by the kidneys, returning essentially to normal values in 72 hours after exposure. Fecal elimination, which is about half that of the urinary, returned to normal similarly in 72 hours. Rats stored relatively more Mo from ingested MoO_3 than from $CaMoO_4$; storage was least from MoS_2. Significant storage above normal of Mo in bone was noted in all cases.

Tissue distribution of Mo in guinea pigs following inhalation was modest (a few p.p.m.) from extremely high (150 to 300 mg./cu. meter) exposure levels of MoS_2, MoO_3 dust and fume, and $CaMoO_4$. In line with other metabolic studies, distribution was least with MoS_2.[474] Within 2 weeks following exposure, tissue Mo had decreased to about 20 to 93 per cent of original values, the highest retention occurring in bone, indicating a relatively slow removal of tissue-deposited Mo.

The distribution of microgram quantities of injected radioactive Mo^{99} in dogs is reported to be selectively concentrated in the liver and kidney, with high concentrations in the endocrine glands—pancreas, pituitary, and especially the thyroid and adrenals. Brain, white marrow, and fat contained negligible amounts.[483]

Mechanism. The antagonism of Cu for Mo is well known.[478] Similarly, addition of inorganic sulfate to diets high in Mo is known to alleviate the growth-depressing effects of Mo. S in liver and bone of rabbits on Mo diets was lower, and urinary S higher, suggesting decreased retention of S.[484] Because S is usually an indicator of protein metabolism, the higher excretion of S indicated a lowered protein metabolism in Mo-fed animals. Nitrogen-balance studies showed more than a 2-fold greater N loss in high-Mo rabbits. That the action of Mo on S metabolism, however, may be indirect, and that it may act directly on the enzyme, intestinal xanthine oxidase, finds support in the isolation of an Mo salt in soy flour that is essential to the maintenance of this enzyme in the rat. It would thus appear that, although Mo is essential to the action of certain enzymes, higher Mo levels may inhibit the action of these essential enzymes.

At least this view finds some support in the work of Monty *et al.*,[485] who

[481] J. D. Burke, L. R. Arrington, and G. K. Davis, *Blood,* **8,** 1105 (1953).

[482] Ferguson, *et al., J. Agr. Sci.,* **33,** 40 (1943).

[483] A. Bru, *et al., Compt. rend.,* **237,** 279 (1953).

[484] Anon., *Nutrition Revs.,* **18,** 54 (1960).

[485] A. W. Halverson, J. H. Phifer, and K. J. Monty, *J. Nutrition,* **71,** 95 (1960).

found that high intake of Mo in rats resulted in a substantial reduction in activity of sulfide oxidase in the liver. The reduced activity of this enzyme leads to accumulation of sulfide in the tissues, and subsequent formation of highly undissociated CuS, thus removing Cu from metabolic activity. This is a probable explanation for the induction of Cu deficiency by molybdate.

Industrial Experience. No reports of industrial exposures to Mo and its compounds appear to have been published in the industrial hygiene literature.

6. *Hygienic Standards of Permissible Exposure*

A threshold limit of 5 mg. of Mo/cu. meter of air for soluble Mo compounds and 15 mg. of Mo/cu. meter for insoluble forms of Mo has been set by the American Conference of Governmental Industrial Hygienists.[486] The levels were estimated from the work of Fairhall et al.[474] in animals. No information has come to the attention of the Committee on Threshold Limits as to the suitability of these limits. Because of the adequacy of Cu intake in most individuals, and in the absence of information to the contrary, the levels would appear appropriate for the prevention of Mo poisoning.

NICKEL, NI

1. *Source and Production*

With continuous increases in world production of Ni to a new high of 314,000 short tons in 1957, the supply of Ni was adequate for both civilian and defense needs for the first time since 1950. Activity, however, continued in exploring new sources, developing new mines, and expanding and increasing the efficiency of smelting and refining operations, developing new and greater uses, and developing Ni-base alloys capable of withstanding extremely high temperatures. A number of new processes for treating Ni–Fe ore, producing ferronickel from low-grade ores, and separating Ni and Co from ores were developed. Canada furnishes more than three fourths of present Ni output; New Caledonia and Cuba are second and third; domestic production remains small. The Office of Defense Mobilization has an expansion goal to provide the United States with an annual supply of Ni of 440 million pounds by 1961; schedules in Canada and Cuba called for 1.5-fold expansion by 1961.

There are three principal classes of Ni ores: sulfide $(Ni,Fe)_9S_8$, occurring with $CuFeS_2$ (Canada), silicate (New Caledonia and Cuba), and arsenide ores, NiAs. The last is of little commercial importance. Sulfide ores contain small amounts of Co, Se, Te, Ag, Au, and 5 of the Pt metals.

The process of extraction varies with the nature of the ore. Copper–nickel concentrates are treated by methods similar to those used in the production of copper, to obtain a mixture of copper and nickel oxides, which may be reduced

---

[486] Comm. on Threshold Limits, ACGIH, *A.M.A. Arch. Ind. Health,* **20,** 266 (1959).

directly to Monel metal, an alloy containing 28 per cent copper and 67 per cent nickel. If nickel alone is desired, a separation is effected at an earlier stage, while the metals are in the form of sulfides, by fusion with coke and sodium sulfate. The nickel sulfide, which forms a lower layer, is reduced by means of coke in a suitable furnace. The final purification is electrolytic.

In the Mond process, the mixed nickel and copper sulfides are converted to oxides and reduced by heating with water gas at 350 to 400°C. The cooled mixture of nickel and copper is then subjected to the action of carbon monoxide at 60°C. The nickel unites with the carbon monoxide to form volatile nickel carbonyl, $Ni(CO)_4$, which is decomposed by passage over nickel pellets heated to 180°C., very pure nickel being deposited upon the pellets. Leakage of the apparatus may lead to exposure to both carbon monoxide and nickel carbonyl. The latter is a nearly colorless liquid, which boils at 45°C., forming a very toxic gas with a peculiar sooty odor, detectable in a concentration of 1 vol. in 2,000,000 of air.

2. Uses and Industrial Exposures

Pure Ni (99.4 per cent containing some Co) is used as anodes in plating for corrosion resistance, thermal conductivity, in radio and electronic tube parts, magnetostriction oscillators, and in coinage (25 per cent Ni, 75 per cent Cu). An Ni–Cd accumulator (battery) is gaining acceptance. More than 3000 Ni alloys constitute the chief and widely varied uses of Ni. Principal alloys are Ni–Fe for alloy steels in heavy machinery, communications, automotive and electric equipment parts; Ni–Cu (Monel) for food processing, chemical and petroleum equipment, condenser plates, and coinage; Ni–Al for automotive and aircraft parts; Ni–Cr for gas-turbine and jet-engine parts; Ni–Cu–Zn for flatware, jewelry, telephone equipment, plumbing fixtures, architectural trim; Ni–Cr–Fe for stainless-steel cooking utensils, chemical and petroleum equipment. Other alloys of Ni with Mn, Mo, and Si find special uses. Millions of pounds of nickel salts are used annually for plating baths, in enamel ground coats (NiO), and in the production of Ni catalysts (sulfate, nitrate, formate). These and other salts (carbonyl and complexes with ammonia) play a large role in metallurgy of Ni.

No well-authenticated cases of acute or chronic industrial intoxication from Ni or its compounds, other than $Ni(CO)_4$, have been reported. Numerous cases of dermatitis, however, have been observed among nickel platers.

3. Physical and Chemical Properties

The physical and chemical properties of Ni and some of its compounds are listed in Table 28. Because of the special industrial hygiene significance and importance of $Ni(CO)_4$, this Ni compound is presented separately under Metal Carbonyls, page 1107.

Ni forms two series of oxides and hydroxides, di- and trivalent, but only the divalent (nickelous) forms are basic. The divalent salts are not oxidizable to the

TABLE 28

Physical and Chemical Properties of Ni and Some of Its Compounds

	Chemical symbol	At. or mol. wt.	Sp. gr.	M.p., °C.	B.p., °C.	Solubility
Nickel	Ni	58.7	8.90	1455	2900	Insol. hot or cold H_2O Sol. dil. HNO_3 Sl. sol. HCl, H_2SO_4
Nickel oxide	NiO	74.69	7.45	1990		Insol. hot or cold H_2O Sol. acids, NH_4OH, KCN
Nickel formate	$Ni(CHO)_2 \cdot 2H_2O$	184.76	2.154	Dec.		13 mg./liter Sol. acids, NH_4OH
Nickel sulfate	$NiSO_4 \cdot 6H_2O$	262.85	2.07	Tr. 53.3	$(-6H_2O)280$	625 g./liter(20°) 3407 g./liter(100°) V. sol. alc., NH_4OH

trivalent form. The complex salts of Ni are rather unstable and thus give some of the reactions of divalent Ni.

4. Analytical Determination

There is no entirely satisfactory chemical method for the quantitative estimation of small amounts of Ni; the methods employing various oximes,[487] although rather sensitive, suffer from the formation of microcrystalline Ni-complex precipitates. The colorimetric method of Fairhall[488] using dithiooxalate, although sensitive, suffers from the unavailability of the reagent, which is difficult to prepare. Ni may be determined spectrographically with a sensitivity of 0.01 μg. of Ni on the electrode per 2 mg. of ash.

5. Physiological Responses

Ni salts are recognized to be highly toxic following access to the blood stream; there is conflicting evidence on their toxicity by mouth. Colloidal Ni or $NiCl_2$ is lethal to dogs when intravenously administered in a single dose of 10 to 20 mg./kg.;[489] the cardiac and respiratory nerve centers are affected; gross changes consist of edema, hemorrhage, and degeneration of the heart muscle, brain, lung, liver, and kidney.[490] Lethal subcutaneous doses of soluble Ni salts are reported[490] to be 7 to 8 mg./kg., for rabbits, 9 to 16 mg./kg. for cats. Although the older pharmacological literature indicates a rather high oral tolerance for either Ni metal powder (1 to 3 g./kg., dogs) or nickel salts when incorporated in diets (4 to

[487] M. Kenigsberg and I. Stone, *Anal. Chem.*, **27**, 1339 (1955); E. B. Sandell, *Colorimetric Determination of Traces of Metals*. 3rd ed., Interscience Publishers, New York–London, 1959, p. 665.

[488] L. T. Fairhall, *J. Ind. Hyg.*, **8**, 528 (1928).

[489] J. Caujoulle and G. Canal, *J. pharm. chim.*, **29**, 391, 410 (1939).

[490] H. W. Armit, *J. Hyg.*, **7**, 525 (1907), **8**, 565 (1908).

12 mg./kg. daily for 200 days in dogs and cats), a later report[491] revealed severe damage to myocardium and liver parenchyma of rabbits drinking water containing $NiSO_4$ equivalent to 0.54 mg./kg. for 160 days. Possible explanation of these findings may be that absorption of Ni from water occurs to a greater extent than when Ni is associated with and probably bound to food components. There is no evidence that the Ni absorbed from foods cooked in Ni or Ni-alloy containers is of hygienic significance[492] although milligram amounts may be daily taken into the body.

Hueper[493] has reported evidence which strongly suggests that finely dispersed nickel may elicit cancerous reactions in tissues (rat) with a minimal latent period of 6 months.

Industrial Experience. Apart from the possible malignant effects of $Ni(CO)_4$ (which see) dermatitis constitutes the only other serious Ni exposure hazard. Numerous cases of dermatitis have been reported among Ni platers.[494] The "nickel itch," which is variable in nature, usually begins with a sensation of burning and itching in the hand, followed by erythema and a nodular eruption on the web of the fingers, wrists, and forearms. The nodules may become pustules or may ulcerate. The acute stage is sometimes accompanied by fever. Recovery usually occurs after a week, although in a few cases the dermatitis has persisted for weeks. The role of degreasing agents has been considered in its causation. It has long been recognized that some individuals are highly susceptible to Ni sensitization, and in such persons the wearing of Ni-alloy or plated objects is sufficient to induce the response.[495] Ni dermatitis appears to have two components, a simple dermatitis localized to the area of contact, and chronic eczema or neurodermatitis without apparent connection to such contact.[496] Ni sensitivity once acquired is likely to persist. A complement-fixation test has been recommended as a diagnostic aid in Ni eczema.[497] It is not yet resolved whether "cross-sensitization" of Ni and Co is a true cross-sensitization or whether it is attributable to contamination of each metal with the other used in patch testing; until recently (1958) "spectrographically pure" Ni and Co salts were unavailable for testing.

Experimental sensitization to Ni salts has been developed in animals.[498] Ni

[491] Y. M. Grushko, V. A. Donskov, and V. S. Kolesnik, *Farmikol. i Toksitol.,* **16,** 47 (1953), (abstract, translation).

[492] K. R. Drinker, L. T. Fairhall, G. B. Ray, and C. K. Drinker, *J. Ind. Hyg.,* **6,** 308 (1924).

[493] W. C. Hueper, *Texas Repts. Biol. Med.,* **10,** 167 (1952).

[494] A. Mullschitzky, *Wien. med. Wochschr.,* **89,** 717 (1939); N. Wedroff, *Arch. Gewerbepathol. Gewerbehyg.,* **6,** 179 (1935); H. E. Stockinger and Schittenhelm, *Z. ges. exptl. Med.,* **45,** 58 (1925); R. P. White, *The Dermatogoses, or Occupational Diseases of the Skin.* 4th ed., Lewis, London, 1934, p. 193.

[495] L. Goldman, *Arch. Dermatol. Syphilol.,* **28,** 688 (1933).

[496] H. T. H. Wilson, *Practitioner,* **177,** 303 (1956); C. D. Calnan, *Brit. J. Dermatol.,* **68,** 229 (1956); G. C. Wells, *Brit. J. Dermatol.,* **68,** 237 (1956).

[497] J. Wendlberger, and E. Frolich, *Arch. Hyg. u. Bakteriol.,* **138,** 430 (1954).

[498] H. Haxthausen, *Arch. Dermatol. u. Syphilis,* **174,** 17 (1936).

salts have been ranked high among the substances producing sensitization.[499] Some investigators consider the high temperature near nickel-plating baths a predisposing factor.[500]

6. Hygienic Standards of Permissible Exposure

With the exception of $Ni(CO)_4$ (which see), no limit has been set for Ni or its salts in air.

NIOBIUM, Nb, formerly Columbium, Cb

1. Source and Production

Nb does not occur free but is associated usually with Ta in a variety of minerals scattered throughout the world. Principal Nb-producing countries are Nigeria, Belgian Congo, and Malaya. Tantalite and columbite $(Fe,Mn)(Ta, Nb)_2O_6$, are common source minerals; other Nb-containing minerals such as pyrochlore, fergusonite, and euxenite are highly complex and contain many lanthanons and U and Th. World production of Nb-Ta concentrates in 1958 totaled 5 million pounds, down from a high of 11.5 million pounds in 1955 owing to lessened consumer demand. Domestic industrial consumption of Nb-bearing mineral concentrates as metal was 786 thousand pounds in 1958, down from 1.22 million pounds the previous year; on the other hand, domestic production increased to a new high of 428 thousand pounds. South Dakota, New Hampshire, Idaho, Connecticut, and Maine are sources of columbite ore in the United States.

Nb and Ta are extracted together from their ores. A method employed by a metallurgic company for winning Nb from its ores is as follows: The ore is fused with caustic soda at red heat to form sodium niobo-tantalate. Following washing, the melt is digested with hot concentrated HCl to form tantaloniobic acid, and to dissolve impurities of FeMn. The insoluble tantaloniobic acid is dissolved in HF with enough KF to form niobium potassium oxyfluoride and tantalum potassium fluoride. The solution is filtered and cooled to collect tantalum potassium fluoride. The niobium solution is concentrated and treated with alkali to yield crude niobium oxide. Pure Nb is obtained by heating a stoichiometric mixture of NbO_2 and NbC in vacuo.

2. Uses and Industrial Exposures

Principal use of Nb has been as ferro Nb or ferro Nb-Ta in the manufacture of stainless steels. Recently Nb metal of unusual purity and Nb alloys possessing high strength at high temperatures are finding applications in jet engines, guided missiles, and atomic reactors that may exceed the demand for stainless steels. Nb

[499] A. Rostenberg, Jr., and M. B. Sulzberger, Arch. Dermatol. Syphilol., 35, 433 (1937); H. H. Johnson, Arch. Dermatol. Syphilol., 43, 575 (1941).

[500] C. DuBois, Schweiz. med. Wochschr., 61, 278 (1931); F. M. R. Bulmer and E. A. MacKenzie, J. Ind. Hyg., 8, 517 (1926).

is an electrolytic-valve metal used in construction of electrolytic rectifiers and condensers; when made the anode in an acid solution, it blocks passage of the electric current by formation of an oxide film; as a cathode, it conducts current. Use in railway locomotive, automobile, and marine engines, and in stationary power generators has been forecast.

Industrial exposures consist of those related to milling of Nb ore and its associated elements Mn and Fe, and HF gas used in its extraction. Forging and other fabrication techniques of Nb metal and alloys present a fume and dust problem.

3. Physical and Chemical Properties

Nb has an atomic weight of 92.91 (see Table 29), and occurs in group 5 of the periodic table together with Ta, which it resembles closely in chemical and physi-

TABLE 29

Physical and Chemical Properties of Nb and Some of Its Compounds

	Chemical symbol	At. or mol. wt.	Sp. gr.	M.p., °C.	B.p., °C.	Solubility
Niobium	Nb	92.91	8.57	2500 ± 50	3700	Insol. hot or cold H_2O Sol. hot H_2SO_4 Sl. sol. HF Insol. HCl, HNO_3, aq. reg.
Niobium pentoxide	Nb_2O_5	265.82	4.47	1520		Insol. cold H_2O Sol. conc. H_2SO_4, HCl, HF, alkalies Insol. NH_3
Niobium pentachloride	$NbCl_5$	270.2	2.75	194	240.5	Dec. cold H_2O Sol. HCl, alc.
Potassium niobate	$4K_2O \cdot 3Nb_2O_5 \cdot 16H_2O$	1462.68				4.25 kg./liter(20°)

cal properties. In pure form Nb is extremely ductile unless allowed to associate at elevated temperatures with common gases, N_2, H_2, or O_2; free of these gases the workability of Nb is unusual. Nb is extremely ductile; sintered bars produced by powder metallurgy can be worked at room temperature into plate, foil, sheet rod. or wire without annealing. Pure Nb oxidizes rapidly at high temperatures, but by alloying, both strength and oxidation resistance at these temperatures are increased. Nb alloys are among the most corrosion-resistant known. The metal has a low neutron cross-section, useful in atomic energy applications. Nb also has the lowest work function (energy required to remove an electron from the surface) of any pure refractory metal.

Nb has a valence of $+5$ in all of its important compounds but some Nb compounds with $+4$ and $+3$ valence are known. Nb is extremely resistant to attack by many chemical agents, particularly at ordinary temperatures. Nb scarcely tarnishes after many years exposure to the weather. Strips heated for long periods change color from yellow at 200°C. to blue at 300°C. to black (whence its name) at 390°C. when the white oxide develops. F_2 reacts with Nb at room temperatures but elevated temperatures are required for action with Cl_2 and Br_2. Potassium niobate, formed from solution of niobic acid ($Nb_2O_5XH_2O$) in concentrated KOH, is important in the final purification of Nb material.

4. Analytical Determination

A spectrochemical procedure has been employed by Rankama and Joensuu[501] for the determination of Nb in rocks. Following necessary concentration with organic chelating acids such as tartaric, salicylic, or phenyl arsonic, and ignition, Nb was determined spectrochemically using the cathode layer technique. Using Nb 4058, a lower limit of 10 μg. of Nb/g. could be detected. Chemical methods for the determination of Nb are extremely complex. Grimaldi[502] has described a modified Nb thiocyanate spectrophotometric procedure that is claimed to be applicable to the p.p.m. range (in ores). The method as applied to rocks is long and laborious and undoubtedly could be considerably shortened if interfering elements were absent. No method for the determination of Nb in specimens of biological origin appears to have been developed or reported.

5. Physiological Response

Only range-finding toxicity studies have been reported for Nb; Schubert[503] found 50 mg./kg. of Nb as sodium niobate produced severe chronic intoxication in rats; DuBois et al.[504] found a rather high toxicity for a soluble salt of Nb, the chloride (14 mg. of Nb/kg.), by the intraperitoneal route in rats; potassium niobate was approximately 6-fold less toxic on the basis of Nb. Orally potassium niobate was practically nontoxic for rats; 3000 mg./kg. or 1140 mg. of Nb/kg. represented the oral LD_{50}.

Metabolism. Metabolism studies by Scott et al.[505] with carrier-free Nb^{93} showed that uptake and retention by the soft tissues, and elimination, are greater than for most fissioned materials. Moreover, a higher proportion was removed via the kidneys.

Mechanism. $NbCl_5$ and potassium niobate produced 50 per cent inhibition of adenosine triphosphatase (ATP) activity at 4×10^{-4} and $5.8 \times 10^{-4}M$, re-

[501] K. Rankama and O. Joensuu, *Compt. rend. soc. géol. Finlande,* **19,** 8 (1946).

[502] F. S. Grimaldi, *Anal. Chem.,* **32,** 119 (1960).

[503] J. Schubert, *J. Lab. Clin. Med.,* **34,** 313 (1949).

[504] K. W. Cochran, J. Doull, M. Mazur, and K. P. DuBois, *Arch. Ind. Hyg. Occupational Med.,* **1,** 637 (1950).

[505] K. G. Scott, R. Overstreet, L. Jacobson, J. G. Hamilton, H. Fisher, J. Crowley, *et al., U.S. Atomic Energy Comm.,* **MDDC-1275,** 1947.

spectively, and thus were more potent inhibitors of the activity of this enzyme than were the chlorides of Sr, Y, Zr, and La, tested under similar conditions.[506] It would thus appear that Nb can seriously interfere with Ca as an activator of ATP activity.

6. Hygienic Standards of Permissible Exposure

No hygienic standard for air has been set for Nb or any of its compounds.

PLATINUM-GROUP METALS—PLATINUM, Pt, PALLADIUM, Pd, IRIDIUM, Ir, OSMIUM, Os, RHODIUM, Rh, RUTHENIUM, Ru

1. Source and Production

Of the Pt-group metals, Pt is of first industrial importance, Pd second, but owing to changes in technology and increasing usages, the relative importance of the metals in the group is changing. The Pt metals are found uncombined in the earth alloyed with each other or associated with Au, Cu, Fe, Ni, or chromite; some metals of the group are associated with As, Sb, S. Native Pt averages about 80 per cent Pt, and native osmiridium about 35 per cent Os and 30 per cent Ir, the other members of the group accounting for most of the remainder. The principal minerals of the Pt group found in the Cu–Ni sulfide-type deposit are sperrylite ($PtAs_2$), cooperite (PtS), braggite (Pt,Pd,Ni,S), laurite (RuS_2), and stibiopalladinite (Pd_3Sb). The average current production of Pt metals, both placer and lode, is about 60 per cent Pt, 30 per cent Pd, and 10 per cent Ir, Os, Rh, and Ru together.

Canada, the Union of South Africa, and the USSR are the leading producers of Pt metals (1958). Domestic mine production, relatively insignificant, fell off 23 per cent, and with secondary metals recovery furnished about 14 per cent of domestic requirements. World production dropped below that of the three preceding years.

Crude placer Pt is recovered by methods similar to those for Au, except that the Pt metals do not respond to amalgamation as does Au. Most is recovered by dredging, but small hand operations are used in less developed areas. The Cu, Ni, and S ores are mined by large-scale underground methods, concentrated by flotation, and smelted and refined. Individual metals are separated and refined by a complex chemical treatment based on solution in aqua regia, and precipitation as the double ammonium chlorides. In some instances, separation procedures are considered closely guarded information. The precipitates are calcined, yielding a spongy mass of pure metal, which is compacted by melting in a high-temperature oxyacetylene or electric furnace. The bulk of the refined metals has a purity of 99.95 to 99.99 per cent; higher purity is required for some purposes. A cationic exchange technique has been reported to give near quantitative separations of base-metal contaminants from anionic complexes of Pt and Rh.

Secondary sources in substantial quantities come from refining and reworking scrap and used pieces.

[506] C. K. Butler, *Ind. Eng. Chem.*, **48**, 711 (1956).

2. Uses and Industrial Exposures

As pure metals, alloyed, combined, or clad, the Pt metals are used for jewelry, chemical and electric industries, in dentistry, and various small miscellaneous purposes. Chemical uses absorb most of the Pt, accounting for 57 per cent in 1958 compared with 70 per cent in 1957. Of the total Pt group, 37 per cent went into catalysts for producing high-octane gasoline, nitric and sulfuric acids, petrochemicals, and pharmaceuticals. Pd usage increased 8 per cent owing to increased usage by the drug and chemical industries. Almost 66 per cent of Pd was used by the electronic or electric industries; electric industries, leading users of Pt metals, accounted for 43 per cent of the total.

TABLE 30

Physical and Chemical Properties of the Pt-Group Metals

	Chemical symbol	At. or mol. wt.	Sp. gr.	M.p., °C.	B.p., °C.	Solubility
Platinum	Pt	195.09	21.37	1773.5	4300	Insol. hot or cold H_2O Sol. aq. reg., fused alkalies
Palladium	Pd	106.7	11.4(22.5°)	1549.4	ca. 2200	Insol. hot or cold H_2O Sol. aq. reg., hot HNO_3, H_2SO_4 Sl. sol. HCl
Iridium	Ir	192.2	22.421	2454	>4800	Insol. hot or cold H_2O Amor. sol. aq. reg. Insol. acids, aq. reg.
Rhodium	Rh	102.91	12.4	1966 ± 3	>2500	Insol. hot or cold H_2O Sol. H_2SO_4 + HCl, hot conc. H_2SO_4 Sl. sol. acids, aq. reg.
Osmium	Os	190.2	22.48	2700	>5300	Insol. cold H_2O Sl. sol. HNO_3, aq. reg. Insol. NH_3
Ruthenium	Ru	101.1	12.6	>1950		Insol. hot or cold H_2O Sol. fused alkalies Sl. sol. aq. reg. Insol. acids

Pt and Pt alloys find numerous uses stemming from their resistance to corrosion and oxidation, particularly at high temperatures, high electric conductivity, and excellent catalytic properties. Some of the more important specific uses include catalytic gauze for NH_3 oxidation; support catalyst for high-octane gasoline; gas ignitors; spark-plug electrodes; electric contacts; spinnerets for extruding synthetic fibers; equipment for handling molten glass, for drawing Fiberglas, dental, and medical devices; laboratory ware; and jewelry. Pd has many similar uses, but is used to a greater extent in low-current electric contacts.

The minor Pt metals find use primarily as alloying metals to improve the properties of Pt and Pd. Iridium crucibles are used because of their high-temperature resistance and Pd–Au alloys are used as thermal-fuse elements for protecting electric furnaces because of their narrow melting range and corrosion-resistance properties.

3. *Physical and Chemical Properties*

The Pt-group metals belong to 2 triads of Group VIII, the transition group of metals, whose atomic weights differ between the triads by a factor of almost 2. The physical and chemical properties of the corresponding metals in each triad nevertheless are very similar (see Table 30). For example, the ductility and chemical properties of Pt and Pd are very much alike. The other pairs are Ir and Rh, Os and Ru. The Pt-metal alloys that find commercial use are those in which Pt metals alloy with each other, but other alloys with Ni and W are also important. Pt is a tin-white metal, malleable and ductile, with a coefficient of expansion the same as that of glass. Pt does not oxidize at any air temperature but reacts with halogens, CN, S, and caustic alkalis. Ir, in the residue from the solution of Pt by aqua regia, is so named because of the varied colors of its salts; green, red, and violet. Ir is the densest of all metals except Os. Its most important salt is $IrCl_4$. Os is a bluish white, hard, crystalline metal and highly infusible. When heated in air, it is oxidized to OsO_4 which has pungent, irritating, and toxic properties, but is readily reduced by organic matter. From the odor of the oxide its name was derived (osme = odor).

Some of the characteristic compounds of the Pt metals are given in Table 31.

4. *Analytical Determination*

At the present stage of development and application, determination of the Pt metals by emission spectrography would appear to be the method of choice.[507] Because of their relative rarity in soils, rocks, etc., a preliminary concentration procedure is often required. Present separation procedures of Pd, Rh, and Ir from Pt consist of hydrolytic precipitation from bromate solution according to Gilchrist and Wickers.[508] Spectrographically, Pt may be determined with a sensitivity considerably greater than 0.1 μg. on the electrode, using the 2659.4A or the 2998.0 line. Co line 3013.6 may be used as the internal standard. Pd is determinable in microgram amounts, using the 3029.9 line which is relatively free of interference but less sensitive than the 3404.6 line. For Rh, the 3263.1 line is used and for Ir, the 2924.8 line. Os and Ru, as the oxides, are considerably more volatile than others of the Pt group, but if account is taken of this, Os and Ru may be determined spectrographically using Os line 3058 and Ru line 3436.

[507] G. H. Ayres and H. J. Belknap, *Anal. Chem.,* **29,** 1536 (1937).

[508] R. Gilchrist and E. Wickers, *J. Am. Chem. Soc.,* **57,** 2565 (1935).

HERBERT E. STOKINGER

TABLE 31

Typical Compounds of the Platinum-Group Metals

Metal	No.	Oxidation states	Oxides	Coordination compounds	Nitro compounds	Ammines	Carbonyls	Carbonyl halides
Platinum	78	2,3,4	PtO_2	$H_2(PtCl_4)$ $H_2(PtCl_6)$[a]	$Na_2Pt(NO_2)_6$	$Pt(NH_3)_6Cl_4$ $Pt(NH_3)_4Cl_2$		$Pt(CO)Cl_2$ $Pt(CO)_2Cl_2$ $2PtCl_2 \cdot 3CO$
Palladium	46	2(3)4	PdO $Pd_2O_3 x H_2O$	$H_2(PdCl_4)$ $H_2(PdCl_6)$[a]	$Na_2Pd(NO_2)_4$	$Pd(NH_3)_2Cl_2$ $Pd(NH_3)_4Cl_2$		$Pd(CO)Cl_2$
Iridium	77	1,2,3,4,6	IrO_2 IrO_3	$H_3(IrCl_6)$ $H_2(IrCl_4)$[a]	$Na_3Ir(NO_2)_6$	$Ir(NH_3)_3Cl_2$ $Ir(NH_3)_6Cl_3$	$Ir(CO)_3$ $Ir(CO)_4$	$Ir(CO)_2Cl_2$ $Ir(CO)_3Cl$
Rhodium	45	3	Rh_2O_3	$H_3(RhCl_6)$	$Na_3Rh(NO_2)_6$	$(Rh(NH_3)_6)_2Cl_6$	$Rh(CO)_4$ $Rh(CO)_3$ $Rh_4(CO)_{11}$	$Rh(CO)_2Cl$
Osmium	76	2,3,4,6,8	OsO_2 OsO_4	$H_3(OsCl_6)$ $H_2(OsCl_6)$[a] $K_2(OsCl_5NO)$[a]		$OsO_2(NH_3)_4Cl_2$	$Os(CO)_5$ $Os_2(CO)_9$	$Os(CO)_3Cl_2$ $Os(CO)_4Cl_2$
Ruthenium	44	(1)3,4,5,6,7,8	RuO_2 RuO_4	$H_3(RuCl_6)$ $H_2(RuCl_6)$[a] $K_2(RuCl_5NO)$[a]		$Ru(NH_3)_6Cl_3$	$Ru(CO)_5$ $Ru(CO)_x$	$Ru(CO)Br$ $Ru(CO_2)Cl_2$

[a] Coordination compound with halogen.

A number of colorimetric procedures are available for the Pt metals; these are best used following some concentration procedure such as those described by Sandell.[509] The colorimetric methods for Pt and others of this group are suggested by Sandell, who gives indications of the relative interference of each metal in the particular method. A colorimetric method claimed to be highly selective for trace amounts of Pd has been reported.[510] A new class of organic reagents for the spectrophotometric determination of trace amounts of Os has been reported by Steele and Yoe.[511] A colorimetric method for the determination of Ir using o-dianisidine is reported;[512] the absorbance follows Beer's law from 2 to 20 μg.

5. Physiological Properties

Two members of the Pt-group metals, Pt and Os, exhibit striking and characteristic toxic responses in animals and man marked by irritation to skin, eyes, and respiratory tract; as far as limited reports are available, Pd and others of the Pt group do not seem to share these irritative characteristics. No significant amount of information is available on the toxicity of the three Pt metals, Ir, Rh, and Ru.

Platinum. Four studies of the effects of exposure to soluble, complex Pt salts of the form Na_2PtCl_6 describe essentially the same findings, which are designated variously as Pt allergy, platinosis, or Pt asthma.[513] The respiratory syndrome starts with repeated sneezing, followed by profuse running of the nose with a watery mucous discharge. This is followed by tightness of the chest, shortness of breath, cyanosis, and a wheeze. A relative lymphocytosis commonly is present. Dermal manifestations may be acute, consisting of itching and urticaria of exposed areas; subacute, in which the urticaria is replaced by typical contact dermatitis; and chronic, in which there is a secondary eczema and persistent contact dermatitis. Roberts[513] considers the incidence of platinosis to be 100 per cent; in about 40 per cent the condition is asymptomatic but with conjunctivitis, inflammation of the nose and throat, and lymphocytosis. Some exposed individuals experience both dermal and respiratory effects. Predisposing symptoms are related to skin types, light- and red-haired individuals with fine-textured skin being more susceptible.

The exposures occur in Pt refineries, which involve dusts and sprays of complex salts of Pt and related metals. Sieving spongy Pt gives rise to a dust of metallic Pt. Hunter *et al.*[513] measured the Pt concentrations in the air at the various stages in four refineries; highest concentrations (1700 μg./cu. meter and 960

---

[509] E. B. Sandell, *Colorimetric Determination of Traces of Metals.* Interscience Publishers, New York–London, 1959, p. 721 ff.

[510] N. C. Sogani and S. C. Bhattacharyya, *Anal. Chem.*, **29**, 397 (1957).

[511] E. L. Steele and J. H. Yoe, *Anal. Chem.*, **29**, 1622 (1957).

[512] S. S. Berman, F. E. Beamish, and W. A. E. McBryde, *Anal. Chim. Acta*, **15**, 363 (1956).

[513] W. Massmann and H. Opitz, *Zentr. Arbeitsmed. u. Arbeitsschutz*, **4**, 1 (1954); A. E. Roberts, *Arch. Ind. Hyg. Occupational Med.*, **4**, 549 (1951); D. Hunter, R. Milton, and K. M. A. Perry, *Brit. J. Ind. Med.*, **2**, 92 (1945); S. R. Karasek and M. Karasek, *Rept. Illinois State Comm. Occupational Dis.*, **1911**, 97.

μg./cu. meter, respectively) were found around the crushing of $(NH_4)_2$ $PtCl_6$ and at the sieving of spongy Pt. With the exception of crushing and packing the complex salts of Pt and Pd, at which concentrations were 50 μg./cu. meter, the air about all other operations, including the purification of Rh, were generally less than 10 μg./cu. meter. Review of the occupational histories and the environmental data convinced Hunter et al. that the complex salts of Pt were the cause of the syndrome; finely divided Pt metal apparently can be inhaled without ill effects (Massmann and Opitz[153]). The intradermal test for Pt sensitivity, however, was found unsatisfactory in their hands, Roberts[513] suspects that platinosis produces a low-grade pulmonary fibrosis, on the basis of chest roentgenograms, but finds no evidence that Pt has carcinogenic activity. Using a scratch test, Roberts found 42 per cent of the laboratory and refinery personnel were sensitive to $PtCl_6$ at dilutions of 1×10^{-3} and higher. Roberts provides an outline of prophylactic, active, and hyposensitization treatment. Removal from Pt exposure results in almost immediate relief of asthma; the dermatitis usually clears in a day or two, but may be more persistent.

A case of contact dermatitis from Pt and related metals, believed to be the first reported, was found by Sheard.[514] According to jewelers, contact dermatitis from Pt jewelry is probably not so uncommon as would be inferred from this first report. The case in question was determined to arise from Pt and/or Pd and Ir as a result of patch tests on an individual wearing 1 ring of 90 per cent Pd and 10 per cent Ru, and another of 90 per cent Pt and 10 per cent Ir; 1 ring was Rh-plated. A patch test was negative for Rh; no separate test was made for Pd or Ir.

Palladium. Pd and its salts appear to be without appreciable toxicity as far as limited information indicates. Subcutaneous administration to rats of 24 mg. of Pd/kg. in a buffered solution caused no injury;[515] the daily administration of 5 mg. of Pd to rabbits during a 2-month period caused no disturbance of health.[516] Patch tests of 7 subjects with buffered $PdCl_2$ caused no reaction in 24 hours; if not neutralized, however, such salts might be expected to be irritant to the skin.

No industrial intoxication from Pd or its salts has been reported.

Osmium. In the metal form Os is innocuous[517] but osmic acid, OsO_4, is extremely irritant and toxic. OsO_4 develops slowly at room temperature when powdered Os is exposed to air. The gentle heat required for annealing Pt or for making pen nibs results in the evolution of poisonous fumes.[518] Illustrative of the persistently irritating property of OsO_4 is the intense conjunctivitis and corneal ulceration that slowly develop following instillation of 1 drop of a 1 per cent solution of osmic acid into the conjunctival sac of a rabbit. The acute phase begins to

[514] C. Sheard, Jr., *Arch. Dermatol.*, **71**, 357 (1955).

[515] S. F. Meek, G. O. Harrold, and C. P. McCord, *Ind. Med.*, **12**, 447 (1943).

[516] M. Kauffman, *Munch. med. Wochschr.*, **60**, 525 (1913).

[517] A. I. G. McLaughlin, R. Milton, and K. M. A. Perry, *Brit. J. Ind. Med.*, **3**, 183 (1946).

[518] B. M. Hoke, *Brass World*, **20**, 242 (1924), *Chem. Abstr.*, **18**, 2673 (1924); F. R. Brunot, *J. Ind. Hyg.*, **15**, 137 (1933).

subside in 10 days. All rabbits died that were exposed to OsO_4 vapor for 30 minutes at rather uncertain concentrations (Brunot[518]) ; their lungs were congested, hemorrhagic, and edematous, voluminous, and doughy; the bronchi stood out prominently in a cut section because of the black staining of the epithelium by the reduced OsO_4, probably OsO; the trachea, epiglottis, and the interior of the larynx were jet black with deposited OsO. In other respects, the lesions resembled those from Cl_2. The very slight exposure of the investigator, which occurred from opening four 250-mg. ampoules and placing them in an exposure chamber, an operation requiring 2 minutes, was followed 30 minutes later by smarting of the eyes and lacrimation. This progressed until at 3 hours reading was difficult and the street lights appeared surrounded with haloes. This is a characteristic ocular effect resulting from exposure to OsO_4. Even small amounts, if inhaled over a considerable period, cause headache, insomnia, pharyngeal and laryngeal distress, and digestive disturbance.[517,519]

Exposures to Os compounds can occur from the spray thrown up from the aqua regia used to dissolve the Pt metals. Samples of air taken near the reaction pots showed from 133 to 640 μ. of Os/cu. meter as determined colorimetrically after $SnCl_2$ reduction by McLaughlin et al.[517] Case histories of 7 men exposed to Os are given. Ventilation over the reaction pots stopped further appearance of symptoms, which subsided in 24 hours. No chronic or cumulative effects were noted.

No reports specifically implicating Ru, Rh, and Ir as industrially hazardous agents have appeared.

6. Hygienic Standards of Permissible Exposure

No threshold limit has been set by an official agency in the United States for any of the Pt-group metals or any of their compounds.

STRONTIUM, Sr

1. Source and Production

The principal source minerals of Sr are celestite ($SrSO_4$) and strontiantite ($SrCO_3$), which are widespread in the rocks and waters of the earth's surface. World production amounted to about 13,000 short tons of Sr minerals in 1958. In recent years, the United States has depended on imports of Sr minerals, although several deposits in California, Texas, Arkansas, Utah, and Ohio are known or have been worked. Chief deterrent to domestic mining of Sr minerals is competition from foreign sources and small consumption owing to lack of development of new uses.

Sr minerals are produced on a small scale by simple methods. In England and Mexico, the ore is mined by open-pit methods and hand-picked. Users of celestite for chemical purposes require a minimum of 92 per cent $SrSO_4$, and a maximum of

[519] F. Flury and F. Zernik, Schädliche Gase. Springer, Berlin, 1931, p. 253.

4 per cent $CaSO_4$ and $BaSO_4$. A commonly used purification method is digestion of finely ground $SrSO_4$ in hot soda-ash solution to produce insoluble $SrCO_3$. This is treated with the appropriate acid to make the desired salt. Another method reduces $SrSO_4$ with powdered coal to yield a soluble SrS; this is leached with water and filtered; then treated with CO_2 to obtain the carbonate, or with nitric acid for the nitrate. Sr metal can be produced by electrolysis of fused $SrCl_2$ and NH_4Cl, or by thermal reduction of SrO with Al.

2. *Uses and Industrial Exposure*

Most of the Sr minerals are converted to Sr compounds. Sr compounds impart a characteristic red color to flame, which is the basis of their use in several pyrotechnical applications such as tracer bullets, marine distress-signal rockets and flares, tactical military signal flares, highway and railroad warning fuses, fireworks, and other pyrotechnical devices. The principal compound for these uses is $Sr(NO_3)_2$. $SrCO_3$ is used in ceramics and Zn refining; other Sr compounds are used in greases, corrosion inhibitors, depilatories, medicines, plastics, and luminous paints. Small quantities of Sr metal are used as a getter to remove traces of gas from vacuum tubes.

Industrial hazards are low, not only because of limited production and applications, but also because the toxicity of Sr, and substances associated with its production and handling, is relatively low. Although Ba is associated with Sr minerals, it is in low concentration and its toxicity is dependent on solubility; $BaCO_3$ and BaS produced in the preparation of Sr compounds are among the more toxic forms of Ba, however. Sr metal must be handled with due regard for contact with moisture and air; and the major use of Sr compounds in pyrotechnics creates its own special handling problems. The major hazard of exposure to Sr is from general environmental pollution from radioactive fallout of Sr^{90}.[520]

3. *Physical and Chemical Properties*

The physical and chemical properties of Sr and some of its compounds are listed in Table 32.

Sr is a characteristic alkaline-earth element with a $+2$ valence. As a rule, Sr salts are less soluble than those of Ba. Sr metal will displace H_2 from water.

4. *Analytical Determination*

Spectrographic methods have been employed to determine Sr in ashed bone using Cu electrodes and Cr as the internal standard at 4616A. The Sr line of 4607A. is used.[521,522] A flame photometric determination of Sr in the range of 1 p.p.m. or less, possessing certain advantages over emission spectrography, has been applied

[520] J. L. Kulp, A. P. Schubert, and E. J. Hodges, *Science*, **129**, 1249 (1959).

[521] R. M. Hodges, *et al.*, *J. Biol. Chem.*, **185**, 519 (1950); G. V. Alexander, R. E. Nusbaum. and N. S. MacDonald, *J. Biol. Chem.*, **218**, 911 (1956), **188**, 137 (1951).

[522] R. M. Forbes and H. H. Mitchell, *A.M.A. Arch. Ind. Health*, **16**, 489 (1957).

TABLE 32

Physical and Chemical Properties of Sr and Some of Its Compounds

	Chemical symbol	At. or mol. wt.	Sp. gr.	M.p., °C.	B.p., °C.	Solubility
Strontium	Sr	87.63	2.6	757(f. p.)	1500	Dec. hot or cold H_2O
						Sol. acids, alc., liq. NH_3
Strontium carbonate	$SrCO_3$	147.64	3.7	1497(60 atm.)	$(-CO_2)1340$	11 mg./liter(18°)
						650 mg./liter(100°)
						Sol. acids, NH_4 salts
Strontium nitrate	$Sr(NO_3)_2$	211.65	2.986	570		401 g./liter(0°)
						1 kg./liter(90°)
						V. sol. NH_3
						Sl. sol. HAc
Strontium sulfate	$SrSO_4$	183.7	3.96	1580 dec.	Dec.	113 mg./liter(0°)
						114 mg./liter(30°)
						140 g./liter H_2SO_4 (70°)
						Sl. sol. acids
						Insol. alc., dil. H_2SO_4

to whiskey.[523] Advantages are that the lower volatilization temperatures limit emissions of other cations to more sensitive lines and lines of Sr can be more readily resolved.

5. *Physiological Response*

Sr compounds have a low to moderate toxicity as indicated by range-finding tests[524] and subchronic feeding tests in rats.[522] On the other hand, the radioisotopic Sr^{90} is highly radiotoxic because of its characteristic deposition in growing bone.[520] The intraperitoneal LD_{50} of five Sr compounds (nitrate, iodide, bromide, lactate, and salicylate) ranged from 400 to 1000 mg./kg. (88 to 247 mg. of Sr/kg.). The salicylate was the most toxic, the lactate and bromide the least. $SrCl_2$ showed a minimal lethal dose for rats of 123 mg. of Sr/kg.,[525] and 590 mg. of Sr/kg. for rabbits, administered intravenously;[526] Sr acetate, an intravenous LD_{50} of 105 mg. of Sr/kg. for rats, 168 mg. of Sr/kg. for mice.[527] Sr acetate was considerably less toxic than Ca acetate for mice; at equal millimoles of compound/kg., the mortality from Sr acetate was 5 per cent as against 70 per cent for Ca acetate; less distinction was noted in rats regarding the toxicities of the two compounds.

[523] M. J. Pro and A. P. Mathers, *J. Assoc. Offic. Agr. Chemists*, **39**, 225 (1956).

[524] K. W. Cochran, J. Doull, M. Mazur, and K. P. DuBois, *Arch. Ind. Hyg. Occupational Med.*, **1**, 637 (1950).

[525] D. Loeser and A. L. Konwiser, *J. Lab. Clin. Med.*, **15**, 35 (1930).

[526] A. Faccini, *Arch. farmacol. sper.*, **62**, 187 (1936).

[527] V. V. Cole, B. K. Harned, and R. Hafkesbring, *J. Pharm. Exptl. Therap.*, **71**, 1, (1941).

Ingestion of doses of $Sr(NO_3)_2$ of 50 mg./kg. (1030 p.p.m.) are well tolerated for 8 weeks by either weanling or adult rats, as evidenced by normal amounts of food intake, normal weight gains, total bone ash, or Ca and P composition of the skeleton.[522]

The ingestion by mice of 16,000 p.p.m. Sr lactate in the drinking water for 402 days produced an immediate stunting of growth, which later was recovered according to Alexander et al.[521] At this level Sr actively inhibited calcification of growing bone. Repeated intraperitoneal injections of $SrCl_2$ to the extent of Sr deposition as 7 per cent of inorganic content of bone did not produce rickets in rats.

The chief sign of acute Sr toxicity is respiratory failure; death occurred within an hour after intravenous injection (Ca death is associated with cardiac failure). The only consistent change in the heart from Sr is a slight slowing in rate as determined in animals.[527]

Metabolism. Orally administered $Sr(NO_3)_2$ accumulated in the skeleton of rats in proportion to the dose, as determined in 4- and 8-week feeding tests. Skeletal storage averaged 2.7 per cent of the dose, and the percentage was uninfluenced by sex or age or by the level fed from 30 to 1030 p.p.m. $Sr(NO_3)_2$.[522] Young (6-week) rats had 7.7 p.p.m. Sr in bone ash before administration of Sr; adults had 11.8 p.p.m. The bone-retention factor "discrimination factor" (number of Sr atoms per 1000 Ca atoms divided by the same ratio for the diet) averaged 0.20, indicating a preferential retention of Ca to that of Sr. Alexander et al.[521] confirmed the discrimination factor for the rat but showed varying factors for different species (guinea pigs, 2.5). Sr was noted in all samples of bone analyzed (rabbit, rat, guinea pig, horse, cow). These findings and those of Hodges et al.[521] showing small amounts (1.5 to 2.9 p.p.m.) of Sr in human bone varying in age from fetuses to 75 years indicate the ubiquitous source of Sr in the diet and the skeletal deposition of Sr. These results were confirmed independently by others.[528] Pigmented parts of the eye contain more Sr than other parts according to results of activation analysis;[529] no marked accumulation of Sr in soft tissues of human cadavers was found by the same method of analysis.

Investigations of the distribution of Sr^{90} from nuclear detonations and its uptake in man show that the average burden of Sr^{90} in the adult human skeleton throughout the world in 1958 was 0.19 micromicrocurie of Sr^{90}/g. of Ca, and was independent of age above 20 years.[520] Highest levels of Sr^{90} (micromicrocurie/g. of Ca) occur in North America up through the fourth year of life. The behavior of Sr^{90} in man is similar to that in animals.[530] Preliminary evidence indicates fluoride may decrease the amount of Sr^{90} retained by the rat.[531]

[528] K. K. Turekian and J. L. Kulp, *Science,* **124,** 405 (1956).

[529] E. Snowden, *Biochem. J.,* **70,** 712, 716 (1958).

[530] C. L. Comar, R. H. Wasserman, S. Ullberg, and G. A. Andrews, *Proc. Soc. Exptl. Biol. Med.,* **95,** 386 (1957).

[531] J. C. Muhler, G. K. Stookey, and M. J. Wagner, *Proc. Soc. Exptl. Biol. Med.,* **102,** 644 (1959).

Acute Sr toxicity is partially antagonized by pentobarbital.[527] BAL has no effect on the retention of Sr^{90}, or on the uptake, distribution, or total retention of Sr.[532]

6. Hygienic Standards of Permissible Exposures

No threshold limit value has been set for Sr or any of its compounds by an official agency in the United States.

TANTALUM, Ta

1. Source and Production

Ta occurs in certain complex minerals that contain Nb (which see). Columbite-tantalite $(FeMn)(NbTa)_2O_5$, and pyrochlore minerals are the commercial source of Ta. Tantalite is produced principally in Brazil, Australia, and South Rhodesia; domestically, South Dakota, New Hampshire, Idaho, Connecticut, and Maine have mineral deposits of tantalite. At present, of the world production of tantalite minerals of about 5 million pounds, United States production is about 10 per cent. Domestic production tends to exceed demand.

Ta and Nb are extracted together by fusing at red heat with NaOH until sodium tantaloniobate is formed; then washed and digested with hot concentrated HCl to form tantaloniobic acid and to dissolve impurities of Fe and Mn. The solid tantaloniobic acid is dissolved in HF and sufficient KF added to form $TaKF_4$, which is obtained as a crystal. Pure Ta metal is obtained by electrolysis of the fused $TaKF_4$. Ta collects on the iron cathode as a crystalline powder, which is formed into ingots at high pressure in a vacuum furnace at sintering temperature.

2. Uses and Industrial Exposures

Chief use for Ta in the past has been in chemical equipment and corrosion-resistant tools, but substitutes have replaced Ta in many applications. Recently, a new line of solid Ta capacitors with high capacity per unit volume, low dissipation, stability, long life, and ruggedness has been developed. Ta has been used in the construction of certain types of reactors. An expanded mesh anode made from Pt-clad Ta for Rh electroplating has been developed. Ta_2O_5 is used as a nonradioactive tracer in glass research.

Industrial exposures to Ta are rather limited. Fluoride exposure constitutes the only hazard anticipated during its extraction. Ta metal exposures in the fabrication of metal ingots or metal parts constitute a hazard, as do the preparation and handling of $TaCl_5$.

3. Physical and Chemical Properties

The physical and chemical properties of Ta and some of its compounds are listed in Table 33.

[532] W. E. Kisieleski, W. P. Norris, and L. A. Woodruff, *U. S. Atomic Energy Comm.*, **ANL-5584,** 91 (1956).

TABLE 33

Physical and Chemical Properties of Ta and Some of Its Compounds

	Chemical symbol	At. or mol. wt.	Sp. gr.	M.p., °C.	B.p., °C.	Solubility
Tantalum	Ta	180.95	Met. 16.6 Powd. 14.491	2996 ± 50	ca. 4100	Insol. hot or cold H_2O Sol. HF, fused alkalies Insol. acids
Tantalum pentoxide	Ta_2O_5	441.76	8.735	1470 dec.		Insol. hot or cold H_2O Sol. fused $KHSO_4$ Insol. acids
Tantalum pentachloride	$TaCl_5$	385.17	3.68(27°)	221	242	Dec. in H_2O Sol. abs. alc., H_2SO_4
Potassium fluotantalate	K_2TaF_6	392.15	5.24	720 ± 10		ca. 5 g./liter(0°) 600 g./liter(100°)

Ta is highly unreactive toward many substances at ordinary temperatures; important exceptions are the strong alkalies, fuming sulfuric acid, hydrofluoric acid, and concentrated sulfuric acid above 150°C. Ta is inert toward most gases below 200° but absorbs H_2. It becomes brittle on contact with dilute HF (aq.), or when deposited cathodically in electrolysis. When heated in H_2 above 250°C. Ta can absorb more than 700 volumes of gas, and it becomes very brittle. F_2 attacks Ta at room temperature, Cl_2 at 250°C., and Br_2 at 300°C., but Ta is inert to I_2 below red heat. Ta powder reacts directly with C and B at high temperatures. Ta wire heated to incandescence is converted to the oxide. Commercial Ta powder is not pyrophoric. Ta is dissolved most readily in a mixture of HF and HNO_3 acids.

The chemical inertness of Ta limits the number of ordinary compounds. TaC, used in conjunction with WC and Co for hard-carbide cutting tools and wear-resistant parts, is an important compound of Ta. K_2TaF_6 is one of the more important salts of Ta. Tantalic acid ($Ta_2O_5n\ H_2O$) forms a series of tantalates, for example, potassium metatantalate ($K_2O \cdot Ta_2O_5$), orthotantalate $4K_2O \cdot 3Ta_2O_5 \cdot 16H_2O$, and pertantalate, K_3TaO_8.

4. Analytical Determination

Because the spectral sensitivity of Ta in the emission spectrograph is low, a chemical enrichment procedure using tartaric, salicylic, and phenylarsonic acids[533] should be employed followed by x-ray spectrophotometric determination.[534] All currently devised colorimetric methods are designed for metallurgic work in which Ta is associated with Nb, a condition not necessarily obtained in industrial exposures. A colorimetric method for determining trace quantities of Ta, in which interfering substances are masked with EDTA, employs phenylfluorone after extraction from acid solution with methylisobutyl ketone.[535]

[533] K. Rankama, *Bull. comm. géol. Finlande,* No. **133** (1944).
[534] B. J. Mitchell, *Anal. Chem.* **30,** 1894 (1958).
[535] C. L. Luke, *Anal. Chem.,* **31,** 904 (1959).

5. Physiological Response

Except for two oblique references[536] to exposures in which Ta as the carbide was associated with Co, WC, and TiC, the report by Schepers[537] on massive (100 mg.) intratracheal introduction of Ta_2O_5 in guinea pigs' lungs constitutes one of the few sources of information on the biological effects of natural Ta. Ta_2O_5, particle diameter 3 μ and below, produced transient bronchitis, interstitial pneumonitis with hyperemia and residual focal hypertrophic emphysema, and organizing pneumonitis about the metal deposits, but was not fibrogenic. Within 1 month focal reactions in the lung appeared that paralleled in severity the amount of deposited dust. After 1 year almost complete recovery from the acute changes had occurred. The author warns against classifying Ta_2O_5 as innocuous on the basis of the single exposure, the single-year observation period, the bronchial epithelial hyperplasia, and the hypertrophic focal emphysema.

A Russian worker,[538] extending his clinical investigation of industrial exposures to Ta (and Nb) to animals, reported that Ta by mouth is almost completely excreted via the intestine with no evidence, as shown by radiotracers, of resorption. The industrial dusts studied were found to be insoluble in physiological media and consequently showed very slight toxicity. Effects on the lung varied according to the Ta compound inhaled; potassium fluotantalate (K_2TaF_7) gave the most pronounced effects, Ta_2O_5 and Ta, next in order. General effects consisted of thickening of the alveolar septa and blood-vessel walls, and proliferation of histiocytes. Tantalum metal has been used in surgery, and appears relatively inert in contact with tissue.[539]

Metabolism. Trace amounts (11 microcuries) of carrier-free radioactive Ta^{182} by intramuscular injection in rats was 99 per cent absorbed and excreted within 32 days, in a ratio of 4 in the urine to 1 in the feces, thus differing considerably in retention and excretion from Ta^{182} administered in microgram amounts with carrier.[540]

Industrial Experience. Clinical studies made on 22 Russian chemical workers and welders handling both Ta and Nb showed little evidence of poisoning apart from radiologic signs of early pulmonary fibrosis and, in 1 or 2 cases, chronic atrophic rhinitis.[538] (The abstract makes no reference to air concentrations or types of compounds to which workers were exposed.)

6. Hygienic Standards for Permissible Exposures

No limit for Ta or any of its compounds has been established by an official agency in the United States.

[536] C. W. Miller, M. W. Davis, A. Goldman, and J. P. Wyatt, *Arch. Ind. Hyg. Occupational Med.*, **8**, 453 (1953); S. G. Sjoberg, *Nord. Med.*, **43**, 117 (1950).

[537] G. W. H. Schepers, *A.M.A. Arch. Ind. Health*, **12**, 121 (1955).

[538] J. L. Egorov, *Gigiena Truda i Professional. Zabolevaniya*, **1**, 16 (1957), (abstr.).

[539] A. R. Koontz and R. C. Kimberly, *Ann. Surg.*, **137**, 833 (1953).

[540] K. G. Scott, *Med. Health & Phys. Quart. Rept.*, **UCRL-960**, Sept., 1950.

THALLIUM, Tl

1. Source and Production

Tl is widely distributed in the earth. Igneous rocks are estimated to contain 30 g./ton. Commercial sources, however, are flue dusts, either from pyrites (FeS_2) burners or from Pb and Zn smelters and refiners, as a by-product of Cd production at the rate of a few 1000 pounds per year. In the dusts, Tl occurs largely as a sulfate, which is extracted with hot water or dilute acid. Crude Tl is precipitated as a metal onto Zn or Al, as the chloride, followed by recrystallization from dilute H_2SO_4. Electrolysis of carbonate, sulfate, or perchlorate solutions isolates Tl metal.

2. Uses and Industrial Exposures

Principal use of Tl, as the sulfate, is as a rodent poison. Tl_2CO_3 is a good fungicide because of its solubility. Some additional uses for Tl were developed during World War II based on the property of the bromoiodide crystals of transmitting radiation of very long wavelength, which is useful in equipment for detection and signaling where visible radiation must be absent. Bromoiodide crystals of thallium find uses as lenses, plates, and prisms in optical systems of spectrometers. Tl oxysulfide finds use in photoelectric cells because of greater sensitivity than Se to light of long wavelength. A saturated solution of Tl malonate and formate is used to separate mineralogic specimens. A Tl–Hg alloy (8.5 per cent Tl) finds application in low-range glass thermometers. In addition to these uses, Tl has limited uses in high-density liquids, special glasses, Se rectifiers, insect proofing, and as a phosphor activator. Tl, as metal or as compound, reportedly has potentialities as an antiknock substance in gasoline.

Industrial exposures to Tl and its compounds constitute a severe hazard in themselves in the handling of pyrites, flue dusts, and other sources of Tl-bearing materials. Chief ancillary exposures in the production of Tl and its compounds are Pb, Cd, and As. Tl exposures occur also in handling of Tl compounds and preparations, such as economic poisons, or solutions used in mineralogical analysis.

3. Physical and Chemical Properties

The physical and chemical properties of Tl and some of its compounds are listed in Table 34.

Tl metal is somewhat softer than Pb and when freshly melted resembles Sn in whiteness. Tl, however, oxidizes rapidly, turning first gray then brownish black from an oxide coating. For this reason Tl rods are often paraffin-coated. The metallurgic properties of Tl are little known because little use of the metal has been made.

The chief valence of Tl is +1; a valence of +3 is known but the compounds are less numerous and less stable. The metal is very reactive, but, in bulk,

TABLE 34

Physical and Chemical Properties of Tl and Some of Its Compounds

	Chemical symbol	At. or mol. wt.	Sp. gr.	M.p., °C.	B.p., °C.	Solubility
Thallium	Tl	204.39	11.85	303.5	1460	Insol. hot or cold H_2O Sol. HNO_3, H_2SO_4 Sl. sol. HCl
Thallous oxide	Tl_2O	424.78		300	1080(600 mm.) (−0)1865	V. sol. Dec. to TlOH Sol. acids, alc.
Thallic oxide	Tl_2O_3	456.78	Amor. 9.65 (21°) Hex. 10.19 (20°)	717 ± 5	(−20)875	Insol. hot or cold H_2O Sol. acids Insol. alkalies
Thallous carbonate	Tl_2CO_3	468.79	7.11	273		40.3 g./liter(15°) 272 g./liter(100°) Insol. abs. alc., ether, acetone
Thallous sulfate	Tl_2SO_4	504.85	6.77	632	Dec.	48.7 g./liter(20°) 155.7 g./liter(99.7°)
Thallous sulfide	Tl_2S	440.85	8.0	443	Dec.	0.2 g./liter(20°) Sol. acids Insol. alkalies
Thallous acetate	$Tl(CH_3COO)$	263.43	3.68	110		V. sol. cold H_2O, alc.

reactions are slowed or inhibited by the formation of coatings over the surface. Mixed steam and air will react to produce TlOH solution. Tl reacts with hot alcohol to produce an ethoxide. Halogen acids react only to a limited extent because of the insolubility of the halides. Nitric acid will react with Tl to produce either $TlNO_3$ or $Tl(NO_3)_3$ according to the acid concentration; hot H_2SO_4 reacts to give the sulfate.

4. Analytical Determination

The spectrochemical method has been used by Truhaut[541] for the determination of small amounts of Tl in biological specimens; the Tl line 5350.5A. is used with the Ag line 5209.1A as internal standard; 0.02 μg. of Tl on the electrode with an error of 20 per cent is reported. Ca must be removed prior to arcing as Ca 5349.5A interferes. This may be done simply by addition of excess NH_4OH, which also removes phosphate. Downs et al.,[542] however, utilize the Tl line

[541] R. Truhaut, *Recherches sur la Toxicologie du Thallium.* Institut National Securité pour la Prevention des Accidents du Travial, Paris, 1959; F. F. Heyroth, *Public Health Repts.* U. S. Suppl. No. **197**, 1947.

[542] W. L. Downs, J. K. Scott, L. T. Steadman, and E. A. Maynard, *Am. Ind. Hyg. Assoc. J.*, **21**, 399, 1960.

2767.9A., employ a Bi line 2898.0A. as internal standard, and determine Tl to 0.2 μg. on the electrode with an error of ± 10 per cent when determinations are made in triplicate. No chemical separation of Tl is necessary by this procedure. Micro-quantities of Tl may also be determined colorimetrically by the use of thionalide (2-mercapto-N-2-naphthylacetamide), or by extraction with 2-octane (Stravinoha et al.)[543] A polarographic method for the determination of Tl in urine has been developed by Winn, Godfrey, and Nelson.[544]

5. *Physiological Response*

A monograph on the toxicology of Tl, consisting of 272 pages and more than 260 references to the world literature on the subject, by Truhaut[541] was published in 1959. Heyroth[541] had previously summarized the medical literature relating to Tl poisoning to 1947. Tl is one of the more toxic elements, both acutely and chronically, in animals and in man, regardless of the route of intake. Toxicity is largely independent of valence state. Truhaut reported[541] the acute LD_{50} of Tl_2SO_4 for the adult French male mouse by oral, subcutaneous, and intravenous routes to be 46, 41, and 33 mg. of Tl/kg respectively. Downs et al.[542] found somewhat greater toxicity of thallous acetate for the rat, guinea pig, rabbit, and dog by the intravenous, intraperitoneal, and oral routes (4 to 29 mg. of Tl/kg.), the guinea pig being the most susceptible, the rat the least susceptible. Trivalent Tl in the form of Tl_2O_3 was somewhat less acutely toxic to these species by these routes (9 to 92 mg. of Tl/kg.). An exception was the oral toxicity of Tl_2O_3 for the guinea pig, which was highest of all (3 to 5 mg. of Tl/kg.).

Repeated daily ingestion of Tl^{1+} acetate incorporated in the diet was tolerated at 0.001 per cent level by rats; 0.003 per cent was lethal to male, but not female, rats (15 weeks); on the other hand, Tl_2O_3 at a 0.0035 per cent level in the diet was lethal to the majority of rats in 15 weeks, but the 0.002 per cent level was lethal only to female rats.[542]

The most characteristic symptom of Tl intoxication is the development of alopecia. Symptoms of acute poisoning are variable but are referable to the gastrointestinal tract and nervous system. Symptoms develop slowly, from several hours to 2 days. Digestive disturbances include metallic taste, salivation, stomatitis, nausea, vomiting, and abdominal pain. Vasomotor disturbances appear as puffiness of the cheek, eyelids, and lips. Tingling pain in the extremities, muscular weakness, delirium, convulsions, and coma may ensue. Death may occur in a few days or after several weeks. Recovery is slow, requiring 6 or more months. Sequelae in the form of sensory and nervous disorders, such as ataxia and chorei-

[543] R. Berg, E. S. Fahrenkamp, and W. Roebling, *Mikrochemie Molisch Festschr.,* **1936,** 44; E. B. Sandell, *Colorimetric Determination of Traces of Metals.* Interscience, Publishers, New York–London, 1959, p. 834; W. B. Stavinoha, J. B. Nash, and G. A. Emerson, *Federation Proc. Abstr.,* **391,** April, 1960.

[544] G. S. Winn, H. L. Godfrey, and K. W. Nelson, *Arch. Ind. Hyg. Occupational Med.,* **6,** 14 (1952).

form movements, blindness and paralyses of the special senses, psychoses, peripheral neuritis, and renal damage, are common.

Symptoms of chronic intoxication resemble those seen in acute forms, but are milder, and in addition may include such manifestations as incoordination, paralysis of the extremities, encephalitis, endocrine disorders, psychoses, and baldness. Loss of hair usually occurs 10 to 14 days after exposure but may occur earlier depending on dosage.

Metabolism. Tl is absorbed readily through the digestive tract and skin. Following absorption, Tl is generally distributed throughout the body. From the data of Downs *et al.*[542] on rats, fed separately thallous acetate and thallic oxide for from 2 to 3 months, the kidney contained largest amounts of Tl, bone next, followed by liver, lung, spleen, and brain, which comprised the tissues analyzed. Tl was not detectable to a significant amount in unexposed controls. Truhaut[541] found general distribution of Tl in all of 26 tissues of rats and rabbits chronically fed Tl metal for 2 weeks. Under these conditions, bone had scarcely detectable amounts, whereas hair and the hypophysis frequently contained more Tl than the kidney. Tl passes the placental barrier as well as the blood-brain barrier. Tl accumulates in the germinating zone of the hair follicle. The general distribution of Tl has been demonstrated also by radiotracer techniques using Tl^{204}.[541,545] The older literature on metabolism of Tl, reviewed by Heyroth,[541] showed sufficient Tl may be in the milk of lactating rats to slow growth of young, but not to kill them. Heyroth quotes the opinion that there is little likelihood of individuals being poisoned by eating game birds poisoned by Tl. His statement that tissues containing as little as 5 μg. of Tl/g. are suggestive of Tl intoxication, although rather unspecific, is probably the right order of magnitude in the light of recent work.

Excretion of Tl follows a rather unusual course, although rapidly and widely distributed and appearing early in the urine and feces, 2 months are required before Tl disappears from the urine in chronic intoxication, and 3 weeks in acute poisonings. Dogs excreted 60 per cent of a single, orally administered dose in a period of 36 days (Heyroth[541]). The not uncommon pattern of urinary excretion of metals (e.g., V) is 60 per cent elimination of the dose in 24 hours. Truhaut[541] has confirmed the long residence time of Tl in the body and has pointed out that the fecal route of excretion is almost as important (in the rabbit) as urinary excretion. During a 23-day period following exposure, 11 mg. of Tl was excreted in the feces as against 14.6 mg. in the urine, constituting respectively 24.4 and 31.9 per cent of the dose. In man, 15.4 per cent of the dose was excreted in the urine in 5.5 days and the rate of excretion of the remaining body Tl burden was 3.2 per cent per day;[546] distribution of Tl^{204} in 57 tissue sites of a man following oral administration is given.

[545] R. Lie, R. G. Thomas, and J. K. Scott, *Health Phys.*, **12**, 334 (1960); N. Thyresson, *Acta Dermato-Venereol.* **31**, 3 (1951).

[546] R. K. Barclay, W. C. Peacock, and D. A. Karnofsky, *J. Pharmacol. Exptl. Therap.*, **107**, 178 (1953).

Mechanism. Truhaut[541] has made an extensive study of the action of Tl salts on more than a dozen organ, cellular, and enzyme systems, but as yet no specific enzyme system has been found that is inhibited by Tl at concentrations that have proved injurious in the body. Among Truhaut's findings: (*a*) Tl salts inhibit the respiration of yeast at 0.01 mM; SH compounds did not prevent this action. Similar though lessened inhibitory effects of Tl were found for homogenates of liver, kidney, and heart; succinic acid tended to suppress the inhibitory effects of Tl. (*b*) Tl acts as an excitor of the preganglionic parasympathetic nervous system; actions of adrenaline and acetyl choline were diminished by Tl. On this basis, Truhaut is of the opinion that Tl oxidatively destroys adrenaline. (*c*) On the isolated intestine, Tl reduces contractions, increases amplitude, and is thus considered to be, like Ba, a general muscle poison, although a relaxant, not a stimulator of muscle contraction. (*d*) Tl is an antimitotic agent, presumably inhibiting the development of keratin, and thus explaining its role in producing alopecia. Inhibition of mitosis of the cells of the intestinal mucosa of the mouse had been previously reported.[547] In this connection, Downs *et al.*[542] quote Opdyke, who studied the cellular changes in the skin and hair in animals poisoned with Tl: there appears to be total atrophy of the hair follicles with replacement by scar or collagen or fat, a general thickening of collagen, disappearance of lymph spaces, and apparent pressure on blood vessels and follicles such as is seen in atrophic scleroderma. The sebaceous glands appear hyperactive, with excess fat, indicating ultimate destruction of the glands. Microincineration showed striking depletion of Ca and Mg in the basal and granular layers of the epidermis in the region of the hair follicles. Cystine (1 to 20 per cent) added to the diet of rats delayed the development of alopecia, and increased the survival period,[548] indicating dietary factors have a distinct role in the mechanism of Tl poisoning.

Clinical and Industrial Experience. Several hundred cases of Tl poisoning in man have been recorded but only a few of industrial exposure, all of them chronic, have been reported. As late as 1955, cases of industrial Tl poisoning in the manufacture of rodenticides were reported from Sweden by Glomme and Sjostrom,[549] and from Germany by Egen.[550] Because the early symptoms are rather nonspecific, and the sequelae often severe and incapacitating, it has been suggested that Tl poisoning be added to the list of compensable industrial diseases. Six cases of Tl intoxication were reported for the first time in the United States in 1958 among men working with solutions of organic Tl compounds for the separation of industrial diamonds.[551] The intoxication resulted purely from skin contact, as no Tl in the workroom air was demonstrable. Diagnostic difficulties were

---

[547] G. Marras and L. Nanetti, *Arch. Sci. biol. Bologna,* **38,** 472 (1954).

[548] P. Gross, E. Runne, and J. W. Wilson, *J. Invest. Dermatol.,* **10,** 119 (1948).

[549] J. Glomme and B. Sjostrom, *Svenska Läkartidn.,* **52,** 1436 (1955); *Ind. Hyg. Foundation Am.* Abstr., **20,** No. 207 (1956).

[550] B. Egen, *Zentr. Arbeitsmed. Arbeitschutz,* **5,** 141 (1955).

[551] E. M. Richeson, *Ind. Med. and Surg.,* **27,** 607 (1958).

pointed out, as well as the inadequacy of definitive laboratory aids for a differential diagnosis. No relation between urinary excretion rate of Tl, the exposure, and the appearance of symptoms was established; BAL did not seem to benefit 2 cases in which it was tried.

A brief review and summary of thallitoxicosis as a recurring problem in the United States was made by Conley,[552] in which reference is made to cases in California of Mexicans poisoned by Tl-containing tortillas, of Tl poisoning in Turkey from bread made from flour contaminated with a Tl rodenticide, as well as more than 60 cases of Tl poisoning in children in Texas from eating Tl-containing insect and rat poisons.

Prior to the review of Conley, a summary of all cases of human thallitoxicosis to the year 1934 was made by Munch.[553] Twelve cases of industrial origin were disclosed, none of them fatal. From clinical use of Tl, however, 692 poisoning cases were revealed, with 31 deaths. Toxicological literature showed 53 poisoning, from which 10 died; from rodenticidal and entomological use 21 poisonings and 5 deaths; in all, 778 Tl poisoning cases with a 6 per cent death rate. In many cases of mild chronic poisoning lymphocytosis and eosinophilia were usual findings.

Therapy. It is apparent from the above that no therapeutic agent has been found satisfactory, to date; BAL has been tested to no apparent avail in a few cases. Results are claimed to be good when a "stabilized" H_2S preparation was used in guinea pigs.[554] More recently, Stavinoha et al.[555] have reported that mercaptopropane and monothioglycerol exert a moderate protective action in mice when administered 6 hours after Tl.

6. Hygienic Standards of Permissible Exposure

A threshold limit of 0.1 mg. of Tl/cu. meter has been set for soluble Tl compounds, by the American Conference of Governmental Industrial Hygienists. It would now appear from the work of Downs et al.[542] that the reference to soluble compounds is unnecessary, and that the 0.1 mg./cu. meter limit should be applied to all Tl compounds, including the metal. No documented evidence of the suitability of the present limit for Tl has reached the Committee on Threshold Limits, but Truhaut has indicated his belief that the limit is a satisfactory one.[556]

THORIUM, Th

1. Source and Production

Monazite, a phosphate mineral of Th containing certain of the rare earths (Ce, Nd, Pr, La) from the Union of South Africa, continues to be the main source

[552] B. E. Conley, *J. Am. Med. Assoc.,* **165,** 1566 (1957).

[553] J. C. Munch, *J. Am. Med. Assoc.,* **102,** 1929 (1934).

[554] S. Moeschlin and B. Demiral, *Schweiz. med. Wochschr.,* **82,** 57 (1952).

[555] W. B. Stavinoha, G. A. Emerson, and J. B. Nash, *Toxicol. Appl. Pharmacol.,* **1,** 638 (1959).

[556] R. Truhaut, personal communication (1959).

of supply of Th for the free world. Domestic production uses small amounts from Florida, Idaho, South Carolina, Tennessee, and Colorado. Th concentrates are obtained by dredging, concentrating by jigging, and separating by a combination of electrostatic and high-intensity separators. Ultrapure nuclear-grade ThO_2 is made by solvent extraction of cracked thorite ore ($ThSiO_4$); Th metal and alloys are made from the same process. Monazite ore is reduced by a caustic soda process to yield $Th(NO_3)_4$, ThO_2, ThF_4, $Th(SO_4)_2$ and $ThCl_4$, the principal Th salts of commerce.

2. Uses and Industrial Exposures

Nonenergy uses of Th in 1957 were first, by a large margin, Mg alloys (100,000 lb. estimated), gas-mantle manufacture (40,000 lb.), chemical and medical products (4000 lb.), electronic products (1000 lb.), and refractories and polishing compounds (unestimated). New alloys capable of withstanding highest temperatures of any metal of equal density, made of 3 per cent Th, 0.7 per cent Zr, the rest Mg, account for much of the increase in Mg alloys; Th alloys have found use in Air-Force missiles, supersonic bombers, and Navy satellites. In addition to being the least expensive airframe material, Mg–Th alloys have weight-saving advantages over Al in some jet-engine castings with excellent uniformity. A thoriated-tungsten alloy (3 to 6 per cent Th) is finding use in welding. Th metal is also utilized as a deoxidant in preparing Mo and its alloys; and in electronic tubes and lamps for controlling starting voltages and maintaining stability. Nonenergy uses of Th for 1957 have been given in detail.[557] Use of thoria in ceramics is being studied.

Energy uses of Th are for breeder reactors, temporarily out of fashion, but which will probably again find favor in the 1960's. The Consolidated Edison Test Reactor in New York is a pressurized-water, Th-converter, power reactor, operating with zircalloy-clad fuel plates of U-235 alloy and a Zr-clad Th blanket. Steps in the manufacture of Th ingots from $Th(NO_3)_4 \cdot 4H_2O$ as well as details of a new process for manufacturing high purity Th are given in the minerals yearbook.[558]

Industrial exposures occur during handling of the various Th salts in the fabrication of Th ingots from the nitrate, in handling of Th salts in various industrial uses, in the fume from welding with thoriated-tungsten electrodes, and in the casting and machining of Th-alloy parts.

3. Physical and Chemical Properties

The physical and chemical properties of Th and some of its compounds are listed in Table 35.

Thorium is a white, relatively soft, ductile metal that burns readily in air. In its analytical reactions it resembles the rare earths, and can be separated

---

[557] W. C. Lilliendahl, *Metal Progr.*, **71**, 104 (1957).
[558] *Minerals Yearbook, U. S. Bur. Mines*, **1957** (1958).

TABLE 35

Physical and Chemical Properties of Th and Some of Its Compounds

	Chemical symbol	At. or mol. wt.	Sp. gr.	M.p., °C.	B.p., °C.	Solubility
Thorium	Th	232.05	11.6	1845	4500	Insol. hot or cold H_2O Sol. HCl, H_2SO_4, aq. reg. Sl. sol. HNO_3
Thorium dioxide	ThO_2	264.12	10.03	3050	4400	Insol. hot or cold H_2O Sol. hot H_2SO_4 Insol. alkalies, dil. acids
Thorium nitrate tetrahydrate	$Th(NO_3)_4 \cdot 4H_2O$	522.22		Swells		V. sol. H_2O, alc. acids

from the common metals as an oxalate or fluoride. Th forms one series of tetravalent salts, many of which exist as various crystalline hydrates. Water and aqueous alkali do not appreciably attack Th metal, but it slowly reacts with acids to form soluble salts. Th ion is colorless; hydrolysis of salts is common but hydroxides or basic salts are not precipitated unless solutions are highly dilute.

Th^{232} is the parent of a series of radioactive elements extending through various Ra isotopes, Th,[228] Po, Bi, and finally stable $Pb.^{208}$ Industrial hazards associated with the radioactivity of Th and its daughters are discussed in Volume I, pages 739 to 742.

4. Analytical Determination

Analysis of Th in samples of industrial hygiene interest is difficult. Successful application of chemical methods for low concentrations require preliminary concentration procedures involving ion exchange or coprecipitating agents. Interferences of commonly encountered ions are serious. Recoveries at best are of the order of 80 per cent. Because of the highly nonvolatile nature of Th, spectrographic methods are insufficiently sensitive; moreover, Th spectrum is extremely complex with no lines of high transition probabilities. One of the better methods for the determination of submicrogram quantities of Th in urine using morin is that of Perkins and Kalkwarf.[559] Albert et al.[560] give details of a similar method using thorin. A chrome–azurol–S reagent has been compared with morin and thorin for the determination of Th in urine;[561] morin was found to be the most sensitive reagent.

[559] R. W. Perkins and D. R. Kalkwarf, Anal. Chem., **28**, 1989 (1956).

[560] R. Albert, P. Kevin, J. Fresco, J. Harley, W. Harris, and M. Eisenbud, A.M.A. Arch. Ind. Health, **11**, 234 (1955).

[561] G. A. Wolford, D. A. Sutton, R. S. Morse, and S. Tatras, Am. Ind. Hyg. Assoc. J., **19**, 464 (1958).

5. *Physiological Response*

An annotated bibliography of 21 references on Th toxicology to 1959 is available.[562] An alphabetically arranged bibliography of more than 500 references on the biological effects of Th appeared in 1960.[563] The literature on the radiotoxic effects of Th, chiefly thorotrast and thorium X, is increasing yearly.

The acute toxicity of Th salts by all routes examined is very low; the acute intraperitoneal LD_{50} of $Th(NO_3)_4 \cdot 4H_2O$ was calculated as 1220 mg./kg. for mature female rats; for weanlings, greater than 2000 mg./kg.[564] Dietary levels of 3 per cent nitrate were required before growth depression occurred in rats; a dog showed growth depression (15%) on 1 g. of nitrate/kg. daily, in a 46-day study. No other changes, histological, hematological, or otherwise were noted. Short-term inhalation exposures of dogs of from 2 to 10 weeks to the nitrate, oxide, fluoride, and oxalate separately, at mean exposure concentrations of 76, 51, 11, and 26 mg./cu. meter of compound respectively, at mean particle diameters around 1 μ showed abnormal leucocytes as the only evidence of toxicity. The leucocytes were identified as: (a) monocytes with atypical nuclei; (b) abnormal forms of lymphocytes such as seen in toxic states; and (c) old hypersegmented polymorphonuclear leucocytes.[565] A 1-year exposure of laboratory animals to ThO_2 at 5 mg./cu. meter (about 50 times the M.P.C.) confirmed the chemical inertness of Th salts by inhalation.[565]

Metabolism. Th is characterized by very low absorption from the gastrointestinal tract, less than 0.001 per cent absorbed at doses of 500 to 800 mg./kg., 0.05 per cent at doses of 5 mg./kg.[566] Following intravenous administration, a soluble Th salt is removed from the blood slowly; 30 per cent of a small dose still remains in the blood after 6 hours. Most of the Th goes to the reticuloendothelial system, the liver, spleen, and bone marrow[506a] When Th salts are intratracheally injected they remain in the lung; when injected in the muscle, they remain at the injection site.[566] Retention of Th from inhalation of ThO_2 dust for 1 year, at approximately 50 times the M.P.C. at particle diameter near 1 μ, left 98 per cent of the body burden of Th in the lung and pulmonary lymph nodes, 2 per cent in bone, and less than 0.1 per cent in the rest of the body.[565] The relative immobility of tissue-deposited Th is shown by the finding of 97.7 per cent in the pulmonary lymph nodes, 2.1 per cent in the lung, 0.003 per cent in the femur, and fractions of this latter amount in the kidney, spleen, and liver 6 to 7 years after inhalation exposure had been terminated.[565]

[562] R. E. Allen, ed., *Thorium—A Bibliography of Published Literature.* TID-3044, Suppl. 1, U. S. Atomic Energy Comm., Oak Ridge, Tenn.

[563] E. D. Hutchinson, *U. S. Atomic Energy Comm.,* Res. Div. Rept. **UR-561,** Jan., 1960.

[564] W. L. Downs, J. K. Scott, E. A. Maynard, and H. C. Hodge, *U. S. Atomic Energy Comm.,* Res. Div. Rept. **UR-561,** Dec., 1959.

[565] H. C. Hodge, E. A. Maynard, and L. J. Leach, *U. S. Atomic Energy Comm.,* Res. Div. Rept. **UR-562,** Jan., 1960.

[566] P. R. Salerno and P. A. Mattis, *J. Pharmacol. Exptl. Therap.,* **101,** 31 (1951).

Industrial Experience. The work experience with Th and its compounds has been exceptionally good. Although the work force has not been large, neither workers in plants making incandescent Welsbach mantles (Th 99 per cent, Ce 1 per cent from the nitrates) for more than 70 years, nor workers refining thoria from monazite have experienced any effects attributable to chemical toxicity or radiation injury. Moreover, exposures during the earlier years were largely unregulated and unquestionably exceeded the M.P.C. for Th_{232} (2×10^{-12} microcuries/cc.). A recent industrial hygiene and medical survey of a Th refinery that had been in operation for 30 years, which revealed exposures well in excess of current standards, revealed "no evidence of overt industrial diseases."[566a] Employee force was 84, 60 of whom worked with thorium and thoron; one half had spent 10 years or more in the thorium process. Six hundred and ninety-three former employees on whom there were records showed no single disease indicating a relation to their work.

Investigation[567] of the exposure hazards associated with welding with thoriated-tungsten electrodes revealed that although the thoria released exceeded the M.P.C. at a distance of 6 in. from the arc, all air samples at the breathing zone of the welder were well below the M.P.C. (less than 0.9 α disintegrations/minute/cu. meter). No special ventilation or protection was recommended. Thus the amounts of thoria released were such as to concern only radiotoxicity, not chemical toxicity.

Medical Exposure. With increasing frequency since 1955 there have been reports in the medical literature of severe radiation damage and cancers of the bone, blood vessels, liver, and other organs as a result of thorotrast or thorium X, administered as a radiopaque medium 11 to 20 years previously.[568] More reports of this unfortunate use of Th will certainly appear. Although this is not of occupational health interest, it serves to point up most forcibly the severely damaging capacity of thorium oxide preparations when introduced in the body spaces. Looney[568] has given a comprehensive review of current information on this problem.

Therapy. EDTA (ethylenediaminetetraacetic acid) is believed to merit consideration in helping rid the body of Th from accidental overexposure.[569] Used in 4 patients whose exposure was complicated by burns and exposure to uranium,

[566a] J. K. Scott, W. F. Neuman, and J. F. Bonner, *J. Pharmacol. Exptl. Therap.*, **106**, 286 (1952).

[567] A. J. Breslin and W. B. Harris, *Am. Ind. Hyg. Assoc. Quart.*, **13**, 191 (1952).

[568] C. M. Gros, et al., *bull assoc. franc. étude cancer*, **42**, 566 (1955); H. Spiess, *Deut. med. Wochschr.*, **81**, 1054 (1956); M. Schraier, et al., *Prens. med. Arg.*, **45**, 1805 (1958); H. Telsuk and W. A. Nordin, *Arch. Pathol.*, **60**, 493 (1955); J. C. Roberts and K. E. Carlson, *Arch. Pathol.*, **62**, 1 (1956); H. Brody and M. Cullen, *Surgery*, **42**, 600 (1957); F. J. Rosenbaum, *Deut. med. Wochschr.*, **84**, 428 (1959); W. B. Looney, *Am. J. Roentgenol. Radium Therapy Nuclear Med.* (1960).

[569] W. N. Young and H. A. Tebrock, *Ind. Med. and Surg.*, **27**, 229 (1958).

EDTA aided removal of Th as a soluble undissociated complex, although bioassay for Th was rendered more difficult by the complex.

6. Hygienic Standards of Permissible Exposure

No limit has been set for Th and its compounds on the basis of its chemical toxicity because of its toxicological inertness. Various limits on the basis of its radiotoxicity have been set from time to time; the presently suggested M.P.C. is 2×10^{-12} microcuries/cc. This is based on the assumption that the M.P.C. constant exposure could be endured for 50 years without exceeding the limiting dose.

7. Flammability

The powdered metal ignites at $270°C.$, ThH_4 at $260°C.$ There is no limiting O_2 pressure that will prevent Th ignition, as ignition will occur in pure CO_2. A limiting concentration of 6 per cent O_2 will prevent ThH_4 from igniting. The minimal explosive concentrations for Th and ThH_4 are 75 and 80 oz. per 1000 cu. ft. (see also Vol. I, pages 560, 562, 565, and 575).

TIN, Sn

1. Source and Production

There are 9 Sn-bearing minerals but only cassiterite (SnO_2) is of commercial importance. All other minerals are complex sulfides; stannite (Cu_2FeSnS_4) and teallite ($PbZnSnS_2$) are of commercial significance only in Bolivia. Although the United States consumes more than 50 per cent of the free-world production, no Sn ore or concentrate of marketable grade was produced in 1958 in the United States. World production, chiefly from Southeast Asia, Malaya, and Indonesia, has been limited by the International Tin Agreement; in 1958, 104,500 tons were produced compared with 152,200 tons the previous year.

Cassiterite is recovered chiefly by bucketline dredging. The gravel is disintegrated into coarse and fine; riffled sluices separate the heavy minerals from the waste fines. Some dredging is done under water in off-shore deposits in Indonesia. Some cassiterite is recovered by hydraulic mining and, where labor is cheap and placer Sn occurs, hand-mining. Cassiterite concentrates often contain 75 per cent Sn (cassiterite, 78.6 per cent Sn). Preparation of high-grade concentrates from lode mines is difficult and recoveries are poor because of underdeveloped technology.

In the roasting of sulfide ores, most of the S and As are removed as SO_2 and As oxides, the sulfides of Fe, Cu, Bi, and Zn are converted to oxides, and Pb to $PbSO_4$. The roasted concentrates are leached with dilute acid, which removes the oxides of Bi, Cu, and Zn; W is removed by heating with Na_2CO_3 and water leaching; and Pb, Sb, Ag, and Bi are removed by a chloridizing roast followed by acid leach.

In smelting, SnO_2 is readily reduced at a low red heat by C or reducing gases (H_2, CO, or hydrocarbons), whereby losses occur when SnO_2 combines with silicate slag to form $SnSiO_3$ or with basic slags to form stannate. Also at the temperatures required to make the slag fusible, 1100 to 1200°C., considerable volatilization of Sn occurs, which requires fume entrapment. The slag is resmelted for further recovery of Sn.

Most smelted Sn requires refining to remove metal impurities. This may be done by heat or electrolytically. In the heat treatment, Sn is heated in a reverberatory furnace to just above its melting point, where metals with higher melting points (e.g., Fe) remain in the dross; Pb and Bi remain in the Sn, but As, Sb, and Cu are partly removed in the dross. Boiling and tossing (poling) agitate the Sn with air causing metal impurities to oxidize and go to the surface, when they are skimmed. The refined Sn is cast into pigs. Electrolytic refining produces a purer grade of Sn, but is little used for lack of demand. There are many grades of Sn ranging from 99.9810 per cent Sn (electrolytic) to 99.0 per cent Sn.

Secondary Sn, recovered from tinplate and other scrap, amounted to 22,800 long tons in 1958 in the United States.

2. Uses and Industrial Exposures

The main outlet of Sn is its use as a protective coating; next in importance is its use as an alloying metal. Tinplate, solder, bronze, and brass account for more than 80 per cent of Sn consumed. Almost 90 per cent of tinplate is used for cans, of which 60 per cent is used for food packaging, 40 per cent for nonfood products. A significant quantity of Sn is used as platings, solders, alloys, and chemical compounds in the automotive and aircraft industries: as Sn-coated wire, phosphor–bronze alloys, Zn–Sn coatings for hydraulic brake parts, landing-gear equipment, and Cd–Sn alloy for plating engine parts. Modern pewter alloys replace the Pb of the old pewter with Sb(4 to 6.5 per cent), Cu(1 to 1.5 per cent), and the balance Sn. For low-temperature soldering, Sn-base soft solders have no serious competitors.

Stannous chloride is a reducing agent employed as a discharge in calico printing. Hydrated stannic acid, sodium metastannate, hydrated stannic chloride, and ammonium stannic chloride are used as mordants in dyeing and in the weighting of silk. Stannous fluoborate, $Sn(BF_4)_2$, forms the basis of a commercially important plating bath. Trialkyl Sn salts are used as fungicides and to a limited extent as an anthelmintic.

Industrial procedures involve exposure to silica, Pb and As in mining the sulfide ores of Sn; and in the roasting and smelting processes, exposure to Bi and Sb as well. Similarly, the preparation and use of Sn alloys and solders present an exposure to these heavy metals as well as to Sn.

3. Physical and Chemical Properties

The physical and chemical properties of Sn and some of its compounds are listed in Table 36.

TABLE 36

Physical and Chemical Properties of Sn and Some of Its Compounds

	Chemical symbol	At. or mol. wt.	Sp. gr.	M.p., °C.	B.p., °C.	Solubility
Tin	Sn	118.70	5.75	231.89	2260	Insol. hot or cold H_2O Dec. HCl, H_2SO_4, dil. HNO_3, aq. reg., hot KOH, NaOH
Stannous oxide	SnO	134.70	6.446(0°)	Dec. 700–950		Insol. hot or cold H_2O Dec. acids, fixed alkalies, hydroxides Sl. sol. NH_4Cl
Stannous chloride	$SnCl_2$	189.61	3.393(245°) Liq.	246	623	839 g./liter(0°) 2.698 kg./liter(15°) Sol. alc., ether, acetone, pyridine, ethyl acetate
Stannous chloride dihydrate	$SnCl_2.2H_2O$	225.65	2.71(15.5°)	37.7	Dec.	1.187 kg./liter Dec. hot H_2O Sol. alc., ethyl acetate, acetic acid
Stannous sulfate	$SnSO_4$	214.77		Dec. <360 to $-SO_2$		330 g./liter(25°) Sol. ether, dil. H_2SO_4
Triethyl tin chloride	$(C_2H_5)_3SnCl$	241.34	1.428(8°)	10(15.5)	208–210	Insol. cold H_2O Sol. organic solvents

Sn is a soft, pliable, white, silvery metal adaptable to all types of cold-working such as rolling, extrusion, and spinning. It is readily joined with low-melting solders. Sn retains its brightness well even during long exposures to the atmosphere. Alloying elements such as Sb, Bi, Cu, and Cd increase its hardness.

Sn exhibits two valence forms, +2 and +4; the most important compounds are those in which Sn is +2. Sn is amphoteric, reacting with both strong acids and bases, but is relatively unreactive to nearly neutral solutions. The presence of O_2 greatly accelerates reaction in solution; in the absence of O_2 the high hydrogen overvoltage of Sn causes a surface film of H_2 that retards reaction with acids. The metal is normally covered with an invisible film of SnO_2. Halogen acids attack Sn when hot and concentrated, and hot H_2SO_4 will dissolve Sn, particularly in the presence of oxidizing agents. Hot HNO_3 converts Sn to hydrated SnO_2. Oxalic acid is the most corrosive organic acid toward Sn. Strong alkalies dissolve Sn with the formation of stannates. Acid-reacting salts, like $AlCl_3$, attack Sn in the presence of an oxidizing agent.

4. Analytical Determination

Sn may be determined by spectrographic,[570] fluorimetric, or colorimetric methods. The last method involves the use of aromatic dithiols and suffers from

[570] J. Cholak and R. V. Story, *Ind. Eng. Chem. Anal. Ed.*, **10**, 619 (1938).

instability of these reagents.[571] It has been applied to the determination of Sn in foods. The fluorimetric method of Coyle and White[572] is probably the best of available methods for Sn. The reagent is sensitive to 0.02 μg. of Sn in the fluorometer.

5. Physiological Response

A review (17 pages with 65 references) of the toxicology of Sn and its compounds by Barnes and Stoner appeared in 1959.[573] Interpretations of the toxicity of simple Sn salts is difficult because of the complication of acidity and irritating properties of the solutions. When parenterally administered, Sn salts are toxic, cause spasms and fatal paralysis, and induce hyperemia, vascular changes, and bleeding in the central nervous system, the liver, heart, and other organs. The lethal intravenous dose of a compound of Sn with citric acid, for rabbits, corresponds to about 100 mg. of Sn/kg. of body weight. Daily subcutaneous injection of rabbits with sodium stannous tartrate, a less acid salt, produced death at all levels from 1.6 to 12.5 mg./kg./day in from 6 days at the highest dose to 255 days at the lowest; doses below 1.6 mg./kg./day were not lethal at any time. Cats were more sensitive than rabbits or dogs and developed paralysis with anesthesia of the limbs.

When taken orally, a very large dose of metallic Sn will cause vomiting. Three cats survived in normal health for 390 to 612 days on a normal diet to which was added daily up to 40 mg. of Sn/kg. as a simple inorganic salt, and guinea pigs did well for 4 months on a diet containing 770 p.p.m. Sn salts. Powdered Sn given to rats and pellets of Sn given to hens were noninjurious. The diet of Americans was found to contain on the average 17.14 mg. of Sn/day, in 1940, much of it derived from canned goods. Buchanan and Schryver,[574] after having reviewed alleged cases of poisoning from canned goods, concluded that there is little likelihood of chronic poisoning from this source.

When inhaled as a dust or a fume, Sn leads to a benign pneumoconiosis, without symptoms of interference with pulmonary function. More than 150 such cases have now been described in the literature as a result of x-raying workers in Sn foundries. Sn provides a very radiopaque shadow. In a lung that contained 110 mg. per cent Sn wet weight as SnO_2, no tissue reaction to the SnO_2 was found. The deposited dust appeared nodular, the particles being mostly extracellular. No necrosis, foreign-body giant-cell reaction, or collagen formation was seen.

SnH_4 is reported to be more toxic than AsH_3 to mice and to guinea pigs. The effects, however, differed in that no hemolysis resulted; mainly the central nervous system was involved.

[571] E. B. Sandell, *Colorimetric Determination of Traces of Metals*. Interscience Publishers, New York–London, 1959, p. 861; D. Dickinson and R. Holt, *Analyst*, **79**, 104 (1954).
[572] C. F. Coyle and C. E. White, *Anal. Chem.*, **29**, 1486 (1957).
[573] J. M. Barnes and H. B. Stoner, *Pharmacol. Revs.*, **11**, 211 (1959).
[574] G. S. Buchanan and S. B. Schryver, *Brit. Food J.*, **11**, 101 (1909).

Metabolism of Inorganic Sn. Evidence has now accumulated to indicate that absorption of inorganic Sn from the alimentary tract of man, dog, cat, rat, and rabbit is small. Only small amounts of Sn from metallic Sn, its inorganic salts, and from such compounds as sodium Sn tartrate pass into the tissues. Even with the last compound, which is soluble, more than 90 per cent of the dose was found in the feces. The small amounts that are absorbed are found mainly in the liver and kidneys, with traces in other tissues. The absorption of Sn on protein inhibited its digestion by proteolytic enzymes.

When Sn salts are injected, Sn is widely distributed, but as the tissue concentrations gradually decrease, Sn levels in the liver and spleen increase, as if the Sn were colloidally transported. It is in these two tissues that pathological changes are seen. Ultimately Sn is excreted in the urine; and the bile is a minor pathway. The excretion of Sn in the urine and feces balances the intake provided intake does not exceed 130 mg. of Sn/day;[574] some accumulates in the tissues when the intake is greater. The average Sn level in the blood of "unexposed" Americans in 1940 was 14 μg. per cent as found by Kehoe, Cholak, and Story;[575] practically all of it was in the cells. The urine averaged 11 μg./liter. The Sn content of human organs was found by the same investigators to be the following expressed as μg. of Sn per 100 g. of wet tissue: kidney, 20; heart, 22; liver, 60; spleen, 22; lung, 45; muscle, 11; long bone, 80; rib, 50; stomach, 50; intestine, 16; brain, 0.0.

Organic Sn Compounds. Considerable study of the physiological effects has been given organo Sn compounds of the type $RSnX_3$, R_2SnX_2, R_3SnX, and R_4Sn despite very limited use as fungicides. Interest in their biological action was stimulated by a tragic group of poisonings in France in which 100 died from their oral use for the treatment of skin disorders. Physiologically each type reacts differently. Little is known of the toxicity of *monoalkyl Sn* derivatives except that the toxicity of monoethyl Sn trichloride is low. The toxicity of the *dialkyl Sn* compounds is characterized by their irritative action and production of a lesion of the bile duct and liver, which may result in death from hepatic failure or peritonitis. The lesions have been described in detail.[576] All of the dialkyls from methyl to hexyl are toxic to rats; dioctyl Sn is nontoxic to rats, mice, and guinea pigs in oral doses up to 400 mg./kg. given for 3 or 4 successive days; similarly, no toxicity was observed when it was incorporated in the diet of rats for 4 months at 200 p.p.m. Toxic effects have been produced in the rat following a single dose of 920 mg./kg.; 690 mg./kg. in the mouse. Intraperitoneal toxicity is about 10-fold oral toxicity. Highest levels of Sn were found in the liver and smaller amounts in the kidney, from injection of dialkyl Sn. Chronic administration did not alter the type of lesion that occurred from single doses. Lower homologs caused severe dermal lesions in animals but dipropyl and higher homologs caused relatively little damage. One of the main pathways by which dialkyl Sn exerts its action in the body is believed to be through the inhibition of alpha-keto oxidase activity, leading to the accumula-

---

[575] R. A. Kehoe, J. Cholak and R. V. Story, *J. Nutrition*, **19**, 582 (1940), **20**, 85 (1940).
[576] J. M. Barnes and P. M. Magee, *J. Pathol. Bacteriol.*, **75**, 267 (1958).

tion of pyruvate.[573] BAL prevents these biochemical effects, the toxicity but not the production of the biliary lesion.

Trialkyl Sn Compounds differ from dialkyl derivatives in that their toxicity is manifested by an apparent brain damage, although their site of action is unknown. Progressive weakness and paralysis with convulsions in some species (rabbits) follow administration. Complete recovery can occur, however. In the rat, triethyl Sn is equally as toxic by mouth as by injection; in the hen it is not toxic orally. Ten to 20 p.p.m. triethyl Sn in the diet of rats is sufficient to induce a progressive weakness but recovery usually occurred even from severe poisoning, although some chronically poisoned animals developed tremors from which they did not recover. In these responses the rat paralleled the responses seen in Frenchmen poisoned with triethyl Sn. The toxicity by mouth rapidly diminishes in progressing from tripropyl to trihexyl Sn. Trioctyl Sn was nontoxic by mouth. The characteristic lesion of trialkyl Sn is interstitial *edema* of the white matter, which is reversible. Neither trimethyl Sn nor the higher members of the series produced *edema*. BAL has no effect on trialkyl Sn derivatives. Triethyl Sn is the most active inhibitor known of oxidative phosphorylation ($1 \times 10^{-7}M$); this is consistent with its known activity in mammals. Despite these and other biochemical observations of the mechanism of action of trialkyl Sn, it is not possible to explain its action in the intact animal.

Tetraalkyl Sn acts toxicologically similar to trialkyl Sn; workers handling tetramethyl and tetraethyl Sn suffered from the same symptoms as those exposed to triethyl Sn. This is explained by the discovery that tetraalkyl Sn derivatives are inert physiologically until degraded to trialkyl Sn by a liver enzyme.[571]

Industrial Experience. Twelve cases of benign pneumoconiosis from SnO_2 have been reported to 1954, 5 of them in the United States. A case, reported by Spencer and Wycoff,[578] was a retired worker, who for 22 years had been a smelter and bagger of SnO_2, but disclaimed any chest symptoms until several years later when he developed dyspnea. He had a vital capacity 70 per cent of normal and a maximal breathing capacity 61 per cent of predicted. The nonprogressive nature of the pneumoconiosis from SnO_2 is emphasized.

Dermal lesions consisting of two types (a) acute, localized burns or (b) subacute irritation, have been reported among process workers handling tributyl Sn, by Lyle.[579]

6. *Hygienic Standard of Permissible Exposure*

No limit of safe exposure has been set for Sn or any of its compounds by an official agency in the United States.

[577] J. E. Cremer, *Biochem. J.*, **68**, 685 (1958).
[578] G. E. Spencer and W. C. Wycoff, *Arch. Ind. Hyg. Occupational Med.*, **10**, 295 (1954).
[579] W. H. Lyle, *Brit. J. Ind. Med.*, **15**, 193 (1958).

TITANIUM, Ti

1. Source and Production

Ti is estimated to be the fourth most plentiful structural metal in the earth. The principal useful ores of Ti are ilmenite, theoretically FeO TiO_2, and rutile, TiO_2. The former is found in beach sands (India and Florida) and in rock deposits associated with Fe (Adirondack Mountains and Quebec). Rutile is less abundant; its chief source is certain Australian beach sands. The ores vary in TiO_2 composition around the world from 39 to 96 per cent.

Following a rapid expansion in 1956, a marked decline in world production occurred in 1958, chiefly due to oversupply of rutile and termination of purchases under United States Government contracts. All producers of Ti pigments, however, increased capacity. Production of Ti sponge metal dropped to about 10 per cent of the industry's capacity; total output for 1958 was 4,600 short tons or about $1/4$ the 1957 peak. A total of 210,500 tons of welding rods containing titaniferous materials were produced in 1958, 21 per cent below 1957; 37 per cent contained rutile only, 24 per cent ilmenite only, 13 per cent rutile and manufactured TiO_2, 14 per cent TiO_2 only, and 12 per cent slag only.

A vast amount of research activity continues on improved metallurgical procedures for production and uses of Ti, its compounds, and alloys. A comprehensive review of the consumable-electrode vacuum arc-melting technique for melting Ti was reported.[580] Development of several new alloys was reported; one containing 8 per cent Al, 8 per cent Zr, and 1 per cent Ta and Nb retained its strength at 1100°F. for long periods. New methods were described for hot-spinning, hot-forming, cold extrusion and explosive-forming of Ti.[581] Various techniques have been reported for machining Ti and its alloys, which indicate increasing facility in Ti metal-processing, and interest in broadening the horizons of fabrication and design.

Ti is mined commercially from rock and sand deposits by open-pit and dredging methods. Mechanical beneficiation methods are used for concentrating the major Ti minerals, ilmenite and rutile, to meet industry specifications. Ti-rich slag (70 per cent TiO_2) is produced by smelting ilmenite with coal and lime at 1500 to 1700°C. in an electric furnace. TiO_2 of pigment grade is made by triturating ground ilmenite with concentrated H_2SO_4, reducing the Fe present with Fe scrap, and precipitating the TiO_2 by hydrolysis; As_2O_3 and alkali-earth oxides are among the substances used to control particle size. TiO_2 may also be made by oxidation of $TiCl_4$. To produce ductile Ti by the Kroll process, Ti minerals are chlorinated in the presence of C to produce $TiCl_4$, which is reduced with Mg in an inert atmosphere to produce Ti sponge metal. Ti sponge is solidified either by arc- or induction-melting to produce Ti ingots.

[580] H. Gruber, *J. Metals*, **10**, 193, (1958).
[581] *Minerals Yearbook*, Vol. I, *U. S. Bur. Mines*, 1958, p. 1085 (see references).

2. Uses and Industrial Exposures

Ilmenite is used principally for making pigments, as is Ti slag. Rutile is used for making Ti metal (36 per cent), welding-rod coatings (51 per cent), and the remainder for alloys, carbides, ceramics, and Fiberglas. Ti metal is used principally in commercial and military aircraft and in missiles. For these applications, Ti is used as tubes, fittings, fire walls, cowlings, skin sections, and jet compressors, which amount to from 700 to 2500 lb./plane. Industrially, Ti metal is used as a protective surface on mixers, such as in the pulp-paper industry, or against chlorides or acids. Ti metal tubing is used as a flexible casing for control wires in atomic reactors. TiO_2 as a pigment constitutes a major use; Pigments are used in ceramics, roofing, siding. TiO_2 is also used in plastics as a pigment, as a starting product for Ti chemicals, and as gems. Use of TiO_2 in curing concrete has been investigated, as has a new fibrous potassium titanate as a thermal insulator, and Ti esters (e.g., tetraisopropyl titanate) as heat-resistant surface coatings in paints and plastics. TiC is a component of cobalt-cemented carbides used as cutting tools. $TiCl_4$ is used by the military as a smoke screen.

Industrial exposures consist chiefly of dust and fume of Ti and TiO_2 from electric furnace operations; and of $TiCl_4$ and its partial hydrolysis product, presumably $TiOCl_2$, associated with the chlorination of the oxide. Some metal and metal-oxide fume and dust exposures are associated with various machining and metal-fabrication operations. Ti powders are highly pyrophoric and liquid Ti burns in air. Adequate ventilation and lubrication should be provided during sawing, grinding, and polishing of Ti metal.

3. Physical and Chemical Properties

The physical and chemical properties of Ti and some of its compounds are listed in Table 37.

TABLE 37

Physical and Chemical Properties of Ti and Some of Its Compounds

	Chemical symbol	At. or mol. wt.	Sp. gr.	M.p., °C.	B.p., °C.	Solubility
Titanium	Ti	47.90	4.5(20°)	1800	>3000	Insol. cold H_2O Dec. hot H_2O Sol. dil. acids
Titanium dioxide (rutile)	TiO_2	79.90	4.26	1640 dec.		Insol. hot or cold H_2O Sol. H_2SO_4, alkalies Insol. acids
Titanium tetra-chloride	$TiCl_4$	189.73	Liq. 1.726	−30	136.4	Sol. cold H_2O Dec. hot H_2O Sol. dil. HCl, alc.
Titanium sulfate	$Ti_2(SO_4)_3$	384.00				Insol. hot or cold H_2O Sol. dil. acids Insol. alc., ether, conc. H_2SO_4

Present commercial production of Ti by magnesium reduction of TiCl$_4$ yields actually an alloy of Ti (99.5 per cent pure) that differs in hardness, strength, and elongation from the pure Ti prepared by thermal decomposition of TiI$_4$. Ti shows unusual creep behavior; at intermediate temperatures high stresses must be imposed in order to induce creep, a behavior in marked contrast to other structural metals. Commercial Ti shows excellent corrosion resistance to cold and hot chloride solutions and to hypochlorite; chromic acid has little effect; but it is attacked by sulfuric and hydrochloric acids, but not nitric acid. Ti is unusually resistant to organic acids except hot solutions of oxalic and trichloroacetic acids. Thus, Ti has good resistance to acidic food stuffs. Ti is resistant to the action of sea water.

The chief valence form of Ti is +4, but +3 (titanous) and +2 are known, including a peroxy-type compound. Oxy compounds exist, such as TiOCl$_2$ called titanyl, but the existence of TiO^{2+} ion is doubtful. Ti^{4+} compounds undergo hydrolysis to yield hydrous titanium oxide, which is capable of control and forms the basis of the TiO$_2$ pigment industry. Intermediate valence stages disproportionate to higher and lower forms, which property is used in producing reduced Ti halides. Bibliographies and detailed information on Ti and its compounds are available.[582]

As with the inorganic compounds, numerous organic derivatives of Ti are known, but as yet are of little industrial significance. The most common types are the alkyl and aryl titanates of the general formula Ti(OR)$_4$ where R stands for alkyl or aryl radical. Complex organic coordination compounds of Ti are known.

4. Analytical Determination

Samples collected from air or biological specimens may be analyzed by a colorimetric procedure employing salicylic acid.[583] The method is claimed to be specific for Ti and free from disturbances by other cations. A more sensitive colorimetric method applicable to air and certain biological samples (urine) depends on the absorbance of Ti thiocyanate complex with tri-n-octylphosphine oxide in an organic phase (Young and White[583]). When the acid extraction medium is modified to use $5M$ H$_2$SO$_4$ containing $2M$ Na$_2$SO$_4$ the method is suitable for the determination of Ti in urine in the 0.1 to 10 μg. range with an error over the working range of ±0.3 μg. Added Ti was not recoverable from blood or spleen by the method (Levinskas[583]) (see also reference 590 under Physiological Response).

Ti may be determined spectrographically. The most sensitive lines are Ti 3349.04, 3398.64, and 3989.76. None are outstandingly sensitive, so that amounts below 2 or 3 p.p.m. are not determinable without prior chemical concentration procedures.

[582] J. Barksdale, *Titanium, Its Occurrence, Chemistry and Technology*. Ronald Press, New York, 1949; G. Skinner, H. L. Johnston, and C. Beckett, *Titanium and Its Compounds*. H. L. Johnston Enterprises, Columbus, Ohio, 1954.

[583] M. Schlenk, *Helv. Chim. Acta*, **19**, 1127 (1936); J. P. Young and J. C. White, *Anal. Chem.*, **31**, 393, 1959; G. Levinskas, American Cyanamid Co., April, 1960, private communication.

5. Physiological Response

The physiological history of TiO_2 is one of inertness. Lehmann, and Zerget[584] found no evidence, from histological study of organs of animals fed TiO_2 for periods up to 16 months, of any tissue change or of appreciable quantities of Ti in the tissues, and concluded that Ti is not absorbed from the gastrointestinal tract by the blood. Similar evidence for physiological inertness of TiO_2 was found by Vernetti Blina[585] from animals exposed by either the respiratory, digestive, or subcutaneous routes. On the other hand, a more recent Russian report,[586] seen in abstract form only, indicated exposure to an aerosol of TiC by animals for 5 months resulted in lung changes resembling silicosis. The silicotic-like lesion did not result in animals exposed for several years to Carboloy dust, a mixture of WC, TiC, TaC, and Co.[587]

$TiCl_4$ exposures can be injurious. Dogs exposed to $TiCl_4$ dust (presumably chiefly a mixture of HCl and $TiOCl_2$) intermittently for a few hours showed respiratory distress characterized by a severe bronchitis and some edema,[588] collapse, and death. Lungs showed focal congestion and hemorrhage indicative of effects of HCl. Repeated exposures to $TiCl_4$ vapor at an average Ti concentration of 8.4 p.p.m. (range 1.6 to 17.1 p.p.m.) produced results described as closely paralleling the reactions caused by silica. Further tests of injections of hydrolized $TiCl_4$ into the peritoneal cavities of guinea pigs to determine whether the lesions were progressive, as in silicosis, revealed only a tissue reaction characteristically produced by inert dusts.[588]

Disks of Ti metal implanted in muscle of dogs and left in situ for 7 months were inert; the wound healed and the metal was encapsulated with fibrous tissue.[589]

TiO_2, used as a protective film on exposed parts of the body as a prevention of flash burns during World War II, was without consequence indicating no capacity of TiO_2 to produce contact dermatitis, allergic sensitization, or appreciable dermal absorption.

The Ti content of normal human organs (Europe) varied from 1.5 to 10 μg. per cent in 17 organs examined; blood contained 3 μg. per cent on the average. Ti was determined on tissue extracts precipitated with nitrosophenylhydroxylamine and estimated colorimetrically with H_2O_2, a method that has the advantage of freedom from interfering ions.[590]

Ti inhibits serum alkaline phosphatase markedly (53 per cent) at low con-

---

[584] K. B. Lehmann and L. Zerget, Chem. Ztg., 82, 793 (1927).

[585] L. Vernetti Blina, Riforma med., 44, 1516 (1928).

[586] O. Ya Mogilevskaya, et al. Chem. Abstr. 50, 9990i (1956).

[587] H. E. Stokinger, data.

[588] Personal communication.

[589] O. E. Beder and G. Eade, Surgery, 39, 470 (1956).

[590] L. C. Maillard and J. Ettori, Compt. rend. soc. biol., 122, 1951 (1936); Compt. rend., 202, 594, 1459, 1621 (1936).

centrations (0.46 mM) in vitro, inhibits slightly yeast invertase and amylase, but has no effect on trypsin even at high concentrations.[591]

Industrial Experience. The Italian investigator, Vernetti Blina, reported[585] no significant pulmonary alterations among workmen employed in enclosed workshops with TiO_2 dust. In the United States no reported cases of effects of TiO_2 exposures have appeared. Similarly, no pulmonary difficulties have occurred among workers in $TiCl_4$ operations where there was an enlightened industrial hygiene program.[588]

6. Hygienic Standards of Permissible Exposure

A threshold limit of 15 mg./cu. meter has been set for TiO_2 dust by the American Conference of Governmental Industrial Hygienists; no limit has been set for $TiCl_4$ mist or other Ti compounds. The limit of 15 mg./cu. meter is set as a maximal limit of permissible dustiness for innocuous substances.

7. Flammability and Explosibility

Dust clouds of Ti metal powder, average particle diameter 10.5 μ, containing 0.08 per cent H, 0.82 per cent O, and 0.062 per cent N by weight, were determined experimentally to ignite at 330°C. in air but not to ignite at temperatures below 850°C. in CO_2 or N_2 gas.[592] TiH_2 containing 2.83 per cent H, 0.24 per cent O, and 0.071 per cent N, containing 70 per cent theoretical H for TiH_2, ignited at 440°C. under similar conditions of test. Neither dust layers nor clouds ignited in an atmosphere of He or A up to 850°C., but layers of both powders ignited in a mixture of 50 per cent air and 50 per cent He at 850°C. Layers of both powders could be ignited by weak electric condenser discharge sparks (see also Vol. I, pages 560, 562, 565, 567, 570, and 575).

A serious explosion has been reported[593] from friction of commercially pure Ti samples that had been undergoing corrosion tests in red fuming nitric acid.

TUNGSTEN, W

1. Source, Production, and Uses

Commercial sources of W of importance are scheelite ($CaWO_4$) and wolframite (Fe,Mn)WO_4. Scheelite, when pure, contains 80.6 per cent WO_3, the commonest impurity being MoO_3. The percentages of FeO and MnO in wolframite vary considerably; hubnerite is the term applied to ore containing more than 20 per cent MnO, ferberite, to ore containing more than 20 per cent FeO; intermediate samples are called wolframite. The richest W deposits are in the China-Malaya area; the second richest area comprises most of the western states in the United States, Alaska, and Mexico. Most domestic ores mined contain less than 1 per cent

[591] B. S. Gould, *Proc. Soc. Exptl. Biol. Med.,* **34**, 381 (1936).

[592] I. Hartman, *U. S. Bur. Mines Rept. Invest.* No. **3202**, 1951.

[593] P. M. Ambrose, *et al., U. S. Bur. Mines Inform. Circ.* No. **7711**, 1955.

WO_3; these are concentrated to 50 to 70 per cent WO_3 at the mine by ore-dressing or hydrometallurgy. Impurities present in the concentrates determine largely the type of use; As, P, S, and Mo are among the most important impurities.

Following a critical shortage of W in 1951, a plan of stockpiling was launched to increase and conserve supplies, chiefly by restricting the quantity of high-W tool steel. This plan was so effective that by 1958 domestic mine production fell far below that of 1957, and consumption of W products, including imports, scrap, and scheelite used in steel, decreased 32 per cent. World production of W simultaneously was curtailed in almost every area.

W is mined in the United States by opencut operations similar to those for gold, but most W in the United States comes from underground workings. More primitive mining methods exist elsewhere. Primitive methods of hand-sorting and hand-jigging frequently require cleaning and retreatment of the ore after shipment.

Modern W mills are of the gravity or gravity-flotation type, which attain recoveries of WO_3 as high as 70 to 90 per cent. Production of finished concentrates may involve acid-leaching (HCl) to remove P, roasting to remove As or S, or magnetic separation to improve grade (e.g., wolframite is separated from scheelite by magnetic separation). Calcite ($CaCO_3$) often complicates flotation procedures used with scheelite because of like densities. The type of concentrate produced must be suited to the needs, whether for direct addition to a steel charge, to produce ferrotungsten, to produce W metal powder, or for W compounds. Cu, As, Sb, Bi, P, S, Pb, Mo, Sn, Zn, and Mn are the most common impurities in concentrates used for direct charging for steel. Some scheelite concentrates require digestion and precipitation to remove combined Cu and Mo. Scheelite is "nodulized" to minimize dust losses. Ferrotungsten is made in electric furnaces with C or Si as reducing agent, or by alumino- or silicothermic methods. Production of the W metal- (powder W) chemical group, which uses more than half of the total W, requires solution of the W from the concentrates whence it may be precipitated as tungstic acid or left in solution as Na_2- or K_2WO_4. Both forms require additional purification. The more important from the industrial hygiene standpoint is treatment with NH_4OH to form $(NH_4)_2WO_4$, followed by heating to form ammonium paratungstate. Because of the undesirable effects of associated impurities noted above, elaborate steps are taken to minimize their presence in the final products. Arsenic, for example, cannot be tolerated higher than 0.02 per cent; P, not more than 0.05 per cent.

There are two divisions of the W metal-powder industry, H reduction and C reduction. H reduction is used for metal powder for most WC and for all filament wire; C reduction is resorted to when W powder is used for welding rods, coating oil-well tools, and for some WC. In H reduction, tungstic acid or ammonium paratungstate in Fe or Ni boats is placed in tubes through which H flows, usually in a 2-stage reduction. C-reduced W is made similarly from tungstic acid but lamp-black, natural, or manufactured gas is the reducer.

For the production of W wire, rod, and sheet, H-reduced powder is compressed, sintered, heated to incipient fusion by passage of electric current, and swaged, drawn, or rolled.

WC is produced from either H- or C-reduced powder by heating with lampblack, or WC crystals may be prepared directly from the ore. A mixed WTiC may be produced. WC tools are produced by sintering a mixture of WC and Co, or Ni, or both in dies of the desired shape. TiC and TaC are both often blended with WC before sintering with Co.

In addition to the metal, ferrous metal-alloy and carbide uses noted above, W compounds are used in dyes, inks, and ceramic frits. An important use of the tungstates is in fluorescent lamps. Recently a new W–Ta refractory alloy with very low impurities was announced.[594] It is capable of withstanding temperatures up to 5200°F. and has many other desirable properties. The W–Ta alloy was made possible by an electron-beam furnace that reduces nonmetallic impurities (C, N, and O) to a few p.p.m.

2. Industrial Exposures

The hazardous exposures would appear to be to substances associated with the production and uses of W, its alloys, and compounds rather than to W itself, although precisely what role, if any, W plays in the exposures is far from clear. In the WC cutting-tool industry exposures are to Co fume and dust as well as to WC, TiC, TaC, and NiC. The industrial hygiene aspects of the cemented tungsten-carbide industry have been described.[595] In the production of W metal, hazardous exposures to associated metals in the ore are chiefly to As, Sb, Mn, Pb, Cu, Bi, Sn, and Mo.

3. Physical and Chemical Properties

The physical and chemical properties of W and some of its compounds are listed in Table 38.

The chemistry of W and its compounds is very similar to that of Mo. W exists in several states of oxidation, 0, +2, +3, +4, +5, and +6; the most stable is +6, the lower valence states being relatively unstable; bivalent W exists only as halogen compounds. Like Mo, W has a strong tendency to form complexes exemplified by a large series of heteropoly acids formed with oxides of P, As, V, and Si among others (e.g., phosphotungstic acid, $H_3PW_{12}O_{40} \cdot 14H_2O$). In addition, compounds of W exist in which W occurs in more than 1 valence state ($2WN \cdot W(NH_2)_2$). W forms a series of oxyhalides (e.g., $WOCl_4$, WO_2Cl_2, and $WOBr_4$).

4. Analytical Determination

A colorimetric thiocyanate method for the quantitative determination of W has been applied to urine, blood, and other tissues by Aull and Kinard.[596] In their

[594] Chem. Eng. News., M, 55 (1960).

[595] L. T. Fairhall, R. G. Keenan, and H. P. Brinton, Public Health Repts. U. S., 64, 485 (1949).

[596] J. C. Aull and F. W. Kinard, J. Biol. Chem., 135, 119 (1940).

TABLE 38

Physical and Chemical Properties of W and Some of Its Compounds

	Chemical symbol	At. or mol. wt.	Sp. gr.	M.p., °C.	B.p., °C.	Solubility
Tungsten	W	183.86	19.3	3370	5900	Insol. hot or cold H_2O Sol. HF + HNO_3 Insol. HF, KOH V. sl. sol. H_2SO_4
Tungsten trioxide	WO_3	231.92	7.16	1473		Insol. hot or cold H_2O Sol. hot alkalies, HF Insol. acids
Tungstic acid (ortho)	H_2WO_4	249.94	5.5	$(-\frac{1}{2}H_2O)100$		Insol. cold H_2O Sl. sol. hot H_2O Sol. alkalies, HF, NH_3 Insol. most acids
Sodium tungstate	$Na_2WO_4 \cdot 2H_2O$	329.95	3.23	$(-2H_2O)100$ Anh. 698		410 g./liter(0°) 825 g./liter(20°) 1235 g./liter(100°) Sl. sol. NH_3 Insol. acids, alc.
Ammonium p-tungstate	$(NH_4)_6W_7O_{24} \cdot 6H_2O$	1887.78		$(-4H_2O)100$		28 g./liter(15°) 45 g./liter(22°) Insol. alc.
Tungsten carbide	WC	195.93	15.63	2780 ± 50	6000	Insol. cold H_2O Sol. HNO_3 + HF, aq. reg.

hands[597] the method had a lower limit of 0.5 mg. of the total sample (10 p.p.m. tissue concentration) with a range of recovery of 95 to 105.8 per cent, and thus is limited to specimens that have been subjected to high levels of exposure. The method requires refinement for use in evaluating the W levels encountered in human exposures. The spectrographic method is not sufficiently sensitive to determine W in the 1-p.p.m. range without chemical concentration procedures. If, however, W is collected on a carrier precipitate of Ti developed from tannin, antipyrine, and cinchonine, a detection limit of 0.7 p.p.m. W is obtained. The W line 2948 is used.[598]

5. Physiological Response

The LD_{50} of sodium tungstate, $Na_2WO_4 \cdot 2H_2O$, for 66-day-old rats subcutaneously injected after 24 hours' starvation was between 223 and 255 mg./kg., equivalent to 140 to 160 mg. of W.[599] A significant postprandial effect reducing the lethality was noted, as was a decided age effect; younger animals were more re-

[597] F. W. Kinard and J. C. Aull, *J. Pharmacol. Exptl. Therap.*, **83**, 53 (1945).

[598] S. H. Wilson and M. Fields, *Analyst*, **69**, 12 (1944).

[599] F. W. Kinard and J. van de Erve, *Am. J. Med. Sci.*, **199**, 688 (1940).

sistant to the injected tungstate; no 30-day-old rats died at a dose of 150 mg. of W/kg., whereas rats of 170 days and 365 days died uniformly at the same dose.

Of three W compounds incorporated separately in the diet and fed to rats in a 70-day study, sodium tungstate was most toxic (2 per cent W), tungstic oxide was intermediate (3.96 per cent W), and ammonium p-tungstate was least toxic (5 per cent W), as determined by 100 per cent mortality in each instance.[600] Tungstic oxide was lethal to rats at 0.5 per cent W in the diet, as was sodium tungstate, but ammonium p-tungstate was nonlethal at this level although it resulted in slight weight loss (4 or 5 per cent). At the 0.1 per cent level tungstic oxide and sodium tungstate gave a slight weight loss.

W metal powder fed for 70 days to weanling rats of both sexes at levels of 2, 5, and 10 per cent of the diet (Purina dog chow) resulted in no effect on growth rate of the male rats, but caused a 15 per cent reduction in weight gain in the females from that of controls.[601] Females consumed 75 g. and males 104 g. of W during the 70 days. The particle size was not reported. Patients are reported[602] to have consumed 25 to 80 g. of W in single doses without ill effects.

No evidence was found for the fibrogenic activity of WC alone or in combination with the other carbides and metal in cemented tungsten carbide when the technique of Miller and Sayers was used.[603] Evidence for inertness of WC and W for animals was given similarly by Fredrick and Bradley, and by Harding.[604] W metal powder (size not stated) intratracheally injected in larger doses (150 mg. in three divided doses) in guinea pigs produced transient focal interstitial pneumonitis and bronchiolitis with almost complete recovery after 1 year; residual pathological changes consisted of an occasional focal peribronchial, peribronchiolar, and perivascular fibrocellular reaction with some bronchiolitis obliterans and atrophic emphysema.[605] When Co was similarly administered with WC in a ratio of 1:3, however, an acute inflammatory response occurred, which was later replaced by focal pneumonitis and residual bronchial epithelial hyperplasia and metaplasia.[606]

Metabolism. W tends to be deposited in the bone and spleen with lesser amounts going to the kidney and liver, and possibly the skin, from oral intake by rats for 100 days of either W metal, tungstic oxide, sodium tungstate, or ammonium p-tungstate,[601] as determined by the method of Aull and Kinard.[596] Trace amounts were found in the lung, muscle, testis, and blood; trace amounts by the method employed were those below 10 μg./g. (10 p.p.m.), a quantity now considered exceptionally high for a trace element.

[600] F. W. Kinard and J. van de Erve, *J. Pharmacol. Exptl. Therap.,* **72,** 196 (1941).

[601] F. W. Kinard and J. van de Erve, *J. Lab. Clin. Med.,* **28,** 1541 (1943).

[602] R. Kruger, *Münch. med. Wochschr.,* **59,** 1910 (1912).

[603] K. D. Lindgren and A. Svensson, *Acta Med. Scand.,* **145,** 20 (1953).

[604] W. G. Fredrick and W. R. Bradley, 7th Ann. Meeting AIHA, 1946; H. E. Harding, *Brit. J. Ind. Med.,* **7,** 76 (1950).

[605] G. W. H. Schepers, *A.M.A. Arch. Ind. Health,* **12,** 134 (1955).

[606] G. W. H. Schepers, *A.M.A. Arch. Ind. Health,* **12,** 140 (1955).

Mechanism. Because tungstate is isomorphic with molybdate, Na_2WO_4 antagonizes the normal metabolic action of molybdate in its role as metal carrier for xanthine dehydrogenase.[607] Dietary Na_2WO_4 equivalent to a W:Mo ratio of 100:1 completely inhibited the deposition of intestinal xanthine oxidase in the rat, and markedly reduced xanthine dehydrogenase and Mo in the liver. Na_2WO_4 produced an Mo deficiency in the chick and one half the uric acid normally excreted by the chick was replaced by xanthine and hypoxanthine. All effects of WO_4^{2-} were reversed by small amounts of MoO_4^{2-}, thus indicating a true antagonism.

Industrial Experience. There are reports of numerous cases of a pneumoconiosis among workers in the cemented WC-tool industry, both in the United States[608] and abroad.[603] Whether the pneumoconiosis is due to WC, or the mixture of carbides and Co, or to Co has not been resolved. No cases, however, of pneumoconiosis among workers exposed strictly to W or its compounds have been reported.

6. Hygienic Standards of Permissible Exposure

No limit has been set for W or any of its compounds for air of industrial workrooms by an official agency in the United States.

URANIUM, U

1. Source and Production

U is distributed more abundantly in the earth's crust than the total of Sb, Bi, Cd, Au, Hg, and Ag and averages about 3 g./ton of rock. Although there are more than 100 U-bearing minerals, carnotite ($K_2O \cdot 2U_2O_3 \cdot V_2O_5 \cdot 3H_2O$), pitchblende (a mineral complex of $UO_3 \cdot UO_2$, PbO, Th, Y, etc.), tobernite ($Cu(UO_2)_2P_2O_8 \cdot 12H_2O$), autunite, and a few others are U-bearing minerals presently of commercial interest. The Colorado plateau areas of Colorado, Arizona, Utah, and New Mexico are the largest domestic sources of U; other (foreign) sources of importance are the Northwest Territory in Canada, and the Belgian Congo in Africa. Canada surpasses the United States as leading world producer of U. Domestic annual production rate in 1958 from 23 mills was 15,000 tons of U concentrate; domestic annual requirement forecast from 1960 to 1966 was 35,000 tons U_3O_8. Some 3.5 million tons of uraniferous ore of average grade of 0.27 per cent U_3O_8 was mined in the United States from almost 1000 mines in 1956, but there was a significant trend toward consolidation of minor holdings (less than 50 tons/year). From the industrial hygiene standpoint this trend is important because larger operations provide better health protection for the

[607] E. S. Higgins, D. A. Richert, and W. W. Westerfeld, *J. Nutrition,* **59,** 539 (1956).

[608] C. W. Miller, M. W. Davis, A. Goldman, and J. P. Wyatt, *Arch. Ind. Hyg. Occupational Med.,* **8,** 453 (1953).

miners. About 5750 persons were employed by 624 shippers in 1958; approximately 1825 persons were employed in mills.

U ore is processed by acid or carbonate leaching to produce a concentrate, after which U is recovered by classified procedures to produce (a) the orange UO_3, (b) conversion of UO_3, to UF_4, and (c) conversion of part of the UF_4 to UF_6 for the U^{235} gaseous diffusion plants, the remainder to U metal for Plutonium manufacture.

2. Uses and Industrial Exposures

The bulk of all U produced is used for nuclear weapons; smaller quantities are used in research, particularly for reactor development, industrial power problems, and propulsion devices. Civilian nuclear power plants with 80 megawatt electric capacity were in operation in 1958 and a number of nuclear-powered submarines were operating, under construction, or were authorized; 50 test reactors were operating, 44 were under construction, and 17 planned. In June, 1958, the AEC removed the ban on nonnuclear uses of U, which prior to the restriction had used about 200 tons U_3O_8 annually for ceramics and glass, and in photography. Study is being given to applications of U as nuclear explosives in excavating, mining, power, and isotope production.

Hazardous exposures in the uranium industry begin in the mining. Hazards are of two types, chemical and radiological; of the two, radiation is the more hazardous. In addition to the alpha-particle radiation hazard from U in the ore, the most hazardous elements are Rn gas and its particulate daughters, RaA and RaC′, chiefly, all alpha emitters (see also Vol. I, page 725). In the mines some beta and gamma exposures from RaB and RaC and Ra also occur but are of relatively minor importance. Effective ventilation control measures have reduced the radiation exposures to satisfactory levels in the larger mines, but far less satisfactory radiation-exposure conditions exist in small mines without benefit of ventilation. Some mine waters are high in Rn and thus should not be used for wet drilling.

The chemical hazards in the mines are to U and V if the ore is carnotite, Pb and Th chiefly, but As and Mn are also present, if pitchblende. Crystalline silica (40 to 60 per cent of ore) is a hazard common to all metal mining in the West.

Hazards in the milling of U to produce a concentrate are relatively minor because a wet process is used. Somewhat greater exposures occur in the production of the U metal, from dusts of the intermediates UO_3 and UF_4, and to the gaseous UF_6 (UO_2F_2 and HF) from accidents or leaks. Hazards from the production of U metal briquettes or in the hot-rolling of U rods are relatively small.[609] An evaluation of surface-contamination control of U-rolling operations has been made by both Blackwell and Hyatt.[609] U metal is pyrophoric; chips from cleaning the briquette readily ignite. Radiation hazards from the U^{235}-enriched U are high but

[609] C. D. Blackwell, J. Am. Ind. Hyg. Assoc., **20**, 92 (1959); E. C. Hyatt, J. Am. Ind. Hyg. Assoc., **20**, 82 (1959).

are recognized. An over-all evaluation of various U fabrication procedures is given by Harris and Kingsley,[610] and a discussion of the environmental exposure to U compounds in terms of air, urinary, and medical findings is reported by Lippmann.[611]

3. Physical and Chemical Properties

The physical and chemical properties of U and some of its compounds are listed in Table 39.

TABLE 39

Physical and Chemical Properties of U and Some of Its Compounds

	Chemical symbol	At. or mol. wt.	Sp. gr.	M.p., °C.	B.p., °C.	Solubility
Uranium	U	238.07	18.7	ca. 1133	Ignites	Insol. hot or cold H_2O Sol. acids Insol. alkalies
Uranium dioxide	UO_2	270.07	10.9	2176		Insol. hot or cold H_2O Sol. HNO_3, conc. H_2SO_4
Uranyl oxide	UO_3	286.07	7.29	Dec.		Insol hot or cold H_2O Sol. min. acids
Triuranium octoxide	U_3O_8	842.21	8.30	Dec.		Insol. hot or cold H_2O Sol. HNO_3, H_2SO_4
Uranium tetra-fluoride	UF_4	314.07		ca. 1000		Insol. cold H_2O Sol. conc. acids and alkalies Insol. dil. acids and alkalies
Uranium hexa-fluoride	UF_6	352.07	4.68(20.7°)	69.2 (2 atm.)	56.2 (764.6°)	Sol. cold H_2O Dec. alc., ether Sol. CCl_4, $CHCl_3$ Insol. CS_2
Uranyl nitrate	$UO_2(NO_3)_2 \cdot 6H_2O$	502.18	2.807(13°)	60.2 Dec. 100		8 g./liter(14°) 33 g./liter(100°) Sol. min. acids, alkalies, oxalates

U is a weak metal, is only moderately ductile at room temperature, and its strength declines rapidly at elevated temperatures. Its hardness is markedly affected by impurities.

U is highly reactive and forms numerous intermetallic compounds. The behavior of U in the electromotive series resembles that of Be. Because U is so

[610] W. B. Harris and I. Kingsley, A.M.A. Arch. Ind. Health, **19**, 540 (1959).

[611] M. Lippmann, Arch. Ind. Health, **20**, 211 (1959).

reactive, casting is a particularly difficult problem requiring protection from contact with air or moisture.

In the dry state U forms compounds in which the valence is $+3$, $+4$, $+5$ or $+6$. In aqueous media U^{3+} and U^{5+} are unstable: U^{3+} readily oxidizes, and U^{5+} disproportionates to U^{4+} and U^{6+} which latter is the most stable form and exists as the oxygen-containing cation UO_2^{++} in acid solution. The uranyl ion has a green-yellow fluorescence. Although U forms a great variety of compounds in which U is either a cation or an anion, the industrially most important compounds are the dioxide, UO_2, the trioxide, UO_3, the octoxide, U_3O_8, the tetrafluoride, UF_4, the hexafluoride, UF_6, and uranyl nitrate hexahydrate, $UO_2(NO_3)_2 \cdot 6H_2O$. The uranyl ion forms soluble complexes with various inorganic and organic anions (e.g., uranyl carbonate and uranyl proteinate) but the exact composition of the complexes cannot be specified.

5. Physiological Response

A simple, rapid colorimetric method utilizing ferrocyanide for the determination of U in dust samples ranging upward from 80 μg. is described by Blackwell.[609] For small samples from air or samples of biological origin, the fluorophotometric method is recommended. It is suitable for the determination of U in the parts per billion range if preliminary protein isolation and electrolysis procedures are used.[609] Polarographic determination of U is particularly valuable for the determination of trace amounts of the hexavalent form in the presence of tetravalant U as it requires no prior separation procedures.[609] A spectrochemical method for the determination of U in all types of samples, with modifications suitable for each, is also described by Blackwell.[609]

5. Physiological Response

The pharmacology and toxicology of U compounds of industrial (atomic energy) interest are presented in four volumes comprising about 2300 pages, which represent the most thorough and extensive study ever given a hazardous substance.[612] The following is a rather inadequate attempt to highlight some of the more important aspects of U toxicity from the vast amount of information therein.

Acute and Subacute Chemical Toxicity—Animals. Oral toxicity of U compounds is rather low; the maximal dosage just failing to be lethal for rats in 30-day feeding tests was about 0.5 per cent U compound in the diet for 6 *soluble* compounds tested; 20 per cent U compound for 3 insoluble U compounds (UO_2 U_3O_8, and UF_4). No amount of insoluble U compounds acceptable to the rat was lethal, but levels of from 1 to 4 per cent soluble U compound produced 50 per cent mortality in 30 days. Compared to toxicity by inhalation, toxicity by ingestion was far less, ranging from 30-fold less for UO_4, to 3300-fold less for UF_4 and U_3O_8 for the rat; for the dog, differences were even greater.

[612] Voegtlin and Hodge, eds., *Pharmacology and Toxicology of Uranium Compounds.* McGraw-Hill, New York, 1953.

Acute intravenous toxicity of soluble U compounds (uranyl nitrate) is extremely high; the approximate LD_{50} for rabbits was 0.1 mg. of U/kg.; guinea pigs, 0.3 mg. of U/kg.; rats, 1 mg. of U/kg.; and mice, 10–20 mg. of U/kg. (from 0.2 to 40 mg. of compounds/kg.).

Acute intraperitoneal toxicity of soluble U compounds for the rat is considerably less than intravenous toxicity; approximate LD values ranged from 40 mg. of UO_2F_2/kg. to 400 mg. of UCl_4/kg. About a 2-fold increase in toxicity with increasing age occurred.

Percutaneous toxicity. All soluble U compounds are lethal when applied in a single dose to the skin of rabbits, either in various vehicles or in some cases (UCl_4, UCl_5) without vehicle; the insoluble salts UO_4, UF_4, UO_2, and U_3O_8 were not lethal by the same route and caused no signs of poisoning.

Ocular toxicity. Soluble U compounds may be lethal when placed in the conjunctival sac of the rabbit eye either in a vehicle or without vehicle; insoluble salts resulted in no mortality. UO_3 was remarkable in that it was uniformly fatal but failed to produced any local signs of poisoning.

Inhalation toxicity. Dusts and mists of respirable particle size of UF_6, UO_2F_2, and UCl_4, and $UO_2(NO_3)_2 \cdot 6H_2O$ were generally fatal to most laboratory species when exposed daily for 1 month at 20 mg. of compound/cu. meter; 2.5 mg./cu. meter was fatal to some species; 0.2 mg./cu. meter was fatal to an occasional animal; and 0.05 mg./cu. meter resulted in no histological damage to any species. At 20 mg./cu. meter, UF_4, UO_2, and "high grade" ore were occasionally fatal to some species; 2.5 mg./cu. meter was essentially nonfatal, with mild or no renal damage. On a relative toxicity scale UF_6 was the most toxic U compound, followed by UO_2F_2, UCl_4, and $UO_2(NO_3)_2 \cdot 6H_2O$; UO_2 and U_3O_8 were the least toxic.

Carnotite ore dust (20 per cent U, 5 per cent V) at a daily average concentration of 84 mg./cu. meter and particle diameter (MMD) of 1.39 μ was uniformly lethal to rabbits (92 per cent mortality) but resulted in only 10 per cent mortality of rats, in a 32-day intermittent daily exposure. Other criteria of injury were consistent with these findings. The degree and nature of the injury resembled that of UO_3 dust.[613]

Signs of Acute Chemical Toxicity. The kidney is the organ most directly affected by U. Effects on the kidney precede in time and degree effects on the liver, which are consequent to the acidosis and azotemia with attendant cachexia induced by renal dysfunction. Tests of renal injury are changes in urinary catalase, phosphatase, and protein, which exhibit a sharp peak 2 to 5 days after acute injury. Catalase normally not present in urine appears in response to as little as 0.02 mg. of uranyl acetate/kg. in the rabbit. Tests of altered functional capacity of the kidney, phenol-red removal from blood, urinary amino acid N to creatinine ratio, and various clearance tests show maximal change 3 to 6 days

[613] H. B. Wilson, H. E. Stokinger, and G. E. Sylvester, *Arch. Ind. Hyg. Occupational Med.*, **7**, 301 (1953).

after exposure to U. Return to normal of these functions occurs despite continued exposure provided exposure is not so severe as to result in fatality. Only chloride clearance and amino acid/creatinine ratio remain elevated for long periods after severe exposure, indicating prolonged interference with renal tubular function.

Chronic Chemical Toxicity. The soluble U compounds UO_2F_2, UCl_4, and $UO_2(NO_3)_2 \cdot 6H_2O$ are tolerated by dogs for 1 year when incorporated in the diet at a level of 0.2 mg./kg./day; 10 g./kg./day of UO_2 was tolerated. $UO_2(NO_3)_2 \cdot 6H_2O$ caused an adverse effect on growth at the end of 1 year at a level of 0.2 g./kg./day; borderline effects resulted from UF_4 at 5 g./kg./day. Most of the dogs showed little change in the NPN or urea N, although some showed abnormal urinary protein and sugar values; renal cortical changes were histologically observable in the dogs showing the urinary changes. No U compound produced significant alteration in the cellular blood picture. In male rats, 0.05 per cent $UO_2(NO_3)_2 \cdot 6H_2O$ in the diet produced no alteration in weight gain over a 2-year period; 0.1 per cent caused just detectable growth depression for male rats, 0.5 per cent for female rats; 2 per cent did not lessen life-span. A tolerance to acute $UO_2(NO_3)_2 \cdot 6H_2O$ poisoning was demonstrated in proportion to the chronic dosage fed.

Inhalation Toxicity. Soluble U compounds are clearly differentiated in toxicity from insoluble in 1-year inhalation studies: 0.2 mg. of U/cu. meter as $UO_2(NO_3)_2 6H_2O$, UF_6, or UCl_4 was toxic, whereas 10 mg. of U/cu. meter as UO_2 was a concentration tolerated for 1 year by the dog, the most sensitive species, and 1 mg. of U/cu. meter was without detectable effect in any animal by any test employed.

Experimental Cancer Production. Sarcomas resulted in rats injected with metallic U in the femoral marrow and in the chest wall;[614] it is unknown whether the sarcomas were due to metallocarcinogenic or radiocarcinogenic action.

Metabolism. A provisional synthesis of metabolism steps involved in the fate of U^{6+} in the animal body indicates that U^{6+} enters the blood stream as UO_2^{2+} wherein it forms two complexes: a nondiffusible complex with plasma proteins and a diffusible bicarbonate complex. The two forms are in equilibrium with each other. The diffusible bicarbonate complex binds U so tightly at the pH of soft tissues that no U is deposited; because some U is deposited in bone, reduced pH values dissociate the soluble bicarbonate–U complex fixing at this site. The bicarbonate–U passes through the kidney glomerulus, the bicarbonate blood level regulating the rate of filtration, because the protein–U complex is nonfilterable. It is convertible to bicarbonate complex, however, so that ultimately all U passes the glomerulus as a bicarbonate complex on all but overwhelming U doses. On reaching the renal tubules, the bicarbonate–U complex dissociates because of lowered pH. The dissociated UO_2^{2+} precipitates on the renal tubular cell walls, inhibiting vital enzyme function, and injures the cell. If alkaline reserve is high,

---

[614] W. C. Hueper and J. H. Zuefle, *et al., J. Natl. Cancer Inst.,* **13**, 291 (1952).

however, and significant amounts of bicarbonate remain in the urine leaving the kidney, little U will deposit and injure the kidneys.

Quantitatively, approximately 20 per cent of U^{6+} in the blood stream is deposited immediately in the kidney, followed by a 60 per cent mobilization of the dose to the urine in 24 hours.[615] About 10 to 30 per cent of the dose is deposited immediately in the bone, which is followed by a slow mobilization. No significant excretion of intravenous U^{6+} occurs via the gastrointestinal tract. No significant deposition of U^{6+} occurs in the liver, or in other soft tissues.

U^{4+} metabolism differs from that of U^{6+} in that significantly less is deposited in the kidney and as much as 50 per cent may be deposited in the liver to be mobilized later to the gastrointestinal tract. Somewhat less U^{4+} is deposited in bone than U^{6+}. Urinary excretion is slower for U^{4+} than U^{6+} and fecal excretion is significant at 2 to 4 days.

Mechanism of Action. Uranyl ion complexes with many substances of biological interest containing phosphate, carboxyl, or hydroxyl groups. Complexing with phosphate occurs more readily at pH values below 7, which exist in certain organs of the body (proximal convoluted tubes of the kidney). The number of complexing uranyl ions has been determined to be 18 for serum albumin, 8 for egg albumin. Among the phosphates the ratio is 3:1 for hexametaphosphate, 4:1 for deoxyribonucleic acid. Interpretation of the considerable amount of data is consistent with the hypothesis that the U-complexing loci of the cell surface are polymers of phosphate such as adenosine triphosphate; and that the U inhibits carbohydrate metabolism of the cell, thus injuring it, by inactivating critical phosphate-containing coenzymes required for carbohydrate metabolism. Other bivalent cations common to the body (Mg^{2+}, Ca^{2+}, and Zn^{2+}) can compete for these loci along with UO_2^{2+} and thus alter the injurious action of U according to conditions of concentration, pH, and other factors.[612,616]

Clinical Experience—Toxicity. Individuals injected with small doses (0.4 to 4 mg.) of $UO_2(NO_3)_2 \cdot 6H_2O$ were found approximately 10-fold more resistant to the effects of U on the kidney than were rabbits, one of the most sensitive species, as judged by urinary catalase, phosphatase, and amino acid N:creatinine ratio, the most sensitive indicators of early mild response of the kidney to U. The amount of U excreted in 24 hours was 76 per cent of the injected dose, 79 per cent at infinite time. Fecal excretion was negligible. In the case of soluble compounds, the critical organ for radiation damage was calculated to be the kidney rather than the bone. The kidney burden of U was the same as that in bone, however; the biological half-life in bone and kidney was found to be 300 days.[617]

[615] W. F. Neuman, R. W. Fleming, and A. L. Dounce, *et al., J. Biol. Chem.,* **173,** 737 (1948).

[616] A. Rothstein, R. C. Meier, and T. G. Scharff, *Am. J. Physiol.,* **173,** 41 (1953).

[617] S. R. Bernard and E. G. Struxness, ORNL Rept. No. **2304,** Oak Ridge, Tenn., June, 1957; A. J. Luessenhop, J. C. Gallimore, W. H. Sweet, E. G. Struxness, and J. Robinson, *Am. J. Roentgenol. Radium Therapy Nuclear Med.,* **79,** 83 (1958).

Industrial Experience—Chemical Toxicity. Except for accidental acute exposures, no evidence of chronic toxicity, either chemical or radiation, was obtained in any worker for any U compound during the first 6 years of the atomic-energy program.

The two accidental exposures involved the same compound, gaseous UF_6 and the hydrolysis products UO_2F_2 and HF. On one accident, sudden rupture of a tank of UF_6 and steam lines resulted in the death of 2 persons in the path of the UF_6 cloud: one who was exposed for 5 minutes died 10 minutes later; the other, after a rapid escape, died 70 minutes later. Live steam complicated the exposure, as some persons had third degree burns. Of 3 other persons seriously injured, 2 were in the vicinity of the accident, the third outside nearby; 10 to 14 days hospitalization was required for recovery. The 13 other exposed individuals required only dispensary care. Detailed case findings are given in Volume 2 of reference 612.

In the second case of accidental exposure to UF_6, important findings were related to the eyes (chemical conjunctivitis and corneal necrosis), the respiratory tract (increased density of bronchovascular markings and hilus shadows), and the urinary tract (increased amount of urinary solids). In 5 days, with treatment, the corneal epithelium had almost completely regenerated and at time of discharge visual acuity was normal; in 10 days the chest was clear; in 13 the hemorrhages of the larynx had disappeared. The urinary signs cleared with improvement of the patient. Some mental derangement, restlessness, and nervous tension accompanied the early responses, but these disappeared in a week (Vol. 2, reference 612).

Industrial Experience—Radiation Hazard. Evaluation of the radiation health hazards in all phases of U fabrication, as well as practical methods of control, are given by Harris and Kingsley;[610] hazards associated with rolling normal and enriched U are similarly treated by Hyatt and by Blackwell.[609] Data on U concentrations in air and in urine, and medical findings, are presented by Lippmann[611] for two U-refineries for a 2-year period. External radiation hazards, which are chiefly beta emissions, can be controlled if attention is given to prescribed procedures; control of exposures to U is gradually being tightened. No dermatitis problems relating to U itself have been found in a $2^1/_2$-year study of a plant processing U ore to form UF_4.[618] EDTA has been employed to aid removal of U (and Th) in 4 accidentally exposed cases, with ostensible value.[619]

6. Hygienic Standards of Permissible Exposure

A limit of 50 μg. of U/cu. meter of air has been adopted for *soluble* U compounds and 250 μg. of U/cu. meter of air for *insoluble* U compounds by the American Conference of Governmental Industrial Hygienists. The levels have been developed from the most thorough evaluation of data from extensive animal inhalation toxicity studies. The selection of the limit for soluble compounds was based on the considerations that: (*a*) the amount of U present in lung, kidney, or

---

[618] R. L. Kile and J. A. Quigley, *A.M.A. Arch. Dermatol.,* **79,** 49 (1959).

[619] W. N. Young and H. A. Tebrock, *Ind. Med. and Surg.,* **27,** 220 (1958).

that to be deposited in bone from tissue sources will be less than that required to be a radiation hazard; and (*b*) the amount that fails to injure the kidney, the most sensitive target organ for chemical toxicity of U, can be considered toxicologically noninjurious. The selection of the limit for insoluble U compounds is based on the amount of U in the air that will not result in a U burden in the lung or pulmonary lymph nodes that will exceed the radiation tolerance corresponding to 25 μg. of U/g. of tissue. The limit of 250 μg. of U/cu. meter for an 8-hour repeated daily exposure is believed to provide a factor of safety of 5- to 10-fold.[612]

Other standards of permissible exposure based on purely radiation-hazard concentrations, as expressed μg. of U/cu. meter with no specification as to compound type, are: 78 (NBS); 270 (ICRP); 75 for adults, 2.6 for minors (Federal Register).

7. *Flammability and Explosibility*

U is pyrophoric and will react with carbon dioxide. A U-metal dust cloud will ignite at ordinary temperatures. The explosive concentration is 60 oz. per 1000 cu. ft. (see also Vol. I, pages 560, 562, 565, and 575). Complete coverage of U-metal scrap with oil is essential for the prevention of fires; burning under supervision of all finely divided U metal before accumulation occurs is essential. Graphite chips and asbestos blankets should be used for fighting fires; speed is essential.[610] Combustibles should not be stored near $UO_2(NO_3)_2 \cdot 6H_2O$ because of excess nitric acid in the product. Cylinders of solid UF_6 should be warmed with extreme caution for release of gaseous contents in order to prevent fracture of container.

VANADIUM, V

1. *Source and Production*

V, although widely distributed, exists in such small quantities throughout the earth that known deposits commercially feasible to mine are confined almost wholly to four countries, the United States, Peru, North Rhodesia, and southwest Africa. The center of V mining in the United States is the Colorado plateau region. Peru, which ranks second to the United States as a producer of V, has deposits in the Andes at altitudes of 14,000 to 16,000 feet. The chief V ores in U-bearing sandstones of Colorado, Utah, and Arizona are carnotite ($K_2O \cdot 2UO_3 \cdot V_2O_5 \cdot 3H_2O$), roscoelite ($CaO \cdot 3V_2O_5 \cdot 9H_2O$), and vanadinite (Pb(PbCl), $(V_2O_4)_3$). Peru, which has the largest known V deposit, also has the only known source of patronite ($V_2S_5 + S$), which usually bears some Fe, Ni, Mo, P, and C. V ores of Rhodesia and southwest Africa are Pb, Cu, Zn-containing (mottramite, descloisite), in addition to vanadinite. In all, more than 65 V ores have been described, but all except 5 or 6 are of secondary origin formed by oxidation or weathering.

The United States produced 6830 short tons of contained V in 1958, which was 150 per cent more than the remaining world production that year, from which were produced nearly 2800 short tons of V_2O_5. On March 31, 1962, commitments for V purchase by the AEC terminate.

United States mining methods for V are relatively simple because size of deposit limits the need for extensive equipment. Typical equipment includes portable 2-drill compressors, jackhammer drills, gasoline hoists, a few mine cars, an air hoist or two, and in the larger mines ventilation equipment. Mining methods for a site near Rifle, Colorado, are given in detail by Burwell.[620] In Peru, V ore is mined by open-pit. In the United States, V is recovered from its ores by direct reduction and chemical treatment, such as roasting with or without reagents (soda ash or salt) followed by a leach to remove V. The process used by one of the larger mills has been described.[621] The resulting product contains 85 to 92 per cent V_2O_5, which is converted largely to ferrovanadium, which contains 35 to 55 per cent V, but some oxide is used for addition to steels. Ductile V 99.8 or 99.9 per cent purity is available; it is obtainable in massive form for remelting, as well as in ingots, bars, sheet, and foil. Vanadium pentoxide (V_2O_5) production decreased 23 per cent in 1958 from 1957, amounting to about 5000 short tons. Ferrovanadium production likewise dropped to approximately half that in 1957, but the quantity of domestic and foreign vanadium ores and concentrate used at domestic plants for V_2O_5 and ferrovanadium was about 13.7 million pounds in 1958, a decrease of only 7 per cent from 1957.

2. Uses and Industrial Exposures

About 81 per cent (1000 short tons) V used in 1958 went for ferrovanadium; of this, about 83 per cent was used for high-speed and other alloy steels; the oxide, about 70 tons; ammonium metavanadate, NH_4VO_3, 90 tons; and an equal amount for various other uses. V_2O_5 was used in welding electrode coatings, and as an additive to special steels. NH_4VO_3 and V_2O_5 are used as catalysts in glass and ceramic glazes and in research. For tool and structural steels only a small amount of V is used; in high-speed steels, the V content is between 0.5 and 2.5 per cent; alloy tool steels contain from 0.2 to 1 per cent V. Except in some carbon steels where V is used alone, V is usually combined with Cr, Ni, Mn, B, and W.

Industrial exposures related to the domestic production of V start with mining. When the mined ore is carnotite, serious exposures occur, not to V, which is relatively minor, but to radon and its daughter products associated with the U and Ra in the carnotite. Some SiO_2 exposures are encountered. In the milling of domestic V, exposures to V, U, and other ore constituents are relatively minor because much of the process is wet and comparatively good industrial hygiene is practiced in most plants. Some dust and fume exposures occur near the production of polyvanadate scale, and in the barreling operations, but they are relatively minor. Somewhat greater exposures to V are encountered in the milling of the patronite

[620] B. Burwell, *U. S. Bur. Mines Inform. Circ.* No. **6662**, 1932.
[621] F. F. Kett, *Vancoram Rev.*, **4**, No. 2, 3, 18 (1945).

ore of Peru, according to Vintinner *et al.*;[622] at most operations air concentrations considerably exceed the recommended limit of 0.5 mg. of V/cu. meter. Numerous reports call attention to health hazards from V-containing residual oil ash of high V-content oil. Potentially hazardous exposures to V occur during the cleaning of the oil-fired burners.

3. Physical and Chemical Properties

The physical and chemical properties of V and some of its compounds are listed in Table 40.

TABLE 40

Physical and Chemical Properties of V and Some of Its Compounds

	Chemical symbol	At. or mol. wt.	Sp. gr.	M.p., °C.	B.p., °C.	Solubility
Vanadium	V	50.95	5.96	1710	3000	Insol. hot or cold H_2O Sol. HNO_3, H_2SO_4, HF, aq. reg. Insol. HCl, alkalies
Vanadium pentoxide	V_2O_5	181.90	3.357(18°)	Dec. 480		V. sl. sol. cold H_2O Sol. acetic anhydride, ethyl acetate, acetone
Ammonium metavanadate	NH_4VO_3	116.99	2.326	Dec.		5.2 g./liter(15°) 69.5 g./liter(96°) Insol. alc., ether, NH_4Cl
Sodium metavanadate	$NaVO_3$	121.95		630		211 g./liter(25°) 388 g./liter(75°)
Vanadyl sulfate	$VOSO_4$	163.02				V. sol. cold H_2O

V is a grey-colored metal and relatively soft when pure. V has marked resistance to nonoxidizing acids and sea water. V forms an alloy with Fe (ferrovanadium) in which there is complete liquid solubility. V forms a very hard and stable carbide, V_4C_3, in carbon- and most alloy steels. As a carbide former, V is stronger than W or Cr. V oxidizes rapidly in the air at temperatures above 675°C.

The chemistry of V is exceeding complex; the exact nature of many of its simple ions in aqueous solution is still not known; some of the valence types are unstable; some are amphoteric. Table 41 shows the characteristic states of oxidation of V and a number of typical compound forms. By United States industrial practice VO^{2+} is vanadyl, but this leads to ambiguity for the forms $VOCl_2$ and $VOCl_3$. Ambiguity is avoided by referring to these forms as vanadium oxydichloride and vanadium oxytrichloride. The terms vanadite and hypovanadate are also applied to these oxy salts. Only vanadium pentoxide (V_2O_5), ammonium metavanadate (NH_4VO_3), and sodium vanadite ($Na_2V_4O_9$) are regular items of commerce, but the toxicity of a number of other V compounds has been determined:

[622] F. J. Vintinner, *et al., A.M.A. Arch. Ind. Health,* **12,** 635 (1955).

sodium orthovanadate (Na_3VO_4), sodium pyrovanadate ($Na_4V_2O_7$), sodium tetravanadate ($Na_2V_4O_{11}$), sodium hexavanadate ($Na_4V_6O_{15}$), as well as a number of complex vanadium–arsenic and vanadium–antimony compounds.

TABLE 41

Oxidation States of Vanadium

Oxidation state	Oxide	Typical compounds	Color in aqueous solution	Characteristics
2	VO, basic	$V(OH)_2$, VCl_2, VSO_4	Yellow to green	Powerful reducing agents; isomorphic with compounds of Cr^{2+}, Fe^{2+}, Mg^{2+}
3	V_2O_3, basic	$V(OH)_3$, VCl_3, VOCl, V_2S_3, VN	Green	Reducing agents; often amorphous with compounds of Al^{3+}, Cr^{3+}, Fe^{3+}; forms alums, complex cyanides, thiocyanates
4	VO_2, basic or weakly acidic	VCl_4 VOCl_2$, $VOSO_4$, $Na_2V_4O_9$	Blue Black	
5	V_2O_5, amphoteric	VF_5, $VOCl_3$, $NaVO_3$, $Na_4V_2O_7$	Yellow to red Light yellow to yellow	Stable forms
		$(NH_4)_2VO_4$, $NH_4H_2V_2O_2$	Yellow to orange	By action of H_2O_2

4. Analytical Determination

Air samples containing V may be conveniently analyzed by the phosphotungstic colorimetric method.[623] V may also be determined colorimetrically with 8-quinolinol in the range of 1 to 50 μg. with an average error of ±.32 μg. by the method of Talvitie;[624] the method is applicable to all types of samples but is particularly designed for analysis of biologic specimens; it is more sensitive and elaborate than the phosphotungstic method. A rapid, semiquantitative method applicable to urine specimens only, has been developed by Rockhold and Talvitie.[625] V may be determined spectrochemically according to procedures described by Daniel et al.;[626] using the 3184A. line, a sensitivity of 0.005 μg. of V on the electrode is obtainable.

5. Physiological Response

Acute Toxicity—Animals. There is a rather large older literature on the acute toxicity of V compounds for animals,[627] but much of it is unprecise by

[623] F. D. Snell and C. T. Snell, *Colorimetric Methods of Analysis,* Vol. II. 3rd ed., Van Nostrand, 1949, p. 455.

[624] N. A. Talvitie, *Anal. Chem.,* **25,** 604 (1953).

[625] W. T. Rockhold and N. A. Talvitie, *Clin. Chem.,* **2,** 188 (1945).

[626] E. P. Daniel, E. M. Hewston, and M. W. Kies, *Ind. Eng. Chem. Anal. Ed.,* **14,** 921 (1942).

present standards, or pertains to V compounds of no present industrial interest. Later work has shown the toxicity of V salts to be high parenterally, low by mouth, and intermediate by respiratory tract. The intraperitoneal LD_{50} for mice of $NaVO_3 \cdot H_2O$ is about 11 mg. of V/kg., and slightly greater for rats, on commercial laboratory diets; intramuscularly, the MLD of the metavanadate for dogs on laboratory chow and milk is 3.7 mg. of V/kg., according to Mitchell and Floyd.[628] Franke and Moxon[629] many years earlier had found intraperitoneally injected $NaVO_3$ in doses of 4 to 5 mg. of V/kg. to be lethal to young rats; the composition of the diet was not specified. Sjoberg[630] found V_2O_5 dust to be lethal to rabbits following a single, 7-hour exposure to 205 mg. of V_2O_5/cu. meter; the animals died of pulmonary edema, presumably from the combined non-specific irritancy and the specific toxicity of V_2O_5. The pH of dilute aqueous solutions of V_2O_5 approximates 3.

Chronic Toxicity. By mouth, the toxicity varies greatly, depending on the composition of the diet. Daniel and Lillie,[631] using a whole wheat, milk, meatscrap, mineral diet in which $NaVO_3 \cdot H_2O$ was incorporated, found levels above 23 p.p.m. V caused diminished weight gains, and gross pathological changes, but all rats survived a 10-week experimental period except those fed 368 p.p.m. V. Similarly, Franke and Moxon,[632] using a "control wheat diet" containing casein, vitamins, and salt mixture, found the metavanadate to be perceptibly toxic to weanling rats at about the same level (25 p.p.m. V), and definitely toxic at twice this level, in a 200-day study. Later, Stokinger et al.[633] showed that by alternating a suboptimal, casein, synthetic diet with an optimal diet (Purina chow) toxicity alternately appeared and disappeared in rats fed $NaVO_3 \cdot H_2O$ at a level of 100 p.p.m. V. Dietary V_2O_5 is less toxic than $NaVO_3 \cdot H_2O$ by a factor of 2, but is toxic at 100 p.p.m. V to rats fed the casein diet, whereas it required 1000 p.p.m. V in the optimal diet to produce definite signs of toxicity. Only at 2000 p.p.m. V, however, did deaths occur from dietary V_2O_5.

V_2O_5 was ingested for a lifetime, 2.5 years, by rats and dogs at levels of 10 and 100 p.p.m. V with no significant change in any of the criteria measured, except for reduced hair cystine values.[633] Growth of rats fed 100 p.p.m. V actually ex-

[627] F. Proescher, H. A. Seil, and A. W. Stillians, *Am. J. Syphilis,* **1,** 347 (1917); L. Larmuth, *J. Anat. Physiol.,* **11,** 251 (1877); J. Priestly, *Phil. Trans.,* **166,** 495 (1876); D. E. Jackson, *Am. J. Physiol.,* **29,** 23 (1911); P. Ricciardi, *Lavori Congr. med. int.,* **19,** 401 (1910); F. Ballotta, *Med. lavoro,* **22,** 250 (1931).

[628] W. G. Mitchell and E. P. Floyd, *Proc. Soc. Exptl. Biol. Med.,* **85,** 206 (1954).

[629] K. W. Franke and A. L. Moxon, *J. Pharmacol. Exptl. Therap.,* **58,** 454 (1936).

[630] S. G. Sjoberg, *Vanadium Pentoxide Dust.* Stockholm, 1950, *Arch. Ind. Hyg. Occupational Med.,* **3,** 631 (1951).

[631] E. Daniel and R. D. Lillie, *Public Health Repts. U. S.,* **53,** 765 (1938).

[632] K. W. Franke and A. L. Moxon, *J. Pharmacol. Exptl. Therap.,* **61,** 89 (1937).

[633] H. E. Stokinger, W. D. Wagner, J. T. Mountain, F. R. Stockell, O. J. Dobrogorski, and R. G. Keenan, in press.

ceeded that of controls for the first 7 months of feeding. Thereafter, the average group weight was slightly lower than the controls but more rats survived for longer periods in the V-fed group than among the controls, showing that chronic V toxicity varies with animal age.

In a study to test experimentally in animals the suitability of the threshold limit for V_2O_5 of 0.5 mg. of V/cu. meter recommended by the Russians, it was found that dogs, rats, guinea pigs, and rabbits tolerated V_2O_5 dust exposure at this level for 6 months of daily 6-hour exposures without evidence of histological change referable to inhalation of the dust.[633] Brief inhalation exposures of rabbits to high concentrations (several mg. of V_2O_5/cu. meter) resulted, however, in marked conjunctivitis, acute laryngotracheitis, acute bronchopneumonia, and in some cases fatty degeneration of the liver.[630] After 24 days there was still some residual chronic tracheitis; atelectatic, postpneumonia areas of the lungs; and areas with dilated alveoli. It was considered from these and other observations that there was little probability of V_2O_5 dust producing pneumoconiosis. In sub-stantiation of this conclusion, Sjoberg[630] found no fibrotic changes in the lungs of rabbits exposed to V_2O_5 dust for 8 months. Indeed, Daniel and Lillie[631] had shown earlier that some tolerance to V is developed by animals; increasing doses were tolerated in amounts far greater than were lethal on the first or second dose.

Physiological Changes. The chief action from large doses of V is reported by Proescher *et al.*[627] to be on the vascular system, as manifested by peripheral con-striction of the vessels of the lungs, spleen, kidneys, and intestines. The vasocon-striction is local in origin. Counteraction by epinephrine indicates a nervous as well as a muscular effect of V. The central nervous system is affected, however, only by fatal doses, which produce depression. Lymph flow is increased. Weight gain by more efficient utilization of food is a common response from well-tolerated doses. Most of the descriptions of effects of V poisoning scattered through the older literature deal with conditions of exposure unrelated to those encountered by man and will not be recounted. Serial microscopic changes from 2 days to 8 weeks in tissues of rats poisoned by feeding $NaVO_3$ have been reported in detail by Daniel and Lillie.[631] One of the more prominent findings was acute desquamative enteritis. In the liver there was congestion, and neutral fat droplets in the cells of the periportal zones. The lungs showed marked congestion, with some hemorrhages into the alveoli. Renal changes were slight and consisted of swelling and finely granular degeneration of the epithelium of the convoluted tubules, generally con-fined to the proximal portion. Changes of a congestive character are described for the adrenals, spleen, bone marrow, brain, and spinal cord. Spinal ganglia were normal as were the pancreas, skeletal muscle, testis, and epididymus.

Skin absorption occurs from approximately saturated (20 per cent) solution of $NaVO_3$; its application to rabbit skin causes irritation. At a 10 per cent concen-tration, $NaVO_3$ acts as a primary irritant to human skin. NH_4VO_3 and V_2O_5 in 0.5 and 0.8 per cent solutions, respectively, representing approximately saturated solutions, do not irritate the skin.[633]

Acute Experimental Toxicity in Man. Complex vanadate salts have been administered to man orally, intramuscularly, and intravenously. A normal, young, adult male injected intramuscularly with $Na_2V_4O_{11}$, 5.6 mg. of V as the first dose, followed on the third and fourth days with 11.2 mg. of V each, responded by showing elevations in the following urinary constituents: urea and purine nitrogen, and neutral sulfate; total nitrogen, uric acid, ethereal and neutral sulfate, and phosphorus showed insignificant increases. Fecal phosphorus was increased (27 per cent), but total nitrogen and sulfate were not significantly changed. When the tetravanadate was taken by mouth in 12 daily doses of 6.9 mg. of V each, 12 per cent was excreted in the urine and most of the remainder was recovered in the feces (Proescher *et al.*[627]). When the tetravanadate was intravenously given to 2 men in 6 daily doses amounting to 18 and 24 mg. per week respectively, 93 and 89 per cent of the administered total dose was recovered, 9 per cent in the feces, and it was concluded that 10 per cent is stored in the body.[634]

The fatal dose of V for man has been given variously as 60 to 120 mg., presumably of compound. From the foregoing, it is apparent that amounts of from 24 to 80 mg. of V in daily divided doses are tolerated by man. At the turn of the century V was in vogue as a therapeutic agent for tuberculosis, chlorosis, and diabetes in doses of 1 to 8 mg.

Metabolism. Although frequent reference to the metabolism of V may be found in the older toxicological literature,[627] the methods of analysis used appear inadequate for accurate, quantitative estimation of the amounts involved; accordingly, metabolism data based on more recent, reliable analytical methods will be presented.

Diammonium oxytartratovanadate fed to normal adult males in daily doses of 150 to 200 mg./day (21 to 32 mg. of V) appeared in the urine in from 6 to 31 μg. of V/liter, indicating an absorption of 0.02 to 0.1 per cent of this soluble V complex.[635] $VOSO_4 \cdot 2H_2O$ fed to adult male rats in daily doses of from 650 to 1250 μg. (160 to 310 μg. V) resulted in a mean absorption of about 0.5 per cent, with considerable variation, as judged by urinary V values.[636] The EDTA complex of V gave a somewhat improved absorption of about 1 per cent, but the biological activity of V-EDTA is very low (see section on Therapy, page 1181).

Talvitie and Wagner[637] have shown that the kidneys are the major route of excretion of intraperitoneally or intravenously administered $NaVO_3 \cdot H_2O$ in rats and rabbits. Renal excretion is rapid; about 60 per cent of a subtoxic dose is eliminated in the first 24 hours and the per cent of excreted V is essentially independent of the number of doses administered. Accordingly, urinary V values rapidly reflect changes in absorbed V. The intestinal excretion amounts to 10 or 12

[634] N. L. Kent and R. A. McCance, *Biochem. J.,* **35,** 837 (1941).

[635] G. L. Curran, private communication, 1958.

[636] J. T. Mountain, unpublished results, Toxicologic Services, Occ. Health Field Hdqts., Cincinnati, 1959.

[637] N. A. Talvitie and W. D. Wagner, *Arch. Ind. Hyg. Occupational Med.,* **9,** 414 (1954).

per cent of the dose. An equivalent amount is deposited in the skeleton, but trace amounts occur in all tissues. Blood levels of V are measurable for at least 6 hours after absorption.

Distribution of V in tissues of rats fed V at dietary levels of from 25 to 1000 p.p.m. V, as V_2O_5 and $NaVO_3\cdot H_2O$, showed that V from the latter was deposited in the tissues in about twice the amounts, at a given dietary level, as that from V_2O_5; but the amount of V deposited in the bone from the more soluble metavanadate represented only 0.07 per cent of the dose received. Taking the deposition of V in the liver as 1, bone contained 14, kidney 7, spleen and lung each 4, times as much V at the end of 20 weeks.[633] Dietary V_2O_5 at 25 p.p.m. V resulted in no measurable accumulation in any tissue of the rat at 20 weeks, but just measurable accumulation occurred in tissue V from $NaVO_3\cdot H_2O$. Retention of V, on the other hand, was relatively high, at least in bone and lung; 55 and 39 per cent were still present in the bone and lung, respectively, 4 months after the end of exposure.

Deposition of V in the femoral epiphyses of dogs fed V_2O_5 in their diets at levels of 10 and 100 p.p.m. V for 2.5 years showed a mean value of only 1.4 μg./g. (range 1.2 to 1.5) and 7.6 μg./g. (range 7.3 to 7.8), respectively. Urinary excretion of V at this time was about 0.2 mg./liter in dogs on the 10 p.p.m. level, 1.1 mg./liter at the 100 p.p.m. V level. The urine was obtained by catheterization. The normal diet showed no measurable amounts of V. At 7 months, dog plasma:cell V ratio averaged 3.7 (range 3.6 to 3.8), the plasma showing 0.26 μg. of V/g. at the 100 p.p.m. V level in the diet.[633]

Deposition of V in tissues of rats, following 6 months of daily, 6-hour inhalations of V_2O_5 dust, 0.2 μ M.M.D., at a level of 0.5 mg. of V/cu. meter, amounted to 30 μg. of V/g. in the lung, 0.8 μg./g. in the kidney, 0.6 μg./g. in the spleen, and but 0.14 μg. of V/g. in the liver.[433] Elimination from these organs was relatively slow; 40 days after the exposure terminated, the lung still contained 10 per cent of the amount at the end of exposure, the kidney 60 per cent, the spleen 50 per cent, and the liver was unchanged. Demonstrable evidence of V exposure was still found in the lung, kidney, and spleen 20 weeks postexposure. Mean daily urinary excretion of V by dogs inhaling the V_2O_5 dust for 6 months was 140 μg. of V/liter compared with 5 μg. of V/liter for control dogs.

Mechanism. V exhibits a variety of biological activities, the mechanisms of some of which are known. V in trace amounts, in the livers of rabbits on high cholesterol diets, reduces plasma cholesterol and liver phospholipids, and diminishes the size of aortic plaques (Mountain, Stockell, and Stokinger[638]). V also reduces plasma cholesterol in normal males on ordinary diets (Curran et al.[638]). The mechanism by which this occurs is by inhibition of conversion of hydroxymethylglutaric to methylcrotonic acid and by preventing the utilization of

[638] J. T. Mountain, F. R. Stockell, and H. E. Stokinger, *Proc. Soc. Exptl. Biol. Med.*, **92**, 582 (1956); G. L. Curran, D. L. Azarnoff, and R. E. Bollinger, *J. Clin. Invest.*, **38**, 1251 (1951); D. L. Azarnoff and G. L. Curran, *J. Am. Chem. Soc.*, **79**, 2968 (1956); C. Lewis, *A.M.A. Arch. Ind. Health*, **19**, 419, 497 (1959).

mevalonic acid (Azarnoff and Curran[638]), important steps in the preliminary stages of cholesterol synthesis. Lowered plasma cholesterol values appeared to be the rule among V workers in the Colorado Plateau, according to Lewis.[638]

V in trace amounts also increases the oxidation of fatty acids of liver phospholipids in vitro (Bernheim and Bernheim[639]), which may account for decreased liver phospholipids noted above. Evidence has also been presented by Snyder and Cornatzer[639] that small amounts of V, fed to animals, result in inhibition of phospholipid synthesis. Concurrently, V caused an alteration in S metabolism as shown by a reduction in the —SH content of the liver, and an increased turnover of protein sulfur. In this connection, cystine content of the hair of animals exposed to V compounds, as well as that of the fingernails of man, are reduced by small exposures to V that produce no other clinical signs (Mountain, Stockell, and Stokinger[639]).

V inhibits the sulfhydryl-dependent succinic dehydrogenase of liver, but not of kidney, of animals exposed to small amounts of V,[633] which possibly accounts for the capacity of V to increase body weight of animals. Barron and Kalnitsky[639] have shown that certain dithiols, but not BAL, can reactivate V-inhibited succinoxidase, thus providing a possible explanation why BAL is ineffective in the therapy of V poisoning (see section on Therapy, page 1181).

Some evidence has been reported[640] that V fed to rats as $NaVO_3 \cdot 4H_2O$ retards the synthesis of coenzyme A by blocking the decarboxylation of pantothenylcysteine, which process involves pyridoxal phosphate as the prosthetic group.

V has been shown[641] to be the most potent of the transition elements in catalyzing the oxidation of pressor amines derived from the decarboxylation of amino acids; $10^{-5}M$ V increased oxidation 13 per cent and $10^{-1}M$, 125 per cent in vitro. Similarly, V was the most potent of 18 metals in catalyzing the oxidation, mediated by monoamine oxidase, of the important catechol amines, 5-hydroxytryptamine (serotonin), adrenaline, noradrenaline, and others.[642] Pentavalent and tetravalent V compounds were equally effective in the in vitro studies. The importance of these studies are twofold: (a) in suggesting that V might be the metal activator for monoamine oxidases, (b) in possibly accounting for the intense vascular effects produced by V in the kidney, spleen, and intestine, and increased peristalsis, allayed by adrenaline. V-treated dogs, however, responded with a *decreased* urinary excretion of the oxidized form of the catechol amine, 5-hydroxytryptamine;[643] no evidence was found for an increased oxidation of this amine at any V dose tested.

[639] F. Bernheim and M. L. C. Bernheim, *J. Biol. Chem.,* **127**, 353 (1939); F. Snyder and W. E. Cornatzer, *Nature,* **182**, 462 (1958); J. T. Mountain, F. R. Stockell, and H. E. Stokinger, *A.M.A. Arch. Ind. Health,* **12**, 494 (1955); E. S. G. Barron and G. Kalnitsky, *Biochem. J.,* **41**, 346 (1947).

[640] E. Mascitelli-Coriandoli and C. Citterio, *Nature,* 1641 (1959).

[641] H. M. Perry, S. Teitelbaum, and P. L. Schwartz, *Federation Proc.,* **14**, 113 (1955).

[642] G. M. Martin, E. P. Binditt, and N. Ericksen, *Nature,* **186**, 885 (1960).

[643] C. E. Lewis, *A.M.A. Arch. Ind. Health,* **20**, 455 (1959).

V also mobilizes Fe in the liver and spleen; at dietary levels of 150 p.p.m. V and below, Fe is moved into the spleen and liver; at levels from 250 to 1000 p.p.m. V, Fe is moved out of these organs.[633]

Rygh,[644] using metal-free diets, has reported that small amounts of V favor deposition of Ca in bone and have a sparing action on vitamin C in scorbutic animals.

Industrial Experience. Industrial exposures to V compounds are restricted to relatively few types: the mining and milling of V, in which latter the chief exposure is to V_2O_5 mixed with H_2SO_4. Sjoberg[630] has given a detailed description of plant process for production of V_2O_5, the types of exposure, symptomatology of the disease, course of the disease, its prognosis and complications, diagnosis, treatment, and prophylaxis. Case histories of 36 affected Swedish workers are appended. As with all other cases reported in the literature, concentrations greatly exceeded the recommended limit of 0.5 mg. of V/cu. meter, sometimes attaining values 150 times the limit about some operations. The onset of symptoms, which are chiefly referable to the respiratory tract, does not generally occur until after a few days or a week of repeated exposures to V dust. The affection is principally acute with relapses, although constant coughing occurs in some cases. The duration of the disease varies from a few days to more than 2 months, with an average of 13.5 days. All 36 workers had symptoms. Clinical observations consisted of irritation of the conjunctiva and nasal mucosa, with nasal catarrh in some, and moderate pathological changes in the mucous membrane, consisting of acute and chronic hyperplasia, in many cases of the allergic type. Dryness and irritation of the throat predominated, associated with chronic atrophic changes. Wheezing and dyspnea were considered the most characteristic signs. Five cases of bronchopneumonia or lobar pneumonia occurred among the long-term employees.

A follow-up study[645] 8 years after the first exposure revealed that 6 of the 36 workers still had bronchitis with rhonchi, resembling asthma, bouts of dyspnea, and fatigue, despite considerable improvement in working conditions indicating, as Symanski[646] pointed out some years before, that V_2O_5 overexposure can lead to a chronic condition.

Similar affections referable to the eyes and respiratory tract, but of a less serious nature, are reported by Vintinner *et al.*[647] among Peruvian ore roasters of patronite. Again, V levels in the air greatly exceeded the recommended limit. On the other hand, no effects except lowered serum cholesterol levels were seen among V-processing workers in Colorado who were exposed to V levels of from 0.1 to 0.3 mg./cu. meter.

[644] O. Rygh, *Bull. soc. chim. biol.,* **31,** 1403 (1949).

[645] S.-G. Sjoberg, *Acta Med. Scand.,* **154,** 381 (1956).

[646] H. Symanski, *Arhiv. hig. rada,* **5,** 360 (1954).

[647] F. J. Vintinner, R. Vallenas, C. E. Carlin, R. Weiss, C. Macher, and R. Ochoa, *A.M.A. Arch. Ind. Health,* **12,** 635 (1955).

Sjoberg[630] found 9 cases of skin diseases among 36 exposed individuals. Eruptions of the face showed highest prevalence (7 cases), in which itching papules or dry seborrhealike patches occurred in chin and cheeks, and around the edge of the face mask. Four cases showed eruptions on the hands, usually between the fingers; some eruptions occurred on the wrists, legs, and dorsum of the feet. Patch tests made with 2 per cent $NaVO_3$ revealed only 1 distinct eczematous reaction persisting for a 2-year period, thus indicating an allergy of the skin to V.

At least six reports[648] from the United States and Europe have called attention to a mild, chronic, inflammatory type of change in the respiratory tract from cleaning oil-fired burners or gas turbines. In some instances (Fallentine and Frost[648]) sulfuric acid accompanying the V_2O_5 in the soot was felt to contribute to the respiratory-tract irritation. Petroleum crudes contain from 0 to 80 per cent vanadium oxides in the ash, depending upon source. V is believed to exist in the crude oil as a complex of mesoporphyrin IX dimethyl ester, soluble or dispersible in the oil but insoluble in water.[649]

In a report drawing attention to two practical considerations, Tara et al.[650] suggest that (a) V and its compounds be added to the list of dangerous chemicals, and (b) there is a need for selection and medical attention of dockers. Four of 12 dockers, rebagging Ca vanadate during a 1.5-day period, came down with a variety of bronchitic manifestations, including dyspnea with hemoptysis.

Therapy. Ascorbic acid, in doses of 125 mg./kg. 20 minutes prior to an LD_{70} dose of $NaVO_3 \cdot H_2O$, gave almost complete protection against the lethal effects in mice;[628] $CaNa_2$-EDTA, as well as ascorbic acid, proved antidotal in rats and dogs. EDTA was effective in antagonizing the lethal action in dogs when given intraperitoneally in doses of 100 mg./kg. after the first sign of poisoning became evident and at 2 and 4 hours later. No clinical test of the efficacy of these drugs in man appears to have been reported.

BAL was found not to be useful in antidoting the lethal action of NH_4VO_3 in rabbits and cats; BAL, however, reversed the blood pressure effects of V, causing a rise, but no change, in respiration.[651]

6. Hygienic Standards of Permissible Exposure

Threshold limits of 0.5 mg. of V_2O_5 dust/cu. meter of air, and 0.1 mg./cu. meter for V_2O_5 as fume, have been adopted by the American Conference of Governmental Industrial Hygienists. Evidence for the suitability of the limit for

[648] L. C. McTurk, C. H. W. Hirs, and R. E. Eckardt, *Med. Bull. Std. Oil N. J.*, **15**, 2, 96 (1955); S.-G. Sjoberg, *A.M.A. Arch. Ind. Health*, **11**, 505 (1955), *Nord. Hyg. Tidskr.*, **3**, 45 (1954); B. Fallentine and J. Frost, *Ind. Hyg. Occupational Med.*, **10**, 175 (1954) (abstr.); N. Williams, *Brit. J. Ind. Med.*, **9**, 50 (1952); R. C. Browne, *Brit. J. Ind. Med.*, **12**, 57 (1955); E. C. Fear and F. H. Tyrer, *Trans. Assoc. Ind. Med. Offic.*, **7**, 153 (1958).

[649] D. A. Skinner, *Ind. Eng. Chem.*, **44**, 1159 (1952).

[650] S. Tara, A. Cavigneaux, and Y. Deplace, *Arch. maladies profess. méd. travail et sécurité sociale*, **14**, 378 (1953).

[651] T. Dalhamn, S. Forssman, and S.-G. Sjoberg, *Acta Pharmacol. Toxicol.*, **9**, 11 (1953).

V_2O_5 dust is good. It was originally suggested by Roschin[652] on the basis of animal studies of short duration, and its suitability was subsequently confirmed in more extensive animal studies by Stokinger et al.[633] Lewis[638] has more recently reported no toxic manifestations from V dusts among workmen exposed to from 0.1 to 0.3 mg. of V/cu. meter. The lower limit for V_2O_5 fume is based on the recognized greater toxicity of fume compared with dusts of larger particle size.

ZINC, Zn

1. Source and Production

Zn is widely distributed and occurs in small amounts in almost all igneous rocks. Sphalerite (zinc blende) ZnS is the principal Zn mineral. Depending on Fe content, natural specimens range in color from light tan to black; above a ratio of Fe:Zn of 1:5 the mineral is called marmatite; above 5:6 the sphalerite structure ceases to exist. Next to Fe, Cd is the most common impurity in sphalerite; when associated with Zn as CdS it is called greenochite. Cd is about $1/_{200}$ as abundant as Zn. Ga and Ge also occur in sphalerite (low-temperature formation), Sn and In occur in traces from high-temperature deposits. Pb minerals are commonly associated with Zn minerals; the ratio Zn:Pb varies widely, from 1:7 to 5:1. Other commonly associated minerals are calcite ($CaCO_3$), dolomite ($CaCO_3 \cdot MgCO_3$), pyrite (FeS_2), quartz (SiO_2), chalcopyrite ($CuFeS_2$), and barite ($BaSO_4$). Other oxidized forms of Zn minerals, such as ZnO, $ZnSO_4 \cdot 7H_2O$, $ZnCO_3$, $Zn_4Si_2O_7(OH)_2 \cdot H_2O$, and $(Zn,Mn)O \cdot Fe_2O_3$, can be thought of as alterations from the sulfide, and are of minor importance.

United States mines produced 412,000 tons recoverable Zn in 1958, 23 per cent less than in 1957 and the smallest annual amount since 1933. Oversupply, increased world production, declining consumption, cessation of government stockpiling, and lower Zn prices contributed to lessened domestic production.

A variety of mining methods are used, which vary with the type of ore body. Underground methods yield most production but some open-pit mining is done. Almost all loading and transportation are handled by power, at present; electric or diesel units supply the power. Production of Zn concentrates is done by crushing and grinding followed by gravity separation, flotation, or magnetic methods, or combinations of them, depending on the complexity of the ore. Considerable metal loss occurs on concentrating; losses of 8 to 20 per cent occur with the sulfide and 15 to 90 per cent of the oxidized Zn. To improve recoveries of the latter, a caustic-leach electrolytic process is used; in this, Zn is extracted from the ores by NaOH solution, the resulting electrolyte is purified with Zn dust and lime, and the Zn is electrodeposited. Zn smelting refers to treatment whereby Zn ores or concentrates are reduced to refined metal. Sulfide Zn concen ates are roasted to eliminate S; in the process Zn is converted to the oxide and small amounts of $ZnSO_4$. The roast may either be leached for electrolytic deposition of Zn, or combined with coke or

[652] I. V. Roschin, Gigiena i Sanit., 11, 49 (1952).

coal and retorted at about 1100°C. Residues from leaching and distillation are shipped to a lead smelter for further processing if they contain economic amounts of metals (Pb, Au, and Ag). Domestic smelters produce Zn of various grades; electrolytic plants produce special high-grade or high-grade slab Zn; slab Zn from horizontal retort plants is mostly prime western grade, although smaller amounts of other grades are produced; vertical retort plants produce regular high grade; all other grades are produced as the market warrants.

2. Uses and Industrial Exposures

Zn as metal in slab or pig was used to the extent of 868,300 tons in 1958; ore and concentrates for pigments and salts, 94,900 tons; as scrap for alloys, Zn dust pigments, and salts, 178,900 tons; other Zn concentrate is used for galvanizing, the largest single use (40 per cent of total slab production). Second largest use of slab Zn is in die-casting: as Zn alloy, of which parts as small as zipper elements or as large as automobile radiator grilles are made. The automobile industry uses a great number of Zn die-castings, as do the electric appliance, light machine, tool, hardware, and toy industries. Large quantities of Zn are used for brass, particularly in wartime for cartridges and shell cases, but developments in weapons and ammunition are lessening the need for Zn. Zn, in sheet and rolled forms as an alloy, is used in dry cells; jar caps (whence comes Cd in our diet); weather stripping; photoengraving plates; roofing; ship's hulls; pipelines; and heavy plates for cathodic protection of steam boilers. Zn pigments and salts, chief of which are ZnO, lithopone (ZnS + BaSO$_4$), ZnCl$_2$, and ZnSO$_4$, have innumerable uses; the major ones are in rubber goods, linoleum, paints, ceramics, cosmetics, textiles, paper filler, and pharmaceuticals. Leaded ZnO is used in paints, varnishes, coated fabrics, and textiles; ZnCl$_2$, for wood preservation, dry-battery cells, refining oil, soldering flux; ZnSO$_4$, in rayon, fertilizers, glue, textile dyes, electrogalvanizing solutions, economic poisons, pure chemicals, and soap. About 30,000 tons of Zn dust is used as a reducing agent in the synthesis of dye intermediates, of Na and Zn hydrosulfite, in metallic paints, and cladding of iron and steel products (sherardizing). Organic Zn compounds of the type R$_2$Zn have limited application in organic synthesis.

Industrial hazards arise from exposure to Zn fume, notorious producer of metal-fume fever, but associated hazards in the metallurgy of Zn, of more serious consequence, arise from the presence of As, Cd, Mn, Pb, and possibly Cu and Ag. The frequent presence of As in Zn is a source of exposure to arsine (AsH$_3$) whenever Zn is dissolved in acids or alkalies; many cases of intoxication by AsH$_3$ have occurred in the pickling of galvanized iron or from the use of powdered impure Zn as a reducing agent in dyeing. It is possible also that effects attributed to exposure to Zn fume may in part be attributable to those of Cd. Cd occurring uniformly as a contaminant in Zn is a continuous source of trace amounts of Cd in man.

3. *Physical and Chemical Properties*

The physical and chemical properties of Zn and some of its compounds are listed in Table 42.

TABLE 42

Physical and Chemical Properties of Zn and Some of Its Compounds

	Chemical symbol	At. or mol. wt.	Sp. gr.	M.p., °C.	B.p., °C.	Solubility
Zinc	Zn	65.38	7.14	419.47	907	Insol. hot or cold H_2O
						Sol. acids, alkalies, acetic acid
Zinc oxide	ZnO	81.38	5.47	>1800		1.6 mg./liter(29°)
						Sol. acids, dil. acetic acid, NH_4OH
Zinc sulfide (sphalerite)	ZnS	97.45	4.102(25°)	Tr. 1020		0.65 mg./liter(18°)
						V. sol. acids
Zinc chloride	$ZnCl_2$	136.29	2.91(25°)	262	732	4.32 kg./liter(25°)
						6.15 kg./liter(100°)
						1 kg./liter alc.(12.5°)
						V. sol. ether
						Insol. NH_3
Zinc sulfate	$ZnSO_4$	161.44	3.74(15°)	Dec. 740		865 g./liter(80°)
						808 g./liter(100°)
						Sl. sol. alc.
Dimethyl zinc	$(CH_3)_2Zn$	95.45	1.386(10.5°)	−42.2	46	Dec. cold H_2O, alc., acids
						Sol. ether

Zn is a silvery metal of low to intermediate hardness; rolled Zn, 99.94 per cent purity, has a scleroscope hardness of 13 to 15; cast Zn of the same purity is slightly softer. The effect of small amounts of common impurities is to increase corrosion resistance to solutions, but not in atmosphere. Ordinary Zn is too brittle to roll at ordinary temperatures, but becomes ductile at elevated temperatures; brittleness is thought to be associated with impurities such as Sn.

Zn has a standard electrode potential of +0.761 and is thus electropositive to most structural metals except Al and Mg. This property is the basis of many important uses of Zn, for example, in batteries and electrogalvanizing of steel. Zn is attacked by moist air, CO_2, and SO_2, resulting chiefly in a coating of hydrated basic carbonate of variable composition; some H_2O_2 may be formed in the process. Zn is resistant to attack by dry F_2, Cl_2, and Br_2 but combines rapidly in presence of water vapor. Zn is attacked by acid gases and acids. Zn is an active reducing agent for many ions such as Fe^{+++}, MnO_4^-, and CrO_4^-. Hot caustic solutions form zincates of uncertain composition. Zn vapor reduces CO_2 to CO, the amount depending on the temperature; above 1100°C., the retort temperature of Zn distillation, essentially all CO_2 is reduced to CO, in the presence of excess C. Zn and S will explode when mixed as a powder and warmed. A protective coating of ZnS is formed on masses of Zn by either S or H_2S. When Zn or one of its alloys is burned, melted, or heated to temperatures above 930°F., Zn metal oxide fume of particle diameter 1 μ and below is formed.

4. Analytical Determination

Samples of fumes of Zn, ZnO in air may be collected by means of an impinger, electrostatic precipitator, molecular-membrane filter, or by passing air through a cotton plug moistened with dilute HNO_3. Analysis may be made colorimetrically, using dithizone or di-β-naphthylthiocarbazone, or polarographically.[653]

5. Physiological Response

A brief review by Vallee (6 pages, 28 references) touching on many of the salient features of Zn and its biological significance to man appeared in 1957.[654] Sufficient evidence has accumulated to show that Zn occurs in the body in two different protein combinations: (a) as a metalloenzyme in which Zn is an integral part of an important enzyme system, such as carbonic anhydrase for the regulation of CO_2 exchange, and (b) as a metal–protein complex in which Zn is loosely bound to a protein, which acts as its carrier and transport mechanism in the body. The possible relation of these Zn–protein combinations in a variety of altered metabolic conditions and diseases is pointed out, and will be discussed in more detail here in the appropriate sections.

Toxicity. Zn salts are astringent, corrosive to the skin, and irritating to the gastrointestinal tract. Because of the last, when ingested they act as emetics. Zn ion, however, is ordinarily too poorly absorbed to induce acute systemic intoxication. After large doses have been ingested, fatal collapse may occur as a result of serious damage to the buccal and gastroenteric mucous membranes. Mass poisonings have been repeatedly recorded[655] from drinking acidic beverages made in galvanized containers; fever, nausea, vomiting, stomach cramps, and diarrhea occurred in 3 to 12 hours following ingestion. The emetic concentration range in water is from 675 to 2280 p.p.m.; the threshold concentration of taste for Zn salts approximates 15 p.p.m.; 30 p.p.m. soluble Zn salts impart a milky appearance to water, and 40 p.p.m., a metallic taste.[656] The lethal dose of Zn ion administered orally to mice is 57 mg./kg.[657] When parenterally administered, Zn depresses the central nervous system, causing tremors and paralysis of the extremities; subcutaneous injection of Zn lactate or valerate in a dose equivalent to 57 mg. of

[653] J. Cholak, D. Hubbard, and R. Burkey, *Ind. Eng. Chem. Anl. Ed.*, **15**, 754 (1943).

[654] B. L. Vallee, *A.M.A. Arch. Ind. Health*, **16**, 147 (1957); *Physiol. Revs.*, **39**, 443 (1959).

[655] G. E. Callender and C. J. Gentkow, *Military Surgeon*, **80**, 67 (1939); J. W. Sale and C. H. Badger, *Ind. Eng. Chem.*, **16**, 164 (1924); *U. S. Natl. Offic. Vital Statist.*, Communicable Diseases Summary, for Sept. 11, 1954.

[656] J. J. Hinman, *J. Am. Water Works Assoc.*, **30**, 484 (1938).

[657] H. Jaeger, *Arch. exptl. Pathol. Pharmakol.*, **159**, 139 (1931).

Zn/kg. killed a cat after 3 days.[657] Orally, soluble Zn salts are more than 100 times less toxic than corresponding salts of Cd, with which Zn is commonly contaminated.

Drinker, Thompson, and Marsh[658] gave from 175 to 1000 mg. of ZnO/day for periods of from 3 to 53 weeks to dogs and cats, and it was tolerated; glycosuria occurred in the dogs, and fibrous degeneration of the pancreas in some of the cats was found at autopsy. No manifest injury occurred in rats from administration of from 0.5 to 34.4 mg. ZnO/day for periods of 1 month to 1 year. Similar lack of response from $ZnCO_3$ is reported. On the other hand, Waltner and Waltner[659] reported that feeding the same salt induced anemia and osteoporosis in rats; 2 per cent metallic Zn in the diet of rats, however, resulted in no injury. Zn acetate fed to rats for 4 months in doses of 10 to 15 mg. daily, and 50 mg. of Zn malate fed to cats for 10 days to 2 months caused no intoxication, according to Salant.[660] Sutton and Nelson[661] found that 0.1 per cent Zn was tolerated in the diet of rats, but that more than 0.5 per cent reduced their capacity to reproduce, and 1 per cent inhibited growth, caused severe anemia, and death. Zn salts in the diet are somewhat more toxic to pigs.[662]

Metabolism. Human intake of Zn is about 10 to 15 mg./day, which is mainly excreted through the intestines. The feces contain about 10 mg. and the urine about 0.4 mg. The normal Zn content of human tissues varies from 10 to 200 μg./g. of fresh tissue. Most organs, including the pancreas, contain around 20 to 30 μg./g.[654] The liver, bone, and voluntary muscle contain from 60 to 180 μg. of Zn/g. of tissue. Exceptionally large amounts are present in the prostate (860 $\pm$ 100 μg./g.) and the retina (500 to 1000 μg./g.). The total amount of Zn in the human adult is estimated to be 2 g. No tissue preferentially stores Zn. Whole blood of man contains about 900 μg. per cent of Zn, of which the serum contains from 80 to 160 μg. per cent. Leukocytes contain 3 per cent of total blood Zn. Blood Zn values exhibit no seasonal or diurnal variations nor do they differ between the sexes. Diabetics have normal Zn blood levels.

Only very small amounts of Zn were absorbed and stored in the tissues of dogs, cats, and rats fed Zn compounds for long periods;[658] chief sites of storage were the liver and pancreas. Intravenously injected radioactive Zn^{65} showed that liver, pancreas, and kidney stored large amounts of Zn^{65}, but that the muscular and mucosal layers of the small intestine contained relatively large amounts; more than 50 per cent of the dose was excreted by mice in the feces and 2 per cent in the urine in 170 hours.[663]

[658] K. R. Drinker, P. K. Thompson, and M. Marsh, *Am. J. Physiol.*, **80**, 31, 65 (1927), **81**, 284 (1927).

[659] K. Waltner and K. Waltner, *Arch. exptl. Pathol. Pharmakol.*, **141**, 123 (1929); **146**, 310 (1929).

[660] W. Salant, *J. Ind. Hyg.*, **2**, 72 (1920).

[661] W. R. Sutton and D. E. Nelson, *Proc. Soc. Exptl. Biol. Med.*, **36**, 211 (1937).

[662] R. E. Grimmett, et al., *N. Z. J. Agr.*, **54**, 216 (1937).

[663] G. E. Sheline, I. L. Chaikoff, H. B. Jones, and M. L. Montgomery, *J. Biol. Chem.*, **147**, 409 (1943), **149**, 138 (1943), *J. Exptl. Med.*, **78**, 151 (1943).

Zn is apparently transported in the blood stream as a loosely bound metal globulin complex;[664] blood serum also contains a firmly bound Zn–protein component, which amounts to about 50 per cent of that of the loosely bound fraction.

Ethylenediaminetetraacetate (EDTA) administered to rats intraperitoneally greatly increased the urinary excretion of Zn. Of the various tissues analyzed, only the pancreas and ileum showed a decreased concentration of Zn after EDTA.[665]

Mechanism. Zn is an active component of carbonic anhydrase; removal of Zn results in irreversible inactivation of the enzyme, which is responsible for rapid exchange of CO_2.[654] Zn is also a component of carboxypeptidase that splits terminal amino groups from peptides; 1 atom of Zn is combined to 1 molecule of enzyme. Four dehydrogenases contain Zn that is essential for their action: alcohol dehydrogenase of yeast and liver, lactic acid dehydrogenase, and glutamic dehydrogenase; 2 to 4 moles of Zn are contained per enzyme molecule. Their presence in liver and retina may explain the high concentration of Zn at these sites.

Serum Zn level is lowered in experimental CS_2 poisoning,[666] in pneumonia, bronchitis, erysipelas, in pyelonephritis, and in untreated pernicious anemia.[654]

Industrial Experience. Batchelor, Fehnel, Thomson, and Drinker found no acute or chronic illness in their examination of 24 men who had been exposed to ZnO at levels of 0.3 to 1.64 mg./cu. ft. for from 2 to 35 years.[667] The mean Zn content of the blood of the exposed men was only slightly greater than that of unexposed men. Exposure to mists or fumes of Zn salts may give rise to irritation of the respiratory or the gastrointestinal tracts, but the evidence for these effects is inconclusive. It has been stated that ZnO as dust may block the sebaceous-gland ducts and give rise to a papulopustular eczema in men engaged in packing ZnO in barrels.[668] Sensitivity to ZnO is extremely rare.[669] Unlike ZnO, $ZnCl_2$ has a caustic action, may result in ulceration of the fingers, hands, and forearms of those who use it as a flux in soldering. Affections of the skin have been noted among men handling railway ties impregnated with $ZnCl_2$. Zn stearate has been implicated as a causative agent in pneumoconiosis.[670]

Metal-Fume Fever. Inhalation of fumes of ZnO results in a malaria-like illness with onset some hours after the exposure.[671] Metal-fume fever is not confined to the inhalation of ZnO but may follow exposure to metal fumes of Sb, As, Be, Cd, Co, Cu, Fe, Pb, Mg, Mn, Hg, Ni, and Sn. ZnO fume is a more frequent

[664] I. Vikbladh, *Scand. J. Clin. & Lab. Invest.*, **3**, Suppl. No. 2, 1 (1951).

[665] M. J. Millar, *et al.*, *Nature,* **174**, 881 (1954).

[666] A. E. Cohen and L. D. Scheel, *et al.*, *Am. Ind. Hyg. Assoc. J.*, **20**, 303 (1959).

[667] R. P. Batchelor, J. W. Fehnel, R. M. Thomson, and K. R. Drinker, *J. Ind. Hyg.*, **8**, 322 (1926).

[668] J. A. Turner, *Public Health Repts. U. S.*, **36**, 2727 (1921); J. G. Downing, *J. Ind. Hyg.*, **17**, 147, 150 (1935).

[669] H. E. Freeman, *J. Am. Med. Assoc.*, **119**, 1016 (1942).

[670] U. Uotila and L. Noro, *Folia Med. Naples*, **40**, 245 (1957).

[671] C. C. Sturgis, P. Drinker, and R. M. Thompson, *J. Ind. Hyg.*, **9**, 88 (1927); P. Drinker, R. M. Thompson, and J. L. Finn, *J. Ind. Hyg.*, **9**, 98, 187, 331 (1927); **10**, 13 (1928).

cause, however. The symptoms include chills and fever, which rarely exceeds 102°F., nausea and sometimes vomiting, dryness of the throat, cough, fatigue, yawning, weakness, and aching of the head and body. After a few hours, the victim sweats profusely, and the temperature begins to fall. The condition lasts a day and is never fatal. Occasionally glucose is found in the urine; albuminuria is rare. Mental confusion and convulsions may be present. The vital capacity (lung) may be reduced, a condition which may persist for 15 hours. In 36 of 100 cases observed by Natvig,[672] the condition recurred weekly or more frequently. Leucocytosis (12,000 to 16,000 leucocytes/cu. mm.) persists for 12 hours after the temperature has returned to normal.[671] While the condition persists there is a measure of immunity. Workers are more susceptible on Mondays, and on week-days following a holiday, than on other workdays. A postulated mechanism that seems most reasonable today is that of Lehmann,[673] who suggested that the inhaled Zn-fume particles liberated modified protein from the lung into the blood stream. The subsequent distribution and absorption of the modified protein results in the characteristic response resembling that from the injection of a foreign protein. A recent review of the problem of metal-fume fever from inhaling ZnO, with reports of cases, is given by Rohrs.[674]

6. Hygienic Standards of Permissible Exposure

A limit of 15 mg. of Zn/cu. meter air accepted by the American Conference of Governmental Industrial Hygienists was suggested by Drinker, Thomson, and Finn[671] as the maximal concentration that can be inhaled daily without the production of metal-fume fever. In laboratory experiments, a concentration of 45 mg./cu. meter was tolerated for 20 minutes by man.

7. Flammability

Powdered Zn presents a hazard of explosion. If stored in damp places, there is danger of spontaneous combustion. Residues from reduction reactions may start a fire if thrown into combustible waste (see also Vol. I, pages 560, 562, 565, 567, 570, and 575).

ZIRCONIUM, Zr

1. Source and Production

Zr is associated with other metals in many minerals, but is recovered only from zircon ($ZrO_2 \cdot SiO_2$) and baddeleyite (brazilite) (ZrO_2). Hafnium, Hf, is invariably associated with Zr. Zircon occurs in all igneous rocks but is more common in granite, sylnite (complex silicates), and diorite (alkaline-earth silicates). Zircon is a common constituent of river gravels and beach sands, whence it is

---

[672] H. S. Natvig, *J. Ind. Hyg. Toxicol.*, **19**, 227A (1937) (abstr.).

[673] K. B. Lehmann, *Arch. Hyg.*, **72**, 358 (1910).

[674] L. C. Rohrs, *A.M.A. Arch. Int. Med.*, **100**, 44 (1957).

recovered as a coproduct of ilmenite ($FeO \cdot TiO_2$), rutile (TiO_2), and monazite (Th and lanthanon phosphates). Baddeleyite usually occurs in phenolite and is also found in river gravels and beach sands; commercial deposits are known only in Brazil. Zr ore is produced in five countries: United States, Australia, Brazil, French West Africa, and India; the principal United States sources is Florida sands. World reserves of Zr are tremendous; United States reserves alone are greater than the foreseeable demand for the next 100 years.

World production of Zr ores and concentrates was about 110,000 short tons in 1958, of which the United States produced about one third. About two dozen companies in the United States supply $ZrSiO_4$, zirconia (ZrO_2), Zr alloys, and ductile Zr. For production of Zr, Zr-containing sands are subjected to electro-magnetic separators that isolate $ZrSiO_4$ of about 99 per cent purity. The ore can be decomposed by heating to volatilize SiO_2, by sintering and digesting with acid, by fluxing with alkali, or by carbiding followed by direct chlorination. $ZrCl_4$ is the starting compound of the Kroll process, which utilizes molten Mg for reduction to Zr in an atmosphere of A or He. The van Arkel iodide process, the only other method of producing sufficiently pure Zr metal to be ductile, employs Ca to reduce ZrO_2 to Zr; this is combined with I_2 at 200° C. and is heated to 1300°C. to yield pure Zr.

2. Uses and Industrial Exposures

Of the approximately 54,000 tons of zircon used in the United States in 1958, 8000 tons were used for the production of Zr metal and ferro alloys, and somewhat less for refractories. Foundry sands and facings took most of the remainder; abrasives, ceramics, paints, and chemicals utilized small quantities. Consumption of Zr sponge is mainly for military purposes; substantial amounts found use in private power reactors; almost 25 tons was used in industry unrelated to nuclear power (i.e., photographic flash bulbs). Zirconia refractories are used by the aluminum, brass, and glass industries; a small amount was used to prepare Zr metal powder. Zr, a powerful deoxidizer (scavenger), is used because of this as an additive to iron and steel. Zr compounds are being investigated for use in ceramics. Zirconia has been used as a reflective surface agent on satellites. Basic studies of binary alloys of Zr have been made with Sb, Be, B, Nb, Ga, In, Ni, Ag, Th, Ti, U, and Zn; and ternary alloys of Zr with Fe and Sn, Ta and Nb, Si and B, and Cu and Cr.[675]

Exposures consist or dust and fume of Zr and compounds, as well as SiO_2 during milling and in various uses.

3. Physical and Chemical Properties

The physical and chemical properties of Zr and some of its compounds are listed in Table 43.

Zr alloys with almost all metals except Hg and those of the alkali and alkaline-

[675] L. E. Tanner, Armour Research Foundation, Rept. No. **ARF-2068-5**, Office of Tech. Serv., Washington, D. C., 1958.

TABLE 43

Physical and Chemical Properties of Zr and Some of Its Compounds

	Chemical symbol	At. or mol. wt.	Sp. gr.	M.p., °C.	B.p., °C.	Solubility
Zirconium	Zr	91.22	6.4	1857	>2900	Insol. hot or cold H_2O Sol. HF, aq. reg. Sl. sol. acids
Zirconia	ZrO_2	123.22	5.6	2715		Insol. hot or cold H_2O Sol. H_2SO_4, HF
Zirconium tetrachloride	$ZrCl_4$	233.05	2.8	Subl. 300		Dec. to $ZrOCl_2$ Sol. alc., ether, conc. HCl
Zirconium iodide	ZrI_4	598.9				S. d. Dec. alc. Sol. acids, ether Sl. sol. benzene, CS_2

earth groups. Zr alloys resist neutron bombardment and corrosion and have a low neutron-capture cross-section; traces of Hf destroy this property. Sponge Zr is pyrophoric when pure, and violently explosive when impure. Zr in bulk form is not reactive at ordinary temperatures but it oxidizes readily at 700°C., and combines with Cl at lower temperatures. The most common valence of Zr is +4 but in rare instances it exhibits a valence of +3 and +2. Zr has a maximal covalency of 8, but covalencies of 5, 6, and 7 are common. Zr probably does not exist in compounds as a monatomic ion, but is frequently found as the central atom of complex anions and cations. Zr forms three categories of compounds: (a) complex Zr cations, (b) complex Zr anions, and (c) Zr in nonionized groups. The simplest Zr cations are ZrO^{2+}, $ZrCl_2^{2+}$; an example of anions of the hypothetic H_4ZrO_4 and H_2ZrO_3 is hexafluorozirconate, ZrF_6^{2-}. Zr alkoxides ($Zr(OR)_4$) are examples of nonionic Zr compounds. Noncomplex Zr compounds exist generally only in the gaseous state, for example, $ZrCl_4$. $ZrCl_4$ reacts violently with water to form $ZrOCl_2$. Important sources of information on the chemistry of Zr are to be found in Tanner,[675] Blumenthal,[676] and Miller.[676]

4. Analytical Determination

A spectrographic method is the one choice for Zr because, if Hf is present, procedures for chemical analysis give substantially the same results for both elements. The spectral lines proving most useful for Zr determination are at wavelengths 2571.39 and 3391.75. Modifications have been made in the Bricker-Waterbury colorimetric method to adapt it to routine analyses of Zr in air for the estima-

---

[676] W. T. Blumenthal, The Chemical Behavior of Zirconium. van Nostrand, Princeton, 1958; G. L. Miller, Zirconium. 2nd ed., Academic Press, New York, 1958.

tions of sample portions of 1 to 10 μg. of Zr.[677] The method is not applicable to Zr determination in urine unless Zr is first isolated.

5. *Physiological Response*

Acute Toxicity. The toxicity of Zr salts is very low by the oral route; the LD$_{50}$ dose of the nitrate, chloride, sulfate, and acetate of Zr, and sodium zirconyl sulfate for rats ranged from 2.5 to 10 g./kg. (0.853 to 2.290 g. of Zr/kg.) (see Table 44). Intraperitoneally the same Zr compounds were considerably more

TABLE 44

LD$_{50}$ Values of Zirconium Compounds for Rats

Compound	LD$_{50}$, mg. Zr/kg.
Oral administration	
Zirconyl acetate	1,660
Zirconium carbonate, hydrate	>10,000
Zirconyl chloride	990
Zirconyl nitrate	853
Zirconium sulfate	1,253
Sodium zirconyl sulfate	2,290
Intraperitoneal administration	
Zirconyl acetate	122
Zirconyl chloride	113
Zirconium citrate	1,710
Zirconium gluconate	247
Zirconyl nitrate	426
Zirconium sulfate	63
Sodium zirconyl sulfate	939

toxic acutely, in most instances (0.175 to 4.1 g./kg.) ;[678] least toxic was the sodium zirconyl sulfate, in which Zr is in the anionic form. ZiSiO$_4$ is physiologically inert in single or repeated intraperitoneal doses in guinea pigs.[679] Zr gluconate was moderately toxic (LD$_{50}$, 247 mg. of Zr/kg.) acutely by intraperitoneal administration in rats, but the citrate was relatively innocuous by the same route (1.71 g. of Zr/kg.).[680] Hydrated Zr carbonate (3ZrO$_2$·CO$_2$·H$_2$O) was also physically inert orally in rats in doses up to 10 g./kg., as it was intraperitoneally in doses up to 1.5 g./kg.[681]

[677] E. E. Campbell, *Am. Ind. Hyg. Assoc. J.*, **20**, 281 (1959).
[678] K. W. Cochran, J. Doull, M. Mazur, and K. P. DuBois, *Arch. Ind. Hyg. Occupational Med.*, **1**, 637 (1950).
[679] H. E. Harding, *Brit. J. Ind. Med.*, **5**, 75 (1948).
[680] L. T. McClinton and J. Schubert, *J. Pharmacol. Exptl. Therap.*, **94**, 1 (1948).
[681] J. W. E. Harrison, B. Trabin, and E. W. Martin, *J. Pharmacol. Exptl. Therap.*, **102**, 179 (1951).

Animals acutely poisoned by Zr compounds show progressive depression and decrease in activity until death. The time of death varies from a few hours to a few days following administration of the compounds; few deaths occur later than 5 days after administration. No signs characteristic of Zr poisoning have been observed.

Subacute Toxicity. Toxicity of ZrO_2 dust by inhalation is of a very low order; neither 75 mg. of Zr/cu. meter inhaled by laboratory animals daily for 1 month or 11 mg. of Zr/cu. meter for 2 months produced any significant alterations in growth rate, mortality, hematological values, biochemical constituents, or in morphological structure.[682] On the other hand, inhalation of $ZrCl_4$ mist at a level of 6 mg. of Zr/cu. meter was perceptibly toxic in 2 months, as shown by slight decreases in hemoglobin and erythrocyte count of dogs, and some increase in mortality over that of controls. Apparently the toxicity of Zr compounds is not altered by the 3 or 4 per cent Hf ordinarily present in the Zr compounds, as toxicity was not observed in the 2-month study of ZrO_2 that was relatively free of Hf.

Chronic Toxicity. Orally $3ZrO_2 \cdot CO_2 \cdot H_2O$ incorporated in a standard diet at a level of 4 per cent and fed to rats for 17 weeks was without adverse effect on all criteria examined (growth rate, blood and urine constituents, mortality.[681] The carbonate likewise produced no evidence of sensitization following 12 weekly applications in ointment form.

ZrO_2 dust and $ZrCl_4$ mist inhaled at respirable particle sizes for 1 year at 3.5 mg. of Zr/cu. meter likewise failed to have any adverse effect on laboratory animals.[683]

Intensive exposure of rats to $ZrSiO_4$ dust for 7.5 months produced radiographic lung shadows but no cellular reaction as determined histologically.[684]

Metabolism. Deposition and retention of Zr in the lung and pulmonary lymph node following inhalation of ZrO_2 dust are typical of that of an insoluble substance;[682] however, deposition of the soluble $ZrOCl_2$ in the lung following inhalation of the $ZrCl_4$ mist was equally great and showed a similar distribution in other body tissues; Zr deposition in the bone was less than 1 per cent of that in the lung, and soft tissue content was a fraction of that in bone. Retention of Zr in all tissues of the rat is high; little, if any, decrease in deposited Zr occurs in 6 months.

Tracer studies of oral absorption of $Zr^{89}Cl_4$, with the stable chloride as carrier, show less than 0.001 per cent of the dose passed from the gastrointestinal tract to the blood stream.[685]

Pharmacological Action. A study[686] of certain pharmacological actions of

[682] C. J. Spiegl, *et al.*, Univ. Rochester, Atomic Energy Comm. Project, Rept. **UR-460**, July, 1956.

[683] H. C. Hodge, *et al.*, unpublished results, Univ. Rochester Atomic Energy Comm. Project, 1955.

[684] H. E. Harding and T. A. L. Davis, *Brit. J. Ind. Med.*, **9**, 70 (1952).

[685] J. G. Hamilton, U. S. Atomic Energy Comm., Rept. No. **MDDC-1001**, 1944.

[686] J. van Niekirk, *Arch. exptl. Pathol. Pharmakol.*, **184**, 686 (1937).

pure zirconyl chloride, $ZrOCl_2$ (free from Hf) on smooth and striated muscle in rabbits and cats indicated that the normal peristaltic action in rabbits was slowed at 1 mM concentration but the effect was less noticeable in cats. The effect was reversible on washing with Ringer's solution. $ZrOCl_2$ at 0.006 mM shortened the amplitude of the movement of the auricle and ventricle of the rabbit heart and constricted the coronary arteries. The latter effect appeared irreversible. No effect on blood pressure, however, was noted below 1 mole% concentration of $ZrOCl_2$ in injected rabbits; 3 to 7 mole% was required to produce appreciable lowering of the blood pressure. No electrocardiographic changes were noted.

Effect on Enzymes. $Zr(NO_3)_4$ produced a slight (10.5 per cent) inhibition of amylase activity at 0.53 mM concentration in vitro, a 20 per cent inhibition of yeast invertase at 0.7 mM, and a 50 per cent inhibition of serum alkaline phosphatase at 0.9 mM, but no inhibition of trypsin at 1 mM concentration.[687] The rather low inhibitory effects produced on these enzymes at the Zr concentrations used indicates no specific effect of Zr on these particular enzymes.

Industrial Experience. A study of 22 workers exposed to the fumes of the Zr reduction process for from 1 to 5 years revealed no striking abnormalities felt to be referable to Zr exposure.[688] Moreover, guinea pigs exposed continuously at several of the steps in the Zr process for from 2 to 6 months showed no pulmonary changes attributable to the exposure on histologic examination.

Clinical Experience. The new and unique clinical entity in man, of Zr granulomas of allergic epithelioid origin, first described by Rubin et al. in 1956,[689] has since been repeatedly reported both clinically[690] and experimentally.[691] Grossly they consist of dusky reddishbrown, discrete papules 1 to 4 mm. in diameter, closely set in the domes of the axillae, sparsely at the periphery. Histologically, they are granulomas of the corium composed of epithelioid cells and Langhans' giant cells surrounded by lymphocytes in a typical tuberculoid pattern. Cases arise from use of a deodorant containing an organic Zr salt, commonly NaZr lactate. Removal of the sensitizing Zr salt results in healing.

6. Hygienic Standards of Permissible Exposure

A threshold limit of 5 mg. of Zr/cu. meter of air for exposure to Zr compounds was set by the American Conference of Governmental Industrial Hygienists on the basis of the lack of effect of ZrO_2 dust and $ZrCl_4$ mist ($ZrOCl_2$) in animals exposed

[687] B. S. Gould, *Proc. Soc. Exptl. Biol. Med.,* **34,** 381 (1936).

[688] C. E. Reed, *A.M.A. Arch. Ind. Health,* **13,** 578 (1956).

[689] L. Rubin, A. H. Slepyan, L. F. Weber, and I. Neuhauser, *J. Am. Med. Assoc.,* **162,** 65, 953 (1956).

[690] Editorial, *Lancet,* May 31, 1958.

[691] J. T. Prior, H. Rustad, and G. A. Cronk, *J. Invest. Dermatol.,* **29,** 449 (1957); J. T. Prior and G. A. Cronk, *A.M.A. Arch. Dermatol.,* **80,** 447 (1959); R. M. Williams and G. B. Skipworth, *A.M.A. Arch. Dermatol.,* **80,** 273 (1959); W. B. Shelley and H. J. Hurley, *Brit. J. Dermatol.,* **70,** 75 (1958).

for 1 year to approximately 3.5 mg. of Zr/cu. meter, as well as the recognized low toxicity of Zr compounds by most routes.

7. *Flammability and Explosibility*

Finely divided or spongy Zr metal is easily ignited in air by a spark or a blow. Fires so started cannot be extinguished by ordinary means and require quenching by smothering with some pulverized mineral carbonate, such as dolomite. Serious explosion has occurred from moist exposure of metal scrap containing Zr as a contaminant. Zr has a minimal explosive concentration of 40 oz. per 1000 cu. ft.; ZrH_4, 85 oz./cu. ft. Zr metal dust dispersed with U and UH_4 as a cloud ignites under certain conditions (see also Vol. I, pages 560, 562, 565, 567, 570 and 575).

Acknowledgment

The author wishes to acknowledge with appreciation the contribution of Mr. Robert G. Keenan, Chief of Analytic Services, Occupational Health Field Headquarters, PHS, who has aided greatly in reviewing the material on analytical methods.

CHAPTER XXVIII

The Aliphatic (Open Chain, Acyclic) Hydrocarbons

HORACE W. GERARDE, M.D., PhD

I. Saturated Aliphatic Hydrocarbons (Paraffins, Alkanes, Methane Series), General Formula: C_nH_{2n+2}

1. Source and Uses

The paraffin hydrocarbons are derived almost exclusively from petroleum. Except for the normal (straight chain) paraffin hydrocarbons, very few of the paraffins above C 8 have been positively identified in petroleum. The lower members of the series (methane, ethane, propane, and butanes) are gases, the members above C_4H_{10} (pentanes to hexadecanes) are liquids, and above $C_{16}H_{34}$ are solids at room temperature. They are used extensively, most often as complex mixtures, as fuels, refrigerants, propellants, solvents for paints, pesticides, protective coatings and plastics, in dry cleaning, in degreasing operations, and as lubricants. The most important hydrocarbon mixtures containing paraffins are listed in Table 1.

The paraffins above octane are not sufficiently volatile to warrant serious consideration as vapor hazards at room temperature unless confined to a tank or other enclosure where vapor-saturated air is encountered.

TABLE 1

Principal Mixtures Containing Paraffin Hydrocarbons

Mixture	Boiling range, °C.	Principal paraffins
Natural gas	Gas at room temp.	C_1, C_2
LPG ("bottled gas")	Gas at room temp.	C_3, C_4
Petroleum ether	20–60	C_4 to C_6
Petroleum benzin	40–90	C_5 to C_7
Petroleum naphtha	65–120	C_6 to C_8
Gasoline	36–210	C_5 to C_{10}
Mineral spirits	150–210	C_7 to C_9
Kerosine (coal oil)	170–300	C_9 to C_{16}
Jet and turbo fuels	40–300	C_5 to C_{16}
Lubricating oils	300–700	C_{17} and up

2. *Physical and Chemical Properties*

The physical and chemical properties of the first eight members of the paraffin hydrocarbon series are given in Table 2.

3. *Determination in the Atmosphere*

Portable electrical devices, such as the methane detector and combustible gas indicators, when properly calibrated are sufficiently accurate for most purposes. Additional methods include the interferometer (see Vol. I, page 179), gas chromatographic analysis, and the Haldane or Orsat gas apparatus.

4. *Physiological Response and Permissible Atmospheric Concentrations*

The first two members of the series, methane and ethane, are pharmacologically "inert," belonging to a group of gases called "simple asphyxiants." These gases can be tolerated in high concentrations in inspired air without producing systemic effects. If the concentration is high enough to dilute or exclude the oxygen normally present in the air, effects produced will be due to oxygen deprivation or asphyxia. Pharmacologically, the hydrocarbons above ethane can be grouped with the general anesthetics in the large class known as the central nervous system depressants, which includes such well-known chemicals as ethyl alcohol, diethyl ether, and acetone. The vapors of these hydrocarbons are mildly irritating to mucous membranes, the irritation increasing in intensity from pentane to octane. The liquid paraffin hydrocarbons are fat solvents and primary skin irritants. Repeated or prolonged skin contact will dry and defat the skin, resulting in irritation and dermatitis. Direct contact of liquid hydrocarbons with lung tissue (aspiration) will result in chemical pneumonitis, pulmonary edema, and hemorrhage. For permissible limits, see specific hydrocarbons.

II. Specific Paraffin Hydrocarbons

METHANE, CH_4

Methane has no appreciable physiological action except when it lowers the partial pressure of oxygen in the air enough to cause systemic effects due to oxygen deprivation. It has no warning odor. The chief danger in the coal mining industry is explosion hazard. However, because of its low density, methane may accumulate in the upper strata of poorly ventilated areas to produce an asphyxiating atmosphere.

ETHANE, C_2H_6

No systemic effects are produced by breathing an atmosphere containing ethane in concentrations below 50,000 p.p.m. (5 per cent).

TABLE 2

Physical and Chemical Properties of C_1 to C_8 Paraffin Hydrocarbons

Property	Methane	Ethane	Propane	n-Butane	n-Pentane	n-Hexane	n-Heptane	n-Octane
Formula	CH_4	C_2H_6	C_3H_8	C_4H_{10}	C_5H_{12}	C_6H_{14}	C_7H_{16}	C_8H_{18}
Molecular weight	16.04	30.07	44.09	58.12	72.15	86.17	100.20	114.22
Density of liquid (25°/4°C.)			0.4928[a]	0.5730[a]	0.62139	0.65481	0.67949	0.69855
F.p., °C.	−182.48	−183.23	−187.65	−138.33	−129.723	−95.320	−90.595	−56.798
B.p., °C. (760 mm.)	−161.49	−88.63	−42.07	−0.5	36.073	68.740	98.426	125.665
Vapor density (air = 1)	0.55	1.04	1.52	2.01	2.49	2.97	3.52	3.94
Vapor pressure (mm. Hg, 25°C.)	Gas	Gas	8.8 atm. (20°)	1823	500 (24.34°)	150 (24.81°)	47.7	10.45 (20°)
n_D (25°C.)					1.35466	1.37226	1.38517	1.39580
Per cent in saturated air (25°C., 760 mm.)	100	100	100	100	66.0	19.7	6.3	1.4
Density of air saturated with vapor (25°C., 760 mm., air = 1)	0.55	1.04	1.52	2.01	1.98	1.39	1.18	1.04
Solubility in water (20°C.)	0.09 cc./g.	0.04724 vol. in 1 vol.	0.065 vol. in 1 vol. (17.8°)	0.15 vol. in 1 vol. (17°)	0.036 g. in 100 ml. (16°)	0.0023 wt. per cent (20°)	0.005 wt. per cent (15.5°)	0.0014 wt. per cent (16°)
Solubility in ethyl alcohol, (20°C.)	0.60 cc./g.	2.3344 vol. in 1 vol.	7.90 vol. in 1 vol. (16.6°)	18.83 vol. in 1 vol. (17°)	Miscible	50 g. in 100 ml.	100 g. in 100 ml.	Slightly soluble
Solubility in ether (20°C.)	0.91 cc./g.		9.26 vol. in 1 vol. (16.6°)	29.8 vol. in 1 vol. (18°)	Miscible	Soluble	Miscible	Soluble
Flash point (°F.)	Gas	Gas	Gas	Gas	−40	−15	25	56
Specific dispersion					97.9	97.9	97.9	97.9
P.p.m. ≈ 1 mg./liter	1524	813	557	421	340	284	244	214
Mg./cu. meter ≈ 1 p.p.m.	0.656	1.230	1.804	2.376	2.94	3.52	4.10	4.67
Flammable limits (% by vol. in air)	5.00–15.00	3.10–12.45	2.10–9.50	1.86–8.41	1.42–7.80	1.18–7.43	1.10–6.70	0.96–4.66
Suggested max. permissible limit	1.00 vol. %[a]	0.62 vol. %[b]	0.42 vol. %[b]	0.37 vol. %[b]	0.2 vol. %[b]	700 p.p.m.	300 p.p.m.	200–300 p.p.m
Attendant warning properties	None	None	None	Odor 0–1	Odor 3	Odor 3	Odor 3–4	Odor 4

[a] At saturation pressure.

[b] Based upon 20 per cent of the lower flammable limit rather than upon physiological effects.

PROPANE, C_3H_8

Brief exposures to 10,000 p.p.m. (1 per cent by volume) cause no symptoms in man. Odor is not detectable below 20,000 p.p.m. (2 per cent). A concentration of 100,000 p.p.m. (10 per cent) is not noticeably irritating to the eyes, nose, or respiratory tract, but will produce slight dizziness in a few minutes.

BUTANE, C_4H_{10}

A 10-minute exposure to 10,000 p.p.m. (1 per cent) of butane gas results in drowsiness but no other evidence of systemic effect. The odor of butane is not detectable below a concentration of 5000 p.p.m. (0.5 per cent).

PENTANE, C_5H_{12}

Pentane is the lowest member of the series that is a liquid at room temperature and pressure. It causes narcosis in 5 to 60 minutes at a concentration range of 90,000 to 120,000 p.p.m. (9–12 per cent by volume) in air.[1] Only a narrow margin exists between the concentrations which cause narcosis and death in mice. In human studies, a 10-minute exposure to 5000 p.p.m. (0.5 per cent) did not cause mucous membrane irritation or other symptoms. The odor of pentane at this concentration is readily detectable. The threshold limit for pentane is 1000 p.p.m.

HEXANE, C_6H_{14}

Hexane is three times as toxic to mice as pentane.[1] Narcosis is produced in mice at concentrations of approximately 30,000 p.p.m. (3 per cent); convulsions and death resulted from exposures of equal duration to 35,000 to 40,000 p.p.m. (3.5 to 4 per cent). In man, 2000 p.p.m. (0.2 per cent) hexane produced no symptoms during a 10-minute exposure, whereas 5000 p.p.m. (0.5 per cent) caused dizziness and a sensation of giddiness.[2] The threshold limit for hexane is 500 p.p.m.

HEPTANE, C_7H_{16}

Heptane, in concentrations of 10,000 to 15,000 p.p.m. (1 to 1.5 per cent), produces narcosis in mice within 30 to 60 minutes.[1] At higher concentrations, 15,000 to 20,000 p.p.m. (1.5 to 2 per cent), a 30- to 60-minute exposure caused convulsions and death in mice. Slight vertigo developed in men exposed for 6 minutes to 1000 p.p.m. (0.1 per cent) and for 4 minutes to 2000 p.p.m. (0.2 per cent).[2] A 4-minute exposure to 5000 p.p.m. (0.5 per cent) heptane caused marked vertigo, inability to walk a straight line, hilarity and incoordination. It is significant that these signs and symptoms of systemic effects were produced in the absence of evidence or complaints of mucous membrane irritation. A 15-minute

[1] H. Fuhner, *Biochem. Z.,* 115, 235 (1921).
[2] F. A. Patty and W. P. Yant, *U. S. Bur. Mines Rept. Invest.,* No. 2979 (1929).

exposure to heptane at this concentration produced a state of intoxication characterized by uncontrolled hilarity in some individuals and in others a stupor lasting for 30 minutes after the exposure. These symptoms were frequently intensified or first noticed at the moment of entry into an uncontaminated atmosphere. These individuals also complained of loss of appetite, slight nausea, and a taste resembling gasoline for several hours after exposure to heptane. The threshold limit for heptane is 500 p.p.m.

OCTANE, C_8H_{18}

Octane in concentrations of 6600 to 13,700 p.p.m. (0.66 to 1.37 per cent) caused narcosis in mice within 30 to 90 minutes. No deaths or convulsions resulted from these exposures to concentrations below 13,700 p.p.m. (1.37 per cent). The threshold limit for octane is 500 p.p.m.

III. Hydrocarbon Mixtures

A. NATURAL GAS, LPG, PETROLEUM ETHER, PETROLEUM BENZIN AND PETROLEUM NAPHTHA

The physiological response resulting from exposure to these hydrocarbon mixtures can be judged from their composition (Table 1) and the information given above for the physiological response to the specific hydrocarbons, methane to octane. Similarly, an approximation can be made for the permissible atmospheric concentration on the basis of the flammability and the threshold limits of the individual components in these mixtures.

B. GASOLINE AND MINERAL SPIRITS (STODDARD SOLVENT, VARSOL)

Gasoline is undoubtedly the most extensively used hydrocarbon mixture. It must be emphasized that this is a complex mixture of hydrocarbons containing a small proportion of nonhydrocarbon additives (antiknock agents, antioxidants, rust inhibitors, dyes, etc.). The revolution in the method of gasoline manufacture in the last few years has been accompanied by changes in hydrocarbon composition. High quality gasoline today consists of highly branched paraffins ("alkylate"), branched and internally unsaturated cyclic olefins (see page 1207), and aromatics (see Chapter XXX) boiling over a range from below 100°F. to somewhat over 400°F. Widely varying amounts of individual constituents are contained in typical gasoline blends, depending on such factors as the origin of the blending streams, seasonal requirements, and intended use.

The changes in the hydrocarbon composition of gasoline have not significantly altered its basic pharmacology and toxicology. An acute exposure to gasoline vapors will elicit the symptoms and signs of intoxication described for exposure to heptane. The atmospheric concentrations and the duration of exposure to elicit these responses will differ with the composition of the gasoline. The concentrations of gasoline hydrocarbons that cause mucous membrane irritation will vary with the degree of branching of the paraffins and the content of alkyl derivatives of benzene and olefins. The foregoing does not apply to gasolines that contain a significant concentration of benzene (benzol), which appears to be unique among hydrocarbons in its effect on blood-forming tissue (see Capter XXX).

The relationship between air concentrations of the volatile fraction (distilling below 230°F.) of unleaded straight-run gasoline and response in man is summarized in Table 3.[4] Massive or overwhelming exposures may cause sudden collapse, coma, and death. In fatal cases there is usually some blood vessel damage, with small hemorrhages into the body organs. Bronchitis, pulmonary edema, and cellular damage of the kidneys, liver, and spleen have been described.[3]

TABLE 3

Human Response to Gasoline Vapors Distilling Below 230°F.

Concentration, p.p.m.	Exposure time	Response
550	1 hr.	No effects
900	1 hr.	Slight dizziness and irritation of eyes, nose, and throat
2,000	1 hr.	Dizziness, mucous membrane irritation, and anesthesia
10,000	10 min.	Nose and throat irritation in 2 min.; dizziness in 4 min.; signs of intoxication in 4 to 10 min.

There are a number of reports in the clinical literature which clearly indicate that severe acute hydrocarbon intoxication may cause central nervous system sequelae. The length of coma and the presence of convulsions or epileptiform seizures at the time of exposure are indications of the severity of the acute exposure. There is little doubt that in most cases of acute hydrocarbon intoxication in man recovery is complete. Overwhelming amounts of hydrocarbons carried by the blood to nerve or brain tissue may cause chemical irritation. This is followed by the chain of events of the inflammatory reaction that may ultimately leave scarring in a small but delicate area of the brain. This may be the focal point for cerebral dysrhythmia resulting in convulsions or seizures months after the initial severe acute exposure. Cases of this type reported in the clinical literature are summarized by Machle.[4]

Repeated exposures to relatively low concentrations of gasoline vapor in air with brief periods of higher concentrations are common for garage workers and filling station attendants. There is no conclusive evidence that this type of exposure has any deleterious effect on the health of these individuals. A potential source of severe exposure exists in the cleaning and repairing of storage tanks on land or in tankers. It is apparent from the above information that an atmosphere containing a concentration of 20 per cent of the lower flammable limit (approxi-

[3] R. H. Wilson, J. Am. Med. Assoc., **139**, 906 (1949).
[4] W. Machle, J. Am. Med. Assoc., **117**, 1965 (1941).

mately 2000 p.p.m.) of gasoline hydrocarbons is *not safe for a man to enter for even a brief time*. The practice of permitting men to work without respiratory protection in these atmospheres should be strongly condemned. The threshold limit for gasoline is 500 p.p.m. The hydrocarbons in gasoline are primary skin irritants and will dry and defat the skin if contact is repeated or prolonged.

Mineral spirits (Stoddard Solvent, Varsol, and numerous other synonyms) consist of mixtures of straight and branched chain paraffins, naphthenes (cycloparaffins), and alkyl derivatives of benzene, boiling in the range of 305 to 410°F. They are used extensively for dry cleaning and degreasing, and as thinners for paints and other finishes. Pharmacologically and toxicologically, these mixtures are qualitatively comparable with heptanes and octanes.

C. KEROSINE, JET FUELS, AND TURBO PROP FUELS

These hydrocarbon mixtures are similar chemically, consisting of aliphatic, olefinic, naphthenic (cycloparaffinic), and aromatic hydrocarbons. The principal components are aliphatics ranging from C_5 to C_{16}. Because of their relatively low vapor pressure, inhalation toxicity is unlikely under ordinary conditions of use. Exposure to mists will cause mucous membrane irritation. Contact with lung tissue will cause chemical pneumonitis. Prolonged or repeated contact with skin will result in drying and dermatitis. The threshold limit for these hydrocarbon mixtures has not been established.

D. LUBRICATING OILS

These hydrocarbon mixtures consist of hundreds of individual hydrocarbons belonging to the paraffinic, cycloparaffinic (naphthenes), aromatic, and polyaromatic series. They have a low order of toxicity. Because of their low vapor pressure inhalation of vapors is not encountered in normal use. Exposure to mists will cause mucous membrane irritation and a chemical pneumonitis from direct contact of the liquid or aerosol with pulmonary tissue. Frequent and prolonged contact with the skin will lead to skin irritation and dermatitis. Skin cancer in man has been reported following severe, prolonged (many years) exposure to cutting oils.[5] These cutaneous reactions can be prevented by good personal hygiene consisting of minimizing skin contact, prompt removal of oil from the skin with soap and water, and wearing clean work garments. The threshold limit for these hydrocarbon mixtures has not been established.

IV. Unsaturated Aliphatic Hydrocarbons: Olefins (Alkenes, Ethylene Series), General Formula C_nH_{2n}; Diolefins (Alkadienes), General Formula C_nH_{2n-2}; Acetylene, CH≡CH

A. SOURCES AND USES

The olefins of industrial importance (ethylene, propylene, butylenes) are not found in petroleum or natural gas but are formed as by-products of the cracking

---

[5] Hunter, D., *The Diseases of Occupations*, Little, Brown, Boston, 1955, p. 728.

of petroleum fractions. They are used for synthesis of polymers (resins, plastics, rubber) and as starting materials for synthesis of numerous chemicals.

The most important diolefins are 1,3-butadiene and isoprene (2-methyl-1,3-butadiene) which are the chief constituents of synthetic rubbers. Most of the 1,3-butadiene today is made from the catalytic dehydrogenation of n-butylenes. Isoprene is obtained commercially by steam-cracking higher hydrocarbons from refinery streams.

Acetylene is prepared by the action of water on calcium carbide and by cracking of natural gas and liquid hydrocarbons (petrochemical acetylene). The gas is usually compressed in cylinders partly filled with acetone which dissolves 25 times its volume of acetylene per atmosphere of pressure. Acetylene is used in welding, metal-cutting, and as a starting material for synthesis of a wide variety of chemicals.

B. PHYSICAL AND CHEMICAL PROPERTIES

Conjugated dienes such as 1,3-butadiene and isoprene form organic peroxides on standing which are explosive when concentrated and heated. Proper precautions should be taken in distilling these hydrocarbons.

The physical and chemical properties of the commercially important unsaturated aliphatic hydrocarbons are summarized in Table 4.

C. DETERMINATION IN THE ATMOSPHERE

In general, analytical methods for the unsaturated aliphatic hydrocarbons are similar to those described for the paraffins. In addition, the reactivity of the olefins with sulfuric acid forms the basis for their estimation by absorption in a gas analysis pipet. Butadiene may be determined by the iodine pentoxide method. The following procedures may be used for acetylene: Combustible gas indicator equipped with a screen to prevent flash back, Interferometer, Ilosvay, and the Acetylide test.[6]

D. PHYSIOLOGICAL RESPONSE AND PERMISSIBLE ATMOSPHERIC CONCENTRATION

The lower members of the olefin series (ethylene, propylene, butylenes), the diolefins and acetylene are simple asphyxiants or weak anesthetics. Ethylene and propylene induce surgical anesthesia when used at a concentration of 60 per cent or more. Anesthetic potency increases with increased chain length in the ethylene series. Amylene (pentene), a liquid at room temperature, has also been used for surgical anesthesia. Propylene trimer (C_9H_{18}), propylene tetramer ($C_{12}H_{24}$), diisobutylene (C_8H_{16}), and triisobutylene ($C_{12}H_{24}$) are not sufficiently volatile to warrant serious consideration as vapor hazards at room temperature unless confined to a tank or other enclosure where vapor saturated air is encountered.

[6] *Natl. Bur. Standards U. S.*, Summary Tech. Rpt. 2273, Sept., 1958.

TABLE 4

Physical and Chemical Properties of Some Unsaturated Aliphatic Hydrocarbons

Property	Ethylene	Propylene	1-Butene	1,3-Butadiene	Isoprene	Acetylene
Formula	C_2H_4	C_3H_6	C_4H_8	C_4H_6	C_5H_8	C_2H_2
Molecular weight	28.05	42.08	56.10	54.09	68.11	26.04
B.p., °C. (760 mm.)	−103.71	−47.70	−6.26	−4.41	33.5	−83.6
M.p., °C.	−169.15	−185.25	−185.35	−108.915	−120.00	−81.8
n_D (25°C.) and satn. pressure		0.5053	0.5888	0.6149	0.8095	
Density of gas (air = 1)	0.97	1.45	1.94	1.87	2.35	0.90
Solubility in water	25.6 vol. in 100 vol. (0°C.)	44.6 vol. in 100 vol.	Slightly soluble	Very low	Very low	100 vol. in 100 vol. (18°C.)
Solubility in ethyl alcohol	360 vol. in 100 vol.	1250 vol. in 100 vol.	Soluble	Very soluble	Soluble	600 vol. in 100 vol. (18°C.)
Flammable[a] limits (% by vol.)	2.75–28.6	2.00–11.10	1.98–9.65	2.0–11.5		2.5–80.00
Suggested max. permissible[b] limit (% by vol.)	0.55	0.40	0.40	0.40		0.50
Attendant warning properties	Faint sweet odor	Weak odor	Weak odor	Faint odor	Faint odor	Faint odor

[a] See Volume I, Chapter XVI.
[b] Based on 20 per cent of the lower flammable limit rather than on physiological effects.

V. Specific Unsaturated Aliphatic Hydrocarbons

A. OLEFINS (ALKENES, ETHYLENE SERIES)

ETHYLENE, $CH_2=CH_2$ (Ethene)

Ethylene is a gas at room temperature with a slightly sweet odor. It has been used at concentrations of 75 to 90 per cent in oxygen as an anesthetic. The hazards in the industrial use of ethylene are flammability and the possibility of causing asphyxia by lowering the oxygen content of the working atmosphere. Its maximum permissible limit in workroom air should not exceed 5500 p.p.m., which is 20 per cent of the lower flammable limit. The threshold limit for ethylene has been established at 1000 p.p.m.

1 mg/liter $\backsimeq$ 872 p.p.m. and 1 p.p.m. $\backsimeq$ 1.15 mg./cu. meter at 25°C and 760 mm. Hg.

PROPYLENE, $CH_3CH=CH_2$ (Propene, Methylethylene)

Propylene has recently gained industrial importance as the starting material for the synthesis of polypropylene, phenol, acetone, glycerine, acrolein and isopropyl alcohol. Its pharmacological properties resemble ethylene, and it also has been used as an anesthetic. The flammable range is 2.00 to 11.10 per cent by volume of air. The suggested maximum permissible limit is 4000 p.p.m., one-fifth of the lower flammable limit. The threshold limit has not been established.

1 mg./liter $\backsimeq$ 581 p.p.m. and 1 p.p.m. $\backsimeq$ 1.72 mg./cu. meter at 25°C., 760 mm. Hg.

BUTYLENES: 1-Butene, $CH_3CH_2CH=CH_2$ (Ethylethylene); 2-Butene, $CH_3CH=CHCH_3$ (sym-Dimethylethylene); Isobutylene, $CH_3C(CH_3)CH_2$ (2-Methylpropene, asym-Dimethylethylene)

The butylene isomers are similar in their pharmacological activity as asphyxiants and weak anesthetics. They are about 4.5 times as toxic as ethylene. Unless encountered in sufficient concentrations to cause asphyxia, these olefins do not appear to warrant serious consideration for their effects on the health of workmen exposed to low concentrations for prolonged periods or to higher concentrations for relatively short periods of time. The flammable range of n-butene is 1.98 to 9.65 per cent by volume in air. This places the maximum permissible concentration for workroom atmospheres at 4000 p.p.m. on the basis of flammability. The threshold limit for the butylenes has not been established.

1 mg/liter $\backsimeq$ 436 p.p.m. and 1 p.p.m. $\backsimeq$ 2.30 mg./cu. meter at 25°C., 760 mm. Hg.

B. DIOLEFINS

1,3-BUTADIENE, $CH_2=CH-CH=CH_2$ (Divinyl, Biethylene)

In high concentrations, 1,3-butadiene is an anesthetic that can cause respiratory paralysis and death. Rabbits exposed to concentrations ranging from 200,000 to 250,000 p.p.m. once a day for 15 to 21 days had no deleterious effects from these exposures. During the actual inhalation of these high concentrations of the hydrocarbon, the animals were lightly anesthetized, there was a loss of the pupillary reflex, and noisy breathing and rales were noted. Two human volunteers breathed 8000 p.p.m. for 8 hours with no effects other than slight irritation of the eyes and upper respiratory tract. No cases of serious illness arising from its industrial use have been published. Contact of liquid butadiene with the skin will cause a freezing burn. The lower flammability limit for butadiene is 20,000 p.p.m. (2 per cent). The threshold limit has been established at 1000 p.p.m.

1 mg./liter $\backsimeq$ 452 p.p.m. and 1 p.p.m. $\backsimeq$ 2.22 mg./cu. meter at 25°C., 760 mm. Hg.

ISOPRENE, $CH_2=C(CH_3)CH=CH_2$ (2-Methyl-1,3-Butadiene)

Isoprene at a concentration of 20,000 p.p.m. (2 per cent) did not cause narcosis in mice exposed for a 2-hour period. Deep narcosis resulted from exposure to 35,000 to 45,000 p.p.m. (3.5–4.5 per cent) and death followed exposure to 50,000 p.p.m. (5 per cent). On the basis of these limited animal studies, it appears that isoprene is a more potent anesthetic than butadiene. Qualitatively, its pharmacological activity is very similar to butadiene. The threshold limit for isoprene has not been established.

1 mg./liter $\backsimeq$ 358 p.p.m. and 1 p.p.m. $\backsimeq$ 2.79 mg./cu. meter at 25°C., 760 mm. Hg.

ACETYLENE, $CH\equiv CH$

The inhalation of 100,000 p.p.m. (10 per cent) acetylene has a slightly intoxicating effect on man. Marked intoxication occurs at 200,000 p.p.m. (20 per cent), incoordination at 300,000 p.p.m. (30 per cent), and unconsciousness in 5 minutes on exposure to 350,000 p.p.m. (35 per cent). There is no evidence that repeated exposure to tolerable levels of acetylene has any deleterious effects on health. Phosphine (as much as 0.06 per cent) has been reported as an impurity in some commercial grades of acetylene. This and other impurities must be considered in evaluating cases of exposure to acetylene.

The flammable range for acetylene ranges from 2.5 to 80 per cent by volume in air. Based on the lower explosive limit of 2.5 per cent, acetylene should not be permitted to exceed 0.5 per cent (5000 p.p.m.) in the working environment. The threshold limit for acetylene has not been established.

1 mg./liter $\backsimeq$ 937 p.p.m. and 1 p.p.m. $\backsimeq$ 1.065 mg./cu. meter at 25°C., 760 mm. Hg.

CHAPTER XXIX

The Alicyclic Hydrocarbons

HORACE W. GERARDE, M.D., PhD

I. General Considerations

A. NOMENCLATURE, SOURCES, AND INDUSTRIAL USES OF ALICYCLIC HYDROCARBONS

The *alicyclic* hydrocarbons are saturated or unsaturated cyclic hydrocarbons having the chemical properties of the aliphatic series, as the name implies. The saturated alicyclic hydrocarbons are called *cycloparaffins, cycloalkanes,* or *naphthenes.* The term "naphthenes" is commonly used in the petroleum industry and is not to be confused with the aromatic hydrocarbon *naphthalene.* The cyclic hydrocarbons with one double bond are called *cycloalkenes,* or *cycloolefins;* the terms *cyclodiolefins* and *cycloalkadienes* are applied to ring structures having two double bonds. Naturally occurring alicyclic hydrocarbons from biological sources are commonly known as *terpenes.*

The simplest cycloparaffin, cyclopropane, is extensively used as an inhalation anesthetic. It is synthesized by reduction of dihalogenated propane precursors. The 5 and 6 carbon cycloparaffins, cyclopentane and cyclohexane, and their alkyl derivatives are found in crude petroleum. Some crude oils contain very high proportions of these compounds. The cycloparaffins are found as constituents in petroleum products, the lower molecular weight series in liquid fuels and solvents, and the higher homologs in lubricants. Cyclopentane, cyclohexane, and their methyl derivatives are available commercially as single compounds and are used as solvents and starting materials for synthesis in the chemical industry. Decalin (decahydronaphthalene) is produced by the catalytic hydrogenation of naphthalene. It is used as a solvent, in paint and varnish removers, and as a turpentine substitute in shoe polishes, oil paints, and floor lacquers.

The cycloolefins (cycloalkenes) and cyclodiolefins (cycloalkadienes) are produced commercially by recovery from hydrocarbon streams originating from high temperature cracking of petroleum fractions and as by-products of the coke-oven industry. The most important commercial cyclodiolefins are cyclopentadiene and methylcyclopentadiene and the dimers of these compounds. They are used as intermediates and starting materials for synthesis in the chemical industry.

The terpene hydrocarbons occur in the essential or volatile oils derived from many plants, especially the Coniferae and Rutaceae. Commercially, the most important terpene is a mixture known as turpentine, consisting principally of alpha and beta pinene. The turpentines are extensively used as solvents in the manufacture of surface coatings and in numerous chemical products.

B. TOXICOLOGY

Toxicologically, the alicyclic hydrocarbons resemble their open chain relatives, the aliphatic hydrocarbons. In general, they are anesthetics and central nervous system depressants with a relatively low order of acute toxicity. They do not tend to accumulate in body tissues so that cumulative toxicity from repeated exposure to low atmospheric concentrations is improbable. An overwhelming acute exposure resulting in prolonged unconsciousness, anoxia, and convulsions may result in central nervous system sequelae. This has been described following exposure to volatile aliphatic hydrocarbons.

The liquid alicyclic hydrocarbons will dehydrate and delipidize the skin on repeated or prolonged contact and cause dermatitis. Direct contact of liquid alicyclic hydrocarbons with lung tissue (aspiration) will cause pulmonary edema, pneumonitis, and hemorrhage. The vapors in sufficient concentrations will cause irritation of the mucous membranes. The atmospheric concentrations eliciting these responses vary with the specific hydrocarbon. In general, the saturated hydrocarbons are less irritating than the corresponding unsaturated compounds. There is no evidence that the alicyclic hydrocarbons are specific blood-forming tissue toxicants.

II. Specific Compounds

CYCLOHEXANE, C_6H_{12} (Hexahydrobenzene)

1. Sources, Uses, and Industrial Exposures

Cyclohexane is found in considerable proportions in certain crude petroleums from which it can be separated by fractionation. It is also prepared by hydrogenation of benzene. It is used as a solvent and a starting material and intermediate for synthesis in the chemical industry.

2. Physical and Chemical Properties

The physical and chemical properties of cyclohexane are given in Table 1.

3. Determination in the Atmosphere

The general nonspecific methods for hydrocarbon analysis may be used for cyclohexane, namely, the combustible gas indicator, interferometer, spectrometer, infrared absorption, gas chromatography, and combustion to carbon dioxide. The refractivity of 1 per cent vapor in air at 25°C. and 760 mm. Hg is 15.1×10^{-6}.

TABLE 1

Physical and Chemical Properties of Some Industrial Alicyclic Hydrocarbons and Hydrocarbon Mixtures

Property	Cyclohexane	Methyl cyclo-hexane	Decalin		Turpentine	(Endo) Dicyclopentadiene
			Cis	Trans		
Mol. wt.	84.6	98.18	138.25	138.25	Approx. 133	132.20
B.p., °C.	80.738	100.93	194.6	185.5	153-175	170
F.p., °C.	6.554	-126.597	-43.26	-31.47	-50--60	33
n_D (25°C.)	1.42354	1.42056	1.48113 (20°C.)	1.46994 (18°C.)	1.459-1.47 (20°C.)	1.5050
Vapor pressure	103.67 mm. (26.347°C.)	43 mm. (25°C.)	1 mm. (22.5°C.)	10 mm. (47.2°C.)		10 mm., 47.6°C.
Density (25°/4°C.)	0.77389	0.76501	0.8963 (20°/4°C.)	0.8699 (20°/4°C.)	0.86-0.88	0.9766 (35°C.)
Vapor density (air = 1)	2.9	3.4	4.8	4.8	4.6	4.55
Per cent in satd. air 760 mm.	13.66 at 26.3°	5.65 at 25°				
Density of satd. vapor-air mixt. at 760 mm. (air = 1)	1.23 at 26°	1.14 at 25°				1.04 (47.6°C.)
Specific dispersion	96.1	97.8				
Solubility in water at 20°C.	Insoluble	Insoluble	Insoluble	Insoluble	Insoluble	Insoluble
Solubility in ethyl alcohol	Miscible		Soluble	Soluble	Miscible	Soluble
Solubility in ethyl ether	Miscible		Soluble	Soluble	Miscible	Soluble
Flammable limits (% by vol. in air)	1.33-8.35	1.15-?			0.69-?	
Flash point (closed cup)	1°F.	25°F.	136°F.	136°F.	95°F.	

1 mg./liter ⟂ 291 p.p.m. and 1 p.p.m. ⟂ 3.44 mg./cu. meter at 25°C., 760 mm. Hg.

4. Physiological Response

Acute Toxicity. The minimum lethal oral dose of cyclohexane for rabbits is 5.5 to 6 g./kg. of body weight.[1] Inhalation of 26,600 p.p.m. (2.66 per cent) caused the death of rabbits after 1 hour of exposure. Death did not result from a single 8-hour exposure to 18,500 p.p.m. (1.85 per cent). Atmospheric concentrations of 12,600 p.p.m. (1.26 per cent) produced evidence of lethargy, narcosis, increased respiration rate, and convulsions. Exposure to 3330 p.p.m. (0.333 per cent) caused no visible effects in rabbits. Additional acute inhalation toxicity data using other animals are summarized in Table 2.[2]

TABLE 2

Comparative Effects of a Single Exposure to 18,000 p.p.m. (1.8 Per Cent) Cyclohexane Vapor in Air

	Minutes to produce effect		
Animal	Trembling	Disturbed equilibrium	Complete recumbency
Mouse	5	15	25
Guinea pig	Slight		
Rabbit	6	15	30
Cat		11	18–25

Chronic Toxicity. Rabbits exposed 8 hours/day, 5 days/week for 26 weeks, to 434 p.p.m. (0.043 per cent) cyclohexane showed no evidence of ill effects during or after exposure and there was no indication of gross or microscopic injury to the tissues in these animals.[1] A monkey received 50 daily 6-hour exposures to 1243 p.p.m. (0.12 per cent) cyclohexane without showing any signs of deleterious effects during or after the exposures. The tissues of this animal were normal on microscopic examination. Minor microscopic changes were observed in the liver and kidneys of rabbits following 50 daily 6-hour exposures to 786 p.p.m. (0.076 per cent) cyclohexane. No deaths or signs of injury were observed in rabbits following 50 daily 6-hour exposures to 3330 p.p.m. (0.333 per cent) cyclohexane. Ten daily 6-hour exposures to cyclohexanes ranging from 7400 (0.74 per cent) to 18,500 p.p.m. (1.85 per cent) caused some fatalities in a group of rabbits exposed.

The pathological changes observed in the tissues of the animals described in these experiments are not specific for cyclohexane. Generalized endothelial injury involving most of the tissues is characteristic of acute injury by a number of

[1] J. F. Treon, W. E. Crutchfield, Jr., and K. V. Kitzmiller, *J. Ind. Hyg. Toxicol.,* **25,** 199, 323 (1943).

[2] F. Flury and F. Zernik, *Schädliche Gase.,* Springer, Berlin, 1931.

irritating hydrocarbons and other organic chemicals. No specific or general toxic effects on the cellular elements of the peripheral blood of the animals were observed in the experiments described with cyclohexane. Degenerative changes in the liver and kidney were found in addition to the endothelial injury.

Absorption and Excretion. Cyclohexane is absorbed by inhalation. A small fraction is exhaled and a portion excreted in the urine unchanged, but most of the cyclohexane that gets into the blood is metabolized and excreted in the urine in the form of glucuronides and ethereal sulfates.

Tests Indicating Exposure. In addition to air analysis, body fluids (blood and urine) could be analyzed for cyclohexane or its metabolites. The decrease in the ratio of inorganic to total sulfates in the urine is somewhat proportional to the concentration of cyclohexane in the inhaled air. The change is not as rapid, nor as complete as in exposure to benzene. In the absence of other chemicals eliminated as glucuronides or ethereal sulfates, glucuronic acid excretion or the change in urinary sulfate ratio could form the basis for a test for exposure to cyclohexane. To establish the validity of these tests a correlation would have to be established between air concentration of cyclohexane and the change in glucuronic acid excretion or urinary sulfate ratio in workmen.

5. *Hygienic Standard of Permissible Exposure*

The threshold limit for cyclohexane has been established at 400 p.p.m.

6. *Odor and Warning Properties*

Cyclohexane has a bland odor compared to those of aromatic hydrocarbons. A concentration of 300 p.p.m. is detectable by odor and is somewhat irritating to the eyes and mucous membranes.

METHYLCYCLOHEXANE, $C_6H_{11}CH_3$ (Hexahydrotoluene)

1. *Sources, Uses, and Industrial Exposures*

Methylcyclohexane is found in considerable proportions in certain crude petroleums from which it can be separated by fractionation and distillation. It is also prepared by hydrogenation of toluene. It is used as a solvent and a starting material and intermediate for synthesis in the chemical industry.

2. *Physical and Chemical Properties*

The physical and chemical properties of methylcyclohexane are given in Table 1.

3. *Determination in the Atmosphere*

The general nonspecific methods for hydrocarbon analysis may be used for cyclohexane, namely, the combustible gas indicator, interferometer, spectrometer, infrared absorption, gas chromatography, and combustion to carbon dioxide. The refractivity of 1 per cent vapor in air at 25°C. and 760 mm. Hg is 17.5×10^{-6}.

1 mg./liter $\backsimeq$ 249 p.p.m. and 1 p.p.m. $\backsimeq$ 4.02 mg./cu. meter at 25°C., 760 mm. Hg.

4. *Physiological Response*

Acute Toxicity. The oral minimum lethal dose for rabbits is 4.0 to 4.5 g./kg. of body weight.[1] Rabbits exposed to 15,000 p.p.m. (1.5 per cent) methylcyclohexane died in 70 minutes. Effects observed preceding death were: conjunctival congestion, salivation, labored breathing, narcosis, and convulsions.

Chronic Toxicity. All rabbits died following exposure to 10,000 p.p.m. (1 per cent) methylcyclohexane, 6 hours/day, 5 days/week for 2 weeks; one-fourth of the group of rabbits exposed to 7300 p.p.m. (0.73 per cent) for a similar time died. Four weeks' exposure of rabbits to 5600 p.p.m. (0.56 per cent) caused no deaths. The only sign of intoxication was slight lethargy which was not observed at half this concentration. Likewise, concentrations of 1162 p.p.m. (0.116 per cent) and 241 p.p.m. (0.024 per cent) produced no evidence of injury in rabbits after 10 weeks of daily exposure. A monkey exposed the same length of time to 373 p.p.m. (0.037 per cent) showed no signs of illness. No evidence of injury was observed on histological examination of the tissues of rabbits exposed to 1162 p.p.m. (0.116 per cent) and only minor evidence of liver and kidney injury after 300 hours' (10 weeks) exposure to 3330 p.p.m. (0.33 per cent). In general, the pathological changes in tissues of animals exposed to methylcyclohexane are similar to effects described for cyclohexane.

Absorption and Excretion. Methylcyclohexane is absorbed by inhalation. A small fraction is exhaled and another fraction excreted in the urine unchanged, but most of the hydrocarbon that gets into the blood is metabolized and excreted in the urine in the form of conjugates with glucuronic acid or sulfuric acid. Total urinary glucuronic acid was higher after exposure to methylcyclohexane than after inhalation of a comparable quantity of cyclohexane. The urinary sulfates were altered less by exposure to methyl cyclohexane than by exposure to cyclohexane.

Tests Indicating Exposure. In addition to air analysis, blood and urine could be analyzed for methylcyclohexane after separation of the hydrocarbon from these fluids. It would appear that glucuronic acid excretion in the urine might form the basis for a test for exposure to methylcyclohexane. This has not been established, and, of course, exposure to other chemicals eliminated as glucuronides would have to be taken into account.

5. *Hygienic Standard of Permissible Exposure*

The threshold limit for methylcyclohexane has been established at 500 p.p.m.

6. Odor and Warning Properties

There are no quantitative data on the relationship between air concentrations of methylcyclohexane and odor intensity or effects on the eyes and mucous membranes of the nasal passages and throat. In general the alicyclic hydrocarbons are less irritating than the aromatics.

TURPENTINES

1. Sources, Uses, and Industrial Exposures

Gum turpentine is the steam-volatile portion of the resin that exudes from incisions cut in trunks of living pine trees. Chemically it contains 58–65 per cent α-pinene, 30 per cent of an isomeric β-pinene, 2 per cent of monocyclic terpenes, and 2 per cent terpene alcohols. Wood turpentine, obtained from waste woodchips or sawdust, contains about 80 per cent α-pinene, traces of β-pinene, 15 per cent monocyclic terpenes, and 1.5 per cent terpene alcohols. Sulfate turpentine is a by-product of the kraft paper industry. It varies in composition since the less stable β-pinene is affected by the pulping process. The turpentines are extensively used in the manufacture of surface coatings, as a solvent for oils, fats and resins, in shoe creams, oil lacquers, and polishing waxes, and in the preparation of chemical products such as camphor, linoleum, ink, soap, and resins.

2. Physical and Chemical Properties

The physical and chemical properties of gum turpentine are given in Table 1.

3. Determination in the Atmosphere

The methods applicable to hydrocarbons in general may be used for turpentine, namely, the combustible gas indicator, the interferometer, the spectrometer, gas chromatography, infrared absorption, and combustion to carbon dioxide. Turpentine hydrocarbons may also be determined by absorbing in suitable scrubbing media and developing a color with a solution of vanillin in hydrochloric acid. Another method with a sensitivity of approximately 0.02 mg. is based on color development in concentrated sulfuric acid, which also serves as the collection fluid.

1 mg./liter ⇌ approximately 179 p.p.m. and 1 p.p.m. ⇌ approximately 5.57 mg./cu. meter at 25°C., 760 mm. Hg.

4. Physiological Response

Acute Toxicity. Lehman[3] reported immediate mucous membrane irritation, particularly of the eyes, and slight convulsions in cats exposed to 540 to 720 p.p.m. (3 to 4 mg./liter) for a few hours. A concentration of 1440 p.p.m. (8 mg./liter) produced disturbances in equilibrium, tonic convulsions in 30 to 60 minutes, and paralysis in 150 to 180 minutes. Death resulted from a 45- to 60-minute exposure

[3] K. B. Lehman, *Arch. Hyg.*, **83**, 239 (1914).

to 2880 p.p.m. (16 mg./liter) of turpentine vapors. Men exposed to concentrations of 720 to 1100 p.p.m. complain of eye irritation, headache, dizziness, nausea, chest pain, and visual disturbances.[4] Albuminuria and hematuria have been reported in men exposed to turpentine vapors.[5] This laboratory evidence of renal injury is absent several weeks after termination of exposure to turpentine.

Chronic Toxicity. No evidence of injury was found in cats exposed to 155 to 180 p.p.m. turpentine vapors 3.5 hours/day for 8 days.[3] Smyth and Smyth[6] found no injury in guinea pigs after prolonged exposure to 750 p.p.m. No adverse effects on the blood have been reported due to exposure to turpentine.

There is some evidence that some individuals develop a hypersensitivity to turpentine after prolonged or repeated exposure to it. Wood turpentine is more irritating to the skin than gum turpentine, according to McCord.[7] This is attributed to the impurities present, which include formic acid, formaldehyde, and phenols.

Absorption and Excretion. The principal portal of entry of turpentine in industrial use is inhalation. Part of it is eliminated unchanged in the expired air and in the urine, but most of it is metabolized and excreted in the urine conjugated with glucuronic acid.

Tests Indicating Exposure. In addition to air analysis for turpentine hydrocarbons, body fluids may also be analyzed after separation from interfering biological chemicals. The ultimate analysis would require a method with sufficient sensitivity to detect the small amounts that would be present. Total glucuronic acid in the urine may also give some indication of the degree of exposure, but cognizance must be taken of possible other sources of glucuronides in the urine. The odor of the urine of men exposed to turpentine has been described as reminiscent of the odor of violets.

5. Hygienic Standard of Permissible Exposure

The threshold limit for turpentine has been set at 100 p.p.m.

6. Odor and Warning Properties

The odor of 200 p.p.m. turpentine in air is readily noticeable; this concentration is moderately irritating to the eyes and mucous membranes of the nasal passages.

DECALIN, $C_{10}H_{18}$ (Decahydronaphthalene)

1. Sources, Uses, and Industrial Exposures

Decalin is produced by the catalytic hydrogenation of naphthalene. Its principal uses are as a solvent for oils, fats, resins, and waxes and as a substitute for

[4] K. B. Lehman and F. Flury, *Toxicology and Hygiene of Industrial Solvents.* Trans. by E. King and H. F. Smyth, Jr., Williams & Wilkins, Baltimore, 1943.

[5] E. M. Chapman, *J. Ind. Hyg. Toxicol.,* **23**, 277 (1941).

[6] H. F. Smyth and H. F. Smyth, Jr., *J. Ind. Hyg.,* **10**, 261 (1928).

[7] C. P. McCord, *J. Am. Med. Assoc.,* **86**, 1978 (1926).

turpentine in oil paints and lacquers. It is used more extensively for these applications in countries other than the United States (Europe, England, and the Latin-American nations).

2. Physical and Chemical Properties

The physical and chemical properties of decalin are given in Table 1.

3. Determination in the Atmosphere

The general methods for hydrocarbon analysis may be used for the determination of decalin in air, namely, the interferometer, combustion apparatus, infrared absorption procedures, and gas chromatography.

1 mg./liter ≈ 177 p.p.m. and 1 p.p.m. ≈ 5.65 mg./cu. meter at 25°C., 760 mm. Hg.

4. Physiological Response

Toxicity. Dermatitis has been reported, but no serious systemic poisoning, in painters working with decalin.[1]

Three guinea pigs were exposed by Cardani[8] to vapors of decalin at a concentration of 319 p.p.m. (1.8 mg./liter) for 8 hours a day. One died after 1 day, the second after 21 days, and the third after 23 days of exposure. On gross and microscopic examination of the tissues of these animals the principal positive pathological findings were lung congestion, kidney and liver injury. No significant deviations were found in the peripheral blood. Guinea pigs died in 10 days following 2 daily applications of decalin to 6 cm.[2] of shaved skin. Tissue injury was the same as described following inhalation of decalin vapors.

Absorption and Excretion. Decalin is metabolized and excreted in the urine conjugated with glucuronic acid. The urine of guinea pigs dosed orally with decalin has a brownish green color. This color was not found in guinea pigs exposed to decalin by inhalation or by percutaneous administration. Colored urine has been reported in workers exposed to a mixture of decalin and tetralin.

Tests Indicating Exposure. Determination of the concentration of decalin in the atmosphere and in body fluids, such as blood and urine, may be made by suitable sensitive methods. Some indication may be obtained by urine analysis for decalin metabolites and glucuronic acid. This test is subject to the limitations described earlier.

5. Hygienic Standard of Permissible Exposure

The threshold limit has not been established for decalin. A value of 25 p.p.m. appears to be a reasonable figure, based on the toxicity studies that have been published.

[3] A. Cardani, *Med. lavoro,* **33,** No. 10, 169 (1942).

6. Odor and Warning Properties

Quantitative studies on the relationship of air concentrations of decalin and odor intensity and mucous membrane irritation have not been published. Qualitatively the odor resembles that of tetralin.

DICYCLOPENTADIENE, $C_{10}H_{12}$

1. Sources, Uses, and Industrial Exposure

Dicyclopentadiene is produced commercially by recovery from hydrocarbon streams originating from high temperature cracking of petroleum fractions, and as a by-product of the coke-oven industry. It is formed from the spontaneous dimerization of cyclopentadiene which is used as a starting material for synthesis in the chemical industry.

2. Physical and Chemical Properties

The physical and chemical properties of dicyclopentadiene are given in Table 1.

3. Determination in the Atmosphere

There are no known specific methods published for the determination of dicyclopentadiene. The general procedures used for hydrocarbons and olefins may be used, viz. combustible gas indicator, the interferometer, spectrometer, infrared absorption, combustion to carbon dioxide and gas chromatography.

1 mg./liter $\approx$ 185.2 p.p.m. and 1 p.p.m. $\approx$ 5.40 mg./cu. meter at 25°C., 760 mm. Hg.

4. Physiological Response

Acute Toxicity. Smyth and co-workers[9] reported death in 4 out of 6 rats exposed for 4 hours to 2000 p.p.m. of dicyclopentadiene and an LD_{50} of 6.72 ml./ kg. by percutaneous administration to rabbits. To ascertain the potential of dicyclopentadiene for affecting the blood-forming tissue, Gerarde[10] dosed rats subcutaneously with 1.0 ml. of dicyclopentadiene/kg. of body weight daily for 14 days. Another group of rats received a single dose of 5.0 ml./kg. of body weight subcutaneously and was sacrificed 96 hours after dosing. A leucocytosis was found in both groups of animals dosed with dicyclopentadiene. In contrast, benzene-dosed animals (serving as positive controls) developed severe leukopenias (500 to 2000 cells/mm.³) after receiving the equivalent dose of benzene. The principal positive pathological findings in rats dosed orally with dicyclopentadiene were generalized congestion, hyperemia and focal hemorrhage in many tissues, including kidney, intestine, stomach, bladder, and particularly the lungs. These effects are typical of irritating hydrocarbons when administered orally in large doses.

[9] H. F. Smyth, Jr., C. P. Carpenter, C. S. Weil, and U. C. Pozzani, *A.M.A. Arch. Ind. Hyg. Occupational Med.,* **10,** 66 (1954).

[10] H. W. Gerarde, personal investigations. To be published.

Chronic Toxicity. There are no published data on the effects of repeated exposure to vapors of dicyclopentadiene.

Absorption and Excretion. Dicyclopentadiene is absorbed by inhalation and partly eliminated unchanged in the lung and in the urine. The largest portion of the hydrocarbon absorbed in the blood would probably be metabolized and excreted in the urine. Because chemically it is a terpenelike molecule, dicyclopentadiene is probably excreted conjugated with glucuronic acid.

Tests Indicating Exposure. Determination of dicyclopentadiene in the air and possibly in the body fluids (blood and urine) in addition to the detection of metabolites conjugated with glucuronic acid may be used as measures of exposure.

5. Hygienic Standard of Permissible Exposure

The threshold limit for dicyclopentadiene has not been established. Data are needed both on human response to various atmospheric concentrations of dicyclopentadiene and on repeated inhalation studies in animals. Based on the limited toxicity data available and careful extrapolation from similar chemicals, a value of 100 p.p.m. seems reasonable.

6. Odor and Warning Properties

No quantitative studies have been published on the relationship of the atmospheric concentration of dicyclopentadiene and odor intensity or mucous membrane irritation. It has an odor which is not distinguishable from the odor of other hydrocarbons with closely related chemical structure, such as the terpenes.

CHAPTER XXX

The Aromatic Hydrocarbons

HORACE W. GERARDE, M.D., PhD

I. General Considerations

A. NOMENCLATURE, SOURCES, AND USES OF AROMATIC HYDROCARBONS

The aromatic hydrocarbons contain one or more benzene rings. They are classified into three principal groups, depending on the number of benzene rings and the type of linkage between the rings in the molecule. These groups are: (*1*) benzene and its aliphatic and alicyclic derivatives, (*2*) polyphenyls—two or more noncondensed rings, and (*3*) polynuclear—two or more condensed rings.

From the industrial hygiene standpoint, the most important aromatic hydrocarbons belong to the first group, consisting of benzene and its aliphatic and alicyclic derivatives. In general, these compounds are liquids with a significant vapor pressure at room temperature which, combined with their toxicity, may present potential working hazards. Included in this group are benzene (benzol), toluene, ethylbenzene, styrene, the xylenes, cumene, *p-tert*-butyltoluene, and tetralin.

Benzene and its alkyl derivatives are obtained as by-products of the coke-oven industry and the petroleum industry. From coke-oven operations they are recovered from the gases and the coal tars. As petrochemicals they are separated from crude oil by fractionation, distillation, or solvent extraction, or converted chemically by dehydrogenation of naphthene fractions, alkylation of benzene with olefins, or produced from paraffins by catalytic cyclization and aromatization.

Benzene is extensively used as a solvent in the chemical and drug industry, as a starting material and intermediate in the synthesis of numerous chemicals, and as a constituent of motor fuels. Toluene and the xylenes are used as solvents for synthetic rubber, paint and lacquers, and are also constituents of motor fuels. *p*-Xylene is the intermediate in the synthesis of synthetic fibers. Toluene is the starting material in the manufacture of the explosive trinitrotoluene. Ethylbenzene is used principally as the source of styrene, which is the constituent of polymers such as synthetic rubber, plastics, and packaging films. Cumene and *p-tert*-butyltoluene are used for synthesis in the chemical industry. Cumene is also a constituent of certain aromatic solvents and motor fuels. Tetralin is prepared by the hydrogenation of naphthalene and is used as a constituent of solvents, motor fuels, and lubricants.

The polyphenyls of industrial importance are diphenyl (biphenyl) and the terphenyls (triphenyls). Diphenyl is produced by thermal dehydrogenation of benzene; terphenyls are by-products in the manufacture of diphenyl. Diphenyl is one of the most thermally stable of known organic compounds. This forms the basis for its use either alone or mixed with diphenyl oxide as a low-pressure, high-temperature, heat-transfer medium. Dowtherm A is a eutectic mixture of 73.5 per cent diphenyl ether (diphenyl oxide) and 26.5 per cent diphenyl. The terphenyls are also used as heat-storage and heat-transfer agents. Because of their low vapor pressure and low order of toxicity these hydrocarbons do not present industrial hygiene hazards and do not warrant detailed discussion in this chapter.

Naphthalene is the most extensively used polynuclear hydrocarbon in industry. It is the most abundant single constituent of coal tar and is available from this source in technical grades at a low price. It is important as a starting material and intermediate in the synthesis of numerous chemicals and is used also as a moth repellent and insecticide.

B. TOXICOLOGY OF THE AROMATIC HYDROCARBONS

The liquid aromatic hydrocarbons are primary irritants which on repeated or prolonged contact with the skin will cause dermatitis due to their dehydrating and defatting action. Contact of the liquid hydrocarbons with lung tissue (aspiration) will cause severe pulmonary edema, pneumonitis, and hemorrhage. Because of their low surface tension a small volume of liquid will cover a large area so that extensive lung injury can result from the aspiration of a few cubic centimeters of a liquid hydrocarbon. The vapors are more irritating to the mucous membranes than equivalent concentrations of the aliphatic and alicyclic hydrocarbons.

Systemic injury can result from the inhalation of vapors of the aromatic hydrocarbons. It is well established that benzene (benzol) is an insidious toxicant which has a specific destructive effect on the blood-forming tissue. The preponderance of the evidence from animal experimentation indicates that the alkyl derivatives do not possess this property and that benzene appears to be unique among hydrocarbons as a myelotoxicant. Because the name "alkylbenzenes" suggests to the uninformed that these compounds are similar to benzene, the term "phenylalkanes" has been suggested to dissociate them from benzene.[1] Pharmacologically, the phenylalkanes can be classified with the central nervous system depressants.

II. Specific Aromatic Hydrocarbons

BENZENE, C_6H_6 (Benzol)

1. *Sources, Uses, and Industrial Exposures*

Benzene (not to be confused with petroleum benzin, a hydrocarbon mixture discussed in Chapter 28) is obtained as a by-product of the coke-oven industry

---

[1] H. W. Gerarde, *A.M.A. Arch. Ind. Health,* **19,** 403 (1959).

and the petroleum industry. From coke-oven operations it is recovered from the gases and coal tar. As a petrochemical it is obtained by dehydrogenation of naphthene fractions or by cyclization and aromatization of paraffin hydrocarbons. Benzene is used extensively as a solvent in the chemical and drug industries, as a starting material and intermediate in the synthesis of numerous chemicals, and as a constituent of motor fuels.

2. Physical and Chemical Properties

See Table 1.

3. Determination in the Atmosphere

In the presence of other gases the most satisfactory method of determining benzene is the nitration procedure described by Schrenk and co-workers.[2] The interferometer (Vol. I, page 179), a properly calibrated and serviced benzol indicator, and the M.S.A. aromatic hydrocarbon detector may be used for air analysis.

1 mg./liter ≎ 313 p.p.m. and 1 p.p.m. ≎ 3.19 mg./cu. meter at 25°C., 760 mm. Hg.

4. Physiological Response

Acute Toxicity. Acute poisoning by benzene is due to its narcotic action and in many respects resembles that caused by other low molecular weight petroleum hydrocarbons. Flury[3] gives the following figures for a single exposure for man: 3000 p.p.m.—endurable for 0.5 to 1 hour; 7500 p.p.m.—dangerous after 0.5 to 1 hour; 20,000 p.p.m.—fatal after 5 to 10 minutes. The inhalation of a high concentration of benzene may cause exhilaration followed by drowsiness, fatigue, vertigo, nausea, and headache. With higher concentrations or longer exposure times, convulsions followed by paralysis and loss of consciousness may result. An initially rapid respiration soon diminishes in rate and circulatory collapse may follow. Death may ensue quickly from respiratory paralysis after severe exposure. Dautrebande[4] found that dogs inhaling benzene initially developed hypertension. This was soon followed by paralysis of the vasomotor system due to the effect of benzene on the smooth muscle of the blood vessels. High concentrations of benzene are irritating to the mucous membranes of the eyes, nose, and respiratory tract. Liquid benzene is irritating to the skin and direct contact of liquid benzene with the lung (aspiration) will cause severe pulmonary edema and hemorrhage which may be fatal depending on the volume aspirated.

Brief exposures to high concentrations in the air may cause chronic benzene intoxication. It is not good practice to encourage or to condone exposure to 100 p.p.m. or more for even brief periods without suitable respiratory protection. A

[2] H. H. Schrenk, S. J. Pearce, and W. P. Yant, *U. S. Bur. Mines Rept. Invest.* No. **3287** (1935).

[3] F. Flury, *Arch. exptl. Pathol. Pharmakol.,* **138,** 65 (1928).

[4] L. Dautrebande, *Arch. intern. pharmacodynamie,* **44,** 394 (1933).

TABLE

Physical and Chemical Properties of

| Property | Benzene | Toluene | Xylenes | | |
			o-	m-	p-
Mol. wt.	78.11	92.13	106.16	106.16	106.16
B.p., °C.	80.103	110.623	144.414	139.102	138.348
F.p., °C.	5.533	−94.991	−25.175	−47.872	13.263
n_D (25°C.)	1.49790	1.49405	1.50282	1.49455	1.49319
Vapor pressure	100 mm. (26.085°C.)	30 mm. (26.04°C.)	10 mm. (32.11°C.)	10 mm. (28.26°C.)	10 mm. (27.30°C.)
Density (25°/4°C.)	0.87368	0.86220	0.87583	0.85985	0.85666
Vapor density (air = 1)	2.7	3.2	3.7	3.7	3.7
Per cent in satd. air, 760 mm.	13.15 (26°C.)	3.94 (26°C.)	1.32 (32°C.)	1.32 (28.3°C.)	1.32 (27.3°C.)
Density of satd. vapor—air mixture at 760 mm. (air = 1)	1.22 (26°C.)	1.09 (26°C.)	1.03 (32°C.)	1.03 (28°C.)	1.03 (27°C.)
Specific dispersion	189.6	184.4	180.3	181.1	181.8
Solubility in water (20°C.)	0.082 g. per 100 ml. (22°C.)	0.047 g. per 100 ml. (16°C.)	Insoluble	Insoluble	Insoluble
Solubility in ethyl alcohol	Miscible	Miscible	Very soluble	Very soluble	Very soluble
Solubility in ethyl ether	Miscible	Miscible	Very soluble	Very soluble	Very soluble
Flammable limits (% by vol. in air)	1.35–7.90	1.17–7.10	1.09–6.40	1.09–6.40	1.08–6.60
Flash point (closed cup)	12°F.	40°F.	63°F.	77°F. (t.o.c.)	77°F.

brief exposure may be warranted in case of an emergency, but should not be permitted in addition to the routine exposure to the threshold limit of 25 p.p.m.

Chronic Toxicity. The chronic effects of inhaling small amounts of benzene over a prolonged period of time are of the greatest importance in the industrial use of this chemical. The symptoms, signs and blood changes in chronic benzene poisoning may not invariably present a typical picture of severe anemia and leukopenia. Greenburg and his associates[5] did complete blood studies on 102 workmen exposed to benzene in the rotogravure industry. A positive diagnosis of

[5] L. Greenberg, M. R. Mayers, L. Goldwater, and A. R. Smith, *J. Ind. Hyg. Toxicol.,* **21,** 395 (1939).

1

Some Aromatic Hydrocarbons

	Ethyl-benzene	Cumene	Styrene	p-tert-Butyl-toluene	Naphthalene	Tetralin
	106.16	120.19	104.14	148.24	128.16	132.20
	136.187	152.393	145.2	192.8	217.9	207.2
	−94.950	−96.028	−30.628		80.22	−30.0
	1.49319	1.48874	1.5441	1.4892	1.58218 (at 99.6°)	1.54614 (at 20.2°)
	10 mm. (25.90°C.)	10 mm. (38.33°C.)	4.3 mm. (15°C.)	0.65 mm. (25°C.)	Approx. 0.082 mm. (25°C.)	
	0.86258	0.85748	0.9021	0.8575	1.145 (20°/4°C.)	0.971 (20°/4°C.)
	3.7	4.2	3.6		4.4	4.6
	1.32 (26°C.)	1.32 (38.3°C.)	0.57 (15°C.)		0.01 (25°C.)	
	1.03 (26°C.)	1.03 (38°C.)	1.02 (15°C.)	1.00479 (25°C.)	1.00 (25°C.)	
	174.6	166.2	265			
	0.014 g. per 100 ml. (15°C.)	Insoluble	0.31 g. per 100 ml. (25°C.)	Insoluble	3 mg. per 100 ml.	Insoluble
	Miscible	Soluble	Miscible	Miscible	4.2 g. per 100 ml. (20°C.)	Very soluble
	Miscible	Soluble	Miscible	Miscible	Very soluble	Very soluble
	0.99–6.70	0.88–6.50	1.10–6.10		0.88–5.9	
	63°F	102°F.	86°F.	155°F.(t.o.c.)	176°F.	ca. 172°F.

chronic benzene poisoning varying in severity was made in 74 men in this group. It is significant that in this group were found men with clinical pictures of benzene poisoning whose blood was normal, and serious blood abnormalities in the complete absence of signs or symptoms of benzene poisoning. A few men developed blood abnormalities in a 60-day period after removal from the source of benzene exposure. The percentages of the positive diagnoses of benzene poisoning detected by any single laboratory blood test in this group are given in Table 2. The bone marrow in chronic benzene poisoning may appear normal, aplastic, or hyperplastic. Signs and symptoms may include headache, dizziness, fatigue, loss of appetite, irritability, nervousness, nosebleed, and other hemorrhagic manifestations.

TABLE 2

Detection of Benzene Intoxication by Blood Tests

Test	Per cent positive
Mean corpuscular volume of erythrocytes (above 94 cu.μ)	64.9
Erythrocyte count (below 4.5 million)	63.5
Platelet count (below 100,000)	41.9
Hemoglobin (less than 13 g. per 100 ml. blood)	40.5
Leucocyte count (below 5000)	40.5

Absorption and Excretion. Industrial benzene poisoning results almost exclusively from the inhalation of benzene in the atmosphere. Saturation of the circulating blood occurs rapidly, with 70 to 80 per cent saturation being reached within 30 minutes (see Vol. I, pages 156–158, and Figs. 2 and 3 for absorption and elimination curves for benzene). However, it requires several days for complete saturation of the body tissues and fluids. Benzene is rather insoluble in blood and the equilibrium constant or coefficient of distribution of benzene between room air and the blood of dogs breathing this air is 6.58, that is, milligrams of benzene per liter of blood/milligrams of benzene per liter of air. This may also be stated as follows: at equilibrium, the average concentration of benzene in the blood is 2.1 mg./liter for each 100 p.p.m. of benzene in the inhaled air. As the blood circulates, however, it comes to equilibrium with the tissues and the fatty tissues store quantities of benzene in this manner.[6] Elimination involves the same process in reverse because most of the benzene is eliminated through the lungs, being picked up by the blood in the capillaries and carried to the lungs, where equilibrium with the alveolar air is rapidly established, as previously explained.

Small amounts of benzene are absorbed through the skin wherever the liquid touches the skin. It is not probable that systemic poisoning can arise from immersing the hands in benzene. However, well-recognized results of skin contact with benzene include skin defatting with erythema, dry scaling, and even secondary infections. In this respect, benzene is not considered to be as severe as toluene, xylene, and many other solvents, including certain of the petroleum distillates.

Some benzene is eliminated unchanged in the urine. Some is oxidized in the body to phenols and diphenols which, in turn, are conjugated in the liver with sulfate ions and excreted in the urine, thus increasing the ratio of ethereal sulfates to total sulfates in the urine. This is the basis of a test to determine the severity of the over-all daily exposure to benzene.

Tests Indicating Exposure. The exposure of an individual at any particular moment may be rather accurately determined by analysis of the benzene in a sample of arterial blood[7] and computation of the corresponding concentration in the air from the distribution coefficient of benzene between blood and room air.[6]

[6] H. H. Schrenk, W. P. Yant, S. J. Pearce, F. A. Patty, and R. R. Sayers, *J. Ind. Hyg. Toxicol.,* **23**, 20 (1941).

[7] S. J. Pearce, H. H. Schrenk, and W. P. Yant, *U. S. Bur. Mines Rept. Invest.* No. **3302** (1936).

The over-all or integrated exposure may be best arrived at by determining the ratio of inorganic to organic sulfates in a sample of urine collected near the end of the day's exposure, or within one hour after cessation of the day's exposure.[8,9] The ratio of inorganic sulfates to total sulfates in the urine is normally 85 per cent or above. Upon exposure to benzene this ratio is decreased, and the decrease is related quantitatively to the severity of the exposure to the point where nearly all sulfates are eliminated as organic sulfates. Ratios of less than 70 per cent inorganic are abnormal and may be indicative of benzene exposure and warrant investigation of exposure sources with a view toward correction. Ratios of 60 per cent or less inorganic indicate dangerous exposures warranting immediate correction. Concurrent exposure to carbon tetrachloride, or any other material having an adverse effect upon the liver, may decrease the sulfate response so that ratios on the order of 70 per cent inorganic would indicate exposures of great significance. As pointed out in the original paper,[8] the sulfate test must be made while the worker is on the job, and it is not to be considered a method for diagnosing benzene poisoning, because the changes are merely indicative of exposure, not of poisoning nor of damage. It is therein that the great preventive value of this test lies; the indication occurs in advance of any demonstrable harmful effects, but forewarns of their coming if exposures are not reduced. However, in complete disregard of these well-known facts, case histories of benzene poisoning too often carry enlightening comments to the effect that from a prognostic viewpoint repeated blood studies on the hospitalized patient are of greater value than concurrent repeated urinary sulfate determinations!

It seems logical to make periodic hematological studies of the blood of all workers coming in contact with any benzene vapors, perhaps every 30 days, and urine sulfate tests, perhaps weekly. Air analyses for benzene in suspected atmospheres should be made frequently. If these three control measures are consistently performed and heeded, as a guide for elimination of significant exposures, it is believed that no fear need be entertained regarding the careful use of benzene in industrial processes.

5. Hygienic Standard of Permissible Exposure

The threshold limit for benzene has been established at 25 p.p.m.

6. Odor and Warning Properties

Benzene has a distinctive odor which should be familiar to all industrial hygienists. The warning properties, however, are inadequate since 100 p.p.m. has an irritation rating of zero and an odor intensity between 1 and 2.

[8] W. P. Yant, H. H. Schrenk, R. R. Sayers, A. A. Horvath, and W. H. Reinhart, J. Ind. Hyg. Toxicol., 18, 69 (1936).

[9] W. P. Yant, H. H. Schrenk, and F. A. Patty, J. Ind. Hyg. Toxicol., 18, 349 (1936).

TOLUENE, $C_6H_5CH_3$ (Toluol, Methylbenzene)

1. Source, Uses, and Industrial Exposure

Toluene is obtained as a by-product of the coke-oven industry and is also a petrochemical. From coke-oven operations it is recovered from the gases and coal tar. In the petroleum industry it is obtained by dehydrogenation of naphthene fractions or by cyclization and aromatization of paraffin hydrocarbons. Toluene is used extensively as a solvent in the chemical and drug industry, as a starting material and intermediate for synthesis of numerous chemicals, as a constituent of motor fuels, and as a thinner for paints, varnishes, enamels, and lacquers.

2. Physical and Chemical Properties

The physical and chemical properties of toluene are given in Table 1.

3. Determination in the Atmosphere

The M.S.A. Aromatic Hydrocarbon Detector and the benzol indicator can be used if properly calibrated for toluene. The indicator registers total combustibles and vapors other than toluene must be considered. The interferometer can be used to determine toluene alone, or in vapor mixtures where the nature of the components and their relative proportions are known. Colorimetric chemical methods have greater sensitivity, such as the nitration procedure developed by Yant and his associates.[10] Toluene may be detected by absorption of the vapor in concentrated sulfuric acid containing formaldehyde. This method is not specific for toluene. Fabre and co-workers[11] have described a method for determining toluene and benzene in air.

1 ml./liter $\backsim$ 266 p.p.m. and 1 p.p.m. $\backsim$ 3.76 mg./cu. meter at 25°C., 760 mm. Hg.

4. Physiological Response

Acute Toxicity. Toluene is a more powerful narcotic and is more toxic acutely than benzene. Smyth and Smyth[12] found animals severely affected, but no deaths resulted after 18 daily 4-hour exposures to 1250 p.p.m. toluene, whereas a few daily 4-hour exposures to 4000 p.p.m. caused fatalities in the exposed animals. A comparison of the acute toxicities of certain aromatic hydrocarbons for mice can be made from the values given by Lazarew[13] shown in Table 3. Some of the early toxicity studies on toluene and xylene are unreliable because of the contamination of these hydrocarbons with benzene. Toluene of high purity is now available commercially.

[10] W. P. Yant, S. J. Pearce, and H. H. Schrenk, *U. S. Bur. Mines Rept. Invest.* No. **3323** (1936).

[11] R. Fabre, R. Truhaut, and M. Peron, *Ann. pharm. franç.,* **8,** 613 (1950).

[12] H. F. Smyth and H. F. Smyth, Jr., *J. Ind. Hyg.,* **10,** 261 (1928).

[13] N. W. Lazarew, *Arch. exptl. Pathol. Pharmakol.,* **143,** 223 (1929).

TABLE 3

Acute Toxicity of Some Aromatic Hydrocarbons—Mice

	Minimum concentrations of vapors that cause			
	Prostration		Death	
Compound	Mg./liter	Vol. %[a]	Mg./liter	Vol. %[a]
Benzene	15	0.47	45	1.41
Toluene	10–12	0.27–0.32	30–45	0.80–1.20
o-Xylene	15–20	0.35–0.46	30	0.69
m-Xylene	10–15	0.23–0.35	50	1.15
p-Xylene	10	0.23	15–35	0.35–0.81
Ethylbenzene	15	0.35	45	1.04

[a] Vol. % × 10,000 = parts per million.

Controlled exposures of human beings to concentrations of 50 to 800 p.p.m. indicate that exposure to a concentration of 200 p.p.m. for a period of 8 hours produces mild fatigue, weakness, confusion, and paresthesias of the skin.[14] The fatigue persisted for hours and moderate insomnia and restlessness resulted. The same symptoms were more pronounced with 300 p.p.m. With 400 p.p.m. mental confusion was added to the list of symptoms. With 600 p.p.m. extreme fatigue, mental confusion, exhilaration, nausea, headache, and dizziness resulted by the end of 3 hours. After 8 hours, the mental confusion, weakness, dizziness, and nausea were pronounced. The pupils were dilated and accommodation to light was impaired. The subjects lost coordination and had a staggering gait. These effects persisted for hours and the subjects complained of insomnia. Fatigue and nervousness were still present on the second day. With 800 p.p.m. the same symptoms were more pronounced and after effects, characterized by severe nervousness, muscular fatigue, and insomnia, lasted for several days. Exposures to 50 and 100 p.p.m. failed to present distinct symptoms or after effects.

Chronic Toxicity. The experiments described above, totaling 15 exposures over a period of 3 months to concentrations of toluene ranging from 50 to 800 p.p.m., did not cause definite changes in the white cell blood count. Greenburg and his co-workers[15] studied a group of 106 painters in an airplane factory. These men were exposed to concentrations of toluene ranging from 100 to 800 p.p.m., although the average exposures of a few men were higher: 900, 1000, and 1100 p.p.m. The paints and finishes contained relatively small proportions of other solvents including acetone, ethyl alcohol, ethyl acetate, butyl alcohol, butyl acetate, petroleum naphtha, and xylene. These finishes also contained zinc chromate, magnesium silicate, titanium oxide, iron pigments, zinc oxide, nitrocellulose, and resins. Some of the brush paints contained aluminum, cadmium, and barium pig-

[14] W. F. von Oettingen, P. A. Neal, and D. D. Donahue, *J. Am. Med. Assoc.*, **118**, 579 (1942).

[15] L. Greenberg, M. R. Mayers, H. Hermann, and S. Moskowitz, *J. Am. Med. Assoc.*, **118**, 573 (1942).

ments. The toluene exposure exceeded that of all other vapors combined, and only the toluene was quantitated in the study. The exposure period ranged from a few weeks to 5 years. None of these men experienced any symptoms throughout the study other than occasional dermatitis. Five men had perforated nasal septums (probably due to zinc chromate) and 32 were found to have enlargement of the liver (as determined by palpation) as the only physical evidence of toxic effects. The hematological findings were: erythrocytes—a small percentage of the workers slightly below normal; hemoglobin—slightly above normal; leucocytes—differential: normal, absolute count: about 20 per cent of the exposed workers were above 5000 as compared with 7.7 per cent in a control group; mean corpuscular volume of erythrocytes—23.6 per cent were above 100 cubic microns as compared with 7.2 per cent in the control group. The platelet and reticulocyte count, basophilic aggregation, sedimentation rate, coagulation time, hematocrit, erythrocyte fragility, and serum bilirubin were all reported to be within normal limits.

Absorption and Excretion. As in the case of benzene, industrial poisoning by toluene probably results only from inhalation. However, since toluene is slowly absorbed through the skin and is also irritating to the skin, it is only common sense to avoid skin contact wherever possible. Since the solubility of toluene in water and blood is low, the circulating blood rapidly comes to equilibrium with toluene vapor in the alveolar air, as is the case with benzene. Von Oettingen, Neal, and Donahue[14] found 7.3 mg. of toluene per liter of blood in men exposed to 300 p.p.m. toluene vapor in air. This corresponds to a coefficient of distribution of 6.5 (milligrams of toluene per liter of blood/milligrams of toluene per liter of room air) which is about the same as the figure for benzene.

Part of the absorbed toluene is eliminated in the exhaled breath, but a large percentage is oxidized to benzoic acid, conjugated with glycine, and excreted as hippuric acid in the urine according to the following chemical reaction:

$$C_6H_5COOH + NH_2CH_2COOH \rightarrow C_6H_5CONHCH_2COOH + H_2O$$

| Benzoic acid | Glycine | Hippuric acid | Water |

The amount of hippuric acid normally excreted in the urine varies with individuals and with the diet, but has been reported to be approximately 0.7 g. per day.[16] The amount of hippuric acid excreted in a 24-hour period by men exposed to toluene was found to be proportional to the concentration of toluene in the air.[14] Within the range of 100 to 600 p.p.m. toluene vapor in air for an 8-hour exposure, the amount of hippuric acid found in the urine was approximately 1.2 g. above normal per 100 p.p.m. (0.376 mg./liter) toluene vapor in air.

The amount of toluene absorbed and the portion excreted as hippuric acid are not known but may be roughly estimated as follows. The subjects were probably

[16] P. B. Hawk and O. Bergeim, *Practical Physiological Chemistry.* 11th ed., Blakiston, Philadelphia, 1944.

breathing at a rate of approximately 10 liters per minute or $8 \times 60 \times 10 = 4800$ liters per 8-hour exposure period. We may safely assume that only about 70 per cent of this could have been absorbed (see Vol. I, page 147) regardless of how rapidly saturation of the blood with toluene and oxidation of the toluene occurred. We then would expect as a maximum $4800 \times 0.70 \times 0.376 = 1263$ mg. of toluene to have reached the alveoli from the exposure to 100 p.p.m. The actual proportion absorbed would depend upon several factors and is unknown. However, since it has been shown that approximately 1.2 g. of hippuric acid results from such an exposure, we can compute the percentage of toluene converted to hippuric acid:

$$1 \text{ mole toluene} \rightarrow 1 \text{ mole benzoic acid} \rightarrow 1 \text{ mole hippuric acid}$$

| 92.066 g. | 179.2 g. |
| X g. | 1.2 g. |

$$92.066/179.2 = X/1.2 \qquad X = (1.2 \times 92.066)/179.2 = 0.611 \text{ g.}$$
$$(0.611 \text{ g.}/1.263 \text{ g.}) \times 100 = 48 \text{ per cent}$$

Therefore it appears that approximately 48 per cent of the available toluene (portion reaching the respiratory tissue) was converted to hippuric acid in the body and excreted in the urine. It is probable that most of the remainder was excreted unchanged in the exhaled air. Ethereal sulfates do not appear in the urine in significant amounts.

Tests Indicating Exposure. Toluene in the blood may be used as an indication of exposure if the exposure period has been sufficiently long to approach equilibrium. At equilibrium, the average toluene concentration per liter of blood is 2.4 mg. for each 100 p.p.m. toluene in the environmental air. Since each 100 p.p.m. toluene per 8-hour day produces an excretion of approximately 1.2 g. hippuric acid, the total daily excretion of hippuric acid appears to be an index of exposure in concentrations less than 800 p.p.m. It must be remembered, however, that normal dietary constituents such as certain fruits and vegetables raise the normal level of hippuric acid excretion.

5. *Hygienic Standard of Permissible Exposure*

The threshold limit for toluene has been set at 200 p.p.m.

6. *Odor and Warning Properties*

The odor intensity of 200 p.p.m. toluene is about 3 on immediate exposure, but olfactory fatigue follows rapidly.

STYRENE, $C_6H_5CH{=}CH_2$ (Vinylbenzene, Phenylethylene)

1. *Sources, Uses, and Industrial Exposure*

Styrene is produced commercially by the dehydrogenation of ethylbenzene. It also occurs naturally in the sap of styraceous trees. It is used principally as the monomer for polystyrene, synthetic rubbers, and film-packaging plastics, and as starting material and intermediate for synthesis of other chemicals.

2. *Physical and Chemical Properties*

The physical and chemical properties of styrene are given in Table 1.

3. *Determination in the Atmosphere*

Rowe, Atchison, Luce, and Adams[17] have presented techniques for the sampling and analysis of styrene by ultraviolet, infrared, and nitration methods. The choice between these methods will depend upon circumstances, as discussed in the original paper. For control purposes the interferometer, because of its rapidity and ease of manipulation, is probably the method of choice even though it is not specific. Because of the relatively high indication given by threshold concentrations of styrene, the 25-cm. interferometer is quite satisfactory. Combustible gas indicators also may be used.

1 mg./liter ⥲ 235.5 p.p.m. and 1 p.p.m. ⥲ 4.26 mg./cu. meter at 25°C., 760 mm. Hg.

4. *Physiological Response*

Acute Toxicity. Under ordinary room conditions styrene does not vaporize sufficiently to reach vapor concentrations that kill animals (rats and guinea pigs) in a few minutes;[18] 10,000 p.p.m. was dangerous to life in 30 to 60 minutes, 2500 p.p.m. was dangerous to life in 8 hours, while 1300 p.p.m. was the highest amount causing no serious systemic disturbance in 8 hours. All animals exposed to these concentrations had evidence of eye and nasal irritation. Animals exposed to 2500 p.p.m. ·or greater showed varying degrees of weakness and stupor, followed by incoordination, tremors, and unconsciousness. Unconsciousness occurred in 10 hours with 2500 p.p.m., in 1 hour with 5000 p.p.m., and in a few minutes at 10,000 p.p.m.

Chronic Toxicity. Rats exposed to 1300 p.p.m. styrene, 7 to 8 hours per day, 5 days a week, for 6 months had definite signs of eye and nasal irritation and appeared unkempt, although they made a normal gain in weight and presented no significant microscopic tissue changes or changes in the blood picture.[18] Twelve rabbits exposed to 1300 p.p.m. for a similar period presented, with the exception of one unexplained death, a similar result. Of nearly 100 guinea pigs exposed to this concentration, 10 per cent died from lung irritation within a few exposures, while the remainder survived the 6-month exposure period with no significant gross or microscopic findings. When another group of guinea pigs was similarly

[17] V. K. Rowe, G. J. Atchison, E. N. Luce, and E. M. Adams, *J. Ind. Hyg. Toxicol.*, **25**, 348 (1943).

[18] H. C. Spencer, D. D. Irish, E. M. Adams, and V. K. Rowe, *J. Ind. Hyg. Toxicol.*, **24**, 295 (1942).

exposed to 650 p.p.m. styrene, the weight gain and gross appearance of the exposed animals were similar in all respects to the controls.

Absorption and Excretion. Absorption is mainly through the respiratory tract, and in both animals[18] and man[19] styrene is metabolized to benzoic acid, conjugated with glycine, and excreted in the urine as hippuric acid. From 50 to 90 per cent of orally administered styrene has been recovered as hippuric acid in the urine. As long as styrene is present in the circulating blood, some of it is excreted in the exhaled air.

Tests Indicating Exposure. In addition to air analyses, tests for styrene can be conducted on blood and urine of individuals exposed. This requires separation of the hydrocarbon from the body fluid chemicals which might interfere, and sufficient sensitivity to detect small amounts of styrene. Hippuric acid analyses may be carried out on the urine. This is subject to the limitations described in the test indicating exposure to toluene.

5. *Hygienic Standard of Permissible Exposure*

The threshold limit for styrene has been set at 100 p.p.m.

6. *Odor and Warning Properties*

Styrene vapor in concentrations of 200 to 400 p.p.m. has a transient irritating effect on the eyes and mucous membranes of the nose and an odor intensity of 3 to 4 initially which diminishes to 1 to 2.

ETHYLBENZENE, $C_6H_5(C_2H_5)$

1. *Source, Uses, and Industrial Exposures*

Ethylbenzene is a petrochemical prepared by dehydrogenation of naphthene fractions, alkylation of benzene, or from paraffins by catalytic cyclization and aromatization. It is used principally as the source of styrene, but is also found in hydrocarbon mixtures used as motor fuels and solvents.

2. *Physical and Chemical Properties*

The physical and chemical properties of ethylbenzene are given in Table 1.

3. *Determination in the Atmosphere*

In general, the procedures described for xylene may be used. Gas chromatography may permit separation of ethylbenzene from other hydrocarbons which give the general chemical reactions of ethylbenzene.

1 mg./liter $\rightleftharpoons$ 230 p.p.m. and 1 p.p.m. $\rightleftharpoons$ 4.35 mg./cu. meter at 25°C., 760 mm. Hg.

[19] C. P. Carpenter, C. B. Shaffer, C. S. Weil, and H. F. Smyth, Jr., *J. Ind. Hyg. Toxicol.*, **26,** 69 (1944).

4. *Physiological Response*

Acute Toxicity. The acute effects in mice are greater than those of benzene and comparable with toluene and *m*-xylene (see Table 3). The acute effects in guinea pigs, according to Yant, Schrenk, Waite and Patty,[20] are given in Table 4. Death in guinea pigs appeared to result primarily from effects on the central nervous system.

TABLE 4

Acute Effects of Exposure of Guinea Pigs to Ethylbenzene Vapor

Effects	Concentration, per cent by volume
Kills in a few minutes	a
Dangerous to life in 30 to 60 minutes	1.0
Dangerous to life after several hours	0.5
Maximum amount for one hour without serious symptoms	0.3
Maximum amount for several hours without serious disturbance	0.1

ᵃ Not produced by 1 per cent, the highest concentration obtained in a closed chamber by extended recirculation of air at 23°C. over wicks wet with ethylbenzene.

Men exposed to 1000 p.p.m. ethylbenzene experienced eye irritation which rapidly diminished in intensity on continued exposure. A concentration of 2000 p.p.m. caused immediate, severe eye irritation, lacrimation, and irritation of the mucous membranes of the nose. The irritation diminished with continued exposure but dizziness became apparent in 6 minutes, when the exposure was terminated. Exposure to a concentration of 5000 p.p.m. ethylbenzene causes intolerable irritation of the eyes and mucous membranes of the nose.

Chronic Toxicity. Wolf and co-workers[21] exposed rats, rabbits, guinea pigs, and monkeys to concentrations of ethylbenzene from 400 p.p.m. to 2200 p.p.m., 7 to 8 hours per day, 5 days a week, for as long as 6 months. The guinea pigs, rabbits, and monkeys were not affected, as judged by any of the criteria of injury, which included hematological study. A slight increase in the average weights of the kidneys and livers was found in the rats exposed to 400 p.p.m. for 186 days. Gerarde[22] dosed rats subcutaneously with 1 ml. of ethylbenzene/kg. of body weight daily for 2 weeks and found no decrease in the total femoral marrow nucleated cell count. These animals developed a leucocytosis instead of the severe leucopenias found in benzene-dosed animals which served as positive controls.

Absorption and Excretion. Absorption is chiefly by inhalation. A small proportion of the ethylbenzene that gets into the blood stream is exhaled un-

²⁰ W. P. Yant, H. H. Schrenk, C. P. Waite, and F. A. Patty, *Public Health Repts. U. S.,* **45,** 1241 (1930).

²¹ M. A. Wolf, V. K. Rowe, D. D. McCollister, R. L. Hollingsworth, and F. Oyen, *A.M.A. Arch. Ind. Health,* **14,** 387 (1956).

²² H. W. Gerarde, *A.M.A. Arch. Ind. Health,* **13,** 468 (1956).

changed, but most of it is found in the urine as metabolites because of oxidation of the side chain.

Tests Indicating Exposure. In addition to determining the concentration of ethylbenzene in the atmosphere, the presence of the hydrocarbon in the blood and urine would serve as an indication of exposure. This would require sensitive methods not affected by the chemicals normally present in these body fluids. Since ethylbenzene is excreted as a glucuronide, glucuronic acid analysis of the urine might be a measure of exposure. It is subject to the limitations described in the test indicating exposure to toluene.

5. *Hygienic Standard of Permissible Exposure*

The threshold limit for ethylbenzene has been established at 200 p.p.m.

6. *Odor and Warning Properties*

The odor intensity is about 2. A concentration of 200 p.p.m. has an eye irritation intensity of 1 or 2; a concentration of 1000 p.p.m. has a transient irritation intensity of 3 and causes profuse lacrimation.

XYLENE, $C_6H_4(CH_3)_2$ (Xylol, Dimethylbenzene)

1. *Source, Uses, and Industrial Exposures*

Xylene exists in three isomeric forms, *o*-, *m*-, and *p*-, di-methylbenzenes. Commercial xylene is a mixture of the three isomers with the meta (1,3-dimethylbenzene) isomer the chief constituent. The sources of xylenes are essentially the same as described for toluene, and the uses are similar, except as modified by higher boiling point. The para isomer, 1,4-dimethylbenzene, is a starting material for synthetic fiber production.

2. *Physical and Chemical Properties*

The physical and chemical properties of the xylenes are given in Table 1.

3. *Determination in the Atmosphere*

The M.S.A. Aromatic Hydrocarbon Detector and the combustible gas indicator and interferometer may be used for the quantitative analysis of the xylenes in air. Infrared and ultraviolet absorption analysis after collection of the xylenes by appropriate methods may also be used (see Chapter VII). In the absence of interfering substances such as benzene and toluene, the xylenes may be nitrated and measured colorimetrically.

1 mg./liter $\backsim$ 230 p.p.m. and 1 p.p.m. $\backsim$ 4.35 mg./cu. meter at 25°C., 760 mm. Hg.

4. *Physiological Response*

Acute Toxicity. The toxicological studies conducted on the xylenes are far fewer in number and less complete than the work reported on benzene and toluene.

It appears that the acute toxicity of the xylenes is greater than the acute toxicity of toluene or benzene. The xylene isomers differ in their acute toxicity although there is not complete unanimity among animal experimentalists about the relative toxicity.[23] The effects observed, mechanism of action, and pathological changes in acute xylene intoxication are very similar to what has been described for toluene.

Chronic Toxicity. The most extensive repeated vapor inhalation studies on xylene have been conducted by Fabre and Truhaut.[24] Rats and rabbits exposed to 690 p.p.m. (3 mg./liter) of mixed xylenes (composition not given) 8 hours per day, 6 days per week, for 130 days developed no significant changes in the peripheral blood. Rabbits exposed for 8 hours per day, 6 days per week, for 55 days to 1150 p.p.m. (5.0 mg./liter) of xylene mixture developed a decrease in the number of red blood cells and leucocytes and an increase in the platelet count. These investigators emphasize that these changes are much less severe than the effects produced by the same amount of benzene.

Absorption and Excretion. Absorption of xylene takes place chiefly through the lungs. It must be remembered, when considering solvents of relatively low vapor pressure, that volatility is not as important a factor where mists are encountered, as in the spraying of finishes thinned with xylene. The absorption of xylene through the skin is probably not of industrial significance, but skin irritation from xylene is more serious than from either benzene or toluene. The fate of xylene in the human body has not been determined, although it has been shown that in dogs[25] xylene is oxidized to toluic acid, conjugated with glycine, and excreted as toluric acid in a manner analogous to the fate of toluene, and it is logical to expect such a course in the human body as well.

Tests Indicating Exposure. In addition to atmospheric measurements to evaluate the extent of exposures, xylene determinations could be conducted on blood or urine samples by adapting methods used for air analysis. The urine could also be analyzed for toluric acid or other xylene metabolites.

5. Hygienic Standard of Permissible Exposure

The threshold limit for xylenes is 200 p.p.m.

6. Odor and Warning Properties

The initial odor of 200 p.p.m. has an intensity of approximately 3 and an irritation value of 1. As in most other instances, olfactory fatigue occurs rapidly and the odor is no longer detected at this concentration.

[23] W. S. Spector, ed., *Handbook of Biological Data.* Saunders, Philadelphia, 1956.
[24] R. Fabre and R. Truhaut, *Atti cong. intern. med. lavoro,* 11th Congr. Napoli, 1954.
[25] O. Schultzen and B. Naunyn, *Arch. Physiol.,* **1867,** 349; cited by W. F. von Oettingen, U. S. Public Health Bull. No. **255** (1940).

CUMENE, $C_6H_5CH(CH_3)_2$ (Isopropylbenzene, Cumol, 2-Phenylpropane)

1. Sources, Uses, and Industrial Exposures

Cumene is normally found in many petroleum distillates and is frequently a significant constituent of commercial petroleum solvents in the boiling range of 150 to 160°C. It is recovered by fractionation of petroleum and is widely used as a diluent or thinner for paints and enamels, as a solvent, and in organic synthesis. It has recently become a commercial source of phenol and acetone.

2. Physical and Chemical Properties

The physical and chemical properties of cumene are given in Table 1.

3. Determination in the Atmosphere

The methods for the determination of cumene vapor in the air are similar to those described for xylene (see page 1233).

1 mg./liter $\backsim$ 203.5 p.p.m. and 1 p.p.m. $\backsim$ 4.92 mg./cu. meter at 25°C., 760 mm. Hg.

4. Physiological Effects

Acute Toxicity. The acute toxicity of cumene is greater than that of benzene or toluene. The minimum lethal concentration of cumene as determined by single 7-hour exposures of white mice is 10 mg./liter, 2000 p.p.m.[26] On a mg./liter basis this is twice the toxicity of toluene and over three times that of benzene. On a volume basis the ratio is even higher, the LD_{50} concentrations of cumene, toluene, and benzene being approximately 2000, 5000, and 10,000 p.p.m., respectively. Cumene is a depressant to the central nervous system, its narcotic action being characterized by slow induction and long duration. Damage to the spleen and fatty change in the liver were consistent findings. Neither renal nor pulmonary irritation was observed.

Chronic Toxicity. The most extensive repeated vapor inhalation studies on cumene have been carried out by Fabre and Truhaut.[24] Rats and rabbits exposed to about 500 p.p.m. (2.5 mg./liter) of cumene 8 hours a day, 6 days per week, for 150 days developed no significant changes in the peripheral blood. Hyperemia and congestion was found in the lungs, liver, and kidneys of the rats exposed. Repeated oral feeding in rats by Wolf and associates[21] and subcutaneous administration by Gerarde[22] showed no evidence of injury to the blood-forming tissues due to cumene.

Absorption and Excretion. Robinson, Smith, and Williams[27] found an increase in glucuronides in the urine of rabbits dosed orally with cumene. The side chain is oxidized, but apparently not to the extent that benzoic acid is formed and eliminated as hippuric acid.

[26] H. W. Werner, R. C. Dunn, and W. F. von Oettingen, *J. Ind. Hyg. Toxicol.*, **26**, 264 (1944).

[27] D. Robinson, J. N. Smith, and R. T. Williams, *Biochem. J.*, **56**, xi (1954).

Tests Indicating Exposure. The general principles described for detecting exposure to toluene are applicable to cumene.

5. Hygienic Standard for Permissible Exposure

The threshold limit for cumene has not been established. Based on the chronic inhalation studies reported in the literature, 50 to 100 p.p.m., if it elicits no mucous membrane effects in man, should be a safe working atmosphere.

6. Odor and Warning Properties

No quantitative data are available on relationship of concentration to odor intensity or mucous membrane irritation. Cumene has a sharp aromatic odor, but it is not sufficiently distinctive to make it possible to distinguish it from other common solvent chemicals. The low vapor pressure of cumene limits the exposure arising from its use in the cold in open vats or containers, but mechanical ventilation or enclosed processes are advised.

*p-tert-*BUTYLTOLUENE, $CH_3C_6H_4C(CH_3)_3$ (1-Methyl-4-*tert*-butylbenzene)

1. Sources, Uses, and Industrial Exposure

p-tert-Butyltoluene is a petrochemical which has been suggested for use as a stable, high-purity, moderately high-boiling solvent for the preparation of resins and as a primary intermediate for synthesis in the chemical industry.

2. Physical and Chemical Properties

The physical and chemical properties of *p-tert*-butyltoluene are given in Table 1.

3. Determination in the Atmosphere

Hine and associates,[28] using an ultraviolet spectrophotometric method, have determined the concentration of *p-tert*-butyltoluene in 1-liter grab samples of air. It was possible to detect concentrations as low as 2 p.p.m. in this volume of air sample. The general methods described for the analysis of aromatic hydrocarbons can probably also be used.

1 mg./liter $\backsimeq$ 165 p.p.m. and 1 p.p.m. $\backsimeq$ 6.05 mg./cu. meter at 25°C., 760 mm. Hg.

4. Physiological Response

Acute Toxicity. The LC_{50} values for female rats following single exposures ranged from 165 p.p.m. after 8 hours of exposure to 934 p.p.m. for a 1-hour exposure.[28] The LC_{50} for the 4-hour exposure was 248 p.p.m. for female rats and mice. Rats exposed to 1500 p.p.m. showed signs of almost immediate respiratory

---

[28] C. H. Hine, H. Ungar, H. H. Anderson, J. K. Kodama, J. K. Critchlow, and N. W. Jacobsen, *A.M.A. Arch. Ind. Hyg. Occupational Med.*, **9**, 227 (1954).

distress with extreme dyspnea. With the exception of a slight dyspnea, there was no evidence of intolerance in rabbits exposed to the vapors of 1000 p.p.m. of *p-tert*-butyltoluene.

The LD_{50} for the percutaneous route of administration, using rabbits as the test animal, ranged from 13.8 to 27.8 ml. with an average of 19.6 ml./kg. of body weight.

Chronic Toxicity. There was no evidence of unusual behavior in rats exposed 50 times to 25 p.p.m. or 25 times to 50 p.p.m. other than a slightly lowered respiratory rate while in the exposure chamber. The duration of the daily exposures varied from 1 to 7 hours. There were also no apparent signs of toxicity in a group of rats exposed to 25 or 50 p.p.m. for 26 weeks, except in 3 rats exposed for 7 hours daily to 50 p.p.m. One of these died after 42 exposures, another developed severe flexor spasticity of both forelegs after 59 exposures, and the third rat showed signs of weakness in the hind legs after 71 exposures.

The principal positive pathological findings in animals exposed to *p-tert*-butyltoluene were generalized hyperemia and congestion, pulmonary edema and hemorrhage, edema and necrosis in the white matter of the brain and spinal cord, and fatty changes in the liver and kidneys. The white and red cell counts in exposed animals were lower than normal but are of questionable significance since the values obtained in unexposed control animals were also abnormally low.

Absorption and Excretion. Metabolic studies on *p-tert*-butyltoluene have not been reported in the literature. It is probable that its absorption, excretion, and metabolism are qualitatively similar to toluene.

Tests Indicating Exposure. In addition to air analysis the concentration of *p-tert*-butyltoluene in blood and urine could be determined by adaptation of the ultraviolet spectrophotometric method described by Hine and associates,[28] after extraction of the hydrocarbon with isooctane or cyclohexane.[29]

5. *Hygienic Standard of Permissible Exposure*

The threshold limit for *p-tert*-butyltoluene has been set at 10 p.p.m.

6. *Odor and Warning Properties*

p-tert-Butyltoluene has a distinctive odor, to which a tolerance is soon acquired. Olfactory recognition is immediate at concentrations of 5 p.p.m., and moderate eye irritation is experienced at 80 p.p.m. for a 5-minute exposure.

NAPHTHALENE, $C_{10}H_8$ (Tar Camphor)

1. *Sources, Uses, and Industrial Exposures*

Naphthalene is a by-product of the coke-oven industry and is recovered from coal tar. It is sold in the form of scales, powder, and the familiar "moth balls" which are widely used as a moth repellent. It is also used extensively in chemical

[29] H. W. Gerarde, *A.M.A. Arch. Ind. Health*, **20**, 262 (1959).

and dye manufacturing, for carbureting illuminating gas, as a disinfectant, and in preserving wood and other materials.

2. Physical and Chemical Properties

The physical and chemical properties of naphthalene are given in Table 1.

3. Determination in the Atmosphere

Naphthalene vapor in air may be determined by weighing, after adsorption on silica gel or activated charcoal, or after collection by condensation at low temperature. It may be determined with the spectrometer, the interferometer, by the picrate method[30] or by nitration and subsequent analysis in the same manner as toluene if interfering vapors are absent.

1 mg./liter $\simeq$ 191 p.p.m. and 1 p.p.m. $\simeq$ 5.24 mg./cu. meter at 25°C., 760 mm. Hg.

4. Physiological Response

Acute Toxicity. Rabbits dosed with 1 g. of naphthalene per kilogram of body weight daily, develop changes in the lens after 3 doses and definite opacity of the lens after 20 doses.[31] A case of a human cataract[32] has been reported which may be the counterpart of the cataracts produced experimentally in rabbits. The inhalation of naphthalene vapor may cause headache, confusion, nausea, and profuse perspiration. Severe exposures to vapors of naphthalene may cause vomiting, optic neuritis, and hematuria. Naphthalene is irritating to the skin and hypersensitivity in certain individuals has been reported.

Chronic Toxicity. No chronic toxic effects have been reported as a result of industrial exposure to naphthalene vapor.[33] There are no published reports on repeated inhalation studies with experimental animals.

Absorption and Excretion. Naphthalene is readily absorbed when inhaled. No information is available concerning skin absorption. In animals, it is metabolized and excreted in the urine conjugated with glucuronic acid, sulfuric acid, and in the form of 1-α-naphthylmercapturic acid.

Tests Indicating Exposure. Air analysis for naphthalene, and the detection of the urinary metabolites may form the basis for evaluating the extent of human exposure to naphthalene.

5. Hygienic Standard for Permissible Exposure

The threshold limit for naphthalene has not been established. Based on the author's experience with similar compounds and the toxicological studies described in the literature, a concentration of 25 p.p.m. is suggested as a tentative limit. This

[30] M. B. Jacobs, *Analytical Chemistry of Industrial Poisons, Hazards, and Solvents.* Interscience Publishers, New York, 1941.

[31] M. C. Bourne, *Physiol. Revs.,* **17,** 16 (1937).

[32] R. J. Meyer, *New Engl. J. Med.,* **252,** 622 (1955).

[33] *A.P.I. Toxicological Review of Naphthalene,* 1960.

level corresponds to 130 mg./cu. meter, which is about 25 per cent of the concentration of naphthalene vapor in saturated air at 25°C. The vapor pressure of naphthalene at this temperature is approximately 0.082 mm. Hg.

6. Odor and Warning Properties

Naphthalene has a familiar, characteristic, well-known odor. The initial odor of 25 p.p.m. has an intensity rating of approximately 3.

TETRALIN, $C_{10}H_{12}$ (Tetrahydronaphthalene)

1. Source, Uses, and Industrial Exposures

Tetralin is produced by the catalytic hydrogenation of naphthalene. Its principal uses are as a solvent for oils, fats, waxes, resins, asphalt, and rubber, and as a substitute for turpentine in shoe polish, oil paints, and lacquers. Tetralin has also been used as a larvicide for mosquitoes. It is used more extensively for these applications in Europe, England, and in the Latin-American countries than in the United States.

2. Physical and Chemical Properties

The physical and chemical properties of tetralin are given in Table 1.

3. Determination in the Atmosphere

General methods of analysis, such as, the interferometer, combustion apparatus, infrared and ultraviolet absorption procedures, and gas chromatography may be employed.

1 mg./liter $\backsim$ 185 p.p.m. and 1 p.p.m. $\backsim$ 5.41 mg./cu. meter at 25°C., 760 mm. Hg.

4. Physiological Response

Toxicity. Dermatitis has been reported in painters working with tetralin.[34] These individuals also complained of headache, malaise, and irritation of the eyes, throat, and mucous membranes of the nasal passages. Severe exposure has resulted in the elimination of a green-colored urine in man.

The most extensive repeated inhalation studies have been conducted by Cardani[35] with guinea pigs. These animals were exposed to about 275 p.p.m. tetralin, 8 hours a day, for approximately 3 weeks. The principal positive pathological findings were severe changes in the kidney and liver and chemical pneumonitis. No significant deviations were found in the peripheral blood.

Absorption and Excretion. Tetralin is absorbed by inhalation, metabolized in part, and excreted in the urine in the form of glucuronides.

[34] F. Flury and F. Zarnik, *Schädliche Gas.* Springer, Berlin, 1931.
[35] A. Cardani, *Med. lavoro,* **33,** 145 (1942).

Test Indicating Exposure. The degree of exposure may be determined by measuring the concentration of tetralin in the atmosphere and body fluids, such as blood and urine, using suitable sensitive methods. Some indication may be obtained by urine analysis for tetralin metabolites and glucuronic acid.

5. Hygienic Standard of Permissible Exposure

The threshold limit has not been established for tetralin. A value of 25 p.p.m. appears to be a reasonable level, based on the toxicity studies conducted and by analogy with naphthalene.

6. Odor and Warning Properties

No quantitative studies have been reported on the relation of air concentration of tetralin and odor intensity or mucous membrane irritation. The odor qualitatively is similar to that of naphthalene.

CHAPTER XXXI

Halogenated Hydrocarbons: I. Aliphatic

DON D. IRISH

I. General Considerations

The halogenated hydrocarbons are among the most widely available industrial chemicals. The aliphatic chlorinated compounds are used in greatest volume, not only because of good properties but also because of low cost. From within the group as a whole, we can choose compounds with excellent solvent properties suited to a particular process requirement. We can choose a specific volatility most appropriate to the particular use. We can choose one with little or with no flammability. While we find a wide range of toxicity among this class of compounds, we can certainly choose one with very low or at least readily controllable health hazard.

Toxicity is, of course, not the prime factor in choosing a compound for industrial use. If the performance and economics are favorable, the health hazard can be properly controlled provided it is well understood.

The excellent solvent properties of the chlorinated aliphatics account for the largest volume of use. Their wide range of volatility allows adaptation to many extraction or cleaning operations, some requiring rapid volatilization, others better suited to slower volatilization. They have frequently been substituted for more flammable materials where they had similar or better solvent properties. In many organic fluid mixtures used as solvents, as fumigants, or other uses where flammability may be a problem, certain halogenated hydrocarbons have been added to suppress flammability. Some of the halogenated aliphatics have found use as fire extinguishers. This includes compounds containing chlorine, bromine and fluorine, or combinations thereof.

Certain mixed halogen compounds containing fluorine and having a high volatility and low toxicity have been widely used as refrigerants. In more recent years, the same or similar compounds have been used as aerosol propellants. Several of the more complex halogenated hydrocarbons have been found to be exceptionally effective in insect control. Some of these are now widely used for the control of household insects as well as agricultural pests.

A. TOXICOLOGY

Not so many years ago, warning labels simply stated "contains halogenated hydrocarbons" or "contains chlorinated hydrocarbons." The assumption appeared to be that they could all be considered in the same class. In examining the physiological response to different halogenated hydrocarbons, it becomes obvious that they cannot be lumped together in considering the hazard, either quantitatively or qualitatively. We could certainly not class dichlorodifluoromethane, which is among the most physiologically inert of organic substances, with a material such as allyl chloride, which is quite toxic, either acutely or chronically. Qualitatively, the most significant response may be liver or kidney injury as is observed with carbon tetrachloride, or it may be almost entirely an anesthetic action as in the case of ethyl chloride.

From a single acute exposure, narcosis is probably the most common physiological response from many of the halogenated hydrocarbons. However, the effect may be deep anesthesia from one compound and only weak narcosis or inebriation from another. On the other hand, respiratory irritation may dominate the picture as it does with allyl chloride. It seems necessary to consider each compound individually.

Many of the halogenated hydrocarbons are known to potentiate adrenalin. While not all halogenated hydrocarbons have been studied for this tendency, it is well to assume that any one of them may have such an effect until proved otherwise. In any case, it is certain that adrenalin should not be administered to a person who has had a serious exposure to halogenated hydrocarbons, unless it is clearly known that the particular compounds concerned do not potentiate adrenalin. In cases of severe acute poisoning, cardiac arrhythmia may occur, even from endogenous adrenalin.

B. PROPER USE OF VOLATILE ORGANIC LIQUIDS

The user of a halogenated solvent (or any other material for that matter) is interested in the likelihood of injury or health hazard. The toxicity of the material is not a measure of that health hazard, although it is a very important factor in determining the hazard. Other factors that may affect a given situation are operating temperature, ratio of exposed liquid surface to available ventilation, evaporation rate, pattern of air flow, and work pattern of the operator. The hazard then, is from a particular operation using the substance, not solely from the toxicity of the substance itself. The proper use of a substance, therefore, implies a thorough knowledge of its physical properties and of all the requirements of a particular operation as well as the toxicological response from exposure. It also implies the competence to comprehend and intelligently apply all of these factors.

C. INDUSTRIAL HYGIENE STANDARDS

The description of toxic properties involves concepts that may be foreign to many people who have the task of designing facilities and planning operations so that attendant health hazards will be controlled. The toxicologists and industrial hygienists have attempted to be helpful by offering control figures, first known as "maximum allowable concentrations," to guide the engineers. Several designations have been used for these limits, such as "MAC" ("maximum allowable concentration" or "maximum acceptable concentration"), threshold limits, and industrial hygiene standards. Unfortunately, many people accepted these figures as an index of relative toxicity or, worse yet, as an index of hazard.

The definition of these standards has not always been clearly indicated. Some individuals still consider them as a sort of "average." This is obviously unsound as it would be possible to have a peak concentration above the acute lethal concentration and yet the average for the day might be well below the standard. The American Standards Association most clearly expressed its definition. It gives a limiting concentration "for exposures not exceeding eight hours daily with the understanding that variations should fluctuate below this value."

Such standards are of great value if clearly understood and competently used. An industrial hygienist or plant engineer must understand the basis for the particular standards he is using. He must be aware of the physiological consequences of exceeding that particular standard. In the case of one substance, it may be serious tissue injury. In another, it may be transient misfunction, such as inebriation or eye irritation. Full awareness of the significance of such reactions in relation to exposure to each individual substance is necessary to intelligently use such standards.

To illustrate, a standard of 1000 p.p.m. for dichlorodifluoromethane is set because there really is no excuse to have higher concentrations in the atmosphere and not because there is any probability of physiological injury at levels even several times that amount. A standard is set at 100 p.p.m. for styrene in order to avoid discomfort, but physiological injury will not occur at several times that concentration. It is obvious that the intelligent industrial hygienist is not going to close down a plant on the basis of occasional peaks which may greatly exceed the standard for either one. In the case of materials, such as carbon tetrachloride, trichloroethylene, and perchloroethylene, the level set as a standard by the ACGIH is still a level that may cause slight injury in experimental animals. It is, therefore, obvious that humans should not be exposed for 8 hours a day to a level maintained as high as the standard for any one of these materials. Under such circumstances, using the standard as the top level, any fluctuations should be around a median of perhaps half the standard. A standard cannot replace a good knowledge of the facts and competent judgement in applying them.

D. TIME

In considering the significance of an exposure condition, one cannot depend solely on the concentration in the work atmosphere. The periods of exposure to varying concentrations must be considered in order to understand the potential exposure of an individual when analyzing his specific job. Unfortunately, it has been impossible to devise any formula by which a fluctuating concentration and varying intervals of time could be integrated into a number that had any significant relationship to the probability of physiological injury. The so-called "time-weighted average" has been proposed as an approach to this problem. Unfortunately, a workman might die of an acute lethal exposure even though the time-weighted average for the day was well below the standard. Time of exposure is of great importance but it cannot be logically reduced to such a simple mathematical expression.

If the industrial hygienist were to use the standard essentially as a ceiling, the normal fluctuations in concentration would then be limited to a much narrower range and the time-weighted average would be very useful in analyzing the fluctuations below the standard.

E. PROBABILITY OF AIR DISPERSION

The degree to which a substance will be vaporized or otherwise dispersed in the air is dependent upon the properties of the substance and the conditions under which it must be used. All of these factors must be taken into account in considering a particular operation using a certain substance. The best evaluation of a health hazard can be made by actual measurement of air concentrations. As an alternative in considering new operations, it is acceptable to use a carefully considered extrapolation from actual experience with comparable unit operations. A great oversimplification of this approach was suggested a number of years ago and has been proposed every year or two since. The industrial hygiene standard was to be divided by the vapor pressure, thus deriving a new "hazard" standard. Just how this was to be used in a practical engineering problem is uncertain. In any case, the problem of interpretation or use is merely confounded by mathematical manipulation of a standard which is not a physical constant but simply a suggested ceiling for operation control. These standards are in no sense a measure of relative toxicity. No simple mathematical calculation can substitute for a thorough understanding of all factors contributing to an exposure plus good judgment in their application to the particular practical problem at hand.

It should also be recognized that vapor pressure is not a good measure of the probability of air dispersion. Factors such as heat of vaporization may profoundly affect the rate of dispersion. Rate of evaporation is probably closest to a practical measure and may be useful in estimating the requirements of the design of a new plant or in planning a new operation. However, the best evaluation of health hazard associated with a process requires a competent survey of actual operations. The factors leading to dispersion of a substance into the air where it may be breathed are of very great importance and should be taken into consideration.

F. INDIVIDUAL COMPOUNDS

Information thought to be essential for the practical evaluation of the hazards associated with industrial use is given separately for each halogenated hydrocarbon in this chapter. The treatment is not in any sense exhaustive; for a more detailed discussion of the voluminous scientific literature, the reader should consult some of the various reviews or briefs.[1-4]

The physical constants used in this chapter were taken largely from the *Handbook of Chemistry and Physics* published by the Chemical Rubber Publishing Co., Cleveland, Ohio, although, in a number of instances, the physical constants were taken from specific publications on the particular substance concerned. The vapor pressure was calculated from the tables by Dreisbach.[5]

Per cent in saturated air and the density of saturated air were calculated from the vapor pressure.

G. SAMPLING AND ANALYSIS

No attempt will be made to outline specific analytical methods or to indicate their advantages for particular compounds. There are a number of the analytical techniques mentioned here that may be used for many of the different halogenated hydrocarbons. An individual intending to use a particular method should verify it for his own particular use.

There are several techniques applicable to a quick field determination that are both very simple and portable. Others, however, are used in the laboratory for the analysis of samples taken in the field.

H. COLORIMETRIC

Several types of reagents may be obtained on an adsorbent in narrow glass tubes through which air can be drawn and produce a color reaction. The apparatus is very simple and light, and the results read directly. Special tubes are available for a number of different compounds in this class. There may be others available in the future. This is not a highly exact quantitative method but is certainly a simple and quick method for field survey. Other colorimetric methods are applicable, particularly those using the Fujiwara reaction.

I. FLAME COLOR

There are two types of apparatus using the development of color in a flame when halogens are in contact with heated copper. One is the simple Halide Leak Detector which has been used for many years by the refrigeration people. This will give a very rough quantitative approximation if the color changes are related to known concentrations of the specific material of interest to the individual who

[1] W. F. von Oettingen, The halogenated hydrocarbons, toxicity and potential dangers, *U. S. Public Health Serv. Publ.* No. **414**, (1955).

[2] *Chemical Safety Data Sheets.* Manufacturing Chemists Assoc., Washington, D. C.

[3] *American Standard Maximum Acceptable Concentrations of Vapors, Dusts and Gases.* Z-37 Committee of the American Standards Assoc., New York.

[4] *Hygienic Guide Series.* American Industrial Hygiene Assoc., Detroit, Mich.

[5] R. S. Dreisbach, *Pressure-Volume-Temperature Relations of Organic Compounds.* Handbook Publishers, Sandusky, Ohio, 1952.

will use the detector. Each person should become acquainted with his own ability to determine the color level. Since it depends upon the presence of an open flame, it has the limitation that it cannot be used where there is any fire hazard. Sensitivity to various halogenated hydrocarbons may vary greatly. Some of them, of course, are not detected at all. Fluorinated hydrocarbons containing no other halogens are not detected by the Halide Leak Detector.

The flame photometer works on a similar principle but is a much more sensitive and accurate method of determination. It is very useful in analyzing for a wide range of halogenated hydrocarbons. Such an apparatus should be carefully calibrated by the individual using it and must be kept in good working order. A simple, available, and portable apparatus using this principle is the Davis Halide Meter. As would be expected, this instrument also is not useful in determining fluorinated hydrocarbons that do not contain other halogens.

J. COMBUSTION AND HALOGEN DETERMINATION

One of the old stand-by methods for determining halogenated hydrocarbons is the combustion of the material in a hot tube yielding the halogen and the halogen acid which are then absorbed in water solution. The halogen may be determined by Volhard titration or by turbidimetric, amperometric, colorimetric, or conductivity measurement. The measurement of conductivity changes has been adapted for the continuous monitoring of a plant atmosphere.

K. INFRARED

The infrared spectrometer is particularly useful where it is necessary to separately determine individual constituents of a mixture. It has been used for the direct determination of the halogenated material in air. It is also very useful for determining the material after it has been sampled by adsorption on silica gel or by collection in a cold trap. Infrared analysis has the advantage of specific identification of the materials as well as quantitative measurement.

L. GAS CHROMATOGRAPHY

One of the newer methods of analysis that has come into recent usage is gas chromatography. This technique is particularly useful in the separation of mixtures. It has not been used for the direct analysis of air, although it might be so used if it could be made sensitive enough to determine concentrations of importance to industrial hygiene. It is very useful, however, in the separation of individual materials in a mixture that has been sampled by adsorption on silica gel or in a cold trap. The fractions from the gas chromatographic apparatus may be analyzed by infrared, or mass spectrograph, or by other common analytical methods.

M. MASS SPECTROGRAPH

This apparatus is also very useful when it is desirable to identify and quantitate the different materials in a mixture. Some of the newer types of apparatus are sensitive enough to analyze air samples directly. Instruments now available would not be considered portable enough for many types of industrial hygiene work. This type of instrument may be used for the analysis of samples which are taken by adsorption on silica gel, condensation in a cold trap, or as grab samples in vacuum bottles.

N. INTERFEROMETER

This instrument is quite sensitive but not at all specific. It has been used for taking quick spot samples where it is desirable to determine peak concentrations. It requires good calibration and the results may be somewhat confusing when it is used with mixtures of materials. Because of its lack of specificity, it has not been used as extensively in recent years.

O. THERMOCONDUCTIVITY

Thermoconductivity has been used in measuring high concentrations of halogenated hydrocarbons. It is not sensitive enough to analyze for concentrations in the range of significance to industrial hygiene. This is true, at least, of the more toxic materials, which must be determined in concentrations of 1 p.p.m. to a few hundred p.p.m.

P. GENERAL

It should be remembered that in sampling or analysis for industrial hygiene purposes there is no apparatus or method which is in any sense foolproof. Every instrument mentioned here must be carefully calibrated and checked by the individual who is using it. The industrial hygienist must be well acquainted with the method and have proved its usefulness under his particular conditions of application. Some of the newer spectrographic methods mentioned have not been widely used in the field of industrial hygiene. They have a great deal of usefulness, however, particularly in separating mixtures and specifically identifying ingredients. They cannot replace the simple, rapid, field methods but they are exceedingly valuable in giving a more definitive check, both qualitatively and quantitatively. Perhaps more important than any sampling or analytical equipment is the skill of the industrial hygienist in knowing when and where to sample so that he may determine the actual potential for exposure in a specific operation.

There is a trend in the direction of continuous monitoring of many types of operation. This is most valuable in detecting fluctuations that may be missed in spot sampling. The apparatus may be expensive, but in many instances it will pay for itself many times over in improved quality of operation. One type of continuous recording apparatus using combustion and conductivity measurement has been in use in several different operations throughout a number of years and has been successful in monitoring the atmosphere for halogenated hydrocarbons. The infrared spectrograph has now been made sensitive enough so that it also can be combined with a continuous recording apparatus and may be set to determine a specific compound with or without other compounds present.

Q. ANALYSIS OF BIOLOGICAL MATERIAL

Stewart et al.[5a-d] in recent investigations have shown the feasibility of analysis of expired air and blood for several of the chlorinated hydrocarbons. Carbon tetrachloride, trichloroethylene, tetrachloroethylene, and methyl chloroform have been studied. Direct analysis of expired air by the infrared method was the most meaningful. This may be a useful tool in the clinic giving qualitative and to some degree quantitative indications of exposure. These methods are of particular value in research for measuring the rate at which the material is cleared from the body in expired air. Further investigations by this group are in the process of publication. An interesting use of these methods in research has been reported by Chenoweth et al.[5e]

II. Specific Compounds

METHYL CHLORIDE, CH_3Cl (Monochloromethane)

1. *Uses and Industrial Exposures*

Methyl chloride has been used as a refrigerant, aerosol propellant, solvent, and chemical intermediate, particularly in methylation reactions.

Exposure is largely in industrial operations. The most serious problem of exposure in past years was from its use as a refrigerant. It is of less significance as a refrigerant today. As it is a gas at normal temperatures, it must be used in closed or well-ventilated systems.

2. *Physical and Chemical Properties*

Physical state: colorless gas
Molecular weight: 50.49
Melting point: $-97.7°C.$
Boiling point: $-24.22°C.$
Vapor density: 1.785 (air $= 1$)
Solubility: 400 ml. in 100 ml. of water at 20°C.; 3500 ml. in 100 ml. of ethanol at 20°C.; soluble in ethyl ether, chloroform, and acetone

[5a] R. D. Stewart, H. H. Gay, D. S. Earley, C. L. Hake, and A. W. Schaffer, *Am. Ind. Hyg. Assoc. J.,* August 1961 (to be published).

[5b] R. D. Stewart, H. H. Gay, D. S. Earley, and C. L. Hake, *Arch. Environ. Health,* **2,** 516 (1961).

[5c] R. D. Stewart, D. S. Earley, T. R. Torkelson, and C. L. Hake, *Nature,* **184,** 192 (1959).

[5d] R. D. Stewart, T. R. Torkelson, C. L. Hake, and D. S. Earley, *J. Lab. and Clin Med.,* **56,** 148 (1960).

[5e] M. B. Chenoweth, D. N. Robertson, D. S. Earley, and R. Golhke, *Anesthesiology,* 1961 (to be published).

1 mg./liter $\approx$ 484 p.p.m. and 1 p.p.m. $\approx$ 2.09 mg./cu. meter at 25°C., 760 mm. Hg

3. *Physiological Response*

Chronic and subacute exposures predominantly effect the nervous system. Symptoms observed are ataxia, staggering gait, weakness, tremors, vertigo, difficulty in speech, and blurred vision. In severe acute poisoning, gastro-intestinal disturbances, such as nausea, vomiting, abdominal pain, and diarrhea are observed. Acute animal experiments have indicated pulmonary congestion. Histopathological changes of the internal organs are not observed.

In a practical sense, the major problem encountered in mild exposure is the "drunkenness" or inebriation. The resulting incoordination and impaired judgment may lead to unsafe manual manipulation. The man may injure himself or endanger others by mechanical misoperation. As this condition may persist for some time, he is a hazard to himself and others if he drives his car after such exposure.

The symptoms may be delayed some time in onset. They may also continue for some hours after the exposure has stopped. There are indications that with long, severe chronic exposure, symptoms may persist for longer than a few hours. Most experience, however, would indicate complete recovery in a matter of hours.

The above practical experience on mild exposure in humans is from The Dow Chemical Co.[5f] From this same experience, we concluded that exposure to fluctuating concentrations essentially below 100 p.p.m. was well tolerated.

TABLE 1

Acute Effects of Methyl Chloride Vapor

Single exposure	Concentration (p.p.m.)
Kills most animals in a short time	150,000–300,000
Dangerous in 30–60 min.	20,000–40,000
Maximum for 60 min. without serious effect	7000
Maximum for 8 hrs.	500–1000

Acute Vapor Exposure. Studies by Sayers *et al.*[6] gave data from acute exposure of guinea pigs. Flury and Zernik[7] reported data from acute exposure of several animals. They also reported limited data from chronic exposure of mice and guinea pigs. They observed injury and deaths from exposure to approximately 3000 p.p.m. for 15 min./day for a varying number of days (between 3 and 100). For the detailed data, the reader should consult the original publication. Table 1 summarizes the effects of acute exposure.

[5f] D. D. Irish, unpublished data, Biochemical Research Lab., The Dow Chemical Co.
[6] R. R. Sayers, W. P. Yant, B. G. H. Thomas, and L. B. Berger, *U. S. Public Health Bull.* No. **185** (1929).
[7] F. Flury and F. Zernik, *Schädliche Gase.* Springer, Berlin, 1931.

Chronic Vapor Exposure. Smith and von Oettingen[8] reported data from chronic exposure of animals. They exposed their animals 6 hours/day, 6 days/week. Guinea pigs, mice, dogs, rabbits, and rats showed injury at 1000 p.p.m. over varying periods up to 175 days. At 500 p.p.m., rats showed no effects but the other animals, including a monkey, showed some response. At 300 p.p.m., no effects were observed on any of the animals. They suggested that the "maximum allowable concentration" should be well below 500 p.p.m.

This correlates reasonably well with the clinical experience, which would indicate the level of exposure should be below 100 p.p.m. based on central nervous system effects.

Experience in Man. Kegel et al.[9] and McNally[10] reported clinical cases of acute poisoning. The exposure resulted from leaking refrigerators. The use of methyl chloride in refrigerators is rare in the United States today. The symptoms reported are indicated in the summary.

Hansen et al.[11] observed the effects of excessive exposure after a spill. Fifteen workers manifested signs of dizziness, blurred vision, incoordination, and gastrointestinal complaints. Recovery was complete in 10 to 30 days.

A more recent report by Klimkova-Dentschova[12] indicates continued use as a refrigerant in other countries. He observed in detail the neurological picture in 100 workers. The report stated: "Involvement of the internal organs (kidney, optic disturbances) was absent even where nervous and mental changes indicated a severe form of poisoning." Levels of exposure were not indicated.

Absorption, Excretion, and Metabolism. The metabolic fate of methyl chloride is not certain. Smith[13] was unable to demonstrate methanol in the blood or significant increases of formic acid in the urine. He found no hematological or biochemical changes.

Sperling et al.[14] indicated that intravenously injected methyl chloride rapidly disappeared from the blood but only about 5 per cent appeared in the expired air in 1 hour and only small amounts in the bile and urine. This slow excretion by the lungs is surprising and may account for the slow recovery from symptoms.

From chronic exposure, methyl chloride shows much lower toxicity than chloroform or carbon tetrachloride and much greater toxicity than methylene chloride.

[8] W. W. Smith and W. F. von Oettingen, *J. Ind. Hyg. Toxicol.,* **29,** 47 (1947).

[9] A. H. Kegel, W. D. McNally, and A. S. Pope, *J. Am. Med. Assoc.,* **93,** 353 (1929).

[10] W. D. McNally, *J. Ind. Hyg. Toxicol.,* **28,** 94 (1946).

[11] H. Hansen, N. K. Weaver, and F. S. Venable, *A.M.A. Arch. Ind. Hyg. Occupational Med.,* **8,** 328 (1953).

[12] E. Klimkova-Dentschova, *Rev. Czechoslov. Med.,* **3,** 1 (1957).

[13] W. W. Smith, *J. Ind. Hyg. Toxicol.,* **29,** 185 (1947).

[14] F. Sperling, F. J. Macri, and W. F. von Oettingen, *A.M.A. Arch. Ind. Hyg. Occupational Med.,* **1,** 215 (1950).

4. Hygienic Standards of Permissible Exposure

The threshold limit of methyl chloride was established by the American Conference of Governmental Industrial Hygienists, in April, 1959, at 100 p.p.m. (210 mg./cu. meter).

5. Flammability

The flammability limits of methyl chloride are 8.25–18.7 vol. %.[15] The ignition temperature is 634°C.[16]

6. Odor and Warning Properties

Methyl chloride has no odor or other warning property. This, and the fact that the material is a gas at normal temperature, increases the seriousness of the hazard.

METHYL BROMIDE, CH₃Br (Monobromomethane)

1. Uses and Industrial Exposures

The largest single use for methyl bromide is in the fumigant field. It is used to fumigate the soil, a wide range of commodities, grain, warehouse, and mill. The principal problems here are associated with the fumigating operator and the control of other people who may enter the fumigated area.

Methyl bromide is used as a chemical intermediate. One of the principal chemical uses is as a methylating agent. A number of publications indicate that it has been used as a refrigerant but such use is not significant in the United States. It has found use as a fire-extinguishing agent, particularly in automatic equipment for the control of engine fires on aircraft. Because of the toxicity of this material, its fire-extinguishing use will probably be limited to such specialized applications. It has been used as a fire extinguisher in Europe and a number of reports of injury from this use will be found in the literature.

As methyl bromide is a gas at ordinary temperatures and has no warning properties, a dangerous concentration may rapidly accumulate in a work area without warning to the operator. In industrial operations regularly using methyl bromide, it is advisable to have some kind of warning or monitoring system for continuous analysis of the air. In fumigation operations, the commonly used flame-type of Halide Leak Detector is probably satisfactory provided the personnel have proper protective equipment.

2. Physical and Chemical Properties

Physical state: colorless gas
Molecular weight: 94.95
Specific gravity: 1.732 (0/0°C.)

[15] G. W. Jones, *Chem. Revs.*, **22** (1938).
[16] H. H. Nuckolls, *Underwriters Labs. Rept.* No. **1418** (1926).

Melting point: —93.66°C.

Boiling point: 4.6°C.

Vapor density: 3.27 (air = 1)

Solubility: 0.09 g. per 100 ml. water at 20°C.; soluble in ethyl ether, ethanol, chloroform, carbon disulfide, benzene, and carbon tetrachloride.

1 mg./liter ≎ 257 p.p.m. and 1 p.p.m. ≎ 3.89 mg./cu. meter at 25°C., 760 mm. Hg

3. *Physiological Response*

Unless the concentration is high enough to cause rapid narcosis and death from respiratory failure, the most striking response to exposure at high concentrations will be lung irritation resulting in congestion and edema. These symptoms are observed in both animals and man and often develop into a typical confluent bronchial pneumonia. At lower levels of exposure, this lung condition may account for delayed deaths. If it leads to secondary infection, the delay may be a matter of days.

At threshold concentrations, this lung condition is not observed. The response is almost entirely referable to the nervous system and usually shows up only after prolonged and repeated exposures. Excitation and even convulsions have been observed in animals; but if they survived repeated exposures, the later signs were a paralysis of the extremities. The paralysis of the extremities is most typical of threshold toxic response from repeated exposures over a long period of time. Animals that have been seriously paralyzed have recovered although the recovery is somewhat slow. Human experience indicates that there is a high probability of complete recovery although the time necessary may be quite long.

Due to the high volatility (methyl bromide is a gas at normal temperature) one can readily attain high concentrations in a work atmosphere. Because methyl bromide essentially has no warning properties, such high concentrations can be attained without recognition. These factors and the fact that methyl bromide is quite toxic would indicate that it is a material with potentially high hazard. It should be used only by individuals who are well acquainted with proper methods of handling and fully cognizant of the consequences of exposure to excessive amounts.

Serious skin burns may be observed from contact, especially under clothing or shoes.

Acute Exposure to Vapors. The 8-hour survival dose for rats is approximately 1 mg./liter (260 p.p.m.). They survive 5200 p.p.m. for 6 minutes and 2600 p.p.m. for 24 minutes. The 6-hour survival dose for rabbits is approximately 2 mg./liter (520 p.p.m.). They survive 5200 p.p.m. for 6 minutes and 2600 p.p.m. for 1 hour. This data is taken from Irish et al.[17]

[17] D. D. Irish, E. M. Adams, H. C. Spencer, and V. K. Rowe, *J. Ind. Hyg. Toxicol.*, **22**, 218 (1940).

These authors studied rats, rabbits, guinea pigs, and monkeys. They described the response of most animals as typically one of lung irritation. If the exposure was severe enough, this resulted in lung edema and usually a typical confluent bronchial pneumonia.

The symptoms observed in humans from acute exposure have been reported by a number of authors whose reports have been reviewed by von Oettingen.[18] The early symptoms may be a feeling of illness, headache, nausea, and vomiting. Tremors and even convulsions are observed much as they were in animals. Lung edema and an associated cyanosis was observed.

It was indicated by von Oettingen that if the patient survived an acute exposure for the first two or three days, the probability of complete recovery was very high.

Chronic Vapor Exposure. In chronic or repeated exposure to relatively low concentrations of methyl bromide, the picture differs from that of the single exposure to high concentrations. Unless the concentration is high enough to cause lung irritation, the response observed from repeated exposures will be essentially one of paralysis of the extremities as observed by Irish *et al.*[17] Rats, rabbits, guinea pigs, and monkeys responded similarly with some quantitative differences. At 0.42 mg./liter (100 p.p.m.), rats showed a varying pulmonary response from essentially normal lungs to quite severe pneumonia. Guinea pigs failed to show any significant pulmonary changes at this level. They survived up to 98 exposures and showed no histopathological changes. A monkey exposed at this level showed severe convulsions. At 0.25 mg./liter (66 p.p.m.), rats showed essentially no response from a 6-month period of repeated exposures. Guinea pigs survived with similar lack of response. Rabbits, however, developed a characteristic paralysis and some pulmonary damage. Monkeys exposed at this level developed a paralysis comparable to that seen in rabbits. At 0.13 mg./liter (33 p.p.m.), rabbits still showed pulmonary damage and paralysis. Monkeys appeared normal. At 0.065 mg./liter (17 p.p.m.), all the animals survived without indications of response to the exposure.

For a review of the human experience in chronic exposure, the reader should refer to the summary by von Oettingen.[18] The response in humans is not unlike that which was reported for animals. Individuals developing severe nervous system effects may show a very slow recovery. Individual cases have been reported to have taken months.

Interesting experience with chronic exposure to methyl bromide was carefully studied during the early 1940's by several drug companies who were packaging methyl bromide to be used for fumigation of clothing. During the investigations on people exposed in this operation, studies of blood bromide were made and were related to the probability of toxic response. It was indicated that concentrations

[18] W. F. von Oettingen, The toxicity and potential dangers of methyl bromide with special reference to its use in the chemical industry, fire extinguishers, and in fumigation, *Nat. Insts. Health Bull.* No. **185**, (1946).

of over 10 mg. % in the blood were indicative of the probability of difficulties. Levels of 15 or 20 mg. % would lead to quite severe response. If blood bromide is to be used in determining the exposure to methyl bromide, it should be kept in mind that the figures can be very misleading if there are other sources of bromine. The most common is inorganic bromide medications.

Skin and Eyes. Watrous[19] described difficulties in handling methyl bromide in the drug industry. Repeated splashes on the skin resulted in severe skin lesions. Severe cases showed "vesicles or blebs." In a less rigorous exposure, severe itching dermatitis was observed.

In the experience of the present author,[5f] methyl bromide will cause difficulty when it is held near the skin by clothes. This is a special problem with shoes. In the case of shoes, it is suspected that a fairly high concentration of vapor near the floor may actually be absorbed into the leather and cause skin irritation. Spills onto or into the shoes may cause very severe burns.

When handling methyl bromide, care should be taken that splashes do not get onto the clothing or shoes. Where there is a fairly high vapor concentration, as in an accidental spill, care should be taken that it does not concentrate in the shoes or the protective clothing.

Absorption, Excretion, Metabolism, and Mode of Action. Methyl bromide is readily absorbed through the lungs. There have been suggestions that it could be significantly absorbed through the skin, but a report of Butler[20] indicates that absorption through the skin is probably not of any significance. It may be concluded that experience so far has not shown absorption through the skin to be an important factor in methyl bromide intoxication. Certainly, the major problem is one of absorption by the lungs.

It is probable that excretion is most predominantly by the lungs as methyl bromide. There is a significant amount of methyl bromide, however, that is metabolized in the body and it appears as inorganic bromide, which is excreted in the urine.

The mechanism of action of methyl bromide is by no means clear. A number of suggestions have been made but certainly nothing conclusively indicated. It would seem to the present author that methyl bromide, which is a very active methylating agent, could methylate any one of a number of substances in the biological system and thus have significant consequences.

4. *Hygienic Standards of Permissible Exposure*

The threshold limit of methyl bromide was established by the American Conference of Governmental Industrial Hygienists, in April, 1959, at 20 p.p.m. (80 mg./cu. meter).

[19] R. M. Watrous, *Ind. Med.*, **11**, 575 (1942).
[20] E. C. B. Butler, K. M. A. Perry, and J. R. F. Williams, *British J. Ind. Med.*, **2**, 30 (1945).

5. Flammability[21]

Methyl bromide is practically nonflammable. Flame propagation is in the narrow range of 1 per cent (13.5 to 14.5 per cent by volume in air). The ignition temperature is 537°C.

6. Odor and Warning Properties

Methyl bromide, itself, has practically no odor or irritating effect and, therefore, no warning, even at physiologically hazardous concentrations. Some mixtures of methyl bromide used in the fumigation field contain chloropicrin or similar irritant material which aids in giving warning of significant concentrations of methyl bromide.

METHYL IODIDE, CH_3I (Monoiodomethane)

1. Uses and Industrial Exposures

Methyl iodide is used principally as a chemical intermediate. It has been proposed as a fire extinguisher[22] and insecticidal fumigant.

2. Physical and Chemical Properties

Physical state:　colorless liquid
Molecular weight:　141.95
Specific gravity:　2.279 (20°/4°C.)
Melting point:　−66.1°C.
Boiling point:　42.50°C.
Vapor density:　4.9 (air = 1)
Vapor pressure:　400 mm. Hg (25°C.)
Refractive index:　1.5293 (21.0°C.)
Per cent in "saturated" air:　53 (25°C.)
Density of "saturated" air:　3.04 (air = 1)
Solubility:　1 vol. per 125 vol. water at 15°C.; soluble in ethanol and ethyl ether

1 mg./liter ≎ 172 p.p.m. and 1 p.p.m. ≎ 5.8 mg./cu. meter at 25°C., 760 mm. Hg

3. Physiological Response

As methyl iodide has not been widely or extensively used, the toxicological investigations and experience have been limited. It appears to be primarily a central nervous system depressant. There are also indications of lung irritation from acute exposures.

[21] *Chemical Safety Data Sheet SD-35*. Manufacturing Chemists Assoc., Washington, D. C., 1949.
[22] *Paint Technol.*, **13**, 90 (1948).

Vapor Exposure. Bachem[23] studied the response of mice to methyl iodide. His work is summarized in Table 2.

TABLE 2

Physiological Response to Various Concentrations of Methyl Iodide—Mice[23]

Concentration		Response
mg./liter	p.p.m.	
454.4	78,693	Rapid narcosis; death after 10 minutes' exposure
105.1	18,109	Death after 30 minutes' exposure
42.6	7,340	After 15 to 30 minutes, side position, no complete narcosis; death 1 hour after beginning of exposure
21.3–31.6	3,670–5,373	Death after 2 to 2¹/₂ hours of exposure
0.43–4.26	73.4–734	Death of all animals within 24 hours
0.31	53.8	No marked toxic symptoms

Chambers *et al.*[24] reported the lethal concentration for rats from a 15-minute exposure to be 22 mg./liter in air (3790 p.p.m.). The rats died within a period of 11 days. They showed lung irritation and pulmonary edema.

Buckell[25] reported the LC_{50} for mice to be 5 mg./liter in air from a 57-minute exposure.

Garland and Camps[26] reported 2 cases of exposure of humans in industrial operations to the vapors of methyl iodide. The first case was quoted from Jaquet reported in 1901. This case showed symptoms of vertigo, diplopia, and ataxia. There was evidence of urinary iodine. This individual developed delirium and serious mental disturbances. The second case was one observed by Garland and Camps. He was found to be drowsy, unable to walk, and with slurred incoherent speech. He showed iodine in the urine. Death occurred from 7 to 8 days after exposure.

Skin Contact and Oral Administration. Buckell[25] determined the LD_{50} for rats when given both subcutaneously and orally to be 0.15 to 0.22 mg./kg. of body weight. This author also reported that it would produce a "vestibular burn" if closely held to the human skin. It was indicated that clothing contaminated with methyl iodide should be removed immediately.

Absorption and Metabolism. Very little has been determined with regard to the absorption and metabolism of methyl iodide. From the little work that has been done, it would appear that the material is readily hydrolyzed to give inorganic iodine, which appears in the tissues and in urine.

[23] C. Bachem, *Arch. Exptl. Pathol. Parmakol.*, **122**, 69 (1927).

[24] W. H. Chambers, E. H. Krackow, C. C. Comstock, F. P. McGrath, S. V. Goldberg, L. H. Lawson, and J. K. MacNamee, *U. S. Army Chem. Corps Med. Div.* Research Rept. No. **23** (1950).

[25] M. Buckell, *Brit. J. Ind. Med.*, **7**, 122 (1950).

[26] A. Garland and F. E. Camps, *Brit. J. Ind. Med.*, **2**, 209 (1945).

METHYLENE CHLORIDE, CH_2Cl_2 (Dichloromethane)

1. *Source, Uses, and Industrial Exposures*

Methylene chloride is used as a solvent in a number of extraction processes, particularly where its high volatility is desirable. It has high solvent power for cellulose esters, fats, oils, resins, and rubber. It has been very useful in paint stripping operations.

Due to its high volatility (comparable to diethyl ether), high concentrations may be rapidly attained in poorly ventilated areas. This property should be recognized in planning an operation using methylene chloride. It should also be remembered that formulations for paint-stripping operations may contain other solvents as well as methylene chloride.

2. *Physical and Chemical Properties*

Physical state: colorless liquid
Molecular weight: 84.94
Specific gravity: 1.325 (20°/4°C.)
Melting point: —96.7°C.
Boiling point: 40.1°C.
Vapor density: 2.93 (air = 1)
Vapor pressure: 440 mm. Hg (25°C.)
Refractive index: 1.4237 (20°C.)
Per cent in "saturated" air: 55 (25°C.)
Density of "saturated" air: 2.06 (air = 1)
Solubility: 2 g. per 100 ml. water at 20°C.; soluble in ethanol, ethyl ether, and acetone
1 mg./liter ≈ 288 p.p.m. and 1 p.p.m. ≈ 3.48 mg./cu. meter at 25°C., 760 mm. Hg

3. *Physiological Response*

Methylene chloride is by far the least toxic of the four chlorinated methanes. The toxic effect is predominantly narcosis, although apparently not effective enough for good anesthesia. It does not cause significant organic injury.

The principal problem from use is the "drunkenness" that may cause inept operation, which may result in injury to the man himself or others around him.

The symptoms of excessive exposure may be dizziness, nausea, tingling or numbness of the extremities, sense of fullness in the head, sense of heat, stupor, or dullness, lethargy, and drunkenness. Exposure to very high concentrations may lead to rapid unconsciousness. Prompt removal from exposure usually results in complete recovery.

The high volatility should be balanced by proper enclosure or ventilation. Methylene chloride should be a particularly appropriate solvent for operations where a high rate of vaporization is required.

Acute Vapor Exposure. Svibely *et al.*[27] reported that the LD_{50} for mice was approximately 50 mg./liter or 15,000 p.p.m. for an 8-hour exposure. Survival of all animals was observed at approximately 11,000 p.p.m.

Lehmann and Schmidt-Kehl[28] reported levels of narcosis in cats; 32 mg./liter or 9000 p.p.m. caused "displacement of equilibrium" in 20 minutes but no narcosis. At 37.5 mg./liter or 10,000 p.p.m., light narcosis occurred in 220 minutes and deep narcosis at 293 minutes. They reported that cats and rabbits tolerated 6 to 7 mg./liter for 8 to 9 hours per day for 4 weeks with no significant observable changes.

Chronic Vapor Exposure. The most definitive work on methylene chloride was that of Heppel and associates.[29] They reported no pathology or growth depression in dogs, puppies, rats, guinea pigs, or rabbits exposed to 5000 p.p.m. 7 hours per day 5 days a week for 6 months. They could not detect depression of the central nervous system by ordinary observation. At 10,000 p.p.m. they observed light to moderate narcosis. Several animals died, apparently from pulmonary congestion.

Heppel and Neal[30] reported experiments with young male rats in an activity cage. At a concentration of 5000 p.p.m. the animals showed definite reduction in activity. As this technique has not been further studied, it is difficult to determine its significance. These authors proposed a "maximum allowable limit" of 500 p.p.m. In practical use, this has been found adequate.

Skin and Eye Contact. Methylene chloride is mildly irritating to the skin on repeated contact. The problem may be accentuated by the chemical being sealed to the skin by shoes or tight clothing. The situation is most severe with paint remover formulations that form a "skin" or film. Some irritation may be contributed by other constituents of the formulation.

Liquid methylene chloride splashed in the eye will be painful and irritating, but is not likely to cause serious injury.

Observations in Man. Experience in use has largely confirmed the laboratory toxicological data. Collier[31] reported 2 cases that recovered. The effects reported were "attributed to its anesthetic action and are largely subjective . . . headache, giddiness, stupor, irritability, numbness and tingling in the limbs."

Moskowitz and Shapiro[32] reported 4 cases of severe exposure, one of which was fatal.

The principal problem from excessive exposure will be the "drunkenness" and incoordination, which may lead to unsafe operation.

[27] J. L. Svibely, B. Highman, W. C. Alford, and W. F. von Oettingen, *J. Ind. Hyg. Toxicol.,* **29,** 382 (1947).

[28] K. B. Lehmann and L. Schmidt-Kehl, *Arch. Hyg. u. Bakteriol.,* **116,** 131 (1936).

[29] L. A. Heppel, P. A. Neal, T. L. Perrin, M. L. Orr, and V. T. Poterfield, *J. Ind. Hyg. Toxicol.,* **26,** 8 (1944).

[30] L. A. Heppel and P. A. Neal, *J. Ind. Hyg. Toxicol.,* **26,** 17 (1944).

[31] H. Collier, *Lancet,* **1,** 594 (1936).

[32] S. Moskowitz and H. Shapiro, *A.M.A. Arch. Ind. Hyg. Occupational Med.,* **6,** 116 (1952).

Absorption, Excretion, and Metabolism. von Oettingen and associates[33] studied the absorption and excretion of methylene chloride. They reported that it was rapidly absorbed and largely excreted by the lungs. Its metabolism has not been studied. The relatively high chemical stability of this compound would suggest that it might not be metabolized to a significant degree, but that it would be excreted unchanged via the lungs. This might account for its low toxicity.

4. Hygienic Standards of Permissible Exposure

The threshold limit of methylene chloride was established by the American Conference of Governmental Industrial Hygienists, in April, 1959, at 500 p.p.m. (1750 mg./cu. meter).

5. Flammability[34]

The autoignition temperature of methylene chloride is 1224°F. (624°C.). There is no flash point or fire point. Under practical conditions, methylene chloride can be considered essentially nonflammable.

The flammability in oxygen is: lower, 15.5 per cent; upper, 66 per cent.

6. Odor and Warning Properties

Methylene chloride has a not unpleasant sweetish odor at concentrations above 300 p.p.m. This cannot be considered a good warning as one could become adapted to it. The odor can be detected, however, at concentrations well below those having definite physiological actions other than odor.

CHLOROFORM, CHCl$_3$ (Trichloromethane)

1. Uses and Industrial Exposures

Chloroform was widely used for many years as an anesthetic. Due to resultant liver injury and cardiac sensitization, this use has been greatly reduced. It has some use as a chemical intermediate and is probably most used as a solvent. Although the use of chloroform as a solvent in industry is not extensive, it may often be used as a constituent in solvent mixtures. Chloroform has some insecticidal value, although it is not widely used as a fumigant.

2. Physical and Chemical Properties

Physical state: colorless liquid
Molecular weight: 119.39
Specific gravity: 1.49845 (15°/4°C.)
Melting point: —63.5°C.
Boiling point: 61.26°C.
Vapor pressure: 200 mm. Hg (25°C.)

[33] W. F. von Oettingen, C. C. Powell, N. E. Sharpless, W. C. Alford, and L. J. Pecora, *Natl. Insts. Health Bull.* No. **191** (1949).
[34] H. F. Coward and G. W. Jones, *U. S. Bur. Mines Bull.* No. **503** (1952).

Vapor density: 4.1 at 25°C. (air = 1)

Solubility: 1.0 g. per 100 ml. water at 15°C.; soluble in ethanol, ethyl ether, benzene, acetone, and CS_2

1 mg./liter ⊃ 206 p.p.m. and 1 p.p.m. ⊃ 4.89 mg./cu. meter at 25°C., 760 mm. Hg

3. Physiological Response

High concentrations of chloroform result in narcosis and anesthesia. The most outstanding effect from acute exposure is depression of the central nervous system. Responses associated with exposure to concentrations below anesthetic or preanesthetic level are typically inebriation and excitation passing into narcosis. Vomiting and gastrointestinal upsets may be observed. Exposures to high concentrations may result in cardiac sensitization to adrenalin. In cases of more chronic or repeated exposure to chloroform, liver injury is most typical. This is not unlike the effect of carbon tetrachloride. Although injury to the kidney is not as common as that to the liver, it may be observed from either acute or chronic exposure.

Although there is extensive literature on the use of chloroform as an anesthetic, there is surprisingly little controlled quantitative toxicological investigation, particularly of subacute and chronic exposure. There has also been relatively little in the way of clinical reports from industrial exposure. It is the present author's opinion that chloroform is much more toxic and represents more of an industrial hazard than is usually indicated. It should be considered much more toxic than methylchloride, perchloroethylene, or tricholorethylene and more nearly comparable to carbon tetrachloride.

Acute Vapor Exposure. Man and animals withstand very high concentrations of chloroform for a short period of time. Table 3 gives indications of the response to be expected in man.[35]

The response to acute exposure has been indicated in the summary as central nervous system depression, liver or kidney injury, and possible cardiac sensitization.

Chronic Vapor Exposure. There have been almost no quantitative toxicological studies of the response from chronic exposure to chloroform. Considering the long history of chloroform, there is surprisingly little clinical literature on chronic exposure. A recent publication by Challen, Hickish, and Bedford appeared in 1958.[36]

These authors studied an industrial operation where chloroform was being used. Groups exposed to concentrations varying between 77 and 237 p.p.m. exhibited definite symptoms. Apparently, there were also some high peak concentrations for very short periods of time. Symptoms were gastrointestinal distress and depression. Another group with shorter service was exposed to concentrations

---

[35] K. B. Lehmann and F. Flury, *Toxicology and Hygiene of Industrial Solvents.* Trans. by E. King and H. F. Smith, William & Wilkins, Baltimore (1943).

[36] P. J. R. Challen, D. E. Hickish, and J. Bedford, *Brit. J. Ind. Med.*, 15, 243 (1958).

from 21 to 71 p.p.m. They also showed symptoms of a comparable nature. Tests of both groups were made to determine possible liver injury, but the liver function tests used did not clearly demonstrate liver injury. It should be remembered, however, that liver function tests are notoriously insensitive to anything but severe liver injury. It is quite possible, as indicated by these authors, that there may have been mild liver injury in these cases. The recommendation of these authors that atmospheric exposure should be kept well below 50 p.p.m. is entirely in order.

TABLE 3

Physiological Response to Various Concentrations of Chloroform in Man[35]

Concentration		Response
mg./liter	p.p.m.	
70–80	14,336–16,384	Narcotic limiting concentration
20	4096	Vomiting, sensation of fainting
7.2	1475	Dizziness and salivation after a few minutes
5	1024	Dizziness, intracranial pressure, and nausea after 7 minutes
5	1024	Definite aftereffects; fatigue and headache still felt hours later
1.9	389	Endured for 30 minutes without complaint
1–1.5	205–307	Lowest amount that can be detected by smell

Skin. Oettel[37] has indicated that chloroform may cause skin irritation. A burning sensation may be noted, followed by erythema and hyperemia and finally vesication. Such a reaction would not be expected from ordinary contact with the material as a solvent. It would be expected to have a drying effect on the skin as many of the fat solvents do. In the eyes, the material would be expected to be quite painful but not to produce permanent injury.

Absorption, Excretion, and Metabolism. Chloroform is very rapidly absorbed from the lungs and is widely distributed in the body. von Oettingen *et al.*[38] exposed dogs to 15,000 p.p.m.; they reached equilibrium in 80 to 100 minutes. Concentrations at that time in mg. per 100 g. were indicated as 31.94 in the heart, 25.55 in the liver, 47.96 in the brain, and 49.05 in the blood. Most of the chloroform is excreted as such by the lungs. It has been indicated that a portion may be metabolized and excreted as chloride in the urine. Good biochemical and metabolic studies to indicate the nature and amount of any metabolism that does occur have not been carried out with chloroform.

4. *Hygienic Standards of Permissible Exposure*

The threshold limit of chloroform was established by the American Conference of Governmental Industrial Hygienists, in April, 1959, at 50 p.p.m. (240 mg./cu. meter).

[37] H. Oettel, *Arch. Exptl. Pathol. Pharmakol.*, **183**, 655 (1936).

[38] W. F. von Oettingen, C. C. Powell, N. E. Sharpless, W. C. Alford, and L. J. Pecora, *Natl. Insts. Health Bull.* No. **191** (1949).

5. Flammability

Chloroform is essentially not flammable.

6. Odor and Warning Properties

Chloroform has a sweetish odor. Lehmann[35] had indicated that the lowest concentration that could be detected was 200 to 300 p.p.m. This might be considered as some warning from exposure to acutely hazardous amounts. It is by no means a low enough level of detection to be considered a warning from chronic exposure.

BROMOFORM, $CHBr_3$ (Tribromomethane)

1. Uses and Industrial Exposures

Bromoform has been used as a chemical intermediate and in the medicinal field.

2. Physical and Chemical Properties

Physical state: colorless liquid
Molecular weight: 252.77
Specific gravity: 2.890 (20°/4°C.)
Melting point: 6 to 7°C.
Boiling point: 149.5°C.
Vapor density: 8.7 (air = 1)
Vapor pressure: 5.6 mm. Hg (25°C.)
Refractive index: 1.5980 (19°C.)
Per cent in "saturated" air: 0.7 (25°C.)
Density of "saturated" air: 1.05 (air = 1)
Solubility: 0.3 g. per 100 ml. water at 30°C.; soluble in ethanol, ethyl ether, and benzene
1 mg./liter ≏ 97 p.p.m. and 1 p.p.m. ≏ 10.34 mg./cu. meter at 25°C., 760 mm. Hg

3. Physiological Response

Relatively little is known of the toxicity of bromoform in industrial uses. It is indicated as causing depression of the central nervous system which results in narcosis. There are also indications of liver injury.

Flury and Zernik[7] reported narcosis in 8 minutes and death in one hour from exposure of dogs to 580 mg./liter of air (29,000 p.p.m.). It is a little difficult to determine the significance of this, as a saturated atmosphere of bromoform is about 7,000 p.p.m. It would be probable that these authors exposed their dogs to something approaching a saturated atmosphere.

Much of the experience in poisoning cases in humans has been from the oral administration of the material. This was summarized by W. F. von Oettingen in 1955.[1]

Bromoform should be suspected of being rather highly toxic and until more definitive toxicological experiments or experience has been reported, it should be handled with a great deal of caution. There is not enough information to suggest any level as a permissible concentration.

4. Hygienic Standards of Permissible Exposure

No standard has been suggested for bromoform.

5. Flammability

Bromoform is not flammable.

6. Odor and Warning Properties

Bromoform has a sweetish, chloroformlike odor. As the material is probably quite highly toxic, this is not to be considered a safe warning property, although the actual threshold of detection of the odor has not been determined.

IODOFORM, CHI_3 (Triiodomethane)

1. Uses and Industrial Exposures

Iodoform is used as a chemical intermediate and for medicinal purposes.

2. Physical and Chemical Properties

Physical state: yellow solid
Molecular weight: 393.78
Specific gravity: 4.008 (20°/4°C.)
Melting point: 119°C.
Boiling point: 210°C., subl. explodes
Vapor density: 13.6 (air = 1)
Refractive index: 1.800 (20°C.)
Solubility: 0.01 g. per 100 ml. water at 25°C.

3. Physiological Response

Most of the problems associated with iodoform have been related to its topical application as an antiseptic material and to its oral administration. Absorption of significant amounts of this material may result in central nervous system depression and injury to the heart, liver, and kidneys. von Oettingen, in 1955,[1] reviewed the toxicological problems related to medicinal use. This compound is not at present of great significance in industry and will, therefore, not be discussed in detail.

4. *Hygienic Standards of Permissible Exposure*

No standard has been proposed for iodoform.

CARBON TETRACHLORIDE, CCl₄ (Tetrachloromethane)

1. *Uses and Industrial Exposures*

Carbon tetrachloride is used widely and for a great many purposes, largely because of its easy availability and low cost. It is used as a chemical intermediate, as a fumigant, as a fire-extinguishing agent, and as a suppressant of the flammability of mixtures of more flammable materials. Its most serious toxicological problems have been introduced, however, by its use as a solvent.

Carbon tetrachloride in years past was one of the most used and at the same time most misused of the chlorinated aliphatic solvents. It was cheap and easily available; therefore, it was found in the home, garage, and the shop. It might be found in pop bottles, tin cans, drinking glasses, or any odd container. It has been sprayed with a paint spray gun, swabbed by hand with a rag or paint brush, and used in all kinds of open-bucket cleaning operations. Considering the fact that carbon tetrachloride is seriously toxic when inhaled, it is not surprising that such misuse has led to numerous cases of toxic injury and a number of fatalities.

The serious misuse of this material in open-bucket cold cleaning operations has been greatly reduced in the last few years in many industries. Carbon tetrachloride, when used under proper closed conditions or proper ventilation, is a useful material. However, it is probably still readily available in odd and sometimes unlabeled containers in the home and small shop and, in this manner, may represent a serious problem to those people not well acquainted with its properties.

The use of carbon tetrachloride as a fire extinguisher has not resulted in toxic hazards in any way comparable to those seen from its use as a solvent.

Carbon tetrachloride is also used in fumigant mixtures, particularly in fumigating grain. It is an active insecticide and is also very effective in suppressing the flammability of more flammable fumigants.

2. *Physical and Chemical Properties*

Physical state: colorless liquid
Molecular weight: 153.84
Specific gravity: 1.585 (25°/4°C.)
Freezing point: −22.8°C.
Boiling point: 76.75°C.
Vapor density: 5.3 (air = 1)
Vapor pressure: 113 mm. Hg (25°C.)
Refractive index: 1.46305 (15°C.)
Per cent in "saturated" air: 15 (25°C.)
Density of "saturated" air: 1.64 (air = 1)

Solubility: 0.08 g. per 100 g. water at 20°C.; miscible with alcohol, diethyl ether, and benzene

1 mg./liter ⊅ 159 p.p.m. and 1 p.p.m. ⊅ 6.29 mg./cu. meter at 25°C., 760 mm. Hg

3. *Physiological Response*

The response from exposure to high concentrations of carbon tetrachloride will be depression of the central nervous system. If the concentration is not high enough to lead to rapid loss of consciousness, other indications of nervous system effect such as dizziness, vertigo, headache, depression, mental confusion, and incoordination, will be observed. Many individuals will also show gastrointestinal responses such as nausea, vomiting, abdominal pain, and diarrhea. Some individuals will respond with nausea and vomiting to surprisingly low concentrations. Some individuals may be nauseated by the "faintest smell of the stuff." Whether this is a conditioned reflex or a direct nervous system effect is difficult to determine. Functional and destructive injury of the liver and kidney may occur from a single acute exposure but it is much more likely to occur from repeated chronic exposures. In a case of long-term chronic exposure to low concentrations, the kidney and liver injuries completely dominate the picture. The milder the exposure, the more tendency for the injury to be predominantly in the liver. At threshold concentrations, the injury of the liver will appear mostly as malfunction and/or enlargement. Many enlarged livers are observed in animals at the threshold of response. The detection of an enlarged liver in humans should be considered of importance, although it will be recognized that enlargement of the liver may occur from a great many different causes.

The 25 p.p.m. proposed as an industrial hygiene standard for carbon tetrachloride should be considered a ceiling, and the time-weighted average of fluctuations in concentrations below that ceiling should be around 10 p.p.m. Occasional short-term peaks above the ceiling should be kept at an absolute minimum and should not be superimposed upon a significant continuous level of exposure. It has been recognized that the concurrent intake of significant amounts of alcohol with exposure to carbon tetrachloride may greatly increase the probability of injury.

Acute Vapor Exposure. Adams et al.[39] reported the response of laboratory animals to single exposures to various concentrations of carbon tetrachloride. The maximum time-concentrations in air survived by rats were as follows: 12000 p.p.m. for 15 minutes, 7300 p.p.m. for 1.5 hours, 4600 p.p.m. for 5 hours, and 3000 p.p.m. for 8 hours.

The maximum time-concentrations in air having no adverse effects upon male rats were as follows: 3000 p.p.m. for 6 minutes, 800 p.p.m. for 30 minutes, and 50 p.p.m. for 7 hours.

[39] E. M. Adams, H. C. Spencer, V. K. Rowe, D. D. McCollister, and D. D. Irish, *A.M.A. Arch. Ind. Hyg. Occupational Med.,* **6,** 50 (1952).

Comparable data were reported for rabbits by Lehmann.[40]

The data from these two sources would indicate that rabbits and guinea pigs show a somewhat greater tolerance for carbon tetrachloride than rats do. This is not the experience with chronic exposures. In any case, the acute data given for the different animals are in the same range.

The response to a single exposure of carbon tetrachloride has been reviewed in the summary. The responses observed in animals and man are reasonably comparable. There seems to be a higher probability of significant kidney response in man than is observed in animals. Qualitatively, such injury is observed in both animals and man. Histopathological and biochemical studies of acutely injured animals will show marked hepatic injury, as has been demonstrated by increased plasma prothrombin clotting time, an increase of serum phosphatase, an increase in liver weight, an increase of total lipid content of the liver, and by central fatty degeneration of the liver. Renal injury was not apparent in the acute exposure of rats in the studies of Adams.[39] Quite significant kidney injury has been reported, however, from what were thought to be single exposures in humans.

Chronic Vapor Exposure. While responses referable to the nervous system or gastrointestinal tract may still be observed in chronic exposure, they are much less important factors. These effects may not be noticed at all following a long period of chronic exposure to low concentrations. The organic and functional injury of the internal organs becomes predominant, particularly of the liver and the kidney. It is noticeable in the literature of the last ten years that carbon tetrachloride has become the classical drug for producing liver injury.

A review of the literature on carbon tetrachloride up to a few years ago is given by von Oettingen.[1]

The most comprehensive toxicological investigation in animals is that of Adams *et al.*[39] These investigators studied rats, guinea pigs, rabbits, and monkeys, which were given repeated 7-hour per day exposures 5 days per week.

At a concentration of 400 p.p.m. (2.52 mg./liter), rats and guinea pigs suffered severe intoxication. Less than half of them lived for 127 exposures during a period of 173 days. There was an increase in liver weight up to twice that of the controls and a moderate increase in kidney weight. Histological examination of the tissues showed central fatty degeneration with cirrhosis of the liver and slight parenchymatous degeneration of the tubular epithelium of the kidneys. Animals examined within 2 weeks demonstrated advanced liver and kidney changes by that time.

At a concentration of 200 p.p.m. (1.26 mg./liter), rats and guinea pigs still showed a definite response, and high mortality. Biochemical and histological studies were comparable but less severe than in the 400-p.p.m. exposure.

At a concentration of 100 p.p.m. (0.63 mg./liter), rats, rabbits, guinea pigs, and monkeys tolerated 146 to 163 exposures without evidence of adverse effect on

[40] K. B. Lehmann, *Arch. Hyg.*, **74**, 1 (1911).

gross appearance, behavior, growth, etc. They all showed histopathological changes. The changes were borderline in the monkey.

At a concentration of 50 p.p.m. (0.32 mg./liter), rats, guinea pigs, and rabbits tolerated up to 134 exposures in 187 days, showing increased liver weight and moderate fatty degeneration of the liver. Monkeys tolerated 198 exposures in 277 days without evidence of gross, histopathological, or biochemical change.

At a concentration of 25 p.p.m. (0.16 mg./liter), rats, guinea pigs, and rabbits tolerated 137 exposures in 191 days. They showed a slight increase in liver weight and some fatty changes but no cirrhosis; otherwise there was no difference from normal animals.

At a concentration of 10 p.p.m. (0.063 mg./liter), rats and guinea pigs showed a slight to moderate fatty degeneration and increased liver weight; rabbits were normal.

At a concentration of 5 p.p.m. (0.032 mg./liter), rats showed no effect. Guinea pigs also showed essentially no effect, although there was slight increase in liver weight in the females.

These authors proposed that a standard of 25 p.p.m. would be acceptable if the average of the air analysis did not exceed 10 p.p.m.

There have been innumerable clinical investigations of carbon tetrachloride exposure. This experience parallels in general the animal investigations. From general industrial experience, it would appear probable that the standards suggested by Adams et al.[39] are satisfactory.

Oral Ingestion. Availability of carbon tetrachloride in odd containers around the home and shop has made oral ingestion a serious problem. The LD_{50} for rats is reported by McCollister et al.[41] as 2.92 g./kg.

Skin and Eyes. As carbon tetrachloride is a good fat solvent, it will remove the fats from the skin and, in so doing, will cause a dry disagreeable feeling and may lead to secondary infection. Contact with the eyes may cause a transient disagreeable irritation but does not lead to serious injury.

Absorption through the skin from the vapor phase was studied by McCollister et al.[42] By using radioactive carbon tetrachloride, they were able to detect small amounts in the blood from exposure of the skin to concentrations of 485 and 1150 p.p.m. in the air. While traces were absorbed through the skin under these conditions, their conclusions were that "absorption through the intact skin would appear to be of no practical significance in considering the hazard to the health of industrial workers exposed to concentrations of at least as high as 1150 p.p.m. in the air."

Absorption, Excretion, and Metabolism. Early studies to measure absorption and excretion of carbon tetrachloride have been handicapped by analytical difficulties. Studies were made only at very high concentrations.

[41] D. D. McCollister, R. L. Hollingsworth, F. Oyen, and V. K. Rowe, *A.M.A. Arch. Ind. Health,* **13,** 1 (1956).

[42] D. D. McCollister, W. H. Beamer, G. J. Atchison, and H. C. Spencer, *J. Pharmacol. Exptl. Therap.,* **102,** 112 (1951).

McCollister *et al.*[42] studied the absorption distribution and elimination of carbon tetrachloride by using radioactive carbon. This allowed them to study these factors at concentrations which were physiologically significant from a chronic exposure point of view. They exposed monkeys to concentrations of 46 p.p.m. (0.290 mg./liter) of carbon tetrachloride which was C^{14}-labeled. Approximately 30 per cent was absorbed. The equivalent of at least 51 per cent of the carbon tetrachloride absorbed during an inhalation period was estimated to have been eliminated in the expired air within 1800 hours. The remainder was excreted largely in the urine and feces. Approximately 4.4 per cent was eliminated as carbon dioxide. Some 94.3 per cent of the radioactivity in the urine was as a nonvolatile, unidentified intermediate. Small amounts occurred in the urea and carbonates.

It will be noted from the work of McCollister *et al.*[42] that the excretion of carbon tetrachloride from the system was relatively slow. This slow disappearance may account for the toxic effect of small amounts of carbon tetrachloride inhaled at repeated short interval exposures. The distribution would indicate highest concentrations in fat, decreasing in liver, bone marrow, blood, and other internal organs.

Ball and Kay[43] have reported a drop in serum esterase activity as a "sensitive index of exposure."

Many reports have appeared in the literature concerning the protective action of a wide range of vitamins, amino acids, etc. The most consistent report concerns the protective action of the sulfur-containing amino acids, particularly methionine. When the diet is low in the sulfur-containing amino acids, the liver is much more sensitive to damage by many agents. Supplementing the diet up to the normal level does give protection from liver damage. Beyond the normal level, however, it does not seem to be of any advantage. This has been reviewed by Shaffer *et al.*[44] This has also been studied by Drill *et al.*[45,46] Another review is by L. L. Miller.[47]

There have been publications indicating that liver necrosis induced by carbon tetrachloride in certain mice would be followed by hepatomas. This work was discussed by Eschenbrenner and Miller[48] of the National Cancer Institute. Another discussion of the production of hepatomas is given by Stowell *et. al.*[49] of the University of Kansas.

There have been a great many publications concerned with the possible mechanism of the toxic action of carbon tetrachloride. Many different enzyme systems of the liver have been studied and the pattern of inhibition or noninhibition is both conflicting and confusing. The most interesting suggestion that has appeared so far seems to be that of Christe and Judah.[50]

[43] W. L. Ball and K. Kay, *A.M.A. Arch. Ind. Health,* **14,** 319 (1956).

[44] C. B. Shaffer, C. P. Carpenter, and C. Moses, *J. Ind. Hyg. Toxicol.,* **28,** 87 (1946).

[45] V. A. Drill, T. A. Loomis, and J. Belford, *J. Ind. Hyg. Toxicol.,* **29,** 180 (1947).

[46] V. A. Drill and T. A. Loomis, *J. Pharmacol. Exptl. Therap.,* **90,** 137 (1947).

[47] L. L. Miller, *Occupational Med.,* **5,** 194 (1948).

[48] A. B. Eschenbrenner and E. Miller, *Gen. Natl. Research Inst.,* **6,** 325 (1946).

[49] R. E. Stowell, C. S. Lee, K. K. Tsuboi, and A. Villasana, *Cancer Research,* **11,** 345 (1951).

[50] G. F. Christe and J. D. Judah, *Proc. Roy. Soc. London,* **B-142,** 241 (1954).

These authors suggest that the mitochondria of the cell are attacked by carbon tetrachloride. The physical disruption of the mitochondria appears to be associated with a disorganization of the enzyme systems involved but not in a way that would indicate specific inhibition. This may account for the erratic results indicated in many publications on specific enzyme systems.

In contrast, Brody et al.[50a] proposed that carbon tetrachloride does not act on the liver cell, but rather that the effects are promulgated via an action upon the central nervous system. This results in blood vessel constriction and release of catechol amines by the adrenal medulla, which in turn causes the observed liver damage. Recknagle et al.[50b] suggested that the primary lesion was inhibition or destruction of a hepatic triglyceride secreting mechanism. Smuckler et al.[50c] suggested that the primary lesion was a defect in protein synthesis.

4. Hygienic Standards of Permissible Exposure

The threshold limit of carbon tetrachloride was established by the American Conference of Governmental Industrial Hygienists, on April, 1959, at 25 p.p.m. (160 mg./cu. meter). This limit should be considered a ceiling with averages around 10 p.p.m.

5. Flammability

Carbon tetrachloride is not flammable. It has no flash or fire point. It is not explosive and will not support combustion.

6. Odor and Warning Properties

Carbon tetrachloride has a sweetish odor. It is not considered particularly disagreeable by most people, although some people may be nauseated by small amounts. The odor is one to which the average individual will become readily adapted. The odor would certainly not be considered a satisfactory warning of excessive exposure. The threshold of detection of the odor of carbon tetrachloride is approximately 79 p.p.m. and the odor is strong at 176 p.p.m.[51]

CARBON TETRABROMIDE, CBr₄ (Tetrabromomethane)

1. Uses and Industrial Exposures

Carbon tetrabromide is used as a chemical intermediate.

2. Physical and Chemical Properties

Physical state: colorless solid
Molecular weight: 331.67

[50a] T. H. Brody, D. N. Calvert, and A. F. Schneider, J. Pharmacol. Exptl. Therap., **131,** 341 (1961).

[50b] R. O. Recknagel and B. J. Lombardi, J. Biol. Chem., **236,** 564 (1961).

[50c] E. A. Smuckler, O. A. Iseri, and E. P. Benditt, Biochem. Biophys. Communi., **5,** 270, (1961).

[51] Chemical Safety Data Sheet SD-3. Manufacturing Chemists Assoc., Washington, D. C., 1946.

Specific gravity: 3.42 (20°C.)

Melting point: α, 48.4°C.; β, 90.1°C. (slight decomposition on melting)

Boiling point: 189.5°C. (slight decomposition)

Vapor density: 11.4 (air = 1)

Refractive index: 1.59998 (99.5°C.)

Solubility: 0.024 g. per 100 ml. water at 30°C.; soluble in ethanol, ethyl ether, and chloroform

3. Physiological Response

Carbon tetrabromide is a highly toxic material. Acute exposure to high concentrations will cause upper respiratory irritation and injury to the lungs, liver, and kidneys. The response to chronic exposure at very low concentrations will be almost entirely liver injury. The material is a potent lacrymator, even at these very low levels.

Chronic Vapor Exposure. There is no information published in the literature on the effect of chronic exposure to carbon tetrabromide in air. Chronic exposures of rats to vapors of carbon tetrabromide have been studied.[5f] Exposure was 7 hours/day, 5 days/week for 6 months. If the concentration in air was determined by the usual technique of combustion and determination of halogen, the concentration without effect was 0.1 p.p.m. by volume. If the concentration in air was determined by polarographic method, the concentration was 0.3 to 0.5 p.p.m. by volume. Higher concentrations than this caused poor growth, and fatty and degenerative changes in the liver.

If one is to monitor the atmosphere for control of exposure to carbon tetrabromide, he should note the difference in results obtained by the two methods indicated.

Oral. The lethal dose by oral administration was found to be between 1 and 3g./kg. of body weight in the rat.

Eye and Skin Irritation. In the eye, undiluted material causes severe irritation and permanent corneal damage. If the material is promptly washed from the eye, pain and irritation will be noted but the corneal damage will be temporary.

Skin contact causes relatively slight irritation. If the material is bandaged tightly to the skin, it may cause hyperemia and a moderate edema. From observations made when the material was bandaged onto the skin in fair amounts, there was no indication of toxic absorption.

4. Hygienic Standards of Permissible Exposure

No standard has been adopted. A tentative standard of 0.1 p.p.m. is suggested on the basis of animal work in the Biochemical Research Laboratory of The Dow Chemical Co.[5f]

5. Flammability

Carbon tetrabromide is nonflammable.

6. *Warning Properties*

Carbon tetrabromide has significantly lacrymatory effects upon the eye at low concentrations. This is a reasonably good warning property.

METHYLENE CHLOROBROMIDE, CH_2ClBr

(Monobromomonochloromethane)

1. *Uses and Industrial Exposures*

Methylene chlorobromide is used as a fire-extinguishing agent and, to some extent, as a chemical intermediate.

2. *Physical and Chemical Properties*

Physical state: colorless liquid
Molecular weight: 129.40
Special gravity: 1.991 (19°/4°C.)
Freezing point: —88°C.
Boiling point: 68 to 69°C.
Vapor density: 4.4 (air = 1)
Vapor pressure: 155 to 160 mm. Hg (25°C.)
Refractive index: 1.4850 (26°C.)
Per cent in "saturated" air: 21 (25°C.)
Density of "saturated" air: 1.72 (air = 1)
Solubility: Insoluble in water; soluble in organic solvents
1 mg./liter $\backsim$ 189 p.p.m. and 1 p.p.m. $\backsim$ 5.3 mg./cu. meter at 25°C., 760 mm. Hg

3. *Physiological Response*

Methylene chlorobromide is one of the less toxic halo methanes. It falls roughly in a class with methylene chloride but is somewhat more toxic. The pattern of response to this material is central nervous system depression. There appears to be very little organic injury following either acute or chronic exposure except lung irritation from acute exposure at high levels.

Primary concern in industrial use would be with the nervous system depression that may lead to disturbances of vision and coordination. Such changes may lead to poor manual manipulation and, therefore, unsafe operation.

Acute Vapor.   Svirbely *et al.*[52] reported the LD_{50} by inhalation for mice with a 7-hour exposure as 12.03 mg./liter of air (2273 p.p.m.).

Comstock *et al.*[53] reported the concentration of ethylene chlorobromide causing anesthesia. Concentrations as low as 3000 p.p.m. produced light narcosis.

[52] J. L. Svirbely, B. Highman, W. C. Alford, and W. F. von Oettingen, *J. Ind. Hyg. Toxicol.,* **29**, 382 (1947).

[53] C. Comstock, R. W. Fogleman, and F. W. Oberst, *A.M.A. Arch. Ind. Hyg. Occupational Med.,* **7**, 526 (1953).

Transient pulmonary edema was observed at concentrations below 27,000. At higher concentrations, interstitial pneumonitis resulted in delayed deaths. Delayed deaths were observed after exposure to 20,000 p.p.m. Deaths during exposure occurred only above 27,000 p.p.m.

Matson and Dufour[54] reported limited acute studies on guinea pigs. At concentrations of 0.8 to 1 per cent in air, the animals survived 1-hour exposures, recovering in 2 days. At 2-hour exposures, 1 out of 3 guinea pigs died. Autopsy findings showed lung injury from the 1-hour exposure. At a concentration of 2 to 2.4 per cent, animals recovered after a $1/2$-hour exposure. After a 1-hour exposure, 1 animal out of 3 died. After a 2-hour exposure, 2 out of 3 died. The principal toxicological observation from the exposure was lung injury.

Chronic Vapor. Repeated exposures were reported by Svirbely et al.[52] They exposed animals 7 hours/day, 5 days/week for a period up to 14 weeks to a concentration of 1000 p.p.m. Rats, rabbits and dogs survived these exposures and showed no evidence of toxic response. Growth was normal and there were no histopathological changes.

Torkelson et al.[54a] indicate that rats will survive without injury up to 400 p.p.m. in air for 7 hours/day, 5 days/week for 6 months. They will show some injury at 1000 p.p.m.

Oral. Highman et al.[55] 1948, reported on the administration of monochloromonobromomethane by stomach tube to mice. No changes were noted after the administration of 500 mg./kg. Single doses of 3000 and 4500 mg./kg. were followed by fatty degeneration of the liver and kidneys. These same authors observed no liver or kidney injury in animals exposed to its vapors dispersed in air. They commented that the difference in liver injury could be due to a different pathway of absorption from the gastrointestinal tract than from the lungs.

A single oral dose to rats of 1 g./kg. of body weight or less has no apparent effect. A dose of 3 g./kg. of body weight by mouth kills most animals within 24 hours.

Skin and Eyes. Methylene chlorobromide, when applied repeatedly to the open skin of rabbit, will result in some hyperemia and exfoliation. When bandaged on, it will rapidly produce moderate irritation and hyperemia. It will, therefore, be expected that this material would have a rather mild effect from ordinary contact. It is well to avoid wetting clothing or shoes.

Absorption, Excretion, and Metabolism. Svirbely et al.[52] determined the inorganic bromide of blood serum and urine in dogs exposed to 1000 p.p.m. of methylene chlorobromide in air. These animals were exposed daily, 7 hours/day, 5 days/week. During the third week, the blood serum inorganic bromide had increased from a normal of 5 to 10 mg. % up to levels over 200 mg. %. During the

[54] A. F. Matson and R. E. Dufour, *Underwriters Lab. Bull. Research,* No. **42** (Aug., 1948).

[54a] T. R. Torkelson, F. Oyen, and V. K. Rowe, *Am. Ind. Hyg. Assoc. J.,* **21,** 275 (1960).

[55] B. Highman, J. L. Svirbely, W. F. von Oettingen, W. C. Alford, and L. J. Pecora, *A.M.A. Arch. Pathol.,* **45,** 299 (1948).

thirteenth and fourteenth weeks, the concentration was over 300 mg. of inorganic bromide per 100 ml. of blood. The same authors determined the blood concentration of volatile bromide expressed as milligrams of methylene chlorobromide per 100 ml. of blood. Taken immediately at the end of the exposure, concentrations between 5 and 9 mg. of methylene chlorobromide per 100 ml. were observed. At periods of 17 to 65 hours after the end of the last exposure, no volatile bromide was observed in one dog and concentrations less than 1 mg. in the other. It would appear that methylene chlorobromide as such appears in the blood during exposure to vapors in air but disappears very rapidly on cessation of exposure. Apparently, a quite significant amount of material is hydrolyzed to yield inorganic bromide.

Female rats exposed to 400 p.p.m. 5 days/week for 8 hours/day showed a blood bromide of 70 mg. %. Dogs similarly exposed showed a blood bromide of 59 mg. %.[54a]

4. Hygienic Standards of Permissible Exposure

The threshold limit of methylene chlorobromide was tentatively established by the American Conference of Governmental Industrial Hygienists in April, 1959, at 400 p.p.m. (2100 mg./cu. meter).

5. Flammability

Methylene chlorobromide has no flash or fire point. It is an effective fire-extinguishing agent.

6. Odor and Warning Properties

Methylene chlorobromide has a distinctive odor at 400 p.p.m. It is a good warning at well below any acutely hazardous concentration. The odor is distinctive at the acceptable concentration and so gives some warning but it is not disagreeable enough to drive anyone from the area and workmen may well tolerate a level well above the acceptable level for chronic exposure.

ETHYL CHLORIDE, CH_3CH_2Cl (Monochloroethane)

1. Uses and Industrial Exposures

Ethyl chloride has been used as a chemical intermediate, as an anesthetic, and to a limited degree as a refrigerant. The most serious problems have been with its use as an anesthetic. Very little physiological difficulty has been encountered in industry. The major problems are fire and explosion.

2. Physical and Chemical Properties

Physical state: colorless gas
Molecular weight: 64.52
Specific gravity: 0.9214 (0°/4°C.)
Melting point: −138.7°C.

Boiling point: 12.2°C.

Vapor density: 2.23 (air = 1)

Solubility: 0.57 g. per 100 ml. water at 20°C., 48 g. per 100 ml. ethanol at 21°C., soluble in ethyl ether

Flash point: —58°F. (closed cup) ; 45°F. (open cup)

1 mg./liter ⊃⊂ 379 p.p.m. and 1 p.p.m. ⊃⊂2.64 mg./cu. meter at 25°C., 760 mm. Hg

3. *Physiological Response*

The physiological problems introduced by ethyl chloride are much like those of methyl chloride except that the threshold of response to ethyl chloride is very much higher than that of methyl chloride. The principal problem in industrial use will be that typical of an anesthetic material where the "drunkenness" and incoordination may lead to inept operation and, therefore, the possibility of an injury.

Acute Vapor.   The acute toxicity for animals was reported by Sayers, et al.[56] in 1929 (see Table 4).

The narcotic concentrations for man were reported by Lehmann and Flury[35] (see Table 5).

TABLE 4

Response of Guinea Pigs to Ethyl Chloride Vapor in Air[56]

Concentration, %	Exposure time, min.	Response
23–24	5–10	Unconscious, some deaths
15.3	40	Some deaths in 30 min.; some survived 40 min.
9.1	30	Survived—histopathological changes in lungs, liver, etc.
5	40	Survived—lung congested
4	122	Survived—returned to normal
	270	Survived—histopathological changes in the lungs, liver, and kidneys
	540	Some deaths
2	270	Survived—returned to normal
	540	Survived—histopathological changes in liver and kidneys
1	810	Survived—returned to normal

Chronic Vapor Exposure.   There has been essentially no experience with chronic vapor exposure to ethyl chloride. No chronic studies on animals have been reported. From the use of this material in industry, where the concentration must be maintained at a level low enough to avoid the fire and explosion hazards, one would not expect any significant chronic toxicity problems.

[56] R. R. Sayers, W. P. Yant, B. H. Thomas, and L. B. Burger, *U. S. Public Health Bull.* No. **185** (1929).

Absorption, Excretion, and Metabolism. Ethyl chloride is readily absorbed through the lungs and excreted by the lungs. Thorough investigation of absorption, excretion, or metabolism has not been reported. The indications are that ethyl chloride is excreted by the lungs and is not metabolized to any significant degree.

Perhaps the most serious problem from severe acute exposure, other than the anesthetic effect, is the possibility of potentiation of adrenalin. The resultant cardiac problems may be the most serious consequences of accidental exposure to very high concentrations.

TABLE 5

Narcotic Ethyl Chloride Concentrations in Man[35]

Concentration		Response
mg./liter	p.p.m.	
105.6	40,000	After 2 inhalations, stupor, irritation of eyes, and stomach cramps
88.7	33,600	After 30 seconds, quickly increased toxic effect
66.0	25,000	Lack of coordination
52.8	20,000	After 4 inhalations, dizziness and slight abdominal cramps
50.4	19,000	Weak analgesia after 12 minutes.
34.3	13,000	Slight symptoms of poisoning

Skin Exposure. Due to the fact that ethyl chloride is a gas at normal room temperatures, the liquid spilled on the skin may cause rapid cooling and possibly frost bite.

A great deal of the general information on the physiological response to ethyl chloride has been obtained in investigations with this material as an anesthetic.

4. Hygienic Standards of Permissible Exposure

The threshold limit of ethyl chloride was established by the American Conference of Governmental Industrial Hygienist in April, 1959, at 1000 p.p.m. (2600 mg./cu. meter).

5. Flammability

The flammable limits of ethyl chloride are 3.75 to 15.40 per cent by volume. The ignition temperature is 519°C.

6. Odor and Warning Properties

Ethyl chloride has an ethereal, somewhat pungent odor. As it requires very high concentrations to have serious physiological effects, this may be considered a fair warning property. The threshold of detection of the odor has not been determined.

ETHYL BROMIDE, CH_3CH_2Br (Monobromoethane)

1. Uses and Industrial Exposures

The principal use of ethyl bromide is as a chemical intermediate. Although it has been proposed occasionally as an anesthetic, it has not been used to any extent for that purpose.

2. Physical and Chemical Properties

Physical state: colorless liquid
Molecular weight: 108.98
Specific gravity: 1.4505 (25°/4°C.)
Melting point: −119°C.
Boiling point: 38.4°C.
Vapor density: 3.76 (air = 1)
Vapor pressure: 475 mm. Hg (25°C.)
Refractive index: 1.42386 (25°C.)
Per cent in "saturated" air: 62.5 (25°C.)
Density of "saturated" air: 2.7 (air = 1)
Solubility: 0.91 g./100 ml. water at 20°C.; soluble in ethanol and ethyl ether
1 mg./liter ≏ 224.3 p.p.m. and 1 p.p.m. ≏ 4.46 mg./cu.meter at 25°C., 760 mm. Hg

3. Physiological Response

The primary response from exposure to ethyl bromide is central nervous system depression, as in the case of ethyl chloride. In contrast, however, ethyl bromide causes irritation of the lungs and definite injury to the liver and kidneys. Chronic quantitative studies have not been made and there has been relatively little experience in industry with the use of ethyl bromide. It would, therefore, be suggested that some caution be given even within the limits of the present standard proposed.

Acute Vapor Exposure. Sayers et al.[56] reported acute studies on the response of guinea pigs to ethyl bromide (see Table 6).

Limited human experience has indicated, in addition to central nervous system depression, the possibility of lung congestion and degeneration of the liver and kidney tissues.

Chronic Exposure. No information is available on quantitative chronic exposure of animals to the vapors of ethyl bromide. The possibility of some significant chronic problems should be recognized from the nature of the tissue injury observed in acute exposure.

Absorption, Excretion, and Metabolism. Relatively little study has been given to this compound. It has been indicated, however, that it may be hydrolyzed to a significant degree in the body, resulting in the formation of inorganic bro-

TABLE 6

Physiological Response of Guinea Pigs to Ethyl Bromide in Air[56]

Concentration, %	Exposure time, min.	Response
14	10	Unconscious—death in several days
5	98	Unconscious—died 1 hour later
6	10	Survived—lung injury
2.4	90	Died in 18 hours
	30	Some delayed deaths; pathological changes in lungs, liver, and spleen
	10	Dizzy—survived—slight congestion in lungs and liver
1.2	270	Some deaths—histopathological changes
	90	Survived—histopathological changes
	55	Survived—slight histopathological changes
0.65	270	Some deaths
	180	Survived—histopathological changes
0.07	810	One death and histopathological changes
	540	Survived—normal

mides. Even the limited information available would indicate that it is much less toxic than methyl bromide, possibly due to the fact that it is an ethylating rather than a methylating agent.

4. Hygienic Standards of Permissible Exposure

The threshold limit of ethyl bromide was established by the American Conference of Governmental Industrial Hygienists, in April, 1959, at 200 p.p.m. (890 mg./cu. meter).

5. Flammability

The flammable limits of ethyl bromide are 6.75 to 11.25% by vol. in air.[57]

6. Odor and Warning Properties

Ethyl bromide has a definite though not particularly distinctive odor. The threshold of odor response has not been determined. Until more definite information is available, the odor should not be considered as a good warning property as the tolerable concentration for ethyl bromide is 200 p.p.m.

ETHYL IODIDE, CH_3CH_2I (Monoiodoethane)

1. Uses and Industrial Exposures

Ethyl iodide is used largely as a chemical intermediate. In the past, it has had limited medicinal use.

[57] H. F. Coward and G. W. Jones, *U. S. Bur. Mines Bull.* No. **279** (1939).

2. *Physical and Chemical Properties*

Physical state: colorless liquid
Molecular weight: 155.98
Specific gravity: 1.9245 (25°/4°C.)
Melting point: −108.5°C.
Boiling point: 72.2°C.
Vapor density: 5.4 (air = 1)
Vapor pressure: 137 mm. Hg (25°C.)
Refractive index: 1.5076 (25°C.)
Per cent in "saturated" air: 18 (25°C.)
Density of "saturated" air: 1.8 (air = 1)
Solubility: 0.4 g./100 ml. water at 20°C.; soluble in ethanol, ethyl ether, benzene, and chloroform
1 mg./liter ≎ 156.7 p.p.m. and 1 p.p.m. ≎ 6.38 mg./cu. meter at 25°C., 760 mm. Hg

3. *Physiological Response*

Most of the information on the toxicity of ethyl iodide has been obtained because of the interest in this material for the treatment of fungus infections. It has been administered to man as a vapor which is inhaled and absorbed by the lungs. It will cause central nervous system depression and, it has been indicated, it may affect the kidneys, thyroid, lungs and the liver.

Flury and Zernik[7] have reported the physiological effect of various concentrations on mice. These are shown in Table 7.

TABLE 7

Physiological Response to Ethyl Iodide Vapors in Air (Mice)[7]

Concentration		Duration of exposure, hrs.	Response
mg./liter	p.p.m.		
1.87	290	3	Fatal
0.94	150	24	Fatal
0.75	120	?	Tolerated

The best discussion of the response of man to the inhalation of the vapors of ethyl iodide are given in the publications of Bloomgarth et al.[58] and Schwartz.[59]

While no standard has been established for ethyl iodide and there is very little basis to establish a standard, it should be recognized from the table of Flury and Zernik that mice tolerate a concentration of 120 p.p.m. A concentration of 150 p.p.m. for 24 hours is fatal. It would, therefore, be advisable to keep concentra-

[58] H. L. Bloomgarth, D. R. Gilligan, and J. H. Schwartz, *J. Clin. Invest.*, **9**, 635 (1931).
[59] J. H. Schwartz, *Arch. Dermatol. and Syphilol.*, **40**, 962 (1939).

tions well below 100 p.p.m. for relatively short periods of exposure. Such exposures should not be repeated frequently.

4. *Hygienic Standards of Permissible Exposure*

No standard has been established for ethyl iodide.

5. *Odor and Warning Properties*

Ethyl iodide has an ethereal odor but it would not be considered an adequate warning of excessive exposure.

ETHYLIDENE DICHLORIDE, CH₃CHCl₂ (1,1-Dichloroethane)

1. *Uses and Industrial Exposures*

Ethylidene dichloride has been used as a chemical intermediate. Formerly used as an anesthetic, it is of no importance in this field today.

2. *Physical and Chemical Properties*

Physical state: colorless liquid
Molecular weight: 98.97
Specific gravity: 1.174 (20°/4°C.)
Melting point: —96.7°C.
Boiling point: 57.3°C.
Vapor density: 3.4 (air = 1)
Vapor pressure: 234 mm. Hg (25°C.)
Refractive index: 1.41655 (20°C.)
Per cent in "saturated" air: 30.8 (25°C.)
Density of "saturated" air: 1.73 (air = 1)
Solubility: 0.5 g. per 100 ml. water at 20°C.; soluble in ethanol and ethyl ether
Flash point: 57°F. (open cup)
1 mg./per liter $\backsim$ 247 p.p.m. and 1 p.p.m. $\backsim$ 4.05 mg./cu. meter at 25°C., 760 mm. Hg

3. *Physiological Response*

Relatively little has been published on this compound. Exposure will result in central nervous system depression. It is also indicated as causing liver injury.

Acute Vapor Exposure.   Most of the information on ethylidene dichloride in acute exposure has come from studies on its possible use as an anesthetic. Lazarew, in 1929, reported the minimum fatal concentration to mice to be 0.00070 mole/liter of air or 70 mg./liter for a 24-hour exposure. Smyth[60] quoted unpublished work in his laboratory indicated that rats would survive 8 hours at 4000 p.p.m. but were killed at 16,000 p.p.m.

[60] H. F. Smyth, Jr., Donald E. Cummings Memorial Lecture, *Am. Ind. Hyg. Assoc. Quart.*, **17**, 129 (1956).

Chronic Vapor Exposure. The only chronic vapor work found was quoted by Smyth[60] from unpublished work in his laboratory. He indicated that repeated inhalation by rats and dogs showed a chronic toxicity somewhat other than that of carbon tetrachloride. His observations would show that the most significant injury from chronic inhalation was in the liver. He recommended 100 p.p.m. as a threshold limit but suggested that more information would be desirable in setting a final standard.

4. *Hygienic Standards of Permissible Exposure*

The 1961 A.C.G.I.H. standard for ethylidene dichloride is 100 p.p.m. (400 mg./cu. meter).

5. *Flammability*

The ignition temperature of ethylidene dichloride is 856°F.

6. *Odor and Warning Properties*

Ethylidene dichloride has been stated to have a "chloroformlike" odor. The level at which this odor is detectable has not been determined. It is probably not a satisfactory warning.

ETHYLENE DICHLORIDE, CH_2ClCH_2Cl (1,2-Dichloroethane)

1. *Uses and Industrial Exposures*

Ethylene dichloride is used as an industrial solvent in cleaning and extraction processes. It is also widely used as a fumigant, usually in combination with other fumigants, and is used in antiknock gasoline.

The principal contact with this material in industry is from its use as a solvent. In its use as a fumigant, the principal contact is during the process of fumigation by the operators and the aeration of the fumigated space after the completion of fumigation.

2. *Physical and Chemical Properties*

Physical state: colorless liquid
Molecular weight: 98.97
Specific gravity: 1.2529 (20°/4°C.)
Melting point: −35.3°C.
Boiling point: 83.5°C.
Vapor density: 3.4 (air = 1)
Vapor pressure: 87 mm. Hg (25°C.)
Refractive index: 1.44432 (20°C.)
Per cent in "saturated" air: 11.5 (25°C.)
Density of "saturated" air: 1.27 (air = 1)
Solubility: 0.9 g./100 ml. water at 20°C.; soluble in ethanol and ethyl ether
Flash point: 18.3°C. (open cup) ; 13°C. (closed cup)

1 mg./liter $\backsimeq$ 247 p.p.m. and 1 p.p.m. $\backsimeq$ 4.05 mg./cu. meter at 25°C., 760 mm. Hg

3. *Physiological Response*

At very high concentrations, ethylene dichloride is irritating to the eyes, nose, and throat. The symptoms are largely related to central nervous system depression or gastrointestinal upset, that is, mental confusion, dizziness, nausea, and vomiting. At subacute levels, similar symptoms of central nervous system depression and gastrointestinal upset are observed. Definite liver, kidney, and adrenal injuries may occur at these levels. From chronic exposure to lower concentrations, some indications of central nervous system depression are still observed. Nausea and vomiting are still quite common in humans. The symptom of nausea and vomiting from ethylene dichloride is quite striking and similar to that often observed from carbon tetrachloride. The pathological picture from chronic exposure is injury of the liver, kidneys, and adrenals. This general pattern has been consistent in animal investigative work and in experience from human exposure.

Acute Vapor Exposure. Animal response from acute exposure to ethylene dichloride was reported by Sayers *et al.*[61] This work was done on guinea pigs.

Heppel *et al.*[62] reported single exposures to rabbits, guinea pigs, hogs, cats, raccoons, mice, and rats. While cats and raccoons seemed to be somewhat more resistant, there were otherwise no very radical differences in sensitivity to acute exposure.

Spencer *et al.*[63] reported an investigation of the response of rats. The maximum time-concentrations in air survived by rats from a single exposure were as follows: 12 minutes at 20,000 p.p.m., 1 hour at 3000 p.p.m., and 7 hours at 300 p.p.m. The maximum time-concentrations in air having no adverse effect on female rats were as follows: 6 minutes at 12,000 p.p.m., 1.5 hours at 1000 p.p.m., and 7 hours at 200 p.p.m. The responses of other species reported by other authors fall within the same general range.

These acute exposures produced considerable depression of the central nervous system. Concentrations below 12,000 p.p.m. produced various degrees of drunkenness but not unconsciousness or death within the duration of the exposure. At concentrations of 3000 p.p.m. or more, definite depression was observed in the form of inactivity or stupor. Organic changes were observed in the form of increased weight of liver and kidneys. Biochemical changes indicated liver misfunction. There were histopathological change in the kidneys, liver, and adrenals.

[61] R. R. Sayers, W. P. Yant, C. P. Waite, and F. A. Patty, *Public Health Repts. U. S.*, Reprint No. **1349**, (Jan. 31, 1930).

[62] L. A. Heppel, Paul A. Neal, T. L. Perrin, K. M. Endicott, and V. T. Porterfield, *J. Pharmacol. Exptl. Therap.*, **1**, 53 (1945).

[63] H. C. Spencer, V. K. Rowe, E. M. Adams, D. D. McCollister, and D. D. Irish, *A.M.A. Arch. Ind. Hyg. Occupational Med.*, **4**, 482 (1951).

Heppel *et al.*[64] demonstrated in dogs the occurrence of a clouding of the cornea that was reversible. After several exposures, the dogs became resistant to this effect. These authors studied many different species but observed this effect only in the fox and the dog. This has not been observed in humans exposed to ethylene dichloride.

There have been some clinical cases of acute exposure to ethylene dichloride reported in the literature. In general, they would confirm the picture observed in the animals. Menschick[65] reported 4 acute cases. The picture was dominated by an "hepato-renal syndrome." The author also lists 27 cases of poisoning by inhalation collected from the literature.

Chronic Vapor. Heppel *et al.*[66] published investigations of the effect of daily inhalations of ethylene dichloride by a number of animals. These authors exposed their animals 7 hours/day for 5 days/week. At a concentration of 1000 p.p.m., guinea pigs survived 2 exposures, rats survived from 3 to 14 exposures. Rabbits survived from 2 to 64 exposures. These three species were all susceptible. Dogs, cats, and monkeys survived a period from 23 to 55 days and showed a very low mortality. The rats, rabbits, guinea pigs, cats, and monkeys showed fatty changes in the liver. Only 1 out of 6 dogs showed liver changes. At 400 p.p.m., rats, rabbits, and guinea pigs still showed a high mortality. There were variable indications of liver injury. Dogs survived up to 177 exposures. The smaller animals were continued for up to 70 exposures. At 200 p.p.m., guinea pigs and rats still showed a higher than normal mortality. Experiments were run up to 125 exposures. The rats showed only mild pulmonary congestion. There were some indications, although variable, of histopathological changes in guinea pigs. Rabbits and monkeys appeared normal during these experiments. At 100 p.p.m., rats, guinea pigs, and mice "survived many exposures and developed no demonstrable lesions." The number of exposures was not indicated.

A comparable chronic study was carried out by Spencer *et al.*[63]; they exposed their animals 7 hours/day, 5 days/week. They likewise showed high mortality at 400 p.p.m. in their rats and guinea pigs in periods of 14 to 56 days of exposure. The animals showed loss of weight, a slight increase in weight of the liver and kidneys, and only relatively slight histopathological changes. Guinea pigs showed more definite histopathological changes in both the liver and kidneys. At 200 p.p.m., rats survived 151 exposures in 212 days without apparent adverse effect. Growth was normal and there was no indication of organic injury. Guinea pigs survived comparable exposures, but about half of the animals showed some histological changes. At 100 p.p.m., rats, guinea pigs, rabbits, and monkeys were given 120 exposures in 168 days without apparent effect.

[64] L. A. Heppel, Paul A. Neal, K. M. Endicott, and V. T. Porterfield, *A.M.A. Arch. Ophthalmol.*, **32**, 391 (1944).

[65] H. Menschick, *Arch. Gewerbepathol. Gewerbehyg.*, **15**, 241 (1957).

[66] L. A. Heppel, P. A. Neal, T. L. Perrin, K. M. Endicott, and V. T. Porterfield, *J. Ind. Hyg. Toxicol.*, **28**, 113 (1946).

The work from these two publications would indicate that animals tolerated a concentration of 100 p.p.m. for 7 hours/day, 5 days/week over many months.

Experience in Man. There have been some reports of chronic exposure of humans in industry. McNally and Fosvedt[67] reported on 2 mild cases that had from 2 to 5 months of exposure. Their symptoms were central nervous system depression and gastrointestinal upset with nausea and vomiting. They recovered when removed from exposure.

Oral. McCollister *et al.*[68] reported the LD_{50} of ethylene dichloride for rats was 0.68 g./kg. of body weight. This would indicate that ethylene dichloride is several times more toxic than carbon tetrachloride when taken in a single oral dose. The carbon tetrachloride LD_{50} indicated by the same authors was 2.98.

Ethylene dichloride does present a problem from oral ingestion. A significant number of the poisoning cases reported in the literature were from oral ingestion. The response is not unlike that which will be observed from acute vapor exposure, that is, central nervous system depression, gastrointestinal upset, and injury to the liver, kidneys, and the lungs.

Skin. Ethylene dichloride is absorbed through the skin though it takes quite large doses to cause serious systemic poisoning. Rabbits will survive a dose of 1.26 g./kg. of body weight, the LD_{50} was calculated to be 2.8 g./kg. of body weight. When ethylene dichloride is held close to the skin within a cuff, such as is used in skin-absorption experiments, quite severe irritation and moderate edema and necrosis may be observed. Ordinary contact with ethylene dichloride on the skin where it is not held by the clothing does not cause serious difficulties. Repeated or prolonged contact, however, may cause a rough, red, dry skin due to extraction of fatty materials. It may result in cracking and a "chapped skin."

Eyes. When ethylene dichloride is splashed in the eyes, it may result in pain, irritation, and lachrymation. If it is promptly removed by washing no significant injury should occur. If not removed, serious damage may be the result.

Absorption, Excretion, and Metabolism. Ethylene dichloride is readily absorbed via the lungs when breathed or via the gastrointestinal tract when taken by mouth. To a less extent, it is absorbed through the skin and at a large enough dose causes a systemic toxic response. It required a dose of around 2 g./kg. of body weight held in contact with a large area of the body for a period of 24 hours in order to cause the death of rabbits. It does not seem probable that under industrial conditions, sufficient exposure can be obtained to allow skin absorption to become a very significant factor in systemic intoxication.

The extent or nature of the metabolism of ethylene dichloride has not been established.

[67] W. D. McNally and G. Fosvedt, *Ind. Med.,* **10,** 373 (1941).

[68] D. D. McCollister, R. L. Hollingsworth, F. Oyen, and V. K. Rowe, *A.M.A. Arch. Ind. Health,* **13,** 1 (1956).

4. *Hygienic Standards of Permissible Exposure*

The threshold limit of ethylene dichloride was established by the American Conference of Governmental Industrial Hygienists, in April, 1959, at 100 p.p.m. (400 mg./cu. meter).

5. *Flammability*

The explosive limits of ethylene dichloride are 6.2 to 15.9% by vol. in air. The ignition temperature is 413°C.

6. *Odor and Warning Properties*

Ethylene dichloride has a sweetish, not particularly disagreeable, odor. The odor is barely detectable at 50 p.p.m. in air, is definite but not unpleasant at 100 p.p.m. While it is pronounced at 200 p.p.m., it still would not be considered unpleasant. Even though the odor may be definite enough to act as a warning of acutely hazardous concentrations, it is probably not sufficiently striking to be considered a significant warning of hazardous chronic exposure. This is particularly true as one can become adapted to the odor at low concentrations.

ETHYLENE DIBROMIDE, CH_2BrCH_2Br (1,2-Dibromoethane)

1. *Uses and Industrial Exposures*

Ethylene dibromide is extensively used in antiknock gasoline and in fumigant mixtures. It is also used to some extent as a fire extinguisher, chemical intermediate, special solvent, and gage fluid.

It will be encountered in industry in manufacture and handling. The most contact may be in fumigant applications.

The high-boiling point, low vapor pressure, and high heat of vaporization greatly reduce the probability of high concentrations accumulating rapidly in the work atmosphere.

2. *Physical and Chemical Properties*

Physical state: colorless liquid
Molecular weight: 187.88
Specific gravity: 2.1701 (25°/4°C.)
Melting point: 9.97°C.
Boiling point: 131.6°C.
Vapor density: 6.5 (air = 1)
Vapor pressure: 12 mm. Hg (25°C.)
Refractive index: 1.53789 (20°C.)
Per cent in "saturated" air: 1.5 (25°C.)
Density of "saturated" air: 1.08 (air = 1)
Solubility: 0.43 g. per 100 ml. water at 30°C.; soluble in ethanol and ethyl ether

1 mg./liter ≎ 130 p.p.m. and 1 p.p.m. ≎ 7.68 mg./cu. meter at 25°C., 760 mm. Hg

3. *Physiological Response*

Exposure to high concentrations of the vapors of ethylene dibromide will result in central nervous system depression, although the anesthetic action is weak. Deaths from acute exposure at high concentrations are usually due to pneumonia developed as a result of an injury to the lungs. Following acute exposures, in general, injury will be observed in the lungs, liver, and kidneys. Chronic exposure over a long period of time to levels significantly above the threshold will result in a response very similar to that seen from acute exposure. The injury will be to the lungs, liver, and kidneys.

It should be particularly noted that the difference between the concentration causing severe injury and death and that which is tolerated for long-term exposure is not great. The 25 p.p.m. threshold should be considered as a ceiling and variation in exposure should stay well below this level.

Acute Vapor Exposure. A quantitative expression of acute vapor toxicity of ethylene dibromide is found in the data of Rowe *et al.*[69]

The maximum survival exposures of rats to ethylene dibromide vapor in air are as follows: 3000 p.p.m. for 6 minutes, 400 p.p.m. for 30 minutes, and 200 p.p.m. for 2 hours. Guinea pigs survive 400 p.p.m. for 2 hours and 200 p.p.m. for 7 hours. The maximum exposures without adverse effect on female rats are 800 p.p.m. for 6 minutes, 100 p.p.m. for 2.5 hours, and 50 p.p.m. for 7 hours.

The pathological changes following acute exposures described by Rowe *et al.*[69] were congestion, edema, hemorrhages, and inflammation of the lungs. The liver showed cloudy swelling and central lobular fatty degeneration and necrosis. The kidneys showed slight interstitial congestion and edema with slight cloudy swelling of the tubular epithelium in some cases.

Chronic Vapor Exposure. Chronic vapor exposure in animals has been reported by Rowe *et al.*[69] 1952. Animals were exposed 7 hours/day, 5 days/week for periods up to 6 months. At 100 p.p.m. in air, both rats and rabbits showed poor condition and some deaths within the first week or two of exposure. The rats showed injury to the lungs, liver, and spleen. The rabbits showed definite liver injury. At 50 p.p.m., the rats showed a fairly high mortality (approximately 50 per cent) due to pneumonia and infection of the upper respiratory tract which may be related to the lung effect of ethylene dibromide. A number of them, however, lived through the full 6-month period. They revealed lung-, liver-, and kidney-weight increases and some histopathological changes in the lungs. Guinea pigs subjected to 57 7-hour exposures in 80 days showed some depression of growth but no increase in mortality. Lung, liver, and kidney weights were apparently increased. There were slight histopathological changes in the liver and

[69] V. K. Rowe, H. C. Spencer, D. D. McCollister, R. L. Hollingsworth, and E. M. Adams, *A.M.A. Arch. Ind. Hyg. Occupational Med.*, **6**, 1958 (1952).

kidneys. An exposure of 25 p.p.m. was well tolerated by rats, guinea pigs, rabbits, and monkeys. The rats showed the highest mortality due to pneumonia and infections of the upper respiratory tract.

There have not been clinical reports of injury to man by breathing vapors of ethylene dibromide. Kochmann[70] carried out some animal investigations and suggested that ethylene dibromide could be a hazardous material. Kochmann proposed that a concentration of 50 p.p.m. could be dangerous to man.

Oral Toxicity. Rowe et al.[69] reported the single oral dose for several species of animals. They determined the LD_{50} in g./kg. as follows:

Female mice	0.420
Male rats	0.148
Female rats	0.117
Guinea pigs	0.110
Chicks	0.079
Female rabbits	0.055

Skin and Eyes. Thomas and Yant[71] indicated that ethylene dibromide could be readily absorbed through the skin in toxic amounts.

Rowe et al.[69] reported quantitative measurements of the toxic dose absorbed through the skin in rabbits for a contact period of 24 hours. A dose of 0.21 g./kg. body weight was survived by 14 out of 15 animals; 1.1 g./kg. body weight killed 5 of 5 animals. Intermediate mortalities were observed at intermediate concentrations.

Ethylene dibromide is definitely irritating and injurious to the skin. Rowe et al. reported that a 1 per cent solution of ethylene dibromide in butylcarbitol acetate applied 10 times in 14 days to a rabbit's ear caused slight irritation characterized by erythema and exfoliation. The same repeated applications, when bandaged on to the shaved abdomen, showed erythema and edema progressing to necrosis and sloughing of the superficial layers of the skin. Pflesser[72] observed a case of prolonged contact with the skin when liquid ethylene dibromide was accidentally spilled into the shoes. There was reddening, blistering, and burning pain.

In our experience,[5f] serious skin injury can occur from clothing and particularly shoes wet with ethylene dibromide. This is true not only when material is spilled inside the shoe but also when it wets the outside as it will penetrate leather.

Calingaert and Shapiro[73] have observed that ethylene dibromide will penetrate through several types of protective clothing, particularly neoprene rubber and several types of plastic gloves. Nylon was found to be the most resistant

[70] M. Kochmann, *Münch. med. Wochschr.*, **75**, 1334 (1928).

[71] B. G. H. Thomas and W. P. Yant, *Public Health Repts. U. S.*, Reprint No. **1139**, 370 (1927).

[72] G. Pflesser, *Arch. Gewerbepathol. Gewerbehyg.*, **8**, 591 (1938).

[73] G. Calingaert and H. Shapiro, *Ind. Eng. Chem.*, **40**, 332 (1948).

material but lacking in good physical characteristics. They proposed a combination of neoprene and nylon.

Rowe et al.[69] reported experiments on the effects of ethylene dibromide in the eye of rabbits. Undiluted material caused obvious pain and conjunctival irritation, clearing in 48 hours. A slight but superficial necrosis of the cornea was observed. Nevertheless, healing was prompt and complete. A 10 per cent solution in propylene glycol was tested and produced a more severe reaction than did the undiluted material. The eye healed without scarring or apparent injury. It is obvious that, in handling ethylene dibromide, the eye should be protected. If the material gets into the eye, it should be washed out promptly.

Absorption, Excretion, and Metabolism. Apparently ethylene dibromide is readily and rapidly absorbed from the lungs when breathed as a vapor, from the gastrointestinal tract when taken by mouth, or through the skin when applied topically. Leuze[74] reported the distribution of bromide in the tissues.

Lucas[75] indicated that inorganic bromide was observed in the urine and livers of animals exposed to ethylene dibromide.

It is quite possible that an appreciable amount of ethylene dibromide is hydrolyzed to give inorganic bromide. However, no quantitative study has been made of the relation between the inorganic bromide in blood or urine and the exposure.

4. Hygienic Standards of Permissible Exposure

The threshold limit of ethylene dibromide was established by the American Conference of Governmental Industrial Hygienists, in April, 1959, at 25 p.p.m. (190 mg./cu. meter).

5. Flammability

Ethylene dibromide is essentially nonflammable.

6. Odor and Warning Properties

The odor of ethylene dibromide is not unpleasant and has been termed "chloroformlike." The odor is not detectable at a low enough concentration to be considered a good warning of excessive exposure.

METHYL CHLOROFORM, CH_3CCl_3 (1,1,1-Trichloroethane)

1. Uses and Industrial Exposures

Until five or six years ago, methyl chloroform was used largely as a chemical intermediate. Recently, it has been used as a solvent. It has been considered as an anesthetic because of its low toxicity, particularly because of its lack of injury to the liver and kidneys. However, it was not effective enough as an anesthetic to develop significant use in this area.

[74] E. Leuze, *Arch. Exptl. Pathol. Pharmakol.*, **95**, 145 (1922).
[75] G. H. Lucas, *J. Pharmacol.*, **34**, 223 (1928).

The principal industrial exposure is from its use as a solvent. Methyl chloroform has excellent solvent characteristics, and the boiling point is very close to that of carbon tetrachloride. As carbon tetrachloride had caused serious difficulties in its use as a "bucket solvent" and in other cold cleaning operations, there was a strong demand for a safer substitute. Methyl chloroform was well suited to this purpose because of its similar physical properties and a much lower toxicity.

2. *Physical and Chemical Properties*

Physical state: colorless liquid
Molecular weight: 133.42
Specific gravity: 1.3249 (26°/4°C.)
Boiling point: 74.1°C.
Vapor density: 4.6 (air = 1)
Vapor pressure: 127 mm. Hg (25°C.)
Refractive index: 1.43765 (21°C.)
Per cent in "saturated" air: 16.7 (25°C.)
Density of "saturated" air: 1.6 (air = 1)
Solubility: insoluble in water; soluble in ethanol and ethyl ether
1 mg./liter $\eqsim$ 183 p.p.m. and 1 p.p.m. $\eqsim$ 5.46 mg./cu. meter at 25°C., 760 mm. Hg

3. *Physiological Response*

The response from acute or chronic exposure to excessive amounts of methyl chloroform is due to depression of the central nervous system. It has little capacity to produce organic injury from either single or repeated exposures.

Humans exposed to approximately 500 p.p.m. showed no response although the odor was noticeable. At approximately 1000 p.p.m., the odor was strong but there appeared to be no significant response in exposures of up to 70 minutes. At approximately 2000 p.p.m., there was definite disturbance of equilibrium. The major problems from excessive exposure will usually be drunkenness and incoordination.

Only 2 fatal cases are known, and they each occurred in a tank where the concentration may well have been close to saturation. Exposure in one case was estimated at close to 30 minutes; the other is not known.

The low toxicity and the fact that the volatility is close to that of carbon tetrachloride have encouraged use in place of the latter. The proposed industrial hygiene standard of 500 p.p.m. has been in use for a number of years and appears to be satisfactory.

Acute Vapor Exposure.   Adams *et al.*[77] reported on the response of animals to acute exposure. Maximum time-concentrations in air survived by rats were as

[76] W. B. Crummett and V. A. Stenger, *Ind. Eng. Chem.,* **48,** 434 (1956).

[77] E. M. Adams, H. C. Spencer, V. K. Rowe, and D. D. Irish, *A.M.A. Arch. Ind. Hyg. Occupational Med.,* **1,** 225 (1950).

follows: 6 minutes at 30,000 p.p.m., $1^1/_2$ hours at 15,000 p.p.m., and 7 hours at 8000 p.p.m.

Maximum time-concentrations in air with no detectable injury in rats were as follows: 18 minutes at 18,000 p.p.m., 5 hours at 8000 p.p.m. It should be noted that the maximum level with no detectable injury is very close to the maximum level survived.

This information has been confirmed by Torkelson et al.,[78] using the inhibited solvent.

Chronic Vapor Exposure. Adams et al.[77] also reported repeated exposures of animals for 7 hours/day, 5 days/week for 1 to 3 months. At 10,000 p.p.m. rats showed staggering gait and weakness in 10 minutes. By 3 hours, they showed loss of color, irregular respiration, and semiconsciousness. Survivors had completely recovered by the following morning. For those that succumbed, death seemed to be due to either cardiac or respiratory failure. At 5000 p.p.m., there was definite but mild narcotic effect within 1 hour. There was reduced activity. They survived for 31 exposures over 41 days without apparent injury. Rabbits showed slight retardation of growth at 5000 p.p.m.

At 3000 p.p.m., rabbits and monkeys showed no response over a 2-month period. Guinea pigs showed a barely significant retardation of growth at 650 p.p.m.

Torkelson et al.[78] confirmed and extended these results, using the inhibited material. The chlorinated solvents require the addition of an inhibitor to prevent corrosion of metals, particularly aluminum. The 1,1,1-trichloroethane studied by Torkelson contained 2.4 to 3 per cent dioxane, 0.12 to 0.3 per cent butanol, and small amounts of ethylene dichloride, water, etc.

The 1,1,1-trichloroethane is considerably less toxic than the 1,1,2-trichloroethane. Due to an error in the literature, several authors have indicated the reverse. Torkelson[78] has explained the circumstances and documented the literature.

Ethyl Browning,[79] in a review of the solvents, discussed the 1,1,2- and the 1,1,1-trichloroethane. She quoted Dr. Hecht of the Bayer Company to the effect that the β(1,1,2-isomer) was much more toxic than the α(1,1,1-isomer), and that the statement to the contrary made by Lehmann and Flury was in error.

Oral. Torkelson et al.[78] reported the LD_{50} on several animals as follows: male rats 12.3 g./kg., female rats 10.3 g./kg., female mice 11.24 g./kg., female rabbits 5.66 g./kg., and male guinea pigs 9.47 g./kg.

Skin and Eyes. The skin shows only slight reddening and scaliness from contact. The reaction is somewhat increased on repeated exposures. Applied under a cuff for 24 hours, 3.9 g./kg. was survived by all animals; 15.8 g./kg. failed to kill any animals. Skin irritation is mild and there is no significant problem from skin absorption.

[78] T. R. Torkelson, F. Oyen, D. D. McCollister, and V. K. Rowe, *Am. Ind. Hyg. Assoc. J.,* **19,** 353 (1958).

[79] E. Browning, *Brit. J. Ind. Med.,* **16,** 23 (1959).

Human Experience. Experimental exposure of humans was reported by Torkelson *et al.*[78] This report and the 2 known fatal cases are discussed in the preceding summary. Experience with the use of methyl chloroform in many solvent applications has been very favorable.

Absorption, Excretion, and Metabolism. Methyl chloroform is very stable in the body. A large part of a dose injected intravenously is excreted rapidly by the lungs unchanged. Hake[80] has shown that a very small amount is metabolized to chloroethanol and excreted in the urine as the glucuronate.

The high stability and rapid excretion may well account for the low toxicity and quick recovery from anesthetic concentrations.

4. Hygienic Standards of Permissible Exposure

The threshold limit of methyl chloroform was established by the American Conference of Governmental Industrial Hygienists, in April, 1959, at 500 p.p.m. (2700 mg./cu. meter).

5. Flammability[76]

The flammable characteristics of methyl chloroform are similar to those of trichloroethylene. It has no flash point or fire point by ASTM procedures for Tag closed cup and Cleveland open cup tests. Limits of flammability of vapors of inhibited 1,1,1-trichloroethane have been found to be 10 to 15.5 per cent in air with hot wire ignition. A considerable amount of energy is required for ignition. It will not sustain combustion.

6. Odor and Warning Properties

While methyl chloroform has a typical sweetish odor, it is not striking enough to be considered a good warning. The odor is noticeable at concentrations well below those known to cause any other physiological response.

VINYL TRICHLORIDE, CH₂ClCHCl₂ (1,1,2-Trichloroethane)

1. Uses and Industrial Exposures

Vinyl trichloride has been used as a chemical intermediate and solvent. Significant industrial usage has not developed.

2. Physical and Chemical Properties

Physical state: colorless liquid
Molecular weight: 133.42
Specific gravity: 1.443 (20°/4°C.)

[80] C. L. Hake, T. B. Waggoner, D. N. Robertson, and V. K. Rowe, *Arch. Environ. Health,* **1,** 101 (1960).

Melting point: −36.7°C.
Boiling point: 113.5°C.
Vapor density: 4.6 (air = 1)
Vapor pressure: 25 mm. Hg (25°C.)
Refractive index: 1.4711 (20°C.)
Per cent in "saturated" air: 3.3 (25°C.)
Density of "saturated" air: 1.1 (air = 1)
Solubility: 0.44 g. in 100 g. water at 20°C.; soluble in ethanol and ethyl ether
1 mg./liter ⇌ 183 p.p.m. and 1 p.p.m. ⇌ 5.46 mg./cu. meter at 25°C., 760 mm. Hg

3. *Physiological Response*

The principal physiological response to vinyl trichloride is depression of the central nervous system. Lehmann and Flury indicated injury to the lungs, liver, and the kidneys. They were dealing solely with the acute lethal dose.

There has been some confusion in the literature between the 1,1,2-trichloro-ethane and the 1,1,1-trichloroethane. Torkelson *et al.*[78] have explained the circumstances and documented the literature. There was apparently an error in quoting the early work of Lazarew.

Carpenter *et al.*[81] reported an acute lethal concentration (50) for rats to be 2000 p.p.m. from a 4-hour exposure followed by a 14-day observation period.

Very comparable measurements were made by Adams *et al.*[77] on the 1,1,1-trichloroethane. They reported the approximate lethal concentration (50) for a 3-hour exposure to be 18,000 p.p.m. and for a 7-hour exposure to be 14,000 p.p.m.

It would be concluded from this information that the toxicity of the 1,1,2-isomer is manyfold that of the 1,1,1-isomer for an acute exposure. Unpublished animal investigations[5f] would indicate that the toxicological response from chronic exposure would be qualitatively and quantitatively comparable to carbon tetrachloride.

Wright and Shaffer[82] reported that the oral lethal dose for dogs was 0.5 cl./kg. Unpublished data[5f] indicate an LD_{50} for rats of 0.1 to 0.2 gm./kg. Liver and kidney pathology are seen at considerably lower doses.

Very little is known of the fate of this compound in the body. Barrett *et al.*[83] were unable to show the excretion of any trichloroacetic acid in the urine of dogs exposed to 1,1,2-trichloroethane.

4. *Hygienic Standards of Permissible Exposure*

No threshold limit for vinyl trichloride has been established.

[81] C. P. Carpenter, H. F. Smyth, and U. C. Pozzani, *J. Ind. Hyg. Toxicol.*, **31**, 343 (1949).

[82] W. H. Wright and J. M. Shaffer, *Am. J. Hyg.*, **16**, 325 (1932).

[83] H. M. Barrett, J. G. Cunningham, and J. H. Johnston, *J. Ind. Hyg. Toxicol.*, **21**, 479 (1939).

ACETYLENE TETRACHLORIDE, CHCl₂CHCl₂ (1,1,2,2-Tetrachloroethane)

1. Uses and Industrial Exposures

Acetylene tetrachloride is used principally as a solvent for cleaning and extraction processes and to some extent as a chemical intermediate.

2. Physical and Chemical Properties

Physical state: colorless liquid
Molecular weight: 167.86
Specific gravity: 1.5869 (25°C.)
Melting point: —42.5°C.
Boiling point: 146.3°C.
Vapor density: 5.79 (air = 1)
Vapor pressure: 6 mm. Hg (25°C.)
Refractive index: 1.4918 (25°C.)
Per cent in "saturated" air: 0.79 (25°C.)
Density of "saturated" air: 1.04 (air = 1)
Solubility: very slightly soluble in water; soluble in ethanol and ethyl ether
1 mg./liter ⩰ 145.8 p.p.m. and 1 p.p.m. ⩰ 6.86 mg./cu. meter at 25°C., 760 mm. Hg

3. Physiological Response

Acetylene tetrachloride is considered to be among the more toxic of the chlorinated hydrocarbons. The most significant injury from subacute or chronic exposure is to the liver. The first indication is a greatly enlarged and palpable liver, which may progress to fatty degeneration and cirrhosis. Injury to the kidneys may also be observed.

This compound also causes some central nervous system depression as do many chlorinated hydrocarbons. Exposure may cause dizziness, nervousness, and incoordination. In very severe acute exposures, unconsciousness and death from respiratory failure may be seen. Central nervous system depression is not a striking part of the response to usual industrial exposure because of the low volatility and other injurious effects predominating at lower levels.

Respiratory irritation may be observed and may lead to pulmonary damage. A significant irritation in the gastrointestinal tract is also observed and may result in nausea, vomiting, and gastric pain.

There have not been significant quantitative laboratory studies of the response to chronic exposure. An estimate of acceptable concentration is based upon limited clinical experience.

Acute Vapor Exposure. While there are a number of scattered animal experiments with this material, a clean cut quantitative study of animal response from acute exposure has not been reported. Smyth[60] quoted unpublished work by his laboratory indicating that rats were found to survive a 4-hour exposure at 500 p.p.m. but would not survive 4 hours at 1000 p.p.m.

Lehmann and Schmidt-Kehl[84] reported 3 p.p.m. to have a noticeable odor and 13 p.p.m. to be tolerated for 10 minutes. Concentrations of 186 p.p.m. inhaled for 30 minutes or 335 p.p.m. inhaled for 10 minutes had a disagreeable and marked odor, causing upper respiratory irritation and central nervous system effects.

Chronic Vapor Exposure. Lehmann and Flury[85] reported some vapor experiments on cats and rabbits. They were exposed to a concentration of from 100 to 160 p.p.m. for 8 to 9 hours daily for 4 weeks. No typical organ changes were found.

These observations are rather surprising when it is considered that in industrial experience reports would indicate that injury to man has occurred at much lower concentrations. Elkins[86] reported that he had been informed of illness in workmen where repeated tests of the work environment had indicated concentrations below 10 p.p.m.

Gurney[87] reported a number of cases of chronic exposure to acetylene tetrachloride. He studied 277 individuals of whom 75 had symptoms, 55 had enlarged livers. There were nearly as many with enlarged livers among those who did not show symptoms as there were among those who did. Coyer[88] reported on 6 cases, 1 of which was fatal.

It is evident from clinical reports on exposure to acetylene tetrachloride that the principal effect involves the liver. Symptoms referable to gastrointestinal injury may also be observed.

Oral. Barsoum and Saad[89] reported that an oral dose of 0.7 g./kg. body weight in dogs is a toxic dose.

Wright and Schaffer[82] reported that a lethal dose for dogs was 0.3 ml./kg. given orally.

Sherman[90] reported on 8 humans who were given 3 ml. of tetrachloroethane by mistake. It was given with 30 g. of magnesium sulfate and water. Within $1\frac{1}{2}$ to $2\frac{1}{2}$ hours, they were all comatose. Reflexes were absent and the pulse barely perceptible. Respiration was shallow and rapid. They all recovered and showed no aftereffects.

Skin. Schwander[91] indicated that acetylene tetrachloride could be absorbed through the intact skin.

[84] K. B. Lehmann and L. Schmidt-Kehl, *Arch. Hyg. u. Bakteriol.,* **116**, 131 (1936).

[85] H. B. Lehmann and F. Flury, *Toxicology and Hygiene of Industrial Solvents.* Trans. by F. King and H. F. Smyth, William & Wilkins, Baltimore, 1943.

[86] H. B. Elkins, *Chemistry of Industrial Toxicology.* Wiley, New York, 1950.

[87] R. Gurney, *Gastroenterology,* **1**, 1112 (1943).

[88] H. A. Coyer, *Ind. Med.,* **13**, 320 (1944).

[89] G. S. Barsoum and K. Saad, *Quart. J. Pharm. Pharmacol.,* **7**, 205 (1934).

[90] J. B. Sherman, *J. Trop. Med. Hyg.,* **56**, 139 (1953).

[91] P. Schwander, *Arch. Gewerpathol. Gewesbehyg.,* **7**, 109 (1936).

Absorption, Excretion, and Metabolism. Studies have shown that acetylene tetrachloride is readily absorbed via the lungs or gastrointestinal tract. Some authors have indicated absorption by the skin. It is apparently readily excreted by the lungs. Its metabolism has not been thoroughly investigated.

The only significant study of metabolism of this compound was by Barrett *et al.*[83] They indicated that the Fujiwara test colors produced with symmetrical tetrachloroethane and trichloroethylene were similar. This was due to the conversion of the tetrachloro to the trichloro, splitting out HCl. They indicated that there were traces of tetrachloroethane in the urine which might account for the reaction given. They concluded that, if the tetrachloroethane were broken down to trichloroethylene in the body, it would be excreted as trichloroacetic acid. Although that possibility was not completely eliminated, it was felt to be improbable.

4. *Hygienic Standards of Permissible Exposure*

The threshold limit for acetylene tetrachloride was established by the American Conference of Governmental Industrial Hygienists, in April, 1959, at 5 p.p.m. (35 mg./cu. meter).

5. *Flammability*

There is no flash point for acetylene tetrachloride. It is not flammable or explosive.

6. *Odor and Warning Properties*

Acetylene tetrachloride has a mild sweetish odor similar to several other chlorinated hydrocarbons. According to Lehmann and Schmidt-Kehl,[84] a noticeable odor is detected at 3 p.p.m. As the odor is not particularly striking and the acceptable level is given as 5 p.p.m., this would not appear to be of any significance as a warning property.

ACETYLENE TETRABROMIDE, $CHBr_2CHBr_2$ (1,1,2,2-Tetrabromoethane)

1. *Uses and Industrial Exposures*

Because of its high density acetylene tetrabromide is used as a gage fluid, for balancing equipment, and for ore separation. It has had some use as a special solvent.

2. *Physical and Chemical Properties*

Physical state: colorless to yellow liquid
Molecular weight: 345.7
Specific gravity: 2.9638 (20/4°C.)
Melting point: 0.13°C.
Boiling point: decomposition, 239 to 242°C.

Vapor density: 11.92 (air = 1)

Refractive index: 1.63795 (20°C.)

Solubility: 0.065 g. per 100 ml. water at 30°C.; soluble in ethanol, ethyl ether, and chloroform

3. Physiological Response

Acetylene tetrabromide is a central nervous system depressant. If given in sufficiently large doses, it may cause narcosis and coma, and eventually death from respiratory failure. There is lung irritation and pathological changes are observed in the liver and kidneys.

As the vapor pressure is exceedingly low at room temperatures, the material may be easily handled with reasonable precautions and ordinary ventilation.

Acute Vapor Exposure. Merzbach,[92] and Glaser and Frisch[93] exposed animals to vapors of acetylene tetrabromide. It is a bit difficult to determine the significance of their findings as Gray[94] indicated that the concentrations reported by these authors were, in most instances, well above a saturated atmosphere.

Gray reported exposure of animals to the vapors of acetylene tetrabromide at near saturation in a static chamber. Rabbits were exposed for up to $2^1/_2$ hours without deaths. Rats were exposed for up to 3 hours without deaths. Guinea pigs exposed $^1/_2$ hour survived. One out of 2 survived from an exposure of 1 hour. All exposed succumbed to exposures of $1^1/_2$ hours or over. In the guinea pigs that died, injury was seen in the liver and kidneys. The guinea pigs recovered consciousness from the exposure but died in periods of from 1 to 5 days later.

Chronic Vapor. Gray[94] exposed mice, rats, guinea pigs, and rabbits for 15 minutes daily for 47 to 92 days to a "saturated atmosphere." One mouse out of 48 died of an unknown cause; the rest survived without adverse effects. Four out of 36 albino rats were ill from lobar pneumonia, a complication that was considered unrelated to the exposure. The rest of the animals survived without injury. The rabbits were observed taking significant amounts of the condensed acetylene tetrabromide orally by licking it from the body. They were, therefore, not considered significant to the experiment. It should also be recognized that a certain amount of oral intake would be possible in the case of the other animals. The chronically exposed animals showed no significant pathological change.

It was found in the Biochemical Research Laboratory of The Dow Chemical Co. that the concentration in air could not be maintained above approximately 14 p.p.m. by analysis. The theoretically saturated concentration should be about 79 p.p.m.

Rats, guinea pigs, rabbits, and a monkey were exposed 7 hours/day, 5 days/week for periods ranging from 100 to 106 days. The average concentration was 14 p.p.m. by analysis. All animals survived and appeared normal. There was growth

---

[92] L. Merzbach, *Z. ges. exptl. Med.,* **63,** 383 (1928).

[93] E. Glaser and S. Frisch, *Arch. Hyg.,* **101,** 48 (1929).

[94] M. G. Gray, *A.M.A. Arch. Ind. Hyg. Occupational Med.,* **2,** 407 (1950).

depression in guinea pigs. All animals showed an increase in liver weight. Histopathological changes were observed in the liver and lungs.

At 4 p.p.m., some animals showed slight histopathological changes in the liver and some in the lungs. At 1.1 p.p.m., all animals were normal.

Oral. Gray[94] reported the LD_{50} for rabbits and guinea pigs as approximately 0.4 g./kg. of body weight. Experiments at the Biochemical Research Laboratory of The Dow Chemical Co. showed that rats survived 0.6 g./kg. and succumbed to 1.6 g./kg.

Skin. Ordinary contact with the open skin does not result in any skin reaction if it is washed off in a reasonable period of time. If the material is bandaged onto the skin and allowed to remain there for a period of hours, a slight redness will appear. In 24 hours, there will be some edema and blistering. It can be concluded that ordinary contact would not represent a skin problem. However, clothing or shoes that may contain the material should be removed and cleaned before reuse.

Aerosols. Arkhangel-Skaya and Yanushkevich[95] reported on the exposure of rats to aerosols of acetylene tetrabromide. A concentration varying from 3.7 to 4.2 mg./liter for a single exposure of 2 hours gave only slight and ill-defined symptoms of toxicity. Concentrations varying from 5.9 to 7.2 mg./liter caused excitation followed by sleepiness. Repeated daily exposure to aerosol concentrations of 3.7 to 4.2 mg./liter resulted in death of the animals.

4. Hygienic Standards of Permissible Exposure

The threshold limit of acetylene tetrabromide was established by the American Conference of Governmental Industrial Hygienists, in April, 1959, at 1 p.p.m. (14 mg./cu. meter).

5. Flammability

Acetylene tetrabromide has no flash or fire point.

6. Odor and Warning Properties

The odor of acetylene tetrabromide is sweetish and has been compared with chloroform or camphor. The odor is not distinctive enough to be considered a good warning property, although it may give sufficient warning to avoid serious acute exposure.

PENTACHLOROETHANE, $CHCl_2CCl_3$

1. Uses and Industrial Exposures

Pentachloroethane has been used as a solvent and chemical intermediate.

2. Physical and Chemical Properties

Physical state: liquid
Molecular weight: 202.31

[95] I. N. Arkhangel-Skaya and R. I. Yanushkevich, *Materialy p.o Voproser Gigieny Truda i Klin. Profess. Boleznei Sbornik,* **5,** 190 (1956); abstr. *Ind. Hyg. Digest,* **21,** 16, #1264 (1957).

Specific gravity: 1.6712 (25°/4°C.)
Melting point: —29°C.
Boiling point: 162°C.
Vapor density: 7.0 (air = 1)
Vapor pressure: 3.4 mm. Hg (25°C.)
Refractive index: 1.50250 (24°C.)
Per cent in "saturated" air: 0.45 (25°C.)
Density of "saturated" air: 1.03 (air = 1)
Solubility: insoluble in water; soluble in ethanol and ethyl ether
1 mg./liter $\backsim$ 120.8 p.p.m. and 1 p.p.m. $\backsim$ 8.27 mg./cu. meter at 25°C., 760 mm. Hg

3. Physiological Response

Pentachloroethane has a strongly narcotic effect which has been indicated as even greater than that of chloroform. Exposure to this material may also result in injury to the liver, lings, and kidneys. It has a local irritating effect on the eyes and the upper respiratory tract.

Lehmann and Flury[85] indicated that cats could inhale 1 mg./liter of air (121 p.p.m.) 8 to 9 hours daily for 23 days without showing serious poisoning symptoms. However, they showed significant pathological changes in the liver, lungs, and kidneys. Dogs exposed to the vapor for 3 weeks showed fatty degeneration of the liver and injury to the kidneys and lungs.

Barsoum and Saad[89] reported the oral lethal dose in dogs to be 1.75 g./kg. of body weight.

4. Hygienic Standards of Permissible Exposure

No standards have been proposed for pentachloroethane. It is probably not possible to set a reliable standard because of the limited toxicological information and experience reported on this material. It is obvious that a safe concentration would be well below the 121 p.p.m. that was shown to cause pathological changes.

5. Flammability

Pentachloroethane is essentially not flammable.

6. Odor and Warning Properties

Pentachloroethane has a sweetish odor, not unlike chloroform. The threshold at which the odor is detected has not been determined and, therefore, its value as a warning is not known.

HEXACHLOROETHANE, CCl_3CCl_3

1. Uses and Industrial Exposures

Hexachloroethane has been used as a chemical intermediate, as an insecticide and as a parasiticide in animals.

2. Physical and Chemical Properties

Physical state: solid—rhombic crystals
Molecular weight: 236.76
Specific gravity: 2.091 (20°C.)
Melting point: sublimes at 187°C.
Vapor density: 8.16 (air = 1)
Solubility: 0.005 g. per 100 ml. water at 22°C.; soluble in ethanol and ethyl ether

1 mg./liter ≳ 103.3 p.p.m. and 1 p.p.m. ≳ 9.68 mg./cu. meter at 25°C., 760 mm. Hg

3. Physiological Response

The primary physiological response to hexachloroethane reported is that of depression of the central nervous system. As the material is a solid and has a rather low vapor pressure, the hazards of breathing the vapor in industrial handling are relatively low. There has been some experience in industry to indicate that an excessive amount of dust in the air can cause irritation. Irritation has been observed from fumes of the material when handled hot. In general, it would not be considered a significant industrial hazard if handled with reasonable cleanliness.

Barsoum and Saad[89] indicated that an intravenous dose of 325 mg./kg. in dogs resulted in death. A corresponding dose for pentachloroethane was 100 mg. and for chloroform, 90 mg. This would indicate that hexachloroethane is intrinsically less toxic than the other materials.

Most of the experience with this material has come about because of its use as a parasiticide in animals.

Jondorf et al.[96] reported that a dose of 0.5 g./kg. of body weight in a rabbit was slowly metabolized. Approximately 5 per cent appeared in the urine in a period of 3 days and from 14 to 24 per cent in the expired air. These authors used chromatographic and isotopic dilution techniques to determine the nature of the metabolites in the urine. They were reported as per cent of the dose given:

Trichloroethanol	1.3
Dichloroethanol	0.4
Trichloroacetic acid	1.3
Dichloroacetic acid	0.8
Monochloroacetic acid	0.7
Oxalic acid	0.1

The radioactive carbon appeared in the expired air as carbon dioxide, hexachloroethane, tetrachloroethylene, and symmetrical tetrachloroethane. It is interesting to note that no trichloroethylene was formed.

[96] W. R. Jondorf, D. V. Parke, and R. T. Williams, *Biochem. J.*, **65**, 14P (1957).

4. Hygienic Standards of Permissible Exposure

No standard has been established for hexachloroethane.

5. Flammability

Hexachloroethane is not flammable.

PROPYL CHLORIDE, $CH_3CH_2CH_2Cl$ (1-Chloropropane)

1. Uses and Industrial Exposures

Propyl chloride has found little significant use in industry. It has been studied as an anesthetic and antiparasiticide but has not found significant use.

2. Physical and Chemical Properties

Physical state: colorless liquid
Molecular weight: 78.54
Specific gravity: 0.8910 (20°/4°C.)
Melting point: −122.8°C.
Boiling point: 46.4°C.
Vapor density: 2.71 (air = 1)
Vapor pressure: 350 mm. Hg (25° C.)
Refractive index: 1.38838 (20°C.)
Per cent in "saturated" air: 44.5
Density of "saturated" air: 1.75 (air = 1)
Solubility: 0.27 g. per 100 ml. water at 20°C.; soluble in ethanol and ethyl ether
Flash point: < 0°F.
1 mg./liter ⊃311.5 p.p.m. and 1 p.p.m. ⊃ 3.21 mg./cu. meter at 25°C., 760 mm. Hg

3. Physiological Response

As this material has not been of importance industrially, toxicological studies have been from the point of view of its use as a parasiticide or as an anesthetic. It has been of no significance in either of these uses. At high concentrations, depression of the central nervous system is observed. Injury of the liver and kidneys are likely to be observed if the time of the administration is long or repeated.

4. Hygienic Standards of Permissible Exposure

None have been established for propyl chloride.

5. Flammability

Propylchloride is flammable. The explosive limits are 2.6 to 11.1% by vol.

ISOPROPYL CHLORIDE, $CH_3CHClCH_3$ (2-Chloropropane)

1. *Uses and Industrial Exposures*

Isopropyl chloride has been used as a solvent and chemical intermediate and to some extent as an anesthetic.

2. *Physical and Chemical Properties*

Physical state: colorless liquid
Molecular weight: 78.54
Specific gravity: 0.859 (20°C.)
Melting point: —117°C.
Boiling point: 35.3°C.
Vapor density: 2.7 (air = 1)
Vapor pressure: 523 mm. Hg (25°C.)
Per cent in "saturated" air: 68.7 (25°C.)
Density of "saturated" air: 2.18 (air = 1)
Solubility: 0.31 g. per 100 ml. water at 20°C.; soluble in alcohol and diethyl ether
1 mg./liter $\backsim$ 311 p.p.m. and 1 p.p.m. $\backsim$ 3.21 mg./cu. meter at 25°C., 760 mm. Hg

3. *Physiological Response*

Isopropyl chloride has a potent anesthetic action and has been proposed for this usage. It will also cause some histopathological changes in the liver and kidneys.

von Oettingen in 1955[1] reviewed the literature on the use of this material as an anesthetic. Vomiting and cardiac arrhythmia have been observed as a result of its usage. It has not been widely used as an anesthetic.

Work in the Biochemical Research Laboratory of The Dow Chemical Co. has indicated something of its vapor toxicity. Rats, rabbits, mice, guinea pigs, and monkeys exposed 7 hours/day, 5 days/week for a total of 127 exposures over a period of 181 days survived a level of 1000 p.p.m. in air by volume with normal growth and appearance. Histological changes in the liver and kidneys were observed in some of the species tested. It would be obvious that an industrial hygiene standard should be well below 1000 p.p.m.

Oral. Guinea pigs will survive 3 g./kg. of body weight but will succumb to 10 g./kg. of body weight.

Skin Irritation. On the open surface of the skin, isopropyl chloride caused very little irritation. If it is bandaged on to the skin, some erythemia and superficial irritation may result.

Eye Irritation. The liquid material splashed into the eyes may cause painful but transient irritation. It should be washed out promptly. No serious injury would be expected.

4. *Hygienic Standards of Permissible Exposure*

None has been established for isopropyl chloride.

5. *Flammability*

Isopropyl chloride is highly flammable.

PROPYL BROMIDE, $CH_3CH_2CH_2Br$ (1-Bromopropane)

1. *Uses and Industrial Exposures*

Propyl bromide has had but minor use in industry. It has been studied as an anesthetic but has not found significant use.

2. *Physical and Chemical Properties*

Physical state: colorless liquid
Molecular weight: 123
Specific gravity: 1.3539 (20°/4°C.)
Melting point: —109.85°C.
Boiling point: 71.0°C.
Vapor density: 4.25 (air = 1)
Vapor pressure: 143 (25°C.)
Refractive index: 1.43411 (20°C.)
Per cent in "saturated" air: 19.3 (25°C.)
Density of "saturated" air: 1.66 (air = 1)
Solubility: 0.25 g. per 100 ml. water at 20°C.; soluble in alcohol and ethyl ether
1 mg./liter $\asymp$ 198.8 p.p.m. and 1 p.p.m. $\asymp$ 5.03 mg./cu. meter at 25°C., 760 mm. Hg

3. *Physiological Response*

Propyl bromide has a depressing action on the central nervous system and, for that reason, has been considered as a possible anesthetic. Exposure of animals to anesthetic concentrations may result in injury to the lungs and liver. As this material has not been of industrial significance, it will not be discussed in detail.

4. *Hygienic Standards of Permissible Exposure*

None has been established.

PROPYLENE DICHLORIDE, $CH_2ClCHClCH_3$ (1,2-Dichloropropane)

1. *Uses and Industrial Exposure*

Propylene dichloride is used as a solvent, chemical intermediate, and fumigant.

2. *Physical and Chemical Properties*

Physical state: colorless liquid
Molecular weight: 112.99
Specific gravity: 1.1593 (20/20°C.)
Freezing point: −100°C.
Boiling point: 96.8°C.
Vapor density: 3.9 (air = 1)
Vapor pressure: 50 mm. Hg (25°C.)
Refractive index: 1.437 (25°C.)
Per cent in "saturated" air: 3.4 (25°C.)
Density of "saturated" air: 1.09 (air = 1)
Solubility: 0.27 g. per 100 ml. water at 20°C.; soluble in alcohol and ethyl ether
Flash point: 60°F.
1 mg./liter ≎ 216.5 p.p.m. and 1 p.p.m. ≎ 4.62 mg./cu. meter at 25°C., 760 mm. Hg

3. *Physiological Response*

Propylene dichloride causes some central nervous system depression and, following chronic exposure, injury to the liver and kidneys. Some indications of adrenal injury in animal experiments have also been reported. There have not been reports of injury from the industrial use of this material.

The principle toxicological laboratory studies on this material were published by Heppel and co-workers.[97,98]

At a concentration of 1000 p.p.m. in air, dogs were exposed 7 hours/day, 5 days/week. Some deaths were observed after 24 exposures. There were some deaths in guinea pigs after 22 exposures. Rats succumbed in as few as 7 exposures. Rabbits survived this concentration for some period of time. The same authors exposed rats, guinea pigs, and dogs to 400 p.p.m. They were exposed 7 hours/day, 5 days/week for periods of 128 to 140 exposures. These investigators saw no ill-effects and no histological changes in these animals.

At concentrations of 1000 p.p.m. or more, the histological changes observed were in the kidneys, liver, and to some degree in the adrenal glands.

A group of C_3H strain mice exposed to 400 p.p.m. showed rather high mortality. Hepatomas were observed in some of the animals that survived.

Oral Toxicity. The lethal dose of propylene dichloride to guinea pigs is between 2 and 4 g./kg. of body weight. Repeated oral feedings of doses as low as 0.2 g./kg. of body weight will be survived for some period of time, but the animal will show definite injury.

[97] L. A. Heppel, P. A. Neal, B. Highman, and V. T. Porterfield, *J. Ind. Hyg. Toxicol.,* **28**, 1 (1946).

[98] L. A. Heppel and E. G. Peake, *J. Ind. Hyg. Toxicol.,* **30**, 189 (1948).

Skin Irritation. Propylene dichloride on the open skin causes only mild irritation. Single short contacts will probably be without any effects. The intensity of the reaction is greatly increased when bandaged on the skin or when held close to the skin by clothing.

Eye Irritation. Propylene dichloride causes some pain and irritation when splashed into the eye. It would not be expected to cause serious or permanent injury. It should be washed out immediately with water.

4. *Hygienic Standards of Permissible Exposure*

The threshold limit of propylene dichloride was established by the American Conference of Governmental Industrial Hygienists, in April, 1959, at 75 p.p.m. (350 mg./cu. meter.)

5. *Flammability*

The flammable limits of propylene dichloride are 3.4 to 14.5 per cent in air. The ignition temperature is 557 to 570°C.

VINYL CHLORIDE, $CH_2=CHCl$ (Monochloroethene)

1. *Uses and Industrial Exposures*

Vinyl chloride is used as a chemical intermediate primarily as a monomer in plastic manufacture. The fire and explosion hazard is by far the dominant problem in handling vinyl chloride.

2. *Physical and Chemical Properties*

Physical state: gas
Molecular weight: 62.5
Specific gravity: 0.9121 (20°/4°C.)
Melting point: −153.71°C.
Boiling point: −13.8°C.
Vapor density: 2.15 (air = 1)
Vapor pressure: 2580 mm. Hg (20°C.)
Solubility: slightly soluble in water; soluble in ethanol and ethyl ether
Flash point: −78°C. (open cup)
1 mg./liter ≍ 391 p.p.m. and 1 p.p.m. ≍ 2.56 mg./cu. meter at 25°C., 760 mm. Hg

3. *Physiological Response*

Vinyl chloride appears to be a material of relatively low toxicity. The principal response seems to be one of central nervous system depression, which may result in symptoms of dizziness and disorientation that are somewhat similar to the response from ethyl chloride exposure. There is the possibility of some lung irritation occurring from chronic exposure as some edema is observed in acute vapor exposure. Most investigators did not observe kidney or liver damage. One group of authors indicated some hyperemia of the liver and kidneys from acute exposure. It is concluded that the material has essentially a narcotic effect, with some lung irritation and a possibility of organ injury. There has been quite extensive use of this material in the chemical industry but no clinical reports of injury.

Acute Vapor Exposure. Patty et al.[100] reported the response of guinea pigs to single exposures to varying concentrations. The maximum time-concentrations in air for a single exposure survived by guinea pigs were as follows: 5 to 7 per cent for 1 hour and $2^1/_2$ per cent for 8 hours. The maximum time-concentration in air for a single exposure without serious disturbance was $1^1/_2$ per cent for 1 hour and 0.5 per cent for 8 hours. It will be noted that it requires a high concentration to cause death.

The same authors report that a concentration of $2^1/_2$ per cent of vinyl chloride in the air for a period of 3 minutes will cause dizziness and disorientation in exposed men. They also observed a faintly pleasant odor at this concentration. Animals that died from acute exposures showed edema of the lungs and hyperemia of the kidneys and liver.

Men exposed to vinyl chloride detected a slight odor at 4100 p.p.m. This is the lowest concentration at which an odor was detected. At 6600 p.p.m. for $^1/_2$ hour, they noticed dizziness, sleepiness, and distinct odor.

Vinyl chloride has been investigated as a possible material for anesthetic purposes by Peoples and Leake[101] and by Oster et al.[102] It was concluded that vinyl chloride was unsatisfactory for use as an anesthetic because of its circulatory and cardiac effects. These studies were at 10 to 20 vol. % and are hardly significant for industrial exposure.

Chronic Vapor Exposure. Essentially no investigations have been published on the response to chronic vapor exposure. Schaumann[85] exposed mice and rats and found that they tolerated a level sufficient for "light narcosis" for periods of 4 hours daily for 5 to 8 consecutive days or for 1 hour daily for 4 weeks without showing kidney or liver injury.

Torkelson et al.[102a] (to be published) report repeated exposures of animals for 7 hours/day, 5 days/week. At 500 p.p.m., rats showed increased liver weight and micropathology. At 200 and 100 p.p.m., rats showed increased liver weight, but no changes could be observed in dogs or guinea pigs. All species tolerated 50 p.p.m. for 6 months. Repeated exposures for 1 hour/day at 200 or 100 p.p.m. were tolerated without observable effect.

[99] *Chemical Safety Data Sheet SD-56.* Manufacturing Chemists Assoc., Washington, D. C., 1954.

[100] F. A. Patty, W. P. Yant, and C. P. Waite, *Public Health Repts. U. S.,* Reprint No. **1405, 45,** No. 34 (Aug., 1930).

[101] S. A. Peoples and C. D. Leake, *J. Pharmacol. Exptl. Therap.,* **48,** 284 (1933).

[102] R. H. Oster, C. J. Carr, J. C. Krantz, and M. J. Sauerwald, *Anesthesiology,* **8,** 358 (1947).

Skin. The boiling point of vinyl chloride is so low that if the liquid material were spilled on the skin, there is a possibility of severe cooling and possibly frost bite.

Absorption, Excretion, and Metabolism. Very little is known of the metabolism of this compound. It is apparently readily absorbed by the lungs and rapidly excreted by the lungs. The little information that is available would indicate that a large part is excreted by the lungs unchanged.

4. Hygienic Standards of Permissible Exposure

The threshold limit of vinyl chloride was established by the American Conference of Governmental Industrial Hygienists, in April, 1959, at 500 p.p.m. (1300 mg./cu. meter).

Torkelson *et al.*[102a] suggest 100 p.p.m.

5. Flammability[99]

The explosive limits of vinyl chloride are: lower, 4% and upper, 22% (by vol. in air). The autoignition temperature is 472.22°C.

6. Odor and Warning Properties

Vinyl chloride has a mild, sweetish odor. The odor is not an adequate warning property for excessive exposure.

VINYLIDENE CHLORIDE, $CH_2=CCl_2$ (1,1-Dichloroethylene)

1. Uses and Industrial Exposures

Vinylidene chloride is used as a chemical intermediate, particularly as a monomer in the production of plastics.

2. Physical and Chemical Properties

Physical state: clear colorless liquid
Molecular weight: 96.95
Specific gravity: 1.218 (20°/4°C.)
Freezing point: −122.5°C.
Boiling point: 31.7°C.
Vapor density: 3.34 (air = 1)
Vapor pressure: 591 mm. Hg (25°C.)
Refractive index: 1.427 (20°C.)
Per cent in "saturated" air: 78
Density of "saturated" air: 2.8 (air = 1)
Solubility: insoluble in water; soluble in organic solvents
Flash point: −15°C. (open cup)
1 mg./liter $\backsimeq$ 252 p.p.m. and 1 p.p.m. $\backsimeq$ 3.97 mg./cu. meter at 25°C., 760 mm. Hg

[102a] T. R. Torkelson, F. Oyen, and V. K. Rowe, *Am. Ind. Hyg. Assoc. J.*, **22**, 354 (1961).

3. Physiological Response

Exposure to high concentrations results primarily in central nervous system depression and the associated symptoms of drunkenness which may progress to unconsciousness. Chronic exposure to low concentrations results primarily in injury in the liver and kidney. In a very general way, vinylidene chloride may be considered both qualitatively and quantitatively comparable to carbon tetrachloride.

There are essentially no publications on the toxicology of vinylidene chloride. The information included in this discussion is taken entirely from unpublished work of the Biochemical Research Laboratory of The Dow Chemical Co.[5f]

Acute Vapor Exposures. High concentrations, in the order of 4000 p.p.m., rapidly produce drunkenness which may progress to unconsciousness if exposure is continued. If the exposure is of short duration, complete recovery from the anesthetic effect may be expected. Animal investigations indicate that the maximum single exposure permitting a reasonably high probability of no injuries is above 1000 p.p.m. for up to 1 hour and 200 p.p.m. for up to 8 hours.

Chronic Vapor Exposure. Animals exposed five days a week, 8 hours per day for several months will show some injury of the liver and kidney at concentrations of 100 p.p.m. and 50 p.p.m. There was minimal but significant injury to the liver and kidney, even at a concentration as low as 25 p.p.m.

This information is quite comparable to that reported by Adams *et al.*[39] on carbon tetrachloride. It would seem logical that, until further information is available, we should handle this material as we would carbon tetrachloride. It should be recognized, however, that it has a much higher vapor pressure than carbon tetrachloride.

Eye Contact. Vinylidene chloride is moderately irritating to the eyes. It will cause pain, conjunctival irritation, and some transient corneal injury. Permanent damage is not likely.

This information is obtained on inhibited vinylidene chloride. A high concentration of the phenolic inhibitor by itself may cause serious and permanent eye injury. A contaminated eye should be flushed immediately with large quantities of flowing water.

Skin Contact. Liquid vinylidene chloride is irritating to the skin after direct contact of only a few minutes. The inhibitor content of the vinylidene chloride may be partly responsible for this irritation. Where leaks occur, vinylidene chloride will evaporate and the phenolic inhibitor may accumulate till it reaches a concentration capable of causing local burns. Particular caution should be used with regard to contaminated clothing which should be removed immediately and thoroughly cleaned before reuse.

Absorption, Excretion, and Metabolism. Nothing is known of the metabolism of vinylidene chloride in the body. It is apparently readily absorbed when the vapor is breathed by the lungs.

4. Hygienic Standards of Permissible Exposure

No standard for vinylidene chloride has been established. It would be suggested that the concentration in the work atmosphere should be below 25 p.p.m.

5. Flammability

The explosive limits of vinylidene chloride (% by vol. with air) are 7–16. The auto ignition temperature is 570°C.

Vinylidene chloride in the presence of air or oxygen, with the inhibitor removed, forms a complex peroxide compound at temperatures as low as −40°C. The peroxide is violently explosive. Reaction products formed with ozone are particularly dangerous.[103]

6. Odor and Warning Properties

Vinylidene chloride has a characteristic sweet smell that resembles carbon tetrachloride or chloroform. Most persons can detect a mild but definite odor at 1000 p.p.m. in air. Some people can detect it at 500 p.p.m. Vapors containing decomposition products have a disagreeable odor and can be detected at concentrations considerably less than 500 p.p.m. Neither odor nor irritating properties of vinylidene chloride are adequate to warn of excessive exposure.

cis-1,2-DICHLOROETHYLENE, CHCl=CHCl

1. Source, Uses, and Industrial Exposures

The cis- and trans-isomers of 1,2-dichloroethylene have had use as solvents and chemical intermediates. Neither of the isomers has developed wide industrial usage.

2. Physical and Chemical Properties

Physical state: liquid
Molecular weight: 96.95
Specific gravity: 1.2743 (25°/4°C.)
Melting point: −80.5°C.
Boiling point: 60.25°C.
Vapor density: 3.34 (air = 1)
Vapor pressure: 208 mm. Hg (25°C.)
Per cent in "saturated" air: 27.4 (20°C.)
Density of "saturated" air: 1.63 (air = 1)
Solubility: 0.35 g. per 100 ml. water at 20°C.; soluble in ethanol and ethyl ether

[103] R. C. Reinhardt, *Chem. Eng. News*, **25**, 2136 (1947).

1 mg./liter ⇌ 252 p.p.m. and 1 p.p.m. ⇌ 3.97 mg./cu. meter at 25°C., 760 mm. Hg

trans-1,2-DICHLOROETHYLENE, CHCl=CHCl

1. *Physical and Chemical Properties*

Physical state: liquid
Molecular weight: 96.95
Specific gravity: 1.2489 (25°/4°C.)
Melting point: −50°C.
Boiling point: 48.35°C.
Vapor density: 3.34 (air = 1)
Vapor pressure: 324 mm.Hg.(25°C.)
Refractive index: 1.44234 (20°C.)
Per cent in "saturated" air: 42.5 (25°C.)
Density of "saturated" air: 2 (air = 1)
Solubility: 0.63 g. per 100 ml. water at 20°C.; soluble in ethanol and ethyl ether

1 mg./liter ⇌ 252 p.p.m. and 1 p.p.m. ⇌ 3.97 mg./cu. meter at 25°C. and 760 mm. Hg

2. *Physiological Response*

Neither *cis*- nor *trans*-1,2-dichloroethylene has found wide usage; toxicological studies have been relatively limited. The major response to both of these materials is one of central nervous system depression. Liver and kidney injury has not appeared to be a major factor. As only acute studies have been made, some caution should be exercised in the use of any proposed acceptable concentration for chronic or repeated exposure.

Acute Vapor.   Smyth[60] stated that the *cis*-isomer did not kill or anesthetize rats in 4 hours at 8000 p.p.m. At 16,000 p.p.m., they were anesthetized in 8 minutes and killed in 4 hours. He also states that the *trans*-isomer was twice as toxic and anesthetic as the *cis*-isomer. Smyth states that the important effect of inhalation is narcosis, and that the 200-p.p.m. threshold limit is probably low enough to prevent definite narcosis.

Lehmann and Flury[85] reported that disturbance of equilibrium and prostration occur in approximately the same length of time from similar concentrations of the *cis*- and *trans*-isomers. Comparable times of slight narcosis and deep narcosis occurred with the *cis*-isomer at concentrations of about half that of the *trans*-isomer.

Chronic Exposure.   Lehmann and Flury[85] reported the results of exposure of cats and rabbits to vapor concentrations of 0.16 to 0.19 per cent in air. Animals exposed to the *cis*-isomer, at this concentration, showed loss of appetite, decrease in body weight, and pathological changes in the lungs, liver, and kidneys. Animals exposed to the *trans*-isomer at the same concentrations showed loss of appetite and some respiratory irritation but no histopathological changes in the organs.

It would seem that more toxicological research would be necessary before definitive conclusions could be drawn.

3. Hygienic Standards of Permissible Exposure

The threshold limit for *cis*- and *trans*-1,2-dichloroethylene was established by the American Conference of Governmental Industrial Hygienists, in April, 1959, at 200 p.p.m. (790 mg./cu. meter).

TRICHLOROETHYLENE, CHCl=CCl₂ (Acetylene Trichloride)

1. Uses and Industrial Exposures

Trichloroethylene is widely used as an industrial solvent, particularly in metal degreasing and extraction processes. It has some use as a chemical intermediate and also as an anesthetic. By far the most important contact with trichloroethylene in industry is with the vapors of the material when it is used as an industrial solvent. In some industrial processes, such as vapor degreasing or hot extraction, the material may be used at elevated temperatures that increase the problem of escape of vapors into the working atmosphere.

2. Physical and Chemical Properties

Physical state: colorless liquid
Molecular weight: 131.4
Specific gravity: 1.45560 (25°/4°C.)
Freezing point: —86.8°C.
Boiling point: 87°C.
Vapor density: 4.54 (air = 1)
Vapor pressure: 77 mm. Hg (25°C.)
Refractive index: 1.4777 (20°C.)
Per cent in "saturated" air: 10.2 (25°C.)
Density of "saturated" air: 1.35 (air = 1)
Solubility: 0.1 g. per 100 ml. water at 20°C.; soluble in ethanol and ethyl ether
1 mg./liter ≏ 185.8 p.p.m. and 1 p.p.m. ≏ 5.38 mg./cu. meter at 25°C., 760 mm. Hg

3. Physiological Response

The predominant physiological response from exposure to trichloroethylene is one of central nervous system depression. This is particularly true as a response from acute exposure. Visual disturbances, mental confusion, fatigue, and sometimes nausea and vomiting are observed. Nausea and other gastrointestinal disturbances are not as striking with trichloroethylene as they are with carbon tetrachloride or ethylene dichloride.

The dangers in industry may be accentuated by visual disturbances and incoordination, which may lead to poor manual manipulation and, therefore, unsafe mechanical operations. While nervous system depression is a dominating problem from exposure to trichloroethylene, some indications of injury to the liver and kidneys may also be observed. While such organic injury has been observed both in experimental animals and in the clinic, it is qualitatively and quantitatively of less significance than in the case of materials like carbon tetrachloride.

Acute Vapor Exposure. The response of laboratory animals to single exposures to various concentrations of trichloroethylene has been reported by Adams et al.[104]

The maximum time-concentrations in air for a single exposure survived by rats were as follows: 18 minutes at 20,000 p.p.m., $1^1/_2$ hours at 6400 p.p.m., and 8 hours at 3000 p.p.m. Light anesthesia was seen at concentrations of 4800 p.p.m. or more, but it was not observed at 3000 p.p.m. Clinical experience from acute exposure has come in a great degree from the use of trichloroethylene as an anesthetic. Death may result from respiratory failure or cardiac arrest. The cardiac response may, in most instances, be due to the potentiation of endogenous epinephrine. Much of the literature on anesthetic use of trichloroethylene has come from Britain.

Chronic Vapor Exposure. Animal investigative work was reported by Adams et al.[104] They exposed several species of animals 7 hours/day, 5 days/week for approximately 6 months. At 3000 p.p.m. by volume in air, rats and rabbits both showed an increase in liver and kidney weight. At 400 p.p.m., rats showed an increase in liver and kidney weights and the male rats showed significantly less growth. Guinea pigs showed an increase in liver weight and the growth of the exposed males was less than the controls. Rabbits showed a slight increase in liver weight. An exposed monkey showed no response at 400 p.p.m. At 200 p.p.m., rats showed no response. Guinea pigs showed less growth than controls. Rabbits showed no response. Monkeys showed no effect from this concentration. At a concentration of 100 p.p.m., guinea pigs showed no significant response. The maximum concentration tolerated for the 6-month period were as follows: monkeys, 400 p.p.m.; rats and rabbits, 200 p.p.m.; guinea pigs, 100 p.p.m.

The principal response observed from chronic exposure at low levels was growth depression and liver and kidney changes. It would appear probable that the principal problem from chronic exposure of humans will be complaints of nervous system disfunction. One would, therefore, expect complete recovery when removed from exposure.

[104] E. M. Adams, H. C. Spencer, V. K. Rowe, D. D. McCollister, and D. D. Irish, *A.M.A. Arch. Ind. Hyg. Occupational Med.*, **4**, 469 (1951).

Stuber[105] has reported on a number of cases of exposure to trichloroethylene. Effects on the trigeminal nerve and upon vision are emphasized. Stuber did not observe either kidney or liver damage. The observations of Ahlmark and Forssman[106] and also those of Frant and Westendorp[107] would indicate largely functional complaints.

Baker[108] exposed dogs to trichloroethylene vapor, "varying from 500 to 3000 p.p.m. repeatedly for periods varying from 2 to 8 hrs. daily often for 5 days weekly." Some dogs were exposed to a total of up to 162 hours. He observed histological changes in a number of areas of the brain. The most striking and severe changes were observed within the cerebellum and involved primarily the Purkinje cell layer. It was implied that all of the chronically exposed dogs showed the same histological change and no relation to specific exposure was given. The ataxia reported clinically following trichloroethylene exposure correlates well with this injury to the cerebellum.

Browning[79] has reviewed the literature on several solvents. She concludes that trichloroethylene is probably more toxic than the 200-p.p.m. standard would indicate. She quoted the literature as indicating that concentrations of 20 to 80 p.p.m. might result in symptoms. She also quoted recent literature which indicated liver injury from trichloroethylene exposure, but she is apparently not convinced that the evidence is conclusive.

Oral Toxicity. There have been indications that a fatal dose by mouth would be in the range of 3, 4, or 5 ml./kg. of body weight. There have been a significant number of clinical cases reported from oral intake. It is obvious that this material is readily absorbed by the gastrointestinal tract and the symptoms and signs of poisoning are comparable to those observed from inhalation.

Skin and Eye Contact. Trichloroethylene is irritating in the eyes and should be washed out promptly. It would not be expected to cause permanent injury from ordinary contact. Trichloroethylene is only mildly irritating to the skin from ordinary open contact. From continued use of the material in contact with the skin, a defatting takes place, producing a rough, chapped skin, which may result in erythema and possibly secondary infection. A much more severe skin response would be expected when the material is held close to the skin under tight clothing, such as shoes. Frant and Westendorp[107] reported their experiments where individuals exposed themselves to severe contact with the liquid and checked for excretion of trichloroacetic acid in the urine. They concluded that there was no significant skin absorption.

Experience in Man. Kleinfeld and Tabershaw[109] reported several fatal cases

[105] K. Stuber, *Arch. Gewerbepathol. Gewerbehyg.,* **2,** 398 (1931).

[106] A. Ahlmark and S. Forssman, *A.M.A. Arch. Ind. Hyg. Occupational Med.,* **3,** 386 (1951).

[107] R. Frant and J. Westendorp, *A.M.A. Arch. Ind. Hyg. Occupational Med.,* **1,** 308 (1950).

[108] A. B. Baker, *J. Neuropathol. Exptl. Neurol.,* **17,** 649 (1958).

[109] M. Kleinfeld and I. R. Tabershaw, *A.M.A. Arch. Ind. Hyg. Occupational Med.,* **10,** 134 (1954).

of trichloroethylene poisoning in the United States. Four of the cases were exposed to significantly high concentrations of vapor in industrial operations. These patients had reported symptoms from a number of exposures over some period of time. Death occurred several hours after the last exposure, which may have been fairly severe. On autopsy, they showed no anatomical abnormalities. It was assumed that they died from cardiac arrhythmia, probably due to the potentiation of endogenous epinephrine. The fifth case accidentally drank trichloroethylene. He showed severe injury in both the liver and kidneys.

Absorption, Excretion, and Metabolism. Barrett, Cunningham, and Johnston[83] showed that the administration of trichloroethylene to animals was followed by the excretion of trichloroacetate in the urine.

Soucek et al.,[110] in 1952, indicated that an average of 56 per cent of the inspired trichloroethylene was retained. The remainder was excreted by the lungs as trichloroethylene. Even after 5 hours, blood concentration did not reach saturation. An average of 16 per cent was excreted as trichloroacetic acid in the urine and a larger amount appeared as unknown metabolites. Soucek,[111] in 1953, indicated that the amount of trichloracetic acid was not in simple relationship to the amount retained. He concluded that this excretion could not be utilized as a simple test for exposure to trichloroethylene.

Butler[112] indicated a remarkable series of metabolic changes in trichloroethylene. He showed that trichloroacetic acid, urochloralic acid, and trichloroethanol, small amounts of chloroform, and monochloroacetic acid may be produced. Bardodej and Vyshocil[113] proposed a chart of the various pathways by which trichloroethylene might be metabolized to the several different metabolites.

A number of investigators have been interested in the possible relationship of the excretion of trichloroacetate in the urine and exposure to trichloroethylene, having in mind the possibilities of a clinical measure of exposure.

Ahlmark and Forssman have several publications on the excretion of trichloroacetate in relationship to exposure. Their general conclusion in 1951[106] was that excretion of less than 20 mg./liter was not likely to result in any complaints. Complaints from some of the people were noted when they excreted from 40 to 75 mg./liter. Almost all people excreting 100 mg./liter or more would have complaints; and if it were 200 mg./liter or greater, the complaint would be pronounced and the individuals often lost time from work. Frant and Westendorp[107] made similar studies. Their data would indicate a somewhat variable relationship between the excreted trichloroacetate and inhaled trichloroethylene. Another clinical

[110] B. Soucek, J. Teisinger, and E. Pavelkova, *Pracovní lékařství,* **4,** 31 (1952); abstr. *A.M.A. Arch. Ind. Hyg. Occupational Med.,* **6,** 461 (1952).

[111] B. Soucek and E. Pavelkova, *Pracovní lékařství,* **5,** 62 (1953); *Chem. Abstr.,* **49,** 4181 (1955).

[112] T. C. Butler, *J. Pharmacol. Exptl. Therap.,* **97,** 84 (1949).

[113] Z. Bardodej and J. Vyshocil, *A.M.A. Arch. Ind. Health,* **13,** 581 (1956).

study was reported by Grandjean[113a]. Forssman and Holmquist[114] studied the relationship between absorption and excretion. They indicated that a significant amount of trichloroethylene was retained by the body for some period of time after exposure. Only a small quantity was exhaled (21 to 28 per cent) from a short exposure and 32 to 69 per cent from a long exposure. The remainder is excreted or converted in some other way. Only a small part (from 1.2 to 7.8 per cent) was recovered as trichloroacetic acid in the urine.

The total matter might be summed up as follows. The excretion of trichloroacetate is a reasonable index of exposure to trichloroethylene. The relation both in quantity and time of exposure between the inhaled trichloroethylene and the excreted trichloroacetate is quite variable. Good monitoring of air exposure seems still to be the best control of exposure. Trichloroacetate analysis in the urine, however, should be very valuable supplemental information, particularly in studying an individual in the clinic.

Bardodej et al.[115] have made some interesting suggestions with regard to the reason for the apparent potentiation of toxic effect observed when trichloroethylene and ethanol are absorbed concurrently. They indicated that the trichloroethylene or the metabolism of trichloroethylene to chlorohydrate may inhibit liver aldehyde dehydrogenase, thus blocking the metabolism of ethanol beyond the aldehyde stage.

4. Hygienic Standards of Permissible Exposure

The threshold limit for trichloroethylene was established by the American Conference of Governmental Industrial Hygienists, in April, 1961, at 100 p.p.m. (520 mg./cu. meter).

It should be noted that animals showed a toxic response at 200 p.p.m.; also, Browning[79] questioned the level of 200 p.p.m.

In 1961 the A.C.G.I.H. lowered their limit to 100 p.p.m.

5. Flammability

The ignition temperature of trichloroethylene is 410°C.

6. Odor and Warning Properties

Trichloroethylene has a typical odor which has been characterized as ethereal or chloroformlike. The odor threshold has not been reported for man but the odor is detectable at levels below the acceptable levels for industrial operations. It could not, however, be considered an effective warning.

[113a] E. Grandjean, R. Munchinger, V. Turrian, P. Haas, H. K. Knoepfel, and H. Rosenmund, *Brit. J. Ind. Med.*, **12**, 131 (1955).

[114] S. Forssman and C.-E. Holmquist, *Acta. Pharmacol. Toxicol.*, **9**, 235 (1953).

[115] Z. Bardodej, M. Krivaucova, and F. Pokorny, *Pracovní léukařství*, **7**, 263 (1955); *Chem. Abstr.*, **49**, 16269 (1955).

TETRACHLOROETHYLENE, $CCl_2=CCl_2$ (Perchloroethylene)

1. *Uses and Industrial Exposures*

Tetrachloroethylene is used as an industrial solvent for a number of purposes, particularly dry cleaning and degreasing. It has been used as an anthelmintic in humans and animals. It also finds some use as a chemical intermediate and for other industrial purposes.

The principal contact with this material in industry is with its use as a solvent.

2. *Physical and Chemical Properties*

Physical state: colorless liquid
Molecular weight: 165.85
Specific gravity: 1.6226 (20°/4°C.)
Melting point: −23.35°C.
Boiling point: 121.2°C.
Vapor density: 5.7 (air = 1)
Vapor pressure: 19 mm. Hg (25°C.)
Refractive index: 1.50534 (20°C.)
Per cent in "saturated" air: 2.5 (25°C.)
Density of "saturated" air: 1.11 (air = 1)
Solubility: insoluble in water; soluble in ethanol, ethyl ether, chloroform, and benzene
Flash point: none
1 mg./liter $\backsimeq$ 147.4 p.p.m. and 1 p.p.m. $\backsimeq$ 6.78 mg./cu. meter at 25°C., 760 mm. Hg

3. *Physiological Response*

The major response to tetrachloroethylene at high concentrations is central nervous system depression. It is not, however, sufficiently effective to be considered a useful anesthetic. Irritation of the eyes, nose, and throat may also be observed at high concentrations. There are some indications of nausea and gastrointestinal upset, but these are not nearly as striking in the case of perchloroethylene as they are in carbon tetrachloride or ethylene dichloride exposure. Humans exposed to 106 p.p.m. were very much aware of the odor and of a very slight irritation of the eyes. They became rapidly acclimated, however, and there was no indication of interference with motor coordination or normal alertness. At 260 p.p.m., they observed slight dizziness and some sleepiness. At 280 p.p.m., they were lightheaded. The eyes were irritated and there was definite impairment of motor coordination.

The liver and kidney injuries are not as severe or striking as they are from a material like carbon tetrachloride. However, changes in the liver and kidneys are seen following chronic exposure. Carpenter[116] observed this finding in his animals.

[116] C. P. Carpenter, *J. Ind. Hyg. Toxicol.*, **19**, 323 (1937).

Similar observations were made by Rowe et al.[117] Dizziness, inebriation, and inco-ordination have been observed in industrial exposure. The possible hazard to the man himself or to his associates because of poor mechanical coordination should be considered in exposure to perchloroethylene.

Relatively few incidents of industrial problems from tetrachloroethylene exposure have been reported. The most recent was that of Coler and Rossmiller in 1953.[118] They reported on 7 cases of exposure to tetrachloroethylene, with a range of concentration between 232 to 385 p.p.m. One man had a severe gastric hemorrhage and cirrhosis of the liver. Six others had symptoms of headache, nausea, lightheadedness, dizziness, tiredness, hang-over, and a feeling of intoxication. Of these, 3 exhibited signs of liver disfunction—all 3 had an altered Bromsulphalein test, 2 showed altered serum protein patterns, 1 had a 3+ cephalin cholesterol reaction. This clinical report would seem to be confirmation of the liver injury reported in animals.

Browning[79] reviewed the toxicity of several solvents. She indicated that in the case of perchloroethylene that recent investigations would indicate that confidence in "its relative non-toxicity may be unfounded." She concludes that if perchloroethylene is substituted for trichloroethylene, it should be used with the same precautions. The literature quoted would indicate the possibility of liver and kidney injury.

Acute Vapor Exposure. The response to single exposures was reported by Rowe et al.[117] They stated that 2000 p.p.m. were tolerated for up to 14 hours, 3000 p.p.m. were tolerated for 4 hours. Unconsciousness was observed in the animals within a few minutes at concentrations of 6000 p.p.m. or more, and after several hours at 3000 p.p.m. But unconsciousness was not observed at 2000 p.p.m. At these high-level single exposures, the predominant response was one of depression of the nervous system. There were slight changes in the liver, characterized by a slight increase in weight, slight increase in total lipid, and slight cloudy swelling.

Chronic Vapor Exposure. The first chronic study of tetrachloroethylene vapors was carried out by Carpenter.[116] He exposed rats 8 hours/day, 5 days/week for periods up to 7 months to concentrations of 70, 230, and 470 p.p.m. All his animals survived with growth comparable to his controls. At 70 p.p.m., no pathological effects were observed. At 230 p.p.m., he observed some pathological changes in both the liver and the kidneys. At 470 p.p.m., the pathological indications of injury to the liver and kidneys were more striking.

Rowe et al.[117] in 1952 reported the results of chronic vapor exposures of several species of animals. At 1600 p.p.m. in air, rats showed a drowsy, depressed condition for the first week and later a definite "irritation." Enlargement of the

[117] V. K. Rowe, D. D. McCollister, H. C. Spencer, E. M. Adams, and D. D. Irish, *A.M.A. Arch. Ind. Hyg. Occupational Med.*, **5**, 566 (1952).

[118] H. R. Coler and H. R. Rossmiller, *A.M.A. Arch. Ind. Hyg. Occupational Med.*, **8**, 227 (1953).

liver and kidneys was noticeable. Guinea pigs showed a loss in body weight, and increase in liver weight with moderate histological changes. At 400 p.p.m., for 130 exposures of 7 hours/day over a period of 183 days, rats showed no evidence of adverse effect. Guinea pigs showed definite increase in liver and kidney weights and slight fatty degeneration of the liver. Rabbits and monkeys showed no evidence of injury. At 200 p.p.m., guinea pigs still showed an increase in liver weight, increase in total liver lipid, and slight to moderate histopathological changes in the liver. At 100 p.p.m., female guinea pigs showed an increase in liver weight. Histologically, they appeared normal.

Guinea pigs in these studies had been shown to be particularly susceptible, showing changes even at 100 p.p.m. However, due to the fact that human experience had been favorable, these authors were inclined to accept the previously used maximum allowable concentration of 200 p.p.m. as satisfactory. They suggested, however, that this should be considered a ceiling and fluctuations should be around an average of 100 p.p.m.

Oral Toxicity. The toxic level for many different species of animals was studied by Schlingman and Gruhzit.[119] Lamson et al.[120] studied the pharmacology and toxicology of tetrachloroethylene.

They indicated that tetrachloroethylene was poorly absorbed from the intestinal tract of dogs. They did not observe pathological changes in dogs from single oral doses. Schlingman and Gruhzit did, however, observe changes in both the liver and the kidney in a number of different species of animals used in their investigations.

Skin and Eyes. Incidental contact on the open skin will not cause serious injury. Tetrachloroethylene may cause skin response after repeated and prolonged contact. The skin may become reddened and rough, giving the appearance of chapping. This condition can be much accentuated by the solvent being held close to the skin by the clothing, particularly in shoes or under areas where the clothing is tight. In the eyes, the material may be irritating, causing lacrimation and burning. However, it would not be expected to cause any permanent injury.

Absorption, Excretion, and Metabolism. Tetrachloroethylene is readily absorbed when breathed as a vapor. Its absorption through the skin is probably not significant, although thorough studies have not been made on this subject. Little is known of the metabolism of tetrachloroethylene in the body. It is partly excreted as such by the lungs but a certain amount may be converted to a water-soluble metabolite, as reported by Barrett et al.[83]

4. Hygienic Standards of Permissible Exposure

The threshold limit of tetrachloroethylene was established by the American Conference of Governmental Industrial Hygienists, in April, 1961, at 100 p.p.m. (670 mg./cu. meter).

[119] A. S. Schlingman and O. M. Gruhzit, *J. Am. Vet. Assoc.*, **71**, 189 (1927).
[120] P. D. Lamson, B. H. Robbins, and C. B. Ward, *Am. J. Hyg.*, **9**, 430 (1929).

It should be noted that one species of animal has shown injury at 200 p.p.m. and minimal effects at 100 p.p.m. This limit should, therefore, be considered a ceiling with fluctuations around 100 p.p.m.

In 1961 the A.C.G.I.H. lowered their limit to 100 p.p.m.

5. *Flammability*

Tetrachloroethylene is nonflammable and nonexplosive. It will not support combustion.

6. *Odor and Warning Properties*

Perchloroethylene has a not unpleasant ethereal or aromatic odor. Carpenter[116] indicated that this odor was detectable at 50 p.p.m. Although this odor is detectable at levels below the acceptable level for chronic exposure, it cannot be considered a good warning property, as an individual may become adapted to the odor in a reasonably short time. It has the value, however, that it would be recognized by an individual entering the area from an area of uncontaminated air.

ALLYL CHLORIDE, CH_2=CH—CH_2Cl (3-Chloropropene-1)

1. *Uses and Industrial Exposures*

The major use of allyl chloride is as a chemical intermediate.

2. *Physical and Chemical Properties*

Physical state: liquid, colorless
Molecular weight: 76.53
Specific gravity: 0.9376 at 20°/4°C.
Freezing point: —136.4°C.
Boiling point: 45.0°C.
Vapor density: 2.64 (air = 1)
Vapor pressure: 368 mm. Hg (25°C.)
Refractive index: 1.4155 (20°C.)
Per cent in "saturated" air: 48 (25°C.)
Density of "saturated" air: 1.87 (air = 1)
Solubility: miscible with ethanol, ethyl ether, and chloroform
Flash point: —20°F. (open cup)
1 mg./liter ≎ 320 p.p.m. and 1 p.p.m. ≎ 3.13 mg./cu. meter at 25°C., 760 mm. Hg

3. *Physiological Response*

Allyl chloride is very irritating to the eyes and upper respiratory tract. It is only weakly narcotic and this would be of very little significance in the usual industrial exposure. In acute exposures, the two primary responses are lung irritation and kidney injury. Injury to the liver is less significant in the acute exposures. Similar injury may be observed in chronic exposure, but the liver injury becomes more important. As the concentrations decrease, the problem of lung irritation is somewhat less. At very low concentrations for chronic exposure, the primary problem is liver and kidney injury.

Allyl chloride is a highly toxic and irritating material and should be handled with due respect.

Acute Vapor. Acute vapor studies on animals were reported by Adams et al.[121] Maximum exposure time-concentrations in air for a single exposure survived by rats were as follows: 3 hours at 290 p.p.m., 1 hour at 2900 p.p.m., and 15 minutes at 29,300 p.p.m. Comparable figures for guinea pigs were: 8 hours at 290 p.p.m., 3 hours at 2900 p.p.m., and $1/2$ hour at 29,300 p.p.m.

The primary lesions observed were in the kidneys and the lungs. The kidney injury was severe. There was congestion throughout the kidney with hemorrhage and marked parenchymatous degeneration. The pulmonary injury was severe, particularly at higher concentrations. It consisted of marked congestion with frequent hemorrhages in the alveolar spaces.

Chronic Vapor Exposure. Chronic vapor work has not been published. The following information is taken from two sources. Unpublished work in the Biochemical Research Laboratory of The Dow Chemical Co.[5f] and the information assembled by the Shell Chemical Corp. in its industrial bulletin.[122] Rats, guinea pigs, and rabbits exposed 7 hours/day, 5 days/week to a concentration of 8 p.p.m. of allyl chloride survived 28 exposures without apparent gross ill effect. However, on histopathological examination, severe damage to the liver and kidneys of essentially all the animals was observed. A comparable experiment was run at 3 p.p.m. The animals survived as many as 127 or 134 exposures over a period of 194 days. There was no evidence of ill effect in most of the animals. Dogs were also included in this experiment.

Although a concentration of 5 p.p.m. has been suggested as an industrial standard of permissible exposure, it would seem desirable to the author to lower this level to 2 p.p.m. with an understanding that fluctuations should stay below this level.

Oral. Smyth and Carpenter[123] reported an oral LD_{50} of 0.7 g./kg. of body weight as determined on rats.

Skin. Allyl chloride is mildly irritating to the skin but may be absorbed through the skin in amounts sufficient to cause systemic intoxication in animals. Smyth and Carpenter[123] reported an LD_{50} of 2.2 g./kg. determined for a 24-hour skin absorption on rabbits. The Shell Chemical Corporation's Industrial Bulletin states that, "The absorption of the liquid through the skin of human beings is attended by deep-seated pain in the contact area."[122]

Eyes. Concentrations of allyl chloride of 50 or 100 p.p.m. are irritating to the eyes. High concentrations of vapor may cause very severe irritation and pain.

[121] E. M. Adams, H. C. Spencer, and D. D. Irish, *J. Ind. Hyg. Toxicol.,* **22,** 79 (1940).

[122] Shell Chemical Corp., *Ind. Hyg. Dept., Bull.* No. **SD-57-80,** New Work.

[123] H. F. Smyth, Jr., and C. P. Carpenter, *J. Ind. Hyg. Toxicol.,* **30,** 63 (1948).

Liquids splashed in the eye would be expected to cause severe irritation and should be washed out immediately with water.

Clinical Experience. There has been very little clinical experience with allyl chloride. The most complete discussion of such experience was published by the Shell Chemical Corp.[122]

4. Hygienic Standards of Permissible Exposure

The threshold limit of allyl chloride was established by the American Conference of Governmental Industrial Hygienists, in April, 1959, at 5 p.p.m. (15 mg./cu. meter).

This standard is probably too high. It is suggested that the concentration of the work atmosphere should be below 2 p.p.m.

5. Flammability

The upper and lower explosive limits of allyl chloride are 11.0% vol. (approx.), and 3.28% vol., respectively.

6. Odor and Warning Properties

Shell Chemical Corp. published the following information in their bulletin in 1958.[122] Odor threshold for half of the people is from 3 to 6 p.p.m. Odor threshold for essentially all people is 25 p.p.m. Eye irritation occurs between 50 and 100 p.p.m. Nose irritation and pulmonary discomfort may be observed at levels below 25 p.p.m.

The above property might be considered a warning of levels hazardous from acute exposure. The material is not detected at a low enough concentration to be considered a warning of levels hazardous from chronic exposure.

CHLOROPRENE, $CH_2{=}CCl{-}CH{=}CH_2$ (2-Chlorobutadiene)

1. Uses and Industrial Exposures

Chloroprene is used as a chemical intermediate, largely as a monomer for the manufacture of a synthetic rubber.

2. Physical and Chemical Properties

Physical state: colorless liquid
Molecular weight: 88.54
Specific gravity: 0.9583 (20°/20°C.)
Boiling point: 59.4°C.
Vapor density: 3.0 (air = 1)
Vapor pressure: 215.4 mm. Hg (25°C.)
Refractive index: 1.4583 (20°C.)
Per cent in "saturated" air: 28 (25°C.)
Density of "saturated" air: 1.57 (air = 1)

Solubility: slightly soluble in water; soluble in alcohol and diethyl ether

1 mg./liter $\backsim$ 276.5 p.p.m. and 1 p.p.m. $\backsim$ 3.62 mg./cu. meter at 25°C.. 760 mm. Hg

3. Physiological Response

The primary responses to chloroprene appear to be central nervous system depression and significant injury to the lung, liver, and kidneys. The reported depilatory action upon the scalp hair, while not serious from the point of view of injury, is still of concern from a cosmetic point of view. Although the quantitative information available is not sufficient to draw any very definitive conclusions with regard to the hazard of this material, von Oettingen's suggestion of 0.3 mg./liter (90 p.p.m.) would indicate a significantly toxic material but not a highly hazardous one.

Eye.   Conjunctivitis and necrosis of the cornea have been reported. von Oettingen in his book of 1955[1] quoted Roubal (1942) to this effect. Apparently, nervousness and irritability are typical responses to exposure. A more thorough discussion of the clinical picture can be found as reported by Nystrom[124] and Ritter and Carter.[125]

Acute Vapor.   von Oettingen et al.[126] indicated that a 1-hour exposure of mice to chloroprene vapor at a concentration of 3 mg./liter of air (829.2 p.p.m.) was fatal to all of the animals exposed. A concentration of 1 mg./liter (277 p.p.m.) killed none of the animals.

They also reported that the LC_{100} for an 8-hour exposure for rabbits is 7.5 mg./liter of air and for cats is 2.5 mg./liter.

Chronic Vapor Exposure.   von Oettingen quoted Nystrom[124] to the effect that rats would tolerate a daily exposure of 8 hours to concentrations of 0.3 mg./liter of air for 13 weeks. A concentration of 1.2 mg./liter was fatal after daily exposures for 6, 9, and 13 weeks. This would indicate that a tolerated level in plant operation should be well below the 0.3 mg./liter. The level that does not cause any histopathological change should be known before determining a level for control of industrial atmospheres.

Oral.   von Oettingen et al.[126] reported the LD_{100} for rats to be 0.67 g./kg. of body weight. Death occurred in periods from 5 hours to 4 days after the administration.

Skin.   von Oettingen et al.[126] indicated some systemic toxicity from repeated topical application of chloroprene to the skin of rats. Ritter and Carter[125] indicated systemic poisoning from the topical application of chloroprene to guinea pigs. It is uncertain how completely the animals were prevented from breathing the vapors when it was applied to the skin.

[124] A. E. Nystrom, *Acta Med. Scand.,* (*Suppl.* No. 219) **132**, 1 (1948).

[125] W. L. Ritter and A. S. Carter, *J. Ind. Hyg. Toxicol.,* **30**, 192 (1948).

[126] W. F. von Oettingen, W. C. Hueper, W. Deichmann-Grubler, and F. H. Wiley, *J. Ind. Hyg. Toxicol.,* **18**, 240 (1936).

Perhaps the most surprising effect from topical application is the loss of hair. As the hair grows readily when the individual is removed from contact, it would appear that the action is directly on the hair. Apparently, this is observed both in humans in industrial operations and in animals exposed in the laboratory. Ritta and Carter question whether this action is of chloroprene itself or an intermediate in the polymerization of chloroprene; but it is agreed that the depilatory action does occur in workers that are handling chloroprene in the polymerization process.

4. Hygienic Standards of Permissible Exposure

The threshold limit of chloroprene was established by the American Conference of Governmental Industrial Hygienists, in April, 1959, at 25 p.p.m. (90 mg./cu. meter).

5. Flammability

Chloroprene is flammable.

FLUORINE-CONTAINING HALOGENATED HYDROCARBONS

I. General

There are some aspects of the fluorine-containing halogenated hydrocarbons which can be more readily handled if they are considered as a group. Only those compounds that appear to have some commercial significance will be considered individually.

There is a wide range of compounds not of commercial significance about which there are a number of publications. A group of publications by Pattison and his associates covers a great many fluorinated compounds. These researchers have published over twenty papers on this subject. A tabulation of these papers and a review of the work was presented by Pattison[127] in 1957.

These authors studied the effect of carbon chain length on the toxicity of these compounds and found a very startling difference between those with an odd number of carbon atoms and those with an even number. The following explanation is quoted from Pattison's paper. "Any compound which can form fluoracetic acid by some simple biochemical process is toxic and gives rise to symptoms in animals similar to those produced by fluoroacetic acid itself."

Robbins[128] studied 46 different fluoro hydrocarbons and ethers; many containing other halogens. He determined by 10-minute exposure of mice the anesthetic concentration and fatal concentration for each of 50 different compounds.

Range-finding toxicological data are given on a number of fluorinated compounds by Smyth, Carpenter, and Pozzani.[129]

[127] F. L. M. Pattison, *Chem. in Can.*, **9**, No. 8, 27 (1957).
[128] B. H. Robbins, *J. Pharmacol. Exptl. Therap.*, **86**, 197 (1946).
[129] H. F. Smyth, C. P. Carpenter, and U. C. Pozzani, *J. Ind. Hyg. Toxicol.*, **31**, 434 (1949).

Krantz and his associates have published several papers on the anesthetic effect of fluorinated hydrocarbons and ethers. These have been published in *Anesthesiology* or the *Journal of Pharmacology*. There are several publications in the *British Medical Journal* for 1957 on the use of bromochlorotrifluoroethane as an anesthetic. Studies on the anesthetic effect of some fluorinated hydrocarbons were also presented by Artusio and Poznak in *Federation Proceedings*, 1958.

It is apparent from the number of people who have worked on the anesthetic effect of these materials that there are a number of them that do act on the central nervous system. A material like dichlorodifluoromethane produces tremors in dogs and monkeys at 20 per cent by volume in air, but the animal recovers completely with no indications of pathological changes. A similar reaction in cats requires a concentration as high as 70 per cent. This certainly represents the extreme of low toxicity in the series. In contrast, we have a material like perfluoroisobutylene, which is an extremely toxic material.

Several of the materials that are exceedingly low in toxicity have been used as refrigerants. In more recent years, these same materials have found wide use as propellants in small "aerosol bombs." About everything conceivable can be obtained in this form from whipped cream or shaving cream to insecticides or paints. In more recent years, the fluorinated hydrocarbons, particularly those containing bromine, are becoming most interesting as fire-extinguishing agents.

The substitution of fluorine for hydrogen appears to produce a more stable compound and the presence of fluorine on the same carbon appears to stabilize other halogens. This high chemical stability appears to be related to the low toxicity of these materials. However, one should not assume that the substitution of fluorine will always reduce the toxicity as some fluorinated hydrocarbons are highly toxic.

II. Specific Compounds

VINYL FLUORIDE, CH_2=CHF (Monofluoroethylene)

1. *Uses and Industrial Exposures*

Vinyl fluoride has been used primarily as a chemical intermediate.

2. *Physical and Chemical Properties*

Physical state: colorless gas
Molecular weight: 46.04
Boiling point: —51°C.
Vapor density: 1.58 (air = 1)
1 mg./liter $\backsim$ 532 p.p.m. and 1 p.p.m. $\backsim$ 1.88 mg./cu. meter 25°C., 760 mm. Hg

3. Physiological Response

Toxicological research was reported by Lester and Greenberg.[130] They exposed rats to high concentrations of this material. They observed slight intoxication at concentrations of 30 per cent by volume in air or more. At 60 per cent, the animals lost the postural reflex. At 70 per cent, they lost the righting reflex.

They exposed animals at 80 per cent for $12^{1}/_{2}$ hours. These animals still retained the corneal reflex. Recovery was immediate when the animals were moved to fresh air. During the exposure to 80 per cent vinyl fluoride, the animals showed a slight intoxication and some dyspnea.

4. Hygienic Standards of Permissible Exposure

None has been proposed for vinyl fluoride.

5. Flammability

Vinyl fluoride is highly flammable.

VINYLIDENE FLUORIDE, $CH_2=CF_2$ (1,1-Difluoroethylene)

1. Uses and Industrial Exposures

Vinylidene fluoride has been used as a chemical intermediate.

2. Physical and Chemical Properties

Physical state: colorless gas
Molecular weight: 64.03
Boiling point: $-70°C$.
Vapor density: 2.2 (air $= 1$)
1 mg./liter $\backsimeq$ 382 p.p.m. and 1 p.p.m. $\backsimeq$ 2.62 mg./cu. meter at 25°C., 760 mm. Hg

3. Physiological Response

Toxicological work has been published by Lester and Greenberg.[130] They exposed rats to this material. They observed a slight intoxication at concentrations of 40 per cent by volume in air and up. At 80 per cent by volume in air, the animals showed an unsteady gait but no loss of reflex. When this exposure was continued for 19 hours, there were no progressive signs of intoxication. When these animals were killed and examined, there was no evidence of pulmonary irritation.

4. Hygienic Standards of Permissible Exposure

None has been proposed for vinylidene fluoride.

5. Flammability

Vinylidene fluoride is flammable.

[130] D. Lester and L. A. Greenberg, *A.M.A. Arch. Ind. Hyg. Occupational Med.*, **2,** 335 (1950).

MONOCHLORODIFLUOROMETHANE, CHClF$_2$

1. Uses and Industrial Exposures

Monochlorodifluoromethane has been used as a propellant.

2. Physical and Chemical Properties

Physical state: colorless gas
Molecular weight: 86.48
Melting point: —146—147°C.
Boiling point: —40°C.
Vapor density: 2.98 (air = 1)
1 mg./liter ≈ 282.6 p.p.m. and 1 p.p.m. ≈ 3.54 mg./cu. meter at 25°C., 760 mm. Hg

3. Physiological Response

Monochlorodifluoromethane is an odorless gas of very low toxicity. It is, however, more toxic than difluorodichloromethane. Booth and Bixby[131] indicated that guinea pigs showed definite nervous system response to a concentration of 16 per cent by volume in air, over a period of 55 minutes. They showed tremors and convulsive movements but recovered on removal. At 40 per cent, they showed tremors and were helpless over a period of 2^1/$_2$ hours but recovered and were observed for 25 days. At a concentration of 58 per cent, death occurred in 8 minutes.

4. Hygienic Standards of Permissible Exposure

None has been suggested for monochlorodifluoromethane.

5. Flammability

Monochlorodifluoromethane is not flammable.

DICHLORODIFLUOROMETHANE, CCl$_2$F$_2$

1. Uses and Industrial Exposures

Dichlorodifluoromethane is used as a refrigerant and propellant.

2. Physical and Chemical Properties

Physical state: colorless gas
Molecular weight: 120.92
Specific Gravity: 1.486 (30°C.)
Melting point: —160°C.
Boiling point: —28°C.
Vapor density: 4.1 (air = 1)
Solubility: 5.7 ml. per 100 ml. of water; soluble in ethanol and ethyl ether

---

[131] H. S. Booth and E. M. Bixby, *Ind. Eng. Chem.*, **24**, 637 (1932).

1 mg./liter $\simeq$ 202.3 p.p.m. and 1 p.p.m. $\simeq$ 4.95 mg./cu. meter at 25°C., 760 mm. Hg

3. Physiological Response

Dichlorodifluoromethane is an odorless gas of very low toxicity. Sayers et al.[132] indicated that dogs and monkeys would tolerate 20 per cent by volume in air. These animals showed generalized tremors, salivation, and lachrymation, but recovered on removal with no evidence of pathological change.

Brenner[133] indicated that cats would tolerate a concentration of 70 per cent for some period of time but would show tremors comparable to that seen by Sayers et al.[132] in dogs and monkeys at lower concentrations. It required this very high concentration to produce this reaction in cats.

Lester and Greenberg[130] investigated the response of rats. They saw no observable effect at 20 per cent. They observed twitching and tremors at concentrations of 30 to 40 per cent, loss of reflex at 50 per cent and above. At 70 and 80 per cent, corneal reflex was absent and the animals were in deep anesthesia. A few of the animals were exposed as long as 4 to 6 hours at 80 per cent and the animals suffered no permanent effects.

Chronic Exposure. Sayers and his associates[132] exposed dogs, monkeys, and guinea pigs 7 to 8 hours daily for a total of 12 weeks at a concentration of 20 per cent in air. They observed the tremors and salivation that had been noted in the acute experiments at such concentrations but no further signs of poisoning. Outside of a few cases of pneumonia, which occurred in the controls also, their animals survived without injury.

None of the authors observed any histopathological changes, even at the exceedingly high concentrations to which their animals were exposed. It would seem that this material is one of extraordinarily low toxicity, presenting very little practical hazard.

4. Hygienic Standards of Permissible Exposure

The threshold limit of dichlorodifluoromethane was established by the American Conference of Governmental Industrial Hygienists, in April, 1959, at 1000 p.p.m. (2950 mg./cu. meter).

This level was approved, not because injury would be expected at higher concentrations, but, apparently, because this is as high a level as has been accepted for any organic vapor and there is certainly no reason why working conditions cannot be controlled well within such levels by practical engineering control.

5. Flammability

Dichlorodifluoromethane is not flammable.

[132] R. R. Sayers, W. P. Yant, J. Chornyak, and H. W. Shoaf, *U. S. Bur. Mines Dept. Invest.* No. **3013** (1930).
[133] C. Brenner, *J. Pharmacol.,* **59,** 176 (1937).

TRICHLOROMONOFLUOROMETHANE, CCl_3F
(Fluorotrichloromethane)

1. Uses and Industrial Exposures

Trichloromonofluoromethane has been used as a refrigerant and propellant.

2. Physical and Chemical Properties

Physical state: colorless gas or liquid
Molecular weight: 137.38
Specific gravity: 1.4944 at (17.2°C.)
Melting point: −111°C.
Boiling point: 24.9°C.
Vapor density: 4.7 (air = 1)
Refractive index: 1.3865 (18.5°C.)
Solubility: insoluble in water; soluble in ethanol and ethyl ether
1 mg./liter ≎ 178 p.p.m. and 1 p.p.m. ≎ 5.61 mg./cu. meter at 25°C., 760 mm. Hg

3. Physiological Response

Trichloromonofluoromethane is a colorless and odorless gas or liquid—it may occur either as a gas or liquid as the boiling point is very near room temperature. The material seemed to be definitely more toxic than dichlorodifluoromethane. Nuckolls[134] indicated that this chemical would cause some slight symptoms if concentrations were around 1 per cent by volume in air. At around 5 per cent in a period of about 2 hours, stupor and incoordination were observed but the animals recovered with no pathological changes. At concentrations of around 10 per cent, he observed serious responses in one hour and unconsciousness in 2 hours. The animals recovered but showed lung injury.

Lester and Greenberg[130] suggested that an allowable concentration for a short single exposure to fluorotrichloromethane should be 5 per cent by volume in air. They observed loss of reflex at concentrations of 6 per cent and above and death from concentrations of 10 per cent and above for exposures of around 30 minutes.

4. Hygienic Standards of Permissible Exposure

The threshold limit of trichloromonofluoromethane was established by the American Conference of Governmental Industrial Hygienists, in April, 1959, at 1000 p.p.m. (5600 mg./cu. meter).

5. Flammability

Trichloromonofluoromethane is not flammable.

[134] A. A. Nuckolls, *Underwriters Labs. Rept.* No. **2375** (1933).

MONOBROMOMONOCHLORODIFLUOROMETHANE, $CBrClF_2$

1. Uses and Industrial Exposures

Monobromomonochlorodifluoromethane has been used as a fire extinguisher.

2. Physical and Chemical Properties

Physical state: colorless gas
Molecular weight: 165.37
Freezing point: —160°C.
Boiling point: —4°C.
Vapor density: 5.7 (air = 1)
1 mg./liter $\eqsim$ 147.9 p.p.m. and 1 p.p.m. $\eqsim$ 6.76 mg./cu. meter at 25°C., 760 mm. Hg

3. Physiological Response

Chambers et al.[135] reported for this compound an approximate lethal concentration of 2200 mg./liter of air (326,000 p.p.m.) for a 15-minute exposure of rats. The same authors indicated that when this material was decomposed at 800°C. in a hot iron tube, the approximate lethal concentration was 50 mg./liter or 8000 p.p.m. The animals showed mild pulmonary irritation.

Data from the Biochemical Research Laboratory of The Dow Chemical Co. would indicate that rats will survive for 15 minutes at a concentration of approximately 300,000 p.p.m. They will survive an exposure of 7 hours to a concentration of approximately 100,000 p.p.m.

4. Hygienic Standards of Permissible Exposure

None has been established for monobromomonochlorodifluoromethane.

5. Flammability

Monobromomonochlorodifluoromethane is not flammable; it is a good fire extinguisher.

DIBROMODIFLUOROMETHANE, CBr_2F_2

1. Uses and Industrial Exposures

Dibromodifluoromethane has been used as a fire extinguisher.

2. Physical and Chemical Properties

Physical state: colorless gas or liquid
Molecular weight: 209.83
Melting point: —146°C.

---

[135] W. H. Chambers, E. H. Krachow, F. P. McGrath, S. B. Goldberg, L. H. Lawson, and J. K. McNamee, U. S. Army Chem. Corps. Med. Div. Research Rept. No. **23** (1950).

Boiling point: 23°C.
Vapor density: 7.2 (air = 1)
1 mg./liter ≈ 116.5 p.p.m. and 1 p.p.m. ≈ 8.58 mg./cu. meter at 25°C., 760 mm. Hg

3. Physiological Response

Chambers et al.[135] reported an approximate lethal concentration of 470 mg./liter of air, or 55,000 p.p.m. by volume in air, for a 15-minute exposure of rats. If the material is decomposed at 800°C. in contact with iron, the approximate lethal concentration will be 16 mg./liter or 1870 p.p.m. Single acute exposures at 4000 p.p.m. for 15 minutes produced significant pulmonary damage, irritation, and edema.

Comstock et al.[136] reported on chronic exposure to vapors of this material. The exposures were 6 hours/day, 5 days/week. Both dogs and rats survived without effect a concentration of 3000 mg./cu. meter of air or 350 p.p.m. by volume in air for a 7-month period.

4. Hygienic Standards of Permissible Exposure

The threshold limit of dibromodifluoromethane was established by the American Conference of Governmental Industrial Hygienists, in April, 1959, at 100 p.p.m. (860 mg./cu. meter).

5. Flammability

Dibromodifluoromethane is not flammable; it is a good fire extinguisher.

MONOBROMOTRIFLUOROMETHANE, $CBrF_3$

1. Uses and Industrial Exposure

Monobromotrifluoromethane has been used as a fire extinguisher.

2. Physical and Chemical Properties

Physical state: colorless gas
Molecular weight: 148.9
Melting point: —166°C.
Boiling point: —58.67°C.
Vapor density: 3.8 (air = 1)
1 mg./liter ≈ 164 p.p.m. and 1 p.p.m. ≈ 6.09 mg./cu. meter at 25°C., 760 mm. Hg

3. Physiological Response

Chambers[135] reported that the approximate lethal concentration for this compound was 5070 mg./liter of air, or 834,000 p.p.m. by volume in air, for a 15-

---

[136] C. C. Comstock, J. Kerschner, and F. W. Oberst, U. S. Army Chem. Corps Med. Div. Research Rept. No. 180 (1953).

minute exposure of rats. If the material was decomposed at 800°C. in an iron tube, the approximate lethal concentration was 90 mg./liter or 14,000 p.p.m. for a 15-minute exposure of rats. The response was primarily respiratory damage with irritation and edema of the lungs. There was congestion of the liver, spleen, and kidney but no changes in the circulatory system.

Comstock et al.[136] reported experiments on the chronic exposure of rats and dogs. All animals survived 6 hours/day, 5 days/week for 18 weeks at a concentration of 140,000 mg./cu. meter of air without observable effect.

4. Hygienic Standards of Permissible Exposure

The threshold limit of monobromotrifluoromethane was established by the American Conference of Governmental Industrial Hygienists, in April, 1959, at 1000 p.p.m. (6100 mg./cu. meter).

5. Flammability

Monobromotrifluoromethane is not flammable; it is a good fire extinguisher.

DICHLOROTETRAFLUOROETHANE, $CClF_2CClF_2$

1. Uses and Industrial Exposures

Dichlorotetrafluoroethane has been used as a refrigerant and propellant.

2. Physical and Chemical Properties

Physical state: colorless gas
Molecular weight: 170.93
Boiling point: 3.6°C.
Vapor density: 5.9 (air = 1)
Refractive index: 1.3092 (0°C.)
Solubility: insoluble in water; soluble in ethanol and diethyl ether
1 mg./liter $\backsimeq$ 143.1 p.p.m. and 1 p.p.m. $\backsimeq$ 6.99 mg./cu. meter at 25°C., 760 mm. Hg

3. Physiological Response

Dichlorotetrafluoroethane has a sweetish, chloroformlike odor. It is of very low toxicity. At high concentrations, the typical responses are tremors and convulsions. There are essentially no histopathological changes.

Vapor Exposure. Yant et al.[137] reported on exposure of animals to 20 per cent by volume in air. Dogs showed tremors and convulsions. They survived a single 8-hour exposure but succumbed to a single exposure of 16 hours or more. They also succumbed to 4 daily 8-hour exposures. At the same concentration, guinea pigs survived single exposures up to 24 hours. Dogs survived a concentration of 15 per cent for single exposures up to 24 hours. They also survived 21 daily ex-

[137] W. P. Yant, H. H. Schrenk, and F. A. Patty, *U. S. Bur. Mines Rept. Invest.* No. **3185** (1932).

posures of 8 hours. These animals developed a tolerance for the material after the first few exposures. There was no sign of gross pathological or cellular change. A slight congestion was observed in several organs. Yant and his co-workers reported an increase in red blood cells, hemoglobin, and younger forms of polymorphonuclear leucocytes.

Nuckolls[134] reported on exposure of guinea pigs to concentrations from 1 to 4.7 per cent. Exposure time was 5 minutes to 2 hours. The animals showed symptoms of exposure but all recovered with no evidence of pathological effect.

4. *Hygienic Standards of Permissible Exposure*

The threshold limit of dichlorotetrafluoroethane was established by the American Conference of Governmental Industrial Hygienists, in April, 1959, at 1000 p.p.m. (7000 mg./cu. meter).

5. *Flammability*

Dichlorotetrafluoroethane is not flammable.

TETRACHLORODIFLUOROETHANE (Symmetrical—1,1,2,2-Tetrachloro-1,2-difluoroethane) CCl_2FCCl_2F (Asymmetrical—1,1,1,2-Tetrachloro-2,2-difluoroethane) CCl_3CClF_2

1. *Physical and Chemical Properties*

	Symmetrical	*Asymmetrical*
Physical state	Colorless liquid or solid	Colorless solid
Molecular weight	203.85	203.85
Melting point	24.65°C.	40.6°C.
Boiling point	92.8°C.	91.5°C.
Vapor density	7.0 (air = 1)	7.0 (air = 1)

1 mg./liter ☌ 120 p.p.m. and 1 p.p.m. ☌ 8.33 mg./cu. meter at 25°C., 760 mm. Hg

2. *Physiological Response*

Greenberg and Lester[138] reported on both the symmetrical and the asymmetrical forms of tetrachlorodifluoroethane.

Oral Ingestion. Both compounds were administered in doses of 2 g./kg. of body weight/day for 23 to 33 days. The animals survived without any apparent observable effect or pathological change. The authors indicated that the materials were probably not readily absorbed from the gastrointestinal tract, as most of the material was recovered from the gastrointestinal tract or the feces.

[138] L. A. Greenberg and D. Lester, *A.M.A. Arch. Ind. Hyg. Occupational Med.*, **2**, 345 (1950).

Inhalation. Rats exposed to the asymmetrical compound for $1^{1}/_{2}$ hours at 1 per cent by volume in air showed only a slight intoxication. At $1^{1}/_{2}$ per cent, there was loss of the corneal reflex. Concentrations of 2 and 3 per cent were fatal in 1 to $2^{1}/_{2}$ hours. There was no evidence of pathological changes in the lungs. Rats were exposed to 0.1 per cent concentration of the vapor of this compound for 18 hours/day for 17 days. At the end of that time, there was no pathological involvement of the lungs, liver, or spleen. Rats exposed to 0.5 per cent for similar periods showed no effects. At 1 per cent, there was evidence of slight intoxication but no pathological involvement of the lungs.

The symmetrical compound produced anesthetic effects at about the same levels as the asymmetrical compound, but caused significant injury to the lungs. Rats exposed to 3 per cent concentration of vapor died in 40 to 60 minutes with severe pulmonary hemorrhage. Rats exposed 18 hours/day for 16 days to 0.1 per cent of the symmetrical compound showed no effect. At 0.5 per cent, they were unconscious on removal. They died in periods of 4 to 36 hours later. All of them showed severe lung response.

These authors suggested an allowable concentration of 0.1 per cent by volume in air.

3. Hygienic Standards of Permissible Exposure

None has been established for tetrachlorodifluoroethane.

5. Flammability

Tetrachlorodifluorethane is not flammable.

DIBROMODIFLUOROETHANE, CH_2Br—$CBrF_2$
(1,2-Dibromo-1,1-difluoroethane)

1. Uses and Industrial Exposures

Dibromodifluoroethane has been considered as a fire-extinguishing agent.

2. Physical and Chemical Properties

Physical state: colorless liquid
Molecular weight: 223.87
Specific gravity: 2.242 (12.2°/4°C.)
Melting point: −56.5°C.
Boiling point: 93°C.
Vapor density: 7.7 (air = 1)
Vapor pressure: 59.7 mm. Hg (25°C.)
Per cent in "saturated" air: 7.85 (25°C.)
Density of "saturated" air: 1.55 (air = 1)
Solubility: insoluble in water

1 mg./liter ⟂ 109.3 p.p.m. and 1 p.p.m. ⟂ 9.16 mg./cu. meter at 25°C., 760 mm. Hg

3. *Physiological Response*

The response to dibromodifluoroethane is primarily depression of the central nervous system. Tremors are observed at anesthetic concentrations. There is some irritation of the lungs but no other organ pathology.

Vapor Exposure. Lester and Greenberg[130] exposed rats for 18 hours to a number of concentrations. Concentrations of 0.5 per cent by volume in air and above were fatal. Concentrations of 0.25 per cent and below were survived, but the animals showed tremors and a loss of reflex. A 4-hour exposure at 1 per cent caused death in 1 out of 3 rats. They showed pulmonary damage. These authors concluded that the maximum allowable concentration for single exposures of human beings should be 0.1 per cent.

Chambers[135] gave the approximate lethal concentration for this compound for an exposure period of 15 minutes as 210 mg./liter of air or 23,000 p.p.m. by volume in air. When the material was decomposed by exposing it to 800°C. in an iron tube, the approximate lethal concentration was 110 mg./kg., or 12,000 p.p.m. for the 15-minute exposure.

4. *Hygienic Standards of Permissible Exposure*

The threshold limit of dibromodifluoroethane was established by the American Conference of Governmental Industrial Hygienists, in April, 1959, at 100 p.p.m. (860 mg./cu. meter).

5. *Flammability*

Dibromodifluoroethane is not flammable.

CHAPTER XXXII

Halogenated Hydrocarbons: II. Cyclic

DON D. IRISH

CHLOROBENZENE, C_6H_5Cl (Monochlorobenzene)

1. *Uses and Industrial Exposure*

Chlorobenzene is used as a solvent and chemical intermediate.

2. *Physical and Chemical Properties*

Physical state: colorless liquid
Molecular weight: 112.56
Specific gravity: 1.1066 (20°C.)
Melting point: —44.9°C.
Boiling point: 132.0°C.
Vapor density: 3.88 (air = 1)
Vapor pressure: 11.8 mm. Hg (25°C.)
Refractive index: 1.5216 (25°C.)
Per cent in "saturated" air: 1.55 (25°C.)
Density of "saturated air: 1.05 (air = 1)
Solubility: 0.049 g. per 100 ml. water at 20°C.; soluble in alcohol, benzene and diethylether
1 mg./liter ⇌ 217 p.p.m. and 1 p.p.m. ⇌ 4.60 mg./cu. meter at 25°C.; 760 mm. Hg
Flash point: —90°F. (closed cup)

3. *Physiological Response*

Chlorobenzene has been used extensively in industry for a good many years but there has been a relatively small amount of information published on its toxicity. It is a central nervous system depressant and will cause symptoms typical of its anesthetic effect. At the same time, degeneration of the liver and kidneys may be observed. The histological changes may progress as the concentration increases or the time of exposure is lengthened. The liver injury may progress to necrosis and parenchymous degeneration. Toxicological studies and experience would indicate that monochlorobenzene will not cause the type of blood changes seen with benzene exposure.

Acute Vapor. The only acute vapor study in animals was reported by Flury and Zernik[1] (see Table 1).

TABLE 1

Physiological Response to Various Concentrations of Chlorobenzene—Cats[1]

| Concentration | | |
mg./liter	p.p.m.	Response
37	8000	Severe narcosis after $1/2$ hour; death 2 hours after removal from exposure
17	3700	Death after 7 hours
11–13	2400–2900	Unsteadiness after 1 hour; tremor, twitching; if removed within 7 hours no serious injury
5.5	1200	Definite narcotic symptoms
1–3	220–660	Tolerated for 1 hour

Chronic Vapor Exposure. Good quantitative chronic vapor exposure experiments have not been reported in the literature. The following information was taken from data in the Biochemical Research Laboratory of The Dow Chemical Co.[2] Rats, rabbits, and guinea pigs were exposed 7 hours/day, 5 days/week for a total of 32 exposures over a period of 44 days. At a concentration of 1000 p.p.m. in air, there were histopathological changes in the lungs, liver, and kidneys. There was a slight depression of growth. While there was no mortality in rats or rabbits, the guinea pigs did show a higher-than-normal mortality. At 475 p.p.m., the same species of animals survived. There was a slight increase in liver weight and slight liver histopathology. The blood was essentially normal in all of these animals. At a concentration of 200 p.p.m., all of the animals appeared to be normal.

Oral. Data from the same source gives the oral toxicity to animals. The LD_{50} for rats was 2.91 g./kg. of body weight, the LD_{50} for rabbits was 2.83 g./kg. of body weight. Doses were given repeatedly to rats 5 days/week for a total of 137 doses over a period of 192 days. A dose of 0.0144 g./kg. of body weight/day was survived without any observable effect. At 0.144 g./kg., there was a slight dip in growth from which they recovered and caught up with the control animals. At both 0.144 g./kg. and 0.288 g./kg., there were significant increases in liver and kidney weight and slight liver pathology. Blood and bone marrow were normal.

Skin. Slight reddening of the skin was observed from application of monochlorobenzene either on the uncovered or covered skin. Continuous contact for a week may result in moderate erythema and slight superficial necrosis. In these studies, there were no indications of toxic absorption.

Eye. There was evidence of moderate pain and slight transient conjunctival irritation. No corneal injury was observed. The irritation usually cleared within 48 hours.

---

[1] F. Flury and F. Zernik, *Schädliche Gase*, Springer, Berlin, 1931.

[2] D. D. Irish, unpublished data, Biochemical Research Lab., The Dow Chemical Co.

Absorption, Excretion, and Metabolism. The metabolism and excretion of this compound have been studied and reported by Spencer and Williams,[3] and by Smith, Spencer, and Williams.[4]

These authors administered monochlorobenzene to rabbits in a dose of 0.5 g./kg. of body weight. They indicated that 27 per cent of the administered dose was excreted unchanged in the expired air. Twenty-five per cent of the material appeared in the urine as the glucuronide, 27 per cent as the ethereal sulfate and 20 per cent as the mercapturic acid. The total is 99 per cent, which is an exceptionally good recovery for such a study.

4. Hygienic Standards of Permissible Exposure

The threshold limit of chlorobenzene was established by the American Conference of Governmental Industrial Hygienists, in April, 1959, at 75 p.p.m. (350 mg./cu. meter).

5. Odor and Warning Properties

Monochlorobenzene has an aromatic odor. The threshold of odor detection has not been determined.

o-DICHLOROBENZENE, $C_6H_4Cl_2$ (1,2-Dichlorobenzene)

1. Uses and Industrial Exposures

o-Dichlorobenzene is used as a solvent, fumigant, insecticide, and chemical intermediate.

2. Physical and Chemical Properties

Physical state: colorless liquid
Molecular weight: 147.01
Specific gravity: 1.2973 (25°C.)
Melting point: —17.6°C.
Boiling point: 180.48°C.
Vapor density: 5.07 (air = 1)
Vapor pressure: 1.56 mm. Hg (25°C.)
Refractive index: 1.5476 (25°C.)
Per cent in "saturated" air: 0.2 (25°C.)
Density of "saturated" air: 1.01 (air = 1)
Solubility: insoluble in water; soluble in ethanol, benzene, and diethyl ether
Flash point: 155°F. (open cup) ; 151°F. (closed cup)
1 mg./liter $\approx$ 166.3 p.p.m. and 1 p.p.m. $\approx$ 6.01 mg./cu. meter at 25°C. and 760 mm. Hg

[3] B. Spencer and R. T. Williams, *Biochem. J.,* **47,** 279 (1950).
[4] J. M. Smith, B. Spencer, and R. T. Williams, *Biochem. J.,* **47,** 284 (1950).

3. Physiological Response

The toxicological effect of o-dichlorobenzene is primarily injury to the liver and secondarily to the kidneys. Short exposure at high concentrations may result in depression of the central nervous system although this material is but weakly anesthetic. The primary problem in industrial exposure is that of liver and kidney injury.

Acute Vapor Exposure. Hollingsworth et al.[5] reported that rats survived 2 hours at a concentration of 977 p.p.m. in air and succumbed to an exposure of 7 hours at the same concentration. Rats survived a single 7-hour exposure to 539 p.p.m. These animals showed drowsiness, unsteadiness, and eye irritation. While they survived the exposure, definite organic injury was observed. There was an increase in weight of liver and kidneys. Microscopic examination showed marked central lobular necrosis in the liver and cloudy swelling of the tubular epithelium of the kidneys.

Chronic Vapor. The only chronic vapor exposure studies reported are those of Hollingsworth et al.[5] They exposed animals 7 hours/day, 5 days/week to the vapors of o-dichlorobenzene. At a concentration of 93 p.p.m. in air, rats, guinea pigs, and rabbits survived for periods of 6 to 7 months. The exposed animals showed no effect in growth, mortality, organ weight, hematology, or histopathology. These authors concluded that 75 p.p.m. was a reasonable standard for control of industrial exposure.

Oral. Hollingsworth et al.[5] fed guinea pigs o-dichlorobenzene in solution in olive oil. All guinea pigs survived 0.8 g./kg. of body weight but they all succumbed to 2.0 g./kg. of body weight. These authors also fed o-dichlorobenzene by stomach tube, 5 days/week for a total of 138 doses in 192 days. The material was suspended in gum arabic water solution. This was fed to rats. There was no effect on growth or mortality at a dose of 376 mg./kg. of body weight/day. There was a moderate increase in average liver weight and a slight increase in average kidney weight. There were slight histopathological changes in the liver. At a dose of 188 mg./kg. of body weight/day, there was a slight increase in liver and kidney weight but no apparent histopathology. No adverse effect could be observed at 18.8 mg./kg. of body weight/day.

Eyes. The same authors studied undiluted o-dichlorobenzene in the eye of the rabbit. The response was moderate pain and slight conjunctival irritation. There was no serious injury and the irritation cleared in a few days.

Industrial Human Experience. Hollingsworth et al. also reported that studies in the workroom atmosphere where large quantities of o-dichlorobenzene were handled showed concentrations ranging from 1 to 44 p.p.m. with an average of 15 p.p.m. The odor of o-dichlorobenzene was not noticeable at these levels. Thorough physiological examination in the clinic of all workmen in these areas failed to show any indication of injury from the exposure.

[5] R. L. Hollingsworth, V. K. Rowe, F. Oyen, T. R. Torkelson, and E. M. Adams, *A.M.A. Arch. Ind. Health,* **17,** 180 (1958).

4. Hygienic Standards of Permissible Exposure

The threshold limit of o-dichlorobenzene was established by the American Conference of Governmental Industrial Hygienists, in April, 1959, at 50 p.p.m. (300 mg./liter), and by the Manufacturing Chemists' Association Safety Data Sheet, on October, 1953, at 75 p.p.m.

5. Flammability

The explosive limits of o-dichlorobenzene are 2 to 9 per cent by volume in air.

6. Odor and Warning Properties

The odor of o-dichlorobenzene is detectable by the average person at 50 p.p.m. in air. Eye and nose irritation are not noted at this level. The odor becomes strong and the irritation noticeable at higher concentrations of around 100 p.p.m. It has fair warning properties at this level but the possibility of adaptation should be recognized.

p-DICHLOROBENZENE, $C_6H_4Cl_2$ (1,4-Dichlorobenzene)

1. Uses and Industrial Exposures

p-Dichlorobenzene has been widely used as an insecticide and disinfectant, as well as a chemical intermediate.

2. Physical and Chemical Properties

Physical state: colorless or white crystals
Molecular weight: 147.01
Specific gravity: 1.248 (55°C.)
Melting point: 53°C.
Boiling point: 174°C.
Vapor density: 5.07 (air = 1)
Solubility: insoluble in water; soluble in benzene, alcohol, and diethyl ether
Flash point: 153°F. (open cup)
1 mg./liter $\backsimeq$ 166.3 p.p.m. and 1 p.p.m. $\backsimeq$ 6.01 mg./cu. meter at 25°C., 760 mm. Hg

3. Physiological Response

The physiological response to p-dichlorobenzene is primarily injury to the liver and secondarily to the kidneys. Central nervous system depression will be observed in concentrations which are very disagreeable because of the strong odor and the irritation to the eyes and nose. Individuals who are exposed to excessive concentrations of p-dichlorobenzene for some period of time may show weakness, dizziness, and loss of weight; vomiting may occur. Jaundice may develop and the liver injury may progress as far as cirrhosis and death.

Vapor Exposure. The most extensive vapor study in animals was reported by Hollingsworth *et al.*[6] in 1956. They exposed animals 5 days/week, 7 hours/day for extended periods of time. Rats, guinea pigs, and rabbits exposed to 798 p.p.m. in air showed definite toxic reactions. Exposures ran from a few up to as many as 69; occasionally an animal died. They showed tremors, weakness, loss of weight, eye irritation, and an unkempt appearance. Some of them became unconscious. Histopathologically, the liver showed cloudy swelling and central necrosis. There was also a slight cloudy swelling of the tubular epithelium of the kidneys in some animals. The rabbits showed some lung changes.

At a concentration of 341 p.p.m., rats and guinea pigs survived for a period of 6 months. There was slight growth depression in male guinea pigs, a slight increase in average liver weight in male rats, and slight pathological changes in the liver of male guinea pigs. At a concentration of 158 p.p.m., animals survived exposures running from 137 to 219 days. No adverse effects on growth or mortality were observed in rats, mice, rabbits, or monkeys. There was a slight growth depression of the guinea pigs. The liver weights were slightly increased in male and female rats and in female guinea pigs. There were some questionable histopathological changes in the liver. At a concentration of 96 p.p.m., rats, guinea pigs, rabbits, mice, and a monkey were exposed for periods up to 6 or 7 months. No adverse effects on any species of animals were observed.

Oral. The same authors fed *p*-dichlorobenzene to rats as a 20 per cent solution in olive oil. Animals survived single doses of 1 g./kg. of body weight, but all animals succumbed to a dose of 4 g./kg. of body weight. Guinea pigs were fed a 50 per cent solution and survived 1.6 g./kg. of body weight as a single dose and succumbed to 2.8 g./kg. of body weight.

Repeated oral administration over a long period of time was also studied. Rats were fed 5 days/week for a total of 138 doses in 192 days. At a daily dose of 376 mg./kg., an increase in liver weight and a slight increase in kidney weight were observed. Microscopic examination of the liver revealed slight cirrhosis and focal necrosis. At 188 mg./kg., a slight increase in average weight of the liver and kidneys occurred. At 18.8 mg./kg. of body weight/dose, no effects could be observed. Similar studies were made on rabbits which showed changes at 1000 mg./kg. of body weight/dose and showed less intense but definite injury at 500 mg./kg.

Eyes. Solid particles, vapor, or fumes of *p*-dichlorobenzene are very painful to the eyes and nose. In order for vapors to be painful, it is usually necessary for the material to be heated or to be dispersed in such a way that there is a very large area for evaporation in a poorly ventilated area. It is painful to most people in concentrations between 50 and 80 p.p.m. and the discomfort becomes quite severe at 160 p.p.m.

[6] R. L. Hollingsworth, V. K. Rowe, F. Oyen, H. R. Hoyle, and H. C. Spencer, *A.M.A. Arch. Ind. Health,* **14,** 138 (1956).

Skin. Solid p-dichlorobenzene has very little irritating effect upon the skin. It does produce a burning sensation when held in close contact for excessive periods of time. Fumes from the surface of hot p-dichlorobenzene may irritate the skin slightly when the contact is repeated or prolonged. There is no evidence of absorption through the skin.

Industrial Experience. Hollingsworth[6] reported extensive industrial experience in handling p-dichlorobenzene. He reported on 58 men who had worked continuously or intermittently on operations involving the handling of p-dichlorobenzene from periods of 8 months to 25 years—an average of 4.75 years. Early investigations showed concentrations ranging from 10 to 550 p.p.m. with an average of 85 p.p.m. A later study separated the job area into two ranges; one running from 100 to 725 p.p.m. with an average of 380 p.p.m., and another with concentrations from 5 to 275 p.p.m. with an average of 90 p.p.m. A third survey was made after there had been some major changes in operations. The concentration varied from 50 to 170 p.p.m. with an average of 105 p.p.m. Under these conditions, there were still some complaints of eye and nose irritation by the workmen. Studies were made that showed conditions running from 15 to 85 p.p.m. with an average of 45 p.p.m. Under these conditions, there were no complaints. Men in these areas throughout the different periods of study have been thoroughly examined in the clinic. There was no evidence of organic injury, hematological effects, or eye changes.

There have been a number of reports of exposure of humans in the literature. They have been reviewed by von Oettingen in 1955[7] and by Hollingsworth in 1956.[6] Berliner[8] reported cataract formation following exposure to a "mothproofing agent" containing p-dichlorobenzene. Considering the publications of a number of authors, particularly Pike,[9] indicating that p-dichlorobenzene did not cause cataract in the eye and the reports by Hollingsworth[6] on industrial experience, it would appear that Berliner's findings may have been due to other materials present in the particular "mothproofing agent."

Absorption, Excretion, and Metabolism. The material is apparently well absorbed by the gastro-intestinal tract and from the lungs but not appreciably through the skin.

Williams[10] quoted Baumann as indicating that it did not form mercapturic acid. He noted that this was likely to be true as the acetylcysteine group normally conjugates in the para position to the halogen and this position is occupied in the case of p-dichlorobenzene.

[7] W. F. von Oettingen, The halogenated hydrocarbons, toxicity and potential dangers, *U. S. Public Health Serv. Publ.* No. **414** (1955).

[8] M. L. Berliner, *A.M.A. Arch. Ophthalmol.,* **22**, 1023 (1939).

[9] M. H. Pike, *J. Mich. Med. Soc.,* **43**, 581 (1944).

[10] R. Tecwyn Williams, *Detoxification Mechanism,* Chapman and Hall, London, 1949.

Azouz, Parke, and Williams,[11,12] studied the metabolism in rabbits. A single oral dose of 0.5 g./kg. of body weight of p-dichlorobenzene as a 25 per cent olive oil solution was given. Thirty-five per cent of the dose was recovered as 2,5-dichlorophenol and 6 per cent as 2,5-dichloroquinol. The conjugates excreted in the urine were 36 per cent glucuronides and 27 per cent ethereal sulfates.

4. *Hygienic Standards of Permissible Exposure*

The threshold limit of p-dichlorobenzene was established by the American Conference of Governmental Industrial Hygienists, in April, 1959, at 75 p.p.m. (450 mg./cu. meter).

5. *Odor and Warning Properties*

p-Dichlorobenzene has a very distinctive aromatic odor. The threshold of detection will vary from 15 to 30 p.p.m. in air. The odor becomes very strong at concentrations between 30 and 60 p.p.m. It is painful to the eyes and nose at concentrations of 80 to 160 p.p.m. Above 160 p.p.m., it is intolerable to any person who has not worked in it long enough to have had some adaptation. This odor and irritating effect are good warnings to prevent overexposure to p-dichlorobenzene. It should be recognized, however, that a person may become sufficiently accustomed to the odor to tolerate high concentrations.

CHLORINATED DIPHENYLS, $C_{12}H_{(10-n)}Cl_n$

1. *Use and Industrial Exposure*

The chlorinated diphenyls have uses comparable to those discussed under chlorinated napthalenes. These materials are sometimes used mixed with the chlorinated napthalenes. The problems associated with the use of these materials are somewhat comparable to chlorinated naphthalenes.

2. *Physical and Chemical Properties*

	Cl_3 equivalent	Cl_5 equivalent
Physical state	Liquid	Viscous liquid
Specific gravity	1.378–1.388 (25/25°C.)	1.538–1.548 (25/25°C.)
Distillation range	325–360°C.	365–390°C.
Vapor pressure	30 mm. Hg (200°C.); 4 mm. (150°C.)	9 mm. Hg (200°C.); 1.3 mm. (150°C.).
Refractive index	1.627–1.629 (20°C.)	1.639–1.641 (20°C.)
Solubility	Insoluble in water; soluble in oils and organic solvents	
Flash point (open cup)	176–180°C.	
At 745 mm. Hg and 25°C.	1 mg./liter ≎ 96.9 p.p.m.	1 mg./liter ≎ 76.5 p.p.m.

[11] W. M. Azouz, D. V. Parke, and R. T. Williams, *Biochem. J.,* **54,** xii (1954).
[12] W. M. Azouz, D. V. Parke, and R. T. Williams, *Biochem. J.,* **59,** 410 (1955).

3. Physiological Response

Much of the literature that refers to chlorinated diphenyls is referring to the mixture of chlorinated diphenyl with chlorinated naphthalene. There is enough information on the chlorinated diphenyls themselves to indicate that they will cause liver injury when absorbed into the body and that they will cause an acneform dermatitis when applied to the skin.

Vapor Exposure. Drinker *et al.*[13] exposed rats 16 hours/day to a concentration which averaged 0.57 mg./cu. meter of air. After 6 weeks, there was slight liver damage which advanced during the next 2 months. These authors concluded that it was difficult to determine the difference in toxicity of the chlorinated diphenyls and chlorinated naphthalenes, but that the chlorinated diphenyl was certainly capable of causing injury in very low concentration and was probably more hazardous than the chlorinated naphthalenes. It should be noted that they were using a chlorinated diphenyl containing 65 per cent chlorine. This is equivalent to a little over 7 chlorine atoms per molecule.

Treon *et al.*[14] exposed cats, guinea pigs, mice, rabbits, and rats to vapors of two different chlorinated diphenyl mixtures. One contained essentially three chlorine equivalents and had a chlorine concentration of 42 per cent. The other, contained approximately 5 chlorine equivalents and had a chlorine concentration of 54.3 per cent.

Experiments with the 42 per cent chlorine-containing mixture were run at a concentration of 8.6 μg./liter of air. The animals were exposed 7 hours/day for 17 exposures over 24 days. This concentration is "approaching saturation." The guinea pigs showed poor growth but there was no indication of effect upon the other animals. Exposure to a concentration of 6.83 μg./liter, 7 hours/day for 84 exposures over 122 days had essentially no effect on the animals.

Exposing animals to the mixture containing 54.3 per cent chlorine at a concentration of 5.4 μg./liter of air, 7 hours/day for 83 exposures over 121 days resulted in liver cell injury and increased liver weight in the rats. At a concentration of 1.5 μg./liter for 150 exposures over 213 days, the rats showed distinct histological changes in the liver. These authors concluded that as they were approaching the saturation point in their exposures and required heat to vaporize the material in order to obtain this concentration, these materials did not probably represent a significant vapor problem from cold operations.

This is reasonably well in accord with reports of Drinker. However, Drinker showed a slightly higher toxicity since he was using a chlorinated diphenyl with 65 per cent chlorine.

Treon and his colleagues proposed that 2 mg./cu. meter would be considered an allowable concentration of the chlorinated diphenyl containing 42 per cent

[13] C. K. Drinker, M. F. Warren, and G. A. Bennett, *J. Ind. Hyg. Toxicol.,* **19,** 283 (1937).

[14] J. F. Treon, F. P. Cleveland, J. W. Cappel, and R. W. Atchley, *Am. Ind. Hyg. Assoc. Quart.,* **17,** 204 (1956).

chlorine and that an allowable concentration of 1 mg./cu. meter would be reasonable for the chlorinated diphenyl containing approximately 55 per cent chlorine.

Oral Toxicity. Drinker[13] reported oral toxicity studies on chlorinated diphenyl containing 65 per cent chlorine. Feeding rats at 0.5 g./day, the first animal died in 9 days; 4 more by the end of the month. Liver lesions were found in these animals similar to those seen from vapor exposure.

Miller[15] fed chlorinated diphenyl to animals. His diphenyl contained 42 per cent chlorine and was approximately equivalent to 3 atoms of chlorine per molecule. Two doses, each dose 0.05 ml. (69 mg.), were given to guinea pigs 1 week apart. Death occurred in 11 to 29 days. The livers showed metamorphosis and central atrophy. Rats received 25 daily doses of 0.1 ml. (139 mg.), These animals survived the administration and were killed at 10-day intervals from 30 to 90 days after the beginning of the feeding. They showed the typical liver injury seen in the other animals.

Skin. Miller[15] also applied the same chlorinated diphenyl to the skin of guinea pigs, rats, and rabbits. Guinea pigs received 11 daily skin applications of $1/_{40}$ of 1 ml. (34.5 mg.) of undiluted chlorinated diphenyl. The animals died at various intervals up to 21 days after the first application. Histopathological examination of the liver showed fatty degeneration and central atrophy. The kidney was essentially normal. There was a thickening of the epidermis at the site of application.

Rats were given 25 daily applications of the same dose, $1/_{40}$ ml. (34.5 mg.) of undiluted chlorinated diphenyl. They survived the application. The animals were killed at 10 day intervals beginning with 30 days and ending at 90 days after the initial treatment. Histopathological examination of the liver showed only slight changes. The treated skin was much thickened and hair follicles were swollen.

Rabbits received skin applications at 2-day intervals. They were given 86 mg./day for the first 7 applications and 172 mg. for the last 8 applications of undiluted chlorinated diphenyl. Death occurred between 17 and 98 days. Fatty degeneration and central atrophy of the liver were more striking in these animals than in the previous animals. The treated skin showed a thinning of the prickle cell layer and a relative thickening of the outer cornified layers.

Human Experience. A number of the references recorded under chlorinated naphthalenes also discussed chlorinated diphenyls. It is difficult to determine the role played by the chlorinated diphenyls as, in most instances, they were mixed with chlorinated naphthalene. Jones and Alden[16] reported on human cases working with chlorinated diphenyls. They observed a typical acneform dermatitis among workers handling this material. As these workers were manufacturing the material from crude coal tar distillates, these authors were inclined to feel that other chlorinated aromatic materials present which were considered to be more unstable might be the more important cause of the dermatological response. The workers

[15] J. W. Miller, *Public Health Repts. U. S.*, **59**, 1085 (1944).
[16] J. W. Jones and H. S. Alden, *Arch. Dermatol. and Syphilol.*, **33**, 1022 (1936).

did not get a typical acneform dermatitis from patch tests. This is not surprising as a number of authors recognize that it is necessary to have a long-term exposure to this material in order to get this typical response. It is quite possible that high chlorinated aromatic materials from the crude coal tar distillates may also cause acneform dermatitis. It is quite probable, however, that the dermatitis can also be observed following exposure to the highly chlorinated diphenyls.

4. Hygienic Standards of Permissible Exposure

The threshold of chlorinated diphenyls was established by the American Conference of Governmental Industrial Hygienists, in April, 1959, at 1.0 mg./cu. meter for chlorodiphenyl (42 per cent chlorine) and 0.5 mg./cu. meter for chlorodiphenyl (54 per cent chlorine).

CHLORINATED NAPHTHALENES, $C_{10}H_{(8-n)}Cl_n$

1. Uses and Industrial Exposures

The chlorinated naphthalenes do not occur in industry as individual compounds. They will, therefore, be treated as a group. Any one product is a mixture but will have a specific range of chlorine substitution. Those with three or less chlorine equivalents are much less toxic than those with four or more.

These materials are used in electric wire insulation. They have also been used as additives to special lubricants.

The first problems were encountered in the use of these materials in wire insulation. Skin problems may arise from contact with the cold material. The hot material and, particularly, the vapors, may cause serious systemic poisoning as well as acne.

Injury to farm animals has occurred when lubricants containing the chloronaphthalenes were ingested. This may occur when cattle feed becomes contaminated from the machinery in which it is processed, or from similar lubricants used in farm machinery.

2. Physical and Chemical Properties

Physical state: Waxy solids
Properties vary with the degree of chlorine substitution
Solubility: insoluble in water; soluble in organic solvents

3. Physiological Response

When the higher chlorinated naphthalenes are absorbed they cause severe injury to the liver characterized as acute yellow atrophy. The systemic injury appears to be exclusively one of liver injury. Injury may occur from ingestion or from breathing hot vapors. Some authors have indicated that chlorinated naphthalenes may also be absorbed through the skin.

The other most striking response to contact with the chlorinated naphthalenes is chloracne. This usually occurs from long-term contact with the material or from a much shorter contact with hot vapors. The reaction is usually slow to appear and may take months to return to normal.

The third problem that has arisen from industrial use of chlorinated naphthalene has been hyperkeratosis in cattle. Special lubricants containing the chloronaphthalene were used in machinery for pelleting feeds. Externally, there is a thick keratinized layer on the skin of the neck, shoulders, and the withers. The gall bladder, liver, pancreas, and kidneys are affected. The condition is often fatal. While the most common cause has been from grease that contaminated pelleted feeds from the pelleting machinery, the same condition can occur from the ingestion of special oils or greases used on farm machinery if they contain the chloronaphthalene.

Vapor Exposure. Drinker et al.[13] exposed rats to vapors of chlorinated naphthalenes. As the material has a high boiling point and had to be heated to vaporize, it is possible that a certain amount of material may be condensed either on the cage or on the animal so that there may have been some oral intake. Animals were exposed for 16 hours/day. In a case of lower chlorinated material represented largely by a mixture of tri- and tetrachloronaphthalene, an average concentration of 1.31 mg./cu. meter of air essentially produced no effect other than a possible slight enlargement of the liver. The same material at a concentration of 10.97 mg./cu. meter resulted in a slight liver injury. Exposure was repeated daily for up to $4\frac{1}{2}$ months.

Similar exposure to the vapors of a mixture of penta- and hexachloronaphthalene at an average concentration of 1.16 mg./cu. meter produced definite liver injury. At a concentration of 8.88 mg./cu. meter, there was some mortality, poor growth, and severe liver injury. The higher chlorinated material showed a greater toxicity than the lower chlorinated material.

Clinical cases of systemic injury from this material have been reported by a number of authors. Sometimes the illness was preceded by skin lesions and sometimes it was not. It is implied that the systemic poisoning was largely due to breathing vapors. The following references are pertinent.[17-23]

Oral. Drinker et al.[13] fed a mixture of penta- and hexachloronaphthalene at a rate of 3 g./day to rats for 1 month. Nine out of 10 animals died. They all lost weight, were sick, and showed severe liver injury. When a mixture of tetra- and pentachloronaphthalene was fed at a dose of 0.5 mg./day for 2 months they observed some sickness, some mortality, and definite liver injury.

[17] L. Greenberg, M. R. Mayers, and A. R. Smith, *J. Ind. Hyg. Toxicol.,* **21**, 29 (1939).

[18] M. R. Mayers and A. R. Smith, *N. Y. State Ind. Bull.,* **21**, 30 (1942).

[19] N. G. McLetchie and D. Robertson, *Brit. Med. J.,* **1**, 641 (1942).

[20] E. C. Riley, *N. Y. State Ind. Bull.,* **22**, 80 (1943).

[21] L. Greenberg, *J. Ind. Med.,* **12**, 520 (1943).

[22] L. Greenberg, *J. Ind. Med.,* **22**, 404 (1943).

[23] L. H. Cotter, *J. Am. Med. Assoc.,* **125**, 273 (1944).

Skin. The most commonly observed problem from the use in handling of chlorinated naphthalenes is the chloracne. This has been observed on the skin of workers from a number of operations where it has been used. Most of the clinical reports listed previously under vapor exposure also include studies of chloracne. A detailed discussion of dermatological problems has been published by Good and Pensky.[24] The clinical aspects of this skin problem were also discussed by Schwartz.[25] Another extensive report was given by the special bulletin of the State of Pennsylvania.[26]

Adams *et al.*[27] studied the nature of the response of rabbit's skin to this type of material. They concluded that the response takes the form, first, of epithelial hyperplasia, second, inflammatory and degenerative changes, and, finally, re-generative processes.

This work is particularly interesting when we consider the nature of the hyperkeratosis observed in cattle. This is described in some detail in Leaflet No. 355, U. S. Dept. of Agriculture, 1954. There is apparently a hyperplasia and a keratinization. It is noticeable in the neck, shoulders, and withers. There are also growths on the gums of the animal. This is also discussed by Bell,[28] and was first described by Olafson.[29]

This condition in animals is supposedly due to ingestion. The injury to the liver and obvious systemic effect certainly would indicate that it is an important factor. It is possible, however, considering the reaction of the tissues of the mouth, cheek, and neck, that some of the material may act by direct contact with the mouth and skin.

Absorption and Metabolism. Cleary *et al.*[30] published an article on the absorption and metabolism of the chlorinated naphthalenes. They indicated that the material was readily absorbed from an olive oil solution. They found no evidence of storage in the lung, liver, skin, or kidneys. They did not find any excretion via the urine and concluded that the chlorine might be removed from the ring and excreted as chloride. They observed a rinse in urinary ethereal sulfate. There was no change in neutral sulfur.

4. Hygienic Standards of Permissible Exposure

Drinker[13] proposed the following limits for the chlorinated naphthalenes: trichloro—49.9 per cent chlorine (10.0 mg./cu. meter); tetra and penta—56.4 per cent chlorine (1.0 mg./cu. meter); penta and hexa—62.6 per cent chlorine (0.5 mg./cu. meter). The A.C.G.I.H. 1961 threshold for trichloronaphthalene is 5 mg./cu. meter; for pentachloronaphthalene is 0.5 mg./cu. meter.

[24] C. K. Good and N. Pensky, *Arch. Dermatol. and Syphilol.,* **48,** 251 (1943).

[25] L. Schwartz and S. M. Peck, *N. Y. State J. Med.,* **43,** 1711 (1943).

[26] Commonwealth of Penn. Dept. Labor & Ind., *Spec. Bull.* No. **43** (1936).

[27] E. M. Adams, D. D. Irish, H. C. Spencer, and V. K. Rowe, *Ind. Med. Ind. Hyg. Section,* **1,** 1 (1941).

[28] W. B. Bell, *Vet. Med.,* **48,** 135 (1953).

[29] P. Olafson, *Cornell Vet.,* **37,** 195 (1947).

[30] R. V. Cleary, J. Maier, and G. H. Hitchings, *J. Biol. Chem.,* **127,** 403 (1939).

HALOGENATED HYDROCARBON INSECTICIDES

I. General

Several halogenated hydrocarbons that have become of great importance as agricultural chemicals have been developed in the last 15 or 20 years. Most of these do not fit into the chemical types discussed as industrial chemicals and they are, therefore, set aside as a small group by themselves. Most of them are much more complex in structure than the simple halogenated aliphatic or aromatic materials. Their only use of any significance is as pesticides.

These materials have a very low vapor pressure. The problems of handling them are most usually observed in connection with their application as insecticides. When they are applied as a dust or a liquid spray, the finer particles in the air dispersion are readily breathed and may be absorbed by the lungs, the coarser materials are trapped in the upper respiratory area and swallowed. There are also some problems in the manufacture and packaging of these materials within the industry.

There is a very extensive toxicological literature on most of these materials. A great deal of it, however, is concerned with the important problem of residues, which may appear in food or feed that has been sprayed with the material. This is a broad public health problem but one that is not particularly appropriate to this book, which is concerned with industrial problems. The present discussion will point out only a very brief consideration of the most important toxicological problems, particularly, those that will be of importance in the industrial handling and use of the material.

Individuals interested in more detailed discussion of these products should consult reviews or original publications in this field. The writings of the following authors will be of particular importance: W. J. Hayes, United States Public Health Service; A. J. Lehman, The Federal Food and Drug Administration; F. Princi, The Kettering Laboratory at Cincinnati; R. Metcalf, The University of California; J. M. Barnes, Medical Research Council, England.

DDT [Dichlorodiphenyltrichloroethane, 1,1,1-Trichloro-2,2-bis(*p*-chlorophenyl)ethane]

$$Cl-\!\!\bigcirc\!\!-\overset{\displaystyle |}{\underset{\displaystyle CCl_3}{C}}\!\!-\!\!\bigcirc\!\!-Cl$$

1. Uses and Industrial Exposures

DDT is a very widely used insecticide. It has not been a serious hazard in its manufacture and packaging or in spraying and dusting in agricultural use. Toxic effects can be observed from industrial and agricultural use but can be avoided by reasonable cleanliness. The major concern has been with chronic absorption and particularly as a residue in food and feed.

2. Physical and Chemical Properties

Physical state: white crystalline solid
Molecular weight: 354.5
Specific gravity: 1.55 (25°C.)
Melting point: 108.5 to 109°C.
Solubility: 78 g. per 100 ml. benzene; 116 g. per 100 ml. cyclohexane; 45 g. per 100 ml. carbon tetrachloride; insoluble in water

3. Physiological Response

The chief effect of DDT is on the nervous system. There are indications of apprehension, excitement, tremors, and finally convulsions followed by paralysis and death. This material is not highly toxic from an acute exposure. The greatest concern is with absorption and storage during a long period of chronic or repeated exposures. The most noticeable response to long-term chronic exposure, even at a fairly low concentration, is tremors. There have been some indications of changes in the liver tissue and very slight changes in the kidneys.

No attempt will be made to cover the tremendous volume of literature that has been published on the toxicology of DDT. The industrial problem does not justify extensive treatment. The few references given here may give a sufficient clue to aid the individual who wishes to go more deeply into the toxicology of this compound.

Air Dispersion. As the principal concern has been the taking of DDT by mouth, there has been relatively less investigation of the problem of breathing air dispersions of the dust or mist. Neal and associates[31] exposed dogs for 3 hours/ day to a concentration of 12.44 mg. of DDT (as a 10 per cent dust)/liter of air for a period of 4 weeks without observing any effect. They also exposed animals to a mist of DDT for 48 minutes/day for 4 weeks at concentrations of 388.7 mg. of DDT/liter of air without effect.

Cameron and Burgess[32] exposed several species of animals to a concentration that approximated 1000 p.p.m. weight-volume in air (mg./cu. meter) for a period of 2 hours daily. The animals showed signs of intoxication and deaths occurred after 4 to 10 exposures.

Neal[33] reported exposures of animals to aerosols of DDT. Dogs, rats, and guinea pigs were exposed to an initial concentration of 54.4 mg. of DDT/liter of air for a period of 45 minutes. No indications of toxicity were observed. He also reported that the oil used as a solvent had some effect on the response. Mice tolerated 6.22 mg. of DDT/liter of air without manifestations of toxic symptoms if the solution contained 6 per cent sesame oil. If the concentration of oil was

[31] P. A. Neal, W. F. von Oettingen, W. W. Smith, R. B. Malmo, R. C. Dunn, H. E. Morann, T. R. Sweeney, D. W. Armstrong, and W. C. White, Public Health Rept. Suppl. **177**, (1944).

[32] G. R. Cameron and F. Burgess, *Brit. Med. J.*, **1**, 865 (1945).

[33] P. A. Neal, *Soap Sanit. Chemicals*, **21**, No. 9, 99 (1945).

9.5 per cent, toxic effects were observed at the same concentration of DDT. Neal also reported experiments on humans exposed to an aerosol of DDT.

Acute Oral. Woodard et al.[34] investigated the acute oral toxicity of DDT. They reported some deaths in rats at 140 mg./kg. Rabbits survived 260 mg./kg. and some died at 400 mg./kg. Mice survived 399 mg./kg. and some died at 448 mg./kg. Guinea pigs survived 178 mg./kg. and some died at 224 mg./kg. Some animals survived as high a dose as was given.

Lehman[35] reported the approximate LD_{50} for rats as 250 mg./kg.

Chronic Oral. There have been many chronic oral studies made with DDT fed to many species of animals. Fitzhugh and Nelson[36] fed rats for 2 years on a diet containing DDT. At concentrations of 600 and 800 p.p.m. by weight in the diet, animals showed moderate tremors, particularly during the early exposures. An occasional animal showed tremors at 400 and rarely at 200 p.p.m. Concentrations of 400 p.p.m. and above showed a higher than normal mortality. Animals fed on a concentration of 400 p.p.m. and higher showed an increase in weight of the liver. If the concentrations were 600 or 800 p.p.m., there was an indication of increase in weight of the kidney. Histopathological studies showed moderate liver damage at concentrations of 200 p.p.m. and above. There were some slight indications of liver damage, even at 100 p.p.m.

Treon[37] indicated that rats maintained from 18 months to 2 years on a diet containing 25 p.p.m. by weight of DDT showed an increase in liver weight. At 12.5 p.p.m. in the diet for a similar period, he reported no effect. Dogs on a diet containing 30 p.p.m. of DDT for a period of 15.7 months showed no effect.

Laug et al.[38] indicated that hepatic cell alterations were seen at 5 p.p.m. in the diet and higher but not at 1 p.p.m. They observed an accumulation of DDT in body fat, even at 1 p.p.m. in the diet.

Skin. Experiments by Draize et al.[39] indicated that dusts and solutions of DDT may cause a slight to moderate erythema on the skin. Using dry dusts of 5 per cent DDT in talc, they were unable to find any indications of systemic toxic effect due to absorption through the skin. There was no indication of systemic toxic effect from absorption following the application of the 10 per cent solution of DDT in corn oil in doses up to 940 mg. of DDT/kg. Solutions of 30 and 25 per cent DDT in dimethylphthalate and dibutylphthalate caused no deaths but toxic symptoms were observed from a dose of 9.4 ml./kg. for a 24-hour exposure.

[34] G. Woodard, A. A. Nelson, and H. O. Calvery, *J. Pharmacol. Exptl. Therapy.*, **82**, 152 (1944).

[35] A. J. Lehman, *Assoc. Food & Drug Officials U. S. Quart. Bull.*, **15**, 122 (1951).

[36] O. G. Fitzhugh and A. A. Nelson, *J. Pharmacol. Exptl. Therap.*, **89**, 18 (1947).

[37] J. F. Treon and F. P. Cleveland, *J. Agr. Food Chem.*, **3**, 402 (1955).

[38] E. P. Laug, A. A. Nelson, O. G. Fitzhugh, and F. M. Kunze, *J. Pharmacol. Exptl. Therap.*, **98**, 268 (1950).

[39] J. H. Draize, A. A. Nelson, and H. O. Calvery, *J. Pharmacol. Exptl. Therap.*, **82**, 159 (1944).

These same authors ran 90-day repeated applications. They found that doses as low as 0.5 ml. of a 30 per cent solution of DDT/kg./day (150 mg./kg./day of DDT) in rabbits, rats, and guinea pigs may cause death in some cases after 30 days. Animal species showed a wide variation in susceptibility.

Absorption, Excretion, and Metabolism. Pearce et al.[40] showed that absorbed DDT was stored in fat as the DDT and as the DDE (2,2-bis-(p-chlorophenyl),1,1-dichloroethylene).

Jensen et al.[41] demonstrated that the primary metabolite of DDT was DDA, which is the acetate of DDT. It is primarily excreted in the bile and appears in the feces. DDA is also excreted in the urine.

Human Response. There have been many reports and studies on human response to DDT. An excellent epidemiological study was reported by Fowler.[42] A sizable population in the state of Mississippi was studied, both before and after a number of years of widespread application of DDT, both on crops and in the control of mosquitoes. It could be concluded, in general, that the health condition had improved over this period of time. This could be due to improved sanitation but it should also be recognized that DDT controls a number of insect carriers of disease. There was no indication of deleterious effect on the population from the use of DDT. There were rare cases of acute poisoning due to misuse of the compound.

Hayes[43,44] has written two very good reviews on DDT. Hayes indicates that acute poisoning in man has occurred but that chronic poisoning in man has not been confirmed. A dose of 10 mg./kg. produced illness in some men but not all; 285 mg./kg. has been taken without fatal results, although with toxic response. The tolerated chronic dose in man is not known, but from animal experiments, 2.5 to 5 mg./kg. body weight/day might produce mild illness. Human volunteers took 0.5 mg./kg./day for several months with no detectable effect. He commented that the liver changes reported were real but that their significance in the toxicological picture was doubtful. He stated that there was "more occupational disease caused by the solvents in DDT formulations than by DDT itself."

Ortelee[45] reported a well-planned study of 40 men exposed to DDT during manufacture and formulation. Their exposure was followed by analysis of DDT concentration in urine and compared with the excretion by men with known oral intake of DDT. He concludes that it is unlikely that any illness will occur from DDT at the current dietary level as the men studied showed no effect from ex-

[40] G. W. Pearce, A. M. Mattson, and W. J. Hayes, Jr., *Science*, **116**, 254 (1952).

[41] J. A. Jensen, C. Cueto, W. E. Dale, C. F. Rothe, G. W. Pearce, and A. M. Mattson, *J. Agr. Food Chem.*, **5**, 919 (1957).

[42] F. E. L. Fowler, *J. Agr. Food Chem.*, **1**, 469 (1953).

[43] W. J. Hayes, *Am. J. Public Health*, **45**, 478 (1955).

[44] W. J. Hayes, Jr., "Pharmacology and Toxicology of DDT," in P. Muller, ed., *DDT Insecticides*, Vol. 2, Birkhauser, Basle, 1955, p. 11.

[45] M. F. Ortelee, *A.M.A. Arch. Ind. Health*, **18**, 433 (1958).

posure for up to 6.5 years, during which they absorbed an average of 200 times that absorbed by the general population from their diet.

4. Hygienic Standards of Permissible Exposure

The threshold limit of DDT was established by the American Conference of Governmental Industrial Hygienists, in April, 1959, at 1 mg./cu. meter of air. The residue tolerance for specific agricultural commodities is 7 p.p.m. by weight.[46]

KELTHANE [4,4'-Dichloro-α-(trichloromethyl)benzhydrol]

1. Uses and Industrial Exposures

Kelthane is used as an insecticide.

2. Physiological Response

The only information available on Kelthane is from animal experimentation. At high levels, the animals show a general weakness, coma, and death. Tremors were not observed as they usually are with DDT, which is a closely related compound. Histopathological changes are limited to the liver and kidneys and are relatively mild in nature. It is reported to cause some suppression of adrenal cortical activity.

All the information indicated here is from the publication of R. Blackwell Smith et al.[47]

Acute Oral. LD$_{50}$ for rats of a 20 per cent solution in corn oil of the technical Kelthane was 809 mg./kg. for males and 684 mg./kg. for females. The LD$_{50}$ for male rabbits was 1810 mg./kg. and for dogs was greater than 4000 mg./kg.

Chronic Oral. Dogs fed 1 year on a diet containing Kelthane survived levels of 300 p.p.m. or less without effect. At a level of 900 p.p.m. 2 out of 4 dogs died. There was no evidence of histopathological changes.

Rats fed a diet containing Kelthane for 2 years survived 1000 p.p.m. by weight in the diet. Females, at levels of 250 p.p.m. or higher, and males at 500 p.p.m. and higher, showed depression of growth.

Skin. Solutions in dimethylphthalate cause erythema and superficial destruction. Emulsions were more severely irritating and showed marked tissue destruction.

When applied to the skin in a cuff, the material may be absorbed through the skin. The LD$_{50}$ for the rabbit of a 30 per cent solution in dimethylphthalate is

---

[46] L. Ingle, *Arch. Ind. Hyg. Occupational Med.*, **6**, 354 (1952).

[47] R. B. Smith, Jr., P. S. Larson, K. J. Finnegan, H. B. Haag, G. R. Henniger, and F. Cobey, *J. Toxicol. Appl. Pharmacol.*, **1**, 119 (1959).

2.1 g./kg. body weight. Repeated skin applications daily, 5 days a week, for 13 weeks of a 30 per cent solution in dimethylphathalate, at doses of 1 ml./kg. and up, caused death. Similar repeated applications of an emulsion of 18.5 per cent active material caused some deaths in doses of 0.1 ml./kg. and up. A wettable powder containing 18.5 per cent active ingredient with 2 parts of water applied repeatedly caused deaths in doses of 0.5 g./kg. and up.

3. Hygienic Standards of Permissible Exposure

None has been established for Kelthane.

HEXACHLOROCYCLOHEXANE [gamma isomer of 1,2,3,4,5,6-hexachloro-cyclohexane (Lindane); mixed isomers of 1,2,3,4,5,6-hexachlorocyclo-hexane (Benzene hexachloride)]

1. Uses and Industrial Exposures

Hexachlorocyclohexane is used as an insecticide.

2. Physical and Chemical Properties

(See Table 2.)

TABLE 2

Physical and Chemical Properties of Hexachlorocyclohexane Isomers

				Solubility		
Isomer	Melting point	Vapor pressure (20°C.)	p.p.m. in water	Ethanol (g. per 100 g.)	Ether (g. per 100 g.)	Benzene (g. per 100 g.)
Alpha	157.8–158	0.02	10	1.8	6.2	9.9
Beta	309	0.005	5	1.1	1.8	1.9
Gamma	112.5	0.03	10	6.4	20.8	28.9
Delta	138 –139	0.02	10	24.4	35.4	41.1
Epsilon	217 –218	—	—	—	—	—

3. Physiological Response

One of the confusing aspects of dealing with hexachlorocyclohexane is the fact that it occurs as five isomers. Four of these are found in sufficient quantity to be considered of importance. A crude mixture of these four isomers is available under the common name, benzene hexachloride. The most effective isomer is the gamma and this, highly purified, is available under the common name, lindane.

The primary response to lindane exposure is stimulation of the central nervous system resulting in hyperexcitability and convulsions. The response seen from benzene hexachloride exposure is somewhat similar, although the delta isomer does depress the central nervous system and thus antagonizes to some degree the stimulating action of the gramma isomer. Quantitatively, the acute toxicity of lindane is somewhat greater than that of the technical benzene hexachloride. Chronically, however, the technical benzene hexachloride is somewhat more toxic than lindane.

Histopathological changes are seen in the liver and kidneys and, to some degree, in the lungs following chronic exposure to these materials.

These substances do not present particularly difficult problems in use and would not be considered greatly hazardous. They are approximately in the same category as DDT, although their chronic toxicity is less.

Acute Oral. Lehman,[35] reports the oral LD_{50} of lindane for rats as 125 mg./kg. body weight. The response is one of hypersensitivity followed by convulsions.

The Council on Pharmacy and Chemistry of the A.M.A.[48] reported an estimated LD_{50} of approximately 600 mg./kg. for the technical benzene hexachloride. It should be recognized, however, that this will vary somewhat with the proportions of the different isomers in the mixture.

Chronic Oral Toxicity. Lehman[49] indicated that the highest tolerated daily dose of lindane over a 2-year period without effect when given to rats was 5 mg./kg.

Fitzhugh et al.[50] indicated that rats could tolerate long term feeding of diets containing 800 p.p.m. by weight of lindane but showed some liver enlargement at this level. A comparable response was observed from the technical benzene hexachloride at a level of 100 p.p.m. in the diet.

The primary response from chronic oral feeding is injury to the liver. Following high doses, indication of changes in the kidney may be seen.

Air Dispersion. Queen[51] reported on the use of lindane by vaporization into the air. He exposed canaries to an air concentration of 0.34 mg./liter of air. He observed death following exposures of from 6 to 16 days.

There is also a possibility of absorption of lindane or technical benzene hexachloride when used as a dust or as a spray. Quantitative information on the levels tolerated were not found.

Skin. Lehman[52] reported that the approximate LD_{50} for rabbits from absorption of lindane in the dry form was greater than 4000 mg./kg. Severe symptoms

[48] Council on Pharmacy and Chemistry, *J. Am. Med. Assoc.,* **147**, 571 (1951).

[49] A. J. Lehman, *Assoc. Food & Drug Officials U. S. Quart. Bull.,* **18**, 3 (1954).

[50] O. G. Fitzhugh, A. A. Nelson, and J. P. Frawley, *J. Pharmacol. Exptl. Therap.,* **100**, 59 (1950).

[51] W. A. Queen, *Assoc. Food & Drug Officials U. S. Quart. Bull.,* **17**, 127 (1953).

[52] A. J. Lehman, *Assoc. Food & Drug Officials U. S. Quart. Bull.,* **14**, 3 (1952).

were observed at this level but the animals survived. Moderate skin irritation is observed from application of the dry material. When lindane was applied as a 2 per cent solution in dimethylphthalate, it was determined that the approximate LD_{50} was greater than 188 mg./kg. When lindane was prepared as a 1 per cent solution in vanishing cream base, the approximate LD_{50} was 50 mg./kg. No information was reported on the technical benzene hexachloride.

Human Experience. Danopoulos,[53] reported on clinical cases of exposure to technical benzene hexachloride in Greece. Exposure was apparently very severe. They used a 40 per cent dry powder or the same powder mixed with either water or a petroleum solvent. The material was sprinkled on the ground, walls, over the clothing, bedding, and bodies of the people. Seventy-nine persons were affected, 18 seriously. Five were treated in a clinic. All survived. Symptoms indicated response of the gastrointestinal system and the central nervous system. There were indications of electrocardiographic changes and hematological changes.

Heiberg and Wright[54] reported a case in the United States. A woman was "washing" two calves with a benzene hexachloride solution. Her arms and hands were wet to the elbows and her clothing partially soaked. She showed a severe response with convulsions but survived. A concentration of 4.95 mg. of benzene hexachloride per 100 ml. of urine was found the day following admission.

Absorption, Excretion, and Metabolism. Either lindane or the technical benzene hexachloride may be absorbed from the gastrointestinal tract or from the lungs, or through the skin.

van Asperen and Oppernoorth[55] reported that 1 mg. of lindane injected subcutaneously in the mouse was eliminated in 4 days. Two hundred gammas injected intravenously disappeared in 24 hours. They found none in the urine and the feces. Traces were found in a number of organs.

Davidow and Frawley[56] gave the distribution in various tissues. It was highest in fat. Chemical changes which may take place in benzene hexachloride in the body are not known.

4. Hygienic Standards of Permissible Exposure

The threshold limit of lindane was established by the American Conference of Governmental Industrial Hygienists, in April, 1959, at 0.5 mg./cu. meter of air. The residue tolerance on specified agricultural commodities is 10 p.p.m. by weight for lindane[57] and 5 p.p.m. by weight for benzene hexachloride.

[53] E. Danopoulos, K. Melissinos, and G. Katsas, *A.M.A. Arch. Ind. Hyg. Occupational Med.*, **8**, 582 (1953).

[54] O. M. Heiberg and H. N. Wright, *A.M.A. Arch. Ind. Health*, **11**, 457 (1955).

[55] K. van Asperen and F. J. Oppernoorth, *Nature*, **173**, 1000 (1954).

[56] B. Davidow and J. P. Frawley, *Proc. Soc. Exptl. Biol. Med.*, **76**, 780 (1951).

[57] Nat. Agr. Chem. Assoc. News, **17**, No. 3 (1959).

CHLORDANE (1,2,4,5,6,7,8,8-Octachloro-2,3,3a,4,7,7a-hexahydro-4,7-methanoindane)

1. *Uses and Industrial Exposure*

Chlordane is used as an insecticide.

2. *Physical and Chemical Properties*

Physical State: a viscous amber liquid; 60 to 75 per cent pure, the rest related compounds; chlorine content, 64 to 67 per cent
Solubility: insoluble in water; soluble in organic solvents

3. *Physiological Response*

Response to the absorption of chlordane is not unlike that of the other members of this group of chlorinated insecticides. The primary acute response is in the central nervous system. The symptoms are irritability and tremors, leading to convulsions and finally death. If the material is taken orally, nausea, vomiting, and diarrhea are likely to be observed. There is some local irritation of the gastro-intestinal tract. Poisoning from chronic exposure also shows indications of effects on the central nervous system. Cellular changes in the liver may occur. Edema of the lungs and irritation of the gastrointestinal tract have been reported.

Vapor.   No work has been reported on dusts and mists of chlordane.   Frings and O'Tousa[58] reported that mice exposed to vapors from technical chlordane showed severe toxicological response.

Ingle[59] made similar studies and decided that the toxicity of the vapor was due to the presence of hexachlorocyclopentadiene in the technical grade chlordane. It is indicated that more recent preparations of chlordane contain much less of this material.

Acute Oral.   Ingle[46] reported the LD_{50} for the rat to be 250 mg./kg. body weight when dissolved in corn oil. Lehman[52] gave the LD_{50} for the rat as 457 mg./kg. body weight.

Chronic Oral.   Ingle[60] reported 2-year chronic oral feedings in rats. There was retardation of growth at concentrations of 150 and 300 p.p.m. by weight in the diet, but no effect at 5, 10, and 30 p.p.m. Liver damage was marked at 150 and 300 p.p.m. It was slight at 30, minimal at 10, and absent at 5 p.p.m. There was no injury to the kidneys at 5, 10, or 30 p.p.m. but there was marked injury at 150

[58] H. Frings and J. E. O'Tousa, *Science,* **111,** 658 (1950).
[59] L. Ingle, *Science,* **118,** 213 (1953).
[60] L. Ingle, *Arch. Ind. Hyg. Occupational Med.,* **6,** 357 (1952).

and 300 p.p.m. The lung showed marked damage at 300 p.p.m., mild injury at 150, and none at lower concentrations.

Lehman[49] reported that the maximum tolerated dose for rats over a 2-year period was 0.125 mg./kg. day. Ambrose[61] reported similar chronic studies in which he observed growth depression at 320 p.p.m. but normal growth at 160 p.p.m. and less. He observed enlargement of the liver at 80 p.p.m. but no effect at 10 p.p.m.

A discussion of the human experience with this material was documented by the Council of Pharmacy.[62]

4. Hygienic Standards of Permissible Exposure

The threshold limit of chlordane was established by the American Conference of Governmental Industrial Hygienists, in April, 1959, as 2 mg./cu. meter of air. The residue tolerance for specified agricultural commodities is 0.3 p.p.m. by weight.[57]

HEPTACHLOR (1,4,5,6,7,8,8-Heptachlor-3a,4,7,7a-tetrahydro-4,7-methanoindene)

1. Uses and Industrial Exposures

Heptachlor is used as an insecticide.

2. Physical and Chemical Properties

Physical state: white crystalline solid
Molecular weight: 373.22
Melting point: 95 to 96°C.

3. Physiological Response

Heptachlor is another chlorinated derivative of methanoindene, similar to chlordane. The toxic response observed from this material is very similar to chlordane.

Acute Oral. Lehman[35] reported that when heptachlor was fed to rats in a single dose, the approximate LD_{50} was 90 mg./kg. body weight. The principal responses were tremors and convulsions.

[61] A. M. Ambrose, H. E. Christensen, D. J. Robbins, and L. J. Rather, *A.M.A. Arch. Ind. Hyg. Occupational Med.,* **7,** 197 (1953).
[62] Council on Pharmacy and Chemistry, *J. Am. Med. Assoc.,* **158,** 1364 (1955).

Chronic Oral. Radomski and Davidow[63] reported that the largest concentration that rats would survive for 6 months was 30 p.p.m. by weight of heptachlor in the diet. These authors indicated that, even at this level, a very significant amount of heptachlor was stored in the fat of the fed animal.

Skin. Lehman[52] reported that when the dried powder of heptachlor was applied to the skin, the approximate LD_{50} was 2000 mg./kg. body weight. If it were applied as a 20 per cent solution in dimethylphthalate, the approximate LD_{50} was less than 780. He observed no skin irritation from either material.

When the heptachlor was applied repeatedly, the approximate LD_{50} was less than 20 mg./kg. body weight/day. There were no survivors after 14 doses at the level of 28 mg./kg.

Metabolism. The reports of Davidow and Radomski[64] and of Radomski and Davidow[63] indicate that the epoxide of heptachlor is formed in the body and that it is stored as such in the fat.

4. *Hygienic Standards of Permissible Exposure*

The A.C.G.I.H. tentative threshold value for 1961 is 0.25 mg./cu. meter. The residue tolerance for specified agricultural commodities is 0.1 p.p.m. by weight.[57]

ALDRIN (1,2,3,4,10,10-Hexachlor-1,4,4a,5,8,8a-hexahydro-endo-exo-dimethanonaphthalene)

1. *Uses and Industrial Exposures*

Aldrin is used as an insecticide.

2. *Physical and Chemical Properties*

Aldrin occurs as four isomers.

3. *Physiological Response*

The administration of aldrin appears to affect the central nervous system predominantly. Indications of hyperirritability, convulsions, and/or coma are observed. Nausea and vomiting may also be associated with poisoning with this material, and are typical of the acute response. Responses from chronic administration may be anorexia, loss of weight, headache, and nervousness. These symptoms are quite general and, of course, are in no way specific to this compound.

---

[63] J. L. Radomsky and B. Davidow, *J. Pharmacol. Exptl. Therap.*, **107**, 266 (1953).

[64] B. Davidow and J. L. Radomski, *J. Pharmacol. Exptl. Therap.*, **107**, 3, 259 (1953).

Histopathological changes have not been clearly defined. Some authors have indicated injury to the liver and kidneys. Others have indicated that this is not usually seen.

Air Dispersions. No quantitative investigative work has been done with air dispersions of aldrin. Some clinical studies have been made on people packaging or handling this material. Princi and Spurbeck[65] studied workers who were exposed to the material in packaging. Any absorption would be by breathing as a dust or possibly through skin contact. The exposure was to a mixture of chlordane, aldrin, and dieldrin. Analysis of the air varied from 5 to 57 mg./cu. meter determined as chlorine and calculated as aldrin. By special absorption technique, the actual aldrin was determined to be between 1 and 2.6 mg./cu. meter. No evidence of physiological response from the exposure was found.

Nelson[66] studied a group of workers exposed to dusts of aldrin. He recorded complaints of headache, dizziness, nausea, and vomiting. He found no evidence of liver injury in these individuals.

It is evident that aldrin can be absorbed when it is breathed as a dust. When considering the toxicity by mouth, it will be recognized that proper protection from air-borne dust in industrial operations is necessary.

Acute Oral. Lehman[35] gave the approximate LD_{50} for rats from aldrin as 67 mg./kg. body weight. Characteristic responses to a toxic dose were tremors and convulsions. Death may be delayed for several days.

Treon and Cleveland[67] reported an LD_{50} for rats of a solution of aldrin in peanut oil to be 45.9 mg./kg. of body weight. The toxic dose will vary with the solvent in which it is given.

Ball *et al.*[68] indicated that the toxicity of aldrin would vary with the quality of the product and the nature of the formulation. He gave LD_{50} figures from 10.6 mg./kg. up to 59.6 mg./kg. for various preparations.

Chronic Oral Toxicity. Treon and Cleveland[67] fed rats for a period of 2 years at concentrations of 2.5, 12.5, and 25 p.p.m. by weight in the diet. There was no increase in mortality or decrease in growth at any of the levels fed. At 12.5 p.p.m. and above, some animals showed increased liver weights and some degenerative hepatic cell changes. Dogs fed 3 p.p.m. in the diet for extended periods of time showed an increase in liver weight and minor histopathological changes in the liver. Dogs fed 1 p.p.m. showed no gross or microscopic abnormalities. The period of feeding extended to over 15 months. Ball, *et al.*[68] fed rats on concentrations of 5, 10, and 20 p.p.m. in the diet. He saw no response for a 6-week period in which he was feeding the pure material. The diet was then changed so as to feed comparable doses of aldrin as a commercial wettable powder. At 20 p.p.m., the growth

[65] F. Princi and G. H. Spurbeck, *Arch. Ind. Hyg. Occupational Med.*, **3**, 64 (1951).

[66] E. Nelson, *Rocky Mt. Med. J.*, **50**, 483 (1953).

[67] J. F. Treon and F. P. Cleveland, *J. Agr. Food Chem.*, **3**, 402 (1955).

[68] W. L. Ball, K. Kay, and J. W. Sinclair, *A.M.A. Arch. Ind. Hyg. Occupational Med.*, **7**, 292 (1953).

showed an initial drop and then a gain to a level higher than the normal. The authors suspect that other materials may be contained in the preparation. They also observed no signs of neuromuscular involvement in the first 6 weeks, but after adding the wettable powder they did observe hyperexcitability and other nervous system responses at levels of 10 and 20 p.p.m. These authors also indicated some disturbance of the estrous cycle of rats at concentrations of 10 and 20 p.p.m.

Kitselman[69] fed dogs doses of 0.02, 0.06, and 2 mg./kg. of body weight/day for a period up to a year. Histopathological studies on the animals fed 0.02 and 0.06 mg./kg. of body weight showed parenchymatous degeneration of the liver and kidneys.

Skin. Only minor erythema is observed from skin contact with aldrin. It should be recognized that commercial preparations for agricultural use may contain other more irritating ingredients.

Treon and Cleveland[67] applied aldrin as a crystalline powder maintained under a rubber sleeve for 24 hours. The minimum lethal dose was 0.6 to 1.25 g./kg. When the material was applied 2 hours/day, 5 days/week for 10 weeks, the minimum lethal dose in mg./kg./day for the dry powder was 35 to 123. In vegetable oil, it was 10 to 26 mg./kg. In ultrasene, it was less than 4.8 mg./kg.

Lehman[52] reported that a 4 per cent aldrin solution in dimethylphthalate had an approximate LD_{50} of less than 150 mg./kg. body weight. It produced no skin irritation but the animals showed severe convulsions and death.

Metabolism. Bann *et al.*[70] indicate that aldrin is largely converted to dieldren in the body and is stored as such.

4. *Hygienic Standards of Permissible Exposure*

The threshold limit of aldrin was established by the American Conference of Governmental Industrial Hygienists, in April, 1959, at 0.25 mg./cu. meter of air. The residue tolerances for specified agricultural commodities are 0, 0.1, 0.25, and 0.75 p.p.m. by weight.[46]

DIELDRIN (1,2,3,4,4a,10-Hexachloro-6,7-epoxy-1,4,4a,5,6,7,8,8a-octahydro-1,4,5,8-endo-exo-dimethanonaphthalene)

1. *Uses and Industrial Exposures*

Dieldrin is used as an insecticide.

[69] C. H. Kitselman, *J. Am. Vet. Med. Assoc.*, **123**, 28 (1953).

[70] J. M. Bann, T. J. DeCino, N. W. Earle, and Y. P. Sun, *J. Agr. Food Chem.*, **4**, 937 (1956).

2. *Physiological Response*

Dieldrin is the epoxide of aldrin. Aldrin is converted to dieldrin in the body. In a general way, the toxicology is similar. Dieldrin is slightly more toxic than aldrin. Treon and Cleveland[67] report the LD_{50} for rats when given in peanut oil to be 38.3 mg./kg. When the material was applied to the dry skin, the minimum lethal dose for a 24-hour exposure was between 0.25 and 0.36 g./kg. When dieldrin was fed for a 2-year period to rats, a concentration of 2.5 p.p.m. was survived; the animals had a slight increase in mortality and an increase in liver weights. Dogs survived a feeding period of approximately 15 months at 1 p.p.m. in the diet but showed an increase in liver weights.

Anyone interested in more details in regard to toxicological factors concerning dieldrin will find that this material is discussed in most of the toxicological investigations referred to under aldrin.

3. *Hygienic Standards of Permissible Exposure*

The threshold limit of dieldrin was established by the American Conference of Governmental Industrial Hygienists, in April, 1959, at 0.25 mg./cu. meter of air. The residue tolerances on specified agricultural commodities are 0, 0.1, 0.25, and 0.57 p.p.m. by weight.[57]

TOXAPHENE, $C_{10}H_{10}Cl_{10}$ (Chlorinated Camphene)

1. *Uses and Industrial Exposures*

Toxaphene is used as an insecticide. The principal exposures will be from dusts or mists when spraying, or from skin contact as it may be absorbed from solutions.

2. *Physical and Chemical Properties*

Physical state: amber wax
Melting point: 70 to 95°C.
Solubility: soluble in water and ethanol

3. *Physiological Properties*

Toxaphene, when absorbed, acts as a stimulant on the brain and spinal cord, resulting in generalized convulsions. Death is usually due to respiratory failure. The symptoms resemble those resulting from absorption of camphor. Internal hemorrhage and high temperatures have been reported following intake in farm animals. Animals that have had chronic oral intake have shown degenerative changes in the liver parenchyma and renal tubules. A general discussion of the response to toxaphene can be found in the report of the Council on Pharmacy and Chemistry of the A.M.A.[71]

[71] Council on Pharmacy and Chemistry, *J. Am. Med. Assoc.*, **149**, 1135 (1952).

Exposure to Mist. The Council on Pharmacy and Chemistry[71] reported that exposures to an oil mist containing 0.2 mg./per 100 ml. of air/minute for periods up to 158 minutes resulted in high mortality. They estimated an approximate LC_{50} as 0.2 mg./per 100 ml. of air/minute for 2 hours.

Acute Oral. Lehman[35] gives the approximate LD_{50} for rats as 69 mg./kg. body weight. The response to this acute dose was hypersensitivity, tremors, and convulsions. Death usually occurs within 24 hours but may be delayed several days.

The A.M.A. Council on Pharmacy and Chemistry[71] reported that the oral LD_{50} could vary from 15 to 375 mg./kg., depending upon the species of animal used. They indicated that a fatal dose in man was estimated to be from 2 to 7 g.

Lackey[72] reported the LD_{50} for dogs, when given orally as a solution in corn oil, to be 25 mg./kg. body weight. When toxaphene was given as a solution in kerosene, the animal tolerated a much larger dose. The author indicated that it was poorly absorbed from kerosene solution. A dose of 10 mg./kg. in the dog caused convulsions.

Chronic Oral. Lackey[72] reported that a dose of 5 mg./kg./day to dogs caused convulsions after a few days administration. He fed 4 mg./kg./day for 106 days. The dogs showed some reaction immediately after being dosed but seemed to recover to normal between doses. At the end of the period, the animals showed a reversible hydropic degeneration and fatty degeneration of the liver. The kidneys showed marked degeneration of the tubular epithelium.

Lehman[49] gave the tolerated chronic oral dose for rats over a 2-year period as 5 mg./kg. body weight/day. This dose was survived without apparent effect.

Skin. The A.M.A. Council on Pharmacy and Chemistry[71] indicated a mild to moderate skin irritation from toxaphene. It was also indicated that it could be absorbed through the skin. Radeleff[73] dipped animals in an emulsion of toxaphene. When used as a dip, an 8 per cent emulsion caused the death of both sheep and goats. One goat survived a dip in 4 per cent emulsion. Calves sprayed with 4 per cent emulsion survived. Calves sprayed with an 8 per cent emulsion died. Injuries to the lungs, liver, and kidneys were observed. It should be recognized that some mist may be breathed when animals are sprayed.

Human Experience. McGee and Reed[74] reported on 6 cases of poisoning from toxaphene that recovered. They also reported 4 fatal cases of children who took toxaphene orally. They observed congestion and edema in the lungs, dilatation of the heart, and petechial hemorrhages in the brain.

4. Hygienic Standards of Permissible Exposure

The threshold limit of toxaphene was established by the American Conference of Governmental Industrial Hygienists, in April, 1959, at 0.5 mg./cu. meter of air.

[72] R. W. Lackey, *J. Ind. Hyg. Toxicol.,* **31,** 117 (1949).
[73] R. D. Radeleff, *Vet. Med.,* **44,** 436 (1949).
[74] J. C. McGee and H. L. Reed, *J. Am. Med. Assoc.,* **149,** 1124 (1952).

The residue tolerance for specified agricultural commodities is 5 or 7 p.p.m. by weight.[57]

HEXACHLOROCYCLOPENTADIENE, C_5Cl_6

1. *Uses and Industrial Exposures*

Hexachlorocyclopentadiene is used as chemical intermediate in the manufacture of aldrin.

2. *Physical and Chemical Properties*

Physical state: liquid
Molecular weight: 273
Freezing point: 9 to 10°C.
Boiling point: 234°C.
Vapor density: 9.4 (air = 1)
Vapor pressure: 0.080 mm. Hg (25°C.)
1 mg./liter ≈ 89.6 p.p.m. and 1 p.p.m. ≈ 11.17 mg./cu. meter at 25°C., 760 mm. Hg

3. *Physiological Response*

Hexachlorocyclopentadiene is highly irritating to mucous membranes, causing lacrimation, sneezing, and salivation.   Degenerative changes were seen in the brain, heart, and adrenal glands; also degeneration and necrosis in the liver and the kidney tubules. There were severe pulmonary hyperemia and edema from breathing the air dispersion.

All information on this compound is taken from Treon *et al.*[75]

Acute Vapor Exposure.   The maximum time-concentrations by volume in air survived by guinea pigs was 0.25 hour at 20.2 p.p.m., 1 hour at 7.2 p.p.m., 3.5 hours at 3.1 p.p.m., and 7 hours at 1.5 p.p.m.

Levels survived by rats were 0.25 hour at 20.2 p.p.m., 0.5 hour at 7.2 p.p.m., 1 hour at 3.1 p.p.m., and 7 hours at 0.33 p.p.m.

Chronic Vapor Exposure. Guinea pigs, rabbits, and rats were given 150 7-hour exposures over 216 days. They survived at 0.15 p.p.m. by volume in air. Guinea pigs survived a concentration of 0.34 p.p.m., but rats and mice succumbed after 30 7-hour exposures.

[75] J. F. Treon, F. P. Cleveland, and J. Cappel, *A.M.A. Arch. Ind. Health*, **11**, 459 (1955).

The response and pathological changes were discussed in the summary. The animals showed mild liver and kidney injury even at 0.15 p.p.m.

Oral. The approximate lethal dose by single oral administration of the 93.3 per cent compound was between 0.42 and 0.62 g./kg. for rats and rabbits.

The animals showed diarrhea, lethargy, and retarded respiration. Rabbits showed diffuse degenerative changes in the brain, heart, liver, and adrenals. There was necrosis of the epithelium of the renal tubules and the lungs showed severe hyperemia and edema.

Skin. The material was extremely irritating to the skin. By skin absorption, the lethal dose was between 0.43 and 0.61 g./kg.

4. *Hygienic Standards of Permissible Exposure*

None has been established for hexachlorocyclopentadiene, but the exposure should be below 0.15 p.p.m.

5. *Odor and Warning Properties*

A faint odor is detected at 0.15 p.p.m. by volume in air. A pronounced, pungent odor is observed at 0.33 p.p.m.

CHAPTER XXXIII

Phenols and Phenolic Compounds

WILLIAM B. DEICHMANN and MORENO L. KEPLINGER

PHENOL, C_6H_5OH (Hydroxybenzene, Carbolic Acid)

1. Source, Uses, and Industrial Exposures

Phenol is one of the many aromatic compounds present in coal tar. It is separated from other substances by fractional distillation (170 to 230°C.) and by other methods of purification until "gray phenic acid" or a pure grade of phenol has been obtained. Synthetic processes developed for the production of phenol include fusion of sodium benzenesulfonate with sodium hydroxide and hydrolysis of chlorobenzene.

Phenol is used in the production or manufacture of a large variety of aromatic compounds, including explosives, fertilizers, coke, illuminating gas, lampblack, paints, paint removers, rubber, asbestos goods, wood preservatives, synthetic resins, textiles, drugs, pharmaceutical preparations, perfumes, bakelite, and other plastics (phenol-formaldehyde resins). Phenol also finds use in the petroleum, leather, paper, soap, toy, tanning, dye, and agricultural industries.[1]

With rare exceptions, human exposure in industry has been limited to accidental contact of phenol with the skin or to inhalation of phenol vapors. Following the introduction of phenol spray by Lister in 1867, the compound became very popular and was used extensively for a number of years. Its medicinal uses are now limited chiefly to its application as an agent for relieving itching, as a disinfectant for septic wounds, as a cauterizing agent, as an insecticide, and as an agent in the treatment of certain systemic disorders.[2–12]

[1] T. F. Mancuso, Ohio State Med. J., 51, 672 (1955).

[2] Bacelli, cited from H. C. Wood, Jr., J. Am. Med. Assoc., 32, 1249 (1899).

[3] U. Conforti, Policlinico Rome, No. 44, 1381 (1909).

[4] A. Ellinger in A. Heffter, Handbuch der experimentellen Pharmakologie, Springer, Berlin, Vol. I, 1. 1923, p. 893.

[5] H. Herding, Bull. med., 53, 281 (1939).

[6] R. Kobert, Lehrbuch der Intoxikationen. Ende, Stuttgart, 1906.

[7] L. Lewin, Gifte und Vergiftungen. Stilke, Berlin, 1929.

[8] B. Mistretta, Policlinico Rome, Sez. prat., 44, 2287 (1937).

[9] A. Pisani, Gazz. med. Lombarda, 87, 99 (1928).

[10] M. Sein, Indian Med. Gaz., 74, 270 (1939).

[11] C. Sironi, Gazz. ospedali e clin., 51, 395 (1930).

[12] H. Zangger in F. Flury and H. Zangger, Lehrbuch der Toxikologie. Springer, Berlin 1928. Occupation and Health. International Labor Office, Geneva, 1934.

2. Physical and Chemical Properties

Physical state: white, crystalline mass or hygroscopic, translucent, needle-shaped crystals

Molecular weight: 94.11

Specific gravity: 1.072

Melting point: 41°C.

Boiling point: 182°C.

Vapor density: 3.24 (air = 1)

Vapor pressure: 0.3513 mm. Hg (25°C.)

Refractive index: 1.54 (45°C.)

Per cent in "saturated" air: 0.046% by volume (25°C.)

Density of "saturated" air: 1.00104 (air = 1)

Solubility in water and common solvents: Phenol added to water (25°C.) forms a true solution when present in concentrations up to 8 per cent, and also in concentrations ranging from about 71 to 97 per cent, in terms of both weight and volume.   The compound is soluble to more than 50 per cent[13] in ethyl alcohol, chloroform, ethyl ether, ethyl acetate, toluene, glycerol, and olive oil. Its solubility in mineral oil is about 0.2 per cent (25°C.), in petroleum ether 5.5 per cent (31°C.), and in rabbit fat 40 per cent (34°C.).[14] According to Pilcher and Sollmann,[15] one part of crystallized phenol dissolves in 8 to 9 parts of petrolatum, 20 to 21 parts of gasoline, 23 to 24 parts of solid petrolatum, and 45 to 50 parts of liquid petrolatum.

Flash point: (closed cup) 80°C. (175°F.), (open cup) 85°C. (185°F.)

1 mg/liter ⪰ 260 p.p.m. and 1 p.p.m. ⪰ 0.00384 mg/liter at 25°C., 760 mm. Hg.

3. Determination in the Atmosphere

Phenol in the air can be collected in an absorbing solution of dilute alkali such as sodium hydroxide or 0.5 per cent sodium bicarbonate. The quantity of phenol is then determined with diazotized p-nitroaniline reagent,[16] Folin-Ciocalteu reagent, 2,6-dibromoquinonechloroimide, or p-aminodimethylaniline sulfate.[17] The last-named reagent is recommended when the concentration in air is very low (2 to 10 parts per billion). The phenol also may be absorbed in spectrograde alcohol with direct determination by ultraviolet spectrophotometry.

It should be emphasized that in order to determine the phenol content of any material a specific analytical method should be employed by a competent chemist. Most of the qualitative tests and many of the quantitative procedures used in the past were not specific for phenol.

[13] N. A. Lange and G. M. Forker, eds., *Handbook of Chemistry*. Handbook Publishers, Sandusky, Ohio, 1956.

[14] W. B. Deichmann, S. Witherup, and M. Christian, unpublished observations.

[15] J. D. Pilcher and T. Sollmann, *J. Pharmacol. Exptl. Therap.*, **6**, 377 (1914).

[16] H. B. Elkins, *The Chemistry of Industrial Toxicology*. Wiley, New York, 1950.

[17] M. M. Braverman, S. Hockheiser, and M. B. Jacobs, *Am. Ind. Hyg. Assoc. Quart.*, **18**, 132 (1957).

Briefly, biological material to be analyzed is extracted with ether and the amount of "free" and "conjugated" phenol is determined spectrophotometrically utilizing the color developed with various reagents such as those mentioned above. Simple, gross, qualitative tests can be made with various color-producing reagents, but separation by paper chromatography gives more specific results.

A critical review of analytical methods with their usefulness and limitations in estimating "free" and "conjugated" phenol in blood, organs, urine, saliva, sweat, feces, etc., was published in 1942.[18] In addition to the analytical procedures recommended therein, the reader is advised to consider also those that have been published more recently by Schmidt,[19,20] Tucker,[21] Chirkov,[22] Baernstein,[23] Lykken et al.,[24] Glick,[25] Armstrong et al.,[26] and Tompsett.[27]

4. Physiological Response

Regardless of the mode of administration, the signs of acute illness induced by phenol in experimental animals, resemble those observed in man. However, in man, phenol usually exerts (directly and indirectly) a predominant action upon the higher centers resulting in sudden collapse. In other mammals the predominant effects are exerted upon motor centers in the spinal cord inducing marked twitchings and severe convulsions. Following absorption of a toxic dose the heart rate first increases, then becomes slow and irregular. The blood pressure increases slightly at first, then falls markedly. There are some salivation, marked dyspnea, and usually a decrease in body temperature.

Prolonged oral or subcutaneous administration can cause damage to the lungs, liver, kidneys, heart, and genitourinary tract.[4,12,28-33] In animals, prolonged inhalation of vapors (30 to 60 p.p.m.) has induced respiratory difficulties, lung damage, loss of weight, and paralysis.[34]

[18] W. Deichmann and L. Schafer, Am. J. Clin. Pathol., 12, 129 (1942).

[19] E. G. Schmidt, J. Biol. Chem., 145, 533 (1942), 150, 69 (1943).

[20] E. G. Schmidt, J. Biol. Chem., 179, 211 (1949).

[21] I. W. Tucker, J. Assoc. Offic. Agr. Chemists, 25, 779 (1942).

[22] S. K. Chirkov, J. Appl. Chem. U.S.S.R., 17, 31 (1944).

[23] H. D. Baernstein, J. Biol. Chem., 161, 685 (1945).

[24] L. Lykken, R. S. Treseder, and V. Zahn, Ind. Eng. Chem., 18, 103 (1946).

[25] D. Glick, ed., Methods of Biochemical Analysis, Vol. I. Interscience Publishers, New York–London, 1954, pp. 27–52.

[26] M. D. Armstrong, K. N. F. Shaw, and P. E. Wall, J. Biol. Chem., 218, 293 (1956).

[27] S. L. Tompsett, Clin. Chem., 4, 237 (1958).

[28] M. Biebl, Beitr. pathol. Anat. u. allgem. Pathol., 84, 257 (1930); Z. ges exptl. Med., 87, 436 (1933), 93, 515 (1934).

[29] V. G. Heller and L. Pursell, J. Pharmacol. Exptl. Therap., 63, 99 (1938).

[30] W. B. Deichmann and P. Oesper, Ind. Med., 9, 296 (1940).

[31] L. Wachholz, Deut. med. Wochschr., 21, 146 (1895).

[32] O. Wandel, Arch. exptl. Pathol. Pharmakol., 56, 161 (1907).

[33] W. Hesselbach, Inaugural Dissertation, Halle, 1890.

[34] W. B. Deichmann, K. V. Kitzmiller, and S. Witherup, Am. J. Clin. Pathol., 14, 273 (1944).

TABLE 1

Toxicity of 2 to 7 Per Cent Aqueous Solutions of Phenol for Adult Experimental Animals

Species	Route	Dose killing approx. 50% of animals, g./kg.	Ref.
Mouse	Subcut.	0.3–0.35	Tollens,[35] Duplay and Cazin[36]
Rat	Subcut.	0.45	Deichmann and Witherup[37]
	Oral	0.53[a]	Deichmann and Witherup[37]
	Cut.	2.5	Deichmann and Witherup[37]
Guinea pig	Subcut.	0.68	Duplay and Cazin[36]
Rabbit	Intrav.	0.18	Deichmann and Witherup[37]
	Subcut.	0.5–0.6	Tauber[38] and Tollens[35]
	Oral	0.6	Clarke and Brown[39]
	Oral	0.4–0.6	Deichmann and Witherup[37]
	Intraper.	0.5–0.6	Deichmann and Witherup[37]
Cat	Subcut.	0.09	Tollens[35]
	Oral	0.1	Macht[40]
Dog	Oral	0.5	Macht[41]
Monkey	Toxicity is of similar order to that for rabbit		Smith, Elvove, and Frazier[41]

[a] LD$_{50}$.

TABLE 2

Comparative Toxicity of Aqueous Preparations of Phenol in Different Concentrations[a] Applied to the Abdominal Skin of Rabbits[37]

Dose of 2 g./kg., administered as	Number of rabbits	Deaths, %
Emulsion: 10 g. phenol and 90 g. water	10	100
Emulsion: 25 g. phenol and 75 g. water	10	90
Emulsion: 50 g. phenol and 50 g. water	10	90
Solution: 75 g. phenol and 25 g. water	10	80
Solution: 90 g. phenol and 10 g. water	10	50
Solution: 95 g. phenol and 5 g. water	17	53
Melted phenol reagent heated to 40°C.	15	30

[a] Standard dose for all concentrations, 2 g. phenol per kilogram of rabbit.

Lethal doses for experimental animals are presented in Tables 1 and 2.

Pathology. The pathological changes produced by phenol in animals vary with the route of absorption, vehicle employed, concentration, and duration of

[35] K. Tollens, *Arch. exptl. Pathol. Pharmakol.*, **52**, 220 (1905).

[36] S. Duplay and M. Cazin, *Compt. rend.*, **112**, 672 (1891).

[37] W. B. Deichmann and S. Witherup, *J. Pharmacol. Exptl. Therap.*, **80**, 233 (1944).

[38] Tauber and S. Tauber, *Z. Physiol. Chem.*, **2**, 366 (1878–79); *Arch. exptl. Pathol. Pharmakol.*, **36**, 197, 211 (1895).

[39] T. W. Clarke and E. D. Brown, *J. Am. Med. Assoc.*, **46**, 782 (1906).

[40] D. I. Macht, *Bull. Johns Hopkins Hosp.*, **26**, 98 (1915).

[41] M. I. Smith, E. Elvove, and W. H. Frazier, *Public Health Repts. U.S.*, **45**, 2509 (1930).

exposure. Local damages to the skin include eczema, inflammation, discoloration, papillomas, necrosis, sloughing, and gangrene. Following oral ingestion the mucous membranes of the throat and esophagus may show swelling, corrosions, and necroses, with hemorrhage and serous infiltration of the surrounding areas.[42] In a severe intoxication the lungs may show hyperemia, infarcts, bronchopneumonia, purulent bronchitis, and hyperplasia of the peribronchial tissues. There can be myocardial degeneration and necrosis. The hepatic cells may be enlarged, pale, and coarsely granular with swollen, fragmented, and pyknotic nuclei.[43] Prolonged administration of phenol may cause, in the kidney, parenchymatous nephritis,[34] hyperemia of the glomerular and cortical region, cloudy swelling, edema of the convoluted tubules, and degenerative changes of the glomeruli. Blood cells become hyaline, vacuolated, or filled with granules. Muscle fibers show marked striation.

Absorption, Excretion and Metabolism. Phenol is readily absorbed through the intact and abraded skin and from the stomach, enteric tract, uterus, intraperitoneal cavity, and subcutaneous tissues of man and animals.[37] Vapors of phenol are readily absorbed into the pulmonary circulation. Figure 1 presents the comparative rates of absorption from the gastroenteric tract of rabbits.

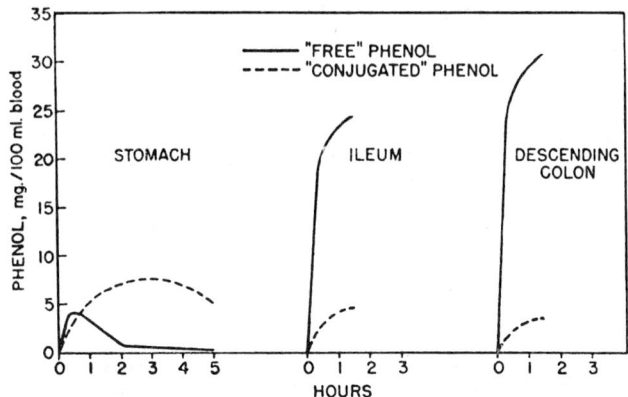

Fig. 1. Comparative rate of absorption of phenol from the stomach, ileum, and descending colon of the rabbit.[44] 0.3 g. phenol/kg., as a 5 per cent aqueous solution, was placed into ligatured sections of the gastroenteric tract.

After absorption into the body, most of the phenol is oxidized and conjugated with sulfuric, glucuronic, and other acids. It is excreted in the urine as "free" and as "conjugated" phenol.[45] Traces of "free" phenol are eliminated with the feces and

[42] A. Lesser, *Arch. pathol. Anat. u Physiol. Virchow's,* **83,** 230 (1881).

[43] P. Binet, *Rev. med. Suisse. romande,* **15,** 561 (1895), **16,** 449 (1896); cited from W. F. von Oettingen, *Natl. Inst. Health Bull.* No. **190** (1949).

[44] W. B. Deichmann, S. Witherup, and M. Dierker, *J. Pharmacol. Exptl. Therap.,* **105,** 265 (1952).

[45] W. Deichmann, *Federation Proc.,* **2,** 77 (1943); *Arch. Biochem.,* **3,** 345 (1944).

expired air. Figures 2 and 3, respectively, show graphically the fate of a lethal and a sublethal dose of phenol.[45,46]

Additional information regarding the absorption[44,47-50] and metabolism[46,51-62] of phenol can be found in the works of other investigators.

Mode of action. The primary site of stimulation in the central nervous system is the spinal cord. The twitchings are due to reflex stimulation,[63] while the clonic convulsions probably are due to an increased excitation of the motor mechanism of the anterior horn cells.[64] Local application of phenol solutions to the cerebellar cortex[65] did not produce the effects seen when phenol was applied to the spinal cord. When applied directly to the cerebral cortex, a dilute solution increased reflex excitability while more concentrated solutions destroyed the tissues.[66] The heart apparently is slowed by a direct myocardial depression not by stimulation of the vagus.[67] The slight rise in blood pressure appears to be due to peripheral vasoconstriction. The more prominent fall of pressure is caused by local cardiac depression and failure of the vasomotor centers.

[46] E. Baumann, *Arch. ges. Physiol.*, **13**, 285 (1876); *Z. physiol. Chem.*, **2**, 335, (1878–79), **3**, 149, 250 (1879), **6**, 183 (1882), **10**, 123 (1886); *Arch. Anat. u. Physiol., Physiol. Abt.*, **1879**, 245.

[47] W. B. Drichmann, T. Miller and J. B. Roberts, *Arch. Ind. Hyg. Occupational Med.*, **2**, 254 (1950).

[48] D. E. Jackson, *Experimental Pharmacology and Materia Media.* Mosby, St. Louis, 1939.

[49] H. Nicolai, *Klin. Wochschr.*, **18**, 123 (1939), **20**, 80 (1941).

[50] T. Sollmann, P. J. Hanzlik, and J. D. Pilcher, *J. Pharmacol. Exptl. Therap.*, **1**, 409 (1910).

[51] L. Brieger, *Z. physiol. Chem.*, **2**, 241 (1878–79), **3**, 134 (1879).

[52] F. Muller, *Berlin. klin. Wochschr.*, **24**, 405, 433, 436 (1887).

[53] W. F. Rogers, Jr., M. P. Burdick, and G. R. Burnett, *J. Lab. Clin. Med.*, **45**, 87 (1955).

[54] W. Deichmann and L. Schafer, *Am. J. Clin. Pathol.*, **12**, 129 (1942).

[55] H. G. Bray *et al., Biochem. J.*, **52**, 422 (1955).

[56] S. Suzaki, N. Takahaski, and F. Egami, *Biochim. et Biophys. Acta*, **24**, 444 (1957).

[57] R. H. de Meio and M. Wizerkaniuk, *Biochim. et Biophys. Acta*, **20**, 428 (1956).

[58] J. D. Gregory and F. Lipmann, *J. Biol. Chem.*, **229**, 1081 (1957).

[59] E. Becher, S. Litzner, W. Täglich, and F. Doenecke, *Münch. med. Wochschr.*, **1925** 1676, **1927** 1656; *Klin. Wochschr.*, **1926** 147; *Z. klin. Med.*, **104**, 182, 195 (1926); *Z. physiol. Chem.*, **47**, 173 (1906).

[60] W. J. Darby, R. H. de Meio, M. L. C. Bernheim, and F. Bernheim, *J. Biol. Chem.*, **158**, 67 (1945).

[61] R. H. de Meio and R. I. Arnold, *J. Biol. Chem.*, **156**, 577 (1944); *J. Pharmacol. Exptl. Therap.*, **84**, 64 (1945); *Rev. soc. arg. biol.*, **17**, 570 (1941), **18**, 158 (1942).

[62] W. L. Lipschitz and E. Bueding, *J. Biol. Chem.*, **129**, 333 (1939).

[63] S. Baglioni and M. Magnini, *Arch. fisiol.*, **6**, 240 (1909); abstr. *Biochem. Z.*, **10**, 97 (1910).

[64] M. Magnini, *et al., Arch. fisiol.*, **8**, 111, 157, 166 (1910); abstr. *Biochem. Z.*, **10**, 831 (1910).

[65] K. Lobker, *Deut. med. Wochschr.*, **15**, 219 (1889).

[66] J. W. C. Gunn, *J. Pharmacol. Exptl. Therap.*, **29**, 297 (1926).

[67] G. Haas and E. F. Schlesinger, *Arch. exptl. Pathol. Pharmakol.*, **104**, 56 (1924).

The toxic effects of phenol are related directly to the amount of "free" phenol in the blood.

Cause of Death. In an acute intoxication, death is usually due to respiratory failure.

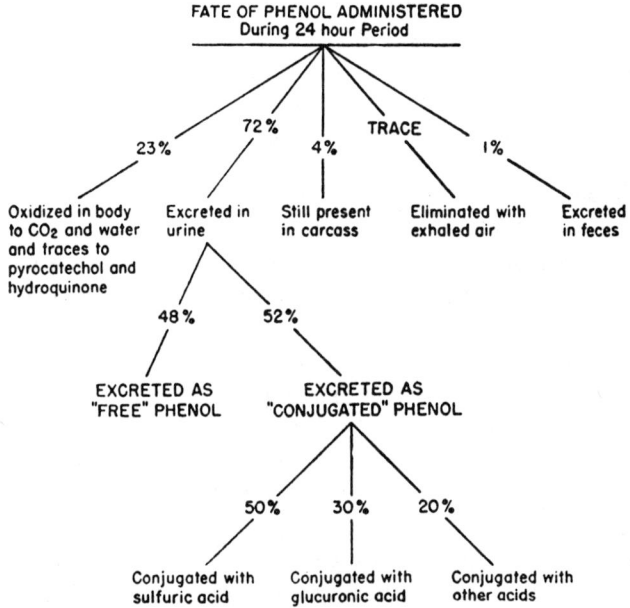

Fig. 2. Metabolism of phenol in a rabbit given a sublethal oral dose (0.3 g./kg.).[45]

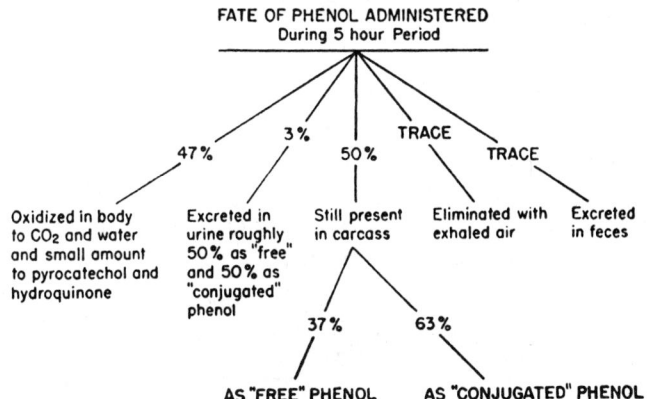

Fig. 3. Metabolism of phenol in a rabbit given a lethal oral dose (0.5 g./kg.).[45]

Acute Effects in Man. Fatalities from poisoning by phenol occurred much more frequently in the past decades than in recent years, although accidental poisonings still occur, particularly in the home.

An oral dose of 1 g. of phenol may be lethal to man; however, in exceptional cases, patients have survived the ingestion of 65 g. of pure phenol or 120 g. of the crude product. Roughly 50 per cent of all reported cases have terminated fatally.[6,7]

The swallowing of phenol causes intense burning of the mouth and throat followed by marked abdominal pain. The breath has the odor of phenol, the face is pale and usually covered with cold sweat, the pupils may be contracted or dilated, and cyanosis is usually marked. Collapse, manifested by muscular weakness and unconsciousness, occurs in many cases a few minutes after the poison is swallowed. The pulse is usually weak and slow, but occasionally it is racing. Respiration may be increased in rate in the early stage of poisoning, but later it decreases in both rate and magnitude. The temperature of the body may fluctuate above or below normal. Reflex activity is lost. General tremors, tonoclonic convulsions, or twitchings of isolated muscles of the face or limbs are occasionally observed, but they are never marked. Death usually results from respiratory failure.[7,12,68-71]

Chronic Effects in Man. In Lister's day,[72] cases of chronic or subacute phenol poisoning were not uncommon among surgeons and their assistants. As Kobert[6] remarks, many of those doctors must have possessed great tolerance considering that they used, applied, and inhaled sprays of carbolic acid for years without illness or apparent discomfort. No doubt the intermittent character of their exposure was a factor in their favor. Data which give us an insight into the quantities of phenol apparently tolerated by man were reported in 1881 by Falkson,[73] who stated that after inhaling these vapors for several hours his own urine contained phenol in amounts up to 2 g. per day, while the urine of patients who absorbed the compound from the inspired air as well as from the skin or from open wounds contained up to 5 g. per day.

At present, chronic phenol poisoning is reported most infrequently. According to Zangger,[12] severe chronic poisoning in man is characterized by systemic disorders such as digestive disturbances, including vomiting, difficulty of swallowing, ptyalism, diarrhea, and anorexia; by nervous disorders, with headache, fainting, vertigo, and mental disturbances; and possibly by an eruption on the skin. The disease is usually fatal when there is extensive damage to the liver and kidneys. Prolonged cutaneous exposure to preparations of phenol may result in ochronosis.

[68] E. Salkowski, *Arch. pathol. Anat. u Physiol. Virchow's*, **73**, 409 (1878); *Arch. ges. Physiol.*, **5**, 335 (1872); *Z. physiol. Chem.*, **7**, 161 (1882–83), **13**, 264 (1889).

[69] T. Sollmann, *A Manual of Pharmacology*. Saunders, Philadelphia, 1927.

[70] F. P. Underhill, *Toxicology or the Effects of Poisons*. Blakiston, Philadelphia, 1928.

[71] A. R. Cushny, C. W. Edmunds, and J. A. Gunn, *Pharmacology and Therapeutics*. Lea & Febiger, Philadelphia, 1940.

[72] J. Lister, *Lancet*, **2**, 95, 353, 668 (1867).

[73] R. Falkson, *Arch. klin. Chir. Langenbecks*, **26**, 204 (1881).

5. Hygienic Standard of Permissible Exposure

The recommended maximum atmospheric concentration (M.A.C.) for 8 hours of exposure is 5 parts of vapor per million parts of air, by volume (p.p.m.).[74]

According to Irish, concentrations in the air (in the presence of steam) of 5 p.p.m. or less are detectable by most individuals.

Beinhart[75] indicates that concentrations of about 4 p.p.m. will impart a decided phenol taste to vegetables and fruit grown within a radius of more than a mile from a plant where such vapors escape.

Threshold limits[76] for the detection of odors in aqueous solutions are as follows: phenol, 25 mg./liter; chlorophenols, 0.001 to 0.0005 mg./liter; cresol, 0.0025 mg./liter; chlorinated cresols, 0.001 to 0.0002 mg./liter; thymol, 0.05 mg./liter; resorcinol, 40.0 mg./liter; creosote, 0.125 mg./liter; chlorinated creosote, 0.01 mg./liter; naphthol, 7.0 mg./liter; chlorinated naphthol, 0.5 mg./liter. Hydroquinone imparted no odor to the water.

6. Flammability

There is a moderate fire hazard. Extinguish with water, carbon dioxide, or dry chemicals.

7. Odor and Warning Properties

There is a distinct, aromatic, somewhat sickening sweet and acrid odor with a sharp and burning taste.

8. Comments on the Local Use of Phenol in Medicine

The once favorite local application of dressings or compresses saturated with dilute (1 to 5 per cent) aqueous solutions of phenol is no longer practiced. Many cases of gangrene have resulted from this type of therapy, and as a result fingers and toes required amputation. Gangrene has been observed less frequently upon the trunk.

An efficient antipruritic is obtained by incorporation of 1 to 2 per cent of phenol into a salve, ointment, a vegetable oil, or calamine lotion. However, preparations of this type should not be covered by a tight bandage after application to the skin.

Addition of camphor to phenol brings about a marked reduction of causticity and disturbs but little the analgesic action of phenol. Claims that the causticity is totally eliminated in mixtures containing more than 30 per cent each of phenol and camphor are exaggerations The causticity remaining can be traced in part to

[74] American Conference of Governmental Industrial Hygienists, *A.M.A. Arch. Environ. Health*, **1**, 142 (1960).

[75] E. G. Beinhart, *Science*, **103**, 207 (1946).

[76] M. S. Nesmeyanova, *Gigiena i Sanit.*, No. 7, 11 (1953), cited from *Chem. Abstr.*, **48**, 913d (1954).

an aqueous phenol phase. This is formed because of the equilibration of the phenol between the phenol–camphor and the tissue fluids or perspiration of which no wound or skin remains free for any length of time.

9. Prevention of Poisoning by Phenol

The manufacture and use of phenol in industry need not be considered an industrial hazard as long as management installs effective safety measures and employees are instructed and supervised in the handling of this compound. New employees should be questioned as to any known susceptibility to phenol or to closely related materials. Those affected with hepatic or kidney diseases should not be exposed to phenol for any length of time, because even intermittent exposure to vapors of phenol may become dangerous, particularly when the material is handled at elevated temperatures. Liquid phenol is combustible and mixtures of air containing 3 to 10 per cent phenol are explosive.

Measures of safety should include at least the following:

1. Publication and distribution of a manual outlining individual steps to be followed in the manufacture, handling, storage, and transport of phenol.

2. Effective ventilation.

3. Proper disposal of phenol waste with precautions against the possible pollution of air, streams, and underground waters.

4. Cleaning of tanks should not be attempted without proper gear: forced air supply, a rescue harness and life line, hose mask, boots, apron and gloves of rubber, and a "watcher" stationed at the entrance of the tank.

5. Continuous vigilance on the part of the hygienist or physician for signs and symptoms of chronic (local or systemic) intoxication.

10. Treatment of Phenol Poisoning

Phenol spilled upon the skin is removed most efficiently by flooding the affected area with water. This must be done promptly; otherwise absorption will occur.

Experimental observations[14,37] have shown that the rate of absorption from the skin of animals depends primarily upon the size of the area involved, and the duration of contact. The concentration of the solution applied was a less important factor. If a preparation of phenol is spilled upon the skin, it is imperative that all parts of the body and clothing that have become wetted by the washings be flushed until all traces of phenol have been removed. It is well to cover phenol burns with wet dressings, such as compresses of saturated sodium sulfate. Healing of cutaneous phenol burns on animals was not expedited by treatment with various vegetable oils or glycerol. Application of a solution of 50 per cent camphor in ethyl alcohol irritated the skin of rabbits and consequently retarded healing.

Speed is equally essential in the treatment of oral poisoning. If a patient is conscious and can be induced to vomit easily, 15 to 30 ml. of castor oil or some other vegetable oil may well be administered. If vomiting is not induced readily and promptly, gastric lavage should be initiated without delay, preferably with an aqueous solution of 40 per cent of Bacto-Peptone or milk; if these are not available, water may be used. Washing should be continued, employing 300 to 400 ml. of liquid at a time, until the odor of phenol is no longer detectable.

After the stomach has been washed out, an oral dose of 15 to 50 ml. of castor or another vegetable oil may be administered to advantage. As Jackson[48] and others[14,77] have pointed out, oils retard the absorption of phenol and tend to reduce the local damage. (Animal experiments have shown that phenol is less toxic when administered in castor oil than in other vegetable oils.)[39] However, caution is in order in the face of serious alimentary damage, and it may be inadvisable to give a cathartic such as castor oil if much time has elapsed since the ingestion of phenol.

Symptomatic treatment consists primarily of the administration of circulatory stimulants. These may augment the systemic poisoning temporarily, but they counteract stasis and therefore tend to prevent thrombosis and necrosis. Transfusions also may be indicated. In collapse, heat should be applied.

The treatment of chronic phenol poisoning is symptomatic after the patient has been removed from the site of exposures.

11. Factors Relating to the Legal Aspect of Poisoning by Phenol

The gross appearance of the lesions is suggestive, but not sufficiently specific to exclude other substances. Microscopic examination of tissues that have had brief contact with phenol is likely to be entirely fruitless because of the excellent "fixing" properties of phenol solutions. Chemical analyses will provide positive or negative evidence in a case of suspected poisoning. Stomach contents (in oral poisoning) and blood and urine are best suited for analysis in the surviving patient. The relative concentrations of "free" and "conjugated" phenol will furnish useful information in relation to quantities ingested and absorbed, time interval between poisoning and drawing of blood, and the prognosis. In fatal cases any tissue or excretion may be employed for analysis.

Analysis of tissues should be carried out as soon as feasible after death because glycolysis will alter the respective concentrations of "free" and "conjugated" phenol. Rising and Lynn[78] suggest ethyl alcohol as a preservative for samples of viscera containing phenol. These recommendations are supported by observations,[14] which indicated that only 10 to 15 per cent of phenol was lost during a period of 12 months when rat tissues (containing originally 25 mg. of phenol) were kept in ethyl alcohol, at 7°C., in paraffin-stoppered containers. Fifty per cent of the original content of phenol was lost from corresponding tissues preserved without alcohol under otherwise identical conditions.

Embalming destroys the evidence of poisoning by phenol because embalming fluids usually contain this or very closely related compounds.

[77] R. Kobert, *Lehrbuch der Intoxikationen.* Enke, Stuttgart, 1906.

[78] L. W. Rising and E. V. Lynn, *J. Am. Pharm. Assoc., Sci. Ed.,* **21,** 138 (1932).

12. *Phenols in Water Supply*

The presence of phenol and related compounds in a public water supply generally presents a serious problem to the engineer of a water plant. The treatment of a public water supply for the purpose of removing phenolic tastes and odors is not the answer to the problem. The phenols must be eliminated at the source of entry into the water.[79]

Chase[79] recommends chlorine dioxide (1 to 10 lb. per million gallons of water) to remove phenolic tastes. Caution is in order, however, since a high dosage of chlorine dioxide produces a breaking or sloughing off of the iron and manganese in the distribution system, and the resulting red water may cause as many, if not more, complaints as undesirable tastes and odors.

Not all tastes are due to phenol or phenols. According to Kinney,[80] organic chemicals other than phenols are more responsible for the production of undesirable tastes and odors in some of our waters. Some of the nonphenols give phenolic tastes, and some of these chemicals apparently are not removed by chlorination. Hence, developing limits on phenols and requiring that phenols be treated to meet certain limits will not necessarily assure taste-free water. For methods now employed in the removal of phenols from industrial wastes consult the summary presented by Pettit.[81] This author emphasizes that none of the present methods of treatment offers a complete solution.

13. *Action of Phenol and Related Compounds in Tumor Formation*

Interesting and valuable contributions to this subject were provided by Boutwell, Rusch, and Bosch[82-85] and others.[86] Their reports deal with the action of phenol and about fifty phenol derivatives containing one to five halogen atoms, hydroxyl, carbonyl, carboxyl, and nitro radicals, as well as β-naphthyl, phenolacetate, anisole, certain cresols, and other related compounds.

Pretreatment with a single application of about 75 μg. of 9,10-dimethyl-1,2-benzanthracene (DMBA) was not essential for the induction of tumors.[85] Phenol and certain other materials alone induced both papillomas and carcinomas, but the time required for the production of tumors was extended when pretreatment with DMBA was excluded. The maximum response was produced with a 10 per cent solution of phenol, while a less marked, but positive, effect was caused by a

[79] D. E. Chase, *J. Am. Water Works Assoc.*, **45**, 493 (1953).

[80] J. E. Kinney, *J. Am. Water Works Assoc.*, **45**, 499 (1953).

[81] G. A. Pettit, *J. Am. Water Works Assoc.*, **45**, 496 (1953).

[82] R. K. Boutwell, H. P. Rusch, and B. Bosch, *Proc. Am. Assoc. Cancer Research*, **2**, 6 (1955).

[83] H. P. Rusch, D. Bosch, and R. K. Boutwell, *Acta Unio Intern. Contra Cancrum*, **11**, 699 (1955).

[84] R. K. Boutwell, K. P. Rusch, and B. Booth, *Proc. Am. Assoc. Cancer Research*, **2**, 96 (1956).

[85] R. K. Boutwell and D. K. Bosch, *Cancer Research*, **19**, 413 (1959).

[86] M. H. Salaman and O. M. Glendenning, *Brit. J. Cancer*, **11**, 433 (1957).

5 per cent solution. A more concentrated (20 per cent) solution caused death from systemic toxicity before tumors were produced.

By using reagent grade, USP, and purified phenol, it was shown that this action was due to phenol itself, not to a contaminant.

Certain inferences concerning the chemical structure required for tumor formation certainly may be made from the results obtained with the derivatives of phenol.

Monohalogenated phenols, or methylphenols, or dimethylphenols, were as potent in promoting papillomas as phenol itself (with the exception of 2,6-dimethylphenol, which was inactive). The addition of a nitro, carbonyl, carboxyl, or second phenolic group essentially eliminated the carcinogenicity, although resorcinol showed some activity.

It appears that there must be at least one unsubstituted position ortho to the phenolic group for papilloma-promoting activity.

There seems to be no correlation between the known biological effects and the carcinogenicity of phenol or its derivatives.

There is no specific evidence of human cancer attributable to phenol or related compounds; however, the carcinogenicity to mice certainly emphasizes the precautions necessary in handling these materials.

PYROCATECHOL, $C_6H_4(OH)_2$ (Catechol, o-Dihydroxybenzene, 1,2-Benzenediol, Pyrocatechin)

1. Source, Uses, and Industrial Exposures

Pyrocatechol may be obtained by the fusion of o-phenolsulfonic acid with alkali, by heating o-chlorophenol with a solution of sodium hydroxide at 200°C. in an autoclave, or by cleavage of the methyl ether group of guaiacol (obtained from beechwood tar) with hydriodic acid.[87]

Pyrocatechol is used for various purposes, but particularly as an antioxidant[88-90] in the rubber, chemical, photographic, dye, fat,[91] and oil[92] industries. It is also employed in cosmetics[93] and in some pharmaceuticals.

Cases of industrial or accidental poisoning have been rare.

2. Physical and Chemical Properties

Physical state: colorless, crystalline solid; sublimes readily; volatile in steam.

[87] R. Q. Brewster, Organic Chemistry. Prentice-Hall, New York, 1948.
[88] J. Lavollay and J. L. Parrot, Compt. rend., 215, 496 (1942).
[89] S. Manskaya and Emelyanova, Biokhimiya, 5, 432 (1940).
[90] G. T. Martin et al., Am. J. Physiol., 136, 66 (1942).
[91] C. H. Lea, J. Soc. Chem. Ind. London, 63, 107 (1944).
[92] K. Biltz and W. Simon, Monatschr. Textil Ind., 56, 195 (1941).
[93] Cosmetics and Applied Preparations. Bur. Investigation, Am. Med. Assoc., Chicago,

Molecular weight: 110.11
Specific gravity: 1.344 (4°C.)
Melting point: 105°C.
Boiling point: 245.6°C. (decomposes at 240 to 245°C.)[94]
Vapor density: 3.79 (air $= 1$)
Soluble in water, alcohol, and ether
Flash point: (closed cup) 127.2°C. (261°F.)
1 mg./liter $\backsimeq$ 222.3 p.p.m. and 1 p.p.m. $\backsimeq$ 0.00450 mg./liter at 25°C., 760 mm. Hg

3. *Determination in the Atmosphere*

Apparently methods for the determination of catechol in air have not been developed.

The methods of Baernstein[23] or Tompsett[95] can be used to determine pyrocatechol in urine and in other biological materials.

4. *Physiological Response*

Phenol-like signs of illness are induced in experimental animals given toxic or lethal doses. Unlike phenol, large doses of pyrocatechol can cause a predominant depression of the central nervous system and a prolonged rise of blood pressure.[96,97] The repeated absorption of sublethal doses by animals may also induce methemoglobinemia, leukopenia, and anemia.

Pyrocatechol is more toxic than phenol. The *Approximate Lethal Oral Dosages* (in grams per kilogram) for various experimental animals are: dog, 0.3;[98] rabbit, 0.2;[99] cat, 0.1; and guinea pig, 0.16.[100] Most rats and guinea pigs die when given a single subcutaneous injection of about 0.22 g./kg., while the lethal intravenous dose for dogs is about 0.04 g./kg.[101]

Pathology.   Dietering[102] reported degenerative changes in the tubuli of the kidney, with red blood cells and fibrin clots in the lumina.

Absorption, Excretion, and Metabolism.   Catechol is readily absorbed from the gastroenteric tract and through the intact skin of mice.[103] It seems quite

[94] *International Critical Tables of Numerical Data, Physics, Chemistry and Technology,* Vol. 7. McGraw-Hill, New York, 1930.

[95] S. L. Tompsett, *J. Pharm. Pharmacol.,* **10**, 20 (1958).

[96] G. Barger and H. H. Dale, *J. Physiol.,* **41**, 19 (1910).

[97] C. H. H. Harald, M. Nierenstein, and H. E. Roaf, *J. Physiol.,* **41**, 308 (1910).

[98] G. Colasanti and I. Moscatelli, *Boll. ed. atti accad. med. Roma,* (1887–88).

[99] F. Heyroth and E. Largent, personal communication.

[100] A. Masing, Inaugural Dissertation, Dorpat, 1882.

[101] W. Gibbs and H. A. Hare, *Dubois Arch. Physiol.,* **1890** 352.

[102] H. Dietering, *Arch. exptl. Pathol. Pharmakol.,* **188**, 493 (1938).

[103] F. Sander, Inaugural Dissertation, Koln, 1933; cited from W. F. von Oettingen, *Natl. Inst. Health Bull.* No. **190** (1949).

probable that this compound can be absorbed from the lungs, but substantiating data are not available.

After absorption, part of the catechol is oxidized with polyphenol oxidase to o-benzoquinone.[104] Another fraction conjugates in the body with hexuronic, sulfuric, and other acids. A small amount is excreted in the urine as "free" pyrocatechol. The "conjugated" fraction hydrolyzes easily in the urine[23,105] with the liberation of the "free" compound; this is oxidized with the formation of dark-colored substances that impart to the urine a "smoky" appearance.

Mode of Action. Apparently, pyrocatechol acts by mechanisms similar to those reported for phenol. The rise of blood pressure appears to be due to peripheral vasoconstriction.[97]

Cause of Death. Death apparently is initiated by respiratory failure.

Effects Observed in Man. Cases of industrial or accidental poisoning have been rare. Contact with the skin has been known to cause an eczematous dermatitis, while absorption through the skin in a few instances has resulted in symptoms of illness resembling closely those induced by phenol, except for certain central effects (convulsions), which were more marked.[106]

5. Hygienic Standard of Permissible Exposure

Standards of permissible exposures or a maximum allowable concentration of catechol vapors in air have not been established.

6. Flammability

Extinguish with water, carbon dioxide, and dry chemicals.

7. Odor and Warning Properties

Faint, characteristic odor.

RESORCINOL, $C_6H_4(OH)_2$ (m-Dihydroxybenzene, Resorcin)

1. Source, Uses, and Industrial Exposures

Resorcinol is usually prepared by fusing sodium meta-benzenedisulfonate with sodium hydroxide. It can be made in several other ways, by destructive distillation of brazilin or by the fusion of galbanum, ammoniac, sagapenum, asafetida, or acroides with caustic potash.[107]

It is used in tanning, in photography, and in the manufacture of explosives,

---

[104] W. G. C. Forsyth and V. C. Quainel, *Biochim. et Biophys. Acta,* **25,** 155 (1957).

[105] F. Vorsatz, *Collegium,* **1942** 424; *Chem. Abstr.* **37,** 6927 (1943).

[106] A. R. Cushny, C. W. Edmunds, and J. A. Gunn, *Pharmacology and Therapeutics.* Lea & Febiger, Philadelphia, 1940.

[107] E. F. Cook and E. W. Martin, *Remington's Practice of Pharmacy.* Mack, Easton, Pa., 1948.

dyes, cosmetics, organic chemicals, and antiseptics.[108,109] It has also been suggested as an aerial bactericide.[110,111]

Industrial exposures are rather rare, but could occur in any industry utilizing the compound.

2. Physical and Chemical Properties

Physical state: white, needle-shaped crystals or rhombic tablets and pyramids, which turn pink upon exposure to light and air

Molecular weight: 110.11

Specific gravity: 1.272 (15°C.)

Melting point: 110.7°C.

Boiling point: 276°C.

Vapor density: 3.79 (air = 1)

Per cent in "saturated" air: 2.64% by volume (25.1° C.)

Density of "saturated" air: 1.0739 (air = 1)

Soluble in water, alcohol, glycerol, and ether[112]

Flash point: (closed cup) 127.2°C. (261°F.)

1 mg./liter $\backsimeq$ 222.3 p.p.m. and 1 p.p.m. $\backsimeq$ 0.00450 mg./liter at 25°C., 760 mm. Hg.

3. Determination in the Atmosphere

Analytical methods for the estimation of resorcinol[113-115] have not been adapted to its determination in air or in body fluids.

4. Physiological Response

The signs of intoxication resemble those induced by phenol, but the antipyretic action of resorcinol is more marked. Resorcinol is less toxic than pyrocatechol. The *Approximate Lethal Oral Dose* of resorcinol, in aqueous solution, for rabbits is 0.75 g./kg.[99] and for rats and guinea pigs 0.37 g./kg.[116] Dogs were killed by intravenous injections of 0.7 to 1.0 g./kg.[101]

Pathology. Very little information is available concerning the pathological changes in animals. Pathology reported for man includes marked siderosis of the

[108] P. A. Ark, *Phytopathology,* **30,** 1 (1940).

[109] A. Laurie, *Agr. News Letter* (Public Relations Dept., E. I. du Pont de Nemours & Co.), **9,** 22 (1941).

[110] C. C. Twort and A. H. Baker, *J. Hyg.,* **42,** 266 (1942).

[111] A. E. Williamson and H. B. Gotaas, *Ind. Med.,* **11,** 40 (1942).

[112] *Merck Index,* Merck, Rahway, N. J., 1940.

[113] F. M. Garfield, *J. Assoc. Offic. Agr. Chemists,* **25,** 897 (1942).

[114] A. O. Songina, *Chem. Abstr.,* **37,** 53 (1943).

[115] P. Torti, *Boll. chim. farm.,* **81,** 28 (1942); *Chem. Abstr.,* **38,** 3220 (1944).

[116] L. Brieger, *Z. physiol. Chem.,* **2,** 241 (1878–79), **8,** 134 (1879).

spleen[117] and marked tubular injury in the kidney. Becker[118] reported fatty changes and anemia of the liver, degenerative changes in the kidney, fatty changes of the heart muscle, moderate enlargement and pigmentations of the spleen, and edema and emphysema of the lungs.

Absorption, Excretion, and Metabolism. Resorcinol is readily absorbed from the gastroenteric tract and, in a suitable solvent, is readily absorbed through the human skin. The compound is excreted in the urine, as are other phenols, in a free state and conjugated with hexuronic, sulfuric, or other acids.

Mode of Action. Resorcinol causes local damage by direct irritation of the tissues. It causes systemic effects in much the same manner as pyrocatechol.

Cause of Death. Death apparently is initiated by respiratory failure.

Effects Observed in Man. The cutaneous application of solutions or salves[119] containing from 3 to 25 per cent of this compound may result in local hyperemia, itching, dermatitis, edema, corrosion, and the loss of the superficial layers of the skin. These changes, if they are severe, may be associated with some or all of the following effects: enlargement of regional lymph glands, restlessness, methemoglobinemia, cyanosis, convulsions, tachycardia, dyspnea, and death.[7,106,120,121] Ingestion of resorcinol induces similar signs and symptoms. Thus a child, after accidentally swallowing 4 g., complained of dizziness and somnolence. The ingestion of 8 g., in another case, induced an almost immediate hypothermia, fall in blood pressure, and decrease in the rate of respiration, with tremors, icterus, and hemoglobinuria. Recovery was noted 2 hours after the poisoning.[7] Other cases are on record in which similar doses apparently had no ill effects.[122]

5. Hygienic Standards of Permissible Exposure

Because of the potential toxicity of this compound, it should be handled and used with caution. Safe concentrations have not been established for its use as an aerial bactericide.

6. Flammability

Extinguish with water, carbon dioxide, or dry chemicals,

7. Odor and Warning Properties

Resorcinol has a faint, characteristic odor and a sweetish followed by a bitter taste.

[117] J. Brudzinski, *Therap. Monatsh.*, **13**, 517 (1899); cited from W. F. von Oettingen, *Nalt. Inst. Health Bull.* No. **190** (1949).

[118] J. Becker, *Samml. Vergiftungsfällen*, **4**, 7 (1933); cited from W. F. von Oettingen, *Natl. Inst. Health Bull.* No. **190** (1949).

[119] E. A. Strakosch, *Arch. Dermatol. u. Syphilis*, **48**, 384 (1943).

[120] M. Haenelt, *Münch. med. Wochschr.*, **72**, 386 (1925).

[121] L. Schwartz and L. Tulipan, *Occupational Diseases of the Skin.* Lea & Febiger, Philadelphia, 1939.

[122] V. Surbeck, *Deut. Arch. klin. Med.*, **32**, 515 (1883).

8. Treatment of Poisoning

Resorcinol should be thoroughly washed from the skin with ample volumes of water. If it is swallowed, the patient should be forced to vomit or gastric lavage should be performed. Saline cathartics will remove unabsorbed portions from the enteric tract. Additional treatment is symptomatic. If a patient has malfunction of the kidneys, he should be observed with extra care.

HYDROQUINONE, $C_6H_4(OH)_2$ (p-Dihydroxybenzene, p-Hydroxyphenol)

1. Source, Uses, and Industrial Exposures

Hydroquinone is usually prepared by the reduction of quinone with sulfurous acid.

Its industrial uses are similar to those of pyrocatechol and resorcinol.[123-133] It is a reducing agent used extensively as a photographic developer and as an antioxidant or stabilizer of certain materials which polymerize in the presence of oxidizing agents.[87] Many derivatives of hydroquinone have been used as bacteriostatic agents. Certain hydroquinone derivatives, specifically 2,5-bis(ethyleneimino)hydroquinone, have been reported to be good antimitotic and tumor-inhibiting agents.[134,135]

Harmful effects are likely to occur to workers who inhale the dust or vapors or whose skin or eyes come in contact with the dust or vapors for prolonged periods.

2. Physical and Chemical Properties

Physical state: crystallizes from water in hexagonal prisms
Molecular weight: 110.11
Specific gravity: 1.332 (15°C.)
Melting point: 172°C.
Boiling point: 285°C. (730 mm. Hg)
Vapor density: 3.81 (air = 1)

[123] U. P. Basu, *Ann. Biochem. and Exptl. Med. Calcutta*, **1**, 165 (1941); *Chem. Abstr.*, **37**, 3559 (1943).

[124] C. Golumbic and H. A. Mattill, *J. Am. Chem. Soc.*, **63**, 1269 (1941).

[125] C. E. Hartt, *Hawaiian Sugar Planter's Assn. Printed Reports Ann. Mtg. Rept. Comm. in Charge Expt. Sta.*, **62**, 88 (1942); cited from *Chem. Abstr.*, **37**, 6150 (1943).

[126] H. A. Hollender and P. H. Tracy, *J. Dairy Sci.*, **25**, 249 (1942).

[127] A. Overman, *J. Biol. Chem.*, **142**, 441 (1942).

[128] W. T. Sumerford, A. B. Huff, and O. K. Coleman, *J. Am. Pharm. Assoc., Sci. Ed.*, **33**, 150 (1944).

[129] H. Sullmann, *Helv. Chim. Acta*, **26**, 1114 (1943).

[130] R. Waite, *J. Dairy Research*, **12**, 178 (1941).

[131] K. Weber, *Radiologica*, **1**, 223 (1937).

[132] A. Weissberger, D. S. Thomas, and J. E. LuValle, *J. Am. Chem. Soc.*, **65**, 1489 (1943).

[133] K. T. Williams, E. Bickoff, and B. Lowrimore, *Oil & Soap*, **21**, 161 (1944).

[134] A. Marxer, *Experientia*, **11**, 184 (1955).

[135] P. Loustalot, B. Schär, and R. Meier, *Experientia*, **11**, 186 (1955).

Vapor pressure: 4.0 mm. Hg (150°C.)[152]
Density of "saturated" air: 1.011 (150°C.) (air = 1)
Solubility in water: 7 per cent at 25°C.
Soluble in hot water, alcohol, and ether
Flash point: (closed cup) 165°C. (329°F.)
1 mg./liter ≈ 222.3 p.p.m. and 1 p.p.m. ≈ 0.00450 mg./liter at 25°C., 760 mm. Hg

3. Determination in the Atmosphere

Air contaminated with hydroquinone in concentrations of 0.1 to 2.0 μg. per liter can be collected by passing the air through isopropyl alcohol in a midget impinger. A color is produced by the addition of phloroglucinol in potassium hydroxide. Spectrophotometric determination of the color at 520 mμ gives quantitative results.[136]

Hydroquinone in biological materials can be determined by the methods reported by Baernstein[23] or by others.[137-140]

4. Physiological Response

A toxic concentration of hydroquinone in the tissues of experimental animals induces signs of illness which resemble, in many respects, those induced by phenol. In acute poisoning, there is an increased motor activity, hypersensitivity to external stimuli, hyperactive reflexes, dyspnea, and cyanosis. These signs are followed by marked clonic convulsions and later by complete exhaustion, hypothermia, paralysis, loss of reflexes, coma, and death. Marked formation of methemoglobin occurs. Subacute poisoning may be characterized by hemolytic icterus, anemia, leukocytosis, reticulocytosis, increased cell fragility, hypoglycemia, depigmentation of fur, and marked cachexia.

Hydroquinone is more toxic than phenol. The *Approximate Lethal Oral Dosages* of this compound in aqueous solution are 0.2 g./kg. for the rabbit[106] and 0.08 g./kg. for the cat.[141] The lethal intravenous dose for the dog is about 0.09 g./kg.,[101] and the lethal subcutaneous dose for mice is about 0.16 g./kg.[142]

Pathology. Very little information concerning the pathology in animals is available. The changes reported for man include hyperemia of abdominal organs that are rich in pigments, pathological changes in the liver and kidney, and bronchopneumonia in the lungs.[142,143]

[136] F. L. Oglesby and J. H. Sterner, *Ind. Med.*, **15**, 483 (1946).
[137] W. Deichmann, *J. Lab. Clin. Med.*, **28**, 770 (1943).
[138] E. Ergriwe, *Z. anal. Chem.*, **125**, 241 (1943).
[139] J. G. Stott, *J. Soc. Motion Picture Engrs.*, **39**, 37 (1942).
[140] J. F. Treon and W. Crutchfield, Jr., *Ind. Eng. Chem.*, **14**, 119 (1942).
[141] H. Oettel, *Arch. exptl. Pathol. Pharmakol.*, **183**, 319 (1936).
[142] S. Busatto, *Deut. Z. ges. gerichtl. Med.*, **31**, 285 (1939); *Arch. antropol. criminale psichiat. e med. legale*, **60**, 620 (1940).
[143] I. Zeidman and R. Deutl, *Am. J. Med. Sci.*, **210**, 328 (1945).

Absorption, Excretion, and Metabolism. Hydroquinone is absorbed more rapidly than phenol from the gastroenteric tract (rats).[144] The compound may be absorbed through the skin, but this is questionable.[103] Analysis of the blood or urine[23,137-140] will demonstrate the absorption of this compound.

Little is known concerning the metabolism of hydroquinone except that it is oxidized to the more toxic quinone.[4,142,145-149] (Hydroquinone will also oxidize to quinone when exposed to light or air.) Hydroquinone and quinone are partially excreted as such, and in conjugation with hexuronic, sulfuric, and other acids.

Mode of Action. Hydroquinone appears to act on the body by first being oxidized to quinone.

Cause of Death. Death apparently is initiated by a type of respiratory failure. The oxygen-carrying capacity of the blood is so markedly reduced that anoxia results.

Effects Observed in Man. Hydroquinone causes signs and symptoms of an illness resembling that induced by pyrocatechol or resorcinol. Ingestion of 1 g. by an adult (smaller quantity by a child) may cause tinnitus, nausea, dizziness, a sense of suffocation, an increased rate of respiration, vomiting, pallor, muscular twitchings, headache, dyspnea, cyanosis, delirium, and collapse. The urine is usually green or brownish green in color and continues to darken on standing.[7] Fatal cases have been reported after ingestion of 5 to 12 g.[150] Certain "health teas" have been prepared from leaves of blueberry, red whortleberry, cranberry, or bearberry. Their ingestion should be avoided because the leaves may contain hydroquinone in a concentration (sometimes exceeding 1 per cent) capable of producing irritation of the intestinal mucosa and systemic poisoning.[7] Cases of dermatitis have resulted from skin contact with hydroquinone. Lapin[151] reports such findings after application of an "antiseptic oil" which apparently contained traces of hydroquinone as an antioxidant.

Anderson,[152] Velhagen,[153] Sterner,[154] and others[155] have reported cases of keratitis and discoloration of the conjunctiva among men exposed to concentrations ranging from 10 to 30 mg. of vapor or dust of hydroquinone per cubic meter of air. Anderson and Oglesby[156] recently reported corneal changes, particularly alteration

[144] M. L. Mareque and A. D. Marenzi, *Compt. rend. soc. biol.*, **127**, 153 (1938).

[145] R. Honorato and R. E. Ortuzar, *Rev. med. y aliment. Santiago, Chile*, **5**, 223 (1943).

[146] R. Labes, *Arch. exptl. Pathol. Pharmakol.*, **146**, 44 (1929), **152**, 111 (1930).

[147] P. Marquardt, *Arch. exptl. Pathol. Pharmakol.*, **201**, 234 (1943).

[148] S. Sato and M. Ugai, *Okayama-Igakkai-Zasshi*, **51**, 829 (1939).

[149] E. N. Speranskaya-Stepanova, *J. Physiol. U.S.S.R.*, **29**, 334 (1940).

[150] I. Zedman and R. Deute, *Am. J. Med. Sci.*, **210**, 328 (1945).

[151] J. H. Lapin, *Am. J. Diseases Children*, **63**, 89 (1942).

[152] B. Anderson, *A.M.A. Arch. Ophthalmol.*, **38**, 812 (1947).

[153] K. Velhagen, cited from L. Schwartz and L. Tulipan, *Occupational Diseases of the Skin*, Lea & Febiger, Philadelphia, 1939.

[154] J. H. Sterner, personal communication.

[155] A. I. Dashevskiy and F. F. Marmorshteyn, *Vestnik Oftalmol.*, **19**, Nos. 1-2, 50 (1941).

[156] B. Anderson and F. Oglesby, *A.M.A. Arch. Ophthalmol.*, **59**, 495 (1958).

of the curvature, which were manifested long after exposure, and after the stain and pigment had disappeared.

5. Hygienic Standards of Permissible Exposure

Sterner[154] recommends that the concentration of dust of hydroquinone in air be kept below 5 mg./cubic meter, which is approximately 1 p.p.m. The A.C.G.I.H. threshold limit for 1961 is 2 mg./cu. meter

6. Odor and Warning Properties

Sweet taste.

7. Safe Handling Procedures

In view of its potential toxicity, hydroquinone should be handled with caution. Contact with the eyes or inhalation of its dust or vapors, liberated particularly at elevated temperatures, must be avoided. Although quinone vapors are considerably more toxic than hydroquinone dust, one circumstance which makes a greater hazard from a dust such as hydroquinone is that the extent of eye injury depends largely upon the concentration of the compound in the fluids bathing the eye. The irritating vapors of quinone become diluted and washed away with the tears, while dust particles of hydroquinone are apt to remain for a longer period of time and, in dissolving, produce localized areas of high concentration.

QUINONE, OC_6H_4O (p-Benzoquinone, 1,4-Benzoquinone)

1. Source, Uses, and Industrial Exposures

Quinone can be prepared by oxidation starting with aniline or by the reduction of hydroquinone with bromic acid.[87]

The compound has found wide application in the dye, textile, chemical, tanning, and cosmetic industries primarily because of its ability to transform certain nitrogen-containing compounds into a variety of colored substances.

Severe local damage to the skin and mucous membranes may occur following contact with solid quinone, solutions of quinone, and with quinone vapors condensing upon exposed parts of the body (particularly moist surfaces).

2. Physical and Chemical Properties

Physical state: large, yellow, monoclinic prisms
Molecular weight: 108.09
Specific gravity: 1.318 (20°C.)
Melting point: 112.9°C.
Boiling point: (sublimes)
Vapor pressure: (considerable; sublimes readily upon gentle heating)
Soluble in hot water, alcohol, ether, and alkalies
1 mg./liter ⇌ 226.4 p.p.m. and 1 p.p.m. ⇌ 0.00442 mg./liter at 25°C., 760 mm. Hg

3. Determination in the Atmosphere

Analytical methods for the estimation of quinone have not been adapted to air or to biological materials.[157-162]

4. Physiological Response

Absorption of large doses of quinone from the gastroenteric tract or from subcutaneous tissues of animals induced local changes, crying, clonic convulsions, respiratory difficulties, drop of blood pressure, and death by paralysis of the medullary centers. Asphyxia appears to play an important role in the terminal picture, both because of pulmonary damage resulting from excretion of quinone into the alveoli and because of certain not too well-defined effects of quinone upon hemoglobin.[163] The urine of severely poisoned animals may contain protein, blood, casts, and "free" and "conjugated" hydroquinone.

Pathology. See Hydroquinone.

Absorption, Excretion, and Metabolism. Quinone is readily absorbed from the gastroenteric tract and subcutaneous tissues. It is partially excreted unchanged; but the bulk is eliminated in conjugation with hexuronic, sulfuric, and other acids.

Mode of Action. Quinone apparently has a direct effect on the medulla and the oxygen-carrying capacity of the blood.

Cause of Death. Paralysis of the medullary centers is caused by lethal doses of quinone.

Effects Observed in Man. The local changes may include discoloration, severe irritation, erythema, swelling, and the formation of papules and vesicles. Prolonged contact may lead to necrosis. Vapors condensing upon the eyes are capable of inducing serious disturbances of vision.[152,164-168] According to Sterner, the injury usually extends through the entire layer of the conjunctiva and is characterized by a deposit of pigment. The staining, varying from diffuse brown

[157] F. P. Dann, *Proc. Soc. Exptl. Biol. Med.*, **42**, 663 (1939).

[158] L. Rosenthaler, *Pharm. Acta Helv.*, **14**, 93 (1939).

[159] J. Rzymkowski, *Z. Elektrochem.*, **31**, 371 (1925).

[160] E. Schulek and P. Rozsa, *Magyar Chem. Folyóirat*, **46**, 155 (1940); *Magyar Chem. Folyóirat*, **47**, 75 (1941); cited from *Chem. Abstr.*, **35**, 5821, 7993 (1941).

[161] B. Singh and S. Singh, *J. Indian Chem. Soc.*, **16**, 346 (1939); *Chem. Abstr.*, **34**, 3204 (1940).

[162] N. R. Trenner and F. A. Bacher, *J. Biol. Chem.*, **137**, 745 (1941).

[163] S. Liu, *Biochem. Z.*, **195**, 248 (1928).

[164] A. Hamilton, *Industrial Poisons in the United States.* Macmillan, New York, 1925.

[165] W. C. Hueper, *Occupational Tumors and Allied Diseases.* Thomas, Springfield, Ill., 1942.

[166] R. L. Mayer, *Klin. Wochschr.*, **7**, 1958 (1928); *Arch. Dermat. u. Syphilis*, **158**, 266 (1929).

[167] N. Takizawa, *Proc. Imp. Acad. Tokyo*, **16**, 309 (1940).

[168] R. P. White, *The Dermatergoses or Occupational Affections of the Skin.* 4th ed., Lewis, London, 1934.

to globules of brownish black, is located primarily in the zones extending from the canthi medially to the edges of the cornea. All layers of the cornea are involved in the injury, with a resultant discoloration that may be white and opaque or brownish-green and translucent. Alteration of the cornea can occur after the pigment has disappeared.[156] Ulceration of the cornea has resulted from one brief exposure to a high concentration of the vapor of quinone, as well as from repeated exposures to moderately high concentrations. Following discontinuation of exposure, recovery occurred promptly and spontaneously, and appeared to be nearly complete. There were no systemic effects or changes in the composition of the blood or urine.[154] Cases of oral poisoning apparently have not been reported.[4]

5. *Hygienic Standards of Permissible Exposure*

Sterner recommends that the concentration of quinone vapor in air be kept below 0.1 p.p.m. (0.44 mg./cu. meter).[154] The A.C.G.I.H. limit for 1961 is 0.1 p.p.m. (0.4 mg./cu. meter).

6. *Odor and Warning Properties*

Quinone has a pungent odor. The vapors are irritating enough to cause sneezing.

7. *Prevention of Poisoning*

Control of the exposure is largely a matter of adequate ventilation.

PYROGALLOL, $C_6H_3(OH)_3$ (Pyrogallic Acid, Pyro, 1,2,3-Trihydroxybenzene)

1. *Source, Uses, and Industrial Exposures*

Pyrogallol is prepared by heating dried gallic acid at about 200°C. Carbon dioxide splits off leaving pyrogallol.[107]

Its usefulness in various industries[169-171] is based primarily upon its property of being easily oxidized in alkaline solutions (even by atmospheric oxygen) so that such solutions become potent reducing agents. It is used specifically as a developer in photography and for maintaining anaerobic conditions for bacterial growth.

2. *Physical and Chemical Properties*

Physical state: white, or nearly white, needle- or leaf-shaped crystals or crystalline powder
Molecular weight: 126.11
Specific gravity: 1.453 (4°C.)
Melting point: 133°C.

[169] F. Bergel, *Chem. & Ind. London*, **14**, 127 (1944).
[170] W. W. Scheumann and J. H. Haslam, *Ind. Eng. Chem.*, **34**, 485 (1942).
[171] K. Ziegler and P. Herte, *Ann.*, **551**, 127 (1942).

Boiling point: 309°C. (decomposes at 293°C.)
Soluble in water (1:2), alcohol (1:1.5), and ether (1:2) at 25°C.[112]
1 mg./liter ⇌ 194 p.p.m. and 1 p.p.m. ⇌ 0.00515 mg./liter at 25°C., 760 mm. Hg

3. Determination in the Atmosphere

Specific methods for the determination of pyrogallol in air are not available. The content of pyrogallol in the urine can be determined by the method described by Tompsett.[27]

4. Physiological Response

Signs of intoxication include vomiting, hypothermia, fine tremors, weakness, muscular incoordination, diarrhea, loss of reflexes, coma, and asphyxia.[7]

Because of its marked reducing action, pyrogallol has a tremendous affinity for the oxygen of the blood. Heyroth and Largent[99] observed that the intravenous injection of 0.3 g./kg. into a rabbit provided a sufficient quantity of pyrogallol to unite with all of the oxygen of the blood, thereby causing the death of the animal. There was extensive destruction and fragmentation of the erythrocytes.

The urine of poisoned animals may contain casts, glucose, hemoglobin, methemoglobin, urobilin, and other compounds that cause discoloration.

Repeated absorptions of toxic but sublethal concentrations into the tissues of animals have been found to cause severe anemia, icterus, nephritis, and uremia.

The *Approximate Lethal Dosages* of pyrogallol in aqueous solution for various animal species, under varying conditions of administration, are as follows: rabbit, 1.1 g./kg. (orally);[99] rabbit or guinea pig, 1.0 g./kg. (subcutaneously);[4] dog or cat, 0.35 g./kg. (subcutaneously); and dog 0,09 g./kg. (intravenously).[101]

Pathology. Pathological changes in animals caused by pyrogallol include edema and hyperemia of the lungs, and moderate fatty degeneration, round cell infiltration and necrosis of the liver. The kidney may show hyperemia, necrosis of the epithelium, granular pigmentation, and glomerular nephritis.[172,173] The nuclei of striated muscle disappear, the striation of the muscle is lost, and the muscle plasma swells, coagulates, and decomposes. In the heart there is separation of fibers of the myocardium, interfibrillary hemorrhages, infiltration of the endocardium, lesions of the endothelium, and fibrinous deposits in the valves.[174] Changes of the bone marrow and myeloid changes in the spleen were noted after chronic administration of this compound.[175]

Absorption, Excretion, and Metabolism. Pyrogallol is readily absorbed from the gastroenteric tract and from parenteral sites of injection. Little is absorbed

---

[172] E. Bauman and E. Herter, Z. physiol. Chem., 1, 244 (1877).

[173] M. A. Afanassiew, Arch. pathol. Anat. u. Physiol. Virchow's, 98, 472 (1884).

[174] A. Natanson, Inaugural Dissertation, Dorpat, 1888; cited from A. Ellinger, Heffter's Handbuch d. experimentellen Pharmakologie Vol. I, 1. Springer, Berlin, 1923, p. 891.

[175] A. von Domarus, Arch. exptl. Pathol. Pharmakol., 58, 335 (1908).

through the intact skin. The bulk of absorbed pyrogallol is readily conjugated with hexuronic, sulfuric, or other acids and excreted within 24 hours via the kidneys.[176] A fraction is excreted unchanged.

Mode of Action. Most of its pharmacological effects can be traced directly to its potent reducing action.

Cause of Death. Death is initiated by respiratory failure. Pyrogallol ties up the oxygen in the blood.

Effects Observed in Man. Cases of human poisoning have not been frequent. Cases reported in the older literature[4] include one man who ingested an aqueous solution containing 8 g. of pyrogallol and who recovered after suffering an acute intoxication; another, who ingested 15 g. of this compound, died despite prompt vomiting. When applied upon the human skin in the form of a salve, it can cause local discoloration, irritation, eczema, and even death.[177] Repeated contact with the skin can cause sensitization.[178] The symptoms observed in acute intoxications in man resemble closely the signs of illness displayed by experimental animals.

5. Hygienic Standard of Permissible Exposure

None established for pyrogallol.

6. Odor and Warning Properties

Pyrogallol is practically odorless.

7. Prevention and Treatment of Poisoning

Avoid skin contact. Wash affected areas thoroughly with water. If pyrogallol has been swallowed, induce vomiting or administer gastric lavage. Further treatment is symptomatic.

Phloroglucinol (1,3,5-trihydroxybenzene) and *benzenetriol* (1,2,4-trihydroxybenzene) are of little industrial or hygienic importance.

CRESOL, $CH_3C_6H_4OH$ (Cresylic Acid, Cresol U.S.P., Tricresol, Methylphenol)

Pure cresol is a mixture of ortho-, meta-, and para-isomers (*o*-cresol, *m*-cresol, and *p*-cresol). Crude cresol (commercial cresol) is a mixture of aromatic compounds containing about 20 per cent of *o*-cresol, 40 per cent of *m*-cresol, and 30 per cent of *p*-cresol with small amounts of phenol and xylenols.

1. Source, Uses, and Industrial Exposures

Crude cresol is obtained by distilling "gray phenic acid" at a temperature within the ranges of about 180 to 205°C. *o*-Cresol may be separated from the

---

[176] D. Vitali, *Jahresber, Tierchem.,* **29,** 827 (1894); cited from W. F. von Oettingen. *Natl. Inst. Health Bull.* No. **190** (1949).

[177] A. Neisser, Z. *klin. Med.,* **1,** 88 (1880).

[178] E. Zurhelle and S. K. de Boer, *Arch. Dermat. u. Syphilis,* **183,** 130 (1942).

crude or purified mixture by repeated fractional distillation and crystallization or by double distillation *in vacuo*. *m*- and *p*-Cresols are separated by treating the mixture with sulfuric acid. This yields the crystallized disulfonic acid of *p*-cresol, and *p*-cresol upon hydrolysis. The liquid fraction, when hydrolyzed with steam at 130°C., yields *m*-cresol, which may be further purified by double distillation *in vacuo*. Each of the cresols can be prepared synthetically by diazotization of the specific toluidine or by fusion of the corresponding toluenesulfonic acid with sodium hydroxide.

The cresols have found wide application in synthetic resin, explosive, petroleum,[179] photographic, paint,[180,181] and agricultural[182,183] industries. They have been used for years as antiseptics, disinfectants,[184-189] and insecticides.[190,191]

The *o*-, *m*-, and *p*-cresols are marketed individually ("practical grade," purity 98 or 99 per cent), or in a mixture as cresylic acid, or as crude cresol. Preparations containing cresols include: (*a*) saponated (or compound) cresol solution, which is a clear 50 per cent solution of cresol, USP, in a mixture of vegetable oil, sodium and/or potassium hydroxide, and water; (*b*) Lysol, a proprietary preparation, which is essentially the same as saponated cresol solution, but its composition is not constant and it may contain various coal-tar derivatives, acids, bases, or salts; (*c*) Carbolineum, the composition of which varies, but usually containing naphthalene and pyridine and sometimes acridine and anthracene, in addition to the cresols and phenol; (*d*) Saprol, a mixture of crude cresols dissolved in petroleum solvents, which upon addition of water yields a turbid solution; (*e*) Creoline, a water-soluble mixture, prepared by the addition of sulfuric acid to mixtures of cresol and phenol.

Industrial exposure has occurred in industries utilizing cresol. Fatalities can occur from prolonged and/or extensive skin contact. Because of its low vapor pressure and disagreeable odor, cresol usually does not present an acute inhalation hazard. Data are not available concerning vapor concentrations that are toxic to man.

[179] T. Kennedy, *J. Inst. Petrol.*, **27**, No. 207, 15 (1941).

[180] F. Moll, *Farben-Chem.*, **11**, 101 (1940).

[181] G. H. Young, G. W. Gerhardt, W. K. Schneider, *Ind. Eng. Chem.*, **35**, 432 (1943).

[182] M. Kondo and Y. Kasahara, *Ber. Ōhara Inst. landwirtsch. Forsch. Kurashiki Japan*, **8**, 325 (1941).

[183] T. Manley *et al.*, *Report of the Research Com., West Va. Gladiolus Soc.*, Gladiolus Suppl., **6**, No. 1, 10 (1942).

[184] A. Bos, *Tijdschr. Diergeneesk.*, **70**, 55 (1943).

[185] C. M. Brewer, *J. Assoc. Offic. Agr. Chemists*, **23**, 557 (1940).

[186] H. J. Henk, *Deut. Parfüm. Ztg.*, **27**, 120 (1941).

[187] G. Klust, *Chem. Zentr.*, **1**, 2468 (1941).

[188] R. Puget, *Ann. hyg. publ. ind. et sociale*, **18**, 319 (1940).

[189] M. Waldhecker, *Münch. med. Wochschr.*, **88**, 949 (1941).

[190] P. H. Berry, *Pharm. J.*, **148**, 112 (1942).

[191] A. J. Salle and H. L. Guest, *Proc. Soc. Exptl. Biol. Med.*, **55**, 26 (1944).

2. Physical and Chemical Properties

Physical state: crude cresol, colorless liquid, turns brown upon exposure to air or light; o-cresol, colorless crystalline compound; m-cresol, yellowish liquid; p-cresol, white crystalline compounds

Readily soluble or miscible with organic solvents, vegetable oils, or ether and alcohol (o-cresol at 30°C., p-cresol at 36°C.)

Other pertinent physicochemical properties are presented in Table 3.

1 mg./liter ⋍ 226.4 p.p.m. and 1 p.p.m. ⋍ 0.00442 mg./liter at 25°C., 760 mm. Hg

TABLE 3
Physicochemical Properties of Cresols

	Commercial cresol	o-Cresol	m-Cresol	p-Cresol
Molecular weight	108.13	108.13	108.13	108.13
Specific gravity	1.030–1.045	1.048	1.034	1.035
Melting point, °C.	Liquid	30.9	12.0	34.8
Boiling point, °C.	185–205	191	202.7	201.9
Vapor density	3.72	3.72	3.72	3.72
Vapor pressure, mm. Hg at 25°C.	—	0.2453	0.1528	0.1080
Refractive index	1.5353^{25}	1.5372^{40}	1.5425^{18}	1.5^{35}
Per cent in "saturated" air at 25°C.	—	0.0323	0.0201	0.0142
Density of "saturated" air at 25°C. (air=1)	—	1.00089	1.00054	1.00039
Solubility in water, %	1.0	2.5	0.5	1.8
Solubility in mineral oil, %	—	2.0	2.5	0.7
Flash point (closed cup), °C. (°F.)	43.5 (110.3)[a]	81 (178)	86 (187)	86 (187)

[a] Closed or open cup not indicated.

3. Determination in the Atmosphere

The amount of cresol in contaminated air can be determined in essentially the same manner as that described for phenol. The cresol is absorbed in dilute alkali and determined colorimetrically with diazotized p-nitroaniline reagent[16] or absorbed in spectrograde alcohol with direct determination by ultraviolet spectrophotometry. The latter method seems to be generally preferred.[192]

4. Physiological Response

The predominant signs of local and systemic intoxication produced by these compounds are very similar to those produced by phenol.[120,193,194] Acute exposures by all routes of absorption may cause muscular weakness, gastroenteric disturbances, severe depression, collapse, and death. While the effects are primarily

[192] D. W. Fassett, personal communication.
[193] J. Hasenbach, Zentr. Chir., 68, 67 (1941).
[194] F. Beran, Nachrbl. deut. Pflanzenschutzdienst Berlin, 20, 33 (1940).

on the central nervous system, edema of the lungs and injury of the kidneys, liver, pancreas, and spleen also may occur. Repeated exposures may result in digestive disturbances, damage to the liver and kidneys, and skin eruptions. Cresol has a marked corrosive action on tissues, producing burns and dermatitis.

Campbell[195] exposed mice to air saturated with vapors of cresylic acid. Single brief exposures did not seem to be harmful, while repeated exposures caused fatalities. Rats were reported to survive 8 hours of inhalation of air substantially saturated with vapors at room temperature.[196]

m-Cresol is generally considered the least toxic; however, reports differ as to whether o- or p-cresol is the more toxic of the two (Table 4). For all practical purposes the three isomers can be considered as having essentially the same degree of toxicity.

TABLE 4

Lethal Dosages of Cresols for Experimental Animals

Species	Route	o-Cresol	m-Cresol	p-Cresol	Ref.
		Dose, g./kg.			
Mouse	Subcut. (dil. aq. suspensions)	0.35	0.45	0.15	Tollens[35]
Rat	Oral (10% soln. in olive oil)	1.35	2.02	1.8	Deichmann and Witherup[37]
Rabbit	Intrav. (0.5% aq. soln.)	0.2	0.28	0.16	Deichmann and Witherup[37]
	Subcut. (dil. aq. suspensions)	0.45	0.5	0.3	Meili[197], Tollens[35]
	Oral (10% soln. in olive oil)	0.8	1.1	1.1	Deichmann and Witherup[37]
Cat	Subcut. (20% aq. suspensions)	0.6	0.15	0.08	Deichmann and Witherup[37]

Pathology. The pathological changes induced by these compounds are similar to those caused by phenol. These include irritation, corrosion, hemorrhages, and destruction of cellular protoplasm of the gastroenteric tract (following oral administration), kidney tubule damage, nodular pneumonia, and congestion of the liver with pallor and necrosis of hepatic cells.[32]

In man[198] it was reported that the lungs showed hyperemia, emphysema, edema, and bronchopneumonia with petechial hemorrhages in the pleura. The liver showed turbidity, inflammatory reactions, and fatty degeneration; the kidney showed parenchymatous and hemorrhagic nephritis; the myocardium was degenerated, and there were small hemorrhages in the epicardium and endocardium.

Absorption, Excretion, and Metabolism. Cresol is absorbed through the skin open wounds, and mucous membranes of the gastroenteric and respiratory tracts. The rate of absorption through the skin depends more upon the size of the area exposed than on the concentration of the material applied.[37]

[195] I. Campbell, *Soap Sanit. Chemicals*, **17**, No. 4, 103 (1941).

[196] H. F. Smyth, Jr., *Am. Ind. Hyg. Assoc. Quart.*, **17**, 2 (1956).

[197] W. Meili, Dissertation, Bern, 1891.

[198] E. Kathe, *Arch. pathol. Anat. u. Physiol. Virchow's*, **185**, 132 (1906).

The major route of excretion of the cresols is with the urine, but considerable amounts may be excreted with bile[32] and traces with the exhaled air.

The information available as to the normal metabolism and the rate of absorption, detoxication, and excretion of o-, m-, and p-cresols indicates that these compounds behave very much like phenol.[37,199-201] They are oxidized and conjugated with sulfuric and glucuronic acids. According to a recent paper by Bray, Thorpe, and White[202] cresols are excreted by the rabbit primarily as oxygen conjugates; 60 to 72 per cent as ether glucuronides and 10 to 15 per cent as ethereal sulfates. The authors could show by paper chromatography that o- and m-cresols are hydroxylated to a small extent and that p-cresol gives rise to the formation of some p-hydroxybenzoic acid. From the urine of animals fed o- and m-cresols they isolated 2,5-dihydroxytoluene, and from those fed p-cresol, p-cresylglucuronide.

The determination of cresols in biological material, particularly urine, may be done by extracting the urine with ether and then distilling at 100°C. Phenol and all three isomers of cresol are recovered. The p-cresol can be separated by treatment of the mixture with 1-nitroso-2-naphthol. The ortho and meta isomers can be separated from phenol and p-cresol by a reaction with 4-aminoantipyrine.[27] o-Cresol may be separated from m-cresol by paper chromatography.[203]

When evaluating analyses it should be remembered that the cresols, particularly p-cresol, are normally present in human urine. It has been reported that the normal human excretes in the urine from 16 to 39 mg. of p-cresol. per day.[27]

Mode of Action. Very meager data are available concerning the exact mechanism of action of cresols; however, they appear to act in much the same manner as phenol.

Cause of Death. In most cases death appears to have been caused by respiratory failure.

Effects Observed in Man. Cutaneous application can cause local tissue corrosion, burns, and dermatitis. Certain individuals are hypersensitive to cresol.

The predominant signs of intoxication closely resemble those induced by phenol. Suitable quantitative analytical methods[204-208] have not been readily available and, consequently, many cases of poisoning induced by Lysol[209] or the cresols have been reported as cases of phenol poisoning.[4] This is of little conse-

[199] R. D. Embody *et al., Trans. Am. Fisheries Soc.,* **70**, 304 (1940).

[200] M. Hunaki, *Mitt. med. Akad. Kioto,* **29**, 99 (1940).

[201] M. E. Klinger and J. F. Norton, *Ind. Hyg. Dig.,* **9**, 355 (1945).

[202] H. G. Bray, W. V. Thorpe, and K. White, *Biochem. J.,* **46**, 275 (1950).

[203] R. L. Hossfeld, *J. Am. Chem. Soc.,* **73**, 852 (1951).

[204] W. Bielenberg and L. Fischer, *Brennstoff-Chem..* **22**, 278 (1941); *Oel u. Kohl ver. Erdoel u. Teer,* **37**, 496 (1941).

[205] N. R. Campbell and D. H. Hey, *Nature,* **153**, 745 (1944).

[206] W. Deichmann, *Ind. Eng. Chem.,* **16**, 37 (1944).

[207] V. P. Maevskaya, *Zavodskaya Lab.,* **8**, 812 (1939).

[208] H. Muhmann, *Pharm. Acta Helv.,* **15**, 141 (1940).

[209] E. F. Koster, *Ohio State Med. J.,* **39**, 840 (1943).

quence from the point of view of industrial hygiene, since the treatment of poisoning and the precautions required for the safe handling and use of cresol and its various preparations are the same as those recommended for phenol. The degree of systemic illness that may be induced by one of the commercial preparations depends primarily upon its content of cresol.

5. Hygienic Standard of Permissible Exposure

The recommended M.A.C. (8 hours) is 5 parts of vapors (*all* isomers) per million parts of air, by volume (p.p.m.).[210]

6. Odor and Warning Properties

Cresols have an odor much like that of phenol or creosote. With concentrations in the air as low as 5 p.p.m. the odors are easily recognized.

7. Safe Handling Procedures

The prevention and treatment of poisoning by cresol should be considered the same as that described for phenol.

CREOSOTE (Creosotum, Creosote Oil, Brick Oil)

1. Source, Uses, and Industrial Exposures

Creosote is obtained by distillation (205 to 220°C.) of the tar obtained from beech and other woods, as well as from coal (200 to 250°C.). It has also been prepared from the residue of olives.[211] Creosote obtained from beechwood is composed almost entirely of guaiacol ($C_6H_4OHOCH_3$) and creosol ($C_6H_3OHCH_3$-OCH_3), while creosote obtained from coal tar contains, in addition to these, phenol, cresols, pyrol, pyridine, and other aromatic compounds. Purification of the crude preparation is accomplished by distillation and extraction from suitable oils.[212,213]

Creosote has found application as an antiseptic,[214] disinfectant[215-222] anti-

[210] American Conference of Governmental Industrial Hygienists, *A.M.A. Arch. Environ. Health,* **1**, 141 (1960).

[211] G. de B. Camps, *Anales real. acad. farm.,* **2**, 353 (1941).

[212] P. A. Bobrov, *Trans. Viatka Sci. Res. Inst.,* **2**, 87 (1926); *Chem. Abstr.,* **22**, 1434 (1928).

[213] G. P. Krivokhatskii, *Mitt. Kirov. forsttech. Akad. U.S.S.R.,* No. 54, 16 (1939).

[214] C. Richet and H. Cardot, *Compt. rend.,* **165**, 491 (1917).

[215] E. F. Armstrong, *Chem. & Ind.,* **1941**, 668; cited from *Chem. Abstr.,* **35**, 8305 (1941).

[216] P. da R. Azevedo, *Anais. assoc. quim. Brasil,* **2**, 97 (1943).

[217] R. L. Datta et al., *Soap Perfumery & Cosmetics,* **12**, 583 (1939).

[218] W. E. Dove and S. W. Simmons, *J. Econ. Entomol.,* **35**, 582 (1942).

[219] E. J. Fellows, *J. Pharmacol. Exptl. Therap.,* **60**, 178, 183 (1937).

[220] E. J. Fellows, *Proc. Soc. Exptl. Biol. Med.,* **42**, 103 (1939).

[221] M. S. Hudson and R. H. Baechler, *Proc. Am. Wood Preservers Assoc.,* **1940** 74.

[222] C. J. Ramsburg, *Am. Inst. Mining Met. Engrs. Contribs.* No. **122** (1942).

pyretic, astringent, styptic, and germicide, and as a therapeutic agent.[223-226] It is also used as a lubricant for die molds, as an agent for waterproofing, as a wood preservative, as an animal dip, as a constituent of fuel oil,[227-231] and in the manufacture of chemicals.

Workers most likely to be exposed are carpenters, railroad workers, farmers, tar distillers, glass- and steel-furnace attendants, engineers, and truckers of diesel fuel oil.

Injuries to the skin or eyes have occurred mainly among men engaged in dipping or in "pickling" and handling "sleepers," mine timbers, and woods for floors and other purposes. Jonas calls attention to burns induced by fine particles of sawdust from creosote-treated lumber. He observed that men with fair skin were very sensitive, while colored workers demonstrated a remarkable resistance. The burns were reduced to a minimum on rainy days, probably because of the decreased dispersion of both the wood particles and creosote. Engels[232] considers the use of creosote-treated timber in mines inadvisable because of the fire hazard and because of the contamination of the air.

2. Physical and Chemical Properties

Physical state: yellowish or colorless, flammable, oily liquid (often brownish colored because of impurities or oxidation)

Molecular weight: varies with purity

Specific gravity: 1.07 to 1.08

Boiling point: 195 to 400°C.

Slightly soluble in water; miscible with alcohol, ether, fixed or volatile oils; soluble in glycerin and in solutions of fixed alkali hydroxides

Flash point: (closed cup) 73.9°C. (165°F.), (open cup) 85°C. (185°F.)

3. Determination in the Atmosphere

Specific methods apparently are not available.

4. Physiological Response

The signs of intoxication produced in animals resemble those for man (see below). When administered orally to various animal species the *Approximate Lethal Dosages* of creosote are 0.1 g./kg. for the pigeon and 0.6 to 0.8 g./kg. for the

[223] E. J. Fellows, *Am. J. Med. Sci.*, **197**, 683 (1939).

[224] J. G. Sampson and G. Limkako, *Philippine J. Sci.*, **23**, 515 (1923).

[225] M. E. Stevens *et al.*, *Can. Med. Assoc. J.*, **48**, 124 (1943).

[226] R. R. Wade, *Chem. Abstr.*, **19**, 3316 (1925).

[227] E. B. Davies, *Fuel Econ. Rev.*, **21**, 82 (1942).

[228] A. J. Gibbs-Smith, *Petrol. Times*, **47**, 496 (1944).

[229] C. A. McDonnell and P. J. Tracy, Brit. Patent 548,125 (1942).

[230] F. Newman, *Steam Engr.*, **12**, 348 (1943).

[231] M. Stuart, *S. African Mining Eng. J.*, **42**, 27 (1931).

rabbit, cat, and dog.[7] Cattle have been fatally poisoned by licking creosote from treated telephone poles.[232-234]

The carcinogenicity of creosote oils has been studied quite thoroughly using mice.[235-237] It was found that the high carcinogenic potency could not be correlated with the content of benzpyrene. This led to the conclusion that something other than (or in conjunction with) benzpyrene caused the skin cancers.

Pathology. Following oral ingestion the resultant lesions are those of intense irritation and congestion of the entire gastroenteric tract.

Absorption, Excretion, and Metabolism. Creosote is rapidly absorbed from the gastroenteric tract and through the skin. It appears to be excreted in the urine mainly in conjugation with sulfuric, hexuronic, and other acids.[164,220,223] Oxidation also occurs with the formation of compounds that impart a "smoky" appearance to the urine. Traces are excreted by way of the lungs. Analytical methods[238-241] for the estimation of creosote apparently have not been applied to analysis of biological material.

Mode of Action. The mechanism of action has not been well defined.

Cause of Death. Death from large doses of creosote appears to be due largely to cardiovascular collapse.

Effects Observed in Man. Fatalities have occurred 14 to 36 hours after the ingestion of about 7 g. by adults or 1 to 2 g. by children.[7] The symptoms of systemic illness included salivation, vomiting, respiratory difficulties, thready pulse, vertigo, headache, loss of pupillary reflexes, hypothermia, cyanosis, and mild convulsions. The repeated absorption of therapeutic doses from the gastroenteric tract may induce signs of chronic intoxication, characterized by disturbances of vision and digestion (increased peristalsis and excretion of bloody feces). In isolated cases of "self-medication," hypertension[242] and also general cardiovascular collapse[7] have been described.

Contact of creosote with the skin or condensation of vapors of creosote upon the skin or mucous membranes may induce an intense burning and itching with local erythema, grayish yellow to bronze pigmentation,[243,244] papular and vesicular

[232] W. Engels, *Chem. Ztg.,* **55,** 285 (1931).
[233] G. Hanlon, *Australian Vet. J.,* **14,** 73 (1938).
[234] K. Kasai, *Chem. Abstr.,* **2,** 2583 (1908).
[235] W. E. Poel, A. G. Kammer, L. J. Sullivan and C. B. Willingham, *Proc. Am. Assoc. Cancer Research,* **2,** 30 (1955).
[236] W. E. Poel and A. C. Kammer, *J. Natl. Cancer Inst.,* **18,** 41 (1957).
[237] W. Lijinsky, U. Saffiotti and P. Shubik, *J. Natl. Cancer Inst.,* **18,** 687 (1957).
[238] H. Degner, *Chem. Abstr.,* **26,** 801 (1932).
[239] L. Ekkert, *Pharm. Zentralhalle,* **73,** 487 (1932), **75,** 49 (1934); *Chem. Abstr.,* **26,** 5380 (1932), **28,** 1955 (1934).
[240] W. Franke, *Braunkohlenarch.,* No. **36,** 1 (1932).
[241] K. L. Nilstead, *J. Assoc. Offic. Agr. Chemists,* **21,** 543 (1938).
[242] S. X. Robinson, *Illinois Med. J.,* **74,** 278 (1938).
[243] Hudelo et al., *Bull. soc. franç. dermatol. syphilig.,* **34,** 144 (1927).
[244] A. D. Jonas, *J. Ind. Hyg. Toxicol.,* **25,** 418 (1943).

eruptions, gangrene,[245,246] and in isolated instances cancer.[168,246-250] A review of this phase of the problem is given by Hueper.[165] Heinz bodies have been noted[165] in the blood of a patient one year after his exposure to creosote. Jonas[244] and Goldenberg[252] made similar observations following percutaneous absorption of this preparation. Eye injuries can include keratitis,[253] conjunctivitis, and abrasion of the cornea. According to Jonas permanent corneal scars result in about one third of such cases. Photosensitization has been reported by Schwartz and Tulipan[121] and severe systemic illness by Lewin.[7]

5. Hygienic Standard of Permissible Exposure

Maximum allowable concentrations for creosote have not been established.

6. Flammability

Creosote is flammable and burns with a luminous, smoky flame.

7. Odor and Warning Properties

It has a penetrating smoky odor and a burning, caustic taste.

8. Prevention of Poisoning by Creosote

Protective measures include adequate ventilation, use and frequent change of protective garments, wearing of goggles, application of a heavy layer of petroleum jelly[244] or lanolin–castor oil ointment upon the skin of the face, thorough cleansing of all parts of the body that have become exposed accidentally, and thorough washing of the face and hands after every working period. Jonas recommended a preparation containing calcium salts, benzocaine, sulfur, and a vegetable oil base for the treatment of creosote burns. For inflamed eyes he suggested washing with aqueous boric acid or, if pain is a factor, the application of Metaphen ophthalmic ointment. (Coal tar ointments will intensify the local damage.[250]) Caution is in order when old creosote-treated lumber is handled or sawed because it retains a considerable portion of the oil for periods up to 25 or 30 years.[254,255]

[245] G. Michel et al., Rev. med. de l'est, **63,** 775 (1935).
[246] D. Schapiro, Vrachebnoe Delo, **11,** 631 (1928).
[247] S. Cabot, N. Shear, and M. J. Shear, Am. J. Pathol., **16,** 301 (1940).
[248] M. Knallinsky, Rev. arg. dermatosif., **23,** 313 (1939).
[249] R. D. Sall et al., J. Natl. Cancer Inst., **1,** 45 (1940).
[250] L. Schwartz, Ind. Med., **11,** 387 (1942).
[251] N. Lenson, New Engl. J. Med., **254,** 520 (1956).
[252] E. Y. Goldenberg, Vrachebnoe Delo, **10–11,** 663 (1939).
[253] G. T. Birdwood, Brit. Med. J., **2,** 18 (1938).
[254] N. A. Richardson, Chem. & Ind. London, **1934,** 710.
[255] H. von Schrenk, A. L. Kammerer, and H. Schmitz, Proc. Am. Wood Preservers Assoc., **1936,** 167.

PENTACHLOROPHENOL, C_6Cl_5OH, and SODIUM PENTACHLOROPHEN-ATE, C_6Cl_5ONa (Santophen 20, Penta, Dowicide 7, Penchlorol, P.C.P., Cuprinol, Evrisan; Santobrite)

The solubilities of pentachlorophenol and its sodium salt differ, but their toxic effects are the same. Therefore, for practical purposes, the following information pertains to both compounds.

1. Source, Uses, and Industrial Exposures

Pentachlorophenol is prepared by the chlorination of phenol.

"Many industrial products of organic origin are subject to microbiological attack at some stage of their preparation or use. Materials to which pentachlorophenol and its sodium salt have been successfully applied include wood and other cellulosic products, starches, adhesives, proteins, leather, oils, paints, latex, and rubber. Pentachlorophenol is equally important for the control of termites in wood and insulating board, the control of powder post beetles and other wood-boring insects, the control of slime and algae, the manufacture of herbicides, and the inhibition of fermentation in materials."[256,257] For recent work in the control of fungi see the report by Verrall and Mook.[258]

After prolonged or repeated contact with the skin, the dust or solutions of pentachlorophenol can cause dermatitis and systemic intoxication. Low concentrations of the dust (as little as 0.3 mg. per cubic meter) irritate the mucous membranes of the nose, throat, and eyes. Fatalities have occurred from absorption of solutions through the skin.[259-263]

2. Physical and Chemical Properties

(These properties pertain to pentachlorophenol only.)

Physical state: white, monoclinic, crystalline solid (technical grade dark gray to brown)

Molecular weight: 266.35

Specific gravity: 1.978 (22°C.)

Melting point: 188 to 189°C.

Boiling point: 310°C. (decomposes)

Vapor pressure: 0.00011 mm. Hg (20°C.)

[256] Pentachlorophenol-Technical, *Monsanto Tech. Bull.* No. **O-25** (1955).

[257] Santobrite, *Monsanto Tech. Bull.* No. **O-23** (1955).

[258] A. F. Verrall, and P. V. Mook, *U. S. Dept. Agr. Tech. Bull.* No. **1046** (1951).

[259] R. Truhaut, P. L. Epee, and E. Poussemart, *Arch. maladies profess. méd. travail et securité sociale,* **13**, 567 (1952).

[260] R. Truhaut, G. Vitte, and E. Poussemart, *Arch. maladies profess. méd. travail et securité sociale,* **13**, 570 (1952).

[261] S. Nomura, *J. Sci. Labour Tokyo,* **29**, 274 (1953).

[262] D. Gordon, *Med. J. Australia,* **2**, 485 (1956).

[263] J. A. Menon, *Brit. J. Med.,* **1**, 1156 (1958).

Per cent in "saturated" air:　0.0000145% by volume (20°C.)

Density of "saturated" air:　1.0000011 (air = 1)

The solubilities of these materials are of particular interest, since they are generally used in solution (Table 5).

1 mg./liter ⇌ 91.9 p.p.m. and 1 p.p.m. ⇌ 0.01088 mg./liter at 25°C., 760 mm. Hg

TABLE 5

Solubility of Pentachlorophenol and Its Sodium Salt

Solvent	Solubility at 20°C., % (w/v)
Pentachlorophenol	
Organic solvents	
Methanol	57.0
Ethanol	53.0
Ethanol (95%)	47.5
Diethylene glycol	27.5
Acetone	21.5
Xylene	14.0
Benzene	11.0
Ethylene glycol	6.0
Carbon tetrachloride	2.0
Oils of vegetable origin	
Pine oil	32.0
Corn oil	14.8
Linseed oil (raw)	13.9
Tung oil	11.2
Turpentine	3.0
Water	14[a]
Sodium Pentachlorophenate	
Water[b]	15.0 at 4°C.
	20.0 at 9.5°C.

[a] Parts per million.

[b] Higher concentrations tend to cause formation of hydrates having lower solubility.

3. Determination in the Atmosphere

An electrostatic precipitator or glass fiber filter paper can be used to collect dust or fumes of pentachlorophenol from contaminated air.[264] The compound can then be determined by the spectrophotometric method of Deichmann and Schaefer,[265] which was developed for the estimation of pentachlorophenol in tissue and water. This procedure is based on the formation and determination of a reddish yellow pigment formed by the action of fuming nitric acid on pentachloro-

[264] Hygienic Guide Series, *Am. Ind. Hyg. Assoc. Quart.*, **18**, 274 (1958).

[265] W. B. Deichmann, and L. J. Schaefer, *Ind Eng. Chem.*, **14**, 310 (1942).

phenol. More recently (1958), Erne[266] published a method that appears to give more specific results with biological material. In this procedure, urine or other body fluids were first extracted with ether. Pentachlorophenol was separated by partition chromatography, extracted with aqueous pyrophosphate, purified further by paper chromatography, eluted with carbon tetrachloride, and determined spectrophotometrically in the ultraviolet region (320 mμ). This latter method will detect 1 or 2 μg./g.

A rapid method for determining concentrations from 1 to 100 p.p.m. of sodium or copper pentachlorophenate in water was recommended by Haskins. [267] Detailed methods for the analysis of other materials are covered in the *Monsanto Technical Bulletin No. 0-24*.[268]

4. *Physiological Response*

When absorbed in sufficient quantity into the tissues of dogs, rabbits, rats, and guinea pigs, pentachlorophenol and sodium pentachlorophenate[269,270] produce an acute toxic state characterized by accelerated respiration, moderately elevated blood pressure, hyperpyrexia, hyperglycemia, glycosuria, and hyperperistalsis (vomiting was observed in dogs after a subcutaneous dose). The urinary output is increased at first, but diminished later. There is a rapidly developing motor weakness which, in fatal cases, terminates in asphyxial convulsions with cardiac and muscular collapse. Rigor mortis is immediate and most profound.

A gradual loss of weight, but no significant changes in the hemoglobin content of the blood, is noted in rabbits given repeated, slightly sublethal doses of the salt. The number of erythrocytes and various types of leucocytes remains within normal limits. Rectal temperature and blood sugar values rise markedly after the administration of a single, slightly sublethal dose, but they fail to show such elevations when the doses are repeated.

The general and local effects produced by the application of solutions of pentachlorophenol to the skin of animals vary with the period of exposure and the vehicle employed. Irritation of the skin and marked local damage followed by complete recovery is the usual result of cutaneous application of single or repeated doses of pentachlorophenol in fuel oils. Single and repeated applications of aqueous solutions of the salt may cause mild or no local damage. Table 6 presents acutely lethal doses in experimental animals.

Pathology. In animals, the post-mortem evidences of injury consist largely of extensive damage to the vascular system, with heart failure, and with involvement of the parenchymatous tissues as a result of the heart failure (congestion and

[266] K. Erne, *Acta Pharmacol. Toxicol.*, **14**, 158 (1958).

[267] W. T. Haskins, *Public Health Repts. U.S.*, **66**, 1047 (1951).

[268] Analytical Methods for Pentachlorophenol, *Monsanto Tech. Bull. No.* **0-24** (1955).

[269] F. Vallier, L. Roche, and A. Brune, *Compt. rend. soc. biol.*, **148**, 374 (1954).

[270] F. Vallier, L. Roche, and A. Brune, *Compt. rend. soc. biol.*, **148**, 690 (1954).

TABLE 6

Acute Toxicity of Pentachlorophenol and Its Sodium Salt in Experimental Animals[271-273]

Species	Route	Solvent	Dose killing approx. 50% of animals, mg./kg.	Ref.
		Pentachlorophenol		
Rabbit	Cut.	11% in olive oil	(450 was not fatal)	271
		5% in Stanolex fuel oil	60	272
		5% in Shell fuel oil	130	272
		1.8% in pine oil	40	272
	Subcut.	5% in olive oil	70	271
	Oral	5% in Stanolex fuel oil	70	271
		11% in olive oil	100	271
Rat	Oral	0.5% in Stanolex fuel oil	27	271
		1% in olive oil	78	271
		Sodium pentachlorophenate (in terms of pentachlorophenol)		
Rabbit	Cut.	10% aqueous	250	271
	Subcut.	10% aqueous	100	271
	Oral	2% aqueous	275	273
	Intraven.	1% aqueous	22	271
Rat	Subcut.	2% aqueous	66	271
	Oral	2% aqueous	210	271
		2, 25% aqueous	125–200	
		Copper pentachlorophenate		
Rat	Oral	2 to 20% suspension in Tween 20	600	

edema), and the injury to small blood vessels (edema and proliferation of endo-thelium, dilatation of arterioles and capillaries, and hemorrhage).

Absorption, Excretion, and Metabolism. Pentachlorophenol and its sodium salt can be absorbed through the skin and from the gastroenteric tract.

Complete data on the distribution of pentachlorophenol in the tissues were obtained from two rabbits. Each was given one oral dose of the sodium salt (0.25 per cent aqueous solution) equivalent to 37 mg. of pentachlorophenol per kilogram; it was then placed in a metabolism cage for 24 hours, after which it was sacrificed for analysis of the tissues. Of the administered quantities of pentachlorophenol, 92 and 91 per cent, respectively, were recovered. The bulk of the material was found in the urine, and about 4 and 7 per cent, respectively, were present in the gastro-enteric tract. The remainder was well distributed throughout the tissues (Table 7). Obviously, in the rabbit, the kidney is the principal route of elimination of penta-chlorophenol from the body, which suggests that urinary concentrations may offer

[271] W. B. Deichmann, W. Machle, K. V. Kitzmiller, and G. Thomas, *J. Pharmacol, Exptl. Therap.,* **76,** 104 (1942).

[272] R. A. Kehoe, W. B. Deichmann, and K. V. Kitzmiller, *J. Ind. Hyg. Toxicol.,* **21,** 160 (1939).

[273] W. Machle, W. B. Deichmann, and G. Thomas, *J. Ind. Hyg. Toxicol.,* **25,** 192 (1943).

means of determining the severity of human exposure. Only traces of pentachlorophenol are excreted in conjugated form in the urine of the rat and rabbit.

Mode of Action. Apparently, pentachlorophenol causes a radical uncoupling of oxidation and phosphorylation cycles in the tissues. This produces a markedly increased basal metabolic rate and a marked temperature increase.[264,265]

TABLE 7

Distribution of Pentachlorophenol in the Tissues of a Rabbit 24 Hours after Oral Administration of One Dose of Sodium Pentachlorophenate Equivalent to 94 Mg. of Pentachlorophenol[271]

Tissue or fluid	Total Weight of tissue or fluid, g.	Pentachlorophenol		
		In 10 g. sample analyzed, mg.	In total tissue or fluid, mg.	Recovered, % of amount ingested
Urine and feces	199.1	3.40	66.21	70.6
Stomach and intestine (wall and contents)	575.7	0.11	6.33	6.76
Muscle	931.0	0.065	6.05	6.42
Bones	382.6	0.075	2.86	3.06
Skin	260.9	0.08	2.08	2.22
Blood	222.6	0.08	1.78	1.90
Liver and gall bladder	100.0	0.14	0.70	0.743
Kidneys	17.7	0.185	0.25	0.267
Heart, lungs, and testes	25.9	0.06	0.16	0.171
Central nervous system	12.8	0.075[a]	0.09	0.096
Total recovery			86.51	92.23

[a] Entire sample of 12.8 g. analyzed.

In vitro tests have shown that 10^{-6} to 10^{-3} (or greater) molar concentrations of pentachlorophenol uncouple oxidative phosphorylation, inhibit mitochondrial and myosin adenosine triphosphatase, inhibit glycolytic phosphorylation, inactivate respiratory enzymes, and cause gross damage to mitochondria.[274]

Cause of Death. Hyperpyrexia and cardiac failure apparently are the cause of death in pentachlorophenol poisoning.

Effects Observed in Man. If not handled with proper precautions, pentachlorophenol and its sodium salt are capable of inducing discomfort and local or systemic effects.

Manufacturing, handling, and use have shown that irritation of the skin is likely to occur as a result of relatively brief, single exposures to solutions containing more than approximately 10 per cent of the material. Solutions as dilute as about 1 per cent may cause irritation if contacts are prolonged or repeated frequently. However, solutions containing 0.1 per cent, or less, are not likely to cause any adverse local effects.[275]

[274] E. C. Weinbach, *Proc. Natl. Acad. Sci. U. S.*, **43**, 393 (1957); cited from *Chem Abstr.*, **51**, 132326 (1957).

[275] D. D. McCollister, personal communication.

"Fine dusts and sprays of pentachlorophenol or sodium pentachlorophenate will cause painful irritation to the eyes and upper respiratory tract, but this very fact lessens the hazard. In our opinion, the dust of the materials has excellent warning properties, so that one will suffer prohibitive painful irritation in the nose before [one] is subjected to concentrations which will produce any adverse systemic effects. We have found in our own plants that atmospheric concentrations appreciably greater than 1.0 mg./cu. meter of air will cause this pain in the un-initiated person. Concentrations as high as 2.4 mg./cu. meter can be tolerated by those who are accustomed to the material. When concentrations are appreciably below 1.0 mg./cu. meter, we have experienced no noticeable effects, even among persons unaccustomed to the material."[275]

Severe effects, including several fatalities, have been reported in Japan,[260] France, Puerto Rico, and Australia.[262] These have resulted from the practically uncontrolled use of pentachlorophenol as an oil solution for the destruction of weeds or termites.[276] Truhaut et al.[259,260] and Menon[263] reported fatalities of men who dipped planks of wood into 1.5 to 3 per cent solutions of a mixture of 80 per cent pentachlorophenol and 20 per cent sodium pentachlorophenate. The observed symptoms of intoxication included loss of appetite, respiratory difficulties, anesthesia, hyperpyrexia, sweating, dyspnea, and a rapidly progressive coma. Pentachlorophenol was found in the liver, blood, and urine. Autopsy revealed an inflamed gastric mucosa in three patients, edema and congestion of the lungs in two, and edema in the brain in one.[263]

Examination of these reports reveals that systemic effects were noted particularly in uninstructed workers who used knapsack sprayers and who drenched their backs as well as walked with bare feet through puddles of the material or who, in dipping operations, disregarded nearly every safe handling precaution.

5. Hygienic Stardard for Permissible Exposure

The recommended maximum atmospheric concentration for 8 hours is 0.5 mg. per cubic meter of air (0.046 p.p.m.).[277,278] This recommendation is based on toxicological observations on animals.[271-273]

6. Odor and Warning Properties

There is a rather marked, characteristic odor.

7. Instructions for Handling Pentachlorophenol

Like many other compounds, pentachlorophenol and its sodium salt must be handled with caution.[45] The dust of these materials is irritating to the skin and mucous membranes and, if inhaled, will induce violent coughing and sneezing.

[276] E. Kelly, personal communication.
[277] American Conference of Governmental Industrial Hygienists, A.M.A. Arch. Ind. Health, **14**, 188 (1956).
[278] Safe Handling and Toxicity of Penta, Monsanto Tech. Bull. No. **0-100** (1956).

Obviously, skin contact or inhalation must be avoided. Skin contact with solutions must be equally avoided in order to prevent absorption and possible systemic intoxication. For the same reason, even the ingestion of traces should be prevented. Individuals suffering from kidney and liver diseases have a lowered resistance and should not be permitted to suffer any exposure.

Good housekeeping in the manufacturing plants has kept undesirable effects at a minimum; therefore, the following precautionary measures are directed primarily at those who use these compounds.

If exposure to the dust cannot be eliminated completely, then protect nose and mouth by a respirator or folded gauze and the eyes by goggles. Wear protective clothing, including gloves, and replace immediately any that becomes soaked with solutions of pentachlorophenol. All clothing worn during one spraying operation should be laundered before re-use. Washing with soap and water must be routine practice before eating, drinking, or smoking. At the end of each day, a workman should shower and change into clean clothing.

8. *Treatment of Local Effects and Systemic Intoxication*

If the material gets in the eyes, they should be washed in flowing water for a period ranging from 5 to 30 minutes. If tissue damage is apparent, a boric acid solution and cortisone ophthalmic drops should be applied alternately at hourly intervals until healing is well apparent.

Little can be done to relieve the irritation of the mucous membranes of the nose or throat. It is essential, of course, to remove the affected person from the site of exposure. If coughing is a problem, any soothing cough syrup may be used to advantage.

If a dermatosis is produced by the material, repeated and prolonged soaking in boric acid or Burow's solution (1 part to 9 parts of water) is recommended until inflammation and swelling have subsided. This treatment is to be followed by the application of bland sterile ointments until the skin has healed.

If a solution of pentachlorophenol or sodium pentachlorophenate is accidentally swallowed, induce vomiting immediately. This first aid treatment should be followed by thorough gastric lavage with water or milk and the administration of a saline cathartic.

Treatment of a systemic intoxication should be directed toward allaying anxiety, insuring adequate intake of oxygen, promoting loss of heat (ice packs if hyperpyrexia becomes marked), and replacing lost fluids and electrolytes.

CHLOROPHENOLS (Other Than Pentachlorophenol)

There are many mono-, di-, tri-, and tetrachlorophenols. Obviously, each compound cannot be discussed separately in a report of this type. Therefore, some of the physical and chemical properties are given for at least one representative in each group (di-, trichlorophenols, etc.).

The toxicity is discussed for the chlorophenols as a whole. Individual differences are emphasized.

1. Source, Uses, and Industrial Exposures

Most of the chlorophenols are prepared by the chlorination of phenol.

The tetrachlorophenols (and their sodium salts) have been used as fungicides and wood preservatives. The trichlorophenols are used as fungicides and bactericides. 2,4-Dichlorophenol is used in organic synthesis. 2,4-Dichlorophenoxyacetic acid is used as a weed killer.

2. Physical and Chemical Properties

Some of the pertinent chemical and physical properties of the chlorinated phenols are presented in Table 8.

3. Determination in the Atmosphere

Methods for the determination of chlorophenols in the air apparently are not available.

For the determination of monochlorophenols in biological materials, Karpow (1893) suggested distillation of acidulated urine and precipitation of the chlorophenol in the distillate with bromine water.[279]

4. Physiological Response

In rats, oral, subcutaneous, and intraperitoneal lethal doses of the chlorophenols produce similar signs of poisoning. Oral administration, however, results in fatal poisoning in smaller dosage and in a shorter period of time than subcutaneous administration.

Restlessness and an increased rate of respiration appear a few minutes after administration of o- and m-chlorophenols and are followed a few minutes later by a rapidly developing motor weakness. Tremors, clonic convulsions (which can be induced by noise or touch), dyspnea, and coma set in promptly and continue until death. Similar signs are produced by p-chlorophenol, but the convulsions are more severe. 2,4- and 2,6-Dichlorophenols and 2,4,6- and 2,4,5-trichlorophenols produce these signs also, but decreased activity and motor weakness do not appear quite so promptly. The tremors are much less severe, but in this case, also, they continue until a few minutes before death. Tetrachlorophenols take an intermediate place between the lower homologs and pentachlorophenol. The signs they produce are similar to those caused by the mono-, di-, and trichlorophenols, except that tremors and convulsions probably are due to asphyxia or hypoglycemia and result from a different mechanism than those noted with the lower chlorinated phenols. Deichmann[45] reported that no hyperpyrexia was produced by these chlorophenols in rats and rabbits, but Farquharson et al.[280] reported hyperpyrexia from injections of tri- and tetrachlorophenols.

[279] G. Karpow, Arch. sci. biol. St. Petersburg, **2**, 304 (1893).
[280] M. E. Farquharson, J. C. Gage, and J. Northover, Brit. J. Pharmacol., **13**, 20 (1958).

TABLE 8

Physicochemical Properties of Some Chlorophenols[13]

Compound[a]	Empirical formula	Physical state	Molecular weight	Specific gravity	Melting point	Boiling point	Solubility, parts per 100 parts		
							Water	Alcohol	Ether
o-Chlorophenol[a]	ClC_6H_4OH	Colorless liquid	128.56	1.241 (18.2°/15°)	α, 7; β, 0; γ, 4.1	175–6	2.85[b]	s.	s. alk.
m-Chlorophenol	ClC_6H_4OH	Needles	128.56	1.268 (25°)	32–33	214	2.6[b]	s.	s.
p-Chlorophenol	ClC_6H_4OH	Needles	128.56	1.306 (20°/4°)	41–43	217	2.71[b]	v.s.	v.s.
2,4-Dichlorophenol	$Cl_2C_6H_3OH$	Needles/benzene	163.01	1.383 (60°/25°)	45	209–10	0.45[b]	v.s.	v.s.
2,3,5-Trichlorophenol	$Cl_3C_6H_2OH$	Needles/aqueous alcohol	197.46		55	249–50	sl.s. h.	s.	s.
2,4,5-Trichlorophenol	$Cl_3C_6H_2OH$	Colorless/ petroleum ether	197.46		61–63	252	i.	s.	s.
2,4,6-Trichlorophenol	$Cl_3C_6H_2OH$	Needles	197.46	1.490 (75°/4°)	68–69	246	0.09[c]	v.s.	v.s.
2,3,4,6-Tetrachlorophenol	Cl_4C_6HOH	Needles	231.90	1.6(60°/4°)	69–70	164 (23 mm.)	v.sl.s.	v.s.	v.s.

[a] Vapor pressure 1.393 (18–80°C.).
[b] 20°C.
[c] 25°C.

The lethal dosages (LD_{50}) of some of these compounds are presented in Table 9.

The higher chlorinated phenols produce contraction of the isolated rat phrenic nerve diaphragm preparation and stimulate oxygen uptake of rat brain homogenate.[280]

From the work of several investigators[45,280,281] it can be concluded that the general effect of increasing the chlorination of phenol is a reduction of the convulsant action but an increase in the inhibition of oxidative phosphorylation. (Pentachlorophenol does not produce convulsions.)

TABLE 9

Lethal Dosages of Some Chlorophenols to Rats

(Doses expressed as g./kg.; compounds administered in olive oil)

Compound	I.P. MLD[280]	Oral LD_{50}[45]	S.C. LD_{50}[45]
o-Chlorophenol	0.23	0.67	0.95
m-Chlorophenol	0.36	0.57	1.39
p-Chlorophenol	0.28	0.67	1.03
2,4-Dichlorophenol	0.43	0.58[a]	1.73[a]
2,4,5-Trichlorophenol	0.36	0.82[a]	2.26[a]
Tetrachlorophenol	0.13	0.14[a]	0.21[a]

[a] Fuel oil was the solvent.

Pathology. In the rat, monochlorophenols caused marked injury to the kidneys with red blood cell casts in the tubules, fatty infiltration of the liver, and hemorrhages in the intestines.[43]

Absorption, Excretion, and Metabolism. The compounds are readily absorbed from the gastroenteric tract and from parenteral sites of injection.

The monochlorophenols are excreted as conjugates of sulfuric and glucuronic acids. The urine darkens after standing.[282]

Mode of Action. It has been postulated that the convulsant action of the lower chlorinated phenols is due to the undissociated molecule.[283] The interference with oxidative phosphorylation evidently caused by tri-, tetra-, and pentachlorophenols may be due to the dissociated chlorophenate ion.

Effects Observed in Man. There is essentially no information in the literature concerning the toxicity of the chlorophenols to man. Dermatoses in man caused by tetrachlorophenol and its sodium salt have been reported; these included papulofollicular lesions, comedones, sebaceous cysts, and marked hyperkeratosis.[283,284]

5. Hygienic Standards for Permissible Exposure

No standards for permissible exposure have been set.

[281] H. Bechold, and P. Ehrlich, *Z. physiol. Chem.*, **47**, 173 (1906).

[282] W. F. von Ottingen, *Natl. Inst. Health Bull.* No. **190** (1949).

[283] M. G. Butler, *Arch. Dermatol. and Syphilol.*, **35**, 251 (1937).

[284] K. O. Stingily, *Southern Med. J.*, **33**, 1268 (1941).

6. Odor and Warning Properties

The chlorophenols have a distinct odor.

RELATED COMPOUNDS

There are many phenol derivatives that are of some importance in industrial toxicology. Because of the large number of such compounds and their varied uses a full treatment of each compound does not seem to be merited. Therefore, the discussion of these compounds is limited to a brief presentation of toxicity data.

BROMO- and IODOPHENOLS

1. Physiological Response in Animals

These compounds are rapidly absorbed from the gastroenteric tract. The approximate (oral) LD_{50} in the rat in mg./kg. of pentabromophenol is slightly more than 200; of 2,4,6-tribromophenol, somewhat less than 2000; and of 2,4,6-triiodophenol, from 2000 to 2500. The compounds were administered as sodium salts in 15 to 40 per cent aqueous solutions. In rats and guinea pigs the subcutaneous LD_{50}'s of o-bromophenol are 1.5 and 1.8 g./kg., respectively.

In general, the symptoms produced by all the compounds except pentabromophenol were the same as those produced by pentachlorophenol and consisted of an increased respiratory rate and amplitude followed by loss of muscle tone, collapse, and death. The signs and symptoms from pentabromophenol included increased respiratory rate and amplitude with general body tremors, occasional convulsions, and death.

Evidence of some cumulative toxicity was obtained with 2,4,6-tribromophenol and 2,4,6-triiodophenol. An increase of one- to threefold in the LD_{50} was obtained from five daily doses of these compounds.[285]

Herdt, Loomis, and Nolan[286] studied the effects of three potential molluscacides, pentabromophenol and sodium and copper pentachlorophenate. These were given in drinking water to three young bulls at a dosage of 7.6 mg./kg./day for 5 weeks. No significant signs of intoxication and no micropathological changes were noted. The authors believe that these halogenated phenols can be used with safety as molluscacides in the field, provided reasonable precautions are taken in their application.

2. Pathology in Animals

The pathologic changes were most marked in the lungs. All compounds produced mild to severe congestion and petechial hemorrhages. In addition, large doses of 2,4,6-triiodophenol produced severe inflammation of the mucous membrane of the pylorus and fundus of the stomach with corrosion and hemorrhages.

[285] E. F. Stohlman, *Public Health Repts. U. S.*, **66**, 1303 (1951).
[286] J. R. Herdt, L. N. Loomis and M. O. Nolan, *Public Health Repts.*, **66**, 1313 (1951).

These were also produced, but in a considerably less severe form, by the larger doses of 2,4,6-tribromophenol. Smaller doses of these two compounds or larger doses of the others produced only moderate inflammation or congestion of the mucosa of the stomach and intestines.

o-PHENYLPHENOL, $C_6H_5C_6H_5CH$ (Dowicide I)

1. Uses

This compound is being used as a fungicide, germicide, household disinfectant, intermediate for dyes, and preservative in water-oil emulsions such as are used in the rubber, textile, and metal-working industries. The water-soluble sodium salt of o-phenylphenol is used for protecting water-extendable paints against decomposition prior to use and is employed as a preservative for proteins and other types of decomposable adhesives. More recently o-phenylphenol has found use in the treatment of citrus fruit to prevent mold, in dish-washing formulations, in a fungistatic wax for coating vegetables, and for dipping paper and similar containers employed for the storage of foods.[287]

2. Physiological Response in Animals

Hodge, Maynard, Blanchet, Spencer, and Rowe[287] have found that o-phenylphenol has a low acute oral toxicity for male rats; the LD_{50} was found to be 2.7 g./kg. When tested on 200 human subjects as a 5.0 per cent solution in sesame oil and as a 0.1 per cent aqueous solution of the sodium salt, o-phenylphenol failed to cause either primary skin irritation or skin sensitization. Male and female rats (25 of each sex per group) that were maintained for two years on diets containing 0.02 or 0.2 per cent o-phenylphenol showed no adverse effects as judged by gross appearance, growth, hematology, rate of mortality, organ weights, and histopathological changes in various tissues. Similar groups of rats maintained for two years on a diet containing 2 per cent of the material deviated from the controls by exhibiting slight retardation of growth, histopathological kidney changes (marked tubular dilation), and the presence of small amounts of o-phenylphenol in tissues of the kidney. Dogs that received oral doses of 0.02, 0.2, and 0.5 g./kg./day of o-phenylphenol for a period of one year showed no adverse effects as judged by body weights, hematological values, organ weights, and histopathological changes in various tissues.

DITERTIARYBUTYLMETHYLPHENOL (Di-tert-Butylmethylphenol DBMP, 4-Methyl-2,6-di-tert-butylphenol, 2,6-Di-tert-butyl-p-cresol, Di-tert-butylhydroxytoluene, Deenax, Paranox, DBPC Antioxidant, Ionol)

1. Uses

Di-tert-butylmethylphenol (DBMP) is an antioxidant which in small amounts prevents the deterioration of a variety of materials, including fats, oils,

[287] H. C. Hodge, E. A. Maynard, H. J. Blanchet, Jr., H. C. Spencer, and V. K. Rowe, J. Pharmacol. Exptl. Therap., 104, 202 (1952).

waxes, resins, and plastic films. To mention a few of the potential uses, it is expected that this material will ultimately be incorporated into most edible vegetable or animal fats and oils, into baked and fried foods, and into waxes or plastic films used for coating food wrappers or containers.

2. *Physical Properties*

Physical state: slightly yellow, crystalline solid
Melting point: 70°C.
Boiling point: 265°C.

3. *Physiological Response in Animals*

When absorbed in toxic concentrations into the tissues of unanesthetized animals, DBMP induced signs of intoxication resembling those seen after absorption of a toxic dose of a parasympathetic drug (salivation, a mild degree of miosis, unsteadiness, restlessness, hyperexcitability, diarrhea, and tremors).[288] When given intravenously to a dog under pentobarbital anesthesia, DBMP (25 mg./kg.) induced a prompt reduction of blood pressure. Atropine sulfate partially antagonized this depressor effect. Large doses of DBMP produced a gross disturbance of sodium, potassium, and water balance in the rabbit.[289] It was concluded that the increase in sodium and aldosterone excretion was due to pyelonephritis, and that death was due to potassium depletion.

From the results of chronic toxicity studies using dogs and rats[288] it was concluded that DBMP is a relatively innocuous compound.

DODECYLTHIOPHENOL, $C_6H_4(C_{12}H_{25})SH$

The acute toxicity of this material is of a low order. An intramuscular dose of 20 g./kg. is not fatal in the rat; lethal oral doses range from 20 to 30 g./kg. When applied to the skin of the rat, rabbit, and guinea pig, the material induces loss of hair in 6 to 12 days. When applied to the human skin, the material may induce local eczema but apparently does not cause loss of hair.[290]

[288] W. B. Deichmann, J. J. Clemmer, R. Rakoczy and J. Bianchine, *A.M.A. Arch. Ind. Health*, **11**, 93 (1955).

[289] F. A. Denz and J. G. Llaurado, *Brit. J. Exptl. Pathol.*, **38**, 515 (1957).

[290] A. Enders and A. Moench, *Arzneimittel Forsch.*, **2**, 587 (1952).

CHAPTER XXXIV

Alcohols

JOSEPH F. TREON

METHYL ALCOHOL, CH₃OH (Methanol)

1. Source, Uses, and Industrial Exposures

Synthetic methyl alcohol made from carbon oxides and hydrogen has practically replaced that distilled from wood.

Methyl alcohol has been known also as carbinol, Columbian spirit, wood alcohol, and wood spirit. It is used extensively as an industrial solvent, in the lacquer industry, in the preparation of celluloid, films, plastics, textile soaps, wood stains, artificial leather, and nonshatterable glass. It is used in enamels, stains, dyes for straw hats, paint and varnish removers, cleaning and dewaxing preparations, embalming fluids, and antifreeze mixtures. It is also used as an intermediate and as an extracting medium in organic synthesis.[1] Seventy-two occupations that offer exposure to methyl alcohol have been reported by the United States Department of Labor.

Industrial injuries or fatalities have been reported from the inhalation of high concentrations of methyl alcohol by persons engaged in varnishing beer vats,[2] varnishing metal,[3] varnishing the engine room of a submarine,[4] shellacking lead pencils,[5] shellacking hogsheads,[6] coloring cloth and other articles in solutions of dyes in methyl alcohol,[7-9] stiffening hats,[10,11] and manufacturing shoes.[12]

[1] I. Mellan, *Industrial Solvents*. Reinhold, New York, 1939, p. 202.

[2] C. A. Wood, *J. Am. Med. Assoc.*, **59**, 1962 (1912).

[3] E. Browning, *Med. Research Council Ind. Health Research Board,* Rept. No. **80,** H. M. Stationery Office, London (1937).

[4] S. L. Ziegler, *J. Am. Med. Assoc.*, **77**, 1160 (1921).

[5] *N. Y. State Dept. Labor Bull.* No. **86** (1917).

[6] A. B. Hale, *J. Am. Med. Assoc.*, **37**, 1447, 1450 (1901).

[7] A. Hamilton, *Industrial Poisons in the United States.* Macmillan, New York, 1925, p. 427.

[8] *Natl. Research Council Can.* Bull. No. 15, 20 (1930).

[9] J. M. Robinson, *J. Am. Med. Assoc.*, **70**, 148 (1918).

[10] C. Baskerville, *Wood Alcohol: A Report on Chemistry, Technology, and Pharmacology of and the Legislation Pertaining to Methyl Alcohol,* Appendix VI, Vol. 2. rept., N. Y. State Factory Investigating Commission, 1913, p. 917.

[11] F. Buller and C. A. Wood, *J. Am. Med. Assoc.*, **43**, 1117 (1904).

[12] E. R. Hayhurst, *Occupational Survey of Ohio*, 1915.

The concentration of methyl alcohol was 22 to 25 p.p.m.[13] in well-ventilated rooms in which a mixture of methyl alcohol and acetone was employed to impregnate fused collars. At a distance of 6 ft. from the site at which artificial flowers were being dipped, there was 200 p.p.m. of methyl alcohol in the air. Vapors were also noticeable at a distance of 75 ft. from the point of dipping and drying.[5] Concentrations of 50 to 6000 p.p.m. of methyl alcohol in workrooms have been reported.[8] On the assumption that there was one change of air per hour, Loewy[14] estimated the concentrations of methyl alcohol in various workrooms in which violins, artificial flowers, shoes, and hats were being processed, through calculations based on the volume of the rooms and the quantities of alcohol vaporized therein. He arrived at values of less than 50 p.p.m. in 7 rooms, 50 to 100 p.p.m. in 5 rooms, and 265 to 622 p.p.m. in 4 rooms.

In the manufacture of photographic film, methyl alcohol is kept essentially within a closed system, but, during the loading of mixers and the changing of filters, Sterner[15] found concentrations of methyl alcohol ranging from 200 to several thousand p.p.m., the latter values occurring for only short periods of time. The daily average of the concentrations to which operators were exposed was probably between 400 and 500 p.p.m. Sterner believes these latter values would not ordinarily result in any serious effect or even moderate discomfort, since numbers of men known by him to have been exposed to such conditions while handling millions of gallons of this solvent failed to show any evidences of methyl alcohol intoxication. Concentrations of methyl alcohol ranging from 165 to 635 p.p.m. were found in four plants employing duplicating machines,[16] while the concentration in a fifth plant varied between 40 and 50 p.p.m. McAllister[17] found methanol in concentrations of 400 to 800 p.p.m. in the breathing zone of workers when four different makes of duplicators were operating simultaneously but the values decreased to 155 to 420 p.p.m. when three or less machines were in operation.

2. *Physical and Chemical Properties*

Physical state: colorless, volatile liquid
Molecular weight: 32.042[18]
Specific gravity: 0.792 (20/4°C.)[1]
Refractive index: 1.329 (20°C.)[1]
Melting point: —97.8°C.[19]
Boiling point: 64.5°C.[19]

[13] L. Greenburg, M. R. Mayers, L. J. Goldwater, and W. J. Burke, *J. Ind. Hyg. Toxicol.*, **20**, 148 (1938).

[14] A. Loewy, *Vierteljahrsschr. gerichtl. Med.*, **48**, Suppl., 93 (1914).

[15] J. H. Sterner, personal communication.

[16] A. E. Goss and G. H. Vance, *Ind. Hyg. Newsletter*, **8**, No. 9, 15 (1948).

[17] R. G. McAllister, *Am. Ind. Hyg. Assoc. Quart.*, **15**, 26 (1954).

[18] "Atomic Weights, (International) 1941," *J. Am. Chem. Soc.*, **63**, No. 3 (1941).

[19] *International Critical Tables of Numerical Data, Physics, Chemistry, and Technology*, Vol. I. McGraw-Hill, New York, 1926, p. 177.

Vapor density: 1.11 (air $= 1$)

Vapor pressure: 160 mm. Hg (30°C.),[1] 125 mm. Hg (25°C. calculated)

Per cent in "saturated" air: 21.05 (30°C.)

Density of "saturated" air: 1.02 (air $= 1$)

Solubility: miscible with water, alcohols, ketones, esters, and halogenated hydrocarbons, and partially miscible with benzene[1]

Flash point: 54°F.

1 mg./liter $\backsim$ 764 p.p.m. and 1 p.p.m. $\backsim$ 1.31 mg./cu. meter at 25°C., 760 mm. Hg

3. Determination in the Atmosphere

The Denigès[20] test, which is the basis for most accepted methods for the determination of methyl alcohol in air, is based upon the oxidation of methyl alcohol by potassium permanganate to formaldehyde and the subsequent measurement of the color produced with fuchsin in sulfurous acid (Schiff's reagent). The sensitivity and precision of this test were increased by Chapin,[21] who employed Elvove's[22] modification of Schiff's reagent. Chapin's method can be used in the presence of ethyl alcohol. Wright[23] found rosaniline a better coupling agent than fuchsin because it is quite stable and more sensitive than fuchsin.

By absorbing methyl alcohol from the air in a bubbler containing water and subsequently employing fuchsin, concentrations as low as 5 p.p.m.[24] can be detected, while 30 p.p.m. can be determined quantitatively. Jephcott[24] found the development of color was dependent upon the reaction of the medium. Rosaniline was employed by Ackerbauer,[25] who used a series of scrubbers to eliminate chlorine, sulfur dioxide, formic acid, and acetic acid from air containing them while the air was flowing at the rate of 1 liter every 25 minutes for 5 or 10 liters. He claimed that the error of determination of concentrations as low as 6 p.p.m. of methyl alcohol in the presence of formaldehyde was not more than 5 per cent.

The following interfering substances have been eliminated by suitable chemical methods followed by distillation: formaldehyde, terpenes, phenol, carbohydrates,[21] pectin,[23] and glycerol.[21,23] Chapin[21] found amyl alcohol, acetone, formic acid, and acetic acid are not apt to interfere.

4. Determination in Tissues and Blood

A method based upon the colorimetric measurement, at 585 mμ, of dichromate reduced to trivalent chromium by methyl alcohol, has been used by Hine et al.[26]

[20] E. Denigès, Compt. rend., **150**, 832 (1910).

[21] R. M. Chapin, Ind. Eng. Chem., **13**, 543 (1921).

[22] E. Elvove, Ind. Eng. Chem., **9**, 295 (1917).

[23] L. O. Wright, Ind. Eng. Chem., **19**, 750 (1927).

[24] C. M. Jephcott, Analyst, **60**, 588 (1935).

[25] C. F. Ackerbauer and R. J. Lebowich, J. Lab. Clin. Med., **28**, 372 (1942).

[26] C. N. Hine, T. E. Shea, Jr., and W. R. Olsdorf, Federation Proc., **6**, 338 (1947).

TABLE 1. Physiological Effects upon Animals of the Inhalation of Methyl Alcohol

Animal	Concentration		Duration of exposure, hr.	Signs of intoxication	Outcome	Reference
	p.p.m.	mg./liter				
Cat	132,000	173.0	5–5.5	Narcosis	Died	29
	65,700	86.0	4.5	On side	50% died	29
	33,600	44.0	6	Incoordination	50% died	29
	18,300	24.0	6	None, but salivation	Survived	29
Mouse	72,600	95.0	54	Narcosis	Died	30
	72,600	95.0	28	Narcosis	Died	30
	54,000	70.7	54	Narcosis	Died	30
	48,000	62.8	24	Narcosis	Survived	30
	10,000	13.1	230	Ataxia	Survived	30
	152,800	200.0	94 min.	Narcosis		31
	101,600	133.0	91 min.	Narcosis		31
	91,700	120.0	95 min.	Narcosis	Over-all	31
	76,400	100.0	89 min.	Narcosis	mortality	31
	61,100	80.0	134 min.	Narcosis	45%	31
	45,800	60.0	153 min.	Narcosis		31
	30,600	40.0	190 min.	Narcosis		31
	173,000	227.0			Died	28
	139,000	182.0		Highest concentration endurable		28
Rat	60,000	78.5	2.5	Narcosis, convulsions		32
	31,600	41.4	18–20		Died	32
	22,500	29.5	8	Narcosis		32
	13,000	17.0	24	Prostration		32
	8,800	11.5	8	Lethargy		32
	4,800	6.3	8	None		32
	3,000	4.0	8	None		32
	50,000	65.4	1	Drowsiness	Survived	33
Dog	37,000	48.4	8	Prostration, incoordination		32
	13,700	17.9	4	None		32
	2,000	2.6	24	None		32
	10,000	13.1	3 min. 8 times each day at hourly intervals for 100 days	None	Survived	34
Dogs and pups	450–500	0.59–0.65	8 hr./day, 7 days/week, for 379 days	None	Survived	34
Monkey, rabbit, rat	40,000	52.4	4	Illness	Death	36
	40,000	52.4	1 daily		Delayed death	36
	10,000	13.1	18 daily		Death	36
	10,000	13.1	7 daily for several weeks		Delayed death	36
	1,000	1.3	41		Death	36

to determine methyl alcohol in blood tissues and expired air. The colorimetric measurement of the diffuse violet color, produced by the reaction products of methanol, permanganate, and manganese dioxide with chromotropic acid, is the basis of a method by Ozburn[27] for the determination of methyl alcohol in blood and body fluids.

5. Physiological Response

Animal Symptomatology. Exposure of animals to concentrations of methyl alcohol in air may induce the following signs of intoxication: increased rate of respiration, a state of nervous depression followed by excitation, irritation of the mucous membranes, ataxia, partial paralysis, agony, prostration, deep narcosis, convulsions, decrease in rectal temperature, loss in weight, and death due to respiratory failure. The narcotic effect of methyl alcohol is weaker than that of ethyl alcohol, but the toxic effect from accumulated doses of methyl alcohol, owing to slow elimination, is greater than that of ethyl alcohol. There is not a very wide difference between the concentration necessary to produce narcosis and that which is lethal. Temporary or permanent visual disturbances and blindness may result from repeated exposure to intermediate concentrations. Numerous investigators have shown that the toxicity of methyl alcohol is due to its inherent properties or to those of its metabolites and not to some contaminating substance.

The minimum lethal concentration of methyl alcohol in the air breathed by different animals over short periods of time varies widely with the species and according to the investigator (Table 1). The highest value is that reported by Bachem,[28] who found that mice tolerated 139,000 p.p.m., but died if exposed for an unstated period of time to 173,000 p.p.m. Mice exposed to air containing 48,000 p.p.m. for 3.5 to 4 hours daily up to a cumulative total of 24 hours were in a state of narcosis, but survived, whereas they succumbed in coma when correspondingly exposed for 54 hours to air containing 54,000 p.p.m.[30] Cats survived a 6-hour exposure to 18,300 p.p.m. without any signs of intoxication other than an initial salivation, but when exposed for 6 hours to 33,600 p.p.m. they suffered incoordination and 50 per cent died.[29] Rats became drowsy when exposed for 1 hour to

[27] E. T. Ozburn, *U. S. Naval Med. Bull.*, **46,** 1170 (1946).

[28] C. Bachem, *Arch. exptl. Pathol. Pharmakol.*, **122,** 69 (1927).

[29] R. Witte, Dissertation, Würzburg, 1931.

[30] H. Weese, *Arch. exptl. Pathol. Pharmakol.*, **135,** 118 (1928).

[31] L. M. Mashbitz, R. M. Sklianskaya, and F. M. Urieva, *J. Ind. Hyg. Toxicol.*, **18,** 117 (1936).

[32] A. Loewy and R. von der Heide, *Biochem. Z.*, **65,** 230 (1914).

[33] R. Müller (cited in Loewy and von der Heide), *Z. angew. Chem.*, **23,** 351 (1910).

[34] R. R. Sayers, W. P. Yant, H. H. Schrenk, J. Chornyak, S. J. Pearce, F. A. Patty, and J. G. Linn, *J. Ind. Hyg. Toxicol.*, **26,** 255 (1944).

[35] R. R. Sayers, W. P. Yant, H. H. Schrenk, J. Chornyak, S. J. Pearce, F. A. Patty, and J. G. Linn, *U. S. Bur. Mines Rept. Invest.* No. **3617** (1942).

[36] C. P. McCord, *Ind. Eng. Chem.*, **23,** 931 (1931).

49,700 p.p.m., but survived without any aftereffect;[37] they died when exposed for 18 to 20 hours to a concentration of 31,600 p.p.m.[32] The lowest fatal concentration for animals was reported by McCord[36] in relation to monkeys, some of which died after a few 18-hour exposures on successive days to the concentration of 1000 p.p.m. Tyson and Schoenberg[38] also found the monkey more susceptible than the guinea pig, rabbit, or dog. It is not surprising that such divergent results have been reported if one recognizes such experimental variables as the different means employed for vaporizing the liquid, the differences in the methods of exposing the animals (some of the inhalation chambers had inadequate air exchange), the qualitative and quantitative variability of the analytical procedures for the determination of the actual concentrations in the air breathed by the animals, and the variations in the age and state of health of the experimental animals. However, despite the variability of the recorded data, it seems probable that it would be dangerous for men to be exposed to the vapors of methyl alcohol in concentrations of the order of 30,000 to 50,000 p.p.m. for as much as 30 to 60 minutes.

A valuable experiment conducted upon dogs for 379 consecutive days by Sayers and his associates[35] has revealed that repeated exposure of this animal to 450 or 500 p.p.m. for 8 hours daily is innocuous. Following preliminary observation, the dogs were exposed to a vapor-air concentration maintained within the stated limits as demonstrated analytically by the method of Chapin. The animals showed no unusual behavior, impairment of vision, or loss of weight, and all survived. Ophthalmoscopic examinations disclosed no remarkable abnormalities. There were no significant changes in the formed elements or the chemical constituents of the blood of the animals, nor were there gross or microscopic abnormalities in their tissue at necropsy. The concentration of methyl alcohol in the blood of these dogs at the end of an 8-hour exposure generally varied between 10 and 15 mg. per 100 ml. of blood, but occasional concentrations as high as 52 mg. were found.

A second report by Sayers and his associates,[34] who exposed two dogs to 10,000 p.p.m. for about 3 minutes in each of 8 periods per day at hourly intervals on 100 consecutive days, reveals no noteworthy effects attributable to methyl alcohol poisoning. Median values of 6.5 mg. and 14 mg. per 100 ml., respectively, were obtained for the methyl alcohol concentration in the blood of these two animals.

Pathology in Animals. The following pathological changes found in the tissues of animals exposed to inhalation of methyl alcohol are quite similar to those observed in animals following ingestion of this compound. In the eyes of the dog Tyson and Schoenberg[39] found hyperemia of choroid, and edema of the ocular tissue with early signs of degeneration of the ganglionic cells of the retina

[37] K. B. Lehmann and F. Flury, *Toxikologie und Hygiene der technischen Lösungsmittel.* Springer, Berlin, 1938, p. 149.

[38] H. H. Tyson and M. J. Schoenberg, *Arch. Ophthalmol.,* **44,** 275 (1915).

[39] H. H. Tyson and M. J. Schoenberg, *J. Am. Med. Assoc.,* **63,** 915 (1914).

and nerve fibers. Scott and his associates[40] also found that the vessels of the choroid of poisoned animals were markedly congested, the entire retina was edematous, and the ganglion cells were degenerated. Occasionally there were degenerative changes and fibrosis of the optic nerve. Although Weese[30] observed degenerative alterations in retinas of mice, he did not attribute them to the effects of methyl alcohol.

Petechial hemorrhages in the lungs of dogs were seen by Tyson and Schoenberg.[39] Lehmann and Flury[37] reported the occurrence of pulmonary edema. Rabbits developed patchy bronchopneumonia,[41] and the lungs of poisoned mice were extensively so involved if they survived for 24 hours, according to Weese.[30] In cases of milder poisoning, Scott and associates[40] found edema, congestion, and desquamation of alveolar epithelium, and in more advanced cases there was terminal pneumonic consolidation.

The only pathological change found by Tyson and Schoenberg[39] in the livers of monkeys, dogs, and rabbits was a slight darkness (congestion). The livers of the poisoned rabbits were increased in size, friable, and involved in an albuminous degeneration (cloudy swelling), fatty degeneration, with also an increased amount of connective tissue in those animals subjected to repeated exposures or examined some time after exposure had been terminated.[41] The cell nuclei of the livers of mice were unaltered, but Weese[30] found mild fatty infiltration of the liver parenchyma. Lehmann and Flury[37] reported the occurrence of mild fatty infiltration of the liver as the consequence of brief, severe exposure and more severe fatty changes in association with prolonged or frequently repeated exposures. Scott and associates[40] found in the livers of rats, monkeys, and rabbits parenchymatous degeneration with focal necroses.

In the kidney Tyson and Schoenberg[39] found only dark purple congestion. Albuminous and fatty degeneration is reported by Eisenberg.[41] Damage to the kidneys of mice, characterized principally by fatty infiltration, was described by Weese.[30] Parenchymatous degeneration of the epithelium lining of the convoluted tubules was seen by Scott and associates.[40] Fatty infiltration of the kidney following brief high exposures and more severe fatty changes in repeated exposures were reported by Lehmann and Flury.[37]

No alteration in the hearts of mice was observed by Weese.[30] Cardiac dilatation with vascular engorgement was reported by Lehmann and Flury.[37] The muscle was dark and the cavities were empty when examined by Tyson and Schoenberg.[39] Eisenberg[41] noted fatty degeneration of the heart muscle with occasional fragmentation and segmentation of the muscle cells. Edema, granular degeneration, and, in some instances, necrosis of heart muscle fibers were described by Scott and associates.[40]

Degenerative injuries of the central nervous system have been described by Lehmann and Flury.[37] The meninges of dogs showed marked congestion, accord-

[40] E. Scott, M. K. Helz, and C. P. McCord, *Am. J. Clin. Pathol.*, **3,** 311 (1933).

[41] A. A. Eisenberg, *Am. J. Public Health,* **7,** 765 (1917).

ing to Tyson and Schoenberg.[39] In the rabbit, disintegration of nerve cells of the cerebrum, with actual atrophy and diffuse fibrosis, was described by Eisenberg.[41] Pathological changes in the central nervous system were manifested by capillary congestion, edema, and patchy degeneration in the neurons.[40] The cellular degeneration occurred more often in the spinal cord than in the brain.

Hyperplasia of the lymph nodes was reported by Scott and co-workers.[40] The spleen was a dark indigo blue.[39] Pin-point hemorrhages and congestion of the gastric mucosa were believed by Tyson and Schoenberg[39] to be characteristic of poisoning from the inhalation of methyl alcohol.

Absorption, Distribution, and Excretion. The distribution of methyl alcohol within the tissues of dogs exposed to 4000 and 15,000 p.p.m. in air over periods ranging from 12 hours to 5 days was found to be rapid.[42] The quantities found in the various tissues were correlated to their water content; although the differences were not great, the highest concentrations were found in the blood, eye fluid, bile, and urine, and the lowest in the bone marrow and fatty tissue.

Loewy and von der Heide[32] showed that the lipoid solubility of methyl alcohol is slight and that fat rats absorb less proportionally than the lean. The entire carcasses of rats exposed for 8 hours to 4500, 8500, or 22,500 p.p.m. contained 0.65, 2.0, and 4.3 g. of methanol/kg. of body weight, respectively.

One to 7 mg. of methyl alcohol/g. of blood (100 to 700 mg. per 100 ml.) was found by Haggard and Greenberg[43] in the blood of rats following oral administration of 4 g. of methyl alcohol/kg. of body weight. Seventy per cent of the methyl alcohol lost by the animals was eliminated in the expired air. The amount eliminated in unit time was determined by the concentration in the blood and the volume of the pulmonary ventilation. This was demonstrated by the exponential curve that resulted when the concentration of methyl alcohol in the blood was plotted against the time elapsing after the administration of methyl alcohol, under varying conditions of pulmonary ventilation. The rate of elimination of methyl alcohol from the blood was increased when pulmonary ventilation was increased by carbon dioxide or 2,4-dinitrophenol. (Newman and Tainter[44] have shown that the increased elimination of methyl alcohol from the blood, following intramuscular injection of dinitrophenol, is due to pulmonary ventilation and not to oxidation, by the simple expedient of determining the rate of methyl alcohol elimination on the part of dogs so injected, under conditions of enforced rebreathing of the expired air, as compared with free breathing of ordinary atmosphere.)

Leaf and Zatman[44a] have studied the metabolism of methanol under carefully controlled experimental conditions in man following ingestion and inhalation. Dosages of 71 to 84 mg./kg. orally resulted in blood levels of 4.7 to 7.6 mg. per 100 ml. of blood 2 to 3 hours afterward. The urine/blood concentration ratio was

[42] W. P. Yant and H. H. Schrenk, *J. Ind. Hyg. Toxicol.,* **19,** 337 (1937).
[43] H. W. Haggard and L. A. Greenberg, *J. Pharmacol.,* **66,** 479 (1939).
[44] H. W. Newman and M. L. Tainter, *J. Pharmacol.,* **57,** 67 (1936).
[44a] G. Leaf and L. J. Zatman, *Brit. J. Ind. Med.,* **9,** 19 (1952).

found to be relatively constant at about 1.30, similar to that found for ethanol in man by Haggard.[43] This suggests that the urinary methanol concentration may be a reliable index of the concentration in body water during the excretory period. Urine concentrations reached a peak rapidly in about an hour and declined exponentially reaching the blank or control value in 13 to 16 hours.

These excretion curves can be represented approximately by an equation

$$\log C = \log C_0 - kt$$

where C_0 and C are urinary concentrations at zero time and at t hours respectively. The constant k was very similar in the three individuals studied (0.094 to 0.108). With these small dosages, the amount of the dose accounted for in the urine only averaged 0.7 per cent of the total with amounts in a few expired air studies only slightly larger, indicating that metabolism (probably by oxidation) was accounting for the major part of the dose.

Inhalation of from 500 to 1100 p.p.m. methanol for periods of 3 to 4 hours gave urine concentrations of about 1 to 3 mg. per 100 ml. of urine at the end of exposure that seemed to be well correlated with time and exposure levels.

The ingestion of 15 ml. of ethanol at the same time as 4 ml. of methanol gave a marked elevation in the peak concentration of methanol excretion in urine, although the time to return to a control level of methanol was unaltered. This indicated an inhibition of methanol oxidation by ethanol. An even more dramatic effect was noted when 10-ml. doses of ethanol were given hourly.

The extent of the oxidation of methyl alcohol to formaldehyde or formic acid, and the role that these substances play in determining the toxicity of methyl alcohol have not been completely established. However, there is evidence in support of their occurrence as metabolic products. Although it has been established that formaldehyde can remain intact for but a short time because of its reaction with proteins, Keeser[45] found it present for a short time in vitreous humor, spinal fluid, and abdominal fluids of rabbits poisoned by methyl alcohol. Keeser[46] also showed that methyl alcohol was oxidized in vitro by the freshly prepared vitreous humor of calves, but that hexamethylenetetramine could be formed if ammonium carbonate were present, and the resulting amount of formaldehyde were materially decreased. The evidence is somewhat stronger for the formation of formic acid. Following inhalation of vapors of methyl alcohol by dogs, Tyson and Schoenberg[39] found an increase in the electroconductivity of blood due to an increase in H-ion content, which was substantiated by alkaline titration. The excretion of formic acid for several days following administration of methyl alcohol was reported by Pohl[47] and Hunt.[48] Formic acid was excreted in the urine of rabbits during inhalation of methyl alcohol, according to Bachem.[28] A few milligrams of formate were

[45] E. Keeser, *Deut. med. Wochschr.*, **57**, 398 (1931).

[46] E. Keeser, *Arch. exptl. Pathol. Pharmarkol.*, **160**, 687 (1931)

[47] J. Pohl, *Arch. exptl. Pathol. Pharmakol.*, **31**, 281 (1893).

[48] R. Hunt, *Bull. Johns Hopkins Hosp.*, **13**, 213 (1902).

found in the blood, muscle, kidney, and lung of a 7-kg. animal by Pohl[47] on the day following the administration of 25 ml. of methyl alcohol. These findings indicate there is very little storage of formate ion in the body. Bastrup[49] found that rabbits that had been given a single oral dose of methyl alcohol (2 to 10 g./kg. of body weight) excreted 0.1 to 1.1 per cent of it as formate and 13 to 20 per cent of it as methyl alcohol in the urine within 47 to 143 hours. Dogs, however, excreted in the urine 5 to 15 per cent of an oral dose of methyl alcohol (1 to 2 g./kg.) as formate and 5 to 8 per cent as unchanged methyl alcohol. Lund[50,51] likewise found that the dog excreted more formate than the rabbit after the ingestion of methanol. The rabbit excreted 10 per cent of the dose (2.38 g./kg.) unchanged in the urine within 95 hours but the urinary formate was only questionably above the normal amount. In the case of 2 dogs given the dosages of 1.70 and 1.97 g./kg., respectively, 6.0 and 8.7 per cent were recovered in urine as methanol and 22.4 and 23.7 per cent were found in the urine as formate within 100 hours.

When kept in prolonged contact with the skin, liquid methyl alcohol induces a moderate feeling of local warmth, slight local hyperemia, and eventually a dryness and brittleness of the involved skin.[37] Apparently, methyl alcohol can penetrate the skin in sufficient quantity to cause fatal intoxication, although the available information on this point is somewhat inadequate and is certainly inconclusive with respect to the dosage required. McCord[36] is authority for the statement that the application of 0.5 ml./kg. of body weight upon the skin of a monkey produced illness and death, but the details of the experiment are not given and the means by which inhalation of vapor was prevented were not described. Yant and his associates[52] drenched the entire bodies of unshaven dogs for several hours in such a manner as to eliminate any inhalation of the vapor. According to Schrenk,[53] the unpublished data for these experiments reveal, from the amount of methyl alcohol in the blood, that so far as dogs are concerned the possibility of poisoning by this method was not great and certainly was very much less than from inhalation. Ocular disturbances and blindness in man have been reported by Campbell[54] and Woods[55] from repeated rubbing of the skin with methyl alcohol under conditions that did not prevent inhalation of the vapor. As may be seen, the available evidence as to the magnitude of the hazard from the percutaneous absorption is somewhat inadequate, but such as it is, it suggests that prolonged or frequently repeated exposure by this means should be avoided.

Effects upon Man. From the time of Buller and Wood[11] in 1904 until the present time, numerous cases of blindness and death due to the drinking of a few ounces of methyl alcohol have been reported. Articles by Jacobson and his asso-

[49] J. T. Bastrup, *Acta pharmacol.*, **3**, 303 (1947).
[50] A. Lund. *Acta. pharmacol.*, **4**, 99 (1948).
[51] A. Lund, *Acta, pharmacol.*, **4**, 108 (1948).
[52] W. P. Yant, H. H. Schrenk, and R. R. Sayers, *Ind. Eng. Chem.*, **23**, 551 (1931).
[53] H. H. Schrenk, personal communication.
[54] J. A. Campbell, *J. Ophthalmol. Otol. Laryngol.*, **21**, 756 (1915).
[55] H. Woods, *J. Am. Med. Assoc.*, **60**, 1762 (1913).

ciates,[56] Kaplan and Levreault,[57] and Voegtlin and Watts[58] record 24 deaths from drinking methyl alcohol. Twenty-three cases of poisoning following the ingestion of methanol by men in Korea were reported by Keeney and Mellinkoff.[59] Six of these died but the others sustained no permanent injury, although they manifested nausea, epigastric pain, vomiting, headache, dizziness, delirium, varying degrees of transitory blindness, acidosis, and acetonuria. Hughes[60] reported the loss of vision of 3 workers following the ingestion of methanol. In acute cases, the use of intravenous injection of sodium bicarbonate or sodium lactate and glucose in physiological saline has been recommended by Johnstone,[61] Jacobson et al.,[56] and Voegtlin.[58] These investigators have also recommended the use of emetics, a high fluid intake, cardiac and respiratory stimulants, and oxygen or artificial respiration. Chew et al.[62] and Roe[63,64] have also recommended the alkali treatment for methyl alcohol poisoning but disagree on the efficacy of the use of ethyl alcohol for poisoning from methyl alcohol. The studies of Leaf and Zatman[44a] appear, however, to support the suggestion by Roe that ethanol may be a useful therapeutic agent in methanol poisoning.

Keeney and Mellinkoff[59] suggested that intravenous injection of glucose may be an adjuvant in the treatment of ketosis. Suprunov[65] has recommended the administration of vitamin C and thiamine in cases of poisoning by methyl alcohol, since he found a reduced amount of these vitamins in the tissues of rabbits following subcutaneous injections of sublethal amounts of methyl alcohol.

Although the individual responses of man to methyl alcohol may vary considerably, industrial exposures are not very hazardous if concentrations are maintained within the upper limit of 200 p.p.m. by proper ventilation. Under varying conditions of severity and duration of exposure to the vapor of methyl alcohol the signs of intoxication may include: irritation of all the mucous membranes, headache, roaring in the ears, tiredness, insomnia, nystagmus, trembling, vertigo, unsteady gait, dyspnea, nausea, vomiting, colic, constipation, dilated pupils, clouded vision, diplopia, blindness, itching of the skin, eczema, and dermatitis.[4,9,10,14,66]

General physical, ocular, and hematological examinations by Greenberg and his associates[13] of 19 workers who had been repeatedly exposed to 22 to 25 p.p.m.

[56] B. M. Jacobson, H. K. Russell, J. J. Grimm, and E. C. Fox, *U. S. Naval Med. Bull.,* **44,** 1099 (1945).

[57] A. Kaplan and G. V. Levreault, *U. S. Naval Med. Bull.,* **44,** 1107 (1945).

[58] W. L. Voegtlin and C. E. Watts, *U. S. Naval Med. Bull.,* **41,** 1715 (1943).

[59] A. H. Keeney, and S. M. Mellinkoff, *Ann. Int. Med.,* **34,** 331 (1951).

[60] J. P. Hughes, *J. Am. Med. Assoc.,* **156,** 234 (1954).

[61] R. T. Johnstone, *Occupational Diseases.* Saunders, Philadelphia, 1941, p. 169.

[62] W. B. Chew, E. H. Berger, O. A. Brines, and M. J. Capron, *J. Am. Med. Assoc.,* **130,** 61 (1946).

[63] O. Roe, *Acta Med. Scand.,* **126,** 182, 253 (1946); *Chem. Abstr.,* **41,** 2805 (1947).

[64] O. Roe, *Quart. J. Studies Alc.,* **11,** 107 (1950).

[65] A. T. Suprunov, *Farmakol. i. Toksikol.,* **9,** 49 (1946); *Chem. Abstr.,* **41,** 3221 (1947).

[66] F. Flury and F. Zernik, *Schädliche Gase.* Springer, Berlin, 1931.

of methyl alcohol and 40 to 45 p.p.m. of acetone revealed no significant abnormalities. Likewise, a survey by Yant and associates[52] of 36 men employed in the manufacture of methyl alcohol and of 24 drivers of trucks using methyl alcohol as an antifreeze disclosed no harmful effects. The latter group of investigators found the concentration of methyl alcohol to be higher in the urine and blood in the evening than in the morning, but the amounts were all less than those found in the urine and blood of dogs poisoned by the inhalation of methyl alcohol.

Elkins[67] has measured the concentration of methanol in the atmosphere of three plants and concurrently determined the concentration of methanol in the urine of the workers. The results were as follows:

Average methanol in air (p.p.m.)	Methanol in urine (mg.%)
125	0.3, 0.6
440	0.35, 0.65, 0.9
2800	1.6, 4.4

The concentration of methanol in the blood was determined at the time of death of 6 men following the ingestion of methanol. Keeney and Mellinkoff[59] found the average concentration was 71.1 mg. per 100 ml. of blood with the values ranging from 15.6 to 150. In 5 other fatal cases Lund[68] reported the following values:

	Mg.%	
	Methanol	Formic acid
Blood	74–110	9–68
Urine	140–240	216–785
Liver	106	60–99

Lund[68] found no noteworthy increase in methanol or formic acid in the blood or urine of humans following the ingestion of a nonfatal estimated dose of 10 to 20 ml. of methanol. However 48 hours following the ingestion of nonfatal estimated doses of about 50 to 80 ml. of methanol, formic acid could be demonstrated in the blood (2.6–7.6 mg.%), as well as an increased excretion in the urine (54–205 mg.%).

Bennett et al.[69] described the symptoms, laboratory and physical findings, pathological alterations, and treatment following an epidemic of acute methyl alcohol poisoning (in Atlanta) that resulted from the ingestion of bootleg whiskey (mortality was 6.2 per cent among 323 cases). These authors suggest that depression of carbon dioxide production is probably related to interference with controlling enzyme systems including succinic dehydrogenase.

[67] H. B. Elkins, personal communication.
[68] A. Lund, Acta pharmakol., **4**, 205 (1948).
[69] I. L. Bennett, F. H. Casey, G. L. Mitchell, and M. N. Cooper, Medicine, **32**, 431 (1935).

Human Pathology. Examination by Province *et al.*[70] of the tissues of 5 persons fatally poisoned by the ingestion of methyl alcohol disclosed the occurrence of catarrhal gastritis, acute enteritis, focal necrosis of the liver with infiltration by polymorphonuclear leucocytes, pulmonary edema, and early degeneration of the neurons of the brain. Chew *et al.*[62] at necropsy of 5 other cases found cerebral edema, hypostatic pulmonary congestion, fatty infiltration of the liver, and passive congestion of all organs. On examining 11 similar cases Tonning[71] found superficial necrosis of the stomach accompanied by mucous distensions of epithelial cells, parenchymatous degeneration of the liver, engorgement of the pulmonary vessels, edema and hyperemia of the brain accompanied by occasional punctate hemorrhages and accumulation of brown pigment in the neurons, and irregular staining of the ganglion cells of the retina accompanied by eccentric nuclei, fraying, vacuolation, and autolysis. Necrosis of the pancreas of a woman succumbing from ingestion of methyl alcohol was the most characteristic finding of Branch.[72] The fatal cases from ingestion of methanol that were studied by Keeney and Mellinkoff[59] revealed gastritis, pulmonary congestion, edema, and patchy necrosis, uniformly, with bronchial desquamation in 2 of 6 necropsies and bronchopneumonia in 1 case. Severe congestion of the glomerular tufts and cloudy swelling of the convoluted tubules, hepatic infiltration, congestion and parenchymal hemorrhage of the pancreas, and congestion of the spleen were also observed.

Upon examination of the ganglion cells of the retina of humans dying of methanol, Roe[73] found severe degenerative cells. Roe attributes diminution in vision to acidosis, since he failed to obtain degenerative changes in the ganglion cells of rats and rabbits, which have a greater alkaline reserve than man.

6. *Hygienic Standard of Permissible Exposure*

The threshold limit for methyl alcohol has been established at 200 p.p.m.[74-77]

7. *Flammability*

The ignition temperature is 867°F. (see Vol. I, page 528). The lower and upper flammable limits are, respectively, 6.72 and 36.50 per cent by volume[78] (see Vol. I, page 516).

[70] W. D. Province, R. A. Kritzler, and F. P. Calhoun, *Bull. U. S. Army Med. Dept.,* **5,** 114 (1946).

[71] D. J. Tonning, *Nova Scotia Med. Bull.,* **24,** 1 (1945).

[72] A. Branch, *Can. Med. Assoc. J.,* **51,** 428 (1944).

[73] O. Roe, *Acta Ophthalmol.,* **26,** 169 (1948).

[74] *Am. Standards Assoc.,* Allowable concentration of methanol, Z37.14–1944.

[75] American Conference of Governmental Industrial Hygienists' Threshold Limit Values for 1959, *A.M.A. Arch. Ind. Health,* **20,** 266 (1959).

[76] *Hygienic Guide.* Am. Ind. Hyg. Assoc., Chicago, Ill., December 1957.

[77] H. F. Smyth, Jr., *Am. Ind. Hyg. Assoc. Quart.,* **17,** 129 (1956).

[78] G. W. Jones, *Chem. Revs.,* **22,** 1 (1938).

8. *Odor and Warning Properties*

Methyl alcohol does not have suitable warning odor or irritating properties except at high concentrations. Witte[29] found an initial salivation by cats when exposed to 18,300 p.p.m. (24 mg./liter). It becomes unendurable in concentrations of 50,000 p.p.m. (65 mg./liter).[37] Scherberger, Happ, Miller, and Fassett[79] state that a level of 2000 p.p.m. of methyl alcohol is barely detectable by odor.

ETHYL ALCOHOL, C_2H_5OH (Ethanol)

1. *Source, Uses, and Industrial Exposures*

Greater quantities of ethyl alcohol are synthesized from ethylene and sulfuric acid than are fermented from such agricultural products as molasses, starch, and cellulose.[80,81] Anhydrous ethanol is manufactured on a large scale by azeotropic distillation.[80]

Ethyl alcohol or ethanol is also called alcohol, grain alcohol, ethyl hydrate, spirit of wine, or cologne spirits. Mellan[82] lists the quantity of ethanol used in the manufacture of synthetic rubber, as an intermediate in the manufacture of certain chemicals and medicinals, as an antifreeze, as fuel, as a solvent or processing agent for various purposes including explosives, plastics, synthetic resins, nitrocellulose, lacquers, pharmaceuticals, cosmetics, adhesives, inks, and preservatives. Numerous denaturants for ethyl alcohol have been listed by Zangger.[83] Mellan[84] lists denaturants authorized by the U. S. Treasury Department.

On the assumption that there was one change of air per hour in workrooms in which violins, artificial flowers, shoes, and hats were being processed, Loewy[85] estimated the prevailing concentrations of ethyl alcohol in the air through calculations based on the volume of the rooms and the quantities of alcohol vaporized therein. He arrived at values of 200 to 1000 p.p.m. in six rooms, 1000 to 5000 p.p.m. in seven rooms, and more than 5000 p.p.m. in three rooms. The hazards from industrial exposure to ethyl alcohol are low.

2. *Physical and Chemical Properties*

Physical state: colorless, volatile liquid
Molecular weight: 46.070[86]
Specific gravity: 0.7904 (20/20°C.)[82]

[79] R. F. Scherberger, G. P. Happ, F. A. Miller, and D. W. Fassett, *Am. Ind. Hyg. Assoc. J.*, **19**, 494 (1958).

[80] I. Mellan, *Industrial Solvents*. Reinhold, New York, 1950, p. 454.

[81] A. K. Doolittle, *The Technology of Solvents and Plasticizers*. Wiley, New York, 1954, p. 617.

[82] I. Mellan, *Industrial Solvents*. Reinhold, New York, 1950, pp. 460–466.

[83] H. Zangger, *Arch. Gewerbepathol. Gewerbehyg.*, **2**, 205 (1931).

[84] I. Mellan, *Industrial Solvents*. Reinhold, New York, 1950, p. 461.

[85] A. Loewy and R. von der Heide, *Biochem. Z.*, **86**, 125 (1918).

[86] "Atomic Weights (International), 1957," *J. Am. Chem. Soc.*, **80**, 4122 (1958).

Refractive index: 1.3633 (15°C.)[82]

Melting point: —116°C.[82]

Boiling point: 78.4°C.[82]

Vapor density: 1.59 (air = 1)

Vapor pressure: 50 mm. Hg (25°C.)[87]

Per cent in "saturated" air: 6.58 (25°C.)

Density of "saturated" air: 1.04 (air = 1)

Solubility: miscible in all proportions with water and with most organic solvents.

Flash point: 65°F. (95 per cent alcohol); 57°F. (absolute alcohol)[82]

1 mg./liter $\backsimeq$ 532 p.p.m. and 1 p.p.m. $\backsimeq$ 1.88 mg./cu. meter at 25°C., 760 mm. Hg

3. Determination in the Atmosphere

Although no method for the determination of ethyl alcohol has been generally adopted, there are several satisfactory but nonspecific methods available. Jacobs[88] states that ethyl alcohol in the air can be determined by measurement of the specific gravity of its aqueous solution if the sample is bubbled through water until the latter contains more than 0.1 per cent alcohol by volume. The immersion refractometer can be employed if more than 0.5 per cent by volume of alcohol is present in water. Haggard and Greenberg[89] determined the amount of alcohol in air by passing it over iodine pentoxide and measuring the liberated iodine by titration with sodium sulfite. Hydriodic acid, which is also liberated by the pentoxide, is reacted with iodate to liberate additional iodine, which is also determined by titration with sodium sulfite.

The oxidation of alcohol to acetic acid by potassium bichromate according to the procedure of Nicloux[90] has been applied to the determination of alcohol in blood and animal tissues and should be applicable to the determination of alcohol in air. After distillation of the acetic acid formed by this reaction, Gettler and Tiber[91] titrated it with 0.05N alkali using phenolphthalein. The quantity of bichromate consumed in the Nicloux reaction was determined iodometrically by McNally[92] or measured by employing methyl orange to determine when an excess of ferrous sulfate was present.[93]

---

[87] *International Critical Tables of Numerical Data, Physics, Chemistry, and Technology,* Vol. III. McGraw-Hill, New York, 1928, p. 27.

[88] M. B. Jacobs, *Analytical Chemistry of Industrial Poisons, Hazards, and Solvents.* Interscience Publishers, New York, 1941, p. 482.

[89] H. W. Haggard and L. A. Greenberg, *J. Pharmacol.,* **52,** 137 (1934).

[90] M. Nicloux, *Compt. rend. soc. biol.,* **3,** 841 (1896).

[91] A. O. Gettler and A. Tiber, *Arch. Pathol. Lab. Med.,* **3,** 218 (1927).

[92] W. D. McNally, *Toxicology.* Industrial Medicine, Chicago, 1937, p. 648.

[93] R. N. Harger, *J. Lab. Clin. Med.,* **20,** 746 (1935).

Ethyl alcohol can be determined in the presence of methyl alcohol by the use of alkaline and acid permanganate.[94] Ethyl alcohol has been determined in the presence of other alcohols by oxidation of the alcohols to fatty acids, which are subsequently determined by partition between isopropyl ether and water.[95] Acetone can be destroyed in the presence of ethyl alcohol by paraformaldehyde.[96]

4. Determination in Tissues and Blood

Most methods for the determination of ethyl alcohol in tissues or blood are based upon measurement of the quantity of dichromate used up in oxidizing the alcohol. Ferrous sulfate has been employed by some investigators to titrate the excess dichromate. Harger[93] employed methyl orange as an indicator and Chaikelis and Floersheim[97] used Congo red. Rochat[98] removed the excess dichromate with Mohr's salt and subsequently titrated with permanganate. McNally and Coleman[99] titrated the excess dichromate with sodium thiosulfate in the presence of potassium iodide. Gingras and Gaudry[100] based their procedure upon the measurement, at 600 mμ, of the green-colored chromic sulfate. Henry et al.[101] oxidized the alcohol to acetaldehyde with dichromate, subsequently determining the aldehyde colorimetrically by its reaction with p-hydroxydiphenyl.

5. Physiological Response

Animal Symptomatology. The toxicological response of animals to ethyl alcohol following its ingestion or application on the skin is given in Table 2, while the physiological response to its inhalation is presented in Table 3.

Animals exposed to ethyl alcohol in air may manifest the following signs of intoxication: slight irritation of the mucous membranes, excitation followed by ataxia, drowsiness, prostration, narcosis, twitching, general paralysis, dyspnea, and occasionally death associated with respiratory failure. Ethyl alcohol has a stronger narcotic effect than methyl alcohol. Although the actual concentration of ethyl alcohol found in the tissues of fatally poisoned animals is less than that of methyl alcohol under corresponding circumstances, the much more rapid oxidation, and hence the slower accumulation of ethyl alcohol in the body, renders ethyl alcohol much less hazardous than methyl alcohol. Ethyl alcohol is oxidized to

---

[94] Hepter, *Z. Nahr. Genussm.*, **26**, 342 (1913).

[95] C. H. Werkman and O. L. Osburn, *Ind. Eng. Chem. Anal. Ed.*, **3**, 387 (1931).

[96] R. D. Stanley, *J. Assoc. Offic. Agr. Chemists*, **22**, 594 (1939).

[97] A. S. Chaikelis and R. D. Floersheim, *Am. J. Clin. Pathol.* Tech. Suppl., **10**, 180 (1946), *Biol. Abstr.*, 21, 5367 (1947).

[98] J. Rochat, *Helv. Chim. Acta*, **29**, 819 (1946), *Analyst*, **72**, 450 (1947).

[99] W. D. McNally and H. M. Coleman, *J. Lab. Clin. Med.*, **29**, 429 (1944), *Chem. Abstr.*, **38**, 5858 (1944).

[100] R. Gingras and R. Gaudry, *Laval méd.*, **9**, 661 (1944), *Chem. Abstr.*, **39**, 954 (1945).

[101] R. J. Henry, C. F. Kirkwood, S. Berkman, R. D. Housewright, and J. J. Henry, *J. Lab. Clin. Med.*, **33**, 241 (1948).

TABLE 2

Toxicological Response to Ethyl Alcohol

Animal	Response
Single Dose	
Rat (orally)	LD_{50} 13.7 ml./kg.[102]
Rabbit (orally)	MLD 12.5 ml./kg.[103]
Rat (I.P.)	MLD 8.0 ml./kg.[104]
Guinea pig (I.P.)	LD_{50} 5.6 ml./kg.[105]
Rabbit (I.V.)	MLD 9.4 ml./kg.[106]
Cat (I.V.)	MLD 3.9 ml./kg.[107]
Skin absorption	
Rabbit	LD_{50} >9.4 ml./kg.[108]
Skin irritation	
Rabbit	Slight dryness of skin[108]

carbon dioxide and water, but small amounts of the unoxidized material remain in the blood and are excreted in the urine and expired air for several hours after exposure. Westerfield and associates[109] have shown the transitory presence of acetaldehyde in the oxidation of ethyl alcohol by the body. As in the case of methyl alcohol, there is no difference in the toxicity of natural and synthetic ethyl alcohol.

The physiological effects of the inhalation of ethyl alcohol in various concentrations in air by various animal species are given in Table 3. Dogs subjected to alcohol vapors for 42 minutes showed no signs of intoxication other than slight staggering.[110] The mouse is able to tolerate 23,000 to 25,000 p.p.m.[111-113] but dies when exposed for a short time to 29,000 p.p.m.[111-112] Guinea pigs survived exposure to 13,300 p.p.m. but succumbed when the concentration was 21,900 p.p.m.[114] Loewy and von der Heide[114] showed that the survival time of rats, which were more susceptible than guinea pigs, was roughly inversely proportional to the concentration, being 6.5, 10–15, and 22 hours in relation to concentrations of 44,000, 22,000 and 12,700 p.p.m., respectively. Rats died when exposed for several days to

[102] H. F. Smyth, Jr., *J. Ind. Hyg. Toxicol.*, **23**, 253 (1941).
[103] J. C. Munch and E. W. Schwartze, *J. Lab. Clin. Med.*, **10**, 985 (1925).
[104] L. Lendle, *Arch. exptl. Pathol. Pharmakol.*, **132**, 214 (1928).
[105] H. F. Smyth, Jr., *J. Ind. Hyg. Toxicol.*, **23**, 259 (1941).
[106] A. J. Lehman and H. W. Newman, *J. Pharmacol. Exptl. Therap.*, **61**, 103 (1937).
[107] D. I. Macht, *J. Pharmacol. Exptl., Therap.*, **16**, 1 (1920).
[108] J. F. Treon, unpublished work.
[109] W. W. Westerfield, E. Stotz, and R. L. Berg, *J. Biol. Chem.*, **149**, 237 (1943).
[110] Gréhant and E. Quinquaud, *Compt. rend. soc. Biol.*, **5**, 426 (1883).
[111] C. Bachem, *Arch. exptl. Pathol. Pharmakol.*, **122**, 69 (1927).
[112] K. B. Lehmann and F. Flury, *Toxikologie und Hygiene der technischen Lösungsmittel.* Springer, Berlin, 1938, p. 152.
[113] H. Weese, *Arch. exptl. Pathol. Pharmakol.*, **135**, 118 (1928).
[114] A. Loewy and R. von der Heide, *Biochem. Z.*, **86**, 125 (1918).

TABLE 3

Physiological Effects upon Animals of the Inhalation of Ethyl Alcohol

Animal	Concentration		Duration of exposure, hr.	Signs of intoxication	Outcome	Reference
	p.p.m.	mg./ liter				
Mouse	29,370	55.2	Short (?)		Died	111
	22,980	43.2	Short (?)	Tolerated	Survived	111
	31,900	70.0	0.33	Ataxia		112
	29,300	55.0	7.0	Narcosis	Died	112
	23,940	45.0	1.25	Narcosis		112
	13,300	25.0	1.33	Ataxia		112
	25,000	48.1	125 over several days	Narcosis	Survived	113
	16,700	31.4	24 over several days	Narcosis	Survived	113
	8,350	15.7	29 × 7.2	Ataxia	Survived	113
Guinea pig	45,000	84.6	3.75	Incoordination		85
	44,000	82.7	7.5	Deep narcosis		85
	50,170	94.3	10.2	Deep narcosis	Died	85
	19,260	36.2	3.75	None		85
	20,000	37.6	6.5	Incoordination		85
	21,900	41.2	9.8	Deep narcosis	Died	85
	9,080	17.1	5.25	None		85
	12,850	24.2	8.75	Incoordination		85
	13,300	25.0	24.0	Light narcosis		85
	6,400	12.0	8.0	None	Survived	85
	3,000ᵃ	5.6	64 × 4	None	Survived	116
Rat	32,000	60.1	8.0		Some died	117
	16,000	30.1	8.0		Some died	117
	45,000	84.6	3.75	Deep narcosis		85
	44,000	82.7	6.5	Deep narcosis	Died	85
	19,260	36.2	2.0	Light narcosis		85
	21,960	41.2	9.8	Deep narcosis	Died	85
	18,200	34.2	1.0	Excitation		85
	18,200	34.2	1.75	Incoordination		85
	22,800	42.9	8.0	Deep narcosis		85
	22,100	41.5	15.0	Deep narcosis	Died	85
	10,750	20.2	0.5	None		85
	10,750	20.2	2.0	Incoordination		85
	12,400	23.3	8.5	Deep narcosis		85
	12,700	23.8	21.75	Deep narcosis	Died	85
	5,660	10.6	1.75	Incoordination		85
	6,400	12.3	12.0	Light narcosis	Survived	85
	3,260	6.1	6.0	None		85
	3,260	6.1	8.0	Drowsiness		85
	4,580	8.6	21.13	Ataxia	Survived	85

ᵃ Volatilized from alcohol containing 0.5 per cent benzene.

[115] D. I. Macht, *J. Pharmacol.*, **16,** 1 (1920).

[116] H. F. Smyth and H. F. Smyth, Jr., *J. Ind. Hyg.*, **10,** 261 (1928).

[117] H. F. Smyth, Jr., *Am. Ind. Hyg. Assoc. Quart.*, **17,** 129 (1956).

saturated vapors of ethyl alcohol, according to Macht.[115] Smyth and Smyth[116] exposed guinea pigs for 4 hours per day 6 days a week on 64 exposure days without any untoward effects.

Animal Pathology. Weese[113] reports that reversible fatty infiltration of the liver occurs following repeated exposure to high concentrations. Mertens[118] exposed rabbits to air saturated with alcoholic vapors for periods ranging from 25 to 365 days and thereby induced cirrhosis of the liver as a common lesion. Petri[119] found hemorrhagic perivascular infiltrates in the tissues of animals subjected to high dosage.

Human Pathology. Petri[119] records that Fahr listed the following as the most commonly encountered lesions resulting from prolonged ingestion of toxic quantities of alcohol: fatty infiltration of the liver and heart muscle, chronic leptomeningitis, and chronic gastritis. Brezina,[120] without much proof, accepts certain cardiac disturbances as effects of the inhalation of the vapors of warm alcohol.

Absorption and Excretion. Loewy and von der Heide[114] found that a state of diminished excitability on the part of rats exposed to the vapors of alcohol was associated with the presence in their carcasses of concentrations of alcohol ranging from 0.16 to 0.27 g./kg. of their total weight. Corresponding concentrations of 1 g./kg. were associated with the induction of a state of narcosis, and concentrations ranging from 3.1 to 5.8 g./kg. were found in the bodies of fatally poisoned animals. Guinea pigs, which are more resistant than rats, are severely intoxicated but may survive in spite of the presence of concentrations of alcohol within the latter range. Rats and guinea pigs oxidized 66.5 to 98.9 per cent of the absorbed alcohol[114] when exposed for about 2 hours to air containing 27,600 to 41,400 p.p.m. of ethyl alcohol. Chickens exposed by Carpenter[121] to alcohol vapors for 2 to 29 hours usually had the highest concentration in the blood, although occasionally higher concentrations were found in the brain. The lowest concentrations of alcohol were found in the fat. According to Carpenter,[121] when the concentration of alcohol in the blood of chickens was more than 2.5 g./kg., or when that in the whole body was more than 1.7 g./kg., the animals showed signs of abnormal behavior. Concentrations of 3.7 to 5.6 g./kg. of body weight proved fatal. The concentration of alcohol in the blood of dogs was found to be about 60 per cent higher than that in the body as a whole.[122] This percentage is in excellent agreement with that of Carpenter. The concentration of alcohol in the blood of dogs breathing an undetermined concentration of alcohol rose from 0.8 g./kg. of blood after 2 hours to 4.0 after 6 hours.

[118] H. Mertens, *Arch. intern. pharmacodynamie*, **2**, 127 (1896).

[119] E. Petri, in F. Henke and O. Lubarsch, *Handbuch der speziellen pathologischen Anatomie und Histologie*, Vol. X. Springer, Berlin, 1930, p. 276.

[120] E. Brezina, *Internationale Übersicht über Gewerbekrankheiten*. Springer, Berlin, 1929, p. 83.

[121] T. M. Carpenter, *J. Pharmacol.*, **37**, 217 (1929).

[122] H. W. Haggard and L. A. Greenberg, *J. Pharmacol.*, **52**, 167 (1934).

Following the ingestion by dogs of 3.3 g. of alcohol per kilogram of body weight, Haggard and Greenberg[123] found that the concentration of alcohol in the urine in relation to that in the arterial blood corresponded closely to the relative solubility in vitro of alcohol in urine and blood (1.14 to 1). These observations were interpreted as suggesting that alcohol passes through the kidneys by simple diffusion. During the period of absorption from the gastroenteric tract, the concentration of ethyl alcohol found in a peripheral vein (femoral) was somewhat lower than that in the arterial blood. The same investigators found that during a 16-hour period following ingestion of alcohol by dogs, 2.1 to 4.3 per cent of the total alcohol ingested was eliminated by the kidneys, the rate of elimination within this range being a function of the concurrent urinary volume. The ratio of the concentration of ethyl alcohol in the arterial blood to that in the alveolar air was found to be 1142 to 1 in correspondence with the distribution of alcohol in vitro in the media involved. The total amount eliminated by dogs in the expired air in a period of 8 hours following ingestion of 4 g./kg. amounted to 4 per cent of the total amount ingested.

In a further article Haggard and Greenberg[122] state that the rate of oxidation of ethyl alcohol is proportional to the amount in the body. Marshall and Fritz[124] employed a sensitive enzymic method for the determination of the concentration of ethanol in the plasma of dogs following the intravenous injection of ethanol. With very low concentrations of alcohol in the plasma (5–10 mg.%) the disappearance of ethanol followed an exponential curve, indicating that rate of disappearance was proportional to the concentration present. The rate of disappearance at higher levels was greater than at lower levels but it was not determined whether this was due to an increased rate of oxidation because of the hourly biological variation in metabolism. Aull et al.[125] observed variation of the rate of metabolism among rats given ethanol intraperitoneally in the dosage of 2.85 or 3.0 g./kg. of body weight. Upon the basis that the rate of metabolism of ethanol is independent of the concentration, Aull et al.[125] stated that alcohol disappeared at the rate of 270 mg./kg./hr. when based upon the declining concentration of alcohol in the blood with time and disappeared at the rate of 293 mg./kg./hr. when based upon the difference between the dosage administered and the amount found in the carcass by analytical determination. These later investigators state that the rat metabolizes ethanol more slowly than the mouse and more rapidly than the dog. Jacobsen,[126] who has reviewed the metabolism of ethyl alcohol, stated that until recently the evidence indicated that in man ethanol was metabolized independently of the concentration, but more recent studies by Hjelt in Finland and Goldberg in Sweden show a dependence of the metabolic rate upon the concentration of alcohol present in man. Jacobsen[126] suggested that ethyl alcohol can

[123] H. W. Haggard and L. A. Greenberg, *J. Pharmacol.*, **52**, 150 (1934).

[124] E. K. Marshall, Jr., and W. F. Fritz, *J. Pharmacol. Exptl. Therap.*, **109**, 431 (1953).

[125] J. S. Aull, Jr., W. J. Roberts, Jr. and F. W. Kinard, *Am. J. Physiol.*, **186**, 380 (1956).

[126] E. Jacobsen, *Pharmacol. Revs.*, **4**, 107, (1952).

proceed to acetaldehyde by means of two different enzymes, namely, alcohol dehydrogenase and catalase, and that the shape of the disappearance curve for ethanol will depend on the ratio between the effects of these two systems. If the former prevails the curve will be rectilinear and independent of the concentration, but if the latter takes prominence then the rate would depend upon the concentration.

Jacobsen reported that the rate of oxidation of ethanol to acetaldehyde is much slower than the rate of oxidation of the aldehyde to acetic acid According to Jacobsen, Antabuse (tetraethylthiuram disulfide) causes an increase in acetaldehyde concentration and a reduction in the formation of carbon dioxide. Although this compound interferes very little with the maximum capacity of an organism to metabolize acetaldehyde, it increases the tissue concentration of acetaldehyde necessary for the metabolism of a certain amount of acetaldehyde per amount of time.[126] However, Casier and Polet[127] maintain that tetraethylthiuram disulfide prevents the oxidation of ethanol rather than the accumulation of acetaldehyde. These later investigators propose that the disagreeable effects resulting from ethanol and this disulfide are not caused by the accumulation of acetaldehyde but rather by the combined effects of the alcohol and metabolites of the disulfide. Casier and Polet found no accumulation of radioactive acetaldehyde in mice treated beforehand with tetraethylthiuram disulfide, but instead they observed an increase in volatile reducing substances. Acetate from alcohol was proved by the use of ethyl alcohol labeled with deuterium.[126] Acetic acid from alcohol joins the "acetylic pool" of normal metabolism.

Although the enzyme system or systems that oxidize acetaldehyde to acetic acid (second step of oxidation of ethanol) have not been as well defined as those for the first step of oxidation of ethanol (ethanol to acetaldehyde), this later step may involve the flavoproteins or the dehydrogenases.[126]

Jacobsen[126] reported that the liver is the most important organ capable of oxidizing ethanol. This was shown by rate of disappearance of alcohol from the liver with concurrent measurement of oxygen consumption and the formation of carbon dioxide and acetic acid. Experiments on hepatectomized dogs revealed that very little, if any, ethanol was oxidized.

Eggleston and Smith[128] observed a depression of chloride excretion in the urine of man during a period of diuresis following the ingestion of ethyl alcohol. Lolli et al.[129] suggested that the water lost in diuresis was insufficient in quantity to account for thirst following alcohol intoxication. They suggested that there is a redistribution of intracellular to extracellular water in the body, since rats given ethyl alcohol by mouth or injection showed an increase in extracellular water.

[127] H. Casier and H. Polet, *Radioisotopes in Scientific Research, Proc. First (UNESCO) International Conference. Pergamon,* New York, 1958, page 481.

[128] M. G. Eggleston and I. G. Smith, *J. Physiol.,* **104,** 435 (1946).

[129] G. Lolli, M. Rubin, and L. A. Greenberg, *Quart. J. Studies Alc.,* **5,** 5 (1944).

Ethyl alcohol penetrates the skin of animals, but not at a rate sufficient to induce serious effects. Deichmann[130] found values of 0.13 and 0.04 mg. of alcohol, respectively, per 100 ml. of blood, at 0.5- and 1-hour intervals following one application of 35 ml./kg. of body weight upon the belly of a rabbit protected against inhalation of the vapor. Boughton[131] made 187 daily applications of 10 drops of 50 per cent solution of ethyl alcohol upon the facial skin of rats without injury to the skin, hair, and eyes other than temporary irritation.

Newman and Lehman[132] demonstrated that dogs habituated to drinking alcohol showed better neuromuscular coordination under the influence of a given concentration of alcohol in the blood than did control dogs. Furthermore, since the brains of habituated rats contained a slightly higher concentration of alcohol than did those of control rats under corresponding condition of dosage, these investigators believe that acquired tolerance in animals is primarily an adaptation of the cells of the central nervous system.

Effects upon Man. The well-known effects of chronic alcoholism from the excessive use of alcoholic beverages are not matters of concern in relation to occupational hazards, except to the extent that they may influence the effects of exposure to other substances and environmental conditions. In any case, they do not enter into the present discussion.

The following nonindustrial cases of idiosyncrasy of children to alcohol vapors have been reported: vomiting, unconsciousness, and nystagmus by James;[133] narcosis by Kalt;[134] and death by Leschke.[135]

Although ethyl alcohol is relatively innocuous if proper ventilation is maintained, prolonged exposures to too high a concentration may produce: irritation of the mucous membrane, irritation of the upper respiratory tract, headache, nervousness, dizziness, tremors, fatigue, nausea, and narcosis.[114,136,137] The effects of ethyl alcohol upon the power of concentration and alertness should be remembered in relation to the prevention of industrial accidents. Lehmann and Flury[112] are authority for the statement that intoxication has been seen among human beings subjected to inhalation of the vapors of hot alcohol.

In terms of symptomatology in relation to dosage, there is no doubt that a tolerance is acquired after repeated exposure to alcohol. However, no proof has been submitted of physiological adaptation in man in terms of metabolic changes or of resistance to cellular injuries. Loewy and von der Heide[114] (Table 4) have shown that the symptoms are less severe and the time required to produce them is

[130] W. Deichmann, personal communication.
[131] L. L. Boughton, *J. Am. Pharm. Assoc. Sci. Ed.*, **33**, 111 (1944).
[132] H. W. Newman and A. J. Lehman, *J. Pharmacol.*, **62**, 301 (1938).
[133] V. C. James, *Brit. Med. J.*, **1**, 539 (1931).
[134] A. Kalt, *Schweiz. Korresp.*, **1906**, 725.
[135] E. Leschke, *Münch. med. Wochschr.*, **79**, 751 (1932).
[136] F. Koelsch, *Zentr. Gewerbehyg. Unfallverhüt*, **9**, 203 (1921).
[137] E. Roth, in R. Abel, *Handbuch der prakt. Hygiene*, Gustav Fischer, Jena, **2**, 232 (1913)

TABLE 4

Symptoms Induced in Man by Inhalation of Ethyl Alcohol[114]

Average concentration mg./liter	p.p.m.	Duration of exposure, min.	Symptoms
			A. Subject unaccustomed to alcohol
2.59	1380	39	None after 28 min; after 33 min , headache and slight numbness
6.28	3340	100	Sensation of warmth and coldness, nasal irritation, headache, and numbness
16.62	8840	64	Initial intolerable odor and difficulty in breathing, soon overcome, conjunctival and nasal irritation, feeling of warmth, headache, drowsiness, fatigue
			B. Subject accustomed to alcohol
9.45	5030	120	Slight headache after 20 min
11.50	6120	120	Odor intense, slight pressure in left temple
13.14	6990	109	Headache, conjunctival irritation, feeling of warmth, drowsiness, and fatigue

greater in subjects accustomed to alcohol than in those unaccustomed to it. A subject unaccustomed to alcohol complained of headache after 33 minutes in an atmosphere of 1380 p.p.m. of ethyl alcohol; a feeling of warmth after 11 minutes, and numbness after 50 minutes of exposure to 3340 p.p.m.; intense stinging of eyes and drowsiness after 10 minutes, and a feeling as of intracranial pressure, numbness, and drowsiness after 29 minutes of exposure to a concentration of 8840 p.p.m. Subjects accustomed to alcohol exhibited only slight headache after 20 minutes of exposure to 5030 p.p.m.; occasional sensations of intracranial pressure after 120 minutes of exposure to 6120 p.p.m.; and a feeling of warmth and drowsiness after 90 minutes when subjected to 6990 p.p.m. of ethyl alcohol in the air.

A more recent extensive study of the effects of inhalation of ethyl alcohol by man has been carried out by Lester and Greenberg.[137a] In this study, subjects were exposed to known concentrations of alcohol and observed for subjective and objective symptoms, ventilation rate, per cent absorption, and blood levels of alcohol at varying times after exposure. These authors question the previous conclusions of Loewy and von der Heide with regard to the conclusion that more than 0.1 per cent in the air would be hazardous in workrooms. Since the subjects in the latter study were only exposed for 2 hours and since no blood levels were taken, it was difficult to interpret the symptoms that were produced at higher levels of exposure.

Symptoms noted by Lester and Greenberg were as follows. At 10–20 mg./liter (about 5000–10,000 p.p.m.) there was some coughing and smarting of the eyes and nose with disappearance of the symptoms within a few minutes. While the atmosphere was not exactly comfortable, it was felt that there would be no difficulty in tolerating these levels. At 30 mg./liter (about 15,000 p.p.m.) there was continu-

[137a] D. Lester and L. A. Greenberg, *Quart. J. Studies Alc.*, **12**, 167 (1951).

ous lachrymation and coughing. Above 40 mg./liter (about 20,000 p.p.m.) it was impossible to tolerate the atmosphere even for a short period of time.

The subjective symptoms increased when the rate of ventilation was increased by two or three times the resting level. Measurements were made of the per cent absorption as related to changes in the ventilation rate from 7 to 25 liter min. and at concentrations of alcohol varying from 11 to 19 mg./liter. The average absorption found was about 62 per cent and seemed to be independent of concentration and rate of ventilation.

Studies on blood alcohol concentrations were made by exposing subjects to concentrations in the range of from 13 to 16 mg./liter for periods from 3 to 6 hours with ventilation rates varying from resting ventilation to as high as 25 liter min. The highest blood level obtained in any of the subjects was 47 mg. of alcohol per 100 ml. of blood after a 6-hour exposure to 16 mg./liter of alcohol at a high ventilation rate of 22 liter/min. When the ventilation rate was reduced to 15 liter/min. with an alcohol concentration in the air of 15 mg./liter, there was definitely a plateau reached in the blood level of about 7 to 8 mg. per 100 ml. of blood. It was obvious that under these conditions the rate of intake of alcohol was equal to the rate of metabolism.

TABLE 5

Effects of Ethyl Alcohol in the Blood

Effect	Alcohol concentration in blood, %
Beginning of uncertainty	0.06–0.08
Slow comprehension	0.10
Stupor	0.12–0.15
Drunkenness	0.16
Severe intoxication	0.2 –0.4
Death	0.4 –0.5

They point out that, based on the results of these studies, a 70-kg. man exposed to 1000 p.p.m. of alcohol would have to breathe at a rate greater than 117 liter/min. in order to obtain any continuous rise in alcohol concentration. They conclude from this that since continuous hard work may involve a ventilation rate of the order of 30 liter/min., concentrations of alcohol in the air not exceeding 7 mg./liter (about 3500 p.p.m.) seems to be a more realistic standard. Fassett[137b] feels that his observations under practical conditions of exposure are in accord with Lester and Greenberg regarding irritant levels and lack of subjective symptoms at levels higher than 1000 p.p.m.

Determination of the content of ethyl alcohol in the blood, spinal fluid, urine, or even the breath is useful in determining the amount absorbed, when such data

---

[137b] D. W. Fassett, personal communication.

are considered in conjunction with accurate information on the time and duration of the exposure. Although there is considerable variation among individuals and in the same individual at different times, as well as some divergence in the data of various authorities, the clinical effects shown in Table 5 are commonly associated with the indicated concentrations in the blood.[112]

6. *Hygienic Standard of Permissible Exposure*

The threshold limit of ethyl alcohol has been established at 1000 p.p.m.[138-140]

7. *Flammability*

The ignition temperature of 95 per cent alcohol is 738°F. (392°C.)[141] (see Vol. I, page 527). The lower and upper flammable limits of alcohol are, respectively, 3.28 and 18.95 per cent by volume in air[142] (see Vol. I, page 516).

8. *Odor and Warning Properties*

Concentrations of 6000 to 9000 p.p.m. have an intense odor and may be practically intolerable at first, but one becomes acclimated to them after a short time. Concentrations of this order of magnitude, however, should not be permitted. The minimum identifiable odor of ethyl alcohol of 350 p.p.m., as determined by Scherberger, Happ, Miller, and Fassett,[143] is well below 1000 p.p.m., which is the value of the hygienic standard.

ι-PROPYL ALCOHOL, $CH_3CH_2CH_2OH$ (Propanol-1)

1. *Source, Uses, and Industrial Exposures*

n-Propyl alcohol is recovered as a by-product of methanol synthesis by high pressure, and in the propane-butane oxidation process.[144]

n-Propyl alcohol, although not used as extensively as certain other alcohols, may be used as a solvent for: vegetable oils; natural gums and resins such as soft copals, rosin, and shellac; certain synthetic resins; ethyl cellulose, and polyvinyl butyral. Propanol is used in lacquers and dopes, cosmetics, dental lotions, cleaners, polishes, and pharmaceuticals.[144,145] Ill effects from industrial usage of n-propyl alcohol have not been reported.

[138] American Conference of Governmental Industrial Hygienists' Threshold Limit Values for 1959, *A.M.A. Arch. Ind. Health,* **20,** 266 (1959).

[139] *Hygienic Guide.* Am. Ind. Hyg. Assoc., Chicago, Ill., March 1956.

[140] H. F. Smyth, Jr., *Am. Ind. Hyg. Assoc. Quart.,* **17,** 129 (1956).

[141] N. J. Thompson, *Ind. Eng. Chem.,* **21,** 134 (1929).

[142] G. W. Jones, *Chem. Revs.,* **22,** 1 (1938).

[143] R. F. Scherberger, G. P. Happ, F. A. Miller and D. W. Fassett, *Am. Ind. Hyg. Assoc. Quart.,* **19,** 494 (1958).

[144] I. Mellan, *Industrial Solvents.* Reinhold, New York, 1950, pp. 466, 467.

[145] A. K. Doolittle, *The Technology of Solvents and Plasticizers.* Wiley, New York, 1954, p. 627.

2. *Physical and Chemical Properties*

Physical state: colorless, volatile liquid
Molecular weight: 60.097[146]
Specific gravity: 0.804 (20/4°C.)[147]
Refractive index: 1.3850 (20°C.)[145]
Melting point: −126.1°C.[145]
Boiling point: 97.3°C.[145]
Vapor density: 2.08 (air = 1)
Vapor pressure: 20.8 mm. Hg (25°C.)[148]
Per cent in "saturated" air: 2.7 (25°C.)
Density of "saturated" air: 1.028 (air = 1) (25°C.)
Solubility: miscible in all proportions with water[145] and with alcohol and ether
Flash point: 59°F. (15°C.)

1 mg./liter ≏ 408 p.p.m. and 1 p.p.m. ≏ 2.45 mg./cu. meter at 25°C., 760 mm. Hg

3. *Physiological Response*

Animal Symptomatology. The toxicological response of animals to *n*-propyl alcohol following its ingestion or application on the skin is given in Table 6.

Animals exposed to sufficient concentrations of the vapors of *n*-propyl alcohol may manifest the following signs of intoxication: irritation of the mucous membranes, ataxia, lethargy, prostration, narcosis, and death. Rats survived when exposed for 2 hours to the saturated vapor by Smyth *et al.*[149] However, these investigators obtained death in 2 of 6 rats exposed for 4 hours to the concentration of 4000 p.p.m.[149]

Weese[150] found that mice survived when they were exposed intermittently to the narcotic vapor of 7874 p.p.m. in air (19.3 mg./liter) for a total period of 95 hours, but died if exposed for 160 minutes to 13,120 p.p.m. (32.2 mg./liter) or for 120 minutes to 19,680 p.p.m. (48.2 mg./liter).

Groups of mice (2 in each) were exposed by Starrek[151] for decreasing lengths of time (480, 240, 135, 120, 90, and 60 minutes) to increasing concentrations of *n*-propyl alcohol in the atmosphere [3250 p.p.m. (8 mg./liter), 4100 p.p.m. (10 mg./liter), 8150 p.p.m. (20 mg./liter), 12,250 p.p.m. (30 mg./liter), 16,300 p.p.m. (40 mg./liter), and 24,500 p.p.m. (60 mg./liter)]. The length of time required for

[146] "Atomic Weights (International), 1957," *J. Am. Chem. Soc.,* **80,** 4122 (1958).
[147] Beilstein, *Handbuch der organischen Chemie,* Vol. I. 4th ed., Springer, Berlin, 1918, p. 350.
[148] I. Mellan, *Industrial Solvents.* Reinhold, New York, 1939, p. 215.
[149] H. F. Smyth, Jr., C. P. Carpenter, and C. S. Weil, *Arch. Ind. Hyg. Occupational Med.,* **10,** 61 (1954).
[150] H. Weese, *Arch. exptl. Pathol. Pharmakol.,* **135,** 118 (1928).
[151] E. Starrek, Dissertation, Wurzburg, 1938.

TABLE 6

Toxicological Response to n-Propyl Alcohol

Animal	Response
Single dose	
Rat (orally)	LD_{50} 1.9 g./kg.[149]
Rabbit (orally)	MLD 3.5 ml./kg.[103]
Mouse (orally)	LD_{50} 4.5 g./kg.[150]
Rat (I.P.)	MLD 4.0 ml./kg.[104]
Rabbit (I.V.)	MLD 4.0 ml./kg.[106]
Cat (I.V.)	MLD 1.6 ml./kg.[107]
Skin absorption	
Rabbit (single application, 24 hrs.)	LD_{50} 5.0 ml./kg.[149]
Rabbit (single application, 24 hrs.)	MLD >9.4 ml./kg.[108]
Rabbit (multiple application, 38 ml./kg./day for 30 days over a period of 6 weeks)	LD_{33}[108]
Skin irritation	
Rabbit (single application)	None[149]
Rabbit (single application)	Slight erythema and dryness[108]
Rabbit (multiple application)	Dryness, superficial desquamation[108]

the appearance of ataxia, prostration, and deep narcosis was inversely proportional to the concentration to which the mice were exposed. Ataxia appeared in 10 to 14 minutes at 24,500 p.p.m. and in 90 to 120 minutes at 3250 p.p.m. Prostration was evident in 19 to 23 minutes at the former concentration and in 165 to 180 minutes at the latter. Deep narcosis was manifest in 60 minutes at 24,500 p.p.m. and in 240 minutes at 4100 p.p.m. Only 1 of 12 mice that showed signs of intoxication died. Mice exposed for 480 minutes to 2050 p.p.m. (5 mg./liter) showed no reaction.

Absorption and Excretion. n-Propyl alcohol was found in the blood of a dog for 275 minutes following the oral administration of 16.1 g. of this alcohol. Acetone, derived from isopropyl alcohol, was found in the blood of a dog for 540 minutes following the oral administration of 15.8 g. of isopropyl alcohol. Hence, n-propyl alcohol is oxidized and eliminated faster than isopropyl alcohol.[152] Basing his opinion upon the concentrations in the blood, the same investigator[152] concluded that n-propyl alcohol was oxidized and eliminated from the dog considerably more rapidly than was ethyl alcohol. Rabbits given n-propyl alcohol by the intravenous route also oxidized it more rapidly than ethyl alcohol.[153]

4. Flammability

n-Propyl alcohol vapor is flammable within the range of 2.15 to 13.50 per cent by volume in air. The ignition temperature is 822°F. (see Vol. I, pages 516 and 530).

---

[152] M. Neymark, *Skand. Arch. Physiol.*, **78**, 242. (1938).
[153] S. M. Berggren, *Skand. Arch. Physiol.*, **78**, 249 (1938).

ISOPROPYL ALCOHOL, $CH_3{\cdot}CHOH{\cdot}CH_3$ (Propanol-2)

1. *Source, Uses, and Industrial Exposures*

Most isopropyl alcohol is synthesized from propylene, which results from cracking petroleum hydrocarbons. However, a small percentage is produced by the reaction of natural gas hydrocarbons.[154]

Isopropyl alcohol is also known as propanol-2,2-propanol, isopropanol, secondary propyl alcohol, dimethyl carbinol, Perspirit, Petrohol, or Avantine. Isopropyl alcohol is used in liniments, skin lotions, cosmetics, permanent wave preparations, pharmaceuticals, and hair tonics. It is employed as a solvent in perfumes, in extraction processes, as a preservative, in nitrocellulose lacquer formulations, and in many dye solutions. It is an ingredient of antifreezes, liquid soaps, and window cleaners. It is used as a coupling agent in oil emulsions and as an extracting agent for sulfonic acids from petroleum oils.[154] It has been used in the "fused collar" industry.[155] It is a raw material for the manufacture of acetone and various isopropyl derivatives. It has gained widespread use as a rubbing alcohol. However, it should not be taken internally. Isopropyl alcohol does not constitute an industrial hazard.

2. *Physical and Chemical Properties*

Physical state: colorless, volatile liquid
Molecular weight: 60.097[146]
Specific gravity: 0.7874 (20/20°C.)[154]
Refractive index: 1.3776 (20°C.)[154]
Melting point: —85.8°C.[154]
Boiling point: 82.4°C.[154]
Vapor density: 2.08 (air = 1)
Vapor pressure: 44 mm. Hg (25°C.)[154]
Per cent in "saturated" air: 5.8 (25°C.)
Density of "saturated" air: 1.06 (air = 1)
Miscible with water and most organic solvents
Flash point: 53°F.

1 mg./liter ≎ 408 p.p.m. and 1 p.p.m. ≎ 2.45 mg./cu. meter at 25°C., 760 mm. Hg

3. *Determination in the Atmosphere*

Isopropyl alcohol that has been collected in water may be oxidized to acetone by chromic acid, and the acetone subsequently measured iodimetrically.[156] The use of sodium nitroprusside for determining the acetone seems feasible.[157] After

[154] I. Mellan, *Industrial Solvents*. Reinhold, New York, 1950, pp. 467–482.

[155] D. E. Donley, *J. Ind. Hyg. Toxicol.*, **18,** 571 (1936).

[156] M. B. Jacobs, *The Analytical Chemistry of Industrial Poisons, Hazards and Solvents*. Interscience Publishers, New York, 1941, p. 487.

[157] G. Kleyer, Pharm. Ztg., **72,** 1262 (1927).

absorbing isopropyl alcohol on silica gel, Hahn[158] recovered it by steam distillation, converted it to isopropyl nitrite, and subsequently titrated the liberated nitrous acid according to the procedure of Knipping and Ponndorf.[159]

4. Physiological Response

The toxicological response of the rat and rabbit to isopropyl alcohol is given in Table 7.

TABLE 7

Toxicological Response to Isopropyl Alcohol

Animal	Response
Single dose	
Rat (orally)	LD_{50} 5.84 g./kg.[160]
Rabbit (orally)	LD 10.0 ml./kg.[161]
Skin absorption	
Rabbit (single application, 24 hrs.)	LD_{50} 16.4 ml./kg.[160]
Rabbit (single application, 24 hrs.)	LD_{20} >9.4 ml./kg.[162]
Rabbit (multiple application, 45 ml./kg./day for 30 days over a period of 6 weeks)	LD_0[162]
Skin irritation	
Rabbit (single application)	Slight erythema[162]
Rabbit (multiple application)	Slight erythema, dryness, superficial desquamation[162]

Animal Symptomatology. Isopropyl alcohol in large amounts is more toxic[163] and more narcotic[150,164] than ethyl alcohol, but less so than *n*-propyl alcohol.[165] There is apparently little or no accumulation in the body. On the basis of symptomatology[166–168] and concentration of alcohol in the blood,[167] there is some evidence that a slight tolerance is acquired.

Animals subjected to vapor of isopropyl alcohol have manifested the following signs of intoxication: irritation of the mucous membranes, ataxia, prostration, deep narcosis, and death.

[158] E. Hahn, *Biochem. Z.*, **292,** 148 (1937).
[159] H. W. Knipping and W. Ponndorf, *Z. physiol. Chem.*, **160,** 25 (1926).
[160] H. F. Smyth, Jr., and C. P. Carpenter, *J. Ind. Hyg. Toxicol.*, **30,** 63 (1948).
[161] J. C. Munch and E. W. Schwartze, *J. Lab. Clin. Med.*, **10,** 985 (1926).
[162] J. F. Treon, unpublished data.
[163] A. J. Lehman and H. F. Chase, *J. Lab. Clin. Med.*, **29,** 561 (1944).
[164] A. J. Lehman, H. Schwerma, and E. Richards, *J. Pharmacol.*, **82,** 196 (1944).
[165] D. I. Macht, *J. Pharmacol.*, **16,** 1 (1920).
[166] H. C. Fuller and O. B. Hunter, *J. Lab. Clin. Med.*, **12,** 326 (1927).
[167] A. J. Lehman, H. Schwerma, and E. Richards, *J. Pharmacol.*, **85,** 61 (1945).
[168] J. Pohl, *Biochem. Z.*, **127,** 66 (1921).

Rats were unaffected except for slight intoxication when exposed by Macht[165,169] intermittently over a period of a week (total number of hours of exposure not given) to air apparently saturated with vapor of isopropyl alcohol. Smyth found that rats survive when exposed for 4 hours to the concentration of 12,000 p.p.m. but exposure for 8 hours to this concentration resulted in death among one half the group.[170] Mice subjected by Weese[150] to 10,900 p.p.m. isopropyl alcohol in air (26.8 mg./liter) for about 4 hours/day until they had accumulated 123 hours of exposure were narcotized but survived. Mice died if exposed to 12,800 p.p.m. (31.4 mg./liter) for 200 minutes or 19,200 p.p.m. (47.1 mg./liter) for 160 minutes. It is wise to assume that high concentrations of isopropyl alcohol would be dangerous.

The length of time required for the development of ataxia, prostration, and deep narcosis on the part of mice exposed by Starrek[151] to vapors of isopropyl alcohol was inversely proportional to the concentration. Ataxia was manifest in 12 to 26 minutes at 24,500 p.p.m. (60 mg./liter), but it occurred with progressively decreasing rapidity at concentrations of 16,300 p.p.m. (40 mg./liter), 12,250 p.p.m. (30 mg./liter), 8150 p.p.m. (20 mg./liter), and 4100 p.p.m. (10 mg./liter), until at 3250 p.p.m. (8 mg./liter) 180 to 195 minutes were required. Prostration appeared in 37 to 46 minutes at 24,500 p.p.m. and in 340 to 350 minutes at 3250 p.p.m. The onset of deep narcosis ranged from 100 minutes at 24,500 p.p.m. to 460 minutes at 3250 p.p.m. Only 1 of 12 mice exposed in these experiments succumbed. Mice exposed for 480 minutes to 2050 p.p.m. (5 mg./liter) gave no evidence of reaction.

Animal Pathology. Reversible fatty changes in the livers of mice, following repeated inhalation of isopropyl alcohol, have been reported by Weese.[150] No gross or microscopic abnormalities of the brain, pituitary, lung, heart, liver, spleen, kidneys, or adrenals of rats given 0.5 to 10.0 per cent of isopropyl alcohol in their drinking water for 27 weeks were observed by Lehman and Chase.[163]

Absorption and Excretion by Animals and Man. Numerous investigators have found acetone in the urine of men and animals following oral administration of isopropyl alcohol. According to Fuller and Hunter,[166] small amounts of acetone were found in the urine of men 2 to 4 days after the last of three daily doses of 20 or 30 ml. of 50 per cent aqueous isopropyl alcohol. Acetone was found by Kemal[171] in the urine of men after ingestion of 0.1 g. of isopropyl alcohol. This investigator found a relatively small quantity of acetone, as compared with the amount of isopropyl alcohol, in the urine of men for 24 hours and 48 hours, respectively, following the ingestion of 5 or 10 g. of isopropyl alcohol. McCord, Switzer, and Brill[172] reported acetone in the urine of 2 patients in a comatose condition after the ingestion of isopropyl alcohol. The alcohol level in the blood of these subjects,

[169] D. I. Macht, *Arch. intern. pharmacodynamie,* **26,** 285, (1922).

[170] H. F. Smyth, Jr., *Cummings Memorial Lecture.* Am. Ind. Hyg. Assoc., Philadelphia, April 25, 1956.

[171] H. Kemal, *Biochem. Z.,* **187,** 461 (1927).

[172] W. M. McCord, P. K. Switzer, and H. H. Brill, Jr., *Southern Med. J.,* **41,** 639 (1948).

who recovered, was 150 and 450 mg.%, respectively. Morris and Lightbody[173] gave 6 ml. of isopropyl alcohol per kilogram of body weight by mouth to each of 6 rabbits, and found acetone in the urine of 5 of them during the following 72 hours, but none thereafter. Kemal[174] gave a progressively increasing daily dose of isopropyl alcohol (5–90 ml.) by stomach tube to 3 dogs. From the thirteenth day on, when doses of 65 ml. or more of isopropyl alcohol were given, 48 to 71 mg. of acetone and 119 to 148 mg. of isopropyl alcohol, per 100 ml. of urine, were found in daily volumes of urine ranging from 1070 to 2250 ml. A method for the determination of both isopropyl alcohol and acetone in the urine has been described by Cook and Smith.[175] The isopropyl alcohol of one aliquot is oxidized to acetone and the total acetone weighed as the mercuric sulfate complex of Denigés. In another aliquot, the original acetone present is distilled into hydroxylamine hydrochloride and the resulting hydrochloric acid titrated using methyl orange as an indicator.

In the experiments of Lehman and associates,[167] this alcohol was metabolized more slowly than was ethyl alcohol at two levels of concentration by cats, rabbits, and pigeons; but in dogs the rate of disappearance of isopropyl alcohol from the blood was more rapid than that of ethyl alcohol when the concentrations were high and slower when they were low. The rate of decrease of isopropyl alcohol in the blood of dogs for a period of 4 to 24 hours following intravenous infusion of isopropyl alcohol (0.64 to 3.84 ml./kg.) or oral administration (0.93 to 3.75 ml./kg.) has been found to vary with the concentration of alcohol in the blood.[164]

Acetone and isopropyl alcohol have been found in the expired air of animals and man following intake of isopropyl alcohol. In a period of 12 hours following the oral administration of 2.37 g. of isopropyl alcohol to rabbits, 0.251 g. of acetone (equivalent to 0.258 g. of isopropyl alcohol) and 0.0281 g. of isopropyl alcohol were found by Pohl[168] in the expelled air. These amounted to 10.9 and 1.2 per cent, respectively, of that administered. The administration of adrenaline, histamine, or oxyphenylethylamine to dogs did not alter significantly the rate of their oxidation of isopropyl alcohol.[168] For a period of a few hours following the ingestion of 720 mg. of aqueous isopropyl alcohol by man, Hahn[158] found the expired air contained isopropyl alcohol and acetone.

The distribution of isopropyl alcohol and acetone in the tissues of dogs 4 hours after the oral administration of 90 ml. of isopropyl alcohol was determined by Kemal[174] (these dogs had been given progessively increasing amounts of the alcohol for the previous 59 days). In general, the concentrations of isopropyl alcohol found in the tissues and body fluids decreased in the following order: brain, urine, heart, kidney, and blood. The relationship of the concentrations of acetone in the tissues and urine was not as clearly defined as that of isopropyl alcohol. Except in the blood, where the value for acetone approached that for

[173] H. J. Morris and H. D. Lightbody, *J. Ind. Hyg. Toxicol.*, **20,** 428 (1938).

[174] H. Kemal, *Z. physiol. Chem.*, **246,** 59 (1937).

[175] C. A. Cook and A. H. Smith, *J. Biol. Chem.*, **85,** 251 (1929).

isopropyl alcohol, the concentration of isopropyl alcohol in the tissues and urine was about twice that for acetone.

By measuring the electric current necessary to induce clonic convulsions in rabbits, rats, and cats, Chu et al.[176] found that isopropyl alcohol had anticonvulsant properties. These authors measured the increase in acetone in the blood and concluded that it paralleled the anticonvulsant action. However, Schaffarzick[177] measured the blood levels of both acetone and isopropyl alcohol in rats and rabbits, and concluded that this property was related to the isopropyl alcohol rather than the acetone.

Single or repeated applications of isopropyl alcohol upon the skin of rats, rabbits, dogs, or human beings induced no untoward effects.[169,178,179]

5. Hygienic Standard of Permissible Exposure

The threshold limit of isopropyl alcohol has been established at 400 p.p.m.[180,181]

6. Flammability

The ignition temperature of isopropyl alcohol is 853°F. The lower limit of flammability is 2.02 per cent by volume in air and the upper limit is 11.80 per cent (see Vol. I, pages 516 and 530).

7. Warning Properties

Mild irritation of the eyes, nose, and throat was induced in human subjects exposed by Nelson and associates[182] for 3 to 5 minutes to 400 p.p.m. of isopropyl alcohol. Although the effects of exposure to 800 p.p.m. were not severe, most subjects found the atmosphere objectionable. From the viewpoint of comfort, these subjects found 200 p.p.m. to be the highest concentration acceptable for an 8-hour exposure. Scherberger et al.[183] stated that the minimum concentration with identifiable odor of isopropyl alcohol is 200 p.p.m.

[176] N. Chu, R. L. Driver, and P. J. Hanzlik, J. Pharmacol., 92, 291 (1948).

[177] R. W. Schaffarzick, Proc. Soc. Exptl. Biol. Med., 74, 211 (1950).

[178] H. Boruttau, Deut. med. Wochschr., 47, 747 (1921).

[179] L. L. Boughton, J. Am. Pharm. Assoc. Sci. Ed., 33, 111 (1944).

[180] W. A. Cook, Ind. Med., 14, 936 (1945).

[181] American Conference of Governmental Industrial Hygienists' Threshold Limit Value for 1959, A.M.A. Arch. Ind. Health, 20, 266 (1959).

[182] K. W. Nelson, J. F. Ege, Jr., M. Ross, L. E. Woodman, and L. Silverman, J. Ind. Hyg. Toxicol., 25, 282 (1943).

[183] R. F. Scherberger, G. P. Happ, F. A. Miller, and D. W. Fassett, Am. Ind. Hyg. Assoc. Quart., 19, 494 (1958)

n-BUTYL ALCOHOL, $C_2H_5 \cdot CH_2 \cdot CH_2OH$ (Butanol-1)

1. Source, Uses and Industrial Exposures

n-Butyl alcohol is synthesized commercially from acetaldehyde or ethanol. Large quantities also are made by fermentation of carbohydrates.[184,185]

n-Butyl alcohol is also known as n-butanol, butanol-1, 1-butanol, butyl hydroxide, propylcarbinol, butyric alcohol, or hydroxybutane. n-Butyl alcohol is employed as a solvent for paints, lacquers (nitrocellulose), coatings, natural resins (congo, dammar, elemi, manila, sandrac, and shellac), gums, vegetable oils (castor and linseed oil), synthetic resins (ureas, alkyds, phenolics, ethyl cellulose, polyvinol butyral), dyes, alkaloids, and camphor. It is used as an extractant in the manufacture of antibiotics, hormones, and vitamins. It is an intermediate in the manufacture of butyl acetate, dibutyl phthalate and dibutyl sebacate.[184,185] The production or, in some cases, use of the following substances may offer exposure to n-butyl alcohol: artificial leather, butyl esters, rubber cement, dyes, fruit essences, lacquers, motion picture and photographic films, raincoats, perfumes, pyroxylin plastics, rayon, safety glass, shellac, varnish, and waterproofed cloth.[184-189]

Concentrations of 5 to 100 p.p.m. of butyl alcohol in the air in six plants manufacturing raincoats and waterproofed cloths for sleeping pads have been reported by Tabershaw and associates.[187] Over a period of 10 years Sterner et al.[189] conducted a study of men exposed to n-butyl alcohol in a baryta-coating operation in the manufacture of photographic paper. Initially the concentrations at the breathing zone exceed 200 p.p.m. but these were reduced to 100 p.p.m. by lowering the concentration of butyl alcohol in the coating suspension and by the introduction of additional ventilation.

2. Physical and Chemical Properties

Physical state: colorless volatile liquid
Molecular weight: 74.124[190]
Specific gravity: 0.8109 (20/20°C.)[185]
Refractive index: 1.3974 (25°C.)[184]
Melting point: −79.9°C.[184]

[184] I. Mellan, *Industrial Solvents*. Reinhold, New York, 1950, pp. 482–488.

[185] A. K. Doolittle, *The Technology of Solvents and Plasticizers*. Wiley, New York, 1954, pp. 640–642.

[186] D. G. Cogan and W. M. Grant, *Arch. Ophthalmol.*, **33**, 106 (1945).

[187] I. R. Tabershaw, J. P. Fahy, and J. B. Skinner, *J. Ind. Hyg. Toxicol.*, **26**, 328 (1944).

[188] U. S. Dept. Labor, Bur. Labor Statistics Bull., No. **41**, 38 (1942).

[189] J. H. Sterner, H. C. Crouch, H. F. Brockmyre, and M. Cusack, *Am. Ind. Hyg. Assoc. Quart.*, **10**, 53 (1949).

[190] "Atomic Weights (International), 1957," *J. Am. Chem. Soc.*, **80**, 4122 (1958).

Boiling point: 117.7°C.[185]
Vapor density: 2.56 (air = 1)
Vapor pressure: 6.5 mm. Hg (25°C.)[184]
Per cent in "saturated" air: 0.86 (25°C.)
Density of "saturated" air: 1.01 (air = 1)
Solubility: 7.7 per cent in water (20°C.)[185]
Flash point: 84°F.
1 mg./liter ≈ 330 p.p.m. and 1 p.p.m. ≈ 3.03 mg./cu. meter at 25°C., 760 mm. Hg

3. Determination in the Atmosphere

No specific analytical method is available, but under circumstances involving the presence of n-butyl alcohol alone, its oxidation by chromic acid may be measured quantitatively. The determination of n-butyl alcohol in the presence of acetone and ethyl alcohol in an aqueous solution is described by Christensen and Fulmer.[191] In measuring industrial air concentrations, Tabershaw and his associates[187] employed the method of Ficklen,[192] which measures iodimetrically the amount of chromate necessary to oxidize n-butyl alcohol to butyric acid. This is similar to a method referred to previously, in the case of ethyl alcohol. Hoch[193] employed a commercial type of equipment (M-6 Vaportester, Davis Emergency Equipment Co.) for determining the concentration of butyl alcohol in air. Sterner and associates[189] passed air through silica gel in U-tubes. The butanol was distilled with water and collected in an ice bath. The amount of butanol in the distillate was determined by means of a Zeiss liquid interferometer.

4. Physiological Response

Animal Symptomatology. The toxicological response of animals to n-butyl alcohol when it is ingested or maintained upon the skin is given in Table 8.

Animals exposed to n-butyl alcohol in the air may manifest the following signs of intoxication: restlessness, irritation of mucous membranes, ataxia, prostration, and narcosis. Following absorption of sufficient quantities, death, associated with respiratory failure, occurs. At high concentrations, n-butyl alcohol is more narcotic than is n-propyl alcohol[199, 200] but slightly less so than is sec-butyl alcohol.[199]

[191] L. M. Christensen and E. I. Fulmer, *Ind. Eng. Chem. Anal. Ed.*, **7**, 180 (1935).

[192] J. B. Ficklen, *Manual of Industrial Health Hazards.* Service to Industry, West Hartford, Conn., 1940, p. 50.

[193] S. M. Hoch, personal communication.

[194] H. F. Smyth, Jr., C. P. Carpenter, and C. S. Weil, *Arch, Ind. Hyg. Occupational Med.*, **4**, 119 (1951).

[195] J. C. Munch and E. W. Schwartze, *J. Lab. Clin. Med.*, **10**, 985 (1925).

[196] L. Lendle, *Arch. exptl. Pathol. Pharmakol.*, **132**, 214 (1928).

[197] D. I. Macht, *J. Pharmacol. Exptl. Therap.*, **16**, 1 (1920).

[198] J. F. Treon, unpublished data.

[199] E. Starrek, Dissertation, Würzburg, 1938.

[200] H. Weese, *Arch. exptl. Pathol. Pharmakol.*, **135**, 118 (1928).

TABLE 8

Toxicological Response to n-Butyl Alcohol

Animal	Response
Single dose	
Rat (orally)	LD$_{50}$ 4.36 g./kg.[194]
Rabbit (orally)	LD 4.25 ml./kg.[195]
Rat (I.P.)	MLD 1.2 ml./kg.[196]
Cat (I.V.)	MLD 0.24 g./kg.[197]
Skin absorption	
Rabbit (single application, 24 hrs.)	LD$_{67}$ 9.4 ml./kg.[198]
Rabbit (multiple application, 42–55 ml./ kg./day for 1 to 4 consecutive days)	LD$_{100}$[198]
Rabbit (multiple application; 20 ml./kg./ day for 30 days over a period of 6 weeks)	LD$_0$[198]
Skin irritation	
Rabbit (single application)	Slight erythema[198]
Rabbit (multiple application)	Hemorrhages and necrosis[198]

Two rabbits exposed by Gardner[201] to unstated varying concentrations of n-butyl alcohol in the air for an unstated length of time per day on each of 55 days out of a total period of 71 days, showed no signs of intoxication other than an increased restlessness, some slight irritation of the mucous membranes, a mild anemia, and a terminal leucocytosis. Guinea pigs exposed to 100 p.p.m. butyl alcohol in the atmosphere (0.303 mg./liter) for 4 hours/day on 64 days (1 day/week omitted) gained in weight but exhibited some decrease in the number of red blood cells and a relative and absolute lymphocytosis.[202] Smyth[203] found rats survived when exposed for 4 hours to the concentration of 8000 p.p.m. According to Weese,[200] mice subjected to 130 hours of total exposure (unstated number of hours per day for several days) to the concentration of 8000 p.p.m. (24.3 mg./liter of air) were narcotized repeatedly but gained in weight and survived. No evidence of intoxication was observed among mice exposed to 3300 p.p.m. (10 mg./liter) or 1650 p.p.m. (5 mg./liter) for 420 min.[199] However, both Starrek[199] and Weese[200] observed deaths among mice when apparently exposed to a mist of n-butyl alcohol (the concentrations were greater than the highest vapor concentration obtainable under ordinary conditions).

Animal Pathology. Gardner[201] reported that a mild bronchial irritation, associated with some enlargement of bronchial lymph nodes, was the only demonstrable lesion resulting from the prolonged exposure of rabbits to air containing the vapors of this alcohol. Smyth and Smyth[202] found hemorrhagic areas in the

[201] H. A. Gardner, *Paint Mfrs. Assoc. U. S. Tech. Circ.* No. **250**, 89 (1925).

[202] H. F. Smyth and H. F. Smyth, Jr., *J. Ind. Hyg.*, **10**, 261 (1928).

[203] H. F. Smyth, Jr., *Cummings Memorial Lecture.* Am. Ind. Hyg. Assoc., Philadelphia, April 25, 1956.

lungs, early degenerative lesions in the livers, and cortical and tubular degeneration in the kidneys of exposed guinea pigs. Weese[200] described reversible fatty changes in the livers of mice.

Absorption and Excretion by Animals. n-Butyl alcohol is oxidized more rapidly than is ethyl alcohol in the tissues of the rabbit.[204] According to Sander,[205] cited by von Oettingen,[206] n-butyl alcohol is absorbed through the skin of animals.

Effects upon Man. n-Butyl alcohol is potentially more toxic than any of the lower homologs, but the practical hazards associated with its industrial production and use (at ordinary temperatures) are appreciably diminished by its relatively low volatility. Exposure of human beings to vapors of this alcohol may induce the following symptoms: irritation of the nose, throat, and eyes; the formation of translucent vacuoles in the superficial layers of the cornea; headache; vertigo; and drowsiness.[187,205,207] Contact dermatitis, involving the fingers and hands, also may occur.

Twenty-eight of 34 women employed in a plant in which raincoats were being manufactured were found to have from 10 to 1000 vacuoles in the corneal epithelium, in association with pain, itching, swelling, and epiphora, but only occasional redness. These signs were more severe upon awakening in the morning than during the day.[186] The significance of a reported case of injury to the liver following exposure to a mixed solvent containing n-butyl alcohol is quite questionable in relation to this compound.[208]

Hoch[193] reported that, in the manufacture of vitamins, n-butyl alcohol, as well as sec- and isobutyl alcohols, have been employed in variable quantities without giving rise to any evidence of systemic intoxication.

In the 10-year study of workers exposed to n-butyl alcohol, Sterner[189] and associates clinically determined and evaluated the hemoglobin, the erythrocyte count and cell volume, leucocyte count, differential leucocyte count, icterus index, and sedimentation rate. They conducted a complete urine analysis and made serial chest x-rays. The eyes were examined by an ophthalmologist. Records of absence due to illness among the men exposed to n-butyl alcohol were compared to absence among all men in the entire plant. The tests for thymol-barbitone precipitation and cephalin-cholesterol flocculation were applied during the later part of the study. During the early portion of this study when the concentration was at 200 p.p.m. or above, occasional individuals developed increasing corneal inflammation associated with burning sensation, blurring of the vision, lacrimation, and photophobia, beginning at the middle of the week and growing more severe toward the end of the week. Slight to moderate corneal edema with injection and mild edema of the conjunctiva were reported by the ophthalmologist. In every instance the

[204] S. M. Berggren, *Skand. Arch. Physiol.,* **78,** 249 (1938).
[205] F. Sander, Inaugural Dissertation, Köln, 1933.
[206] W. F. von Oettingen, *U. S. Public Health Bull.* No. **281,** 123 (1943).
[207] E. Krüger, *Arch Gewerbepath. Gewerbehyg.,* **3,** 798 (1932).
[208] G. E. C. Burger and B. H. Stockmann, *Zentr. Gewerbehyg. Unfallverhüt.,* **9,** 29 (1932).

condition subsided over the weekend. At this period the mean erythrocyte count was slightly reduced. During the greater portion of this study when the concentration averaged 100 p.p.m. no systemic effects were observed and total absence or absence due to illness was no greater than among the total plant personnel. Only rare complaints of irritation or disagreeable odor were made when the concentration was at the 100 p.p.m. level.

5. Hygienic Standard of Permissible Exposure

The threshold limit of n-butyl alcohol has been established at 100 p.p.m.[187,189,203,209,210]

6. Flammability

The ignition temperature of n-butyl alcohol is 653°F. (345°C.)[211] (see Vol. I, page 526). The lower limit of flammability is 1.45 per cent by volume and the upper limit is 11.25 per cent (Vol. I, page 516).

7. Odor and Warning Properties

Nelson et al.[182] reported mild irritation of the nose, throat, and eyes of subjects briefly exposed to 25 p.p.m. and stated that exposure to 50 p.p.m. was objectionable because it produced pronounced irritation of the throat in all subjects and mild headaches in some instances. According to Tabershaw,[187] exposure in excess of 50 p.p.m. is associated with irritation of the eyes. However, the more extensive 10-year study by Sterner and associates[189] revealed little or no irritation or complaints among workers when the average concentration was 100 p.p.m. Scherberger, Happ, Miller, and Fassett[212] determined that the minimum concentration with identifiable odor of butyl alcohol was 15 p.p.m.

sec-BUTYL ALCOHOL, C_2H_5·CHOH·CH_3 (Butanol-2)

1. Source, Uses, and Industrial Exposures

sec-Butyl alcohol is synthesized from butylene obtained in the cracking of petroleum. The butylene is reacted serially with sulfuric acid and then steam.[213]

sec-Butyl alcohol is known also as butanol-2, 2-butanol, and methyl ethyl carbinol. This secondary alcohol is employed to some extent as a solvent of lacquers, enamels, vegetable oils, gums, and natural resins. It finds usage in hydraulic brake fluids, industrial cleaning compounds, polishes, paint removers, and pene-

[209] American Conference of Governmental Industrial Hygienists' Threshold Limit Values for 1959, A.M.A. Arch. Ind. Health, 20, 266 (1959).

[210] K. B. Lehmann and F. Flury, Toxikologie und Hygiene der technischen Lösungsmittel. Springer, Berlin, 1938, p. 157.

[211] N. J. Thompson, Ind. Eng. Chem., 21, 134 (1929).

[212] R. F. Scherberger, G. P. Happ, F. A. Miller, and D. W. Fassett, Am. Ind. Hyg. Assoc. Quart., 19, 494 (1958).

[213] I. Mellan, Industrial Solvents. Reinhold, New York, 1950, pp. 488–493.

trating oils. It is employed in the preparation of ore-flotation agents, fruit essences, perfumes, and dyestuffs. A large portion of sec-butyl alcohol is converted into methyl ethyl ketone.[213,214]

2. Physical and Chemical Properties

Physical state: colorless, volatile liquid
Molecular weight: 74.124[190]
Specific gravity: 0.8077 (20/20°C.)[214]
Refractive index: 1.3971 (20°C.)[214]
Freezing point: −114.7°C.[214]
Boiling point: 98.8°C.[214]
Vapor density: 2.56 (air = 1)
Vapor pressure: 23.9 mm. Hg (30°C.)[213]
Per cent in "saturated" air: 3.14 (30°C.)
Density of "saturated" air: 1.05 (air = 1)
Solubility: 20.1 per cent in water (20°C.)[214]
Flash point: 70°F.
1 mg./liter ≅ 330 p.p.m. and 1 p.p.m. ≅ 3.03 mg./cu. meter at 25°C., 760 mm. Hg

3. Determination in the Atmosphere

The nonspecific determination of the ethylenic hydrocarbon by bromination following dehydration of the alcohol has been reported by Crane and associates.[215]

4. Physiological Response

Toxicological Response to sec-Butyl Alcohol

(single oral dose)

Rat	LD$_{50}$ 6.48 g./kg.[216]
Rabbit	MLD 6.0 ml./kg.[195]

The signs of intoxication in animals exposed to vapors of sec-butyl alcohol in the air are similar to those induced by n-butyl alcohol. The vapor of sec-butyl alcohol is slightly more narcotic[199,200] and lethal[200] than n-butyl alcohol. Based upon the effects of intraperitoneal injections into mice, Butler and Dickison[217] found the optically isomeric sec-butyl alcohols were equal in anesthetic activity.

Five of 6 rats died following exposure for 4 hours to the concentration of 16,000 p.p.m. (48.5 mg./liter) sec-butyl alcohol by Smyth and associates.[216] Mice

---

[214] A. K. Doolittle, *The Technology of Solvents and Plasticizers,* Wiley, New York, 1954, pp. 644, 645.

[215] O. Crane, B. Löfström, and R. Winbladh, *Ing. Vetenskaps Akad. Handl.,* No. **147** (1938).

[216] H. F. Smyth, Jr., C. P. Carpenter, C. S. Weil, and U. C. Pozzani, *Arch. Ind. Hyg. Occupational Med.,* **10,** 61 (1954).

[217] T. C. Butler and H. L. Dickison, *J. Pharmacol.,* **69,** 225 (1940).

subjected repeatedly by Weese[200] to the concentration of 5330 p.p.m. (16.2 mg./liter) sec-butyl alcohol for a total of 117 hours were narcotized, but survived. Concentrations of 10,670 p.p.m. (32.3 mg./liter) for 225 minutes and 16,000 p.p.m. (48.5 mg./liter) for 160 minutes were fatal for mice.[200]

Groups of mice (2 in each) were exposed by Starrek[199] for decreasing lengths of time (300, 190, 75, 60, 45, and 40 minutes) to increasing concentrations of sec-butyl alcohol in the air: 3300 p.p.m. (10 mg./liter), 6600 p.p.m. (20 mg./liter), 9900 p.p.m. (30 mg./liter), 13,200 p.p.m. (40 mg./liter), 16,500 p.p.m. (50 mg./liter), and 19,800 p.p.m. (60 mg./liter). As in the case of other alcohols used by Starrek, the durations of exposure necessary to induce ataxia, prostration, or deep narcosis were inversely proportional to the concentration. Thus at 3330 p.p.m., ataxia, prostration, and narcosis became evident in 51 to 100 minutes, 120 to 180 minutes, and 300 minutes, respectively; whereas at 19,800 p.p.m. these signs appeared in 7 to 8 minutes, 12 to 20 minutes, and 40 minutes, respectively. No deaths occurred among any of these 12 mice. No signs of intoxication were observed in mice exposed for 420 minutes to 1650 p.p.m. (5 mg./liter).

5. Flammability

The ignition temperatures of sec-butyl alcohol in air and oxygen are respectively 763°F. (406°C.)[218] and 711°F. (377°C.).[219]

ISOBUTYL ALCOHOL, $(CH_3)_2$:$CH \cdot CH_2OH$

1. Source, Uses and Industrial Exposures

Isobutyl alcohol is obtained commercially as a by-product of high pressure methanol synthesis and by the oxidation of natural gas hydrocarbons.[220]

Isobutyl alcohol is known also as 2-methyl-1-propanol, 2-methyl propanol-1, and isopropyl carbinol. This alcohol is used to some extent in lacquers, paint removers, cleaners, and hydraulic fluids. It is employed in the manufacture of isobutyl esters, which serve as solvents, plasticizers, flavorings, and perfumes.[221,222]

2. Physical and Chemical Properties

Physical state: colorless, volatile liquid
Molecular weight: 74.124[223]
Specific gravity: 0.8032 (20/20°C.)[221]
Refractive index: 1.3959 (20°C.)[221]

[218] M. G. Zabetakis, A. L. Furno, and G. W. Jones, Ind. Eng. Chem., 46, 2173 (1954).
[219] G. S. Scott, G. W. Jones, and F. E. Scott, Anal. Chem., 20, 238 (1948).
[220] I. Mellan, Industrial Solvents. Reinhold, New York, 1950, p. 488.
[221] A. K. Doolittle, The Technology of Solvents and Plasticizers. Wiley, New York, 1954, p. 647.
[222] D. Steinkoff, Zentr. Arbeitsmed. Arbeitsschutz, 2, 13 (1952).
[223] "Atomic Weights (International), 1957," J. Am. Chem. Soc., 80, 4122 (1958).

Freezing point: $-108°C$.[221]
Boiling point: $108.3°C$.[221]
Vapor density: 2.56 (air = 1)
Vapor pressure: 12.2 mm. Hg $(25°C.)$[224]
Per cent in "saturated" air: 1.61 $(25°C.)$
Density of "saturated" air: 1.03 (air = 1)
Solubility: 10 per cent in water $(20°C.)$[221]
Flash point: 82°F. $(27.7°C.)$
1 mg./liter $\backsim$ 330 p.p.m. and 1 p.p.m. $\backsim$ 3.03 mg./cu. meter at 25°C., 760 mm. Hg

3. Determination in the Atmosphere

Although the reactions are not specific for isobutyl alcohol, the oxidation of isobutyl alcohol by chromic acid to the aldehyde and the subsequent reaction of the latter with o-nitrobenzaldehyde may serve as a means of detecting isobutyl alcohol qualitatively.[225]

4. Physiological Response

Toxicological Response to Isobutyl Alcohol

Single oral dose

Rat LD_{50} 2.46 g./kg.[216]

Skin absorption (single, 24-hr. application)

Rabbit LD_{50} 4.24 g./kg.[216]

Animal Symptomatology. Like the other alcohols, isobutyl alcohol is primarily narcotic in action. It is somewhat less lethal when inhaled in high concentrations than is normal or *sec*-butyl alcohol.[226] Weese[226] subjected mice to a concentration of 2125 p.p.m. (6.44 mg./liter) for 223 hours in a series of intermittent exposures, each of which was of 9.25 hours' duration, without any untoward effects. Mice were narcotized repeatedly in a series of intermittent exposures totaling 136 hours at 6400 p.p.m. (19.3 mg./liter), but survived. Exposure of mice to concentrations of 10,600 p.p.m. (32.2 mg./liter) for 300 minutes, and 15,950 p.p.m. (48.3 mg./liter) for 250 minutes, resulted in fatal poisoning.

Smyth and associates[216] reported uniform survival among rats exposed for 2 hours to the saturated vapor of isobutyl alcohol in air but observed 2 deaths among a group of 6 rats when they were exposed for 4 hours to the concentration of 8000 p.p.m.

Effects upon Man. Slight erythema and hyperemia, without the formation of wheals, were observed by Oettel[227] following the application of isobutyl alcohol to

[224] I. Mellan, *Industrial Solvents.* Reinhold, New York, 1939, pp. 226–236.
[225] H. H. Weber and W. Koch, *Chem. Ztg.,* **57,** 73 (1933).
[226] H. Weese, *Arch. exptl. Pathol. Pharmakol.,* **135,** 118 (1928).
[227] H. Oettel, *Arch. exptl. Pathol. Pharmakol.,* **183,** 641 (1936)

the skin of man. According to Schwartz and Tulipan,[228] isobutyl alcohol may be a skin irritant.

Irritation of the eyes and throat, formation of vacuoles in the superficial layers of the cornea, and loss of appetite and weight were reported among workers subjected to an undetermined, but apparently high, concentration of isobutyl alcohol and butyl acetate arising from the lacquering of cables under crowded conditions, ineffectual ventilation, and oppressive heat (89–107°F.). Rectification of these conditions removed the symptoms.[222]

Fassett[229] has studied isobutyl alcohol under the same conditions of exposure as for n-butyl alcohol. No evidence of eye irritation was noted with repeated 8-hour exposures to levels on the order of 100 p.p.m. Animal data showed that it was very similar to n-butyl alcohol in its effect. It does, however, have a vapor pressure higher than n-butyl alcohol; therefore, under comparable conditions of exposure higher concentrations will be encountered.

5. *Flammability*

The ignition temperature of isobutyl alcohol is 813°F. (434°C.)[230] The lower limit of flammability of this alcohol is 1.68 per cent by volume in air.[231]

tert-BUTYL ALCOHOL, $(CH_3)_3COH$

1. *Uses and Industrial Exposures*

tert-Butyl alcohol, known also as 2-methyl-propanol-2 and trimethylcarbinol, has but little use in industry. It is employed for the removal of water from substances, in the extraction of drugs, in the manufacture of perfumes, and in the recrystallization of chemicals.[232] According to Schwartz and Tulipan,[233] it is used for intermediates, fruit essences, cellulose esters, plastics, and lacquers.

2. *Physical and Chemical Properties*

Physical state: colorless, volatile liquid
Molecular weight: 74.124[223]
Specific gravity: 0.783 (25/25°C.)[232]
Refractive index: 1.3878[232]
Melting point: 25.6°C.[232]
Boiling point: 82.4°C.[232]
Vapor density: 2.56 (air = 1)

[228] L. Schwartz and L. Tulipan, *A Textbook of Occupational Diseases of the Skin.* Lea & Febiger, Philadelphia, 1939, p. 717.

[229] D. W. Fassett, personal communication.

[230] N. J. Thompson, *Ind. Eng. Chem.,* **21,** 134 (1929).

[231] G. W. Jones, *Chem. Revs.,* **22,** 1 (1938).

[232] I. Mellan, *Industrial Solvents.* Reinhold, New York, 1950, pp. 493–495.

[233] L. Schwartz and L. Tulipan, *A Textbook of Occupational Diseases of the Skin.* Lea & Febiger, Philadelphia, 1939, p. 717.

Vapor pressure: 42.0 mm. Hg (25°C.)[234]

Per cent in "saturated" air: 5.53 (25°C.)

Density of "saturated" air: 1.09 (air = 1)

Solubility: miscible in water; soluble in alcohols, esters, ethers, aromatic and aliphatic hydrocarbons

Flash point: 52°F.

1 mg./liter ⟋ 330 p.p.m. and 1 p.p.m. ⟋ 3.03 mg./cu. meter at 25°C., 760 mm. Hg

3. Determination in the Atmosphere

Although no specific method is available for the determination of *tert*-butyl alcohol in the air, the possibility of developing a method from known color reactions seems feasible. Several alcohols including *tert*-butyl give color reactions with aldehydes, for example, anisaldehyde, furfural, piperonal, salicylaldehyde, sucrose, vanillin, and cinnamaldehyde in the presence of sulfuric acid.[235]

4. Physiological Response

Toxicological Response to tert-Butyl Alcohol

Single oral dose

Rat	LD$_{50}$ 3.5 g./kg.[236]
Rabbit	MLD 6.0 ml./kg.[237]

Animal Symptomatology. The signs of intoxication on the part of animals exposed to vapors of *tert*-butyl alcohol are similar to those induced by the other butyl alcohols. It has a stronger narcotic action upon mice than has normal or isobutyl alcohol.[238]

Effects upon Man. Oettel[227] observed no reaction other than slight erythema and hyperemia following the application of *tert*-butyl alcohol to human skin. Schwartz and Tulipan[233] are authority for the statement that it may be a skin irritant. The A.C.G.I.H. threshold limit for 1961 is 100 p.p.m.

5. Flammability

The ignition temperatures of *tert*-butyl alcohol in air and oxygen are, respectively, 892°F. (478°C.) and 860°F. (460°C.).[239] The lower limit of flammability at 25°C. is 2.35 per cent by volume; the upper limit of flammability at 55°C. is 8 per cent by volume.[239]

[234] I. Mellan, *Industrial Solvents.* Reinhold, New York 1939, p. 236.

[235] L. Ekkert, *Pharm. Zentralhalle,* **69,** 289 (1928).

[236] R. W. Schaffarzick, *Science,* **116,** 663 (1952).

[237] J. C. Munch and E. W. Schwartze, *J. Lab. Clin. Med.,* **10,** 985 (1925).

[238] H. Weese, *Arch. exptl. Pathol. Pharmakol.,* **135,** 118 (1928).

[239] W. J. Huff, *U. S. Bur Mines Rept. Invest.* No. **3669** (1942).

AMYL ALCOHOLS, $C_5H_{12}O$

1. *Source, Uses and Industrial Exposures*

There are eight structural isomers of amyl alcohol. There are four primary amyl alcohols: (*a*) 1-pentanol, *n*-amyl alcohol, *n*-butylcarbinol, or pentan-1-ol; (*b*) 2-methyl-1-butanol, primary active amyl alcohol, *sec*-butylcarbinol, methyl ethylcarbinol, or 2-methyl-butan-1-ol; (*c*) 3-methyl-1-butanol, isoamyl alcohol, primary isobutylcarbinol, 2-methylbutan-4-ol, or 3-methylbutan-1-ol; and (*d*) 2,2-dimethyl-1-propanol, *tert*-butylcarbinol, or neopentyl alcohol. There are three secondary amyl alcohols: (*a*) 2-pentanol, secondary active amyl alcohol, methyl propylcarbinol, 1-methyl-1-butanol, or pentan-2-ol; (*b*) 3-pentanol, diethylcarbinol, 1-ethyl-1-propanol, or pentan-3-ol; and (*c*) 3-methyl-2-butanol, methyl isopropylcarbinol, *sec*-isoamyl alcohol, 2-methyl-3-butanol, or 3-methylbutan-2-ol. There is one tertiary amyl alcohol, known as 2-methyl-2-butanol, *tert*-amyl alcohol, amylene hydrate, dimethylethylcarbinol, or 2-methylbutan-2-ol. The preferable names are given in Table 9.

Three of these, namely, 2-methyl-1-butanol, 2-pentanol, and 3-methyl-2-butanol, possess an asymmetric carbon atom; hence each may exist as two optical isomers in addition to the optically inactive form. 2-Methyl-1-butanol is formed from the fermentation of an optically active substance; it exists in commerce largely as the *d*-amyl alcohol.[240]

Commercial amyl alcohols are secured from different sources. Refined fusel oil or fermentation amyl alcohol is obtained from by-product fusel oil, which arises during the production of ethyl alcohol by fermentation. This consists largely of 3-methyl-1-butanol with minor amounts of 2-methyl-1-butanol and other alcohols.[241] Fusel oil is essentially 7 parts of 3-methyl-1-butanol and 1 part of the active form of 2-methyl-1-butanol (*d*-amyl alcohol) with small amounts of pyridine, furfurals, and esters.[242] Amyl alcohols are also made synthetically from pentane by chlorination and hydrolysis. Synthetic amyl alcohol consists of about 70 per cent of three of the primary alcohols (described above) and three other amyl alcohols. 3-Pentanol is available in high purity by synthetic methods and 3-methyl-1-butanol is available as a pure product by separation from fusel oil.[241]

"Pentasol" is a mixture of several amyl alcohols.[243] A commercial product containing about 80 per cent of 2-pentanol and 20 per cent of 3-pentanol is marketed as *sec*-amyl alcohol.[243]

In the past the handling and use of fusel oils have been the principal means of industrial exposure. Amyl alcohols are used in the manufacture of lacquers,

[240] F. C. Whitmore, *Organic Chemistry*. Van Nostrand, New York, 1937, p. 125.
[241] A. K. Doolittle, *The Technology of Solvents and Plasticizers*. Wiley, New York, 1954, pp. 653–655.
[242] I. Mellan, *Industrial Solvents*. Reinhold, New York, 1939, p. 236.
[243] L. L. Fieser and M. Fieser, *Organic Chemistry*. Heath, Boston, 1944, p. 124.

TABLE 9

Physical Properties of Amyl Alcohols

Isomer	Formula	Specific gravity	Melting or freezing point, °C.	Boiling point, °C.	Refractive index n_D	Vapor pressure, mm. Hg (°C.)	Solubility of alcohol in water (°C.)
		Primary alcohols					
1-Pentanol	$CH_3—CH_2—CH_2—CH_2—CH_2OH$	0.817 (20/20°)[248]	−78.5[248]	137.9[248]	1.414 (13°)[248]		Slightly soluble
2-Methyl-1-butanol	$CH_3—CH_2—CH—CH_2OH$ $\;\;CH_3$	0.816 (20/4°)[248]		128[248]		3.4 (25°)[242]	3.6% (30°)
3-Methyl-1-butanol	CH_3 $HC—CH_2—CH_2OH$ CH_3	0.8129 (15/4°)[241]	−117.2[241]	131.4[241]	1.4014 (20°)[241]	2.8 (20°)[241]	2.0% (14°)[241]
2,2-Di-methyl-1-propanol	CH_3 $CH_3—C—CH_2OH$ CH_3		53[248]	114[248]			Slightly soluble
		Secondary alcohols					
2-Pentanol	$CH_3—CH_2—CH_2—CH—OH$ $\;\;\;\;CH_3$	0.809[248]	Glassy in liquid air[250]	119.5[248]	1.4972 (20°)[248]		In 6 vol. H_2O[249]
3-Pentanol	$CH_3—CH_2—CH—CH_2—CH_3$ $\;\;\;\;OH$	0.8169 (20/20°)[241]	Glassy in liquid air[250]	115.9[241]	1.4078 (20°)[241]	2 (20°)[241]	4.1% (20°)[241]
3-Methyl-2-butanol	$CH_3—CH—CH—CH_3$ $\;\;\;\;CH_3\;\;OH$	0.819[248]		114[248]			2.8% (30°)
		Tertiary alcohol					
2-Methyl-2-butanol	CH_3 $CH_3—CH_2—C—OH$ CH_3	0.809 (20/4°)[248]	−11.9[248]	101.8[248]	1.406 (20°)[248]		In 8 parts H_2O[249]

chemicals, rubber, plastics, fruit essences, and explosives,[242,244-246] They are used as a component of paint strippers and hydraulic fluids and as an intermediate of ore-flotation agents and other amyl derivatives.[241] 3-Methyl-1-butanol is used principally as an intermediate in the preparation of pharmaceuticals and isoamyl derivatives.[241] The United States Department of Labor listed twenty occupations that offer exposure to amyl alcohol.[246]

2. Physical and Chemical Properties

Except for 2,2-dimethyl-1-propanol, which is a crystalline solid, the amyl alcohols when purified are colorless liquids with a mild odor. They are solvents for camphor, alkaloids, natural and synthetic resins, iodine, phosphorus, sulfur, benzyl abietate, copal ester, ester gum, etc.[242] They are miscible with organic solvents but are only slightly soluble in water. The physical properties are shown in Table 9. The molecular weight is 88.151.[247]

1 mg./liter $\eqsim$ 278 p.p.m. and 1 p.p.m. $\eqsim$ 3.60 mg./cu. meter at 25°C., 760 mm. Hg.

3. Determination in the Atmosphere

Amyl alcohol might be determined by either of two nonspecific methods employed for the determination of fusel oil in distilled spirits: (a) oxidation by chromic acid or (b) measurement of the color produced from coupling with various aldehydes in the presence of sulfuric acid.

Titration of valeric acid distilled from a reaction mixture with sulfuric acid, and chromate was employed by Allen and Chattaway[251] and is given in the AOAC Methods.[252] Furfural,[253] salicylaldehyde,[254,255] benzaldehyde, p-dimethyl-aminobenzaldehyde, and vanillin,[255] in the presence of sulfuric acid, have been utilized for distilled spirits. Korenman[256] used furfural to determine the amount of

[244] L. Carozzi, *Occupation and Health*. International Labor Office, Geneva, 1930, p. 115.

[245] E. B. Ley and F. J. Vintinner, *The Toxicology and Prevention of Industrial Disease*. 3rd ed., U. S. War Dept., Eighth Service Command, 1944, p. 13.

[246] *U. S. Dept. Labor Bur. Labor Statistics Bull.* No. **41**, 33 (1942).

[247] "Atomic Weights (International), 1957." *J. Am. Chem. Soc.*, **80**, 4122 (1958).

[248] *International Critical Tables of Numerical Data, Physics, Chemistry, and Technology*, Vol. I. McGraw-Hill, New York, 1926, p. 193.

[249] Beilstein, *Handbuch der organischen Chemie*, Vol. I. 4th ed., Springer, Berlin, 1918, p. 383.

[250] J. Timmermans and Mme. Hennault-Roland, *J. Chim. phys.*, **29**, 529 (1932).

[251] A. H. Allen and W. Chattaway, *Analyst*, **16**, 102 (1891).

[252] *Official and Tentative Methods of Analysis of the Association of Official Agricultural Chemists*. 5th ed., Assoc. Offic. Agr. Chemists, Washington, D. C., 1940, p. 175.

[253] H. P. Basset, *Ind. Eng. Chem.*, **2**, 389 (1910).

[254] A. Komarowsky, *Chem. Ztg.*, **27**, 1086 (1903).

[255] W. B. D. Penniman, D. C. Smith, and E. I. Lawshe, *Ind. Eng. Chem. Anal. Ed.*, **9**, 91 (1937).

[256] I. M. Korenman, *Arch. Hyg.*, **109**, 108 (1932).

amyl alcohol in the air. Quantities of the order of 0.6 to 2.0 mg. in 3.5 liters of air were determined with an error of ± 10 per cent.

4. Physiological Response

Animal Symptomatology. The toxicological response of animals to certain of the amyl alcohols following ingestion or application on the skin follows:

Toxicological Response to Amyl Alcohol

Single oral dose

3-Methyl-1-butanol, rabbit	LD 4.25 ml./kg.[257]
2-Pentanol, rabbit	LD 3.5 ml./kg.[257]
3-Pentanol, rat	LD_{50} 1.87 g./kg.[258]
2-methyl-2-butanol, rat	LD_{50} 1.00 g./kg.[259]
2-methyl-2-butanol, rabbit	LD 2.5 ml./kg.[257]

Skin absorption (24-hr. contact)

3-pentanol	LD_{50} 2.52 ml./kg.[258]

Except for one report by Smyth,[260] in which rats survived when exposed to amyl alcohol in the concentration of 8000 p.p.m. (close to saturation), animals have not been exposed experimentally to the vapors of any of the amyl alcohols, but most investigators have found that amyl alcohol administered otherwise is more narcotic and lethal than the lower homologs. Baer[261] and Munch and Schwartze[257] compared amyl alcohol with the lower homologs by their oral administration to rabbits and found the latter less toxic. Joffroy and Serveaux[262] and Lehman and Newman[263] obtained similar results following intravenous administration of alcohols to rabbits; Macht[264] arrived at the same conclusion from corresponding experiments upon cats. Intraperitoneal injections of the alcohols into rats, by Lendle,[265] likewise showed that amyl alcohol was more toxic than the lower homologs.

According to Munch and Schwartze[257] the narcotic action of a series of amyl alcohols, in the case of rabbits, decreases in the following order: 2-pentanol (a secondary alcohol), 2-methyl-2-butanol (a tertiary alcohol), and 3-methyl-1-butanol (a primary alcohol); the toxicity (orally to rabbits) of the same three isomers decreases in the following order:[257] tertiary, secondary, and primary (see

[257] J. C. Munch and E. W. Schwartze, *J. Lab. Clin. Med.,* **10**, 985 (1925).

[258] H. F. Smyth, Jr., C. P. Carpenter, C. S. Weil, and U. C. Pozzani, *Arch. Ind. Hyg. Occupational Med.,* **10**, 61 (1954).

[259] R. W. Schaffarzick, *Science,* **116**, 663 (1952).

[260] H. F. Smyth, Jr., *Cummings Memorial Lecture.* Am. Ind. Hyg. Assoc., Philadelphia, April 25, 1956.

[261] G. Baer, *Arch. Anat. u. Physiol. Anat. Abt.,* 1898, 283.

[262] A. Joffroy and R. Serveaux, *Arch. méd. exptl. anat. pathol.,* **7**, 569 (1895).

[263] A. J. Lehman and H. W. Newman, *J. Pharmacol.,* **61**, 103 (1937).

[264] D. I. Macht, *J. Pharmacol.,* **16**, 1 (1920).

[265] L. Lendle, *Arch. exptl. Pathol. Pharmakol.,* **132**, 214 (1928).

Table 9). Lendle[265] confirmed the latter results in his experiments on rats, reporting upon the toxicity of three alcohols in decreasing order: 2-methyl-2-butanol (tertiary), 2-pentanol (secondary) and 1-pentanol (primary). Starrek[266] found 1-pentanol (*n*-amyl) less toxic than 3-methyl-1-butanol (isoamyl)—both primary alcohols—when administered subcutaneously.

Gross Pathology in Animals. Strauss[267] reported the occurrence of liver injury in the rabbit following repeated oral administration of amyl alcohol.

Absorption and Excretion by Animals. Apparently dogs and cats exhale more unchanged amyl alcohol than do rabbits or rats (see Table 10). A dog weighing

TABLE 10

Excretion of Amyl Alcohols by Rats in Per Cent of Total Given

Compound	In expired air	In urine	Total excreted
Primary alcohols			
1-Pentanol	0.88	0.29	1.2
2-Methyl-1-butanol	5.6	2.0	7.6
3-Methyl-1-butanol	0.86	0.22	1.1
3-Methyl-1-butanol (levo form)	0.97	0.27	1.2
Secondary alcohols			
2-Pentanol	6.2 (42.3)[a]	1.3 (2.4)[a]	52.2
3-Pentanol	0.3 (51.2)[a]	0.1 (4.7)[a]	56.3
3-Methyl-2-butanol	8.3 (49.1)[a]	2.9 (2.5)[a]	62.8
Tertiary alcohol			
2-Methyl-2-butanol	26.4	8.9	35.3

[a] Ketone excreted.

11 kg., when injected subcutaneously with a dose of 0.1 ml. of 2-methyl-2-butanol (*tert*-amyl alcohol) exhaled 65 per cent of it within 5.75 hours. Another dog given a slightly larger dose of the same alcohol intravenously exhaled 52 per cent within 6 hours.[268] Cats also exhaled a large quantity unchanged.[268] When rabbits were injected intravenously with two levels of dosage, one approximately the same as that given to dogs and the other half as much, they exhaled, in the first instance, 21 per cent of the alcohol in unchanged form within 4 hours, and in the second, 22 per cent within 3 hours. Over a period of 3 hours a dog eliminated 55 per cent, whereas a rabbit expelled 17 per cent. A rat given 1 g./kg. of the same isomer eliminated 26.4 per cent unchanged in the expired air within 50 hours.[269]

Following oral administration of a primary alcohol (3-methyl-1-butanol), a secondary alcohol (2-pentanol), and the tertiary alcohol (2-methyl-2-butanol) to

[266] E. Starrek, Dissertation, Würzburg, 1938.
[267] Strauss, *Compt. rend. soc. biol.*, **4**, 54 (1887).
[268] J. Pohl, *Arch. exptl. Pathol. Pharmakol. Suppl.*, **1908**, 427.
[269] H. W. Haggard, D. P. Miller, and L. A. Greenberg, *J. Ind. Hyg. Toxicol.*, **27.** 1 (1945).

rabbits, small amounts of conjugated glucuronic acid were found in the urine.[270,271] Dogs excreted less glucuronic acid than rabbits following administration of the secondary alcohol, and none at all when given the tertiary alcohol.[270,271]

The rate of elimination (or oxidation) of the amyl alcohols from the body decreases in the following order: primary, secondary, tertiary. Haggard and his associates[269] gave each of the structural isomers, except 2,2-dimethyl-1-propanol, and also the levo form of 2-methyl-1-butanol, in the dose of 1 g./kg. to rats, intraperitoneally. These alcohols disappeared from the blood following their administration, after the following intervals of time: primary, 3.5 to 9 hours; secondary, 13 to 16 hours; and tertiary, 50 hours. The concentrations (mg.%) found in the blood 1 hour after administration were as follows: primary, 14 to 55; secondary, 51 to 65; and tertiary, 12.5. Unlike the large amounts of unchanged tertiary alcohol exhaled by dogs, Pohl[268] found only traces of alcohol in the expired air of a dog following the intravenous administration of 2-methyl-1-butanol, a primary alcohol.

Guggenheim and Löffler[272] found that isovaleric acid was formed during the perfusion of the rabbit liver with isoamyl alcohol. Haggard and his associates[269] showed the transitory presence of small amounts of valeraldehydes in the blood following the administration of the primary alcohols. They believe the aldehyde is oxidized to valeric acid. The secondary alcohols were oxidized to the ketones, which were present in the blood in measurable quantities. The ketones were present about twice as long as the secondary alcohols. No volatile metabolites were found following the administration of *tert*-amyl alcohol. The conversion of alcohol to aldehyde is dependent upon the action of the liver, since in partially hepatectomized animals this conversion was largely inhibited. Although there was a definite decrease in the conversion of secondary amyl alcohols to ketone in partially hepatectomized mice, the inhibition was not nearly as great as in the case of the primary alcohol.

Haggard *et al.*[269] also found the following concentrations (mg.%) of the respective alcohols present in jugular blood at the time of death of mice from respiratory failure: 1-pentanol, 76; 2-methyl-1-butanol, 110; levo form of 2-methyl-1-butanol, 76; 3-methyl-1-butanol, 76; 2-pentanol, 86; 3-pentanol, 87; 3-methyl-2-butanol, 90; and 2-methyl-2-butanol, 191.

From the increasing amounts of 2-methyl-2-butanol (tertiary) required to narcotize a dog in successive experiments, evidence of habituation (tolerance) was obtained.[273] Similar observations upon a rabbit yielded no such evidence.[273]

Mice were narcotized by immersion in dilute aqueous solutions of amyl alcohol under conditions that prevented ingestion and inhalation of the alcohol.[274]

[270] O. Neubauer, *Arch. exptl. Pathol. Pharmakol.*, **46**, 133 (1901).
[271] H. Thierfelder and J. V. Mering, *Z. physiol. Chem.*, **9**, 511 (1885).
[272] M. Guggenheim and W. Löffler, *Biochem. Z.*, **72**, 325 (1916).
[273] J. Biberfeld, *Biochem. Z.*, **92**, 198 (1918).
[274] Schwenkenbecher, *Arch. Anat. u. Physiol. Anat. Abt.*, **1904**, 121.

Effects upon Man. Without regard to any specific isomer most investigators have found that the inhalation of amyl alcohol vapors by man caused marked irritation of the eyes and respiratory tract, headache, and vertigo;[244-246,275-279] dyspnea, and cough;[275,278,280] nausea, vomiting, and diarrhea.[245,246,275,280,281] Double vision, deafness, delirium, and occasionally fatal poisoning, preceded by severe nervous symptoms, have been ascribed by Flury and Zernik[275] and by Eyquem[280] to the effects of the absorption of amyl alcohol. Coma, glycosuria, and sometimes methemoglobinemia are represented by Fuchter[282] and Underhill[279] as characteristic of amyl alcohol intoxication.

A few cases of industrial poisoning appear to have been caused by amyl alcohol, although in each reported instance some other solvent was present to becloud the issue. A neurasthenic brewery manager who inhaled the vapors (primary active amyl and isoamyl) from fermentation vats exhibited psychic stimulation, insomnia, and chromatopsia.[281] Eyquem[280] reported the following symptoms among workers engaged in producing smokeless powder: cough, irritation of the eyes, colic, diarrhea, vomiting, palpitation of the heart, nervous symptoms, headache, vertigo, disturbances of vision, forgetfulness, insomnia, somnolence, and weakness. One fatal case occurred. Although other alcohols and ether were employed, the signs increased as the use of amyl alcohol (probably fusel oil) increased. Two fatal cases from the use of a lacquer containing amyl alcohol and probably tetrachloroethane, for coating the inside surface of a tank, were reported by Zangger.[283] A lacquerer exposed to amyl alcohol and amyl acetate had digestive symptoms and secondary anemia, according to Baader.[284]

An increase of urobilin in the urine of lacquer sprayers exposed to a solvent containing amyl and butyl alcohols, amyl and butyl acetates, and acetone was reported by Burger and Stockmann.[285]

Nonindustrial cases of poisoning from drinking fusel oil, characterized by coma, glycosuria, and methemoglobinuria, were seen by Fuchter.[282] A patient died following an enema of 35 g. of amylene hydrate (Jacobi and Speer,[286] cited by Lewin[281]). Anker[287] reported the recovery of a woman who intentionally drank

[275] F. Flury and F. Zernik, *Schädliche Gase.* Springer, Berlin, 1931, p. 352.

[276] *Natl. Research Council Can. Bull.* No. **15,** 21 (1930).

[277] K. W. Nelson, J. F. Ege, Jr., M. Ross, L. E. Woodman, and L. Silverman, *J. Ind. Hyg. Toxicol.,* **25,** 282 (1943).

[278] L. Resnick, *Eye Hazards in Industry.* Columbia Univ. Press, New York, 1941, p. 251.

[279] F. P. Underhill, *Toxicology.* 2nd ed., Blakiston, Philadelphia, 1928, p. 212.

[280] Eyquem, *Ann. hyg. publ. et méd. légale,* **3,** 71 (1905).

[281] L. Lewin, *Gifte und Vergiftungen.* Stilke, Berlin, 1929, p. 407.

[282] T. B. Fuchter, *Am. Med.,* **2,** 210 (1901).

[283] H. Zangger, *Arch. Gewerbepathol. Gewerbehyg.,* **4,** 117 (1933).

[284] E. Baader, *Verhandl. deut. Ges. inn. Med.,* **45,** 318 (1933).

[285] G. E. C. Burger and B. H. Stockmann, *Zentr. Gewerbehyg. Unfallverhüt,* **9,** 29 (1932).

[286] W. Jacobi and E. Speer, *Therap. Halbmonatsh.,* **34,** 445 (1920).

[287] M. Anker, *Therap. Monatsh.,* **6,** 623 (1892).

about 27 g. of amylene hydrate. Recovery was preceded by coma, dyspnea, irregular pulse, and dilation then contraction of the pupils.

No effects upon the nerves of the skin of men, and no local wheal formation, erythema, or hyperemia were observed by Oettel[288] following the application of all isomers except 1-pentanol. Schwartz and Tulipan[289] state that isoamyl, normal, secondary, and tertiary amyl alcohols may be irritating to the skin.

5. Hygienic Standard of Permissible Exposure

The threshold limit of isoamyl alcohol has been established at 100 p.p.m.[290]

6. Flammability

The available data upon the flash points, limits of flammability, and ignition temperatures for the various isomers of amyl alcohol are given in Table 11.

TABLE 11
Flammability of Amyl Alcohols[a]

Isomer	Common alcohol name	Flash point, °F.	Lower limit of flammability, vol. %	Ignition temperature, °C.
Primary alcohols				
1-Pentanol	Primary amyl	100[291]	1.19[292]	391[291]
2-Methyl-1-butanol	Primary active amyl			
3-Methyl-1-butanol	Isoamyl	114[291]	1.2[292]	343[291]
2,2-Dimethyl-1-propanol	*tert*-Butyl carbinol			
Secondary alcohols				
2-Pentanol	Secondary active amyl	94[291]		343[291]
3-Pentanol	Diethylcarbinol			
3-Methyl-2-butanol	*sec*-Isoamyl	102.9[291]		
Tertiary alcohol				
2-Methyl-2-butanol	*tert*-Amyl	67[291]		437[a]

[a] See Vol. I, pages 516 and 525.

7. Warning Properties

According to Nelson et al.,[277] the following concentrations of 3-methyl-1-butanol (isoamyl alcohol) caused irritation of the respective mucous membranes of the majority of persons subjected to exposure for a few minutes: eyes, 150

[288] H. Oettel, *Arch. exptl. Pathol. Pharmakol.*, **183**, 641 (1936).

[289] L. Schwartz and L. Tulipan, *Occupational Diseases of the Skin.* Lea & Febiger, Philadelphia, 1939, p. 716.

[290] American Conference of Governmental Industrial Hygienists' Threshold Limit Values for 1959, *A.M.A. Arch. Ind. Health*, **20**, 266 (1959).

[291] Committee on Flammable Liquids, *Fire Hazard Properties*. Natl. Fire Protec. Assoc., Boston, 1941, pp. 6, 17.

[292] G. W. Jones, *Chem. Revs.*, **22**, 1 (1938).

p.p.m. (0.54 mg./liter); nose, 150 p.p.m.; and throat, 100 p.p.m. (0.36 mg./liter). From the standpoint of comfort these subjects agreed that the highest concentration for an 8-hour exposure should be less than 100 p.p.m.

METHYLAMYL ALCOHOL, $(CH_3)_2 CH \cdot CH_2 \cdot CHOH \cdot CH_3$
(Methylisobutylcarbinol)

1. Source, Uses, and Industrial Exposures

Methylamyl alcohol is prepared commercially by reduction of mesityl oxide[293] in acetic acid solution with platinum, or over reduced copper on asbestos at 120°C.

Methylamyl alcohol, which is a hexyl alcohol, is also known as 4-methyl-2-pentanol or methylisobutylcarbinol. It is used in brake fluids, as a frothing agent in ore flotation, and in the manufacture of ore-flotation agents, lubricant additives, solvents, plasticizers, lacquers, and in organic synthesis.[294,295]

2. Physical and Chemical Properties

Physical state: colorless liquid
Molecular weight: 102.178[296]
Specific gravity: 0.8079 (20/20°C.)[294]
Refractive index: 1.4113 at 20°C.[294]
Melting point: —90°C.[293]
Boiling point: 132.0°C.[294]
Vapor density: 3.52 (air = 1)
Vapor pressure: 3.52 mm. Hg (20°C.)[295]
Per cent in "saturated" air: 0.46 (20°C.)
Density of "saturated" air: 1.01 (air = 1)
Solubility: 1.7 per cent in water (20°C.);[294] soluble in ethyl alcohol, hydrocarbons, and most organic solvents
Flash point: 130°F. (55°C.)[294]
1 mg./liter ≍ 239.3 p.p.m. and 1 p.p.m. ≍ 4.17 mg./cu. meter at 25°C., 760 mm. Hg

3. Physiological Response

The following toxicological properties of methylamyl alcohol were reported by Smyth, Carpenter, and Weil.[297]

[293] R. E. Kirk and D. F. Othmer, eds., *Encyclopedia of Chemical Technology*, Vol. 1. Interscience Encyclopedia, New York, 1947, p. 316.

[294] A. K. Doolittle, *The Technology of Solvents and Plasticizers*. Wiley, New York, 1954, pp. 400–401, 661–664.

[295] I. Mellan, *Industrial Solvents. Reinhold*, New York, 1950, pp. 503–507.

[296] "Atomic Weights (International), 1957," *J. Am. Chem. Soc.*, **80**, 4122 (1958).

[297] H. F. Smyth, Jr., C. P. Carpenter, and C. S. Weil, *Arch. Ind. Hyg. Occupational Med.*, **4**, 119 (1951).

Toxicological Response to Methylamyl Alcohol

Single oral dose

Rat   LD$_{50}$ 2.59 g./kg.

Skin absorption
(24-hr. contact)

Rabbit   LD$_{50}$ 3.56 ml./kg.

Skin irritation

Rabbit   Trace of capillary injection

Inhalation

Rat   Saturated vapor, 2 hr., no deaths
Rat   2000 p.p.m., 8 hr., 5 of 6 died

2-ETHYLBUTYL ALCOHOL, C$_2$H$_5$·CH(C$_2$H$_5$)CH$_2$OH

1. *Source, Uses, and Industrial Exposures*

2-Ethylbutyl alcohol can be prepared commercially by the aldol condensation of acetaldehyde and butyraldehyde.[298]

2-Ethylbutyl alcohol is a hexyl alcohol known also as 2-ethyl-1-butanol, 2-ethylbutanol-1, or 2-ethyl-n-butyl alcohol. 2-Ethylbutyl alcohol is used as a solvent for printing inks, as a component in lacquers, and in the manufacture of surface active agents, synthetic lubricants, and lubricant additives.[299]

2. *Physical and Chemical Properties*

Physical state: colorless liquid
Molecular weight: 102.178[296]
Specific gravity: 0.8328[299]
Refractive index: 1.4208 (20°C.)[299]
Melting point: <-50°C.[298]
Boiling point: 149.4°C.[299]
Vapor density: 3.52 (air = 1)
Vapor pressure: 1.80 mm. Hg (20°C.)[300]
Per cent in "saturated" air: 0.22 (20°C.)
Density in "saturated" air: 1.005 (20°C.)
Solubility: 0.43 per cent in water (20°C.)[299]
Flash point: 135°F. (57°C.)[299]

1 mg./liter $\backsim$ 239.3 p.p.m. and 1 p.p.m. $\backsim$ 4.17 mg./cu. meter at 25°C., 760 mm. Hg

[298] R. E. Kirk and D. F. Othmer, eds., *Encyclopedia of Chemical Technology*, Vol. 1, Interscience Encyclopedia, New York, 1947, pp. 316–317.

[299] A. K. Doolittle, *The Technology of Solvents and Plasticizers*. Wiley, New York, 1954, pp. 400, 665, 666.

[300] I. Mellan, *Industrial Solvents. Reinhold*, New York, 1950, pp. 507–509.

3. *Physiological Response*

Animal Symptomatology. The following toxicological data obtained on animals are available for 2-ethylbutyl alcohol:

Toxicological Response to 2-Ethylbutyl Alcohol

Single dose

Rat (orally)	LD$_{50}$ 1.85 g./kg.[301]
Rat (I.P.)	LD$_{50}$ 0.80 g./kg.[302]
Guinea Pig (I. P.)	LD$_{50}$ 0.45 g./kg.[302]

Skin absorption

Rabbit (24-hr. contact)	LD$_{50}$ 1.26 ml./kg.[301]
Guinea pig	LD$_{50}$ <5.0 ml./kg.[301]

Skin irritation

Rabbit	Capillary injection[301]
Guinea pig	Moderate irritation[302]

Inhalation

Rat	Concentrated vapor, 8 hr., no deaths

Absorption and Excretion. Rabbits excreted 40 per cent of an ingested dose of 2-ethylbutyl alcohol as urinary glucuronide (diethyl acetyl glucuronide).[303] Kamil, Smith, and Williams[303] have shown that the 2-substituted primary alcohols are oxidized rapidly resulting in very little conjugation with glucuronic acid if there is a methyl group at the 2 position (isobutanol and 2-methyl-butanol) but, on the other hand, that 2-ethyl-substituted alcohols give rise to alpha ethyl fatty acids (which are not readily oxidized in vivo) that are readily conjugated with glucuronic acid (2-ethylbutyl alcohol and 2-ethylhexyl alcohol). The fatty acid resulting in vivo from 2-ethylbutyl alcohol is not completely resistant, since a small amount of methyl *n*-propyl ketone is also excreted in rabbit urine following the ingestion of 2-ethylbutyl alcohol.[303]

2-OCTANOL, $CH_3(CH_2)_5CHOH \cdot CH_3$ (Capryl Alcohol)

1. *Source, Uses, and Industrial Exposures*

2-Octanol is produced commercially by heating a soap of castor oil with sodium hydroxide.[304]

This secondary octyl alcohol is also known as capryl alcohol, *sec*-octyl alcohol, and methyl-*n*-hexylcarbinol. The content of methyl hexyl ketone ranges from 2 per cent maximum in the chemical grade to 14 per cent maximum in the

---

[301] H. F. Smyth, Jr., C. P. Carpenter, C. S. Weil, and U. C. Pozzani, *Arch. Ind. Hyg. Occupational Med.*, **10**, 61 (1954).

[302] D. W. Fassett, personal communication.

[303] I. A. Kamil, J. N. Smith, and R. T. Williams, *Biochem. J.*, **53**, 137 (1953).

[304] R. E. Kirk and D. F. Othmer, eds., *Encyclopedia of Chemical Technology*, Vol. 1, Interscience Encyclopedia, New York, 1947, pp. 317–318.

solvent grade.[305] It is used in lacquers, enamels, perfumes, organic synthesis, and as an antifoaming agent. It is used as a solvent and in the manufacture of plasticizers.[304,305]

2. Physical and Chemical Properties

Physical state: colorless liquid
Molecular weight: 130.232[306]
Specific gravity: 0.8232 (20/20°C.)[305]
Refractive index: 1.4260 (20°C.)[307]
Melting point: −38.6°C.[304]
Boiling point: 178–179°C.[304]
Vapor density: 4.49 (air = 1)
Vapor pressure: 1 mm. Hg (32.8°C.) ;[308] 40 mm. Hg (98°C.)[307]
Per cent in "saturated" air: 0.132 (32.8°C.)
Density of "saturated" air: 1.001 (air = 1)
Solubility: insoluble in water (<0.1 per cent) ;[305] soluble in ether, ethanol, aromatic and aliphatic hydrocarbons
Flash point: 150°F. (65.5°C.) for solvent grade; 180°F. (82°C.) for chemical grade[305]
1 mg./liter $\approx$ 187.8 p.p.m. and 1 p.p.m. $\approx$ 5.32 mg./cu. meter at 25°C., 760 mm. Hg

3. Physiological Response

The following toxicological information is available on 2-octanol:

Toxicological Response to 2-Octanol

	Single oral dose
Rat	LD$_{50}$ >3.2 g./kg.[307]
	Skin absorption
Guinea pig	LD$_{50}$ >0.5 g./animal[307]
	Skin irritation
Guinea pig	Slight[307]

2-ETHYLHEXANOL, $CH_3(CH_2)_3CH(C_2H_5)CH_2OH$

1. Source, Uses, and Industrial Exposures

2-Ethylhexyl alcohol is synthesized on a commercial scale.[309]
2-Ethylhexanol, known also as 2-ethylhexyl alcohol and 2-ethyl-1-hexanol,

[305] I. Mellan, *Industrial Solvents.* Reinhold, New York, 1950, pp. 512–514.
[306] "Atomic Weights (International), 1957," *J. Am. Chem. Soc.,* **80,** 4122 (1958).
[307] D. W. Fassett, personal communication.
[308] D. R. Stull, *Ind. Eng. Chem.,* **39,** 517 (1947).
[309] R. E. Kirk and D. F. Othmer, eds., *Encyclopedia of Chemical Technology,* Vol. 1. Interscience Encyclopedia, New York, 1947, p. 318.

is commercially the most important of the higher alcohols. Its principle use is as an intermediate in the manufacture of plasticizers, wetting agents, and the lacquer solvent, 2-ethylhexyl acetate.[310] It is also used as a solvent for nitrocellulose, urea resins, enamels, alkyd varnishes and lacquers. It is used as a defoaming and wetting agent. 2-Ethylhexanol finds use in ceramics, paper coatings, rubber latex, and textiles.[309,311]

2. *Physical and Chemical Properties*

Physical state: colorless liquid
Molecular weight: 130.232[306]
Specific gravity: 0.8340 (20/20°C.)[309,311,312]
Refractive index: 1.4313 (20°C.)[309,311]
Melting point: < -76.0°C.[309,312]
Boiling point: 184.8°C.[311]
Vapor density: 4.49 (air = 1)
Vapor pressure: 0.05 mm. Hg (20°C.)[311]
Per cent in "saturated" air: 0.0066 (20°C.)
Density of "saturated" air: <1.001 (air = 1)
Solubility: 0.1 per cent in water; soluble in ether, ethanol, and other organic solvents
Flash point: 185°F. (84°C.)[312]
1 mg./liter ≎ 187.8 p.p.m. and 1 p.p.m. ≎ 5.32 mg./cu. meter at 25°C., 760 mm. Hg

3. *Physiological Response*

Toxicological Response to 2-Ethylhexanol

Single dose

Rat (orally)	LD_{50} 3.2–6.4 g./kg.[312]
Rat (orally)	LD_{50} 3.2 g./kg.[313]
Mouse (orally)	LD_{50} 3.2–6.4 g./kg.[312]
Rat (I.P.)	LD_{50} 0.8–1.6 g./kg.[312]
Rat (I.P.)	LD_{50} 0.65 g./kg.[313]
Mouse (I.P.)	LD_{50} <0.40 g./kg.[312]
Mouse (I.P.)	LD_{50} 0.78 g./kg.[313]

Skin absorption

Guinea pig	LD_{50} >10 ml./kg.[312]

Skin irritation

Guinea pig	Moderate[312]

Inhalation

Rat	235 p.p.m., 6 hr., all survived[312]

[310] I. Mellan, *Industrial Solvents*. Reinhold, New York, 1950, pp. 511–512.

[311] A. K. Doolittle, *The Technology of Solvents and Plasticizers*. Wiley, New York, 1954, pp. 676–678.

[312] D. W. Fassett, personal communication.

[313] H. Hodge, *Proc. Soc. Exptl. Biol. Med.*, **53**, 20 (1943).

Absorption and Excretion. Kamil, Smith, and Williams[314] found that nearly 90 per cent of a dose of 2-ethylhexanol, when ingested by rabbits is excreted in the urine conjugated with glucuronic acid (α-ethylhexanonylglucuronide).

Effects on Man. Large quantities have been used in industry without any apparent reports of injury.

NONYL ALCOHOL, $C_9H_{19}OH$

1. *Source, Uses, and Industrial Exposures*

Nonyl alcohol has numerous potential isomers. By means of the "oxo process" (the addition of carbon monoxide and hydrogen to an olefin in the presence of a catalyst) a primary nonyl alcohol that consists largely of trimethyl hexanols, with some dimethyl heptanols, may be prepared. This material is probably rich in 3,5,5-trimethyl hexanol ($(CH_3)_3CCH_2CH(CH_3)CH_2CH_2OH$). Another commercial nonyl alcohol is a secondary alcohol rich in diisobutyl carbinol, ($(CH_3)_2CHCH_2)_2CHOH$, (2,6-dimethyl-4-heptanol).

Diisobutyl carbinol is employed as a reaction medium in a process for the manufacture of hydrogen peroxide. It is used as a solvent for coating compositions of urea or melamine resins, and as a defoaming agent and for the preparation of lubricant additives and plasticizers.[315]

2. *Physical and Chemical Properties*

The physical and chemical properties of 3,5,5-trimethyl hexanol, diisobutyl carbinol (studied for its toxicological properties by Smyth *et al.*) and a nonyl alcohol rich in trimethyl hexanols (which was examined by the writer for its toxicological properties) are given in Table 12. The properties of another nonyl alcohol containing 2 per cent of 2-propyl heptanol (examined by D. W. Fassett[319]) are also presented in Table 12.

3. *Physiological Response*

The toxicological response to the various nonyl alcohols are listed in Table 13.

Pathology in animals. Fatal intoxication following the nonyl alcohol rich in trimethyl hexanol (listed as superscript c in Table 13) resulted in degenerative changes in the neurons in all parts of the brain and brain stem. Hepatocellular degeneration involved the liver cords and was pronounced in the central areas. Renal

[314] I. A. Kamil, J. N. Smith, and R. T. Williams, *Biochem. J.*, **53**, 137 (1953).

[315] A. K. Doolittle, *The Technology of Solvents and Plasticizers*. Wiley, New York, 1954, p. 680.

[316] "Atomic Weights (International), 1957," *J. Am. Chem. Soc.*, **80**, 4122 (1958).

[317] I. Heilbron, *Dictionary of Organic Compounds*, Vol. 4. Oxford Univ. Press, New York, 1953, p. 606.

[318] I. Heilbron, *Dictionary of Organic Compounds*, Vol. 2. Oxford Univ. Press, New York, 1953, p. 262.

[319] D. W. Fassett, personal communication.

TABLE 12

Physical Properties of Some Nonyl Alcohols[a]

	3,5-Trimethyl hexanol	Diisobutyl carbinol	Nonyl alcohol rich in trimethyl hexanols	Nonyl alcohol containing 2% 2-propyl heptanol
Physical state	Liquid	Liquid	Colorless liquid	Liquid
Molecular weight	144.259[316]	144.259[316]		
Specific gravity	0.8236 (25/4°C.)[317]	0.809 (21/4°C.)[318]	0.845 (15.6/15.6°C.)	0.8274 (20/4°C.)[319]
Refractive index (n)	1.4300 (25°C.)[317]	1.4229 (20°C.)[318]	1.438	1.43347 (20°C.)[319]
Melting point (°C.)				−5[319]
Boiling point (°C.)	193–194[317]	178.1[315]	187–208	192[319]
Vapor density (air = 1)	4.97	4.97		
Vapor pressure, mm.		0.21 (20°C.)[315]		40 (129°C.)[319]
Per cent in "saturated" air		0.027		
Density of "saturated" air		1.001		
Solubility in water	Insoluble	0.06% (20°C.)[315]	Insoluble	Insoluble

[a] 1 mg./liter ≈ 169.6 p.p.m. and 1 p.p.m. ≈ 5.90 mg./cu. meter at 25°C, 760 mm. Hg.

TABLE 13

Toxicological Response to Nonyl Alcohol

Animal	Response	Reference
	Single dose	
Mouse (orally)	LD_{50} 6.4–12.8 g./kg.[a]	319
Rat (orally)	LD_{50} 3.2–6.4 g./kg.[a]	319
Rat (orally)	LD_{50} 3.56 g./kg.[b]	320
Rat (orally)	LD 1.4–1.75 ml./kg.[c]	321
Rabbit (orally)	LD 1.4–2.1 ml./kg.[c]	321
Mouse (I.P.)	LD_{50} 0.80–1.60 g./kg.[a]	319
Rat (I.P.)	LD_{50} 0.80–1.60 g./kg.[a]	319
Rabbit (I.V.)	LD 0.015–0.021 g./kg.[c]	321
	Skin absorption	
Guinea pig	LD_{50} >10 ml./kg.[a]	319
Rabbit (24 hr.)	LD_{50} 5.66 ml./kg.[b]	320
Rabbit (24 hr.)	LD_{50} <3.6 ml./kg.[c]	321
	Skin irritation	
Guinea pig	Moderate[a]	319
Rabbit	None[b]	320
Rabbit	Erythema, slight necrosis[c]	321
	Repeated ingestion	
Rabbit	Dosage of 0.148 g./kg. given on each of 67 days over a period of 83 days resulted in normal growth, uniform survival, and no signs of intoxication[c]	321
	Repeated application on skin	
Rabbit	Contact of 5 ml. (1.6–2.0 g./kg.) of nonyl alcohol for 1 hr./day with the skin on each of 50 days over a period of 75 days resulted in retarded growth and erythema but no mortality[c]	321
	Inhalation	
Rat	730 p.p.m., 6 hr., no deaths[a]	319
Rat	215 p.p.m., 6 hr., no deaths[a]	319
Rat	Saturated vapor, 8 hr., no deaths[b]	320

[a] Nonyl alcohol employed by D. W. Fassett.
[b] Diisobutyl carbinol.
[c] Alcohol rich in trimethyl hexanol.

damage was manifested by severe degeneration and frequent necrosis of the epithelium of the proximal and loop tubules, as well as the epithelium of the glomeruli. There were slight degenerative changes in the myocardium.[322]

[320] H. F. Smyth, Jr., C. P. Carpenter and C. S. Weil, *J. Ind. Hyg. Toxicol.*, **31**, 60 (1949).
[321] J. F. Treon, unpublished data.
[322] K. V. Kitzmiller, personal communication.

DECYL ALCOHOL, $CH_3(CH_2)_8CH_2OH$

1. Source, Uses, and Industrial Exposures

n-Decyl alcohol, 1-decanol, or n-nonylcarbinol is a primary alcohol that is manufactured commercially by the catalytic reduction of coconut oil and of coconut oil fatty acids or their esters under pressure.[323] Decyl (oxo) alcohol, which contains mixed isomers of decanol, is manufactured by the "oxo process."

Surface active agents and detergents are manufactured by the sulfonation of n-decyl alcohol. n-Decyl alcohol is also used as an antifoaming agent and as an intermediate in the manufacture of perfumes.[323]

TABLE 14

Toxicological Response to Decyl Alcohol

Animal	Response	Reference
	Single dose	
Mouse (orally)	LD_{50} 6.4–12.8 g./kg.[a]	326
Rat (orally)	LD_{50} 12.8–25.6 g./kg.[a]	326
Rat (orally)	LD_{50} 9.80 g./kg.[b]	327
Mouse (I.P.)	LD_{50} 0.80–1.6 g./kg.[a]	326
Rat (I.P.)	LD_{50} 0.80–1.6 g./kg.[a]	326
	Skin absorption	
Guinea pig	LD_{50} >10 ml./kg.[a]	326
Rabbit	LD_{50} 3.56 ml./kg.[b]	327
	Skin irritation	
Guinea pig	Moderate[a]	326
Rabbit (24 hr.)	Severely[b]	327
	Inhalation	
Rat	905 p.p.m., 6 hr., no deaths[a]	326
Rat	65 p.p.m., 6 hr., no deaths[a]	326
Rat	Saturated vapor, 8 hr., no deaths[b]	327

[a] Mixture of n-decyl and sec-decyl alcohol.

[b] Mixed isomers (decyl (oxo) alcohol).

2. Physical and Chemical Properties

Physical state: colorless to light yellow liquid
Molecular weight: 158.286[324]
Specific gravity: 0.829 (20/20°C.)[323]
Refractive index: 1.4638 (20°C.)[323]

[323] R. E. Kirk and D. F. Othmer, eds., *Encyclopedia of Chemical Technology*, Vol. 1. Interscience Encyclopedia, New York, 1947, p. 318.

[324] "Atomic Weights (International), 1957," *J. Am. Chem. Soc.*, **80**, 4122 (1958).

Melting point: $-6°C$.[323]
Boiling point: $231°C$.[323]
Vapor density: 5.46 (air $= 1$)
Vapor pressure: 1 mm. Hg (69.5°C.)[325]
Solubility: insoluble in water; soluble in ether, alcohol, acetone, benzene, and glacial acetic acid

1 mg./liter $\approx$ 154.5 p.p.m. and 1 p.p.m. $\approx$ 6.47 mg./cu. meter at 25°C., 760 mm. Hg

3. *Physiological Response*

The toxicological responses to decyl alcohols are given in Table 14.

LAURYL ALCOHOL, $C_{11}H_{23}CH_2OH$ (Dodecanol)

1. *Source, Uses, and Industrial Exposures*

Lauryl alcohol was previously prepared commercially by the reduction of ethyl laurate (from coconut oil) with sodium and absolute alcohol, but presently is manufactured by the catalytic hydrogenation of coconut oil fatty acids or their esters under high pressure.[328]

Lauryl alcohol is used in the manufacture of detergents.[328]

2. *Physical and Chemical Properties*

Physical state: crystalline solid
Molecular weight: 186.340[329]
Specific gravity: 0.8309 (24/4°C.)[328]
Melting point: 24°C.[328]
Boiling point: 255–259°C.[328]
Vapor density: 6.43 (air $= 1$)
Vapor pressure: 1 mm. Hg (91°C.)[330]
Solubility: insoluble in water; soluble in alcohol and ether

1 mg./liter $\approx$ 131.3 p.p.m. and 1 p.p.m. $\approx$ 7.62 mg./cu. meter at 25°C., 760 mm. Hg

[325] D. R. Stull, *Ind. Eng. Chem.*, **39**, 517 (1947).

[326] D. W. Fassett, personal communication.

[327] H. F. Smyth, Jr., C. P. Carpenter, and C. S. Weil, *Arch. Ind. Hyg. Occupational Med.*, **4**, 119 (1951).

[328] R. E. Kirk and D. F. Othmer, eds., *Encyclopedia of Chemical Technology*, Vol. 1. Interscience Encyclopedia, New York, 1947, p. 319.

[329] "Atomic Weights (International), 1957," *J. Am. Chem. Soc.*, **80**, 4122 (1958).

[330] T. E. Jordan, *Vapor Pressure of Organic Compounds*. Interscience Publishers, New York–London, 1954, p. 80.

3. Physiological Response

Toxicological Response to Lauryl Alcohol

Single dose

Rat (orally)	LD_{50} >12.8 g./kg.[331]
Rat (orally)	LD_{50} >36.0 ml./kg.[332]
Rabbit (orally)	LD_{50} >36.0 ml./kg.[332]
Rat (I.P.)	LD_{50} 0.80–1.6 g./kg.[331]

Skin absorption

Guinea pig	LD_{50} >10 ml./kg.[331]

Skin irritation

Guinea pig	Practically none[331]

Animal Pathology. Seven rats and 7 rabbits, which survived a dosage of either 24 or 36 ml. of technical lauryl alcohol/kg. of body weight, demonstrated no significant gross or microscopic changes. One rat, which died 6 days after the oral administration of 36 ml./kg., exhibited fatty degeneration of the liver and confluent bronchcpneumonia.[333]

CETYL ALCOHOL, $C_{16}H_{33}OH$

1. Source, Uses, and Industrial Exposures

Cetyl alcohol, which was first obtained from spermaceti, is prepared commercially by the catalytic reduction of fats containing palmitic acid and by the saponification of spermaceti wax.[334]

Cetyl alcohol, which is also known as hexadecyl alcohol, palmityl alcohol, "ethal and ethol," is used in cosmetics, perfumes, lotions, toilet articles, medicinals, and in manufacturing detergents.[334]

2. Physical and Chemical Properties

Physical state: white, crystalline solid
Molecular weight: 242.448[329]
Specific gravity: 0.8176 (50/4°C.)[335]
Refractive index: 1.4283 (78.9°C.)[334]
Melting point: 49.3°C.[334]
Boiling point: 344°C.;[334] 190°C. (15 mm. Hg)[334]
Vapor pressure: 1 mm. Hg (122.7°C.)[330]

[331] D. W. Fassett, personal communication.
[332] J. F. Treon, unpublished data.
[333] F. E. Shaffer, personal communication.
[334] R. E. Kirk and D. F. Othmer, eds., *Encyclopedia of Chemical Technology*, Vol. 1. Interscience Encyclopedia, New York, 1947, p. 320.
[335] D. W. Fassett, personal communication.

Vapor density: 8.360 (air = 1)

Solubility: insoluble in water; soluble in alcohol, ether, and chloroform

1 mg./liter $\backsimeq$ 100.9 p.p.m. and 1 p.p.m. $\backsimeq$ 9.91 mg./cu. meter at 25°C., 760 mm. Hg

3. *Physiological Response*

The alcohol studied in reference 335 was a synthetic liquid C_{16} alcohol that may have contained impurities.

Toxicological Response to Cetyl Alcohol

Single dose

Mouse (orally)	LD_{50} 3.2–6.4 g./kg.[335]
Rat (orally)	LD_{50} 6.4–12.8 g./kg.[335]
Mouse (I.P.)	LD_{50} 1.6–3.2 g./kg.[335]
Rat (I.P.)	LD_{50} 1.6–3.2 g./kg.[335]

Skin absorption

Guinea pig	LD_{50} <10 g./kg.[335]

Skin irritation

Guinea pig	Slight[335]

Inhalation

Rat	2.22 mg./liter (calculated), 6 hr., all died within 2 days[335]
Rat	0.41 mg./liter (calculated), 6 hr., all survived[335]

DIACETONE ALCOHOL, $CH_3COCH_2COH(CH_3)_2$

1. *Source, Uses, and Industrial Exposures*

Diacetone alcohol is prepared by condensing acetone in the liquid phase in the presence of alkali and alkaline earth hydroxides.[336] The technical grade contains up to 15 per cent acetone.[337]

Diacetone alcohol is known also as diacetonyl alcohol, diacetone, 4-hydroxy-4-methyl-2-pentanone, dimethyl acetonyl carbinol, and pyranton A. It is a solvent for nitrocellulose, cellulose acetate, cellulose esters, epoxy resins, hydrocarbons, oils, fats, resins, gums, and dyes.[337,338] It is used in hydraulic fluids, metal-cleaning compounds, the manufacture of photographic film, making artificial silk and leather, and in coating compositions for paper and textiles.[336,337] Unless neutral or slightly alkaline it will decompose to acetone.[336]

---

[336] R. E. Kirk and D. F. Othmer, eds., *Encyclopedia of Chemical Technology*, Vol. 1. Interscience Encyclopedia, New York, 1947, p. 320.

[337] I. Mellan, *Industrial Solvents*. Reinhold, New York, 1950, pp. 609–610.

[338] A. K. Doolittle, *The Technology of Solvents and Plasticizers*. Wiley, New York, 1954, pp, 518–520.

2. Physical and Chemical Properties

Physical state: colorless liquid when pure, becomes yellow on aging
Molecular weight: 116.162[339]
Specific gravity: 0.9406 (20/20°C.)[338]
Refractive index: 1.4226 (20°C.)[338]
Freezing point: −42.8°C.[338]
Boiling point: 169.2°C.[338]
Vapor density: 4.01 (air = 1)
Vapor pressure: 0.97 mm. Hg (20°C.);[338] 1.5 mm. Hg (25°C.)[340]
Per cent in "saturated" air: 0.13 (20°C.); 0.2 (25°C.)
Density of "saturated" air: 1.005 (air = 1)
Solubility: miscible in all proportions in water and miscible with most organic solvents
1 mg./liter $\backsim$ 216.5 p.p.m. and 1 p.p.m. $\backsim$ 4.75 mg./cu. meter at 25°C., 760 mm. Hg

3. Determination in the Atmosphere

Diacetone alcohol may be differentiated qualitatively from acetone, since with 2,4-dinitrophenyl hydrazine, the former gives a red, and the latter a yellow, precipitate.[341]

4. Physiological Response

Physiological Response to Diacetone Alcohol

	Single dose
Rat (orally)	LD$_{50}$ 4.0 g./kg.[342]
	Skin absorption
Rabbit (24-hr. contact)	LD$_{50}$ 14.5 ml./kg.[342]
	Skin irritation
Rabbit	Comparable to acetone[342]

Animal Symptomatology. Walton and his associates[341] observed that intravenous injections of diacetone alcohol into mice induced narcosis more rapidly than acetone, and that diacetone alcohol was about twice as toxic as acetone. Following intravenous, intramuscular, or oral administration to rabbits, diacetone alcohol depressed the respiration markedly, decreased the blood pressure, induced narcosis, and caused death by respiratory failure.[341] The progressive decrease in the blood pressure of dogs injected repeatedly with diacetone alcohol, noted by these investigators, led them to believe that increased susceptibility had resulted from the repetition of the injections.

[339] "Atomic Weights (International), 1957." *J. Am. Chem. Soc.,* **80,** 4122 (1958).
[340] G. S. Gardner, *Ind. Eng. Chem.,* **32,** 226 (1940).
[341] D. C. Walton, E. F. Kehr, and A. S. Lovenhart, *J. Pharmacol.,* **33,** 175 (1928).
[342] H. F. Smyth, Jr., and C. P. Carpenter, *J. Ind. Hyg. Toxicol.,* **30,** 63 (1948).

Keith[343] observed a temporary decrease in the hemoglobin content and numbers of erythrocytes in the peripheral blood of rats for 1 to 4 days following the oral administration of a sublethal dose of diacetone alcohol.

Smyth and Carpenter[342] placed diacetone alcohol in the drinking water of groups of rats for 30 days. The least daily dosage that resulted in any evidence of micropathological alteration was 0.04 g./kg., whereas no effects were observed at the level of 0.01 g./kg.

Mice, rats, rabbits, and cats subjected for 20 minutes to inhalation of air containing 2100 p.p.m. (10/mg.liter) of diacetone vapors manifested restlessness, symptoms of irritation, coryza, symptoms of excitation, then sleepiness (unpublished work of E. Gross, cited by Lehmann and Flury[344]). Smyth and Carpenter[342] found 1500 p.p.m. did not kill rats in 8 hours.

Animal Pathology. Walton *et al.*[341] have reported 1 case of acute kidney damage among several rabbits. Gross, cited by Lehmann and Flury,[344] also found damage in the kidneys of exposed rabbits. Keith[343] described hepatic lesions, following the oral administration of a sublethal dose to rats, characterized by vacuolization and granulation of the parenchymal cells, which reached the maximum stage in about 24 hours.

5. Hygienic Standard of Permissible Exposure

The threshold limit of diacetone alcohol has been established at 50 p.p.m.[345]

6. Flammability

The flash point of pure diacetone alcohol is 131°F.[346] The technical grade, which may contain as much as 15 per cent acetone, has a flash point of 40 to 57°F.[347]

7. Odor and Warning Properties

Silverman, Schulte, and First[348] observed irritation of the eyes, nose, and throat in most subjects exposed to 100 p.p.m. Although some found the odor and taste unpleasant, this concentration was not intolerable.

[343] H. M. Keith, *Arch. Pathol.,* **13,** 707 (1932).

[344] K. B. Lehmann and F. Flury, *Toxikologie und Hygiene der technischen Lösungsmittel.* Springer, Berlin, 1938, 248.

[345] American Conference of Governmental Industrial Hygienists' Threshold Limit Values for 1959, *A.M.A. Arch. Ind. Health,* **20,** 266 (1959).

[346] Committee on Flammable Liquids, *Fire Hazard Properties.* Natl. Fire Protect. Assoc., Boston, 1941, pp. 6, 17.

[347] G. W. Jones, *Chem. Revs.,* **22,** 1 (1938).

[348] L. Silverman, H. F. Schulte, and M. W. First, *J. Ind. Hyg. Toxicol.,* **28,** 262 (1946).

BENZYL ALCOHOL, $C_6H_5CH_2OH$

1. Source, Uses, and Industrial Exposures

Commercially, benzyl alcohol is manufactured from benzyl chloride by refluxing with sodium or potassium carbonate.[349]

Benzyl alcohol is known also as phenyl carbinol or phenyl methanol. Benzyl alcohol has been used as a lacquer solvent and a plasticizer. It is employed in the manufacture of perfumes, pharmaceuticals, and dyestuffs.[350,351] It is useful as a solvent in cosmetics and inks. Its esters are utilized in soaps, perfumes, and flavors.[349]

2. Physical and Chemical Properties

Physical state: colorless liquid

Molecular weight: 108.141[339]

Specific gravity: 1.0472 (20/20°C.)[350]

Refractive index: 1.5399 at 20°C.[350]

Freezing point: —15.3°C.[350]

Boiling point: 205.3°C.[350]

Vapor density: 3.72 (air = 1)

Vapor pressure: 0.15 mm. Hg (25°C.)[351]

Per cent in "saturated" air: 0.02

Density of "saturated" air: 1.0005 (air = 1)

Solubility: 3.5 per cent in water (20°C.);[350] soluble in alcohol, ether, and aromatic hydrocarbons

Flash point: 96°C.[351]

1 mg./liter $\backsimeq$ 226.1 p.p.m. and 1 p.p.m. $\backsimeq$ 4.42 mg./cu. meter at 25°C., 760 mm. Hg

3. Determination in the Atmosphere

Callaway and Reznek[352] determined the amount of benzyl alcohol in an aqueous solution by measuring the refractive index, or by measuring the amount of benzoic acid resulting from oxidation with a saturated potassium permanganate solution. A paper by Mohler and Hämmerle[353] describes a method in which benzyl

[349] R. E. Kirk and D. F. Othmer, eds., *Encyclopedia of Chemical Technology*, Vol. 2. Interscience Encyclopedia, New York, 1948, pp. 483–486.

[350] A. K. Doolittle, *The Technology of Solvents and Plasticizers*. Wiley, New York, 1954, p. 672.

[351] I. Mellan, *Industrial Solvents*. Reinhold, New York, 1950, p. 86, 519.

[352] J. Callaway, Jr. and S. Reznek, *J. Assoc. Offic. Agr. Chemists*, **16**, 285 (1933).

[353] H. Mohler and W. Hämmerle, *Z. anal. Chem.*, **122**, 202 (1941), *Chem. Abstr.*, **36**, 4970 (1942).

alcohol in bitter-almond water was determined by means of its absorption spectrum in the ultraviolet region. They report results within the limits of an error of 5 per cent. The maxima at 267, 264, 258, and 252 mμ are recommended for quantitative estimation. This method should be adaptable to air analysis.

4. *Physiological Response*

Toxicological Response to Benzyl Alcohol

Single oral dose

Rat   LD$_{50}$ 3.10 g./kg.[354]

Skin irritation

Rabbit   Slight erythema[354]

Inhalation (satd. vapor, 200–300 p.p.m.)

Rat   2 hr., LC$_0$[354]
Rat   4 hr., LC$_{33}$[354]
Rat   8 hr., LC$_{100}$[354]

Animal Symptomatology. Fassett[354a] found that orally the approximate LD$_{50}$ in rats was between 1600 and 3200 mg./kg. for the undiluted liquid. Intraperitoneally, the approximate LD$_{50}$ was 400–800 mg./kg. in rats and guinea pigs. The undiluted material applied to depilated skin of guinea pigs for a period of 24 hours caused moderately strong primary irritation, and there was evidence of systemic symptoms with death at less than 5 ml./kg. Systemic symptoms were still noted when it was applied under the same conditions as a 20 per cent solution in acetone. It was not a skin sensitizer in guinea pigs. No fatalities or symptoms were found in rats when exposed 6 hours to a calculated concentration of 61 p.p.m. nor were symptoms produced by exposures obtained by bubbling air through the liquid heated to 100 and 150°C.

In a comparison of the lethal doses of various alcohols injected subcutaneously into mice, Starrek[355] indicated that benzyl alcohol is 2.5 to 3 times as toxic as *n*-butyl or isopropyl alcohol. Poisoned mice suffered respiratory stimulation, respiratory and muscular paralysis, convulsions, and narcosis. Macht[356] also observed convulsions when lethal doses were injected into small animals and reported the occurrence of a decrease in the blood pressure of animals following various modes of injection. Gruber[357] observed a decrease in the arterial blood pressure of rabbits, cats, and dogs following intravenous injection of benzyl alcohol, but failed to find such decrease in the case of dogs following the oral administration of 0.1 to 1.0 ml./kg. of body weight. Doses of 0.2 ml./kg., or more, ad-

---

[354] H. F. Smyth, Jr., C. P. Carpenter, and C. S. Weil, *Arch. Ind. Hyg. Occupational Med.,* **4,** 119 (1951); H. F. Smyth, Jr., personal communication.

[354a] D. W. Fassett, personal communication.

[355] E. Starrek, Dissertation, Würzburg, 1938.

[356] D. I. Macht, *J. Pharmacol.,* **11,** 263, 419 (1918).

[357] C. M. Gruber, *J. Lab. Clin. Med.,* **9,** 15, 92 (1923).

ministered to dogs by stomach tube induced emesis and defecation. This was apparently due to irritation of the gastric mucosa, since no such effects resulted from smaller doses given by this route or from the injection of larger doses.[357] Diuresis, more pronounced in the rabbit than in the dog, resulted from the administration of benzyl alcohol by various means.[358] The blood sugar of fasting animals was increased somewhat by prolonged administration of benzyl alcohol.[359] The administration of benzyl alcohol stimulated the rate of elimination of an injected pigment by the liver.[359] Macht[356] believed death was caused by paralysis of the respiratory center, but Gruber[357] believed that cardiac paralysis might precede that of the respiratory center.

Animal Pathology. The injection of 5 or 10 per cent benzyl alcohol in oil of sweet almond in the region of the auditory meatus of cats caused temporary degeneration of the small facial nerves.[360] Local necrosis of tissue following the accidental injection of pure benzyl alcohol in preparation for a circumcision has been described by Macht.[361]

Absorption and Excretion. The animal (and human) organism readily oxidizes benzyl alcohol to benzoic acid, which, after conjugating with glycine, is rapidly eliminated as hippuric acid in the urine. Rabbits, given 1 g. of benzyl alcohol subcutaneously, eliminated 300 to 400 mg. of hippuric acid within the following 24 hours.[362] Within 6 hours after the oral administration of 0.40 g. of benzyl alcohol per kilogram of body weight, rabbits eliminated 65.7 per cent of the dose as hippuric acid in the urine.[363] Within 6 hours after taking 1.5 g. of benzyl alcohol orally, human subjects eliminated 75 to 85 per cent of the dose in the urine as hippuric acid.[364]

Effects upon Man. Seven cases of illness in association with the use of a lacquer containing 5 per cent benzene, 10 per cent benzyl alcohol, acetone, denatured alcohol, butyl tartrate, and cellulose acetate, were reported by de Gaulejac and Dervillée.[365] It was believed that benzene (which was present in the air in a poorly ventilated room to the extent of 0.3 mg./liter) and benzyl alcohol were the chief causes of the violent headaches, vertigo, nausea, gastric pains, vomiting, diarrhea, and loss of weight. These signs of intoxication, which appeared after $1^1/_2$ to 2 months of exposure, disappeared upon removal from the lacquer exposure.

5. *Flammability*

The ignition temperature of benzyl alcohol is 426°C. (see also Vol. I, pages 526 and 533).

[358] C. M. Gruber, *J. Lab. Clin. Med.*, **10**, 284 (1924).
[359] I. Hosino, *Zikken Syokakibyogaku (Exptl. Gastroenterol.)*, **15**, 117 (1940), *Japan. J. Med. Sci.* II, *Biochem.*, **4**, No. 4, Abstr. (in English), 104 (1941).
[360] D. Duncan and W. H. Jarvis, *Anesthesiology*, **4**, 465 (1943).
[361] D. I. Macht, *J. Pharmacol. Proc.*, **13**, 509 (1919).
[362] J. A. Stekol, *J. Biol. Chem.*, **128**, 199 (1939).
[363] S. L. Diack and H. B. Lewis, *J. Biol. Chem.*, **77**, 89 (1928).
[364] I. Snapper, A. Grünbaum, and S. Sturkop, Biochem. Z., **155**, 163 (1925).
[365] R. de Gaulejac and P. Dervillée, *Ann. méd. légale criminol. et police sci.*, **18**, 146 (1938).

β-PHENYLETHYL ALCOHOL, $C_6H_5CH_2CH_2OH$

1. Source, Uses, and Industrial Exposures

β-Phenylethyl alcohol is produced from benzene and ethylene oxide by the Friedel-Crafts reaction.[366]

β-Phenylethyl alcohol is known also as phenylethyl alcohol, 2-phenylethanol, and benzyl carbinol. Along with citronellol and geraniol this alcohol forms the basis of rose-type perfumes. It is used to enhance other perfumes.[366]

2. Physical and Chemical Properties

Physical state: liquid
Molecular weight: 122.168[367]
Specific gravity: 1.0235 (15/15°C.)[368]
Refractive index: 1.5240 (20°C.)[368]
Melting point: −25.8°C. (extremely difficult to crystallize; tends to super-cool)[366]
Boiling point: 219.5–220°C.[366]
Vapor density: 4.21 (air $= 1$)
Vapor pressure: 1 mm. Hg (58°C.);[369] 10 mm. Hg (100°C.)[368]
Per cent in "saturated" air: 0.13 (58°C.)
Solubility: 2 per cent in water;[366] 1 part in 1.5 parts of 50 per cent ethanol[366]
Flash point: 216°F. (102°C.)[368]
1 mg./liter $\backsim$ 200.2 p.p.m. and 1 p.p.m. $\backsim$ 5.00 mg./cu. meter at 25°C., 760 mm. Hg

3. Determination in the Atmosphere

β-Phenylethyl alcohol may be identified as the p-nitrobenzoate (m.p. 106–107°C.) and as β-phenylethyl-p-nitrobenzylphthalate (m.p. 84.3°C.). It also forms styrene on treatment with alkali.[366]

4. Physiological Response

Toxicological Response to β-Phenylethyl Alcohol

Single dose

Mouse (orally)	LD$_{50}$ 0.8–1.5 g./kg.[368]
Guinea pig (orally)	LD$_{50}$ 0.4–0.8 g./kg.[368]
Mouse (I.P.)	LD$_{50}$ 0.2–0.4 g./kg.[368]
Guinea pig (I.P.)	LD$_{50}$ 0.4–0.8 g./kg.[368]

[366] R. E. Kirk and D. F. Othmer, eds., *Encyclopedia of Chemical Technology*, Vol. 2. Interscience Encyclopedia, New York, 1948, pp. 486–489.

[367] "Atomic Weights (International), 1957," *J. Am. Chem. Soc.*, **80**, 4122 (1958).

[368] D. W. Fassett, personal communication.

[369] T. E. Jordan, *Vapor Pressure of Organic Compounds*. Interscience Publishers, New York–London 1954, p. 78.

Skin absorption

Guinea pig   LD$_{50}$ 5–10 ml./kg.[368]

Skin irritation

Guinea pig   Slight[368]

CYCLOHEXANOL, H$_2$C(CH$_2$)$_4$CHOH

1. Source, Uses, and Industrial Exposures

Cyclohexanol may be prepared by (a) the catalytic hydrogenation of phenol at elevated temperatures and pressures or (b) catalytic oxidation of cyclohexane in the liquid phase at 100–250°C.[370]

Cyclohexanol is also known as hexahydrophenol, Hexalin, hydrophenol, cyclohexyl alcohol, Hydralin, Adronal, and Anol. Cyclohexanol is used as a solvent for oils, alkyd resins, shellac, alcohol-soluble phenolics, ethylcellulose, basic chrome and acid dyes, metallic soaps, gums, and natural resins. It is used in dry-cleaning, textile cleaning, laundry and household preparations. Cyclohexanol is used in leather lacquers, paint and varnish removers, polishes, rubber cements, and as an intermediate in the preparation of plasticizers and other chemicals.[371,372]

2. Physical and Chemical Properties

Physical state: colorless liquid

Molecular weight: 100.162[367]

Specific gravity: 0.9493 (20/4°C.)[371]

Refractive index: 1.4656 (20°C.)[371]

Freezing point: 25.1°C.[371]

Boiling point: 161.1°C.[371]

Vapor density: 3.46 (air = 1)

Vapor pressure: 3.5 mm. Hg (34°C.);[373] 2.5 mm. Hg (30°C.) (extrapolated)

Per cent in "saturated" air: 0.33 (30°C.)

Density of "saturated" air: 1.01 (air = 1)

Solubility: 3.6 per cent in water (20°C.);[371] miscible in most proportions with most aliphatic, aromatic, hydrogenated, and chlorinated solvents

Flash point: 63°C. (closed cup)[372]

1 mg./liter ≎ 244.1 p.p.m. and 1 p.p.m. ≎ 4.10 mg./cu. meter at 25°C. and 760 mm. Hg

[370] R. E. Kirk and D. F. Othmer, eds., *Encyclopedia of Chemical Technology*, Vol. 4, Interscience Encyclopedia, New York, 1949, p. 769.

[371] A. K. Doolittle, *The Technology of Solvents and Plasticizers*. Wiley, New York, 1954, pp. 658–659.

[372] I. Mellan, *Industrial Solvents*. Reinhold, New York, 1950, pp. 519–521.

[373] G. S. Gardner and J. E. Brewer, *Ind. Eng. Chem.*, **29**, 179 (1937).

3. *Determination in the Atmosphere*

The concentration of cyclohexanol in the air may be determined colorimetrically by measuring the intensity of the straw color produced by the reaction with catechol and concentrated sulfuric acid.[374] The error of analysis of an aqueous solution of cyclohexanol containing no other alcohol is ±0.009 mg. in the range of 0.05 to 0.25 mg.

4. *Physiological Response*

Toxicological Response to Cyclohexanol

Single oral dose

Rabbit   MLD 2.2–2.6 g./kg.[375]

Skin absorption

Rabbit   MLD 12.4–22.7 g./kg.[375]

Animal Symptomatology. Pohl[376] was unable to observe any effects when a dog was exposed to air saturated with cyclohexanol for 10 minutes/day on 7 successive days. However, exposure of animals to sufficiently high concentrations

TABLE 15

Physiological Response of Animals Subjected to Vapors of Cyclohexanol[374]

Number of animals exposed	Concentration		Duration of exposure, hr.[a]	Percentage mortality	Signs of intoxication
	mg./ liter	p.p.m.			
4 rabbits	4.93	1229	150	50	Narcosis, lethargy, incoördination, conjunctival congestion and irritation, and salivation
4 rabbits	4.00	997	330	50	Narcosis, lethargy, conjunctival congestion and irritation, lacrimation, salivation, few convulsive movements
1 monkey	2.78	693	300	0	Lethargy, conjunctival irritation
4 rabbits	1.09	272	300	0	Slight conjunctival irritation
4 rabbits	0.58	145	300	0	None

[a] 6 hrs./day, 5 days/week.

of cyclohexanol in air for 6 hours/day, 5 days/week (see Table 15) induces an intoxication characterized by conjunctival congestion and irritation, lacrimation, salivation, lethargy, incoordination, narcosis, and mild convulsions.[374] Under

[374] J. F. Treon, W. E. Crutchfield, Jr., and K. V. Kitzmiller, *J. Ind. Hyg. Toxicol.*, **25**, 323 (1943).

[375] J. F. Treon, V. E. Crutchfield, Jr., and K. V. Kitzmiller, *J. Ind. Hyg. Toxicol.*, **25**, 199 (1943).

[376] J. Pohl, *Zentr. Gewerbehyg. Unfallverhüt*, **12**, 91 (1925).

suitable conditions this intoxication may terminate in death. Unlike benzene, this compound, when absorbed by experimental animals over considerable periods of time, has shown no tendency to bring about injury to the cellular elements of the peripheral blood.[374] A transient decrease in the blood pressure of rabbits was observed by Sato[377] following the intravenous injection of this compound.

Animal Pathology. The oral administrations of lethal doses of cyclohexanol (2.6 g. or more/kg. of body weight) to rabbits caused severe vascular damage and extreme toxic effects with massive coagulation necrosis of myocardium, lung, liver, kidney, and brain. Animals that survived somewhat smaller doses developed toxic degenerative lesions and vascular damage of much lesser degree. The changes observed following the cutaneous application of cyclohexanol were very similar, in general, to those caused by oral administration, and so afforded corroborative evidence of the absorption of the compound through the skin.[375]

Toxic degenerative changes were found in the brain, heart, liver, and kidneys of rabbits exposed repeatedly to concentrations of cyclohexanol in air ranging from 997 to 1229 p.p.m. (4.00 to 4.93 mg./liter). Similar but less severe changes were seen in the myocardium, liver, and kidneys of rabbits exposed to vapors of this compound at the concentration of 272 p.p.m. in air (1.09 mg./liter). Rabbits exposed to a concentration of 145 p.p.m. (0.58 mg./liter) suffered little or no injury, slight degenerative changes being barely demonstrable in the liver and kidneys.[374]

Absorption and Excretion. When applied upon the intact skin of rabbits in single large doses, cyclohexanol induced tremors, narcosis, hypothermia, and deaths.[375] The application of 10 ml. of cyclohexanol upon the intact skin of a rabbit for 1 hour on each of 10 successive days induced narcosis, tremors, athetoid movements, and hypothermia, together with necrosis, exudative ulceration, and thickening of the skin in the area of contact.[375] A number of animals subjected to these applications were fatally poisoned. Only temporary erythema and superficial sloughing of the skin of rabbits occurred when an ointment consisting of potassium oleate and cyclohexanol, up to 15 per cent by weight, was applied in 5-g. portions for 1 hr./day over a period of 15 days.[375]

Pohl[376] found glucuronic acid in the urine of a dog following the oral administration of cyclohexanol. Following oral administration of cyclohexanol to rabbits, this compound is excreted in the urine in conjugation with sulfuric and glucuronic acids.[375,378] Similar results were obtained when animals were subjected to repeated inhalation of the compound.[374] In general, the increased elimination of conjugated sulfates could be correlated with the increase in the concentration of cyclohexanol in the air.[374] On the other hand, rabbits exposed to the lowest concentration used, 145 p.p.m. cyclohexanol in the air, showed no increase in the conjugation of urinary sulfates but excreted five times the normal amount of glucuronic acid.[374]

---

[377] K. Sato, *Japan. J. Med. Sci. IV Pharmacol. Trans. Abstr.,* **3,** No. 1, (1928).
[378] Y. Sasaki, *Acta Schol. Med. Univ. Imp. Kioto,* **1,** 413 (1917).

Di Prisco[379] found a decrease of 1 to 2 per cent in the reduced glutathione of the blood of rabbits subjected to the inhalation of vapors of cyclohexanol for 10 to 15 minutes upon alternate days over a period of 21 days.

Effects upon Man. Browning[380] has reported 1 case of suspected intoxication, characterized by vomiting, coated tongue, and slight tremors, in a worker engaged in spraying leather with a preparation that contained butyl acetate and cyclohexanol. The evidence in this case was not adequate to convict cyclohexanol as the offending agent. On the other hand, it is apparent that headache and conjunctival irritation have resulted from prolonged exposure to excessive concentrations.

5. Hygienic Standard of Permissible Exposure

The threshold limit of cyclohexanol has been established at 100 p.p.m.[381,382,382a] The A.C.G.I.H. limit for 1961 is 50 p.p.m.

6. Flammability

The ignition temperature of cyclohexanol is 572°F. (see Vol. I, page 526).

7. Odor and Warning Properties

Irritation of the eyes, nose, and throat were induced in human subjects exposed by Nelson *et al.*,[383] for 3 to 5 minutes to 100 p.p.m. of cyclohexanol. The majority of these subjects believed that the highest permissible concentration for an 8-hour exposure, from the standpoint of comfort, should be less than 100 p.p.m. At the concentration of 272 p.p.m., which produced irritation of the eyes and nose in rabbits, the odor of the compound was recognized at once by human subjects.

METHYLCYCLOHEXANOL, $CH_3C_6H_{10}OH$

1. Source, Uses, and Industrial Exposures

Methylcyclohexanol is manufactured by the hydrogenation of *m*- and *p*-cresols.[384] There are three structural isomers of this compound (ortho, meta, and para) each of which may occur as *cis* and *trans* geometric isomers. The commercial product consists essentially of a mixture of two cyclic secondary alcohols (meta and para isomers).[384]

[379] L. DiPrisco, *Minerva med.*, **1932**, 423, (II).

[380] E. Browning, *Toxicity of Industrial Organic Solvents.* H. M. Stationery Office, London, 1953, pp. 236–239.

[381] W. A. Cook, *Ind. Med.*, **14**, 936 (1945).

[382] American Conference of Governmental Industrial Hygienists' Threshold Limit Values for 1959, *A.M.A. Arch. Ind. Health*, **20**, 266 (1959).

[382a] H. F. Smyth, Jr., *Cummings Memorial Lecture.* Am. Ind. Hyg. Assoc., Philadelphia, April 25, 1956.

[383] K. W. Nelson, J. F. Ege, Jr., M. Ross, L. E. Woodman, and L. Silverman, *J. Ind. Hyg. Toxicol.*, **25**, 282 (1943).

[384] I. Mellan, *Industrial Solvents.* Reinhold, New York 1950, pp. 521–522.

Methylcyclohexanol is also known as hexahydrocresol, methylhexalin, hexahydromethylphenol, methyl Adronal, methyl Anol, and Sextol. Methylcyclohexanol dissolves gums, oils, resins, and waxes. This cyclic alcohol serves as a solvent in lacquers, a blending agent in textile soaps, and an antioxidant in lubricants.[384,385]

2. Physical and Chemical Properties

Physical state: colorless liquid
Molecular weight: 114.189[386]
Specific gravity: 0.913 (25/15.5°C.)[387]
Refractive index: 1.461 (20°C.)[385]
Freezing point: −50°C.[384]
Boiling point: 173.0–175.3°C.[384]
Vapor density: 3.94 (air = 1)
Vapor pressure: 1.5 mm. Hg (30°C.)[387]
Per cent in "saturated" air: 0.20 (30°C.)
Density of "saturated" air: 1.01 (air = 1)
Solubility: 3–4 per cent in water (20°C.);[384] miscible with common solvents, plasticizers, and gum solutions
Flash point: 154°F. (67.8°C.) (closed cup)[385]
1 mg./liter ≏ 214.2 p.p.m. and 1 p.p.m. ≏ 4.67 mg./cu. meter at 25°C., 760 mm. Hg

3. Determination in the Atmosphere

The concentration of methylcyclohexanol in air may be determined colorimetrically by measuring the intensity of the straw color produced by the reaction with catechol and sulfuric acid.[388] The error of analysis of an aqueous solution of methylcyclohexanol containing no other alcohol is ± 0.009 mg. in the range of 0.05 to 0.25 mg.

4. Physiological Response

Toxicological Response to Methylcyclohexanol

Single oral dose

Rabbit MLD 1.75–2.0 g./kg.[389]

Skin absorption

Rabbit MLD 6.8–9.4 g./kg.[389]

[385] A. K. Doolittle, *The Technology of Solvents and Plasticizers.* Wiley, New York, 1954, pp. 249, 672.

[386] "Atomic Weights (International), 1957," *J. Am. Chem. Soc.,* **80,** 4122 (1958).

[387] The Barrett Division of Allied Chemical and Dye Corp., personal communication.

[388] J. F. Treon, W. E. Crutchfield, Jr., and K. V. Kitzmiller, *J. Ind. Hyg. Toxicol.,* **25,** 323 (1943).

[389] J. F. Treon, W. E. Crutchfield, Jr., and K. V. Kitzmiller, *J. Ind. Hyg. Toxicol.,* **25,** 199 (1943).

Inhalation

Rabbit	LC₀ 50 × 6 hr., 503 p.p.m.[388]
Rabbit	LC₀ 50 × 6 hr., 232 p.p.m.[388]
Rabbit	LC₀ 50 × 6 hr., 121 p.p.m.[388]

Animal Symptomatology. According to Pohl[390] there were no signs of intoxication in a dog exposed for 10 minutes daily on 7 consecutive days to air saturated with methylcyclohexanol. However, in other investigations of the compound, rabbits, subjected to 503 p.p.m. methylcyclohexanol in air for 6 hours/day 5 days/week over a total of 10 weeks, developed salivation, conjunctival congestion and irritation, and lethargy.[388] Corresponding conditions of exposure to lower concentrations (121 and 232 p.p.m.) resulted in no observed effects. No evidence of specific or general change in the cellular elements of the peripheral blood of animals, exposed repeatedly to any of these concentrations, was found.[388] On the basis of the comparative results of the intraperitoneal injection of the three isomeric forms into mice, Fillipi[391] concluded that *o*-methylcyclohexanol is more toxic than the other two isomers.

Animal Pathology. The oral administration of lethal doses of methylcyclohexanol to rabbits (2.0 g. or more/kg. of body weight) induced severe acute toxic parenchymal and vascular changes in the heart, liver, and kidneys, and toxic vascular damage in the lungs. As a general rule, these lesions were accompanied by cerebral edema and congestion. Diffuse degenerative changes in the liver were the only histopathological evidences of intoxication in the case of animals given sublethal oral doses of the compound.[389]

Comparable toxic lesions were found in the tissues of animals subjected to inhalation of the vapors in air and to percutaneous absorption of methylcyclohexanol. The severity of the toxic changes varied with the severity of the experimental conditions, assuming borderline or questionable significance (when compared with the incidental variations in the histological pattern in the tissues of control animals) in the case of rabbits exposed to the least concentration of methylcyclohexanol in air (121 p.p.m., 0.56 mg./liter).[388]

Absorption and Excretion. The application of large doses of methylcyclohexanol upon the intact skin of rabbits induced fatal poisoning characterized by tremors, narcosis, and hypothermia.[389] The application of 10 ml. of methylcyclohexanol upon the intact skin of a rabbit for 1 hour on each of 6 successive days resulted fatally. At various stages of the experiment, weakness, tremors, deep anesthesia, and local petechiae, gross hemorrhage, and thickening of the skin were seen.[389] The application upon the intact skin of rabbits of 5 g. portions of a mixture of methylcyclohexanol (up to 15 per cent by weight) in potassium oleate, for 1 hour/day over a period of 15 days, produced only temporary erythema and superficial sloughing of the skin.

[390] J. Pohl., *Zentr. Gewerbehyg. Unfallverhütt,* **12,** 91 (1925).

[391] E. Fillipi, *Arch. farmacol. sper.,* **18,** 178 (1914).

Glucuronic acid has been found in the urine of a dog following the oral administration of methylcyclohexanol.[390] In the case of rabbits, under similar conditions of administration, conjugation of the compound or a metabolite with both glucuronic and sulfuric acids has been demonstrated by analysis of the urine.[389] Some conjugation product with sulfuric acid was found also in the urine of animals exposed to 232 and 503 p.p.m. methylcyclohexanol in air.[388] The rate of excretion of glucuronic acid in the urine of rabbits is correlated directly with the concentration of methylcyclohexanol in the air to which they have been subjected.[388] Twice the normal quantity of glucuronic acid was found in the urine of rabbits subjected to 121 p.p.m. methylcyclohexanol in the air.[388]

Effects upon Man. Headache and irritation of the ocular and upper respiratory membranes may result from prolonged exposure to excessive concentrations of the vapor of methylcyclohexanol in air.

After examining several workers who had been exposed to a cellulose solvent containing methylcyclohexanol, Browning[392] concluded that a few of them had slightly but significantly diminished total numbers of leucocytes in the peripheral blood streams, while one had a slight relative lymphocytosis.

5. *Hygienic Standard of Permissible Exposure*

The threshold limit of methylcyclohexanol has been established at 100 p.p.m.[393–395]

6. *Odor and Warning Properties*

Methylcyclohexanol vapor in air can be detected and recognized by its odor when present to the extent of 500 p.p.m., a concentration capable of causing upper respiratory irritation.

ALLYL ALCOHOL, CH$_2$:CHCH$_2$OH

1. *Source, Uses, and Industrial Exposures*

Allyl alcohol is prepared (a) by high temperature chlorination of propylene yielding allyl chloride, which is subsequently hydrolyzed to the alcohol and (b) from glycerol by dehydration and reduction.

Allyl alcohol (2-propen-1-ol, propenol-3, vinyl carbinol) is used in the preparation of allyl resin and plastics, in the preparation of pharmaceuticals, and in the chemical industry.

[392] E. Browning, *Toxicity of Industrial Organic Solvents.* H. M. Stationery Office, London, 1953, pp. 239–240.

[393] W. A. Cook, *Ind. Med.,* **14,** 936 (1945).

[394] American Conference of Governmental Industrial Hygienists' Threshold Limit Values for 1959, *A.M.A. Arch. Ind. Health,* **20,** 266 (1959).

[395] H. F. Smyth, Jr., *Cummings Memorial Lecture.* Am. Ind. Hyg. Assoc., Philadelphia. April 25, 1956.

2. *Physical and Chemical Properties*

Physical state: colorless, pungent liquid
Molecular weight: 58.081[386]
Specific gravity: 0.8476 (25/4°C.)
Melting point: < -129°C.
Boiling point: 96.9°C.[400]
Vapor density: 2.00 (air = 1)
Vapor pressure: 23.8 mm. Hg (25°C.)
Refractive index: 1.4111 (25°C.)
Per cent in "saturated" air: 3.13 (25°C.)
Density of "saturated" air: 1.031 (air = 1) (25°C.)
Solubility: miscible with water, alcohol, and ether
Flash point: 22°C. (closed cup) ; 32°C. (open cup)
1 mg./liter ∞ 422 p.p.m. and 1 p.p.m. ∞ 2.37 mg./cu. meter at 25°C., 760 mm. Hg

3. *Determination in the Atmosphere*

Dunlap *et al.*[396] collected allyl alcohol in air by drawing it through water. To this aqueous sample was added 0.01N bromine in acetic acid, in the presence of a catalyst, mercuric acetate. Excess bromine was reduced by iodide and the iodine titrated with 0.01N thiosulfate according to the method of Reid and Beddard.[397]

4. *Physiological Response*

The toxicological response of animals to allyl alcohol is shown in Table 16.

Several investigators have exposed experimental animals to allyl alcohol in respiratory chambers (see Table 17).

Animal Symptomatology. Allyl alcohol readily penetrated the animal skin, producing systemic intoxication with only slight alteration to the skin. Transitory conjunctival erythema and corneal opacity were observed in rabbit eyes.[396]

Sufficiently high concentration of the vapor of allyl alcohol causes lacrimation, pulmonary irritation, diarrhea, and coma among rats. In the repeated exposure of rats to the vapor, Dunlap *et al.*[396] reported that retardation of growth appeared at 20 p.p.m., an increase in the weight of the lungs at 40 p.p.m., and an increase in the weight of the kidneys at 60 p.p.m. The vapor of allyl alcohol was toxic to rats when exposed on one occasion, but on repeated exposure the effects seemed to be those of repeated insult rather than those of accumulation in the tissue. For example, one exposure of 8 hours to the concentration of 76 p.p.m. resulted in the death of one of the group of rats (LC_{50}) but the repeated exposure (7 hr./day) to the concentration of 60 p.p.m. resulted in only an LC_{10}. A similar

[396] M. K. Dunlap, J. K. Kodama, J. S. Wellington, H. H. Anderson, and C. H. Hine, *A.M.A. Arch. Ind. Health,* **18,** 303 (1958).
[397] V. W. Reid and J. W. Beddard, *Analyst,* **79,** 456 (1954).

TABLE 16

Toxicological Response to Allyl Alcohol

Animal	Response	Reference
	Single dose	
Mouse (orally)	96[a]	396
Mouse (I.P.)	60[a]	396
Rabbit (orally)	71[a]	396
Rat (orally; 111–143 g.)	105[a]	396
Rat (orally; 170–252 g.)	99[a]	396
Rat (orally)	64[a]	398
	Skin absorption	
Rabbit	89[a]	396
Guinea Pig	53[b]	399
	Skin irritation	
Rabbit	Slight erythema in 1 of 12	396
Guinea Pig	Comparable to acetone	399
	Repeated ingestion	
Rat	Daily dosage of 9.7 mg./kg. given daily in the diet over the period of 30 days resulted in reduced appetite, mortality, and microscopic lesions of the tissues	398
Rat	Daily dosage of 4 mg./kg. given similarly had no effect	398
Rat	Daily dosage of 29 mg./kg. in drinking water over period of 90 days produced an increase in weights of liver and kidneys but no other effect	396
Rat	Daily dosage of 12 mg./kg. given similarly had no effect	396

[a] LD_{50} (mg./kg.).
[b] LD_{50} (ml./kg.).

result was observed among the rats that ingested a repeated dose of allyl alcohol. Based upon growth, behavior, mortality, gross appearance, final average body and organ weights, Torkelson et al.[400] found no evidence of ill effect among animals exposed repeatedly to 7 p.p.m. of allyl alcohol in the atmosphere. However, Torkelson et al. reported pathological alterations at 7 p.p.m.; these findings were not in agreement with those of Dunlap et al.,[396] who reported no pathological alterations among animals subjected repeatedly to an atmosphere containing 20 p.p.m. Torkelson et al.[400] found no effect at 2 p.p.m. when based upon those criteria measured at 7 p.p.m. and, in addition, blood nonprotein nitrogen and urea nitrogen.

Animal Pathology. Single exposures by the various routes among the different species of animals resulted in pulmonary and visceral congestion. The livers showed differing degrees of injury from congestion of the periportal sinusoids to

[398] H. F. Smyth, Jr., C. P. Carpenter, and C. S. Weil, *Arch. Ind. Hyg. Occupational Med.*, **4**, 199 (1951).
[399] H. F. Smyth, Jr., and C. P. Carpenter, *J. Ind. Hyg. Toxicol.*, **30**, 63 (1948).

TABLE 17

Response of Animals to Inhalation of Allyl Alcohol.

Concentration (p.p.m.)	Animal	Duration of exposure (hr.)	Outcome	Signs of intoxication	Reference
1000	Monkey	4	Death		401
1000	Rabbit	4	LC_{100}		401
1000	Rat	4	LC_{100}		401
1060	Rat	1	LC_{50}		396
1000	Rat	1	LC_{67}		399
500	Rat	1	Survived		402
250	Rat	4	Some deaths		403
165	Rat	4	LC_{50}		396
76	Rat	8	LC_{50}		396
200	Rat	2 x 7	LC_{100}		401
200	Rabbit	18 x 7	LC_{100}		401
150	Rat	60 x 7	LC_{100}	Gasping, depression, nasal discharge, eye irritation, 1 case of corneal opacity	396
100	Rat	55 x 7	LC_{60}	As above, but less intense	396
60	Rat	60 x 7	LC_{10}	Eye irritation, persistent, but gasping only during first few exposures	396
40	Rat	60 x 7	Survived	Irritation during first few exposures	396
20	Rat	60 x 7	Survived	None	396
7	Rat Guinea pig Rabbit	28 x 7	Survived	None	400
5	Rat	60 x 7	Survived	None	396
2	Rat Guinea pig Rabbit Dog	134 x 7	Survived	None	400
1	Rat	60 x 7	Survived	None	396

periportal necrosis, central pallor to central necrosis. The kidneys were grossly swollen and discolored but microscopically revealed only heme casts and cloudy swelling.[396]

Repeated ingestion of 42 mg./kg./day by rats resulted in a decrease in the peritoneal fat and well-localized areas of necrosis with regeneration in the liver. Nothing remarkable was observed at lower levels.[396]

When rats were exposed to the vapor in the concentration of 100 p.p.m., their

[400] T. R. Torkelson, M. A. Wolf, F. Oyen, and V. K. Rowe, *Am. Ind. Hyg. Assoc. J.*, **20**, 224 (1959).

[401] C. P. McCord, *J. Am. Med, Assoc.*, **98**, 2269 (1932).

[402] H. F. Smyth, Jr., *Am. Ind. Hyg. Assoc. Quart.*, **17**, 129 (1956).

[403] C. P. Carpenter, H. F. Smyth, Jr., and U. C. Pozzani, *J. Ind. Hyg. Toxicol.*, **31**, 343 (1949).

livers were slightly hemorrhagic; the lungs were pale and spotted, but the kidneys were normal. Microscopically, only slight congestion of the lungs and liver was observed. At 20 p.p.m. and below, no gross or microscopic lesions were found.[396]

Torkelson et al.[400] reported mild reversible changes in the liver (dilation of the sinusoids, cloudy swelling, and necrosis) and kidneys (necrosis of the epithelium of the convoluted tubules and proliferation of the interstitial tissue) of experimental animals subjected repeatedly to 7 p.p.m. These investigators found no alterations in the tissues of animals exposed repeatedly to allyl alcohol in the concentration of 2 p.p.m.

Absorption and Excretion by Rats. During the period of 15 to 120 minutes after the administration of a single oral dosage of allyl alcohol (120 mg./kg.) to rats, the mean concentrations of this alcohol in the portal vein were between 9 and 15 μg./ml.[404]

Allyl alcohol is apparently oxidized readily since within a few minutes after the intravenous injection of rats with the dosage of 30 mg./kg. the vena caval blood contained an average concentration of about 24 μg./ml.; within 15 minutes the concentration was about 4 μg./ml. and within an hour the alcohol had almost disappeared from the blood. During constant intravenous infusion the allyl alcohol disappeared at the rate of about 23 mg./hr.[404]

Pharmacodynamics and Mode of Action. A rapid drop in the carotid blood pressure, a definite reduction in respiratory rate and amplitude, and an increase in hemoconcentration (hematocrit), but no increase in the concentration of histamine in the plasma, followed the intravenous injection of 40 mg./kg. into a dog.[404]

Severe inhibition of oxygen uptake was demonstrated in liver slices of rats given an oral administration of 120 mg. of allyl alcohol/kg., but only slight inhibition occurred at the dosage of 60 mg./kg. The kidney slices from the above rats manifested an increase in oxygen uptake but the oxygen uptake of the cerebral cortex was not altered. Both phenoxybenzamine and hexamethonium, when given to rats prior to the oral administration of allyl alcohol in the dosage of 120 mg./kg., afforded partial protection to the liver against (a) inhibition of oxygen uptake by the liver and (b) increase in water content. Phenoxybenzamine more than doubled the mean time of death for rats given allyl alcohol orally but it did not alter mortality. These results indicate that the hepatic injury from alcohol is not the primary cause of death.[404]

Liver dehydrogenases (in vitro) were inhibited only by excessively high concentrations ($2 \times 10^{-1} M$ to $1M$), which are unlikely to be obtained in vivo.[404]

Cause of Death. Kodama and Hine[404] state that the primary cause of death from allyl alcohol is probably cardiovascular failure, which probably follows depression of the respiratory and vasomotor centers and a decrease in effective blood pressure.

Effects on Man. In contrast to the slight irritation produced on the skin of animals, Dunlap et al.[396] reported that skin irritation is quite common with this

[404] J. K. Kodama and C. H. Hine, *J. Pharmacol. Exptl. Therap.,* **124,** 97 (1958).

compound and that absorption through the skin leads to deep pain, probably due to muscle spasm. Intravenous injection of or infiltrating the site with calcium gluconate has been found to relieve the pain. Eye contamination from the liquid may cause a severe chemical burn.[396] Torkelson *et al.*[400] corroborate delayed eye irritation and muscular spasm resulting from skin contact with allyl alcohol, in industry.

Unlike the vapors of the saturated lower aliphatic alcohols, the vapor of allyl alcohol does not possess particular narcotic properties but rather is an irritant to the mucous membranes and to the lungs.

Dunlap *et al.*[396] reported that when the air was moderately contaminated with allyl alcohol, men frequently manifested lacrimation, retrobulbar pain, photophobia, and some blurring of vision. They reported that no permanent or irreversible severe scarring of the cornea or loss of corneal substances had been observed following exposure to these vapors. Smyth[402] reported that the vapor of allyl alcohol temporarily blinded one man through delayed corneal necrosis.

No evidence of liver damage or disturbed kidney function was noted among a group of employees working for 10 years with allyl alcohol.[396]

5. Hygienic Standard of Permissible Exposure

The threshold limit of allyl alcohol has been established at 5 p.p.m.[402,405] Torkelson *et al.*[400] suggest that the concentration should not exceed 5 p.p.m. at any time and that the over-all, time-weighted, daily average of 7 to 8 hours should not exceed 2 p.p.m. The A.C.G.I.H. limit for 1961 is 2 p.p.m.

6. Flammability

The lower and upper explosive limits of allyl alcohol have been established at 2.5 and 18.0 per cent, respectively[400] (see also Vol. I, pages 516, 525, 533, and 538).

7. Odor and Warning Properties

Allyl alcohol has a pungent odor and is a potent lacrimator.
Threshold, odor: <0.78 p.p.m.[396]
Eye irritation, slight: 6.25 p.p.m.[396]
Eye irritation, moderate: 25.0 p.p.m.[396]
Threshold, nasal irritation: <0.78 p.p.m.[396]
Nasal irritation, moderate: 12.5 p.p.m.[396]
Threshold, pulmonary discomfort: >25.0 p.p.m.[396]
Threshold, CNS effects: >25.0 p.p.m.[396]
Five out of 10 human volunteers reported a definite odor but no irritation when exposed to 2 p.p.m. of allyl alcohol for 1 to 3 minutes.[400]

[405] American Conference of Governmental Industrial Hygienists' Threshold Limit Values for 1959, *A.M A. Arch. Ind. Health*, **20**, 266 (1959).

FURFURYL ALCOHOL, $C_4H_3O \cdot CH_2OH$

1. Source, Uses, and Industrial Exposures

Furfuryl alcohol is prepared by high pressure catalytic hydrogenation of furfural.[406]

Furfuryl alcohol is also called 2-furyl carbinol and 2-furanmethanol. It is a solvent for cellulose ethers and esters, ester gum, coumarone resins, and natural resins. It is used in the manufacture of dark-colored thermosetting resins and phenolic resins and as a solvent in the manufacture of abrasive wheels. It is employed in the textile industry as a solvent for dyes and a dispersant for difficultly soluble dyes.[406–408] Under ordinary usage in the industrial plant the use of furfuryl alcohol for about 20 years has resulted in no impairment of health. More recently furfuryl alcohol has been employed as a liquid propellant.[409]

2. Physical and Chemical Properties

Physical state: colorless liquid; turns amber due to autoxidation and intermolecular dehydration during storage;[408] turns black in presence of air[406]

Molecular weight: 98.103[410]

Specific gravity: 1.1287 (20/4°C.)[408]

Refractive index: 1.484 (20°C.)[408]

Freezing point: −14.6°C.[406,408]

Boiling point: 171°C., 750 mm. Hg[406,408]

Vapor density: 3.38 (air = 1)

Vapor pressure: 1 mm. Hg (31.8°C.)[411]

Per cent in "saturated" air: 0.13 (31.8°C.)

Density of "saturated" air: 1.003 (air = 1)

Solubility: miscible in water in all proportions above 21°C.;[408] miscible with most organic solvents but immiscible with petroleum hydrocarbons and most oils[407]

Flash point: 167°F.

1 mg./liter ⇌ 249.4 p.p.m. and 1 p.p.m. ⇌ 4.01 mg./cu. meter at 25°C., 760 mm. Hg

3. Determination in the Atmosphere

Furfuryl alcohol has been collected from air by passing it through glacial

---

[406] R. E. Kirk and D. F. Othmer, eds., *Encyclopedia of Chemical Technology*, Vol. 6. Interscience Encyclopedia, New York, 1951, pp. 997, 1002–1003.

[407] I. Mellan, *Industrial Solvents*. Reinhold, New York, 1950, p. 523.

[408] A. K. Doolittle, *The Technology of Solvents and Plasticizers*. Wiley, New York, 1954, p. 658.

[409] K. H. Jacobson, W. E. Rinehart, H. J. Wheelwright, Jr., M. A. Ross, J. L. Papin, R. C. Daly, E. A. Greene, and W. A. Groff, *Am. Ind. Hyg. Assoc. J.*, **19**, 91 (1958).

[410] "Atomic Weights (International), 1957," *J. Am. Chem. Soc.*, **80**, 4122 (1958).

[411] D. R. Stull, *Ind. Eng. Chem.*, **36**, 517 (1947).

acetic acid. Comstock and Oberst[412] then added 0.05N pyridine bromine sulfate solution to the acetic acid. After reaction of the bromine with the furane moiety (1 hour in the dark) 5 per cent KI was added and the solution was titrated with 0.1N $Na_2S_2O_3$.

4. *Physiological Response*

Toxicological Response to Furfuryl Alcohol

Single dose

Rat (orally)	LD_{50} 0.275 g./kg.[413]
Rabbit (I.V.)	LD_{50} 0.650 g./kg.[413]
Rabbit (S.C.)	LD 0.600 g./kg.[414]

Inhalation

Mouse	LC_0, 10 min., 700 p.p.m.[415]
Rat	LC_{17}, 4 hr., 700 p.p.m.[412]
Rat	LC_{25}, 8 hr., 700 p.p.m.[412]
Rat	LC_{50}, 4 hr., 233 p.p.m.[409]
Rat	LC_0, 30 × 6 hr., 19 p.p.m.[412]
Mouse	LC_0, 15 × 6 hr., 19 p.p.m.[412]

Animal Symptomatology. When rats were exposed to furfuryl alcohol in the atmosphere in the concentration of 700 p.p.m. the symptoms noted were initial excitement, followed by eye irritation and drowsiness. The eyes became red within 8 minutes. Rats and mice exposed to the average concentration of 19 p.p.m. (12–29 p.p.m.) in the atmosphere exhibited restlessness for the first 5 or 10 minutes; the animals were drowsy during the remaining portion of the 6-hour periods of exposure.[412]

Erdman[414] reported that small doses of furfuryl alcohol stimulated respiration in both man and rabbit. Larger doses were reported to depress respiration, to lower the body temperature and to produce nausea, salivation, diarrhea, dizziness, and diuresis. Sensory nerves were paralyzed by dilute solutions of furfuryl alcohol.[416]

Fine and Wills[417] observed that furfuryl alcohol (*a*) had a negative inotropic effect on the heart without a significant chronotropic one, (*b*) decreased the tonus and contractility of intestinal smooth muscle, and (*c*) depressed the central nervous system, producing changes in the electroencephalographic record similar to those produced by certain anesthetic agents. Fine and Wills[417] produced severe falls in blood pressure and temporary apnea when the intravenous dosage ex-

[412] C. C. Comstock and F. W. Oberst, *Chem. Corps Med. Labs. Research Rept.*, No. **139**, Oct., 1952.

[413] J. Gajewski and W. Alsdorf, *Federation Proc.*, **8**, 294 (1949).

[414] E. Erdmann, *Arch. exptl. Pathol. Pharmacol.*, **48**, 233 (1902) (quoted from reference 417).

[415] NDRC, Univ. of Chicago, unpublished report, Feb. 21, 1942 (quoted from reference 412).

[416] M. Okubo, *J. Pharm. Soc. Japan*, **539**, 39 (1937) (quoted from reference 417).

[417] E. H. Fine and J. H. Wills, *Arch. Ind. Hyg. Occupational Med.*, **1**, 625 (1950).

ceeded 0.500 to 0.600 g./kg. of body weight. Death occurred from respiratory paralysis at dosages from 0.800 to 1.400 g./kg. The central depressant action of lethal dosages of furfuryl alcohol was combated effectively with pentamethylenetetrazol, amphetamine, or ephedrine. Fine and Wills[417] suggested that furfuryl alcohol probably is distributed equally throughout most of the body and has little specific action on enzyme systems localized in special structures in the body.

Animal Pathology. When rats were exposed repetitively to furfuryl alcohol in the atmosphere in the concentration of 19 p.p.m., changes were observed only in the respiratory tract; they consisted of moderate diffuse congestion without any significant cellular changes.

5. Hygienic Standard of Permissible Exposure

The threshold limit of furfuryl alcohol has been established at 5 p.p.m.[409] The A.C.G.I.H. in 1961 set a limit of 50 p.p.m.

6. Flammability

The ignition temperature of furfuryl alcohol is 736°F. (see also Vol. I, pages 516, 528, and 534).

7. Odor and Warning Properties

The median detectable concentration of furfuryl alcohol in the atmosphere is reported by Jacobson *et al.*[409] to be 8 p.p.m.

TETRAHYDROFURFURYL ALCOHOL, $C_4H_7OCH_2OH$

1. Source, Uses, and Industrial Exposures

Tetrahydrofurfuryl alcohol is prepared by liquid-phase hydrogenation of furfuryl alcohol over a nickel catalyst.[418]

Tetrahydrofurfuryl alcohol or tetrahydro-2-furancarbinol is used in the preparation of esters, plasticizers, and as a chemical intermediate. It is a solvent for cellulose acetate, cellulose nitrate, ethyl cellulose, furfuryl alcohol, polymers, styrene, phenol-aldehyde resins, and vinyl acetate.[418]

2. Physical and Chemical Properties

Physical state: colorless liquid; slowly discolors on exposure to air in the presence of iron or copper[418]

Molecular weight: 102.135[410]
Specific gravity: 1.0495 (20/4°C.)[419]
Boiling point: 177.5°C., 743 mm. Hg[418]
Vapor density: 3.522 (air = 1)

[418] R. E. Kirk and D. F. Othmer, eds., *Encyclopedia of Chemical Technology*, Vol. 6. Interscience Encyclopedia, New York, 1951, pp. 997, 1003–1004.

[419] D. W. Fassett, personal communication.

Refractive index: 1.4505 (20°C.)[418]

Solubility: completely miscible in water;[418] soluble in acetone, ethyl ether, and ethanol

Flash point: 167°F.[419]

1 mg./liter ≏ 239.5 p.p.m. and 1 p.p.m. ≏ 4.18 mg./cu. meter at 25°C., 760 mm. Hg

3. Physiological Response

The toxicological response of animals to tetrahydrofurfuryl alcohol is given in Table 18.

TABLE 18

Toxicological Response to Tetrahydrofurfuryl Alcohol

Animal	Response
	Single dose
Rat (orally)	LD$_{50}$ 1.6–3.2 g./kg.[419]
Guinea pig (orally)	LD$_{50}$ 0.8–1.6 g./kg.[419]
Rat (I.P.)	LD$_{50}$ 0.4–0.8 g./kg.[419]
Guinea pig (I.P.)	LD$_{50}$ 0.4–0.8 g./kg.[419]
	Skin absorption
Guinea pig	LD$_{50}$ <5 ml./kg.[419]
	Skin irritation
Guinea pig	Moderate[419]
	Skin sensitization
Guinea pig	None of 5 guinea pigs sensitized
	Inhalation
Rat	2 out of 3 deaths following 6-hr. exposure at 12,650 p.p.m. calculated
Rat	0 out of 3 deaths in 6 hr. at 655 p.p.m.; loss of co-ordination, prostration, and vasodilation in ears and feet noted at high levels

4. Flammable Limits

The limits of flammability of tetrahydrofurfuryl alcohol are 1.5–9.7 per cent (72.5–122°C.).[418] The ignition temperature is 540°F. (see Vol. I, page 531).

β-CHLOROETHYL ALCOHOL, CH$_2$Cl·CH$_2$OH

1. Source, Uses, and Industrial Exposures

β-Chloroethyl alcohol may be produced by passing chlorine and ethylene simultaneously into water.[420]

[420] R. E. Kirk and D. F. Othmer, eds., *Encyclopedia of Chemical Technology*, Vol. 3. Interscience Encyclopedia, New York, 1949, p. 852.

β-Chloroethyl alcohol, also known as 2-chloroethanol-1, ethylene chlorohydrin, and glycol chlorohydrin, is used to produce ethylene glycol and ethylene oxide.[420] Huntress[421] states that β-chloroethyl alcohol is employed for the separation of butadiene from hydrocarbon mixtures, in dewaxing and removing naphthenes from mineral oil, in the refining of rosin, in the extraction of pine lignin, and as a solvent for cellulose acetate, cellulose ethers, and various resins. Ambrose[422] reports that it is an effective agent in hastening the early sprouting of potatoes and has been proposed for treating seeds for the inhibition of biological activity.

Several deaths have been reported by Koelsch,[423] Middleton,[424] Cavallazzi,[425] Dierker and Brown,[426] and Goldblatt and Chiesman,[427] from the manufacture or other industrial exposure to β-chloroethyl alcohol by inhalation and/or percutaneous contact. In the fatal case reported by Dierker and Brown,[426] the deceased had been using β-chloroethyl alcohol for 2 hours to clean trays upon which rubber strips were stored. Analysis of the air revealed a concentration of about 1 mg./liter (304 p.p.m.). Percutaneous absorption was a significant factor in the death reported by Middleton.[424] Nine cases of nonfatal intoxication during the manufacture of β-chloroethyl alcohol were reported by Goldblatt and Chiesman.[427] The average concentration of β-chloroethyl alcohol in the plant at the time of these nonfatal cases was 18 p.p.m. Smyth and Carpenter[428] warn that rubber gloves offer little protection, since dangerous amounts of β-chloroethyl alcohol or its aqueous solution rapidly penetrate through rubber.

2. Physical and Chemical Properties

Physical state: colorless liquid
Molecular weight: 80.519[429]
Specific gravity: 1.2045 (20/20°C.)[420]
Refractive index: 1.4417 (20°C.)[420]
Melting point: −62.6°C.[420]
Boiling point: 128.7°C.[420]
Vapor density: 2.78 (air = 1)

[421] E. T. Huntress, *The Preparation, Properties, Chemical Behavior and Identification of Organic Chlorine Compounds*. Wiley, New York, 1948, p. 705.

[422] A. M. Ambrose, *Arch. Ind. Hyg. Occupational Med.*, **2**, 591 (1950).

[423] F. Koelsch, *Zbl. Gewerbehyg.*, **14**, 312 (1927), cited from E. Browning, *Toxicity of Industrial Organic Solvents*. Chemical Publishing Co., New York, 1953, p. 249.

[424] E. L. Middleton, *J. Ind. Hyg. Toxicol.*, **12**, 265 (1930).

[425] D. Cavallozzi, *Samml. Vergiftungsfällen*, **12**, 79 A-910 (1942), cited from E. Browning, *Toxicity of Industrial Organic Solvents*. Chemical Publishing Co., New York, 1953, p. 249.

[426] H. Dierker and P. Brown, *J. Ind. Hyg. Toxicol.*, **26**, 277 (1944).

[427] M. W. Goldblatt and W. E. Chiesman, *Brit. J. Ind. Med.*, **1**, 207 (1944).

[428] H. F. Smyth, Jr., and C. P. Carpenter, *J. Ind. Hyg. Toxicol.*, **27**, 93 (1945).

[429] "Atomic Weights (International), 1957," *J. Am. Chem. Soc.*, **80**, 4122 (1958).

[430] T. E. Jordan, *Vapor Pressure of Organic Compounds*. Interscience Publishers, New York–London, 1954, p. 65.

Vapor pressure: 4.9 mm. Hg (20°C.) ;[420] 10 mm. Hg (30.3°C.)[430]
Per cent in "saturated" air: 0.644 (20°C.) ; 1.32 (30.3°C.)
Density of "saturated" air: 1.011 (20°C.) ; 1.022 (30.3°C.)
Solubility: soluble in water in all proportions[420]
Flash point: 135°F. (closed cup)[420]
1 mg./liter $\backsimeq$ 303.8 p.p.m. and 1 p.p.m. $\backsimeq$ 3.29 mg./cu. meter at 25°C., 760 mm. Hg

When β-chloroethyl alcohol is heated to 184°C., it decomposes into ethylene chloride and acetaldehyde; when heated with water to 100°C., β-chloroethyl alcohol decomposes into glycol and aldehyde.[423]

3. Determination in the Atmosphere

Uhrig[431] describes a method for the determination of large quantities of this compound. An aqueous solution of this chlorinated alcohol is refluxed with potassium hydroxide. In the presence of nitric acid, excess standardized silver nitrate is titrated with ammonium thiocyanate, using ferric ammonium sulfate as indicator.

4. Physiological Response

Toxicological Response to β-Chloroethyl Alcohol

(single dose)

	LD_{50}(mg./kg.)
Rat (orally)	95[432]
Guinea pig (orally)	110[432]
Rat (orally)	72[433]
Rat (I.P.)	56[433]

Animal Symptomatology. Rats maintained on diets containing β-chloroethyl alcohol in concentrations ranging from 0.01 to 0.08 per cent for at least 220 days grew normally but those fed on diets from 0.12 to 0.24 per cent had retarded growth. The tissues of the rats at all levels were without histopathological alteration.[430]

β-Chloroethyl alcohol readily penetrates the skin. Smyth and Carpenter[428] found the toxicity of this compound to be greater by percutaneous absorption than by ingestion. These investigators applied β-chloroethyl alcohol to a poultice that was maintained in contact with the skin of guinea pigs. They reported an LD_{50} of 0.070 ml./kg. when the undiluted chemical was applied, with the majority of the deaths among the guinea pigs occurring within 24 hours. When contact was limited to 2 hours, the LD_{50} was about 0.3 ml./kg. When β-chloroethyl alcohol was applied as a 10 per cent solution in water, the LD_{50}, in terms of the chemical was about 1.14 ml./kg.[428] Ambrose[422] also reported that undiluted and aqueous solutions of β-chloroethyl alcohol were lethal when applied to the skin of rats. Rabbits died following a few repeated applications of the undiluted compound.[422] Gold-

[431] K. Uhrig, *Ind. Eng. Chem. Anal. Ed.,* **18**, 469 (1946).
[432] H. F. Smyth, Jr., J. Seaton, and L. Fischer, *J. Ind. Hyg. Toxicol.,* **23**, 259 (1941).
[433] M. W. Goldblatt, *Brit. J. Ind. Med.,* **1**, 213 (1944).

blatt[433] reported that the application of 0.03 to 0.09 ml. of this material to the skin of mice resulted in fatalities.

Table 19 lists the experimental data available in relation to exposure of animals to air containing β-chloroethyl alcohol.

TABLE 19

Physiological Response of Animals to Inhalation of β-Chloroethyl Alcohol

Dose (mg./liter)	Animal	Duration of exposure (hr.)	Outcome	Reference
18.0	Guinea pig	0.25	Death	434
7.0	Mouse	2.0	Death	433
5.0	Guinea pig	0.9	Survived	433
4.5	Mouse	0.5	Death	433
4.0	Rat	0.5	Death	433
4.0	Mouse	0.25	LC_{67}	433
3.6	Guinea pig	1.0	Death	434
3.0	Guinea pig	1.8	Death	433
3.0	Mouse	1.0	Death	433
3.0	Guinea pig	0.5	Survived	433
3.0	Rat	0.25	Survived	433
2.5	Cat	4×3	Death	434
1.2	Mouse	2	LC_{17}	426
1.0	Mouse	2	Survived	433
0.11	Rat	4	$\cong LC_{50}$	435

Three rats, which were exposed to β-chloroethyl alcohol in the average concentration of 3.4 mg./liter for 15 minutes on each of 11 days, died after 3, 6, and 11 exposures, respectively.[433]

Ambrose[422] exposed rats to air that was bubbled through aqueous solutions of β-chloroethyl alcohol maintained at 40°C. Rats exposed for 1 hour to air passed through 12.5, 25, or 50 per cent aqueous β-chloroethyl alcohol died 1 or 2 hours after the exposure. Rats that were exposed, during three periods each of 1-hour duration over a total period of less than 2 days, to air bubbled through a 6.25 per cent aqueous solution of this alcohol died following the third exposure. A total of four exposure periods, each of 2 hours' duration, on 2 consecutive days to air bubbled through a 3.13 per cent solution resulted in morbidity, depression, paralysis, and mortality among rats.

Inhalation of β-chloroethyl alcohol by animals resulted in nasal irritation, incoordination, convulsions, prostration, and respiratory failure. True narcosis was absent.[433]

[434] F. Koelsch, *Zbl. Gewerbehyg. N. F.,* **4,** 312 (1927), cited from K. B. Lehmann and F. Flury, *Toxicology and Hygiene of Industrial Solvents.* Transl. by E. King and H. F. Smyth, Jr., Williams & Wilkins, Baltimore, 1948, p. 215.

[435] C. P. Carpenter, *J. Ind. Hyg. Toxicol.,* **31,** 343 (1949).

Goldblatt[433] reported that intravenous injection (but not inhalation) of β-chloroethyl alcohol induced a decrease in blood pressure and inhibition of respiration in cats, but that vagal action and cardiovascular reflexes were not altered. This compound is an inhibitor of the perfused frog's heart. β-Chloroethyl alcohol inhibited both the tone and rhythm of the smooth muscle. In contact with the nerve it induced complete nerve block, which is reversible.[433]

Pathology in Animals. Upon microscopic examination of the liver, kidneys, and lungs of a mouse that died following the inhalation of β-chloroethyl alcohol, Dierker and Brown[426] reported edema, capillary engorgement, and interstitial hemorrhages of all these organs. Goldblatt[433] believed that the kidneys were the earliest focus of stress, since following repeated exposure to the vapor there were large numbers of hemorrhages, mainly at junctional areas between the cortex and medulla, and there was a complete disintegration of the cells of the convoluted tubules. Congestion, formation of pigment, and fatty degeneration were observed in the liver. Pulmonary hemorrhages and collapse were noted.

Effects on Man. In the case of fatalities of humans exposed to β-chloroethyl alcohol the following signs of intoxication were usually reported: nausea, vomiting, incoordination of the legs, vertigo, weakness, weak, irregular pulse, and respiratory failure.[424,426,427] Vomiting of bile, profuse perspiration, headache, visual disturbance, decreased blood pressure, hematuria, and spastic contracture of the hands were reported by Goldblatt and Chiesman.[427]

Microscopic examination of the tissues by Dierker and Brown[426] revealed edema, swelling, and vacuolation of the hepatic cells. Necrosis and engorgement of the hepatic blood vessels with interstitial hemorrhages were also observed. The tubules of kidneys showed cloudy swelling; there was intense engorgement of the tubules and some parenchymal cells in the tubules. Goldblatt and Chiesman[427] stressed cerebral edema. These investigators stated that fatty degeneration of the liver, and edema, collapse, and extravasation of the lungs were insufficient to account for death. Unfortunately, kidney sections were unavailable.

In nonfatal cases, Goldblatt and Chiesman[427] reported nausea, epigastric pain, repeated vomiting, occasionally with bile, signs of circulatory shock, headaches, confusion, vertigo, incoordination, slight albuminuria, polyuria, cough, and erythema of the skin.

5. Hygienic Standards of Permissible Exposure

The threshold limit of β-chloroethyl alcohol has been established at 5 p.p.m.[436]

6. Flammability

The lower and upper explosive limits of β-chloroethyl alcohol are 4.9 and 15.9 per cent, respectively.[437]

[436] American Conference of Governmental Industrial Hygienists' Threshold Limit Values for 1959, *A.M.A. Arch. Ind. Health*, **20**, 266 (1959).

[437] A. K. Doolittle, *The Technology of Solvents and Plasticizers*. Wiley, New York, 1954, p. 248.

CHAPTER XXXV

Glycols

V. K. ROWE

ETHYLENE GLYCOL, $C_2H_6O_2$ or $HOCH_2CH_2OH$ (1,2-Ethanediol)

1. *Source, Uses, and Industrial Exposure*

Ethylene glycol is synthesized commercially by first preparing ethylene oxide from ethylene by direct oxidation with oxygen or air or by the chlorohydrin process and then hydrolyzing the ethylene oxide to the glycol. It can also be prepared by the hydrolysis of ethylene dichloride or ethylene dibromide, and by other methods that can be employed in the laboratory but are of no commercial significance.[1]

Large amounts of ethylene glycol are used in antifreeze mixtures for motor vehicles. It is used in hydraulic fluids, heat exchangers, and as a solvent. It is also used in large amounts as a chemical intermediate in the production of ethylene glycol dinitrate, glycol esters, and certain resinous products.

Contact with the skin and eyes is most likely to occur in industrial handling. Inhalation may be a problem if the material is handled hot or if a mist is generated by heat or by violent agitation. Swallowing is not expected unless the material is stored in unmarked containers or is put in mislabeled containers.

2. *Physical and Chemical Properties*

Ethylene glycol is a colorless, odorless, viscous, hygroscopic liquid with a bittersweet taste. It has the following physical properties:[1]

Molecular weight: 62.07
Specific gravity: 1.1133 (20/4°C.) ; 1.1155 (20/20°C.)
Freezing point: —13°C.
Boiling point: 197.6°C. (760 mm. Hg)
Vapor pressure: 0.06 mm. Hg (20°C.)
Refractive index: 1.4316 (20°C.)
Per cent in "saturated" air: 0.0131 (25°C.) ; 0.0079 (20°C.)

[1] G. O. Curme, Jr. and F. Johnston, eds., *Glycols, American Chemical Society Monograph, Series 114.* Reinhold, New York, 1952.

Solubility: miscible with water, lower aliphatic alcohols, aldehydes, and ketones; practically insoluble in hydrocarbons and similar compounds

Flash point: 240°F. (TOC)

1 p.p.m. ≎ 2.74 mg./cu. meter at 25°C., 760 mm. Hg

3. Determination in the Atmosphere

Although numerous methods are available for the determination of ethylene glycol[1] when it is present in substantial amounts, few methods are applicable to the determination of small amounts such as may be of industrial hygiene significance. The physical methods employing such instruments as the interferometer or combustible gas analyzer are not sufficiently sensitive to be useful. Spectrographic procedures employing infrared absorption undoubtedly can be used to advantage where appropriate instruments are available. Ultraviolet absorption would seem to offer little promise but mass spectrophotometry and vapor phase chromatography should be applicable. According to Bergner and Sperlich,[2] detection of ethylene glycol and other glycols in mixtures can be accomplished by chromatographic procedures.

The chemical method most likely to be useful in most laboratories depends upon trapping the material in water, oxidation to formaldehyde with periodate and the colorimetric determination of the formaldehyde formed by the use of sodium chromotropate in concentrated sulfuric acid. Other methods for the determination of formaldehyde formed by periodate oxidation are also applicable.[3,4] It should be remembered that the oxidation of ethylene glycol with periodate yields formaldehyde, an end product common to other glycols that have adjacent hydroxyl groups, such as 1,2-propylene glycol and glycerol. Another chemical method that can be used is described by Duke and Smith.[5] It is based upon the reaction of alcoholic hydroxyl groups with ammonium hexanitrato cerate to form a red product. The optical density of the color can be quantitated. It should be noted, however, that this method is not specific for ethylene glycol.

4. Physiological Response

Summary. Ethylene glycol presents negligible hazards to health in industrial handling, except possibly where it is being used at elevated temperatures. It is low in acute oral toxicity, is not significantly irritating to the eyes or skin, is not readily absorbed through the skin, and its vapor pressure is sufficiently low so that toxic concentrations cannot occur in the air at room temperatures. The principal hazard to health of ethylene glycol is associated with the ingestion of large quanti-

[2] K. G. Bergner and H. Sperlich, *Z. Lebensm. Untersuch. u. Forsch.,* **97,** 253 (1953), *Chem. Abstr.,* **48,** 2273 (1954).

[3] B. Worshowsky and P. J. Elving, *Ind. Eng. Chem.,* **18,** 253 (1946).

[4] R. C. Reinke and E. N. Luce, *Ind. Eng. Chem.,* **18,** 244 (1946).

[5] F. R. Duke and G. F. Smith, *J. Ind. Eng. Chem. Anal. Ed.,* **12,** 201 (1940).

ties in single doses or of small quantities repeatedly over a prolonged period of time.

Most of the older published articles dealing with the toxicity of ethylene glycol and the other common glycols have been ably reviewed.[6-10]

Single-Dose Oral. The toxicity of ethylene glycol for animals has been determined by numerous investigators. The findings of Laug *et al.*[7] and Smyth *et al.*[11] on small animals are representative. Laug reports the LD_{50} for rats, guinea pigs, and mice to be 5.50, 7.35, and 13.1 ml./kg., respectively. Smyth found the LD_{50} for rats and guinea pigs to be 8.54 and 6.61 g./kg., respectively. Bornmann[12] found the LD_{50} for mice to be 13.70 mg./kg. Earlier, Page[13] reported that dogs fed 9 ml./kg. were unaffected. The single oral dose lethal for humans has been estimated at 1.4 ml./kg. or about 100 ml. per person.[7] It is apparent that ethylene glycol is much more acutely toxic for humans than for the animals studied.

Repeated-Dose Oral. The most comprehensive study of the repeated dose oral toxicity of ethylene glycol is that reported by Morris, Nelson, and Calvery.[9] They maintained rats for 2 years on diets containing 1 and 2 per cent of ethylene glycol and observed a shortening of the life span, calcium oxalate bladder stones, severe renal injury particularly of the tubules, and centrolobular degeneration of the liver. More recently, Bornmann[12] has reported administering ethylene and other glycols to mice in their drinking water in concentrations of 1, 2, 5, 10, and 20 per cent but details are not available to the writer, other than that the material had a strong diuretic effect, and caused narcosis and central depression of the heart and respiration.

There seems to be little doubt that the primary effect of repeated small oral doses of ethylene glycol is severe kidney injury. Unfortunately, studies have not been made in which dosage has been decreased progressively to the level that caused no effect.

Repeated Injections. Recently, Paterni, Dotta, and Sappa[14] have reported on studies in which they administered, subcutaneously, to rats daily doses of 4 ml. of ethylene glycol diluted with water. Progressive hemolytic anemia and a

[6] W. F. von Oettingen, *U. S. Public Health Bull.* No. **281**, 1943.

[7] E. P. Laug, H. O. Calvery, H. J. Morris, and G. Woodard, *J. Ind. Hyg. Toxicol.,* **21**, 173 (1939).

[8] E. M. K. Geiling and P. R. Cannon, *J. Am. Med. Assoc.,* **111**, 919 (1938).

[9] H. J. Morris, A. A. Nelson, and H. O. Calvery, *J. Pharmacol. Exptl. Therap.,* **74**, 266 (1942).

[10] E. Browning, *Toxicity of Industrial Organic Solvents.* Chemical Publishing, New York, 1953.

[11] H. F. Smyth, Jr., J. Seaton, and L. F. Fischer, *J. Ind. Hyg. Toxicol.,* **23**, 259 (1941).

[12] G. Bornmann, *Arzneimittel Forsch.,* **4**, 643, 710 (1954), **5**, 38 (1955), *Chem. Abstr.,* **49**, 7131 (1955).

[13] I. H. Page, *J. Pharmacol.,* **30**, 313 (1927).

[14] L. Paterni, F. Dotta, and M. Sappa, *Folia Med. Naples,* **39**, 242 (1956), *Chem. Abstr,* **50**, 11524 (1956).

variety of changes in the leucocytes were seen. Other findings included renal lesions, changes in the liver and spleen, and iron deposits in all organs. Bornmann[12] also gave the material parenterally to mice and observed a hemolytic effect but concluded that death was due to the narcotic effect and to renal insufficiency and not to the hemolytic effect.

Skin Irritation. Ethylene glycol produces no significant irritant action upon the skin. A slight macerating action on the skin may result from very severe, prolonged exposures, which is comparable to that caused by glycerine under similar conditions.

Skin Absorption. Hanzlik *et al.*[15] have shown in animal studies that toxic amounts of ethylene glycol can be absorbed through the skin. The data, however, are erratic and difficult to quantitate. Volkmann[16] reports a case of what was believed to be ethylene glycol poisoning resulting from the massive application of an eczema remedy containing ethylene glycol. A comatous condition accompanied by miosis and slowed pulse occurred 4 hours after application but no oxalate was found in the urine.

For industrial hygiene purposes, it would seem prudent to avoid prolonged and repeated contact with skin, particularly contacts that involve extensive areas of skin.

Eye Irritation. Carpenter and Smyth[17] report that ethylene glycol failed to cause appreciable irritation when introduced into the eyes of rabbits. No cases of injury to human eyes have been reported nor would any be expected.

Inhalation. According to Browning,[10] Flury and Wirth exposed rats for 28 hours during 5 days to an atmosphere essentially saturated (0.5 mg./liter) with ethylene glycol. No deaths occurred but the animals reportedly exhibited slight narcosis. The most extensive inhalation studies reported are those of Wiley, Hueper, and von Oettingen,[18] who exposed rats and mice to concentrations of 0.35 to 0.40 mg./liter (140 to 160 p.p.m.) 8 hours a day during 16 weeks without producing injury. Since these concentrations represent essentially saturated atmospheres at warm room conditions, it may be concluded that the hazard from even repeated exposure to vapors under ordinary room conditions is of no practical significance. It should be noted, however, that a hazard due to inhalation may exist in circumstances where the material is being handled hot or where agitation or other mechanical operations may create a fog or mist in the air. Troisi[19] reports the results of an investigation of complaints among women workers in an electrolytic condenser factory and attributes the trouble to the inhalation of ethylene

[15] P. J. Hanzlik, W. S. Lawrence, J. K. Fellows, F. P. Luduena, and G. L. Laqueur, *J. Ind. Hyg. Toxicol.,* **29,** 325 (1947).

[16] E. Volkmann, *Hippokrates,* **21,** 549 (1950), *Chem. Abstr.,* **47,** 9567 (1953).

[17] C. P. Carpenter and H. F. Smyth, Jr., *Am. J. Ophthalmol.,* **29,** 1363 (1946).

[18] F. J. Wiley, W. C. Hueper, and W. F. von Oettingen, *J. Ind. Hyg. Toxicol.,* **18,** 123 (1936).

[19] F. M. Troisi, *Brit. J. Ind. Med.,* **7,** 65 (1950).

glycol vapor. The work involved coating aluminum and paper with a mixture containing 40 per cent ethylene glycol, 55 per cent boric acid, and 5 per cent ammonia at 105°C. Nine of 38 women exposed to the vapors suffered frequent attacks of loss of consciousness and nystagmus and 5 of these showed an absolute lymphocytosis. Further examination revealed 5 additional cases of nystagmus among the other 29 workers. Proper enclosure of the system to prevent inhalation of vapors resulted in complete recovery of all affected individuals, although the 2 most severely affected were removed to other work.

Metabolism. It has been reported recently by Deichmann and Gerarde[20] that Gessner and Williams have disclosed results of metabolism studies on C^{14}-ethylene glycol. They are: (a) Rats excreted oxalate in the urine, 0.25 per cent to 1 per cent of the dose, when the dose was 1 g./kg. or more; with a smaller dose, 0.1 g./kg., oxalate excretion was negligible. (b) Guinea pigs failed to excrete in the urine detectable amounts of oxalate when given doses of 1.3 and 2.7 g./kg. (c) A cat excreted oxalate in the urine, 9 per cent of a dose of 1 g./kg. (d) The proportion of administered C^{14} (as ethylene glycol) excreted in the urine of rats varied with dosage, being 21 per cent at a dose of 0.1 g./kg. and 78 per cent at a dose of 7.5 g./kg. (e) The distribution of radioactivity 24 hours after the administration of the glycol to rats was highest in the bones (2 to 10 per cent); and highest in the muscle of rabbits (3.4 per cent), although the liver contained 1.4 and the bone 1.7 per cent of the dose. (f) Two metabolic pathways are suggested, the major one leading to carbon dioxide through glycolaldehyde, glycolic acid, or glyoxylic acid, and the minor one leading to oxalic acid; the significance of the minor pathway appears to depend on dosage and species.

Mode of Action. Ethylene glycol in large doses is a depressant of the central nervous system. Deaths from large single doses are likely to be due to this action. Nonfatal, acutely toxic doses may well exert their effect primarily upon the kidney and brain and to a lesser degree upon the liver. The role of the metabolites is uncertain. Chronic effects resulting from prolonged and repeated exposure are most likely to be centered in the kidneys, although subjective symptoms of nystagmus, loss of appetite, "dopiness," and periods of unconsciousness have been reported.

Human Experience. When one considers the huge volumes of ethylene glycol that are handled and used industrially, there have, indeed, been few instances of adverse effects. Only one episode that can be considered to be of industrial origin is reported[19] and this involves inhalation of vapors from heated material. One case of alleged absorption through the skin involved a medicinal product.[16]

Numerous poisonings have occurred among humans who have ingested substantial amounts of ethylene glycol, usually as antifreeze. Judging from the cases where the amount ingested could be well estimated, it appears that the lethal dose for humans is about 1.4 ml./kg. It is apparent, therefore, that the human is far more susceptible to acute ethylene glycol poisoning than most animals.

[20] W. B. Deichmann and H. W. Gerarde, *J. Occupational Med.,* **1,** 465 (1959) (abstr.).

The cause of acute deaths has been debated. Renal failure, central nervous system depression, and brain injury have been suggested as the cause. It would seem from the evidence available that large single doses are more likely to cause death by depression (narcotic effect) followed by respiratory and/or cardiac failure; and that delayed deaths from smaller single doses would be more likely to be due to a combination of the organic injury to the kidneys and to the brain. This latter effect is stressed by Pons and Custer[21] in their review of 18 human cases. Grant[22] has also noted cerebral injury from acute poisoning and believes it may have caused permanent injury. Ross[23] describes a fatal case in which $1/4$ to $1/2$ pint of an antifreeze solution was ingested; acute meningoencephalitis followed by anuria and death from renal failure resulted after 12 days.

Morini[24] describes three stages of poisoning by ethylene glycol. The first stage is drunkenness typical of ethanol intoxication; the second stage consists of the residual symptoms of drunkenness along with beginning symptoms of oxalic acid poisoning; and the third stage is characterized by loss of control of sphincters, psychological disturbances, marked cyanosis, and renal failure. He believes the toxic dose to be about 100 g. per person, the fatal dose to be but slightly more.

5. Hygienic Standards of Permissible Exposure

Based upon the work of Wiley, Hueper, and von Oettingen,[18] it is believed that a vapor concentration of 100 p.p.m. of ethylene glycol in the air would be safe for prolonged and repeated 8-hour exposures in humans.

DIETHYLENE GLYCOL, $C_4H_{10}O_3$ or $(HOCH_2CH_2)_2O$ (2,2'-Oxydiethanol, Bis(2-hydroxyethyl) ether)

1. Source, Uses, and Industrial Exposure

Diethylene glycol is produced commercially as a by-product of ethylene glycol production. It can also be produced directly by reaction between ethylene glycol and ethylene oxide.[25] Diethylene glycol is used in permanent antifreeze formulations as a constituent of brake fluids, lubricants, mold release agents, and inks; as a humectant in tobacco; as a softening agent for textiles; as a plasticizer for cork, adhesives, paper, packaging materials, and coatings; as an intermediate in the production of the explosive, diethylene glycol dinitrate; and as an intermediate in the production of certain resins and diethylene glycol esters and ethers.

Diethylene glycol presents practically no hazard from the standpoint of industrial handling. It is quite stable chemically, does not present a hazard due to flam-

[21] C. A. Pons and R. P. Custer, *Am. J. Med. Sci.*, **211**, 544 (1946).
[22] A. P. Grant, *Lancet,* **263**, 1252 (1952).
[23] I. P. Ross, *Brit. Med. J.*, **1**, 1340 (1956).
[24] I. Morini, Minerva med., **1**, 72 (1954), *Ind. Hyg. Dig.*, **20**, 210 (1956).
[25] G. O. Curme, Jr. and F. Johnston, eds., *Glycols, American Chemical Society Monographs, Series 114.* Reinhold, New York, 1952.

mability, except at high temperatures or where mist may be involved. It is not appreciably irritating to the eyes and skin, is not absorbed through the skin in appreciable amounts except possibly under adverse conditions where extensive and prolonged skin contact occurs. Its vapor pressure at room temperatures is so low that toxic concentrations of vapor are impossible. It should be noted, however, that a hazard from repeated prolonged inhalation may exist in operations involving heated material or where mists or fogs are generated. However, any reasonable industrial hygiene control would eliminate this possibility. Although the principal hazard to health presented by diethylene glycol is that of ingestion of substantial amount, this should not occur in industrial handling unless the material is put in unlabeled or mislabeled containers.

2. Physical and Chemical Properties

Diethylene glycol is a colorless, essentially odorless, viscous, hygroscopic liquid. Initially, it has a sweetish taste but its aftertaste is bitter. It has the following properties:[25]

Molecular weight: 106.12

Specific gravity: 1.116 (20/4°C.) ; 1.1184 (20/20°C.)

Freezing point: —8.0°C.

Boiling point: 245°C. (760 mm. Hg)

Vapor pressure: less than 0.01 mm. Hg (20°C.)

Refractive index: 1.446 (25°C.) ; 1.4472 (20°C.)

Per cent in "saturated" air: 0.0013 (20°C.)

Solubility: miscible with water, lower aliphatic alcohols, acetone; quite soluble in benzene, toluene, and carbon tetrachloride

Flash point: 280 to 290°F. (TOC)

1 p.p.m. $\approx$ 4.35 mg./cu. meter at 25°C., 760 mm. Hg

3. Determination in the Atmosphere

Methods applicable to the determination of small amounts of diethylene glycol, such as may be of industrial hygiene significance, are few. Physical methods applying such instruments as the interferometer or combustible gas analyzer, are not sufficiently sensitive to be useful. Spectrographic procedures employing infrared absorption undoubtedly can be used to advantage where appropriate instruments are available. Ultraviolet absorption would seem to offer little promise but mass spectrophotometry and vapor phase chromatography should be useful. According to Bergner and Sperlich,[26] diethylene glycol can be identified in the presence of other glycols by chromatographic procedures. A chemical method, described by Duke and Smith,[27] based upon the reaction of alcoholic hydroxyl groups with ammonium hexanitrato cerate to form a red product has been found

---

[26] K. G. Bergner and H. Sperlich, *Z. Lebensm. Untersuch. u. Forsch.,* **97,** 253 (1953), *Chem. Abstr.,* **48,** 2273 (1954).

[27] F. R. Duke and G. F. Smith, *Ind. Eng. Chem. Anal. Ed.,* **12,** 201 (1940).

useful. The optical density of the color can be quantitated. It should be noted, however, that this method is not specific for diethylene glycol.

Recently, Amlinskaya and Erikh[28] reported a method for the determination of diethylene glycol in aqueous solution. It involves oxidizing with dry solid potassium dichromate in sulfuric acid and titrating the excess dichromate with sodium thiosulfate. This method also would appear to be limited by its nonspecificity.

4. Physiological Response

Summary. Diethylene glycol presents negligible hazards to health in industrial handling except possibly where it is being used at elevated temperatures. It is low in acute oral toxicity, it is not irritating to the eyes or skin, it is not readily absorbed through the skin, and its vapor pressure is sufficiently low so that toxic concentrations of vapor can not occur in air at room temperatures. The principal hazard to health of diethylene glycol is associated with the ingestion of large quantities in single doses.

Prior to 1937, the toxicological information on diethylene glycol and other glycols was rather incomplete. However, in 1937 more than 100 deaths were caused by the ingestion of an elixir consisting of sulfanilamide and diethylene glycol. As a result of this tragedy, a large number of investigations were conducted in an effort to clarify the toxicological picture in regard to the various glycols.

Single-Dose Oral. Most of the older published articles dealing with the toxicity of diethylene glycol and the other common glycols have been ably reviewed.[29-33] The toxicity of diethylene glycol for animals has been determined by numerous investigators. The findings of Laug et al.[31] and Smyth et al.[34] on small animals are representative. Laug[31] reports the single dose oral LD_{50} values for rats, guinea pigs, and mice to be 14.8, 7.76, and 23.7 ml./kg., respectively, and Smyth[34] reports similar values of 20.76 g./kg. for rats and 13.21 g./kg. for guinea pigs. Laug[31] reports the symptomatology for rabbits, dogs, mice, and guinea pigs to be quite similar; first noted were thirst, diuresis, roughened coat, and refusal of food, followed days later by suppression of urine, proteinuria, prostration, dyspnea, a bloated appearance, coma, lowering of body temperature, and death. As a result of the elixir of sulfanilamide episode the single oral dose lethal for humans has been estimated by Calvery and Klumpp[35] to be about 1 ml./kg.

[28] M. A. Amlinskaya and V. N. Erikh, *Trudy Vsesoyuz. Nauch. Issledovatel. Inst. Khim. Pererabotki Gazov.*, **6**, 213 (1951), *Chem. Abstr.*, **49**, 9445 (1955).

[29] W. F. von Oettingen, *U. S. Public Health Bull.* No. **281**, 1943.

[30] E. Browning, *Toxicity of Industrial Organic Solvents.* Chemical Publishing, New York, 1953.

[31] E. P. Laug, H. O. Calvery, H. J. Morris, and G. Woodard, *J. Ind. Hyg. Toxicol.*, **21**, 173 (1939).

[32] H. J. Morris, A. A. Nelson, and H. O. Calvery, *J. Pharmacol. Exptl. Therap.*, **74**, 266 (1942).

[33] E. M. K. Geiling and P. R. Cannon, *J. Am. Med. Assoc.*, **111**, 919 (1938).

[34] H. F. Smyth, Jr., J. Seaton, and L. Fischer, *J. Ind. Hyg. Toxicol.*, **23**, 259 (1941).

[35] H. O. Calvery and T. G. Klumpp, *Southern Med. J.*, **32**, 1105 (1939).

Repeated-Dose Oral. The most comprehensive study of the repeated dose oral toxicity of diethylene glycol is that reported by Fitzhugh and Nelson,[36] who maintained rats for 2 years on diets containing 1, 2, and 4 per cent diethylene glycol. Unfortunately, however, the purity of the diethylene glycol was not described and, hence, some of the effects noted may not, in fact, be attributable to the chemical diethylene glycol. At the dietary level of 1 per cent of diethylene glycol, they observed slight growth depression, a few bladder stones identified as calcium oxalate, slight kidney damage, and infrequent liver damage. At the 2 per cent dietary level, slight growth depression, a number of bladder stones and bladder tumors, moderate kidney damage, and slight liver damage were noted. At the 4 per cent dietary level, there was marked growth depression, slight mortality, frequent bladder stones and bladder tumors, marked kidney damage, and moderate liver damage. There seems to be little doubt that the bladder stones are directly attributable to the material fed. However, the tumors may well have been a secondary result of mechanical irritation of the bladder by the stones and not the direct effect of a chemical on the cells of the bladder. As a result of this study, and an observation previously reported from the U. S. Food and Drug Administration by Morris, Nelson, and Calvery,[32] it was concluded that, in the ethylene glycol series, the chronic oral toxicity for rats decreased with an increase in molecular weight of the glycols.

More recently, German workers[37-39] have studied the effects of diethylene and other glycols. Wegener[37] gave rats 1 ml. of a 20 per cent aqueous solution of diethylene glycol per 100 g. of body weight daily over a period of 12 weeks and concluded that it had no influence on the reproductive ability of the animals or on their offspring. Bornmann[38] administered the material to rats in concentrations of 1, 2, 5, 10, and 20 per cent in their drinking water and found the material to have a narcotic effect and to cause central paralysis of the respiratory and cardiac centers. Loesser *et al.*[39] found that concentrations of 5 to 20 per cent in the drinking water of rats caused weight loss and death but that 1 and 2 per cent had no such effect. They also report finding 2-naphthol in the urine and bile of animals treated with diethylene glycol; this is difficult to rationalize.

There seems to be little doubt that further studies of the repeated oral toxicity of pure diethylene glycol are needed before the true chronic toxicity of the material can be established.

Eye Irritation. Carpenter and Smyth[40] report that diethylene glycol failed to

[36] O. G. Fitzhugh and A. A. Nelson, *J. Ind. Hyg. Toxicol.,* **28,** 40 (1946).
[37] H. Wegener, *Arch. exptl. Pathol. Pharmakol.,* **220,** 414 (1953), *Chem. Abstr.,* **48,** 2919 (1954).
[38] G. Bornmann, *Arzneimittel Forsch.,* **4,** 643, 710 (1954), **5,** 38 (1955), *Chem. Abstr.,* **49,** 7131 (1955).
[39] A. Loesser, G. Bornmann, L. Grosskinsky, G. Hess, R. Kopf, K. Ritter, A. Schmitz, E. Stürner, and H. Wegener, *Arch. exptl. Pathol. Pharmakol.,* **221,** 14 (1954), *Chem. Abstr.,* **48,** 4698 (1954).
[40] C. P. Carpenter and H. F. Smyth, Jr., *Am. J. Ophthalmol.,* **29,** 1363 (1946).

cause appreciable irritation when introduced into the eyes of rabbits. No cases of injury to human eyes have been reported nor would any be expected.

Inhalation. The effects of inhaling vapors of diethylene glycol have not been determined and there would seem to be very little reason to determine them. The vapor pressure of diethylene glycol at ordinary temperatures is sufficiently low that possible concentrations would seem to offer no hazard, even if breathed continuously over prolonged periods of time. A hazard from inhalation might result from industrial operations that involve the handling of hot material or where agitation or other mechanical operations create a fog or mist in the air. It would seem only wise to avoid prolonged and repeated inhalation of such atmospheres.

Skin Irritation. Diethylene glycol produces no significant skin irritation; however, prolonged contact over an extended period of time may produce a macerating action comparable to that caused by glycerol.

Skin Absorption. Hanzlik et al.[41] have shown that commercial diethylene glycol of unknown purity can be absorbed in toxic amounts through the skin of rabbits. The data, however, are erratic and therefore difficult to quantitate. It would seem that the hazard to health from skin absorption in ordinary industrial operations would be quite small. It would seem wise, however, to avoid prolonged and repeated contact and any contact involving extensive areas of skin.

Metabolism. Repeated administration for a week to dogs did not lead to consistent increases in urinary oxalate. However, the urinary oxalate was increased in rats maintained on water containing diethylene glycol. Wiley, Hueper, Bergen, and Blood[43] were unable to demonstrate the presence of oxalic acid in the urine following large doses of diethylene glycol to rabbits and dogs. These apparent discrepancies may well be attributable to the purity of the material studied.

Mode of Action. Diethylene glycol in large doses appears to be a depressant to the central nervous system. Deaths from large single doses which occur within 24 hours are believed to result from this action. Acutely toxic doses, not immediately fatal, may exert their effect primarily upon the kidney and, to a lesser extent, upon the liver. Deaths or serious injuries are associated primarily with renal insufficiency caused by the swelling of the convoluted tubules and a plugging of the tubules with debris. Chronic effects resulting from prolonged and repeated exposure to the commercial product, at least, are most likely to be centered in the kidney and to a lesser degree in the liver. In metabolism studies with the dog, Haag and Ambrose[42] found that a large portion of the diethylene glycol administered was excreted in the urine unchanged.

Human Experience. The human experience in the industrial handling and use of diethylene glycol has been excellent. Following the "elixir of sulfanilamide

[41] P. J. Hanzlik, W. S. Lawrence, J. K. Fellows, F. P. Luduena, and G. L. Laqueur, *J. Ind. Hyg. Toxicol.*, **29**, 325 (1947).

[42] H. B. Haag and A. M. Ambrose, *J. Pharmacol. Exptl. Therap.*, **59**, 93 (1937).

[43] F. H. Wiley, W. C. Hueper, D. S. Bergen, and F. R. Blood, *J. Ind. Hyg. Toxicol.*, **20**, 269 (1938).

tragedy" in which more than 100 deaths were attributed to the ingestion of diethylene glycol, a great number of articles have been published dealing with the clinical and experimental aspects of diethylene glycol poisoning. These have been well summarized by Geiling and Cannon.[33] A few cases have also been reported from other uses of diethylene glycol in medicinals. In general, pathology observed in human victims resembles closely that which has been described previously for laboratory animals and consists primarily of degeneration of the kidney with lesser lesions in the liver. Death in practically all of these cases was due to renal insufficiency.

5. *Hygienic Standards of Permissible Exposure*

Because of the low vapor pressure of diethylene glycol and the fact that it is actually quite low in toxicity, there seems to be no need to establish an industrial hygiene standard for inhalation. However, if a guide to the amount of air contamination that could be tolerated under unusual circumstances were desired, it would seem logical to conclude that 100 p.p.m. calculated as the vapor, although partly present as a fog, would be very unlikely to cause any difficulty. This figure is based upon the fact that oral and skin absorption studies indicate that diethylene glycol is quantitatively less toxic than ethylene glycol, and the work available on ethylene glycol indicates that 100 p.p.m. would be without serious hazard. Hence, a similar figure for diethylene glycol would not be out of line.

TRIETHYLENE GLYCOL, $C_6H_{14}O_4$ or $(CH_2OCH_2CH_2OH)_2$ (2,2'-(ethylenedioxy)diethanol, Triglycol, Bis(2-hydroxyethoxy)ethane)

1. *Source, Uses, and Industrial Exposure*

Triethylene glycol, like diethylene glycol, is produced commercially as a by-product of ethylene glycol production; its formation being favored by a high ethylene oxide to water ratio.[44]

Triethylene glycol is used for many of the same applications as diethylene glycol but it has two distinct properties of importance; it is less volatile and less toxic. It is used to a limited extent as an air disinfectant in air-conditioning systems, particularly in hospitals and public buildings. Much has been written about this application. It is used as a humectant and softening agent, as a plasticizer, as a dehydrating agent for natural gas, and as a selective solvent. It is a valuable intermediate for the manufacture of plasticizers, resins, emulsifiers, demulsifiers, lubricants, explosives, and many others.

The industrial handling and use of triethylene glycol should present no significant problem from ingestion, skin contact, or vapor inhalation. Its low oral

[44] G. O. Curme, Jr., and F. Johnston, eds., *Glycols, American Chemical Society Monograph Series 114.* Reinhold, New York, 1952.

toxicity suggests that it may be considered safe for many applications where intake is limited. Similarly, its negligible skin irritation and absorption properties make it suitable for use to some extent in preparations intended to be applied over appreciable areas of the body. Furthermore, it is stable chemically, does not present a hazard due to flammability, except possibly at high temperatures or where fogs or mists are involved.

2. Physical and Chemical Properties

Triethylene glycol is a colorless to pale straw-colored, essentially odorless, viscous, hygroscopic liquid. It has the following properties:[44]

Molecular weight: 150.17

Specific gravity: 1.1254 (20/20°C.)

Freezing point: —4.3°C.

Boiling point: 287.4°C. (760 mm. Hg); 198°C. (50 mm. Hg); 162°C. (10 mm. Hg)

Vapor pressure: 0.001 mm. Hg (20°C.)

Refractive index: 1.4559 (20°C.)

Per cent in "saturated" air: approximately 0.00013 at 20°C.

Solubility: miscible with water, most common aliphatic alcohols, ketones, esters, and low molecular weight halogenated hydrocarbon solvents

Flash point: 330°F. (TOC)

1 p.p.m. ≎ 6.14 mg./cu. meter at 25°C., 760 mm. Hg

3. Determination in the Atmosphere

There would seem to be no need for determining atmospheric concentrations of triethylene glycol for industrial hygiene purposes. If analysis of the atmosphere were to be made, however, the methods noted under Diethylene Glycol probably would be useful. If additional information is desired, it is suggested that Curme and Johnston[44] be consulted.

4. Physiological Response

Summary. Triethylene glycol is very low both in acute and chronic oral toxicity, is not irritating to the eyes or skin, and the inhalation of amounts that conceivably could cause injury does not seem likely.

Single-Dose Oral. Latven and Molitor,[45] Smyth *et al.*,[46] and Laug *et al.*[47] have studied the single dose oral toxicity of triethylene glycol and found it to be less toxic than diethylene glycol. Smyth *et al.*[46] report the oral LD_{50} for rats and guinea pigs to be 22.06 and 14.66 g./kg., respectively. Laug *et al.*[47] found the LD_{50}

[45] A. R. Latven and H. Molitor, *J. Pharmacol. Exptl. Therap.*, **65**, 89 (1939).

[46] H. F. Smyth, Jr., J. Seaton, and L. Fischer, *J. Ind. Hyg. Toxicol.*, **23**, 259 (1941).

[47] E. P. Laug, H. O. Calvery, H. J. Morris, and G. Woodard, *J. Ind. Hyg. Toxicol.*, **21**, 173 (1939).

values for rats, guinea pigs, mice, and rabbits to be 16.8, 7.9, 18.7, and 8.4 ml./kg., respectively.

Repeated-Dose Oral. The most comprehensive study of the repeated dose oral toxicity of triethylene glycol is that reported by Fitzhugh and Nelson.[48] These investigators fed the material at concentrations of 1.0, 2.0, and 4.0 per cent in the diet of rats for 2 years without producing adverse effects. These dosage levels are equivalent to as much as 3 to 4 g./kg./day without effect. Earlier, Lauter and Vrla[49] reported that rats could tolerate 3 per cent in their drinking water for 30 days without effect, but 5 per cent caused ill effects. These dosages are equivalent to about 5 and 8 g./kg./day.

Lauter and Vrla[49] described the material they used as "commercial grade" but unfortunately Fitzhugh and Nelson[48] give no indication of the quality of the material they studied. Since commercial grade triethylene glycol may contain several per cent of diethylene glycol, the possibility that the toxic effect seen by Lauter and Vrla[49] may have been caused by diethylene glycol rather than by triethylene glycol cannot be overlooked.

From these findings, it is apparent that triethylene glycol is very low in repeated dose oral toxicity, far less than ethylene or diethylene glycols.

Eye Irritation. Latven and Molitor[45] tested triethylene glycol for its effect upon the rabbit eye and found it to be similar to glycerin and diethylene glycol and less irritating than propylene glycol.

Carpenter and Smyth[50] report that triethylene glycol failed to cause appreciable irritation when introduced into the eyes of rabbits. No cases of injury to human eyes have been reported nor would any be expected.

Skin Irritation. Triethylene glycol produces no significant irritation of the skin. However, prolonged contact over an extended period of time may result in a macerating action similar to that caused by glycerin.

Skin Absorption. No studies have been reported dealing with the skin absorption of triethylene glycol. Although it is possible that, under conditions of very severe prolonged exposures, some of the material may be absorbed through the skin, it is extremely doubtful that a quantity sufficient to produce an appreciable systemic injury would be absorbed.

Intramuscular Toxicity. Lauter and Vrla[49] administered single doses of triethylene glycol to rats intramuscularly by injection and found the LD_{50} to be approximately 8.4 g./kg.

Intraperitoneal Toxicity. Karel et al.[51] administered single doses of triethylene glycol to rats intraperitoneally by injection and found the acute LD_{50} dose to be 8.15 g./kg.

[48] O. G. Fitzhugh and A. A. Nelson, *J. Ind. Hyg. Toxicol.*, **28**, 40 (1946).

[49] W. M. Lauter and V. L. Vrla, *J. Am. Pharm. Assoc.*, **29**, 5 (1940).

[50] C. P. Carpenter and H. F. Smyth, Jr., *Am. J. Ophthalmol.*, **29**, 1363 (1946).

[51] L. Karel, B. H. Landing, and T. S. Harvey, *J. Pharmacol. Exptl. Therap.*, **90**, 338 (1947).

Inhalation. Interest in the toxicity of triethylene glycol when inhaled was initiated by the observation by Robertson[52] in 1943 and later in 1947[53] that triethylene glycol was an effective air sterilizer.

During the studies on effectiveness, numerous persons were exposed, and, according to Jennings *et al.*[54] and Harris and Stokes,[55] none were adversely affected. The developments in the field of air sterilization have been well reviewed by Polderman.[56]

Also in 1947, Robertson *et al.*[57] reported extensive experiments with monkeys and rats showing that prolonged inhalation of saturated vapors, as in air disinfection (about 1 p.p.m.), was without any physiological effect. It is generally concluded that indefinitely long exposures to air at room temperatures substantially saturated with vapors of triethylene glycol are harmless.

Metabolism. McKennis and co-workers[57a] recently have studied the fate of C^{14}-labeled triethylene glycol in rats and of unlabeled material in rabbits. They found in both species that a high per cent of small doses was eliminated in the urine unchanged and possibly as the mono- and dicarboxylic acid derivatives of triethylene glycol. In the studies with rats, little if any C^{14}-oxalate or C^{14}-triethylene glycol in conjugated form was found in the urine. Small portions of the administered C^{14} activity were found in the feces (2 to 5 per cent) and in expired air (1 per cent). Recoveries of the administered C^{14} activity ranged from 91 to 98 per cent.

Human Experience. The human experience in the handling and use of triethylene glycol has been uneventful and without reported cases of any adverse effects.

5. *Hygienic Standards of Permissible Exposure*

It does not seem that a hygienic standard for triethylene glycol is necessary.

POLYETHYLENE GLYCOLS

1. *Source, Uses, and Industrial Exposure*

The polyethylene glycols are prepared commercially by adding ethylene oxide to ethylene glycol, diethylene glycol, or water in the presence of caustic or other

[52] O. H. Robertson, *Harvey Lectures, Ser.* **38**, 227 (1943).

[53] O. H. Robertson, *Wisconsin Med. J.*, **46**, 311 (1947).

[54] B. H. Jennings, E. Biggs, and F. C. W. Olson, *Heating, Piping, Air Conditioning*, **16**, 538 (1944).

[55] T. N. Harris and J. Stokes, Jr., *Am. J. Med. Sci.*, **209**, 152 (1945).

[56] L. D. Polderman, *Soap Sanit. Chemicals*, 133 (July 1947).

[57] O. H. Robertson, C. G. Loosli, T. T. Puck, H. Wise, H. M. Lemon, and W. Lester, Jr., *J. Pharmacol. Exptl. Therap.*, **91**, 52 (1947).

[57a] H. McKennis, Jr., R. A. Turner, L. B. Turnbull, E. R. Bowman, W. W. Muelder, M. P. Neidhardt, C. L. Hake, R. Henderson, H. G. Nadaeu, and S. Spencer, *Toxicol. Appl. Pharmacol.*, 1962, in press.

catalysts. The molecular weights of the product can be controlled by the proportions of the reactants used.[58]

The polyethylene glycols are used as water-soluble lubricants, plasticizers, softening agents, mold release agents, solvents, dispersing agents, binders, and where appropriate in pharmaceuticals and cosmetics. They are also used as chemical intermediates in the production of polyethylene glycol ethers and esters, which have many and varied uses, and in the production of various resins.

Industrial exposure to the polyethylene glycols is almost entirely limited to topical contact.

2. *Physical and Chemical Properties*

The polyethylene glycols may be represented by the formula $H(OCH_2CH_2)_n$-OH. Those having average molecular weights of 200, 300, 400, and 600 are viscous, nearly colorless, odorless, water-soluble liquids having very low vapor pressures. The polyethylene glycols having average molecular weights of 1000 and more are nearly colorless, water-soluble, waxy solids at room temperature. All of the unstabilized polyglycols are inherently susceptible to oxidative degradation, which occurs with increased rapidity as temperature increases and as the availability of oxygen increases. Their physical and chemical properties are given in Table 1.

TABLE 1

Physical and Chemical Properties of Polyethylene Glycols

Material designation[a]	Physical state	Molecular wt. range	Specific gravity (25/25°C.)	Freezing range, °C.	Refractive index (25°C.)	Flash point, °F.[b]
200[c,d]	Liquid	190–210	1.125	Supercools	1.459	340–360
300[c,d]	Liquid	285–315	1.125	−15--−6	1.463	385–415
400[c,d]	Liquid	380–420	1.125	4–8	1.465	435–460
600[c,d]	Liquid	570–630	1.125	20–25	1.466	475–480
1000[c,d]	Solid	956–1050	1.117	36–40		490–510
1450[d]	Solid	1300–1600	1.210	43–46		490
1540[c]	Solid	1300–1600	1.21	43–46		510
2000[d]	Solid	1900–2300	1.211	47–50		510
4000[c]	Solid	3000–3700	1.204	53–56		520
4000[d]	Solid	4200–4800	1.212	54–57		515
6000[c]	Solid	6000–7500		60–63		520
6000[d]	Solid	7000–8000	1.212	56–59		515
9000[d]	Solid	9000–10,000	1.212	60–64		520

[a] Generally designates average molecular weights.

[b] Cleveland open cup.

[c] *Physical Properties of Synthetic Organic Chemicals.* Union Carbide Chemicals Co., New York, N.Y. 1958.

[d] *Dow Polyethylene Glycols.* The Dow Chemical Co., Midland, Mich., 1959.

[58] G. O. Curme, Jr. and F. Johnston, eds., *Glycols, American Chemical Society Monograph, Series 114.* Reinhold, New York, 1952.

3. Determination

The determination of the polyglycols in aqueous solution can be accomplished by the method of Duke and Smith[59] based upon the reaction of alcoholic hydroxyl groups with ammonium hexanitrato cerate to form a red product. The optical density of the color can be quantitated. Infrared spectrophotometry also can be employed to advantage where appropriate instruments are available. Shaffer and Critchfield[60] have described a method for quantitatively determining the solid polyethylene glycols in biological materials.

4. Physiological Response

Summary. The polyethylene glycols present practically no hazards to health in industrial handling and use. They are not significantly irritating to the eyes, skin, or mucous membranes; they are exceptionally low in oral toxicity; and their vapor pressures are so low that there is no hazard from inhalation.

Single-Dose Oral. All of the polyethylene glycols are very low in single-dose oral toxicity. It is noteworthy that toxicity decreases as molecular weight increases. Representative figures are given in Table 2.

TABLE 2

Single Dose Oral Toxicity of Polyethylene Glycols
(Approximate LD_{50} values, g./kg.)

Material	Rats M	Rats B	Rats F	Guinea pigs M	Guinea pigs B	Guinea pigs F	Rabbits M	Rabbits B	Rabbits F	Mice
200[a]		34						20		
200[b]	34		28			17	14			34
300[a]		39			20			21		
300[b]	30		29	21					21	31
400[a]		44			16			27		
400[b]	33		32			21	22			36
600[b]	33		30			28	19			36
1,000[a]		42			22					
1,000[b]	45		32			41			>50	>50
1,540[a]		51								
2,000[b]	45		>50	>50			>50			>50
4,000[a]		59			51		76			
4,000[b]	>50		>50			46	>50			>50
6,000[a]		>50			>50					
6,000[b]	>50		>50			>50	>50			>50
9,000[b]	>50		>50	>50					>50	>50
10,000[a]		>50								

Note: M = males; B = both sexes; F = females.

[a] H. F. Smyth, Jr., C. P. Carpenter, and C. S. Weil, *J. Am. Pharm. Assoc. Sci. Ed.*, **39**, 349 (1950).

[b] Biochemical Research Laboratory, The Dow Chemical Co., unpublished data.

[59] F. R. Duke and G. F. Smith, *Ind. Eng. Chem. Anal. Ed.*, **12**, 201 (1940).

[60] C. B. Shaffer and H. Critchfield, *Ind. Eng. Chem.*, **19**, 32 (1947).

Repeated-Dose Oral. Smyth and co-workers[61,62] have summarized the extensive feeding studies they have conducted with the polyethylene glycols. In their most recent article,[62] they conclude that the polyethylene glycols having average molecular weights of 400, 1540, and 4000 caused no adverse effect upon dogs when fed in their diet for 1 year at the level of 2 per cent. When fed to rats for 2 years as a part of their diet, polyethylene glycols 1540 and 4000 had no effect at a level of 4 per cent and polyethylene glycol 400 had no effect at a level of 2 per cent.[61] The other polyethylene glycols have been studied by dietary feeding techniques using rats, but for shorter periods of 3 to 4 months.[62,63] It appears that several per cent of these materials can be tolerated in the diet of rats without appreciable adverse effects, indicating that they are exceptionally low in repeated-dose oral toxicity (see Table 3). The hazard from their ingestion would seem to be slight.

Eye Irritation. Carpenter and Smyth[64] have reported that the polyethylene glycols do not cause appreciable irritation to the eyes of rabbits. This has been confirmed.[65] No cases of injury to human eyes have been reported nor would any be expected.

Skin Irritation and Sensitization. Although early reports by Smyth *et al.*[66] reported that skin sensitization was observed among a few human subjects and in guinea pigs tested with certain polyethylene glycols, later studies[61,65] show that currently produced materials are without irritating or sensitizing properties. This has been borne out by their very wide application without difficulty in cosmetics.

Skin Absorption—Single Doses. As concluded by Smyth *et al.*,[61] the size of the single dose of the polyethylene glycols required to kill by skin penetration is so large as to defy the establishment of LD_{50} values. Unpublished studies,[63] employing essentially the technique of Draize *et al.*,[67] have shown that single doses of 20 g./kg. of the various polyethylene glycols ranging from 200 through 9000 were without toxic effects.

Skin Absorption—Repeated Doses. The studies reported by Luduena *et al.*[68] in 1947 indicated that toxic amounts of polyethylene glycols 200, 400, 1500, and 4000 were quite readily absorbed through the skin of rabbits when applied by inunction 6 days a week for 5 weeks. However, since part of this study was on animals on a deficient diet and since there were no controls, its significance cannot be

[61] H. F. Smyth, Jr., C. P. Carpenter, and C. S. Weil, *J. Am. Pharm. Assoc. Sci. Ed.*, **39**, 349 (1950).

[62] H. F. Smyth, Jr., C. P. Carpenter, and C. S. Weil, *J. Am. Pharm. Assoc. Sci. Ed.*, **44**, 27 (1955).

[63] Biochemical Research Laboratory, The Dow Chemical Co., unpublished data.

[64] C. P. Carpenter and H. F. Smyth, Jr., *Am. J. Ophthalmol.*, **29**, 1363 (1946).

[65] *Dow Polyethylene Glycols,* The Dow Chemical Co., Midland, Mich., 1959.

[66] H. F. Smyth, Jr., C. P. Carpenter, C. B. Shaffer, J. Seaton, and L. Fischer, *J. Ind. Hyg. Toxicol.*, **24**, 281 (1942).

[67] J. H. Draize, G. Woodard, and H. O. Calvery, *J. Pharmacol. Exptl. Therap.*, **82**, 377 (1944).

[68] F. P. Luduena, J. K. Fellows, G. L. Laqueur, and R. L. Driver, *J. Ind. Hyg. Toxicol.*, **29**, 390 (1947).

TABLE 3

Summary of Repeated Oral Dose Toxicity of the Polyethylene Glycols

Mean molecular weight	Species	Sex[a]	Duration of study, mos.	Dosage level (% conc. in diet) Without effect	With effect	First sign of adverse effect noted[b]	Reference
200	Rat	B	3	8	16	LW	62
300	Rat	B	3	4	8	W	62
400	Rat	B	3	8	16	W	62
	Rat	M	24	2	4	W, LW, LP	62
	Rat	F	24	4			62
	Dog	B	12	2			62
600	Rat	B	3	8	16	KW, W	62
1000	Rat	B	3	8	16	W	62
	Rat	B	3	10	15	W	63
1500	Rat	B	3	4	8	W	62
1540	Rat	B	3	4	8	W	62
	Rat	B	24	4	8	LP	62
	Dog	B	12	2 (Possibly 8)			62
2000	Rat	B	4	15			63
4000	Rat	B	3	4	8	W	62
	Rat	B	4	5	10	W, LW	63
	Rat	B	24	4	8	W	62
	Dog	B	12	2 (Possibly 8)			62
6000	Rat	B	3	16	24	KW, W	62
	Rat	M	3	10	15	W	63
	Rat	F	3	15			63
9000	Rat	B	3	15			63

[a] B = both sexes; M = male; F = female.

[b] W = decrease in body weight; LW = increase in liver weight per 100 g. of body weight; KW = increase in kidney weight per 100 g. of body weight; LP = slight histological changes in the liver.

evaluated. This is emphasized by the fact that other well-controlled studies by Smyth et al.[62] using similar materials and by Tusing et al.[69] using a wider spectrum of polyethylene glycols found no adverse effects from larger doses over a longer period of time.

It is concluded from the data available that there is no hazard from the repeated skin application of currently produced polyethylene glycols.

Inhalation. There would not seem to be any hazard from inhalation of the polyethylene glycols.

[69] T. W. Tusing, J. R. Elsea, and A. B. Sauveur, *J. Am. Pharm. Assoc. Sci. Ed.,* **43,** 489 (1954).

Pharmacology. The polyethylene glycols in general appear to be slow-acting parasympathomimeticlike compounds, according to Smyth *et al.*[61] When they are given intravenously, they tend to increase the tendency of the blood to clot and if given rapidly cause clumping of the cells and death occurs from embolism.

Absorption and Excretion. Shaffer and Critchfield[70] in 1947 reported that polyethylene glycols having an average molecular weight of 4000 and 6000 were not absorbed from the rat intestine within 5 hours, whereas lower molecular weight materials (1000 and 1540) were absorbed to a very slight extent. When 1-g. doses of materials having average molecular weights of 6000 and 1000 were given intravenously to human subjects, 96 per cent of the 6000 molecular weight material and 85 per cent of the other were excreted in the urine in 12 hours. When these same two materials in 10-g. doses were given orally to 5 human subjects, none of the 6000 molecular weight material was found in the urine in the following 24 hours, whereas about 8 per cent of the other was found.

Shaffer, Critchfield, and Nair[71] in 1950 reported on studies with human subjects using polyethylene glycol having an average molecular weight of 400. They were able to recover 77 per cent in the urine in 12 hours following the administration of 1 g. intravenously, and to recover 40 to 50 per cent in the urine when a 5- to 10-g. dose of the material was given orally. These authors are of the opinion that ethylene glycol is not a metabolite of polyethylene glycol 400.

Effect on Absorption of Other Materials. Schutz[72] has observed that polyethylene glycol 400 markedly reduces the penetration of some other chemicals through the skin. He found that phenol, dimethylaniline, phenol red, barbitol, salicylic acid, and gamma-hexachlorocyclohexane were very poorly absorbed from solutions of the glycol.

5. Hygienic Standards of Permissible Exposure

No hygienic standard is believed necessary.

PROPYLENE GLYCOL, $C_3H_8O_2$ or $CH_3CHOHCH_2OH$ (1,2-Propanediol, Methyl ethylene glycol)

1. Source, Uses, and Industrial Exposure

Propylene glycol generally is synthesized commercially by starting with propylene, converting to the chlorohydrin, and hydrolyzing to propylene oxide, which is then hydrolyzed to propylene glycol. It can also be prepared by other methods.[73]

Propylene glycol is used in antifreeze formulations, heat exchangers, brake and hydraulic fluids; in the manufacture of resins, polypropylene glycols, propyl-

[70] C. B. Shaffer and F. H. Critchfield, *J. Am. Pharm. Assoc. Sci. Ed.,* **36,** 152 (1947).

[71] C. B. Shaffer, F. H. Critchfield, and J. H. Nair, III, *J. Am. Pharm. Assoc. Sci. Ed.,* **39,** 340 (1950).

[72] E. Schutz, *Arch. exptl. Pathol. Pharmakol.,* **232,** 237 (1957), *Ind. Hyg. Dig. Abstr.,* **22,** 720 (1958).

ene glycol ethers and esters; as a solvent in pharmaceuticals, foods, cosmetics, and inks; as a plasticizer for resins and paper; and as a humectant in textiles and tobacco. It is also used in the vapor form as an air sterilizer for hospitals and public buildings.

Industrial exposures are from direct contact, or from inhalation of vapors and of mists where the material is heated or violently agitated. Other exposure is by ingestion resulting from its use in foods and drugs.

2. Physical and Chemical Properties

Propylene glycol is a colorless, almost odorless, slightly viscous liquid with the following properties:

Molecular weight: **76.10**
Specific gravity: **1.038** (20/20°C.) ; **1.036** (25/25°C.)
Melting point: supercools
Boiling point: 187.2°C. (760 mm. Hg)
Vapor pressure: 0.13 mm. Hg (25°C.)
Refractive index: 1.431 (25°C.)
Per cent in "saturated" air: 0.018 (25°C.)
Solubility: miscible with water, alcohol, ether, and many organic solvents
Flash point: 207°F. (closed cup) ; 215 to 225°F. (open cup)
Flammability limits: 2.62 to 12.55 per cent
1 p.p.m. $\backsim$ 3.11 mg./cu. meter at 25°C., 760 mm. Hg and 1 mg./liter $\backsim$ 322 p.p.m. at 25°C., 760 mm. Hg

3. Determination in the Atmosphere

The determination of propylene glycol in air can be accomplished by several of the methods noted under Ethylene Glycol, although it would seem unnecessary for industrial hygiene purposes. It is suggested that Curme and Johnston[73] be consulted for additional methods.

A method for detection of propylene glycol in body fluids is described by Lehman and Newman.[74]

4. Physiological Response

Summary. The hazards to health in the industrial handling and use of propylene glycol would seem to be negligible. Its systemic toxicity is especially low and, since 1942, it has been considered a proper ingredient for pharmaceutical products.[75] The Food and Drug Administration does not object to its use in food products or in cosmetics.[76] The inhalation of atmospheres containing propylene

[73] G. O. Curme, Jr., and F. Johnston, eds., *Glycols, American Chemical Society Monograph, Series 114*. Reinhold, New York, 1952.

[74] A. J. Lehman and H. W. Newman, *J. Pharmacol. Exptl. Therap.,* **60,** 312 (1937).

[75] Council on Pharmacy and Chemistry of the American Medical Association, *New and Nonofficial Remedies.* Lippincott, Philadelphia, 1949.

[76] Food and Drug Administration, *Food, Drug, Cosmetic Law J.,* **13,** 856 (1958).

glycol vapor presents no hazard to health. Exposures created by operations producing hot vapors, or by high-speed mechanical action in which a fog of propylene glycol is produced, have not been studied. However, it is difficult to visualize how this condition could create a hazard, since the material is so extremely low in systemic toxicity.

Single-Dose Oral. The single dose oral toxicity of propylene glycol has been studied by a number of investigators.[74,77,78] The single dose oral LD_{50} values for rats, rabbits, and dogs have been found to be 32.5, 18.5, and 9.63 ml./kg., respectively. Laug et al.[79] report observing minimal kidney changes from large doses. From one fourth to one half of an oral dose given to rats, dogs, or human beings appears unchanged in the urine within 24 hours.[74,78,80,81]

Repeated-Oral Dose. Seidenfeld and Hanzlik[82] gave groups of rats drinking water containing 1.0, 2.0, 5.0, 10.0, 25.0, and 50.0 per cent propylene glycol over a period of 140 days. Animals receiving water containing 50.0 and 25.0 per cent propylene glycol died in 69 days while those receiving 1.0, 2.0, 5.0, and 10.0 per cent appeared normal throughout the observation period. The average daily intakes for the latter four groups were calculated to be about 1.6, 3.68, 7.7, and 13.2 g./kg./day of propylene glycol, respectively. Histopathological examination of the tissues from these animals revealed no renal or other pathological disturbances. Weatherby and Haag[77] confirmed the fact that rats will tolerate 10.0 per cent propylene glycol in the drinking water without physiological impairment.

Hanzlik and associates[78] found that rats could tolerate up to 30 ml./kg. daily of propylene glycol when fed in the diet over a 6-month period. Morris et al.[83] fed rats 2.45 and 4.9 per cent propylene glycol in the diet, allowing, respectively, average daily intakes of 0.9 to 1.77 ml./kg. over a 24-month period without significant effect on growth rate. Microscopic examination of the tissues revealed very slight liver damage, but no renal pathology.

Whitlock et al.[84] found that a diet containing 30 per cent of propylene glycol was not well tolerated by young rats and that producing females were unable to bring their young to weaning. Glycerin at 30 per cent in the diet was well tolerated. Diets containing 40, 50, or 60 per cent of propylene glycol were lethal after a few days.

[77] J. H. Weatherby and H. B. Haag, *J. Am. Pharm. Assoc.*, **27**, 466 (1938).

[78] P. J. Hanzlik, H. W. Newman, W. Van Winkle, A. J. Lehman, and N. K. Kennedy, *J. Pharmacol. Exptl. Therap.*, **67**, 101 (1939).

[79] E. P. Laug, H. O. Calvery, H. J. Morris, and G. Woodard, *J. Ind. Hyg. Toxicol.*, **21**, 173 (1939).

[80] W. Van Winkle, Jr., *J. Pharmacol. Exptl. Therap.*, **72**, 344 (1941).

[81] H. W. Newman and A. J. Lehman, *Proc. Soc. Exptl. Biol. Med.*, **35**, 601 (1936–37).

[82] M. A. Seidenfeld and P. J. Hanzlik, *J. Pharmacol. Exptl. Therap.*, **44**, 109 (1932).

[83] H. J. Morris, A. A. Nelson and H. O. Calvery, *J. Pharmacol. Exptl. Therap.*, **74**, 266 (1942).

[84] G. P. Whitlock, N. B. Guerrant and R. A. Dutcher, *Proc. Soc. Exptl. Biol. Med.*, **57**, 124 (1944).

Van Winkle and Newman[85] showed that propylene glycol, when given in concentrations of 5 or 10 per cent in the drinking water of dogs for from 5 to 9 months, caused no adverse effects. Criteria employed were liver function, kidney function, and histopathological examination of the visceral organs. Further, they found no alterations in the serum calcium levels of cats and dogs fed large doses of propylene glycol.

Skin Irritation and Absorption. Propylene glycol generally produces no significant irritant action upon the skin. From the results of extensive studies by Warshaw and Herrmann[86] on some 866 human subjects with various dermatological backgrounds, it appears that propylene glycol may cause primary skin irritation in some people, possibly due to dehydration, but the material does not appear to be a sensitizer. Because of the very low systemic toxicity of propylene glycol, it is very doubtful that a quantity sufficient to produce any systemic injury would be absorbed. Propylene glycol has been used widely in preparations for topical application and no evidence of systemic injury to humans has been reported.

Vapor Inhalation. Robertson et al.[87] exposed sizable groups of rats and monkeys for periods of 12 to 18 months to atmospheres saturated with propylene glycol vapor and produced no ill effects. Human beings also have been exposed to saturated and supersaturated atmospheres for prolonged periods in the air-sterilization program without adverse effect.

Eye Contact. Propylene glycol is not injurious to the eyes of rabbits[88] and has not caused any eye irritation in human beings, nor would such be expected.

Pharmacology. Propylene glycol is so low in pharmacological activity that very large doses can be given intravenously, providing it is given slowly. Rapid administration will cause death. The material appears to have a sedative-type of effect and is glycogenic,[78] entering into normal carbohydrate metabolism probably through the intermediate, lactic acid.

5. *Hygienic Standards of Permissible Exposure*

None necessary.

1,3-PROPANEDIOL, $C_3H_8O_2$ or $HOCH_2CH_2CH_2OH$ (Trimethylene glycol)

1. *Source, Uses, and Industrial Exposure*[89]

1,3-Propanediol is prepared as a by-product in the manufacture of glycerine through the saponification of fat.

[85] W. Van Winkle, Jr. and H. W. Newman, *Food Research,* **6,** 509 (1941).

[86] T. G. Warshaw and F. Herrmann, *J. Invest. Dermatol.,* **19,** 423 (1952).

[87] O. H. Robertson, C. G. Loosli, T. T. Puck, H. Wise, H. M. Lemon, and W. Lester, Jr., *J. Pharmacol. Exptl. Therap.,* **91,** 52 (1947).

[88] C. P. Carpenter and H. F. Smyth, Jr., *Am. J. Ophthalmol.,* **29,** 1363 (1946).

[89] G. O. Curme, Jr. and F. Johnston, eds., *Glycols, American Chemical Society Monograph, Series 114.* Reinhold, New York, 1952.

It is used to lower the freezing point of water and as a chemical intermediate. Industrial exposure is limited and of little concern.

2. *Physical and Chemical Properties*[89]

1,3-Propanediol, an isomer of propylene glycol, is of little commercial significance. It is a viscous, colorless, odorless, hygroscopic liquid with a brackish irritating taste. It possesses the following properties:

Molecular weight: **76.10**
Specific gravity: 1.0554 (20/20°C.)
Boiling point: 210 to 211°C. (760 mm. Hg)
Solubility: miscible with water, alcohol, and ether

3. *Determination in the Atmosphere*

Methods applicable to the other glycols having hydroxyl groups that are not vicinal should be useful.

4. *Physiological Response*

Summary. 1,3-Propanediol has not been subjected to extensive toxicological studies. Van Winkle[90] found the material to be about twice as toxic as the 1,2-isomer, to be nonglycogenic, and to cause marked depression in near-fatal doses.

Single-Dose Oral. When given in single oral doses, the LD_{50} for rats might be estimated to be 14 or 15 ml./kg. with a lethal range of from 9 to 18 ml./kg. For cats the LD_{50} was not determined but probably was less than 1 ml./kg.; 3 ml./kg. was always fatal to cats. Cats, for reasons unknown, appear to be particularly sensitive to the material when given in single oral doses.

Repeated-Dose Oral. When the material was fed to rats as a part of their diet for 15 weeks, a concentration of 5 per cent was without grossly apparent toxic effects but no autopsies were performed. Twelve per cent in the diet was not well tolerated, as evidenced by poor growth. A daily dose of 5 ml./kg. given by intubation also caused poor growth; a daily dose of 10 ml./kg. was fatal to all rats within 5 weeks.

Parenteral Toxicity. When given intravenously as a 50 per cent aqueous solution, the LD_{50} was found to be 4 to 5 ml./kg. for rabbits and greater than 3 ml./kg. for cats. When given intramuscularly the LD_{50} for rats was 6 to 7 ml./kg. and for cats more than 3 ml./kg.

5. *Hygienic Standards of Permissible Exposure*

None seems necessary because of the low vapor pressure at room temperatures and the relatively low toxicity.

[90] W. Van Winkle, Jr., *J. Pharmacol. Exptl. Therap.*, **72**, 227 (1941).

DIPROPYLENE GLYCOL, $C_6H_{14}O_3$ or $HO(C_3H_6O)_2H$

1. Source, Uses, and Industrial Exposure

Dipropylene glycol is prepared commercially as a by-product of propylene glycol production. There are three linear isomers possible but these have not been separated and studied and the exact composition of the commercial product is not known. It is also possible to prepare cyclic isomers such as 2,6-dimethyl-1,4-dioxane and 2,5-dimethyl-1,4-dioxane but these are not likely to form under conditions employed commercially.

Dipropylene glycol is used for many of the same purposes as the other glycols but mostly in particular applications where its solubility characteristics (greater hydrocarbon solubility) and lower volatility are useful. It is not used in drugs, pharmaceuticals, or food applications because its toxicological characteristics have not been clearly defined.

Industrial exposure is most likely to be from direct contact and possible inhalation of mist from heated or violently agitated material.

2. Physical and Chemical Properties

Dipropylene glycol is a colorless, odorless, slightly viscous liquid with the following properties:

Molecular weight: 134.18
Specific gravity: 1.0252 (20/20°C.); 1.026 (25/25°C.)
Melting point: supercools
Boiling point: 231.9°C. (760 mm. Hg)
Vapor pressure: <0.01 (20°C.)
Refractive index: 1.439 (25°C.)
Per cent in "saturated" air: <0.0013 (20°C.)
Solubility: miscible with water, methanol, and ether
Flash point: 250 to 280°F. (open cup)
1 p.p.m. ≎ 5.49 mg./cu. meter at 25°C., 760 mm. Hg and 1 mg./liter ≎ 182 p.p.m. at 25°C., 760 mm. Hg

3. Determination in the Atmosphere

Dipropylene glycol undergoes the same general reactions as other polyols. The chemical methods for determination will be the same nonspecific methods described for other glycols. Fair specificity can be obtained through the use of spectrographic methods.

4. Physiological Response

Summary. Although dipropylene glycol is more active physiologically than propylene glycol, it is still very low in toxicity. The industrial handling and use of dipropylene glycol should present no significant problems from ingestion, skin contact, or vapor inhalation. The information available, however, is not considered adequate to allow an evaluation relative to the suitability of this material for use in foods, drugs, or cosmetics.

Single-Dose Oral. Dipropylene glycol is low in acute oral toxicity. The single dose LD_{50} for rats has been reported to be 14.8 g./kg.[91]

Repeated-Dose Oral. Rats were not affected by 5 per cent dipropylene glycol in their drinking water for 77 days.[92] In those animals administered a level of 10 per cent, some died with hydropic degeneration of kidney tubular epithelium and liver parenchyma. These effects were similar to those of diethylene glycol but less severe and less uniformly produced.

Skin Contact. When dipropylene glycol was applied repeatedly for prolonged periods (10 applications in 12 days) to the skin of rabbits it had a negligible irritating action and there was no indication that toxic quantities were absorbed through the intact skin.[93]

Inhalation. Experimental data are not available on the vapor toxicity of dipropylene glycol. However, it is not likely to produce injury because of its low vapor pressure and low systemic toxicity.

Human Experience. No untoward effects have been reported from the use of dipropylene glycol nor would any be expected.

5. Hygienic Standards of Permissible Exposure

Because of the low vapor pressure and the low toxicity of dipropylene glycol, a hygienic standard seems unnecessary.

TRIPROPYLENE GLYCOL, $C_9H_{20}O_4$ or $HO(C_3H_6O)_3H$

1. Source, Uses, and Industrial Exposure

Tripropylene glycol is made commercially as a by-product of propylene glycol production.

It is used much as the other glycols: as an intermediate in the production of ethers, esters, and resins; as a nonvolatile solvent; and as a humectant and a plasticizer.

Industrial exposure is limited almost entirely to direct contact with the liquid. Vapors or mists may be encountered under certain conditions.

2. Physical and Chemical Properties

Tripropylene glycol is a colorless, odorless, slightly viscous liquid with the following properties:

Molecular weight: **192.3**

Specific gravity: 1.019 (25/25°C.)

[91] P. J. Hanzlik, H. W. Newman, W. Van Winkle, Jr., A. J. Lehman, and N. K. Kennedy, *J. Pharmacol. Exptl. Therap.*, **67**, 101 (1939).

[92] H. D. Keston, N. G. Mulinos, and L. Pomerantz, *Arch. Pathol.*, **27**, 447 (1939).

[93] Biochemical Research Laboratory, The Dow Chemical Co., unpublished data.

Melting point: supercools
Boiling point: 268.0°C. (760 mm. Hg)
Refractive index: 1.442 (25°C.)
Vapor pressure: very low—<0.01 mm. Hg (25°C.)
Solubility: miscible with water, alcohol, and ether
Flash point: 285°F.
1 p.p.m. 7.86 mg./cu. meter at 25°C., 760 mm. Hg

3. Determination in the Atmosphere

Tripropylene glycol undergoes the typical reactions of polyols not having vicinal hydroxyl groups. Analysis generally depends upon reaction with the free hydroxyl groups.

4. Physiological Response

There is no toxicological literature readily available on tripropylene glycol. The results of unpublished range finding toxicological studies[94] show the material to be low in single-dose oral toxicity, the LD_{50} for rats being between 3 and 10 g./kg.; it is not irritating to the eyes or skin of rabbits; and is not absorbed through the skin in acutely toxic amounts even from prolonged and repeated contact.

The effect of inhaling vapors or mists has not been investigated, but it is doubtful that a hazardous condition would occur under conditions reasonably to be expected.

5. Hygienic Standards of Permissible Exposure

No data; none believed necessary.

POLYPROPYLENE GLYCOLS

1. Source, Uses, and Industrial Exposure

The polypropylene glycols are prepared commercially by reacting propylene glycol or water with propylene oxide.

They are used as lubricants, solvents, plasticizers, softening agents, antifoaming agents, mold release agents, and as intermediates in the production of resins, surface active agents, and a large series of ethers and esters. They are widely used in hydraulic fluid compositions.

Industrial exposure is most likely to be direct contact with the skin and eyes. Ingestion should not be a problem except from accident. The very low volatility of these materials makes inhalation improbable except perhaps where mists are formed from violent agitation or high temperatures.

[94] Biochemical Research Laboratory, The Dow Chemical Co., unpublished data.

2. *Physical and Chemical Properties*

The polypropylene glycols may be represented by the formula $HO(C_3H_6O)_nH$. They are clear, lightly colored, slightly oily, viscous liquids having very low vapor pressures. All of these materials are quite stable chemically and do not present hazards of flammability except at elevated temperatures. Their physical and chemical properties are given in Table 4.

TABLE 4

Physical and Chemical Properties of Polypropylene Glycols

Material designation[a]	Molecular wt. range	Specific gravity	Pour point, °C.	Refractive index, 25°C.	Flash point, °F.[b]	Fire point, °F.[b]
400[c]		1.007[d]	−45	1.445	390	405
425[e]	400–450	1.0092[f]			420	
750[c]		1.004[d]	−44	1.447	495	525
1025[e]	975–1075	1.0065[f]			450	
1200[c]		1.003[d]	−40	1.448	460	505
2000[c]		1.002[d]	−35	1.450	445	510
2025[e]	1950–2100	1.0061[f]			450	
3000[c]		1.001[d]	−29	1.449	440	505
4000[c]		1.001[d]	−29	1.449	440	510

[a] Average molecular weight.

[b] Open cup.

[c] *Polypropylene Glycols* (No. 125-129-57). The Dow Chemical Co., Midland, Mich., 1957.

[d] 25/25°C.

[e] *Physical Properties of Synthetic Organic Chemicals* (No. F-6136K). Union Carbide Chemicals Co., New York, N.Y. 1957.

[f] 20/20°C.

3. *Determination in the Atmosphere*

Methods that depend upon the reactivity of the free hydroxyl groups can probably be adapted. Infrared spectrophotometry also may be useful.

4. *Physiological Response*

Summary. The low molecular weight polypropylene glycols (200 to 1200) have an appreciable single dose oral toxicity, are mildly irritating to the eyes, are not irritating to the skin, and, although are absorbed through the skin to some extent, skin penetration would not seem to present a serious industrial hazard. The inhalation of mists or vapors from heated material, particularly low molecular weight material, could be hazardous. These materials are not like the low molecular weight polyethylene glycols in their physiological activity; they are rapidly absorbed from the gastrointestinal tract, are potent central nervous system stimulants, and readily cause cardiac arrythmias. The higher molecular weight materials with average molecular weights of 2000 or more are very low in toxicity by all routes and do not have the stimulant effect upon the central nervous system typical of the lower molecular weight materials.

Single-Dose Oral. The single dose oral toxicities of the polypropylene glycols are given in Table 5. The low molecular weight materials (400 to 1200) are

TABLE 5

Single-Dose Oral Toxicities of Various Polypropylene Glycols

| Material (average molecular weight) | Approximate LD_{50} values (g./kg.) | | |
| | Rats | | Guinea pigs, both sexes |
	Male	Female	
400[a]	1.2	0.7	2.3
425[b]	2.91		
750[a]	0.5	0.3	1.7
1025[b]	2.15		
1200[a]	0.6		1.5
2000[a]	10	5	17
2025[b]	9.76		
3000[c]	>40		
4000[c]	>40		

[a] *Polypropylene Glycols* (No. 125-129-57). The Dow Chemical Co., Midland, Mich., 1957.

[b] C. B. Shaffer, C. P. Carpenter, F. H. Critchfield, J. H. Nair, III, and F. R. Frank, *Arch. Ind. Hyg. Occupational Med.*, **3**, 448 (1951).

[c] Biochemical Research Laboratory, The Dow Chemical Co., unpublished data.

rapidly absorbed since excitement and convulsions appear within minutes after administration. With the higher molecular weight material no excitement or convulsions were observed. Autopsy of animals treated with the largest doses 1 to 8 days after exposure revealed nothing remarkable.

Parenteral Toxicity. All of these materials have been given to animals parenterally[95,96] and have been found to have essentially the same relative toxicity, one to another, as by the oral route.

Repeated-Dose Oral.[97] Small groups of male rats were maintained for 100 days on diets containing 0.1 and 1.0 per cent of polypropylene glycol 750 and 0.1, 0.3, 1.0, and 3.0 per cent of polypropylene glycol 2000.

Those animals that received the diet containing 0.1 per cent of P750 were unaffected, as judged by studies of mortality, growth, organ weights, and gross and microscopic examination of the principal internal organs. The animals that received the diet containing 1.0 per cent of P750, when judged by the same criteria, exhibited only a slight increase in the weight of the livers and kidneys without

[95] C. B. Shaffer, C. P. Carpenter, F. H. Critchfield, J. H. Nair, III, and F. R. Franke, *Arch. Ind. Hyg. Occupational Med.*, **3**, 448 (1951).

[96] F. E. Shideman and L. Procita, *J. Pharmacol. Exptl. Therap.*, **103**, 293 (1951).

[97] The Dow Chemical Company, Polypropylene Glycols Form No. 125-129-57 (1957).

histological changes. Hematological studies on this latter group of animals failed to reveal any abnormalities. One per cent of P750 in the diet was well accepted by the rats, and it is especially worthy of note that there was no evidence of any of the pharmacological signs (excitement, tremors, convulsions) seen in the acutely poisoned animals. It is postulated that the material is readily metabolized or eliminated when absorbed in small doses; this probably accounts for its lack of apparent physiological effect.

The animals that received 0.1, 0.3, and 1.0 per cent of P2000 suffered no ill effects as judged by the criteria listed above. Hematological studies were conducted only at the 1.0 per cent level with all values falling in the normal range. Although the growth of those animals maintained on the diet containing 3.0 per cent of P2000 was slightly below normal during most of the test period there were no other changes attributable to the experimental diets.

Skin and Eye Irritation. Tests conducted on rabbits have indicated that these materials are not significantly irritating to the skin even when exposures are prolonged and repeated.[95,97]

Direct contact with the eyes may cause slight transient pain and conjunctival irritation but no corneal damage. The response is similar to that caused by a mild soap.[97]

Skin Irritation and Skin Sensitization. Polypropylene glycol 2000 is neither a skin irritant nor a skin sensitizer. This conclusion is based upon tests with the undiluted material conducted by both continuous and repeated application techniques on a total of 300 human subjects.[97]

Skin Absorption. Acute skin absorption tests conducted by means of a "sleeve" technique similar to that developed by Draize et al.[98] have indicated that the materials are all poorly absorbed through the skin. When single doses of 30.0 ml./kg. were applied for 24 hours, 4 of 5 animals treated with either polypropylene glycols 400, 750, or 1200 survived and all of 6 animals so treated with polypropylene glycol 2000 survived.

Chronic skin absorption studies have been carried out only on polypropylene glycol 2000. In these studies, the material was bandaged onto the shaved abdomens of groups of 5 rabbits each, 24 hours a day, 5 times a week, for 3 months. At a dosage level of 1.0 ml./kg. there were no adverse effects as judged by studies of growth, hematology, weights of organs, and gross and microscopic examination of the lungs, heart, liver, kidney, adrenal, testes, stomach, intestine, and skin taken from the site of the prolonged and repeated exposure. Judged by the same criteria, dosages levels of 5.0 and 10.0 ml./kg. caused slight depression of growth. At the 10.0 ml./kg. level, mortality was increased but since the cause was respiratory infection, the significance is questionable.

Pharmacology. Extensive investigation of the pharmacological activity of these polypropylene glycols has indicated that they are all central nervous system

[98] J. H. Draize, G. Woodard and H. O. Calvery, *J. Pharmacol. Exptl. Therap.*, **82**, 377 (1944).

stimulants. P400, P750, and P1200 are quite potent in this respect, whereas P2000 has but little such activity.[95,96]

Human Experience. No cases of toxicity have resulted from the manufacturing, handling, and use of the polypropylene glycols.

5. *Hygienic Standards of Permissible Exposure*

None would seem necessary.

BUTYLENE GLYCOLS (Butanediols)

1. *Source, Uses, and Industrial Exposure*

1,2-Butanediol is a relatively new product and is produced commercially by the hydration of the corresponding 1,2-butylene oxide in a manner similar to that described for other simple glycols.

According to Curme and Johnston,[99] 1,3-butanediol is prepared commercially by the catalytic reduction of acetaldol, but may be produced by other routes as well. 1,4-Butanediol is produced in Germany by hydrogenation of 2-butyne-1,4-diol but other methods can also be used. 2,3-Butanediol is produced by fermentation, the distribution of optical isomers being dependent upon the species of bacteria used.

The butanediols are not used extensively commercially but there is considerable interest in them as intermediates in the polyester resins. The 1,3-isomer, because of its low toxicity, has been proposed for cosmetic and pharmaceutical applications.

Exposure will be from direct contact in handling and use.

2. *Physical and Chemical Properties*

The commercial preparations of 1,2-, 1,3-, 1,4-, and 2,3-butanediols are clear, viscous liquids miscible with water and alcohol. They all have a molecular weight of 90.1; thus, 1 p.p.m. $\backsim$ 3.68 mg./cu. meter and 1 mg./liter $\backsim$ 272 p.p.m. at 25°C., 760 mm. Hg. The most important physical properties are given in Table 6 but additional properties are given for all but the 1,2-isomer by Curme and Johnston.[99]

3. *Physiological Response*

Summary. 1,2- and 1,3-Butanediol appear to be very low in oral toxicity when administered in both single and repeated doses. In single oral doses, 1,4-butanediol is much more toxic than the 1,3-isomer (about 10 times) and the 2,3-isomer is intermediate. Data on the 1,3- and 1,4-isomers indicate that they are not significantly irritating to the eyes, skin, or mucous membranes, nor are they likely to be absorbed through the skin in hazardous amounts. The undiluted 1,2-

[99] G. O. Curme, Jr., and F. Johnston, eds., *Glycols, American Chemical Society Monograph, Series 114.* Reinhold, New York, 1952.

TABLE 6

Physical and Chemical Properties of Butylene Glycols (Butanediols)

Butane-diol isomer	Boiling point, °C. (760 mm. Hg)	Flash point, °F.	Freezing point, °C.	Refractive index, 20°C.	Vapor pressure, mm. Hg (20°C.)	Specific gravity (20/20°C.)
1,2-[a]	193.5–195			1.4369 (25°C.)		1.0017
1,3-[b]	207.5	250[c]	< −50	1.4401	0.06	1.0059
1,4-[b]	230		16 (m. p.)			1.020
2,3-[b]	182	185[d]	19	1.4377	0.17	1.0093

[a] Biochemical Research Laboratory, The Dow Chemical Co., unpublished data.
[b] G. O. Curme, Jr., and F. Johnston, eds., *Glycols, American Chemical Society Monograph, Series 114.* Reinhold, New York, 1952.
[c] Cleveland open cup.
[d] Tag open cup.

isomer is not significantly irritating to the skin but appears to be irritating to the eyes. Dilution to 10 per cent with water eliminates the effect upon the eyes.

It would not seem that the butanediols would present any appreciable handling hazards other than possible eye irritation from contact with the 1,2-isomer.

1,2-Butanediol. Single dose oral toxicity studies have shown this material to be very low in acute oral toxicity for rats, the LD_{50} being about 16 g./kg. In large doses, the material causes narcosis, irritation of the gastrointestinal tract, profound vasodilatation of the visceral vessels, as well as a marked congestion of the kidneys. No hemorrhage was apparent. Deaths that occurred within a few hours are believed to be due to narcosis and those that were delayed are believed to be due to kidney injury.[100]

Schlüssel,[101] along with his work on the 1,3-isomer, found that young female rats could tolerate a basic diet in which up to 30 per cent of the calories were replaced by 1,2-butanediol, but that 40 per cent replacement caused death in 11 to 29 days.

Intravenous injection of up to 1 g./kg. in an anesthetized dog failed to cause any noticeable response in blood pressure, heart rate, or respiration.[100]

When applied to the eyes of rabbits, the undiluted liquid was painful, irritating, and injurious, whereas a 10 per cent aqueous solution caused no response.[100] The material was not irritating to the skin of rabbits even when exposures were prolonged and repeated and it was not absorbed through the skin in toxic amounts.[100] Rats were unaffected by a single 7-hour exposure to an atmosphere essentially saturated at 100°C. and then cooled to room temperature.

[100] Biochemical Research Laboratory, The Dow Chemical Co., unpublished data.
[101] H. Schlüssel, *Arch. exptl. Pathol. Pharmakol.,* **221,** 67 (1954), *Chem. Abstr.,* **48,** 5315 (1954).

1,3-Butanediol. This material is very low in oral toxicity. When given in single oral doses, Loeser,[102] Fischer *et al.*,[103] and Bornmann[104] all report the LD_{50} to be 23.31 ml./kg. for mice and 29.42 ml./kg. for rats. Smyth *et al.*[105] report the oral LD_{50} for rats to be 22.8 g./kg. When given subcutaneously, the LD_{50} is reported to be 16.51 ml./kg. for mice and 20.06 for rats.[102,104]

1,3-Butanediol also is very low in toxicity when given in repeated oral doses. Loeser[102] reports the feeding of 2 to 3 median lethal doses to rats during a 6-week period without organic damage or growth depression. Fischer *et al.*[103] report that 20 per cent in the drinking water of rats for 44 days was without any effect when judged by studies of growth, hematology, liver, kidney, and bladder. However, Bornmann[104] states that 20 per cent in the drinking water caused slight depression of growth, but no effect at 10 per cent or less. Kopf and co-workers[106] fed rats orally 0.5 or 1.0 ml. of 1,3-butanediol twice a week for 45 to 185 days without any effect and dogs 2.0 ml. of a 50 per cent aqueous solution twice a week for 5 to 6 months without effect. Schlüssel[101] reports that young rats tolerated a basic diet in which 1,3-butanediol accounted for up to 40 per cent of the total calories. Smyth *et al.*[105] fed groups of 10 rats for 90 days on diets containing sufficient 1,3-butanediol to cause a daily ingestion of up to 5.6 g./kg. without any adverse effect as judged by growth, mortality, food consumption, liver and kidney weight changes, and histopathological examination of the liver, kidney, spleen, and testes. Unfortunately, larger doses were not fed.

Carpenter and Smyth[107] found 1,3-butanediol not to be irritating to the rabbit eye. Smyth *et al.*[105] report the material not irritating to the rabbit skin. This is borne out by the work of Loeser,[102] Husing *et al.*,[108] and Fischer et al.,[103] who all conclude that the 1,3-isomer is not irritating to human skin or mucous membranes.

Bornmann[104] states that 1,3-butanediol does not cause hemolysis when given parenterally and that acute intoxication results in deep narcosis.

Smyth *et al.*[105] exposed rats to saturated vapors of 1,3-butanediol for 8 hours without any adverse effects.

1,4-Butanediol. This isomer is about 10 times as toxic when administered to animals as is the 1,3-isomer. The oral LD_{50} has been found to be 2.14 ml./kg.

[102] A. Loeser, *Pharmazie,* **4,** 263 (1949), *Chem. Abstr.,* **43,** 8558 (1949).

[102a] P. K. Gessner, D. V. Parks, and R. T. Williams, *Biochem. J.,* **74,** 1 (1960).

[103] L. Fischer, R. Kopf, A. Loeser, and G. Meyer, *Z. ges exptl. Med.,* **115,** 22 (1949), *Chem. Abstr.,* **44,** 9070 (1950).

[104] G. Bornmann, *Arzneimittel Forsch.,* **4,** 643, 710 (1954), **5,** 38 (1955), *Chem. Abstr.,* **49,** 7131 (1955).

[105] H. F. Smyth, Jr., C. P. Carpenter, and C. S. Weil, *Arch. Ind. Hyg. Occupational Med.,* **4,** 119 (1951).

[106] R. Kopf, A. Loeser, G. Meyer, and W. Franke, *Arch. exptl. Pathol. Pharmakol.,* **210,** 346 (1950), *Chem. Abstr.,* **45,** 5308 (1951).

[107] C. P. Carpenter and H. F. Smyth, Jr., *Am. J. Ophthalmol.,* **29,** 1363 (1946).

[108] E. Husing, R. Kopf, and A. Loeser, *Fette u. Seifen,* **52,** 45 (1950), *Chem. Abstr.,* **44,** 7999 (1950).

for mice[103] and 1.78 g./kg. for rats.[100] Hinrichs *et al.*[109] found the material to cause deep narcosis, constriction of the pupils, total loss of reflexes and kidney injury; they attributed death to paralysis of the vital centers. This has been confirmed.[100]

Application to the eyes of rabbits showed the material to be but slightly irritating; it caused a very slight conjunctival irritation but no corneal injury. Repeated application to the rabbits' skin, both intact and abraded, resulted in no appreciable irritation and no evidence of absorption of acutely toxic amounts.[100] Judging from these observations, there would seem to be no appreciable hazard associated with ordinary industrial handling. Schneider,[110] however, reports finding the material highly toxic on the skin. Perhaps this apparent discrepancy can be attributed to the quality of the test material, a factor not to be ignored and one that has been called to attention by others.[102]

Gessner *et al.*[102a] report that when butane-1,4-diol is fed to rabbits, most of it appears to be destroyed, but small amounts of the corresponding dicarboxylic acid are found in the urine. The intermediates in the *in vivo* destruction have not been identified.

2,3-Butanediol. Toxicological data on this isomer seem to be scanty. Fischer *et al.*[103] report the oral LD_{50} for mice to be 9.0 ml./kg.

4. *Hygienic Standards of Permissible Exposure*

None seems necessary for the butanediols because of their low volatility and low toxicity.

POLYBUTYLENE GLYCOLS[111]

1. *Source, Uses, and Industrial Exposure*

The polybutylene glycols considered herein are prepared commercially by reacting 1,2-butylene glycol or water with 1,2-butylene oxide. Their number designations indicate their average molecular weights. Other polybutylene glycols can, of course, be made using the other butylene oxides.

Uses for these materials are being developed. Industrial exposure is most likely to be from direct contact with the skin and eyes.

2. *Physical and Chemical Properties*

These polybutylene glycols may be represented by the formula $HO-(C_4H_8O)nH$. They are clear, viscous, oily, slightly yellow liquids with sweetish tastes. They are less than 1.0 per cent soluble in water and greater than 25 per cent

[109] A. Hinrichs, R. Kopf, and A. Loeser, *Pharmazie*, **3**, 110 (1948), *Chem. Abstr.*, **42**, 5567 (1948).

[110] W. Schneider, *Pharm. Ind.*, **12**, 226 (1950), *Chem. Abstr.*, **45**, 3998 (1951).

[111] Biochemical Research Laboratory, The Dow Chemical Co., unpublished data.

soluble in methanol and ether. They are quite stable chemically and do not present hazards of flammability except at elevated temperatures.

3. *Physiological Response*

Summary. Limited toxicological information is available on two polybutylene glycols known as Polyglycol B-1000 and Polyglycol B-2000. Both appear to be low in single dose oral toxicity, their oral LD_{50} values for rats being greater than 4.0 g./kg. In such oral doses, however, they do produce marked injury to the kidneys. They are slightly irritating but not damaging to the eyes of rabbits. Prolonged and repeated contact with the skin of rabbits failed to cause any significant topical effect and there was no evidence of absorption of toxic amounts through the skin. It would appear that these materials do not present any appreciable hazards in industrial handling and use. However, until more data become available, care should be taken to avoid ingestion.

MIXED POLYGLYCOLS[112]

More or less recently a number of mixed polyglycols have become available commercially. Those for which there are significant toxicological data available are included here.

1. *Source, Uses, and Industrial Exposure*

The Polyglycol 11 series is prepared by reacting glycerol with different amounts of propylene oxide. The Polyglycol 15 series is prepared by reacting glycerol with different mixtures of propylene oxide and ethylene oxide.

The commercial uses for these materials are typical of the glycols in general: hydraulic fluids, plasticizers, demulsifiers, chemical intermediates, mold release agents, and in the manufacture of certain resins.

Industrial exposure is expected to be from direct contact.

2. *Physical and Chemical Properties*

All of these polyglycols are clear viscous liquids with the properties shown in Table 7.

3. *Physiological Response*

Summary. None of the mixed polyglycols in the 11 or 15 series described above presents any handling hazards of significance. The toxicological information is summarized in Table 8.

[112] Biochemical Research Laboratory, The Dow Chemical Co., unpublished data.

TABLE 7

Physical and Chemical Properties of Polyglycols of the 11 and 15 Series

Commercial designation, Polyglycol	Average molecular weight	Specific gravity (25/25°C.)	Refractive index, 25°C.	Flash point, °F.	Solubility, g. per 100 g. (25°C.)	
					Water	Methanol or ether
11-100	1030	1.026	1.452	435	<0.1	100
11-200	2700	1.018	1.452	435	<0.1	100
11-300	4000	1.017	1.450	440	<0.1	100
11-400	4900	1.017	1.450	445	<0.1	100
15-100	1100	1.070	1.460	510	Misc.	Misc.
15-200	2600	1.063	1.460	470	Misc.	Misc.
15-500	5000	1.051	1.459	475	Misc.	Misc.
15-1000	9000	1.053	1.458	480	Misc.	Misc.

4. Hygienic Standards of Permissible Exposure

The low volatility and low toxicity of these materials would seem to make a hygienic standard unnecessary.

TABLE 8

Summary of Toxicological Information on Certain Mixed Polyglycols

Polyglycol	Rat LD$_{50}$, g./kg.	Eye irritation (rabbits)[b]	Effect on skin		Percutaneous absorption (rabbits)[e]
			Irritation (rabbits)[c]	Sensitization (humans)[d]	
11-100	2.0	None	Slight		None
11-200	>4.0[a]	Trace	None		None
11-300	>4.0[a]	Trace	Slight	None	None
11-400	>4.0[a]	None	None	None	None
15-100	31.6	Trace	Very slight		None
15-200	15 to 20	Trace	Very slight	None[f]	None[g]
15-500	16	Trace	None		None
15-1000	>4.0[a]	None	None	None	None

[a] No deaths at this, the largest dose fed.

[b] One to two drops directly in eye; "trace" = conjunctival irritation but no corneal injury.

[c] Contact 24 hours a day for 12 days.

[d] Repeated insult test on 50 human beings.

[e] None apparent from skin irritation test "c".

[f] Repeated insult test on 50 human beings plus "Swartz" test on 200 human beings.

[g] LD$_{50}$ by "Draize" test = >30 g./kg.

2-METHYL-2,4-PENTANEDIOL, $C_6H_{14}O_2$ or $(CH_3)_2COHCH_2CHOHCH_3$ (Hexylene glycol)

1. Source, Uses, and Industrial Exposure

2-Methyl-2,4-pentanediol is prepared commercially by the catalytic hydrogenation of diacetone alcohol (4-hydroxy-4-methyl-2-pentanone).[113]

It is used as a chemical intermediate, a selective solvent in petroleum refining, a component of hydraulic fluids, a solvent for inks, and as an additive for cement.[113]

Industrial exposure is likely to be from direct contact or from inhalation, particularly if the material is handled hot.

2. Physical and Chemical Properties[113,114]

2-Methyl-2,4-pentanediol is a mild-odored liquid with the following properties:

Molecular weight: 118.17
Specific gravity: 0.9234 (20/20°C.); 0.9216 (20/4°C.)
Boiling point: 198°C. (760 mm. Hg)
Freezing point: sets to glass below −50°C.
Vapor pressure: 0.05 mm. Hg (20°C.)
Refractive index: 1.4263 (20°C.)
Flash point: 210 to 215°F. (COC)
Per cent in "saturated" air: 0.0066 (20°C.)
Solubility: miscible with water and alcohol; soluble in a variety of organic solvents

3. Determination in the Atmosphere

Determination can be accomplished by the usual nonspecific methods applicable to polyols. Infrared spectrophotometry and vapor chromatography may be applicable if specificity is necessary.

4. Physiological Response

Summary. 2-Methyl-2,4-pentanediol is low in single dose oral toxicity, appreciably injurious to the eyes, slightly irritating to the skin, not readily absorbed through the skin, and sufficiently low in vapor pressure at ordinary temperatures so as not to present appreciable hazard from inhalation. Atmospheres essentially saturated at room temperature are detectable by odor and may be slightly irri-

[113] G. O. Curme, Jr. and F. Johnston, eds., *Glycols, American Chemical Society Monograph, Series 114.* Reinhold, New York, 1952.

[114] Shell Chemical Corp., Ind. Hyg. Bull., *Hexylene Glycol* SC:57-101 and SC:57-102, 1958.

tating to the eyes. Atmospheric contamination resulting from handling at elevated temperature causes marked irritation of the eyes and, hence, warning of excessive concentrations. Pharmacologically, the material is a hypnotic. Studies on human subjects show that the material is slowly excreted, largely as the glucuronic acid conjugate.

Single-Dose Oral.[114,115] When given orally, the LD_{50} for mice is 3.8 ml./kg. and for rats approximately 4.79 g./kg. Hypnosis occurred in mice following single doses of 2.0 ml./kg.; with higher doses it was profound. The material caused irritation of the lungs and large intestine, but no gross effects were apparent in the kidney, brain, or heart.

Eye Irritation.[114,115] When introduced into the eyes of rabbits, the undiluted material caused appreciable irritation and corneal injury that was slow to heal.

Skin Irritation and Absorption. A single 24-hour application of 1.84 g./kg. to rabbits caused transitory mild edema and erythema but no deaths.[114] The range finding LD_{50} by cutaneous application to rabbits was found to be 13.3 ml./kg.[114,115]

Inhalation. Rats exposed for 8 hours to air saturated at room temperature all survived.[114,115]

Metabolism. Deichmann and Dierker[116] found that the oral administration of hexylene glycol to rats and rabbits resulted in a substantial increase in the amount of hexuronates in the plasma and in the urine. Jacobsen,[117] in studies on 5 human subjects, found both free and conjugated hexylene glycol in the urine after single or repeated oral doses. When the daily dose was 600 mg. or less, none was detected in the urine. With daily doses up to 5 g./day, substantial amounts of the free hexylene glycol and the conjugated form were found in the urine. Excretion was slow, persisting for up to 10 days after cessation of dosing.

5. *Hygienic Standards of Permissible Exposure*[114]

In the absence of adequate data upon which to establish a hygienic standard, it is suggested that atmospheric concentrations be maintained below those that cause discomfort in the unacclimated individual, probably about 75 p.p.m.

6. *Odor and Warning Properties*[114]

Most human beings exposed 15 minutes to 50 p.p.m. in the air were able to detect the odor and a few noted eye irritation. At a concentration of 100 p.p.m., the odor was plain and some noted nasal irritation and respiratory discomfort; at 1000 p.p.m., there was irritation of the eyes, nose, and throat, and respiratory discomfort.

[115] H. F. Smyth, Jr. and C. P. Carpenter, *J. Ind. Hyg. Toxicol.,* **30,** 63 (1948).

[116] W. B. Deichmann and M. Dierker, *J. Biol. Chem.,* **163,** 753 (1946).

[117] E. Jacobsen, *Acta Pharmacol. Toxicol.,* **14,** 207 (1958).

2-ETHYL-1,3-HEXANEDIOL, $C_8H_{18}O_2$ or $CH_3CH_2CH_2CHOHCH(C_2H_5)CH_2OH$

1. Source, Uses, and Industrial Exposure[118]

2-Ethyl-1,3-hexanediol is produced commercially by the hydrogenation of butyraldol (2-ethyl-3-hydroxy caproaldehyde).

It is used largely as an insect repellent, but it is also used as a solvent for resins and inks, a plasticizer, and a chemical intermediate in the production of polyurethan resins.

Industrial exposure is largely by direct contact. Extensive experience with human beings has been acquired through its extensive use as an insect repellent.

2. Physical and Chemical Properties[118]

2-Ethyl-1,3-hexanediol is an oily, colorless, slightly viscous liquid with the following properties:

Molecular weight: 146.22
Specific gravity: 0.9422 (20/20°C.)
Boiling point: 244.2°C. (760 mm. Hg)
Freezing point: —40°C., sets to glass
Refractive index: 1.4511 (20°C.)
Vapor pressure: <0.01 mm. Hg (20°C.)
Solubility: poor in water (4.2% by wt. at 20°C.); soluble in alcohol and ether
Flash point: 265°F. (COC)

3. Determination in the Atmosphere

Methods applicable to other polyols having terminal hydroxyl groups should be useful.

4. Physiological Response

Summary. 2-Ethyl-1,3-hexanediol is low in single and repeated dose oral toxicity, not appreciably irritating to the human skin, somewhat irritating to mucous membranes, and slowly absorbed through the skin. Once absorbed it causes narcosis but little organic injury. It is considered safe for use as an insect repellent.

Single-Dose Oral. Lehman[119] reports that when 2-ethyl-1,3-hexanediol was fed in single oral doses to various species, the LD_{50} values obtained were as follows: rats, 2.5; mice, 4.2; guinea pigs, 1.9; and chicks, 1.4 g./kg. Smyth et al.[120] report the oral LD_{50} for rats to be 2.71 g./kg. In large doses, the material appears to cause deep narcosis and this is believed to be the cause of death.

[118] G. O. Curme, Jr., and F. Johnston, eds., *Glycols, American Chemical Society Monograph, Series 114.* Reinhold, New York, 1952.

[119] A. J. Lehman, *Assoc. Food & Drug Officials U.S. Quart. Bull.,* **19,** 87 (1955).

[120] H. F. Smyth, Jr., C. P. Carpenter, and C. S. Weil, *Arch. Ind. Hyg. Occupational Med.,* **4,** 119 (1951).

Repeated-Dose Oral. Smyth and co-workers[120] report that rats fed for 90 days on a diet that provided a daily intake of 0.70 g./kg. of the glycol did not grow as well as the controls, but apparently suffered no organic injury. When rats were maintained on a diet that supplied a daily intake of 0.48 g./kg., growth was normal and no detectable adverse effects were noted.

Lehman[119] reports that rats were fed for up to 2 years on diets containing 2.0, 4.0, and 8.0 per cent of 2-ethyl-1,3-hexanediol. Growth was depressed at all levels. At the 8.0 per cent level, all animals were dead within 18 weeks, death being due to inanition. Those at the 4.0 and 2.0 per cent levels survived and autopsy revealed no organic injury attributable to the glycol.

Skin Irritation. 2-Ethyl-1,3-hexanediol is somewhat irritating to the skin of rabbits but human skin appears to be quite resistant.[119] Mucous membranes, however, are quite sensitive to the material.[119] This is confirmed by Carpenter and Smyth[121] in their studies with the rabbit eye.

Skin Absorption.[119] The single dose LD_{50} by skin absorption for rabbits is reported to be greater than 10 ml./kg. However, when rabbits were inuncted daily for 90 days, 2 ml./kg. caused about 50 per cent mortality, and somewhat greater mortality resulted at the dosage level of 4 ml./kg. Continued contact caused appreciable irritation to the skin of rabbits and animals that died exhibited moderate liver and kidney injury. Leucocytosis was observed in only 1 animal treated at the 4 ml./kg. level. Lehman[119] concludes, "The fact that the compound is poorly absorbed through the skin of humans and is nonirritating warrants the conclusion that it may safely be used as a component of an insect repellent product."

5. *Hygienic Standards of Permissible Exposure*

None would seem necessary.

STYRENE GLYCOL, $C_8H_{10}O_2$ or $C_6H_5CHOHCH_2OH$
(1-Phenyl-1,2-ethanediol, Phenyl glycol)

1. *Sources, Uses, and Industrial Exposure*

Styrene glycol is made commercially from styrene oxide by hydrolysis.

Styrene glycol is used largely as a chemical intermediate.

Industrial exposure is likely to be by direct contact with the solid or solutions of the material. Vapors and mists may be encountered under particular conditions.

2. *Physical and Chemical Properties*

Styrene glycol is a white, practically odorless solid with the following properties:

Molecular weight: 138.2
Freezing point: 64°C.
Boiling point: 221°C. at 760 mm. Hg

[121] C. P. Carpenter and H. F. Smyth, Jr., *Am. J. Ophthalmol.*, **29**, 1363 (1946).

Solubility: 40 g. in 100 g. of water at 25°C.; 278 g. in 100 g. of methanol at 25°C.; and 29 g. in 100 g. of ether at 25°C.

1 p.p.m. ⇌ 5.65 mg./cu. meter at 25°C., 760 mm. Hg and 1 mg./liter ⇌ 177 p.p.m. at 25°C., 760 mm. Hg.

3. *Determination in the Atmosphere*

Methods for the determination of small amounts of styrene glycol have not been developed. It is believed, however, that chemical methods applicable to other glycols could easily be developed. It is quite possible also that spectrographic methods, particularly ultraviolet absorption, would be useful.

4. *Physiological Response*[122]

Unpublished range finding toxicological studies[122] indicate that styrene glycol is low in oral toxicity and not appreciably irritating to the skin. It would not be expected to present an appreciable hazard in ordinary industrial handling and use.

Single-Dose Oral. The LD_{50} for guinea pigs appears to be between 2.0 and 2.6 g./kg.

Repeated-Dose Oral. When given by intubation as a solution in olive oil 5 times per week for a month, dosage levels of 0.5 and 1.0 g./kg. were tolerated by rabbits and rats. A dosage level of 1.0 g./kg. caused minor liver injury in the rabbit.

Skin Contact. Prolonged and repeated contact with a 20 per cent solution of styrene glycol in propylene glycol caused no injury to the skin of rabbits nor did it penetrate the skin in toxic amounts.

5. *Hygienic Standards of Permissible Exposure*

No inhalation data. Data available, however, would not indicate the necessity for a standard.

[122] Biochemical Research Laboratory, The Dow Chemical Co., unpublished data.

CHAPTER XXXVI

Derivatives of Glycols

V. K. ROWE

DIOXANE (1,4-Dioxane, *p*-Dioxane, Diethylene-1,4-dioxide)

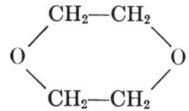

1. *Source, Uses, and Industrial Exposure*

Dioxane can be made by several routes. Probably the most common are by dimerizing ethylene oxide or by dehydration of ethylene glycol. These and other methods are discussed in Curme and Johnston.[1] The material is available in large amounts and is used largely in industry as a solvent for lacquers, plastics, varnishes, paints, dyes, fats, greases, waxes, and resins. When perfectly dry it is stable indefinitely. However it is hygroscopic and because of its ether linkages, it will produce peroxides and other degradation products upon standing in the presence of moisture.

2. *Physical and Chemical Properties*

Dioxane is a colorless liquid with the following properties:
Molecular weight: 88.10
Specific gravity: 1.0353 (20/4°C.), 1.0356 (20/20°C.)
Freezing point: 11.8°C.
Boiling point: 101.3°C. (760 mm. Hg)
Vapor density: approximately 3 (air = 1)
Vapor pressure: 37 mm. Hg (25°C.) (approx.)
Refractive index: 1.4224 (20°C.)
Per cent in "saturated" air: 4.75 (25°C.)
Density of "saturated" air: approximately 1.1 (air = 1)
Solubility: miscible with water and most organic solvents; forms a constant

[1] A. B. Boese, Jr., C. K. Fink, and H. G. Goodman, Jr., in G. O. Curme, Jr., and F. Johnston, eds., *Glycols* (American Chemical Society Monograph, Series 114). Reinhold, New York, 1952.

boiling mixture with water that contains 81.6 per cent dioxane and boils at 87.8°C., 760 mm. Hg

Flash point: 65°F. (TOC)

Flammability limits: 1.97 to 22.25 per cent

1 p.p.m. $\backsim$ 3.6 mg./cu. meter at 25°C., 760 mm. Hg and 1 mg./liter $\backsim$ 278 p.p.m. at 25°C., 760 mm. Hg

3. Determination in the Atmosphere

A satisfactory chemical method for determining low concentration of dioxane vapor in the air has not been developed. It can be determined by means of the interferometer, adsorption, or a combustible gas indicator. It is also reported[2] that dioxane reacts with tetranitromethane with formation of a bright yellow color. It is possible that this reaction can be adapted to determinations in air. Spectrographic techniques also offer promising possibilities.

4. Physiological Response

Summary.   Dioxane is low in single-dose oral toxicity. The liquid is painful and irritating to the eyes, irritating to the skin upon prolonged or repeated contact, and can be absorbed through the skin in toxic amounts. Dioxane vapor has poor warning properties and can be inhaled in amounts that may cause serious systemic injury, principally in the liver and kidney. It is this latter effect that is largely responsible for the hazardous nature of this solvent. Serious and fatal exposures can be experienced without forewarning; illness sometimes becomes apparent hours after exposure.

Single-Dose Oral.   Laug and co-workers[3] studied the single-dose oral toxicity of dioxane rather thoroughly. They determined the LD_{50} values for mice, rats, and guinea pigs to be 5.66, 5.17, and 3.90 g./kg., respectively. Symptoms progressed from weakness, depression, incoordination, and coma, to death. Autopsy revealed hemorrhagic areas in the pyloric region of the stomach, bladders distended with urine, enlarged kidneys, slight proteinuria, but no hematuria. Microscopic changes in the liver and kidneys varied in intensity from animal to animal and generally were of a type seen in diethylene glycol poisoning.

Repeated Oral.   Gross[4] reports that rabbits and guinea pigs fed 0.1 ml./kg. 10 times exhibited dropsical changes in the liver and also that some animals fed 0.5 ml./kg. repeatedly, died after 5, 16, and 20 feedings and sometimes showed typical histological changes.

Skin Irritation.   Dioxane is not considered to be a skin irritant. However, prolonged and repeated contact can cause eczema, as can any effective fat solvent.

[2] E. W. Reid and H. E. Hoffman, *Ind. Eng. Chem.*, **21**, 695 (1929).

[3] E. P. Laug, H. O. Calvery, H. J. Morris, and G. Woodard, *J. Ind. Hyg. Toxicol.*, **21**, 173 (1939).

[4] E. Gross in K. B. Lehmann and F. Flury, *Toxicology and Hygiene of Industrial Solvents.* Trans. by E. King and H. F. Smyth, Jr., Springer, Berlin, 1938.

Skin Absorption. Fairley *et al.*[5] report observing kidney and liver injury in rabbits and guinea pigs as a result of repeated topical application of dioxane.

Single Injection. De Navasquex[6] reports that intravenous injection of dioxane in guinea pigs, rabbits, and cats causes a selective action on the convoluted tubules of the kidney characterized by acute hydropic degeneration. Deaths from dioxane were due to uremia caused by intrarenal obstruction and anuria.

Inhalation. Acute effects. Yant *et al.*[7] exposed guinea pigs for 3 hours to concentrations of from 1000 to 30,000 p.p.m. Gross[4] exposed rats, mice, guinea pigs, and rabbits for 8 hours to concentrations ranging from 4000 to 11,000 p.p.m. At the higher concentrations, marked irritation of the mucous membranes was apparent. Deaths occurring during exposure or shortly afterwards were usually due to respiratory failure because of lung edema, but the animals also exhibited congestion of the brain. Delayed deaths were usually due to pneumonia. Liver and kidney injuries were almost always apparent upon microscopic examination in animals dying days after exposure and in those apparently recovering if they were killed several days after exposure.

Inhalation. Chronic effects. Fairley *et al.*[5] exposed rats, mice, guinea pigs, and rabbits $1^1/_2$ hours a day to concentrations of 10,000, 5000, 2000, and 1000 p.p.m. of dioxane vapor. At the higher levels, mortality was high and deaths were usually due to lung injury. Animals that survived repeated exposures at all levels suffered marked liver and kidney injury.

Gross[4] reports giving cats, rabbits, and guinea pigs (2 of each species) forty-five daily 8-hour exposures to a concentration of 1350 p.p.m. Only 1 cat became ill and had to be sacrificed; it exhibited typical liver and kidney injury. The other animals killed after exposures ceased either showed none or but slight liver and kidney injury. This investigator also exposed a similar group of animals plus 2 mice to 2700 p.p.m. 8 hours per day. Seven of the 10 animals died after 4 to 26 exposures while the rest survived 34 exposures. The signs were irritation of mucous membranes, emaciation, sometimes cramps, narcosis, and albuminuria, and always severe liver and kidney injury. In some cases, the blood urea doubled in concentration.

Mode of Action. Fairley, Linton, and Ford-Moore[8] expressed the opinion that the injury produced by dioxane is a result of its metabolism to diglycolic and oxalic acids. Later, Wiley, Hueper, Bergen, and Blood[9] questioned this hypothesis when they found that dogs and rabbits fed dioxane repeatedly did not excrete in

[5] A. Fairley, E. C. Linton, and A. H. Ford-Moore, *J. Hyg.,* **34,** 486 (1934).

[6] S. de Navasquex, *J. Hyg.,* **35,** 540 (1935).

[7] W. P. Yant, H. H. Schrenk, F. A. Patty, and C. P. Waite, *U. S. Public Health Repts.,* **45,** 2023 (1930).

[8] A. Fairley, E. C. Linton, and A. H. Ford-Moore, *J. Hyg.,* **36,** 341 (1936).

[9] F. H. Wiley, W. C. Hueper, D. S. Bergen, and F. R. Blood, *J. Ind. Hyg., Toxicol.,* **20,** 269 (1938).

the urine more oxalic acid than control animals, whereas animals fed ethylene glycol and ethylene glycol monoacetate did excrete considerably more oxalic acid than did control animals. While there seems to be no question that dioxane exerts its primary toxic action upon the kidneys, the manner in which this is done is not clear.

Human Experience. In 1933, 5 cases of fatal industrial poisoning from dioxane were reported in England.[10,11] These reports are summarized and discussed excellently by Browning.[12] In general, it might be said that men working in a synthetic textile factory inhaled excessive amounts of dioxane. The symptoms were irritation of the upper respiratory passages, coughing, irritation of the eyes, drowsiness, vertigo, headache, anorexia, stomach pains, nausea, vomiting, uremia, coma, and death. Autopsy revealed congestion and edema of the lungs and brain, and marked injury of the liver and kidney. Death was attributed to kidney injury. Blood counts on 3 of the men who died showed no abnormalities other than considerable leukocytosis. The exposure that these men received is still unknown and it is still debatable whether the deaths were caused by chronic exposures or whether they were a result of relatively few intense exposures. De Navasquez[6] believes, on the basis of previous experience in this same factory over several months during which time no trouble was encountered, and on the basis that others in the factory at the same time did not show any serious symptoms, that the persons who died were exposed to high concentrations over a relatively short period of time. It is obvious from this incident that dioxane does not have warning properties adequate to prevent exposure dangerous to life.

5. Hygienic Standards of Permissible Exposure

For a number of years, official agencies and individuals have recommended an industrial hygiene standard of 100 p.p.m. It would seem from the data available that this would give a high degree of safety. It would also seem that 1000 p.p.m. may be suggested as a relatively safe concentration for a single exposure not exceeding $1/2$ hour.

6. Odor and Warning Properties

The odor of dioxane in low concentrations is faint and generally inoffensive and has been described as being somewhat alcoholic. According to Silverman, Schulte, and First,[13] it was concluded from studies on 12 subjects exposed 15 minutes to various concentrations of dioxane that 200 p.p.m. was the highest that they considered acceptable; at 300 p.p.m., it caused irritation of the eyes, nose,

---

[10] H. Barber, *Guy's Hosp. Repts.*, **84**, 267 (1934).

[11] S. A. Henry, Annual Report of Chief Inspector of Factories, H. M. Stationery Office, London, 1934.

[12] E. Browning, *Toxicity of Industrial Organic Solvents*. Chemical Publishing, New York, 1953.

[13] L. Silverman, H. F. Schulte, and W. W. First, *J. Ind. Hyg. Toxicol.*, **28**, 262 (1946).

and throat; and at 500 p.p.m., it was objectionable. Even at higher concentrations, the initial irritation to eyes and respiratory passages is transitory and it is certain that the warning properties of dioxane are completely inadequate to prevent exposure to toxic amounts. Yant *et al.*[7] reported immediate slight burning of the eyes accompanied by lachrymation and slight irritation of the nose and throat from concentration of 1600 p.p.m. for 10 minutes; at 5500 p.p.m. eye irritation and a burning sensation in the nose and throat were noted; and at 10,000 p.p.m. or more, pulmonary irritation occurred.

ETHERS OF MONO-, DI-, AND TRIETHYLENE GLYCOL

1. Source, Uses, and Industrial Exposure

The glycol ethers most commonly encountered industrially are colorless liquids with mild ethereal odors. These monoalkyl ethers are usually produced by reacting ethylene oxide with the alcohol of choice, but also may be made by the direct alkylation of a selected glycol with an alkylating agent such as dialkyl sulfate. The dialkyl ethers may be prepared by reacting the sodium alcoholate of the glycol monoether with an alkyl halide. The yields of these ethers can be varied by changing mole-ratios of reactants and by changing catalysts. For a more comprehensive discussion of the preparation of these materials and references, refer to Boese *et al.* in Curme and Johnston.[14]

The fact that most of these ethers are miscible with water and a large number of organic solvents makes them especially useful as mutual solvents in many oil–water compositions. They are used as solvents for various resins, lacquers, paints, varnishes, dyes, inks, printing pastes, cleaning compositions, liquid soaps, and even cosmetics. They also are used widely as components of hydraulic fluids, and as chemical intermediates.

Industrial exposure may occur by any of the common routes. Excessive exposure to certain of the glycol ethers may occur from inhalation or by absorption through the skin. Ingestion is not likely to be a factor in industrial handling. The toxicological information, experience, and hazards of each individual compound will be discussed separately.

2. Physical and Chemical Properties

The physical and chemical properties of the glycol ethers most commonly encountered are given in Table 1.

3. Determination in the Atmosphere

The choice of the best method for the determination of the glycol ethers will vary with existing conditions. If the ether is known and there are no other inter-

---

[14] A. B. Boese, Jr., C. K. Kink, and H. G. Goodman, Jr., in G. O. Gurme, Jr., and F. Johnston, eds., *Glycols* (American Chemical Society Monograph, Series 114). Reinhold, New York, 1952.

TABLE

Physical and Chemical Properties of Common

| Property | Ethers of glycol | | | | | |
	Methyl	Ethyl	n-Propyl	Iso-propyl	Diethyl	Butyl
Mol. formula	$C_3H_8O_2$	$C_4H_{10}O_2$	$C_5H_{12}O_2$	$C_5H_{12}O_2$	$C_6H_{14}O_2$	$C_6H_{14}O_2$
Mol. wt.	76.1	90.1	104.1	104.1	118.2	118.2
Sp. gr. (25/25°C.)	0.963	0.928	0.909	0.900	0.842	0.900
B. p. (760 mm. Hg)	124.2	134.7	150–152	140–143	121.4	170.6
Vapor pressure, mm. Hg (25°C.)	9.7	5.3	2.9	5.2	12.5	0.88
Refractive index (25°C.)	1.400	1.406	1.412	1.407		1.417
Flash point, °F. (open cup)	120	115	125	145	95	165
Vapor density (air = 1)	2.6	3.0 (approx.)			4.1	4.0 (approx.)
Per cent in satd. air (25°C.)	1.28	0.7	0.38	0.68	1.64	0.11
1 p.p.m. ≏ mg./cu. meter at 25°C., 760 mm. Hg	3.11	3.68	4.29	4.29	4.84	4.84
1 mg./liter ≏ p.p.m. at 25°C., 760 mm. Hg	322	272	234	235	207	207

fering materials present, it can be collected and oxidized with potassium dichromate, as suggested by Werner and Mitchell.[15] The conditions of oxidation giving quantitative results for the methyl, ethyl, n-propyl, and n-butyl ethers of ethylene glycol are described by these authors, but there would seem to be no reason why the basic method could not be adapted to other ethers of various glycols. Nawrocki et al.[16] described an infrared method for the determination of monoalkyl ethers of ethylene glycol in 1944. Improvements in technique, spectrometers, and the ready availability of such machines would seem to make the infrared method the most practical method today for quantitating these materials. Liquid-gas phase chromatography would be useful in resolving mixtures, which could then be identified by infrared or mass spectrophotometry. In instances where a free hydroxyl is available, as in the monoethers, the method of Duke and Smith,[17] based upon the reaction with ammonium hexanitrato cerate, should be applicable but the conditions required have not been defined.

Seikel and Huntress,[18] Elkins et al.,[19] and Morgan[20] have also described

[15] H. W. Werner and J. L. Mitchell, Ind. Eng. Chem. Anal. Ed., **15**, 375 (1943).
[16] C. Z. Nawrocki, F. S. Brackett, and H. W. Werner, J. Ind. Hyg. Toxicol., **26**, 193 (1944).
[17] F. R. Duke and G. F. Smith, J. Ind. Eng. Chem. Anal. Ed., **12**, 201 (1940).
[18] M. K. Seikel and E. H. Huntress, J. Am. Chem. Soc., **63**, 593 (1941).
[19] H. B. Elkins, E. D. Storlazzi, and J. W. Hammond, J. Ind. Hyg. Toxicol., **24**, 229 (1942).
[20] P. W. Morgan, Ind. Eng. Chem. Anal. Ed., **18**, 500 (1946).

1

Ethers of Mono-, Di- and Triethylene Glycol

		Ethers of diethylene glycol					Ether of triethylene glycol
Dibutyl	Phenyl	Methyl	Ethyl	Diethyl	Butyl	Dibutyl	Ethyl
$C_{10}H_{22}O_2$	$C_8H_{10}O_2$	$C_5H_{12}O_3$	$C_6H_{14}O_3$	$C_8H_{18}O_3$	$C_8H_{18}O_3$	$C_{12}H_{26}O_3$	$C_8H_{18}O_4$
174.3	138.2	120.1	134.2	162.2	162.2	218.3	178.2
0.837	1.106	1.018	0.980	0.908	0.952	0.885	1.021
203.6	245	194.1	202.0	188.4	230.0	254.6	255.8
0.12	0.03	0.18	0.13	0.5 (approx.)	0.023	0.015 (approx.)	<0.01
	1.536	1.424	1.425		1.430		1.436
185	275	200	205	180	230	245	275
		4.14	4.62		5.58		
0.016	0.04	0.024	0.017	0.066 (approx.)	0.003	0.002 (approx.)	<0.0013
7.13	5.65	4.91	5.49	6.64	6.64	8.93	7.29
140	177	204	188.2	150.8	150.8	112.0	137.2

methods of qualitative and quantitative analysis that are useful under certain conditions.

ETHYLENE GLYCOL MONOMETHYL ETHER, $CH_3OCH_2CH_2OH$
(2-Methoxyethanol, Methyl Cellosolve, Dowanol EM)

4. *Physiological Response*

Summary. Ethylene glycol monomethyl ether is a volatile liquid with a mild ethereal odor and a bitter taste. It is low in single-dose oral toxicity, moderate in repeated dose oral toxicity, not appreciably irritating to the skin, fairly irritant to mucous membranes, readily absorbed through the skin in toxic amounts, and appreciably toxic when inhaled. Its vapors are irritant in acutely toxic concentrations, but concentrations that may cause serious systemic toxicity upon prolonged and repeated inhalation have negligible warning properties. The material exerts its principal physiological action upon the brain, blood, and kidneys.

Single-Dose Oral. According to Carpenter and co-workers,[21] the single-dose oral toxicity of ethylene glycol monomethyl ether is 3.4 g./kg. for rats, 0.89 g./kg. for rabbits, and 0.95 g./kg. for guinea pigs. In massive doses, the material has a

[21] C. P. Carpenter, U. C. Pozzani, C. S. Weil, J. H. Nair, III, G. A. Keck, and H. F. Smyth, Jr., *A.M.A. Arch. Ind. Health*, **14**, 114 (1956).

narcotic action but at lower dosage levels deaths are delayed and are accompanied by lung edema, slight liver injury, and marked kidney injury. Hematuria may occur from single doses.

Repeated Oral Administration. Gross[22] fed rabbits repeated daily doses of ethylene glycol monomethyl ether and found that 7 doses of 0.1 ml./kg. caused temporary hematuria. Larger doses caused exhaustion, tremors, albuminuria, hematuria, and death. Autopsy revealed severe kidney injury.

Parenteral Administration and Metabolism Studies. Gross[22] gave repeated subcutaneous injections of ethylene glycol monomethyl ether to guinea pigs and rabbits. His results show that with the guinea pig, whereas 7 daily injections of 0.25 ml./kg. caused no symptoms, 5 injections of either 0.5 or 1.0 ml./kg. caused prostration, labored breathing, and death. The response of rabbits was quite similar except that the rabbit appears to be slightly more resistant than the guinea pig to the effects of the material; deaths occurred only after 7 injections of 1.0 ml./kg.

Carpenter *et al.*[21] found that ethylene glycol monomethyl ether in concentrations of more than 25 per cent in 0.75 per cent sodium chloride was hemolytic to rat erythrocytes. When such a 25 per cent solution was given intravenously to rats, the LD_{50} was found to be 2.7 g./kg.; when the undiluted material was given the LD_{50} was 2.2 g./kg. When given intraperitoneally to the rat, the LD_{50} for the undiluted material was 2.5 g./kg.

Wiley, Heuper, Bergen, and Blood[23] injected (site unspecified but believed to be intramuscular) 2 dogs with 6 ml. and 2 rabbits with 2 ml. daily. One rabbit died after 2 injections and 1 dog was anuric after the third (last) injection. They were unable to find any increase in oxalic acid in the urine of any of these animals and they found no increase in methanol or formic acid in the urine of the rabbits. Since under similar conditions the injection of ethylene glycol tripled the urinary excretion of oxalic acid, the injection of ethylene glycol monoacetate doubled it, and since they could find no evidence of the methyl group being liberated, they concluded that this glycol ether was not hydrolyzed to the glycol. Quantitatively, they showed ethylene glycol monomethyl ether to be more toxic than ethylene glycol monoacetate, ethylene glycol diethyl ether, ethylene glycol, or diethylene glycol.

Clinical examination and autopsy of the animals treated with ethylene glycol monomethyl ether revealed anuria, calcified casts in the urine, irritation of the bladder mucosa, hemorrhage in the gastrointestinal tract, lung edema, and liver and testicular injury.

Inhalation Studies. The toxicity of ethylene glycol monomethyl ether when inhaled has been determined for several animal species. Perhaps the most pertinent studies in this area have been conducted by Gross[22] and by Werner and

[22] E. Gross in K. B. Lehmann and F. Flury, *Toxicology and Hygiene of Industrial Solvents.* Trans. by E. King and H. F. Smyth, Jr., Springer, Berlin, 1938.

[23] F. H. Wiley, W. C. Hueper, D. S. Bergen, and F. R. Blood, *J. Ind. Hyg. Toxicol..* **20.** 269 (1938).

co-workers.[24-26] This work has been well summarized by Smyth in Curme and Johnston[14] and by Browning.[27] Gross[22] found that a few repeated exposures to 800 p.p.m. or 1600 p.p.m. produced serious systemic intoxication, characterized for the most part by irritation of the respiratory tract and lungs, hematuria, albuminuria, cylindrical casts in the urine, and severe glomerulitis. Werner et al.[24] exposed mice for 7 hours to various concentrations and found the LC_{50} to be 1480 p.p.m. They concluded that lung and kidney injury was generally the cause of death. Werner et al.[25] exposed groups of rats 7 hours a day, 5 days a week, for 5 weeks to a concentration averaging 310 p.p.m. They noted after 1 week of exposure that there was an increase in the percentage of young granulocytes in the circulating blood. They observed no changes in the kidneys or lungs. Werner et al.[26] exposed 2 dogs to a vapor concentration of 750 p.p.m. of ethylene glycol monomethyl ether 7 hours a day, 5 days a week, for 12 weeks. Again, the most significant changes were in the blood. The hemoglobin concentration, cell volume, and the number of erythrocytes were decreased. The red cells showed an increased hypochromia, polychromatophilia, and microcytosis. The blood picture, in regard to white cells, was characterized by a greater than normal number of immature forms. They[26] point out that the methyl ether, which is significantly less hemolytic in vitro than the other common alkyl ethers of ethylene glycol, produced the greatest alteration in the red cells in the dogs. This is quite a different finding than occurs in mice[24] and rats,[25] where there appears to be a definite correlation between hematological effects and hemolytic potency. The site at which these blood changes occur is obscure. The lack of significant amounts of hemosiderin in the spleen suggests that it is not hemolytic. Although no serious damage to the bone marrow was observed in these studies, it is doubtful these studies were sufficiently prolonged to demonstrate that the effect was not centered in the marrow.

Skin Contact.[28] Ethylene glycol monomethyl ether in repeated and prolonged contact with the skin of rabbits failed to cause any appreciable irritation. However, it readily was absorbed through the skin in toxic amounts. Quantitation by the "sleeve" technique essentially as described by Draize et al.[29] indicated that the LD_{50} was approximately 2 g./kg. The signs of intoxication resulting from

[24] H. W. Werner, J. L. Mitchell, J. W. Miller, and W. F. von Oettingen, *J. Ind. Hyg. Toxicol.*, **25**, 157 (1943).

[25] H. W. Werner, C. Z. Nawrocki, J. L. Mitchell, J. W. Miller, and W. F. von Oettingen, *J. Ind. Hyg. Toxicol.*, **25**, 374 (1943).

[26] H. W. Werner, J. L. Mitchel, J. W. Miller, and W. F. von Oettingen, *J. Ind. Hyg. Toxicol.*, **25**, 409 (1943).

[27] E. Browning, *Toxicity of Industrial Organic Solvents*. Chemical Publishing, New York, 1953.

[28] Biochemical Research Laboratory, The Dow Chemical Co., unpublished data.

[29] J. H. Draize, G. Woodard, and H. O. Calvery, *J. Pharmacol. Exptl. Therap.*, **82**, 377 (1944).

absorption through the skin are essentially the same as those resulting from other routes of administration.

Eye Irritation. When ethylene glycol monomethyl ether was introduced into the eyes of rabbits, it produced immediate pain, conjunctival irritation, and slight transitory cloudiness of the cornea, which cleared within 24 hours.[28] Carpenter and Smyth[30] classify the material along with ethanol as regards its effect on the rabbit eye.

Human Experience. The only fatal case of poisoning in a human due to the ingestion of ethylene glycol monomethyl ether is that recorded by Young and Woolner.[31] The amount of material consumed is somewhat speculative but it is believed that the man consumed about 200 ml. of the material mixed with rum. He was admitted to the hospital in a comatose condition and died 5 hours later without regaining consciousness. The urine contained ethanol but no methanol, thus supporting the conclusion of Wiley *et al.*[23] that the ether is not hydrolyzed. Autopsy revealed hemorrhagic gastritis, marked degeneration of the kidney tubules, and fatty degeneration of the liver. In 1936, Donley[32] described a case of "toxic encephalopathy" suffered by a female who was employed in "fusing" shirt collars by dipping them in a solution composed of ethylene glycol monomethyl ether, isopropanol, and cellulose acetate. She suffered from headache, drowsiness, lethargy, generalized weakness, irregular and unequal pupils, disorientation, and psychopathic symptoms. Two years later Parsons and Parsons[33] described 2 cases of poisoning resulting from the inhalation of vapors of ethylene glycol monomethyl ether again encountered in the manufacture of "permanently starched" collars. The symptoms experienced by these 2 men were weakness, sleepiness, headache, gastrointestinal upset, nocturia, loss of weight, burning of the eyes, and a complete change of personality from one of sharp intelligence to one of stupidity and lethargy. Clinical examination revealed a macrocytic anemia. Both patients apparently recovered completely. As a result of the 2 cases reported,[33] Greenberg and co-workers[34] examined these 2 and 17 other workers employed in the same factory who were using ethylene glycol monomethyl ether. Actually, the solvent being used was 33 per cent ethylene glycol monomethyl ether and 67 per cent denatured ethanol; no benzene was present and the denaturants are not suspect. All revealed abnormalities in the blood pictures suggesting macrocytic anemia and all suffered some degree of excessive fatigue, abnormal reflexes, and tremors. Examination revealed general immaturity of the leukocytes in every case.

Unfortunately, from the industrial hygiene viewpoint, ventilation of the operation causing these effects was improved before Greenberg *et al.*[34] had a

---

[30] C. P. Carpenter and H. F. Smyth, Jr., *Am. J. Ophthalmol.*, **29**, 1363 (1946).

[31] E. G. Young and L. B. Woolner, *J. Ind. Hyg. Toxicol.*, **28**, 267 (1946).

[32] D. E. Donley, *J. Ind. Hyg. Toxicol.*, **18**, 134 (1936).

[33] C. E. Parsons and M. E. M. Parsons, *J. Ind. Hyg. Toxicol.*, **20**, 124 (1938).

[34] L. Greenburg, M. R. Mayers, L. J. Goldwater, W. J. Burke, and S. Moskowitz, *J. Ind. Hyg. Toxicol.*, **20, 134** (1938).

chance to measure the concentrations to which the men were actually exposed. After changes in the ventilation system had been accomplished, concentrations ranged from 25 to 76 p.p.m. Apparently, these findings led these investigators to suggest a threshold limit of 25 p.p.m. even though the concentrations to which the affected persons were exposed were, in all probability, much higher.

The high frequency of mental retardation, neurological symptoms, drowsiness, fatigue, macrocytic anemia, and the abnormal leukocyte picture presented by persons excessively exposed to ethylene glycol monomethyl ether leaves little doubt that the material is a toxic substance and that the effects are centered primarily in the brain, blood, and the kidneys.

5. *Hygienic Standards of Permissible Exposure*

The American Conference of Governmental Industrial Hygienists[35] presently recommends a threshold limit of 25 p.p.m. for ethylene glycol monomethyl ether. While such a value is undoubtedly safe, the data available suggests that it may be unnecessarily stringent.

ETHYLENE GLYCOL MONOETHYL ETHER (2-Ethoxyethanol, Cellosolve Solvent, Dowanol EE)

$$C_2H_5OCH_2CH_2OH$$

4. *Physiological Response*

Summary. Ethylene glycol monoethyl ether is a volatile, colorless liquid with a mild, pleasant, ethereal odor and a bitter taste. It is low in oral toxicity, not significantly irritating to the skin, somewhat irritating to the eyes and mucous membranes, readily absorbed through the skin, and somewhat toxic when inhaled. Its vapors are irritant and disagreeable in acutely toxic concentrations but they are not objectionable at levels considered safe for prolonged and repeated daily exposure. The material exerts its action primarily upon the blood and it is believed that changes in the blood picture will reflect the first evidence of excessive exposure.

Single-Dose Oral. Carpenter and co-workers[21] found the single-dose oral toxicity of ethylene glycol monoethyl ether to be 5.5 g./kg. for rats, 3.1 g./kg. for rabbits, and 1.4 g./kg. for guinea pigs. The value given for rats, 5.5 g./kg., is somewhat higher than the 3.0 g./kg. previously reported by Smyth, Carpenter, and Weil.[36] Laug and co-workers[37] report the following oral LD_{50} values: 3.46 g./kg. for rats, 4.31 g./kg. for mice, and 2.79 g./kg. for guinea pigs. These latter

[35] Conference of Governmental Industrial Hygienists, *Am. Ind. Hyg. Assoc. J.*, **22**, 325 (1961).

[36] H. F. Smyth, Jr., C. P. Carpenter, and C. S. Weil, *Arch. Ind. Hyg. Occupational Med.*, **4**, 119 (1951).

[37] E. P. Laug, H. O. Calvery, H. J. Morris, and G. Woodard, *J. Ind. Hyg. Toxicol.*, **21**, 173 (1939).

authors observed that the animals displayed no immediate signs of distress. However, they did observe hemorrhage of the stomach and intestine, mild liver injury, severe kidney injury, and hematuria in animals that were seriously affected or died. They concluded as a result of their study that ethylene glycol monoethyl ether should not be used in applications where consumption by human beings could be expected.

Repeated-Dose Oral. Gross[22] fed rabbits repeated daily doses of ethylene glycol monoethyl ether and found that 7 doses of 0.1 ml./kg. caused temporary albuminuria, whereas 7 doses of 0.25 ml./kg. caused both albuminuria and hematuria after the seventh feeding. When the dosage was increased to 1 ml./kg., albuminuria and hematuria were observed after the seventh day followed by death on the eighth day due to kidney injury. Two doses of 2 ml./kg. caused exhaustion, refusal to eat, albuminuria, cylinders in the urine, and death believed due to kidney injury.

Smyth et al.[36] report maintaining rats for 90 days on drinking water containing ethylene glycol monoethyl ether. They found that the maximum dose having no effect was 0.21 g./kg./day, that a dose of 0.74 g./kg. reduced growth and appetite, altered liver and kidney weights, and produced microscopic lesions in these organs, and that mortality was increased when the dosage was 1.89 g./kg./day.

Morris, Nelson, and Calvery[38] report upon studies in which ethylene glycol monoethyl ether was fed in the diet of rats for a 2-year period. At the dosage level of 1.45 per cent, equivalent to about 0.9 g./kg./day, they observed only slight kidney damage but did see appreciable tubular atrophy in the testes accompanied by marked interstitial edema in about two thirds of the animals. They did not find any oxalate concretions in the kidneys or bladders, such as have been reported for ethylene glycol.

Parenteral Administration. Gross[22] gave repeated subcutaneous injection of ethylene glycol monoethyl ether to rabbits. The results show that 7 doses of 0.25 ml./kg. produced no apparent effect. However, higher doses produced essentially the same response as observed from oral administration, although the intensity of response seemed to be somewhat greater.

Carpenter et al.[21] found ethylene glycol monoethyl ether in concentrations of more than 18 per cent in 0.75 per cent sodium chloride to be hemolytic to rat erythrocytes. When such an 18 per cent solution was given intravenously to rats, the LD_{50} was found to be 3.3 g./kg. and when the undiluted material was given, the LD_{50} was found to be 2.4 g./kg. When given intraperitoneally to rats, the LD_{50} of the undiluted material was found to be 2.14 g./kg.

Von Oettingen and Jirouch[39] concluded that when given subcutaneously, ethylene glycol monoethyl ether was less toxic than ethylene glycol which, in turn,

[38] H. J. Morris, A. A. Nelson, and H. O. Calvery, *J. Pharmacol. Exptl. Therap.,* **74,** 266 (1942).

[39] W. F. von Oettingen and E. A. Jirouch, *J. Pharmacol. Exptl. Therap.,* **42,** 355 (1931).

was much less toxic than ethylene glycol monobutyl ether. They found the minimum lethal dose to be about 5.0 ml./kg. for the monoethyl ether, 2.5 ml./kg. for ethylene glycol, and 0.5 ml./kg. for ethylene glycol monobutyl ether. They observed also that the ethyl ether had much less effect upon the central nervous system than did the other ethylene glycol ethers. They observed, however, that large doses were capable of causing severe kidney injury.

Skin Contact. Ethylene glycol monoethyl ether, even when in prolonged and repeated contact with the skin of rabbits, failed to cause more than a very mild irritation.[28] However, the material is readily absorbed through the skin of rabbits in toxic amounts. Quantitation by the "sleeve" technique essentially as described by Draize *et al.*[29] indicated that the LD_{50} was 3.6 ml./kg.[21] When the material was applied by inunction, the LD_{50} was 16.3 ml./kg.

Eye Irritation. Carpenter and Smyth[30] classified ethylene glycol monoethyl ether along with ethanol. Other studies[28] confirm this and indicate that when the material is introduced directly into the eye, it will produce immediate pain, some conjunctival irritation, and a slight transitory irritation of the cornea, which clears within 24 hours. These observations would indicate that the material does not present a serious hazard to the eyes, although it may be painful and uncomfortable.

Inhalation Studies. The acute response of guinea pigs to ethylene glycol monoethyl ether in air was studied by Waite, Patty, and Yant.[40] They found that guinea pigs could survive without apparent harm, exposure intensities of 6000 p.p.m. for 1 hour, 3000 p.p.m. for 4 hours, and 500 p.p.m. for 24 hours. More intense exposures caused injury of the lungs, hemorrhage in the stomach and intestines, and congestion of the kidneys. They concluded that air essentially saturated with the vapor of ethylene glycol monoethyl ether at room temperature was sufficiently disagreeable and irritating to the eyes to provide adequate warning to prevent acute poisoning. Gross[22] reports that the majority of animals repeatedly exposed to 1400 p.p.m. of ethylene glycol monoethyl ether 8 hours per day died after 4 to 12 exposures. Cats were found to be most susceptible, dying 2 days after 4 or 5 days of exposure. One of 2 mice died after 9 exposures but the others survived 12 exposures without evident effects. Two rabbits survived 12 exposures, 1 dying 7 days later, while 2 guinea pigs survived 12 exposures without evidence of injury.

Werner and co-workers[24] exposed mice for 7 hours to various concentrations and found the LC_{50} to be 1820 p.p.m. They attributed death to lung and kidney injury.

Werner *et al.*[25] exposed groups of rats 7 hours a day, 5 days a week, for 5 weeks to a concentration averaging 370 p.p.m. of ethylene glycol monoethyl ether vapor and noted only a slight effect upon the cellular elements of the blood. Werner *et al.*[26] exposed 2 dogs to a vapor concentration of 840 p.p.m. of ethylene

[40] C. P. Waite, F. A. Patty, and W. P. Yant, *U. S. Public Health Repts.*, **45**, 1459 (1930).

glycol monoethyl ether 7 hours a day, 5 days a week, for 12 weeks and observed a slight decrease in hemoglobin and red cells. The blood picture, in regard to the white cells, was characterized by a greater than normal number of the immature forms. There was no evidence of kidney injury or of bone marrow injury. There was, however, an increase in the number of calcium oxalate crystals in the urine. The material appeared to be distinctly less toxic than ethylene glycol monomethyl ether or ethylene glycol monobutyl ether.

Human Experience. Human experience in the use of ethylene glycol mono-ethyl ether has been quite uneventful. According to Browning,[27] examinations of workers employed in the manufacture of lacquers and paint by the Factory Department revealed very little evidence of any injury to health from the use of this material.

5. *Hygienic Standards of Permissible Exposure*

The American Conference of Governmental Industrial Hygienists[35] presently recommends a threshold limit of 200 p.p.m. for ethylene glycol monoethyl ether. This appears to be a reasonable figure but the margin of safety it provides is believed to be small.

ETHYLENE GLYCOL MONO-*n*-PROPYL ETHER (2-*n*-Propoxyethanol)

$$C_3H_7OCH_2CH_2OH$$

4. *Physiological Response*

Summary. Ethylene glycol mono-n-propyl ether is a volatile liquid with a mild ethereal odor and a bitter taste. It is moderate in single-dose oral toxicity, the LD_{50} being between 0.5 and 1.0 g./kg. The material is appreciably irritating to the eyes of rabbits, causing injury to the conjunctival membranes and the cornea, and also some iritis; healing appeared to be complete within a week. The material does not appear to be appreciably irritating to the skin but it is readily absorbed through the skin in lethal quantities. It is quite toxic when inhaled in high concentrations. The most striking observation in rats treated with toxic doses by any route was the presence of large amounts of blood in the urine.

Single-Dose Oral. According to Smyth in Curme and Johnston,[14] the LD_{50} for rats of ethylene glycol mono-*n*-propyl ether is 4.89 g./kg. However, other work[28] indicates that the oral LD_{50} for rats is between 0.5 and 1.0 g./kg. The reason for this difference is unknown. Gross[22] states that a single dose of 1 ml./kg. caused the death of a rabbit. In both rabbits and rats, rather large amounts of blood appear in the urine in a matter of hours after feeding 1 g./kg. and death occurs within a matter of a few days after feeding.

Parenteral Administration. Gross in Lehmann and Flury[22] reports that a single subcutaneous injection of 1 ml./kg. in the rabbit caused death with serious kidney injury and blood pigments in the urine. A guinea pig given an intraperitoneal injection of 1 ml./kg. died and showed considerable kidney injury but one given 0.5 ml./kg. survived. Repeated subcutaneous injections of 0.5 ml./kg. were survived by the guinea pig and the rabbit but when the dosage was 1 ml./kg., the guinea pig died several days after the seventh injection.

Eye Contact. When ethylene glycol mono-*n*-propyl ether was introduced into the eyes of rabbits, Gross[22] found it to cause reddening and swelling of the conjunctiva and the lids and corneal damage. In other studies, these same effects were noted but, in addition, some iritis also was observed.[28]

Skin Contact. Gross in Lehmann and Flury[22] reports that the material is not irritating to the skin. Other work[28] confirms this for ordinary exposure but indicates also that if the material is confined to the skin for prolonged periods of time, it may produce appreciable irritation and possibly even a burn. In addition, however, the material may readily be absorbed through the skin in lethal amounts,[28] but no quantitative data are available.

Inhalation Studies. Gross in Lehmann and Flury[22] reports that rabbits tolerated exposures lasting 1 or 3 hours to a concentration of 2400 p.p.m. with only irritation of the mucous membranes and no aftereffect. Others[28] have observed that when rats were given single 7-hour exposures to an atmosphere essentially saturated with ethylene glycol mono-*n*-propyl ether that, although all survived, all showed bloody urines within 2 hours after the exposures terminated together with lung, liver, and kidney injuries. Hematological studies revealed a reduction in packed cell volume but no evidence of hemolysis. Animals exposed for only 4 hours were normal 7 days later, whereas those exposed for 7 hours exhibited severe kidney injury 2 weeks later. In repeated inhalation experiments, Gross[22] found that mice and guinea pigs were unaffected by twelve 8-hour exposures to 600 p.p.m., whereas cats and rabbits died.

Human Experience. There is no report of adverse effects in humans, but this may be accounted for by the fact that the material has not been as widely used as some of the other glycol ethers. Certainly, the material produces effects that should cause one to be very cautious and to avoid exposure wherever possible.

5. Hygienic Standards of Permissible Exposure

There is no data available that would allow the establishment of a hygienic standard for repeated vapor exposure. In view of the rather serious effects produced by this material it would seem only prudent to avoid inhaling the vapors wherever possible. In addition, it would be wise to take precautions to prevent contact with the eyes and prolonged or repeated exposure to the skin.

ETHYLENE GLYCOL MONOISOPROPYL ETHER (2-Isopropoxyethanol)

$$(CH_3)_2CHOCH_2CH_2OH$$

4. Physiological Response[28]

Summary. Ethylene glycol monoisopropyl ether is a volatile liquid with a mild ethereal odor and a bitter taste. It is moderate in single-dose oral toxicity, the

LD$_{50}$ being between 0.5 and 1.0 g./kg. The material is appreciably irritating to the eyes of rabbits causing injury to the conjunctival membranes, the cornea, and also some iritis. Healing appeared to be complete within a week. The material does not appear to be appreciably irritating to the skin but it is readily absorbed through the skin in lethal quantities. It causes severe kidney injury when inhaled in appreciable concentrations. The most striking observation in rats treated with the material by any route was the passing of large amounts of blood in the urine. It appears to be somewhat more toxic than the n-propyl isomer and most other ethers of ethylene glycol.

Single-Dose Oral. The single-dose oral LD$_{50}$ of ethylene glycol monoisopropyl ether for rats is between 0.5 and 1.0 g./kg. It causes severe kidney and liver injury and causes the passage of large amounts of blood in the urine.

Eye Contact. When the material was instilled into the eyes of rabbits it caused marked conjunctival irritation, marked corneal injury, and some iritis. Healing was essentially complete in about 7 days.

Skin Contact. The material is not appreciably irritating to the skin under ordinary conditions of exposure, but if it is confined for prolonged periods, it may cause appreciable irritation, even a burn. In addition, the material may readily be absorbed through the intact skin in lethal amounts.

Inhalation Studies. The results of a limited amount of single dose inhalation work are detailed in Table 2. It appears that ethylene glycol monoisopropyl ether is a highly toxic substance when inhaled, more so than any of the other simple glycol ethers.

Human Experience. No reports of adverse effects from the handling or use of ethylene glycol monoisopropyl ether have been made. However, this may be because of its limited use to date.

5. *Hygienic Standards of Permissible Exposure*

There is no data available that would allow the establishment of a hygienic standard for repeated exposure to vapors. In view of the high toxicity of the material for animals, it would seem only prudent to take particular care to prevent all possible exposure of human beings to this material until such time as its toxicity becomes better evaluated.

ETHYLENE GLYCOL MONOBUTYL ETHER (2-Butoxyethanol, Butyl Cellosolve, Dowanol EB)

$$C_4H_9OCH_2CH_2OH$$

4. *Physiological Response*

Summary. Ethylene glycol monobutyl ether is a volatile colorless liquid with a mild ethereal odor. According to von Oettingen and Jirouch,[39] it at first tastes sour but later causes a burning sensation followed by numbness of the tongue, indicating paralysis of the sensory nerve endings. It is moderately toxic orally,

TABLE 2

Summary of Results of Single Exposures of Rats to Vapors of
Ethylene Glycol Monoisopropyl Ether

Concentration	Duration of exposures (hrs.)	No. dying no. exposed	Comments and observations
Essentially saturated at 100°C. and cooled to room temperature	7	$^4/_4$	After 3 hours, blood in the urine; 1 animal autopsied after exposure had black and severely enlarged kidneys
Essentially saturated	4	$^2/_3$	Passed bloody urine immediately after exposure; excess urine; 3 days after exposure the kidneys were severely necrotic and dark in color
Essentially saturated	1	$^0/_3$	Bloody urine, slight weight loss; severely injured kidneys
Essentially saturated	0.5	$^0/_3$	Bloody urine, slight weight loss, questionable kidney injury
160 p.p.m.	6.7 and 4	$^0/_5$	Bloody urine during exposure; slight weight loss; one rat was autopsied 2 days after exposure and the liver and kidneys were pale in color, urine in bladder was clear; the 4 survivors autopsied 15 days after exposure exhibited evidence of slight to moderate kidney injury
80 p.p.m.	7.0	$^0/_5$	Slight weight loss; much blood passed in urine; one rat was sacrificed immediately after exposure and no gross pathological changes noted; another sacrificed the day after exposure had severely affected kidneys, a pale-colored liver, and bloody urine in the bladder; a rat sacrificed 9 days after exposure showed no evidence of gross changes of the kidney
80 p.p.m.	4.0	$^0/_5$	Questionable evidence of blood in urine, slight weight loss; 1 rat sacrificed immediately after exposure had severely affected kidneys; 1 sacrificed the day after exposure had slight pathology of the kidneys, but no blood in urine; another sacrificed 9 days after exposure appeared grossly to have slightly injured kidneys

appreciably irritating and injurious to the eyes, not significantly irritating to the skin, readily absorbed through the skin in toxic amounts, and moderately toxic when inhaled. Ethylene glycol monobutyl ether is metabolized, at least in part, to butoxyacetic acid and this substance is excreted in the urine of most animal species and of human beings. The material itself and its metabolite are hemolytic agents. The presence of butoxyacetic acid in the urine would suggest that exposure to ethylene glycol monobutyl ether had occurred.

Exposure of human beings to high concentrations of ethylene glycol mono-butyl ether vapors, probably in the range of 300 to 600 p.p.m., for several hours would be expected to cause respiratory and eye irritation, narcosis, and damage to the kidney and liver. Deaths that occur promptly from inhalation in animals are generally attributed to narcosis, but if they are delayed several days they are likely to be caused by pneumonitis and/or kidney injury. The first sign of organic abnormality in man resulting from excessive exposure by any route likely would be an abnormal blood picture characterized by erythropenia, reticulocytosis, granulocytosis, and leucocytosis. Somewhat more intense exposure would be likely to cause fragility of erythrocytes and hematuria.

Single-Dose Oral. Gross in Lehmann and Flury[22] reports that rabbits toler-ated without apparent ill effect single oral doses of 0.1 and 0.5 ml./kg. and that 1.0 and 2.0 ml./kg. caused death within 30 or 22 hours, respectively. Carpenter and co-workers[21] fed single doses of ethylene glycol monobutyl ether to young adults of various species and have calculated the following LD_{50} values: rat, 2.5 g./kg.; mouse, 1.2 g./kg.; rabbit, 0.32 g./kg.; and guinea pig, 1.2 g./kg. They noted that, at least in the rat, large old animals are much more susceptible than weanling animals. In fact, they found the LD_{50} for large old rats to be about 0.55 g./kg. and for weanlings to be 3.0 g./kg. Another group[28] found the LD_{50} for young adult female rats weighing 150 to 200 g. to be 0.47 g./kg. The reason for the differences between the values reported[21,28] for the young adult rat is obscure. Acute or prompt deaths are likely to be due to the narcotic effects of the sub-stance, whereas deaths that were delayed several days usually can be attributed to congested lungs and severely damaged kidneys. Autopsy of animals that died revealed congested lungs, mottled livers, severely congested kidneys, and hemo-globinuria.

Repeated-Dose Oral. Carpenter and co-workers[21] report maintaining rats on diets containing 2.0, 0.5, 0.125, and 0.03 per cent of ethylene glycol monobutyl ether. At the top level, growth depression and increased kidney and liver weights were observed but no hematuria and no histopathological lesions of pertinence were noted. At the 0.5 per cent level, growth depression and increased liver weight were observed. No effects of significance were observed at the two lowest dosage levels.

Parenteral Administration. Gross[22] gave single subcutaneous injections to rabbits and cats. He observed that rabbits tolerated up to 0.1 ml./kg. without reaction, that 0.2 ml./kg. caused temporary slight kidney inflammation, and that doses of 0.4 ml./kg. were fatal. He found that cats tolerated 1 ml./kg. without particular symptoms but that 2 ml./kg. caused death 3 days after injection with signs of kidney injury.

Carpenter *et al.*[21] found that undiluted ethylene glycol monobutyl ether caused hemolysis of rat erythrocytes and that all concentrations above 3 per cent in 0.75 per cent sodium chloride did likewise. When such a solution was injected intravenously in rats, mice, and rabbits, the LD_{50} values were found to be

0.38, 1.1, and 0.50 g./kg. When the undiluted material was injected intravenously, the LD_{50} values found were 0.34 ml./kg. for rats and 0.28 ml./kg. for rabbits. When the undiluted material was given intraperitoneally, the LD_{50} value for rats was found to be 0.55 ml./kg.

Metabolism. Carpenter and co-workers[21] have demonstrated that butoxyacetic acid is a metabolite of ethylene glycol monobutyl ether in the rat, guinea pig, rabbit, dog, rhesus monkey, and man. They have developed a method for estimating its concentration in the urine that can be used to detect the first evidence of absorption of the material by any route. However, it does not appear that the butoxyacetic acid content of the urine can be used as a measure of intensity of exposure. They found 55 mg. of butoxyacetic acid in 16-hour urine samples of dogs exposed to 385 p.p.m. They found 100 and 42 mg. in 24-hour urine samples from 2 dogs exposed to 200 p.p.m. of vapor and 100 and 94 mg. in similar urine samples from 2 dogs exposed to 100 p.p.m. One of 2 monkeys exposed to 100 p.p.m. excreted 30 mg. of butoxyacetic acid in a 48-hour period and very little was found in a similar urine sample from the other animal. Human beings exposed 8 hours to 195 p.p.m. excreted anywhere from 6 to 300 mg. of butoxyacetic acid in a 24-hour period, whereas persons exposed to 98 p.p.m. for 8 hours excreted from 75 to 250 mg. in a 24-hour period. At best, it would seem that the presence of butoxyacetic acid in the urine can be considered evidence of exposure to ethylene glycol monobutyl ether.

Carpenter *et al.*[21] also showed that ethylene glycol monobutyl ether, and to a greater extent, its metabolite, butoxyacetic acid, both increase the osmotic fragility of the erythrocyte. This action appears to be greatest in the rat, mouse, and rabbit and distinctly less in the guinea pig, dog, rhesus monkey, and human.

Inhalation Studies. Werner and co-workers[24] exposed mice for 7 hours to various concentrations and found the LC_{50} to be 700 p.p.m. Death was attributed to lung and kidney injury. Carpenter *et al.*[21] exposed rats of different ages to various concentrations of vapor. They found that 1-year-old rats were more susceptible than young, actively growing rats. At a concentration of 375 p.p.m., the old adults died after 7 hours while the 6-week-old rats survived 8 hours at 500 p.p.m. Gross[22] reports that the inhalation of 520 p.p.m. 8 hours a day for 8 to 12 days caused the death of cats, rabbits, and guinea pigs but had no adverse effects upon mice. Kidney inflammation was present in all animals that died.

Werner *et al.*[25] exposed one group of rats 7 hours a day, 5 days a week, for 5 weeks to concentrations averaging 320 p.p.m., and another group to 135 p.p.m. of ethylene glycol monobutyl ether. They noted, after a week, an increase in the percentage of reticulocytes and young granulocytes in the circulating blood, and a decrease in the hemoglobin concentration and erythrocyte count. The blood picture of animals so affected returned to normal shortly after exposure was terminated.

Werner *et al.*[26] exposed 2 dogs to a vapor concentration of 415 p.p.m. 7 hours a day, 5 days a week, for 12 weeks. Although the animals failed to grow well and displayed the typical reversible blood changes, they did not show evidence of organic injury by histological procedures. However, it was noted that there was an increase in the number of calcium oxalate crystals in the urine and that there was a moderate retention of urea in the blood throughout the exposure period.

Extensive repeated inhalation studies have been conducted by Carpenter et al.[21] The quantitative results are summarized in Table 3. The qualitative

TABLE 3

Summary of Results of Repeated Inhalation Studies on
Ethylene Glycol Monobutyl Ether[21]

Species	Exposure intensity, no. of 7-hr. exposures	Solvent conc., causing some injury[a] (p.p.m.)	Highest vapor concentration (p.p.m.) not causing			
			Injury[a]	Increased erythrocyte fragility	Hemo-globin-uria	Other hemato-logical changes
Rat	30	107	54	<54	107	
Guinea pig	30	203	107	494	494	
Mouse	30	396	200	<112	112	
	60	200	111	<111	111	
	90	401	201	<112	112	
	90 plus 42 days' rest	401	401	401	401	
Dog	90	385	200	100	?	<100
Monkey	90	?	210	<100	?	<100
Human	Two 4-hr. periods separated by a 30-minute lunch period			>195		
Rat	Two 4-hr. periods separated by a 30-min. lunch period			<195		
Rat	4 hrs.			<113		
Human	4 hrs.			>113		

[a] As judged by mortality, growth, kidney and/or liver weight changes, and gross and/or micropathology of organs.

observations are as follows. At high concentrations, the rats exhibited hemorrhage of the lungs, congestion of the viscera, liver injury, hemoglobinuria, and a marked erythrocyte fragility. Females were more sensitive than males. At the lower concentrations, increased fragility of the erythrocytes appears to be the most sensitive criterion of effect. This response was transitory, however, since it was apparent during and shortly after exposure, but not the following day.

The guinea pigs were quite resistant to the effects of ethylene glycol monobutyl ether. At high concentrations, congestion and cloudy swelling of the tubules of the kidneys were observed, but no increase in the fragility of the erythrocytes occurred at any concentration studied. Mice were essentially as resistant as the guinea pigs, with the exception that their erythrocytes were as fragile as those of the rat.

Dogs exposed to high concentrations suffered congestion of the kidneys and lungs, weight loss, increased fragility of the erythrocytes, nasal and eye infections, apathy, anorexia, nausea, and other changes in the circulating blood. The leucocytes were markedly increased whereas the erythrocyte and hemoglobin content was markedly decreased. There was also a marked increase in plasma fibrinogen. At the lowest level of exposure, 100 p.p.m., the dogs exhibited a transitory increase in the leucocyte count, a marked decrease in erythrocyte count, and a low hematocrit. Twenty-four hour urine samples contained 94 to 100 mg. of butoxyacetic acid.

Monkeys exposed to 200 p.p.m. suffered marked reduction in the number of circulating red blood cells and in hemoglobin concentration. The fragility of the erythrocytes was markedly increased and plasma fibrinogen levels were elevated to about 4 times normal values. There was, however, no appreciable organic injury observed at autopsy. At the lowest level, 100 p.p.m., monkeys exhibited typical hematological effects but all had returned to normal by the end of the exposure period. The female monkey excreted 30 mg. of butoxyacetic acid over a 48-hour period after receiving the forty-second exposure.

When human subjects inhaled 200 p.p.m. of ethylene glycol monobutyl ether for 8 hours, there were no objective effects although the presence of butoxyacetic acid in their urines proved that absorption had occurred. The consensus was, however, that this concentration was too high for comfort, with eye, nose, and throat irritation evident. When the concentration was lowered to 100 p.p.m., similar subjective complaints arose from sensitive persons. It appears that this chemical is one of the few materials to which the human is more resistant than the usual experimental animals. This appears to be due, in part, at least, to the fact that humans are more resistant than are most laboratory animals to the hemolytic effects caused by the material itself or its metabolite.

Eye Contact. Carpenter and Smyth[30] have classified ethylene glycol monobutyl ether along with such materials as acetonyl acetone, dioxane, and isopropanol. Other studies[22,28] indicate that when the material is introduced directly into the eye it will produce marked pain, appreciable conjunctival irritation, and slight transitory injury of the cornea, which should heal within a few days.

Skin Contact. Prolonged and repeated contact of ethylene glycol monobutyl ether with the skin of rabbits failed to cause more than a very mild simple irritation. However, the material is readily absorbed through the skin of rabbits in toxic amounts. According to Carpenter et al.[21] the LD_{50} for rabbits was found to be 0.45 ml./kg. when the material was confined to the skin for 24 hours by the method of Draize et al.[29] and 2 ml./kg. when applied to the skin by gentle massage. Evidence of rapid absorption through the skin also was obtained when these investigators showed that erythrocyte fragility was increased an hour after a single 3-minute contact with 0.56 ml./kg. on 4.5 per cent of the total skin surface area.

Human Experience. The human experience in the use of ethylene glycol monobutyl ether has been remarkably free of serious complications. Browning[27] states that there is only one possible injury in industrial use of the material and this was reported by the English Factory Inspection in 1934. A man had two isolated attacks of hematuria at 5-month intervals. The cause, however, was somewhat complicated by the fact that diethylene glycol monobutyl ether was also associated with the process in which he was working. The man also suffered some nasal and eye irritation. Browning[27] also mentions 2 girls, who reported irritation of the mucous membranes and headache as a result of working on a process in which the material was used.

5. Hygienic Standards of Permissible Exposure

The American Conference of Governmental Hygienists[35] presently recommends a threshold limit value of 50 p.p.m. for ethylene glycol monobutyl ether. On the basis of the data on animals and human beings, it would appear that this is a reasonable figure. There is, however, a hazard other than vapor that must not be overlooked when handling this material—that of possible absorption of toxic quantities through the skin. Because of the low vapor pressure of this substance at room temperature, the hazard from skin absorption could well be greater, or contribute substantially to the over-all hazard.

ETHYLENE GLYCOL DIETHYL ETHER (Diethyl Cellosolve)

$$C_2H_5OCH_2CH_2OC_2H_5$$

4. Physiological Response

Summary. Ethylene glycol diethyl ether is a colorless, volatile liquid with a bitter but sweetish taste. Its vapors are irritating to the eyes and mucous membranes. It is low in oral toxicity, appreciably irritating to the eyes, not appreciably irritating to the skin, but it is moderately toxic when inhaled. The material is a weak narcotic agent. In the dog, at least, its administration does not result in an increase in urinary oxalic acid, but it does exert its toxic action principally on the kidney. The irritating nature of the vapors is probably sufficient to warn of concentrations hazardous to life upon single exposure but probably not adequate to prevent exposure to vapor concentrations hazardous upon prolonged and repeated exposure.

Single-Dose Oral. Smyth, Seaton, and Fischer[41] fed single doses of ethylene glycol diethyl ether as a 10 per cent aqueous solution to rats and guinea pigs and found the LD_{50} to be 4.39 g./kg. for the rat and 2.44 g./kg. for the guinea pig.

Repeated-Dose Oral. Gross in Lehmann and Flury[22] reports feeding a dog and a rabbit 1 ml./kg. 6 times within a week without symptoms, but the urine was not investigated. A cat that received 1 ml./kg. 4 times was seriously intoxicated

---

[41] H. F. Smyth, Jr., J. Seaton, and L. Fischer, *J. Ind. Hyg. Toxicol.*, **23**, 259 (1941).

each time and died after the fourth dose. Autopsy revealed no characteristic findings.

Parenteral Administration. Gross[22] also reports that guinea pigs survived 7 subcutaneous injections of 0.5 ml./kg. even though they suffered serious weight loss. When the dose was increased to 1 ml./kg., however, death resulted after 7 injections. After 4 injections, the animals showed temporary narcotic symptoms and prostration was apparent just prior to death. Autopsy revealed kidney injury characterized by parenchymatous and interstitial nephritis.

Wiley and co-workers[23] gave 2 dogs 9.5 ml./day subcutaneously for 7 days and observed no increase in the oxalic acid content of the urine. These doses caused no noticeable effect. However, autopsy revealed injury to the vasculature, liver, brain, testes, and particularly to the kidney.

Skin Contact. Gross[22] reports that studies on guinea pigs, rabbits, and dogs revealed that the subject material does not injure the skin.

Eye Contact. Carpenter and Smyth[30] have classified ethylene glycol diethyl ether along with ethylene glycol monobutyl ether, acetonyl acetone, isopropanol, and dioxane. Hence, the material can be expected to be painful, to produce conjunctival irritation and slight transitory injury of the cornea, which should heal within a few days.

Inhalation Studies. Gross[22] reports that the inhalation of 10,000 p.p.m. for 1 hour caused irritation in the mucous membranes and a suggestion of narcosis. Cats were more sensitive than were rabbits, guinea pigs, or dogs but all survived the exposure. The same authors report that 12 daily 8-hour exposures of mice, guinea pigs, rabbits, and cats to 500 p.p.m. resulted in the death of 1 of 2 rabbits and the 2 cats but no evident injury to the mice and guinea pigs. Microscopic examination of the tissues of both cats showed definite symptoms of kidney injury and in one of them, a serious purulent inflammation of the trachea.

Human Experience. There were no reported instances in which ethylene glycol diethyl ether caused any adverse effect in human beings.

5. *Hygienic Standards of Permissible Exposure*

There are not sufficient data upon which to base a standard for permissible vapor exposure. The indications are that such a limit probably should be somewhat less than 100 p.p.m. The material does not have sufficient warning properties to prevent excessive exposure when contact is prolonged and repeated.

ETHYLENE GLYCOL MONOPHENYL ETHER (2-Phenoxyethanol, Dowanol EP)

$$C_6H_5OCH_2CH_2OH$$

Ethylene glycol monophenyl ether is a clear liquid, soluble in most organic solvents but only very slightly soluble in water. Other properties are given in **Table 1.**

4. *Physiological Response*

Extensive toxicological data on ethylene glycol monophenyl ether are not available. Range-finding studies[28] on the technical or commercially available product show it to be low in single dose oral toxicity, the LD_{50} being between 1.0 and 2.0 g./kg. It is not appreciably irritating to the intact skin and is not readily absorbed through the skin. However, it is severely damaging to the eyes of rabbits. Rats tolerated, without apparent adverse effects, one 7-hour exposure to vapors saturated at 100°C. and cooled to room temperature. Reasonable handling precautions plus particular care to prevent contact with the eyes should prevent any serious toxic effects.

DIETHYLENE GLYCOL MONOMETHYL ETHER (Methyl Carbitol, Dowanol DM)

$$CH_3OCH_2CH_2OCH_2CH_2OH$$

4. *Physiological Response*

Summary. Diethylene glycol monomethyl ether is a colorless liquid with a mild pleasant odor and a bitter taste. It is low in oral toxicity, painful but not seriously injurious to the eyes, and not irritating to the skin. Although it can be absorbed through the skin, severe exposure would be required before serious effects would be expected. Hazardous amounts are not likely to be inhaled under ordinary conditions but where heated material is encountered, care is warranted. No adverse human experience has been reported.

Single-Dose Oral. Smyth, Seaton, and Fischer[41] fed diethylene glycol monomethyl ether as a 50 per cent aqueous solution to rats and guinea pigs and found the LD_{50} values to be 9.21 ml./kg. for rats and 4.16 for guinea pigs. Others[28] have fed the material undiluted to groups of 10 male and 10 female rats and have found the LD_{50} value to be between 6.5 and 7.0 ml./kg. for males and between 5.5 and 6.0 ml./kg. for females. Deaths usually occurred within 48 hours or not at all, and were believed due either to profound narcosis or kidney injury.

Repeated-Dose Oral. Smyth and Carpenter[42] report administering commercial diethylene glycol monomethyl ether to rats for 30 days as a part of their drinking water. They found that the maximum dose having no effect was less than 0.19 g./kg. based upon microscopic study of the tissues. Animals survived the highest dosage level, 1.83 g./kg.

Eye Contact. Diethylene glycol monomethyl ether is somewhat painful to the eyes and is capable of causing only transitory injury.[28,30] Thus, the material presents no serious hazard from eye contact under ordinary industrial handling conditions.

Skin Contact. Diethylene glycol monomethyl ether is not appreciably irritating to the skin.[28] Although the material can be absorbed through the skin of

---

[42] H. F. Smyth, Jr., and C. P. Carpenter, *J. Ind. Hyg. Toxicol.*, **30**, 63 (1948).

rabbits in toxic amounts, extensive and prolonged contact is required to cause serious effects. Studies using the "sleeve" technique of Draize et al.[29] revealed the LD_{50} to be about 20 ml./kg.[28] In view of these observations, there would seem to be no significant hazard from skin contact in ordinary industrial operations.

Inhalation. No data are available upon which to evaluate the hazard from inhalation of the vapors of diethylene glycol monomethyl ether. Because of the low volatility of this material at normal room temperatures and because of its low oral toxicity, the material is believed to present no unusual hazards from inhalation when handled at room temperature. However, vapors of this material generated at elevated temperatures or when breathed repeatedly over a prolonged period may well present a hazard from inhalation.

Human Experience. No adverse experience of humans is reported.

5. Hygienic Standards of Permissible Exposure

Data adequate to allow the establishment of a hygienic standard are lacking.

DIETHYLENE GLYCOL MONOETHYL ETHER (Carbitol, Dowanol DE)

$$C_2H_5OCH_2CH_2OCH_2CH_2OH$$

Diethylene glycol monoethyl ether is a colorless liquid with a very slight pleasant odor and a bitter taste. For additional physical data, see Table 1. Commercial products may contain an appreciable amount of ethylene glycol and such products may be more toxic than the relatively pure ether. This fact must not be overlooked when evaluating the available toxicological data or when choosing a product for a specific use.

4. Physiological Response

Summary. Diethylene glycol monoethyl ether is low in oral toxicity, not appreciably irritating to the eyes or skin, but is readily absorbed through the skin. Its volatility is sufficiently low that acutely hazardous vapor concentrations do not occur at ordinary temperatures. The material appears to be readily oxidized in the body. Experience with human subjects has been uneventful. It is generally agreed that the relatively pure ether does not present any serious industrial hazards. Reasonable precautions are adequate to ensure safe handling.

Single-Dose Oral. Smyth, Seaton, and Fischer[41] fed especially purified diethylene glycol monoethyl ether as a 50 per cent aqueous solution and found the LD_{50} values to be 8.69 g./kg. for the rat and 3.67 g./kg. for the guinea pig. When a 50 per cent aqueous solution of a commercial product was fed, the LD_{50} values found were 9.74 g./kg. for the rat and 4.97 g./kg. for the guinea pig. Gross,[22] working with an industrial product, found that 1 ml./kg. was lethal to cats. It caused disturbance of equilibrium, gastrointestinal inflammation, pneumonia, and kidney injury, but no albuminuria. A dog that received a single dose of 2 ml./kg. as a 20 per cent aqueous solution was unaffected except for slight vomiting 3 hours after feeding. Laug and co-workers[37] found the oral LD_{50} value to be 5.54 g./kg. for rats, 6.58 g./kg. for mice, and 3.87 g./kg. for guinea pigs. They likened the response of animals fed this ether to that caused by ethanol but also noted pneumonia and kidney injury in treated animals. They concluded that this material is not suitable for use in foods or drugs. Others[28] have found the LD_{50} for rats to be 5.4 g./kg. and observed the effects of acutely toxic doses to be characterized by ataxia and depression with little injury to visceral organs apparent grossly.

Repeated-Dose Oral. Smyth and Carpenter[42] report maintaining groups of 5 rats for 30 days on drinking water containing various amounts of purified diethylene glycol monoethyl ether. Dosage levels ranged from 0.21 to 3.88 g./kg./day. They found that 0.49 g./kg. was tolerated without any adverse effect, that 0.87 g./kg. caused reduction in appetite, and that 1.77 g./kg. caused some organic injury.

Morris, Nelson, and Calvery[38] maintained rats for 2 years on a diet containing 2.16 per cent of a purified diethylene glycol monoethyl ether. This probably is equivalent to slightly more than 1.0 g./kg./day. The only adverse effects they noted were a few oxalate concretions in a kidney of 1 animal, slight liver damage, and some interstitial edema in the testes. Since the quality of the material tested was not established, the possibility of the concretions being caused by the presence of small amounts of ethylene glycol in the test sample cannot be overlooked.

Hanzlik, Lawrence, and Laqueur[43] administered the pure ether to rats at the level of 1.0 per cent in their drinking water and to mice at the level of 5.0 per cent in their food for a 2-year period without causing significant adverse effects. They also found that when this same ether containing ethylene glycol was fed, kidney injury typical of the glycol occurred. They concluded that the ether was relatively noninjurious.

Eye Irritation. Diethylene glycol monoethyl ether is slightly painful but causes no more than a trace of irritation of the conjunctival membranes.[28,30,39] Thus, this material probably does not present a serious hazard from eye contact.

Skin Irritation. Diethylene glycol monoethyl ether is not irritant to the skin of rabbits even upon prolonged and repeated contact.[28] Cranch *et al.*[44] found the material to be neither a primary irritant nor a sensitizer and no different than wool fat or glycerine when applied to the skin of 98 human subjects. Furthermore, they report that 70 per cent aqueous material did not retard wound healing. Meininger[45] confirmed the lack of skin irritating effect in humans and was unable to find evidence that it was absorbed through the skin of human subjects. This is unexpected and contrary to the observation of numerous workers in studies on animals and may be due to the lower dosage levels employed.

[43] P. J. Hanzlik, W. S. Lawrence, and G. L. Laqueur, *J. Ind. Hyg. Toxicol.*, **29**, 233 (1947).

[44] A. G. Cranch, H. F. Smyth, Jr., and C. P. Carpenter, *Arch. Dermatol. and Syphilol.*, **45**, 553 (1942).

[45] W. M. Meininger, *Arch. Dermatol. and Syphilol.*, **58**, 19 (1948).

Skin Absorption. Hanzlik *et al.*[46] have reported the results of extensive skin absorption studies with diethylene glycol monoethyl ether. They report the LD_{50} to be 8.5 ml./kg. when applied by inunction for 2 hours to 100 sq. cm. or slightly more. When such applications were repeated daily for 30 days, the LD_{50} was estimated to be 0.32 ml./kg. with the no-effect level between 0.04 and 0.08 ml./kg. They report transient dermatitis and both microscopic injury and impairment of kidney function.

Purified diethylene glycol monoethyl ether has been subjected to repeated skin absorption studies by others.[28] The material used in these studies had a boiling range in °C. at 760 mm. Hg as follows:

First drop—198.7
5%—201.2
50%—202.5
95%—203.7
Dry point—208.0

The material was applied on a 3 × 3 in. cotton pad just heavy enough to hold the dose, this was covered with a 5 × 5 in. impervious film and the whole bandaged onto the clipped abdomen of each rabbit 5 times a week for 3 months. The dosages used were 0.1, 0.3, 1.0, and 3.0 ml./kg./day and 5 animals were used at each dosage. The bandage was held in intimate contact with the skin for 24 hours, thus allowing essentially all to be absorbed each day; there was no leakage of consequence. The criteria used in judging effect were growth, mortality, hematological studies, organ weight studies, blood-urea-nitrogen determinations, and gross and microscopic examination of the treated areas of skin and of the principal organs. The only effects at the top dosage level of 3.0 ml./kg. were an increase in blood-urea-nitrogen and severe kidney injury in 1 of the 4 surviving animals. Minor kidney changes were seen in 2 of the animals and 1 was unaffected. At the dosage level of 1 ml./kg., moderate kidney changes were seen in 3 of the 4 surviving animals and at the lower dosage levels no adverse effects were seen. Thus, the no-effect dosage level for repeated exposures over a 90-day period approximates 0.3 ml./kg./day. This figure is about 10 times larger than that found by Hanzlik[46] and might well reflect a difference in purity of the material tested.

Metabolism. Fellows *et al.*[47] found that diethylene glycol monoethyl ether is largely destroyed by the body or excreted as the glucuronate. They noted that when given to rabbits orally or by injection the urinary content of glucuronic acid increased as it did when propylene glycol was given. Why this should occur with these two materials and not with ethylene and diethylene glycol and glycerol is unknown.

[46] P. J. Hanzlik, W. S. Lawrence, J. K. Fellows, F. P. Luduena, and G. L. Laqueur, *J. Ind. Hyg. Toxicol.*, **29**, 325 (1947).

[47] J. K. Fellows, F. P. Luduena, and P. J. Hanzlik, *J. Pharmacol. Exptl. Therap.*, **89**, 210 (1947).

Human Experience. Insofar as is known, no toxic effects have resulted from the industrial use of diethylene glycol monoethyl ether. However, if the material is intended for uses involving prolonged or repeated contact with the skin, it would seem only prudent to choose a product of low ethylene glycol content. The material is not considered suitable for uses where ingestion may be expected.

Inhalation. Gross in Lehmann and Flury[22] reports that rabbits, cats, guinea pigs, and mice were not injured by 12 daily exposures to an atmosphere essentially saturated with diethylene glycol monoethyl ether.

5. *Hygienic Standards of Permissible Exposure*

Data adequate for the establishment of a hygienic standard for diethylene glycol monoethyl ether are not available. However, in view of the low vapor pressure and the low toxicity of the material, such a standard hardly seems necessary. Reasonable and ordinary precautions to avoid inhalation of vapors or mists would seem adequate to prevent excessive vapor exposure.

DIETHYLENE GLYCOL MONOBUTYL ETHER (Butyl Carbitol, Dowanol DB)

$$C_4H_9OCH_2CH_2OCH_2CH_2OH$$

Diethylene glycol monobutyl ether is a colorless liquid with a mild odor. Additional physical data are given in Table 1. The material generally available commercially is a fairly pure product.

4. *Physiological Response*

Summary. Diethylene glycol monobutyl ether is low in single-dose oral and vapor toxicity, moderately toxic in repeated dose oral toxicity, moderately irritating and injurious to the eyes, not appreciably irritating to the skin, and not absorbed through the skin in acutely toxic amounts except at large dosage levels. The results of the limited repeated dose oral work reported suggests that the material may be rather toxic when inhaled or absorbed through the skin in repeated small doses.

Single-Dose Oral. Smyth, Seaton, and Fischer[41] fed diethylene glycol monobutyl ether as a 50 per cent aqueous solution and found the LD_{50} to be 6.56 g./kg. for rats and 2.00 g./kg. for guinea pigs. Others have found the LD_{50} for rats to be 5.66 g./kg. when it was fed undiluted.[28]

Repeated-Dose Oral. Smyth and Carpenter[42] maintained groups of 5 rats for 30 days on drinking water containing various amounts of the material. Based on the water consumption, the dosage levels ranged from 0.051 to 1.83 g./kg./day. They found that 0.051 g./kg. was the maximum dosage having no effect, 0.094 g./kg. caused a reduction in appetite, and that 0.65 g./kg./day caused histopathological injury to some organs.

Eye Contact. Diethylene glycol monobutyl ether is capable of causing moderate irritation and moderate transient corneal injury.[28,30] This indicates that although the material is not likely to cause impairment of vision, it may cause considerable discomfort.

Skin Contact.[28] Diethylene glycol monobutyl ether is only very slightly irritating to the skin of rabbits even upon prolonged and repeated skin contact. The material is absorbed through the skin but the acute toxicity by this route is slight, the LD_{50} being about 4 g./kg. when exposure was continuous for 24 hours.

Inhalation.[28] Three rats survived a single 7-hour exposure to an atmosphere saturated with the material at 100°C. and then cooled to room temperature. The animals appeared normal throughout the exposure and exhibited only a slight transient weight loss. No autopsies were made. These results indicate that there is little hazard from a single vapor exposure.

Human Experience. There has been no adverse human experience that can be attributed definitely to this material.[27]

5. *Hygienic Standards of Permissible Exposure*

Data for the establishment of a hygienic standard for diethylene glycol monobutyl ether are not available. However, the limited repeated dose data available suggest caution in handling where repeated exposures may be expected.

TRIETHYLENE GLYCOL MONOETHYL ETHER (Ethoxy Triethylene Glycol, Ethoxytriglycol)

$$C_2H_5OCH_2CH_2OCH_2CH_2OCH_2CH_2OH$$

Triethylene glycol monoethyl ether is a colorless, essentially odorless liquid. Other properties are given in Table 1.

4. *Physiological Response*

Extensive toxicological data are not available on triethylene glycol monoethyl ether. Smyth and co-workers[36,42] report that range-finding studies showed the oral LD_{50} to be 10.6 g./kg. for rats, the skin absorption LD_{50} to be 8 ml./kg. for rabbits, and that the material is not injurious to the skin or eyes. They also report maintaining groups of 10 rats for 30 days on drinking water containing sufficient material to result in daily intakes ranging from 0.18 to 3.30 g./kg. They found the maximum intake having no effect to be 0.75 g./kg./day, but did not disclose the nature of the effect at higher dosage levels. The effect noted apparently was not increased mortality, decreased food intake, decreased growth, or microscopic changes in the liver, kidney, spleen, or testes, for none of these criteria were affected at the 3.30 g./kg./day level. It appears from this data that the material would present no appreciable hazard in ordinary handling or use.

ETHERS OF MONO-, DI-, TRI-, AND POLYPROPYLENE GLYCOL

1. Source, Uses, and Industrial Exposure

The ethers of mono-, di-, tri-, and polypropylene glycol generally are prepared commercially by reacting propylene oxide with the alcohol of choice in the presence of a catalyst. They also may be prepared by direct alkylation of the selected glycol with an appropriate alkylating agent such as a dialkyl sulfate in the presence of alkali. Preparation under commercial conditions yields products that are mixtures of the alpha and beta isomers, largely alpha.

The methyl and ethyl ethers of these propylene glycols are miscible with water and a great variety of organic solvents. The butyl ethers have limited water solubility but are miscible with most organic solvents. This mutual solvency in water and oils makes some of these materials exceptionally useful as coupling and dispersing agents, as solvents for lacquers, paints, resins, dyes, oils, and greases. The methyl ether of dipropylene glycol is used in various cosmetics as a solvent and dispersing agent. The butyl ethers of polypropylene glycols 400 and 800 are fly repellents, the latter being known as Crag Fly Repellent.

Industrial exposure may occur by any of the common routes, but the hazards would not seem to be great except under most adverse conditions. The toxicological information, experience, and hazards of each individual compound will be discussed separately.

2. Physical and Chemical Properties

The available physical properties of the most common ethers are given in Table 4. The chemical composition, as far as is known, is described in the following paragraph.

The monoalkyl ethers of propylene glycol appear in two isomeric forms, the alpha or 1-alkyloxy-2-propanol ($ROCH_2CHOHCH_3$), and the beta or 2-alkyloxy-1-propanol ($HOCH_2CHORCH_3$). A commercial product, Dowanol PM, is a mixture of the two isomers consisting of at least 95 per cent alpha, with the balance the beta isomer. The monoalkyl ethers of dipropylene glycol presumably can appear in four isomeric forms. The commercial product, Dowanol DPM, is believed to be a mixture of these but to consist to a very large extent of the isomer in which the alkyl group has displaced the hydrogen of the primary hydroxy group of the dipropylene glycol; the internal ether linkage is between the secondary carbon of the alkyl etherized propylene unit and the primary carbon of the other propylene unit, thus leaving the remaining secondary hydroxy group unsubstituted. They are usually designated by the formula, $ROC_3H_6OC_3H_6OH$. The monoalkyl ethers of tripropylene glycol can appear in eight isomeric forms. The commercial product, Dowanol TPM, however, is believed to be a mixture of isomers consisting largely of the one in which the alkyl group displaces the hydrogen of the primary hydroxyl group of the tripropylene glycol and the internal ether linkages are

TABLE 4

Physical Properties of Monoalkyl Ethers of Mono-, Di-, Tri-, and Polypropylene Glycols

Property	Ethers of propylene glycol				Ethers of dipropylene glycol			Ethers of tripropylene glycol			Ethers of polypropylene glycol	
	Methyl	Ethyl	Isopropyl	n-Butyl	Methyl	Ethyl	n-Butyl	Methyl	Ethyl	Butyl	BPG[a] 400	BPG[a] 800
Mol. formula	$C_4H_{10}O_2$	$C_5H_{12}O_2$	$C_6H_{14}O_2$	$C_7H_{16}O_2$	$C_7H_{16}O_3$	$C_8H_{18}O_3$	$C_{10}H_{22}O_3$	$C_{10}H_{22}O_4$	$C_{11}H_{24}O_4$	$C_{13}H_{28}O_4$		
Mol. wt.	90.1	104.1	118.2	132.2	148.2	162.2	190.2	206.3	220.3	248.3	400 (approx.)	800 (approx.)
Sp. gr. (25/25°C.)	0.919	0.896	0.875	0.879	0.951	0.927		0.965			0.973 (approx.)	0.990 (approx.)
B. p., °C. (760 mm. Hg)	120.1	131–134	139–141	169–172	188.3	193–195	214–217	242.4	250±	255±		
Vapor pressure mm. Hg (25°C.)	10.9	8.2	5.3	1.4	0.36	0.30	0.06	0.022	0.011	0.008	<0.1	<0.1
Refractive index (25/25°C.)	1.402	1.404	1.405	1.415	1.419	1.421	1.418	1.428	1.426	1.424		
Flash point, °F. (open cup)	100	130	140	150	185	205		250				
Per cent in satd. air (25°C.)	1.43	1.08	0.70	0.18	0.047	0.04	0.008	0.003	0.0014	0.001		
1 p.p.m. ⇌ mg./cu. meter at 25°C., 760 mm. Hg	3.68	4.25	4.83	5.40	6.06	6.64	7.77	8.44	9.00	10.1		
1 mg./liter ⇌ p.p.m. at 25°C., 760 mm. Hg	272	235	207	185	165	150.7	128.6	118.5	111	98.6		
Solubility in H₂O g. per 100 g. (25°C.)	∞	∞	∞	4.4	∞	∞	∞	∞			0.2 (20°C.)	0.1 (20°C.)

[a] Butoxypolypropylene glycol.

between secondary and primary carbons. They may be represented by the formula, $ROC_3H_6OC_3H_6OC_3H_6OH$.

3. *Determination in the Atmosphere*

Infrared spectrophotometry would seem to be one of the best methods for the detection of significant amounts of these ethers in the air. Air samples can be drawn through cold carbon bisulfide and the absorbed ether identified and quantitatively determined with sufficient accuracy for industrial hygiene purposes. It is also possible to use other methods, both physical and chemical, as described under the section on ethers of ethylene glycol.

PROPYLENE GLYCOL MONOMETHYL ETHER (Dowanol PM, Ucar Solvent LM)

$CH_3OCH_2CHOHCH_3$

4. *Physiological Response*

Summary. Commercial propylene glycol monomethyl ether is a volatile, colorless liquid with a mild, ethereal odor and a bitter taste. It is low in both single- and repeated-dose oral toxicity, transiently painful to the eyes, not appreciably irritating to the skin, but can be absorbed through the skin in toxic amounts if exposure is extensive and prolonged. The vapors are low in toxicity and the hazard from inhalation also is low because acutely toxic concentrations are essentially intolerable to humans and concentrations that might cause effect from repeated exposures are very disagreeable (irritating to the eyes and mucous membranes and nauseating to some persons). The primary effect of the material is that of an anesthetic agent.

Single-Dose Oral. The single-dose oral LD_{50} for rats for commercial propylene glycol monomethyl ether (Dowanol PM) is reported by Rowe et al.[48] to be 6.6 g./kg. Deaths from massive doses are associated with profound central nervous system depression. Smyth et al.[49] fed special preparations of the two isomers to rats and found the LD_{50} values to be 7.51 g./kg. for the alpha isomer and 5.71 g./kg. for the beta isomer. Shideman and Procita[50] estimate the oral LD_{50} for the dog to be about 10 ml./kg. These investigators attribute death in dogs to respiratory arrest and point out that if the acute effects, which may last for as long as 48 hours depending on dosage, are survived, there are no residual effects.

Repeated-Dose Oral. Rowe et al.[48] report that groups of rats which received 26 doses of 1.0 g./kg. or less over a 35-day period showed no ill effects as judged by appearance, growth, organ weights, and histopathological examination of the

[48] V. K. Rowe, D. D. McCollister, H. C. Spencer, F. Oyen, R. L. Hollingsworth, and V. A. Drill, *Arch. Ind. Hyg. Occupational Med.*, **9**, 509 (1954).

[49] H. F. Smyth, Jr., J. Seaton, and L. Fischer, *J. Ind. Hyg. Toxicol.*, **23**, 259 (1941).

[50] F. E. Shideman and L. Procita, *J. Pharmacol. Exptl. Therap.*, **102**, 70 (1951).

organs. Under the same test conditions, 3.0 g./kg. of this material produced only minor effects in the liver and kidney.

Eye Contact. Repeated application of 1 drop of the undiluted material onto the eyeballs of rabbits for 5 days caused only a mild transitory irritation of the eyelids after each dose.[48] Thus, it should not cause a problem of eye contact under the usual conditions of manufacture and use.

Skin Contact.[48] Propylene glycol monomethyl ether, when tested for skin irritation on rabbits, failed to cause more than a very mild simple irritation and that only after constant contact for several weeks. When applied to rabbits under a "cuff" as advocated by Draize *et al.*,[51] the LD_{50} value was found to be in the range of 13 to 14 g./kg. Depression and incomplete anesthesia are signs commonly associated with the absorption of acutely toxic quantities of this material, especially at dosage levels above 10 ml./kg. When measured doses of propylene glycol monomethyl ether were bandaged repeatedly onto the clipped abdomens of rabbits over a 90-day period, a significant amount of absorption through the skin occurred. At the high dosage levels of 7 and 10 ml./kg., the material caused narcosis and an increased mortality. At the lower levels of 2 and 4 ml./kg., only mild narcosis was apparent. The only organic injury noted in any case was a slight increase in kidney weights of the animals at the top dosage level (10 ml./kg.).

The above skin contact data show that this material is not likely to cause primary irritation. Although it may be absorbed through the skin in toxic amounts, the hazard from skin absorption is very low under the usual conditions of contact.

Inhalation.[48] Acute vapor studies conducted on rats and guinea pigs have shown that: (*a*) at a concentration of approximately 5000 p.p.m., rats and guinea pigs survived single 7-hour exposures; (*b*) at a concentration of approximately 10,000 p.p.m., the LD_{50} for rats was 5 to 6 hours and for guinea pigs it was greater than 7 hours; (*c*) at a concentration of approximately 15,000 p.p.m. (saturated, with some mist present), the LD_{50} for rats was 4 hours and for guinea pigs it was 10 hours; and (*d*) deaths resulting from single inhalation exposures appeared to be due to anesthetic action.

Rabbits and monkeys subjected to 132 exposures of propylene glycol monomethyl ether at 800 p.p.m. (2.91 mg./liter) over a period of 186 days showed no evidence of adverse effects as judged by gross appearance and behavior, growth, final body and organ weights, hematology, and microscopic examination of tissues. Rats and guinea pigs showed no ill effects by the same criteria when they received 130 exposures in a period of 184 days to a concentration of 1500 p.p.m. (5.46 mg./liter). The effects observed at higher concentrations were those of slight growth depression and very slight liver and lung effects. A mild central nervous system depression was noted to result from exposure at the start of the

[51] J. H. Draize, G. Woodard, and H. O. Calvery, *J. Pharmacol. Exptl. Therap.,* **82,** 377 (1944).

experiments at the 3000 p.p.m. level. However, recovery was rapid after cessation of each day's exposure. The animals developed a tolerance after several weeks so that this response was not observed later.

Propylene glycol monomethyl ether in concentrations of the order of 3000 p.p.m. or more had a very strong odor and caused marked nasal irritation that was exceedingly difficult for humans to tolerate. Concentrations of 1000 p.p.m. have quite an objectionable odor and are not likely to be tolerated voluntarily.

The above data on inhalation, both from single and repeated exposures, show that the vapor toxicity of the material is surprisingly low and the hazard from vapor exposure is considerably less than that presented by many common solvents.

The hazard of acute poisoning from single exposures is believed to be relatively minor; toxic vapor concentrations are close to saturation values, they are extremely disagreeable, if not intolerable, and any harmful effects consist of functional depression (anesthesia) of the nervous system. Certainly there is no serious organic injury. The hazard of chronic poisoning, likewise, is quite low. Levels that may be toxic upon repeated exposure probably will not be tolerated voluntarily.

Pharmacology. Shideman and Procita,[50] in studying the pharmacology of the propylene glycol ethers in anesthetized and artificially respired dogs, discovered that in proper dosage propylene glycol monomethyl ether caused auricular fibrillation. The significance of this insofar as practical use is concerned is doubtful for they never observed it in intact animals.

Human Experience. There is no unfavorable human experience known to the author.

5. Hygienic Standard of Permissible Exposure

No hygienic standard for propylene glycol monomethyl ether has been established. However, in view of the data available, it is the author's opinion that vapor concentrations which are not disagreeable to human subjects are safe for prolonged and repeated inhalation. It is further opined that prolonged and repeated skin contact should be avoided.

PROPYLENE GLYCOL MONOETHYL ETHER

$$C_2H_5OCH_2CHOHCH_3$$

4. Physiological Response

Summary. Commercial propylene glycol monoethyl ether is a colorless, volatile liquid with a bitter taste and an ethereal odor. It is low in single-dose oral toxicity, somewhat irritating to the eyes, and not appreciably irritating to the skin. Although it can be absorbed through the skin, this is not likely to be a hazard in ordinary handling and use. Its vapors do not appear to be particularly toxic and concentrations dangerous in short periods (hours) are strongly irritating and narcotic. Data adequate to allow the establishment of safe limits for repeated exposure are not available. The over-all hazard associated with handling and use would seem to be slight.

Single-Dose Oral. Smyth, Seaton, and Fischer[49] fed the two propylene glycol monoethyl ethers, alpha and beta, as 50 per cent aqueous solutions to rats and found the LD_{50}'s to be 7.11 g./kg. and 7.00 g./kg., respectively. Unpublished studies[52] on a commercial product (Dowanol PE) yielded an oral LD_{50} value for rats of 4.9 ml./kg. Marked narcosis and some kidney injury were observed from large doses in both studies.

Repeated-Dose Oral. According to Smyth and Carpenter,[53] who administered the beta isomer to rats in their drinking water for 30 days, a daily dose of 0.68 g./kg. was without adverse effect while a daily dose of 2.14 g./kg. caused reduced growth. Reference to this same article later by Smyth[54] indicates that kidney injury, but no death, also resulted from the 2.14 g./kg. daily dosage level.

Eye Contact.[52] When the commercial product was applied to the eyes of rabbits on 5 consecutive days, it caused conjunctival irritation and some transient cloudiness of the cornea.

Skin Contact.[52] When the material was confined to the skin of rabbits in a manner essentially the same as described by Draize *et al.*,[51] all of 6 animals treated survived 5 ml./kg., 3 of 5 survived 7 ml./kg., and 1 of 5 survived doses of either 10 or 15 ml./kg. The LD_{50} was estimated to be about 9 ml./kg. The material in large doses caused marked central nervous system depression and deaths usually occurred within 48 hours after treatment. No appreciable irritation of the skin resulted.

Inhalation. Gross[55] reports that a mouse, guinea pig, and rabbit tolerated up to 7000 p.p.m. for 1 hour without effect other than irritation of the eyes and respiratory organs. A 2-hour exposure caused more irritation and a rabbit showed signs of kidney injury, transient albumin and red cells in the urine. All of 5 rats exposed for 4 hours to a concentration calculated to be 10,000 p.p.m. survived, but they showed marked irritation of the eyes and nares and they were deeply anesthetized at the end of the exposure.[52] Gross[55] also reported exposing cats, guinea pigs, and rabbits 8 hours a day to a concentration of about 1200 p.p.m. One of 2 cats and 1 of 2 guinea pigs tolerated 12 such exposures without apparent effect but the other cat and guinea pig died after the treatment. The 2 rabbits succumbed after 3 or 9 days; autopsy revealed pneumonia and kidney injury.

Human Experience. No adverse human experience has been reported.

[52] Biochemical Research Laboratory, The Dow Chemical Co., unpublished data.

[53] H. F. Smyth, Jr. and C. P. Carpenter, *J. Ind. Hyg. Toxicol.*, **30**, 63 (1948).

[54] H. F. Smyth, Jr., in G. O. Curme, Jr. and F. Johnston, eds., *Glycols* (American Chemical Society Monograph, Series 114). Reinhold, New York, 1953.

[55] E. Gross in K. B. Lehmann and F. Flury, *Toxicology and Hygiene of Industrial Solvents*. Trans. by E. King and H. F. Smyth, Jr., Springer, Berlin, 1938.

5. *Hygienic Standard of Permissible Exposure*

Data adequate to establish a hygienic standard for repeated exposure are not available.

PROPYLENE GLYCOL MONOISOPROPYL ETHER

$$(CH_3)_2CHOCH_2CHOHCH_3$$

4. *Physiological Response*

Summary.[52] Propylene glycol monoisopropyl ether is a colorless, volatile liquid with a bitter taste and a slight odor. The toxicological information available concerns the commercial product and is not extensive. It is low in single-dose oral toxicity for rats, the LD_{50} being greater than 2 g./kg. and in the neighborhood of 4 g./kg. In the eyes of rabbits, it causes conjunctival irritation, some corneal injury, and iritis which heal within a week. It is only very slightly irritating even upon prolonged and repeated contact with the intact skin and does not seem to be readily absorbed in acutely toxic amounts. Rats that received a single 7-hour exposure to an essentially saturated atmosphere survived but exhibited drowsiness, labored breathing, temporary weight loss, and mild kidney injury at autopsy.

PROPYLENE GLYCOL MONO-*n*-BUTYL ETHER

$$C_4H_9OCH_2CHOHCH_3$$

4. *Physiological Response*

Summary.[52] Propylene glycol mono-*n*-butyl ether is a colorless, slightly volatile liquid with a bitter taste and a slight odor. The toxicological information available is not extensive. The material is low in single-dose oral toxicity, the LD_{50} for rats being 2.2 ml./kg. When studied on rabbits, it was found to be appreciably irritating to the eyes: 1 drop in an eye on 5 consecutive days caused marked conjunctival irritation and corneal cloudiness, which healed within a week. Repeated applications (10 in 14 days) to the skin of rabbits resulted in slight simple irritation and some evidence that toxic amounts were absorbed. Single dose absorption studies conducted essentially as described by Draize *et al.*[51] showed that all of 5 animals receiving 2 ml./kg. survived, 2 of 5 receiving 3 ml./kg. survived and none of 5 receiving 5 ml./kg. survived. When the material was confined under a cuff for 24 hours, as in these studies, rather severe injury to the skin occurred and the animals became deeply narcotized. Deaths from the larger doses occurred within a few hours after application of the material. All deaths occurred within 24 hours after treatment or not at all. Information regarding the effects from inhalation is lacking.

Human Experience. There has been no adverse human experience reported. It would seem, however, that propylene glycol mono-*n*-butyl ether should be handled in such a way that contact with the eyes and prolonged or repeated contact with the skin are prevented.

5. *Hygienic Standard of Permissible Exposure*

Until appropriate data regarding the vapor inhalation toxicity become available, no hygienic standard can be suggested. In the meantime, it would seem prudent to avoid inhaling vapors particularly over a prolonged period or repeatedly.

DIPROPYLENE GLYCOL MONOMETHYL ETHER (Dowanol DPM, Ucar Solvent 2LM)

$$CH_3OC_3H_6OC_3H_6OH$$

4. *Physiological Response*

Summary. Dipropylene glycol monomethyl ether is a colorless liquid, low in volatility, with a mild, pleasant, ethereal odor and a bitter taste. It is low in single-dose oral toxicity, transiently painful but not damaging to the eyes, and is neither appreciably irritating to the skin nor readily absorbed through the skin of rabbits in toxic amounts when exposures are prolonged and repeated.

It caused neither irritation nor sensitization when tested on human subjects. It is low in toxicity by inhalation. The hazards to health associated with the handling and ordinary use of this material would seem to be minimal.

Single-Dose Oral. The single dose oral LD_{50} for male and female rats was found to be 5.50 ml./kg. and 5.45 ml./kg., respectively.[48] The material produces marked central nervous system depression. Shideman and Procita[50] estimated the single dose oral LD_{50} for dogs to be 7.5 ml./kg. They noted that death was due to respiratory failure and usually occurred within 48 hours or not at all.

Eye Contact.[48] Dipropylene glycol monomethyl ether is not appreciably irritating to the eyes. When 1 drop of undiluted material was placed in a rabbit's eye on each of 5 consecutive days, a mild transitory irritation of the conjunctival membranes occurred. Fluorescein staining revealed no corneal damage.

Skin Contact. Irritation.[48] Continuous contact of dipropylene glycol monomethyl ether with the skin of numerous rabbits for 90 days caused only a very slight scaliness, far less in intensity than might have been expected on the basis of the solvent properties of the material. In fact, the response was similar to that produced by water alone under the same conditions. When patch-tested on 250 human beings by accepted techniques, it produced no evidence of primary irritation or sensitization of the skin.

Skin, Contact. Absorption.[48] Single application studies conducted essentially as described by Draize *et al.*[51] revealed that dipropylene glycol monomethyl ether is not absorbed through the skin in acutely dangerous amounts even when massive doses (20 ml./kg.) are held in continuous contact with a large area of the rabbit's skin for a period of 24 hours. Sufficient absorption did occur, however, to result in transient narcosis.

When dipropylene glycol monomethyl ether was applied 5 times a week for 90 days at dosage levels of 1.0, 3.0, 5.0, and 10.0 ml./kg., the following observations were made:

1. Mortality was high at the 10.0 ml./kg. dosage level, slight at 5.0 ml./kg. level, and absent at the 1.0 and 3.0 ml./kg. levels.

2. No adverse body weight occurred at any level except just prior to death in those animals that succumbed, presumably to the narcotic effects of the top dosage levels.

3. No hematological changes occurred at any dosage level.

4. The effect of severe (repeated and prolonged) exposure to the skin was slight, being similar to that caused by distilled water under similar conditions.

5. Observations for gross pathology revealed only gastric distension and occasional gastric irritation in those animals dying at the 10 ml./kg. dosage level.

6. No significant organ weight changes occurred at any dosage level.

7. The blood urea nitrogen concentration was unaffected in the animals surviving the 3.0 and 5.0 ml./kg. dosage level.

8. Histopathological studies conducted on the liver, lung, spleen, adrenal, heart, testes, and stomach of those animals receiving the 5.0 and 10.0 ml./kg. dosage levels revealed no changes. The kidneys of those animals on the 10.0 ml./kg. level showed some granular and some hydropic changes; at the 5.0 ml./kg. level these same kidney abnormalities were observed but they were of no greater intensity than those observed in some of the controls.

Inhalation.[48] Three groups of 3 adult male white rats were given single 7-hour exposures to an atmospheric concentration of dipropylene glycol monomethyl ether calculated to be about 500 p.p.m. This atmosphere was laden with fog and the animals were wet with the material at the end of the exposure. They exhibited only mild narcosis from which they rapidly recovered.

Groups of rats, rabbits, guinea pigs, and monkeys were exposed 7 hours a day, 5 days a week, for periods of from 6 to 8 months to an atmosphere containing about 300 p.p.m. (essentially saturated) of the material. When judged by growth, general appearance, hematological studies, gross observations at autopsy, and organ weight studies, only the rats exhibited a mild transitory narcosis during the first few weeks of exposure and a statistically significant but very slight increase in the weights of their livers. The other species failed to exhibit any abnormalities. The results of histopathological examination revealed no evidence of adverse effect except for minor changes in the livers of the female guinea pigs, rabbits, and monkeys. It should be noted that this concentration of vapor is quite disagreeable to human subjects and it is doubtful if it would be tolerated voluntarily.

Pharmacology. Shideman and Procita[50] in 1951 reported rather extensive studies on dipropylene glycol monomethyl ether. In the intact dog, they found the material to be primarily a depressant of the central nervous system with death being due to respiratory failure. In the anesthetized dog, they observed that intravenous injection caused a precipitous drop in blood pressure. This response was unaffected by the prior administration of atropine and by bilateral vagotomy. They also noted that in the anesthetized and artificially respired dog proper doses of the material intravenously induced auricular, but not ventricular, fibrillation.

Rucknagel and Surtskin[56] in 1952 confirmed the observations of Shideman and Procita. The practical significance of these findings is questionable since it has never been observed in the intact animal, regardless of the dose or mode of administration.

Human Experience. No injury or adverse effects have been reported from the handling and use of dipropylene glycol monomethyl ether.

5. *Hygienic Standards of Permissible Exposure*

In view of the data and experience available, it would appear that if vapor concentrations were controlled to levels not disagreeable to unacclimated human subjects, there would be no hazard from inhalation. In 1961 The Conference of Governmental Industrial Hygienists recommended a threshold limit value of 100 p.p.m. Control of vapors to this level will certainly prevent systemic injury.

DIPROPYLENE GLYCOL MONOETHYL ETHER

$$C_2H_5OC_3H_6OC_3H_6OH$$

4. *Physiological Response*

Summary.[52] Dipropylene glycol monoethyl ether is a colorless liquid, low in volatility, with a mild, pleasant, ethereal odor and a bitter taste. The material is low in single-dose oral toxicity, the LD_{50} for rats being 4 ml./kg. When 1 drop of the liquid was introduced into the eyes of rabbits for 5 consecutive days, it caused a slight transitory conjunctival irritation but no corneal injury. Repeated applications (10 in 14 days) to the skin of rabbits resulted in only a very slight exfoliation and there was no evidence that toxic amounts were absorbed. However, when single doses of the material were confined to the skin for 24 hours under a cuff, essentially as described by Draize et al.,[51] all of 5 animals receiving 5 ml./kg. survived, 3 of 10 receiving 10 ml./kg. survived, and 0 of 5 receiving 15 ml./kg. survived. All animals exhibited transient weight loss. Narcosis was apparent in all animals but was profound at the higher dosage levels. The animals were cold to the touch and the skin beneath the cuff was burned. Death occurred in almost all cases within 1 or 2 days and recovery was apparently complete in the survivors within 3 days. Rats were exposed to atmospheres essentially saturated (calculated to be about 400 p.p.m.) for 7 hours. One of 12 animals so exposed died. The others evidenced some degree of irritation of the eyes and nares, transient weight loss, but recovery appeared complete 24 hours later.

Human Experience. There have been no reports of adverse human experience.

5. *Hygienic Standards of Permissible Exposure*

Until more extensive data regarding the vapor inhalation toxicity become available, no hygienic standard can be suggested. It would seem, however, that

---

[56] D. L. Rucknagel and A. Surtskin, *Proc. Soc. Exptl. Biol. Med.,* **80,** 584 (1952).

under reasonable conditions of handling at room temperature, there would be no appreciable hazard.

DIPROPYLENE GLYCOL MONO-*n*-BUTYL ETHER

$$C_4H_9OC_3H_6OC_3H_6OH$$

4. Physiological Response

Summary.[52] Dipropylene glycol mono-*n*-butyl ether is a colorless, slightly volatile liquid with a bitter taste and a slight odor. When the material was given in single oral doses to rats, the LD_{50} was found to be 2 ml./kg. When 1 drop of the material was introduced into the eye of a rabbit, on 5 consecutive days, it produced only a transient, slight conjunctival irritation but no corneal injury. Repeated applications (10 in 14 days) to the skin of rabbits resulted in very slight simple irritation that readily cleared; based upon appearance, behavior, and weight change, there was no evidence that absorption of toxic quantities had occurred. No quantitative skin absorption studies or inhalation work were done.

Human Experience. There has been no adverse human experience reported and none would be expected under reasonable conditions of handling.

5. Hygienic Standards of Permissible Exposure

Until appropriate data regarding the vapor inhalation toxicity of the material become available, no hygienic standard can be suggested. It would seem, however, that there would be little hazard from inhalation under ordinary conditions of handling and use.

TRIPROPYLENE GLYCOL MONOMETHYL ETHER (Dowanol TPM)

$$CH_3OC_3H_6OC_3H_6OC_3H_6OH$$

4. Physiological Response

Summary. Tripropylene glycol monomethyl ether is a colorless liquid, low in volatility, with slight, pleasant, ethereal odor and a bitter taste. It is low in single-dose oral toxicity, transiently painful but not damaging to the eye, not appreciably irritating to the skin unless exposure is severe, but can be absorbed through the skin in toxic quantities if exposure is prolonged and repeated. Its vapor pressure and toxicity are sufficiently low so that no hazard exists from a single vapor exposure and probably none exists even when exposures are repeated. The hazards to health from ordinary handling and use would seem to be negligible.

Single-Dose Oral. When fed in single doses to rats the LD_{50} for the material appeared to be about 3.3 g./kg. The primary effect of the material appears to be narcotic.[48] Shideman and Procita[50] have estimated the LD_{50} for dogs to be 5 ml./kg. They assert that the primary effect of the material is central nervous system depression with death from large doses due to respiratory failure.

Repeated-Dose Oral. No data available.

Acute Vapor Inhalation. When rats were exposed once for 7 hours to an atmosphere essentially saturated at 25°C. with the vapors, no ill effects were observed. This indicates that the material does not present a hazard from acute vapor exposure at ordinary temperatures.

Repeated Vapor Inhalation. Since no repeated vapor exposures have been conducted, the hazard from repeated vapor inhalation cannot be evaluated. However, since the vapor pressure is such that at ordinary room conditions a saturated atmosphere would contain only about 50 p.p.m. of the material, the hazard from repeated inhalation cannot be considered to be serious.

Eye Contact.[48] Repeated instillation of the liquid into the eye of a rabbit failed to cause serious injury. Evidence of transient pain was observed.

Skin Contact.[48] When tripropylene glycol monomethyl ether was applied repeatedly to the skin of rabbits, it caused only a very mild simple irritation. This would indicate that the material would not be likely to produce skin irritation unless exposures were very severe. When applied to rabbits for 24 hours under a cuff, as advocated by Draize,[51] a single dose of 20 ml./kg. was survived. Such an observation indicates no practical hazard of systemic intoxication from occasional skin contact.

When measured doses were bandaged repeatedly onto the clipped abdomens of rabbits over a 90-day period, a significant amount of absorption through the skin occurred. At the high dosage levels (5 to 10 ml./kg.), the material caused narcosis and kidney injury. At lower dosage levels (below 5 ml./kg.) narcosis was not apparent, but there was some kidney injury even at the lowest dosage level administered (1 ml./kg./day). These results indicate that prolonged and repeated skin contact with appreciable amounts of the material should be avoided.

Pharmacology. Shideman and Procita[50] report that in anesthetized and artificially respired dogs proper dosage of the material caused auricular fibrillation. The significance of this in practical use is doubtful since they did not cause such a response from oral feeding.

Human Experience. No unfavorable experience is known to the writer.

5. Hygienic Standard of Permissible Exposure

No hygienic standard for tripropylene glycol monomethyl ether would seem to be required due to its relatively low toxicity and low volatility.

TRIPROPYLENE GLYCOL MONOETHYL ETHER

$$C_2H_5OC_3H_6OC_3H_6OC_3H_6OH$$

4. Physiological Response

Summary.[52] Tripropylene glycol monoethyl ether is a colorless liquid low in volatility, with a slight pleasant ethereal odor and a bitter taste. When single oral doses of the material were administered to rats, the LD_{50} was found to be 2 ml./kg. When 1 drop of the liquid was introduced into the eye of a rabbit on 5 consecutive days, it produced only a very slight conjunctival irritation and no corneal injury. Repeated applications (10 in 14 days) to the skin of rabbits resulted in very slight simple irritation without evidence that toxic amounts were absorbed. No quantitative skin absorption studies or inhalation studies were conducted.

Human Experience. There has been no adverse human experience reported.

5. *Hygienic Standards of Permissible Exposure*

No hygienic standard for tripropylene glycol monoethyl ether would seem to be required because of its relatively low volatility and low toxicity. There would seem to be little or no hazard associated with ordinary handling and use of this substance.

TRIPROPYLENE GLYCOL MONO-*n*-BUTYL ETHER

$$C_4H_9OC_3H_6OC_3H_6OC_3H_6OH$$

4. *Physiological Response*

Summary.[52] Tripropylene glycol mono-*n*-butyl ether is a colorless liquid, low in volatility, with a slight ethereal odor and a bitter taste. When administered in single oral doses to rats, the LD_{50} was found to be 1.84 ml./kg. When 1 drop of the liquid was instilled into the eye of a rabbit on 5 consecutive days, no apparent irritation was produced. Repeated applications (10 in 14 days) to the skin of rabbits resulted in only a very slight exfoliation. There was, however, a slight weight loss during the course of the experiment indicating that perhaps toxic amounts were being absorbed. It would appear, therefore, that prolonged and repeated skin contact with this substance should be avoided.

Human Experience. There has been no adverse human experience reported.

5. *Hygienic Standards of Permissible Exposure*

Because of the very low vapor pressure of this substance and its relatively low toxicity, it does not seem that a hygienic standard would be necessary. There would seem to be very little hazard associated with the handling and use of this substance.

POLYPROPYLENE GLYCOL BUTYL ETHERS

4. *Physiological Response*

Summary. Carpenter, Critchfield, Nair, and Shaffer[57] have reported on toxicological studies conducted on two polypropylene glycol butyl ethers having molecular weights approximating 400 (BPG 400) and 800 (BPG 800). Both BPG 400 and BPG 800 are low in toxicity when given either orally or intraperitoneally.

---

[57] C. P. Carpenter, F. N. Critchfield, J. H. Nair, III, and C. B. Shaffer, *Arch. Ind. Hyg. Occupational Med.*, **4**, 261 (1951).

The oral LD_{50} values for BPG 400 for male rats, male guinea pigs, and male rabbits were found to be 5.84, 2.46, and 3.30 g./kg., respectively. Similar values for these same species for BPG 800 were 9.16, 6.8, and 23.7 g./kg., respectively. The intraperitoneal LD_{50} values for rats were found to be 0.32 g./kg. for BPG 400 and 0.91 g./kg. for BPG 800. When large single oral doses were given, the principal effects of both materials were gastrointestinal irritation, congestion of internal organs, and death, usually within 24 hours or not at all. The BPG 800 was less distressing to the animals than the BPG 400 and more likely to cause convulsions and lung hemorrhage.

Dietary feeding studies over a 90-day period showed the no-effect dosage level for BPG 400 to be between 0.16 and 0.67 g./kg./day and for BPG 800 to be less than 0.52 g./kg./day. The liver and/or kidney were the first organs affected when subacute dosage levels were given repeatedly.

Neither material is more than very slightly irritating to the rabbit's skin or eyes. BPG 800 is neither an irritant nor a sensitizer of human skin. Neither is readily absorbed through the skin in acutely toxic amounts but repeated inunction of the rabbit's skin for 30 days showed BPG 400 to be moderate in toxicity and BPG 800 to be low in toxicity by this route, the no-effect levels being less than 0.1 ml./kg./day and 1.0 or more ml./kg./day, respectively. BPG 400 is readily absorbed from the gastrointestinal tract whereas BPG 800 is poorly absorbed. BPG 400 was not stored in the bodies of rats fed large doses for 30 days. Neither material presents a hazard from inhalation under reasonable conditions. Rats exposed to atmospheres essentially saturated at room temperatures for 8 hours were unaffected and they suffered only mild effects when exposed for 8 hours to fogs of the material.

It would seem that under ordinary or reasonable conditions, neither BPG 400 nor BPG 800 would present any appreciable hazards to health in industrial handling. It is clear, however, that the BPG 800 is distinctly less toxic than BPG 400.

ETHERS OF BUTYLENE GLYCOL

1. *Source, Uses, and Industrial Exposure*

The methyl, ethyl, and *n*-butyl ethers of butylene glycol considered herein are prepared by reacting the appropriate alcohol with so-called "straight-chain" butylene oxide consisting of about 80 per cent 1,2-isomer and about 20 per cent 2,3-isomer in the presence of a catalyst. They are colorless liquids with slight pleasant odors. The methyl and ethyl ethers are miscible with water but the butyl ether has limited solubility. All are miscible with many organic solvents and oils; thus, they are useful as mutual solvents, dispersing agents, solvents for inks, resins, lacquers, oils, and greases. Industrial exposure may occur by any of the common routes.

2. Physical and Chemical Properties

The physical and chemical properties of these ethers are given in Table 5.

TABLE 5

Physical and Chemical Properties of Ethers of Butylene Glycol

Property	Methyl	Ethyl	n-Butyl
Mol. formula	$C_5H_{12}O_2$	$C_6H_{14}O_2$	$C_8H_{18}O_2$
Mol. wt.	104.1	118.2	146.2
Sp. gr. (25/25°C.)	0.983	0.888	0.877
B. p., °C. (760 mm. Hg)	136	147	180–187
Vapor pressure, mm. Hg (25°C.)	5.5	3.0	0.62
Refractive index (25°C.)	1.408	1.410	1.420
Flash point, °F. (open cup)	110	145	160
Per cent in satd. air (25°C.)	0.72	0.40	0.081
1 p.p.m. ≏ mg./cu. meter at 25°C., 760 mm. Hg	4.25	4.83	5.98
1 mg./liter ≏ p.p.m. at 25°C., 760 mm. Hg	235	207	167
Solubility in H_2O, g. per 100 g. (25°C.)	∞	∞	3.7

3. Determination in the Atmosphere

The same methods as described for the ethers of propylene glycol are applicable.

BUTYLENE GLYCOL MONOMETHYL ETHER

$CH_3OC_4H_8OH$

4. Physiological Response[58]

Summary. Butylene glycol monomethyl ether is low in single-dose oral toxicity for rats, moderately irritating and injurious to the eyes of rabbits, not appreciably irritating to the skin, and not readily absorbed through the skin of rabbits in toxic amounts. Prolonged inhalation by rats of air essentially saturated with vapors causes drowsiness, unsteadiness, and slight injury of the liver and kidneys; when some fog was present, some died.

Single-Dose Oral. When single oral doses of butylene glycol monomethyl ether were fed to rats, doses of 2 g./kg. were survived and doses of 4 g./kg. were fatal. No pathology was noted upon gross observation of the organs of the survivors.

Eye Contact. The undiluted material when instilled into the eyes of rabbits was painful, and caused moderate conjunctival irritation, moderate corneal injury, and some iritis, all of which cleared within a week.

Skin Contact. The material caused only very slight irritation of the skin of rabbits even though exposures were prolonged and repeated. Evidence of absorption through the skin was not apparent.

[58] Biochemical Research Laboratory, The Dow Chemical Co., unpublished data.

Inhalation. An atmosphere essentially saturated (probably **6000** to **7000** p.p.m.) with vapor was not lethal to any of 3 rats exposed for 7 hours but it did cause drowsiness, unsteadiness, temporary weight loss, and some mild kidney and liver changes. An exposure lasting 4 hours caused drowsiness and unsteadiness but no histopathological changes. When 6 rats were exposed for 7 hours to a super-saturated atmosphere (containing some fog), all animals became very drowsy, were unable to stand, breathed with difficulty, and 4 died. No significant pathological changes were seen in the organs of the survivors.

Human Experience. No adverse human experience has been noted.

5. Hygienic Standards of Permissible Exposure

Data adequate to establish a hygienic standard for repeated inhalation are lacking. Although the butylene glycol monomethyl ether considered herein would not appear to present serious hazards in ordinary handling and use, care should be taken to avoid contact with the eyes. Otherwise, ordinary precautions such as the avoidance of prolonged or repeated contact with the skin and the inhalation of high concentrations of vapors for prolonged periods would seem to be adequate to insure safety in industrial handling and use.

BUTYLENE GLYCOL MONOETHYL ETHER

$$C_2H_5OC_4H_8OH$$

4. Physiological Response[58]

Summary. Butylene glycol monoethyl ether is low in single-dose oral toxicity for rats, moderately irritating and injurious to the eyes of rabbits, not appreciably irritating to the skin, and not readily absorbed through the skin in toxic amounts. Air essentially saturated with vapor was only slightly toxic to rats.

Single-Dose Oral. When single oral doses of butylene glycol monoethyl ether were fed to rats, doses of 1 g./kg. were survived, doses of 2 g./kg. caused the death of one of three, and doses of 4 g./kg. caused the death of one of two. Doses of 2 and 4 g./kg. caused injury to the lungs, liver, and kidney, but lower dosage levels did not.

Eye Contact. When the material was instilled into the eyes of rabbits it caused pain, marked conjunctival and corneal injury, and slight iritis.

Skin Contact. The undiluted material was not appreciably irritating to the skin of rabbits even when exposures were prolonged and repeated. Evidence of absorption through the skin was not apparent.

Inhalation. Rats exposed for 7 hours to an atmosphere essentially saturated (probably 3000 to 4000 p.p.m.) with vapor survived without serious effects, although some kidney injury was apparent upon gross examination of the organs. However, when a fog was present, 6 rats exposed for 7 hours exhibited irritation of the eyes and nares, drowsiness, inability to stand, and difficulty in breathing.

Although none died, all were ill and gross examination of the organs revealed injury to the lungs, kidneys, and livers.

Human Experience. No adverse human experience has been noted.

5. Hygienic Standard of Permissible Exposure

Data adequate to establish a hygienic standard for repeated inhalation are lacking. Although the butylene glycol monoethyl ether considered herein would not appear to present serious hazards in ordinary handling and use, care should be taken to prevent contact with the eyes. Otherwise, ordinary precautions such as the avoidance of prolonged and repeated skin contact and the inhalation for prolonged periods of time of high concentrations of vapor would seem adequate to insure safety in industrial handling and use.

BUTYLENE GLYCOL MONO-*n*-BUTYL ETHER

$$C_4H_9OC_4H_8OH$$

4. Physiological Response[58]

Summary. Butylene glycol mono-*n*-butyl ether is low in single-dose oral toxicity for rats. In studies on rabbits, it was found to be moderately irritating and injurious to the eyes, not appreciably irritating to skin when unconfined but moderately injurious when confined, and not readily absorbed through the skin in acutely toxic amounts. A saturated atmosphere was not especially toxic to rats.

Single-Dose Oral. When single oral doses of butylene glycol mono-*n*-butyl ether as a 20 per cent solution in corn oil were fed to rats, doses of 2.0 g./kg. were survived and doses of 4.0 g./kg. were lethal. High doses caused prostration and labored breathing. Autopsy of surviving animals revealed some kidney injury.

Eye Contact. The undiluted material, when instilled into the eyes of rabbits, was painful, irritating to the conjunctival membranes, injurious to the cornea, and caused some iritis. All of these effects disappeared within 7 days after exposure.

Skin Contact. The material was not irritating to the skin of rabbits when applied to the uncovered skin. However, when applied under a bandage for 48 hours, it caused a burn. Acutely toxic amounts were not readily absorbed through the intact skin.

Inhalation. Rats were not noticeably affected by an exposure of 7 hours duration to an atmosphere saturated at 100°C. and then cooled to room temperature. Autopsy, however, showed that some kidney injury had resulted.

Human Experience. No adverse human experience has been reported.

5. Hygienic Standard of Permissible Exposure

Data adequate to establish a hygienic standard for repeated inhalation are lacking. The limited toxicological information available does not indicate that butylene glycol mono-*n*-butyl ether presents any unusual hazards. Ordinary

TABLE 6

Physical Properties of Esters, Diesters, and Ether–Esters of Glycols

Property	Ethylene glycol Mono-acetate	Di-acetate	Di-nitrate	Ethylene glycol ether–esters Methyl ether acetate	Ethyl ether acetate	Butyl ether acetate	Diethylene glycol ether–esters Methyl ether acetate	Ethyl ether acetate	Butyl ether acetate
Mol. formula	$C_4H_8O_3$	$C_6H_{10}O_4$	$C_2H_4O_6N_2$	$C_5H_{10}O_3$	$C_6H_{12}O_3$	$C_8H_{16}O_3$	$C_7H_{14}O_4$	$C_8H_{16}O_4$	$C_{10}H_{20}O_4$
Mol. wt.	104.1	146.2	152	118.13	132.16	160.21	162.2	176.22	204.27
Sp. gr., 20/20°C.	1.106	1.1063	1.4962 (20/15° C.)	1.0067	0.9748	0.9422	1.04	1.0114	0.9810
F. pt. °C.		−39.5	−22.3	−65.1	−61.7	−64.6		−25	−32.2
B. p. °C. (760 mm. Hg)	182	190.8	114–116 (explodes)	145.1	156.4	191.5	194.2	217.4	246.8
Vapor pressure mm. Hg (20°C.)		0.25	0.049	2.0–3.7	1.2–1.7	0.25–0.3	0.12	0.05	<0.01
Flash point, °F. (open cup)	>102	255		140	150	190	180	230	240
Vapor density (air = 1)	3.59	5.04	5.25	4.07	4.72			6.07	
Refractive index (20°C.)	1.4224	1.4159	1.4473	1.4019	1.4058	1.4200			
Per cent in satd. air (25°C.) (approximate)		0.044		0.31–0.60	0.21–0.27	0.044–0.06		0.01	<0.002
1 p.p.m. ⇌ mg./cu. meter at 25° C., 760 mm. Hg	4.25	5.97		4.83	5.40	6.54	6.63	7.20	8.34
1 mg./liter ⇌ p.p.m. at 25°C., 760 mm. Hg	235	167		207	185	157	151	139	120
Solubility in H_2O, g. per 100 g. (20°C.)	16.4	16.4		∞	22.9	1.1	∞	∞	6.5

precautions to prevent contact with the eyes, prolonged and repeated contact with the skin, and inhalation of vapors would seem adequate to prevent the occurrence of systemic toxicity.

ESTERS, DIESTERS, AND ETHER–ESTERS OF GLYCOLS

1. Source, Uses, and Industrial Exposure

The common esters and diesters of the common polyols are prepared commercially by esterifying the particular polyol with the acid, acid anhydride, or acid chloride of choice in the presence of a catalyst. Mono- or diesters result, depending upon the proportions of each reactant employed. The ether–esters are prepared by esterifying the glycol ether in a similar manner. Other methods also can be used.[59]

The acetic acid esters have remarkable solvent properties for oils, greases, inks, adhesives, and resins. They are widely used in lacquers, enamels, dopes, and adhesives to dissolve the plastics or resins. They are also used in lacquer, paint, and varnish removers.

The nitric acid ester, ethylene glycol dinitrate, is widely used as an explosive, usually in combination with nitroglycerin to reduce the freezing point.

Industrial exposures of consequence are most likely to occur through the inhalation of vapors, although excessive contact with the eyes and skin may also occur. With the dinitrate, a serious hazard exists from absorption through the skin.

2. Physical and Chemical Properties

The physical and chemical properties of the materials considered herein are given in Tables 6 and 7.

TABLE 7

Physical Properties of Monomethyl Ether Acetates of
Mono-, Di-, and Tripropylene Glycol

Property	Mono-	Di-	Tri-
Mol. formula	$C_6H_{12}O_3$	$C_9H_{18}O_4$	$C_{12}H_{24}O_5$
Mol. wt.	132.1	190.1	248.2
B. p., °C. (760 mm. Hg)	146	209	258
Refractive index (25°C.)	1.3995		
Sp. gr. (25/4°C.)	0.957		
Solubility in H_2O, g. per 100 g. (25°C.)	>25		
1 p.p.m. ≈ mg./cu. meter at 25°C., 760 mm. Hg	5.40	7.77	10.15
1 mg./liter ≈ p.p.m. at 25°C., 760 mm. Hg	185	128.6	98 5

[59] A. B. Boese, Jr., C. K. Fink, and H. G. Goodman, Jr., in G. O. Curme, Jr., and F. Johnston, eds., *Glycols* (American Chemical Society Monograph, Series 114). Reinhold, New York, 1953.

3. Determination in the Atmosphere

The choice of methods for the determination of the esters, diesters, and ether–esters of various glycols will vary with existing conditions. Nawrocki and co-workers[60] described an infrared method in 1944 for glycol ethers but there is no doubt that it can be adapted also for the esters. Improvements in technique and instruments and their ready availability would seem to make infrared spectroscopy the method of choice. Liquid-gas chromatography would also offer a means not only of resolving mixtures of vapors but also of identifying the components. Mass spectrophotometry may also be used where appropriate instruments are available. Chemical methods such as proposed by Werner and Mitchell[61] for the ethers or by Morgan[62] for esters or ether esters may be useful where spectroscopic equipment is not available.

4. Physiological Response

Summary. Generally speaking, the fatty acid esters of the glycols and glycol ethers, either in the liquid or vapor state, are more irritating to the mucous membranes than those of the parent glycol or glycol ethers. However, once absorbed into the body, the esters are saponified and the systemic effect is quite typical of the parent glycol or glycol ether. Lepkovski, Over, and Evans,[63] in studies with higher fatty acids of glycols, concluded that the fatty acids were liberated and used nutritionally. Furthermore, they observed that severe injury to the tubular epithelium of the kidneys occurred when the esters of ethylene glycol and diethylene glycol were fed, but not when equivalent amounts of fatty esters of propylene glycol, glycerol, ethyl alcohol, methyl alcohol, or the free fatty acids themselves were fed. Shaffer and Critchfield,[64] in studies with polyethylene glycol 400 monostearate, concluded that it was low in toxicity and also utilized nutritionally.

Lest generalizations be too broadly interpreted, it should be noted that the nitric acid esters, ethylene glycol dinitrate and nitroglycerin, are highly toxic and exert a physiological action quite different than that of the parent polyol.

ETHYLENE GLYCOL MONOACETATE (Glycol Monoacetate, Solvent GC)

$$HOCH_2CH_2OOCCH_3$$

4. Physiological Response

Summary. Ethylene glycol monoacetate is a colorless, slightly volatile liquid with a faint fruity odor and a slightly bitter taste. It is low in single-dose oral

[60] C. Z. Nawrocki, F. S. Brackett, and H. W. Werner, *J. Ind. Hyg. Toxicol.,* **26,** 193 (1944).
[61] H. W. Werner and J. L. Mitchell, *Ind. Eng. Chem. Anal. Ed.,* **15,** 375 (1943).
[62] P. W. Morgan, *Ind. Eng. Chem. Anal. Ed.,* **18,** 500 (1946).
[63] S. Lepkovski, R. A. Over, and H. M. Evans, *J. Biol. Chem.,* **108,** 431 (1935).
[64] C. B. Shaffer and F. H. Critchfield, *Federation Proc.,* **7,** 254 (March, 1948).

toxicity, the LD_{50} values reported by Smyth et al.[65] being 8.25 g./kg. for rats and 3.80 g./kg. for guinea pigs when fed as a 50 per cent aqueous solution. It is moderately irritating to the eyes,[66] but not appreciably irritating to the skin of animals or humans.[67] Dogs were apparently unaffected by 12 feedings of 0.1 or 0.5 ml./kg.[67] Seven subcutaneous injections of 0.5 or 1.0 ml./kg. did not injure guinea pigs.[67] Twelve 8-hour exposures to an atmosphere essentially saturated with vapor at room temperature were survived by cats, guinea pigs, and mice, but caused lung irritation and slight kidney injury. One rabbit treated similarly died.[67] Studies by Rosser,[68] quoted by Gross,[67] indicate that cats tolerated a single 6-hour exposure to an atmosphere containing mist (28 mg./liter) but succumbed from two such exposures. Wiley and co-workers[69] gave repeated daily injections of ethylene glycol monoacetate to dogs (8.5 ml./day) and rabbits (3.5 ml./day) and observed a substantial increase in urinary oxalic acid similar to that caused by the administration of an equivalent amount of ethylene glycol. Also noted were degenerative changes in the kidneys, testes, and brain.

Human Experience. There have been no reports of untoward effects in human beings.

5. *Hygienic Standards of Permissible Exposure*

The toxicological data available are inadequate to allow the establishment of a hygienic standard for repeated inhalation.

ETHYLENE GLYCOL DIACETATE

$$CH_3COOCH_2CH_2OOCH_2CH_3$$

4. *Physiological Response*

Summary. Ethylene glycol diacetate is a colorless, slightly volatile liquid with an acetic, esterlike odor and a bitter taste. The single-dose oral toxicity of this material is low, the LD_{50} values being 6.86 g./kg. for rats and 4.94 g./kg. for guinea pigs when fed as a 50 per cent aqueous solution.[65] Keston and co-workers[70] gave the material intravenously and orally to animals and concluded that it did not cause hydropic degeneration of the renal convoluted tubules as do the glycols having ether linkages between the glycol units. Mulinos and co-workers[71] fed the

[65] H. F. Smyth, Jr., J. Seaton, and L. Fischer, *J. Ind. Hyg. Toxicol.*, **23**, 259 (1941).

[66] C. P. Carpenter and H. F. Smyth, Jr., *Am. J. Ophthalmol.*, **29**, 1363 (1946).

[67] E. Gross in K. B. Lehmann and F. Flury, *Toxicology and Hygiene of Industrial Solvents.* Trans. by E. King and H. F. Smyth, Jr., Springer, Berlin, 1938.

[68] E. Rosser in K. B. Lehmann and F. Flury, *Toxicology and Hygiene of Industrial Solvents.* Trans. by E. King and H. F. Smyth, Jr., Springer, Berlin, 1938.

[69] F. H. Wiley, W. C. Hueper, D. S. Bergen, and F. R. Blood, *J. Ind. Hyg. Toxicol.*, **20**, 269 (1938).

[70] H. D. Kesten, M. G. Mulinos, and L. Pomerantz, *Arch. Pathol.*, **27**, 447 (1939), *Chem. Abstr.*, **33**, 4659 (1939).

[71] M. G. Mulinos, L. Pomerantz, and M. E. Lojkin, *Am. J. Pharm.*, **115**, 51 (1943), *Chem. Abstr.*, **37**, 4136 (1943).

material in water over a prolonged period to rats and rabbits and observed that 1 to 3 per cent solutions caused occasional calcium oxalate crystals in the kidneys, and that 5 per cent solutions caused large crystalline deposits and death. They did not observe any hydropic changes in the kidneys of either rats or rabbits, confirming the earlier results of Keston et al.[70]

Human Experience. None reported.

5. Hygienic Standard of Permissible Exposure

None can be established on the very limited toxicological information available. However, it appears that the material should be considered like ethylene glycol.

ETHYLENE GLYCOL MONOMETHYL ETHER ACETATE (2-Methoxyethyl Acetate, Methyl Cellosolve Acetate, Methyl Glycol Acetate)

$$CH_3OCH_2CH_2OOCCH_3$$

4. Physiological Response

Summary. Ethylene glycol monomethyl ether acetate is a colorless, slightly volatile liquid with mild, pleasant, esterlike odor and a bitter taste. It is low in single-dose oral toxicity, not significantly irritating to the eyes or skin, poorly absorbed through the skin, and moderately toxic when inhaled. Its effects are similar to those of ethylene glycol monomethyl ether, being centered in the kidneys and brain.

Single-Dose Oral. The LD_{50} values of 1.25 g./kg. for guinea pigs and 3.93 g./kg. for rats reported by Smyth et al.[65] were determined by feeding 50 per cent aqueous solutions. Later, Smyth and Carpenter[72] reported the LD_{50} for rats to be 3.39 g./kg., but did not specify in what form it was fed.

Eye Contact. The material is but mildly irritating to the eyes.[66]

Skin Contact. The material is not significantly irritating to the skin.[67] However, it can be absorbed through the skin if exposure is prolonged. Using the "cuff" procedure on rabbits as described by Draize et al.,[73] Smyth and Carpenter[72] estimated the LD_{50} to be 5.25 ml./kg.

Repeated-Dose Oral. Gross[67] reports that rabbits died after receiving 3 daily doses of either 0.5 or 1.0 ml./kg. and all showed kidney injury with albumin and granular casts in the urine.

Subcutaneous Injections. Seven injections of 0.5 ml. or 4 injections of 1.0 ml. in guinea pigs caused their death 1 to 5 days after treatment. Kidney injury was apparent.[67]

Inhalation. Gross[67] reports the results of several inhalation studies; they may be summarized as follows: (*1*) Employing an essentially saturated atmos-

---

[72] H. F. Smyth, Jr., and C. P. Carpenter, *J. Ind. Hyg. Toxicol.*, **30**, 63 (1948).

[73] J. H. Draize, G. Woodard, and H. O. Calvery, *J. Pharmacol. Exptl. Therap.*, **82**, 377 (1944).

phere, 22 mg./liter (supersaturated according to present vapor pressure data), mice and rabbits tolerated single exposures for 3 hours with only irritation of the mucous membranes; guinea pigs survived an exposure of 1 hour, but succumbed days later. (2) Cats died after receiving one 9-hour exposure to 2500 p.p.m. and survived one 7-hour exposure to 1500 p.p.m. but did show an increase in blood clotting time and showed changes in the brain. (3) Cats, guinea pigs, rabbits, and mice were given repeated 8-hour exposures to 500 and 1000 p.p.m.; at 500 p.p.m. the cats showed slight narcosis and died, but the others lived. At 1000 p.p.m. deaths occurred in all species. With both concentrations kidney injury occurred. (4) Cats tolerated repeated 4- to 6-hour exposures to 200 p.p.m., but decreases in blood pigments and in numbers of red cells were noted. Smyth et al.[72] report that a single 4-hour exposure to 7000 p.p.m. killed 2 of 6 rats.

Human Experience. No adverse human experience has been reported.

5. Hygienic Standard of Permissible Exposure

The American Conference of Governmental Industrial Hygienists[74] has recommended a threshold limit value of 25 p.p.m. Since the data on this ester are not adequate to permit such a recommendation, this value must be based upon analogy with ethylene glycol monomethyl ether. Since the ester is converted to the ether in the body, this seems justifiable, but as with the ether, the 25 p.p.m. value may be unnecessarily stringent.

ETHYLENE GLYCOL MONOETHYL ETHER ACETATE (2-Ethoxyethol Acetate, Ethyl Glycol Acetate, Cellosolve Acetate)

$$C_2H_5OCH_2CH_2OOCCH_3$$

4. Physiological Response

Summary. Ethylene glycol monoethyl ether acetate is a colorless, slightly volatile liquid with a mild odor and a bitter taste. High concentrations of vapor, however, are irritating to the eyes and nose. The material is fairly low in single-dose oral toxicity, somewhat irritating to the eyes, not appreciably irritating to the skin, poorly absorbed through the skin, and not especially toxic when inhaled in amounts likely to be encountered under ordinary conditions. It is capable of causing central nervous system depression and lung and kidney injury.

Single-Dose Oral. Smyth et al.[65] report LD$_{50}$ values of 5.10 g./kg. for rats and 1.91 g./kg. for guinea pigs when the material was fed as a 50 per cent aqueous suspension in 1 per cent Tergitol 7. Carpenter[75] reports the oral LD$_{50}$ for rabbits to be 1.95 g./kg.

Eye Irritation. Ethylene glycol monoethyl ether acetate is somewhat irritating to the eyes of rabbits.[66]

[74] Conference of Governmental Industrial Hygienists, *Am. Ind. Hyg. Assoc. J.,* **22,** 325 (1961).

[75] C. P. Carpenter, *J. Am. Med. Assoc.,* **135,** 880 (1947).

Skin Irritation. The liquid is not significantly irritating to the skin unless exposure is prolonged or frequently repeated. Absorption of the material through the intact skin can occur but the lethal dose is large. Carpenter[75] reports the LD_{50} for rabbits to be 10.3 g./kg. when the "sleeve" technique of Draize et al.[73] is employed.

Injection. Von Oettingen and Jirouch[76] report the fatal dose for mice to be 5.0 ml./kg. Gross[67] reports that single doses of 0.5 or 1.0 ml./kg. were well tolerated when injected intraperitoneally in guinea pigs. When injected subcutaneously 7 times, doses of 0.5 and 1.0 ml./kg. caused temporary ill effects but no deaths.

Inhalation. According to Gross,[67] a 1-hour exposure to an atmosphere essentially saturated with vapor (<4000 p.p.m.) was survived by guinea pigs. Cats exposed once for 2 to 6 hours to an atmosphere laden with fog survived but 2 such exposures caused vomiting, paralysis, albumin in the urine, and death. Mice, guinea pigs, and a rabbit were unaffected by twelve 8-hour exposures to a concentration of 450 p.p.m., but another rabbit and 2 cats died before the end of the exposure period. Albumin occurred in the urine, and the kidneys of the animals that died were injured.

Carpenter[75] states that rats died as a result of a 2-hour exposure to an atmosphere (probably fog-laden) generated by bubbling the air through the boiling liquid, but survived a 4-hour exposure when the air was bubbled through the liquid at room temperature (1500 p.p.m.). Under the latter conditions deaths resulted from an 8-hour exposure. Dogs survived 120 daily 7-hour exposures to a concentration of 600 p.p.m. without apparent injury. He was unable to detect methemoglobin in the blood, other hematological changes, any effect upon numerous clinical tests, or any histopathological changes in the tissues. This is a bit unexpected when one considers the effects caused by ethylene glycol monoethyl ether and the ether ester on rats, but it may be explained by the difference in susceptibility of the two species and perhaps the dosage. He concludes that ethylene glycol monoethyl ether acetate is in the same range of toxicity as methyl ethyl ketone, propylene dichloride, and tetrachloroethylene, but its hazards are believed to be less because its vapor pressure is substantially lower.

Human Experience. There are no records of adverse human experience that can be attributed to ethylene glycol monoethyl ether acetate. Perhaps the reason for this is that the vapors are objectionable in concentrations necessary to cause adverse effects.

5. *Hygienic Standard of Permissible Exposure*

The Conference of Governmental Industrial Hygienists[74] has recommended a threshold limit value of 100 p.p.m. This level is probably a satisfactory standard, for the systemic effects of the material are in all probability directly related to the metabolic product, ethylene glycol monoethyl ether.

[76] W. F. von Oettingen and E. A. Jirouch, *J. Pharmacol. Exptl. Therap.*, **42**, 355 (1931).

ETHYLENE GLYCOL DINITRATE

$$C_2H_4(ONO_2)_2$$

4. *Physiological Response*

Summary. Ethylene glycol dinitrate is a yellow oily liquid as ordinarily produced, but when pure it is colorless. It is soluble in many solvents, such as carbon tetrachloride, ether, benzene, toluene, and acetone, limited in solubility in common alcohols, and only slightly soluble in water. According to Smyth,[77] it lowers blood pressure by dilating the blood vessels and causes methemoglobin formation. It is readily absorbed through the skin in toxic amounts, more readily than nitroglycerin. It is highly toxic when inhaled. Inhalation of 2 p.p.m. 8 hours a day for 1000 days caused transient blood changes in cats. Anemia, irreversible erythrocyte changes, fatty changes in heart muscle, liver, kidney, and hyperplasia of the bone marrow are associated with chronic poisoning.

Human Experience. Humans consuming alcohol are believed to be especially susceptible to this material and fatalities have been attributed to its handling and use.[78] However, a report by Forssman *et al.*[79] of a study on workers exposed to nitroglycerin and to glycol dinitrate in the Swedish explosive industry indicates that no chronic effects have resulted from repeated exposures to up to 5 mg./cu. meter (0.5 p.p.m.).

5. *Hygienic Standard of Permissible Exposure*

The Conference of Governmental Industrial Hygienists[74] suggests a threshold limit of 0.5 p.p.m. for nitroglycerin and it seems probable that a similar figure for glycol dinitrate would be appropriate, at least until more data become available.

DIETHYLENE GLYCOL MONOMETHYL ETHER ACETATE
[2-(2-Methoxyethoxy)ethyl Acetate, Methyl Carbitol Acetate]

$$CH_3OCH_2CH_2OCH_2CH_2OOCCH_3$$

4. *Physiological Response*

Summary. Diethylene glycol monomethyl ether acetate is a colorless liquid with a faint, not unpleasant odor and a bitter taste. The single-dose oral LD_{50} values reported by Smyth *et al.*[65] are 11.96 g./kg. for rats and 3.46 g./kg. for guinea pigs when the material was fed as a 50 per cent aqueous solution. Carpenter and Smyth[66] found the material to be appreciably irritating to the rabbit eye. There has been no adverse human experience reported nor would any be expected in ordinary industrial handling and use.

[77] H. F. Smyth, Jr., in G. O. Curme, Jr., and F. Johnston, eds., *Glycols* (American Chemical Society Monograph, Series 114). Reinhold, New York, 1953.

[78] W. F. von Oettingen, *Natl. Insts. Health Bull.* No. **186**, 1946.

[79] S. Forssman, N. Nasretiez, G. Johansson, G. Sundell, O. Wilander, and G. Boström, Swedish Employers' Confederation, Stockholm, Sweden.

5. *Hygienic Standard of Permissible Exposure*

None has been suggested nor would one seem necessary in view of the low volatility and the nature of the material.

DIETHYLENE GLYCOL MONOETHYL ETHER ACETATE
[2-(2-Ethoxyethoxy)ethyl Acetate, Carbitol Acetate]

$$C_2H_5OCH_2CH_2OCH_2CH_2OOCCH_3$$

4. *Physiological Response*

Summary. Diethylene glycol monoethyl ether acetate is a colorless liquid with a faint, not unpleasant odor and a bitter taste. The single-dose oral LD_{50} values reported by Smyth et al.[65] are 11.00 g./kg. for rats and 3.93 g./kg. for guinea pigs when the material was administered as a 50 per cent aqueous solution. Carpenter and Smyth[66] found the material to be only very slightly irritating to the rabbit eye. No adverse human experience has been reported nor would any be expected under ordinary conditions of industrial handling and use.

5. *Hygienic Standard of Permissible Exposure*

None has been suggested nor would one seem necessary in view of the low volatility and the nature of the compound.

DIETHYLENE GLYCOL MONOBUTYL ETHER ACETATE [Butyl Carbitol Acetate, 2-(2-Butoxyethoxy)ethyl Acetate]

$$C_4H_9OCH_2CH_2OCH_2CH_2OOCCH_3$$

4. *Physiological Response*

Summary. Diethylene glycol monobutyl ether acetate is a clear liquid with a mild, not unpleasant odor and a bitter taste. In the past, it has been used as an insect repellent but it is no longer so used. The material is low in single-dose oral toxicity, not appreciably irritating to the eyes or skin, but it can be absorbed through the skin in toxic amounts when contact is extensive and prolonged.

Single-Dose Oral. Smyth and co-workers[65] fed the material as a 50 per cent suspension in 1 per cent Tergitol 7 and found the LD_{50} to be 11.92 g./kg. for rats and 2.34 g./kg. for guinea pigs. Others[80] have fed the material emulsified with 5 per cent gum arabic and found it to be more toxic, the LD_{50} being estimated at 7 ml./kg. In doses slightly below the lethal level, it causes marked narcosis. Draize et al.[81] report LD_{50} values for several species as follows: rabbits, 2.8; guinea pigs, 2.7; rats, 7.1; mice, 6.6; and chicks, 5.0 ml./kg.

Eye Contact. Carpenter and Smyth[66] have found the material to be only slightly irritating to the eyes of rabbits.

[80] Biochemical Research Laboratory, The Dow Chemical Co., unpublished data.

[81] J. H. Draize, E. Alvarez, M. F. Whitesell, G. Woodard, E. C. Hagen, and A. A. Nelson, *J. Pharmacol. Exptl. Therap.,* **43,** 26 (1948).

Skin Contact. Repeated and prolonged contact with the skin causes mild erythema and exfoliation in rabbits and humans, particularly when sweating.[81] However, the material is readily absorbed through the skin of rabbits in toxic quantities. Draize *et al.*[81] estimate the single dose LD_{50} by skin absorption for the rabbit to be 5.5 ml./kg. and the repeated dose (90-day) LD_{50} to be 2 ml./kg. In rabbits inuncted repeatedly, they observed hematuria with degenerative changes in the kidney. Hemoglobinuria was severe when the dose was 4 ml./kg.

Human Experience. Diethylene glycol monobutyl ether acetate has been used by many human beings as an insect repellent. This has resulted in rather extensive and intimate contact with the skin and undoubtedly some inhalation of vapors. Hoehn[82] reports one case of a three-year-old child who allegedly suffered nephrosis as a result of such use. As a result and because this conceivably could have occurred, the material was withdrawn from this market. It would seem that the material would present no appreciable hazard in industrial handling and use.

5. *Hygienic Standard of Permissible Exposure*

None has been suggested nor would one seem necessary in view of the low vapor pressure and low toxicity.

PROPYLENE GLYCOL MONOMETHYL ETHER ACETATE
DIPROPYLENE GLYCOL MONOMETHYL ETHER ACETATE
TRIPROPYLENE GLYCOL MONOMETHYL ETHER ACETATE

4. *Physiological Response*

Summary.[80] All three of these compounds are colorless, slightly volatile liquids (see Table 7). They are all low in single-dose oral toxicity for rats. When fed undiluted, single doses of 3 ml./kg. were survived by all animals in each case; doses of 10 ml./kg. caused the death of 3 of 5 fed the propylene glycol derivative, and all of 5 fed the di- and tripropylene glycol derivatives. The propylene glycol derivative is somewhat painful and irritating to the eyes, but the others are not. None of them is appreciably irritating to the skin of rabbits nor are they absorbed through the skin in significant amounts, even when applied repeatedly for a 2-week period of time. It would not seem that any of them would present any serious hazards in ordinary industrial handling and use.

[82] D. Hoehn, *J. Am. Med. Assoc.*, **128**, 513 (1945).

CHAPTER XXXVII

Epoxy Compounds

C. H. HINE, m. d., and V. K. ROWE

I. General Considerations

A. INTRODUCTION

Epoxy compounds are a group of cyclic ethers or alkene (alkylene) oxides in which there is a three-membered ring formed between an oxygen atom and two adjacent carbon atoms, thus:

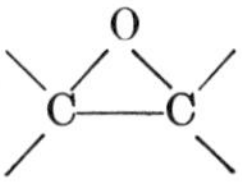

The alpha epoxy compounds in which the epoxy group is in the 1-2 position have considerable academic and industrial importance, whereas the beta, or oxetane compounds, are quite rare. The gamma and delta compounds have been studied extensively in the carbohydrate field. The importance of these cyclic ethers lies principally, however, in the reactivity of the alpha epoxy group, which is seldom found in natural products, although it may be formed by autoxidation of drying oils and unsaturated fatty acids. They have been found to be of considerable use as chemical intermediates in the manufacture of a variety of surface-active agents, special solvents, plasticizers, synthetic resins, cements and adhesives, and fine chemicals. A few of the compounds are useful as end products. The recently developed epoxy resins contain terminal unreacted alpha epoxy groups. The molecular weight of the compounds in this series varies tremendously from ethylene oxide, the simplest, with a molecular weight of 44.05, to the epoxy resin monomers with weights of approximately 4000.

Berthelot, in 1854, first described the preparation and characterization of compounds of the alpha epoxy type, and the chemistry has been reviewed by a number of authors, among them Meerwein[1] and Jungnickel.[2] Because of the

[1] H. Meerwein, in J. Houben, ed., *Methoden der organischen Chemie*, Vol. III. 3rd ed., Thieme, Leipzig, 1930, p. 213.

[2] J. L. Jungnickel, E. D. Peters, A. Polgar, and F. T. Weiss, in J. Mitchell, I. M. Kolthoff, E. S. Proskauer, and A. Weissberger, eds., *Organic Analysis*, Vol. I. Interscience Publishers, New York–London, 1953, p. 127.

strained three-membered ring, the alpha epoxies are the most active of the oxides. They are attacked by almost all nucleophilic substances with a resulting opening of the ring and formation of addition compounds; in this variety of chemicals are included halogen acids, thiosulfate, carboxylic acid, hydrogen cyanide, water, amines, aldehydes, and alcohols.

The most commercially important and widely used epoxy compounds are the epoxy resins. These were first synthesized by P. Caston in Switzerland and S. O. Greenlee in the United States in the 1930's.[3] They are thermosetting materials, which when converted by curing agents become hard and fusible systems with a three-dimensional, cross-linked network. Among their most important properties are:

1. Versatility: Epoxides are compatible with a variety of modifiers, so that the resin systems can be engineered to widely diverse specifications.

2. Good Handling Characteristics: Many can be worked at room temperature. The resins have indefinite shelf life. The ratio of curing agent is not extremely critical.

3. Toughness: This has been alleged to be due to the distance between cross-linking points and the presence of interval aliphatic chains.

4. High Adhesive Properties: The polarity of the aliphatic hydroxyl and ether groups serves to create electromagnetic bonding forces between the molecule and an adjacent surface.

5. Low Shrinkage: The shrinkage is in the order of less than 2 per cent for an unmodified system.

6. Inertness: The ether groups, substituted phenyl rings and aliphatic hydroxyls in a cured system are relatively invulnerable to attack from either caustics or acids.

The epoxy resins are currently marketed by a number of manufacturers and distributors under a variety of trade marks. These include the Bakelite Co. (Bakelite), Ciba Co., Inc. (Araldite), Jones-Dabney (Epi-Rez) Shell Chemical Corp. (EPON) and The Dow Chemical Co. (D.E.R.). They are used as surface coatings, such as appliance primers, and when corrosion resistance is required; in castings and pottings for tooling and encapsulating electronic circuits; as adhesives, especially in metal-to-metal bonding; as textile-treating agents to improve wearability and crease-resistance; and in laminating, especially for structural use in aircraft. For toxicological as well as technical purposes, we should distinguish between cured and uncured epoxy compositions.[4] Unreacted epoxy resins have limited useability. The properties that make them useful are obtained by chemical reaction with curing agents, which convert them into mechanically strong polymers. Both hydroxyl and epoxy groups are involved in cured systems. The epoxy groups are

[3] H. Lee and K. Neville, *Epoxy Resins: Their Application and Technology.* McGraw-Hill, New York, 1957.

[4] C. H. Hine, J. K. Kodama, H. H. Anderson, D. W. Simonson, and J. S. Wellington, *A.M.A. Arch. Ind. Health,* **17,** 129 (1958).

usually cured by reaction with amino, carboxyl, or hydroxyl groups and inorganic acids to give the secondary hydroxyl groups and a bond from the remaining epoxy carbon atom to the nucleus of the donor. Amines are one of the most important types of curing agent. An example of the chemical reaction that occurs with the primary amines is as follows:

$$RNH_2 + \overset{O}{\overset{/\backslash}{CH_2{-}CH}}{\cdot}R' \rightarrow RNH{\cdot}CH_2{\cdot}\overset{OH}{\overset{|}{CH}}{\cdot}R'$$

In addition to amines, acid anhydrides, dibasic acids, and other resins, including amino formaldehyde, phenol formaldehyde, urea formaldehyde, polyamides, and melamine formaldehyde are typical commercial curing agents, attacking both the hydroxyl and epoxy groups. There are four characteristics of epoxy resins that may be followed as guides for structure and uses. These are viscosity, epoxide equivalent (the weight of resin in grams containing one gram equivalent of epoxy), hydroxyl equivalent (the weight of resin in grams containing one equivalent weight of hydroxyl groups), and average molecular weight and molecular weight distribution.

B. METHODS FOR DETERMINATION

Methods available for determining epoxy groups in air for industrial hygiene purposes are not as simple, precise, or reliable as would be desirable. Techniques that have been used include a specially calibrated combustible gas indicator, a gas interferometer, infrared absorption spectroscopy, and chemical methods. The most easily utilizable methods for determination of the alpha epoxies are based on the addition to this group of hydrogen chloride, resulting in a chlorohydrin, according to the following equation:

$$-\overset{|}{\underset{\backslash}{C}}{-}\overset{|}{\underset{/}{C}}{-} + H^+ + Cl^- \rightarrow -\overset{|}{\underset{|}{C}}{-}\overset{Cl}{\overset{|}{\underset{OH}{C}}}{-}$$

The quantity of the alpha epoxide is measured by a Volhard titration, the difference in the amount of acid added and that consumed being a measure of the epoxy group. The reaction is carried out in various solutions because of difference in chemical reactivity. These may be generally classified into three types: (1) water and alcohols, (2) ethers, and (3) pyridine.[2] Appropriate corrections must be made in those circumstances where acid or alkaline vapors are also present.

C. PHYSIOLOGICAL PROPERTIES

No extensive pharmacological study has been carried out as to the mechanism of action of a homologous series of these compounds. Their general physiological activity varies considerably from the highly active, low molecular weight, aliphatic mono- and diepoxides and glycidyl ethers to the practically inert cured resin systems with only a few free epoxy groupings. The majority of observations seem to indicate that physiological activity can be pointed toward three main effects: (1) central nervous system depression, (2) irritation of surface tissues, and (3) radiomimetic action.

1. Central Nervous System: The lower molecular weight monoepoxide hydrocarbons are weakly anaesthetic, but this activity is largely overshadowed by their irritating effect. Some of the monoepoxy ethers show more pronounced central nervous system action. In the case of the phenyl-substituted groups, this may be due to an interneuronal blocking action similar to that seen with mephenesin. In large doses most of the monoepoxides produce a nonspecific depression. Repeated exposures to ethylene oxide cause a species specific reversible paralysis limited to the lower extremities.

2. Irritating Action: Most of the low molecular weight aliphatic epoxies, the diepoxy compounds, and monoepoxy compounds with a reactive group elsewhere in the molecule are extremely irritating to mucous membranes and surface tissues. Inhalation of vapors or aerosols may lead to acute pulmonary edema and chemical pneumonia. In most cases, the warning properties are sufficient to cause avoidance of unsafe exposure. Experimentally, both concentrated and dilute aqueous solutions will cause varying degrees of irritation of the dermis extending to frank necrosis. These compounds not only cause direct skin irritation but may lead to sensitization response as well. Dermatitis is, in fact, the major problem encountered in dealing with these compounds. It affects a great number of persons having prolonged or unusual contact. The reported incidence runs between 10 and 60 per cent. Sensitization is not uncommon and may affect as high as 2 per cent of the exposed population.

3. Radiomimetic Effects: Haddow[5] pointed out that certain diepoxy compounds active as cross-linking agents probably lead to biological alkylation. In a series of homologs of butadiene, he found activity to be inversely related to molecular weight, with a peak at about C_6 and a rapid fall-off of activity above C_9. Monoepoxides were not active. Hendry and his co-workers[6] found both radiomimetic and cytotoxic activity in their series of bisepoxides, a number of these compounds being tumor-inhibitory. With regard to the effect on bone marrow, Williams[7] observed atypical cells in the peripheral blood of certain employees working with epoxy compounds and noticed occasional reversal of the percentages of granulocytes and nongranulocytic cells and depressions in the total white count. His observations, as yet, have not been confirmed by other industrial physicians

[5] A. Haddow, in F. Hamberger and W. H. Fishman, eds., *The Physiopathology of Cancer*, Vol. II. Hoeber, New York, 1953, p. 441.

[6] J. A. Hendry, R. F. Homer, F. L. Rose, and A. L. Walpole, *Brit. J. Pharmacol.*, **6**, 235 (1951).

[7] M. H. C. Williams, Imperial Chemical Industries, Manchester, England, personal communication.

who have reviewed the problem. Experimentally, Kodama et al.[8] found it possible to produce, with certain of the more active compounds, a decrease in the number of nucleated cells in the bone marrow, in the total number of circulating white cells, and in the predominant white cell type in the peripheral blood when these compounds were administered parentally to rats, mice, and dogs. He concluded that certain of these substances resembled benzene in respect to their effect on blood cell production. Petrakis[9] observed changes in the blood of three of eight volunteers administered a low molecular weight aliphatic diglycidyl epoxy resin. A lowering of the white cell count accompanied by decrease in the total numbers of lymphocytes and monocytes occurred. The counts returned to pretreatment levels within a matter of weeks, and it was concluded that these compounds were not comparable to nitrogen mustards or other more active alkylating agents.

Hine et al.[10] and McCammon et al.[11] have produced experimental tumors in animals through repeatedly administering epoxy compounds, either cutaneously or subcutaneously, for prolonged periods. In man no tumors of any type have been suggested as having arisen from exposure to these agents, and as the **oncogenic** activity is low, the hazard in this respect is considered to be minimal. Kotin and Falk[12] postulated that epoxides might be responsible for the cancerogenic activity in a concentrate removed from the polluted air of Los Angeles. Identification of the compound was not furnished. Kotin[13] has also suggested epoxides along with ozonides and peroxides as suspect carcinogens in general air pollution.

The following section summarizes the toxicology and safe handling information for 28 epoxy-containing compounds. Some of these chemicals are in experimental stage; others are used in industry only in small quantities. Only 5 of these have any widespread commercial use. These are ethylene oxide, epichlorohydrin, propylene oxide, the diglycidyl ether of diphenylol propane, and condensation products of the latter compound with additional diphenylol propane. The majority of compounds, including all of those widely used, are of a low order of toxicity and no serious systemic intoxications have resulted from their use. As previously mentioned, all the compounds are dermatitic, either in their own right or when associated with compounds used in their further reaction, and skin irritation and sensitization must be carefully guarded against. The final answer as to the significance of the radiomimetic effect has not been answered. Industrial hygienists and industrial physicians are advised to watch carefully for changes in the blood and skin of exposed personnel.

[8] J. K. Kodama, R. J. Guzman, C. H. Hine, and H. H. Anderson, *Federation Proc.*, **18**, 411 (1959).

[9] N. L. Petrakis, Univ. of California, San Francisco, personal communication.

[10] C. H. Hine, R. J. Guzman, M. M. Coursey, J. S. Wellington, and H. H. Anderson, *Cancer Research*, **18**, 20 (1958).

[11] C. J. McCammon, P. Kotin and H. L. Falk, *Proc. Am. Assoc. Cancer Research*, **2**, 229 (1957).

[12] P. Kotin and H. L. Falk, *Proc. Am. Assoc. Cancer Research*, **2**, 30 (1955).

[13] P. Kotin, *Cancer Research*, **16**, 375 (1956).

Unfortunately, the available toxicity data on these compounds is fragmentary and frequently not reported in similar terms. Where possible, an attempt has been made to convert the expressions to standard meanings. Freedom of interpretation has been utilized by the authors for this purpose where necessary.

II. Specific Compounds

ALLYL GLYCIDYL ETHER (1-Allyloxy-2,3-epoxypropane)

1. Source, Uses, and Industrial Exposure

Allyl glycidyl ether is manufactured through the condensation of allyl alcohol and epichlorohydrin with subsequent dehydrochlorination with caustic to form the epoxy ring. It is a commercial chemical of primary interest as a resin intermediate, and is also used as a stabilizer of chlorinated compounds, vinyl resins, and rubber. It is not classified as a dangerous article and is not regulated by the Bureau of Explosives.

2. Physical and Chemical Properties

Allyl glycidyl ether is a colorless liquid of characteristic, but not unpleasant, odor with the following properties:[14]

Molecular weight: 114.14
Specific gravity: 0.9698 (20/4°C.)
Freezing point: forms glass at −100°C.
Boiling point: 153.9°C. (760 mm. Hg)
Vapor density: 3.32 (25°C.)
Vapor pressure: 4.7 mm. Hg (25°C.)
Refractive index: 1.4348 (20°C.)
Solubility: 14.1 per cent in water; miscible with acetone, toluene, and octane
Flash point: 135°F. (tag open cup)
Per cent in "saturated" air: 0.62 (25°C.)
1 p.p.m. ≏ 4.66 mg./cu. meter at 25°C., 760 mm. Hg and 1 mg./liter ≏ 214 p.p.m. at 25°C., 760 mm. Hg

3. Determination in the Atmosphere

Suggest hydrochloric acid–dioxane method for epoxy groups.[2]

4. Physiological Response

Summary. Allyl glycidyl ether is a central nervous system depressant and also causes acute pulmonary edema. It is classified as slightly toxic after oral ad-

---

[14] Organic Chemicals, *Tech. Publ. SC:52-10,* Shell Chemical Corp., New York, 1952, p. 68.

ministration and percutaneous application, appreciably irritating and injurious to the eyes and skin, capable of causing skin sensitization in human subjects and moderately toxic following vapor exposure.

Response in Animals. Acute toxicity. Administration of the compound to rats and mice orally produced moderate depression and dyspnea within 15 to 19 minutes. Deaths occurred in 4 hours to 5 days. LD_{50} for mice was 0.39 and for rats, 1.60 g./kg. At autopsy hypotonicity of the enteric tract and extensive adhesions of the stomach walls to adjacent tissues were noted. No characteristic microscopic pathology occurred, although occasionally the liver showed focal areas of necrosis.

Percutaneous absorption. Depression occurred progressively during the 7 hours of contact. LD_{50} was 2.55 g./kg. for rabbits. Necropsy showed constricted kidneys and spleens.

Vapor exposures. Exposed mice and rats showed severe irritation of the eyes and respiratory tract accompanied by lachrymation and salivation, dyspnea and severe gasping, and gaseous distention of the abdomen. LC_{50} value for 4-hour exposure in the mouse was 270 p.p.m.; for 8 hours in the rat, 670 p.p.m. Corneal opacities occurred in rats.

Rats exposed at concentrations of 260, 400, 600, and 900 p.p.m. 7 hours a day for 50 exposures showed the following evidence of toxicity. Increased mortality occurred at 600 and 900 p.p.m., and these exposures were terminated after 25 were given. None of the groups on experiment maintained a normal weight gain. An increase in organ/body weight ratio occurred only with the kidneys at 400 p.p.m. Persistent microscopic lesions consisted of varying degrees of bronchopneumonic consolidation, severe emphysema, bronchiectasis, and, occasionally, enlargement of the adrenal glands. Decreased hemoglobin was noted at 400 p.p.m., the only level tested. Considerable ocular irritation and respiratory distress occurred at levels of 400 p.p.m. and above.

Eye irritation. A score of 72 was obtained with the Draize[14a] method. The compound is classified as severely irritating to the eye.

Skin irritation. A score of 4 was obtained on rabbits by the Draize[14a] method. The compound is classified as moderately irritating. Studies on human subjects indicate that skin sensitization occurs readily.

Mode of action. The compound is a pulmonary irritant and central nervous system depressant.

Cause of death. Varies with the route and chronicity of exposure. Central nervous system depression is the predominant symptom after oral or percutaneous administration. Irritation of the respiratory tract with pulmonary edema or secondary pneumonia generally occurred at excessive levels of exposure to vapor.

Effects observed in man. Primary irritation and occasional sensitization noted with this compound. Possible cross-sensitization to other epoxy agents.

[14a] J. H. Draize, G. Woodard, and H. O. Calvery, *J. Pharmacol. & Exptl. Therap.*, **82**, 377 (1944).

5. *Hygienic Standards of Permissible Exposure*

None established. On basis of comfort, 10 p.p.m. is suggested. The A.C.G.I.H. tentative limit value for 1961 is 10 p.p.m.

6. *Odor and Warning Properties*

Sensory threshold not established. Has characteristic odor noticeable at room temperature in less than excessive concentrations.

BUTADIENE DIOXIDE (1,2-3,4-Diepoxybutane, Bioxirane)

$$H_2C \overset{O}{\underset{}{\diagup\diagdown}} \overset{}{\underset{H}{C}} - \overset{}{\underset{H}{C}} \overset{O}{\underset{}{\diagup\diagdown}} CH_2$$

1. *Source, Uses, and Industrial Exposures*

Butadiene dioxide is available in research quantities. It is prepared by chlorination of butadiene followed by epoxidation with peracetic acid, with subsequent hydrolysis of the epoxide group and final reepoxidation with caustic. Suggested as a chemical intermediate, cross-linking agent, and in the preparation of erythritol and pharmaceuticals.

2. *Physical and Chemical Properties*

Butadiene dioxide is a water-white, low-viscosity liquid with the following properties:[15]

Molecular weight: 86.09
Specific gravity: 0.962 (25/4°C.)
Boiling point: 138°C. (760 mm. Hg)
Vapor pressure: 3.9 mm. Hg (20°C.)
Refractive index: 1.435 (20°C.)
Solubility: miscible in water in all proportions
1 p.p.m. $\backsim$ 3.52 mg./cu. meter at 25°C., 760 mm. Hg and 1 mg./liter $\backsim$ 284 p.p.m. at 25°C., 760 mm. Hg

3. *Determination in the Atmosphere*

Suggest pyridinium chloride–chloroform method for epoxy groups.[2]

4. *Physiological Response*

Summary. Butadiene dioxide is severe pulmonary irritant. It is classed as highly toxic on inhalation, moderately toxic following ingestion or skin absorption. It causes severe eye and skin irritation. Experimentally, it is an active radiomimetic substance and produces skin cancers and sarcomas and depression of the hemopoietic system.

[15] B. Phillips, *Peracetic Acid and Derivatives.* Union Carbide Chemicals Co., New York, n. d.

Response in Animals. Acute toxicity. Concentrated vapors in air killed all of 6 rats in 15 minutes. The LC_{50} for a 4-hour exposure to rats is 90 p.p.m. Lachrymation, clouding of the cornea, labored breathing, and congestion of the lungs occurs. Survivors have atrophy of the thymus, involution of the spleen, and decreased weight gain during the recovery period.[16] The LD_{50} value for single oral dose in rats is 0.078 g./kg. Single dose skin penetration studies in rabbits give an LD_{50} value of 0.089 ml./kg.[17]

Eye irritation. Classed as a serious chemical eye injurant.

Skin irritation. The undiluted material causes burns and blister formation on rabbit skin.

Chronic toxicity. Skin applications repeated 3 times weekly for 1 year caused consistent sebaceous gland suppression, intense hyperkeratosis, marked hyperplasia, and a significant number of skin tumors in mice. Repeated subcutaneous injections caused sarcomas in rats.[11] Following 6 intramuscular administrations of 25 mg./kg. to rats, a leukopenia and relative lymphopenia occur.

Mode of action. Locally necrotizing to tissues. Causes extreme irritation of pulmonary tract. Pronounced radiomimetic effect.

Cause of death. Pulmonary edema and shock following extreme local irritation.

Effects Observed in Man. Accidental minor exposure caused swelling of the eyelids, upper respiratory tract irritation, and painful eye irritation 6 hours post exposure.

5. Hygienic Standards of Permissible Exposure

None established. Recommend 1 p.p.m. based on potential hemopoietic and lung irritating effects.

6. Odor and Warning Properties

Lower sensory threshold limits not established. Pronounced nasal and eye irritation at 10 p.p.m., barely recognizable at 5 p.p.m. after 5 minutes' exposure. No significant warning properties at suggested threshold concentration of 1 p.p.m.

BUTYLENE OXIDES

1. Source, Uses, and Industrial Exposures

Butylene oxide is available commercially as the mixed isomers or as the single 1,2-isomer. They are prepared commercially from butylene through the intermediate butylene chlorohydrin. The butylene oxides are used for the production of the corresponding butylene glycols and their derivatives, such as poly-

---

[16] C. H. Hine, J. K. Kodama, and R. J. Guzman, Dept. or Pharmacology & Experimental Therapeutics, Univ. of California Medical Center, San Francisco, unpublished data.

[17] *Toxicology Studies: Butadiene Dioxide (Ep-268),* Industrial Medicine and Toxicology Dept., Union Carbide Corp., New York, 1957.

butylene glycols, mixed polyglycols, and glycol ethers and esters. They are also used to make butanolamines, surface active agents, and gasoline additives, and as acid scavengers and stabilizers for chlorinated solvents.

The butylene oxides are highly flammable and highly reactive chemically, but are less reactive than ethylene or propylene oxide. The liquids are relatively stable but they may react violently with materials having a labile hydrogen, particularly in the presence of catalysts such as acids, alkalies, and certain salts. They are capable of polymerizing exothermically. The same general precautions should be taken when handling the butylene oxides as when handling ethylene oxide.[18]

The hazard to health of the butylene oxides is not as great as for either ethylene oxide or propylene oxide. They are less volatile, more odorous, and less toxic. Excessive exposure, however, can result in eye, skin, and respiratory injury.

2. Physical and Chemical Properties

1,2-Butylene oxide (1,2-epoxybutane) and butylene oxide(s), a mixture of 1,2- and 2,3-epoxybutane, are water-white liquids with the following properties:

	1,2-Butylene oxide	Butylene oxide(s)
Molecular formula	C_4H_8O	C_4H_8O
Structural formula	H_2C—$CHCH_2CH_3$ (epoxide O)	80–90% H_2C—$CHCH_2CH_3$ (epoxide O) and 10–20% CH_3CH—$CHCH_3$ (epoxide O)
Molecular weight	72.1	72.1
Specific gravity (25/25°C.)	0.826	0.824
F. p., °C.	Below −60	Below −50
B. p., °C. (760 mm. Hg)	62.0–64.5	59–63
Refractive index (25°C.)	1.381	1.378
Density of sat'd air (air = 1)	Approx. 0.977	Approx. 1.36
Vapor density	Approx. 177	Approx. 183
Solubility at 25°C., g. per 100 g.		
H₂O	Approx. 8.24	Approx. 9
Other solvents	Miscible with common aliphatic and aromatic solvents	Miscible with common aliphatic and aromatic solvents
Flash point, °F. (closed cup)	−15	5
Flammability limits		1.5–18.3% by volume in air

1 p.p.m. ≏ 2.94 mg./cu. meter at 25°C., 760 mm. Hg and
1 mg./liter ≏ 340 p.p.m. at 25°C., 760 mm. Hg

3. Determination in the Atmosphere

The methods available for the determination of butylene oxide in air for industrial hygiene purposes are the same as those described for propylene oxide and are subject to the same limitations.

[18] Mfg. Chemists' Assoc. Chem. Safety Data Sheet **SD-38**, (1951).

4. *Physiological Response*[19]

Summary. The toxicological information available on the butylene oxides has been derived from studies on the mixed isomers. However, since the mixed isomers consist to such a large degree of the 1,2-isomer, the data so derived are believed applicable also to the commercial 1,2-butylene oxide. The straight chain mixed isomers of butylene oxide are primary irritants of the eyes and skin; however, skin irritation is not likely to follow vapor exposure alone. Although the vapors are moderately anesthetic, the concentrations required are very disagreeable. The hazard from inhalation is not great. Excessive exposure to vapors would be expected to cause irritation of the lungs and its sequellae.

Response in Animals. Single dose oral. When mixed butylene oxide was fed to rats by intubation as a 30 per cent solution in corn oil, the LD_{50} was found to be about 0.5 g./kg. Deaths occurred within a day or not at all.

Eye irritation. When applied to the eyes of rabbits, the liquid oxide was found to be markedly irritating. Pain, conjunctival irritation, and transient corneal injury may be expected from such exposure in human beings.

Skin irritation. The liquid is markedly irritating to the skin of rabbits when confined beneath a covering. When free to evaporate, it does not cause appreciable irritation. A short contact would not be expected to cause more than a mild irritation, but prolonged or repeated exposure would be expected to cause blistering and necrosis.

Skin absorption. Butylene oxide is not absorbed through the skin in amounts likely to cause systemic effects.

Vapor Inhalation. Rats exposed to an atmosphere saturated with butylene oxides at room temperature showed anesthetic effects within minutes. Exposures lasting 12 minutes were lethal whereas those lasting 6 minutes caused some delayed deaths, all of which were due to secondary pneumonia. It appears that the principal hazard from acute vapor exposure is irritation of the lungs followed by edema and pneumonitis. Because of the irritating nature of acutely hazardous concentrations, the hazard of acute exposure would seem to be slight.

Toxicological studies have shown that rats, guinea pigs, and rabbits can tolerate, for prolonged periods, repeated 7-hour exposures to a vapor concentration of 400 p.p.m. It appears from these findings that the mixed butylene oxide is far less toxic when inhaled than ethylene oxide and appreciably less toxic than propylene oxide. Since a vapor concentration of 400 p.p.m. is quite disagreeable to breathe, the hazard of chronic intoxication would seem to be slight *provided persons avoided concentrations detectable by odor.*

Effects Observed in Man. Industrial experience with these materials apparently has been favorable for no references to any toxic effects are published in the readily available literature.

---

[19] The Dow Chemical Co., Midland, Mich., unpublished data.

5. *Hygienic Standards of Permissible Exposure*

On the basis of limited data available, it seems that there would be little possibility of injury if concentrations of butylene oxide vapor were maintained below 400 p.p.m.

6. *Odor and Warning Properties*

The odor of the straight chain mixed isomers of butylene oxide may be described as sweetish, somewhat like butyric acid, and disagreeable.

Limited experience in handling the butylene oxides indicates that, at acutely toxic concentrations, they have a disagreeable odor and are irritating to the eyes and nasal passages. While their odor is disagreeable at concentrations below those likely to cause adverse physiological effects upon repeated exposure, it is doubtful that the odor can be relied upon to prevent excessive exposure.

n-BUTYL GLYCIDYL ETHER (1-*n*-Butoxy-2,3-epoxypropane)

$$\text{C}_4\text{H}_9\text{OCH}_2\text{CH}\overset{\displaystyle O}{\overset{\diagup\diagdown}{-}}\text{CH}_2$$

1. *Source, Uses, and Industrial Exposure*

n-Butyl glycidyl ether is made through the condensation of *n*-butyl alcohol and epichlorohydrin with subsequent dehydrochlorination with caustic to form the epoxy ring. Suggested for use as a viscosity-reducing agent for easier handling of conventional epoxy resins, as an acid accepter for stabilizing chlorinated solvents, and as a chemical intermediate.

2. *Physical and Chemical Properties*

n-Butyl glycidyl ether is a colorless liquid with a slightly irritative odor and the following properties:[20]

Molecular weight: 130.21
Specific gravity: 0.9087 (25/4°C.)
Boiling point: 164°C. (760 mm. Hg)
Vapor pressure: 3.2 mm. Hg (25°C.)
Per cent in "saturated" air: 0.42 (25°C.)
Density of vapor: 3.78 (25°C.)
Solubility: 2 per cent soluble in water at 20°C.
1 p.p.m. ≎ 5.32 mg./cu. meter at 25°C., 760 mm. Hg and 1 mg./liter ≎ 188 p.p.m. at 25°C., 760 mm. Hg

3. *Determination in the Atmosphere*

Suggest hydrochloric acid–dioxane method for epoxy group.[2]

[20] Shell Development Co., Emeryville, Calif., unpublished data.

4. Physiological Response

Summary. *n*-Butyl glycidyl ether is a central nervous system depressant, also causing slight irritation of the respiratory tract. It is classified as slightly toxic after oral administration and practically nontoxic after single percutaneous application and vapor exposure. However, the material is a fairly potent skin sensitizer. Repeated exposures lead to cumulative toxicity of a moderate degree.

Response in Animals. Acute toxicity. Incoordination, ataxia, agitation, and excitement precede marked depression. The LD_{50} following intragastric administration is 1.53 g./kg. in mice and 2.26 g./kg. in rats. The percutaneous LD_{50} in rabbits is 4.93 g./kg. Air saturated with the vapors of the compound did not kill mice exposed for 4 hours. The LC_{50} for an 8-hour exposure in rats was 1030 p.p.m. Intraperitoneal injection caused essentially the same pattern of signs as intragastric administration. The LD_{50} values of 1.14 and 0.70 g./kg. were obtained with rats and mice, respectively. Minimal histological changes consisted of focal inflammation and moderate congestion of central zones of the liver.[21]

Eye irritation. A score of 4 was obtained with the Draize[14a] method in rabbits, indicating mild irritation.

Skin irritation. A score of 2.8 was obtained with the Draize[14a] method in rabbits, indicating mild irritation. Tests upon human subjects employing repeated applications demonstrated an appreciable capacity to cause skin irritation and sensitization.[19]

Chronic toxicity. Repeated applications to the skin of rabbits gave a skin irritation score of 3.3 rated on the Draize scale, indicating moderate irritation. No effects on the number of circulating white cells occur on repeated intramuscular injection of rats. Male rats given 50 7-hour exposures showed no signs of toxicity at 37 or 75 p.p.m. At 150 p.p.m. retardation of growth was noted, at 300 p.p.m. additional signs of chronic toxicity included increased mortality, unkempt appearance, and a significant increase in the kidney/body and lung/body weight ratios.[16]

Mode of action. Depression of the central nervous system and irritation of surface tissues.

Cause of death. Paralysis of the respiratory center and acute pulmonary edema.

Effects observed in man. None reported other than skin irritation and sensitization.

5. Hygienic Standards of Permissible Exposure

None established. Suggest 50 p.p.m. based on comfort, but this may cause response in sensitized persons. The A.C.G.I.H. tentative limit for 1961 is 50 p.p.m.

[21] C. H. Hine, J. K. Kodama, J. S. Wellington, M. K. Dunlap, and H. H. Anderson, *A.M.A. Arch. Ind. Health,* **14,** 250 (1956).

6. Odor and Warning Properties

n-Butyl glycidyl ether possesses satisfactory warning properties. The odor of vapors is not unpleasant but leads to irritation.

C_{16}—C_{18} OLEFIN OXIDE

$$CH_3(CH_2)_{13}CH\text{—}CH_2 \text{ and } CH_3(CH_2)_{15}CH\text{—}CH_2$$
$$\diagdown O \diagup \qquad\qquad \diagdown O \diagup$$

1. Source, Uses and Industrial Exposure

The material is available in multidrum quantities but is still a laboratory chemical. Its suggested uses include intermediate and organic synthesis, a solvent, stabilizer, plasticizer, lubricant additive, a surface-active material, and in a synthesis of modified alkyd resins.

2. Physical and Chemical Properties

C_{16}—C_{18} olefin oxide is a mixed, epoxidized olefin that is a colorless, odorless liquid with the following properties:[22]

Molecular weight: approximately 254
Specific gravity: 0.842 (25/4°C.)
Freezing point: 14 to 15°C.
Boiling point: 200°C. (92 mm. Hg)
Refractive index: 1.4446 (25°C.)
Density of vapor: 7.40 (25°C.)
Solubility: insoluble in water; soluble in hydrocarbons and most common solvents
Flash point: 245°F. (tag open cup)

3. Determination in the Atmosphere

Suggest hydrochloric acid–dioxane method for epoxy group.[2]

4. Physiological Response

Summary. Classified as practically nontoxic.

Response in Animals. Acute toxicity. Only preliminary toxicity data are available on this compound. It is reported that oral administration of 7.5 cc./kg. to the rat is tolerated, as is 2.4 cc. intraperitoneally. Two cc. in a rabbit applied percutaneously is reported to have caused death.[23]

Eye irritation. Not irritating.

[22] *Tech. Bull.* No. **74**, Chemical & Plastics Div., Food Machinery & Chemical Corp., New York, n. d.

[23] *Tech. Bull.* No. **74A**, Chemicals & Plastics Div., Food Machinery & Chemical Corp., New York, n. d.

Skin irritation. Skin fissures, necrosis, and subcutaneous hemorrhage were noted in rabbits 5 days after application of 2 cc. of the material.

Chronic toxicity. No data.

Mode of action. Nothing known.

Cause of death. No specific mechanism.

Effects Observed in Man. No cases of dermatitis or sensitization as a result of handling C_{16}—C_{18} olefin oxide have as yet been reported.

5. Hygienic Standards of Permissible Exposure

None established; suggest 100 p.p.m. based on comfort.

6. Odor and Warning Properties

None.

DICYCLOPENTADIENE DIOXIDE (EP-207)

1. Source, Uses and Industrial Exposures

Available in experimental quantities, this difunctional epoxide is suggested for use as a primary building block for an epoxy resin system. It has also been suggested in the modification of alkyd resins, as a plasticizer, and as a chemical intermediate.

2. Physical and Chemical Properties

Dicyclopentadiene dioxide is a white solid with a mild terpene-like odor having the following properties:[24,25]

Molecular weight: 164

Melting point: 180 to 184°C.

Boiling point: sublimes at 120 to 135°C. (10 mm. Hg)

Density of vapor: 4.77 (25°C.)

Solubility: in water, 1.4 per cent; in methanol, 18.6 per cent; in acetone, 44.7 per cent; and in ether, 18.7 per cent

3. Determination in the Atmosphere

Suggest HCl–dioxane method for epoxy group.[2]

[24] *Tech. Bull.* No. **95**, Chemicals & Plastics Div., Food Machinery & Chemical Corp., New York, n. d.

[25] B. Phillips, *Peracetic Acid and Derivatives.* 2nd ed.. Union Carbide Chemicals Co.. New York, 1957, p. 27.

4. Physiological Response

Summary. The compound is moderately toxic following oral or intraperitoneal entry into the body and practically nontoxic on percutaneous absorption.

Response in Animals. Acute toxicity. Limited toxicity data indicate a tolerated dose to be 34 mg./kg., the lethal range when given orally to rats as 50 to 140 mg./kg., and 11 to 53 mg./kg. intraperitoneally.[25] The oral LD_{50} is reported as 0.21 g./kg.[26] Rabbits tolerated 1 g./kg. percutaneously without effect. The LD_{50} by this route is reported as 8.0 g./kg. In inhalation experiments a dose of 1 g./cu. meter for 1 hour was moderately toxic, and 10 g./cu. meter for 1 hour was lethal to 5 of 6 test animals. Saturated vapors were nonlethal to rats in 8-hour exposure.[27]

Eye irritation. Mildly irritating but no corneal injury in rabbits with test quantities of 0.05 g. A 40 per cent solution was the least concentration to cause barely detectable injury.

Skin irritation. The undiluted material caused redness of short duration. Guinea pig sensitivity tests were negative.

Chronic toxicity. No data.

Mode of Action. Not known.

Cause of Death. Not known.

Effects Observed in Man. No cases of dermatitis or sensitization have been reported.

5. Hygienic Standards of Permissible Exposure

None established.

6. Odor and Warning Properties

The mild terpenelike odor may be masked by a more powerful and distinctive hydrocarbon odor of dicyclopentadiene.

DIGLYCIDYL ETHER [Di (2-epoxypropyl) Ether]

$$CH_2\text{---}CHCH_2\text{---}O\text{---}CH_2CH\text{---}CH_2$$

1. Source, Uses and Industrial Exposure

Diglycidyl ether is made through epoxidation of diallyl ether through chlorohydrination followed by dehydrochlorination with caustic. Suggested for use as a reactive diluent for epoxy resins for easier handling, as a chemical intermediate and a stabilizer of chlorinated organic compounds, and as a textile-treating agent.

---

[26] *Tech. Bull.* No. **95A**, Chemicals & Plastics Div., Food Machinery & Chemical Corp., New York, n. d.

[27] *Toxicology Studies: Dicyclopentadiene Dioxide (EP-207).* Industrial Medicine & Toxicology Dept. Data Sheet, Union Carbide Corp., New York, 1958.

2. *Physical and Chemical Properties*

Diglycidyl ether is a colorless liquid with a pronounced irritant odor, possessing the following properties:[20]

Molecular weight: 130
Specific gravity: 1.262 (25/4°C.)
Boiling point: 260°C. (760 mm. Hg)
Vapor pressure: 0.09 mm. Hg (25°C.)
Per cent in "saturated" air: 0.0121 (25°C.)
Density of vapor: 3.78 (25°C.)

3. *Determination in the Atmosphere*

Suggest hydrochloric acid–dioxane method.[2]

4. *Physiological Response*

Summary. The compound is moderately toxic following ingestion or respiratory exposure; slightly toxic following percutaneous absorption; severely irritating to the skin, eyes, and mucous membranes. Vapor exposure causes chemical pneumonia. Experimentally it produces skin cancers and depression of the hemopoietic system. See Table 1

TABLE 1

Hematopoietic Effects of Repeated Daily Cutaneous Application
of Diglycidyl Ether to Rats

Dose, g./kg.	No. of applications	Drop in WBC	Drop in marrow nucleated cells	Change in WBC differential count
0.030	20	0	0	0
0.060	20	0	+	0
0.125	20	+	+	+
0.250	6[a]	+	+	+
0.500	6[a]	+	+	+

[a] Discontinued because of fractional mortality.

Response in Animals. Acute toxicity. Central nervous system depression, characterized by incoordination, ataxia, and decreased motor activity occurs following intragastric and percutaneous absorption. Intragastric LD_{50} is 0.17 g./kg. in mice and 0.45 g./kg. in rats. Percutaneous approximate lethal dose for 4-hour application is 1.5 g./kg. in rabbits and the LD_{50} dose is 1.0 g./kg. in rats. Vapor exposure causes few immediate effects; however, within 24 hours, depression, clouding of the cornea, increased nasal discharge, and swollen eyelids and feet occur. The LC_{50} in mice for a 4-hour exposure is 86 p.p.m. and for 8 hours 30 p.p.m.; for rats these values are 200 and 68 p.p.m.[21]

Eye irritation. A score of 74 was obtained on the Draize[14a] test, indicating severe ocular irritation. No blindness or permanent defect in the cornea, lens, or iris occurred, however.

Skin irritation. A score of 7.5 was obtained on the Draize[14a] test, indicating severe irritation. Marked erythema, edema, and eschar appeared.

Chronic toxicity. Five repeated applications of 0.2 ml. spread over the back of rabbits caused marked skin irritation characterized by necrosis and penetration through the derma.[21] Applications thrice weekly to the skin of mice for a period of 1 year caused suppression of the sebaceous glands, hyperkeratosis, epithelial hyperplasia, and skin cancers. Repeated subcutaneous injections did not cause sarcomas in rats.[11] Four repeated 4-hour exposures of rats to 20 p.p.m. caused marked respiratory tract irritation, loss in body weight, decrease in the leukocyte count, involution of the thymus and spleen, a hemorrhagic bone marrow, and halving of the count of nucleated cells in the marrow. Twenty-four of 29 rats survived 20 repeated 4-hour exposures to 3 p.p.m. but also showed effects on hemopoiesis. As many as 60 repeated exposures to 0.3 p.p.m. for 4-hour intervals were without effect. Rats sacrificed 1 year after the subacute exposures to 3 p.p.m. did not show any evidence of permanent effects on the hematopoietic system or manifestation of detrimental changes in any of the organs. Dogs, rabbits, and rats administered the compound intramuscularly or intravenously showed evidence of depression of hematopoiesis characterized by one or more of the following: leukopenia, relative lymphopenia or neutropenia (depending on the predominant cell type), hypoplasia of the bone marrow, and decrease in visible lymphoid tissue.[16]

Mode of Action. An acute irritant to skin, eyes, and mucous membranes. It causes respiratory tract irritation and possesses moderate radiomimetic properties.

Cause of Death. Acute pulmonary edema and chemical pneumonitis. Secondary bacterial invasion following acute radiomimetic effects.

Effects Observed in Man. Skin burns, slow in healing. Acute irritation of the eyes and respiratory tract.

5. *Hygienic Standards of Permissible Exposure*

None established. Suggest 1 p.p.m. based on potential hemopoietic effects. The A.C.G.I.H. tentative limit value for 1961 is 10 p.p.m.

6. *Odor and Warning Properties*

Sensory threshold limits not established. Recognizable by odor above 5 p.p.m. Moderately irritating to the eyes at 10 p.p.m. Warning properties not significant at suggested threshold concentration of 1 p.p.m.

DIGLYCIDYL ETHER OF SUBSTITUTED GLYCERINE (EPON 562)

$$CH_2\!\!-\!\!CHCH_2\!\!-\!\!O\!\!-\!\!R\!\!-\!\!O\!\!-\!\!CH_2CH\!\!-\!\!CH_2$$

R = aliphatic radicals, average molecular weight $\sim$ 100

1. Source, Uses, and Industrial Exposures

The diglycidyl ether of substituted glycerine is obtained as a reaction mixture of epichlorohydrin and glycerine. It is used in conjunction with other epoxy resins in the manufacture of adhesives. Since the compound is generally used in conjunction with a curing agent, frequently active amine, the hazardous properties of these substances must also be considered.

2. Physical and Chemical Properties

The diglycidyl ether of substituted glycerine is the reaction product of epichlorohydrin and glycerine. Although the exact molecular arrangement is not known, it is of uniform composition and has the general properties:[3]

Molecular weight: Ca. 300
Specific gravity: 1.023 (20/4°C.)
Refractive index: 1.478 (25°C.)
Solubility: miscible with water, soluble in ketones
Epoxy equivalent*: 140 to 165

3. Determination in the Atmosphere

Suggest hydrochloric acid–dioxane method for the epoxy group.[2]

4. Physiological Response

Summary. Systemic toxicity is low, the compound being practically non-toxic by ingestion and percutaneous absorption. No vapor hazard exists based on results of animal studies and extremely low vapor pressure. Repeated skin contact causes irritation and occasionally sensitization. Slight radiomimetic effects are demonstrable experimentally.

Response in Animals. Acute toxicity. The pharmacologic responses are not remarkable following absorption, depression and slight dyspnea being the only ante-mortem events of significance. A summary of the acute toxicity data appears in Table 2.

Eye irritation. A score of 82 obtained by the Draize[14a] test. The compound is classified as severely irritating.

Skin irritation. A score of 0 obtained by the Draize[14a] test. The compound is classified as nonirritating by this method. Applications of larger amounts or repeated applications actually give rise to severe skin injury.

Chronic toxicity. Repeated 7-hour vapor exposures: Groups of 10 male rats were exposed for 7 hours a day 5 days a week for a total of 50 exposures to air saturated with the vapors of EPON 562. Aside from a slight encrustation of the eyelids in some animals, none of the rats showed any signs of toxicity or irritation attributable to the exposure.

* Grams of resin containing one gram equivalent of epoxide.

TABLE 2

Summary on Acute Toxicity Data on the Diglycidyl Ether
of Substituted Glycerine[4]

Route	Species	Response
Intragastric	Rat	LD$_{50}$, 5.0 g./kg.
Intragastric	Mouse	LD$_{50}$, 1.87 g./kg.
Intragastric	Rabbit	LD$_{50}$, 4.01 g./kg.
Intraperitoneal	Rat	LD$_{50}$, 0.38 g./kg.
Intraperitoneal	Mouse	LD$_{50}$, 0.30 g./kg.
Vapor (8-hr. exposure to "saturated" air)	Mouse	No effect other than slight eye irritation
Vapor (8-hr. exposure to "saturated" air)	Rat	No effect
Percutaneous (12 g./kg. for 2 hrs.)	Rabbit	No mortality, severe local irritation

Repeated skin applications: Rabbits received 20 applications of 0.2 g. total dose of the resin, which was allowed to remain for 1 or 7 hours. Scoring according to the Draize method gave high scores of 8 in both cases, with a final mean of 7.6 to 7.8, indicating the compounds to be highly irritating on repeated applications. In a second experiment, rabbits receiving 1 g./kg. cutaneously on 5 successive days developed severe subcutaneous hemorrhage and skin necrosis; 2 of 4 died. The uncured resin was fed in the diet of rats for 26 weeks at a level of 0.04, 0.2, and 1 per cent. No mortalities occurred at any of the levels but at the highest level there was retardation of weight gain. No significant pathology occurred.

Radiomimetic effects: The effects of repeated exposure on the hemopoietic system will be found in Table 3.

Oncogenic activity: Carcinomas were produced on the skin of mice painted with the material from once to thrice weekly over a period of a year. No tumors were produced on rabbit ears in a similar test. Sarcomas were produced in rats by subcutaneous injection. When fed at a concentration of 0.2 per cent in the diet of A strain mice with spontaneous pulmonary adenomas, there was no effect on the incidence of occurrence.[10]

Mode of Action. The mechanism of systemic effects is not clearly established. Central nervous system depression occurs. Primary action is that of irritation. Radiomimetic effects are probably based on cross-linking of nuclear proteins.

Cause of Death. Depression of the respiratory center and secondary shock.

Effects Observed in Man. Causes primary irritation and occasionally sensitization. In 3 of 8 persons receiving the material intravenously, changes in the blood occurred. These consisted of lowering of the white cell count, decrease in the total number of lymphocytes and monocytes, and a drop in the platelet count. No impairment in liver or kidney function was observed. The effects were reversible.[9]

TABLE 3

Effects on the Hemopoietic System of Exposure to the Diglycidyl Ether
of Substituted Glycerine[16]

Route	Species	Dose	No. of doses	Response
Respiratory	Rat	Saturated vapors	50 (8 hrs. each)	No effect noted
Intramuscular	Rat	0.10 g./kg.	6	No effect
		0.20 g./kg.	6	Depression of white cell count and bone marrow nucleated cell count
Intramuscular	Dog	0.2 g./kg.	2 (1 wk. apart)	Marked depression of white cell count and relative neutropenia, followed by leukocytosis; ulceration and abscess of injection site
Intravenous	Dog	0.2 g./kg.	1	Progressive decline in WBC to count of
		0.05 g./kg.	3	500 WBC; death from overwhelming infection
Intravenous	Rabbit	0.1 g./kg.	2	Decrease in total white count
Percutaneous	Rat	1.0 g./kg.	20	No effect
		2.0 g./kg.	20	No effect
		4.0 g./kg.	20	Depression of bone marrow nucleated cell count (only)

5. Hygienic Standards of Permissible Exposure

None established. Extremely low vapor pressure precludes any significant vapor concentration.

6. Odor and Warning Properties

The compound produces skin irritation shortly after contact. No characteristic odor; no irritation from vapors.

DIGLYCIDYL ETHERS OF BISPHENOL A [2,2-Bis-(4-hydroxyphenyl)propane] and Condensation Products of Their Further Reaction with Bisphenol A

1. *Source, Uses, and Industrial Exposure*

The synthesis of the basic epoxy resin molecule involves the reaction of epichlorohydrin with bisphenol A, the latter requiring two basic intermediates for synthesis, acetone and phenol. Theoretically, the production of the diglycidyl ether of bisphenol A requires 2 moles of epichlorohydrin for each mole of the phenol.

Epoxy resins of higher molecular weight are obtained by reaction in the presence of excess caustic and phenol. This reaction requires consumption not only of the initial epoxy groups in the epichlorohydrin, but of groups formed by dehydrohalogenation.

The properties of epoxy resins make them ideally suited for sealing, encapsulating, making castings and pottings, and formulating light-weight foams. Castings may be used for patterns, molds, and finished products. Patterns are particularly convenient to duplicate when mass-production runs are required, since they may be cast quickly and free surfaces give excellent draws. Molds are used for short production runs, for prototype models, and custom-designed parts. These cast products can be used as bobbins, gears, and metal-working hobs; foam-encased castings are valuable in the aircraft industry. Because of the high degree of fill, they are used for potting all sizes of electric circuits. The tough construction of the potted unit eliminates the need for an exterior totally enclosing case, with consequent saving of space, cost, and weight. For nearly all of the above uses, liquid epoxy resins of low molecular weight are used. Resins with a molecular weight of 300 to 450 are preferred for most of these commercial applications.

The epoxy resins have outstanding adhesive properties, which can be improved selectively by the addition of diluents, resinous modifiers, and fillers. The adhesive systems are usually of the 2-container type, and may be applied either as liquids or as hot melts. Variables considered in the formation and application of these epoxy adhesive systems include a choice of curing agent, degree of cure, curing pressure, surface preparation, and material to be bonded.

These resins are used as binders in the preparation of laminates of paper, polyester cloth, fiber-glass cloth, and wood sheets. The fiber-glass laminate is most important commercially. Essentially two techniques are used to prepare laminates: (*1*) wet layups, which involve the impregnation of the cloth with a 100 per cent solid, nonsolvent-type of containing resin, and immediate application of the cloth to the laminating operation; (*2*) dry layups, which involve impregnation of the cloth with a 100 per cent solid or a solvent-containing resin system in advance of the production operation. The laminates may be formed by contact but are more usually formed under pressure. The properties are dependent on laminating pressure, resin content, and curing cycle. The majority of epoxy production is absorbed in coating formulations. These are of three types: 100 per cent solid coatings, nonesterified solution, and esterified solution coatings. These are formulated either as air-dry or baking-type. A solvent system is generally employed to provide a fluid carrier.

The amines were the first materials to gain general acceptance as curing agents for epoxy resins. Polyfunctional primary aliphatic amines give fast cures and provide over-all properties satisfactory for a wide variety of applications. These materials are usually considerably more active physiologically than the epoxy resins and more volatile, and many cases of skin and eye irritation occur during their use. Other curing agents, including the acid anhydrides and organic acids, have given relatively less problems in handling. A number of diluents are physiologically more active than the resins themselves. Some of these are the epoxy ethers and aliphatic compounds of low molecular weight.

Resin modifiers include phenolic substances, aniline formaldehyde resins, furfural, isocyanates, and silicone resins, all of which may contribute to the handling problem.

Since much of the handling of the products is carried out in small operations without the benefit of good ventilation or housekeeping, exposures are frequent and often give rise to dermatitis.

2. Physical and Chemical Properties

The resins are usually mixtures and may contain homologs of higher weight, isomers, branch-chain homologs, and, occasionally, monoglycidyl ethers. The general formula may be written as indicated for the molecular formula above, where n is the number of repeating units in the resin chain. In the compounds whose general properties are given, n varies from 0 to 9.

Molecular weight (approx.): 350, 900, 2900

n: 0, 2, 9

Specific gravity: 1.168, 1.204, 1.146

Melting point, °C.*: 8–12, 64–76, 127–133

Epoxy equivalent: 190–210, 450–525, 1650–2050

3. Determination in the Atmosphere

The liquid resin would not reach sufficient levels to allow sampling. Dust from the solid resins may be trapped according to standard filter paper and liquid entrapment methods and determined by the hydrochloric acid–dioxane method.[2]

4. Physiological Response

Summary. The liquid members of the group are practically nontoxic perorally, and no physiological systemic effect results from inhalation of their vapors or absorption through the skin. The solid resins are relatively harmless. None of the compounds are more than mildly irritating even on repeated contact, according to animal tests. They do present a major dermatitis problem on use, however, and the liquid type at least causes sensitization.

Response in Animals. Acute toxicity. The pharmacological effects even in lethal doses are not remarkable. Moderate ante-mortem depression occurs, and loss of weight and diarrhea in surviving animals. Toxicity generally decreases with

---

* Durran's mercury method.

increase in molecular weight and epoxy number. Intraperitoneal toxicity is greater than intragastric by a factor of 10. Gross lesions are nonspecific, the chief effect being local irritation. Acute toxicity data is summarized in Table 4.

TABLE 4

Summary of Acute Toxicity Data on Diglycidyl Ethers of Bisphenol A (2,2-Bis(4-Hydroxyphenyl) Propane) and Condensation Products of Further Reaction with Bisphenol A

Characteristics of resin tested		Route	Species	LD$_{50}$ (g./kg.)
Mol. wt.	Epoxy value			
350	190–210	Intragastric	Mouse	15.6
			Rat	11.4
			Rabbit	19.8
		Intraperitoneal	Mouse	4.0
		Intraperitoneal	Rat	2.4
		Respiratory (8-hr.)	Rat	>Saturated[a]
900	450–525	Intragastric	Mouse	>20[a]
			Rat	>30[a]
		Intraperitoneal	Rat	2.40
2900	1650–2050	Intraperitoneal	Rat	2.20

[a] No deaths at these levels of exposure.

Eye irritation. Scores of 2 to 41 obtained by the Draize method. Higher scores probably represent mechanical rather than chemical irritation.

Skin irritation. No irritation produced by single contact. Draize score of 0. Repeated dermal contacts lead to mild or moderate irritation in some test animals.

Chronic toxicity. No systemic toxicity or visceral lesions followed repeated cutaneous contact with any of the compounds.

Compound, molecular weight 350: Feeding 0.2 per cent in the diet of rats for 26 weeks caused no effect other than an increase in the kidney/body weight ratio. One per cent in the diet resulted in retardation of weight gain as well, without discernible lesions. Five per cent dietary intake resulted in death in 2 weeks. When reacted with curing agents and then added to the diet in quantities up to 10 per cent, no effects were seen in rats after a 6-week period.

Compounds, molecular weight 900 and 2900: No effect from feeding up to 5 per cent in the diet.

Radiomimetic effects: Compound, molecular weight 350: No effects on hemopoiesis following repeated intramuscular injection of rats. Skin painting of mice and rabbits for up to 2 years did not cause any tumors. Repeated subcutaneous injection of rats resulted in sarcomas in 25 per cent of the animals. No effect on the spontaneous incidence of pulmonary tumors in strain A mice.

Mode of Action. Central nervous system depressant. Mild irritant.

Cause of Death. Depression of the respiratory center.

Effects Observed in Man. Dermatitis from epoxy resins was first reported by Pletscher *et al.*[28] Clinical aspects have been described by a number of workers including Hine *et al.*,[4] Plüss,[29] and Grandjean.[30] The usual lesion is typical of contact dermatitis, the early manifestations being redness and edema, with weeping followed by crusting and scaling. Following initial contact there is usually an erythematous, discrete area confined to the point of contact. Since this frequently involves the face, it is likely that it is caused by the vapors of the hardener or active diluent, though contact with contaminated gloves or droplets may also play a part, as do occasionally the vapors of the liquid-type epoxy resin. The initial lesion usually persists for from 48 hours to 10 days, after which the erythema fades and gives way to a macular rash followed by scaling.

Sensitization may follow the initial contact, resulting in the development of a papular, vesicular eczema. This is accompanied by considerable itching, and extension beyond the point of original contact. Only occasionally are areas other than the backs of the hands, the forearms, the face and neck involved. Recommended treatment consists of bland ointments and soaps. The worker is withdrawn from further contact, and the lesion usually subsides in 10 to 14 days. However, it may recur on further contact. If the worker is not withdrawn from contact, the dermatitis usually persists for longer periods, but usually does not become more intense. The lesions may assume a brownish color, and scaling is frequently noted.

Grandjean's[30] investigation of European factories indicated that 43 per cent of all workmen involved with epoxy resins suffered dermatitis, and 22 per cent of these were severe, as judged by lost time amounting to 3 days. Malten[31] reported that 26 of 225 workers processing the compounds had become sensitized to resins, hardeners, or both. This probably reflects inadequate hygiene. Only rarely, in our experience, does sensitization occur to the epoxy resins based on bisphenol A, although sensitization to diluents and hardeners is not infrequent. An over-all sensitization rate of approximately 2 per cent may be expected even with good industrial hygiene practice.

Fabrics impregnated with cured resins have been worn by large numbers of persons without any allergic manifestations.

Because of the persistence of dermatitis and the possibility of sensitization, preventive measures are especially important when handling these compounds. The most important measure for combating dermatitis is good personal hygiene. This requires supervisory instruction of personnel and good work habits, and pro-

---

[28] A. Pletscher, R. Schuppli, and R. Reipert, *Z. Unfallmed.,* **47,** 163 (1954).

[29] J. Plüss, *Z. Unfallmed.,* **47,** 83 (1954).

[30] E. Grandjean, *Brit. J. Ind. Med.,* **14,** 1 (1957).

[31] K. E. Malten, Professional eczema in working-up synthetic plastic materials, in particular those derived from unsaturated polyester resins and ethoxyline resins, Doctor's Thesis, Univ. of Amsterdam, 1956.

vision of adequate facilities for removing the material periodically, together with a designated cleanup period prior to "break" and quitting times. In addition, protective devices offer considerable aid in minimizing personal contact. Gloves should be worn, and contamination of the skin should be scrupulously avoided. Protective clothing and personal protective creams are of help. Where possible, only persons with no history of allergic conditions or eczematous eruptions should be selected. Procedures should be carefully supervised to ensure that mixing of volatile agents is carried out in properly ventilated areas. Bench and floor areas should be protected with disposable paper when contamination is likely.

Remedial measures following skin contact should include thorough cleansing with soap and water, followed by a solvent only when absolutely necessary. Accidental eye contamination is unlikely, but treatment consists of the usual measures.

The source of contact should be identified when dermatitis develops, and the improper work condition corrected. Since the solvent curing agents are flammable liquids, fire hydrants and control measures are required. Safety cans should be used for storing flammable solvents and fire extinguishers should be located in the area.

5. *Hygienic Standards of Permissible Exposure*

None established. None required on basis of physical properties and low toxicity.

6. *Odor and Warning Properties*

None; sticky and tacky when handled.

DIPENTENE DIOXIDE (Limonene Dioxide)

$$CH_3$$

S = Cyclohexane

1. *Source, Uses, and Industrial Exposure*

Dipentene dioxide is available in developmental quantities. It has been suggested as diluent for epoxy resins, as an intermediate in the preparation of modified alkyd resins, a plasticizer, lubricant additive, and chemical intermediate. The internal epoxy group is more reactive to ring opening with acids and anhydrides than is the external group. The external group has greater reactivity with amines.

2. *Physical and Chemical Properties*

Dipentene dioxide is a colorless liquid combining the reactivities of an internal and external epoxy group in the same molecule, with a mild menthol-like odor. It has the following properties:[32,33]

Molecular weight: 168

Specific gravity: 1.0287 (20/4°C.)

Freezing point: <60°C.

Boiling point: 228°C. (760 mm. Hg) (by extrapolation)

Vapor density: 7.40

Vapor pressure: 0.02 mm. Hg (20°C.)

Refractive index: 1.4682 (25°C.)

Solubility: slightly soluble in water; miscible with methanol, benzene, carbon tetrachloride, and hexane

Flash point: 118°C. (tag open cup)

1 p.p.m. $\approx$ 6.87 mg./cu. meter at 25°C., 760 mm. Hg and 1 mg./liter $\approx$ 145 p.p.m. at 25°C., 760 mm. Hg

3. *Determination in the Atmosphere*

Hydrochloric acid–dioxane method for the epoxy group.[2]

4. *Physiological Response*

Summary. Slight toxicity by the oral route and practically nontoxic percutaneously.

Response in Animals. Acute toxicity. The tolerated oral dose for rats has been stated to be from 0.4 to 1.55 ml./kg., and the lethal range to lie within 1.5 to 3.8 ml./kg. Intraperitoneal doses were proportionally smaller, with the lethal range lying between 0.4 and 1.2 ml./kg. Tolerated percutaneous dose for rabbits was 1 ml./kg., lethal dose, 6 ml./kg. Lethality occurred in rats after 1-hour exposure to an air concentration of 60 g./cu. meter. Stated to be only slightly toxic on the basis of this study.[34]

Eye irritation. Moderately irritating.

Skin irritation. No data.

Chronic toxicity. No data.

Mode of Action. Not known.

Cause of Death. Nonspecific.

Effects Observed in Man. No cases of sensitization have been reported.

[32] *Tech. Bull.* No. **96,** Chemicals & Plastics Div., Food Machinery & Chemical Corp., New York, n. d.

[33] B. Phillips, *Peracetic Acid and Derivatives.* 2nd ed., Union Carbide Corp., New York, 1958, p. 28.

[34] *Data Sheet* No. **96A,** Chemicals & Plastics Div., Food Machinery & Chemical Corp., New York, n. d.

5. *Hygienic Standards of Permissible Exposure*

None established. Vapor pressure too low to present a hazard.

6. *Odor and Warning Properties*

None reported.

DIPENTENE MONOXIDE (Limonene Monoxide, 1-Methyl-1,2-epoxy-4-isopropenylcyclohexane)

CH₃

CH₂—C=CH₂

S = Cyclohexane

1. *Source, Uses, and Industrial Exposure*

Dipentene monoxide is in the developmental stage. Laboratory samples are available. Suggested uses include solvent stabilizer, plasticizer, lubricant additive, surface active material, corrosion inhibitor, and modifier of alkyd resins.

2. *Physical and Chemical Properties*

Dipentene monoxide is a colorless liquid with a characteristic odor, combining reactivity of an epoxy group with that of an olefinic double bond in a cyclic terpene molecule. It has the following properties:[35]

Molecular weight: 153 (average)
Specific gravity: 0.929 (20/4°C.)
Freezing point: $< -6°C.$
Boiling point: 74 to 76°C. (10 mm. Hg)
Vapor density: 4.45
Refractive index: 1.4697
Solubility: less than 0.3 per cent in water; partially miscible in hexane, completely miscible in methanol, acetone, benzene, carbon tetrachloride, and ether.

Flash point: 152°F. (tag open cup)

1 p.p.m. ≎ 6.26 mg./cu. meter at 25°C., 760 mm. Hg and 1 mg./liter ≎ 160 p.p.m. at 25°C., 760 mm. Hg

3. *Determination in the Atmosphere*

Hydrochloric acid–dioxane method for epoxy group.[2]

4. *Physiological Response*

Summary. The general physiological properties have not been established. The compound is only slightly toxic.

[35] *Tech. Bull.* No. **81**, Chemicals & Plastics Div., Food Machinery & Chemical Corp., New York, n. d.

Response in Animals. Acute toxicity. Limited toxicity information indicates tolerated dose is 2.4 ml./kg. orally to a rat and 0.64 ml./kg. intraperitoneally to this species. Inhalation experiments show that a concentration of 3.9 mg./liter (approximately 600 p.p.m.) resulted in no mortality or weight disturbance in rats.[36]

Eye irritation. Mild to moderate irritation.

Skin irritation. Mild irritation on application of 2 ml. to the skin of rabbits for 24 hours.

Chronic toxicity. No data available.

Mode of action. Not known.

Cause of death. Not known.

Effects observed in man. None reported.

5. Hygienic Standards of Permissible Exposure

None established.

6. Odor and Warning Properties

Characteristic odor.

DODECENE OXIDE (1,2-Epoxydodecene)

1. Source, Uses, and Industrial Exposure

Dodecene oxide is a developmental chemical that has been proposed for general uses as a solvent, stabilizer, plasticizer, lubricant additive, surface-active material, corrosion inhibitor, and reactive diluent for epoxy resins. It is available in multidrum quantities.

2. Physical and Chemical Properties

Dodecene oxide is an oxidized olefin of characteristic odor, colorless, combining a long-chain hydrocarbon structure with a reactive epoxy group. Because the compound is a developmental chemical, its properties may vary slightly with different lots.[37]

Molecular weight: 175

Specific gravity: 0.836 (25/4°C.)

Freezing point: —10 to —12°C.

Boiling point: 215°C. (760 mm. Hg)

Vapor density: · 5.09

[36] *Data Sheet* No. **81A,** Chemical & Plastics Div., Food Machinery & Chemical Corp., New York, n. d.

[37] *Tech. Bull.* No. **73,** Chemical & Plastics Div., Food Machinery & Chemical Corp., New York, n. d.

Refractive index: 1.4347

Solubility: insoluble in water; soluble in hydrocarbons and most common solvents

1 p.p.m. $\backsim$ 7.16 mg./cu. meter at 25°C., 760 mm. Hg and 1 mg./liter $\backsim$ 140 p.p.m. at 25°C., 760 mm. Hg

3. *Determination in the Atmosphere*

Hydrochloric acid–dioxane method for epoxy groups.[2]

4. *Physiological Response*

Summary. Acute toxicity: The compound is slightly toxic on oral ingestion and practically nontoxic on intraperitoneal administration to laboratory animals.

Dodecene oxide is also a general depressant of the central nervous system.

Response in Animals. Acute toxicity. Limited toxicity information indicates the tolerated dose orally in rats is 2.8 ml./kg.; intraperitoneally it is 0.7 ml./kg.[38]

Eye irritation: Moderate erythema, lid edema, and vascularization following minimal contact. Situation reversible, disappearing in 72 hours.

Skin irritation. Erythema, eschar, fissures, and necrosis within six days of application of 2 ml. to rabbits.

Chronic toxicity. No data available.

Mode of action. Not known.

Cause of death. Probable central nervous system depression.

Effects observed in man. None reported.

5. *Hygienic Standards of Permissible Exposure*

None established.

6. *Odor and Warning Properties*

A characteristic odor is described. No sensory threshold data are available.

EPICHLOROHYDRIN (1-Chloro-2,3-epoxypropane, γ-Chloropropylene oxide, α-Epichlorohydrin)

$$
\underset{\displaystyle CH_2\!\!-\!\!CHCH_2Cl}{\overset{\displaystyle O}{\diagup\!\diagdown}}
$$

1. *Source, Uses, and Industrial Exposure*

Epichlorohydrin is available in large-scale commercial quantities through the discovery and development of processes for its production from propylene. It is employed as a raw material for the manufacture of a number of glycerol and glycidol derivatives, in the manufacture of epoxy resins, as a stabilizer in chlorine-containing materials, and as an intermediate in preparation of condensates with polyfunctional substances.

[38] *Data Sheet* No. **73A,** Chemical & Plastics Div., Food Machinery & Chemical Corp., New York, n. d.

2. *Physical and Chemical Properties*

Epichlorohydrin is a colorless, mobile liquid with a characteristic chloroform-like, irritating odor and the following properties:[39]

Molecular weight: 92.53

Specific gravity: 1.1807 (20/4°C.)

Freezing point: —57.2°C.

Boiling point: 116.1°C. (760 mm. Hg)

Vapor density: 3.21

Vapor pressure: 13 mm. Hg (20°C.)

Refractive index: 1.43805 (20°C.)

Solubility: 6.48 per cent in water; miscible with ethers, alcohols, carbon tetrachloride, and benzene.

Flash point: 105°F. (tag open cup)

Per cent in "saturated" air: 1.7 (25°C.)

1 p.p.m. ⇌ 3.78 mg./cu. meter at 25°C., 760 mm. Hg and 1 mg./liter ⇌ 265 p.p.m. at 25°C., 760 mm. Hg

3. *Determination in the Atmosphere*

The modified hydrochloric acid–dioxane method is suitable either by direct sampling into the reagent mixture or by sampling aliquots of a bottle-collected specimen. Infrared absorption in the frequency range 1240 to 1260 cm.$^{-1}$ will identify a characteristic oxirane (α-epoxy) group. A method for the determination in the range 10 to 1000 p.p.m. in aqueous solution involves sampling of the air through two absorbers containing 0.5N alcoholic potassium hydroxide. The absorbent solutions are combined, refluxed to saponify the epichlorohydrin, and the resulting chloride ion titrated potentiometrically with silver nitrate.

4. *Physiological Response*

Summary. Epichlorohydrin is intensely irritating and systemically moderately toxic by the oral, percutaneous and subcutaneous routes as well as by inhalation of the vapors. Death in experimental animals is attributed to its depressant action on the central nervous system, particularly on the respiratory center, and its irritating effect on the respiratory tract. The cumulative toxic action shown in animals is believed to be due to a nephrotoxic action. Skin contact causes acute irritation and the vapors are highly irritating to the mucous membranes of the eye and respiratory tract.

Response in Animals. Acute toxicity. Regardless of the route of administration, toxic doses of epichlorohydrin produce a similar chain of symptoms differing only in their time of onset. There is gradual development of cyanosis followed by muscular relaxation of the extremities. Despite depression of respiration and

---

[39] *Epichlorohydrin: Technical Booklet CS:49-35.* Shell Chemical Corp., New York, reprinted 1953.

skeletal musculature, narcosis is not observed. A summary of acute toxicity data appears in Table 5.

Eye irritation. Markedly irritating to the eye on local contact. Vapors also give rise to eye irritation.

Skin irritation. Undiluted epichlorohydrin is intensely irritating to the depilated skin of laboratory animals. Repeated applications lead to widespread necrosis.

Chronic toxicity. Rats did not gain weight normally when exposed to 32 p.p.m. of vapor 7 hours a day for a total exposure of 90 days. Exposure at 16 p.p.m. produced a significant increase in the size of the kidneys. Alteration in urinary coproporphyrin was detected at the 16 p.p.m. level.

TABLE 5

Summary of Acute Toxicity Data on Epichlorohydrin[a-d]

Route	Species	Dose	Result
Oral	Rats	0.09 g./kg.	LD_{50}
Oral	Guinea pigs	0.178 g./kg.	LD_{50}
Oral	Mice	0.238 g./kg.	LD_{50}
Intravenous	Rats	0.154 g./kg.	LD_{50}
Intravenous	Mice	0.178 g./kg.	LD_{50}
Percutaneous	Rabbits	0.88 ml./kg.	LD_{50}
Percutaneous	Rats	0.5 ml./kg. (3 applications)	LD_{50}
Inhalation	Mice	2370 p.p.m.	0[e]/30
Inhalation	Mice	8300 p.p.m.	20[e]/20
Inhalation	Rats	250 p.p.m. (8 hrs.)	LC_{50}
Inhalation	Rats	500 p.p.m. (4 hrs.)	LC_{50}
Inhalation	Guinea pigs	561 p.p.m. (4 hrs.)	LC_{50}
Inhalation	Rabbits	445 p.p.m. (4 hrs.)	LC_{50}

[a] C. D. Leake, *Univ. Calif. Publs. Pharmacol.*, **2**, 69 (1941).
[b] H. F. Smyth, Jr., and C. P. Carpenter, *J. Ind. Hyg. Toxicol.*, **30**, 63 (1948).
[c] C. P. Carpenter, H. F. Smyth, Jr., and V. C. Pozzani, *J. Ind. Hyg. Toxicol.*, **31**, 343 (1949).
[d] *Ind. Hyg. Bull. SC: 57-86.* Shell Chemical Corp., New York, 1957.
[e] Deaths.

Mode of action. Central nervous system depression and irritation of the respiratory tract.

Cause of death. Depression of the respiratory center.

Effects observed in man. There have been no cases of serious pulmonary injury or systemic toxicity in the manufacture or handling of epichlorohydrin. Because of the potential nephrotoxic effects, persons working with this material on a continuing basis should undergo medical supervision. Cases of sensitization with resulting intolerance to even small quantities of epichlorohydrin occur occasionally in handling this material. Several cases of severe skin burns have resulted from prolonged contact with the liquid.

5. *Hygienic Standards of Permissible Exposure*

No official level has been set. Suggest 5 p.p.m. would be appropriate on the basis of animal studies, comfort, and freedom from respiratory tract irritation.

6. *Odor and Warning Properties*

The odor is generally perceived as a slightly irritating chloroformlike odor. Sensory perception studies have indicated that the mean threshold for odor recognition is approximately 10 p.p.m., and that at 25 p.p.m. it is recognized by the majority of persons. Marked nose and eye irritation occur only at levels exceeding 100 p.p.m. It is concluded that local irritation of the eyes is not severe enough to force workers to evacuate potentially harmful areas. Eye irritation may be accepted as indicating an undesirable atmospheric contamination.

EPOXIDE-201 (α-Dicyclodiepoxycarboxyl-3,4-Epoxy-6-methylcyclohexylmethyl-3,4-epoxy-6-methylcyclohexanecarboxylate)

S = Cyclohexane

1. *Source, Uses, and Industrial Exposure*

Epoxide-201 is a developmental chemical available in drum quantities. Suggested as an epoxy monomer for producing resins that feature high heat-distortion temperatures. It is stated to be a stabilizer for vinyl chloride resins.

2. *Physical and Chemical Properties*

Epoxide-201 is a new low-viscosity, high-activity epoxy resin monomer with the following properties:[40]

Molecular weight: 280.3
Specific gravity: 1.121 (20/20°C.)
Boiling point: 215°C. (5 mm. Hg)
Vapor density: 8.15
Refractive index: 1.4920 (20°C.)
Solubility: 0.3 per cent in water
Flash point: 310°F. (tag open cup)

3. *Determination in the Atmosphere*

Air concentrations are not great enough to be significant.

4. *Physiological Response*

Summary. A slightly toxic compound by oral ingestion. No particular physiological response characteristic.

[40] *Epoxide-201, F-401, 130C.* Union Carbide Corp., New York, 1958.

Response in Animals.   Acute toxicity.   Single dose oral LD$_{50}$ in rats is 4.92 ml./kg. Single dose skin penetration tests in rabbits showed that 10 ml./kg. killed none of 4 test animals. Rats exposed to a mist in air saturated at 170°C. survived a 4-hour exposure, but 5 of 6 were killed by an 8-hour exposure.[41]

Eye irritation.   Slight hazard.

Skin irritation.   The undiluted chemical caused only a faint erythema on rabbit skin. Sensitization tests on guinea pigs were negative.

Chronic toxicity.   Repeated application of the undiluted chemical 3 times a ·week for a year on the skin of mice resulted in tumor formation and occasional skin cancer.

Mode of action.   Not known.

Cause of Death.   Not known.

Effects Observed in Man.   None reported.

5. Hygienic Standards of Permissible Exposure

None established.   Suggest 10 p.p.m.

6. Odor and Warning Properties

None.

ETHYLENE OXIDE (1,2-Epoxyethane, Oxirane, Dimethylene Oxide)

$$\underset{H_2C\overline{}CH_2}{\overset{O}{\triangle}}$$

1. Source, Uses, and Industrial Exposure

Ethylene oxide is synthesized commercially from ethylene, either through the intermediate ethylene chlorohydrin or by direct oxidation with either oxygen or air. It is available in unrestricted quantities and is used largely for the production of ethylene glycol and its derivatives (i.e., polyethylene glycols, mixed polyglycols, and glycol ethers and esters). Substantial amounts are also used in the preparation of such chemicals as monoethanolamine, acrylonitrile, and surface-active agents.[42] Ethylene oxide, usually mixed with carbon dioxide to reduce flammability, is used as a fumigant, but this use is limited because of its tendency to react with certain proteinaceous materials and because of the possibility of undesirable residues. It is used also as a sterilizer, particularly for surgical supplies.

The high chemical reactivity and the general exothermic nature of ethylene oxide reactions present a number of problems in storage, handling, and use.[43,44]

---

[41] *Toxicology Studies: Epoxide-201, F-40423.* Industrial Medicine and Toxicology Dept., Union Carbide Corp., New York, n. d.

[42] G. O. Curme, Jr. and F. Johnston, eds., *Glycols* (American Chemical Society Monograph, Series 114). Reinhold, New York, 1952.

[43] L. G. Hess and V. V. Tilton, *Ind. Eng. Chem.,* **42,** 1251 (1950).

[44] Mfg. Chemists Assoc. *Chem. Safety Data Sheet* **SD-38,** (1951).

Liquid ethylene oxide is relatively stable, but the vapor in concentrations of 3 to 100 per cent is highly flammable and subject to explosive decomposition. Ignition and decomposition are initiated by many common sources of heat, and the pressure rise is sufficiently rapid and extensive to cause violent rupture of containing equipment. It reacts vigorously with materials having a labile hydrogen. Polymerization is catalyzed by a number of materials, such as acids, alkalis, some carbonates, oxides of iron and aluminum, and chlorides of iron, tin, aluminum, and boron. No acetylide-forming metals such as copper or copper alloys should be in contact with ethylene oxide.

The hazards to health associated with the handling of ethylene oxide are those of inhalation of the vapor and of contact of the eyes and skin with the liquid or solutions even as dilute as one per cent.

2. Physical and Chemical Properties

Ethylene oxide is a colorless gas or liquid with the following properties:

Molecular weight: 44.05

Specific gravity: 0.8966 (0/4°C.) ; 0.8711 (20/20°C.)

Freezing point: −112.5°C.

Boiling point: 10.4°C. (760 mm. Hg)

Vapor density: 1.49 (40°C.)

Refractive index: 1.3614 (4°C.)

Solubility: miscible with water, acetone, methanol, ether, benzene, and carbon tetrachloride

Flash point: −4°F.

Flammability limits: 3 to 100 per cent

1 p.p.m. ⪦ 1.80 mg./cu. meter at 25°C., 760 mm. Hg and 1 mg./liter ⪦ 556 p.p.m. at 25°C., 760 mm. Hg

3. Determination in the Atmosphere

The methods available for the determination of ethylene oxide in air for industrial hygiene purposes are not as precise or as reliable as is desirable. For general information the reader should refer to the method described for propylene oxide.[2,45,46]

Recently Critchfield and Johnson[47] have reported a colorimetric method for the determination of small amounts of ethylene oxide in spices. The method depends upon the hydrolysis of ethylene oxide to ethylene glycol and its reaction with sodium periodate to form formaldehyde, which is then determined colorimetrically with sodium chromotrope in concentrated sulfuric acid. Although industrial hygiene experience with this method has not been reported, it would seem to offer promising possibilities.

[45] O. F. Lubatti, *J. Soc. Chem. Ind. London,* **63,** 133 (1944).

[46] R. L. Hollingsworth and B. F. Waling, *Am. Ind. Hyg. Assoc. Quart.,* **16,** 52 (1955).

[47] F. E. Critchfield and J. B. Johnson, *Anal. Chem.,* **29,** 797 (1957).

Gage[48] has reported success with a colorimetric method for the determination of ethylene oxide in the atmosphere based on removal of the ethylene oxide from the air sample by scrubbing with silica gel, elutriation of the oxide from the silica gel with water, oxidation with periodic acid, and determination of the formaldehyde formed colorimetrically by reacting it with sodium arsenite and acetylacetone.

TABLE 6

Summary of Acute Response of Animals to Exposure to Ethylene Oxide[a–d]

P.p.m. by vol. in air	Time, hr.	Animal	Results
250–280	8	Guinea pig	Slight respiratory changes; no deaths
	48	Guinea pig	Occasional death
560–600	7	Guinea pig, cat and dog	No deaths
	8	Guinea pig	Occasional death
	22	Guinea pig and cat	Death during or following exposure
	22	Rabbit, dog	No deaths
710	4	Dog	0/3 in 14 days
1100	5	Rat, guinea pig, rabbit	Moderate injury, no deaths
	8	Guinea pig, dog, rabbit	Slight injury, no deaths
	8	Rat, cat	Death within 24 hours
1300–1400	8	Guinea pig	Majority died in 1 to 8 days
	4	Dogs	3/3 first day
2200	1½	Cat	Injurious, no deaths
	3	Cat	Death within 24 hours
	4	Guinea pig	Injurious, few deaths
	4	Rabbit	Injurious, no deaths
	4	Dog	Death within 24 hours
3000	1	Guinea pig	No deaths
	3	Guinea pig	Death of majority within 1 to 8 days
	8	Guinea pig	Death of majority within 24 hours
7000	⅓	Guinea pig	No evidence of injury
	1	Guinea pig	Death of majority within 1 to 8 days
	2½	Guinea pig	Death within 24 hours
14,000	10 min.	Guinea pig	No evidence of injury
	20 min.	Guinea pig	Majority died in 1 to 8 days
	60 min.	Guinea pig	Death within 24 hours
51,000–64,000	5 min.	Guinea pig	Majority died in 1 to 8 days
	10 min.	Guinea pig	Death in 24 hours

[a] K. H. Jacobsen, E. B. Hackley, and L. Feinsilver, *A.M.A. Arch. Ind. Health,* **13**: 237 (1956).

[b] R. L. Hollingsworth, V. K. Rowe, D. D. McCollister, and H. C. Spencer, *A.M.A. Arch. Ind. Health,* **13**: 217 (1956).

[c] C. P. Waite, F. A. Patty, and W. P. Yant, *U. S. Public Health Repts.,* **45**: 1832 (1930).

[d] F. Flury and F. Zernik, *Schädliche gase.* Springer, Berlin, 1931.

---

[48] J. C. Gage, *Analyst,* **82,** 587 (1957).

4. *Physiological Response*

Summary. Ethylene oxide may be described as a central depressant, an irritant, and a protoplasmic poison. Contact with even dilute solutions may cause irritation and necrosis of the eyes and irritation, blistering, edema, and necrosis of the skin. Excessive exposure to vapor may cause irritation of the eyes, respiratory tract, and lungs, and central depression. Nausea and vomiting are usually delayed and may be followed by convulsive seizures and profound weakness of the extremities and secondary infection of the lungs.

Response in Animals. Acute toxicity. When single doses of ethylene oxide were given intragastrically as a 10 per cent solution in olive oil to groups of 5 rats, all animals survived 0.1 g./kg., while 0.2 g./kg. killed all animals.[50] A 1 per cent aqueous solution[51] had an intragastric LD_{50} of 0.33 g./kg. for rats and 0.27 g./kg. for guinea pigs. High concentrations of ethylene oxide gas are both irritating to the mucous membranes and depressing to the central nervous system. Early symptoms in a number of species include lachrymation, nasal discharge, and salivation. Later, gasping and labored breathing become apparent. Corneal opacity has been observed in certain species, particularly guinea pigs. The delayed effects of acute exposure include nausea, vomiting, diarrhea, edema of the lungs, paralysis (particularly of the hind quarters), convulsions, and death. Prompt deaths are usually due to lung edema; delayed deaths frequently result from secondary infection of the lungs, but general systemic intoxication may also be a factor. The results of single exposures of animals to various concentrations of ethylene oxide in air are summarized in Table 6.

The quantitative aspects of acute exposure are summarized in Figure 1. The area bounded by AB and CD represents intensities of exposure that may be expected to cause more or less severe injury, including death. The broken line, EF, is located to give the industrial hygienist an indication of the maximum intensities of single exposures that probably will be without appreciable injury to human beings. It should be noted that this line is not founded on actual data but is interpreted from the acute data available from the experimental work on animals.

Eye irritation. Vapors of ethylene oxide in high concentration are known to be irritating to the eyes of animals and man. Hess and Tilton,[43] Walker and Greeson,[52] and McLaughlin[53] have reported that liquid ethylene oxide produces severe irritation and corneal injury. The eyes should therefore be protected from ethylene oxide and its solutions.

[49] K. H. Jacobson, E. B. Hackley, and L. Feinsilver, *A.M.A. Arch. Ind. Health*, **13**, 237 (1956).

[50] R. L. Hollingsworth, V. K. Rowe, F. Oyen, D. D. McCollister, and H. C. Spencer, *A.M.A. Arch. Ind. Health*, **13**, 217 (1956).

[51] H. F. Smyth, Jr., J. Seaton, and L. Fischer, *J. Ind. Hyg. Toxicol.*, **23**, 259 (1941).

[52] W. J. C. Walker and C. E. Greeson, *J. Hyg.*, **32**, 409 (1932).

[53] R. S. McLaughlin, *Am. J. Ophthalmol.*, **29**, 1355 (1946).

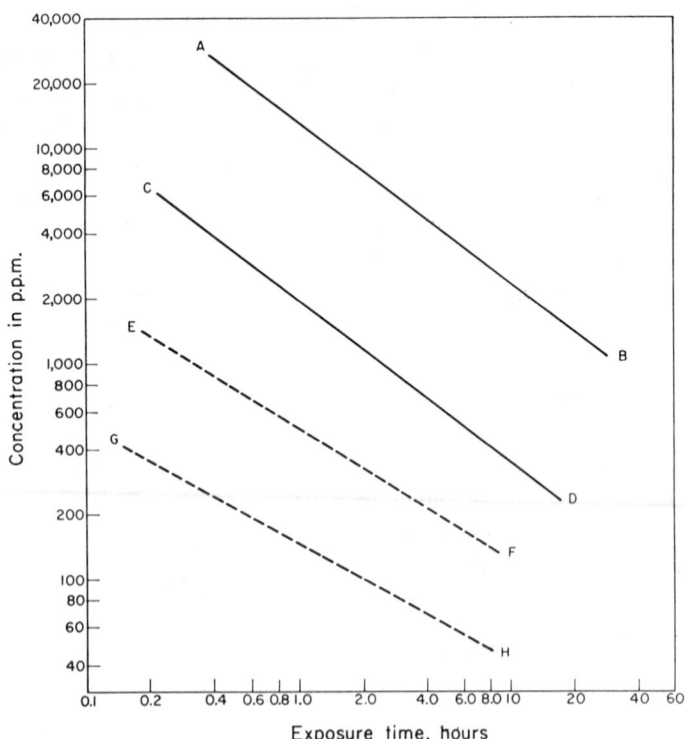

Fig. 1. Inhalation toxicity of ethylene oxide in air. Severe injury or death likely in areas ABCD. EF probable maximum without injury for 1 exposure. GH probable maximum without injury for repeated daily exposures.

Skin irritation. Liquid ethylene oxide apparently is without adverse effect on rabbit and human skin on single mild exposures if the material evaporates rapidly. If large amounts of material are involved, evaporation may cause sufficient cooling to cause "frostbite." When confined to rabbit skin, even dilute aqueous solutions produce irritation, edema, and occasional burns on single short exposures. Sexton and Henson[54,71] have reported that concentrated ethylene oxide and dilute (1 per cent) aqueous solutions produce blisters and cause blebs on prolonged contact with human skin. Such contact ordinarily results from wearing contaminated clothing (for example, shoes or gloves) or may arise from ethylene oxide becoming entrapped under a ring or watchband. The reaction may not appear for several hours after the exposure. They[71] also demonstrated that ethylene oxide can cause sensitization of the skin.

Pharmacology.[19] When freshly prepared aqueous solutions (2 to 5 per cent) were injected intravenously into dogs, the LD_{50} was found to be about 125 mg./kg.

[54] R. J. Sexton and E. V. Henson, *J. Ind. Hyg. Toxicol.*, **31**, 297 (1949).

A dose of 30 mg./kg. or more usually caused vomiting and defecation for about 2 hours, followed by weakness and flaccidity, usually apparent in the hind limbs first. Doses up to 100 mg./kg. in dogs under barbiturate anesthesia caused no apparent changes in blood pressure or cardiac rate. Respiration is adequate until terminal stages, when it becomes labored and cyanosis develops. Tonic extensor spasm may precede respiratory cessation. Since the heart usually beats after all reflexes disappear, death is believed due to respiratory failure.

Chronic effects. Groups of 5 young adult, female, white rats were given olive oil solutions of the test material intragastrically 5 times a week for a total of 22 doses in 30 days. The dosage levels were 0.003, 0.01, 0.03, and 0.1 g./kg. The no-effect dosage level, as judged by gross appearance, growth, blood-urea-nitrogen determinations, hematology, organ weights, and histopathology, was 0.03 g./kg. The dosage level of 0.1 g./kg. caused loss of body weight, gastric irritation, and slight liver damage.[50]

The chronic toxicity of ethylene oxide vapors has been studied extensively by two groups of investigators.[49,50] Their results are summarized in Table 7. Deaths for the most part were due either to primary lung irritation or to secondary infection resulting from the primary irritation. At the highest concentration studied, 841 p.p.m., the animals could tolerate fewer than eight 7-hour exposures.

When animals were exposed repeatedly to vapor concentrations in the range of 350 to 400 p.p.m., some interesting phenomena were observed. Growth depression or weight loss, nasal discharge, diarrhea, and greater or lesser amounts of respiratory irritation occurred constantly. Monkey and rabbits were resistant to respiratory injury, but readily developed a flaccid paralysis of the hindquarters accompanied by severe atrophy of the musculature of the hind legs and back. Mice were the most susceptible species to lung injury and subsequent infection and guinea pigs were most resistant; neither mice nor guinea pigs developed any paralytic effects. Rats were susceptible to both the injury to the lungs and to the paralytic effect. Severely paralyzed rats, rabbits, and monkeys recovered completely upon cessation of exposure. Two of 3 dogs exposed repeatedly to 290 p.p.m. were nauseated, weak in the hindquarters, and slightly anemic.

As the concentration to which animals were exposed was decreased, the effects diminished in intensity until at about 100 p.p.m. one group[49] saw only a minimum hematological effect in dogs and the other group[50] saw only a slight growth depression in male rats. No effects were seen in animals exposed repeatedly to a concentration of 49 p.p.m. These observations led both groups to recommend an industrial hygiene standard of 50 p.p.m.

The results reported by Walker and Greeson[52] indicate that a few short exposures to relatively high concentrations will not be injurious. The results of Koelsch and Lederer[55] indicate a high toxicity not in keeping with the results of most other investigators. These observations should not be ignored, but their sig-

[55] F. Koelsch and E. Lederer, *Zentr. Gewerbehyg. Unfallverhüt.*, **17**, 264 (1930).

TABLE 7

Results of Repeated Exposure of Animals to Vapors of Ethylene Oxide[49,50]

P.p.m.	Exposure Hrs.	No.	Mortality ratio	Species	Pathological findings
841	7	6	10/10	Rat	Gross irritation of the respiratory tract; mice seemed most suscept-ible
		8	8/8	Guinea pig	
		4	1/1	Rabbit	
		1	5/5	Mouse	
		7	1/1	Monkey	
357	7	7	2/20	Rat	Moderate loss of body weight; severe lung injury
			4/10	Mouse	
		33	10/10	Mouse	Secondary respiratory infection the primary cause of death; impaired nervous function at lumbar and sacral level, reversible, no blood changes
		38	10/10	Rat, F	
		38	8/10	Rat, M	
		148	1/2	Rabbit	
		38–94	0/4	Monkey	
		123	0/16	Guinea pig, M&F	Growth depression, increased lung weight in males, and degeneration of testicular tubules; no nervous signs
290	6	6 wk.	0/3	Dog	Vomiting, occasional tremors, tran-sient paraplegia, anemia
204	7	122–157	22/40	Rat	Appreciable number of rats died of secondary respiratory infection; growth depression in rat; Poste-rior paresis in monkey and rabbit; increased lung weight in rat and guinea pig
			1/16	Guinea pig	
			0/4	Rabbit	
			2/10	Mouse	
			0/2	Monkey	
113	7	122–157	0/40	Rat	No findings except growth depression in male rats
			0/16	Guinea pig	
			0/4	Rabbit	
			0/2	Monkey	
100	6	130	3/20	Rat	No clinical signs, no significant find-ings except anemia in dog
			8/30	Mouse	
			0/3	Dog	

nificance is difficult to assess since few animals were used and analyses of chamber concentrations were not made.

Mode of Action. Ethylene oxide in high concentration is narcotic and a primary irritant of the respiratory system. Such irritation creates a condition favorable for the development of secondary infection, frequently the cause of delayed deaths. Moderate exposure to ethylene oxide is unlikely to injure the eyes, liver, kidney, spleen, heart, adrenals, blood, and blood-forming organs. It possesses, however, the peculiar property of causing a delayed and reversible weakness and paralysis of the hindquarters that accompanies, or is caused by, sensory and motor dysfunction of the nervous system, particularly in the sacral and lumbar regions. Ethylene oxide is completely soluble in water and, once dissolved, has a low vapor pressure. Exhalation is not a major route of elimination from the body. Furthermore, its high chemical reactivity tends to prevent its elimination as such. There is no reversible equilibrium and the only limitations to absorption are exposure intensity in terms of concentration and time.

There has been some speculation but no proof regarding the mechanisms through which the systemic effects occur. The most popular theory is that ethylene oxide is absorbed into the cell where it undergoes hydrolysis to ethylene glycol, which in turn causes dysfunction at the cellular level. This would seem to be an oversimplification when dealing with a chemical as reactive as ethylene oxide. It would seem that there must be many vital processes that would be altered by the reaction of ethylene oxide with molecules having labile hydrogen atoms.

Effects Observed in Man. Serious systemic poisoning from ethylene oxide has been rare indeed. Von Oettingen[56] refers to 3 cases reported by Enke, in which headache, vomiting, dyspnea, diarrhea, and lymphocytosis were observed. Hess and Tilton[43] refer to one case reported by Sexton and considered "severe" in which essentially the same symptomatology, including the lymphocytosis noted above, was observed.

Other experience with human beings has been confined to injury to the eyes, and particularly to the skin, including sensitization, without systemic effects.

5. Hygienic Standards of Permissible Exposure

In Figure 1, a broken line, GH, has been drawn to give the industrial hygienist an indication of the maximum intensities of repeated daily exposures that prob-

TABLE 8

Summary of Suggested Permissible Limits for Ethylene Oxide

Exposure	Ethylene oxide (p.p.m.)
Probably safe for daily exposures of up to 7 hours' duration	50
Probably safe for occasional (2/wk.) repeated exposure of up to 4 hrs./day	100
Probably safe for single exposures of several hours duration (up to 7, no more than 1/wk.)	150
Probably safe for repeated exposures of up to 1 hr./day	150
Probably safe for single (no more than 1/wk.) exposures of up to 1 hour's duration	500

ably will be without effect to human beings, and tabulation of exposure intensities believed to be safe is given in Table 8. Whereas the 50 p.p.m. figure for repeated 7-hour exposures is based upon substantial laboratory data, the other figures, at

---

[56] W. F. von Oettingen, Supplement to occupation and health, *Encyclopedia of Hygiene, Pathology and Social Welfare.* International Labour Office, Geneva, 1939.

best, are interpolations of the limited data available and are intended only as a guide for industrial hygienists. These values may well be altered when additional data, such as repeated exposures for short daily periods, became available. The A.C.G.I.H. limit value for 1961 is 50 p.p.m.

6. Odor and Warning Properties

Although ethylene oxide is considered to be an irritant gas, neither irritation nor odor can be relied upon to warn of the presence of vapor concentrations harmful upon prolonged or repeated exposure. The mean detectable concentration[49] is 700 p.p.m. with 95 per cent confidence limits of 317 to 1540 p.p.m. Furthermore, the material is reported to cause olfactory fatigue rather readily.

GLYCIDOL (2,3-Epoxy-1-propanol)

1. Source, Uses, and Industrial Exposure

Glycidol is made through dehydrochlorination of glycerol and monochlorohydrin with caustic. It is commercially available. Suggested for use in preparation of glycerol and glycidyl ethers, esters, and amines, in the pharmaceutical industry, and sanitary chemicals.

2. Physical and Chemical Properties

Glycidol is a colorless, slightly viscous liquid with the following properties:
Molecular weight: 74.05
Specific gravity: 1.115 (20/4°C.)
Boiling point: 160°C. (760 mm. Hg)
Vapor density: 2.15
Vapor pressure: 0.9 mm. Hg (25°C.)
Solubility: completely water-soluble
Per cent in "saturated" air: 0.118
1 p.p.m. $\backsimeq$ 3.03 mg./cu. meter at 25°C., 760 mm. Hg and 1 mg./liter $\backsimeq$ 330 p.p.m. at 25°C., 760 mm. Hg

3. Determination in the Atmosphere

Pyridinium chloride–chloroform method.[2]

4. Physiological Response

Summary. Slightly toxic after ingestion or percutaneous absorption. Moderately toxic upon inhalation. Moderately irritating to surface tissues. Stimulant and depressant of the central nervous system.

Response in Animals. Acute toxicity. Depends in part on route of administration. Depression followed by stimulation with hypersensitivity to sound, vibration of the facial muscles, and involuntary tremors occurred with oral administration. Less frequently epileptiform convulsions and lachrymation appeared in rats. LD_{50} is 0.45 g./kg. in mice and 0.85 g./kg. in rats receiving the compound by this route. Minimal depression preceding death is seen in rabbits receiving the compound percutaneously. The percutaneous LD_{50} is 1.98 g./kg. Irritation of the lungs, emphysema, and pneumonitis are noted after vapor exposure. LC_{50} value for 4-hour exposure in mice is 450 p.p.m. and for 8 hours in rats, 580 p.p.m.

Eye irritation. Severely irritating with a score of 68, according to the Draize[14a] method.

Skin irritation. Moderately irritating with a Draize[14a] score of 4.5 after single application. Repeated daily applications to the skin caused severe irritation after 4 days with a modified Draize score of 5.7 and localized, deeply penetrated areas of skin necrosis.

Chronic toxicity. Fifty exposures of 7 hours each to 400 p.p.m. gave no signs of cummulative toxicity in rats as measured by changes in mortality, growth, organ/body weight ratios, or significant microscopic lesions. Repeated intramuscular injections do not effect hemopoiesis in rats.[21]

Mode of action. Central nervous system stimulant and skin and mucous membrane irritant.

Cause of death. Central nervous system fatigue following central stimulation. Pulmonary irritation and emphysema when inhaled.

Effects Observed in Man. None reported.

5. *Hygienic Standards of Permissible Exposure*

None established; suggest 100 p.p.m. The A.C.G.I.H. tentative limit for 1961 is 50 p.p.m.

6. *Odor and Warning Properties*

Not unusual.

GLYCIDALDEHYDE (2,3-Epoxypropionaldehyde)

$$\underset{\text{CH}_2\text{—CHCH}}{\overset{\overset{\displaystyle O}{\diagup\diagdown}\quad \overset{\displaystyle O}{\|}}{}}$$

1. *Source, Uses, and Industrial Exposure*

Glycidaldehyde is prepared from hydrogen peroxide epoxidation of acrolein. It is suggested as a bifunctional chemical intermediate and as a cross-linking agent for textile treatment, leather tanning, and protein insolubilization.

2. *Physical and Chemical Properties*

Glycidaldehyde is a mobile, colorless liquid with a pungent odor, having the following properties:[57]

[57] *Glycidaldehyde (GDA), Data Sheet S-58:2.* Shell Development Co., Emeryville, Calif., 1958.

Molecular weight: 72.1
Specific gravity: 1.1403
Freezing point: $-61.8°C.$
Boiling point: 112 to 113 (760 mm. Hg) ; 57 to 58 (100 mm. Hg)
Vapor density: 2.58
Refractive index: 1.4200 (20°C.)
Solubility: completely soluble in all common solvents; insoluble in petroleum ether
Flash point: 88°F. (tag open cup)
1 p.p.m. $\backsim$ 2.94 mg./cu. meter at 25°C., 760 mm. Hg and 1 mg./liter $\backsim$ 339 p.p.m. at 25°C., 760 mm. Hg

3. *Determination in the Atmosphere*

Hydrochloric acid–dioxane method for epoxy groups.[2]

4. *Physiological Response*

Summary. Moderately toxic following single exposure by ingestion, inhalation, or percutaneous absorption. Irritating to surface tissues including the respiratory tract. Skin sensitizer. Death from pulmonary edema or secondary shock. Radiomimetic effects demonstrated on hemopoiesis.

Response in Animals. Acute toxicity. Following ingestion, transitory excitement succeeded by depression and labored breathing. LD_{50} in rats is 0.23 g./kg. Respiratory exposure causes marked pulmonary tract irritation and tearing. LC_{50} in rats exposed 4 hours is 251 p.p.m. Percutaneous absorption gives an approximately lethal dose of 0.20 g./kg. in rabbits. None of 3 died at 0.04 g./kg. Severe local injury to the skin occurred at this dose.[20]

Eye irritation. Animal data and human experience not available. On basis of analogy with other low molecular weight aldehydes and glycidyl compounds, it is probably highly irritating.

Skin irritation. Score of 3.6 obtained on Draize test. Classified as moderately irritating on single contact; repeated contacts will result in a high degree of skin irritation.

Chronic toxicity. Rats fail to survive five 4-hour exposures of 80 p.p.m. Marked depression of the white count and nucleated cells of the bone marrow occurred. Further hemopoietic effects in rats receiving 50 mg./kg. intramuscularly included a shift in the differential count and decrease in the numbers of red blood cells.

Mode of action. Causes irritation of surface tissues, depression of the central nervous system. May interfere with glucose utilization.

Cause of death. Pulmonary edema and shock.

Effects observed in man. Marked skin irritation with slow healing followed by bronzing. Sensitization in several cases.

5. Hygienic Standards of Permissible Exposure

None established. Suggest 1 p.p.m. based on comfort and potential hemopoietic effects.

6. Odor and Warning Properties

Pronounced aldehydelike odor at low levels. Voluntary exposure to serious lung-irritating levels unlikely.

ISOPROPYL GLYCIDYL ETHER (1,2-Epoxy-3-isopropoxypropane)

$$CH_3—CH—O—CH_2CH—CH_2$$
with the epoxy oxygen bridging the last two carbons, and CH_3 below the first CH.

1. Source, Uses, and Industrial Exposure

Isopropyl glycidyl ether is manufactured through the condensation of isopropyl alcohol and epichlorohydrin with subsequent dehydrochlorination with caustic to form the epoxy ring. It is available in laboratory quantities. Suggested as a reactive diluent for epoxy resins, stabilizer for organic compounds, and as an intermediate for synthesis of ethers and esters.

2. Physical and Chemical Properties

Isopropyl glycidyl ether is a mobile, colorless liquid with the following properties:[21]

Molecular weight: 116.16
Specific gravity: 0.9186 (20/4°C.)
Boiling point: 137°C. (760 mm. Hg)
Vapor density: 4.15
Vapor pressure: 9.4 mm. Hg (25°C.)
Solubility: 18.8 per cent soluble in water, soluble in ketones, and alcohols
Per cent in "saturated" air: 1.237
1 p.p.m. ≏ 4.74 mg./cu. meter at 25°C., 760 mm. Hg and 1 mg./liter ≏ 211 p.p.m. at 25°C., 760 mm. Hg

3. Determination in the Atmosphere

Hydrochloric acid–dioxane method for epoxy group.[2]

4. Physiological Response

Summary. Slightly toxic following ingestion or respiratory exposure. Practically nontoxic following absorption through the skin. Moderately irritating to the eyes and skin and capable of causing skin sensitization. No radiomimetic effects demonstrated.[16]

Response in Animals. Acute toxicity. Incoordination and depressed motor activity preceding respiratory depression occur on absorption. Intragastric LD_{50} values are 1.30 and 4.20 g./kg., respectively, in mice and rats. Rabbits administered the compound percutaneously showed altered physiology only at high doses, the LD_{50} being 9.65 g./kg. Respiratory exposure produced typical chemical pneumonia in levels exceeding 500 p.p.m. The LC_{50} values are 1500 p.p.m. in the mouse after 4 hours' exposure and 1100 p.p.m. in the rat after 8 hours.[21]

Eye irritation. Moderate eye irritation is obtained on contact, with a Draize[14a] score of 40.

Skin irritation. A score of 4.3 was obtained by the Draize[14a] technique. The compound is classified as moderately irritating. Repeated 1-hour exposures result in some tolerance and a scoring of 2.2 was obtained after 20 applications. This is the least irritating of the simple glycidyl ethers covered in this review.

Chronic toxicity. Rats exposed to 400 p.p.m. for fifty 7-hour periods had a slight retardation of weight gain but no other evidence of cumulative toxicity. Occasional animals had patchy bronchopneumonia at autopsy. No hemopoietic effect appeared in rats receiving repeated intramuscular injections.

Mode of action. The compound is mildly irritating. Its only observed pharmacological property is that of central nervous system depression.

Cause of death. Central depression of respiration.

Effects observed in man. No untoward effects reported other than slight skin irritation on repeated contact.

5. Hygienic Standards of Permissible Exposure

None established. Suggest 100 p.p.m. based on comfort. The A.C.G.I.H. tentative limit for 1961 is 50 p.p.m.

OCTYLENE OXIDE (Mixed 1,2- and 2,3-Epoxyoctane)

$$CH_3(CH_2)_5\overset{O}{\overbrace{CH-CH_2}} \quad \text{and} \quad CH_3(CH_2)_4\overset{O}{\overbrace{CH-CHCH_3}}$$

1. Source, Uses, and Industrial Exposure

Octylene oxide is an experimental compound, available in limited quantities. It has been suggested as an intermediate in organic synthesis, as a solvent stabilizer, plasticizer, lubricant additive, and for other uses generally assigned to epoxy compounds.

2. Physical and Chemical Properties

Octylene oxide is a colorless to pale yellow liquid with a fruity odor. It is a mixed epoxidized olefin combining a medium-chain hydrocarbon structure with a reactive epoxy group and the following general properties.[58]

---

[58] *Tech. Bull.* No. **72,** Chemicals & Plastics Div., Food Machinery & Chemical Corp., New York, n. d.

Molecular weight: 130 (average)
Specific gravity: 0.830 (25/4°C.)
Freezing point: < -50°C.
Boiling point: 156°C. (760 mm. Hg)
Vapor density: 3.78
Refractive index: 1.4160
Solubility: slightly soluble in water; soluble in hydrocarbons and most common solvents
Flash point: 110°F. (tag open cup)
1 p.p.m. $\backsimeq$ 5.32 mg./cu. meter at 25°C., 760 mm. Hg and 1 mg/liter $\backsimeq$ 188 p.p.m. at 25°C., 760 mm. Hg

3. Determination in the Atmosphere

Hydrochloric acid–dioxane method for epoxy group.[2]

4. Physiological Response

Summary. The compound is slightly toxic by oral route and is stated to give no untoward effect from inhalation of reasonable quantities.

Response in Animals. Acute toxicity. Limited. Toxicity data indicate tolerated dose to be 1.6 ml./kg. for rats when administered intraperitoneally. Inhalation studies in rats indicate a concentration of 17.7 mg./liter (approximately 3000 p.p.m.) produced no untoward effect in rats.[59]

Eye irritation. 0.05 ml. in rabbits' eyes produced no significant untoward effects.

Skin irritation. Eschar and bloody fissures were noted 9 days after application of 2 ml. of the material to the skin of rabbits.

Chronic toxicity. No data reported.

Mode of action. Not known.

Cause of death. Not known.

Effects observed in man. No cases of dermatitis or sensitization have been reported.

5. Hygienic Standards of Permissible Exposure

Not established; suggest 50 p.p.m.

6. Odor and Warning Properties

The compound has a slight fruity odor.

PHENYL GLYCIDYL ETHER (Glycidyl Phenyl Ether, 1,2-Epoxy-3-phenoxypropane)

[59] *Data Sheet* No. **72A,** Chemicals & Plastics Div., Food Machinery & Chemical Corp., New York, n. d.

1. Source, Uses and Industrial Exposure

Phenyl glycidyl ether is synthesized by condensation of phenol with epichloro-hydrin, with subsequent dehydrochlorination with caustic to form the epoxy ring. As an acid accepter it is very effective as a stabilizer of halogenated compounds. Its high solvency for halogenated materials offers many possibilities as an intermediate.

2. Physical and Chemical Properties

Phenyl glycidyl ether is a relatively high-boiling, colorless liquid with the following properties:[60]

Molecular weight: 150.17
Specific gravity: 1.1092 (20/4°C.)
Melting point: 3.5°C.
Boiling point: 245°C. (760 mm. Hg)
Vapor density: 4.37
Vapor pressure: 0.01 mm. Hg (20°C.)
Refractive index: 1.5314
Solubility: 0.24 per cent in water; 12.9 per cent in octane; completely soluble in acetone and toluene

3. Determination in the Atmosphere

Hydrochloric acid–dioxane method for epoxy groups.[2]

4. Physiological Response

Summary. Slightly toxic following ingestion, practically nontoxic after percutaneous absorption. Use experience indicates moderate skin irritation on prolonged or repeated contact and skin sensitization.

Response in Animals. Acute toxicity. Incoordination, ataxia, and depressed motor activity preceded frank depression. Animals were comatose just prior to death. Intragastric LD_{50} for mice was 1.40 g./kg., and for rats, 3.85 g./kg. Percutaneous LD_{50} in rabbits, 2.99 g./kg. No deaths were produced in mice exposed for 4 hours, or rats exposed for 8 hours, to saturated vapors at room temperature.[21]

Eye irritation. A score of 8 was obtained by the Draize technique. The compound is classified as mildly irritating.

Skin irritation. A score of 0.7 was obtained by the Draize technique. The compound was classified as mildly irritating on single applications. Repeated applications, however, gave rise to erythema and edema after 4 days of application, and a final rating of 5.2 irritation grade was obtained.

Chronic toxicity. Groups of rats exposed to supersaturated vapors at approximately 15 p.p.m. for 7 hours per day for 50 exposures showed no evidence of toxicity as reflected in mortality, weight gain, organ body weight ratio, microscopic

[60] *Tech. Publ. SC:52-10.* Shell Chemical Corp., New York, 1952, p. 56.

lesions, or histological alteration. No hemopoietic effects on repeated intramuscular injections of rats.

Mode of action. Pharmacological property of the compound following oral or percutaneous absorption is depression of motor activity. This is probably due to the interference with interneuronal transmission. Death in experimental animals is due to a central depression together with paralysis of the respiratory muscles. The compound has little effect on sensory mechanisms.

Cause of death. Paralysis of respiratory muscles.

Effects observed in man. No systemic effects have been reported; however, dermatitis has been experienced in persons repeatedly exposed, and several cases of skin sensitization.

5. *Hygienic Standards of Permissible Exposure*

None established. As saturated vapors approach only 13 p.p.m., it appears that none is required. The A.C.G.I.H. tentative limit for 1961 is 50 p.p.m.

6. *Odor and Warning Properties*

None.

α-PINENE OXIDE (D-2,6,6-Trimethyl-2,3-epoxybicyclo [3.1.1]heptane)

1. *Source, Uses, and Industrial Exposure*

Limited quantities of this compound are available. It has been suggested as a perfumery intermediate, surface-active material, intermediate for organic synthesis, and lubricant.

2. *Physical and Chemical Properties*

α-Pinene oxide is a water-immiscible, colorless liquid with a camphorlike odor. It combines the reactivity of an epoxy group with that of the bicyclic system of α-pinene.[61]

Molecular weight: 146
Specific gravity: 0.936 (20/4°C.)
Freezing point: <60°C.
Boiling point: 61°C. (10 mm. Hg)

[61] *Tech. Bull.* No. **82**, Chemicals & Plastics Div., Food Machinery & Chemical Corp., New York, n. d.

Vapor density: 4.25

Refractive index: 1.4697

Solubility: water-immiscible; soluble in methyl alcohol, hexane, benzene, carbon tetrachloride, and acetone

1 p.p.m. ≈ 5.97 mg./cu. meter at 25°C., 760 mm. Hg and 1 mg./liter ≈ 167 p.p.m. at 25°C., 760 mm. Hg

3. *Determination in the Atmosphere*

Hydrochloric acid dioxane method for the epoxy group.[2]

4. *Physiological Response*

Summary. The compound would be rated as moderately toxic following oral ingestion. General physiological properties have not been established. Mild irritation on skin contact.

Response in Animals. Acute toxicity. Limited toxicity data indicate tolerated doses to be 0.6 cc./kg. administered orally to rats, and 0.06 cc./kg. for administration interperitoneally to this species. Inhalation exposure of rats to 9.7 g./cu. m. (approximately 16,000 p.p.m.) caused no mortality or growth changes.[62]

Eye irritation. Severely irritating.

Skin irritation. Mild irritant on rabbit skin.

Chronic toxicity. No data available.

Mode of action. Not established.

Cause of death. Not established.

Effects observed in man. No untoward effects noted. The compound should be handled with adequate eye protection because of the severe eye-irritating properties.

5. *Hygienic Standards of Permissible Exposure*

No standards set; suggest 50 p.p.m.

6. *Odor and Warning Properties*

Camphorlike odor.

PROPYLENE OXIDE (1,2-Epoxypropane, Propene Oxide, Methyl Oxirane)

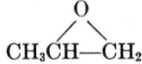

1. *Source, Uses, and Industrial Exposures*

Propylene oxide is synthesized commercially from propylene through the intermediate propylene chlorohydrin. However, it also can be made by direct

[62] *Data Sheet* No. **82A,** Chemicals & Plastics Div., Food Machinery & Chemical Corp., New York, n. d.

oxidation of propylene with either air or oxygen, but this method produces other materials as well. Propylene oxide is available in unrestricted amounts and is used largely for the production of propylene glycol and its derivatives (i.e., polypropylene glycols, mixed polyglycols, and various propylene glycol ethers and esters). Substantial quantities are used also in the preparation of hydroxy propyl celluloses and sugars, surface active agents, isopropanol amine, and a host of other chemicals. It is also used as a fumigant, herbicide, preservative, and in some cases as a solvent.[42]

Propylene oxide is highly reactive chemically, being intermediate between ethylene oxide and butylene oxide. It is also extremely flammable. The liquid is relatively stable but may react violently with materials having a labile hydrogen, particularly in the presence of catalysts such as acids, alkalies, and certain salts. It polymerizes exothermically. No acetylide-forming metals such as copper or copper alloys should be in contact with propylene oxide. The same general handling procedures as described for ethylene oxide[44] should be employed.

The hazard to health of propylene oxide is not as great as that of ethylene oxide but is of the same general character. Adverse physiological effects can follow inhalation of vapors and from contact of the eyes and skin with the liquid or with solutions as dilute as 1 per cent.

2. *Physical and Chemical Properties*

Propylene oxide is a colorless liquid with the following properties:
Molecular weight: 58.03
Specific gravity: 0.8304 (20/20°C.) ; 0.826 (25/25°C.)
Freezing point: −112°C.
Boiling point: 34.2°C. (760 mm. Hg)
Vapor density: 2.0 (air = 1)
Refractive index: 1.363 (25°C.)
Solubility: 40.5 per cent by weight in water at 20°C.; 59 per cent by weight in water at 25°C.; miscible with acetone, benzene, carbon tetrachloride, ether, and methanol
Flash point: <-30°C.
Flammability limits: 2.1 to 38.5 per cent by volume in air
1 p.p.m. $\rightleftharpoons$ 2.376 mg./cu. meter at 25°C., 760 mm. Hg and 1 mg./liter $\rightleftharpoons$ 421 p.p.m. at 25°C., 760 mm. Hg

3. *Determination in the Atmosphere*

The methods available for the determination of propylene oxide in air for industrial hygiene purposes are not as precise or as reliable as is desirable. Propylene oxide in high concentrations in air can be estimated by adsorption techniques, by using a combustible gas indicator or by means of a gas interferometer, but these methods are nonspecific and subject to severe limitations. The method described by Lubatti[45] and studied by Hollingsworth and Waling[46] involves: (a) passing a known volume of air through a dilute sulfuric acid solution containing a high concentration of magnesium bromide, (*b*) titrating with sodium hydroxide to determine the amount of sulfuric acid consumed in converting the propylene oxide to propylene bromohydrin. Similar methods, using dilute hydrochloric acid in a concentrated calcium chloride medium or the pyridinium chloride–chloroform method can also be used.[2] Both of these methods, at best, give only an approximation of the actual amount of propylene oxide present in the air. Vapor phase chromatography offers considerable promise especially when used to determine the amount collected on an absorbing medium or in a vapor trap.

4. *Physiological Response*

Summary. Propylene oxide may be described as primarily an irritant, a mild protoplasmic poison, and a mild depressant of the central nervous system. Contact with even dilute solutions may cause irritation and necrosis of the skin. Excessive

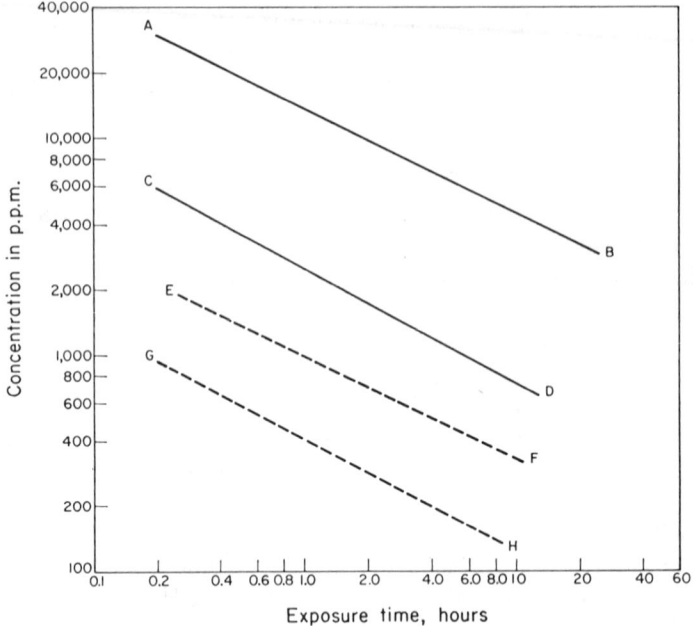

Fig. 2. Vapor toxicity of propylene oxide. Line AB represents the minimum intensities of single exposures causing death in all animals studied.[49,63] Line CD represents the maximum intensities of single exposures causing no detectable injury in rats. Broken line EF represents the intensities of single exposure the author believes to be safe for human subjects. Broken line GH represents the intensities of repeated daily exposures the author believes to be safe for human subjects.

exposure to vapors may cause irritation of the eyes, respiratory tract, and lungs, and central nervous effects characterized by ataxia, incoordination, and general depression. The most likely secondary effect from acute exposure is pulmonary infection. The quantitative aspects of acute exposure are summarized in Figure 2.

The areas bounded by AB and CD represent intensities of exposure that may cause in animals a response ranging from barely detectable injury to certain death. Guinea pigs are most resistant. The broken line, EF, is located to give the industrial hygienist an indication of the maximum intensities of single exposures that probably will be without appreciable effect in human subjects. It should be noted that this line is not founded on actual data but is interpreted from the data obtained on animals. There is no evidence, either from experimental studies or from practical experience, to indicate that propylene oxide can produce systemic intoxication by absorption through the skin.

Response in Animals. Acute toxicity. When propylene oxide as a 5 per cent aqueous solution was administered intragastrically, LD_{50} values of 1.14 g./kg. for rats and 0.69 g./kg. for guinea pigs were obtained.[49] When administered intragastrically as a 10 per cent olive oil solution, rats survived 0.3 g./kg. and succumbed at 1.0 g./kg.[63] The results of single inhalation exposures of animals to various concentrations and for various times are summarized in Table 9.

TABLE 9
Summary of Acute Response of Animals to Vapors of Propylene Oxide[49,63]

Vapor conc. (p.p.m.)	Time (hr.)	Animal	Response: no. dying/no. exposed
900	7	Rat	0/5—no detectable injury
1,330	4	Mouse	1/10 in 1 day
	4	Rat,	0/10 in 14 days
		Dog	0/3 in 14 days
1,800	2	Rat	0/5—no detectable injury
	7	Rat	0/10 in 14 days
	7	Guinea pig	0/10 in 14 days
2,480	4	Dog	3/3 within hours
3,600	1	Rat	0/5—no detectable injury
	2	Rat	4/10
		Guinea pig	0/5
	4	Rat	4/10
		Guinea pig	1/5
	7	Rat	10/10
		Guinea pig	2/5
7,200	0.25	Rat	0/10
	0.5	Rat	2/10
	1.0	Rat	5/10
		Guinea pig	0/10
	2.0	Rat	10/10
		Guinea pig	1/5
	4.0	Guinea pig	10/10
14,400	0.25	Rat	0/15
	0.5	Rat	10/10
		Guinea pig	0/5
	1.0	Guinea pig	5/5

[63] V. K. Rowe, R. L. Hollingsworth, F. Oyen, D. D. McCollister, and H. C. Spencer, *A.M.A. Arch. Ind. Health,* **13,** 228 (1956)

Eye irritation. Vapors of propylene oxide are known to be irritating to the eyes of animals. Liquid propylene oxide and its solutions cause serious local injury in the eyes of rabbits,[64] and 3 cases of corneal burns in humans have been reported.[65]

Skin contact. Undiluted propylene oxide is without adverse effect upon the skin if the material is not confined and can evaporate freely. However, when confined to the skin, as from wearing contaminated clothing or shoes, the material and water solutions as dilute as 10 per cent are likely to cause irritation, blistering, and even burns upon single short exposures. There is some evidence which indicates that solutions more dilute than 10 per cent may be more irritating than the undiluted propylene oxide.

TABLE 10

Summary of Results from Animals That Received
Repeated Seven-Hour Exposures to the Vapors of Propylene Oxide[63]

Vapor concn. (p.p.m.)	Species	Sex	No. exposures received	Mortality: no. drying/ no. exposed	Pathologic findings
457	Rat	M	79	5/10	Eye and respiratory irritation, growth depression, increased mortality, lung injury
		F	138	7/10	
	Guinea pig	M	110	0/8	Eye and respiratory irritation, growth depression, slight liver injury in males, and slight lung injury in females
		F	110	0/8	
	Rabbit	M, F	154	0/2	None
	Monkey	F	154	0/1	None
195	Rat	M, F	138	7/40	None
	Guinea pig	M, F	128	0/16	None except lungs of females; slightly heavy
	Rabbits	M, F	154	0/4	None
	Monkey	F	154	0/2	None
102	Rat	M, F	138	9/40	None
	Guinea pig	M, F	128	0/16	None

Chronic effects. The effect on animals of propylene oxide vapors when inhaled repeatedly has been studied rather extensively. The results are summarized in Table 10. In general, it appears to be about one third as toxic as ethylene oxide. The principal effect appears to be primary irritation of the lungs. This, in the rat, at least, tends to create a medium favorable for the development of secondary infection, frequently the cause of death. Injury to the internal organs, other than the lungs, appears to be insignificant.

Matched groups of 5 young adult female white rats each were fed by intuba-

[64] C. P. Carpenter and H. F. Smyth, Jr., *Am. J. Ophthalmol.,* **29,** 1363 (1946).
[65] R. L. McLaughlin, *Am. J. Ophthalmol.,* **29,** 1355 (1946).

tion olive oil solutions of propylene oxide 5 times a week until a total of 18 doses had been given at levels of 0.1, 0.2, and g./kg./day. Animals receiving a dosage level of 0.2 g./kg. and 0.1 g./kg. showed no effect as judged by gross appearance, growth, blood urea-nitrogen determinations, organ weights, and micropathology of the various organs in comparison with controls receiving olive oil. A dosage level of 0.3 g./kg. caused loss of body weight, gastric irritation, and slight liver injury. Hematological studies of the group receiving 0.1 g./kg. showed no significant changes.

Mode of action. Acute vapor toxicity studies on laboratory animals have indicated that the principal acute systemic toxic effect of propylene oxide vapor is an irritation of the eyes and of the breathing passages and lungs. Early symptoms include lachrymation, nasal discharge, and salivation followed by gasping and labored breathing in all species, and vomiting in dogs. Severe irritation in the lungs may persist for several days and, in some cases, lead to pneumonia. Injury to other organs such as the liver and kidney appears to be insignificant. Propylene oxide has but a relatively weak anesthetic action; this effect became apparent when exposure was to concentrations of 4,000 p.p.m. or more.

Effects Observed in Man. Adverse effects on human beings have apparently been confined to injury to the eyes and skin, and no cases of systemic injury are reported in the readily available literature. Experience has shown that the observance of simple precautions readily permits safe handling and use.

5. *Hygienic Standards of Permissible Exposure*

In Figure 2 a broken line, GH, has been drawn to give the industrial hygienist an indication of the maximum intensities of repeated daily exposure that probably will be safe for human subjects.

A tabulation of exposure intensities believed to be safe is given in Table 11

TABLE 11
Summary of Suggested Permissible Limits for
Vapor Exposure to Propylene Oxide

Exposure	Concentration, p.p.m.
Probably safe for repeated exposures of up to 7 hours' duration (no more than 5/wk.)	150
Probably safe for repeated exposures of up to 4 hours' duration (no more than 5/wk.)	200
Probably safe for repeated exposures of up to 1 hour duration (no more than 5/wk.)	400
Probably safe for repeated exposures of up to 10 minutes' duration (no more than 5/wk.)	1000
Probably safe for single exposure of up to 7 hours' duration (no more than 1/wk.)	400
Probably safe for single exposure of up to 1 hour duration (no more than 1/wk.)	1000

Whereas the 150 p.p.m. figure for repeated 7-hour exposures is based upon substantial laboratory data, the other figures are at best interpolations of the limited data available. They are intended only as guides for industrial hygienists and may well be changed when more data become available. The A.C.G.I.H. threshold limit for 1961 is 100 p.p.m.

6. *Odor and Warning Properties*

The median detectable concentration of propylene oxide vapors is reported to be 200 p.p.m. with 95 per cent confidence limits of 114 to 353 p.p.m.[49] The odor is described as sweet, alcoholic, and like natural gas, ether, or benzene.[49] Neither odor nor irritation can be relied upon to warn of the presence of vapor concentrations not suitable for prolonged and repeated exposure. Odor and/or irritation will warn of the presence of acutely dangerous concentrations.

RESORCINOL DIGLYCIDYL ETHER [1,3-Bis(2,3-epoxypropoxy)benzene]

$$H_2C \overset{O}{\overset{\diagup\diagdown}{-}} \underset{H}{C} - CH_2 - O - \langle \rangle - O - CH_2 - \underset{H}{C} \overset{O}{\overset{\diagup\diagdown}{-}} CH_2$$

1. *Source, Uses and Industrial Exposures*

Resorcinol diglycidyl ether is made through reaction of epichlorohydrin and resorcinol in the presence of caustic. It is available commercially. It is suggested for use as an epoxy resin, as a stabilizer of organic chemicals, as a curing agent for "Thiokol" rubber and for the solubilizing of protein adhesives.

2. *Physical and Chemical Properties*

Resorcinol diglycidyl ether is a colorless solid with a slight phenolic odor and the following properties:[20]
Molecular weight: 222
Specific gravity: 1.2183
Melting point: 32 to 33°C.
Boiling point: 150 to 160°C. (0.05 mm. Hg); 208 to 210°C. (12 mm. Hg)
Vapor density: 7.95
Refractive index: 1.5409 (20°C.)
Epoxy number: 110

3. *Determination in the Atmosphere*

Vapor pressure so low as to not require measurement. While dust conditions have not been encountered, sampling could be carried out by entrapment of airborne particles by standard filter paper or impinger techniques. Subsequent analysis may be carried out through the hydrochloric acid–dioxane method.[2]

4. *Physiological Response*

Summary. Slightly toxic by ingestion. Moderately to highly irritating on skin contact. Moderate radiomimetic activity.

Response in Animals. Acute toxicity. First signs of intoxication following oral administration are minor and include lachrymation and restlessness. This is followed by dyspnea, depression, and ataxia prior to death. The intragastric LD_{50} values for rats, mice, and rabbits, respectively, are 2.57, 0.98, and 1.24 g./kg. Intraperitoneally the LD_{50} values are 0.178 and 0.243 for rats and mice. Air saturated with the vapors is nontoxic to mice and rats after 8 hours of exposure. Rabbits receiving 6 ml./kg. died.[4]

Eye irritation. Score by Draize method, 45, severely irritant.

Skin irritation. Single applications give a score of 5 by the Draize[14a] method. The compound may be considered to be moderately irritating. Repeated applications severely irritating with a leathery appearance of the skin and a Draize[14a] score of 7. Highly purified samples are only moderately irritating.

Chronic toxicity. No evidence of toxicity in rats receiving 50 7-hour exposures to air saturated with the vapors. Repeated skin applications of 1 ml. total dose cause mortality in rabbits.[4] Monkeys receiving 100 to 200 mg./kg. intravenously once monthly show a progressively increasing depression of the white count.[7] Mice receiving repeated applications to the skin for 1 year develop cancers, and rats similarly receiving repeated subcutaneous injections develop sarcomas.[11]

Mode of action. Central nervous system depression. Locally irritating to tissues. Moderate radiomimetic effects.

Cause of death. Depression of the respiratory center and secondary shock.

Effects observed in man. Severe burns on local contact. Sensitization has occurred in a limited number of cases. Questionable decrease in the white count and suggestion of the appearance of atypical monocytic cells in the peripheral blood.

5. *Hygienic Standards of Permissible Exposure*

None established. Atmospheric contamination unlikely, but 1 p.p.m. suggested on basis of possible hemopoietic effects.

6. *Odor and Warning Properties*

Phenolic odor easily perceptible.

STYRENE OXIDE (1-Phenyl-1,2-epoxyethane, 1,2-Epoxyethylbenzene)

1. Source, Uses, and Industrial Exposure[66,67]

Styrene oxide is made commercially from styrene through the intermediate chlorohydrin. It is available in quantity and is used largely as an intermediate in the production of styrene glycol and its derivatives. Substantial amounts are also used as a liquid diluent in the epoxy-resin industry.

Styrene oxide has a flash point of about 80°C. (175°F.) as determined by the open cup procedure. On this basis, it presents a hazard of flammability similar to that encountered with such well-known chemical products as o-cresol, o-dichlorobenzene, naphthalene, phenol, and dimethyl aniline. A definite hazard exists whenever styrene oxide is heated to temperatures at and above the flash point.

Styrene oxide will polymerize exothermally or react vigorously with compounds having a labile hydrogen, including water, in the presence of catalysts such as acids, bases, and certain salts. Although experience has not shown styrene oxide to be as hazardous as ethylene oxide, precautions should be taken to prevent excessive pressure under storage or reaction conditions and to relieve such pressure should it occur.

2. Physical and Chemical Properties

Styrene oxide is a colorless liquid with the following properties:[66,67]
Molecular weight: 120.1
Specific gravity: 1.054 (25/25°C.)
Freezing point: −36.7°C.
Vapor density: 4.30
Boiling point: 194.2°C. (760 mm. Hg)
Vapor pressure: 0.3 mm. Hg (20°C.)
Refractive index: 1.533 (25°C.)
Solubility: 0.28 per cent in water at 25°C.; miscible in methanol, ether, carbon tetrachloride, benzene, and acetone
Flash point: 175°F. (TOC)
Flammability limits: precautions necessary at elevated temperature
1 p.p.m. ≎ 4.91 mg./cu. meter at 25°C., 760 mm. Hg and 1 mg./liter ≎ 203.6 p.p.m. at 25°C., 760 mm. Hg

3. Determination in the Atmosphere.

Methods for the determination of styrene oxide in air are undeveloped. Suggest the pyridinium chloride–chloroform method for epoxy groups.[2]

4. Physiological Response[19,67,68]

Summary. The greatest hazard to health from styrene oxide resides in its ability to cause skin irritation and skin sensitization. This has been demonstrated

[66] B. Phillips, *Peracetic Acid and Derivatives.* 2nd ed., Union Carbide Chemicals Co., New York, 1957.

[67] *Styrene Oxide.* The Dow Chemical Co., Midland, Mich., 1958.

[68] *Toxicology Styrene Oxide (EP-182).* Industrial Medicine & Toxicology Dept., Union Carbide Corp., New York, 1958.

in practice. Other hazards are minimal. The systemic toxicity, as revealed by the feeding of single doses to animals, is quite low; the LD_{50} for guinea pigs and rats is about 2.0 g./kg. Single 24-hour skin exposures in rabbits gave an LD_{50} value of 2.83 g./kg. Rats exposed to air saturated with vapors survived a 2-hour exposure, but 3 of 6 died following a 4-hour exposure.

Eye irritation. Undiluted styrene oxide may cause relatively severe irritation and pain to the eyes, but it is not apt to cause serious burns with permanent loss of vision. Solutions as dilute as 1 per cent may have some irritating action.

Skin irritation. Tests with laboratory animals and human subjects indicate that styrene oxide is capable of causing moderate skin irritation and skin sensitization. These effects may result from single or repeated contact with undiluted material and with solutions as dilute as 1 per cent. Experience indicates that persons who have become hypersensitive may react rather severely to contact with the vapor as well as with the liquid material. There is some evidence that styrene oxide is absorbed slowly through the skin. This absorption could be significant only from exposures that produced extensive and serious local injury to the skin.

5. Hygienic Standards of Permissible Exposure

None established. No data are available upon which a hygienic standard can be based. It is strongly recommended that every precaution be taken when handling or using styrene oxide to prevent contact with the person.

VINYLCYCLOHEXENE DIOXIDE (1-Epoxyethyl-3,4-epoxycyclohexane, EP-206)

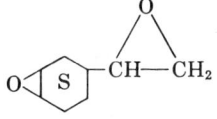

S = Cyclohexane

1. Sources, Uses, and Industrial Exposure

Vinylcyclohexene dioxide is available in developmental quantities as a chemical intermediate and as a monomer for preparation of polyglycols containing unreacted epoxy groups.

2. Physical and Chemical Properties

Vinylcyclohexene dioxide is a clear, water-soluble liquid with a faintly olefinic odor and the following priperties:[69]

Molecular weight: 140
Specific gravity: 1.0986 (20/20°C.)
Freezing point: sets to glass at −55°C.
Boiling point: 227°C. (760 mm. Hg)

[69] *Toxicology Studies: Vinylcyclohexene Dioxide (EP-206).* Industrial Medicine & Toxicology Dept., Union Carbide Corp., New York, 1958.

Vapor density: 4.07
Vapor pressure: <0.1 mm. Hg (20°C.)
Refractive index: 1.4787 (20°C.)
Solubility: 18.3 per cent in water at 20°C.
1 p.p.m. $\approx$ 5.73 mg./cu. meter at 25°C., 760 mm. Hg and 1 mg./liter $\approx$ 174 p.p.m. at 25°C., 760 mm. Hg

3. *Determination in the Atmosphere*

Hydrochloric acid–dioxane method for epoxy groups.[2]

4. *Physiological Response*

Summary. Slightly toxic by ingestion or percutaneous absorption. Moderate radiomimetic activity is present in the molecule. Causes central nervous system depression.

Response in Animals. Acute toxicity. Limited toxicity data indicate the compound has an oral LD_{50} of 2.83 g./kg. for rats and a percutaneous LD_{50} of 0.62 ml. for rabbits. Rats exposed to essentially saturated vapors for 8 hours survived.[69] The LC_{50} for a 4-hour exposure of rats is 800 p.p.m.[16] Vasodilatation and unsteady gait occur during exposure. Death occurs during or soon after exposure. Pathology consists of congestions of the lungs and liver and occasionally testicular atrophy.

Eye irritation. Highly irritating.

Skin irritation. The undiluted chemical causes redness and swelling comparable to a first-degree burn on the skin of rabbits. Twenty repeated skin applications to rabbits causes severe to extreme irritation, with a Draize score of 7.

Chronic toxicity. Repeated skin applications over some months cause skin tumors with some cancers in mice. Repeated subcutaneous injections cause sarcomas in rats.[6,7]

Mode of action. Central nervous system depressant and irritant of surface tissues.

Cause of death. Depression of the respiratory center and acute pulmonary tract irritation.

Effects observed in man. None reported.

5. *Hygienic Standards of Permissible Exposure*

Not established; suggest 1 p.p.m. based on potential effects on hemopoietic system.

6. *Odor and Warning Properties*

Not remarkable.

VINYLCYCLOHEXENE MONOXIDE (1,2-Epoxy-4-vinylcyclohexane)

$$O \triangleleft \underset{\text{S}}{\bigcirc} \negmedspace - CH{=}CH_2$$

S = Cyclohexane

1. Source, Uses, and Industrial Exposure

Vinylcyclohexene monoxide is available in research quantities. It is proposed for use as a chemical intermediate; can be copolymerized with other epoxides to yield polyglycols having unsaturation available for further reaction.

2. Physical and Chemical Properties

Vinylcyclohexene monoxide is a colorless, mobile liquid combining readily with water, alcohols, phenols, and other agents containing active hydrogens and has the following properties:[69]

Molecular weight: 124.18
Specific gravity: 0.9598 (20/20°C.)
Freezing point: sets to a glass below −100°C.
Boiling point: 169°C. (760 mm. Hg)
Vapor density: 3.75
Vapor pressure: 2.0 mm. Hg (20°C.)
Refractive index: 1.4700 (20°C.)
Solubility: 0.5 per cent in water
Per cent in "saturated" air: 0.263
1 p.p.m. ≎ 5.07 mg./cu. meter at 25°C., 760 mm. Hg and 1 mg./liter ≎ 197 p.p.m. at 25°C., 760 mm. Hg

3. Determination in the Atmosphere

No specific method available. Suggest hydrochloric acid–dioxane method for epoxides.[2]

4. Physiological Response

Response in Animals. *Acute toxicity.* LD_{50} in rats given the material by mouth is 2.00 ml./kg. Percutaneous LD_{50} for rabbits is 2.83 g./kg. Rats exposed to saturated vapors for 2 hours survived, but 3 of 6 died after 4 hours' exposure.[70]

Eye irritation. Not marked.
Skin irritation. Moderate.
Chronic toxicity. None reported.
Mode of action. Central depressant and pulmonary irritant.

[70] *Toxicology Studies: Vinylcyclohexene Monoxide (EP-101).* Industrial Medicine & Toxicology Dept., Union Carbide Corp., New York, 1958.
[71] R. J. Sexton and E. V. Henson, *Arch. Ind. Hyg. Occupational Med.,* **2,** 549 (1950).

Cause of death. Depression of the respiratory center.
Effects Observed in Man. None.

5. Hygienic Standards of Permissible Exposure

None established; suggest 50 p.p.m. based on comfort and analogy to other compounds.

6. Odor and Warning Properties

Not known.

CHAPTER XXXVIII

Ethers

C. L. HAKE AND V. K. ROWE

METHYL ETHER (Dimethyl Ether, Methoxymethane)

C_2H_6O or CH_3OCH_3

1. Source, Uses, and Industrial Exposures[1]

Methyl ether is obtained industrially by the catalytic dehydration of methanol. It is also obtained as a by-product in the production of acetic acid from carbon monoxide and methanol, and in the production of methanol by the carbon monoxide–hydrogenation process. The principal use of methyl ether has been that of a liquid refrigerant. It has also been used in a limited amount to freeze meat and fish by direct immersion. Other proposed uses have been as a selective solvent, a propellent for rockets, an anesthetic, an aerosol dispersant, and as a starter for engines. However, none of these uses are extensive.

Industrial exposure is most likely to occur from the inhalation of the gas.

2. Physical and Chemical Properties[1-3]

Methyl ether at room temperature is a colorless gas with an ethereal odor. It is noncorrosive. It has the following properties:

Molecular weight: 46.07
Boiling point: $-23.65°C.$ (760 mm. Hg)
Melting point: $-138.5°C.$
Density: 1.617 (air $= 1$)
Vapor pressure: 1026 mm. Hg $(-20°C.)$; 2128 mm. Hg $(0°C.)$; 3982 mm. Hg $(20°C.)$; 35,400 mm. Hg $(120°C.)$
Refractive index: 1.3441 $(n_D^{-42.5})$
Solubility: in 100 ml. of water, 3700 ml. of gas; more soluble in alcohol; generally soluble in organic solvents with the exception of polyalcohols
Flash point: $-41°F.$

---

[1] Y. Mayor, *Ind. chim. Paris*, **39**, 259 (1952).
[2] *Handbook of Chemistry and Physics.* 40th ed., Chemical Rubber Publishing, Cleveland, 1958.
[3] *The Merck Index.* 7th ed., Rahway, N. J., 1960.

Flammability limits: 3.4 per cent by volume in air, lower limit; 26.7 per cent by volume in air, upper limit

1 p.p.m. ≎ 1.881 mg./cu. meter at 25°C., 760 mm. Hg and 1 mg./liter ≎ 532 p.p.m. at 25°C., 760 mm. Hg

3. Determination in the Atmosphere

Methyl ether can be absorbed from the atmosphere with activated charcoal, titania gel, or silica gel. Use of silica gel has the advantage of ease of elution with ethanol and subsequent analysis by gas chromatography. Air samples may also be analyzed directly by infrared analysis with a gas cell of long path length.

4. Physiological Response

There are no reports in the literature concerning any experimental toxicological research devoted to methyl ether. In 1924, Brown[4] studied the anesthetic effect of this gas on the cat. A mixture of 85 per cent methyl ether–15 per cent air caused profound anesthesia with gradual cessation of respirations. Approximately 20 minutes were necessary for complete recovery after 50 minutes of anesthesia. Brown also reported that methyl ether concentrations of 50 per cent were inhaled by humans in the laboratory, with the observation that the gas is most unpleasant to inhale, being distinctly suffocating, even when taken with a high percentage of oxygen. From this report it appears that the acute toxicity of methyl ether is very low. Like the other alkyl ethers, its principal physiological effect is that of anesthesia. The effect of long-term exposure to this compound is unknown.

5. Hygienic Standards of Permissible Exposure

There have been no hygienic standards of permissible exposure proposed for methyl ether.

ETHYL ETHER (Diethyl Ether, Ether, Sulfuric Ether, Diethyl Oxide, Ethyl Oxide, Ethoxyethane)

$$C_4H_{10}O \quad \text{or} \quad CH_3CH_2OCH_2CH_3$$

1. Source, Uses, and Industrial Exposures[5,6]

Ethyl ether is produced commercially by the dehydration of ethanol or by the hydration of ethylene, both processes being carried out in the presence of sulfuric acid. Because of its solvent powers, it has a host of uses as a commercial solvent. Among these are uses as a solvent for waxes, fats, oils, perfumes, alkaloids, gums, and resins. It is used in a mixture with ethanol as a solvent for nitrocellulose and in the manufacture of guncotton, collodion solutions, and pyroxylin plastics. It is also used as a solvent or extractant in the manufacture of dyes, cellulose acetate

---

[4] W. E. Brown, *J. Pharmacol. Exptl. Therap.*, **23**, 497 (1924).

[5] *The Merck Index.* 7th ed., Rahway, N. J., 1960.

[6] Union Carbide Chemicals Co., *Technical Information, Ethyl Ether*, No. **F-8139**, 1953

rayon, and plastics such as photographic film. It is sometimes used as a primer for gasoline engines and for cleaning fabrics. Ethyl ether was one of the first successful inhalation anesthetics and is still used in large quantities for such purposes (for additional medicinal uses see also under Repeated Doses).

Industrial exposure to ethyl ether is most likely to occur to low concentrations of its vapor. Exposure to deleterious vapor concentrations is unlikely because of its sensory warning properties. Excessive exposure of the skin to the liquid may occur if the material is improperly handled. Splashes of the liquid into the eye also may occur.

The peroxides of ethyl ether become explosive upon heating and subsequent vaporization. The addition of a small amount of water or reducing agent to the ether will lessen the tendency to form peroxides. Peroxides may be removed from ether by shaking with 5 per cent aqueous ferrous sulfate solution.

2. *Physical and Chemical Properties*[5-7]

Ethyl ether is a colorless, highly volatile, hygroscopic, mobile liquid with the characteristic sweetish, pungent odor of the ethers. It is highly flammable. Upon standing in air, and especially in the presence of sunlight, it forms explosive peroxides. It is noncorrosive to most metals. It has the following properties:

Molecular weight: 74.12

Boiling point: 34.6°C. (760 mm. Hg); 17.9°C. (400 mm. Hg); −11.5°C. (100 mm. Hg); −74.3°C. (1.0 mm. Hg)

Melting point: −116.3°C. (stable crystals); −123.3°C. (metastable crystals)

Specific gravity: 0.7146 (20/20°C.); 0.7134 (20/4°C.)

Vapor density: 2.55 (air = 1)

Vapor pressure: 438.9 mm. Hg (20°C.); 921.0 mm. Hg (40°C.)

Per cent in "saturated" air: 68 (25°C., 760 mm. Hg)

Density of "saturated" air: 2.1 (25°C., 760 mm. Hg) (air = 1)

Refractive index: 1.3526 (n_D^{20})

Solubility: in water, per cent by weight, 8.43 at 15°C., 6.9 at 20°C., and 6.05 at 25°C.; miscible with lower aliphatic alcohols, benzene, chloroform, petroleum ether, other fat solvents, and many oils

Flash point: −40°F. (O.C.)

Explosive limits: 1.9 per cent by volume in air, lower limit; 36.5 per cent by volume in air, upper limit

1 p.p.m. ⪬ 3.03 mg./cu. meter at 25°C., 760 mm. Hg and 1 mg./liter ⪬ 330 p.p.m. at 25°C., 760 mm. Hg

3. *Determination in the Atmosphere*

The concentration of the vapors of ethyl ether in air can be determined chemically by the absorption of the material in a sulfuric acid solution of potas-

[7] *Handbook of Chemistry and Physics.* 40th ed., Chemical Rubber Publishing, Cleveland, 1958.

sium dichromate followed by iodometric titration.[8] It can also be absorbed on charcoal or silica gel. Physical methods for its determination utilize heat of combustion, the interferometer, and infrared absorption. It can be detected at fairly low concentrations in the atmosphere by infrared analysis with a gas cell of long path length. A sensitive explosimeter would also be of value for concentrations known to be below explosive limits.

4. Physiological Response

Summary. Without question, the primary physiological response to ethyl ether is that of general anesthesia. Absorption through the lungs occurs quickly upon the onset of exposure and elimination by this route postexposure is also rapid. Absorption through the skin is of no consequence in humans. It may cause a slight irritating effect on the skin, or cause it to become dry and cracked after repeated applications. Exposure of the eyes or mucous membranes to the liquid is irritating and should be avoided.

Single-Dose Inhalation. Concentrations of ethyl ether anesthetic to humans range from 3.6 to 6.5 volume per cent in air. Respiratory arrest may occur from 7 to 10 volume per cent, while air concentrations over 10 per cent are usually fatal[9] to humans. Kärber[10] reported that the lethal concentration for mice was 133.4 mg./liter, or 4.4 volume per cent, during a continuous exposure for 97 minutes. Similar results for the mouse were obtained by Molitor,[11] who calculated the LC_{50} for a continuous 3-hour exposure to be 127.4 mg./liter or 4.2 volume per cent. The lethal concentrations for the rat, dog, and monkey are reported to be 6.4, 10.6, and 7.16 to 19.25 volume per cent,[12] respectively. It would appear that the smaller animal species succumb to a lower concentration of ethyl ether in air than do the larger animal species. However, Robbins[13] found a lethal concentration for dogs of 6.7 to 8 volume per cent. He states that differences in induction mixtures and in the duration of anesthesia lead to the differences in results for lethal concentrations. This is further illustrated by more recent work[14] in which the concentration necessary to produce respiratory arrest in mice was found to be 18 volume per cent. Rapid induction and short duration of anesthesia were used in this experiment.

Cases of human death in industry due to acute inhalation of ethyl ether are rare. Browning[15] cites one such case where ethyl ether was used in perfumery

[8] L. Silverman, *Industrial Air Sampling and Analysis.* Industrial Hygiene Foundation, Pittsburgh, 1947.

[9] F. Flury and O. Klimmer in K. B. Lehmann and F. Flury, *Toxicology and Hygiene of Industrial Solvents.* Trans. by E. King and H. F. Smyth, Jr., Williams & Wilkins, Baltimore, 1943.

[10] Kärber, *Arch. exptl. Pathol. Pharmakol.,* **142,** 1 (1929).

[11] H. Molitor, *J. Pharmacol. Exptl. Therap.,* **57,** 274 (1936).

[12] W. S. Spector, ed., *Handbook of Toxicology,* Vol. 1. Saunders, Philadelphia, 1956.

[13] B. H. Robbins, *J. Pharmacol. Exptl. Therap.,* **53,** 251 (1935).

[14] E. T. Mörch, J. B. Aycrigg, and M. S. Berger, *J. Pharmacol. Exptl. Therap.,* **117,** 184 (1956).

[15] E. Browning, *Toxicity of Industrial Organic Solvents.* Chemical Publishing, New York, 1953.

manufacture. The subject developed acute mania and died in uremic convulsions. Doubtless, there are a few other such cases. Cases of narcosis are more frequent, largely due to the fact that permanent aftereffects are rare. Early symptoms of acute overexposure range from excitement to drowsiness, vomiting, paleness in the face, lowering of the pulse and body temperature, and irregular respiration. Temporary aftereffects of acute exposure are vomiting, salivation, irritation of the respiratory passages, headaches, and depression or excitation.[9] Kidney injury has been listed as a result of acute intoxication but this has been open to question.[15]

A large amount of information on the anesthetic action of ethyl ether is available in the literature concerning anesthesia. Most of the textbooks on pharmacology, medicine, and anesthesiology, and some related annual reviews contain chapters on this subject. They should be consulted for a more detailed account of the use and effects of ethyl ether in anesthesia.

Repeated-Dose Inhalation. Exposures of animals to repeated doses of ethyl ether have not been reported. Repeated exposure of humans in industry was often the result of an intentional exposure called an "ether jag," or of unintentional exposure to nonflammable concentrations. Symptoms consist of loss of appetite, exhaustion, headaches, sleepiness, dizziness, excitation, and psychic disturbances.[9] Clinically, albumin may appear in the urine, and polycythemia in the blood.[15] Nephritis may develop in rare cases.

Repeated Doses, Oral and by Injection. Because of its irritating effect on mucous membranes, humans generally refrain from the oral consumption of ethyl ether. However, ether drinking has taken place,[15] with subsequent repetition leading to the "ether habit" and general debility. It is also used occasionally in cough medicine as an expectorant, and in other medicinal preparations. It has been employed for intravenous anesthesia as a 5 per cent solution, and has been introduced into the common bile duct in concentrated form to dissolve biliary calculi. Sewall[16] reports a fatality from the intravenous injection of ethyl ether in the treatment of vascular diseases.

Effect on Skin and Eyes. Ethyl ether has no deleterious effect on human skin if contact is of short duration. It does have a cooling effect due to rapid evaporation. Repeated exposure of the skin causes cracking and drying due to the extraction of oils. Absorption through the skin is not great enough to cause a deleterious effect. This chemical does cause irritation to the mucous membranes and to the eye in either liquid form or in high concentrations in the air. However, it does not cause permanent damage. Nelson *et al.*[17] reported that human subjects found ethyl ether irritating to the nose, but not to the eyes or throat, at a vapor concentration of 200 p.p.m.

Absorption, Excretion, and Metabolism. Ethyl ether is rapidly absorbed from the inhaled air into the blood stream, from whence it passes rapidly into the brain. Dybring and Skovlund[18] showed that it is also taken up rapidly by fatty

[16] S. Sewall, *Bull. Hosp. Joint Diseases*, **8**, 33 (1947), Chem. Abstr., **46**, 1151 (1952).

[17] K. W. Nelson, J. F. Ege, Jr., M. Ross, L. E. Woodman, and L. Silverman, *J. Ind. Hyg. Toxicol.*, **25**, 282 (1943).

[18] O. Dybring and K. Skovlund, *Acta Pharmacol. Toxicol.*, **13**, 252 (1957).

tissue in rats. The concentration in muscle tissue remains much lower than the concentration in the brain and fatty tissue. Haggard[19] showed in dogs and rabbits that 87 per cent of an absorbed dose is expired in the breath unchanged and 1 to 2 per cent is excreted in the urine. The concentration in the urine does not exceed that in the blood flowing through the kidneys. Upon discontinuance of exposure, the ethyl ether deposited in the fatty tissues remains at a fairly high level until the concentration in the blood has become relatively low. In the rat,[18] the concentration of ethyl ether in the fat was 0.12 mg./g., while in the blood it was 0.03 mg./g., 24 hours after an exposure to 300 mg./liter (100,000 p.p.m.) for 1 hour. Chenoweth et al.,[20] using infrared and mass spectrographic methods of analysis, also found high levels of ether in the fatty depots of the dog after $2^1/_2$ hours of anesthesia. Ether was still detectable in the fat 24 hours after the discontinuance of anesthesia, while the concentration in the blood had reached a similar low level approximately 17 hours earlier. Measurement of these concentrations in 11 organs or tissues and 3 body fluids revealed that the adrenals contain an unexplained, high concentration of the compound during anesthesia.

Although no tracer studies with labeled ethyl ether have been carried out, it is generally felt that the material is stable in vivo and is not metabolized by either man[21] or animal.[22]

Mode of Action. The exact mode of action of ethyl ether in its anesthetic effect has been the subject of much thought and experimentation over the years. However, the basic biochemical alteration existing during anesthesia has not been elucidated. Krantz and Carr,[21] in their book on the pharmacological principles of medical practice, theorize thus for volatile anesthetics in general: "The presence of the narcotic in the central nervous system attacks the 'main line' oxidation in the cells of the central nervous system. The target molecules appear to be either the cytochrome reductase or some flavoprotein linking it with the phosphonucleotide dehydrogenase. The enzyme inactivation is a reversible process."

Human Experience. Almost all of the reported human experience concerning the use of ethyl ether involves its use as an anesthetic agent. Industrial exposure has caused very few deaths or illnesses, and thus very few reports appear in the literature. There is a large margin of safety between the concentration causing nasal irritation[17] and that causing anesthesia, permanent damage, and death.

5. Hygienic Standards of Permissible Exposure

In 1940, Bowditch et al.,[23] published maximum concentrations of various vapors suggested by the State of Massachusetts as a guide to manufacturers and

[19] H. W. Haggard, *J. Biol. Chem.*, **59**, 745 (1924).

[20] M. B. Chenoweth, D. N. Robertson, D. S. Erley, and R. Gohlke, *Anesthesiology*, **23**, 101 (1962).

[21] J. C. Krantz, Jr. and C. J. Carr, *The Pharmacologic Principles of Medical Practice.* 3rd ed., Williams & Wilkins, Baltimore, 1954, pp. 421–484.

[22] R. T. Williams, *Detoxication Mechanisms.* 2nd ed., Wiley, New York, 1959. p. 323.

[23] M. Bowditch, C. K. Drinker, P. Drinker, H. H. Haggard, and A. Hamilton, *J. Ind. Hyg. Toxicol.*, **22**, 251 (1940).

others interested in maintaining satisfactory working conditions. They suggested that this value for ethyl ether be 400 p.p.m. Nelson *et al.*,[17] in 1943, using human subjects, found that a majority of a group of at least 10 persons found 300 p.p.m. objectionable, and estimated that 100 p.p.m. would be the highest concentration satisfactory for an 8-hour exposure. Amor[24] suggested that unsatisfactory conditions would exist in a plant only if the concentration was greater than 500 p.p.m. (1500 mg./cu. meter at 20°C.). He also listed 2000 p.p.m. (6200 mg./cu. meter at 20°C.) as the concentration that might lead to symptoms of illness if exposure were continuous for more than a short time. The recommended threshold limit value as adopted by the American Conference of Governmental Industrial Hygienists[25] in 1960 was 400 p.p.m., or approximately 1200 mg./cu. meter. This value refers to a time-weighted-average concentration for a normal workday.

ISOPROPYL ETHER (Diisopropyl Ether, 2-Isopropoxypropane)

$$C_6H_{14}O \quad \text{or} \quad (CH_3)_2CHOCH(CH_3)_2$$

1. *Source, Uses, and Industrial Exposures*

Isopropyl ether is prepared commercially by the action of sulfuric acid on isopropyl alcohol, and is also obtained as a by-product in the production of isopropyl alcohol from the propylene fraction of cracked gasoline.

Isopropyl ether has many and varied industrial uses.[26,27] It is an excellent solvent for animal, vegetable, and mineral oils, fats, waxes, and ethylcellulose. It is used as a solvent in the manufacture of pharmaceuticals, celluloid articles, smokeless gunpowder, paint and varnish removers, rubber cements, and spotting compositions. It is used as a blending agent in aviation gasolines, to recover acetic acid from dilute solutions, to extract nicotine from tobacco, and in the analysis of fish livers for vitamin content. It is also used as an intermediate in alkylation reactions. It has been tested as a surgical anesthetic but its depressant effects and its low margin of safety eliminated it from this field.[28]

Isopropyl ether, like the other ethers of this series, forms explosive peroxides upon standing in contact with air.[26,29] The development of these peroxides can be retarded by the addition upon storage of oxidation inhibitors such as diphenylamine, α-naphthol, β-naphthol, or hydroquinone. Approximately 0.05 per cent of the inhibitor is added. Water at a concentration of 1.0 per cent by weight is also effective as an inhibitor.

[24] A. J. Amor, *Paint Manuf.*, **20**, 53 (1950).

[25] Conference of Governmental Industrial Hygienists, *A.M.A. Arch. Environmental Health*, **1**, 140 (1960).

[26] Union Carbide Chemicals Co., *Technical Information, Isopropyl Ether*, No. **F-40003**, 1955.

[27] L. Scheflan and M. B. Jacobs, *The Handbook of Solvents*. Van Nostrand, New York, 1953.

[28] L. G. Amiot, *Presse méd.*, **40**, 300 (1932).

[29] E. F. Degering, *J. Chem. Educ.*, **13**, 494 (1936).

Industrial exposure to isopropyl ether is most likely to occur through a direct contact with the skin or eyes due to an accidental spill, or through an inhalation exposure to low concentrations of the material in the atmosphere. Acute exposures to physiologically significant concentrations are rare because of the flammability associated with such concentrations.

2. Physical and Chemical Properties[29,30-32]

Isopropyl ether is a colorless, flammable liquid with an odor like camphor and ethyl ether. Upon standing in air it forms explosive peroxides. It has the following properties:

Molecular weight: 102.17

Boiling point: 68.3°C. (760 mm. Hg); 3°C. (50 mm. Hg); —23°C. (10 mm. Hg)

Melting point: —60°C.

Specific gravity: 0.7258 (20/4°C.)

Vapor density: 3.5 (air = 1)

Vapor pressure: 119.4 mm. Hg (20°C.)

Per cent in "saturated" air: 21 (25°C., 760 mm. Hg)

Density of "saturated" air: 1.5 (25°C., 760 mm. Hg) (air = 1)

Refractive index: 1.3682 (n_D^{20})

Solubility: in water, 0.2 to 0.9 per cent by weight at 20°C.; miscible with alcohol, ether, most organic solvents, and also acetic acid

Flash point: 15°F. (T.O.C.); —18°F. (C.C.)

Explosive limits: 1.4 to 2 per cent by volume in air, lower limit; 7.9 per cent by volume in air, upper limit

1 p.p.m. ≎ 4.18 mg./cu. meter at 25°C., 760 mm. Hg and 1 mg./liter ≎ 240 p.p.m. at 25°C., 760 mm. Hg

3. Determination in the Atmosphere

The concentration of isopropyl ether in the atmosphere may be determined by adsorption on silica gel or charcoal and the use of physical methods such as interferometer, heat of combustion, infrared, and gas chromatograph techniques. It can also be determined by a modification of the bichromate oxidation method used for ethyl ether.[33] Machle et al.[34] describe a special combustion technique for its analysis.

[30] The Merck Index. 7th ed., Rahway, N. J., 1960.

[31] Shell Chemical Corp., Safety Data Sheet, Isopropyl Ether, No. SC:57-104, 1959.

[32] G. W. Jones in F. A. Patty, ed., Industrial Hygiene and Toxicology, Vol. I. 2nd revised ed. Interscience Publishers, New York–London, 1958.

[33] Methods of Association of Official Agricultural Chemists. 6th ed., 1945, through M. B. Jacobs, The Analytical Chemistry of Industrial Poisons, Hazards, and Solvents. 2nd ed., Interscience Publishers, New York, 1949.

[34] W. Machle, E. W. Scott, and J. Treon, J. Ind. Hyg. Toxicol., 21, 72 (1939).

4. *Physiological Response*

Summary. As with ethyl ether, the primary physiological response to isopropyl ether is that of general anesthesia. Even though it is somewhat more toxic than ethyl ether, the vapor concentration necessary for this anesthetic effect is much higher than the concentration that causes irritation and unpleasant odors. This compound is only slightly irritating to the skin and is capable of producing only minor injury to the eyes. It is not absorbed through the skin in harmful amounts.

Single-Dose Inhalation. Machle et al.[34] exposed animals (monkey, rabbit, and guinea pig) to a vapor concentration of 6.0 per cent of isopropyl ether in air. All died due to respiratory failure. At a vapor concentration of 3.0 per cent all animals survived a 1-hour exposure but all showed signs of anesthesia, with the monkey being most susceptible. At a vapor concentration of 0.3 per cent for 2 hours there was no visible indication of anesthetic action. Smyth[35] found the lethal vapor concentration for rats to be 1.6 per cent for a 4-hour exposure.

Repeated-Dose Inhalation. Machle et al.[34] repeatedly exposed the animals referred to above for various periods of time. The animals exposed to a vapor concentration of 3.0 per cent were given 10 exposures. Recoveries from the anesthesia were prompt and seemed not to be additive. However, weight loss and blood changes were apparent during and for several weeks after exposure. Animals exposed to a vapor concentration of 1.0 per cent for 1 hour daily, although exhibiting signs of intoxication and depression, revealed no significant weight or blood changes during or after 20 exposures. At 0.3 per cent for 2 hours and 0.1 per cent for 3 hours daily, there was no deleterious effect noted during or after 20 exposures.

Single-Dose Oral. The minimum lethal dose for rabbits[34] was found to be between 7 and 9 ml./kg. (5 to 6.5 g./kg.). A rapid, intense intoxication was produced. Death was due to respiratory failure caused by depressant action.

Effect on Skin and Eyes. Machle et al.[34] tested the effect of isopropyl ether on the skin of rabbits. Single exposures to the liquid for one hour produced no deleterious effect. However, repeated exposure for 10 days caused dermatitis. Isopropyl ether is irritating and capable of producing minor injury to the rabbit eye.[26] Skin absorption in deleterious amounts seems to be no problem.[26] The compound is irritating to the mucous membranes in liquid form or in high vapor concentration.[36]

Absorption, Excretion, and Metabolism. From the work cited by Machle et al.,[34] it is apparent that isopropyl ether is rapidly absorbed by the blood from the lungs or gastrointestinal tract. Although no studies have been carried out, it is probable that the major portion of a dose is eliminated through the lungs after cessation of exposure, and that the compound is not metabolized in vivo.

Human Experience. Silverman et al.[36] reported that humans exposed to a vapor concentration of 500 p.p.m. isopropyl ether for 15 minutes while watching a

[35] W. S. Spector, ed., *Handbook of Toxicology*, Vol. I. Saunders, Philadelphia, 1956.
[36] L. Silverman, H. F. Schulte, and M. W. First, *J. Ind. Hyg. Toxicol.*, **28**, 262 (1946).

movie noted no irritating properties. However, 35 per cent of the subjects objected to the unpleasant odor of this solvent at a vapor concentration of 300 p.p.m.[36] At 800 p.p.m. for 5 minutes most subjects reported irritation of the eyes and nose, and the most sensitive reported respiratory discomfort.[31]

There have been few cases reported of death or serious illness attributed to isopropyl ether. Its chief hazards are flammability and explosions due to peroxide formation, rather than toxicity.

5. Hygienic Standards of Permissible Exposure

Cook,[37] citing the animal work by Machle et al.,[34] suggested a maximum atmospheric concentration for industry of 500 p.p.m. The American Conference of Governmental Industrial Hygienists still recommends this value (500 p.p.m., approximately 2100 mg./cu. meter) as the threshold limit for 1960.[38] It recommends this value as the time-weighted-average concentration for a normal workday.

n-BUTYL ETHER (n-Dibutyl Ether, 1-Butoxybutane, Butyl Ether)

$$C_8H_{18}O \quad or \quad CH_3(CH_2)_3O(CH_2)_3CH_3$$

1. Source, Uses, and Industrial Exposures

n-Butyl ether is prepared from butanol with the use of special catalysts. It is also available as a by-product in the manufacture of butyl esters.

n-Butyl ether has various industrial uses in organic synthesis, as an extracting agent, and as a solvent for esters, gums, hydrocarbons, alkaloids, oils, organic acids, and resins. Industrial exposure could occur through inhalation of its vapors or through contact with the liquid on the skin or eyes due to an accidental spill or careless handling.

n-Butyl ether, like the other ethers, also tends to form explosive peroxides, especially when anhydrous.

2. Physical and Chemical Properties[39-42]

n-Butyl ether is a colorless, flammable liquid with a mild ethereal odor and the following properties:

Molecular weight: 130.22
Boiling point: 142.4°C. (760 mm. Hg)

[37] W. A. Cook, Ind. Med., **14,** 936 (1945).

[38] Conference of Governmental Industrial Hygienists, A.M.A. Arch. Environmental Health, **1,** 140 (1960).

[39] The Merck Index. 7th ed., Rahway, N. J., 1960, p. 181.

[40] Union Carbide Chemicals Co., Physical Properties of Organic Chemicals. New York, 1959.

[41] A. I. Vogel, J. Chem. Soc. London, **1948,** 616.

[42] L. Scheflan and M. B. Jacobs, The Handbook of Solvents. Van Nostrand, New York, 1953, p. 168.

Melting point: $-95.2°C$.

Specific gravity: 0.7694 ($20/20°C$.)

Vapor density: 4.48 (air $= 1$)

Vapor pressure: 4.8 mm. Hg ($20°C$.)

Per cent in "saturated" air: 0.9 ($25°C$., 760 mm. Hg)

Density of "saturated" air: 1.1 ($25°C$., 760 mm. Hg) (air $= 1$)

Refractive index: 1.4010 (n_D^{20})

Solubility: in water, 0.03 to 0.05 per cent by weight at $20°C$.; miscible with alcohol, benzene, acetone, and most organic solvents

Flash point: $100°F$. (C.C.)

1 p.p.m. $\backsim$ 5.33 mg./cu. meter at $25°C$., 760 mm. Hg and 1 mg./liter $\backsim$ 188 p.p.m. at $25°C$., 760 mm. Hg

3. Determination in the Atmosphere

n-Butyl ether can be determined qualitatively and quantitatively by silica gel absorption, elution with ethanol, and gas chromatography of the eluate. It can also be determined by a sensitive explosimeter or by infrared analysis in a gas cell of long path length.

4. Physiological Response

Summary. n-Butyl ether, although not highly toxic orally, seems to have a greater toxicity by inhalation than the lower ethers of the series. It is also more irritating to the skin, nose and eyes. However, little work has been done from which definite conclusions can be drawn.

Animal Toxicity. The only report in the literature concerning the toxicity of n-butyl ether is that of Smyth *et al.*[43] in their report on range-finding studies. The acute oral LD_{50} for the rat was found to be 7.4 g./kg. When the material was applied under an impervious cuff to the shaved skin of 4 rabbits, the LD_{50} was calculated to be 10.08 ml./kg., although the range was quite large. In inhalation studies, it was found that rats survived for 30 minutes when exposed to concentrated vapors. Two of 6 rats died after a 4-hour exposure to 4000 p.p.m. of the vapor. Primary skin irritation studies on rabbits indicated rather severe irritation but no necrosis. Eye injury was not apparent after applying the undiluted material to rabbit eyes.

Human Experience. Silverman *et al.*[44] exposed at least 12 human subjects to low concentrations of n-butyl ether for 15 minutes. A majority of the subjects found 200 p.p.m. irritating to the eyes and nose, although the odor was not objectionable, even at 300 p.p.m. They estimated that 100 p.p.m. should be the highest concentration to which a workman should be exposed for an 8-hour day.

No reports have been found in the literature concerning serious injury or death to humans due to n-butyl ether exposure.

[43] H. F. Smyth, Jr., C. P. Carpenter, C. S. Weil, and U. C. Pozzani, *Arch. Ind. Hyg Occupational Med.,* **10**, 61 (1954).

[44] L. Silverman, H. F. Schulte, and M. W. First, *J. Ind. Hyg. Toxicol.,* **28**, 262 (1946).

5. Hygienic Standards of Permissible Exposure

There has been no threshold limit recommended for n-butyl ether.

OTHER SATURATED ALKYL ETHERS

There is a small group of saturated alkyl ethers other than those treated individually, which are used industrially or have proposed industrial uses, and about which some physiological information is available. Information regarding their physical and chemical properties is given in Table 1; information regarding their physiological properties is given in Table 2. Since these physiological properties vary greatly, anyone interested in a more complete study should consult the literature cited.

DIVINYL ETHER (Vinyl Ether, Divinyl Oxide, Ethenyloxyethene)

$$C_4H_6O \quad \text{or} \quad CH_2CHOCHCH_2$$

1. Source, Uses, and Industrial Exposures

Divinyl ether is prepared by the action of caustic on dichloroethyl ether, which is formed by the Williamson synthesis from chloroethyl alcohol and ethylene dichloride.[45]

Divinyl ether is used principally as an anesthetic agent, where it is especially suitable for brief surgical procedures or induction. The anesthetic, vinyl ether, USP, contains about 4 per cent alcohol and not more than 0.025 per cent of a preservative, phenyl α-naphthylamine.

Industrial exposure in the preparation of the anesthetic could result from low concentrations of vapor. Due to the high flammability, it is unlikely that exposures of a deleterious nature would exist if the fire or explosion hazard is controlled.

2. Physical and Chemical Properties[46-48]

Divinyl ether is a colorless, volatile, flammable liquid with a characteristic, disagreeable odor. On exposure to light, air or acid fumes, it decomposes to formaldehyde, formic acid and peroxide. It polymerizes readily to a solid glasslike mass when no inhibitor is present. It has the following properties:

Molecular weight: 70.09
Boiling point: 28.3°C. (760 mm. Hg)
Specific gravity: 0.774 (20/20°C.)
Vapor density: 2.4 (air = 1)

[45] H. R. Fleck, *Chem. Prod.*, **19**, 144 (1956).

[46] *The Merck Index.* 7th ed., Rahway, N. J., 1960.

[47] L. Scheflan and M. B. Jacobs, *The Handbook of Solvents.* Van Nostrand, New York, 1953.

[48] T. E. Jordan, *Vapor Pressure of Organic Compounds.* Interscience Publishers, New York–London, 1954.

TABLE 1

Physical and Chemical Properties of Other Saturated Alkyl Ethers

Ether	Mol. formula	Mol. wt.	B. p., °C. (760 mm. Hg)	M. p., °C.	Sp. gr.	Vapor pressure, mm. Hg	Refr. index, n_D^{20}	Sol. in H_2O (g. per 100 ml.)	Flash point, °F.	Miscellaneous
Hexyl	$CH_3(CH_2)_5O(CH_2)_5CH_3$	186.3	226	−43	0.7942 (20/20°C.)	0.07 (20°C.)	1.42041	0.01	170 (OC)	Liquid
Methyl propyl	$CH_3OCH_2CH_2CH_3$	74.1	38.8		0.738 (20/4°C.)	520 (28°C.)	1.36019	3.05		Flammable liquid
Methyl isopropyl	$CH_3OCH(CH_3)_2$	74.1	32.5 (777 mm.)		0.7347 (20/20°C.)	635 (26°C.)				Flammable liquid
Ethyl butyl	$C_2H_5OC_4H_9$	102.2	91.5		0.749 (20/4°C.)	44.4 (20°C.)	1.3875			Liquid
Di-(2-ethylhexyl)	$[CH_3(CH_2)_3CH(C_2H_5)CH_2]_2O$	242.5								
2,2-Dimethoxypropane (ketal)	$CH_3C(OCH_3)_2CH_3$	104.1	80	−47	0.850 (20/20°C.)	100 (26.14°C.)	1.376	18	20	Colorless liquid
1,1,3-Trimethoxybutane	$(CH_3O)_2CHCH_2CH(OCH_3)CH_3$	148.2								

TABLE 2

Physiological Properties and Uses of Other Saturated Alkyl Ethers

| Ether | Anesthetic | | Oral LD50, g./kg. (rat) | Vapor exposure[a] | Primary skin irritation[b] | Eyes[c] | Odor | Remarks | Uses |
	Surg. anes., ml./kg.	Resp. arrest, ml./kg.							
Hexyl[d-f]			30.9	8 hr.	5	1	Mild	Rabbit skin absorption, LD50 = 6.9 ml./kg.	Reaction medium, component of foam breaker
Methyl propyl[d,g,h]	Dog, 0.84	2.1					Ethereal	Caused no deleterious effect in dog, monkey, or rat	Anesthetic
Methyl isopropyl[i]	Dog, 1.12	2.62					Ethereal	Caused no deleterious effect in dog, monkey, or rat	Anesthetic
Ethyl butyl[e,j]			1.87	5 min.	2	2		Inhalation; all of 6 rats survived 1000 p.p.m. for 4 hr.	
Di-(2-ethylhexyl)[f]			33.9	4 hr.	3	1			
2,2-Dimethoxypropane[k] (ketal)			>1.0	12 min.	1	4	Irritating	Inhalation results in anesthesia	Dehydrating agent, solvent
1,1,3-Trimethoxybutane[f]			1.48	2 hr.	2	2		Inhalation : 1 of 6 rats died after 4 hr. at 2000 p.p.m.	

[a] Maximum time for no death in rats when exposed to concentrated vapor obtained by passing air through compound at room temperature.
[b] Grade of primary irritation hazard to rabbit skin as explained by H. F. Smyth, Jr., C. P. Carpenter, and C. S. Weil, J. Ind. Hyg. Toxicol., 31, 60 (1949).
[c] Grade of hazard to rabbit eye as explained by C. P. Carpenter and H. F. Smyth, Jr., Am. J. Ophthalmol., 29, 1363 (1946).
[d] L. Scheflan and M. B. Jacobs, The Handbook of Solvents. Van Nostrand, New York, 1953.
[e] A. I. Vogel, J. Chem. Soc. London, 1948, 616.
[f] H. F. Smyth, Jr., C. P. Carpenter, C. S. Weil, and U. C. Pozzani, Arch. Ind. Hyg. Occupational Med., 10, 61 (1954).
[g] The Merck Index. 7th ed., Rahway, N. J., 1960.
[h] J. C. Krantz, Jr., W. E. Evans, Jr., C. J. Carr, and D. V. Kibler, J. Pharmacol. Exptl. Therap., 86, 138 (1946).
[i] J. C. Krantz, Jr., C. J. Carr W. E. Evans, Jr., and R. Musser, J. Pharmacol. Exptl. Therap., 87, 132 (1946).
[j] H. F. Smyth, Jr., C. P. Carpenter, and C. S. Weil, Arch. Ind. Hyg. Occupational Med., 4, 119 (1951).
[k] The Dow Chemical Co., unpublished data.

Vapor pressure: 430 mm. Hg (20°C.)

Per cent in "saturated" air: 56 (25°C., 760 mm. Hg)

Density of "saturated" air: 1.8 (25°C., 760 mm. Hg) (air = 1)

Refractive index: 1.3989 (n_D^{20})

Solubility: 0.53 g. dissolves in 100 ml. of water at 37°C., 0.4 g. at 20°C.; miscible with alcohol, ether, oils, and most organic solvents

Flash point: < -22°F. (C.C.)

Explosive limits: 1.7 per cent by volume in air, lower limit; 27.0 per cent by volume in air, upper limit

Autoignition temperature: 680°F.

1 p.p.m. ≏ 2.86 mg./cu. meter at 25°C., 760 mm. Hg and 1 mg./liter ≏ 349 p.p.m. at 25°C., 760 mm. Hg

3. Determination in the Atmosphere

Specific methods for the determination of divinyl ether in the atmosphere have not appeared in the literature. However, grab-bag sampling with subsequent infrared analysis with a gas cell of long path length should provide a highly specific method. Absorption on silica gel, elution with ethanol, and analysis of the eluate by gas chromatography also should be of value. A sensitive explosimeter could be used only when the vapor concentrations are known to be below the lower explosive limit.

4. Physiological Response

Summary.[49,50] Divinyl ether has been studied extensively in animals and humans from the standpoint of anesthesia. As with ethyl ether, the primary physiological effect of divinyl ether is that of general anesthesia. Although it is somewhat more toxic than ethyl ether, the concentration in the blood necessary for anesthesia is much lower. Its rapidity of action is much greater than that of ethyl ether; therefore induction and recovery are more rapid. It has a distinct disadvantage for prolonged anesthesia since it appears to have appreciable capacity to cause liver damage. Recent studies by Cavaliere *et al.*[51] indicate that vinyl ether causes a sharp increase in bile products in the dog. This was not found during ethyl ether anesthesia.

Further details of the physiological response to divinyl ether may be obtained from most textbooks on pharmacology or anesthesiology.

5. Hygienic Standards of Permissible Exposure

There have been no hygienic standards of permissible exposure to divinyl ether proposed.

[49] L. S. Goodman and A. Gilman, *The Pharmacological Basis of Therapeutics.* 2nd ed., Macmillan, New York, 1955, p. 68.

[50] T. Sollmann, *A Manual of Pharmacology.* 8th ed., Saunders, Philadelphia, 1957, p. 915.

[51] R. Cavaliere, B. Giovanella, and G. Moricca, *Med. sper.,* **31**, 253 (1957), *Chem. Abstr.,* **52**, 12228 (1958).

OTHER UNSATURATED ALKYL ETHERS

There are several other unsaturated alkyl ethers that have received some study regarding their toxicological or physiological effects, and that are used industrially. Several of these contain the vinyl radical and are useful as monomers for polymers or are useful in some other manner in the field of industrial polymers. These others are listed in Tables 3 and 4, along with their known physical, chemical, and physiological properties, and some uses.

For details the references cited should be consulted.

CHLOROMETHYL ETHER (Chloromethyl Methyl Ether, Monochloromethyl Ether)

$$C_2H_5OCl \quad \text{or} \quad ClCH_2OCH_3$$

1. Source, Uses, and Industrial Exposures[52]

Chloromethyl ether is prepared by bubbling HCl through a heated solution of formaldehyde and methanol. It can also be prepared by the direct chlorination of methyl ether.

Chloromethyl ether is used in industry as a methylating agent. Industrial exposure to chloromethyl ether should not be allowed to occur because of its extreme toxicity. In case of an accident, exposure to the vapors of this compound would probably occur since it has a low boiling point. Skin or eye contact will cause severe burns and necrosis.

2. Physical and Chemical Properties[52]

Chloromethyl ether is a colorless liquid that has the following properties:
Molecular weight: 80.5
Boiling point: 59.5°C. (759 mm. Hg) ; 58°C. (743 mm. Hg)
Melting point: −103.5°C.
Specific gravity: 1.070 (20/4°C.)
Solubility: decomposes in water and hot alcohol; soluble in 95 per cent ethanol and acetone
1 p.p.m. $\backsimeq$ 3.29 mg./cu. meter at 25°C., 760 mm. Hg and 1 mg./liter $\backsimeq$ 304 p.p.m. at 25°C., 760 mm. Hg

3. Determination in the Atmosphere[52]

Chloromethyl ether can be determined in the atmosphere by combustion and analysis of the trapped chloride by Volhard titration. Silica-gel absorption should be of value in obtaining samples. Infrared spectroscopy and gas chromatography are of value for chloromethyl ether determination.

[52] Biochemical Research Laboratory, The Dow Chemical Co., unpublished data.

TABLE 3

Physical and Chemical Properties of Other Unsaturated Alkyl Ethers

Ether	Mol. formula	Mol. wt.	B. p., °C. (760 mm. Hg)	M. p., °C.	Sp. gr.	Vapor pressure, mm. Hg (20°C.)	Refr. index, n_D^{20}	Sol. in H_2O (g. per 100 ml.)	Flash point, °F.	Miscellaneous
Allyl	$(CH_2CHCH_2)_2O$	98.1	94.3–95		0.808 (20/4°C.)			0.3	20 (O.C.)	Colorless liquid
Ethyl vinyl	$C_2H_5OCHCH_2$	72.1	35.5	−115	0.755 (20/20°C.)	428	1.3763	0.9		Colorless liquid
Isopropyl vinyl	$(CH_3)_2CHOCHCH_2$	86.2	55.7	−140	0.754 (20/20°C.)	195	1.3845	0.64		Pale yellow liquid
Butyl vinyl	$C_4H_9OCHCH_2$	100.2	93.8	−113	0.780 (20/20°C.)	42	1.3997	0.04	30 (Q.C.)	Liquid
2-Ethylhexyl vinyl	$C_8H_{17}OCHCH_2$	156.3	177.5	−100	0.810 (20/20°C.)	0.90	1.4232	0.01		Liquid
2,6,8-Trimethylnonyl vinyl	$C_{12}H_{25}OCHCH_2$	212.4								Liquid
2-Methoxyethyl vinyl	$CH_3OC_2H_4OCHCH_2$	102.1	108.8	−83	0.897 (20/20°C.)	18		8.8		Liquid
2-Butoxyethyl vinyl	$C_4H_9OC_2H_4OCHCH_2$	125.2								Liquid
Methyl isopropenyl	$CH_3OC(CH_3)CH_2$	88.0	35.8							Colorless liquid

TABLE 4

Physiological Properties and Uses of Other Unsaturated Alkyl Ethers

Ether	Anesthetic Surg. anes.	Anesthetic Resp. arrest	Oral LD50, g./kg. (rat)	Vapor exposure[a]	Primary skin irritation[b]	Eyes[c]	Odor	Remarks	Uses
Allyl[d,e]			0.32		2	4	Horse-radish	Rabbit skin absorption, LD50 = 0.6 ml./kg.	Organic synthesis
Ethyl vinyl[d,f-h]	Mouse, 6 vol. % Dog, 0.56 cc./kg.	16 vol. % 1.66 cc./kg.					Divinyl ether	16,000 p.p.m. for 4 hr. killed 2, 3, or 4 of 6 rats	Organic synthesis, plasticizers, adhesives
Isopropyl vinyl[d,g,i]	Dog, 0.50 cc./kg.	3.08 cc./kg.					Divinyl ether		Organic synthesis manufacture of polymers
Butyl vinyl[d,g,j]			10.30	5 min.	3	2	Ethereal	Rabbit skin absorption, LD50 = 4.24 ml./kg. Inhalation: all 6 rats survived 8000 p.p.m. for 4 hr.	Organic synthesis
2-Ethylhexyl vinyl[d,g,j]			1.35	4 hr.	3	1		Rabbit skin absorption, LD50 = 3.56 ml./kg.	Organic synthesis
2,6,8-Trimethylnonyl vinyl[j]			1.22	4 hr.	3	1		Rabbit skin absorption, LD50 = 5.0 ml./kg.	
2-Methoxyethyl vinyl[g,k]			3.90	1 hr.	1	2		Rabbit skin absorption, LD50 = 7.13 ml./kg. Inhalation: 4 of 6 rats survived 8000 p.p.m. for 4 hr.	
2-Butoxyethyl vinyl			3.10	4 hr.	2	5		Rabbit skin absorption, LD50 = 3.00 ml./kg. Inhalation: 4 of 6 rats survived 2000 p.p.m. for 8 hr.	
Methyl isopropenyl[k]			>2.0	10 min.	4	2			Chemical intermediate

[a] Maximum time for no death in rats when exposed to concentrated vapor obtained by passing air through compound at room temperature.

[b] Grade of primary irritation hazard to rabbit as explained by Smyth et al.[e]

[c] Grade of eye hazard to rabbit as explained by Carpenter et al.[k]

[d] L. Scheflan and M. B. Jacobs, The Handbook of Solvents. Van Nostrand, New York, 1953.

[e] H. F. Smyth, Jr., C. P. Carpenter, and C. S. Weil, J. Ind. Hyg. Toxicol., 31, 60 (1949).

[f] E. T. Mörch, J. B. Ayerigg, and M. S. Berger, J. Pharmacol. Exptl. Therap., 117, 184 (1956).

[g] Union Carbide Chemicals Co., Technical Information, Vinyl Ethers, No. F-6701-A, 1948.

[h] C. P. Carpenter, H. F. Smyth, Jr., and U. C. Pozzani, J. Ind. Hyg. Toxicol., 31, 343 (1949).

[i] J. C. Krantz, Jr., C. J. Carr, G. Lu, and M. J. Fassel, J. Pharmacol. Exptl. Therap., 105, 1 (1952).

[j] H. F. Smyth, Jr., C. P. Carpenter, C. S. Weil, and U. C. Pozzani, Arch Ind. Hyg. Occupational Med., 10, 61 (1954).

[k] C. P. Carpenter and H. F. Smyth, Jr., Am. J. Ophthalmol., 29, 1363 (1946).

4. *Physiological Response*[52]

Summary. Although chloromethyl ether is only moderately toxic when given orally, it is very severely irritating to the eyes and skin. The vapors of the material are highly irritating and painful to the eyes and nose even in concentrations of the order of 100 p.p.m., which can be tolerated by animals for several hours. Vapor concentrations that are rapidly fatal are irrespirable for humans. Illness or death that results from exposure to vapors of chloromethyl ether will occur from lung edema or secondary pneumonia.

Animal Toxicity. Chloromethyl ether was fed orally in single doses to rats. Doses of 0.3 g./kg. allowed survival, whereas doses of 1.0 g./kg. caused death. An approximate LD_{50} of 0.5 g./kg. was calculated. When tested in the rabbit eye as a 1 per cent solution in propylene glycol, severe irritation and necrosis developed. Rabbit skin tests using the undiluted material resulted in severe hyperemia, edema, denaturation, and even complete destruction of the skin.

Groups of 6 rats were exposed for periods of $1/2$, 1, and 4 hours to vapor concentrations of from 100 to 10,000 p.p.m. The results indicated that 2000 p.p.m. of vapor is dangerous to life if the exposure lasts over 30 minutes, while 100 p.p.m. is dangerous to life if the exposure lasts 4 hours. The vapors were highly irritating to the mucous membranes at all concentrations. Deaths resulting from vapor exposure were usually from pneumonia and were delayed, occurring several days or even weeks after exposure. The role of formaldehyde in the vapor toxicity of chloromethyl ether is unknown. It may be significant because the compound decomposes to formaldehyde in the presence of moist air.

Human Experience.[52] An industrial exposure to rather high concentrations of chloromethyl ether occurred. The subject had symptoms of a sore throat, fever, and chills and was not able to work for 8 days, at which time recovery appeared complete. Another subject who received only very slight exposure had difficulty in breathing for several days.

5. *Hygienic Standards of Permissible Exposure*

No hygienic standards have been proposed for chloromethyl ether.

DICHLOROETHYL ETHER [*sym*-Dichloroethyl Ether, β,β'-Dichloroethyl Ether, 2,2'-Dichlorodiethyl Ether, 1-Chloro-2-(β-chloroethoxy) Ethane, Bis(2-chloroethyl) Ether]

$$C_4H_8OCl \quad \text{or} \quad ClCH_2CH_2OCH_2CH_2Cl$$

1. *Source, Uses, and Industrial Exposures*[53-56]

Dichloroethyl ether is obtained as a by-product in the production of ethylene glycol from ethylene chlorohydrin. It is used extensively in industry as a solvent

[53] L. Scheflan and M. B. Jacobs, *The Handbook of Solvents.* Van Nostrand, New York, 1953.

[54] *The Merck Index.* 7th ed., Rahway, N. J., 1960.

[55] F. A. Patty, ed., *Industrial Hygiene and Toxicology,* Vol. II. Interscience Publishers. New York, 1949.

[56] H. Allen, *Chem. Prod.,* **19,** 482 (1956).

for special lacquers, resins, and oils; to assist in the scouring of textiles; as a de-waxing agent for lubricating oils; as a chemical intermediate; to replace sodium hydroxide in kier boiling; as a penetrant and wetting agent; in agriculture in 1 per cent aqueous solution as an insecticidal soil fumigant; and it is sometimes used as a dry-cleaning agent.

Except in the case of accidental exposure to high concentrations of the vapor, the chief hazard of industrial exposure is the possibility of repeated exposures to low concentrations of the vapor, which may cause a form of chemical bronchitis.

2. *Physical and Chemical Properties*[53,55]

Dichloroethyl ether is a clear, colorless, stable liquid with a pungent, fruity odor. It has the following properties:

Molecular weight: 143.02
Boiling point: 178.5°C. (760 mm. Hg) ; 178°C. (744 mm. Hg)
Melting point: −51.7°C.
Specific gravity: 1.222 (20/4°C.)
Vapor density: 4.93 (air = 1)
Vapor pressure: 0.73 mm. Hg (20°C.) ; 1.4 mm. Hg (25°C.)
Per cent in "saturated" air: 0.18 (25°C., 760 mm. Hg)
Density of "saturated" air: 1.007 (25°C., 760 mm. Hg) (air = 1)
Refractive index: 1.457 (n_D^{20})
Solubility: 1.07 g. in 100 ml. of water; miscible with alcohol, ether, and most organic solvents; immiscible with paraffin hydrocarbons
Flash point: 185°F. (O.C.) ; 131°F. (C.C.)
1 p.p.m. ≎ 5.85 mg./cu. meter at 25°C., 760 mm. Hg and 1 mg./liter ≎ 171 p.p.m. at 25°C., 760 mm. Hg

3. *Determination in the Atmosphere*

Dichloroethyl ether can be determined in the atmosphere by trapping the compound in alcoholic KOH bubblers, hydrolyzing by refluxing for 2 hours, and determining the chloride by Volhard titration.[56] Other available ways of determining dichloroethyl ether are the vapor pressure, infrared, and interferometer methods. Silica-gel absorption is another valuable tool for the trapping of vapors. The dichloroethyl ether can be removed by warming, combusted, and determined by chloride titration.

4. *Physiological Response*

Summary. Dichloroethyl ether vaporizes to an irritant gas, which can cause a delayed response resulting in lung lesions. However, the vapor concentration necessary to cause damage in the respiratory tract is easily detected by its odor, and is intolerable in most cases. The compound is injurious to the eyes, both in liquid and concentrated vapor forms. It has no deleterious effect on the intact skin. However, it is rapidly absorbed in lethal amounts by the skin; therefore, skin contact should be avoided. Oral consumption also should be avoided.

Single-Dose Oral. Smyth and Carpenter[57] in 1948 reported the single-dose oral LD_{50} for rats to be 75 mg./kg. They regarded ethylene chlorohydrin as a chemical having a hazard in the same order of magnitude. More recently, the *Handbook of Toxicology*, edited by Spector,[58] reports the following oral LD_{50} values, giving unpublished data from Smyth as reference: rat, 105 mg./kg.; mouse, 136 mg./kg.; and rabbit, 126 mg./kg. Further work[59] using the rat has given an LD_{50} value of 150 mg./kg., with 19/20 confidence limits of 110 to 210 mg./kg.

There have been no studies reported on the repeated oral feeding of dichloroethyl ether.

Inhalation. Schrenk, Patty, and Yant[60] studied the effect of inhalation of dichloroethyl ether vapors upon the guinea pig. Concentrations of 500 to 1000 p.p.m. caused immediate, severe eye and nasal irritation as evidenced by lachrymation and scratching of the nose. After $1^1/_2$ to 3 hours at these concentrations, respiratory disturbances were evident, with death taking place after 5 to 8 hours of continuous exposure. The chief lesions in these animals were in the respiratory organs, with some effects noted in the brain, kidneys, and liver. Even if exposure at these concentrations was stopped after 90 minutes, many of the animals died later from lung lesions. The authors, therefore, suggested that an exposure of 30 to 60 minutes duration to dichloroethyl ether vapor at a concentration of 500 to 1000 p.p.m. was dangerous to the life of the guinea pig. Ten to 15 hours of continuous exposure to 105 and 260 p.p.m. eventually caused death. However, if the exposure time to 100 p.p.m. was limited to 1 hour, no serious disturbances resulted, even though eye and nose irritation was still evident. The latter was indicated even at 35 p.p.m., although there were no other signs of adverse effects and no deaths after $13^1/_2$ hours of continuous exposure. Smyth and Carpenter[57] exposed rats to 1000 p.p.m. vapor for 45 minutes. Three of 6 animals died within 14 days. This group also reported deaths in rats when they were exposed to 250 p.p.m. vapor for a single 4-hour period.[61] They classified this compound in the "definite hazard" group.

Repeated exposures of rats and guinea pigs to an average concentration of 69 p.p.m. of dichloroethyl ether vapor have been carried out.[59] There were 93 7-hour exposures, 5 per week, over a period of 130 days. No adverse effect was observed as judged by gross appearance, behavior, mortality, hematological values, and gross

[57] H. F. Smyth, Jr., and C. P. Carpenter, *J. Ind. Hyg. Toxicol.*, **30,** 63 (1948).

[58] W. S. Spector, ed., *Handbook of Toxicology*, Vol. I., Saunders, Philadelphia, 1956.

[59] Biochemical Research Laboratory, The Dow Chemical Co., unpublished data.

[60] H. H. Schrenk, F. A. Patty, and W. P. Yant, *Public Health Repts. U. S.*, **48,** 1389 (1933).

[61] C. P. Carpenter, H. F. Smyth, Jr., and U. C. Pozzani, *J. Ind. Hyg. Toxicol.*, **31,** 343 (1949).

and microscopic examination of the tissues at autopsy. However, all of the animals showed a varying degree of growth depression, which was significant in the males of both species. Organ weights also showed significant variance from controls; however, these variances were not consistent among species or sex. It was concluded that repeated exposure to 69 p.p.m. of dichloroethyl ether produced no serious injury and that the abnormalities observed reflected only mild physiological stress.

Effects on Skin and Eyes. Smyth and Carpenter[57] reported that dichloroethyl ether has a mild primary irritating effect on the skin. Allen,[56] in his review of the safety hazards of dichloroethyl ether, stated that, "The pure liquid has no irritative effect or indeed any other effect on the skin, a factor greatly in its favour." This has essentially been confirmed by studies with rabbits,[59] which showed that a 10 per cent solution of the compound has very little or no effect on the intact or abraded skin after repeated applications. However, in the same studies, when using the pure material, acutely toxic amounts rapidly penetrated the skin and caused death within a day. Using graded doses of a 10 per cent solution of dichloroethyl ether in propylene glycol, the LD_{50} for 24-hour skin absorption in rabbits was calculated to be 90 mg./kg., with 19/20 confidence limits of 55 to 145 mg./kg. Smyth and Carpenter[57] reported an LD_{50} for guinea pigs of 0.3 ml./kg., using the poultice method of testing. From these results it is obvious that, although dichloroethyl ether is not a primary skin irritant, contact with the skin should be avoided because of its rapid absorption and deleterious consequences.

Carpenter and Smyth[62] reported an injury grade of 4 for dichloroethyl ether when applied to rabbit eyes. This indicates appreciable irritation and injury. Further studies[59] with rabbits have revealed that both the pure material and a 10 per cent solution in propylene glycol cause moderate pain, conjunctival irritation, and corneal injury, which generally heals within 24 hours. Immediate irrigation of the eye with flowing water greatly reduces the deleterious effects. From these results it is evident that exposure of human eyes to dichloroethyl ether could cause serious damage, although probably not permanent loss of vision.

Mode of Action. From the work of Schrenk, Patty, and Yant,[60] it is evident that the physiological action of dichloroethyl ether is that of a primary respiratory irritant. Although the compound is capable of causing narcosis at high concentrations, the delayed deaths at lower concentrations have been due to severe irritation of the respiratory tree, with a resultant respiratory collapse. The congestions in the brain, liver, and kidneys are probably secondary effects. Allen[56] makes the interesting comparison of the chemical structure relationship of dichloroethyl ether to mustard gas, dichloroethyl sulfide. The replacement of the S atom by the O atom in dichloroethyl ether abolishes the blistering action of mustard gas on the skin. However, many of the respiratory irritant effects are retained.

[62] C. P. Carpenter and H. F. Smyth, Jr., *Am. J. Ophthalmol.,* **29,** 1355 (1946).

Human Experience. Schrenk, Patty, and Yant[60] exposed human volunteers to dichloroethyl ether. Brief exposure to concentrations above 550 p.p.m. were very irritating to the eyes and nasal passages and were considered intolerable. They also caused coughing, retching, and nausea. At 260 and 100 p.p.m. the irritating effects were still present to some extent but were not considered intolerable. The nauseous odor was still detectable at 35 p.p.m. but there was no irritation.

There have been no cases of toxic effects reported from industry, either from acute or chronic exposure to dichloroethyl ether.

5. Hygienic Standards of Permissible Exposure

Based upon the work of Schrenk, Patty, and Yant,[60] Cook[63] suggested 15 p.p.m. as the maximum allowable concentration for daily exposure. This value (15 p.p.m., approximately 90 mg./cu. meter) has been adopted as the threshold limit by the American Conference of Governmental Industrial Hygienists, and is still the recommended value.[64] Fortunately, most people can detect the chemical at its threshold limit by its odor. However, this criterion should not be relied upon for monitoring purposes.

DICHLOROISOPROPYL ETHER [Bis(β-chloroisopropyl) Ether, β,β'-Dichlorodiisopropyl Ether]

$$C_6H_{12}OCl_2 \quad \text{or} \quad CH_3(CH_2Cl)CHOCH(CH_2Cl)CH_3$$

1. Source, Uses, and Industrial Exposures

Dichloroisopropyl ether is obtained as a by-product in the commercial production of propylene glycol. It has some commercial use as a solvent for fats, greases, and waxes, and as a cleaning and spotting agent. It is also used in textile processes, where it assists the action of soap solutions without excessive loss by evaporation. It has been used in paint and varnish removers, spotting agents, cleaning solutions, and as an intermediate in the manufacture of dyes, resins, and pharmaceuticals. Industrial exposure of greatest significance is likely to be exposure to the vapors of this compound, but contact with the skin and eyes can also be expected.

2. Physical and Chemical Properties[65,66]

Dichloroisopropyl ether is a colorless liquid with the following properties:
Molecular weight: 171.07
Boiling point: 187.3°C. (760 mm. Hg)

[63] W. A. Cook, *Ind. Med.,* **14,** 936 (1945).

[64] Conference of Governmental Industrial Hygienists, *A.M.A. Arch. Environmental Health,* **1,** 140 (1960).

[65] L. Scheflan and M. B. Jacobs, *The Handbook of Solvents.* Van Nostrand, New York, 1953.

[66] Biochemical Research Laboratory, The Dow Chemical Co., unpublished data.

Melting point: −96.8 to −101.8°C.

Specific gravity: 1.1122 (20/20°C.) ; 1.096 (25/25°C.)

Vapor density: 5.9 (air = 1)

Vapor pressure: 0.71 to 0.85 mm. Hg (20°C.) ; 44.5 mm. Hg (100°C.)

Per cent in "saturated" air: 0.12 (25°C., 760 mm. Hg)

Density of "saturated" air: 1.05 (25°C., 760 mm. Hg) (air = 1)

Refractive index: 1.4451 (n_D^{25})

Solubility: 0.17 g. in 100 ml. of water; miscible with most organic solvents and oils

Flash point: 185°F. (O.C.)

1 p.p.m. ≎ 7.00 mg./cu. meter at 25°C., 760 mm. Hg and 1 mg./liter ≎ 143 p.p.m. at 25°C., 760 mm. Hg

3. Determination in the Atmosphere

Methods used for the determination of dichloroethyl ether in the atmosphere should be applicable to the determination of dichloroisopropyl ether.

4. Physiological Response

Summary. The physiological response to dichloroisopropyl ether is not entirely clear. It does not seem to be quite as toxic orally or by inhalation as dichloroethyl ether, and its deleterious effects are found in the liver and kidneys rather than in the lungs. Although absorbed readily through the skin, it has little effect on it, or upon the eyes.

Single Oral Dose. Smyth et al.[67] fed rats single oral doses of dichloroisopropyl ether and obtained an LD$_{50}$ of 0.24 g./kg. Other studies[66] indicate similar results. None of 5 animals died at 0.2 g./kg., but 4 of 6 died at 0.4 g./kg. All of the 5 rats fed 0.8 g./kg. died. Smyth[68] found the LD$_{50}$ for guinea pigs to be 0.45 g./kg. These results indicate that dichloroisopropyl ether is fairly toxic orally, but less toxic than dichloroethyl ether.

Repeated Oral Dose.[66] Rats were fed 22 doses of dichloroisopropyl ether in olive oil by stomach tube during a period of 31 days. Even the lowest dose administered, 0.01 g./kg., caused a decrease in growth rate when compared to the controls. However, there were no differences in organ weights or hematological values at this level. At the highest dosage level, 0.20 g./kg., both the liver and kidney weights (per unit of body weight) were increased when compared to the controls. Spleen weights also increased but there was no indication of effects on the blood. Intermediate dosage levels showed only a decrease in growth rate. It is evident from this work that repeated, low oral doses of dichloroisopropyl ether, although not rapidly lethal, have a deleterious effect on rats.

   [67] H. F. Smyth, Jr., C. P. Carpenter, and C. S. Weil, *Arch. Ind. Hyg. Occupational Med.,* 4, 119 (1951).

   [68] W. S. Spector, ed., *Handbook of Toxicology,* Vol. I., Saunders, Philadelphia, 1956.

Inhalation. Smyth et al.[67] exposed 6 rats to a vapor concentration of 1000 p.p.m. for 4 hours. One animal died within a period of 14 days. In other studies,[66] rats exposed to an atmosphere believed to be essentially saturated with dichloro-isopropyl ether exhibited signs of immediate eye irritation and some incoordination; the maximum exposure time causing no deaths was 1 hour. When rats were exposed to 700 p.p.m., deaths occurred after 6 hours of exposure. Autopsy revealed slight lung irritation and moderate to severe liver damage. All of 10 rats survived a 6-hour exposure to 350 p.p.m. but 2 of 5 died after an 8-hour exposure. These animals exhibited moderate lung congestion and some liver necrosis. One of 4 animals died after an 8-hour exposure to 175 p.p.m. It appears that the length of exposure to dichloroisopropyl ether vapors is a highly significant factor in its capacity to cause death.

There have been no studies reported involving repeated exposure of animals to dichloroisopropyl ether.

Effects on Skin and Eyes. Dichloroisopropyl ether evidently has little effect on rabbit skin. After 20 applications to the ear there was no response. Moderate scaliness was produced after 20 applications to the rabbit belly by the poultice technique.[66] Smyth et al.[67] also reported that this compound has no primary irritating effect on rabbit skin; however, a dose of 3.0 ml./kg. was found to be the LD_{50} for skin penetration.

Smyth et al.[67] found that liquid dichloroisopropyl ether caused no damage to the eye beyond irritation. The concentrated vapors are also irritating to the eyes and mucous membranes.

Human Experience. There have been no reports of overexposure of humans to dichloroisopropyl ether.

5. *Hygienic Standards of Permissible Exposure*

No hygienic standards of permissible exposure have been proposed for dichloroisopropyl ether.

OTHER HALOGENATED ALKYL ETHERS

Very little toxicological research has been reported on other halogenated alkyl ethers that are used industrially. Smyth et al.[69] conducted range-finding studies on 2-chloroethyl vinyl ether and found that it was moderately toxic orally to rats (LD_{50} = 0.25 g./kg.), but that it had little effect on the skin and in the eyes of rabbits. One of 6 rats died after a 4-hour exposure to 500 p.p.m. of the vapor in air. This compound is used as an intermediate in chemical reactions and in the manufacture of polymers.

Range-finding studies were also conducted on 2-chloro-1,1,2-trifluoroethyl methyl ether.[70] Its oral toxicity to rats was very low (LD_{50} = 5.13 g./kg.).

[69] H. F. Smyth, Jr., C. P. Carpenter, and C. S. Weil, *J. Ind. Hyg. Toxicol.,* **31,** 60 (1949).

[70] H. F. Smyth, Jr., C. P. Carpenter, and C. S. Weil, *Arch. Ind. Hyg. Occupational Med.,* **4,** 119 (1951).

Although it had no primary irritant effect on the skin of the rabbit, it penetrated the skin easily, as evidenced by an LD_{50} of 0.2 ml./kg. when applied to the skin by the "cuff" technique. It was quite injurious to the rabbit eye. Rats lived for a maximum time of only 5 minutes when exposed to concentrated vapors in the atmosphere. From these data it appears that this toxic compound is readily absorbed through the skin and lungs but not from the gastrointestinal tract.

Many halogenated ethers, particularly fluorinated ethers, have been tested as anesthetics in a search for nonflammable agents. Buu-Hoï listed 7 fluorinated compounds in his review.[71] 2,2,2-Trifluoroethyl vinyl ether (called Fluoromar) has met with some success as an anesthetic but it presents no outstanding advantages over ethyl ether.[71] Mixed halogenated ethers have also been studied as anesthetic agents. One of the most promising of these seems to be 1,1-difluoro-2,2-dichloroethyl methyl ether (methoxyfluorane).[72,73] Increased fluorination of diethyl ether progressively diminishes its anesthetic potency so that perfluoroethyl ether is devoid of anesthetic properties.[74] On the other hand, a partially fluorinated compound, di-(2,2,2-trifluoroethyl) ether, in addition to some anesthetic action, is a powerful convulsant. Krantz et al.[74] reported its trial in man as a substitute for electroshock therapy. The threshold concentration for the convulsive syndrome in rats was found to be 21 p.p.m.

Many of the halogenated ethers have been tested for insecticidal activity. Pattison et al.[75] report that 2-fluoro-1',2',2',2'-tetrachlorodiethyl ether shows outstanding activity as a systemic insecticide. It has an intraperitoneal LD_{50} for mice of 48 mg./kg. These authors, from their toxicological studies of mixed halogenated ethers of various chain lengths, propose that the ether link is ruptured in vivo. Buckle and Saunders[76] believe that the toxicity of these compounds can be explained by β-oxidation. For a further discussion of the toxicity of these compounds see Pattison.[77]

ANISOLE (Phenyl Methyl Ether, Methoxybenzene)

$$C_7H_8O \quad \text{or} \quad C_6H_5OCH_3$$

1. Source, Uses, and Industrial Exposures

Anisole generally is made by reacting methyl chloride with sodium phenate. It is used as a chemical intermediate and in perfumery. In industrial handling,

[71] N. P. Buu-Hoï, Les dérivés organique du fluor d'intérêt pharmacologique. To be published.

[72] A. Van Poznak and J. F. Artusio, Jr., Toxicol. Appl. Pharmacol., 2, 374 (1960).

[73] J. F. Artusio, Jr., A. Van Poznak, R. E. Hunt, F. M. Tiers, and M. Alexander, Anesthesiology, 21, 512 (1960).

[74] J. C. Krantz, Jr., E. B. Truitt, Jr., A. S. C. Ling, and L. Speers, J. Pharmacol. Exptl. Therap., 121, 362 (1957).

[75] F. L. M. Pattison, W. C. Howell, and R. G. Woolford, Can. J. Chem., 35, 141 (1957).

[76] F. J. Buckle and B. C. Saunders, J. Chem. Soc. London, 1949, 2774.

[77] F. L. M. Pattison, Toxic Aliphatic Fluorine Compounds. Elsevier, New York, 1959.

contact with the eyes and skin and inhalation of vapors may occur but the consequences of such exposure would not be expected to be serious.

2. *Physical and Chemical Properties*[78]

Anisole is a colorless liquid having a low volatility and a distinctive odor. It has the following properties:

Molecular weight: 108.1
Boiling point: 154°C.
Freezing point: —37.4°C.
Specific gravity: 0.989 (25/4°C.)
Vapor density: 3.7 (air = 1)
Refractive index: 1.514 (25°C.)
Vapor pressure: 3.1 mm. Hg (25°C.)
Per cent in "saturated" air: 0.41
Density of "saturated" air: 1.1 (air = 1)
Solubility: insoluble in water; soluble in alcohol and ether
1 p.p.m. $\backsimeq$ 4.46 mg./cu. meter at 25°C., 760 mm. Hg and 1 mg./liter $\backsimeq$ 226.2 p.p.m. at 25°C., 760 mm. Hg

3. *Determination in the Atmosphere*

Collection followed by analysis employing ultraviolet spectrophotometry, as suggested under Phenyl Ether, is recommended. When in isooctane, anisole has strong absorption bands at 271 and 277.5 mμ, and 0.05 mg./100 ml. can be measured readily.

Vapor-phase chromatography and mass spectrometry also can be used.

4. *Physiological Response*

The toxicological information available on anisole is very limited. Spector[79] reports that Binet found the lethal dose for the rat to be 3500 to 4000 mg./kg. when given subcutaneously and 100 to 900 mg./kg. when given intraperitoneally. One would not expect the material to be more than slightly toxic and it seems doubtful that it would present any unusual hazard in industrial use.

Metabolism. Coombs and Hele[80] in 1926 reported that the administration of anisole to the dog caused an increase in the excretion of ethereal sulfate. Williams[81] interpreted these observations to indicate cleavage of the ether linkage with subsequent conjugation of the formed phenol with sulfuric acid. Later, however, he revised this conclusion,[82] probably as a result of the work of Bray *et al.*[83] The

[78] R. R. Dreisbach and R. A. Martin, *Ind. Eng. Chem.*, **41,** 2875 (1949).
[79] W. S. Spector, ed., *Handbook of Toxicology*, Vol. 1. Saunders, Philadelphia, 1956.
[80] H. I. Coombs and T. S. Hele, *Biochem. J.*, **20,** 606 (1926).
[81] R. T. Williams, *Detoxification Mechanisms.* Chapman and Hall, London, 1947.
[82] R. T. Williams, *Detoxification Mechanisms.* 2nd ed., Wiley, New York, 1959.
[83] H. G. Bray, S. P. James, W. V. Thorpe, and M. R. Wasdell, *Biochem. J.*, **54,** 547 (1953).

work of Bray *et al.*[83] with the rabbit suggests that the increase in ethereal sulfate observed by Coombs and Hele is probably due to para-hydroxylation and conjugation with sulfuric acid. These authors showed the major metabolite of anisole to be *p*-hydroxyphenyl methyl ether, which was excreted unconjugated (2 per cent) and conjugated with glucuronic acid (48 per cent) and sulfuric acid (29 per cent).

Aryl-substituted anisoles such as *m*- and *p*-nitro-, *p*-chloro-, *p*-methoxy-, and *p*-cyanoanisoles are reported by Bray, Craddoch, and Thorpe[84] to be demethylated to form the phenols, which are excreted largely as the glucuronates and ethereal sulfates. Axelrod[85,86] has found an ether-splitting enzyme system in the microsomes of rat liver cells, which would account for the observation of Bray *et al.*[84]

5. *Hygienic Standard of Permissible Exposure*

Data adequate to suggest a hygienic standard are lacking. However, the absence of recorded adverse effects indicates that there probably is little need for such a figure.

PHENETOLE (Phenyl Ethyl Ether, Ethoxybenzene)

$$C_8H_{10}O \quad \text{or} \quad C_6H_5OC_2H_5$$

1. *Source, Uses, and Industrial Exposures*

Phenetole can be made by the direct ethylation of phenol by reacting ethyl chloride with sodium phenate. It is used as a chemical intermediate and in perfumery. In industrial handling, contact with the eyes and skin and inhalation of vapors may occur but the consequences of exposure are not serious.

2. *Physical and Chemical Properties*

Phenetole is a colorless liquid having a low volatility and a distinctive odor. It has the following properties:

Molecular weight: 122.1
Boiling point: 170°C.
Freezing point: −29.5°C.
Specific gravity: 0.960 (25/4°C.)
Vapor density: 4.2 (air = 1)
Refractive index: 1.505 (25°C.)
Vapor pressure: 1.7 mm. Hg (25°C., 760 mm. Hg)
Per cent in "saturated" air: 0.224
Density of "saturated" air: 1.01 (air = 1)
Solubility: insoluble in water; soluble in alcohol and ether
Flash point: 160°F. (O.C.)

[84] H. G. Bray, V. M. Craddoch, and W. V. Thorpe, *Biochem. J.*, **60**, 225 (1955).
[85] J. Axelrod, *J. Pharmacol. Exptl. Therap.*, **115**, 259 (1955).
[86] J. Axelrod, *Biochem. J.*, **63**, 634 (1956).

1 p.p.m. $\backsimeq$ 4.99 mg./cu. meter at 25°C., 760 mm. Hg and 1 mg./liter $\backsimeq$ 200.2 p.p.m. at 25°C., 760 mm. Hg

3. Determination in the Atmosphere

Collection followed by analysis employing ultraviolet spectrophotometry, as suggested under Phenyl Ether, is recommended. When in isooctane, phenetole has strong absorption bands at 271 and 277.5 mμ and 0.05 mg./100 ml. can be measured readily.

Vapor-phase chromatography and mass spectrometry also can be used.

4. Physiological Response

Summary. Toxicological information on phenetole is limited. That available, however, indicates that the material is quite low in acute toxicity. It would seem unlikely that the material would present any unusual hazards in industrial handling.

Acute Oral.[87] Phenetole has been found to be low in single dose oral toxicity for the guinea pig; doses of 3.0 g./kg. allowed survival of all animals and 10.0 g./kg. caused the death of all animals fed.

Skin Contact.[87] Phenetole is only very slightly irritating to the skin of rabbits.

Injection. Spector[88] reports that Binet found the minimum lethal dose for phenetole to be between 3.5 and 4.0 g./kg. when it was given subcutaneously to rats.

Metabolism. According to Williams,[89] Kossel (1880, 1883), Lehmann (1889), and Kuhling (1887) have shown that phenetole, like anisole, is hydroxylated in the para position and excreted as the glucuronide and ethereal sulfate.

Human Experience. No report of adverse effects from phenetole has been noted in the literature.

5. Hygienic Standard of Permissible Exposure

Data adequate to set a hygienic standard are lacking. The lack of reports of adverse effect in the past indicates that there probably is little need for such a figure.

GUAIACOL (*o*-Methoxyphenol, 1-Hydroxy-2-methoxybenzene, Methylcatechol)

$$C_7H_8O_2 \quad \text{or} \quad CH_3OC_6H_4OH$$

1. Source, Uses, and Industrial Exposures

Guaiacol is produced commercially through the destructive distillation of selected hardwoods. It is used as an antioxidant, an antigumming agent in hydro-

[87] Biochemical Research Laboratory, The Dow Chemical Co., unpublished data.

[88] W. S. Spector, ed., *Handbook of Toxicology,* Vol. 1. Saunders, Philadelphia, 1956.

[89] R. T. Williams, *Detoxification Mechanisms.* 2nd ed., Wiley, New York, 1959.

carbon solvents, an antiskinning agent in surface coatings, to control oxidation in printing inks, and as a chemical intermediate. It also has antiseptic properties, and has been used medically as an expectorant and antipyretic agent.

Direct contact with the eyes and skin would be the most likely industrial exposure. Since its vapor pressure is low and ordinary use does not involve elevated temperatures, vapor exposure of significance would not seem likely.

2. Physical and Chemical Properties

Guaiacol is a clear, colorless to pale yellow liquid with a mild aromatic odor. It has the following properties:

Molecular weight: 124.1

Boiling point: 205°C. (760 mm. Hg)

Melting point: 32°C.

Specific gravity: 1.097 (25/25°C.)

Vapor density: 4.27 (air = 1)

Vapor pressure: 100 mm. Hg (144°C.); 10 mm. Hg (92°C.); 5 mm. Hg (79°C.); 1 mm. Hg (52.4°C.); 0.103 mm. Hg (25°C.) (by extrapolation)

Per cent in "saturated" air: 0.0135 (25°C., 760 mm. Hg)

Density of "saturated" air: 1.005 (air = 1)

Solubility: soluble in alkalies and glycerine; miscible with alcohols, ether, esters, benzene, toluene, glacial acetic acid

Flash point: 180°F. (approximate)

1 p.p.m. $\backsim$ 5.07 mg./cu. meter at 25°C., 760 mm. Hg and 1 mg./liter $\backsim$ 197 p.p.m. at 25°C., 760 mm. Hg

3. Determination in the Atmosphere

Guaiacol vapor can be trapped by passing the samples of contaminated air through methanol or other suitable organic solvent, or through 0.1 to 1N sodium hydroxide. Guaiacol as the sodium salt absorbs strongly in the ultraviolet at a wavelength of 289 mμ and as little as 0.01 mg./100 ml. of solution can be determined when using a 10-cm. cell. Vapor-phase or paper chromatography and mass spectrometry may be useful analytical tools under specific circumstances.

4. Physiological Response

Summary. Guaiacol appears to be about one third as toxic as phenol and to have pharmacological properties quite similar to phenol. It causes muscular weakness, cardiovascular collapse, and paralysis of the vasomotor centers. Medical experience indicates that toxic quantities can be absorbed through the skin quite readily. The material is not especially irritating to the skin but prolonged contact may cause injury, particularly if the skin is abraded. It is severely injurious to the eyes, however. It does not seem likely that the inhalation of vapors would constitute a serious industrial hazard, unless, of course, the material were handled hot or fumes were generated.

Single-Dose Oral. When guaiacol as a 20 per cent solution in corn oil was administered by intubation to rats, the LD_{50} was found to be approximately 1.5 g./kg. (1.0 to 2.0 g./kg.). Deaths occurred within hours after feeding or not at all. Autopsy of surviving animals 24 hours and 2 weeks after feeding revealed no significant internal injury.[90] Solis-Cohen and Githens[91] state that when given by mouth the lethal dose for rabbits was 4 g. and for cats, 60 drops. When given to humans, the material causes irritation, burning pain with vomiting, and diarrhea that may be bloody. A tolerance develops when repeated small doses are given.

Injection. Solis-Cohen and Githens[91] report that when given subcutaneously, the fatal dose for pigeons is 0.2 g., for rabbits 2.5 g., and for guinea pigs and rats, 0.9 g./kg.

Eye Contact.[90] Undiluted guaiacol was severely injurious to the eyes of rabbits, 1 drop causing severe corneal necrosis and severe injury to the conjunctival membranes. One drop of a 10 per cent solution in propylene glycol, however, was only mildly irritating.

Skin Contact. Several repeated 24-hour exposures of intact rabbit skin to undiluted guaiacol were required to cause serious irritation. Injury to abraded skin occurred much more readily. Solis-Cohen and Githens[91] state that guaiacol causes burning and partial loss of sensation similar to that caused by phenol. They point out that contact with impure preparations may result in dermatitis with vesication.

Absorption of acutely toxic amounts of the material did not occur from rather severe repeated exposures to the rabbit skin. However, human experience indicates that it is readily absorbed and that doses larger than 2 g./man may result in chills, sudden fall in temperature, weakness, and even collapse and death due to respiratory failure.[91]

Human Experience. Industrial experience to date has not indicated that guaiacol presents any unusual handling hazards. However, medical experience indicates that the material may be more hazardous to human beings than to lower animals. Solis-Cohen and Githens[91] report that a 9-year old girl died after swallowing 5 ml. and that doses of more than 2 g. are hazardous when applied to the skin. Goodman and Gilman[92] report that its medical use as an expectorant has shown that large doses can cause cardiovascular collapse and that clinical doses sometimes cause gastrointestinal irritation. It would seem from this experience and from the pharmacological action of guaiacol that it behaves in human subjects much like phenol.

5. Hygienic Standard for Industrial Exposure

It would not seem that a hygienic standard for guaiacol would be required. It would be considered wise, however, to avoid vapor concentrations that were

[90] Biochemical Research Laboratory, The Dow Chemical Co., unpublished data.

[91] S. Solis-Cohen and T. S. Githens, *Pharmacotherapeutics*. Appleton, New York, 1928.

[92] L. S. Goodman and A. Gilman, *The Pharmacological Basis of Therapeutics*. 2nd ed., Macmillan, New York, 1955.

irritating to the eyes or respiratory passages. The prevention of topical contact and swallowing would seem quite important in industrial operations.

HYDROQUINONE ETHERS (Hydroquinone Dimethyl Ether (1,4-Dimethoxybenzene); Hydroquinone Monomethyl Ether (p-Methoxyphenol); Hydroquinone Monobenzyl Ether (p-Benzyloxyphenol))

1. Source, Uses, and Industrial Exposures

The monomethyl and dimethyl ethers of hydroquinone are produced commercially by the methylation of hydroquinone with dimethyl sulfate.[93] Hydroquinone monobenzyl ether is prepared from hydroquinone and benzyl bromide, the reaction being carried out in an alcoholic potassium hydroxide solution.[94]

Hydroquinone dimethyl ether is used as an intermediate in the manufacture of certain naphthol dyes and as a weathering agent for paints and plastics. Because of its fragrant odor it is also used to some extent in soap perfumes, lotions, and creams. The monomethyl ether is used as a stabilizing agent in textile lubricating oils and other formulations, and as a chemical intermediate. The monobenzyl ether has been used in the rubber industry as an antioxidant, and medically as a melanin pigment inhibitor for dermal conditions characterized by melanosis.

An exposure to the dusts of these hydroquinone ethers could occur as a result of industrial handling. Since there have been no reports from industry of deleterious effects upon the skin caused by these compounds, normal habits of skin cleanliness must be sufficient. However, from the animal studies, it would appear that hydroquinone monomethyl ether could easily cause skin irritation; human studies with monobenzyl ether indicate that dermatitis could be a problem upon repeated exposure of the skin.

2. Physical and Chemical Properties

The hydroquinone ethers listed above are light-colored solids at room temperature. They are all quite stable upon storage. Hydroquinone dimethyl ether has the fragrance of sweet clover, and the monomethyl ether has an odor of caramel and phenol. These hydroquinone ethers have the following properties:

Hydroquinone ether	Dimethyl	Monomethyl	Monobenzyl
Molecular formula	$C_8H_{10}O_2$	$C_7H_8O_2$	$C_{13}H_{12}O_2$
	$CH_3OC_6H_4OCH_3$	$CH_3OC_6H_4OH$	$C_6H_5CH_2OC_6H_4OH$
Molecular weight	138.16	124.16	200.23
Boiling point	210–212°C.	243°C.	
Melting point	55–57°C.	53°C.	122.5°C.
Specific gravity	1.053		
	(55/55°C.)		

[93] W. H. Shearon, Jr., L. G. Davy, and H. Von Bramer, *Ind. Eng. Chem.*, **44**, 1730 (1952).
[94] *The Merck Index*. 7th ed., Rahway, N. J., 1960.

Solubility	Very slightly sol. in water; sol. in acetone and ben- zene	4 g. in 100 ml. of H_2O (25°C.)	Slightly sol. in cold water, sol. in hot water, ace- tone, and other organic solvents
1 p.p.m. ⇌ mg./cu. meter at 25°C., 760 mm. Hg	5.65	5.08	8.19
1 mg./liter ⇌ p.p.m. at 25°C., 760 mm. Hg	177	197	122

3. Determination in the Atmosphere

There have been no methods described in the literature for the determination of the hydroquinone ethers in air. However, trapping their dusts with filter paper devices and their vapors with organic solvents, and subsequent analysis by chromatography, spectroscopy, or methods suitable for the detection of the aromatic ring should enable one to make an accurate atmospheric analysis.

4. Physiological Response

Summary. Hydroquinone dimethyl ether. This compound is very low in acute oral toxicity to animals, and therefore its acute ingestion in rather large amounts should present no problem in humans. Studies using repeated oral doses were not of sufficient length to draw final conclusions, but it is probable that there would be no effect from the repeated ingestion of the small amounts expected to occur during industrial handling. This compound may cause irritation of the eyes if contacted in undiluted form.

Hydroquinone monomethyl ether. The monomethyl ether of hydroquinone is more toxic to rats when given orally than is the dimethyl ether. The presence of an unsubstituted hydroxyl group on the monomethyl ether evidently is the cause of its increased toxicity. Yet the compound is considered to be only slightly toxic when ingested since its acute oral LD_{50} value for rats was found to be approximately 1.6 g./kg. Industrial handling should cause no problems as far as oral ingestion is concerned. This compound causes considerable damage to the rabbit eye if contacted in undiluted form; therefore eye protection is necessary. Prolonged contact has caused some injury to rabbit skin and absorption through the skin when the compound is in solution has also been noted. Cleanliness must be observed during industrial handling.

Hydroquinone monobenzyl ether. This material is of interest because of its ability to inhibit melanin formation in the skin and to cause depigmentation on local contact. This quality was first noted because of its use as an antioxidant in the rubber used for making gloves. Its use in rubber also elicited considerable contact dermatitis. Therefore skin contact must be controlled. Humans have consumed rather large amounts of this compound orally without deleterious effects. Studies of its effect on the eyes have not been reported.

Single-Dose Oral. Hodge et al.[95] found the acute oral LD_{50} values in the

[95] H. C. Hodge, J. H. Sterner, E. A. Maynard, and J. Thomas, *J. Ind. Hyg. Toxicol.,* **31,** 79 (1949).

rat to be about 8.5 g./kg. for hydroquinone dimethyl ether, about 1.6 g./kg. for the monomethyl ether, and about 0.38 g./kg. for hydroquinone itself. There have been no reports concerning the acute toxicology of hydroquinone benzyl ether. However, the oral LD_{50} value of its sodium salt is greater than 3.2 g./kg. for rats.[96]

Single-Dose Intraperitoneal. Hodge et al.[95] also studied the acute intraperitoneal toxicity of the monomethyl and dimethyl ethers of hydroquinone. Table 5

TABLE 5

Acute Intraperitoneal Toxicity of Hydroquinone Ethers
(LD_{50} in mg./kg.)

Species	Monomethyl ether	Dimethyl ether	Hydroquinone
Mouse	725	2000	160–170
Rat	430	1100	170–175
Rabbit	720–970		125–150

lists the LD_{50} values obtained for these two compounds, plus hydroquinone, in three species. It is evident that increased methylation reduces the acute toxicity of hydroquinone considerably, and also that the rat is a species more susceptible to the deleterious effects of the monomethyl ether than the mouse and rabbit. The symptoms of acute intraperitoneal poisoning were those of paralysis and anoxia at lower levels and narcosis at high levels. The acute intraperitoneal toxicity of hydroquinone monobenzyl ether has not been reported.

Repeated-Dose Oral. Hodge et al.[95] reported maintaining rats for 7 weeks on diets containing either the monomethyl or the dimethyl ether of hydroquinone. Slight growth depression resulted in males when the concentration of the monomethyl ether was 0.1 per cent with moderate depression at 2 per cent and marked depression at 5 per cent. In females, a diet containing 0.5 per cent of the dimethyl ether produced questionable effects on growth, with more definite depression when the concentration was 2 per cent or greater. Since there were no other deleterious effects noted at these levels, the authors concluded that the flavor may have reduced the palatability of the ration and thus reduced food intake. Support for this theory is gained by a comparison with the rabbit-feeding studies wherein these compounds at a level of 10 per cent in the diet caused little or no growth depression. However, it must also be remembered that the rat is the species more susceptible to poisoning by intraperitoneal injection of these compounds. There were no deleterious effects noted in dogs fed these compounds in amounts up to 6 g./day; 12 g./day caused only minor body weight losses. The repeated oral feeding studies on hydroquinone monobenzyl ether are even less complete. Denton et al.[97] report that no toxic changes were observed in guinea pigs ingesting this

[96] D. W. Fassett, Laboratory of Industrial Medicine, Eastman Kodak Co., unpublished data.

[97] C. R. Denton, A. B. Lerner, and T. B. Fitzpatrick, *J. Invest. Dermatol.*, **18**, 119 (1952).

compound at a level of 160 mg./kg. daily for 2 months. The only change noted was a depigmentation of the hair. Peck and Sobotka[98] fed the same species approximately 12 g. over a period of 5 months with neither deleterious effects nor depigmentation of skin or hair.

Repeated-Dose Intraperitoneal. No reports concerning the repeated-dose intraperitoneal toxicity of the monomethyl and dimethyl ethers of hydroquinone have been made. Lerner and Fitzpatrick[99] reported that no toxic changes were observed in mice that were injected intraperitoneally with 600 mg./kg. of hydroquinone monobenzyl ether daily for 3 weeks.

Eye Irritation. Hydroquinone monomethyl ether was found to be moderately irritating when applied undiluted to the rabbit eye.[100] Conjunctival irritation, slight to moderate corneal damage, and slight iritis were produced. Immediate irrigation with water reduced the injury appreciably. No experimental work has been reported on the effects on eyes of the dimethyl and monobenzyl ethers of hydroquinone. Since it is suspected that they may cause similar damage to the eye protection is indicated when these compounds are handled industrially.

Skin Irritation and Absorption. Hodge et al.[95] studied the effects of the monomethyl and dimethyl ethers of hydroquinone on rabbit skin. Both were applied at 10 per cent concentration in sun-tan lotion base. Some erythema and escharification developed. The rabbits failed to gain any weight over a 17-day period of application. The use of sun-tan lotion base clouds the conclusion that can be drawn from this study. Further rabbit studies with the monomethyl ether[100] have shown that the undiluted material can cause considerable necrosis if exposure is not terminated in one day. Prolonged contact can cause a severe burn in the rabbit. There was also an indication in this study that the material is absorbed in toxic amounts when in solution, especially through abraded skin. Fassett[96] has found that neither of these two compounds is a skin sensitizer in guinea pigs and that 40 per cent solutions in olive oil and acetone caused only slight or moderate irritation to guinea pig skin on a single 24-hour contact.

Considerable study has been devoted to the effect of hydroquinone monobenzyl ether on human skin since Schwartz and co-workers[101,102] reported that a group of Negro workers developed depigmentation of the hands following the wearing of rubber gloves containing this substance. Peck and Sobotka[98] obtained depigmentation of skin, but not of hair, in guinea pigs upon topical application. This has been confirmed by Fassett.[96] The human studies will be discussed in a later paragraph.

[98] S. M. Peck and H. Sobotka, *J. Invest. Dermatol.*, **4**, 325 (1941).

[99] A. B. Lerner and T. B. Fitzpatrick, *J. Am. Med. Assoc.*, **152**, 577 (1953).

[100] Biochemical Research Laboratory, The Dow Chemical Co., unpublished data.

[101] E. A. Oliver, L. Schwartz, and L. H. Warren, *J. Am. Med. Assoc.*, **113**, 927 (1939).

[102] L. Schwartz, E. A. Oliver, and L. H. Warren, *Public Health Repts. U. S.*, **55**, 1111 (1940).

Metabolism. No direct studies involving the metabolism of these hydro-quinones have been reported. Astill *et al.*[103] concluded that hydroquinone mono-methyl ether is not demethylated by the rat but that it is excreted as a glucuronide. Fassett[96] has predicted that, because of its low chronic toxicity, hydroquinone monobenzyl ether is probably readily metabolized in the same manner as phenol.

Mode of Action. No biochemical studies on the mode of action of these compounds when taken internally have been reported. Schwartz *et al.,*[102] Peck and Sobotka,[98] and Lea[104] have studied the mechanism of action of hydroquinone monobenzyl ether with regard to depigmentation of the skin. The mechanism is not entirely clear, but depigmentation appears to result from some alteration of the tyrosine–tyrosinase reaction or from inhibition of the autoxidation of dihy-droxyphenylalanine.

Human Experience. No problems have been reported with the industrial handling of either the mono- or dimethyl ethers of hydroquinone. Unlike the monobenzyl ether, there have been no cases of dermatitis reported in their com-mercial use. Denton *et al.*[97] and Blank and Miller[105] refer to many cases of der-matitis and leukoderma that were evidently caused by skin contact with rubber that contained hydroquinone monobenzyl ether as an antioxidant. Lerner[99] re-ports a 13 per cent incidence of contact dermatitis in cases where this compound has been used for depigmentation purposes. He has reviewed the use of this com-pound for such treatment and feels it is useful under conditions of correct exposure. Kelly *et al.*[106] found that cancer patients tolerated ingestion of up to 10 g./day of this compound without toxic symptoms. Its use for depigmentation is therefore limited by its ability to cause contact dermatitis rather than by its effects when absorbed internally.

5. Hygienic Standards of Permissible Exposure

There have been no such standards proposed for these hydroquinone ethers and it appears that they are unnecessary.

EUGENOL (1-Allyl-3-methoxy-4-hydroxybenzene, Allyl Guaiacol)

ISOEUGENOL (1-Propenyl-3-methoxy-4-hydroxybenzene)

1. Source, Uses, and Industrial Exposures

Eugenol is a constituent of many essential oils. It occurs in clove oil (80 to 95 per cent), pimento oil (80 per cent), cinnamon leaf oil (95 per cent), and bay oil (60 per cent). It is usually extracted from these natural sources with caustic, the

[103] B. D. Astill, D. W. Fassett, and R. L. Roudabush, *Biochem. J.*, **75**, 543 (1960).

[104] A. J. Lea, *Nature*, **167**, 906 (1951).

[105] I. H. Blank and O. G. Miller, *J. Am. Med. Assoc.*, **149**, 1371 (1952).

[106] K. H. Kelly, H. R. Bierman, and M. B. Shimkin, *Proc. Soc. Exptl. Biol. Med.*, **79**, 589 (1952).

other natural oils and volatile impurities are removed by steam distillation, the salt is neutralized, and then purified by steam distilling it from the nonvolatile impurities. Isoeugenol is made from eugenol by treating with caustic to effect the allylic transformation.[107]

Eugenol has been used as an antipyretic but it is relatively ineffective.[108] It is most widely used as a flavoring agent and in perfumery to impart a carnationlike odor.[108] It has also been used in medicine for the study of mucous secretion and gastric cytology, without gastric resection or gastroenterostomy,[109,110] and it has been shown to have anthelmintic properties.[111] It is also used as a chemical intermediate to produce isoeugenol, which in turn may be used to produce vanillin.[107]

Industrial exposure to either of these materials is most likely to occur by a direct contact of the skin and eyes with the oil or solutions of the oil. Excessive exposure to vapors does not seem likely in view of the low volatility and the pungent odor of these materials when in high concentrations.

2. *Physical and Chemical Properties*

Eugenol and isoeugenol are colorless to slightly yellow liquids. Eugenol has a strongly aromatic odor characteristic of clove and a strong spicy taste. They have the following properties:

	Eugenol	Isoeugenol
Molecular formula	OH C_6H_3(OCH$_3$) CH$_2$CHCH$_2$	OH C_6H_3(OCH$_3$) CHCH$_2$CH$_3$
Molecular weight	164.15	164.15
Boiling point, °C., (760 mm. Hg)	253.2	266
Melting point, °C.	−9.2	
Specific gravity	1.064 (25°C.)	1.091 (15°C.)
Vapor pressure	100 mm. Hg (182.2°C.)	16 mm. Hg (142°C.)
	10 mm. Hg (123°C.)	
	1 mm. Hg (78.4°C.)	
	0.03± mm. Hg (25°C.)	0.02± mm. Hg (25°C.)
		(extrapolated)
Per cent in "saturated" air (approx.)	0.004	0.0026
Refractive index (20°C.)	1.541	1.573
Solubility	Both are slightly soluble in water, soluble in alkalies, and miscible with ether, alcohol, and chloroform	

1 p.p.m. of either ≈ 6.71 mg./cu. meter at 25°C., 760 mm. Hg and
1 mg./liter of either ≈ 149 p.p.m. at 25°C., 760 mm. Hg

[107] P. Z. Bedoukian, *Perfumery Synthetics and Isolates.* Van Nostrand, New York, 1951.

[108] S. Solis-Cohen and T. S. Githens, *Pharmacotherapeutics.* Appleton, New York, 1928.

[109] H. A. Sober, F. Hollander, and E. K. Sober, *Proc. Soc. Exptl. Biol. Med.*, **73**, 148 (1950).

[110] F. V. Lauber and F. Hollander, *Gastroenterology*, **15**, 481 (1950), *Chem. Abstr.*, **44**, 9051 (1950).

[111] G. Valette, R. Cavier, and J. Dehelmas, *Ann. pharm. franç.*, **11**, 649 (1953), *Chem. Abstr.*, **48**, 4181 (1954).

3. Determination in the Atmosphere

No information regarding the collection and determination of eugenol or iso-eugenol in the air has been published. It could be expected, however, that the materials could be trapped by passing the air through $1N$ caustic solution or a suitable organic solvent. Analysis of such solutions by ultraviolet spectrophotometry would be expected to yield highly accurate results even at concentrations as small as 0.005 mg./ml.[112] Other spectrographic and chromatographic methods[113] should also be applicable.

4. Physiological Response

Summary. There is but a limited amount of toxicological information available on eugenol, and none on isoeugenol. Most of that available has been developed in conjunction with the development of eugenol as a mucigogue. Sober *et al.*[109] have reported the oral LD_{50} of eugenol for the rat to be 1.93 g./kg. They observed that paralysis occurred initially in the hindquarters and in the lower jaw, the forelegs being unaffected unless general prostration or coma developed. Deaths were attributed to circulatory collapse. Animals that survived the acute effects remained lethargic, showed kidney injury as manifested by urinary incontinence and sometimes hematuria, and exhibited malfunction of the hind legs for several days. Lauber and Hollander[110] report giving eugenol in 10 oral doses of 0.2 g./kg. each to dogs over a 3-week period without ill effects. Single doses of 0.25 g./kg. sometimes caused vomiting and single doses of 0.5 g./kg. were sometimes fatal. Signs of adverse effect were similar to those observed in the rat.

According to Spector,[114] Binet found the lethal dose for rats to be about 5 g./kg. when given subcutaneously and 0.8 to 1.0 g./kg. when given intraperitoneally.

Human Experience. No cases of adverse effects in industrial handling have been reported.

5. Hygienic Standard of Permissible Exposure

No hygienic standard has been proposed nor does it seem that one should be necessary under reasonable conditions of industrial handling.

BUTYLATED HYDROXYANISOLE (BHA)

$$C_{11}H_{16}O_2 \quad \text{or} \quad CH_3OC_6H_3(OH)C(CH_3)_3$$

1. Uses

BHA is used as an antioxidant for edible fats and oils, essential oils, waxes, coating materials, vitamin A, and so forth. The levels ordinarily used in edible

---

[112] F. Meyer and E. Meyer, *Arch. Pharm.,* **290**, 109 (1957), *Chem. Abstr.,* **51**, 11660 (1957).

[113] K. Hayashi and Y. Hashimoto, *Pharm, Bull. Tokyo,* **4**, 494 (1956), *Chem. Abstr.,* **51**, 13320 (1957).

[114] W. S. Spector, ed., *Handbook of Toxicology,* Vol. 1. Saunders, Philadelphia, 1956.

fats are of the order of 50 to 100 p.p.m. It may be used in conjunction with butylated hydroxytoluene (BHT) or *n*-propyl gallate (NPG). These materials act to reduce rancidity and improve stability by providing reactive hydrogen atoms, which stop the process of free radical formation. Citric acid is frequently added as a metal-scavenging agent in such mixtures.[115]

2. Physical and Chemical Properties

Butylated hydroxyanisole (BHA) is a waxy white solid with a slight odor and a peppery taste. It consists of about 95 per cent 3-*tert*-butyl-4-hydroxyanisole and 5 per cent 2-*tert*-butyl-4-hydroxyanisole and has the following properties:

Molecular weight: 180
Melting point: 48 to 55°C.
Boiling point: 264 to 270°C. (733 mm. Hg)
Solubility: insoluble in water; soluble in fats, acetone, and propylene glycol

3. Determination in the Atmosphere

While there have been few occasions to determine air concentrations during the use of BHA, this can be done by collecting it in dilute caustic solution or in organic solvents and then employing colorimetric methods suitable for phenolic materials, ultraviolet spectrophotometry, or polarographic methods for its analysis.[116]

4. Physiological Response

Summary. The solid, concentrated solutions and heated vapor of BHA are slightly irritating to the skin or eyes. BHA is not a skin sensitizer in guinea pigs or humans. It is low in single-dose oral toxicity and not especially toxic when given to animals in repeated oral doses. The disagreeable taste of foods containing amounts that might cause adverse toxicological effects would probably prevent the accidental ingestion of hazardous quantities of BHA. No health problems have been encountered during its manufacture or industrial use.

Single-Dose Oral. The single-dose oral toxicity is very low. When the material was given to unfasted rats as a corn oil solution or as a water emulsion, the LD_{50} was found to be between 4 and 5 g./kg. When given to fasted rats in propylene glycol solution, the LD_{50} was somewhat lower, about 2.5 g./kg.[117-119]

Repeated-Dose Oral. A number of long-term feeding studies have been carried out. Wilder and Kraybill[117,119] found that rats could tolerate up to 0.12

[115] *Manufacturer's Bulletin.* Eastman Chemical Products, Kingsport, Tenn., 1959.

[116] D. W. Fassett, Laboratory of Industrial Medicine, Eastman Kodak Co., unpublished data.

[117] O. H. M. Wilder and H. R. Kraybill, *Federation Proc.,* **8,** 165 (1949).

[118] J. H. Sterner, S. Ames, and D. W. Fassett, *Federation Proc.,* **8,** 334 (1949).

[119] O. H. M. Wilder and H. R. Kraybill, *Report of American Meat Institute Foundation.* Univ. of Chicago, Chicago, 1948.

per cent in their diet (1200 p.p.m.) for a period of 21 months. No histopathological or carcinogenic effects were found to be associated with its ingestion nor were there any effects on reproduction. Attempts to feed higher levels resulted in refusal to eat, probably due to the unpleasant odor and taste of the diet. There was no evidence of storage of BHA in fatty tissues of these rats. Brown, Johnson, and O'Halloran[120] also found no evidence of toxic effects in rats fed up to 0.1 per cent (1000 p.p.m.) in the diet for 2 years, at 0.5 per cent (5000 p.p.m.), slight initial growth changes and slight liver weight increases without histopathological changes were present.

Graham et al.[121] fed rats a diet consisting of about 75 per cent bread. The bread contained, among other chemical additives, BHA and other antioxidants in an amount 50 times greater than the normal use level. No toxic effects were discernible after a year's feeding. The daily dosage level of BHA was from 3.3 to 7.0 mg./kg. of body weight.

Hodge and Fassett[122] fed dogs up to 100 mg./kg. for a period of over a year. No pathological changes resulted from the treatment and there was no evidence of storage of BHA in brain, liver, kidney, or fat. Dogs fed an antioxidant mixture containing BHA and hydroquinone over long periods also showed no toxic effects.[119]

Wilder, Ostby, and Gregory[123] maintained groups consisting of 4 cocker spaniel pups each for 15 months on diets that contributed a daily dosage of 0, 5, 50, and 250 mg./kg. of BHA. Those animals receiving the highest dosage level ate poorly, probably because of objectionable taste, grew poorly, and exhibited signs of toxicity characterized by the presence of sugar in the urine and definite liver injury. When judged by general appearance, behavior, growth hematological studies, urine examination, and gross and microscopic examination of the tissues, all the other groups of animals were normal.

Metabolism. The metabolism of BHA in the rabbit has been studied by Dacre and Denz[124] and in the rat by Astill, Fassett, and Roudabush.[125] The metabolic pattern is similar in these species, the principal excretion product in the urine being a glucuronide ester. The excretion is relatively rapid and complete even at high dose levels. Recent studies have shown a similar fate in humans.[126] Wilder et al.[123] showed that in the dog BHA is conjugated and excreted both as the glucuronide and as the ethereal sulfate.

[120] W. D. Brown, A. R. Johnson, and M. W. O'Halloran, *Australian J. Exptl. Biol. Med. Sci.*, **37,** 533 (1959).

[121] W. D. Graham, H. Teed, and H. C. Grice, *J. Pharm. and Pharmacol.*, **6,** 534 (1954).

[122] H. C. Hodge and D. W. Fassett, University of Rochester and Eastman Kodak Co., unpublished data.

[123] O. H. M. Wilder, P. C. Ostby, and B. R. Gregory, *J. Agr. Food. Chem.*, **8,** 504 (1960).

[124] J. C. Dacre and F. A. Denz, *Biochem. J.*, **64,** 777 (1956).

[125] B. D. Astill, D. W. Fassett, and R. L. Roudabush, *Biochem. J.*, **75,** 543 (1960).

[126] B. D. Astill, J. Mills, and D. W. Fassett, Laboratory of Industrial Medicine, Eastman Kodak Co., unpublished experiments, 1960.

5. *Hygienic Standard for Permissible Exposure*

It does not seem likely that such a standard is necessary for industrial exposure. The United States Food and Drug Administration[127] has established 200 p.p.m. as the maximum amount of BHA that can legally be present in fats and oils intended for human consumption.

VANILLIN (3-Methoxy-4-hydroxybenzaldehyde, Vanillic Aldehyde, Methylprotocatechuic Aldehyde)

ETHYL VANILLIN (3-Ethoxy-4-hydroxybenzaldehyde, Vanillal, Ethylprotocatechuic Aldehyde, Burbonal)

1. *Source,[128] Uses,[128] and Industrial Exposures*

Vanillin occurs widely in nature, especially in the vanilla plant, and usually in the form of its glucosides. Vanillin is formed in the vanilla bean by hydrolysis of the glucoside by endogenous enzymes. Vanilla extract is prepared by extracting the cured vanilla bean with 50 to 60 per cent ethanol. Vanillin, as such, is not produced commercially by extraction from natural sources. It has been produced synthetically by various routes, usually starting directly with eugenol, isoeugenol, guaiacol, or similar materials.[128] More recently, however, considerable quantities are being produced from sulfite liquor wastes generated in the wood-processing industry.

Information regarding the production of ethyl vanillin is lacking in the readily available literature. However, there have been numerous patents issued that describe the various procedures. A few of these are cited[129–133] for the benefit of those who are concerned.

Both vanillin and ethyl vanillin are used as flavoring agents and in perfumery, the ethyl vanillin being about four times as strong in flavor. Both are also used as chemical intermediates.

Industrial exposure is generally by direct contact. Neither material presents any serious hazards to health in industrial handling.

2. *Physical and Chemical Properties[128]*

Vanillin and ethyl vanillin are white solids having similar and strongly characteristic odors and tastes and the following properties:

[127] G. P. Larrick, *Federal Register,* **24,** 9369 (1959).

[128] P. Z. Bedoukian, *Perfumery Synthetics and Isolates.* Van Nostrand, New York, 1959.

[129] E. Mather and W. E. Hamer, Brit. Pat. 453,482 (Sept. 4, 1936), *Chem. Abstr.,* **31,** 1039 (1937).

[130] E. Mather and W. Hamer, U. S. Pat. 2,199,748 (May 7, 1940), *Chem. Abstr.,* **34,** 5854 (1940).

[131] F. Boedecker and H. Volk, U. S. Pat. 2,062,205 (Nov. 24, 1936), *Chem. Abstr.,* **31,** 701 (1937).

[132] G. A. Kirkhgof, Russ. Pat. 44,929 (Nov. 1935), *Chem. Abstr.,* **32,** 2961 (1938).

[133] C. N. Geneff, U. S. Pat. 2,154,979 (Apr. 18, 1939), *Chem. Abstr.,* **33,** 5415 (1939).

	Vanillin	Ethyl vanillin
Molecular formula	$C_8H_8O_3$	$C_9H_{10}O_3$
	$(CH_3O)C_6H_3(OH)CHO$	$(C_2H_5O)C_6H_3(OH)CHO$
Molecular weight	152.1	166.2
Boiling point, °C., (760 mm. Hg)	285	
Melting point, °C.	82–83.5	77–78
Specific gravity	1.056	
Vapor pressure	10.0 mm. Hg (154°C.)	
	1.0 mm. Hg (107°C.)	
	0.0022 mm. Hg (25°C.)	
Per cent in "saturated" air	0.00029	
Solubility		
Water	1 per cent (25°C.)	1 per cent (14°C.)
	6 per cent (80°C.)	
Alcohol, ether	Soluble	Soluble
Glycerin, propylene glycol	Soluble	Soluble
1 p.p.m. $\approx$ mg./cu. meter at 25°C., 760 mm. Hg	6.22	6.78
1 mg./liter $\approx$ p.p.m. at 25°C., 760 mm. Hg	160.8	147.2

3. Determination in the Atmosphere

Vanillin and ethyl vanillin can be trapped from the air by scrubbing with methanol. Methanol solutions can be analyzed by ultraviolet spectrophotometry. Strong absorption bands occur from vanillin at 308 and 278 mμ and from ethyl vanillin at 310 and 278 mμ. Concentration of either in the range of 0.01 mg./100 ml. of methanol can be measured accurately when a 10-cm. cell is used.

There are methods described in the literature for the detection of these materials in foods[134,135] and for the differentiation of vanillin and ethyl vanillin from each other and from related materials.[136–140]

4. Physiological Response

Summary. Vanillin and ethyl vanillin are both sufficiently low in systemic toxicity to present no significant hazards from industrial operation. Information relative to their local effect upon the eyes and skin and their effect when inhaled is lacking. Both materials are pharmacologically active, causing depressed blood pressure, increased respiratory rate, and death due to cardiovascular collapse.

[134] H. Böhme and O. Winkler, *Z. Lebensm. Untersuch. u. Forsch.*, **99**, 22 (1954), *Chem. Abstr.*, **48**, 13111 (1954).

[135] K. G. Bergner and H. Sperlich, *Deut. Lebensm. Rundschau*, **6**, 134 (1951), *Chem. Abstr.*, **45**, 8156 (1951).

[136] W. R. Gailey, *Chemist Analyst*, **39**, 59 (1950).

[137] A. Bevenue and K. T. Williams, *Chemist Analyst*, **41**, 5 (1952).

[138] H. W. Chenoweth, *Ind. Eng. Chem. Anal. Ed.*, **12**, 98 (1940).

[139] H. Nechamkin, *Ind. Eng. Chem. Anal. Ed.*, **15**, 268 (1943).

[140] F. Meyer and E. Meyer, *Arch. Pharm.*, **290**, 109 (1957), *Chem. Abstr.*, 11660 (1957).

Acute Oral. Deichmann and Kitzmiller[141] administered to rabbits by intubation single doses of vanillin and ethyl vanillin as 4 or 5 per cent solutions in milk. They found that doses of 3.0 g./kg. or more of either material were likely to be lethal. Typical signs of intoxication for both materials were increased respiration, lachrymation, dyspnea, collapse, and death in coma. No convulsions were observed.

Others[142] have administered single doses of vanillin as a suspension in corn oil to rats by intubation and have found the LD_{50} to be 2.0 g./kg., with 95 per cent confidence limits of 1.6 to 2.5 g./kg.

Repeated Oral Administration. Deichmann and Kitzmiller[141] report administering approximately 300 mg./kg. of either vanillin or ethyl vanillin intragastrically to rats twice a week for 14 weeks without signs of illness. They also fed groups of 16 rats each, diets containing either vanillin or ethyl vanillin. Those that received approximately 20 mg./kg./day for 126 days of either material showed no evidence of adverse effect. Those groups that received approximately 64 mg./kg./day for 70 days of either material appeared normal and active but growth was depressed and histopathological examination revealed injury of varying severity in the myocardium, kidney, liver, lungs, spleen, and stomach.

In a recent study[142] matched groups of 10 male and 10 female rats, 4 to 6 weeks of age, were maintained for 91 days on diets containing up to 50,000 p.p.m. of vanillin, equivalent to about 2500 mg./kg./day. When judged by appearance, behavior, growth, mortality, final body and organ weights, terminal hematological examination, and histological studies, no adverse effects were detected when the diet contained 3000 p.p.m. vanillin, equivalent to as much as 150 mg./kg./day. At the next higher dosage level, 10,000 p.p.m., mild adverse effects were noted and at 50,000 p.p.m. growth was depressed and the liver, kidneys, and spleen were enlarged.

Although the experimental techniques employed in these two studies[141,142] were somewhat different, it does not seem that such differences would explain the differences in toxicity observed. Thus, it would seem desirable to resolve these differences before final conclusions are reached regarding the repeated oral toxicity of vanillin.

Injection. When either vanillin or ethyl vanillin was administered subcutaneously to rats in single doses as a 4 per cent solution in milk, doses of 1.8 g./kg. or more were likely to be lethal.[141] Caujolle, Meynier, and Moscarella[143,145] gave both materials to dogs by slow intravenous infusion and found the lethal doses to be 1.32 g./kg. for vanillin and 0.76 g./kg. for ethyl vanillin.

Caujolle and Meynier[146] in a similar study on dogs with *o*-vanillin and ethyl

[141] W. Deichmann and K. V. Kitzmiller, *J. Am. Pharm. Assoc.*, **29**, 425 (1940).
[142] Biochemical Research Laboratory, The Dow Chemical Co., unpublished data.
[143] F. Caujolle, D. Meynier, and C. Moscarella, *Compt. rend.*, **236**, 2549 (1950).
[145] F. Caujolle, D. Meynier, and C. Moscarella, *Compt. rend.*, **237**, 765 (1953).
[146] F. Caujolle and D. Meynier, *Compt. rend.*, **238**, 2576 (1954).

o-vanillin found the lethal doses to be 0.43 and 0.36 g./kg., respectively. They also studied the comparative acute intraperitoneal toxicities of vanillin, ethyl vanillin, and their ortho isomers on the mouse and guinea pig. The results are given in Table 6.

TABLE 6

LD$_{50}$ Values from Single Intraperitoneal Injection

Animal	Vanillin (g./kg.)	Ethyl vanillin (g./kg.)	*o*-Vanillin (g./kg.)	Ethyl *o*-vanillin (g./kg.)
Mouse	0.78	0.75	0.40	0.36
Guinea pig	1.19	1.14	0.40	0.32

They observed that all of the materials, particularly *o*-vanillin, were strong convulsive agents and respiratory excitants.

Pharmacology. Deichmann and Kitzmiller,[141] employing rabbits anesthetized with sodium barbital, found that lethal doses of vanillin administered intraperitoneally caused a sudden drop in blood pressure accompanied by a sudden doubling of the respiration rate. The respiration rate rapidly returned to normal and remained normal, whereas the blood pressure continued to fall, finally causing death. With sublethal doses, the effects were the same but not so severe. Ethyl vanillin produced essentially effects of the same type but they occurred more gradually.[141] Essentially the same observations were made by Caujolle and co-workers[143,145] using dogs.

Metabolism. According to Williams,[147] vanillin when fed to rabbits in 2-g. doses was excreted as glucurovanillin (14 per cent) and vanillic acid (70 per cent), of which one third was conjugated with glucuronic and sulfuric acids.

Although no direct studies have been reported on ethyl vanillin, it would be predicted that it would be metabolized in a manner similar to that of vanillin.

Human Experience. Deichmann and Kitzmiller,[141] in their review of the older literature, point out that a case of mass poisoning originally attributed to vanilla-flavored food was, in reality, due to food poisoning of microbiological origin. Aside from this, no adverse effects have been reported. It seems probable that the strong odor and taste of vanillin or ethyl vanillin effectively prevent excessive exposure and the use of excessive quantities as a flavoring agent.

5. *Hygienic Standard for Industrial Exposure*

None has been proposed and none seems necessary.

PHENYL ETHER (Diphenyl Oxide, Phenoxybenzene, Diphenyl Ether)

$$C_{12}H_{10}O \quad \text{or} \quad (C_6H_5)_2O$$

[147] R. T. Williams, *Detoxication Mechanisms.* 2nd ed., Wiley, New York, 1959.

1. Source, Uses, and Industrial Exposures

Phenyl ether is made commercially as a coproduct with phenol when the latter is made by the chlorobenzene process. It is widely used as a heat-transfer agent and is the major component of Dowtherm A heat-transfer medium. It is used in increasing amounts as a chemical intermediate in the production of surface active agents and high-temperature lubricants. It has a geraniumlike odor and is used in perfumery.

In industrial handling, contact with the eyes and skin and exposure to vapors and mists may occur but the consequences of such exposure are not serious.

2. Physical and Chemical Properties

Phenyl ether is a colorless liquid having a low volatility and a geraniumlike odor. It has the following properties:

Molecular weight: 170.2
Boiling point: 257°C. (760 mm. Hg)
Freezing point: 27°C.
Specific gravity: 1.070 (20/4°C.)
Vapor density: 5.86 (air = 1)
Refractive index: 1.579 (25°C.)
Vapor pressure: 0.0213 mm. Hg (25°C., 760 mm. Hg)
Per cent in "saturated" air: 0.0028
Density of "saturated" air: 1.0014 (air = 1)
Solubility: insoluble in water; miscible with alcohol and ether
Flash point: 205°F. (O.C.)
Spontaneous ignition temp.: 1195°F.
1 p.p.m. $\rightleftharpoons$ 7.0 mg./cu. meter at 25°C., 760 mm. Hg and 1 mg./liter $\rightleftharpoons$ 143.6 p.p.m. at 25°C., 760 mm. Hg

3. Determination in the Atmosphere

Phenyl ether can be trapped from the air by scrubbing with methanol or isooctane. Such a solution can be analyzed by ultraviolet spectrophotometry. Experience has shown that concentrations as low as 0.05 mg. of phenyl ether per 100 ml. of methanol can be measured accurately at wavelengths of either 271 or 278 mμ when using a 10-cm. cell.

A procedure based upon the color formed when aromatic compounds react with piperonal chloride in trifluoroacetic acid has been described by Sawicki, Stanley, and Houser.[148] Phenyl ether produces a red color with a maximum absorption at a wavelength of 537 mμ.

4. Physiological Response

Summary. Information regarding the physiological effects of phenyl ether is not extensive. It appears to be low in oral toxicity, not appreciably irritating to

[148] E. Sawicki, T. Stanley, and T. R. Houser. *Chemist Analyst.* **47.** 69, 77 (1958).

the skin, and its vapors do not present a toxicological problem, but may be a nuisance due to disagreeableness.

Acute Oral. Phenyl ether is low in acute oral toxicity for both rats and guinea pigs. When the material was given in single doses by means of a stomach tube, doses of 4.0 g./kg. were required to produce the death of all animals of both species treated. Doses of 1.0 g./kg. were survived by guinea pigs and doses of 2.0 g./kg. were survived by all the rats treated.[149] Pecchiai and Saffiotti[150] report the oral LD$_{50}$ for rats to be 3.99 g./kg. They also report injury to the liver, spleen, kidney, thyroid, and intestinal tract in surviving animals.

Repeated Oral. Pecchiai and Saffiotti[150] apparently have conducted some repeated feeding studies with phenyl ether and report some organic injury, but sufficient details are not available to allow a discussion of the results.

Skin Irritation.[149] Skin irritation tests conducted upon rabbits indicate that the undiluted material is somewhat irritating if exposures are prolonged or repeated. Effects from such an exposure are characterized by erythema and exfoliation, which clears promptly upon cessation of exposure. When the material is diluted, as in perfume compositions, it does not appear to present any hazard of skin irritation.

Inhalation.[149] The vapor toxicity of diphenyl ether has not been determined directly. However, some experiments on materials that consisted largely of this compound have indicated that vapor concentrations that can occur at ordinary room temperatures present no hazard of systemic injury. It should be noted that such concentrations are quite low because of the low vapor pressure of the material and, furthermore, that such concentrations have an odor which may be disagreeable.

Metabolism. Stroud[151] in 1940 concluded that phenyl ether, when injected into the rabbit, was excreted in the urine as *p*-hydroxyphenyl phenyl ether and that none was excreted as the unchanged ether. In 1953 Bray *et al.*[152] confirmed the observations of Stroud that in the rabbit no cleavage of the ether linkage occurs. They also showed the major metabolite to be the *p*-hydroxylated ether, which is excreted unconjugated (15 per cent), and conjugated with glucuronic acid (63 per cent) and sulfuric acid (12 per cent). Another metabolite was also isolated from the rabbit urine and fairly well identified as di(*p*-hydroxyphenyl) ether.

Human Experience. Experience in manufacturing, handling, and selling this material over many years has indicated that the material does not present an appreciable hazard to health as ordinarily used.

[149] Biochemical Research Laboratory, The Dow Chemical Co., unpublished data.

[150] L. Pecchiai and U. Saffiotti, *Med. lavoro,* **48**, 247 (1957), *Chem. Abstr.,* **51**, 15040 (1957).

[151] S. W. Stroud, *Nature,* **146**, 166 (1940).

[152] H. G. Bray, S. P. James, W. V. Thorpe, and M. R. Wasdell, *Biochem. J.,* **54**, 547 (1953).

5. Hygienic Standards of Permissible Exposure

In order to avoid complaints of disagreeable odor and, in some cases, nausea, it probably is necessary to control vapor concentrations to less than 1 p.p.m.

PHENYL ETHER—BIPHENYL EUTECTIC MIXTURE (Dowtherm A)

1. Source, Uses, and Industrial Exposures

The composition is prepared by mixing together the required amounts of phenyl ether and biphenyl. It is used widely in the liquid phase or in the vapor phase as a heat-transfer agent at temperatures below 750°F. Industrial exposure is most likely to be to vapors or mists or to the liquid. The vapor or mists present the greatest problem because in most installations the material is heated and under pressure and hence any small leak gives rise to concentrations in the air that are likely to be very objectionable, frequently nauseating.

2. Physical and Chemical Properties

Dowtherm A heat-transfer medium is a colorless to straw-colored liquid with a distinctive objectionable odor. It is a eutectic mixture consisting of 73.5 per cent phenyl ether and 26.5 per cent biphenyl by weight. It has the following properties:

Molecular weight (av.) : 166
Boiling point: 257.4°C.
Freezing point: 12°C.
Specific gravity: 1.06 (25/25°C.)
Vapor pressure: 144 psig (750°F.) ; approximately 0.08 mm. Hg (25°C.) (extrapolated)
Solubility: insoluble in water; soluble in alcohol, ether, and benzene
Flash point: 255°F. (O.C.)
1 p.p.m. $\backsimeq$ 6.79 mg./cu. meter at 25°C., 760 mm. Hg and 1 mg./liter $\backsimeq$ 147.3 p.p.m. at 25°C., 760 mm. Hg.

3. Determination in the Atmosphere

Dowtherm A can be trapped from air by scrubbing with methanol or isooctane. Ultraviolet spectrophotometry appears to offer the simplest method of analysis of such samples. In methanol, the biphenyl absorbs strongly at 247 mμ and the phenyl ether at 271 and 278 mμ. It is recommended, however, that since the band at 271 mμ is more specific, it be employed if possible. Standards should be prepared with Dowtherm A. A sensitivity of 0.05 mg./100 ml. can be expected.

4. Physiological Response

Summary. The data from oral administration reveal a comparatively low systemic toxicity, sufficiently low that health problems would not be expected in the normal handling and use of Dowtherm A. Neither would a particular problem be anticipated in the food industry, where there might conceivably be accidental contamination of food. The acute toxicity is low so that the probability of harm from ingestion is remote, and the low solubility in aqueous media plus the outstanding odor and taste of Dowtherm A would cause contamination to be readily detected and excessive ingestion unlikely.

The hazard to health from inhalation is low since no adverse effects were detected in animals exposed to vapor concentrations that were quite painful, and approaching saturation. These data, together with considerable experience in use, indicate that vapor concentrations that are not painful or particularly disagreeable present no health hazards.

Skin irritation of more or less mild degree may be expected only from prolonged and repeated contacts. No adverse effect should result from occasional short contacts. The liquid may cause pain and transient irritation of the eyes.

Acute Oral. Rats and guinea pigs were fed by intubation single doses of Dowtherm A as manufactured. Guinea pigs also were fed Dowtherm A that had been in commercial use in a boiler for 5 years. The doses of the unused material permitting survival and causing the death of all rats were 2.0 and 4.4 g./kg., respectively; for guinea pigs these doses were 0.3 and 4.4 g./kg. When the used material was fed the corresponding dosages were 1.0 and 4.4 g./kg., indicating no change in acute oral toxicity as a result of prolonged commercial usage.[153]

Pecchiai and Saffiotti[154] report the LD_{50} for Dowtherm A when administered as a 25 per cent solution in olive oil to be 5.66 ±1.28 g./kg. for the rat. They observed degenerative changes in the liver and kidneys in surviving animals.

Repeated Oral. Small groups of rats were given repeated doses by stomach tube of either unused Dowtherm A or the material that had been in use for five years. In each case doses of 0.5 to 1.0 g./kg. were given 5 days a week until 132 doses were administered. Those rats receiving 0.5 g./kg. daily showed only a very slight depression of growth and slight weight increases of the liver and kidneys; there were no histopathological changes. Those rats receiving 1.0 g./kg. daily showed a moderate depression of growth, with slight to moderate histopathological changes in the kidneys, and slight changes in the liver and spleen. The kidneys showed degeneration of tubules and hyaline cast formation. A striking observation in those rats receiving 1.0 g./kg. was the gross liver enlargement, up to 188 per cent normal, with no morphological abnormalities. There were no changes in the blood cells nor was there any evidence of irritation in the gastrointestinal tract. The results of repeated feeding experiments lasting for 4 weeks indicated that doses of 1.5 g./kg. and larger resulted in deaths, and that doses of the order of 0.1 g./kg. had no detectable adverse effect. Large doses causing death after 2 to 4 doses produced but slight to moderate histopathological change in the kidneys and minor changes in the liver.[153]

Pecchiai and Saffiotti[154] report feeding daily doses of 50 and 100 mg. in the diet (equivalent to 0.25 and 0.5 g./kg./day) and observing moderate degenerative

---

[153] Biochemical Research Laboratory, The Dow Chemical Co., unpublished data.

[154] L. Pecchiai and U. Saffiotti, *Med. lavoro*, **48**, 247 (1957), *Chem. Abstr.*, **51**, 15040 (1957).

changes to the livers and kidneys in about 2 months. These changes were neither increased nor extended in rats fed for periods up to 13 months.

Skin Contact. When applied daily to the ear of the rabbit, undiluted Dowtherm A produced but a slight irritation manifested by hyperemia, edema, exfoliation, hair loss, and enlargement of hair follicles. When applied daily to the rabbit's abdominal skin, with a cloth bandage, undiluted Dowtherm A produced hyperemia and blistering in 3 days. There have been a very few persons who displayed an idiosyncrasy toward Dowtherm A. These cases are so rare that they are of no practical importance.

Eye Contact.[153] Eye contact with liquid Dowtherm A causes pain and conjunctival irritation but no necrosis in the rabbit.

Inhalation.[153] Concentrations of vapor sufficiently high to cause toxic effects from a single exposure of up to 7 hours duration are not attainable.

A group of 12 rats, 8 guinea pigs, and 1 monkey was given repeated 8-hour exposures to an atmospheric concentration of 0.182 mg./liter of Dowtherm A, partly as a vapor and partly as a fine mist. Most of the rats and guinea pigs became ill and many of them died between the twenty-second and thirty-fifth exposure. The remaining animals were autopsied after the thirty-seventh exposure. Most of the effects observed were those of emaciation and starvation as the rats and guinea pigs refused to eat even when not exposed. The study of the organ weights of the animals indicated that there possibly may have been a slight increase in the weight of the liver on a g. per 100 g. body weight basis. The results of blood studies made on the exposed rats showed that no significant changes occurred. It is noteworthy that during the first few exposures the monkey vomited in the chamber and showed other signs of nausea. However, after the feeding time was changed from immediately before exposure to after the exposure, much of the nausea was avoided. At the end of the experiment, the animal appeared to be in good condition. Hematological studies and blood pressure measurements made during the course of the experiment revealed no evidence of adverse effect.

In a subsequent experiment, 15 rats, 9 guinea pigs, 4 rabbits, and 2 monkeys were exposed 7 hours a day, 5 days a week, for a period of 6 months to a vapor concentration of 7 to 10 p.p.m. (0.05 to 0.07 mg./liter). These animals showed no adverse effect as judged by studies of mortality, growth, organ weights, and hematological and histopathological observations. The vapors in these concentrations are extremely nauseating and even painful to the eyes and upper respiratory passages of human subjects.

Metabolism. Both biphenyl and phenyl ether are converted in the body largely to the 4-hydroxy compounds, and are excreted in the urine conjugated with glucuronic and sulfuric acids (see section on Phenyl Ether).

Human Experience. Dowtherm A has been used rather widely in diverse fields for a considerable number of years without causing serious difficulty due to toxicity. There have been occasional cases of mild skin irritation. and complaints in regard to the unpleasantness of the odor.[155]

[155] W. L. Badger, *Ind. Chemist,* **13,** 343 (1937).

5. *Hygienic Standards of Permissible Exposure*

In order to avoid complaints of disagreeable odor and, in some cases, nausea, it probably is necessary to control vapor concentration to less than 1.0 p.p.m. and perhaps to less than 0.1 p.p.m. for some persons. The exact level will depend upon the individuals concerned.

6. *Odor and Warning Properties*

Experience with human subjects has indicated that concentrations of vapors of Dowtherm A ranging from 0.1 to 1 p.p.m. are readily detected by the olfactory organs and are unpleasant, sometimes producing nausea. It would appear, therefore, that warning properties of Dowtherm A are excellent and they will serve to prevent excessive exposure.[153, 155]

BIS(PHENOXYPHENYL) ETHER

1. *Source, Uses, and Industrial Exposures*

Bis(phenoxyphenyl) ether can be prepared by condensation of halogenated phenyl ether and hydroxyphenyl ethers. The choice of the isomers in the starting materials determines the character of the finished product.

These compounds have been found to be useful synthetic high-temperature lubricants and hydraulic fluids.

Industrial exposure is most likely to be from direct contact with the liquid or fume.

2. *Physical and Chemical Properties*

Bis(phenoxyphenyl) ether, mixed isomers, is a viscous water-clear liquid with practically no odor. Bis(p-phenoxyphenyl) ether is a crystalline solid. They have the following properties:

	Mixed isomers[156]	p-Isomer[157]
Molecular formula	$C_{24}H_{18}O_3$	$C_{24}H_{18}O_3$
Structural formula	$C_6H_5OC_6H_5OC_6H_5OC_6H_5$	
Molecular weight	354.4	354.4
Boiling point, °C., (760 mm. Hg)	443	444
(1 mm. Hg)	225	
Melting point, °C.	None	110
Pour point, °F.	21	
Density, g./ml. (26°C.)	1.179	
(320°C.)	0.971	
Flash point, °F.	505	

[156] Biochemical Research Laboratory, The Dow Chemical Co., unpublished data.

[157] M. V. Shelanski and K. L. Gabriel, Wright Air Force Development Center, *Tech. Rept.* No. **59-124,** June, 1959.

Solubility
   Water   Insoluble   Insoluble
   Alcohol   3 per cent
   Benzene and other nonpolar
      solvents   Miscible   Miscible

3. Determination in the Atmosphere

The material probably can be removed from the air by scrubbing through a nonpolar solvent. Analysis of the resultant solution can probably be accomplished by ultraviolet spectrophotometry.

4. Physiological Response

Summary. Bis(phenoxyphenyl) ethers seem to be very low in toxicity by any of the routes likely to be encountered in industrial handling.

Single-Dose Oral.[156] The mixed isomers as a 20 per cent solution in corn oil were fed to rats by intubation. Doses of 4 g./kg. caused no deaths but did cause diarrhea and slight liver and kidney injury apparent 24 hours after feeding.

Repeated-Dose Oral.[156] Matched groups of 5 male and 5 female young adult rats each were maintained for 31 days on diets containing 0, 0.1, 0.3, 1, 3, and 10 per cent of the mixed isomers of bis(phenoxyphenyl) ether. The well-being of these animals was judged by gross appearance and behavior, food consumption, growth, mortality, hematological determinations, final body and organ weights, and gross and microscopic examination of the tissues. On the basis of these criteria the females on the diet containing 0.1 per cent of the test material were unaffected, whereas the males exhibited abnormal livers characterized by cloudy swelling in the central lobular region and slight necrosis. At the 0.3 per cent level the females showed only slightly depressed growth and the males exhibited liver injury. As dosage increased, evidence of adverse effects became more prevalent in both sexes with poor growth, and injury to the liver, kidney, spleen, and testes. No hematological changes occurred even at the highest dosage level.

Eye Contact.[156] When a drop of the undiluted mixed isomer was placed in the rabbit eye it caused only very slight pain, which disappeared very quickly.

Skin Contact—Rabbits. The mixed isomers failed to cause any apparent irritation even when confined to the skin of rabbits for 14 days. Furthermore, there was no evidence observed to indicate that toxic amounts were absorbed.[156] Two 48-hour exposures to the crystalline *p*-isomer, 14 days apart, also failed to cause any irritation.[157]

Skin Contact—Human Subjects. A repeated-insult test essentially as described by Shelanski and Shelanski[158] was conducted by Klauder[159] with the undiluted mixed isomers of bis(phenoxyphenyl) ether on 50 human subjects of mixed racial background. It resulted in neither primary irritation, fatiguing

---

[158] H. A. Shelanski and M. V. Shelanski, *Proc. Sci. Sect. Toilet Goods Assoc.,* **19** (1953).

[159] Biochemical Research Laboratory, The Dow Chemical Co., unpublished data obtained by J. V. Klauder, Consulting Dermatologist, Philadelphia.

reactions, nor allergenic responses in any of the subjects. Shelanski and Gabriel[157] employed the method of Schwartz and Peck[160] to study the crystalline *p*-isomer on 300 human subjects. They observed no primary irritation or evidence of allergenicity.

Human Experience. No cases of adverse effects due to the bis(phenoxyphenyl) ethers have been reported nor would any be expected from ordinary industrial handling and use.

5. Hygienic Standard for Industrial Exposure

None has been proposed and none would seem necessary.

CHLORINATED PHENYL ETHERS (Monochlorodiphenyl Oxide, Dichlorodiphenyl Oxide, etc., through Hexachlorodiphenyl Oxide)

1. Source, Uses, and Industrial Exposures[161]

All of the chlorinated phenyl ethers can be made by the direct chlorination of phenyl ether. The mono- and dichloro materials are of interest as chemical intermediates. The higher chlorinated materials have been used in the electrical industry and have been considered as constituents of high-pressure greases and as plasticizers; their high toxicity, however, has seriously hampered their use. Industrial exposure has been very limited.

2. Physical and Chemical Properties[161]

The chlorinated phenyl ethers vary from water-clear oily liquids of varying viscosities to white to yellowish waxy semisolids as the equivalents of chlorine they contain increase from 1 to 6. Properties of three of the more common chlorophenyl ethers follow:

Chlorinated phenyl ether	Monochloro-	Dichloro-	Hexachloro-
Molecular formula	$C_{12}H_9ClO$	$C_{12}H_8Cl_2O$	$C_{12}H_4Cl_6O$
Molecular weight	204.5	238.9	376.9
Boiling point, °C. (8 mm. Hg)	153	168.2	230–260
Specific gravity	1.19	1.32	1.60
	(25/25°C.)	(25/25°C.)	(20/60°C.)
Refractive index (25°C.)	1.5868	1.5980	1.621
Vapor pressure (25°C.) (extrapolated)	0.007	0.0006	Very low
Percent in "saturated" air (25°C.)	0.0009	0.00008	Low
Solubility			
Water	Insoluble	Insoluble	0.1 g. per 100 g.
Methanol	Soluble	Soluble	11 g. per 100 g.
Ether, aromatic hydrocarbons, and			
most chlorinated solvents	Miscible	Miscible	Miscible
Flash point, °F. (O.C.)	258.8	298.4	None
Ignition temperature, °F.	334.4	464	1163

[160] L. Schwartz and M. S. Peck, *U. S., Public Health Repts.,* **59,** 546 (1944).
[161] Biochemical Research Laboratory, The Dow Chemical Co., unpublished data.

1 p.p.m. ⇌ mg./cu. meter at 25°C., 760 mm. Hg	8.34	9.78	15.4
1 mg./liter ⇌ p.p.m. at 25°C., 760 mm. Hg	119.6	102.3	64.7

3. *Determination in the Atmosphere*

The chlorinated phenyl ethers can be determined by direct combustion technique followed by absorption and titration of the liberated halogen. They also may be scrubbed from the air by appropriate solvents and determined spectrophotometrically using standard ultraviolet methods.

4. *Physiological Response*[161]

Summary. The chlorinated phenyl ethers are a class of synthetic organic compounds that present potential industrial handling hazards. In general, the toxicity of these materials increases with the degree of chlorination. The mono-, di-, and trichlorophenyl ethers do not present serious handling hazards, whereas the higher chlorinated commercial products do. Toxic effects, which may result from exposure to the higher chlorinated derivatives, may appear either as an "acneform" dermatitis, a systemic intoxication, or both. The more common manifestation of excessive exposure to these materials is the skin condition, which develops at the site of exposure, resembles common acne in appearance, and may be accompanied by intensive itching. This condition generally is caused by prolonged or repeated skin contact. Systemic intoxication has not been observed among men handling the chlorinated phenyl ethers so far as the authors are aware. However, toxicological work on laboratory animals has indicated that the effect is cumulative, and that liver damage will result if the material is taken into the body repeatedly. Repeated inhalation of vapors or fumes from the higher chlorinated materials would be most likely to cause such an effect. Special handling precautions should be used, particularly when the higher chlorinated products are handled routinely or when heated.

Single-Dose Oral.[161] The various chlorinated phenyl ethers were fed to guinea pigs by intubation. The results are summarized in Table 7 and show both

TABLE 7

Chlorinated Phenyl Ethers
Summary of Single-Dose Oral Feeding Studies on Guinea Pigs

	After 4 days		After 30 days	
Material	Lethal dose (g./kg.)	Survival dose (g./kg.)	Lethal dose (g./kg.)	Survival dose (g./kg.)
1 X	0.7	0.2	0.6	0.1
2 X	1.3	0.4	1.0	0.05
3 X	2.2	0.4	1.2	0.2
4 X	3.0	0.4	0.05	0.0005
5 X	3.4	1.8	0.1	0.005
6 X	3.6	0.4	0.05	0.005

the marked delayed effect and also the high toxicity of those materials containing 4 or more equivalents of chlorine.

Repeated-Dose Oral.[161] Rabbits were fed by intubation 5 times a week for 4 weeks unless death intervened. The results obtained with the various chlorinated products are summarized in Table 8. In one instance noted, specially purified

TABLE 8

Chlorinated Phenyl Ethers
Results of Repeated Oral Feeding of Rabbits

Material	Dose (g./kg.)	No. of doses	No. of days	Effect
1 X	0.1	19	29	None
2 X	0.1	19	29	Mild liver injury
3 X	0.1	5	12	Death
	0.05	20	29	Slight liver injury
	0.01	20	29	No effect
4 X	0.05	4	10	Death
	0.005	20	29	Severe liver injury
5 X	0.05	8	21	Death
Pentachlorophenyl ether[a]	0.1	20	29	Moderate liver injury No growth
	0.01	20	29	Slight liver injury
	0.001	20	29	No effect
6 X	0.005	8	10	Death
	0.001	20	28	Severe liver injury
	0.0001	20	28	No effect

[a] Highly purified pentachlorophenyl ether.

pentachlorophenyl ether was studied and the results indicated that it was appreciably less toxic than the regular reaction product. Presumably this reflects the presence of some higher chlorinated products in the regular commercial grade material. This probably accounts for the high toxicity of the tetrachloro compound also.

The organic injury in all cases was centered in the liver and was characterized by congestion and varying degrees of fatty degeneration. The other organs did not appear to be injured.

Topical Contact.[161] The effects of the various chlorinated phenyl ethers were studied on rabbits.

All the materials examined produced an irritation when applied to the skin. The more severe exposures (except in the case of hexachlorophenyl ether) produced a necrosis and sloughing. This acute reaction was followed by a hyperplastic reaction of the epithelium. Milder exposures of a more chronic nature, which failed to produce necrosis, did, however, produce the hyperplastic reaction in the epithelium. Progressive chlorination of the phenyl ethers modified the type of reaction produced. The ability to cause the necrotic type of reaction increased with chlorine content, apparently reaching the maximum at four equivalents of chlorine, then declining. Hexachlorophenyl ether produced a marked irritation but never definite necrosis. The ability to induce the epithelial hyperplasia increased throughout the series, the hexachlorophenyl ether being the most active. There was not a marked difference between the tetra-, penta-, and hexachloroderivatives, but there seemed to be a sharp increase in the ability to induce hyperplasia as the degree of chlorination increased from 3 to 4 or more equivalents of chlorine.

Hexachlorophenyl ether does not appear to be absorbed rapidly through the skin, but when exposures were prolonged and repeated there was definite evidence of toxic effects, weight loss and liver injury. It should be noted, however, that the methods employed in these skin application studies were not such as to exclude the possibility of some ingestion.

Human Experience.[161] Experience with human subjects has not been extensive but is sufficient to show conclusively that exposure to even small amounts of hexachlorophenyl ether may result in appreciable acneform dermatitis. No cases of systemic toxicity are known to the authors.

5. *Hygienic Standards for Industrial Exposure*

No hygienic standards for the chlorinated phenyl ethers have been proposed. In the case of those materials having 4 equivalents of chlorine or more it would be the authors' recommendation that special handling procedures be employed to prevent all possible exposure, particularly repeated exposure. The A.C.G.I.H. threshold limit for 1961 of Chlorinated diphenyl oxide is 0.5 mg./cu. meter.

CELLULOSE ETHERS (General)

1. *Source, Uses, and Industrial Exposures*

Methyl- and ethylcellulose are made by reacting cellulose, cotton linters, or wood pulp with sodium hydroxide and either methyl or ethyl chloride. Hydroxyethylcellulose is made by reacting alkali-cellulose with ethylene oxide. Hydroxypropyl methylcellulose is made by reacting alkali-cellulose with methyl chloride and propylene oxide. Carboxymethylcellulose is made by reacting alkali-cellulose with monochloroacetic acid in the presence of excess alkali to obtain the sodium salt, the form most commonly encountered.

Commercial ethylcellulose is the only one of these ethers that is not water-soluble. It is soluble in the common alcohols and many organic solvents. It is used to form films for packaging, in lacquers, in molded plastic articles, and as a binder in many compositions such as pills.

The other cellulose ethers considered herein are soluble in water. They are available commercially in various viscosity ranges and thermal gel points. They are used very widely in many diverse fields as thickening, dispersing, suspending, binding, and sticking agents, and film formers. They are perhaps best known for their many applications in the food, cosmetic, and pharmaceutical industries.

Industrial exposures of all types occur but are of little concern because of the very low toxicity of the cellulose ethers.

2. *Physical and Chemical Properties*

The cellulose ethers—methylcellulose, hydroxypropyl methylcellulose, carboxymethylcellulose, sodium salt, ethylcellulose, and hydroxyethylcellulose—are all formed by reacting alkali-cellulose of predetermined average molecular weight with various materials to form the respective ether.

Each anhydroglucose unit of the cellulose polymer has three free hydroxyl groups that can be etherified. The degree to which this is effected, and the nature of the substituent group influences markedly the physical properties, particularly solubility, of the product. The molecular weight of the alkali-cellulose markedly affects the viscosity of the final product. All of the ethers are odorless, tasteless, and very stable chemically. Properties peculiar to the particular ethers will be discussed individually.

3. *Determination in the Atmosphere*

All of the cellulose ethers may be trapped from the air by filtration through a membrane filter. The water-soluble ethers can also be trapped in cold water. Ethylcellulose can be trapped in organic solvents.

The determination of methylcellulose and ethylcellulose can be accomplished in various media by employing the method of Samsel and McHard[162] for methoxyl and ethoxyl groups. This method is the basis for ASTM methods D1347-56 for methylcellulose and D914-50 for ethylcellulose. The colorimetric methods of Samsel and DeLap[163] and of Kanzaki and Berger[164] for methylcellulose and of Samsel and Aldrich[165] for ethylcellulose also are very useful.

The method of Morgan[166] is generally the basis for determining the hydroxyalkyl ethers of cellulose. It is based upon the hydrolysis of the ether with hydriodic acid, which yields the alkyl iodide and the corresponding olefin. Measurement of the olefin formed can be accomplished by absorbing in Wijs solution with subsequent determination of the iodine number.[167]

Hydroxylpropyl methylcellulose may be determined by employing the method of Samsel and McHard[162] to determine the methoxyl content and the method of Lemieux and Purves[168] as modified[169] for determining hydroxypropyl content.

Carboxymethycellulose can be determined by the anthrone colorimetric method described by Black.[170] It can also be estimated by hydrolyzing with acid

[162] E. P. Samsel and J. S. McHard, *J. Ind. Eng. Chem. Anal. Ed.,* **14,** 750 (1942).

[163] E. P. Samsel and R. A. DeLap, *Anal. Chem.,* **23,** 1795 (1951).

[164] G. Kanzaki and E. Y. Berger, *Anal. Chem.,* **31,** 1383 (1959).

[165] E. P. Samsel and J. C. Aldrich, *Anal. Chem.,* **29,** 574 (1957).

[166] P. W. Morgan, *Ind. Eng. Chem. Anal. Ed.,* **18,** 500 (1946).

[167] The Dow Chemical Co., unpublished data.

[168] R. U. Lemieux and C. B. Purves, *Can. J. Research,* **B-25,** 485 (1947).

[169] The Dow Chemical Co., *Analytical Method* No. **200,** Nov. 17, 1955.

[170] H. C. Black, *Anal. Chem.,* **23,** 1792 (1951).

and determining the resulting glycolic acid by the method of Calkins.[171] ASTM method No. D1439-58T may also be adaptable.

4. Physiological Response

Summary. The cellulose ethers are all very low in toxicity when administered by normal routes. They are not irritating to the skin or other delicate membranes of the body. When swallowed they are not absorbed to any appreciable degree and appear unchanged in the feces. No inhalation studies have been conducted, but exposure of humans to the dust in manufacturing operations over many years has not led to any known adverse effects. Parenteral administration of the water-soluble ethers has led to serious adverse effects in animals and it would seem unwise to administer them by such routes to human subjects.

5. Threshold Limit for Industrial Exposure

No limit has been suggested for any of the cellulose ethers nor does it seem that any is necessary.

METHYLCELLULOSE (Cellulose Methyl Ether, Methocel MC)

1. Source, Uses, and Industrial Exposures

See Cellulose Ethers.

2. Physical and Chemical Properties[172]

Methylcellulose is a white solid, soluble in cold water but insoluble in hot water and most organic solvents. Its aqueous solutions are colorless, clear, odorless, tasteless, neutral in reaction, and they gel at a temperature of about 50°C. Its methoxyl content varies from 26.0 to 33.0 per cent, which corresponds to a "degree of substitution" averaging 1.78. It is generally available in viscosities ranging from 10 cp. to 8000 cp. material when measured in a 2 per cent solution at 20°C. The highest concentrations pourable range from 18 per cent for the 10 cp. material to 2 per cent for the 8000 cp. product.

3. Determination in the Atmosphere

See Cellulose Ethers.

4. Physiological Response

Summary. As a result of extensive toxicological studies conducted both upon animals and human subjects, it may be concluded that methylcellulose is quite innocuous when swallowed or when in contact with the skin or other delicate membranes of the body. On the other hand, the evidence is equally as strong that it should not be administered parenterally.

[171] V. P. Calkins, *Ind. Eng. Chem. Anal. Ed.*, **15**, 762 (1943).
[172] The Dow Chemical Co., *Methocel Handbook*, Form No. **125-239-60**, 1960.

Repeated-Dose Oral. Deichmann and Witherup[173] gave the material at the rate of about 0.44 g./rat/day, part in their food and part in their drinking water, for a period of 8 months without observing any evidence of adverse effect. Bauer, Lehman, and Yonkman[174] fed rats up to 6.2 g./kg./day of methylcellulose for 6 months and observed no adverse effect. Bauer and Lehman[175] report maintaining rats through three generations on a diet containing 5 per cent methylcellulose without any adverse effects, including reproductive function. They also note that methylcellulose did not appear to be absorbed from the intestinal tract or to be hydrolyzed to methanol and cellulose. Machle, Heyroth, and Witherup[176] fed human subjects single doses of 5 or 10 g. of methylcellulose and recovered essentially all of it in the feces.

Parenteral Administration. Heuper, Martin, and Thompson[177] in 1942 pointed out that the intravenous administration of 400 cp. methylcellulose solution was effective in restoring blood volume associated with shock. They considered it to have certain advantages over some other macromolecular substances, but also certain limitations. Heuper[178–180] reported that when methylcellulose was given intravenously to the dog and rabbit it caused hematological reactions, which he designated as "hematologic macromolecular syndrome." He noted that aside from the effect upon the circulating blood, the inability of the body to degrade the substance led to its retention and accumulation in the liver, spleen, lymph nodes, kidney, and vascular walls.

Katzenstein and co-workers[181,182] repeatedly administered low viscosity (15 cp.) methylcellulose intravenously to dogs and observed serious effects characterized by a progressive decrease in the volume of urine and the presence in the urine of albumin, casts, and both red and white blood cells. Nonprotein nitrogen levels of the blood rose sharply but blood pressure was not increased. Generalized necrotizing vascular disease was frequently observed and death was generally due to renal failure.

Palmer and co-workers[183] injected groups of rats intraperitoneally with methylcellulose (400 cp.) twice weekly for 15 weeks. In intact animals they noted grossly enlarged spleens, changes in the bone marrow and cellular elements of the

[173] W. Deichmann and S. Witherup, *J. Lab. Clin. Med.*, **28**, 1725 (1943).

[174] R. Bauer, A. J. Lehman, and F. F. Yonkman, *Federation Proc.*, **3**, 65 (1944).

[175] R. O. Bauer and A. J. Lehman, *J. Am. Pharm. Assoc. Sci. Ed.*, **40**, 257 (1951).

[176] W. Machle, F. F. Heyroth, and S. Witherup, *J. Biol. Chem.*, **153**, 551 (1944).

[177] W. C. Heuper, G. J. Martin, and M. R. Thompson, *Am. J. Surgery*, **56**, 629 (1942).

[178] W. C. Heuper, *Arch. Pathol.*, **33**, 1 (1942).

[179] W. C. Heuper, *Arch. Pathol.*, **33**, 267 (1942).

[180] W. C. Heuper, *Am. J. Pathol.*, **20**, 737 (1944).

[181] R. Katzenstein, M. C. Winternitz, and J. Meneely, *Yale J. Biol. and Med.*, **16**, 561 (1944).

[182] R. Katzenstein, M. C. Winternitz, and J. Meneely, *Yale J. Biol. and Med.*, **16**, 571 (1944).

[183] J. G. Palmer, E. J. Eichwald, G. E. Cartwright, and M. M. Wintrobe, *Blood*, **8**, 72 (1953).

blood, ascites, and infiltration of the spleen, liver, and kidneys with "storage-cell" macrophages. When the methylcellulose was given to previously splenectomized rats similar histological lesions resulted but none of the hematological changes were observed.

Heuper[184] reported recently that when methylcellulose was injected subcutaneously in the rat, it failed to cause a neoplastic response, whereas a number of other water-soluble polymers tested in a similar manner did cause such a response.

5. Threshold Limit for Industrial Exposure

None would seem necessary.

HYDROXYPROPYL METHYLCELLULOSE (Propylene glycol ether of methyl cellulose, Methocel HG)

1. Source, Uses, and Industrial Exposures

See Cellulose Ethers.

2. Physical and Chemical Properties

The pertinent properties of the various hydroxypropyl methylcelluloses are given in Table 9. Aqueous solutions are clear, smooth and viscous. They resemble

TABLE 9

Properties of Hydroxypropyl Methylcellulose

Commercial designation	Methoxyl content (average D.S.[a])	Propylene glycol ether content (average D.S.[a])	Thermal gel point, °C.
Methocel 60 HG	1.75	0.24	55–60
Methocel 65 HG	1.68	0.14	60–65
Methocel 70 HG	1.52	0.10	66–72
Methocel 90 HG	1.25	0.20	about 84

[a] D.S. refers to degree of substitution. Maximum D.S. is 3.0 since there are only three hydroxyl groups available for substitution per anhydrous glucose unit.

methylcellulose in most of their properties but possess higher thermal gel points; specific types possess additional properties, such as increased surface activity and solubility in organic solvents. The character of the gels obtained on heating also varies.

3. Determination in the Atmosphere

See Cellulose Ethers.

[184] W. C. Heuper, A.M.A. Arch. Pathol., **67,** 589 (1959).

4. *Physiological Response*

Summary. The hydroxypropyl methylcelluloses have not been studied as extensively as has methylcellulose. However, the data available and the similarity in chemical nature strongly suggest that they can be expected to behave physiologically like methylcellulose, that is, essentially inert upon direct contact, essentially innocuous when swallowed, and unsuitable for parenteral administration.

Repeated-Dose Oral. Hodge and co-workers [185] studied the oral toxicity of a high-gel-point methylcellulose, Methocel HG, now designated commercially as Methocel 65 HG. They fed groups of 50 male and 50 female rats each for 2 years on diets containing 1, 5, and 20 per cent of the test material. The only effect observed was growth retardation in the males fed the 20 per cent diet. They also fed doses of 0.1, 0.3, 1.0, and 3.0 g./kg./day to groups of 2 dogs each for 1 year without observing any adverse effects. A dog fed 25 g./kg./day for 30 days suffered no apparent ill effects. Another dog fed 50 g./kg./day exhibited some diarrhea, slight weight loss, and slight depression in red blood cell count but no evidence of histological injury was found.

Knight and co-workers[186] fed the same material as Hodge *et al.*[185] to 25 human subjects. Each individual received 3 graduated doses ranging from 0.6 to 8.9 g. each at least 1 week apart. They found that essentially all of the ingested material was eliminated in the feces within 96 hours following single doses of 3.0 to 8.9 g. Recovery of the smaller doses was not attempted. They concluded that the high-gel methylcellulose was similar to methylcellulose in physiological properties.

McCollister and Oyen[187] fed Methocel 2602 (now designated as Methocel 60 HG), to rats as a part of their diet for 121 days. They found that dietary levels of 3 per cent or less caused no adverse effects, that 10 per cent caused slight retardation in the growth of male rats, and that 30 per cent in the diet caused marked growth depression and excessive mortality in both males and females. No specific organic injury occurred even at the highest dose and the adverse effects were considered to be due to poor nutrition caused by the excessive bulk of such high-level diets.

Further dietary feeding studies[188] on rats have been carried out with other hydroxypropyl methylcelluloses, namely, Methocel 70 HG and Methocel 90 HG. The results obtained are almost identical with those described for Methocel 60 HG.

Parenteral Administration. Hodge *et al.*[185] administered Methocel HG intraperitoneally in aqueous solution to rats and mice and found the LD_{50} to be about 5 g./kg. for both species.

[185] H. C. Hodge, E. A. Maynard, W. G. Wilt, Jr., H. J. Blanchett, Jr., and R. E. Hyatt, *J. Pharmacol. Exptl. Therap.*, **99**, 112 (1950).

[186] H. F. Knight, Jr., H. C. Hodge, E. P. Samsel, R. E. DeLap, and D. D. McCollister, *J. Am. Pharm. Assoc. Sci. Ed.*, **41**, 427 (1952).

[187] D. D. McCollister and F. Oyen, *J. Am. Pharm. Assoc. Sci. Ed.*, **43**, 664 (1954).

[188] D. D. McCollister, F. Oyen, and G. K. Greminger, *J. Pharm. Sci.*, **50**, 615 (1961).

5. *Threshold Limit for Industrial Exposure*

None seems necessary.

CARBOXYMETHYLCELLULOSE (Cellulose Glycolic Acid, Sodium Salt; CMC)

1. *Source, Uses, and Industrial Exposures*

See Cellulose Ethers.

2. *Physical and Chemical Properties*[189]

Carboxymethylcellulose, sodium salt, is a white powder that readily dissolves in both hot or cold water to form smooth viscous solutions. It usually contains 0.3 to 1.4 glycolic acid units per anhydrous glucose unit of the cellulose molecule.

3. *Determination in the Atmosphere*

See Cellulose Ethers.

4. *Physiological Response*

Summary. Carboxymethylcellulose is very low in oral toxicity when fed either in single or repeated doses. It is not irritating to the skin or delicate membranes of the body. In the rat and man it passes through the gut unchanged. When given parenterally it causes undesirable effects characterized by deposition in various organs and in the walls of the blood vessels. It has been reported to cause sarcomas when injected subcutaneously in massive repeated doses.

Single-Dose Oral. Rowe et al.[190] reported being unable to feed carboxymethylcellulose and its sodium and aluminum salts in single doses large enough to cause any apparent illness in rats, guinea pigs, and rabbits. Shelanski and Clark,[191] however, succeeded in feeding larger doses of the sodium salt and found the single-dose oral LD_{50} to be 27 g./kg. for rats and 16 g./kg. for guinea pigs. These workers also studied the fate of the sodium salt in the rat and recovered approximately 90 per cent of a single oral dose in the feces.

Repeated-Dose Oral. The first published information regarding the physiological properties of carboxymethylcellulose appeared in 1941. In reports of Brown and Houghton[192] and of Werle,[193] it was concluded that the material is harmless when swallowed. Rowe and co-workers[190] reached the same conclusion. They maintained groups of rats for 201 to 250 days on diets containing 5 per cent of either the free acid, the sodium salt, or the aluminum salt without observing any adverse effects.

[189] Hercules Powder Co., Hercules Cellulose Gums, CMC, Form 835, Copyright, 1958.
[190] V. K. Rowe, H. C. Spencer, E. M. Adams, and D. D. Irish, *Food Research*, **9**, 175 (1944).
[191] H. A. Shelanski and A. M. Clark, *Food Research*, **13**, 29 (1948).
[192] C. J. Brown and A. A. Houghton, *J. Soc. Chem. Ind.*, **60**, 254 (1941).
[193] E. Werle, *Chem. Ztg.*, **65**, 320 (1941); *Chem. Abstr.*, **37**, 2807 (1943).

Shelanski and Clark[191] fed groups of rats doses of 1.0 g./kg./day for 25 months without causing any detectable adverse effects as judged by appearance, growth, urine examinations, hematological studies, fertility studies, and histological examination of numerous tissues. No neoplasms were found in any of the experimental animals. They also fed groups of guinea pigs 1.0 g./kg./day for 1 year and groups of dogs 1.0 g./kg./day for 6 months without causing any adverse effects as judged by appearance, growth, and histological examination of numerous organs.

Ziegelmayer and co-workers[194] reported that rats were unaffected by diets containing up to 14 per cent of carboxymethylcellulose and that the material passes through the digestive tracts of rats and humans unchanged. They observed that, in 3 humans fed 20 to 30 g./day, a depression of protein digestion and an increase in fat digestion occurred. In vegetarian dogs only about one half of that fed could be recovered in the feces and it is postulated that this may be the result of degradation by the bacterial flora of the intestines.

Topical Contact. Rowe *et al.*[190] found carboxymethylcellulose to be nonirritating to the skin of rabbits even upon prolonged and repeated contact. Shelanski and Clark,[193] employing the method of Schwartz *et al.*[195] on 200 human subjects, concluded that the material was neither a primary irritant nor a skin sensitizer. They also found it nonirritating when in contact with the membranes of the external genitalia and the vagina.

Parenteral Administration. Heuper[196] gave carboxymethylcellulose intravenously to dogs and found that single doses caused only mild transitory shifts in the cellular elements of the blood and that repeated doses caused a decrease in hemoglobin and an increase in sedimentation rate. He found the material stored in the Kupffer cells, the reticulum cells of the spleen, the endothelial cells of the glomeruli, and on the walls and large branches of the aorta. In general it appeared that carboxymethylcellulose was tolerated more readily than methylcellulose under similar conditions.

Lusky and Nelson[197] reported in 1957 that sarcomas were produced at the site of repeated subcutaneous injection of aqueous solutions of carboxymethylcellulose. It should be noted, however, that massive doses were given and therefore there must have been considerable local trauma.

5. *Threshold Limit for Industrial Exposure*

None seems necessary.

[194] W. Ziegelmayer, A. Columbus, W. Klausch, and R. Wieske, *Arch. Tierernähr.,* **2,** 35 (1951); *Chem. Abstr.,* **46,** 5680 (1952).

[195] L. Schwartz, L. W. Spolyar, F. U. Gastineau, J. E. Dalton, A. B. Loveman, M. B. Sulzberger, E. P. Cope, and R. L. Baer, *J. Am. Med. Assoc.,* **115,** 906 (1940).

[196] W. C. Heuper, *Am. J. Pathol.,* **21,** 1021 (1945).

[197] L. M. Lusky and A. A. Nelson, *Federation Proc.,* **16,** 318 (1957).

ETHYLCELLULOSE (Cellulose Ethyl Ether, Ethocel)

1. *Source, Uses, and Industrial Exposures*

See Cellulose Ethers.

2. *Physical and Chemical Properties*

Ethylcellulose is not only a white granular solid as normally sold, but it is also available as a clear film or sheet. It is insoluble in water but soluble in common alcohols and numerous organic solvents. Its solubility depends upon its ethoxyl content, which normally ranges from 2.25 to 2.58 degrees of substitution. The melting point range is from 165 to 195°C.

3. *Determination in the Atmosphere*

See Cellulose Ethers.

4. *Physiological Response*

Summary. Ethylcellulose is very inert physiologically.

Repeated Oral Administration. Deichmann and Witherup[198] maintained a group of 80 rats for 8 months on a diet containing 1.2 per cent of ethylcellulose, which amounted to an average dose of 182 mg./rat/day. They found no evidence of adverse effects in any of the rats as judged by appearance, behavior, growth, and gross and microscopic examination of the tissues.

Parenteral Administration. Heuper[199] introduced 500 mg. of powdered ethylcellulose subcutaneously into 25 rats and did not observe a greater than normal incidence of tumors.

5. *Threshold Limit for Industrial Exposure*

None seems necessary.

HYDROXYETHYLCELLULOSE (Natrosol, Cellosize)

1. *Source, Uses and Industrial Exposures*

See Cellulose Ethers.

2. *Physical and Chemical Properties*

Hydroxyethylcellulose is a white powder that dissolves in water to yield clear viscous solutions. It is essentially insoluble in most fats and oils.

3. *Determination in the Atmosphere*

See Cellulose Ethers.

[198] W. Deichmann and S. Witherup, *J. Lab. Clin. Med.,* **28,** 1725 (1943).
[199] W. C. Heuper, *A.M.A. Arch. Pathol.,* **67,** 589 (1959).

4. *Physiological Response*

Summary.   Hydroxyethylcellulose, like the other water-soluble cellulose ethers, is very low in oral toxicity but causes undesirable effects when given parenterally.

Repeated-Dose Oral.   Smyth and co-workers[200] maintained groups of rats for 2 years on diets containing 5, 1, and 0.2 per cent of hydroxyethylcellulose without any adverse effects. Criteria employed included growth, food intake, life span, frequency of extraneous infections, body measurements, kidney and liver weights, litters, hematological examinations, occurrence of neoplasms, and histological examination of numerous organs.

Parenteral Administration.   Hueper[201] gave up to 55 intravenous injections of hydroxyethylcellulose to dogs without causing injury other than that typical of the other water-soluble cellulose ethers. He noted transitory changes in the blood picture and the deposition of the material on the intima of the blood vessels.

5. *Threshold Limits for Industrial Exposure*

None seems necessary.

[200] H. F. Smyth, Jr., C. P. Carpenter, and C. S. Weil, *J. Am. Pharm. Assoc. Sci. Ed.,* **36,** 335 (1947).

[201] W. C. Heuper, *Arch. Pathol.,* **41,** 130 (1946).

CHAPTER XXXIX

Ketones

V. K. ROWE AND M. A. WOLF

I. General Considerations

A. SOURCE, USES, AND INDUSTRIAL EXPOSURES

1. Source of Ketones

Historically, some of the first commercially available ketones were produced by fermentation of grain or by the destructive distillation of wood. Some are still being made in this manner. In recent years, however, more and more of the commercial ketones are made by organic synthesis. In general, the usual method involves the dehydration or the oxidation of an appropriate secondary alcohol using suitable catalysts and conditions. Certain of the more complex ketones, however, are made by special reactions. The petrochemical industry is becoming an increasingly important source of the raw materials for making ketones.

2. Uses

In industrial operations, ketones find three major uses: as solvents for a wide variety of materials, as raw materials or intermediates in organic syntheses, and for special uses such as in perfumes. Their use as solvents, however, accounts for most of the ketones produced. They may be encountered in the manufacture of smokeless powder and other explosives, lacquers, varnishes, lacquer and varnish removers, plastics, rubber, artificial silk and leather, lubricating oils, cosmetics, pharmaceuticals, perfumes, and many organic chemicals. They are used widely as solvents for dyes, oils, fats, tars, waxes, and many natural and synthetic resins and gums. They are to be found in many consumer items such as synthetic coatings, dopes, and adhesives.

3. Industrial Exposures

The ketones are generally quite stable chemically but all of them are flammable to some degree. It is important, therefore, that due consideration be given to this property in their industrial handling.

TABLE
Physical and Chemical

Compound	Acetone	Methyl ethyl ketone	Methyl n-propyl ketone	Methyl n-butyl ketone	Methyl isobutyl ketone
Synonym	Dimethyl ketone	2-Butanone, MEK	2-Pentanone	2-Hexanone	4-Methyl-2-pentanone, Hexone
Structural formula	CH_3 $\mid$ $C{=}O$ $\mid$ CH_3	CH_3 $\mid$ CH_2 $\mid$ $C{=}O$ $\mid$ CH_3	CH_3 $\mid$ CH_2 $\mid$ CH_2 $\mid$ $C{=}O$ $\mid$ CH_3	CH_3 $\mid$ CH_2 $\mid$ CH_2 $\mid$ CH_2 $\mid$ $C{=}O$ $\mid$ CH_3	$(CH_3)_2$ $\mid$ CH $\mid$ CH_2 $\mid$ $C{=}O$ $\mid$ CH_3
Mol. formula	C_3H_6O	C_4H_8O	$C_5H_{10}O$	$C_6H_{12}O$	$C_6H_{12}O$
Mol. wt.	58.08	72.06	86.13	100.16	100.16
B. p., °C.	56.1	79.6	102.2	127.5	115.8
M. p., °C.	−95.6	−86.6	−83.5	−56.9	−83.5
Sp. gr.[a]	0.7911	0.8072 (25/25°C.)	0.8064 (20/4°C.)	0.8072 (25/4°C.)	0.8020
Refractive index (20°C.)	1.3589	1.3814 (15°C.)	1.3895	1.3969 (17.4°C.)	1.3959
Vapor pressure[b]	226.3	100	16.0	3.8	7.5
Per cent in satd. air	29.8	13.2	2.1	0.5	1.0
Vapor density (air = 1)		2.41			
Evaporation rate (ether = 1)	1.9	2.7		8.1	5.6
Flash point, °F.[c]	0	24	45	73	64
Flammability range (% by vol.)					
Lower limit	2.15	1.8	1.55	1.22	1.35
Upper limit	13.0	12.0	8.15	8.0	7.6
Solubility in:					
Water—g. per 100 g.	∞	25.57	5.51	1.64	1.91
Org. solv.	∞	∞	∞	∞	∞
1 p.p.m. ⇌ mg./cu. meter[d]	2.37	2.94	3.52	4.10	4.10
1 mg./liter ⇌ p.p.m.[d]	422	340	284	244	244
Warning properties[e]					
Odor	1–2	2–3	3	3	3
Irritation	0–1	1–2	1–2	2	2
Suggested max. practical working concentration (p.p.m.)	1000	250	200	100	100

1
Properties of Ketones

Methyl n-amyl ketone	Ethyl butyl ketone	Dipropyl ketone	Methyl n-hexyl ketone	Ethyl sec-amyl ketone	Diisobutyl ketone
2-Heptanone	3-Heptanone	Butyrone, 4-Heptanone	2-Octanone	5-Methyl-3-heptanone	2,6-Dimethyl-4-heptanone
CH_3–CH_2–CH_2–CH_2–CH_2–$C{=}O$–CH_3	CH_3–CH_2–CH_2–CH_2–$C{=}O$–CH_2–CH_2–CH_3	CH_3–CH_2–CH_2–$C{=}O$–CH_2–CH_2–CH_3	CH_3–CH_2–CH_2–CH_2–CH_2–CH_2–$C{=}O$–CH_3	CH_3–CH_2–$CH{-}CH_3$–CH_2–$C{=}O$–CH_2–CH_3	$(CH_3)_2$–CH–CH_3–$C{=}O$–CH_2–CH–$(CH_3)_2$
$C_7H_{14}O$	$C_7H_{14}O$	$C_7H_{14}O$	$C_8H_{16}O$	$C_8H_{16}O$	$C_9H_{18}O$
114.18	114.18	114.18	128.21	128.21	142.2
150.6	147.6	143.7	172.9	160.5	168.1
−26.9	−39	−32.6	−16		−5.9
0.8166	0.8164	0.8174 (20/4°C.)	0.8360 (25/25°C.)	0.850 (0/4°C.)	0.8089
1.4073	1.3994	1.4073	1.4161		1.421 (15°C.)
1.6	1.4	1.2	1.2	2.0	2.4
0.21	0.18	0.16	0.16	0.26	0.32
17.4					30.8
120		120	155	135 (open cup)	120 (open cup)
0.43		Insoluble	0.09	Insoluble	Very slight
∞	∞	∞	∞	∞	∞
4.66	4.66	4.66	5.24	5.24	5.81
214	214	214	191	191	172
3		3	3	4	4
2		2	2	2	2
100	100	100	75–100	75	50

TABLE 1

Compound	Diacetone alcohol	Acetonyl acetone	Trimethyl-nonanone	Methyl isopropenyl ketone	Mesityl oxide
Synonym	4-Hydroxy-4-methyl-2-penta-none	2,5-Hexane-dione, α,-β-diacetyl-ethane	2,6,8-Tri-methyl-4-nonanone	2-Methyl-1-butene-3-one	Methyl iso-butenyl ketone, 4-methyl-3-pentene-2-one
Structural formula	$(CH_3)_2$ $\mid$ COH $\mid$ CH_2 $\mid$ C=O $\mid$ CH_3	CH_3 $\mid$ C=O $\mid$ CH_2 $\mid$ CH_2 $\mid$ C=O $\mid$ CH_3	CH_3 $\mid$ $CHCH_3$ $\mid$ CH_2 $\mid$ C=O $\mid$ CH_2 $\mid$ $CHCH_3$ $\mid$ CH_2 $\mid$ $CHCH_3$ $\mid$ CH_3	CH_2 $\parallel$ CCH_3 $\mid$ C=O $\mid$ CH_3	$(CH_3)_2$ $\mid$ C $\parallel$ CH $\mid$ C=O $\mid$ CH_3
Mol. formula	$C_6H_{12}O_2$	$C_6H_{10}O_2$	$C_{12}H_{24}O$	C_5H_8O	$C_6H_{10}O$
Mol. wt.	116.16	114.14	184.19	84.06	98.14
B.p., °C.	169.2	194	207–228	97.7	129.55
M.p., °C.	−42.8 (f.p.)	−9		−53.7	−46.4
Sp. gr.[a]	0.9406	0.9737 (20/4°C.)	0.8165	0.8550	0 8569
Refractive index (20°C.)	1.4242	1.423	1.4273	1.4220	1.444
Vapor Pressure[b]	1.2	1.6	<0.5 (est.)	42	9.5
Per cent in satd. air	0.16	0.21	<0.07 (est.)	5.5	1.25
Evaporation rate (ether = 1)	60	230			8.4
Flash point, °F.[c]	144 (open cup)	174	195 (open cup)	49	90
Flammability range (% by vol.)					
Lower limit					
Upper limit					
Solubility in:					
Water—g. per 100 g.	∞	∞	Insoluble	4.7	Very slight
Org. solv.	∞	∞	∞	∞	∞
1 p.p.m. ⇌ mg./cu. meter[d]	4.74	4.66	7.54	3.44	402
1 mg./liter ⇌ p.p.m.[d]	211	214	133	291	249
Warning properties[e]					
Odor	3	3		5	2
Irritation	2	2		4	2
Suggested max. practical working concentration (p.p.m.)	50	75		0.3	25

[a] Specific gravity is given at 20/20°C. unless otherwise noted.
[b] Vapor pressure in mm. Hg at 25°C.
[c] Closed cup.
[d] At 25°C., 760 mm. Hg.

(*continued*)

	3-Butyn-2-one	3-Pentyn-2-one	Aceto-phenone	Isophorone	Cyclo-hexanone	Methyl-cyclo-hexanone
			Phenyl methyl ketone	3,5,5-Trimethyl-2-cyclohexene-1-one	Pimelic ketone, Anon, Sextone	Methylanon, Sextone B

Structural formulas:

3-Butyn-2-one:
```
CH
‖
C
|
C=O
|
CH₃
```

3-Pentyn-2-one:
```
CH₃
|
C
‖
C
|
C=O
|
CH₃
```

Acetophenone: C₆H₅–C(=O)–CH₃ (phenyl ring with C=O and CH₃)

Isophorone: CH_3 and $(CH_3)_2$ substituted 2-cyclohexene-1-one ring

Cyclohexanone: cyclohexane ring with C=O

Methylcyclohexanone: methyl-substituted cyclohexanone (two isomers shown)

	3-Butyn-2-one	3-Pentyn-2-one	Aceto-phenone	Isophorone	Cyclo-hexanone	Methyl-cyclo-hexanone
	C_4H_4O	C_5H_6O	C_8H_8O	$C_9H_{14}O$	$C_6H_{10}O$	$C_7H_{12}O$
	68.04	82.05	120.2	138.21	98.14	112.17
	83 (approx.)	133 (approx.)	202	215.2	155.6	160–170
		−28.7	20.5 (f.p.)	−8.1 (f.p.)	−45	
	0.8793	0.910	1.03	0.9229	0.9478 (20/4°C.)	0.921 (20/0°C.)
	1.4024	1.141	1.5339	1.4789 (21.5°C.)	1.4500	1.4458
			0.45	0.44	4.5	3.0 (approx.)
			0.06	0.06	0.60	0.4 (approx.)
				200	40.6	
			180	184	143	130
				0.8	1.1 (212°F.)	
				3.8		
			0.55	Very slight	Slight	Very slight
			∞	∞	∞	∞
	2.78	3.35	4.95	5.65	4.02	4.58
	360	298	204	177	249	218
	4	4		2	1–2	1–2
	4	4		2	2	2
				10	100	100

[e] Degree	Odor intensity	Irritation intensity
0	No odor	No irritation
1	Very faint	Faint
2	Faint	Moderate
3	Easily noticeable	Strong
4	Strong	Intolerable
5	Very strong	

Industrial exposure to the common ketones is most likely to be through inhalation of vapors or through contact with the liquids. Swallowing is not likely unless they are stored in improperly labeled containers.

B. PHYSICAL AND CHEMICAL PROPERTIES

The physical and chemical properties of ketones are listed in Table 1.

C. DETERMINATION IN THE ATMOSPHERE

The determination of ketones in the atmosphere has been given little study. A number of authors have described chemical methods for the quantitative determination of ketones in the atmosphere. However, only the method for the estimation of isophorone as described by Kacy and Cope[1] is considered to be specific. Essentially it consists of the spectrographic measurement of the blue color developed when air contaminated with isophorone is scrubbed through glacial acetic acid containing molybdic acid. The method developed by Messinger and modified and described by Jacobs[2] and Patty et al.[3] depends upon the reaction of the ketone or ketones with iodine in an alkaline solution to produce iodoform and acetic acid. After acidifying the solution, the excess free iodine is then titrated with sodium thiosulfate, using starch as an indicator. This reaction, however, proceeds in the presence of most methyl ketones and thus is suitable only for estimating the total concentration of such ketones present. It is not suitable for aliphatic ketones not having a methyl group adjacent to the carboxyl groups or for cyclic ketones.

The method of Morasco,[4] as described by Jacobs,[2] is less sensitive than the Messinger method but can be used for any of the ketones. It depends upon the reaction between the ketone and hydroxylamine hydrochloride to form the ketoxime liberating hydrochloric acid which is titrated with sodium hydroxide in the presence of methyl orange.

The method described by Greenberg and Lester[5] for acetone and acetone bodies in air or biological fluids depends upon the reaction of 2,4-dinitrophenylhydrazine with acetone to form the corresponding phenylhydrazone. The hydrazone is extracted with CCl_4, which is in turn extracted with an alkaline solution that develops color and absorbs light in the 400 to 420 mμ wavelength range. This method, according to Oglesby,[6] is less dependent on time and temperature than most other methods and is readily adaptable to the determination of acetone in the atmosphere, urine, and blood.

Of the physical methods of analysis, vapor phase chromatography, infrared and mass spectrometry are the methods of preference.

[1] H. W. Kacy and R. W. Cope, Am. Ind. Hyg. Assoc. Quart., 16, 55 (1955).

[2] M. Jacobs, The Analytical Chemistry of Industrial Poisons, Hazards, and Solvents. Interscience Publishers, New York, 1949.

[3] F. A. Patty, W. P. Yant, and H. H. Schrenk, U. S. Public Health Repts., 50, 1217 (1935).

[4] M. Morasco, Ind. Eng. Chem., 18, 701 (1926).

[5] L. A. Greenberg and D. Lester, J. Biol. Chem., 154, 177 (1944).

[6] F. L. Oglesby, Tennessee Eastman Kodak Co., personal communication.

The choice of the physical method or methods to use will depend upon the analytical problem at hand. Contaminated atmospheres containing a ketone or a mixture of ketones of known structure as well as other contaminants present a real problem to the analyst. Vapor phase chromatography, mass spectrometry, or infrared techniques may each be used. However, the analysis of air suspected of being contaminated with one or more ketones of unknown structure will present a more difficult problem—one that may well require a combination of the physical methods in order both to identify the ketone or ketones present and to establish the level of contamination in the atmosphere.

It is of interest that both the infrared and the mass spectrometry analytical techniques may be adapted for the continuous monitoring of work atmospheres. Vapor phase chromatography is not suited for such use at the present time.

D. PHYSIOLOGICAL RESPONSE

Industrial exposure to the commonly used ketones has been occurring to an appreciable degree for many years. The lack of reports in the literature of serious injury indicates that these substances do not present serious hazards to health, probably because they have fairly effective warning properties. The vapors of the saturated aliphatic ketones generally are classed as narcotic, but concentrations required to cause frank narcosis are irritating to the eyes and respiratory passages and, hence, are not voluntarily breathed. Lower concentrations, however, which can be breathed without discomfort, may cause impairment of judgment and thereby create a secondary hazard. Generally, toxicity, irritation, and narcotic potency of the aliphatic ketones increase with increasing molecular weight; irritation and toxicity also increase with unsaturation. In liquid form, the common ketones are painful and irritating to the eyes and prolonged and/or repeated contact with the skin may cause detrimental effects. These latter effects are probably due, in part at least, to a defatting action that causes the skin to chap, rendering it susceptible to invasion by other materials. There is little hazard from absorption through the skin with the common ketones. Generally, the hazard of swallowing is slight providing they are kept in properly labeled containers.

Since a few of the common ketones, particularly the cyclic ones, differ in their effect, it is suggested that the discussions of the individual materials be consulted before drawing conclusions as regards their hazard to health.

1. Mode of Action

In general, the common ketones usually exert a narcotic-type action, with death apparently due to their depression of the respiratory center. The initial indications of exposure are those of irritation of the eyes, nose, and throat. This is followed by the development of drowsiness, loss of control, coma, and death. Recovery has been shown to occur in laboratory animals if they are removed to fresh air at any stage short of deep coma, indicating that the ketones are metabolized or excreted rapidly from the body. Excretion may be considered to be primarily through the lungs and by the kidneys. Little is known of the metabolism of the ketones. It is known, however, that ketones in small amounts occur normally in the blood and urine. Depending upon diet, endocrine balance, and other factors, they may accumulate to a considerable extent. Wick, Sherrill, and MacKay[7] report maximum amounts as high as 200 mg. per liter in the blood of human subjects after 4 days of fasting.

Because of the rapid excretion of the saturated aliphatic ketones, their inhalation usually causes only minor systemic effects. The effects most usually seen are those of lung irritation characterized by emphysema, limited tissue congestion usually centering in the liver, kidneys, and sometimes the brain. Irritation of the intestinal tract has been reported also.

Generally, the effects of repeated exposures to the vapors of these ketones have been reported to be those seen on single exposure. The few recorded instances of repeated vapor exposure of humans have shown that the usual effects are those of headache, nausea with vomiting, dizziness, and drowsiness. A feeling of fatigue or excitement may occur, also.

E. HYGIENIC STANDARDS OF PERMISSIBLE EXPOSURE

The hygienic standards of permissible exposure of the ketones are listed in Table 1.

II. Specific Compounds

ACETONE (Dimethyl Ketone, 2-Propanone)

1. *Source, Uses, and Industrial Exposures*

Acetone is produced commercially by destructive distillation of wood; by distillation of calcium acetate; by fermentation of corn products using selected bacteria; and by catalytic oxidation of isopropyl alcohol, cumene, or natural gas.

Acetone is a low-cost industrial solvent and chemical intermediate that is used widely on a large scale. It may be encountered in the manufacture of smokeless powder and explosives; in the lacquer and varnish industry; in the plastics industry; in the manufacture of rubber; in the chemical industry both as a solvent and as an intermediate for chloroform, ketones, iodoform, and sulfonal; in the dyeing industry; in the leather industry in the form of a solvent for cements; in the manufacture of artificial silk and leather; as a solvent for acetylene; and in the production of lubricating oils. It is used as a solvent for many oils and fats, as a remover of stains in clothing, and as a volatile carrier solvent in many applications. It also finds wide usage as a solvent in laboratories.

[7] A. N. Wick, J. W. Sherrill, and E. M. MacKay, *Proc. Soc. Exptl. Biol. Med.*, **45**, 437 (1940).

Due to the high volatility of acetone, inhalation is the most likely avenue of contact in industrial handling that may lead to health problems. Eye and particularly skin contact may occur. Swallowing is not likely to occur because of its sharp and bitter taste. Because of its low flash point, the fire and explosion hazard of acetone is a major factor in its handling.

2. *Physical and Chemical Properties*

Acetone (see Table 1) is a water-clear, very volatile liquid with a peculiar pungent, aromatic, non-residual odor, and a pungent taste. It is considered to be extremely flammable.

3. *Physiological Response*

Summary. Acetone presents a low degree of hazard to health under conditions of industrial handling. It is low in acute oral toxicity, is but very slightly irritating to the skin even after rather severe exposure, and is moderately irritating to the eyes. It may be slowly absorbed through the skin when confined but this appears to be of little practical significance. It is but slightly toxic when inhaled, causing narcosis when inhaled in very high concentration but no serious systemic injury. The principal hazards to health of acetone are associated with the inhalation of the vapors at very high concentrations and with repeated and prolonged extensive skin contact.

Single-Dose Oral. The toxicity by ingestion of acetone for animals is low. Walton *et al.*[8] have reported a lethal and narcotic dose for rabbits of 10 and 7 ml./kg., respectively. Albertoni,[9] as cited by Browning,[10] established a narcotic and lethal dose for dogs of 4 and 8 g./kg., respectively.

Repeated-Dose Oral. Albertoni,[9] as cited by Browning,[10] has reported that acetone taken by mouth by humans in doses of 15 to 20 g. daily for several days produced no ill effects other than slight drowsiness.

Single-Dose Injection. Walton *et al.*[8] have reported that the lethal dose for acetone when given by intravenous injection was 6 to 8 ml./kg. for rats and 4 ml./kg. for rabbits and that the narcotic dose was 2 ml./kg. Rabbits were depressed but not made unconscious when given 5 ml./kg. intramuscularly.

Skin Irritation. Acetone may produce local dermatitis due to its defatting action on the skin. This develops, however, only after repeated prolonged contact. An occasional short exposure should cause no skin irritation.

Skin Absorption. Lazarew and co-workers, as reported by Browning,[10] estimated the amount of acetone absorbed when the foot of the animal was immersed by measuring the amount exhaled and that present in the blood. They regard the danger of absorption through the skin to be small and very unlikely to occur in

[8] D. Walton, E. Kehr, and A. Loevenhart, *J. Pharmacol. Exptl. Therap.*, **33**, 175 (1928).

[9] P. Albertoni, *Arch. exptl. Pathol. Pharmakol.*, **18**, 219 (1884).

[10] E. Browning, *Toxicity of Industrial Organic Solvents.* Chemical Publishing, New York, 1953.

industry. Cesaro and Pinerolo[11] and Oglesby, Williams, Fassett, and Sterner[12] agree.

Eye Irritation. Carpenter and Smyth[13] report that acetone in small amounts caused moderate irritation when introduced onto the eyes of rabbits. Larson and co-workers[14] demonstrated mild edema and Gomer[15] suggested that the injurious effect of acetone on the eye was caused by dehydration of the sclera which resulted in gelatinous flocculation and opacity of the sclera. Complete healing of contaminated eyes is to be expected.

TABLE 2
Summary of Results of Single Exposures of Animals
to the Vapors of Acetone

Animal	Conc. of acetone		Hours exposed	Effects
	mg./liter	p.p.m.		
Mice	110	46,000	1	Fatal
	48	20,256	1.5	Narcotic
	40	16,880	3.0	Narcotic
Rats	300	126,600	1.75–2.25	Fatal
	100	42,200	4.5–5.5	Fatal
	100	42,200	1.75–2.0	Loss of corneal reflex
Guinea pigs	95	40,000	4–8	Dangerous to life
	47	20,000	8–9	Loss of reflexes

Nelson *et al.*[16] report that unacclimated persons experience eye irritation from vapor concentrations of 500 p.p.m. of acetone and that 200 p.p.m. is as much as such persons can tolerate. However, the extensive studies of Oglesby *et al.*[12] made in actual plant operations with acclimated persons show that with such people eye irritation is not a factor until concentrations become greater than 2500 p.p.m.

Inhalation. The effects of the vapors of acetone on various laboratory animals as recorded by various authorities are given in Table 2.[17–19]

[11] A. N. Cesaro and A. Pinerolo, *Med. lavoro,* **38**, 384 (1947), *Ind. Hyg. Dig.,* **12**, Abstr. No. 896 (1948).

[12] F. L. Oglesby, J. E. Williams, D. W. Fassett, and J. H. Sterner, Eastman Kodak Co., Rochester, N. Y., unpublished paper presented at the Annual Meeting of the American Industrial Hygiene Association, Detroit, 1949.

[13] C. P. Carpenter and H. F. Smyth, Jr., *Am. J. Ophthalmol.,* **29**, 1363 (1946).

[14] P. S. Larson, J. K. Finnegan, and H. B. Haag, *J. Pharmacol. Exptl. Therap.,* **116**, 119 (1956).

[15] J. J. Gomer, *Z. Hyg. Infektrouskrankh.,* **130**, 680 (1960), *Ind. Hyg. Dig.,* **15**, Abstr. No. 168 (1951).

[16] K. Nelson, J. F. Ege, Jr., M. Ross, L. E. Woodman, and L. Silverman, *J. Ind. Hyg. Toxicol.,* **25**, 282 (1953).

[17] H. W. Haggard, L. A. Greenburg, and J. M. Turner, *J. Ind. Hyg. Toxicol.,* **26**, 133 (1944).

[18] F. Flury and W. Wirth, *Arch. Gewerbepathol. Gewerbehyg.,* **5**, 1 (1934).

[19] H. Specht, J. W. Miller, P. J. Valaer, and R. R. Sayers, *Natl. Insts. Health Bull. No.* **176**, 36 (1940).

Kagan[20] and Flury and Wirth[18] studied the effects on cats of repeated exposures to acetone vapors. They used doses of 3 to 5 mg./liter (1265 to 2110 p.p.m.) and observed no ill effects other than slight irritation of the eyes and nose. Kagan[20] observed that repeated exposures resulted in an increased tolerance to acetone vapors.

Kagan,[20] in experiments on himself, found that it was impossible to inhale acetone concentrations of 22 mg./liter (9300 p.p.m.) for longer than 5 minutes because of the acute irritation of the throat.

Nelson et al.[16] have reported observing eye, nasal, and throat irritation in unacclimated volunteers subjected to 500 p.p.m. of acetone. Oglesby et al.[12] concluded, as a result of their extensive studies on thousands of workers exposed to acetone in the manufacture of cellulose acetate yarn, that 200 to 400 p.p.m. was detectable only upon immediate contact; that after a short time 700 p.p.m. cannot be detected; and that 2500 to 3000 p.p.m. causes, at most only minor irritation of the eyes and nose.

Mode of Action. Acetone, because of its solubility in water, is readily absorbed into the blood stream and thus is transported rapidly throughout the body. Kagan[20] found that a man breathing an estimated concentration of 22 mg./liter (9300 p.p.m.) for 5 minutes absorbed 71 per cent of the inhaled acetone; 2 men breathing 11 mg./liter (4650 p.p.m.) for 15 minutes absorbed 76 and 77 per cent; 23 to 29 per cent was carried out with the expired air. Since the alveolar air during these short exposures would be expected to be relatively exhausted of acetone, due to its great solubility in blood and body fluids, the per cent of acetone found in the exhaled breath should be a measure of the physiological dead space in the respiratory tract. The average figure for this space, as given by Best and Taylor,[21] is 24 per cent and may vary for individuals and for depth of respiration. Kagan's percentage of acetone expired agrees satisfactorily with Best's figure for dead-air space and we may thus conclude that essentially all the acetone actually reaching the effective area of the lungs was absorbed, as would be expected. Those ketones less soluble in water are naturally less readily absorbed.

Briggs and Schaffer[22] found that the coefficient of distribution of acetone between alveolar air and blood or water was 1:333, expressed in mg./liter. Thus, a workman breathing 1000 p.p.m. (2.3 mg./liter) acetone in air would reach equilibrium when he had attained a blood concentration of approximately 0.77 g./liter. Since the body is about 70 per cent water, if this relation holds true in the tissues throughout the body, there would be an accumulation of approximately 40 g. in the entire body, for a man of average weight. After this level of saturation was reached, the only acetone absorbed would be to replace any amount metabolized or excreted and sufficient to equilibrate water consumed. That this equilibrium is never actually reached even after several days of continuous exposure has been demonstrated by Haggard, Greenberg, and Turner,[17] who found the value 330 for

[20] E. Kagan, *Arch. Hyg.,* **94,** 41 (1924).

[21] C. H. Best and N. B. Taylor, *The Physiological Basis of Medical Practice.* 6th ed., Williams & Wilkins, Baltimore, 1955, p. 364.

[22] A. P. Briggs and P. A. Schaffer, *J. Biol. Chem.,* **48,** 413 (1921).

the coefficient of distribution. These investigators have shown that when men are exposed to moderate amounts of acetone vapor for 8 hours daily with 16 hours away from exposure, the residual amount at the end of the 16-hour period is so small that the accumulation over a period of days is only slightly more than at the end of the first 8 hours, which was on the order of 20 to 30 per cent of the equilibrium figure. A man exercising moderately for 8 hours while inhaling an atmosphere containing 2100 p.p.m. acetone accumulated 330 mg. of acetone per liter of blood and exhibited no distinct symptoms.

It appears likely that analysis of the blood or urine of the exposed persons at the end of the exposure day and week would be helpful indications of the extent of exposure and maximum absorption. Samples of blood or urine taken for the purpose of establishing the absorption of ketones should be taken as soon as possible after termination of exposure and the time elapsed between the termination of exposure and sampling should be noted and considered.

Acetone is excreted rapidly mainly by the lungs. However, some may be excreted through the skin in cases of excessive exposures and in the urine. Parmeggiani and Sassi[23] showed that the excretion of acetone in humans is rapid for 8 hours after a single oral dose but was not complete in 24 hours. They reported that, under conditions of light work and normal diuresis, the ratio of excretion of acetone was approximately 40 to 70 per cent in the breath, 15 to 30 per cent in the urine, and 10 per cent of the total excreted through the skin.

Available data upon the metabolism of acetone suggest that much of it is split to formate and acetate. Price and Rittenberg[24] and Sakami and Lafaye,[25] employing C^{14} tracer techniques, agree that at least 50 per cent of small oral doses is excreted through the lungs as $C^{14}O_2$ in 14 to 24 hours. The finding of C^{14}-methyl in methionine and choline, β-labeled carbon in serine and 1- and 6-labeled carbon in glycogen strongly suggests that degradation involved "formate." The appearance of C^{14}-acetylated materials and 2,5-labeled glycogen is evidence that labeled "acetate" was also formed. Sakami and Rudney[26] believe that acetone is also converted to acetoacetate and to the 3-carbon intermediates of glycolysis.

Human Experience. Cases of acute acetone poisoning are few if one considers the huge volume of acetone that is used. Although occasional nonfatal industrial poisonings have been reported, they have resulted from the inhalation of high (anesthetic) concentrations of acetone. In general, the characteristic effects were ketosis, narcosis of variable degree, and inflammation of the gastrointestinal tract accompanied by vomiting.[10] Ketosis due to acetone is not to be confused with "ketosis" occurring in diabetic acidosis.

[23] L. Parmeggiani and C. Sassi, *Med. lavoro,* **45,** 431 (1954), *Chem. Abstr.,* **49,** 6509 (1955).

[24] T. D. Price and D. Rittenberg, *J. Biol. Chem.,* **185,** 449 (1950).

[25] W. Sakami and J. M. Lafaye, *J. Biol. Chem.,* **187,** 369 (1950).

[26] W. Sakami and H. Rudney, *Brookhaven Symposia in Biol.,* No. **5,** 176 (1952).

Reports of poisonings due to repeated exposures have usually involved a solvent consisting of acetone in combination with other materials. Headache, irritation of the upper respiratory passages and eyes, gastritis and digestive disturbances with resultant loss of weight, anemia, menstrual irregularities, liver enlargement, dryness of skin on the hands, excitement, and fatigue have all been attributed to such exposures.[10,27,28] It is doubtful that these observations can be attributed to acetone.

Oglesby and co-workers[12] have conducted extensive studies on human subjects exposed repeatedly to acetone vapors over a period of up to 15 years. Concentrations averaged up to about 2000 p.p.m. The study includes environmental, medical, laboratory, and statistical studies and represent 21 million man-hours of experience. They concluded that no injury had occurred to any individual. In a recent communication from Dr. Fassett,[29] he states that experience over the 10 years since 1948 has yielded no data which contradicts previous conclusions.

It is obvious from these data that the toxicity of acetone is low; the lethal dose for humans cannot be estimated.

4. Hygienic Standards of Permissible Exposure

The American Conference of Governmental Industrial Hygienists[30] has recommended a threshold limit of 1000 p.p.m. (2400 mg./cu. meter) for acetone. This level seems appropriate, particularly in view of the vast experience of Oglesby et al.[12] with human subjects.

METHYL ETHYL KETONE (2-Butanone, M.E.K.)

1. Source, Uses, and Industrial Exposures

M.E.K. has been produced for many years as a by-product in the distillation of wood. However, in recent years the production has been that of synthesis from 2-butanol and from butylene oxide because of the purer product obtained.

M.E.K. has wide use as a solvent for such materials as cellulose acetate and nitrate, some gums and some resins, lacquers, and varnishes. It is finding increased use in pharmaceutical and cosmetic manufacture and synthetic rubber. It is used in combination with other solvents in dewaxing operations. Like acetone, M.E.K. is used extensively by laboratory workers.

Industrial exposures to M.E.K. are mainly those of inhalation and skin and eye contact. Swallowing is not expected to be a problem in industrial handling. Skin absorption, while it may occur, is not considered to present a problem.

[27] S. Sessa and F. M. Troisi, Folia. Med. Naples, 30, 129 (1947), Chem. Abstr., 41, 7136 (1947).

[28] A. R. Smith and M. R. Mayers, Ind. Bull. N. Y. State Dept. Labor, 23, 174 (1944).

[29] D. W. Fassett, Eastman Kodak Co., personal communication.

[30] American Conference of Governmental Industrial Hygienists, Am. Ind. Hyg. Assoc. J., 22, 325 (1961).

2. *Physical and Chemical Properties*

Methyl ethyl ketone (see Table 1) is a water-clear, highly volatile, flammable liquid with an odor resembling that of acetone.

3. *Physiological Response*

Summary. M.E.K. presents a low degree of hazard in industrial handling. It has a low acute oral toxicity. The liquid may produce moderate skin irritation if the exposures are frequent and prolonged. Such effects may occur also if M.E.K. is confined to the skin. Eye contact with the liquid may result in irritation and transient corneal injury. The vapors may cause marked irritation of the eyes and mucous membranes so that harmful exposures should not occur unless the person is unable to extricate himself from the area. The fire hazard of M.E.K. is considered to be low even though it is highly flammable because of the intolerably irritating nature of the vapors to humans at levels well below the lower limit of flammability.

The threshold for odor is reported to be less than 25 p.p.m. and the threshold for eye and nose irritation about 200 p.p.m. for 50 per cent of unacclimated individuals.[31]

Single Dose Oral. The single dose oral toxicity of methyl ethyl ketone has been determined to be 3.3 g./kg. of body weight for rats. Massive doses may be expected to result in irritation of the lungs and narcosis.[31]

Skin Irritation. Moderate skin irritation has been observed as the result of a 24-hour exposure of rabbits in which the methyl ethyl ketone was confined to the skin.[31] Human experience has shown that dermatitis can result if excessive repeated prolonged skin contact occurs. Minor skin contacts have been shown to cause no evidence of irritation.[31]

Although absorption of methyl ethyl ketone through the skin may occur, it has been shown that it is low in toxicity by this route, the LD_{50} value being greater than 8 ml./kg. of body weight for rabbits.[31]

For industrial hygiene purposes care should be exercised to avoid excessive or repeated skin contact.

Eye Irritation. The liquid is highly irritating to the eyes.[31] Larson and co-workers[32] have reported that methyl ethyl ketone has a greater capacity to cause edema of the cornea than acetone.

Inhalation. The toxicity of methyl ethyl ketone when inhaled has been determined on both rats and guinea pigs. Patty and co-workers[33] have reported that a 10 per cent concentration (100,000 p.p.m.) in air caused no deaths of guinea pigs

---

[31] Shell Chemical Corp., *Ind. Hyg. Bull. Toxicity Data Sheet,* Methyl ethyl ketone, SC 57-109, 1959.

[32] P. S. Larson, J. K. Finnegan, and H. B. Haag, *J. Pharmacol. Exptl. Therap.,* **116,** 119 (1956).

[33] F. A. Patty, H. H. Schrenk, and W. P. Yant, *U. S. Public Health Repts.,* **50,** 1217 (1935).

exposed for a few minutes. An exposure of 1 hour to a 1 per cent concentration (10,000 p.p.m.) was without serious disturbance. At this level, irritation of the eyes and nose occurred soon after the start of the exposure and narcosis was observed in 4 to 5 hours. Smyth[34] has reported that a 2-hour exposure of rats to 2000 p.p.m. caused no deaths. Four of 6 rats exposed to 4000 p.p.m. for a 2-hour period died.

Nelson et al.[35] have reported that 350 p.p.m. of methyl ethyl ketone in the air caused irritation of the eyes, nose, and throat in human subjects. This is well below the levels tolerated without effects by animals.

Mode of Action. According to animal experiments, methyl ethyl ketone in high concentrations causes narcosis, emphysema of the lungs, and congestion of the liver and kidneys. Because it is less soluble than acetone in blood, it is more rapidly excreted through the lungs.

Human Experience. There have been no recorded instances of human illness due to the use of methyl ethyl ketone.[36] Dermatitis has occurred. Elkins,[37] reporting on industrial exposures in Massachusetts, noted that human exposures to 700 p.p.m. of methyl ethyl ketone in the air were without evidence of permanent ill effects. At concentrations of 500 p.p.m. vomiting and nausea occurred; however, it was found that the solvent contained 10 per cent 2-nitropropane. He indicated that concentrations of 300 p.p.m. or higher usually give rise to complaints of headaches, throat irritation, and similar symptoms. Smith and Mayers[38] reported that 2 girls exposed to a mixture of acetone (330 to 496 p.p.m.) and methyl ethyl ketone (398 to 561 p.p.m.) became unconscious or fainted. One girl noted a gastric upset prior to the central nervous system effects. Recovery was complete.

These authors[38] also reported that dermatoses among workers handling methyl ethyl ketone are not uncommon. Several workers exposed both to the liquid and to the vapors at 300 to 600 p.p.m. were found to complain of numbness of the fingers and arms. Two men exposed only to the vapors were reported to have developed dermatoses of the face, thus indicating that vapors as well as the liquid may cause skin problems.

4. Hygienic Standards of Permissible Exposure

The American Conference of Governmental Industrial Hygienists[39] have recommended a threshold limit of 200 p.p.m. for methyl ethyl ketone. This appears to be a reasonable level that should prevent serious toxic effects in humans.

[34] H. F. Smyth, Jr., *Am. Ind. Hyg. Assoc. Quart.,* **17,** 129 (1956).

[35] K. Nelson, J. F. Ege, Jr., M. Ross, L. E. Woodman, and L. Silverman, *J. Ind. Hyg. Toxicol.,* **25,** 282 (1943).

[36] E. Browning, *Toxicity of Industrial Organic Solvents.* Chemical Publishing, New York, 1953.

[37] H. B. Elkins, *The Chemistry of Industrial Toxicology.* 2nd ed., Wiley, New York, 1959.

[38] A. R. Smith and M. R. Mayers, *Ind. Bull. N. Y. State Dept. Labor,* **23,** 174 (1944).

[39] American Conference of Governmental Industrial Hygienists, *Am. Ind. Hyg. Assoc. J.,* **22,** 325 (1961).

METHYL n-PROPYL KETONE (Ethyl Acetone, 2-Pentanone)

1. Source, Uses, and Industrial Exposures

Methyl n-propyl ketone is made primarily by the oxidation of 2-pentanol. It is used as a solvent either alone or in combination with other solvents in much the same places acetone is used.

In industrial handling, exposures most likely to occur are those of inhalation of the vapors and those of skin and eye contact. Swallowing is not expected to present a problem.

2. Physical and Chemical Properties

Methyl n-propyl ketone (see Table 1) is a water-clear liquid with an odor resembling acetone but having a more ethereal character.

3. Physiological Response

Summary. Methyl n-propyl ketone may be considered to present a low degree of hazard in industrial handling. Its toxicological properties have received little investigation except for those involved in breathing of its vapors. It probably will be low in acute oral toxicity, cause dermatitis if skin contact is prolonged and frequently repeated, and cause moderate transient irritation and perhaps corneal injury when in contact with the eye. It is possible that methyl n-propyl ketone may be absorbed through the skin but it should be low in toxicity by this route. The vapors do not present a serious hazard to health from inhalation unless the adequate warning properties are ignored.

Eye Irritation. Yant, Patty, and Schrenk[40] reported that 1500 p.p.m. of methyl n-propyl ketone had a strong odor and was markedly irritating to the eyes and nasal passages of humans.

Inhalation. Yant, Patty, and Schrenk[40] studied the acute effects on guinea pigs of exposure to the vapors of methyl n-propyl ketone. They report that 30,000 to 50,000 p.p.m. of this material was dangerous to the life of guinea pigs after exposures of 30 to 60 minutes. The maximum concentration that caused no serious disturbances in 1 hour was 5000 p.p.m.; for 8 hours without serious disturbances, 2000 p.p.m., and for several hours with but slight or no symptoms, 1500 p.p.m.

Specht et al.[41] also studied the acute toxicity to guinea pigs of the vapors of this material. They reported similar results. In addition to the symptoms reported by Yant et al.,[40] Specht and co-workers noted that the vapors of methyl n-propyl ketone caused progressive general narcosis characterized by depression of body temperature, respiration rate, and heart beat rate, and loss of corneal, auditory, and equilibratory reflexes. Symptoms of general congestion of the tissues were seen at

[40] W. P. Yant, F. A. Patty, and H. H. Schrenk, *U. S. Public Health Repts.*, **51**, 392 (1936).

[41] H. Specht, J. W. Miller, P. J. Valaer, and R. R. Sayers, *Natl. Insts. Health Bull.* No. **176**, 1940.

the levels tested (2500, 5000, 10,000, 20,000 and 40,000 p.p.m.). Smyth[42] observed that a 4-hour exposure to 2000 p.p.m. of methyl n-propyl ketone killed part of a group of rats, thus indicating that rats may be more susceptible than guinea pigs to the vapors.

Yant and co-workers[40] showed that 1500 p.p.m. of this ketone caused complaints of irritation of the nasal passages and that this level was strongly odorous to humans.

Mode of Action. Methyl n-propyl ketone is irritating to the mucous membranes and has a pronounced narcotic effect. Excessive exposure results in depression of such functions as rectal temperature, respiration, and heart rate.

Human Experience. There are no reported instances of poisonings due to methyl n-propyl ketone. Its warning properties appear adequate to prevent voluntary exposure to dangerous vapor concentrations.

4. *Hygienic Standards of Permissible Exposure*

The American Conference of Governmental Industrial Hygienists[43] have recommended a threshold limit for methyl n-propyl ketone of 200 p.p.m. This appears to be a reasonable level, one that is low enough to prevent definite narcosis.

METHYL n-BUTYL KETONE (2-Hexanone)

1. *Source, Uses, and Industrial Exposures*

Methyl n-butyl ketone is produced commercially by the catalyzed reaction of acetic acid and ethylene under pressure followed by distillation to purify the material.

Methyl n-butyl ketone is an effective solvent for nitrocellulose, resins, oils, fats, and waxes. It is used as a solvent for lacquers and in lacquer and varnish removers.

Industrial exposures are mainly to vapors. Skin and eye contact may occur with the liquid as well, but such exposures should not present a serious problem. Absorption through the skin and swallowing are not expected to present a problem in industrial handling.

2. *Physical and Chemical Properties*

Methyl n-butyl ketone (see Table 1) is a water-clear liquid with an odor resembling that of acetone, but more pungent.

3. *Physiological Response*

Summary. Methyl n-butyl ketone presents no unusual hazards to health in industrial handling. It is low in single dose oral toxicity. It is capable of causing

---

[42] H. F. Smyth, Jr., *Am. Ind. Hyg. Assoc. Quart.*, **14**, 129 (1956).

[43] American Conference of Governmental Industrial Hygienists, *Am. Ind. Hyg. Assoc. J.*, **22**, 325 (1961).

mild transient eye irritation, but should cause essentially no irritation to the skin unless contact is repeated and prolonged. Absorption of this ketone through the skin can occur but it is low in toxicity by this route. The inhalation of the vapors of this material may result in eye and upper respiratory tract irritation followed by narcosis if the concentration is high. However, the inhalation of toxic amounts of methyl n-butyl ketone should not occur because of its warning properties at low levels.

The principal hazard in industrial handling is that of inhalation.

Single Dose Oral. Smyth and co-workers[44] have reported that the acute oral toxicity of methyl n-butyl ketone is low. The LD$_{50}$ value for rats is given as 2.59 g./kg. of body weight. Thus, this ketone should not present a hazard from swallowing in industrial handling. Some systemic injury might result, however, if large amounts were swallowed accidentally due to mistaken identity.

Skin Irritation. Smyth et al.[44] indicated in their tests on rabbits that no significant irritation occurred. Because of its fat solvent action, this ketone may be expected to defat the skin with resultant dermatitis if repeated prolonged skin contact should occur.

Skin Absorption. These workers[44] showed that the acute toxicity by skin absorption to rabbits was low. They established an LD$_{50}$ value of 5.99 ml./kg. of body weight for this species.

Eye Irritation. Smyth et al.[44] reported that methyl n-butyl ketone is capable of causing mild irritation, possibly some transient minor corneal injury.

Inhalation. Only the effects of single exposures to rats and guinea pigs as well as to man have been studied. Smyth et al.[44] reported that in exposure to saturated atmospheres, more than 30 minutes were required to cause the death of rats. Exposures to concentrations of 4000 p.p.m. caused no deaths in rats exposed for 4 hours, whereas exposure to concentrations of 8000 p.p.m. for 4 hours resulted in the death of all the rats exposed.

Schrenk and co-workers,[45] based upon their studies with guinea pigs, suggested the following points of reference:

	Concentration (p.p.m.)
Kills in a few minutes	Greater than 20,000
Dangerous to life in 30–60 min.	10,000 to 20,000
Dangerous to life after several hours	4000 to 6000
Maximum concentration for 60 min. without serious disturbances	3000
Maximum concentration for several hours without serious disturbances	1500
Maximum concentration for several hours resulting in only slight or no symptoms	1000

[44] H. F. Smyth, Jr., C. P. Carpenter, C. S. Weil, and U. C. Pozzani, *Arch. Ind. Hyg. Occupational Med.,* **10,** 61 (1954).

[45] H. H. Schrenk, W. P. Yant, and F. A. Patty, *U. S. Public Health Repts.,* **51,** 624 (1936).

In general, the sequence of effects observed were those of eye and nasal irritation followed by narcosis and death. Pathological examinations revealed congestion and hemorrhage in the lungs, slight congestion of the brain, and moderate congestion of the liver and kidneys.

Schrenk et al.[45] exposed human volunteers for several minutes to concentrations of 1000 p.p.m. of methyl n-butyl ketone. At this level, moderate eye and nasal irritation occurred. This was the level shown to be essentially without symptoms to guinea pigs, even after several hours exposure.

Mode of Action. Specht et al.[46] found that methyl n-butyl ketone when inhaled caused progressive narcosis in guinea pigs, characterized by depression of body temperature, respiratory rate, and heart rate. Corneal, auditory, and equilibratory reflexes were abolished. Continued exposure resulted in coma and death.

Human Experience. None reported.

4. *Hygienic Standards of Permissible Exposure*

The American Conference of Governmental Industrial Hygienists[47] has recommended a threshold limit of 100 p.p.m. (410 mg./cu. meter) for methyl n-butyl ketone. This seems a reasonable level, one that certainly should prevent serious effects from exposure.

METHYL ISOBUTYL KETONE (Hexone, 4-Methyl-2-pentanone)

1. *Source, Uses, and Industrial Exposures*

Methyl isobutyl ketone is produced commercially by the selective catalytic hydrogenation of the double bond in mesityl oxide.

Methyl isobutyl ketone finds its chief utility as a solvent in the lacquer industry, for nitrocellulose and cellulose ethers, for various oils, fats, waxes, natural and synthetic gums, and resins. It has also been proposed as a denaturant for alcohol.

Industrial exposures are mainly those of inhalation of vapors and skin and eye contact. Swallowing and absorption through the skin are not expected to present a problem in industrial handling. The fire hazard is sufficient to require the taking of proper precautions.

2. *Physical and Chemical Properties*

Methyl isobutyl ketone (see Table 1) is a water-clear liquid with a pleasant ketonelike odor.

3. *Physiological Response*

Summary. Methyl isobutyl ketone presents a low degree of hazard to health in industrial handling. It has a low single dose oral toxicity. When in contact with

[46] H. Specht, J. W. Miller, P. J. Valaer, and R. R. Sayers, *Natl. Insts. Health Bull.* No. **176,** 1940.

[47] American Conference of Governmental Industrial Hygienists, *Am. Ind. Hyg. Assoc. J.,* **22,** 325 (1961).

the eyes it may cause transient irritation and swelling. Repeated prolonged skin contact may result in a dermatitis. Short or occasional contacts should cause essentially no irritation unless confined to the skin as by clothing or shoes. Methyl isobutyl ketone is apparently not absorbed through the skin in acutely toxic amounts. Inhalation of this ketone, because of its warning properties, does not constitute a health problem. However, its vapors when breathed in high concentration can cause narcosis, even death.

The principal hazard to health in industrial handling will probably be that of inhalation of vapors.

Single Dose Oral. The LD$_{50}$ value of methyl isobutyl ketone for rats has been reported[48] to be 2.08 g./kg. of body weight. Because this ketone is rapidly eliminated from the body no pathological effects are expected.

For industrial hygiene purposes, no special precautions need be taken. It should be recognized, however, that the swallowing of large amounts of methyl isobutyl ketone accidentally or willfully may cause serious systemic effects.

Skin Irritation. In skin tests on rabbits, liquid methyl isobutyl ketone has been shown to cause transient erythema as the result of a single prolonged contact especially when confined to the skin. Repeated daily applications of 10 ml. of this ketone to the skin of rabbits resulted in drying and scaliness of the skin.[48]

Skin Absorption. There are no specific data at hand. However, it can be inferred from the skin tests reported above that methyl isobutyl ketone is probably low in toxicity by this route.

Eye Contact. Undiluted methyl isobutyl ketone when in contact with the eyes of rabbits has been reported to cause swelling and irritation, which subside in a few days.[48]

Inhalation. Mice exposed for 30 minutes to 19,500 p.p.m. of methyl isobutyl ketone became anesthetized. They recovered consciousness rapidly when exposed to fresh air. At higher concentrations (above 20,000 p.p.m.) deep anesthesia was induced with subsequent death of most of the animals. No gross pathology other than congestion of the lungs was seen in the mice that died.[48] Smyth[49] found rats survived a 4-hour exposure to methyl isobutyl ketone at 2000 p.p.m.; death occurred as the result of a 4-hour exposure to 4000 p.p.m.

Specht,[50] working with guinea pigs, found that high concentrations of methyl isobutyl ketone (16,800 p.p.m.) caused immediate symptoms of eye and nasal irritation followed by salivation, drunkenness, narcosis, and death. The breathing of concentrations of 1000 p.p.m. caused a low grade narcosis in 6 hours. Death resulted after a 4-hour exposure to 10,000 p.p.m.

Human volunteers have reported that the vapors of methyl isobutyl ketone at concentrations of 200 p.p.m. caused definite eye irritation. This level was de-

[48] *Shell Chemical Corp., Ind. Hyg. Bull. Toxicity Data Sheet,* Methyl isobutyl ketone, SC 57-113, 1957.

[49] H. F. Smyth, Jr., *Am. Ind. Hyg. Assoc. Quart.,* **17,** 129 (1956).

[50] H. Specht, *U. S. Public Health Repts.,* **53,** 292 (1938).

clared to have an objectionable odor.[51] Another source[48] reports the following results when unconditioned humans were exposed 5 minutes to the vapors of this ketone:

Threshold of odor for all volunteers	Less than 100 p.p.m.
Threshold for eye irritation for 50 per cent of the volunteers	200 to 400 p.p.m.
Threshold for nasal irritation for 50 per cent of the volunteers	400 p.p.m.

Limited studies have been reported on the chronic effects of inhalation of methyl isobutyl ketone.[48] In these studies, mice were given 20-minute daily exposures to 20,000 p.p.m. for 15 days. One death occurred after each of the first 6 days; 3 after the ninth, and one after the tenth day. The 4 remaining mice survived the 15 exposures.

Mode of Action. Specht[50] and Specht *et al.*[52] showed from guinea pig studies that methyl isobutyl ketone when breathed caused narcosis characterized by depression of body temperature, and respiratory and heart rate, as well as inhibition of corneal, auditory, and equilibratory reflexes. Gross and microscopic pathology was very slight. Death was the result of the narcotic action.

Human Experience. Very few instances of human poisonings due to methyl isobutyl ketone have been recorded. Elkins[53] reports that workers exposed to about 100 p.p.m. of this material complained of headache and nausea. A tolerance apparently developed during the working week but was lost over the weekend. In another plant where the exposure was similar only complaints of irritation of the respiratory tract were raised. When exposures were reduced to 20 p.p.m., the complaints were largely eliminated.

There have been no fatalities attributable to methyl isobutyl ketone recorded in the literature.

4. *Hygienic Standards of Permissible Exposure*

The American Conference of Governmental Industrial Hygienists in 1961[54] recommended a threshold limit for methyl isobutyl ketone of 100 p.p.m. (410 mg./cu. meter). Based upon the human experience reported by Elkins,[53] this level may be too high. It certainly is low enough to prevent definite narcosis, but perhaps not low enough to prevent objectionable subjective symptoms.

[51] L. Silverman, H. Schulte, and M. First, *J. Ind. Hyg. Toxicol.*, **28**, 262 (1946).

[52] H. Specht, J. W. Miller, P. J. Valaer, and R. R. Sayers, *Natl. Insts. Health Bull.* No. **176**, 1940.

[53] H. B. Elkins, *The Chemistry of Industrial Toxicology.* 2nd ed., Wiley, New York, 1959.

[54] American Conference of Governmental Industrial Hygienists, *Am. Ind. Hyg. Assoc. J.*, **22**, 325 (1961).

METHYL n-AMYL KETONE (2-Heptanone)

1. Source, Uses, and Industrial Exposures

Methyl n-amyl ketone is produced commercially primarily by the catalytic dehydrogenation of 2-heptanol.

In industrial handling, the main avenue of contact with methyl n-amyl ketone is that of inhalation. Skin and eye contact may occur but little is known of the effects of such contact. Swallowing is not likely to occur.

2. Physical and Chemical Properties

Methyl n-amyl ketone (see Table 1) is a liquid of low volatility and with a marked fruity odor similar to that of isoamyl acetate. It is said to have a pearlike flavor.

3. Physiological Response

Summary. Methyl n-amyl ketone has received little attention from a toxicological viewpoint. Inhalation of the vapors of this ketone have been shown to be capable of serious narcotic effects if concentrations are high enough. It is expected that the health hazard from inhalation is low in degree, however, because of the expected strong odor and irritation to the eyes and nose of these dangerous levels.

Inhalation. Specht et al.[55] conducted single inhalation studies on guinea pigs. From their data, it appears that a concentration of 1500 p.p.m. was irritating to the mucous membranes, 2000 p.p.m. was strongly narcotic, and that 4800 p.p.m. caused narcosis and death in from 4 to 8 hours of exposure. Prolonged exposure to all of these concentrations caused varying degrees of irritation to the mucous membranes, depression of rectal temperature, respiratory rate, and heart-beat rate.

Human Experience. No reports of poisonings due to the handling of methyl n-amyl ketone have appeared in the literature.

4. Hygienic Standards of Permissible Exposure

Although no threshold limit has been suggested for this ketone by the American Conference of Governmental Industrial Hygienists, it would seem that 100 p.p.m. should be low enough to prevent narcosis and could serve as a guide for practical control.

ETHYL BUTYL KETONE (3-Heptanone)

1. Source, Uses, and Industrial Exposures

Ethyl butyl ketone is produced commercially by the hydrogenation of the mixed alcohol condensation product of propionaldehyde and methyl ethyl ketone, or by the catalytic dehydrogenation of 3-heptanol.

---

[55] H. Specht, J. W. Miller, R. J. Valaer, and R. R. Sayers, *Natl. Insts. Health Bull.* No. **176,** 1940.

Ethyl butyl ketone finds use primarily as a solvent for many organic materials and as an intermediate in the making of other organic products. As a solvent, it may be expected to appear in operations where other similar ketones are used.

In industrial handling, the principal sources of exposure are those of inhalation of the vapors and of skin and eye contact. Swallowing should not be a problem.

2. *Physical and Chemical Properties*

Ethyl butyl ketone (see Table 1) is a water-clear, slightly volatile liquid with a typical ketone-type odor.

3. *Physiological Response*

Summary. As with most of the ketones, ethyl butyl ketone presents a low degree of hazard to health when handled industrially. It is low in acute oral toxicity and in toxicity by skin absorption. When in contact with the skin and eyes, it may cause, at most, only mild irritation. Due to its defatting action to the skin, a dermatitis may result from frequently repeated or excessively prolonged skin exposures. The vapors of this ketone, while toxic in high concentrations, are not likely to present a health problem when handling at room temperature, because of the ketone's low volatility. The problem of flammability should not be great unless the material is handled at elevated temperatures.

Single Dose Oral. Smyth, Carpenter, and Weil[56] have reported the LD$_{50}$ value for ethyl butyl ketone to be 2.76 g./kg. of body weight for rats.

Skin Irritation. Ethyl butyl ketone in skin tests on rabbits has been reported to cause, at most, only mild irritation.[56] It is possible, due to its defatting action to the skin, that a dermatitis could result if frequently repeated or excessively prolonged skin contact should occur.

Skin Absorption. Smyth et al.[56] using a modified Draize technique for the study of skin absorption, showed that the LD$_{50}$ value for rabbits for a 24-hour exposure to ethyl butyl ketone was greater than 20 ml./kg. of body weight.

Eye Irritation. In eye tests conducted on rabbits, Smyth and co-workers[56] showed that ethyl butyl ketone may cause mild irritation when in contact with the eyes.

Inhalation. Toxicological information concerning the effect of inhalation of ethyl butyl ketone is limited. Smyth et al.[56] have reported that when rats were exposed for 4 hours to 2000 p.p.m. of ethyl butyl ketone no deaths occurred; at 4000 p.p.m. all the rats died.

Human Experience. No cases of ill effects to humans due to exposures to ethyl butyl ketone have been reported in the literature.

4. *Hygienic Standards of Permissible Exposure*

The limited inhalation data preclude the suggesting of a threshold limit value for ethyl butyl ketone. Until more is known about this material, a level of 100

[56] H. F. Smyth, Jr., C. P. Carpenter, and C. S. Weil, *J. Ind. Hyg. Toxicol.*, **31**, 60 (1949).

p.p.m. (that suggested for methyl n-amyl ketone) may be considered as a guide for engineering control.

DIPROPYL KETONE (Butyrone, 4-Heptanone)

1. Source, Uses, and Industrial Exposures

Dipropyl ketone may be made by the fermentation of wood in the presence of alkali with subsequent distillation of the ketone from the fermentation products. The commercially available dipropyl ketone contains some impurities.

Dipropyl ketone has excellent solvent properties. It finds utility as a solvent for nitrocellulose, raw and blown oils, and for resins. It is used in the manufacture of lacquer and resin finishes.

The principal industrial exposures are those of inhalation of the vapors, and those of skin and eye contact. Swallowing and absorption through the skin probably will not present a health problem.

2. Physical and Chemical Properties

Dipropyl ketone (see Table 1) is a water-clear liquid with low volatility and a pleasant ketone odor.

3. Physiological Response

Summary. There are no published data on the toxicity of dipropyl ketone.

4. Hygienic Standard of Permissible Exposure

On the basis of the bench marks for closely related ketones, a practical working guide of 100 p.p.m. may be postulated as a maximum level for prolonged and repeated exposure to dipropyl ketone.

METHYL n-HEXYL KETONE (2-Octanone)

1. Source, Uses, and Industrial Exposures

Methyl n-hexyl ketone is prepared by the distillation of sodium ricinoleate with caustic soda. It is used as an antiblushing agent for nitrocellulose lacquers, as a chemical reagent, and as a constituent in perfumes to impart an apricot, peach, or plum fragrance in a sweet-type essence or as a carowaylike fragrance in bitter-type essence.

The principal exposure likely to be encountered in industrial handling is that of vapor inhalation. Skin and eye contact may present a minor problem from exposure. Swallowing or absorption through the skin is not expected to present a health hazard in industrial handling. Flammability should not be a problem unless the material is handled at elevated temperatures.

2. Physical and Chemical Properties

Methyl n-hexyl ketone (see Table 1) is a water-clear, low-volatility liquid with an applelike odor and a camphorlike taste.

3. Physiological Response

Summary. Only limited toxicological information has been published on methyl n-hexyl ketone. The available data show that this product presents a low degree of hazard from inhalation.

Inhalation. The effects of single exposures of guinea pigs to the vapors of methyl n-hexyl ketone have been reported by Specht, Miller, Valaer, and Sayers.[57] They indicate that symptoms of eye and nasal irritation developed immediately upon exposure to an essentially saturated atmosphere (1300 p.p.m.). This was followed by muscular weakness after approximately 10 hours and coma after approximately 12 hours of exposure. An exposure of 1 hour was considered to be the maximum to this concentration without serious disturbances to the guinea pigs.

Mode of Action. The data presented by Specht *et al.*[57] suggest that the vapors of methyl n-hexyl ketone, upon single exposures, may act as a narcotic as well as an irritant. As there were no deaths in the animals at the levels studied, the cause of death was not established.

4. Hygienic Standard of Permissible Exposure

Insufficient toxicological data are available to permit the recommendation of a threshold limit value for methyl n-hexyl ketone. A figure of 75 to 100 p.p.m. could probably be used as a practical working guide until enough information is accumulated to provide a basis for a recommendation.

ETHYL sec-AMYL KETONE (5-Methyl-3-Heptanone)

1. Source, Uses and Industrial Exposures

Ethyl sec-amyl ketone is derived primarily from French lavender oil by isolation and purification, but it is also produced synthetically.

Besides its use as a solvent for various purposes and as an organic intermediate, ethyl sec-amyl ketone has been suggested as a synthetic apricot and peach essence in perfumes.

The principal industrial exposure is that of inhalation of vapors. Skin and eye contact may occur but should cause no more than minor health problems unless exposures are frequently repeated or excessive in nature. Neither swallowing nor absorption through the skin of acutely toxic amounts is expected in industrial handling. The problem of flammability should not be great unless this ketone is handled at elevated temperatures.

[57] H. Specht, J. W. Miller, P. J. Valaer, and R. R. Sayers, *Natl. Insts. Health Bull.* No. **176**, 1940.

2. *Physical and Chemical Properties*

Ethyl sec-amyl ketone (see Table 1) is a water-clear, low-volatility liquid with an agreeable, aromatic, penetrating, fruity odor.

3. *Physiological Response*[58]

Summary. Ethyl sec-amyl ketone has a low degree of hazard to health in industrial handling. It has a low acute oral toxicity. Eye contact may result in mild irritation accompanied by mild transient corneal injury. Occasional skin contacts with the liquid ethyl sec-amyl ketone should cause no irritation. However, continuous or frequently repeated skin contact may result in some irritation. The inhalation of the vapors may cause irritation to the eyes, nose, and throat so that overexposure should be avoided. However, ethyl sec-amyl ketone, besides being irritating, may cause narcosis with resultant death if highly contaminated air is inhaled.

The principal hazard to health in the handling of ethyl sec-amyl ketone is that of inhalation of its vapors.

Single Dose Oral. The acute oral toxicity on several species has been reported. The LD_{50} values for mice, rats, and guinea pigs were found to be 3.8, 3.5, and 2.5 g./kg. of body weight, respectively. It was noted that local irritation of the gastric mucosa and slight discoloration of the liver, kidneys, and spleen developed.

Injection. When given intraperitoneally in single doses, the LD_{50} for mice was found to be 406 mg./kg. of body weight.

Skin Irritation. Skin irritation tests, using laboratory animals, have shown that ethyl amyl ketone is only a mildly irritating compound. However, continuous or frequently repeated skin contact with the liquid may lead to drying and cracking of the skin.

Eye Contact. Eye contact tests on laboratory animals demonstrated that ethyl amyl ketone is capable of producing mild irritation and some transient corneal injury. Prompt flushing of contaminated eyes reduces the injury.

Inhalation. The results of single dose inhalation studies show that mice and rats, when exposed for 4 hours to saturated air at 25°C. (approximately 3000 p.p.m.), quickly developed signs of irritation of the respiratory tract and of the eyes. Three of the 6 mice exposed and none of the 6 rats exposed died. An exposure of 8 hours to approximately 6000 p.p.m. (saturated air at 35°C.) caused the same symptoms and death of all the mice, but of only 4 of the 6 rats exposed. Varying degrees of ataxia, prostration, and respiratory distress followed by hypnosis were noted. Those animals that survived did not appear to be greatly affected.

The sensory response of unconditioned humans exposed 5 minutes to the vapors of ethyl sec-amyl ketone showed that: 5 p.p.m. will be detected by its odor by most people; 25 p.p.m. has a strong odor and may cause mild irritation of the nose in sensitive people; 50 p.p.m. was considered to be the threshold limit for eye, nose, and throat irritation by the majority of the people; and 100 p.p.m. has a strong local irritation and may cause headache and possibly nausea.

Mode of Action. High concentrations of the vapors of ethyl sec-amyl ketone have a narcotic effect. Nausea, headache, incoordination and a slowing of respiration and heart beat may be expected from excessive exposures. Recovery should be complete and uncomplicated if the terminal stages of anesthesia are not reached.

Human Experience. No incidents of illness caused by the industrial handling of ethyl sec-amyl ketone have been recorded. Experience has shown that workers may complain of odor and transient eye irritation when this ketone is handled in poorly ventilated areas. Prolonged exposure to high concentrations may produce transient nausea and headaches. Limited experience would indicate that concentrations of vapor causing these transient responses are incapable of producing serious systemic effects.

4. *Hygienic Standard of Permissible Exposure*

The limited animal data and human experience are not sufficient to permit the recommending of a threshold limit for ethyl sec-amyl ketone. Based upon human comfort, a level of 75 p.p.m. has been suggested as a guide to a practical working level.[58] This level seems a reasonable one and should prevent definite narcotic effects.

DIISOBUTYL KETONE (2,6-Dimethyl-4-heptanone)

1. *Source, Uses, and Industrial Exposures*

Commercially, diisobutyl ketone is produced primarily by the hydrogenation of phorone.

Diisobutyl ketone has found widespread use in industry as a solvent for nitrocellulose, milled crepe rubber, vinylite, and synthetic coatings, and as a dispersant for organosol-type resins. It is also used as an intermediate in the synthesis of inhibitors, dyes, pharmaceuticals, and insecticides.

Industrial handling of this ketone may lead to vapor exposures and to skin and eye contact with the liquid. Eye contact, swallowing, and absorption are not likely to present a health problem. Flammability of this material is low. It should not present a problem unless it is handled at elevated temperatures.

2. *Physical and Chemical Properties*

Diisobutyl ketone (see Table 1) is a water-clear, low-volatility liquid.

3. *Physiological Response*

Summary. Diisobutyl ketone has a low degree of hazard in industrial handling. It has a low acute oral toxicity. When in contact with the eyes, it should cause no more than minor irritation. Skin contact with the liquid may result in mild irritation or even a dermatitis depending upon the type and duration of contact. The vapors, in high concentrations, exert both an irritant and narcotic ac-

[58] Shell Chemical Corp., *Ind. Hyg. Bull., Toxicity Data Sheet,* SC:57–99, 1958.

tion. Humans find concentrations below those causing narcosis in animals to be sufficiently irritating and discomforting to be offensive. Hence, overexposure to concentrations of diisobutyl ketone that may produce serious effects are not likely.

The principal hazard to health in industrial handling is that of vapor inhalation.

Single Dose Oral. Diisobutyl ketone has been shown by Smyth, Carpenter, and Weil[59] to have a low acute oral toxicity. They report an LD_{50} value for rats of 5.8 g./kg. of body weight.

Skin Irritation. Using rabbits, the above authors[59] found that liquid diisobutyl ketone caused only a mild irritation. Because of its high fat solvency it is suspected that frequently repeated skin contacts may cause a dermatitis.

Skin Absorption. Smyth and co-workers[59] reported the LD_{50} value for rabbits by skin absorption to be greater than 20 ml./kg. of body weight. An exposure of 24 hours was used. This indicates a low degree of toxicity by this route.

Eye Irritation. When the undiluted liquid was instilled into the eyes of rabbits, essentially no irritation developed.[59,60] Diisobutyl ketone thus should present no great hazard from eye contact.

Inhalation. McOmie and Anderson[61] reported that rats and guinea pigs survived single exposures of from $7^1/_2$ to 16 hours to essentially saturated vapors of diisobutyl ketone. Smyth et al.[59] reported a higher degree of toxicity to rats. They found that an 8-hour exposure to a concentration of 2000 p.p.m. (essentially a saturated atmosphere) caused the deaths of 5 of the 6 animals tested.

Silverman, Schulte, and First[62] studied the sensory response of humans to various levels of diisobutyl ketone vapors. They report that concentrations higher than 25 p.p.m. were found to have an objectionable odor and that eye irritation occurred at 50 p.p.m., but no nose and throat irritation was observed at this level. Carpenter, Pozzani, and Weil[63] in studies on 3 human volunteers found that, at levels of 50 and 100 p.p.m., a 3-hour exposure caused slight irritation of the eyes, nose, and throat. A concentration of 100 p.p.m. was considered to be "unsatisfactory," whereas 50 p.p.m. would be "satisfactory," if not ideal.

In studies on the effect of repeated exposures to the vapors of diisobutyl ketone, McOmie and Anderson[61] found that mice survived twelve 3-hour exposures to a saturated atmosphere. Carpenter et al.,[63] in a more extensive study, repeatedly exposed guinea pigs and rats to concentrations ranging from 125 to 1650 p.p.m. These animals were given thirty 7-hour exposures. At the 125 p.p.m. level, no effects could be found in either the rats or guinea pigs. At the 250 p.p.m.

[59] H. F. Smyth, Jr., C. P. Carpenter, and C. S. Weil, *J. Ind. Hyg. Toxicol.*, **31**, 60 (1949).

[60] C. P. Carpenter and H. F. Smyth, Jr., *Am. J. Ophthalmol.*, **29**, 1363 (1946).

[61] W. A. McOmie and H. H. Anderson, Univ. California Publ. Pharmacol., **2**, 217 (1949).

[62] L. Silverman, H. Schulte, and M. First, *J. Ind. Hyg. Toxicol.*, **28**, 262 (1946).

[63] C. P. Carpenter, U. C. Pozzani, and C. S. Weil, *Arch. Ind. Hyg. Occupational Med.*, **8**, 377 (1953).

level, the female rats developed significantly increased liver and kidney weights, whereas the male guinea pigs exhibited a decreased liver weight. At the intermediate levels of 530 and 920 p.p.m. only rats were exposed and the only significant finding was that of increased liver and kidney weights. Increased mortality and kidney and liver injury were observed only in those animals exposed at the 1650 p.p.m. level.

Mode of Action. Diisobutyl ketone is an irritant and causes narcosis at high concentrations. The cause of death, while not established, was probably due to respiratory failure induced by narcotic action on the respiratory center. Lower concentrations of vapors are primarily irritating to the eye and upper respiratory passages. Repeated exposures to concentrations that may be irritating may cause nausea and dizziness. In addition, such exposures may cause liver and kidney effects but it does not seem likely that any such effects would be serious.

Human Experience. There are no records in the literature of human illnesses due to the handling of diisobutyl ketone.[64]

4. Hygienic Standard of Permissible Exposure

Carpenter, Pozzani, and Weil[63] have recommended a tentative threshold limit value of 50 p.p.m. for diisobutyl ketone based upon the human comfort level. This level is also recommeded by the American Conference of Governmental Industrial Hygienists.[65] It appears low enough to prevent adverse effects.

2,6,8-TRIMETHYL-4-NONANONE (Trimethylnonanone)

1. Source, Uses, and Industrial Exposures

Trimethylnonanone is produced commercially by catalytic condensation of methyl isobutyl ketone, using calcium carbide followed by catalytic hydrogenation of the reaction product at 280°C.

Trimethylnonanone is used primarily as a dispersant for vinyl chloride resins. It finds some use as a solvent for other materials.

Industrial exposures are mainly those of inhalation of vapors, and skin and eye contact. Swallowing and absorption through the skin are not expected to present a problem in industrial handling. Trimethylnonanone, because of its relatively high flash point, is not considered a flammable solvent.

2. Physical and Chemical Properties

2,6,8-Trimethyl-4-nonanone (see Table 1) is a colorless liquid with a pleasant fruity odor.

[64] E. Browning, *Brit. J. Ind. Med.*, **16**, 23 (1959).

[65] American Conference of Governmental Industrial Hygienists, *Am. Ind. Hyg. Assoc. J.*, **22**, 325 (1961).

3. Physiological Response

Summary. Trimethylnonanone presents a low degree of hazard to health in industrial handling. It has a low single dose oral toxicity. When in contact with the eye, it should cause no significant irritation. An occasional skin contact may cause mild erythema; a dermatitis may result from excessive skin contact or when trimethylnonanone is confined to the skin as by clothing and shoes. It is low in toxicity by skin absorption. Inhalation of the vapors of this ketone does not constitute a health problem unless it is handled at elevated temperatures.

The principal hazard to health in industrial handling will probably be that of skin contact.

Single-Dose Oral. The LD_{50} value of trimethylnonanone for rats has been reported[66] to be 8.47 g./kg. of body weight. The physiological and pathological effects of this ketone have not been reported.

For industrial hygiene purposes, no special precautions need be taken. It should be recognized, however, that the swallowing of excessive amounts of this ketone accidentally or willfully may cause some systemic effects.

Skin Irritation. In skin tests on rabbits, trimethylnonanone has been shown to cause mild erythema when contact is prolonged.[66] It is suspected that this ketone may produce a mild dermatitis if confined to the skin as in clothing or shoes or if the contact is frequently repeated.

Skin Absorption. Smyth et al.[66] have reported an LD_{50} value by skin absorption for rabbits of 11 ml./kg. of body weight for trimethylnonanone. Thus, this ketone may be considered to have a low degree of hazard to health by this route.

Eye Contact. Smyth et al.[66] have reported that trimethylnonanone caused no significant effects when the undiluted ketone was applied to the eyes of rabbits.

Inhalation. There is little published information upon the effects of the inhalation of the vapor of trimethylnonanone. Smyth et al.[66] reported that an exposure of 4 hours to an essentially saturated atmosphere of this ketone caused no deaths of rats.

These meager data would indicate that trimethylnonanone should present a low degree of hazard to health from single vapor exposure. Its relatively high boiling point would support this conclusion also.

Human Experience. There are no published data of poisonings to humans by trimethylnonanone.

4. Hygienic Standards of Permissible Exposure

There are no data upon which to base a suggested threshold limit value for trimethylnonanone. Wisdom would indicate the use of reasonable care and caution to avoid inhalation, particularly where this ketone is handled at elevated temperatures or in confined places.

[66] H. F. Smyth, Jr., C. P. Carpenter, and C. S. Weil, *Arch. Ind. Hyg. Occupational Med.,* **4,** 119 (1951).

ACETONYL ACETONE (2,5-Hexanedione, α,β-Diacetyl Ethane)

1. *Source, Uses, and Industrial Exposures*

Acetonylacetone is made commercially from acetoacetic acid using alkylated iodine as a catalyst or by hydrolysis of 2,5-dimethyl furan. It finds uses both as a solvent for a wide variety of materials and as an intermediate in chemical synthesis.

In industrial handling, the primary avenue of contact will be through inhalation of the vapors. Skin and eye contact may occur but are not expected to present a serious problem. Neither swallowing nor absorption through the skin of toxic amounts is expected to be a problem in industrial handling. While acetonylacetone will burn, it does not flash until heated to 174°F.

2. *Physical and Chemical Properties*

Acetonylacetone is a water-clear, low-volatility liquid that becomes yellow on standing. See Table 1, page 1722.

3. *Physiological Response*

Summary. The limited toxicological information on acetonylacetone suggests that it presents a low degree of hazard to health in industrial handling. When in contact with the eye, moderate to marked irritation and injury may occur. Although no data are at hand, prolonged or frequently occurring skin contacts may be expected to cause dermatitis due to its fat-solvent action. The vapors of acetonylacetone are irritating to humans at low levels, hence, there is little likelihood of overexposure to concentrations that may lead to serious effects if the warning properties are heeded.

Single-Dose Oral. Smyth and Carpenter[67] report the range finding oral LD_{50} for rats to be 2.7 g./kg. Hence, there is little hazard from ingestion in industrial handling unless the material were swallowed accidentally or willfully.

Eye Irritation. Carpenter and Smyth[68] have shown that acetonylacetone has the capacity to cause moderate to marked eye irritation and some transient corneal injury. This would not suggest an unusual hazard to health from eye contact.

Skin Irritation. Although there are no laboratory data published regarding the topical effects of acetonylacetone, experience indicates that it does not present any unusual problems other than that it stains the skin an orange-brown color. It would be expected to cause irritation if exposure was prolonged or frequently repeated.

Skin Absorption. Smyth and Carpenter[67] report an LD_{50} of 6.6 ml./kg. for guinea pigs. Although this indicates that skin penetration can occur, it is doubtful if it is of practical significance in reasonable industrial handling.

[67] H. F. Smyth, Jr. and C. P. Carpenter, *J. Ind. Hyg. Toxicol.*, **26**, 269 (1944).
[68] C. P. Carpenter and H. F. Smyth, Jr., *Am. J. Ophthalmol.*, **29**, 1363 (1946).

Inhalation. The data on inhalation are scanty and difficult of interpretation. Smyth and Carpenter[67] report that the maximum time rats could tolerate a saturated atmosphere without deaths was 1 hour. On the other hand, Specht *et al.*[69] exposed guinea pigs for 8 hours to an atmosphere which they describe as saturated at 25°C. and which they state contained 400 p.p.m. acetonylacetone. They noted slight irritation of the eyes and nose and a slight decrease in respiration rate but no apparent loss of reflexes and no changes in rectal temperature or pulse rate. However, since the vapor pressure of acetonylacetone is reported to be 1.3 mm. Hg at 20°C., a saturated atmosphere at even this temperature would be about 1700 p.p.m. Thus, if the atmosphere to which Specht *et al.*[69] exposed their animals was saturated to the degree that Smyth's was, then there must be a marked difference in susceptibility of rats and guinea pigs to acetonylacetone. It seems more logical, however, to suspect that the guinea pigs were exposed to an atmosphere somewhat less than saturated.

Until this apparent discrepancy is resolved, it would seem wise to assume that prolonged exposure to concentrations attainable at room temperatures could result in more or less serious effects.

Human Experience. Authentic reports of adverse effects from handling acetonylacetone are lacking.

4. *Hygienic Standard of Permissible Exposure*

The limited data and experience available do not permit the recommendation of a threshold limit value for acetonylacetone. Elkins[70] notes that 75 p.p.m. is used as a guide for practical working levels by the Imperial Chemical Industries in England. This level would appear to be low enough to avoid serious systemic effects.

DIACETONE ALCOHOL (4-Hydroxy-4-methyl-2-pentanone)

1. *Source, Uses, and Industrial Exposures*

Diacetone alcohol is prepared by the action of barium or calcium hydroxides on acetone. It finds widespread use as a solvent for cellulose acetate, nitrocellulose, other cellulosic materials, coatings, dyes, dopes, cements, lacquers, oils, resins, tars, and waxes. It is found in the artificial leather and silk industry, in some antifreezes, in hydraulic brake fluids, in nongrain stains, metal-cleaning compounds, wood preservatives, photographic films, and as a preservative in pharmaceutical preparations.

In the industrial handling of diacetone alcohol, inhalation of the vapors and skin and eye contact are the most likely exposures to occur. Because of the warning properties of diacetone alcohol, overexposure leading to serious injury is not likely

[69] H. Specht, J. W. Miller, P. J. Valaer, and R. R. Sayers, *Natl. Insts. Health Bull.* No. **176**, 1940.

[70] H. B. Elkins, *The Chemistry of Industrial Toxicology.* 2nd ed., Wiley, New York, 1959.

unless the person is entrapped. Swallowing and absorption of the material through the skin are not expected to present a problem. It is readily flammable.

2. Physical and Chemical Properties

Diacetone alcohol (see Table 1) is a flammable liquid with a faint pleasant odor.

3. Physiological Response

Summary. Diacetone alcohol presents only a low degree of hazard to health in industrial handling. It has a low acute oral toxicity. The liquid defats the skin and thus may produce dermatitis upon prolonged or frequently occurring contacts. The liquid when in contact with the eyes may cause moderate to marked irritation. Absorption of diacetone alcohol occurs readily from the lungs. However, its toxicity by skin absorption is slight. Diacetone alcohol vapors, while capable of causing narcosis and systemic injury at high concentrations, produce transitory local irritation of the eyes, nose, and throat at concentrations well below those levels required to produce injury to health. Hence, diacetone alcohol has good warning properties so that acute systemic injury, even mild narcosis, should not occur under the normal conditions found in industrial operations.

Single Dose Oral. The various workers have found diacetone alcohol to have a low acute oral toxicity. Smyth and Carpenter[71] found the LD_{50} value for rats to be 4.0 g./kg. of body weight. In a similar study[72] rats fed 2 ml./kg. developed transient liver injury and succumbed to 4 ml./kg. Flury and Klimmer[73] report that rabbits exhibited narcotic effects when fed 2.4 to 4.0 ml./kg. and succumbed when fed 5 ml./kg.

Repeated Dose Oral. Smyth and Carpenter[71] report that a dosage level of 0.01 g./kg./day of diacetone alcohol administered in the drinking water of rats over a period of 30 days was the maximum level not causing any effect; organic injury occurred when the dose was 0.04 g./kg. Flury and Klimmer [73] noted that in rabbits, 2 ml. given 12 times daily caused moderate narcosis, injury to kidneys, and death in three-fourths of the rabbits treated.

Injection.[73] The effects of intravenous and intramuscular injections to rabbits have been studied. By the intravenous route, the narcotic dose was found to be 1.0 to 1.5 ml./kg. and the minimum fatal dose to be 3.25 ml./kg. A minimum fatal dose for rabbits via the intramuscular route was found to be in the range of 3 to 4 ml./kg. Repeated subcutaneous injections of 0.08 ml. of diacetone alcohol resulted in somnolence followed by recovery in rats.

[71] H. F. Smyth, Jr., and C. P. Carpenter, *J. Ind. Hyg. Toxicol.*, **30**, 63 (1948).
[72] Shell Chemical Corp., *Ind. Hyg. Bull. Toxicity Data Sheet,* Diacetone alcohol, SC 57-84, 1957.
[73] F. Flury and O. Klimmer, in K. B. Lehmann and F. Flury, *Toxicology and Hygiene of Industrial Solvents.* Trans. by E. King and H. F. Smyth, Jr., Springer, Berlin, 1938.

Skin Irritation. Diacetone alcohol is not highly irritating to the skin. However, it has been reported to be capable of causing dermatitis upon prolonged or frequently repeated contact.[72]

Skin Absorption. Smyth and Carpenter[71] using rabbits have reported an LD_{50} value for skin absorption of 14.5 ml./kg. of body weight. These data show that skin contact with liquid diacetone alcohol is not particularly hazardous.

Eye Irritation. Carpenter and Smyth[74] present evidence that liquid diacetone alcohol is capable of causing moderate to marked eye irritation and transient corneal damage. Thus, this material presents no unusual hazards from eye contact.

Inhalation. According to Smyth and Carpenter[71] rats survived an 8-hour exposure to an atmosphere saturated with the vapors of diacetone alcohol. It is reported[73] that mice, rats, rabbits, and cats exposed for 1 to 3 hours to 2100 p.p.m. exhibited restlessness, irritation of the membranes, excitement, and later somnolence. In view of the vapor pressure of acetone alcohol at ordinary room temperatures, information regarding the conditions under which the concentration of 2100 p.p.m. was achieved would be helpful in evaluating the significance of these observations.

Silverman, Schulte, and First[75] in studies with humans found that 100 p.p.m. of diacetone alcohol caused definite eye and throat irritation. This level was also reported to cause a definite bad taste and an objectionable odor. Nasal irritation was not encountered until levels higher than 100 p.p.m. were used.

In a similar study[72] using unconditioned humans exposed for 15 minutes, a level of 400 p.p.m. was found to cause pulmonary discomfort in addition to eye, nasal, and throat irritation. The results at 100 p.p.m. were essentially the same as reported[75] previously except that eye irritation was also noted. It was concluded that 50 p.p.m. was not objectionable.

Mode of Action. Diacetone alcohol causes irritation and pulmonary discomfort at high concentrations. However, the primary effect is that of narcosis. Death may be expected to result from respiratory failure due to the depression of the respiratory centers. In addition, injury to the kidneys and liver as well as a decrease in blood pressure may occur.[72]

Human Experience. There are no reports in the literature of toxic effects to humans resulting from the industrial handling of diacetone alcohol.

4. Hygienic Standard for Permissible Exposure

There is not adequate toxicological data available upon which to establish a safe level for prolonged and repeated vapor exposure. The American Conference of Governmental Industrial Hygienists,[76] however, suggests a threshold limit of 50

[74] C. P. Carpenter and H. F. Smyth, Jr., *Am. J. Ophthalmol.*, **29**, 1363 (1946).

[75] L. Silverman, H. Schulte, and M. First, *J. Ind. Hyg. Toxicol.*, **28**, 262 (1946).

[76] American Conference of Governmental Industrial Hygienists, *Am. Ind. Hyg. Assoc. J.*, **22**, 325 (1961).

p.p.m. This is probably based upon comfort. Until appropriate data becomes available, 50 p.p.m. can be considered to be a reasonable guide.

METHYL ISOPROPENYL KETONE (2-Methyl-1-butene-3-one)

1. Source, Uses, and Industrial Exposures

Methyl isopropenyl ketone is made commercially either by the catalytic hydration of isopropenyl acetylene or by the condensation of acetylene with acetone to form dimethyl ethynyl carbinol, which is then reacted with acetic acid to form dimethyl acetyl carbinol acetate. This intermediate is then pyrolyzed to methyl isopropenyl ketone. It can also be made by reacting butanone with formaldehyde in the presence of caustic to yield 2-methyl-3-keto-butanol, which is in turn cracked to the 2-methyl-1-butene-3-one.

Methyl isopropenyl ketone finds some use as a solvent but the greater bulk of it is used as a monomer in commercial copolymers.

The industrial handling of this ketone may lead to serious problems of health from eye and skin contact. Because of its strong lachrymatory properties, the vapors should not present a health problem from single exposures. However, these warning properties are not considered sufficient to prevent overexposure from repeated contacts. Swallowing should not present a problem in the usual conditions of industrial handling. Serious systemic injury may occur, however, if relatively small amounts of this material should be swallowed accidentally. Methyl isopropenyl ketone is a moderately flammable material.

2. Physical and Chemical Properties

Methyl isopropenyl ketone (see Table 1) is a chemically reactive, clear, colorless liquid with a very pungent odor. It is a strong lachrymator. Because of its chemical reactivity, it must be stored at subzero temperatures to prevent polymerization.

3. Physiological Response

Summary. Methyl isopropenyl ketone, unlike the saturated aliphatic ketones, presents a moderate to high degree of hazard to health during industrial handling. It has a moderate to high degree of toxicity by single oral exposure. Eye contact with the liquid may result in severe damage resulting possibly in some permanent impairment of vision unless flushing of the contaminated eye is started promptly. Methyl isopropenyl ketone is moderately irritating to the skin upon contact. The vapors of this ketone, even in low concentrations, cause burning of the eyes and profuse lachrymation. Also the vapors in high concentration are capable of causing irritation of the skin and even a burn if exposure is prolonged. The vapors are also highly toxic as well as irritating to the respiratory passages. The lachrymatory and irritating properties, although potent, are not strong enough to prevent the breathing of harmful amounts when exposures are repeated and prolonged. Methyl isopropenyl ketone is flammable.

The principal hazard in the handling of methyl isopropenyl ketone is considered to be that of vapor inhalation. However, eye contact with both the liquid and vapor and skin contact with the liquid may be quite hazardous.

Single-Dose Oral. Smyth *et al.*[77] have reported an LD_{50} value for methyl isopropenyl ketone of 0.18 g./kg. of body weight for rats. Unpublished data[78] indicate a lethal range for guinea pigs of from 0.06 to 0.25 g./kg. of body weight. These data indicate a moderate to high degree of toxicity from swallowing.

For industrial purposes, no special precautions need be taken to avoid ingestion. However, it should be borne in mind that serious systemic injury may occur from relatively small amounts of methyl isopropenyl ketone. Therefore, containers of this material should be properly labeled at all times.

Skin Irritation. Skin tests on rabbits[78] have shown that methyl isopropenyl ketone may be moderately irritating to the skin especially when confined to the skin.

Human experience[78] has shown this ketone to be a definite skin irritant. Of importance is the fact that the development of blistering and pain may not occur immediately upon contact. Several minutes may elapse before such manifestations arise.

Skin Absorption. According to Smyth *et al.*[77] methyl isopropenyl ketone has a moderate to high toxicity by this route. The LD_{50} value for rabbits was calculated to be 0.23 g./kg. of body weight.

Eye Irritation. In eye tests conducted on rabbits,[77] liquid methyl isopropenyl ketone was found to cause severe irritation and damage to the eye. This evidence suggests that contact with the human eye may result in permanent impairment of vision.

For industrial purposes, eye contact with liquid methyl isopropenyl ketone must be prevented. Suitable eye protection in the form of chemical workers' goggles should be worn where eye contact with the liquid is possible. Also in many instances where only vapors are encountered, it will be necessary to wear chemical workers' goggles because vapor concentrations that can be tolerated by the respiratory system may well produce eye irritation and lachrymation.

Inhalation. Methyl isopropenyl ketone has been shown to have a high degree of toxicity from single vapor exposures. Smyth *et al.*[77] reported that an exposure of 2 minutes to an essentially saturated atmosphere of this ketone killed all the rats tested. A 4-hour exposure to 125 p.p.m. resulted in the death of five of the six rats exposed.

The results of unpublished work[78] show that rats responded as follows to single exposures to the given nominal vapor concentrations:

---

[77] H. F. Smyth, Jr., C. P. Carpenter, and C. S. Weil, *Arch. Ind. Hyg. Occupational Med.*, **4**, 119 (1951).

[78] Biochemical Research Laboratory, The Dow Chemical Co., unpublished data.

524 p.p.m. All survived a 90-minute exposure but exhibited marked irritation of the eyes and nose

1455 p.p.m. Exposures of 30 minutes or more are dangerous to life and very irritating

2910 p.p.m. Dangerous to life in a few minutes

The animals that died during exposure or shortly afterwards were cyanotic and died in convulsions. Those that died a day or more after exposure died of respiratory irritation, either directly or indirectly.

The unpublished results[78] of repeated vapor exposures of rats, guinea pigs, and rabbits may be summarized as follows: Twenty 7-hour exposures in 28 days to a concentration of 30 p.p.m. caused marked irritation of the nose and eyes of all species, mortality and weight loss in the rats, and depressed growth in guinea pigs. In the rats, the lungs were severely affected and there were some changes in their kidneys and spleens; a distinct leucocytosis was also observed. The organs of the other species were not significantly affected.

Rats, guinea pigs, and rabbits also were given up to 100 7-hour exposures in up to 140 days to a concentration of 15 p.p.m. The rats sustained a slightly increased mortality, leucocytosis, and slight kidney injury characterized by a slight increase in weight and slight tubular injury. Otherwise, all the animals appeared normal other than for eye and nasal irritation.

Vapor concentrations below 15 p.p.m. have not been studied on animals.

Mode of Action. Based upon animal tests, the principal effect of vapor exposure is due to the irritant effect upon the eyes and respiratory tract.

Human Experience. There are no published reports of poisonings to humans attributable to methyl isopropenyl ketone. This probably is due to the fact that it

TABLE 3

Summary of Response of Man to Short Exposures
of Methyl Isopropenyl Ketone[78]

Vapor concentration		Effect
mg./liter	p.p.m.	
0.001	0.291	Questionable odor, no irritation
0.0025	0.77	Definite immediate odor, not unpleasant; could be detected by the eyes after a few minutes but not actually irritating
0.005	1.45	Immediate slight odor; eye irritation apparent and would probably be too severe to withstand for a full workday
0.01	2.91	Immediate slight odor; definite unpleasant eye irritation after 6 to 8 minutes
0.05	14.5	Immediate somewhat strong odor, but not unpleasant; eye irritation apparent after about 2 minutes, rapidly becoming quite strong (eyes shut involuntarily)

has fairly effective warning properties, its lachrymatory and irritating effects on the upper respiratory tract. The effects of various concentrations are given in Table 3. It should also be noted that irritation resulting from methyl isopropenyl ketone may be delayed and persist for some time after exposure. Effects due to skin contact are local and are usually slow in developing after contact.

4. *Hygienic Standards of Permissible Exposure*

Based upon animal data and human experience, a maximum concentration for repeated exposures of 0.3 p.p.m. of methyl isopropenyl ketone is suggested. This probably is low enough to prevent local or systemic injury. It is further suggested that processes involving this material be enclosed as completely as possible, that exhaust ventilation be supplied wherever vapors may be encountered in irritating amounts, and that appropriate protective equipment be readily available for emergency use.

MESITYL OXIDE (Methyl Isobutenyl Ketone, Isopropylidene Acetone, 4-Methyl-3-pentene-2-one)

1. *Source, Uses, and Industrial Exposures*

Mesityl oxide may be made by dehydration of diacetone alcohol using a trace of iodine as a catalyst.

It is considered an effective solvent for many uses. It is found extensively in the lacquer and leather industries. It has been used as a solvent for cellulosic materials, for polyvinyl chloride, high molecular weight copolymer resins, stains, and roll-coating inks. It has found utility in ore flotation, as an organic intermediate, and was at one time suggested as an insect repellent.

In the course of industrial handling, skin and eye contact may occur. Skin contact can result in both skin irritation and absorption of acutely toxic amounts. The inhalation of vapors in high concentrations may cause discomfort, even serious systemic injury. Swallowing is not expected to present a health problem in industrial handling.

2. *Physical and Chemical Properties*

Mesityl oxide (see Table 1) is a water-clear liquid with a strong odor suggestive of peppermint. Upon standing, it may darken in color. The commercial product tends to contain traces of aldehydes.

3. *Physiological Response*

Summary. Mesityl oxide presents a low degree of hazard to health in industrial handling. The liquid is capable of causing marked irritation and transient corneal injury when in contact with the eye. An occasional short skin contact should produce no irritation. Sustained skin contact may produce a dermatitis. In addition, mesityl oxide is capable of causing systemic injury due to absorption if skin contact is protracted and extensive. Also, it is entirely possible that frequent intermittent contact with small quantities may cause some systemic injury. The vapors of this ketone, in high concentrations, are narcotic and may cause injury to the lungs, liver, and kidneys. However, such concentrations are not likely to be tolerated voluntarily because of the strong odor and irritation to the eyes and nose. Because it has a distinct odor and causes irritation of the mucous membranes at vapor levels considered safe for prolonged and repeated exposure, there would seem to be little hazard from inhalation if its warning properties are not ignored.

Eye Contact. Carpenter and Smyth[79] found that liquid mesityl oxide caused marked irritation and corneal injury when in contact with the eye. This is substantiated by later workers.[80] For industrial purposes, suitable eye protection, such as safety glasses with side shields, should be worn wherever eye contact may occur.

Skin Irritation. Liquid mesityl oxide has been reported to produce dermatitis when in sustained contact with the skin.[80]

Skin Absorption. When 0.5 ml. of mesityl oxide was applied to the skin of mice, marked irritation developed in a few minutes. The mice became ataxic and narcotized within 15 minutes; and death of all the mice occurred in 3 to 9 hours. In a similar test on a set of 10 mice, treated with 0.1 ml. per mouse, only 1 mouse died.[80]

For industrial purposes, precautions should be taken to prevent direct skin contact with the liquid.

Inhalation. Specht et al.[81] exposed guinea pigs to vapor concentrations of 2,300, 5,000, and 10,000 p.p.m. for periods up to eight hours. They observed a progressive narcosis proportional to the intensity of exposure. Signs of intoxication were characterized by respiratory irritation, decrease in body temperature and respiratory and heart rate, loss of reflexes, coma, and death in all cases. They noted that the bodies of pigs had a vile odor, particularly apparent at autopsy, and also that the bodies of those dying sometimes became rigid even before death.

Smyth and co-workers[82] studied the effects resulting from both single and repeated exposures of rats and guinea pigs to various concentrations of mesityl oxide. Their conclusions in regard to single exposures are given in Table 4.

Groups of 10 rats and 10 guinea pigs given ten 8-hour exposures at a concentration of 500 p.p.m. suffered marked mortality, congestion and cloudy swelling of the kidneys, and some injury of the livers and lungs. At concentrations of 250, 100, and 50 p.p.m., similar groups of animals received thirty 8-hour exposures. The frequency and intensity of the organic injury decreased as exposure intensity

---

[79] C. P. Carpenter and H. F. Smyth, Jr., *Am. J. Ophthalmol.,* **29,** 1363 (1946).

[80] Shell Chemical Corp., *Ind. Hyg. Bull., Toxicity Data Sheet* SC 57-106, 1957.

[81] H. Specht, J. W. Miller, P. J. Valaer, and R. R. Sayers, *Natl. Insts. Health Bull.* No. **176,** 1940.

[82] H. F. Smyth, Jr., J. Seaton, and L. Fischer, *J. Ind. Hyg. Toxicol.,* **24,** 46 (1942).

decreased until at the lowest concentration, 50 p.p.m., no adverse effects were observed.

Human sensory perception of the vapors of mesityl oxide have been studied by several groups.[80,83] Silverman et al.[83] reported that 25 p.p.m. caused eye irritation; 50 p.p.m., nasal irritation, an objectionable odor and a bad taste in humans. Somewhat different values were established when unconditioned persons were subjected to 5-minute vapor exposures.[80] One half the people reported detecting the odor of mesityl oxide at 12 p.p.m. and all detected 25 p.p.m. Eye irritation was experienced at concentrations of 50 to 100 p.p.m.; nasal irritation and pulmonary discomfort was experienced by essentially half of the volunteers at 25 p.p.m.

TABLE 4

Acute Effects of Exposure of Rats and Guinea Pigs to Mesityl Oxide

Effect	Concentration of vapor, p.p.m.
Death in a few minutes	13,000 (saturated air)
Dangerous to life in 30 to 60 minutes	5,000
Maximum concentration for 60 minutes without serious disturbance	1,000
Maximum concentration for several hours without serious disturbances	200
Maximum concentration for several hours with slight or no symptoms	100

Mode of Action. Mesityl oxide is considered a strong narcotic with some irritative effect. When exposures to sublethal concentrations of vapor occur, some congestion, primarily of the kidneys, and to a lesser degree of the liver and lungs, may develop. Smyth et al.[82] postulate that these effects are secondary to the anesthetic action of mesityl oxide upon the circulatory and respiratory systems. Deaths are generally attributed to its narcotic action.

Human Experience. No ill effects from the industrial handling of mesityl oxide have been reported.

4. Hygienic Standards of Permissible Exposure

The American Conference of Governmental Hygienists[84] has recommended a threshold limit for mesityl oxide of 25 p.p.m. This level also has been suggested by others.[83] It appears low enough to prevent definite symptoms of narcosis and is probably low enough to prevent organic injury, but may not be low enough to prevent discomfort in unacclimated individuals.

[83] L. Silverman, H. Schulte, and M. First, *J. Ind. Hyg. Toxicol.*, **28**, 262 (1946).
[84] American Conference of Governmental Industrial Hygienists, *Am. Ind. Hyg. Assoc. J.*, **22**, 325 (1961).

3-BUTYN-2-ONE

1. *Physical and Chemical Properties*

3-Butyn-2-one (see Table 1) is a water-clear liquid, chemically reactive, and very odorous.

2. *Physiological Response*[85]

Summary. 3-Butyn-2-one is highly toxic orally, very irritating to the skin, eyes, and mucous membranes, readily absorbed through the skin in lethal amounts, and very dangerous to handle without proper knowledge and equipment.

The toxicological information given in the following paragraphs was derived from range-finding studies.

Single Dose Oral. When rats were fed the material by intubation as a 0.3 per cent solution in corn oil, the LD_{50} was found to be between 6.3 and 12.6 mg./kg. Tremors, diarrhea, and depression were signs seen prior to death.

Eye Contact. Undiluted material is extremely painful and damaging to the eyes of rabbits. Furthermore, it is readily absorbed through the eye (conjunctival membranes) in lethal amounts. Even a 1 per cent solution in propylene glycol causes such damage as to suggest complete loss of vision.

Skin Contact. Undiluted 3-butyn-2-one and strong solutions (10 per cent) are extremely damaging to the skin and are fatal if confined for a few hours. Dilute solutions (1 per cent) are also very damaging, particularly to abraded skin.

Skin Absorption. 3-Butyn-2-one is absorbed through the skin readily. Using the "cuff" technique essentially as described by Draize et al.[86] and applying the material as a 10 per cent solution in propylene glycol, the LD_{50} was found to be 40 to 50 mg./kg. Deaths occurred with typical signs within hours.

Inhalation. Groups of rats were exposed to an atmosphere essentially saturated with vapor by passing the entire air supply through a fritted glass disk immersed in the liquid held at 23°C., and to nominal vapor concentrations of 200, 100, 50, 25, and 10 p.p.m. The results are given in Table 5.

Immediate severe eye and respiratory irritation was noted at all concentrations studied. Lung congestion occurred in all cases. Liver and kidney injury was not observed to any appreciable extent in those animals receiving the most acute exposures but was apparent in those receiving the longer exposures even though the animals survived.

Human Experience. Human experience has been very limited but sufficient to suggest that the human subject is at least as sensitive to the material as the animals.

3. *Hygienic Standards of Permissible Exposure*

Data adequate to establish a permissible limit for repeated exposure are lacking. However, the data available are sufficient to indicate that the material is

---

[85] Biochemical Research Laboratory, The Dow Chemical Co., unpublished data.

[86] J. H. Draize, G. Woodard, and H. O. Calvery, *J. Pharmacol. Exptl. Therap.*, **82**, 377 (1944).

TABLE 5

Summary of Range-Finding Vapor Inhalation
Studies on Rats with 3-Butyn-2-one

Vapor concentration, p.p.m.	Exposure time, hr.	No. dying/no. exposed
Saturated		
	0.1	3/3
200	0.5	4/4
	0.2	2/4
	0.1	3/4
	0.05	1/4
100	2.0	4/4
	1.0	4/4
	0.5	3/4
	0.2	0/4
50	4.0	4/4
	2.0	4/4
	1.0	3/4
	0.5	2/4
	0.2	0/4
25	1.0	4/4
	0.5	2/4
	0.2	0/4
10	7.0	4/4
	4.0	2/4
	2.0	0/4

more toxic than methyl isopropenyl ketone. Precautions for safe handling should be designed to prevent all exposure. Appropriate impermeable clothing and appropriate respiratory protection should be worn or be readily available whenever exposure is possible.

3-PENTYN-2-ONE

1. *Physical and Chemical Properties*

3-Pentyn-2-one (see Table 1) is a water-clear liquid, chemically reactive, and very odorous.

2. *Physiological Response*[87]

Summary. 3-Pentyn-2-one is quite toxic orally, very irritating to the skin and mucous membranes, readily absorbed through the skin in lethal amounts, and dangerous to handle without proper knowledge and equipment.

The toxicological information given in the following paragraphs was derived from range-finding studies.

[87] Biochemical Research Laboratory, The Dow Chemical Co., unpublished data.

Single Dose Oral. When rats were fed the material by intubation as a 1.26 per cent solution in corn oil, the LD_{50} was found to be between 63 and 126 mg./kg. Deaths usually occurred within 2 days after tremors and convulsions occurred. Autopsy of surviving animals indicated severe liver and kidney damage and marked irritation of the stomach mucosa.

Eye Contact. Undiluted material or strong solutions (10 per cent) are severely painful and injurious to the eyes. Contact may be expected to cause impairment of vision.

Skin Contact. The undiluted material or a 10 per cent solution was severely irritating to the skin; if 1 to 2 ml. was confined to the skin deaths occurred within a few hours. Quantitation employing the Draize[88] technique showed the LD_{50} for rabbits to be between 6 and 12 mg./kg. Thus, the material is far more toxic when applied to the skin than when given intragastrically.

Inhalation. A group of 3 rats all died after receiving a 6-minute exposure to an atmosphere essentially saturated with vapor. "Saturation" was accomplished by bubbling all the air through a fritted glass disk submerged in the liquid at 23°C. Immediate irritation of the eyes and nose was evident. Deaths occurred over a 5-day period. No autopsies were conducted.

Human Experience. Human experience has been very limited but sufficient to suggest that the human subject is at least as sensitive to the material as the animals.

3. *Hygienic Standards of Permissible Exposure*

Data adequate to establish a permissible limit for repeated exposure are lacking. Precautions for safe handling should be designed to prevent all exposure. Appropriate impermeable clothing and appropriate respiratory protection should be worn or be readily available whenever exposure is possible.

ACETOPHENONE (Phenylmethyl Ketone, Acetyl Benzene, Hypnone)

1. *Source, Uses, and Industrial Exposures*

Acetophenone may be made by a number of industrial processes. Perhaps one of the most widely used methods is that of reacting benzene and acetyl chloride to form acetophenone. It may be made by reacting acetic and benzoic acid under the proper conditions, by combining benzaldehyde and diazomethane, or by the oxidation of ethyl benzene.

Because of its odor, acetophenone has been used in the perfume industry. In addition, it is used extensively as an intermediate in the manufacture of organic chemicals. In the past, it enjoyed use as an anesthetic and as an analgesic agent in medicine.

[88] J. H. Draize, G. Woodard, and H. O. Calvery, *J. Pharmacol. Exptl. Therap.,* **82,** 377 (1944).

The industrial handling of acetophenone may result in skin and eye exposures if handled carelessly. The inhalation of vapors should present no problem unless the material is heated excessively. Swallowing or absorption through the skin in toxic amounts is not expected to occur. It is stable under the normal conditions of storage and does not form flammable mixtures with air at room temperature.

2. *Physical and Chemical Properties*

Acetophenone (see Table 1) is a low-melting solid or a colorless, low-volatility liquid depending upon the temperature. It has a persistent odor not unlike that of orange blossoms and jasmine.

3. *Physiological Response*

Summary. Acetophenone presents no unusual hazards to health in industrial handling. It does have definite narcotic and analgesic properties when swallowed or injected intravenously or subcutaneously into the body. However, it has a moderate to low degree of toxicity by these routes. Skin contact with the liquid may result in marked irritation, even a burn if confined. Eye contact can be expected to produce up to marked irritation and transient corneal injury depending upon the extent and duration of contact. Its vapors are not expected to present a problem unless acetophenone is heated.

The main hazard in industrial handling is that of skin and eye contact.

Single-Dose Oral. Smyth and Carpenter[89,90] have presented data showing that acetophenone has a moderate to low acute oral toxicity. They have reported range-finding LD_{50} values of 3 g./kg. in 1944 and 0.9 g./kg. in 1948. For industrial purposes, no special precautions need be taken to avoid swallowing. However, it should be remembered that narcotic effects may result from swallowing.

Repeated-Dose Oral. Smyth and Carpenter[90] fed acetophenone in the diet of rats for 30 days. Under such conditions, they observed no effect at the highest level fed, 102 mg./kg./day.

Injection. Considerable work has shown that subcutaneous or intravenous injection of acetophenone will produce hypnotic effects. Typical of such work is that of Quevauviller,[91] who found that, when injected intraperitoneally, mice could survive 0.7 g./kg. and that clear-cut hypnotic action developed rapidly at levels of 0.4 or 0.5 g./kg. The minimum fatal dose was found to be 1.6 g./kg. whereas the LD_{50} was 1.07 g./kg.

Skin Irritation. Acetophenone has been reported to be capable of causing dermatitis to humans following skin contact. Katz[92] reported this material to be a known skin irritant. Experiments on rabbits confirm that acetophenone is capable of causing irritation, perhaps a mild burn, if contact is prolonged or frequently repeated.[89]

[89] H. F. Smyth, Jr., and C. P. Carpenter, *J. Ind. Hyg. Toxicol.,* **26,** 269 (1944).
[90] H. F. Smyth, Jr., and C. P. Carpenter, *J. Ind. Hyg. Toxicol.,* **30,** 63 (1948).
[91] A. Quevauviller, *Compt. rend. soc. biol.,* **140,** 367 (1946).
[92] A. E. Katz, *Spice Mill,* **69,** 46 (1946).

Skin Absorption. Smyth and Carpenter[89] showed that the LD_{50} value for guinea pigs by skin absorption was greater than 20 ml./kg. of body weight. Thus, acetophenone may be considered to be low in toxicity by this route.

Eye Irritation. Animal data reported by Smyth and Carpenter[90] indicate that acetophenone may cause eye irritation and possibly transient corneal injury. Hence, acetophenone presents no unusual hazard from eye contact.

Inhalation. Because of the low vapor pressure of acetophenone, its toxicity by inhalation has been given little study. Smyth and Carpenter[89] found in range-finding studies that an 8-hour exposure of rats to a saturated atmosphere produced no deaths.

Mode of Action. Acetophenone, when given by various routes, produces anesthesia and analgesia in animals. However, its action is erratic. Some of the treated animals went into a state of deep coma, which was followed by death. This finding resulted in the loss of interest in acetophenone as a medicinal.

According to Fairhall,[93] Thierfelder and Klenk[94] found that acetophenone is metabolized to a large extent (91.7 per cent) to benzoic acid, which appears in the urine as hippuric acid. Small amounts of other metabolites also were found. Apparently the phenyl ring is left intact.

Human Experience. There are no known records of systemic effects due to the handling of acetophenone. This is probably due to its low volatility.

4. Hygienic Standard of Permissible Exposure

Because of the low volatility of acetophenone, a problem is not expected. A recommended threshold limit has not been suggested. Adequate ventilation to remove its vapors should be provided in those operations where acetophenone is heated to elevated temperatures.

ISOPHORONE (Trimethyl Cyclohexenone, Isoacetophenone, 3,5,5-Trimethyl-2-cyclohexene-1-one)

1. Source, Uses, and Industrial Exposures

Isophorone is prepared commercially by two methods. Acetone is used in both as the intermediate. It is either passed over calcium oxide, hydroxide or carbide or their mixtures at 350°C. and atmospheric pressure, or it is heated at 200 to 250°C. under pressure. The isophorone is separated from the resultant products by distillation.

Isophorone is an excellent solvent for many oils, fats, gums, and resins. It finds widespread use as a solvent for lacquers, nitrocellulose, and vinyl-resin copolymers. Because of its chemical structure, it also finds use as a chemical intermediate.

[93] L. T. Fairhall, *Industrial Toxicology*. 2nd ed., Williams & Wilkins, Baltimore, 1957.
[94] H. Thierfelder and E. Klenk, *Z. physiol. Chem.*, **141**, 13 (1924).

In the industrial handling of isophorone, inhalation of the vapors is the most likely mode of contact. However, skin and eye contact with the liquid may occur also. Because of the odor and taste of isophorone, swallowing is not expected unless by accident due to storage in unlabeled or mislabeled containers. It is stable chemically and does not present a problem of flammability unless it is handled at elevated temperatures.

2. *Physical and Chemical Properties*

Isophorone (see Table 1) is a water-clear, low-volatility liquid with a peppermintlike odor and a cooling taste.

3. *Determination of Isophorone in Air*

Kacy and Cope[95] developed on analytical procedure for estimation of small amounts of isophorone. Essentially it consists of the spectrographic measurement of the blue color developed when air contaminated with isophorone is scrubbed through glacial acetic acid containing phosphomolybdic acid (see also methods suggested in the general discussion of ketones).

4. *Physiological Response*

Summary. Except for the effect of inhalation of the vapors of isophorone, there is little recorded upon the physiological activity of isophorone. It is considered to have a low degree of hazard to health by inhalation even though it has a moderately high degree of toxicity by this route. Its vapors are quite irritating and have an unpleasant odor to unacclimated humans so that overexposure is not likely if its warning properties are heeded. It is possible that humans may lose their ability to detect the presence of isophorone if vapor exposures are prolonged.

Inhalation. In 1940 and in 1942, Smyth and co-workers[96,97] reported the results obtained when animals allegedly were exposed to vapors of isophorone. The vapor concentrations reported are impossible to attain under the conditions employed. Later investigation lead to the conclusion that the material used in these studies was an impure commercial product containing appreciable amounts of material(s) more volatile than isophorone, a fact not known to the investigators at the time. Since the concentration of vapors within the exposure chamber was measured by means of an interferometer calibrated against pure isophorone, they assumed the vapors present in the chamber were this material. Because the composition of the material tested is unknown, it seems best not to consider the results reported in these studies in evaluating the hazards to health of isophorone. For this reason, care should also be taken in interpreting the conclusions regarding isophorone expressed in other texts.

[95] H. W. Kacy and R. W. Cope, *Am. Ind. Hyg. Assoc. Quart.*, **16,** 55 (1955).
[96] H. F. Smyth, Jr., and J. Seaton, *J. Ind. Hyg. Toxicol.*, **22,** 477 (1940).
[97] H. F. Smyth, Jr., J. Seaton, and L. Fischer, *J. Ind. Hyg. Toxicol.*, **24,** 46 (1942).

Silverman, Schulte, and First,[98] using unconditioned human volunteers distracted with movies, found that a short exposure to the vapors of isophorone was definitely irritating to the eyes, nose, and throat at a level of 25 p.p.m.

5. Hygienic Standard of Permissible Exposure

The American Conference of Governmental Industrial Hygienists[99] has recommended a threshold limit value of 25 p.p.m. for repeated prolonged chemical exposure to isophorone. In view of the foregoing discussion and the observations of Silverman et al.,[98] it would seem prudent to employ a somewhat lower level for engineering purposes. Tentatively, 10 p.p.m., as recommended by Silverman et al.,[98] is suggested.

CYCLOHEXANONE (Pimelic Ketone, Sextone, Anone, Hexanon)

1. Source, Uses, and Industrial Exposures

Cyclohexanone generally is made by catalytic oxidation of cyclohexanol. Distillation of pimelic acid salts will also yield cyclohexanone.

Like many of the ketones, cyclohexanone is used both as a solvent and as a chemical intermediate. It is a solvent for cellulose esters and ethers, dyes, resins, lacquers, shellac, oils, and fats. Thus, it is found in the textile, lacquer, and leather industries. Its solvent properties make it desirable as a degreaser, as a spotting agent for removing stains in the dry-cleaning and textile industries, as a solvent in paint removers, for printing inks, and in the plastic industry.

In the handling of cyclohexanone industrially, contact with the skin and eye and inhalation of the vapors are most likely to occur. Swallowing and absorption of toxic amounts through the skin are not likely to be a problem unless excessive exposures are encountered.

It is chemically very stable and should not present a problem of flammability unless handled at elevated temperatures.

2. Physical and Chemical Properties

Cyclohexanone (see Table 1) is a colorless, slightly volatile, liquid with a peculiar ketone-type odor similar to peppermint.

3. Determination in the Atmosphere

In addition to the methods discussed at the beginning of this chapter, Treon et al.[100,101] have described a method for the determination of cyclohexanone in

[98] L. Silverman, H. F. Schulte, and M. W. First, *J. Ind. Hyg. Toxicol.*, **28**, 262 (1946).

[99] American Conference of Governmental Industrial Hygienists, *Am. Ind. Hyg. Assoc. J.*, **22**, 325 (1961).

[100] J. F. Treon, W. E. Crutchfield, Jr., and K. V. Kitzmiller, *J. Ind. Hyg. Toxicol.*, **25**, 199 (1943).

[101] J. F. Treon, W. E. Crutchfield, Jr., and K. J. Kitzmiller, *J. Ind. Hyg. Toxicol.*, **25**, 323 (1943).

the atmosphere. It depends upon measuring the intensity of the pink color produced when cyclohexanone (and methylcyclohexanone) reacts with m-dinitrobenzene.

4. Physiological Response

Summary. Cyclohexanone is considered to be low in degree of hazard to health under the usual conditions of industrial handling. Although cyclohexanone is capable of causing narcosis and death at high concentrations, it is only slightly volatile; hence, such concentrations are not likely to occur unless it is handled at elevated temperatures. Cyclohexanone has strong warning properties at low concentrations, thus overexposure to concentrations that may cause systemic injury are not likely to be tolerated voluntarily by most humans. It has a low acute oral toxicity. An occasional skin contact with the liquid probably should cause no irritation. However, because it is capable of defatting the skin, prolonged or frequently repeated skin contact may logically be expected to result in irritation or dermatitis. Eye contact with the liquid may result in marked irritation and some transient corneal injury.

The principal hazard to health in the handling of cyclohexanone is that of inhalation of the vapors.

Single Dose Oral. Several authors have reported on the intragastric toxicity of cyclohexanone to laboratory animals. Jacobi, Hayashi, and Szubinski[102] found the minimum lethal dose for mice to be 1.3 to 1.5 g./kg. of body weight. They observed that the treated mice developed paresis of the hind quarters, narcosis, and deep, slow respiration before death. Treon, Crutchfield, and Kitzmiller[100] found the LD_{100} values for rabbits to be 1.6 to 1.9 g./kg. They also observed an increased excretion of organic sulfates and of glucuronic acids in the urine and, at high dosages, some lung damage was observed.

Injection Data. A number of workers have reported on the effects of injection, by various routes, of cyclohexanone. Intraperitoneal injection of 0.5 ml./ mouse has been reported to cause excitation, paresis of hind quarters, marked hypothermia, and convulsions followed by death.[103] One of the metabolic products these workers found was adipic acid, presumably due to the oxidation of cyclohexanone. Caujolle and co-workers,[104-106] working with anesthetized dogs, found that 630 mg./kg. of cyclohexanone, when given intravenously, caused death in 60 minutes. They observed accelerated respiration, vasodepression, and hypotension in the dogs so treated. Deichmann and Dierker[107] established that cyclohexanone greatly increased the hexuronate content of the plasma in rats and rabbits.

[102] C. Jacobi, Hayashi, and Szubinski, *Arch. exptl. Pathol. Pharmakol.,* **50,** 199 (1903).

[103] E. Fillipe, *Arch. farmacol. sper.,* **18,** 178 (1914).

[104] F. Caujolle, P. Couturier, G. Roux, and Y. Gase, *Compt. rend.,* **236,** 633 (1953).

[105] F. Caujolle and G. Roux, *Compt. rend.,* **239,** 680 (1954).

[106] F. Caujolle, G. Roux, and P. Thomas, *Trav. soc. pharm. Montpellier,* **14,** 329 (1954), *Ind. Hyg. Dig.,* **19,** Abstr. 953 (1955).

[107] W. B. Deichmann and M. Dierker, *J. Biol. Chem.,* **163,** 753 (1946).

Skin Irritation. There are no published data on the effect of skin contact with cyclohexanone. However, as it is an excellent fat solvent, it is reasonable to assume that prolonged or frequently repeated contacts with the skin may possibly cause some irritation, even dermatitis. It is probable, however, that an occasional contact with the skin will result in no appreciable irritation.

Skin Absorption. Treon *et al.*[100] working with rabbits, found the LD_{100} value by skin absorption to be in the range of from 10.2 to 23.0 g./kg. of body weight. They noted tremors, narcosis, and hypothermia prior to death. Thus, the absorption of cyclohexanone through the skin produces the same effects as by other routes but the dosage required is larger.

This ketone, even though it may cause dermatitis because of its defatting action to the skin, may be considered to have a low degree of hazard from skin contact.

Eye Irritation. When liquid cyclohexanone was placed in the eyes of rabbits, it caused marked irritation and some corneal injury.[108] Hence, liquid cyclohexanone may be expected to cause marked irritation possibly some transient corneal injury when in contact with the human eye.

Inhalation. Specht, Miller, Valaer, and Sayers[109] observed that a 6-hour exposure of guinea pigs to 4000 p.p.m. of cyclohexanone caused typical narcotic symptoms, lacrymation, salivation, depression of body temperature and respiratory heart rates, and opacity of the corneas. Recovery from the effects of narcosis was slow.

Smyth[110] found that a 4-hour exposure of rats to a concentration of 4000 p.p.m. permitted all the rats to survive, whereas a 4-hour exposure to 8000 p.p.m. resulted in death due to anesthesia.

Flury and Klimmer[111] report that E. Gross exposed mice, guinea pigs, and cats to 3800 p.p.m. cyclohexanone and observed signs as noted above and no abnormalities in the urine.

Treon, Crutchfield, and Kitzmiller[101] gave monkeys and rabbits fifty 6-hour exposures to 190 p.p.m. with no detectable effects other than very slight kidney and liver injury. At 309 p.p.m., the animals developed very slight eye irritation. At 773 p.p.m., the animals were observed to salivate in addition to exhibiting eye irritation. At the highest level, 3082 p.p.m., light narcosis, labored breathing, incoordination, and a slightly increased mortality were also seen. However, no hematological changes or pathology specific to cyclohexanone were detected. As in the single-dose oral studies, increased organic sulfate and glucuronic acid were found in the urine of the rabbits.

[108] C. P. Carpenter and H. F. Smyth, Jr., *Am. J. Ophthalmol.,* **29,** 1363 (1946).

[109] H. Specht, J. W. Miller, P. J. Valaer, and R. R. Sayers, *Natl. Insts. Health Bull.* No. **176,** 1940.

[110] H. F. Smyth, Jr., *Am. Ind. Hyg. Assoc. Quart.,* **17,** 129 (1956).

[111] F. Flury and O. Klimmer, in K. B. Lehmann and F. Flury, *Toxicology and Hygiene of Industrial Solvents. Trans.* by E. King and H. F. Smyth, Jr., Springer, Berlin, 1938.

Nelson and co-workers[112] studied the sensory response of humans exposed for a single short period to the vapors of cyclohexanone. They reported that 50 p.p.m. was definitely objectionable. Definite eye, nose, and throat irritation was recorded at 75 p.p.m. A level of 25 p.p.m. was thought by most of the volunteers to be the highest concentration satisfactory for an 8-hour exposure.

Mode of Action. Cyclohexanone is both an irritant and a narcotic agent. Death is thought to be due to respiratory failure. At high dosage levels, the organic sulfate and glucuronic acid output in the urine is increased.

Human Experience. No serious industrial poisonings have been reported. Browning[113] states that no ill effects were observed in a number of workers employed in operations where cyclohexanone was used as a degreasing solvent. Drowsiness, however, was observed in some 20 workers who were using a mixture of 25 per cent cyclohexanone, cyclohexyl acetate, and methyl cyclohexanone.

5. Hygienic Standard of Permissible Exposure

The American Conference of Governmental Hygienists[114] has recommended a threshold limit value of 50 p.p.m. for cyclohexanone. This level should prevent definite narcosis. However, based on the work of Nelson *et al.*,[112] this level may be somewhat high and if comfort is to be attained, the concentration in the air may well have to be maintained below 50 p.p.m.

METHYLCYCLOHEXANONE (Methylanon, Sextone B)

1. Source, Uses, and Industrial Exposures

Primarily, methylcyclohexanone is made by hydrogenation of cresol. Hence, the commercially available methylcyclohexanone is primarily a mixture of the meta and para isomers. The ratio of the isomers depends upon the sample of cresol used.

Methylcyclohexanone is frequently used as a cosolvent with cyclohexanone. However, it is not as good a solvent as cyclohexanone for many purposes. Commercially, the major share of methylcyclohexanone is used as a solvent in the manufacture of lacquers and varnishes and in plastics. In addition, it is used in the leather industry and as a rust remover.

Like cyclohexanone, the main industrial exposure is that of inhalation of the vapors although skin and eye contact with the liquid may occur. Swallowing and absorption of the liquid through the skin are not expected to present a problem.

[112] K. Nelson, J. F. Ege, Jr., M. Ross, L. E. Woodman, and L. Silverman, *J. Ind. Hyg. Toxicol.*, **25**, 282 (1953).

[113] E. Browning, *Toxicity of Industrial Organic Solvents*. Revised ed., Chemical Publishing, New York, 1953.

[114] American Conference of Governmental Industrial Hygienists, *Am. Ind. Hyg. Assoc. J.*, **22**, 325 (1961).

It is possible, however, that systemic injury may occur if large amounts of methylcyclohexanone are swallowed. Hence, containers of this material should be clearly labeled so that accidental swallowing is not likely.

Chemically, methylcyclohexanone is stable. It is flammable at temperatures readily attainable in many industrial operations.

2. Physical and Chemical Properties

Methylcyclohexanone (see Table 1) is an almost colorless, low-volatility liquid with an acetone or peppermintlike odor. Upon storage, it may darken, especially if exposed to light.

3. Determination in the Atmosphere

In addition to the methods discussed at the beginning of this chapter, Treon et al.[115,116] have described a method for the determination of methylcyclohexanone in the air. It depends upon measuring the intensity of the pink color produced when methylcyclohexanone (and cyclohexanone) react with *m*-dinitrobenzene.

4. Physiological Response

Summary. Methylcyclohexanone is not considered a hazardous material even though it is capable of causing narcosis and death at high vapor concentrations. Such exposures are not likely because of its irritant and warning odor at levels well below those causing serious effects. It has a moderate to low acute toxicity from swallowing and a low degree of toxicity from absorption through the skin.

Single Dose Oral. Treon, Crutchfield, and Kitzmiller[115] found the LD$_{100}$ value for rabbits to be 1.0 to 1.2 g./kg. of body weight. In these studies they noted an increased excretion of organic sulfates and glucuronic acid in the urine. These findings indicate a moderate to low degree of toxicity for methylcyclohexanone from swallowing.

Injection. Using chloral treated dogs, Caujolle et al.[117] injected methylcyclohexanone intravenously as a solution in olive oil. They reported a lethal dose range of 270 to 370 mg./kg. depending upon the isomer studied. Flury and Klimmer[118] report that E. Gross noted accelerated respiration but no circulatory effects after injecting 0.1 ml. subcutaneously.

Skin Irritation. Flury and Klimmer[118] also report no effect from contact of methylcyclohexanone with the skin of rabbits, guinea pigs, or humans. However,

[115] J. F. Treon, W. E. Crutchfield, Jr., and K. V. Kitzmiller, *J. Ind. Hyg. Toxicol.*, **25**, 199 (1943).

[116] J. F. Treon, W. E. Crutchfield, Jr., and K. V. Kitzmiller, *J. Ind. Hyg. Toxicol.*, **25**, 323 (1943).

[117] F. Caujolle, P. Couturier, G. Roux, and Y. Gase, *Compt. rend.*, **236**, 633 (1953).

[118] F. Flury and O. Klimmer, in K. B. Lehmann and F. Flury, *Toxicology and Hygiene of Industrial Solvents. Trans.* by E. King and H. F. Smyth, Jr., Springer, Berlin, 1938.

as with other such materials, prolonged or frequently repeated contact would be expected to cause some irritation.

Skin Absorption. Treon *et al.*[115] studied the skin-penetrating properties of methylcyclohexanone. Using rabbits, they reported an LD_{100} value of 4.9 to 7.2 g./kg. Marked hypothermia, tremors, and narcosis were observed in the animals exposed to the high doses. It is apparent, therefore, that methylcyclohexanone may be considered to present a low degree of hazard from skin contact. For industrial purposes, reasonable precautions to avoid prolonged or frequently occurring skin contacts should be adequate to avoid skin difficulties.

Eye Contact. The effect of liquid methylcyclohexanone on the eye has not been studied using laboratory animals. Based upon chemical similarity to cyclohexanone, it is expected that it will be painful, irritating, and somewhat injurious. The vapors in high concentration are irritating.

Inhalation. Flury and Klimmer[118] report that mice, guinea pigs, and rats suffered marked pain and irritation of the mucous membranes, became incoordinate in 15 minutes, and prostrate in 30 minutes when exposed to 3500 p.p.m. of methylcyclohexanone vapor. At 2500 p.p.m., rabbits and cats became sleepy within an hour, suffered respiratory irregularities, and were poorly coordinated after an hour but recovered completely. Mice were markedly irritated by a vapor concentration of 450 p.p.m. and tried violently to escape.

Studies reported by Treon *et al.*[116] show that, while methylcyclohexanone is an irritant at low concentrations, and is somewhat narcotic at high concentrations, lethal levels cannot be reached at ordinary temperatures.

Treon *et al.*[116] gave monkeys and rabbits fifty 6-hour exposures to various concentrations of methylcyclohexanone in the air. A level of 1139 p.p.m. caused lethargy, salivation, lachrymation, and eye irritation. Only eye irritation was noted at a level of 514 p.p.m. and the "no-effect" level was 182 p.p.m. At the higher levels, an increased excretion of organic sulfates and glucuronic acid was observed.

Mode of Action. The main effects of the inhalation of high concentrations of methylcyclohexanone are narcosis and eye and nasal irritation. Metabolically increased levels of organic sulfate and glucuronic acid may occur.

Human Experience. No records of toxic effects from the handling of methylcyclohexanone have been reported.

5. *Hygienic Standard of Permissible Exposure*

The American Conference of Governmental Industrial Hygienists[119] has recommended a threshold limit value of 100 p.p.m. for methylcyclohexanone. This appears a low enough level to prevent narcotic effects but may not be low enough to prevent complaints of eye and respiratory irritation.

[119] American Conference of Governmental Industrial Hygienists, *Am. Ind. Hyg. Assoc. J.,* **22,** 325 (1961).

CHAPTER XL

Organic Acids, Anhydrides, Lactones, Acid Halides and Amides, Thioacids

DAVID W. FASSETT, M.D.

I. General Considerations

A. CHEMICAL CLASSIFICATION

Organic acids and their derivatives cover a wide range of substances, many of which are very important in industry. For the purposes of this chapter, they can be classified in three ways:

1. According to the nature of the acid group or derivative: e.g., carboxylic, RCOOH; sulfuric, RSO_4H; sulfonic, RSO_3H; sulfinic, RSO_2H; acid halides, RCOCl, RSO_2Cl; acid amides, $RCONH_2$, RSO_2NH_2; thioacids, RCOSH, RCSSH. While there are other important acid groups, such as phosphoric or arsonic, these will not be considered here.

2. According to the general nature of the substituent to which the acid group is attached: e.g., R can be aliphatic (saturated or unsaturated), alicyclic, aromatic, heterocyclic, etc. The substituent groups can have one or more acid groups attached, e.g., dicarboxylic or disulfonic. There can also be mixtures of different types of acid groups on the same substituent, i.e., $HO_3SRCOOH$.

3. According to the specific groups on the substituent: e.g., R as described above can be further modified by the presence of other groups such as OH, NH_2, Cl, SH, etc. The latter groups can also vary in their position in relation to the acid grouping.

Because of the great variety of possible combinations, a large number of organic acids or derivatives are known and the nomenclature may be confusing. As a general rule, however, the acid group takes preference (e.g., propionic acid, propionyl chloride, propionamide, β-chloropropionic acid, etc.) and will usually appear last in the name, being preceded by the substituent group.

B. INDUSTRIAL APPLICATIONS

These compounds are used in nearly every type of chemical manufacturing. Some of the more important uses are as follows:

1. *Aliphatic Carboxylic Acids* (e.g., acetic, propionic, lactic, butyric, etc.).

These may be used in preparation of cellulose resins, in manufacture of various esters for solvent purposes, in food applications, in cosmetics, and as chemical intermediates. The anhydrides of the lower members of the series are especially useful in syntheses.

2. *Dicarboxylic Acids.* Certain of these have wide uses in food products, for example, citric and tartaric. Oxalic acid is used as a rust remover, bleach, etc. Others, such as maleic, are of importance in alkyd resins.

3. *Sulfonic Acids.* These are of major importance in the manufacture of detergents and dyes. Sulfate esters are also used as detergents.

4. *Acid Halides.* Because of their chemical reactivity, they find wide use as intermediates.

5. *Acid Amides.* These have their principal uses in the formation of resins, in certain packaging applications, as solvents, and as chemotherapeutic agents.

6. *Rosin Acids.* These are used as sizes and for other purposes in the manufacture of paper.

7. Certain thio derivatives, such as mercaptoacetic acid, are used in hairwave preparations.

C. PHYSIOLOGICAL EFFECTS

Because of the variety of chemical structures in the organic acid group, several types of toxic effects may occur. These are generally in the nature of primary irritant effects, occasional sensitization of the skin, or enzyme inhibitory action. Occasionally, chronic organ damage is seen.

1. *Primary Irritant Effect.* This is the combined result of the degree of acid dissociation, water solubility, and other factors influencing the penetration of skin and mucous membranes. The vapor pressure will, of course, control the degree of hazard from vapor contact. The type of burn produced by formic, acetic, oxalic, and other short chain acids resembles that of a mineral acid. They are readily absorbed through the lung. Oxalic acid produces a rapid onset of severe generalized symptoms and renal injury if swallowed. Table 1 illustrates the importance of these factors for the undiluted form of the carboxylic acids. Although less information is available on the sulfur acids, the same principles will apply.

Formic acid has the highest acid dissociation constant of the simple unsubstituted monocarboxylic fatty acids and probably the most severe local effect. It is followed closely by acetic acid, with propionic and butyric only slightly less severe. Trichloroacetic acid has a more severe local effect than acetic acid, related to the marked increase in acid dissociation. The presence of a second carboxyl in oxalic acid also greatly increases the acidity and to some degree the corrosive action. The fact that it is a solid and not as water soluble may counteract to some extent the effect of the increased ionization.

In the case of propionic acid, substitution of a chlorine or hydroxy group in the alpha position (α-chloropropionic and lactic acids) causes a decided increase in dissociation of the carboxylic acid and an increase in local action. While the

TABLE 1

Some Organic Acids, Their Physical Constants, and Damage to Tissue

Acid	Melting point, °C.	Boiling point, °C.	Dissociation constant	Solubility in H_2O	Damage to tissue
Formic	8.4	100.5	1.76×10^{-4}	Very	Severe
Acetic	16.6	118.1	1.8×10^{-5}	Very	Severe
Oxalic	189	Sublimes	6.5×10^{-2}	Moderate	Very severe
Trichloroacetic	57.5	197.5	2×10^{-1}	Very	Very severe
Propionic	−22.0	141.1	1.4×10^{-5}	Very	Moderately severe
α-Chloropropionic	—	186	1.5×10^{-3}	Very	Severe
Lactic	26	Decomposes	1.4×10^{-4}	Very	Severe
Butyric	−7.9	163.5	1.5×10^{-5}	Very	Moderately severe
Caprylic	16	237	About 10^{-5}	0.25% (100°C.)	Very slight
Stearic	69.4	383	About 10^{-5}	0.03% (25°C.)	None
Salicylic	159	Sublimes	1×10^{-3}	0.18% (20°C.)	Severe

common saturated fatty acids from acetic to stearic all have similar dissociation constants, the lack of water solubility above about C_8 does not allow the usual acidic properties to become evident. A saturated solution of stearic acid in water, for example, does not change the color of indicator papers. This lack of water solubility and ability to penetrate skin or mucous membranes probably explains the lack of irritation of the higher fatty acids, the limited amounts in solution being neutralized promptly.

On the other hand, salicylic acid has a rather low water solubility, but because of a specific ability to penetrate skin and a large dissociation constant it can cause severe local burns.

The presence of unsaturation, aldehydes, keto groups, OH groups, and nitro groups also tend to increase the chances of increased local damage to tissues. The acid halides dissociate into the respective organic acid and halogen acid in the presence of water and produce severe local effects as a general rule. The anhydrides also hydrolyze readily to the original organic acid and produce either the same or an increased local irritation.

2. *Specific Sensitization Effects.* While sensitization to either carboxylic or sulfur acids or amides is very rare, it does occur with anhydrides, acid halides, and certain substituted organic acids (e.g., iodoacetic acid). This is apparently explained by the reactivity of these groups allowing them to combine with proteins, forming an antigen. This is more frequent with certain aromatic sulfonyl chlorides, maleic anhydrides, and phthalic anhydrides. They are not as potent skin sensitizers, however, as phenylenediamines or hydrazines.

3. *Enzyme Inhibition.* The outstanding examples of enzyme inhibitors are iodoacetic acid and fluoroacetic acid. Lundsgaard discovered that muscle contracting anaerobically derives its energy by a glycolytic process in which lactic acid accumulates. The understanding of the nature of this was advanced when iodo-

acetic poisoned muscle was shown not to produce lactic acid.[1] This led to other discoveries of major importance in the chemistry of respiration. The present opinion is that this effect is probably produced by a combination of iodoacetic acid with the SH groups of triose phosphate dehydrogenase (e.g., $RSH + ICH_2COOH \rightarrow RSCH_2COOH + HI$).[2] Dixon has also discussed this type of reaction in connection with the mechanism of action of lachrymators such as ethyl iodoacetate.[3] These effects can be shown either in vivo or in vitro. Iodoacetamide has a similar, but more widespread, action. Similar mechanisms may be the cause of the toxic effects of monochloro- and monobromoacetic acid.

In contrast, fluoroacetic acid has no action in vitro on the isolated enzyme systems sensitive to iodoacetic acid. The substance, nevertheless, has extraordinary toxicity, having lethal effects in some species at 50 μg./kg. It has been found to be the cause of the toxic effects of a South African plant known as "Giffblaar." The symptoms are those of acute convulsive or cardiac effects, coming on after a variable latent period. The study of homologs by Saunders[4] indicated that straight chain acids containing even numbers of carbon atoms were toxic (i.e., could be broken down to FCH_2COOH) while odd carbon chains were not active. It was also shown that fluoroethanol was active, presumably because of conversion to FCH_2COOH.

The discovery of an accumulation of citrate in poisoned tissues led to the demonstration of the formation of fluorocitrate from the condensation of fluoroacetate CoA (coenzyme A) and oxaloacetate. Fluorocitrate inhibits aconitase and thus the remainder of the Krebs cycle.[5] Peters refers to this as a "lethal synthesis." No very effective antidotes have been found, probably because of the stability of the fluorocarbon bond, and the fact that the synthesis takes place inside cells in the mitochondria. Chenoweth et al.[6] have found that glycerol monoacetate has some antidotal properties in animals when given in relatively large amounts.

4. *Metabolism.* The oxidation of fatty acids is now known to proceed by conversion to thio esters of coenzyme A following which they are oxidized by the β-oxidation–condensation process to two carbon fragments which condense to form acetoacetic acid. This procedure goes on with odd-numbered carbon chains as well as even-numbered ones. However, in the case of odd number acids, the removal of C_2 units leads ultimately to the formation of a propionyl CoA. The propionyl is converted to succinic and oxidized through the citric acid cycle.[2] If for some reason the β-oxidation is blocked, as in the case of 2,2-dimethyl stearic

[1] E. Lundsgaard, *Biochem. Z.*, **233**, 322 (1932).

[2] J. Fruton and S. Simmonds, *General Biochemistry*. 2nd ed., Wiley, New York, 1958.

[3] M. Dixon, *Biochemical Society Symposia No. 2*. Cambridge Univ. Press, Cambridge, England, 1948.

[4] B. Saunders and G. Stacey, *J. Chem. Soc.*, **1948**, 1773, **1949**, 773.

[5] R. Peters in F. F. Nord, ed., *Advances in Enzymology*, Vol. XVIII. Interscience Publishers, New York–London, 1957, p. 113.

[6] M. G. Chenoweth, A. Kandel, L. B. Johnson, and D. R. Bennett, *J. Pharmacol. Exptl. Therap.*, **102**, 31 (1951).

acid, there may be oxidation to a dicarboxylic acid. In this particular case, dimethyladipic acid was formed and excreted in the urine.[7]

The anhydrides and acid halides react readily with water to form the original acids. They also act as acylating agents for free amino groups. Once they have been hydrolyzed, their fate would be the same as the original acidic group.

The fatty acid amides are generally hydrolyzed to the corresponding acid and ammonia (e.g., acetamide, propionamide, oleic acid amide, stearamide) and are inert. Acrylamide, however, has produced central nervous system damage in humans and in experimental animals, which is at least partly reversible. The typical delay in onset and the reversibility of the effects strongly suggest some type of enzyme disturbance, the nature of which is as yet unknown. Possibly, this reflects another example of a toxic synthetic process. For the details of metabolism of organic acids and amides, see the series of papers by Bray et al.,[8] C. H. Fiske,[9] and R. T. Williams.[10]

Certain N-substituted amides, such as dimethyl formamide and dimethyl acetamide, may produce chronic liver or renal damage. The details of their metabolism are not clearly understood at present. They are readily absorbed through the skin. The metabolism of aromatic carboxylic acids is influenced considerably by the nature and position of substitutes on the ring. Generally, the monocarboxylic acid group is either conjugated with glycine, as in the case of benzoic acid → hippuric acid; or conjugates with glucuronic acid to form an ester glucuronide, for example, benzoic acid → benzoylglucuronide, or is excreted unchanged. The dicarboxylic aromatic acids, such as phthalic, isophthalic, or terephthalic, are probably excreted unchanged. The alicyclic acids, such as cyclohexyl acetic acid, are said to be dehydrogenated to benzoic acid, which then appears in the urine as hippuric acid. If the acid group is not attached to the ring as in cyclohexyl propionic acid, some conjugation with glycine may occur. The aromatic acid amides are generally hydrolyzed to the corresponding acid, conjugated with glycine, and excreted in the urine. The aliphatic and aromatic sulfonic and probably sulfuric acids tend to be excreted unchanged. In contrast to the carboxylic acid amides, the sulfonic acid amides tend to be stable in the body. In the case of sulfanilamide, for example, the p-amino group is acetylated while the SO_2NH_2 group remains intact.

The metabolism of thioacids such as RCOSH is apparently unknown. Several general principles are apparent from the above review of the physiological effects and metabolism of organic acids and their derivatives: (a) In the absence of modifying factors, the organic acids are generally free from serious cumulative effects and are well handled by the body. Their local irritant effects are of primary

[7] S. Bergström, B. Borgström, N. Tryding, and G. Westöö, Biochem. J., 58, 604 (1954).

[8] H. G. Bray, W. V. Thorpe, et al., Biochem. J., 44, 39, 618 (1949); 45, 45, 467 (1949); 47, 294 (1950).

[9] C. H. Fiske, J. Biol. Chem., 55, 191 (1923).

[10] R. T. Williams, Detoxication Mechanisms. 2nd ed., Wiley, New York, 1959.

importance and, if present, are usually apparent only at higher concentrations. (b) Sensitization reactions occur only under special circumstances and are usually related to the probability of some type of protein binding. (c) A variety of specific and potent toxic agents can be produced by certain simple modifications in structure. These are most noticeable thus far in certain smaller molecules of importance in the metabolic cycle, that is, acetic versus fluoroacetic or propionamide versus acrylamide.

II. Specific Compounds

A. ALIPHATIC MONOCARBOXYLIC ACIDS

FORMIC ACID, HCOOH (Methanoic Acid)

1. Source

Carbon monoxide and sodium hydroxide react to give sodium formate. This is acidified and formic acid produced by distillation.[11]

2. Uses and Industrial Exposures

Formic acid is used for producing formate esters, as an acidifying and reducing agent, in wool, leather, rubber, and electroplating industries, and has been used as a fumigant. Over 17,000,000 pounds were produced in 1958.[12]

3. Physical and Chemical Properties

Physical state: colorless liquid
Molecular weight: 46.03
Melting point: 8.40°C.
Boiling point: 100.7°C.
Specific gravity: 1.214 (20°C.)
Vapor density: 1.59 (air = 1)
Vapor pressure: 43 mm. Hg (25°C.)
Density of saturated air: 1.03 (air = 1)
Soluble in alcohol
Flash point: 156°F. (open cup)

4. Determination in the Atmosphere

Sample in alkaline water solutions, reduce to formaldehyde, and determine colorimetrically.[13] Can also be determined by the method of Miller et al.[14]

[11] R. N. Shreve, The Chemical Process Industries. McGraw-Hill, New York, 1956.

[12] Synthetic organic chemicals, U. S. production and sales, 1958, U. S. Tariff Comm. Repts. No. 205, Washington, D. C., 1959.

[13] L. T. Fairhall, Industrial Toxicology. 2nd ed., Williams & Wilkins, Baltimore, 1957.

[14] F. M. Miller, R. F. Scherberger, H. Brockmyre, and D. W. Fassett, Am. Ind. Hyg. Assoc. Quart., 17, 221 (1956).

5. *Physiological Response*

The principal hazard is that of severe primary damage to the skin, eye, or mucosal surfaces.[15,16] Sensitization is apparently rare, but may have occurred in a person previously sensitized to formaldehyde.[17] Von Oettingen[18] has recently summarized a number of reports of accidental injury from formic acid, which indicates that the effects in man are the same as for other relatively strong acids. Baldi[16] found no evidence of delayed or chronic effects in exposed workman. Sollman[19] found that concentrations up to 0.2 per cent in the rat diet over prolonged periods produced no toxic effects, and that it was similar to acetic acid under these circumstances—at 0.5 per cent, there was less food consumption and slower growth.

The lack of any evidence of cumulative effects is of interest in view of reports that formic acid formation is involved in the toxic effects of methanol. Many recent studies of the biochemistry of "active C_1" groups[2] have established beyond question that formaldehyde and formate groups are both readily converted to the labile methyl groups present in methionine, choline, and in other compounds derived from methionine.

The limitation of possible intake by the irritant qualities may also account for the absence of chronic effects.

6. *Threshold Limit Value*

None established. Baldi[16] suggests 20 p.p.m. A more conservative level would be 5–10 p.p.m.

7. *Odor*

Pungent, penetrating.

ACETIC ACID, CH_3COOH (Ethanoic Acid)

1. *Source*

Acetic acid is produced up to about 8–10 per cent concentration by bacterial oxidation of alcohol. More concentrated solutions are produced synthetically by oxidation of propane or butane, by reaction of methanol and CO, and by oxidation of ethanol from acetylene.[11]

2. *Uses and Industrial Exposures*

Acetic acid is one of the most widely used organic acids. Important in foods, cellulose and vinyl resins, esters, manufacture of the anhydride, and as a solvent

[15] P. S. Larson, et al., *J. Pharm. Exptl. Therap.,* **116,** 119 (1956).

[16] G. Baldi, *Med. lavoro,* **44,** 483 (1953).

[17] A. J. Weil, and H. E. Rogers, *J. Invest. Dermatol.,* **17,** 227 (1951).

[18] W. F. von Oettingen, *A.M.A. Arch. Ind. Health,* **20,** 517 (1959).

[19] T. Sollman, *J. Pharm. Exptl. Therap.,* **16,** 463 (1921).

and intermediate.[11] Five hundred forty-six million pounds were produced in 1958.[12]

3. Physical and Chemical Properties

Physical state: white liquid
Molecular weight: 60.05
Melting point: 16.6°C.
Boiling point: 118.1°C.
Density: 1.049 (20°C.)
Vapor density: 2.7 (air = 1)
Vapor pressure: 15 mm. Hg (25°C.)
Refractive index : 1.37182 (20°C.)
Density of saturated air: 1.02 (air = 1)
Soluble in water and alcohol
Flash point: 104°F. (closed cup)
1 mg./liter $\simeq$ 408 p.p.m. and 1 p.p.m. $\simeq$ 2.554 mg./cu. meter at 25°C., 760 mm. Hg

4. Determination in the Atmosphere

Collect in alkali and back titrate. This has serious limitations at low levels because of the effect of atmospheric CO_2. A better method has been described.[14]

5. Physiological Response

The only important effects are those resulting from the direct effect of the acid on the skin, eye, and mucous membranes, or exposed teeth. Studies of the effect of acetic acid on the guinea pig skin[20] indicate that concentrations from 80 per cent to glacial produce severe burns; from 50 to 80 per cent, moderate to severe burns; and below 50 per cent, relatively mild injury. No injury is noted at 5 to 10 per cent concentrations. The effects of eye contact are thought to be similar. Oral ingestion of glacial acetic acid is very serious, and the details of reports of accidental human injury have recently been summarized by von Oettingen.[21] Skin sensitization to acetic acid is rare but has occurred.[17,22]

Animal studies show the oral LD_{50} in the rat (for the neutralized acid) to be 3.3 g./kg.[23] Smyth found similar values[24] for the oral LD_{50} and also studied the effect of acute inhalation and skin and eye contact. Sixteen thousand p.p.m. for 4 hours gave a mortality of 1 out of 6. Eye and skin injury were severe. Ghiring-

[20] D. W. Fassett, unpublished data, Laboratory of Industrial Medicine, Eastman Kodak Co.

[21] W. F. von Oettingen, *A.M.A. Arch. Ind. Health*, **21**, 28 (1960).

[22] F. L. Oglesby, personal communication.

[23] G. Woodard, S. Lange, K. Nelson, and H. Calvery, *J. Ind. Hyg. Toxicol.*, **23**, 78 (1941).

[24] H. F. Smyth, Jr., C. P. Carpenter, and C. S. Weil, *Arch. Ind. Hyg. Occupational Med.*, **4**, 119 (1951).

helli[25] found the LC_{50} for a 1-hour exposure in guinea pigs and mice to be about 5000 p.p.m. Symptoms of irritation of the eyes and respiratory tract occurred above 100 p.p.m. It was of interest that guinea pigs exposed to 5000 p.p.m. showed no change in their alkali reserve—evidence of the rapid metabolism of acetic acid.

In spite of the great importance of acetic acid, very few studies have been reported on the health of exposed persons. Parmeggiani[26] studied 5 workers exposed 7–12 years to high concentrations (80–200 p.p.m. at peak concentrations). The principal findings were blackening and hyperkeratosis of the skin of the hands, conjunctivitis (but no corneal damage), bronchitis, and pharyngitis, and erosion of the exposed teeth (incisors and canines). The latter finding appears to be similar to that described by Lynch and Bell,[27] who carefully examined the teeth of 126 workers exposed to inorganic acid vapors and mists. Of the 126 people exposed, 45 showed definite evidence of erosion. This interesting report and accompanying photographs should be consulted by anyone interested in the dental effects of exposure to acids.

Except for these local effects, no evidence of cumulative toxicity has been found. This is not unexpected in view of the well-known role of acetic acid and the acetyl group in carbohydrate and fat metabolism.

6. Threshold Limit Value

10 p.p.m. (25 mg./cu. meter).[28]

7. Flammability

Lower explosive limit is 4 per cent.

8. Odor

Penetrating; can be detected below 10 p.p.m.[14]

PROPIONIC ACID, CH_3CH_2COOH (Methylacetic Acid)

1. Source

Propionic acid is produced by reaction of CO with olefins or ethyl alcohol.

2. Uses and Industrial Exposures

Propionic acid is used as mold inhibitor in manufacture of esters and cellulose resins. About 23 million pounds were produced in 1958.[12]

3. Physical and Chemical Properties

Physical state: colorless liquid
Molecular weight: 74.08

[25] L. Ghiringhelli, *Med. lavoro*, **48,** 559 (1957).
[26] L. Parmeggiani and C. Sassi, *Med. lavoro*, **45,** 319 (1954).
[27] J. Lynch and J. Bell, *Brit. J. Ind. Med.*, **4,** 84 (1947).
[28] American Conference of Governmental Industrial Hygienists, *Am. Ind. Hyg. Assoc. J*, **22,** 325 (1961).

Melting point: −22°C.
Boiling point: 141.1°C.
Specific gravity: 0.993 (20°C.)
Vapor density: 2.56 (air = 1)
Vapor pressure: 10 mm. Hg (40°C.)
Density of saturated air: 1.02 (air = 1)
Soluble in alcohol
Flash point: 140°F.

4. Determination in the Atmosphere

Can be determined by methods used for acetic acid.[14]

5. Physiological Response

The chief effects are those of local damage to the skin, eye, or mucosal surfaces on contact with the concentrated acid. The oral LD_{50} of a 1 per cent water solution was greater than 400 mg./kg. in the rat. Intraperitoneally, it was 200–400 mg./kg. A 50 per cent solution in acetone was a severe skin irritant.[20]

Graham[29] has studied the effect of prolonged feeding of the sodium salt to rats at 50 times the level used in bread. A temporary growth depression was the only effect seen. Sodium propionate is not irritating when used clinically as a dusting powder for control of fungus infections.[30]

The metabolism of propionic acid (given as such or produced as a result of the oxidation of odd numbered fatty acids) is known to proceed by conversion to succinic and then through the normal processes.[2] No cumulative effects are known from industrial exposures, and it is generally recognized as safe in connection with food uses.

BUTYRIC ACID, $CH_3CH_2CH_2COOH$ (Butanoic Acid, Ethylacetic Acid)

1. Source

Butyric acid is produced by oxidation of n-butyraldehyde, also by fermentation.

2. Physical and Chemical Properties

Physical state: colorless liquid
Molecular weight: 88.10
Melting point: −4.7°C.
Boiling point: 163°C.
Specific gravity: 0.964 (20°C.)
Vapor density: 3.04 (air = 1)
Vapor pressure: 0.84 mm. Hg (20°C.)

[29] W. D. Graham and H. C. Grice, J. Pharm. Pharmacol., 6, 534 (1954).
[30] T. Sollmann, A Manual of Pharmacology. Saunders, Philadelphia, 1957.

Refractive index: 1.39906 (20°C.)
Slightly soluble in water; soluble in alcohol
Flash point: 170°F.

3. Determination in the Atmosphere

Same methods as for acetic acid.[14]

4. Physiological Response

The effects are similar to those of propionic acid. The oral LD_{50} of a 1 per cent water solution is greater than 400 mg./kg. in the rat. It is a moderately strong primary irritant in the guinea pig.[20]

Smyth[31] found the oral LD_{50} in the rat to be 2.94 g./kg. An 8-hour inhalation of saturated vapor caused no deaths. In a subsequent paper[32] the LD_{50} is said to be 8.79 g./kg. By contact with rabbit skin it was 6.35 ml./kg.

No cumulative effects are known. It occurs normally in fats and undergoes normal metabolic processes.

5. Odor and Warning Properties

Pungent odor like rancid butter which is detectable at very low levels.

ISOBUTYRIC ACID, $(CH_3)_2CHCOOH$ (2-Methylpropanoic Acid, α-Methylpropionic Acid)

1. Source

Isobutyric acid can be prepared by oxidation of isobutyraldehyde produced by the oxo process.

2. Uses and Industrial Exposures

Isobutyric acid is used as an intermediate in preparation of esters.

3. Physical and Chemical Properties

Physical state: colorless liquid
Molecular weight: 88.10
Melting point: −47.0°C.
Boiling point: 154.4°C.
Density: 0.949 (20°C.)
Vapor density: 3.04 (air = 1)
Vapor pressure: 1 mm. Hg (14.7°C.)
Refractive index: 1.39300

[31] H. F., Smyth, Jr., C. P. Carpenter, and C. S. Weil, *A.M.A. Arch. Ind. Hyg. Occupational Med.*, **4**, 119 (1951).

[32] H. F. Smyth, Jr., C. P. Carpenter, C. S. Weil, and U. C. Pozzani, *A.M.A. Arch. Ind. Hyg. Occupational Med.*, **10**, 61 (1954).

Soluble in water (20 g. per 100 ml.) and in alcohol
Flash point: 170°F.

4. Determination in the Atmosphere

See acetic acid.

5. Physiological Response

The primary irritant effects are similar to butyric acid. The oral LD_{50} in the rat (undiluted) is 400 to 800 mg./kg. A 10 per cent water solution has an oral LD_{50} in mice of more than 800 mg./kg. Isobutyric acid is a moderately severe primary irritant to the skin and eye in concentrated form. It is not a skin sensitizer.[20]

It is unlikely that any cumulative effects will be found in man or animals since isobutyric acid occurs normally in the metabolism of valine.[2] Valine is deaminated to form α-ketoisovaleric acid which loses CO_2 → isobutyryl-CoA → β-hydroxyisobutyryl-CoA. CO_2 is removed → propionyl-CoA. Thus isobutyric is converted to a propionyl group and enters into the glycogenic process.

6. Odor

Similar to butyric acid.

n-VALERIC ACID, $CH_3(CH_2)_3COOH$

1. Source and Uses

n-Valeric acid occurs normally in valerian and can be distilled. It also can be made by oxidation of amyl alcohol. It is used in perfumes, flavors, and medicines.

2. Physical and Chemical Properties

Physical state: clear liquid
Molecular weight: 102.14
Melting point: —59°C.
Boiling point: 184 to 187°C.
Specific gravity: 0.939 (20°C.)
Vapor pressure: 1 mm. Hg (42°C.)
Soluble in water (3.7 g. per 100 ml.) and in alcohol

3. Determination in the Atmosphere

See acetic acid.

4. Physiological Response

The LD_{50} in the rat of a 1 per cent water solution was greater than 400 mg./kg.[20] n-Valeric acid is a strong skin irritant in undiluted form. It is metabolized by splitting into acetic acid and pyruvic acid.[2] No cumulative effects are known.

5. *Odor*

n-Valeric acid has an unpleasant odor.

ISOVALERIC ACID, $(CH_3)_2CHCH_2COOH$

1. *Source and Uses*

Isovaleric acid occurs naturally in tobacco, valeriana, and hop oil. It can be made by oxidation of isoamyl alcohol. It is used in flavors, perfumes, and medicinals.

2. *Physical and Chemical Properties*

Physical state: colorless liquid
Molecular weight: 102.14
Melting point: −37°C.
Boiling point: 173 to 175°C.
Specific gravity: 0.931 (20/20°C.)
Vapor pressure: 1 mm. Hg (34°C.)
Soluble in water (4.2 g. per 100 ml.) and in alcohol

3. *Determination in the Atmosphere*

See acetic acid.

4. *Physiological Response*

The oral LD_{50} in rats is less than 3200 mg./kg. The symptoms were weakness, retraction of the abdomen, and vasodilation. The undiluted liquid was a strong skin irritant in the guinea pig.[20] No cumulative effects are known. It is metabolized in a normal manner for similar fatty acids.[2]

5. *Odor*

Isovaleric acid has an unpleasant odor.

CAPROIC ACID, $C_5H_{11}COOH$ (Hexanoic Acid, Pentanecarboxylic Acid)

1. *Source*

Caproic acid is produced from fermentation of butyric acid or by distillation of fatty acids.

2. *Uses and Industrial Exposures*

Caproic acid is used as a chemical reagent, in manufacture of flavors, driers, resins, and drugs.[33]

[33] A. Rose and E. Rose, *The Condensed Chemical Dictionary*. 5th ed., Reinhold, New York, 1956.

3. *Physical and Chemical Properties*

Physical state: oily liquid
Molecular weight: 116.16
Melting point: $-5.4°C$.
Boiling point: 205°C.
Specific gravity: 0.9295 (20/20°C.)
Vapor pressure: 1 mm. Hg (72°C.)
Refractive index: 1.41635
Slightly soluble in water (0.4 per cent); soluble in alcohol
Flash point: 215°F.

4. *Physiological Response*

The oral LD_{50} in rats is given as 6.44 g./kg. The skin LD_{50} in rabbits is 0.6 ml./kg. An 8-hour inhalation of saturated vapor caused no deaths. Skin and eye irritation was relatively severe.[32] A previous paper[34] gave a value for the skin LD_{50} in the guinea pig of 5 ml./kg.

5. *Odor*

Caproic acid has an odor like limberger cheese.

2-METHYLPENTANOIC ACID, $CH_3CH_2CH_2CH(CH_3)COOH$
(α-Methylvaleric Acid)

1. *Physical and Chemical Properties*

Physical state: liquid
Molecular weight: 116.16
Boiling point: 193.5°C.
Density: 0.928 (20°C.)
Soluble in water (0.57 g. per 100 ml.) and in alcohol

2. *Physiological Response*

The oral LD_{50} in the rat for the undiluted acid was 1600 to 3200 mg./kg. The symptoms were those of weakness, vasodilation, and respiratory difficulty. Contact with the rabbit eye and guinea pig skin caused severe injury.[20]

4-METHYLPENTANOIC ACID, $CH_3CH(CH_3)CH_2CH_2COOH$
(Isocaproic Acid)

1. *Physical and Chemical Properties*

Physical state: liquid
Molecular weight: 116.16

[34] H. F. Smyth, Jr. and C. P. Carpenter, *J. Ind. Hyg. Toxicol.*, **26**, 269 (1944).

Melting point: $-35°C.$
Boiling point: 110°C., 20 mm. Hg
Density: 0.925 (20°C.)
Slightly soluble in water; soluble in alcohol

2. Physiological Response

The oral LD_{50} in the rat (given as the undiluted acid) was greater than 3200 mg./kg. The symptoms were those of weakness, vasodilation, respiratory distress, excessive salivation, and in the higher doses bloody urine.

Application of isocaproic acid to guinea pig skin and rabbit eye caused severe damage.[20]

2-ETHYLBUTYRIC ACID, $(C_2H_5)_2CHCOOH$ (Diethylacetic Acid)

1. Uses

Diethylacetic acid is used in formation of esters and as a chemical intermediate.

2. Physical and Chemical Properties

Physical state: colorless liquid
Molecular weight: 116.16
Melting point: $<-15°C.$
Boiling point: 195°C.
Specific gravity: 0.9225 (20/20°C.)
Vapor pressure: 0.08 mm. Hg (20°C.)
Refractive index: 1.41788 (10°C.)
Soluble in water and alcohol
Flash point: 210°F.

3. Determination in the Atmosphere

The method of Miller[14] is probably suitable.

4. Physiological Response

The oral toxicity in animals is low. The oral LD_{50} in rats was 2.2 g./kg; the skin LD_{50} in rabbits, 0.5 ml./kg, indicating more possibility of injury by skin absorption. An 8-hour saturated vapor inhalation produced no deaths in rats. Skin irritation was slight in rabbits, but eye injury was severe. No human injuries have been reported.[32]

5. Odor

Similar to butyric acid, but less powerful.

HEPTANOIC ACID, $CH_3(CH_2)_5COOH$ (Enanthic Acid, Oenanthic Acid, Heptylic Acid, Heptoic Acid)

1. Source and Uses

Heptanoic acid is found in fused oils and can be prepared by oxidation of heptaldehyde. It is used in organic syntheses.

2. Physical and Chemical Properties

Physical state: colorless liquid
Molecular weight: 130.18
Melting point: —7.5°C.
Boiling point: 223.0°C.
Specific gravity: 0.922 (15°C.)
Vapor pressure: 1 mm. Hg (78.0°C.)
Slightly soluble in water; soluble in alcohol and ether

3. Determination in the Atmosphere

See acetic acid.

4. Physiological Response

Mice receiving 125 mg./kg./day died in 2 to 4 days.[20] Heptanoic acid is converted to glycogen, probably by splitting into two C_2 fragments plus a propionyl group.[35] It is a primary irritant in concentrated solutions.

CAPRYLIC ACID, $CH_3(CH_2)_6COOH$ (Octanoic Acid)

1. Source

Caprylic acid is produced by saponification and distillation of cocoanut oil.[33]

2. Uses and Industrial Exposures

Caprylic acid is used as a fungicide, flavoring agent, and in preparation of plasticizers.

3. Physical and Chemical Properties

Physical state: colorless liquid or solid
Molecular weight: 144.21
Melting point: 17°C.
Boiling point: 235 to 237°C.
Specific gravity: 0.910 (20/20°C.)
Vapor pressure: 1 mm. Hg (92°C.)
Very slightly soluble in water; soluble in alcohol and ether

[35] H. J. Deuel, Jr., The Lipids, Vol. III. Interscience Publishers, New York–London, 1957.

4. *Physiological Response*

The effects are those of relatively mild irritation. It occurs normally and goes through the normal metabolic pathways for similar fatty acids. Feeding 1 to 5 per cent of the free acid or glycerides in the diet of dogs caused some diarrhea.[36]

5. *Odor*

Caprylic acid has an unpleasant and irritating odor. It can be detected at 0.008 p.p.m.[37]

2-ETHYLHEXANOIC ACID, $CH_3(CH_2)_3CH(C_2H_5)COOH$
(3-Heptanecarboxylic Acid)

1. *Uses*

The metallic salts such as lead, manganese, cobalt, zinc, and calcium are used in paints and varnishes as driers. The aluminum salt is a gelling agent for hydrocarbons such as gasoline. In 1958, 1,466,000 pounds of 2-ethylhexanoic acid were produced.[12]

2. *Physical and Chemical Properties*

Physical state: colorless liquid
Molecular weight: 144.21
Melting point: $<0°C.$
Boiling point: 223 to 225°C.
Specific gravity: 0.903 (20°C.)
Vapor density: 4.9 (air = 1)
Vapor pressure: 0.03 mm. Hg (20°C.)
Soluble in water (0.2 g. per 100 ml.)
Flash point: 260°F.

3. *Determination in the Atmosphere*

Probably can be determined by the method of Miller.[14]

4. *Physiological Response*

Ethylhexanoic acid appears to be of generally low toxicity. The oral LD_{50} in rats was 3.0 g./kg. The skin LD_{50} in the guinea pig was about 6.3 ml./kg. Inhalation of saturated vapor 8 hours caused no deaths in rats. It was a severe eye irritant on direct contact with the liquid and showed some primary irritation of the skin.[38]

There appear to be no reports of human injury except for one case of corneal injury with prompt healing.[39]

[36] H. L. Wikoff, *et al., Am. J. Digest. Diseases,* **13,** 228 (1946).
[37] H. L. Wikoff, *et al., Chem. Abstr.,* **49,** 13308b (1955).
[38] H. F. Smyth, Jr. and C. P. Carpenter, *J. Ind. Hyg. Toxicol.,* **26,** 269 (1944).
[39] R. S. McLaughlin, *Am. J. Ophthalmol.,* **29,** 1355 (1946).

PELARGONIC ACID, $CH_3(CH_2)_7COOH$ (Nonanoic Acid)

1. *Physical and Chemical Properties*

Physical state: colorless, oily liquid
Molecular weight: 158.23
Melting point: 12.5°C.
Boiling point: 253 to 254°C.
Specific gravity: 0.906 (20°C.)
Very slightly soluble in water; soluble in alcohol and ether

2. *Physiological Response*

A 10 per cent solution in corn oil given orally to rats did not produce deaths up to 3200 mg./kg. Pelargonic acid is a strong skin irritant to the guinea pig. No symptoms could be produced by inhalation in rats.[20]

UNDECYLIC ACID, $CH_3(CH_2)_9COOH$ (Undecanoic Acid)

1. *Source*

Hydrogenation of undecylenic acid.

2. *Uses and Industrial Exposures*

Undecylic acid is used in organic synthesis and has been studied as an insect repellent.

3. *Physical and Chemical Properties*

Physical state: colorless crystals
Molecular weight: 186.29
Melting point: 29 to 30°C.
Boiling point: 228°C. at 160 mm. Hg
Specific gravity: 0.891 (30°C.)
Insoluble in water; soluble in alcohol and ether

4. *Physiological Response*

No data could be found on this compound. There are no reports of human injury. Undecylic acid is converted to glycogen.

UNDECYLENIC ACID, $CH_2{=}CH(CH_2)_8COOH$ (10-Undecenoic Acid)

1. *Source*

Destructive distillation of castor oil.

2. *Uses and Industrial Exposures*

Undecylenic acid is used as a therapeutic agent for fungus infections of the skin, in perfumes, and as a flavoring agent.

3. *Physical and Chemical Properties*

Physical state: solid or liquid
Molecular weight: 184.27
Melting point: 24.5°C.
Boiling point: 295°C. decomposes
Specific gravity: 0.907 (24°C.)
Vapor pressure: 10 mm. Hg (160°C.)
Insoluble in water; soluble in alcohol and ether
Flash point: 295°F.

4. *Physiological Response*

In its use as a therapeutic agent for fungus infections, occasional primary irritation may be found.[30] sensitization is rare; the oral toxicity has been studied in rats.[40] The oral LD_{50} is 2.5 g./kg. Up to 0.4 g./kg. daily in the diet of rats caused no toxicity even when continued for periods of 6 to 9 months. Newell *et al.*[41] found that 2.5 per cent in the diet affected rat growth markedly, but that the effect was much less marked at 0.5 per cent. Undecylenic acid has been used orally in humans in the therapy of psoriasis.

5. *Odor*

The odor of undecylenic acid is suggestive of perspiration.

LAURIC ACID, $CH_3(CH_2)_{10}COOH$ (Dodecanoic Acid)

1. *Source*

Distillation of cocoanut oil acids.

2. *Uses and Industrial Exposures*

Lauric acid is used in alkyd resins, detergents, as an intermediate in the formation of peroxide catalyst, and in chemical synthesis.

3. *Physical and Chemical Properties*

Physical state: colorless solid
Molecular weight: 200.31
Melting point: 44.1°C.
Boiling point: 225°C., 100 mm. Hg
Specific gravity: 0.871 (50°C.)
Vapor pressure: 1 mm. Hg (121.0°C.)
Insoluble in water; soluble in benzene and alcohol

[40] R. Tislow, S. Margolin, E. J. Foley, and S. W. Lee, *J. Pharm. Exptl. Therap.*, **98,** 31 (1950).
[41] G. W. Newell, A. K. Petretti, and L. Reiner, *J. Invest. Dermatol.*, **13,** 145 (1949).

4. Physiological Response

The glycerides of lauric occur naturally in many fats. The free acid is not particularly irritating to the skin, and it is not considered a skin sensitizer.

No toxic effects were found from feeding rats lauric acid at a 10 per cent level in the diet for 18 weeks.[42] No effects from lauric acid glycerides were found at a level of 25 per cent in the rat diet for 2 years.

Lauric acid appears to be metabolized, as do the longer fatty acids, and incorporated into depot fat, rather than going through the glucogenic cycle as do some of the shorter chain acids. Studies for carcinogenicity were negative.[43]

5. Odor

Slight odor, like oil of bay.

MYRISTIC ACID, $CH_3(CH_2)_{12}COOH$ (Tetradecanoic Acid)

1. Source

Fractional distillation of cocoanut acids.

2. Uses and Industrial Exposure

Myristic acid is used in soaps, cosmetics, flavors, and perfumes.

3. Physical and Chemical Properties

Physical state: colorless leaflets
Molecular weight: 228.36
Melting point: 54.2°C.
Boiling point: 250.5°C., 100 mm. Hg
Specific gravity: 0.853 (70°C.)
Insoluble in water; soluble in benzene, alcohol, and ether

4. Physiological Response

No hazards are known in industrial handling or in subsequent uses. Studies for carcinogenic action were negative.[43]

PALMITIC ACID, $CH_3(CH_2)_{14}COOH$ (Hexadecanoic Acid)

1. Source

Hydrolysis of natural fats; present in most commercial stearic acid.

2. Uses

Palmitic acid is used in manufacture of soaps and oils.

[42] O. G. Fitzhugh, P. J. Schouboe, and A. A. Nelson, *Toxicol. Appl. Pharmacol.*, **2**, 59 (1960).

[43] P. Shubik and J. L. Hartwell, Compounds which have been tested for carcinogenic activity, *U. S. Public Health Serv. Publ.* No. **149**, Suppl. 1 (1957).

3. *Physical and Chemical Properties*

Physical state: colorless crystals
Molecular weight: 256.42
Melting point: 62.8°C.
Boiling point: 271.5°C., 100 mm. Hg
Specific gravity: 0.849 (70°C.)
Insoluble in water; soluble in alcohol and ether

4. *Physiological Response*

Palmitic acid is one of the major saturated natural fatty acids. There are no known hazards in industrial handling or in its usage in products. It is not carcinogenic.[43]

OLEIC ACID, $CH_3(CH_2)_7CH=CH(CH_2)_7COOH$

1. *Source*

Derived from the hydrolysis and distillation of triolein.

2. *Uses*

Oleic acid is used in soap manufacture; also manufacture of detergents.

3. *Physical and Chemical Properties*

Physical state: yellowish, oily liquid
Molecular weight: 282.45
Melting point: 16°C.
Boiling point: 285°C., 100 mm. Hg
Specific gravity: 0.891 (20°C.)
Insoluble in water; soluble in alcohol and ether

4. *Physiological Response*

Oleic acid is the most abundant of all natural fatty acids. There are no known hazards in its use in industry. Like other fatty acids, it may cause marked hemolysis if injected intravenously.[44]

STEARIC ACID, $CH_3(CH_2)_{16}COOH$ (Octadecanoic Acid)

1. *Source*

From saponification of tallow.

2. *Uses*

Stearic acid is used in soaps, driers, cosmetics, in rubber manufacturing; usually contains about 45 per cent palmitic acid.

[44] N. C. Jefferson and H. Necheles, *Proc. Soc. Exptl. Biol. Med.*, **68**, 248 (1948).

3. *Physical and Chemical Properties*

Physical state: crystalline
Molecular weight: 284.47
Melting point: 69.6°C.
Boiling point: 291°C., 110 mm. Hg
Specific gravity: 0.847 (69.3°C.)
Slightly soluble in water; alcohol, and ether

4. *Physiological Properties*

The extensive use of stearic acid in industry has not been accompanied by any reports of injury.

5. *Odor*

Stearic acid has a slight odor and taste.

LINOLEIC ACID, $CH_3(CH_2)_4CH=CHCH_2CH=CH(CH_2)_7COOH$ (9,12-Octadecadienoic Acid)

1. *Source*

From linseed oil, or from dehydration of castor oil.

2. *Uses*

Linoleic acid is of major importance as a drying oil in paints and varnishes. The production of linoleic salts was 770,000 pounds in 1958.[12]

3. *Physical and Chemical Properties*

Physical state: straw-colored liquid
Molecular weight: 280.44
Melting point: −5°C.
Boiling point: 230°C., 16 mm. Hg
Specific gravity: 0.905 (20°C.)
Insoluble in water; soluble in most organic solvents

4. *Physiological Response*

Linoleic and linolenic (the corresponding 9, 12, and 15 unsaturated acid) are considered as "essential" fatty acids in the diet. Linoleic acid seems to be relatively free of skin irritant effects in its use in paints and varnishes.

Spontaneous combustion of rags soaked with linseed oil and linoleic acid salts presents a fire hazard.

5. *Odor*

Linoleic acid has a characteristic odor.

RICINOLEIC ACID, $CH_3(CH_2)_5CH(OH)CH_2CH{=}CH(CH_2)_7COOH$

1. Source

Ricinoleic acid is produced in saponification of castor oil, where it occurs as high as 80 per cent of the fatty acid content.

2. Uses

Ricinoleic acid is used for producing turkey red oils, sebacic acid, and in textile finishing.

3. Physical and Chemical Properties

Physical state: yellow viscous liquid
Molecular weight: 298.46
Melting point: 5 to 16°C.
Boiling point: 250°C., 15 mm. Hg
Density: 0.945 (15°C.)
Insoluble in water; soluble in alcohol and ether

4. Physiological Response

Ricinoleic acid is the effective cathartic agent produced when castor oil is hydrolyzed in the intestinal tract. There have been no reports of health hazards from its industrial use.

B. UNSATURATED MONOCARBOXYLIC ALIPHATIC ACIDS

METHACRYLIC ACID, $CH_2{=}C(CH_3)COOH$ (α-Methacrylic Acid, 2-Methylpropenoic Acid)

1. Source

From acetone cyanhydrin plus sulfuric acid.

2. Uses

Methacrylic acid is used in preparation of polymers and in syntheses.

3. Physical and Chemical Properties

Physical state: unstable crystalline or colorless liquid which will polymerize
Molecular weight: 86.09
Melting point: 16°C.
Boiling point: 163°C.
Specific gravity: 1.0153 (20°C.)
Vapor density: 2.97 (air = 1)
Vapor pressure: 1 mm. Hg (25.5°C.)
Refractive index: 1.43143

Soluble in warm water and alcohol
Flash point: 170°F.

4. Physiological Response

Similar to acrylic acid.

5. Odor

Acrid, repulsive.

ACRYLIC ACID, ($CH_2=CH-COOH$) (Propenoic acid, Ethylenecarboxylic Acid)

1. Source

Oxidation of acrolein or hydrolysis of acrylonitrile.

2. Uses

Acrylic acid is used in manufacture of plastics and in syntheses. Six hundred thousand pounds were produced in 1958.[12]

3. Physical and Chemical Properties

Physical state: unstable, colorless liquid which will polymerize
Molecular weight: 72.06
pK_a: 4.25[148]
Melting point: 12.3°C.
Boiling point: 141.9°C.
Density: 1.062 (16°C.)
Vapor Density: 2.5 (air = 1)
Vapor pressure: 3.1 mm. Hg (20°C.)
Refractive index: 1.4224
Miscible with water and alcohol
Flash point: 130°F. (open cup)

4. Physiological Response

Acrylic acid is a severe skin, eye, and respiratory irritant in concentrated solutions or as a liquid. No cumulative toxic reactions are known. It is probably formed in normal metabolic processes.[2] The oral LD_{50} in the rat is 2520 mg./kg.; the skin LD_{50} in the rabbit is about 950 mg./kg.[45,46] It should be handled with the usual precautions for other strong acids.

5. Odor

Acrid.

[45] W. Spector, ed., *Handbook of Toxicology*, Vol. I. Saunders, Philadelphia, 1956.
[46] H. F. Smyth, Jr., unpublished data.

CROTONIC ACID, $CH_3CH=CHCOOH$ (*trans*-2-Butenoic Acid)

1. Uses

Crotonic acid is used in manufacture of resins, plasticizers, and drugs.

2. Physical and Chemical Properties

Physical state: colorless crystals
Molecular weight: 86.09
pK_a: 4.69[148]
Melting point: 72°C.
Boiling point: 189°C.
Specific gravity: 0.964 (79.7°C.)
Vapor density: 2.97 (air = 1)
Vapor pressure: 0.19 mm. Hg (20°C.)
Soluble in water (8.3 g. per 100 ml.), alcohol, ligroin, and toluene

3. Physiological Response

The *trans* form of crotonic acid is a solid at room temperature (the *cis* form is less stable and exists as a liquid above 15°C.) and is capable of acting as a severe skin, eye, and respiratory irritant. The irritation decreases rapidly with dilution. It is present as a constituent of croton oil.

No cumulative effects would be expected at nonirritating exposure levels since it is known to be converted to the β-hydroxybutyryl-CoA by the enzyme known as crotonase present in liver and other tissues. It is also normally produced in metabolism of fatty acids by the action of the acyl CoA dehydrogenases.[2] Smyth et al.[38] found the oral LD_{50} in rats to be 1 g./kg. by skin contact; in the guinea pig, 600 mg./kg. The intraperitoneal LD_{50} in the rat is 100 mg./kg. and in the guinea pig, 60 mg./kg.[20] Crotonic acid should be regarded as having the same hazards as other strong acids.

C. HALOGENATED ACETIC ACIDS

CHLOROACETIC ACID, $ClCH_2COOH$ (Monochloroacetic Acid)

1. Source

Chlorination of acetic acid.

2. Uses and Industrial Exposures

Chloroacetic acid is used in syntheses. It also has bacteriostatic properties but is considered too hazardous for use in foods. Thirty-eight million pounds were produced in 1958.[12]

3. Physical and Chemical Properties

Physical state: colorless crystals
Molecular weight: 94.50

pK$_a$:   2.86[148]

Melting point: 50–63°C.

Boiling point: 189.4°C.

Specific gravity: 1.58 (20/20°C.)

Vapor density: 3.25 (air = 1)

Vapor pressure: 1 mm. Hg (43.0°C.)

Very soluble in water; soluble in benzene, alcohol, and ether

4. *Determination in the Atmosphere*

The method for acetic acid by Miller *et al.*[14] should be useful—also methods used for determining organic halogens.

5. *Physiological Response*

The toxic effects of monochloroacetic acid (and other halogenated acetic acids) have been studied extensively (see review by Dalgaard-Mikkleson *et al.*[47]).

Woodard *et al.*[48] pointed out that the acute oral toxicity of the sodium salt of the monochlor compound was much greater than for the corresponding di- or trichloroacetic acids. The oral LD$_{50}$ in rats was 76 mg./kg.; in mice, 255 mg./kg.; and in guinea pigs, 80 mg./kg. The symptoms were also different and consisted of apathy, loss of weight, and death in 1 to 3 days. Acetic acid, dichloro- and trichloroacetic acids had LD$_{50}$'s in the rat of 3.3 and 4.5 g./kg. and caused narcosis and either death or complete recovery in 36 hours. Similar results were noted by Dalgaard-Mikkelson *et al.*[47] Fuhrman *et al.*[49] studied the effects of chronic administration to rats compared with iodo- and bromoacetate. At 0.1 per cent in the diet, rats grew more slowly over the 200-day period of testing. No specific lesions were found, although the livers showed increased glycogen. Lower concentrations were inactive.

The mechanism of the toxic action is probably the result of a reaction with SH groups of essential enzymes such as triosephosphate dehydrogenase. Monochloroacetic acid can produce severe local reactions of the skin, eye, or respiratory tract, as might be expected from its ionization constant. Watrous has stated that it is a lung and skin irritant.[50] Screening tests indicated absorption through the skin.[20]

Skin contact and inhalation of dust or vapor should be avoided. All persons exposed to repeated contact or handling appreciable quantities should be under careful medical supervision.

DICHLOROACETIC ACID, CHCl$_2$COOH

1. *Source*

Chlorination of acetic acid in presence of iodine.

[47] S. Dalgaard-Mikkelson *et al.*, *Acta Pharmacol. Toxicol.*, **11**, 13 (1955).

[48] G. Woodard *et al.*, *J. Ind. Hyg. Toxicol.*, **23**, 78 (1941).

[49] F. A. Fuhrman *et al.*, *Arch. intern. pharmacodynamie*, **102**, 113 (1955).

[50] R. M. Watrous, *Brit. J. Ind. Med.*, **4**, 111 (1947).

2. *Uses*

Dichloroacetic acid is used as an intermediate.

3. *Physical and Chemical Properties*

Physical state: colorless liquid; also crystalline
Molecular weight: 128.95
Melting point: −4°C.
Boiling point: 193 to 194°C.
Specific gravity: 1.5724 (13°C.)
Vapor density: 4.40 (air = 1)
Vapor pressure: 1 mm. Hg (44°C.)
Refractive index: 1.4659 at 22°C.
Soluble in water (8.6 g. per 100 ml.) and alcohol

4. *Determination in the Atmosphere*

See chloroacetic acid.

5. *Physiological Response*

It is definitely less toxic on acute ingestion than monochloroacetic acid. Woodard et al.[48] found the oral LD_{50} of the sodium salt in the rat to be 4.5 g./kg.; in the mouse, 5.5 g./kg. Smyth, Carpenter and Weil[31] obtained an oral LD_{50} in the rat of 2.8 g./kg. The LD_{50} by rabbit skin was 0.5 ml./kg. Saturated vapor inhalation for 8 hours caused no deaths in rats. Severe skin and eye injury are produced.

It has a powerful keratolytic effect and has been used in dermatology for removal of skin lesions.[51]

TRICHLOROACETIC ACID, CCl_3COOH

1. *Source*

Chlorination of acetic acid with catalysts.

2. *Uses*

Trichloroacetic acid is used in syntheses and as a reagent for albumin; it also has medicinal and herbicidal uses.

3. *Physical and Chemical Properties*

Physical state: colorless deliquescent crystals
Molecular weight: 163.40
pK_a: 0.65[148]
Melting point: 57.5°C.
Boiling point: 197.5°
Specific gravity: 1.6298 (61°C.)

[51] G. A. Rau, *Clin. Med. and Surg.*, May 1937.

Vapor density: 5.65 (air = 1)
Vapor pressure: 1 mm. Hg (51°C.) ; 10 mm. Hg (88°C.)
Soluble in alcohol and ether

4. Determination in the Atmosphere

See monochloroacetic acid.

5. Physiological Response

Woodard et al.[48] found the oral LD_{50} in the rat (sodium salt) to be 3.3 g./kg. The symptoms were those of narcosis. Trichloroacetic acid produces very severe damage to tissues by direct contact and is a strong acid (see Table I, page 177 for ionization constant). Spiegl et al.[52] found that 0.0035 ml. applied to the rabbit cornea caused immediate, severe coagulation necrosis. It is excreted in the urine after exposure to trichloroethylene.[53]

Ahlmark and Forssman[54,54a] state that urine concentrations of less than 20 mg./liter of trichloroacetic acid will not usually be associated with trichloroethylene exposures of a degree giving rise to subjective symptoms.

6. Odor

Sharp, pungent.

BROMOACETIC ACID, BrCH₂COOH

1. Source

Bromination of acetic acid.

2. Uses

Bromoacetic acid is used in organic synthesis.

3. Physical and Chemical Properties

Physical state: colorless deliquescent crystals
Molecular weight: 138.96
pK_a: 2.86[148]
Melting point: 50°C.
Boiling point: 208°C.
Specific gravity: 1.934 (20°C.)
Soluble in alcohol

[52] C. J. Spiegl, R. E. Howe, and L. La France, U. S. Atomic Energy Commission Report MDDC-1715, 1948.

[53] H. E. Elkins, The Chemistry of Industrial Toxicology. 2nd ed., Wiley, New York, 1959.

[54] A. Ahlmark and S. Forssman, Arch. Ind. Hyg. Occupational Med., 3, 386 (1951).

[54a] R. Frant and J. Westendorp, Arch. Ind. Hyg. Occupational Med., 1, 308 (1950).

4. Determination in the Atmosphere

See acetic acid.

5. Physiological Response

It produces severe local reactions on skin, eye, or mucosa, similar to mono-chloroacetic acid. The chronic toxicity has been studied by Dalgaard-Mikkelson et al. in pigs[47] and the acute effects in dogs.[55]

The oral LD$_{50}$ in the rat is 100 mg./kg., and the subcutaneous fatal dose in mice is 150 mg./kg. Repeated subcutaneous injections of 20 to 30 mg./kg. to rats and rabbits are fatal in about a week.[56-58]

The pigs studied by Dalgaard-Mikkelson et al. were fed diets to which a 10 per cent aqueous monobromoacetic acid had been added in varying amounts. Daily doses of 10 to 54 mg./kg. produced fatal poisoning in 28 to 105 days, characterized by gastroenteritis, jaundice, and muscle weakness. Histological examination showed marked degenerative changes in liver, muscle, myocardium and kidney. No effects on fertility were present. Doses of the order of 2 to 6 mg./kg./day for about a year did not produce obvious toxic effects. Organically bound bromine was found in kidney, liver, and muscle in proportion to the size of the daily dose at the time of sacrifice (i.e., at 15 mg./kg./day, the liver showed 290 μg./g.; the kidney, 510 μg./g.; and muscle, 200 μg./g.).

Studies of the acute effects in the dog[55] of a buffered solution of bromoacetic acid given intravenously in doses of 8 mg./kg. or orally at 50 mg./kg. caused a delayed onset of vomiting (about 1 hour) and diarrhea, with mild to severe muscular weakness. Transient T-wave changes were seen in the electrocardiogram.

Bromoacetic acid should be handled so as to prevent skin or eye contact and and inhalation of dust or vapor. Exposed persons should be carefully followed by medical examinations.

IODOACETIC ACID, ICH₂COOH

1. Source

Iodination of acetic acid.

2. Uses

Iodoacetic acid is used in synthesis and in biochemical research.

3. Physical and Chemical Properties

Physical state: colorless crystals
Molecular weight: 185.96

[55] A. H. Anderson, S. Dalgaard-Mikkelson, and S. A. Kvorning, Acta Pharmacol. Toxicol., 11, 33 (1955).
[56] F. Neuss, Arch. exptl. Pathol. Pharmakol., 160, 551 (1931).
[57] J. I. Morrison, J. Pharm. Exptl. Therap., 86, 336 (1946).
[58] K. Hikiji, Arch. exptl. Pathol. Pharmakol., 168, 1 (1932).

pK$_a$:　3.12[148]
Melting point:　82°C.
Boiling point:　decomposes
Soluble in alcohol

4. Determination in the Atmosphere

See acetic acid.

5. Physiological Response

The local effects are similar to monochloroacetic acid. The systemic toxic effects are probably produced by reaction with SH groups (see introduction to this chapter).

The toxicity has been reviewed by Dalgaard-Mikkelson *et al.*[47] Morrison found the oral LD$_{50}$ in mice (buffered solution) to be 83 mg./kg.[57] Iodoacetic acid was fatal to rats subcutaneously at 120 mg./kg.[58a] Orally in the diet of the rat, 50–70 mg./kg./day has been found to be fatal in 10 to 40 days.[59,60]

Marcus and Frerichs[61] report 2 cases of severe contact dermatitis, one in a chemist making the material and the other in a laboratory technician. The lesions were severe and bullous in the chemist and vesicular in the technician. While primary irritation was the probable cause, sensitization could not be ruled out. In view of the protein binding, the halogenated acetic acids should be regarded as probable sensitizers.

Iodoacetic acid should be handled with the same care as in the case of monochloro- or monobromoacetic acid.

FLUOROACETIC ACID, CH$_2$FCOOH (Monofluoroacetic Acid, Fluoroethanoic Acid); CH$_2$FCOONa (Sodium Fluoroacetate)

1. Source

The sodium salt of fluoroacetic acid is prepared by reaction of ethyl chloroacetate and potassium fluoride, forming ethylfluoroacetate which is hydrolyzed with sodium hydroxide.

2. Uses

Fluoroacetic acid is used as a rodenticide, but requires highly trained personnel because of its high toxicity.

3. Physical and Chemical Properties

Physical state:　colorless solid (the sodium salt is a white powder, soluble in water)

[58a] K. Klingshoffer, *J. Biol. Chem.,* **126**, 201 (1938).
[59] L. Laszt and F. Verzar, *Arch. ges. Physiol. Pfluger's,* **236**, 693 (1935).
[60] L. Laszt and F. Verzar, *Arch. ges. Physiol. Pfluger's,* **237**, 483 (1936).
[61] M. D. Marcus and J. B. Frerichs, *J. Am. Med. Assoc.,* **142**, 805 (1950).

Molecular weight: 78.04
pK$_a$: 2.66[148]
Melting point: 33°C.
Boiling point: 165°C.
Soluble in alcohol

4. Determination in the Atmosphere

Collection of dust on membrane filters and determination by the appropriate modification of the method of Ramsey and Clifford[62] for food analysis should be useful.

5. Physiological Response

The high degree of toxicity and the mode of action have been reviewed earlier in the chapter (see page 1772). For details of the toxicity of monofluoroacetic acid and related compounds, the review by Chenoweth should be consulted.[63] Its manufacture and use require extreme precaution and special training.[64]

6. Threshold Limit Value

0.1 mg./cu. meter.[28]

TRIFLUOROACETIC ACID, CF$_3$COOH

1. Uses

Trifluoroacetic acid is used in syntheses; also has solvent properties for resins.

2. Physical and Chemical Properties

Physical state: colorless, fuming liquid
Molecular weight: 114.03
pK$_a$: 0.23[148]
Melting point: −15.3°C.
Boiling point: 72.4°C.
Specific gravity: 1.53514 (0°C.)
Vapor density: 3.92 (air = 1)
Vapor pressure: 191 mm. Hg (37°C.)
Refractive index: 1.2850
Density of saturated air: 1.74 (37°C.) (air = 1)

3. Determination in the Atmosphere

Acetic acid method of Miller is known to be useful.[14]

[62] L. L. Ramsey and P. A. Clifford, U. S. Food and Drug Administration, Washington, D. C., *J. Assoc. Offic. Ag. Chemists*, **32**, 788 (1949).
[63] M. B. Chenoweth, *J. Pharm. Exptl. Therap.*, **97**, 383 (1949).
[64] R. L. Jenkins and H. C. Koehler, *Chem. Ind. London*, **62**, 232 (1948).

4. *Physiological Response*

The local effect on skin, eye, or mucous membranes is similar to trichloro-acetic acid but probably more intense, with more pronounced penetration of tissue. Contact with guinea pig or rat skin at concentrations of 20 per cent or higher caused marked coagulation, necrosis, or complete dissolution of tissue. Ten per cent solutions caused moderately severe reactions. Solutions of 2 to 5 per cent were only moderate irritants. The oral LD_{50} in the rat and mouse (2–5 per cent solutions in water) was about 200 to 400 mg./kg. Intraperitoneally, the rat LD_{50} was less than 100 mg./kg. Inhalation of about 100 mg./liter (calculated concentrations) caused rapid death in rats with symptoms of local and respiratory damage. No deaths were produced nor were delayed effects noted at levels of about 4 to 20 mg./liter for single 6-hour exposures in rats.[20]

The precautions should be the same as for other strong acids such as trichloroacetic acid. Because of its high volatility, special attention should be given to adequate ventilation or respiratory protection.

5. *Odor*

Pungent.

D. MISCELLANEOUS HALOGENATED ALIPHATIC ACIDS

Table 2 indicates certain physical properties and acute oral LD_{50}'s of halogenated aliphatic acids up to C_6 in length. The LD_{50}'s were found in the paper by Morrison.[57] The acids were neutralized to pH 7. The species used was the mouse.

The influence of an α- or β-position of the halogen on toxicity is seen in chloropropionic acid. The α-position is associated with a much greater toxicity.

TABLE 2
Physical Properties of Some Halogenated Aliphatic Acids

Acid	Molecular weight	Melting point, °C.	Boiling point, °C.	Solubility in water	Ionization constant	Oral LD_{50}, mg./kg.
α-Chloropropionic	108.53	<25	186	Very	1.4×10^{-3}	980
β-Chloropropionic	108.53	41	204	Soluble	8.59×10^{-5}	>2000
Bromoacetic	138.96	50	208	Very	1.38×10^{-3}	100
α-Bromopropionic (*dl* form)	152.99	25.7	203.5	Very	1.08×10^{-3}	250
β-Bromopropionic	152.99	62.5		Soluble	9.8×10^{-5}	>2000
α-Bromobutyric	167.01	−4	212	6.7 g./100 ml.	pK^a 2.99[148]	310
α-Bromoisobutyric	167.01	48	198	Very		>2000
α-Bromovaleric	181.04		67, 10 mm. Hg	Slightly		380
α-Bromoisovaleric	181.04	44	230	70–80 g./100 ml.		>2000
α-Bromo-*n*-caproic	195.07	<25	240			590

It is of interest that the α-position also conferred greater potency in the nitrile series (i.e., α-hydroxy $>$ β-hydroxy propionitrile). These differences are also associated with a much larger acidic ionization constant for the α- as compared to the β-position.

It is likely that this difference in toxicity is not the direct result of the ionization constant, but that the latter simply happens to be associated with some other property of the molecule that alters its activity.

The effect of carbon chain length on oral toxicity can be seen by referring to the α-bromo derivatives (e.g., bromoacetic, α-bromopropionic, α-bromobutyric, α-bromovaleric, and α-bromocaproic). There is a steady increase in the LD_{50} with chain length. This is not accounted for by molecular weight difference, but probably results from differences in type and rate of metabolism.

The branching of the carbon chain in the α-bromo series had a marked effect, with the isobutyric and valeric acids being much less toxic than the normal acids. This certainly points to some major difference in metabolism, the nature of which does not appear to be known.

E. MISCELLANEOUS SUBSTITUTED ALIPHATIC MONOCARBOXYLIC ACIDS

GLYCOLIC ACID, HOCH₂COOH (Hydroxyacetic Acid)

1. *Source*

Prepared by hydrolysis of chloroacetic acid or by oxidation of glycol.

2. *Uses*

Glycolic acid is available in tank-car quantities and has wide uses in leather, textile, electroplating, adhesive, and metal-cleaning industries as well as in syntheses.

3. *Physical and Chemical Properties*

Physical state: crystalline solid (the 70 per cent technical grade is a light, straw-colored liquid, specific gravity 1.27, melting point, $-10°C$.)

Molecular weight: 76.05
Melting point: 78 to 79°C.
Boiling point: decomposes
Soluble in alcohol

4. *Determination in the Atmosphere*

See acetic acid.

5. *Physiological Response*

Glycolic acid is a stronger acid than acetic acid ($pK_a = 3.83$ as compared to acetic acid where $pK_a = 4.76$). It produces very severe burns of the skin or eye in the 70 per cent technical solution.[39] At 20 per cent the reaction is greatly reduced. The oral LD_{50} of a 20 per cent water solution in the rat is 1600 to 3200 mg./kg. Inhalation of air passed through the 70 per cent solution at room temperature caused no fatalities or symptoms in rats over a 6-hour exposure. It is not a skin sensitizer in guinea pigs.[20] No cumulative effects are known or expected since it is probably a normal metabolite and can be converted to glycine through the intermediate oxidation to glyoxylic acid.[2] However, no glycogen is formed in the rat.[65]

Glycolic acid is not converted to oxalic acid.[66] Dogs and cats can tolerate large daily amounts of the sodium salt over long periods of time.[67]

AMINOACETIC ACID, NH_2CH_2COOH (Glycine)

1. Source

Occurs in sugar cane and gelatin. Can be prepared by reacting monochloroacetic acid with ammonium hydroxide.

2. Uses

Aminoacetic acid is used as a buffering agent and in syntheses.

3. Physical and Chemical Properties

Physical state: crystalline; has sweet taste
Molecular weight: 75.07
Melting point: 232 to 236°C., decomposes
Specific gravity: 1.575 (50°C.)
Soluble in water; insoluble in alcohol and ether

4. Physiological Response

Glycine is a normal constituent of proteins. Its isoelectric point is 5.97, so that it is essentially neutral at the pH existing on body surfaces. No unusual precautions are needed in handling in industrial uses.

SULFOACETIC ACID, $HOSO_2CH_2COOH \cdot H_2O$

1. Physical and Chemical Properties

Physical state: solid, hygroscopic
Molecular weight: 140.11
pK_a: 4.05[148]
Melting point: 86°C.

[65] R. H. Barnes and A. Lerner, *Proc. Soc. Exptl. Biol. Med.,* **52,** 216 (1943).

[66] A. J. Milhorat and V. Toscani, *J. Biol. Chem.,* **114,** 461 (1936).

[67] H. F. Smyth, Jr., in G. Curme and F. Johnston, eds., *Glycols,* ACS Monograph No. 114. Reinhold, New York, 1952.

Boiling point: 245°C., decomposes
Soluble in water and alcohol

2. *Determination in the Atmosphere*

See acetic acid.

3. *Physiological Response*

The LD_{50} by mouth in rats is 3.16 g./kg.; by skin contact in rabbits, 1.57 g./kg.[68] Sulfoacetic acid produces severe skin and eye injury in concentrated form. Inhalation of saturated vapor for eight hours caused no deaths in rats. It is excreted as such by rats after subcutaneous injection.[68a]

Prevention of skin and eye contact with the solid or concentrated solutions is necessary in industrial handling.

PERACETIC ACID, CH_3COOOH (Peroxyacetic Acid)

1. *Source*

Reaction of acetic acid and hydrogen peroxide in the presence of sulfuric acid.

2. *Uses*

Peracetic acid is used as a bleach, catalyst, oxidant, and in epoxidation.

3. *Physical and Chemical Properties*

Physical state: colorless liquid (40 per cent solution in H_2O)
Molecular weight: 76.05
Melting point: —30°C.
Boiling point: 105°C. (approximately)
Specific gravity: 1.15 (20°C.)
Soluble in water, alcohol, and ether
Flash point: 110°C.

4. *Physiological Response*

Peracetic acid is a strong skin and eye irritant.

5. *Flammability*

Peracetic acid explodes above 110°C. Decomposes with generation of oxygen at lower temperatures. Reacts vigorously with organic materials.[69]

6. *Odor*

Acrid.

[68] H. F. Smyth, Jr., C. P. Carpenter, and C. Weil, *J. Ind. Hyg. Toxicol.*, **31**, 60 (1949).
[68a] G. A. Maw, *Biochem. J.*, **55**, 37 (1953).
[69] A. G. White and E. Jones, *J. Soc. Chem. Ind. London*, **69**, 206 (1950).

LACTIC ACID (dl), $CH_3CHOHCOOH$ (α-Hydroxypropionic Acid)

1. Source

Lactic acid is produced by fermentation or by synthetic methods from sulfite pulp liquor.

2. Uses

Lactic acid is available in edible grades (22 to 44 per cent concentration) and in 50 to 80 per cent concentration for plastic and USP grades. Used in foods as an acidulant, in adhesives, plastics, and textiles. Over 5,000,000 pounds produced in 1958.[12]

3. Physical and Chemical Properties

Physical state: hygroscopic solid or liquid
Molecular weight: 90.08
pK_a: 3.86[148]
Melting point: 16.8°C.
Boiling point: 122°C., 14 mm. Hg
Specific gravity: 1.249 (15°C.)
Infinitely soluble in water, alcohol, and ether

4. Physiological Response

The more concentrated solutions can cause severe burns of skin or eye. Smyth *et al.* found the oral LD_{50} in the rat to be 3.73 g./kg.; in the guinea pig, 1.81 g./kg.[70] No cumulative effects are known or expected because of its normal occurrence in the diet and in metabolic processes. However, massive oral doses given daily (1.5 g./kg.) to rats produced weight loss, anemia, and increased blood CO_2. The same was true of acetic acid. The reaction was less when the acids were given in the diet.[71] Precautions should be taken against skin and eye contact with the concentrated solutions.

SORBIC ACID, $CH_3(CH)_4COOH$ (2,4-Hexadienoic Acid)

1. Source

Formation of a trimer from acetaldehyde and air oxidation.

2. Uses

Sorbic acid is used as a fungicide in edible foods.

   [70] H. F. Smyth, Jr., J. Seaton, and L. Fischer, *J. Ind. Hyg. Toxicol.*, **23**, 259 (1941).
   [71] Z. Wysokinska, *Roczniki Państwowego Zakladu Hig.*, **3**, 273 (1952); *Chem. Abstr.*, **49**, 5638c (1955).

3. *Physical and Chemical Properties*

Physical state: crystalline
Molecular weight: 112.12
Melting point: 134.5°C.
Boiling point: 228°C. decomposes
Vapor pressure: 10 mm. Hg (120°C.) ; 50 mm. Hg (153°C.)
Soluble in hot water; very soluble in alcohol and ether
Flash point: 260°F. (open cup)

4. *Physiological Response*

Smyth and Carpenter[72] found the oral LD_{50} in the rat to be 7.36 g./kg. The sodium salt had a similar LD_{50}; 30 daily doses of about 50 mg./kg. in the food of rats caused some reduced growth.

The sodium salt caused no effects at a level of 1 g./kg./day. Fryklöf[73] reported that, in a small number of subjects tested, sorbic acid was a primary irritant at 0.15 per cent concentration in water under the conditions of a closed patch test. Ointments containing it were said to cause itching of the face, and 3 subjects were thought to have skin sensitivity.

Unpublished studies[20] did not show primary irritation or sensitization of the guinea pig skin when 2 samples were applied in 0.1 molar concentration. No reports of injury in industrial handling have been noted.

THIOGLYCOLIC ACID, $HSCH_2COOH$ (Mercaptoacetic Acid, Thiovanic Acid)

1. *Source*

Prepared by reacting chloroacetic acid with potassium sulfhydrate.

2. *Uses*

Thioglycolic acid is used for hair waving and as an analytical reagent. Production in 1958 was 2.1 million pounds (thioglycolic acid plus derivatives).[12]

3. *Physical and Chemical Properties*

Physical state: colorless liquid
Molecular weight: 92.11
pK_a: 3.68[148]
Melting point: —16.5°C.
Boiling point: 123°C., 29 mm. Hg
Specific gravity: 1.325 (20°C.)
Vapor pressure: 10 mm. Hg (18°C.)
Soluble in alcohol

[72] H. F. Smyth, Jr. and C. P. Carpenter, *J. Ind. Hyg. Toxicol.*, **30**, 63 (1948).
[73] L. E. Fryklöf, *J. Pharm. and Pharmacol.*, **10**, 719 (1958).

4. *Determination in the Atmosphere*

Thioglycolic acid produces a blue color with ferric iron, which may be useful for air analysis.[13]

5. *Physiological Response*

Thioglycolic acid has a high degree of acute toxicity. The oral LD_{50} of the undiluted acid in rats is <50 mg./kg., and fatalities were produced by application of a 10 per cent solution to guinea pig skin at <5 ml./kg. The symptoms were weakness, gasping respirations, and convulsions. The concentrated acid is a severe skin and eye irritant, and it is absorbed through the skin. The primary irritation decreases considerably with dilution; at 2 to 10 per cent, the reactions are moderate. It was not a potent skin sensitizer in guinea pigs.[20]

In its use in cold hair-waving preparations, it may be present as a salt (*e.g.*, NH_4) at a pH of about 9 or 9.5. Under these conditions the R—S—S—R groups of keratin are reduced to RSH and the thioglycolic acid is converted to $(—SCH_2-COOH)_2$. Skin irritation occasionally occurs in professional hair dressers but is apparently relatively rare in home use. Specific sensitization appears rare.[30]

The cause of the rather high degree of acute toxicity and details of metabolism appear to be unknown, but it seems probable that a specific action on certain enzymic thiol groups will be found.*

Skin and eye contact should be avoided and adequate ventilation provided. It has good warning properties, but there is rapid fatigue of odor detection. Persons exposed to appreciable quantities in industry should be under careful medical supervision.

6. *Odor*

Pronounced, disagreeable odor.

β-MERCAPTOPROPIONIC ACID, $HSCH_2CH_2COOH$

1. *Source*

Reaction of β-propiolactone with sodium hydrogen sulfide.

2. *Uses*

Mercaptopropionic acid has been tried in cold-wave preparations and in syntheses.[30]

3. *Physical and Chemical Properties*

Physical state: liquid (clear)
Molecular weight: 106

---

* Freeman *et al.*[73a] reported some aspects of the absorption, distribution, and excretion of S^{35}-labeled sodium thioglycolate. It was rapidly absorbed through the skin. Sixty per cent or more was excreted in the rabbit urine in twenty-four hours as inorganic sulfate or neutral sulfur.

pK$_a$: 4.34[148]
Melting point: 16.8°C.
Boiling point: 111.5°C., 15 mm. Hg
Specific gravity: 1.218 (21°C.)
Soluble in water, alcohol, benzene, and ether

4. Physiological Response

The oral LD$_{50}$ in rats of the undiluted liquid was 50 to 400 mg./kg. The symptoms were similar to thioglycolic acid—weakness and convulsions. Deaths occurred within 30 minutes after dosing. The liquid appears to be absorbed directly through the guinea pig skin with deaths from as low as 5 ml./kg. Severe delayed eschar formation was noted. A 10 per cent solution in water applied as a compress to guinea pig skin also produces severe damage and eschars. No deaths were produced at this concentration. It is not a skin sensitizer in the guinea pig.

Adequate ventilation and prevention of contact with body surfaces are needed in handling in industry.[20]

Voss[74] has studied the effects of a large number of mercapto acids and amides on the skin of guinea pigs and humans. The β-form did not sensitize human subjects, although the α-isomer gave an incidence of 10 per cent sensitization. This appears to be another instance of the importance of α- vs. β-substitution in enhancing a biological effect.

Schroeder[75] has studied the effects of various mercapto acids on hypertensive dogs and rats. β-Mercaptopropionic acid caused a drop in blood pressure in hypertensive, but not in normal, animals.

5. Odor

Strong and acrid, sulfidelike odor.

THIOLACETIC ACID, CH$_3$COSH

1. Source and Uses

Reaction of glacial acetic acid and phosphorus pentasulfide and distillation. Thiolacetic acid is used in syntheses and as a chemical reagent.

2. Physical and Chemical Properties

Physical state: yellow, volatile liquid
Molecular weight: 76.11
Melting point: <-17°C.
Boiling point: 81.8°C., 630 mm. Hg
Specific gravity: 1.074 (10°C.)
Soluble in water, alcohol, and ether

[73a] M. V. Freeman, J. H. Draize, and P. K. Smith, *J. Pharm. and Exptl. Therap.,* **118,** 304 (1956).

[74] J. G. Voss, *J. Invest. Dermatol.,* **31,** 273 (1958).

[75] H. A. Schroeder, *Science,* **114,** 441 (1951).

3. *Physiological Response*

Little toxicity data were found. Thiolacetic acid is a lachrymator, and the vapors are irritating to the nose and throat. The intraperitoneal LD_{50} in the mouse was 60 to 125 mg./kg., with clonic convulsive movements and gasping respirations noted.[76]

In another study, the oral LD_{50} of a dilute water solution was 200 to 400 mg./kg. in the mouse and rat and below 100 mg./kg. intraperitoneally. There was a rapid onset of weakness and unconsciousness and death. When the undiluted material was applied to the skin of the guinea pig, it was a strong irritant, and deaths were produced by this route at less than 1 ml./kg. Inhalation produced rapid deaths in the rat down to as low as 700 p.p.m. for 2 hours.[20] Skin contact and inhalation of vapor should be avoided.

4. *Odor*

Thiolacetic acid has a pungent odor resembling both acetic acid and hydrogen sulfide.

F. ALIPHATIC DICARBOXYLIC ACIDS

The aliphatic dicarboxylic acids (also citric acid, a tricarboxylic acid) are of importance because of their use in foods, beverages, and drugs, and in manufacturing processes. A number of them present no hazard from low level chronic exposure and are normally present in metabolic processes. Primary irritant effects of varying severity are present as with the monocarboxylic acids. Sensitization is rare and certainly less of a problem than with the anhydrides.

These materials are all solids at room temperature so that contact is usually in the form of dust or crystals.

Table 3 gives some of the pertinent physical properties.

OXALIC ACID, HOOC—COOH · 2H$_2$O

1. *Source and Uses*

Oxalic acid can be made from carbon monoxide and sodium hydroxide.[11] It is used as a bleach, metal polish, rust remover, and in syntheses. In 1958 14 million were produced.[12]

2. *Determination in the Atmosphere*

See acetic acid.

3. *Physiological Response*

Oxalic acid can cause severe burns of the skin, eye, or mucous membranes either as a dust, or in solution. Klauder[77] reports that solutions of 5 to 10 per cent

---

[76] Sloan-Kettering Institute, unpublished report No. 7600.
[77] J. V. Klauder, L. Shelanski, and K. Gabriel, *A.M.A. Arch. Ind. Health,* **12,** 412 (1955).

TABLE 3

Physical Properties of Polycarboxylic Acids

Acid	Formula	Molecular weight	Melting point, °C.	Boiling point, °C.	Solubility in H_2O g. per 100 ml.	Solubility in Other	Specific gravity	Acid dissociation pK_{a1}	pK_{a2}	Other
Oxalic	HOOC.COOH.2H₂O	126.07	187, anhydride 101.5, dihydrate	150 sublimes	9.5 (15°C.)	Alcohol, ether	1.653 (18.5/4)	1.46	4.40	
Malonic	CH₂(COOH)₂	104.06	135.6	Decomposes	74		1.631 (15/4)	2.80	5.85	
Succinic	HOOCCH₂CH₂COOH	118.09	189	235 (−H₂O)	6.8 (20°C.)	Alcohol, ether	1.572 (25/4)	4.17	5.64	
Malic (dl)	HOOCCH(OH)CH₂COOH	134.09	128.5	150 decomposes	144	Alcohol	1.601 (20/4)	3.4	5.05	
Thiomalic (mercaptosuccinic)	HOOCCH(SH)CH₂COOH	150.2	149–150			Alcohol				
Tartaric	HOOCCH(OH)COOH.H₂O	168.10	205	100 (−H₂O)	Soluble	Alcohol	1.697 (20/4)	2.95	4.16	
Adipic	HOOC(CH₂)₄COOH	146.14	153	265 (10 mm.)	1.4 (15°C.)	Alcohol, ether (slightly)	1.360 (25/4)	4.43	5.52	V.P. = 1 mm., 159.5°C.
Pimelic	HOOC(CH₂)₅COOH	160.17	106	272 (100 mm.)	5	Alcohol		4.47	5.42	
Azelaic	HOOC(CH₂)₇COOH	188.22	106.5	286.5 (100 mm.)	0.2 (15°C.)	Alcohol	1.029 (20/4)	4.54	5.52	V.P. = 1 mm., 178.3°C.
Sebacic	HOOC(CH₂)₈COOH	202.24	134.5	294.5 (100 mm.)	0.1 (17°C.)	Alcohol, ether	1.207 (25/4)	4.55	5.52	
Citric	HOOCCH₂C(OH)(COOH)-CH₂COOH	192.12	153	Decomposes	133	Alcohol	1.542 (18/4)	3.08	4.75	
Maleic (cis)	HOOCCH=CHCOOH	116.07	130.5	135 decomposes	78.8 (25°C.)	Alcohol	1.590 (20/4)	1.83	6.58	
Fumaric (trans)	HOOCCH=CHCOOH	116.07	sublimes at 200		0.7	Alcohol	1.635 (20/4)	3.00	4.52	
Itaconic (methylene succinic)	HOOCC(=CH₂)CH₂COOH	130.10	161, dec.	268 sublimes	8.3 (20°C.)	Alcohol	1.632	3.84	5.55	

are irritating on prolonged exposure. There have been a number of cases of fatalities recorded following ingestion, even with doses as low as 5 g. The symptoms come on rapidly and consist of shock, collapse, and convulsions. Marked renal damage and deposition of calcium oxalate in the lumen of the renal tubules occurs.[78]

Oxalate ions occur normally in metabolic processes. The quantities in food are harmless. Rats fed 1.2 per cent showed no toxic effects.[79] The precautions are the same as for other strong acids.

MALONIC ACID, $CH_2(COOH)_2$ (Methanedicarboxylic Acid)

1. Source and Uses

Malonic acid is produced by reaction of monochloroacetic acid with potassium cyanide and subsequent hydrolysis. It is used as an intermediate in production of barbiturates.

2. Determination in the Atmosphere

See acetic acid.

3. Physiological Response

Malonic acid is a relatively strong acid and can damage skin and mucous membranes, although not as severely as oxalic acid. The LD_{50} by intraperitoneal injection in the rat and mouse is about 1500 mg./kg.[45] While it is thought to inhibit succinic dehydrogenase, no toxic effects in humans have been attributed to this action.

SUCCINIC ACID, $HOOCCH_2CH_2COOH$

1. Source and Uses

Succinic acid is derived from fermentation of ammonium tartrate. It is used in the syntheses of medicinals, and in the manufacture of perfumes, dyes, and lacquers.

2. Determination in the Atmosphere

See acetic acid.

3. Physiological Response

The free acid may cause some irritation but is otherwise free of toxic effects and is a normal metabolite. Studies of succinic acid and its sodium and magnesium salts by Friend and Gold[80] did not demonstrate any systemic toxic effects. Large

[78] H. Többen, Arch. pathol. Anat. u. Physiol., **302**, 246 (1938).
[79] O. G. Fitzhugh and A. A. Nelson, J. Am. Pharm. Assoc., **36**, 217 (1947).
[80] V. L. Friend and H. Gold, J. Am. Pharm. Assoc., **36**, 50 (1947).

oral doses produce nonspecific vomiting and diarrhea. No injuries in industrial use have been reported.

4. *Odor*

No odor; very acid taste.

MALIC ACID (*dl*), HOOCCH (OH) CH₂COOH (Hydroxysuccinic Acid)

1. *Source and Uses*

Malic acid occurs naturally in many fruits. Made synthetically by catalytic oxidation of benzene to maleic acid, with reduction to malic by steam. Used in manufacture of wine and food and in medicinals.

2. *Determination in the Atmosphere*

See acetic acid.

3. *Physiological Response*

Malic acid is a relatively strong acid and can produce some irritation of skin and mucous membranes in concentrated form. No cumulative effects are known, and there are no reports of injury in industrial handling. The oral LD_{50} in mice and rats is about 1600 to 3200 mg./kg. (1 per cent water solution). The symptoms are weakness, retraction of the abdomen, cyanosis, and respiratory distress. It is a strong skin irritant in the guinea pig. No differences were noted in the *dl* or *l* forms of malic acid.[20]

THIOMALIC ACID, HOOCCH(SH)CH₂COOH (Mercaptosuccinic Acid)

1. *Uses*

Thiomalic acid is used as a chemical intermediate, in rubber additives, and in biochemical research.

2. *Physiological Response*

Thiomalic acid has been reported to be a heavy metal antidote.[81] The LD_{50} in mice was reported to be about 2900 mg./kg. for the sodium salt (route was not stated). It was said to inhibit or reverse mercurial diuresis in a manner similar to BAL or Unithiol (sodium-2,3-dimercaptopropane sulfonate).

In another laboratory the LD_{50} in rats orally was 800 to 1600 mg./kg. (10 per cent solution of the free acid). The symptoms were weakness, retraction of the abdomen, depressed respiration, and cyanosis. The LD_{50} by a 24-hour skin contact with the moist solid was greater than 2 g./kg. in the guinea pig. Severe skin damage resulted under these conditions. It was not a skin sensitizer in this species.[20]

[81] N. M. Kostygov, *Pharmacol. Toxicol. U.S.S.R.*, **1**, 274 (1958).

Voss[74] reports that it caused sensitization of the skin under experimental conditions in some human subjects.

3. Odor

Thiomalic acid has a sulfidelike odor.

TARTARIC ACID, HOOC(CHOH)$_2$COOH.H$_2$O (Racemic Acid, Dihydroxysuccinic Acid)

1. Source

From reaction of maleic anhydride and hydrogen peroxide or from residues in wine manufacture.

2. Determination in the Atmosphere

See acetic acid.

3. Physiological Response

Tartaric acid is a relatively strong acid and can cause local irritation. Elsbury[82] has given an interesting account of a marked dental erosion in the exposed surfaces of teeth in girls packaging tartaric acid powder. Air analyses showed an average of 1.1 mg./cu. meter of free tartaric acid. Clinical erosion was obvious in 6 months; by 3 years, there was major destruction of the tooth structure. Health hazards from tartaric acid have also been discussed by Barsotti,[83] who examined 156 workers in a large tartaric acid factory. Levels in air ranged from very low to 32 mg./cu. meter. Thirty persons showed some erosion of teeth, but this was felt to be of borderline significance. Small chronic skin ulcers, gastric disturbance, and febrile attacks similar to metal-fume fever were noted.

ADIPIC ACID, (CH$_2$CH$_2$COOH)$_2$

1. Source and Uses

Oxidation of cyclohexanol by nitric acid.[11] Adipic acid is used in the manufacture of nylon.

2. Physiological Response

Enders[84] reports that it is only slightly toxic on acute or chronic exposures in rabbits and rats. It is slowly excreted in the urine and probably partly oxidized. No reports of injury in industrial handling were found.

[82] W. B. Elsbury, R. C. Browne, and J. Boyes, *Brit. J. Ind. Med.*, **8**, 179 (1951).
[83] M. Barsotti, C. Sassi, and G. Ghetti, *Med. lavoro*, **45**, 239 (1954).
[84] A. Enders, *Arch. exptl. Pathol. Pharmakol.*, **197**, 706 (1941).

PIMELIC ACID, HOOC(CH$_2$)$_5$COOH

1. *Source*

Pimelic acid has been found in castor oil and may be one of the precursors in the synthesis of biotin.[2]

2. *Physiological Response*

Pimelic acid appears to be of low acute toxicity. The oral LD$_{50}$ in the rat is greater than 3200 mg./kg. with no obvious signs of toxicity. It is not a skin irritant nor is it absorbed through the guinea pig skin.[20]

AZELAIC ACID, (CH$_2$)$_7$(COOH)$_2$ (Nonandioic Acid)

1. *Source and Uses*

Oxidation of oleic acid by ozone. Azelaic acid is used in large quantities in plasticizers, alkyd resins, lacquers, and in syntheses.

2. *Physiological Response*

See adipic acid.[84]

SEBACIC ACID, (CH$_2$)$_8$(COOH)$_2$

1. *Source*

Sebacic acid is derived from butadiene or from dry distillation of castor oil.[33]

2. *Physiological Response*

See adipic acid.[84]

CITRIC ACID, HOOCCH$_2$C(OH)(COOH)CH$_2$COOH

1. *Source and Uses*

Citric acid is derived by fermentation processes. It is used in very large amounts as a flavoring or acidifying agent in beverages, foods, and medicines— also as a sequestering agent and as a chemical intermediate.

2. *Physiological Properties*

Citric acid is a normal metabolite in the body. No major problems are present in industrial handling. Some hypocalcemic effects have been reported arising out of the calcium-binding effects during transfusions of large volumes of citrated blood.[85]

[85] J. D. Bunker *et al, J. Am. Med. Assoc.*, **157**, 1361 (1955).

MALEIC ACID, (:CHCOOH)₂ (Butanedioic Acid, Toxilic Acid)

1. *Source and Uses*

Maleic acid is a by-product of phthalic acid manufacture, and is also made by the oxidation of benzene. It is used in synthetic resins, textiles, and syntheses.

2. *Physiological Response*

The toxicity of maleic acid has recently been reviewed by Henson.[86] It is a strong acid (pK_{a_1} is nearly that of oxalic acid) and can produce marked irritation of skin and mucous membranes. Winter and Tullius have studied the effects on the rabbit eye and find that severe effects can result from concentrations as low as 5 per cent.[87] Fitzhugh and Nelson have studied the chronic toxicity in rats and report some toxic effects as low as 0.5 per cent. The differences from controls were not marked, however, and the pathology was nonspecific.[79] Dye[88] studied the effect on growth of young rats by subcutaneous injection in sesame oil. Daily doses of 0.5 to 2 mg./rat for 60 days were tolerated, but 5–10 mg./rat produced deaths, retarded growth, and patchy hair distribution. Comparisons were made with other substances affecting plant growth such as indoleacetic acid.

There are no reports of cumulative toxic effects in man, and the hazard in industrial use is essentially that of primary irritation of exposed surfaces.

FUMARIC ACID, (:CHCOOH)₂

1. *Source and Uses*

Fumaric acid is derived from molasses fermentation, isomerization of maleic acid, or oxidation of benzene. It is used in polyester and alkyd resins, lacquers, inks, foods, and in syntheses.

2. *Determination in the Atmosphere*

See acetic acid.

3. *Physiological Response*

Fumaric acid, in contrast to maleic, is a much weaker acid and is relatively insoluble in water. It is a normal metabolite. Studies by Fitzhugh and Nelson[79] show that rats tolerate about three times as much in the diet as in the case of maleic acid. Levey[89] has made a careful study of the acid and sodium salt. They were found to be less toxic orally than tartaric acid. Humans tolerated 500 mg./day for a year. The degree of local irritation is mild compared to maleic acid. No problems in industrial handling are known.

86 E. V. Henson, *J. Occupational Med.*, **1**, 339 (1959).
87 C. A. Winter and E. J. Tullius, *Am. J. Ophthalmol.*, **33**, 387 (1950).
88 W. S. Dye *et al.*, *Growth*, **8**, 1 (1944).
89 S. Levey *et al.*, *J. Am. Pharm. Assoc.*, **35**, 298 (1946).

ITACONIC ACID, HOOCC($=CH_2$)CH$_2$COOH (Methylenesuccinic Acid)

1. Source and Uses

Itaconic acid is produced by carbohydrate fermentation. It is used in copolymerization, oil additives, resins, and in syntheses.

2. Determination in the Atmosphere

See acetic acid.

3. Physiological Response

Itaconic acid seems to be of low toxicity, although it may inhibit utilization of succinic acid when fed in relatively large doses to rats.[90] One per cent in the diet of rats caused slow growth. A large single oral dose causes an increased output of succinic acid in rabbit urine. It is probably a mild irritant. No industrial injuries have been reported.

G. ACID ANHYDRIDES

The acid anhydrides have higher boiling points than the corresponding acids or acid halides and are more stable toward hydrolysis than the latter. Their physiological effects generally resemble those of the corresponding acids, but they are more potent eye irritants in the vapor phase, producing a somewhat persistent conjunctivitis. They are probably slowly hydrolyzed on contact with body tissues, but it is possible that some of their irritant effects may be due to the intact molecule. They can react with amino groups of proteins.[2] Therefore, it is not surprising that occasional sensitization occurs. Except for irritant and sensitizing phenomena, cumulative toxic effects are not seen in industrial handling.

They can usually be determined by methods similar to those used for acids.

The properties of certain anhydrides are listed below.

ACETIC ANHYDRIDE, (CH$_3$CO)$_2$O (Ethanoic Anhydride)

1. Source

Oxidation of propane and butane, oxidation of acetaldehyde, reaction of ketene plus acetic acid.

2. Uses

The major use of acetic anhydride is acetylation of cellulose in the production of cellulose acetate. It is also used in alkyd resins. Nine hundred and sixty-five million pounds were produced in 1958.[12]

3. Physical and Chemical Properties

Physical state: colorless liquid
Molecular weight: 102.09

[90] A. N. Booth, J. Taylor, R. H. Wilson, and F. De Eds, *J. Biol. Chem.*, **195**, 697 (1952).

Melting point: −73.1°C.
Boiling point: 140.8°C.
Density: 1.08712 (15°C.)
Vapor density: 3.6 (air = 1)
Vapor pressure: 10 mm. Hg (36°C.)
Refractive index: 1.39038
Density in saturated air: 1.04 (air = 1)
Soluble in water (13 g. per 100 ml.—decomposes), alcohol, and ether; reacts with alcohols
Flash point: 150°F.

4. Determination in the Atmosphere

See acetic acid—also can be determined by method of Diggle and Gage.[91]

5. Physiological Response

Acetic anhydride is known to be a severe eye and skin irritant. McLaughlin[39] reported 16 cases of corneal burns in humans, 1 of which resulted in loss of vision. Three healed slowly, the remainder rapidly.

The oral LD_{50} in rats was 1.78 g./kg.; inhalation of 2000 p.p.m. by rats for 4 hours caused deaths. It caused severe eye burns in rabbits.[31] No cumulative effects are known; specific skin sensitization occasionally occurs.

6. Threshold Limit Value

20 mg./cu. meter or 5 p.p..m.[28]

7. Flammability

Explosive limit: 2.67 to 10.13 per cent by volume.

8. Odor

Acetic anhydride has a pungent odor and is a lachrymator.

PROPIONIC ANHYDRIDE, $(CH_3CH_2CO)_2O$

1. Source

Propionic anhydride is produced by methods similar to those for acetic anhydride.

2. Use

Propionic anhydride is produced by methods similar to those for acetic in production of alkyd resins, and as an intermediate.

[91] W. M. Diggle and J. C. Gage, *Analyst*, **78**, 473 (1953).

3. *Physical and Chemical Properties*

Physical state: colorless liquid
Molecular weight: 130.14
Melting point: $-45°$C.
Boiling point: 167 to 169°C.
Specific gravity: 1.0119 (20°C.)
Vapor density: 4.5 (air $= 1$)
Vapor pressure: 1 mm. Hg (20°C.)
Decomposes in water
Soluble in alcohol but may react with it; soluble in ether

4. *Determination in the Atmosphere*

See acetic anhydride.

5. *Physiological Response*

The effects are similar to those of acetic anhydride with eye irritation from the vapor being a prominent symptom. Skin irritation also is present. Smyth[68] found the oral LD_{50} in rats to be 2.36 g./kg.; the skin LD_{50} in rabbits, 10 ml./kg.; and that breathing air saturated with vapor for more than 1 hour produced deaths in rats.

6. *Odor*

Propionic anhydride has a pungent odor.

BUTYRIC ANHYDRIDE, $(CH_3CH_2CH_2CO)_2O$

1. *Source*

Butyric anhydride is prepared by methods similar to acetic anhydride.

2. *Uses*

Butyric anhydride is used in production of cellulose acetate butyrate, drugs, and tanning agents.

3. *Physical and Chemical Properties*

Physical state: colorless liquid
Molecular weight: 158.19
Melting point: $-75°$C.
Boiling point: 199.5°C., 760 mm. Hg
Specific gravity: 0.968 (20/20°C.)
Vapor density: 5.5 (air $= 1$)
Vapor pressure: 0.3 mm. Hg (20°C.)
Decomposes in water

Soluble in ether
Flash point: 190°F.

4. Determination in the Atmosphere

See acetic anhydride.

5. Physiological Response

Butyric anhydride causes persistent eye irritation if contact with the vapor is repeated and prolonged. It is also a skin irritant. The oral LD_{50} in the rat is reported to be 8.8 g./kg. and the skin LD_{50} in the rabbit, 6.4 g./kg.[45]

6. Odor

Butyric anhydride has a pungent odor.

MALEIC ANHYDRIDE, OCOCH=CHCO (Toxilic Anhydride)

1. Source

Maleic anhydride is produced by reaction of benzene vapor and air with vanadium catalyst—also from phthalic anhydride or butenes.

2. Uses

Maleic anhydride is used in alkyd resin ester and polyester resins, in drying oils, and agricultural chemicals. In 1958, 52 million pounds were produced.[12]

3. Physical and Chemical Properties

Physical state: crystalline
Molecular weight: 98.06
Melting point: 53°C.
Boiling point: 202°C., sublimes
Specific gravity: 0.934 (20°C.)
Insoluble in water; very slightly soluble in alcohol
Flash point: 218°F.

4. Physiological Response

Maleic anhydride can produce severe eye and skin burns. Winter and Tullius[87] studied the effect on the rabbit eye and reported that either the powder or solutions produced persistent congestion and vascularization of the cornea. Very severe injury to the rabbit eye was also reported by Carpenter and Smyth.[92] McLaughlin[39] studied 14 cases of eye burns in humans.

The oral LD_{50} in the rat was found to be 400 to 800 mg./kg. The skin LD_{50} in the guinea pig was greater than 20 g./kg., although severe burns were produced.

[92] C. P. Carpenter and H. J. Smyth, Jr., *Am. J. Ophthalmol.*, **29**, 1363 (1946).

It was not a skin sensitizer in the guinea pig in one test; however, this has occurred in humans.[20]

Strict precautions should be taken to prevent contact of the solid or solution with the skin or eyes.

CITRACONIC ANHYDRIDE, $OCOC(CH_3)=CHCO$ (Methylmaleic Anhydride)

1. Uses

Citraconic anhydride is used as a reagent for alkalies, alcohols, and amines.

2. Physical and Chemical Properties

Physical state: colorless liquid
Molecular weight: 112.08
Melting point: 8 to 10°C.
Boiling point: 213°C.
Specific gravity: 1.25 (15°C.)
Decomposes in water
Soluble in alcohol

3. Physiological Response

The oral LD_{50} in the rat is 2.6 g./kg.; the skin LD_{50} in the guinea pig, 1 ml./kg. Eye injury in the rabbit was severe.[38]

CROTONIC ANHYDRIDE, $(CH_3CH=CHCO)_2O$

1. Uses

Crotonic anhydride is used in syntheses.

2. Physical and Chemical Properties

Physical state: colorless liquid
Molecular weight: 154.16
Melting point: 72°C.
Boiling point: 246 to 248°C., 766 mm. Hg
Specific gravity: 1.040 (20°C.)
Decomposes in water
Soluble in ether

3. Physiological Response

Contact with the liquid will produce severe skin and eye burns. Delayed vesicant or bullous responses may occur in the skin following contact of contaminated clothing with skin. There may be little sensation of pain at the time of contact.[93]

4. Odor

Crotonic anhydride has a slight odor.

PHTHALIC ANHYDRIDE, $C_6H_4(CO)_2O$

1. Source

Phthalic anhydride is produced by air oxidation of naphthalene or *o*-xylene.

2. Uses

Phthalic anhydride is used in production of plasticizers for vinyl and acetate resins, alkyd resins, manufacture of phenophthaleins and dyes. Over 237 million pounds of phthalic anhydride esters were produced in 1958.[12]

3. Physical and Chemical Properties

Physical state: crystalline
Molecular weight: 148.11
Melting point: 130.8°C.
Boiling point: 284°C., sublimes
Specific gravity: 1.527 (4°C.)
Very slightly soluble in water; soluble in alcohol; slightly soluble in ether

4. Physiological Response

Phthalic anhydride is a potent skin, eye, and upper respiratory irritant and can cause skin and possibly pulmonary sensitization. In spite of this, enormous quantities are handled yearly with relatively little difficulty. Baader[94] and Menschick[95] have published studies on exposed people in factories making phthalic acid and anhydride. In both of these studies the clinical findings were principally conjunctivitis, bloody nasal discharge, atrophy of the nasal mucosa, hoarseness, cough, occasional bloody sputum, bronchitis, and emphysema. Four cases of bronchia asthma were attributed by Menschick to this exposure. A variety of alterations in the blood were reported, but these seem slight and of questionable significance. Some lowering of blood pressure and minor signs of central nervous system excitation were reported. Skin sensitization and eczematous responses were noted and, in some instances, urticaria. Baader states that air concentrations of 30 mg./cu. meter cause definite conjunctival irritation, and that some signs of mucous membrane irritation were noted at 25 mg./cu. meter. Some air pollution problems in the vicinity of the factory were studied.

[93] D. W. Fassett, personal observations.
[94] E. W. Baader, *Arch. Gewerbepathol. Gewerbehyg.*, **13**, 419 (1955).
[95] H. Menschick, *Arch. Gewerbepathol. Gewerbehyg.*, **13**, 454 (1955).

It is apparent that phthalic anhydride handling requires adequate local and general ventilation, careful education of personnel, and medical supervision with placement and follow-up examinations.

Other studies on the toxicity of phthalic anhydride have been made by Friebel[96] and Jacobs.[97] The acute oral LD_{50} in the rat is 800 to 1600 mg./kg. and less than 100 mg./kg. intraperitoneally in the guinea pig.[20]

5. Threshold Limit Value

A suggested value would be <25 mg./cu. meter.[98]

6. Flammability

Explosive limits: 1.7 per cent at 140°C. (lower); 10.5 per cent at 193°C. (upper).

7. Odor

Phthalic anhydride has a characteristic choking odor.

H. LACTONES

The lactones (inner esters of hydroxy acids) are of increasing importance industrially and have many uses as intermediates, solvents, and are used in the perfume industry. The lactone portion of cardiac glycosides is an unsaturated α-lactone and is essential in their physiological activity. Information on the toxic properties of lactones is incomplete, although a number of them appear to be of low acute toxicity. A striking exception is β-propiolactone.

β-PROPIOLACTONE, $OCH_2CH_2C=O$

1. Source

Reaction of ketene with formaldehyde with zinc chloride as a catalyst.

2. Uses

β-Propiolactone has a wide variety of uses in organic syntheses. It is used as a viricidal agent in plasma and tissue grafts.

3. Physical and Chemical Properties

Physical state: liquid
Molecular weight: 72.1
Melting point: —33.4°C.
Boiling point: 56 to 57°C., 15 mm. Hg

[96] H. Friebel, *Arch. Gewerbepathol. Gewerbehyg.*, **14**, 465 (1956).
[97] J. L. Jacobs, *Proc. Soc. Exptl. Biol. Med.*, **43**, 74 (1940).
[98] *Quart. Safety Summary*, **27**, 22 (1956).

Density: 1.1460 (20°C.)
Soluble in water; miscible with alcohol; reacts with alcohol
Flash point: 165°F.

4. Physiological Response

The acute oral and intraperitoneal toxicity is high in rats (orally, the undiluted material has an approximate LD_{50} of 50–100 mg./kg. and about the same intraperitoneally). The symptoms come on rapidly and consist of twitching, gasping, convulsions, and collapse. The LD_{50} by skin application is less than 5 ml./kg. in the guinea pig, indicating considerable absorption. It is a strong primary irritant in the undiluted form.[20]

A number of studies have been made to determine the safety of use in treating plasma and tissue grafts. While β-propiolactone causes liver necrosis and renal tubular damage when given by itself intravenously, if it is allowed to react with proteins before injection, the toxicity is said to be very much reduced.[99–103] Some studies have shown it to be a skin carcinogen or cocarcinogen in mice.[104,105] Intracutaneous injection in mice, however, did not produce tumors in the experiments of Boyland and Sargent.[106] Eckardt[107] refers to it as a skin carcinogen in the mouse.

In view of its high toxicity and possible carcinogenicity, skin contact and inhalation should be prevented and exposed persons should be under careful medical supervision.

γ-BUTYROLACTONE, $CH_2CH_2CH_2CO$
$\underline{\qquad\quad O \qquad\quad}$

1. Source

γ-Butyrolactone can be prepared from γ-hydroxybutyric acid.

2. Uses

γ-Butyrolactone is used as a solvent, an intermediate, paint remover, resin solvent, and in petroleum processing.

[99] F. W. Hartman and A. R. Kelly, *Federation Proc.*, **12**, 390 (1953).

[100] A. R. Kelly, C. E. Rupe, J. J. Tazuma, and F. W. Hartman, *Federation Proc.*, **13**, 434 (1954).

[101] D. H. Basinski and D. G. Remp, *Federation Proc.*, **14**, 178 (1955).

[102] F. W. Hartman, G. A. LoGrippo, and A. R. Kelly, *Federation Proc.*, **15**, 518 (1956).

[103] A. R. Kelly, F. W. Hartman, and G. A. LoGrippo, *Gastroenterology*, **28**, 24 (1955).

[104] F. J. C. Roe and M. H. Salaman, *Brit. J. Cancer*, **9**, 177 (1955).

[105] F. J. C. Roe and O. M. Glendenning, *Brit. J. Cancer*, **10**, 357 (1956).

[106] E. Boyland and S. Sargent, *Brit. J. Cancer*, **5**, 433 (1951).

[107] R. E. Eckardt, *Industrial Carcinogens*. Grune and Stratton, New York, 1959.

3. *Physical and Chemical Properties*

Physical state: liquid
Molecular weight: 86.09
Melting point: —44°C.
Boiling point: 90 to 92°C., 17 mm. Hg
Density: 1.1286 (15°C.)
Miscible with water and alcohol
Flash point: 209°F.

4. *Physiological Response*

γ-Butyrolactone probably has a low acute toxicity. One report gives the lethal dose in mice intraperitoneally as about 1600 mg./kg. and the maximum tolerated dose/day by this route in the mouse as 15 mg./day for 10 days.[108]

The oral LD_{50} in the rat and mouse is 800 to 1600 mg./kg. The intraperitoneal LD_{50} was 200 to 400 mg./kg. with deaths delayed as long as 2 weeks. The symptoms were those of weakness, unconsciousness, and increased depth of respiration. It appears to be readily absorbed through guinea pig skin with some irritation produced. The LD_{50} by skin contact is less than 5 ml./kg. It is not a skin sensitizer in this species.[20] No data or reports on cumulative effects were found. The metabolism is unknown. Because of possible delayed effects, exposures by skin or inhalation should be controlled.

5. *Odor*

γ-Butyrolactone has an unpleasant odor.

γ-VALEROLACTONE, $CH_3CHCH_2CH_2CO$
$$\lfloor\!-\!-\!-O\!-\!-\!-\rfloor$$

1. *Source*

Isomerization of allylacetic acid.

2. *Uses*

γ-Valerolactone is used as a solvent, in dye baths as a complexing agent, in brake fluids, and in cutting oils.

3. *Physical and Chemical Properties*

Physical state: liquid
Molecular weight: 100.12
Melting point: —37°C.
Boiling point: 207°C.
Specific gravity: 1.0518
Vapor pressure: 6 mm. Hg (73°C.)

[108] *Summary Tables of Biological Tests,* **7,** 687 (1955), National Research Council.

Miscible with water and alcohol

Flash point: 205°F.

4. *Physiological Response*

Deichmann reported that it has a low oral toxicity; the LD_{50} is about 9200 mg./kg. in the rat and 2600 mg./kg. in the rabbit.[109] Similar results have also been noted.[20] The oral LD_{50} in the mouse and rat was 3200 to 6400 mg./kg. The symptoms produced were simply weakness. No delayed deaths were noted. It is not a skin irritant or sensitizer. The results of skin absorption studies varied with two samples, but it is probably not readily absorbed through the skin. The metabolism is not known; no human injuries have been reported. Because of possible skin absorption, skin contact should be avoided.

HEXANOIC ACID ε-LACTONE CH₃CH(CH₂)₃COO

1. *Physiological Response*

The acute toxicity of hexanoic acid ε-lactone is low. Smyth[32] reported the oral LD_{50} in the rat as 4.29 g./kg.; the LD_{50} by skin contact in the rabbit as 5.99 ml./kg. Inhalation of saturated vapor for 8 hours caused no deaths in rats. Little skin, but considerable eye damage resulted in the rabbit.

I. ACID HALIDES (RCOCl or RSO₂Cl, etc.)

There has been very little investigation of the specific toxic action of the organic acid halides, although their hazardous properties are generally well known to chemists. When they are soluble in water, they hydrolyze very rapidly, in some cases even fuming in moist air (i.e., acetyl chloride or bromide) and give rise to the corresponding organic and inorganic acids, as, for example,

$$RCOCl + HOH \rightarrow RCOOH + HCl$$

Similar reactions go on with alcohol or ammonia, for example,

$$RCOCl + HOR' \rightarrow RCOOR' + HCl$$

$$RCOCl + HNH_2 \rightarrow RCONH_2 + HCl$$

in which cases esters or amides plus the inorganic acids are produced.

These acid halides are well known to produce severe primary irritation of the eyes, skin, or respiratory tract. Delayed vesication is not uncommon. Sensitization may also occur, for example, methyl *p*-toluene sulfonyl chloride. The boiling points of acid halides are usually considerably lower than the corresponding acids; therefore, higher vapor concentrations may be encountered under comparable circumstances. The delayed deep vesicles produced by some of the compounds suggest that in some instances penetration of surfaces by the intact molecule may occur,

[109] W. Deichmann, *J. Ind. Hyg. Toxicol.*, **27**, 263 (1945).

followed by a slow hydrolysis or reaction with proteins, probably at —OH, —NH₂, or —SH groups.

While cumulative toxic effects are theoretically possible if the original organic acid is also known to be cumulative in action (i.e., monochloroacetyl chloride), the irritant properties of the halide prevent voluntary exposure to significant amounts and make difficult any experimental investigation of this aspect of their toxicity. Table 4 gives some comparisons of the important effects of halide formation on the physical properties.

TABLE 4

Effects of Acid Halide Formation on Physical Properties

Acid and acid halide	Melting point, °C.	Boiling point, °C.	Solubility in water, g. per 100 ml.	Remarks
Formic acid	8.4	100.5	Soluble	Major damage to all tissues
Methyl chloroformate		71.4	Decomposes	Chlorides of formic acid are un-
$\overset{\text{O}}{\underset{\parallel}{\text{CH}_3\text{O}-\text{C}-\text{Cl}}}$				known, but chlorides of carbonic acid derivatives are in use; severe burns and inhalation hazard
Acetic acid	16.6	118	Soluble	Severe tissue damage
Acetyl bromide	−97	76	Decomposes	Severe tissue damage; inhalation hazard
Propionic acid	−22	141	Soluble	Severe burns
Propionyl chloride	−94	80	Decomposes	Severe burns
Phenyl acetic acid	77	266	1.66	
Phenylacetyl chloride		95 (12 mm.)	Decomposes	
p-Toluene sulfonic acid	106	140 (20 mm.)		Strong irritant
p-Toluene sulfonyl chloride	69	146 (15 mm.)		Strong irritant and vesicant
Oxalic acid	101	150	Soluble	Severe tissue damage
Oxalyl chloride	−12	64	Decomposes	Severe burns; inhalation hazard

When handling these materials, there should be education of personnel in the nature of the hazard, provision of personal protective equipment (safety goggles, gloves), adequate ventilation, and provision for flushing eyes or skin with water.

J. ACID AMIDES

The simple amides of carboxylic acids (C₂ or above) are solids with higher boiling points than their respective acids. They have a variety of uses (especially acetamide) in organic syntheses, surfactants, stabilizers, and soldering fluxes. The higher amides, such as stearic and oleic, are used as release agents in the manufacture of plastics and films.

TABLE 5

Physical Properties of Amides

Amide	Formula	Molecular weight	Melting point °C.	Boiling point °C.	Vapor pressure, mm. Hg	Density	Solubility in water	Other
Formamide	$HCONH_2$	45.04	2.6	210 (decomposes)	10 (109°C.)	1.134 (20°C.)	Infinite	
Acetamide	CH_3CONH_2	59.07	81	222	10 (105°C.)	1.159 (20°C.)	98 g. per 100 ml.	Mousy odor
Propionamide	$CH_3CH_2CONH_2$	73.09	79	213	10 (105°C.)	1.042 (20°C.)	Soluble	
Butyramide	$CH_3CH_2CH_2CONH_2$	87.12	116	216		1.032 (20°C.)	16.3 g. per 100 ml.	
Valeramide	$CH_3(CH_2)_3CONH_2$	101.2	114–116.			1.023	Soluble	
Lauramide	$CH_3(CH_2)_{10}CONH_2$	199.3	103. (freezing point)				Insoluble	
Palmitamide	$CH_3(CH_2)_{14}CONH_2$	255.4	106	236 (12 mm.)			Insoluble	
Oleamide	$C_{17}H_{33}CONH_2$	281.5	76				Insoluble	
Stearamide	$CH_3(CH_2)_{16}CONH_2$	283.5	109	251 (12 mm.)			Insoluble	

The substituted amides such as dimethylformamide are liquids with valuable solvent properties. Aromatic N-derivatives of acetamide form a large class of important dye and medicinal intermediates. The amides of aromatic carboxy or sulfonic acids comprise a very large group of useful materials of importance in the manufacture of sulfonamide drugs, insect repellents, and dyes. Unsaturated amides such as acrylamide are reactive monomers used in certain acrylic resins.

Some physical properties of certain typical amides are listed in Tables 5 through 8.

1. Simple Aliphatic Carboxylic Amides

The physiological effects of the simple carboxylic amides in Table 5 are (with the possible exception of formamide) marked by a lack of apparent serious hazard by any type of contact. The lack of cumulative or other toxic effect is probably

TABLE 6

Physical Properties of Unsaturated Aliphatic Amides

Amide	Formula	Molecular weight	Melting point, °C.	Vapor pressure, mm. Hg	Solubility in water
Acrylamide	$CH_2{=}CHCONH_2$	71.08	85	2 (87°C.) 10 (117°C.)	205 g. per 100 ml.
Methacrylamide	$CH_2{=}C(CH_3)CONH_2$	85	110		Infinite
N,N-Dimethyl acrylamide	$CH_2{=}CHCON(CH_3)_2$	99	<20	2 (46°C.)	Soluble
N-Isopropyl acrylamide	$CH_2{=}CHCONHCH(CH_3)_2$	113	60	2 (83°C.)	Soluble
N-tert-butyl acrylamide	$CH_2{=}CHCONHC(CH_3)_3$	127.2	128		0.7 g. per 100 ml.

explained by their relatively rapid hydrolysis to the corresponding acid or in some cases their excretion in the urine unchanged. The details of the metabolism have been studied extensively in rabbits by Bray et al.[8] and also in cats by Fiske.[9] There appears to be good evidence for hydrolysis, both in vivo and in vitro. The site of hydrolysis is probably the liver, where there are relatively nonspecific amidases. The rate of hydrolysis increases rapidly as the molecular weight increases. Actually, acetamide was hydrolyzed more slowly than formamide, with about 70 per cent of the dose appearing unchanged in the urine over a 4-day period. Decomposition products are sometimes irritating.

Formamide is absorbed directly through the guinea pig skin with an LD_{50} of <5 ml./kg.; only slight irritation was noted, however. The effects of vapor inhalation do not appear to have been studied. No human injury has been reported.

The relative lack of irritant effects on skin and mucous membranes indicates that hydrolysis is probably not occurring there to any extent, otherwise the lower

TABLE 7

Physical Properties of N-Substituted Aliphatic Amides

Amide	Formula	Molecular weight	Melting point, °C.	Boiling point, °C.	Vapor pressure, mm. Hg	Density	Solubility in Water	Other
N,N-Dimethylformamide	HCON(CH₃)₂	73	−61	153	3.7 (25°C.)	0.953 (15.6/15.6)	Infinite	Flash point: 153°F. (open cup)
N,N-Diethylformamide	HCON(CH₃)₂	101.2		177	1.0 (25°C., approx.)	0.908	Infinite	
N-Phenylacetamide (acetanilide)	C₆H₅NH(COCH₃)	135.17	114	305			Soluble in hot water	
N,N-Dimethylacetamide	CH₃CON(CH₃)₂	87.12		166	9 (60°C.) 2 (35°C.)	0.9366	Infinite	
N,N-Diethylacetamide	CH₃CON(C₂H₅)₂	91	<20	182–186	2 (35°C.)	0.920		Flash point: 170°F.

TABLE 8

Physical Properties of Amides of Aromatic Carboxylic and Sulfonic Acids

Amide	Formula	Molecular weight	Melting point, °C.	Boiling point, °C.	Density	Solubility in water
Benzamide	$C_6H_5CONH_2$	121.13	130	290	1.341 (4°C.)	1.4 g. per 100 ml.
α-Phenylacetamide	$C_6H_5CH_2CONH_2$	135.2	156–160	280–290 decomposes		Soluble in hot water
Benzene sulfonamide	$C_6H_5SO_2NH_2$	157.2	156			0.43 g. per 100 ml.
p-Aminobenzene sulfonamide (sulfanilamide)	$NH_2C_6H_4SO_2NH_2$	172.2	163			0.4 g. per 100 ml.
o-Toluamide	$CH_3C_6H_4CONH_2$	135.2	139	May explode on heating[125]		Insoluble
m-Toluamide		135.2				
N,N-Diethyl-m-toluamide (85–95% meta)	$CH_3C_6H_4CO(C_2H_5)_2$	191.	<25	288–292 decomposes	0.9962– 0.9986	Insoluble
Salicylamide	$2\text{-}OHC_6H_4CONH_2$	137.1	140–142	270 decomposes		Slightly
Nicotinamide	$N{=}CHC(CONH_2){=}CHCH{=}CH_2$	122.1	129–130			100 g. per 100 ml.

molecular weight acids produced would cause local damage. Skin sensitization also appears to be rare, indicating that they probably do not specifically bind to these proteins.

2. Unsaturated Aliphatic Amides

In contrast to the innocuous nature of the amides in Table 5, which we just discussed, certain of the unsaturated and N-substituted amides in Table 6 exhibit a tendency to produce toxic effects on the central nervous system, liver, or kidneys. While they are relatively free from sensitizing properties, they may in some instances be local irritants and may have considerable ability to penetrate through intact skin.

ACRYLAMIDE and Related Compounds

Acrylamide has been reported to cause unusual neurotoxic effects in a variety of animal species as well as in a few men occupationally exposed.[110-112] The syndrome is characterized (particularly in the cat) by ataxia, incoordination, and weakness of the extremities (especially hind quarters) with full retention of sensory responses and normal or hyperactive reflexes. Changes in personality and "visual hallucinations" appear in cats.

The symptoms are always delayed in onset for a day or two after even the larger doses. As the daily dose is decreased, the time for onset becomes longer. In the cat, for example, a daily dose (intravenous or intraperitoneal) of 1 mg./kg. will take about 125 days to produce the picture, while only two daily doses of 50 mg./kg. are required. It seems to require a cumulative dose on the order of 80 to 130 mg./kg. for induction of the full effect. If the exposure is terminated as soon as symptoms appear, the symptoms regress and full recovery eventually takes place. If exposure continues, weight loss, weakness, and muscular atrophy may progress and become irreversible. No histological effects are seen in the nervous system or elsewhere. The location of action is probably subcortical and in the midbrain, according to Kuperman.[111]

The signs in humans were somewhat similar to those in the cat. Initially there were complaints of drowsiness, fatigue, and tingling of the fingers. The principal complaints were a stumbling, propulsive type of gait with a sense of unsteadiness, especially when the eyes were closed. Neurological findings were vague and equivocal. All cases recovered on cessation of exposure at the first sign of the effects, although in one case this took about a year.[112] The exposures causing the effect were difficult to determine because of the uncertain factor of skin penetration. Thus far the mechanism of action and metabolism have not been solved. Con-

---

[110] *American Cyanamid Company New Products Bulletin,* Vol. II, Collective. New York, 1952.

[111] A. S. Kuperman, *J. Pharm. Exptl. Therap.,* **123,** 180 (1958).

[112] H. H. Golz, personal communication, 1955.

version to a more toxic material by some metabolic process seems a good probability in view of the delay in onset and the need for a certain, fairly constant, total threshold dose before symptoms occur. Effects can be produced by oral or skin contact as well as by injection. In rabbits a 10 per cent water solution is not a primary irritant although neurotoxic symptoms developed a week after 10 daily applications in doses of 1 ml./kg.[110] In humans, a 1 per cent water solution causes dermal irritation. Polyacrylamides, free of monomer, are inert.

Acrylamide should be handled only under carefully controlled conditions, with all sources of contact (especially skin) held as low as possible. In addition, exposed persons should be under careful medical supervision, with examinations at reasonably frequent intervals to detect any early neurological symptoms. Abnormal physical findings would not be expected at early stages of intoxication.

The following compounds have not been studied as intensively as acrylamide, although in two instances it seems probable that much weaker but similar actions are present.

METHACRYLAMIDE

Intraperitoneal injections of a 10 per cent aqueous solution in the cat failed to produce any neurological symptoms in several series of repeated injections, starting with 30 mg./kg./day and progressing to 120 mg./kg./day. The total cumulative dose in a period of about three weeks was 900 mg./kg. Prolonged observation following treatment failed to show any signs of neurotoxicity or of any other action.[20]

Application of 1 g. of the moist solid to rabbit skin on an area of 12 sq. cm. for 4 hours caused only minor primary irritation.[113]

The removal of any apparent neurological effect by substitution of an α-methyl group in acrylamide is of considerable interest, since the two monomers have many similar chemical properties. No reports of human injury or illness in handling have been reported.

N,N-DIMETHYLACRYLAMIDE

Doses of 35 mg./kg./day for 10 days given intraperitoneally as a 10 per cent water solution in a cat produced weight loss but no neurological effect. The dose was increased to 70 mg./kg./day; and after a total cumulative dose of 540 mg./kg., some difficulty in walking, tremors, and slight spasticity of the legs were seen. The typical acrylamide syndrome was never apparent.

The oral LD_{50} in the rat is 200 to 400 mg./kg. No delayed neurological signs were noted. Weakness, secretions around the mouth and eyes, and convulsions were produced. N,N-Dimethylacrylamide is readily absorbed through guinea pig skin when applied by gauze pad, producing fatalities at less than 0.5 ml./kg. It was not a skin sensitizer in this species.[20]

[113] *Methacrylamide Bulletin.* Rohm and Haas, Philadelphia, 1955.

N-ISOPROPYLACRYLAMIDE

A total cumulative dose of 840 mg./kg. intraperitoneally in a cat over a period of 18 days produced paralysis of the hind quarters and head tremors. No visual hallucinations, hypermetria, or loss of righting were present. Treatment was discontinued and full recovery took place in about five weeks. In the rat, the oral LD$_{50}$ is about 350 mg./kg. Crude solutions may be skin irritants. N-Isopropylacrylamide is not readily absorbed through the skin. Polymers are inert.[20]

Other Acrylamide Derivatives

Investigation has shown that similar symptoms can be produced by repeated feeding of high doses of methylol acrylamide to rats, but that the characteristic effect was not seen with N-tert-butylacrylamide, N-tert-octylacrylamide, or N,N'-methylenebisacrylamide.[114,115]

3. N-Substituted Aliphatic Amides

The physiological properties of one of the N-substituted amides, N,N-dimethylformamide (see Table 7), have been summarized in the American Industrial Hygiene Association's Hygienic Guide.[116] N,N-Dimethylformamide is considered to be moderately hazardous by inhalation, is moderately irritating to skin, and is a definite hazard by skin absorption.[117] Its metabolism is uncertain although some may be excreted unchanged. There is experimental evidence of liver and kidney damage in rabbits and cats and to a lesser extent in rats when the compound is given by injection or inhalation (rats tolerated 420 p.p.m. daily for long periods, while cats were affected by 100 p.p.m.). There does not appear to have been injury in exposed persons or dogs exposed to low levels (50 p.p.m.) by inhalation. The odor is unpleasant—fishy—and probably serves as a warning. The current tentative threshold limit is 20 p.p.m. (60 mg./cu. meter).

N,N-Diethylformamide (DEF) and N,N-dimethylacetamide (DMA) have also been studied experimentally.[20] They resemble dimethylformamide in being readily absorbed through the skin and in having some tendency to cumulative effects. These are more marked with dimethylacetamide than with either dimethyl- or diethylformamide. Marked weight loss occurred in rats with repeated intraperitoneal injections of dimethylacetamide when compared to the two formamide derivatives. Horn[118] has noted liver damage in some rats and dogs exposed to repeated inhalations of DMA of 100–200 p.p.m.

[114] American Cyanamid Company New Products Bulletin, Vol. III, Collective. New York, 1954.

[115] B. C. Shaffer, personal communication.

[116] American Industrial Hygiene Association Hygienic Guide, Dimethyl formamide, Am. Ind. Hyg. Assoc. J., 18, 279 (1957).

[117] W. Massmann, Brit. J. Ind. Med., 13, 51 (1956).

[118] H. J. Horn, personal communication.

No threshold limits have been proposed for DEF or DMA, although it would seem reasonable to use the value proposed for DMF (20 p.p.m.) as a rough guide. It is especially important to recognize the possibility of significant skin absorption and to prevent repeated or prolonged skin contact with the concentrated liquids or their solutions. Exposed persons should be under adequate medical supervision, and air concentrations should be evaluated. It is probable that methods described for DMF will be suitable.[119,120]

N-PHENYLACETAMIDE (Acetanilide)

N-Phenylacetamide (see Table 7) is a representative of one of the large series of important N-aromatic derivatives of acetamide. It can also be regarded as an acetylated aniline. By varying the aromatic substituent, many useful compounds are known and used as intermediates in manufacture of dyes and medicinals. N-Phenylacetamide is also used as a preservative for peroxide solutions and as a rubber accelerator. Because of its long and extensive use as an antipyretic and analgesic drug, there have been many studies of its toxicity in animals and humans.[30]

The p-ethoxy derivative is currently used medically even more widely. The acute toxicity of N-phenylacetamide is very low—rats and monkeys tolerating doses of 200 to 400 mg./kg. orally per day for many weeks.[30] With very large doses, however, methemoglobin and some hyperplasia of the bone marrow can be produced.[121] It is well known to produce cyanosis in some humans when taken repeatedly, possibly due to formation of sulfhemoglobin.[122] Large doses in acute poisoning produce methemoglobin, however.

The metabolism has been studied by several workers. It is thought to be oxidized by liver microsomes to N-acetyl-p-aminophenol, which is apparently the active analgesic substance. It is readily excreted in the urine as sulfate or glucuronate conjugate of the phenol. Minor amounts may be deacetylated to p-aminophenol and aniline.[123,124]

In spite of the occasional occurrence of cyanosis in humans during therapeutic use, this rarely occurs in occupational exposures.[125] It is not mentioned by Watrous in his review of the health hazards of the pharmaceutical industry,[50] nor has it been noted in the author's experience. Specific sensitization dermatitis is very rare.

No threshold limits for air concentrations have been proposed.

[119] F. Bergman, Anal. Chem., 24, 1367 (1952).

[120] W. Massmann, Arch. Gewerbepath. Gewerbehyg., 14, 91 (1955).

[121] A. G. Karczmar, Federation Proc., 8, 305 (1949).

[122] T. B. Reynolds and A. G. Ware, J. Am. Med. Assoc., 149, 1538 (1952).

[123] L. A. Greenberg et al., J. Pharm. Exptl. Therap., 90, 68, 150 (1947).

[124] B. B. Brodie and J. Axelrod, J. Pharm. Exptl. Therap., 94, 24, 76 (1948).

[125] P. G. Stecher, ed., The Merck Index of Chemicals and Drugs, 7th ed., Merck & Co., Rahway, N. J., 1960.

4. Amides of Aromatic Carboxylic and Sulfonic Acids

The amides of aromatic carboxylic and sulfonic acids provide another series of useful intermediates (see Table 8). α-Phenylacetamide has been used as a penicillin precursor. The sulfonamides are used as chemotherapeutic agents. Diethyl-m-toluamide is an insect repellent. Salicylamide has been used as an analgesic similar to aspirin; and the amide of nicotinic acid is, of course, an essential food element. The physiological properties can be interpreted in terms of their metabolism which in some cases has received extensive study.[8,10]

The simple unsubstituted aromatic carboxylic amides, such as benzamide and α-phenylacetamide, are hydrolyzed to the corresponding acid, which is then conjugated with glycine and excreted in the urine, for example, as hippuric acid or phenylaceturic acid.

In case of methyl or nitro substituents on the ring, the metabolic pattern appears to be a combination of hydrolyses of the amide as above, plus the customary alterations of the ring substituents. In the case of the toluamides, the methyl group is oxidized to the acid, so that the compound may appear in the urine as a glycine or glucuronic acid conjugate or both. The nitro groups were reduced to amino groups and then acetylated; or in some cases, the ring was further oxidized forming an additional hydroxyl substituent.

o-Toluamide was of particular interest since it was considerably more acutely toxic than the others, causing a deep reversible narcosis lasting several hours at doses of 250 mg./kg. by mouth. Slight narcosis only was noted with m-toluamide. The metabolism of the ortho compound was not clearly established but it probably was excreted as an o-hydroxymethyl benzoic acid or amide.[8]

N,N-Diethyltoluamide has recently come into wide use as an insect repellent. Ambrose[126] has carried out an extensive study of its acute and chronic toxicity in rats and rabbits. Aside from moderate primary irritation of the skin and eyes from concentrated solutions, no specific toxic action was found.

High levels in the diet (1 per cent) of rats produced depression of weight gain, but no histological changes were noted. Inhalation of vapor or aerosols produced only local upper respiratory irritation. No study of the metabolic fate was found, although this would appear to be of considerable interest. No cases of human injury have been reported.

Salicylamide has had some use as an analgesic, and its metabolism is similar to aspirin.

Nicotinamide is excreted as the N-methyl derivative in man, rats, and dogs. It is free from the marked cutaneous flushing effect of nicotinic acid in industrial handling.[50]

K. NAPHTHENIC ACIDS

1. Source and Uses

The naphthenic acids are derived from various cycloparaffins in petroleum, probably by oxidation. The commercial acids are usually viscous liquid mixtures

---

[126] A. M. Ambrose, Toxicol. Appl. Pharmacol., **1**, 97 (1959).

of general type formulas $C_nH_{2n-2}O_2$ to $C_nH_{2n-10}O_2$. They can be classified as low-boiling fractions C_8–C_{10} acids, e.g., methylcyclopentyl acetic acid, and as high-boiling fractions—C_{14} to C_{19} acids—with an average of 2.6 rings. In 1958, 16 million pounds were produced.[12] The crude acids have strong odors, but this is reduced on refining. The molecular weights vary from about 180 to 350.

The principal uses are in the preparation of paint driers, where the metallic salts, such as lead, cobalt, manganese, act as oxidizing agents. Their solubility in oil is an advantage in this use and is in contrast to the insolubility of similar salts of straight chain aliphatic acids in oil. Copper naphthenate is used as a fungicide especially ror rope, wood, hemp, etc.

2. Determination in Air

No published methods were found. It would seem probable that freeze-out samples followed by gas chromatography would be feasible.

3. Physiological Response

This has not received extensive study.[127] Rockhold[128] states that the acute oral toxicity to rats is low. A fraction derived from crude kerosene acids had an LD_{50} of about 3 g./kg.; that from crude mixed acid, 5.2 g./kg. The acute toxicity of the metal (cobalt, copper, calcium, lead, manganese, and zinc) naphthenates was also quite low, the oral LD_{50} in rats being from 4 to more than 6 g./kg. Phenyl mercury naphthenate was more toxic, having an oral LD_{50} of 0.4 g./kg.

Lead naphthenate, given to rats as 20 daily oral doses of 0.25 ml. of a 1 per cent solution (as Pb), showed no signs of toxicity nor of significant storage of lead.

It seems probable that their principal effect is that of mild primary irritation when encountered in high concentrations. If their metabolism is similar to the cyclohexyl acids discussed by Williams,[10] they are probably conjugated with glycine.

No reports of human injury were found nor have threshold limits for air concentrations been proposed.

L. AROMATIC ACIDS

The aromatic carboxy and sulfonic acids comprise one of the largest and most important groups of industrial chemicals. They are used in the synthesis of dyes, elastomers, medicinals, pesticides, and plastics. The total 1958 production of cyclic intermediates, of which these acids form a significant percentage, was about 6.6 billion pounds.[12] Their commercial importance is inversely proportional to the attention they have received from industrial hygienists and toxicologists. The Threshold Limits List for 1959[28] mentions only one material of this class—2,4-

[127] API Toxicological Review, *Naphthenic Acids*. American Petroleum Institute, New York, 1948.
[128] W. T. Rockhold, *A.M.A. Arch. Ind. Health*, **12**, 477 (1955).

TABLE 9
Aromatic Carboxy Acids (Benzene)

Acid	Molecular weight	Melting point, °C.	pK_a	Solubility in H_2O (g. per 100 ml.)	Acute oral toxicity in rats, approximate LD50 (mg./kg.)	Acute effects on skin	Metabolic fate and remarks
Benzoic	122.13	123	4.17	0.21	1700	Mild irritant, allergy rare[19]	Conjugated with glycine, excreted as hippuric acid; cumulative effects absent
o-Toluic	136.15	104	3.89	0.12	400–3200 (about 750 mg./kg. orally in rabbit)	Slight irritant in guinea pig, not absorbed	Conjugated with glucuronic acid[8]
m-Toluic	136.15	109	4.28	0.09	>3200	Not an irritant in guinea pig, not absorbed	Conjugated with glycine and glucuronic acid; some also excreted unchanged
p-Toluic	136.15	179	4.35	0.03	400–3200	Not an irritant in guinea pig; not absorbed	Conjugated with glycine and glucuronic acid; some also excreted unchanged
p-tert-Butylbenzoic	178.1	161			600 in mice[130]	Mild irritant, not absorbed through skin	Metabolism unknown, probably readily metabolized
o-Chlorobenzoic	156.7	139	2.89	0.21	>500 orally in rabbit	Moderate irritant	Excreted as such, but to lesser extent as glucuronide ester[131]
m-Chlorobenzoic	156.57	157	3.82	0.04	>500 orally in rabbit	Irritant	Largely converted to glycine and glucuronide conjugates
p-Chlorobenzoic	156.57	241	4.03	0.007	>200 orally in rabbit		Similar to m-chlorobenzoic acid
o-Nitrobenzoic	167.13	146	2.21	0.68			Excreted principally unchanged, about 10–20% reduced to the aminobenzoic acid and acetylated[8]
m-Nitrobenzoic	167.13	141	3.46	0.31	>200 orally in rabbit	Moderate irritant in guinea pig; possible absorption	Same as for o-nitrobenzoic acid
p-Nitrobenzoic	167.13	241	3.40	0.024	>200 orally in rabbit >400 orally in rat		Same as for o-nitrobenzoic acid
o-Hydroxybenzoic (salicylic)	138.13	160	3.00	0.18	Oral lethal dose of salicylates in adults about 20–30 g.; wide variation[133]	Strong irritant and keratolytic	Excreted as such, also conjugated with glycine and as an oxidation product (gentisic acid)
m-Hydroxybenzoic	138.13	204	4.12	0.92	2800 mg./kg. I.P. in guinea pig		Excreted principally as a glycine conjugate, plus small amounts of the glucuronic acid conjugate
p-Hydroxybenzoic	138.13	216	4.54	0.79	3000 mg./kg. I.P. in guinea pig	Slight irritant in solid form; no absorption	Excreted largely as the free acid plus smaller amounts of the glycine conjugate
2,5-Dihydroxybenzoic (gentisic)	154.13	205	2.93	Soluble	800–1600	Moderate irritant in solid form; no absorption; not a sensitizer	Excreted largely as free acid, but also conjugated on the 5-hydroxyl; no cumulative effects; humans tolerate relatively large doses orally[30]

Compound	Mol. wt.	M.P. (°C)			Toxicity (LD50 mg./kg.)	Skin irritant	Remarks
3,4,5-Trihydroxybenzoic (gallic)	170.13	220 (dec.)	4.40	1.16	>500 I.P. in mice	Skin irritant	Probably conjugated; less toxic than tannic acid[133]
o-Aminobenzoic (anthranilic)	— 137.14	146	5.00	0.35	Low toxicity in mammals; does not form methemoglobin	Not irritating	0.2% in rat diet said to produce bladder papillomas in some animals on long feeding.[134] Conjugated with glucuronide acid.[10]
m-Aminobenzoic	137.14	172	4.82	0.59	250-500 I.P. in mice[135]	Not irritating	No tumors produced; similar metabolism to ortho, but some combination with glycine also.
p-Aminobenzoic	137.14	188 (dec.)	4.92	0.34	2850 orally in mice[45] 1 g./kg. in dogs causes severe gastroenteritis[138]	Not irritating	No tumors produced in rats; excreted as p-acetaminobenzoic acid; antagonist to sulfonamide drugs[134]
o-Benzoylbenzoic	226.23	127		Soluble in hot water	Maximum daily dose tolerated I.P. in mice = 31 mg./kg.	Moderate irritant	Unknown
Phthalic	166.14	207 (dec.)	2.98	0.62	>3200 (mono K salt)	Slight irritant	Probably excreted as such[10] 1-4% in rat diet for 1 year had no effect[136]
Isophthalic	166.14	345	3.46	0.013	>500 I.P. in mice	Slight irritant; not absorbed; not a sensitizer	Same metabolism as for phthalic acid
Terephthalic	166.14	Sublimes	3.51	Insoluble	>6400	Slight irritant (guinea pig)	Same metabolism as for phthalic acid
4-Hydroxyisophthalic	182.13	310	0.3		>3200 in rat 1600-3200 in mice	Not absorbed	0.2% in diet of rats for 26 weeks produced no pathology[137]
Phenylacetic (α-toluic)	136.15	76	4.31	1.66	Low toxicity	Slight irritation	Conjugated with glutamine in man[10]
Phenoxyacetic	152.15	98	3.12	1.2	>200 in rabbits	Slight irritation	Excreted as free acid by rabbits[139]
o-Monochlorophenoxyacetic	186.60	145		0.5 (82°C.)	>800	Slight irritation	Same as for phenoxyacetic acid[139]
p-Monochlorophenoxyacetic	186.60	158			>800; in rabbits 200 mg./kg. caused marked hematuria	Slight irritation	Same as for phenoxyacetic acid[139]
2,4-Dichlorophenoxyacetic (2,4-D)	221.05	139		Very low	Relatively low toxicity for mammals;[30] 280 mg./kg. subcutaneously in mice[140]	Slight irritation	Relatively low acute and chronic toxicity;[141] possibly some neurological effects in dogs at high dose levels.[142] Threshold limit is 10 mg./cu. meter.[38]
2,4,5-Trichlorophenoxyacetic (2,4 5-T)	255.50	149		Very low	About 100 orally in dogs[142]	Slight irritant	Somewhat similar to 2,4-D

[129] K. A. Baird, J. Allergy, 16, 195 (1945). [130] Industrial Hygiene Bulletin, Tertiary butyl benzoic acid, No. SC:57-117. Shell Chemical Corp., New York, 1959. [131] H. G. Bray et al., Biochem.. J., 50, 583 (1952). [132] M. Gleason, R. Gosselin, and H. Hodge, Clinical Toxicology of Commercial Products, Williams & Wilkins, Baltimore, 1957. [133] H. J. Robinson and O. E. Graessle, J. Pharm. Exptl. Therap., 77, 63 (1943). [134] B. Eckman and J. P. Strombeck, Acta Pathol. Michobiol. Scand., 26, 447 (1949a). [135] Summary Tables of Biological Tests, 6, 53 (1954), National Research Council. [136] H. Hodge, unpublished data. [137] H. O. J. Collier and G. B. Chesher, Brit. J. Pharmacol, 11, 20 (1956). [138] C. C. Scott and E. B. Robbins, Proc. Soc. Exptl. Biol. Med., 49, 212 (1942). [139] S. Levey and H. B. Lewis, J. Biol. Chem., 168, 213 (1947). [140] N. L. R. Bucher, Proc. Soc. Exptl. Biol. Med., 63, 204 (1946). [141] E. V. Hill and H. Carlisle, J. Ind. Hyg. Toxicol., 29, 85 (1947). [142] V. A. Drill and T. Hiratzka, Arch. Ind. Hyg. Occupational Med., 7, 61 (1953).

TABLE 10

Aromatic Carboxy Acids (Naphthalene)

Acid	Molecular weight	Melting point, °C.	pK_a	Solubility in H_2O (g. per 100 ml.)	Acute oral toxicity in mice, approximate LD_{50} (mg./kg.)	Acute effects on skin	Metabolic fate and remarks
α-Naphthoic (1-naphthoic)	172.19	162	3.7	Very slightly in hot water	>500 I.P.	Moderate irritant	Probably excreted as ester glucuronide and unchanged; considerable species variation[10]
β-Naphthoic (2-naphthoic)	172.19	184	4.15	0.007	>500 I.P.	Moderate irritant	Excreted unchanged in the dog; in rabbits is conjugated with glycine
β-Hydroxynaphthoic (2-hydroxy-1-naphthoic)	188.17	156 (dec.)		Slightly in hot water	125–250 I.P.	Moderate irritant; heated vapor irritating	
3-Hydroxy-2-naphthoic	188.17	216		Slightly in hot water	800–1600	Moderate irritant (guinea pig); not absorbed; not a sensitizer	Probably readily conjugated
3-Amino-2-naphthoic	181.14	216			1600–3200	Not irritating (guinea pig); not a sensitizer	Probably readily conjugated
α-Naphthaleneacetic (2-naphthaleneacetic)	186.21	131		Soluble in hot water	>500 I.P.		

TABLE 11

Aromatic Sulfonic Acids (Benzene)

Acid	Molecular weight	Melting point, °C.	pKₐ	Solubility in H₂O (g. per 100 ml.)	Acute oral toxicity in mice, approximate LD₅₀ (mg./kg.)	Acute effects on skin	Metabolic fate and remarks
Benzenesulfonic	158.17	44	0.7	Very	400–3200	Severe skin irritation (guinea pig)	Probably excreted as such
Sodium benzenesulfonate	180.16			Very	>3200	Only slight irritation (guinea pig)	Probably excreted as such
p-Toluenesulfonic	172.19	106		Very	400–3200	Severe skin and eye irritant	Probably excreted as such
Xylenesulfonic (2,5-dimethylbenzenesulfonic)	222.27	85		Soluble	500 I.P.	Severe skin and eye irritant	Probably excreted as such
p-Hydroxybenzenesulfonic (p-phenolsulfonic)	174.17			Soluble		Skin irritant	
p-Aminobenzenesulfonic (Sulfanilic)	173.20	280 (dec.)	3.32	1.08	>3200	Slight irritant (guinea pig); not absorbed	
m-Aminobenzenesulfonic	173.20		3.80	.67	>3200	Slight irritant (guinea pig); not absorbed	

TABLE 12
Aromatic Sulfonic Acids (Naphthalene)

Acid	Molecular weight	Melting point, °C.	Solubility in H_2O (g. per 100 ml.)	Acute oral toxicity in rats, approximate LD_{50} (mg./kg.)	Acute effects on skin	Metabolic fate and remarks
Naphthalene sulfonic·$2H_2O$ (1-naphthalene sulfonic·$2H_2O$)	244.26	90	Very soluble			
β-Naphthalene sulfonic·H_2O (2-naphthalene sulfonic·H_2O)	226.26	124	77	400–3200	Slight irritant (guinea pig); not absorbed	
1,5-Naphthalene disulfonic disodium salt·$2H_2O$	368.30		102	>3200, >3200 (mice)	Slight irritant; not a sensitizer	
1-Naphthol-4-sulfonic (Nevile and Winther's acid)	224.23	170 (dec.)	Very soluble			
2-Naphthol-6-sulfonic (Schäffer's acid)	224.23	125	Very soluble			
2-Naphthol-3,6-disulfonic (R acid)	304.30		Very soluble			
4,5-Dihydroxy-2,7-naphthalene disulfonic sodium salt (chromotropic acid)	400.30		Very soluble	>3200	Slight irritant (guinea pig); absorption (?)	
Naphthionic (1-naphthylamine-4-sulfonic acid)	232.25	170 (dec.)	.03	>500 I.P. (mice)	Slight irritant	Not carcinogenic if pure, but may be contaminated with α-naphthylamine[143]
1,5-Naphthionic·H_2O (α-naphthylaminesulfonic acid) (Cleve's acid)	241.26	190	0.1	Similar to naphthionic	Similar to naphthionic	Similar to naphthionic
1-Amino-2-naphthol-4-sulfonic (1,2,4 acid)	239.26		Very slightly soluble	>3200	Slight irritant (guinea pig); not absorbed; has caused dermatitis in humans	Similar to naphthionic
7-Amino-1-naphthol-3,6-disulfonic mono-sodium salt (2-R acid)	341.31			>3200	Slight irritant (guinea pig); not absorbed	
1-Amino-8-naphthol-3,6-disulfonic (H acid)	319.31		Slight			
1-Naphthylhydrazine-4-sulfonic	239.3		Soluble	>1600	Moderate irritant (guinea pig) not absorbed; probably is a skin sensitizer	

[143] T. S. Scott and M. H. C. Williams, *Brit. J. Ind. Med.*, **14**, 150 (1957).

TABLE 13

Aromatic Sulfonic Acids (Anthracene)

Acid	Molecular weight	Solubility in H_2O (g, per 100 ml.)	Acute oral toxicity in rats, approximate LD_{50} (mg./kg.)	Acute effects on skin	Metabolic fate and remarks
1-Anthraquinone sulfonic sodium salt	310.26	0.5	Similar to 2-anthraquinone sulfonic	Similar to 2-anthraquinone sulfonic	
2-Anthraquinone sulfonic sodium salt (β acid)	310.26	Soluble	>3200 >500 I.P. (mice)[144]	Slight irritant (guinea pig); no absorption	
1,8-Anthraquinone disulfonic dipotassium salt (chi acid)	444.53	Slightly soluble	>3200	Slight irritant (guinea pig); no absorption	Large oral doses in rats caused retraction of abdominal wall, rough coat and diarrhea
2,6-Anthraquinone disulfonic	253.04	3.9 (Na salt)			
1,5-Anthraquinone disulfonic (rho acid)	253.04	Very soluble (Na salt)			

[144] Summary Tables of Biological Tests, 4, 316 (1952), National Research Council.

TABLE 14

Miscellaneous Acids

Acid	Molecular weight	Melting point, °C.	Solubility in H₂O (g. per 100 ml.)	Acute oral toxicity in rats, approximate LD_{50} (mg./kg.)	Acute effects on skin	Metabolic fate and remarks
2,6-Dihydroxy-4-carboxy-pyridine (Citrazinic acid)	155.11	Decomposes	Very slight	>3200	Slight skin irritation; not absorbed; not a sensitizer	
Abietic	302.44	137–166	Insoluble	1600–3200 I.P. >3200 orally (mice) Low oral toxicity	Not a skin irritant	
Methane sulfonic	96.11	20 B.p. 167 at 10 mm. (dec.)	Very soluble	200–400 <50 I.P. (rat) 100–200 orally (guinea pig)	Severe skin irritant; not absorbed	No effect in rats after 6 hours of inhaling vapor from heated liquid: excreted as such in rats after 1 g./kg. subcutaneously[145]
Taurine (2-aminoethane-sulfonic acid)	125.14	328 (dec.)	Very soluble	LD_{50} > 1 g. per kg. subcutaneously	Slightly irritating	Excreted as such[145]
Sodium lauryl sulfate	288.38		Very soluble	Low oral toxicity[146]	Not irritating in low concentrations	Rats tolerate 1% in diet; decreased growth at 4%[147]

[145] G. A. Maw, *Biochem. J.*, **55**, 42 (1953).

[146] G. Woodard and H. O. Calvery, *Proc. Sci. Sect. Toilet Goods Assoc.*, No. **3**, 1 (1945).

[147] G. Fitzhugh and A. A. Nelson, *J. Am. Pharm. Assoc. Sci. Ed.*, **37**, 29 (1948).

[148] E. A. Braude and F. C. Nachod, *Determination of Organic Structures By Physical Methods.* Academic, New York, 1955.

dichlorophenoxy acetic acid (2,4-D). Little toxicological information is available for most of them, although some have received remarkably detailed metabolic studies.

The absence of any systematic investigation of this group probably reflects the fact that in general they have not proved to be very hazardous. As a group they are almost entirely free from serious systemic or cumulative effects. Even rings with amino or nitro groups (which would be methemoglobin formers in the absence of the acid grouping) are often relatively inert. The explanation of this probably derives from the rapid excretion in the urine, either unchanged or conjugated with glycine or glucuronic acid and from a change in activity of these functional groups. The presence of the acid in many cases seems to reduce or abolish the ability of the substance to penetrate through the skin or cause the development of skin sensitivity.

The primary irritant effects vary greatly, and there are many anomalies in the relationship of acid ionization constants and local damage to body surfaces. The fact that they are often crystalline solids, and in some cases have low water solubility, may help to reduce the local effect. On the other hand, they are not infrequently used in such a manner that contact with heated vapor or sublimates may occur, and in these instances there may be irritation of the upper respiratory tract and eyes and skin.

There have been few attempts to measure occupational exposures to these materials, although analyses in air should not be difficult. Because of the strong absorbance of the ring structures in the ultraviolet and the reactivity of many of the ring substituents (i.e., NH_2 and OH), analytical methods should be easy to develop. For those acids readily soluble in water and with reasonably good ionization constants, simpler methods such as those of Miller et al.[14] should suffice.

In Tables 9 through 14, the aromatic acids will be grouped as carboxylic or sulfonic derivatives of benzene, naphthalene, or anthracene. The general order of acute toxic effect, effects on skin, and metabolism will be indicated if known. If data are absent, an opinion of the hazard based on general experience may be used, or in some instances, a lower limit for the approximate lethal dose may be inferred from data described in metabolic studies. The source of most of the data in these tables is reference 20.

CHAPTER XLI

Esters

DAVID W. FASSETT, M.D.

I. General Considerations

A. CLASSIFICATION AND CHEMICAL PROPERTIES

Esters are compounds formed when the hydrogen of an acid is replaced by an organic grouping. They can be prepared from either organic or inorganic acids, and are named in a manner similar to salts of acids, namely,

$$\text{acetic acid} + \text{ethyl alcohol} \rightarrow \text{ethylacetate } (CH_3\overset{\overset{\displaystyle O}{\|}}{C}-O-C_2H_5) + H_2O$$

The formation of an ester is known as esterification. It is often a reversible reaction not quite going to completion. The reverse reaction is referred to as hydrolysis. Esterification is usually promoted by acidic catalysts, while hydrolysis of the ester usually proceeds faster under alkaline conditions. If the velocity of the esterification reaction is divided by the velocity of hydrolysis reaction, a constant, K_s, can be derived, which is useful as a measure of the stability of the ester.

$$(\text{Conc.}_{\text{ester}})(\text{Conc.}_{H_2O})/(\text{Conc.}_{\text{alcohol}})(\text{Conc.}_{\text{acid}}) = K_s$$

Ester specifications frequently give the maximum amount of free acid present.

Esters can be prepared by several types of reactions, all of which are similar in principle.[1]

1. The reaction of an alcohol plus an acid

$$RCOOH + HOR' \rightarrow RCOOR' + H_2O$$

2. The reaction of an alkyl halide plus an acid salt

$$RCOONa + ClR' \rightarrow RCOOR' + NaCl$$

3. The reaction of an alcohol with an acid halide

$$RCOCl + HOR' \rightarrow RCOOR' + HCl$$

4. The reaction of an alcohol with an acid anhydride

$$(RCO)_2O + HOR' \rightarrow RCOOR' + RCOOH$$

[1] L. Fieser and M. Fieser, *Organic Chemistry*. 3rd ed., Reinhold, New York, 1956.

5. Conversion of one ester to another by reaction with another alcohol; this is referred to as an alcoholysis or ester interchange

$$RCOOCH_3 + R'OH \rightarrow RCOOR' + CH_3OH$$

While there are a very large number of possible types of esters, for the present purposes they can be regarded as being of the following chemical types (glycol esters are considered in another chapter).

1. Esters of inorganic acids, for example, dimethyl sulfate, methyl-p-toluene sulfonate, triethyl phosphate, triphenyl phosphate, organic nitrates and nitrites (see Chapter XLVI) ethyl silicate, trimethyl borate.

2. Esters of aliphatic acids:

(a) Mono-, di-, or tricarboxylic esters, for example, ethyl acetate, diethyl oxalate, triethyl citrate; (b) esters in which the alcohol or acid is unsaturated or contains substituents, for example, vinyl acetate, ethyl acrylate, ethyl iodoacetate.

3. Esters of aromatic acids:

(a) Mono- or dicarboxylic, for example, methyl benzoate, dibutyl phthalate, dimethyl terephthalate; (b) Esters with substituents on the ring or on the alcohol, for example, methyl-p-hydroxybenzoate, dimethoxyethyl phthalate

Most esters are liquids at room temperature. In the case of the simple esters of aliphatic acids, the boiling points are much lower than the corresponding acid. They tend to have very low water solubility, except for a few of those with low molecular weights. They are frequently flammable with low flash points. They often have pleasant odors in contrast to the acids from which they are derived, for example, ethyl butyrate vs. butyric acid. A great variety of esters are normally present in fruits, flavors, and vegetables. Because of the striking difference in physical properties of esters from those of their corresponding acids, esterification is frequently used as a means of isolating and protecting an acid group.

The esters of monocarboxylic aromatic acids have moderate volatility and pleasant odors, while those of dicarboxylic acids are essentially nonvolatile and often have no characteristic odor. The relative position of the ester groups on the ring may alter the physical properties, for example, dimethyl-o-phthalate is a liquid, while dimethyl terephthalate is a solid. The aromatic carboxylic esters usually have a much higher flash point than aliphatic esters. Benzyl esters are usually more irritating than corresponding aliphatic esters.

B. INDUSTRIAL USES OF ESTERS

Esters are comparable to organic acids in their industrial importance and are often the end product of the use of an organic acid. Some idea of their importance can be gathered from the 1958 production figures for some of the principal groups of esters.[2]

[2] U. S. Tariff Commission Report #205, Synthetic organic chemicals—U. S. production and sales, Washington, D. C., 1958.

1. Plastics and resins (phthalic alkyd, polyester, vinyl and cellulose acetate) : about 1.54 billion pounds.

2. Plasticizers (phosphates, phthalates, and aliphatic acid esters) : 418 million pounds.

3. Lacquer solvents (ethyl and butyl acetates make up more than half of the total) : 265 million pounds.

4. Flavor and perfume (methyl and amyl salicylate, methyl anthranilate, benzyl and phenethyl acetate, linalyl and terpinylacetate, and ethyl butyrate account for more than 90 per cent) : 5.3 million pounds.

5. Medicinals (complete figures for the usage of esters for the preparation of medicinals could not be ascertained from the data available. Some of the more important ones—excluding aspirin—are as follows: p-aminobenzoates, methyl and propyl-p-hydroxy benzoates, diethyl malonate) : 1.7 million pounds.

The types of industries using esters are automotive and aircraft, food processing, chemical and drug manufacturing, soap and cosmetic, surface coating, textiles, and leather.[3] In other words, they are to be encountered in most chemical industries.

The chief exposures will occur during use for the more volatile esters in lacquers, during handling of some of the resin monomers, and, to a minor degree, from the heating of plasticized resins.

C. ANALYTICAL METHODS FOR DETERMINING ESTERS IN AIR

Of principal interest are the methods for simple aliphatic esters such as formates and acetates, methods for acrylic esters, for halogenated esters such as ethyl chloroformate, methods for the phthalate plasticizer esters, and those for the phosphate esters.

Methods for the formates and acetates are described by Jacobs and others.[4–6]

The most widely used method for esters depends on the following reaction:

$$\text{RCOOR}' + \text{NaOH} \rightarrow \text{RCOONa} + \text{R}'\text{OH}$$

This is a simple saponification reaction in which the hydrolysis is carried out by a known excess of sodium hydroxide and the amount of excess alkali is determined by back titration with acid. The usual procedure is to collect the sample in fritted scrubbers using ethanol or isopropanol as a solvent. The sample may also be collected on silica gel.[7] Samples of the order of 100–200 liters of air are necessary to obtain adequate sensitivity.

[3] I. Mellan, *Industrial Solvents.* 2nd ed., Reinhold, New York, 1950.

[4] M. B. Jacobs, *The Analytical Chemistry of Industrial Poisons, Hazards, and Solvents,* Vol. I. 2nd ed., Interscience Publishers, New York, 1949.

[5] F. H. Goldman and M. B. Jacobs, *Chemical Methods of Industrial Hygiene.* Interscience Publishers, New York–London, 1953.

[6] N. Strafford, C. R. N. Strouts, and W. V. Stubbings, *The Determination of Toxic Substances in Air.* W. Heffer, Cambridge, England, 1956.

[7] J. A. Houghton, Research Center, Liberty Mutual Insurance Co., Hopkinton, Mass., personal communication.

A colorimetric method has been described by Strafford et al.[6] This is based on the method of Hestrin,[8] which depends on the reaction of the ester with alkaline hydroxylamine forming acethydroxamic acid. The addition of $FeCl_3$ forms a purple color. This reagent will also react with acid anhydrides or acid chlorides, but it will not react with the acids themselves. It is described as being suitable for methyl, ethyl, butyl, and isoamyl acetates. They suggest collection in two scrubbers in series using ethanol as a solvent. The color developed is measured with the aid of Ilford No. 604 filters.

Ethyl acrylate, ethyl methacrylate, and methyl methacrylate can be determined by collection in scrubbers in ethanol, refluxing with NaOH, and back titrating.[7]

Pozzani et al.[9] measured ethyl acrylate in animal exposures by the use of a Zeiss portable interferometer calibrated as described by Stamm and Whalen.[10] One scale division equaled 22 p.p.m. of ethyl acrylate vapor.

Treon et al.[11] measured methyl and ethyl acrylates by a method similar to that of Werner and Mitchell for ethylene glycol ethers.[12] The sample was collected in fritted glass scrubbers in a solution of dichromate and sulfuric acid. After heating to oxidize the acrylates, the excess chromate was determined iodimetrically.

Deichmann[13] determined methyl methacrylate in air by passing air through acid permanganate, adding an equivalent amount of oxalate, and back titrating the excess with permanganate.

No methods could be found in the literature for the halogenated esters. These could probably be collected on silica gel or charcoal, decomposed, and the halogen determined as in the case of other organic halogens. The chloroformates readily decompose in water, releasing HCl, and might be titrated. Sampling with CCl_4 in a midget impinger, followed by measurement of the absorbance in the infrared at 5.6 μ (1777 cm.$^{-1}$) using a 5–10 mm. cell appears to be very promising and of high sensitivity. As low as 0.2 p.p.m. can be determined in relatively small air samples.[14]

The phthalate plasticizers can usually be measured with a high degree of sensitivity by collecting in scrubbers using spectrographic grade 3-A alcohol as a solvent, and measuring the absorbance in the UV at 225 mμ. Interferences will, of course, be present if other aromatic substances are present in the sample. A 1-cu. ft. air sample will allow detection of 0.02 p.p.m. of dimethoxy ethyl phthalate or

[8] J. Hestrin, *J. Biol. Chem.*, **180**, 249 (1949).

[9] U. C. Pozzani, C. S. Weil, and C. P. Carpenter, *J. Ind. Hyg. Toxicol.*, **31**, 311 (1949).

[10] R. F. Stamm and J. J. Whalen, *J. Ind. Hyg. Toxicol.*, **29**, 203 (1947).

[11] J. R. Treon, H. Sigmon, H. Wright, and K. V. Kitzmiller, *J. Ind. Hyg. Toxicol.*, **31**, 317 (1949).

[12] H. W. Werner and J. L. Mitchell, *Ind. Eng. Chem. Anal. Ed.*, **15**, 375 (1943).

[13] W. Deichmann, *J. Ind. Hyg. Toxicol.*, **23**, 343 (1941).

[14] W. E. Bissinger, Barberton Laboratories, Columbia-Southern Chemical Corp., Barberton, Ohio, personal communication.

methylphthalyl ethyl glycolate. It seems probable that similar methods will work with all esters containing aromatic rings.[15]

There are a number of possible methods for the determination of phosphate esters, depending on the type of material esterified with the phosphoric acid. In the case of simple aliphatic esters, such as triethyl or tributyl phosphate, the material can be collected in scrubbers using alcohol or some similar solvent, followed by hydrolysis, separation of the phosphate radical, and determination by methods suitable for phosphates. In the case of aromatic phosphate esters, such as triphenyl phosphate, the analysis can be carried out by absorption in scrubbers in 3-A spectrographic grade alcohol and determination of the absorbance in the UV at 260 mμ. In the case of triphenyl phosphate, a 1-cu. ft. air sample will allow detection of 3.5 mg./cu. meter. Triphenyl phosphate can also be determined by absorbing in gas-washing bottles in 3-A alcohol, evaporating to dryness, hydrolyzing with concentrated sulfuric acid, and determining the phosphate by a colorimetric procedure with ammonium molybdate. This method is not as useful as the UV absorption method.[15]

D. PHYSIOLOGICAL EFFECTS OF ESTERS

While there are confusing reports in the literature regarding the effects of esters, there appear to be about four basic types of action, which can be correlated fairly well with chemical structure. In a few instances the mechanism of action can be related to known biochemical facts, although the lack of data on metabolic changes often prevents any rational interpretation or organization of the toxicological data. The four groups listed below are probably incomplete, but seem to represent the principal discernible types of response. The esters of nitric acid and of glycols are considered elsewhere in this volume and have not been included in this classification.

1. Anesthesia and Primary Irritation

The simple aliphatic esters used as lacquer solvents produce anesthesia when inhaled in sufficiently high concentrations. As expected from Meyer-Overton theories, the more water-soluble lower molecular weight esters, such as methyl and ethyl formate, and methyl and ethyl acetate, are less potent anesthetics than butyl and amyl acetates.

Their anesthetic potency is weaker than most chlorinated hydrocarbons and usually less than ethyl ether, although they are more active than ethanol, acetone, and aliphatic hydrocarbons such as pentane. Because of their relatively good solubility in plasma, they readily pass through the alveoli. Those with higher water solubility presumably have higher blood–air distribution coefficients and reach saturation more slowly. They are thought to be hydrolyzed readily, either

---

[15] F. A. Miller and R. F. Scherberger, Laboratory of Industrial Medicine, Eastman Kodak Co., Rochester, N. Y., unpublished data.

by simple chemical hydrolysis or by esterases present in liver and plasma, and then readily metabolized as corresponding alcohols or acids. While the details of metabolism are lacking, they are probably partly excreted in exhaled air and urine and partly metabolized.

Most of the aliphatic esters possess some degree of irritant action on exposed surfaces. The formates are especially irritating to the eyes and respiratory tract. This may well be related to their relatively rapid hydrolysis upon contact with water, forming the original alcohol plus formic acid.[16] Even the higher aliphatic esters have some degree of local irritant effect, which serves in some measure to reduce the probability of voluntarily inhaling enough to produce serious anesthetic effects, although these have occurred from massive exposures inside tanks, etc.[17]

The irritant effects of the methyl and ethyl formates and acetates commence at lower levels than for the corresponding alcohols. For example, ethyl acetate may be irritating at levels of 400–800 p.p.m. while ethanol is practically devoid of irritation at several thousand p.p.m. However, butyl acetate begins to be irritating at about the same level as butanol. The irritant effect of esters of higher molecular weight is a function of the whole molecule, not apparently related to hydrolysis. Most of the esters are more irritating than chlorinated hydrocarbons at the same level.

Except for irritant effects on the eyes or respiratory tract, cumulative effects are not marked in either man or animals. The pathology noted in animals after either acute or chronic exposure is limited to vascular congestion, respiratory irritation, and loss of weight at high levels of daily exposure in animals.

The evaluation of the significance of older reports of effects on the blood or gastrointestinal system of humans is difficult since the exposures were nearly always to mixtures of solvents, often of unknown composition.

The local effects on the skin are about the same as for other volatile solvents; defatting and cracking may occur. Specific sensitization dermatitis is negligible.

Considering the large quantities used and the fact that the exposure levels are potentially high, the saturated aliphatic esters are among the least toxic of industrial solvents.

2. Lachrymation, Vesication, and Lung Irritation

Several very potent lachrymators and vesicants are known, of which ethyl-chloro-, ethylbromo-, and ethyliodoacetate are outstanding. They appear to produce this effect by reason of the reactivity of the halogen atom and its combination with certain SH groups.[18] They may also possess the general toxic effects of the corresponding acids (see Chapter XL on Organic Acids). Ethyl fluoroacetate possesses nearly the original potency of fluoroacetic acid, indicating

[16] W. F. von Oettingen, *Arch. Ind. Health*, **20**, 517 (1959).

[17] E. Browning, *Toxicity of Industrial Organic Solvents*, Chemical Publishing, New York, 1953.

[18] M. Dixon, *Biochemical Soc. Symposia* No. **2**, 1948.

a rapid hydrolysis in the body to the latter compound.[19] Methyl-*p*-toluene sulfonate is a vesicant and skin sensitizer. Dimethyl sulfate and methyl, ethyl, and isopropyl chloroformates can produce delayed pulmonary edema similar to that produced by phosgene. Dimethyl carbonate is also active in this respect to animals. Unsaturation (e.g., acrylates) may be associated with lachrymatory action.

3. Cumulative Organic Damage to the Nervous System

The outstanding example is the neuropathy produced in a variety of species by triorthocresylphosphate. The weakness and paralysis are usually reversible if exposure ceases in the early stages; demyelination may occur. The mechanism of action is not entirely clear, although it may be associated with some effects on cholinesterase. Certain other phosphate compounds cause similar effects (diisopropyl fluorophosphorate and *N,N'*-diisopropyl phosphorodiamidic fluoride). This is not a general property of phosphate esters, however, and many of them are devoid of this action.

Some phosphate esters are readily absorbed through the skin (triorthocresyl phosphate) without any evident local irritant effects, while others are quite irritating (tributyl phosphate).

4. Physiologically Inert Materials

Practically all the common aliphatic and aromatic esters (some phosphate esters excepted) used as plasticizers are inert. At the most, minor degrees of irritation may follow inhalation of heated vapor or extensive, prolonged skin exposure to certain types. While sensitization has been reported, continuing experience indicates that it must be very rare and possibly due to occasional impurities or decomposition products. Many of them are so inert that it is impractical to determine an acute LD_{50}. Specific pathology is usually absent even when the compound is fed in massive quantities to the point of inadequate nutrition. Oily, liquid stools are sometimes seen at high levels, indicating lack of absorption. In other cases, the apparent lack of toxicity may indicate conversion by hydrolysis to normal metabolites, or to materials readily excreted in the urine, although in most cases little actual data are available.

The ester-type resins are similar to many other resins in being completely inert. They are unabsorbed from the gut, do not cause skin irritation or sensitization, and do not show specific pulmonary reactions upon inhalation as dust.

[19] F. L. M. Pattison, *Toxic Aliphatic Fluorine Compounds*. Elsevier, Princeton, 1959.

TABLE 1

Physical and Chemical Properties of Formates

Compound	Formula	Mol. wt.	Sp. gr., °C. (g./ml.)	M. p., °C.	B. p., °C.	Vapor pressure, mm. Hg. (°C.)	Vapor density (air = 1)	Density of sat. air (air = 1)	Flash point, °F.	Solubility in H₂O (g. per 100 ml.)	Conversion factors	
											1 mg./ liter (p.p.m.)	1 p.p.m. (mg./cu. meter)
Methyl formate	HCOOCH₃	60.05	0.9631 (25)	−100.4	31.5	600(26)	2.07	1.83	−2	30	408	2.55
Ethyl formate	HCOOC₂H₅	74.08	0.9236 (25/4)	−80.5	54.3	200(21)	2.56	1.25	−4	11.8	330	3.03
n-Propyl formate	HCOO(CH₂)₂CH₃	88.10	0.9006 (20/4)	−92.9	81.3	85(25)	3.04	1.23	27	2.79	278	3.60
n-Butyl formate	HCOOC₄H₉	102.13	0.8885 (20/4)	−90.0	106.8	30(25)	3.5	1.09	64	Slightly soluble	240	4.17
n-Amyl formate	HCOOC₅H₁₁	116.16	0.8926 (15/4)	−73.5	130.4	9.6(25, approximately)	4.0	1.04	79	Slightly soluble	211	4.74
Benzyl formate	HCOOCH₂C₆H₅	136.14	1.081 (20/4)	3.6	203.4	10(84)	4.7			Insoluble	180	5.56
Cyclohexyl formate	HCOOC₆H₁₁	128.10			162.5		4.4		124		191	5.24
Methyl cyclohexyl formate	HCOOC₆H₁₀CH₃	142.12			176–180		4.9		64		172	5.81

II. Specific Compounds

A. ESTERS OF CARBOXYLIC ACIDS

1. Monocarboxylic—Saturated—Aliphatic

FORMATES

The physical and chemical properties of some formic acid esters are given in Table 1.

The literature on the physiological properties of formate esters has recently been reviewed by von Oettingen.[16] Little information of value has appeared in the

TABLE 2

Physiological Response to Inhalation of Formates

Compound	Species	Concentration p.p.m.	mg./liter	Duration of exposure	Results[20,21]
Methyl formate	Guinea pig	50,000	128	30 min.	Lethal
	Guinea pig	25,000	64	60 min.	Lethal
	Guinea pig	10,000	25	3–4 hr.	Lethal
	Guinea pig	3,500	9	8 hr.	No deaths
	Man	1,500	3.8	1 min.	No symptoms
Ethyl formate	Rat	Conc. vapor		5 min.	No deaths[22]
	Rat	8,000		4 hr.	5 deaths of 6 exposed
	Rat	4,000		4 hr.	No deaths[23]
	Mouse	10,000	32	20 min.	Eye irritation, dyspnea[21]
	Mouse	5,000	16	20 min.	Eye irritation, dyspnea[24]
	Cat	10,000	32	80 min.	Eye irritation, narcosis, death
	Cat	5,000	16	20 min.	Eye irritation, salivation
	Dog	10,000	32	4 hr.	Pulmonary edema, death
	Man	330	1		Eye and nose irritation[24]
n-Butyl formate	Cat	10,000	44	60 min.	Irritation, narcosis and death[24]
	Dog	10,000	44	60 min.	Irritation, narcosis, and recovery[24]
	Man	10,000	44	<1 min.	Intolerable irritation

[20] H. H. Schrenk, W. P. Yant, J. Chronyak, and F. A. Patty, *Public Health Repts. U. S.*, **51**, 1329 (1936).

[21] K. B. Lehmann and F. Flury, *Toxicology and Hygiene of Industrial Solvents*. Trans. by E. King and H. F. Smyth, Jr., Williams & Wilkins, Baltimore, 1943.

[22] H. F. Smyth, Jr., C. P. Carpenter, C. S. Weil, and U. C. Pozzani, *A.M.A. Arch. Ind. Hyg. Occupational Med.*, **10**, 61 (1954).

[23] H. F. Smyth, Jr., *Am. Ind. Hyg. Assoc. Quart.*, **17**, 129 (1956).

[24] F. Flury and F. Zernik, *Schädliche Gase*, Springer, Berlin, 1931.

literature since the first edition of this volume. The experimental data consists principally of single exposure inhalation experiments with a variety of species and some sensory responses of man to brief inhalations of known concentrations. Table 2 presents some of the data on methyl formate, ethyl formate, and *n*-butyl formate. No specific data of value could be found for propyl, amyl, benzyl, cyclohexyl, or methyl cyclohexyl formates. With the exception of methyl formate, there appear to be no reports of injury from the use of aliphatic saturated formates. The principal difference between these esters and the higher esters such as acetates or butyrates concerns their marked irritant properties at higher concentrations, probably due to their ready hydrolysis on moist surfaces.

METHYL FORMATE

Methyl formate has had some use as a larvicide and a fumigant as well as being used for solvent purposes. The report by Gettler[25] cites the case of a 19-month-old child who developed serious symptoms and died about 20 or 30 minutes following an application of 1 oz. of liquid insecticide mixture containing methyl formate. This mixture had been applied on the scalp under a bathing cap as a treatment for pediculosis capitis. Pathological findings were not specific; however, some methyl formate and also methanol were recovered in the brain and other organs, indicating that this material may have been absorbed directly through the skin under these circumstances.

von Oettingen reviews a report by Duquenois and Revel[26] in which workers were exposed to vapors of a solvent mixture containing methyl and ethyl formate and also methyl and ethyl acetate. A variety of central nervous symptoms was reported, including some visual disturbances which were temporary in nature. It is difficult to assess the significance of this report since specific air concentrations do not appear to have been determined. The current threshold limit for methyl formate is 100 p.p.m.

ETHYL FORMATE

In addition to the values given in Table 2, the oral LD_{50} has been determined by Smyth[22] to be 4.3 g./kg. It did not appear to penetrate the rabbit skin, the LD_{50} being greater than 20 ml./kg. Injury to the skin from direct contact was mild while contact with the eye produced moderate irritation. Duquenois and Revel[26] considered ethyl formate to be definitely less irritating than methyl formate and less apt to produce respiratory or nervous system effects. The present threshold limit is 100 p.p.m.

[25] A. O. Gettler, *Am. J. Clin. Pathol.*, **10**, 188 (1940).
[26] T. Duquenois and T. Revel, *J. pharm. chim. Ser. 8*, **19**, 599 (1935).

BUTYL FORMATE

Butyl formate appears to be similar in its properties to ethyl formate. Its lower vapor pressure reduces the possibility of encountering very high concentrations. No threshold limit has been proposed.

ACETATES

1. *Uses and Industrial Exposures*

The saturated aliphatic acetates are important lacquer solvents, especially ethyl and butyl acetate. Isopropyl and isobutyl acetate are finding increasing use. The total production of all butyl acetates in 1958 was reported to be 85 million pounds, of which isobutyl acetate represented 12 million and *n*-butyl acetate 63 million. Ethyl acetate production in the same year was 86 million pounds; amyl acetates, 9 million pounds; and methyl acetate, 6 million pounds.[2] These make up the principal solvents for lacquers but are usually combined with some type of aromatic solvent such as toluene and xylene. Glycol esters or alcohols are also frequently used along with the other solvents, although they are usually not one of the major components in terms of concentration.

In addition to solvent uses, a number of the aliphatic acetates are used as flavors or in perfumes. Benzyl, phenethyl, geranyl, citronellyl, and linalyl acetates are of considerable importance in perfumes.

The acetates currently in use for solvent purposes are of relatively high purity and uniformity, (see Table 3) compared with those in use 30 years ago, during which time much of the animal data presented in Table 4 were developed. The absence of any obvious evidence of systemic toxicity in spite of enormous current usage might indicate that some of the reports of the early 1930's of cumulative symptoms in man (reviewed by Browning[17] and von Oettingen[32]) may have been the result of impurities. The much more frequent use of well-designed spray booths in recent years has also greatly reduced the hazard.

2. *Physical and Chemical Properties*

The physical and chemical properties of acetate esters are given in Table 3.

3. *Physiological Response*

The physiological effects of higher concentrations are similar in both man and experimental animals and consist of signs of irritation of the eyes, nose, and throat, followed by a relatively slow and gradual onset of narcosis with slow recovery after exposure ceases (see Table 4). This is in contrast to the abrupt onset of anesthesia with the chlorinated hydrocarbons. The characteristic vomiting and nausea that invariably accompany a light anesthesia with methylene chloride, for example, do not seem to be typical of the acetates. On the other hand, there is more tendency to acidosis, especially with methyl acetate, presumably from its hydrolysis in vivo.[27]

[27] F. Flury and W. Wirth, *Arch. Gewerbepathol. Gewerbehyg.*, **5,** 1 (1934).

DAVID W. FASSETT

Compound	Formula	Mol. wt.	Sp. gr., °C. (g./ml.)
Methyl acetate	CH_3COOCH_3	74.08	0.9272 (20/4)
Ethyl acetate	$CH_3COOC_2H_5$	88.10	0.901 (20/4)
n-Propyl acetate	$CH_3COOC_3H_7$	102.13	0.8884 (20/4)
Isopropyl acetate	$CH_3COOCH(CH_3)_2$	102.13	0.8732 (18/4)
n-Butyl acetate	$CH_3COOC_4H_9$	116.16	0.8824 (18/4)
sec-Butyl acetate	$CH_3COOCH(CH_3)C_2H_5$	116.16	0.8701 (20/4)
Isobutyl acetate	$CH_3COOCH_2CH(CH_3)_2$	116.16	0.8747 (20/4)
n-Amyl acetate	$CH_3COOC_5H_{11}$	130.18	0.8756 (20/4)
sec-Amyl acetate	$CH_3COOCH(CH_3)C_3H_7$	130.18	0.86
Isoamyl acetate (banana oil)	$CH_3COO(CH_2)_2CH(CH_3)_2$	130.18	0.876 (15/4)
sec-Hexyl acetate (2-methylamyl acetate)	$CH_3COOCH(CH_3)CH_2CH(CH_3)_2$	144	
Hexyl acetate (1,3-dimethyl-butyl acetate)	$CH_3COOCH(CH_3)CH_2(CH_3)_2$	144.1	
Heptyl acetate	$CH_3COOC_7H_{15}$	158.24	0.876 (16/16)
2-Ethylhexyl acetate (octyl acetate)	$CH_3COOCH_2CHC_2H_5C_4H_9$	172.3	0.872

3

Properties of Acetates

M. p., °C.	B. p., °C.	Vapor pressure, mm. Hg (°C.)	Vapor density (air = 1)	Density of sat. air (air = 1)	Flash point, °F.	Solubility in H_2O (g. per 100 ml.)	Conversion factors	
							1 mg./ liter (p.p.m.)	1 p.p.m. (mg./ cu. meter)
−98.7	57	235 (25)	2.55	1.48	14	32	330	3.03
−82.4	77	100 (25)	3.04	1.02	28	8.6	278	3.60
−92.5	101.6	35 (25)	3.5	1.11	58	1.89	240	4.17
−73.4	88.9	73 (25)	3.5	1.24	40	3	240	4.17
−76.8	124–126	15 (25)	4.0	1.06	84	1	211	4.75
	112–113	24 (25)	4.0	1.09	66	3	211	4.75
−98.85	117.2	20 (25) 40 (39)	4.0	1.08	72	0.6	211	4.75
−71	148.8	5 (25)	4.5	1.02	77	0.18	188	5.32
	133	9 (25)	4.5	1.04	89	Slightly soluble	188	5.32
	143	6 (25)	4.5	1.03	92	0.25	188	5.32
−63.8	146.3	3.8 (20)			43		170	5.9
140–147					113	0.08	170	5.9
	191.5	12 (73)	5.5			Insoluble	155	6.4
−93	198.6	0.4 (20)	6.0		180	0.55	142	7.0

TABLE 3

Compound	Formula	Mol. wt.	Sp. gr., °C. (g./ml.)
Methyl aceto-acetate	$CH_3COCH_2COOCH_3$	116.1	1.0785 (20/20)
Ethyl acetoace-tate	$CH_3COOCH_2COOC_2H_5$	130.1	1.025
Methyl diaceto-acetate	$(CH_3CO)_2CHCO_2CH_3$	158.2	
Butyl aceto-acetate	$CH_3COCH_2COOCH_2CH_2CH_3$	158.2	0.9694
Methoxy butyl-acetate (butoxyl)	$CH_3CO_2C_4H_8OCH_3$	146.2	0.956
Cyclohexyl acetate	$CH_3COOC_6H_{11}$	142.10	0.963
Benzyl acetate	$CH_3COOCH_2C_6H_5$	150.17	1.057 (17)
Phenethyl acetate	$C_6H_5CH_2CH_2OOCCH_3$	164.2	1.031
Geranyl acetate	$CH_3-C(CH_3)=CH(CH_2)_2CH(CH_3)=CHCH_2OOC-CH_3$	196.28	0.9174 (15/15)
Linalyl acetate	$(CH_3)_2C=CH(CH_2)_2-C(CH_3)(OOCCH_3)-CH=CH_2$	196.28	0.895 (20/4)
Citronellyl acetate	$CH_2=C(CH_3)(CH_2)_3-CH(CH_3)(CH_2)_2OOCCH_3$	198.30	0.893

(*continued*)

M. p., °C.	B. p., °C.	Vapor pressure, mm. Hg (°C.)	Vapor density (air = 1)	Density of sat. air (air = 1)	Flash point, °F.	Solubility in H₂O (g. per 100 ml.)	Conversion factors	
							1 mg./ liter (p.p.m.)	1 p.p.m. (mg./ cu. meter)
−80	170	0.7 (20)	4.0		180	38	211	4.7
−80	180–181	0.8 (20)	4.5		185	13	188	5.3
			5.45				155	6.4
	213.9	0.19 (20)	5.45		185	2.8	155	6.4
	135–173	3 (30)	5.05		170	6.46	168	5.9
	170–177	7 (30)	4.9		136 (c.c.) 165 (c.c.)		172	5.8
−51.5	216	1.9 (60)	5.2	1.01	216	Insoluble	163	6.13
	226		5.7			Insoluble	149	6.7
	242					Insoluble		
	220					Insoluble		
	172 (34 mm.)					Insoluble		

TABLE 4. Physiological Response to Inhalation of Saturated Acetates

Compound	Species	Concentration p.p.m.	Concentration mg./liter	Duration of exposure	Results
Methyl acetate	Cat	31,000	95	2–3 hr.	Lethal[17,21,24]
	Cat	22,000	65		Lethal
	Cat	19,000	56		Narcosis
	Cat	10,000	32	22 hr.	Narcosis
	Cat	5,000	16	20 min.	Eye irritation; salivation
	Cat	6,600	20	6 hr./day for 8 days	Weight loss; weakness; slow recovery
	Mouse	21,000	63	3 hr.	Lethal
	Mouse	11,000	32	3 hr.	Narcosis; lethal
	Mouse	5,000	16	20 min.	No effect
	Man	10,000	32	Short time	Irritation persisting after exposure stopped
Ethyl acetate	Mouse	8,600	31	3 hr.	Narcosis; lethal[24]
	Mouse	5,000	18	3–4 hr.	Narcosis; recovery
	Cat	12,000	43	5 hr.	Lowest narcotic concentration[27]
	Cat	20,000	72	45 min.	Deep narcosis; recovered[24]
	Cat	43,000	155	15 min.	Deep narcosis; lethal
	Guinea pig	4,300	16	6 hr./day for 7 days	Loss of weight;[27] minor changes in blood
		2,000	7	6 hr./day for 65 days	No effects[28]
n-Propyl acetate	Man	400		30 min.	Irritation of nose and throat[29]
	Cat	24,500	102	30 min.	Narcosis and death[24]
Isopropyl acetate	Cat	5,300	22	6 hr./day for 5 days	Eye irritation; salivation
	Rat	Conc. vapor		>30 min.	Lethal[22]
	Rat	32,000		4 hr.	Lethal; 5/6
	Man	200			Eye irritation[30]
n-Butyl acetate	Guinea pig	14,000	67	4 hr.	Eye irritation; narcosis; lethal[31]
	Guinea pig	7,000	33	13 hr.	Eye irritation; deep narcosis; recovery
	Guinea pig	3,300	16	13 hr.	Eye irritation; no other symptoms
	Mouse	7,400	35	3 hrs.	Narcosis; recovery[27]
	Cat	17,500	83	30 min.	Narcosis; lethal to some[32]
	Cat	12,000	55	30 min.	Narcosis; recovered
	Cat	900		65 daily exposures	Weakness[25]
	Cat	4,200	20	6 hr. for 6 days	Weakness; loss of weight; minor blood changes[27]
Isobutyl acetate	Man	3,300	16	Brief	Marked irritation to eyes and nose[31]
	Man	200–300		Brief	Mild irritation on eyes and nose[29]
	Rat	21,000		150 min.	Narcosis; lethal; 6/6[33]
	Man	3,000		5 hr.	No symptoms
sec-Amyl acetate	Guinea pig	10,000	53	6 hr.	Eye and nose irritation; narcosis; lethal[14]
	Guinea pig	5,000	27	13 hr.	Eye and nose irritation; narcosis; recovered
	Guinea pig	2,000	11	13 hr.	Eye and nose irritation; no narcosis; recovered

Compound	Species	Conc.	27–53	Brief	Effects
Isoamyl acetate	Man	5,000–10,000	27–53	Brief	Mild to marked eye and nose irritation
	Dog	5,000	27	1 hr.	Nasal irritation; drowsiness[24]
	Cat	7,200	40	24 hr.	Light narcosis; delayed death due to pneumonia
	Cat	4,000	21	20 min.	Irritation of eyes and nose
	Cat	1,900	10	8 hr. for 6 days	Irritation; weakness; loss of weight
	Cat	500–1,000	2.7–5.3	2–3 hr. for 4–6 exposures	No effects[17,21]
	Man	950	5	30 min.	Irritation of nose and throat; headache; weakness[24]
sec-Hexyl acetate (2-methyl amyl acetate)	Rat	Conc. vapor		8 hr.	Nonlethal[22]
Hexyl acetate (1,3-dimethyl butyl acetate)	Rat	Conc. vapor		2 hr.	Nonlethal[22]
3-Heptyl acetate (acetic acid, 3-heptanol ester)	Rat	4,000		4 hr.	Lethal; 2/6
	Rat	Conc. vapor		8 hr.	Nonlethal[22]
2-Ethylhexyl acetate (octyl acetate)	Rat	Conc. vapor		15 min.	Nonlethal[35]
2-Ethylisohexyl acetate	Rat	5,900	45	6 hr.	Nonlethal; light narcosis; vasodilatation[33]
Methyl acetoacetate	Rat	Conc. vapor		8 hr.	Nonlethal[36]
Ethyl acetoacetate	Rat	Conc. vapor		8 hr.	Nonlethal[37]
Methyl diacetoacetate	Rat	Conc. vapor		8 hr.	Nonlethal[36]
Butyl acetoacetate	Rat	Conc. vapor		8 hr.	Nonlethal[22]
3-Methoxy butylacetate	Rat	Conc. vapor		8 hr.	Nonlethal[22]
Cyclohexyl acetate	Cat	1,700	10	10 hr.	Deep narcosis and death[38]
	Cat	860	5	9 hr.	
	Cat	1,600	9.5	8 hr. for 5 days	Irritation plus light narcosis
	Cat	637	3.7	8 hr. for 30 days	No symptoms
	Dog	637	3.7	8 hr. for 30 days	No symptoms
	Man	516	3	Brief	Irritation of eyes and throat
Benzyl acetate	Mouse	212	1.3	7–13 hr.	Dyspnea; narcosis; lethal[32]
	Cat	180	1.1	8–9 hr. for 7 days	Irritation; gradual weakness; loss of appetite and weight; drowsiness; lethal

[22] H. F. Smyth and H. F. Smyth, Jr., J. Ind. Hyg., 10, 163, 261 (1928). [33] K. W. Nelson, J. F. Ege, M. Ross, L. E. Woodman, and L. Silverman, J. Ind. Hyg. Toxicol., 25, 282 (1943). [30] L. Silverman, H. F. Schulte, and M. First, J. Ind. Hyg. Toxicol., 28, 262 (1946). [17] R. R. Sayers, H. H. Schrenk, and F. A. Patty, Public Health Repts., U. S., 51, 1229 (1936). [35] W. F. von Oettingen, Arch. Ind. Health, 21, 28 (1960). [24] D. W. Fassett, Laboratory of Industrial Medicine, Eastman Kodak Co., Rochester, N. Y., unpublished data. [21] F. A. Patty, W. P. Yant, and H. H. Schrenk, Public Health Repts., 51, 811 (1936). [36] H. F. Smyth, Jr, and C. P. Carpenter, J. Ind. Hyg. Toxicol., 26, 269 (1944). [37] H. F. Smyth, Jr, and C. P. Carpenter, J. Ind. Hyg. Toxicol., 30, 63 (1948). [38] H. F. Smyth, Jr., C. P. Carpenter, and C. S. Weil, J. Ind. Hyg. Toxicol., 31, 60 (1949). [32] K. B. Lehmann, Arch. Hyg., 78, 260 (1913).

In the author's experience, no anesthetic symptoms result in man from ethyl or butyl acetate at levels of 400 to 600 p.p.m. in exposures of 2 or 3 hours' duration. While butyl acetate may produce slight eye irritation in some people at 200–300 p.p.m., it does not appear to produce the characteristic temporary corneal edema caused by butanol at such levels.[39] No changes upon physical examination nor in blood or other laboratory findings have been found in persons exposed to these solvents. No skin sensitization and only minor dryness of the skin has been noted.[33]

An eye splashed with butyl acetate healed promptly in 48 hours, while one with isopropyl acetate took slightly longer.[40]

There appears to be no information as to the effect on humans of the aceto-acetates in Table 4, but from the known facts concerning metabolism of aceto-acetic acid, they are in all probability converted to normal metabolic products as are most of the simple aliphatic esters.

Both cyclohexyl and benzyl acetates appear to produce narcosis and lethal effects in animals at levels considerably below those of the other esters (see Table 4). They are of low volatility, however; and except for local irritation, no effects have been reported in man. Snapper[41] reports that benzyl acetate is rapidly hydrolyzed to benzyl alcohol, which in turn is oxidized to benzoic acid and excreted as hippuric acid.[42,43]

While no specific toxicity information could be found for phenethyl acetate, it seems probable that it would be readily hydrolyzed to phenethyl alcohol. This is known to be oxidized to phenyl acetic acid, which, in man, is excreted as the glutamine conjugate, phenyl acetylglutamine. This is known to be normally present in human urine in relatively large amounts (250 mg./day).[43,44]

The open chain olefinic terpene acetates (geranyl, linalyl, and citronellyl acetates) have also had little toxicological attention, although the metabolism of the alcohols is known. Assuming that these terpene esters are hydrolyzed to the alcohols, they probably are then oxidized in part to an acid known as Hildebrandt acid, $CH_3(COOH)C=CHCH_2CH_2C(CH_3)=CHCOOH$. In the case of geraniol, some reduced Hildebrandt acid (at the 2–3 double bond) is also formed. It seems probable that linalool (in which the OH group is tertiary) might resist oxidation and be excreted as a conjugate. This does not appear to have been proved, however.

Citronellol does not become cyclized (as does the aldehyde citronellal) and is excreted as a dicarboxylic acid, similar in structure to Hildebrandt acid.[43]

None of these perfume acetates seems to have given rise to any health problems in manufacturing, handling, or in their end uses. Skin sensitization to them appears to be rare.

[39] J. H. Sterner, H. C. Crouch, H. F. Brockmyre, and M. Cusack, *Am. Ind. Hyg. Assoc. Quart.*, **10**, 53 (1949).

[40] R. S. McLaughlin, *Am. J. Ophthalmol.*, **29**, 1361 (1946).

[41] I. Snapper, A. Grünbaum, and S. Sturkop, *Biochem. Z.*, **155**, 163 (1925).

[42] H. G. Bray, W. V. Thorpe, and M. R. Wasdell, *Biochem. J.*, **48**, 88 (1951).

[43] R. T. Williams, *Detoxication Mechanisms*. 2nd ed., Wiley, New York, 1959.

[44] W. H. Stein, A. C. Paladini, C. H. W. Hirs, and S. Moore, *J. Am. Chem. Soc.*, **76**, 2848 (1954).

4. *Hygienic Standards of Permissible Exposure*

The current threshold limits for some of the acetates are as follows:

Methyl acetate	200 p.p.m. (610 mg./cu.meter)
Ethyl acetate	400 p.p.m. (1400 mg./cu. meter)
n-Butyl acetate	200 p.p.m. (950 mg./cu. meter)
Amyl acetate	200 p.p.m. (1050 mg./cu. meter)
Propyl acetate	200 p.p.m. (840 mg./cu. meter)
sec-Hexyl acetate	100 p.p.m. (590 mg./cu. meter)

Baldi,[45] in reviewing industrial experiences with amyl, butyl, and propyl acetates, suggests that the concentrations (in terms of the acetate concentration) be kept below 400 to 500 p.p.m.

PROPIONATES, LACTATES, AND BUTYRATES

These esters are used as solvents, in perfumes, and flavors. Their solvent use is not as extensive as the acetates, although the propionates, lactates, and methyl and ethyl butyrate have been used for this purpose. Use in perfumes has been mentioned for butyl and amyl propionate, ethyl butyrate, and for methyl, ethyl, and isobutyl isobutyrates.

The physical and chemical properties are given in Table 5.

Little toxicity data are available, and there appears to be no problem in their industrial handling. No cumulative effects would be expected for most of them in view of their hydrolysis to materials that are either normally in the diet or that are readily converted to such materials. Some experimental data were found on the following compounds.

METHYL PROPIONATE

Treon[11] reported the oral LD_{50} in rabbits to be 2.5 to 3.2 g./kg. with the production of ataxia, gasping respirations, with hypothermia at lethal dose levels.

ETHYL PROPIONATE

The oral LD_{50} in the rabbit was 3.2 to 3.9 g./kg.[11] The symptoms were similar to methyl propionate. These symptoms may represent some acidosis due to the release of quantities of propionic acid at rates faster than can be handled by normal metabolic processes.

ETHYL β-ETHOXYPROPIONATE

Smyth[46] reported the oral LD_{50} in the rat to be about 5.0 g./kg. and the LD_{50} on rabbit skin to be about 10 ml./kg. An 8-hour inhalation of saturated vapor

---

[45] G. Baldi, *Med. lavoro*, **44**, 473 (1953).

[46] H. F. Smyth, Jr., C. P. Carpenter, and C. S. Weil, *A.M.A. Arch. Ind. Health & Occupational Med.*, **4**, 119 (1951).

TABLE 5

Physical and Chemical Properties of Propionates and Butyrates

Compound	Formula	Mol. weight	Sp. gr. °C (g./ml.)	M.p., °C	B.p., °C	Vapor pressure, mm. Hg (°C.)	Vapor density (air = 1)	Density of sat. air (air = 1)	Flash point, °F.	Solubility in H₂O (g. per 100 ml.)	1 mg./liter (p.p.m.)	1 p.p.m. (mg./cu. meter)
Methyl propionate	$CH_3CH_2COOCH_3$	88.10	0.915 (20/4)	−87.0	80	100(29)	3.0	1.26	−2	6	284	3.5
Ethyl propionate	$CH_3CH_2COOC_2H_5$	102.13	0.891 (20/4)	−73.0	99	40(27)	3.5	1.13	12	2.4	246	4.1
n-Amyl propionate	$CH_3CH_2COO(CH_2)_4CH_3$	144.21	0.8761 (15/4)	−73.1	164–166		5.0			Insoluble	170	5.9
n-Butyl propionate	$CH_3CH_2COOC_4H_9$	130.18	0.8828 (15)	−89.55	145.4		4.5		110	Very slightly soluble	188	5.3
Ethyl lactate	$CH_3CHOHCOOC_2H_5$	118.13	1.031 (20/4)	−25.0	154	5(30)	4.1	1.02	115	Very soluble	207	4.8
Butyl lactate	$CH_3CH(OH)COO(CH_2)_3CH_3$	146.18	0.9803 (22/4)	−43.0	155–195	0.4(20)	5.04	1.0	160	3.4	168	5.97
Ethyl β-ethoxy-propionate	$CH_3CH_2OCH_2CH_2COOC_2H_5$	146.0	0.948 (25/25)		165–172		5.0				168	5.9
Methyl butyrate	$CH_3CH_2CH_2COOCH_3$	102.13	0.8721 (20/20)	<−95.0	102.3	40(30)	3.52	1.13	57	1.56	240	4.2
Ethyl butyrate	$CH_3CH_2CH_2COOC_2H_5$	116.16	0.879 (20/4)	−93.3	121.3	20(28)	4.0	1.08	85	0.75	211	4.7
Methyl 3-hydroxybutyrate	$CH_3CHOHCH_2COOCH_3$	118.1	1.0559 (20/20)		174.9	0.85(20)	4.1		180	Very soluble	207	4.8
Methyl isobutyrate	$(CH_3)_2CHCOOCH_3$	102.13	0.8903 (20/20)	−84.0	92.0	50(24)	3.5	1.2	60	Slightly soluble	240	4.1
Isobutyl isobutyrate	$(CH_3)_2CHCOOCH_2CH(CH_3)_2$	144.21	0.8557 (20/20)	−81.0	147.0	10(38)	5.0	1.05	120	0.5	174	5.8

caused no deaths. Ethyl β-ethoxypropionate was a slight skin and moderate eye irritant.

BUTYL LACTATE

The subcutaneous lethal dose in mice was 12 g./kg. with symptoms of dyspnea, prostration, and loss of reflexes being noted.[21] No symptoms were produced at 5 g./kg.

METHYL 3-HYDROXYBUTYRATE

This material produced severe eye damage in the rabbit by application in the undiluted form.[47]

METHYL ISOBUTYRATE

Methyl isobutyrate has been found to be only slightly toxic. The undiluted material administered orally to rats killed at doses of 16,000 mg./kg., while intraperitoneal doses killed at 3200 mg./kg. Symptoms in the rats included labored respiration, vasodilitation, slight roughening of the coat, and in some instances, muscular twitching. Deaths were delayed as long as 2 days.

When rats were exposed to atmospheric concentrations of 175 mg./liter, or roughly 42,000 p.p.m. (calculated), for a period of 1 hour and 3 minutes, 2 out of 3 exposed animals died. In a repeated test at a lower concentration, rats subjected to 25 mg./liter, or roughly 6400 p.p.m. (calculated), for a period of 6 hours exhibited signs of loss of coordination and prostration; however, none of the animals died during the 6-hour exposure. Fourteen days later the animals had all gained the normal amount of weight and had no apparent residual effects from exposure to the compound.[33]

ISOBUTYL ISOBUTYRATE

Isobutyl isobutyrate is considered to be practically nontoxic orally and is classified as being only slightly toxic intraperitoneally. It was lethal to both rats and mice orally at 12,800 mg./kg. and did not kill any animals at 6400 mg./kg. When injected into the abdominal cavity, it killed rats at 6300 mg./kg. and mice at 1600 mg./kg.

When held in contact with the skin for 24 hours, the compound caused only slight skin irritation. This same test indicated that the material may be absorbed through the skin. No animals were killed by doses up to 10 cc./kg., but those receiving the highest dose came up to original weight only after the 2-week recovery period, thus indicating some interference with normal growth.

Inhalation of vapors at a concentration of about 5000 p.p.m. killed 2 out of 3

[47] C. P. Carpenter and H. F. Smyth, Jr., *Am. J. Ophthalmol.,* **29,** 1363 (1946).

rats in 6 hours. Symptoms showed prostration and complete narcosis. Similar exposures to 600 p.p.m. caused no death or symptoms in 6 hours.[33,48,49]

No threshold limits have been suggested for any of the members of this group.

GLYCEROL ESTERS

While the glycerol esters of fatty acids are of primary interest because of their importance as foods, a number of them also are of significance as industrial chemicals. The lower molecular weight esters, such as triacetin and tributyrin, have uses as solvents and plasticizers, while the mono- and diglycerides are useful because of their surface active properties. Tristearin has a number of uses such as in soap, candles, adhesives, polishes, and sizes.[50]

There are no health problems in the industrial handling of these materials nor in their end uses in foods, etc. They are hydrolyzed in the stomach and small intestine by lipases, which are relatively specific for glycerol esters. The esterases that act on simple esters such as ethyl butyrate hydrolyze most glycerides only slowly or not at all.[51] However, the lower molecular weight esters, such as mono-, di-, or triacetin, may also be hydrolyzed by simple esterases in serum and other tissues.

While the even-numbered fatty acid esters are far more prevalent in nature, the odd-numbered glycerol esters appear to be readily handled by mammalian metabolic processes, being split by esterases and lipases as usual with the acids being broken down into C_2 units with a terminal C_3 unit.

Since lipases are probably present in both the lung[51] and skin,[52] some hydrolysis may take place at these sites. The amount of free fatty acids released at these sites must be relatively small, however, since no major irritant effects occur on skin contact or inhalation, even when the acids are relatively highly ionized as would be the case with triacetin or tripropionin. The rate at which hydrolysis can occur is probably limited by the accumulation of fatty acids.

MONOACETIN (Acetin, Glyceryl Monoacetate)

$$C_3H_7O_2 \cdot OOCCH_3$$

1. Uses and Industrial Exposures

Monoacetin is a colorless or pale yellow hygroscopic liquid with a characteristic odor. It is used in the manufacture of smokeless powders and dynamite, in

[48] H. E. Parish and E. F. Knipling, *J. Econ. Entomol.*, **35**, 70 (1942).

[49] M. B. Jacobs, *Am. Perfumer Essent. Oil Rev.*, **54**, 303, 471 (1949).

[50] A. Rose and E. Rose, *The Condensed Chemical Dictionary*. 5th ed., Reinhold, New York, 1956.

[51] J. Fruton and S. Simmonds, *General Biochemistry*. 2nd ed., Wiley, New York, 1958.

[52] S. Rothman, *Physiology and Biochemistry of the Skin*. Univ. of Chicago Press, Chicago, 1954.

tanning, and as a solvent. It has been proposed as an antidote for fluoroacetate poisoning.[53]

2. Physical and Chemical Properties

Molecular weight: 134.13
Specific gravity: 1.206 (20/4°C.)
Boiling point: 258°C.
Freezing point: −78°C.
Vapor pressure: 3 mm. Hg (130°C.)
Soluble in water

3. Physiological Response

The physiological properties of monoacetin have recently been reviewed by von Oettingen.[32] Li et al.[54] determined the LD_{50} in rats and mice by subcutaneous injections using a 50 per cent water solution buffered with phosphate to pH7. In mice the solution was diluted 1:5 or 1:10. The LD_{50} in rats was 5.5 ml./kg. and in mice, 3.5 ml./kg. Symptoms appear rapidly and consist of prostration, labored respirations, and in some cases, terminal convulsions. The survivors make an uneventful recovery. Autopsy shows marked pulmonary congestion and right heart dilatation.

Li and Anderson[55] have also studied the intravenous toxicity in dogs and rabbits. The LD_{50} in these species are 5 ml./kg. and 4 ml./kg., respectively. The symptoms are those of severe dyspnea, tremors, retraction of the neck, and terminal convulsions, occurring 2 to 20 minutes after injection. The pathological findings were hemorrhagic areas in the lung.

Studies were also made in vitro of hemolysis of red cells. No hemolysis resulted in 24 hours at concentrations of 1/4000 up to 1/500. The pH of a 50 per cent water solution of monacetin was 3.88; at a 1:14 dilution the pH was 4.5.

DIACETIN (Glyceryl Diacetate)

$$C_3H_5(OH)(CH_3COO)_2$$

1. Uses and Industrial Exposures

Diacetin is a colorless hygroscopic liquid used to a limited extent as a plasticizer and softening agent. It usually is a mixture of its isomers.

2. Physical and Chemical Properties

Molecular weight: 176.7
Specific gravity: 1.178 (15/15°C.)

[53] M. G. Chenoweth, A. Kandel, L. B. Johnson, and D. R. Bennett, J. Pharm. Exptl. Therap., 102, 31 (1951).
[54] R. C. Li, P. T. Sah, and H. H. Anderson, Proc. Soc. Exptl. Biol. Med., 46, 26 (1941).
[55] R. C. Li and H. H. Anderson, J. Pharm. Exptl. Therap., 72, 26 (1941).

Boiling point: 259°C. (approximately)
Freezing point: −30°C.
Vapor pressure: about 40 mm. Hg (175°C.)
Soluble in water

3. *Physiological Response*

Diacetin has been studied by Li *et al.*[54] The subcutaneous LD_{50} in the rat is 4 ml./kg. and in the mouse 2.5 ml./kg. The intravenous lethal dose in the dog is 3 ml./kg. and in the rabbit is 1.5 ml./kg. The symptoms are similar to those from monoacetin.

TRIACETIN (Glyceryl Triacetate)

$$(CH_3COO)_3C_3H_5$$

1. *Uses and Industrial Exposures*

Triacetin is more widely used than the other acetins. It is a cellulose plasticizer, fixative in perfumes, solvent, and is used in the therapy of fungus infections. It is a colorless, essentially odorless liquid.

2. *Physical and Chemical Properties*

Molecular weight: 218.20
Specific gravity: 1.156 (25/4°C.)
Boiling point: 258°C.
Melting point: −78°C.
Vapor pressure: 1 mm. Hg (100°C.)
Flash point: 290°F.
Soluble in water

Studies by Li *et al.*[54] show that triacetin has the same type of effect as the other acetins. The subcutaneous LD_{50} in the rat was 2.8 ml./kg.; mouse, 2.3 ml./kg. The intravenous lethal dose in the dog was 1.5–2.0 ml./kg.; in the rabbit, 0.75 ml./kg. The symptoms were similar to the other acetins. Triacetin has a slightly lower pH (3.88) in dilute water solutions than mono- or diacetin and will produce slight hemolysis at concentrations of 1000 to 2000 in vitro.

Hughes[56] found that it produced no injury to the rabbit eye after a continuous 6-minute irrigation.

Other studies on triacetin have shown that it has a low toxicity orally in the rat (6.4 to 12.8 g./kg.) and mouse (3.2–6.4 g./kg.). The symptoms were principally weakness and ataxia. It was not a skin irritant and not absorbed through guinea pig skin. It was not a skin sensitizer in this species. Inhalation of concentrations (produced by using heated vapor) averaging 250 p.p.m. 6 hours a day, 5 days a week, for 13 weeks produced no symptoms or histopathology in rats. No changes were seen in liver and kidney weights, blood counts, or urine analysis. Exposure to

[56] W. F. Hughes, Jr., *Bull. Johns Hopkins Hosp.*, **82**, 338 (1948).

saturated vapor (plus some mist) for 6 hours a day for 5 days caused no symptoms or histopathology.[33] Experiments by Cox[57] showed that rats made very good weight gains when triacetin replaced fat in the diet to the extent of 55 per cent of the total diet.

While triacetin may have some irritant effects when injected, it appears to be innocuous when swallowed, inhaled, or in contact with skin.

Triacetin has recently been used in the treatment of fungal skin infections.[58, 58a] Esterases in fungi or in serum act at pH >4 to slowly release acetic acid *in situ*. The extent of hydrolysis is automatically limited by the increased acidity and lowering of pH.

TRIPROPRIONIN (Glyceryl Triproprionate)

$$(CH_3CH_2COO)_3C_3H_5$$

1. Physical and Chemical Properties

Molecular weight: 260.29
Boiling point: 155°C., 15 mm. Hg

2. Physiological Response

The acute toxicity of tripropionin has been studied by Hodge.[59] In mice, the intraperitoneal LD_{50} is about 1.7 ml./kg. About 10 minutes after injection of lethal doses, there was a sudden onset of excitement, followed by stupor and death in about 1 hour. The oral LD_{50} in rats is about 15 ml./kg. The symptoms were considerably different from those in mice given parenteral injections and consisted of lethargy and narcosis. Survivors showed some diarrhea. The symptoms on intraperitoneal injection are suggestive of some in vivo hydrolysis releasing propionic acid.

In another laboratory also the oral toxicity of tripropionin in the rat was found to be very low (LD_{50} 6.4–12.8 g./kg.).[33] The only symptoms seen were weakness and ataxia. Similar findings were noted in the mouse and guinea pig. On intraperitoneal injection in mice, rats, and guinea pigs, the results were similar to those of Hodge.[59] Rats exposed to heated vapor (about 750 p.p.m.) 6 hours a day for 90 days showed no symptoms or abnormalities in laboratory data or histopathology. It is not a skin irritant or sensitizer in the guinea pig.[33]

TRIBUTYRIN

$$(C_4H_7COO)_3C_3H_5$$

1. Physical and Chemical Properties

Molecular weight: 302.36
Specific gravity: 1.0350 (20/4°C.)

[57] W. M. Cox, *J. Biol. Chem.*, **103,** 777 (1933).
[58] S. A. M. Johnson and J. L. Tuura, *Arch. Dermatol.,* **74,** 73 (1956).
[58a] S. G. Knight, *J. Invest. Dermatol.,* **28,** 363 (1957).
[59] H. C. Hodge, *Proc. Soc. Exptl. Biol. Med.,* **49,** 277 (1942).

Melting point: $< -75°C$.
Boiling point: $315°C$.
Insoluble in water

2. *Physiological Response*

The oral LD_{50} in the rat is about 13 g./kg. Weakness, ataxia, and vasodilitation in the ears and feet were noted. In the mouse the LD_{50} is even higher. There is no percutaneous absorption or irritation in the guinea pig. Inhalation of 78 p.p.m. for 6 hours caused temporary (30 minutes) hyperpnea, but no fatalities or other symptoms in rats.[33]

TRIISOBUTYRIN

The oral LD_{50} in rats is 1.6 to 3.2 g./kg. The symptoms were only those of weakness and ataxia. It is not a skin irritant in the guinea pig. Inhalation of 145 p.p.m. for 6 hours by rats caused no fatalities and only transient increased depth of respiration.[33]

GLYCERIDES OF LAURIC ACID

The oral toxicity of the glycerides of lauric acid has been studied recently by Fitzhugh, Schouboe, and Nelson[60] as a result of a question regarding their safety when used in foods. Prolonged feeding of high levels (25 per cent) in the rat diet was not productive of any toxic symptoms. They apparently are metabolized in the same manner as other fats and occur normally in foods such as milk.[61]

MONO- AND DIGLYCERIDES

For many years, mono- and diglycerides isolated from normal sources or prepared by synthesis have been used as emulsifiers in foods. A review of the evidence as to their safety has been published by the Food Protection Committee of the National Research Council.[61]

While animal data had indicated that they were utilized as other fats, further evidence bearing on their safety was obtained by the demonstration that triglycerides are normally converted to mono- and diglycerides in the human intestine.[62] Acetylated mono- and diglycerides also appear to be utilized as foods.[62]

HIGHER MOLECULAR WEIGHT ALIPHATIC ESTERS

The esters of the higher molecular weight fatty acids have their principal uses as plasticizers (butyl laurate, myristate, and stearate) and as components of

[60] O. G. Fitzhugh, P. J. Schouboe, and A. A. Nelson, *Toxicol. Appl. Pharmacol.,* **2,** 59 (1960).

[61] The safety of mono- and diglycerides for use as intentional additives in foods, Nat. Acad. Sci., Nat. Research Council Publ. No. **251,** 1952.

[62] H. Kuhrt, E. A. Welch, W. P. Blum, E. S. Perry, W. H. Weber, and E. S. Nassett, *J. Am. Oil Chemists Soc.,* **29,** 271 (1952).

flavors (ethyl caproate, heptanoate, caprylate, pelargonate, and caprate). The physical properties are given in Table 6.

None of these is known to produce any ill effects either in industrial handling or in their end uses. They are thought to be readily hydrolyzed to the corresponding alcohols and acids and then further metabolized by routes reviewed in the

TABLE 6
Physical Properties of Ethyl Esters of Some Saturated Fatty Acids[a]

Ester	Mol. wt.	Sp. gr., °C. (g./ml.)	M. p., °C. (b = f. p.)	B. p., °C. B. p., °C.(mm. Hg)
Ethyl valerate (C$_5$)	130.18	0.877 (20)	−91.0 (b)	144.6[736]
Ethyl caproate (C$_6$)	144.21	0.889 (0)	−67.5 (b)	166–167[760]
Ethyl heptanoate (C$_7$)	158.24	0.886 (0)	−66.3 (b)	188.6[760]
Ethyl caprylate (C$_8$)	172.26	0.884 (0)	−43.2 (b)	208.5[760]
Ethyl pelargonate (C$_9$)	186.29	0.869 (15)	−36.7 (b)	216–219[760]
Ethyl caprate (C$_{10}$)	200.31	0.857 (30)	−19.9 (b)	243–245[760]
Ethyl undecanoate (C$_{11}$)	214.29		−147.0 (b)	140[20]
Ethyl laurate (C$_{12}$)	228.37	0.867 (19)	−1.8 (b)	163[25]
Ethyl myristate (C$_{14}$)	256.42	0.857 (25)	12.3	139[4]
Isopropyl myristate (C$_{14}$)	270.42			
Ethyl palmitate (C$_{16}$)	284.47	0.858 (25)	25.0	185[10]
Ascorbyl palmitate (C$_{16}$)	414.0		116.0	
Ethyl stearate (C$_{18}$)	312.52		33.9	152[0.18]
Butyl stearate (C$_{18}$)	340.57	0.875 (25/25)	27.5	220–225[25]

[a] Adapted from *The Lipids*, Vol. I, by H. J. Deuel, Jr., Interscience Publishers, New York, 1951.

chapters on organic acids and alcohols. If methyl groups are inserted on the lower fatty acid (C$_4$ to C$_7$), they are still readily metabolized. However, if the substituent is ethyl or propyl, for example, 2-ethyl butyric, they tend to be excreted unchanged or as an ester glucuronide.[43] They may undergo ω-oxidation to dicarboxylic acids (see review by Weitzel[63]).

[63] G. Weitzel, *Z. physiol. Chem.*, **287**, 254 (1951).

Deuel *et al.*[64] have studied the effects of the ethyl esters of fatty acids on the production of ketone bodies in the rat.

Oettel[65] reports no skin damage from ethyl pelargonate and ethyl heptanoate when applied to human skin for periods from 5 minutes to 5 hours.

Ethyl laurate was said to produce a diffuse myocarditis in cats when fed at very high levels (35 to 40 per cent) with diets lacking choline. When choline or methionine was given the myocarditis was absent.[66]

Isopropyl myristate has been studied by Platcow and Voss[67] as a possible vehicle for parenteral injection. Its oral toxicity in mice was >100 ml./kg. Repeated intraperitoneal injections of 5 ml./kg. in rats daily over 20 days caused some deaths after 5 days. In the remainder, no growth effects or toxic symptoms were noticed. It was not a sensitizer on intracutaneous injection in the guinea pig. No eye irritation was produced on direct contact. It is undoubtedly hydrolyzed to normal metabolic products.

Fitzhugh and Nelson[68] fed ascorbyl palmitate to rats over long periods of time. The 5 per cent level caused poor growth and some bladder stones were noted. At 0.25 per cent levels, no effects were noted from either the dextro or levo forms.

Butyl stearate was studied by Smith.[69] The oral LD_{50} in rats was greater than 32 g./kg. Two-year feeding studies in the rat at 1.25 and 6.25 per cent produced no effects on growth, mortality, or fertility.

Stetten and Salcedo[70] studied the effect of the chain length of the ethyl esters of fatty acids (C_4 to C_{18}) on the development of fatty liver in rats deficient in choline. Young rats were fed the ethyl esters at a 35 per cent level in the diet for 2 weeks and the liver fatty acids determined. The liver fats were normal for ethyl stearate, but increased rapidly as the chain lengths decreased to 16 or 14 carbon atoms. No severe fatty livers were seen when ethyl esters of less than 12 carbon atoms were fed. Gastric hemorrhages were noted with ethyl butyrate and some renal hemorrhage with ethyl caprylate. The myocarditis caused by ethyl laurate was not found when choline was provided.

2. Monocarboxylic—Unsaturated—Aliphatic

1. Uses

There are two types of esters in this series; those with unsaturation in the

$$\text{acid portion (acyl } R\overset{\displaystyle O}{\overset{\displaystyle \|}{-C}}-) \text{ and those in which the unsaturation is in the alcohol}$$

[64] H. J. Deuel, Jr., L. F. Hallman J. S. Butts, and S. Murray, *J. Biol. Chem.*, **116**, 621 (1936).

[65] H. Oettel, *Arch Exptl. Pathol. Pharmacol.*, **183**, 641 (1936).

[66] H. D. Kesten, J. Salcedo, Jr., and De W. Stetten, Jr., *J. Nutrition,* **29**, 171 (1945).

[67] E. L. Platcow and E. Voss, *J. Am. Pharm. Assoc.*, **43**, 690 (1954).

[68] O. G. Fitzhugh and A. A. Nelson, *Proc. Soc. Exptl. Biol. Med.*, **61**, 195 (1946).

[69] C. C. Smith, *Arch. Ind. Hyg. Occupational Med.*, **7**, 310 (1953).

[70] De W. Stetten, Jr., and J. Salcedo, Jr., *J. Nutrition*, **29**, 167, (1945).

$$O$$
$$\|$$
portion of the ester (R—C—OCH=CH$_2$). The various esters of acrylic acid are examples of the former and the vinyl esters, of the latter. They are among the most important of industrial chemicals and have their principal uses in production of thermoplastic, vinyl, and acrylic resins.

The total production of vinyl resins in 1958 was reported to be 870 million pounds; the production of vinyl acetate monomer was 190 million pounds.[2] In most vinyl resins, vinyl acetate is copolymerized with vinyl chloride, with the latter being used in larger quantities.

The acrylic resins were produced in smaller quantities, being somewhat less than 190 million pounds in 1958.[2]

Some long chain unsaturated fatty acid esters have uses as solvents, lubricants, and plasticizers.

2. *Physical and Chemical Properties*

The physical and chemical properties of examples of vinyl and acrylic resins are given in Table 7.

Most of these are liquids with relatively low boiling points and high vapor pressures. They generally have sharp, unpleasant and persistent odors. Ethyl acrylate can be readily detected in concentrations as low as 8 p.p.m.

3. *Methods of Preparation*

The acrylate esters can be prepared by reacting ethylene chlorohydrin with sodium cyanide and a subsequent reaction with an alcohol and sulfuric acid, for example,

$$HOCH_2CH_2CL \xrightarrow{NaCN} HOCH_2CH_2CN \xrightarrow[H_2SO_4]{CH_3OH} CH_2{=}CHCOOCH_3$$
$$\text{methyl acrylate}$$

They also can be produced by oxo processes, for example, acetylene + CO + ethanol → ethyl acrylate.

The methacrylate esters can be produced from acetone cyanohydrin, an alcohol, and sulfuric acid,[1,71] for example,

$$CH_3{-}C(OH)CN \xrightarrow[H_2SO_4]{ROH} CH_2{=}C{-}\overset{\overset{O}{\|}}{C}{-}OR + (NH_4)_2SO_4$$
$$\underset{CH_3}{|} \qquad\qquad \underset{CH_3}{|}$$

Vinyl acetate is made by reacting acetic acid with acetylene in the presence of a catalyst such as a mercuric salt.

$$CH{\equiv}CH + HOOC{\cdot}CH_3 \rightarrow CH_2{=}CH\ OOC\ CH_3$$

Polymerization of vinyl and acrylate monomers occurs very readily and is an exothermic reaction, making it necessary to handle them initially only in the

[71] R. N. Shreve, *The Chemical Process Industries,* McGraw-Hill, New York, 1956.

TABLE 7

Physical and Chemical Properties of Monocarboxylic, Unsaturated Aliphatic Esters

Compound	Formula	Mol. wt.	Sp. gr. (g./ml.)	M. p., °C.	B. p., °C.	Vapor pressure, mm. Hg (°C.)	Vapor density (air = 1)	Flash point, °F.	Solubility in H₂O (g. per 100 ml.)	Conversion factors: 1 mg./liter (p.p.m.)	1 p.p.m. (mg./cu. meter)
Methyl acrylate	$CH_2CHCO_2CH_3$	86.09	0.95	< −75	80	68 (20)	3.0	27 (c.c.)	5.2	284	3.5
Ethyl acrylate	$CH_2CHCO_2C_2H_5$	100.11	0.924	< −75	99.5	30 (20)	3.5	48	1.5	245	4.1
Butyl acrylate	$CH_2CHCO_2C_4H_9$	128.17	0.898	−65	145	4 (20)	4.4	105	Insoluble	191	5.2
2-Ethylbutyl acrylate	$CH_2CHCO_2CH_2C(C_2H_5)HC_2H_6$	156.2	0.896	−70	82 (10 mm.)	1.7 (20)	5.4	125	Insoluble	157	6.4
2-Ethylhexyl acrylate	$CH_2CHCO_2CH_2CH(C_2H_5)C_4H_9$	184.3	0.887	−90	130 (50 mm.) / 217 (760 mm.)	1.0 (20) / 15 (100)	6.35	180	Insoluble	133	7.5
Methyl methacrylate	$CH_2C(CH_3)COOCH_3$	100.1	0.936	−50	101.0	40 (25)	3.45	85	Slightly soluble	245	4.1
Ethyl methacrylate	$CH_2C(CH_3)CO_2C_2H_5$	114.07	0.911	< −75	118		3.94	95	Insoluble	215	4.6
n-Butyl methacrylate	$CH_2C(CH_3)COCC_4H_9$	142.19	0.895		163		4.9	150	Insoluble	172	5.8
Isobutyl methacrylate		142.19			60 (20 mm.)	1.8 (20)	4.9		Insoluble	172	5.8
2-Ethyl isohexyl-methacrylate	$CH_2C(CH_3)CO_2C_8H_{17}$	198			113 (20 mm.)	<1 (20)	6.8		Insoluble	124	8.1
Methyl crotonate	$CH_3CH=CHCOOCH_3$	100.11	0.946		118	18 (25)	3.5	105	2.4	245	4.1
Ethyl crotonate	$CH_3CH=CHCO_2C_2H_5$	114.11	0.92	45	139		3.9	36	Insoluble	215	4.6
2-Ethylhexyl crotonate	$CH_3CH=CHCO_2C_8H_{17}$	198.30	0.888		204	10 (119)	6.8	245	Insoluble	124	8.1
Allyl formate	$HCOOCH_2CH=CH_2$	86.09	0.948		82–84		3.0		Slightly soluble	284	3.5
Allyl acetate	$CH_3COOCH_2CH=CH_2$	100.12	0.928		103–104		3.5		Slightly soluble	245	4.1
Allylidene diacetate	$CH_2=CHCH(OCOCH_3)_2$	158.15	1.0749	−36.6	107 (50)		5.46	180		155	6.5
Isopropenyl acetate	$CH_3COOC(CH_3)=CH_2$	100.12			97		3.45	60		245	4.1
Vinyl acetate	$CH_3COOCH:CH_2$	86.09	0.9342	< −84	73	115 (25.3)	3.0	−8	2.5	284	3.5
Vinyl butyrate	$CH_3(CH_2)_2COOCH=CH_2$	114.15			116		3.9			215	4.6

presence of inhibitors such as hydroquinone. They are distilled off in the reactive form at the time of polymerization. Special precautions are necessary in their handling because of a high degree of flammability and because of the possibility of high pressures and temperatures being developed as a result of the exothermic polymerization reactions. A useful summary of storage and handling precautions has been given by Riddle.[72] Chemical Safety Data Sheets are available for methyl and ethyl acrylate and vinyl acetate.[73,74] The vinyl resins are especially important in films and coatings and the acrylates in transparent molding compositions and adhesives.

Another useful class of ester resins can be prepared by condensing partially hydrolyzed polyvinyl acetate with aldehydes. These are known by the general name of polyvinyl acetals or more specifically, as polyvinyl formal, polyvinyl butyral, etc. A typical unit in polyvinyl formal is:

$$-CH_2CH-CH_2CH-CH_2-CH-CH_2CH-$$

Polyvinyl butyral is widely used as an interlayer in safety glass in automobiles, etc.

The crotonate esters are similar to methacrylates in structure except that the methyl group is in the beta position. They do not polymerize as readily as the methacrylates,[1] although under some conditions they may polymerize with vinyl acetates to form transparent high-melting resins.[75]

Some long chain unsaturated fatty acid esters have uses as solvents, lubricants, and plasticizers (butyloleate, methoxyethyl oleate, and methyl ricinoleate). These can be made by the usual methods of esterification of the acids with alcohols.

4. Physiological Response

The introduction of a double bond into either the acyl (acid) or alcohol portion of the ester results in considerable modification of its physiological effects. These unsaturated esters acquire marked irritant, and in some cases, lachrymatory, properties. The acute LD_{50} by various routes tends to be lower than for the corresponding saturated ester. Treon[11] has pointed out that methyl and ethyl acrylate are in general 10 times as toxic as methyl and ethyl propionate. The symptoms following ingestion include signs of collapse, severe respiratory difficulty, and central nervous system stimulation. By inhalation they are usually marked irritants, causing salivation, conjunctival irritation, and marked respira-

[72] E. H. Riddle, *Monomeric Acrylic Esters*. Reinhold, New York, 1954.

[73] *Mfg. Chemists Assoc. Chem. Safety Data Sheet* No. **SD-75,** Vinyl acetate, 1959.

[74] *Mfg. Chemists Assoc. Chem. Safety Data Sheet* No. **SD-79,** Methyl and ethyl acrylates, 1960.

[75] *Tech. Data Sheet,* Methyl crotonate, Eastman Chemical Products, Kingsport, Tenn., 1959.

tory irritation or pulmonary edema at high concentrations. Prolonged contact with the skin or eye may cause severe damage. The pathology reported from single lethal or sublethal doses by mouth or inhalation[9,11] is not particularly characteristic except for the almost universal finding of marked pulmonary congestion or hemorrhage. Such findings as cloudy swelling of the liver and kidney have been noted at very high levels, but probably represent only nonspecific effects of the very high tissue concentrations of such irritants.

It is of interest that no pathology in the liver or kidney was found by Treon[11] upon repeated inhalation of methyl or ethyl acrylate at levels where local irritant effects were minimal, nor was any cumulative effect noted on repeated oral ingestion of amounts as high as a tenth of the oral LD_{50}.

The cause of the lachrymatory action of the esters containing unsaturated acids is undoubtedly the result of some mechanism such as that discussed by Dixon[18] in which the neighboring keto group has a polarizing effect on the double bond, making it capable of adding nucleophilic agents such as SH groups. These esters are not general enzyme inhibitors, but probably do have some specificity for certain enzyme SH groups, which may be associated with their effects on the eye.

It is unknown whether similar reasoning regarding mechanisms of action should be applied to esters having unsaturation in the alkyl group, but on theoretical grounds it seems inappropriate.

Table 8 illustrates the effect of unsaturation. There is no question but that striking differences exist in the case of acrylates versus propionates and in the vinyl and allyl acetates versus ethyl and propyl acetates. There is, however, little difference between the isomer of allyl acetate (isopropenyl acetate) and isopropyl acetate. The position of the double bond in relation to the carbonyl group is apparently one of the determining factors in the biological action. The substituent on either of the carbons of the double bond is important. Ethyl acrylate, for example, is several times as toxic as ethyl methacrylate. Moving the methyl group to the beta position, as in ethyl crotonate, increases the toxicity somewhat orally but probably not by inhalation.

The metabolism of these unsaturated esters appears not to have been studied. Some literature reviewed by Williams[43] indicates that while acrylate is oxidized by liver and kidney enzymes, it probably is not an intermediate in propionate metabolism. On the other hand, crotonyl and methacrylyl groups are known to occur during the metabolism of butyryl and isobutyryl molecules.[51,76]

Some of the data (mostly from range-finding studies by Smyth et al.) on toxicity in animals are summarized in Table 9.[22,35-37,46]

No comprehensive study in man at known exposure levels has been reported, although the primary irritant and lachrymatory effects are well known. No reports were found of injury to man from long-term exposure to low level concentrations. While there is fatigue of the ability to detect odors, it is unlikely that high

---

[76] M. J. Coon, *Federation Proc.*, **14**, 762 (1955).

TABLE 8

Effects of Unsaturation of Acid or Alcohol on Monocarboxylic, Unsaturated Aliphatic Esters

Ester	Formula	Physiological Effects
Ethyl propionate	$CH_3CH_2COOCH_2CH_3$	Oral LD_{50} (rabbit): 3,500 mg./kg.
Ethyl acrylate	CH_2=$CHCOOCH_2CH_3$	Oral LD_{50} (rabbit): 400 mg./kg.[11] Oral LD_{50} (rat): 1,000 mg./kg.[9] Inhalation LC_{50} (rat) 4 hr.: <1,000 p.p.m. (4 mg./liter)
Methyl isobutyrate	$CH_3CH(CH_3)COOCH_3$	Oral LD_{50} (rat): 16,000 mg./kg.[33] Inhalation LC_{50} (rat) 6 hr.: >25 mg./liter
Methyl methacrylate	CH_2=$C(CH_3)COOCH_3$	Oral LD_{50} (rat): 8,400 mg./kg.[13] Inhalation LC_{50} (rat) 8 hr.: 15 mg./liter
Ethyl acetate	$CH_3COOCH_2CH_3$	Oral LD_{50} (rat): 5,600 mg./kg.[77] Inhalation LC_{50} (rat) 8 hr.: 1,600 p.p.m.
Vinyl acetate	CH_3COOCH=CH_2	Oral LD_{50} (rat): 2,900 mg./kg. Inhalation LC_{50} (rat) 4 hr.: 4,000 p.p.m.
n-Propyl acetate	$CH_3COOCH_2CH_2CH_3$	Inhalation LC_{50} (cat) 30 min.: 24,000 p.p.m.[24]
Allyl acetate	CH_3COOCH_2CH=CH_2	Inhalation LC_{50} (rat) 1 hr.: 1,000 p.p.m.[3]
Isopropyl acetate	$CH_3COOCH(CH_3)_2$	Oral LD_{50} (rat): 3,000 mg./kg. Inhalation LC_{50} (rat) 4 hr.: 32,000 p. p. m.
Isopropenyl acetate	$CH_3COOC(CH_3)$=CH_2	Oral LD_{50} (rat): 3,000 mg./kg.[37] Inhalation LC_{50} (rat): concentrated vapor tolerated 30 min.
Ethyl methacrylate	CH_2=$C(CH_3)COOC_2H_5$	Oral LD_{50} (rat): 15,000 mg./kg.[13] Inhalation LC_{50} (rat): 15 mg./liter
Ethyl crotonate	CH_3CH=$CHCOOC_2H_5$	Oral LD_{50} (rat): 3,000 mg./kg.[35] Inhalation saturated vapor (rat) 8 hr.: no deaths

levels would be tolerated voluntarily. Because of their generally high vapor pressure, a high degree of hazard exists in the event of spills or leakage in confined spaces.

Threshold limits have been proposed for ethyl acrylate (25 p.p.m. or 100 mg./cu. meter) and methyl acrylate (10 p.p.m. or 35 mg./cu. meter).[78] Precautions such as those given in reference 74 should be used for all volatile unsaturated esters.

[77] W. Spector, ed., *Handbook of Toxicology*, Vol. I. Saunders, Philadelphia, 1956.

[78] American Conference of Governmental Industrial Hygienists, *Am Ind. Hyg. Assoc. J.* **22**, 325 (1961).

TABLE 9

Physiological Response to Monocarboxylic, Unsaturated Aliphatic Esters

Compound	Acute oral LD50 (rats) (g./kg.)	Skin toxicity LD50 (rabbit) (mL./kg.)	Inhalation by rats p.p.m.	Time	Mortality	Sat. vapor time for no deaths	Skin irritation (rabbit)	Eye effect (rabbit)
Methyl acrylate[34]	0.2 (rabbit)[11]	1.3	1,000	4 hr.	3/6		Moderately severe[23]	Moderate
Ethyl acrylate	0.4 (rabbit)[11] 1.0[30]	1.95	4,000 2,000 1,000 50,000 540 70	4 hr. 4 hr. 4 hr. 15 min. 19 days 30 hr.	6/6 5/6 0/6 6/6 12/18 2/29		Slight (severe if confined)	Moderate
2-Hydroxy ethylacrylate	1.0	1.0	500	4 hr.	5/6	1 hr.	Slight	Severe
2-Ethoxy ethylacrylate	1.0	1.0	500	4 hr.	5/6	1 hr.	Slight	Severe
Ethoxy propylacrylate	0.8	1.4	250	4 hr.	1/6	1 hr.	Moderate	Moderate
n-Butyl acrylate	3.7	3.4	1,000	4 hr.	5/6	30 min.	Slight	Moderate
2-Ethyl butylacrylate	6.5	5.5				4 hr.	Moderate	Slight
2-Ethylhexyl acrylate	5.6 6.4–12.8[33]	8.5 >10 (guinea pig)[33]				8 hr.	Moderately severe[23] (under cuff) (guinea pig)	Slight
Methyl methacrylate[13,79]	8.4 6–7 (rabbit)	>10	3,750	8 hr.	Approx. LC50		Moderate	
Ethyl methacrylate[13]	14.8 (rat) 4–6 (rabbit)	>10	13,500 3,300	3 hr. 8 hr.	 Approx. LC50		Moderate	
n-Butyl methacrylate[13]	>20 (rat) 7–10 (rabbit)	>10	>880	8 hr.				
Isobutyl methacrylate[33]	6.4–12.8 (rat)	>20 (guinea pig)	3,600	6 hr.	0/3		Moderate	Moderate
2-Ethyl isohexyl methacrylate[33]	>12.8	>20 (guinea pig)	14	6 hr.	0/3		Moderate	Slight
2-Butyl octyl methacrylate	25.8	10–20 (guinea pig)				8 hr.	Slight	Slight
Methyl crotonate[33]	>3.2 1.6–3.2 (mice)		19,000	6 hr.	0/3		Moderate	Slight
Ethyl crotonate	3.0	>10 (guinea pig)	4,000	4 hr.	6/6	8 hr.	Like morpholine	Severe
Vinyl crotonate	6.5		2,000	4 hr.	0/6	1 hr.	Slight	Slight
2-Ethyl hexyl crotonate[23]	>3.2 (mice)	>20 (guinea pig)	2,500	6 hr.	0/3	6 hr.	Moderate	Slight
Methoxyethyl oleate[69]	>3.2 (mice)	(a 1-year feeding study in rats showed that 0.05 to 0.25% caused growth depression; higher levels caused some renal calculi)						
Allyl formate[16]	16 (approx.) about 15 mg./kg. (mice I.P.)	(liver damage produced in a number of species)	1,000	1 hr.	3/6			
Allyl acetate	0.13	1.1	8	4 hr.	4/6		Slight	Moderate
Allylidine diacetate	0.25	0.32					Severe	Severe
Isopropenyl acetate	3.0					30 min.	Slight	Slight
Vinyl acetate	2.9	2.5	4,000	4 hr.	3/6		Slight	Slight
Vinyl butyrate	8.5		4,000	4 hr.	3/6	30 min.	Slight	Slight

| Vinyl 2-ethylhexanoate (2-ethyl hexoic acid vinyl ester) | 4.29 | | | | Slight |
| 1-Propene-1,3-diol diacetate | 0.15 | 0.67 | | | Severe |

| | | 16 | 4 hr. | 4/6 | 8 hr. | Moderate Severe |

C. R. Spealman, R. J. Main, H. B. Haag, and P. S. Larson, *Ind. Med.*, **14**, 292 (1945).

TABLE 12

Physical and Chemical Properties of Saturated, Aliphatic Di- and Tricarboxylic Acid Esters

Compound	Formula	Mol. wt.	Sp. gr., °C. (g./ml.)	M. p., °C.	B. p., °C.	Vapor pressure mm. Hg (°C.)	Flash point, °F.	Solubility in H$_2$O (g. per 100 ml.)
Diethyl oxalate	$(COOC_2H_5)_2$	146.14	1.0842 (15)	-40.6	185.4	10 (84) 1 (47)	168	Slightly soluble
Ethylidine diacetate	$(CH_3COO)_2CHCH_3$	146.14	1.061 (12/4)		167-169	17 (69)		Slightly soluble
Diethyl malonate	$CH_2(COOC_2H_5)_2$	160.17	1.055 (20/4)	-49.8	198.9	10 (81)	200	2.08 (20°C.)
Diethyl glutarate	$C_2H_5OOC(CH_2)_3COOC_2H_5$	188.22	1.025 (20/4)	-24.1	237			0.88 (20°C.)
Diethyl succinate	$(CH_2COOC_2H_5)_2$	174.19	1.0402 (20/4)	-21.0	217.7	1 (55)	230	2.8
Dipropyl succinate	$CH_2COO(CH_2CH_2CH_3)_2$	202.24	1.001 (20/4)	-10.4	250.8			Insoluble
Dibutyl succinate		230.3	0.9760 (20/4)	-29.25	274.5			Insoluble
Di-n-butyl malate	$CH_3(CH_2)_3OCOCH(OH)CH_2COO(CH_2)_3CH_3$	246.3	1.038 (20/4)		170-171			Slightly soluble
Dibutyl adipate	$(CH_2CH_2COOC_4H_9)_2$	258.35	0.9652 (20/4)	-37.5	183 (14 mm.)		241	Insoluble
Dibutoxyethyl adipate	$(C_4H_9OC_2H_4OCOCH_2CH_2)_2$	346	0.997		208 (4 mm.)	<0.17 (150)	370	Insoluble
Di-2-ethyl butyl adipate	$((CH_2)_2COOC_6H_{13})_2$	314.0						Insoluble
Di-2-ethyl hexyl adipate	$C_4H_8(COOCH_2CH(C_2H_5)C_4H_9)_2$	371.0	0.9268	<-60.0	210-218 (5 mm.)	2.4 (200)	384	<0.01
Diisooctyl adipate	$((CH_2)_2COOC_8H_{17})_2$	371.0	0.928	<-70.0	205-220 (4 mm.)	<0.12 (150)	410	Insoluble
Di-2-hexyl oxyethyl adipate (dihexoxy ethyl adipate)	$(CH_2)_4(COOC_2H_4OC_6H_{13})_2$	402.0	0.9642 (25)	-6.0				Insoluble

(continued)

DAVID W. FASSETT

TABLE 12 (continued)

Compound	Formula	Mol. wt.	Sp. gr., °C. (g./ml.)	M.p., °C.	B.p., °C.	Vapor pressure mm. Hg (°C.)	Flash point, °F.	Solubility in H_2O (g. per 100 ml.)
Didecyl adipate (isodecyl adipate)	$C_{10}H_{21}OCO(CH_2)_4COOC_{10}H_{21}$	427	0.915 (pour point)	-64	245 (5 mm.)		424	Insoluble
Dibutyl azelate	$(CH_2)_7(COOC_4H_9)_2$	300			336			Insoluble
Di-2-ethyl hexyl azelate	$(CH_2)_7(COOCH_2CH(C_2H_5)C_4H_9)_2$	413	0.919 (20)	-65	376	5 (237)	414	Insoluble
Dibutyl sebacate	$C_4H_9OCO(CH_2)_8COOC_4H_9$	314.45	0.933 (15)	-11	344-345	3 (180)	356	Insoluble
Triethyl citrate	$(C_2H_5OCOCH_2)_2C(OH)COOC_2H_5$	276	1.136	-55	294	1 (127)	311	6.5
Tribu yl citrate	$(C_4H_9)_3C_6H_5O_7$	360.44	1.045 (20/20)	-85	233 (17 mm.)	1 (170)	365	<0.002
Acetyl triethyl citrate	$(C_2H_5OCOCH_2)_2C(OCOCH_3)COOC_2H_5$	318	1.135	-50	225 (5 mm.)	1 (132)	370.4	0.72
Acetyl tributyl citrate	$(C_4H_9OCOCH_2)_2C(OCOCH_3)COOC_4H_9$	402	1.048	-80		1 (173)	399.2	Insoluble

TABLE 13

Physiological Response of Saturated Di- and Tricarboxylic Acid Esters

Compound	LD50 Species	LD50 Route	LD50 g./kg.	Skin LD50 Species	Skin LD50 ml./kg.	Rat Inhalation Toxicity Sat. vap. or p.p.m.	Time, hr.	Mortality	Irritant effect Species	Irritant effect Skin	Irritant effect Eye
Diethyl oxalate	Rat[33]	Oral	0.4-1.6	Guinea pig	>10.0				Guinea pig	Slight	
Diethyl malonate	Rat[33]	Oral	>1.6	Guinea pig	>10.0				Guinea pig	Slight	
Ethylidine diacetate	Rat[33]	Oral	>1.6	Guinea pig	>10.0				Guinea pig	Slight	
Diethyl glutarate	Rat[33]	Oral	>1.6	Guinea pig	>10.0				Guinea pig	Moderate	
Diethyl succinate	Rat[46]	Oral	8.53			Sat. vap.	8	None	Guinea pig	Slight	Slight
	Rat[22]	Oral	>3.2	Guinea pig	>10.0						
Dipropyl succinate	Rat[36]	Oral	6.5			Sat. vap.	8	None		None	
Dibutyl succinate	Rat[22]	Oral	8.0						Guinea pig	Slight	Slight
Di-2-hexoxyethyl succinate	Rat[22]	Oral	4.3	Rabbit	12.3	Sat. vap.	8	None			
Diethyl adipate	Rat[22]	Oral	>1.6	Guinea pig	>10.0				Guinea pig	Slight	
Dibutyl adipate	Rat[46]	Oral	12.9	Rabbit	20.0	Sat. vap.	8	None		Slight	Slight
Di-2-ethylbutyl adipate	Rat[27]	Oral	5.6	Rabbit	17.0	Sat. vap.	8	None		Slight	Slight
Dibutoxyethyl adipate	Rat[47]	I.P.	0.6							Slight	Slight

Compound	Route			Animal						Animal		
1,1-Di(2-methoxyethoxy)-ethane[g]	Rat	Oral	3.26	Rabbit			8		Rabbit	Slight	Slight
Glyoxal tetrabutyl acetal[d]	Rat	Oral	8.9	Rabbit		Conc. vapor	8	0/6	Rabbit	Slight	Slight
Chloroacetal[b] (chloroacetaldehyde diethylacetal)	Rat	Oral	0.05–0.4	Guinea pig	<10 (loss of weight)	Conc. vapor		0/6	Guinea pig	Slight	Severe
Keto acetal[b] (4,4-dimethoxy-2-butanone)	Mouse	Oral	1.6–3.2		>5				Guinea pig	Slight	
	Mouse	I.P.	3.2–6.4								
Crotonaldehyde acetal[h]	Mouse	I.P.	0.06–0.25								
Chloral hydrate[i,j] (trichloroacetaldehyde monohydrate)	Rat	Oral	0.8								
	Rabbit	Oral	1.3 (LD100)								
	Cat	Oral	0.5 (LD50)								
	Dog	Oral	1.1 (LD50)								

[a] F. L. Weaver, A. R. Hough, B. Highman, and L. T. Fairhall. *Brit. J. Ind. Med.*, **8**, 279 (1951). [b] D. W. Fassett, unpublished data, Laboratory of Industrial Medicine, Eastman Kodak Co., Rochester, N. Y. [c] H. F. Smyth, Jr., and C. P. Carpenter, *J. Ind. Hyg. Toxicol.*, **30**, 63 (1948). [d] H. F. Smyth, Jr., C. P. Carpenter, and C. S. Weil, *J. Ind. Hyg. Toxicol.*, **31**, 60 (1949). [e] C. W. LaBelle and H. Brieger, *A.M.A. Arch. Ind. Health*, **12**, 623 (1955). [f] P. Knoefel, *J. Pharmacol.*, **50**, 88 (1934). [g] H. F. Smyth, Jr., C. P. Carpenter, C. S. Weil, and U. C. Pozzani, *A.M.A. Arch. Ind. Health Occupational Med.*, **10**, 61 (1954). [h] Sloan-Kettering Institute Screening Data, New York, N. Y. [i] W. Spector, ed., *Handbook o Toxicology*, Vol. I. Saunders, Philadelphia, 1956. [j] J. Adams, *J. Pharmacol. Exptl. Therap.*, **78**, 340 (1943).

TABLE 10. Physical and Chemical Properties of Aromatic and Heterocyclic Aldehydes

Compound	Formula	Molecular weight	Specific gravity, g./ml.	Melting point, °C.	Boiling point, °C.	Vapor pressure, mm. Hg (°C.)	Vapor density (air = 1)	Flash point, °F.	Solubility in H₂O, g. per 100 ml.	Conversion factors	
										1 mg./liter (p.p.m.)	1 p.p.m. (mg./cu. meter)
Benzaldehyde	C₆H₅CHO	106.1	1.046	−26	178	1 (26) 40 (90)	3.66	165	Slightly soluble	231	4.3
p-Hydroxybenzaldehyde	HOC₆H₄CHO	122.12		116	Sublimes	1 (121) 10 (170)	4.2		1.38	200	5.0
o-Hydroxybenzaldehyde (salicylaldehyde)	HOC₆H₄CHO	122.12	1.153	−7	197	1 (33) 10 (74)	4.2		Slightly soluble	200	5.0
p-Tolualdehyde (p-methylbenzaldehyde)	CH₃C₆H₄CH	120.1	1.020		204		4.1		Slightly soluble	204	4.9
p-Methoxybenzaldehyde (anisaldehyde)	CH₃OC₆H₄CHO	136.14	1.119	0	248	1 (73) 10 (118)	4.7		Slightly soluble	180	5.6
p-Aminobenzaldehyde	NH₂C₆H₄CHO	121.13		71.5			4.2		Soluble	202	5.0
3,4-Dihydroxybenzaldehyde (protocatechualdehyde)	(HO)₂C₆H₃CHO	138.12		153			4.8		5	177	5.6
3-Methoxy-4-hydroxybenzaldehyde (vanillin)	(CH₃O)C₆H₃(OH)CHO	152.14	1.056	82	285	1 (107) 10 (154)	5.2		1	161	6.2
3,4-Dimethoxybenzaldehyde (Veratric aldehyde)	(CH₃O)₂C₆H₃CHO	166.17		42	281	1 (155)	5.7		Slightly soluble	147	6.8
Piperonal	3,4-OCH₂OC₆H₃CHO	150.14		36	263	1 (87) 10 (132)	5.2		Slightly soluble	163	6.1
Furfural (2-Furfuraldehyde)	OCH=CHCH=CCHO	96.09	1.1563	−37	162	1 (19) 15 (60)	3.31	155	8.3	255	3.9
Phenylacetaldehyde (α-tolualdehyde)	C₆H₅CH₂CHO	120.15	1.023	33	195	10 (78)	4.1		Slightly soluble	204	4.9
Cinnamaldehyde (phenylacrolein)	C₆H₅CH=CHCHO	132.16	1.048	−8	246	1 (76) 10 (120)	4.6		Slightly soluble	185	5.4

3. Monocarboxylic—Aliphatic Esters Containing Halogens

CHLOROFORMATES (Chlorocarbonates)

A number of chloroformates are of industrial importance in certain types of organic syntheses. They are generally made by reacting phosgene with an alcohol, for example,

$$COCl_2 + C_2H_5OH \rightarrow C_2H_5O\overset{\displaystyle O}{\overset{\|}{-}}CCl \text{ (ethyl chloroformate)}$$

They are corrosive, flammable, colorless liquids that have many of the properties of acid chlorides (see Table 10). They tend to decompose spontaneously, give rise to HCl and other products. They hydrolyze in the presence of water or moist air.

TABLE 10

Physical and Chemical Properties of Chloroformates

Ester	Formula	Mol. wt.	Sp. gr. (g./ml.)	B.p., °C.	F.p., °F.	Vapor density (air =1)	Solubility in H_2O
Methyl chloro-formate	ClCOOCH₃	94.50	1.223	71		3.3	Slightly soluble; decomposes
Ethyl chloro-formate	ClCOOC₂H₅	108.53	1.135	93	61	3.7	Insoluble; decomposes
Isopropyl chlorofor-mate	ClCOOCH(CH₃)₂	122.5	1.08	103	60	4.2	Insoluble; decomposes
n-Propyl chlorofor-mate	ClCOO(CH₂)₂CH₃	122.5	1.090	114		4.2	Insoluble; decomposes
n-Butyl chlorofor-mate	ClCOO(CH₂)₃CH₃	136.6	1.074	140		4.7	Insoluble; decomposes
Chloromethyl chlorofor-mate	ClCOOCH₂Cl	128.9	1.465	107		4.5	Decomposes
Trichloro-methyl chlorofor-mate (Di-phosgene)	ClCOOCCl₃	199.82	1.656	128 (vapor pressure = 10 mm. Hg 20°C.)			

Storage containers need to be vented because of the gaseous decomposition products.

While chloroformates have had little systematic study in experimental animals, their intensely irritating and lachrymatory properties are well known to chemists. One of these, chloromethyl chloroformate, was known as "K Stoff" in

World War I. According to Vedder[80] a concentration of 10 p.p.m. (0.0528 mg./liter) caused lachrymation and 1 mg./liter (190 p.p.m.) was lethal in 10 minutes. He rated its irritant potency as 5 times that of chlorine and about one half that of phosgene.

Eye irritation, irritation of the upper respiratory tract, and surface burns have been the principal findings in exposed humans. Pulmonary edema would undoubtedly occur with higher levels of exposure. The eye irritation may persist some time after exposure ceases. Skin sensitization may occur. No effects on general health nor in the laboratory findings (blood studies and liver function tests) of persons manufacturing chloroformates were noted.[81] Some preliminary toxicity studies have been made on two of these materials.

Ethyl and methyl chloroformate have oral LD_{50}'s in the rat of less than 50 mg./kg. They cause severe damage to the guinea pig skin and are probably absorbed through the skin.[33]

Isopropyl chloroformate has an oral LD_{50} in rats and mice of less than 100 mg./kg. Respiratory distress was usually evident. Inhalation of 122,000 p.p.m. (calculated) by rats caused immediate fatalities with pulmonary edema. Inhalation of 6450 p.p.m. caused the death of all rats in about 50 minutes, with respiratory distress, prostration, and convulsions. Application to the guinea pig skin from a medicine dropper caused deep necrosis and eschar formation. Contact with the rabbit eye causes permanent corneal opacity.[33]

These materials should be handled so as to prevent all contact as far as possible, and with due regard for their corrosive and flammable nature. Air-supplied respirators and facilities for flushing the eyes and skin with water should be provided. Special attention should be given the eyes and respiratory tract in medical examinations.

HALOGENATED ACETATES

The halogenated acetates are of importance chiefly as intermediates in various types of organic synthesis. They can be prepared by a number of reactions, usually starting with the reaction of a halogen with acetic acid. They are generally colorless liquids of rather high density (see Table 11). They are insoluble in water, but are more stable toward hydrolysis than the chloroformates. However, ethyl iodoacetate may break down to some extent with heating, releasing iodine.

The physiological properties of ethyl chloro-, bromo-, and iodoacetates are similar and are marked by the intensely irritating character of the vapor to the eyes, producing lachrymation. The potency is in the order of $I > Br > Cl$. They are also highly toxic compounds in general and can produce pulmonary edema.

Lachrymators of this type probably owe their action to some type of specific action with certain enzyme SH groups, according to Dixon.[18] The presence of the

[80] E. B. Vedder, *Medical Aspects of Chemical Warfare*. Williams & Wilkins, Baltimore, 1925.

[81] E. R. Plunkett, Columbia-Southern Chemical Corp., Barberton, Ohio, personal communication.

TABLE 11

Physical and Chemical Properties of Halogenated Acetates

Ester	Formula	Mol. wt.	Sp. gr. (g./ml.)	B.p., °C.	F.p., °F.	Vapor density (air = 1)	Solubility in H_2O
Ethyl chloro-acetate	$CH_2ClCOOC_2H_5$	122.6	1.158	145	120	4.2	Insoluble; decomposes
Ethyl bromo-acetate	$CH_2BrCOOC_2H_5$	167.0	1.506	160	110	5.8	Insoluble
Ethyl iodoace-tate	$CH_2ICOOC_2H_5$	214.0	1.817	179 (vapor pressure = 0.54 mm. Hg, 20°C.)		7.4	Insoluble
Ethyl fluoroace-tate	$CH_2FCOOC_2H_5$	106.2		116		3.7	Insoluble
2-Fluoroethyl acetate	$CH_3COOCH_2CH_2F$	106.2		117		3.7	Insoluble

ester group seems to activate the halogen. This reaction is very rapid and probably does not require hydrolysis to the free acid, which is, in fact, not a lachrymator although it also may combine with SH groups. The effect on the eyes is rapidly reversible and at lower concentrations ceases in a few minutes after exposure stops. Most of the information about these materials was developed during World War I and has been summarized in the official reports of the War Department,[82] by Vedder,[80] and Prentiss.[83]

ETHYL CHLOROACETATE

Ethyl chloroacetate is a relatively weak lachrymator compared to the bromo or iodo esters. It is, however, highly irritating and toxic. The oral LD_{50} in the rat is less than 50 mg./kg. It is a moderately strong irritant to the guinea pig skin and is probably absorbed percutaneously.[33]

ETHYL BROMOACETATE

Ethyl bromoacetate is a potent lachrymator in humans at levels as low as 0.003 mg./liter. At 0.04 mg./liter, the effect on the eye is intolerable. Lethal effects are produced at 2.3 mg./liter in 10 minutes in animals.[82] It can produce a sensitization dermatitis in humans.[84]

ETHYL IODOACETATE

Lachrymation is present in humans at 0.14 p.p.m. (0.0014 mg./liter) and becomes intolerable at 0.015 mg./liter (about 1.5 p.p.m.). Ethyl iodoacetate is

[82] U. S. War Department, *Chemical Aspects of Gas Warfare*, Vol. XIV. 1926.

[83] A. M. Prentiss, *Chemicals in War*. 1st ed., McGraw-Hill, New York, 1937.

[84] D. W. Fassett, Laboratory of Industrial Medicine, Eastman Kodak Co., Rochester, N. Y., personal observations.

highly toxic for dogs at 1.6 mg./liter. Vedder[80] calls attention to 2 cases of fatal pulmonary edema in workmen accidentally exposed to sudden high concentrations. The autopsy findings showed pulmonary edema.

Chronic low-level exposures to the above esters do not appear to have been studied. It is probable that in spite of the highly toxic nature of these molecules, their warning properties prevent any significant prolonged intake by inhalation.

Handling of these materials should never be undertaken without careful training and constant supervision of all procedures. It is very hazardous to store any quantity of these substances in densely populated areas of a plant or community. All people handling them should be subject to repeated medical evaluation.

FLUORINATED ESTERS

Since the discovery of the action of fluoroacetic acid (see Chapter XL on Organic Acids) a very large number of derivatives, including many esters have been made for experimental study. This information has recently been reviewed in a monograph by Pattison[19] and should be consulted by anyone interested in this problem. Information of great significance for proof of the current theories of fatty acid metabolism has been obtained; and in the cases of the esters, much valuable information as to hydrolysis mechanisms in vivo.

ETHYL FLUOROACETATE $CH_2FCOOC_2H_5$

Ethyl fluoroacetate, has an LD_{50} of 6 to 10 mg./kg. in mice, very similar to fluoroacetic acid. If the fluorine is in an alkyl ethyl group (as in 2-fluoroethyl acetate), it is only slightly less toxic, having an LD_{50}'s of 18 mg./kg. (these active fluoro compounds have almost identical $LD_{50's}$ by any route of administration). In this case, hydrolysis gives rise to fluoroethanol, which is rapidly converted to fluoroacetic acid in vivo.

If the acid portion of the ester has an odd number of carbons, the toxicity is much less, since the hydrolysis leads to an odd carbon acid which is metabolized by C_2 degradation to a terminal fluoropropionate. Fluoropropionate does not possess the specific high toxicity of fluoroacetic acid. The same reasoning applies to the alkyl portion of the ester. While the literature gives little information on irritant properties, it appears as though esters such as ethyl fluoroacetate are pleasant smelling and lack the intense local effects on the eye of the bromo- and iodoacetate esters.

Any handling of fluorinated esters (and also fluorinated acids, alcohols, aldehydes, or ketones) should be done only after study of the literature and preparation for all possible means of preventing intake into the body.

OTHER CHLORO ESTERS

Some data have been published by Smyth[46] on two other halogenated esters, both of which appear to have a somewhat lower degree of toxicity than the above materials, although they are still very potent in their action.

2-CHLOROETHYL ACRYLATE

2-Chloroethyl acrylate was reported to have an oral LD_{50} in the rat of 0.18 g./kg. Inhalation of 125 p.p.m. for 4 hours killed none of 6 rats while 250 p.p.m. for 4 hours killed 6 of 6. Skin and eye irritation were severe in the rabbit.

2-CHLOROALLYLIDENE 3,3-DIACETATE

2-Chloroallylidene 3,3-diacetate had an oral LD_{50} in the rat of 0.32 g./kg. and a skin LD_{50} in the rabbit of 0.98 ml./kg. Skin irritation was moderately severe in the rabbit, while eye irritation was severe. The compound was evidently highly toxic by inhalation, causing deaths in 3 of 6 rats after a 4-hour exposure to 8 p.p.m.

4. Saturated Aliphatic Di- and Tricarboxylic Acid Esters

The esters of the saturated dicarboxylic acids are usually high-boiling liquids. They have little or no odor except for those with the lower molecular weights, where the odor may be sharp and pungent. They tend to have high flash points and, in the higher molecular weights, may be viscous. Their uses are principally as plasticizers or as special lubricants, except for the lower molecular weight substances, such as diethyl oxalate or malonate, which have their principal uses as solvents for resins or as intermediates. Some of their physical and chemical properties are listed in Table 12.

The industrial use of these substances has not been associated with any particular problems of toxicity. Their low vapor pressure limits the chance of inhalation except under conditions where aerosols might be present. A possible exception is diethyl oxalate, which is said to have a mild benzenelike effect.[17,85]

Little appears to be known of their metabolism, although some hydrolysis of the ester linkage seems likely with a subsequent fate similar to that of the corresponding acid. The absence of any pronounced acute or chronic effects for these simple aliphatic dicarboxylic esters is probably a reflection of their ease of metabolism or excretion. Absorption from the gut is probably very low for the higher molecular weight materials.

Toxicity studies have been mostly in the nature of acute screening tests and are summarized in Table 13. A few of these have received extensive study as a result of usage in connection with foods or food packaging. No systematic industrial hygiene or chemical studies have been reported for any of these esters with the exception of diethyl oxalate. An occasional contact dermatitis may be seen but, in general, this is a minor problem. No methods for air analyses could be found. However, the methods previously described for other esters would appear to be feasible.

Some additional comments can be made on these specific compounds that have received more detailed study.

[85] H. Desoille, L. Truffert, Assouly, and Bidegarray, Arch. maladies profess. méd travail et securité sociale, **8**, 265 (1947).

DIETHYL OXALATE

Diethyl oxalate has the lowest molecular weight (146) of any of the industrially important members of this series of dicarboxylic acid esters. Its vapor pressure, however, is so low (1 mm. Hg at 47°C. and 10 mm. Hg at 84°C.) as to limit the quantity of material that can be taken in by the respiratory route. The one report of significance to industrial hygienists is that by Desoille et al.[85] which was reviewed by Browning.[17] Desoille and his associates reported that complaints of weakness, headache, and nausea, together with evidence of some slight anemia and leucopenia, neutropenia, and eosinophilia were encountered in workmen who had been exposed for several months to concentrations of 0.76 mg./liter. It was stated that the symptoms regressed when exposure ceased. Exposure of guinea pigs produced a slight anemia.

In spite of the continued use of diethyl oxalate as a solvent and for miscellaneous other purposes as an intermediate, etc., no further reports have been found of injuries from this material. Calculations from vapor pressures show that a value of 0.76 mg./liter corresponds to about one half the amount present in a saturated atmosphere at 25°C. It would appear as though Desoille's report must be accepted with some reservations as to the accuracy of the concentrations. On theoretical grounds, it seems possible that if there were rapid hydrolysis of this material in vivo the oxalic acid produced might cause some symptoms of hypocalcemia provided the exposures were high or in the event that there was skin absorption.

The rat oral LD_{50} is 0.4 to 1.6 g./kg. with disturbance of respiration and muscle twitching noted as prominent symptoms. Massive renal oxalate deposits and dilitation of tubules were seen after an oral dose of 400 mg./kg. in the rat. Doses of 10 ml./kg. on the guinea pig skin caused no fatalities and little irritation.[33]

While the evidence for the toxicity of diethyl oxalate remains obscure, it would seem reasonable to use strict precautions to avoid prolonged inhalation or skin contact with this substance.

DIETHYL MALONATE

Diethyl malonate has low acute toxicity in the rat; the oral LD_{50} is > 1.6 g./kg. The only symptom noted was weakness. It produces no symptoms or local irritation to guinea pig skin in doses of 10 ml./kg.[33]

No reports of injury to humans handling this material have been found. Its physical properties are fairly similar to those of diethyl oxalate (vapor pressure is 1 mm. Hg at 40°C.). In view of the fact that hydrolysis would produce malonic acid it would seem reasonable to use adequate caution against repeated or prolonged inhalation or skin contact. Malonic acid (as mentioned in Chapter XL on Organic Acids) is a relatively strong acid and is capable of acting as an inhibitor of certain isolated enzyme systems such as succinic dehydrogenase. It is of interest in this connection, however, as pointed out by Williams,[43] that the calcium salt of malonic acid has a much greater solubility than calcium oxalate. In view of these facts, toxic reactions would appear to be probable only under extraordinary conditions of high temperatures and possibly extensive skin contact.

ETHYLIDINE DIACETATE

Ethylidine diacetate showed the same oral toxicity and skin effects as diethyl malonate.[33]

DIETHYL GLUTARATE

Diethy glutarate also has low oral toxicity in the rat (oral $LD_{50} = >1.6$ g./kg.), although vasodilitation was noted. It was a moderate skin irritant in the guinea pig, but no toxic symptoms resulted from single doses of 10 ml./kg.[33]

SUCCINATE ESTERS

Reference to the table on physiological properties (Table 13) shows that none of the three succinates mentioned (ethyl, propyl, butyl) nor 2-hexoxyethyl) has a high degree of toxicity by oral ingestion; that none of them is particularly irritating to the skin or eye; nor is it likely that they are absorbed through the skin. It appears as though the somewhat greater toxicity of the hexoxyethyl compound might be related to the difference in metabolism associated with the disposal of the oxy containing group.

There are no reports of human injury from handling these materials.

ADIPATE ESTERS

These esters have extensive use as plasticizers. Twenty adipate esters are listed in the plasticizer section of the *Modern Plastics Encyclopedia*.[90] None of these materials possesses a very high degree of acute toxicity, and their irritant effect on skin or eye is very slight.

In the case of the 2-ethylhexyl and the decyl adipates, the oral lethal dose is so high as to be practically indeterminate. The oily stools produced by a number of esters with higher molecular weights would seem to indicate a lack of absorption from the gut.

The relatively high acute toxicity of dibutoxyethyl adipate, compared to closely related materials such as di-2-ethylhexyl adipate, is of interest. For example, dibutoxyethyl adipate has an intraperitoneal LD_{50} in the rat of 0.6 ml./kg. compared to greater than 47 ml./kg. for di-2-ethylhexyl adipate.[87] It seems possible that this result is due to hydrolysis of the ester, giving rise to the butyl ether of ethylene glycol (butyl cellosolve), which has a known high toxicity in this species. The protocols of this experiment, however, do not indicate that there

[90] C. A. Breskin, *Modern Plastics Encyclopedia*, Vol. 31, No. 1a. Bristol, Conn., 1955.

was any hemoglobinuria nor that there were any unusual findings in the kidney. Therefore, this explanation of the toxicity is uncertain.

The other oxy compounds, di-2-hexyloxyethyl adipate and di-2(2-ethylbutoxy)ethyl adipate, were somewhat more toxic, but the difference was not as striking as in the case of butoxyethyl adipate.

Di-2-ethylhexyl adipate has been fed to rats at levels of 0.5, 2, and 5 per cent in the diet for a period of a month with findings of a definite growth effect at 5 per cent, but not at the lower levels. No effects were found at any levels in the blood or urine, nor in histopathology. The dog was fed 2 g./kg. in the diet for a period of 2 months with only transient loss of appetite and with no changes in the blood, urine, or histopathology.[87] Di-2-ethylhexyl adipate has also been studied by Smyth, Carpenter, and Weil.[46] The material was given to rats in food in doses of 0.16 to 4.74 g./kg./day. Deaths were produced at a 4.74 g./kg. level; no effects no growth, appetite, liver and kidney weight, or histopathology were found at 0.16 g./kg.

Didecyl adipate has very low acute toxicity in the rat and mouse upon oral administration and produces no effect on the skin. A 6-hour inhalation of air that had been bubbled through a column of the liquid at 100°C. produced no symptoms or fatalities in rats.[33]

AZELATE ESTERS

Several azelates are used as plasticizers. The available toxicity data indicated that they also are inert as far as any acute exposures are concerned (see Table 13).

SEBACATE ESTERS

Fourteen sebacate esters are listed as being plasticizers in the *Modern Plastics Encyclopedia*.[90] Of those listed, dibutyl sebacate has received the most extensive study. Smith[69] fed dibutyl sebacate to rats at levels of 0.01 to 6.25 per cent over a period of 2 years. No effects were noted on growth, mortality, pathology, or reproductive ability. A very slight growth depression was noted when the material was fed at high levels of weanlings.

Lack of toxicity is apparently explained by the splitting of this material into its components and their normal metabolism.[69] In vitro tests with pancreatin show that it is split at least as rapidly as triolein. Its acute effects by oral ingestion and by skin contact are very slight.

Di-2-ethylhexyl sebacate has similar properties to dibutyl sebacate upon single oral doses. It is not a skin irritant nor absorbed through the skin. Rats exposed for 6 hours to air bubbled through a column of liquid at 100°C. showed no mortality and no symptoms.

CITRATE ESTERS

A variety of citrate esters also have uses as plasticizers. In addition, the isopropyl and stearyl citrates are said to have uses as emulsifiers and as flavor-preserving agents.[89]

The toxicity of triethyl and tributyl, acetyl triethyl and acetyl tributyl citrates have been studied by Finklestein and Gold.[88]

The symptoms produced by single oral doses of either of the triethyl esters are similar in both rats and cats and include signs of weakness, depression, and finally hyperirritability with convulsions and respiratory failure. The onset of symptoms was quite rapid, although in some cases the symptoms continued for 2 days. In contrast, single oral doses of the butyl esters produced no marked effects, although some oily material was noted in the stool.

The ethyl esters were fed to rats at levels of 0.5, 1, and 2 per cent for a period of 6 weeks. No notable effects were seen on weight gain, blood count, blood chemistry, urinalysis, or histopathology. Cats tolerated an oral dose of 0.25 cc./kg. of the triethyl citrate and 0.5 cc./kg. of the acetyl triethyl citrate daily for a period of 8 weeks but showed mild symptoms of poisoning after the fourth and fifth doses, consisting of weakness, ataxia, and depression.

The tributyl and acetyl tributyl citrates were fed to rats at levels of 5 and 10 per cent in the diet for a period of 6 weeks. The 5 per cent level had no effect on growth while the 10 per cent eve showed some growth depression. Diarrhea was noted with the higher concentration. No effect was seen on the blood counts and histopathology. Cats were given 5 cc./kg. of the butyl ester through a stomach tube daily for 2 months. There were no effects on the appearance or behavior of the animals nor on the urine, blood chemistry, or blood count. There was a decline in weight of about 30 per cent which may have been associated with the diarrhea produced by the compound.

Some additional pharmacological studies were carried out to determine the cause of the effect of ethyl esters. While this was not proved conclusively, there was some evidence that the type of effects produced may have resulted from the binding of calcium by the release of citrate ion with resultant hypocalcemia. Similar effects have also been noted by others.[33] Triethyl citrate has been studied by inhalation.[33] The LD_{50} for a 6-hour period is somewhere between 1300 and 3500 p.p.m. for the rat. To obtain these levels it was necessary to vaporize these materials at elevated temperatures (200 and 240°C.). It is possible that some decomposition occurred at these temperatures. At the lower level, 296 p.p.m., rats tolerated 6-hour daily exposures for a period of 62 days with no reported symptoms. At the higher concentrations, the symptoms were those of gasping, weakness, and post-mortem examination showed pleural infusion; some pulmonary edema was probably present. The ethyl esters have no effect on the skin of the guinea pig and are not skin sensitizers.[33]

Isopropyl and stearyl citrates have been studied by Deuel et al.[89] The isopropyl compound consisted of 70 per cent diester with the rest mono and tri. The acute lethal doses in rats and dogs are given in Table 13. They have low acute oral toxicity in both rats and dogs. Isopropyl citrate was fed for 2 years to rats at levels of 0.28 to 2.8 per cent of the diet with no effect on growth, mortality, or histology. Rabbits were fed for a 6-week period at levels of 9.2 per cent in the diet, and dogs were fed for 12 weeks at levels of 0.06 per cent in the diet with no effects.

The stearyl citrate compound was fed for 2 years to rats at levels of 0.5 to 10 per cent of the diet with no effects. Rabbits were similarly fed for 6 weeks at 2 and 10 per cent levels; dogs were fed 12 weeks with a 3 per cent level in the diet with no apparent effect on the functions tested. No symptoms were mentioned similar to those with the triethyl citrate compounds following acute doses, and there was nothing to suggest any hypocalcemic phenomenon. The assumption was made that all the products went through normal metabolic pathways.

5. Unsaturated Aliphatic Dicarboxylic Acid Esters

While these compounds (see Table 14) have had some use as plasticizers, only one (dibutyl maleate) is listed in the *Modern Plastics Encyclopedia*;[90] they have also had uses as intermediates.

The available toxicity data consist chiefly of the screening data given in various papers by Smyth *et al.*[22,35-37,46] Table 15 gives the physiological properties of some of these materials. In general they possess a higher degree of acute toxicity than the corresponding saturated aliphatic dicarboxylic acids and have a somewhat greater tendency to cause either skin or eye irritation. Their low vapor pressures greatly reduce the hazard from inhalation, however. No cumulative effects are known.

6. Aromatic Monocarboxylic Acid Esters

The physical and chemical properties of a number of aromatic monocarboxylic acid esters of commercial importance are given in Table 16. Although many of them have not been studied extensively, some toxicological properties are given in Table 17.

The simple aliphatic esters of benzoic acid (methyl, ethyl, isopropyl, butyl, *n*-hexyl) are liquid. They have uses as solvents but are more important in flavors or perfumes. Benzyl benzoate, in addition, has some use as a miticide and plasticizer. There have been no reports of difficulties in industrial handling. As can be seen from the table on physiological properties, their acute lethal doses would indicate a low order of toxicity, and their vapor pressures make it unlikely that significant amounts would be inhaled under ordinary conditions. In the two compounds for which data are available (isopropyl benzoate and hexyl benzoate), there was little evidence of skin absorption. The undiluted materials may be either slight or moderate skin irritants.

While benzoic acid itself has had extensive study with regard to its metabolic fate, the metabolism of these esters does not appear to have been worked out. In view of their low toxicity they undoubtedly are either hydrolyzed and then metabolized according to the normal pattern for the alcohol and acid produced, or it

TABLE 14

Physical and Chemical Properties of Unsaturated Aliphatic Dicarboxylic Acid Esters

Compound	Formula	Mol. wt.	Sp. gr., °C. (g./ml.)	M. p., °C.	B. p., °C.	Vapor pressure, mm. Hg (°C.)	Flash point, °F.	Solubility in H_2O
Dimethyl maleate	$(:CHCOOCH_3)_2$	144.12	1.1513 (20/4)	−19	200	10 (84)	235	Insoluble
Diethyl maleate	$(:CHCOOC_2H_5)_2$	172.18	1.064 (25.2)	−11.5	225 105–106 (14 mm.)	10 (99.4) 1 (57.3)	250	Insoluble
Di-n-propyl maleate		200.24				6 (114)		
Dibutyl maleate	$C_4H_9OCOCH=CHCOOC_4H_9$	228.28	0.995 (20)	<−80	281	10 (139)	285	Insoluble
Di-2-ethylhexyl maleate		340.5	0.9436	−60		10 (164)	365	
Diallyl maleate		196.21				9 (127)		
Diethyl fumarate		172.18	1.0529	1–2	219	1 (53)	220	
Diisopropyl fumarate		200.24				10 (108)		
Dibutyl fumarate		228.29				8 (140)		

TABLE 15

Physiological Response to Unsaturated Aliphatic Dicarboxylic Acid Esters

Compound	LD$_{50}$			Skin LD$_{50}$		Rat inhalation toxicity			Irritant effect		
	Species	Route	g./kg.	Species	ml./kg.	Sat. vap. or p.p.m.	Time, hr.	Mortality	Species	Skin	Eye
Dimethyl maleate									Rabbit		Moderate
Diethyl maleate	Rat[37]	Oral	3.2	Rabbit	5.0	Sat. vap.	8	None	Rabbit	Slight	Slight
Di-2-methoxyethyl maleate	Rat[46]	Oral	3.3	Rabbit	1.9	Sat. vap.	8	None	Rabbit	Slight	Moderate
Monohydroxyethyl maleate	Rat[22]	Oral	2.5			Sat. vap.	8	None	Rabbit	Slight	Severe
Mon(hydroxyethoxyethyl) maleate	Rat[22]	Oral	2.8	Rabbit	7.1	Sat. vap.	8	None	Rabbit	Slight	Severe
Di-isopropyl maleate	Rat[22]	Oral	2.1			Sat. vap.	8	None	Rabbit	Slight	Slight
Mono-2-hydroxypropyl maleate	Rat[22]	Oral	3.7	Rabbit	8.5	Sat. vap.	8	None	Rabbit	Slight	Severe
Dibutyl maleate	Rat[22]	Oral	3.7	Rabbit	10.1	Sat. vap.	8	None	Rabbit	Slight	Slight
	Mouse[91]	I.P.	>0.5								
Di(1,3-dimethylbutyl) maleate	Rat[22]	Oral	7.5	Rabbit	11.9	Sat. vap.	8	None	Rabbit	Slight	Slight
Di-2-ethylhexyl maleate	Rat[37]	Oral	14.2	Rabbit	15.0	Sat. vap.	4	None	Rabbit	Slight	Slight
	Rat[87]	I.P.	>5.0								
Diallyl maleate	Rat[37]	Oral	0.3	Rabbit	1.2	Sat. vap.	8	None	Rabbit	Moderate	Slight
	Mouse[91]	I.P.	>5.0								
Didecyl maleate	Rat[22]	Oral	3.3						Rabbit	Moderate	Slight
Di-isopropyl fumarate	Rat[46]	Oral	8.5	Rabbit	10.0	Sat. vap.	8	None	Rabbit	Moderate	Slight
Dibutyl fumarate	Mouse[92]	I.P.	>0.5	Rabbit	16.0	Sat. vap.	8	None	Rabbit	Moderate	Slight
Di-2-ethylhexyl fumarate	Rat[22]	Oral	29.0	Rabbit	>20.0	Sat. vap.	8	None	Rabbit	Moderate	Slight

[91] *Summary Tables of Biol. Tests.* **7**, 788 (1955).

[92] *Summary Tables of Biol. Tests.* **7**, 782 (1955).

TABLE 16

Physical and Chemical Properties of Aromatic Monocarboxylic Acid Esters

Compound	Formula	Mol. wt.	Sp. gr., °C. (g./ml.)	M. p., °C.	B. p. °C.	Vapor density (air = 1)	Flash point, °F.	Solubility in H_2O (g. per 100 ml.)
Methyl benzoate	$C_6H_5COOCH_3$	136.14	1.0937 (15/4)	−12.5	199.6	4.7	181	0.0157 (30°C.)
Ethyl benzoate	$C_6H_5COOC_2H_5$	150.17	1.0509 (15/4)	−34.6	212.6	5.2	200	0.08 (20°C.)
Isopropyl benzoate	$C_6H_5COOCH(CH_3)_2$	164.20	1.0162 (15/4)		218.5	5.7	210	Insoluble
Butyl benzoate	$C_6H_5COOC_4H_9$	178.22	1.000 (20/4)	−22.4	250.3	6.2	225	Insoluble
n-Hexyl benzoate	$C_6H_5COO(CH_2)_5CH_3$	206.29			160 (20 mm.)	7.1		
Benzyl benzoate	$C_6H_5COOCH_2C_6H_5$	212.24	1.114 (18)	21	323–324	7.31	298	Insoluble
Ethyl-p-aminobenzoate	$NH_2C_6H_4COOC_2H_5$	165.19		88–90				0.04
2-Diethylaminoethyl-p-aminobenzoate (Procaine)		236.3		61				
Methyl-p-hydroxy-benzoate	$HOC_6H_4COOCH_3$	152.14		131	270–280 decomposes			
Propyl-p-hydroxy-benzoate	$HOC_6H_4COOC_3H_7$	180.20		96–97				
Methyl salicylate	$HOC_6H_4COOCH_3$	152.14	1.180	98	219	5.2	219	0.5 (30°C.)
Amyl salicylate (iso-pentyl salicylate)	$HOC_6H_4CO_2C_5H_{11}$	208.25	1.065 (15)		265			Insoluble
Phenyl salicylate	$HOC_6H_4COOC_6H_5$	214.21	1.250	43	173 (12)			0.015 (12°C.)
Resorcinol mon acetate	$HOC_6H_4OCOCH_3$	152.14			283 (approx.) decomposes			Insoluble
Resorcinol monobenzoate	$C_6H_5COOC_6H_4OH$	214.1		132	140 (0.15 mm.)			Insoluble
Methyl anthranilate	$2\text{-}NH_2C_6H_4COOCH_3$	151.17		24–25				
n-Propyl gallate	$3,4,5\text{-}(HO)_3C_6H_2COOCH_2CH_2CH_3$	212.21		150				0.035 (25°C.)
		306.45						

TABLE 17

Physiological Response to Aromatic Monocarboxylic Acid Esters

Compound	LD$_{50}$ Species	LD$_{50}$ Route	LD$_{50}$ g./kg.	Skin LD$_{50}$ Species	Skin LD$_{50}$ ml./kg.	Rat inhalation toxicity Sat. vap. or p.p.m.	Time, hr.	Mortality	Irritant effect Species	Irritant effect Skin	Irritant effect Eye
Methyl benzoate	Rat[22]	Oral	3.4			Sat. vap.	8	None		Moderate	Slight
Ethyl benzoate	Rat[22]	Oral	6.5			Sat. vap.	8	None		Moderate	Slight
Isopropyl benzoate	Rat[37]	Oral	3.7	Rabbit	20	Sat. vap.	4	None		Slight	Slight
Butyl benzoate	Rat[22]	Oral	5.1			Sat. vap.	8	None		Moderate	Slight
n-Hexyl benzoate	Rat[46]	Oral	12.3	Rabbit	21	Sat. vap.	8	None		Slight	Slight
Benzyl benzoate	Rat[93]	Oral	1.7	Rabbit	4						
	Mouse[94]	I.P.	>0.5								
Vinyl benzoate	Rat[22]	Oral	3.3			Sat. vap.	8	None		Moderate	Slight
Ethyl-p-aminobenzoate (Benzocaine)									Local anesthetic action on skin and mucous membranes; skin sensitization may occur		
2-Diethylamino ethyl-p-aminobenzoate (Procaine)	Mouse	Oral	900.0								
	Rat[77]	I.P.	225.0								
Methyl-p-hydroxy-benzoate[95,96]	Dog[97]	Oral	3.0								
Propyl-p-hydroxy-benzoate[95,96]	Dog[98]	Oral	6.0								
Methyl salicylate	Guinea pig[77]	Oral	0.7							Severe	Severe
	Rabbit	Oral	2.8								
	Dog	Oral	2.1								
	Human[86]	Oral	0.5 (adult)								
Amyl salicylate	Dog[97]	I.V.	0.5-0.8								
Resorcinol monobenzoate[33]	Rat	Oral	0.8-1.6	Guinea pig	>20				Guinea pig	Slight	
	Rat	I.P.	0.4-0.8								
n-Propyl gallate[33]	Rat	Oral	2.5-4.0						Guinea pig	Slight	
	Mouse	Oral	2.5-3.1								
	Cat	Oral	0.4-0.8								
Lauryl gallate[33]	Mouse	Oral	1.6-3.2						Guinea pig	(Skin sensitizer)	

93 J. Draize, E. Alvarez, and M. F. Whitesell, J. Pharmacol. Exptl. Therap., 93, 26 (1948).

94 Summary Tables Biol. Tests, 7, 783 (1955).

95 W. Metzger et al., J. Am. Med. Assoc., 155, 352 (1954).

96 M. Siegel, Antibiotics & Chemotherapy, 3, 478 (1953).

97 T. Sollmann and P. J. Hanzlik, Fundamentals of Experimental Pharmacology, J. W. Stacy, Cincinnati, 1940.

98 T. Sabalitschka, Z. angew. Chem., 42, 936 (1929).

seems probable that in some cases the ring might be hydroxylated and the product excreted as a glucuronide or sulfate ester.

BENZYL BENZOATE

Benzyl benzoate has been studied by Draize et al.[93] for its safety when used as a repellent or scabicide. The oral LD_{50} was between 1.0 and 1.7 g./kg. in several species. The LD_{50} in the rabbit skin was about 4 ml./kg. Repeated skin applications in the rabbit resulted in an LD_{50} at about 2 ml./kg./day for 90 days. Little skin irritation was noted. There was some inanition in the rabbits on the 90-day skin test, but the pathological findings were not particularly significant.

ETHLY-p-AMINOBENZOATE (Benzocaine)

Benzocaine has wide use medicinally because of its local anesthetic action on skin and mucous membranes. It has also had some use in sun-screening agents. The closely related ester, 2-diethylaminoethyl-p-aminobenzoate (Procaine) has been in use for many years as an injectable local anesthetic in dentistry and surgery. Details of the toxicological actions will not be reviewed here, since they are described in many pharmacology textbooks, for example, Sollmann's A Manual of Pharmacology.[99] The general toxicity of this material seems to be quite low. Although the metabolism of ethyl-p-aminobenzoate does not appear to have been studied in detail, it is well established that the diethylaminoethyl-p-aminobenzoate is readily hydrolyzed into its components, p-aminobenzoic acid and 2-diethyl-aminoethanol.[43] These esters are of particular interest to the toxicologist in that occasional local and sometimes general sensitization may occur. In the·case of the diethylaminoethyl compound, this may include anaphylactic type of reactions upon injection.

METHYL- AND PROPYL-p-HYDROXYBENZOATE

Methyl- and Propyl-p-Hydroxybenzoate have wide use as stabilizers in materials such as cosmetics because of their bacteriostatic properties. They also have low acute toxicities and appear to be relatively free from any skin irritant or skin sensitizing properties.

METHYL SALICYLATE

Methyl salicylate occurs naturally in oil of wintergreen and is used as a flavoring agent. It is also used as a counterirritant in some ointments and lina-ments. Its toxicity in humans is well known since it has been the cause of many accidental poisonings in children and occasionally in adults. The oral lethal dose in the adult human is approximately 0.5 g./kg. This is somewhat similar to the

---

[99] T. Sollmann, A Manual of Pharmacology, Saunders, Philadelphia, 1957.

oral lethal dose in guinea pigs. The symptoms produced are generally similar to those of other salicylates, and there is evidence that considerable hydrolysis of the ester occurs in the intestinal tract.[43] Marked gastrointestinal irritation, central nervous system symptoms, and hyperpyrexia may be noted. A toxic effect on the optic nerve from the methanol liberated during hydrolysis does not appear to have been described, perhaps because the quantity of methanol liberated is not sufficiently great and is overshadowed by the effects of the other portions of the molecule. It is possible that some of the toxic action is due to the intact molecule. In some species, such as the rabbit, the material may be partly excreted as sulfate or glucuronic acid conjugate on the free hydroxyl group. The conjugation appears to take place before the hydrolysis of the methyl ester.[43]

PHENYL SALICYLATE

Phenyl salicylate has had some use because of a theoretical intestinal antiseptic action. This is thought to arise from the fact that it is hydrolyzed in the gut to phenol and salicylic acid. While this does in fact occur to a considerable extent, the products are absorbed fairly rapidly and excreted in the usual manner for phenols and salicylic acid.

RESORCINOL MONOBENZOATE

Resorcinol monobenzoate is used as an ultraviolet stabilizer in certain types of plastic films. It appears to have a low acute toxicity in the rat and is not absorbed through the skin of the guinea pig; it is only a slight primary irritant in the concentrated form.[33]

METHYL ANTHRANILATE

Methyl anthranilate is an ester occurring naturally in grapes, and the synthetic material is used in flavors and some perfumes.

Its acute oral toxicity in the rat is low—about 3 to 5 g./kg.[33,100] It was tolerated in the diet of rats in chronic feeding studies at 0.3 per cent but not at 1.0 per cent, at which level some slight histological changes in the kidney and increased liver and kidney weights were seen. It is only a slight skin irritant in the rabbit and guinea pig, but may cause eye irritation in concentrated form. Seven-hour exposure of rats to atmospheres saturated at 100°C. caused no deaths and only transient weight loss.[100]

The metabolism of this ester does not appear to have been established, but it appears probable that it would in part be hydrolyzed and the acid then excreted as an ester glucuronide.[43]

[100] V. K. Rowe, Biochemical Research, Dow Chemical Co., Midland, Mich. unpublished data.

TABLE 18

Physical and Chemical Properties Aromatic Dicarboxylic Acid Esters

Compound	Formula	Mol. wt.	Sp. gr. °C.	M. p., °C.	B. p., °C.	Vapor pressure mm. Hg. (°C.)	Flash point, °F.	Solubility in H₂O (g. per 100 ml.)
Dimethyl phthalate	$C_6H_4(COOCH_3)_2$	194.18	1.189 (25/25)	0.0, freezes	282	<0.01 (20)	325	0.5
Diethyl phthalate	$C_6H_4(COOC_2H_5)_2$	222.23	1.123 (25/4)	−40.5	296.1	0.05 (70)	335	Insoluble
Dimethoxy ethyl phthalate	$C_6H_4(COOC_2H_4OCH_3)_2$	282.0	1.171 (20)		190–210	0.01 (20) 0.25 (150)	379.4	0.85
Dibutoxy ethyl phthalate		366.0	1.063		210	<0.01 (20)	208	<0.03
Diallyl phthalate	C_6H_4-1,2-$(COOCH_2CH:CH_2)_2$	246.27	1.120 (20/20)	−77	290	2.4 (150)	166	<0.01
Dibutyl phthalate	$C_6H_4(COOC_4H_9)_2$	278.34	1.0465 (21)	−35	340	2.0 (150) 0.1 (115)	315	0.45 (25°C.)
Diisobutyl phthalate		278.3	1.040	−50	327		322	Insoluble
Dihexyl phthalate	$C_6H_4(COO.CH_2.CH(C_2H_5)C_2H_5)_2$	334.0	0.990			5 (210)		Insoluble
Di-n-octyl phthalate		391.0	0.978	−25	220 (4 mm.)	<0.2 (150)	219	Insoluble

Name	Formula	Mol. wt.	Sp. gr.	M.p. / pour pt.	B.p.		Flash pt.	Solubility
Di-2-ethylhexyl phthalate (octyl phthalate)	$C_6H_4(CO_2CH_2CH(C_2H_5)C_4H_9)_2$	391.0	0.985 (20/20)	−50	386.9 (5 mm.)	<0.01 (20)	425	Insoluble
Di-2-ethylhexyl isophthalate		391.0			400	1 (223)		Insoluble
Diisooctyl phthalate	$C_6H_4(COOC_8H_{17})_2$	391.0	0.981 (25)	−4	239 (5 mm.)	1.0 (200)	410	Insoluble
Dinonyl phthalate		419.0	0.965		413	1 (205)	420	Insoluble
Butyl octyl phthalate	$C_6H_4(COO(CH_2)_3CH)(COO(CH_2)_7CH_3)$	334.0		−50	340		370.4	
Dicyclohexyl phthalate	$C_6H_4(COOC_6H_{11})_2$	330.0	1.20 (25/25)	62−65 (pour pt.)	220−228 (4 mm.)	<0.06	440.6	Insoluble
Methyl phthalyl ethyl glycolate		266.24	1.220 (25/25)	<−35	189 (5 mm.)		193°C.	0.053
Ethyl phthalyl ethyl glycolate		280.27	1.220 (25/25)	20	320		386	<0.08
Butyl phthalyl butyl glycolate	$C_4H_9OOCC_6H_4COOCH_2COOC_4H_9$	336.37	1.097 (25/25)	<−35	219 (5 mm.)		199°C.	0.0012%
Dimethyl ter-phthalate	$C_6H_4(COOCH_3)_2$	194.18	1.04	140	285 (sublimes)	16 (100) 140 (150) 78 (200)	146°C. (o.c.)	0.3 in hot H_2O
Di-2-ethylhexyl tetrahydro-phthalate	$C_6H_8(COOCH_2-CH(C_2H_5)(CH_2)_3CH_3)_2$	395	0.969	−50 (pour pt.)	219 (5 mm.)		350	

ETHYL GENTISATE

Esters of gentisic acid, such as ethyl gentisate, have not had extensive toxicological study, but appear to have relatively low toxicity. They are undoubtedly readily excreted, probably by etheral sulfate conjugation.[43,100a]

n-PROPYL GALLATE AND LAURYL GALLATE

The acute and chronic toxicities of n-propyl gallate have been studied by Orten et al.[101] in connection with its use as an antioxidant for edible fats. The oral toxicity in rats is low (acute $LD_{50} = 3.8$ g./kg.). The symptoms produced at these levels are gasping respirations and terminal convulsions. Levels of 1.1 to 2.3 per cent in the diet were not tolerated by rats, definite stunting of growth resulting. Some tubular damage in the kidneys was noted at these high levels, although this may have been the result of severe inanition. No effects on growth of rats were seen after 1 year on levels of 0.001 to 0.1 per cent. After 2 years the number of surviving animals was too small to determine any significant difference, but no pronounced effect was seen. No blood or histological changes were found that could be attributed to the treatment.

Guinea pigs fed 0.01 per cent in the diet for 14 months also showed no toxic effects. Dogs tolerated 0.01 per cent in the diet for over a year with no signs of toxicity.

The metabolism of n-propyl gallate has not been worked out.* Ferric chloride tests on the urine were negative, suggesting that it may undergo the usual esterification reactions of other phenolic antioxidants.

Lauryl (dodecyl) gallate has been studied by Allen and De Eds[102] and by Van Sluis.[103] It has a low acute oral toxicity (about 4 g./kg.) and is tolerated up to levels of 0.5 per cent in the diet of rats. It appears to have some skin irritant properties[102]; when applied topically, is a skin sensitizer in the guinea pig.[33] Its metabolism is unknown.*

7. Aromatic and Cyclic Dicarboxylic Acid Esters

The aromatic dicarboxylic acid esters are among the most important of industrial chemicals and are used as plasticizers for a variety of films in which they may be present as high as 30 per cent or more by weight. They are used particularly with cellulose and vinyl resins and give the required properties of toughness and flexibility. As can be seen from Table 18 on physical and chemical properties,

---

* Recent publications by J. C. Dacre (*J. New Zealand Inst. Chem.*, **24**, 161 (1960)) and A. N. Booth et al. (*J. Biol. Chem.*, **234**, 3014 (1959)) indicate that in rats and rabbits the ester linkage is cleared with subsequent 4-O-methylation and/or conjugation with glucuronic acid.

[100a] N. E. Clarke and R. E. Mosher, *Circulation*, **7**, 337 (1953).

[101] J. M. Orten, A. A. Kuyper, and A. H. Smith, *Food Technol.*, **2**, 308 (1948).

[102] S. C. Allen and F. De Eds, *J. Am. Oil Chemists Soc.*, **28**, 304 (1951).

[103] K. J. H. Van Sluis, *Food Manuf.*, **26**, 99 (1951).

they are in most cases liquids with very high boiling points and very low vapor pressures. The low vapor pressure is important in contributing to their stability in plastics.

Two of these (dimethyl and dibutyl phthalate) have had use as insect repellents. Dimethyl terephthalate (the terephthalates have the carboxy groups in the 1–4 position instead of the ortho position) is used as an intermediate in the manufacture of certain polyesters.

These phthalate esters are in general among the most inert of all industrial materials. They are rarely the cause of skin difficulties. They are not absorbed through the skin, and their low vapor pressures usually preclude the inhalation of a toxicologically significant amount. A number of them have been studied, both for their single acute lethal doses and in chronic feeding experiments because of their use in food packaging. Many of these materials have high oral lethal dosages and can be tolerated in relatively high levels in the diet. In some cases this is apparently the result of poor absorption. As is pointed out by Williams,[43] these esters are very likely to be hydrolyzed in vivo giving phthalic acid and an alcohol. Since phthalic acid is of low toxicity and is thought to be excreted quantitatively, the toxicity of these materials may depend to a considerable extent on the nature of the alcohol released on hydrolysis (see section on methyl phthalyl ethyl glycolate). This is frequently the case with other esters as well.

The pathology produced by these materials is usually nonspecific unless the alcoholic portion released on hydrolysis in vivo has activity. Comments on some of these specific materials follow. Physiological properties are shown in Table 19.

DIMETHYL PHTHALATE

Dimethyl and dibutyl phthalates were introduced during World War II as effective mosquito and other insect repellents. Dimethyl phthalate was more widely used for this purpose and there were few toxic symptoms reported. Accidental ingestion may cause symptoms of gastrointestinal irritation with some coma and hypotension.[99]

Draize[93] also studied dimethyl phthalate for oral and skin toxicity. Its oral LD_{50} in a variety of species ranged from about 2 to 8 ml./kg. Large oral doses produce an anesthetic reaction. Two-year feeding studies in female rats at levels of 2 to 8 per cent in the diet showed only slight growth effects at 4 and 8 per cent. Some chronic nephritic changes were seen at the 8 per cent level, but not at lower levels.[104] The skin LD_{50} in the rabbit was > 10 ml./kg. The 90-day LD_{50} by repeated skin application in the rabbit was > 4 ml./kg. No skin irritation or sensitization resulted. The pathology in these experiments was limited to slight renal changes of uncertain significance.

DIETHYL PHTHALATE

Diethyl phthalate has a somewhat lower LD_{50} than some of the other phthalates, but is generally regarded as having little acute or chronic toxic properties. It

[104] A. J. Lehman, *Assoc. Food & Drug Offic., U. S. Quart. Bull.,* **19,** 87 (1955).

TABLE

Physiological Response to

Compound	LD$_{50}$			Skin LD$_{50}$	
	Species	Route	g./kg.	Species	Dose
Dimethyl phthalate	Mouse	Oral	7.2	Rabbit	>10 ml./kg.[104]
	Mouse	I.P.	3.6		
	Rat	Oral	6.9		
	Guinea pig	Oral	2.4		
Diethyl phthalate[77]	Rabbit	Oral	1.0		
	Mouse	I.P.	2.8		
	Guinea pig	S.C.	3.0		
Dimethoxy ethyl phthalate[33]	Rat	Oral	>4.4	Guinea pig	>10 ml./kg.
	Mouse	Oral	3.2–6.4		
	Guinea pig	Oral	1.6–3.2		
Dibutoxy ethyl phthalate[36]	Rat	Oral	8.4		
Diallyl phthalate[77]	Rat	Oral	1.7	Rabbit	3.4 ml./kg.
	Rabbit	Oral	1.7		
	Mouse	I.P.	0.7		
Dibutyl phthalate[69]	Rat	Oral	8.0	Rabbit	>20 ml./kg.[104]
	Rat	I.M.	>8.0		

has been widely used as a plasticizer in cellulosic materials and seems to be devoid of any major irritant or sensitizing effects on the skin. Exposure to heated vapor may produce some transient irritation of the nose and throat. There are no reports of cumulative effects in its occupational use.

DIMETHOXY ETHYL PHTHALATE

This material has a low acute oral toxicity and is not a skin irritant or skin sensitizer. Inhalation of high concentrations produced by exposure to heated vapor

19

Aromatic Dicarboxylic Acid Esters

	Inhalation toxicity				Irritant effect		
Species	Concentration	Time, hr.	Mortality		Species	Skin	Rabbit eye
Cat[21]	9.3 mg./liter (1213 p.p.m.)	6.5	Lethal[77]		Guinea pig; zation[33]	no sensiti-	Slight[47,93,105]
Cat[21]	2 mg./liter		Nasal irritation				
Rat[33]	511 p.p.m.; air through liquid at 150°C.	6	0/3; vasodilatation of ears and feet			Slight	
Cat[21]	10 mg./liter (1120 p.p.m.) mist	5	Nasal irritation				
Rat[33]	air through liquid at 200°C. (5500 p.p.m.)	1.5	3/3; nasal irritation		Guinea pig	None	None
Rat	1595 p.p.m.	6	3/3; nasal irritation				
Rat	770 p.p.m.	6	0/3; no symptoms				
Rat	145 p.p.m. (air through liquid at 100°C.)	6/day for 62 days	0/3; no symptoms				
					Rabbit Human[98]	Slight Irritant	Slight[106]
Cat[21]	1 mg./liter	5.5	Nasal irritation			Not irritating[99]	

may cause signs of nasal irritation; but at lower levels, such as 145 p.p.m., no symptoms are produced in rats even upon prolonged repeated exposures.[33]

DIBUTOXY ETHYL PHTHALATE (Dibutyl Cellosolve Phthalate)

This ester has a low acute oral lethal dose and appears to be relatively well tolerated when fed in the diet of rats for a 30-day period at levels up to 0.5 g./kg./day. At higher levels, such as 1.7 g./kg./day, some reduced growth, appe-

TABLE 19

Compound	LD₅₀			Skin LD₅₀	
	Species	Route	g./kg.	Species	Dose
Diisobutyl phthalate[33]	Mouse	Oral	>12.8	Guinea pig	>10 ml./kg.
	Rat	Oral	20–25		
Dihexyl phthalate[22]	Rat	Oral	30	Rabbit	>20 ml./kg.
Di-2-ethylhexyl phthalate	Rat[33]	Oral	>26		
	Rat[107]	Oral	30	Guinea pig	>10 ml./kg.[33]
	Rabbit	Oral	34		
Di-n-octyl phthalate[33]	Mouse	Oral	>13	Guinea pig	>5 ml./kg.
Di(2-ethylisohexyl)phthalate[33]	Rat	Oral	>26	Guinea pig	>10 ml./kg.
	Mouse	Oral	>26		
	Rat	I.P.	>13		
Di(2-ethylhexyl)isophthalate[33]	Rat	Oral	>3.2	Guinea pig	>20 ml./kg.
Dinonyl phthalate[108]	Rat	Oral	>2		
Methyl phthalyl ethyl glycolate	Rat[36]	Oral	9.0		
	Rat[33]	Oral	3.2–6.4	Guinea pig	>10 ml./kg.
Ethyl phthalyl ethyl glycolate[109]	Rat	Oral	10% in diet (all dead in 7–15 days)		
Butyl phthalyl butyl glycolate[36]	Rat	Oral	14.6		
Dimethyl terphthalate[33]	Rat	Oral	>3.2	Guinea pig	>5 g./kg.
	Rat	I.P.	>3.2		
Di(2-ethylhexyl) tetrahydro-phthalate[46]	Rat	Oral	114		

[105] L. Karel, *Federation Proc.*, **6**, 342 (1947).

[106] W. A. McOmie, *Federation Proc.*, **5**, 191 (1946).

[107] C. B. Shaffer, C. P. Carpenter, and H. J. Smyth, Jr., *J. Ind. Hyg. Toxicol.*, **27**, 130 (1945).

[108] D. K. Harris, *Brit. J. Ind. Med.*, **10**, 255 (1953).

[109] H. C. Hodge, *Arch. Ind. Hyg. Occupational Med.*, **8**, 289 (1953).

(*continued*)

Inhalation toxicity				Irritant effect		
Species	Concentration	Time, hr.	Mortality	Species	Skin	Rabbit eye
				Guinea pig; no irritation or sensitization		
Rat	Sat. vap.	8	No deaths	Rabbit	Slight	Slight
Rat	8 p.p.m. (100°C.)	6	0/3	Guinea pig Slight[33] No sensitization		Slight[33,47]
						Slight
				Guinea pig Slight		Slight
				Guinea pig Slight		Slight
Rat	Sat. vap. (28°C.)	6/day for 12 days	No effect	Rat Human	None None	
Rat[33]	2500 p.p.m. (320°C.)	6	0/3			
Rat	290 p.p.m.	6/day for 62 days	No effect	Guinea pig None		None
				Guinea pig Slight No sensitization		Slight

tite, and histological changes were noted. It is not clear whether the pathology produced was the result of hydrolysis to butoxy ethanol.[36]

DIALLYL PHTHALATE

Diallyl phthalate seems to be somewhat more toxic than the other phthalates and is said to be an irritant to the skin in humans.[98] This probably results from the presence of unsaturation of the alcohol portion of the ester.

DIBUTYL PHTHALATE

Dibutyl phthalate has a low acute and chronic toxicity. Smith[69] found that all rats tolerated 0.25 per cent and that some animals tolerated 1.25 per cent for a year. Some rats succumbed during the first week of the experiment at this higher level. No histopathology was found; it appears to be readily hydrolyzed by pancreatic lipase.

Fairhall[110] refers to a report by Cagianut[111] in which a chemical operator swallowed 10 g. of dibutyl phthalate by error. The symptoms were nausea and dizziness, with lachrymation, photophobia, and conjunctivitis. Some albuminuria was noted. The eye symptoms conceivably could have been the result of the sudden release of large quantities of butanol by in vivo hydrolysis of the ester. The patient made a prompt, uneventful recovery.

DIISOBUTYL PHTHALATE

Diisobutyl phthalate is also very inert and has an even higher oral LD_{50} than dibutyl phthalate. Daily doses of as high as 1600 mg./kg. were injected intraperitoneally in rats for a period of 30 days without producing any effects on growth, symptoms, or histopathology. No effects were noted in blood count, platelets, or clotting times.[33] It was well tolerated in the diet of rats over a period of several months up to levels of 5 per cent, at which point growth retardation was noted.[87] No toxic effects were seen in dogs given 2 cc./kg./day for several months.

DI-2-ETHYLHEXYL PHTHALATE

As has been mentioned in the introduction to this chapter, di-2-ethylhexyl phthalate (also commonly known as dioctyl phthalate) is one of the most widely used plasticizers and has received considerable toxicological investigation. The acute and chronic toxicities in various species by various routes have been studied by Shaffer et al.,[107] Hodge,[112] Carpenter,[113] and Harris.[114]

The acute oral LD_{50} in rats is in the neighborhood of 30 to 34 g./kg. The principal effect produced was that of soft liquid feces similar to those produced by other oily or insoluble materials such as mineral oil. The LD_{50} intraperitoneally in the rat is about 24 to 30 g./kg.; the principal finding is the presence of unabsorbed milky emulsion and some nonspecific changes in the liver. Dioctyl phthalate apparently is poorly absorbed from the intraperitoneal cavity. It appears to be very poorly absorbed through the skin of the rabbit and doses as high as 25 ml.kg. were

[110] L. T. Fairhall, *Industrial Toxicology*. 2nd ed., Williams & Wilkins, Baltimore, 1957.

[111] B. Cagianut, *Schweiz. med. Wochschr.*, **84**, 1243 (1954).

[112] H. C. Hodge, *Proc. Soc. Exptl. Biol. Med.*, **53**, 20 (1943).

[113] C. P. Carpenter, C. S. Weil, and H. F. Smyth, Jr., *Arch. Ind. Hyg. Occupational Med.*, **8**, 219 (1953).

[114] R. S. Harris, H. C. Hodge, E. A. Maynard, and H. J. Blanchet, Jr., *A. M. A. Arch Ind. Health*, **13**, 259 (1956).

needed to produce fatalities. Inhalation of the vapor–mist mixture produced by bubbling air through a column of plasticizer maintained at 170°C. could be tolerated for 2 hours without producing fatalities. In a 4-hour period, however, all rats had succumbed. On the basis of these experiments, it was felt that the hazard to exposed individuals would be very low under ordinary circumstances.

Di-2-ethylhexyl phthalate produces no irritant effect on the rabbit eye nor is there any appreciable irritant effect on rabbit skin. Patch tests with the undiluted material in human subjects show no irritation or sensitization. This has generally been borne out in subsequent experience.

The metabolism of di-2-ethylhexyl phthalate has been studied by Shaffer et al.[107] In rats and rabbits there was no evidence of conjugation of the carboxy groups of the phthalate, although in the case of rabbits sufficient phthalate was excreted in the urine to account for between 26 and 65 per cent of the total dose with considerable variation in different animals. In dogs, however, the excretion of phthalate accounted for only 2 to 4.5 per cent of the total dosage. In the case of 2 human subjects, taking single doses of 5 and 10 g. of di-2-ethylhexyl phthalate, only 4.5 per cent of the total dose could be accounted for as phthalic acid in the urine in the succeeding 24-hour period. A great majority of this was excreted between a 5- and 7-hour period after dosing. It is of interest that no symptoms resulted from the 5-g. dose and only mild gastric disturbances and some loose stools from the larger dose.

In the case of rabbits, there was definite increase in fatty acids found in their urine after a single oral dose of 2 g./kg.; but it was stated that a large proportion of these acids were made up of shorter chain lengths than 2-ethylhexoic acid. This does not entirely fit in with other studies discussed by Williams,[43] in which it was pointed out that in the rabbit 2-ethylhexanol is very largely excreted as 2-ethyl-hexanoyl glucuronide, as are a number of other branch-chained alcohols. However, the methods used by Shaffer were apparently not designed specifically to look for this particular metabolic product.

There have been two principal studies on the chronic oral toxicity of di-2-ethylhexyl phthalate.[113,114] The results of feeding studies in rats over a 2-year period were essentially similar in these two reports. There was no effect on any of the usual criteria for chronic toxicity when the levels were below 0.13 per cent in the diet. In both studies some retardation of growth and some statistically significant increases in weight of liver and kidneys were noted at levels of 0.4 to 0.5 per cent in the diet. As is often the case in such studies the increase in weight of these organs was not accompanied by any definite histopathology. This presumably results from a kind of compensatory hypertropy or hyperplasia in response to the intake of very large quantities of a foreign material, influencing these organs that are involved in metabolism and excretion. There was no unusual incidence of tumors nor were there any effects on the blood or fertility of the animals. Studies by Carpenter et al.[113] also included a 1-year oral feeding to guinea pigs at levels up to 0.13 per cent of di-2-ethylhexyl phthalate. Except for a possible change in liver weight of the females treated with 0.13 per cent, no deleterious effects were noted. Harris[114] found no definite effects in dogs even at 5 g./kg. in the diet for 14 weeks.

A more elaborate study on dogs[113] in which the dosages were given once a day in capsules for a year also indicated a very low order of toxicity. During the majority of the year, the treated animals were given 0.06 ml./kg./day. One animal, which had been used in pilot studies, was given an increased dose of 0.09 ml./kg./day for a total of 169 doses. In none of these studies was there any effect on body weight nor on liver and kidney weights at autopsy. The only pathology mentioned was in the dog receiving 0.09 ml./kg./day, and this consisted of some fatty vacuolization and congested areas in the liver with some congestion and cloudy swelling in the kidney. None of these changes was noted in the dogs receiving the slightly lower dosage of 0.06 ml./kg. In view of the fact that this was in a single animal and that the changes were relatively nonspecific the significance of this finding is somewhat dubious. Bromsulphalein tests of liver function, prothrombin times, and plasma cholinesterase were determined. Although no specific details were given, apparently none of these findings were thought to be significant.

In retrospect, it would seem as though the only conclusive effects at all in either of these chronic studies were the changes in body weight and liver and kidney weights in rats exposed at levels of 0.4 to 0.5 per cent in the diet (corresponding to levels of 0.2 to 0.4 g./kg./day).

It seems clear that there is no reason for predicting that the use of this material in industry would be associated with health hazards, nor that it would be unsafe for use in food-packaging materials, especially for nonfatty foods.

Other isomers of the octyl phthalates, such as *n*-octyl, isooctyl, and 2-ethyl isohexyl, appear to be similar to di-2-ethylhexyl phthalate. It is of interest that the di-2-ethylhexyl isophthalate is also of very low acute toxicity.

METHYL PHTHALYL ETHYL GLYCOLATE

This material has been studied by Smyth and Carpenter[36] for acute and subacute oral toxicity. It has a low oral toxicity in the rat and no effect was noted when it was fed to this species for 30 days at levels of 0.48 g./kg./day. Some micropathology was said to be noted (type not specified) at a level of 1.9 g./kg./day.

Hodge[115] fed the material to rats for 2 years at levels up to 1.5 per cent without effects on growth, life span, blood, urine, or histopathology. At 5 per cent there was definite weight loss, hematuria; the experiment was terminated at about 5 months. The kidney was enlarged and showed many tubules plugged with oxalate crystals. There were marked changes in the architecture of the kidney due to dilitation of tubules. The effect was not present at lower levels.

Dogs tolerated levels of 1.0 g./kg./day (given by mixing in the diet) and showed no effects on their appearance, appetite, body weight, urine, blood, or

[115] H. C. Hodge, Univ. of Rochester, School of Medicine and Dentistry, Rochester, N. Y., 1955, unpublished report.

microscopic cell structure. The deposition of oxalate in the kidney was not present in this species at these levels.

ETHYL PHTHALYL ETHYL GLYCOLATE

Hodge[109] studied this material in rats and dogs. Rats ate up to 0.5 per cent in the diet for 2 years with no effects, but at 5 per cent marked renal changes were noted as in the case of methyl phthalyl ethyl glycolate. These changes were present after 10 weeks as well as at the end of the feeding at this level (55 to 72 weeks) at which point there was 100 per cent mortality. Dogs showed no significant effects or histopathology at doses of 0.25 g./kg./day for a year.

BUTYL PHTHALYL BUTYL GLYCOLATE

Smyth[36] studied the subacute oral toxicity of this material for rats. The results were essentially similar to his findings with methyl phthalyl ethyl glycolate, and it was tolerated at about 0.45 g./kg./day (0.9 per cent in diet). Some reduced growth and micropathology were noted at high levels of 1.56 g./kg./day in this 30-day study.

DIMETHYL TERPHTHALATE

In this compound the carboxy groups are in the 1–4 position. It has a low acute oral toxicity and is not absorbed through the skin nor is it a skin irritant or sensitizer. Feeding high levels (5 per cent) in the diet of rats for 28 days caused loss of weight, marked reduction of food consumption, and high mortality. No effects on the blood were found, and the cause of death was not established. The average daily dose was 3.75 g./kg.[33]

DI-2-ETHYLHEXYL TETRAHYDROPHTHALATE

This material is of interest because of the fact that the ring is partially reduced. It has very low acute oral toxicity, and has been fed to rats for 90 days[46] without effect at a dietary level of 0.2 g./kg./day. Some liver and kidney weight changes were noted at 0.84 g./kg./day, and an effect on growth at 2.67 g./kg./day.

DIMETHYL TETRAHYDROPHTHALATE

The dimethyl ester has a higher acute oral toxicity in the rat (0.7 g./kg.) than the 2-ethylhexyl ester. It is not a skin irritant and has apparently little skin absorption.[37]

B. Carbonic Acid and Ortho Esters

The esters of carbonic acid and ortho esters are usually relatively low boiling liquids, although this varies considerably with the molecular weight (see Table

DAVID W. FASSETT

TABLE 20

Physical and Chemical Properties of Carbonic Acid and Ortho Esters

Compound	Formula	Mol. wt.	Sp. gr., °C. (g./ml.)	M. p., °C.	B. p., °C.	Flash point, °F.	Solubility in H₂O
Dimethyl carbonate	CO(OCH₃)₂	90.08	1.0694 (20/4)	−0.5	90–91		Insoluble
Diethyl carbonate	(C₂H₅)₂CO₃	118.13	0.9751 (20/4)	−43	125.8	25	Insoluble
Di-n-propyl carbonate	CO(OCH₂CH₂CH₃)₂	146.18	0.9411 (20/4)		168.2		Very slightly soluble
Diisopropyl carbonate	(C₃H₇)₂CO₃	146.18	0.921 (20/4)		147.2		
Di-n-butyl carbonate	CO(OCH₂CH₂CH₂CH₃)₂	174.24	0.9244 (20/4)		207 (720 mm.)		Insoluble
Diisobutyl carbonate	CO(OCH₂CH(CH₃)₂)₂	174.24	0.919 (15/4)		190.3		Insoluble
Ethylene carbonate	(CH₂O)₂CO	88.0	1.3218 (39/4)	36.4	248	320	
Propylene carbonate	C₃H₆CO₃	102.0	1.2057 (20/4)	−49.2 (freezes)	241.7	270	
Triethyl orthoformate	HC(OC₂H₅)₃	148.21	0.895 (20/20)		145	86	
Triethyl orthoacetate	CH₃C(OC₂H₅)₃	162.23			141		
Triethyl orthopropionate	CH₃CH₂C(OC₂H₅)₃	176.26			40 (8 mm.)		

20). In the case of the cyclic esters, such as ethylene and propylene carbonate, the boiling points are higher. These materials have miscellaneous uses but principally as intermediates or solvents. Very little attention has been given to their toxicological properties, and only screening-type data are available on a few of them.

DIMETHYL CARBONATE[33]

The undiluted liquid has an oral LD_{50} in the mouse and rat of between 6.4 and 12.8 g./kg. Intraperitoneally, the LD_{50} is in the neighborhood of 800 to 1600 mg./kg. The symptoms produced were those of weakness, ataxia with gasping, and unconsciousness. When the undiluted liquid was applied to the skin of a guinea pig with a cuff and covered with a gauze pad and rubber cuff, the LD_{50} was > 10 ml./kg. Some weight loss was noted, and it is possible that some skin absorption took place. The degree of irritation was relatively slight.

The material appears to be relatively hazardous by inhalation since values as low as 8000 p.p.m. caused rapid onset of gasping, loss of coordination, frothing from the mouth and nose, and pulmonary edema with death of all rats in a period of 2 hours. It seems possible that this material may be similar to dimethyl sulfate and act as a methylating agent in tissues.

ETHYLENE CARBONATE[22]

The oral LD_{50} is 10.4 g./kg.; the skin LD_{50} in the rabbit is > 20 ml./kg. Inhalation of the concentrated vapor for 8 hours by rats produced no deaths. It was only a slight skin irritant in the rabbit and produced moderate irritation in the rabbit eye.

PROPYLENE CARBONATE[22]

The oral LD_{50} in the rat is 29 g./kg.; the LD_{50} for skin in the rabbit is > 20 ml./kg. An 8-hour concentrated vapor inhalation produced no deaths in rats. It is only a slight skin irritant in the rabbit and produces moderate eye irritation in this species.

TRIETHYL ORTHOFORMATE[33]

The oral LD_{50} in the rat is between 3.2 and 6.4 mg./kg. Symptoms produced are those of dyspnea and weakness; it does not appear to be absorbed through guinea pig skin. The LD_{50} by skin in this species is > 10 ml./kg.; no skin irritation was noted.

TRIETHYL ORTHOACETATE[33]

The oral LD_{50} in the rat is between 6.4 and 12.8 g./kg. The intraperitoneal LD_{50} in the rat is apparently even higher—between 12.8 and 25.6 g./kg. It is not absorbed through guinea pig skin; the LD_{50} in this species is > 10 ml./kg. and only very slight skin irritation is caused under these circumstances.

TABLE 21

Physical and Chemical Properties of Phosphate and Phosphite Esters

Compound	Formula	Mol. wt.	Sp. gr., °C (g./ml.)	M. p., °C	B. p., °C	Vapor pressure, mm. Hg °C.	Flash point, °F.	Solubility in H₂O (g. per 100 ml.)
Trimethyl phosphate	$(CH_3)_3PO_4$	140.08	1.220 (15)	-47.1	196			100 (25°C.)
Triethyl phosphate	$(C_2H_5)_3PO_4$	182.6	1.0686 (25/4)	-56.4	216		240	100 (25°C., decomposes)
Tripropyl phosphate	$(CH_3(CH_2)_2O)_3PO$	224.24	1.0121					
Triisopropyl phosphate	$((CH_3)_2CHO)_3PO$	224.24	0.9867					
Tri-n-butyl phosphate	$(C_4H_9)_3PO_4$	266.32	0.973	<-80	289	127 (177)	380	0.6
Tributyl phosphite	$(CH_3(CH_2)_3O)_3P$	250.32			124-126 (9 mm.)			Insoluble
Triisobutyl phosphate	$(CH_3(CH_2)_3O)_3PO$	266.32	0.9681		117 (5 mm.)	0.01 (25)		Insoluble
Tri-2-ethylhexyl phosphate	$(C_4H_9CH(C_2H_5)CH_2O)_3PO$	435.0	0.9260 (25)	-74 (pour pt.)		1.9 (200) 5.0 (220)	404.6	<0.01% by weight
Tri-β-chlorethyl phosphate	$(Cl(CH_2)_2O)_3PO$	285.5	1.425	<-55 (pour pt.)		0.5 (145)	451	Insoluble
Tri-o-cresyl phosphate	$(CH_3C_6H_4)_3PO_4$	368.36	1.170		410	10 (200) 0.5 (185) 0.25 (170) 0.1 (155)		Sparingly soluble
Cresyl diphenyl phosphate	$(CH_3C_6H_4O)(C_6H_5O)_2PO$	337.0	1.208 (25)	-40	390	5.0 (220)	449.6	Insoluble
Tri-p-cresyl phosphate (commercial mixture of isomers)	$(CH_3C_6H_4O)_3PO$	368.36	1.165	-28 (pour pt.)		10 (265)	410 (c.c.)	Slightly soluble

Compound	Formula					
Triphenyl phosphate	$(C_6H_5O)_3PO$	326.29	1.286	49–50	11 (245)	Insoluble
Triphenyl phosphite	$(C_6H_5O)_3P$	310.29	1.184	22–24	360	Insoluble
2-Ethylhexyl diphenyl phosphate	$(CH_3(CH_2)_3CH(C_2H_5)CH_2O)(C_6H_5O)_2PO$	362.40	1.0884	–30	0.5 (190)	Insoluble

TABLE 22

Physiological Response to Phosphate Esters[a]

Compound	Acute LD₅₀ (approx.)			Central nervous system effects		Cholinesterase changes	Irritant effects
	Species	Route	ml. or g./kg.	Species	Dose and type of effect		
Trimethyl phosphate[120]	Rat	Oral	1.65	Rabbit	0.3 ml./kg./day × 6 days orally and 2.0 ml./kg./day × 20 doses on skin Flaccid and spastic paralysis;	Not reported	None on rabbit skin
	Rabbit	Oral	1.05				
	Guinea pig	Oral	<1.4	Cat	0.1 ml./kg./day × 79 days subcutan. loss of weight and weakness		
Triethyl phosphate[33,121]	Rat	Oral	>0.8	Rat	400 mg./kg. I.P. × 37 days; peritoneal irritation; ascites; anesthesia; no paralysis	Weak in vitro rat brain inhibition[122]	Slight irritant on guinea pig skin
	Rat	I.P.	0.95				
	Mouse	Oral	1.5				
	Guinea pig	Oral	>0.8	Rat	Inhalation of 28,000 p.p.m. × 6 hr.—3/3 deaths; weakness: gasping respirations		
Tri-2-chloroethyl phosphate[121]	Rat	I.P.	0.28	Rat	Oral or I.P., LD₅₀—convulsions; 80–125 mg./kg. I.P. × 37 days; no paralysis; hemorrhagic effect at high dose level[123]	Weak in vitro rat brain inhibition[122]	None on guinea pig
	Rat	Oral	0.2–0.4	Rat			
Triallyl phosphate[124,125]	Mouse	I.P.	0.25–0.50	Mouse	0.5 g./kg. I.P.—5/5 deaths, ataxia and dyspnea; no paralysis; no delayed deaths	Not reported	Not reported but probably irritating
	Mouse	I.V.	0.071				
Tributyl phosphate[33,35,47]	Rat	Oral	3.0	Rat	Large oral or I.P. doses cause weakness, dyspnea, pulmonary edema, twitching; no paralysis in survivors;	Weak in vitro inhibition human RBC and plasma C.E.[126]	Strong skin and respiratory irritant; inhalation of 123 p.p.m. × 6 hr.—0/3 deaths in rats[33]
	Rat	I.P.	0.8–1.6	Guinea pig	1 cc./kg./day × 4 days; 2/3 deaths		

(continued)

TABLE 22. (continued)

Compound	Acute LD₅₀ (approx.) Species	Route	ml. or g./kg.	Central nervous system effects Species	Dose and type of effect	Cholinesterase changes	Irritant effects
Triisobutyl phosphate[33]	Rat	Oral	3.2-6.4	Rat	Same as tri-n-butyl phosphate; no paralysis in survivors	Not reported	Inhalation of 122 p.p.m. × 6 hr.—0/3 deaths in rats
	Rat	I.P.	0.8-1.6				
Tri-2-ethyl isohexyl phosphate[33]	Rat	Oral	>25.0		No paralysis	Not reported	Moderate skin irritant in guinea pig
Tri-2-ethyl hexyl phosphate[33,36,47,87]	Rat	I.P.	6.4-12.8	Mouse and rat	Delayed deaths; no paralysis	Not reported	Not a skin irritant in guinea pig; no irritation in rabbit eye
	Rat	Oral	37.0				
	Mouse	I.P.	30.0				
	Mouse	Oral	>12.8	Rat	Tolerated up to 0.4 g./kg./day in diet for 30 days	Not reported	Not a skin irritant in humans
	Rat	I.P.	3.2-6.4				
2-Ethylhexyl diphenyl phosphate[127]	Rat	Oral	>24	Rat	2-yr. feeding—rapid loss of weight and death at 5%; retarded growth at 1%; no effects at 0.125%; no paralysis or CNS symptoms	Not reported	
	Rabbit	Skin	>13				
Triphenyl phosphate[116,128]	Rat	Oral	>6.4	Rat	No effect	Weak inhibition of RBC acetylcholine esterase, but not plasma esterases in man	Not a skin irritant; not absorbed through skin appreciably
	Cat	S.C.	0.1-0.2	Chicken	No effect		
	Chicken	Oral	>2.0	Cat	Flaccid paralysis; delayed onset; no cholinergic symptoms		
Tri-o-cresyl phosphate[116,129]	Rat	Oral	3-10	Cats, rabbits, chickens, dogs, monkeys, calves, and man	Flaccid or spastic paralysis; minimum oral paralytic dose in man about 10-30 mg./kg.	Probably inhibits chiefly pseudocholine esterases[130]	Not a skin irritant; 0.1-0.4% of dose applied to human skin is absorbed[131]
	Cat	S.C.	0.1-0.2				
	Chicken	Oral	0.1-0.2				
Tri-p-cresyl phosphate[116]	Rat	Oral	>12.8		No paralysis or neurological symptoms with pure ester	No effects on choline esterases[129,132]	Not a skin irritant
	Cat	S.C.	>1.0				
	Chicken	Oral	>2.0				
Cresyl diphenyl phosphate[33]	Rat	Oral	6.4-12.8		No paralysis; some dyspnea at high doses	Not reported	Not a skin irritant; not absorbed through guinea pig skin
	Mouse	Oral	6.4-12.8				
	Guinea pig	Oral	1.6-3.2				

ᵃ The compounds described here have uses principally as plasticizers, solvents, and intermediates, NOT as insecticides.

[120] W. B. Deichmann and S. Witherup, J. Pharmacol. Exptl. Therap., 88, 338 (1946). [121] M. I. Smith, Ntl. Insts. Health Bull. No. 165, 1936. [122] K. Dubois, Univ. of Chicago, Chicago, personal communication. [123] D. W. Fassett and R. L. Roudabush, Arch. Ind. Hyg. Occupational Med., 6, 525 (1952). [124] Summary Tables of Biol. Tests, 5, 337 (1953). [125] Summary Tables of Biol. Tests, 6, 138 (1954). [126] J. C. Sabin and F. N. Hayes, Arch. Ind. Hyg. Occupational Med., 6, 74 (1952). [127] J. F. Treon, F. R. Dutra, and F. P. Cleveland, Arch. Ind. Hyg. Occupational Med., 8, 170, 268, 281, 284 (1953). [128] W. L. Sutton, C. J. Terhaar, F. A. Miller et al., A.M.A. Arch. Environmental Health, 1, 33 (1960). [129] C. H. Hine, M. K. Dunlap, E. G. Rice, M. M. Coursey, R. M. Gross, and H. H. Anderson, J. Pharmacol. Exptl. Therap., 116, 227 (1956). [130] K. P. Strickland, R. H. S. Thompson, and G. R. Webster, Biochem. J., 62, 512 (1956). [131] H. C. Hodge and J. H. Sterner, J. Pharmacol. Exptl. Therap., 79, 225 (1943). [132] M. M. Coursey, M. K. Dunlap, and C. H. Hine, Proc. Soc. Exptl. Biol. Med., 96, 673 (1957).

TRIETHYL ORTHOPROPIONATE[33]

The oral LD_{50} in the rat is 6.4 to 12.8 g./kg.; the intraperitoneal LD_{50} in this species has the same value. Application to the guinea pig skin shows that it is not absorbed, and the LD_{50} by this route is > 10 ml./kg. Only very slight skin irritation was noted.

C. Phosphate and Phosphite Esters

The phosphate esters (particularly the aromatic type) are of considerable industrial importance as plasticizers. *The Modern Plastics Encyclopedia*[90] lists twenty phosphorous ester plasticizers. In their use as plasticizers they also contribute some fire-retarding properties. Some of the lower molecular weight phosphate esters may also be used as solvents or intermediates, for example, triethyl phosphate. Tributyl phosphate and trioctyl phosphate have been used as antifoaming agents.

Tricresyl phosphate has wide use as a gasoline additive and certain mixed triaryl phosphates are used for special lubricants. Some phosphite esters, such as triphenyl phosphite, have been employed as stabilizers or antioxidants in the rubber and plastics industry.

The majority of these materials are very high-boiling liquids, although some (e.g., triphenyl phosphate) exist as low-melting solids. Their vapor pressures are quite low thus reducing the possibility of inhalation of significant quantities. The triphosphate esters are usually stable to hydrolysis, while the triphosphite esters are less stable. The mono- or diesters are acidic and may hydrolyze readily. Table 21 gives some of their physical and chemical properties.

While a few of these esters have received intensive toxicological study, in most cases only fragmentary information is available. The phosphorus esters that have come into use as insecticides are covered in Chapter XLII and will not be discussed here.

1. Physiological Response

The interest of toxicologists in this series of compounds was aroused by the classical studies of M. I. Smith, R. D. Lillie, et al.[116-118] made in the course of an investigation of a widespread epidemic of poisoning by tri-o-cresyl phosphate used in the manufacture of "Jamaica ginger." While many studies have been published since the occurrence of this disaster in 1930, anyone interested in this topic should read the account by Smith, Engle, and Stollman.[116]

At that time some 10,000 to 15,000 people developed a neuromuscular disturbance characterized by flaccid paralysis, from which recovery was slow, or with

[116] M. I. Smith, E. W. Engel, and E. F. Stohlman, *Ntl. Insts. Health Bull.* No. **160** 1932.
[117] M. I. Smith, E. Elvove, and W. H. Frazier, *Public Health Repts. U. S.* (2nd Rept.), **45**, 2509 (1930).
[118] M. I. Smith and R. D. Lillie, *Arch. Neurol. Psychiat.*, **26**, 976 (1931).
[119] H. V. Smith and J. M. K. Spaulding, *Lancet*, **2**, 1019 (1959).

TABLE 23

Physiological Response to Phosphite Esters

Compound	Acute LD50 (approx.)			Central nervous system effects		Cholinesterase changes	Irritant effects
	Species	Route	ml. or g./kg.	Species	Dose and type of effect		
Triethyl phosphite[133]	Mouse	I.P.	>0.5				
Tri-n-propyl phosphite[134]	Mouse	I.P.	0.25–0.5				
Triisopropyl phosphite[135]	Mouse	I.P.	0.5				
Dibutyl phosphite[37]	Rat	Oral	3.2		None mentioned	Not reported	Slight skin, but severe eye injury
Tributyl phosphite[35]	Rat	Oral	3.0		No paralysis reported; no deaths after inhalation of saturated vapor for 8 hr.	Not reported	Slight eye irritant similar to morpholine on skin
Tri-sec-butyl phosphite[136]	Mouse	I.P.	>0.5				
Trihexyl phosphite[134]	Mouse	I.P.	>0.5				
Triphenyl phosphite[33,137,138]	Rat	Oral	1.6–3.2		Tremors, diarrhea, vasodilatation by all routes; no paralysis in survivors	Inhibition in whole blood in mice[33]	Skin and eye irritant; absorbed through guinea pig skin[33]
	Rat	I.P.	0.8–1.6				
	Mouse	I.P.	0.05–0.1				
Tri-o-cresyl phosphite[116]	Rat	S.C.	10.0	Rat	Tremors, but no paralysis	Inhibition in fowl plasma in vivo[129]	
	Chicken	Oral	>0.8	Cat	Flaccid paralysis—extensor rigidity		
	Cat	S.C.	0.1	Monkey			
	Monkey	S.C.	1.0	Chickens	Less effect		
Triphenyl phosphine[33]	Rat	Oral	0.8–1.6		No paralysis; weakness and ataxia only symptoms	Not reported	Slight skin irritant in guinea pig
	Rat	I.P.	1.6–3.2				
	Mouse	Oral	0.8–1.6				
	Mouse	I.P.	0.8–1.6				

[133] *Summary Tables of Biol. Tests*, **5**, 286 (1953).
[134] *Summary Tables of Biol. Tests*, **5**, 340 (1953).
[135] *Summary Tables of Biol. Tests*, **9**, 130 (1957).
[136] *Summary Tables of Biol. Tests*, **9**, 129 (1957).
[137] R. B. Aird, W. E. Cohen, and S. Weiss, *Proc. Soc. Exptl. Biol. Med.*, **45**, 306 (1940).
[138] M. I. Smith, R. D. Lillie, E. Elvove, and C. F. Stohlman, *J. Pharmacol. Exptl. Therap.*, **49**, 78 (1933).

permanent disability. In spite of the general knowledge of the toxic properties of tri-*o*-cresyl phosphate, there have been eight or nine epidemics since that time, the most recent occurring in 1959 in Morocco, involving several thousand persons.[119]

The search for the causative agent by Smith and others resulted in the accumulation of valuable toxicological information on closely related phosphate esters; the discovery of the powerful insecticidal action of certain pyrophosphates and sulfur-containing phosphate esters has resulted in the synthesis and study of many thousands of compounds. The mode of action of the insecticide type depends usually to a considerable extent on the ability to inhibit cholinesterase; this is described in detail in Chapter XLII.

A study of the physiological properties of the phosphate and phosphite esters in Tables 22 and 23 shows that they can be considered as having several different types of toxic effects, which are as follows:

1. Production or organic damage to the central nervous system with the resultant flaccid or spastic paralysis.

2. Production of nervous system stimulant or convulsive effects or in some cases an anestheticlike action.

3. An effect (usually relatively weak) on true or pseudocholinesterase with the latter predominating.

4. Irritant action on skin and respiratory surfaces.

5. A group that appear to have no major toxic effects.

Some of the compounds that fall into these classes are listed in Table 24. Not all the phosphate esters studied are in this table, since in some cases insufficient data were available to classify them as to type of action.

In the group causing paralysis, tri-*o*-cresyl phosphate has been studied most extensively and is the one that has been stated to be the causative agent in various epidemics of poisoning in man. In all cases these were associated with the ingestion of a material containing the ortho isomer, although in most cases this was present as only a small proportion of the total ester.

Reviews of the findings in human cases of tri-*o*-cresyl phosphate poisoning are given in papers by Hunter *et al.*,[139] Susser and Stein,[140] and Smith and Lillie.[117,118] The clinical picture has been very similar in all of these accounts.

Immediately after ingestion either there may be no symptoms or there may be some gastrointestinal disturbance, which lasts but a few hours or a day or two. This does not occur in all cases and probably depends on the amount taken at one time. After a latent period of about 5 to 28 days, sharp, cramplike pains may occur in the calves with some numbness in the hands and feet. In a few hours after this there may be increasing weakness of the legs and feet and the development of bilateral footdrop. At this time the cramping pains may disappear. In another few days, there may be progression to involve the fingers or wrists. This process

---

[139] D. Hunter, K. Perry, and R. B. Evans, *Brit. J. Ind. Med.*, **1**, 227 (1944).
[140] M. Susser and A. Stein, *Brit. J. Ind. Med.*, **14**, 111 (1957).

TABLE 24

Classification of Toxic Action of Phosphate and Phosphite Esters

Paralysis	Convulsive or central nervous system stimulant action	Inhibition (weak) of cholinesterase	Irritants	Relatively inert
Trimethyl phosphate	Tri-β-chloroethyl phosphate	Triethyl phosphate	Triallyl phosphate	Tri-2-ethylhexyl phosphate
Triphenyl phosphate	Tributyl phosphate	Tri-β-chloroethyl phosphate	Tributyl phosphate	Tri-2-ethylisohexyl phosphate
Tri-o-cresyl phosphate	Triisobutyl phosphate	Triphenyl phosphate	Triisobutyl phosphate	2-Ethylhexyl diphenyl phosphate
Triphenyl phosphite	Triphenyl phosphite	Tri-o-cresyl phosphate	Dibutyl phosphite	Tri-p-cresyl phosphate (free from o-isomer)
Tri-o-cresyl phosphite	Triethyl phosphate (anesthetic)	Triphenyl phosphite	Triphenyl phosphite	p-Cresyl diphenyl phosphate
		Tri-o-cresyl phosphite	Tri-o-cresyl phosphite	Tributyl phosphite

does not seem to extend above the elbows and the larger muscles of the thigh are only infrequently involved.

There is some difference of opinion as to the extent to which sensory loss may be present. Susser et al.[1] find some sensory loss and also some loss of pain, temperature, and vibration sense corresponding roughly to the areas of motor weakness. On the other hand, Hunter et al. [139] feel there are no sensory changes of any consequence or any loss of control of sphincters. There is general agreement, however, that the sensory changes are minor in comparison to the extent of dysfunction in the motor system.

The minimum paralytic dose in man is uncertain, but it can be estimated to be somewhere between 10 and 30 mg./kg. of tri-o-cresyl phosphate in an adult, based on studies by Smith.[116,117]

The progress of these cases varies considerably but in general the muscular weakness may increase over a period of several weeks or even months and then become more or less stationary. Any sensory changes noted regress rapidly particularly in the milder cases. Various studies have been made as to the prognosis of these cases and have been reviewed by Susser.[140] Fatalities are quite rare and occur principally in those who have taken larger quantities in a short period of time. The extent to which the paralysis will be permanent probably also depends on the quantity taken in over a short period, rather than being the result of a cumulative effect from the intake of very small quantities over very long periods. The per cent of cases having permanent residual effects is probably on the order of 25 or 30 per cent.

The pathological changes are well known and have been reproduced in cats, rabbits, chickens, dogs, and monkeys, as well as in calves; rats, mice, and guinea pigs are resistant. The pathology in the human is very similar to that in various experimental animals, although there is some lack of autopsy material because of the infrequency of fatal results. The pathology described by Smith and Lillie[118] in 6 human cases was characterized by some involvement of the anterior horn cells, fatty degeneration of the white substance of the cord, tigrollysis of nerve cells, displacement of the nucleous to the periphery, and especially a demyelination with marked fragmentation and fatty degeneration of the myelin sheath. No other pathology of any consequence was noted in the human cases, although in certain animal species Lillie commented on minor fatty changes in the liver and some degenerative changes in the kidney. Barnes and Denz[141] studied the demyelination produced by tri-o-cresyl phosphate (TOCP) as well as other phosphate esters. The demyelination is not solely or directly related to the cholinesterase inhibitory activity of these substances.

2. *Mode of Action*

There have been a number of attempts to review the metabolism and cellular mode of action of phosphate ester demyelination. It seems clear that these esters are in many instances not hydrolyzed by the body, as shown by the absence of an increase of excretion of phenols.[116] Also see the study with P^{32}-labeled TOCP by Hodge and Sterner.[131] The latter study dealt with skin absorption and it was evident that 0.1 to 0.4 per cent of the total dose was absorbed through the skin. The distribution of labeled phosphorus in the tissues was not what one would expect if it had been hydrolyzed. Appreciable concentrations were found in the brain.

While triphenyl and tri-o-cresyl phosphite[138] produce a typical flaccid paralysis in cats, dogs, and monkeys (see Table 23), it appears likely that these materials may be hydrolyzed readily. The chicken appears to be somewhat more resistant to tri-o-cresyl phosphite than to tri-o-cresyl phosphate.

Triphenyl phosphite is a definite skin and eye irritant and is readily absorbed through guinea pig skin. It produces definite reduction of whole blood cholinesterase in mice; and the symptoms of tremors, diarrhea, etc., have somewhat more resemblance to those seen with cholinesterase inhibitors in species such as the rat. The experiments of Aird, Cohen, and Weiss,[137] who used P^{32}-labeled triphenyl phosphite (0.3 ml./kg.) intraperitoneally in cats, indicated considerable hydrolysis and only a small amount of labeled phosphorus in the central nervous system. The authors conclude that the presence of small concentrations of phosphorous acid in the nervous tissue may be related to the degenerative changes, and that these are not characteristic of phenols.

The papers by Strickland, Thompson, and Webster[130] and Earl, Thompson, and Webster[142] should be consulted for a review of the possible cellular mecha-

---

[141] J. M. Barnes and F. A. Denz, *J. Pathol. Bacteriol.*, **65**, 587 (1953).

[142] C. J. Earl, R. H. S. Thompson, and G. R. Webster, *Brit. J. Pharmacol.*, **8**, 110 (1953).

nisms of action of materials such as tri-*o*-cresyl phosphate. There seems to be no question but that poisoned animals in susceptible species such as hens may show marked depression of the pseudo-cholinesterases in nervous tissue. However, other, more powerful cholinesterase inhibitors may show no paralysis whatever. None of these compounds produces the typical symptoms associated with cholinesterase inhibition shown by the type of phosphate esters described in Chapter XLII.

The depression of pseudo-cholinesterase does not appear to have been accompanied by any change in glucose or pyruvate oxidation or in trypsin, brain amine oxidase, pancreatic lecithinase, and brain cephalinase. There is some reduction of tributyrinase activity in the spinal cords of hens poisoned with TOCP. It is possible that the primary effect is on the nerve cell itself and the demyelination processes are secondary. Another possible mode of action is through a defect on the lipid biosynthesis in the nervous tissue.[143]

3. *Industrial Hygiene Investigations*

In spite of the extensive industrial use of the aryl phosphate esters, there have been relatively few reports of neurological symptoms in workers handling these substances. Hunter *et al.* described 3 occupational cases of polyneuritis from manufacturing various aryl phosphates;[139] it was their opinion that TOCP was the causative agent. Tabershaw and Kleinfeld[144] noted that men manufacturing aryl phosphates (including TOCP and triphenyl phosphate) showed some plasma cholinesterase depression, although no definite evidence of neuromuscular difficulties attributable to this exposure were found. Both of these exposures involved inhalation and probable extensive skin contact. The concentrations of aryl phosphates ranged from about 0.2 up to 3.4 mg./cu. meter. Tabershaw found no correlation of cholinesterase effect with degree and duration of exposure nor with minor gastrointestinal or neuromuscular symptoms.

A case of permanent paralysis in a man engaged in manufacturing the meta and para isomers of tricresyl phosphate was described by Bidstrup and Bonnell.[145] While the final product contained only 1 per cent of TOCP, as much as 6 to 10 per cent was present as a contaminant during manufacturing. No exposure data were given, although the patient had worked a total of 5 months in the occupation before symptoms of anorexia, nausea, and aching of the legs had developed. Cholinesterase determinations were not done until several months after the onset of these symptoms, by which time they were within normal limits for both plasma and red cells.

A recent study was made by Baldridge *et al.*[146] of the exposure of humans to a triaryl phosphate used in hydraulic systems of aircraft carriers. The men exposed

[143] L. Austin, *Brit. J. Pharmacol.*, **12**, 356 (1957).

[144] I. R. Tabershaw and M. Kleinfeld, *A.M.A. Arch. Ind. Health*, **15**, 541 (1957).

[145] P. A. Bidstrup and J. A. Bonnell, *Chem. & Ind. London*, **1954**, 674.

[146] H. D. Baldridge, D. J. Jenden, C. E. Knight, T. J. Preziosi, and J. R. Tureman, *A.M.A. Arch. Ind. Health*, **20**, 258 (1959).

in this operation were studied carefully for clinical signs of neurological disturbances and also for cholinesterase effects. The total duration of the exposure was relatively limited, being only about three months. The air concentrations were very low, averaging only 0.1 p.p.m. No cholinesterase changes or symptoms were noted in this short, low-level exposure. There were a few incidents of extensive skin exposure lasting 1 or 2 hours. However, the material was carefully removed by thorough washing. Toxicological studies by Carpenter *et al.*[147] revealed the typical paralysis of tri-*o*-cresyl phosphate. The *o*-cresyl content of the cresylic acid from which it was manufactured was said to be below 1.5 per cent.

Sutton *et al.*[128] recently published a study covering clinical, toxicological, and industrial hygiene aspects of triphenyl phosphate (TPP). Clinical studies, including true and pseudo-cholinesterase determinations were made on a group of about 14 men who were exposed to triphenyl phosphate vapor, mist, or dust over a period of from 8 to 10 years. Particle-size measurements on the dust indicated that 90 per cent of the particles were less than 1 μ in diameter. The average concentration was probably on the order of 3.5 mg./cu. meter; although, on occasion, levels as high as 40 mg./cu. meter were encountered. There were no signs of clinical illness related to this exposure; the only positive finding was a slight, but statistically significant, reduction in red-cell cholinesterase. The paralysis in cats previously reported by M. I. Smith was confirmed and while TPP was found to inhibit cholinesterases in vivo and in vitro it was not a potent anticholinesterase agent. It seems clear from this study that triphenyl phosphate presents almost no industrial hazard as contrasted to tri-*o*-cresyl phosphate.

A tentative threshold limit for tri-*o*-cresyl phosphate of 0.1 mg./cu. meter has been suggested.[78]

TRIMETHYL PHOSPHATE

Trimethyl phosphate is the only simple alkyl phosphate that has been reported to produce a flaccid paralysis.[120] Its mechanism of action is not clear, and there have been no reports of industrial injuries from it, although it has little industrial use at present.

TRIETHYL AND TRI-2-CHLOROETHYL PHOSPHATES

These materials present an interesting contrast with triethyl phosphate producing an anestheticlike picture with considerable muscle relaxation in relatively high dosages. While in vitro inhibition of brain cholinesterase occurs,[122] it appears to have no cumulative action, and there are no reports of industrial injury. Tri-2-chloroethyl phosphate produces prolonged epileptiform convulsions in the rat at levels of 0.28 g./kg. intraperitoneally.[33,87,121] The convulsions are violent and may proceed at intervals for several hours; they are not present at lower doses and are

---

[147] H. M. Carpenter, D. J. Jenden, N. R. Schulman, and J. R. Tureman, *A.M.A. Arch. Ind. Health*, **20**, 234 (1959).

less apt to occur after oral administration. In vitro tests with brain show only a weak cholinesterase inhibition.[122] The effects are not similar to those that might have been expected if β-chloroethyl alcohol had been produced by hydrolysis.[121]

The material does not appear to be absorbed through the skin and is not a skin irritant. There are no reports of injury from industrial use.

TRIBUTYL AND TRIISOBUTYL PHOSPHATES

These materials possess a definite central nervous system excitory action, which does not usually progress to a state of convulsions. Muscle twitching, weakness, and terminal pulmonary edema are present. They are respiratory irritants. Although their vapor pressures are low, if they are used in boiling water, the steam produced may be quite irritating to the eyes and throat. Heated vapor from the hot concentrated material also may be irritating. It is unknown whether this is due to hydrolysis, but it probably results from the intact molecule. Tributyl phosphate shows weak in vitro inhibition of human red-blood-cell and plasma cholinesterase.[126]

TRI-2-ETHYLHEXYL PHOSPHATE AND 2-ETHYLHEXYL DIPHENYL PHOSPHATE

Both of these long-chain alkyl derivatives appear to be inert, both acutely and chronically, in the rat. However, they do not appear to have been studied in species sensitive to neurological effects such as the cat, chicken, or monkey.[33,87,127]

THE ALKYL PHOSPHITE ESTERS

Most of these compounds have had only very preliminary screening studies by oral injection for acute effects (see Table 23). No paralytic symptoms were mentioned in these studies. However, they were carried out in the mouse and rat so that it is unlikely that any conclusions could be drawn regarding this type of action. Some members of the series are probably more irritating than the corresponding phosphate esters due perhaps to their tendency to hydrolysis.

TRIPHENYL PHOSPHINE

While it might have been predicted that this substance would be quite toxic, as suggested by Hine,[129] screening data would indicate that it is relatively inert in the rat and mouse.[33]

HANDLING PRECAUTIONS FOR PHOSPHATE AND PHOSPHITE ESTERS

In view of the variety of toxicological properties of these esters, it is essential that all handling of phosphate esters in industry be accompanied by a careful study of the potential hazard of the material being used. Avoidance of skin contact and inhalation of mist, dust, or vapor (especially if heated liquids are handled) will usually be desirable and, in some cases, mandatory. All persons handling these substances should have regular, careful medical examinations with particular reference to the occurrence of any effects on the central nervous system. Cholinesterase determinations may be helpful in some instances in establishing the fact that some absorption has occurred, although there is little evidence at present that they will be helpful in establishing the extent of exposure nor in predicting the likelihood of toxic symptoms.

If contamination of the skin or clothing occurs, the material should be removed as rapidly as possible and followed by thorough washing with soap and water. Any individual who has an extensive exposure should be carefully followed medically for a long period to be certain that neurological effects do not occur. Since the esters described in this chapter do not produce typical cholinergic effects of the insecticidal type of phosphate ester, there is probably no logical reason for the usual antidotes of atropine or PAM (pyridine-2-aldoxime methiodide) in the event of acute symptoms following exposure. There appears to have been no study of the possible effects of agents such as PAM in reversing the neurological damage of the aryl phosphates and phosphites.

The toxicity of thermal decomposition products has not been studied extensively, although in the opinion of Treon et al.,[148] they are probably not particularly hazardous.

Finally, it should be stated that any toxicological data on this group of compounds should be interpreted with caution. Predictions as to human effects are probably best made with sensitive, nonrodent species, such as cats, chickens, dogs, or monkeys.

D. Sulfuric and Sulfonic Acid Esters

Several esters of sulfuric and sulfonic acids are of considerable importance in organic syntheses as methylating or ethylating agents. The physical properties of several of these materials are listed in Table 25. They are usually colorless or slightly yellowish oily liquids or, as in the case of the toluene sulfonates, low-melting solids. They have high boiling points and relatively low vapor pressures. Some of them tend to hydrolyze readily in the presence of moisture.

These materials have a physiological property in common in that they all are capable of acting as severe tissue irritants. This may be due in part to their ability to methylate certain amino acids, although it has been speculated that part of the action may be due to hydrolysis in vivo giving rise to sulfuric or sulfonic acids. In some instances, the lack of warning property makes them especially hazardous.

Some methane sulfonates, such as 1,4-bis(methane sulfonoxy)butane, are currently being used in cancer chemotherapy. They are probably less active as

[148] J. F. Treon, J. W. Cappel, F. P. Cleveland, E. E. Larson, R. W. Atchley, and R. T. Denham, Am. Ind. Hyg. Assoc. Quart., **16**, 187 (1955).

TABLE 25

Physical and Chemical Properties of Sulfuric and Sulfonic Acid Esters

Compound	Formula	Mol. wt.	Sp. gr., °C. g./ml.	M. p., °C.	B. p., °C.	Flash point, °F.	Solubility in H₂O (g. per 100 ml.)
Dimethyl‡ sulfate	(CH₃O)SO₂	126.14	1.3322 (20/4)	−26.8	188.5 (decomposes) 69–70 at 10 mm.	182	2.8 (hydrolyzes)
Diethyl sulfate	(C₂H₅O)₂SO₂	154.19	1.180 (20/4)	−24.0	208 (decomposes) 88–90° at 9 mm. 20° at 0.2 mm.	250	0.7 (hydrolyzes)
Methyl chlorosulfonate (methyl sulfuryl chloride)	CH₃OSO₂Cl	130.57	1.492 (10)	−70.0	133–135 decomposes		Decomposes
Ethyl chlorosulfonate (ethyl sulfuryl chloride)	C₂H₅OSO₂Cl	144.59	1.379 (0)		152–153 62–63° at 32 mm.		Decomposes
Methyl-p-toluene sulfonate	CH₃C₆H₄SO₂OCH₃	186.23	1.230 (25/25)	27–28			
Ethyl-p-toluene sulfonate	CH₃C₆H₄SO₂OC₂H₅	200.26	1.17	31–33		316	Insoluble, (hydrolyzes slowly)
1,4-bis(methane sulfonoxy)butane (Myleran, Busulfon)		246.31		114–118			

primary irritants but have a highly selective ability to depress certain proliferative tissues, such as the bone marrow, lymphoid tissue, and some neoplastic cells. This is thought to be the result of a specific alkylating action in the proteins of these tissues.

While in some instances only very scanty experimental data are available, the acute effects on humans are well known. The properties of some of these specific compounds are as follows.

DIMETHYL SULFATE

Dimethyl sulfate is a colorless oily liquid, which is nearly odorless or which may have a faint onionlike odor. It has a fairly heavy vapor density (4.35), but its boiling point is quite high and the vapor pressure at 25°C. is on the order of 0.1 mm. Hg. It can be decomposed very readily by alkalies and also hydrolyzes readily in the presence of water or moisture with the production of sulfuric acid and methanol.

The vesicant and pulmonary actions of dimethyl sulfate were described as early as 1902 by Weber.[149] It was used to a limited extent as a chemical warfare agent in World War I,[83] and there have been a number of reports of human injury in its industrial use. Some of the older literature is reviewed in *Occupation and Health*,[150] and a number of more recent reports have been reviewed by Fairhall.[110]

Two typical cases have been reported by Littler and McConnell.[151] In the first case a 20-year old chemist sustained a splash of dimethyl sulfate on the skin and clothing. He immediately washed the affected areas with water and then in turn applied 5N caustic and ammonia. He had no immediate symptoms but in about 4 hours he noted marked swelling of the eyes and prepuce and shortly after this was found to have a temperature of 99.2°F. and a pulse of 116/minute. By this time he had also developed marked evidence of bronchospasm and rales in the chest. Thirteen hours after the exposure large vesicles had developed at the site of the skin splash. His cough was severe and the sputum included pieces of necrotic trachial mucosa. The next morning subcutaneous emphysema was noted, which indicated a perforation of the trachea or bronchi. His white cell count was 25,000 with 90 per cent polys; hemoglobin, 120 per cent; and red cell count, 6.09 million. Albumin and red cells were noted in the urine. Slight improvement began about 2 days after the exposure with subsiding of the marked swelling around the eyes and lowering of the white count. A repetition of the bronchitis was noted about 10 days after the occurrence but soon subsided. The skin burns healed very slowly.

[149] Weber, *Arch. exptl. Pathol. Pharmakol.*, **47**, 113 (1902).

[150] *Occupation and Health*, **1**, 565 (1930), International Labour Office, Geneva, Switzerland.

[151] T. R. Littler and R. B. McConnell, *Brit. J. Ind. Med.*, **12**, 54 (1955).

This case illustrates the fact that marked, generalized symptoms can occur from a moderate skin contact (even though promptly and vigorously treated) where the opportunity for inhalation of vapor must have been at the most a matter of a few minutes.

The second case was a 28-year old male chemist who had only a very minimal vapor exposure, the day preceding the onset of symptoms, as he emptied a vessel containing dimethyl sulfate. He had no symptoms on that day, but on the following day, at 8 A.M., he walked past the area where he had been working and did not notice any unusual vapor in the air. By 10 A.M. he noticed that he had lost the upper half of his visual fields and that his face and eyelids were beginning to swell. On the advice of another worker he breathed ammonia fumes, but this did not appear to be of any benefit. About 10 hours after the onset of the eye difficulty, he developed pulmonary edema, increased swelling of the face and eyelids, edema of the soft palate, and swelling and vesication of the prepuce. His temperature was 99 to 100°F., the pulse showed bradycardia at 50 (his normal pulse rate was 75 to 85). He was treated with oxygen and local therapy was given to the burns. In due course he made a good recovery. This case is also interesting in that he showed definite signs of analgesia; this has been commented on previously by Auer.[152,153] Soldiers who had been gassed with dimethyl sulfate during World War I were noted to have analgesia and could be operated upon without pain. This apparently was not present in the first case, however, and is variable.

This second case also illustrates other features of dimethyl sulfate, namely, that the vapor alone can cause very severe damage to all exposed surfaces, and that the duration of onset of pulmonary symptoms may be delayed by several hours.

Two other typical cases have been described by Mohlau.[154] In this report 2 males, aged 53 and 55, had opened a tank of dimethyl sulfate and had not noticed any immediate effects except for very slight momentary irritation of the eyes and nose. However, that evening they developed cyanosis, marked coughing, pulmonary edema, and central nervous system signs, which consisted of delirium. In these cases analgesia was not present and the patients were in considerable discomfort and also developed photophobia and headache. Both men recovered; however, in one instance, there was a marked disturbance of color vision. Photophobia persisted for a long time.

In an experiment in which a single rabbit was exposed to dimethyl sulfate vapor (concentration undetermined but sufficient to cause death), the histological findings showed marked edema of the lung and an intense degenerative change in the liver and kidneys.

A number of experimental studies have been made to determine the acute effects of inhalation of dimethyl sulfate on various species. Some of these are

[152] J. Auer, *Proc. Soc. Exptl. Biol. Med.,* **15,** 104 (1918).
[153] J. Auer, *J. Exptl. Med.,* **35,** 97, (1922).
[154] F. D. Mohlau, *J. Ind. Hyg.,* **2,** 238 (1920).

summarized in Table 26. It is evident that dimethyl sulfate is an extremely potent toxic agent and that severe symptoms or fatalities may occur in susceptible species after a few minutes' inhalation of as low as 13–20 p.p.m. The symptoms observed in experimental animals are those of irritation of the eyes and respiratory tract and signs of central nervous system effects.

Ghiringhelli et al.[155,156] have investigated the pathology in the mouse, rat, and guinea pig after inhalation of 75 p.p.m. of dimethyl sulfate for periods varying from 17 to 26 minutes. This is approximately the LC_{50} in these species. The pathology was similar to that described by Mohlau[154] and consisted of marked pulmonary edema, pulmonary emphysema, peribronchitis, and fatty degeneration and necrotic areas in the liver. These workers attempted to determine the role of

TABLE 26

Physiological Response to Inhalation of Dimethyl Sulfate

Species	Concentration, p.p.m.	Time	Mortality
Rat	75	18 min.	LC_{50}[155,156]
	30	4 hr.	5/6[46]
	15	4 hr.	Survived[23]
Cat	175	11 min.	Death after several days[24]
	78	11 min.	Death in 1.5 weeks
	20	11 min.	Death in 1.5 weeks
Monkey	26	40 min.	Death in 1.5 weeks[24]
	13	20 min.	Severe symptoms; recovery in 4 weeks
Guinea pig	75	24 min.	LC_{50}[155,156]
	26	40 min.	Death in 3 weeks[24]
	13	20 min.	Severe symptoms; recovery in 4 weeks
Mouse	75	26 min.	LC_{50}[155,156]

methyl alcohol liberation in the production of the symptoms. The maximum concentration found was 1.87 mg.% in guinea pig urine 18 minutes after inhalation of air containing 76 p.p.m. of dimethyl sulfate. There was a drop of about 22 per cent in the alkali reserve of the animals in the days following exposure. The amount of methanol excerted was only a relatively small fraction of that which would have been expected if all the dimethyl sulfate had been hydrolyzed.

Ghiringhelli also studied blood catalase in guinea pigs exposed to similar concentrations. No changes were found in vivo in blood catalase nor was there any appreciable inhibition of crystalline catalase incubated in vitro with dimethyl sulfate. There was no change in the ultraviolet spectrum. Evidently its toxic effect was not due to methylation of this enzyme.

Ghiringhelli believes it is probable that the local effects on the eye and skin are in part due to the liberation of sulfuric acid and that the other more generalized

[155] G. L. Ghiringhelli and G. Sironi, Med. lavoro, 49, 690 (1958).
[156] G. L. Ghiringhelli, U. Columbo, and A. Monteverde, Med. lavoro, 48, 634 (1957).

effects, central nervous system effects, etc., may be due to the intact molecule. He suggests that the primary effect is probably one of methylation of certain important groups, enzymes, etc.

The acute oral toxicity of the liquid has been reported as being on the order of 50 mg./kg. in the rabbit[77] and about 440 mg./kg. in the rat.[46] The liquid produces very severe eye injury in the rabbit.

The threshold limit for dimethyl sulfate has for some time been suggested to be about 1 p.p.m. (5 mg./cu. meter). The basis for this limit is probably an extrapolation from various animal experiments, since there does not appear to have been any comprehensive study of the tolerance of humans for such a level for an 8-hour period. Smyth[23] suggests that while this level may be low enough to protect against pulmonary injury, some bronchial irritation might be present. In view of the fact that extremely severe symptoms have been reported with as low as 13 p.p.m. for a period of 20 minutes, it seems somewhat unlikely that 1 p.p.m. would be tolerated for a full 8-hour day. There is no good evidence for any chronic effects of dimethyl sulfate although subacute effects involving lesser degrees of primary irritation may have occurred.

Suggestions for an analytical method for the determination of dimethyl sulfate have been made by Fairhall and include absorption in alcohol and potassium hydroxide, dilution, and subsequent precipitation with barium sulfate.[110]

The safe use of dimethyl sulfate depends on education of all operators and all chemical workers on its properties and the provision of completely closed systems for handling as far as practical. All areas where spills, or leaks, or transfer of material may occur should have adequate ventilation. Air-supplied respirators and clothing impervious to the vapor should be available in order to enter any contaminated area. Adequate facilities for washing the eye or skin should be provided at convenient locations. In the event of contact with the skin or eye, the area should be immediately flushed with copious amounts of water and continued for at least 15 minutes; then medical attention should be sought. Exposed surfaces, such as floors, may be decontaminated by flushing with dilute alkali or ammonia and water.

DIETHYL SULFATE

Diethyl sulfate is also a colorless, oily liquid with a high-boiling point and low vapor pressure. It is used for ethylation processes for organic synthesis reactions.

Its physiological effects have not received intensive study. However, Smyth and Carpenter[37,47] report that the oral LD_{50} in the rat is about 880 mg./kg. and that rats tolerate saturated vapor for 2 hours without production of deaths. Inhalation of 250 p.p.m. for 24 hours caused 0/6 deaths in rats while 500 p.p.m. for 4 hours caused 6/6 deaths. It produced severe eye injury in the rabbit and necrosis of the rabbit skin when applied in the undiluted state.

Weber[149] felt that it would produce the same effects as dimethyl sulfate but that the lesions would be less severe. No reports of human injury could be found.

METHYL CHLOROSULFONATE (Methyl Sulfuryl Chloride)

Methyl chlorosulfonate is a colorless, oily liquid with a relatively high specific gravity and a relatively low boiling point (133). Its vapor density is quite high (4.5, air = 1). It is used as a methylating agent and an intermediate.

The principal information about its physiological action comes from its use as a chemical warfare agent in World War I.[83] It produces pulmonary edema in a manner similar to chlorine and phosgene but, in addition, is an intense lachrymatory agent in concentrations as low as 0.008 mg./liter (1.5 p.p.m.). The irritant effect on the upper respiratory tract is intolerable at 0.05 mg./liter, and a concentration of 2.0 mg./liter is fatal upon 10 minutes' exposure.

No reports could be found of injury during industrial use, although this undoubtedly could occur. No threshold limit has been suggested. The precautions should be similar to those suggested for dimethyl sulfate.

ETHYL CHLOROSULFONATE (Ethyl Sulfuryl Chloride)

Ethyl chlorosulfonate is a colorless, heavy liquid which is used as an intermediate.

Its physiological properties are similar to those of methyl chlorosulfonate.[83] It is an intense lachrymator in concentrations as low as 1 p.p.m. Concentrations of 0.05 mg./liter are intolerable, and concentrations of 1 mg./liter are highly dangerous.

The precautions in handling should be the same as those for dimethyl sulfate. No threshold limits have been suggested.

METHYL-p-TOLUENE SULFONATE

Methyl-p-toluene sulfonate is an extremely useful methylating agent in organic synthesis and exists as a low-melting solid or liquid.

While there has been very little experimental study of this compound, it is well known by chemists to cause a potent vesicant action. Observation of a number of these cases[84] shows that in most instances contact with the solution or solid may give rise to no symptoms for several hours. Several hours later, there may be the development of vesication at the site of contact, frequently with little accompanying pain. Very large bullae may be formed in some instances. Generalized toxic symptoms do not seem to occur and these burns heal usually with some pigmentation, which gradually fades. Probably because of the low vapor pressure, there have been few instances of pulmonary irritation. However, one instance was noted of urticarial reaction from heated vapor; skin sensitization also occurs.

ETHYL-p-TOLUENE SULFONATE

Ethyl-p-toluene sulfonate has also been used as an ethylating agent and has physical properties similar to methyl-p-toluene sulfonate. Information as to its physiological effects is lacking although it undoubtedly might produce, to some degree, the effects of the methyl derivative.

1,4-BIS (METHANESULFONOXY) BUTANE

While this compound has no industrial use except as a drug, it is of interest to toxicologists because it is an example of a highly potent "alkylating agent" and is in use in the chemotherapy of certain chronic leukemias. Its toxic properties have been studied extensively in connection with this use and have recently been summarized by Sternberg et al.[157] The intraperitoneal LD_{50} (suspended in peanut oil) in the rat is 18 mg./kg. The immediate symptoms produced are those of convulsions. In the mouse the intraperitoneal LD_{50} is 125 mg./kg. Maximum effects on the bone marrow are seen on about the fourth day, by which time a marked aplasia is present with as high as 70 per cent depletion of all the nucleated elements. Remarkable atrophy of all lymphoid tissue, including that in the spleen, occurs; there may be damage to the intestinal mucosa. In its use in therapy of humans, the principal effect seems to be on bone marrow or on leukemic cells, although some side effects such as alopecia or discoloration of the skin have been noticed with certain derivatives. Studies on the distribution of alkylating agents by Smith et al.[158] would indicate that the material disappeared rapidly from the blood, and that most of the radiosulfur was concentrated in the liver, kidneys, and small intestines of the rat immediately after injection. In 24 hours there was no selective concentration in any particular organ and 60 to 85 per cent of the injected radioactivity was excreted into the urine chiefly in the form of methane sulfonic acid.

Roberts and Warwick[159] have also studied the metabolism of this compound and believe that it may react with cysteine to form a cyclic compound, which is thought to be S-β-alanyl tetrahydrothiophenium ion. Isolation of this compound appears to have been accomplished. These authors feel a simpler compound, such as ethyl methane sulfonate, is excreted as N-acetyl-S-ethylcysteine.

Similar derivatives such as 1,4-bis(methanesulfonoxy)nonane and 1,4-bis-(methanesulfonoxy)dimethyl butane have also been used for chemotherapy and have similar toxic properties. The intraperitoneal LD_{50}'s in the mouse of a series of such derivatives have been reported by Carlson and Morgan.[160]

There appear to have been no published reports of the direct irritant effects

[157] S. S. Sternberg, F. S. Philips, and J. Scholler, Ann. N. Y. Acad. Sci., **68**, 811 (1958).

[158] P. K. Smith, M. V. Nadkarni, E. G. Trams, and C. Davison, Ann. N. Y. Acad. Sci., **68**, 834 (1958).

[159] J. J. Roberts and G. P. Warwick, Biochem. J., **72**, 3p (1959).

[160] W. W. Carlson and C. A. Morgan, Proc. Soc. Exptl. Biol. Med., **85**, 211 (1954).

on the skin or mucous membranes of such compounds, but an instance has been observed of slowly healing skin burns in an individual who was splashed with an alkylmethane sulfonoxy derivative.[33]

E. Silicate Esters

Methyl and ethyl silicates have had industrial uses in protective coatings and as preservatives and waterproofing agents for stone and concrete. Their physical and chemical properties are listed in Table 27.

TABLE 27

Physical and Chemical Properties of Silicate Esters

	Methyl Silicate (tetramethylorthosilicate)	Ethyl Silicate (tetraethylorthosilicate or tetraethoxysilane)
Formula	$Si(OCH_3)_4$	$Si(OC_2H_5)_4$
Physical state	Liquid	Liquid
Molecular weight	152.2	208.2
Specific gravity, g./ml.	1.028	0.933 (20/4)
Boiling point, °C.	121 (759 mm.)	168.2
Vapor pressure, mm. Hg	12 (25°C.)	1.5 (25°C.)
Vapor density (air = 1)	5.2	7.2
Flash point, °F.		125
Solubility	Soluble in alcohol	Soluble in alcohol

METHYL SILICATE

Methyl silicate has been used in the ceramic industry for closing pores, including those in concrete and cement, for coating metal surfaces, and as a bonding agent in paints and lacquers.

Methyl silicate appears to be a moderately toxic material in experimental animals.[161] The minimum oral lethal dose in the rat is about 700 mg./kg. while the intraperitoneal lethal dose is on the order of 100 mg./kg. Animals are said to show kidney damage regardless of the route of administration.

One report states that methyl silicate vapor under certain conditions of humidity or liquid methyl silicate may cause a necrosis of the cornea, which may be progressive.[162] No threshold limit has been suggested for this material.

Precautions should include avoiding the inhalation of vapor and contact of the liquid or vapor with the eyes.

ETHYL SILICATE

Ethyl silicate is used as a preservative for stone, brick, concrete, and plaster. It is used in waterproofing, weatherproofing, and acid-proofing work, in bricks, in

[161] W. G. Fredrick, Detroit Dept. of Health, Detroit, Mich. personal communication.
[162] Chem. Age London, **55**, 208 (1946).

heat- and chemical-resistant paints, and in protective coatings. Nearly three million pounds were produced in 1958.[2] It is a moderately high-boiling liquid with a low vapor pressure.

The physiological properties of ethyl silicate have been studied by Smyth and Seaton,[163] Kasper, McCord, and Fredrick,[164] Rowe, Spencer, and Bass,[165] and by Pozzani and Carpenter.[166] The studies by Smyth et al. would indicate that the principal acute symptoms are those of irritation of the eyes and respiratory tract followed by tremors, weakness, narcosis, and death. Some kidney damage was noted. Inhalation of 400 p.p.m. by rats for 7 hr./day for 30 days[166] caused definite mortality and also some pathology in the lung, liver, and kidney. Exposure under similar conditions to 88 p.p.m. did not cause injury.

Rowe et al.[165] found some kidney damage in rats when the rats repeatedly inhaled 125 p.p.m. Severe symptoms and death result in rats and guinea pigs inhaling from 1500 to 2300 p.p.m. for a period of 4 hours. The maximum amount tolerated by a guinea pig for several hours is about 500 p.p.m.

Humans can detect about 85 p.p.m. by odor; 250 p.p.m. is irritating to the eye and nose; 3000 p.p.m. is intolerable. In view of its good warning properties and the lack of toxic effects in animals at 100 p.p.m. (the suggested threshold limit[78]), this concentration is probably low enough to prevent lung or kidney injury.[23] The threshold limit of 100 p.p.m. does not seem to have been confirmed by any reported studies on humans exposed over long periods of time to such concentrations.

Precautions should be taken in industrial handling to avoid repeated or prolonged contact with higher concentrations of the vapor and to avoid contact of the vapor or liquid with the eyes.

OTHER SILANE DERIVATIVES

Some other silane derivatives have been studied by Smyth et al.:[22] vinyltriethoxysilane, vinyltrichlorosilane, amyltriethoxysilane, tris(2-chloroethoxy)silane, and phenyltrichlorosilane. Of these, tris(2-chloroethoxy) silane was the most toxic with an oral LD_{50} in the rat of 0.19 g./kg. and a skin LD_{50} in the rabbit of 0.089 ml./kg. It was an eye irritant. Vinyltrichlorosilanes were skin and eye irritants.

[163] H. F. Smyth, Jr., and J. Seaton, J. Ind. Hyg. Toxicol., **22,** 288 (1940).

[164] J. A. Kasper, C. P. McCord and W. G. Fredrick, Ind. Med., **6,** 660 (1937).

[165] V. K. Rowe, H. C. Spencer, and S. L. Bass, J. Ind. Hyg. Toxicol., **30,** 332 (1948).

[166] U. C. Pozzani and C. P. Carpenter, Arch. Ind. Hyg. Occupational Med., **4,** 465 (1951).

CHAPTER XLII

Organic Phosphates

LLOYD W. HAZLETON AND ROBERT J. WEIR

I. General Considerations

The organic phosphates are most widely recognized in their usage as insecticides, and the bulk of this chapter will be devoted to discussion of chemicals intended for this use. Their biocidal properties present appreciable toxicological problems from the standpoint of manufacture and use. Organic phosphates are also used as gasoline additives, hydraulic fluids, cotton defoliants, and as industrial intermediates where their highly toxic effect is neither desirable nor always apparent. In dealing with this group of chemicals, it is clear that the toxicity of the organic phosphate pesticide is not universal to the organic phosphates as a class. One must only consider those that occur naturally in the body, such as the phospholipids, phosphonucleotides, phosphoproteins, etc., to observe that some of the class members are relatively nontoxic.

The nomenclature of the organic phosphates frequently is confusing. Trade names, generic names, and manufacturers' experimental designations add to the confusion. For the purpose of this chapter, class names will follow those outlined by Negherbon.[1]

A. SYMPTOMS IN ANIMALS

The universal signs of intoxication of the organic phosphates that are insecticidal appear to result from the inhibition of the cholinesterases (these esterases hydrolyze acetylcholine, butyrylcholine, benzoylcholine, acetyl-β-methylcholine, etc., depending on the species).

Following excessive exposure, the signs of toxicity reflect stimulation of the autonomic and central nervous systems, resulting from inhibition of acetylcholinesterase and consequent accumulation of acetylcholine. Prolongation and intensification of the acetylcholine action results in two degrees of response, depending upon dosage and specific action of the inhibitor. The initial action is on smooth muscles, cardiac muscle, and exocrine glands and, in general, is comparable to stimulation of the postganglionic parasympathetic nerves. This phase of action results in the early signs of toxicity resembling those of muscarine and, hence, is

---

[1] W. O. Negherbon, *Handbook of Toxicology*, Vol. III. Saunders, Philadelphia, 1959.

referred to as the *muscarinic* action of acetylcholine. The action, and hence the signs, can be counteracted by atropine. The most common early signs are intestinal cramps, tightness in the chest, blurred vision, headache, diarrhea, decrease in blood pressure, and salivation.

The second stage of intoxication results from stimulation of the peripheral motor system and of all autonomic ganglia. Experimentally, these actions can be counteracted by curare and ganglionic blocking agents and in other respects resemble the classical action of nicotine, hence are referred to as the *nicotinic* action of acetylcholine. The complexity of toxic action during the second stage includes neuromuscular and ganglionic blockade; thus, curare therapy would be contra-indicated. Ultimately the toxic manifestations of poisoning are referable to stimulation and/or paralysis of the somatic, autonomic, and central nervous systems. A more detailed understanding of this complex action and the influence of adequate atropinization may be obtained from Goodman and Gilman.[2] A basic review in terms of health problems has been published by Hazleton.[3]

The signs of toxicity outlined above do not apply to animals that receive small doses over a long period of time. In this case, the correlation of signs of toxicity and inhibition of the cholinesterase(s) is not clear-cut. With many organic phosphates, inhibition in rats may be so complete in plasma and red blood cells that the cholinesterase activity is immeasurable with present techniques, and the brain activity may be markedly inhibited, while the animals appear normal in all respects. With other members of the organic phosphate class, chronic exposure in rats produces marked inhibition of plasma and red cell cholinesterase and moderate inhibition of brain activity, and results in diarrhea and tremors as the only toxic signs. These examples point out the variability of signs of toxicity, which depend on the toxicant, vehicle, route of administration, dosage, species studied, specific enzyme system on which the toxicant acts (true or pseudocholinesterase), metabolic conversion products, reversibility of the inhibition, etc. These variables and their ultimate discussion are beyond the scope of this book and only those factors that have bearing on industrial hygiene will be covered here.

B. GROSS PATHOLOGY IN ANIMALS

The greatest bulk of the organic phosphate insecticides do not produce morphological alterations in animals. Some of the more recent additions to the group are chlorinated and are suspected of being capable of producing liver and kidney damage similar to that produced by the halogenated hydrocarbons. There is very little literature to support the supposition, as these members have not received extensive evaluation to date.

A few members of the group have been observed to produce demyelination of

[2] L. S. Goodman and A. Gilman, *The Pharmacological Basis of Therapeutics*. 2nd ed., Macmillan, New York, 1955.

[3] L. W. Hazleton, *J. Agr. Food Chem.*, **3**, 312 (1955).

the peripheral nerves, anterior horn cell degeneration, and fatty degeneration of spinal cord white matter. This results in a syndrome resembling "jake leg" or "ginger jake" paralysis, which has been described for Jamaica ginger poisoning. Paralytic effects of TOCP (tri-o-cresyl phosphate) have been described by Smith et al.[4,5] More recent reports on this phenomenon after administration of TOCP are reported by Durham et al.,[6] Barnes and Denz,[7] Hine et al.,[8] and Frawley et al.[9] Durham et al.[6] also studied Chlorthion (O,O-dimethyl-O-3-chloro-4-nitrophenyl thionophosphate), DDVP (dimethyl 2,2-dichlorovinyl phosphate), Systox (O,O-diethyl-O-2-[ethylmercapto]-ethyl thionophosphate), Diazinon (O,O-diethyl-O-[2-isopropyl-6-methyl-4-pyrimidyl] phosphorothioate), OMPA (octamethyl pyrophosphoramide), EPN (ethyl-p-nitrophenyl benzenethionophosphonate), malathion (O,O-dimethyl S-[1,2-dicarboethoxyethyl] dithiophosphate), and Isopestox (bis-[monoisopropylamino]-fluorophosphine oxide). Of these, only Isopestox was found to produce the demyelination syndrome. Later, Frawley et al.[9] showed that EPN also produced demyelination, contrary to findings of Smith et al.[4] and Barnes and Denz;[7] and Austin and Davies[10] demonstrated that DFP (diisopropyl fluorophosphate) also produced demyelination in animals. The symptomatic observations have been confirmed with histological evidence in all cases (DFP, EPN, TOCP, and Isopestox).

C. EXPOSURE IN MAN

The most common, and most important, route of industrial exposure to the organic phosphates is by accidental spillage on the skin. Most of the materials later discussed in detail are rapidly absorbed through the skin. Percutaneous absorption frequently is unnoticed since dermal irritation rarely occurs, unless the solvent systems of the formulated materials possess this irritative property.

The second most frequent exposure route is through the respiratory tract. Intoxication may occur with some of the more highly toxic members of the group, such as TEPP (tetraethyl pyrophosphate), demeton (Systox), parathion (O,O-diethyl O-p-nitrophenyl thiophosphonate), and Phosdrin (alpha isomer of 2-carbomethoxy-1-methylvinyl dimethyl phosphate), but, in general, it is agreed that exposure is due to particulate matter rather than to vapor.[3] The organic phosphates as a group have extremely low vapor pressures; despite this, Kay and co-workers[11]

[4] M. I. Smith, E. Ehrve, and W. H. Frazier, *Public Health Repts. U. S.,* **45,** 2509 (1930).

[5] M. I. Smith and R. D. Lillie, *A.M.A. Arch. Neurol. Psychiat.,* **26,** 976 (1931).

[6] W. F. Durham, T. B. Gaines, and W. J. Hayes, Jr., *A.M.A. Arch. Ind. Health,* **13,** 326 (1956).

[7] J. M. Barnes and F. A. Denz, *J. Pathol. Bacteriol.,* **65,** 587 (1953).

[8] C. H. Hine, E. F. Gutenburg, M. M. Coursey, K. Seligman, and R. M. Gross, *J. Pharmacol. Exptl. Therap.,* **113,** 28 (1955).

[9] J. P. Frawley, R. E. Zwickey, and H. N. Fugat, *Federation Proc.,* **15,** 424 (1956).

[10] L. Austin and D. R. Davies, *Brit. J. Pharmacol.,* **9,** 145 (1954).

[11] K. Kay, L. Monkman, J. P. Windish, T. Doherty, J. Park, and C. Racicat, *Arch. Ind. Hyg. Occupational Med.,* **6,** 252 (1952).

give analytical values for parathion (vapor pressure 3.78×10^{-5} mm. Hg) in the air of treated orchards for up to 3 weeks after application. Summerford et al.[12] correlated blood cholinesterase level with symptomatology in plant personnel, orchard workers, and other groups during and after a spray season. Inhalation exposure was not distinguished from dermal exposure. Fatal or near fatal illnesses have resulted from brief, massive exposure to parathion due to gross carelessness rather than repeated exposure.

Oral exposure is rarely a problem except for accidental ingestion by children and in the case of suicide. The more toxic members of the organic phosphate group may be an ingestion problem in manufacturing and spraying operations if good personal hygiene practices are not followed.

The organic phosphate insecticides share the biological action of inhibiting cholinesterase(s). While this is generally recognized to be not the sole toxic action, it does provide the toxicologist and industrial hygienist with an excellent tool for the measurement of exposure of animals or workers to the toxicant. This measurement of exposure serves as a warning of impending toxicity and is useful in prophylactic programs. Beyond this its reliability, either diagnostic or prognostic, is of little value.

For practical reasons, the most widely used method for determination of cholinesterase inhibition appears to be that of Michel[13] or some modification of it. The manometric method of Ammon[14] is also reliable. The colorimetric method of Metcalf,[15] and the modified method for whole blood described by Fleisher and Pope[16] are also useful, and the field kits such as described by Edson[17] are based on this method.

Using the electrometric technique, Wolfsie and Winter[18,19] have evaluated the plasma and red blood cell cholinesterase levels of men and women who were not exposed to organic phosphate insecticides (ingestion of residues from treated crops could not be eliminated). Mean values were as follows:

Red blood cell   0.67–0.86 Δ pH units/hr.
Plasma   0.70–0.97 Δ pH units/hr.

In the prophylactic program, both red-blood-cell and plasma values should be obtained. Measurement should be made frequently and with regularity. The workers' previous values should be available for inspection and comparison. In addition to knowledge of the specific action of the compound, judgment and ex-

[12] W. T. Summerford, W. J. Hayes, Jr., J. M. Johnston, K. Walker, and J. Spillane, *Arch. Ind. Hyg. Occupational Med.*, **7**, 383 (1953).

[13] H. O. Michel, *J. Lab. Clin. Med.*, **34**, 1564 (1949).

[14] R. Ammon, *Arch. ges. Physiol. Pflüger's*, **233**, 57 (1933).

[15] R. L. Metcalf, *J. Econ. Entomol.*, **44**, 883 (1951).

[16] J. H. Fleisher and E. J. Pope, *Arch. Ind. Hyg. Occupational Med.*, **9**, 323 (1954).

[17] E. F. Edson, *World Crops*, **1** (Aug., 1958).

[18] J. H. Wolfsie and G. D. Winter, *Arch. Ind. Hyg. Occupational Med.*, **6**, 43 (1952).

[19] J. H. Wolfsie and G. D. Winter, *Arch. Ind. Hyg. Occupational Med.*, **9**, 396 (1954).

perience are essential to adequate interpretation of results. Marked or severe depression of either the plasma or red-blood-cell value may be considered strong evidence of exposure, whether accompanied by gross signs or not. The red-cell level is more significant, since it represents the true neurohormone esterase level for humans, acetylcholinesterase. Plasma enzyme inhibition is less specific but may be important as a diagnostic aid in acute exposure since it usually, but not always, responds more rapidly and at lower dosage. Whole blood analysis is least reliable since it may give a composite effect and may mask the individual level of either the plasma or cells.

In the event exposure occurs, the symptoms in man are qualitatively similar to those described for animals and include headache, vertigo, blurred vision, lacrimation, salivation, sweating, muscular weakness and ataxia, dyspnea, diarrhea, abdominal cramps, vomiting, coma, pulmonary edema, and death.

D. TREATMENT

The onset of symptoms is rapid and maximum effects may develop within a few hours. It is thus important that medical care be obtained without delay. Since the early symptoms of headache, malaise, etc., are easily confused with other diseases, it is important that workers exposed to the organic phosphates be instructed to report any such indications.

Adequate atropinization is essential to relieve the muscarinic effects, and to provide central respiratory stimulant action. An average adult may require from 12 to 24 mg. total dose of atropine intravenously during the first 24 hours. Since this is far in excess of the usual therapeutic dose, the physician unacquainted with the mutually antagonistic action of this drug and the organic phosphate may be hesitant to employ such large doses. A general rule is that atropine should be administered until visible effects of atropinization are observed. Since, as pointed out above, the muscarinic effects are only a part of the action produced by heavy exposure, it is essential that the patient be treated symptomatically with artificial respiration, postural drainage, warmth, etc. Prognosis depends largely upon the exposure, type of compound, and adequacy of treatment. Care should be exercised until the patient is obviously free of any signs of toxicity. A detailed survey of this subject is presented by Gordon and Frye.[20]

Since the toxicity of the organic phosphates is due to the inhibition of cholinesterase, the reactivation of these enzymes would offer great promise as a therapeutic measure. This action is apparently achieved by pyridine-2-aldoxime methiodide (PAM, PAM-2, 2-PAM, P-2-AM), diacetyl monoxime (DAM), and other oximes. Namba and Hiraki[21] report both experimental and clinical investigations of PAM. In cases of parathion poisoning, they recommend intravenous doses of 1 g. of PAM or more, if indicated.

[20] A. S. Gordon and C. W. Frye, *J. Am. Med. Assoc.*, **159**, 1181 (1955).
[21] T. Namba and K. Hiraki, *J. Am. Med. Assoc.*, **166**, 1834 (1955).

Grob and Johns[22] give a step-by-step outline of combined therapy for organic phosphate intoxication. After removal from exposure, a patent airway and artificial respiration should be established. The therapeutic regimen includes atropine, 2 to 4 mg. intravenously, repeated frequently until muscarinic symptoms disappear. PAM or DAM is recommended in doses of 2000 mg. intravenously.

Until these agents are more adequately studied and evaluated, both alone and in combination with atropine, their ultimate place in therapy will be uncertain. In the meantime, the guiding principle is adequate therapy since neither atropine nor the oximes offers serious hazards of overdosage in the presence of severe cholinesterase inhibition.

II. Specific Compounds

CHLORTHION (O,O-Dimethyl O-3-chloro-4-nitrophenyl Thionophosphate; Bayer 22/190)

1. Source, Uses, and Industrial Exposures

Chlorthion controls a wide range of agricultural and household insects, including houseflies, mosquitoes, and roaches.

2. Physical and Chemical Properties[23]

Physical state: yellowish brown liquid

Molecular weight: 295.5

Density: 1.433 (20°C.)

Boiling point: 112°C. (0.04 mm. Hg)

Vapor pressure: 7.0×10^{-6} mm. Hg (30°C.)

Refractive index: $1.5680\ n_D{}^{20°C.}$

Solubility: 1:25,000 in water; readily miscible with benzene, toluene, alcohols, ethers, and oils; unstable in alkaline solutions

Flash point and flammable limits: not flammable at normal temperature; thermal decomposition point about 150°C.

3. Determination in the Atmosphere

See parathion (pure material for color standards available from Chemagro Corporation).

[22] D. Grob and R. J. Johns, *J. Am. Med. Assoc.*, **166**, 1855 (1955).

[23] *Pesticide Official Publication and Condensed Data on Pesticide Chemicals.* Assoc. of Am. Pest Control Officials, College Park, Md., 1955. See also 1957 Supplement, Reference 57. Data in the text have been gathered also from unpublished sources and manufacturers data sheets and may vary from these basic references.

4. *Physiological Response*

Acute. The acute oral LD_{50} in rats is 1500 mg./kg., while the intraperitoneal LD_{50} is 750 mg./kg.[24] The dermal application of approximately 1400 mg./kg. to rabbits produced no signs of toxicity.

Subacute and Chronic. Daily administration of Chlorthion in the diet of groups of rats at 50, 100, and 200 mg./kg./day resulted in cholinesterase inhibition at all levels. The two lower dosages produced 50 per cent cholinesterase inhibition. Forty per cent mortality resulted at 100 mg./kg./day over the 60-day feeding period. Two hundred mg./kg./day produced 75 per cent cholinesterase inhibition and was not tolerated for more than 5 to 10 days. The 50 mg./kg./day dosage was tolerated without mortality for 60 days.[1,24]

Pharmacological Effect. Chlorthion is an active inhibitor in vitro. Low mammalian toxicity is ascribed to slow absorption.[1] As a typical member of the organic phosphate group, no pathology has been described in animals receiving repeated high dosages.

5. *Hygienic Standard of Permissible Exposure*

No threshold limit or pesticide tolerance has been established for Chlorthion.

DDVP (Dimethyl 2,2-Dichlorovinyl Phosphate)

1. *Source, Uses, and Industrial Exposures*

DDVP is used almost exclusively as a quick, knockdown agent for the control of houseflies. Its action depends on its fumigant action as it is a poor stomach and contact poison.

2. *Physical and Chemical Properties*[23,24a]

Physical state: oily liquid
Molecular weight: 221.0
Density: 1.415 (25°C.)
Boiling point: 84°C. (1 mm. Hg)
Vapor pressure: 0.01 mm. Hg (30°C.)
Refractive index: $1.451\ n_D^{25°C.}$
Solubility: miscible with alcohol and most nonpolar solvents; 1.0 per cent in water and 0.5 per cent in glycerine at room temperature
Flash point and flammable limits: practically nonflammable

[24] K. P. DuBois, J. Doull, J. Deroin, and O. K. Cummings, *Arch. Ind. Hyg. Occupational Med.*, **8**, 350 (1953).

3. Determination in the Atmosphere

A micro method, which could be adapted to air analysis, has been published by Giang et al.[25]

4. Physiological Response

Acute. The acute oral LD_{50} is 80 mg./kg. in male rats and 55 mg./kg. in female rats.[26] The acute dermal LD_{50} is 107 mg./kg. in male rats and 75 mg./kg. in female rats.

Subacute and Chronic. Edson[27] has evaluated the cumulative effects of DDVP by administering to rats repeated one half LD_{50} doses in rapid succession over a short period of time. The reversible nature of the cholinesterase inhibition produced is unique among organic phosphates.

5. Hygienic Standard of Permissible Exposure

No threshold limit or pesticide tolerance has been set for DDVP.

DIAZINON (*O,O*-Diethyl *O*-[2-isopropyl-6-methyl-4-pyrimidyl] phosphorothioate; G-24480)

1. Source, Uses, and Industrial Exposures

Diazinon is a broad-spectrum insecticide and acaricide. It has also received much use in the control of cockroaches, particularly those resistant to chlorinated hydrocarbon pesticides.

2. Physical and Chemical Properties[23]

Physical state: the pure material is a colorless liquid; the technical material is a pale to dark brown liqiud

Molecular weight: 340.4

Density: 1.116–1.118 (25°C.)

Boiling point: 83–84°C. (0.002 mm. Hg)

Vapor pressure: 1.4×10^{-4} mm. Hg (20°C.)

Refractive index: 1.4978 to 1.4981 $n_D{}^{20°C.}$

[24a] *Pesticide Official Publication, 1957 Supplement.* Assoc. of Am. Pest Control Officials, College Park, Md.

[25] P. A. Giang, F. T. Smith, and S. A. Hall, *J. Agr. Food Chem.*, **4**, 621 (1956).

[26] A. M. Mattson, J. T. Spillane, and G. W. Pearce, *J. Agr. Food Chem.*, **3**, 319 (1955).

[27] E. F. Edson, personal communication.

Solubility: 0.004 per cent in water at room temperature; miscible with alcohol, xylene, acetone, and petroleum oils

Flash point and flammable limits: practically nonflammable

3. Determination in the Atmosphere

A method of analysis of Diazinon has been described by Harris[28] and presumably could be adapted to air analysis.

4. Physiological Response

Acute. The acute oral LD_{50} of 95 per cent technical Diazinon in rats is 100 to 150 mg./kg. The LD_{50} of a 23 per cent wettable powder is 264.5 mg./kg. on the basis of active ingredient.[29,30] This discrepancy was later explained by Gysin[31] as resulting from the formation of a number of possible isomerization and decomposition products resulting from the technical form. Presumably the responsible form is monothiono-TEPP.[32] Formulation prevented the degradation.

Subacute and Chronic. Male and female rats receiving 100 and 1000 p.p.m. (weight) technical Diazinon in the diet for 4 weeks showed no gross signs of intoxication, alteration of growth, or gross pathology at autopsy. Red blood cells and brain cholinesterase levels were significantly inhibited at both dosages. Plasma activity at both dosages was comparable to the control.

Rats received 10, 100, and 1000 p.p.m. (weight) active Diazinon as a wettable powder in the diet for 72 weeks with no apparent gross signs of toxicity. Dogs received orally various doses of active Diazinon as a wettable powder for 46 weeks. No pathology, gross or microscopic, was observed at the lowest dosage (4.6 mg./kg./day) in 2 weeks. After 12 weeks cholinesterase inhibition was complete at the lowest dosage. At a dosage of 9.3 mg./kg./day for 5 weeks signs of toxicity and complete cholinesterase inhibition were observed. Withdrawal of Diazinon at the highest dosage resulted in reversal of signs and regeneration of cholinesterase activity to normal limits after 2 weeks.

5. Hygienic Standard of Permissible Exposure

No threshold limit has been set for Diazinon, but a tolerance of 0.75 p.p.m. (weight) has been established by the Food and Drug Administration on a number of vegetable crops, indicating that this daily oral dosage is considered to produce no measurable response in man.

[28] H. J. Harris, *A Tentative Ultraviolet Method for Analysis of Diazinon in Spray Residues.* Geigy Chemical Corp., Yonkers, N. Y., 1953.
[29] R. B. Bruce, *J. Agr. Food Chem.,* **3,** 1017 (1955).
[30] R. B. Bruce, *Federation Proc.,* **13,** (1954).
[31] H. Gysin and A. Margot, *J. Agr. Food Chem.,* **6,** 900 (1958).
[32] H. Gysin, personal communication.

DIPTEREX (*O*,*O*-Dimethyl-2, 2, 2-trichloro-1-hydroxyethyl phosphonate; Bayer L 13/59)

1. Sources, Uses, and Industrial Exposures

Dipterex has been used to control houseflies. It has shown promise in agriculture against lepidopterous and dipterous insects and mites.

2. Physical and Chemical Properties[23]

Physical state: white to pale yellow crystalline solid
Molecular weight: 275.5
Density: 1.73 (20°C.)
Melting point: 78 to 80°C.
Boiling point: 120°C. (0.4 mm. Hg)
Vapor pressure: volatile
Refractive index: 1.3439 (10 per cent aqueous solution) $n_D{}^{20°C.}$
Solubility: soluble in water to 13 to 15 per cent at 25°C.; soluble in alcohols, benzene, toluene, chloroform; slowly unstable in water; decomposition speeded by heat or alkali; decomposition product DDVP
Flash point and flammable limits: practically nonflammable

3. Determination in the Atmosphere

A method suitable for air analysis of Dipterex has been published by Giang et al.[33]

4. Physiological Response

Acute. The acute oral and intraperitoneal LD_{50} in rats has been reported by DuBois[34] as 450 and 225 mg./kg., respectively.

Subacute and Chronic. Daily intraperitoneal dosages of 100 mg./kg. (more than one fourth of the LD_{50} dose) produced 40 per cent mortality of treated rats in 60 days. Duration of toxic action was brief; complete recovery occurred in a few hours after dosing.

5. Hygienic Standard of Permissible Exposure

No threshold limit or pesticide tolerance has been set for Dipterex.

[33] P. A. Giang, W. F. Barthel, and S. A. Hall, *J. Agr. Food Chem.*, **2**, 1281. (1954).
[34] K. P. DuBois and C. J. Cotter, *Arch. Ind. Hyg. Occupational Med.*, **11**, 53 (1955).

EPN (Ethyl-*p*-nitrophenyl Benzenethionophosphonate)

1. Sources, Uses, and Industrial Exposures

EPN has been used as an insecticide and acaricide. It has shown a broad spectrum of activity against mites and insects. Present use in agriculture is limited.

2. Physical and Chemical Properties[38a]

Physical state: white crystalline solid; the technical material is a dark amber liquid

Molecular weight: 323.3

Density: 1.268 (25°C.)

Melting point: 36°C.

Vapor pressure: 3.0×10^{-4} mm. Hg (100°C.)

Refractive index: 1.5978 $n_D^{30°C.}$

Solubility: practically insoluble in water; soluble in most of the common organic solvents; stable at ordinary temperatures and in neutral and acid media

3. Determination in the Atmosphere

The method of Averell and Norris[35] for parathion is applicable to EPN. This method has been modified by Gunther and Blinn[36] and more recently by Wilson et al.[37]

4. Physiological Response

Acute. The acute oral LD_{50} for pure EPN is 42 and 14 mg./kg. for male and female rats, respectively; for the technical materials, values of 28 to 33 and 7 to 13 mg./kg. are given by Hodge et al.[38]

Subacute and Chronic. In 2-year chronic feeding studies in rats, doses of 150 p.p.m. for males and 75 p.p.m. for females produced no effect. Doses of 2.0 mg./kg./day were administered to male and female dogs for 1 year without effect. EPN has been found to potentiate the effects of malathion (details under malathion).

[35] P. R. Averell and M. V. Norris, *Anal. Chem.*, **20**, 753 (1948).

[36] F. A. Gunther and R. C. Blinn, *Advances in Chem. Ser.*, **1**, 75 (1950).

[37] C. W. Wilson, R. Baier, D. Genung, and J. Mullowney, *Anal. Chem.*, **23**, 1487, (1951).

[38] H. C. Hodge, E. A. Maynard, L. Hurwitz, V. DiStefano, W. L. Downs, C. K. Jones, and H. J. Blanchet, Jr., *J. Pharmacol. Exptl. Therap.*, **112**, 29 (1954).

[38a] All threshold limit values are quoted from "Threshold Limit Values for 1959," adopted at the 21st Annual Meeting of the American Conference of Governmental Industrial Hygienists, Chicago, April 25–28, 1959, *A.M.A. Arch. Ind. Health*, **20**, 266 (1959).

5. *Hygienic Standard of Permissible Exposure*

The threshold limit for EPN as a dust, fume, or mist is 0.5 mg./cu. meter of air.[38a] A pesticide tolerance of 3.0 p.p.m. (weight) has been established for fruits and vegetables by the Food and Drug Administration.

ISOPESTOX (Bis-[monoisopropylamino]-fluorophosphine Oxide; Mipafox)

$$\begin{array}{ccccccc} CH_3 & H & H & O & H & H & CH_3 \\ & \diagdown & | & | & \| & | & | & \diagup \\ & & C-N-P-N-C \\ & \diagup & & | & & \diagdown \\ CH_3 & & & F & & & CH_3 \end{array}$$

1. *Source, Uses, and Industrial Exposures*

Isopestox was introduced as an effective systemic insecticide and acaricide. After being placed on the market, it was found responsible for near fatalities and paralysis of several workers in England, which resulted in its being withdrawn from commerce.

2. *Physical and Chemical Properties*[23]

Physical state: crystalline solid
Molecular weight: 182.2
Density: 1.2 (25°C.)
Melting point: 60 to 65°C.
Boiling point: 125 to 126°C. (2 mm. Hg)
Vapor pressure: 0.001 mm. Hg (5°C.)
Solubility: soluble in water and polar organic solvents; slightly soluble in petroleum oils.

3. *Determination in the Atmosphere*

Presumably the general method of Giang and Hall[39] for enzymic determination of organic phosphorous insecticides could be applied to measure Isopestox air contamination.

4. *Physiological Response*

Acute. The acute oral LD_{50} in various species has been reported to range from 25 to 100 mg./kg.[23,40,41] Near lethal doses produce severe neurotoxic signs.

Subacute and Chronic. In chickens, rabbits, and humans, flaccid paralysis has been demonstrated and appears to result from demyelinization of nerves, as

[39] P. A. Giang and S. A. Hall, *Anal. Chem.,* **23,** 1830 (1951).

[40] H. Martin, *Guide to the Chemicals Used in Crop Protection.* 2nd ed., Univ. of Western Ontario, 1955.

described for TOCP (tri-o-cresyl phosphate).[41-43] In chickens, a single dose of 1.0 mg./kg. orally produced the response in 10 to 14 days. Like other organic phosphates, Isopestox reduces the activity of cholinesterase(s), both in vitro and in vivo. The signs of toxicity, except for the demyelination syndrome, are typical of the class.

5. *Hygienic Standard of Permissible Exposure*

No threshold limit or pesticide tolerance has been set for Isopestox.

MALATHION (*O,O*-Dimethyl *S*-[1,2-dicarboethoxyethyl] dithiophosphate; Experimental Insecticide No. 4049 [American Cyanamid Co.])

1. *Source, Uses, and Industrial Exposures*

Malathion is regarded as the least toxic of the class and is generally considered as a wide spectrum insecticide for fruits, vegetables, and ornamental plants.

2. *Physical and Chemical Properties*[23]

Physical state: yellow to dark brown oily liquid, dependent on purity
Molecular weight: 330
Density: 1.23 (25°C.)
Melting point: 2.85°C.
Boiling point: 156 to 157°C. (0.7 mm. Hg)
Vapor pressure: 4.0×10^{-5} mm. Hg (30°C.)
Refractive index: 1.495 $n_D^{25°C.}$
Solubility: 145 p.p.m. (weight) in water; miscible with alcohols, ethers, and vegetable oils
Flash point and flammable limits: not flammable at normal temperature; thermal decomposition above the boiling point

3. *Determination in the Atmosphere*

The colorimetric method of Norris *et al.*[44] is applicable to analysis for malathion in air.

[41] W. E. Ripper, *Compt. rend. IIIme congr. intern. de phytopharmacie*, Paris, 1952.
[42] D. R. Davies, *J. Pharm. and Pharmacol.*, **6**, 1 (1954).
[43] D. R. Davies, *Proc. Roy. Soc. Med.*, **45**, 570 (1952).
[44] M. V. Norris, W. A. Vail, and P. R. Averell, *J. Agr. Food Chem.*, **2**, 570 (1954).

4. *Physiological Response*

Acute. The acute LD_{50} for various species and routes has been compiled by Golz and Shaffer.[45] The acute oral LD_{50} has been reported to vary widely, but it is generally agreed to range around 1400 mg./kg. for female rats when administered in vegetable oils. Acute vapor inhalation exposure does not appear to be too great a problem due to the low vapor pressure and low inherent toxicity of malathion. Single dermal application of 2460 to 6150 mg./kg. (90 per cent grade) to rabbits produces some toxicity.

Subacute and Chronic. Subacute and chronic exposures of rats[45,46] indicate that an oral dosage of 100 p.p.m. (weight) can be tolerated without effect on cholinesterase(s) activity. While 5000 p.p.m. (weight) has slight effect on survival, food consumption, and growth, some rats have survived 20,000 p.p.m. (weight) in the diet for 2 years. Frawley *et al.*[47,48] demonstrated that the simultaneous administration of malathion and EPN resulted in mortality greater than expected from the administration of the components alone (potentiation). Slight potentiation was also found in subacute studies of these materials when cholinesterase activity was studied as the criterion of evaluation. DuBois and Coon[49] found that EPN interfered with the enzymic hydrolysis of malathion and thus produced the potentiating effect by disrupting the detoxication mechanism of malathion.

5. *Hygienic Standard of Permissible Exposure*

The threshold limit for malathion as a dust, fume, or mist is 15 mg./cu. meter of air.[38a] The present pesticide tolerance set for malathion by the Food and Drug Administration is 8.0 p.p.m. (weight).

METHYL PARATHION (O,O-Dimethyl O-p-nitrophenyl Thionophosphate; Metacide)

1. *Source, Uses, and Industrial Exposures*

Methyl parathion is closely related to parathion in its chemistry and toxicology. It controls aphids, boll weevils, and mites especially well, although its spectrum for control of insects is nearly as broad as parathion.

[45] H. H. Golz and C. B. Shaffer, *Malathion, Summary of Pharmacology and Toxicology.* Central Medical Dept., American Cyanamid Co., New York, revised January, 1955.

[46] L. W. Hazleton and E. G. Holland, *Arch. Ind. Hyg. Occupational Med.*, **8**, 399 (1953).

[47] J. P. Frawley, E. C. Hagan, O. G. Fitzhugh, H. N. Fuyat, and W. I. Jones, *J. Pharmacol. Exptl. Therap.*, **119**, 147 (1957).

[48] J. P. Frawley, H. N. Fuyat, E. C. Hagan, J. R. Blake, and O. G. Fitzhugh, *J. Pharmacol. Exptl. Therap.*, **121**, 96 (1957).

[49] K. P. DuBois and J. M. Coon, *Arch. Ind. Hyg. Occupational Med.*, **6**, 9 (1952).

2. *Physical and Chemical Properties*[23]

Physical state: white crystalline solid in pure form; brown liquid crystallizing at 29°C. as the technical material

Molecular weight: 263.3

Density: 1.358 (20°C.)

Melting point: 35 to 36°C. (pure)

Vapor pressure: 0.5 mm. Hg (109°C.)

Refractive index: $1.5515 \; n_D^{35°C.}$

Solubility: 50 p.p.m. (weight) in water at 25°C.; soluble in most aromatic solvents; slightly soluble in paraffin hydrocarbons

Flash point and flammable limits: not flammable at normal temperatures.

3. *Determination in the Atmosphere*

The method of Averell and Norris[35] for parathion is applicable to methyl parathion. This method has been modified by Gunther and Blinn[36] and more recently by Wilson *et al.*[37] The method is applicable to the analysis of air for the presence of methyl parathion.

4. *Physiological Response*

Acute. The acute oral LD_{50} for methyl parathion in the rat ranges between 9 and 25 mg./kg.,[23] depending on the purity of the material studied. Methyl parathion is toxic by all routes but has been described as especially hazardous via the eye.

Subacute and Chronic. DuBois and Coon[49] state that methyl parathion is approximately as toxic as parathion to rats but is a much less potent cholinesterase inhibitor. This material, like parathion, is a poor cholinesterase inhibitor in vitro but is converted in the mammalian liver to the toxic form,[50] which is probably the oxy analog.

5. *Hygienic Standard of Permissible Exposure*

No threshold limit in air has been established, but a pesticide tolerance of 1.0 p.p.m. (weight) has been established by the Food and Drug Administration, indicating this dosage in the diet of man is safe.

PARATHION (*O,O*-Diethyl *O-p*-nitrophenyl Thiophosphonate; E-605; Compound 3422)

[50] R. L. Metcalf, *Organic Insecticides.* Interscience Publishers, New York–London, 1955.

1. Source, Uses, and Industrial Exposures

Parathion is one of the best known of the class and, despite relatively high toxicity, it has received extensive use in agriculture as a broad spectrum insecticide.

2. Physical and Chemical Properties[23]

Physical state: technical grade clear, medium to dark brown liquid

Molecular weight: 291.27

Density: 1.265 (25°C.)

Melting point: 6.1°C.

Boiling point: 157 to 162°C. (0.6 mm. Hg); 375°C. (760 mm. Hg)

Vapor pressure: 0.00003, 0.00066, and 0.0028 mm. Hg at 24.0, 54.5, and 70.7°C., respectively

Refractive index: 1.53668 $n_D{}^{20°C.}$

Solubility: 24 p.p.m. (weight) in water at 25°C.; miscible with most organic solvents

Flash point and flammable limits: flash point at 120 to 160°C. until flammable impurities of technical material are removed; boils with decomposition at 215°C. and residue supports combustion when temperature reaches 221°C.

3. Determination in the Atmosphere

The methods of Averell and Norris,[35] Gunther and Blinn,[36] and, more recently, Wilson et al.[37] may be applied to analysis of parathion in air. A method of analysis in biological material has been reported by Hazleton and Holland.[51]

4. Physiological Response

Acute. According to Hazleton and Holland,[51] the oral LD_{50} for parathion is 3.5 and 12.5 mg./kg. for female and male rats, respectively. Essential aspects of pharmacology, acute toxicology, and antidotes for parathion appeared in 1948 in reports by DuBois et al.,[52] Hagan and Woodard,[53] and Hazleton and Godfrey.[54]

Subacute and Chronic. Parathion was not found to be stored in tissues of rats in subacute feeding studies.[51] No effect was noted in growth, food consumption, gross or microscopic morphology, or survival when 100 p.p.m. (weight) parathion was fed to male rats for a period of 2 years. Cholinesterase inhibition was observed at lower levels. Parathion was found by Gardocki and Hazleton[55] to be excreted in the urine as *p*-nitrophenol. This is a useful method, together with cholinesterase determinations, for measurement of exposure to parathion. The toxicity to humans

[51] L. W. Hazleton and E. G. Holland, *Advances in Chem. Ser.,* **1**, 31 (1950).

[52] K. P. DuBois, J. Doull, and J. M. Coon, *Federation Proc.,* **7**, 216 (1948).

[53] E. C. Hagan and G. Woodard, *Federation Proc.,* **7**, 224 (1948).

[54] L. W. Hazleton and E. Godfrey, *Federation Proc.,* **7**, 226 (1948).

[55] J. F. Gardocki and L. W. Hazelton, *J. Am. Pharm. Assoc. Sci. Ed.,* **40**, 491 (1951).

can be evaluated from the studies of Grob *et al.*[56] and acute human intoxication incidences; a picture similar to that observed in laboratory animals is apparent.

5. Hygienic Standard of Permissible Exposure

The threshold limit for parathion as a dust, fume, or mist is 0.1 mg./cu. meter of air.[38a] A pesticide tolerance of 1.0 p.p.m. (weight) has been established by the Food and Drug Administration, indicating that this level is a safe daily dosage for oral consumption by man.

PHOSDRIN (Alpha Isomer of 2-Carbomethoxy-1-methylvinyl Dimethyl Phosphate; OS2046)

$$CH_3O \quad O \qquad CH_3$$
$$\diagdown \underset{\|}{P}-O-\underset{|}{C}=CHCOOCH_3$$
$$CH_3O \diagup$$

1. Source, Uses, and Industrial Exposures

Phosdrin is useful in the control of aphids, mites, thrips, and lepidopterous larvae on a wide variety of crops.[57]

2. Physical and Chemical Properties[58]

Physical state: light yellow to orange liquid
Molecular weight: 224.1
Density: 1.23 (20°C.)
Boiling point: 106–107.5°C. (1.0 mm. Hg)
Vapor pressure: 0.0029 mm. Hg (21°C.)
Refractive index: 1.4493 $n_D^{25°C.}$
Solubility: miscible with water, alcohols, ketones, chlorinated and aromatic hydrocarbons; slightly soluble in aliphatic hydrocarbons.
Flash point: 175°F. by the tag open cup

3. Determination in the Atmosphere

Standard micro and macro methods for determination of Phosdrin are available from the Shell Corporation.[58] A recent preliminary method of analysis for the determination of Phosdrin was reported by Zweig at the 136th Meeting of the American Chemical Society, October, 1956. The method is based on incubation of the water soluble Phosdrin with a standard amount of enzyme acetylcholinesterase, addition of a standard amount of acetylcholine, and the measurement of unhydrolyzed acetylcholine remaining by means of a color reaction with alkaline hydroxylamine and ferric chloride.

[56] D. Grob, W. L. Garlick, and A. M. Harvey, *Bull. Johns Hopkins Hosp.*, **87**, 106 (1950).

[57] *Pesticide Official Publication, 1957 Supplement.* Assoc. of Am. Pest Control Officials, College Park, Md.

[58] *Bull. SC: 59–37, Summary of Basic Data for Phosdrin Insecticide.* Shell Chemical Corp., New York.

4. Physiological Response

Acute. The acute oral toxicity of Phosdrin for rats is 6.0 to 7.0 mg./kg. The dermal LD_{50} for rabbits is approximately 34 mg./kg. The LC_{50} in inhalation studies of 1-hour duration is approximately 14 p.p.m. for female rats.

Subacute and Chronic. In chronic studies the minimal lethal dose for rats has been established as between 100 and 200 p.p.m. in the diet (weight). Lower doses affect tissue cholinesterase levels. Dosages below 5.0 p.p.m. have no effect on cholinesterase activity.[59]

5. Hygienic Standard of Permissible Exposure

No threshold limit has been set for Phosdrin insecticide, but a tolerance of 1.0 p.p.m. (weight) has been established by the Food and Drug Administration, indicating that this level is safe for oral consumption by man.

RONNEL (*O,O*-Dimethyl-*O*[2,4,5-trichlorophenyl] Phosphorothioate; ET-57; Korlan)

1. Source, Uses, and Industrial Exposures

Ronnel is a systemic organic phosphate that has been shown to be highly effective in the control of insects affecting cattle as well as plants. As a systemic livestock pest control agent, it is effective against cattle grub, and occasional reports indicate utility against sheep keds, sheep nasal botfly, chicken lice and mites, dog fleas and ticks.

2. Physical and Chemical Properties[60, 61]

Physical state: white to light tan crystalline solid
Molecular weight: 321.56
Melting point: 40.97°C.
Vapor pressure: 0.0008 mm. Hg (25°C.)
Solubility: insoluble in water; soluble in most organic solvents

3. Determination in the Atmosphere

There is no method presently available in the literature. For analysis of Ronnel, contact the Dow Chemical Co., Agricultural Chemicals Division, Midland, Mich.

[59] J. K. Kodama, C. H. Hine, and M. S. Morse, *Arch. Ind. Hyg. Occupational Med.*, **9**, 54 (1954).

[60] *Korlan Information Sheet.* The Dow Chemical Co., May 1, 1958.

[61] *Trolene (Dow-ET-57) The First Animal Insecticide.* Agricultural Chemical Dept. Information Bull. No. 108, Dow Chemical Co., 1957.

4. *Physiological Response*

Acute. Ronnel has a low toxicity to warm-blooded animals. The acute oral LD_{50} of Ronnel in rats is 3000 mg./kg. The LD_{50} in rabbits is around 1000 mg./kg., and in chickens between 4000 and 5000 mg./kg.

Subacute and Chronic. Ronnel is also a weak inhibitor of cholinesterase. It affects the pseudo-esterase of the plasma predominantly rather than the true acetylcholinesterase of the red blood cells upon both single and repeated oral doses.[62]

5. *Hygienic Standard of Permissible Exposure*

No threshold limit or official pesticide tolerance has been established for Ronnel.

SCHRADAN (Octamethyl Pyrophosphoramide; OMPA; Bis-[bis-dimethylamino] Phosphonous Anhydride)

$$(CH_3)_2N\ \ \ O\ \ \ \ \ \ \ O\ \ \ N(CH_3)_2$$
$$\diagdown \ \ \underset{\|}{P}\!-\!O\!-\!\underset{\|}{P} \ \diagup$$
$$(CH_3)_2N \diagup \ \ \ \ \ \ \ \ \ \diagdown N(CH_3)_2$$

1. *Sources, Uses, and Industrial Exposures*

Schradan is a systemic insecticide primarily used in Europe for the control of sucking insects such as mites, aphids, etc. Crop use in this country is limited to walnuts.

2. *Physical and Chemical Properties*[23]

Physical state: viscous, dark brown liquid
Molecular weight: 286.3
Density: 1.1343 (25°C.)
Melting point: below −10°C.
Boiling point: 135 to 137°C. (1 mm. Hg)
Vapor pressure: 0.0003 mm. Hg (25°C.)
Refractive index: 1.4612 $n_D{}^{25°C.}$
Solubility: miscible with water, soluble in ethanol, acetone, chloroform, and benzene; insoluble in heptane and petroleum ether

3. *Determination in the Atmosphere*

A method suitable for the analysis of Schradan in air has been reported by Hartley *et al.*[63]

[62] D. C. McCollister, F. Oyen, and V. K. Rowe, *J. Agr. Food Chem.*, **7**, 689 (1959).
[63] G. S. Hartley, D. F. Heath, J. M. Hulme, D. W. Pound, and M. Whittaker, *J. Sci. Food Agr.*, **2**, 303 (1951).

4. Physiological Response

Acute. In acute studies, the LD_{50} has been reported by Lehman[64] to be 13.5 mg./kg. orally in rats. DuBois *et al.*[65] reported the oral LD_{50} in rats to be 10 mg./kg. and the acute intraperitoneal LD_{50} to be 8.0 mg./kg. Frawley[66] listed oral LD_{50}s of 35.5 mg./kg. for female rats and 13.5 mg./kg. for male rats, thus indicating some sex variation. Reports from these and other investigators indicate that there is very little, if any, species variation to the acute response.

Subacute and Chronic. The chronic toxicity in rats has been studied by Barnes and Denz.[67] At 50 p.p.m. (weight) in the diet for 1 year, male rats showed toxic signs, growth suppression, and marked cholinesterase depression without producing tissue pathology. At dosages of 10 and 50 p.p.m. (weight) whole blood cholinesterase activity was reduced but brain activity was unaffected. Further studies in male rats at 1.0 p.p.m. (weight) showed that the acetylcholinesterase (true cholinesterase) in the red cell was unaffected. A dosage of 0.3 p.p.m. produced no effect. This was confirmed by Edson *et al.*,[68] who fed 0.25 p.p.m. (weight) to rats without effect. Studies conducted on humans indicate that a level of 0.6 p.p.m. (weight) was without effect in 6 male and 6 female subjects. One subject was administered 2.4 p.p.m. (weight) in the total diet, which produced marked plasma and red cell cholinesterase depression.[68]

5. Hygienic Standard of Permissible Exposure

No threshold limit has been set for Schradan, but a tolerance of 0.75 p.p.m. (weight) has been established by the Food and Drug Administration for Schradan on walnuts. This would indicate that a dosage of 0.75 p.p.m. (weight) is safe in the daily diet of man.

SYSTOX (O,O-Diethyl-O-2-[ethylmercapto]-ethyl Thionophosphate; E-1059)

1. Sources, Uses, and Industrial Exposures

Systox is a systemic insecticide, which also has contact action. It is particularly useful for the control of sucking insects such as mites and aphids.

[64] A. J. Lehman, *Assoc. Food & Drug Officials U. S. Quart. Bull.*, **15**, 122 (1951).

[65] K. P. DuBois, J. Doull, and J. M. Coon, *J. Pharmacol. Exptl. Therap.*, **99**, 376 (1950).

[66] J. P. Frawley, E. C. Hagan, and O. G. Fitzhugh, *J. Pharmacol, Exptl. Therap.*, **105**, 156 (1952).

[67] J. M. Barnes and F. A. Denz, *Brit. J. Ind. Med.*, **11**, 11 (1954).

[68] E. P. Edson, K. P. Fellowes, and F. MacL. Carey, *Medical Department Report*, Fisons Pest Control Ltd., England, 1954.

2. *Physical and Chemical Properties*[23]

Physical state: pale yellow to light brown liquid
Molecular weight: 258
Density: 1.1183 (20°C.)
Boiling point: 134°C. (2 mm. Hg)
Vapor pressure: 0.001 mm. Hg (33°C.)
Refractive index: 1.4875 $n_D^{20°C.}$
Solubility: 0.01 per cent in water; miscible with most organic solvents

3. *Determination in the Atmosphere*

The method of Gardner and Heath[69] presumably can be applied to the analysis of Systox in air.

4. *Physiological Response*

At this time it is not practical to give the LD_{50} values for Systox without further qualification, as it is a mixture of isomers. This is further complicated by an apparent literature discrepancy as to the structure of the isomers. The isomer (formula above) is referred to as Systox by Deichmann[70] and Martin and Miles,[71] while Barnes and Denz[67] apply this term to the mixture and designate the above structure as the P=S isomer. They designate the second isomer as P=O, an exchange of S and O positions, while Deichmann[70] designates the second isomer as Iso Systox and replaces the O with S, in effect producing a dithio compound.

Based on the above terminology, the Barnes and Denz[67] data for the mixture are as follows: the oral LD_{50} for rats is 4.0 mg./kg. for females and 10 mg./kg. for males.

5. *Hygienic Standard of Permissible Exposure*

No threshold limit has been set for Systox, but a tolerance of 0.75 p.p.m. (weight) has been established by the Food and Drug Administration, indicating that this amount can be tolerated by man without effect.

TEPP (Tetraethyl Pyrophosphate)

$$H_5C_2O \diagdown \underset{\underset{H_5C_2O}{\diagup}}{\overset{\overset{O}{\|}}{P}}-O-\underset{\underset{OC_2H_5}{\diagdown}}{\overset{\overset{O}{\|}}{P}}\diagup OC_2H_5$$

1. *Source, Uses, and Industrial Exposures*

TEPP is a relatively insoluble, nonsystemic insecticide used in the control of some aphids, spider mites, mealybugs, leafhoppers, and thrips. It is particularly

---

[69] K. Gardner and D. F. Heath, *Anal. Chem.*, **25**, 1849 (1953).

[70] W. B. Deichmann, *Federation Proc.*, **13**, 346 (1954).

[71] H. Martin and J. R. W. Miles, *Guide to the Chemicals Used in Crop Protection.* Science Service, Dominion of Canada Dept. of Agriculture, London, Ontario, 1952.

useful close to harvest because of its rapid degradation and absence of residue in a short period after application.

2. Physical and Chemical Properties[23]

Physical state: amber liquid
Molecular weight: 290.2
Density: 1.1810 (25°C.)
Boiling point: 104 to 110°C. (0.08 mm. Hg)
Vapor pressure: volatile
Refractive index: 1.4170 to 1.4180 $n_D^{25°C.}$
Solubility: miscible with water resulting in rapid hydrolysis (half-life 6.8 hours at 25°C., pH 7.0); miscible with most organic solvents except kerosene of low aromatic content.

3. Determination in the Atmosphere

Methods for analysis of TEPP have been accepted by the Association of Official Agricultural Chemists[72] and presumably could be adapted to air analysis.

4. Physiological Response

Acute. TEPP is highly toxic by ingestion, skin absorption, and by way of the eye. The LD_{50} for all species in general is 50 mg./kg. or less, according to Harris.[73] He also reports general acute toxicity by inhalation: the LC_{50} is 200 p.p.m. (volume) following a 1-hour exposure. In dermal studies, Harris gives the LD_{50} for animals in general as 200 mg./kg. following a 24-hour exposure.

Subacute and Chronic. Chronic exposure to sublethal amounts lowers the blood cholinesterase level. Some experience has been gained with the use of TEPP as a therapeutic agent in myasthenia gravis. Single dosages of 5.0 mg. or 3.6 mg. daily for 2 days, or 7.2 mg. every 3 hours orally for 3 to 5 doses, produce symptoms in normal subjects, as did somewhat larger doses in myasthenia gravis patients. Symptoms appeared within 30 minutes after final dose and were typical of organic phosphate poisoning. The estimated lethal dose in man is 20 mg. intramuscularly or 100 mg. orally.

5. Hygienic Standard of Permissible Exposure

The threshold limit for TEPP as a dust, fume, or mist is 0.05 mg./cu. meter of air.[38a]

[72] W. Horwitz, *Official Methods of Analysis of the Association of Official Agricultural Chemists.* 8th ed., Association of Official Agricultural Chemists, Washington, D. C., 1955, pp. 86–87.
[73] J. S. Harris, *Agr. Chem.,* **2,** 27 (1947).

TRITHION (*O,O*-Diethyl *S-p*-Chlorophenylthiomethyl Phosphorodithioate; R-1303)

$$C_2H_5O \diagdown \overset{\overset{S}{\parallel}}{P}-SCH_2S\langle=\rangle Cl$$
$$C_2H_5O \diagup$$

1. Source, Uses, and Industrial Exposures

Trithion is a nonsystemic insecticide, miticide, and ovicide with a relatively long residual action.

2. Physical and Chemical Properties[57]

Physical state: light amber liquid
Molecular weight: 342.9
Density: 1.265 to 1.285 (25°C.)
Vapor pressure: very low
Refractive index: $1.590–1.597 \ n_D{}^{25°C.}$
Solubility: not appreciably soluble in water, but miscible with vegetable oils and most organic solvents.

3. Determination in the Atmosphere

Air may be analyzed for Trithion by the method described by Patchett.[74]

4. Physiological Response

Acute. The acute oral LD_{50} of Trithion for male albino rats is 17.2 to 28 mg./kg.

Subacute and Chronic. Exposure of rats and dogs to a near saturated vapor (in air) of Trithion for 4 weeks resulted in plasma cholinesterase depression in dogs while there was no reduction in enzyme activity in rats. The animals appeared normal throughout the exposure. It would appear that Trithion is not volatile enough to be a hazard by the inhalation route. Subacute studies in rats and dogs have been performed by Weir and Fogleman.[75] Dosages of 100 p.p.m. by weight in the diet of rats produced tremors and reduced body weight gains. Cholinesterase activity was markedly depressed, but no gross or microscopic morphological changes occurred. Dosages of 5.0 p.p.m. (weight) produced no effect. Dosages of 1.0 p.p.m. (weight) in dogs produced no effect while higher dosages, up to 30 p.p.m. (weight) produced plasma and red-cell cholinesterase depression without overt signs or pathology.

[74] G. G. Patchett, *Determination of R-1303 Spray Residues in Oranges, Lemons and Alfalfa.* Stauffer Chemical Co., Richmond, Calif., 1956.
[75] R. J. Weir and R. W. Fogleman, unpublished data.

5. *Hygienic Standard of Permissible Exposure*

No threshold limit has been established for Trithion, but a tolerance of 0.8 p.p.m. (weight) has been established by the Food and Drug Administration, indicating that this dosage is safe in the lifetime diet of man.

III. Further Reference.

Because of the great number of organic phosphates that have been studied, it is impossible to enumerate the specific information on all of them in this chapter. For further reference, it is recommended that the following general material be explored by the reader:

G. M. Kosolapoff, *Organophosphorus Compounds.* Wiley, New York, 1950.

R. L. Metcalf, *Advances in Pest Control Research,* Vol. I. Interscience Publishers, New York–London, 1957.

R. L. Metcalf, *Advances in Pest Control Research,* Vol. II. Interscience Publishers, New York–London, 1958.

F. A. Gunther and R. C. Blinn, *Analysis of Insecticides and Acaricides,* Vol. 6. Interscience Publishers, New York–London, 1955.

L. W. Hazleton, Pesticide Toxicity, *J. Agr. Food Chem.,* **3,** 312 (1955).

E. R. de Ong, *Chemistry and Uses of Pesticides.* 2nd ed., Reinhold, New York, 1956.

W. O. Negherbon, *Handbook of Toxicology,* Vol. III. Saunders, Philadelphia, 1959.

L. W. Hanna, *Hanna's Handbook of Agricultural Chemicals.* 2nd ed., Hanna, Forest Grove, Ore., 1958.

R. L. Rudd and R. E. Genelly, *Pesticides: Their Use and Toxicity in Relation to Wildlife.* State of California Dept. of Fish and Game, Game Bull. No. 7, Sacramento, 1956.

Pesticide Official Publication and Condensed Data on Pesticide Chemicals. Assoc. of Am. Pesticide Control Officials, College Park, Md., 1955.

Pesticide Official Publication and Condensed Data on Pesticide Chemicals. Assoc. of Am. Pesticide Control Officials, College Park, Md., 1957.

CHAPTER XLIII

Aldehydes and Acetals

DAVID W. FASSETT, M.D.

I. General Considerations

The aldehydes comprise one of the most important classes of industrial chemicals. They undergo a great variety of chemical reactions and can be prepared economically from readily available raw materials such as natural gas. The uses of aldehydes are so numerous that it is impossible to do more than indicate some of the major applications. A single substance may be used in many different applications as, for example, formaldehyde. Three important classes of use and quantities produced in 1958 are listed below.[1]

A. USES OF ALDEHYDES

1. Resins

Formaldehyde is the outstanding material in this class, as indicated by the quantities of formaldehyde-containing resins produced in 1958.[1]

Phenol formaldehyde	449 million pounds
Melamine formaldehyde	110 million pounds
Urea formaldehyde	238 million pounds

2. Intermediates

The aldehydes are used as intermediates for the synthesis of alcohols, acids, and other chemicals; they are also used in rubber, tanning, and paper industries, and in agriculture.

Formaldehyde	1,358 million pounds

(a considerable proportion of this is included in the synthesis of resins)

Acetaldehyde	46 million pounds
Chloral	49 million pounds
Hexamethylene tetramine	17 million pounds
Sodium and zinc formaldehyde sulfoxylate	7 million pounds
Benzaldehyde	2 million pounds
o-Chlorobenzaldehyde	198 thousand pounds

[1] Synthetic Organic Chemicals—U. S. Production and Sales, 1958, *U. S. Tariff Comm. Rept.* No. **205**, Washington, D.C., 1959.

3. Flavors and Perfumes

Anisaldehyde	318 thousand pounds
Citronellal	250 thousand pounds
Hydroxy citronellal	217 thousand pounds
Piperonal	228 thousand pounds
Geranial	61 thousand pounds
Veratraldehyde	8 thousand pounds

Many aldehydes are of importance in the manufacture of medicinals and dyes, and formaldehyde itself has wide use as a deodorizing, bactericidal agent and as a hardening agent for proteins.

While relatively few of these are used as solvents, furfural has important applications in the solvent refining of mineral oil, as a solvent for various types of resins, as well as other uses as an intermediate.

B. PHYSICAL AND CHEMICAL PROPERTIES OF ALDEHYDES

Various types of aldehydes are listed in Tables 1, 2, 3, 4, and 5. They vary greatly; some lower members are gases and the higher aromatic materials are high-melting solids. In general, however, the majority of the aliphatic aldehydes are liquids, usually with characteristic odors.

A useful reference to the properties of various derivatives of aldehydes and other organic compounds has been published recently.[2] In the lower members of the saturated aliphatic series, the boiling points are about 30 to 40°C. lower than those of the corresponding alcohols. Many of these aldehydes have low flash points.

Another important characteristic is the tendency of aldehydes to polymerize, with the result that frequently they are usable only in the presence of an inhibitor. In some cases they may react violently with oxygen; for example, acetaldehyde can produce explosive reactions.

C. CHEMICAL REACTIONS OF ALDEHYDES

These cannot be reviewed in any detail here since they are too numerous and are well covered in organic chemistry textbooks.[3] It may be of some use, however, in understanding the variety of possible applications of aldehydes to mention briefly some of the more important reactions.

1. Reduction to alcohols or oxidation to acids

2. Polymerization reactions; for example, formaldehyde and acetaldehyde are converted to paraformaldehyde and paraldehyde, respectively

3. Addition of sodium bisulfite

$$\overset{\text{H}}{\underset{\text{(excess)}}{\text{RC}=\text{O}}} + \text{NaHSO}_3 \rightarrow \overset{\text{H}}{\underset{\overset{\diagdown}{\text{SO}_3\text{Na}}}{\text{RC}-\text{OH}}}$$

[2] *Handbook of Chemistry and Physics*. Supplement Tables for identification of organic compounds), Chemical Rubber Publishing Co., Cleveland, 1960.

[3] L. Fieser and M. Fieser, *Organic Chemistry*. 3rd ed., Reinhold, New York, 1956.

4. Formation of acetals

$$R\text{—CHO} + R'\text{OH} \rightarrow R\text{—CH}\underset{\underset{\text{Hemiacetal}}{\text{OH}}}{\overset{\text{OR}'}{\diagup\diagdown}} + R'\text{—OH} \rightarrow R\text{—CH}\underset{\text{OR}'}{\overset{\text{OR}'}{\diagup\diagdown}}$$

5. Formation of secondary alcohols with the Grignard reagent

$$\text{R-Mg-X} + R'\text{—CHO} \rightarrow R'\text{—CHOH—R}$$

6. Condensation with active methylene groups; for example, ethyl malonate
7. Aldol condensation to form unsaturated aldehydes

$$2RCH_2CHO \xrightarrow{\text{HO}^-} RCH_2CHOHCHRCHO \xrightarrow{-H_2O} RCH_2CH\!=\!CRCHO$$

8. Reaction with hydroxylamine to form **oximes**

$$RCHO + H_2NOH \rightarrow RCH\!=\!NOH + H_2O$$

9. Reaction with hydrazines to form hydrazones

$$R\text{—CHO} + H_2N\text{—}NH_2 \rightarrow RCH\!=\!N\text{—}NH_2 + H_2O$$

10. Reaction with amines to form Schiff bases

$$R\text{—CHO} + H_2NR' \xrightarrow{-H_2O} RCH\!=\!NR'$$

11. Cannizzaro reactions

$$2C_6H_5CHO \xrightarrow{\text{OH}^-} C_6H_5OH + C_6H_5COOH$$

Nearly all of the wide variety of reactions listed above are important in large-scale production as well as in the chemical laboratory.

D. SYNTHESIS OF ALDEHYDES

Aldehydes are synthesized by a variety of methods but one of the oldest is the oxidation of alcohols, from which process the name aldehyde is said to be derived (*alcohol dehydr*ogenated). In this manner methyl and ethyl alcohols can be converted to formaldehyde and acetaldehyde, respectively. The oxidation of natural gas can also be used as a means of synthesizing formaldehyde. Acetaldehyde can be prepared by reaction of water with acetylene.

One of the more recent, important methods of producing aldehydes is based on the hydroformylation of alkenes.[3]

$$RCH\!=\!CH_2 + CO + H_2 \xrightarrow[\text{pressure}]{\text{metal catalyst}} RCH_2CH_2CHO + \underset{\overset{|}{CHO}}{RCH\text{—}CH_3}$$

A wide variety of aliphatic aldehydes can be produced by this reaction from simple raw materials such as natural gas, propane, steam, and hydrogen.

Two of the important unsaturated aldehydes are acrolein, which can be produced by the dehydration of glycerol, and crotonaldehyde, which can be formed through an aldol condensation of two molecules of acetaldehyde. Aromatic aldehydes can be produced from coal-tar sources.

E. METHODS FOR DETERMINATION OF ALDEHYDES IN AIR

While the reactivity of aldehydes makes it possible to devise many methods for their determination, only a few have been widely used. One of the most common methods and possibly one of the most satisfactory is that based on the reaction of the aldehyde with bisulfite forming the addition product, and the determination of the excess bisulfite by iodometric processes.[4] Another method is that of using the color developed with the Schiff reagent.[4] Formaldehyde can be determined with considerable sensitivity by the purple color formed with chromotropic acid.[5]

Unsaturated aldehydes such as acrolein can be determined by direct microiodometric titration or by colorimetry with Schiff reagent.[4] Very few methods for the determination of aromatic aldehydes in air appear to have been published. It seems likely, however, that many of the same reactions, depending on the properties of the aldehyde group, would be useful. In view of the high absorption of the aromatic ring in ultraviolet light, it would seem probable that methods based on this principle would be very useful and sensitive.

The problems in the analysis of the aldehydes are concerned with their intense chemical reactivity and tendency to polymerize. In addition, physiologically significant concentrations are often so low that methods of high sensitivity and freedom from interference are necessary. Methods applicable to aldehydes in air-pollution work have been described by Jacobs[6] and in the Air Pollution Manual of the American Industrial Hygiene Association.[7] It seems certain that more modern techniques such as gas chromatography, infrared spectrometry, and polarography will be useful in the determination of aldehydes.

F. PHYSIOLOGICAL EFFECTS OF ALDEHYDES

There have been few comprehensive reviews of the toxicity of aldehydes. The literature on formaldehyde up to 1945 was thoroughly reviewed by the U. S. Public Health Service.[8] Skog has studied the toxicity of several lower aliphatic

---

[4] M. B. Jacobs, *The Analytical Chemistry of Industrial Poisons, Hazards, and Solvents,* Vol. I. 2nd ed., Interscience Publishers, New York, 1949.

[5] C. E. Bricker and H. R. Johnson, *Ind. Eng. Chem. Anal. Ed.,* **17,** 400 (1945).

[6] M. B. Jacobs, *Chemical Analysis of Air Pollutants.* Interscience Publishers, New York–London, 1960 (Vol. X of *The Analytical Chemistry of Industrial Poisons, Hazards, and Solvents*).

[7] *Air Pollution Manual* (Part I, Evaluation). American Industrial Hygiene Assoc., Detroit, 1960.

[8] *Formaldehyde—Its Toxicity and Potential Dangers.* Suppl. No. **181,** Public Health Repts., U. S. Government Printing Office, Washington, D. C., 1945.

aldehydes.[9,10] Salem and Cullumbine[11] and Pattle and Cullumbine[12] have studied certain saturated and unsaturated aldehydes in relation to air-pollution problems. The literature pertaining to the role of aldehydes in air pollution has been covered up to 1957 in an annotated bibliography prepared by the Kettering Laboratory of the University of Cincinnati.[13]

In spite of the wide use of aldehydes, comprehensive studies of the health of individuals exposed to known atmospheric concentrations are lacking. There have been almost no long-term studies in experimental animals. Much of the available information deals with the effects of single, acute exposures; this has been summarized in the tables on physiological properties. Industrial experience has not indicated the presence of serious cumulative effects, although primary irritant reactions and contact dermatitis are occasionally seen.

There appear to be about four basic types of effects of aldehydes:

1. *Primary Irritation of the Skin, Eyes, and Mucosa of the Respiratory Tract.* While this property is possessed to some degree by nearly all of the aldehydes, it is most characteristic and important in those with lower molecular weights, those with unsaturation in the aliphatic chain, or in some cases halogenated substituents. The general and parenteral toxicities of these molecules appear to be related primarily to the irritant properties, although in some cases, such as fluoroacetaldehyde or fluorobutanal, their metabolic conversion to the corresponding fluorinated acids gives them an extraordinarily high degree of toxicity. The irritant properties of the dialdehydes have not been intensively studied, but in some instances concentrated solutions can be severe irritants to the skin or especially to the eyes. The acetals and the aromatic aldehydes in general have a lower degree of primary irritant action although there are some exceptions. Furfural has irritant properties but it is not nearly as active as acrolein or formaldehyde.

2. *Sensitization.* Sensitization is well known to occur on contact with liquid solutions of formaldehyde and probably has occurred with various other aldehydes. A direct sensitization to vapor appears to be relatively rare. Sensitization to addition products such as the bisulfites almost never occurs. Because of the bifunctional nature of the dialdehydes, they should theoretically be capable of acting as sensitizers but thus far there have been few reports of this occurring in actual handling.

Sensitization to the unsaturated aldehydes may occur, but it is usually very difficult to separate the primary irritant and sensitizing components of the action. Sensitization to the acetals and aromatic aldehydes appears to be infrequent. Sensitization to resins and polymers containing formaldehyde may occur, but it ap-

---

[9] E. Skog, *Acta Pharmacol. Toxicol.*, **6**, 299 (1950).

[10] E. Skog, *Acta. Pharmacol. Toxicol.*, **8**, 275 (1952).

[11] H. Salem and H. Cullumbine, *Toxicol. Appl. Pharmacol.*, **2**, 183 (1960).

[12] R. E. Pattle and H. Cullumbine, *Brit. Med. J.*, **11**, 913 Suppl. (1956).

[13] *Annotated Bibliography—The Effects of Atmospheric Pollution on the Health of Man, 1957.* The Kettering Laboratory, Cincinnati, 1957.

pears rare and may be associated with the presence of excess formaldehyde, or with some breakdown process releasing formaldehyde. Sensitization of the pulmonary tract or asthmatic-like symptoms are rarely caused by the inhalation of aldehydes.

3. *Anesthesia.* Two materials that have unquestionable anesthetic properties are chloral hydrate and paraldehyde. The former may act because of its conversion to trichloroethanol in the body, and the latter by depolymerization to acetaldehyde. Although when administered experimentally a number of the aliphatic aldehydes will give anesthetic-like symptoms in large parenteral or oral doses, in industrial exposures this action is overshadowed by the primary irritant action, which prevents voluntary inhalation of any significant quantities. The small quantities that can be tolerated by inhalation are usually so rapidly metabolized that no anesthetic symptoms occur. Some nausea, vomiting, headache, and weakness have been reported by Wilkinson[14] in chemists exposed to high concentrations of isovaleraldehyde (2-methyl butyraldehyde). None of these, however, could be classed as a definite anesthetic reaction.

4. *Organic Pathology.* The principal pathology experimentally produced in animals exposed to aldehyde vapors is that of damage to the respiratory tract and pulmonary edema. Multiple hemorrhages and alveolar exudate may be present, although these effects are usually much less dramatic than with gases such as phosgene. The effects produced with ketene, acrolein, crotonaldehyde, chloracetaldehyde, etc., are, however, much more pronounced and similar to those of phosgene, chlorine, etc. High dosages of materials such as methylal and furfural have been reported to cause various changes in the liver, kidneys, and central nervous system, but there appears to have been no confirmation of this type of action in human industrial exposures. In general, the aldehydes are remarkably free of actions that lead to definite cumulative organic damage to tissues (other than those which may be associated with primary irritation or sensitization).

G. METABOLISM OF ALDEHYDES

The metabolism of aldehydes has recently been reviewed by Williams.[15] It seems clear that the simple aliphatic aldehydes are readily handled by the same general mechanisms that have been reviewed in the case of organic acids. Both formaldehyde and acetaldehyde are present in normal metabolism and the importance of acetaldehyde in the action of alcohol is well known. The fact that the general toxicity of aldehydes appears to decrease with increasing molecular weight is probably related to the primary irritant action of the lower molecular weight substances, which makes them appear to be more potent. The use of the "tagging" with fluorine[16] (see also fluoroacetic acid, Chapter 40) added conclusive proof of

[14] J. F. Wilkinson, *J. Hyg.,* **40,** 555 (1940).

[15] R. T. Williams, Detoxication Mechanisms. 2nd ed., Wiley, New York, 1959.

[16] F. L. M. Pattison, *Toxic Aliphatic Fluorine Compounds.* Elsevier, Princeton, N. J., 1959.

the conversion of aldehydes into acids. The same general alternation of properties occurs, that is, the even-numbered omega-substituted fluoroaldehydes are very toxic, while the odd-numbered ones are much less toxic.

A number of interesting studies have been done with paraldehyde, demonstrating the depolymerization to acetaldehyde. It has been shown, for example, that about 80 per cent of the dose is destroyed and about 20 per cent is eliminated via the lungs, with only minute traces excreted in the urine. This depolymerization of paraldehyde probably occurs in the liver, since dogs in which liver damage was produced by chloroform showed prolongation of hypnotic action and an increase in the excretion through the lungs. Paraldehyde can also furnish the two-carbon fragment necessary for acetylation of materials such as sulfanilamide.

The metabolism of chloral hydrate has been studied by a number of investigators, and it is known to be converted to trichloroethanol and to trichloroacetic acid in both dogs and man. Some of the alcohol derivative is excreted as the glucuronide. Bromal hydrate is metabolized in a different manner and is more toxic.[15]

The aromatic aldehydes, such as benzaldehyde, are oxidized to the corresponding acids. This probably occurs at a relatively slow rate in the liver, but it is usually complete except where substituents such as hydroxy groups make it capable of being excreted by alternate metabolic pathways such as sulfate or glucuronic-acid conjugations on the hydroxy group.[15]

The rapid metabolism of the aliphatic and aromatic aldehydes by normal pathways undoubtedly accounts for the apparent safety of the large number of these substances that are ingested by humans and animals as a result of their natural occurrence in foods or from their addition as flavoring agents.

While the mechanism of the primary irritant action with tissues has not been explained in detail, it probably is associated with the reactivity of aldehydes with proteins and amino acids. For example, with aliphatic aldehydes, methylol or hydroxy methyl derivatives may be formed from reaction with amino groups.[17]

$$R-NH_2 + HCHO \rightarrow R-NHCH_2OH$$

$$R-NH_2 + 2HCHO \rightarrow R-N(CH_2OH)_2$$

Through subsequent reactions cyclic compounds may be formed, such as tricarboxymethyl trimethylene triamine from the reaction of glycine with formaldehyde.

In the case of the aromatic aldehydes, for example, benzaldehyde, the products of reaction with amino groups appear to be Schiff bases ($C_6H_5CH=N-R$). Various types of cross-linking reactions can also occur with either the mono- or dialdehydes resulting in alteration in protein structures. A discussion of the reaction of formaldehyde with glycine and other amino acids has been given by French and Edsall.[18]

Dixon[19a] has discussed the lachrymatory effect of unsaturated materials with

[17] J. Fruton and S. Simmonds, *General Biochemistry*. 2nd ed., Wiley, New York, 1958.

[18] F. V. French and J. T. Edsall, *Advances in Protein Chem.*, **2**, 278 (1945).

[19] (a) M. Dixon, *Biochemical Society Symposia* No. 2. Cambridge Univ. Press, Cambridge, England, 1948; (b) M. Dixon and E. C. Webb, *Enzymes*, Academic Press, New York, 1958.

TABLE 1
Physical and Chemical Properties of Saturated Aliphatic Aldehydes

Compound	Formula	Molecular weight	Specific gravity, g./ml.	Melting point, °C.	Boiling point, °C.	Vapor pressure, mm. Hg (°C.)	Vapor density (air = 1)	Flash point, °F.	Solubility in H_2O (g. per 100 ml.)	Conversion factors 1 mg./liter (p.p.m.)	Conversion factors 1 p.p.m (mg./cu. meter)
Formaldehyde (gas)	HCHO	30.0	0.815—20°C.	-92	-19.5	10 (-88.0)	1.075	572 (ignit. temp.) 7.0-73% (exp. limits by vol.)	Very soluble	815	1.2
Formaldehyde (37% in H_2O with 0-15% methanol)			1.075-1.081		98 (approx.)			122 (with 15% methanol)	Very soluble		
Sodium formaldehyde bisulfite	$CH_2O \cdot HSO_2Na \cdot H_2O$	152.11							Very soluble		
Sodium formaldehyde sulfoxalate	$CH_2O \cdot HSO_2Na \cdot 2H_2O$	154.13		65 (approx.)	Decomposes				Soluble		
Paraformaldehyde	$(CHO)_x$	$(30)_x$		25 (solid)	Decomposes			158	Slightly soluble, decomposes		
Acetaldehyde	CH_3CHO	44.05	0.788 (16/4)	-123.5	21	740 (20)	1.52	-36 4.1-55% (exp. limits by vol.)	Soluble	556	1.8
Acetaldehyde sodium bisulfite	$CH_3CHO \cdot NaHSO_2$	148.12									
Fluoroacetaldehyde	FCH_2CHO	62.05			64-65		2.14			394	2.5
Hydroxy acetaldehyde (glycolaldehyde)	$HOCH_2CHO$	60.05	1.366 (100)	95-97			2.1		Very soluble	408	2.6
Chloroacetaldehyde (40% in water)	$ClCH_2CHO$	78.5	1.190	-16.3	90-100		2.7	190	Very soluble	309	3.2
Trichloroacetaldehyde (chloral)	CCl_3CHO	147.4	1.512	-57.5	98	35 (20)	5.1	167	Soluble (forms hydrate)	166	6.0
Paraldehyde	$(C_2H_4O)_3$	132.2	0.9943 (20/4)	12.6	124 (752 mm. Hg.)		2.0	111.2	10.5		
Propionaldehyde (propanal)	CH_3CH_2CHO	58.1	0.807	-81	48.8	687 (45) 300 (25, approx.)	2.0	15-19	20	422	2.3
n-Butyraldehyde	$CH_3(CH_2)_2CHO$	72.1	0.817	-99	76	92 (20)	2.48	15	4.0	340	2.9
Isobutyraldehyde	$(CH_3)_2CHCHO$	72.1	0.7938	-66	62	170 (20, approx.)	2.48	-11	11	340	2.9

Continuation of preceding table (physical properties):

Compound	Formula	Mol. wt.	Sp. gr.	F.p.	B.p.				Solubility		
3-Hydroxybutanal (aldol)	$CH_3CHOHCH_2CHO$	88.1	1.098	<0	85 (decomposes)	21 (20)	3.0	181 (c.c.)	Very soluble	278	3.6
4-Fluorobutanal	$F(CH_2)_3CHO$	76.1				50 (49)	2.4		Slightly soluble	322	3.1
n-Valeraldehyde (pentanal)	$CH_3(CH_2)_3CHO$	86.1	0.819	−91.5	103	50 (25, approx.)	3.0		Slightly soluble	284	3.5
Isovaleraldehyde	$(CH_3)_2CHCH_2CHO$	86.1	0.7845	−51	92.5	50 (25, approx.)	3.0		Slightly soluble	284	3.5
2-Methylbutyraldehyde	$CH_3CH_2CH(CH_3)_2CHO$	86.1				50 (25, approx.)	3.0			284	3.5
n-Hexaldehyde (caproaldehyde, hexanal)	$CH_3(CH_2)_4CHO$	100.2	0.8335	−56	128	10 (20)	3.5	90	Insoluble	245	4.1
2-Ethylbutyraldehyde (α-ethylbutyraldehyde)	$(C_2H_5)_2CHCHO$	100.2	0.818	−89	116 (80–135)	14 (20)	3.5	70	0.3	245	4.1
n-Heptylaldehyde (heptanal, enanthaldehyde)	$CH_3(CH_2)_5CHO$	114.2	0.850	−45	154	3 (25)	3.9		Slightly soluble	215	4.6
2-Ethylhexylaldehyde (2-Ethylcaproaldehyde, 2-ethylhexanal)	$CH_3(CH_2)_3(C_2H_5)CHO$	128.2		<−100	163 (155–185)	2 (22)	4.4	125	0.04	191	5.2

TABLE 2

Physiological Properties of Saturated Aliphatic Aldehydes

Compound	LD₅₀ Species	LD₅₀ Route	LD₅₀ g./kg.	Skin LD₅₀ Species	Skin LD₅₀ ml. or g./kg.	Inhalation Species	Inhalation p.p.m.	Inhalation Time, hrs.	Inhalation Mortality	Irritant Species	Irritant Skin	Irritant Eye
Formaldehyde	Rat[a]	Oral	0.8			Rat[b]	250	4	LC₅₀	Rabbit[c]	Moderate	Severe
	Guinea pig	Oral	0.26			Rat[d]	815	0.5	LC₅₀	Man	Moderate	
	Rat[d]	S.C.	0.42			Cat[e]	650	8	LC₅₀ approx.			
	Rat[f]	Oral	0.1–0.2			Cat	200	3.5	All survived			
Sodium formaldehyde bisulfite[f]	Mouse	Oral	3.2–6.4	Guinea pig	>20							
	Mouse	I.P.	1.6–3.2									
Sodium formaldehyde sulfoxylate[g]	Mouse[g]	I.P.	>0.5									
	Rat[h]	Oral	>1.0									
	Rat[h]	I.V.	>2.0									
Paraformaldehyde[f]	Rat	Oral	>1.6			Rat[i]	Sat. vap.	2 min.	LC₁₀₀	Man		
Acetaldehyde	Rat[i]	Oral	1.93			Rat	16,000	4	0/6			Moderate
	Rat[d]	S.C.	0.64			Rat[d]	20,000	30 min.	LC₅₀			
						Cat[j]	13,600	0.25	1/1			

TABLE 2 (continued)

Compound	LD50 Species	LD50 Route	LD50 g./kg.	Skin LD50 Species	Skin LD50 ml. or g./kg.	Inhalation Species	Inhalation p.p.m.	Inhalation Time, hrs.	Inhalation Mortality	Irritant Species	Irritant Skin	Irritant Eye
Acetaldehyde sodium bisulfite[f]	Rat	Oral	>3.2	Guinea pig	>20	Cat[j]	4,100	3–5	0/1	Guinea pig	Slight	
	Guinea pig	Oral	1.6–3.2			Cat[j]	256	5	0/1			
Fluoroacetaldehyde[k]	Mouse	I.P.	0.006									
Glycolaldehyde[h] (hydroxyacetaldehyde)	Rabbit	S.C.	4.0									
Chloroacetaldehyde[f]	Rat	Oral	0.05–0.4	Guinea pig	0.1–1.0					Guinea pig	Severe	Severe
Trichloroacetaldehyde[f] (chloral)	Rat	Oral	0.05–0.4	Guinea pig	1.0–10					Guinea pig	Severe	Severe
Paraldehyde[h]	Rat	Oral	1.65									
	Rabbit	Oral	5.0									
	Dog	Oral	3.0–4.0									
Propionaldehyde[d, f, i]	Rat	Oral	1.4	Rabbit	5.0	Rat	8,000	4	5/6	Rabbit		Severe
	Rat	Oral	0.8–1.6	Guinea pig	10–20	Rat	60,000	0.3	3/3	Guinea pig	Severe	
	Rat	S.C.	0.8			Rat	26,000	0.5	LC50			
Ethoxypropionaldehyde[l]	Rat	Oral	0.9	Rabbit	1.0	Rat	500	4	6/6	Rabbit	Slight	Severe
α,β-Dichloropropionaldehyde[i]	Rat	Oral	0.16	Rabbit	0.078	Rat	Conc. vap.	2 min.	6/6	Rabbit	Severe	Severe
n-Butyraldehyde[d, f, i]	Rat	Oral	5.9	Guinea pig	>20	Rat	16	4	4/6	Guinea pig	Moderate	Severe
	Rat	I.P.	0.8			Rat	8,000	4	1/6	Rabbit		
	Rat	S.C.	10.0			Rat	60,000	0.5	LC50			
Isobutyraldehyde[f, m]	Rat	Oral	3.7	Rabbit	7.1	Rat	8,000	4	1/6	Rabbit	Slight	Severe
	Rat	I.P.	1.6–3.2	Guinea pig	>20					Guinea pig	Moderate	
	Rat	I.P.	1.6–3.2									
β-Hydroxybutyraldehyde[n] (aldol, acetaldol)	Rat	Oral	2.2	Rabbit	10	Rat	4,000	4	2/6	Rabbit	Slight	Slight
				Rabbit	0.14	Rat	Sat. vap.	0.5	No deaths			
4-Fluorobutanal[k]	Mouse	I.P	0.002									
n-Valeraldehyde[f]	Mouse	Oral	6.4–12.8	Guinea pig	>20	Rat	48,000	1.2	3/3	Guinea pig	Severe	Severe
	Rat	Oral	3.2–6.4			Rat	1,400	6	0/3	Rabbit		
Isovaleraldehyde[f]	Rat	Oral	>3.2	Guinea pig	>10					Guinea pig	Moderate	
				Guinea pig	>20							
2-Methylbutyraldehyde[f]	Mouse	Oral	3.2–6.4			Rat	67,000	0.3	3/3	Guinea pig	Moderate	Moderate
	Rat	Oral	6.4–12.8			Rat	3,800	6.0	0/3	Rabbit	Moderate	Severe
						Rat	1,013	6.0	0/3			
n-Hexaldehyde[m] (hexanal)	Rat	Oral	4.9			Rat	Conc. vap.	1	0/6	Rabbit	Slight	Slight
						Rat	2,000	4	1/6			
2-Ethylbutyraldehyde[m]	Rat	Oral	3.98			Rat	Conc. vap.	5 min.	0/6	Rabbit	Slight	Slight
						Rat	8,000	4	5/6			

Compound	Species	Route					
n-Heptaldehyde[o, p]	Mouse	Oral	25				
	Mouse	I.P.	>0.5				
2-Ethylhexylaldehyde[f, i] (α-ethylcaproaldehyde)	Rat	Oral	3.73				
	Rat	Oral	>3.2				
	Rabbit		5.04				
	Guinea pig		>20				
	Rat			25,000	13 min.	3/3	
	Rat			4,000	4	1/6	
	Rat			2,000	23 min.	3/3	
	Rat			145	6	0/3	
	Rabbit			Moderate			Slight
	Rabbit			Moderate			
	Guinea pig			Moderate			Moderate

[a] H. F. Smyth, Jr., J. Seaton, and L. Fischer, *J. Ind. Hyg. Toxicol.*, **23**, 259 (1941).

[b] C. P. Carpenter, H. F. Smyth, Jr., and U. C. Pozzani, *J. Ind. Hyg. Toxicol.*, **31**, 343 (1949).

[c] C. P. Carpenter and H. F. Smyth, Jr., *Am. J. Ophthalmol.*, **29**, 1363 (1946).

[d] E. Skog, *Acta. Pharmacol. Toxicol.*, **6**, 299 (1950).

[e] F. Flury and F. Zernik, *Schädliche Gase*. Springer, Berlin, 1931.

[f] D. W. Fassett, unpublished data, Laboratory of Industrial Medicine, Eastman Kodak Co., Rochester, N. Y.

[g] *Summary Tables of Biological Tests*, **8**, 743 (1956).

[h] W. Spector, ed., *Handbook of Toxicology*, Vol. I. Saunders, Philadelphia, 1956.

[i] H. F. Smyth, Jr., C. P. Carpenter, and C. S. Weil, *A.M.A. Arch. Ind. Health Occupational Med.*, **4**, 119 (1951).

[j] N. Iwanoff, *Arch. Hyg.*, **73**, 307 (1910–1911).

[k] F. L. M. Pattison, *Toxic Aliphatic Fluorine Compounds*. Elsevier, Princeton, N. J., 1959.

[l] H. F. Smyth, Jr., and C. P. Carpenter, *J. Ind. Hyg. Toxicol.*, **30**, 63 (1948).

[m] H. F. Smyth, Jr., C. P. Carpenter, C. S. Weil, and U. C. Pozzani, *A.M.A. Arch. Ind. Health Occupational Med.*, **10**, 61 (1954).

[n] H. F. Smyth, Jr., C. P. Carpenter, and C. S. Weil, *J. Ind. Hyg. Toxicol.*, **31**, 60 (1949).

[o] E. Boyland, *Biochem. J.*, **34**, 1196 (1940).

[p] *Summary Tables of Biological Tests*, **8**, 487 (1956).

adjacent aldehyde groups. There is some possibility that this type of effect may be an indication of an attack on the SH group in enzymes present in nerve endings. The mechanism of this effect has not been established with certainty; however, it is clear that the presence of unsaturation greatly increases the primary irritant activity of the aldehyde. In the same manner halogen substitution of aldehydes may greatly increase the local tissue irritation.

In the oxidation of aldehydes, combination with enzyme SH groups may be needed in the dehydrogenation process, [19b] that is,

$$\underset{\substack{|\\ \text{H}}}{R\text{C}}{=}\text{O} + \text{HSR}'_{\text{enz}} \rightarrow \underset{\substack{|\\ \text{OH}}}{\overset{\substack{\text{H}\\|}}{R\text{C}}}{-}\text{SR}'_{\text{enz}} \xrightarrow{-2\text{H}} \overset{\substack{\text{O}\\||}}{R\text{C}}{-}\text{SR}'_{\text{enz}} \rightarrow \overset{\substack{\text{O}\\||}}{R\text{C}}\text{OH} + \text{R}'_{\text{enz}}\text{SH}$$

II. Specific Compounds

A. SATURATED ALIPHATIC ALDEHYDES

The physical and chemical properties of the saturated aliphatic aldehydes are given in Table 1. Table 2 lists the physiological responses of animals to these compounds.

FORMALDEHYDE

Inhalation of formaldehyde by animals causes prompt and severe irritation of the eyes and respiratory tract. Skog[9] found edema and hemorrhages of the rat lung, and signs of hyperemia and perivascular edema in the liver and kidneys. Changes similar to these were reported by Salem and Cullumbine.[11] It is of interest that the same type of symptoms (although less intense) could be produced by subcutaneous injection. Most survivors recovered fully in 2 or 3 days. The LC_{50} for a 30-minute exposure is about 800 p.p.m. in the rat.

Relatively large quantities can be tolerated orally in animals and humans. Yonkman et al.[20] reported no significant toxic effects in humans from the daily ingestion of from 22 to 200 mg. over a period of 13 weeks. Higher doses caused moderate irritation of the upper digestive tract.

The characteristic effects in humans of inhalation of formaldehyde gas are as follows:[21] It can be detected by odor well below 1 p.p.m. by nearly all persons. No discomfort is noted until about 2 or 3 p.p.m., when a very mild tingling sensation in the eyes, nose, and posterior pharynx may be felt. Some tolerance occurs so that repeated 8-hour exposures at this level are possible. At 4 or 5 p.p.m., the discomfort increases rapidly and some mild lachrymation occurs in most people. This level can be tolerated fairly well for periods of perhaps 10 to 30

[20] F. F. Yonkman, A. J. Lehman, C. C. Pfeiffer, and H. F. Chase, J. Pharmacol. Exptl. Therap., **72**, 46 (1941).

[21] D. W. Fassett, personal observations, Laboratory of Industrial Medicine, Eastman Kodak Co., Rochester, N. Y.

minutes by some, but not by all people. Longer exposures at this level are decidedly unpleasant.

Concentrations of 10 p.p.m. are borne with difficulty for only a few minutes. Profuse lachrymation occurs in all people even if acclimated to lower levels. Between 10 and 20 p.p.m., it becomes difficult to take a normal inspiration voluntarily; the burning of the nose and throat becomes more severe and extends to the trachea. Coughing is noted. The lachrymation subsides promptly, but the nasal and respiratory irritation may persist for an hour or so after exposure ceases. The levels at which serious inflammation of the bronchi and lower respiratory tract would occur in man are unknown; but based on the above findings, it is likely that 5- or 10-minute exposures to levels of 50 to 100 p.p.m. might be expected to cause very serious injury.[22]

One of the more recent reports on the general health of individuals exposed to formaldehyde is that by Harris.[23] Clinical data are given on 25 men who were engaged in the manufacture of urea–formaldehyde and phenol–formaldehyde resins. The exposure periods varied from about 5 to 18 years and as a general rule the concentrations were well below 10 p.p.m. The only positive finding clearly related to the formaldehyde exposure was that of dermatitis in 4 of the men.

Harris also studied the occurrence of dermatitis in another group of 150 to 200 men in a factory manufacturing urea and formaldehyde resins, glues, molding powders, etc. He points out that three different types of clinical lesions are to be seen. In the first type, a worker may develop a sudden eczematous reaction of the face, neck, scrotum, flexor surfaces of the arms, eyelids, etc., which may come on only a few days after commencing work. Another eczematous type of reaction, which may not appear for a number of years, starts in the digital areas, back of the hands, wrists, forearms, and parts of the body that are exposed to friction from clothing. The third type of reaction may include a combination of the first two types. It is of interest that 77 per cent of the cases lost no time and were successfully treated while continuing on the job.

There have been many descriptions of formaldehyde dermatitis in the literature, but these generally are related to situations in which there is a direct contact with either liquid solutions or solid materials or resins containing free formaldehyde. Peck and Palitz[24] described some cases of sensitization to facial tissues containing urea–formaldehyde resins. Rostenberg et al.[25] described cases occurring in nurses handling thermometers that had been placed in 10 per cent formaldehyde for sterilization. Rostenberg[25] believes that it is probable that eczematous skin reactions to aldehydes will usually be quite specific for a given compound, but that urticarial reactions may be much less so.

[22] Personal observations—Laboratory of Industrial Medicine, Eastman Kodak Co., Rochester, N. Y., 1936–1960.

[23] D. K. Harris, *Brit. J. Ind. Med.*, **10**, 255 (1953).

[24] S. Peck and L. Palitz, *J. Am. Med. Assoc.*, **160**, 1226 (1956).

[25] A. Rostenberg, D. Bairstow, and T. W. Luther, *J. Invest. Dermatol.*, **19**, 459 (1952).

In the author's experience, skin sensitization to formaldehyde in the vapor state is very rare; no instances of authentic pulmonary sensitization have occurred. Individuals who have already developed an eczematous skin sensitization may, however, show a flareup of the skin reaction upon encountering formaldehyde in the vapor state. Rappeport and Hoffman[26] have reported a typical urticarial reaction in a patient exposed to formaldehyde for a period of 4 years in the course of research work in histology. This person proved to be sensitive to aldehyde groups in general and showed reactions to all aliphatic saturated aldehydes from C_1 to C_{18} and also acrolein. He was negative to crotonaldehyde, furfural, and aromatic aldehydes.

The present threshold limit for formaldehyde is 5 p.p.m.[27] The level of 10 p.p.m. originally proposed some years ago is undoubtedly too high in that very few individuals could become acclimated to such levels. A level of 5 p.p.m. will produce mild nose and throat irritation and some lachrymation in unacclimated persons. Acclimation to the lachrymatory effect occurs in some, but not all, people. Based on studies of relatively large numbers of individuals over long periods of time, it can be stated that levels of 5 p.p.m. will not be continuously tolerated by the average individual over an 8-hour workday. Levels of 2 or 3 p.p.m. will, however, be tolerable to the average individual with little signs of discomfort or lachrymation and with no effects on general health.[21]

All persons handling solids containing free formaldehyde, or concentrated solutions, or exposed to the gas, should be carefully instructed in the need for protective measures. Adequate ventilation should be provided to maintain concentrations well below 5 p.p.m. and preferably below 3 p.p.m. First-aid treatment for eye and skin splashes is the same as that for other chemical irritants; but in the case of eye splashes, extremely prompt action is required to prevent coagulation of the surface of the cornea. Careful consideration should be given to placement of individuals with previous skin, eye, or respiratory difficulties. Safety data sheets and Hygienic Guides are available on this chemical.[28,29]

PARAFORMALDEHYDE

Paraformaldehyde is a solid polymer of formaldehyde, which will depolymerize readily, especially upon heating or in the presence of alkalies or acids, liberating formaldehyde. It is a combustible solid and usually has a strong odor of formaldehyde.

[26] B. Z. Rappaport and M. M. Hoffman, *J. Am. Med. Assoc.*, **116**, 2656 (1941).

[27] American Conference of Governmental Industrial Hygienists, *Am. Ind. Hyg. Assoc. J.*, **22**, 325 (1961).

[28] *Chemical Safety Data Sheet SD-1, Formaldehyde.* Manufacturing Chemists Assoc., Washington, D. C., 1960.

[29] American Industrial Hygiene Association Hygienic Guide, Formaldehyde, *Am. Ind. Hyg. Assoc. Quart.*, **16**, 336 (1955).

Contact of the solid material with the skin or mucous membranes will cause the same type of symptoms as formaldehyde but the symptoms may not appear quite as rapidly. A chemical safety data sheet is available for paraformaldehyde.[30]

ACETALDEHYDE

Because of the explosive hazards of acetaldehyde, it is usually handled in industry under closed systems and exposures are not apt to be continuous or at high levels. It can be readily detected well below 50 p.p.m.[31] Some individuals can notice it below 25 p.p.m. At 50 p.p.m. a majority of volunteers exposed for 15 minutes showed some signs of eye irritation, and at 200 p.p.m. all subjects had red eyes and transient conjunctivitis. Eye irritation and, to a lesser extent, nose and throat irritation, are the only signs noted during usual industrial exposures. There appear to have been very few studies of men actually exposed while at work, although there have been many reports of the pharmacology of acetaldehyde in relation to the effects of alcohol. One interesting study is that by Asmussen, Hald, and Larsen.[32] Human subjects were given intravenous infusions to raise the blood level to 0.2 to 0.7 mg.% (about 10 times normal). At these levels an increase in heart rate and respiratory ventilation occurs, and a "hangover" sensation is noted.

Skog's studies with rats show that the LC_{50} is about 20,000 p.p.m. for a 30-minute exposure. At these levels, the animals develop pronounced excitement, followed by an anestheticlike state after about 15 minutes; survivors recover rapidly. The principal finding at autopsy is pulmonary edema.[9]

Although the current threshold limit of 200 p.p.m. is probably sufficiently low to prevent serious acute lung injury,[33] there is little in the literature to confirm its validity for an 8-hour day. While there is no question but that anesthetic symptoms can be produced at very high exposure levels or by parenteral injection, it seems almost impossible for this to occur under working conditions because of the irritant character of these vapors. A chemical safety data sheet and a hygienic guide are available on acetaldehyde.[34,35]

BISULFITE AND HYDROSULFITE ADDITION PRODUCTS OF ALDEHYDES

Sodium formaldehyde bisulfite and sodium formaldehyde sulfoxylate are both solids and are used as reducing agents. Sodium formaldehyde sulfoxylate was used

[30] *Chemical Safety Data Sheet SD-6, Paraformaldehyde.* Manufacturing Chemists Assoc., Washington, D. C., 1960.

[31] L. Silverman, H. F. Schulte, and M. W. First, *J. Ind. Hyg. Toxicol.*, **28**, 262 (1946).

[32] E. Asmussen, J. Hald, and V. Larsen, *Acta Pharmacol.*, **4**, 311 (1948).

[33] H. F. Smyth, Jr., *Am. Ind. Hyg. Assoc. Quart.*, **17**, 129 (1956).

[34] *Chemical Safety Data Sheet SD-43, Acetaldehyde.* Manufacturing Chemists Assoc., Washington, D. C., 1952.

[35] American Industrial Hygiene Association Hygiene Guide, Acetaldehyde, *Am. Ind. Hyg. Assoc. Quart.*, **16**, 273 (1955).

at one time as an antidote for mercury poisoning, but was not particularly effective.

These addition products possess little of the irritating qualities of the aldehyde. Skin sensitization to these materials is rare, and they can be handled without unusual precautionary measures. They have a relatively low oral or parenteral toxicity in experimental animals. The same applies to acetaldehyde sodium bisulfite.

HIGHER ALIPHATIC ALDEHYDES

The higher aliphatic aldehydes (C_3 or higher) have received a certain amount of preliminary toxicological study, as indicated in Table 2, and in recent years there has been increasing industrial experience with these materials as a result of their availability from petro-chemical reactions. These aldehydes of higher molecule weight are characterized by lower general toxicity, particularly by oral administration. They appear to be relatively well tolerated by inhalation although local reactions on skin and eyes may still be quite pronounced. Sensitization may occur, but is not as troublesome as in the case of formaldehyde.

Skog[9] has noted the occurrence of anesthesia in rats at high levels of inhalation for propionaldehyde and butyraldehyde. The survivors recovered promptly. Autopsies showed principally evidence of bronchial and alveolar inflammation. While subcutaneous injection of rats with large doses of butyraldehyde caused hemoglobinuria, this was not seen in rats on inhalation.

Salem and Cullumbine[11] have studied the acute toxicity of aldehydes in mice, guinea pigs, and rabbits. These studies included propionaldehyde, butyraldehyde, isobutyraldehyde, and n- and isovaleraldehyde. All animals exposed to high levels by inhalation developed pulmonary edema as a cause of death. The toxicity decreases with increasing chain length of the aldehyde. Branched-chain aldehydes are similar to straight-chain aldehydes in potency and type of effect.

There have been a few clinical reports of individuals exposed to higher aldehydes. Wilkinson[14] reported an instance in which several chemists engaged in distilling isovaleraldehyde (2-methyl butyraldehyde) developed some signs of chest discomfort, nausea, vomiting, and headaches. Exposure levels were not measured although the odor was very pronounced, and it is possible that fairly high levels were present. None of these cases was severely affected, and all recovered in a few days without any particular after effects.

A chemical safety data sheet[36] is available on the properties and safe handling of normal and isobutyraldehydes. The precautions for the higher aldehydes are essentially those for most other reactive organic compounds, and should include adequate ventilation in areas where high exposures are expected, fire and explosion precautions, and proper instruction of employees in use of respiratory, eye, and

[36] *Chemical Safety Data Sheet SD-78, Butyraldehydes.* Manufacturing Chemists Assoc., Washington, D. C., 1960.

skin protection. No cumulative effects have been reported nor would any be expected because of the conversion of these compounds to normal metabolites.

HALOGENATED AND OTHER SUBSTITUTED ALDEHYDES

As has been mentioned previously, halogenation generally greatly increases the local irritant action and general toxicity of aldehydes. Chloroacetaldehyde is one of the more widely used halogenated derivatives and has a tentative threshold limit of 1 p.p.m. (3 mg./cu. meter).[27] While no detailed studies of exposed persons could be found, the animal data in Table 2 would indicate that it possesses a rather high degree of acute toxicity and severe primary irritant action on skin. Fluoro-acetaldehyde and fluorobutanal are included in this discussion primarily because of the interest in their conversion to fluoroacetic acid. In the recent monograph by Pattison,[16] a series of fluorinated aldehydes was synthesized and studied for toxic effects. As expected by current theories on fatty-acid metabolism, the even-numbered compounds were extremely toxic while the odd-numbered possessed a relatively low degree of toxicity.

Trichloroacetaldehyde (chloral) has a rather high degree of local irritant action and general toxicity in experimental animals.[37] Dichloropropionaldehyde also is highly toxic and produces severe local irritant reactions.[38]

Substitution of a hydroxy group, such as in glycol aldehyde (hydroxyacetaldehyde) or β-hydroxybutyraldehyde (acetaldol) does not confer any toxic properties on the molecules. On the other hand, ethoxypropionaldehyde appears to have a considerably greater toxicity than propionaldehyde, especially by inhalation.

Since halogen and possibly other types of substituents may markedly increase the toxicity, the use of these compounds requires very close control of exposures and careful examination of exposed persons.

B. UNSATURATED ALIPHATIC ALDEHYDES

A number of examples of industrially important compounds of this class are listed in Tables 3 and 4. Three of these are of considerable industrial importance (ketene, acrolein, and crotonaldehyde), and the nature of their acute irritant effects is well known in both man and experimental animals. There have been, however, few studies of long-term exposures to lower, nonirritant concentrations. The presence of unsaturation greatly increases the acute physiological effects as compared to the corresponding saturated compounds. This is well illustrated by considering the approximate lethal concentrations in the rat for certain pairs of saturated and unsaturated aldehydes, as shown in Table 5.

[37] D. W. Fassett, unpublished data, Laboratory of Industrial Medicine, Eastman Kodak Co., Rochester, N. Y.

[38] H. F. Smyth, Jr., C. P. Carpenter, and C. S. Weil, *A.M.A. Arch. Ind. Health Occupational Med.*, **4**, 119 (1951).

TABLE 3

Physical and Chemical Properties of Unsaturated Aliphatic Aldehydes

Compound	Formula	Molecular weight	Specific gravity, g./ml.	Melting point, °C.	Boiling point, °C.	Vapor pressure, mm. Hg. (°C.)	Vapor density (air = 1)	Flash point, °F.	Solubility in H₂O, (g. per 100 ml.)	Conversion factors	
										1 mg./liter (p.p.m.)	1 p.p.m. (mg./cu. meter)
Ketene	$CH_2=C=O$	42.04		−150	−56		1.45		Decomposes	582	1.7
Acrolein (2-propenal) (acrylic aldehyde)	$CH_2=CHCHO$	56.1	0.8389	−87	52.7	214 (20)	1.94	<0	22	437	2.3
Methacrylaldehyde (methacrolein)	$CH_2=C(CH_3)CHO$	70.1	0.837		68		2.4	5	6.4	349	2.8
2-Ethyl-3-propyl acrolein	$C_3H_7CH=C(C_2H_5)CHO$	126.19	0.8518		175	1 (20)	4.4	155	0.07	194	5.2
Crotonaldehyde (β-Methyl acrolein)	$CH_3CH=CHCHO$	70.1	0.852	−69	103 (99–104)	19 (20)	2.4	128	18	349	2.8
Methyl-β-ethyl acrolein	$C_2H_5CH=C(CH_3)CHO$	98.14	0.854		137		3.1		Insoluble	250	4.0
Mucochloric Acid	$CHOCCl=CClCOOH$	168.97		125–127					Slightly soluble (hot)		
Citral (geranial)	$(CH_3)_2C=CH(CH_2)_2C(CH_3)=C—CHO$ (CH₃ H)	152.23	0.8898		229 (decomposes)	5 (91–95)			Insoluble	161	6.2
Citronellal (rhodinal)	$(CH_3)_2C=CH(CH_2)_2CH(CH_3)CH_2CHO$	154.25	0.856		204	11 (88)			Slightly soluble	159	6.3

TABLE 4

Physiological Properties of Unsaturated Aliphatic Aldehydes

Compound	LD50 Species	Route	g./kg.	Skin LD50 Species	ml. or g./kg.	Inhalation Species	p.p.m.	Time, hrs.	Mortality	Irritant Species	Skin	Eye
Ketene[a,b,c]						Mouse[a]	120	3	16/20			
						Mouse	70	4	20/20			
						Rat	250	150 min.	4/4			
						Rat	120	10 days	0/4			
						Cat	370	8-12	1/2			
						Cat	230	15 days	0/1			
						Rat[b]	53	100 min.	2/2			
						Rat	1	7 × 14	0/1			
						Cat	23	4 × 2	0/2			
						Cat	1	7 × 55	0/1			
						Cat	1	7 × 55	0/2			
						Monkey	1	7 × 55	0/1			
Acrolein[d,e,f,g]	Rat	Oral	0.046	Rabbit	0.43	Rat[c]	200	5 min.	LC100	Rabbit	Severe	
						Rat[e]	8	4	1/6			
						Cat[f]	690-1150	2	3/3			
						Cat	18-92	3-4	0/2			
						Cat	11	3-10	0/2			
						Rat[d]	130	30 min.	LC50			
						Man[g]	150	10 min.	Fatal			
						Man	5	1 min.	Intolerable			
						Man	0.25	5 min.	moderate irritation			
Methacrylaldehyde[h] (methacrolein)	Rat	Oral	0.14	Rabbit		Rat	250	4	5/6	Rabbit	Slight	Severe
2-Ethyl-3-propyl acrolein[i]	Rat[i]	Oral	3.0	Guinea pig	>20	Rat	Conc. vap.	8	0/6	Rabbit	Slight	Severe
Crotonaldehyde[d,i,j] (β-methyl acrolein)	Rat[j]	Oral	0.3	Rabbit	0.38	Rat	Conc. vap.	1 min.	0/6	Rabbit	Slight	Severe
	Rat	Oral	0.16	Guinea pig	0.5-1.0	Rat	1,400	30 min.	LC50			
	Rat	I.P.	0.07	Rabbit	0.15-0.2							
	Rat[d]	S.C.	0.14									
Mucochloric acid[k]	Rat	Oral	0.05-0.1	Guinea pig	<5					Guinea pig	Severe	Severe
	Rat	I.P.	0.01-0.025									
Methyl-β-ethyl acrolein[l] (2-methyl-2-pentene-1-al)	Rat	Oral	4.3	Rabbit	4.5	Rat	2,000	4	3/6	Rabbit	Slight	Severe

[a] H. A. Wooster, C. C. Lushbaugh, and C. E. Redemann, J. Ind. Hyg. Toxicol. 29, 56 (1947).
[b] J. F. Treon, H. E. Sigmon, K. V. Kitzmiller, F. F. Heyroth, W. J. Younker, and J. Cholak, J. Ind. Hyg. Toxicol. 31, 209 (1949).
[c] G. R. Cameron and A. Neuberger, J. Pathol. Bacteriol. 45, 653 (1937).
[d] E. Skog, Acta Pharmacol. Toxicol. 6, 299 (1950).
[e] H. F. Smyth, Jr., C. P. Carpenter, and C. S. Weil, A.M.A. Arch. Ind. Health Occupational Med. 4, 119 (1951).
[f] N. Iwanoff, Arch. Hyg. 73, 307 (1910-1911).
[g] A. M. Prentiss, Chemicals in War, 1st ed., McGraw-Hill, New York, 1937.
[h] H. F. Smyth, Jr., C. P. Carpenter, and C. S. Weil, J. Ind. Hyg. Toxicol. 31, 60 (1949).
[i] H. F. Smyth, Jr., and C. P. Carpenter, J. Ind. Hyg. Toxicol. 26, 269 (1944).
[j] H. C. Hodge, unpublished data, Univ. of Rochester, Rochester, N. Y.
[k] D. W. Fassett, unpublished data, Laboratory of Industrial Medicine, Eastman Kodak Co., Rochester, N. Y.
[l] H. F. Smyth, Jr., C. P. Carpenter, C. S. Weil, and U. C. Pozzani, A.M.A. Arch. Ind. Health Occupational Med. 10, 61 (1954).

TABLE 5
Effect of Unsaturation on the Inhalation Toxicity of Aldehydes

Compound	Formula	LC_{50} (p.p.m.), rats	Time of exposure
Acetaldehyde	CH_3CHO	20,000	30 min.
Ketene	$CH_2\!=\!CHO$	130	30 min.
Propionaldehyde	CH_3CH_2CHO	26,000	30 min.
Acrolein	$CH_2\!=\!CHCHO$	130	30 min.
Isobutyraldehyde	$(CH_3)_2CHCHO$	>8,000	4 hr.
Methacrolein	$CH_2\!=\!C(CH_3)CHO$	250	4 hr.
n-Butyraldehyde	$CH_3(CH_2)_2CHO$	60,000	30 min.
Crotonaldehyde	$CH_3CH\!=\!CHCHO$	1,400	30 min.

An analogous situation was found in the case of unsaturated versus saturated esters (see Table 8, page 1879). In the case of aldehydes, however, the effects of unsaturation are more striking.

The symptoms in animals following inhalation of acrolein and crotonaldehyde have been described by Skog,[9] and were marked by the immediate onset of severe respiratory irritation, and evidence of primary irritation of the eyes and upper respiratory tract. In contrast to the corresponding saturated aldehydes, propionaldehyde and butyraldehyde, no anesthetic symptoms were noted following inhalation. Pathological changes were principally those of pulmonary edema and marked damage to the bronchial epithelium. These effects were less marked with crotonaldehyde than with acrolein. Acrolein is about 10 times as toxic as crotonaldehyde in terms of the quantities required to produce pulmonary edema and death in experimental animals.

Ketene has been studied by Cameron and Neuberger[39] in mice, rats, and guinea pigs. Concentrations of over 100 p.p.m. invariably proved fatal with death occurring rapidly with symptoms of marked eye, nose, throat, and pulmonary irritation. Autopsies showed that there was dilatation of the right heart and major damage to the alveolar walls with pulmonary edema. The epithelium of the bronchi and bronchioles, however, was not particularly affected. Similar findings were noted by Treon et al[40] and by Wooster et al.[41] following acute exposures; the conclusion was reached that ketene resembled phosgene closely in its ability to cause delayed pulmonary edema. Treon's studies also included repeated inhalations at 1 p.p.m. in the rat, cat, and monkey. On autopsy all animals showed evidence of lung damage, with varying degrees of interstitial fibrosis, cellular infiltration, atelectasis, and emphysema.

[39] G. R. Cameron, and A. Neuberger, *J. Pathol. Bacteriol.*, **45**, 653 (1937).

[40] J. F. Treon, H. E. Sigmon, K. V. Kitzmiller, F. F. Heyroth, W. J. Younker, and J. Cholak, *J. Ind. Hyg. Toxicol.*, **31**, 209 (1949).

[41] H. A. Wooster, C. C. Lushbaugh, C. E. Redemann, *J. Ind. Hyg. Toxicol.*, **29**, 56 (1947).

The effects of acrolein in man have been summarized by Prentiss.[42] Levels as low as 0.25 p.p.m. may cause some irritation, while a level of 1 p.p.m. is practically intolerable and is capable of causing lachrymation and marked eye, nose, and throat irritation within a period of 5 minutes.[43] Smyth[33] is of the opinion that the current threshold limit of 0.5 p.p.m.[27] can be interpreted from human sensory data, and that it is low enough to prevent pulmonary edema. In view of the fact that concentrations of 0.25 p.p.m. are moderately irritating and that studies of human tolerance to such levels are lacking, the accuracy of this value may be questioned.

It is unlikely that any chronic systemic effects will be found with either ketene, acrolein, or crotonaldehyde other than those that might be associated with primary irritation and local pulmonary effects. All these materials should be capable of being handled by the normal metabolic processes at levels below those that cause irritation to exposed tissues.

Mucochloric acid provides an example of the combined effect of halogen substitution and unsaturation on the properties of an aldehyde. This material is a solid that can produce severe burns of the skin or eye, and is also a potent skin sensitizer in man and experimental animals. It is capable of penetrating the intact skin of the guinea pig and can cause loss of weight and lowered hemoglobin and red-cell counts on repeated application to the skin of guinea pigs at relatively high dose levels.[37]

Citral and citronellal are examples of complex unsaturated aldehydes, which are components of natural flavors in lemon and orange oils. They are relatively mild irritants compared to the materials above. Although little is known about their toxicity in experimental animals, their metabolism has recently been reviewed by Williams.[15] Citral A (geranial) has two double bonds in the 2-3 and 6-7 positions. This material is converted in part to the so-called Hildebrandt acid in which a double omega oxidation has taken place.

$$CH_3(COOH)C=CHCH_2CH_2C(CH_3)=CHCOOH$$
Hildebrandt acid

$$(CH_3)_2C=CHCH_2CH_2C(CH_3)=CHCHO$$

$$CH_3(COOH)C=CHCH_2CH_2CH(CH_3)CH_2COOH$$
Reduced Hildebrandt acid

In the case of citronellal (rhodinal) the aldehyde is cyclized and converted to p-menthane-3, 8-diol, which is excreted as the glucuronide in the rabbit. It is possible that this cyclization of citronellal was not actually a metabolic reaction but simply a chemical one, which took place in the stomach in the presence of gastric hydrochloric acid. Williams[15] refers to a report by Leach and Lloyd,[44] which indicates that under some circumstances citral has caused damage to vascular endothelium of rabbits and monkeys. It was suggested that this might

[42] A. M. Prentiss, *Chemicals in War*. 1st ed., McGraw-Hill, New York. 1937.

[43] W. P. Yant, H. H. Schrenk, F. A. Patty, and R. R. Sayers, *U. S. Bur. Mines Repts. Invest.*, No. **3027** (1930).

[44] E. H. Leach, and J. P. F. Lloyd, *Proc. Nutrition Soc. Engl. and Scotland,* **15,** xv (1956).

have been the result of a competitive reaction with retinine (Vitamin A aldehyde). The significance of this report is unknown at the present time.

While there are other interesting correlations that might be made between chemical structure and primary irritant activity, it seems probable that the postulated mechanism of action given by Dixon[19a] is correct. The presence of an aldehyde group adjacent to a double bond has a polarizing effect on the latter, which then makes it capable of adding nucleophilic groups such as SH groups. If the SH groups in certain enzyme systems are attacked, it seems reasonable that this might be related to the physiological response. There are, however, certain difficulties with this explanation, as pointed out by Dixon.[19a] The lachrymatory action of such materials is usually very transient and ceases immediately upon removal of the irritant. He speculates that the nerve endings may respond to a change in the relative amount of the SH compound present, but further evidence seems necessary on this point. It is interesting that, if an exposure to a lachrymator of this type is sufficiently prolonged, a point is reached at which lachrymation no longer occurs, which might suggest a complete saturation of some reactive site.

The precautions for handling reactive unsaturated aldehydes such as ketene, acrolein, methacrolein, and crotonaldehyde should be the same as those for other highly active eye and pulmonary irritants, as, for example, phosgene. One manufacturer's bulletin gives a comprehensive discussion of the safe-handling precautions for acrolein.[45]

C. ALIPHATIC DIALDEHYDES

A number of dialdehydes have become available commercially (see Table 6); and while not all their properties are completely known, some toxicological data are summarized in Table 7. These materials have many of the same properties as the monoaldehydes, but because of their bifunctional nature may provide different types of useful cross-linking reactions. They tend to polymerize readily and are sometimes available only in a water solution in the presence of polymerization inhibitors.

Of these compounds glyoxal has probably had the most widespread use up to the present time. Henson[46] states that glyoxal vapors do not irritate the skin or mucous membranes. In the author's experience, glyoxal vapors are somewhat irritating to the eyes and nasal mucosa, although this may be less pronounced than in the case of formaldehyde. A 30 per cent solution in water produces severe irritation of guinea pig skin, and is quite toxic to rats ($LD_{50} = 0.2$ to 0.4 g./kg. orally and <100 mg./kg. intraperitoneally). Williams[15] discusses some evidence that glyoxal is the metabolite responsible for the toxic effects produced by large doses of ethylene glycol. Water solutions of succinaldehyde, glutaraldehyde, and 3-methyl glutaraldehyde also are relatively strong irritants to the skin or eyes.

[45] Industrial Hygiene Bulletin, *Acrolein, Its Chemistry and Applications.* Shell Chemical Corp., New York,

[46] E. V. Henson, *J. Occupational Med.*, **1**, 457 (1959).

TABLE 6
Physical and Chemical Properties of Aliphatic Dialdehydes

Compound	Formula	Molecular weight	Specific gravity, g./ml.	Melting point, °C.	Boiling point, °C.	Vapor pressure, mm. Hg (°C.)	Vapor density (air = 1)	Solubility in H_2O (g. per 100 ml.)	Conversion factors: 1 mg./liter (p.p.m.)	1 p.p.m. (mg./cu. meter)
Glyoxal (butanedial)	O=CH—CH=O	58.0	1.14	15	51	220 (20, approx.)	2.0	Very soluble	422	2.4
Succinaldehyde	O=CH(CH₂)CH=O	86.1	1.064		169	1 (25, approx.)	3.0	Soluble	284	3.5
Glutaraldehyde	O=CH(CH₂)₃CH=O	100.1			187		3.4		245	4.1
β-Methyl Glutaraldehyde	CH₃CH(CH₂CH=O)₂	114.1					4.0		215	4.7
Adipaldehyde (hexanedial)	CHO(CH₂)₄CHO	114.14				9 (92–94)	4.0	Slightly soluble	215	4.7

TABLE 7
Physiological Properties of Aliphatic Dialdehydes

Compound	LD₅₀ Species	Route	LD₅₀ g./kg.	Skin LD₅₀ Species	Skin LD₅₀ ml. or g./kg.	Inhalation Species	p.p.m.	Time, hrs.	Mortality	Irritant Species	Skin	Eye
Glyoxal (30% in H_2O)[a,b]	Rat	Oral	2.02	Guinea pig	5–10					Guinea pig	Severe	
	Rat	Oral	0.2–0.4	Rabbit	6.6							
	Guinea pig	Oral	0.76									
	Rat	I.P.	<0.1									
Succinaldehyde (25% in H_2O)[a,c]	Rat	Oral	0.3	Rabbit	1.0	Rat	Conc. vap. (ca. 15 mg./liter)	6	0/3	Rabbit		Severe
Hexa-2,4-dienal[d]	Rat	Oral	0.7	Rabbit	0.27	Rat	2000	4	1/6	Guinea pig	Severe	
Glutaraldehyde[c]	Rat	Oral	0.82	Rabbit	0.64					Rabbit	Severe	Severe
Glutaraldehyde bis-sodium bisulfite[a]	Rat	Oral	1.6–3.2	Guinea pig	>20					Rabbit	Moderate	Severe
	Mouse	Oral	1.6–3.2							Guinea pig	Moderate	
	Rat	I.P.	0.4–0.8									
3-Methyl glutaraldehyde[a,d]	Rat	Oral	0.78	Rabbit	0.3	Rat	Conc. vap.	8	0/6	Rabbit	Severe	Severe
	Rat	Oral	0.1–0.2	Guinea pig (4% in H_2O)	>20	Rat	Conc. vap.	6	0/3	Guinea pig	Moderate	Severe
	Rat	I.P.	0.005–0.010									
3-Methyl glutaraldehyde bis-sodium bisulfite[a]	Rat	Oral	16.–3.2	Guinea pig	>20					Guinea pig	Moderate	
	Mouse	Oral	0.8–1.6									
	Rat	I.P.	0.2–0.4									
α-Hydroxyadipaldehyde[d]	Rat	Oral	17	Rabbit	>20	Rat	Conc. vap.	8	0/6	Rabbit	Slight	Slight

[a] D. W. Fassett, unpublished data, Laboratory of Industrial Medicine, Eastman Kodak Co., Rochester, N. Y.
[b] H. F. Smyth, Jr., in G. Curme and F. Johnston, eds., Glycols, ACS Monograph No. 114. Reinhold, New York, 1952.
[c] H. F. Smyth, Jr., unpublished data, Mellon Institute, Pittsburgh, Pa.
[d] H. F. Smyth, Jr., C. P. Carpenter, C. S. Weil, and U. C. Pozzani, A.M.A. Arch. Ind. Health Occupational Med. 10, 61 (1954).

TABLE 8. Physical and Chemical Properties of Acetals

Compound	Formula	Molecular weight	Specific gravity g./ml.	Melting point °C.	Boiling point °C.	Vapor pressure, mm. Hg. (°C.)	Vapor density (air = 1)	Flash point, °F.	Solubility in H₂O, (g. per 100 ml.)	Conversion factors	
										1 mg./ liter (p.p.m.)	1 p.p.m. mg./cu. meter
Methylal (formal, dimethoxymethane)	CH₂(OCH₃)₂	76.1	0.8630	−104.8	43	400 (25, approx.)	2.6	0	33	322	3.1
Diethoxymethane (diethyl formal, ethylal)	CH₂(OC₂H₅)₂	104.2	0.824	−67	89	60 (25, approx.)	3.6		7.0	235	4.2
Dichloroethyl formal (bis-2-chloroethyl formal)	CH₂(OCH₂CH₂Cl)₂	173.0	1.234		218.1	0.1 (20)	6.0	230	0.78	141	7.1
1,1-Dimethoxyethane (dimethyl acetal)	CH₃CH(OCH₃)₂	90.1	0.850	−58	82–83	61 (20)	3.1	40	∞	272	3.7
Acetal (1,1-diethoxyethane, diethyl acetal)	CH₃CH(OC₂H₅)₂	118.2	0.825		107–112	20 (19.6)	4.1	−5 (o. c.) 1.6–10.4% Exp. limits	5.5	207	4.8
Chloroacetal (chloroacetaldehyde diethylacetal)	CH₂ClCH(OC₂H₅)₂	152.6	1.026		157	20 (62)	5.3		Slightly soluble	160	6.2
Acetaldehyde dibutyl acetal	CH₃CH(OC₄H₉)₂	174					6.0			141	7.1
Crotonaldehyde acetal		144.2					5.0			170	5.9
Chloral hydrate	CCl₃CH(OH)₂	165.4	1.9081	61–63	98 (decomposes)				Soluble		
Ketoacetal (4,4-dimethoxy-2-butanone)		132.09					4.5		Soluble	185	5.4

TABLE 9. Physiological Properties of Acetals

Compound	LD₅₀			Skin LD₅₀		Inhalation toxicity				Irritant effect		
	Species	Route	g./kg.	Species	ml. or g./kg.	Species	p.p.m.	Time, hrs.	Mortality	Species	Skin	Eye
Methylal[a] (formal, dimethoxymethane)	Guinea pig	S.C.	>5			Guinea pig Mouse Mouse	150,000 18,000 11,000	2 7 7 × 15	Fatal LC₅₀ 6/50	Guinea pig		Moderate
Ethylal[b] (diethoxymethane, diethylformal)	Rat	Oral	>3.2	Guinea pig	>10					Guinea pig		Slight
Dichloroethylformal[c]	Rat	Oral	0.065	Guinea pig	0.17					Rabbit	Slight	Slight
Dimethylacetal[d, e] (1,1-dimethoxyethane)	Rat	Oral	6.5	Rabbit	20	Rat[d] Rat Rat[d] Rat[e] Rat[e]	120 60 Conc. vapor 16,000 3,000	4 4 5 min. 4 4	6/6 0/6 0/6 3/6 LC₅₀ 0/6	Rabbit	Slight	Moderate
Acetal[d, f] (diethylacetal or 1,1-diethoxyethane)	Rat	Oral	4.6	Rabbit	10	Rat	Conc. vapor 3,000	5 min.	0/6	Rabbit	Slight	Slight
Dibutylacetal[g] (1,1-dibutoxy ethane, acetaldehyde dibutylacetal)	Rat Rat	I.P. Oral	0.9 8.8			Rat	4,000 Conc. vapor	4 8	2/6 0/6	Rabbit	Moderate	Slight

Compound	Animal	Route	LD50 (g/kg)	Inhal. species	Conc.	Form	No.	Deaths	Skin species	A	B	C	D
Di-2-ethylhexyl adipate	Rat[46]	Oral	9.1										
	Rat[87]	Oral	20–50										
	Rat	I.P.	>47										
	Mouse	I.P.	About 150	Rabbit	16.3	Sat. vap.	8	None		Slight	Slight	Slight	Slight
Di-2-hexyloxyethyl adipate	Rabbit	I.P.	>38.0										
Di-2-(2-ethylbutoxy)ethyl adipate	Rat[22]	Oral	4.3	Rabbit	12.3							Slight	Slight
Didecyl adipate	Rat[22]	Oral	3.25	Rabbit	4.24	Sat. vap.	8	None				Moderate	Slight
	Rat[33]	Oral	>25.6	Guinea pig	>10.0	Sat. vap.	6	0/3	Guinea pig	Slight	Slight		
	Mouse[33]	Oral	12.8–25.6										
	Rat	I.P.	>25.0										
	Mouse[33]	I.P.	>25.0										
Dibutyl azelate	Mouse[33]	Oral	>12.8	Guinea pig	>10.0				Guinea pig	Slight	None		
	Mouse	I.P.	>12.8										
Di-2-ethylhexyl azelate	Rat[33]	Oral	6.4–12.8	Guinea pig	>10.0	Sat. vap.	6	0/3		None	None		
	Mouse	Oral	>25.6										
	Rat	I.P.	>25.0										
	Mouse	I.P.	>25.0										
Dibutyl sebacate	Rat[69]	Oral	16–32										
Di-2-ethylhexyl sebacate	Rat[33]	Oral	12.8–25.6	Guinea pig	>10.0	Sat. vap.	6	0/3		None	None		
	Mouse	Oral	12.8–25.6										
	Rat	I.P.	>25.0										
	Mouse	I.P.	>25.0										
Triethyl citrate	Rat[88]	Oral	7.0	Guinea pig	>10.0	3500	6	2/3		None	None		
	Rat	Oral	3.2–6.4			1700	6	0/3					
	Cat[88]	Oral	.3.5			1300	6	3/3					
						1700	6	0/6					
						(guinea pig)							
Acetyl triethyl citrate	Rat[88]	Oral	7.0	Guinea pig	>10.0					None			
	Rat[33]	Oral	3.2–6.4										
	Mouse	Oral	6.4–12.8										
	Cat[88]	Oral	7.0										
Tributyl citrate	Rat[88]	Oral	>30 ml./kg.										
	Cat	Oral	>50 ml./kg.										
Acetyl tributyl citrate	Rat[88]	Ora	>30 ml./kg.										
	Rat[33]	Oral	>25 ml./kg.										
	Cat[88]	Oral	>50 ml./kg.										
Isopropyl citrates (71% mono with rest di and tri)	Rat[69]	Oral	3.6 (water solution)										
	Rat	Oral	>20.0 (oil solution)										
	Dog	Oral	>2.25										
Stearyl citrates (75% di with rest mono and tri)	Rat[69]	Oral	>5.4 (oil solution)										
	Dog	Oral	>5.0										

86 *The Merck Index of Chemicals and Drugs.* 6th ed, Merck & Co., Rahway, N. J., 1952.   87 H. C. Hodge, University of Rochester, N. Y., unpublished data.   88 M. Finklestein and H. Gold, *Toxicol. Appl. Pharmacol.,* **1,** 283 (1959).   89 H. J. Deuel, S. N. Greenberg, C. E. Valvert, R. Baker, and H. R. Fischer, *Food Research,* **16,** 258 (1951).

Their lower vapor pressures, however, may make them less hazardous from the point of view of inhalation under comparable conditions of use.

While the extent of precautions needed under various types of industrial use is not known with certainty, it would appear reasonable to handle these dialdehydes with the same general precautions as used for formaldehyde and other lower molecular weight monoaldehydes.

D. ACETALS OR KETALS

The acetals or ketals are produced by reactions of aldehydes with alcohols. They are coming into increasing industrial use and may be utilized as solvents, chemical intermediates, plasticizers, or they may be used to generate aldehydes in the presence of acid. These materials have some of the properties of ethers and are stable under neutral or slightly alkaline conditions, but hydrolyze readily in the presence of acids to generate aldehydes (see Table 8). This latter reaction makes them capable of hardening natural adhesives, such as glue or casein.[47]

The hazards of the use of acetals in industry are not known with certainty, but a number of them have received a certain amount of experimental study, some of which is summarized in Table 9. The physiological properties of the simple unsubstituted acetals are characterized by an etherlike anesthetic action, and by a relatively low degree of primary irritation compared to the parent aldehyde. A number of these have been studied for their anesthetic properties although at present they are not used for this purpose. Bacq and Dallemagne[48,49] and Knoefel[50-52] have investigated methylal, acetal, and a number of other similar materials. The toxicity of methylal has recently been investigated by Weaver et al.[53] These authors reviewed some of the older literature and pointed out that a number of attempts have been made to use methylal as an anesthetic, and that Bacq and Dallemagne[48,49] had investigated this intensively, both in dogs and in humans. Apparently, anesthesia could be produced in humans, but the onset was slower than with ether and the effect more transitory.

The experiments of Weaver et al.[53] were concerned principally with the effects of inhalation of various concentrations on guinea pigs and mice. At extremely high levels, 153,000 p.p.m., anesthesia occurred in 20 minutes, and death occurred in about 2 hours. At these levels, definite evidence of irritation was noted in the guinea pig, including squinting, lachrymation, sneezing, and nasal discharge. Other pronounced signs of eye and respiratory-tract irritation were also noted at lower

[47] *The Physical Properties of Synthetic Organic Chemicals.* Carbide and Carbon Chemicals Corp., F-6136, P-18212, New York,

[48] Z. M. Bacq, and M. J. Dallemagne, *Arch intern. pharmacodynamie,* **69,** 127 (1943).

[49] Z. M. Bacq, and M. J. Dallemagne, *Arch. intern. pharmacodynamie,* **69,** 235 (1943).

[50] P. Knoefel, *J. Pharmacol.,* **50,** 88 (1934).

[51] P. Knoefel, *J. Pharmacol.,* **53,** 440 (1935).

[52] P. Knoefel, *J. Pharmacol.,* **55,** 235 (1935).

[53] F. L. Weaver, A. R. Hough, B. Highman, and L. T. Fairhall, *Brit. J. Ind. Med.,* **8,** 279 (1951).

levels, and the LC_{50} in mice for a 7-hour exposure was found to be about 18,000 p.p.m. Most of the deaths occurred during the course of exposure.

Experiments were also carried out with repeated inhalations in the case of mice. A group of 50 mice received fifteen 7-hour exposures at concentrations of approximately 11,000 p.p.m. Only minor irritation was noted at this level, although lack of coordination appeared after about 3 or 4 hours of exposure. Recovery was usually complete in 1 hour after removal from the chamber. Six deaths occurred in the 50 animals during the 22-day exposure period. Repetition of these experiments at 14,000 p.p.m. showed more evidence of irritation and a greater degree of anesthesia. About 30 per cent of the group of mice succumbed during a 17-day exposure.

Attempts were made to determine the metabolism of methylal in these animals by testing for formaldehyde and formic acid in vitreous humor and urine. No evidences of these metabolic products were found. However, in view of the rather marked irritation occurring during inhalation and the necrosis following subcutaneous injections in guinea pigs, it would seem possible that hydrolysis to formaldehyde does take place. This is readily metabolized so that it would be difficult to detect it under these conditions.

Histopathological studies were made on the guinea pigs and mice exposed by inhalation. In the case of guinea pigs exposed to very high levels, the animals sacrificed 16 to 74 hours after the beginning of exposure showed moderate to severe fatty degeneration of the liver and kidney, and an extensive bronchopneumonia. Other guinea pigs sacrificed 23 hours after three successive 7-hour exposures showed similar changes in the lungs, liver, and kidneys. However, guinea pigs exposed to five daily 7-hour inhalations at levels of about 45,000 p.p.m. showed no significant changes. Mice having about 15 seven-hour exposures at levels up to 14,000 p.p.m. showed occasional evidence of pulmonary edema and slight fatty changes in the kidney.

No changes were found in the optic nerves or retinas of mice that could be attributed to methylal. Occasionally some corneal blebs were seen, but these could not be attributed with certainty to the methylal exposure.

These authors conclude that the threshold for production of toxic effects in guinea pigs and mice is of the order of 11,000 p.p.m. They extrapolate from this to the conclusion that 1000 p.p.m. might be safe for an 8-hour working day. The present threshold limit is 1000 p.p.m.[27] No studies of workers exposed to such concentrations over long periods of time have been reported, and the validity of this level would appear to be uncertain at the present time.

Safe-handling precautions should include the use of adequate ventilation to be certain that the average concentrations are well below 1000 p.p.m., and avoidance of excessive or prolonged skin contact. Methylal should be handled with due regard for its flammable properties.

Ethylal produced only minor symptoms of weakness in the rat; even at high dose levels no typical anesthesia was noted.[37] It is of interest that the halogenated compound, dichloroethyl formal, possessed a high degree of toxicity in the rat orally or in the guinea pig by skin contact. It also was highly potent upon inhalation in the rat, giving 100 per cent fatalities with levels as low as 120 p.p.m. It was stated to be only a slight skin or eye irritant in the rabbit.[54] It is obvious that this halogenated material should be handled with considerable care.

Dimethylacetal in animal experiments seems to be somewhat similar to methylal. The pathological effects have not been reported, however.[21,55]

Acetal also appears to have anesthetic properties.[56] Knoefel[50] believes that it is probably rapidly hydrolyzed in the stomach. Hydrolysis would give rise to either a hemiacetal or to acetaldehyde and ethyl alcohol.

The introduction of a halogen as in chloroacetal also greatly increased toxic properties upon oral administration in rats. (Note the similarity to the toxicity increase in the case of dichloroethylformal.) This suggests the hydrolysis of this compound to give rise to chloracetaldehyde and ethanol.[37] The influence of unsaturation on an acetal is indicated by the high degree of intraperitoneal toxicity in the mouse for crotonaldehyde acetal.[57]

Ketoacetal is interesting in that the presence of the keto group in the beta position to the acetal grouping did not appear to enhance the toxicity.[37] While not enough compounds have been studied to predict the effect of unsaturation on aldehyde groups adjacent to acetal groups, from the fragmentary data available the same principles would apply as mentioned in the case of aldehydes.

In the view of the lack of specific information, it would seem well to regard substituted acetals as potentially capable of being hydrolyzed to the component alcohols and aldehydes, and to take precautions to avoid excessive skin contact or inhalation.

E. AROMATIC AND HETEROCYCLIC ALDEHYDES

The physical and biological properties of a number of aromatic and heterocyclic aldehydes are summarized in Tables 10 and 11. A number of these aldehydes occur naturally as components of essential oils or plant products. They are widely used in perfumes and as flavoring agents.

While there have been few studies of their toxicity, a considerable amount of work has been devoted to a study of their metabolic fate, and this has been summarized by Williams.[15]

The metabolism of these aromatic aldehydes follows the pattern established for aromatic acids. In general the aldehyde grouping in compounds such as benzaldehyde is converted to the acid, probably by liver aldehyde dehydrogenases. While this may occur at a relatively slow rate, it is usually complete. If the

[54] H. F. Smyth, Jr., C. P. Carpenter, and C. S. Weil, *J. Ind. Hyg. Toxicol.,* **30**, 63 (1948).

[55] C. W. LaBelle and H. Brieger, *A.M.A. Arch. Ind. Health,* **12**, 623 (1955).

[56] E. Browning, *Toxicity of Industrial Organic Solvents.* Chemical Publishing Co., New York, 1953.

[57] Sloan-Kettering Institute Screening Data, New York,

TABLE 11. Physiological Properties of Aromatic and Heterocyclic Aldehydes

Compound	LD_{50} Species	Route	g./kg.	Skin LD_{50} Species	ml. or g./kg.	Irritant effect Species	Skin	Eye
Benzaldehyde[a]	Rabbit	S.C.	5.0					
Salicylaldehyde (2-hydroxybenzaldehyde)[a]	Rat	S.C.	0.9					
p-Aminobenzaldehyde[a]	Mouse	I.P.	0.9					
p-Acetamidobenzaldehyde[b]	Rat	Oral	>3.2	Guinea pig	>1.0	Guinea pig	Slight	
p-Dimethylaminobenzaldehyde[b]	Mouse	Oral	0.8–1.6					
	Mouse	I.P.	0.2–0.4					
p-n-Propylbenzaldehyde[a]	Rat	Oral	4.2					
	Mouse	Oral	1.8					
p-(n-Propoxy)benzaldehyde[a,c]	Rat	Oral	1.6	Rabbit	9.0			
	Mouse	Oral	1.8					
m-Nitrobenzaldehyde[b,d]	Rat	Oral	0.05–0.4	Guinea pig	>1.0	Guinea pig	Moderate	
	Mouse	I.P.	>0.5					
2,4-Dihydroxybenzaldehyde[b]	Rat	Oral	0.4–3.2	Guinea pig	>1.0	Guinea pig	Slight	
	Mouse	I.P.	>0.5					
2,5-Dimethoxybenzaldehyde[e]	Rat	Oral	0.8–1.6					
	Rat	I.P.	0.1–0.2					
2-Hydroxy-5-chlorobenzaldehyde[b]	Rat	Oral	0.8–1.6	Guinea pig	10–20	Guinea pig	Moderate	
2-Hydroxy-5-bromobenzaldehyde[b]	Rat	Oral	0.8–1.6	Guinea pig	>20	Guinea pig	Slight	
	Rat	I.P.	0.2–0.4					
p-Hexoxybenzaldehyde[f]	Mouse	I.P.	>0.5					
Piperonal[f]	Mouse	I.P.	>0.5					
Cinnamaldehyde[g]	Mouse	I.P.	0.2					
Furfural[b,h]	Rat	Oral	0.05–0.1	Guinea pig	<10	Guinea pig	Slight	
	Rat	I.P.	0.02–0.05					
	Mouse	Oral	0.5					
	Dog	Oral	0.65					

a W. Spector, ed., *Handbook of Toxicology*, Vol. I. Saunders, Philadelphia, 1956.
b D. W. Fassett, unpublished data, Laboratory of Industrial Medicine, Eastman Kodak Co., Rochester, N. Y.
c J. H. Draize, E. Alvarez, and M. F. Whitesell, *J. Pharmacol. Exptl. Therap.*, **93**, 26 (1948).
d *Summary Tables of Biological Tests*, **6**, 144 (1954).
e *Summary Tables of Biological Tests*, **6**, 215 (1954).
f Sloan-Kettering Institute Screening Data, New York, N. Y.
g *Summary Tab'es of Biological Tests*, **7**, 687 (1955).
h E. Boyland, *Biochem. J.*, **34**, 1196 (1940).

aromatic ring contains phenolic groups, as in the case of p-hydroxybenzaldehyde, the compound may be excreted partially as the glucuronide and partially as the free acid or as the conjugated acid. Reduction of the aromatic aldehyde group to an alcohol has not been observed. While benzaldehyde itself is not excreted in any appreciable amounts as an ester glucuronide, some substituted aldehydes, such as veratraldehyde (3,4-dimethoxybenzaldehyde), are oxidized to the corresponding acid and excreted as ester glucuronides. Nitrobenzaldehydes are oxidized to nitro-benzoic acids, which are either excreted as such or in conjugation with hippuric acid or as acetamido benzoic acids. p-Dimethylaminobenzaldehyde may undergo partial demethylation.

The literature references to the aromatic aldehydes do not give many details of the type of toxic reactions found. It may therefore be of interest to mention the few that have been noticed in the course of screening tests.[37]

p-Acetamidobenzaldehyde: the only symptoms noted were those of moderate weakness in rats receiving up to 3200 mg./kg. orally.

p-Dimethylaminobenzaldehyde: weakness, ataxia, unconsciousness, and trem-ors were noted in mice receiving up to 1600 mg./kg. orally or up to 400 mg./kg. intraperitoneally. Repeated intraperitoneal injection in mice at levels of 100 to 200 mg./kg. caused weakness and ataxia, but there was no significant reduction of hemoglobin during such treatment. No difference was found between the pure and technical grade samples.

p-Nitrobenzaldehyde: in doses of 50 or 400 mg./kg. orally in the rat, the symptoms were those of prostration and cyanosis.

2,4-Dihydroxybenzaldehyde: rats receiving 50 to 3200 mg./kg. orally showed weakness, tremors, and violent convulsions.

2-Hydroxyl-5-chlorobenzaldehyde: the symptoms in mice (either orally or intraperitoneally) were those of weakness, ataxia, gasping respirations, and uncon-sciousness. Skin irritation was more marked with this compound than with the corresponding 5-bromo compound.

2-Hydroxy-5-bromobenzaldehyde: oral or intraperitoneal administration in mice or rats caused weakness, ataxia, and unconsciousness.

Furfural is extensively used in industry for the solvent refining of lubricating oils, resins, and other organic materials. It is also used in connection with rubber manufacturing; it is a constituent of some insecticidal preparations and rubber cements, and is widely used as a reagent. The details of its chemical reactions can be found in the book by Dunlop and Peters.[58] This book also has a section that deals with the physiological properties of several of the furanes, including furfural, furfural alcohol, and tetrahydrofurfural alcohol. The toxicity of furfural has also been published in the American Petroleum Institute Toxicological Review series.[59] Furfural can be derived from pentosans present in straws and bran by hydrolysis

[58] A. P. Dunlop and F. N. Peters, *Furans*. Reinhold, New York, 1953.

[59] API Toxicological Review, *Furfural*. American Petroleum Institute, Dept. of Safety, New York, 1948.

and dehydration with sulfuric acid. While the vapor is quite irritating, it has relatively low volatility (1 mm. Hg at 19°C.), so that the inhalation of toxicologically significant quantities is unlikely. It is a relatively strong skin irritant and is said to be capable of producing a dermatitis in man. The hazard of handling furfural resins is similar to that of other aldehyde resins.[60]

In animals, furfural has a relatively high degree of toxicity upon oral or intraperitoneal administration. The oral LD_{50} in the rat is between 50 and 100 mg./kg. and the intraperitoneal between 20 and 50 mg./kg.; the symptoms following oral or intraperitoneal administration[37] appear to be those of weakness, ataxia, and unconsciousness. No signs of central nervous system stimulation were seen, such as those reported in the American Petroleum Institute review of the toxicity in dogs, cats, and rabbits. The inhalation exposure of cats to very high levels of furfural (2800 p.p.m.) for half an hour resulted in death due to pulmonary edema.[59]

A report by Korenman and Resnik[61] states that levels of 1.9 to 14 p.p.m. caused complaints of eye and throat irritation and headache.

The metabolism of furfural has recently been reviewed by Williams.[15] The aldehyde group is converted to an acid and this in turn is conjugated with glycine. In dogs and rabbits, furoic acid, furoyl glycine, and furfuracryluric acid are excreted. Not much is known about the metabolism of furfural derivatives, although 5-nitrofurfural is excreted in rats as 5-nitro-2-furoic acid.

While no studies on the health of workmen exposed over long periods of time have been published, Dunlop and Peters[58] state that other than an occasional allergic skin manifestation no injury due to exposure to furfural has been reported. The current threshold limit of 5 p.p.m.[27] is based on its primary irritant property at this concentration.

The precautions in handling furfural should include adequate ventilation and provision of skin and eye protection. In case of contact with skin or eyes, the chemical should be removed promptly by washing with copious amounts of water. While no unusual medical examinations are indicated, it would seem well to consider carefully the placement of individuals who had shown previous tendencies toward contact dermatitis.

[60] L. Schwartz, L. Tulipan, and D. J. Birmingham, *Occupational Diseases of the Skin*. 3rd ed., Lea & Febiger, Philadelphia, 1957.

[61] J. Korenman and J. B. Resnik, *Arch. Hyg.*, **104**, 344 (1930).

CHAPTER XLIV

Cyanides and Nitriles

DAVID W. FASSETT, M.D.

I. General Considerations

While cyanides are among the most toxic of all industrial chemicals, they are produced in large quantities, used in many different applications, but yet are the cause of few serious accidents or deaths. This is due in part to the fact that the word cyanide is synonymous with a highly poisonous substance, and a certain amount of care in handling is thereby insured. The good record is also due in no small part to the fact that manufacturers have generally provided adequate labeling and instructions for safe handling and first-aid treatment.

It is essential that all personnel working with processes involving cyanides or nitriles be specially trained so that they are fully aware of the hazard and so that they will follow faithfully all rules laid down for safe handling. It is also essential that special training be given in the specific first-aid measures, and that adequate specific antidotes be available for first aid and for use by physicians.

A. SYMPTOMS PRODUCED BY CYANIDES AND NITRILES

For purposes of the toxicologist cyanides and nitriles can be thought of as belonging to three classes: (*1*) hydrogen cyanide, cyanogen, simple salts of hydrogen cyanide (such as sodium, potassium, calcium, and copper cyanide); (*2*) halogenated compounds such as cyanogen chloride or bromide; (*3*) nitriles, such as acetonitrile (methyl cyanide), acrylonitrile, and isobutyronitrile.

Hydrogen cyanide itself and its simple soluble salts, such as those in group *1*, are among the most rapidly acting of all known poisons. A few inhalations of higher concentrations of HCN vapor or the ingestion of amounts as low as 50–100 mg. of sodium or potassium cyanide may be followed by almost instantaneous collapse and cessation of respiration. At much lower dosages, the earliest symptoms may be simply those of weakness, headaches, confusion, and occasionally nausea and vomiting. The respiratory rate and depth will usually be increased at the beginning and at later stages become slow and gasping. Blood pressure is usually normal, especially in the mild or moderately severe cases, although the pulse rate is usually more rapid than normal. It is characteristic that the heart beat may continue for some time even after respirations have ceased. If cyanosis is present, it usually indicates that respiration has either ceased or been very inadequate for a few minutes.

The halogenated materials, cyanogen bromide and cyanogen chloride, are also highly toxic and possess some of the same properties as hydrogen cyanide and its soluble salts. However, at low concentrations, these materials behave more like the highly irritating vesicant gases, such as phosgene, and cause severe lachrymatory effects and both acute and delayed pulmonary irritation and pulmonary edema.

Nitriles such as acrylonitrile, isobutyronitrile, propionitrile, etc., can cause the same general symptoms as hydrogen cyanide, but the onset of symptoms is apt to be slower. They are also apt to be more active as primary irritants on the skin or eye, and they are frequently absorbed rapidly and completely through the intact skin. Skin absorption, however, is not absent with hydrogen cyanide nor even its soluble salts, and this may be a prominent factor in preventing recovery unless all the material is removed from the skin.

While acetonitrile has long been known to have specific action on the thyroid gland, it is generally considered to be much less toxic than other nitriles. However, it recently has been demonstrated by Amdur[1] and Pozzani et al.[2] that inhalation of relatively high concentrations of acetonitrile may produce a severe delayed reaction with some features not typical of hydrogen cyanide intoxication.

A number of other related materials, such as cyanamide, calcium cyanamide, cyanates, isocyanates, isonitriles, thiocyanates, ferri- and ferrocyanides, and cyanoacetates, do not have all the typical toxic properties of cyanides and nitriles and may act by different mechanisms. Most of these (a notable exception being certain isocyanates) are of a somewhat lower order of toxicity, although there is great variability in their activity.

B. MODE OF ACTION AND METABOLISM OF NITRILES

Cyanide exerts its typical dramatic effect because it is capable of inactivating certain biological heavy metal catalysts by forming very stable complexes with the metal in these catalysts. Cytochrome oxidase is probably the most important of these since it occupies a fundamental position in the respiratory process and is involved in the ultimate electron transfer to molecular oxygen. Since cytochrome oxidase is present in practically all cells that are functioning under aerobic conditions, and since the cyanide ion is very easily diffusable to all parts of the body, it is capable of suddenly bringing to a halt practically all cellular respiration.

There are probably a number of other materials, such as hydrogen sulfide and azides, that have similar properties of complexing with metals in these

[1] Marvin L. Amdur, *J. Occup. Med.,* **1,** 627 (1959).

[2] U. C. Pozzani, C. P. Carpenter, P. E. Palm, C. S. Weil, and J. H. Nair, *J. Occup. Med.,* **1,** 634 (1959).

biological catalysts. Dixon and Webb[3] call attention to the fact, however, that cyanide is a relatively nonspecific inhibitor of enzyme systems and list 42 enzyme reactions that it can inhibit. Many, but not all of these, contain iron or copper as part of the enzyme molecule. Cytochrome oxidase is, however, the most sensitive of all these enzymes and actually will show 50 per cent inhibition at concentrations of cyanide as low as 10^{-8} molar. Cyanide is thought to be capable of acting by a number of different mechanisms in addition to combining with the essential metal in the enzyme itself. It may actually remove metals from the enzyme or it may combine with carbonyl groups in some prosthetic group of an enzyme or in the substrate of an enzyme; or it can act as a reducing agent and break up disulfide links.

Cyanide does not combine appreciably with either the oxidized or reduced form of hemoglobin in the blood, although it will combine with the 2 per cent or so of methemoglobin (in which the iron is in the Fe^{+++} form) normally present. In the cytochrome oxidase system the iron is in the oxidized form and combines in an especially stable manner with cyanide.

Thus, while there are a number of enzyme actions with cyanide, there seems little doubt but that the most sensitive and fundamental action is that of an inhibition of cytochrome oxidase. This in turn prevents the oxidation of reduced cytochrome C, thus stopping the utilization of molecular oxygen by cells.[3]

The most specific pathological finding in acute cases is the bright red color of venous blood. This is striking, visible evidence of the inability of the tissue cells to utilize oxygen, as a result of which the venous blood is only about one volume per cent lower in oxygen content than arterial blood—in contrast to the usual A-V difference of 4–5 volumes per cent. As recovery takes place, the A-V oxygen difference returns to normal.

In the case of aliphatic nitriles, it is probable that hydrolysis takes place with the release of the CN group. The rapidity with which this process occurs may determine the relative toxicity of the various aliphatic nitriles.[4] The fate of the remaining portion of aliphatic nitriles is unknown, but possibly it is converted to an organic acid.

In the case of aromatic nitriles, Williams[4] says that when the CN group is directly attached to the aromatic ring it probably is hydrolyzed to a carboxy group and ammonia. More recent work indicates that cyanobenzene is slowly excreted, principally as phenolic conjugates, and only about 10 per cent as benzoic acid.[4a] In cases where the CN group is separated from the ring by CH_2 groups, it is split off as a CN group, which presumably would exert a cyanide-type effect and be converted to SCN. Again, the rapidity with which such a process occurs would determine the extent of cyanide-like activity.

[3] M. Dixon and E. C. Webb, *Enzymes*, Academic Press, New York, 1958.
[4] R. T. Williams, *Detoxication Mechanisms*, 2nd ed., Wiley, New York.
[4a] J. N. Smith and R. T. Williams, *Biochem. J.*, **46**, 243 (1950).

If the concentration of cyanide ion is not so great as to cause death, then it is gradually released from its combination with the ferric iron of cytochrome oxidase or methemoglobin, converted to thiocyanate ion (SCN^-), and excreted in the urine.

It has been known since the publication of Lang[5] that a widely distributed enzyme known as rhodanese (transulfurase) would enable thiosulfate ions to react with cyanide ions to form sulfite and thiocyanate ions

$$S_2O_3^{2-} + CN^- \rightarrow SO_3^{2-} + SCN$$

This enzyme has recently been crystallized from beef liver by Sörbo.[6] The urine normally contains only small amounts of thiosulfate. Thiosulfate is formed by enzymes present in the liver and kidney as part of the metabolism of 1-cysteine,[7] which may account for the ability of the body to tolerate limited amounts of cyanide without any apparent toxic symptoms. While thiocyanate is capable of producing some toxic symptoms in relatively large dosages, it is not as acutely hazardous as cyanide and is readily excreted in the urine.

A number of years ago, Voegtlin[8] suggested that preliminary intravenous treatment with cystine might protect against the toxic action of cyanide. Wood and Cooley[9] present evidence that cyanide may also react with sulfur amino acids, such as cystine, and form 2-imino-4-thiazolidine carboxylic acid, which is then excreted in the urine. The quantitative importance of this process is unknown at the present time, and it seems probable that the thiocyanate detoxication mechanism is more important.

Serum thiocyanate may be elevated in the case of acute poisoning. Confusion may occur, however, if the individual is a smoker since this also can give rise to increased thiocyanate levels. Cyanide also can be determined directly in the blood.[10,10a] While blood cyanide and thiocyanate levels are useful in following acute poisoning, there is no clear indication that they are of value in following the exposure of individuals under more normal working conditions.

C. TREATMENT OF POISONING BY CYANIDES AND NITRILES

While specific and effective antidotes for poisonous substances are very rare, fortunately they are available for some cyanides and nitriles. The effectiveness of the antidote derives from two basic facts: (*a*) Methemoglobin binds extremely firmly with any free cyanide ion. Therefore, if additional methemoglobin can be

[5] K. Lang, *Biochem. Z.*, **259**, 243 (1933).

[6] B. Sörbo, *Acta. Chem. Scand.*, **7**, 1129, 1137 (1953).

[7] B. Sörbo, *Biochem. et Biophys. Acta*, **24**, 324 (1957).

[8] C. Voegtlin, J. M. Johnson, and H. A. Dyer, *J. Pharmacol. Exptl. Therap.*, **27**, 467 (1956).

[9] J. L. Wood and S. L. Cooley, *J. Biol. Chem.*, **218**, 1 (1956).

[10] M. Feldstein and N. C. Klendshoj, *J. Lab. Clin. Med.*, **44**, 166 (1954).

[10a] R. B. Bruce, J. W. Howard, and R. F. Hanzal, *Anal. Chem.*, **27**, 1346 (1955).

formed in the blood, this will trap circulating cyanide ions that are either being continually absorbed from the stomach, through the skin or possibly returning to the blood stream from the tissues. Since the formation of 10 or 20 per cent methemoglobin usually involves no great risk, this can provide a large amount of cyanide-binding substance. (b) Since the amounts of thiosulfate formed in the body are relatively limited, the introduction of excess thiosulfate ions appears to increase the activity of the transulfurase enzyme and thus increase the rate of conversion of cyanide to the less toxic thiocyanate.

The combined use of methemoglobin-forming agents plus thiosulfate is capable of protecting experimental animals against at least 20 lethal doses of cyanide. A detailed account of this therapy can be found in reviews by Chen, Rose, and Clowes,[11] Wolfsie and Shaffer,[12] and Wolfsie.[13] In practice the antidotes are administered by the combined use of artificial respiration (where necessary) and the simultaneous inhalation of amyl nitrite vapor from ampules crushed in a handkerchief and held by an assistant close to the nose of the victim. Several ampules may be used in the course of the first half hour. This procedure alone may suffice for some of the milder cases provided the sources of absorption are also removed (e.g., residual material on skin).

In more serious cases, if there is no response to the above, it may be necessary to produce methemoglobin in greater amounts and to use thiosulfate. This can be done by intravenous injection of 0.3 g. of sodium nitrite (10 ml. of a 3 per cent solution at a rate of 2.5–5 ml./min.), followed at once by 12.5 g. of sodium thiosulfate intravenously (50 ml. of a 25 per cent solution at the same rate as sodium nitrite). The sodium nitrile and thiosulfate therapy should be repeated in an hour at half the original dose if symptoms recur or persist. In milder cases, where the patient is conscious and not having much respiratory difficulty, recovery may occur without any specific therapy.

In order to use effectively such first-aid and medical therapy, it is necessary that all personnel and all physicians and nurses be thoroughly familiar with the toxic effects of cyanide and with the specific first-aid therapy. Because of the rapidity of onset, it is necessary to have first-aid kits in convenient locations in all areas where these materials are to be used.[12] The essential ingredients of each kit are:

2 boxes (2 dozen) of ampules each containing 5 minims (0.3 ml.) of amyl nitrite
2 ampules of sterile sodium nitrite solution (10 ml. of a 3 per cent solution in each)
2 ampules of sterile sodium thiosulfate solution (50 ml. of a 25 per cent solution in each)
1 10-ml. and 1 50-ml. sterile glass syringe with sterile intravenous needles
1 ampule file
1 tourniquet
12 gauze pads

[11] K. K. Chen, C. L. Rose, and G. H. A. Clowes, *J. Indiana Med. Assoc.*, **37**, 344 (1944).
[12] J. H. Wolfsie and B. C. Shaffer, *J. Occup. Med.*, **1**, 281 (1959).
[13] J. H. Wolfsie, *Arch. Ind. Hyg. Occup. Med.*, **4**, 1 (1951).

1 small bottle of 70 per cent alcohol
1 stomach tube
2 1-pint (473 ml.) bottles of 1 per cent sodium thiosulfate solution

Specific training of employees in artificial respiration and in the use of amyl nitrite ampules is extremely important.

D. CHRONIC POISONING FROM CYANIDE EXPOSURE

Hardy et al.[14] and Wolfsie and Shaffer[12] discuss the possibility of chronic poisoning from cyanide exposure. Hardy suggests that in some individuals the thiocyanate excretion may be inadequate and that increased thiocyanate levels may produce either goiters or symptoms of thiocyanate intoxication, or both.

One of her cases, however, came from an area of known endemic goiter so that there may have been other factors present. If chronic poisoning does occur, it appears to be relatively rare. Further studies on this problem appear to be needed. It is of interest in this connection that a long-term study of the chronic toxicity to rats of foods containing hydrogen cyanide at levels up to 300 p.p.m. produced no typical evidence of chronic toxicity even though definite increases in thiocyanate concentrations could be found in the tissues of these animals.[15] It is also of interest that no changes were found in the thyroid even though rats are sensitive to goiterogenic agents.

E. FIRE AND EXPLOSION HAZARDS

Some materials in this chapter are flammable, or in some cases, explosive. The hazard is increased by the release of HCN under the influence of heat, moisture or acid. Fires involving nitriles or materials capable of generating HCN are always potentially very hazardous.

II. Specific Compounds

HYDROGEN CYANIDE, HCN, (Prussic Acid, Hydrocyanic Acid)

1. Source

While hydrogen cyanide can be prepared by treating cyanide salts with dilute sulfuric acid, it is now manufactured largely by the reaction of ammonia, air, and methane in the presence of a platinum catalyst. It has been stated that more than 150,000,000 pounds were produced in 1957, but over half of this was used for the synthesis of acrylonitrile.[12,16]

[14] H. L. Hardy, W. McK. Jeffries, M. M. Wasserman, and W. R. Wadell, *New Engl. J. Med.*, **242**, 968 (1950).

[15] J. W. Howard and R. F. Hanzal, *J. Agr. Food Chem.*, **3**, 325 (1955).

[16] R. N. Shreve, *The Chemical Process Industries*, McGraw-Hill, New York, 1956.

2. Uses and Industrial Exposures

Hydrogen cyanide has wide usage, which may involve many different types of exposure. The chief uses are in fumigation of ships, buildings, orchards, and various foods; in electroplating; in mining; in the production of various resin monomers such as acrylates, methacrylates, and hexamethylenediamine; and in the production of other nitriles. It also has many uses as a chemical intermediate, and may be generated in such operations as blast furnaces, gas works, and coke ovens.

3. Physical and Chemical Properties

Physical state: colorless liquid with characteristic odor
Molecular weight: 27.03
Melting point: $-13.4°C.$
Boiling point: $25.7°C.$
Refractive index: 1.2619 (20°C.)
Vapor density: 0.94 (air $= 1$)
Vapor pressure: 807.23 mm. Hg (27.22°C.)
Per cent in "saturated" air: 100 (25.7°C.)
Soluble in alcohol and ether, miscible with water
Flash point: (closed cup) $-17.8°C.$
1 mg./cu. meter $\approx$ 0.9 p.p.m. at 25°C., 760 mm. Hg

4. Determination in the Atmosphere

Methods have been described by Robbie and Leinfelder[17] and Jacobs.[18] A variety of field methods and test papers is available; their limitations have recently been discussed.[12] Detector tubes, operating on the length of stain principle, are useful, but results should be interpreted with caution in the presence of reactive materials such as hydrogen sulfide, styrene, nitric acid, hydrochloric acid, and various organic vapors.

5. Physiological Response

The typical symptoms and mode of action have been described previously. Relatively little gross or microscopic pathology can be seen following fatal inhalation of hydrogen cyanide. While there may be scattered hemorrhages and scattered congestion, these are probably the result of anoxia. Venous blood may appear a brighter red color than normal. Hydrogen cyanide vapor is absorbed extremely rapidly through the respiratory tract; the liquid and possibly the concentrated vapor are absorbed directly through the intact skin.

There is relatively close agreement between the lethal concentrations in various species and in man. Of the usual experimental animals, the dog is most sensi-

[17] W. A. Robbie and P. J. Leinfelder, *J. Ind. Hyg. Toxicol.*, **27**, 136 (1945).

[18] M. B. Jacobs, *Analytical Chemistry of Industrial Poisons, Hazards, and Solvents,* 2nd ed., Interscience Publishers, New York–London, 1949.

tive. The responses to various concentrations of hydrogen cyanide in animals and in man are listed in Tables 1 and 2.

TABLE 1

Physiological Response to Various Concentrations of Hydrogen Cyanide in Air—Animals[19, 20]

	Concentration		
Animal	mg./liter	p.p.m.	Response
Mouse	1.45	1300	Fatal after 1 to 2 min.
Mouse	0.12	110	Fatal after $^3/_4$ hr. exposure
Mouse	0.05	45	Fatal after $2^1/_2$ to 4 hr. exposure
Cat	0.350	315	Quickly fatal
Cat	0.20	180	Fatal
Cat	0.14	125	Markedly toxic in 6 to 7 min.
Dog	0.350	315	Quickly fatal
Dog	0.125	115	Fatal
Dog	0.1	90	May be tolerated for hours; death after exposure
Dog	0.07–0.04	65–35	Vomiting, convulsions, recovery; may be fatal
Dog	0.035	30	May be tolerated
Guinea pig	0.350	315	Fatal
Guinea pig	0.23	200	Tolerated $1^1/_2$ hrs. without symptoms
Rabbit	0.350	315	Fatal
Rabbit	0.13	120	No marked toxic symptoms
Monkey	0.14	125	Distinctly toxic after 12 min.
Rat	0.12	110	Fatal after $1^1/_2$ hrs. exposure

TABLE 2

Physiological Response to Various Concentrations of Hydrogen Cyanide in Air—Man[19, 20]

Response	Concentration	
	mg./liter	p.p.m.
Immediately fatal	0.3	270
Fatal after 10 min.	0.2	181
Fatal after 30 min.	0.15	135
Fatal after $^1/_2$ to 1 hr. or later, or dangerous to life	0.12–0.15	110–135
Tolerated for $^1/_2$ to 1 hr. without immediate or late effects	0.05–0.06	45–54
Slight symptoms after several hours	0.02–0.04	18–36

6. Precautions in Handling

Detailed precautions for handling in industry may be obtained from manufacturers' bulletins and from the Manufacturing Chemists Association Safety Data Sheet.[21] An American Industrial Hygiene Association Hygienic Guide is also available for hydrogen cyanide.[22] It is essential that all personnel be adequately

[19] H. C. Dudley, T. R. Sweeney, and J. W. Miller, *J. Ind. Hyg. Toxicol.,* **24,** 255 (1942).

[20] F. Flury and F. Zernik, *Schädliche Gase,* Springer, Berlin, 1931.

trained in recognition of odors of hydrogen cyanide and in the application of proper emergency first-aid measures. All sources of vapor or liquid exposure should be carefully studied, and adequate local exhaust ventilation used where necessary. Special attention needs to be given to the location of the outlets of such exhausts and for their adequate maintenance.

First-aid kits, described previously, should be properly and conveniently located; and all employees in potentially hazardous areas should be under constant supervision in case of emergency.

Since hydrogen cyanide is highly toxic to all species living in water, special attention should be given to the possibility of water pollution.[23]

7. Threshold Limit Value

The current threshold limit value suggested by the American Conference of Governmental Industrial Hygienists for hydrogen cyanide is 10 p.p.m. or 11 mg./cu. meter.[24]

8. Flammability

Hydrogen cyanide is definitely flammable and will burn in the air with a bluish flame. The flammable limits are from 5.6 to 40 per cent by volume in air.

9. Odor and Warning Properties

Hydrogen cyanide has a characteristic odor, which can be recognized by trained individuals at 2–5 p.p.m.[25] The sense of smell is, however, easily fatigued; and there is wide individual variation in the minimum odor threshold.

SODIUM CYANIDE, NaCN

1. Source

Sodium cyanide can be prepared by heating sodium amide with carbon or by melting sodium chloride and calcium cyanamide together in an electric furnace. A more recent method involves the dehydration of formamide by heat and a catalyst with the formation of hydrocyanic acid. HCN may then be converted to sodium cyanide by treatment with caustic soda.[16]

2. Uses and Industrial Exposures

Some of the more important uses of sodium cyanide are in the extraction of gold and silver from ores, the heat-treating of metals, electroplating, in various

[21] Mfg. Chemists Assoc. Chem. Safety Data Sheet, **SD-67,** Hydrocyanic Acid, 1957.

[22] Am. Industrial Hygiene Association Hygienic Guide, Hydrogen Cyanide, Am. Ind. Hyg. Assoc. Quart., **17,** 347 (1956).

[23] Charles E. Renn, Sewage and Ind. Wastes, **27,** 297 (1955).

[24] American Conference of Governmental Industrial Hygienists, Am. Ind. Hyg. Assoc., **22,** 325 (1961).

[25] A. M. Prentiss, Chemicals in War, McGraw-Hill, New York, 1937.

organic reactions, and in the manufacture of adiponitrile.[16] It is also used as a pesticide.

3. Physical and Chemical Properties

Physical state: white crystalline solid, deliquescent
Molecular weight: 49.02
Melting point: 564°C.
Boiling point: 1496°C.
Vapor pressure: 1.0 mm. Hg (817°C.), 10.0 mm. Hg (983°C.)   (these temperatures may be encountered in metal-treating processes using cyanide)
Readily soluble in water; slightly soluble in alcohol

4. Physiological Response

Sodium cyanide produces all the typical symptoms of other sources of cyanide ion. It can produce acute symptoms by inhalation and by skin absorption as well as by ingestion. The fatal dosage by oral ingestion will vary considerably depending on whether or not food is present in the stomach, etc. It is probably on the order of 1–2 mg./kg. in man as it is in a variety of experimental animals.[26-27]

The symptoms and therapy are the same as those described previously in the introduction of this chapter. It is important to remove all remaining dust or solutions containing sodium cyanide from the skin in the event of acute exposures.

5. Threshold Limit Value

The threshold limit value suggested by the American Conference of Governmental Industrial Hygienists for cyanide dust is 5 mg./cu. meter expressed as CN.[24]

6. Odor

The solid may have a light odor of hydrogen cyanide, especially if moisture is present.

POTASSIUM CYANIDE, KCN

1. Source

Potassium cyanide may be produced by methods similar to those for sodium cyanide.

2. Uses and Industrial Exposures

Similar to sodium cyanide.

[26] M. Gleason, R. Gosselin, and H. Hodge, *Clinical Toxicology of Commercial Products,* Williams & Williams, Baltimore, 1957.

[27] T. Sollmann, *A Manual of Pharmacology,* Saunders, Philadelphia, 1957.

3. *Physical and Chemical Properties*

Physical state: white crystalline solid, deliquescent
Molecular weight: 65.11
Melting point: 636°C.
Specific gravity: 1.560
Readily soluble in water: at 25°C., 1000 g. dissolves 716 g. of KCN; slightly soluble in alcohol: at 19.5°C., 100 g. dissolves 0.875 g. of KCN

4. *Physiological Response*

Similar to sodium cyanide. Streicher[28] has studied the effect of temperature on the toxicity of potassium cyanide in mice. The LD_{50} following oral administration at 23–25°C. is said to be 6.02 ± 3.3 mg./kg. When the mice were kept at temperatures of 4°C., the LD_{50} was 2.86 ± 1.6 mg./kg., indicating that toxicity increased with a decrease in environmental temperature.

Liebowitz and Schwartz[29] discuss the recovery of a patient with no specific therapy after a suicidal ingestion of an unusually large amount of potassium cyanide (3–5 g.).

5. *Odor*

Similar to HCN.

CALCIUM CYANIDE, $Ca(CN)_2$

1. *Source*

Calcium cyanide may be prepared by fusing calcium cyanamide ($CaCN_2$) with sodium chloride to give a crude mixture of calcium cyanide and sodium cyanide.

2. *Uses and Industrial Exposures*

Used as a fumigant and pesticide.

3. *Physical and Chemical Properties*

Physical state: amorphous white powder
Molecular weight: 92.12
Readily soluble in water (with gradual liberation of HCN); soluble in alcohol

4. *Physiological Response*

Physiological properties similar to those for other cyanide salts.

5. *Threshold Limit Value*

The current threshold limit value as suggested by the American Conference of Governmental Industrial Hygienists for cyanide dusts is 5 mg./cu. meter expressed as CN.[24]

[28] E. Streicher, *Proc. Soc. Exptl. Biol. Med.*, **76**, 536 (1951).
[29] D. Liebowitz and H. Schwartz, *Am. J. Pathol.*, **18**, 965 (1950).

6. Odor and Warning Properties

The solid may have a rather definite odor of hydrogen cyanide.

CALCIUM CYANAMIDE, CaCN₂

1. Source

The process of making calcium cyanamide involves three raw materials— coke, coal, and limestone—plus nitrogen. The limestone (calcium carbonate) is burned with coal to produce calcium oxide. The calcium oxide is then allowed to react with amorphous carbon in the furnace at about 2000°C. with the formation of calcium carbide (CaC_2). Finely powdered calcium carbide is heated at about 1000°C. in an electric furnace into which pure nitrogen is passed. It is then removed and uncombined calcium carbide removed by leaching.[16]

2. Uses and Industrial Exposures

Calcium cyanamide has its major use as a fertilizer. However, it has a number of other uses such as an herbicide and a defolliant for cotton plants. It is finding increased use as a chemical intermediate. For example, it is being used to produce dicyandiamide, which in turn can be polymerized to form the widely used resin monomer, melamine. The conversion to calcium cyanide and hence into a variety of other uses is also important commercially.

3. Physical and Chemical Properties

Physical state: white crystalline solid
Molecular weight: 80.11
Melting point: 1300°C. (sublimes > 1150)
Specific gravity: 2.3
Decomposes in water, liberating ammonia

4. Physiological Response

The principal exposures to calcium cyanamide dust (other than during manufacturing processes) result from its application as a fertilizer. The character of the toxic effect seems to be principally that of a transient vasomotor disturbance of the upper portion of the body. Irritation of the exposed mucous membranes and skin can occur, but this is probably related to the caustic content. It is thought that in the presence of body fluids calcium cyanamide will react with carbon dioxide to form calcium carbonate and cyanamide ($CN-NH_2$). Cyanamide is not converted to cyanide and its method of action is unknown. Glaubach[30] suggests that cyanamide may react with the sulfur groups of glutathione and thus influences catalytic oxidation-reduction processes. Apparently, there is a wide variation in the sensitivity to the vasomotor effect and some evidence that it is increased by

[30] S. Glaubach, Arch. exptl. Pathol. Pharmakol., 117, 247 (1926).

the simultaneous intake of alcohol.[20] The possible Antabuse-like effect of cyanamide has been discussed recently by Hald *et al.*[31] Reports from older literature[20] of polyneuritis following acute exposures do not appear to be confirmed. DeLarrard and Lazarini[32] discuss five cases of poisoning in farmers characterized by dermatitis, vasomotor changes, and dyspnea.

The acute toxicity of calcium cyanamide is low; the oral lethal dose is said to be 40–50 g. in an adult.[27]

5. Threshold Limit Value

None proposed.

DIMETHYL CYANAMIDE, $(CH_3)_2NCN$

1. Physical and Chemical Properties

Physical state: colorless liquid
Molecular weight: 70.10
Melting point: —41.0°C.
Boiling point: 162–164°C.
Density: 0.8768 (30°C.)
Vapor density: 2.41
Vapor pressure: 40 mm. Hg (80°C.)
Flash point (closed cup): 160°F.

2. Physiological Response

Fassett[33] found the oral LD_{50} in rats and guinea pigs to be 50–100 mg./kg. The symptoms were weakness, ataxia, gasping respirations, and unconsciousness. It was readily absorbed through the guinea pig skin (LD_{50} less than 5 ml./kg.) with little skin irritation. It was not a severe eye irritant to the rabbit. No skin sensitization was produced in the guinea pig.

The compound appears to be hazardous, especially by skin absorption.

CYANOGEN, N:CC:N

1. Source

Cyanogen can be prepared by slowly dropping potassium cyanide solution into copper sulfate solution or by heating mercury cyanide.[34]

[31] J. Hald, E. Jacobson, and V. Larson, *Acta Pharmacol. Toxicol.*, **8**, 329 (1952).

[32] J. DeLarrad and H. J. Lazarini, *Arch. maladies Profess. méd. travail. et sécurite sociale,* **15**, 282 (1954).

[33] D. W. Fassett, unpublished data. Eastman Kodak Co., Rochester 4, N. Y.

[34] F. M. Turner, *The Condensed Chemical Dictionary*, 4th ed., Reinhold, New York, 1950.

2. *Uses and Industrial Exposures*

Cyanogen has been used as a fumigant and may be encountered in situations in which there is heating of nitrogen-containing carbon bonds and in blast-furnace gases, etc.

3. *Physical and Chemical Properties*

Physical state: colorless gas
Molecular weight: 52.04
Melting point: −34.4°C.
Boiling point: −27.17°C.
Specific gravity: 1.8064 (air = 1)
Soluble in water: (450 cc. per 100 ml. of water at 20°C.) ; ethyl alcohol: 2300 cc. per 100 ml. of alcohol at 20°C.) ; ethyl ether: (500 cc. per 100 ml. of ether at 20°C.)
1 mg./liter ≏ 469.6 p.p.m. and 1 p.p.m. ≏ 2.127 mg. cu. meter at 25°C., 760 mm. Hg

4. *Determination in the Atmosphere*

Cyanogen may be determined in the presence of HCN by first scrubbing out the HCN with silver nitrate solution and then estimating the cyanogen by the ferrocyanide or thiocyanate methods.[18]

5. *Physiological Response*

The effect of cyanogen is similar in nature to that of other cyanides. It is thought to be converted in the body partly to hydrogen cyanide and partly to cyanic acid (HOCN). It is stated to be somewhat more irritating than hydrogen

TABLE 3

Toxicity of Cyanogen in Air for Various Animal Species[20]

Animal	Concentration		Length of time, hr.	Response
	mg./liter	p.p.m.		
Mouse	0.5	235	15 min.	Recovered
Mouse	5.5	2,600	12 min.	Fatal
Mouse	31.5	15,000	<1 min.	Fatal
Rabbit	0.21	100	4	Practically no effect
Rabbit	0.42	200	4	Slight symptoms
Rabbit	0.63	300	3.5	Severe symptoms; delayed death
Rabbit	0.84	400	1.8	Fatal
Cat	0.1	50	4	Severe symptoms but recovered
Cat	0.21	100	2–3	Fatal
Cat	0.42	200	1/2	Fatal
Cat	4.26	2,000	13 min.	Fatal

cyanide; quantitatively, it appears to be less potent in a variety of species (see Table 3).

The effects on man are similar to those of animals, although it appears to be more irritating than hydrogen cyanide. Quantitative data on symptoms at various exposure levels appear to be lacking. Therapy and precautions are the same as for hydrogen cyanide.

6. *Flammability*

Flammable within the range of 6.60 to 42.60 per cent by volume in air. Burns with peach-blossom red flame.

7. *Odor*

Pungent odor.

CYANOGEN CHLORIDE, CNCl

1. *Source*

Cyanogen chloride is produced by the action of chlorine on moist sodium cyanide suspended in carbon tetrachloride and kept cooled to —3°C., followed by distillation.[34]

2. *Uses and Industrial Exposures*

Cyanogen chloride is used in organic synthesis.[34] It is used as a warning agent in fumigant gases.

3. *Physical and Chemical Properties*

Physical state: colorless liquid or gas
Molecular weight: 61.48
Melting point: —6°C.
Boiling point: 13.8°C.
Density of liquid: 1.218 (4/4°C.)
Vapor density: 2 (air = 1)
Vapor pressure: 1000 mm. Hg at 20°C.
Soluble in water: (2500 cc. in 100 ml. of H_2O at 20°C.); ethyl alcohol: (10,000 cc. in 100 ml. of alcohol at 20°C.); ether: 5000 cc. in 100 ml. of ether at 20°C. Dissolves readily, soluble in all organic solvents. Tends to form polymers upon storage

1 mg./liter $\approx$ 398 p.p.m. and 1 p.p.m. $\approx$ 2.51 mg./cu. meter at 25°C., 760 mm. Hg

4. *Determination in the Atmosphere*

A colorimetric method is described by Jacobs.[18]

5. Physiological Response

Cyanogen chloride possesses the same general type of toxicity and mode of action as hydrogen cyanide, but is much more irritating even in very low concentrations. It can cause a marked irritation of the respiratory tract with a hemorrhagic exudate of the bronchi and trachea and pulmonary edema. Because of the high degree of irritant properties resembling those of phosgene, it is improbable that anyone would voluntarily remain in areas with a high enough concentration to exert a typical nitrile effect. Tables 4 and 5 indicate the relationship of concentration to symptoms produced in various animal species and in man.

TABLE 4

Effects of Cyanogen Chloride Inhalation on Various Animal Species[20]

| Animal | Concentration | | Length of time, min. | Response |
	mg./liter	p.p.m.		
Mouse	0.2	80	5	Tolerated by some animals
Mouse	0.3	120	3.5	Fatal to some animals
Mouse	1.0	400	3	Fatal
Rabbit	3.0	1200	2	Fatal
Cat	0.1	40	18	Delayed fatalities after 9 days
Cat	0.3	120	3.5	Fatal
Cat	1.0	400	<1	Fatal
Dog	0.05	20	20	Recovered
Dog	0.12	48	6 hrs.	Fatal
Dog	0.3	120	8	Severe injury, recovered
Dog	0.8	320	7.5	Fatal
Goat	2.5	1000	3	Fatal after 70 hrs.

TABLE 5

Effects of Varying Concentrations of Cyanogen Chloride in Air on Man[20,25]

| Concentration | | Response |
mg./liter	p.p.m.	
0.4	159	Fatal after 10 min.
0.12	48	Fatal after 30 min.
0.05	20	Intolerable concentration, 1-min. exposure
0.005	2	Intolerable concentration, 10-min. exposure
0.0025	1	Lowest irritant concentration, 10-min. exposure

If the patient is conscious, first-aid and medical treatment should generally be directed toward the relief of any pulmonary symptoms. The patient should immediately be put at bed rest with the head slightly elevated and a medical examination carried out as quickly as possible. Oxygen should be administered if there is any dyspnea or evidence of pulmonary edema. If the patient has been trapped in an area so that the exposure was prolonged, it is possible that both the cyanide effects and pulmonary edema may develop.

This situation has been studied experimentally by Jandorf and Bodanski.[35] Dogs were exposed in pairs to concentrations of cyanogen chloride varying from 2.3 to 3.6 mg./liter for periods of 1 to 2 minutes. One member of each pair received artificial respiration and inhalation of amyl nitrite (0.3 cc. administered by a nose cone) beginning 45–90 seconds after the end of exposure. This was continued until the animals resumed spontaneous respiration or until a fatality occurred. At these high levels of exposure, 8 per cent of the untreated animals recovered, whereas 77 per cent of the amyl nitrite-treated animals survived. When the exposures were increased to concentrations ranging from 3.1 to 5.9 mg./liter for 1.5–2.5 minutes, there was no benefit from the amyl nitrite therapy. Similar results were obtained with mice. Therefore, in cases with symptoms of both the nitrile-type effects and pulmonary edema, the combined therapy with oxygen plus amyl nitrite inhalations and artificial respiration seems to be indicated.

Effects of cyanogen chloride on experimental animals have also been studied by Aldridge and Evans.[36] These authors also point out the combined effect of pulmonary edema and the interference of the cellular metabolism by the cyanide radical.

The metabolism of cyanogen chloride has been studied by Aldridge.[37] Cyanogen chloride is apparently converted to cyanide ion in vivo by a reaction with hemoglobin and glutathione, which eventually liberates the CN ion.

6. Theshold Limit Value

No official threshold limit value have been published. The value should certainly be less than 0.5 p.p.m.

7. Odor and Warning Properties

Cyanogen chloride has a pungent odor detectable at 2.5 mg./cu. meter (1 p.p.m.).[25]

CYANOGEN BROMIDE, CNBr

1. Source

Cyanogen bromide may be prepared either by the action of bromine on potassium cyanide or by the interaction of sodium bromide, sodium cyanide, sodium chlorate, and sulfuric acid.[34]

2. Uses and Industrial Exposures

Cyanogen bromide is used in organic synthesis, as a fumigant and pesticide, and in gold-extraction processes. It has also been used in connection with cellulose technology.

[35] B. J. Jandorf and O. Bodansky, J. Ind. Hyg. Toxicol., 28, 125 (1946).
[36] W. N. Aldridge and C. L. Evans, Quart. J. Exp. Physiol., 33, 241 (1946).
[37] W. N. Aldridge, Biochem. J., 48, 271 (1951).

3. *Physical and Chemical Properties*

Physical state: colorless crystals (needles or cubes)
Molecular weight: 105.93
Specific gravity: 2.015 (20/4°C.)
Melting point: 52°C.
Boiling point: 61.6°C.
Vapor density: 3.62 (air = 1)
Vapor pressure: 92.0 mm. Hg (20°C.)
Density of "saturated" air: 1.32 (20°C.)
Per cent in "saturated" air: 12.1 per cent (20°C.)
Soluble in water with hydrolysis, and in alcohol and ether
  1 mg./liter ≎ 230.9 p.p.m. and 1 p.p.m. ≎ 4.33 mg./cu. meter at 25°C., 760
mm. Hg

4. *Determination in the Atmosphere*

A colorimetric method is described by Jacobs.[18]

5. *Physiological Response*

Cyanogen bromide is stated to be similar to cyanogen chloride in its effect. The systemic toxicity may be greater with the irritant properties somewhat less than cyanogen chloride. Tables 6 and 7 indicate the response to various concentrations of cyanogen bromide in animals and man.

TABLE 6

Response of Animals to Various Concentrations of Cyanogen Bromide in Air[20]

Concentration		Response	
mg./liter	p.p.m.	Mice	Cats
1	230	Fatal	Fatal
0.3	70	Paralysis after 3-min. exposure	Paralysis after 3-min. exposure
0.15–0.05	35–12	—	Severe injury; fatal on prolonged inhalation

TABLE 7

Response of Man to Various Concentrations of Cyanogen Bromide in Air[2,25]

Concentration		Response
mg./liter	p.p.m.	
0.4	92	Fatal after 10 min.
0.085	20	Intolerable concentration, 1-min. exposure
0.035	8	Intolerable concentration, 10-min. exposure
0.006	1.4	Lowest irritant concentration, 10-min. exposure

6. Threshold Limit Value

No official limit suggested; should certainly be less than 0.5 p.p.m.

7. Odor and Warning Properties

Has a penetrating odor and a bitter taste.

ACRYLONITRILE, CH_2:CHCN (Vinyl Cyanide, Propenenitrile)

1. Source

Acrylonitrile can be prepared by two methods. In one, ethylene oxide is reacted with hydrogen cyanide to form ethylene cyanohydrin (β-hydroxypropionitrile ($HOCH_2CH_2CN$), which is then dehydrated in the presence of a catalyst to form acrylonitrile. A somewhat similar synthesis involves the treatment of ethylene chlorohydrin ($HOCH_2CH_2Cl$) with sodium cyanide to form ethylene cyanohydrin. A second method involves the partial oxidation of natural gas to acetylene which is then reacted with hydrogen cyanide to form acrylonitrile.[16]

2. Uses and Industrial Exposures

Acrylonitrile is one of the major industrial intermediates. The United States Tariff Commission's report states that in 1958 approximately 180 million pounds were produced in the United States.[38] A considerable proportion of this was used in synthetic fiber production and the remainder in various plastics, synthetic rubber, and in organic synthesis.

3. Physical and Chemical Properties[39,40]

Physical state: clear, colorless, volatile liquid (some technical grades slightly yellowish)

Molecular weight: 53.06

Specific gravity: 0.8004 (25/4°C.)

Freezing point: —83°C.

Boiling point: 77.3°C.

Refractive index: 1.3885 (25°C.)

Vapor density: 1.9 (air = 1)

Vapor pressure: 110–115 mm. Hg (25°C.)

Per cent in "saturated" air: 14.5 (25°C.)

Density of "saturated air": 1.13 (25°C.) (air = 1)

Solubility: in water, 7.3 per cent by weight; soluble in all common organic solvents; forms azeotropes with water–benzene; soluble in isopropyl alcohol

[38] U. S. Tariff Comm. Rept. No. 205, Synthetic Organic Chemicals—U. S. Production and Sales, 1958, Washington, D. C.

[39] E. R. Blout and H. Mark, Monomers, Interscience Publishers, New York–London, 1951.

[40] H. C. Dudley and P. A. Neal, J. Ind. Hyg. Toxicol., 24, 27 (1942).

Flash point: 0 ±2.5°C.[39]

1 mg/liter ≎ 460.5 p.p.m. and 1 p.p.m. = 2.168 mg./cu. meter at 25°C., 760 mm. Hg

4. Determination in the Atmosphere

Acrylonitrile can be determined by polarography following collection in 95 per cent ethyl alcohol[41] or by ultraviolet spectrophotometry following collection in water.[42] A chemical method suitable for field testing has been described by Gisclard, et al.[43] which involves collection in dilute alkaline potassium permanganate and subsequent colorimetry. It can also be determined by collection in a suitable solvent followed by micro-Kjeldahl analysis, provided ammonia or other nitrogen-containing vapors are not present.

None of these methods is entirely specific with the possible exception of polarography, and there is considerable likelihood of encountering interfering substances under usual conditions of use.

5. Physiological Response

Acrylonitrile can be readily absorbed by mouth, through intact skin, or by inhalation. It has long been known to possess a high degree of toxicity and to possess many of the characteristics of poisoning by the cyanide ion. Extensive studies of the toxic effects in a variety of species in experimental animals have

TABLE 8

Physiological Response to Various Concentrations of Acrylonitrile in Air—Animals

| Animal | Concentration | | Response |
	mg./liter	p.p.m.	
Rat	1.38	636	Fatal after 4-hr. exposure
Rat	0.28	129	Slight transitory effect
Rat	0.21	97	Slight transitory effects
Rabbit	0.56	258	Fatal during or after exposure
Rabbit	0.29	133	Marked transitory effects
Rabbit	0.21	97	Slight transitory effects
Cat	0.60	276	Markedly toxic
Cat	0.33	152	Markedly toxic, sometimes fatal
Guinea pig	1.25	576	Fatal during or after exposure
Guinea pig	0.58	267	Slight transitory effect
Dog	0.24	110	Fatal to three fourths of the dogs
Dog	0.213	98	Convulsions and coma; no death
Dog	0.12	55	Transitory paralysis; 1 dog died
Dog	0.063	29	Very slight effects

[41] W. I. Bird and C. H. Hale, Anal. Chem., 24, 586 (1952).

[42] H. L. Brieger, F. Rieders, and W. A. Hodes, Arch. Ind. Hyg. Occup. Med., 6, 128 (1952).

[43] J. B. Gisclard, D. B. Robinson, and P. J. Kuczo, Am. Ind. Hyg. Assoc. J., 19, 43 (1958).

been made.[40, 42, 44] Some typical values for which various toxic effects occur are listed in Table 8.[44]

Brieger et al.[42] have reviewed the literature regarding human exposures and have evaluated the effects of acrylonitrile in dogs, rats, and monkeys. The level of cyanide ion in blood appears to be correlated with the degree of poisoning, and the symptoms are similar to those due to cyanide. Wilson[45] discussed the problem of toxicity encountered in industrial exposures and believed that the effects are similar to hydrogen cyanide, although some phases of the toxicity may be different. He felt that acute liver damage might be produced by acrylonitrile, but this does not seem to have been noted in subsequent experiences in industry.

Ghiringhelli[46] also suggested that the compound may have toxic actions not related to its ability to release cyanide ion. No study appears to have been published on levels of thiocyanate in blood under conditions of chronic exposure, and there is no definite evidence of thiocyanate-type symptoms such as those described by Hardy.[14]

The toxicity by other routes of administration has been studied. Orally the LD_{50} in the mouse is on the order of 35 mg./kg.; in the rat, about 78 mg./kg.; and in the guinea pig, about 90 mg./kg.[47] Rats ingesting 0.1 per cent in water for a period of 13 weeks showed retarded growth and emaciation; some effects were also noted with as low as 0.05 per cent in water in 2-year feeding studies. Acrylonitrile is known to be a severe skin and eye irritant.[47]

Education of personnel and first-aid and medical therapy should be the same as that for hydrogen cyanide. Because of its ability to penetrate the intact skin, special attention should be given to removing all the liquid from contaminated skin area. In the case of eye contact, irrigation with large amounts of water for 15 minutes should be carried out and medical attention sought. A number of excellent bulletins are available describing the hazards and proper handling precautions in industry.[48-50]

6. Threshold Limit Value

The threshold limit value is 20 p.p.m. (45 mg./cu. meter).[24]

7. Flammability

Acrylonitrile forms explosive mixtures with air in the range of 3.05 to 17.0 ±0.5 per cent by volume. The ignition temperature is 481°C. It is considered a severe fire and explosion hazard.[49]

[44] H. E. Dudley, T. R. Sweeney, and J. W. Miller, *J. Ind. Hyg. Toxicol.*, **24**, 255 (1942).

[45] R. H. Wilson, *J. Am. Med. Assoc.*, **124**, 701 (1944).

[46] G. L. Ghiringhelli, *Med. del. lavoro*, **45**, 305 (1954).

[47] *American Cyanamid Co. Bull.*, Toxicology of acrylonitrile, 1954.

[48] *Am. Petrol. Inst.*, API Toxicology Review, Acrylonitrile, 1948.

[49] American Industrial Hygiene Association Hygienic Guide, Acrylonitrile, *Am. Ind. Hyg. Assoc. Quart.*, **18**, 78 (1957).

[50] *Mfg. Chemists' Assoc. Chemical Safety Data Sheet* **SD-31**, Acrylonitrile, 1949, Washington, D. C.

8. *Odor and Warning Properties*

Acrylonitrile has a characteristic unpleasant odor somewhat resembling that of pyridine. It can be detected by trained observers at or below the threshold limit, but the sense of smell fatigues rapidly and is unreliable as an index of exposure.

METHACRYLONITRILE, $CH_2:C(CH_3)CN$

1. *Source*

Methacrylonitrile can be derived from isobutyraldehyde.

2. *Uses and Industrial Exposures*

It is used as a monomer.

3. *Physical and Chemical Properties*

Physical state: colorless liquid
Molecular weight: 67.09
Melting point: −35.8°C.
Boiling point: 90.3°C.
Specific gravity: .8001 (20/4°C.)
Vapor density: 2.31 (air = 1)
Vapor pressure: 40 mm. Hg (12.8°C.); 65 mm. Hg (25°C.); 100 mm. Hg (32.8°C.)
Density of saturated air: 1.17 (air = 1)
Solubility in water: 2.5 per cent (20°C.)
Flash point: 55°F. (open cup)
1 p.p.m. ⊃⊂ 2.74 mg./cu. meter and 1 mg./liter ⊃⊂ 365 p.p.m.

4. *Physiological Response*

McOmie[51] has made a comparative study of methacrylonitrile and acrylonitrile. The approximate LC$_{50}$ for mice exposed one hour was 630 p.p.m. (1700 mg./cu. meter); for 4 hours it was about 400 p.p.m. None out of 6 mice was killed by an 8-hr. exposure to 75 p.p.m. The animals showed respiratory paralysis and convulsions.

Methacrylonitrile was found to penetrate the rabbit skin readily, causing fatalities in doses of 2 to 4 ml./kg. One of the rabbits was treated with 20 mg./kg. of sodium nitrite intravenously and was revived, indicating a typical nitrile effect.

Fassett[33] found the oral LD$_{50}$ in mice to be 20–25 mg./kg. and in rats 25–50 mg./kg. Symptoms were those of weakness, tremors, cyanosis, and convulsions. There was no damage to the rabbit cornea. It was absorbed readily through guinea pig skin with no skin irritation. It was not a skin sensitizer in this species.

[51] W. A. McOmie, *J. Ind. Hyg. Toxicol.,* **31,** 113 (1949).

Inhalation of 9880 p.p.m. for 2 hours killed 3 out of 3 rats. It should be handled the same as other toxic nitriles.

5. *Odor and Warning Properties*

Methacrylonitrile has a slight odor resembling cyanide.

ACETONITRILE, CH_3CN (Methyl Cyanide, Ethanenitrile)

1. *Source*

Methyl cyanide is prepared by heating acetamide with glacial acetic acid.[34]

2. *Uses and Industrial Exposures*

The industrial uses of acetonitrile are increasing as an intermediate for chemical reactions, as a solvent, and as an extractant for animal and vegetable oils.[2]

3. *Physical and Chemical Properties*

Physical state: colorless liquid
Molecular weight: 41.05
Specific gravity: 0.7768 (25/4°C.)
Melting point: −41°C.
Boiling point: 81.6°C.
Refractive index: 1.34596 (16.5°C.)
Vapor pressure: 73 mm. Hg (20°C.)
Per cent in "saturated" air: 9.6
Vapor density: 1.42 (air = 1)
Density of "saturated" air: 1.04 (air = 1)
Solubility: Infinitely soluble in water; readily miscible with alcohol, ether, and acetone
Flash point: 55°F.
1 mg./liter ⩴ 595.3 p.p.m. and 1 p.p.m. ⩴ 1.68 mg./cu. meter at 25°C., 760 mm. Hg

4. *Determination in the Atmosphere*

A method for determining acetonitrile in air under laboratory conditions has recently been described.[2] This method makes use of a portable interferometer, calibrated by a modification of a method previously described by Jacobs.[18] Since the chemical method depends on the hydrolysis of acetonitrile and subsequent titration of the ammonia, interference might be expected if the latter were present.

5. Physiological Response

Until recently acetonitrile had been investigated by toxicologists chiefly because of its relationship to thyroid metabolism. For example, Marine et al.[52] showed that a progressive bilateral exophthamus could be produced in rabbits with a daily intramuscular injection of 0.05 cc. of acetonitrile, and that this reaction could be inhibited by feeding of fresh vegetables. The degree of exophthalmus was related to the thyroid hyperplasia, and could be prevented by previous administration of iodine. Most recent investigations of the acute toxicity[2,53,54] indicate that acetonitrile has a relatively low acute toxicity.

The oral LD_{50} in the rat, for example, varies from about 1.7 to 8.5 g./kg., depending on the conditions of the experiment. The guinea pig appears to be more sensitive than the rat with an oral LD_{50} of 0.18 g./kg. (This was also noted by Fassett.[33])

It is absorbed through the intact skin of rabbits with an LD_{50} of 1.25 ml./kg. and through the skin of guinea pigs.[33] The LC_{50} for a single, 8-hr. inhalation in male rats is 7500 p.p.m. with females being somewhat less sensitive. The rabbit and guinea pig are somewhat more sensitive. In the case of dogs, no fatalities occurred up to and including 8000 p.p.m. for a 4-hr. exposure; deaths occurred at levels of 16,000 and 32,000 p.p.m. Symptoms in animals appear to be those of prostration, followed by convulsive seizures. Autopsy findings indicate pulmonary hemorrhage and vascular congestion. At the lower dosage levels, the deaths always appeared to be delayed.[2]

Repeated inhalation studies have also been made on a variety of species. Rats exposed 7 hours a day to acetonitrile vapor for a period of 90 days showed no specific effects at 166 or 330 p.p.m. At 665 p.p.m., a pulmonary inflammatory change and minor changes in the kidney and liver were noted in some animals. No mention was made of any effect on the thyroid.[2]

While these animals excreted some thiocyanate, apparently the amount was not proportional to the acetonitrile inhaled. Dogs and monkeys were exposed to acetonitrile vapor for 7 hours a day, 5 days a week, for 91 days. The mean concentration was approximately 350 p.p.m. The symptoms produced were not remarkable and only some minor variations in weight, hematocrit, and hemoglobin were reported. At autopsy some cerebral hemorrhage was noted in the monkeys and some evidence of focal emphysema and proliferation of alveolar septa in the lung. Rather marked pigment-bearing macrophages were consistently noted in monkeys. A similar picture in the lungs was noted in dogs. Small amounts of cyanide and thiocyanate ion were present, but the significance of this seems somewhat uncertain at the lower levels of exposure.[2]

Human volunteers were studied at levels of 40, 80, and 160 p.p.m. of acetonitrile vapor for periods of 4 hours. No specific subjective responses were noted.

[52] D. Marine, S. H. Rosen, and A. Cipra, Proc. Soc. Exptl. Biol. Med., **30**, 649 (1933).
[53] L. Cuny and D. Quivy, Compt. rend. soc. biol., **132**, 429 (1939).
[54] H. F. Smyth, Jr., and C. P. Carpenter, J. Ind. Hyg. Toxicol., **30**, 63 (1948).

There was no consistent change in the blood-cyanide level or urinary thiocyanate. From these various studies, the authors conclude that, at least in the case of dogs, the fatal concentrations are associated with the formation of cyanide in vitro; but that in some instances it may be that direct action of acetonitrile is responsible. They point out the important fact that determination of blood cyanide or urinary thiocyanate should not be relied on as evidence for brief inhalations of lower concentrations of acetonitrile vapor. Because acetonitrile vapor is definitely lower in acute toxicity than other nitriles, they suggest that a threshold limit value of 40 p.p.m. would probably not be unduly hazardous.[2]

In 1955, Grabois[55] described a fatality and several cases of accidental poisoning in workers exposed to methyl cyanide vapor. A comprehensive discussion of this incident has been given by Amdur.[1]

The fatality occurred in a 23-year-old man who had been engaged for 2 days in hand-painting the interior of a tank with a resin containing 30–40 per cent methyl cyanide as well as other substances, such as diethylenetriamine and a mercaptan. Methyl cyanide was the major volatile component. About four hours after leaving the job, the subject complained of chest pain, vomited, and had a massive hematemesis, followed by convulsions. About 9 hours later, he was admitted to the hospital in a comatose state, with an ashen-gray color and irregular and infrequent respirations; he expired about an hour after admission with convulsive seizures and marked rigidity of the neck. A post-mortem examination disclosed only generalized vascular congestion. Examination of the blood and various organs showed high levels of cyanide ion (μg.%: blood, 796; urine, 215; kidney, 204; spleen, 318; lungs, 128; liver, 0).

Two additional cases were hospitalized with severe symptoms consisting of nausea and vomiting, respiratory depression, extreme weakness, and a semicomatose state. Both of these men were treated with oxygen, fluids, and whole blood intravenously, as well as with ascorbic acid and sodium thiosulfate. Amyl nitrite was not used. One of the cases developed a transient weakness of the flexor muscles of the arms and wrists, and both developed urinary frequency, associated in one instance with albuminuria and in the other with the passage of a small oxalate-type urinary calculus. These cases showed also elevated blood-cyanide levels and somewhat increased serum-thiocyanate levels. All other exposed workers were evaluated; and in some instances, increased blood-cyanide and thiocyanate values were obtained with minor degrees of the symptoms described previously. It is of interest that none of these individuals developed any enlargement of the thyroid or alteration in the thyroid function.

Amdur[1] is of the opinion that because of the very marked delay in onset of symptoms and because of their character, the cause was not primarily that of the cyanide ion, but that of the excessive accumulation of thiocyanate. While there is some resemblance between these symptoms and those of thiocyanate intoxication, it is difficult to overlook the very high blood levels of cyanide found in the most

[55] B. Grabois, *N. Y. State Dept. Labor Monthly Rev., Div. Ind. Hyg.*, **34**, 1 (1955).

severely ill individuals and in some of the other cases. An alternate explanation might be that acetonitrile is hydrolyzed to give the CN ion rather more slowly than other similar nitriles; and that the final clinical picture is a result of the intact molecule or perhaps a combined effect of the molecule and gradual, increasing hydrolysis to give CN.

The hemorrhagic effect might well have been the result of the extensive primary irritation of the respiratory tract.

In view of the recent report of severe intoxication from exposure to high concentrations of acetonitrile vapor, it seems important that the same protective measures should be applied as in the case of other nitriles, especially education of personnel, and proper ventilation and protective measures. Certainly in the event of expected high concentrations such as those inside tanks, etc., precautions should include supplied-air respirators and complete skin protection.

In view of the uncertainty as to the specific mode of action of acetonitrile, the effectiveness of combined use of amyl nitrite and sodium thiosulfate cannot be considered to have been established. Certainly, the use of oxygen and other supportive therapy plus careful removal of any liquid from the skin seems indicated. However, it would not appear unduly hazardous to use both amyl nitrite and sodium thiosulfate in the event of high exposures with serious symptoms.

6. *Threshold Limit Value*

No specific values have been established. It has been suggested that 40 p.p.m. might be a tentative level[24] based on animal and limited human experimental data.

7. *Odor and Warning Properties*

Acetonitrile has an ethereal odor with a burning, sweetish taste. Minimum odor threshold level is not known; it is probable that the odor sensation is rapidly fatigued, which would make it an unreliable index of exposure.

PROPIONITRILE, C_2H_5CN (Ethyl Cyanide, Propanenitrile)

1. *Source*

Propionitrile is made by heating barium–ethyl sulfate and potassium cyanide, with subsequent distillation.[34]

2. *Uses and Industrial Exposures*

Propionitrile is used in organic synthesis.[34]

3. *Physical and Chemical Properties*

Physical state: colorless liquid
Molecular weight: 55.08
Specific gravity: 0.7770 (25/4°C.)
Boiling point: 97.1°C.

Melting point: −103.5°C.
Refractive index: 1.3659 (24°C.)
Vapor density: 1.9 (air = 1)
Vapor pressure: 40 mm. Hg (22°C.)
Density of "saturated" air: 1.05 (22°C.) (air = 1)
Soluble in water up to 12 per cent at 40°C.; infinitely soluble in alcohol; slightly soluble in ether

4. *Physiological Response*

Propionitrile has a high degree of toxicity and is thought to produce its action by fairly rapid metabolism to the cyanide ion. The fate of the other portion of the molecule is somewhat uncertain.[4] Smyth[56] indicates that the oral single dose LD_{50} in the rat is about 39 mg./kg. The LD_{50} by skin absorption in the rabbit was 0.21 ml./kg. A 2-min. inhalation of saturated vapor killed all rats. A 4-hr. exposure to 500 p.p.m. gave a mortality of 2 out of 6 rats. Application to the skin and eye of rabbits did not result in severe damage. Fassett[33] found that the approximate oral LD_{50} in the rat was 50–100 mg./kg. and in the guinea pig 25–50 mg./kg. Intraperitoneally, the values for the rat were 25–50 mg./kg. and for the guinea pig 10–25 mg./kg. The material was only slightly irritating to the skin and had an LD_{50} by skin in the guinea pig of less than 5 ml./kg. One and a half hours' exposure to 9500 p.p.m. killed all rats.

The subcutaneous LD_{50} in the guinea pig was found by Ghiringhelli[61] to be 18 mg./kg., and 70 per cent of the dose was accounted for as SCN. This material is obviously highly toxic and requires the same care in handling as other toxic nitriles.

n-BUTYRONITRILE, $CH_3CH_3CH_2CN$ (Butanenitrile, *n*-Propyl Cyanide)

1. *Physical and Chemical Properties*

Physical state: colorless liquid
Molecular weight: 69.10
Refractive index: 1.3816 (24°C.)
Density: 0.796 (15°C.)
Melting point: −112.6°C.
Boiling point: 118°C.
Vapor pressure: 10 mm. Hg (15.4°C.) ; 40 mm. Hg (38.4°C.)
Vapor density: 2.4 (air = 1)
Density of "saturated" air: 1.07 (38.4°C.) (air = 1)
Slightly soluble in water ; soluble in alcohol

[56] H. F. Smyth, C. P. Carpenter, and C. S. Weil, *Arch. Ind. Hyg. Occup. Med.,* **4,** 119 (1951).

2. Physiological Response

Fassett[33] found the oral LD_{50} in rats to be 50–100 mg./kg.; intraperitoneally, less than 50 mg./kg. The symptoms were weakness, tremors, vasodilatation, labored respiration, and terminal convulsions—similar to other active nitriles.

Similar symptoms were produced in mice. The LD_{50} by skin contact in the guinea pig was 0.1–0.5 ml./kg. Skin and eye irritation was slight. Inhalation of vapor readily produced fatalities in rats with symptoms of nitrile toxicity.

n-Butyronitrile is considered a highly hazardous material and full precautions should be used to prevent skin contact or inhalation of the vapor.

The first-aid and medical therapy should be the same as for HCN.

ISOBUTYRONITRILE, $(CH_3)_2CHCN$ (2-Methyl Propanenitrile, Isopropylcyanide)

1. Source

Isobutyronitrile can be derived from isobutyraldehyde.

2. Uses and Industrial Exposures

Isobutyronitrile is used in organic synthesis.

3. Physical and Chemical Properties

Physical state: colorless liquid
Molecular weight: 69.10
Density: 0.773
Boiling point: 107°C.
Vapor density: 2.4 (air = 1)
Slightly soluble in water; soluble in alcohol and ether

4. Physiological Response

Fassett[33] found the oral LD_{50} in rats to be 50–100 mg./kg.; in mice, 5–10 mg./kg. The symptoms were those of weakness, vasodilatation, tremors, and convulsions similar to those caused by other nitriles. SCN was present in the urine. The LD_{50} by skin contact in the guinea pig was less than 5 ml./kg. with only slight irritation noticed.

Vapor inhalation at a calculated concentration of 5500 p.p.m. for about 1 hour killed all rats with similar symptoms to those seen after oral dosages. Isobutyronitrile is considered highly hazardous and full precautions should be taken to prevent skin contact or inhalation of vapor.

First-aid and medical therapy should be the same as for HCN.

β-HYDROXYPROPIONITRILE, $HOCH_2CH_2CN$ (Ethylene Cyanohydrin, Hydracrylonitrile, 3-Hydroxypropanenitrile, Glycol Cyanohydrin)

1. Source

β-Hydroxypropionitrile can be prepared by reacting ethylene oxide with hydrogen cyanide or by reacting ethylene chlorohydrin with sodium cyanide.

2. *Uses and Industrial Exposures*

Its major use is in the synthesis of acrylonitrile.

3. *Physical and Chemical Properties*

Physical state: colorless or straw-colored liquid
Molecular weight: 71.08
Density of liquid: 1.059 (0/4°C.)
Melting point: —46°C.
Boiling point: 221°C., 724 mm. Hg
Refractive index: 1.4241 (25°C.)
Vapor pressure: 0.08 mm. Hg (25°C.); 1 mm. Hg (58.7°C.); 10 mm. Hg (102°C.)
Vapor density: 2.45 (air = 1)
Soluble in water and alcohol
Flash point: <80°F.

4. *Physiological Response*

Smyth[57] found the oral LD_{50} in rats to be 10 g./kg. Saturated vapor inhalation for 8 hours produced no effect. Sunderman[58] reports that Hamblin found the minimum lethal dose in rabbits to be 0.9–1.4 g./kg. The oral LD_{50} in mice was 1.8 g./kg. Single applications to the skin caused moderate local irritation but no toxicity up to 3.8 g./kg. in the rabbit. In 15 repeated applications to the rabbit skin, there was no injury. β-Hydroxypropionitrile was applied to guinea pig skin (0.5 ml./guinea pig) on a gauze pad 1-inch square. No effect was noted in 24 hours. Rats and guinea pigs were exposed to vapor in an 8-liter chamber. Dry air at a rate of 0.9 liters/min. was passed through 250 ml. in a 5-in. sintered glass tube. No effect was produced in rats or guinea pigs by a 1-hour exposure.[58]

Fassett[33] found that the oral LD_{50} in the rat was between 3200 and 6400 mg./kg. with about the same values intraperitoneally. Little evidence of skin irritation was noted, and there was no significant skin absorption.

This material seems to be of a very low order of toxicity compared to some nitriles. Apparently, when the hydroxy group is in the beta position relative to the nitrile group, the compound is not hydrolyzed in the body to release cyanide. When the hydroxyl group is in the alpha position adjacent to the CN group, the extreme toxicity of nitriles is retained.[57,58] In view of the very low vapor pressure and the lack of significant toxicity in animals and the lack of reports of human injury, the customary training of personnel and handling precautions would seem sufficient.

LACTONITRILE, $CH_3CH(OH)CN$ (2-Hydroxypropanenitrile, Acetaldehyde Cyanohydrin, Ethylidene Cyanohydrin)

1. *Physical and Chemical Properties*

Physical state: colorless or straw-colored liquid

[57] H. F. Smyth, Jr., *J. Ind. Hyg. Toxicol.*, **26**, 269 (1944).
[58] F. W. Sunderman and J. F. Kincaid, *Arch. Ind. Hyg. Occup. Med.*, **8**, 371 (1953).

Molecular weight: 71.08
Melting point: −40°C.
Boiling point: 103°C., 50 mm. Hg
Refractive index: 1.4058 (18.4°C.)
Specific gravity: 0.992 (0/18.4°C.)
Vapor density: 2.45
Vapor pressure: 10 mm. Hg (74°C.)
Flash point: 76.7°C. (170°F.)
Readily miscible with water, acetone, alcohol, and other organic solvents

2. *Physiological Response*

Lactonitrile is reported[59] to be an extremely toxic compound by oral administration and skin or eye contact. The acute oral LD_{50} (species not mentioned)
was 21 mg./kg. with deaths occurring as low as 10 mg./kg. As little as 0.05 ml. of
the undiluted compound applied to the eye was fatal to all animals within a
period of 5 minutes. The LD_{50} by skin application was less than 1 ml./kg. with all
deaths occurring within a period of one hour.

It is unknown whether or not lactonitrile produces its effects by virtue of
hydrolysis to give the cyanide ion or whether it acts as an intact molecule.
Methods for determination in blood have been reported.[10a,61] Extreme care is
necessary in handling this material with particular attention to the education of
personnel and availability of prompt first-aid and medical treatment.

GLYCOLONITRILE, $HOCH_2CN$

1. *Uses*

Glycolonitrile is used as an organic intermediate.

2. *Physical and Chemical Properties*

Physical state: (anhydrous glycolonitrile) colorless, odorless oil with
sweetish taste
Molecular weight: 57.05 (theoretical)
Boiling point: 183°C. (slight decomposition)
Density: 1.104 (19°C.)
Vapor density: 1.96
Vapor pressure: 1 mm. Hg (63°C.) ; 14 mm. Hg (102°C.)
Soluble in water, ethanol, and ether

3. *Physiological Response*

Glycolonitrile has been reported[59] to be extremely toxic, similar to lactonitrile. A method has been reported for determination in blood.[10a]

The same handling precautions are necessary as for other nitriles.

ACETONE CYANOHYDRIN, $(CH_3)_2C(OH)CN$ (α-Hydroxyisobutyronitrile, 2-Hydroxy-2-methylpropanenitrile, Isopropylcyanohydrin)

1. Source

Acetone cyanohydrin is formed by the reaction of acetone plus hydrogen cyanide and may contain 0.2 per cent free hydrogen cyanide.

2. Physical and Chemical Properties

Physical state: colorless liquid
Molecular weight: 85.10
Melting point: −19°C.
Boiling point: 82°C., 23 mm. Hg
Refractive index: 1.3996
Density: 0.932 (19°C.)
Vapor density: 2.95 (air = 1)
Vapor pressure: 0.8 mm. Hg, 20°C.
Very soluble in water, alcohol, and ether; very slightly soluble in petroleum ether
Decomposes rapidly in alkali → HCN

3. Physiological Response

Sunderman and Kincaid[58] have reviewed the toxicity and hazards. The oral LD_{50} in mice is 15 mg./kg. Death occurs readily when applied to rabbit skin at 100 mg./kg., but not at 50 mg./kg. Fatalities and severe symptoms were reported. (Case No. 1: Splash on skin at 10 A.M. followed by nausea, unconsciousness, convulsions, with death at 4:30 P.M. Case No. 2: Ingestion of alcohol and acetone cyanohydrin was followed by death, in spite of thiosulfate and nitrite use. Case No. 3: Packing pump exposures were followed by unconsciousness and then recovery. Symptoms included palpitation, headache, and vomiting.)

Skin penetration was studied in the guinea pig by application on a gauze pad for 24 hours; the LD_{50} was 0.15 ml./kg.

Inhalation of saturated vapor was followed by death of all rats in 2 minutes. Treatment of the unconscious rats with nitrites and thiosulfate saved 2 out of 2 rats, whereas the untreated rats died.

Acetone cyanohydrin toxicity has also been described by Krefft,[58a] who discusses 2 fatal cases of poisoning in humans from inhalation and skin contact. A study was made of skin penetration in guinea pigs under carefully controlled conditions. Guinea pigs absorb this material rapidly and show the typical symptoms of nitrile poisoning. Potassium cyanide was included in this study and also was shown to penetrate the intact skin of this species.

[58a] S. Krefft, *Arch. Gewerbepathol. Gewerbehyg.*, **14**, 110 (1955).

This material is considered very hazardous and should only be handled under conditions that prevent any inhalation of vapor or skin contact.

The first-aid and medical therapy should be the same as for HCN.

SUCCINONITRILE, CN—CH₂—CH₂CN (Butanedinitrile, Ethylene Cyanide)

1. Physical and Chemical Properties

Physical state: colorless, waxy solid
Molecular weight: 80.1
Melting point: 57.7°C.
Density: 0.9880 g./cc. (60°C.)
Vapor density: 2.8 (air = 1)
Vapor pressure: 6 mm. Hg (125°C.)
Soluble in water (12.8 g. in 100 ml. of H_2O) ; soluble in alcohol and benzene
Flash point: 270°F.

2. Physiological Response

The acute toxicity appears somewhat lower than materials such as propiono- or butyronitrile. The oral LD_{50} in rats is 450 mg./kg. The effects on the skin of rabbits of a 95 per cent water solution were those of mild irritation. Continued contact of the solution with rabbit skin for 18 hours produced fatalities, indicating a probable hazard by skin absorption. A 24-hour exposure of mice to vapor from a 95 per cent solution caused no symptoms.[59] Whether this compound acts as a typical nitrile is unknown. Adequate precautions should be taken against skin or eye contact. The inhalation hazard is uncertain.

ADIPONITRILE, CN(CH₂)₄CN (Tetramethylene Dicyanide, Adipyl Dinitrile)

1. Source and Uses

Adiponitrile is derived from butadiene and used as an intermediate for hexamethylenediamine in nylon manufacturing.[60]

2. Physical and Chemical Properties

Physical state: colorless liquid
Molecular weight: 108.1
Melting point: 2.3°C.
Boiling point: 295°C.
Density: 0.965 (20/4°C.)
Vapor density: 3.73

[59] *Am. Cyanamid Co. New Products Bull.,* revised ed., Collective Vol. **I,** 1, 1952, New York.
[60] W. N. Aldridge, *Analyst,* **69,** 262 (1944).

Vapor pressure: 2 mm. Hg (119°C.)
Slightly soluble in water; soluble in alcohol
Flash point: 199.4°F. (open cup)

3. Physiological Response

Ghiringhelli[61] reports that this nitrile is quite toxic, with an LD_{50} subcutaneously in the guinea pig of about 50 mg./kg.; and that it is hydrolyzed to HCN in the body giving rise to SCN in the urine. Seventy-nine per cent of the dose was eliminated as SCN in the urine.

The symptoms of acute poisoning following the ingestion of a few ml. by a young man were said to be weakness, mental confusion, vomiting, rapid respiration, and tachycardia and convulsions, which were relieved by injection of sodium thiosulfate and glucose. In exposed guinea pigs thiosulfate was a more effective treatment than nitrites. No effect was seen on the blood of guinea pigs from repeated doses (3–30 mg./kg. subcutaneously—6 days/week for 40 to 70 days). Skin penetration was suggested by the increase in SCN in the urine of guinea pigs after application to depilated skin. Greater quantities were absorbed when the skin was excoriated.

Ceresa and De Blasiis[62] report that hexamethylenediamine may have been the cause of a hemolytic anemia previously reported in nylon workers exposed to crude adiponitrile. However, this was not found by Ghiringhelli[61] in a study of workers exposed to hexamethylenediamine.

The same care and precautions should be used as for other reactive nitriles. The vapor hazard at room temperature seems low. Skin contact should be avoided.

4. Odor and Warning Properties

There are none.

β-DIMETHYLAMINOPROPIONITRILE, $(CH_3)_2NCH_2CH_2CN$

1. Physical and Chemical Properties

Physical state: colorless, mobile fluid
Molecular weight: 98.15
Melting point: —44.2°C.
Boiling point: 172°C.
Density: 0.8617 (30°C.)
Vapor density: 3.4 (air = 1)
Vapor pressure: 10 mm. Hg (57°C.)
Density of "saturated" air: 1.03 (air = 1)

[61] G. L. Ghiringhelli, *Med. del. lavoro,* **46**, 221, 229 (1955); **47**, 192 (1956); **49**, 683 (1958).
[62] C. Ceresa and M. De Blasiis, *Med. del. lavoro,* **41**, 78 (1950).

Miscible with water, alcohol, and other solvents
Flash point: 147°F. (closed cup)

2. *Physiological Response*

Preliminary data[63] indicate a low order of acute toxicity in mice and rats. The oral LD_{50} in mice was 1.5 g./kg. The vapor was thought to be hazardous and care in handling was suggested.

β-ISOPROPYLAMINOPROPIONITRILE, $(CH_3)_2CH_2NHCH_2CH_2CN$

1. *Physical and Chemical Properties*

Physical state: liquid
Molecular weight: 112.18
Melting point: $< -20°C$.
Boiling point: 87°C., 17 mm. Hg
Density: 0.864 (25°C.)
Vapor density: 3.9 (air $= 1$)
Vapor pressure: 2 mm. Hg (60°C.)
Miscible with water and other solvents
Flash point: $>105°F$. (open cup)

2. *Physiological Response*

Similar to β-dimethylaminopropionitrile.[63] The oral LD_{50} in mice is 2175 mg./kg.

β-METHOXYPROPIONITRILE, $CH_3OCH_2CH_2CN$

1. *Physical and Chemical Properties*

Physical state: colorless liquid
Molecular weight: 85.1
Melting point: $-62.9°C$.
Boiling point: 160°C.
Density: 0.9299 (30°C.)
Vapor density: 2.9 (air $= 1$)
Vapor pressure: 10 mm. Hg (55°C.)
Solubility in water: 33.5 g. per 100 g.; miscible with alcohol, toluene, and other solvents
Flash point: 149°F. (closed cup)

2. *Physiological Response*

Similar to β-dimethylaminopropionitrile.[63] The oral LD_{50} in mice is 3.2 g./kg.

[63] *Am. Cyanamid Co. New Products Bull.*, Collective Vol. **II**, 1952, New York.

β-ISOPROPOXYPROPIONITRILE, $(CH_3)_2CH_2OCH_2CH_2CN$

1. *Physical and Chemical Properties*

Physical state: liquid
Molecular weight: 113.16
Melting point: $-67°C$.
Boiling point: 177°C.
Density: 0.883 (25°C.)
Vapor density: 3.9 (air $= 1$)
Soluble in water: 6.4 g. per 100 g. of water; miscible with acetone, benzene, and other solvents

2. *Physiological Response*

Similar to β-dimethylaminopropionitrile.[63] The oral LD_{50} in mice is 4450 mg./kg. β-Isopropoxypropionitrile is a skin irritant in rabbits.

β-CHLOROPROPIONITRILE, $ClCH_2CH_2CN$

1. *Physical and Chemical Properties*

Physical state: colorless liquid
Molecular weight: 89.53
Melting point: $-51°C$.
Boiling point: 132°C., 200 mm. Hg
Density: 1.1363 (25°C.)
Vapor density: 3.1 (air $= 1$)
Vapor pressure: 5 mm. Hg (46°C.)
Soluble in water: 4.5 g. per 100 g. at 25°C.; miscible with acetone, carbon tetrachloride, benzene, and other solvents
Flash point: 168°F. (closed cup)

2. *Physiological Response*

β-Chloropropionitrile is reported to be highly toxic.[59] The oral LD_{50} in mice is 9 mg./kg.; in rats, 100 mg./kg. Symptoms are those of deep anesthesia with no demonstrable pathology. Exposure to the vapor of 0.01 ml. in a 1-liter beaker killed all mice in 18 hours. It is probably absorbed through the intact skin. The mechanism of action, however, appears unknown. The marked increase in toxicity associated with the beta-chloro in contrast to the beta-hydroxy substitution and the atypical symptoms suggest a different mode of action. It is also of interest that the substitution of a CH_3 group (as in n-butyronitrile) results in the retention of the typical symptoms and potency of an active nitrile.

Substitution of an amino group (see β-aminopropionitrile) causes an even more extraordinary change in response; namely, that of an alteration of growth of mesodermal tissues at low levels in the diet. If the other hydrogen of the amino group is replaced by a second propionitrile (see β-β'-iminodipropionitrile) group, the effect changes to one of marked central nervous system damage.

The remarkable variety of toxicological effects produced by this series of compounds indicates that they should be handled with caution and all exposed persons closely followed medically.

β-AMINOPROPIONITRILE, NH₂CH₂CH₂CN

1. *Physical and Chemical Properties*

Physical state: the free base is a liquid; the hydrochloride is a crystalline solid

Molecular weight: 70

Boiling point: 79–81°C., 16 mm. Hg

Refractive index: 1.4396

Vapor pressure: 2 mm. Hg (38–40°C.)

2. *Physiological Response*

β-Aminopropionitrile has been studied extensively since the isolation of its glutamyl derivative as the probable causative factor in the toxic effect of sweet peas.[64] It was proved soon afterward that β-aminopropionitrile itself was fully active.[65,66] The disease produced by ingestion of large quantities of sweet peas in man is known as lathyrism and is characterized by paralysis of the legs, and other central nervous system symptoms. In young rats, and various avian species, it produces severe skeletal deformities and aneurysms leading to rupture of the aorta. The effective doses to turkey poults may be as low as 0.01% in the diet.[67]

In the rat somewhat higher concentrations may be necessary (0.1–0.2 per cent).[68] The mechanism of the effect is unknown, but it is thought to be by some action on growth of certain mesodermal tissues. It is not due to one of its major metabolites, cyanoacetic acid,[69] and both the free amino group and the cyano group seem essential for activity. It is not produced if the amino group is in the alpha position nor if placed in the gamma position in butyronitrile. On the contrary, aminoacetonitrile appears fully potent.

Some other related compounds found not to produce growth effects were propionitrile, KCN, bis(β-cyanoethyl)amine (β,β'-iminodipropionitrile), ethylene

[64] E. D. Schilling, *Federation Proc.*, **13**, 290 (1954).

[65] T. E. Backhuber, J. J. Lalich, D. M. Angevine, E. D. Schilling, and F. M. Strong, *Proc. Soc. Exptl. Biol. Med.*, **89**, 294 (1955).

[66] S. Wawzonek, I. V. Ponseti, R. S. Shepard, and L. E. Wiedenmans, *Science,* **121,** 63 (1955).

[67] B. D. Barnett, H. R. Bird, J. J. Lalich, and F. M. Strong, *Proc. Soc. Exptl. Biol. Med.*, **94,** 67 (1957).

[68] W. Dasler, *Proc. Soc. Exptl. Biol. Med.*, **85,** 485 (1954).

[69] J. J. Lalich, *Science,* **128,** 206 (1958).

cyanohydrin, β-methylaminopropionitrile, β-dimethylaminopropionitrile, and trimethylenediamine.[65,66]

β,β'-IMINODIPROPIONITRILE, $HN(CH_2CH_2CN)_2$ (bis(β-cyanoethyl)amine)

1. Physical and Chemical Properties

Physical state: colorless liquid
Molecular weight: 123.2
Melting point: —5.50°C.
Boiling point: 173°C., 10 mm. Hg
Density: 1.0165 (30°C.)
Vapor density: 4.2 (air = 1)
Vapor pressure: 1 mm. Hg (140°C.)
Soluble in water, ethanol, acetone, and benzene
Flash point: >176°F.

2. Physiological Response

The LD_{50} is greater than 3000 mg./kg. when β,β'-iminodipropionitrile is given orally to mice. Central nervous system damage was apparent in 3 days and persisted for prolonged periods. The same symptoms were noted after skin application. Damage to the lens of the eye was noted after oral dosage, but not after skin contact. The inhalation hazard is unknown.[27,63] Injection of 1–2 g./kg. in rats, mice, birds, and fish was followed in 2–10 days by a great increase in motor activity, changes in behavioral patterns, backward walking, and head twitching, similar to results caused by lysergic acid diethylamide, except for the delay in onset and permanence of symptoms. Marked histological damage was found in the brain.[70]

MALONONITRILE, $CH_2(CN)_2$ (Malonicdinitrile, Methylenecyanide, Propanedinitrile)

1. Physical and Chemical Properties

Physical state: colorless solid
Molecular weight: 66.06
Melting point: 32.1°C.
Boiling point: 220°C.
Density: 1.049 (34°C.)
Soluble in water: 13 g. per 100 ml. of water; soluble in alcohol and ether

2. Physiological Response

Stern et al.[71] found that 14 mg./kg. subcutaneously in rats produced severe symptoms of dyspnea, cyanosis, and convulsions and was a nearly fatal dose.

---

[70] H. A. Hartman and H. F. Stich, *Federation Proc.*, **16**, 358 (1957).

[71] J. Stern, C. Weil, M. Malherbe, and R. H. Green, *Biochem. J.*, **52**, 114 (1952).

Studies of tissue homogenates exposed to malononitrile showed that cyanide and thiocyanate are produced, along with an inhibition of respiration, and an increase of aerobic glycolysis resembling the action of cyanide.

Based on the above facts, the precautions and medical therapy should be the same as for cyanide. Skin contact and inhalation of dust or vapor should be prevented.

CYANOACETIC ACID, CNCH₂COOH (Malonic Mononitrile, Cyanoethanoic Acid)

1. *Source*

Reaction of sodium chloroacetate and potassium cyanide.

2. *Uses and Industrial Exposures*

As a chemical intermediate.

3. *Physical and Chemical Properties*

Physical state: white crystals
Molecular weight: 85.06
Melting point: 66°C.
Boiling point: 108°C., 15 mm. Hg
Soluble in water and alcohol

4. *Physiological Response*

While no studies were found concerning industrial hazards, cyanoacetic acid has been studied with reference to its possible role in the production of the symptoms of lathyrism by β-aminopropionitrile.[69] Injection of C_{14}-labeled β-aminopropionitrile in rats showed that 25–30 per cent could be recovered as cyanoacetic acid. In order to evaluate this metabolite, rats were given drinking water containing 200 mg. of cyanoacetic acid per 100 ml. daily for 7 weeks. No toxic effects of any sort were noted, indicating that cyanoacetic acid is not responsible for the skeletal deformities, etc., produced by feeding β-aminopropionitrile.

CYANOACETAMIDE, CNCH₂CONH₂ (Propionamide Nitrile, Nitrilomalonamide)

1. *Physical and Chemical Properties*

Physical state: white powder
Molecular weight: 84.08
Melting point: 119°C.
Boiling point: decomposes
Soluble in water: 15 g. per 100 g. of water; soluble in ethanol. 2 g. per 100 g. of ethanol

2. *Physiological Response*

Fassett[33] noted that the oral LD_{50} in rats was greater than 3200 mg./kg. and greater than 800 mg./kg. intraperitoneally. Contact with the skin of guinea pigs caused slight irritation with no evidence of toxic symptoms by skin absorption. Valdecasas[72] states that it has a very low toxicity.

METHYL CYANOACETATE, CH_3OOCCH_2CN

1. *Physical and Chemical Properties*

Physical state: liquid
Molecular weight: 99.09
Melting point: —22.5°C.
Boiling point: 203°C.
Specific gravity: 1.123 (15/4°C.)
Vapor density: 3.4 (air = 1)
Insoluble in water; soluble in alcohol and ether

2. *Physiological Response*

Fassett[33] found the oral LD_{50} in the guinea pig to be 400–800 mg./kg., and the same value intraperitoneally. Some toxic effects following skin contact were noted. While there are no reports of injury to humans handling the material, care should be used to avoid skin contact and inhalation of vapor, especially heated vapor.

ETHYL CYANOACETATE, $CH_2(CN)COOC_2H_5$ (Cyanoacetic Acid Ethyl Ester)

1. *Physical and Chemical Properties*

Physical state: colorless liquid
Molecular weight: 113.12
Melting point: —22.5°C.
Boiling point: 205–208°C.
Specific gravity: 1.0560 (25°/4°C.)
Vapor pressure: 1 mm. Hg (68°C.)
Slightly soluble in water; soluble in alcohol and ether

2. *Physiological Response*

Fassett[33] found the oral LD_{50} in rats to be greater than 400 and less than 3200 mg./kg. The LD_{50} by skin contact in the guinea pig was greater than 5 ml./kg. No skin irritation was noted, although some effects were probably produced by skin absorption. Ghiringhelli[61] obtained an LD_{50} subcutaneously in the guinea pig of about 1100 mg./kg.

[72] F. G. Valdecasas, *Arch. inst. farmacol. exptl. Madrid,* **5,** 64 (1953) *Chem. Abstr.,* **48,** 13084c (1954).

3. Odor

Mild, pleasant odor.

METHYLCYANOFORMATE, CN COOCH₃ (Cyanomethylcarbonate, Methylcyanomethanoate)

1. Physical and Chemical Properties

Physical state: colorless liquid
Molecular weight: 85.03
Boiling point: 97°C.
Density: 1.08

2. Physiological Response

Methylcyanoformate is said to act like HCN[20] but to be more active at lower concentrations. A dog recovered from a 10- to 20-minute exposure to 29 p.p.m. (0.1 mg./liter). Cats were severely affected and developed pulmonary damage by short exposures to 3–18 p.p.m. Mice succumbed to 15-minute exposures of 86 p.p.m. (0.3 mg./liter).

ETHYLCYANOFORMATE, CN·COOCH (Cyanoethylcarbonate, Ethylcyanomethanoate)

1. Physical and Chemical Properties

Physical state: colorless liquid
Molecular weight: 99.05
Boiling point: 116°C.
Density: 1.013 (20/4°C.)
Insoluble in water, soluble in alcohol

2. Physiological Response

Similar to, but slightly less potent than, methylcyanoformate.[20]

METHYLISOCYANIDE, CH₃NC (Methylcarbylamine)

1. Physical and Chemical Properties

Physical state: colorless liquid
Molecular weight: 41.05
Melting point: —45°C.
Boiling point: 59.6°C.
Density: 0.756 (4°C.)
Vapor density: 1.42 (air = 1)
Soluble in water: 10 g. per 100 ml. at 15°C.; soluble in alcohol

2. Physiological Response

The isocyanides do not appear to have been investigated intensively.[20,73] Flury and Zernik say methylisocyanide is more toxic than HCN, although ethyl isocyanide is eight times less toxic.[20,74] The isocyanides are generally regarded by chemists as hazardous. The applicability of the usual antidotes for HCN to isonitriles appears somewhat uncertain.

3. Odor

Has a strong odor.

CYANURIC CHLORIDE, N:CClN:CClN:CCl (Trichloro-s-Triazine, Tricyanogen Chloride)

1. Uses and Industrial Exposures

Cyanuric chloride is used as a chemical intermediate.

2. Physical and Chemical Properties

Physical state: colorless crystal
Molecular weight: 184.4
Melting point: 145.8°C.
Boiling point: 190°C.
Slightly soluble in water; soluble in alcohol

3. Physiological Response

Cyanuric chloride is a lachrymator and respiratory irritant similar to cyanogen chloride. The oral LD_{50} in mice was 1000 mg./kg. and in rats, 425 mg./kg.[59] Deaths were delayed with evidence of corrosive damage to the intestinal tract. Repeated application to the skin caused increasing injury. A repeated oral dose of 37 mg./kg. daily for 5 weeks caused no injury in rabbits.

Fassett[33] noted similar effects with an oral LD_{50} in the mouse of 400–800 mg./kg., but less than 10 mg./kg. intraperitoneally in this species. Slight initial skin irritation was noted in the guinea pig, which later developed a hard eschar. Eye damage was severe.

Cyanuric chloride should be handled with full precautions against skin or eye contact and inhalation of dust avoided.

4. Odor

Cyanuric chloride has a pungent odor.

[73] S. Frankel, *Die Arzneimittels Synthese,* Springer, Berlin, 1927.
[74] Y. Henderson and H. Haggard, *Noxious Gases,* Reinhold, New York, 1943.

BROMOBENZYLCYANIDE, C₆H₅CHCNBr (Bromobenzylnitrile)

1. *Physical and Chemical Properties*

Molecular weight: 182.03
Melting point: 25°C.
Boiling point: 225°C.
Density: 1.47 (solid)
Vapor density: 6.6 (air = 1)
Vapor pressure: 0.01 mm. Hg (20°C.)

2. *Physiological Response*

Bromobenzylcyanide is a highly potent lachrymator. The CN group is probably released and converted to SCN.[4] Like some other potent lachrymators, it probably acts by a progressive reaction with SH groups.[75]

Prentiss[25] gives the physiological effects of various levels in air as follows:

Lowest detectable level: 0.09 mg./cu. meter
Lowest irritant concentration: 0.15 mg./cu. meter
Intolerable concentration: 0.8 mg./cu. meter (10 min.)
Lethal concentration: 900.0 mg./cu. meter (30 min.) ; 3500.0 mg./cu. meter (10 min.)

TOLUENE-2,4-DIISOCYANATE, CH₃C₆H₃(NCO)₂ (Tolylene 2,4-diisocyanate, TDI)

1. *Uses and Industrial Exposures*

TDI is used in manufacture of polyurethane foams, foam-type insulation, etc.

2. *Physical and Chemical Properties*

Physical state: white liquid
Molecular weight: 174
Boiling point: 250°C.
Specific gravity: 1.21 (28°C.)
Vapor density: 6.0 (air = 1)
Vapor pressure: 1 mm. Hg (80°C.)
Insoluble in water; soluble in acetone, ethyl acetate, toluene, and kerosene
1 p.p.m. ⪰ 7.12 mg./cu. meter and 1 mg./liter ⪰ 140.5 p.p.m.

3. *Determination in the Atmosphere*

Colorimetric methods are available.[76]

[4] M. Dixon, *Biochemical Society Symposia No. 2*, Cambridge Univ. Press, Mass., 1948.
[76] American Ind. Hyg. Assoc. Hygienic Guide, Toluene 2,4-Diisocyanate, *Am. Ind. Hyg. Assoc. Quart.*, **18**, 370 (1957).

4. Physiological Response

Since the introduction of this material in the manufacture of synthetic foams, there have been a number of reports of severe pulmonary effects in man, usually beginning after a latent period with repeated exposures and characterized by an acute asthmalike reaction.[76] Direct pulmonary irritant effects are noted in animals,[77] although the specific picture of a sensitization-type asthma has not been reproduced in animals. Skin sensitization in animals has been demonstrated. This does not appear to have been a major problem in humans.

Merewether[78] points out that the isocyanates are very reactive substances and are known to react with various groupings in proteins and thus should be capable of forming antigens. The reaction is probably with free amino groups. In the case of phenyl isocyanate, the reaction leads to the formation of a phenylhydantoic acid.[79] There seems little doubt that the pulmonary reaction is at least partly based on some type of delayed or sensitization-type reaction.

Because the pulmonary effects can be produced in man at very low levels in air, TDI should only be used in areas with adequate general and local ventilation or with air-supplied respirator equipment. Skin contact should be avoided. Persons with chronic respiratory disease or respiratory allergies should not be exposed.

Other isocyanates may have irritant properties on the eyes or respiratory tract.

5. Threshold Limit Value

0.02 p.p.m. (0.14 mg./cu. meter).[24] May not protect persons having a previous specific sensitization.

6. Odor Threshold

0.4 p.p.m. in about half of subjects.

SODIUM DICYANAMIDE, $NaN(CN)_2$

1. Uses and Industrial Exposures

Sodium dicyanamide is used as a chemical intermediate. It is said to repel moths and to have some insecticidal properties.

2. Physical and Chemical Properties

Physical state: colorless crystals
Molecular weight: 89.04
Melting point: 315°C. (decomposes)
Soluble in water: 26.5 g. per 100 g. of water at 30°C.; soluble in methanol: 4.2 g. per 100 g. of methanol at 30°C.

[77] J. A. Zapp, Jr., *Arch. Ind. Health,* **15,** 324 (1957).

[78] E. R. A. Merewether, *Industrial Medicine and Hygiene,* Vol. III, Butterworth, London, 1956.

[79] J. S. Fruton, *General Biochemistry,* 2nd ed., Wiley, New York, 1959.

3. Physiological Response

The oral LD_{50} in mice is stated to be 1000 mg./kg. and the intraperitoneal LD_{50}, 610 mg./kg. It is not absorbed in significant amounts through the intact skin of rabbits, although it apparently penetrates the abraded skin of this species.[63]

4. Flammability

Heavy metal salts are said to explode on heating.[63]

DICYANDIAMIDE, $NH_2 \cdot C(:NH) NHCN$ (Cyanoguanidine)

1. Physical and Chemical Properties

Physical state: crystalline solid
Molecular weight: 84.08
Melting point: 209–211°C.
Boiling point: decomposes
Specific gravity: 1.40 (14°C.)
Soluble in water: 23 g. per 100 g. of water; slightly soluble in alcohol and ether

2. Physiological Response

Hald et al.[31] found the oral LD_{50} in mice to be greater than 4 g./kg. when given with alcohol and greater than 3 g./kg. in rabbits.

SODIUM CYANATE, NaOCN

1. Physical and Chemical Properties

Physical state: colorless solid
Molecular weight: 65.9
Specific gravity: 1.937 (20°C.)
Soluble in water

2. Physiological Response

Birch and Schütz[81] noted that the LD_{50} in rats intramuscularly was 310 mg./kg. Lower doses caused drowsiness. Larger doses caused drowsiness with intermittent clonic convulsions, terminating in tonic convulsions. Loss of weight and apathy was caused by repeated intramuscular doses of 50–100 mg./kg. in rats and rabbits.

Increased urinary output and diarrhea were also present.

No details of metabolism were found, but presumably the toxic effect is produced by the OCN ion and not by breakdown products. Care should be used to avoid inhalation of dust and prolonged or repeated skin contact.

[80] R. W. Berliner, Am. J. Physiol., 160, 325 (1950).

POTASSIUM CYANATE, KOCN

1. Uses and Industrial Exposures

Potassium cyanate is used as a chemical intermediate, weed killer, and in agriculture.

2. Physical and Chemical Properties

Physical state: white solid
Molecular weight: 81.1
Melting point: 315°C.
Specific gravity: 2.056 (20°C.)
Decomposes in hot water
Soluble in alcohol

3. Physiological Response

The LD_{50} in rats and mice by oral doses is about 1000 mg./kg. Dogs given 400 mg./kg. intraperitoneally show severe or fatal symptoms (vomiting, defecation, urination, lachrymation, salivation, rapid respiration, tremors, and convulsions).[59]

Birch and Schütz[81] have described somewhat similar symptoms with the sodium salt (see Sodium Cyanate).

The degree of hazard appears less than with cyanides and some nitriles, but care should be used to avoid inhalation of dust and prolonged and repeated skin contact.

POTASSIUM FERRICYANIDE, $K_2Fe(CN)_6$

1. Source

Oxidation of ferrocyanide.

2. Uses and Industrial Exposures

As a chemical reagent and in metallurgy, photography, and pigments.

3. Physical and Chemical Properties

Physical state: red solid
Molecular weight: 298.97
Specific gravity: 1.8109
Soluble in water

4. Physiological Response

Only slightly toxic (see Potassium Ferrocyanide). Converted rapidly to Ferrocyanide.[80]

POTASSIUM FERROCYANIDE, $K_4Fe(CN)_6 \cdot 3H_2O$

1. Source

Potassium ferrocyanide can be produced from gas plant by-products or from alkaline earth cyanides.

2. Uses and Industrial Exposures

Chemical reagent, metallurgy, graphic arts.

3. Physical and Chemical Properties

Physical state: lemon-yellow solid
Molecular weight: 422.39
Melting point: loses water at 60°C.
Specific gravity: 1.85 (17°C.)
Soluble in water

4. Physiological Response

Appears only slightly toxic.[33] Fassett found the oral LD_{50} in rats to be 1600–3200 mg./kg. The handling hazard is slight. No dermatitis was observed in workers handling ferro- or ferricyanide over a number of years. Sollmann[27] states that ferri- and ferrocyanide are not toxic as such, because the CN group is tightly bound. Dogs tolerate 35 cc./kg. of a 7.5 per cent solution intravenously (2625 mg./kg.) of crystalline ferrocyanide. It is rapidly excreted by glomerular filtration, similar to creatinine.[80]

The intravenous injection of 0.25 g. (calculated as anhydrous ferrocyanide) has been proposed as a test of glomerular filtration in humans.[27] Poisoning from oral ingestion seems to have been questionable.

NITROPRUSSIDE, $NA_2[FE(NO)(CN)_5] \cdot H_2O$ (Sodium Nitroferricyanide)

1. Uses and Industrial Exposures

Used as an analytical reagent. Also has been tried in hypertension.

2. Physical and Chemical Properties

Physical state: red crystals
Specific gravity: 1.72
Soluble in water: 40 g. per 100 ml.
Soluble in alcohol

3. Physiological Response

Nitroprusside is said to be decomposed in vivo to liberate cyanide. Five mg./kg. by mouth produces a fall in blood pressure similar to nitrites. It is of interest that methemoglobin is not formed. There is evidence that the cyanide liberated is converted to SCN as in the case of other nitriles.[27]

[81] K. M. Birch and F. Schütz, *Brit. J. Pharmacol.*, **1**, 186 (1946).

CHAPTER XLV

Aliphatic and Alicyclic Amines

WILLIAM L. SUTTON, M.D.

I. General Considerations

A. PHYSICAL AND CHEMICAL PROPERTIES

The aliphatic amines are derivatives of ammonia in which one or more of the hydrogen atoms are replaced by an alkyl radical. They are strongly alkaline in character. They give an alkaline reaction in aqueous solution and form salts with acids. In general the more common and widely used amines are gases or fairly volatile liquids with a pronounced odor similar to ammonia, but more fishlike. The lower amines are very soluble in water, the gaseous members of the series being commonly supplied as water solutions. The higher molecular weight amines are less volatile, odorless, and only partially soluble in water. Branching of the alkyl chain tends to enhance volatility, whereas hydroxy substitution as in the alkanolamines decreases volatility. Many of the lower aliphatic amines have low flash points and fall into the category of flammable liquids or gases. The physical properties of some of the aliphatic and alicyclic amines are given in Tables 1 and 2. The data have been obtained from standard references, 1–3 and in some cases from the manufacturer's literature.

The aliphatic amines are conveniently classified as primary, secondary, and tertiary amines according to the number of substitutions on the nitrogen atom. If only one radical is substituted, the amine is a primary amine, even though the alkyl substituent may have a secondary or tertiary structure. A further convenient subdivision that will be used in this chapter is as follows:

1. Monoamines: (a) primary; (b) secondary; (c) tertiary; (d) unsaturated; (e) alicyclic

2. Polyamines

3. Alkanolamines

The alkyl amines behave as bases in organic solvents and aqueous solutions. Base strength is expressed as the negative logarithm of the dissociation constant

[1] *Handbook of Chemistry and Physics,* 41st ed., Chemical Rubber Publishing, Cleveland, 1960.

[2] The Condensed Chemical Dictionary, 5th ed., Reinhold, New York, 1956.

[3] *Merck Index,* 7th Ed., Merck, Rahway, N. J., 1960.

TABLE 1 Physical and Chemical Properties of Aliphatic and Alicyclic Monoamines

Name	Formula	Mol. wt.	M.p., °C.	B.p., °C.	Density, g./ml.	Solubility in H₂O (g. per 100 ml.)	Vapor pressure, mm. Hg (°C.)	Vapor density (air = 1)	Flash point, °F.	Conversion units	
										1 mg./liter (p.p.m.)	1 p.p.m. (mg./cu. meter)
Methylamine	CH₃NH₂	31.06	−92.5	−6.5	0.7691(−70/4)	V. sol.	2 atm. (25°C.)	1.07	34 (30% soln.)	788	1.27
Dimethylamine	(CH₃)₂NH	45.08	−96.0	7.4	0.6804(0/4)	V. sol.	2 atm. (10°C.)	1.55	54 (25% soln.)	542	1.84
Trimethylamine	(CH₃)₃N	59.11	−124	3.5		V. sol.	760(2.9)	2.04	38 (25% soln.)	414	2.42
Ethylamine	CH₃CH₂NH₂	45.08	−80.6	16.6	0.6836(20/20)	Complete	400(2.0)	1.55	<0.0	542	1.84
Diethylamine	(CH₃CH₂)₂NH	73.14	−50	55.5	0.7108(18/4)	Complete	195(20)	2.52	5	334	2.99
Triethylamine	(CH₃CH₂)₃N	101.19	−115.3	89.5	0.7229(25/4)	1.5	53.5(20)	3.49	25	242	4.14
Propylamine	CH₃CH₂CH₂NH₂	59.11	−83	48.7	0.718(20/20)	Sol.	400(31.5)	2.04	<20	414	2.42
Di-n-propylamine	(CH₃CH₂CH₂)₂NH	101.19	−39.6	110.7	0.7384(20/4)	Sol.	30(25)	3.49	45	242	4.14
Isopropylamine	(CH₃)₂CHNH₂	59.11	−101.2	34	0.694(15/4)	Complete	460(20)	2.04	<0.0	414	2.42
Diisopropylamine	((CH₃)₂CH)₂NH	101.19	−96.3	83.4	0.720(20/20)	Sl. sol.	70(20)	3.49	30	242	4.14
n-Butylamine	CH₃(CH₂)₃NH₂	73.14	−50.5	77.8	0.740(20/4)	Complete	72(20)	2.52	45	334	2.99
Di-n-butylamine	(CH₃(CH₂)₃)₂NH	129.24	<−50	159.6	0.7613(20/20)	Sol.	1.9(20)	4.46	125	189	5.29
Tri-n-butylamine	(CH₃(CH₂)₃)₃N	185.34	<−70	214	0.775(20/20)	Insol.	20(100)	6.39	175	132	7.58
Isobutylamine	(CH₃)₂CHCH₂NH₂	73.14	−85.5	68	0.736	Complete	100(18.8)	2.52	<20	334	2.99
n-Amylamine	CH₃(CH₂)₄NH₂	87.16	−55	104	0.7614(20/4)	Sol.		3.01	30	281	3.56
Isoamylamine	(CH₃)₂CHCH₂CH₂NH₂	87.16		95	0.7505(20/4)	Sol.		3.01		281	3.56
n-Hexylamine	CH₃(CH₂)₅NH₂	101.19	−19	132.7	0.767	1.2	6.5(20)	3.49	105	242	4.14
2-Ethylbutylamine	(CH₃CH₂)₂CHCH₂NH₂	101.19		125	0.776(20/20)			3.49	70	242	4.14
n-Heptylamine	CH₃(CH₂)₆NH	115.22	−23	158.3	0.777(20/4)	Sl. sol.		3.97	130	212	4.71
Di-n-heptylamine	(CH₃(CH₂)₆)₂NH₂	213.4						7.35		115	8.73
2-Ethylhexylamine	CH₃(CH₂)₃(CH₃CH₂)CHCH₂NH₂	129.24		169.2	0.7894(20/20)	0.25		4.46	140	189	5.29
Di(2-ethylhexyl)-amine	(CH₃(CH₂)₃(CH₃CH₂)CHCH₂)₂NH	241.45		280.7	0.8062(20/20)	Insol.	<0.01(20)	8.33	270	101	9.88
Octadecylamine	CH₃(CH₂)₁₇NH₂	269.5		232.0		Insol.		9.29		91	11.02
Allylamine	CH₂:CHCH₂NH₂	57.09		53.2	0.761(20/4)	Complete		1.97	20	428	2.33
Diallylamine	(CH₂:CHCH₂)₂NH	97.16	−88.4	110.4	0.7627(10/4)	8.6		3.35	60	252	3.97
Triallylamine	(CH₂:CHCH₂)₃N	137.22	<−70	149.5	0.800(20/4)	0.25		4.73	103	178	5.61
Cyclohexylamine	C₆H₁₁NH₂	99.17		134	0.8191(20/4)	Sl. sol.		3.42	90	247	4.06
Dicyclohexylamine	(C₆H₁₁)₂NH	181.31	20	254–256	0.913–0.919 (15/15)	Sl. sol.		6.25	110	135	7.42
N,N-Dimethylcyclo-hexylamine	C₆H₁₁N(CH₃)₂	127.22	<−77	159	0.8490(20/20)	1.1	3(25)	4.39	110	192	5.20

TABLE 2

Physical and Chemical Properties of Aliphatic Polyamines

Name	Formula	Mol. wt.	M.p., °C.	B.p., °C.	Density, g./ml.	Solubility in H_2O g. per 100 ml.	Vapor pressure, mm. Hg (°C.)	Vapor density (air = 1)	Flash point, °F.	1 mg./liter (p.p.m.)	1 p.p.m. (mg./cu. meter)
Ethylenediamine	$NH_2CH_2CH_2$	60.10	8.5	116.1	0.8994(20/4)	Sol.	10(21.5)	2.07	110	407	2.46
N,N-Diethylethylenediamine	$(CH_3CH_2)_2NCH_2CH_2NH_2$	116.20		145.2	0.8211	V. sol.	4.1(20)	4.01	115	210	4.75
Trimethylenediamine	$NH_2(CH_2)_3NH_2$	74.13	−23.5	135.5	0.884(25/4)	Sol.		2.56		330	3.03
1,2-Propanediamine	$CH_3CH(NH_2)CH_2NH_2$	74.13	−37.2	120.9	0.864(20/20)	Complete	8.0(20)	2.56	92	330	3.03
Tetramethylenediamine	$NH_2(CH_2)_4NH_2$	88.15	27	158		V. sol.		3.04		277	3.60
1,3-Butanediamine	$CH_3CH_2CH(NH_2)CH_2NH_2$	88.15		142–150	0.85			3.04	125	277	3.60
Pentamethylenediamine	$NH_2(CH_2)_5NH$	102.18	9	178–180	0.9174(0/4)	Sol.		3.52		239	4.18
Hexamethylenediamine	$NH_2(CH_2)_6NH_2$	116.21	39–40	196		Sl. sol.		4.01		210	4.75
Diethylenetriamine	$(NH_2CH_2CH_2)_2NH$	103.17	−39	207.1	0.9586(20/20)	Complete	0.2(20)	3.56	215	237	4.22
Triethylenetetramine	$NH_2(CH_2CH_2NH)_2CH_2CH_2NH_2$	146.24		277.4	0.9818(20/20)	Complete	<0.01(20)	5.04	290	167	5.98
Tetraethylenepentamine	$NH_2(CH_2CH_2NH)_3CH_2CH_2NH_2$	189.30		340.3	0.9980(20/20)	Complete	<0.01(20)	6.53	92	129	7.74

(pK_b). The pK_a value (14 — pK_b) is most frequently used, the stronger bases having the higher values. The pK_a values for some of the more common amines are given in Table 3. In general, primary amines are stronger bases than ammonia,

TABLE 3

Base Strengths of Amines[a]

Amine	pK_a	Amine	pK_a
Methylamine	10.64	n-Butylamine	10.61
Dimethylamine	10.61	Di-n-butylamine	11.31
Trimethylamine	10.71	Tri-n-butylamine	10.89
Ethylamine	10.75	Allylamine	9.53
Diethylamine	11.00	Cyclohexylamine	10.79
Triethylamine	10.74	Ethylenediamine	10.08
n-Propylamine	10.59	Hexamethylenediamine	11.11
Di-n-propylamine	10.91	Ethanolamine	9.44
Isopropylamine	10.63	Diethanolamine	8.88
Diisopropylamine	11.05	Triethanolamine	7.77

[a] pK_a values in water.

and secondary amines are stronger bases than tertiary amines. As the length of the chain increases to 4 or 5 carbon atoms, the base strength tends to decrease. Diamines such as ethylenediamine also behave as strong bases.[4] The alkanolamines are weaker bases than the corresponding unsubstituted amines.

Amine salts contrast with the corresponding bases in their physical properties. They are odorless, nonvolatile solids that are insoluble in hydrocarbon solvents, which dissolve typically organic amines, but are generally readily soluble in water. Because of these more convenient properties, the amines are often used or handled in salt form.

B. MANUFACTURE AND USES

The aliphatic amines have had increasingly wide use in the past decade. Their largest use has been for chemical intermediates in the production of pharmaceutical agents, soaps, emulsifiers, dyestuffs, rubber products, flotation agents, finishing agents, and ion exchange resins. They have been used as, and for the production of, corrosion inhibitors and stabilizing agents. A recent and important use of industrial hygiene interest has been their employment as catalytic agents or hardeners in polymer formation, particularly with epoxy resins (see Chapter XXXVII). They are used in photographic processing, as petroleum additives, as reactants in polymer formations, as additives to paint strippers, paints, and leather-tanning processes. The total United States production of acyclic amines in 1958 was approximately 260 million pounds.[5]

[4] H. K. Hall, Jr., J. Phys. Chem., **60**, 63 (1956).

[5] U. S. Tariff Comm. Rept. No. **205**, 2nd ser., Synthetic Organic Chemicals, U. S. Production and Sales, 1958 (1959).

The amines are prepared in a variety of ways, the most important of which are (1) alkylation of ammonia by reacting the appropriate alkyl halide with ammonia, usually under heat and pressure and (2) reduction of unsaturated nitrogen compounds, particularly the catalytic hydrogenation of the corresponding nitrile. Where primary amines, unmixed with higher substituted amines, are desired, various processes are used, such as reduction of the corresponding oxime, the Hoffman reaction (reacting sodium hypochlorite with the corresponding amide in the presence of sodium hydroxide), or the reaction of the alkylcarboxylic acid with hydrazoic acid in the presence of an acyl azide and sulfuric acid. The methylamines are prepared by a special method in which methanol is reacted with ammonium chloride in the presence of a catalyst at elevated temperatures. The technical production of ethanolamines is by the reaction of ethylene oxide with ammonia.

C. ABSORPTION, EXCRETION, AND METABOLISM

There has been relatively little study on the metabolism of industrially important aliphatic amines, although more interest has been directed toward the pharmacologically important substituted amines. A number of aliphatic amines have been identified as normal constituents of mammalian and human urine. These include methylamine, dimethylamine, trimethylamine, ethanolamine, ethylamine, and isoamylamine as well as the catechol amines (hydroxytyramine and norepinephrine) and histamine and piperidine.[6] Rechenberger states that man excretes on the average approximately 10 mg. of volatile alkyl amine nitrogen/day.[7] Davies found as many as 8 aliphatic and ring-substituted primary amines in 24-hour urine samples studied by paper chromatography. These were excreted in amounts of approximately 20 to 100 μg./day.[8] The origin of these amines is not clear, although it has been repeatedly suggested that they may arise from absorption of primary amines formed by decarboxylation of amino acids through the action of bacteria in the gut. The sympathomimetic catechol amines, adrenalin, noradrenalin, and hydroxytyramine, and the heterocyclic-substituted ethylamine, histamine, are naturally occurring amines with a wide species distribution and considerable pharmacological importance.[9]

The amines are well absorbed from the gut and respiratory tract. The simple aliphatic amines can produce lethal effects by percutaneous absorption and the LD_{50} by this route is often approximately that determined orally. Rechenberger (see Table 4) recovered little methylamine, propylamine, or n-butylamine after oral administration of the hydrochlorides to man, whereas a high percentage of dimethylamine and diethylamine and intermediate amounts of ethylamine and isobutylamine were recovered from the urine. These findings correlate well with

[6] R. T. Williams, Detoxication Mechanisms, 2nd ed., Wiley, New York, 1959.

[7] J. Rechenberger, Z. physiol. Chem., **265**, 275 (1940).

[8] D. F. Davies, J. Lab. Clin. Med., **43**, 620 (1954).

[9] H. Blaschko, Pharmacol. Revs., **4**, 415 (1952).

what is known about amine metabolism and enzymic deamination by amine oxidases.[9,10]

Monoamine oxidase and diamine oxidase (histaminase) occur widely in animal tissues, being most concentrated in the liver, kidney, and intestinal mucosa. It is assumed that they play an important role in the "detoxication" of amines not

TABLE 4

Urinary Excretion of Amines after Oral Administration to Man[7]

Amine (HCl salt)	Dose, g.	Excretion, %
Methylamine	10.0	1.85
Dimethylamine	8.0	91.5
Ethylamine	2.0	32.0
Diethylamine	5.0	86.2
n-Propylamine	6.0	9.5
n-Butylamine	3.5	1.95
Isobutylamine	3.0	14.9

normally present as well as in the metabolism of pharmacologically important substituted amines. Monoamine oxidase will catalyze the deamination of primary, secondary, and tertiary amines according to the following over-all reaction:

$$2\ RCH_2NR'R'' + O_2 + 2\ H_2O \rightarrow 2\ RCHO + 2\ NH_2\ R'R'' + H_2O_2$$

Diamine oxidase deaminates one end of the diamine molecule as follows:

$$R'\ CH_2NH_2 + O_2 + H_2O \rightarrow R'CHO + NH_3 + H_2O_2$$

(R' contains the other basic group).

In these reactions the ammonia that is eventually formed is converted to urea. The hydrogen peroxide is acted on by catalase and the aldehyde formed is probably converted to the corresponding carboxylic acid by the action of aldehyde oxidase. The rate of oxidation of monoamines is faster with primary amines than with secondary. Tertiary amines and branched chains are more slowly oxidized than straight chains. The rate of oxidation varies with the number of carbon atoms in the carbon chain, methylamine not being attacked at all by monoamine oxidase whereas ethylamine is slowly oxidized. The rate of oxidation increases to a maximum at 5 or 6 carbon atoms and falls off with further increase in chain length. Compounds of longer chain lengths, for example, octadecylamine, may inhibit the enzyme. Short-chain diamines are oxidized by diamine oxidase (histaminase). The 4-carbon diamine, putrescene (tetramethylenediamine, 1,4-butanediamine), is most rapidly oxidized. The rate drops off with increasing chain lengths to a minimum around C_{10}. Monoamine oxidase which shows a lack of

[10] D. Richter, *Biochem. J.*, **32**, 1763 (1938).

affinity for the short diamines does oxidize the longer chains with a maximum at 13 methylene groups.[11]

Although the amine oxidases are enzymes of considerable biological importance in the inactivation of amines occurring naturally in the human body and probably for oxidative deamination of the terminal amino groups of foreign amines, this is not the only means of metabolism. Methylamine, for example, is not oxidized by amine oxidases yet it is rapidly absorbed and is not excreted in the urine to any appreciable extent. Trimethylamine is partly metabolized to ammonia and subsequently urea but is also partly oxidized to trimethylamine oxide by a specific enzyme, trimethylamine oxidase.[6] Cadaverine (pentamethylenediamine), which may be formed naturally from the amino acid lysine, appears to be cyclized to piperidine since oral administration of cadaverine to rabbits causes a several-fold increase in the normal piperidine excretion. Ring-substituted primary amines such as benzylamine and β-phenylamine are metabolized by deamination. Histamine is acetylated and methylated as well as deaminated. Many of the biologically and pharmacologically important secondary and tertiary substituted amines are metabolized by dealkylation, which may be carried out through the function of an enzyme system that is different from monoamine oxidase and is located in the microsomes of liver cells.[6]

D. PHYSIOLOGICAL AND PATHOLOGICAL EFFECTS IN ANIMALS

The results of acute toxicity tests in animals are summarized in Tables 5 and 6. The majority of the information is taken from the extensive range finding toxicity data lists from Smyth's laboratory.[12] The remainder are unpublished observations by Fassett.[13] For convenience Smyth's grading systems for eye and skin irritation have been retained; they are presented in abbreviated form at the end of Table 5.

From the industrial hygiene point of view, the most important action of the amines is the strong local irritation produced by contact with liquids, solutions, or vapors. Animals exposed to concentrated vapors exhibit signs and symptoms of mucous membrane and respiratory tract irritation. Single exposures to near lethal concentrations and repeated exposures to sublethal concentrations result in tracheitis, bronchitis, pneumonitis, and pulmonary edema. For most of the amines listed in Tables 5 and 6, a single skin application will cause deep necrosis and a drop applied to a rabbit's eye results in severe corneal damage or complete eye destruction. These effects are undoubtedly related to the alkalinity of the bases, even though a perfect correlation between pK_a values and the degree of skin or

[11] H. Blaschko and J. Hawkins, *Brit. J. Pharmacol.*, **5**, 625 (1950).

[12] (a) H. F. Smyth and C. P. Carpenter, *J. Ind. Hyg. Toxicol.*, **26**, 269 (1944), (b) **30**, 63 (1948), (c) H. F. Smyth, C. P. Carpenter, and C. S. Weil, *J. Ind. Hyg. Toxicol.*, **31**, 60 (1949), (d) *Arch. Ind. Hyg. Occupational Med.*, **4**, 119 (1951), (e) H. F. Smyth, C. P. Carpenter, C. S. Weil, and U. C. Pozzani, *Ibid.*, **10**, 61 (1954).

[13] D. W. Fassett, Laboratory of Industrial Medicine, Eastman Kodak Co., Rochester, N. Y., unpublished observations.

TABLE 5

Aliphatic Monoamines—Acute Animal Toxicity

Amine	Acute oral toxicity, LD$_{50}$, rats, g./kg.	Skin toxicity, LD$_{50}$ ml./kg.	Inhalation toxicity, rats				Skin[a] irritation, rabbit	Eye[b] effect, rabbit
			Concentration		Mortality	"Saturated" vapor, time for no deaths		
			p.p.m.	Time, hr.				
Methylamine	0.1–0.2 (10% soln.)	0.1 ml. survived[c] 40% 1.0 died					40% soln., 0.1 ml., necrosis[c]	40% soln., corneal damage
Ethylamine	0.40 0.4–0.8 (70% soln.)	0.39[d]	8,000	4	2/6	2 min.[e]	Grade 1, Necrosis from 70%[c]	Grade 9
Diethylamine	0.54	0.82[d]	4,000	4	3/6	5 min.[e]	Grade 4	Grade 10
Triethylamine	0.46	0.57[d]	1,000	4	1/6		Grade 2	Grade 9
n-Propylamine	0.2–0.4	0.05 ml. survived[c] 0.1 ml. died					Necrosis[c]	1 drop, severe eye damage
Isopropylamine	0.82	0.55[d]	8,000	4	6/6	2 min.[e]	Grade 6, Necrosis[c]	Grade 10
	<0.1 (40% soln.)	0.5[c]	4,000	4	0/6			
Diisopropylamine	0.77							
Di-n-propylamine	0.2–0.4	10 ml./kg. died[c]	1,000	4	2/6	5 min.[e]	Grade 1 Necrosis[c]	Grade 8 1 drop, corneal damage
n-Butylamine	0.5 0.2–0.4 (10% soln.)	0.5[c]	4,000 2,000	4 4	Deaths Survived	2–5 min.[e]	Necrosis[c]	Grade 9
Di-n-butylamine	0.5	1.01[c]	500 250	4 4	6/6 0/6		Grade 5	Grade 9
n-Hexylamine	0.67	0.42[d]	500	4	2/6	1 hr.	Grade 6	Grade 8
2-Ethylbutylamine	0.39	2.0[d]	1,000 500	4 4	6/6 0/6	15 min.	Grade 6	Grade 9

Compound								
Di-n-heptylamine	0.2–0.4 (undil.)	5 ml./kg. died[e]					Edema, necrosis[c]	1 drop, moderately severe
2-Ethylhexylamine	0.45		250 / 125	4 / 4	6/6 / 0/6	1 hr.	Grade 6	Grade 9
Di(2-ethylhexyl)amine	1.64					8 hrs.	Grade 5	Grade 8
Cyclohexylamine	0.4–0.8 (5% soln.)	1–5 ml./kg. LD₅₀[e]	12,000 / 1,000	6 / 6	3/3 / 0/3		Necrosis[c]	1 drop, 50%, destroyed eye
N,N-Dimethylcyclohexyla-mine	0.2–0.4 (undil.)	10–20[e]	2,670	1½	3/3		Necrosis[c]	1 drop (undil.), corneal damage
	0.8–1.6 (10%)		150	6	0/3			

[a] Skin irritation grades (adapted from H. F. Smyth, C. P. Carpenter, and C. S. Weil, *J. Ind. Hyg. Toxicol.*, **31**, 60 (1949)).
Grade 1. No reaction from undiluted.
Grade 5. Erythema, edema, slight necrosis from undiluted.
Grade 6. More necrosis from undiluted, less than edema from 10%.
Grade 7. Less than edema from 1%.
Grade 8. Less than edema from 0.1%.
Grade 10. Less than edema from less than 0.01%.

[b] Eye irritation grading system (adapted from C. P. Carpenter and H. F. Smyth, *Am. J. Ophthamol.*, **29**, 1363 (1946).

Grades
Grade 1. 0.5 ml. undiluted gives 0–1 point.
Grade 5. 0.005 ml. undiluted gives 0–5
 (0.02 ml. undiluted gives >5).
Grade 6. >40% solution gives 0–5
 (0.005 ml. undiluted gives >5).
Grade 8. >5% solution gives 0–5
 (15% solution gives >5).
Grade 10. >1% solution gives >5.

Points
Opaque cornea:
 $<\frac{1}{2}$ area = 4 points
 $>\frac{1}{2}$ area = 6
Keratoconus = 6
Corneal necrosis:
 5% of area = 1
 13–37% of area = 3
 63–87% of area = 5
 88–100% of area = 6

[c] Guinea pig.
[d] Rabbit.
[e] All died.

TABLE 6

Aliphatic Polyamines—Acute Animal Toxicity

Amine	Acute oral toxicity, LD_{50}, rats, g./kg.	Skin toxicity, LD_{50} ml./kg.	Inhalation Toxicity, Rats				"Saturated" vapor, time for no deaths, hr.	Skin[a] irritation, rabbit	Eye[b] effect, rabbit
			Concentration						
			p.p.m.	Time, hr.	Mortality				
Ethylenediamine	1.16	0.73[d]	4,000	8	6/6			Grade 6	Grade 8
			2,000	8	0/6				
N,N-Diethylethylenediamine	2.83	0.82[d]					8	Grade 6	Grade 10
1,2-Diaminopropane	2.23	0.50[d]					4	Grade 6	Grade 8
1,3-Diaminobutane	1.35	0.43[d]					8	Grade 6	Grade 9
Diethylenetriamine	2.33	1.09[d]					8	Grade 6	Grade 8
	1.8	0.17[c]							
Triethylenetetramine	4.34	0.82[d]					4	Grade 6	Grade 5
Tetraethylenepentamine	3.99	0.66[d]					8	Grade 6	Grade 4

[a] See footnote a, Table 5.
[b] See footnote b, Table 5
[c] Guinea pig.
[d] Rabbit.

eye irritation cannot be observed. The acute oral toxicities range from moderately high to slight. Some of the effects observed result from the local corrosive action of the bases on the gastrointestinal tract. The salts are less irritating and, therefore, less toxic orally. They also exhibit reduced irritation of the skin and eye when applied as solutions (see Table 7).

TABLE 7

Relative Toxicity of Amine Bases and Their Hydrochloride Salts[13] in Aqueous Solutions

| | Base[a] | | Hydrochloride Salt[b] | |
| | | | Approxi-mate oral LD_{50}, | |
Amine	Approximate Oral LD_{50}, rats, g./kg.	Eye irritation, rabbit, 1 drop	rats, g./kg.	Eye irritation, rabbit, 1 drop
Methylamine	0.1–0.2 (40%)	Inmediate, severe (40%)	1.6–3.2 (40%)	Mild, normal 24 hrs. (40%)
Ethylamine	0.4–0.8 (70%)	Inmediate, severe (70%)	>3.2 (10%)	Moderate, normal at 14 days (70%)
n-Propylamine	0.2–0.4 (10%)	Immediate, severe (undiluted)	3.2–6.4 (25%)	Mild, normal 24 hrs. (crystals)
n-Butylamine	0.2–0.4 (10%)	Immediate, severe (undiluted)	1.6–3.2 (10%)	Moderate, normal at 14 days (crystals)
Ethylenediamine	0.7–1.4 (85%)	Severe (undiluted)	1.6–3.2 (10%)	Mild, normal at 14 days (crystals)

[a] Bases = vol./vol.

[b] Salts = wt./vol.

Interest in the pharmacology of the simple aliphatic amines was initially stimulated by their structural relationship to adrenalin (epinephrine), a catechol amine. Barger and Dale introduced the term "sympathomimetic" to describe effects similar to those produced by epinephrine, including blood pressure elevation, contraction of smooth muscle, salivation, and dilatation of the pupil.[14] Using intravenous doses of aliphatic amine hydrochlorides, they found increasing pressor response with increasing straight-chain length up to C_6. Branched chain members were less active. There was decreasing sympathomimetic activity above C_7 with increasing cardiac depression. Studies on the influence of structure on epinephrinelike activity show that primary amines have somewhat more pressor activity than secondary and tertiary; straight chains are more active than branched; a second amine group in the chain decreases activity; and an amine group on the second carbon gives maximum activity.[15–17] It has generally been

[14] G. Barger and H. H. Dale, *J. Physiol.*, **41**, 19 (1910–1911).

[15] W. H. Hartung, *Chem. Revs.*, **9**, 389 (1931).

[16] E. E. Swanson and K. K. Chen, *J. Pharmacol. Exptl. Therap.*, **88**, 10 (1946).

[17] M. F. W. Dunker and W. H. Hartung, *J. Am. Pharmacol. Assoc., Sci. Ed.*, **30**, 619 (1941).

observed that when amines are administered repeatedly, cardiac stimulation is replaced by vasodilatation and cardiac depression.[18] Convulsions are a frequent finding after fatal or near fatal doses.

Pharmacology of the simple amines has also been studied in connection with the normal role of amine oxidases. Simple aliphatic amines may be both inhibitors and substrates of amine oxidase. Despite extensive investigation, the *in vivo* role of these interesting enzyme systems is still unclear. Aliphatic amines possess the ability to cause the release of histamine and potentiate its action.[19,20] The monoamines produce a typical histaminelike "triple response" (white vaso-constriction, red flare, wheal) in human skin at concentrations needed to release histamine from guinea pig lung. Maximum histamine release in the series $C_nH_{2n+1}NH_2 \cdot HCL$ is at C_{10}. Straight chain diamines show increasing release from C_6 to a maximum at about C_{14}. Potent histamine releasors, such as compound 48/80 and octylamine, cause decrease in blood pressure, tachycardia, headache,

TABLE 8

Toxicity of Amines to Paramecia[20]
(Concentration to Immobilize 50 Per Cent in Five Minutes)

Monoamines		Diamines	
No. of carbons in chain	Concentration[a]	No. of carbons in chain	Concentration[a]
C_4	400.0	C_6	300.0
C_6	300.0	C_8	200.0
C_8	20.0	C_{10}	60.0
C_{10}	2.0	C_{12}	20.0
C_{12}	0.6	C_{13}	10.0
C_{18}	2.1	C_{15}	1.0

[a] Wt./vol. of HCl salts $\times 10^{-5}$.

itching, erythema, urticaria, and facial edema when administered intravenously in man, as does histamine. It is possible that amine histamine-releasing agents could produce bronchoconstriction and wheezing by inhalation since histamine aerosol has this effect.[21] This mechanism has been proposed as a possible explana-tion for the bronchoconstriction observed in byssinosis.[22]

In addition to increasing pressor activity, histamine release, and the rate of oxidation by amine oxidase, increasing chain length results in increased toxicity to paramecia, as shown in Table 8.[20] Acute oral, cutaneous and eye toxicities of the bases do not show marked change with increasing chain length, presumably

[18] R. P. Ahlquist, *J. Pharmacol. Exptl. Therap.*, **85**, 283 (1945).
[19] J. L. Mongar and H. O. Schild, *Brit. J. Pharmacol.*, **8**, 103 (1953).
[20] J. L. Mongar, *Brit. J. Pharmacol.*, **12**, 140 (1957).
[21] T. Sollmann, *A Manual of Pharmacology*, 8th ed., Saunders, Philadelphia, 1957.
[22] A. Bouhuys, S.-E. Lindell, and G. Lundin, *Brit. Med. J.*, **1**, 324 (1960).

because these are influenced to a major degree by the acute local irritant properties, which show only slight variation (Table 5). There is a distinct tendency for an increase in the toxicity of an amine in passing from the primary to the secondary and tertiary, although this is not marked.[23,24] The monoamines of higher molecular weight generally exhibit increasing vapor toxicity (Table 5). Addition of a hydroxyl group tends to decrease toxicity. Unsaturation (allylamines) increases toxicity.

In the relatively few pathological studies that have been reported, changes in the lungs, liver, kidneys, and heart have been observed. For example, Brieger and Hodes produced pulmonary edema with hemorrhage and bronchopneumonia, nephritis, liver degeneration, and muscular degeneration of the heart in rabbits exposed repeatedly to the ethylamines.[23] With the exception of the myocardial degeneration, similar effects occur in animals exposed to ethylenediamine.[25] Myocardial damage has been described after vapor exposures to allylamines.[26] Tabor and Rosenthal[61] have shown that spermine (diamino-propyltetramethyl-enediamine) has a high degree of nephrotoxicity. Monoamines (C_1–C_{10}), diamines (C_4–C_{10}), and diethylenetriamine, triethylenetetramine, and tetraethylenepentamine were inactive. Ethylenediamine, ethyleneimine, 1,3-diaminopropane and 1,2-diaminopropane produced proteinuria and tubular damage of a lesser degree than spermine. No carcinogenic effects of aliphatic amines have been described.

Skin sensitization in animals has been reported for only a few aliphatic amines: cyclohexylamine,[27] diethylenetriamine, and n-hydroxyethyldiethylenetriamine.[28] Negative results have been obtained with di-n-propylamine, butylamine, n-hexylamine, and di-n-heptylamine in guinea pigs.[13]

E. EFFECTS IN MAN

The recorded effects in man are largely those related to the local action of the amines. Exposure to the vapors of the volatile amines produces eye irritation with lacrimation, conjunctivitis, and corneal edema resulting in "halos" around lights.[24,29,30] Inhalation causes irritation of the mucous membranes of the nose and throat and lung irritation with respiratory distress and cough. After exposure to some polyamines, asthmatic symptoms (wheezing) have been observed and tentatively attributed to respiratory tract sensitization.[31] The vapors may also produce primary skin irritation and dermatitis.[29] Direct local contact with the

[23] H. Brieger and A. M. Hodes, *Arch. Ind. Hyg. Occupational Med.*, **3**, 287 (1951).

[24] P. J. Hanzlik, *J. Pharmacol. Exptl. Therap.*, **20**, 435 (1923).

[25] U. C. Pozzani and C. P. Carpenter, *Arch. Ind. Hyg. Occupational Med.*, **9**, 233 (1954).

[26] C. H. Hine, J. K. Kodama, R. J. Guzman, and G. S. Loguvam, *Arch. Environ. Health*, **1**, 343 (1960).

[27] F. S. Mallette and E. VonHaam, *Arch. Ind. Hyg. Occupational Med.*, **5**, 311 (1952).

[28] A. K. Ingberman and R. K. Walton, *Ind. Eng. Chem.*, **49**, 1105 (1957).

[29] L. B. Bourne, F. J. M. Milner, and K. B. Alberman, *Brit. J. Ind. Med.*, **16**, 81 (1959).

[30] A. J. Amor, *Mfg. Chemist*, **20**, 540 (1949).

[31] C. U. Dernehl, *Ind. Med. and Surg.*, **20**, 541 (1951).

liquids is known to produce severe and sometimes permanent eye damage,[32] and skin burns. It has been noted that after skin contact the immediate discomfort is often less than might be expected on the basis of the severity of the skin burn that follows. Cutaneous sensitization has been repeatedly observed from some members of the series, notably the ethyleneamines.[29,31,33,34] Systemic symptoms may also result from inhalation. These include headache,[29,31] nausea, faintness, and anxiety.[35] Davies[8] states that intravenous injection of isoamylamine produces apprehension and flushing and Rechenberger[7] found an unpleasant degree of nausea and salivation from 40 mg./kg. of isobutylamine orally. These systemic symptoms are transient and are probably related to the pharmacodynamic action of the amines. The uncontrolled use of the amines in industry may be associated with lung and mucous membrane irritation from vapor exposure, skin and eye burns from the liquids, and sensitization reactions from the ethyleneamines. Experience has shown that with full use of standard industrial hygiene techniques, and with some attention to detail, these effects can be avoided.[29,36,37]

F. THRESHOLD LIMIT VALUES, MAXIMUM PERMISSIBLE EXPOSURES

Because (1) industrial experience has been relatively brief, (2) few chronic animal studies have been conducted, and (3) observations on the effects in man have seldom been accompanied by quantitation of the exposures, threshold limit values have been recommended for only a few of the aliphatic and alicyclic amines. These values appear in Table 9 with an indication of the apparent basis for the recommendation. Although these values generally appear reasonable, they are based largely on human sensory response, short-term animal exposures, or analogy and should, therefore, be considered as tentative values, useful for bench marks but requiring further evaluation and confirmation. Dernehl's experience[31] indicates the possibility of symptoms occurring in certain individuals at very low vapor concentrations. The typical odor of the amines and their mucous membrane irritation afford warning properties, which should not be entirely relied upon for exposure control.

G. DETERMINATION IN AIR

Relatively few methods for the determination of aliphatic amines in air have been published. These have largely consisted of collection in standard acid followed by titration of the excess acid with standard alkali using an appropriate

[32] C. P. Carpenter and H. F. Smyth, *Am. J. Ophthalmol.*, **29**, 1363 (1946).

[33] E. Grandjean, *Z. Präventivmed.*, **2**, 77 (1957).

[34] L. E. Savitt, *A.M.A. Arch. Dermatol.*, **71**, 212 (1955).

[35] R. M. Watrous and H. N. Schulz, *Ind. Med. and Surg.*, **19**, 317 (1950).

[36] D. J. Birmingham, *A.M.A. Arch. Ind. Health*, **19**, 365 (1959).

[37] C. H. Hine, J. S. Kodama, H. H. Anderson, D. W. Simonson, and J. S. Wellington, *A.M.A. Arch. Ind. Health*, **17**, 129 (1958).

TABLE 9

Suggested Threshold Limit Values

Amine	P.P.M.	Basis
n-Amylamine	25[a]	Analogy to ethylamine
n-Butylamine	5[b]	Analogy to ethylamine;[c] unpublished industrial experience
Cyclohexylamine	20[d]	Animal experiments;[e] unpublished industrial experience
Dicyclohexylamine	20[d]	Like cyclohexylamine
Diethylamine	25[b]	Repeated animal exposures[f]
Diisopropylamine	10[d]	Repeated animal exposures[g]
Ethylamine	25[b]	Repeated animal exposures[f]
Ethylenediamine	10[b]	Repeated animal exposures;[h] human sensory response
Isopropylamine	5[b]	Analogy to ethylamine; unpublished industrial experience
Triethylamine	25[b]	Repeated animal exposures[f]

[a] *Handbook of Organic Industrial Solvents.* Industrial Hygiene Subcommittee of the National Association of Mutual Casualty Companies, Chicago, 1958.

[b] American Conference of Governmental Industrial Hygienists, *A.M.A. Arch Ind. Health*, **20**, 266 (1959).

[c] H. F. Smyth, *Am. Ind. Hyg. Assoc. Quart.*, **17**, 129 (1956).

[d] H. B. Elkins, *The Chemistry of Industrial Toxicology*, 2nd ed., Wiley, New York, 1959.

[e] R. M. Watrous and H. N. Schulz, *Ind. Med. and Surg.*, **19**, 317 (1950).

[f] H. Brieger and A. M. Hodes, *Arch. Ind. Hyg. Occupational Med.*, **3**, 287 (1951).

[g] J. F. Treon, H. Sigmon, K. V. Kitzmiller, and F. F. Heyroth, *J. Ind. Hyg. Toxicol.*, **31**, 142 (1949).

[h] U. C. Pozzani and C. P. Carpenter, *Arch. Ind. Hyg. Occupational Med.*, **9**, 233 (1954).

indicator.[35,38,39] Although this approach is generally applicable, it suffers from interference by other bases and from relatively low sensitivity. Variations on the micro-Kjeldahl method have been recommended and used for some amines.[40,41] The ninhydrin method described by Davies *et al.*[42] has been successfully used for the determination of primary amines in air, with adequate sensitivity for industrial hygiene purposes.[43]

[38] U. C. Pozzani and C. P. Carpenter, *Arch. Ind. Hyg. Occupational Med.*, **9**, 223 (1954).

[39] E. Grandjean, *Brit. J. Ind. Med.*, **14**, 1 (1957).

[40] M. G. Jacobs, *Analytical Chemistry of Industrial Poisons, Hazards and Solvents*, Interscience Publishers, New York, 1949.

[41] American Industrial Hygiene Association Hygienic Guide Series; Diethylene Triamine and Diethylamine, *Am. Ind. Hyg. Asso. J.*, **21**, 266, 268 (1960).

[42] D. F. Davies, K. M. Wolf, and H. M. Perry, *J. Lab. Clin. Med.*, **41**, 802 (1953).

[43] R. F. Scherberger, F. A. Miller, and D. W. Fassett, *Am. Ind. Hyg. Asso. J.*, **21**, 471 (1960).

II. Specific Compounds

A. ALIPHATIC AND ALICYCLIC MONOAMINES

The physical properties of the monoamines are given in Table 1. Acute animal toxicity data appear in Table 5.

METHYLAMINES

The methylamines are supplied commercially as aqueous solutions in concentrations from 25 to 40 per cent. Although they are widely used as intermediates in the chemical and pharmaceutical industries and for dehairing hides in leather manufacture, there is little published information on their toxicity or their effects in man. The Manufacturing Chemists' Association states that methylamines are irritating to the lungs, upper respiratory tract, and eyes, and are characterized by a fish-like odor at concentrations less than 100 p.p.m.[44] Toxicity data for animals as summarized by Hartung[15] appear in Table 10.

TABLE 10

Toxicity of Methylamines for Rabbits[15]

Amine	Dose	Route	Response
Methylamine	0.3–0.4 g.	Intravenous	Not fatal
	2 g.	Subcutaneous	Not fatal
Dimethylamine	0.6 g. (salt)		Fatal
	4 g.	Oral	Fatal
Trimethylamine	6 g.	Subcutaneous	Minimum lethal dose
	0.4 g./kg.	Intravenous	Minimum lethal dose
	0.8 g./kg.	Subcutaneous	Minimum lethal dose

The odor of *methylamine* is faint but readily detectable at less than 10 p.p.m., becomes strong at from 20 to 100 p.p.m. and intolerably ammoniacal at 100 to 500 p.p.m. Olfactory fatigue occurs readily. Brief exposures to 20 to 100 p.p.m. produce transient eye, nose, and throat irritation. No symptoms of irritation are produced from longer exposures at less than 10 p.p.m.[13]

ETHYLAMINES

Brieger and Hodes[23] exposed rabbits repeatedly to measured concentrations of *ethylamine, diethylamine*, and *triethylamine*. The three amines produced lung, liver, and kidney damage at 100 p.p.m. Triethylamine produced definite degenerative changes in the heart at 100 p.p.m., whereas this was an inconstant finding with ethylamine and diethylamine. Fifty p.p.m. of the three amines was sufficient

[44] *Mfg. Chemists' Assoc. Chemical Safety Data Sheet* **SD-57**, Methylamines, 1955.

to produce lung irritation and corneal injury (delayed until 2 weeks with ethylamine). Their findings are summarized in Table 11. Ethylamine, diethylamine

TABLE 11

Response of Rabbits to Inhalation of Ethylamines[23]
(Seven Hours a Day, Five Days a Week, for Six Weeks)

Amine	Concentration	Result
Ethylamine	100 p.p.m.	Lung irritation, kidney damage
	50 p.p.m.	Lung irritation, some myocardial degeneration, eye irritation, corneal erosions, and edema
Diethylamine	100 p.p.m.	Lung irritation, liver and kidney damage
	50 p.p.m.	Lung irritation, slight liver and kidney changes, eye irritation, corneal erosions, and edema
Triethylamine	100 p.p.m.	Lung irritation, liver and kidney damage, heart muscle degeneration, and edema
	50 p.p.m.	Lung irritation, slight liver damage, eye irritation, corneal erosions, and edema

Note: Six rabbits in each exposure group. Experiments with diethylamine and triethylamine performed twice.

and triethylamine produce fractional mortalities in rats from single 4 hour exposures at calculated concentrations of 8000, 4000, and 1000 p.p.m., respectively (Table 5). Although Smyth[12] recorded no skin reaction in rabbits from undiluted ethylamine, Fassett found severe skin irritation with extensive necrosis and deep scarring from 0.1 ml. of 70 per cent ethylamine base held in contact with guinea pig skin for 24 hours and prompt necrotic skin burns from 70 per cent dropped on guinea pig skin.[13] Eye irritation and corneal edema have been reported from exposures to the three amines in industry.[30] The threshold limits of 25 p.p.m.[45] are probably sufficiently low to prevent significant eye and mucous membrane irritation. In view of the definite pathological changes produced in animals at 50 and 100 p.p.m. these threshold limits require further validation.

PROPYLAMINES

n-Propylamine (1-aminopropane) and di-n-propylamine show approximately the same degree of acute toxicity in animals. The approximate oral LD_{50} for each is 0.2 g./kg. in rats and 0.8 to 1.6 g./kg. in mice (10 per cent solutions of the bases). Both cause severe injury to the rabbit eye and guinea pig skin.[13] Five of 5 rats died during exposure to 3200 p.p.m. (calculated) of n-propylamine, but no deaths occurred after 1600 p.p.m. for 8 hours. One of 10 rats died and others exhibited reduced growth during fifty 7-hour exposures to 400 p.p.m. Weight loss, corneal opacities, and deaths occurred in all animals exposed repeatedly to 800 p.p.m.[26]

[45] American Conference of Governmental Industrial Hygienists, *A.M.A. Arch. Ind. Health*, **20**, 266 (1959).

Isopropylamine (2-aminopropane) exhibits strong local irritant properties. The results of animal studies appear in Table 5. It has good warning properties.[46] At 5 to 10 p.p.m. its ammonialike odor is definite. Between 10 and 20 p.p.m. the odor becomes strong and nose and throat irritation result from short exposures.[13] The high volatility, low flash point, and relatively low ignition temperature produce a high potential fire and explosion hazard.

Di-isopropylamine has been studied by Treon et al.[47] Deaths occurred in all animals from a single exposure to 2200 p.p.m. and in some animals from repeated exposures at 261 p.p.m. Rabbits and guinea pigs were more susceptible than cats and rats. Corneal opacities and lung irritation occurred at all concentrations. Workers exposed to concentrations between 25 and 50 p.p.m. experienced nausea and temporary impairment of vision.[47]

BUTYLAMINES

Hanzlik found that the normal butylamines produce symptoms similar to other alkylamines when administered systemically to rats.[24] The symptoms were restlessness, excitability, increased pulse and respiratory rates, dyspnea, convulsions, and death. n-Butylamine, di-n-butylamine, and tri-n-butylamine caused essentially the same kind and degree of systemic toxicity. He states that the bases produced no irritant effects on human skin.[24] However, others have found that n-butylamine and di-n-butylamine are severe skin and eye irritants (see Table 5).

n-Butylamine (1-aminobutane) at measured concentrations of 3000 to 5000 p.p.m. produces an immediate irritant response, labored breathing, and pulmonary edema with death of all rats in minutes to hours. Ten and 50 per cent v/v aqueous dilutions and the undiluted base produce severe skin and eye burns in animals. The immediate skin and eye reactions are not appreciably altered by prolonged washing or attempts at neutralization when these are commenced within 15 seconds after application. Direct skin contact with the liquid causes severe primary irritation and deep second degree burns (blistering) in man. The odor of butylamine is slight at less than 1 p.p.m., noticeable at 1 to 2 p.p.m., moderately strong at 2 to 5 p.p.m., strong at 5 to 10 p.p.m., and strong and irritating at concentrations exceeding 10 p.p.m. Workers with daily exposures of from 5 to 10 p.p.m. complain of nose, throat, and eye irritation and headaches. Concentrations of 10 to 25 p.p.m. are unpleasant to intolerable for more than a few minutes' exposure. Daily exposures to less than 5 p.p.m. (most often between 1 and 2 p.p.m.) produce no complaints or symptoms.[13] The suggested threshold limit value of 5 p.p.m. can be interpreted from these observations.

Secondary butylamine (2-aminobutane) and tertiary butylamine (2-amino-isobutane, trimethylaminomethane) have not been studied. The data attributed

[46] Mfg. Chemists' Assoc. Chemical Safety Data Sheet SD-72, Isopropylamine, 1959.

[47] J. F. Treon, H. Sigmon, K. V. Kitzmiller, and F. F. Heyroth, J. Ind. Hyg. Toxicol., 31, 142 (1949).

to Flury and Zernik[48] is apparently the information on di-n-butylamine and tri-n-butylamine developed by Hanzlik. Isobutylamine (1-amino-2-methylpropane) is said to be sympathomimetic, cardiac depressant, and convulsant.[24] Rechenberger observed unpleasant nausea and striking salivation in man after oral administration of 40 mg./kg.[7]

AMYLAMINES

In the absence of specific toxicity data, a threshold limit value of 25 p.p.m. has been suggested for n-amylamine (1-aminopentane, pentylamine) on the basis of respiratory tract and mucous membrane irritation expected by analogy with ethylamine.[49] First and second degree skin burns have been observed from direct liquid contact.[13]

Isoamylamine (1-amino-3-methylbutane) shows pressor activity in man. It produces flushing and apprehension when injected intravenously. The reaction is mild at 83 mg./kg.[8, 10] It stimulates salivary and lacrimal secretion and smooth muscle.[13] Hartung reports that 250 mg./kg. is not toxic to rabbits, 1.5 g. of the sulfate kills rats and 1.8 g. of the hydrochloride kills rabbits.[15]

HEXYLAMINES

n-Hexylamine (1-aminohexane) is an active pressor agent,[14] and an irritant for the skin, eyes, mucous membranes, and respiratory tract (see Table 5). The vapor toxicity is somewhat higher than the butylamines and skin irritation is at least as great.

2-Ethylbutylamine (1-amino-2-ethylbutane) exhibits acute toxicity for animals that is roughly equivalent to that of n-hexylamine (Table 5).

HEPTYLAMINES

Dunker et al.[50] determined the acute intraperitoneal toxicity of the four primary aminoheptanes for mice with results as listed in Table 12. 2-Aminohep-

TABLE 12

Acute Intraperitoneal Toxicity of the n-Heptylamines for Mice
(Aqueous Solutions Neutralized with 1 N HCl[50])

Amine	LD$_{50}$, mg./kg.
1-Aminoheptane	100
2-Aminoheptane	60
3-Aminoheptane	70
4-Aminoheptane	110

[48] F. Flury and F. Zernik, *Schädliche Gase,* Springer, Berlin, 1931, p. 420.

[49] *Handbook of Organic Industrial Solvents,* Industrial Hygiene Subcommittee of the National Association of Mutual Casualty Companies, Chicago, 1958.

[50] F. W. Dunker and W. Hartung, *J. Am. Pharmacol. Assoc., Sci. Ed.,* **30,** 623 (1941).

tanc has been used as a nasal vasoconstrictor. The sulfate salt has an intraperitoneal LD_{50} for rats of 42 mg./kg. It produces a sustained elevation in blood pressure in dogs. Some depressant activity and vasodilatation appear after repeated doses. In man 2 mg./kg. by mouth results in palpitation, dry mouth, and headache with slight rise in blood pressure.[51]

Di-n-heptylamine has an approximate oral LD_{50} for rats (undiluted) of 0.2 to 0.4 g./kg. The approximate oral LD_{50} for mice (5 per cent in corn oil) is 0.2 to 0.4 g./kg. Deaths occur within a few minutes with dyspnea and convulsions. One drop of the undiluted amine causes strong irritation of the eye and surrounding tissues and permanent corneal damage. It is a strong primary skin irritant.[13]

HIGHER ALKYLAMINES (C_8–C_{18})

Little toxicity information is available on amines with alkyl chains containing 8 to 18 carbons. The evidence suggests that these higher alkylamines should be strong local irritants for skin, eyes, and mucous membranes. Their higher boiling points should decrease the hazard from vapor exposures.

2-Aminooctane produces elevation of blood pressure in dogs at 1 mg./kg. The minimum lethal injected dose is 0.135 g./kg. in mice. Lethal doses result in dyspnea, excitation, convulsions, and death in respiratory paralysis.[52]

2-Ethylhexylamine (C_8) exhibits a high degree of acute vapor toxicity for rats and is a potent skin and eye irritant (Table 5). Di(2-ethylhexyl)amine (dioctylamine) shows a lower degree of oral toxicity for rats and is slightly less irritating to the skin and eye. A single concentrated vapor exposure produces no deaths in 8 hours (Table 5).

Laurylamine (dodecylamine) is classified by Fleming et al. with compounds that produce severe burns and vesication of the skin.[53]

Octadecylamine has been studied by Deichmann et al.[54] in connection with its possible use as an anticorrosive agent in live steam which could be used to cook food. Rats fed levels of 0 to 500 p.p.m. in the diet for 2 years showed no detectable effects on growth, food consumption, hematology, or microscopic pathology. At 3000 p.p.m. there was anorexia, weight loss, and some histological changes in the gastrointestinal tract, mesenteric nodes, and liver.[54] The acute oral LD_{50} for mice and rats is approximately 1 g./kg. It is a known primary skin irritant.[55]

ALLYLAMINES

The allylamines are the only unsaturated alkylamines that have been studied. Unlike the saturated lower alkylamines, the secondary and tertiary allylamines

[51] D. F. Marsh, J. Pharmacol. Exptl. Therap., **94**, 225 (1948).

[52] H. Morin, Thérapie, **7**, 57 (1952); Chem. Abstr., **47**, 1850i (1953).

[53] A. J. Fleming, C. A. D'Alonzo, and J. A. Zapp, Modern Occupational Medicine, Lea & Febiger, Philadelphia, 1954.

[54] W. B. Deichmann, J. L. Radomski, W. E. MacDonald, R. L. Kascht, and R. L. Erdman, A.M.A. Arch. Ind. Health, **18**, 483 (1958).

[55] H. H. Anderson and G. H. Hurwitz, Arch. exptl. Pathol. Pharmakol., **219**, 119 (1953).

are less toxic than monoallylamine. Their effect on vascular tissues appears to be unusual.[21] Hart found that all but 1 or 2 of 30 mice exposed to concentrations of 1.27 mM/liter of allylamine or 0.88 mM/liter of diallylamine died during the course of 10 minutes' inhalation. All survivors died within the next 48 hours. Symptoms were irritation, flushing of the ears, irregular respiration, cyanosis, convulsions, coma, and death. In rats, 0.05 cc. of allylamine and 0.1 cc. of diallylamine produced severe necrosis and death when applied to 1 sq. cm. area of shaved abdominal skin. The experimenters experienced transient irritation of the mucous membranes of the nose, eyes, and mouth with lacrimation, coryza, and sneezing after accidental exposure to the allylamine vapors. Hart and Leake felt that the characteristic ammoniacal odors and irritant properties at low concentrations would afford adequate warning. They recommended good ventilation where the allylamines were present and care to prevent skin contact.[56,57]

The toxicity of the allylamines as determined by Hine[26] is summarized in Table 13. He found decreasing acute oral and percutaneous toxicity from mono-

TABLE 13

Toxicity of Allylamines for Rats[26]

	Monoallylamine	Diallylamine	Triallylamine
Oral LD$_{50}$, mg./kg.	106	578	1310
Percutaneous LD$_{50}$ (rabbit, mg./kg.)	35	356	2250
Inhalation LC$_{50}$, p.p.m.[a]			
4-hour	286	2755	828
8-hour	177	795	554
Repeated Inhalation, 7 hours × 50, p.p.m.[b]			
Change in liver or kidney weights	5	200	100
Reduced growth	10	200	200
Deaths	40	200	200[c]

[a] Calculated.
[b] Measured.
[c] $^1/_{15}$ occurred at 100 p.p.m.

allylamine to triallylamine. However, triallylamine showed increased relative toxicity as compared to diallylamine on inhalation. Both mono- and diallylamines were severely irritating to skin and eyes while triallylamine was mildly irritating. Acute exposures to the vapors produced symptoms and findings of respiratory tract irritation. The pathological changes observed in rats following repeated inhalation included chemical pneumonias and some liver and kidney damage. The most prominent pathological effect described was myocarditis, which occurred

[56] E. R. Hart, *Univ. of Calif. Pub. in Pharmacology*, **1**, 213 (1938–1941).
[57] E. R. Hart and C. Leake, *J. Pharmacol. Exptl. Therap.*, **66**, 18 (1939).

after repeated inhalation[26] of all three allylamines. Experimental human exposures gave the following results:

1. Monoallylamine: recognizable at 2.5 p.p.m., mucous membrane irritation and chest discomfort in some persons at 2.5 p.p.m., intolerable to most at 14 p.p.m.

2. Diallylamine: recognizable but not unpleasant at 2 to 9 p.p.m., mucous membrane irritation and chest discomfort in a few subjects at 22 p.p.m., not intolerable at 70 p.p.m.

3. Triallylamine: recognizable at 0.5 p.p.m., mucous membrane irritation or chest discomfort in some at 12.5 p.p.m., increasingly frequent symptoms to 50 p.p.m., irritant symptoms more severe at 75 to 100 p.p.m., with unpleasant systemic symptoms including nausea, vertigo, and headache.

Calandra[58] conducted a 1-year chronic vapor-inhalation study with mono-allylamine in which rats, rabbits, and dogs were exposed for 8 hours a day, 5 days a week, to 5 and 20 p.p.m. No adverse effects on growth, behavioral reactions, or abnormal blood or urine changes were observed. Deaths from pneumonia occurred in 3 of 6 rabbits exposed to 20 p.p.m. Lung changes consistent with chronic irritation were found at both exposure levels. However, no myocardial damage was found in rabbits or dogs and only a few rats showed slight changes, which were not considered different than could be expected in unexposed rats. Periodic liver and kidney function tests, transaminase determinations, and electrocardiographic examinations on the dogs did not reveal any abnormalities. Congestive changes in the liver and kidneys were noted in dogs at both exposure levels.[58]

Threshold limit values have not been proposed for the allylamines. The data available indicate that their warning properties are sufficiently good to make voluntary exposure in concentrations that are significant for short-term exposure unlikely.

CYCLOHEXYLAMINES

Cyclohexylamine produced convulsant deaths in rabbits when injected in olive oil at doses of 0.5 g./kg. When administered daily for 82 days in the drinking water at 100 mg./kg., pathological findings or weight loss appeared in rabbits, guinea pigs, and rats.[59]

Watrous and Schulz[35] exposed rabbits, guinea pigs, and rats to cyclohexylamine vapors, 7 hours a day, 5 days a week, at average concentrations of 1200, 800, and 150 p.p.m. At 1200 p.p.m. all animals except 1 rat showed extreme irritation and died after a single exposure. Fractional mortality occurred after repeated exposures at 800 p.p.m. At 150 p.p.m. 4 of 5 rats and 2 guinea pigs survived 70 hours of exposure, but 1 rabbit died after 7 hours. The chief effects were irritation of the respiratory tract and eye irritation with the development of corneal opacities. No convulsions were observed.

[58] J. C. Calandra, Industrial Bio-Test Laboratory, report to Shell Chemical Co., July, 1959.
[59] T. C. Carswell and H. L. Morrill, *Ind. Eng. Chem.,* **11,** 1247 (1937).

These investigators report 3 cases of transitory systemic toxic effects from acute accidental industrial exposures. The symptoms were light headedness, drowziness, anxiety and apprehension, and nausea. Slurred speech, vomiting, and pupillary dilatation occurred in 1 case. Operators exposed to 4 to 10 p.p.m. had no symptoms.[35] In human patch tests a 25 per cent solution produced severe skin irritation and possible skin sensitization.[60] Tests in guinea pigs do not give evidence of skin sensitization.[13]

Dicyclohexylamine appears to be somewhat more toxic. Symptoms and death appear earlier in rabbits after injection of 0.5 g./kg. Doses of 0.25 g./kg. are just sublethal, causing convulsions and temporary paralysis. It is a skin irritant.[59]

N,N-Dimethylcyclohexylamine is finding use in polyurethane plastics, textiles, and as a chemical intermediate. Acute toxicity tests (see Table 5) show that it is somewhat less irritating than cyclohexylamine and less toxic on oral and intraperitoneal administration to rats and mice. The symptoms produced by effective doses are similar: weakness, tremor, salivation, gasping, and convulsions. Inhalation of both compounds causes respiratory tract irritation. Repeated skin applications of the diluted amine (1 per cent) do not give evidence of sensitization in the guinea pig.[13]

B. ALIPHATIC POLYAMINES

The physical properties of the aliphatic polyamines are given in Table 2. Acute animal toxicity data appear in Table 6.

ALIPHATIC DIAMINES

Introduction of a second amine group into the alkyl radical tends to decrease systemic toxicity. The shorter chain diamines have a sympatholytic effect upon the blood pressure rather than a sympathomimetic effect. Ethylenediamine, tetramethylenediamine (1,4-butanediamine, putrescene) and pentamethylenediamine (1,5-pentanediamine, cadaverine) cause depression of the blood pressure in animals. The longer chain diamines may exhibit sympathomimetic activity, for example, hexadecylmethylenediamine.[9] The histamine-releasing activity of the diamines is slight at C_4 (tetramethylenediamine) and increases to a maximum at C_{10} (1,10-decanediamine). There is increasing toxicity to paramecia from C_6 to C_{15}.[20] The diamines are strong bases and exhibit skin and eye irritant properties similar to the monoamines. In some cases (ethylenediamine, hexamethylenediamine) they exhibit skin sensitization properties not experienced with the corresponding monoamines. They are absorbed through the skin. The acute percutaneous toxicity is often approximately equivalent to that of the corresponding monoamine. Some renal tubular damage and proteinuria is produced by intraperitoneal injections in rats of 1,3-propanediamine and 1,2-propanediamine. Simi-

---

[60] F. S. Mallette and E. von Haam, *Arch. Ind. Hyg. Occupational Med.*, **5**, 311 (1952).

lar effects are not produced by tetramethylenediamine, pentamethylenediamine, hexamethylenediamine, or decamethylenediamine.[61]

Hexamethylenediamine (1,6-hexanediamine) causes anemia, weight loss, and degenerative microscopic changes in the kidneys and liver and to a lesser degree in the myocardium of guinea pigs after repeated doses.[61a] Conjunctival and upper respiratory tract irritation have been observed in workers handling hexamethylenediamine. One worker, out of the 20 studied, developed acute hepatitis followed by dermatitis which was attributed to hexamethylenediamine. No anemia was observed. Air concentrations varied from 2 and 5.5 mg./cu. meter during normal operations to 32.7 and 131.5 mg./cu. meter during autoclave operations in two plants.[62]

ETHYLENEDIAMINE

Ethylenediamine is a hygroscopic, fuming liquid that is used as a chemical intermediate in the preparation of dyes, inhibitors, resins, and pharmaceuticals. The results of acute animal toxicity studies are summarized in Table 6. Observations on animals show that the vapors are irritating to the eyes, mucous membranes, and respiratory tract and that the liquid causes severe skin corrosion and corneal injury.[12d,13,34] Repeated exposures of rats to measured concentrations of ethylenediamine vapors produced depilation, and lung, kidney, and liver damage at 484 p.p.m. with lesser degrees of injury at 225 and 132 p.p.m. No injury was observed at 125 p.p.m. continued for thirty, 7-hour exposures (Table 14). Renal

TABLE 14

Response of Rats to Thirty, Seven-Hour Exposures to Ethylenediamine Vapors[25]

Nominal concentration, p.p.m.	1000	500	250	125
Observed concentration, p.p.m.	484	225	132	59
Toxic deaths	30/30	16/20	0/26	0/30
Mean days to death	11.4	17.2		
Pathology	Lung, liver and kidney damage	Increased liver and kidney weights	Depilation	None

tubular damage and proteinuria are produced in rats from intraperitoneal doses of 300 mg./kg.[61] Dermatitis occurred in a high proportion of exposed operating personnel manufacturing mixed ethylene amines. It is probable that both primary irritation and sensitization occur. Respiratory irritation and asthmatic symptoms may follow exposures to low vapor concentrations.[31] Ethylenediamine is known to cause severe eye damage in man.[32] Voluntary vapor inhalation for 5 to 10 seconds produced tingling of the face and irritation of the nasal mucosa at 200 p.p.m. and

[61] C. W. Tabor and S. M. Rosenthal, *J. Pharmacol. Exptl. Therap.*, **116**, 139 (1956).
[61a] D. Ceresa and M. DeBlasis, *Med. del. lavoro*, **41**, 78 (1950).
[62] G. Gallo and L. Ghiringhelli, *Med. del. lavoro*, **49**, 688 (1958).

severe nasal irritation at 400 p.p.m.[25] A threshold limit value of 10 p.p.m. has been suggested.[45] This might be too high for sensitive individuals.

N,N-Diethylethylenediamine is less volatile than ethylenediamine. Concentrated vapors (not measured), produced no deaths in rats following an 8-hour exposure. Skin and eye effects in animals approximate those of ethylenediamine (Table 6).

DIETHYLENETRIAMINE (2,2'-Diaminodiethylamine)

Diethylenetriamine is a skin, eye, and respiratory tract irritant. It is known to produce skin sensitization and probable pulmonary sensitization. The acute toxic effects for animals as determined by Smyth are listed in Table 6. Hine and associates found the oral LD_{50} for rats to be 1.08 g./kg. The intraperitoneal LD_{50} was 0.074 g./kg. Application to rabbit skin produced maximum irritation.[37] Rats exposed to concentrated vapors (Table 6) and to 300 p.p.m.[34] showed no effects. Solutions of 15 to 100 per cent produced severe corneal injury, whereas a 5 per cent solution produced only minor injury. Human skin sensitization has been repeatedly observed, particularly during the use of diethylenetriamine as a catalyst for epoxy resins. It has been stated that, in view of the relatively high frequency of cutaneous and pulmonary sensitization, great care must be used in handling diethylenetriamine. If a definite odor of diethylenetriamine can be detected, process control may be inadequate.[41]

The substituted diethylenetriamines, N(hydroxyethyl)diethylenetriamine and N(cyanoethyl)diethylenetriamine are less toxic orally and intraperitoneally than diethylenetriamine and are less irritating to the skin of rabbits on single or repeated applications.[37]

TRIETHYLENETETRAMINE

In common with the other ethylene amines, triethylenetetramine causes skin sensitization as well as primary irritation. Exposure to the hot vapor results in respiratory tract irritation and itching of the face with erythema and edema.[29] Grandjean[33] was unable to detect triethylenetetramine in the workroom air where dermatitis was occurring. He concluded that the control problem was primarily one of preventing direct skin contact. Successful control requires good personnel training and scrupulous handling techniques.[33]

Tetraethylenepentamine has essentially the same degree of acute toxicity for animals as triethylenetetramine (Table 6). It would presumably exhibit the same important capacity for producing allergic responses in man as do the other ethylene amines.

C. ALKANOLAMINES AND ALKYLALKANOLAMINES

1. General Considerations

Properties and Uses. The alkanolamines (amino alcohols) are hydroxy substituted primary, secondary, and tertiary amines. Under appropriate conditions they enter into reactions characteristic of both alcohols and amines. Their solutions are alkaline. They form salts readily with inorganic and organic acids. The salts formed with fatty acids are technically important in emulsifying agents and special soaps. They find wide use in the chemical and pharmaceutical industries as intermediates for the production of emulsifiers, detergents, solubilizers, cosmetics, drugs, and textile-finishing agents. The physical properties of some of the more common amino alcohols appear in Table 15.

Metabolism. The metabolism of the amino alcohols has received little attention. Ethanolamine is naturally formed in mammals from serine and is a normal constituent of mammalian urine.[63] Forty per cent of N^{15}-labeled ethanolamine appears as urea in 24 hours when given to rabbits, suggesting it is deaminized. It is also methylated to choline and converted to serine and glycine. Monomethylaminoethanol and dimethylaminoethanol are intermediates in the conversion to choline. Some 33 per cent of diethylaminoethanol injected into man in 1-g. doses is excreted unchanged. The transformation of the remaining portion is unknown. It could be de-ethylated to ethanolamine and thus enter the normal metabolic pathways.[6]

Pharmacology. The pharmacological properties of dimethylaminoethanol have been studied more extensively than the other amino alcohols because of its potential therapeutic usefulness as a central nervous system stimulant.[64] Pfeiffer and co-workers found that large doses of the tartrate salt result in depression and pulmonary edema in rats. Intravenous injection in anesthetized dogs produces transient fall in blood pressure with moderate doses whereas large doses (greater than 30 mg./kg.) cause a pressor effect. It exhibits a low order of acute toxicity (LD_{50} 3.1 g./kg. intraperitoneally in mice) and convulsions do not result from single doses. On chronic administration to rats in doses of 500 mg./kg./day central nervous system stimulation appears with a lowered threshold for audiogenic seizures. Occasional deaths from maximal convulsions occur after 3 to 4 weeks. In man oral doses of 10 to 20 mg. of the base (as the tartrate salt) produce mild mental stimulation. At 20 mg./day there is a gradual increase in muscle tone and an apparent increased frequency of convulsions in susceptible individuals. Large doses produce insomnia, muscle tenseness, and spontaneous muscle twitches.[64] Triethanolamine is said to be a powerful vasodilator.[65] Smyth's laboratory[66] has found that intravenous injections of mono-, di-, and triethanolamines in dogs resulted in increased blood pressure, diuresis, salivation, and pupillary dilatation. These symptoms resemble those produced by the pharmacologically active aliphatic amines. Larger doses produced sedation, coma, and death following depression of blood pressure and cardiovascular collapse. Monoethanolamine was the most effective and triethanolamine the least effective.

[63] J. N. Luck and A. Wilcox, *J. Biol. Chem.*, **205**, 859 (1953).

[64] C. C. Pfeiffer, E. H. Jenney, W. Gallagher, R. P. Smith, W. Bevan, K. F. Killam, E. K. Killam, and W. Blackmore, *Science,* **126**, 610 (1957).

[65] E. Browning, *Toxicity of Industrial Organic Solvents,* Chemical Publishing, New York, 1953.

[66] H. F. Smyth, Mellon Institute of Industrial Research, Rept. No. **19-5**, 1956.

TABLE 15

Physical and Chemical Properties of Alkanolamines and Alkylalkanolamines

Name	Formula	Mol. wt.	B.p., °C.	Flash point, °F.	Solubility in water	Approximate oral LD_{50}, rats, g./kg.	Approximate percutaneous LD_{50}, ml./kg.	Source of toxicity data
Ethanolamine	$HOCH_2CH_2NH_2$	61.08	170.5	200	Complete	2.74		12d
Diethanolamine	$(HOCH_2CH_2)_2NH$	105.14	268	280	96.4% w/w	1.82		12d
Triethanolamine	$(HOCH_2CH_2)_3N$	149.19	360	375	Complete	9.11		12d
3-Amino-1-propanol	$HOCH_2CH_2CH_2NH_2$	75.11	168(500 mm. Hg)	>175	Misc.	0.8–1.6	Absorbed	13
Isopropanolamine	$CH_3CH(OH)CH_2NH_2$	75.11	159.9	160	Complete	4.26	1.64	12c
Triisopropanolamine	$(CH_3CH(OH)CH_2)_3N$	191.27	306.5	320	V. sol.	6.50		12b
2-Methylaminoethanol	$CH_3NHCH_2CH_2OH$	75.11	159.5	165	Complete	2.34	Absorbed	12e
2-Dimethylaminoethanol	$(CH_3)_2NCH_2CH_2OH$	89.14	134.6	105	Complete	2.34	1.37	12d
2-Ethylaminoethanol	$CH_3CH_2NHCH_2CH_2OH$	89.14	167.1	160	Complete	1.48	0.36	12e
2-Diethylaminoethanol	$(CH_3CH_2)_2NCH_2CH_2OH$	117.19	162.1	140	Complete	1.3	1.0	12b
2-Dibutylaminoethanol	$(CH_3(CH_2)_3)_2NCH_2CH_2OH$	173.29	229.7	220	0.4% w/w	1.07	1.68	12e
N-Methyl-2,2'-iminodiethanol	$CH_3N(CH_2CH_2OH)_2$	119.16	247.2	260	Complete	4.78	5.99	12e
N-Ethyl-2,2'-iminodiethanol	$CH_3CH_2N(CH_2CH_2OH)_2$	133.19	252.2	280	Complete	4.57		12e
1-Dimethylaminopropanol-2	$(CH_3)_2NCH_2CH(OH)CH_3$	103.16	125.8	95	Complete	1.89		12e

Toxicity for Animals. The toxicological properties of the alkanolamines generally resemble those of the corresponding alkylamines. The most pronounced effects in animals are those related to the local irritant effects of the concentrated alkaline liquids or solutions. Their salts show reduced local irritant activity and the tertiary alkanolamines are less irritating than the primary compounds. The *n*-alkyldialkanol compounds are less irritating and less toxic orally than the dialkylalkanolamines. The liquids and alkaline solutions produce severe eye injury in animals. The eye irritation scores in rabbits as determined in the extensive series studied in Smyth's laboratory usually lie between 5 and 9.[12,32] Primary irritation of the skin varies from slight to moderately severe. This is enhanced by repeated application or by application under an occlusive dressing. They are absorbed through the skin and when held in contact with the skin of small animals may cause death in doses that are less than those producing death by mouth (see Table 15). The acute oral toxicities for laboratory animals are generally low. Concentrated, unneutralized solutions of the more soluble alkanolamines cause intense gastrointestinal irritation with hemorrhage and congestion of the intestinal tract. Adhesions of the visceral organs are a frequent finding in survivors. Neutralization and increasing dilution reduce the oral toxicity. In the series studied by Smyth, single exposures of rats to "saturated" vapors seldom produced deaths in less than 8 hours. Except for monoethanolamine, repeated vapor exposures in animals have not been reported. The generally low vapor pressures reduce the inhalation hazard in industry. Human injuries by inhalation have not been recorded, despite the wide use of some of the alkanolamines. No threshold limit values have been established for these materials.

2. Specific Compounds

ETHANOLAMINE (2-Aminoethanol)

Monoethanolamine has had wide use in industry, yet reports of injury in man are lacking. Its physical and chemical properties are enumerated in Table 16. It is a normal constituent of human urine. The excretion rate in 8 males has been found to vary between 4.8 and 22.9 mg./day with a mean of 0.162 mg./kg./day. Eleven females excreted larger amounts varying from 7.7 to 34.9 mg./day with a mean of 0.492 mg./kg./day. The excretion rates in animals were approximately as follows: cats, 0.47 mg./kg./day; rats, 1.46 mg./kg./day; rabbits, 1.0 mg./kg./day. From 6 to 47 per cent of monoethanolamine administered to rats may be recovered in the urine.[63]

Smyth obtained the following results in a 90-day subacute oral toxicity study in rats:[12] maximum daily dose with no effect, 0.32 g./kg.; dose at which altered liver or kidney weight appeared, 0.64 g./kg.; dose at which microscopic pathology and deaths appeared, 1.28 g./kg. The acute oral LD_{50} was 2.74 g./kg. Treon and associates have studied the inhalation toxicity of monoethanolamine.[67] The con-

---

[67] J. F. Treon, F. P. Cleveland, E. E. Larson, J. Cappel, *The Response of Animals to Airborne Monoethanolamine,* presented at The 1958 Industrial Health Conference of the American Industrial Hygiene Association, Atlantic City.

TABLE 16

Physical and Chemical Properties of the Ethanolamines[1]

	Monoethanolamine (2-aminoethanol)	Diethanolamine (2,2,'- iminodiethanol)	Triethanolamine (2,2,'2"- nitrilotriethanol)
Formula	$HOCH_2CH_2NH_2$	$(HOCH_2CH_2)_2NH$	$(HOCH_2CH_2)_3N$
Molecular weight	61.08	105.14	149.19
Physical state	Clear liquid	Crystalline solid	Viscous liquid
Melting point, °C.	10.5	28	21.2
Boiling point, °C., 760 mm. Hg	170.5	268	360
Flash point, °F., open cup	200	280	375
Specific gravity	1.0179(20/20°C.)	1.0919(30/20°C.)	1.1258(20/20°C.)
Vapor pressure, 20°C.	0.4 mm. Hg	<0.01 mm. Hg	<0.01 mm. Hg
Vapor density (air = 1)	2.1	3.6	5.1
Solubility in water	Complete	96.4% by wt.	Complete
Conversion units (approx.)			
1 mg./liter to p.p.m.	401	232	164
P.P.M. to mg./cu. meter	2.5	4.3	6.1
Odor	Mildly ammoniacal	Ammoniacal	Characteristic
Uses	CO_2 extractor, alkaline conditioning agent, intermediate for soaps, detergents, textile agents	Absorbent for gases; solubilizer for 2,4-D; softener, emulsifier intermediate for detergents	Plasticizer, neutralizer alkaline dispersions, lubricant additive, corrosion inhibitor, to form soaps, detergents

ditions were such that an unknown proportion of the ethanolamine base was converted to the carbonate in the exposure chamber. Dogs and cats survived exposures to concentrations of 2.47 mg./liter for 7 hours on each of 4 consecutive days. Four of 6 guinea pigs died following exposure to concentrations of 0.58 mg./liter for 1 hour. Rats, rabbits, and mice were less susceptible than guinea pigs but more susceptible than cats or dogs. Sixty of 61 animals survived exposure to concentrations of 0.26 to 0.27 mg./liter for 7 hours on each of 5 consecutive days and 25 of 26 animals survived 25, 7-hour exposures (over a period of 5 weeks) to concentrations of 0.26 mg./liter (104 p.p.m.). The observed symptoms were primarily those of respiratory tract irritation. Eye irritation was negligible, presumably due to the formation of the carbonate under the conditions of the experiment. Pathological changes in those animals exposed to higher concentrations were chiefly those of pulmonary irritation, with some nonspecific degenerative changes in the liver and kidneys. Survivors of the lower concentrations had normal pathological findings. Weeks[68] reports that dogs, rats, and

[68] M. H. Weeks, U.S. Chemical Warfare Laboratories, Spec. Publ. **2-10**, Army Chemical Center, Maryland, 1958.

guinea pigs survived inhalation of 12 to 25 p.p.m. for 90 days, whereas fractional mortality occurred in 24 to 30 days at 100 p.p.m. (dogs) and 66 to 75 p.p.m. (rodents). Skin irritation and lethargy occurred at 5 and 12 p.p.m. He found the median detectible (odor) concentration for humans to be 3 to 4 p.p.m.[68] Ethanolamine instilled in the rabbit eye produces severe injury.[32] Browning states that when undiluted monoethanolamine is applied to human skin on gauze for $1\frac{1}{2}$ hours only marked redness and infiltration of the skin are produced.[65] Monoethanolamine can be determined in air titrametrically by collection in water and titration with standardized acid.

DIETHANOLAMINE (2,2'-Iminodiethanol)

The physical and chemical properties of diethanolamine are listed in Table 16.

The acute and subacute oral toxicity of diethanolamine for rats is somewhat greater than that for monoethanolamine. The acute oral LD_{50} is 1.82 g./kg. The maximum daily dose having no effect over a 90-day period is 0.02 g./kg. A daily dose of 0.17 g./kg. over the same period produces microscopic pathology and deaths and 0.09 g./kg. causes changes in liver or kidney weights. The undiluted liquid and 40 per cent solutions produce severe eye burns whereas 15 per cent produces minor damage. Ten per cent solutions applied to rabbits' skin cause redness. Higher concentrations cause increasing injury.[12,32]

TRIETHANOLAMINE (2,2',2''-Nitrilotriethanol)

The physical and chemical properties of triethanolamine are listed in Table 16.

Triethanolamine is generally considered to have low acute and chronic toxicity. A. J. Lehman states that if deleterious effects were to occur in man from triethanolamine, these would probably be acute in nature and due to its alkalinity rather than its intrinsic toxicity.[69] Kindsvatter[70] found the acute oral LD_{50} in rats and guinea pigs to be 8 g./kg. The symptoms observed were confined to the effects on the gastrointestinal tract. He felt the toxic effects were probably from the alkaline irritation, since larger doses of the neutralized material produced no symptoms when given in doses (10 g./kg.) at which the free base would cause 100 per cent mortality. Repeated feeding produced only slight reversible pathology in the liver and kidneys. Applications to the skin gave evidence of skin absorption.[70] Smyth also found low acute oral toxicity, the LD_{50} for rats being 9.11 g./kg.[12] In a 90-day subacute feeding experiment with rats, the maximum dose producing no effect was 0.08 g./kg. daily. Microscopic lesions and deaths occurred at 0.73 g./kg. and 0.17 g./kg. produced alterations in liver and kidney weights. Applications of 5 or 10 per cent solution to rabbit and rat skin did not

[69] A. J. Lehman, *Assoc. Food & Drug Officials U. S. Quart. Bull.,* **14,** 82 (1950).
[70] V. H. Kindsvatter, *J. Ind. Hyg. Toxicol.,* **22,** 206 (1940).

produce irritation.[65] No industrial injuries from triethanolamine have been recorded. It appears to be free of skin sensitization effects in its extensive use in cosmetic formulations.

In view of the low vapor pressure (Table 16) significant exposure by inhalation appears unlikely and the chief risk in industry would be from direct local contact of the skin and eyes with the undiluted, unneutralized liquid.

CHAPTER XLVI

Aliphatic Nitro Compounds, Nitrates, Nitrites

WILLIAM L. SUTTON, M.D.

ALIPHATIC NITRO COMPOUNDS

I. General Considerations

A. PHYSICAL AND CHEMICAL PROPERTIES

Nitro compounds are characterized by the linkage —C—NO$_2$. The nitroparaffins (nitroalkanes) have the general formula $C_nH_{2n+1} \cdot NO_2$. They occur as primary, secondary, and tertiary nitro compounds and as mono- and polynitroparaffins. The commercially available nitroparaffins, nitromethane, nitroethane, 1-nitropropane, and 2-nitropropane, are said to be of increasing interest in industry. They are oily liquids with low water solubility. Their boiling points and flash points are higher than the corresponding hydrocarbons. Although the flash points are relatively high, under certain conditions of temperature, confinement, chemical reaction, and shock explosion can result. Nitromethane is classified as a potentially explosive chemical and some serious explosions have occurred.[1] Some of the polynitroparaffins such as di- and trinitromethane and trinitroethane are more explosive. However, they have little industrial use at present.

B. MANUFACTURE AND USE

The mononitroparaffins are manufactured commercially by the vapor-phase reaction between nitric acid vapor and propane at elevated temperature and pressure. The nitroparaffins are separated by subsequent distillation. They find use in industry as solvents for cellulose esters and other resins and for fats, oils, waxes, and dyes. They are intermediates in organic synthesis of certain pharmaceuticals, dyes, insecticides, and textile chemicals and for possible explosive formulations. They have been used for fuels and fuel additives.

C. SYMPTOMS IN ANIMALS

The nitroparaffins act chiefly as moderate irritants when inhaled. Exposed animals give evidence of restlessness, discomfort, and signs of respiratory-tract

---

[1] National Board of Fire Underwriters, *Nitroparaffins and Their Hazards*, New York, 1959.

irritation followed by eye irritation, salivation, and later central nervous system symptoms consisting of abnormal movements with occasional convulsions. 2-Nitropropane appears to be more irritating than nitroethane, which is more irritating than nitromethane. Anesthetic symptoms are generally mild, and appear late; most animals that manifest anesthesia die later. This is more marked with nitroethane and nitropropane than with nitromethane inhalation. Incomplete information on the polynitroparaffins indicates that an increased number of nitro groups results in increased irritant properties.[2,3] The chlorinated nitroparaffins are more irritating than the unchlorinated compounds.[4] This reaches a severe degree with trichloronitromethane (chloropicin). Unsaturation of the hydrocarbon chain in the nitroolefins also results in an increase in the irritant effects.[5]

The nitroparaffins fail to show significant pharmacological effects on blood pressure or respiration.[6,7] Oral doses result in symptoms similar to those produced by inhalation except for the additional evidence of gastrointestinal tract irritation. They are less potent methemoglobin formers than the aromatic nitrocompounds. As with other methemoglobin formers, cats are more susceptible than other animals. The formation of Heinz bodies from the inhalation of 2-nitropropane has been described.[8]

D. PATHOLOGY IN ANIMALS

Animals dying following inhalation of the nitroparaffins show general visceral and cerebral congestion. After exposure to high concentrations there is pulmonary irritation and edema, the latter inadequate to be the sole cause of death. Inhalation of 2-nitropropane results in general vascular endothelial damage in all tissues as well as pulmonary edema and hemorrhage, brain and liver damage. Microscopically the liver shows severe parenchymal degeneration and focal necrosis.[8] Sublethal concentrations of nitromethane produce severe liver changes in dogs consisting of infiltration with chronic inflammatory cells, fatty changes, congestion, and some hemorrhage and necrosis.[7] Toxic damage to the kidneys and heart is less prominent. Inhalation of trichloronitromethane produces severe injury of the respiratory tract consisting of inflammation and necrosis of the bronchi and edema and congestion in the alveoli. The chlorinated nitroparaffins and nitroolefins produce gastrointestinal tract irritation and damage when given by mouth. The monochloronitroparaffins are not markedly irritating to the skin or eyes but

[2] W. Machle, E. W. Scott, and J. Treon, *J. Ind. Hyg. Toxicol.*, **22**, 315 (1940).

[2a] E. W. Scott and J. Treon, *Ind. Eng. Chem., Anal. Ed.*, **12**, 189 (1940).

[3] H. J. Horn, *Arch. Ind. Hyg. Occupational Med.*, **10**, 213 (1954).

[4] W. Machle, E. W. Scott, J. F. Treon, F. F. Heyroth, and K. V. Kitzmiller, *J. Ind. Hyg. Toxicol.*, **27**, 95 (1945).

[5] W. B. Deichmann, M. L. Keplinger, and G. E. Lanier, *A.M.A. Arch. Ind. Health*, **18**, 312 (1958).

[6] W. Machle and E. W. Scott, *Proc. Soc. Exp. Biol. Med.*, **53**, 42 (1943).

[7] J. H. Weatherby, *A.M.A. Arch. Ind. Health*, **11**, 102 (1955).

[8] J. Treon and F. R. Dutra, *Arch. Ind. Hyg. Occupational Med.*, **5**, 52 (1952).

the dichloro compounds and particularly the nitroolefins are strong skin and eye irritants.

E. ABSORPTION, EXCRETION, AND METABOLISM

The nitroparaffins are absorbed through the lung and from the gastrointestinal tract. Applications to the skin give no evidence of sufficient absorption to produce systemic injury. The studies by Scott[9-11] have demonstrated that the mononitroparaffins disappear from the blood rapidly after inhalation or oral administration. A portion is excreted unchanged in the expired air and part is metabolized with the formation of nitrite and nitrate, which can be found in the blood and urine. Nitrite can be found in the blood after the administration of nitroethane, the nitropropanes, and nitrobutanes but not after nitromethane or 2-nitro-2-methylpropane.[9-11] The over-all reactions *in vivo* are assumed to be as follows:

$$R\text{-}CH_2NO_2 \rightarrow R\text{-}CHO + H^+ + NO_2^- \rightarrow NO_3^-$$

The chloronitroparaffins do not show appreciable skin absorption but the nitroolefins are rapidly absorbed by all routes including the skin. Their metabolism does not appear to have been studied.

F. EFFECTS IN MAN

Published reports of human experience are scanty. Anorexia, nausea, vomiting, and intermittent diarrhea and headaches in men exposed to 20 to 45 p.p.m. of nitropropane during a dipping process ceased to appear when methylethyl ketone was substituted.[12] Nasal irritation, burning eyes, dyspnea, cough, chest oppression, and dizziness in men handling crude TNT have been attributed to their tetranitromethane exposure.[13] Headache, methemoglobinemia, and a few deaths have also been attributed to similar exposures. There has been more human experience with trichloronitromethane (chloropicrin) because it was used as a war gas often in mixtures with chlorine or phosgene. Chloropicrin is a lacrimator which produces coughing nausea, and vomiting and severe injury of the respiratory tract resulting in pulmonary edema. It has been noted that individuals who have been injured with chloropicrin appear to become more susceptible so that concentrations not producing symptoms in others cause them distress.[14] No specific injuries from the nitroolefins have been recorded. It is felt that these compounds might occur in irritant "smogs" of the Los Angeles type as a result of the combination of nitrogen oxides and olefins from automobile exhaust.[5]

[9] W. Machle, E. W. Scott, and J. Treon, *J. Ind. Hyg. Toxicol.*, **24,** 5 (1942).

[10] E. W. Scott, *J. Ind. Hyg. Toxicol.*, **24,** 226 (1942).

[11] E. W. Scott, *J. Ind. Hyg. Toxicol.*, **25,** 20 (1943).

[12] J. B. Skinner, *Ind. Med.*, **16,** 441 (1947).

[13] R. F. Sievers, E. Rushing, H. Gray, and A. R. Monaco, *Public Health Repts. U. S.*, **62,** 1048 (1947).

[14] E. B. Vedder, *The Medical Aspects of Chemical Warfare*, Williams & Wilkins, Baltimore, 1925.

G. THRESHOLD LIMIT VALUES

Threshold limit values have been suggested for several of the aliphatic nitro compounds (see Tables 1 and 7).[15] Except for 2-nitropropane and chloropicrin (trichloronitromethane) these are based on animal studies.

H. DETERMINATION IN AIR

Nitromethane can be determined by means of the color developed when it is heated with vanillin in an ammoniacal solution.[2] Scott and Treon[2a] have described a colorimetric method for the primary mononitroparaffins utilizing the red color developed by ferric chloride in an acidified sample that has been treated with sodium hydroxide. Tetranitromethane has been measured by collection in reagent grade methanol followed by reading at 240 mμ on a Beckman spectrophotometer and comparison to reference calibration curves.[3] Treatment of a sample collected in ethyl alcohol with pyridine and alcoholic benzidine produces a color that can be compared with freshly prepared standards.[16] Other methods for the nitroparaffins may be useful.[8,17] The chlorinated nitroparaffins can be determined by colorimetric procedures.[4]

II. Specific Compounds

A. NITROPARAFFINS

The physical and chemical properties of the more common nitroparaffins appear in Table 1 along with recommended threshold limit values and acute oral lethal doses for rabbits. Details of the toxicity for the other nitroparaffins are lacking. Scant information indicates that dinitromethane and nitropentane have actions similar to the more common nitroparaffins.[2]

NITROMETHANE, CH$_3$NO$_2$

Nitromethane is obtained from the nitration of propane. It can also be produced by the reaction of methane with oxides of nitrogen under pressure. Increased interest has been exhibited in its use as a fuel additive and as an intermediate for fuels and propellants. Nitromethane is classified as potentially explosive. Under ordinary conditions it is relatively stable but it can be detonated by mechanical and air-pressure impact or by heat, particularly when confined or mixed with certain other materials. The force of such an explosion is large and the damage from two separate tank-car explosions was extreme.[1]

Animal toxicity has been studied by Machle and co-workers and by Weatherby.[2,7] This information is summarized in Table 2. Animals exposed to 30,000 p.p.m. in air for longer than 1 hour developed pronounced nervous system

[15] American Conference of Governmental Industrial Hygienists, *A.M.A. Arch. Ind. Health,* **20,** 266 (1959).

[16] V. B. Vouk and O. A. Weber, *Brit. J. Ind. Med.,* **9,** 32 (1952).

[17] I. R. Cohen and A. P. Altshuller, *Anal. Chem.,* **31,** 1638 (1959).

TABLE 1

Properties of the Mononitroparaffins and Tetranitromethane

Name	Formula	Mol. wt.	B.p., °C.	Sp. gr.	Solubility in H_2O, % by vol.	Vapor pressure, mm. Hg (°C.)	Vapor density (air=1)	Flash point °F.	Conversion units		Oral lethal dose, rabbits, g./kg.[a]	Threshold limit value,[b] p.p.m.
									1 mg/liter (p.p.m.)	1 p.p.m. (mg./cu. meter)		
Nitromethane	CH_3NO_2	61.04	101	1.139 (20/20°C.)	9.5	27.8 (20)	2.11	112	400.7	2.495	0.75–1.0	100
Nitroethane	$CH_3CH_2NO_2$	75.07	114.8	1.052 (20/20°C.)	4.5	15.6 (20)	2.58	106	325.7	3.07	0.50–0.75	100
1-Nitropropane	$CH_3CH_2CH_2NO_2$	89.09	132	1.003 (20/20°C.)	1.4	7.5 (20)	3.06	120	274.7	3.64	0.25–0.50	
2-Nitropropane	$CH_3CH(NO_2)CH_3$	89.09	120	0.992 (20/20°C.)	1.7	12.9 (20)	3.06	103	274.7	3.64	0.50–0.75	50
1-Nitrobutane	$CH_3CH_2CH_2CH_2NO_2$	103.12	151	0.9774 (15.6/15.6°C.)	0.5	5(25)	3.6		237.1	4.21	0.50–0.75	
2-Nitrobutane	$CH_3CH_2CH(NO_2)CH_3$	103.12	139	0.9728 (15.6/15.6°C.)	0.9	8 (25)	3.6		237.1	4.21	0.50–0.75	
Tetranitromethane	$C(NO_2)_4$	196.04	125.7	1.650 (13/4°C.)	Insoluble	13 (approx., 25)	6.8		124.7	8.02		1

[a] W. Machle, E. W. Scott, and J. Treon, *J. Ind. Hyg. Toxicol.*, **22**, 315 (1940).
[b] American Conference of Governmental Industrial Hygienists, A.M.A. *Arch. Ind. Health*, **20**, 266 (1959).

TABLE 2
Acute Toxicity of Nitromethane[2,7]

Route	Animal	Dose	Mortality
Oral	Dog	0.125[a]	0/2
		0.25–1.5[a]	12/12
	Rabbit	0.75–1.0[a]	Lethal dose
	Mouse	1.2[a]	1/5
		1.5[a]	6/10
Subcutaneous	Dog	0.5–1.0[b]	MLD
Intravenous	Rabbit	0.8[a]	2/5
		1.0[a]	2/6
		1.25–2.0[a]	9/9
Inhalation	Rabbit	30,000 p.p.m., <2 hrs.	0/6
		2 hrs.	2/2
		10,000 p.p.m., 6 hrs.	2/2
		1–3 hrs.	0/4
		5,000 p.p.m., 6 hrs.	1/2
		3 hrs.	0/2
		500 p.p.m., 140 hrs.	0/2
	Guinea pig	30,000 p.p.m., 1–2 hrs.	4/4
		30 min.	1/2
		15 min.	0/2
		10,000 p.p.m., 3–6 hrs.	4/4
		1 hr.	0/2
		1,000 p.p.m., 30 hrs.	2/2
		500 p.p.m., 140 hrs.	0/2
	Monkey	1,000 p.p.m., 48 hrs.	1/1
		500 p.p.m., 140 hrs.	0/1

[a] Dose, g./kg.
[b] Dose, ml./kg.

symptoms. At 10,000 p.p.m. nervous system symptoms did not appear until after 5 hours. During exposure to lower concentrations there was slight irritation of the respiratory tract without evidence of eye irritation. This was followed by mild narcosis, weakness, and salivation. Rabbits, guinea pigs, and monkeys all survived repeated exposures for a total of 140 hours at concentrations of 500 p.p.m. A single monkey exposed to 1000 p.p.m. for eight 6-hour exposures died. No remarkable changes in either blood pressure or respiration followed intravenous injection of nitromethane in anesthetized dogs. The histopathological changes observed following acute poisoning by all routes were chiefly confined to the liver and kidneys with the liver showing the most prominent injury. Subcapsular damage, focal necrosis, both periportal and midzonal fatty infiltration, congestion, and edema were observed. Three of 10 rats given 0.25 per cent nitromethane in their drinking water for 15 weeks and 4 of 10 rats given 0.1 per cent died during the course of the experiment. The surviving animals failed to gain weight normally. Histopathological examination showed mild but definite liver abnormalities. Methemoglobinemia has not been observed in rabbits or rats. Nitromethane is apparently metabolized by a different mechanism than nitroethane and nitropropane in that negligible amounts of nitrites are found in the blood following intravenous injection of 1 mM in rabbits. Skin application of nitromethane does not produce irritation or death in animals.

The most important observations are that inhalation of nitromethane produces mild irritation and toxicity before narcosis occurs and that liver damage can result from repeated administration. The odors of nitroparaffins are easily detectable and concentrations below 200 p.p.m. are disagreeable to most observers.[2] The odor and sensory symptoms are not dependable warning properties. The suggested threshold limit value of 100 p.p.m. can be interpreted from the acute and repeated vapor exposures of animals. No injuries in man have been reported.

TETRANITROMETHANE, $C(NO_2)_4$

Tetranitromethane is a colorless oily fluid with a distinct odor. It is explosive and is more easily detonated than TNT. Its explosive power is less than TNT except when mixed with hydrocarbons, the mixtures being more powerful explosives and very sensitive to shock. Accidental explosions have occurred in handling and manufacture. It occurs as a contaminant of crude TNT. It can be prepared by the nitration of acetylene with nitric acid to form trinitromethane, the mixture of trinitromethane and nitric acid being converted to tetranitromethane by sulfuric acid at elevated temperatures. It is of interest for use as a propellant and as a fuel additive.

A summary of the data on the response of animals to inhalation of various concentrations of tetranitromethane appears in Table 3. In all reported experiments, exposed animals have exhibited similar symptoms, chiefly those of respira-

TABLE 3
Response to Various Concentrations of Tetranitromethane

Animal	Concentration, p.p.m.	Duration of exposure	Response
1 cat	100	20 min.	Death in 1 hr.[a]
1 cat	10	20 min.	Death in 10 days[a]
5 cats	7–25	$2^1/_2$–5 hrs.	Death in 1–$5^1/_2$ hr.[b]
2 cats	3–9	6 hrs. × 3	Severe irritation[b]
2 cats	0.1–0.4	6 hrs. × 2	Mild irritation[b]
20 rats	1230	1 hr.	All died in 25–50 min.[c]
20 rats	300	$1^1/_2$ hrs.	All died in 40–90 min.[c]
20 rats	33	10 hrs.	All died in 3–10 hrs.[c]
19 rats	6.35	6 mos.	11 deaths[c]
2 dogs	6.35	6 mos.	Mild symptoms[c]

[a] F. Flury and F. Zernik, *Schädliche Gase*. Springer, Berlin, 1931.

[b] R. F. Sievers, E. Rushing, H. Gay, and A. R. Monaco, *Public Health Repts. U. S.*, **62**, 1048 (1947).

[c] H. J. Horn, *Arch. Ind. Hyg. Occupational Med.*, **10**, 213 (1954).

tory tract irritation. The first signs are increased preening, change in the respiratory pattern, and evidences of eye irritation followed by rhinorrhea, gasping and salivation. The symptoms progress to cyanosis, excitement, and death at higher concentrations. Methemoglobinemia occurred in exposed cats. It should be noted that Sievers *et al.*[13] exposed their animals to tetranitromethane from crude trinitrotoluene and although the concentrations recorded for tetranitromethane were determined by sampling and analysis, other unknown contaminants could have been present. These investigators found that animals exposed to 3 to 9 p.p.m. for 1 to 3 days developed pulmonary edema. Lower concentrations (0.1 to 0.4 p.p.m.) produced only mild irritation. Horn[3] exposed 2 dogs and 19 rats to 6.35 p.p.m. for 6 hours a day, 5 days a week, for 6 months. Eleven of the 19 rats died in the course of the experiment with evidence of pulmonary irritation, edema, and pneumonia. Some initial anorexia was observed in the dogs. Repeated examinations did not reveal anemia, Heinz bodies, methemoglobinemia, or biochemical disturbances.

The results of pathological examinations on animals dying from acute exposures were all similar. There was marked lung irritation with destruction of epithelial cells, vascular congestion, pulmonary edema, and emphysema with tracheitis and bronchopneumonia. Nonspecific changes in the liver and kidney were observed in some animals. Rats surviving 6.35 p.p.m. for 6 months developed pneumonitis and bronchitis of a moderate degree whereas those dying developed more severe pneumonia. Histopathological examination of 2 dogs surviving the same concentration for 6 months revealed no evidence of injury.[3]

A few deaths and intoxications that have occurred during handling of heated contaminated TNT have been attributed to tetranitromethane.[2] Symptoms experienced in the laboratory production of tetranitromethane were irritation of eyes, nose, and throat from acute exposures and, after more prolonged inhalation, headache and respiratory distress.[18] Skin irritation does not result from repeated contact in man or animals.

Tetranitromethane can be recognized by its characteristic acrid biting odor. It can be determined in air by the methods used by Horn[3] or Vouk and Weber.[16] Sievers *et al.*[13] felt that the safe working level should be below 5 p.p.m. The American Conference of Governmental Hygienists suggests a threshold limit value of 1 p.p.m.[15] Human experience is lacking but the animal experiments suggest that this level may not be low enough to prevent all irritation. Elkins has selected 0.5 p.p.m. for the maximum allowable concentration.[19]

NITROETHANE, $CH_3CH_2NO_2$

Nitroethane is an oily liquid with a somewhat pleasant odor. It represents a lesser explosive hazard than nitromethane and tetranitromethane. Unconfined

---

[18] K. F. Hager, *Ind. Eng. Chem.*, **41**, 2168 (1949).

[19] H. B. Elkins, *The Chemistry of Industrial Toxicology*, 2nd ed., Wiley, New York, 1959.

quantities are not exploded by heat or shock. Since under appropriate conditions of confinement or contamination with other materials explosions could result, safe handling procedures for nitroethane have been recommended in detail.[1]

The response of rabbits and guinea pigs to inhalation of nitroethane as determined by Machle and co-workers[2] appears in Table 4. It is a moderate respiratory tract irritant. There was more respiratory tract irritation and less narcosis with nitroethane than was observed with nitromethane. Except for this, the symptomatology and pathological findings were similar to nitromethane. They found no evidence of skin irritation or skin absorption. Scott found increasing nitrite concentrations in the blood of rabbits during inhalation of nitroethane.[11]

TABLE 4
Response to Inhalation of Nitroethane[2]
(Two rabbits and two guinea pigs in each experiment)

Concentration, p.p.m.	Time, hr.	Mortality	
		Rabbit	Guinea pig
30,000	1.25	2	2
	1	1	1
	0.5	1	None
10,000	3	2	1
	1	1	None
5,000	3	2	None
	2	1	None
2,500	3	None	None
1,000	2	1	None
	6	None	None
500	30	None	
	140	None	None[a]

[a] One monkey exposed, not fatal.

On the basis of this information Cook suggested a maximum allowable concentration of 200 p.p.m.[20] The American Conference of Governmental Industrial Hygienists and Elkins have selected 100 and 50 p.p.m., respectively.[15,19] No injuries or observations on men exposed to measured concentrations in industry have been recorded.

2-NITROPROPANE, $CH_3CH(NO_2)CH_3$

The physical and chemical properties and the acute oral toxicity for rabbits of 2-nitropropane appear in Table 1. The nitropropanes are less of an explosive hazard than the nitromethanes.

Treon and Dutra[8] exposed rats, guinea pigs, rabbits, and cats to measured concentrations of 2-nitropropane. Considerable difference between species was

[20] W. A. Cook, *Ind. Med.*, **14**, 936 (1945).

observed. Cats were the most susceptible and guinea pigs the least (see Table 5). High concentrations of 2-nitropropane produced dyspnea, cyanosis, prostration, some convulsions, lethargy, and weakness proceeding to coma and death. Some animals surviving the acute exposure died 1 to 4 days later. Cats, rabbits, rats, guinea pigs, and monkeys were exposed repeatedly to 328 and 83 p.p.m. for 7 hours a day. Cats died following several days exposure to 328 p.p.m. but rabbits, rats, and guinea pigs survived 130 exposures and a monkey survived 100. No signs or symptoms were observed in any animals during 130 exposures to 83 p.p.m.

High concentrations of 2-nitropropane caused pulmonary edema and hemorrhage, selective disintegration of brain cells, and hepatocellular damage, with general vascular endothelial injury in all tissues. Cats that died following several exposures to 328 p.p.m. had microscopic evidence of focal necrosis and parenchymal degeneration in the liver and slight to moderate degeneration of the heart and kidneys. The lungs showed pulmonary edema, intraalveolar hemorrhage, and interstitial pneumonitis. The other species exposed at this concentration did not

TABLE 5
Response to Inhalation of 2-nitropropane[3]

Animal	Highest tolerable concentration (p.p.m.)			Lowest lethal concentration (p.p.m.)		
	1 hr.	2.25 hrs.	4.5 hrs.	1 hr.	2.25 hrs.	4.5 hrs.
Rat	2353	1372	714[a]	3865	2633	1513
Guinea pig	9523	4313	2381		9607	4622[b]
Rabbit	3865	2633	1401	9523	4313	2381
Cat	787	734	328	2353	1148	714

[a] Time, 7 hrs.
[b] Time, 5.5 hrs.

exhibit these findings. Except for 1 cat, no microscopic tissue changes were observed in the animals exposed to 83 p.p.m. Inhalation of 2-nitropropane induced methemoglobin formation in cats and to a lesser extent in rabbits. Cats developed 25 to 35 per cent methemoglobin when exposed to 750 p.p.m. for $4\frac{1}{2}$ hours and about 15 to 25 per cent methemoglobin during repeated daily 7 hour exposures to 280 p.p.m. Heinz bodies appeared in the erythrocytes of cats and rabbits at even lower concentration.

Workers dipping forms into a solvent mixture of xylene and 2-nitropropane at 110 to 120°F. developed anorexia, nausea with vomiting, and occasionally diarrhea. They felt well the following morning but the symptoms recurred during exposure. Concentrations of 20 to 45 p.p.m. of 2-nitropropane were found in the work area (xylene concentrations not given). Substitution of methyl ethyl ketone for 2-nitropropane in the solvent mixture relieved the symptoms.[12] A spraying operation utilizing a lacquer with 20 per cent 2-nitropropane did not produce symptoms in personnel exposed not more than 4 hours a day for 3 days a week.

Concentrations found ranged from 10 to 30 p.p.m.[12,19] On this basis Skinner[12] suggested 25 p.p.m. as the maximum allowable concentration.

Treon and Dutra[8] on the basis of their animal experiments suggested that 50 p.p.m. might be considered tolerable. The American Conference of Governmental Industrial Hygienists has accepted 50 p.p.m.[15] Human experience in industry indicates that persons exposed repeatedly to concentrations averaging 50 p.p.m. should be carefully observed medically. 2-Nitropropane cannot be detected by odor at 83 p.p.m.[8]

1-NITROPROPANE, $CH_3CH_2CH_2NO_2$

The acute inhalation toxicity of 1-nitropropane is not greatly different from that of 2-nitropropane (see Tables 1 and 6). Exposures to 5000 p.p.m. of 1-nitro-

TABLE 6
Response to Inhalation of 1-Nitropropane[2]

Concentration, p.p.m.	Time, hr.	Mortality[a]	
		Rabbit	Guinea pig
10,000	3	2	2
10,000	1	None	1
5,000	3	2	2

[a] Two rabbits and 2 guinea pigs in each experiment

propane for 3 hours killed rabbits and guinea pigs, whereas the lowest lethal concentrations of 2-nitropropane for these animals after a 2.25-hour exposure were 4313 and 9607 p.p.m., respectively.[2,8] The symptoms and gross pathological changes observed in the exposed animals were similar to those exposed to nitroethane. Most subjects exposed experimentally to 1-nitropropane for short periods found concentrations exceeding 100 p.p.m. irritating.[21] No human experience in industry has been reported. Since the limited animal data suggest approximately equivalent acute toxicity to 2-nitropropane, the threshold limit value for 1-nitropropane should probably be the same. Cook selected 100 p.p.m. and Elkins has suggested 25 p.p.m.[19,20]

NITROBUTANES

The toxicology of 1-nitrobutane and 2-nitrobutane has not been studied beyond that reported by Machle et al.[2] (Table 1). The effects following oral administration in rabbits were similar to those produced by the other nitroparaffins and the lethal dose range was the same as for 2-nitropropane and nitroethane. As with these materials, no skin irritation or systemic symptoms were observed after 5 daily open applications to rabbit skin. Less nitrite can be recovered from rabbit

[21] L. Silverman, H. F. Schulte, and M. W. First, *J. Ind. Hyg. Toxicol.*, **28**, 262 (1946).

TABLE 7
Physical and Chemical Properties of the Chlorinated Mononitroparaffins

Name	Formula	Mol. wt.	B. p. °C.	Sp. gr.	Solubility in H_2O, ml. per 100 ml. (20°C.)	Vapor pressure, mm. Hg (25°C.)	Vapor density (air = 1)	Flash Point, °F.	Conversion units 1 mg./liter (p.p.m.)	Conversion units 1 p.p.m. (mg./cu. meter)	Oral lethal dose, rabbits, g./kg.	Threshold limit value, p.p.m.
Trichloronitromethane (chloropicrin)	CCl_3NO_2	164.39	112	1.651 (20/4°C.)	Insoluble	16.9 (20°C.)	5.7		148.8	6.72		0.1
1-Chloro-1-nitroethane	$CH_3CHClNO_2$	109.5	127.5	1.286 (20/20°C.)	0.4	11.9	3.6	133	237	4.21	0.10–0.15	
1,1-Dichloro-1-nitroethane	$CH_3CCl_2NO_2$	143.9	124	1.4271 (20/20°C.)	0.25	16	5.0	168	169.9	5.89	0.15–0.20	10
1-Chloro-1-nitropropane	$CH_3CH_2CHClNO_2$	123.5	139.5–143.3	1.209 (20/20°C.)	0.5	5.8	4.3	144	198	5.05	0.05–0.10	20
2-Chloro-2-nitropropane	$CH_3CHClNO_2CH_3$	123.5	133.6	1.197 (20/20°C.)	0.5	8.5	4.3	135	198	5.05	0.5–0.75	

blood following intravenous injection of the nitrobutanes than after an injection of equivalent doses of the nitropropanes or nitroethanes. It would be expected that the nitrobutanes would present hazards qualitatively similar to 2-nitropropane.

CHLORINATED MONONITROPARAFFINS

The acute animal toxicity of 1-chloro-1-nitroethane, 1,1-dichloro-1-nitroethane, 1-chloro-1-nitropropane, and 2-chloro-2-nitropropane have been studied by Machle and co-workers.[2,4] Their formulas, physical and chemical properties, oral lethal doses, and threshold limit values appear in Table 7. They have limited industrial use, being employed chiefly in organic chemical synthesis. 1,1-Dichloro-1-nitroethane has been used as a fumigant for stored produce. The chlorinated compounds are more toxic by mouth and inhalation than the unsubstituted mononitroparaffins. Of the four, 2-chloro-2-nitropropane is the least toxic for rabbits and 1-chloro-1-nitropropane the most toxic. 1,1-Dichloronitroethane is considerably more irritating to skin and mucous membranes than 1-chloro-1-nitropropane. It exhibits greater toxicity by inhalation (see Tables 8 and 9). Both materials are

TABLE 8
Response to Inhalation of 1,1-Dichloronitroethane[4]

Average concentration, p.p.m.	Duration of exposure	Mortality[a]	
		Rabbit	Guinea pig
4910	30 min.	2	2
985	3$^1/_2$ hrs.	2	1
594	2$^1/_2$ hrs.	1	None
254	1 hr.	None	None
169	2 hrs.	1	1
100	6 hrs.	2	2
60	2 hrs.	None	None
52	18 hrs., 40 min.	2	None
34	4 hrs.	None	None
25	204 hrs.	None	None

[a] Two rabbits and 2 guinea pigs in each experiment.

lung irritants. They cause pulmonary edema and death within 24 hours following exposure to high concentrations. The chief site of injury is the lungs but damage is also observed in the heart muscle, liver, and kidneys after lethal exposures. Although 1-chloro-1-nitroethane and 2-chloro-2-nitropropane have not been studied in detail, it would be expected that their inhalation toxicity would be qualitatively and quantitatively similar to 1-chloro-1-nitropropane. No human experience has been recorded. On the basis of the animal experiments, threshold limit values of 10 and 20 p.p.m. have been proposed for 1,1-dichloro-1-nitroethane and 1-chloro-1-nitropropane, respectively.[15]

TABLE 9
Response to Inhalation of 1-Chloro-1-nitropropane[4]

Average concentration, p.p.m.	Duration of exposure	Mortality[a]	
		Rabbit	Guinea pig
4950	60 min.	2	1
2574	2 hrs.	2	None
2178	1 hr.	None	1
1069	1 hr.	None	None
693	2 hrs.	None	None
393	6 hrs.	1	None

[a] Two rabbits and 2 guinea pigs in each experiment.

TRICHLORONITROMETHANE, CCl_3NO_2 (Chloropicrin)

Chloropicrin has been used in organic syntheses, as an insecticide and soil fumigant, and as a war gas. It is a colorless, slightly oily liquid with a very intense odor and definite lacrimatory effect. Data on exposures of man and animals to various concentrations of chloropicrin, largely obtained during World War I, have been summarized in Tables 10 and 11.[14,22,23] Chloropicrin is both a lacrimator and a lung irritant. It is intermediate in toxicity between chlorine and phosgene. Nominal lethal concentrations (mg. per liter) after 30 minutes' exposure are: chlorine, 3.0; chloropicrin, 0.8; phosgene, 0.36. Flury and Zernik state that exposure to 4 p.p.m. for a few seconds renders a man unfit for combat and 15 p.p.m. for approximately the same period of time results in respiratory tract injury.[22] Whereas chlorine in fatal concentrations produces more injury of the upper respiratory tract, trachea, and larger bronchi than in the alveoli and phosgene acts primarily on the alveoli, chloropicrin produces more injury to medium and small bronchi than to the trachea and large bronchi. The alveolar injury is less than with phosgene but pulmonary edema occurs and is the most frequent cause of early deaths. Exposure to chloropicrin produces more coughing than does phosgene and there is less delay in onset of pulmonary edema. Late deaths may occur from secondary infections, bronchopneumonia, or bronchiolitis obliterans. During World War I chloropicrin was noted for its tendency to cause nausea and vomiting. It has been stated that individuals injured by inhalation of chloropicrin become more susceptible so that concentrations of the gas not producing symptoms in others cause them distress. Chloropicrin is also a potent skin irritant.

The American Conference of Governmental Industrial Hygienists has recently reduced their suggested threshold limit value from 1 to 0.1 p.p.m. This 0.1 p.p.m. value affords a wider margin of safety. It is below the levels detectable by odor or irritation.

[22] F. Flury and F. Zernik, *Schädliche Gase*, Springer, Berlin, 1931.
[23] A. M. Prentiss, *Chemicals in War*, McGraw-Hill, New York, 1937.

TABLE 10

Response to Various Concentrations of Trichloronitromethane—Animals

Animal	Concentration mg./liter	p.p.m.	Duration of exposure, min.	Response
Dog	1.05	155	12	Became ill
	0.8–0.95	117–140	30	Death of 43% of the animals; survival of remainder
Mouse	0.85	125	15	Death in 3 hrs. to 1 day
Cat	0.51	76	25	Death usually in 1 day
Mouse	0.34	50	15	Death after 10 days
Dog	0.32	48	15	Tolerated
Cat	0.32	48	20	Death after 8 to 12 days
	0.26	38	21	Survived 7 days
Mouse	0.17	25	15	Tolerated

TABLE 11

Response to Various Concentrations of Trichloronitromethane—Man

Concentration mg./liter	p.p.m.	Duration of exposure, min.	Response
2.0	297.6	10	Lethal concentration
0.8	119.0	30	Lethal concentration
0.1	15.0	1	Intolerable
0.050	7.5	10	Intolerable
0.009	1.3		Lowest irritant concentration
0.0073	1.1		Odor detectable
0.002–0.025	0.3–3.7	3–30 sec.	Closing of eyelids according to individual sensitivity

NITROOLEFINS

Interest in the nitro derivatives of straight chain olefins stems largely from study of air pollution. They are probably formed by the reaction of oxides of nitrogen with olefins from automobile exhaust. Their concentration in Los Angeles air during irritant "smog" accumulations has been estimated to range from 10 to 15 p.p.b. For this reason Deichmann, Keplinger, and Lanier conducted studies on the acute effects of a series of straight chain nitroolefins on experimental animals.[5] Part of their data appears in Table 12.

All members of the series showed a relatively high degree of toxicity and local irritation. Exposure to the vapor caused marked irritation of the eyes, entire respiratory tract, and skin. The exposed animals exhibited lacrimation, nasal secretion, vasodilatation, and dyspnea progressing to cyanosis, hyperexcitability, convulsions, coma, and death. There was rapid absorption from the gastrointestinal

TABLE 12
Acute Effects of Nitroolefins[5]

Name	Formula	Vapor exposure (5 hrs.)		Oral toxicity, rats, Single dose, undiluted (approx. lethal dose)		Intraperitoneal toxicity, rats, Single dose, undiluted (approx. lethal dose)		Percutaneous toxicity, rabbits, open, 5-hr. single dose (approx. lethal dose)	
		Concentration, p.p.m.	Survival time, rats, 47% humidity	g./kg.	mM/kg.	g./kg.	mM/kg.	g./kg.	mM/kg.
2-Nitro-2-butene	$CH_3C(NO_2):CHCH_3$	1400	100 min.	0.28	2.8	0.08	0.8	0.62	6.1
2-Nitro-2-pentene	$CH_3C(NO_2):CHCH_2CH_3$	240	240 min.	0.28	2.4	0.08	0.7	0.94	5.4
		55	Survived						
3-Nitro-3-pentene	$CH_3CH_2C(NO_2):CHCH_3$	268	280 min.	0.42	3.7	0.05	0.4	0.62	8.2
2-Nitro-2-hexene	$CH_3C(NO_2):CH(CH_2)_2CH_3$	515	50–85 min.	0.42	3.3	0.12	0.9	1.40	7.3
		152	Survived						
3-Nitro-3-hexene	$CH_3CH_2C(NO_2):CHCH_2CH_3$	557	30–70 min.	0.42	3.3	0.08	0.6	0.94	10.9
		50	Survived						
2-Nitro-2-heptene	$CH_3C(NO_2):CH(CH_2)_3CH_3$	308	3–18 hrs.	0.94	6.6	0.28	2.0	0.94	6.6
		135	Survived						
3-Nitro-3-heptene	$CH_3CH_2C(NO_2):CH(CH_2)_2CH_3$	54	24 hrs.	0.62	4.3	0.28	2.0	1.40	9.8
2-Nitro-2-octene	$CH_3C(NO_2):CH(CH_2)_4CH_3$	47	Survived	1.4	9.0	0.28	1.8	0.62	4.0
3-Nitro-3-octene	$CH_3CH_2C(NO_2):CH(CH_2)_3CH_3$	142	18–24 hrs.	0.62	4.0	0.18	1.2	0.94	6.0
		72	Survived						
3-Nitro-2-octene	$CH_3CH:C(NO_2)(CH_2)_4CH_3$	141	18–72 hrs.	0.62	4.0	0.18	1.2	0.62	4.0
		44	Survived						
2-Nitro-2-nonene	$CH_3C(NO_2):CH(CH_2)_5CH_3$	64	Survived	2.1	12.3	0.28	1.6	0.62	3.6
3-Nitro-3-nonene	$CH_3CH_2C(NO_2):CH(CH_2)_4CH_3$	59	24 hrs.	2.1	12.3	0.42	2.5	0.42	2.5
		10	Survived						

tract, peritoneal cavity, and skin. Signs and symptoms of intoxication appeared promptly and were similar by all routes of exposure. Respiratory tract injury occurred in all cases. Open application to the skin (for 5 hours) resulted in intense local irritation, pain, erythema, edema, and later necrosis. One drop in the eye produced marked irritation and corneal damage. Pathological changes were most marked in the lungs. These included hemorrhages, edema, and emphysema. Rabbits, rats, and chicks were exposed to calculated concentrations of the nitro-olefins at different chamber humidities. Increasing the chamber humidity did not have marked effect upon survival time at equivalent concentrations except in the case of 3-nitro-3-hexene, where survival time was increased by increasing humidity. Increasing humidity did, however, increase the amount of edema and serous fluid in the thoracic cavity. Inhalation toxicity showed no definite relationship to chain length, but the acute oral and intraperitoneal toxicities decreased with increasing length of the carbon chain.

ALIPHATIC NITRATES

I. General Considerations

A. PROPERTIES, MANUFACTURE, AND USES

The aliphatic nitrates are nitric acid esters of monovalent and polyvalent aliphatic alcohols. The nitrate group has the structure —$CONO_2$. The nitric acid esters of the lower mono-, di-, and trivalent alcohols are liquids and those of the quadrivalent alcohols (erythritol tetranitrate and pentaerythritol tetranitrate) and the hexavalent alcohol (mannitol hexanitrate) are solids. They are generally insoluble or very slightly soluble in water but are more soluble in alcohol and other organic solvents. They are prepared by the esterification of the corresponding alcohol with nitric acid. Because of their explosive nature, their major use has been as industrial and military explosives. Several members of the group have been used as therapeutic agents. Methyl and ethyl nitrate are not used very extensively. n-Propyl nitrate has recently had increasing use in the propellant field. Nitroglycerin (glyceryl trinitrate) is widely used as an industrial explosive and as a vasodilating agent in medicine. In 1953 nearly 800 million pounds of industrial explosives were manufactured and sold in the United States. Of this, approximately 670 million pounds were high explosives.[24]

B. PHARMACOLOGY, SYMPTOMATOLOGY, AND MODE OF ACTION

The chief effects of the aliphatic nitrates are dilatation of blood vessels and methemoglobin formation. The vascular dilatation accounts for the characteristic lowering of blood pressure and headache. The individual members in the series differ in intensity and duration of these effects. Animals given effective doses by mouth or parenterally exhibit such signs and symptoms as marked depression in

[24] R. N. Shreve, *The Chemical Process Industries*, 2nd ed., McGraw-Hill, New York, 1956.

blood pressure, tremors, ataxia, lethargy, alteration in respiration (usually hyperpnea) cyanosis, prostration, and convulsions. Death, when it occurs, is either from respiratory or cardiac arrest. Animals surviving the acute exposure recover promptly.

Nitroglycerin and erythritol tetranitrate (ETN) are capable of producing approximately the same degree of hypotension in man but the effect of erythritol tetranitrate is more prolonged and requires a larger dose. The maximum blood pressure depression from nitroglycerin occurs at approximately 4 minutes whereas that from ETN occurs at approximately 20 minutes. Pentaerythritol tetranitrate (PETN) is less effective as a hypotensive agent than ETN. Methyl nitrate causes little depression in blood pressure.[25] The outstanding symptom produced in man is headache. This is usually described as very severe and throbbing, and is often associated with flushing, palpitation, and nausea and, less frequently, with vomiting and abdominal discomfort. Temporary tolerance develops from continued or repeated daily exposures. Doses resulting in headaches in man are: nitroglycerin, 18 mg. (skin); ethylene glycol dinitrate, 35 mg. (skin); ETN, 45 mg. (oral).[25] PETN in doses of 64 mg. orally does not produce headache.[49] Thus PETN is the least effective and nitroglycerin the most effective. Other pharmacological consequences of vasodilatation are increased pulse rate, an increase in cardiac stroke volume, variable cardiac dilatation and cardiac output, and a shift in blood distribution with increased stasis and pressure in the pulmonary arteries.[25]

For some members of the series, the ease of hydrolysis to the alcohol and nitrite and the degree of blood pressure lowering are parallel. Krantz and co-workers, however, feel that there is little evidence that hydrolysis to produce nitrite is necessary for hypotensive action.[26-28] They present evidence that the intact nitrate molecule (e.g., isomannide dinitrate) can act directly to lower blood pressure. It appears that this effect of the nitrate esters does not depend exclusively on the liberation of nitrite groups. Dilatation can occur without measurable nitrite in the blood, or when the amount measured is not sufficient to account for the effect observed. The *in vivo* formation of nitrite is commonly assumed to be the explanation for the methemoglobin-forming properties of the aliphatic nitrates.[29] The mechanism of formation of nitrite is not clear.[30] It is possible that reduction to nitrite occurs before hydrolysis as follows:[31]

$$RONO_2 \xrightarrow{+2H} RONO \xrightarrow{+H_2O} ROH + NO_2^-$$

[25] W. F. von Oettingen, The effects of aliphatic nitrous and nitric acid esters on the physiological functions with special reference to their chemical constitution, *National Inst. Health Bull.* No. **186**, 1946.

[26] J. C. Krantz, C. J. Carr, and S. E. Forman, *Proc. Soc. Exp. Biol. Med.*, **42**, 472 (1939).

[27] J. C. Krantz, C. J. Carr, S. E. Forman, and N. Cone, *J. Pharmacol. Exptl. Therap.*, **70**, 323 (1940).

[28] M. Rath and J. C. Krantz, *J. Pharmacol. Exptl. Therap.*, **76**, 33 (1942).

[29] O. Bodansky, *Pharmacol. Revs.*, **3**, 144 (1951).

[30] R. T. Williams, *Detoxication Mechanisms*, 2nd ed., Wiley, New York, 1959.

[31] P. Rofe, *Brit. J. Ind. Med.*, **16**, 15 (1959).

Ethyl nitrate, ethylene glycol dinitrate, nitroglycerin, propyl nitrate, and amyl nitrates are known to cause methemoglobin formation in experimental animals. Ethyl nitrate is a weak methemoglobin former. Nitroglycerin is a moderately active methemoglobin former but ethylene glycol dinitrite is considerably more effective (approximately 4 times). Ethylene glycol mononitrate, on the other hand, is not very active.[25]

Ethyl, propyl, and amyl nitrates, and ethylene glycol dinitrate and nitroglycerin induce Heinz body formation in animals. Although ethyl nitrate induces Heinz body formation, ethyl nitrite does not. Ethylene glycol dinitrate is more effective than nitroglycerin, which is more effective than ethyl nitrate. Toxicity data are not sufficiently complete to permit further comparisons of the effectiveness of the nitrates in producing Heinz body formation. The precise nature of these small, rounded inclusion bodies in the red blood cells, described by Heinz in 1890, is not clear. They have been observed in man and animals after absorption of a variety of chemical compounds, the most prominent of which are the aromatic nitrogen compounds, inorganic nitrites, and the aliphatic nitrates. Their appearance is commonly associated with anemia and the production of methemoglobin. Some evidence indicates that they are proteins in nature, possibly hemoglobin degradation products. Red blood cells containing the inclusion bodies have a shorter life span and are removed from the circulation by the spleen. Special stains are required to satisfactorily demonstrate their presence. In the case of the aliphatic nitrates, erythrocytes containing Heinz bodies disappear from the circulating blood more slowly than methemoglobin.[31-33]

The alkyl nitrates are all absorbed from the gastrointestinal tract.[25] PETN is relatively slowly absorbed by this route. Ethylene glycol dinitrate and nitroglycerin are absorbed through the skin but the absorption of ETN and PETN by this route is slow or absent. The nitric acid esters of the monovalent alcohols are rapidly absorbed from the lung. The absorption of ETN through the lungs is slower than with ethylene glycol dinitrate. Nitroglycerin is more slowly absorbed from the lungs than ethylene glycol dinitrate and PETN is more slowly absorbed than ETN.

Trimethylenetrinitramine (cyclonite, RDX) has been included in this section because it is also used as a high explosive. The nitramines contain the grouping —$NHNO_2$. Little is known about the action of the nitramines. However, neither methyl nitramine nor trimethylenetrinitramine exhibits nitrite or nitratelike actions.[34]

C. PATHOLOGY IN ANIMALS

Pathological examinations of animals dying following acute intoxication have either been negative or have revealed only slight nonspecific pathologic changes

[32] J. B. Hughes and J. F. Treon, *Arch. Ind. Hyg. Occupational Med.*, **10**, 192 (1954).

[33] J. F. Treon, F. P. Cleveland, and J. Duffy, *A.M.A. Arch. Ind. Health*, **11**, 290 (1955).

[34] W. F. von Oettingen, D. D. Donahue, H. Yagoda, A. R. Monaco, and M. R. Harris, *J. Ind. Hyg. Toxicol.*, **31**, 21 (1949).

consisting of congestion of internal organs. Hueper and Landsberg[35] have described degenerative vascular and parenchymatous lesions in the heart, kidneys, lungs, brain, and testes following several months administration of large doses of erythritol tetranitrate to young rats. They felt that these changes were induced by inadequate nutrition of the tissues following hypoxia and stagnation of the organs' blood supply associated with vasodilatation. On the other hand, Donahue and Monaco were unable to confirm these findings after chronic administration of ETN and PETN at lower doses.[25] Evidence of injury to these organ systems in men chronically exposed to nitroglycerin, ETN, and PETN is lacking.

D. INDUSTRIAL EXPERIENCE

No human injuries from the monovalent alcohol esters of nitric acid have been recorded. An apparent increased number of sudden deaths among explosives workers exposed to ethylene glycol dinitrate has been observed but not adequately explained.[25] The occurrence of severe headaches in persons handling nitroglycerin is so frequent that the headaches have acquired such names as "power headache," "dynamite head," etc.[36,37] Hypotension and a tendency to peripheral vascular collapse have been described in workers exposed to nitroglycerin and ethylene glycol dinitrate.[38] McConnell and his associates have summarized the experience with occupational disease and industrial hygiene in government-owned ordnance explosives plants in the United States during World War II.[39] These included chemical and explosives manufacturing, munitions and munition loading, and proving and testing areas. Trinitrotoluene and tetryl were common to most areas but nitroglycerin, pentaerythritol tetranitrate, and trimethylenetrinitramine (RDX) were also handled. In 915,000 man-years of experience no fatalities due to the aliphatic nitrates or RDX occurred. Most of the illness and 21 deaths were attributed to trinitrotoluene. Some of the 93,000 cases of mild illness or dermatitis were attributed to PETN, nitroglycerin, or RDX manufacture.[39]

E. DETERMINATION IN AIR, MAXIMUM ALLOWABLE CONCENTRATIONS

Foulger has used the color developed by phenol disulfonic acid for the determination of ethylene glycol dinitrate. This method can also be applied to nitroglycerin.[40] Yagoda and Goldman[41] utilized a colorimetric method for the determination of traces of aliphatic nitrate esters in air. The method can be used for nitroglycerin or PETN. Oxides of nitrogen and ammonium nitrate interfere. Their method is based on the nitration of m-xylene in sulfuric acid by the nitric

---

[35] W. C. Hueper and J. W. Landsberg, *Arch. Pathol.*, **29**, 633 (1940).

[36] G. Ebright, *J. Am. Med. Assoc.*, **62**, 201 (1914).

[37] A. M. Schwartz, *New Eng. J. Med.*, **255**, 541 (1946).

[38] A. J. Fleming, C. A. D'Alonzo, and J. A. Zapp, *Modern Occupational Medicine*, Lea & Febiger, Philadelphia, 1954.

[39] W. J. McConnell, R. H. Flinn, and A. D. Brandt, *Occupational Med.*, **1**, 551 (1946).

[40] J. H. Foulger, *J. Ind. Hyg. Toxicol.*, **18**, 127 (1936).

[41] H. Yagoda and F. H. Goldman, *J. Ind. Hyg. Toxicol.*, **25**, 440 (1943).

acid formed from the ester. The nitroxylene is isolated by steam distillation and determined colorimetrically. In the absence of interfering nitrogen compounds in air, the Kjeldahl method for determination of nitrogen could be applied.[25]

Except for nitroglycerin no threshold limit values have been established for the aliphatic nitrates. The threshold limit value of 0.5 p.p.m. (5 mg./cu. meter) for nitroglycerin is based largely on human experience. A threshold limit value of 25 p.p.m. (110 mg./cu. meter) for *n*-propyl nitrate has been tentatively proposed on the basis of animal studies.[15]

II. Specific Compounds

METHYL NITRATE, CH_3ONO_2

1. Uses

Methyl nitrate is a liquid that explodes on heating. It has been used as a rocket propellant.

2. Physical and Chemical Properties

Molecular weight: 77.04
Boiling point: 65°C. (explodes)
Specific gravity: 1.217 (15°C.)
Vapor density: 2.66 (air = 1)
Solubility: slightly soluble in water, soluble in alcohol and ether
1 mg./liter $\asymp$ 317 p.p.m., 1 p.p.m. $\asymp$ 3.15 mg./cu. meter at 25°C., 760 mm. Hg

3. Physiological Response

No injuries in industry have been reported. Doses of 12.5 mg./kg. have practically no effect on the blood pressure and pulse rate of rabbits. Doses of 52 mg./kg. have a slight transient effect.   Methyl nitrate is considerably less effective in these respects than nitroglycerin. The minimal dose causing headache in man is between 117 and 470 mg. As has been observed with the other nitric acid esters, fractional doses produce tolerance that lasts for several days.[25]

ETHYL NITRATE, $CH_3CH_2ONO_2$

1. Uses

Ethyl nitrate is a colorless flammable liquid with a pleasant odor.
It has been used in organic syntheses of drugs, perfumes, and dyes.

2. Physical and Chemical Properties

Molecular weight: 91.07
Specific gravity: 1.105 (20/4°C.)
Vapor density: 3.14 (air = 1)

Solubility: 1.3 g. in 100 ml. of water at 55°C.; miscible in all portions in alcohol and ether

1 mg./liter $\backsimeq$ 269 p.p.m. and 1 p.p.m. $\backsimeq$ 3.72/cu. meter at 25°C., 760 mm. Hg

3. Physiological Response

No industrial intoxications have been recorded from ethyl nitrate. It is said that ethyl nitrate has anesthetic properties and on inhalation causes headache, narcosis, and vomiting.[25] In cats, 400 mg./kg. in olive oil intraperitoneally produces unconsciousness, increased respiratory rate, dilatation and fixation of pupils, and death in 90 minutes; 300 mg./kg. intraperitoneally is followed by recovery. Moderate methemoglobinemia and Heinz body formation are observed after doses of 125 to 250 mg./kg.[25] Thus the effects of ethyl nitrate resemble those of the other aliphatic nitrates that have been studied.

n-PROPYL NITRATE, $CH_3CH_2CH_2ONO_2$

1. Source, Uses, and Industrial Exposures

n-Propyl nitrate is a pale yellow liquid with a sweet, sickening odor. It is being used as a fuel ignition promoter, in liquid rocket propellants, and as an organic intermediate. It is prepared by continuous nitration of propyl alcohol with nitric acid, usually in the presence of urea and ammonium nitrate or sulfuric acid.

2. Physical and Chemical Properties

Molecular weight: 105.09
Boiling point: 110.5°C.
Specific gravity: 1.058 (20/4°C.)
Vapor pressure: approximately 16 mm. Hg, 25°C.
Vapor density: 3.62 (air = 1)
Solubility: very slightly soluble in water, soluble in alcohol and ether

1 mg./liter $\backsimeq$ 233 p.p.m. and 1 p.p.m. $\backsimeq$ 4.30 mg./cu. meter at 25°C., 760 mm. Hg

3. Physiological Response

Pharmacological and acute toxicity studies in animals have shown that n-propyl nitrate produces hypotension, presumably by its direct action on vascular muscle, and methemoglobinemia. It has relatively low acute oral and intravenous toxicity for rats, cats, and dogs.[42,43] Its percutaneous toxicity is low. Inflammation and thickening of the skin result from repeated application. Inhalation of sufficient concentration is capable of producing cyanosis, methemoglobinemia, lethargy, convulsions, and death in rodents and dogs. Repeated inhalation

[42] D. B. Hood, Haskell Laboratory for Toxicology and Industrial Medicine, E. I. du Pont de Nemours & Co., Rept. No. 21-53, 1953.

[43] E. F. Murtha, D. E. Stabile, and J. H. Wills, J. Pharmacol. Exptl. Therap., 118, 77 (1956).

produces increasing methemoglobin concentrations. Some dogs with exposures to 260 p.p.m. for 26 weeks developed hemoglobinuria and mild hemolytic anemia, but not methemoglobinemia.[44] Some of the symptoms and signs resulting from repeated absorption of *n*-propyl nitrate become less severe or disappear with continuing exposure.[42,44] Other symptoms observed in exposed animals are hypotension, lethargy, weakness with vomiting, excitement, and convulsions. Un-

TABLE 13

Response to *n*-Propyl Nitrate—Animals[42,43]

Animal	Dose, g./kg.	Route	Effect
Rat	7.5	Oral	Approximate lethal dose (sample I)
Rat	5.0	Oral	Approximate lethal dose (sample II)
Rat	1.0	Oral	Weakness, incoordination, cyanosis
Rats	1.5 × 10	Oral	Weakness, cyanosis, weight loss (first week)
Rabbit	11, 17	Skin	Essentially none
Rabbit	0.2–0.25	Intravenous	Approximate LD_{50}
Dog	0.005	Intravenous	Slight fall in blood pressure
	0.050	Intravenous	Hypotension, cyanosis
	0.2–0.25	Intravenous	Death in respiratory arrest
Cat	0.1–0.25	Intravenous	$6/7$ died in 1 minute
	0.025–0.075	Intravenous	Hypotension, methemoglobinemia, survived

TABLE 14

Response to Inhalation of Various Concentrations of
n-Propyl Nitrate—Animals[44]

Animal	Concentration, p.p.m.	Duration of exposure	Mortality[a]
Rats	7134	4 hrs.	$1/10$
	5816	4 hrs.	$1/10$
	3235	8 wks.[b]	$5/20$
	2110	26 wks.	$9/20$
Mice	7134	4 hrs.	$15/20$
	5816	4 hrs.	$2/10$
	3235	8 wks.	$6/29$
Guinea pig	3235	8 wks.	$0/10$
	2110	26 wks.	$0/10$
Hamster	3235	8 wks.	$0/10$
Dog	2000	11 hrs.	$3/3$
	900	6 days	$1/1$
	560	6 wks.	$1/2$
	260	26 wks.	$0/3$

[a] Symptoms observed in all animals at all concentrations except for hamsters and guinea pigs.

[b] Six hours a day, 5 days a week.

[44] W. E. Rinehart, R. C. Garbers, E. A. Greene, and R. M. Stoufer, *Am. Ind. Hyg. Assoc., Quart.*, **19**, 80 (1958).

diluted n-propyl nitrate instilled into rabbits' eyes results in mild transient inflammation without evidence of corneal damage.

Pathological examinations have revealed either no changes or moderate non-specific tissue changes compatible with anoxia. These consist of congestion of the brain, liver, and myocardium and occasionally increased pigment deposition in liver and spleen. Animals allowed to recover for 2 weeks after exposure show no pathological findings. The results of animal toxicity studies are summarized in Tables 13 and 14.

Rinehart and co-workers[44] found that 3 dogs recovered from 26 weeks of repeated exposures to 260 p.p.m., which produced hemoglobinuria, mild anemia, and slight depression during the first 2 or 3 weeks of exposure. They state that the odor of n-propyl nitrate is fairly obvious and detectable at concentrations of 50 p.p.m. or above. Their data indicate that dogs are considerably more susceptible to inhalation of n-propyl nitrate than rodents. They suggest that a reasonable maximum acceptable concentration for man may be 50 p.p.m. or less. The American Conference of Governmental Industrial Hygienists has suggested a tentative threshold limit value of 25 p.p.m. (110 mg./cu. meter).[15] This value appears acceptable pending clinical and hematological observations on men with repeated measured exposures.

Hood[42] also studied the acute and subacute animal toxicity of isopropyl nitrate. She found that the effects are essentially the same as those produced by n-propyl nitrate with a suggestion of slightly increased toxicity for isopropyl nitrate. Animals exposed to the combustion products of both nitrates developed signs and symptoms of irritation and pulmonary edema.

AMYL NITRATE, $C_5H_{11}O_3N$

1. *Uses*

A mixture of amyl nitrate isomers used as a diesel fuel improver has been studied by Treon, Cleveland, and Duffy.[33] This was a clear, slightly yellowish liquid, with a sickening, sweet odor at 1.4 mg./liter.

2. *Physical and Chemical Properties*

Molecular weight: 133.15
Vapor density: 0.997 (20/4°C.)
Boiling range: 150–155°C. (unstable)
Vapor pressure: (isoamyl nitrate) 5 mm. Hg, 28.8°C.
Solubility: 0.3 ml. in 100 ml. of water
Flash point 107.6°F. (closed cup)
1 mg./liter ≈ 184 p.p.m. and 1 p.p.m. ≈ 5.44 mg./cu. meter at 25°C., 760 mm. Hg

3. *Physiological Response*

Treon et al.[33] exposed cats, guinea pigs, rabbits, rats, and mice to measured concentrations of amyl nitrates in air. Some of their results are given in Table 15.

Mice responded variably to amyl nitrate inhalation and cats were most resistant. The signs and symptoms observed were tremors, ataxia, alterations in respiration, lethargy, cyanosis, convulsions, coma, and deaths. All exposures, except 262 p.p.m., produced signs or symptoms in cats, guinea pigs, and rats and some alterations in respiration were observed in rabbits and mice at the lowest concentration, 262 p.p.m. A cat exposed to 599 p.p.m. developed methemoglobin levels up to 59.5 per cent after the seventh exposure. Cats exposed to concentrations ranging from 1700 to 3700 p.p.m. of the amyl nitrates showed Heinz body formation, which reached a peak several days after exposure and disappeared slowly after a period of 1 to 3 weeks.

Animals dying during exposure had diffuse degenerative changes in the liver, kidneys, and brain with hyperemia and edema of the lungs. Those sacrificed at varying intervals after exposures had normal findings on pathological examination.

TABLE 15

Response to Inhalation of Various Concentrations of Amyl Nitrates[33]

Concentration		Duration, hr.	Mortality[a]			
mg./liter	p.p.m.		Guinea pigs	Rabbits	Rats	Mice
19.9	3730	7	$^2/_2$	$^2/_2$	$^3/_4$	$^5/_5$
19.17	3593	3.5	$^0/_2$	$^2/_2$	$^1/_4$	$^4/_5$
17.22	3227	1	$^0/_2$	$^0/_2$	$^0/_4$	$^0/_5$
16.39	3072	3×1	$^2/_2$	$^2/_2$	$^4/_4$	$^5/_5$
14.8	2774	3.5	$^0/_2$	$^0/_2$	$^0/_4$	$^2/_4$
13.6	2549	0.33	$^0/_2$	$^0/_2$	$^0/_4$	$^0/_5$
12.7	2380	2×7	$^2/_2$	$^2/_2$	$^0/_4$	$^5/_5$
12.3	2305	1	$^0/_2$	$^0/_2$	$^0/_4$	$^5/_5$
9.6	1807	7	$^0/_2$	$^1/_2$	$^0/_4$	$^4/_4$
9.09	1703	3×7	$^2/_2$	$^1/_2$	$^2/_4$	$^5/_5$
8.6	1612	7	$^0/_2$	$^0/_2$	$^0/_4$	$^0/_5$
3.2	599	$9 \times 7 + 6.25$	$^0/_2$	$^0/_2$	$^0/_4$	$^0/_2$
1.4	262	20×7	$^0/_2$	$^0/_2$	$^0/_3$	$^0/_5$

[a] No cats died following any of these exposures.

Persons exposed in the laboratory during these studies developed nausea and headache. No other illness was observed.[33]

No observations on men exposed in industry have been recorded and no maximum allowable concentrations have been proposed. The effects of mixed amyl nitrates in animals are qualitatively similar to those of the other alkyl mononitrates. The acute and subacute inhalation toxicity for guinea pigs, rats, and mice is greater than that of n-propyl nitrate. The higher boiling point and lower vapor pressure of the amyl nitrates tend to reduce this differential as far as the industrial hazard is concerned.

ETHYLENE GLYCOL DINITRATE, $C_2H_4(ONO_2)_2$

1. Uses

Ethylene glycol dinitrate is a yellow liquid. It explodes with heat or impact and has an explosive force approximately equivalent to nitroglycerin. Its major use has been for high explosive formulations, particularly in nonfreezing dynamites. It is produced by the nitration of ethylene glycol.

2. Physical and Chemical Properties

Molecular weight: 152.07
Specific gravity: 1.483 (8°C.)
Boiling point: explodes at 114–116°C.
Vapor density: 5.24 (air = 1)
Vapor pressure: 0.045 mm. Hg. (20°C.)
Solubility: insoluble in water, soluble in alcohol and dilute alkali
1 mg./liter $\backsim$ 161 p.p.m. and 1 p.p.m. $\backsim$ 6.24 mg./cu. meter at 25°C., 760 mm. Hg

3. Physiological Response

Ethylene glycol dinitrate is absorbed through the skin, the lungs, and the gastrointestinal tract. von Oettingen has summarized the toxicity data in his review of nitrate esters (Table 16). It causes hypotension, methemoglobinemia, and Heinz body formation. Inhalation of the low level of 2 p.p.m., 2 hours a day, for 1000 days by cats caused transient blood changes and 21 p.p.m. for the same

TABLE 16
Response to Ethylene Glycol Dinitrate[25]

Species	Dose	Route	Response
Cat	100 mg./kg.	Subcutaneous	Certain fatal dose
Rabbit	400 mg./kg.	Subcutaneous	Certain fatal dose
Cat	60 mg./kg.	Subcutaneous	45% methemoglobin, Heinz bodies
Rabbit	12.5 mg./kg.		Hypotension
Cat	21 p.p.m. × 1000	Inhalation	Marked blood changes only
Cat	2 p.p.m. × 1000	Inhalation	Temporary blood changes
Man	1.8–3.5 cc. of 1% solution	Skin	Minimal dose for headache

period resulted in marked blood changes.[25] Rats and guinea pigs are quite tolerant, surviving 6 months' exposure to 500 μg./liter (80 p.p.m.) with the only effect being slight drowsiness and some Heinz body formation.[45] Chronic poisoning in animals is associated with anemia as well as methemoglobinemia and Heinz body formation. Ethylene glycol dinitrate is absorbed through the skin more rapidly than nitroglycerin. It is more toxic for cats and rabbits and produces a greater

[45] W. Stein, *Arch. Gewerbepathol. Gewerbehyg.*, **15**, 19 (1956).

methemoglobin and Heinz body response. The pathological findings observed in experimental animals with chronic poisoning are fatty changes in the heart muscle, liver, and kidney, and pigment deposition in the liver and spleen characteristic of anemia. In man the cardiovascular effects overshadow all others.

Acute exposure in man results in headache, nausea, vomiting, hypotension, and tachycardia. The much quoted reports of sudden deaths in explosives workers that have been attributed to exposure to ethylene glycol dinitrate have never been explained satisfactorily.[25] Blood pressure abnormalities (presumably hypotension) have been observed in appreciable numbers of men engaged in the production of dynamite, with exposures to both ethylene glycol dinitrate and nitroglycerin.[38] Similar findings were observed in 265 Italian workers producing antifreeze dynamite.[46] No methemoglobinemia, anemia, or Heinz bodies were observed although the exposure was sufficient to produce headache, palpitation, nausea, and peripheral vasodilatation in some of the workers. Evidence of anemia, hypotension, or electrocardiographic abnormalities was lacking in 47 explosives workers with chronic exposure to 20 mg./cu. meter (3.2 p.p.m.). Low levels of methemoglobin (1 to 6 per cent) and some alcohol intolerance were observed. According to von Oettingen, a temporary tolerance to ethylene glycol dinitrate headache appears from repeated exposures as it does with nitroglycerin.[25]

Careful attention should be given to preventing skin contact and to the provision of adequate skin protection if transient symptoms are to be avoided.[38] A maximum allowable concentration has not been established. However, on the basis of industrial experience, animal studies, and by analogy to nitroglycerin, 0.5 p.p.m. is suggested as a bench mark at which chronic effects would be unlikely but transient symptoms may occur.

NITROGLYCERIN, $C_3H_2(ONO_2)_3$ (Glyceryl Trinitrate, Trinitroglycerin)

1. Uses

Nitroglycerin is a colorless, oily liquid which explodes violently from shock or when heated to about 260°C. It is used as an explosive in dynamite, cordite, or blasting gelatin and as a pharmaceutical agent in the treatment of angina pectoris.

2. Physical and Chemical Properties

Molecular weight: 227.1
Specific gravity: 1.601
Melting point: 13°C.
Vapor density: 7.8 (air = 1)
Vapor pressure: 0.00025 mm. Hg, 20°C.

Slightly soluble in water, partly soluble in alcohol; miscible with ether and chloroform

[46] I. Maccherini and E. Camarri, *Med. del. Lavoro,* **50,** 193 (1959).

1 mg./liter $\backsimeq$ 108 p.p.m. and 1 p.p.m. $\backsimeq$ 9.29 mg./cu. meter at 25°C., 760 mm. Hg

3. *Physiological Response*

Nitroglycerin is absorbed through the skin, lungs, and mucous membranes. In view of the skin absorption and low vapor pressure, skin exposures are particularly significant. Absorption of relatively small amounts results in an intense throbbing headache, often associated with nausea, and occasionally with vomiting and abdominal pain. Larger amounts may result in hypotension, depression, confusion, occasionally delirium, methemoglobinemia, and cyanosis. Aggravation of these symptoms and the occurrence of maniacal manifestations after alcohol ingestion have been repeatedly observed. A temporary tolerance to headache develops from repeated exposures but this is usually lost after a few days without exposure. Workers have utilized this phenomenon by placing a small amount of nitroglycerin in their hat bands to ensure continued absorption and to prevent "Monday headache." The occurrence of typical headache, hypotension, palpitation, and flushing have been observed in explosives workers,[36,38,46] dynamite handlers,[37] and men handling cordite.[47] The headache is presumably due to cerebral vasodilatation and resembles clinically that produced by histamine. Temporary relief can be obtained from adrenalin or from the administration of ergotamine tartrate. Fatalities from industrial intoxication are uncommon. Medical studies of explosives workers with combined nitroglycerin and ethylene glycol dinitrate exposures have not given evidence of chronic intoxication or injury despite the occurrence of transient symptoms. An extensive study of 276 workers with long exposure to nitroglycerin and ethylene glycol dinitrate in three Swedish explosives factories gave no evidence of permanent deterioration in health. The average air concentrations of nitroglycerin–ethylene glycol dinitrate for most operations were below 5 mg./cu. meter, usually 2 to 4 mg./cu. meter. In the group with exposures to concentrations generally below 3 mg./cu. meter, symptoms such as fatigue and alcohol intolerance were less frequent but there was little difference in the frequency of headaches. Headache and fatigue were the predominant symptoms. Vomiting, dyspnea, and alcohol intolerance were less frequent and chest pain was least frequent.[48] The occurrence of "anginoid crises" or other acute heart emergencies has not been thoroughly substantiated but the suggestive reports point to the need for further documentation of this important possibility.

A threshold limit value of 0.5 p.p.m. (5 mg./cu. meter) has been suggested.[15] Headaches and other symptoms will occur at this level but chronic injury is unlikely. Skin absorption is probably important and careful attention is required to reduce this exposure.

[47] J. S. Weiner and M. L. Thomson, *Brit. J. Ind. Med.*, **4**, 205 (1947).
[48] S. Forssman, N. Masreliez, G. Johansson, G. Sundell, O. Wilander, and G. Boström, *Arch. Gewerbepathol. Gewerbehyg.*, **16**, 157 (1958).

PENTAERYTHRITOL TETRANITRATE, $C(CH_2ONO_2)_4$

1. Uses

Pentaerythritol tetranitrate (PETN) is the most sensitive of the bursting charge explosives and is loaded as such only in detonators. It is also used as a booster charge, in plastic demolition explosives, or bursting charges when desensitized with TNT or wax. It is prepared by the nitration of pentaerythritol with nitric acid. The solid explodes at 205 to 215°C. and tends to decompose above 150°C.

2. Physical and Chemical Properties

Molecular weight: 316.15
Melting point: 140–141°C.
Insoluble or slightly soluble in water, partially soluble in alcohol, soluble in acetone

3. Physiological Response

PETN is absorbed slowly from the gastrointestinal tract and lung but not appreciably through the skin. Its physiological effects are similar to the other aliphatic nitrates. It is considerably less effective than nitroglycerin. Doses of 5 mg./kg. by mouth in dogs result in a fall in blood pressure but no effect is observed in man after 64 mg. orally. The daily oral administration of 2 mg./kg. for 1 year caused no effects on growth, hematology, or pathology in rats. Patch tests in 20 persons gave no evidence of skin irritation or sensitization.[49] Although some cases of mild illness and dermatitis have been attributed to contact with PETN in ordnance plants,[39] it is apparent that pentaerythritol tetranitrate is relatively nontoxic and the controls and good housekeeping necessary to prevent explosions from this shock sensitive material should be adequate to prevent injurious effects in workers.

TRIMETHYLENETRINITRAMINE (cyclotrimethylenetrinitramine, cyclonite, hexogen, T_4, RDX)

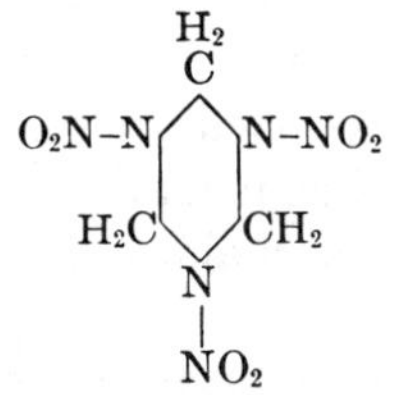

$$O_2N-N \diagdown \diagup N-NO_2$$

with the ring structure showing H_2C at top, H_2C and CH_2 at the sides, and $N-NO_2$ at the bottom.

1. Uses

Trimethylenetrinitramine is a highly explosive solid which found extensive use in military high explosives, generally in combination with TNT. It is pre-

[49] W. F. von Oettingen, D. D. Donahue, A. H. Lawton, A. R. Monaco, H. Yagoda, and P. J. Valaer, Toxicity and potential dangers of penta-erythritol-tetranitrate (PETN), *U. S. Public Health Bull.* No. **282**, 1944.

pared by the nitration of hexamethylenetetramine (hexamine). The most efficient process utilizes acetic acid, acetic anhydride, ammonium nitrate, hexamine, and nitric acid.[24]

2. Physical and Chemical Properties

Melting point: 200–203°C.
Insoluble in water, soluble in acetone

3. Physiological Response

RDX does not exhibit pharmacological effects similar to the nitrites or nitrates. Chronic intoxication is characterized by the occurrence of repeated convulsions. It is slowly absorbed from the stomach and apparently from the lungs but there is no evidence of skin absorption.

The minimum lethal dose as determined in rats by single oral doses of a 4 per cent solution was approximately 200 mg./kg. There was 1 death in 35 rats given 15 mg./kg. daily for 10 weeks by mouth. Deaths occurred in 17 of 35 rats given 50 mg./kg. daily and 15 of 35 on daily doses of 100 mg./kg. The animals lost weight, became increasingly irritable and vicious, and developed frequent convulsions. Gross pathology in those dying during exposure showed lung and gastrointestinal tract congestion. Those surviving had no pathological changes. Seven dogs given 50 mg./day, 6 days a week for 6 weeks showed no blood changes and no methemoglobinemia. A few hours after the first dose they became excited and irritable. As dosing continued reflexes became hyperactive and within the first week the animals had generalized convulsions characterized by hyperexcitability and increased activity followed by clonic movements and salivation, then tonic convulsions and collapse. There was weight loss in all animals and death in one. No microscopic pathology was observed.[34]

Epileptiform seizures have occurred in workers manufacturing trimethylenetrinitramine (T_4) in Italy.[50] The convulsions occurred either without warning or after 1 or 2 days of insomnia, restlessness, and irritability. They were generalized tonic-clonic convulsions resembling in all clinical respects the seizures seen in epilepsy but occurring in individuals without a previous history of seizures. They were most frequent in persons doing the drying, sieving, and packing where the dust could be inhaled. The attacks disappeared when the workers were removed from contact with trimethylenetrinitramine. The seizures were followed by temporary post convulsive amnesia, malaise, fatigue, and asthenia but there was eventually complete recovery.

Similar evidence of systemic intoxication was not observed in a major RDX manufacturing plant during World War II. In this operation there was little or no dusting since the material was handled in a moist state. However, primary irritant and sensitization dermatitis, particularly of the face and eyelids, was encountered during the nitration operation. Studies indicated that an unidentified component

[50] M. Barsotti and G. Crotti, *Med. lavoro*, **40**, 107 (1949).

in the fumes from the reaction mixture was responsible.[51] Although McConnell[39] attributed some dermatitis to RDX manufacture, this probably was due to intermediates since significant dermatitis was not observed in individuals handling the final purified material.[51] This observation is corroborated by von Oettingen's finding that patch testing with the moistened solid did not produce irritation.[34]

No threshold limit value has been proposed and quantitative exposure data for man or animals are inadequate to suggest safe levels. It is evident that fume exposure and skin contact should be prevented during nitration and processes controlled to prevent dust inhalation in order to avoid irritation, dermatitis, and convulsions.

ALKYL NITRITES

I. General Considerations

The alkyl nitrites are aliphatic esters of nitrous acid. The nitrite group has the structure —CONO. Except for methyl nitrite, which is a gas, the lower members of the series are volatile liquids. In general they are insoluble or only very slightly soluble in water but are soluble or miscible with alcohol and ether in most proportions. They tend to decompose to oxides of nitrogen with exposure to light or heat. Violent decomposition can occur. As a group, they tend to be flammable and potentially explosive. They are oxidizing materials which present the possibility of violent reactions from contact with readily oxidized compounds. The physical and chemical properties of the alkyl nitrites are given in Table 17.

The aliphatic nitrites have been of interest mainly because of their pharmacological properties and therapeutic use, but they are also used to a limited extent in industry as intermediates in chemical syntheses. n-Butyl nitrite has been used in the manufacture of rare-earth azides. n-Propyl nitrite, isopropyl nitrite, and tertiary butyl nitrite have been used as jet propellants and for the preparation of fuels. They are usually prepared by the action of sodium nitrite on a mixture of the alcohol and sulfuric acid.

The pharmacological and toxicological effects of the aliphatic nitrites are chiefly characterized by vasodilatation resulting in a fall in blood pressure and tachycardia. Methemoglobin is produced by larger doses. In these respects the alkyl nitrites resemble closely the inorganic nitrites (sodium nitrite) and the aliphatic nitrates. Inhalation by animals and man results in smooth muscle relaxation, vasodilatation, increased pulse rate, and decreased blood pressure progressing to unconsciousness with shock and cyanosis. Headache is often a prominent symptom and may be due to meningeal congestion and vascular dilatation. The development of tolerance has been observed in the therapeutic use of amyl nitrite for angina pectoris. This disappears after a week or so of "nonexposure." The branched chain compounds are more effective than the straight chains in lowering blood pressure. Isopropyl nitrite is considerably more

---

[51] J. H. Sterner, Eastman Kodak Co., personal communication, 1961.

effective than *n*-propyl nitrite and isobutyl nitrite more than normal butyl. The secondary and tertiary butyl compounds also have a more pronounced hypotensive effect than normal butyl nitrite. Methyl nitrite is more effective than are ethyl- and propyl nitrites and amyl nitrite is more effective than ethyl nitrite. As far as the duration of the hypotensive effect is concerned, methyl and ethyl nitrites are most effective, *n*-propyl is the least effective of the lower alkyl nitrites, and the iso derivatives of propyl and butyl nitrite are more effective than the normal compounds.[25] The hypotensive effects are very transient. Amyl nitrite, for example, produces a rapid fall in blood pressure, which lasts only a few minutes after inhalation. Krantz, Carr, and associates have conducted extensive studies on the pharmacology of the alkyl nitrites.[26-28] They found that when dogs were exposed by administering the vapor of 0.3 cc. through an aspirating bottle into the trachea, the degree of hypotension produced decreased from *n*-hexyl (58 per cent fall in blood pressure) through *n*-heptyl (47 per cent) and *n*-octyl (30 per cent) to *n*-decyl (16 per cent). Alkyl nitrites with 11 to 18 carbon atoms in their chain showed slight or no effect on blood pressure under these conditions. However, if injected they produced hypotension. With chains longer than 2-ethyl-*n*-hexyl-1-nitrite, the duration of action became shorter. Cyclohexyl nitrite produced a fall in blood pressure equivalent to ethyl nitrite or amyl nitrite but the duration was longer. In man it produced severe headache. They felt that the major effects are related to the relaxing action of the nitrites on smooth muscle. The mechanism of this is not clear.

Methemoglobin formation has been repeatedly observed following administration to man and animals. The aliphatic nitrites act as direct oxidants of hemoglobin. One molecule of nitrite and 2 molecules of hemoglobin can react to form 2 methemoglobin molecules under appropriate conditions. Side reactions to form nitrosohemoglobin and nitrosomethemoglobin may occur. The amount of methemoglobin formed in cats is directly proportional to the intravenous dose.[29] The longer chain compounds induce more methemoglobin formation relative to their hypotensive effect.[26] The therapeutic usefulness of methylene blue in acute intoxications accompanied by methemoglobinemia remains controversial even though support for its effectiveness in severe methemoglobinemia continues to appear.[29] Although methemoglobinemia is a prominent effect of nitrite absorption, the action of the alkyl nitrites on the vascular system is the major determinant in their toxicity.

The lower aliphatic nitrites are promptly absorbed from the lung. Amyl nitrite is ineffective by mouth since it is destroyed in the gut. It is less effective by injection than by inhalation. Octyl nitrite (2-ethyl-*n*-hexyl-1-nitrite) is not absorbed through the mucous membranes and is ineffective sublingually. It appears that the nitrites are hydrolyzed *in vivo* to nitrite and the corresponding alcohol, which is then partly oxidized and partly exhaled unchanged.

Although the inorganic nitrites, particularly sodium nitrite, have produced many accidental poisonings by ingestion, industrial intoxications from alkyl

TABLE 17 Physical and Chemical Properties of the Alkyl Nitrites

Name	Formula	Molecular weight	Boiling point, °C.	Physical state	Specific gravity
Methyl nitrite	CH_3ONO	61.04	−12	Gas	0.991 (15°C.)
Ethyl nitrite	CH_3CH_2ONO	75.07	17	Colorless liquid	0.900 (15.5°C.)
n-Propyl nitrite	$CH_3CH_2CH_2ONO$	89.09	57	Liquid	0.935
Isopropyl nitrite	$(CH_3)_2CHONO$	89.09	45	Pale yellow oil	0.844 (25/4°C.)
n-Butyl nitrite	$CH_3(CH_2)_2CH_2ONO$	103.12	78.2	Oily liquid	0.9114 (0/4°C.)
Isobutyl nitrite	$(CH_3)_2CHCH_2ONO$	103.12	67	Colorless liquid	0.8702 (20/20°C.)
sec-Butyl nitrite	$CH_3CH_2CH(CH_3)ONO$	103.12	68	Liquid	0.8981 (0/4°C.)
tert-Butyl nitrite	$(CH_3)_3CONO$	103.12	63	Yellow liquid	0.8941 (0/4°C.)
n-Amyl nitrite	$CH_3(CH_2)_4ONO$	117.15	104	Pale yellow liquid	0.8528 (20/4°C.)
Isoamyl nitrite	$(CH_3)_2CHCH_2CH_2ONO$	117.15	97–99	Transparent liquid	0.872
n-Hexyl nitrite	$CH_3(CH_2)_5ONO$	131.17	129–130	Liquid	0.8851 (20/4°C.)
n-Heptyl nitrite	$CH_3(CH_2)_6ONO$	145.20	155	Yellow liquid	0.8939 (0/4°C.)
n-Octyl nitrite	$CH_3(CH_2)_7ONO$	159.23	174–175	Greenish liquid	0.862 (17°C.)

TABLE 18 Comparative Data on Aliphatic Nitro Compounds, Nitrates, and Nitrites

Chemical group	Skin absorption	Irritation	Vascular dilatation	Methemoglobin formation	Industrial experience
Aliphatic nitro compounds —CNO_2					
Nitroparaffins	None	Moderate	None	Positive	Irritation, systemic symptoms
Chlorinated nitroparaffins	None	Marked	None	Unknown	Lung injury
Nitroolefins	Positive	Marked	Unknown	Not observed	None
Aliphatic nitrates —$CONO_2$	Positive	None	Marked	Positive	Systemic symptoms, possible deaths
Aliphatic nitrites —$CONO$	Unknown	None	Marked	Positive	Systemic symptoms, 1 fatality
Trimethylenetrinitramine —$CNHNO_2$	None	None	None	None	Convulsions

nitrites have rarely occurred. Since quantitative data on the inhalation of various concentrations of the aliphatic nitrites by experimental animals or by man in industry are not available, no maximum allowable concentrations have been established. Methods for the determination of nitrites in biological fluids and in air have been described.[25]

II. Specific Compounds

Except for ethyl nitrite and amyl nitrite, specific information on the effects of the alkyl nitrites in man is lacking. However, the pharmacological properties as determined in animals are so uniform within the group that information on ethyl and amyl nitrite can be taken as illustrative of the effects and potential hazards of the other members of the series (see Table 17).

ETHYL NITRITE, CH_3CH_2ONO

Ethyl nitrite is a volatile, flammable, colorless liquid. It has a boiling point of 17°C., and a specific gravity of 0.900 at 15.5°C. Its flash point is —31°F. and the explosive limits in per cent by volume in air are 3.01 to 50. The autoignition temperature of the liquid is 195°F. Thus, it has a high potential fire and explosion hazard. It decomposes readily to form oxides of nitrogen.

Inhalation of ethyl nitrite by dogs results in as much as 60 mm. Hg depression in the blood pressure. This lasts approximately 2 minutes after a single inhalation. Methemoglobin is formed but Heinz bodies have not been found.[25] Mice and cats exposed for 15 minutes to 15 p.p.m. did not show recognizable effects. Industrial intoxications characterized by headache, tachycardia, and methemoglobinemia have occurred. A fatality has been described following the inhalation of ethyl nitrite after accidental breakage of a 4-liter bottle containing spirits of ethyl nitrite (24 per cent ethyl nitrite in alcohol).[22]

AMYL NITRITE, $(CH_3)_2CHCH_2CH_2ONO$ (Isoamyl Nitrite)

Amyl nitrite is a light yellow, transparent liquid with a pleasant, fragrant, fruity odor. Its physical properties are: molecular weight, 117.15; boiling range, 97 to 99°C.; specific gravity, 0.872. It decomposes upon exposure to air and sunlight. It is flammable and explosive. It is prepared by the addition of sodium nitrite to a mixture of isoamyl alcohol and sulfuric acid followed by distillation.

Amyl nitrite was introduced to medicine in 1859 and has received considerable pharmacological investigation since that time. Its major use was for the treatment of angina pectoris through its vasodilatory effect on the coronary arteries. However, this effect is transient and it has largely been replaced by nitroglycerin and longer acting nitrates. It continues to find use in the emergency treatment of acute cyanide intoxication (see Chapter XLIV, Cyanides and Nitriles). The symptoms following inhalation of large doses by man are flushing of the face, pulsatile headache, disturbing tachycardia, cyanosis (methemoglobinemia), weakness, confusion, restlessness, faintness, and collapse, particularly if the individual is standing. These symptoms are usually of short duration. Industrial intoxications have not been reported.[25]

Summary

Although the aliphatic nitro compounds and the nitrates and nitrites have several features in common (nitrogen-oxygen grouping, explosiveness, methemoglobin formation) there are significant differences. Some of their attributes are summarized in Table 18. The esters of nitric and nitrous acid, with the nitrogen linked to the carbon through oxygen, are very similar in their pharmacological effects. Both produce methemoglobinemia and vascular dilatation, with hypotension and headache. These effects are transient. None of the series has appreciable irritant properties. Pathological changes occur in animals only after high levels of exposure and are generally nonspecific and reversible. The nitric acid esters of the monovalent and lower polyvalent alcohols are absorbed through the skin. Information is not available on the skin absorption of the alkyl nitrites. Members of both groups are well absorbed from the mucous membranes and lungs. Heinz body formation has been observed with the nitrates but not with the nitrites.

The nitroparaffins, like the nitrates and nitrites, cause methemoglobinemia in animals. Heinz body formation parallels this activity within the series. Although some members are metabolized to nitrate and nitrite, there is no significant effect on blood pressure or respiration. As with the lower nitrates and nitrites, anesthetic symptoms are observed in animals during acute exposures but these occur late. The prominent effect is irritation of the skin, mucous membranes, and respiratory tract. This is most marked with the chlorinated nitroparaffins and the nitroolefins. In addition to the respiratory tract injury, cellular damage may be observed in the liver and kidneys. Except for the nitroolefins, skin absorption is negligible.

The nitramine, trimethylenetrinitramine, has entirely different activity. It is a convulsant for man and animal. Skin absorption, irritation, vasodilatation, methemoglobin formation, and permanent pathological damage after repeated doses are either insignificant or absent.

Transient illness has been associated with the industrial use or manufacture of these materials but fatalities have been rare and chronic intoxication has been uncommon. Some members of each group present extremely high fire and explosion hazards.

CHAPTER XLVII

Aromatic Nitro and Amino Compounds*

DONALD O. HAMBLIN, M.D.

I. General Considerations

Nitro and amino derivatives of the aromatic series constitute a large and varied group of compounds of great commercial importance. They are widely used as intermediates in the synthesis of "coal tar" or "aniline" dyes, pharmaceuticals, and accelerators and antitoxidants for the rubber industry. Several are classed as fur "dyes," although they actually are not dyestuffs because they have no tinctorial value in their unoxidized form. In addition, these compounds find more limited uses in the production of paints, varnishes, shoe polishes, perfumes, fungicides, pesticides, plastics, petroleum products, and synthetic resins.

This series is characterized chemically by the substitution of an amino (NH_2) radical or nitro (NO_2) radical for a hydrogen atom of the benzene ring or one of its homologs (toluene, xylene). The naphthalene and anthracene rings may be regarded chemically as analogs of benzene. The nitro or amino radicals may be substituted on these rings almost at will at any position in the ring, along with the halogens (most frequently chlorine) and certain of the alkyl radicals, chiefly the methyl and ethyl groups (CH_3, C_2H_5). The simplest of these series of compounds and those that are typical toxicologically are aniline and mononitrobenzene (oil of mirbane). From these basic compounds and their homologs stem a long and interesting succession of products.

These compounds are, for the most part, similar in chemical properties, as well as in pharmacological or toxicological effects. These effects are modified to a greater or lesser degree by the nature of the substituted radicals. The entire series, therefore, lends itself well to a discussion of a few characteristic toxic properties. It should be emphasized that the nitro and amino derivatives of benzene and its

* *Author's Note:* This chapter, originally published in 1949, was based upon the literature deemed to be relevant and pertinent, and upon the author's personal observations up to that time. Since its publication, there have come to the writer's attention approximately 250 additional references or communications on this subject matter. The revision of this chapter has been written with the aim of including, either by reference within the text to this new literature, or by appending a more comprehensive bibliography. A relatively small percentage of the references are thought to have been of true significance. However, during the approximately 10 years that have elapsed, our over-all experience and consequent knowledge of this particular field have served to clarify some of the more obscure areas.

homologs and analogs differ quite markedly toxicologically from their parent compounds (i.e., benzene, toluene, xylene, naphthalene, etc.).

A. MODES OF ABSORPTION

It was pointed out above that the substitution of various radicals and atoms on the ring modifies the toxic properties in varying degrees. It should be borne in mind, in addition, that the physical properties of these compounds markedly influence the magnitude of hazard to the exposed worker. The vapor pressure or volatility of a given compound largely determines to what extent there is a hazard by absorption through the respiratory tract. Similarly, the fat-solvent properties of a compound largely determine its hazard of absorption through the skin. In general, those compounds that are soluble in the common organic solvents, such as alcohol, ether, chloroform, and so forth, are fat-soluble or fat-solvent and insoluble in water. Such compounds usually penetrate the intact skin readily. Conversely, compounds that are water-soluble and insoluble in organic solvents usually do not penetrate the skin appreciably.

Therefore, compound A, with a high vapor pressure and high fat solubility, may be much more of a hazard to exposed workmen than compound B, even though B, if administered directly by the oral route or parenterally, is of a much higher order of systemic toxicity. Aniline is an oily, aromatic, fat-soluble compound, which readily penetrates the intact skin, shoes, clothing, and leather gloves, so that a small area of contamination in clothing or gloves will produce evidence of poisoning if left in contact with the skin for several hours. Aniline hydrochloride, on the other hand, which is a white crystalline solid and is water-soluble and insoluble in the common organic solvents, presents no hazard from skin absorption. This is true, although when administered parenterally or orally to animals, it exhibits toxicity almost identical with that of aniline oil.

B. TOXICOLOGY

This group of compounds, with few exceptions, is characterized by a common and outstanding property: the ability to form methemoglobin in man. Obviously, this cannot be the sole property, toxicologically or pharmacologically, of the various substituent derivatives of aniline and nitrobenzene. The added effects of varying substituents usually do not stand out clearly; so that, with the exceptions to be referred to later, the physician in industry will most often be confronted with methemoglobinemia as the presenting sign.

A great deal of work has been done in an effort to elucidate the mechanism by which these compounds exert their effect. With this class of organic compounds, as well as others, it is manifestly impossible to predict with accuracy what pharmacological changes may result from the substitution of any one radical or even for the substitution of the identical radical in a different ring position.

Since methemoglobin formation plays an important role in the effects of these compounds, methemoglobinemia will be discussed.

The chemistry of hemoglobin and its derivatives is complex, at best, and many of its phases still remain obscure. Reduced to the simplest terms, methemoglobin is the oxidation product of natural hemoglobin and may be considered a true oxide, in which one atom of oxygen combines with one of iron.[1] In other words, heme (an iron-porphyrin compound) combined with globin (protein) constitute hemoglobin when the iron is in the Fe^{++} ferrous state. In this state, hemoglobin easily transports oxygen in a readily dissociable phase. Oxidation of the iron of the heme to the ferric (Fe^{+++}) state constitutes methemoglobin, and its oxygen is so bound that it is unavailable for tissue exchange. In this form, oxygen is in fact so firmly held that exposure to vacuum liberates none.

It has been postulated by Lemberg and Legge[2] that, with 4 hemes available in hemoglobin, 21 classes of intermediates are possible if the free hemes are combined with oxygen, carbon monoxide, or both. Carrying this further, Darling and Roughton,[3] with regard to the hemoglobin and methemoglobin equilibrium, propose the existence of the following intermediates: $(Fe^{++})_3 (Fe^{+++})_1X$; $(Fe^{++})_2 (Fe^{+++})_2X$; $(Fe^{++})_1 (Fe^{+++})_3X$; $(Fe^{+++})_4X$.

This chapter obviously is not the place for more than a cursory discussion of methemoglobinemia. Only those phases that appear essential to an elementary understanding of induced methemoglobinemia in warm-blooded animals and man will be briefly mentioned. For a comprehensive and excellent review of methemoglobinemia and compounds forming it, the reader is referred to Bodansky.[4]

Methemoglobinemia remains a rather puzzling phenomenon, in spite of intensive study that it has received from numerous investigators. Many compounds, other than the aryl amino-nitro group, are methemoglobin formers. Among the direct formers are the nitrites, quinones, and methylene blue. Indirectly, nitrates, such as bismuth subnitrate, when ingested orally, apparently by the action of intestinal bacteria, may yield nitrites with resultant methemoglobin formation. Idiopathic or familial methemoglobinemia need not be considered here.

There is considerable evidence that many of the aryl nitroamino compounds do not directly form methemoglobin, but that such an effect is obtained only from their metabolites. For example, dimethylaniline does not form methemoglobin in vitro,[5] and there is a considerable lag in methemoglobin appearance, following administration to nearly all species in vivo.[6] The marked difference in species response to methemoglobin-forming compounds has been adequately demonstrated by Lester,[7] as well as that there appears to be a rather marked difference of response within species to a given dose.[8]

[1] C. H. Best and N. B. Taylor, *Physiological Basis of Medical Practice*, 5th ed., Williams & Wilkins, Baltimore, 1950, p. 57.

[2] R. Lemberg and J. W. Legge, *Hematin Compounds and Bile Pigments*, Interscience Publishers, New York, 1941, p. 228.

[3] R. C. Darling and F. J. W. Roughton, *Am. J. Physiol.*, **137**, 56 (1942).

[4] O. Bodansky, *Pharmacol. Revs.*, **3**, 144 (June, 1951).

[5] W. Heubner, *Arch. exptl. Pathol. Pharmakol.*, **72**, 239 (1913).

[6] W. W. Cox and W. B. Wendel, *J. Biol. Chem.*, **143**, 331 (1942).

[7] D. Lester, *J. Pharmacol. Exptl. Therap.*, **77**, 154 (1943).

[8] M. V. Bredow and F. Jung, *Arch. exptl. Pathol. Pharmakol.*, **200**, 335 (1942).

Cox and Wendel[6] have demonstrated that single doses of the compounds listed in Table 1, administered either intravenously or by stomach tube, regularly produced methemoglobin in dogs.

TABLE 1

Methemoglobin-Forming Compounds—Dog[6]

Compound	Single dose, mg./kg.	Route[a]
Acetanilid	200	T
o-Aminophenol	20	V
p-Aminophenol	20	V
Aniline	50	V and T
Dimethylaniline	50	T
Hydroxylamine	5	V
α- or β-Naphthylamine	200	T
p-Nitroaniline	15	V
Nitrobenzene	200	T
Nitroglycerin	10	V
Sodium dichromate	60	V
Sodium nitrite	30	V

[a] T = stomach tube administration; V = intravenous administration.

They further state that single doses of H acid (8-amino-1-naphthol-3,6-disulfonic acid) (30,V), hydroquinone (30,V), o-nitrophenol (700,T), and p-nitrotoluene (50,T) caused no accumulation of methemoglobin. Dinitrophenol, sodium chlorate, sodium ferricyanide, sodium sulfanilamide gave negative results even after repeated administrations of large doses. They found that the rates of accumulation, as well as those of disappearance of methemoglobin, varied widely. While intravenous sodium nitrite yields a maximal concentration of methemoglobin in about 45 minutes, nitrobenzene did not become maximally effective for 12 to 15 hours.

To point up the species difference, in response to methemoglobin-formers, Lester[9] found that the cat was most sensitive. If this species is listed as 100, the sensitivities of other species are as follows for acetanilid: man, 56; dog, 29; rat, 5; rabbit, 0; monkey, 0. For acetophenetidin, the sensitivities were as follows: cat, 100; man, 63; dog, 35; rat, 5.

Bodansky[4] has sharply pointed up the wide variation in methemoglobin-forming capacity in the cat for different members of the aryl nitro-amino group. Many investigators have used molecular ratios (molar ratio of methemoglobin formed to dose of test compound) as the unit of comparison for dosage schemes to illustrate this wide variation in the methemoglobin-forming capacity of the cat. Bodansky states it will be noted that the molecular ratio is greater than 1 for a

[9] D. Lester, *J. Pharmacol. Exptl. Therap.*, **77**, 154 (1943).

considerable number of compounds. He has assembled a most informative table, illustrating the wide variation (see Table 2).

Such wide variations, of course, led to speculation as to what metabolic changes within the organism may explain them. It would appear that the compounds themselves or their metabolites must vary widely in their capacity to ox-

TABLE 2

Methemoglobin-Forming Capacity of Aryl Amino and Nitro Compounds in the Cat[a]

Compound	Molecular ratio
Aniline	2.5[b]
	2.7[c]
Acetanilid	1.0[d]
m-Phenylenediamine	1.4[e]
Acetophenetidin	0.14[d]
o-Aminophenol	6.8[f]
Nitrosobenzene	8.6[f]
p-Aminophenol	3.6[g]
p-Aminophenol	1.3[h]
Phenylhydroxylamine	34.0[h]
Nitrobenzene	0.86[i]
o-Dinitrobenzene	1.9[j]
	3.7[h]
m-Dinitrobenzene	7.1[h]
	7.8[i]
	6.4[j]
p-Dinitrobenzene	55[j]
	198[h]
Trinitrobenzene	4.8[i]
o-Nitrotoluol	0.05[i]
m-Nitrotoluol	0.04[i]
p-Nitrotoluol	Very slight[i]
2,4-Dinitrotoluol	1.4[i]
2,6-Dinitrotoluol	0.55[i]
2,4,6-Trinitrotoluol	1.7[i]
m-Chloronitrobenzene	2.3[e]
m-Aminonitrobenzene	3.0[e]
2,4-Dinitrochlorobenzene	0.6[e]
p-Nitro-o-toluidine	3.7[k]

[a] Reproduced with permission of O. Bodansky and William & Wilkins Co.[4]
[b] H. Herken, *Arch. exptl. Pathol. Pharmakol.*, **202**, 70 (1943).
[c] G. Schwedtke, *Arch. exptl. Pathol. Pharmakol.*, **188**, 121 (1937).
[d] D. Lester, *J. Pharmacol. Exptl. Therap.*, **77**, 154 (1943).
[e] F. Jung, *Arch. exptl. Pathol. Pharmakol.*, **204**, 133 (1947).
[f] C. Petersen, *Arch. exptl. Pathol. Pharmakol.*, **198**, 675 (1941).
[g] G. Schwedtke and L. Sing, *Arch. exptl. Pathol. Pharmakol.*, **188**, 138 (1937).
[h] B. V. von Issekutz, *Arch. exptl. Pathol. Pharmakol.*, **193**, 551 (1939).
[i] M. V. Bredow and F. Jung, *Arch. exptl. Pathol. Pharmakol.*, **200**, 335 (1942).
[j] W. Heubner and L. Sing, *Arch. exptl. Pathol. Pharmakol.*, **188**, 143 (1937).
[k] M. Reiter, *Arch. exptl. Pathol. Pharmakol.*, **205**, 327 (1948).

idize hemoglobin. For example, *p*-aminophenol, nitrosobenzene, and phenylhydroxylamine have been advanced as the metabolites of both aniline and nitrobenzene, responsible for methemoglobin formation. The reader is referred to the work of Heubner *et al.*,[10] Lipschitz,[11] and Ellinger.[12] Whatever the mechanism, metabolically, certain compounds are methemoglobin-formers par excellence, such as *p*-dinitrobenzene, phenylhydroxylamine, *p*-aminopropiophenone,[13] and nitrosobenzene. Their marked effectiveness has been attributed to their intermediary metabolites, without very detailed substantiating evidence. Bodansky concludes there is at present no conclusive evidence concerning the nature of the intermediary compound or compounds that oxidize hemoglobin to methemoglobin.

Recently, Evans *et al.*[14] succinctly summarized their views on the action of aromatic amino and nitro compounds as follow: Such "compounds probably are not, in themselves, cyanosis producers, but biochemical redox processes create derivatives which hinder oxygen transport to tissues by forming Hb complexes. Enzyme systems catalyze oxidation of the amine group and reduction of the nitro group to known cyanopathic nitroso- and hydroxylamine derivatives (Fig. 1). The phenylhydroxylamines are probably the most potent cyanosis producers in the redox series.[4,15] Detoxification by rearrangement of the phenylhydroxylamine to ortho- and para-aminophenols of lower potential for excretion through the kidneys probably is responsible for restoration of the oxygen transport balance."

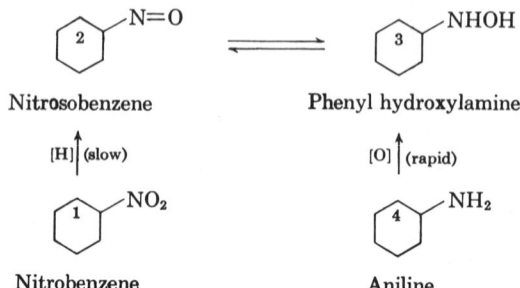

Fig. 1. Cyanosis precursor mechanism (after Evans, Charsha, and Linch[14]).

Quite aside from the difference among species, the differences between metabolites, redox systems, steric relationships, not to mention enzymes and substrates, there is a great deal of evidence pointing to an ultimate equilibrium in vivo between hemoglobin and methemoglobin. Most compounds of this series that have

[10] W. Heubner, R. Meier, and H. Rhode, *Arch. exptl. Pathol. Pharmakol.*, **100**, 149 (1923).

[11] W. Lipschitz, *Z. physiol. Chem.*, **109**, 189 (1920).

[12] O. Ellinger, *Z. physiol. Chem.*, **111**, 86 (1920).

[13] J. M. Vanderbelt, C. Pfeiffer, M. Kaiser, and M. Sibert, *J. Pharmacol. Exptl. Therap.*, **80**, 31 (1944).

[14] E. E. Evans, R. C. Charsha, and A. L. Linch, *A.M.A. Arch. Ind. Health*, **18**, 422 (1958).

[15] W. F. von Oettingen, *U. S. Public Health Bull.* No. **271**, 1941.

been investigated, regardless of species, including man, come to a point of equilibrium, Hb⇌MetHb, beyond which, in spite of further dosage, no appreciable increase in methemoglobin concentration can be obtained.[16-19] Bodansky points out that, both in man and in several animal species, small amounts of methemoglobin appear to be normally present; he postulates that there normally exists an equilbrium in blood between hemoglobin and methemoglobin, which is normally shifted far to the right. This shift, he believes, is regulated by various oxidizing and reducing substances (largely unidentified), and that such a concept helps to explain the difference in degree of methemoglobin formation in various species, as well as the differing rates of reduction of methemoglobin to hemoglobin.

It is of further interest that, while one might assume if 40 per cent of hemoglobin has been converted to carboxyhemoglobin or 40 per cent methemoglobin, that 60 per cent available hemoglobin would continue to transport dissociable oxygen; apparently this is not the case. While the essential effect of a methemoglobin-forming compound is similar to that of carbon monoxide in rendering a portion of the hemoglobin unavailable for oxygen transport, the presence of carboxyhemoglobin and methemoglobin renders residual oxyhemoglobin less capable of dissociation but not to the same degree.[20]

Bodansky concluded, from the dissociation curves of Darling and Roughton, that a content of 23 per cent of carboxyhemoglobin shifted the dissociation curve of the residual oxyhemoglobin in human blood about as far to the left as did a content of 43 per cent methemoglobin.

The effect of various substituents in the ring structure cannot so far be correlated, with any accuracy, as to the nature of the substituted radical, nor can the position of one or more radicals in relation to another be predicted, insofar as methemoglobin-forming capacity is concerned. The additional pharmacological effects of such substituents, that is, the production of effects other than methemoglobin formation, have not been extensively investigated and remain, to a much greater extent, obscure.

Sollmann[21] indulges in some interesting generalizations as to the effect of specific substituents in connection with the antipyretic-analgesic drugs, such as acetanilid and acetophenetidin. These and closely related drugs are derivatives of p-aminophenol.

$C_6H_4OHNH_2$	$C_6H_5NHCOCH_3$	$C_2H_5OC_6H_4NHCOCH_3$
p-Aminophenol	Acetanilid	Acetophenetidin

He considers that antipyretic action resides in the benzene ring, but that benzene itself is not antipyretic because of its inability to react readily with body cells. He

[16] B. B. Clark, E. J. Van Loon, and R. W. Morrissey, *J. Ind. Hyg. Toxicol.*, **25**, 1 (1943).

[17] D. Lester, L. A. Greenberg, and E. Shukovsky, *J. Pharmacol. Exptl. Therap.*, **80**, 80 (1944).

[18] M. Reiter, *Arch. exptl. Pathol. Pharmakol.*, **205**, 327 (1948).

[19] C. Petersen, *Arch. exptl. Pathol. Pharmakol.*, **198**, 675 (1941).

[20] V. A. Drill, *Pharmacology in Medicine*, McGraw-Hill, New York, 1954, p. 56.

[21] T. Sollmann, *A Manual of Pharmacology*, 8th ed., Saunders, Philadelphia, 1957, p. 722.

further states that "this capacity of reacting may be given to it by substituting for one of its H atoms an OH group, as in phenol, C_6H_5OH; or still more strongly an NH₂ group, as in aniline, $C_6H_5 \cdot NH_2$ (phenylhydrazine, $C_6H_5NH \cdot NH_2$ being stronger); or by both as in p-aminophenol, $C_6H_4OHNH_2$."

He continues that the aromatic alcohols and amines are all strongly anti-pyretic but they also cause collapse. The diminution of this "collapse" action, he continues, may be accomplished by replacing the H, either of the OH or of the NH₂, by other radicals. He further states that reduction of toxicity is greater if the substitution is made in the NH₂ group and is greatest if the substituted radical is alkyl and less if "acidyl." However, if several H atoms are substituted, the toxicity is less if both an "acidyl" and an alkyl radical are introduced than if both radicals be of the same kind. Later, he states that aniline acetate preserves the original toxic action of aniline, whereas this is greatly weakened in acetanilid, and that the substitution of an H by I, Br, or Cl does not modify the antipyretic or toxic action of the original substance.

The difficulty in predicting the differences between pharmacological effects produced by varying substituents and their steric relationships is sharply pointed up by a comparison of dinitrobenzene with dinitrophenol and dinitrocresol.

2,4-Dinitrophenol   p-Dinitrobenzene   2-Methyl-4,6-dinitrophenol

p-Dinitrobenzene is a very powerful methemoglobin-former, and this appears to be its primary effect when absorbed by man. The dinitrophenols and dinitrocresols, on the other hand, are poor methemoglobin formers but exert profound metabolic disturbances when absorbed. These effects will be briefly mentioned later.

C. POISONING IN MAN BY THE ARYL NITRO-AMINO COMPOUNDS

By and large, in chemical plants engaged in the production of dyestuffs and their intermediates, rubber accelerators, etc., exposures to nitrobenzene and aniline are most frequent because these are the basic compounds from which a large percentage of the intermediates are formed. Aniline is formed by the reduction of nitrobenzene. Aniline is highly reactive and enters into numerous syntheses.

Derivatives of aniline are important in the pharmaceutical, dye, resin, pigment, agricultural, rubber chemical, and many other fields. The schematic drawings in Fig. 2 serve as examples of some reactions and uses of aniline.

Clinically, upon appreciable absorption of aniline, nitrobenzene, and most of their immediate homologs, methemoglobinemia is the outstanding effect. These compounds appear to be relatively free of any other outstanding side effects. Both

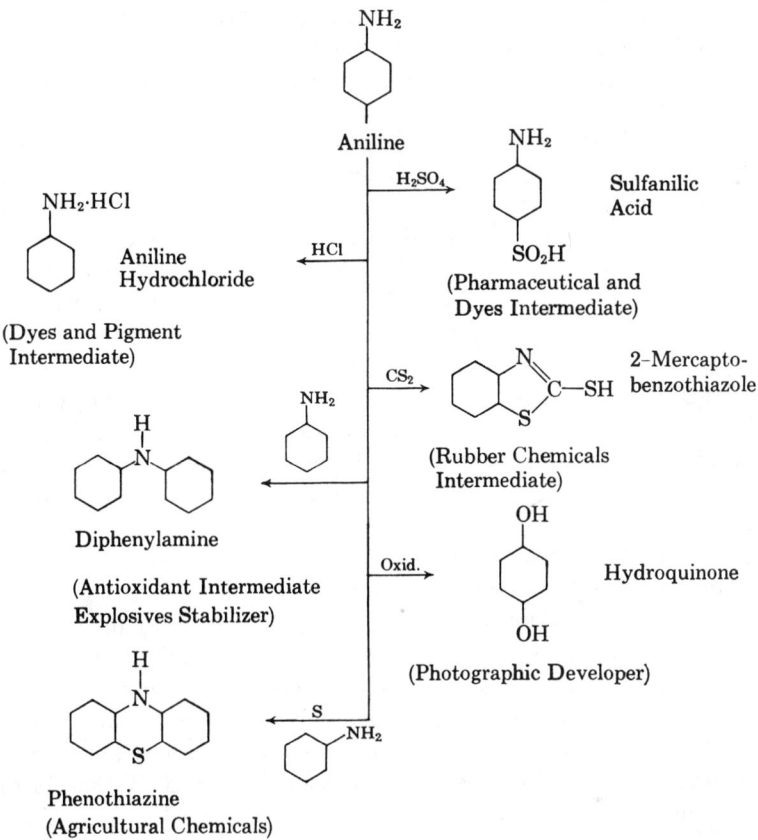

Fig. 2. Some reactions and uses of aniline.

of these compounds are fat-soluble, penetrate the intact skin rapidly, and their vapors are quickly absorbed upon inhalation.

Following absorption of a methemoglobin-former of this class, the rate of appearance of methemoglobinemia, as would be expected, is dependent upon the nature of the exposure and its intensity. If absorption follows contamination of a leather shoe or glove with aniline oil, for example, several hours may elapse before methemoglobinemia is apparent. Similarly, a considerable delay may be expected if absorption has taken place slowly through inhalation of a low concentration of vapor. On the other hand, when, for example, a pipeline failure occurs, drenching a man from head to foot with aniline or nitrobenzene, little time elapses before the typical "blue lip" or "huckleberry pie" faces are seen. The intensity of cyanosis is only a rough guide to the degree of methemoglobinemia present, because of pigmentary differences.

It has been a common observation within the industries handling such compounds routinely that the onset of methemoglobinemia is indeed insidious. The patient initially usually feels very well and perhaps, to some degree, euphoric. This is not surprising, when one considers the close pharmacological relationship of aniline and its homologs to the antipyretic–analgesic group characterized by acetanilid and acetophenetidin.

As methemoglobinemia develops in its intensity, the first symptom is frequently a headache. If methemoglobinemia progresses, the headache may develop to great intensity. The only safe or reliable index is a quantitative determination of the methemoglobin content of venous blood by an accepted method.

Within rather wide limits, cyanosis is usually grossly recognizable when the methemoglobin concentration reaches 15 per cent or more. As the methemoglobin concentration increases to 40 per cent or so, the patient usually feels well, has no complaint, and is quite insistent there is nothing wrong with him. From 40 to 70 per cent, weakness and a sense of dizziness or ataxia may be observed. Dyspnea on exertion becomes apparent, to a greater or lesser degree, as does an accompanying increase in pulse rate. However, with the patient at rest in bed, there are very few complaints; and we have never observed convulsive phenomena of any sort, even when the methemoglobin concentration has reached 76 per cent. At this level, which is the highest we have recorded, there is an increase in pulse rate to 110 or so, and some, slight increase in the respiratory rate. With marked cyanosis, the appearance of the patient is quite alarming and, for most of us, continues to be, even though we have seen numerous cases of similar gross appearance. Headache and a general sense of weakness seem to be promptly relieved by placing the patient within an oxygen tent. This, of course, does not alter the methemoglobin content of the blood. The relief is transient and the headache usually returns promptly if the supply of oxygen is stopped. Without attempts at specific therapy, we have found that patients who absorbed aniline, nitrobenzene, p-nitroaniline, etc., returned to normal in approximately 24 hours, provided all sources of further absorption were completely eliminated. Perhaps thorough cleansing of the patient is the most important single measure to be taken for the patient's welfare. This means thorough cleaning from head to foot with soap and warm water, scrubbing finger and toenails; and, if there has been a dust exposure, repeated irrigation of the nasal passage with warm saline. This, coupled with rest in bed, seems to be all that is actually required because of the reversibility of the methemoglobin to hemoglobin equilibrium. The factors influencing the return of this equilibrium to the right are described by Bodansky;[4] first, metabolism of the offending compound to a metabolite that does not oxidize hemoglobin; and second, reduction of methemoglobin to hemoglobin by largely unidentified but constant enzyme systems.

On the other hand, this assumption, based on rather extensive experience over many years, may be unwarrantedly complacent. Bodansky and Gutmann,[22,23] by using p-aminopropiophenone, were able to kill dogs with methemoglobin

[22] O. Bodansky and H. R. Gutmann, *J. Pharmacol, Exptl. Therap.*, **90**, 45 (1947).

[23] H. R. Gutmann and O. Bodansky, personal communication, 1946.

concentrations of greater than 90 per cent. We have never encountered this compound as an in-plant hazard, although we have had rather severe poisonings from phenylhydroxylamine, and still have not observed in man concentrations beyond approximately 75 per cent methemoglobin. A patient with such marked hypoxia is, of course, critically ill and must be so regarded even though recovery without specific therapy is the rule.

Therapy, such as the administration of ascorbic acid, methylene blue, and possibly glucose, has since been recommended on apparently sound experimental and clinical bases. Ascorbic acid, in the treatment of congenital methemoglobinemia, was reported by Lian et al.[24] as apparently hastening the reversion of methemoglobin to hemoglobin. It is less efficacious than methylene blue.[22]

Wendel[25] demonstrated the value of methylene blue in bringing about a rapid and even dramatic reduction of methemoglobin to hemoglobin. This work was initially received with some skepticism because methylene blue is, itself, a methemoglobin former. Wendel explained his reasoning as to how methemoglobin might be reduced by methylene blue, as follows:

"Since each molecule of dye injected effects conversion of many molecules of methemoglobin to hemoglobin, this reaction, too, is catalytic. Here, however, the catalysis is one of reduction, and leuco methylene blue would appear to be the effective reductant. Two possible sources of the leuco methylene blue in the body are reduction of methylene blue in the erythrocytes by enzymes systems present there and reduction of methylene blue in other tissues. Preliminary experiments suggest that the rate of formation of leuco methylene blue in the erythrocytes may not be sufficiently rapid to account for all the methemoglobin reduced. Thus, it would appear that leuco methylene blue formed in the more actively metabolizing tissues and returned as such to the erythrocytes may play a role in reducing methemoglobin to hemoglobin. Experiments designed to test this possibility are in progress."

This skepticism was commented upon by Crisler[26] and editorially in the *Journal of the American Medical Association*,[27] and joined by many of us working with methemoglobin-formers in industry. This skepticism appears to have been ill-founded and methylene blue is now widely used in the treatment of severe methemoglobinemia. Bodansky explains that, according to Nadler et al.,[28] it has been demonstrated that methylene blue is not a good methemoglobin-former in vivo in man. Bodansky's explanation of its action follows:

"When methylene blue is injected into man, two reactions ensue: (1) a direct oxidation of hemoglobin to methemoglobin, and (2) an opposing reaction, the reduction of methemoglobin to hemoglobin, for which coenzyme and coenzyme factor are necessary. Apparently, the latter reaction is much more effective, so that the equilibrium state between the two reactions is pitched at a point of very low methemoglobin formation. This equilibrium is arrived at whether methylene blue is injected in a person with no methemoglobinemia or in one with a high concentration of methemoglobin."

[24] C. Lian, P. Frumasan, and Sassier, *Bull. mém. méd. hôp. Paris*, **55**, 1194 (1939).
[25] W. B. Wendel, *J. Clin. Invest.*, **18**, 179 (1939).
[26] G. Crisler, *Am. J. Physiol.*, **110**, 580 (1935).
[27] Editorial, *J. Am. Med. Assoc.*, **141**, 534 (1949).
[28] J. E. Nadler, H. Green, and A. Rosenbaum, *Am. J. Med. Sci.*, **188**, 15 (1934).

Brooks,[29] in 1935, from her work with rabbits on cyanide antidotes, noted that intravenous glucose seemed to hasten the reversion of methemoglobin to hemoglobin. Cox and Wendel,[6] working with dogs, found no such effect with similar nitrite-induced methemoglobinemia. We believed that the use of intravenous glucose in normal solutions was preferable to the possibly undemonstrated hazards of intravenous methylene blue, and so began using it in cases in man showing 40 per cent or more of methemoglobin. Due to the impossibility of having adequate controls in man and because of the normal rapid rate of reversion of otherwise untreated methemoglobinemia in man, it has not been clearly shown that intravenous glucose is without value.

Mangelsdorff,[30] in a concise and excellent paper, presents comparison of no medication with intravenous glucose and intravenous methylene blue. In this he mentions the possible harmful effects of methylene blue, such as hemolysis, altered oxygen capacity of red cells, and electrocardiographic changes.[28,31-34] With one exception, we are in complete agreement with his conclusions, which we know have been based on careful and extensive observations. "In cases showing methemoglobinemia above 30%, three methods of treatment are advocated: (a) watchful waiting; (b) intravenously administered dextrose; and (c) intravenously injected methylene blue.

"Because a methemoglobinemia of moderate degree returns to normal usually within 20 hours, no drastic treatment is necessary.

"Methylene blue or other reducing compounds should only be used in severe cases of methemoglobinemia and then with caution."

The exception is that he apparently considers that glucose is not a reducing compound, although this has not been unequivocally demonstrated.

Fleming et al.,[35] in their textbook, also reported the use of glucose and oxygen as standard procedure in the treatment of methemoglobinemia, presumably in du Pont plants. Bodansky is in agreement that, in most cases of induced methemoglobinemia in which no more than 40 per cent of the blood pigment is oxidized, watchful waiting appears adequate, provided the patient is at rest. He further agrees that, when high levels of methemoglobinemia are reached, perhaps 60 per cent or above, methylene blue appears to be indicated. He advocates methylene blue in a dosage of 1 to 2 mg./kg. of body weight as a 1 per cent saline or aqueous solution. However, he deprecates the administration of oxygen and intravenous glucose as being without sound physiological basis. One would be hard put to it to demonstrate a truly sound physiological basis, except for the

[29] M. M. Brooks, Am. J. Physiol., **114**, 160 (1935).

[30] A. F. Mangelsdorff, M. D., A.M.A. Arch. Ind. Health, **14**, 148 (1956).

[31] E. Huyghebaert, Arch. intern. pharmocodynamie, **29**, 405 (1924).

[32] S. S. Spicer and E. C. Thompson, J. Ind. Hyg. Toxicol., **31**, 206 (1949).

[33] W. B. Wendel and M. L. Hefley, Proc. Soc. Exptl. Biol. Med., **31**, 973 (1934).

[34] G. Crisler, Am. J. Physiol., **110**, 580 (1935).

[35] A. J. Fleming, C. A. D'Alonzo, and J. A. Zapp, Modern Occupational Medicine, Lea & Febiger, Philadelphia, 1954, p. 342.

reports of Brooks,[29] previously referred to, and practical, although largely empirical, experience with this clinical approach.

Occasionally, as Bodansky states, one finds the transfusion of whole blood advocated for the treatment of severe or moderately severe induced methemoglobinemia. This, in view of the usual prompt reversion of methemoglobin to hemoglobin, appears to the writer to be wholly unwarranted and basically unsound.

At any rate, as has been previously stated, intelligent evaluation of the clinical progress of induced methemoglobinemia requires the quantitative determination of the methemoglobin content of the blood by an approved method. The method most commonly employed and most generally available is that described by Evelyn and Malloy.[36] This requires, of course, the use of a well-standardized photoelectric colorimeter. In our early determinations, we were fortunate in having access to a very sensitive and accurate instrument, a modified General Electric Recording Spectrophotometer. This method was described by Hamblin and Mangelsdorff.[37] The beauty of this method is the wide spectral range that is covered in a single reading recorded as a logarithmic curve from 400 to 700 mμ. Thus, deviant chromogens would be readily apparent.

Evans et al.,[14] following the in vitro work of Jackson and Thompson,[38] believe they have demonstrated hemoglobin complexes other than methemoglobin in the blood of patients who have absorbed certain of the aryl nitro-amino compounds. They believe these hemoglobin complexes are precursors to true methemoglobin. They further imply that, upon recognition of such precursors, the prompt administration of methylene blue will prevent formation of true methemoglobin. This is an interesting observation but its clinical value remains obscure to this writer.

At any rate, with the concentration of methemoglobin known to the observer, the clinical course can be intelligently followed. In spite of some earlier reports to the contrary,[39] with aniline and its immediate homologs the severity of the clinical picture remains throughout directly proportionate to the concentration of methemoglobinemia. As the percentage of methemoglobin increases, the patient becomes sicker; and as it decreases, he is better.

The secondary effects, other than formation of methemoglobin, are for the most part so obscured that they cannot be readily distinguished. Lester et al.[17] dosed rats with p-aminophenolhydrochloride and β-phenylhydroxylamine and the rats died, but the levels of methemoglobin were insufficient to cause death by anoxia. They therefore concluded that methemoglobinemia plays no important part in the acute toxicity of the two compounds used. They also state that: "Indications are given that this fact applies also to aniline and acetanilid." This is a rather sweeping conclusion and apparently is not relevant to aniline-induced

[36] K. A. Evelyn and H. T. Malloy, J. Biol. Chem., **126**, 655 (1938).
[37] D. O. Hamblin and A. F. Mangelsdorff, Ind. Hyg. Toxicol., **20**, 523 (1938).
[38] H. Jackson and R. Thompson, Biochem. J., **57**, 619 (1954).
[39] Price-Jones and Boycott, Guy's Hosp. Repts., **63**, 309 (1901).

methemoglobinemia in man since no significant aspects of the clinical picture in man—with a possible exception of analgesia—need be explained other than hypoxia. We find Lester's statement perpetuated in a recent textbook, *Industrial Toxicology*, by Fairhall.[40]

Hemolysis is frequently reported as a secondary or accompanying effect of the absorption of methemoglobin formers of this class. We have never observed evidences of hemolysis as concomitant of poisoning by aniline and its immediate homologs. We did blood counts regularly for many months when we were observing numerous cases of methemoglobinemia, and completely abandoned this procedure as unrewarding. At this time, we were also unable to demonstrate increased urobilin, urobilinogen, or other urinary evidence of increased hemolysis.

Bodansky also is apparently not impressed with hemolysis as a prominent characteristic of induced methemoglobinemia. He cites equivocal evidence from Van Loon and associates,[41] based upon the administration of acetanilid and acetophenetidin to dogs. He further states that these drugs are not directly hemolytic, but that the possibility exists that their metabolic products may render cells more sensitive to hemolysis. He also states that Jung[42] was unable to demonstrate any decreased resistance to hemolysis, using methemoglobinemic erythrocytes in various NaCl solutions.

Inclusion of Heinz bodies have been described by many writers, and particularly well by Hughes and Treon,[43] as regular concomitants of induced methemoglobinemia from the aryl nitro or amino group. These bodies have eluded this observer and his associates, possibly because of faulty staining techniques. At any rate, their recognition appears to be of little practical clinical significance.

Similarly, chiefly in German literature, we find reference to liver damage, bone marrow changes, respiratory tract inflammation, involvement of the gastro-intestinal system, etc. Good examples are the writings of Holstein[44] and Otto.[45] Otto also emphasizes electrocardiographic changes as being of great significance. Conceivably, such changes may be concomitant of severe anoxia but probably otherwise do not occur. These observations are very puzzling to those of us who have worked for some time with patients poisoned by aniline and its immediate homologs, and who believe this is an essentially simple, reversible process with no residual pathology of any significance. Such tissue changes may be properly attributable to trinitrotoluene and dinitrobenzene, but they occur only after prolonged and intensive exposure.

By and large, then, methemoglobinemia is the predominant effect of this group of compounds. As was stated previously, because of widely varying structures and substituents, this property is not the sole pharmacological effect.

[40] L. T. Fairhall, *Industrial Toxicology*, 2nd ed., Williams & Wilkins, Baltimore, 1957.
[41] E. J. Van Loon, B. B. Clark, and D. Blair, *J. Lab. Clin. Med.*, **29**, 942 (1944).
[42] F. Jung, *Klin. Wochschr.*, **19**, 1016 (1940).
[43] J. P. Hughes and J. F. Treon, *Arch. Ind. Hyg. Occupational Med.*, **10**, 192 (1954).
[44] E. Holstein, *Arbeitsmedizin*, **27**, 7 (1953).
[45] H. Otto, *Arbeitsmedizin*, **27**, 46 (1953).

Bodansky, in Table 2, has clearly demonstrated the wide variation in methemoglobin-forming properties among compounds representative of this group, as expressed as molecular ratios. Cox and Wendel[6] demonstrated that o-nitrophenol, p-nitrotoluene and dinitrophenol appear incapable of producing methemoglobin in dogs. Dinitrochlorobenzene is a poor methemoglobin former but a very powerful sensitizing agent for skin and mucous membrane. These examples illustrate wide variations in pharmacological properties.

In addition, we repeat, many compounds—solely because of their physical properties—cease to be of major toxicological significance in industrial usage, even though when administered to animals orally or parenterally a relatively high degree of toxicity is demonstrated.

It is manifestly impossible to attempt to mention each of the hundreds of compounds of commercial importance, let alone the thousands known in the laboratory, which constitute this series.

There follow brief descriptions of some of the characteristics that illustrate further these variants.

D. VARIATIONS IN TOXICOLOGICAL EFFECTS

Dinitrobenzene, for example, even though it is a solid, is a methemoglobin-former par excellence. Even though of comparatively low vapor pressure, it appears to be much more toxic than aniline and nitrobenzene. Comparable exposure to or absorption of this compound results in a more intense methemoglobinemia, which is much less readily reversible than that from an equivalent exposure to its parent compound, nitrobenzene. This is perhaps explained as suggested by Lipschitz[11] that, in vitro, blood may reduce dinitrobenzene to nitrosophenylhydroxylamine, with the further assumption that the body is unable to break down this compound further than nitroaniline, which, in itself, is highly toxic in addition to causing a more persistent methemoglobinemia than dinitrobenzene and may, following prolonged exposures, result in a toxic hepatitis which may occasionally progress into acute yellow atrophy.[46-48]

Dinitrochlorobenzene has the outstanding characteristic of being an almost universal skin sensitizer. Repeated and minute contacts with this compound may produce anything from a few itching papular vesicles to a generalized exfoliative dermatitis.[49-52] One of the most useful compounds for experimental sensitization is 2,4-dinitrochlorobenzene, which combines with the amino groups of the lysine component of the epidermal proteins.[53] As a hazard in the production of dinitro-

[46] M. Kiese, Arch. exptl. Pathol. Pharmakol., 206 (5/6), 505 (1949).

[47] M. Kiese, Arch. exptl. Pathol. Pharmakol., 208 (1), 43 (1949).

[48] M. Kiese, Arch. exptl. Pathol. Pharmakol., 206 (4), 361 (1949).

[49] K. Landsteiner, A. Rostengerg, and M. S. Sulzberger, J. Invest. Dermatol., 2, 25 (1939).

[50] J. Edmund, Acta Allergol., 6, 118 (1953).

[51] H. Haxthausen, Acta Dermato-Venereol., 31, 659 (1951).

[52] W. S. Jeter and P. M. Seebohm, Proc. Soc. Exptl. Biol. Med., 80, 694 (1952).

[53] D. M. Pillsbury, W. B. Shelley, and A. M. Kligman, Dermatology, 158 (1956).

chlorobenzene, methemoglobinemia seems to be of little importance. In producing many tons of this compound, we recall no instances in which it has given rise to clinical cyanosis.

Dinitrophenol is rather aberrant in its behavior. Dinitrophenol is again not important as a methemoglobin-former, but profoundly affects metabolism. Absorption of this compound in toxic quantities leads to a marked elevation of the basal metabolic rate and rises in temperature to as high as 110°F., with perhaps additional nervous system effects. Likewise, liver damage and kidney damage have been reported, as well as destructive changes in the thyroid.[54-58] It will be recalled that, in 1933, 2,4-dinitrophenol[59] was advocated in this country as an agent for the easy treatment of obesity. It will be recalled how disastrously toxic this compound proved to be in many cases, not only in severe acute poisonings and deaths, but in a most unfortunate delayed effect, the formation of cataracts of the lens.[60-61]

Dinitro-o-cresol is pharmacologically rather similar to dinitrophenol. Bidstrup and Payne[62] have made an excellent report of 8 fatal cases of poisoning with this compound occurring in Great Britain, where dinitro-o-cresol was widely used as a selective weed killer in cereal crops. For details, the reader is referred to this well-documented presentation.

Aniline hydrochloride, a simple addition product of aniline, is not a hazard of significance commercially. It is a white, crystalline solid with a very low vapor pressure, which apparently is not readily absorbed through the intact skin, and which can be handled with impunity. It is, of course, toxic if administered either orally or parenterally and its effects are very similar to those of aniline. Precautions should be taken to prevent inhalation of its dusts.

The addition products of this series, usually the hydrochloride or the sulfate, like aniline hydrochloride, with few if any exceptions, behave similarly.

m-Toluylenediamine (2,4-diaminotoluene) and m-phenylenediamine, reduction products of the highly toxic compounds, dinitrotoluene and dinitrobenzene, cease to be of any practical importance as industrial hazards since they are, when pure, colorless crystalline water-soluble compounds, which are not fat-soluble. Both compounds exhibit very low vapor pressures. Industrially, they have never, insofar as we are aware, given rise to methemoglobinemia nor to any other toxic effects more troublesome than a deep staining of the skin. However, when administered either orally or parenterally, both compounds are reported to be highly toxic.[63] In repeated small doses, it was found that toluylenediamine can

[54] H. Magne, A. Mayer, and L. Plantefol, *Ann. physiol. physicochim. biol.*, **7**, 269 (1931).

[55] N. Alwall and G. Mansfeld, *Arch. exptl. Pathol. Pharmakol.*, **185**, 93 (1937).

[56] L. Lutz and G. Baume, *Compt. rend. soc. biol.*, **80**, 483 (1917).

[57] M. L. Tainter and W. C. Cutting, *Arch. Pathol.*, **18**, 881 (1934).

[58] I. Peissakowitsch and P. Kostenko, *Arch. Gewerbepathol. Gewerbeghyg.*, **6**, 160 (1935).

[59] M. L. Tainter, A. B. Stockton, and W. C. Cutting, *J. Am. Med. Assoc.*, **101**, 1472 (1933).

[60] W. D. Horner, R. B. Jones, and W. W. Boardman, *J. Am. Med. Assoc.*, **105**, 108 (1935).

[61] W. W. Boardman, *J. Am. Med. Assoc.*, **105**, 108 (1935).

[62] P. L. Bidstrup and D. J. Payne, *Brit. Med. J.*, 16, (1951).

[63] W. Gibbs and E. T. Reichert, *Anat. Physiol. Suppl.*, 259 (1892); *Am. Chem. J.*, **16**, 443 (1894).

be used to produce jaundice at will. The mechanism by which such damage is produced is summarized by Greene and Schaal,[64] who offer the following three explanations: (a) that the toluylenediamine icterus is caused by damage to the epithelium of the larger biliary ducts, resulting in passage of the bile into the lymph spaces, the thoracic duct, and the blood; (b) that it may be of hemolytic and hyperfunctional origin as indicated by the greater activity of Kupffer's cells; and (c) that biliary thrombi may be formed, blocking the central biliary ducts, causing a static icterus.

p-Phenylenediamine (known as Ursol in the fur-dyeing industry) also is water-soluble and not appreciably soluble in fats, and is of little importance as a toxic hazard in industry. When pure, it is a colorless crystalline compound, but, when exposed to air and moisture, it is oxidized rather rapidly and progresses to red, brown, and finally black. It is widely employed in the dyeing of furs when a deep black is desired. Its oxidation may also be arrested at the brown or reddish stages, but such an arrest of oxidation is difficult to control. This compound, which is really an intermediate and not a dyestuff, has done more than any other compound to bring "aniline" dyes as a whole into unmerited opprobrium. In its intermediate stages of oxidation it frequently is a skin sensitizer and may produce contact dermatitis of varying intensities. Thus, carelessly dyed furs, which have not been completely oxidized and after-treated, have been the cause of litigation following dermatoses of varying severity in sensitized wearers of furs. It is stated by Mayer[65] and others that the oxidation product that is the offender is quinonediimine. In addition, p-phenylenediamine has, in industrial exposures, caused true allergenic bronchial asthma.[66] Systemic poisoning from industrial exposures to this compound is unknown, although it is reported as being highly toxic when administered orally or parenterally to experimental animals.

p-Aminophenol is very similar in its properties to p-phenylenediamine, both chemically and pharmacologically. It also finds considerable use as a fur dye, chiefly in obtaining varying shades of brown. It, likewise, may cause contact dermatitis in its various stages of oxidation and has also been reported as causing bronchial asthma.

TNT (trinitrotoluene), many million tons of which were produced during World Wars I and II—with the resultant exposure of many thousands of workers —is a highly toxic compound, but one which, like all others of the aromatic series, may be produced without injury to the exposed individuals if rigid hygienic measures are enforced. That it is perfectly feasible to manufacture TNT without harmful effects is illustrated by reports of Cone[67] and by reports of plant physicians whom the author visited personally during the recent war. Because of poor hygiene, however, serious poisoning and fatalities unfortunately have occasionally

[64] H. H. Greene and W. Schaal, *Beitr, pathol. Anat. u. allegem. Pathol.*, **89**, 61 (1932).
[65] R. L. Mayer, *Arch. Dermatol. u. Syphilis*, **156**, 331 (1928).
[66] H. Reichel, *Vergifungsfallen*, **5**, A21 (1934).
[67] T. E. Cone, Jr., *Ind. Hyg. Toxicol.*, **26**, 260 (1944).

occurred. These fatalities may, in some degree, be due to idiosyncrasies, since exposures equal in duration and severity bring about wide variations in clinical responses. McConnell and Flinn[68] report for the United States Ordnance Department that, in the period of $3^1/_2$ years during the past war, 22 fatalities occurred, which, however, is at a rate of only 3 fatalities per 100,000 operating employees, a record, that speaks well for the program of hygiene enforced. In this series of 22 fatal cases, these writers report, 8 died of toxic hepatitis, 13 of aplastic anemia, and 1, who recovered partially from hepatitis, died from aplastic anemia or a combination of anemia and hepatitis. The number of cases of toxic jaundice and aplastic anemia who did not die is not known, but very few cases of aplastic anemia recover from such poisoning. Hepatitis was observed to occur more frequently among the younger age group, the average age being 30 years; aplastic anemia more frequently among the older age group, the average age being 45 years. These writers report that, unfortunately, the early symptoms of poisoning are indefinite and not sufficiently marked to cause patients to report for medical aid until their condition has become advanced. Most of the employee groups were examined at from 1- to 2-month intervals during exposure. It is reported that the most significant laboratory findings were the elevation of the icteric index or a decrease in hemoglobin or total cell pack for anemia. It is obvious, however, that by the time a significant elevation of icteric index has occurred liver damage of some severity has already occurred, which makes this a rather unsatisfactory test. In the Lake Ontario Ordnance Works we thought that a prothrombin time (modified Quick test) showed some promise of giving earlier indications of disturbed liver function and believe that this is well worth trying. Sievers et al.,[69] in a case study of TNT workers in a bomb- and shell-loading plant, reported no cases of severe TNT intoxication, but state that cyanosis was observed among 68 per cent of the men and 36 per cent of the women. They state that cyanosis was only suggestive in the majority of subjects. They quote other authors,[70-72] who have made similar observations in TNT workers, and further state that the cyanosis observed evidently was not due entirely to methemoglobinemia, and report that the highest concentration observed was 6 to 7 per cent. They conclude that no correlation was found to exist between the degree of cyanosis observed and the amount of methemoglobin found in the blood. We believe that these observations are open to question, as has been pointed out previously, and that the methods employed for the determination of methemoglobin may have been at fault. Also, as has been pointed out previously, in our experience with the nitro and amino series, 6 to 7 per cent methemoglobin, let alone 2 per cent methemoglobin, is not grossly detectable. Mild hypochromic anemia was frequently observed, as well as a slightly, but signifi-

[68] W. J. McConnell and R. H. Flinn, *Ind. Hyg. Toxicol.*, **28**, 76 (1946).

[69] R. F. Sievers, A. H. Lawton, F. Shoog, P. A. Neal, and W. F. von Oettingen, *U. S. Public Health Bull.* No. **291**, 1945.

[70] C. Voegtlin, C. W. Hooper, and J. M. Johnson, *Hygienic Lab. Bull.* No. **126**, 1920.

[71] P. N. Panton, *Lancet*, **2**, 77 (1917).

[72] T. J. Putnam and W. Herman, *J. Ind. Hyg.*, **1**, 238 (1919).

cantly, shortened coagulation time in exposed men. No positive urobilins or urobil-inogens in urine were encountered. They concluded that the health of exposed workers could be satisfactorily controlled by accurate hemoglobin determinations, total cell volume determinations (Wintrobe hematocrit) and an icteric index, done each month. It is also of interest that these observers concluded that the adminis-tration of ascorbic acid at a rate of 100 mg. daily, as well as "shotgun" vitamins regularly administered, had no recognizable value in the prevention or therapy of TNT poisoning. With this we are in complete agreement.

Tetryl (tetranitromonomethylaniline), an explosive (a booster), was manu-factured by the millions of tons during both wars. As an industrial hazard, even though it contains 4 nitro groups, it is of very minor significance. Hatch and Probst[73] state that the consensus appears to be that systemic poisoning from this compound does not occur, although some earlier writers report the appearance of systemic effects, none of which, however, are severe. Tetryl, however, causes a high incidence of contact dermatitis, nasal irritation, and epistaxis. One of the most troublesome effects of tetryl, although of no significance, is a yellow staining of the skin and hair, which cosmetically and psychologically disturbs many individuals.

o-Toluidine appears to approximate aniline as a methemoglobin former. However, in our experience it has an additional property of causing transient hematuria. Following absorption of this compound, one occasionally may observe sufficient red blood cell content in the urine to render a specimen grossly sus-pect. However, more often, the urinary sediment, on microscopic examination, shows numerous erythrocytes without having grossly altered its appearance. To our relief, cystoscopies of a few of these men failed to show any alterations in the bladder mucosa, but rather, red blood cells in urine descending both ureters. Further, the urinary sediment has not shown casts, although this transient hema-turia must be of renal origin, presumably glomerular. This hematuria has invari-ably proved to be transient, mild, and of a duration not longer than a week.

E. BLADDER TUMORS

The term "aniline tumor" of the bladder is one that is found so often in litera-ture that apparently it is here to stay. The term "aniline tumor," however, is a misnomer, since no convincing evidence has ever been adduced to demonstrate that aniline itself has ever been the cause of bladder tumors, although there is much evidence to the contrary. Understandably, this misnomer apparently came about from early German observers, for example, Rehn, in 1895,[74] because it was common practice to produce several dyestuff intermediates within the same shop or working area, tended by the same crew of men, with resultant widely mixed ex-posure to a multiplicity of compounds. Apparently, more often than not, β-naph-thylamine and benzidine were among the intermediates produced and handled, along with other analogs and homologs of aniline.

[73] H S. Hatch and E. W. Probst, *Ind. Med.*, **14,** 189 (1945).

[74] L. Rehn, *Arch. Klin. Chir.*, **50,** 588 (1895), *Ver. duet. ges Chir.*, **34,** I., 220 (1905).

There is no question but that β-naphthylamine or its metabolites directly cause bladder tumors, both benign and malignant, in man as well as in experimental animals.[75–77] As late as 1949, Gehrmann, Foulger, and Fleming,[78] because of their inability to cause experimental bladder tumors in dogs with any compound

Aniline   o-Toluidine   Xylidine β-Naphthylamine   Benzidine

other than β-naphthylamine, believed this to be the sole demonstrated carcinogenic agent of this group. As Walpole, Williams, and Roberts[79] point out, there is likewise no satisfactory evidence that any of the simpler aromatic amines, such as the toluidines and xylidines, are, in fact, carcinogenic.

Benzidine, which has been under suspicion since the 1920's as a carcinogen, seems to have been widely accepted as such in Europe, but not on the basis of experimental evidence in the United States. Maguigan (unpublished lectures) has long maintained that benzidine is a carcinogen among dyestuff intermediates workers. Benzidine tumors, experimentally, have been very difficult to demonstrate. Spitz, Maguigan, and Dobriner[80] demonstrated its carcinogenicity in dogs and rats, but at sites other than the bladder.

Case et al.[81] have statistically reviewed hospital records, case reports, death certificates, etc., where the chemical industry was mentioned as the occupation from 1921 to 1950, and have come up with evidence, which would appear to be largely irrefutable, concerning the marked increase of bladder tumors as the cause of death among this group. The following tabulation is taken from Goldblatt:[82]

"From the data thus obtained the following results emerged:

Nominal roll of workers engaged in (i) manufacture, (ii) use or (iii) purification of one or more of aniline, 1-naphthylamine, 2-naphthylamine and benzidine	4622
Expected number of death-certificates (cause: cancer of the bladder) in the three groups allowing for age and date of entry into the industry	3–5
Number of cases of bladder tumour in the three groups	at least 341
Number of these 341 cases which had contact with one or more of 1-naphthylamine, 2-naphthylamine and benzidine	298

[75] E. E. Evans, J. Urol., **38**, 212 (1937).

[76] G. H. Gehrmann, J. Urol., **31**, 126 (1934).

[77] H. D. Wolfe, J. Urol., **38**, 216 (1937).

[78] C. G. Gehrmann, J. H. Foulger, and A. J. Fleming, Proc. Intern. Congr. Ind. Med., 9th Congr. London, **1948**, 472 (1949).

[79] A. L. Walpole, M. H. C. Williams, and D. C. Roberts, Brit. J. Ind. Med., **9**, 255, 1952.

[80] S. Spitz, W. H. Maguigan, and K. Dobriner, Cancer, **3**, 789, 1950.

[81] R. A. M. Case, M. E. Hosker, D. B. McDonald, and J. T. Pearson, Brit. J. Ind. Med., **11**, 75, 1954.

[82] M. W. Goldblatt, Brit. Med. Bull., 136 (1958).

Including aniline (4) and magenta (9)   311
Number of these 311 who appeared in nominal roll, i.e., worked in the
    industry for more than six months   262
Number of death-certificates (cause: cancer of the bladder) among the 262   127

"Thus, the over-all risk of dying from bladder cancer in the manufacture of synthetic dyestuffs was about 30 times that in the general population. It was also concluded that there was a definite hazard of bladder cancer in the manufacture of auramine (a diphenylmethane dyestuff) and of magenta (a triphenylmethane dyestuff). It may be recalled that the earliest cases of occupational bladder cancer described by Rehn (1895) were among workers in magenta manufacture. Case and Pearson (1954) found no evidence to suggest that the use or manufacture of aniline during 1910–52 has been a cause of bladder cancer.

"From Case's analysis the number of fatal tumours already found among the manufacturing workers was 243 out of 2,466 during 1915–50, and the calculated forecast is that a further 243 fatal tumours will be found, even if no further exposure takes place."

Similarly, statistics compiled by Uebelin and Pletscher[83] are of interest and bear out the contention that β-naphthylamine and benzidine are the two compounds among the dyestuff intermediates are the true carcinogens. They have analyzed, statistically, the cases of 100 workers in the dye department of a Basel enterprise, who developed tumors of the bladder from 1924–1953, and have tabulated their results as follow:

Of these 100 patients:

20 had worked with only benzidine.

25 had worked with only β-naphthylamine.

20 had worked with both benzidine and β-naphthylamine.

6 had worked with benzidine and at times other aromatic amines.

17 had worked with β-naphthylamine and at times other aromatic amines.

9 had worked with β-naphthylamine, benzidine, and at times other aromatic amines.

3 had worked with only other aromatic amines.

They conclude that benzidine and β-naphthylamine were chiefly responsible for the appearance of the industrial bladder tumors. Other aromatic amines seemed to play no part. Benzidine also was probably responsible for tumors in other organs.

Latent Period. The average latent period for all bladder tumors was 18.6 years with a range of 1 to 44 years; for early diagnosed tumors 17.0 years (range 4 to 44 years). The latent period was often much longer than the period of exposure.

Age at Which Contact Began. Except for the years between 15 and 20, this did not seem to have any relation to the latent period. The average latent period in 7 cases was 30.5 years with a range of 22 to 52. In 2 cases examined cystoscopically this was 28.0 years (25 and 31 years, respectively).

[83] F. Uebelin and A. Pletscher, *Schweiz. med. Wochschr.*, **84**, 917 (1954).

The authors[83] further emphasized the importance of annual cystoscopic examinations for all workers with these compounds. Good housekeeping techniques for prevention of bladder tumors are described. In discussing prevention the authors point out that, since 1944, Ciba has not used the dry benzidine base but the moist hydrochloride and since 1948, instead of β-naphthylamine, has used 1-sulfonic acid β-naphthylamine.

Even though aromatic amines, other than benzidine and β-naphthylamine, appear in the statistical compilations of Case et al.[81] and Uebelin,[83] the role of simple aromatic amines, such as aniline, toluidines and xylidines, remains obscure even though they would apparently account, according to Hergt (personal communication to Walpole et al.,[79]) for bladder tumors found in fuschine (magenta) workers.

Also, Case et al.[81] statistically implicate magenta (a triphenylmethane dyestuff) and auramine (a diphenylmethane dyestuff) manufacture as giving rise to an increased incidence of bladder tumors. Case et al. consider this evidence concerning magenta to be only strongly suggestive. Case also demonstrates that α-naphthylamine appears to require a much longer exposure than do benzidine or β-naphthylamine before bladder papilloma appear. This, in all probability, was due to the varying β-content of α-naphthylamine, which, in earlier days, was said to have been present as an impurity, to the extent of 5 to 15 per cent.

Walpole et. al.,[84] in two papers, reported carcinoma in dog bladder, after feeding 4-aminodiphenyl for 2 years and 9 months, in 2 dogs out of 30. Later, Deichmann et al.[85] confirmed these findings.

4-Aminodiphenyl

In 1955, Melick et al.[86] reported the occurrence of bladder tumors, both benign and malignant, in 11.1 per cent of 171 workers producing 4-aminodiphenyl. The duration of exposure to this compound in these 19 workers varied from $1^1/_4$ to 19 years. Efforts have been made to pinpoint the exact metabolite of 4-aminodiphenyl responsible, but this remains largely of academic interest. At any rate, the production of this compound for use as a rubber antioxidant in a single plant in the United States was promptly abandoned.

It was also discovered by Wilson et al.[87] and later confirmed by Morris and Eyestone[88] that a promising pesticide, 2-acetylaminofluorene, was also a potent bladder carcinogen; fortunately, as a consequence, it was never marketed.

[84] A. L. Walpole, M. H. C. Williams, and D. C. Roberts, Brit. J. Ind. Med., 11, 105, 1954.
[85] W. B. Deichmann, J. L. Radomsky, W. A. D. Anderson, M. M. Coplan, and F. M. Woods, Ind. Med. and Surg., 27, 25 (1958).
[86] W. F. Melick, H. M. Escue, J. J. Naryka, R. A. Mezera, and E. P. Wheeler, J. Urol., 74, 760 (1955).
[87] R. H. Wilson, F. De Eds, and A. J. Cox, Jr., Cancer Research, 1, 595 (1941).
[88] H. P. Morris and W. H. Eyestone, J. Natl. Cancer Inst., 13, 1139 (1953).

2-Acetylaminofluorene

In summary, then, it would appear to have been unequivocally demonstrated that of compounds of commercial importance, β-naphthylamine, benzidine, 4-aminodiphenyl, and 2-acetyl-aminofluorene are important as causative agents of bladder tumor. It would appear equally clear that aniline, the toluidines, and the xylidines have never clearly been demonstrated to be carcinogens. The present trend in literature to substitute the term "aromatic amine tumors" for "aniline tumors" is not helpful, since aniline and its immediate homologs are the first compounds that come to mind as representative of aromatic amines.

In terminating this discussion, it is of interest that Walpole, Williams, and Roberts have become intrigued by the possibility that o-hydroxyamines are the effective carcinogenic metabolites, but this hypothesis seems to be based on rather tenuous grounds. On this basis, they postulate that o-toluidines may be capable of producing bladder tumors in man and dog, although this still remains purely speculative.

F. PREVENTION AND CONTROL OF BLADDER TUMORS IN THE DYESTUFF INDUSTRY

The British dyestuff industry is to be complimented on having had the interest and foresight to promote the study of Case and his co-workers, and to furnish financial support for further research on papilloma, which was already in progress under the direction of Professor Passey and Dr. Bonser, in the Department of Experimental Pathology and Cancer Research of Leeds University. A description of this very intelligent approach to the problem is set forth in a booklet, *Papilloma of the Bladder in the Chemical Industry*, issued by the Association of British Chemical Manufacturers in 1953.[89] As a result of these findings, the production of β-naphthylamine in the British Isles has ceased, and presumably will not be resumed. Dyestuff manufacturers in the United States and elsewhere, who have not already done so, would do well to take heed of what has transpired in Great Britain.

In 1957, Scott and Williams[90] published an excellent and comprehensive paper, "The Control of Industrial Bladder Tumours," which, likewise, is highly recommended for the consideration of manufacturers of organic chemicals everywhere. Also, in 1956, Crabbe et al.[91] reported the successful application of the Papanicolaou staining technique to the urine of dyestuff workers. It should also be

[89] Association of British Chemical Manufacturers, *Papilloma of the Bladder in the Chemical Industry*, 1953.

[90] T. S. Scott and M. H. C. Williams, *Brit. J. Ind. Med.*, **14**, 150 (1957).

[91] J. G. S. Crabbe, W. C. Cresdee, T. S. Scott, and M. H. C. Williams, *Brit. J. Ind. Med.*, **13**, 270 (1956).

noted that the detection of abnormal cells in urine is enhanced by the concentration method of Rofe.[92,93]

G. DYESTUFFS

Some of the simpler dyestuffs may properly be considered members of the nitro-amino aromatic series. These are among the dyestuffs, structurally, which are truly substituted o- or p-nitrophenols or the phenolamines, in which the chromophore is the NO_2 group and the auxochromes are the OH and NH_2 groups.[97] Examples are:

Salicyl yellow　　　　　　　　Sudan yellow

Also, the monoazo dyes are simple and are composed, for the most part, of two or more aryl nitro-amino radicals, coupled by the azo linkage. Examples are:

Disperse black 3　　　　　　　　Disperse red 41

As the structure of the dyestuffs becomes more complex, the chemical and physical resemblance to these simple aromatic amines and nitro compounds is lost. In general, dyestuffs are very complex and are comparatively high in molecular weight. Thus, by the time the vat dyes are reached, the resemblance to the simple amines and nitro compounds has largely disappeared. Example:

Indanthrene dark blue BT (B)

The dyestuffs as a group present a relatively low order of systemic toxicity; this statement is based on more than 100 years of manufacturing experience of these dyestuffs. Actually, very few of them have been systematically investigated for toxic properties. However, no group of compounds can have been produced in high tonnage over so long a period of time without having manifested recognizable toxic effects upon those engaged in the handling, blending, and dye application of these compounds, and be considered hazardous. A few of them have proved to be sources of contact dermatitis, but these are usually mild manifestations, such as may be encountered in "hat band" dermatitis, which usually arises from Bismarck

[92] J. Rofe, J. Clin. Pathol., **8**, 25 (1955).
[93] J. Rofe, Brit. J. Ind. Med., **14**, 164 (1957).

brown or chrysoidine. In dyestuff plants, where dyes are dried, blended, and packaged, contact dermatitis or any other observable deleterious effects are indeed rare.

Several dyestuffs have proved of value therapeutically; among them are chrysoidine and chrysarobine, which were extensively used in the alleviation of psoriasis, and scarlet red, a simple azo dye, which is of value in stimulating the proliferation of epithelium. Methylene blue has proved to be of value in the treatment of methemoglobinemia and for many years was used as a urinary antiseptic, although of dubious value. Crystal violet and methyl violet were, prior to the advent of sulfonamides, used occasionally, intravenously, as the last resort in septicemia.

TABLE 3

Dyes Certified in 1940 as Food Colors

Trade name	F & D color	No.	Colour index[97] no.
Brilliant Blue FCF	Blue	1	
Indigotine (Indigo Cermine)	Blue	2	1180
Guinea Green B	Green	1	666
Light Green SF Yellowish	Green	2	670
Fast Green FCF	Green	3	
Orange I	Orange	1	150
Orange SS	Orange	2	
Ponceau 3R	Red	1	80
Amaranth	Red	2	184
Erythrosine	Red	3	773
Ponceau SX	Red	4	
Oil Red XO	Red	32	
Acid Violet 6B	Violet	1	697
Naphthol Yellow S	Yellow	1	10
Naphthol Yellow S, potassium salt	Yellow	2	10
Yellow AB	Yellow	3	22
Yellow OB	Yellow	4	61
Tartrazine	Yellow	5	640
Sunset Yellow FCF	Yellow	6	

Among the few potentially grave offenders are o-aminoazotoluene and 4-dimethylaminoazobenzene (butter yellow). With these dyes, cancer of the liver has been produced in rats and mice, although never in man.[94-96] Use as a butter or margarine colorant was discontinued years ago. Other azo colors have caused tumors in rats and mice. The Federal Food and Drug Administration, by 1940, accepted 19 dyes for certification as food colors; they are listed in Table 3.

[94] J. A. Miller and E. C. Miller, Advances in Cancer Research, 1, 339 (1953).
[95] J. A. Miller, E. C. Miller, and G. C. Finger, Cancer Research, 17, 387 (1957).
[96] J. A. Miller, E. C. Miller, and G. C. Finger, Cancer Research, 17, 387 (1957).
[97] J. Soc. Dyers Colourists, Colour Index, 3, 1956.

However, in the last 2 or 3 years, 3 of these colors have been reexamined, toxicologically, in experimental animals; and, as a result, all 3 of them have been removed from the F and D list. These are: Orange #1, Orange SS, and Oil Red XO. At the present time, FDA has announced its intention of delisting 4 others. These are: Naphthol Yellow S; Naphthol Yellow S, potassium salt; Yellow AB; and Yellow OB.

In addition, 69 colors, classified as D & C, are acceptable for certification for use in drugs and cosmetics but not in food. An additional 29 synthetic dyes, classified as Ext. D & C, are certified for use in externally applied drugs and cosmetics.[98]

H. SUMMARY

We have tried to point out that the most characteristic effect of the absorption of nitro-amino compounds of the aromatic series in the human is methemoglobinemia, which, in most instances, is a readily reversible reaction. Most of the methemoglobin formers in this series appear to exert an analgesic effect with or without the "collapse" action referred to by Sollmann.[21] In addition, however, we have tried to make it clear that there are numerous deviations from this pattern, which may vary from a high incidence of bladder tumors, as with β-naphthylamine, or profound metabolic disturbances with dinitrophenol, to lack of appreciable systemic toxicity with tetryl.

I. ENVIRONMENT AND THE WORKMAN

In conclusion, we should like to emphasize that, with proper plant hygiene, all of the members of this series of compounds can be produced and handled without hazard to workers so engaged. It is, of course, inevitable that an occasional accident, such as the breaking of a valve or of a pipeline or an overflow, may bring about an acute poisoning. However, in this day of modernized industry, there is no acceptable excuse for the production of compounds hazardous to health in a dirty, sloppy, uncontrolled working environment.

Proper hygiene consists primarily of cleanliness in the work area, including its atmosphere, and of every individual employee. This can be brought about with the cooperation of an intelligent and sympathetic management in various ways:

1. Through competent engineering, which is directed not only at good yields and low production costs, but also toward protection of the worker from exposure. This must be equally as important an objective in engineering as low production costs if the objective is to be reached.

2. Where totally closed systems are not feasible, adequate ventilation at the source of exposure must be insisted upon.

3. Daily changes of freshly laundered clothing must be provided for workers engaged in the production of these compounds and a shower bath must be compulsory at the end of each work period.

[98] M. B. Jacobs, *The Chemistry and Technology of Food and Food Products,* Interscience Publishers, New York, 1951 p. 102.

4. Every individual employed in a hazardous area should pass through the medical department for inspection at the end of each shift, since the onset of acute methemoglobinemia is symptomless.

5. Individuals showing the slightest gross evidence of cyanosis should be kept under observation until an accurate methemoglobin determination can be made and as much longer as appears necessary.

6. Workmen should never be permitted to enter a vessel that has contained compounds of this series until a determination of the degree of atmospheric contamination has been quantitatively established by the plant hygienist.

7. Workers exposed to these compounds should be given a complete physical examination, including hemoglobin determinations, red, white, and differential counts, and complete urinalysis, every 6 months or oftener. In cases of TNT, dinitrobenzene, and dinitrophenol exposures, workers should be examined monthly and inspected daily.

8. In exposures to β-naphthylamine, benzidine, and 4-aminodiphenyl, cystoscopic examination of the bladder mucosa should be performed every 6 to 12 months. Likewise, the Papanicolaou examination of urinary sediment, after concentration by the method of Rofe, should be borne in mind as a diagnostic procedure of great value.

9. Those suffering from cardiovascular or renal disease of significant severity or individuals suffering from lowered vital capacity for any reason should not be employed. Likewise, those suffering from blood dyscrasia, as well as chronic alcoholics, should never be employed in areas in which this series of compounds is produced.

II. Specific Compounds

o-AMINOPHENOL, $NH_2C_6H_4OH$ (o-Hydroxyaniline)

1. *Uses*

Dye intermediate and fur and hair "dye."

2. *Physical and Chemical Properties*

Physical state: colorless rhombic needles or plates
Molecular weight: 109.12
Melting point: 173°C.
Boiling point: sublimes
Solubility: water—1.7 at 0°C.; alcohol—4.3 at 0°C.; ether—very soluble
Odor and warning properties: none

3. *Toxicity*

Not readily absorbed through intact skin, but may prove to be a sensitizing agent with resultant contact dermatitis. Inhalation of dust should be avoided since, if inhaled in excessive amounts, it may cause methemoglobinemia. In rare instances, o- and p-aminophenol have caused a bronchial asthma.

p-AMINOPHENOL, NH₂C₆H₄OH (p-Hydroxyaniline)

1. Uses

Identical with o-aminophenol.

2. Physical and Chemical Properties

Physical state: white leaflets
Molecular weight: 109.12
Melting point: 184°C.
Boiling point: sublimes
Solubility: water—1.1 at 0°C.; alcohol—4.5 at 0°C.; ether—slightly soluble
Odor and warning properties: none

3. Toxicity

Identical with o-aminophenol.

m-AMINOPHENOL, NH₂C₆H₄OH (m-Hydroxyaniline)

1. Uses

Of little commercial importance. Used chiefly in the synthesis of dyes and occasionally as a fur "dye."

2. Physical and Chemical Properties

Physical state: colorless prisms
Molecular weight: 109.12
Melting point: 122°C.
Solubility: water—2.6 at 0°C.; alcohol and ether—very soluble
Odor and warning properties: none

3. Toxicity

Similar to the ortho and para compounds.

ANILINE, C₆H₅NH₂ (Aminobenzene, Phenylamine)

1. Uses

Manufacture of dyestuffs, other dyestuff intermediates, rubber accelerators and antioxidants; also as an intermediate in the manufacture of pharmaceuticals, photographic developers, plastics, and ion exchange resins.

2. Physical and Chemical Properties

Physical state: liquid
Molecular weight: 93.12
Density: 1.022 (20/4°C.)
Melting point: —6.2°C.
Boiling point: 184.4°C.
Density of vapor: 3.22 (air = 1)
Vapor pressure: 15 mm. Hg (77°C.)
Refractive index: 1.5863 (20°C.)
Solubility: water—3.4 at 20°C.; alcohol, ether, benzene, chloroform, and carbon tetrachloride—soluble
Flash point: 168°F.
Autoignition temperature: 1418°F.
Maximum allowable concentration: 5 p.p.m.
Odor and warning properties: characteristic, peculiar odor and burning taste

3. Toxicity

See Section I, B.

4. Hygienic Standards of Permissible Exposure

The threshold limit was established by the American Conference of Governmental Industrial Hygienists in 1958 at 5 p.p.m. (19 mg./cu. meter).

BENZIDINE, $NH_2C_6H_4C_6H_4NH_2$ (4,4'-Diaminobiphenyl, p,p'-Bianiline)

1. Uses

Benzidine is used in the synthesis of dyes and dye intermediates.

2. Physical and Chemical Properties

Physical state: white or slightly reddish crystals, powder, or leaflets
Molecular weight: 184.23
Density: 1.250 (20/4°C.)
Melting point: 116.5°C.
Boiling point: 401.7°C.
Solubility:water—0.04 at 12°C.; 0.94 at 100°C.; alcohol—soluble; ether—2.2

3. Toxicity

Benzidine is of little importance as a methemoglobin former and of a low order of toxicity in industrial exposures. It remains on the suspect list as a possible cause of bladder tumors, although available information indicates that bladder tumors cannot be caused by administration of this compound to experimental animals as in the case of β-naphthylamine.

p-CHLOROANILINE, ClC₆H₄NH₂ (2-Chlorophenylamine)

1. Uses

p-Chloroaniline is used in the synthesis of dyestuffs and other intermediates.

2. Physical and Chemical Properties

Physical state: rhombic prisms
Molecular weight: 127.57
Density: 1.427 (19/4°C.)
Melting point: 70°C.
Boiling point: 231°C.
Solubility: water, alcohol and ether—soluble
Odor and warning properties: characteristic sweet odor

3. Toxicity

p-Chloroaniline is absorbed through the intact skin and may cause methemo-globinemia. It is less hazardous than aniline or mononitrobenzene in industrial exposures. Relatively low vapor pressure, but precaution should be taken to avoid inhalation of vapors.

o-CHLOROANILINE, ClC₆H₄NH₂ (2-Chlorophenylamine)

1. Uses

Same as p-chloroaniline.

2. Physical and Chemical Properties

Physical state: liquid
Molecular weight: 127.57
Density of liquid: 1.2125 (20/4°C.)
Melting point (solid states): α, 14°C.; β, 3.5°C.
Boiling point: 208.8°C.
Refractive index: 1.5895 (20°C.)
Solubility: water—insoluble; alcohol—miscible; ether—soluble
Odor and warning properties: characteristic sweet odor

3. Toxicity

Absorbed through the intact skin. According to Lehmann,[99] methemoglo-binemia does not follow its absorption but rather kidney and, to a lesser extent, liver damage may result. Relatively low vapor pressure, but precaution should be taken to avoid inhalation of vapors.

[99] K. B. Lehmann, *Arch. Hyg. Bakteriol.*, **110**, 12 (1933).

m-CHLOROANILINE, ClC₆H₄NH₂ (3-Chlorophenylamine)

1. *Uses*

Same as *p*-chloroaniline.

2. *Physical and Chemical Properties*

Physical state: liquid
Molecular weight: 127.57
Melting point: −10.4°C.
Boiling point: 229.8°C.
Refractive index: 1.59424 (20°C.)
Solubility: water—insoluble; alcohol—miscible; ether—soluble
Odor and warning properties: characteristic sweet odor

3. *Toxicity*

Readily absorbed through the intact skin. May cause methemoglobinemia and possible liver and kidney damage. Relatively low vapor pressure, but precaution should be taken to avoid inhalation of vapors.

4-CHLORO-1,2-DINITROBENZENE, C₆H₃Cl(NO₂)₂

1. *Uses*

Commercial chlorodinitrobenzene is usually a mixture of the six possible isomers, which are closely related chemically and physically to each other. The physical constants given refer to 4-chloro-1,2-dinitrobenzene and illustrate general properties common to this series. It is used in the manufacture of dyestuffs, other dye intermediates, and in the explosive roburite.

2. *Physical and Chemical Properties*

Physical state: monoclinic prisms, needles
Molecular weight: 202.56
Melting point: α, 36.3°; β, 37.1°; γ, 38.8°; δ, 28°C.
Boiling point: 315°C.
Solubility: water—insoluble; alcohol—soluble in hot, difficulty in cold; ether—soluble

3. *Toxicity*

In industrial exposures 4-chloro-1,2-dinitrobenzene is of little importance as a systemic poison, although it is almost a universal sensitizer, causing contact dermatitis in from 60 to 80 per cent of individuals having even minute contact with it. This contact dermatitis may vary from a few itching, vesicular papules to a generalized exfoliative dermatitis.

3-CHLORO-1,2-DINITROBENZENE, $C_6H_3Cl(NO_2)$

1. *Uses*

Identical with 4-chloro- 1,2-dinitrobenzene.

2. *Physical and Chemical Properties*

Physical state: crystals from alcohol or ether
Molecular weight: 202.56
Melting point: 78°C.
Boiling point: 315°C.
Refractive index: 1.6867 (16.5°C.)
Solubility: water—insoluble; alcohol and ether—soluble

3. *Toxicity*

Identical with 4-chloro-1,2-dinitrobenzene.

2-CHLORO-1,4-DINITROBENZENE, $C_6H_3Cl(NO_2)_2$

1. *Uses*

Identical with 4-chloro- 1,2-dinitrobenzene.

2. *Physical and Chemical Properties*

Physical state: light yellow crystals
Molecular weight: 202.56
Melting point: 64°C.
Solubility: water—insoluble; alcohol and ether—soluble

3. *Toxicity*

Identical with 4-chloro-1,2-dinitrobenzene.

4-CHLORO-1,3-DINITROBENZENE, $C_6H_3Cl(NO_2)_2$

1. *Uses*

Identical with 4-chloro-1,2-dinitrobenzene.

2. *Physical and Chemical Properties*

Physical state: crystals
Molecular weight: 202.56
Melting point: α, 53.4°C.; β, 43°C.
Boiling point: 315°C. (762 mm. Hg)
Solubility: water—insoluble; alcohol—readily soluble in hot, difficulty in cold; ether—soluble

3. *Toxicity*

Identical with 4-chloro-1,2-dinitrobenzene.

5-CHLORO-1,3-DINITROBENZENE, $C_6H_3Cl(NO_2)_2$

1. *Uses*

Identical with 4-chloro-1,2-dinitrobenzene.

2. *Physical and Chemical Properties*

Physical state: colorless needles
Molecular weight: 202.56
Melting point: 55°C.
Solubility: water—insoluble; alcohol and ether—soluble

3. *Toxicity*

Identical with 4-chloro-1,2-dinitrobenzene.

DIETHYLANILINE, $C_6H_5N(C_2H_5)_2$ (*n*-Phenyldiethylamine)

1. *Uses*

n-Phenyldiethylamine is used in dyestuffs, and in the synthesis of other intermediates and pharmaceuticals.

2. *Physical and Chemical Properties*

Physical state: colorless or yellow or brown flammable oil
Molecular weight: 149.23
Density: 0.93507 (20/4°C.)
Melting point: —38.8°C.
Boiling point: 215.5°C.
Refractive index: 1.54105 (22°C.)
Solubility: water—1.44 at 12°C.; alcohol and ether—soluble

3. *Toxicity*

n-Phenyldiethylamine is quantitatively less toxic than aniline, but very similar in its effects. It is readily absorbed through the intact skin and precautions must be taken to avoid inhalation of its vapors.

DIMETHYLANILINE, $C_6H_5N(CH_3)_2$ (*n*-Phenyldimethylamine)

1. *Uses*

Synthesis of dyestuffs, other dyestuff intermediates, as a solvent, as an aid in methylation, and as a reagent.

2. Physical and Chemical Properties

Physical state: yellow liquid
Molecular weight: 121.18
Density of liquid: 0.9557 (20/4°C.)
Melting point: 2.5°C.
Boiling point: 192.5°C.
Density of vapor: 4.17 (air = 1)
Refractive index: 1.55819 (20°C.)
Solubility: water—slightly soluble; alcohol and ether—soluble
Flash point: 145°F. (closed cup), 170°F. (open cup)
Autoignition temperature: 700°F.
Maximum allowable concentration: 5 p.p.m.

3. Toxicity

Same as diethylaniline.

4. Hygienic Standards of Permissible Exposure

The threshold limit was established by the American Conference of Governmental Industrial Hygienists, in 1958 at 5 p.p.m. (25 mg./cu. meter).

o-DINITROBENZENE, $C_6H_4(NO_2)_2$ (1,2-Dinitrobenzene)

1. Uses

Dinitrobenzene in the meta, para, and ortho isomers is of commercial importance and is usually manufactured as a mixture of the three isomers. It is used in the synthesis of dyestuffs, of other dyestuff intermediates, in explosives, and as a camphor substitute in celluloid production.

2. Physical and Chemical Properties

Physical state: colorless to yellow monoclinic plates
Molecular weight: 168.11
Density: 1.565 (17/4°C.)
Melting point: 117 to 118°C.
Boiling point: 319°C. (773 mm. Hg)
Vapor density: 5.79 (air = 1)
Solubility: water—0.01 (cold), 0.38 at 100°C.; alcohol—3.8 at 25°C.; chloroform—27.1 at 18°C.; benzene—5.0 at 18°C.; methyl alcohol—soluble
Flash point: 302°F. (closed cup)

3. Toxicity

o-Dinitrobenzene is a powerful methemoglobin former and on prolonged exposure may lead to liver damage. It is readily absorbed through the intact skin and its vapors are highly toxic. It is reported to cause a secondary anemia on absorption, but this has not been a consistent finding during our observations over a period of years.

4. Hygienic Standards of Permissible Exposure

The threshold limit was established by the American Conference of Governmental Industrial Hygienists in 1958 at 1 mg./cu. meter (specific compound not indicated by ACGIH).

p-DINITROBENZENE, $C_6H_4(NO_2)_2$ (1,4-Dinitrobenzene)

1. Uses

Identical with o-dinitrobenzene.

2. Physical and Chemical Properties

Physical state: colorless to yellow monoclinic needles
Molecular weight: 168.11
Density: 1.625 (20/4°C.)
Melting point: 173°C.
Boiling point: 299°C. (777 mm. Hg)—sublimes
Solubility: water—0.18 at 100°C.; alcohol—0.4 at 20°C.; chloroform—1.82 at 18°C.; benzene—2.3 at 18°C.
Maximum allowable concentration: 5 p.p.m.

3. Toxicity

Identical with o-dinitrobenzene.

4. Hygienic Standards of Permissible Exposure

The threshold limit was established by the American Conference of Governmental Industrial Hygienists in 1958 at 1 mg./cu. meter (specific compound not indicated by ACGIH).

m-DINITROBENZENE, $C_6H_4(NO_2)_2$ (1,3-Dinitrobenzene)

1. Uses

Identical with o-dinitrobenzene.

2. Physical and Chemical Properties

Physical state: colorless to yellow rhombic needles or plates
Molecular weight: 168.11

Density: 1.571 (0/4°C.)
Melting point: 89.57°C.
Boiling point: 302.8°C. (770 mm. Hg)
Solubility: water—0.0469 at 15°C., 0.32 at 100°C., alcohol—2.6 at 20°C.; ether—6.7 at 15°C.; benzene, toluene, chloroform, and ethyl acetate—soluble
Flash point: 302°F. (closed cup)
Maximum allowable concentration: 5 p.p.m.

3. Toxicity

Identical with o-dinitrobenzene.

4. Hygienic Standards of Permissible Exposure

The threshold limit was established by the American Conference of Governmental Industrial Hygienists in 1958 at 1 mg./cu. meter (specific compound not indicated by ACGIH).

2,3-DINITROPHENOL, $C_6H_3OH(NO_2)_2$

1. Uses

The isomers of dinitrophenol are usually not separated but are prepared as mixtures; however, the mixtures or the individual isomers are so similar toxicologically and chemically that they need not be considered separately. They are used in the synthesis of dyestuffs, picric acid, and picramic acid. They are also used in the preservation of timber and in the manufacture of the photographic developer, Amidol.

2. Physical and Chemical Properties

Physical state: yellow monoclinic prisms
Molecular weight: 184.11
Melting point: 144°C.
Solubility: water—slightly soluble; alcohol—very soluble in hot; ether—very soluble

3. Toxicity

This highly toxic compound is readily absorbed through the intact skin; its vapors are absorbed through the respiratory tract. For details see Section I, D on 1,2,4-dinitrophenol (page 2120).

2,4-DINITROPHENOL, $C_6H_3OH(NO_2)_2$

1. Uses

Identical with 2,3-dinitrophenol.

2. *Physical and Chemical Properties*

Physical state: yellow rhombic crystals or needles
Molecular weight: 184.11
Melting point: 114, 115°C.
Solubility: water—0.56 at 18°C., 4.3 at 100°C.; alcohol—3.9 at 19°C.; ether—3.065 at 15°C.; chloroform and benzene—soluble
Odor and warning properties: bitter taste

3. *Toxicity*

Identical with 2,3-dinitrophenol.

2,5-DINITROPHENOL, $C_6H_3OH(NO_2)_2$

1. *Uses*

Identical with 2,3-dinitrophenol.

2. *Physical and Chemical Properties*

Physical state: yellow needles
Molecular weight: 184.11
Melting point: 104°C.
Solubility: water—slightly soluble; alcohol—soluble in hot; ether—easily soluble

3. *Toxicity*

Identical with 2,3-dinitrophenol.

2,6-DINITROPHENOL, $C_6H_3OH(NO_2)_2$

1. *Uses*

Identical with 2,3-dinitrophenol.

2. *Physical and Chemical Properties*

Physical state: yellow rhombic crystals
Molecular weight: 184.11
Melting point: 63.5°C.
Solubility: water—slightly soluble in cold, more soluble in hot; alcohol—readily soluble in hot; ether—readily soluble; benzene—soluble

3. *Toxicity*

Identical with 2,3-dinitrophenol.

3,4-DINITROPHENOL, $C_6H_3OH(NO_2)_2$

1. *Uses*

Identical with 2,3-dinitrophenol.

2. *Physical and Chemical Properties*

Physical state: colorless needles
Molecular weight: 184.11
Melting point: 134°C.
Solubility: alcohol and ether—very soluble

3. *Toxicity*

Identical with 2,3-dinitrophenol.

3,5-DINITROPHENOL, $C_6H_3OH(NO_2)_2$

1. *Uses*

Identical with 2,3-dinitrophenol.

2. *Physical and Chemical Properties*

Physical state: monoclinic prism
Molecular weight: 184.11
Melting point: 122 to 123°C.
Solubility: alcohol and ether—very soluble; chloroform and benzene—
soluble

3. *Toxicity*

Identical with 2,3-dinitrophenol.

4,6-DINITRO-*o*-CRESOL, $C_6H_2(CH_3)OH(NO_2)_2$
(2-Methyl-4,6-dinitrophenol)

1. *Uses*

The two dinitrocresol isomers for which physical constants are presented are those of greatest commercial importance. Their uses are similar to those given for dinitrophenol.

2. *Physical and Chemical Properties*

Physical state: yellow prisms
Molecular weight: 198.13
Melting point: 85.8°C.

Solubility: water—slightly soluble; alcohol—10.82 at 15°C.; ether—very soluble; acetone—soluble

3. *Toxicity*

Resembles dinitrophenol.

4. *Hygienic Standards of Permissible Exposure*

The threshold limit was established by the American Conference of Governmental Industrial Hygienists in 1958 at 0.2 mg./cu. meter.

2,6-DINITRO-*p*-CRESOL, $C_6H_2(CH_3)OH(NO_2)_2$
(4-Methyl-2,6-dinitrophenol)

1. *Uses*

Identical with 4,6-dinitro-*o*-cresol.

2. *Physical and Chemical Properties*

Physical state: long yellow prisms
Molecular weight: 198.13
Melting point: 81°C.
Solubility: water—slightly soluble; alcohol—soluble; ether—very soluble

3. *Toxicity*

Identical with 4,6-dinitro-*o*-cresol.

DIPHENYLAMINE, $(C_6H_5)_2NH$ (Phenylaniline)

1. *Uses*

Diphenylamine is used in the synthesis of dyestuffs, other dyestuff intermediates, and explosives.

2. *Physical and Chemical Properties*

Physical state: colorless monoclinic leaflets
Molecular weight: 169.22
Density: 1.159 (20/4°C.)
Melting point: 53°C.
Boiling point: 302°C.
Solubility: water—0.03 at 25°C.; alcohol—44; ether—very soluble; methyl alcohol—57.5
Odor and warning properties: floral odor

3. *Toxicity*

Similar to that of aniline but much less toxic and less readily absorbed through the skin and respiratory tract.

α-NAPHTHYLAMINE, $C_{10}H_7NH_2$ (1-Naphthylamine)

1. Uses

α-Naphthylamine is used in the synthesis of dyestuffs and other dyestuff intermediates and as a developer for naphthol AS colors.

2. Physical and Chemical Properties

Physical state: yellow rhombic needles
Molecular weight: 143.18
Density: 1.123 (25/25°C.)
Melting point: 50°C.
Boiling point: 301°C.
Vapor density: 4.93 (air = 1)
Refractive index: 1.6703 (51.2°C.)
Solubility: water—0.17; alcohol and ether—very soluble
Flash point: 315°F. (closed cup)

3. Toxicity

As an industrial hazard α-naphthylamine is of little importance except that commercially it may contain up to 10 per cent β-naphthylamine. Because of this beta content, it may theoretically give rise to bladder tumors in prolonged exposures.

β-NAPHTHYLAMINE, $C_{10}H_7NH_2$ (2-Naphthylamine)

1. Uses

β-Naphthylamine is used in dyestuffs and other dyestuff intermediates.

2. Physical and Chemical Properties

Physical state: leaflet form
Molecular weight: 143.18
Density: 1.061 (98/4°C.)
Melting point: 110.2°C.
Boiling point: 306.1°C.
Refractive index: 1.64927 (98.4°C.)
Solubility: water—insoluble; alcohol, ether, and benzene—soluble

3. Toxicity

β-Naphthylamine is readily absorbed through the skin and respiratory tract. It may cause mild methemoglobinemia but it is a highly dangerous compound because it causes bladder tumors that may occur following from 1 to 12 years' exposure.

m-NITROANILINE, NO₂C₆H₄NH₂ (1-Amino-3-nitrobenzene)

1. *Uses*

Synthesis of dyestuffs and other intermediates.

2. *Physical and Chemical Properties*

Physical state: yellow rhombic needles
Molecular weight: 138.12
Density: 1.430 (20/4°C.)
Melting point: 111.8°C.
Boiling point: 286°C.
Solubility: water—0.089 at 25°C.; alcohol—6.10 at 25°C.; ether—5.67 at 20°C.
Odor and warning properties: burning sweet taste

3. *Toxicity*

m-Nitroaniline is a powerful methemoglobin former with attendant hemolytic effect. On prolonged and excessive exposures it may also cause liver damage. It is readily absorbed through the intact skin and its vapors are highly toxic as well.

p-NITROANILINE, NO₂C₆H₄NH₂ (1-Amino-4-nitrobenzene)

1. *Uses*

p-Nitroaniline is used in the synthesis of dyestuffs and other intermediates.

2. *Physical and Chemical Properties*

Physical state: yellow monoclinic needles
Molecular weight: 138.12
Density: 1.424 (20/4°C.)
Melting point: 147.5°C.
Boiling point: 331.73°C.
Solubility: water—0.08 at 19°C. or 2.2 at 100°C.; alcohol—4.61 at 20°C.; ether—4.39 at 20°C.

3. *Toxicity*

p-Nitroaniline is a powerful methemoglobin former with attendant hemolytic effect. On prolonged and excessive exposures it may also cause liver damage. It is readily absorbed through the intact skin and its vapors are highly toxic as well.

4. *Hygienic Standards of Permissible Exposure*

The threshold limit was established by the American Conference of Governmental Industrial Hygienists, in 1958 at 1 p.p.m. (6 mg./cu. meter).

o-NITROANILINE, $NO_2C_6H_4NH_2$ (1-Amino-2-nitrobenzene)

1. *Uses*

o-Nitroaniline is of slight importance commercially. It is used as a dye intermediate.

2. *Physical and Chemical Properties*

Physical state: orange rhombic needles
Molecular weight: 138.12
Density: 1.442 (20/4°C.)
Melting point: 71.5°C.
Boiling point: 284.11°C.
Solubility: water—0.126 at 25°C.; alcohol—15.8 at 15°C., 27.87 at 25°C.; ether—very soluble

3. *Toxicity*

Practically identical with the meta and para isomers.

NITROBENZENE, $C_6H_5NO_2$ (Oil of Mirbane)

1. *Uses*

Nitrobenzene is one of the most important and basic intermediates, widely used in the preparation of other intermediates and other organic syntheses. It is also used to mask unpleasant odors because of its musklike smell.

2. *Physical and Chemical Properties*

Physical state: liquid
Molecular weight: 123.11
Density: 1.19867 (25/4°C.)
Melting point: 5.7°C.
Boiling point: 210.9°C.
Density of vapor: 4.1 (air = 1)
Refractive index: 1.55291 (20°C.)
Solubility: water—0.19 at 20°C., 0.8 at 80°C.; alcohol and ether—very soluble; benzene and oils—soluble
Flash point: 190°F. (closed cup)
Maximum allowable concentration: 1 p.p.m.
Odor and warning properties: oil of bitter almond odor

3. *Toxicity*

Nitrobenzene is a powerful methemoglobin former; in this respect probably more potent than aniline per unit of weight or vapor concentration. See preceding text for discussion.

4. *Hygienic Standards of Permissible Exposure*

The threshold limit was established by the American Conference of Governmental Industrial Hygienists, in 1958 at 1 p.p.m. (5 mg./cu. meter).

o-NITROPHENOL, $NO_2C_6H_4OH$

1. *Uses*

o-Nitrophenol is used in the synthesis of dyestuffs and other intermediates.

2. *Physical and Chemical Properties*

Physical state: light yellow monoclinic needles or prisms
Molecular weight: 139.11
Density: 1.657 (20°C.)
Melting point: 45°C.
Boiling point: 214.5°C.
Solubility: water—0.21 at 20°C., 1.08 at 100°C.; alcohol—46.0 at 25°C.; ether—very soluble; alkali—soluble
Odor and warning properties: aromatic odor and sweet taste

3. *Toxicity*

o-Nitrophenol is a methemoglobin former, but less so than aniline and mononitrobenzene. It may be absorbed through the intact skin and its vapors through the respiratory tract.

m-NITROPHENOL, $NO_2C_6H_4OH$

1. *Uses*

Identical with o-nitrophenol.

2. *Physical and Chemical Properties*

Physical state: colorless to yellow monoclinic form
Molecular weight: 139.11
Density: 1.485 (20°C.)
Melting point: 96°C.
Boiling point: 174°C. (70 mm. Hg)
Solubility: water—1.35 at 25°C.; ether—very soluble; benzene and alkali—soluble

3. *Toxicity*

Identical with o-nitrophenol.

p-NITROPHENOL, NO₂C₆H₄OH

1. Uses

Identical with o-nitrophenol.

2. Physical and Chemical Properties

Physical state: colorless to yellowish monoclinic prisms
Molecular weight: 139.11
Density: 1.479 (20°C.
Melting point: 114°C.
Boiling point: 279°C. (decomposes)
Solubility: water—1.6 at 25°C., 26.9 at 90°C.; alcohol—189.5 at 25°C.; ether—very soluble; benzene and alkali—soluble

3. Toxicity

Identical with o-nitrophenol

o-NITROTOLUENE, NO₂C₆H₄CH₃

1. Uses

o-Nitrotoluene is used in the synthesis of dyestuffs and other intermediates and explosives.

2. Physical and Chemical Properties

Physical state: yellow liquid
Molecular weight: 137.13
Density: 1.163 (20/4°C.)
Melting point: α, 10.6°C.; β, 4.1°C.
Boiling point: 222.3°C.
Vapor pressure: 1.6 mm. Hg (60°C.)
Refractive index: 1.54739 (20.4°C.)
Per cent in "saturated" air: 0.21 (60°C.)
Solubility: water—0.0652 at 30°C.; alcohol, ether, benzene, and chloroform —soluble

3. Toxicity

o-Nitrotoluene is a methemoglobin former of an apparently low grade which is not important as an industrial hazard. It may be absorbed through the intact skin and through the respiratory tract.

4. *Hygienic Standards of Permissible Exposure*

The threshold limit was established by the American Conference of Governmental Industrial Hygienists in 1958 at 5 p.p.m. (30 mg./cu. meter) (specific compound not indicated by ACGIH).

p-NITROTOLUENE, $NO_2C_6H_4CH_3$

1. *Uses*

Identical with *o*-nitrotoluene.

2. *Physical and Chemical Properties*

Physical state:　colorless rhombic needles
Molecular weight:　137.13
Density:　1.286 (20°C.)
Melting point:　51.3°C.
Boiling point:　238°C.
Density of vapor:　4.72 (air = 1)
Vapor pressure:　1.3 mm. Hg (65°C.)
Refractive index:　1.5346 (62.5°C.)
Per cent in "saturated" air:　0.17 (65°C.)
Solubility:　water—0.0442 at 30°C.; alcohol—soluble; ether—very soluble; benzene—soluble
Flash point:　223°F. (closed cup)

3. *Toxicity*

Identical with *o*-nitrotoluene.

4. *Hygienic Standards of Permissible Exposure*

The threshold limit was established by the American Conference of Governmental Industrial Hygienists in 1958 at 5 p.p.m. (30 mg./cu. meter) (specific compound not indicated by ACGIH).

m-NITROTOLUENE, $NO_2C_6H_4CH_3$

1. *Uses*

Identical with *o*-nitrotoluene.

2. *Physical and Chemical Properties*

Physical state:　liquid
Molecular weight:　137.13
Density:　1.157 (20/4°C.)

Melting point: 15.5°C.
Boiling point: 231°C.
Vapor pressure: 1.0 mm. Hg (60°C.)
Refractive index: 1.5475 (20°C.)
Per cent in "saturated" air: 0.13 (60°C.)
Solubility: water—0.0498 at 30°C.; alcohol, ether, and benzene—soluble

3. *Toxicity*

Identical with *o*-nitrotoluene.

4. *Hygienic Standards of Permissible Exposure*

The threshold limit was established by the American Conference of Governmental Industrial Hygienists, in 1958 at 5 p.p.m. (30 mg./cu. meter) (specific compound not indicated by ACGIH).

p-NITROSODIMETHYLANILINE, $NOC_6H_4N(CH_3)_2$

1. *Uses*

p-Nitrosodimethylaniline is used in the synthesis of dyestuffs and as an accelerator in the vulcanization of rubber.

2. *Physical and Chemical Properties*

Physical state: green triclinic crystals
Molecular weight: 150.18
Melting point: 85°C.
Solubility: water—insoluble; alcohol and ether—soluble

3. *Toxicity*

Apparently not a methemoglobin former, but is highly irritating to the skin both as a primary irritant and as a sensitizing agent.

m-PHENYLENEDIAMINE, $C_6H_4(NH_2)_2$ (1,3-Diaminobenzene)

1. *Uses*

m-Phenylenediamine is chiefly used in the synthesis of dyestuffs and of other dyestuff intermediates.

2. *Physical and Chemical Properties*

Physical state: colorless rhombic needles
Molecular weight: 108.14
Density: 1.1389 (5°C.); 1.107 (58°C.)

Melting point: 62.8°C.
Boiling point: 287°C.
Refractive index: 1.63390 (57.7°C.)
Solubility: water—35.1 at 25°C.; alcohol and ether—soluble

3. Toxicity

Industrially presents no recognized hazard.

p-PHENYLENEDIAMINE, $C_6H_4(NH_2)_2$ (1,4-Diaminobenzene)

1. Uses

p-Phenylenediamine is used in the synthesis of dyestuffs and other intermediates and as a fur "dye."

2. Physical and Chemical Properties

Physical state: colorless monoclinic crystals
Molecular weight: 108.14
Melting point: 139.7°C.
Boiling point: 267°C.
Solubility: water—3.8 at 24°C., 669 at 107°C.; alcohol, ether, and chloroform—soluble

3. Toxicity

Systemic toxicity in industrial exposures apparently is not recognizable. However, it may cause contact dermatitis through sensitization and may cause bronchial asthma. Such asthma, of course, ceases promptly with complete withdrawal from exposure.

TETRYL, $(NO_2)_3C_6H_2N(NO_2)CH_3$ (Trinitrophenylmethylnitramine, N-Methyl-N-2,4,6-tetranitroaniline)

1. Uses

Tetryl is used as an explosive.

2. Physical and Chemical Properties

Physical state: yellow monoclinic crystals
Molecular weight: 287.15
Density: 1.57 at 19°C.
Melting point: 130°C.
Boiling point: 187°C., explodes

Solubility: water—insoluble; alcohol—0.422 at 18°C.; ether—very soluble; benzene and acetic acid—soluble

Maximum allowable concentration: 1.5 mg./cu. meter

3. Toxicity

Systemic toxicity has never been recognized from this compound. Irritating to the mucous membranes of the upper respiratory tract.

2,3,4-TRINITROTOLUENE (β), $C_6H_2CH_3(NO_2)_3$

1. Uses

2,3,4-Trinitrotoluene is used as an explosive.

2. Physical and Chemical Properties

Physical state: crystals
Molecular weight: 227.13
Melting point: 112°C., explodes at 290 to 310°C.
Solubility: water—insoluble; alcohol and ether—slightly soluble

3. Toxicity

Of little importance as a methemoglobin former but may cause aplastic anemia and liver damage. For further details, see Section I.

2,4,5-TRINITROTOLUENE (γ), $C_6H_2CH_3(NO_2)_3$

1. Uses

Identical with 2,3,4-trinitrotoluene.

2. Physical and Chemical Properties

Physical state: pale yellow prisms
Molecular weight: 227.13
Melting point: 104°C.
Boiling point: 290°C. (decomposes), 288 to 293°C. (explodes)
Solubility: water—insoluble; alcohol—very slightly soluble in cold, much more readily in hot; ether—readily soluble; hot glacial acetic acid, benzene, and acetone—soluble

Maximum allowable concentration: 1.5 mg./cu. meter

3. Toxicity

Identical with 2,3,4-trinitrotoluene.

2,4,6-TRINITROTOLUENE (α, or TNT), $C_6H_2CH_3(NO_2)_3$

1. *Uses*

Identical with 2,3,4-trinitrotoluene.

2. *Physical and Chemical Properties*

Physical state: monoclinic prisms, crystals (colorless)
Molecular weight: 227.13
Melting point: 80.35, 81.1°C.
Boiling point: 240°C. (explodes)
Vapor pressure: 0.046 mm. Hg (82°C.)
Solubility: water—0.02 at 15°C., 0.15 (hot); alcohol—0.1 g. in 8 cc. at 18°C.; ether—0.1 g. in 4 cc. at 18°C.; chloroform—readily soluble, 0.1 g. in 0.4 cc. at 18°C.; carbon tetrachloride—0.1 g. in 7 cc. at 18°C.; and benzene, toluene, acetone—readily soluble
Maximum allowable concentration: 1.5 mg./cu. meter

3. *Toxicity*

Identical with 2,3,4-trinitrotoluene.

4. *Hygienic Standards of Permissible Exposure*

The threshold limit was established by the American Conference of Governmental Industrial Hygienists in 1958 at 1.5 mg./cu. meter (specific compound not indicated by ACGIH).

2,3,5-TRINITROTOLUENE (ε), $C_6H_2CH_3(NO_2)_3$

1. *Uses*

Identical with 2,3,4-trinitrotoluene.

2. *Physical and Chemical Properties*

Physical state: yellow rhombic crystals
Molecular weight: 227.13
Melting point: 97°C.
Boiling point: explodes at 333 to 337°C.
Solubility: alcohol—soluble; ether—insoluble
Maximum allowable concentration: 1.5 mg./cu. meter

3. *Toxicity*

Identical with 2,3,4-trinitrotoluene.

2,3,6-TRINITROTOLUENE, $C_6H_2CH_3(NO_2)_3$

1. *Uses*

Identical with 2,3,4-trinitrotoluene.

2. *Physical and Chemical Properties*

Physical state: monoclinic needles (from alcohol)
Molecular weight: 227.13
Melting point: 111°C.
Boiling point: explodes at 327 to 335°C.
Solubility: alcohol—soluble in 9 parts boiling alcohol or 11 parts of cold
Maximum allowable concentration: 1.5 mg./cu. meter

3. *Toxicity*

Identical with 2,3,4-trinitrotoluene.

3,4,5-TRINITROTOLUENE, $C_6H_2CH_3(NO_2)_3$

1. *Uses*

Same as other isomers.

2. *Physical and Chemical Properties*

Physical state: yellow-green monoclinic prisms or needles (alcohol)
Molecular weight: 227.13
Melting point: 132.0°C.
Boiling point: explodes 305 to 318°C.
Solubility: alcohol—100 parts 95 per cent alcohol dissolves 1 part at 15°C.
Maximum allowable concentration: 1.5 mg./cu. meter

3. *Toxicity*

Same as other isomers.

p-TOLUIDINE, $CH_3C_6H_4NH_2$ (p-Methylaniline)

1. *Uses*

p-Toluidine is used in the synthesis of dyestuffs and other intermediates and in the preparation of ion exchange resins.

2. *Physical and Chemical Properties*

Physical state: leaflets
Molecular weight: 107.15

Melting point: 45°C.
Boiling point: 200.3°C.
Refractive index: 1.55324 (59.1°C.)
Density: 1.046 (20/4°C.), 0.973 (50/50°C.)
Solubility: water—0.74 at 21°C.; alcohol—156 at 30°C.; ether—soluble
Odor and warning properties: aromatic, winelike odor; burning taste

3. Toxicity

Very similar to aniline, but in addition causes a transient hematuria which, in our experience, has been unaccompanied by any further signs of kidney damage and which is cleared completely upon removal from further exposure.

o-TOLUIDINE, $CH_3C_6H_4NH_2$ (o-Methylaniline)

1. Uses

Same as p-toluidine.

2. Physical and Chemical Properties

Physical state: colorless liquid
Molecular weight: 107.15
Density: 1.004 (20/4°C.)
Melting point: α, 24.4°C.; β, 16.3°C.
Boiling point: 199.84°C.
Refractive index: 1.57276 (20°C.)
Solubility: water—1.5 at 25°C.; alcohol and ether—soluble

3. Toxicity

Same as p-toluidine.

4. Hygienic Standards of Permissible Exposure

The threshold limit was established by the American Conference of Governmental Industrial Hygienists in 1958 at 5 p.p.m. (22 mg./cu. meter).

III. Determination in the Atmosphere

These compounds may exist in the air of industrial workplaces in the form of either dusts or vapors. Specific methods for air sampling or for analysis, however, appear in the literature for only a few of them. Studies of industrial environments where these substances may exist as contaminants deal with trace quantities, so that techniques of efficient trapping and sensitive methods for analysis are required.

The dusts, of course, may be collected by impingement in a suitable vehicle, or may be electrically precipitated where there are no explosion hazards. The vapors may be trapped by scrubbing through weak acid solutions or specific solvents. Condensation employing Dry Ice and acetone and a condensing chamber of three or more bulbs is preferred. The condensate may then be dissolved in the solvent of choice. Methods employing the collection of aromatic nitro and amino compound vapors on silica gel or activated charcoal may not prove satisfactory.

TABLE 4

Ultraviolet Absorption[a]

Compound	Solvent	E (1%, 1 cm.)	Wavelength
m-Aminophenol	95% ethanol	204	285 mμ
Aniline	Cyclohexane[b]	192	291
Aniline	Cyclohexane[b]	211	288
Aniline	Cyclohexane[b]	204	283.9
Aniline	Cyclohexane[b]	185	281
o-Dinitrobenzene	Cyclohexane[b]	388	250
m-Dinitrobenzene	95% ethanol	1025	233
p-Dinitrobenzene	95% ethanol	876	260
Diphenylamine	95% ethanol	137	284
o-Nitroaniline	95% ethanol	408	408–409
o-Nitroaniline	95% ethanol	390	278
m-Nitroaniline	95% ethanol	109	374–375
p-Nitroaniline	95% ethanol	151	374
p-Nitroaniline	95% ethanol	66	230
Nitrobenzene	Cyclohexane	753	253
o-Nitrophenol	95% ethanol	231	347
o-Nitrophenol	95% ethanol	446	273
m-Nitrophenol	95% ethanol	154	329
m-Nitrophenol	95% ethanol	444	270
p-Nitrophenol	95% ethanol	802	315
p-Nitrosodimethylaniline	95% ethanol	388	271.6
p-Nitrosodimethylaniline·HCl	95% ethanol	288	233
m-Phenylenediamine	95% ethanol + water	100	288
p-Phenylenediamine·HCl	95% ethanol	191	309
Sulfanilic acid	70% ethanol	225	247
o-Toluidine	Cyclohexane	209	281–289
m-Toluidine	Cyclohexane	172	286–294
p-Toluidine	Cyclohexane	563	294

Calculations for sensitivity: If it is assumed that the minimum detectable amount transmits 90% of the light, then:

$$\log I_0/I = \log (100/90) = 0.045$$
$$E \text{ (1%, 1 cm.) for } m = \text{dinitrobenzene} = 1025$$
$$0.045/1025 = 5 \times 10^{-5} \% = 0.5 \text{ p.p.m.}$$
$$E \text{ (1%, 1 cm.) for } m\text{-phenylenediamine} = 100$$
$$0.045/100 = 45 \times 10^{-5} \% = 4.5 \text{ p.p.m.}$$

[a] Unpublished data, Stamford Research Laboratories, American Cyanamid Co.
[b] Cyclohexane must be entirely free of impurities such as benzene.

The chemical analysis for many of these compounds, such as aniline, toluidine, phenylenediamine, the nitrobenzenes, the nitrotoluenes, picric acid, and other nitrophenols have been reviewed by Jacobs.[100] However, those chemical analytical methods that involve diazotization and coupling do not avoid the opportunity for side reactions and other interferences that reduce the accuracy. The official British[101] method for aniline depends on the formation of the hydrochlorite complex and a blue coloration following the addition of ammonia and phenol. Jacobs[100] lists a table on the "Colorimetric Detection of Certain Dyestuffs and Intermediates," which is an aid in identification.

Trinitrotoluene and tetryl can be satisfactorily collected and analyzed by the method of Goldman and Rushing.[102] The collecting solvent they employ is diethylaminoethanol. Ficklen[103] has reported on methods for collecting and analyzing aniline, nitrobenzene, and toluidine.

The analysis of these compounds existing as vapors is greatly facilitated by the use of physical testing instruments. Instruments employing infrared and ultraviolet techniques, visible spectroscopy, and refractometric tools such as the interferometer are available. The indices of refraction of this group of compounds are sufficiently high to make possible determinations in air of concentrations between 5 and 20 p.p.m., making the interferometer a practical tool.

Spectrometric methods (see Chapter VII) are also applicable; and trace analysis of aromatic nitro and amino vapors may be accomplished with ultraviolet absorption techniques. In collecting vapors, for application of this method, the solvents of choice are 95 per cent ethyl alcohol or, preferably, cyclohexane. Table 4 presents, for various compounds and solvents, the extinction coefficients for 1 per cent solutions by weight in a 1-cm. cell at the specified wavelengths. Assuming a conservative limit to be 10 per cent absorption for a 1-cm. cell, the range of sensitivities will for such solutions be between 0.5 and 4.5 p.p.m., barring any interferences and assuming Beer's law to be obeyed. A 10 per cent absorption is quite conservative as the sensitivity of most ultraviolet spectrometers is of the order of 0.5 to 1 per cent.

General References

Methemoglobinemia

Angeleri, C., (Methemoglobinemia in a group of workers exposed to the absorption of aromatic amines), "Comportamento della metaemoglobinemia in un gruppo di operai esposti all' assorbimento di amine aromatische," *Med. lavoro*, **40**, 313 (1949).
Cheramy, P., (Methemoglobinizing poisons in industry), "Les poisons methemoglobinisants dans l'industrie," *semaine Hôp.*, **25**, 3084 (1949).

[100] M. B. Jacobs, *Analytical Chemistry of Industrial Poisons, Hazards, and Solvents,* Interscience Publishers, New York, 1941.
[101] *Grt. Brit. Dept. Sci. Ind. Research,* Methods for the detection of toxic gases in industry —aniline vapor, No. **11**, H. M. Stationery Office, London, 1939.
[102] F. H. Goldman and D. E. Rushing, *J. Ind. Hyg. Toxicol.,* **25**, 164, 195 (1943).
[103] J. B. Ficklen, *Manual of Industrial Health Hazards,* Science Press, Lancaster, Pa., 1940.

Gaertner, H., (Therapeutic properties, mechanism of action and application of alkyarylamine: bibliographic review), *Postepy Hig., Med. Doswiadczaine*, **8**, 411 (1954).

Gibson, Q. H., Methaemoglobin and sulphaemoglobin, *Biochem. J.*, **57**, 111 (1954).

Holstein, E., (Industrial diseases caused by aromatic nitro and amino compounds), "Berufskrankheiten durch aromatishche Nitro- und Aminoverbindungen," *Arbeitsmedizin*, **27**, 7 (1953).

Kiese, M., (Pathological chemistry of blood pigment), "Zur Pathochemie des Blutfarbstoffs," *Deut. Arch. klin. med.*, **195**, 442 (1949), *Excerpta Med. Sect.*, **3**, No. 6514 (1950).

Kiese, M., and M. Soetbeer, (Kinetics of methemoglobin formation. 3. Formation by nitrosobenzene in vitro) "Kinetik der Hämiglobinbildung. 3. Hämiglobinbildung durch Nitrosobenzol in vivo," *Arch. Exptl. Pathol. Pharmakol.*, **210**, 305 (1950), *Excerpta Med. Sect. II*, **4**, No. 4303 (1951, abstr.).

Kiese, M., D. Reinwein, and H. D. Waller, (4. Methemoglobin formation by hydroxylamine and nitrosobenzene in red cells in vitro), "Die Hämiglobinbildung durch Hydroxylamine und Nitrosobenzol in roten Zellen in vitro," *Arch. Exptl. Pathol. Pharmakol.*, **210**, 393 (1950), *Chem. Abstr.*, **46**, 11260c-f (1952).

Kiese, M., and H. Dannenberg, (Kinetics of methemoglobin formation. 5. Demonstration of nitrosobenzene in red cells after formation of methemoglobin by phenylhydroxylamine), "Kinetik der Hämiglobinbildung. 5. Nachweis von Nitrosobenzol in roten Zellen bei der Hämiglobinbildung durch Phenylhydroxylamin," *Arch. Exptl. Pathol. Pharmakol.*, **211**, 102 (1950), *Chem. Abstr.*, **46**, 11260c-f (1952).

Kiese, M., and W. Münch, (6. Formation of methemoglobin by hydroxylamine), "6. Hämiglobinbildung durch Hydroxylamin," *Arch. Exptl. Pathol. Pharmakol.*, **211**, 115 (1950), *Chem. Abstr.*, **46**, 11260c-f (1952).

Kiese, M., and H. D. Waller, (7. The metabolic process involved in red cell formation of methemoglobin through the phenylhydroxylamine-nitrosobenzene cycle), "Die Stoffwechselvorgänge in roten Zellen bei der Hämiglobinbildung durch den Kreisprozess Phenylhydroxylamin-Nitrosobenzol," *Arch. Exptl. Pathol. Pharmakol.*, **211**, 345 (1950), *Chem. Abstr.*, **47**, 1057id (1953).

Kiese, M., and D. Reinwein, (8. The oxidation of hemoglobin by phenylhydroxylamine and oxygen), "8. Die Oxydation von Hämiglobin durch Phenylhydroxylamin und Sauerstoff," *Arch. Exptl. Pathol. Pharmakol.*, **211**, 392 (1950), *Chem. Abstr.*, **47**, 10571d (1953).

Kiese, M., and H. Dannenberg, (Kinetics of methemoglobin formation. 10. Reduction of nitrosobenzene by red cells), "Kinetik der Hämiglobinbildung. 10. Reduktion von Nitrosobenzol in roten Zellen," *Arch. Exptl. Pathol. Pharmakol.*, **211**, 410 (1950), *Chem. Abstr.*, **47**, 10571d (1953).

Lawford, D. J., E. King, and D. G. Harvey, On the metabolism of some aromatic nitro compounds by different species of animals. 2. The elimination of various nitrocompounds from the blood of different species of animal, *J. Pharm. and Pharmacol.*, **6**, 619 (1954).

Löher, R. M., (Liver damage due to nitro and amino compounds), "Leberschäden durch Nitro und Aminoverbindugen," *Arbeitsmedizin*, **27**, 25 (1953).

Mangelsdorff, A. F., Methemoglobinemia—recognition, treatment and prevention, *Ind. Med.*, **21**, 395 (1952).

Nilzen, A., Some endocrine aspects of skin sensitization and primary irritation. II. Observations on the influence of a renalectomy on cutaneous irritation and sensitization, *J. Invest. Dermatol.*, **19**, 337 (1952).

Otto, H., (Heart and circulation effects from occupational poisonings by benzene and its aromatic nitro and amino compounds), "Das Verhalten von Herz und Kreislauf bei beruflichen Vergiftungen durch Benzol sowie seine aromatischen Nitro- und Aminoverbindungen," *Arbeitsmedizin*, **27**, 46 (1953).

Parmentier, R., (Study of cellular lesions induced by different phenols and aromatic amines), "Etude des lesions cellulaires provoquees pars divers phenols et amines aromatiques," *Rev. belge pathol. et med. Exptl.*, **22**, 1 (1952).

Piotrowski, J., Attempted application of biochemical indexes of absorption of aniline, nitrobenzene and benzene in dye workers, *Med. Pracy.* **5**, 299 (1954).

Pletscher, A., H. Thoelen, and R. Richterich, (Experimental studies on the effect of aromatic amines), "Experimentelle Untersuchungen über die Wirkung aromatischer Amine," *Helv. Physiol. et Pharmacol. Acta.*, **11**, 171 (1953).

Pletscher, A., F. Uebelin, and H. Buess, (Chronic action of aromatic and nitro compounds), "Zur chronischen Wirkung der Aromatischen Amino- und Nitroverbindungen," *Zt. Unfallmed. Berufskr.*, **45**, 40 (1952).

Pletscher, A., (Effects of chronic exposure to aromatic nitro and amino compounds), *Rev. Acc. Trav. Mal. Prof.*, **1**, 40 (1952), *Arch. Ind. Hyg. Occupational Med.*, **6**, 284 (1952) (abstr.).

Pletscher, A., F. Uebelin, and H. Buess, (Injuries to health of solvent workers), "Gesundheitsschäden bei Lösungsmittel-Arbeitern," *Zt. Unfallmed. Berufskr.*, **46**, 39 (1953), *Arch. Ind. Hyg. Occupational Med.*, **9**, 178 (1954) (abstr.).

Reiter, M. and K. Leusser, (Effect of methylsubstitution on methemoglobin formation and the general toxicity of aniline), "Einfluss der Methylsubstitution auf die Hämiglobin (Methämoglobin)-Bildung und die Allgemeintoxicität des Anilins." *Arch. Exptl. Pathol. Pharmakol.*, **214**, 158 (1952).

Sroka, K. H., (Treatment of industrial diseases caused by nitro and amino compounds of benzene and its homologs), "Zur Behandlung der Berufskrankheiten durch Nitro- und Amidokörper des Benzols und seiner Homologen," *Seifen-Öle-Fette-Wachse*, No. **5**, 101 (1952).

Vigliani, F. O., (Nutritional problems of workers exposed to the action of toxic substances), "Problemi di alimentazione per i lavoratori esposti all 'azione di sostanze tossiche," *med. lavoro*, **45**, 423 (1954), *Am. Ind. Hyg. Assoc. Quart.*, **15**, 305 (1954), *Arch. Maladies Profess. méd travail et sécurite sociale*, **16**, 90 (1955).

Aminophenol

Bernstein, S., and R. W. McGilvery, The enzymatic conjugation of *m*-aminophenol, *J. Biol. Chem.*, **198**, 195 (1952), *Excerpta Med. Sect. II*, **7**, No. 1191 (1954) (abstr.).

Bray, H. G., R. C. Clowes, and W. W. Thorpe, The metabolism of aminophenols, *o*-formamidophenol, benzoxazole, 2-methyl- and 2-phenyl-benzoxazoles and benzoxazolone in the rabbit, *Biochem. J.*, **51**, 70 (1952).

Robinson, D., J. N. Smith, and R. T. Williams, Studies in detoxication. 40. The metabolism of nitrobenzene in the rabbit. *o*-, *m*-, and *p*-Aminophenols and metabolites of nitrobenzene, *Biochem. J.*, **50**, 228 (1951).

Zini, F., (Glucuronic acid conjugation of aminophenol slices by organs of rats), "Sulla glicurono-coniugazione dello' o-amino-fenolo da fettine d'organi di ratto," *Sperimentale*, **102**, 40 (1952), *Chem. Abstr.*, **46**, 8766b, (1952).

Aniline

Bartalini, E., and L. Ghiringhelli, (Methemoglobinemia and coproporphyrinuria in experimental aniline poisoning) "Metaemoglobinemia e coproporfinuria nell' intossicazione sperimentale da anilina," *med. lavoro*, **44**, 88 (1953), *Ind. Hyg. Dig.*, **17**, No. 824, (1953) (Abstr.), *Excerpta Med. Sect. II*, **7**, No. 2275 (1954).

Borisenko, A. N., Effects of nicotinic acid in acute aniline poisoning, *Farmakol. i. Toksikol.*, **16**, 34 (1953), *Russ. Chem. Abstr.*, **47**, 12655de (1953), *Ind. Hyg. Dig.*, **17**, No. 59 (1954).

Carpenter, C. P., H. F. Smyth, Jr., and U. C. Pozzani, The assay of acute vapor toxicity, and the grading and interpretation of results on 96 chemical compounds, *J. Ind. Hyg.*, **31**, 343 (1949).

Carozzi, L., Intoxication from insecticides, disinfectants and herbicides, *Folia Med.*, **33**, 97 (1950), *Chem. Abstr.*, **45**, 2591b (1951).

Casciano, A. D., Acute methemoglobinemia due to aniline; report of a case, *J. Med. Soc. New Jersey*, **49**, 141 (1952).

Coignet, R., (Curious history of a sunday cyanosis), "Curieuse histoire d'une cyanose dominical," *Marseille Med.*, **91**, 575 (1954).

Comstock, C. C., and F. W. Oberst, Inhalation toxicity of aniline, furfuryl alcohol and their Mixtures in rats and mice, *Chem. Corps Med Lab.* (Army Chem. Center, Maryland), **CMLRE-ML-52**, Med. Lab. Research Rept. No. **139**, 1 (1952).

Conn, L. W., R. G. Horton, J. S. Wiles, and H. C. Khalouf, The penetration of aniline through polyethylene films proposed for protective clothing, *Chem. Corps Med. Lab.* (Army Chem. Center, Maryland) Med. Lab. Research Rept. No. **169**, (1953).

Ekman, B. and J. P. Strömbeck, Studies of aniline metabolism in rats on varying diets, *Acta Physiol. Scand.*, **25**, 377 (1952), *Excerpta Med. Sect. II*, **6**, No. 5156 (1953).

Ekman, B. and J. P. Strömbeck, Effect of aniline on the urinary bladder in rats, *Acta Pathol. Microbiol. Scand.*, **26**, 472 (1949), *Arch. Ind. Hyg.*, **1**, 127 (1950).

Erbslöh, J., (Aniline poisoning from marking ink in the newborn), "Anilin-Vergiftung durch stempelfarben bei Neugeborenen," *Samml. Vergiftungsfallen*, **14**, 321 (1953), *Excerpta Med. Sect. II*, **8**, No. 581 (1955).

Friehoff, F. J., and K. H. Löbermann, (Toluidine blue in the treatment of toxic hemoglobinemia), "Toluidinblau bei der Behandlung von toxischen Hämoglobinämien," *Therap. Gegenw.*, **91**, 446 (1952), *Excerpta Med. Sect. II*, **7**, No. 567 (1954).

Ghiringhelli, L., and C. Molina, (Methemoglobin in acute poisoning by aniline in experimental animals and in man; its relation to cyanosis, anemia and Heinz bodies). "La metaemoglobinemia nell' intossicazione acuta da anilina nell' animale da esperimento e nell' uomo, Suoi rapporti colla cianosi l'anemia e i corpi di Heinz," *Med. lavoro*, **42**, 125 (1961), *Ind. Hyg. Dig.* **16**, No. 408 (1952), *Chem. Abstr.*, **46**, 2185i (1952).

Ghiringhelli, L., (Relation of methemoglobinemia to catalase in experimental aniline poisoning), "Rapporti tra metaemoglobinemia e catalasi nell' intossicazione sperimentale da anilina," *Med. lavoro*, **43**, 272 (1952), *Excerpta Med. Sect. II*, **6**, No. 5767 (1953).

Goldblatt, M. W., Research in Industrial Health in the Chemical Industry, *Brit. J. Ind. Med.*, **12**, 1 (1955).

Hackley, E. B., C. C. Comstock, and F. W. Oberst, Chronic inhalation toxicity of aniline to experimental animals, *Federation Proc.*, **12**, 327 (1953).

Hahn, H. J. A., Two cases of poisoning by marking ink, *J. Roy. Naval Med. Serv.*, **40**, 39 (1954), *Excerpta Med. Sect. II*, **8**, No. 1100 (1955).

Hill, D. L., Excretion of diazotizable metabolites in man after aniline exposure, *Arch. Ind. Hyg. Occupational Med.*, **8**, 347 (1943).

Jacobson, S. M., Acute aniline poisoning, *West V. Med. J.*, **48**, 298 (1952).

Jasinski, B., (Acute aniline poisoning, its clinical-hematologic course and its treatment), "Akute Anilinvergiftung, ihr klinisch-hämatologischer Ablauf und ihre Behandlung," *Schweiz. Med. Wochschr.*, **78**, 1282 (1948), *Arch. Ind. Hyg.*, **1**, 126 (1950).

Lundberg, A., Acute methaemoglobinaemia caused by aniline dye intoxication; report of a case, *Acta Paediat.*, **43**, 83 (1954).

MacMath, I. F., and J. Apley, Cyanosis from absorption of marking ink in newborn babies, *Lancet*, **2**, 895 (1954).

Merlevede, E., (Vitamin C as antidote to aromatic hydrocarbons), "La vitamine C comme antidote des hydrocarbures aromatiques," *Acta Med. Leg. Soc. Louvain*, **3**, 27 (1950), *Excerpta Med. II*, **4**, No. 1060, 1951.

Montanaro, O., E. G. Saraco, and C. J. Paglilla, (Poisoning caused by cutaneous absorption of aniline), "Intoxicacion por absorcion cutanea de anilina," *Semana Med. Buenos Aires*, **101**, 7 (1952).

Pickup, J. D., and J. Eeles, Cyanosis in newborn babies caused by aniline dye poisoning, *Lancet,* **2,** 118 (1953).

Pujol, A., (Vesical cancer possibly due to aniline dyes), "Cancer vesical de probable origen anilinico," *Rev. Arg. Urol.,* **23,** 101 (1954).

Rocchietta, S., (Accidental aniline poisoning in children), "Intossicazioni accidentali da anilina nei bambini," *Boll. chim. farm.,* **92,** 88 (1953).

Rodeck, H., and H. Westhaus, (Aniline poisoning by laundry inks and stamp dyes in newborn; report of a group poisoning of 41 newborn), "Die Anilinvergiftung durch Wäschetinten und Stempfelfarben bei Säuglingen; Ein Bericht an Hand einer Gruppenvergiftung von 41 Säuglingen," *Arch. Kinderheilk.,* **145,** 77 (1952).

Spicer, S. S., Species differences in susceptibility to methemoglobin formation, *J. Pharmacol. Exptl. Therap.,* **99,** 185 (1950), *Excerpta Med. Sect. II,* **4,** No. 949 (1951).

Spicer, S. S., and P. A. Neal, The effect of hypoxia on the in vivo formation of methemoglobin by aniline and nitrite, *J. Pharmacol. Exptl. Therap.,* **95,** 438 (1940).

Spicer, S. S., and E. C. Thompson, Heinz body formation in vivo, a property of methylene blue, *J. Ind. Hyg.,* **31,** 206 (1949).

Zakabunina, M. S., Effects of aniline spread in minimal doses on rabbit skin, *Farmakol. i. Toksikol.,* **16,** 40 (1953), *Chem. Abstr.,* **47,** 12653b (1953), *Ind. Hyg. Dig.,* **17,** No. 58 (1954).

Benzidine

Baker, R. K., and J. G. Deighton, The metabolism of benzidine in the rat, *Cancer Research,* **13,** 529 (1953).

Barsotti, M., and E. C. Vigliani, Bladder lesion from aromatic amines; statistical considerations and prevention, *Arch. Ind. Hyg. Occupational Med.,* **5,** 234 (1952).

Barsotti, M., and E. C. Vigliani, Bladder lesions from amines. Statistical considerations and preventive measures, *Proc. Internat. Congr. Ind. Med. 9th Congr., London, 1948,* 484 (1949).

Engelbertz, P., and E. Babel, The use of the benzidine test method with potassium 1,2-naphthoquinone-4-sulfonate for the determination of health hazards in benzidine plants, *Zentralbl. Arbeitsmed.,* **4,** 40 (1952).

Gehrmann, G. H., J. H. Foulger, and A. J. Fleming, Occupational cancer of the bladder, *Proc. Internat. Congr. Ind. Med. 9th Congr., London, 1948,* 472 (1949), *Chem. Abstr.,* **45,** 7275a 65 (1951).

Goldblatt, M. W., Vesical tumours induced by chemical compounds, *Brit. J. Ind. Med.,* **6,** (1949).

Meigs, J. W., L. J. Sciarini, and W. A. Van Sandt, Skin penetration of diamines of the benzidine group, *Arch. Ind. Hyg. Occupational Med.,* **9,** 122 (1954).

Meigs, J. W., R. M. Brown, and L. J. Sciarini, A study of exposure to benzidine and substituted benzidines in a chemical plant; a preliminary report, *Arch. Ind. Hyg. Occupational Med.,* **4,** 533 (1951).

Scott, T. S., The incidence of bladder tumours in a dyestuffs factory, *Brit. J. Ind. Med.,* **9,** 127 (1952).

Thoelen, H., and A. Pletscher, (Effect of calcium, glucose, and fructose on subacute experimental benzidine poisoning), "Die Wirkung von Calcium, Glukose und Fructose bei subakuter experimenteller Benzidinvergiftung," *Helv. Physio. et Pharmacol. Acta,* **12,** 293 (1955).

Dinitrochlorobenzene

Andreasen, E., and H. Haxthausen, Lymph node response to cutaneous sensitization with dinitrochlorobenzene, *Acta Pathol. Microbiol. Scand.,* **29,** 345 (1951).

Edmund, J., Localized allergic reaction in the cornea of guinea-pigs, *Acta Allergol.*, **6**, 118 (1953).

Depaoli, M., and M. Dogliotti, (Leucocytic blood picture in allergic reaction to 2-4-dinitro-chlorobenzene), "La Formula leucocitaria nella reazione allergica da 2-4-dinitrochloro-benzolo," *Minerva dermatol.*, **29**, 410 (1954), *Arch. maladies profess. méd. travail et sécurité sociale*, **16**, 264 (1955).

Frey, J. R., and H. P. Bächtold, (Effect of cortisone—adrenocortical preparation—and various drugs on contact eczema of guinea pig produced by dinitrochlorobenzene), "Versuche zur Beeinflussung des Dinitrochlorobenzol-Kontakt-Ekzems am Meerschweinchen mit Cortison und Verschiedenen Pharmaca," *Bull. schweiz. Akad. med. wiss.*, **8**, 180 (1952).

Grimmer, H., and S. Rust, (Experimental investigations on animals of the effect of deep trich-ophytosis on epidermal sensitization by dinitrochlorobenzene), "Tierexperimentelle Untersuchungen über den Einfluss der tiefen Trichophytie auf die epidermale Sensibilisie-rung durch Dinitrochlorobenzol," *Arch. Dermatol. u. Syphilis*, **194**, 663 (1952).

Jeter, W. S., M. M. Tremaine, and P. M. Seebohm, Passive transfer of delayed hypersensitivity to 2,4-dinitrochlorobenzene in guinea pigs with leucocytic extracts, *Proc. Soc. Exptl. Biol. Med.*, **86**, 251 (1954).

Korossy, S., and I. Boggyan, (Significance of dental foci for dinitrochlorobenzene sensitization), "Zur Frage der Bedeutung dentaler Herde," *Dermatologica*, **104**, 168 (1952).

Nilzen, A., Some aspects of epidermal testing of guinea pigs sensitized and not sensitized to 2,4-dinitrochlorobenzene, *Acta Dermat.-Venereol. Suppl. 29*, **32**, 231 (1952).

Seebohm, P. M., M. M. Tremaine, and W. S. Jeter, The effect of cortisone and adenocortico-tropic hormone on passively transferred delayed hypersensitivity to 2,4-dinitrochloro-benzene, *J. Immunol.*, **73**, 44 (1954), *Chem. Abstr.*, **48**, 11610d (1954).

Werner, H., and U. Wetzel, (Manifestations of chloronitrobenzene poisoning), "Zur Sympto-matik der Chlornitrobenzolvergiftung," *Arztl. Wochschr.*, **7**, 1210 (1952), *Ind. Hyg. Dig.*, **17**, No. 588 (1953), *Excerpta Med. Sect. II*, **7**, No. 1109 (1954).

Tabachnick, M., and H. N. Eisen, Quantitative measurement of dinitrophenyl amino acids formed by the reaction in vivo of 2,4-dinitrochlorobenzene and epidermis, *Federation Proc.*, **14**, 290 (1955).

2-Chloro-4-nitroaniline, toxicity, mice, *Summary Tables Biol. Tests*, **6**, 138 (1954).

4-Chloro-2-nitroaniline, toxicity, mice, *Summary Tables Biol. Tests*, **6**, 138 (1954).

1,2-Dichloro-4,5-dinitrobenzene, toxicity, mice, *Summary Tables Biol. Tests*, **6**, 53 (1954).

3-Chloro-4-nitrophenol, toxicity, mice, *Summary Tables Biol. Tests*, **6**, 60 (1954).

Dinitrobenzene

Carsten, M. E., and H. N. Eisen, The interaction of dinitrobenzene derivatives with bovine serum albumin, *J. Am. Chem. Soc.*, **75**, 4451 (1953).

Carsten, M. E., and H. N. Eisen, The specific interaction of some dinitrobenzenes with rabbit antibody to dinitrophenylbovine globulin, *J. Am. Chem. Soc.*, **77**, 1273 (1955).

Danopoulos, E., and K. Melissinois, (Remarks on clinical and hematological observations of collective poisoning by dinitrobenzene and trinitrotoluene), "Remarques cliniques et hematologiques a propos d'une intoxication collective par le dinitrobenzene et trinitro-toluene," *Arch. maladies profess. méd. travail et séaurité sociale*, **13**, 458 (1952).

Eisen, H. N., and S. Belman, Studies of hypersensitivity to low molecular weight substances. II. Reaction of some allergenic substituted dinitrobenzenes with cysteine or cystine of skin proteins. *J. Exptl. Med.*, **92**, 533 (1953).

Dinitro-o-Cresol

Barnes, J. M., The effect of dinitro-*o*-cresol on the deposition of liver glycogen in the rat, *Biochem. J.*, **54**, 148 (1953), *Arch. Ind. Hyg. Occupational Med.*, **9**, 265 (1954).

Bidsrup, P. L., Clinical aspects of poisoning by dinitro-o-cresol, *Proc. Roy. Soc. Med.*, **45**, 574 (1952).

Bidsrup, P. L., J. A. L. Bonnell, and D. G. Harvey, Prevention of acute dinitro-o-cresol (D.N.O.C.) poisoning, *Lancet*, **1**, 794 (1952).

Bruce, J., and R. Mackay, Estimation of dinitro-o-cresol in blood, *Lancet*, **1**, 1115 (1952).

Bruusgaard, A., Dangers of intoxication from use of insecticides with special consideration of dinitro-o-cresol and parathion, *Tidsskr. Norske Laegeforen.*, **72**, 357 (1952).

Corti, A. L., (Fatal poisoning with dinitro-o-cresol) "Intoxication mortal con dinitro-o-cresol," *Rev. farm. Buenos Aires*, **95**, 157 (1953), *Excerpta Med. Sect. II*, **7**, No. 5533, (1954), *Chem. Abstr.*, **48**, 4698h (1954).

Fairhall, L. D., Dinitro-o-cresol, *Occupational Health*, **12**, 132 (1952).

Galley, R. A. E., *et al.*, Discussion on agricultural poisons, *Proc. Roy. Soc. Med.*, **45**, 567 (1952).

Harvey, D. G., Estimation of dinitro-o-cresol in blood, *Lancet*, **1**, 796 (1952).

Harvey, D. G., and P. L. Bidstrup, Routine estimation of dinitro-o-cresol in blood, *Lancet*, **1**, 1213 (1952).

Harvey, D. G., The toxicity of the dinitro-cresols, Pt. II. The formation and toxic properties of some nitro-compounds derived from metal and para cresols, *J. Pharm. and Pharmacol.*, **8**, 497 (1953).

Jensen, H. L., and K. Gunderson, Biological decomposition of aromatic nitro compounds, *Nature*, **175**, 341 (1955).

King, E., and D. G. Harvey, The absorption and excretion of 4,6-dinitro-o-cresol (DNOC) I. Blood dinitro-o-cresol levels in the rat and rabbit following different methods of absorption. II. The elimination of 4,6-dinitro-o-cresol by man and animals, *Biochem. J.*, **53**, 185, 196 (1953), *Arch. Ind. Hyg. Occupational Med.*, **8**, 485 (1953), *Ind. Hyg. Dig.*. **17**, No. 589, 590 (1953).

Locker, A., and K. H. Spitzky, (Effect of dinitrocresol on respiration of normal and damaged cells and tissues), "Die Wirkung von Dinitrokresol auf die Atmung Normaler und geschädigter Zellen und Gewebe," *Z. Ges. exptl. Med.*, **118**, 155 (1951), *Excerpta Med. Sect. II*, **5**, No. 4085 (1952).

Parker, V. H., J. M. Barnes, and F. A. Denz, Some observations on the toxic properties of 3:5 dinitro-o-cresol, *Brit. J. Ind. Med.*, **8**, 226 (1951), *Arch. Ind. Hyg. Occupational Med.*, **5**, 593 (1952).

Smith, J. N., R. H. Smithies, and R. T. Williams, Studies in detoxication. 48. Urinary metabolites of 4:6-dinitro-o-cresol in the rabbit, *Biochem. J.*, **54**, 225 (1953).

Stoner, H. B., C. J. Threlfall, and H. N. Green. The effect of 3:5-dinitro-o-cresol on the organic phosphates of muscle, *Brit. J. Exptl. Pathol.*, **33**, 398 (1952).

Dinitrophenol

Bardino, M., (Rigidity of cadavers after administration of 2,6-dinitrophenol), "Comportamento della rigidita cadaverica in seguito a somministrazione di dinitrofenolo 1-2-6," *Boll. soc. ital. biol. sper.*, **28**, 325 (1952), *Chem. Abstr.*, **48**, 274b (1954).

Derobert, L., A. Desclaux, A. Hadengue, and R. Le Breton, (2-4-dinitrophenol poisoning. III. Action of cortisone), "Le 'Intoxication par le dinitrophenol 1-2-4 (thermol): action de la cortisone," *Ann. med. légale et criminol. police sci. et toxicol.*, **32**, 246 (1952), *Chem. Abstr.*, **47**, 3470h (1953), *Ind. Hyg. Dig.*, **17**, No. 480 (1953).

Ershoff, B. H., Comparative effects of B vitamins and liver on the dinitrophenol toxicity in the rat, *J. Nutrition*, **42**, 271 (1950), *Excerpta Med. Sect. II*, **4**, No. 2355 (1951).

Dinitrophenol residues, *Federal Register*, **19**, 6740 (1954).

Hicks, S. P., Brain metabolism in vivo: distribution of lesions caused by azide, malononitrile, plasmocid (quinoline derivative) and dinitrophenol poisoning in rats, *Arch. Pathol.*, **50**, 454 (1950).

Piette, M., Blood and bone marrow poisons, *Document. Biol. Pract.*, **5**, 202 (1952), *Chem. Abstr.*, **46**, 8761g (1952).

Simon, E. W., Mechanisms of dinitrophenol toxicity, *Biol. Revs. Cambridge Phil. Soc.*, **28**, 453 (1953).

Swamy, S. A., Suicidal poisoning by dinitrophenol, *J. Indian Med. Assoc.*, **22**, 504 (1953), *Excerpta Med. Sect. II*, **7**, No. 6097 (1954).

Ther, L., and M. Meuller, (Kinetics of glycogen phosphorylation in adrenal insufficiency and dinitrophenol intoxication), "Kinetik der Glykogenphosphorylierung bei Nebennieren- insuffizienz und Dinitrophenol Vergiftung," *Arch. Exptl., Pathol.*, **209**, 194 (1950), *Excerpta Med. Sect. II*, **4**, No. 1802 (1951).

Aisenberg, A. C., and V. R. Potter, Effect of fluoride and dinitrophenol on acetate activation in kidney and liver homogenates, *J. Biol. Chem.*, **215**, 737 (1955).

Barnes, J. M., and J. I. Duff, Action of 2:4-dinitrophenol on striated muscle, *J. Physiol.*, **124**, 37P (1954).

Beevers, H., 2,4-Dinitrophenol and plant respiration, *Am. J. Botany*, **40**, 91 (1953).

Bodine, J. H., Combined action of 2,4-dinitrophenol-DNP and ethyl carbamate on oxygen uptake of embryonic cells, *Physiol. Zoöl.*, **23**, 63 (1950), *Chem. Abstr.*, **47**, 5461h (1953).

Bodine, J. H., and W. L. West, Reagents modifying stimulating action of 2,4-dinitrophenol, *Proc. Soc. Exptl. Biol. Med.*, **85**, 322 (1954).

Castor, C. W., and Beierwaltes, The effect of dinitrophenol on protein-bound iodine in man; a preliminary report, *Univ. Mich. Med. Bull.*, **21**, 101 (1955).

Coussens, R., and L. Massart, The influence of cetylammoniumbromide on the activity of dinitrophenol, *Arch. Internat. Pharmacodynamie*, **93**, 456 (1953).

Eaton, M. D., and M. E. Perry, Further observations on the effect of 2,4-dinitrophenol on the growth of influenza virus, *J. Infectious Diseases*, **93**, 269 (1953).

Frommel, E., Radouco, and F. Vallette, (Effect of antiepileptic medication on hyperthermia in- duced by dinitrophenol in Guinea pig), "L'influence de la médication antiépileptique sur l'hyperthermie provoquée par le dinitrophenol chez le cobaye," *Helv. Physiol. et Pharma- col. Acta*, **10**, 288 (1952).

Frunder, H., (Changes of pH in living tissue in disturbances of carbohydrate metabolism), "Über pH-Anderungen im lebenden Gewebe bei gestörtem Kohlenhydratstoffwechsel," *Arch. ges. Physiol.*, **252**, 520 (1950), *Excerpta Med. Sect. II*, **4**, No. 1804 (1951).

Ginetti, Y., (Hydrogen bonds of para and meta nitrophenols), "Liasons hydrogene des nitrophenols para et meta," *Naturwissenschaften*, **41**, 333 (1954).

Giotti, A., and N. F. Buffoni, (Studies on isolated atrial activity in various experimental con- ditions and under the action of various groups of drugs. Action of 2,4-dinitrophenol), "Studi sull 'attivita degli atri isolati in varie condizioni sperimentali e sotto l'azione di vari gruppi di farmaci. L'azione del 2,4-dinitrofeno, *Arch. ital. Sci. Farmacol.*, **5**, 271 (1955).

Lardy, H. A., and H. Wellman, The catalytic effect of 2,4-dinitrophenol on adenosinetriphos- phate hydrolysis by cell particles and soluble enzymes, *J. Biol. Chem.*, **201**, 357 (1953), *Excerpta Med. Sect. II*, **7**, No. 1817 (1954).

Maruyama, K., Effects of 2,4-dinitrophenol and azide on the latent apyrase of the toad embryo, *Arch. Biochem. Biophys.*, **52**, 485 (1954).

Mudge, G., H. W. Neuberg, and S. W. Stanbury, The effect of 2,4-dinitrophenol on the for- mation of phosphoenolpyruvate by liver mitochondria, *J. Biol. Chem.*, **210**, 966 (1954).

Parker, V. H., Enzymic reduction of 2:4-dinitrophenol by rat tissue homogenates, *Biochem. J.*, **51**, 363 (1952).

Robinson, J. R., Effect of 2,4-dinitrophenol on osmoregulation in the isolated kidney slices, *Nature*, **166**, 989 (1950), *Excerpta Med. Sect. II*, **4**, No. 3192 (1951).

Rothlin, E., M. Taeschler, and A. Cerletti, Action of dinitrophenol and lantoside C on the canine heart-lung preparation, *Circulation Research*, **3**, 32 (1955).

Sammartino, U., and A. Amici, (Effects of nitro derivatives of phenol and naphthol on blood catalase and intracellular respiration. IV. Effects of 2,4-, 2:5- and 2:6-dinitrophenol on intracellular respiration studied by the Warburg method. V. Effects of 2:4-dinitro-1-naphthol and 1:6-dinitro-2-naphthol on intracellular respiration as studied by the Warburg method), "Azione del nitroderivativi del fenolo e del naftolo sulla catalasi del sangue e sulla respirazione intracellulare. IV. Influenza del 2-4 dinitrofenolo, del 2-5 dinitrofenolo, del 2-6 dinitrofenolo sulla respirazione intracellular studiata con il metodo di Warburg. V. Influenza del 2-4 dinitro-1-naftolo e dell' 1-6-dinitro-2-naftolo sulla respirazione intra-.cellular studiata con il metodo di Warburg," *Boll. soc. ital. biol. sper.*, **27**, 987, 990 (1951), *Excerpta Med. Sect. II*, **5**, Nos. 3789 and 3323 (1952).

Schwartz, H. S., and S. B. Barker, Dinitrophenol and rat brain metabolism. *Am. J. Physiol.*, **171**, 765 (1952).

Siekevitz, P., and V. R. Potter, The adenylate kinase of rat liver mitochondria, *J. Biol. Chem.*, **200**, 187 (1953), *Excerpta Med. Sect. II*, **7**, No. 1186 (1954).

Siekevitz, P., and V. R. Potter, Effect of 2,4-dinitrophenol and of fluoride on oxidations in normal and tumor tissues, *Cancer Research*, **13**, 513 (1953).

Simpson, J. R., and W. C. Evans, The metabolism of nitrophenols by certain bacteria, *Biochem. J.*, **55**, xxiv (1953).

Slater, E. C., and S. W. Lewis, Stimulation of respiration by 2:4-dinitrophenol, *Biochem. J.*, **58**, 337 (1954).

Vantaggi-Cozzari, L., Action of 2,4-dinitrophenol on alkaline phosphatase, *Boll. soc. ital. biol. sper.*, **28**, 1922 (1952), *Chem. Abstr.*, **47**, 10569h (1953).

Walton, R. P., L. I. Goldberg, et. al., Effects of hyperpyrexia on the heart in situ; studies with dicumarol, dinitrophenol and external heat, *Am. J. Physiol.*, **169**, 78 (1952).

Witter, R. F., E. H. Newcomb, and E. Stotz, Studies of the mechanism of action of dinitrophenol, *J. Biol. Chem.*, **202**, 291 (1953).

Wollenberger, A., and M. L. Karsh, Effect of cardiac glycoside on contraction and energy-rich phosphate content of heart poisoned with dinitrophenol, *J. Pharmacol. Exptl. Therap.*, **105**, 477 (1952).

Turba, F., and G. Gundlach, (Separation of dinitrophenols from dinitrophenyl-aminoacids and peptides), "Abtrennung des Dinitrophenols von Dinitrophenyl-Aminosäuren und Peptiden," *Biochem. Z.*, **326**, 322 (1955).

Tyler, D. B., Some factors affecting the action of 2,4-dinitrophenol on the oxygen uptake of excised rat brain, *J. Biol. Chem.*, **184**, 711 (1950), *Excerpta Med. Sect. II*, **4**, No. 2026 (1951).

Vaughan, S. L., Drug-induced blood dyscrasias, *N. Y. State J. Med.*, **55**, 2457 (1955).

Shils, M. E., and L. J. Goldwater, Effect of diet on the susceptibility of the rat to poisoning by 2,4-dinitrotoluene, *Arch. Ind. Hyg. Occupational Med.*, **8**, 262 (1953).

4-Biphenylamine, toxicity, mice, *Summary Tables Biol. Tests*, **6**, 54 (1954).

Carcinogenic Aromatic Amines

Walpole, A. L., M. H. C. Williams, D. Roberts, and J. A. Hendry, Experiments in carcinogenesis with methyl derivatives of 4-aminodiphenyl, *Acta. Unio Intern. contra Cancrum*, **10**, 174 (1954).

Walpole, A. L., M. H. C. Williams, and D. Roberts, Bladder tumors induced in rats of two strains with 3:2'-dimethyl-4-aminodiphenyl, *Brit. J. Cancer*, **9**, 170 (1955).

Bonser, G. M., J. G. S. Crabbe, J. W. Jull, and L. N. Pyrah, Induction of epithelial neoplasms in the urinary bladder of the dog by intravesical injection of a chemical carcinogen, *J. Pathol. Bacteriol.*, **68**, 561 (1954).

Boyland, E., Different types of carcinogens and their possible modes of action, *Cancer Research*, **12**, 77 (1952).

Buu-Hoi, N. P., (On the physico-chemical importance of the mode of action of carcinogens), "Zur physikalischchemischen Deutung des Wirkungsmechanismus von krebserregenden Verbindungen," *Arch. Geschwulstforsch.*, **6**, 19 (1953).

Clayson D. B., A working hypothesis for the mode of carcinogenesis of aromatic amines, *Brit. J. Cancer*, **7**, 460 (1953).

Di Maio, G., Affections of the bladder due to aromatic amines, *Proc. Intern. Congr. Ind. Med., 9th Congr., London, 1948*, **476** (1949), *Chem. Abstr.*, **45**, 2274h (1951).

Scott, W. W., and H. L. Boyd, A Study of the Carcinogenic Effect of Beta-Naphthylamine on the Normal and Substituted Isolated Sigmoid Loop Bladder of Dogs, *J. Urol.*, **70**, 914–925 (Dec. 1953).

Carcinogenic Aromatic Amines, *Brit. M. J.*, **2**, 28–29 (July 1953). Editorial.

Truhaut, R., (Occupational Cancer. Bladder Carcinogens) "Contribution a l'etude des cancers professionals. Les cancerigenes vesicaux." *Sem. D. Hop. Paris*, **25**, 3078–3084 (Oct. 10, 1949).

Paranitroaniline

Baldi, G., and A. Raule, (Industrial poisoning by paranitroaniline. Clinical case with neuritic symptomatology), "Intossicazione professionale da paranitroanilina. Un caso clinico a sintomatologia neuritica," *Med. lavoro*, **45**, 584 (1954).

Gupta, M. N., Acute paranitraniline poisoning in dock laborers, *Proc. Soc. Study Ind. Med. India*, **5**, 32 (1953), *Arch. Ind. Hyg. Occupational Med.*, **10**, 84 (1954), *Ind. Hyg. Dig.*, **18**, No. 162 (1954).

Rieders, F., and H. Brieger, Mechanism of poisoning from wax crayons, *J. Am. Med. Assoc.*, **151**, 1490 (1953).

Kiese, M., (Toxicity of some new sweetening agents—2-alkoxy derivatives of *m*-nitraniline), "Zur Toxizität neuer Süszstoffe (1-Alkoxy-2-amino-4-nitrobenzole)," *Arch. Exptl. Pathol. Pharmakol.*, **207**, 446 (1949), *Excerpta Med. Sect. II*, **3**, No. 4643 (1950).

Nitrobenzene

Dollinger, A., Peroral poisoning with nitrobenzene or aniline in the newborn, *Monatschr. Kinderheilk.*, **97**, 91 (1949), *Chem. Abstr.*, **43**, 8060e (1949).

Izzhizu, S., and K. Akiyama, Heinz bodies in toxic anemia with methemoglobin formation, *J. Sci. Labour Rodo Kagaku*, **29**, 527 (1953), *Chem. Abstr.*, **48**, 6585h (1954).

Naevestad, R., Intoxication by DDT and nitrobenzene, *Tidsskr. Norske Laegeforen.*, **67**, 261 (1947), *Chem. Abstr.*, **43**, 8525i (1949).

Parkes, W. F., and D. W. Neill, Acute nitrobenzene poisoning with transient aminoaciduria, *Brit. Med. J.*, **1**, 653 (1953), *Chem. Abstr.*, **47**, 11513d (1953), *Ind. Hyg. Dig.*, **17**, No. 481 (1953), *Excerpta Med. Sect. II*, **7**, No. 2273 (1954).

Nitrophenols

4-Nitro-*m*-cresol, toxicity, mice, *Summary Tables Biol. Tests*, **6**, 54 (1954).

4-Nitroso-*m*-cresol, toxicity, mice, *Summary Tables Biol. Tests*, **6**, 55 (1954).

6-Nitro-*m*-cresol, toxicity, mice, *Summary Tables Biol. Tests*, **6**, 55 (1954).

p-Nitrosophenol, toxicity, mice, *Summary Tables Biol. Tests*, **6**, 148 (1954).

2, 4-6-Trinitro-*m*-cresol, toxicity, mice, *Summary Tables Biol. Tests*, **6**, 55 (1954).

Judah, J. D., Mode of action of the nitrophenols, *Proc. Roy. Soc. Med.*, **45**, 574 (1952).

Osgood, E. E., Hypoplastic anemias and related syndromes caused by drug idiosyncrasy, *J. Am. Med. Assoc.*, **152**, 816 (1953).

Raule, A., and C. Sassi, (A case of sulfhemoglobinemia caused by occupational poisoning by paranitrophenol), "Un caso di solfoemglobinemia da intossicazione professionale da paranitrofenolo," *Osped. maggiore*, **41**, 474 (1953).

Robinson, D., J. N. Smith, and R. T. Williams, Studies in detoxication. 39. Nitro compounds. a. The metabolism of o-, m-, and p-nitrophenols in the rabbit. b. The glucuronides of the mononitrophenols and observations on the anomalous optical rotations of triacetyl-β-o-nitrophenyl glucuronide and its methyl ester, *Biochem. J.*, **50**, 221 (1951).

Simon, E. W., The relative toxicity of nitrophenols to various organisms, *Ind. Hyg. Dig.*, **17**, No. 54 (1953), *Ann. Appl. Biol.*, **39**, 416 (1952).

Phenylenediamine

Avezzu, G., and P. Ottaviani, (Some aspects of experimental poisoning with p-phenylenediamine and its mechanism), "Su alcuni aspetti e sul meccanismo patogenetico della intossicazione sperimentale da parafenilendiamino," *Arch. Med. Int. Parma*, **4**, 205 (1952), *Excerpta Med. Sect. II*, **7**, No. 2276 (1954).

Bonomi, U., (Diaminophilia (behavior in presence of phenylenediamine) of mast cells), "Sulla diaminofilia delle mastzellen," *Arch. Sci. Biol. Bologna*, **36**, 314 (1952).

Davis, J. E., Paraphenylene diamine anemia in rats, *Federation Proc.*, **11**, 337 (1952).

Rajka, G., On hypersensitive to "para group," *Acta Allergol.*, **5**, 11 (1952), *Brit. J. Ind. Med.*, **10**, 216 (1953) (abstr.).

Reiches, A. J., Skin reactions to hair dyes, *Arch. Dermatol. and Syphilol.*, **65**, 619 (1952).

Schwartz, L., Paraphenylenediamine in dermatology, *Southern Med. J.*, **46**, 769 (1953).

Stahl, K. E., and F. Jung, (Toxic action on blood of phenylenediamine and tolulyenediamine), Über Blutgiftwirkungen des Phenylenediamins und des Toluylenediamins," *Arch. Exptl. Pathol. Pharmakol.*, **220**, 503 (1953).

Tara, S., J. N. Lamberton, Y. Delplace, and A. Cavigneaux, (Asthma due to paraphenylene-diamine), "Les Asthmes à la Paraphenylenediamine," *Arch. maladies profess. med travail et sécurité sociale*, **13**, 376 (1952).

Zucchi, M., (Psychosis caused by p-phenylenediamine), "Psicosi da parafenilendiamina," *Sistema nervoso*, **6**, 205 (1954).

Heubner, W., B. Wahler, and C. Ziegler, (Methemoglobin formation from acylated phenylhydroxylamine), Über die Bildung von Hämiglobin durch acylierte Phenylhydroxylamine," *Z. physiol. Chem.*, **295**, 397 (1953), *Excerpta Med. Sect. III*, **7**, No. 6408 (1954).

Trinitrotoluene

Bergman, B. B., Tetryl toxicity; summary of 10 years' experience, *Arch. Ind. Hyg. Occupational Med.*, **5**, 10 (1952).

Foulger, J. H., Physiologic effects of certain explosives and chlorinated hydrocarbon solvents, *U. S. Armed Forces Med. J.*, **4**, 1425 (1953), *Arch. Ind. Hyg. Occupational Med.*, **9**, 431 (1954).

Crawford, M. A. D., Aplastic anemia due to trinitro-toluene intoxication, *Brit. Med. J.*, **2**, 430 (1954), *A.M.A. Arch. Ind. Health*, **11**, 442 (1955).

Devereux, J. M., A nutritional factor in TNT poisoning; a study of fifty employees exposed to the toxic influence of handling trinitrotoluene, *Ind. Med. and Surg.*, **24**, 171 (1955).

Di Lauro, S., Bellucci, and F. Sessa, (Thymol turbidity test in chronic experimental poisoning with trinitrotoluene), "La prova del timolo nella intossicazione cronica sperimentale da T.N.T.," *Folia Med. Naples*, **33**, 29 (1950), *Excerpta Med. Sect. II*, **4**, No. 2426 (1951).

Durante, U., (Results of glucose tolerance tests in trinitrotoluene poisoning), "Comportamento della curva glicemica da carico nell'intossicazione da trinitrotoluena," *Folia Med. Naples*, **33**, 526 (1950), *Excerpta Med. Sect. II*, **4**, No. 2754 (1951).

Hayhoe, F. G. J., Aplastic anemia occurring eight years after TNT poisoning, *Brit. Med. J.*, **1**, 1143 (1953), *Ind. Hyg. Dig.*, **17**, No. 712 (1953).

Kuratsune, M., and A. Iwaya, Effect on cyanocobalmin on experimental trinitrotoluene (TNT) poisoning. I. Effect of oral administration of cyanocobalmin, *Igaku to Seibutsugaku,* **29,** 33 (1953), *Chem. Abstr.,* **48,** 2929i (1954).

Masturzo, A., (Blood phosphatase curve and liver and kidney phosphatase activity in experimental intoxication with trinitrotoluene), "Curve fosfatasemche e comportamento della attivita fosfatasica del fegato e del rene nella intossicazione sperimentale la tritolo." *Pathologica,* **42,** 65 (1951), *Excerpta Med. Sect. II,* **6,** 538 (1953).

Pecora, L., (Protein metabolism in experimental poisoning with trinitrotoluene), "Contributo allo studio del ricambio proteico nell' intossicazione sperimentale da tritolo." *Folia Med. Naples,* **33,** 57 (1950), *Excerpta Med. Sect. II,* 4, No. 516 (1951).

Roubal, J., J. Zdrazil, and F. Picha, Polarographic determinations of 2,4,6-trinitrotoluene in the air and of 2,6-dinitro-4-amino toluene in urine, *Ceskoslov. hyg. epidemiol. mikrobiol. Immunol.,* **2,** 300 (1953), *Chem. Abstr.,* **48,** 13548h (1954).

Sessa, T., and S. Di Lauro, (Effects of methionine liver extracts, the vitamin B complex and nicotinic acid on the peripheral blood and the myelogram in experimental chronic poisoning with trinitrotoluene), "L'azione della metionina, degli estratti epatici, del complesso B e dell' acido nicotinico sul sange periferico e sul mielogramma nella intossicazione cronica sperimentale da trinitrotoluolo," *Folia Med. Naples,* **33,** 460 (1950), *Excerpta Med. Sect. II,* **4,** No 2060 (1951).

Shilo, M. E., and L. J. Goldwater, The effect of diet on the susceptibility of rats to poisoning by 2:4:6-trinitrotoluene (TNT), *J. Nutrition,* **41,** 293 (1950), *Excerpta Med. Sect. II,* **4,** No. 523 (1951).

Zambrano, A., Viscosity of blood in chronic experimental (TNT) trinitrotoluene poisoning; the methemoglobinemia of experimental TNT poisoning, *Folia Med. Naples,* **33,** 553, 555, (1950), *Chem. Abstr.,* **47,** 5028h (1953).

Zambrano, A., (Viscosity of blood in trinitrotoluene intoxication), "La viscosita del sangue nell ' intossicazione da tritolo; Metaemoglobinemia; intossicazione cronica sperimentale da T.N.T.," *Folia Med. Naples,* **34,** 219, 424 (1951), *Chem. Abstr.,* **46,** 4114d (1952).

Treon, J. F., W. G. Deichmann, H. E. Sigmon, *et al.,* The toxic properties of xylidine and monomethylaniline. I. The comparative toxicity of xylidine and monomethyl aniline when administered orally or intravenously to animals or applied upon their skin, *J. Ind. Hyg.,* **31,** 1 (1949).

White, J., and P. Moro-Chavez, Acute necrotizing renal papillitis experimentally produced in rats fed mono-*N*-methylaniline, *J. Natl. Cancer Inst.,* **12,** 777 (1952).

Spicer, S. S., B. Highman, and A. R. Monaco, Toxic and pathologic effects of xylidine in the fasting and non-fasting states, *J. Pharmacol. Exptl. Therap.,* **95,** 256 (1949), *Chem. Abstr.,* **43,** 3930g (1949).

4-Ethyl-2,6-dinitrophenol, toxicity, mice, *Summary Tables Biol. Tests,* **6,** 60 (1954).

4-Isopropyl-2,6-dinitrophenol, toxicity, mice, *Summary Tables Biol. Tests,* **6,** 60 (1954).

3-Chloro-4-nitrophenol, toxicity, mice, *Summary Tables Biol. Tests,* **6,** 60 (1954).

α,α,α-Triphenyl-*p*-toluidine, toxicity, mice, *Summary Tables Biol. Tests,* **6,** 64 (1954).

4-Nitro-2-6-xylenol, toxicity, mice, *Summary Tables Biol. Tests,* **6,** 63 (1954).

Dinitro-*o*-cresol salts:

   a. 4,6-Dinitro-*o*-cresol *salt* with l.f.wt. morpholine, toxicity, mice, *Summary Tables Biol. Tests,* **6,** 146 (1954).

   b. 4,6-Dinitro-*o*-cresol salt with l.f.wt. methylamine, toxicity, mice, *Summary Tables Biol. Tests,* **6,** 146 (1954).

   c. 4,6-Dinitro-*o*-cresol acetate, toxicity, mice, *Summary Tables Biol. Tests,* **6,** 146 (1954).

o-Tolidine, toxicity, mice, *Summary Tables Biol. Tests,* **6,** 64 (1954).

Blaisdell, C. T., Comparison of water and acetic acid as decontaminants for aniline, *Chem. Corps Med. Lab. Project No.* **4-61-14-002,** Rept. No. 327, Nov., 1954, p. 1.

Bray, H. G., S. P. James, and W. V. Thorpe, The metabolism of 3:4-dichloronitrobenzene in the rabbit with special reference to the formation of mercapturic acids, *Biochem. J.*, **60**, xxiii (1955).

Fonnesu, A., and C. Severi, Glycogen accumulation in the liver due to 2:4-dinitrophenol, *Brit. J. Exptl. Pathol.*, **36**, 35 (1955).

Holstein, E., (Poisoning by nitro and Amino compounds of the aromatic series), "Beiträge zur Vergiftung durch Nitro- and Aminoverbindungen der aromatischen Reihe," *Arch. Gewerbepathol. Gewerbehyg.*, **13**, 522 (1955).

Druckrey, H., D. Schmähl, and A. Reiter, (Absence of carcinogenic action of three isomers of N-dimethyltoluidine in rats), "Fehlen einer cancerogenen Wirkung bei den drei isomeren N-Dimethyltoluidinen an Ratten," *Arzneimittel Forsch.*, **4**, 365 (1954).

Ekman, B., and J. P. Strombeck, "The effect of some split products of 2,3-azotoluene on the urinary bladder in the rat and their excretion on various diets." *Acta Pathol. Microbiol. Scand.*, **26**, 447 (1949), *Chem. Abstr.*, **43**, 6315df (1949).

Rumpel, W., Causes and prevention of poisoning in chemical plants, *Mitt. chem. Forschungs. inst. Ind. Osterr.*, **4**, 113 (1950), *Chem. Abstr.*, **45**, 2603i (1951).

CHAPTER XLVIII

Heterocyclic and Miscellaneous Nitrogen Compounds

WILLIAM L. SUTTON, M. D.

Part A. Heterocyclic Nitrogen Compounds

Compounds that contain one or more nitrogen atoms in their ring structure are extremely numerous. They occur widely in nature, many in plant alkaloids and other biological materials, and a large number have been synthesized for their pharmaceutical potentialities or for other special uses. Relatively few can be considered here. These have been selected for one or more of these attributes: (1) high toxicity, (2) important industrial use, or (3) the occurrence of human occupational injury. Some are included as examples of a chemical class. Much pharmacological information that is of interest from the point of view of mechanism of action and its relation to chemical structure has been omitted since this is adequately covered in the pharmacological literature.

THREE-MEMBERED RINGS

I. General Considerations

Ethyleneimine and its derivatives have been of interest to chemists and biologists because of their high reactivity and the potentially useful effects produced by low doses. The uses for these materials do not extend beyond limited applications in organic syntheses, polymer and textile technology, and as chemotherapeutic agents in certain malignant diseases. However, in view of the reactivity of ethyleneimine and the chemical properties of its derivatives, other technical applications might be expected. Several series of monofunctional and polyfunctional ethyleneimine derivatives have been studied in connection with their so-called radiomimetic activity (tumor inhibition, hematological depression, lymphatic tissue damage, mutagenic, and antifertility effects).[1-4] Their biological

[1] J. A. Hendry, R. F. Homer, and F. L. Rose, *Brit. J. Pharmacol.*, **6**, 357 (1951).

[2] A. L. Walpole, D. C. Roberts, F. L. Rose, J. A. Hendry, and R. F. Homer, *Brit. J. Pharmacol.*, **9**, 306 (1954).

[3] F. S. Philips and J. B. Thiersch, *J. Pharmacol. Exptl. Therap.*, **100**, 398 (1950).

[4] J. J. Biesele, F. S. Philips, J. B. Thiersch, J. H. Burchenal, S. M. Buckley, and C. C. Stock, *Nature*, **166**, 1112 (1950).

effects are similar to those of the β-chloroethylamines (nitrogen mustards), which have also been classified as biological alkylating agents. These materials have been discussed in detail in a recent informative conference.[5]

Many of the ethyleneimine derivatives have reduced acute toxicity as compared to ethyleneimine. The monofunctional derivatives are less potent in producing the effects characteristic of the group than are the derivatives with two or more ethyleneimine groups. Although the four-membered ring compound, trimethyleneimine, (azetidine, 1,3-propylenimine) does not appear to have been studied, the two derivatives tested by Hendry and co-workers (2,4,6-tris-trimethyleneimino-1,3,5-triazine and $N:N'$-bis-cyclotrimethylene carbamyl hexamethylenediamine) showed little tumor inhibition or cytotoxic action at high doses.[1] The polymers of ethyleneimine and its derivatives have low toxicity. There is no consistent parallel relationship between hematological effects, tumor inhibition, and antifertility activity within the series.[6] Aside from the two ethyleneimines for which threshold limit values have been proposed, only triethylenemelamine (TEM) is discussed in detail, as an example of one of the more active and useful polyfunctional ethyleneimine derivatives.

II. Specific Compounds

ETHYLENEIMINE, NHCH$_2$CH$_2$, (Ethylenimine, Dimethylenimine,

Aziridine, Vinylamine (obsolete))

1. Uses

Ethyleneimine is a mobile, colorless, very volatile fluid with an ammoniacal odor. It is a strong caustic material, which polymerizes easily and behaves much like a secondary amine. It is prepared by the treatment of β-chloroethylamine with sodium hydroxide. Because of its reactivity, it is useful in organic synthesis. A major use is in the manufacture of triethylenemelamine.

2. Physical and Chemical Properties

Molecular weight: 43.07
Boiling point: 55 to 56°C.
Vapor density: 1.48 (air = 1)
Vapor pressure: 160 mm. Hg, 20°C.
Solubility: infinitely soluble in water, soluble in alkali
Flash point: 12°F.
1 mg./liter $\backsimeq$ 569 p.p.m. and 1 p.p.m. $\backsimeq$ 1.8 mg./cu. meter

[5] L. H. Schmidt, Comparative clinical and biological effects of alkylating agent, *Ann. N. Y. Acad. Sci.*, **68**, 657 (1958).

[6] H. Jackson, B. W. Fox, and A. W. Craig, *Brit. J. Pharmacol.*, **14**, 149 (1959).

3. Determination in the Atmosphere

In the absence of interfering alkalies, ethyleneimine can be determined in air by collection in standard acid and titration with sodium hydroxide.[7]

4. Physiological Response

According to the acute toxicity classification of Hodge and Sterner,[8] ethyleneimine is highly toxic by oral administration or percutaneous absorption and extremely toxic by inhalation (see Tables 1 and 2).

TABLE 1

Response to Ethyleneimine—Animals[a,b]

Species	Dose	Route	Response
Rat	15 mg./kg.	Oral	Approximate LD_{50}
Guinea pig	0.014 ml./kg.	Skin (poultice)	LD_{50}, skin necrosis
Rabbit	10 mg./kg.	Oral	Increase in blood urea nitrogen
Rabbit	0.005 ml.	Eye	Severe corneal damage, death
Rat	20 mg./kg. (total in 67 doses)	Subcutaneous	$6/12$ with sarcomata[a] No toxic deaths

[a] A. L. Walpole, D. C. Roberts, F. L. Rose, J. A. Hendry, and R. F. Homer, *Brit. J. Pharmacol.*, **9**, 306 (1954).

[b] C. P. Carpenter, H. F. Smyth, and C. B. Shaffer, *J. Ind. Hyg. Toxicol.*, **30**, 2 (1948).

Effects in Animals. The effects of ethyleneimine in experimental animals have been studied by several groups of investigators.[2,7,9] The high toxicity and the unique effect of ethyleneimine on the kidney were recognized as early as 1898 by Paul Ehrlich. The free imine is a potent skin irritant and vesicant. Application of 0.005 ml. of the imine or 0.5 ml. of a 15 per cent aqueous solution to the eye of rabbits causes corneal damage and death. Silver and McGrath[7] exposed mice to measured concentrations of ethyleneimine in air for 10 minutes and found the LC_{50} to be 3.93 ± 0.42 mg./liter (2236 p.p.m.). The animals exhibited signs of irritation of the eyes and nose during exposure. Deaths were delayed, occurring from about 24 hours after exposure throughout the 10 day observation period. Carpenter, Smyth, and Shaffer[9] exposed rats and guinea pigs to concentrations of ethyleneimine for varying periods of time (see Table 2). They also found that deaths were delayed. The animals exhibited extreme respiratory difficulty after exposures to concentrations above 10 p.p.m. and there was evidence of irritation of the eyes and nose at 100 p.p.m.

Inhalation of ethyleneimine results in delayed lung injury with congestion, edema, and hemorrhage. Kidney damage from absorption of ethyleneimine has been repeatedly observed. Proteinuria, hematuria, and increased blood urea

[7] S. D. Silver and F. P. McGrath, *J. Ind. Hyg. Toxicol.*, **30**, 7 (1948).

[8] H. C. Hodge and J. H. Sterner, *Am. Ind. Hyg. Assoc. Quart.*, **10**, 93 (1949).

[9] C. P. Carpenter, H. F. Smyth, and C. B. Shaffer, *J. Ind. Hyg. Toxicol.*, **30**, 2 (1948).

TABLE 2

Response to Inhalation of Ethyleneimine[a]

Calculated concentration, p.p.m.	Duration of exposure, min.	Fourteen-day mortality (six animals per exposure)	
		Guinea pigs	Rats
10	480	0	
25	60	0[b]	
	480	2	1
50	120	0	0
	480	6	5
100	60	1	0
	120	1	1
	240	6	6
250	60	2	2
	240	6	6
500	15	0	3
	60	6	6
1000	15	0	5
	30	6	6
2000	10	1[b]	1
	15	6	5
4000	5	4	4
	10	6	5

[a] C. P. Carpenter, H. F. Smyth, and C. B. Shaffer, *J. Ind. Hyg. Toxicol.*, **30**, 2 (1948).
[b] Twelve animals exposed.

nitrogen result from inhalation or injection. Micropathological examination reveals renal tubular damage. A decrease in white blood cell count and depression of all blood elements have also been observed. Unlike the polyfunctional ethyleneimine derivatives, ethyleneimine does not have appreciable growth inhibitory activity on the Walker tumor in rats, nor is it significantly cytotoxic. However, it does have a mutagenic effect in the fruit fly, *Drosophila*, and has carcinogenic activity in rats following subcutaneous injection in Carbowax or arachis oil.[2]

Effects in Man. Two workers in Smyth's laboratory developed skin sensitization and a case of slowly healing dermatitis was observed during small-scale production. Two or three cases of nose and throat irritation with conjunctivitis resulted from exposure to unknown concentrations. It was determined experimentally that eye and nose irritation in humans becomes evident at about 100 p.p.m.[9] Danehy and Pflaum report that 2 to 3 minutes' exposure to ethyleneimine vapors in the laboratory did not produce any symptoms until 3 hours later when vomiting occurred followed by irritation of the mouth, throat, and eyes, which subsided in 1 to 2 days.[10] Severe eye burns have resulted from direct contact but no systemic injury or fatalities have been reported.

[10] J. P. Danehy and D. J. Pflaum, *Ind. Eng. Chem.*, **30**, 778 (1938).

5 *Hygienic Standard of Permissible Exposure*

The American Conference of Governmental Industrial Hygienists recommends a threshold limit value of 5 p.p.m. (9 mg./cu. meter).[11] This is based on the acute animal inhalation studies. In view of the action of ethyleneimine on the skin, the serious nature of its systemic effects, and the suggestive but incomplete evidence of carcinogenic and mutagenic activity, exposures should be rigidly controlled and exposed individuals should have careful medical observation.

6. *Odor and Warning Properties*

Ethyleneimine has an ammoniacal odor, which is detectable at 2 p.p.m. However, since eye and nose irritation does not occur at concentrations under about 100 p.p.m. and since the odor is so similar to ammonia that it does not suggest danger, its warning properties are not good. The high volatility, skin absorption, high toxicity, and delay in onset of symptoms accentuate the hazard.

PROPYLENEIMINE, $NHCH_2CHCH_3$ (2-Methylaziridine)

1. *Physiological Response*

Propyleneimine has limited use. The acute effects are similar to ethyleneimine although it is somewhat less toxic for rats and guinea pigs (see Table 3). The effects of repeated exposure and the possibility of carcinogenic and mutagenic activity apparently have not been investigated. No injuries in industry have been recorded beyond the statement that it is known to cause severe eye injury.[9]

TABLE 3

Response to Propyleneimine—Animals[a]

Species	Dose	Route	Response
Rat	19 mg./kg.	Oral	Approximate LD_{50}
Guinea pig	43 ml./kg.	Skin (poultice)	Approximate LD_{50}
Rabbit	0.005 ml. (5% soln.)	Eye	Corneal damage
Rat	500 p.p.m., 2 hrs. 500 p.p.m., 4 hrs.	Inhalation	Survived $^5/_6$ died
Guinea pig	500 p.p.m., $^1/_2$ hr. 500 p.p.m., 2 hrs.	Inhalation	Survived $^3/_5$ died

[a] C. P. Carpenter, H. F. Smyth, and C. B. Shaffer, *J. Ind. Hyg. Toxicol.,* **30,** 2 (1948).

2. *Physical and Chemical Properties*

Molecular weight: 58.10
Boiling point: 63 to 64°C.

[11] American Conference of Governmental Industrial Hygienists, *A.M.A. Arch. Ind. Health.* **20,** 266 (1959).

3. *Hygienic Standard of Permissible Exposure*

A threshold limit value of 25 p.p.m. (60 mg./cu. meter) has been recommended on the basis of the animal exposures and analogy to ethyleneimine.[11]

TRIETHYLENEMELAMINE, $C_9H_{12}N_6$ (2,4,6-Tris(ethylenimino)-*s*-triazine, 2,4,6-Triethylenimino-1,3,5-triazine, Tretamine)

1. *Use*

Triethylenemelamine (TEM) is a crystalline solid, which is prepared from ethyleneimine and cyanuric chloride. Aqueous solutions polymerize readily at room temperature but are stable at reduced temperatures. It was originally of interest as a cross-linking agent in textile technology. Major interest has been directed toward its medical use as a therapeutic agent for leukemia and other malignancies.

2. *Physical and Chemical Properties*

Molecular weight: 204.23
Melting point: Decomposes at 139°C.
Solubility: 40 per cent in water (w/w at 26°C.), less soluble in other common solvents

3. *Physiological Response*

Absorption and Metabolism. TEM is highly toxic by injection, or by oral administration. It is less well absorbed from the gastrointestinal tract than from the peritoneal cavity (see Table 4). Absorption through the skin or by inhalation

TABLE 4

Toxicity of TEM in Animals[a]

Species	Dose, mg./kg.	Route	Response
Mouse	2.8	Intraperitoneal	LD_{50}
	15.0	Oral	LD_{50}
	1.1 (5×)	Intraperitoneal	LD_{50}
Rat	1.0	Intraperitoneal	LD_{50}
	13.0	Oral	LD_{50}
	0.32 (5×)	Intraperitoneal	LD_{50}
Cat	1.0	Intravenous	Lethal in 7–12 days
Dog	1.0	Oral	Lethal in 3–5 days

[a] F. S. Philips and J. B. Thiersch, *J. Pharmacol. Exptl. Therap.*, **100**, 398 (1950).

has not been studied. Blood levels of TEM fall rapidly after intravenous injection in man, the concentration being only 10 per cent of the expected value 2 minutes after administration. There is no persistent selective uptake by any tissue. Eighty per cent of the C^{14} ring-labeled TEM is excreted in the urine within 24 hours, in the form of cyanuric acid. The triazine ring portion probably functions only as a carrier for the ethyleneimine groups, the fate of which are unknown.[12]

Effects in Animals. Like other polyfunctional ethyleneimine derivatives, TEM has been classified as a "radiomimetic" compound. The term is used because the observed effects in animals grossly resemble those produced by acute doses of ionizing radiation. There are delayed gastrointestinal symptoms with diarrhea and characteristic damage to the lymphatic cells of the intestinal tract and elsewhere, bone marrow damage with depression of all blood elements (pancytopenia), tumor inhibition and cytotoxicity (chromosomal aberrations and inhibition of cell division), increased mutation rate (in *Drosophila* and *Neurospora*), and antifertility effects. These effects all appear within a fairly narrow range of effective doses (Table 5). The carcinogenic activity of TEM is less than that of several

TABLE 5

Effects of TEM in Animals

Species	Dose, mg./kg.	Route	Response
Mouse	>125	Intraperitoneal	Weakness, convulsions respiratory failure death in
Rat	>33	Intraperitoneal	<20 hrs.[a]
Mouse	<125	Intraperitoneal	Delayed deaths hematological damage[a]
Rat	<33	Intraperitoneal	
Dog	1	Oral	Lethal, damage to lymphatic tissue, bone marrow[a]
Rat (male)	0.2	Injection	Sterility at 4 weeks recovery[b]
Rat (pregnant female)	0.3 (2 ×)	Intraperitoneal	Fetal deaths, malformed offspring[b]
Rat (Walker tumor)	0.25	Intraperitoneal	Cytotoxic action on tumor[c]
Mouse (stock)	1.25/wk. (52 X)	Subcutaneous	No tumors (0/20)[c]
Mouse (stock)	10 (total dose) arachis oil	Subcutaneous	$^{11}/_{12}$ sarcomata[d]

[a] F S. Philips and J. B. Thiersch, *J. Pharmacol. Exptl. Therap.*, **100**, 398 (1950).
[b] H. Jackson, B. W. Fox, and A. W. Craig, *Brit. J. Pharmacol.*, **14**, 149 (1959).
[c] J. H. Hendry, R. F. Homer, and F. L. Rose, *Brit. J. Pharmacol.*, **6**, 357 (1951).
[d] L. H. Schmidt, *Ann. N. Y. Acad. Sci.*, **68**, 657 (1958).

other ethyleneimine derivatives.[5] Although several attractive hypotheses have been suggested, the precise mode of action at the cellular level is not completely understood.

Effects in Man. The symptoms produced in man from therapeutic administration of TEM are similar to those observed in experimental animals. Oral doses of 1 to 5 mg./kg./day may produce gastrointestinal symptoms and depres-

[12] H. B. Mandel, *Pharmacol. Revs.*, **11**, 743 (1959).

sion of the formed elements in peripheral blood. No injuries in industry have been reported. Threshold limit values have not been suggested. However, it is clear that exposures should be maintained at the lowest possible levels.

FIVE-MEMBERED RINGS

I. General Considerations

Five-membered nitrogen heterocycles have comparatively limited industrial use. There is little information on this group of materials that is of industrial hygiene interest, although numerous compounds of pharmacological and biological importance contain five-membered rings with one or more nitrogen atoms. Pyrrole, for example, is part of the structure of naturally occurring biological pigments such as bilirubin and heme. The pyrrolidine ring occurs in such drugs as atropine and cocaine, pyrazole in antipyrine and its derivatives, imidazoline in some vasodilators, and imidazolidine (tetrahydroimidazoles) in antiepileptic drugs (hydantoin derivatives). The thiazole ring occurs in 2-aminothiazole, sulfathiazole, and vitamin B_1, the tetrahydro form in penicillin, and tetrazole in the central nervous system stimulant and convulsant, 1,5-pentamethylenetetrazole. It can be seen that such derivatives exhibit a wide range of pharmacological activities.

Although the industrial use of the simpler members of this group is limited and the toxicological information is incomplete, individual compounds occasionally receive considerable hygienic interest, as is illustrated by 2-aminothiazole and aminotriazole. Therefore, some information on these materials is presented below.

II. Specific Compounds

PYRROLE, NHCH:CHCH:CH (Azole, Divinylenimine, Imidole)

1. Uses

Pyrrole is a constituent of coal tar and bone oil. It is a colorless liquid when fresh but darkens with exposure to oxygen. It is a very weak base ($pK_b = 13.6$) with an odor resembling chloroform. It finds use as an intermediate, particularly in drug manufacture.

2. Physical and Chemical Properties

Molecular weight: 67.09
Boiling point: 130 to 131°C.
Density: 0.9691 ml. (20/4°C.)
Vapor density: 2.31 (air = 1)
Solubility: slightly soluble in water
Flash point: 102°F.

3. *Physiological Response*

Scanty evidence indicates that pyrrole is generally of low toxicity, both systemically and locally.[13] Injection in mammals produces discoloration of the urine. Intraperitoneal injection of large doses in dogs is said to produce convulsions and liver injury, although this does not occur with other routes of administration.[14] Experience suggests that ordinary handling precautions should be adequate to prevent injury.

PYRROLIDINE, NHCH$_2$CH$_2$CH$_2$CH$_2$ (Tetramethylenimine, Tetrahydropyrrole)

1. *Source and Uses*

Pyrrolidine is a moderately alkaline liquid (pK$_b$ = 11.10) with an ammonia-like odor. It is formed by the reduction of pyrrole. It occurs in tobacco and carrot leaves and is used as a synthetic intermediate.

2. *Physical and Chemical Properties*

Molecular weight: 71.12
Boiling point: 88.5 to 89°C.
Vapor density: 2.45 (air = 1)
Vapor pressure: 128 mm. Hg, 139°C.
Solubility: miscible in water; soluble in alcohol, ether, and chloroform
Flash point: 37°F.

3. *Physiological Response*

Because of its stronger alkalinity, pyrrolidine could be expected to be more irritating to skin and mucous membranes than pyrrole; however, this does not appear to have been studied. Small intravenous doses (less than 1 mg./kg.) in dogs and cats produce an increase in blood pressure and respiratory rate. This pressor activity is reduced by ganglionic blocking agents or sympathectomy. These actions are essentially identical to those observed with similar doses of piperidine.[15] No industrial injuries have been reported from pyrrolidine. Ordinary precautions used for handling any caustic liquid should be observed.

2-AMINOTHIAZOLE, SCH:NCH:CH (2-Thiazolylamine)

1. *Uses*

Aminothiazole came into large use for the synthesis of the sulfa drug, sulfathiazole. It also had brief and limited use as an antithyroid medication. It is a

[13] T. Sollmann, *A Manual of Pharmacology*, 8th ed., Saunders, Philadelphia, 1957.
[14] L. T. Fairhall, *Industrial Toxicology*, 2nd ed., Williams & Wilkins, Baltimore, 1957.
[15] M. F. Lockett, *Brit. J. Pharmacol.*, **4**, 111 (1949).

white to yellowish crystalline solid, which sublimes readily and slowly changes and darkens on exposure to air.

2. *Physical and Chemical Properties*

Molecular weight: 100.14
Melting point: 90°C.
Solubility: slightly soluble in water, alcohol, and ether

3. *Physiological Response*

Deichmann and co-workers[16] found that single oral doses of 2-aminothiazole were moderately toxic and produced gastrointestinal tract irritation and liver damage. The approximate lethal dose was 0.12 g./kg. for guinea pigs and cats. The oral LD_{50} for rats was 0.48 g./kg. Repeated administration of 0.15 g./kg. in 30 oral doses produced some mortalities and liver damage in rabbits. This represents a relatively low degree of accumulation since the approximate lethal dose in rabbits was found to be 0.28 g./kg. Repeated inhalation exposures on 43 days at a concentration of 0.2 mg./liter caused no apparent ill effects in 6 rats and 2 rabbits, whereas 5 of 8 guinea pigs died. Five guinea pigs exposed for a similar period to 0.025 mg./liter developed liver damage with hepatocellular degeneration and fatty change and 2 of 5 died.[16]

Aminothiazole is also known to cause reduced thyroid activity and adenomatous hyperplasia in laboratory animals. Thiourea, which has antithyroid activity, has been detected in the urine of rabbits injected with aminothiazole.

Watrous[17] reports that symptoms developed in a number of workmen exposed to air concentrations of aminothiazole ranging from 0.1 to 3.0 mg./cu. ft. (0.0036 to 0.11 mg./liter) during the production of sulfathiazole. These consisted of brown staining of the skin, transient brown discoloration of the urine during periods of exposure, anorexia, and occasionally nausea and vomiting, and contact dermatitis of the primary irritant type. A smaller number of individuals developed an unusual illness of sudden onset, which lasted several weeks and was characterized by itching and joint or muscle pain. In 2 cases the itching was associated with urticaria and in 1 with a maculopapular rash. Watrous pointed out the resemblance of the symptoms to serum sickness and felt that this represented a special sensitivity to aminothiazole.[17] The use of better equipment for handling the material resulted in elimination of these occurrences. Except for the occurrence of goiters and hypothyroidism in men manufacturing 2-aminothiazole in France, other plants manufactured 2-aminothiazole without observing injury.[16]

[16] W. B. Deichmann, K. V. Kitzmiller, F. F. Heyroth, and S. Witherup, *J. Ind. Hyg. Toxicol.*, **30**, 71 (1948).

[17] R. M. Watrous, *Ind. Med.*, **12**, 832 (1943).

AMINOTRIAZOLE, NHN:CH(NH₂)N:CH (Amizol)

1. Uses

3-Amino-1,2,4-triazole is a crystalline solid that has been used as a herbicide, cotton defoliant, and as a reagent in photography. It proved particularly effective as a weed-control agent in the cranberry industry.

2. Physical and Chemical Properties

Molecular weight: 84.08

Solubility: soluble in water, alcohol, and chloroform; insoluble in ether and acetone

3. Physiological Response

It has been shown that 3-amino-1,2,4-triazole inhibits the enzyme catalase in the liver and kidney of experimental animals. One g./kg. intraperitoneally in rats reduces liver catalase activity to approximately 11 per cent of the control values.[18] However, it has quite low acute toxicity for experimental animals, the single dose LD_{50} being in excess of 10 g./kg. Much lower doses than those necessary to produce marked inhibition of liver and kidney catalase activity produce sharp reduction in the uptake of I^{131} by the thyroid gland, apparently by inhibition of the formation of organically bound iodine; intraperitoneal doses of approximately 5 mg./kg. were effective in rats. Rats fed aminotriazole in their drinking water (12 to 14 mg./day) developed thyroid goiter with considerable increase in thyroid weight and microscopic evidence of colloid loss with glandular hyperplasia.[19] Increased thyroid weight resulted from feeding of 50, 250, and 1250 p.p.m. in the drinking water for 106 days. Rats fed aminotriazole in their diet for two years at levels of 10, 50, and 100 p.p.m. developed abnormal thyroid growth and thyroid tumors.[20] Because of this finding, cranberries that had been improperly treated with the herbicide and, therefore, were contaminated were removed from the market. No human injuries have been reported.

SIX-MEMBERED RINGS WITH ONE HETERO ATOM

I. General Considerations

A. PROPERTIES, MANUFACTURE, AND USES

Pyridine is the parent compound of this series. It is a fully unsaturated 6-membered ring containing 1 nitrogen and 5 carbons. Piperidine is the corresponding fully saturated compound. For almost every type of known benzene compound there is an analogous compound in the pyridine series so that the

[18] W. G. Heim, D. Appleman, and H. T. Pyfrom, *Science*, **122**, 693 (1955).

[19] N. M. Alexander, *J. Biol. Chem.*, **234**, 148 (1959).

[20] U. S. Dept. Health, Educ., and Welfare, News Release, November 17, 1959.

number of potential pyridine derivatives is very extensive. Only a few of these can be considered here.

Pyridine was first isolated from bone oil along with the picolines (methyl-pyridines) and lutidines (dimethylpyridines). These pyridine bases are found in the distillates of many nitrogenous organic materials but are principally isolated from coal tar. They are stable, colorless liquids with characteristic unpleasant odors. The stability of pyridine and its wide solvent power make it a very useful chemical. Pyridine behaves like a tertiary amine and a weak base. The pyridines in general exhibit properties characteristic of aromatic compounds. Piperidine, on the other hand, is a much stronger base and it behaves more like typical aliphatic secondary amines. The pyridine ring occurs naturally in some alkaloids such as nicotine and trigonelline and in the vitamins, nicotinic acid and pyridoxine. Piperidine is frequently found in nature, for example, in the alkaloids of pepper and in coniine. Whereas pyridine has a very wide use, the applications for the picolines, lutidines, collidines, and piperidines are more limited. They are mainly used as special solvents or intermediates in chemical syntheses and in the manufacture of pharmaceuticals and resins.

B. ABSORPTION, EXCRETION, AND METABOLISM

Pyridine and its alkyl derivatives are absorbed from the gastrointestinal tract, intraperitoneal cavity, and lungs. Peritoneal absorption is apparently only slightly more rapid and complete than gastrointestinal absorption since the approximate intraperitoneal LD_{50}'s are generally slightly lower than the acute oral LD_{50}'s. In general the bases are rapidly absorbed through intact skin. Piperidine is well absorbed from the gastrointestinal tract but its skin absorption is less than that seen with the pyridine derivatives.

The metabolic fate of the pyridines is not completely known. Hydroxylation, N-methylation, oxidation, and conjugation reactions have been identified.[21] This is more fully discussed under the specific compounds.

C. SYMPTOMS IN ANIMALS

Pyridine has a narcotic action when administered to experimental animals. Effective doses by any route produce weakness, ataxia, unconsciousness, and salivation, but convulsions are uncommon. Animals that survive the acute episode generally recover. With exposure to pyridine vapors there are symptoms of moderate mucous membrane irritation. Pyridine has relatively low acute toxicity. Most observers have found that the oral LD_{50} for small animals exceeds 1 g./kg.

Pyridine and a series of its derivatives have been studied by Fassett and Roudabush.[22] These were range-finding studies intended to indicate the relative

---

[21] R. T. Williams, *Detoxication Mechanisms*, 2nd ed., Wiley, New York, 1959.

[22] D. W. Fassett and R. L. Roudabush, *Toxicity of Pyridine Derivatives with Relationship to Chemical Structure*, presented to the American Industrial Hygiene Association, Los Angeles, 1953.

TABLE 6
Acute Toxicity of Pyridine Derivatives for Animals[22]

Chemical	Approximate LD$_{50}$, rats, g./kg.		Approximate LD$_{50}$, mice, g./kg.		Approximate LD$_{50}$, guinea pigs, ml./kg., percutaneous	Inhalation toxicity, rats		
	Oral	Intraperitoneal	Oral	Intraperitoneal		Conc., p.p.m.	Time, hr.	Mortality
Pyridine	0.8–1.6	0.8–1.6	0.8–1.6	0.8–1.6	1.0–2.0	4000	6	3/3
2-Methyl	0.4–0.8	0.4	0.4–0.8	0.2–0.4	1.0–2.0	1700	6	0/3
3-Methyl	0.4–0.8	0.1–0.2	0.8–1.6	0.4–0.8	1.0–2.0	8700ᵃ	2	3/3
4-Methyl	0.8	<0.1	0.4–0.8	0.2–0.4	<0.5	8000ᵃ	2	3/3
2,3-Dimethyl	0.4	0.1–0.4	0.4–0.8	0.1–0.2	<2.5	3800ᵃ	2.3	3/3
2,4-Dimethyl	0.2–0.4	0.2–0.4	0.4–0.8	0.2–0.4	1.0–2.5	650	6	0/3
2,6-Dimethyl	0.4–0.8	0.1	0.2–0.4	0.1–0.2	2.5–5.0	7500ᵃ	1.2	3/3
2,4,6-Trimethyl	0.4	0.05–0.1	0.2–0.4	0.1–0.2	1.0–2.0	2500ᵃ	2	3/3
5-Ethyl-2-methyl	0.8–1.6	0.2–0.4	0.8–1.6	0.1–0.2	2.5–5.0	1800ᵃ	3.7	6/6
2-n-Amyl	0.1–0.2	0.1	0.2–0.4	0.1–0.2	5.0–10	400ᵃ	6	0/3
4-n-Amyl	0.2–0.4	0.2	0.2–0.4	0.1–0.2	5.0–10	300ᵃ	6	0/3
2-Vinyl	0.1–0.2	0.1–0.2	0.4–0.8	0.2–0.4	<0.5	160	6	0/3
4-Vinyl	0.1–0.2	0.05	0.2–0.4	0.1–0.2	<0.5	150	6	0/3
2-Amino	0.2	0.025	0.05	0.025–0.05	0.5ᵇ			
2-Amino-3-methyl	0.1	0.01–0.025	0.025–0.05	0.025–0.05	0.2–0.4ᵇ	1200ᶜ	6	3/3
2-Amino-4-methyl	0.2	0.025–0.05	0.1–0.2	0.05–0.1	0.5–1.0ᵇ			
2-Amino-5-methyl	0.2	0.1	0.2–0.4	0.1–0.2	0.4–0.5ᵇ			
2-Amino-6-methyl	0.1	0.025	0.5–1.0	0.001–0.025	0.2–0.4ᵇ			

Note: Where a range is given, all animals died at the high dose and all survived at the lower dose.
ᵃ Concentrated vapor produced by bubbling air through compound at room temperature.
ᵇ Diluted solution used; estimate of LD$_{50}$ range if undiluted.
ᶜ Compound heated to 55°C.

order of magnitude and type of the toxicity of the various derivatives. Some of the results are presented in Table 6. All the compounds were administered as the bases. For oral and intraperitoneal doses either the undiluted liquid or aqueous solutions were used. The skin absorption studies in guinea pigs were done by applying the undiluted liquids or aqueous solutions (as indicated in Table 6) to the intact skin under a rubber cuff, which was left in place for 24 hours. All animals were observed for 2 weeks. The doses given indicate the approximate LD_{50}. Where a range is given, all animals died at the high dose and all survived at the lower dose. Where a single figure is given, approximately one half the animals died at that dose. The vapor concentrations are calculated values.

The determinations of acute oral or intraperitoneal toxicity in rats or mice show that the 2-amino and vinyl pyridines are generally more toxic than the other derivatives. The amyl pyridines and collidine are intermediate in their toxicity. The lutidines and picolines show decreasing toxicity and pyridine is the least toxic. Acute doses of the 2-amino and 2-amino methyl pyridine derivatives produce symptoms quite different from pyridine. There is central nervous system stimulation and irritability. Abrupt clonic convulsive seizures, which are almost always fatal, follow higher doses. The 2-amino pyridines also cause the interesting phenomenon of chromodacryorrhea by stimulation of Harder's gland to release red-colored protoporphyrin.[22a] The vinyl pyridines also cause convulsions but these are less dramatic and are slower in onset than is the case with the amino derivatives. The alkyl derivatives, picolines, lutidines, and collidines, cause symptoms resembling those observed with pyridine. These are largely nonspecific and include weakness, ataxia, diarrhea, and unconsciousness. Delayed deaths following a single dose are uncommon and survivors generally gain weight normally.

The majority of the derivatives tested are primary irritants and penetrate the intact guinea pig skin readily. The amino pyridines, the vinyl pyridines, and 4-picoline penetrate the skin most rapidly and the amyl derivatives and 5-ethyl-2-methyl pyridine are least readily absorbed. Most of these materials cause rather intense skin and eye irritation. Attempts to induce skin sensitization in guinea pigs have generally been unsuccessful except in the case of 4-vinyl pyridine. The amino derivatives, the 2- and 4-vinyl derivatives, and collidine appear to be the most potent primary irritants of the series.

The results of acute inhalation studies appear in Table 7. Rats were exposed to "saturated" vapor concentrations obtained by bubbling dry air through the unheated liquids. The first three compounds listed caused no deaths, principally because of their rather low rate of volatilization. Of the more volatile substances, the most toxic were 5-ethyl-2-methyl pyridine and 4-vinyl pyridine. Collidine, the lutidines, and the picolines exhibited decreasing vapor toxicity and pyridine was the least toxic. The toxicity and irritant properties of the derivatives substituted

[22a] E. J. Towbin, P. E. Fanta, and H. C. Hodge, *Proc. Soc. Exptl. Biol. Med.*, **60**, 228 (1945).

at the 4 position (methyl, ethyl, amyl, vinyl) tend to be greater than the corresponding derivatives substituted at the 2 position. This is true for inhalation exposures as well as other routes of administration. The exposed animals showed the same symptoms as observed with other routes of administration plus the additional evidence of respiratory tract irritation.

Piperidine behaves chemically like a typical aliphatic secondary amine and, like this group of compounds, it exhibits strong primary irritant properties for

TABLE 7

Concentrated Vapor Inhalation of Pyridine Derivatives—Rats[c]

Chemical	Calculated concentration, p.p.m.	Time for 100 per cent mortality, hr.
4-n-Amylpyridine	300	6.0[a]
2-n-Amylpyridine	400	6.0[a]
2-Amino-3-methylpyridine	650	6.0[a]
4-n-Propylpyridine	1,000	6.0[b]
5-Ethyl-2-methylpyridine	1,700	3.7
4-Vinylpyridine	2,000	2.0
Collidine	2,500	2.0
4-Ethylpyridine	2,500	5.0
2,3-Lutidine	3,800	2.3
2,4-Lutidine	4,300	2.0
2-Ethylpyridine	5,400	3.0
2-Vinylpyridine	5,500	1.5
2,6-Lutidine	7,500	1.2
4-Picoline	8,000	2.0
3-Picoline	8,700	2.2
2-Picoline	15,400	1.5
Pyridine	23,200	1.5

[a] No deaths.

[b] One third died 30 minutes after exposure.

[c] D. W. Fassett and R. L. Roudabush, *Toxicity of Pyridine Derivatives with Relationship to Chemical Structure,* presented to the American Industrial Hygiene Association, Los Angeles, 1953.

the skin, mucous membranes, and eyes, and a relatively high degree of acute toxicity. Effective doses produce weakness, ataxia, respiratory distress, and convulsions, as well as local irritation. The simple alkyl piperidine derivatives also show a high degree of toxicity and local irritation.

Nicotine, the N-methyl pyrrolidine derivative of pyridine, is well known to be highly toxic. The acute LD_{50} in small animals by the oral route is approximately 50 mg./kg. and the injected lethal dose is on the order of a few mg./kg. With doses in the lethal range, animals may exhibit central nervous system stimulation with convulsions and respiratory stimulation followed by depression with paralysis and respiratory arrest.

D. PHARMACOLOGY AND MODE OF ACTION

The major actions of pyridine and its simple alkyl derivatives are local irritation and nervous system depression with narcosis. Little is known about their mechanism of action. The amino pyridines produce convulsions somewhat like those caused by strychnine. They also cause contraction of skeletal muscle and smooth muscle producing vasoconstriction and an increase in blood pressure, probably through direct stimulation.[23] Piperidine and 2-propylpiperidine are also active pressor agents when administered intravenously in dogs and cats in doses of 5 to 10 mg./kg. On the other hand, 2,3- and 2,4-dimethyl piperidine have no pressor activity, but depress ganglionic activity somewhat like atropine. Piperidine stimulates respiration initially, causes muscular contraction and, after larger doses, depresses parasympathetic and sympathetic ganglion, resembling in these effects the actions of nicotine. The major action of nicotine is that of primary transient stimulation and a secondary depression of the central nervous system and all sympathetic and parasympathetic ganglia through its direct action on the ganglion cells. It appears that nicotine initially stimulates by depolarizing ganglionic cells and then prevents transmission by competitive blockade of acetylcholine.[24]

E. PATHOLOGY IN ANIMALS

Most interest has been directed toward the transient actions of these pharmacologically active materials, and there has been little study of the cellular damage produced by repeated doses. Emphysema, chronic bronchitis, and fatty degeneration of the liver and kidneys have been observed in animals repeatedly exposed to pyridine vapor.[25] Baxter[26] found that pyridine incorporated into a choline- and casein-deficient diet of rats caused chronic liver and kidney damage with fatty changes and fibrosis. Acute cell necrosis was also observed. Increasing the choline and casein content of the diet resulted in some reduction in the chronic changes.

F. EFFECTS IN MAN

Despite the relatively large industrial use and some medicinal use of pyridine, reports of injurious effects in man have been relatively uncommon. Although repeated administration of oral doses of 0.3 to 1.5 cc. in man generally does not result in toxic effects, 2 cases of kidney and liver injury with 1 death have been reported in subjects receiving 1.8 to 2.5 cc. of pyridine a day for periods up to 2 months. Cases with central nervous system symptoms including headache, dizziness, fatigue, and gastrointestinal disturbances have been reported from exposures to pyridine

[23] F. N. Fastier and M. A. McDowall, *Australian J. Exptl. Biol. Med. Sci.*, **36**, 365 (1958).

[24] W. D. M. Patton and W. L. M. Perry, *J. Physiol.*, **119**, 43 (1953).

[25] E. Browning, *Toxicity of Industrial Organic Solvents*, Chemical Publishing, New York, 1953.

[26] J. H. Baxter, *Am. J. Pathol.*, **24**, 503 (1948).

[27] J. Teisinger, *J. Ind. Hyg. Toxicol.*, **30**, 58 (1948) (abstr.).

[28] A. Meyer, *Zutschift Unfallmed u. Berufskrankh.*, **43**, 144 (1950).

vapor.[25,27,28] Transient symptoms of nausea, headache, insomnia, nervousness, and low back or abdominal discomfort with urinary frequency have been observed in persons with repeated exposures to pyridine in concentrations varying from 15 to 330 p.p.m.[22] Three separate cases of human intoxication from exposure to 2-amino pyridine have been reported.[29,30] One employee who spilled 2-amino pyridine on his clothing during distillation subsequently died. Several cases of transient mucous membrane irritation, systemic symptoms, skin burns, and skin sensitization from exposure to 2- and 4-vinyl pyridine have been observed during manufacture and in laboratory pilot-plant use.[31] No cases of human injury from picolines, lutidines, or collidines have been reported.

Many fatal cases of nicotine poisoning have occurred, usually as a result of accidental or suicidal ingestion of insecticides. A few cases of intoxication have been attributed to spraying nicotine insecticides and to exposures during nicotine extraction.[13]

G. WARNING PROPERTIES, DETERMINATION IN AIR, AND THRESHOLD LIMIT VALUES

Pyridine can be estimated in air by collection in a known volume of standard $0.1N$ sulfuric acid and back titration of the excess with standard alkali. This procedure would also be applicable to determinations of piperidine in the absence of interfering bases. Pyridine can be detected in smaller concentrations by collection in alcohol and determination by ultraviolet spectrophotometry at 256 mμ.[32] The other simple pyridine derivatives also can easily be measured by ultraviolet absorption methods. Infrared spectroscopy has been used for determination of ethyl and methyl pyridines.[33]

Pyridine, the methyl pyridines, and amino pyridines have strong, characteristic, and generally unpleasant odors, which are readily recognizable in low concentrations. Pyridine, for example, is easily detectable at less than 1 p.p.m. and it is very disagreeable to most individuals at 30 p.p.m. However, the odors are generally an unreliable guide at concentrations exceeding a few p.p.m., since acclimatized individuals can readily learn to tolerate obnoxious concentrations and olfactory fatigue occurs quickly.

Threshold limit values have been suggested for pyridine (10 p.p.m., 30 mg./cu. meter) and nicotine (0.5 mg./cu. meter.).[11] Although no threshold limit values have been established for the other members of the series, the information available on their acute toxicity indicates the need for more than ordinary care in avoiding skin and eye contact and inhalation.

[29] R. M. Watrous and H. N. Schulz, *Ind. Med. and Surg.,* **19**, 317 (1950).

[30] L. W. Spolyar, *Ind. Health Monthly,* **11**, 119 (1951).

[31] D. W. Fassett and R. L. Roudabush, Laboratory of Industrial Medicine, Eastman Kodak Co., unpublished observations.

[32] American Industrial Hygiene Association Hygienic Guide, Pyridine, *Am. Ind. Hyg. Assoc. Quart.,* **18**, 372 (1957).

[33] E. A. Coulson and J. L. Hales, *Analyst,* **78**, 114 (1953).

TABLE 8

Physical and Chemical Properties of Pyridine Derivatives

Name	Formula	Molecular weight	Physical state	B. p., °C.	M. p., °C.	Density, g./ml.	Solubility[a] In water, g. per 100 ml.	In other
Pyridine	C_5H_5N	79.10	Liquid	115.3		0.982 (20/4°C.)	∞	∞ in alcohol, ether
2-Methylpyridine (2-picoline)	$(CH_3)C_5H_4N$	93.12	Liquid	128		0.950 (15/4°C.)	V. sol.	∞ in alcohol, ether
3-Methylpyridine (3-picoline)	$(CH_3)C_5H_4N$	93.12	Liquid	143.5		0.9613 (15/5°C.)	∞	∞ in alcohol, ether
4-Methylpyridine (4-picoline)	$(CH_3)C_5H_4N$	93.12	Liquid	143.1		0.9571 (15/5°C.)	∞	∞ in alcohol, ether
2,4-Dimethylpyridine (2,4-lutidine)	$(CH_3)_2C_5H_3N$	107.15	Liquid	157.1		0.9493 (0/4°C.)	20	Sol. in alcohol, ether
2,6-Dimethylpyridine (2,6-lutidine)	$(CH_3)_2C_5H_3N$	107.15	Liquid	143		0.942 (0/4°C.)	∞ (cold)	Sol. in alcohol, ether
2,4,6-Trimethylpyridine (collidine)	$(CH_3)_3C_5H_2N$	121.8	Liquid	172		0.917 (20/4°C.)	3.5 (20°C.)	Sol. in alcohol, ∞ in ether
5-Ethyl-2-methylpyridine (aldehydine)	$(C_2H_5)(CH_3)C_5H_3N$	121.2	Liquid	178.3		0.9215 (20/20° C.)	1.2 (20°C.)	Sol. in alcohol, ether
2-n-Amylpyridine	$(C_5H_{11})C_5H_4N$	149.24	Liquid	82–83 (6 mm. Hg)				
4-n-Amylpyridine	$(C_5H_{11})C_5H_4N$	149.24	Liquid	93–94 (6 mm. Hg)				
2-Vinylpyridine	$(CH_2{:}CH)C_5H_4N$	105.13	Liquid	158–159 (decomp.)		0.9746 (20/20° C.)	2.5 (20°C.)	V. sol in ether
4-Vinylpyridine	$(CH_2{:}CH)C_5H_4N$	105.13	Liquid	121 (150 mm. Hg)		0.988 (20/20°C.)	2.91 (20°C.)	
2-Aminopyridine	$(NH_2)C_5H_4N$	94.11	Solid	204	57.5		Sol.	V. sol. in alcohol
2-Amino-3-methylpyridine	$(CH_3)(NH_2)C_5H_4N$	108.08	Liquid	221.1	33.3			
2-Amino-4-methylpyridine	$(CH_3)(NH_2)C_5H_4N$	108.08	Solid	230.9	99			

[a] ∞ = complete; v. sol. = very soluble; sol. = soluble.

II. Specific Compounds

Because the information on the methyl, ethyl, and amyl pyridines is scanty, they are not specifically discussed below. However, some information on their toxicity for animals appears in Tables 6 and 7, and their physical and chemical properties are listed in Table 8.

PYRIDINE, C_5H_5N

1. Source and Uses

Pyridine is a colorless basic liquid ($pK_b = 8.81$) with a characteristic unpleasant odor. Most of the commercial pyridine is derived from crude coal tar although it can be synthesized from aliphatic compounds. In 1957 some 1,820,000 pounds of pyridine were produced and sold in the United States. It is used as a solvent in many chemical processes and syntheses and as a denaturant for alcohol. It finds use in the dye industry, in manufacture of some explosives, pharmaceuticals, textiles, and paints.

2. Physical and Chemical Properties

Molecular weight: 79.1
Boiling point: 115.3°C.
Specific gravity: 0.982 (20/4°C.)
Vapor density: 2.72 (air $= 1$)
Vapor pressure: 20 mm. Hg, 25°C.
Flash point: 68°F. (closed cup)
Solubility: infinitely soluble in water, alcohol, ether, and chloroform
1 mg./liter $\backsim$ 309 p.p.m. and 1 p.p.m. $\backsim$ 3.23 mg./cu. meter

3. Physiological Response

Pyridine is absorbed from the gastrointestinal tract, through the skin, and by inhalation. Part of the absorbed pyridine is excreted in the urine unchanged and a smaller portion is methylated at the N position. The metabolite, N-methyl pyridinium hydroxide, when injected as N-methyl pyridinium chloride exhibited much higher acute toxicity for mice than pyridine, although it had lesser toxicity on chronic administration.[34] The major effects produced in animals by administration of pyridine by any route are anesthesia and irritation from acute doses, and liver and kidney injury from repeated feeding. The results of toxicity studies in animals are summarized in Table 9.

Most of the effects that have been observed in man are transient and center on the central nervous system and the gastrointestinal tract. These have most often resulted from repeated or intermittent exposures to pyridine vapor. The symptoms

---

[34] J. H. Baxter and M. F. Mason, *J. Pharmacol. Exptl. Therap.*, **91**, 350 (1947).

TABLE 9

Toxicity of Pyridine—Animals

Species	Route	Dose	Response
Rat	Oral	0.8–1.6 g./kg.	Approx. LD_{50}[a,b]
	Oral	1.58 g./kg.	LD_{50}[c]
Mouse	Oral	0.8–1.6 g./kg.	Approx. LD_{50}[a,b]
	Intraperitoneal	0.8–1.6 g./kg.	Approx. LD_{50}[a,b]
	Intraperitoneal	1.2 g./kg.	MLD_{50}[d]
Guinea pig	Intraperitoneal	0.87 g./kg.	Lethal dose[e]
	Skin	1–2 ml./kg.	Approx. LD_{50}[b]
Rat	Inhalation	23,000 p.p.m., 1.5 hrs.	$^6/_6$ died[b]
		3,600 p.p.m., 6 hrs.	$^2/_3$ died[b]
		4,000 p.p.m., 6 hrs.	$^5/_6$ died[c]
Guinea pig	Eye	1 drop, undiluted	Corneal damage[b]
Rabbit	Eye	40% solution	Corneal necrosis[c]
Rat	Oral	0.1% in diet	Deaths, liver and kidney injury[d,f]

[a] All died at 1.6 g./kg. and all survived at 0.8 g./kg.

[b] D. W. Fassett and R. L. Roudabush, *Toxicity of Pyridine Derivatives with Relationship to Chemical Structure*, presented to the American Industrial Hygiene Association, Los Angeles, 1953.

[c] H. F. Smyth, C. P. Carpenter, and C. S. Weil, *Arch. Ind. Hyg. Occupational Med.*, **4.** 119 (1951).

[d] J. H. Baxter and M. F. Mason, *J. Pharmacol. Exptl. Therap.*, **91**, 350 (1947).

[e] E. Browning, *Toxicity of Industrial Organic Solvents*. Chemical Publishing, New York. 1953.

[f] J. H. Baxter, *Am. J. Pathol.*, **24**, 503 (1948).

include headache, dizziness or giddiness, nervousness, insomnia, mental dullness, nausea, and anorexia. In some cases, lower abdominal or back discomfort with urinary frequency has been observed. These transient symptoms, without associated evidence of liver or kidney damage, have occurred in individuals exposed to pyridine concentrations averaging 125 p.p.m., 4 hours a day, for 1 to 2 weeks.[22] Pollock and co-workers have reported serious liver and kidney injury in 2 individuals (one of whom died) after administration of 1.8 to 2.5 cc. of pyridine a day by mouth for up to 2 months in the treatment of epilepsy.[35] The possibility of more permanent central nervous system injury with persistent symptoms is suggested by scattered case reports that do not allow critical appraisal.

4. *Warning Properties and Threshold Limit Value*

The characteristic odor of pyridine is detectable at less than 1 p.p.m. and becomes objectionable to unacclimatized individuals at 10 p.p.m. The odor and mild irritant properties of pyridine do not present enough warning to prevent exposures to concentrations capable of producing symptoms. Maintenance of atmospheric concentrations of pyridine below the recommended threshold limit

L. J. Pollack, I. Findelman, and A. J. Arieff, *Arch. Internal Med.*, **71**, 95 (1943).

value of 10 p.p.m. will prevent systemic symptoms but complaints of unpleasant odor may occur. Because pyridine is absorbed through the skin and may cause irritation, skin contact should be avoided.

2-AMINOPYRIDINE, $NH_2C_5H_4N$ (α-aminopyridine, α-pyridylamine)

1. Uses

2-Aminopyridine is a low melting solid that has been chiefly used in the manufacture of pharmaceuticals. It can be prepared in several ways, the most important in which is direct amination of pyridine with sodium amide.

2. Physical and Chemical Properties

Molecular weight: 94.11
Melting point: 57.5°C.
Boiling point: 204°C.
Solubility: soluble in water, alcohol, benzene, and ether

3. Physiological Response

The chemical properties of the NH_2 group of both 2- and 4-aminopyridine resemble those of an amide or amidine rather than those of an aromatic amine. 3-Aminopyridine, on the other hand, shows chemical properties typical of an aromatic amine. 2- and 4-Aminopyridine are both strong bases, whereas 3-aminopyridine is somewhat less basic. Intravenous injection of 1 mg./kg. of 2-aminopyridine in cats causes an increase in blood pressure and respiratory rate with symptoms of central nervous system stimulation and muscle twitching. Direct application to blood vessels and skeletal muscle results in muscular stimulation and contraction. Lethal doses in experimental animals produce excitement with tremors progressing to convulsions, tetany, and death. The convulsions occur rapidly and are almost always fatal. The same responses are induced by 4-aminopyridine and to a somewhat lesser degree by 3-aminopyridine. The minimum doses for convulsions by intraperitoneal injection in mice are: 2-aminopyridine, 15 mg./kg.; 4-aminopyridine, 3 mg./kg.; 3-aminopyridine, 10 mg./kg. The corresponding intraperitoneal LD_{50}'s in mice are: 35, 9, and 28 mg./kg.[23] 2-Aminopyridine is readily absorbed through the skin and produces convulsive deaths by this route. The 2-aminomethylpyridines also exhibit a high order of toxicity, rapid skin absorption, and moderate skin and eye irritation, and produce striking central nervous system stimulation and convulsions. The amino derivatives of pyridine do not produce significant methemoglobinemia in rats.[22]

Three cases of 2-aminopyridine intoxication in man have been reported. Watrous and Schulz[29] reported the case of a chemical operator engaged in milling 2-aminopyridine without protective equipment; he developed severe pounding headache, nausea, flushing of the extremities, and elevated blood pressure, but recovered fully the next day. Subsequent determinations on this operation revealed air concentrations of 0.02 mg./liter (5.2 p.p.m.). They also cite a more serious intoxication observed by Schmid in which the patient had severe headache and weakness, followed by convulsions and a stuporous state that lasted several days.[29] Fatal intoxication occurred in a chemical operator who spilled 2-aminopyridine during distillation.[30] He continued to work in contaminated clothing for $1^1/_2$ hours. Two hours later he developed dizziness, headache, respiratory distress, and convulsions that progressed to respiratory failure and death. In view of the percutaneous toxicity in guinea pigs it is probable that skin absorption was important in this case.

2-Aminopyridine can be determined by the colorimetric method of Watrous and Schulz.[29] Ultraviolet spectroscopy also can be utilized.

It is evident that acute intoxication can occur from inhalation of the dust or vapor at relatively low concentrations, or by skin absorption following direct contact. Most operations will require controls to prevent inhalation. In cases of accidental skin contact, prompt, thorough skin cleansing and clothing change are necessary.

VINYLPYRIDINES

1. Uses

The vinylpyridines are reactive liquid pyridine derivatives with the vinyl group at the 2, 3, or 4 position. They are monomers for polyvinylpyridine polymers and have been investigated for use in synthetic rubbers, photographic film, ion exchange resins, and other polymers as well as pharmaceuticals. Commercial 2-vinylpyridine and 4-vinylpyridine are supplied with small amounts of polymerization inhibitors added. Both of these materials are colorless, volatile liquids with a pungent unpleasant odor.

Physical and Chemical Properties

	2-Vinylpyridine	4-Vinylpyridine
Formula	$C_5H_4NCHCH_2$	$C_5H_4NCHCH_2$
Molecular weight	105.13	105.13
Boiling point	110°C., 150 mm. Hg	121°C., 150 mm. Hg
Density	0.9746 g./ml., 20°C.	0.988 g./ml. (20/20°C.)
Vapor pressure	10 mm. Hg, 44.5°C.	
Solubility	Approx. 2.5 g. per 100 ml. of water at 20°C., freely soluble in common organic solvents	2.91 g. in 100 ml. of water at 20°C.

Conversion units 1 mg./liter $\approx$ 233 p.p.m. and 1 p.p.m. $\approx$ 4.3 mg./cu. meter

2. Physiological Response

2-Vinylpyridine and 4-vinylpyridine are intermediate between the aminopyridines and pyridine in their acute toxicity for rats and mice (Tables 6 and 7).

They are absorbed from the gastrointestinal tract, the skin, and respiratory tract in these animals. Absorption by these routes results in weakness, ataxia, vasodilatation, respiratory distress, and convulsions. In addition to these symptoms, exposure to the vapor of 2- and 4-vinylpyridine results in nasal and eye irritation with accelerated respiration and respiratory distress. Instillation of the undiluted liquids in rabbit eye causes moderately severe eye irritation. Skin sensitization was produced in some guinea pigs by both of these materials. The inhalation toxicity of 4-vinylpyridine is greater than that of 2-vinylpyridine, and 4-vinylpyridine is definitely more irritating to mucous membranes.[22]

Smyth and co-workers have reported range-finding toxicity studies on 5-ethyl-2-vinylpyridine.[36] It appears to be less toxic and less irritating than 2-vinylpyridine. The oral LD_{50} in rats was 1.23 g./kg. and the dermal LD_{50} in rabbits was 0.89 g./kg. Saturated vapor caused no deaths in 4 hours, whereas a similar exposure period at 8000 p.p.m. resulted in deaths of 2 of 6 rats.

Brief exposures to undetermined concentrations of 2- and 4-vinylpyridine during laboratory use of these materials have caused eye, nose, and throat irritation, and headache, nausea, nervousness, and anorexia. The systemic symptoms are mild and transient resembling those observed with pyridine exposures, although apparently produced by lower concentrations. Direct skin contact with the liquid results in burning pain, followed by fairly severe skin burns in spite of immediate attempts to cleanse the skin. The burns develop a reddish brown color that disappears in the course of about a month. Skin sensitization has been observed with both 2- and 4-vinylpyridine.[22]

3. *Warning Properties, Determination in Air, and Threshold Limit Value*

The odor of 2-vinylpyridine can be detected at levels of approximately 0.3 p.p.m. and is quite strong at 0.5 p.p.m. At higher unmeasured concentrations, both compounds have a very unpleasant nauseating odor. Concentrations can be determined in air by collection in distilled water and measurement of their ultraviolet absorption spectrum. Experience indicates that the unpleasant odor can be tolerated at concentrations that will produce acute symptoms and irritation. No threshold limit values have been proposed for these materials. The acute toxicity information on animals and the experience with intermittent exposures in man show that more than ordinary care is needed to avoid skin contact and inhalation.

NICOTINE, $CH:NCH:CHCH:CCH(CH_2)_3NCH_3$

(3-(1-Methyl-2-pyrrolidyl)pyridine, 1-Methyl-2-γ-(3-pyridyl)-pyrrolidine)

1. *Source and Uses*

Nicotine is a volatile, strongly alkaline liquid that becomes brownish in color and develops a pyridine or tobaccolike odor on exposure to air. The alkaloid is

[36] H. F. Smyth, C. P. Carpenter, C. S. Weil, and U. C. Pozzani, *Arch. Ind. Hyg. Occupational Med.*, **10**, 61 (1954).

obtained from the stems and leaves of tobacco plants where it occurs in concentrations of from 2 to 14 per cent. It is extracted by treatment with alkali and steam distillation or by extraction with solvents. It is most frequently encountered in insecticide preparations. It is usually marketed for this use as a 40 per cent solution of the sulfate. Approximately one million pounds of nicotine are used annually in the United States for agricultural preparations.

2. Physical and Chemical Properties

Molecular weight: 162.23
Boiling point: 247.3°C.
Density: 1.010 g./ml. (20°C.)
Vapor density: 5.61 (air = 1)
Vapor pressure: 1 mm. Hg, 61.8°C.
Solubility: soluble in water, alcohol, and ether
1 mg./liter $\backsim$ 150 p.p.m. and 1 p.p.m. $\backsim$ 6.64 mg./cu. meter

3. Physiological Response

Nicotine is highly toxic. It is absorbed from the gastrointestinal tract, respiratory tract, and the skin. When C^{14}-labeled nicotine is administered to dogs a major portion of the radioactivity can be found in the urine but only approximately 10 per cent is unchanged. Unchanged nicotine in small amounts has been detected in the urine of cigarette smokers.[21] Variable quantities of nicotine have been found in tobacco smoke; the smoke from a cigarette may contain somewhat less than 10 mg./cigarette. Bowman and co-workers[37] gave nicotine to adult male nonsmokers in 10 hourly doses of 3 mg. each day for 3 days. They were unable to identify unchanged nicotine in the urine. They did find that approximately 10 per cent of the administered dose was eliminated as cotinine and smaller amounts as cotinine derivatives. Part of the cotinine probably arises from lactamization of γ-(3-pyridyl)-γ-methylaminobutyric acid.[37] When nicotine was administered to dogs by continuous intravenous administration over an 8-hour period, increasing doses resulted in a linear increase in the rate of urinary excretion of unchanged nicotine. Elimination was essentially complete 16 hours after administration was stopped.[38]

The pharmacological activity and toxicology of nicotine are thoroughly described in standard pharmacological texts.[13] In brief, the major effects are transient stimulation, followed by depression and paralysis of the central nervous system, peripheral autonomic nervous system ganglia, and skeletal muscle nerve endings. These symptoms develop rapidly and are generally of short duration (on the order of a few hours). Death usually occurs within a minute to an hour and is

[37] E. R. Bowman, L. B. Turnbull, and H. McKennis, *J. Pharmacol. Exptl. Therap.,* **127,** 92 (1959).

[38] J. K. Finnegan, P. S. Larson, and H. B. Haag, *J. Pharmacol. Exptl. Therap.,* **91,** 357 (1947)

most often due to paralysis of the respiratory muscles. Artificial respiration and cardiac resuscitation can protect dogs against otherwise fatal doses of nicotine.[39] Nicotine is locally irritating. The free alkaloid is absorbed rapidly through the skin, but absorption of its acid salts is less complete.[40] The response of animals to nicotine is summarized in Table 10.

TABLE 10

Response to Nicotine—Animals

Species	Route	Dose, mg./kg.	Response
Mouse	Oral	24	LD_{50}[a]
	Subcutaneous	16	LD_{50}[a]
	Intravenous	7.1	LD_{50}[b]
	Intraperitoneal	63.5	LD_{50}[b]
Rat	Oral	50–60	Approx. LD_{50}, convulsions, paralysis[c]
	Oral	4 (daily)	No weight loss[d]
Rabbit	Intravenous	9.4	LD_{50}[b]
	Dermal	50	Estimated LD_{50}[e]
Dog	Intravenous	15 (over an 8-hr. period)	Respiratory failure[f]
	Intravenous	5	LD_{50}[b]
Cat	Intravenous	2	LD_{50}[b]

[a] R. L. Metcalf, *Organic Insecticides.* Interscience Publishers, New York–London, 1955.

[b] P. S. Larson, J. K. Finnegan, and H. B. Haag, *J. Pharmacol. Exptl. Therap.*, **95**, 506 (1949).

[c] A. J. Lehman, *Assoc. Food & Drug Officials U. S. Quart. Bull.*, **15**, 122 (1951).

[d] R. H. Wilson and F. DeEds, *J. Ind. Hyg. Toxicol.*, **18**, 553 (1936).

[e] A. J. Lehman, *Assoc. Food & Drug Officials U. S. Quart. Bull.*, **16**, 3 (1952).

[f] J. K. Finnegan, P. S. Larson, and H. B. Haag, *J. Pharmacol. Exptl. Therap.*, **91**, 357 (1947).

Many fatal human cases of nicotine intoxication have occurred, usually as a result of accidental or suicidal ingestion of nicotine insecticides. There were 288 such fatalities reported in the United States in the 5-year period from 1930 to 1934. Intoxications have also been described in persons engaged in nicotine extraction and in spraying insecticides.[13] Considering the relatively enormous use of nicotine, occupational intoxications have been infrequently recorded. The symptoms that have been described in man are quite varied, as would be expected from the complex and phasic pharmacological actions of nicotine. The mild symptoms of nicotine absorption have been experienced by most of those persons who have smoked tobacco for the first time or after a period of abstinence. The common symptoms of moderate intoxication include nausea, vomiting, abdominal pain,

[39] F. E. Franke and J. E. Thomas, *J. Am. Med. Assoc.*, **106**, 507 (1936).

[40] M. N. Gleason, R. E. Gosselin, and H. C. Hodge, *Clinical Toxicology of Commercial Products,* Williams & Wilkins, Baltimore, 1957.

diarrhea, headache, sweating, palpitation, and fatigue. More severe symptoms are faintness, dizziness, weakness, and confusion progressing to prostration with increasing muscular weakness, collapse, and respiratory arrest. Most deaths occur within a few minutes of ingestion and recovery usually occurs if the patient survives 1 to 4 hours. It has been estimated that approximately 60 mg. of nicotine orally would be fatal to most adults.

The chronic toxicity of nicotine has been mainly investigated in connection with the use of tobacco. Some of the effects that have been reasonably attributed to chronic nicotine absorption in smokers are gastrointestinal symptoms, disturbance in heart rhythm, vasoconstriction, and rarely visual impairment. The development of habituation and tolerance to nicotine is well recognized. There is controversy about the existence of many other disturbances that have been attributed to the use of tobacco. Wilson and DeEds[41] were able to produce growth retardation in rats fed a diet containing more than 0.006 per cent nicotine for 300 days. This was only partially attributable to reduced food intake. The animals receiving the diet with 0.006 per cent nicotine ingested 4 mg. of nicotine/kg. daily without showing significant difference in their growth as compared to the controls.[41] Repeated subcutaneous injections in rats has not produced tumors.[42]

Nicotine may be determined quantitatively using the silicotungstic acid method.[43] Various colorimetric procedures are also applicable.[14] A threshold limit value of 0.5 mg./cu. meter has been recommended.[11] This would correspond to a nicotine intake well below that of cigarette smokers, and many times lower than the highest amount that does not alter growth rate in rats.

PIPERIDINE, $(CH_2)_5NH$ (Hexahydropyridine, Pentamethylenimine)

1. Uses

Piperidine is a clear, colorless, strongly basic ($pK_b = 2.88$) flammable liquid with a characteristic aminelike odor. It is obtained commercially by the reduction of pyridine. The most common method for synthesis of piperidine derivatives is by the reduction of the corresponding pyridine derivatives. The chemistry of piperidine has been closely identified with that of the alkaloids, since the piperidine nucleus occurs as such or in the fused states in a wide variety of these plant derivatives. A large number of piperidine compounds have been made for testing and use as local anesthetics and analgesics. Piperidine behaves like a typical aliphatic secondary amine and not like an aromatic amine. It is mainly used in the manufacture of pharmaceuticals, wetting agents, and germicides.

[41] R. H. Wilson and F. DeEds, *J. Ind. Hyg. Toxicol.*, **18**, 553 (1936).

[42] W. C. Hueper, *Arch. Pathol.*, **35**, 846 (1943).

[43] *Association of Official Agricultural Chemists Official and Tentative Methods of Analysis*, 7th ed., Washington, D. C., 1950.

2. *Physical and Chemical Properties*

Molecular weight: 85.15
Boiling point: 106.3°C.
Density: 0.8622 at 20°C.
Vapor density: 3.0 (air = 1)
Vapor pressure: 40 mm. Hg, 29.2°C.
Flash point: 61°F.
Solubility: infinitely soluble in water, alcohol, and ether
1 mg./liter ⊃ 287 p.p.m. and 1 p.p.m. ⊃ 3.5 mg./cu. meter

3. *Physiological Response*

Piperidine has been isolated and identified in animal and human urine.[44] Man excretes about 3 to 20 mg./day. It is probably derived from cadaverine and from the amino acid lysine, which are cyclized to piperidine. Oral administration of cadaverine to rabbits causes a several fold increase in the normal piperidine excretion. When piperidine is injected into hens and rabbits, the major portion is excreted unchanged.[21] It is well absorbed from the gastrointestinal tract and absorption through the skin is sufficient to produce death in small animals. Absorption from the respiratory tract has not been studied.

Piperidine causes moderately severe skin irritation when applied to guinea pig skin, and markedly severe eye damage with permanent corneal injury results from instillation in the rabbit eye. Like many secondary aliphatic amines, it is a pronounced pressor agent. A significant and sustained elevation of blood pressure is observed following intravenous injection in dogs or cats in doses of 5 to 10 mg./kg. Similar doses stimulate respirations and increase the heart rate.[15,44,45] The effects of piperidine resemble those of nicotine and of coniine (2-propyl piperidine) in that smaller doses produce initial stimulation of parasympathetic and sympathetic ganglion and larger doses produce depression. Coniine is the active alkaloid from poison hemlock. Oral administration results in weakness, nausea, vomiting, salivation, labored respirations, muscular paralysis, and asphyxia. It has been stated that 30 to 60 mg./kg. may cause symptoms in man. The information available on other simple alkyl piperidine derivatives is incomplete. Smyth found that 5-ethyl-2-methylpiperidine was highly irritating to the rabbit eye and skin and, like piperidine, was absorbed through the skin (dermal LD_{50} in rabbits, 0.63 cc./kg.; oral LD_{50} in rats, 0.54 g./kg.)[36] Five out of 6 rats exposed to 250 p.p.m. for 4 hours died. This represents a greater degree of toxicity than determined in the same laboratory for 5-ethyl-2-methylpyridine where it was found that the oral LD_{50} in rats was 1.54 g./kg., the percutaneous LD_{50} was 3.8 ml./kg. in rabbits and exposure to 1000 p.p.m. for 4 hours produced deaths in

[44] U. S. von Euler, *Acta Pharmacol. Toxicol.*, **1**, 29 (1945).
[45] T. Koppanyi and E. A. Vivino, *Federation Proc.*, **5**, 186 (1946).

5 out of 6 rats.[46] The 1-ethyl derivative of piperidine is a strong skin irritant and highly toxic for rats when administered undiluted. Piperidine also exhibits a relatively high degree of toxicity when administered undiluted to laboratory animals. The oral and intraperitoneal LD$_{50}$'s are under 50 mg./kg. A 10 per cent solution in water is somewhat less toxic when given orally to rats since some animals survive 100 mg./kg. but not 200 mg./kg. The animals show weakness, respiratory distress, and convulsions.[31]

4. Hygienic Standards of Permissible Exposure

Piperidine can be determined by the colorimetric method of von Euler which employs sodium-β-naphthoquinone-4-sulfonate to develop a red color. This is stable and allows determinations of piperidine in concentrations as low as 5 μg./ml.[44]

No threshold limit value has been proposed. Inhalation studies have not been reported that would allow estimation of safe levels. Piperidine is more irritating and more toxic for small animals than pyridine and is also more volatile. This indicates that process controls would be necessary for most operations. The use of personal protection to prevent accidental skin or eye contact is clearly indicated.

SIX-MEMBERED RINGS CONTAINING MORE THAN ONE HETERO ATOM

I. General Considerations

There is little information pertinent to industrial hygiene or toxicology on the majority of the large number of 6-membered nitrogen heterocycles containing more than 1 hetero atom. Of the diazines, pyrimidine, or 1,3-diazine, is of considerable biological importance since it is the fundamental ring structure in purine and nucleic acids and its derivatives are found in several important classes of pharmaceutical agents. Although there are some observations on the vesicant action of 2,4,6-trichloropyrimidine and the toxicity of 2-aminodiazine and 2-amino-4-methylpyrimidine in connection with their use in the production of therapeutic agents, pertinent data on this group of materials are limited.[47,48] Only the fully saturated 1,4-diazine derivative, piperazine, is considered in detail in this section. The compounds with 3 nitrogens symmetrically placed in the 6-membered rings, the symmetrical triazines, are growing in industrial importance. Of these, melamine, (2,4,6-triamino-s-triazine) has the widest application. Hexamethylene tetramine and morpholine (tetrahydro-1,4-oxazine) have had considerably larger industrial use and minor injuries have been reported in industry. Since only one polycyclic nitrogen heterocycle, quinoline, will be considered, it is included in this section.

[46] H. F. Smyth, C. P. Carpenter, and C. S. Weil, Arch. Ind. Hyg. Occupational Med., 4, 119 (1951).

[47] M. W. Goldblatt, Brit. J. Ind. Med., 2, 183 (1945).

[48] J. F. Treon, H. Wright, K. V. Kitzmiller, and W. J. Younker, J. Ind. Hyg. Toxicol., 30, 79 (1948).

As with the other heterocyclic compounds that have been discussed, the actions of these materials are strongly influenced by the nature of their substituent chemical groups. Their local action is dominated by their alkalinity and their systemic activity tends to be pharmacological in nature. No serious or chronic intoxications have been reported from the use of these materials and threshold limits have not been established.

II. Specific Compounds

PIPERAZINE, NHCH$_2$CH$_2$NH CH$_2$CH$_2$, hydrate—C$_4$H$_{10}$N$_2$·6H$_2$O

(Hexahydropyrazine)

1. *Source and Uses*

Piperazine is a transparent, deliquescent solid. Aqueous solutions of the base are strongly alkaline (pK$_b$ = 4.19); the pH of a 10 per cent solution is approximately 11. It is prepared from N-(2-hydroxyethyl) ethylenediamine by heating in the presence of a metal catalyst. It is used in the preparation of medicinals and to a lesser extent for corrosion inhibitors. The salts are used for the treatment of parasitic infestations in man and animals.

2. *Physical and Chemical Properties*

Molecular weight: 86.14
Melting point: 104°C.
Boiling point: 145°C.
Flash point: 190°F.
Solubility: soluble in water and alcohol

3. *Physiological Response*

The acute oral toxicity of piperazine salts is very low. The single dose LD$_{50}$ of piperazine citrate for mice is 11 g./kg. and that of piperazine phosphate is 20 g./kg.[49] Carpenter and Smyth found that a 5 per cent solution produced severe injury of the rabbit eye.[50] Solutions would also be expected to produce caustic burns from direct contact with skin. Attempts to induce systemic and skin sensitivity in guinea pigs have not been successful.[51]

No injuries in industry have been reported. Side reactions have been observed in individuals taking from 30 to 75 mg./kg./day of various piperazine salts for medicinal purposes. The symptoms have included hives, headaches, nausea and vomiting, diarrhea, lethargy, tremor, incoordination, and muscular weakness. These have been transient and disappear when the medication is stopped.

[49] H. W. Brown, K. Chan, and K. L. Hussey, *Am. Med. Assoc.,* **161,** 515 (1956).
[50] C. P. Carpenter and H. F. Smyth, *Am. J. Ophthalmol.,* **29,** 1363 (1946).
[51] B. Ratner, J. G. Flynn, and K. M. Mayer, *Ann. Allergy,* **13,** 176 (1955).

MELAMINE, N:C(NH₂)N:C(NH₂)N:C(NH₂) (2,4,6-Triamino-s-triazine, Cyanurotriamide)

1. Source and Uses

Melamine is a fine crystalline powder, that is usually prepared by heating dicyandiamide with ammonia under pressure. It can also be prepared by the action of ammonia on urea or cyanuric acid. The major use of melamine is in the manufacture of thermosetting resins, which are used in a wide variety of products. It is also used in the manufacture of adhesives and wet-strength resins for paper, and to a lesser extent for bactericides, fungicides, and flameproofing compounds.

2. Physical and Chemical Properties

Molecular weight: 126.13
Melting point: less than 250°C. (sublimes)
Vapor pressure: 50 mm. Hg, 315°C.
Vapor density: 4.34 (air = 1)
Solubility: slightly soluble in water; very slightly soluble in hot alcohol

3. Physiological Response

Melamine has low toxicity for rats and mice. Single doses of a 10 per cent melamine suspension in corn oil by mouth did not cause deaths in rats at 3.2 g./kg. (the highest dose used), but did cause acute deaths when administered intraperitoneally at this dose, but not at 1.6 g./kg. Mice, however, were killed by oral administration of 1.6 g./kg. and intraperitoneal injection of 0.8 g./kg. A 1 per cent solution in water applied under a rubber cuff to guinea pig skin produced little or no irritation and no apparent absorption. It was not possible to produce skin sensitization in guinea pigs.[31] Lipschitz and Stokey[52] found that melamine has a pronounced diuretic effect in rats and dogs. Large single oral doses of 20 mM/kg. (2.4 g./kg.) in rats caused no effects other than crystalluria and diuresis. The crystalluria was due to the excretion of dimelamine-monophosphate crystals. Following single oral doses in rats and dogs, 50 to 60 per cent of the administered melamine was recovered in 6 hours. In dogs, 60 to 85 per cent was recovered in 24 hours. Histological examination of rabbits and dogs fed 1 mM/kg. (126 mg./kg.) daily for 1 to 4 weeks were negative. Philips and Thiersch[3] also observed crystalline deposits in the renal tubules of rats given 5 successive intraperitoneal doses of 500 mg./kg. No symptoms were observed except for moderate transient weight loss and histological examinations were essentially negative. There were no effects on blood or bone marrow resembling those observed with triethylenemelamine. Although certain methylolmelamines, especially trimethylolmelamine, resemble the nitrogen mustards and other alkylating agents in their cytotoxic and

---

[52] W. L. Lipschitz and E. Stokey, *J. Pharmacol. Exptl. Therap.*, **83**, 235 (1945).

tumor inhibitory effects, melamine gave no evidence of cytotoxicity or tumor inhibition.[53]

Dermatitis has been reported from the manufacture of melamine formaldehyde resins and glues. It is probable that these cases were chiefly due to formaldehyde or intermediate reaction products of formaldehyde and melamine. No instances of melamine intoxication in industry have been reported. Experience indicates that ordinary handling precautions should be adequate to prevent injury.

CYANURIC ACID, N:C(OH)N:C(OH) N:C(OH)

(2,4,6-Trihydroxy-1,3,5-triazine, s-Triazinetroil, Tricyanic acid)

1. Uses

Cyanuric acid is an odorless, crystalline powder. It is used in chemical syntheses and as an intermediate for chlorinated bleaches. It can be used with sodium hypochlorite for chlorination of swimming pools. It is a convenient laboratory source of cyanic acid.

2. Physical and Chemical Properties

Molecular weight: 129.08
Melting point: $>360°C.$, decomposes to cyanic acid (HOCN)
Solubility: 0.25 g. in 100 cc. of water at 17°C.; slightly soluble in alcohol and ether; soluble in hot alcohol and pyridine

3. Physiological Response

Cyanuric acid exists in two tautomeric forms of which the iso or amide form predominates over the enol form. Single dose and feeding toxicity studies in animals shows a very low degree of toxicity[54] (see Table 11). Its toxicity does not resemble that of the cyanides. At the high doses, most of the compound is excreted unchanged. Biochemical or hematological changes were not found in any of the studies. No histopathological damage was identified other than a characteristic dilatation of the collecting tubules in the kidneys of rats fed 6.8 per cent (as the monosodium salt) for 6 months and in the rabbits treated with daily skin applications of 6.8 per cent suspension on 10 per cent of the body surface for 3 months. The monosodium salt and water solutions of cyanuric acid are not irritating. Cyanuric acid does not inhibit tumor growth in rats as do some methylolmelamines.[54]

Some related compounds have been studied. Cyanuric chloride (2,4,6-trichloro-1,3,5-triazine) is a highly toxic lacrimator and an irritant; it is discussed in Chapter XLIV, on nitriles. Sodium dichloroisocyanurate ($Cl_2Na(NCO)_3$) may be used to maintain adequate chlorination in swimming pools or as a bleach

[53] J. A. Hendry, F. L. Rose, and A. L. Walpole, *Brit. J. Pharmacol.*, **6**, 201 (1951).

[54] J. L. Svirbely, Office memorandum, Robert A. Taft Engineering Center, Cincinnati, Feb. 1, 1960.

TABLE 11

Response to Cyanuric Acid—Animals[a]

Species	Administration	Dose	Response
Rat, rabbit	Single oral	10 g./kg.	No symptoms
	Repeated oral	20 g./kg. (4 X)	Decreased appetite
Rat, weanling	30 days in drinking water	400–8000 p.p.m. as monosodium salt	No weight loss No tissue pathology
Rat	6-month feeding	0.68%	No effect
Rat	6-month feeding	6.8%	Weight loss, fractional mortality dilated collecting tubules in kidneys
Dog	6-month feeding	0.68%	No effect
Dog	1-year feeding	6.8%	No symptoms
Rabbit	Daily skin application for 3 months	0.68% soln. 6.8% susp.	No symptoms or irritation. Kidney changes at 6.8%

[a] J. L. Svirbely, Office memorandum, Robert A. Taft Engineering Center, Cincinnati, Feb. 1, 1960.

(available chlorine about 60 per cent). It is more toxic than cyanuric acid. The oral LD_{50} administered as a 10 per cent solution in rats is 1.67 g./kg. Lethal doses produce gastrointestinal tract irritation, liver dysfunction, and congestion in the liver, kidneys, and lungs. Rats and dogs fed 16.6 p.p.m. to 333 p.p.m. in their diet for 6 months showed no signs of toxicity or organ damage. Dichloroisocyanuric acid exhibits a similar degree of acute toxicity. Trichloroisocyanuric acid is slightly more toxic for rats as judged by a single oral dose. These materials are not appreciably irritating to the intact skin but concentrated solutions will irritate abraded skin. The solids are moderately severe eye irritants.[53]

No injuries in humans have been reported. The use of the chloroisocyanuric compounds in swimming pools has not given rise to adverse effects and this use is considered to be safe.[53] Cyanuric acid does not present significant potential for injury in industry except under conditions prompting decomposition to cyanic acid.

MORPHOLINE, $C_2H_4OC_2H_4NH$ (Tetrahydro-1,4-oxazine, Diethylenimide oxide)

1. Uses

Morpholine is a colorless, volatile alkaline liquid with an amine odor. It is prepared by the condensation of ethylene oxide with ammonia or, on large scale, by dehydrating diethanolamine. It is used as a corrosion inhibitor, a neutralizing and scrubbing agent, and for the preparation of derivatives, which are used as plasticizers, synthetic lubricants, emulsifiers, antioxidants, and pharmaceuticals.

2. Physical and Chemical Properties

Molecular weight: 87.12

Boiling point: 128.6°C.

Density: 1.0020 g./ml. (20/20°C.)

Vapor pressure: 8.0 mm. Hg, 20°C.

Flash point: 100°F. (open cup)

Solubility: completely soluble in water; soluble in alcohol, ether, acetone, and benzene

1 mg./liter ⇌ 281 p.p.m. and 1 p.p.m. ⇌ 35.6 mg./cu. meter

3. Physiological Response

Morpholine acts like a secondary amine and resembles this group of compounds in its chemical and toxicological properties. Its solutions are highly alkaline ($pK_b = 9.61$). A 25 per cent aqueous solution has a pH of approximately 11. The base, therefore, is a potent skin and mucous membrane irritant.

TABLE 12

Response to Morpholine—Animals[a]

Species	Dose	Route	Response
Rat	1.6 g./kg.	Oral	LD_{50}
Guinea pig	0.9 g./kg.	Oral	LD_{50}
Rat	0.8 g./kg., daily × 30	Oral	Death in $^{19}/_{20}$, injury of liver, kidney, stomach
Rat	0.16 g./kg., daily × 30	Oral	Death in $^{8}/_{20}$, milder liver and kidney injury
Guinea pig	0.45 g./kg., daily × 30	Oral	Death in $^{16}/_{20}$, injury of liver, kidney, stomach
Guinea pig	0.09 g./kg., daily × 30	Oral	Death in $^{3}/_{20}$, slight changes in liver and kidney
Rabbit	0.9 g./kg., undiluted	Skin, (occlusive dressing)	Death in $^{2}/_{7}$ severe burns penetrating skin
Rabbit	0.9 g./kg., 50% v/v aqueous	Skin, (occlusive dressing)	$^{7}/_{7}$ died before eleventh dose; skin burns, systemic injury
Rat and guinea pig	18,000 p.p.m., 8 hrs. × 6	Inhalation	Fractional mortality, respiratory tract injury
Rat and guinea pig	12,000 p.p.m., 8 hrs. × 1	Inhalation	No deaths, reversible respiratory tract and systemic injury

[a] T. E. Shea, J. Ind. Hyg. Toxicol., 21, 236 (1939).

Smyth[36] found that the single dose oral LD_{50} in rats was 1.05 g./kg. The percutaneous LD_{50} in rabbits was 0.5 ml./kg., indicating skin injury and absorption. A 40 per cent solution and 0.005 ml. of the undiluted base produced corneal injury in the rabbit eye (grade 7). All rats exposed to the saturated vapor died within 4 hours; 1 hour was the maximum time for no deaths. Exposure to 8000 p.p.m. (calculated) for 8 hours resulted in no deaths in 6 rats. Shea[55] found essentially the same degree of acute toxicity in rabbits, rats, and guinea pigs (see Table 12). Single oral doses of undiluted and unneutralized morpholine caused irritation of the intestinal tract with hemorrhage. In subacute feeding experiments, there was also evidence of intense irritation of the intestinal tract and microscopic examination revealed increasing liver and kidney injury with increasing doses. Skin, liver, and kidney injury was also produced by repeated skin application of morpholine base. However, neutralized morpholine in doses up to 3 times the oral LD_{50} for guinea pigs caused only dermal thickening after 30 daily applications in these animals. Exposure of rats and guinea pigs to a measured concentration of 1800 p.p.m. in air resulted in irritation of the eyes, nose, and respiratory tract. Repeated exposures resulted in lung injury largely centered about the bronchi and bronchioles, and kidney and liver sections showed cloudy swelling and congestion.

It is apparent that morpholine acts mainly as a local irritant due to its alkalinity. Its hazards, therefore, are similar to those of other volatile organic bases (see Chapter XLV, on aliphatic amines). Some instances of skin and respiratory tract irritation have been observed in industry, but no chronic effects have been reported.[56] It would be expected that morpholine would possess good warning properties for acute vapor exposures, but this has not been adequately investigated. No threshold limit values have been proposed. Vapor concentrations should be maintained below those producing sensory response in exposed individuals. Skin and eye protection should be provided.

N-SUBSTITUTED MORPHOLINES

Smyth and his co-workers have conducted range-finding toxicity tests on a series of N-substituted morpholines.[36,46,50,57-59] Their results and the physical properties of six of these compounds are shown in Table 13. All exhibited a low degree of oral toxicity and like morpholine, were absorbed through the intact skin. The two amine derivatives, N-aminoethylmorpholine and N-aminopropylmorpholine, showed skin and eye irritant properties similar to morpholine, whereas the other materials were less irritating to the skin and eye. Of the group, the N-methyl and N-ethyl derivatives present the highest potential hazard from vapor exposure and the N-alkyl amine derivatives the highest hazard from local skin and eye contact. No injuries in industry have been reported from these materials.

[55] T. E. Shea, *J. Ind. Hyg. Toxicol.*, **21**, 236 (1939).

[56] M. Clinton, *American Petroleum Institute Toxicological Review on Morpholine*, New York, 1948.

[57] H. F. Smyth and C. P. Carpenter, *J. Ind. Hyg. Toxicol.*, **26**, 269 (1944).

[58] H. F. Smyth and C. P. Carpenter, *J. Ind. Hyg. Toxicol.*, **30**, 63 (1948).

[59] H. F. Smyth, C. P. Carpenter, and C. S. Weil, *J. Ind. Hyg. Toxicol.*, **31**, 60 (1949).

TABLE 13

Physical and Chemical Properties and Acute Toxicity of N-Substituted Morpholines

	N-acetyl	N-aminoethyl	N-amino-propyl	N-ethyl	N-hydroxy-ethyl	N-methyl
Molecular weight	129.16	130.19	144.21	115.17	131.17	101.15
Specific gravity (20/20°C.)	1.1164	0.9915	0.9143	0.9872	1.0740	0.9213
Melting point, °C.	14	25.6	−15	−63	1.6	−65.9
Boiling point, °C.	decomposes	204.2	224.7	138.3	226.3	115.6
Flash point, °F.	235	175	220	100	215	75
Vapor pressure, mm. Hg, 20°C.			0.01	6.1		16.6
LD_{50}, single dose oral, rat, g./kg.	6.13	3.0	5.66	1.78	12.06	2.72
LD_{50} skin, rabbit, ml./kg.	7.5	0.3	1.23		2.5[a]	1.35
Saturated vapor, maximum time for no deaths, hr.	8	8	8	2	8	1
Skin irritation, grade[b]	1	6	6	1	Like acetone	2
Eye irritation, grade[b]	2	9	9	7	5	5

[a] Guinea pig.
[b] See Table VI in Chapter XIV on aliphatic and alicyclic amines.

HEXAMETHYLENETETRAMINE, $(CH_2)_6N_4$ (Hexamine, Methamine, Urotropin)

1. Uses

Hexamethylenetetramine is an odorless, crystalline solid, which is prepared by the condensation of ammonia and formaldehyde in a continuous process. It has been used as an accelerator in the rubber industry, as a curing agent in thermosetting plastics, in the manufacture of resins and pharmaceuticals, and was marketed under various trade names for use as a urinary antiseptic. Large quantities were used in World War II in the manufacture of the explosive, trimethylenetrinitramine (RDX). Approximately 23 million pounds were produced in the United States in 1957. Because it will burn in contact with flame to give a smokeless flame, it has been used as a fuel pellet for camp stoves.

2. Physical and Chemical Properties

Molecular weight: 140.19
Melting point: 263°C. (decomposes)
Boiling point: sublimes
Density: 1.270 (25°C.)
Flash point: 482°F.
Solubility: 150 g. in 100 ml. of water at 20°C.; soluble in alcohol and glycerin; insoluble in ether

3. Physiological Response

It has been said that the occurrence of hexamethylenetetramine dermatitis among workers in the rubber industry was partly responsible for its abandonment in most rubber formulations.[60] On the other hand, dermatitis from the handling of pure hexamethylenetetramine was not observed in a large ordnance plant manufacturing RDX during World War II.[61] The scarcity of recent reports of hexamethylenetetramine dermatitis and the absence of reports of systemic intoxication suggest that it has relatively low toxicity and skin irritation potency.

[60] H. B. Elkins, The Chemistry of Industrial Toxicology, 2nd ed., Wiley, New York, 1959.
[61] J. H. Sterner, Eastman Kodak Co., unpublished observations.

Hexamethylenetetramine is absorbed from the gastrointestinal tract rapidly and appears in the urine in a few minutes reaching a maximum in 1 to 3 hours. Excretion is essentially complete after 24 hours. It dissociates in acid solutions and formaldehyde has been identified in the stomach and in acid urines after oral administration. It has been assumed that the antiseptic effects of hexamethylenetetramine are attributable to this liberation of formaldehyde. Oral doses of up to 5 g./day have been used therapeutically. Some of the side reactions that have been observed are urinary tract irritation, cystitis, hematuria, skin rashes, and digestive disturbances.[13]

The information available indicates that ordinary care in handling should be sufficient to prevent systemic intoxication. Repeated skin contact should be avoided. No threshold limit values have been suggested.

QUINOLINE, $C_6H_4N{:}CHCH{:}CH$ (Benzo(β)pyridine, 1-Benzazine)

1. Uses

Quinoline is a refractive, hygroscopic liquid with a penetrating odor that is not as offensive as that of pyridine. It is a weak base ($pK_b = 8.94$) that resembles pyridine in its general chemical behavior. It is used in the manufacture of dyes, antiseptics, fungicides, and pharmaceuticals, and as a solvent.

2. Physical and Chemical Properties

Molecular weight: 129.15
Boiling point: 237.7°C.
Vapor density: 4.45 (air = 1)
Vapor pressure: 1 mm. Hg, 59.7°C.
Solubility: 6 g. in 100 ml. of water; completely soluble in alcohol and ether

3. Physiological Response

Smyth et al.[46] found that the single dose oral LD_{50} in rats was 0.46 g./kg. and the percutaneous LD_{50} in rabbits was 0.59 ml./kg. The liquid was moderately irritating for the skin and produced rather severe eye injury in rabbits. No deaths occurred in rats exposed to saturated vapor (room temperature) for 8 hours.[46] Other screening toxicity studies have shown essentially the same degree of acute oral and percutaneous toxicity.[31] With doses near the lethal range by either route, the animals exhibited lethargy, respiratory distress, and prostration progressing to coma. No deaths occurred in 3 rats exposed to the concentrated vapor at room temperature (calculated concentration of 17 p.p.m.) for 6 hours. All 3 rats exposed to the vapor produced by heating the compound to 100°C. (calculated concentration of 4000 p.p.m.) died within $5\frac{1}{2}$ hours. These limited animal tests indicate that quinoline has a moderate degree of toxicity by ingestion, skin application, or inhalation. No injuries in industry have been reported.

Part B. Miscellaneous Nitrogen Compounds

The compounds discussed in this section are mainly aliphatic nitrogenous bases and materials with nitrogen to nitrogen bonds. These include the azides (RNNN), diazomethane (CH_2N_2), certain nitroso compounds (RNO), the hydrazines ($RNHNH_2$), and the β-chloroethylamines ($R,R'NCH_2CH_2Cl$). They are all biologically active materials with distinctive pharmacological or pathological effects and a relatively high degree of toxicity.

AZIDES

I. General Considerations

A. PROPERTIES AND USES

The organic and inorganic azides have a well-deserved reputation as dangerous materials. Consequently, their use has been limited. They are derivatives of hydrazoic acid, HN_3, and have the general structural formula R-NNN. Hydrazoic acid is a weakly acidic volatile liquid. The salts of heavy metals such as silver azide (AgN_3) and lead azide (PbN_6) are highly explosive. The percussion sensitivity of PbN_6 led to its important use for primers in munitions. The salts of alkalies and alkaline earths are generally not explosive and, of this group, sodium azide (NaN_3) is most useful. It dissociates fully in aqueous solutions and releases hydrazoic acid with hydrolysis. The alkyl azides (methyl azide, b.p. 20°C.; ethyl azide, b.p. 49°C., vinyl azide, b.p. 26°C.) are stable at room temperature but are apt to explode on rapid heating. They decompose at elevated temperatures to release HN_3. The aryl azides are usually colored, comparatively stable solids. Some explode on percussion and they usually melt with decomposition, some with the evolution of HN_3.

B. PHYSIOLOGICAL RESPONSE

1. Effects in Animals

Considerable interest has been directed toward the effects of the azide anion at the cellular level and, because of its effects on enzyme systems, it has been used as a research tool in biochemistry. Studies on *in vitro* preparations have demonstrated inhibition of cytochrome oxidase and a variety of other enzymes.[62] Because of this, parallels have been drawn to other "metabolic inhibitors" such as cyanide, malonitrile and fluoride.[63] The biochemical effects of azide on intact cells are similar, but not identical, to that of 2,4-dinitrophenol. Although the effects in these systems are complex, there is general agreement that azide causes a dissociation of phosphorylation and cellular respiration.[64]

[62] M. Dixon and E. C. Webb, *Enzymes,* Academic, New York, 1958.
[63] P. Handler, *J. Biol. Chem.,* **161,** 53 (1945).
[64] H. E. Robertson and P. D. Boyer, *J. Biol. Chem.,* **214,** 295 (1955).

Hydrazoic acid and sodium azide are highly toxic materials. The major acute effect in experimental animals is profound hypotension from direct action on the smooth muscle of blood vessels. This action is similar to nitrites and is more powerful. However, unlike the nitrites, the alkyl azides do not produce methemoglobin formation in vivo.[65] Azide stimulates respiration and increases the force of the heart beat. After large doses there may be an increase in blood pressure and generalized convulsions followed by depression and collapse. Injury of myelinated nerve fibers and the gray matter in the central nervous system result from repeated administration.[66,67] It has been suggested that this is due to inhibition of cellular respiration in these critical areas. The azide anion is responsible for the qualitatively similar acute toxicities of other inorganic azides. Inhalation of the vapor or mist of hydrazoic acid causes mucous membrane and respiratory tract irritation with bronchitis and pulmonary edema in addition to the pharmacological effects noted with sodium azide.

A few organic azides have been studied. Werle and Fried[68] found that ethyl and amyl azides were effective hypotensive agents but several aromatic azides were ineffective. Of the acyl azides, the aromatic compounds also depressed the blood pressure, but the aliphatic acyl azides were not tested because of their explosiveness. F. E. Roth and co-workers[69] studied a series of 21 organic azides of which 6 produced greater than 25 per cent fall in mean blood pressure with hypotension lasting for 1 hour or more. The acute toxicity of these, as judged by the single dose oral LD_{50} in mice, was less than that of sodium azide, which was both more toxic, on a mg./kg. basis, and more hypotensive. L. W. Roth and Morphis[70] studied six straight-chain diazides with the general formula $N_3-(CH_2)_n-N_3$. All of these produced a fall in blood pressure in normal and hypertensive animals by the intravenous, oral, or intramuscular routes. There was a peak in the duration and magnitude of the hypotension at C_7. The acute toxicity in mice did not correspond closely to the hypotensive activity. The information on these series is not sufficiently complete to permit further conclusions concerning the relationship of chemical structure to toxicity and hypotensive activity.

2. Effects in Man

There have been scattered reports on the effects of acute hydrazoic acid exposures and medical observations have been made on individuals chronically exposed during the manufacture of lead azide. Sodium azide has been used experimentally for the therapy of hypertension but no human experience with the organic azides has been recorded.

[65] J. D. P. Graham, Brit. J. Pharmacol., 4, 1 (1949).

[66] S. P. Hicks, Arch. Pathol., 50, 545 (1950).

[67] E. W. Hurst, Australian J. Exptl. Biol. Med. Sci., 20, 297 (1942).

[68] E. Werle and R. Fried, Biochem. Z., 322, 507 (1952).

[69] F. E. Roth, J. Schurr, E. Moutis, and W. M. Govier, Arch. intern. pharmacodynamie, 108, 473 (1956).

[70] L. W. Roth and B. R Morphis, Federation Proc., 15, 477 (1956).

II. Specific Compounds

SODIUM AZIDE, NaN₃

1. *Uses*

Sodium azide is a stable, neutral, white crystalline solid that can be prepared from sodium amide and nitrogen monoxide. Hydrazoic acid is released from solutions. It is used as a source of hydrazoic acid and for the preparation of lead azide and pure sodium. It is also used in some organic syntheses.

2. *Physical and Chemical Properties*

Molecular weight: 65.02
Melting point: decomposes
Density: 1.846
Solubility: 40.2 g. in 100 g. of water at 10°C.; slightly soluble in alcohol and benzene

3. *Physiological Response*

Sodium azide is rapidly absorbed from the gastrointestinal tract and from injection sites, but its absorption from the lungs and skin does not appear to have been studied. It has a high degree of toxicity for experimental animals (see Table 14). The symptoms observed in animals after relatively large doses are respiratory stimulation and convulsions, then depression and death. With lower doses the occurrence of convulsions is variable but there is a consistent prompt transient fall in blood pressure. In rabbits an oral dose of 3 to 10 mg./kg. causes a 40 to 60 per cent reduction in blood pressure, which lasts more than 1 hour. At 2 mg./kg. the hypotension is less severe and of shorter duration. Associated hematuria and cardiac irregularities have been observed.[69] As little as 1 mg./kg. intravenously in cats will produce hypotension, which increases when this dose is repeated.[65] Repeated intraperitoneal injections in rats (5 to 10 mg./kg. every 15 to 30 minutes for 3 to 6 hours) result in severe intoxication, and some survivors show injury and demyelination of myelinated nerve fibers in the central nervous system, and testicular damage, but no lesions of liver or kidney.[66] Blindness and attacks of rigidity with abnormal motions are manifestations of the central nervous system damage produced by repeated doses in monkeys.[67]

It has been found that sodium azide produces a larger fall in blood pressure in hypertensive patients than in normotensive individuals. Doses of 0.65 to 1.3 mg. (approximately 0.01 to 0.02 mg./kg.) by mouth produced a prompt fall in blood pressure, which lasted 10 to 15 minutes. The administration of as much as 1.3 mg. of sodium azide 3 times a day for 10 days to 9 normal individuals did not have a sustained effect on blood pressure and no other effects were observed except for a transient pounding sensation in the head. Thirty hypertensive individuals were treated with 0.65 to 3.9 mg. by mouth daily for 1 week to 2½ years. In 15 of these, blood pressure was maintained at near normotensive levels. In 3 patients, who took sodium azide daily for more than a year, there was no evidence of organic damage.[71]

No threshold limit values for sodium azide have been proposed. It is clear from the animal data and the medicinal use of sodium azide that dust concentrations should be kept at low levels if hypotension is to be avoided. It is important to note that the use of sodium azide in organic synthesis may result in significant hydrazoic acid vapor exposure. This is illustrated by the following case that the

TABLE 14

Response to Sodium Azide—Animals

Species	Route of administration	Dose, mg./kg.	Mortality
Rat	Intraperitoneal	25	$^0/_4$ died in 3 hrs.[a]
	Intraperitoneal	33	$^8/_{12}$ died in 3 hrs.[a]
	Intraperitoneal	37	$^5/_5$ died in 3 hrs.[a]
Rat	Subcutaneous	33	$^0/_5$ died in 3 hrs.[a]
	Subcutaneous	35	$^4/_9$ died in 3 hrs.[a]
	Subcutaneous	38	$^8/_8$ died in 3 hrs.[a]
Rat	Oral	40	$^0/_3$ died in 3 hrs.[a]
	Oral	45	$^5/_8$ died in 3 hrs.[a]
	Oral	46	$^3/_3$ died in 3 hrs.[a]
Mouse	Intraperitoneal	28–34	LD_{50}[b]
	Intravenous	19	LD_{50}[b]
	Oral	27	LD_{50}[b]
Mouse	Oral	37.4	LD_{50}[c]
Monkey	Intramuscular	20	Sick but survived[d]
	Intramuscular	10–12 (repeated)	Death after 3 to 4 doses[d]
	Intramuscular	5 (daily)	$^5/_6$ died in 60 to 130 days[d]

[a] L. T. Fairhall, W. V. Jenrette, S. W. Jones, and E. A Pritchard, *Public Health Repts. U. S.,* **58,** 607 (1943).

[b] J. D. P. Graham, *Brit. J. Pharmacol.,* **4,** 1 (1949).

[c] F. E. Roth, J. Shurr, E. Moutis, and W. M. Govier, *Arch. intern. pharmacodynamie,* **108,** 173 (1956).

[d] E. W. Hurst, *Australian J. Exptl. Biol. Med. Sci.,* **20,** 297 (1942).

author has observed. A chemist was acidifying 10 g. of sodium azide in a malfunctioning hood. After a few minutes' exposure, he developed dizziness, weakness, blurred vision, slight shortness of breath, and a feeling that he was going to faint. He was observed to have moderate reduction in blood pressure and bradycardia. There was complete recovery in one hour with no subsequent symptoms. White blood cell counts, blood sugar determination, and the electrocardiogram were normal.

[71] M. M. Black, B. W. Sweifach, and F. D. Speer, *Proc. Soc. Exptl. Biol. Med.,* **85,** 11 (1954).

HYDRAZOIC ACID, HN$_3$ (Azoimide)

1. Source and Uses

Hydrazoic acid is a weak acid ($pK_a = 4.72$). It is a colorless, volatile liquid with a characteristic odor, which has been described as sickening. It is highly explosive. It is obtained by the hydrolysis of acyl azides or from hydrazine and is liberated from aqueous solutions of its salts. It is encountered mainly in organic synthesis or as a by-product of the use or manufacture of other azides. The manufacture of lead azide for use in explosives entails exposure to hydrazoic acid.

2. Physical and Chemical Properties

Molecular weight: 43.03
Boiling point: 37°C.
Solubility: infinitely soluble in water, soluble in alcohol
1 mg./liter ≎ 569 p.p.m. and 1 p.p.m. ≎ 1.76 mg./cu. meter

3. Physiological Response

Hydrazoic acid has essentially the same degree of toxicity for mice when injected intraperitoneally as does sodium azide.[72] The major effects observed are stimulation of respiration, fall in blood pressure, and generalized convulsions followed by depression. Inhalation of the vapor or spray causes acute bronchiolar inflammation in guinea pigs and in some, pulmonary edema. When death is delayed from 3 hours to 3 days, pathological changes are observed only in the lungs. Inhalation and injections fail to produce liver and kidney damage. Fairhall et al. also found that hydrazoic acid vapor was highly toxic and compared the acute toxicity to that of hydrogen sulfide or hydrogen cyanide.[73] Their results appear in Table 15.

Acute effects in man following inhalation of hydrazoic acid vapor are eye irritation, bronchitis, headache, a fall in blood pressure, and weakness or collapse.[65,73] Headache and a feeling of weakness, dizziness, or faintness are the most consistent findings. The pulse rate is variable, from rapid to slow, and the blood pressure is either low or normal. The finding of the normal blood pressure probably represents the transient nature of the hypotension following a single, acute, mild exposure. The headache may last longer than the other symptoms. Recovery usually occurs within an hour after a single exposure. In one instance, a concentration of 3 p.p.m. was found during a repetition of an operation that had produced symptoms after less than an hour's exposure.[31]

Graham and his associates[72] examined workers exposed to hydrazoic acid in the manufacture of lead azide from lead nitrate and sodium azide. Detailed medi-

[72] J. D. P. Graham, J. M. Rogan, and D. G. Robertson, J. Ind. Hyg. Toxicol., **30**, 98 (1948).

[73] L. T. Fairhall, W. V. Jenrette, S. W. Jones, and E. A. Pritchard, Public Health Repts. U. S., **58**, 607 (1943).

TABLE 15

Response to Hydrazoic Acid Vapor—Rats[a]

Concentration, p.p.m.	Duration of exposure, min.	Mortality
849–967	60	$0/14$
1024	60	$3/8$
1081	60	$3/4$
1138	60	$17/18$
1162–1365	60	$16/16$
1566	30	$2/2$
1872	19	$2/2$
2080	16	$2/2$
2900	10	$2/2$

[a] L. T. Fairhall, W. V. Jenrette, S. W. Jones, and E. A. Pritchard, *Public Health Repts. U. S.*, **58**, 607 (1943).

cal examination on 10 workers exposed for 1 to 15 years did not reveal any pathological changes that they could attribute to occupational exposure. However, they did find definite hypotension, which became more pronounced as the work period progressed and returned to normal after the men left the plant. The workmen also experienced throbbing headaches, palpitation, episodes of weakness and unsteadiness, and mild eye and nose irritation. The maximum concentrations in air were between 1.3 and 3.9 p.p.m.

4. Warning Properties, Determination in Air, and Threshold Limit Value

The odor and irritant properties of hydrazoic acid do not present sufficient warning to prevent the occurrence of alarming symptoms. Concentrations in air can be estimated by collection in $0.2N$ potassium hydroxide, neutralization with $0.2N$ nitric acid, followed by spectrophotometric determination of the optical density (at 448 mμ) produced with ferric chloride. In the absence of interfering acids, collection in standard alkali and back titration may be useful.[14] Imperial Chemicals Industries, Ltd. (1952) has suggested that concentrations greater than 1 p.p.m. indicate an unsatisfactory situation for prolonged exposure. On the basis of the limited observations in industry, this suggestion appears to be reasonable.

DIAZOMETHANE, CH_2N_2 (Azimethylene)

1. Source and Uses

Diazomethane is an explosive, yellowish gas with a musty odor. It is highly reactive and has great potential usefulness as a methylating agent in chemical synthesis. However, its high toxicity and explosive hazard have limited its use. Diazomethane can be prepared by the treatment of nitrosomethylurea with potassium hydroxide in ether solution or from methylnitrosourethane and sodium methoxide.

2. Physical and Chemical Properties

Molecular weight: 42.04

Boiling point: $-23°C$., may explode spontaneously or on heating

Solubility: freely soluble in ether, benzene, and alcohol; decomposes in water

1 mg./liter $\backsimeq$ 582 p.p.m. and 1 p.p.m. $\backsimeq$ 1.72 mg./cu. meter

3. Physiological Response

Chemists working with diazomethane have called attention to its high degree of toxicity and have remarked on such effects as skin irritation, chest discomfort, asthmatic symptoms, and the development of hypersensitivity. A 10-minute exposure of cats to a concentration of 175 p.p.m. resulted in hemorrhage and edema of the lungs and death in 3 days.[74] Guinea pigs exposed to unmeasured concentrations showed symptoms of severe respiratory tract irritation, and on autopsy had marked pulmonary edema.[75] Landsteiner and DiSomma were able to produce skin sensitization in guinea pigs with repeated applications of diazomethane in dioxane or cottonseed oil.[76] In view of the high chemical reactivity and alkylating action of diazomethane, it is perhaps not surprising that it has been found by Auerbach to be a mutagen.[77]

Two serious intoxications from the laboratory use of diazomethane have been reported, one of them fatal. A graduate chemist developed violent coughing during brief exposure to diazomethane and for 2 weeks thereafter had repeated attacks of shortness of breath and coughing. His symptoms gradually increased after exposure until on the sixth day his condition was grave, and the clinical findings were compatible with severe respiratory tract irritation and pulmonary edema. His condition improved with treatment. By 2 weeks after the accident, he was free of attacks.[75] LeWinn[78] has credited diazomethane with the fatality of a research chemist engaged in a process employing known pulmonary irritants (phosphorus pentachloride, hydrogen chloride, and acetyl chloride) and diazomethane. The patient developed a cough during a distillation operation. His symptoms progressed to severe respiratory distress with fever. The clinical course was consistent with fulminating pneumonia. He died 4 days after exposure. Autopsy examination revealed generalized acute inflammation of the trachea, bronchi, and bronchioles.[78]

4. Warning Properties, Determination in Air, and Threshold Limit Value

The warning properties of diazomethane are not adequate to prevent serious intoxication or fatalities. Diazomethane can be determined by its reaction with benzoic acid and titration of the excess acid with barium hydroxide.[14] No threshold

[74] F. Flury and F. Zernik, *Schädliche Gase*. Springer, Berlin, 1931.

[75] F. W. Sunderman, R. Connor, and H. Fields, *Am. J. Med. Sci.*, **195**, 469 (1938).

[76] K. Landsteiner and A. H. DiSomma, *J. Exptl. Med.*, **68**, 505 (1938).

[77] C. Auerbach, *Ann. N. Y. Acad. Sci.*, **68**, 731 (1958).

[78] E. B. LeWinn, *Am. J. Med. Sci.*, **218**, 556 (1949).

limit value has been proposed. In view of the potentially lethal effect of unnoticed overexposure and the development of hypersensitivity, it is justly classified as an extrahazardous material. Anyone contemplating the use of diazomethane should be fully aware of its dangerous properties and should employ all precautions to prevent exposure.

NITROSOMETHYLURETHANE, $CH_3N(NO)COOCH_2CH_3$

1. Uses

Nitrosomethylurethane is a sweet-smelling, lightly colored, slightly volatile liquid that has been mainly used for the preparation of diazomethane. It will hydrolyze in water to form methylnitrosamine, which is unstable and decomposes to diazomethane. Nitrosomethylurea has greater use for this purpose since it is conveniently prepared from methylurea and hydrolyzes with alkalies to diazomethane.

2. Physical and Chemical Properties

Molecular weight: 132.12
Boiling point: 65°C. at 13 mm. Hg (decomposes on overheating)
Specific gravity: 1.122 (20/4°C.)
Solubility: slightly soluble in hot water, infinitely soluble in alcohol, ether, and benzene
1 mg./liter $\backsimeq$ 185 p.p.m. and 1 p.p.m. $\backsimeq$ 5.4 mg./cu. meter

3. Physiological Response

There is practically no toxicological information on true nitroso compounds, in which the nitroso group ($-N=O$) is attached to a carbon atom, except for some information on the aromatic series (see Chapter XLVII on aromatic nitrogen compounds). These compounds tend to be skin irritants and skin sensitizers (nitrosodimethyl aniline) and some are methemoglobin formers (p-nitrosotoluene). Skin irritation from p-nitrosophenol and skin sensitization from nitrosodiethyl-N-toluidine have been observed.[31]

Nitrosomethylurea ($NH_2CON(NO)CH_3$) and nitrosomethylurethane are known to cause dermatitis. The reaction to nitrosomethylurea appears to be one of sensitization dermatitis rather than primary irritation.[79,80] Nitrosomethylurethane vapor is highly irritating to the eyes and respiratory tract and direct contact with the liquid causes skin irritation and blistering. These effects may appear after delay of several hours or a day and subside slowly.[80,81] Experience indicates that the greatest care is necessary in handling this material if injury is to be avoided.

[79] W. E. Rosen, *Chem. Eng. News,* **31,** 2132 (1953).
[80] R. M. Watrous, *Brit. J. Ind. Med.,* **4,** 111 (1947).
[81] F. Wrigley, *Brit. J. Ind. Med.,* **5,** 26 (1948).

In some cases it has been necessary to discontinue use since reactions occurred despite precautions.[31]

DIMETHYLNITROSAMINE, (CH₃)₂NNO (*N*-Nitrosodimethylamine)

1. *Source and Use*

Dimethylnitrosamine is a mobile yellowish liquid with a faint characteristic odor. It is neutral and, unlike primary nitrosamines and the nitrosamides, it is stable. It can be prepared by the addition of acetic acid and sodium nitrite to dimethylamine. It has come into industrial prominence in the manufacture of 1,1-dimethylhydrazine.

2. *Physical and Chemical Properties*

Molecular weight: 74.08
Boiling point: 152°C.
Specific gravity: 1.015 (20°C.)
Solubility: completely miscible with water, alcohol, ether, and methylene chloride

1 mg./liter $\approx$ 330 p.p.m. and 1 p.p.m. $\approx$ 3.03 mg./cu. meter

3. *Physiological Response*

At least one death and several cases of poisoning have been attributed to dimethylnitrosamine exposure. It is absorbed from the gastrointestinal tract and lungs, but skin adsorption is slow. When administered to rats, mice, and rabbits, it is distributed uniformly in tissue and metabolized rapidly, with a half-life of approximately 4 hours. Although the liver is the main organ concerned with its metabolism and is the site of selective toxicity, dimethylnitrosamine does not concentrate there. Only a small percentage is excreted unchanged in rat urine after oral and intravenous doses of 50 to 100 mg./kg. A large proportion of C^{14}-labeled dimethylnitrosamine appears in the expired air as C^{14}-labeled carbon dioxide (approximately 60 per cent in 24 hours). Although it has been suggested that a metabolite may be responsible for the toxic liver injury, the metabolites have not been identified.[82-84] In connection with the high reactivity and the carcinogenic activity of this material, it is interesting that Farber and Magee have been able to show that, after injection of C^{14}-labeled dimethylnitrosamine, liver RNA contains guanine that has been methylated in the 7 position.[85]

Barnes and Magee[86] found that dimethylnitrosamine produced deaths in rats, rabbits, mice, guinea pigs, and dogs at oral or intraperitoneal doses on the order of

[82] P. N. Magee, *Biochem. J.*, **64**, 676 (1956).
[83] P. N. Magee, *Biochem. J.*, **70**, 606 (1958).
[84] D. F. Heath and A. Dutton, *Biochem. J.*, **70**, 619 (1958).
[85] P. N. Magee, personal communication, 1960.
[86] J. M. Barnes and P. N. Magee, *Brit. J. Ind. Med.*, **11**, 167 (1954).

25 mg./kg. In rats and guinea pigs, deaths were often delayed for from 4 to 8 days following a single dose, whereas, in rabbits, mice, and dogs, deaths occurred earlier. All animals showed little in the way of symptoms except for general quietness and appearance of illness prior to death. Histological examination revealed severe liver necrosis, which was characteristically centrilobular. There was hemorrhage into the gut and peritoneal cavity in the dogs, rats, and guinea pigs. No significant cellular changes were observed in other organs. It is noteworthy that the earliest detectable liver damage appeared within a few hours after a single dose. Repeated feedings in rats resulted in liver damage with regeneration and fibrosis. The histological changes in the liver had some features in common with, but were distinguishable from, those produced by carbon tetrachloride.[86] These investigators also found that 19 of 20 rats fed 50 p.p.m. of dimethylnitrosamine in their diet developed primary hepatic tumors between 26 and 40 weeks of treatment; 7 of these tumors showed metastatic spread. Six rabbits fed 20 p.p.m. for 10 weeks, 30 p.p.m. for 4 additional weeks, and 50 p.p.m. for 8 additional weeks developed poisoning

TABLE 16

Response to Four-Hour Vapor Inhalation of Dimethylnitrosamine—Animals[a]

Species	Concentration, p.p.m.	Mortality
Mouse	57	LD_{50} (14 days)
	60	$^9/_{10}$ (1–3 days)
	39	$^1/_{10}$ (1st day)
Rat	78	LD_{50} (14 days)
	151	$^{10}/_{10}$ (2–4 days)
	51	$/_{10}$ (3rd day)
Dog	43–144	$^9/_9$ (1–3 days)
	16	$^2/_3$ ($^3/_3$ had evidence of liver injury)

[a] K. Jacobson, H. J. Wheelwright, J. H. Clem, and R. M. Shannon, *A.M.A. Arch. Ind. Health,* **12,** 617 (1955).

and liver injury with deaths occurring between the eleventh and twenty-second weeks. However, no tumors or malignancies developed.[87] More recently, it has been found that rats develop renal tumors after intermittent feeding and, in survivors, even after a single dose around the acute LD_{50}.[85] The liquid does not exhibit significant primary irritant properties for skin or eyes. Jacobson and co-workers found that nitrosodimethylamine vapor caused liver damage, ascites, and disruption of blood coagulation with bleeding in dogs, rats, and mice[88] (see Table 16).

Freund[89] described 2 cases of severe liver damage with jaundice and ascites in chemists manufacturing dimethylnitrosamine. One recovered after a prolonged

[87] P. N. Magee and J. M. Barnes, *Brit. J. Cancer,* **10,** 114 (1956).
[88] K. Jacobson, H. J. Wheelwright, J. H. Clem, and R. N. Shannon, *A.M.A. Arch. Ind. Health,* **12,** 617 (1955).
[89] H. A. Freund, *Ann. Internal Med.,* **10,** 1144 (1937).

illness and the other died. Autopsy revealed acute diffuse centrilobular necrosis of the liver with intense regenerative activity.[89] Two similar (possibly the same) cases were reported by Hamilton and Hardy.[90] Other nonfatal intoxications with liver injury and elevated body temperatures have been described from exposure to dimethylnitrosamine.[86,88] Of the 2 cases reported by Barnes and Magee, cirrhosis was an incidental finding in one and in the other there was evidence of improvement in liver function after 3 months of nonexposure.

4. Warning Properties, Determination in Air, and Threshold Limit Value

The human experience and the experimental vapor exposures in animals show that the vapor pressure of dimethylnitrosamine is sufficient to result in concentrations that can be lethal. Dimethylnitrosamine is not significantly irritating to the eyes or mucous membranes nor does the odor present adequate warning properties. Concentrations in air can be determined by application of the polarographic method of Heath and Jarvis[91] or by collection in water and spectrophotometric determination of the absorption at 335 mμ.[88] On the basis of their animal studies, Barnes and Magee suggested that 250 mg./cu. meter of air would be a clearly dangerous concentration and that the safe level should not be more than 25 mg./cu. meter.[86] The experiments of Jacobson and co-workers indicate that, assuming man might be as susceptible as dogs, a 4-hour exposure to a concentration of 48 mg./cu. meter (16 p.p.m.) would be very dangerous.[88] A threshold limit value for repeated daily exposure has not been proposed.

HYDRAZINES

I. General Considerations

A. PROPERTIES AND USES

Hydrazine (NH_2NH_2) and its derivatives are reactive materials with a variety of biological effects and considerable toxicity. As a group, the alkyl hydrazines are basic liquids and strong reducing agents. They are useful as intermediates in organic syntheses. Hydrazine has been used in photography, metal processing, preservatives, and the preparation of anticorrosives, textile agents, insecticides, and pharmaceuticals. Its methyl derivatives, particularly 1,1-dimethylhydrazine, as well as hydrazine itself, are important rocket fuels. The carboxylic acid derivatives of hydrazine ($RCONHNH_2$), isonicotinic acid hydrazide (isoniazide) and maleic hydrazide, have stimulated interest because of their use as a therapeutic agent for tuberculosis and as an agricultural chemical, respectively. Of the aryl hydrazines, phenylhydrazine is a familiar and useful chemical reagent and intermediate, and has also been used in medicine.

[90] A. Hamilton and H. L. Hardy, *Industrial Toxicology*, 2nd ed., Hoeber, New York, 1949.

[91] D. F. Heath and J. A. E. Jarvis, *Analyst*, **80**, 613 (1955).

B. PHYSIOLOGICAL RESPONSE

1. Response in Animals

The hydrazine derivatives tend to be local irritants, convulsants, hepatotoxins, and hemolytic agents, which are absorbed by all routes of administration. There are, however, important variations among individual members of the series. Hydrazine is a strong skin and mucous membrane irritant, a convulsant, hepatotoxin, and a moderate hemolytic agent. It is absorbed from the lungs, gastrointestinal tract, parenteral injection sites, and through the intact skin. Inhalation results in eye and respiratory tract irritation with lung congestion, bronchitis, and pulmonary edema in some animals. Nervous system symptoms ending in convulsions are prominent after absorption by any route.[92,93] Similar effects are exhibited by 1,1-dimethylhydrazine (*asym*-dimethylhydrazine) but it is less irritating to the skin and it is not as highly toxic by this route. Its oral toxicity is somewhat lower than hydrazine but its acute vapor toxicity is greater. Central nervous system stimulation with terminal convulsions is also observed with *sym*-dimethylhydrazine and methylhydrazine. Methylhydrazine is the strongest convulsant and the most toxic of the methyl derivatives.[94-96] Their properties and acute toxicities can be compared by reference to the information summarized in Table 17.

A large number of other hydrazine derivatives are convulsants including acetylhydrazine ($CH_3CONHNH_2$), benzhydrazide ($C_6H_5CONHNH_2$), and isonicotinic acid hydrazide ($C_5H_5NCONHNH_2$). The monoacylhydrazines have greater convulsant potency and are more toxic than the simple 1,2-diacylhydrazines (acetylhydrazine: intravenous LD_{50} in mice 175 mg./kg.—liver damage; 1,2-diacetylhydrazine: LD_{50} 3000 mg./kg.—no liver damage).[97] The convulsant activity of a series of hydrazides have been studied by Jenny and Pfeiffer.[98]

Liver damage is an important feature of hydrazine toxicity, particularly after chronic exposure. This effect has long been recognized and has frequently been used in experimental studies of liver injury in animals. However, despite predictions to the contrary, hydrazinelike liver injury has not been observed after single doses of the methyl derivatives or after chronic intoxication with 1,1-dimethylhydrazine.[94,99] On the other hand, many other hydrazine derivatives have been

[92] S. Krop, *Arch. Ind. Hyg. Occupational Med.*, **9**, 199 (1954).

[93] C. C. Comstock, L. H. Lawson, E. A. Greene, and F. W. Oberst, *Arch. Ind. Hyg. Occupational Med.*, **10**, 476 (1954).

[94] K. H. Jacobson, J. H. Clem, H. J. Wheelwright, W. F. Rinehart, and N. Mayer, *A.M.A. Arch. Ind. Health*, **12**, 609 (1955).

[95] L. B. Witkin, *A.M.A. Arch. Ind. Health*, **13**, 34 (1956).

[96] S. Rothberg and O. B. Cope, *U. S. Army Chem. Warfare Lab. Rept.* No. **2027,** Army Chemical Center, Maryland, 1956.

[97] H. McKennis, A. S. Yard, J. H. Weatherby, and J. A. Hagy, *J. Pharmacol. Exptl. Therap.*, **126**, 109 (1959).

[98] E. H. Jenny and C. C. Pfeiffer, *J. Pharmacol. Exptl. Therap.*, **122**, 110 (1958).

[99] W. E. Rinehart, E. Donati, and E. A. Greene, *Am. Ind. Hyg. Assoc. J.*, **21**, 207 (1960).

TABLE 17

Physical and Chemical Properties and Toxicity of Hydrazine and Methylhydrazines[b]

	Hydra-zine	Methyl-hydrazine	1,1-Dimethyl-hydrazine	1,2-Dimethyl-hydrazine
Formula	NH_2NH_2	CH_3NHNH_2	$(CH_3)_2NNH_2$	$CH_3NHNHCH_3$
Physical state	Liquid	Liquid	Liquid	Liquid
Molecular weight	32.05	46.07	60.10	60.10
Boiling point, °C.	113.5	87	62.5	81
		(745 mm. Hg)	(717 mm. Hg)	(747 mm. Hg)
Vapor pressure, mm. Hg, (25°C.)	14.4	49.6	156.8	69.9
Solubility medium	Water, alcohol	Water, alcohol, ether	Water, alcohol, ether	Water, alcohol, ether
Oral LD_{50}, rat, mg./kg.	60	32.5	122	160
Intraperitoneal LD_{50}, rat, mg./kg.	59	32	131	163
Skin, LD_{50}, guinea pig, mm.³/kg.	190	56	1680	158
Vapor inhalation, 4-hr. LC_{50}, rat, p.p.m.	570	74	252	280–400
Hazard index, vapor pressure/4-hr. LC_{50}	0.03	0.67	0.62	0.21[a]

[a] 340 used as LC_{50} for this calculation.

[b] K. H. Jacobson, J. H. Clem, H. J. Wheelwright, W. F. Rinehart, and N. Mayer, *A. M. A. Arch. Ind. Health*, **12**, 609 (1955); L. K. Witkin, *ibid.*, **13**, 34 (1956); S. Rothberg and O. B. Cope, *U. S. Army Chem. Warfare Lab. Rept.* No. **2027**, Army Chemical Center, Maryland, 1956.

observed to cause alterations in liver function, fatty infiltration, or cellular damage. These include diisopropylhydrazine, benzhydrazide, phenacethydrazide, acetylhydrazine, phenylhydrazine, acetylphenylhydrazine and the drugs, isoniazide and iproniazide.[97,100–102] The characteristic liver pathology seen after effective doses of the hepatotoxic hydrazines is fatty degeneration, which is most marked in the center of the liver lobules. There may also be damage to other organs, particularly the kidney. In the case of hydrazine, the centrilobular liver damage is predominant, whereas with the phenylhydrazines, the liver damage is more diffuse and kidney injury is prominent. Other pathological changes, such as extramedullary hepatopoiesis, pigment (hemosiderin) deposition in cells, and vascular thrombosis, are secondary to blood cell destruction.

Hydrazine hemolyzes red blood cells when injected intravenously or when given by stomach tube, but blood cell destruction and anemia are less consistent findings after subcutaneous injection or inhalation, usually occurring only after

[100] M. Bodansky, *J. Biol. Chem.*, **58**, 799 (1924).
[101] M. Bodansky, W. L. Marr, and P. Brindley, *Am. J. Clin. Pathol.*, **2**, 391 (1932).
[102] A. Yard and H. McKennis, *J. Pharmacol. Exptl. Therap.*, **114**, 391 (1955).

chronic exposure and severe intoxication.[93] Phenylhydrazine exhibits a powerful and more consistent hemolytic effect by all routes of administration. Many of the phenylhydrazine derivatives that have been studied also show anemiagenic activity, although this tends to be reduced as substitutions are made in the hydrazine radical.[103,104] Diphenylhydrazine and diisopropylhydrazine, however, are active hemolysins.[105] Of the methyl derivatives, monomethylhydrazine is more potent in this respect than are hydrazine or the dimethylhydrazines, since it produces moderately severe intravascular hemolysis after a single 4-hour vapor exposure and the others do not.[94] Hemolytic anemia does appear after chronic inhalation of 1,1-dimethylhydrazine.[99]

Skin sensitization and primary irritation effects of the hydrazines have not been completely studied. However, hydrazine and especially phenylhydrazine are known to be skin irritants and strong skin sensitizers. The methyl hydrazines are less irritating; monomethylhydrazine being more active than dimethylhydrazine, but less active than hydrazine.

2. Metabolism

Several possible metabolic pathways for hydrazine and its derivatives have been postulated including (1) hydrolysis of hydrazides to hydrazine and the carboxylic acid, (2) reaction of hydrazine and hydrazides with natural aldehydes and ketones, (3) acetylation, and (4) splitting of symmetrical disubstituted hydrazines to yield 2 amines. Acethylation has been shown to occur with hydrazine in the rabbit, but not in dogs; and with isonicotinic acid hydrazide in man, monkey, and rats. A number of other hydrazino compounds are metabilized to 1,2-diacylhydrazines in the rabbit.[21,97] A portion of the hydrazine administered to dogs is excreted unchanged in the urine, and it has also been found in rabbit urine after administration of several hydrazides.[97] It is probable that the toxicity of hydrazine derivatives and their ability to cause hepatocellular damage depends in part on the competing rates at which they are converted to hydrazine and monoacylhydrazines (liver toxins, convulsants) or diacylhydrazines (not hepatotoxic). Phenylhydrazine apparently is not acetylated in animals. p-Hydroxyphenylhydrazine and phenylhydrazones of natural keto acids have been identified as metabolites.[21]

3. Effects in Man

Most of the injurious effects of hydrazine in man have been reported by chemists. These have been limited to prompt eye and upper respiratory tract irritation after vapor inhalation, severe eye and skin damage after direct liquid contact, and sensitization-type dermatitis. Phenylhydrazine, on the other hand,

[103] W. F. von Oettingen and W. Deichmann-Gruebler, J. Ind. Hyg. Toxicol., **18**, 1 (1936).

[104] W. F. von Oettingen, The aromatic amino and nitro compounds, their toxicity and otential dangers, U. S. Public Health Bull. No. **271**, 1941.

[105] M. Bodansky, J. Pharmacol. Exptl. Therap., **23**, 127 (1924).

has caused systemic injury, characterized mainly by anemia, and many cases of dermatitis in chemical laboratories, industry, and after medicinal use. Although laboratory evidence of liver damage has been observed in some workers exposed to 1,1-dimethylhydrazine, it appears possible that this was due to their associated exposure to nitrosodimethylamine. These observations on man are considered more fully under the specific compounds.

4. Determination in Air, Warning Properties, Threshold Limit Values

Hydrazine and the methylhydrazines can be determined by potassium iodate titration, potentiometric titration with potassium bromate (except for 1,2-dimethylhydrazine), or a colorimetric molybdenum blue procedure.[106] Collection in standard acid and titration with alkali has also been used for hydrazine.[93] These materials have distinct amine, fishy, or ammoniacal odors which, according to Jacobson et al., have median detectable odor levels of 1 to 10 p.p.m.[94] This should constitute adequate warning, in the average informed worker, of acutely dangerous concentrations. Threshold limit values of 1 p.p.m. (1.3 mg./cu. meter), 0.5 p.p.m. (1 mg./cu. meter), and 5 p.p.m. (22 mg./cu. meter) have been proposed for hydrazine, 1,1-dimethylhydrazine, and phenylhydrazine, respectively.[11]

II. Specific Compounds

HYDRAZINE, NH₂NH₂ (Diamide)

1. Source

Anhydrous hydrazine is prepared from the hydrate obtained from hydrazine sulfate, which is produced by the oxidation of ammonia or urea by sodium hypochlorite. Both anhydrous hydrazine and the hydrate are fuming, strongly basic ($pK_{b1} = 5.52$), colorless liquids. Hydrazine may ignite under various circumstances, such as on contact with rust, and it decomposes violently on contact with oxidizing materials. It attacks cork, rubber, and other organic materials. In order to reduce the flammable hazard and to maintain purity, it is usually stored under nitrogen. Exposures most commonly occur during its use as a chemical intermediate or rocket fuel.

2. Physical and Chemical Properties

Molecular weight: 32.05
Boiling point: 113.5°C.
Density: 1.008 g./cc. (20°C.)
Vapor density: 1.11 (air = 1)
Vapor pressure: 14.4 mm. Hg, 25°C.
Flash point: 100 to 126°F.

[106] L. Feinsilver, J. A. Perregrino, and C. J. Smith, *Am. Ind. Hyg. Assoc. J.*, **20**, 26 (1959).

Explosive limits: 4.7 to 100 per cent

Solubility: completely miscible with water and alcohol; insoluble in hydrocarbons

1 mg./liter $\simeq$ 764 p.p.m. and 1 p.p.m. $\simeq$ 1.3 mg./cu. meter

3. Physiological Response

Hydrazine is well absorbed by all routes of administration including application to the intact skin. Witkin points out that since the median lethal doses by oral, intravenous, and intraperitoneal routes are all roughly equivalent, there must be rapid absorption and slow detoxication and/or excretion.[95] Hydrazino-nitrogen (assumed to be largely unchanged hydrazine) is excreted in the urine after intravenous or subcutaneous administration of hydrazine in dogs. Five to 11 per cent of large doses (50 mg./kg.—twice the LD_{50}) is excreted within the first 4 hours and approximately 50 per cent of 15 mg./kg. doses is excreted within the first 2 days after injection.[107] A small percentage of hydrazine administered to rabbits is excreted as 1,2-diacetylhydrazine but this metabolite cannot be identified in dog urine.[97]

TABLE 18

Acute Toxicity of Hydrazine—Animals[a]

Species	Route of administration	LD_{50}, mg./kg. $\pm$ 1 S.D.
Dog	Intravenous	25
Mouse	Intravenous	57 $\pm$ 7.5
	Intraperitoneal	62 $\pm$ 4.0
	Oral	59 $\pm$ 7.2
Rat	Intravenous	55 $\pm$ 2.7
	Intraperitoneal	59 $\pm$ 3.9
	Oral	60 $\pm$ 3.8
Rabbit	Intravenous	34
	Percutaneous	91

[a] S. Krop, Arch. Ind. Hyg. Occupational Med., 9, 199 (1954); L. B. Witkin, A.M.A. Arch. Ind. Health, 13, 34 (1956).

Several groups of investigators have studied the acute and chronic toxicity of hydrazine for experimental animals.[92-96] Some of their results are summarized in Tables 17, 18, and 19. The effects noted after oral administration, injection, or skin application include anorexia, weight loss, weakness, vomiting, excitement, and convulsions; the chief histological findings are fatty degeneration of the liver, and a lesser degree of kidney inflammation. Inhalation of high concentrations of hydrazine results in respiratory tract irritation with histological evidence of

[107] H. McKennis, J. H. Weatherby, and L. B. Witkin, J. Pharmacol. Exptl. Therap., 114, 385 (1955).

TABLE 19

Toxicity of Hydrazine by Inhalation—Rats[a]

Concentration, p.p.m.	Duration of exposure	Response	Remarks
570	4 hrs.	LC$_{50}$	Dyspnea, convulsions
80–300	4 hrs.	$^{14}/_{30}$ died	Deaths delayed 1–13 days
54	6 hrs. daily	$^{14}/_{16}$ died	Deaths after 4–13 exposures
20	6 hrs. daily	$^{11}/_{13}$ died	Deaths after 13–30 exposures
14	6 hrs. daily	$^{23}/_{30}$ died	Deaths after 1–105 exposures
5	6 hrs. daily	$^{2}/_{10}$ died	Deaths in 28th week, symptoms in all

[a] C. C. Comstock, L. H. Lawson, E. A. Greene, and F. W. Oberst, *Arch. Ind. Hyg. Occupational Med.*, **10**, 476 (1954); K. H. Jacobson, *et al.*, *A.M.A. Arch. Ind. Health*, **12**, 609 (1955).

damage to the lungs, liver, and kidneys. Inhalations of concentrations insufficient to produce significant lung damage can result in vomiting, diarrhea, weight loss, convulsions, and death. Two dogs exposed to 5 p.p.m. for 6 months developed decreased appetite, weight loss, easy fatiguability, and muscular tremors, but survived the full period of exposure and showed very little pathology. Two of 4 dogs survived 194, 6-hour exposures to concentrations averaging 14 p.p.m. All 4 animals developed signs of severe intoxication.[93] There is evidence that the lung, liver, and kidney injury, which may occur after repeated vapor exposures, is reversible within several months after cessation of exposure.[93]

Although it is known the vapors of hydrazine are highly irritating to the eyes, nose, and throat, and direct liquid contact with the skin or eyes produces severe burns, detailed descriptions of systemic effects in man are lacking.[92,93] Evans[108] has described contact dermatitis of the hands appearing in 2 individuals about 5 months after their first contact with hydrazine hydrate. Neither showed any signs of systemic intoxication.[108] Cook has reported dermatitis of the hands and face in workers using hydrazine hydrobromide soldering flux; however, there were no systemic effects.[109]

4. Warning Properties, Determination in Air, and Threshold Limit Value

It appears that the warning properties (irritation and odor) of hydrazine are sufficient to prevent systemic injury following acute vapor exposures. It is not certain that the warning properties would be adequate for prolonged exposures. Jacobson and co-workers[94] found that the median detectable concentration of hydrazine vapor in air for less than 1 minute exposure was 3 to 4 p.p.m. On the

[108] D. M. Evans, *Brit. J. Ind. Med.*, **16**, 126 (1959).
[109] W. A. Cook, *Arch. Gewerbepathol. Gewerbehyg.*, **13**, 616 (1955).

other hand, the suggested threshold limit value, based on Comstock's findings of symptoms or injury in dogs and rats exposed to 5 p.p.m., is 1 p.p.m. or 1.3 mg./cu. meter.[11] Observations correlating air analyses and the presence or absence of symptoms in man are needed. Hydrazine can be determined in air using the methods suggested by Feinsilver and co-workers,[106] or in the presence of ammonia by the method of McKennis and Witkin.[110]

1,1-DIMETHYLHYDRAZINE, $(CH_3)_2NNH_2$ (asym-Dimethylhydrazine, unsym-Dimethylhydrazine (UDMH))

1. Uses

1,1-Dimethylhydrazine is a volatile, flammable, strongly alkaline colorless liquid with a sharp ammoniacal or aminelike odor. It is manufactured from dimethylnitrosamine for use as a rocket propellant fuel.

2. Physical and Chemical Properties

Molecular weight: 60.1
Boiling point: 62.5°C. (717 mm. Hg)
Vapor pressure: 156.8 mm. Hg, 25°C.
Solubility: soluble in water, alcohol, and ether
1 mg./liter $\simeq$ 4.07 p.p.m. and 1 p.p.m. $\simeq$ 2.5 mg./cu. meter

TABLE 20

Acute Toxicity of 1,1-Dimethylhydrazine—Animals[a]

Species	Route of administration	LD_{50}, mg./kg. $\pm$ 1 S.D.
Dog	Intravenous	60
Mouse	Intravenous	250 $\pm$ 19
	Intraperitoneal	290 $\pm$ 38
	Oral	265 $\pm$ 22
Rat	Intravenous	119 $\pm$ 3.8
	Intraperitoneal	131 $\pm$ 5.1
	Oral	122 $\pm$ 10.6

[a] L. B. Witkin, A.M.A. Arch. Ind. Health, **13**, 34 (1956).

3. Physiological Response

The metabolic fate of 1,1-dimethylhydrazine is not known. It is rapidly absorbed from the lungs, gastrointestinal tract and injection sites, and skin application can cause death, although it is less toxic by this route than hydrazine and the other methyl derivatives.[96] The acute and chronic toxicity for experimental animals is summarized in Tables 17, 20, and 21. The outstanding effects after single doses are central nervous system stimulation and convulsions, and the most sig-

[110] H. McKennis and L. B. Witkin, A.M.A. Arch. Ind. Health, **12**, 511 (1955).

TABLE 21

Toxicity of 1,1-Dimethylhydrazine Vapor—Animals[e]

Species	Concentration		Duration,[d] weeks	Mortality
	p.p.m.	mg./cu. meter		
Mouse	140	342	6	$29/30^{a}$
	75	183	7	$8/30^{b}$
Rat	140	342	6	$1/20^{b}$
	75	183	7	$0/30$
Dog	25	61	13	$1/3^{c}$
	5	12.2	26	$0/3$

[a] Deaths occurred within 2 weeks.
[b] Deaths occurred within 5 weeks.
[c] Death occurred after third exposure.
[d] 6 hours per day, 5 days a week.
[e] W. E. Rinehart, E. Donati, and E. A. Greene, Am. Ind. Hyg. Assoc. J., 21, 207 (1960).

nificant results of chronic inhalation are hemolytic anemia and convulsive seizures. Dogs exposed for approximately 3 hours to a vapor concentration of 111 p.p.m. showed salivation, vomiting, respiratory distress, and convulsions. All 3 died on the day of exposure. Similar symptoms were observed in 2 of 3 dogs exposed to 52 p.p.m. for 4 hours; 1 of these died. Three dogs survived 4-hour exposures to 24 p.p.m. Two showed no signs or symptoms, whereas the third developed vomiting and convulsions from which it recovered. None of the surviving animals showed significant changes in hemoglobin, red blood cell count, white blood cell count, sulfabromophthalein retention, or prothrombin time. Post-mortem examination revealed no significant pathological changes except that those animals that had convulsions also had pulmonary edema and patchy pulmonary hemorrhage.[94] Three dogs exposed repeatedly to 25 p.p.m. developed depression, salivation, vomiting, diarrhea, ataxia, convulsive seizures, and hemolytic anemia. There was a decrease in the hematocrit, hemoglobin, and red blood cell counts and hemosiderin was deposited in the cells of the reticuloendothelial system. One dog died on the third day of the 13-week exposure period. No severe toxic signs were observed in 3 dogs exposed to 5 p.p.m. for 6 hours a day, 5 days a week for 26 weeks. Lethargy and weight loss developed after 2 or 3 weeks of exposure. There was evidence of mild anemia after 6 weeks of exposure. Examination, 2 weeks after exposure, showed only hemosiderin deposition in the spleen with no lesions in the other organs.[99] No liver damage has been identified following acute or chronic intoxication with 1,1-dimethylhydrazine. It is less irritating for the skin, eyes, and mucous membranes than hydrazine or monomethylhydrazine.[96]

Shook and Cowart cite two instances in which workers experienced respiratory distress and, later, nausea and vomiting after accidental exposure to 1,1-dimethylhydrazine vapor.[111] Others have observed that acute accidental exposures

[111] B. S. Shook and O. H. Cowart, Ind. Med. and Surg., 26, 333 (1957).

will produce nose and throat irritation, mild conjunctivitis, and nausea. Lower respiratory tract irritation and bronchitis have apparently not been seen.[112] Shook and Cowart[111] reported that 11 individuals exposed to UDMH developed laboratory findings of possible hepatocellular changes (positive cephalin cholesterol flocculation tests) but no clinical symptoms or signs. Since, 1,1-dimethylhydrazine has not been shown to cause liver damage in animals, and others have not observed laboratory evidence of liver injury in persons exposed to sufficient concentrations to develop mucous membrane irritation,[112,113] it appears that some factor other than liver injury from UDMH was responsible.

4. Warning Properties, Determination in Air, and Threshold Limit Value

A fishy or aminelike odor of 1,1-dimethylhydrazine can be detected by most individuals in less than 1 minute at concentrations of 6 to 14 p.p.m. The odor offers adequate warning of exposures to concentrations that would be dangerous for short exposures.[94] Air concentrations can be determined by collection in an acid solution and analysis by potassium iodate titration or a colorimetric molybdenum blue procedure. The molybdenum blue method is sufficiently sensitive to determine microgram quantities of 1,1-dimethylhydrazine.[106] Since dogs exposed to 5 p.p.m. for 26 weeks exhibited mild toxic effects, it has been suggested that 0.5 p.p.m. should be used as a tentative threshold limit value and as a guide to safe handling practices.[11,99]

PHENYLHYDRAZINE, $C_6H_5NHNH_2$

1. Source and Uses

Phenylhydrazine is a colorless, oily liquid at room temperature. It is a weakly basic ($pK_b = 8.79$) reducing agent with a faint aromatic odor. It is prepared by the diazotization of aniline with sodium nitrite and hydrochloric acid followed by treatment with sodium sulfite and liberation of the base with sodium hydroxide. Because it reacts readily with carbonyl compounds, it is a valuable reagent for sugars, aldehydes, and ketones. It is also used in the manufacture of dyes and other organic syntheses, and has been used in medicine.

2. Physical and Chemical Properties

Molecular weight: 108.14
Melting point: 19.5°C.
Boiling point: 243.5°C. (decomposes)
Density: 1.0978 (20/4°C.)
Solubility: miscible with alcohol, ether, chloroform, and benzene; sparingly soluble in water

[112] P. J. Clancy, Aerojet General Corp., unpublished observations.

[113] K. H. Jacobson, U. S. Army Chem. Warfare Lab. Spec. Publ. No. **2-10**, Army Chemical Center, Maryland, 1958, p. 53.

1 mg./liter $\backsimeq$ 226 p.p.m. and 1 p.p.m. $\backsimeq$ 4.4 mg./cu. meter

3. *Physiological Response*

The outstanding effects of phenylhydrazine in man and animals are contact dermatitis (skin sensitization), hemolytic anemia, and liver and kidney injury. It is absorbed from the gastrointestinal tract and through the intact skin. It is said that it is less well absorbed from the lungs but this does not appear to have been carefully studied.[104] Studies with the C^{14}-labeled compound in rabbits show that 30 to 50 per cent of oral doses of 50 mg./kg. is excreted in the urine in 48 hours and 40 to 60 per cent in 4 days. There is continued excretion of the metabolites, *p*-hydroxyphenylhydrazine and the phenylhydrazones of pyruvic acid and α-oxoglutaric acid, at 10 days. In animals killed 4 days after dosing, 5 to 10 per cent of the radioactivity is found in the erythrocytes.[114]

The minimum subcutaneous dose in mice that is lethal for 100 per cent of the animals is 180 mg./kg.[103] These animals show progressive cyanosis, irregular respiration, and increasing dyspnea progressing to death. Degenerative lesions are found in the liver, kidneys, and other organs along with evidence of vascular damage. Subcutaneous injection of around $1/_{10}$ the lethal dose (i.e., 20 mg./kg. in dogs) results in severe anemia, which is due to blood destruction (hemolysis) rather than bone marrow damage. This contention is supported by: (*1*) the appearance of red blood cell pigments in the circulating blood and their deposition in tissues, (*2*) the appearance of immature blood cells, reticulocytes, and pathological red blood cell forms (Heinz bodies), and (*3*) the regular occurrence of bone marrow hyperplasia with increased marrow cell oxygen consumption and mitotic activity. Repeated dosing results in liver damage with hepatic insufficiency and kidney injury as well as anemia. Extramedullary hepatopoiesis is a frequent finding.[13,100,101,103-105]

Experience in effects of phenylhydrazine in man has been obtained from the use of the hydrochloride in the treatment of polycythemia vera (a condition in which there is an overabundance of blood cell elements). The usual oral dose was 0.2 g./day for 3 or 4 doses, then 0.1 g./day until leukocytosis or a decrease in hemoglobin became manifest, usually after a total of 1 to 3 g. had been given. Some of the undesirable side effects noted were jaundice, anorexia, nausea, dermatitis, and vascular thromboses.[13] Stealy and Summerlin[115] report the interesting case of a women who was treated with phenylhydrazine hydrochloride intermittently for 11 years without evidence of liver or kidney damage. She eventually developed peripheral blood changes indicative of response to a toxic hemolytic agent and representative of bone marrow hyperactivity. At autopsy, there was cirrhosis of the liver, myeloid infiltration in the spleen, lymph nodes, and liver,

[114] W. McIsaac, D. Parke, and R. Williams, *Biochem. J.*, **65**, 15P (1957).
[115] C. L. Stealy and H. S. Summerlin, *J. Am. Med. Assoc.*, **126**, 954 (1944).

and splenomegaly.[115] These changes, however, cannot definitely be attributed to phenylhydrazine since they have also been observed in cases of polycythemia not treated in this way.

A development of considerable interest is an *in vitro* test to detect red blood cells deficient in glucose-6-phosphate dehydrogenase by their characteristic pattern of Heinz body formation produced by incubation with acetylphenylhydrazine or phenylhydrazine. This genetically determined defect is responsible for the abnormal sensitivity in some ethnic groups to the hemolytic effect of various drugs and fava beans.[116,117] When phenylhydrazine is given to these sensitive individuals in doses smaller than those usually required to cause hemolysis, hemoglobin levels may drop but return toward normal during medication.[118] However, hemolytic anemia of occupational origin has not been observed in persons known to have this red blood cell defect.[119]

Since the initial observations of Lewin,[120] many cases of contact dermatitis from phenylhydrazine have been reported in industrial employees, laboratory workers, and patients. It is apparent that phenylhydrazine is a potent skin sensitizer that will produce eczematous dermatitis with redness, swelling, and vesiculation in a high proportion of individuals who have repeated skin contact.[104] Hemolytic anemia identical to that seen in patients and animals treated with phenylhydrazine may occur. Fatalities from skin or vapor exposures have not been recorded.

4. Threshold Limit Value

The most important route of exposure is by skin contact and every effort should be made to prevent this. Quantitative data on the inhalation of various concentrations of phenylhydrazine are not available; however, a threshold limit value of 5 p.p.m. (22 mg./cu. meter) has been recommended.[11] If absorption from the respiratory tract is complete, continued exposure to this concentration could theoretically result in the absorption of as much as 0.2 g./day (assuming respiratory rate of 10 cu. meter per 8-hr working day). Since oral doses of this magnitude are effective in producing hemolysis in polycythemia patients, it is possible that continued exposures to these concentrations may be excessive. Consequently, in situations in which air concentrations of phenylhydrazine approach 5 p.p.m., and particularly where there is the additional possibility of skin contact, workers should be under careful medical observation. Phenylhydrazine can be determined in air by collection in cyclohexane, followed by measurement of the absorption in an ultraviolet spectrophotometer at 272 or 278 mμ.[31]

[116] E. Beutler, *Blood,* 14, 103 (1959).
[117] E. Beutler, R. J. Dern, and A. S. Alving, *J. Lab. Clin. Med.,* 45, 40 (1955).
[118] R. J. Dern, E. Beutler, and A. S. Alving, *J. Lab. Clin. Med.,* 45, 30 (1955).
[119] A. Szeinberg, A. Adam, F. Meyers, C. Sheba, and B. Ramot, *A.M.A. Arch. Ind. Health,* 20, 510 (1959).
[120] L. Lewin, *Z. Biol.,* 42, 107 (1901).

β-CHLOROETHYLAMINES

I. General Considerations

A. PROPERTIES AND USES

Intensive study of the biological effects of the β-chloroethylamines was stimulated during World War II by interest in their potential application as chemical warfare agents. Because the β-dichlorodiethylamines $RN(CH_2CH_2Cl)_2$, are similar to dichlorodiethylsulfide (sulfur mustard, mustard gas), $S(CH_2CH_2Cl)_2$, in their chemical structure and vesicant activity they have been called nitrogen mustards. Unlike sulfur mustard, the nitrogen mustards are absorbed through the skin and respiratory tract in sufficient amounts to produce systemic intoxication. Although they have not been used in warfare, studies on the biological properties of the β-dichlorodiethylamines led to their early application in the treatment of neoplastic disease in man.[121] Many of the actions and uses of the notrogen mustards resemble those of ethyleneimine derivatives. This is probably because in aqueous solutions the β-chloroethylamines are transformed into the highly reactive ethyleneimonium intermediates.

$$\begin{array}{cc} R' \\ \diagdown \\ \diagup \mathrm{NCH_2CH_2Cl} \\ R \end{array} \rightleftharpoons \begin{array}{c} R' + \diagup \mathrm{CH_2} \\ \diagdown \mathrm{N} \diagdown | \\ R \diagup \diagdown \mathrm{CH} \end{array} + \mathrm{Cl^-}$$

Nitrogen Mustard   Ethyleneimonium ion

These ions have been shown to react readily with a large number of organic compounds *in vitro*, especially with amino, sulfhydro, and carboxyl groups of proteins, and phosphate groups in nucleic acid and, therefore, can alkylate macromolecules of biological importance.[3,122] For this reason, the β-dichlorodiethylamines have been classified with other biological alkylating agents such as the ethyleneimines, the methane sulfonates, the bisepoxides, the β-lactones, and the sulfur mustards.[5]

The nitrogen mustards are tertiary amines in which the halogen atom and the amine portion have reactivity similar to alkyl halides and alkyl amines. They are oily liquids with limited water solubility but form readily soluble hydrochlorides. They are prepared by the action of thionyl chloride on the appropriate alkanolamine. The β-chloroalkyl tertiary amines with a single β-chloroethyl group, as exemplified by N,N-dibenzyl(β-chloroethyl) amine (Dibenamine), have been used therapeutically as adrenergic blocking agents.[123] These monofunctional β-chloroethylamines have some properties resembling the polyfunctional nitrogen mustards but do not dimerize in aqueous solutions and apparently cannot function as "cross-linking" agents.

[121] A. Gilman and F. S. Philips, *Science*, **103**, 409 (1946).
[122] K. S. Stacey, M. Cobb, S. F. Cousens, and P. Alexander, *Ann. N. Y. Acad. Sci.*, **68**, 682 (1958).
[123] M. Nickerson, *Pharmacol. Revs.*, **1**, 27 (1949).

B. ABSORPTION, METABOLISM, MECHANISM OF ACTION

Oral absorption of the nitrogen mustards is variable and less complete than absorption from subcutaneous injection sites, but they are readily absorbed from the intact skin and lungs. Within 30 seconds after intravenous injection of C^{14}-labeled methyl-bis(β-chloroethyl)amine, over 90 per cent has disappeared from the blood. The promptness with which these agents are transformed *in vivo* is also illustrated by their rapid action. Isolation of the intestine from the circulation for several minutes during intravenous injection affords protection for this sensitive organ. Although many *in vitro* reactions have been described, firm conclusions about the *in vivo* fate of these alkylating agents is not yet possible and conclusive evidence for *in vivo* alkylation of proteins is lacking.[12]

C. BIOLOGICAL EFFECTS

The important characteristic actions of the bis(β-chloroethyl) amines on biological systems are: (*1*) severe local irritation of tissues, vesicant action on the skin and corneal damage in the eye; (*2*) delayed deaths with doses around the LD_{50}, prominent gastrointestinal effects with diarrhea; (*3*) cytotoxic effects, inhibition of cell division; (*4*) bone marrow and lymph node damage, leukopenia; (*5*) tumor inhibition; (*6*) antifertility effects, impairment of menses and spermatogenesis; (*7*) mutagenic activity (*Drosophila*, bacteria); (*8*) carcinogenicity or tumor initiation; and (*9*) pharmacological and neurotoxic activity (large doses). Their proclivity to damage actively proliferating cells (bone marrow, fetal tissue, germinal epithelium, neoplasms) is evident from the effects listed. Because many of these effects resemble those of ionizing radiation, the term "radiomimetic" is frequently used. Not all of the series have been shown to produce all of these effects or to the same degree; however, the differences between the polyfunctional β-chloroethylamines are minimal.

It appears that the structural configuration necessary for the characteristic action against proliferating cells is the presence of at least two β-haloethyl groups per molecule. The third substituent on the tertiary amine can be a wide variety of aromatic or aliphatic groups without abolishing typical activity, even though the potency may be influenced. The secondary amine mustards are less potent than the tertiary amines but they still possess mustard activity whereas the quaternary amines do not. The chlorine, bromine, or iodine compounds are all active but the halogen group must be in the beta position. The bis-β-chloropropyl derivatives also have typical nitrogen mustard actions. The amines with a single β-haloethyl group are less toxic and do not possess full radiomimetic or cytotoxic activity.[3,5,124,125] They do retain the local irritant and vesicant effects. With sublethal doses they show a greater tendency than the polyfunctional compounds to produce

[124] W. P. Anslow, D. A. Karnofsky, B. Val Jager, and H. W. Smith, *J. Pharmacol. Exptl. Therap.*, **91**, 224 (1947).

[125] L. S. Goodman and A. Gilman, *The Pharmacological Basis of Therapeutics*, 2nd ed., Macmillan, New York, 1955.

neurotoxic effects. Dimethyl-β-chloroethylamine, for example, causes prolonged neurological effects such as tremor, uncoordinated movements, ataxia, and derangement of positional reflexes.[126] Other analogs such as Dibenamine have parasympathomimetic activity.[123] Some of these relationships of structure to toxicity and hematopoetic effects are apparent from the data summarized in Table 22.

TABLE 22

Toxicity of β-Chloroethylamines–Mouse[b]

Compound[a]	Route	LD$_{50}$, mg./kg.	Leukopenia
Methyl-bis(β-chloroethyl)amine	Cutaneous	29	+
	Subcutaneous	2.6	
	Oral	20	
Ethyl-bis(β-chloroethyl)amine	Cutaneous	13	+
	Subcutaneous	2.6	
	Intravenous	Approx. 2.0	
Tris(β-chloroethyl)amine	Cutaneous	7	+
	Subcutaneous	2.0	
Isopropyl-bis(β-chloroethyl)amine	Subcutaneous	1.1	+
	Oral	22	
Bis(β-chloroethyl)amine	Subcutaneous	20–33	+
Methyl-β-chloroethylamine	Intravenous	Approx. 100	
Ethyl-β-chloroethylamine	Intravenous	Approx. 100	
Diethyl-β-chloroethylamine	Subcutaneous	100	
	Intravenous	100	

[a] Administered as solutions of the hydrochlorides except for cutaneous application for which the free amine was employed.

[b] M. Nickerson, *Pharmacol. Revs.*, **1**, 27 (1949).

D. EFFECTS IN MAN

All of the effects that have been observed in animals can presumably occur in man. Most of these, with the exception of the carcinogenic and mutagenic effects, have been observed during the therapeutic use of these agents. Nearly all patients develop nausea, vomiting, headache, and diarrhea. A major toxic effect is depression of normal bone marrow function with a decrease in the total number of circulating leukocytes and platelets. Because the β-chloroethylamines have limited specialized use in industry and chemical laboratories, and because their high degree of toxicity and serious effects are rather universally recognized, there have been few reports of injury from these sources.

II. Specific Compounds

The brief data in this section are illustrative of the properties and toxicity of some of the simpler β-chloroethylamines.

[126] E. Boyland, *Biochemical Society Symposia No. 2*, Cambridge Univ. Press, London, 1948.

METHYL-BIS (β-CHLOROETHYL) AMINE, $CH_3N(CH_2CH_2Cl)_2$
(2,2'-Dichloro-N-methyldiethylamine, Methyl-di(2-chloroethyl)amine, HN_2)

1. Physical and Chemical Properties

Physical state: liquid
Molecular weight: 156.07
Boiling point: 75°C. (10 mm. Hg)

2. Physiological Response

This valuable nitrogen mustard produces all the effects described in the previous section. The single dose intravenous LD_{50} for mice, rabbits, rats, and dogs is between 1 and 2 mg./kg. and the subcutaneous LD_{50} is only slightly higher.[124] Intravenous doses of 0.4 mg./kg. in man result in prompt gastrointestinal symptoms and delayed leukocyte depression.[127] It is a potent vesicant and local irritant.

TRIS (β-CHLOROETHYL) AMINE, $N(CH_2CH_2Cl)_3$
(2,2',2''-Trichlorotriethylamine, HN_3)

1. Physical and Chemical Properties

Physical state: liquid
Molecular weight: 204.53
Boiling point: 144°C. (15 mm. Hg)

2. Physiological Response

Tris(β-chlorethyl)amine produces the same effects as methyl-bis(β-chloroethyl)amine. The intravenous LD_{50} in rats is 0.7 mg./kg. and the cutaneous LD_{50} is 4.9 mg./kg.[124] In normal human volunteers single oral doses of 4 to 6 mg. resulted in nausea, vomiting, and diarrhea; 3 mg. daily for 5 or 6 days produced leukopenia.[127]

DIMETHYL-β-CHLOROETHYLAMINE, $(CH_3)_2NCH_2CH_2Cl$
(2-Chloro-N,N-dimethylethylamine)

1. Physical and Chemical Properties

Physical state: liquid
Molecular weight: 107.59
Boiling point: 109 to 110°C. (750 mm. Hg)

2. Physiological Response

This monofunctional β-chloroethylamine lacks the cytotoxic and leukopenic potency of the nitrogen mustards. Its major actions are local irritation, pharmaco-

[127] D. A. Karnofsky, *Ann. N. Y. Acad. Sci.*, **68**, 899 (1958).

logical effects on the autonomic nervous system, and neurotoxicity. The single dose LD_{50} in mice by subcutaneous injection is 200 mg./kg.[126] It has been observed that workers briefly exposed to air concentrations estimated to be between 10 and 100 p.p.m. became severely ill with nausea, vomiting, and dilated pupils.[60]

DIETHYL-β-CHLOROETHYLAMINE, $(CH_3CH_2)_2NCH_2CH_2Cl_2$ (2-Chlorotriethylamine)

1. Physical and Chemical Properties

Physical state: liquid
Molecular weight: 135.64
Boiling point: 51 to 52°C. (16 mm. Hg)

2. Physiological Response

The effects are like dimethyl-β-chloroethylamine. It is not leukotoxic. The LD_{50} in mice by subcutaneous and intravenous injection is 100 mg./kg.[124] It is a strong skin and eye irritant and has produced delayed pulmonary edema after accidental vapor exposure.[60]

CHAPTER XLIX

Potential Exposures in Industry: Their Recognition and Control

FRANK A. PATTY

This chapter endeavors to give significant information about some of the processes, occupations, or industries that are of hygienic interest. The listing is alphabetical and, where information is desired on an industry not listed, significant processes of the industry may be found elsewhere in the chapter.

Abrasive Blasting

There are two types of abrasive-blasting equipment, the automatic and the manually operated. Either type may use sand, steel shot, or artificial abrasives.

Although the degree of potential health hazard is more severe when silica is employed, essentially all abrasive blasting requires adequate exhaust ventilation. The use of steel shot or an artificial abrasive on a casting coated with sand would produce a mixed dust containing free silica within the dangerous particle sizes (less than 5 μ) and in amounts above the maximum allowable concentration for continuous exposure (5 million particles of free silica per cubic foot of air). Even when steel shot or artificial abrasives are used on clean metal surfaces, the amounts of dust produced are so great that exposure of men is undesirable.

In the inspection of automatic equipment air flow into the machine through all openings should be ascertained. The exhausted air should be cleaned by a dust collector that discharges the air outside the building at a point remote from windows or air intakes. The air from the dust collector should not be returned to the plant.

Abrasive-blasting helmets are required to protect workmen stationed within abrasive-blasting rooms. The helmet should be of a type approved for this purpose by the United States Bureau of Mines and should be maintained adequately. The air flow into the helmet should be fixed at not less than 6 c.f.m. to ensure protection. Adjustment of the air flow should not be left to the discretion of the workman. The air supply must be clean and free from oil or carbon monoxide. Dust counts on air samples taken under the helmet, while worker is engaged in blasting, should not exceed one million particles per cubic foot of air.

Good housekeeping is an essential feature of the proper operation of abrasive-blasting equipment. Piles of abrasive near the equipment should be avoided.

As a check on the effectiveness of exhaust ventilation, samples of air for dust-counting purposes should be collected in the breathing zone of operators of automatic blasting machines. If the air from the dust collector is returned to the work-room, it also should be checked by dust count. If counts exceed two or three million particles per cubic foot of air, equipment should be checked for leakage, unless there are obvious sources of dust from adjacent operations.

See also Casting Cleaning.

Wet-Sand Blasting. A "vapor blast" consisting of sand, water, and air is sometimes used. Although the dust arising from such an operation does not compare in magnitude with that from an ordinary sandblast, nevertheless dust counts on the order of 100 million particles per cubic foot of air are common in the vicinity of the blast.

The operation must be conducted inside of exhaust enclosures and when the operators are stationed inside the enclosure they must wear supplied-air abrasive-blasting helmets. Openings to the booth must be protected from the direct blast. The volume of air exhausted from the booth must be sufficient to prevent escape of air, sand, or mist into the room and maintain an inward flow through all openings. If short open drainpipes are provided for the booths, they may become sources of excessive silica dust or mist, requiring control. The filling of sand hoppers and reservoirs also may necessitate control measures.

Where finely powdered quartz is used as the abrasive (microblast, microbrasive blast, and so forth) housekeeping assumes greater importance because the major portion of the abrasive is of "air-float" or respirable size and may be swept from the floor into the air by drafts.

Abrasives Manufacture and Use

1. Artificial Abrasives

The raw materials for the manufacture of silicon carbide include sand, coke, salt, and sawdust, which are fused together in an electric-arc furnace and then broken up, washed, and classified by size. Carbon monoxide and smoke, or fume, arise in considerable quantity from the furnaces and necessitate exhaust ventilation, the extent of which can usually be adjusted satisfactorily by controlling the carbon monoxide. Aluminum oxide abrasives, Alundum and Aloxite, are similarly made by fusing bauxite, crushing the fused mass, then sizing the resulting particles. Bonding materials used with these abrasives are for the most part ceramics, artificial resins, and glue. Infrequently these resins may give rise to dermatitis. Ultraviolet and infrared radiations arising from the electric arcs offer exposures that should be considered. A lung malady named "Shaver's Disease" after discoverer Dr. C. G. Shaver[1] is believed to result from exposure to silica[2] and alumina fumes arising from bauxite furnaces.

[1] C. G. Shaver and A. R. Riddell, *J. Ind. Hyg. Toxicol.,* **29,** 145 (1947).

[2] C. M. Jephcott, J. H. Johnston, and G. R. Finlay, *J. Ind. Hyg. Toxicol.,* **30,** 145 (1948).

In the manufacture of abrasive wheels, abrasive paper, and abrasive cloth, there are numerous dust-producing operations such as crushing, grinding, sifting or screening, edging, facing, and shaving. Although the dusts of all these manufactured abrasives are considered to be of an inert character and of a nuisance nature, they are usually kept well below the frequently stated nuisance dust standard of 50 million particles per cubic foot of air, by light-field count. Whenever such dust concentrations exceed 15 or 20 million particles per cubic foot, the dustiness becomes noticeable and therefore undesirable. A further incentive to control is the abrading action of such dusts on all bearings or moving parts of machinery in use.

The characteristics of the dusts, and their control, mentioned above in connection with the manufacture of wheels, paper, and cloth apply also to situations in which these products are being used. In the main, they can be applied, as well, to the manufacture and use of similar products made of garnet, pumice, and other naturally occurring mineral abrasives other than quartz sand and possibly tripoli.

2. Sand Abrasives

Sand abrasives, wheels, disks, or paper, have been replaced almost entirely by artificial abrasives; but wherever sand-abrasive products are being made or used free silica dust concentrations should be maintained below 5 million particles per cubic foot of air. The use of water does not assure the control of a silica dust exposure. Mist containing finely divided silica may reach excessive concentrations. This is particularly true if the water is recirculated, allowing dust to accumulate in the water or other coolant. It has been observed that, where the humidity is low, the water evaporates from dust particles leaving them suspended in the air.

In the use of abrasive wheels the highest concentrations of dust arise from dry dressing of the wheels.

Acetylene Manufacture

Acetylene is manufactured by the action of water on calcium carbide. The reaction is frequently automatically controlled in enclosed apparatus. The calcium carbide is made by fusing a mixture of lime and coke in an electric furnace. The danger of an explosion of acetylene is the greatest potential hazard in the industry. Some operations are dusty to the extent of polluting the atmosphere surrounding the plant, but such dust is not considered harmful to health. The operations disperse coke, lime, and carbide dusts, of which the lime and carbide dusts are caustic and therefore somewhat irritant to the skin and to the respiratory tract. Carbon monoxide is a by-product and the possibility of its presence in the atmosphere must be recognized. Excessive temperatures are prevalent near the furnaces. Crude acetylene frequently contains traces of ammonia, hydrogen sulfide, and phosphine, which are usually scrubbed out with caustic soda solution. In making tests in atmospheres containing acetylene, ignition sources must be carefully avoided and combustible gas indicators should be used only when they are suitably protected by flash arresters.

When acetylene is to be confined in cylinders, acetone is used as an absorbent for it. The placing of asbestos or siliceous absorbents for the acetone in the cylinders frequently presents a dust problem. Cylinder reclaim involves a possible lead hazard, through grinding on painted surfaces, and exposure to toxic solvents in the use of paint removers.

Acid Manufacture and Recovery

1. Hydrochloric Acid

Hydrochloric acid is made by absorbing hydrogen chloride gas in water, utilizing absorption towers or scrubbers of various designs. The hydrogen chloride may be a by-product of chlorination; or it may be obtained from a bisulfate retort, from burning hydrogen, methane, or water gas in chlorine, or from the decomposition of magnesium chloride in the presence of steam.

Some of the absorption systems have a high efficiency so that little hydrogen chloride gas escapes, while in others the tail gas escaping from the scrubber has sufficient hydrogen chloride in it to cause objectionable concentrations in the surrounding atmosphere. Where little personal attention is required, irritant levels of hydrogen chloride are sometimes condoned, particularly for brief exposures where gas masks are worn. Concentrations in the air can readily be controlled by means of passing the tail gas through one or more additional scrubbers and there seems no logical reason why amounts above 5 p.p.m. in the atmosphere should be tolerated.

2. Nitric Acid

Nitric acid manufacture is more hazardous than hydrochloric acid manufacture in that it requires more personal attention and the oxides of nitrogen have inadequate warning properties in low, toxic concentrations. Nitric acid may be made by heating sodium nitrate with sulfuric acid in retorts, distilling off the resultant nitric acid, and condensing it; by the electric-arc process in which atmospheric oxygen and nitrogen are combined directly to form nitric oxide; or by the ammonia-oxidation process wherein air or oxygen and ammonia are passed over a hot platinum catalyst to yield nitric oxide. Ammonia oxidation may be accomplished at either atmospheric pressure or at pressures up to 100 p.s.i. The concentration of ammonia must be maintained well below the lower flammable limit (15.5 per cent in air; 14.8 per cent in oxygen) in order to avoid an explosion. In each case the NO oxidizes to NO_2 and is absorbed in water to yield nitric acid. The periodic sampling of nitric oxide may present an exposure hazard as does the escape of nitrogen dioxide, which may occur at cracks or at the joints in ceramic pipes. Breaking up nitrogen dioxide-saturated sulfate cake residue from the retorts may also entail excessive exposure to nitrogen dioxide. Nitrogen dioxide in the atmosphere should not be permitted to exceed 5 p.p.m. The escape of ammonia from storage tanks or cylinders at gage glasses, valves, or lines may present a catastrophe hazard of explosion or injury from the irritant action of ammonia on eyes, nose, throat, and lungs.

Nitric acid recovery plants, for the recovery and concentration of spent or weak nitric acid, frequently have local atmospheres ranging above 5 p.p.m. nitrogen dioxide due to liquid or gas leaks at the cemented joints of ceramic pipes. Vent stacks from nitrogen dioxide scrubbers, unless well elevated, are frequent sources of excessive amounts of nitrogen dioxide.

3. Sulfuric Acid

Sulfuric acid is made by either the chamber process or the contact process. In the chamber process sulfur or pyrite ore is burned to sulfur dioxide and this gas is mixed with water and oxides of nitrogen in lead chambers. A nitrosylsulfuric acid is formed, and this is passed through a Glover tower where a countercurrent of hot gases removes the oxides of nitrogen, leaving sulfuric acid, up to 80 per cent, in water.

In the contact process sulfur dioxide, along with an excess of air, is passed through a catalyst such as vanadium, platinum, or chromium–tin. The SO_2 is oxidized to SO_3 with evolution of heat. The sulfur trioxide is absorbed in 80 to 99 per cent sulfuric acid. When sulfur trioxide is absorbed, it is important that it not be allowed to come in contact with water vapor else a persistent fog of finely divided droplets of sulfuric acid is formed. Sulfuric acid of 98 to 99 per cent strength is said to be the most satisfactory absorbent, since concentrations below 98 per cent may fog and those above 99 per cent may permit sulfur trioxide to escape. Cottrell precipitators have been used successfully to prevent the escape of sulfur trioxide fog.

Both sulfur dioxide and sulfur trioxide are irritant. Recommended practice is to keep concentrations of sulfur dioxide below 5 p.p.m. and not to exceed 1 mg./cu. meter sulfuric acid. These concentrations are frequently exceeded around the ordinary plant and, if not too excessive, may be acceptable where men are not stationed in the contaminated area and where respiratory protection is provided for use when needed. Pyrites often contain arsenic which then contaminates the product and may present an arsine hazard where the sulfuric acid is used. Sulfur and pyrite dusts are considered as nuisance dusts.

Aircraft Manufacture, Maintenance, and Repair

Aircraft manufacture embraces many operations that are common to other industries: thus, it may include various potentially harmful conditions. Outstanding among these are the fluoride exposures in magnesium foundry practice, as well as the solvent and paint exposures during the cleaning and painting of the fuselages, especially interiors. Fluoride exposure may cause nasal irritation, eye irritation, epistaxis, and possibly adverse systemic effects. The maximum permissible 8-hour exposure is $2^1/_2$ mg. of total fluoride per cubic meter of air. The fluoride may exist as hydrogen fluoride gas, fluoride fume, dust, mist, or any combination of these. Good process ventilation is recommended for controlling fluorides. Substitution of nontoxic materials, where applicable, is to be preferred.

Any work involving solvent exposure, such as in the application of deicers, or the swabbing of solvents for cleaning purposes, can present a serious hazard unless factors, such as toxicity and volatility of the solvent, and amount and direction of air flow, are considered. Painting of large parts or entire planes requires a well-engineered spray room or booth. Painting of interior compartments requires individual air supply or exhaust and, frequently, effective respiratory protection.

For a discussion of additional exposures incidental to aircraft manufacture, maintenance, and repair see Abrasive Blasting, Aluminum and Magnesium Foundry, Anodizing, Degreasing, Doping, Electroplating, Engine Testing, Grinding and Polishing, Heat Treating, Chlorinated Oils and Waxes, Industrial X-Ray, Metalizing, Metal Cleaning, Quartz Crystal Cutting, Radium and Radium Dial Painting, Soldering, and Spray Painting.

Aluminum Manufacturing

Any health problems associated with the production and fabrication of aluminum are primarily due to the processes and the allied materials used in the processes rather than to aluminum itself. The operations in this industry may be divided into four categories: mining, refining, smelting, and fabricating.

The primary environmental health problem encountered in the mining of aluminum ore (bauxite) is that of nuisance dust. Over the years, exposure to bauxite dust has not produced pneumoconiosis or predisposition to lung disease. Free silica is not associated with the bauxite ores that can be economically used for the production of metal-grade alumina.

Refining of the ore is accomplished by dissolving bauxite in a hot caustic solution, precipitating the purified aluminum hydroxide, and calcining the hydrate to form aluminum oxide. As in mining, the transfer of ore and alumina creates a potential dust problem, which may be readily controlled by engineering methods. The use of a highly alkaline solution poses problems that may range from simple dermatoses to serious chemical burns.

In the smelting operation, aluminum metal is produced electrolytically from a solution of alumina in molten cryolite. Thermal burns are an obvious problem. The fluoride gases and fumes evolved during the electrolytic process present a potential environmental hazard. However, ventilation control measures, used primarily to prevent air pollution, reduce the fluoride concentrations in aluminum smelters below levels of toxic significance. In the smelting operation, nuisance dust problems associated with the transfer of raw materials are also present.

Fabrication of aluminum begins with the alloying of the metal in furnaces. The industrial hygiene problems encountered during the production of aluminum alloys are primarily those associated with alloying metals or the use of fluxes in the metal bath. Although small amounts of toxic metals may occasionally be used in aluminum alloys, the vast majority of alloys contain metals (e.g., Cu, Mg, Zn, Sn) that are not significantly toxic. Fluxing materials, such as chlorine, can cause serious health and corrosion problems unless proper control measures are taken. Dross handling may involve the generation of considerable dust and fume.

The health hazards associated with sand-casting of aluminum are the same as those involved in any other type of sand-casting operation. Silica dust and noise problems are common. The dusting problem, however, is not as severe as that experienced in the casting of metals that melt at higher temperatures than aluminum. Permanent mold and die-casting procedures do not create any significant environmental health problems other than that of radiant heat (see Vol. I, pages 221 and 801).

The cast product may be further subjected to any of the many finishing processes intended to produce a salable end product. The metal may be extruded, rolled, forged, drawn, stamped, sawed, welded, brazed, soldered, riveted, etched, plated, degreased, painted, trimmed, burnished, ground, etc. For the most part, none of these operations produces hazards that are unique to the aluminum industry. In the United States there have been no incidents to suggest that aluminum dusts generated during metal working will cause a pneumoconiosis.

A fabricating operation of aluminum that has some degree of uniqueness as regards industrial hazards is that of welding. The inert-gas-shielded consumable-electrode method of welding aluminum is capable of generating significant quantities of ozone and fume and generally requires ventilation control.

Aluminum dust is combustible and when dispersed in the air may present an explosion hazard (see Vol. I, Chapter XVI, Section Two).

Ammonia Manufacture and Use

Ammonia is a by-product of the distillation of coal. It is manufactured by the action of steam on cyanamide, and by the catalytic combination of nitrogen and hydrogen. It is extensively used in refrigeration, and in the manufacture of chemicals, such as fertilizers, nitric acid, explosives, plastics, and other materials. The workroom atmosphere should be so controlled that extended exposure to 100 p.p.m. does not occur. In the manufacture and use of compressed ammonia gas the prolonged exposure to low concentrations, although undesirable, is not as much of a hazard as is the accidental exposure to much higher concentrations. High concentrations may result from the failure of a valve, or line, or the breakage of a gage glass on a storage tank of the compressed gas. Canister gas masks approved by the United States Bureau of Mines should be provided for emergency use whenever compressed ammonia is used; and where the amounts involved are likely to produce mixtures of more than 3 per cent in the atmosphere, fresh-air hose masks or oxygen rebreathing apparatus should be provided. Fire and explosion hazards are significant. Ammonia escaping from a leak can be ignited readily and gas-air mixtures of 15.5 to 26.6 per cent ammonia can cause violent explosions.

Aniline Manufacture, Distillation, and Handling

Aniline and dimethylaniline are hazardous liquids and exposed workmen should always be instructed how to recognize and avoid the dangers involved. Aniline manufacture by nitration of benzene and subsequent reduction of the nitrobenzene involves potential exposures to nitric acid, nitrogen oxides, benzene, and nitrobenzene, in addition to aniline itself. A chlorination process involves benzene, chlorine, and chlorobenzene exposures. Aniline, although only slightly volatile when used at room temperature, requires either local or good general ventilation; and when used at elevated temperatures, requires either very effective process ventilation, or enclosed processes and general ventilation. The dangers and likelihood of absorption of the aniline or nitrobenzene through the skin are even greater than those from inhalation of the vapors; and shoes, gloves, or any other articles of clothing that have come in contact with any visible amount of these materials should be discarded without delay and the skin should be thoroughly washed.

Even the small amount of aniline used in some formulas for waterproof ink offers exposure in the manufacture, storage, handling, and use of the ink. Poisoning has occurred due to skin contact with unlaundered cloth stamped with such inks. The maximum amount of aniline in air permissible for an 8-hour exposure is 5 p.p.m., and a few drops on the skin or clothing may cause poisoning. Aniline is especially dangerous on shoe soles, where it is likely to be overlooked and remain, to soak through gradually and be absorbed by the feet.

Anodizing

Anodizing is an operation whereby a coating highly resistant to corrosion is formed upon a metal surface. It is accomplished in an electrolytic bath with the metal to be treated forming the anode. A resistant oxide film is formed in this way on aluminum and aluminum alloys. The bath may contain any of a number of materials, with dilute chromic acid and sulfuric acid electrolytes being common. Hydrogen evolved in the process may carry corrosive mists out of the tank unless the tank is correctly ventilated. A trained observer can recognize excessive chromic-acid or sulfuric-acid mists or sprays by nasal irritation. Where any uncertainty exists regarding an exposure, air analyses should be made and corrective measures taken where indicated. It is the usual practice, however, to provide effective ventilation for all anodizing tanks. Permissible limits are discussed under the compounds concerned.

Armature Workers

Lead "tinning" baths and lead pots, when small and equipped with temperature controls, may not present serious lead exposures. However, the larger operations, especially when no temperature control is in use, can offer serious exposures. Wherever there is any question of exposure, the operation should either be provided with mechanically induced air flow to remove or dilute the lead fumes before they can be inhaled, or else air, blood, or urine samples should be collected and analyzed for lead to establish whether ventilation is necessary.

Waterproof varnishes, especially those made from cashew nutshell liquid–formaldehyde resins, may cause dermatitis. This may result from contact with the wet varnish, from the varnish dust dispersed in grinding armatures, or from skin contact with dirty solvent baths for dipping and cleaning old armatures. The solvent bath should not be permitted to become badly contaminated and skin contact must be avoided. In the case of wet or dry varnish, good personal cleanliness is an effective preventive measure, and varnish dust should be controlled by process ventilation. Protective creams and clothing may be useful in some instances.

Art-Metal Casting

Art-metal casts for displays, trophies, medals, and the like are usually cast from white metal or britannia metal, which are alloys of tin, antimony, and copper, often having some zinc and bismuth, occasionally lead, and possibly arsenic. The alloys are melted in pots, temperatures rarely exceeding 300 or 400°C. Burring, filing, belt "sanding," and buffing may be done and, unless controlled, significant exposures to mechanically generated metal dusts may result. At the temperatures ordinarily employed there is little likelihood of an exposure to metal fumes, however. Where uncertainties exist, exposures should be evaluated and, when found excessive, controlled. Other possible exposures in the industry involve plating baths and cyanide dips. Soldering operations are usually at low temperatures to avoid fusing the low-melting-point alloys, and therefore are not conducive to the formation of metal fumes.

Asbestos Workers

Dustiness around asbestos workers is sometimes above the proposed standard of 5 million particles per cubic foot. Perhaps mining offers the greatest exposure to dust, and in fabrication plants the operations, listed in the probable, decreasing order of dustiness, are: screening, picking, stacking, and willowing in the preparation room; carding, dry weaving, spinning, twisting, winding, and warping. The sawing, filing, drilling, and grinding of brake linings is ordinarily well controlled. Grinding, mixing, and bagging asbestos cement and insulating material, and the sawing or beveling of asbestos board, are dusty operations unless properly engineered. The asbestos content of these dusts frequently is not high but diatomaceous earth in considerable quantities may accompany it, thus complicating the exposure with free silica. It is desirable in asbestos exposures to keep the dust count down to 5 million particles, or less, per cubic foot of air.

Asphalt (Mineral Pitch)

The name "asphalt" applies to bitumen or to earth or rocks impregnated with bitumen and to a residue from certain petroleums, coal tar, lignite tar, etc. (artificial asphalt). It is a black solid formed in the earth from slow decomposition of organic matter. It is composed chiefly of hydrocarbons along with varying amounts of oxygen, nitrogen, and sulfur compounds.

Asphalt is used principally for water-excluding coatings upon roads, floors, or roofs. It is employed as a rust preventive in ducts, collectors, tanks, and so forth to protect metals from corrosive (acid) liquids and gases. For such application it is commonly dissolved in benzene.

Vapors and gases from heated asphalt are obnoxious and toxic, containing among other compounds some hydrogen sulfide. Dermatitis from contact with asphalt has been reported as common, with the entire body sometimes affected. Skin cancer has also been reported, but statistics are lacking. Some authorities attribute most of the poisoning that has occurred from the use of asphalt to the solvents employed with it (benzene, phenol, etc.).

Grinding operations and hot processes should have exhaust ventilation. Personal cleanliness is the heart of any program for preventing skin disease.

Automobile Manufacture

The manufacture of automobiles has few operations that are peculiar to the industry. There are many exposures and potential exposures common to other industries.

These are discussed elsewhere under specific headings such as: Abrasive Blasting, Anodizing, Battery Manufacture, Electroplating, Foundry Operations, Forging and Iron Working, Garages, Grinding and Polishing, Heat Treating, Industrial X-Ray, Lead Workers, Metal Cleaning, Metalizing, Motor Testing, Painting and Decorating, Pickling, Plastics and Resins, Radio Manufacture and Repair, Welding, etc.

Body polishing by the use of tripoli polishing compounds is an operation more or less peculiar to the production of highly polished automobile bodies. No harmful effects have been associated with this operation. However, any dust arising from the polishing, and especially from compressed-air blowoff operations, and from the cleaning of the brushes should be controlled. There seems no logical excuse for condoning such dust in concentrations of more than 5 million particles per cubic foot of air. Since the tripoli polish is applied as a slurry, dust control has not been a difficult problem.

Another operation peculiar to the industry is that of filling seams with solder. This operation or, more likely, the subsequent operation of removing excess metal by belt sanding, disk grinding, and buffing, has caused lead absorption in years past. Grinding and buffing of leaded surfaces is now done in exhausted enclosures and air-supplied helmets are used. Careful checks of atmospheric and blood or urinary lead concentrations should accompany these operations wherever exposures are suspected. Temperatures used in applying the lead are not sufficiently high to volatilize excessive amounts of lead, therefore canopy-type hoods are adequate, but the disking operations require well-engineered ventilation to take care of the dust evolved. If the operation is closed in a booth, and only sufficient air exhausted to prevent the spread of dust to other areas of the plant, supplied-air helmets are necessary. The use of blood-lead determinations permits the removal of the employee from work involving exposure to lead dust and fumes *before* intoxication occurs. In a group of plants employing thousands of men working with lead and its oxides, where periodic blood-lead analyses have been used coupled with transfer of employees with confirmed blood-lead levels of above 0.08 mg. lead per 100 g. blood, no definite symptoms of lead intoxication have been encountered.

Contact with oils, emulsions, and solvents, although not peculiar to automobile manufacture, is the most common exposure in the many machining operations that are involved in the making of an automobile.

A statistical study of about 400 cases of dermatitis indicated that where contact with the hands is frequent, mineral seal oil, containing an "improver," caused the highest incidence of dermatitis—on the order of 33 cases per hundred persons exposed per year—while cutting oils, such as are encountered by bar-machine, screw-machine, and other operators, caused the next highest incidence of dermatitis—on the order of 20 per hundred per year of operation—and emulsions, "soluble oils," were third with 10 per hundred per year.

Some of the inhibitors, antiseptics, and "improvers" occasionally used in each of these oils may make matters worse in that they may be themselves either primary irritants or sensitizers. Offensive agents have included nitrobenzene, nitrophenol, cresylic acid, some sulfur compounds, chlorine compounds, aromatic amines, and strong alkalies. Not all sulfur or chlorine additives, however, are objectionable.

It is important to remove metallic particles from recirculated oils, but sterilization or the addition of disinfectants to cutting oils does not appear to be desirable. In the case of emulsifiers or coolants, there appears to be less difficulty where they are discarded and replaced frequently. Elaborate systems for pasteurization have not proved advantageous. There are, however, many who claim that certain bactericidal agents can be added to retard the formation of offensive odors without causing any harmful effect.

The most successful measure for prevention of comedones and folliculitis is effective washing with correct cleaning agents. Unless prevented by supervision, solvents used for cleaning parts and machines will be used for cleaning hands, with definitely unfavorable results. Where low viscosity oils are encountered, some means such as impervious gloves or creams can be used to advantage to prevent defatting the skin or causing sensitization. Creams may have some use both before and following the work period. When the hands and arms frequently become covered with oil, clean waste or rags should be provided. The splash or spray of oils and emulsions, or dispersion of excessive oil fog or mist, should be controlled.

Battery Manufacture

Lead Storage Battery. The manufacture of lead storage batteries involves lead exposures that necessitate continuous and competent industrial hygiene supervision. Mechanization, enclosed processes, and well-designed process ventilation have been advanced to the point where inhalation exposures can be held down to less than 0.2 mg. per cubic meter consistently, if the control program is vigilantly supervised. The industrial hygienist should be thoroughly familiar with the basic information in Chapter XXVI.

There is little information to be gained by a few random breathing-zone samples. The sampling must be extensive enough to indicate the average, as well as the range, of the exposures. Of more practical interest than individual exposures is information on sources and extent of atmospheric contamination, so that it can be tied in with improved engineering control.

Casting or molding operations, if ventilated, ordinarily do not present an exposure control problem. Smelting and reclaim require well-engineered ventilation and rigid housekeeping. Oxide manufacture requires enclosed processes with well-designed ventilation control, as well as due regard for housekeeping. Maintenance or repair jobs in furnaces offer hazardous exposures and also present room contamination problems. Oxide mixing is ordinarily provided with a dust control system that properly ventilates each dust dispersion point, and spilled oxide is removed by a vacuum cleaning system. The operation of hand scooping and weighing presents more of a control problem. Respirators should be used only as a last resort where ventilation control is unsuccessful, but should always be used wherever the exposure exceeds the practical control limit established as a part of a particular control program.

From the time the lead oxide paste is applied to the plates, the control of the exposure becomes more difficult. The paste is applied wet either by hand or machine, and the application itself presents little difficulty, but as soon as the paste becomes dry it is easily dispersed into the air. This introduces a problem wherever heat is applied, or the plates are cleaned, transferred, racked, stacked, trimmed, split, etc. Moreover, the wet paste, if it is allowed to fall on the floor or contaminate equipment, sooner or later becomes dry and adds to the exposure sources. Housekeeping is more than just a name in this control program. Work tables with grids provided with good exhaust have proved very effective in controlling dust dispersion, as well as floor contamination. The plates should be handled manually as little as possible. Ready-charged batteries, the plates of which are handled and shipped dry, ready for service as soon as the electrolyte is added, offer much more of a dust-control problem than plates handled and shipped wet.

Controlling lead exposures in storage battery manufacture is not a piecemeal problem, but one that requires a well-rounded health maintenance program including: (1) competent engineering control; (2) preemployment examinations with blood-lead determinations; (3) periodic surveys of atmospheric contamination; (4) periodic blood-lead determinations of persons in potentially hazardous work places (urinary lead and coproporphyrins may also be determined, but are superfluous if blood leads are determined accurately and with sufficient frequency); (5) education of employees about personal hygiene, good health practices, and safe working procedures; and (6) transfer of any affected person to a position of minimum exposure.

Experience in the battery industry has indicated that if atmospheric lead concentrations are kept below 0.2 mg. per cubic meter and the above outlined program is maintained, there is a reasonable assurance of the prevention of lead poisoning. In one group of plants employing more than 1000 men working with lead and its oxides, the use of blood-lead determinations permitted the removal of the employee from exposure to lead *before* intoxication occurred.

The charging room atmosphere is necessarily subject to contamination by hydrogen and acid mist and, since battery grids often contain from 5 to 10 per cent antimony, production of stibine is at least theoretically possible. Haring and Compton[3] concluded that stibine is generated in perceptible quantities during overcharging. The amount is very small in relation to total battery gases but it increases with the age of the battery. Ventilation sufficient to dilute or limit the sulfuric acid mist in the battery gases to not more than 1 mg. per cubic meter of air must be provided. Under normal conditions this will also control any other gases formed.

Edison Cell. The Edison or "iron nickel" cell uses a spongy iron anode and a compressed flake nickel cathode in a solution of 15 per cent caustic soda containing a small amount of lithium hydrate. Except for the corrosive action of caustic, no significant exposures accompanying the manufacture of the Edison cell have been recognized or described.

Dry Cells. Dry cells are manufactured in many sizes and shapes. The positive pole is a carbon rod, while the negative pole is a sheet of zinc that serves as a container for the battery. A mixture of manganese dioxide, powdered graphite, and ammonium chloride made into a paste is used to pack around the positive center pole. The filled cell is then sealed on the top with a layer of hot pitch or sealing wax. The only significant exposures are to manganese dioxide dust generated from the handling and mixing of manganese dioxide.

[3] H. E. Haring and K. G. Compton, *Trans. Electrochem. Soc.*, **68**, 283 (1935).

Mercuric oxide now extensively used in dry cells may involve exposures to mercury dust and mercury vapor. The exposures are not confined to the manufacturing process but may, under adverse conditions, result where the cells are used. The cells have been known to explode, when shorted, dispersing microscopic mercury droplets which readily disappear in crevices or in dusty places and continue to emit mercury vapor for an indefinite period.

Beer Vat Coating

Steel vats or tanks about 8 feet in diameter and up to 30 or more feet in length are frequently used for the storage of beer. Various types of coating materials are used on the inside of these vats and the application of the coatings involves at least potentially hazardous exposures. Men enter each tank through a manhole to apply plastic coatings in solvent solution by spraying, or to apply waxes with the aid of gas torches.

In the spraying operation hose masks must be used for respiratory protection and, also, explosive atmospheres may be involved. Air sufficient to keep the vapor concentration well below the flammable limit should be supplied at the bottom of one end of the tank and exhausted at the top of the other end, in order to provide effective vapor removal.

In the case of the waxes, where torches are employed, no personal respiratory protection is needed but ventilation must be provided to remove the excessively hot and oxygen-depleted atmosphere. This involves an air supply to the bottom of one end of the tank and an exhaust at the top of the other end of not less than 250 c.f.m. per torch. Portable ducts may be used to direct the air flow and the workman should hold the torch on the downstream side of him.

Bleaching

The bleaching of textiles normally consists of oxidation or reduction reactions to remove color. Strong chemicals such as hypochlorites, formaldehyde, hydrosulfites, ozone, and peroxides may be used.

Mechanical ventilation is required for most processes as the lung irritants, chlorine, sulfur dioxide, and ozone are likely to be evolved. In some instances control of high humidity, due to steam arising from heated vats, is desirable. Process ventilation is more effective, but general ventilation is frequently, and not always advantageously, employed. Suitable clothing for protection against irritant materials is necessary.

Beryllium Processing Operations

Operations in which dust or fumes of beryllium or its compounds may be generated require industrial hygiene supervision.[4-6] Air sampling and analysis are

[4] J. H. Sterner and M. Eisenbud, *Arch. Ind. Hyg. Occupational Med.*, **4**, 123 (1951).

[5] Hygienic Guide Series: Beryllium and Compounds, *Am. Ind. Hyg. Assoc. Quart.*, **17**, 345 (1956).

[6] R. N. Mitchell and E. C. Hyatt, *Am. Ind. Hyg. Assoc. Quart.*, **18**, 207 (1957).

required to establish that atmospheric concentrations of beryllium in the plant do not average above 2 μg. per cubic meter of air for any 8-hour work period. In the plant neighborhood the average monthly concentration at the breathing-zone level should not exceed 0.01 μg. per cubic meter of air.

These low tolerances make it necessary to provide well-engineered process ventilation and air-cleaning installations. Housekeeping should include wet dust-removal methods for machines and floors and any other measures necessary to avoid the spread of contamination. Depending upon the nature and extent of the operations, this may entail coveralls or smocks, shoes or shoe covers, or in more extensive operations may warrant a complete daily change of clothing and facilities for routine shower baths between work and donning street clothing.

Brick and Tile Manufacture

The appraisal of exposures in brick and tile manufacture is concerned principally with free silica. The mineralogical composition of air-borne dust must be ascertained with emphasis on the determination of quartz, tridymite, and cristobalite. Plant operations are generally dusty, releasing to the atmosphere various quantities of clays and shales.

Glazing operations may introduce a lead exposure; and other relatively minor hazards include exposures to excessive temperatures, sulfur dioxide, and carbon monoxide.

Dust appraisal calls for particle enumeration by the standard light-field counting procedure, and particle composition by petrography and x-ray diffraction (silica) and chemical analysis (lead).

Broom Manufacture

In the manufacture of brooms, the brush or seed top of broom corn, after removal of the seed, is used. The scraping and trimming of the brush is done by machines and is a dusty operation unless control ventilation is applied. Bleaching and dyeing operations may require control depending upon the materials employed. The industrial hygienist should find out what materials are in use, know their toxicity and, if necessary, then devise measures to safeguard workmen. Broom handles are frequently lacquered by dipping them in an open vat or tank. Both the dipping and the subsequent drying operations evolve excessive lacquer-solvent vapors unless ventilation control is applied.

Carpentry

Most of the operations in carpentry include accident hazards. Mechanical saws, planers, shapers, and other equipment require a safety program of high caliber, as does work from scaffolds.

Dermatitis and irritations of the upper respiratory tract occur among some workmen. Generally these reactions are due to allergenic substances, but some woods are known to contain irritants. Wood dusts do not cause characteristic, disabling lung disease, but dust control is advisable for dust-producing machines such as power saws in order to prevent fires or dust explosions.

Chemically coated nails are sometimes used to produce strong joinings with the minimum number of nails. The chemical coatings in some cases cause a characteristic dermatitis on the hand used for holding the nails. Other parts of the body may be involved.

Circular saws and planers frequently require noise control methods (see Vol. I, page 685).

Carroting

The carroting of fur by a mixture of mercuric nitrate and nitric acid, now being replaced by less toxic materials, has had mercury poisoning associated with it for two hundred years. The mixture is applied to the tips of the fur on processed rabbit pelts, either by hand or by a revolving brush, in order to improve the felting properties of the fur.

The skins are then dried below 140°F. to produce a white carrot or at 160 to 250°F. for a yellow carrot. The dried pelts may be stored for several months, or immediately sent to the brushing department. After the fur is brushed to smooth it, it is sheared and the skin is shredded. The separated fur is cleaned and sent to the hatting departments.

Bloomfield and DallaValle[7] reported average mercury exposures varying from 7.2 mg. per 10 cu. meters of air for shippers, to 0.6 mg. per 10 cu. meter of air for office workers, machinists, etc. They recommend: (1) segregation of skin-handling operations; (2) local exhaust ventilation for cutting, brushing, and blowing operations; (3) natural or mechanical ventilation to decrease exposures associated with piling and shipping; (4) removal of treated fur from workrooms as quickly as possible and storage in well-ventilated rooms; (5) good housekeeping and sanitation; and (6) arrangement of processes to increase efficiency of operation, with resulting decrease of mercury exposure.

Beal, McGregor, and Harvey[8] report that satisfactory carrot can be obtained by means of hydrolyzing acid-hydrogen peroxide mixtures. Oxidizing acids such as permanganic, iodic, chloric, and so forth may also be used with hydrolyzing acids such as sulfuric or phosphoric. The mixtures are controlled by Beal and McGregor patents. These carroting solutions usually contain chloric and sulfuric acids, or hydrogen peroxide with a hydrolyzing acid. It is claimed that the only hazards involved in the use of these carrots are those of corrosion. Adequate protection of the workers by gloves and guards, and exhaust ventilation capable of removing spray are necessary. It is also advisable to remove fumes evolved during the drying of carroted fur regardless of the type of carroting used. There are said to be no health hazards in any of the succeeding mechanical operations.

[7] J. J. Bloomfield and J. M. DallaValle, *Am. J. Public Health*, **27**, 167 (1937).

[8] G. D. Beal, R. R. McGregor, and A. W. Harvey, *Chem. Eng. News*, **19**, 1239 (1941).

Cement and Concrete

Cement is made from cement rock or a mixture of finely ground limestone, or other form of calcium carbonate, with clay, shale, slate, or blast furnace slags, rarely sandstone, and certain accelerators or retarders. The mixed powders are heated in a kiln, usually rotary, to about 1300 to 1400°C. The kilns are heated by fuel jets of powdered coal, gas, or oil. The cement rock rarely contains more than 6 or 8 per cent quartz; the finished product usually less than 1 per cent, though occasionally it may contain up to 6.5 per cent. The dusts are ordinarily classed as nuisance dusts but their concentration may exceed the most liberal ideas of permissible limits.

There are several dust-producing operations and they are amenable to control measures. Some of these sources are: stone quarrying, crushing, grinding, the rotary kiln, screens, bagging operations, and the loading and unloading of cars. Dust clouds, if uncontrolled, not only are conducive to undesirable working conditions but also constitute an atmospheric pollution nuisance in the community. Electrostatic precipitation as well as centrifugal collectors have been used successfully to remove the dust from the effluent air from kilns and silos, and the amount recovered is said sometimes to exceed 5 per cent of total raw materials. A method that has been successfully applied to the rotary kiln elsewhere (see Sand Refining) is to exhaust the hot kiln gases through baffled water-spray chambers and conduct the water flowing from the sprays through a settling basin from which the sludge may be recovered. The important factors in evaluating exposures are the degree of dust control and the free silica content of dust particles less than 5 μ in diameter. Lung injury from exposure to cement dust, however, has not been demonstrated and the most frequent, harmful result of exposure is that of skin irritation from the alkaline action of cement. In the mixing and use of concrete, the irritant, alkaline action of the wet mixture is similar to that of the cement dust, and both warrant suitable control to prevent prolonged contact with the skin.

Chlorinated Waxes and Oils

Synthetic Chlorowaxes. The synthetic chlorowaxes comprise a number of chlorinated hydrocarbons that are derivatives of naphthalene or diphenyl. Halowax is the trade name of one manufacturer for the chlorowaxes.

The damage that may occur from inhalation, ingestion, and skin absorption of the chlorowaxes will vary with the degree of chlorine saturation of the compounds. The amount of dermatitis and poisoning increases rapidly as the amount of chlorine in the waxes increases. The commonly accepted maximum permissible limit for atmospheric contamination with trichloronaphthalene is 5 mg. per cubic meter of air. Pentachloronaphthalene, one of the chlorowaxes carrying a high percentage of chlorine is restricted to 0.5 mg. per cubic meter of air.

Systemic poisoning from the chlorowaxes is generally characterized by damage to the liver. Serious exposure may produce acute yellow atrophy. The

On pages 2251 and 2252 the expression "chlorowax" has been used incorrectly as an alternate name for chlorinated naphthalene and diphenyl compounds.

CHLOROWAX is a registered trademark of the Diamond Shamrock Corporation for a series of chlorinated paraffin waxes containing 40 to 70% chlorine. (*Continued on p. 2252.*)

skin effects are commonly in the form of acne, which is more widely distributed on the body than common acne.

Among the recommendations for the prevention of chlorowax poisoning are: (1) vapors and dust should be controlled by exhaust ventilation to prevent exposure to amounts of the compounds in excess of the safe limits mentioned above; (2) foremen and workers should be told of the toxicity of the materials that they are handling and instructed to keep skin contacts with the material to a minimum; (3) preemployment and periodical physical examinations should be given with special attention to the skin and to the liver, liver function tests being performed periodically; and (4) the best hygienic conditions should prevail, including the provision of clean work clothing, protective gloves, and protective creams. Those manufacturers who have had the most favorable results in preventing chlorowax poisoning have provided double locker rooms, one for street clothes and one for work clothes, with a shower room in between. The workers should be required to take a shower every day before leaving the factory; it is well to provide supervision to make certain that this practice is complied with.

Air analysis for the chlorowaxes is most commonly performed by passing the air over heated platinum in an electrically heated quartz tube; the waxes decompose and the effluent gas is scrubbed in a column of glass beads moistened with sodium carbonate; the chlorine is converted to sodium chloride which is recovered by washing the beads, and is usually determined nephelometrically. The impinger, using amyl acetate as the collecting medium, has also been employed for collection of samples; afterward the samples are burned, products of combustion collected and chlorides determined.

Chlorinated Oils. There are two general types of oils falling within this class: the first contains additive substances such as carbon tetrachloride to improve cutting properties; the second has chlorine combined in complex organic structure.

The chlorine additive agents may be recovered by distillation and are released from the oils in accordance with vapor pressure laws. The use of oil plus chlorinated hydrocarbon may be hazardous unless exhaust ventilation is provided. Nausea and malaise are common complaints of workers handling the mixtures. The haphazard methods used in preparing the mixtures tend to increase the danger. Commonly the chlorine compound is added in unknown proportion as an antidote for cutting difficulties. In some instances the mixture has been found to contain as much as 40 per cent carbon tetrachloride.

The chlorine in the second type of oil is firmly bound at lower temperatures; upon heating, little decomposition occurs below 200°C.; near and above 250°C. hydrogen chloride is evolved. It is known that temperatures in excess of 300°C. are produced locally from heavy cutting operations. This type of oil probably can be used safely for light precision machining.

Animal experimentation has shown that chlorine compounds in oils may be absorbed through the intact skin to cause liver damage. Dermatitis from chlorinated oils also occurs.

Unlike chlorinated naphthalene and diphenyl compounds, chlorinated paraffin waxes have not been considered to be toxic or to present any special hygiene problems.

Chlorine Manufacture

Chlorine is manufactured by the electrolysis of brine, sodium hydroxide and hydrogen being simultaneously produced. It is not necessary here to emphasize the corrosive action of caustic soda or the flammability of hydrogen, but it is in order to point out the more obscure mercury vapor exposure that may accompany the use of the Castner, the Sorensen, the Krebs, or any other cell in which mercury is used. These cells can be used safely when the exposure possibilities of mercury are recognized and controlled. Wherever mercury is used in quantity, the potential exposure should be evaluated and controlled.

The toxicity of chlorine itself is well recognized; chlorine is no longer considered a desirable or beneficial addition to the atmosphere. It has warning properties that aid in its recognition and control. Concentrations of chlorine in workrooms should not exceed 1 p.p.m.

Compressed Air Work

Work in compressed air has been discussed extensively in Vol. I, Chapter XVII. Discussion here will be limited to the use of helium–oxygen mixtures for the alleviation of "ear block." "Ear block," or more properly tubal or sinus block, occurs with considerable frequency during compression and decompression even when the working pressure is not more than one atmosphere above normal. The condition, which is similar to one encountered in rapid ascent or descent[9] in an airplane, is painful and may be accompanied by deafness, vertigo, tinnitus, and rupture of the eardrum. A mixture of 80 per cent helium and 20 per cent oxygen is administered by means of a face mask.[10] This is most effective when done in the man lock but may be accomplished in the first-aid room near the lock. The similar use of oxygen has been described under oxygen.

The principal objection to administering the gas in the lock is the difficulty of sterilization of facepieces. Only 1 to 3 minutes inhalation of the gaseous mixture is required, and the method is highly successful. A difficulty of the application of this method in the first aid-room prior to compression in the lock is that the lapse of time between inhalation of the gas and start of compression permits partial loss of the gas and its beneficial effect.

The use of the helium–oxygen mixture has been abused and it should never be administered by persons who have a lack of appreciation of sanitation. Medical supervision of personnel is required else persons with colds and respiratory infections that make them unsuited for compressed-air work are likely to take advantage of helium-oxygen inhalation as a means of getting through the lock.

[9] W. R. Lovelace, C. W. Mayo, and W. M. Boothby, *Proc. Staff Meetings Mayo Clinic,* **14,** 91 (1939).

[10] J. W. Crosson, R. R. Jones, and R. R. Sayers, *U. S. Public Health Repts.,* **55,** 1487 (1940).

Cork and Linoleum Industry

The term "cork" is applied to a part of the bark of the cork oak, which is grown principally in southern Europe. There are no recognized hazards peculiar to stripping the bark from the trees.

The cork is graded according to quality and cut into various shapes for marketing. It is said that in some shaping operations about 35 per cent of the cork is converted to dust. Cork dust is of the nuisance variety except for the real danger of dust explosions. Attention should be directed to dust control and elimination, to housekeeping, and to the elimination of static electricity and other sources of ignition.

Linoleum is manufactured by impregnating a woven framework of hemp or jute with oxidized linseed oil combined with vegetable gums, rosin, cork, and pigments. Substitutes may be employed for linseed oil or for cork. Here again, dust explosions from cork or other fillers are the principal hazard. Acrolein and other obnoxious volatile organic vapors or gases are released from linseed oil during oxidation. Lead, manganese, arsenic, and mercury compounds are used in oxidizing the oils, and metal poisoning must be considered. Volatile solvents such as benzene, xylene, methyl alcohol, and turpentine may be encountered also.

Cotton Industry

Some picker-room operations result in very high dust concentrations and the dust may cause acute inflammatory irritation of the nasal pharynx and bronchi. Fever and a bad cough are frequent symptoms. These conditions occur most commonly when low grade cotton is processed. The British[11] consider the respiratory response to be caused by histaminelike substances present in the fine dust. The dust is described as being ultramicroscopic, 0.2μ and less.

Trice[12] recommends that carding machines be housed and provided with exhaust ventilation. Vacuum strippers and grinders are also recommended. It has been found that vacuum cleaning is valuable in reducing the dust load on machines. This prevents redistribution of the dust in the air.

Britten et al.[13] found dust to be of little consequence in cotton cloth plants that they investigated. High temperatures and humidities were found in southern mills but no definite effect on health was established.

Caminita, Schneiter, Kolb, and Neal[14] found a Gram-negative microorganism occurring in large numbers in low-grade stained cotton. The organism was responsible for acute illness, which closely resembled "mill fever" reported among cotton-mill operators.

[11] C. Prausnitz, *Investigations on Respiratory Dust Disease in Operatives in the Cotton Industry,* H. M. Stationery Office, London, 1932.

[12] M. F. Trice, *Respiratory Disturbances Suffered by Cardroom Workers Attributed to Protein in Cotton Dust,* North Carolina State Board of Health, Sept., 1939.

[13] R. H. Britten et al., *U. S. Public Health Bull.* No. **207**, 1933.

[14] B. H. Caminita, R. Schneiter, R. W. Kolb, and P. A. Neal, *U. S. Public Health Repts.,* **58**, 1165 (1943).

Detinning Scrap and the Manufacture of Tin Tetrachloride

Tin scrap is pressed into bundles or bales of about one to two hundred pounds and detinned in large steel chambers. The scrap must be completely dry, and dry chlorine is admitted to the chlorination chamber at ordinary temperatures. The chlorine reacts with the tin, with evolution of heat, to produce liquid tin tetrachloride, which drips to the bottom of the reaction chamber and is removed through pipe lines for purification by distillation. The large quantities of chlorine used necessitate emergency equipment for protection against both low and high concentrations of chlorine in the event of leaks or ruptures of connecting pipe lines. There are opportunities for exposure to chlorine, hydrochloric acid, and tin tetrachloride.

The operation of filling drums with tin tetrachloride and scaling them is accompanied by excessive exposures to tin tetrachloride and hydrogen chloride unless effective control, such as exhaust ventilation, is provided. "Sampling" of the stannic chloride from the reaction cylinder also may be accompanied by exposure to hydrogen chloride and stannic chloride. Tin tetrachloride, with a boiling point of 114.1°C., is quite volatile and in filling drums an essentially saturated vapor-air mixture is displaced from the drum into the room. Concentrations as high as 700 p.p.m. hydrogen chloride have been measured in the room near the drum during filling, when there was no ventilation. No permissible limit for tin tetrachloride has been established, but, since it readily hydrolyzes in the presence of moisture to form hydrochloric acid, it is logical to assume that its physiological effect would not be less than that of hydrogen chloride, and might be more severe in that some of it would tend to reach the alveoli before hydrolysis occurred. Although tin tetrachloride can be handled in iron or steel containers when dry, not so the moist vapor-air mixture: any fans, ducts or enclosures for control of the vapors must either be replaced frequently or be constructed of material resistant to hydrochloric acid.

Doping

Dope for airplane fabrics consists essentially of cellulose esters and volatile solvents. Since large surfaces are involved and the solvents evaporate readily, they contaminate the surrounding atmosphere during application and drying of the dope. Solvents in use today are chiefly alcohols, ketones, esters, and ethers, with moderate amounts of toluene. The dope is often applied in large rooms with only general ventilation. This requires a volume of air movement directly proportional to the amount of dope applied in order to dilute the vapors to acceptable levels. Where humidity-controlled or air-conditioned rooms are used for doping, ventilation is more of a problem. Recirculation through scrubbers has been successfully applied to the control of water-soluble solvent vapors where it was necessary to limit the discharge of air.

A permissible concentration of vapors in the workroom air must be decided upon by the industrial hygienist for each particular situation, based upon the known composition of the dope.

Dye Manufacture and Use

Aniline dyes have an unwarrantedly bad reputation. In their manufacture, raw materials, intermediates, and by-products, such as ammonia, aliphatic amines, acids, sulfur dioxide, nitrogen dioxide, methyl alcohol and other solvents, aniline, m- and p-phenylenediamine, nitrosodiethylaniline, tolylenediamine, toluidine, pyrogallol, quinone, benzidine, and phenylnaphthylamine, warrant careful attention and control to prevent inhalation exposures or skin contact.

The mixing, grinding, and packaging of dyes should be segregated and provided with engineering control measures that prevent objectionable air contamination. Because of the small particle size and the costliness of most dyes, exhaust ventilation must be sparingly applied or else a filter must be used to recover the product.

Drying ovens should be vented to the exterior. Vats evolving steam or hydrogen sulfide, formaldehyde, or other noxious gases should be provided with process ventilation designed to give satisfactory control; reliance should not be placed upon general ventilation alone. Alkalies, chromates, arsenic, some dyes, and most of the materials named above require effective control measures to avoid skin irritation or sensitization from direct contact.

Electroplating

Electroplating is preceded by metal cleaning, which is described elsewhere. Chromium plating is the most common plating operation requiring good exhaust ventilation control. The recommended exhaust volume of 120 to 150 c.f.m. per square foot of tank surface (discussed in Vol. I, Chapter X) may vary considerably depending upon tank location and design, baffles, interfering air movements, current density, and plating efficiency. The plating solution may be conserved by keeping the surface of the solution 8 in. or more below the exhaust slot. A further, remarkable saving in plating solution can be accomplished by the periodic addition of an agent to reduce the surface tension. One such agent, "Zero mist," is said to markedly reduce ventilation requirements.

Plating solutions containing arsenic require very effective exhaust ventilation. Cadmium plating and lead fluoroborate plating operations do not ordinarily present harmful exposures even though unventilated. Copper, zinc, and nickel acid plating operations seldom necessitate mechanical exhaust. Cyanide baths frequently require exhaust and alkaline plating baths sometimes evolve offensive amounts of alkali mists requiring exhaust measures. It is good practice to ventilate all hot baths in order to control excessive humidity and there is some merit in making it general practice to hood and ventilate *all* plating baths as a safeguard against any change in plating procedures that might cause harmful exposures; but such recommendations in many cases cannot be justified by air analyses.

It seems superfluous to say that cyanides should be kept locked and stored where they cannot cause a catastrophe from accidental contact with acids, but they still frequently are stored carelessly.

Ammonia from zinc cyanide plating operations may reach annoying levels, but it is doubtful if it ever exceeds permissible concentrations where the plating department has suitable general ventilation. Concentrations of 5 to 10 p.p.m. ammonia are easily recognized by odor and therefore are sometimes offensive to persons in adjacent departments. Methods of control include foam-producing agents as well as ventilation.

Fertilizer Manufacture

The principal health hazard in the production of fertilizers occurs in the preparation of soluble phosphates. The raw material, fluoroapatite, is a calcium phosphate rock containing as much as 3 or 4 per cent of calcium fluoride. When phosphate rock is acidified to produce soluble phosphates, about half of the fluoride is evolved as silicon tetrafluoride and hydrofluoric acid. The tetrafluoride is produced through the reaction of hydrofluoric acid with silica present in the rock.

It is common practice to reclaim the valuable fluoride gases and to convert them into fluosilicates. Throughout the entire operation there are potential exposures to fluoride gases and dusts. The finished phosphates themselves normally contain about one half of the original fluoride, and are potentially dangerous. The permissible 8-hour fluoride exposure is 2.5 mg. of total fluoride per cubic meter of air.

The exposure should be evaluated by air analyses. One procedure is to collect fluoride dusts with an electrostatic precipitator and to determine gaseous fluoride in the cleaned air. It is also possible to determine dust with a precipitator, and both dust and gas by bubblers. The fluoride gas is then obtained by difference.

The preparation of mixed fertilizers is frequently accompanied by the release of quantities of dust. It is advisable to determine the extent of the exposures by dust-counting technique, and to limit the dust to fifty million particles per cubic foot of air. No health hazard is to be anticipated if this standard of permissibility for nuisance dust is met.

In some plants, phosphate fertilizers are treated with ammonia to provide available nitrogen. Ammonia gas in the air should not exceed 100 p.p.m.

Roasting Phosphate Rock and the Manufacture of Superphosphate

In this industry phosphate rock is fused in an electric furnace with carbon and silica or clay. The gases from this furnace, consisting of elemental phosphorus and carbon monoxide, pass to a condenser where the phosphorus is condensed to a liquid, and the carbon monoxide passes through to be recovered for fuel. The elemental phosphorus may either be collected as such, or vaporized with compressed air and burned to phosphorus pentoxide, which subsequently is absorbed in water to make phosphoric acid. Ground phosphate rock and phosphoric acid are in turn used to make superphosphate. Although the processes are carried out within a closed system, opportunities exist, especially during cleaning and repairing of the system, for exposures to elemental phosphorus, phosphorus oxides, carbon monoxide, and fluorides from calcium fluoride in the phosphate rock. The water that has been in contact with phosphorus is under suspicion as a source of exposure because it contains some phosphorus. All potential exposures should be evaluated periodically.

See also Phosphorus.

Forging and Iron Working

Exposure to high temperatures, sudden changes of temperature, and radiant energy present the principal health problems in forging and iron working.

Ventilation by means of air douches and man-cooling fans, protective asbestos and aluminum clothing, and aluminum radiant-heat shields are commonly used to alleviate the exposures. It is also of value to ventilate the building, as by means of roof ventilators that allow heated air to escape from the structure while supplying air in the occupied zone. Wherever excessive gas and smoke occur they should be removed by local exhaust ventilation.

Excessive infrared radiant energy is undesirable and it may damage the eyes of workers. Heat cataract may develop, depending upon the intensity of radiation, source (point or broad band), length of time exposed, and the age and health of the individual. The danger from radiant bodies multiplies rapidly as the size of the radiant body increases and as the temperature exceeds 2500°F. It has not been possible to set a safe limit based on our knowledge to date. Workers exposed to significantly intense radiation should wear approved protective lenses.

The use of salt in drinking water, or as tablets, is well recognized as a necessity in the prevention of heat sickness (see Vol. I, pages 119 and 807).

Carbon monoxide also is an ever-present danger and excessive fumes and gases from furnaces should be vented outside the building at such a height as to produce adequate dispersion.

Foundry Operations

The foundry environment varies not only with the kind of metal poured, such as steel, iron, brass, bronze, aluminum, or magnesium, but with industries, location, communities, experience and attitudes of management and labor, and other factors. All of these must be taken into consideration in evaluating and controlling the environment. The industry as a whole has been subjected to much criticism in recent years and its environment described as "hot, dusty, smoky, unhealthful, and hazardous." Such a description may apply in isolated instances, but not to the industry; foundrymen are improving the environment and are not only desirous of learning of any possibly harmful conditions existing in their plants, but are ready to listen to any practical, positive means of controlling them. The manager or superintendent who thumps his chest and exhibits his robust physique as evidence that the environment in his foundry is beyond reproach may need to be reminded that the office worker's environment is somewhat different from that of the shakeout man. Foundry management in general not only is desirous of controlling exposures known to be harmful but they want to clean up the plant atmosphere sufficiently to make it a desirable place in which to work.

How far to go in making recommendations for control is a matter to be left to the judgment of the responsible investigator. It is perhaps better to be satisfied, at first at least, with bringing only the more obviously harmful situations under immediate control and to leave the way open for progressive friendly relations than it is to prescribe sufficient control to make a dirty, dusty place into a model of cleanliness throughout. A "clean up" plan that is too ambitious may be received with antagonism and is likely to be placed in the "permanent file." The foundry that had a part in producing the armor for the "Merrimac," for instance, is steeped in lore and tradition. Such older foundries could proudly point to the fact that in nearly one hundred years of operation, during which flasks have always been poured and upset in the general foundry area, no one has died on the job or, so far as is known, been a victim of silicosis. The management of such a foundry may not welcome with open arms all of the newfangled medical and engineering ideas advanced by some itinerant industrial hygienist. Here, suggestions for such changes as the installation of flush toilets, substitution of electrically operated hoists and trolleys to replace a system of belts and pulleys operated by water power for transporting ladles of molten metal, and the segregation and ventilation of floor shakeout operations possibly would be a logical start in improving the environment.

If the foundry happens to be staffed by hardy mountaineers they may prefer to wear nothing more than a pair of shorts in warm weather and they may scoff at any mention of protective clothing or at the idea of wearing respirators in dust clouds ranging upward of one or two hundred million particles per cubic foot of air. Management and employees alike may scorn the "sissy" ideas expressed by a tenderfoot hygienist, especially if he gives way to the urge to don a respirator while making his survey. The obviously hazardous situations should be attacked first and tactful education should start with the management.

On the other hand, if the foundry is more or less completely mechanized, has correctly exhausted, enclosed shakeouts, model core knockouts, dustless molding and core making, a tempered and well-engineered air supply, well-engineered and controlled sand-handling equipment, spacious, well-kept aisles and so on, and all that can be found wrong is that some smoke and fumes are escaping from the pouring operation or the cooling sheds or tunnels and marring the freshly painted girders and walls, the method of approach will be somewhat different. But, even here, the hygienist has an opportunity to be of service to the management and workmen in helping them keep a showplace by advising how the escape of the smoke can be prevented or controlled.

The foundry survey is not merely a matter of collecting breathing-zone samples to evaluate exposures, but should include the location of sources of harmful or annoying exposures and the development of methods for control. The silica exposure has been overrated but is nevertheless the most important factor where dustiness is excessive (on the order of 30 million particles per cubic foot). Sea coal is perhaps the most offensive source of visible dirt and has been the cause of disastrous dust fires and explosions: satisfactory substitutes are being sought and, in a limited degree, applied. Iron fumes may give rise to confusing x-ray photographs of the chest. Cold drafts should be corrected, especially where shake-out or core-knockout operations are located in front of an open window so that the operator is excessively heated in front with radiant heat while his back is chilled with the inrush of cold air.

The control of the foundry environment will be discussed by following the metal from melting to cleaning.

I. Iron and Steel

A. MELTING

The *cupola* may be the source of carbon monoxide exposures and, if no dust control is employed on the stack, dust and iron fumes may be an objectionable source of atmospheric pollution in some neighborhoods, and may cause excessive deposits on the roof of the foundry. A simple type of water-spray dust arrestors with and without fans has proved a fairly satisfactory means of control.

Electric furnaces may give rise to large amounts of iron oxide and various other fumes, such as silicon dioxide and manganese dioxide, depending upon the composition of the steel. Roof ventilation above the furnace is common practice, but local exhaust is more effective. Breathing-zone samples collected with the electrostatic precipitator should be evaluated for toxic fumes and dusts, as well as for free silica (possibly from ferrosilicon[15]). Gases such as nitrogen oxides should be considered also.

If tellurium, selenium, bismuth, or other elements are added to the molten metal, any fumes resulting should be controlled by ventilation.

B. MOLD AND CORE MAKING

Ordinarily there are not many potentially harmful operations involved in making molds and cores. Some points worth investigating, though, are the dusting of flint or silica flour, the spraying of liquid mixtures containing harmful solvents,

[15] T. Bruce, *J. Ind. Hyg. Toxicol.*, **19,** 155 (1937).

chemicals, or silica particles, and the cutting or grinding of baked cores. Chills are sometimes sprayed with shellac or other adhesive material and dusted with sand. This may involve an exposure to solvent vapors and, where the sand contains fines, a dust exposure. Carbon tetrachloride, if used in molds, presents an exposure problem during the making of molds and a more difficult one at pouring stations due to decomposition products.

C. POURING

Smoke, gases, and vapors from the destructive distillation of sea coal mixed into the molding sand, and insignificant amounts of iron fume, arise from pouring operations, and for a considerable period after pouring. The gases, including carbon monoxide, escaping from the mold usually ignite spontaneously or are fired by a torch.

The heat, smoke, and fumes arising from this operation usually need no evaluation other than by the senses. The problem, which is chiefly of a nuisance nature rather than one of health, may involve mildly irritant aldehydes and much smoke that not only makes the atmosphere appear offensive but deposits black residues on all objects in the plant. Where the sand-conditioning process involves the addition of a solvent, this solvent may distill out of the mold at the instant of pouring and necessitate effective ventilation control. A properly constructed side-draft hood with the top extending out as far as possible over the flask and provided with an exhaust volume of 100 c.f.m. per square foot of face will remove most of the smoke, fog, or fume from a central pouring area. Up to more than twice this volume may be required to remove all contamination where conditions are not ideal.

Where flasks are poured on the floor, the general practice has been to provide high ceilings with air outlets as high as possible and inlets at or near floor level to provide dilution and directional general ventilation, preferably mechanical.

Cooling canopies or cooling tunnels, however, are desirable for removing smoke, fumes, and gases from the molds during cooling. Where mold conveyors are used, fairly close-fitting, heavy-gage metal tunnels from pouring station to shakeout, with 30- to 36-in. gravity ventilation stacks, have proved satisfactory if the make-up air supply is adequate. Tunnel air inlets, 8 to 18 in. high, are provided along the sides at floor level and, so long as the foundry is not airbound, smoke and gases can be controlled. Some tunnels have been constructed with removable side panels to permit ready access for maintenance purposes, but this almost invariably has resulted in some of the panels being removed and not replaced, with the result that the smoke pours out of this opening instead of being drawn to the gravity stack. The hinging of these panels might improve the situation, but a more positive method is to make the tunnel of solid welded construction. When necessary to provide an opening for maintenance, a cutting torch may be used and the plate replaced by welding as soon as the repair is completed. Of course if sufficient mechanical exhaust ventilation is applied, rather than gravity stack ventilation, the results are more dependable, and a few missing panels may not seriously interfere with smoke removal. Perhaps the most essential part of the ventilation system is an adequate supply of make-up air delivered to the occupied zone.

D. SHAKEOUT

1. Enclosure

The most effective control for shakeouts is the relatively complete enclosure with sufficient exhaust volume removed at the top to maintain an inflow of about 200 f.p.m. at all openings. The degree of enclosure will be limited by such factors as the kind and extent of mechanization for bringing in and upsetting flasks, size of flasks and castings, and manual or automatic castings discharge. This type of hood is limited to relatively small castings. The exhaust volume for such enclosures can be determined as 200 times the total area of openings in square feet, but must never be less than a minimum volume, which has been suggested by Kane[16] to be 200 c.f.m. per square foot of shakeout grate. This minimum volume is required to remove expanded air and steam.

2. Side Hood

The side hood is applicable to castings of nearly any size. It should be parallel to and larger than the long side of the grate; and the top, which should be reinforced with a bumper, should extend as near as possible to the centerline of the grate. At times it may be expedient to place baffles or side shields at one or both ends of the hood to increase the effect of air flow over the grate. Many factors, such as cross drafts, ratio of sand to metal, temperature of casting, and hood location and design, influence the exhaust volume required (see Vol. I, Chapter X), but 400 to 500 c.f.m. per square foot of grate area is ordinarily required for satisfactory control. It is practically impossible to control shakeout dust if "man-cooler" fans or air-supply ducts are directed toward the shakeout hood.

3. Downdraft

Although downdraft has many appealing features, effective dust control is not one of the major points. Hot castings create intense, opposing undrafts, and the sand covers the effective grate area at the time and place at which a downward flow of air is most necessary. A good part of the air for the exhaust flows in at floor level and so offers no dust control. It is not practical to control by downdraft the upward surge of hot air and its heavy charge of fine dust particles rising directly above red-hot castings.

Such an arrangement has been applied successfully to the shakeout of cooled castings, with volumes as low as 200 c.f.m. per square foot of grate area. At the usual temperatures for shakeout of iron castings volumes of 500 to 600 c.f.m. per

[16] J. M. Kane, *Foundry*, **74,** 86, 104 (1946).

square foot of grate area frequently fail to control the dust, while the high-velocity air tends to aspirate the sand out through the exhaust system.

E. CORE KNOCKOUT

The side-hood arrangement with exhaust volume similar to that for the shakeout is the preferred method of control for the vibrating core knockout, but enclosures can be applied and, when the castings are not too hot, the floor grill is more applicable here than it is to the shakeout. A combination of side hood and downdraft is very effective.

The use of compressed-air jets to blow out the last of the core sand is objectionable in that it disperses excessive amounts of fine silica dust. Hydraulic methods of core removal are reputedly successful.

F. SAND HANDLING AND CONDITIONING

It is a very difficult job to clean up a foundry having mechanical shakeouts and sand-handling equipment but lacking enclosures and correct ventilation. The outstanding points of dust dispersion requiring enclosures and exhaust ventilation are: the shakeout hopper, transfer to return conveyor, transfer to elevator, transfer to and from other belt conveyors, sand screens, tailing pipes, sand mixers, and receiving ports on sand bins. Exhaust ducts must be designed for correct conveying velocities, yet each hood connection must have a low-velocity entrance into the duct in order to avoid picking up large-sized grains. Abrupt hood connections at or near dust-dispersing points waste sand, which may settle out in the exhaust duct. Sea coal, being relatively light and fine, is rapidly removed from the hot sand by the exhaust system and can be reclaimed from collectors in some instances. Its explosion properties should not be overlooked. Shakeout tunnels usually require general ventilation in addition to the control mentioned above, especially where hot sand is transported on an open belt. Depending upon the length and size of the tunnel, it may be practical to sweep the air the length of the tunnel by exhausting at one end, while in others a central exhaust, or a distributed supply along one side and exhaust along the other, may be necessary. In any case there must be a controlled flow and displacement of air rather than dilution and mixing. Conditioning equipment and portable sand riddles that chop and discharge the sand by means of propellers, with sufficient force to throw the sand several feet, are dangerous sources of dust when used with relatively dry sand.

G. CASTING CLEANING

Blast cabinets, Wheelabrators, and tumbling mills must be maintained so that the relation between the area of openings and the volume of exhaust is in the proper balance to prevent the escape of dust into the room; and all unnecessary openings should be eliminated.

Blast rooms must be sufficiently tight to prevent the escape of larger particles through the velocity of their travel, either direct from the blast, or rebound from the work or walls. At the same time they should be provided with a sufficient volume of exhaust to prevent the escape of dust at crevices and to promote visibility. Where the room is large enough so that its ventilation with plant air involves significant heat loss, it may be supplied with some air from out-of-doors.

Air chisels for cleaning castings are frequently the source of excessive dustiness, and therefore dust respirators may be required unless a side or grill exhaust can be provided. Here again compressed-air jets, especially high-pressure ones, are objectionable for blowing off dust.

See also Abrasive Blasting.

H. MAKE-UP AIR

The necessity for make-up air has been discussed in Vol. I, Chapter VIII. In summer, if the air discharged by the cupolas and exhaust stacks is sufficiently clean, make-up air ordinarily comes through the open windows and doors. However, the use of air supplied through insulated ducts, or ducts below floor level, offers attractive cooling possibilities. With windows and doors closed, as in winter, there is necessity for a supply of air sufficient to keep the indoor—outdoor pressure essentially balanced if drafts and the other adverse effects associated with reduced interior pressure are to be avoided.

Various ideas for utilization of waste heat around the foundry in winter by delivering the make-up air through heat exchangers have some merit. Perhaps the most available sources of waste heat are furnace stacks, and casting-cooling conveyors. There are many places where cold air might be used to supply make-up air for process control without having it traverse working spaces, thus avoiding the heat loss from using tempered room air for this purpose. Examples are: shakeouts, cooling tunnels, tumbling mills, blast cabinets, and so on.

I. LAYOUT

It is not possible to have a clean foundry and satisfactory environmental control along with overcrowding. The establishment of good "housekeeping" must necessarily start with the provision of sufficient working space and aisles. Mechanization is an aid to any environmental control program. Conveyorized lines for mold and core preparation, pouring, cooling, and shakeout should be laid out with a view toward minimum-handling, coordinated flow of materials, and ample work space. Those operations that involve heavy lifting, or are stationed in a hot, smoky, or dusty atmosphere should be studied with a view toward simplification or mechanization to an extent that will provide satisfactory environmental conditions for each workman. Automatic pushoffs, vibrating core knockouts, and enclosures for shakeouts and core knockouts that permit the operator to remain outside, protected from heat, noise, and dust, are outstanding examples of mechanized means of improving the environment.

J. HEAT, NOISE, AND DUST CONTROL

In addition to the gas, dust, and fume control measures discussed up to this point, *heat* is a sufficient problem in some areas of the foundry to warrant control by shielding and by ventilation (see Vol. I, pages 221 and 801).

Noise, likewise, is a problem that warrants engineering planning and control efforts. Practical control measures have included: compressed-air release mufflers; liners for tumblers; acoustic-lined enclosures for shakeout and core knockout; spot-welded damping plates on hoppers, tote boxes, bins, and conveyors; separation and isolation of noisy operations; and the application of personal protection measures (see Vol. I, Chapter XVIII). The wise purchaser of new machines and equipment will require the manufacturers to supply information on the intensity and frequency of noise produced by the equipment offered.

Dust control, heat control, and noise control, all three, have been engineered successfully into a single acoustic-lined, ventilated enclosure having heat-shielding glass windows through which workmen view the mechanized operation and manipulate necessary remote controls.

II. Brass and Bronze

Many of the problems previously discussed under the iron and steel foundry apply here. In addition, the melting and pouring operations usually necessitate mechanical ventilation control of zinc fumes, and this exposure must always be considered.

Where phosphor bronze is made by adding white phosphorus to the molten metal, strict safety precautions and good ventilation are necessary. Replacing the white phosphorus with a copper-phosphorus alloy or an ore containing phosphorus might well be given a trial.

If lead is used, the possibility of a lead exposure should be investigated.

III. Aluminum

Sometimes chlorine gas is bubbled through aluminum in the melting furnace to volatilize impurities: control by ventilation is then necessary. Lateral-flow hoods with partial enclosures on the sides are effective, while conventional canopy hoods have not proved satisfactory. Carbon tetrachloride and mixtures containing it, if used in a similar manner, involve a greater potential exposure due not only to the vapors of carbon tetrachloride, but also to its possible decomposition with formation of phosgene gas.

Sometimes an alloy is made by adding about 0.3 per cent mercury to a mixture of molten zinc and aluminum in the melting crucible at about 1000°F. and pouring it at 900°F. Since the boiling point of mercury is 674.4°F. this obviously presents a serious potential mercury exposure even though the mercury is introduced through a tube to the bottom of the crucible. The operation requires effective ventilation control, as well as care to avoid spillage.

Any other exposures are similar to those discussed under iron and steel foundries.

IV. Magnesium

The use of antioxidants, fluxes, or conditioning agents such as sulfur, sulfur dioxide, and fluorides makes it necessary to provide ventilation control for melting, pouring, and mold-cooling operations. Where fluorides are mixed with the molding sand or sprayed on cores, fluoride exposures must be considered and controlled at pouring, core spraying—especially when sprayed hot, shakeout, and core-knockout operations. Depending upon what fluorides are used, the air contamination may be gaseous (hydrogen fluoride), or particulate, or both.

Heat-treat ovens with a reducing atmosphere containing sulfur dioxide, usually on the order of 0.5 per cent, require mechanical ventilation for all openings, and the hoods and ducts must be designed to accommodate the escape of large volumes of the atmosphere when the oven doors are opened for loading and unloading.

Due to the use of a fine-grained sand, even those exposures attendant to the handling of the dry, raw sand must be well controlled.

Galvanizing

Galvanizing is done by dipping cleaned metal into molten zinc. The iron must be cleaned by dilute acids, zinc chloride, molten caustic, abrasive blasting, or barrel tumbling before galvanizing.

Zinc is a physiologically inert element and it should not be considered an industrial poison.[17] As fume (finely divided zinc oxide) it does, however, cause metal-fume fever. This is a transient effect, which is unpleasant but which seldom lasts longer than 24 hours. The fever may be caused by other metallic oxides. Aged fume (partially agglomerated) does not cause fever. The maximum allowable concentration for zinc fume is 15 mg. per cubic meter of air. Skin affections from zinc are uncommon but zinc chloride has been described by several investigators as a skin irritant.

The processes employed for cleaning iron surfaces should be examined carefully. Exposures to the splash of molten caustic, to acid vapors or mists, or to silica dust from abrasive blasting, may require control.

Garages

The principal problem around garages is the exposure to carbon monoxide from exhaust gases. This problem is easily solved by the use of flexible exhaust duct connections for automobiles during test operations (see Vol. I, page 316). Unwarranted, and possibly unsafe, devices such as ozonators for "purifying"

[17] H. Engel, *Occupation and Health,* International Labor Office, Geneva, 1934, p. 1285.

exhaust gases should not be relied upon because they have no beneficial effect upon carbon monoxide, the toxic constituent of exhaust gas. It is essential to provide sufficient general ventilation to dilute and remove exhaust gases generated by cars in motion, so as to keep the average concentration of carbon monoxide below 100 p.p.m. Brief and fleeting concentrations of carbon monoxide up to five times this amount are of no consequence.

The use of trichloroethylene and other solvents is discussed under Metal Cleaning. Spray painting when confined to touch-up work is not a serious problem; for more extensive operations see Spray Painting. When solder is applied to automobile bodies by the use of a torch for the purpose of filling dents or cracks, it does not cause a significant exposure to lead, but in a room where men work, this solder should never be removed by a disk grinder without positive personal respiratory protection. Such disking, unless done in a controlled and segregated area, contaminates the room with sufficient lead dust to present an exposure for an extended time after the actual operation has been finished. Rasping or scraping does not produce a significant amount of respirable dust, but does present a "housekeeping" problem.

The use of gasoline in open containers such as vats, pans, or buckets is hazardous and should be avoided. Sandblast cleaners for spark plugs, if operated for consistently prolonged periods, warrant checking for possible silica dust exposures and lead exposures.

Glass Manufacture and Fabrication[18]

The four main divisions of the glass manufacturing industry are flat glass, container glass, specialty (or technical) glass, and fiber glass. In modern glass technology nearly every element of the periodic table has been utilized, with the attendant hazards found in each. The major portion of all glass batches, however, is still silica sand. In the common glasses the remaining constituents are materials such as limestone, lime, feldspar, salt cake, and soda ash. Minor constituents include arsenic, antimony, carbon, cobalt, rouge, and selenium.

With the major proportion of each batch being sand, it would seem to present a potentially serious silicosis hazard. Actually, in most cases today washed sand is used, from which most of the fine particles have been removed. It is common to find that air-borne dust from the mixed batch will contain only from 1 to 5 per cent of free silica. Silicosis under modern conditions is rare.

However, with certain types of sand the methods of handling can present a silicosis hazard. The unloading of dry sand from boxcars, either by power scoop, or by shovel and wheelbarrow, may produce dangerous quantities of fine silica dust. When wet washed sand is obtained in hopper cars and unloaded by gravity or, after drying, handled mechanically in a totally enclosed system with exhaust ventilation, there is no harmful exposure. The system of obtaining various glass

[18] This section was written by Karl L. Dunn of the Corning Glass Works, Corning, New York.

batch materials in bottom-dump hopper cars has become fairly commonplace in the modern glass plant to the degree that dust exposure from materials other than sand has become a minor consideration. Frequently, one man with automatic handling equipment and weighing devices can handle the batch requirements for a very large, multiple-furnace glass plant.

In the manufacture of optical glass and certain decorative glasses, lead becomes a consideration as far as employee exposure is concerned. Handling of this material, usually in the form of lead oxide or lead silicate, requires the usual personal hygiene procedures, the use of localized exhaust ventilation, and, where necessary, the use of dust respirators. The other major constituents of glass represent a nuisance dust problem rather than an exposure harmful to health. Their only detrimental effect upon health has been dermatitis, where reasonable standards of cleanliness are neglected. The minor constituents have caused some harm to health, with arsenic the principal offender. Perforation of the nasal septum or severe skin effects caused by exposure to arsenic and to highly alkaline constituents of the batch were not uncommon occurrences in the past. Modern methods of handling have eliminated most of this trouble.

Glass workers' cataract has been referred to in the literature and was alleged to be caused as a result of exposure to high intensities of infrared radiation. More recent studies in both American and European glass plants have failed to indicate the existence of this disease in the modern glass industry (see Vol. I, pages 719 and 787). No limitation on tolerance to infrared radiation is recognized at the present time.

The various types of glass manufactured in the modern glass industry are made by two processes—the older pot process and the more modern and more common tank method.

Pot Process. The older, or pot process, is now used principally for the manufacture of high-quality glass, such as optical and mirror glass, and for small-quantity specialty glass, such as railroad, marine, and aviation ware. The tank method is used for high-volume requirements, such as window glass, television tubes, incandescent lamp blanks, fluorescent tubing, and container glass. The pots vary in size up to those capable of holding nearly 2 tons of ingredients, and their manufacture has been responsible in the past for the greater portion of the silicosis in the glass industry. The pots are made from a number of different types of clay combined with flint or silica flour. Pot manufacture was commonly a very dusty industry. It is only in the last 20 to 35 years that dust control has been practiced in the pot manufacturing trade. Pot glass is manufactured in furnaces that, because of waste heat, provide their own industrial gravity ventilation. Dust concentrations in the breathing zone of workers in the furnace halls are usually below 5 million particles per cubic foot of air.

Pot melting of glass necessarily introduces the hazard of hand shoveling and filling of the pots. Optical and specialty glasses frequently contain heavy metals such as lead, barium, and manganese. Close attention must be given to handling of these other toxic materials during this hand-filling process. Because of the introduction of a more efficient tank melting system, the hazards of the pot melting process are rapidly disappearing.

Tank Process. Most of the latest glass tanks provide for enclosed continuous feeding of batch ingredients. This system has reduced radically the exposure of large numbers of people at the batch end of the tank. Numerous dust counts made at the feed end of glass tanks show that typical dust concentrations are on the order of 1 to 2 million particles per cubic foot of air. This is true even with poorly designed feed systems. The heat of the tank provides effective updraft removal of the fine dust that may inadvertently occur because of leakage in the poorly designed systems.

Furnace and Tank Construction and Repair. The blocks and bricks used in the construction of the furnaces and tanks contain free silica in significant amounts. Silica brick contains tridymite as its principal constituent. When assaying the dust exposure, care should be taken to determine tridymite and cristobalite as well as quartz. Previously the wearing of respirators was difficult, if not impossible, because of excessive heat. Respirators are now commonly worn by repairmen, a situation made practical by advances in heat control. Introduction of a system of prefabrication of furnace blocks and parts has reduced the dust concentrations formerly found during tank repair. These were caused by cutting and chipping of blocks and bricks to be fitted into the furnace structure. Large, well-ventilated, mechanized cutting shops now allow for prefabrication of furnace parts to be shipped to the furnace site for construction. Only occasional cutting at the construction site is done under present-day methods.

Glass Fabrication. Glass objects may be formed by blowing, pressing, and casting.

Hand blowing in the modern glass plant is used only for the manufacture of art objects, large objects of low demand incapable of being fabricated by machine, and complicated technical shapes. Hand blowing is the classical method of glass fabrication and was supposed to have been attended by such things as glass workers' cataract, mentioned previously, glass blowers' emphysema, supposedly caused by intense back pressures on the lungs, tuberculosis, and a host of other ailments. Glass blowers' emphysema is nonexistent, as the pressures required for blowing glass are in the order of ounces rather than pounds per square foot. The incidence of tuberculosis and other respiratory diseases is no greater in the hand glass-blowing trade than in the average population. Exposure to high heat was the major difficulty in hand glass fabrication, but modern means of radiant heat control and convective cooling systems have minimized this problem.

The modern glass industry is one of almost complete mechanization. Automatic blowing machines now exist that are capable of producing 2000 incandescent lamp bulb blanks per minute. The same order of speed and efficiency is true of the glass container industry. Glass tubing, window glass, tableware, and high-volume technical glass objects are now formed by highly efficient and intricate glass-forming machinery.

The hazards, then, of the modern glass industry are those of a highly specialized machine trade with the attendant exposures to lubricants, solvents, coolants, and other chemicals.

Because of the high volumes of hot glass flowing through these modern factories, the problem of radiant heat was recognized in the glass industry 20 to 25 years ago. Methods for the control of radiant heat were first introduced into the glass industry and consisted of aluminum-panel radiant shielding. More recently the glass industry, taking care of its own, has introduced infrared reflecting plate glass to provide transparent radiant heat shields to protect the employees. Exhaust ventilation for the control of solvent vapors and the like, and cooled supplied air are commonplace in the modern glass plant.

Annealing. All glass objects, after forming, must undergo a process of annealing to reduce internal stresses in the formed object. This is accomplished in most cases by introducing the objects into long, continuous annealing chambers called "lehrs." The lehrs, because of their size and the quantity of heat generated, introduce again the problem of heat control. This has been accomplished by radiation shielding and convective cooling. The only points of employee exposure in the annealing process are at the charging and discharging ends of the lehr. Cool air is introduced at these points through floor grates and overhead piping, in addition to radiation heat shielding.

Glass-Finishing Processes. After the glass has been formed and annealed, it frequently becomes necessary to give it further treatment for purposes of decoration, labeling, smoothing of rough edges, and the like. The silvering of mirrors introduces the hazards of possible formation of fulminate of silver in the silvering solutions, the contact of the skin with these solutions, and attempts by the employees to remove silver stains by the use of such hazardous chemicals as the cyanides. Such exposures, of course, should be avoided. Glass grinding and polishing is largely accomplished by wet processes with the use of abrasives on large revolving and grinding mills. No hazard other than mechanical is present here because neither the grinding media, usually silicon carbide, nor glass dust are by-products. Glass dust in itself is only a nuisance material, all silica being in the combined or silicate stage. Polishing is accomplished by the use of revolving felt pads with rouge (iron oxide) as the agent. No health hazard is present.

Abrasive blasting is sometimes used, but, with the use of enclosed cabinets with exhaust ventilation and the use of nonsiliceous abrasive materials, no severe hazard exists in such operations.

Application of decorative enamels by spraying or silk-screen processes introduces the possibility of exposure to solvent vapors, which must be controlled by exhaust ventilation.

Hydrofluoric acid etching presents a severe potential hazard, both from the standpoint of skin contact and inhalation of the acid vapor. The use of protective clothing and exhaust ventilation, and adherence to strict measures of personal hygiene, minimize the hazard in acid-etching operations.

The cleaning of glass by various solvent and detergent solutions introduces the same hazards that might be encountered in the machine trades. Control methods for these exposures in the glass trade are identical with those found in other industrial processes.

The processes that are used in cutting and engraving glass in the finishing operations produce some small amount of glass dust but the amounts are far below the nuisance level. As stated previously, the sand that forms the major constituent of glass batch is converted in the pot or tank to silicate. There is no free silica in glass.

Glass Bending. Many of the new shapes of glass used in store fronts, airplanes, etc., are produced by bending the glass in furnaces and kilns. The larger plates are laid in finely ground burned fire clay that has been molded into the proper contours. The molds are often built for each piece.

For mass production, the kiln process is used most frequently and iron molds are replacing the older fire-clay methods. If fire clay is used, the composition of the air-borne dust should be determined and the dust concentration measured. Because of the heat stored in the kiln cars, the continuous kiln process provides ventilation by means of heated air, which rises, and carries the dust up with it and out of high, open windows.

Lampworking. Nearly all laboratory and pharmaceutical and some special decorative ware is formed from glass tubing or special blanks by reheating the tubing or blanks in oxygen gas flames mounted on a work bench. This special fabrication process is called "lampworking" and introduces a special hazard of glare because of the high heat generated. This intense glare or incandescence is caused largely by the sodium doublet in the spectrum. In recent years it has been discovered that this glare can be controlled by providing the lampworker with didymium glass spectacles. Didymium-bearing glass is very light pink in color, almost clear, but has the facility of canceling out the sodium doublet in the spectrum, thereby reducing the major source of glare without reduction in the intensity of the rest of the spectrum.

In the working of the higher melting glasses of high silica content and in the working of fused quartz, it sometimes becomes necessary to introduce in the spectacles coloring agents other than the didymium because in these higher melting glasses other portions of the spectrum likewise present a glare problem.

Heat generated in these laboratory lampworking rooms is nearly all from open flames. General room ventilation of high volume becomes necessary in addition to the introduction of cool supplied air. Shielding of fellow workers from flames on adjacent benches is accomplished by radiation shielding, usually of the infrared-reflecting plate-glass variety.

Glass Fibers. In the manufacture and fabrication or use of glass wool (spun glass), glass fiber particles tend to disperse in the air and settle on the skin. In earlier forms of glass fibers, which were coarser and had inclusions of slugs (larger particles of sharp glass), these slugs and larger particles came in contact with the skin and produced minor irritations. Since moist surfaces tended to aggravate the condition, the use of protective clothing was not always successful. Most workmen soon became hardened to this phenomenon and process ventilation was introduced to control the escape of broken fibers and slugs. Frequent bathing with soap and water tended to remove the particles and was the most successful personal control measure. Recent methods of manufacture of glass fibers have reduced the tendency toward the formation of slugs and larger glass spicules. Pulmonary irritation or injury from the inhalation of glass fibers has not been recognized in persons or demonstrated in animals.

Grain Handlers—Elevators

The cost of dust explosions in the United States up to January 1957 was: 676 persons killed, 1770 injured, and $100,000,000 in property damage (see Vol. I, Chapter XVI, page 549). The control of dust explosions is based upon: (1) prevention or control of dust, (2) elimination of sources of ignition, and (3) construction of buildings so as to minimize damage.

Concentrations of grain dust to be within the explosive range must be high, probably in excess of 35 g. per cubic meter of air: clouds of dust of this intensity are practically opaque. The greatest danger lies within enclosures where the grain is in motion, and in the accumulation of piles of dust on ledges, and so forth, whence it can be dispersed as such a cloud.

Sources of ignition include open flames, static electricity, faulty electrical wiring and electrical equipment, and sparks from pieces of metal. Magnetic separators should be used to remove metal from the grain.

Explosions can be minimized or diverted into less harmful channels by means of lightweight roofs, special windows, and other vents. An inert gas, such as carbon dioxide, is sometimes employed to reduce the oxygen content of the atmosphere in storage bins below 12 per cent, the minimum amount that will permit an explosion.

Grain handlers find that irritation and discomfort occur when fine dust enters the respiratory tract. Flour and meal dusts may form ulcers in the mucous membranes of the nose. "Grain fevers" from the dust or from fungi have also been described. Allergic manifestations from grain dust proteins also occur. The mold, aspergillus, when taken into the lungs, can cause a condition that may be misinterpreted as silicosis or miliary tuberculosis.

The fumigation of grain to destroy weevils may entail exposure to a wide variety of chemicals such as: carbon tetrachloride, carbon disulfide, ethylene dichloride, chloropicrin, ethylene oxide, and hydrogen cyanide. Where seed corn is treated with an insecticide containing mercury[19] and cadmium, excessively harmful exposures may result.

"Grain itch" is common. It is caused by a parasite that infests the grain; cleanliness is the principal means of prevention.

[19] H. F. Schulte, *J. Ind. Hyg. Toxicol.*, **28**, 159 (1946).

Grinding, Buffing, and Polishing

Grinding. The dust produced by grinding operations will be composed of the material being ground and the abrasive. Either or both may be dangerous to health or the dust may be solely of the nuisance variety. Dry grinding must have exhaust ventilation if the dust that is produced contains significant quantities of free silica, asbestos, or a toxic material. Exhaust ventilation should be used in all other cases where nuisance-dust concentrations exceed fifty million particles per cubic foot of air. In attempting to provide a clean environment, some companies now exhaust all fixed-position dry-grinding wheels that operate more than an hour daily.

When natural abrasives containing free silica are used, wheel dressing frequently causes the greater portion of the dust exposure. Attention should be directed to this operation, which occurs in both dry and wet grinding. It should be remembered also that even though the dust from grinding with artificial abrasives on iron is of the nuisance variety, it can be contaminated by sand adhering to the iron castings.

Wet-grinding operations produce less dust than dry, but with high-speed wheels dangerous spray may result. Under some conditions the moisture may vaporize leaving dust particles suspended in the atmosphere. The spray itself also is known to be hazardous under some conditions. With low-speed or moderate-speed wheels exhaust ventilation is normally unnecessary.

Buffing. Buffing compounds may contain free silica or toxic materials. Some of the ingredients may irritate the skin. Any specific operation can be appraised by air analysis. Knowledge of the composition of the buffing compound is desirable, particularly if skin irritation has been suspected. Where irritants are known to be present and it is not practical to eliminate them, protective clothing—or possibly creams—may furnish control, but personal cleanliness is most important.

Polishing. The most commonly used polishing agents, such as rouge or emery, are not harmful to health. If the operation is excessively dusty, exhaust ventilation is desirable.

Hat Manufacture

The exposures to mercury attendant to working with carroted fur have been discussed under Carroting. In the manufacture of hats the mercury exposure problem can be solved by using fur that has not been treated with mercury, or by providing efficient exhaust ventilation as proposed by the United States Public Health Service.[20]

The mercury exposures include mercury vapor, mercury nitrate, and particulate mercury or its compounds; and they occur wherever the fur or felt is

[20] P. A. Neal, R. H. Flinn, T. I. Edwards, W. H. Reinhart, J. W. Hough, J. M. Dalla-Valle, F. H. Goldman, D. W. Armstrong, A. S. Gray, A. L. Coleman, and B. F. Postman, *U. S. Public Health Bull.* No. **263**, 1941.

stored, agitated, subjected to hot water, steam, or drying operations. The average mercury vapor concentration found, in milligrams per cubic meter of air, was: fur storage, mixing, and blowing, 0.5; coning, 0.27; hardening, 0.25; starting, wetting down, and sizing, 0.21; and drying, 0.49. The highest vapor–air concentration recorded was 15.0 mg. per cubic meter of air around kettle fur-dyeing operations at a temperature near the boiling point of water. The highest concentration that has been encountered by the author was in the drying room, which, at 100°F., was unventilated except for one small open window: the concentration was 3.3 mg. of mercury vapor per cubic meter of air. The safe limit of exposure is considered to be 0.1 mg. of total mercury per cubic meter of air.

Many gas-fired appliances, some of which are not properly vented, make carbon monoxide a factor to be considered.

In the manufacture of straw hats, bleaching and cleaning operations involve the use of oxalic acid and sulfur. Lacquer-spraying operations involve the use of highly volatile, flammable thinners and hats with a defective paint job are sometimes sponged off with the thinner. Any exposures or hazards accompanying the use of these materials should be investigated.

Heat-Treating

There are many ways of heat-treating steel for hardening purposes: most common are controlled-atmosphere furnaces and salt and metal baths.

Controlled atmospheres may consist of inert gas containing either carbon dioxide or carbon monoxide. Benzene (C_6H_6) is used in one process. Special atmospheres for nitriding employ ammonia. When carbon monoxide (up to 20 per cent) is used, care must be exercised to prevent its escape into the workroom. The inert-gas producers should be checked for leakage when first used, and after any repairs, to make certain that all of the joints are tight. Good practice dictates the use of exhaust hoods above the furnaces and flame curtains at the openings. It must be ascertained that no explosive mixture is present when the furnaces are lighted. If an excess gas-producing capacity exists, as when only a portion of furnace capacity is used, the gas must be vented safely to the atmosphere outside the plant. Furnaces depending upon gravity stack ventilation should be watched for carbon monoxide leakage during the warm-up period.

Ventilation should be employed on nitriding furnaces when ammonia gas is used. As ammonia is explosive in the range of 15.5 to 26.6 per cent by volume in air, the tanks and lines containing liquid ammonia must be protected from breakage. The release of liquid ammonia could result in an explosive atmosphere coming in contact with one of the many open flames in the department.

It is common practice to provide exhaust ventilation for cyanide baths. This is not because of the production of cyanide fume. The fume consists essentially of sodium carbonate, which is somewhat irritant. Cyanides should be stored under lock and key away from acid carboys

Nitrate baths are exhausted as a precaution against irritant gases. Care should be taken to prevent organic matter from coming in contact with the hot salt because the mixture might be explosive. Neutral salt baths also are frequently exhausted, principally to control the heat during hot weather and to remove vapors corrosive to metal parts.

Lead baths are frequently held at temperatures between 1000 and 1500°F. and therefore require exhaust ventilation.

Where sprinkler systems are used, canopies should be erected above all oil, salt, and metal baths to prevent water from cascading into them. Any workman adjacent to a hot bath when water struck it would be in grave danger.

Oil quench tanks are frequently ventilated to remove the irritating smoke that is evolved during their use. Some tanks have cooling coils to control the temperature of the oil and reduce any fire hazard as well as the tendency to produce smoke.

Induction Furnaces

Converters used for the operation of some high-frequency induction furnaces contain upwards of 50 lb. of mercury in two iron pots, in which adjustable electrodes may be raised or lowered in relation to the pools of sealed-in mercury. This assembly may be sealed in with an iron, ceramic, or plastic cylinder, and cylinder heads. It may be in either two or three parts, with gaskets of rubber or other material to prevent the escape of hydrogen and mercury vapor. An atmosphere of hydrogen is maintained by allowing hydrogen gas to flow slowly through tubing from a cylinder of the compressed gas into the electrode chamber. The hydrogen escapes by bubbling slowly through a mercury trap that maintains the pressure at about $1/2$ in. of Hg above atmospheric. When the furnace is being heated, the electrodes arc and the heat volatilizes the mercury within the electrode chamber. Any leaking hydrogen carries mercury vapor out with it. Leaks may result from the use of faulty gaskets, insecurely fastened heads, or from dispersion of vapor through rubber tubes used as connectors between the electrode chamber and the hydrogen escape trap. Exposures to mercury vapor and dust in the converter area, and especially those of maintenance men who service and clean the converter, should be evaluated and, where necessary, controlled. The dust deposited inside the cages surrounding some of these converters has been found to have a very high mercury content. Although the design of these converters has undergone many improvements since the exposure was first reported by Turner,[21] there are installations in use today that warrant careful investigation and control if mercurialism is to be avoided.

Ionizing Radiation and Industrial Radiography

Radioactive isotopes, industrial radiography installations for the nondestructive inspection of metal, and to some extent also x-ray diffraction apparatus, are

[21] J. A. Turner, *U. S. Public Health Repts.*, **39**, 329 (1924).

serious potential exposure sources of ionizing radiation. The control is relatively simple and consists of shielding with lead or its equivalent so as to prevent exposure to radiation in excess of 300 mrem per week. The ability of concrete or other surfaces to deflect x-rays at any angle must always be considered. Tests should be made with a suitable rate meter, especially at cracks, joints, pipe chases, or other suspected crevices and flaws in protective barriers, to detect any significant leakage. Ceilings and floors of x-ray rooms are frequently found to lack shielding. Mechanical restriction on the angle of projection is an acceptable control method, and so is sufficient distance between the source and possible exposure areas (see Vol. I, Chapter XIX, and recent pertinent publications).[22-24]

Iron and Steel Industry

Iron ore, the life blood of the iron and steel industry, is received substantially free from silica: hence, there is no silicosis hazard at the blast furnaces. However, serious health hazards that may be present during the production of iron are carbon monoxide, hydrogen sulfide, and sulfur dioxide.

By-product coke plants produce many valuable chemicals, which must be carefully handled. Carbon monoxide, ammonia, benzene, carbon disulfide, and so forth, are potentially harmful. Most operations occur in totally enclosed systems so that the principal difficulty is unexpected leakage resulting in momentarily high concentrations. Plant maintenance, intelligent supervision, and thorough training are required to control the chemicals safely. Urine sulfate tests, air analyses, and red and white cell counts are valuable tools in the appraisal of benzene exposures.

Heat is a real problem in the iron and steel industry. Heat sickness is now controlled by means of extra salt (see Vol. I, pages 119 and 807). The alternating heat and cold experienced in winter months is related to the high pneumonia rate found by the United States Public Health Service among iron and steel workers.

Iron and steel foundries have been discussed under Foundry Operations.

Open hearth operations may expose men to carbon monoxide. Furnace repairs also present a silica hazard. Leaded steels were reported by Fehnel[25] to introduce a lead poisoning problem during pouring. Concentrations in crane cabs were as high as 88 mg. per 10 cubic meters of air. Special steels may contain nickel, bismuth, chromium, ferromanganese, tungsten, and molybdenum. Fehnel reported that these metals were not found in the air near the furnace charger nor on the pouring platform. Fluorides may be encountered.

[22] *Natl. Bur. Standards, U. S. Handbook* No. **59,** 1954, Permissible dose from external sources of ionizing radiation.

[23] *Natl. Bur. Standards, U. S. Handbook* No. **69,** 1959, Maximum permissible body burdens and maximum permissible concentrations of radionuclides in air and in water for occupational exposures.

[24] *Radiation Protection Manual.* American Foundrymen's Society, Des Plaines, Ill., 1960.

[25] J. W. Fehnel, *Ind. Med.,* **11,** 358 (1942).

The production of tin plate involves exposure to acids while terneplate also involves lead. Zinc baths have been found contaminated with lead in amounts sufficient to constitute a hazard.

Iron and steel plants also may have plating operations, spray painting, welding, and so forth. Fehnel advises that attention be directed to maintenance shops, for the detection of other significant exposures.

See also Galvanizing, Heat-Treating, and Metal Cleaning.

Lead Workers

The hazard present in exposures to lead and its compounds is now known to be dependent not only upon the air-borne concentration but also the form in which it is present. G. C. Harrold et al.[26] have shown that 1.5 mg. of lead per 10 cubic meters of air in lead-chromate spraying operations is far less than a reasonable safety limit. Whether the explanation for this fact is low solubility as ascribed by Harrold or failure of the lead chromate to reach the lungs (particle size, etc.) is not known. However, the absence of specific data on the safety limits for most lead compounds makes it advisable to adhere to the maximum allowable concentration of 2 mg. per 10 cubic meters of air in most cases.

Molten lead does not produce significant quantities of vapor below 900°F. but lead oxide dross formed on the surface may be thrown into the air. Burning operations are well-known producers of dangerous concentrations. When lead is poured, agitated, or skimmed, the danger from oxides increases.

The appraisal of a lead hazard is best accomplished by a combination of air analysis with blood or urine determinations. Air samples should be collected with the electrostatic precipitator when fume is present, but dust is satisfactorily retained by the impinger. In evaluating the exposure of an individual the samples should be collected from the breathing zone. Respirator filter analysis, when properly supervised, affords a valuable adjunct to other sampling methods (see Chapter XXVI).

The use of blood-lead determinations permits the removal of the employee from work involving exposure to lead dust and fumes *before* intoxication occurs. In a group of plants employing thousands of men working with lead and its oxides, where periodic blood-lead analyses have been used, coupled with transfer of employees with confirmed blood-lead levels of above 0.08 mg. of lead per 100 g. of blood, no definite symptoms of lead intoxication have been encountered.

Leather Industry

The leather industry has no typical, or outstanding, harmful exposures, but many common exposures that require recognition and control. In the tanning of leather there is first the matter of handling the raw hides and skins. One of the

[26] G. C. Harrold, S. F. Meek, G. R. Collins, and T. F. Markell, *J. Ind. Hyg. Toxicol.*, **26**, 47 (1944).

potential exposures is that of infection from contact with organisms such as anthrax. This was a serious problem in the past, especially with skins imported from anthrax-infested areas, but improved methods of treating imported hides have done much to bring the exposure under control. Anthrax is acquired usually by contact of the anthrax bacillus with wounds or abrasions, but the organism may be acquired by inhalation or ingestion. Persons working with raw hides or skins have greater need for medical supervision and inspection than do the average workers because the prevention and control of infections is primarily a medical problem.

Soaking, unhairing, and pickling may involve, in addition to potential exposures to pathogenic organisms, exposures to sodium sulfide, hydrogen sulfide, sulfurous acid, arsenic sulfide, chlorine, 2-naphthol, p-nitrophenol, mercury compounds, dimethylamine, sodium cyanide, lime, formic acid, and ammonia.

The tanning process may involve exposures to chromates, chromic acid, alkalies such as trisodium phosphate and borax, oxalic acid, and formaldehyde.

In the finishing and cementing of leather almost any of the common solvents may be encouraged. Any exposures should be sought out and evaluated. In the process of coating leather sides with a mixture of ethyl acetate and castor oil, after which the hides are hung in the room on racks to dry, very high atmospheric concentrations of ethyl acetate may result. It has been the author's privilege to observe and measure concentrations of ethyl acetate ranging from 375 to 1500 p.p.m. in the breathing zone of such workmen. Although this condition had existed for several months no adverse symptoms or illnesses were observed. This is of interest as an indication of the relatively low toxicity of ethyl acetate.

Gas-heated shaping presses are a potential source of carbon monoxide. Various dyes and stains are likewise potentially harmful. Grinding, sanding, buffing, and polishing operations are dust producers requiring ventilation control. In the leather industry there is much opportunity for the correct application of control measures such as protective clothing, protective creams, segregation of processes, and exhaust ventilation.

Lime

Protection of the eyes and of the skin presents the major environmental control problems in the production of calcium oxide or quick lime. The material is of small particle size and is very irritant to both mucous membranes and the moist skin. It combines with water, with evolution of heat, to form calcium hydroxide, which is nearly as caustic as potassium hydroxide.

Air-slaked lime, which is more or less completely calcium carbonate, has mild causticity and usually it will attack mucous membranes only. If warm, it may cause dermatitis after prolonged exposure.

Quick lime rarely affects the lungs as its irritant action on the upper respiratory tract precludes the necessary exposure. It rapidly produces coughing and sneezing limiting further exposure.

Workmen at lime kilns may be exposed to dangerous quantities of gases: carbon monoxide, carbon dioxide, hydrogen sulfide, and arsine. Severe temperatures are also common.

Roessle and others have shown that among lime workers the mortality rate from tuberculosis is less than the average. Some cases of pneumonia among workers exposed to excessive dust concentrations have been described.

No maximum allowable concentration for dust exposure has been published. However, dust control sufficient to prevent significant irritation of the nose, throat, and eyes is indicated.

Mantle Manufacture

The manufacture of gas mantles, though not the thriving business it once was, still furnishes a means of obtaining excellent illumination from gas, kerosene, and gasoline lamps and lanterns as well as a means of livelihood for many people. No occupational diseases have been found peculiar to the industry. However, since thorium nitrate, a radioactive material, is used to impregnate the mantles, potential exposures to thorium-bearing dusts and thoron gas should be evaluated and controlled.

The Welsbach gas mantle is made by dipping a mantle woven of ramie fiber into an aqueous solution of nitrates—essentially 99 parts thorium nitrate and 1 part cerium nitrate. The mantle is dried, formed, burned to remove the fiber and leave the metal oxides, and coated with collodion to protect it during shipment. Mantle-soaking operations require ventilation to keep the thoron gas within the accepted concentration. In mantle-cutting or mantle-trimming operations and especially in reclaiming operations there is, besides the potential exposure to thoron gas, possibly a more serious exposure to thorium-bearing dusts. There are also insignificant amounts of cerium and beryllium involved. Any dust containing thorium should be controlled (see Vol. I, Chapter XIX).

In lamp manufacture, which often accompanies the mantle manufacture, there may be glazing operations, spray painting, glass etching, soldering, and silvering.

Meat-Packing and Slaughter Industry

The environmental conditions common to the meat-packing industry include extreme dry radiant heat, extreme cold, sudden temperature changes, and high humidity with wet surroundings. Sickness frequency rates[27] indicate the highest rates in decreasing order to be associated with high humidity, sudden temperature change, and extreme cold. The occupations that had the highest frequencies were: cold-meat workers, among white females; and by-products workers, among Negro males. An excess of respiratory diseases was associated with the high frequency

[27] H. P. Brinton. H. E. Seifert, and E. S. Frasier, *U. S. Public Health Repts.,* **54,** 2196 (1939).

rates. Excessive rates for rheumatic diseases also were found. Material exposures associated with the highest rates were hides, glue, and entrails; digestive diseases were most in evidence in this last exposure.

Dermatitis occurs among meat handlers from contact with alkaline cleaning and dehairing baths and resinous dehairing agents; and also as a rash termed "hog itch"—among casing workers.

Brucellosis, which has had a high incidence[28] in the packing industry, may result from eating partially cooked meat in process. Packing-house workers have been known to eat slices of freshly killed pork after partly cooking it in hot water baths or on steam radiators. Skin abrasions and cuts on the hands may be another mode of entry of the causative organisms.

Ammonia refrigeration systems may develop leaks and so present a potential exposure harmful to health and a possible explosion hazard.

Metal-Cleaning Processes

Metal cleaning covers a wide field; an adequate discussion of all its ramifications would require more space than can be allotted here.

The choice of a cleaning method is influenced by many factors, including: composition and structure of the parts to be cleaned, soil or dirt to be removed, and sequence of the cleaning operation. The preceding operation frequently determines the nature of the contaminant and hence the cleaning method, and the next following operation may regulate the degree of cleanliness required. The equipment available, its cost and operating expense, and, above all, the health and safety problems need to be considered. Brief discussions of several different processes of metal cleaning follow.

I. Acid Cleaning

A. PICKLING TANKS

Acids such as sulfuric, hydrochloric, phosphoric, sometimes with chromic or hydrofluoric, are used in water solutions and their splash hazard and corrosive action on skin, clothing, and machinery are well recognized. Bubbles of hydrogen rising from the bath carry an invisible acid mist the amount of which depends upon bath temperature, the acid, its concentration, the metal, its surface area, and whether or not the bath contains an inhibitor. Commercial organic inhibitors have varied, more or less secret compositions, but paper-mill waste, flour-mill waste, and flour (1 oz. per 15 gal.) have been used successfully. Inhibitors may work either as a protective film that clings to clean metal surfaces, slowing down the action on the metal more than on the oxide; or as a producer of foam that inhibits escape of mist; or by lowering the surface tension.

The nature and extent of ventilation control required depends on the rate of acid mist escape; where the exposure is a mild one the addition of an inhibitor may

[28] M. G. Levine, *J. Ind. Hyg. Toxicol.,* **25,** 451 (1943).

make process or slot ventilation unnecessary. Inhibitors should never be added to an automatically timed conveyorized job without adjustment because pickling action is slowed by most inhibitors.

The possibility of exposure to arsine or other metal hydrides, though not common, should not be overlooked: impure acids or the metal being cleaned can be the source of arsenic, phosphorus, selenium, antimony, and so forth.

A good point to remember is that atmospheric acid even below the physiological danger point may corrode intricate and costly metallic parts and increase the plant's general corrosion problems, particularly in humid atmospheres.

Hydrofluoric and chromic acids require good tank-ventilation control and increased precautions to avoid skin contact.

B. BRIGHT DIPS

Acid bright dips are usually mixtures of nitric and sulfuric acids employed to remove tarnish from copper and copper alloys. The extremely corrosive bath is frequently contained in a large crock. Brown nitrogen oxide gases are evolved from the bath and since these gases are very corrosive, as well as dangerously toxic, bright-dip baths require more efficient hooding and exhausting than do most other metal-cleaning operations. Canopy hoods are not satisfactory. Hooding must be arranged so as to direct the gases away from the operator's face and prevent him from even momentarily inhaling the concentrated gas (see Vol. I, Chapter X).

C. PHOSPHORIC ACID CLEANERS

Phosphoric acid solution in water is used alone or with the admixture of alcohols and ethers, including butyl Cellosolve, to remove light rust on steel in preparation for lacquering and enameling. It may be applied by dipping, swabbing, or brushing, and usually when sufficient water is employed the operation does not produce harmful gases, vapors, or mists that require control measures other than protection from splashing and contact with the dilute acid. If mists, fogs, or elevated temperatures are involved, ventilation control is necessary.

D. FERRIC SULFATE PICKLE

Ferric sulfate pickling baths in themselves present no special problem. When the temperature involved is sufficient to produce steam it requires control measures, and if hydrofluoric or other acids are added they then become the problem for control.

E. GAS PICKLING

Gas pickling, utilizing 10 to 40 per cent hydrogen chloride at a temperature of about 1300°F., is used for the pickling of cold-rolled steel strip in preparation for galvanizing. The pickling atmosphere may be produced by burning a mixture of fuel gas, chlorine, and air and adding flue gas. This atmosphere is confined in a furnace through which the steel sheet passes. Any oil on the strip is burned off and the oxides are removed. It is proposed to apply this pickling method previous to the coating of steel with tin, aluminum, and lead, as well as to vitreous enameling and electroplating. Obviously, the pickling atmosphere must be effectively confined by suitable traps or reduced pressure, or else all points of escape must be provided with effective exhaust ventilation so as to maintain an atmosphere of less than 10 p.p.m. hydrogen chloride in working areas. It may be necessary to provide an exhaust hood to remove acid gas from the pickled metal after it leaves the pickling atmosphere.

II. Alkali Cleaning

Alkaline cleaners are used in soak tanks, dipping tanks, and power washers. Alkaline baths may contain caustic soda, soda ash, trisodium phosphate, sodium pyrophosphate, sodium hexametaphosphate, rosin, sodium resinates and other soaps, wetting agents, and emulsifiers. Sometimes solvents such as o-dichlorobenzene, butyl Cellosolve, pine oil, or petroleum distillates are added. Clay is sometimes added as a scrubber.

In general, when alkaline baths contain no solvents other than water, and no electric current is employed, ventilation is not required for the control of alkali mists unless readily attacked metals such as aluminum and zinc are being cleaned. However, it is desirable to remove the steam from heated baths in order to avoid excessive humidity, which contributes to the corrosion of metals and the discomfort of workmen.

The splash hazard of an alkaline bath is mainly dependent on the temperature of the bath and its degree of alkalinity. Caustic soda, upon contact with body tissue, gelatinizes the tissue, forming soluble compounds, and may produce deep and painful destruction of tissue. Even weak alkalies soften the epidermis, emulsify the skin fats, and cause severe skin irritation. All soaps hydrolyze in water to produce free alkali. Trisodium phosphate is somewhat less caustic than caustic soda and is a better detergent. Its cleaning properties are improved by the addition of either tetrasodium pyrophosphate or sodium metasilicate, either of which is still less alkaline than trisodium phosphate.

Paint-stripping operations usually involve baths of a high degree of alkalinity. The operation lends itself well to a conveyorized machine.

III. Emulsion Cleaners

Emulsion cleaners containing kerosene are used in power washers and soak tanks. When used in soak tanks at room temperature no ventilation is required. When the cleaner is sprayed or used hot, the operation should be confined and ventilated.

Emulsion cleaners that contain cresylic acid, phenols, or halogenated hydrocarbons should be provided with ventilation to prevent vapor concentrations in excess of accepted permissible limits. Skin contact should be carefully avoided, impervious rubber gloves should be used, and the face should always be protected by a suitable shield.

IV. Cyanide Bright Dip

Cyanide dip tanks are sometimes used for the removal of tarnish or light oxide films from brass and copper. They should be operated at temperatures below 140°F. and preferably should be ventilated. It is of the greatest importance that they be maintained well on the alkaline side and protected from the accidental addition or accumulation of acid, which might release potentially fatal concentrations of hydrogen cyanide. Overflow or drippings from the bath should not be permitted to mix with acid overflow and drippings, but should be promptly flushed into the sanitary sewer with plenty of water. Cyanide residues should not be dumped into the sewer but may be admitted very slowly, well diluted with water, providing the concentration does not exceed the limits set by official agencies.

V. Burn Off

In preparing for many types of painting operations, particularly sheet metal to be black-enameled, it is possible to replace other means of oil and grease removal with a burn-off operation. This is usually done in an oven heated with open gas burners to a temperature sufficient to ignite and burn any residual oil or grease film. No special problems are involved in venting the products of combustion to the exterior.

VI. Molten Salt Baths

Molten salt baths for heat-treating have some incidental application to the cleaning of metal. For a discussion of this see Heat-Treating.

VII. Molten Caustic Descaling Baths

Molten caustic, either with or without an electric current, is used for cleaning and descaling alloy steel and cast iron. Its use is restricted to those metals and alloys that are not adversely affected by caustic soda at the temperatures employed, usually from 680 to 950°F., in some instances up to 1000°. The advantage claimed for this type of cleaning is a good bond, especially with cast iron, when coating it with lead, zinc, or solder of any kind, and in brazing and vitreous enameling. The method also cleans sand from castings.

Oxides such as those of chromium, nickel, and iron are removed by reduction and the metal is not attacked. Oils are burned off, graphite and carbon are removed, and sand is removed. The oxides and possibly some of the sand collect as sludge on a sludge pan in the bottom of the bath and must be removed periodically. The other contaminants either combine with the molten caustic, float on the surface of the bath, or are volatilized as vapor, smoke, or fume. One patented process employs an electric current through the molten caustic bath. In this electrolytic bath the current is reversed during the cleaning process, it being claimed that certain contaminants are oxidized while the work is in the anodic position and others reduced while it is in the cathodic position.

A. SODIUM HYDRIDE DESCALING

One variation of molten caustic cleaning is the sodium hydride descaling process, in which no current is used and sodium hydride is fed continually to the bath through the action of metallic sodium in order to obtain the scale reduction required. This bath is maintained at about 700°F. and metallic sodium bricks weighing $2^1/_2$ to 5 lb. are added to the molten caustic by admitting them through a partially submerged generator box suspended along one side of the bath. An atmosphere of hydrogen, a mixture of hydrogen and nitrogen (cracked ammonia), or, in fact, any dry gas containing some hydrogen and at the same time free from oxygen, carbon monoxide, and carbon dioxide is maintained in this generator by letting the gas flow in at the bottom and out of a port at the top. The excess hydrogen is burned as it flows from the exit port and some is burned also at the charge hole in the top of the generator box during the time the sodium bricks are being added.

Obviously in either of these descaling processes there are the hazards connected with the use of a bath of molten caustic. The precautions practiced with alkali baths are in order and, in addition, protection from direct burns due to splashes from the hot alkali are important. Moisture entering the bath from any source, such as on or in the work being descaled, through leaks in the roof, leaks or condensation from overhead water pipes, splashes from a quench bath, and so forth, may be vaporized with an explosive violence that can throw particles of the molten caustic a considerable distance. It is necessary to have a remote-control hoist or crane to move the work, and safety shields or enclosures for the bath or for the operator are desirable as additional protection. Overhead water pipes, including automatic sprinkling systems, should be avoided. Safety practices required for handling metallic sodium must be enforced.

Contamination of the environment resulting from the caustic bath is ordinarily not excessive since the caustic is only slightly volatile at the temperature of the bath. Considerable heat arises from the hot bath. Where no electric current is used, and where general ventilation is good, the caustic bath may not require exhaust ventilation other than that necessary to remove any smoke or fumes resulting from the metal being descaled, but it is frequent practice to provide exhaust to remove the heat and any caustic that escapes into the air. The quench tank following the descaling tank requires good exhaust in order to control the steam and alkali mist. Complete enclosures with mechanical exhaust have the advantage of controlling the splash and splatter hazard, as well as any caustic-mist and heat exposures.

VIII. Solvent Degreasing

A. COLD-DIPPING

Cold-dipping is practiced to a certain extent for the cleaning of various objects including internal combustion engines and for the removal of carbon and resinous binders from pistons. The solvents used may vary from a high-flash petroleum distillate to a mixture that includes aliphatic and aromatic chlorinated hydrocarbons, ketones, Cellosolves, creosote, and cresylic acid. No generalities regarding control can be made to apply satisfactorily except that no readily volatile or fast-drying solvent should be used in large, open containers without effective mechanical exhaust ventilation if workmen are to be exposed to its vapors for more than a few minutes a day. Skin contact with these materials should be avoided and a face shield should be used to protect the eyes and face. Covered soak tanks with an adjacent mechanically ventilated work table offer satisfactory vapor control.

Under certain conditions covered soak tanks may be used successfully without ventilation and if the tank has a water solution layer over the surface, that also serves to retard the escape of solvent vapors. Low volatility materials such as mineral spirits and kerosene usually present no inhalation exposure, but the subsequent widely practiced compressed air blow-off operation may produce an objectionable irritant mist unless controlled.

B. SOLVENTS APPLIED BY BRUSHING OR WIPING

When solvents are kept in a safety can or other suitable covered container and applied in small amounts by brushing or wiping, the inhalation hazard far exceeds any fire hazard; and where this kind of operation is found necessary it is better to use petroleum distillates, ketones, and esters rather than chlorinated hydrocarbons. If chlorinated hydrocarbons of established moderate toxicity are found necessary to the operation, exhaust should be provided at the point of usage unless the solvent is applied very sparingly. Soldering and brazing operations are often said to require the use of carbon tetrachloride but in nearly all instances it has been found that where the right flux is used it makes little difference as to what solvent is used or whether any is used on the joint just before soldering.

Where work is conducted without ventilation in confined spaces such as small rooms, vats, tanks, and the like, volatile organic fluids should never be liberally applied by a swab or brush or used in a container that permits much surface contact of the air and liquid. If it is necessary to use volatile solvents in such situations fresh-air hose masks should be employed for personal respiratory protection and the explosion hazard must be reckoned with.

C. PETROLEUM SOLVENT SPRAYS

Spraying with high-flash petroleum distillates such as oleum spirits, mineral spirits, or kerosene is a widely used method of cleaning oils and grease from metal. Solvents with a flash point below 100°F. should not be used for this purpose. The operation should always be provided with suitable mechanical exhaust ventilation, preferably a hood as small as practical with exhaust sufficient to control any mist or vapor, and an exhaust volume of at least 100 c.f.m. per square foot of hood face. The hood may be of conventional spray-booth type or a much smaller one and may or may not be fitted with a fire door and automatic extinguishers. The fire hazard attendant to spraying a high-flash petroleum solvent is no more than that attendant to spraying many lacquers and paints.

D. DEGREASING MACHINES

From the viewpoint of hygiene and safety there are three major questions involving degreasing machines: (1) Is the machine being operated safely? (2) Are excessive amounts of vapor escaping into the room atmosphere? (3) How should such escape be prevented or controlled?

There are many angles to be considered before answering these questions. A degreasing machine is essentially a heated chamber in which to boil a grease solvent, space above the boiling solvent for hot solvent vapors, a condenser above this for cooling and condensing the vapor to liquid, and an extension of the sides above the condenser to minimize air currents inside the machine. The air space above the vapor inside a degreaser contains a vapor–air mixture somewhat richer in solvent vapor nearer the vapor line.

1. Classes of Degreasing Machines

Vapor Degreasers. In vapor degreasers the work is lowered into the vapor above the boiling solvent, and the solvent condensing upon the cool metal surface dissolves and washes away oil and grease films. This action should be continued until the metal reaches the temperature of the solvent vapor; whereupon it becomes dry. The length of time required will depend upon the size, shape, surface area, temperature, and specific heat of the parts to be cleaned. If the material is not allowed to remain long enough to reach the temperature of the vapor, considerable solvent will be dragged out with the dripping parts.

Immersion Degreasers. Where articles to be degreased have intricate shapes or are heavily contaminated with dirt and grease, immersion in boiling solvent gives not only prolonged solvent action but the mechanical scrubbing of the boiling liquid.

The usual practice is to combine immersion and vapor cleaning in a two- or three-chamber machine.

Spray Degreasers. Where insoluble matter is of such a nature or extent as not to be removed by immersion in boiling solvent, spray degreasing may be necessary. In this case the work is first lowered into the vapor to remove oil and grease, then pressure-sprayed with warm solvent, and finally given a vapor rinse to remove all traces of oil and grease. In order to conserve solvent and preserve health it is essential that any spraying be done below the vapor level of the degreaser in such a manner as not to disturb the vapor level, and that baffles or screens be placed so as to prevent the rebound or ricochet of droplets of solvent into the area above the vapor level.

All these methods are applicable to either hand-operated or conveyorized machines. Where the work is of sufficient quantity to justify the cost, conveyorized equipment is more satisfactory, in that solvent loss, with its attendant exposure possibly harmful to health, is easier to control.

2. Operation of Degreasing Machines

Condensers on the safer operating equipment consist of water jackets or a set of pipe coils or a combination of the two extending around the tank at some distance from the top. Some vapor-type degreasers using tetrachloroethylene, without condenser units, depend upon a bimetallic thermostat control to keep the vapor within the tank. The use of a mechanical device as the sole means of keeping the vapor from overflowing the tank is open to criticism; and when the thermostat is placed within a few inches of the top of the tank the device is not sufficiently quick and certain of action to be dependable and safe. The purpose of the condenser is to prevent the escape of the concentrated vapors into the room. The vertical distance between the lowest point at which vapors can escape from the degreaser machine and the highest normal vapor level is called the "free board." The free board should be at least 15 inches and not less than half the width of the machine. That portion of the condenser above the vapor line should be maintained above room temperature and below 110°F. The effluent water should be regulated to this same range, and a temperature indicator or control is desirable.

Solvents. The preferred solvent for water-cooled degreasers is trichloroethylene, but tetrachloroethylene can be used in these machines after they have been adjusted to suit the characteristics of the higher boiling solvent. The machines not equipped with water-cooled condensers are designed to operate with tetrachloroethylene only and should not, under any circumstances, be operated with other more volatile solvents such as trichloroethylene. Carbon tetrachloride, ethylene dichloride, and methyl chloroform also have been used in water-cooled machines. Carbon tetrachloride because of its volatility, greater toxicity, and susceptibility to hydrolysis into acid is not usually considered a suitable solvent for degreasing machines. Ethylene dichloride not only involves the toxicity problem common to some extent to all chlorinated hydrocarbons, but also is flammable. It flashes at 56°F. by the closed-cup method, and vapor–air mixtures ranging from 6.2 to 15.9 per cent by volume ethylene dichloride will explode with violence when ignited. Methyl chloroform necessitates special engineering design and corrosion-resistant material.

Trichloroethylene at room temperature will not burn, but trichloroethylene vapors when heated above 110°F. have a narrow flammable range around 20 per cent by volume. This range increases with temperature and above 135°F. the flammable range is from about 15 to 40 per cent by volume. The ignition temperature is 770°F. These conditions do not ordinarily occur in plant atmospheres but may occur within a degreaser. Trichloroethylene vapors will not explode violently under any circumstances but may burn slowly to form dense smoke and gases such as chlorine, hydrogen chloride, and phosgene. Although tetrachloroethylene vapor will not ignite or burn, oils or greases accumulated during cleaning will, and for that reason sources of ignition, especially overheating with gas or electric heaters, should be avoided during distillation for sludge removal. Also, welding on or in a degreaser when it contains any solvent should be avoided.

Fatalities have resulted from the use of solvent degreasers. Intoxication results chiefly from inhalation of vapors. It is doubtful if systemic industrial poisoning results from absorption of these materials through the skin of the hands, but skin irritation as a result of defatting may result from these as from other fat solvents.

Stabilizers. Commercial degreasing solvents sold under trade names usually contain a small amount of a material known as a stabilizer. The purpose of the stabilizer, frequently organic amines, is to neutralize any free hydrochloric acid that might result from: (*1*) oxidation of the degreasing liquid in the presence of air, (*2*) hydrolysis in the presence of water, or (*3*) pyrolysis under the influence of high temperatures. Some of these stabilizers are highly toxic materials that may cause serious injury to health by inhalation or absorption through the skin and although the amount supplied in commercial degreasing solvents has not, so far as we are aware, caused injury, the addition of the concentrated stabilizer in the plant by unqualified men is a dangerous practice, which should not be encouraged.

Heat Source and Controls. Degreasers may be heated by electricity, gas, or steam—steam being usually preferred. In the case of electricity and gas, thermostatic controls should be provided in the boiling chamber to prevent overheating. Gas-fired units should be provided with a flue from the combustion chamber to remove products of combustion. In the case of steam, the pressure should never exceed 25 lb. for trichloroethylene, and preferably it should not exceed 15 lb.; tetrachloroethylene requires pressure up to 50 lb. All machines should have thermostatic controls located a few inches above the normal vapor level to shut off the source of heat if the vapor rises above the condensing surface.

By-pass steam lines for emergency operation if the safety steam shutoff valve becomes plugged should not be to easy for anyone to operate as to nullify the safety features of the steam shutoff.

Sludge Removal. Sludge and metal chips should be removed as often as necessary to prevent their accumulation. The solvent should be distilled off until the heating surface or element is nearly, but not quite, exposed or until the solvent vapors fail to rise to the collecting trough. After cooling, the oil and solvent should be drained off and the sludge removed. A fire hazard may exist during the cleaning of machines heated by gas or electricity because the flash point of the residual oil may be reached and because trichloroethylene itself is flammable at elevated temperatures. After sludge and solvent removal a degreaser must be thoroughly ventilated mechanically before undertaking any maintenance work involving flames or welding. A person should not be permitted to enter or place his head within a degreaser until after all compartments have been blown free of vapors. When entering a degreaser he should then wear a supplied-air respirator as well as a life line held by an attendant, because in such circumstances not only very high and anesthetic concentrations of vapor may be encountered but also the oxygen content of the atmosphere may be insufficient. Such an atmosphere may cause unconsciousness with little or no warning.

Location of Equipment. Degreasers should be installed in large open departments with good general ventilation but away from drafts such as from open windows, spray booths, space heaters, ventilating-duct openings, or fans. When degreasers are installed in pits, mechanical-exhaust ventilation should be provided at the lowest part of the pit. It is not good practice to locate a degreaser beside a vat that evolves steam, which may enter and condense in the degreaser. Open flames, electric heating elements, and some welding operations within 50 ft. of a degreaser should be vented to the exterior, because if a high concentration of chlorinated solvent vapor comes in contact with a flame, arc, or hot surface corrosive and irritant gases are formed.

Ventilation. A properly constructed and operated degreasing unit located in a room of over 30,000 cu. ft. need not require a direct exhaust system; such a system may cause serious solvent loss and may or may not offer satisfactory control for an improperly constructed or operated machine. Several fatalities have resulted from carelessness during the cleaning of degreasers or the entering of conveyorized machines to make emergency adjustments, but there have been few injuries or fatalities attributed to exposures arising from the normal operation. Two fatalities attributed[29] to prolonged exposure to trichloroethylene vapors arising from degreasing machines occurred where the degreasers were equipped with direct-exhaust systems. This supports the general opinion that individual exhaust systems do not necessarily offer satisfactory control for faulty operating practices. General ventilation is a desirable safeguard even where direct-exhaust systems have been installed. Degreasers should be kept covered when not in use but frequent, abrupt covering and uncovering during operation is worse than leaving the cover off as it tends to fan the vapors out of the machine. For the conveyorized, enclosed machine a relatively small volume of air exhausted just inside the opening at which the operator stands for loading or unloading causes a slight indraft and controls the exposure without increasing solvent loss. A considerably greater volume of exhaust correctly applied outside the machine is likewise effective in control without causing excessive solvent loss. A slot at the top of an open degreaser offers satisfactory control but may increase solvent loss, especially if the volume exceeds 30 c.f.m. per square foot of tank surface.

Common Causes of Solvent Loss and Its Attendant Environment Contamination. Five general and very common causes of solvent loss and atmospheric contamination may be pointed out:

[29] *Ind. Bull. N. Y. State Dept. Labor,* **22,** 122 (1943).

(1) *Air motion* of more than about 50 f.p.m. across an open-top degreaser especially when directed lengthwise. Drafts should be eliminated and degreasers should be covered when not in use.

(2) *Mechanical displacement of solvent vapor.* Overloading may drive vapor out of the machine by physical displacement, especially where there is insufficient clearance between the sides of the rack or the work and the machine. Also when the load is great in relation to the heat input it will cool and condense so much vapor as to cause the vapor level to fall and draw air into the degreaser. The area above the vapor level in a degreaser is filled with a more or less rich solvent-air mixture. When the vapor level is lowered, the volume of this vapor–air mixture is increased by drawing in room air and when the level is brought back to normal some of the mixture is forced out into the room. The heat input should be sufficient to prevent the vapor line from falling below the bottom of the condenser under maximum load. Machines not equipped with water condensers but equipped only with bimetallic thermostats placed well up on the side of the machine may displace an appreciable volume of vapor–air mixture each time the vapor rises to the thermostat. Where the change of vapor level exceeds 6 or 8 in. in the normal operation of the machine excessive solvent loss and air contamination will result. The greater the change in level and the more frequent its occurrence the greater the air contamination.

(3) *Improper operation of condenser.* All too frequently, exposures can be traced to failure to open the valve to start the condenser operating before heat is applied to the boiling compartment. It is desirable practice to provide a safety control to make it impossible to heat the boiling chamber before water is turned on in the condenser. It is also a mistake to turn too much water through the condenser. The effluent water should never be below room temperature and may safely range up to 110°F. with trichloroethylene and 130° with tetrachloroethylene. If the temperature of the condenser water falls below that of the room, water may be condensed from the room air and enter the solvent causing a mixed vapor of water and solvent to form and float above the true-vapor level. This condition, which results in excess solvent loss, is made apparent by a fog or white mist floating in the bath. Water from any other source such as material degreased, or a leaky condenser, will do the same thing. Water is further undesirable because of its tendency to produce corrosive acid by hydrolysis of the solvent. The control of the temperature of the condenser should not be left to guesswork, but the water flow should be adjusted by a key valve, or other means, to give the correct temperature reading on an indicator and the water turned on full at the shutoff valve. This adjustment should be made either for the minimum temperature while operating under full load or for maximum while idling.

(4) *Speed of work.* One of the hardest things to control on hand-operated, open-top machines is the speed of the work, which ordinarily should not exceed 10 to 20 f.p.m. while work is being lowered into or raised out of the machine, as well as when work is moved within it. Speeds above this may displace vapor or mix it with air by agitation.

(5) *Carry out.* The work should always remain in the vapor until it appears dry, otherwise liquid solvent will be carried out to evaporate in the room air. Absorbents such as rope, cloth, or wood should not be a part of the work degreased or a part of the rack or hoist. Solvent may also be carried out in tubes, cups, intricate shapes, and recesses. This kind of work should be racked at an angle or should be tilted or rotated in the vapor zone until all liquid has been drained out.

Metalizing (Metal Spraying)

Metals, in the form of wire, are fed through a spray gun, in which a hydrogen, or acetylene, torch melts the wire. Compressed air, or other gas, is used to atomize and spray the metal onto the surface to be coated. The exposures involve not only air-borne particles of the sprayed metal and its oxides, but also ozone, nitrogen oxides, and ultraviolet light. Obviously the spraying of the more toxic metals offers the more serious potential inhalation exposures.

Many installations are automatic, but, where the gun requires close attention, goggles for protection from ultraviolet rays are necessary.

Process ventilation, with a properly designed hood, is the control method of choice. Many operations are successfully controlled in small enclosures with a relatively small volume but good velocity of air flow.

Ventilation requirements depend on the metal atomized, the velocity and volume of gases from the spray gun, the size and contour of the surface sprayed, the shape and fit of the enclosure, and the size of the enclosure opening.

Milling and Baking

Milling and baking operations involve few exposures harmful to health. Flour dust is the most common exposure and there have been complaints of nasal irritation and asthmatic attacks resulting from the inhalation of flour dusts. When such attacks occur, they would appear to be due to an allergic reaction. "Improvers" that are added to flour are not significant sources of harmful dust. The explosibility of flour dust is of more concern than the possibly harmful effects of inhalation. Well-engineered dust control should be practiced and it should include the prevention of any accumulation of dust where it can be a potential source of an explosive dust cloud.

Bleaching processes employing chlorine, chlorine oxide, or nitrogen trichloride are possible sources of toxic-gas exposures requiring careful handling and the provision of gas masks for emergency use. In the preparation and use of powdered sugar icing, which is frequently mixed and applied by hand, sugar dermatitis involving the nails occasionally results. Lard oil, when applied to pans by means of swabs, sometimes gives rise to skin irritation. Where ovens are allowed to cool between bakes, carbon monoxide in dangerous amounts may be generated during the warm-up period.

Mining

The subject of mining is too broad to be given more than a superficial touch here and the discussion will be confined to generalizations. In nearly all operations underground, dampness and wet surroundings are encountered. Dry drilling, blasting, dry mucking, loading and unloading cars, timbering, and ore crushing are dusty operations necessitating control measures.

In metal mines the metals may constitute an exposure problem as, for instance, in mining lead, zinc, arsenic, or mercury. The sulfide ores of lead and mercury are not readily absorbed, however and therefore have a relatively low toxicity. The inhalation of iron oxide ore in relatively pure state causes a nondisabling mottling of the lungs, a condition termed "siderosis."

In mining lead, zinc, and mercury ores the silica mixed with the ore may exceed 90 per cent of the mined material. Rigid dust control is indicated: wet drilling, wetting of muck piles and the work face, sprinkling of haulage ways, the use of water jets, and well-engineered ventilation.

In coal mining, jack-hammer, dry-cutting, and shot-firing operations produce much dust, as do shoveling and loading operations. These operations are productive of amounts of dust that not only are harmful to health but also have been the source of many coal-dust explosions in mines. Wet drilling and cutting have largely replaced the dry methods; and ventilation, dust traps, spray jets, and other measures have also been applied to the control of what was once one of the dustiest operations in the United States. Work in rock frequently involves exposure to dust high in quartz content, especially in the anthracite mines.

The condition of coal-dust deposits in the lungs has been termed anthracosis and, where a combination of coal and silica dust has produced lung pathology, anthraco-silicosis. A high mortality rate from respiratory diseases has been found in persons exposed to excessive anthracite dust.[30]

In addition to the dust exposures in mining, there are exposures to those injurious gaseous combustion products accompanying the use of explosives, the major ones of which are nitrogen dioxide, carbon monoxide, sulfur dioxide, and hydrogen sulfide. Approved explosives are designed with the proper oxygen balance so that the production of toxic gases is held at a minimum. Nevertheless, it is necessary to provide control measures for these gases where explosives are used underground. The control may consist of ventilation, or sufficient time for natural diffusion and absorption. The use of personal respiratory protection is at times necessary for either gases or dusts.

Naturally occurring gases such as methane, hydrogen, carbon dioxide, and hydrogen sulfide, as well as oxygen deficiency, must be controlled by ventilation. The use of Diesel engines in mines necessitates sufficient ventilation to control the minor amount of aldehydes, carbon monoxide, and nitrogen dioxide present in the exhaust gas.

[30] R. R. Sayers, J. J. Bloomfield, J. M. DallaValle, R. R. Jones, W. C. Dreessen, D. K. Brundage, and R. H. Britten, *U. S. Public Health Bull.* No. **221**, 1936.

In pegmatite and pyrophyllite mines, when considerable quartz is encountered, free silica becomes the criterion for dust control. In sandstone and quartzite quarrying and mining, the dust control obviously must be of a high order. Among mica workers, where little or no silica is involved, there is some question concerning the control standards that should be enforced. There is no logic, however, in permitting dustiness above 50 million particles per cubic foot of air; and there are considerable data[31] indicating the need for more rigid control.

Tremolite talc miners, exposed to rather high concentrations of the asbestine variety of talc dust, have been reported[32] to have a high incidence of advanced fibrosis, although the free silica content of the atmospheric dust was believed less than 1 per cent.

In salt mining, there are exposures to blasting gases and to salt dust, which is irritant to mucous surfaces and even causes some dermatitis.

In all mining work, periodic physical examinations of workers are even more important than they are in work above ground. Sanitation[33] is an important control measure.

Motor Testing

In motor testing, carbon monoxide, noise, and heat furnish the problems for consideration. Carbon monoxide is readily controlled by the correct application of exhaust ventilation; the control of noise is a more difficult engineering problem; while temperature control is frequently more or less incidental to the control of exhaust gases and noise. Where there are individual test chambers provided with exterior controls and instrumentation, the problems are readily solved. Where many engines are operated simultaneously in one room, the exhaust gases are readily removed through flexible connections to exhaust ducts, but noise control is not easy. Walls and ceilings can be deadened with fireproof, sound-absorbing materials, or curtains and irregular space arrangements may be used to absorb sound and avoid reverberation, but the only way of lowering general noise production is by segregating the operations. In some rooms with many closely stationed, simultaneous test operations it has not been practical to bring the general noise level below 100 decibels which, although tolerated by many individuals, is at least very annoying to some. The question of possible ill effects is an open one but it is generally believed that the low-frequency engine-test noises of this level do not have a permanent adverse effect upon the hearing of workmen.

[31] W. C. Dreessen, J. M. DallaValle, T. I. Edwards, R. R. Sayers, H. F. Easom, and M. F. Trice, *U. S. Public Health Bull.* No. **250,** 1940.
[32] W. Siegal, A. R. Smith, and L. Greenburg, *Am. J. Roentgenol. Radium Therapy,* **49,** 11 (1943).
[33] R. R. Sayers, *U. S. Bur. Mines Miners' Circ.* No. **28,** 1924.

Nickel

The element nickel and its compounds, with one exception, are of little significance as industrial poisons. The exception is a toxic gas, nickel carbonyl (see Chapter XXVII).

Nickel is found in nature as the sulfide, which, after concentration and grinding, is roasted. The sulfur is removed as sulfur dioxide, leaving nickel oxide. Sulfur dioxide is not present in the atmosphere of smelters in significant quantities, as the smelting furnaces remove the gas efficiently.

Nickel oxide is reduced to the metal in either open-hearth or electric furnaces. The electric furnaces generally require exhaust ventilation because of excessive fumes. Fluorspar is used as a flux and fluoride concentrations should be recognized and controlled in order to avoid nasal irritation and fluoride storage.

Strong mineral acids are used during the fabrication of nickel, especially nitric acid in pickling processes. Lead is employed as a coating on rods and tubes which are cold-drawn, and there is a potential exposure involved in the application of the lead. Furnaces that have inert atmospheres should be checked for carbon monoxide leakage.

Painting

Paints consist of pigments, binders, thinners, wetting agents, and driers. The pigments may be any of the following: the oxides or other compounds of lead, zinc, titanium, chromium, iron, antimony, or cadmium; metallic zinc, bronze, brass, or aluminum; carbon, talc, china clay, quartz silica, diatomaceous earth, or mica; barium sulfate, calcium carbonate, or calcium sulfate; or organic dyes. The binders include oils, gums, natural and synthetic resins, casein, cellulose esters, gilsonite pitch or tar, dibutyl, diethyl, and diamyl phthalates, and tricresyl phosphate. The thinners include turpentine, aliphatic and aromatic hydrocarbons, esters, alcohols, and ketones. Wetting agents have been relatively innocuous. Driers include lead, cobalt, and manganese compounds.

Paint is applied by brushing, dipping, tumbling, and spraying. The necessity for ventilation control of the solvents during application and drying is common to all methods. Potential exposures include: (1) ingestion of lead and other toxic pigments, and (2) difficulties arising from sensitivity to the irritant effects of zinc chromate, of certain incompletely reacted resins such as cashew shell liquid–formaldehyde resin, and of the thinners, especially wood turpentine, which may cause dermatitis upon contact with the skin. A more serious exposure involves inhalation of excessive amounts of solvent vapor arising from painted surfaces in poorly ventilated areas. In addition to exposures possibly harmful to health, vapor explosions must be considered, especially in the initial period in drying ovens.

In spray painting there are potential exposures to pigments, mists, and solvent vapors during spraying and to solvent vapors during drying. It is desirable to provide all spraying operations with booths engineered to prevent contact of significant vapor or mist with the faces of workmen. Although recommended air-flow velocities for spray booths are on the order of 100 to 200 l.f.m., much depends upon the spray gun, the size and shape and location of the booth, size and shape of the object, the pigments and solvents involved, and other factors. This is discussed more fully in Chapter X. Set rules for air volumes and velocities are not in order, but either the ventilation control should prevent significant vapor, mist, and pigment exposure or the operator should wear a respirator, preferably one of the supplied-air types. The painting of the interiors of cabs, busses, and railroad cars is an exposure difficult to control without resorting to personal respiratory protection. In any type of filter respirator, as in the case of air cleaners for ventilation systems, the important consideration is not the amount of contaminant removed by the filter but the amount that passes through.

Paint Manufacture

The mixing of dry pigments with oils and varnishes constitutes the principal health hazard of this industry. Lead oxides, lead carbonate, and other leaded pigments, as well as cadmium compounds and numerous toxic dry colors, are usually purchased by the paint manufacturer in paper sacks. A dangerous dust exposure occurs when the pigments are added to the mixers unless exhaust ventilation is used. In lieu of such ventilation it is common practice for the workmen to wear approved respirators, as the time of exposure is only a minor portion of each work period.

Paint manufacture is generally conducted in a building of three or four stories so that materials may flow by gravity from operation to operation. Some of the finer products go to stone mills for production of the proper dispersion; other products require roll mills and Banbury mixers. Stones used for grinding pigments and oil generally contain free silica: stone dressing must have exhaust ventilation if silicosis is to be prevented.

After being thinned and tinted, the products are placed in cans by automatic machinery or hand operation, depending on the volumes handled. The chief harmful effect from the paints after the pigments and oils have been blended is dermatitis, but this is rather rare in modern plants where cleanliness is rigidly enforced, especially if coveralls and laundry service are provided.

Turpentine and hydrogenated naphthas are the most common thinners. Hydrogenated naphthas generally consist of saturated cyclic hydrocarbons, but the absence of benzene should not be taken for granted. The saturated cyclic compounds have low toxicity.

Paper Manufacture

Pulping. Wood pulp produced by treatment of wood with steam in digesters is the chief source of paper. In the *sulfate process* (sulfide), the shredded wood is digested with steam in tanks under pressure of about 125 p.s.i. using up to 8.5 lb. of sodium sulfide and 23 lb. of sodium hydroxide per 100 lb. of wood. During cooking, relief gases are released into the atmosphere from the top of the digester or along with the condensate, which contains commercial quantities of "sulfate wood turpentine." These gases contain methyl alcohol, acetone, aldehydes, traces of acetic and formic acids, ammonia, hydrogen sulfide, ammonium sulfide, methyl mercaptan, and dimethyl sulfide. The gases also arise from the pulp and digestion liquor and they persist with the pulp during washing, screening, and some subsequent processes. The liquor is partially evaporated in evaporators, salt cake (sodium sulfate) is added, and the mixture sprayed into recovery furnaces, where the water is removed and the other volatile matter burned, leaving the alkali and sodium sulfide to be reused. The resultant heat is utilized and combustion gases are released from high stacks. These stack gases contain both sulfurous gases and particulate matter that consists of sodium sulfate, sodium carbonate, and sodium sulfide. The odor in and around the plant is exceedingly disagreeable. Although in most instances gases and vapors are in very low concentrations, there are opportunities for harmful exposures and for explosions of flammable gases or vapors.

Various methods have been tried for controlling the escape of odorous gases. These include passing the gases into high stacks, scrubbing them through caustic, passing them through the furnace with or without previous washing with water sprays, and treating them with waste bleach water. Electric precipitation has also been suggested for the stacks.[34]

The *sulfite process* is similar except that the digester liquor is an aqueous solution of sulfurous acid with lime or other base to form bisulfites. The sulfur dioxide is obtained either as a compressed gas or from the burning of sulfur or the roasting of pyrite ores. The relief gas in this process contains high amounts of sulfur dioxide, which must be recovered for economical operation. This is accomplished by separators and coolers.

Bleaching. Bleaching is usually accomplished by the use of chlorine, which may be supplied from cylinders of the compressed gas or from bleaching powder. Chlorine hydrate may form where gaseous chlorine enters the vat and be carried to the surface, where it emits chlorine into the atmosphere. However, the exposure to chlorine is ordinarily not difficult to control by process or general ventilation.

Coating. Paper is coated by various types of coating machines and the materials used include borax, clay, mica, talc, casein, soda ash, dyes, plastics, gums, varnishes, linseed oil, and organic solvents. The principal exposures arising from these operations involve: (*1*) acrolein and other aldehydes resulting from the atmospheric oxidation of linseed oil, and (*2*) solvent vapors from the coating and subsequent drying of the paper, when coating mixtures dissolved in organic solvents are used. When the coating and drying are done in air-conditioned rooms, the environmental-control problems sometimes become difficult.

[34] J. M. DallaValle and H. C. Dudley, *U. S. Public Health Repts.,* **54**, 35 (1939)

Photographic Industry

Dermatitis and skin sensitization are the hazards of the photographic industry. Nasal and bronchial irritation and asthma are also reported from contact with developers and other photographic chemicals. The aminophenols are among the most common sources of skin disease from developers. Other irritating chemicals include caustics, iron salts, mercuric chloride, strong acids, bromides, iodides, pyrogallic acid, and silver nitrate. The last substance is reported to have caused argyria, a condition in which silver is deposited beneath the skin. It is difficult, if not impossible, to remove the deposits completely.

The prevention of dermatitis lies in reducing to a minimum the contact of the chemicals with the skin. Cleanliness, protective clothing, and protective creams are the triumvirate for controlling the hazard.

Plastics and Synthetic Resins

Synthetic resins that have undergone complete condensation cause little difficulty in the cold, but where heat is applied, or an imperfectly combined component is present, skin irritation and sensitization may result. The phenol-formaldehyde and urea–formaldehyde resins owe their irritant and sensitizing properties chiefly to formaldehyde,[35] but phenol and furfural (phenol-furfural resins) may also have some irritant effect. It is advisable to provide process ventilation for any dust-producing or vapor-producing process involving the manufacture, fabrication, or use of plastics whereby formaldehyde is released into the workroom atmosphere. Hexamethylenetetramine, which has a bad reputation as a sensitizer, is harmful because it releases formaldehyde.[36] Cashew nutshell liquid-formaldehyde resin is particularly offensive if any uncombined cashew nutshell liquid remains in the resin. "Oil stop," a waterproof resin made by mixing cashew nutshell liquid to a paste with powdered paraform (a polymer of formaldehyde) and allowing it to "set" or condense, is popular with electricians. Both of the constituents are irritants and sensitizers and there is a high incidence of dermatitis among users who carelessly contaminate their hands and clothing with the mixture.

Plastic glues for the manufacture of plywood, laminated asbestos, fiberboard, glass fabric, and similar products may be made of incompletely condensed resins containing formaldehyde along with acids, alkalies, peroxides, and other irritants and sensitizers. The screening, scaling, and mixing of such powdered glues are productive of dust and conducive to dermatitis.

Many plastics such as vinyl chloride, vinyl acetate, polyethylene, polystyrene, and methyl methacrylate have proved to be more or less inert physiologically. Antioxidants and stabilizers added to some plastics, however, may occasionally cause adverse physiological effects. The majority of the dermatitis

---

[35] A. G. Cranch, *Ind. Med.,* **15,** 168 (1946).
[36] L. Schwartz, *J. Invest. Dermatol.,* **6,** 239 (1945).

cases arising from prolonged contact with plastics of this nature are caused by plasticizers added to eliminate brittleness and to produce flexibility. Some of these plasticizers are susceptible to the effects of heat and moisture and may separate from plastics that are in prolonged contact with the skin and cause either primary irritation or, more likely, sensitization. This presents a use problem rather than a production problem. These plasticizers[36] include derivatives of glycol, glycolic acid, phthalic acid, phosphoric acid, ricinoleic acid, and sebacic acid.

In evaluating and controlling exposures arising from the manufacture, fabrication, or use of plastics, dermatitis is the primary consideration, but eye irritation and even the possibility of lung irritation should be considered. Dusts, especially those involving incompletely reacted materials, and curing agents, especially organic amine vapors, should be controlled by engineering methods; excessively warm and humid atmospheres also should be controlled; and direct skin contact with suspected irritants or sensitizers should be avoided. A supervised program of personal cleanliness and frequent changes of clothing are the best personal control measures wherever dermatitis is involved.

Occasionally persons do not respond to "hardening" and the standard preventive and protective measures. In such cases of unusual susceptibility it is necessary to transfer the workers to other work.

Pottery Industry

Silicosis and lead poisoning are the traditional occupational diseases to potters. Free silica is present as flint in the pottery slip in amounts that make dust control of a high order imperative. Respirable sizes of dust commonly show 40 to 50 per cent free silica in slip houses. The use of lead compounds in decorating ware also necessitates a high degree of dust elimination.

Most of the dangerous silica operations are concentrated in a few departments. The slip house normally has more than half of the total significantly exposed workers. Lead exposure is confined to sprays and dust from ware prior to its being fired.

Jiggering and batting out normally do not involve harmful exposures. Stampers should be provided with exhaust ventilation, however, as should finishers. Dish makers do not have significant dust exposures. This statement also applies to casting shops. Bisque and glost kiln placing and drawing are dust-free operations. Flatware brushing is one of the dustiest occupations outside of the slip house and it requires control. The transfer of raw materials from boxcars to storage bins may involve excessive dust exposures.

It is poor hygienic practice to draw hot air directly from the fire chambers of the kilns into the workrooms to heat them: a system of heat exchangers should be used.

Quartz Crystal Cutting

The manufacture of quartz crystals for radio oscillators results in several processes that may cause harm to the health of workers. Quartz dust, solvents, and x-radiation may be present in damaging quantities.

Both Goss[37] and Schulte[38] have shown that dust counts from cutting operations are closely related to the degree of contamination of the oil. Oil that contains much quartz dust will spread the quartz into the air along with oil spray. Schulte also found, however, that clean oil was not always an insurance of a safe process. Exhaust ventilation is required for adequate control.

Xylene, carbon tetrachloride, and methyl alcohol are commonly used to remove cement from the crystal-mounting bases. These materials may be present in hazardous concentrations. Other alcohols and lacquer thinners are used in lesser amounts.

Etching compounds composed of fluoride salts with weak acids or hydrofluoric acid itself are used to reduce the amount of grinding required. Protective clothing and ventilation are required to protect workmen.

X-rays of low power (35,000 volts) are employed to test quartz crystals. The exposures should be appraised by measuring any stray radiation, and periodic blood counts should be made on operators who are routinely exposed.

Dermatitis, resulting from excessive contact with the solvents and oils, has been of frequent occurrence.

Radio Manufacture

The majority of the potential exposures in radio manufacture have been discussed under Chlorinated Oils and Waxes, Electroplating, Heat-Treating, Metal Cleaning, Plastics and Resins, Quartz Crystal Cutting, Sand Blasting, Soldering, Spray Painting, and Welding.

Where Halowax is used on coils or wires, rigid control is required to avoid dispersing the vapors as well as to avoid skin contact. Paraform introduces both a dust and a formaldehyde gas exposure, either of which may cause skin irritation or sensitization.

Copper cyanide plating operations may be sources of excessive amounts of cyanogen or hydrogen cyanide gases, and zinc cyanide baths may be sources of excess alkaline cyanide mists unless ventilated.

Carbon tetrachloride is frequently found in open containers around radio and wire soldering operations. It may present a harmful exposure and its use is unnecessary.

[37] A. E. Goss, *J. Ind. Hyg. Toxicol.*, **16**, 208 (1944).
[38] H. F. Schulte, *Ind. Med.*, **14**, 68 (1945).

Radium Dial-Painting

Radium dial-painting is treated at some length in Vol. I, Chapter XIX. Radioactive dusts of respirable size probably offer more of an inhalation hazard than does radon, because any particles of radioactive material lodged in the respiratory tract would continue to emit rays for an indefinite period unless removed by ciliary action or other means. Where dial painters are permitted to mix their own paints, radioactive particles can be demonstrated by ultraviolet light on the skin, the clothing, personal belongings, in desk drawers, or any place where the work tools are kept. A factor that may minimize this potential exposure is that the particles of radioactive material are for the most part well above the respirable range in size. The fact that they must be filtered out of air samples collected for radon analysis, however, is a definite indication that air-borne dust particles exist. Centralized, carefully controlled mixing and scrupulous house-keeping evaluated by ultraviolet light are very important control factors. Methods of dust removal must be such as not to present an exposure through redispersion. Drying ovens should be provided with sufficient mechanical ventilation to afford an inward flow through all openings under operating conditions. Radioactive dust should not be permitted to collect in cabinets or ovens whence it may be dispersed by opening or closing the doors. The environment should be monitored and controlled.

Sand Refining

Sand refining for the production of a pure graded product may involve serious exposures to quartz dust. Where the raw product contains fine quartz grains, or the cryptocrystalline variety of quartz, or grains that have been fractured by explosives and drills, there is a serious potential exposure that requires careful control.

Where the refining of such a product involves rotary-kiln drying, screening, elevating, storing, and car loading without the benefit of modern engineering control measures, exposures have been found to range upward of 100 million particles of quartz per cubic foot of air in and around the mills, with a visible cloud of dust that pollutes the atmosphere for a considerable distance leeward of the plant.

This type of plant has been brought under satisfactory control by enclosing and exhausting elevators, screens, chutes, bins, conveyors, and other dust-producing operations; combining the dust-laden exhausted air with the hot rotary drier effluent; and scrubbing both through a homemade collector consisting of water sprays, baffles, and a bed of coke that is also sprayed with water. The heated particles of quartz, although too small to be collected by impingement, are readily collected after each has formed the nucleus of a water droplet in a cooled, supersaturated atmosphere. The collector was the source of huge quantities of "silica flour" that presented a disposal problem until uses for this by-product were developed.

Shipbuilding and Repair

Before a vessel enters a drydock for the purpose of undergoing construction work, repairs, or alterations of any kind, all tanks, compartments, or lines that have contained flammable liquids should be cleaned and freed of flammable vapor to comply with the code of the National Fire Protection Association concerning marine fire hazards. The atmosphere in all unventilated areas or compartments should be checked for harmful or flammable gases and for oxygen deficiency by a qualified chemist or industrial hygienist, who should examine each compartment before workmen are permitted to enter. Tests through long sampling lines from the deck are not reliable.

Tankers that have carried gasoline or volatile crude oils require frequent, periodic checks even after a "gas-free" status has been established. Any rust on the walls or sludge on the floors of compartments may continue to dissipate flammable vapor. The pumping of ballast also may introduce flammables from some inaccessible part of the pipe lines or storage tanks.

When work is conducted in tanks used for transporting leaded gasoline, the lead exposure involved in welding or cutting operations on rust-coated surfaces is not significant; general ventilation, sufficient to control welding fumes, furnishes satisfactory control for lead exposures. The amount of adsorbed or occluded lead in the rust is so small that, except in very dense suspensions of rust in the atmosphere, the exposure is negligible even for prolonged periods. Where extensive and prolonged scratch brushing or other dust-producing work is performed, a respirator to filter out the iron oxide is advisable.

Repairs on pipe lines should be undertaken only after all "hot work" on the hull and in tanks and compartments has been completed. Heat must never be applied to any closed line or section of line; all lines must be opened by cold operation for examination; and, if any welding or torch cutting is to be done on a line that may contain flammables, a blast of air should be blown through it before, and during, the operation. Work of any nature on refrigeration lines must be under the direction of a man who is qualified to recognize and control the hazards involved.

The most generally widespread exposures in shipbuilding and repair are those connected with welding and flame cutting; and since these are discussed under welding, they will not be enlarged upon here more than to point out that there are many opportunities for welding in confined spaces on a ship, areas into which a man must crawl and where there is no ventilation except what is supplied mechanically. Under such circumstances the amount of nitrogen oxides and other gases produced by a gas torch, although normally not a matter for great concern, can reach fatal concentrations in a matter of minutes, and has, where no ventilation was provided. The most favorable reaction temperature for production of nitrogen oxides is said to be 4200°F., at which temperature air passing through the flame may produce as much as 1.75 per cent nitric oxide in the effluent gas. This is the basis of a commercial process for nitrogen fixation.[39] In such circumstances reliance upon respirators is foolhardy unless the respirators are of the supplied-air type. Properly distributed forced ventilation is the control method of choice.

In cutting, burning, or welding operations that involve a potential metal-fume exposure, as in welding or cutting galvanized parts, where the ventilation is sufficient to control the nitrogen oxides, type B fume respirators have been found useful. It might well be pointed out again that canister-type respirators that rely upon activated carbon and soda lime offer little protection against nitrogen oxides.

Burning lead paints off interior surfaces is a hazardous practice requiring effective respiratory protection or ventilation control. One saving feature of such work, however, is that its intermittent nature ordinarily places it in the acute exposure category and precludes any long-continued exposure.

Spray-painting interiors requires not only control of inhalation exposures to thinners, oils, and pigments, but also prevention of the explosion of flammable vapors. Control can be accomplished by the generous and judicious application of forced air supply and exhaust. Where necessary this can be supplemented by air-line or cartridge respirators, depending upon the nature and amount of air contamination. Painting of the hull poses a problem, not alone of respiratory protection, but of being able to see, because uncontrolled paint mist quickly covers goggles or face shields. Where natural or artificially induced air movement cannot be used to advantage to avoid inhaling paint mists, spray nozzles mounted on long pipes have been used; but brush painting has been found the most satisfactory answer to the control problem in many instances.

Riveting interiors with hot rivets where oily or painted surfaces are involved creates an exposure to irritant smoke containing aldehydes and, in the case of lead paints, may present a fume exposure. Ventilation control is advisable.

The compartments of floating dry docks should be tested for flammables periodically and vents should be protected by flash-back arresters.

To be able to decide upon the hazardous nature of cargoes and the proper precautions to be exercised in the handling of them requires a rather broad understanding of industrial toxicology. Oxygen deficiency, resulting from fermentation, Dry-Ice refrigeration, or displacement of air by gases other than carbon dioxide, has probably caused more fatalities on cargo ships than any other aftermath of atmospheric contamination except explosions of flammable vapors. Skin irritation from cargoes is not uncommon and is usually a result of carelessness or a failure to practice elementary preventive measures. For a description of an unusual epidemic see the Preface to Volume I.

Soldering

Hand-soldering operations employing electric irons rarely present a significant lead exposure but the smoke and gas from the flux are sometimes offensive.

[39] Anon., *Chem. Inds.*, **58**, 245 (1946).

Gas-heated irons do not ordinarily present a harmful exposure during use but when the irons are heated in a stove or muffle furnace, any solder falling off the iron into the furnace is soon volatilized into the air. It is therefore important to hood and exhaust any furnace used for heating soldering irons. Iron soldering and small pot tinning need not be provided with process ventilation but are satisfactorily controlled by a reasonable amount of general ventilation. Care should be used, however, to prevent contaminating the surrounding area with lead dross and scrap.

Production torch soldering is usually provided with process ventilation to remove smoke and gas arising from the flux and flame of the torch but the amount of lead volatilized rarely is sufficient to justify such provision. A hood has the advantage of promoting good housekeeping.

Solder dipping where large surfaces are involved requires hooding and exhausting.

Silver soldering, with its higher temperatures, silver–cadmium alloy solder, and often a fluoride flux, warrants better control by ventilation in order to prevent excessive exposures to cadmium fume and fluoride fume.

Stone Industry

Quarrying. Most quarrying operations conducted in the open air are not likely to be accompanied by harmful concentrations of dust unless the stone contains a considerable percentage of free silica, 30 per cent or more. However, dry-drilling operations are usually excessively dusty. Evaluation of individual exposures should be made by dust counts of air samples, and analysis of representative samples of the air-borne dust. Wet drills, and water sprays on power shovels, have been found to provide satisfactory dust control where necessary. Dust traps may be used to advantage on the drills in dry drilling.

Crushing Mills. The primary crushing operation is ordinarily well supplied with natural ventilation and dustiness is not sufficient to require additional control when the free silica content is low and the stone is moist; but if the operation is objectionably dusty, a mist spray can be applied effectively without causing gumming of the fines sufficient to interfere with the operation of conveyor belts, chutes, or buckets. This spray has been used to best advantage at the shovel. Ordinarily, secondary crushers are excessively dusty unless the hopper and belt-loading zone is enclosed and exhausted or the crusher is equipped with sprays. Conveyors, likewise, create dust at their terminals unless enclosed and exhausted, or subjected to water sprays.

Vibrating and rotary screens are the most prolific sources of dust when not enclosed, and frequently produce dust clouds of 100 million to 1 billion particles per cubic foot of air. Even one of the least harmful or offensive dusts, such as relatively pure calcium carbonate, when present in such excessive amounts is not only objectionable from the standpoint of inhalation, but is an accident hazard because it interferes with vision. The situation is especially bad if, as sometimes happens, these screens are located at or near the top of a several story structure with wood floors that have wide cracks that continually sift dust down throughout the structure.

Bagging operations are very dusty unless well engineered. Water-mist spray at dust dispersion points, or suitable enclosures and dust-control systems, are necessary.

Granite cutting has been found to require dust-control measures; maximum permissible concentrations of 10 to 20 million particles per cubic foot have been proposed; 20 is perhaps more generally accepted.

Cement dust, also discussed under Cement, is somewhat alkaline and causes irritation of the intact skin. Opinions differ as to the maximum permissible dustiness, but 50 million particles per cubic foot of air appears to be lenient.

Welding

Eye protection is needed for both intense visible light and ultraviolet radiation from electric welding. The skin also must be shielded to avoid burns. Gas welding does not release significant ultraviolet light. Goggles are necessary, however, to reduce glare and to prevent eye damage from sparks. Flame-resistant clothing is advisable for all types of welding.

When plain steel electrodes are used, long-continued exposures with inadequate ventilation may cause a chronic bronchial cough. The condition clears in a few weeks after exposure ceases. The lungs of welders may have sufficient deposition of iron oxide fume to cause characteristic x-ray findings (siderosis). This condition is benign and is not associated with disability or discomfort. Siderosis complicates the diagnosis of silicosis where there is an associated or subsequent exposure to free silica.

In unventilated spaces nitrogen dioxide and ozone can be generated in harmful quantities: this is especially true of gas shrinking[40] operations.

Coated welding rods release both iron oxide and fume containing the constituents of the coating. Repeated laboratory studies show that harmful amounts of fume are not produced. The only exception occurs in confirmed quarters with no ventilation.

Where the room volume is relatively large, over 50,000 cubic feet with the space per welder over 10,000 cubic feet, and construction is of conventional building materials, dilution and natural ventilation are believed to be sufficient to prevent the accumulation of significant fume levels in the general atmosphere when welding clean carbon steel with steel rods and electrodes. Where space is inadequate, or smoke and fumes become a nuisance, ceiling outlets and a make-up air supply to the occupied zone are advisable.

Welding on surfaces coated with cadmium is dangerous and deaths have occurred where there was no exhaust ventilation. Cadmium oxide may cause a chemical pneumonitis fatal within 24 hours.

[40] F. E. Adley, *J. Ind. Hyg. Toxicol.*, **28, 17** (1946).

Zinc-coated surfaces, during welding, release large amounts of zinc oxide which can cause metal-fume fever. The condition is unpleasant but has no permanent effects after 24 hours. Welding on painted metal is hazardous to a degree dependent upon the composition of the paint. Paints containing lead may cause plumbism.

Welding on aluminum or stainless steel introduces a fluoride exposure from the flux used. In the absence of control measures nasal irritation is common and long-continued exposures might result in fluorosis.

Tuberculosis is no more frequent among welders than among the average population. Animal studies indicate that welding fumes neither predispose to tuberculosis nor reactivate healed lesions. The incidence of pneumonia among welders also approximates that of the general public.

Local exhaust ventilation is advisable for welding processes for the following reasons: (1) to prevent throat and bronchial irritation; (2) to avoid siderosis which may be mistaken for silicosis; (3) to eliminate toxic quantities of lead, beryllium, manganese, cadmium, fluorides, and so forth, when present; (4) to prevent deaths, which occasionally occur from welding in confined spaces. Recirculating the air through a mechanical filter cannot be considered satisfactory ventilation control where nitrogen dioxide is involved.

Spot welding does not produce harmful quantities of gases or rays. When the parts are oil covered, offensive oil smoke may be given off. Burns from flying sparks are possible and in spot-welding stainless steel penetration of metallic particles into the finger tips has been demonstrated by x-ray photographs.

In gas-shielded arc welding objectionable quantities of ozone may be generated by ultraviolet radiation. When rods containing more than 2 per cent thorium are used the thorium exposure should be evaluated. Dust from the grinding of thoriated electrodes should be evaluated. Chlorinated solvent vapors near welding operations, especially gas-shielded, should be avoided because they are readily converted to acid gases by intense ultraviolet radiation.

INDEX